Please Return

607 Pickwick Ct
Mt. Prospect, IL
60056

312-593-6068

Handbook of Materials and Processes for Electronics

Other McGraw-Hill Handbooks of Interest

AMERICAN INSTITUTE OF PHYSICS · American Institute of Physics Handbook
BAUMEISTER AND MARKS · Standard Handbook for Mechanical Engineers
BEEMAN · Industrial Power Systems Handbook
BLATZ · Radiation Hygiene Handbook
BRADY · Materials Handbook
BURINGTON AND MAY · Handbook of Probability and Statistics with Tables
COCKRELL · Industrial Electronics Handbook
CONDON AND ODISHAW · Handbook of Physics
COOMBS · Printed Circuits Handbook
CROFT, CARR, AND WATT · American Electricians' Handbook
ETHERINGTON · Nuclear Engineering Handbook
FINK AND CARROLL · Standard Handbook for Electrical Engineers
GRUENBERG · Handbook of Telemetry and Remote Control
HAMSHER · Communication System Engineering Handbook
HARPER · Handbook of Electronic Packaging
HENNEY · Radio Engineering Handbook
HENNEY AND WALSH · Electronic Components Handbook
HUNTER · Handbook of Semiconductor Electronics
HUSKEY AND KORN · Computer Handbook
IRESON · Reliability Handbook
JASIK · Antenna Engineering Handbook
JURAN · Quality Control Handbook
KLERER AND KORN · Digital Computer User's Handbook
KOELLE · Handbook of Astronautical Engineering
KORN AND KORN · Mathematical Handbook for Scientists and Engineers
KURTZ · The Lineman's and Cableman's Handbook
LANDEE, DAVIS, AND ALBRECHT · Electronic Designers' Handbook
MACHOL · System Engineering Handbook
MAISSEL AND GLANG · Handbook of Thin Film Technology
MARKUS · Electronics and Nucleonics Dictionary
MARKUS · Handbook of Electronic Control Circuits
MARKUS AND ZELUFF · Handbook of Industrial and Electronic Circuits
PERRY · Engineering Manual
SHEA · Amplifier Handbook
SKOLNIK · Radar Handbook
SMEATON · Motor Application and Maintenance Handbook
STETKA · NFPA Handbook of the National Electrical Code
TERMAN · Radio Engineers' Handbook
TRUXAL · Control Engineers' Handbook

Handbook of Materials and Processes for Electronics

CHARLES A. HARPER *editor*
Westinghouse Electric Corporation
Baltimore, Maryland

McGRAW-HILL BOOK COMPANY

New York   St. Louis   San Francisco   Düsseldorf   London
Mexico   Panama   Sydney   Toronto

HANDBOOK OF MATERIALS AND PROCESSES
FOR ELECTRONICS

Copyright © 1970 by McGraw-Hill, Inc. All Rights Reserved.
Printed in the United States of America. No part of this
publication may be reproduced, stored in a retrieval system,
or transmitted, in any form or by any means, electronic,
mechanical, photocopying, recording, or otherwise, without
the prior written permission of the publisher.
Library of Congress Catalog Card Number 76-95803

07-026673-5

1234567890 HDBP 7543210

Sponsoring Editor   Daniel N. Fischel
Director of Production   Stephen J. Boldish
Editing Supervisor   Lila M. Gardner
Designer   Naomi Auerbach
Editing and Production Staff   Gretlyn Blau,
    Teresa F. Leaden, George E. Oechsner

Contributors

BEATTY, ROSEMARY *Electronic Research Center, NASA:* CHAPTER 11. THIN FILMS

BRANDS, E. R. *Autonetics Division, North American Rockwell Corporation:* CHAPTER 5. COATINGS FOR ELECTRONICS

BROACHE, EUGENE W. *Aerospace and Electronics Systems Division, Westinghouse Electric Corporation:* CHAPTER 9. NONFERROUS METALS

CERYAN, JOSEPH G. *Republic Rubber Division, Aeroquip Corporation:* CHAPTER 3. ELASTOMERS

DAVIDSON, EDMUND B. *RCA Laboratories, Radio Corporation of America:* CHAPTER 14. PHOTOFABRICATION

EBERLY, WARREN S. *Steel Division, Carpenter Technology Corporation:* CHAPTER 8. FERROUS METAS

GOETZEL, CLAUS G. *Lockheed Research Laboratory, Lockheed Missiles and Space Company:* CHAPTER 15. MATERIALS FOR THE SPACE ENVIRONMENT

HARPER, CHARLES A. *Systems Development Division, Westinghouse Electric Corporation:* CHAPTER 1. PLASTICS FOR ELECTRONICS

HARRISON, DON E. *Research and Development Center, Westinghouse Electric Corporation:* CHAPTER 6. CERAMICS, GLASSES, AND MICAS

HOOK, HARVEY O. *RCA Laboratories, Radio Corporation of America:* CHAPTER 14. PHOTOFABRICATION

LICARI, JAMES J. *Autonetics Division, North American Rockwell Corporation:* CHAPTER 5. COATINGS FOR ELECTRONICS

LINDEN, ALBERT E. *Advanced Microelectronics Division, Optimax, Inc.:* CHAPTER 12. THICK FILMS

McCORMICK, JOHN E. *Rome Air Development Center, Griffiss Air Force Base:* CHAPTER 13. METALS JOINING OF ELECTRONIC CIRCUITRY

MORATIS, CHRISTY J. *Allis-Chalmers Manufacturing Company:* CHAPTER 6. CERAMICS, GLASSES, AND MICAS

PINKERTON, H. L. *Aerospace and Electronic Systems Division, Westinghouse Electric Corporation:* CHAPTER 10. METALLIC AND CHEMICAL FINISHES ON METALS AND NONCONDUCTORS

RITTENHOUSE, JOHN B. *Advanced Technology, Lockheed–California Company:* CHAPTER 15. MATERIALS FOR THE SPACE ENVIRONMENT

RUNYAN, WALTER R. *Components Group, Texas Instruments, Inc.:* CHAPTER 7. SEMICONDUCTOR MATERIALS

RYAN, ROBERT J. *RCA Laboratories, Radio Corporation of America:* CHAPTER 14. PHOTOFABRICATION

SAMPSON, RONALD N. *Research and Development Center, Westinghouse Electric Corporation:* CHAPTER 2. LAMINATES, REINFORCED PLASTICS, AND COMPOSITE STRUCTURES

SCHUH, ARTHUR G. *Orlando Division, Martin Marietta Corporation:* CHAPTER 4. WIRES AND CABLES

SINGLETARY, JOHN B. *Lockheed Missiles and Space Company:* CHAPTER 15. MATERIALS FOR THE SPACE ENVIRONMENT

WATELSKI, STACY B. *Components Group, Texas Instruments, Inc.:* CHAPTER 7. SEMICONDUCTOR MATERIALS

Preface

In recent years, the role of materials and processes has assumed a rapidly increasing degree of importance in the electronic and electrical industries. Necessity has indeed been the mother of invention, with the demands of more and more complex electronic and electrical systems requiring advances in materials and processing technologies. Based on these advances, the need has become acute for extending basic and practical knowledge of modern materials and processes. It is this need for more broadly communicating the vast new realms of knowledge which dictated the need for this *Handbook of Materials and Processes for Electronics*. The extensive data on materials and processes found in this Handbook, coupled with the extensive electromechanical data found in the companion *Handbook of Electronic Packaging*, should provide an invaluable data bank and reference library for the desktop of all persons with any degree of interest in these subjects, in both the electronic and electrical industries, and in those industries which supply components and materials to the electronic and electrical industries.

It might be well to expand briefly on the above statement regarding the applicability of the presentations in this Handbook to both the electronic and electrical industries. In some instances, these two industrial areas are clearly distinct. In others they are not. In any event, however, the progression of materials and processing technologies very often begins in advanced electronics and aerospace systems, and with time, extends to broad usefulness in the commercial and consumer electrical industries. Many current materials and processes used by the electrical industries have their origin in the advanced electronic and aerospace industries. This progression is becoming more rapid.

The successful progression of materials and processing technologies also requires improvements in bridging the knowledge gap between chemical and electrical scientists—a gap which has long been a problem. It is hoped that this *Handbook of Materials and Processes for Electronics* will help bridge this gap. Certainly a key guideline in the preparation of this Handbook has been to present information in a manner which allows an easy understanding of chemical concepts by those who are not chemical specialists, and an easy understanding of electrical concepts by those who are not electrical specialists. Therefore, this Handbook should be helpful to those in the chemical and material industries who have interests in the electronic and electrical industries, as well as to those in the electronic and electrical industries. Again, the practical presentation plan used in this Handbook parallels that of the earlier mentioned companion *Handbook of Electronic Packaging*.

The chapter content of this Handbook illustrates the breadth of coverage which has been developed. An attempt has been made to conveniently group the chapters. The first five chapters cover important organic based materials, such as plastics, laminates, composites, adhesives, elastomers, coatings, and wire and cable systems. Next come presentations on other nonconductors such as ceramics, glasses, and micas, and on increasingly important semiconductors. Following these come chapters on ferrous and nonferrous metals, and metallic and chemical finishes on metals and nonconductors. The last group of chapters concern what might be considered the more advanced subjects of thin films, thick films, metals joining of electronic circuitry, and photofabrication. Finally, one chapter is devoted to the important subject of materials for the space environment. Each chapter might well be considered the equivalent of a book in itself.

In addition to the broad presentation of data and guidelines, several other features of this Handbook should be noted. First, considerable coverage is given to definitions and basic explanations, which will be an invaluable asset to broad, ready understanding of the material presented. Next, tradeoff guidance is a key element. Such guidance will be very useful in making decisions among various possible approaches to given problem areas. Last, every attempt has been made to provide a very complete and thoroughly cross-referenced index, again, for maximum reader convenience.

Length and coverage of a Handbook of this magnitude are necessarily measured compromises. Inevitably, varying degrees of shortages and excesses will exist, depending on the needs of each individual reader. Then too, the time required to complete such a major work as this necessarily demands that some most recent data may not be fully covered. Further, in spite of the tremendous efforts involved, some errors or omissions may exist. While every effort has been made to minimize such shortcomings, it is my greatest desire to improve each successive edition. Toward this end, any and all reader comments will be welcomed and appreciated.

Charles A. Harper

Contents

Contributors v
Preface vii

1. Plastics for Electronics 1-1
2. Laminates, Reinforced Plastics, and Composite Structures . . . 2-1
3. Elastomers 3-1
4. Wires and Cables 4-1
5. Coatings for Electronics 5-1
6. Ceramics, Glasses, and Micas 6-1
7. Semiconductor Materials 7-1
8. Ferrous Metals 8-1
9. Nonferrous Metals 9-1
10. Metallic and Chemical Finishes on Metals and Nonconductors . . 10-1
11. Thin Films 11-1
12. Thick Films 12-1
13. Metals Joining of Electronic Circuitry 13-1
14. Photofabrication 14-1
15. Materials for the Space Environment 15-1

Index follows Chapter 15.

xi

Chapter 1

Plastics for Electronics

CHARLES A. HARPER
Westinghouse Electric Corporation,
Aerospace and Electronic Systems Division,
Baltimore, Maryland

Introduction	1–2
Thermosetting Plastics	1–2
Alkyds	1–12
Aminos	1–15
Diallyl Phthalates	1–18
Epoxies	1–20
Phenolics	1–21
Polyesters	1–22
Silicones	1–26
Thermoplastics	1–28
ABS Plastics	1–29
Acetals	1–40
Acrylics	1–41
Cellulosics	1–43
Chlorinated Polyether	1–45
Ethylene–Vinyl Acetate	1–45
Fluorocarbons	1–45
Ionomers	1–53
Nylons	1–54
Parylenes	1–58
Phenoxies	1–59
Polyallomers, Polyethylenes, and Polypropylenes	1–60
Polyamide-imides and Polyimides	1–61
Polycarbonates	1–66

Polyesters	1–69
Polyphenylene Oxides	1–71
Polystyrenes	1–74
Polysulfones	1–76
Vinyls	1–78
Glass-fiber-reinforced Thermoplastics	1–79
Plastic Processing Methods and Design Guides	1–82
Embedding Materials and Processes	1–82
Embedding Processes	1–88
Processing Characteristics of Embedding Resins	1–91
Embedding Resins	1–93
Fillers for Embedding Resins	1–102
References	1–105

INTRODUCTION

Plastics represent one of the very important classes of materials used in electronics. Plastic materials are broadly used for dielectric or insulation purposes, as well as for structural or ruggedization purposes. The useful application areas are steadily increasing, owing to continuing advances in plastic technology. The upper temperature limit for plastics has risen substantially in recent years, and the number of high-performance engineering plastics has increased rapidly. In addition, chemical development has led to creation of many electronic grades of plastics, that is, plastics with improved or more stable dielectric properties. Many old industry workhorse standards, not basically suitable for high-performance electronics, can now be obtained in high-performance electronic grades. It is the purpose of this chapter to discuss the types of plastics and bulk properties of plastics, as related to electronic and electrical industry applications. Specific forms of these plastics, such as laminates, films, coatings, etc., will be discussed in other chapters. Embedding materials and processes are discussed at the end of this chapter. Nevertheless, much of the bulk information and data presented in this chapter will apply to all forms in which the specific plastic is found. Some properties, of course, are unique to the individual form. Such properties will be discussed in the other chapters dealing with specific plastic material forms.

Although there are numerous possible classifications for plastics, depending on how one wishes to categorize them, nearly all plastics can be placed in one of two major classifications. These two major plastic material classes are thermosetting materials (or thermosets) and thermoplastic materials. These will be discussed in detail, below. Most thermosets tend to be hard or brittle, and thus most commonly have fillers added to them. Thermoplastics are generally softer and more flexible, with actual softness and flexibility varying over a fairly broad range. Although it has generally not been common to add fillers to thermoplastic materials, the use of filled (especially glass fiber) or reinforced thermoplastics has grown dramatically in recent years, owing to the many improvements in strength made possible by such reinforcement. A later section in this chapter will discuss glass-fiber-reinforced thermoplastics in more detail. Likewise, plastic processing will be discussed in a later section of this chapter.

Definitions for some of the terms associated with plastics and dielectric materials are given in Table 1. The significance of important electrical insulation properties is shown in Table 2.

THERMOSETTING PLASTICS

As the name implies, *thermosetting materials* or *thermosets* are cured, set, or hardened into permanent shape. This curing is an irreversible chemical reaction

TABLE 1 Definition of Terms for Plastic Materials[1]

Accelerator. A chemical used to speed up a reaction or cure. For example, cobalt naphthanate is used to accelerate the reaction of certain polyester resins. The term *accelerator* is often used interchangeably with the term *promoter*. An accelerator is often used along with a catalyst, hardener, or curing agent.
Adhesive. Broadly, any substance used in promoting and maintaining a bond between two materials.
Aging. The change in properties of a material with time under specific conditions.
Aliphatic hydrocarbon. See *Hydrocarbon*.
Alkali. A chemical that gives a base reaction. See *Base*.
Ambient temperature. The temperature of the surrounding cooling medium, such as gas or liquid, which comes into contact with the heated parts of the apparatus.
Arc resistance. The time required for an arc to establish a conductive path in a material.
Aromatic hydrocarbon. See *Hydrocarbon*.
B stage. An intermediate stage in the curing of a thermosetting resin. In this stage, a resin can be heated and caused to flow, thereby allowing final curing in the desired shape. The term *A stage* is used to describe an earlier stage in the curing reaction, and the term *C stage* is sometimes used to describe the cured resin. Most molding materials are in the *B* stage when supplied for compression or transfer molding.
Blowing agent. Chemicals that can be added to plastics and that generate inert gases upon heating. This blowing or expansion causes the plastic to expand, thus forming a foam. Also known as *foaming agent*.
Bond strength. The amount of adhesion between bonded surfaces.
Capacitance (capacity). That property of a system of conductors and dielectrics which permits the storage of electricity when potential difference exists between the conductors. Its value is expressed as the ratio of quantity of electricity to a potential difference. A capacitance value is always positive.
Cast. To embed a component or assembly in a liquid resin, using molds that separate from the part for reuse after the resin is cured. See *Embed* and *Pot*.
Catalyst. A chemical that causes or speeds up the cure of a resin, but that does not become a chemical part of the final product. Catalysts are normally added in small quantities. The peroxides used with polyester resins are typical catalysts.
Chlorinated hydrocarbon. An organic compound having hydrogen atoms and more importantly, chlorine atoms in its chemical structure. Trichloroethylene, methyl chloroform, and methylene chloride are chlorinated hydrocarbons.
Coat. To cover with a finishing, protecting, or enclosing layer of any compound (such as varnish).
Coefficient of expansion. The fractional change in dimension of a material for a unit change in temperature.
Cold flow (creep). The continuing dimensional change that follows initial instantaneous deformation in a nonrigid material under static load.
Compound. Some combination of elements in a stable molecular arrangement.
Contact bonding. A type of adhesive (particularly nonvulcanizing natural rubber adhesives) that bonds to itself on contact although solvent evaporation has left it dry to the touch.
Copolymer. See *Polymer*.
Crosslinking. The forming of chemical links between reactive atoms in the molecular chain of a plastic. It is this crosslinking in thermosetting resins that makes them infusible.

TABLE 1 Definition of Terms for Plastic Materials[1] **(Continued)**

Crystalline melting point. The temperature at which crystalline structure in a material is broken down.

Cure. To change the physical properties of a material (usually from a liquid to a solid) by chemical reaction, by the action of heat and catalysts, alone or in combination, with or without pressure.

Curing agent. See *Hardener*.

Curing temperature. The temperature at which a material is subjected to curing.

Curing time. In the molding of thermosetting plastics, the time it takes for the material to be properly cured.

Dielectric constant (permittivity or specific inductive capacity). That property of a dielectric which determines the electrostatic energy stored per unit volume for unit potential gradient.

Dielectric loss. The time rate at which electric energy is transformed into heat in a dielectric when it is subjected to a changing electric field.

Dielectric loss angle (dielectric phase difference). The difference between 90° and the dielectric phase angle.

Dielectric loss factor (dielectric loss index). The product of dielectric constant and the tangent of dielectric loss angle for a material.

Dielectric phase angle. The angular difference in phase between the sinusoidal alternating potential difference applied to a dielectric and the component of the resulting alternating current having the same period as the potential difference.

Dielectric power factor. The cosine of the dielectric phase angle (or sine of the dielectric loss angle).

Dielectric strength. The voltage that an insulating material can withstand before breakdown occurs, usually expressed as a voltage gradient (such as volts per mil).

Dissipation factor (loss tangent, tan δ, approximate power factor). The tangent of the loss angle of the insulating material.

Elastomer. A material which at room temperature stretches under low stress to at least twice its length and snaps back to original length upon release of stress. See *Rubber*.

Electric strength (dielectric strength or disruptive gradient). The maximum potential gradient that a material can withstand without rupture. The value obtained for the electric strength will depend on the thickness of the material and on the method and conditions of test.

Element. A substance composed entirely of atoms of the same atomic number, e.g., aluminum or copper.

Embed. To encase completely a component or assembly in some material—a plastic, for current purposes. See *Cast* and *Pot*.

Encapsulate. To coat a component or assembly in a conformal or thixotropic coating by dipping, brushing, or spraying.

Exotherm. The characteristic curve of a resin during its cure, which shows heat of reaction (temperature) versus time. Peak exotherm is the maximum temperature on this curve.

Exothermic. Chemical reaction in which heat is given off.

Faying surface. The surface of an object that comes in contact with another object to which it is fastened.

Filler. A material, usually inert, that is added to plastics to reduce cost or modify physical properties.

Film adhesive. Thin layer of dried adhesive. · Also describes a class of adhesives provided in dry-film form with or without reinforcing fabric, which are cured by heat and pressure.

TABLE 1 Definition of Terms for Plastic Materials[1] (Continued)

Flexibilizer. A material that is added to rigid plastics to make them resilient or flexible. Flexibilizers can be either inert or a reactive part of the chemical reaction. Also called a *plasticizer* in some cases.

Flexural modulus. The ratio, within the elastic limit, of stress to corresponding strain. It is calculated by drawing a tangent to the steepest initial straight-line portion of the load-deformation curve and calculating by the following equation:

$$E_B = \frac{L^3 m}{4bd^3}$$

where E_B = modulus — L = span in inches
b = width of beam tested
d = depth of beam
m = slope of the tangent

Flexural strength. The strength of a material in bending, expressed as the tensile stress of the outermost fibers of a bent test sample at the instant of failure.

Fluorocarbon. An organic compound having fluorine atoms in its chemical structure. This property usually lends stability to plastics. Teflon* is a fluorocarbon.

Gel. The soft, rubbery mass that is formed as a thermosetting resin goes from a fluid to an infusible solid. This is an intermediate state in a curing reaction, and a stage in which the resin is mechanically very weak. *Gel point* is defined as the point at which gelation begins.

Glass transition point. Temperature at which a material loses its glasslike properties and becomes a semiliquid.

Glue-line thickness. Thickness of the fully dried adhesive layer.

Hardener. A chemical added to a thermosetting resin for the purpose of causing curing or hardening. Amines and acid anhydrides are hardeners for epoxy resins. Such hardeners are a part of the chemical reaction and a part of the chemical composition of the cured resin. The terms *hardener* and *curing agent* are used interchangeably. Note that these can differ from catalysts, promoters, and accelerators.

Heat-distortion point. The temperature at which a standard test bar (ASTM D 648) deflects 0.010 in. under a stated load of either 66 or 264 psi.

Heat sealing. A method of joining plastic films by simultaneous application of heat and pressure to areas in contact. Heat may be supplied conductively or dielectrically.

Hot-melt adhesive. A thermoplastic adhesive compound, usually solid at room temperature, which is heated to a fluid state for application.

Hydrocarbon. An organic compound having hydrogen atoms in its chemical structure. Most organic compounds are hydrocarbons. Aliphatic hydrocarbons are straight-chained hydrocarbons, and aromatic hydrocarbons are ringed structures based on the benzene ring. Methyl alcohol, trichloroethylene, etc., are aliphatic; benzene, xylene, toluene, etc. are aromatic.

Hydrolysis. Chemical decomposition of a substance involving the addition of water.

Hygroscopic. Tending to absorb moisture.

Impregnate. To force resin into every interstice of a part. Cloths are impregnated for laminating, and tightly wound coils are impregnated in liquid resin using air pressure or vacuum as the impregnating force.

Inhibitor. A chemical added to resins to slow down the curing reaction. Inhibitors are normally added to prolong the storage life of thermosetting resins.

---

* Trademark of E. I. du Pont de Nemours & Co., Inc., Wilmington, Del.

TABLE 1 Definition of Terms for Plastic Materials[1] (Continued)

Inorganic chemicals. Chemicals whose chemical structure is based on atoms other than the carbon atom.
Insulation resistance. The ratio of the applied voltage of the total current between two electrodes in contact with a specific insulator.
Micron. A unit of length equal to 10,000 Å, 0.0001 cm, or approximately 0.000039 in.
Modulus of elasticity. The ratio of stress to strain in a material that is elastically deformed.
Moisture resistance. The ability of a material to resist absorbing moisture, either from the air or when immersed in water.
Mold. To form a plastic part by compression, transfer, injection molding, or some other pressure process.
Monomer. See *Polymer*.
NEMA Standards. Property values adopted as standard by the National Electrical Manufacturers Association.
Noble elements. Those elements that either do not oxidize or oxidize with difficulty, e.g., gold and platinum.
Organic. Composed of matter originating in plant or animal life, or composed of chemicals of hydrocarbon origin, either natural or synthetic. Used in referring to chemical structures based on the carbon atom.
Permittivity. Preferred term for dielectric constant.
pH. A measure of the acid or alkaline condition of a solution. A pH of 7 is neutral (distilled water), pH values below 7 are increasingly acid as pH values go toward 0, and pH values above 7 are increasingly akaline as pH values go toward the maximum value of 14.
Plastic. An organic resin or polymer. See *Resin* and *Polymer*.
Plasticizer. Material added to resins to make them softer and more flexible when cured.
Polymer. A high-molecular-weight compound (usually organic) made up of repeated small chemical units. For practical purposes, a polymer is a plastic. The small chemical unit is called a mer, and when the polymer or mer is a crosslink between different chemical units (e.g., styrene-polyester), the polymer is called a copolymer. A monomer is any single chemical from which the mer or polymer or copolymer is formed. Styrene is the monomer in a styrene-polyester copolymer resin. Polymers can be thermosetting or thermoplastic.
Polymerize. To unite chemically two or more monomers or polymers of the same kind to form a molecule with higher molecular weight.
Pot. To embed a component or assembly in a liquid resin, using a shell, can, or case which remains as an integral part of the product after the resin is cured. See *Embed* and *Cast*.
Pot life. The time during which a liquid resin remains workable as a liquid after catalysts, curing agents, promoters, etc., are added; roughly equivalent to gel time. Sometimes also called *working life*.
Power factor. The cosine of the angle between the voltage applied and the current resulting.
Promoter. A chemical, itself a feeble catalyst, that greatly increases the activity of a given catalyst.
Refractive index. The ratio of the velocity of light in a vacuum to its velocity in a substance. Also the ratio of the sine of the angle of incidence to the sine of the angle of refraction.

TABLE 1 Definition of Terms for Plastic Materials[1] (Continued)

Relative humidity. The ratio of the quantity of water vapor present in the air to the quantity which would saturate it at any given temperature.

Resin. High-molecular-weight organic material with no sharp melting point. For current purposes, the terms *resin, polymer,* and *plastic* can be used interchangeably.

Resistivity. The ability of a material to resist passage of electric current either through its bulk or on a surface. The unit of volume resistivity is the ohm-centimeter, and the unit of surface resistivity is the ohm.

Rockwell hardness number. A number derived from the net increase in depth of impression as the load on a penetrator is increased from a fixed minimum load to a higher load and then returned to minimum load. Penetrators include steel balls of several specified diameters and a diamond cone penetrator.

Rubber. An elastomer capable of rapid elastic recovery.

Shore hardness. A procedure for determining the indentation hardness of a material by means of a durometer. Shore designation is given to tests made with a specified durometer.

Solvent. A liquid substance that dissolves other substances.

Storage life. The period of time during which a liquid resin or adhesive can be stored and remain suitable for use. Also called *shelf life.*

Strain. The deformation resulting from a stress, measured by the ratio of the change to the total value of the dimension in which the change occurred.

Stress. The force producing or tending to produce deformation in a body, measured by the force applied per unit area.

Surface resistivity. The resistance of a material between two opposite sides of a unit square of its surface. Surface resistivity may vary widely with the conditions of measurement.

Thermal conductivity. The ability of a material to conduct heat; the physical constant for the quantity of heat that passes through a unit cube of a material in a unit of time when the difference in temperatures of two faces is 1°C.

Thermoplastic. A classification of resin that can be readily softened and resoftened by repeated heating. Hardening is achieved by cooling.

Thermosetting. A classification of resin that cures by chemical reaction when heated and, when cured, cannot be resoftened by heating.

Thixotropic. Describing materials that are gel-like at rest but fluid when agitated.

Vicat softening temperature. A temperature at which a specified needle point will penetrate a material under specified test conditions.

Viscosity. A measure of the resistance of a fluid to flow (usually through a specific orifice).

Volume resistivity (specific insulation resistance). The electrical resistance between opposite faces of a 1-cm cube of insulating material, commonly expressed in ohm-centimters. The recommended test is ASTM D 257-54T.

Vulcanization. A chemical reaction in which the physical properties of an elastomer are changed by causing it to react with sulfur or other crosslinking agents.

Water absorption. The ratio of the weight of water absorbed by a material to the weight of the dry material.

Wetting. Ability to adhere to a surface immediately upon contact.

Working life. The period of time during which a liquid resin or adhesive, after mixing with catalyst, solvent, or other compounding ingredients, remains usable. See *Pot life.*

1-8 Plastics for Electronics

known as *crosslinking*, which usually occurs under heat. For some thermosetting materials, however, curing is initiated or completed at room temperature. Even then, however, it is often the heat of the reaction, or the exotherm, which actually cures the plastic material. This is true, for instance, of a room-temperature-curing epoxy or polyester compound. The crosslinking that occurs in the curing reaction is brought about by the linking of atoms between or across two linear polymers,

TABLE 2 Significance of Important Electrical Insulation Properties[2]

Property and definition	Significance of values
Dielectric strength All insulating materials fail at some level of applied voltage for a given set of operating conditions. The dielectric strength is the voltage that an insulating material can withstand before dielectric breakdown occurs. Dielectric strength is normally expressed in voltage gradient terms, such as volts per mil. In testing for dielectric strength, two methods of applying the voltage (gradual or by steps) are used. Type of voltage, temperature, and any preconditioning of the test part must be noted. Also, thickness of the piece being tested must be recorded because the voltage per mil at which breakdown occurs varies with thickness of test piece. Normally, breakdown occurs at a much higher volt-per-mil value in very thin test pieces (a few mils thick) than in thicker sections (1/8 in. thick, for example).	The higher the value, the better the insulator. Dielectric strength of a material (per mil of thickness) usually increases considerably with decrease in insulation thickness. Materials suppliers can provide curves of dielectric strength versus thickness for their insulating materials.
Resistance and resistivity Resistance of insulating material, like that of a conductor, is the resistance offered by the conducting path to passage of electric current. Resistance is expressed in ohms. Insulating materials are very poor conductors, offering high resistance. For insulating materials, the term "volume resistivity" is more commonly applied. Volume resistivity is the electrical resistance between opposite faces of a unit cube for a given material and at a given temperature. The relationship between resistance and resistivity is expressed by the equation $\rho = RA/l$, where ρ = volume resistivity in ohm-centimeters, A = area of the faces, and l = distance between faces of the piece on which measurement is made. This is not resistance per unit volume, which would be ohms per cubic centimeter although this term is sometimes erroneously used. Other terms are sometimes used to describe a specific application or condition. One such term is surface resistivity, which is the resistance between two opposite edges of a surface film 1-cm square. Since the length and width of the path are the same, the centimeter terms cancel. Thus, units of surface resistivity are actually ohms. However, to avoid confusion with usual resistance values, surface resistivity is normally given in ohms per square. Another broadly used term is insulation resistance which, again, is a measurement of ohmic resistance for a given condition, rather than a standardized resistivity test. For both surface resistivity and insulation resistance, standardized comparative tests are normally used. Such tests can provide data such as effects of humidity on a given insulating material configuration.	The higher the value, the better for a good insulating material. The resistance value for a given material depends on a number of factors. It varies inversely with temperature, and is affected by humidity, moisture content of the test part, level of the applied voltage, and time during which the voltage is applied. When tests are made on a piece that has been subjected to moist or humid conditions, it is important that measurements be made at controlled time intervals during or after the test condition has been applied, since dry-out and resistance increase occur rapidly. Comparing or interpreting data is difficult unless the test period is controlled and defined.

TABLE 2 Significance of Important Electrical Insulation Properties[2] **(Continued)**

Property and definition	Significance of values
Dielectric constant The dielectric constant of an insulating material is the ratio of the capacitance of a capacitor containing that particular material to the capacitance of the same electrode system with air replacing the insulation as the dielectric medium. The dielectric constant is also sometimes defined as the property of an insulation which determines the electrostatic energy stored within the solid material. The dielectric constant of most commercial insulating materials varies from about 2 to 10, air having the value 1.	Low values are best for high-frequency or power applications, to minimize electric power losses. Higher values are best for capacitance applications. For most insulating materials, dielectric constant increases with temperature, especially above a critical temperature region which is unique for each material. Dielectric constant values are also affected (usually to a lesser degree) by frequency. This variation is also unique for each material.
Power factor and dissipation factor Power factor is the ratio of the power dissipated (watts) in an insulating material to the product of the effective voltage and current (volt-ampere input) and is a measure of the relative dielectric loss in the insulation when the system acts as a capacitor. Power factor is nondimensional and is a commonly used measure of insulation quality. It is of particular interest at high levels of frequency and power in such applications as microwave equipment, transformers, and other inductive devices. Dissipation factor is the tangent of the dielectric loss angle. Hence, the term "tan delta" (tangent of the angle) is also sometimes used. For the low values ordinarily encountered in insulation, dissipation factor is practically the equivalent of power factor, and the terms are used interchangeably.	Low values are favorable, indicating a more efficient system, with lower power losses.
Arc resistance Arc resistance is a measure of an electrical breakdown condition along an insulating surface, caused by the formation of a conductive path on the surface. It is a common ASTM measurement, especially used with plastic materials because of the variations among plastics in the extent to which a surface breakdown occurs. Arc resistance is measured as the time, in seconds, required for breakdown along the surface of the material being measured. Surface breakdown (arcing or electrical tracking along the surface) is also affected by surface cleanliness and dryness.	The higher the value, the better. Higher values indicate greater resistance to breakdown along the surface due to arcing or tracking conditions.

TABLE 3 Application Information for Thermosetting Plastics[1]

Material	Major application considerations	Common available forms	Typical suppliers and trade names*
Alkyds	Excellent dielectric strength, arc resistance, and dry insulation resistance. Low dielectric constant and dissipation factor. Good dimensional stability. Easily molded.	Compression moldings, transfer moldings	Allied Chemical Corp. (Plaskon); American Cyanamid Co. (Glaskyd)
Aminos (melamine formaldehyde and urea formaldehyde)	Available in an unlimited range of light-stable colors. Exhibit hard glossy molded surface, and good general electrical properties, especially arc resistance. Excellent chemical resistance to organic solvents and cleaners and household-type cleaners.	Compression moldings, extrusions, transfer moldings, laminates, film	Allied Chemical Corp. (Plaskon); Monsanto Co. (Resimene); American Cyanamid Co. (Cymel for melamine; Beetle for urea)
Diallyl phthalates (DAP) (allylics)	Unsurpassed among thermosets in retention of properties in high humidity environments. Also, have among the highest volume and surface resistivities in thermosets. Low dissipation factor and heat resistance to 400°F or higher. Excellent dimensional stability. Easily molded.	Compression moldings, extrusions, injection moldings, transfer moldings, laminates	FMC Corp. (Dapon); Allied Chemical Corp. (Diall)
Epoxies	Good electrical properties, low shrinkage, excellent dimensional stability and good to excellent adhesion. Extremely easy to compound, using nonpressure processes, for providing a wide variety of end properties. Useful over a wide range of environments. Bisphenol epoxies are most common, but several other varieties are available for providing special properties.	Castings, compression moldings, extrusions, injection moldings, transfer moldings, laminates, matched-die moldings, filament windings, foam	Shell Chemical Co. (Epon); Jones-Dabney Co. (Epi-Rez); Dow Chemical Co. (D.E.R.); Ciba Products Co. (Araldite); Union Carbide Corp. (ERL); 3M Co. (Scotchcast)

Phenolics	Among the lowest-cost, most widely used thermoset materials. Excellent thermal stability to over 300°F generally, and over 400°F in special formulations. Can be compounded to a broad choice of resins, fillers, and other additives.	Castings, compression moldings, extrusions, injection moldings, transfer moldings, laminates, matched-die moldings, stock shapes, foam	Union Carbide Corp. (Bakelite); Hooker Chemical Corp. (Durez)
Polyesters	Excellent electrical properties and low cost. Extremely easy to compound using nonpressure processes. Like epoxies, can be formulated for either room temperature or elevated temperature use. Not equivalent to epoxies in environmental resistance.	Compression moldings, extrusions, injection moldings, transfer moldings, laminates, matched-die moldings, filament windings, stock shapes	Pittsburgh Plate Glass Co. (Selectron); American Cyanamid Co. (Laminac); Rohm & Haas Co. (Paraplex)
Silicones (rigid)	Excellent electrical properties, especially low dielectric constant and dissipation factor, which change little up to 400°F and over. Nonrigid silicones are covered in elastomers and embedding material sections.	Castings, compression moldings, transfer moldings, laminates	Dow Corning Corp. (DC Resins)

* This listing is only a very small sampling of the many possible excellent suppliers. It is intended only to orient the nonchemical reader into plastic categories. No preferences are implied or intended.

with this crosslinking of atoms making a three-dimensional rigidized chemical structure. A simplified reaction for producing a thermosetting plastic is shown in Fig. 1. Although the cured part can be softened by heat, it cannot be remelted or restored to the flowable state which existed before the plastic resin was cured.

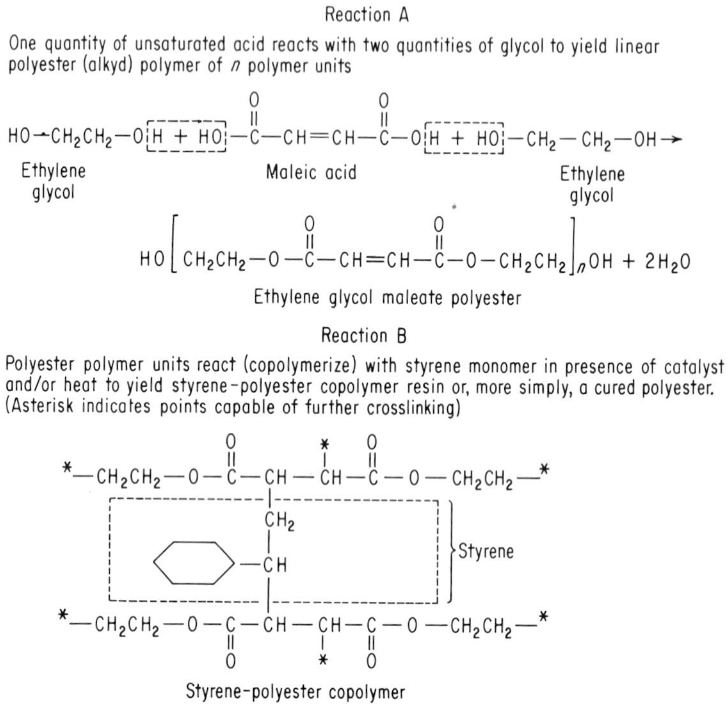

Reaction A

One quantity of unsaturated acid reacts with two quantities of glycol to yield linear polyester (alkyd) polymer of n polymer units

$$HO-CH_2CH_2-O[H + HO]-\overset{O}{\underset{\|}{C}}-CH=CH-\overset{O}{\underset{\|}{C}}-O[H + HO]-CH_2-CH_2-OH \rightarrow$$

Ethylene glycol   Maleic acid   Ethylene glycol

$$HO\left[CH_2CH_2-O-\overset{O}{\underset{\|}{C}}-CH=CH-\overset{O}{\underset{\|}{C}}-O-CH_2CH_2\right]_n OH + 2H_2O$$

Ethylene glycol maleate polyester

Reaction B

Polyester polymer units react (copolymerize) with styrene monomer in presence of catalyst and/or heat to yield styrene-polyester copolymer resin or, more simply, a cured polyester. (Asterisk indicates points capable of further crosslinking)

Styrene-polyester copolymer

Fig. 1 Illustration of a simple chemical reaction which produces a crosslinked thermosetting resin (styrene-polyester copolymer resin).

Design data and information on the various thermosets will be discussed in this section. The major thermosetting plastics, along with the primary design and application considerations for each, and typical commonly used names for these materials are shown in Table 3. More detailed data for each type of thermosetting material is given below.

Alkyds

Alkyds are widely used for molded electrical parts where the general application considerations given in Table 3 apply. They are chemically somewhat similar to polyester resins. Normally, alkyds are solid molding compounds, and polyesters are liquid or paste-form resins. Alkyd molding compounds are commonly available in putty, granular, glass-fiber-reinforced, and rope form. The properties of these forms are shown in Table 4.

Alkyds are easy to mold, and economical to use. Molding dimensional tolerances can be held to within ±0.001 in./in. Postmolding shrinkage is small, especially for glass-filled compounds. Alkyds have long been used in the electrical industry, and development has led to high-performance electronic-grade compounds. Com-

TABLE 4 Typical Properties of Alkyd Molding Compounds[3]

	ASTM test method	Putty	Granular	Glass-fiber reinforced	Rope
Electrical					
Arc resistance, sec	D495-58T	180+	180+	180+	180+
Dielectric constant	D150-54T				
1 Mc		5.4–5.9	5.7–6.3	5.2–6.0	7.4
1 Mc		4.5–4.7	4.8–5.1	4.5–5.0	6.8
Dissipation factor	D150-54T				
60 cycles		0.030–0.045	0.030–0.040	0.02–0.03	0.019
1 Mc		0.016–0.022	0.017–0.020	0.015–0.022	0.023
Dielectric strength, volts/mil	D149-59				
Short-time		350–400	350–400	350–400	360
Step by step		300–350	300–350	300–350	290
Physical					
Specific gravity	D792-50	2.05–2.15	2.21–2.24	2.02–2.10	2.20
Water absorption, 24 hr at 23°C, %	D570-57T	0.10–0.15	0.08–0.12	0.07–0.10	0.05
Heat resistance max., °F					
Long periods (continuous)		250	300	300	300
Short periods (0–24 hr)		300	350	350	350
Short periods (0–1 hr)		325	375	400	400
Heat distortion temp., 264 psi, °F	D648-56	350–400	350–400	>400	>400
Coefficient of linear thermal expansion/°F	D696-44	10–30×10^{-6}	10–30×10^{-6}	10–30×10^{-6}	20×10^{-6}
Thermal conductivity,(g)(cal)/(sec)(cm²)(°C/cm)		15–25×10^{-4}	15–25×10^{-4}	8–12×10^{-3}	10×10^{-4}
Flammability	D635-56T	Nonburning	Self-extinguishing	Nonburning	Self-extinguishing
Mechanical					
Impact strength, Izod, ft-lb/in. of notch	D256-56	0.25–0.35	0.30–0.35	8–12	2.2
Comprehensive strenth, psi	D695-54	20,000–25,000	16,000–20,000	24,000–30,000	28,800
Flexural strength, psi	D790-59T	8,000–11,000	7,000–10,000	12,000–17,000	19,500
Tensile strength, psi	D651-48	4,000–5,000	3,000–4,000	5,000–9,000	7,100
Modulus of elasticity, psi	D790-49T	2.0–2.7×10^6	2.4–2.9×10^6	2.0–2.5×10^6	1.9×10^6
Barcol hardness		60–70	60–70	70–80	72
		MIL-M-14F Type MAG	MIL-M-14F Type MAG	MIL-M-14F Type MAI-60	

parisons of dielectric constant and dissipation factor for general-purpose and electronic-grade alkyd compounds are shown in Fig. 2.

The greatest limitation of alkyds is in extremes of temperature and in extremes of humidity. Silicones and diallyl phthalates are superior here, silicones especially

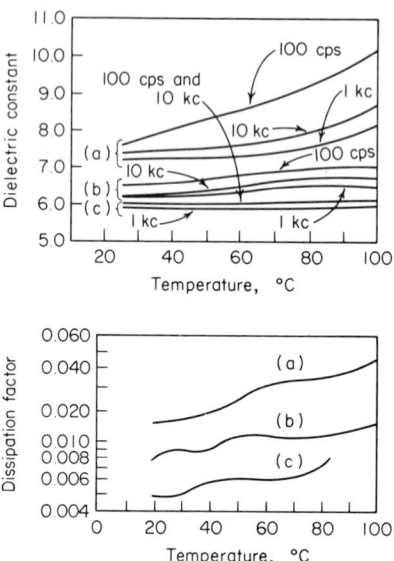

Fig. 2 Dielectric constant and dissipation factor comparisons for (*a*) general-purpose alkyd, (*b*) general-purpose electronic-grade alkyd, and (*c*) high-performance electronic-grade alkyd.[4]

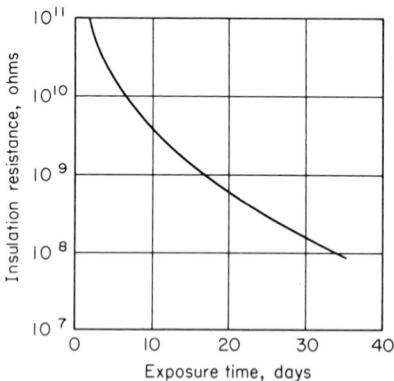

Fig. 3 Effect of 95 percent humidity at 60°C on the 500-volt dc insulation resistance of an alkyd molding compound.[5]

with respect to temperature and diallyl phthalates especially with respect to humidity. The electrical insulation resistance of alkyds decreases considerably in high, continuous-humidity conditions, as shown in Fig. 3. Strength properties, flexural modulus versus temperature, and room-temperature tensile creep at 3,000 psi for alkyds are shown in Fig. 4, as compared with epoxy, phenolic, and diallyl phthalate (DAP) molding materials.

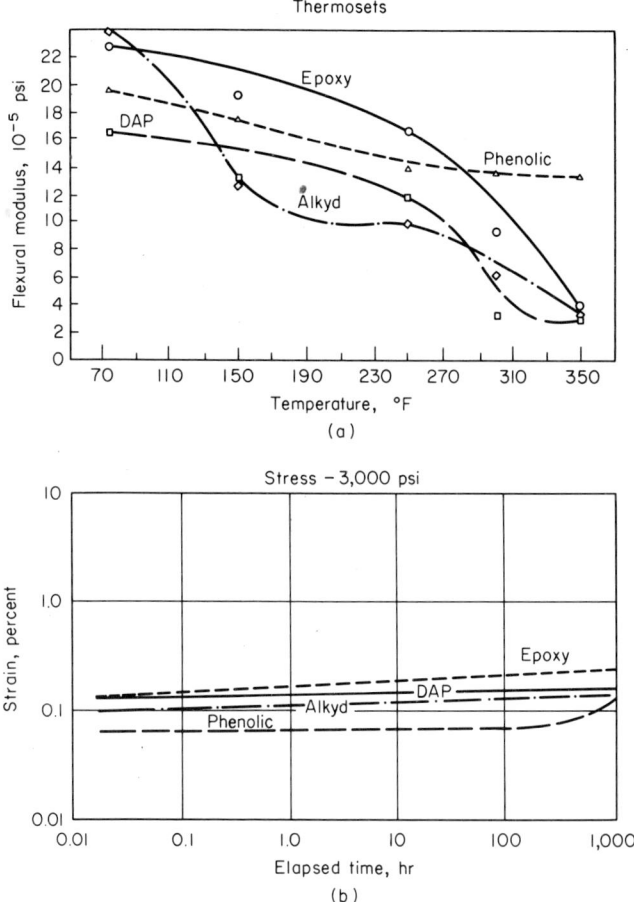

Fig. 4 Flexural modulus versus temperature (*a*) and tensile creep at 73°F and 3,000 psi (*b*) for four thermosetting materials.[6]

Aminos

Amino molding compounds can be fabricated by economical molding methods. They are hard, rigid, and abrasion-resistant, and have high resistance to deformation under load. These materials can be exposed to subzero temperatures without embrittlement. Under tropical conditions, the melamines do not support fungus growth.

Amino materials are self-extinguishing and have excellent electrical insulation characteristics. They are unaffected by common organic solvents, greases and oils, and weak acids and alkalies. Melamines are superior to ureas in resistance to acids, alkalies, heat, and boiling water, and are preferred for applications involving cycling between wet and dry conditions or rough handling. Aminos do not impart taste or odor to foods.

Addition of α-cellulose filler produces an unlimited range of light-stable colors and high degrees of translucency. Colors are obtained without sacrifice of basic material properties.

Melamines and ureas provide excellent heat insulation; temperatures up to the destruction point will not cause parts to lose their shape.

Amino resins exhibit relatively high mold shrinkage, and also shrink on aging.

TABLE 5 Typical Physical Properties of Amino Molding Compounds[7]

	Urea		Melamine					
	Alpha cellulose	Wood flour	Alpha cellulose	Wood flour	Alpha cellulose, modified	Rag	Asbestos	Glass fiber
Specific gravity	1.5	1.5	1.5	1.42	1.43	1.5	1.78	1.94–2.0
Density, g/in.³	24.6	24.6	24.6	23.8	23.5	24.6	29.2	31.8–32.8
Hardness, Rockwell E	94–97	95	110	94	....	100	90	....
Shrinkage,* in./in.								
Molding	0.006–0.009	0.006–0.014	0.008–0.009	0.007–0.008	0.006–0.008	0.003–0.004	0.005–0.007	0.002–0.004
Postmold	0.006–0.012	0.006–0.012	0.009–0.011	0.004–0.007	0.001–0.002	0.004–0.008	0.002–0.003	0.002–0.005
Deflection temperature, °F, at 264 psi	266	270	361	266	266	310	266	400
Heat resistance, continuous, °F	170†	170	210†	250	250	250	300	300
Coefficient of thermal expansion, per °C × 10⁻⁶	22–36	30	20–57	32–50	34–36	25–30	21–43	12–25
Thermal conductivity, (cal)(cm)/(sec²)(cm)(°C) × 10⁻⁴	10.1	10.1	10.1	8.4	....	10.6	13.1	....
Water absorption, 24 hr, at 23°C, percent	0.4–0.8	0.7	0.3–0.5	0.34–0.6	0.3–0.6	0.3–0.6	0.13–0.15	0.09–0.3
Color possibilities	Unlimited	Brown, black	Unlimited	Brown	Brown	Limited	Brown	Natural, gray

* Test specimen: 4-in. diam. × ⅛-in. disk.
† Based on no color change.

TABLE 6 Typical Mechanical and Electrical Properties of Amino Molding Compounds[7]

	Urea		Melamine					
	Alpha cellulose	Wood flour	Alpha cellulose	Wood flour	Alpha cellulose modified	Rag	Asbestos	Glass fiber
Mechanical:								
Tensile strength, 1,000 psi	5.5–7	5.5–10	7–8	5.7–6.5	5.5–6.5	8–10	5.5–6.5	5.9
Compressive strength, 1,000 psi	30–38	25–35	40–45	30–35	24.5–26	30–35	25–30	20–29
Flexural strength, 1,000 psi	11–18	8–16	12–15	6.5–9	11.5–12	12–15	7.4–10	13.2–24
Shear strength, 1,000 psi	11–12		11–12	10–10.5	11.4–12.2	12–14	7–8	13.0–15.6
Impact strength, Izod, ft-lb/in. of notch	0.24–0.28	0.25–0.35	0.30–0.35	0.25–0.38	0.30–0.42	0.55–0.90	0.30–0.40	0.5–6.0
Tensile modulus, 10^6 psi	1.3–1.4		1.35	1.0	1.0	1.4	1.95	
Flexural modulus, 10^6 psi	1.4–1.5	1.3–1.6	1.1	1.0	1.1	1.4	1.8	2.4
Electrical:								
Arc resistance, sec.	80–100	80–100	125–136	70–106	90–120	122–128	120–180	180–186
Dielectric strength, volts/mil								
Short time At 23°C	330–370	300–400	270–300	350–370	350–390	250–340	410–430	170–370
At 100°C	200–270		170–210	290–330	140–190	110–130	280–310	90–350*
Step by step At 23°C	220–250	250–300	240–270	200–240	200–250	220–240	280–300	170–270
At 100°C	110–150		90–130	190–210	90–100	60–90	190–210	60–250*
Slow rate of rise At 23°C	250–260		210–240	240–260	280–290	210–240	270–290	170–210
At 100°C	120–170		90–120	170–200	90	70–80	170–190	70–90
Dielectric constant At 60 cps	7.7–7.9	7.0–9.5	7.9–8.2	6.4–6.6	7.0–7.7	8.1–12.6	10.0–10.2	7.0–11.1
At 10^6 cps	6.7–6.9	6.4–6.9	7.6–8.0	5.6–5.8	5.2–6.0	6.7–6.9	5.3–6.1	6.6–7.9
At 3×10^9 cps							4.9	5.5
Dissipation factor At 60 cps	0.034–0.043	0.035–0.040	0.052–0.083	0.026–0.033	0.192	0.100–0.340	0.100	0.14–0.23
At 10^6 cps	0.029–0.031	0.028–0.032	0.026–0.030	0.034–0.035	0.044–0.12	0.036–0.041	0.039–0.048	0.013–0.016
At 3×10^9 cps							0.032	0.040
Dielectric loss factor At 60 cps	0.28–0.34	0.24–0.38	0.44–0.78	0.17–0.22	0.90–2.4	2.0–5.0	0.5–1.0	1.5–2.5
At 10^6 cps	0.19–0.21	0.18–0.22	0.20–0.33	0.20–0.21	0.19–0.28	0.24–0.26	0.21–0.31	0.09–0.19
Volume resistivity, ohm-cm	$0.5–5.0 \times 10^{11}$		$0.8–2.0 \times 10^{12}$	$6–10 \times 10^{12}$	6×10^{10}	$1.0–3.0 \times 10^{11}$	1.2×10^{12}	$0.9–20 \times 10^{11}$
Surface resistivity, ohms	$0.4–3.0 \times 10^{11}$		$0.8–4.0 \times 10^{11}$	$0.3–5.0 \times 10^{12}$	1.7×10^{12}	$0.7–7.0 \times 10^{11}$	1.9×10^{13}	$3.0–4.6 \times 10^{12}$
Insulation resistance, ohms	$0.2–5.0 \times 10^{11}$		$1.0–4.0 \times 10^{10}$	$1.0–3.0 \times 10^{11}$	$2.0–5.0 \times 10^{9}$	$0.1–3.0 \times 10^{10}$	$1.0–4.0 \times 10^{10}$	$0.2–6.0 \times 10^{10}$

* At 50°C.

1-18   Plastics for Electronics

Cracks develop in urea moldings subjected to severe cycling between dry and wet conditions.

Prolonged exposure to high temperature affects the color of both urea and melamine products.

A loss of certain strength characteristics also occurs when amino moldings are subjected to prolonged elevated temperatures. Some electrical characteristics are also adversely affected: arc resistance of some industrial types, however, remains unaffected after exposure at 500°F.

Ureas are unsuitable for outdoor exposure. Melamines experience little degradation in electrical or physical properties after outdoor exposure, but color changes may occur.

Typical physical properties of amino plastics are shown in Table 5, and typical mechanical and electrical properties in Table 6.

Diallyl Phthalates

Diallyl phthalates, part of the polymer class known as allylics, are among the best of the thermosetting plastics with respect to high insulation resistance and low electrical losses. Excellent properties are maintained up to 400°F or higher, and in the presence of high humidity environments. Also, diallyl phthalate resins are easily molded and fabricated.

There are several chemical variations of diallyl phthalate resins, but the two most commonly used are diallyl phthalate (DAP) and diallyl isophthalate (DAIP). The primary application difference is that DAIP will withstand somewhat higher temperatures than will DAP. Typical properties of DAP and DAIP molding compounds, using various fillers, are shown in Table 7. The retention of electrical

TABLE 7   Typical Properties of Several Diallyl Phthalate Molding Compounds with Various Fillers

Property	Orlon	Dacron	Long glass	Asbestos	Short glass	Short glass*
Tensile strength, psi	6,000	5,000	10,000	5,500	7,000	7,000
Compressive strength, psi	25,000	25,000	25,000	25,000	25,000	28,000
Flexural strength, psi	10,000	11,500	16,000	9,600	12,000	12,000
Flexural modulus, psi $\times 10^{-6}$	0.71	0.64	1.3	1.2	1.2	1.3
Impact strength, Izod, ft-lb/in. of notch	1.2	4.5	6.0	0.4	0.6	0.5
Hardness, Rockwell M	108	108	100	100	105	110
Specific gravity at 25°C	1.31–1.45	1.39–1.62	1.55–1.70	1.55–1.65	1.6–1.8	1.65–1.75
Dielectric constant:						
At 1 kHz	3.7–4.0	3.79	4.2		4.4	4.1
At 1 MHz	3.3–3.6	3.4	4.2	4.5–6.0	4.4	3.4
Dissipation factor:						
At 1 kHz	0.020–0.025	0.008	0.004–0.006	0.04–0.08	0.006	0.004
At 1 MHz	0.015–0.020	0.012	0.008	0.04–0.08	0.008	0.008
Mold shrinkage, in./in.	0.009	0.010	0.002	0.006	0.003	0.003
Postmold shrinkage, in./in.	0.001	0.0006	0.0007	0.001	0.0007	0.0002
Heat-deflection temperature, °F	265	290	392	325	400	500+
Heat resistance, continuous, °F	300–500	350–400	350–400	350–400	350–400	450

* Based on diallyl isophthalate.

resistance properties after humidity conditioning is compared in Fig. 5 for several glass-filled DAP, DAIP, epoxy, and phenolic compounds, and one mineral and glass-filled epoxy.

The excellent dimensional stability of diallyl phthalates has been mentioned above.

This is demonstrated in Fig. 6, which compares diallyl phthalates to other plastic materials at various temperatures.

Likewise, the excellent electrical properties of diallyl phthalates have been stressed. The effect of frequency and temperature on dielectric constant is shown

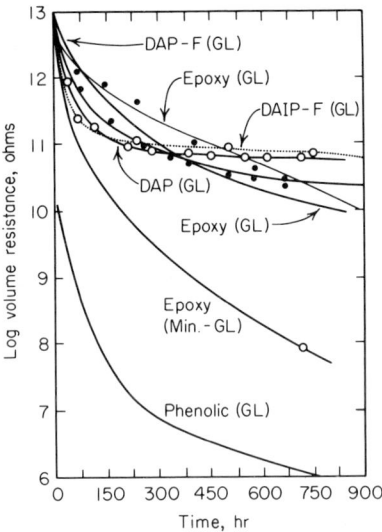

Fig. 5 Volume resistance decrease of several thermosetting materials at 70°C and 100 percent RH.[9] (GL indicates glass filler, and Min-GL indicates mineral and glass filler.)

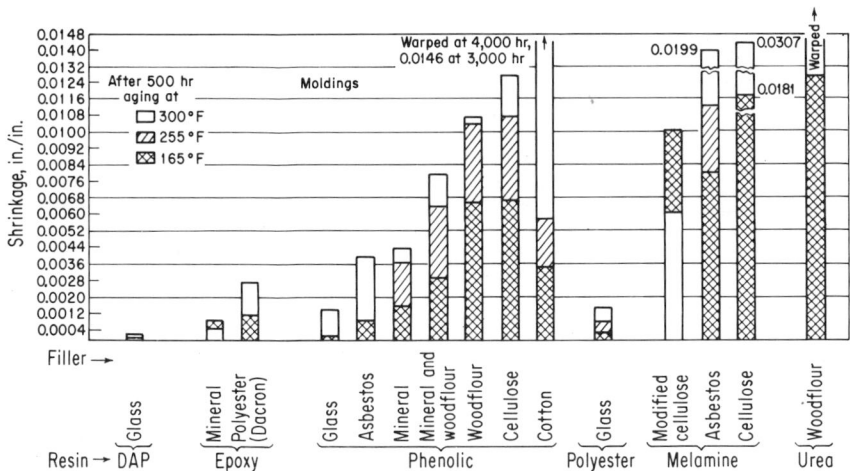

Fig. 6 Shrinkage of various thermosetting molding materials as a result of heat aging.[10]

in Fig. 7, and the effect of frequency and temperature on dissipation factor in Fig. 8. The loss characteristics of DAP, epoxy, and phenolic compounds as a function of temperature are presented in Fig. 9. Diallyl isophthalate, DAIP, is also especially good in retention of dielectric strength, as indicated in Fig. 10.

Epoxies

Epoxies are among the most versatile and most widely used plastics in the electronic packaging field—primarily because of the wide variety of formulations possible, and the ease with which these formulations

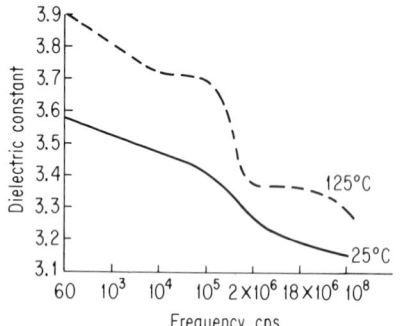

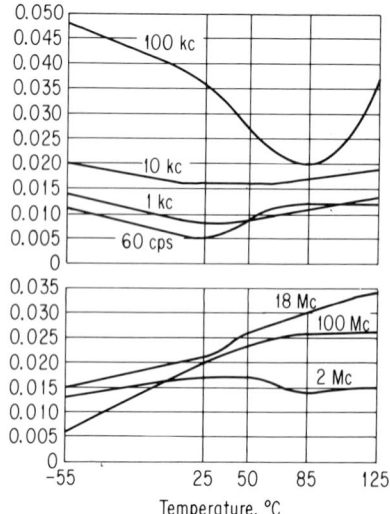

Fig. 7 Effect of frequency and temperature on the dielectric constant of unfilled diallyl phthalate.[11]

Fig. 8 Effect of frequency and temperature on the dissipation factor of unfilled diallyl phthalate.[11]

can be made and utilized with minimal equipment requirements. Formulations range from flexible to rigid in the cured state, and from thin liquids to thick pastes in the uncured state. Conversion from uncured to cured state is accomplished by use of hardeners and/or heat. The largest applications of epoxies in electronic packaging are in embedding applications (potting, casting, encapsulating, and impregnating), and in laminated constructions such as metal-clad laminates for printed circuits and unclad laminates for various types of insulating and terminal boards. Hence, the detailed properties of these embedding compounds and laminates are given in those parts of this book that describe laminates and embedding materials.

Basically, epoxies are available as liquid or solid resins, and as powdered molding compounds. The molding compounds, while broadly used for embedment of electronic assemblies by the transfer molding technique, are also employed for transfer and compression molding of many other types of electrical parts. Typical properties of several epoxy molding compounds are shown in Table 8. Physical and electrical properties are generally very stable in humid environments, although resistance properties in humidity are dependent on the filler used, to some extent, as shown in Figs. 11 and 12. Also, although the general electrical properties of epoxies are good, they are not as good as those of diallyl phthalates as a function of either temperature or humidity. This is shown in Figs. 5 and 9.

In addition to their versatility and good electrical properties, epoxies are also outstanding in their low shrinkage, their dimensional stability, and their adhesive properties. Their shrinkage is often less than 1 percent, and the as-molded dimensions of an epoxy part change little with time or environmental conditions, other than excessive heat. Their excellent performance in this respect is shown in Fig. 6. Because of the low shrinkage and good strength properties of epoxies, cured epoxy parts resist cracking, both upon curing and in thermal shock, better than most other rigid thermosetting materials. Flexural modulus and tensile creep properties are shown in Fig. 4. Based on the excellent bonds obtained with epoxy resins to most substrates, epoxy formulations are broadly used as adhesives. Adhesives are discussed separately in Chap. 2. Even when not specifically used as adhesives, the bonding properties of epoxies often provide a better seal around inserts, terminals, and other interfaces than do most other plastic materials.

Phenolics

Phenolics are among the oldest, best-known general-purpose molding materials. They are also among the lowest in cost and the easiest to mold. There are an extremely large number of phenolic materials available, based on the many resin and filler combinations, and they can be classified in many ways. One common way of classifying them is by type of application or grade. Typical properties for some of these common classifications are shown in Table 9.

Although it is possible to get various grades of phenolics for various applications, as shown in Table 9, phenolics, generally speaking, are not equivalent to diallyl phthalates and epoxies in resistance to humidity, shrinkage, dimensional stability, and retention of electrical properties in extreme environments. Critical electrical property comparisons are shown in Figs. 5 and 9. Phenolics are, however, quite adequate for a large percentage of electrical applications. Furthermore improved grades have been developed, which yield considerable improvements in humid environments (Fig. 13) and at higher temperatures (Fig. 14). In addition, the glass-filled, heat-resistant grades are outstanding in thermal stability up to 400°F and higher, with some being useful

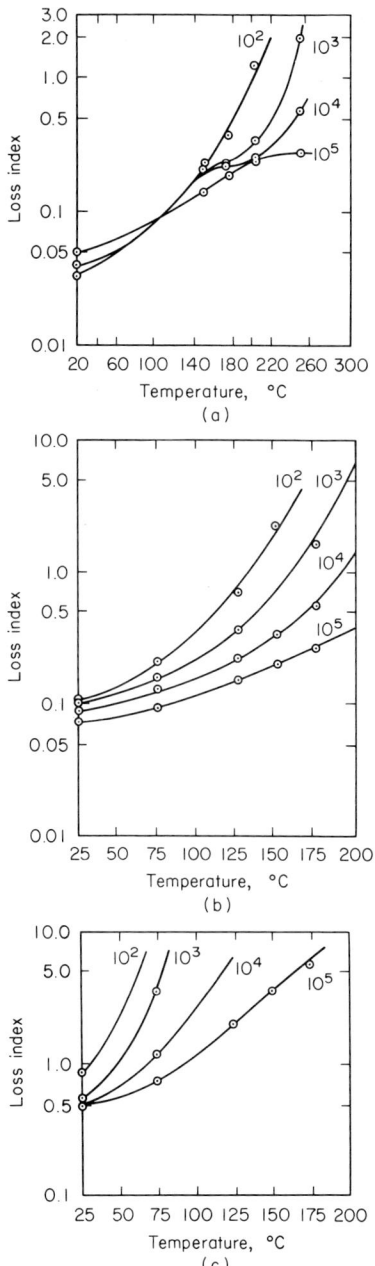

Fig. 9 Loss index versus temperature and frequency for (a) glass-filled diallyl phthalate (DAP) compound, (b) mineral and glass-filled epoxy compounds, and (c) glass-filled phenolic compound.[9]

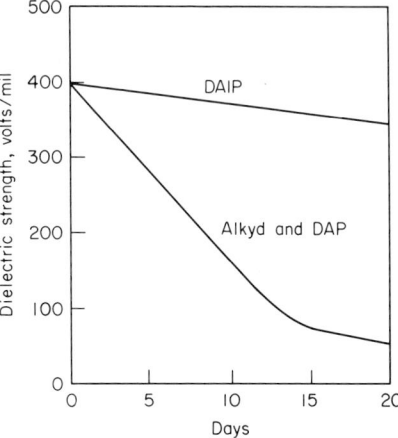

Fig. 10 Effect of heat aging at 400°F on the dielectric strength of diallyl phthalate (DAP), diallyl isophthalate (DAIP), and alkyd molding materials.[12]

TABLE 8 Typical Properties of Epoxy Molding Compounds with Various Fillers

	Glass-fiber filler	Mineral filler
Tensile strength, psi	14,000–30,000	5,000–7,000
Elongation, percent	4	
Tensile modulus, 10^5 psi	30.4	
Compressive strength, psi	25,000–30,000	18,000–25,000
Flexural strength, psi	20,000–26,000	10,000–15,000
Impact strength, Izod, ft-lb/in. of notch	8–15	0.25–0.45
Hardness, Rockwell	M100–M108	M101
Specific gravity	1.8–2	1.6–2.06
Thermal conductivity, $(cal)(cm)/(sec)(cm^2)(°C)$	$7-10 \times 10^{-4}$	$7-18 \times 10^{-4}$
Specific heat, per °C	0.19	
Coefficient of thermal expansion, per °C $\times 10^{-5}$	1.1–3	2.4–5
Heat resistance, continuous, °F	330–500	300–500
Heat-distortion temp., °F	400–500	250–450
Volume resistivity, ohm-cm	3.8×10^{15}	9×10^{15}
Dielectric strength, ⅛-in. (volts/mil)		
Short time	360	330–400
Step by step	340	350
Dielectric constant at 60, 10^3, and 10^6 cps	4–5	4–5
Arc resistance, sec	125–140	150–180
Burning rate	Self-extinguishing	Self-extinguishing
Water absorption, percent*	0.05–0.095	0.1

* Test performed on ⅛-in.-thick piece immersed in distilled water for 24 hr at room temperature.

up to 500°F. Phenolics are relatively stable, physically, as shown in Fig. 4. Shrinkage in heat aging varies over a fairly wide range, depending on filler used. Glass-filled phenolics are the more stable, as shown in Fig. 6.

Polyesters

Polyesters are versatile resins, which handle much like the epoxies. They are available in forms ranging from low-viscosity liquids to thick pastes or putties. The

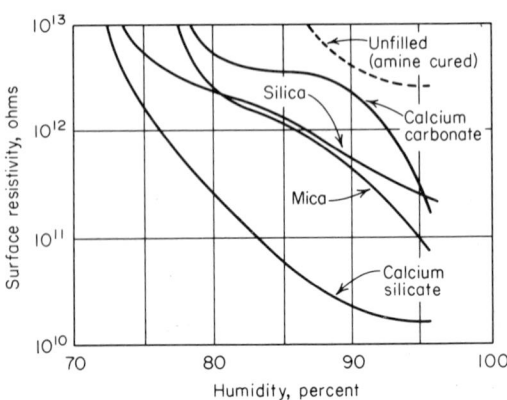

Fig. 11 Effect of humidity on surface resistivity of filled and unfilled epoxy resins at 35°C.[13]

liquids are used for embedding applications and laminated products, much like the epoxies, and the pastes are used for molding applications. Polyester properties and embedding applications are discussed later in the section specifically devoted to embedding. Although both epoxies and polyesters are available in formulations for room-temperature cure and formulations for heat cure, the chemical curing mechanism, or polymerization mechanism, is different for the two types of resins. Likewise, of course, the basic resins are chemically different. It is their physical forms and application forms which make them similar. A simple polyester reaction is shown in Fig. 1.

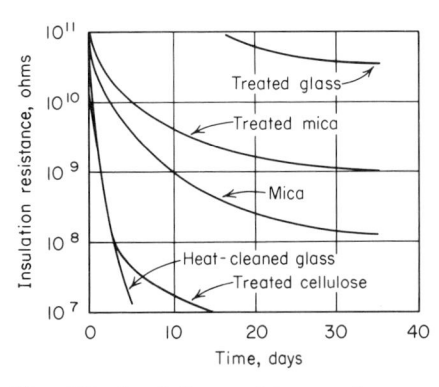

Fig. 12 Insulation resistance of epoxy compounds with various fillers, at 140°F and 95 percent RH. Treatment indicated is a chromic chloride sizing.[5]

The major advantages of polyesters over epoxies are lower cost and appreciably lower electrical losses, especially for the best electrical-grade polyesters. Some important disadvantages of polyesters, as compared to epoxies, are lower adhesion to most substrates, higher polymerization shrinkage, a greater tendency to crack during cure or in thermal shock, and

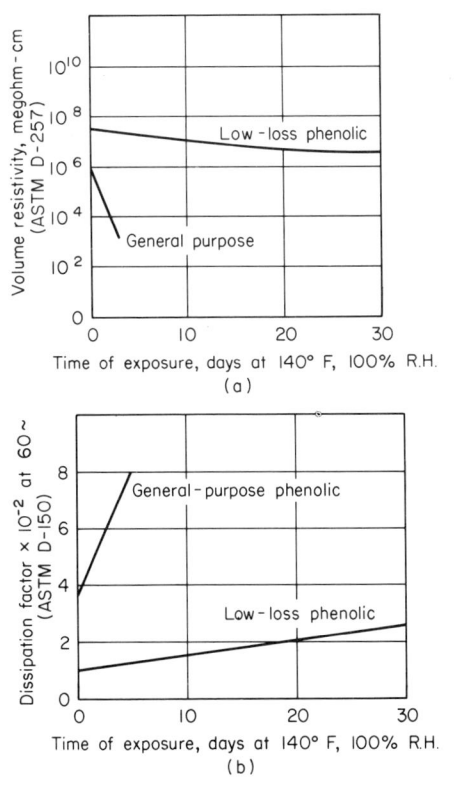

Fig. 13 Comparison of humidity effects on (a) volume resistivity and (b) dissipation factor for a general-purpose and a low-loss phenolic compound.[15]

1-24 Plastics for Electronics

TABLE 9 Typical Properties of Several Types

ASTM test method	Property	Phenolics General-purpose	Phenolics Shock-resistant (cellulose fiber)
D792	Specific gravity	1.33–1.45	1.33–1.43
	Density, lb/in.^{3}	0.048–0.052	0.048–0.051
D955	Mold shrinkage, in./in.	0.004–0.009	0.004–0.009
D635	Flammability	←	
C177	Thermal conductivity, (cal)(sec)(cm^2)/(°C)(cm $\times$ 10^{-4})	4–7	4–7
D570	Water absorption, 24 hr, ⅛-in. thickness, 73°F, percent increase in weight	0.3–0.7	0.3–0.7
D696	Coefficient of thermal expansion, 10^{-5} in./(in.)(°C)	3.0–4.5	3.0–4.5
D648	Heat distortion, temperature, °F	260–340	260–340
	Maximum recommended intermittent service, temperature, °F	300	300
	Maximum recommended continuous service, temperature, continuous loading, °F	250	250
D638	Tensile strength, psi (average)	6,500–10,000	6,500–10,000
D638	Tensile modulus, psi, at 2 in./min	8–12 $\times$ 10^5	8–13 $\times$ 10^5
D790	Flexural strength, psi	8,000–12,000	8,000–12,000
D790	Flexural modulus, psi	8–12 $\times$ 10^5	8–13 $\times$ 10^5
D695	Compressive strength, psi	22,000–35,000	22,000–36,000
D732	Shear yield strength	9,600	....
D638	Elongation, percent (ultimate) at 2 in./min	0.4–0.8	0.4–0.8
D256	Impact strength, Izod, 73°F (notched), ft-lb/in. of notch	0.24–0.33	0.34–8.0
D785	Hardness (Rockwell M scale)	95–120	95–120
D257	Volume resistivity, ohm-cm (50% RH and 23°C)	10^9–10^{13}	10^9–10^{13}
D150	Power factor 24 hr in H$_2$O, 23°C conditioning		
	60 cycles	0.05–0.30	0.05–0.30
	10^6 cycles	0.03–0.07	0.03–0.07
D150	Dielectric constant		
	60 cycles	5.0–12.0	5.0–12.0
	10^6 cycles	4.0–6.0	4.0–6.0
D149	Dielectric strength, step by step, ⅛-in. thickness, volts/mil at 23°C	200–400	200–400
D495	Arc resistance, sec	20–50	20–50
	UV resistance	←	
D543	Resistance to acids	←——————— Resistant to weak
D543	Resistance to bases	←——————— Resistant to weak
D543	Resistance to solvents	←	
	Relative material costs (cents/in.^{3})	1.01–1.06	1.22–1.38

of Phenolic Molding Compounds[14]

		Phenolics		
Heat-resistant (mineral-filled)	Glass-filled	Low-loss	Chemical-resistant	Cast phenolics
1.50–1.95	1.7–1.83	1.50–1.95	1.24–1.50	1.3
0.054–0.070	0.061–0.066	0.054–0.070	0.045–0.054	0.047
0.001–0.004	0.001–0.004	0.001–0.005	0.005–0.009	0.005
	← Self-extinguishing		→	
8–22	2–4	10–14	4–7	4
0.1–0.5	0.05–0.5	0.01–0.05	0.01–0.4	0.02
1.5–4.0	1.0	1.9–2.6	3.0–4.5	5
300–400	375–600	230–350	300–350	95–212
400	450	350	300	
350	400	300	250	
5,500–7,500	7,000–11,000	6,000–7,000	4,000–7,000	3,000–10,500
10–20×10^5	3×10^6	30–50×10^5	1–15×10^5	
8,000–12,000	16,000–22,000	8,000–12,000	7,000–9,000	2,000–15,000
10–20×10^5	3×10^6	30–40×10^5	7–15×10^5	
20,000–35,000	16,500–32,000	25,000–30,000	25,000–33,000	4,000–33,000
	15,000			
0.18–0.50	<1	0.13–0.5	0.2–0.4	
0.27–3.5	0.6–15.0	0.30–0.38	0.20–0.30	0.25–0.6
95–120	90–120	95–120	95–120	20–120
10^9–10^{13}	10^{12}–10^{14}	10^{12}–10^{14}	10^{11}–10^{14}	10^9
0.1–0.3	0.03–0.07	0.03–0.05	0.10–0.20	
0.4–0.8	0.010–0.10	0.005–0.20	0.03–0.06	
7.5–50.0	4.5–6.8	4.5–6.0	6.0–13.0	4.0
5.0–10.0	4.2–6.8	4.2–5.2	4.5–6.0	5.0
200–350	300–400	350–450	225–300	400
30–70	70–180	30–70	20–50	
	← Darkens →		Slightly attacked by strong acids	Good to weak; decomposes with strong acids
acids, attacked by strong acids →				
alkalies, attacked by strong alkalies →			Slightly attacked by strong bases	Poor
Resistant at room temperature →				None to poor
1.21–1.93	2.11	......	1.80	3.5

Plastics for Electronics

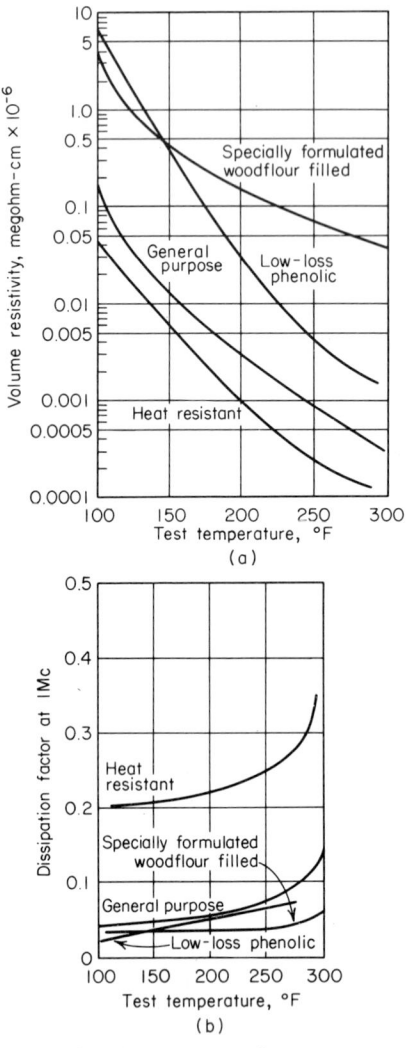

Fig. 14 Comparison of temperature effects on (a) volume resistivity and (b) dissipation factor for several grades of phenolic compounds.[15]

greater change of electrical properties in a humid environment.

Silicones

Silicones are thermosetting polymers which can be classified as elastomers when the cured product is rubberlike, or as embedding materials when the basic plastic form is a castable liquid (either rubberlike or rigid), or as low-pressure transfer molding compounds, when used in the moldable, powdered form. Furthermore, silicone polymers are widely used in electronic and electrical applications in other forms such as laminated products, coatings, and adhesives. Silicone elastomers, laminates, coatings, adhesives, and embedding materials will be discussed later in other chapters that cover those material categories. It is the silicone molding materials that will be discussed at this point.

Silicone molding compounds are useful for many of the molded products required in electronic and electrical applications. These molding compounds are normally either mineral-filled or glass-fiber-filled, and the properties of these two classes of molding compounds are shown in Table 10.

The most important properties of silicones, for electronic applications, are excellent electrical properties which do not change drastically with temperature or frequency over the safe operating temperature range of silicones. This is clearly demonstrated in Fig. 15. Furthermore, silicones are among the best of all polymer materials in resistance to temperature. This is shown in Fig. 16, which compares the relative thermal stability of a group of plastic materials. Useful temperatures of 500 to 700°F are available for silicone materials. Hence, silicones are broadly used for high-temperature electronic applications, especially those applications requiring low electrical losses. Additional data in this respect, for other material forms, are presented in other chapters of this book.

Mechanical properties of silicone molding compounds are affected by temperature aging, as shown in Fig. 17. Although silicones are not as strong as diallyl phthalates, epoxies, and phenolics upon aging at lower temperatures, Fig. 17 shows them to be stronger upon aging above 400°F. Strength at temperature will of course be lower in most cases. Generally, most silicone properties are stable in extreme environments such as humidity and vacuum. Thus, silicones are widely used for both military and space applications. Weight loss at 10^{-6} torr, at several temperatures, is shown for silicone and epoxy materials in Fig. 18. Additional data on stability of silicones in a vacuum environment are given in other chapters of this book.

TABLE 10 Typical Properties of Silicone Molding Compounds[16]

ASTM test method	Property	Mineral filled	Glass-fiber filled
	Specific gravity	1.80–1.95	1.88
	Density, lb/in.^{3}	0.065–0.101	0.068
	Mold shrinkage, in./in.	0.007	0.005
	Flammability	Self-extinguishing	Self-extinguishing
	Thermal conductivity at 500°F, Btu/(hr)(ft^2)(°F)(ft)	2.0–2.1	1.2
D 570	Water absorption in 24 hr, ⅛-in. thickness, at 73°F, % increase in weight	0.05–0.20	0.10–0.12
D 696	Coefficient of thermal expansion, in./(in.)(°C) × 10^5, at 25 to 250°C		
	Perpendicular	2.5–6.0	Not isotropic 1.0
	Parallel	....	12.1
D 648	Heat-distortion temperature, °F	340–900	>900
	Maximum recommended intermittent-service temperature, °F	500–600	700–750
	Maximum recommended continuous-service temperature (continuous loading), °F	500–750	700–750
D 638	Tensile yield strength (avg.) at 73°F, psi × 10^{-3}	2.5–3.5	3.5–6.5
D 790	Flexural strength at 77°F, psi × 10^{-3}	6.5–8.5	18–21
D 790	Flexural modulus at 77°F, psi × 10^{-6}	1.4–1.6	2.5
D 695	Compressive strength at 77°F, psi × 10^{-3}	11–18	10–12.5
D 256	Impact strength, Izod, at 73°F (notched), ft-lb/in. notch	0.30–0.40	10–24
D 785	Hardness, Rockwell M, at 73°F	70–95	83–88
D 257	Volume resistivity at 50% RH and 23°C, ohm-cm	2 × 10^{13}–5 × 10^{15}	Dry, 1.4 × 10^{14} 9 × 10^{-14}
D 150	Power factor at 1 MHz:		
	Dry	0.002	0.004
	After 24 hr immersion	0.003	0.005
D 150	Dielectric constant:		
	At 60 Hz	3.4–6.3	4.35
	At 1 MHz	3.4–6.3	4.26–4.28
D 149	Dielectric strength, 500 volts/sec rate of rise, ⅛-in. thickness, volts/mil	300–465	275–300
D 495	Arc resistance (tungsten electrodes), sec	190–420	230–240
	UV resistance	Excellent	Excellent
	Resistance to acids	To mild acids, excellent	To mild acids, excellent
	Resistance to bases	To mild bases, fair to good	To mild bases, fair to good
	Resistance to solvents	Fair to excellent	Fair to excellent
	Relative material costs, cents/in.^{3}	26–41	24

One other area of specific usefulness is the molding of semiconductor devices. This is particularly true when the device is expected to function or be tested under humid conditions. Ions that dissolve in water or may be transported by moisture vapors may irreversibly alter the characteristics of the device.

One method of determining the presence of ionic constituents in molding compounds is by the *water-extract conductance test*.[19] This procedure involves extracting weighed samples of ground plastic with deionized water for 288 hr at 71°C and measuring the electrical resistivity of the extract solution at 23°C.

Water-extract resistivity data obtained using typical electronic encapsulating grades of silicone, epoxy, and phenolic molding compounds are presented in Table 11. It

1-28 Plastics for Electronics

can be seen that silicones excel in this test. Selection of proper curing agents and control of impurities can yield optimum-quality epoxy formulations in this respect. Also diallyl phthalates offer high water-extract resistivities, since, as with silicones, relatively nonpolar catalyst systems can be used.

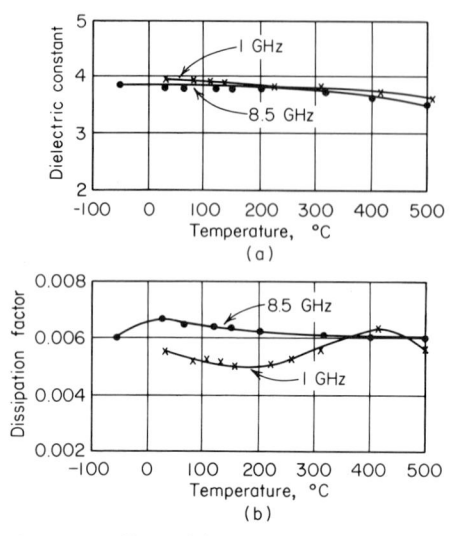

Fig. 15 Effect of heat aging at 300°C on (a) dielectric constant and (b) dissipation factor of a silicone molding compound.[17]

Fig. 16 Thermal stability of a group of plastic materials.[18]

THERMOPLASTICS

Thermoplastics differ from thermosets in that thermoplastics do not cure or set under heat as do thermosets. Thermoplastics merely soften, when heated, to a flowable state whereby under pressure they can be forced or transferred from a heated cavity into a cool mold. Upon cooling in a mold, thermoplastics harden and take the shape of the mold. Since thermoplastics do not cure or set, they can be remelted and rehardened, by cooling, many times. Thermal aging, brought about by repeated exposure to the high temperatures required for melting, causes eventual degradation of the material and so limits the number of reheat cycles. However, thermoplastics will melt to a flowable state upon heating, whereas thermosets do not melt. Essentially all thermoplastics are processed by heating to a soft state and applying pressure, such as injection molding, extruding, and thermoforming. Both thermosets and thermoplastics soften upon heating, to varying degrees. However, fully cured thermosets are not usually processed in this fashion. Rather, they are cured and chemically crosslinked, as mentioned earlier in this chapter. Thermoplastics are straight-chain polymers, rather than infusible, crosslinked polymers.

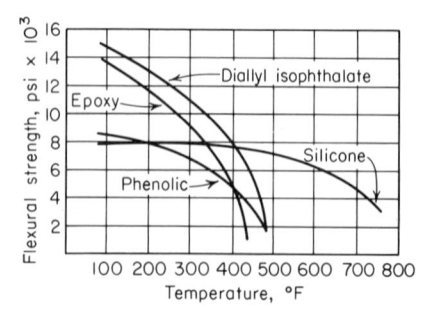

Fig. 17 Effect of temperature on flexural strength of several thermosetting compounds.[17] (All samples were aged 250 hr at temperature and tested at room temperature.)

Design data and information on the

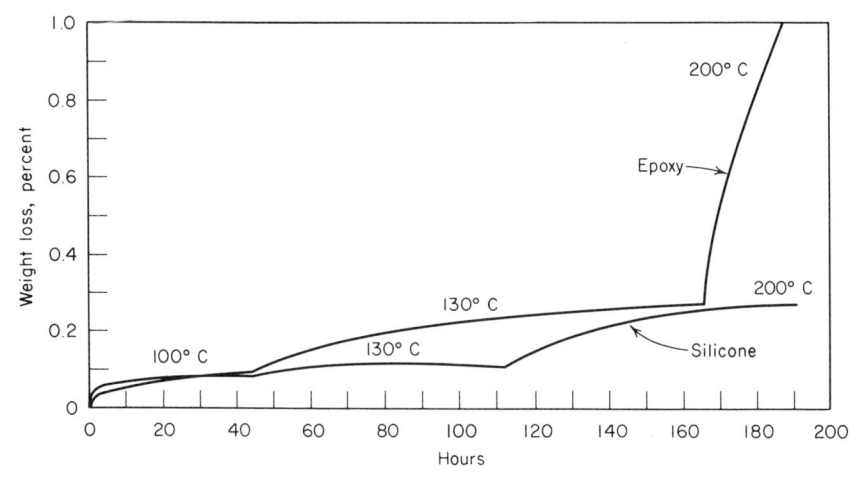

Fig. 18 Weight loss upon vacuum aging at 10^{-6} torr, at temperatures shown, for silicone and epoxy materials.[17]

various thermoplastics will be discussed at this point. The major thermoplastics, along with the primary design and application considerations for each, and typical commonly used names for these materials are shown in Table 12. Basic physical and mechanical data on these thermoplastic materials are presented in Table 13, and basic electrical data in Table 14. A more detailed discussion on each is given below. In addition, further strength data are given in a subsequent portion of this chapter which covers glass-fiber-reinforced thermoplastics. This is of special importance because of the continuous improvements and increasing usage of reinforced thermoplastic materials.

ABS Plastics

ABS plastics are derived from *a*crylonitrile, *b*utadiene, and *s*tyrene. This class possesses hardness and rigidity without brittleness, at moderate costs. ABS materials have a good balance of tensile strength, impact resistance, surface hardness, rigidity, heat resistance, low-temperature properties, and electrical characteristics.

Mechanical properties The most outstanding mechanical properties of ABS plastics are impact resistance and toughness. A wide variety of modifications are available for improved impact resistance, toughness, and heat resistance. Impact resistance does not fall off rapidly at lower temperatures. Stability under load

TABLE 11 Water-extract Resistivities for Silicone, Epoxy, and Phenolic Compounds[17]

Molding compound	Resistivity, ohm-cm	
	Received powder	Molded and postcured 2 hr at 200°C
Silicone............	245,000	245,000
Epoxy.............	4,180	43,000
Phenolic...........	4,580	8,370

TABLE 12 Application Information for Thermoplastics[1]

Material	Major application considerations	Common available forms	Typical suppliers and trade names [a]
ABS (acrylonitrile-butadiene-styrene)	Extremely tough, with high impact resistance. Can be formulated over a wide range of hardness and toughness properties. Special grades available for plated surfaces with excellent pull-strength values. Good general electrical properties but not outstanding for any specific electric applications.	Blow moldings, extrusions, injection moldings, thermoformed parts, laminates, stock shapes, foam	Borg-Warner Corp. (Marbon Cycolac); Monsanto Co. (Lustran); Goodrich Chemical Co. (Abson)
Acetals	Outstanding mechanical strength, stiffness, and toughness properties, combined with excellent dimensional stability. Good electrical properties at most frequencies, which are little changed in humid environments up to 125°C.	Blow moldings, extrusions, injection moldings, stock shapes	Du Pont, Inc. (Delrin); Celanese Corp. (Celcon)
Acrylics (polymethyl-methacrylate)	Outstanding properties are crystal clarity, and resistance to outdoor weathering. Excellent resistance to arcing and electrical tracking.	Blow moldings, castings, extrusions, injection moldings, thermoformed parts, stock shapes, film, fiber	Du Pont, Inc. (Lucite); Rohm and Haas Co. (Plexiglas)
Cellulosics	There are several materials in the cellulosic family, such as cellulose acetate (CA), cellulose propionate (CAP), cellulose acetate butyrate (CAB), ethyl cellulose (EC), and cellulose nitrate (CN). Widely used plastics in general, but not outstanding for electronic applications.	Blow moldings, extrusions, injection moldings, thermoformed parts, film, fiber, stock shapes	Eastman Chemical Co. (Tenite); Dow Chemical Co. (Ethocel-EC); Celanese Corp. (Forticel-CAP)
Chlorinated polyethers	Good electrically, but most outstanding properties are corrosion resistance and good physical and thermal stability by thermoplastic standards.	Extrusions, injection moldings, stock shapes, film	Hercules Powder Co. (Penton)
Ethylene-vinyl acetates (EVA)	Excellent flexibility, toughness, clarity, and stress-crack resistance. Somewhat like a tough synthetic rubber or elastomer. Not widely used in electronics. Comparatively low resistance to heat and solvents.	....	U.S. Industrial Co. (Ultrathene); Du Pont, Inc. (Alathon); Union Carbide Corp. (Bakelite EVA)

Fluorocarbons			
a. Chlorotrifluoroethylene (CTFE)	Excellent electrical properties and relatively good mechanical properties. Somewhat more stiff than TFE and FEP fluorocarbons, but does have some cold flow. Widely used in electronics, but not quite so widely as TFE and FEP. Useful to about 400°F.	Extrusions, isostatic moldings, injection moldings, film, stock shapes	3M Co. (Kel-F); Allied Chemical Corp. (Plaskon CTFE)
b. Fluorinated ethylene propylene (FEP)	Very similar properties to those of TFE, except useful temperature limited to about 400°F. Easier to mold than TFE.	Extrusions, injection moldings, laminates, film	Du Pont, Inc. (Teflon FEP)
c. Polytetrafluoroethylene (TFE)	Electrically one of the most outstanding thermoplastic materials. Exhibits very low electrical losses, and very high electrical resistivity. Useful to over 500°F and to below −300°F. Excellent high-frequency dielectric. Among the best combinations of mechanical and electrical properties but relatively weak in cold-flow properties. Nearly inert chemically, as are most fluorocarbons. Very low coefficient of friction. Nonflammable.	Compression moldings, stock shapes, film	Du Pont, Inc. (Teflon TFE); Allied Chemical Corp. (Halon TFE)
d. Polyvinyl fluoride	Mostly used as a weatherable, architectural facing sheet. Not widely used in electronics.	Extrusions, injection moldings, laminates, film	Du Pont, Inc. (Tedlar)
e. Polyvinylidine fluoride (PVF$_2$)	One of the easiest of the fluorocarbons to process. Stiffer and more resistant to cold flow than TFE. Good electrically. Useful to about 300°F. A major electronic application is wire jacketing.	Extrusions, injection moldings, laminates, film	Pennsalt Chemicals Corp. (Kynar)
Ionomers	Excellent combination of toughness, solvent resistance, transparency, colorability, abrasion resistance, and adhesion. Based on ethylene-acrylic copolymers with ionic bonds. Not widely used in electronics.	Film, coatings, injection moldings	Du Pont, Inc. (Surlyn A); Union Carbide Corp. (Bakelite)
Nylons (polyamides)	Good general purpose for electrical and nonelectrical applications. Easily processed. Good mechanical strength, abrasion resistance, and low coefficient of friction. There are numerous types of nylons; nylon 6, nylon 6/6, and nylon 6/10 are most common. Some nylons have limited use due to moisture-absorption properties. Nylon 6/10 is best here.	Blow moldings, extrusions, injection moldings, rotational moldings, stock shapes, film, fiber	Du Pont, Inc. (Zytel); Allied Chemical Corp. (Plaskon); Union Carbide Corp. (Bakelite)
Parylenes (Polyparaxylylene)	Excellent dielectric properties and good dimensional stability. Low permeability to gases and moisture. Produced as a film on a substrate, from a vapor phase. Such vapor-phase polymerization is unique in polymer processing. Used primarily as thin films in capacitors and dielectric coatings. Numerous polymer modifications exist.	Film coatings	Union Carbide Corp. (Parylene)

TABLE 12 Application Information for Thermoplastics[1] (Continued)

Material	Major application considerations	Common available forms	Typical suppliers and trade names[a]
Phenoxies	Tough, rigid, high-impact plastic. Has low mold shrinkage, good dimensional stability, and very low coefficient of expansion for a thermoplastic. Useful for electronic applications below about 175°F. Useful in adhesive formulations.	Blow moldings, extrusions, injection moldings, film	Union Carbide Corp. (Bakelite Phenoxy)
Polyallomers	Thermoplastic polymers produced from two monomers. Somewhat similar to polyethylene and polypropylene, but with better dimensional stability, stress-crack resistance, and surface hardness than high-density polyethylene. Electronic application areas similar to polyethylene and polypropylene. One of the lightest commercially available plastics.	Blow moldings, extrusions, injection moldings, film	Eastman Chemical Products, Inc. (Tenite)
Polyamide-imides and polyimides	Among the highest-temperature thermoplastics available, having useful operating temperatures between about 400°F and about 700°F or higher. Excellent electrical properties, good rigidity, and excellent thermal stability. Low coefficient of friction. Polyamide-imides and polyimides are chemically similar but not identical in all properties. They are difficult to process, but are available in molded and block forms, and also as films and resin solutions.	Films, coatings, molded and/or machined parts, resin solutions	Du Pont, Inc. (Vespel fabricated blocks, Kapton film, and Pyre-M.L. Resin)
Polycarbonates	Excellent dimensional stability, low water absorption, low creep, and outstanding impact-resistance thermoplastics. Good electrical properties for general electronic packaging application. Available in transparent grades.	Blow moldings, extrusions, injection moldings, thermoformed parts, stock shapes, film	Mobay Chemical Co. (Merlon); General Electric Co. (Lexan)
Polyethylenes and polypropylenes (polyolefins or polyalkenes)	Excellent electrical properties, especially low electrical losses. Tough and chemically resistant, but weak to varying degrees in creep and thermal resistance. There are three density grades of polyethylene: low (0.910–0.925), medium (0.926–0.940), and high (0.941–0.965). Thermal stability generally increases with density class. Polypropylenes are generally similar to polyethylenes, but offer about 50°F higher heat resistance.	Blow moldings, extrusions, injection molding, thermoformed parts, stock shapes, film, fiber, foam	Du Pont, Inc. (Alathon Polyethylene); U.S.I. Chemical Co. (Petrothene Polyethylene); Allied Chemical Corp. (Grex H.D. Polyethylene); Hercules Powder Co. (Hi-Fax

1-32

			H.D. Polyethylene): Hercules Powder Co. (Pro-Fax Polypropylene); Eastman Chemical Co. (Tenite Polyethylene and Polypropylene)
Polyethylene terephthalates	Among the toughest of plastic films with outstanding dielectric strength properties. Excellent fatigue and tear strength and resistance to acids, greases, oils, solvents. Good humidity resistance. Stable to 135–150°C.	Film, sheet, fiber	Du Pont, Inc. (Mylar)
Polyphenylene oxides (PPO)	Excellent electrical properties, especially loss properties to above 350°F, and over a wide frequency range. Good mechanical strength and toughness. A lower-cost grade (Noryl) exists, having somewhat similar properties to PPO, but with a 75–100°F reduction in heat resistance.	Extrusions, injection moldings, thermoformed parts, stock shapes, film	General Electric Co. (PPO and Noryl)
Polystyrenes	Excellent electrical properties, especially loss properties. Conventional polystyrene is temperature-limited, but high-temperature modifications exist, such as Rexolite or Polypenco crosslinked polystyrene, which are widely used in electronics, especially for high-frequency applications. Polystyrenes are also generally superior to fluorocarbons in resistance to most types of radiation.	Blow moldings, extrusions, injection moldings, rotational moldings, thermoformed parts, foam	Dow Chemical Co. (Styron); Monsanto Co. (Lustrex); American Enka Corp. (Rexolite); Polymer Corp. (Polypenco Q-200.5)
Polysulfones	Excellent electrical properties and mechanical properties to over 300°F. Good dimensional stability and high creep resistance. Flame-resistant and chemical-resistant. Outstanding in retention of properties upon prolonged heat aging, as compared to other tough thermoplastics.	Blow moldings, extrusions, injection mold thermoformed parts, stock shapes, film sheet	Union Carbide Corp. (Polysulfone)
Vinyls	Good low-cost, general-purpose thermoplastic materials, but not specifically outstanding electrical properties. Greatly influenced by plasticizers. Many variations available, including flexible and rigid types. Flexible vinyls, especially polyvinyl chloride (PVC), widely used for wire insulation and jacketing.	Blow moldings, extrusions, injection moldings, rotational moldings, film sheet	Diamond Alkali Co. (Diamond PVC); Goodyear Chemical Co. (Pliovic); Dow Chemical Co. (Saran)

* This listing is only a very small sampling of the many possible excellent suppliers. The listing is intended only to orient the nonchemical reader into plastic categories. No preferences are implied or intended.

TABLE 13 Typical Physical and Mechanical Properties of Thermoplastics[19]

Resin material	Coefficient thermal expansion, (in./in) (°C × 10^{-5})	Thermal conductivity, (cal)/ (cm²)(sec) (°C)(cm) × 10^{-4}	Water absorption, 24 hr, %	Rockwell hardness	Flammability, (in./min) 0.125 in.	Specific gravity	Mold shrinkage, in./in.	Clarity	Price range per lb
Acetal	0.25	1.6	0.25	M94, R120	1.1	1.410–1.425	0.022	Translucent to opaque	$0.65
ABS	3–10.5	4–9	0.2–0.5	R80–120	1.0–2	1.01–1.07	0.003–0.007	Opaque	$0.33–0.43
Acrylic	....	1.4	0.3	M84–97	9–1.2	1.18–1.19	0.002–0.006	Transparent	$0.455–0.75
Acrylic high impact	6.5–10.5	4.0	0.2–0.3	M20–67	1.1–1.2	1.11–1.18	0.004–0.008	Translucent to opaque	$0.525–0.70
Cellulose acetate	8–18	4–8	1.7–4.4	R7–122	0–2	1.22–1.34	0.001–0.008	Transparent	$0.40
Cellulose acetate butyrate	11–17	4–8	0.9–2.2	R17–113	0.5–1.5	1.15–1.22	0.003–0.006	Transparent	$0.62
Cellulose propionate	11–16	4–8	1.2–2.8	R15–120	0.5–1.5	1.16–1.23	0.001–0.006	Transparent	$0.62
Chlorinated polyether	8	3.13	0.01	R100	Self-extinguishing	1.4	0.004–0.006	Semitranslucent to opaque	$2.50
Ethyl vinyl acetate	10–20	8	<0.01	R3–7	Slow burning	0.93–0.95	0.01–0.02	Transparent	$0.2775–0.3575
Chlorotrifluoroethylene	5–7	4–6	Nil	R85–112	Nil	2.09–2.14	0.010–0.015	Transparent to opaque	$4.70
Fluorinated ethylene propylene	8.3–10.5	5.9	<0.05	D55	Nonflammable	2.16	0.03–0.05	Transparent to opaque	$5.60–9.60
Polytetrafluoroethylene	5.5 (25–60°C)	6	0.01	D60–65	Nonflammable	2.13–2.18	0.02–0.06	Transparent to opaque	$3.25
Nylon 6	4.6–5.8	5.9	1.5	R107–119	Self-extinguishing	1.13–1.14	0.007–0.011	Transparent to opaque	$0.86–1.19
Nylon 6/6	8.1	5.8	1.3	R118–123	Self-extinguishing	1.13–1.15	0.007–0.015	Translucent to opaque	$0.84–0.875
Nylon 6/10	10	5.5	0.4	R111	Self-extinguishing	1.07–1.09	0.015	....	$1.26
Polyallomer	8–11	2–4	<0.05	R50–85	Slow burning	0.90–0.906	0.01–0.02	Transparent to opaque	$0.28
Polycarbonate	6.7–7	4.6	0.15	M70, R112	Self-extinguishing	1.2	0.005–0.007	Transparent	$0.90–3.05
Polyethylene, low-density	10–20	8	<0.05	R10	Slow burning	0.910–0.925	0.01–0.03	Transparent to opaque	$0.1525–0.29
Polyethylene, medium-density	10–20	8	<0.05	R15	Slow burning	0.926–0.940	0.01–0.035	Transparent to opaque	$0.17–0.235

1-34

Material								
Polyethylene, high-density	10–20	1.9–3.3	<0.01	R30–60	Slow burning	0.941–0.965	0.01–0.04	Translucent to opaque	$0.18–0.32
Polyethylene, high molecular weight	13	8	<0.01	R55	Slow burning	0.93–0.94	0.03	Translucent to opaque	$0.26–0.50
Polyimide	.....	.....	0.32	R85–95	.....	1.43	.....	Opaque	$0.19–0.55
Polypropylene	3.8–9	2.8–4	<0.01	R45–99	Slow burning to nonburning	0.90–1.24	0.008–0.025	Transparent to opaque	$0.145–0.245
Polystyrene	6–8	8	0.03–0.05	M65–80	0.5–2.5	1.05–1.06	0.002–0.006	Transparent	
Polystyrene, high-impact	6.5–8.5	1–3	0.05–0.10	M25–69	.5–2.5	1.04–1.06	0.003–0.005	Translucent to opaque	$0.16–0.27
Polyurethane	10–20	7.4	0.60–0.80	M26, R90	Slow to self-extinguishing	1.11–1.26	0.009	Translucent to opaque	$1.19–1.60
Polyvinyl chloride (flexible)	7–25	3–4	0.15–0.75	.....	Self-extinguishing	1.15–1.80	0.002–0.004	Translucent to opaque	$0.16–0.455
Polyvinyl chloride (rigid)	5–10	3–5	0.07–0.40	R100–120	Self-extinguishing	1.33–1.58	.....	Transparent to opaque	$0.21–0.42
Polyvinyl dichloride (rigid)	7–8	3–4	0.07–0.11	R118	Self-extinguishing	1.50–1.54	0.006–0.007	Translucent to opaque	$0.50–0.53
Styrene acrylonitrile (SAN)	7	3	0.23–0.28	M30–83	0.4–0.7	1.07–1.08	0.003–0.004	Transparent	$0.26–0.30
Ionomer	12–13	5.8	0.1–1.4	D60–65	0.9–1.1	0.94–0.96	0.001–0.005	Transparent	$0.47–0.49
Phenoxy	3.2–3.8	.....	0.13	R113–118	Slow burning, self-extinguishing	1.17–1.34	0.003–0.004	Transparent to opaque	$0.75–1.00
Polyphenylene oxide	5.2	.....	0.06	R120	Self-extinguishing	1.06	0.006–0.008	Transparent to opaque	$1.15
Polysulfone	3.1–10⁻⁵ in./(in.)(°F)	1.8 Btu/(hr) (ft)(°F)(in.)	0.22	M69, R120	Self-extinguishing	1.24–1.25	0.0076	Transparent to opaque	$1.00–1.25

TABLE 13 Typical Physical and Mechanical Properties of Thermoplastics[19] (Continued)

Resin material	Impact strength notched Izod, ft-lb/in., ½" bar	Tensile strength, psi × 10³	Tensile modulus, psi × 10³	Elongation, %	Flexural strength, psi × 10³	Compressive strength, psi × 10³	Compressive modulus, psi × 10³	Heat distortion temperature, °F, 264 psi	Heat resistance, continuous, °F
Acetal	1.1–1.4	8.8–10	400–410	12–75	13–14	18	410	230–255	185
ABS	1.3–10.0	4.5–8.5	200–450	5–200	5–13.5	5–11	120–200	180–245	160–235
Acrylic	0.3–0.4	8.7–11.0	350–450	3–6	14–17	14–17	350–430	167–198	130–195
Acrylic high impact	0.5–2.3	5.5–8	225–330	23–38	8.5–12	7–12	250–360	169–190	140–195
Cellulose acetate	0.5–5.6	2.3–8.1		10–70	2.2–11.5	2.0–10.9		111–209	140–175
Cellulose acetate butyrate	0.4–11	2.6–6.9		40–88	1.8–9.3	2.1–9.4		113–227	140–175
Cellulose propionate	0.7–10.7	1.8–7.3		30–100	2.8–11	2.4–9.6		119–250	140–175
Chlorinated polyether	0.4	6	160	60–160	5		130	185–210	250–275
Ethyl vinyl acetate	No break	20–40	3.0–15	500–1,500					120–170
Chlorotrifluoroethylene	3.5	6	150–190	60–190	8–10	6–12	180	160–170	390
Fluorinated ethylene propylene	No break	2–3.2	60–80	250–350			70	124	400
Polytetrafluoroethylene	No break	2–5	50–100	75–400		4–12	70–90	132	500
Nylon 6	0.9–4	9.5–12.4	200–450	25–300	9–16.6	4–11	347	150–175	250
Nylon 6/6	0.9–2	11.2–13.1	410–480	60–300	14.6	5–13	400	200	250
Nylon 6/10	0.8–3	7–8.5	160–280	50–300	10.5	4–6		145	220
Polyallomer	1.5–12	2.9–4.2	100–170	400–650	4–5			124–133	250
Polycarbonate	2–3	8–9.5	345	60–110	11–13	12.5	350	265–290	250
Polyethylene, low-density	No break	1–2.4	14–38	20–800					140–175
Polyethylene, medium density	No break	1.7–2.8	50–80	80–600					150–180
Polyethylene, high-density	0.5–23	2.8–5	75–200	10–800	1–4	0.8–3.6	50–110	110–125	180–225
Polyethylene, high molecular weight	>20	2.3–5.4	102	525–600	3.5	2.4	110	120	180–225
Polyimide	0.8–1.1	5–14.0		6–7	7–14	12–24		680	500–600
Polypropylene	0.5–15	3.2–5	150–650	3–700	4.5–8	6–10		140–205	250
Polystyrene	0.25–0.40	6–8.1	400–500	1.5–2.5	9–15	11.5–16	300–560	160–215	150–190
Polystyrene, high-impact	0.7–3.5	1.9–4	200–430	10–75	5.5–12.5	8–16		160–205	130–180
Polyurethane	No break	4.1–4	1–3.7	400–650	0.7–1	>20	85		190
Polyvinyl chloride (flexible)	Varied			100–450					150–175
Polyvinyl chloride (rigid)	0.4–22	6–9	200–600	5–40	8–15	10–11	300–400	140–175	160–165
Polyvinyl dichloride (rigid)	1.5–7.0	7.5–9.0	360–450	10–65	14.2–17	13–22		212–235	195–210
Styrene acrylonitrile (SAN)	0.3–0.50	8–12	500–600	1–3.2	17	15–17.5	650	200–218	170–210
Ionomer	5.7–14	3.5–5.5	28–40	300–450					140
Phenoxy	1.5–12	8–9.5	350–410	50–100	12–14.5	15	325	175–188	
Polyphenylene oxide	1.5–1.9	11	380	8–80	15	15.4	380	375	250
Polysulfone	1.3	10.2	360	50–100	15.4		370	345	300

TABLE 14 Typical Electrical Properties of Thermoplastics[19]

Resin material	Volume resistivity, ohm-cm	Dielectric constant, 60 cycles	Dielectric strength, ST,* 1/8-in. thickness, volts/mil	Dissipation or power factor, 60 cycles	Arc resistance, sec
Acetal	$1-10^{14}$	3.7-3.8	500	0.004-0.005	129
ABS	$10^{15}-10^{17}$	2.6-3.5	300-450	0.003-0.007	45-90
Acrylic	$>10^{14}$	3.3-3.9	400	0.04-0.05	No tracking
Acrylic high impact	$10^{16}-10^{17}$	3.5-3.7	450-480	0.04-0.05	No tracking
Cellulose acetate	$10^{10}-10^{12}$	3.2-7.5	290-600	0.01-0.10	50-130
Cellulose acetate butyrate	$10^{10}-10^{12}$	3.2-6.4	250-400	0.01-0.04	
Cellulose propionate	$10^{12}-10^{16}$	3.3-4.2	300-450	0.01-0.05	170-190
Chlorinated polyether	1.5×10^{16}	3	400	0.01	
Ethyl vinyl acetate	1.5×10^{8}	3.16	525	0.003	
Chlorotrifluoroethylene	10^{18}	2.65	450	0.015	>360
Fluorinated ethylene propylene	10^{18}	2.1	500	0.0002	>165
Polytetrafluoroethylene	$>10^{18}$	2.1	400	<0.0001	No tracking
Nylon 6	$10^{14}-10^{15}$	6.1	300-400	0.4-0.6	140
Nylon 6/6	$10^{14}-10^{15}$	3.6-4.0	300-400	0.014	140
Nylon 6/10	$10^{14}-10^{15}$	4.0-7.6	300-400	0.04-0.05	140
Pollyallomer	$>10^{16}$	2.3	500-1,000	0.0001-0.0005	10-120
Polycarbonate	6.1×10^{15}	2.97	410	0.0001-0.0005	Melts
Polyethylene, low-density	$10^{15}-10^{18}$	2.28	450-1,000	0.006	Melts
Polyethylene, medium-density	$10^{15}-10^{18}$	2.3	450-1,000	0.0001-0.0005	Melts
Polyethylene, high-density	$6 \times 10^{15}-10^{18}$	2.3	450-1,000	0.002-0.0003	Melts
Polyethylene, high molecular weight	$>10^{16}$	2.3-2.6	500-710	0.0003	230
Polyimide	$10^{16}-10^{17}$	3.5	400	0.002-0.003	36-136
Polypropylene	$10^{15}-10^{17}$	2.1-2.7	450-650	0.005-0.0007	60-100
Polystyrene	$10^{17}-10^{21}$	2.5-2.65	500-700	0.0001-0.0005	60-90
Polystyrene, high-impact	$10^{13}-10^{17}$	2.5-3.5	500	0.003-0.005	
Polyurethane	2×10^{11}	6-8	850-1,100	0.276	
Polyvinyl chloride (flexible)	$10^{11}-10^{15}$	5-9	300-1,000	0.08-0.15	
Polyvinyl chloride (rigid)	$10^{12}-10^{16}$	3.4	425-1,040	0.01-0.02	
Polyvinyl dichloride (rigid)	10^{15}	3.08	1,200-1,550	0.018-0.0208	100-150
Styrene acrylonitrile (SAN)	10^{15}	2.8-3	400-500	0.006-0.008	
Ionomer	$>10^{16}$	2.4-2.5	1,000	0.001	
Phenoxy	$2.75-5 \times 10^{-5}$	4.1	404-520	0.0012-0.0009	70
Polyphenylene oxide	10^{17}	2.58	400-500	0.00035	75
Polysulfone	5×10^{16}	2.82	425	0.008-0.0056	122

* Short-time.

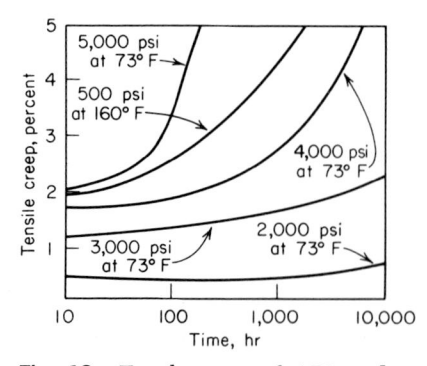

Fig. 19 Tensile creep of ABS under various loads.[20]

is excellent with limited loads, shown in Fig. 19. Heat-resistant ABS is equivalent or better than acetals, polycarbonates, and polysulfones in creep at 3,000 psi at room temperature. The Izod impact strengths at 75°F is in the range of 3 to 5 ft-lb/in. of notch. This figure is gradually reduced to 1 ft-lb/in. of notch at −40°F. When impact failure does occur, the failure is ductile rather than brittle. Modulus of elasticity versus temperature is shown in Fig. 20. Physical properties are little affected by moisture, which greatly contributes to the dimensional stability of ABS materials. ABS plastics are compared with some other materials as to flexural modulus, tensile strength, and creep behavior later in this chapter (Figs. 57, 59, and 61).

Electrical properties Although ABS plastics are used largely for mechanical purposes, they also have good electrical properties that are fairly constant over a wide range of frequencies. These properties are little affected by temperature and atmospheric humidity in the acceptable operating range of temperatures.

The dielectric strength of ABS plastics is about 350 volts/mil. The approximate dissipation factors of the best electrical grades are 0.004 at 60 Hz, 0.005 at kHz, and 0.009 at 11 MHz.

Dielectric constants of these resins are also quite low (2.84 to 3.17) and are relatively independent of frequencies between 60 Hz and 1 MHz.

ABS alloys and electroplating grades Much work has been done to modify ABS plastics by alloying, to improve certain properties, and by modifying, to enhance adhesion of electroplated coatings.

ABS alloyed or blended with polycarbonate combines some of the best qualities of both materials, resulting in an easier-processing, high-heat, and high-impact thermoplastic selling for considerably less than polycarbonate. A comparison of properties is shown in Table 15.

The impact strength of the alloy, 10.7 ft-lb/in. of notch, is well above average for the high-impact engineering thermoplastics, but still not as high as that of polycarbonate. However, unlike polycarbonate, the alloy does not have critical thicknesses with respect to notched impact strength. Its notched Izod value drops

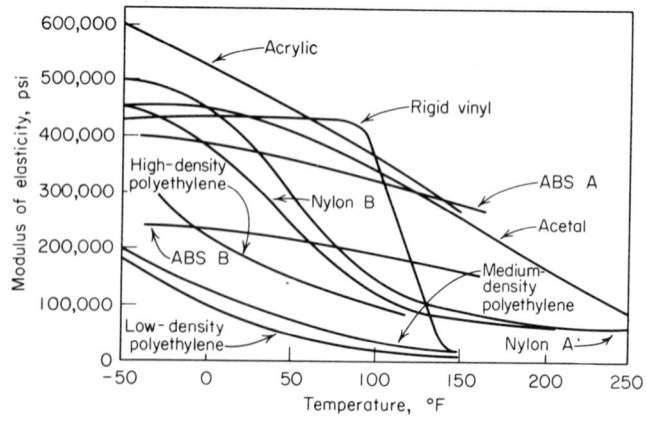

Fig. 20 Modulus of elasticity of several thermoplastic materials as a function of temperature.[21]

TABLE 15 Property Comparisons for ABS-Polycarbonate Alloy and Base Materials[22]

Properties	ASTM No.	Polycarbonate	ABS-Polycarbonate alloy	ABS*
Physical				
Specific gravity	D 792	1.20	1.14	1.04
Thermal conductivity, Btu/(hr)(ft^2)(°F)(in.)		1.3	2.46 per ft	1.55
Coefficient of thermal expansion, 10^{-5} per °F	D 696	3.75	6.12	5.3
Specific heat, Btu/(lb)(°F)		0.30		
Refractive index, n_D	D 542	1.58	Opaque	Opaque
Water absorption in 24 hr, %	D 570	0.15	0.21	0.43
Flammability, in./min	D 635	Self-ext	0.90	1.5
Coefficient of static friction		0.52	0.2	
Mechanical				
Tensile yield strength, psi × 10^{-3}	D 638	9.5	8.2	6.3
Elongation (ultimate), %	D 638	110		
Modulus of elasticity in tension, psi × 10^{-5}	D 638	3.45	3.7	3.1
Flexural Strength, psi × 10^{-3}	D 790	13.5	14.3	9.7
Modulus of elasticity in flex, psi × 10^{-5}	D 790	3.4	4.0	3.2
Impact strength, Izod notched, ft-lb/in	D 256	16	10.7	6.2
Compressive strength, psi × 10^{-3}	D 695	12.5	11.8	
Hardness, Rockwell	D 785	M 70	R 118	R 103
Abrasion resistance, Taber, mg/kHz	D 1044	10		
Electrical				
Volume resistivity, ohm-cm × 10^{-16}	D 257	2.1	2.2	1.05
Dielectric strength, short time, volts/mil	D 149	400	500	
Dielectric constant:	D 150			
At 60 Hz		3.17	2.74	2.84
At 1 MHz		2.96	2.69	2.76
Dissipation factor:				
At 60 Hz		0.009	0.0026	
At 1 MHz		0.010	0.0059	
Arc resistance, sec	D 495	120	96	76
Heat resistance				
Maximum recommended service temperature, °F		250	240	200
Deflection temperature, °F:	D 648			
At 66 psi		280	260	211
At 264 psi		270	245	193

* Grade with maximum toughness and high modulus.

by only 2 to 4 ft-lb/in. in the ⅛- to ¼-in. range. The notched Izod value for a polycarbonate ⅛ in. thick is about 16 ft-lb/in. but only 3 to 4 ft-lb/in. at thicknesses greater than ¼ in. The flexural modulus of the ABS-polycarbonate alloy is about 15 percent greater than that of polycarbonate alone. The alloy remains more rigid than polycarbonate up to about 200°F, as shown in Fig. 21.

The 264-psi heat-deflection temperature of the alloy is 245°F, and the 66-psi value is 260°F. These values are 35 and 30°F, respectively, lower than those of polycarbonate. However, maximum recommended continuous (no load) temperature of the alloy is only 10°F lower than that of polycarbonate.

The good creep resistance of polycarbonate, one of its biggest advantages (shown in Fig. 22), is maintained after alloying.

In addition to the polycarbonate alloy, ABS can be alloyed with other plastics to obtain special properties. Furthermore, ABS has been modified to gain improved adhesion of electroplated metals. Advances have been so great in this area that

1-40 Plastics for Electronics

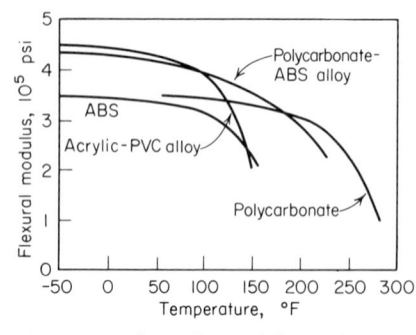

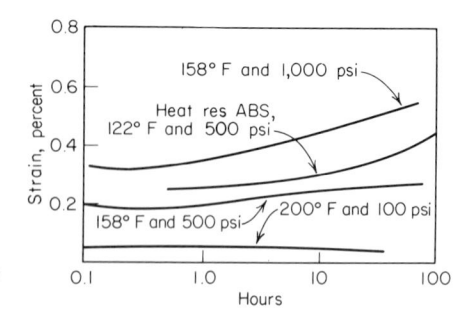

Fig. 21 Flexural modulus of ABS-polycarbonate alloy and base materials.[22]

Fig. 22 Creep resistance of ABS-polycarbonate alloy and heat-resistant ABS.[22]

ABS is perhaps the most widely used material for producing electroplated plastic parts. Electroplated ABS is extensively employed for many electrical and mechanical products, in many forms and shapes. Adhesion of the electroplated metal is excellent.

Acetals

Acetals are among the group of high-performance engineering thermoplastics that resemble nylon somewhat in appearance but not in properties. They are strong and rigid (but not brittle) and have good moisture, heat, and chemical resistance. There are two basic types of acetals, namely, the homopolymers by Du Pont, and the copolymers by Celanese. Table 12 further identifies these materials. The homopolymers are harder, have higher resistance to fatigue, are more rigid, and have higher tensile and flexural strength with lower elongation. The copolymers are more stable in long-term, high-temperature service, and more resistant to hot water. Neither type of acetal is resistant to strong mineral acids, but the copolymers are resistant to strong bases. References are frequently made to acetals, without identification of polymer type. Such references usually imply the homopolymer material.

Mechanical properties The most outstanding properties of acetals are high tensile strength and stiffness, resilience, good recovery from deformation under load, and toughness under repeated impact. They exhibit excellent long-term load-carry-

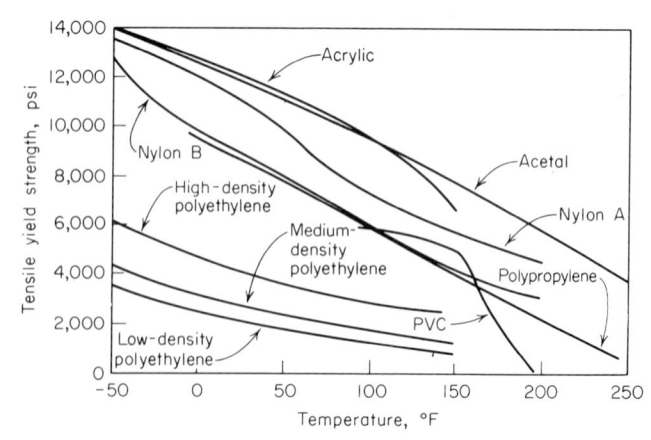

Fig. 23 Tensile yield strength of several thermoplastic materials as a function of temperature.[21]

Thermoplastics   1-41

ing properties and dimensional stability and can be used for precision parts. Acetals have low static and dynamic coefficients of friction and are usable over a wide range of environmental conditions. The plastic surface is hard, smooth, and glossy. A fluorocarbon fiber-filled acetal, Delrin* AF, is available and offers even better low friction and resistance properties. This is discussed in more detail below.

The modulus of elasticity as a function of temperature for acetals and several other thermoplastic materials was shown earlier in Fig. 20. The tensile yield strength of acetals is compared with that of some other thermoplastics in Fig. 23. The deflection under load for Delrin acetal is compared with that of other thermoplastics in Fig. 24. The long-term creep behavior is shown later in this chapter, in Fig. 61. Also presented in subsequent figures, in comparison with other plastics, are data on impact strength (Fig. 55), flexural modulus (Fig. 57), tensile modulus (Fig. 58), and dimensional changes due to absorbed moisture (Fig. 63).

Electrical properties Acetal resins are good insulators, having relatively low dissipation factors and dielectric constants over the operating temperature range for these materials. The electrical properties (Table 14) of acetals are largely retained under exposure to high humidity and water immersion. Critical electrical properties do change with temperature, as shown in Table 16. These properties degrade more drastically above about 200°F. Dissipation factors versus temperature and frequency for acetals and some other plastics are compared in a later section (Figs. 64 and 65).

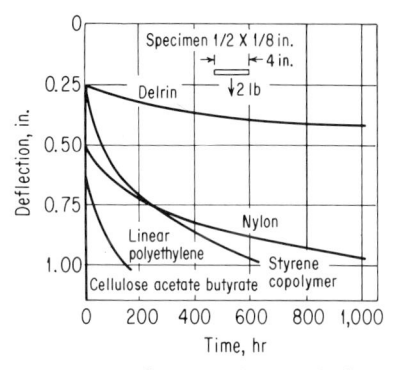

Fig. 24 Deflection of several thermoplastic materials as a function of time at 90 percent RH and 150°F.[23]

Fluorocarbon fiber-filled acetal This acetal modification, mentioned above, is a modified acetal homopolymer developed to meet the need for a thermoplastic injection molding material to be used in moving parts in which low friction and exceptional wear resistance are the principal requirements. This resin consists of oriented TFE fluorocarbon fibers uniformly dispersed in a matrix of acetal resin. The result is an injection molding and extrusion resin that combines the strength, toughness, dimensional stability, and fabrication economy of acetals with the unusual surface and low frictional characteristics of the fluorocarbons.

The outstanding properties of TFE fiber-filled acetal are those associated with sliding friction. Bearings made from this material sustain high loads when operating at high speeds, and show little wear. In addition, such bearings are essentially free of slipstick behavior because their static and dynamic coefficients are almost equal. Comparative properties of the filled and unfilled acetal are given Table 17.

Acrylics

The general properties of acrylics are presented in Tables 12 to 14. Acrylics, or polymethyl methacrylate, provide numerous engineering advantages. They have exceptional optical clarity and basically good weather resistance, strength, electrical properties, and chemical resistance. They do not discolor or shrink after fabrication, have low water-absorption characteristics and a slow burning rate, and will not flash-ignite. For disadvantages, acrylics are attacked by strong solvents, gasoline, acetone, and other similar fluids.

Parts molded from acrylic powders in their natural state may be crystal clear and nearly optically perfect. The total light transmittance is 92 percent, and haze measurements average only 1 percent. Light transmittance and clarity can be modified by addition of transparent and opaque colors, most of these being formu-

---
* Trademark of E. I. du Pont de Nemours & Co., Inc., Wilmington, Del.

lated for long outdoor surface. Acrylics can be injection-molded, extruded, cast, vacuum-and-pressure-formed, and machined, although molded parts for load bearing should be carefully analyzed, especially for long-term loading.

Acrylics are divided into the four main product groups of cast sheet, high-impact sheets, molding powder, and high-impact molding powder. Furthermore, they can be modified to improve heat resistance. Modified acrylics can provide about five times the impact strength of the standard grades. Although their outdoor weatherability is inferior to that of the straight acrylics, their indoor aging characteristics are excellent.

TABLE 16 Effect of Temperature and Frequency on Dissipation Factor and Dielectric Constant of Acetals[24]

Temperature, °F	Dissipation factor		Dielectric constant	
	At 1 kHz	At 100 kHz	At 1 kHz	At 100 kHz
−94	0.0160	0.0130	3.00	2.84
−76	0.0225	0.0200	3.15	2.91
−58	0.0260	0.0300	3.30	3.00
−40	0.0150	0.0360	3.36	3.14
−22	0.0030	0.0300	3.38	3.28
−4.0	0.0011	0.0170	3.40	3.36
14	0.0007	0.0080	3.41	3.39
32	0.0007	0.0036	3.41	4.41
50	0.0007	0.0018	3.41	4.43
68	0.0007	0.0012	3.43	4.43
77	0.0007	0.0010	3.44	4.43
104	0.0009	0.0018	3.52	3.50
140	0.0035	0.0029	3.58	3.55
176	0.0038	0.0060	3.68	3.62

Acrylic-PVC alloys, in sheets of various thicknesses for thermoforming, are also commercially available. They are characterized by high impact resistance. Very high elongation at forming temperatures allows extreme draw ratios and deep parts without hot tearing, not normally possible without the PVC addition. The flexural modulus of such an alloy is shown in Fig. 21.

Mechanical properties Acrylics, although primarily known for their exceptional clarity and excellent light transmission, are strong and rigid. They are among the strongest of the thermoplastics, but only for short-term loadings, owing to creep. Their physical properties are not affected by outdoor weathering, and they do not become exceptionally brittle at low temperatures, although the impact strength of the unmodified acrylic is low. Acrylics exhibit high tensile strength which is comparable to that of acetals, as shown in Fig. 23. However, the strength of acrylics falls off above 150°F. Modulus of elasticity is shown in Fig. 20. Although acrylics are basically stable, dimensional changes do occur in continuously humid environments, as shown in Fig. 25.

Electrical properties Acrylics have no tendency toward arc tracking. Their high arc resistance and excellent tracking characteristic make them a good choice for certain high-voltage applications such as circuit breakers. The dielectric strength is 450 to 500 volts/mil. Acrylics are one of the few plastics that exhibit an essentially linear decrease in dielectric constant and dissipation factor with increase in frequency, as shown in Fig. 26.

Cellulosics

These are a class of thermoplastics that are prepared by various treatments of purified cotton or special grades of wood cellulose. Cellulosics are among the toughest of plastics, are generally economical, and are basically good insulating materials. However, they are temperature-limited, and are not as resistant to ex-

TABLE 17 Property Comparisons for Unfilled and TFE Fiber-filled Acetal[25]

		Average values*	
Property	ASTM No.	Delrin 500 (unfilled)	Delrin AF (TFE filled)
Tensile strength and yield point†, psi $\times 10^{-3}$:			
At $-68°F$	D 638	14.7	9
At $73°F$	D 638	10	6.9
At $158°F$	D 638	7.5	4.7
Elongation,† %:			
At $-68°F$	D 638	13	6
At $73°F$	D 638	15	12
At $158°F$	D 638	330	38
Flexural modulus, psi $\times 10^{-3}$:			
At $73°F$	D 790	410	400
At $170°F$	D 790	190	180
At $250°F$	D 790	90	85
Compressive stress, psi $\times 10^{-3}$:			
1% deformation	D 695	5.2	1.8
10% deformation	D 695	18	13
Impact strength, Izod, ft-lb/in.:			
At $-40°F$	D 256	1.2	0.6
At $73°F$	D 256	1.4	0.7
Coefficient of linear thermal expansion, in./(in.)(°F) $\times 10^5$:			
At $85°F$	D 696	4.5	4.5
At $85-140°F$	. . . . . .	5.5	5.5
Heat-distortion temperature, °F:			
At 264 psi	D 648	255	212
At 66 psi	D 648	338	329
Flammability, in./min	D 635	1.1	0.8
Water absorption after 24 hr immersion, %	D 570	0.25	0.06
Rockwell hardness	D 785	M 94, R 120	M 78, R 118
Specific gravity	D 792	1.425	1.54
Taber abrasion (1,000-g load, CS-17 wheel), mg/kHz	D 1044	20	9
Coefficient of friction (no lubricant)	. . . . . .	0.1–0.3	0.05–0.51

* These values represent those obtained under standard ASTM conditions and should not be used to design parts that function under different conditions. Since they are average values, they should not be used as minimums for material specifications.
† Determined at 0.2 in./min.

treme environments as many other thermoplastics. Furthermore, they are not outstanding in any respect for electronic and electrical applications. Hence, they are not as frequently selected for other than general-purpose use. The four most prominent industrial cellulosics are cellulose acetate, cellulose acetate butyrate, cellulose propionate, and ethylcellulose. A fifth member of this group is cellulose nitrate. Cellulosic materials are available in a great number of formulas and flows and

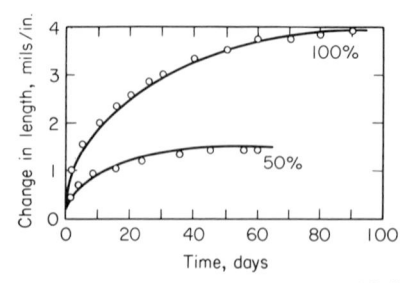

Fig. 25 Changes in length of molded acrylic part during exposure to 50 and 100 percent levels of RH. Parts had been conditioned at 5 percent RH and 73°F.[26]

are manufactured to offer a wide range of properties. They are formulated with a wide range of plasticizers for specific plasticized properties.

Cellulose butyrate, propionate, and acetate provide a range of toughness and rigidity that is useful for many applications, especially where clarity, outdoor weatherability, and aging characteristics are needed. The materials are fast-molding plastics and can be provided with hard, glossy surfaces and over the full range of color and texture.

Butyrate, propionate, and acetate are rated in that order in dimensional stability in relation to the effects of water absorption and plasticizers. The heat deflection of cellulose acetate butyrate at 150°F is shown, compared to that of some other thermoplastics, in Fig. 24. The materials are slow-burning, although self-extinguishing forms of acetate are available. Special formulations of butyrate and propionate are serviceable outdoors for long periods. Acetate is generally considered unsuitable for outdoor uses. From an application standpoint, the acetates generally are used where tight dimensional stability, under anticipated humidity and temperature, is not required. Hardness, stiffness, and cost are lower than for butyrate or propionate. Butyrate is generally selected over propionate where weatherability, low-temperature impact strength, and dimensional stability are required. Propionate is often chosen for hardness, tensile strength, and stiffness, combined with good weather resistance.

Ethylcellulose, best known for its toughness and resiliency at subzero temperatures, also has excellent dimensional stability over a wide range of temperature and humidity conditions. Alkalies or weak acids do not affect this material, but cleaning fluids, oils, and solvents are very harmful.

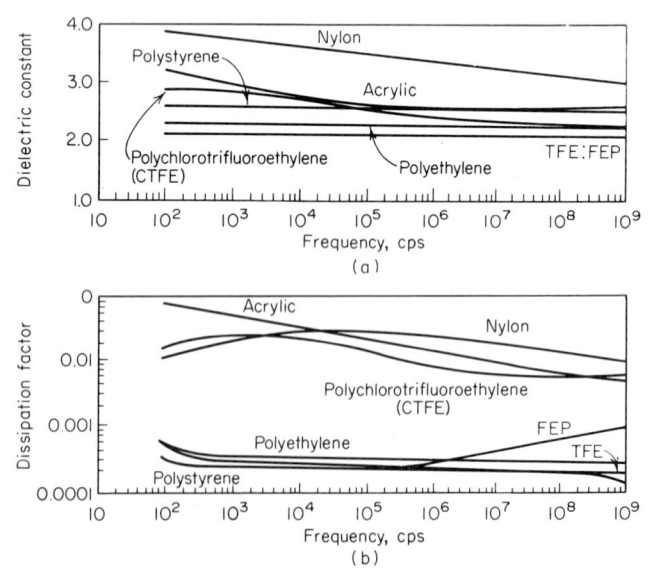

Fig. 26 Effect of frequency on dielectric constant (*a*) and dissipation factor (*b*) of several thermoplastic materials at 23°C.[27]

Chlorinated Polyether

Chlorinated polyether is better known for its use in piping and hardware in the chemical processing industry than for its use in electronic applications. This is primarily due to its outstanding resistance to many chemicals, up to 250°F. Nevertheless, it does have, as mentioned below, properties that are useful for certain electronic and electrical applications. Tables 12 to 14 outline the major physical, electrical, and thermal properties of this material at room temperature.

Mechanical properties Mechanical creep properties are particularly good. The creep remains below 4 percent after 10,000 hr under a sustained 2,000-psi load at 75°F, and below 1 percent after 24 hr under the same load at 125°F. A creep percentage of slightly over 4 percent exists after 5,000 hr under a sustained load of 1,000 psi at 280°F. For most cases up to the conditions just mentioned, the percentage creep increases very little after the first 200 to 400 hr of sustained loading.

Electrical properties The electrical properties of dielectric constant and dissipation factor do not change much in the frequency range of 60 Hz to 50 MHz, nor are they much affected by immersion up to 20 hr in boiling water. Regarding the effect of temperature, however, the dissipation factor does increase considerably in the range of 73 to 250°F. Dielectric constant, however, increases only to about 3.5 at 250°F. The 73°F properties are shown in Tables 12 to 14.

Ethylene–Vinyl Acetates

These materials, commonly known as EVA copolymers, are thermoplastics whose properties approach elastomeric properties in softness and flexibility, but which can be processed like conventional thermoplastics. Hence, they are primarily useful as replacements for plasticized vinyls and rubbers, where the EVA properties are better. Basic limitations of EVA copolymers are low resistance to heat and solvents.

The electrical properties of EVA copolymers are determined somewhat by the specific material composition, with dielectric constant and dissipation factor increasing with increasing comonomer content. The dielectric constant is in the general range of 2.6 to 2.7, and is frequency-dependent. Typical dissipation factor values are 0.015 at 100 kHz and 0.032 at 1 MHz.

Fluorocarbons

Fluorocarbons are very important in electronic and electrical applications, owing to their excellent electrical properties which are relatively unaffected by most extreme environments or operating conditions. Some fluorocarbon classes are more used than others, of course. The most widely used fluorocarbon, perhaps, is polytetrafluoroethylene, or TFE fluorocarbon. This was the original fluorocarbon, and is still known to many as Teflon. Correctly speaking, however, Teflon is the trade name for Du Pont TFE and FEP fluorocarbons. There are now multiple suppliers of fluorocarbon plastics. Table 12 clarifies the fluorocarbon terminology, as does further discussion below. Tables 13 and 14 give the general properties of these materials. Because of the wide use and importance of these materials, additional discussion and data are given below.

It can be considered for practical purposes that there are five types of fluorocarbons, as noted in Table 12. Like other plastics, each type is available in several grades. As mentioned above, the original, basic fluorocarbon, and perhaps still the major one, is TFE fluorocarbon. It has the optimum of electrical and thermal properties, but does have the disadvantage of cold flow or creep under mechanical loading. Stronger, filled modifications exist, as do newer, more cold-flow-resistant grades. FEP fluorocarbon is quite similar to TFE in most properties, except that its useful temperature is limited to about 400°F. FEP is much more easily processed, and molded parts are possible with FEP which might not be possible with TFE.

The CTFE resins, like the FEP materials, are melt-processable and can be injection-, transfer-, and compression-molded or screw-extruded. Compared with TFE, CTFE has greater tensile and compressive strength within its service temperature range. However, at the temperature extremes, CTFE does not perform as well as TFE for parts such as seals. At the low end of the range, TFE has somewhat better physical properties; at the higher temperatures, CTFE is more prone to stress cracking and other difficulties.

Electrical properties of CTFE are generally excellent, but dielectric losses are higher than those of TFE. Chemical resistance is poorer than that of TFE, but radiation resistance is better. CTFE does not have the low-friction and bearing properties of TFE.

Vinylidene floride is another melt-processible fluorocarbon capable of being injection- and compression-molded and screw-extruded. Its 20 percent lower specific gravity compared with that of TFE and CTFE and its good processing characteristics permit economy and provide excellent chemical and physical characteristics. Useful temperature range is −80 to +300°F. Although it is stiffer and has higher resistance to cold flow than TFE, its chemical resistance, useful temperature range, antistick properties, lubricity, and electrical properties are lower. Principal applications for electrical insulation have been wire insulation.

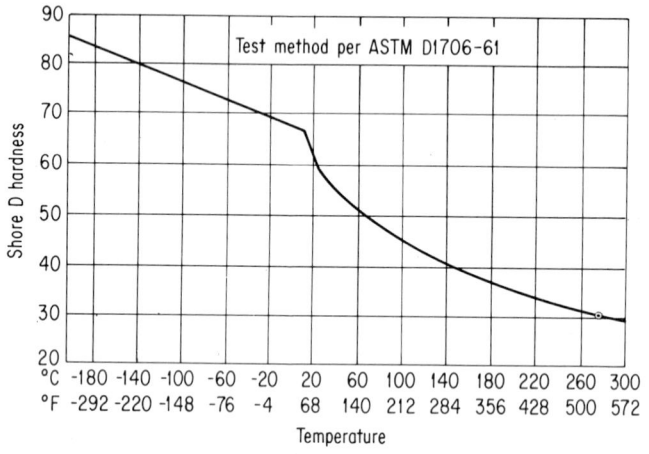

Fig. 27 Hardness of TFE fluorocarbon as a function of temperature.[28]

Mechanical properties As with all plastics, especially thermoplastics, mechanical properties of fluorocarbons vary with temperature. Some of these properties, and their relation to temperature, will be given at this point. In addition, since there are countless data on the many properties, the references listed for this subject area are especially recommended.

TFE fluorocarbon, perhaps the most commonly used fluorocarbon in electronics, is a semisoft plastic, which exhibits some cold-flow properties. TFE and FEP are unique in their ability to retain a useful balance of flexibility and strength over a wide temperature range. TFE plastics are superior in their combination of toughness, flex resistance, and general abuse resistance. The hardness of TFE as a function of temperature is shown in Fig. 27.

Regarding creep properties, it is general practice in plastics engineering to apply the term "creep strain" to total deformation in a creep test. Such usage permits definition of a simple and practical creep modulus. In the most frequently used mechanical design formulas, the material constant required is a modulus, not a strain. Also, for purposes of comparing and selecting materials, rigidity is usually identified with modulus. Therefore, it is advantageous to convert creep-strain curves to modulus curves.

Except for very low strains of about 1 percent or less, the creep moduli of most plastics materials are significantly affected by the levels of applied stress. Fig-

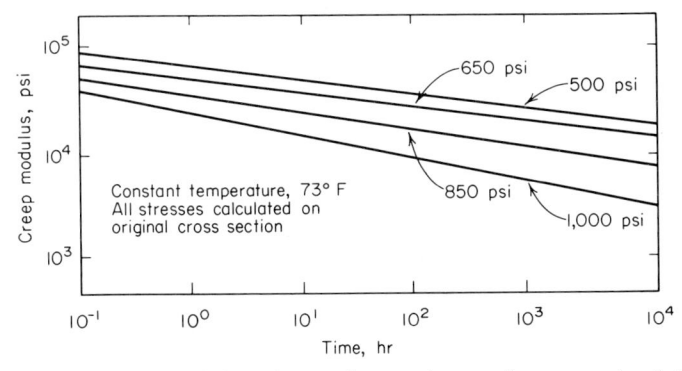

Fig. 28 Creep modulus of TFE fluorocarbon at four stress levels.[29]

ure 28 shows the creep modulus of TFE plotted at four different stresses. The characteristic effect of increasing stress level is to decrease the creep modulus, with the magnitude of the effect increasing with increasing stress level.

Creep properties of plastics materials are also nearly always substantially affected by temperature, especially at room temperature and above. The creep modulus invariably decreases with increasing temperature. Figure 29 presents a plot of the creep modulus of TFE at four different temperatures.

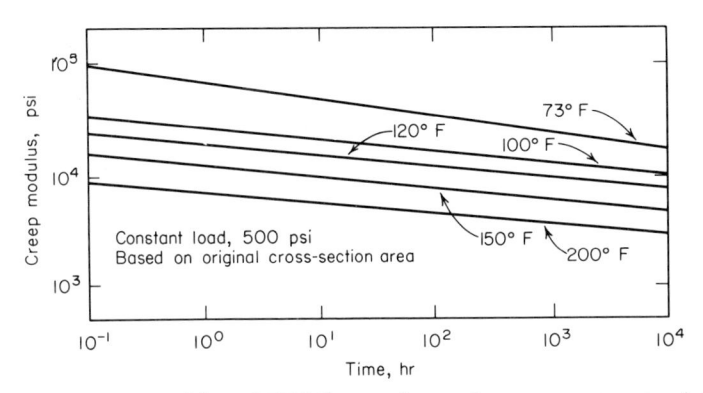

Fig. 29 Creep modulus of TFE fluorocarbon at four temperature levels.[29]

Long-term creep data for over 100 plastic materials are given in Ref. 29, identified at the end of this chapter.

Stress-strain relationship for TFE, FEP, and CTFE are given in Fig. 30, which shows both tensile and compressive loadings. In many instances, recovery from stress is nearly complete, provided the original strain does not exceed the yield strain. Compressive recovery curves for TFE and FEP at 730°F are shown in Fig. 31, and at 212°F in Fig. 32. Tensile strength of these resins, as a function of temperature, is shown in Fig. 33.

Linear thermal expansion of TFE and FEP increases with temperature, and this is shown in Fig. 34. Change of various mechanical properties up to 500°F are given in Table 18. Strength properties at very low temperatures are shown in Table 19. Mechanical properties of fluorocarbon plastics (as well as of other thermoplastics) can be improved by use of glass-reinforcing fillers. This will be discussed later in this chapter.

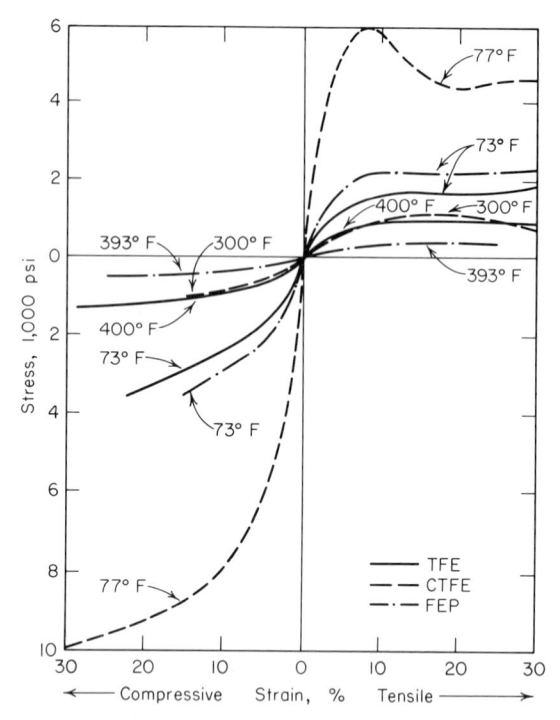

Fig. 30 Stress-strain relationships for three fluorocarbon plastics at several temperatures.[30]

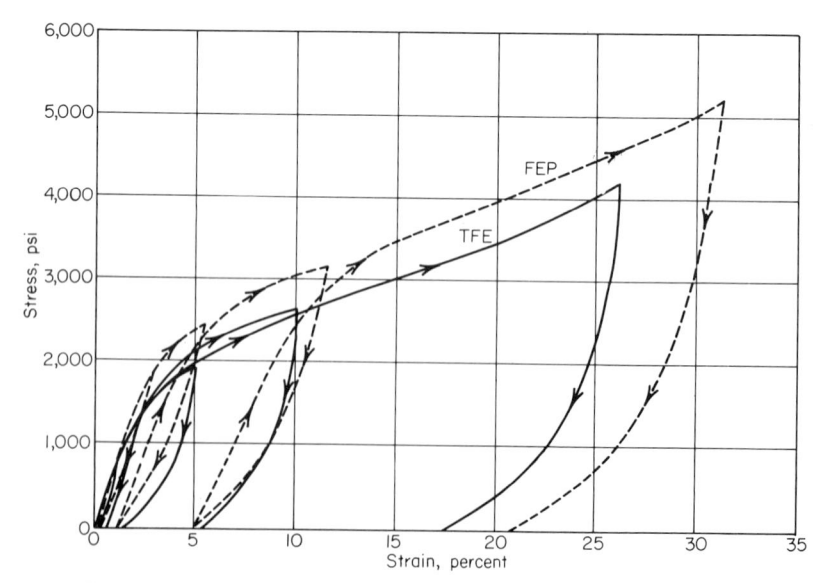

Fig. 31 Compressive recovery curves for TFE and FEP at 73°F.[31]

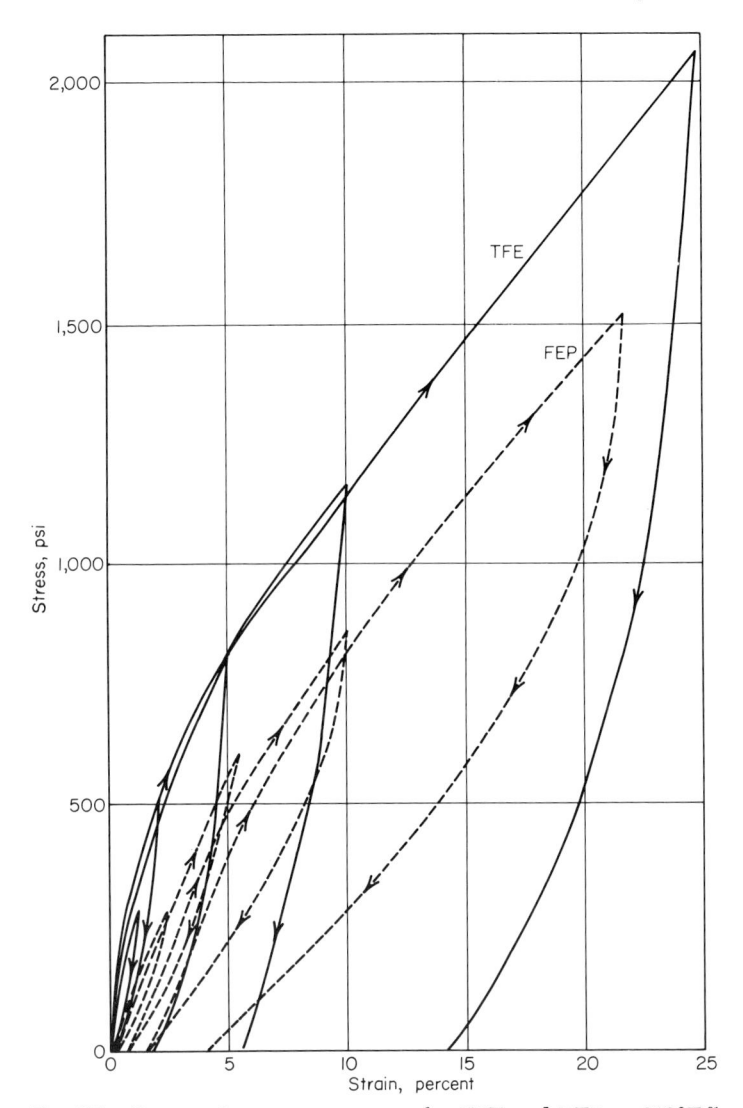

Fig. 32 Compressive recovery curves for TFE and FEP at 212°F.[31]

Electrical properties Fluorocarbons have excellent electrical properties. TFE and FEP fluorocarbons in particular have low dielectric constants and dissipation factors which change little with temperature or frequency. This is shown in Figs. 26 and 35, which also presents data for some other thermoplastics.

The dielectric strength of TFE and FEP resins is high and does not vary with temperature and thermal aging. Initial dielectric strength is very high as measured by the ASTM short-time test. As with most dielectric materials, the value drops as thickness of the specimen increases, as shown in Fig. 36. Further, the dielectric strength is a function of frequency, as noted in Fig. 37. Life at high dielectric stresses is dependent on corona discharge, as shown in Fig. 38. The absence of corona, as in special wire constructions, permits very high voltage stress without

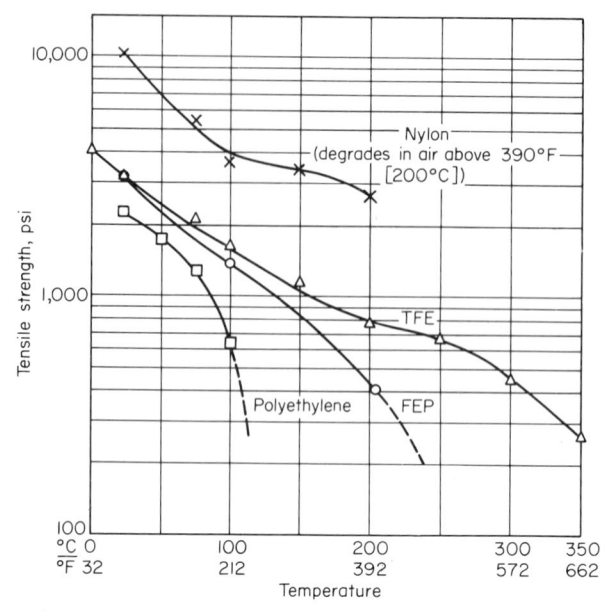

Fig. 33 Tensile strength of TFE, FEP, polyethylene, and nylon as a function of temperature.[31]

damage to either TFE or FEP resins. Changes in relative humidity or physical stress imposed upon the material do not diminish life at these voltage stresses.

Surface arc-resistance of TFE and FEP resins is high, and is not affected by heat aging. When these resins are subjected to a surface arc in air, they do not track or form a carbonized conducting path. When tested by the procedure of ASTM D 495, they pass the maximum time of 300 sec without failure.

The unique nonstick surface of these resins helps reduce surface arc phenomena

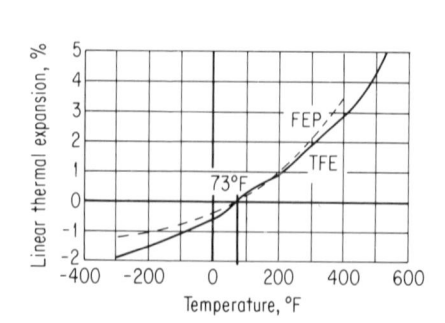

Fig. 34 Linear thermal expansion of TFE and FEP fluorocarbons as a function of temperature.[31]

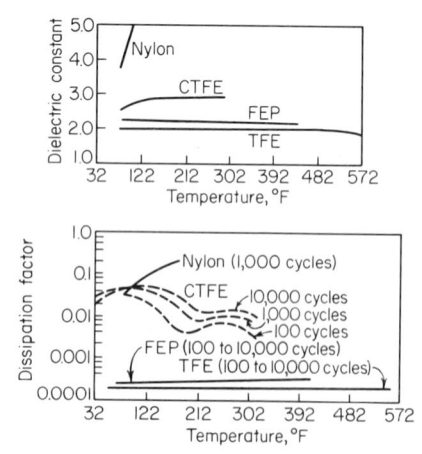

Fig. 35 Effect of temperature on dielectric constant and dissipation factor of fluorocarbons and nylon.[32]

TABLE 18 High-temperature Mechanical Properties of TFE Fluorocarbon Resins[23]

Property	Temperature *			
	72°F (23°C)	212°F (100°C)	400°F (204°C)	500°F (260°C)
Ultimate tensile strength, psi	3,850	2,500	1,500	900
Yield strength, psi	1,050 at 2%	400 at 2%	200 at 4%	200 at 8%
Ultimate elongation, %	300	>400	360	360
Flexural modulus of elasticity, psi	80,700	28,700	....	6,500
Flexural modulus of elasticity (35% glass-reinforced), psi	208,700	113,000	....	25,900
Compressive stress, 1% strain, psi	700	290	100	60
Compressive stress, 5% strain, psi	1,850	800	430	260
Compressive stress, 5% strain (15% glass-reinforced), psi	2,600	1,300	600	320
Linear expansion, in./(in.)(°F)	6.90×10^{-5}	6.90×10^{-5}	8.40×10^{-5}	9.70×10^{-5}
Linear expansion, %	0.0	0.9	2.7	4.3
Linear expansion (35% glass-reinforced), %	0.0	0.15	....	1.0
Coefficient of friction	0.04 over temperature range 80°F to 621°F for static loads			

* These values are typical of those for "Teflon" TFE-fluorocarbon resins in general. Variations may be expected from the values shown depending on the exact type of TFE resin used, methods of molding, and fabrication techniques employed.

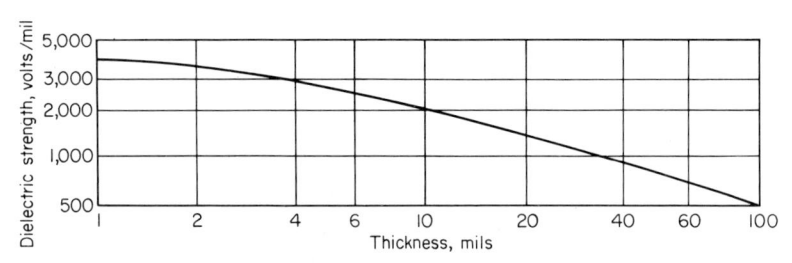

Fig. 36 Short-term dielectric strength of TFE fluorocarbon as a function of thickness.[32]

in two ways: (1) it helps prevent formation of surface contamination, thereby reducing the possibility of arcing; (2) if an arc is produced, the discharge frequently cleans the surface of the resin, increasing the time before another arc.

Volume resistivity ($>10^{18}$ ohm-cm) and surface resistivity ($>10^{16}$ ohm/sq) for both FEP and TFE resins are at the top of the measurable range. Neither resistivity is affected by heat aging or temperatures up to recommended service limits.

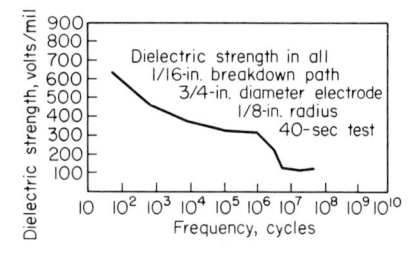

Fig. 37 Dielectric strength of TFE resins as a function of frequency.[32]

Space-environment properties There has been much discussion of the effects of the space environment, especially radiation, on fluorocarbons. The reason for this is partly due to the known effects of radiation on TFE and FEP fluorocarbons

TABLE 19 Strength Properties of TFE and FEP Fluorocarbons at Low Temperature[31]

Mechanical property	°F	°C	TFE Unfilled*	TFE Filled 65% bronze	TFE Filled 15% carbon	TFE Filled 25% asbestos	TFE Filled 25% glass fiber	FEP Unfilled†	FEP Filled, 25% glass fiber
Tensile yield strength, psi × 10⁻³	−423 −320 −297 −200 −110 +77	−253 −196 −183 −129 −79 +25	−17.8 13.3 (12.1) 7.7 4.7 1.7	 6.8 (6.4) (5.1) (3.7) 1.5	 6.3 (5.9) (4.7) (3.4) 1.2	6.6 5.3 (5.0) (3.8) (2.6) 0.5	4.3 3.5 (3.3) (2.3) (1.3) 0.6	23.7 19.0 (18.2) 11.4 5.6 2.0	15.5 13.5 2.0
Ultimate tensile strength, psi × 10⁻³	−423 −320 −297 −200 −110 +77	−253 −196 −183 −129 −79 +25	17.9 14.9 (13.8) 9.1 5.9 4.5	 6.8 (6.5) (5.1) (3.8) 1.5	 6.8 (6.5) (5.1) (3.9) 1.6	6.6 5.4 (5.1) (4.0) (2.8) 0.9	4.2 3.6 (3.5) (2.9) (2.2) 1.0	23.8 18.0 (17.1) 12.2 6.5 4.0	15.5 13.5 (12.9) (10.0) (7.1) 2.2
Tensile modulus, psi × 10⁻⁶	−423 −320 −297 −200 −110 +77	−253 −196 −183 −129 −79 +25	0.62 0.47 0.45 0.30 0.20 0.08				0.46 0.38 (0.36) (0.29) (0.23) 0.08	0.73 0.58 0.56 0.48 0.30 0.07	1.20 1.10 (1.04) (0.80) (0.58) 0.15
Elongation, %	−423 −320 −297 −200 −110 +77	−253 −196 −183 −129 −79 +25	3 7 (7) 13 31 198	0 2 (4) (13) (23) 40	0 1 (2) (6) (11) 20	0 1 (2) (5) (9) 15	1 2 (3) (13) (22) 40	5 7 (8) 15 33 353	3 4 (8) (25) (41) 70
Flexural strength, psi × 10⁻³	−423 −320 −297 −200 −110 +77	−253 −196 −183 −129 −79 +25	28.3 25.8 (24.7) 15.2 6.9 3.7				8.7 6.3 (6.1) (5.2) (4.2) 2.8	36.3 26.7 (24.7) 20.1 9.6 2.9	28.6 23.4 (22.4) (17.5) (13.3) 4.7
Flexural modulus, psi × 10⁻⁶	−423 −320 −297 −200 −110 +77	−253 −196 −183 −129 −79 +25	0.74 0.68 (0.66) 0.46 0.23 0.17				0.41 0.36 (0.34) (0.26) (0.19) 0.04	0.77 0.68 (0.68) 0.57 0.33 0.20	0.94 0.85 (0.82) (0.70) (0.59) 0.38
Compressive strength, psi × 10⁻³	−423 −320 −297 −200 −110 +77	−253 −196 −183 −129 −79 +25	31.9 21.1 (20.2) 16.0 7.4 3.7	30.7 22.5 (21.4) (17.2) (13.0) 5.5	23.2 17.6 (16.3) (12.9) (10.2) 4.1	27.0 24.3 (23.3) (18.3) (13.5) 4.8	27.3 21.3 (20.3) (15.8) (11.5) 3.0	35.9 30.0 (28.8) 23.4 13.3 1.5	37.6 31.9 (30.3) (23.3) (16.7) 4.2
Compressive modulus, psi × 10⁻⁶	−423 −320 −297 −200 −110 +77	−253 −196 −183 −129 −79 +25	0.90 0.80 (0.78) 0.59 0.28 0.10	0.90 0.82 (0.78) (0.59) (0.43) 0.08	0.88 0.84 (0.80) (0.62) (0.46) 0.13	0.77 0.71 (0.67) (0.52) (0.38) 0.10	0.99 0.86 (0.81) (0.64) (0.47) 0.15	1.02 0.92 (0.91) 0.74 0.39 0.09	1.50 1.34 (1.25) (1.01) (0.78) 0.23
Torsion modulus of rigidity, psi × 10⁻³	−423 −320 −297 −200 −110 +77	−253 −196 −183 −129 −79 +25	322 212 (202) 142 70 23				244 104 (100) (76) (56) 16	964 (320) (284) 172 96 16	528 304 (288) (220) (156) 36
Notched Izod impact strength, ft lb/in.	−423 −320 −297 −200 −110 +77	−253 −196 −183 −129 −79 +25	1.40 1.30 (1.35) (1.40) 1.50 1.90	1.25 1.25 (1.30) (1.50) (1.70) 2.05	0.95 0.70 (0.80) (1.20) (1.50) 2.20	0.90 0.80 (0.85) (1.00) (1.15) 1.45	1.00 1.00 (1.05) (1.30) (1.55) 2.00	1.83 1.73 >9.0 >19.5	1.65 1.85 >14.5

NOTE: Filled composition by volume. Interpolated data indicated by parentheses.
* Crystallinity 41–71%.
† Crystallinity 44–51%.

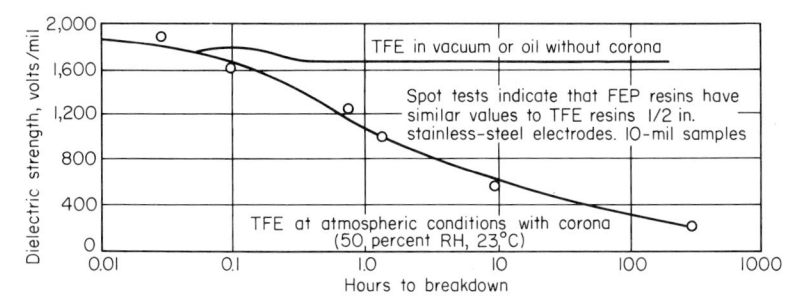

Fig. 38 Insulation life versus continuously applied voltage stress for TFE and FEP fluorocarbons.[32]

in the presence of oxygen. Based on a study made by Du Pont,[33] the following major findings resulted:
1. TFE in air:
   a. Has a damage threshold in the range 2 to 7 × 10^4 rads.
   b. Retains 50 percent of initial tensile strength after a dose of 10^6 rads.
   c. Retains 40 percent of initial tensile strength after a dose of 10^7 rads or more.
   d. Retains ultimate elongation of 100 percent or more (adequate for flexible wire) for doses up to 2 to 5 × 10^5 rads.
2. TFE in vacuum:
   a. Has a damage threshold in or above the range 2 to 7 × 10^5 rads.
   b. Retains 50 percent of initial tensile strength after a dose of 10^7 rads or more.
   c. Retains 40 percent of initial tensile strength after a dose of 8 × 10^8 rads or more.
   d. Retains ultimate elongation of 100 percent or more for doses up to and probably beyond 2 to 5 × 10^6 rads.
3. FEP in air will tolerate a dose 10 to 100 times larger than TFE in air.
4. FEP in vacuum will tolerate a dose up to 10 times larger than TFE in vacuum.
5. Testing in air after vacuum irradiation does not necessarily yield a true measure of radiation tolerance in the absence of oxygen.
6. The radiation tolerance of TFE and FEP is at least as good as that of some common electronic components, transistors in particular.
7. The radiation tolerance of TFE and FEP is much greater than that of man.
8. Since radiation produces electric charge distributions in the resin which decay with time, dielectric properties are a function of dose rate and exhibit transients. The low-frequency loss properties of TFE polymers are drastically affected by x-ray irradiation; the high-frequency loss properties considerably less so. The increases in dielectric constant and dissipation factor depend on the ambient oxygen concentration during exposure and recovery. The dielectric constant and dissipation factor of FEP remain unaffected by x-ray irradiation in vacuum at frequencies from 60 Hz to 100 kHz.
9. Teflon TFE and FEP resins will not evaporate in a vacuum of 10^{-7} mm Hg. Theoretical calculations show that they will not evaporate in any anticipated space vacuum to an extent that would limit their usefulness. Although some outgassing occurs initially, the volatiles are all absorbed atmospheric gases.

Ionomers

Outstanding advantages of this polymer class are combinations of toughness and transparency, and transparency and solvent resistance. Ionomers have high melt strength for thermoforming and extrusion coating processes, and a broad processing temperature range.

Limitations of ionomers include low stiffness, susceptibility to creep, low heat-distortion temperature, and poor ultraviolet resistance unless stabilizers are added, where these properties are important.

Most ionomers are very transparent. In 60-mil sections, internal haze ranges from 5 to 25 percent. Light transmission ranges from 80 to 92 percent over the visible region, and in specific compositions high transmittance extends into the ultraviolet region.

Mechanical properties Basically, commercial ionomers are nonrigid, unplasticized plastics. Outstanding low-temperature flexibility, resilience, high elongation, and excellent impact strength typify the ionomer resins.

Deterioration of mechanical and optical properties occurs when ionomers are exposed to UV light and weather. Some grades are available with UV stabilizers that provide up to one year of outdoor exposure with no loss in mechanical properties. Formulations containing carbon black provide UV resistance equal to that of black polyethylene.

Electrical properties Most ionomers have good dielectric characteristics over a broad frequency range. The combination of these electrical properties, high melt strength, and abrasion resistance qualifies these materials for insulation and jacketing of wire and cable.

Nylons

Also known as polyamides, nylons are strong, tough thermoplastics having good impact, tensile, and flexural strengths from freezing temperatures up to 300°F; excellent low-friction properties; and good electrical resistivities. They are not generally recommended for high-frequency, low-loss applications. Also, since all nylons absorb some moisture from environmental humidity, moisture-absorbtion characteristics must be considered in designing with these materials. They will absorb anywhere from 0.5 to nearly 2 percent moisture after 24 hr water immersion. There are low-moisture-absorption grades, however; and hence moisture-absorption properties do not have to limit the use of nylons, especially for the lower-moisture-absorption grades. Application information on nylons is given in Table 12, and data for three grades are presented in Tables 13 and 14.

Regarding the identifications for the grades of nylon, certain nylons are identified by the number of carbon atoms in the diamine and dibasic acid used to produce that particular nylon. For instance, nylon 6/6 is the reaction product of hexamethylene and adipic acid, both of which are materials containing six carbon atoms in their chemical structure. The commercially available nylons are 6/6, 6, 6/10, 8, 11, and 12. Grades 6 and 6/6 are the strongest structurally; grades 6/10 and 11 have the lowest moisture absorption, best electrical properties, and best dimensional stability; and grades 6, 6/6, and 6/10 are the most flexible. Grades 6, 6/6, and 8 are heat-sealable, with nylon 8 being capable of crosslinking. Another grade, nylon 12, offers advantages similar to those of grades 6/10 and 11, but offers lower cost possibilities owing to being more easily and economically processed. Also a high-temperature type of nylon, known as Nomex,* exists. It is discussed separately in a subsequent section entitled High Temperature Nylon.

In situ polymerization of nylon permits massive castings. Cast nylons are readily polymerized directly from the monomer material in the mold at atmospheric pressure. The method finds application where the size of the part required and/or the need for low tooling cost precludes injection molding. Cast nylon displays excellent bearing and fatigue properties as well as the other properties characteristic of other basic nylon formulations, with the addition of size and short-run flexibility advantages of the low-pressure casting process.

One special process exists in which nylon parts are made by compressing and sintering, thereby creating parts having exceptional wear characteristics and dimensional stability. Various fillers such as molybdenum disulfide and graphite can be incorporated into nylon to give special low friction properties. Also, nylon can be reinforced with glass fibers, thus giving it considerable additional strength. These variations are further discussed in a later part of this chapter dealing with glass-fiber-reinforced thermoplastic materials.

---

* Trademark of E. I. du Pont de Nemours & Co., Inc., Wilmington, Del.

Mechanical properties Modulus of elasticity data for nylons are compared with those for several other thermoplastic materials in Fig. 20. Figure 23 compares similar materials in tensile yield strength. Deflection is shown in Fig. 24.

Nylon, like all thermoplastic materials, exhibits some creep when subjected to stress. Extent of creep depends on stress level, temperature, and time. A part that is subjected to long-time stress must be designed accordingly so that deformation with time is not excessive for the application, and so that fracture will not occur. Typical values of creep versus time under various stress levels for nylon and polycarbonate are shown in Fig. 39. For nylon, most cold flow takes place during the first 24 hr. This is a useful checkpoint for testing parts under load.

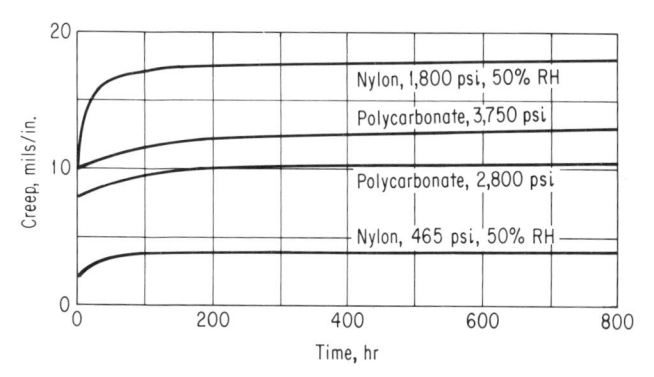

Fig. 39 Creep versus time at room temperature for nylon and polycarbonate.[34]

The creep values shown in Fig. 39 are at relatively low stress levels compared to published tensile stress values. However, these are typical working stresses. In metals, it is common practice to use yield stress or endurance limit and apply a safety factor to arrive at design stress. The same procedure does not apply for plastics. Stress levels must be kept at a fraction of published strength values to ensure proper performance. Published tensile strength of nylon, for example, is 12,000 psi. Stressing a part to 1,800 psi, a safety factor of 6.7, can still result in 16 mils/in. cold flow.

Frictional Properties. Studies of the frictional properties of nylon will not show a constant coefficient of friction, since the values obtained depend upon many variables. Of particular importance are the type of equipment used, the temperature at which the test is conducted, the cleanliness of the material being tested, and the surface finish. The load and speed also affect the values obtained.

Data on coefficients of friction are shown for several conditions in Table 20. Tests have indicated that there is little variation in the coefficient of friction over a temperature spread of 73 to 250°F and speed changes from 8 to 400 ft/min. In any application where friction is critical, it is recommended that measurements be made under simulated operating conditions.

Effect of Temperature on Nylon. One of the major property advantages of nylon resins is that they retain useful mechanical properties over a wide range of temperatures. Properly designed parts can be used successfully from −60 to +400°F.

Both long-term and short-term effects of temperature must, however, be considered. In the short term, there are effects on such properties as stiffness, toughness, etc. There is also the possibility of stress relief and its effect on dimensions. Of most concern in long-term applications at high temperature is gradual oxidative embrittlement, and for such cases the use of heat stabilized resins is recommended.

The important consideration for design work with nylon resins is that exposure to high temperatures in air for a period of time will result in a permanent change

TABLE 20 Coefficients of Friction for Nylon under Various Conditions[35] (Test method: thrust washer)

Zytel* on Zytel		
No lubricant	Static	Dynamic
Maximum............	0.46	0.19
Minimum............	0.36	0.11
Zytel on Delrin		
No lubricant	Static	Dynamic
Maximum............	0.20	0.11
Minimum............	0.13	0.08
Zytel on Steel		
No lubricant	Static	Dynamic
Maximum............	0.74	0.43
Minimum............	0.31	0.17
Normal pressure........	20 psi	(1.41 kg/cm^2)
Sliding speed..........	95 ft/min	(29 meters/min)
Temperature..........	73°F	(23°C)

* Zytel 101 at 2½% moisture. Trademark of E. I. du Pont de Nemours & Co., Inc. Wilmington, Del.

in properties due to oxidation. The permanent change in properties depends on the temperature level, the time exposed, and the composition of nylon used. The effect on the tensile strength, as measured at room temperature and 2.5 percent moisture content, is that a 25 percent strength reduction can occur in three months at 185°F, and a 50 percent reduction can occur in three months at 250°F. Stabilized nylon does not change appreciably at 250°F aging.

High-temperature oxidation reduces the impact strength even more than the static strength properties. For instance, the impact strength and elongation of nylon are reduced considerably after several days at 250°F.

A similar change in properties is encountered on exposure to high-temperature water for long periods of time. In this case a reaction with water takes place. There is no significant reaction up to 120°F. This has been confirmed by molecular-weight measurements on 15-year-old samples.[35]

In boiling water, the tensile strength is slowly reduced until after 2,500 hr it levels off at 6,000 psi (tested at room temperature and 2.5 percent moisture content). The elongation drops rapidly after 1,500 hr; hence, this time has been taken as the limit for the use of basic nylon. Some compositions are especially resistant to hot water exposure, however.

Effect of Moisture on Dimensional Control. Freshly molded objects normally contain less than 0.3 percent of water, since only dry molding powder can be successfully molded. These objects will then absorb moisture when they are exposed to air or water. The amount of absorbed water will increase in any environment until an equilibrium condition based on relative humidity (RH) is reached. Equilibrium moisture contents for two humidity levels are approximately as follows:

	Zytel 101	Zytel 31
50% RH air.....................	2.5%	1.4%
100% RH air (or water)..........	8.5%	3.5%

These equilibrium moisture contents are not affected by temperature to any significant extent. Thus, final water content at equilibrium will be the same whether objects of nylon are exposed to water at room temperature or at boiling temperature.

The time required to reach equilibrium, however, is dependent on the temperature, the thickness of the specimen, and the amount of moisture present in the surroundings. Nylon exposed in boiling water will reach the equilibrium level, 8.5 percent, much sooner than nylon in cool water.

When nylon that contains some moisture is exposed to a dry atmosphere, the loss of water will be the reverse of the changes described above and will take about the same length of time for a corresponding change to occur.

In the most common exposure, an environment of constantly varying humidity, no true equilibrium moisture content can be established. However, moldings of nylon will gradually gain in moisture content in such an environment until a balance is obtained with the midrange humidities. A slow cycling of moisture content near this value will then occur. In all but very thin moldings, the day-to-day or week-to-week variations in relative humidity will have little effect on total moisture content. The long period changes, such as between summer and winter, will have some effect depending on thickness and the relative humidity range. The highest average humidity for a month will not generally be above 70 percent. In cold weather, heated air may average as low as 20 percent relative humidity. Even at these extremes, the change in moisture content of nylon is small in most cases, because of the very low rate of both absorption and desorption.

There are two significant dimensional effects that occur after molding. In some cases these oppose each other so that critical dimensions may change very little in a typical air environment. The first is a shrinkage in the direction of flow due to the relief of molded-in stresses. The second is an increase due to moisture absorption. In applications where dimensions are critical to performance and one of these effects predominates it may be necessary to anneal or moisture condition the parts to obtain the best performance. For example, a part that will be exposed to high temperatures might require annealing, and an object in water service might be moisture-conditioned to effect dimensional changes before use.

The magnitude of the dimensional change due to molded-in stresses depends on molding conditions and part geometry. This change in size can be determined in a given case only by measuring and annealing a few pieces. In general, long dimensions (in the direction of flow) will shrink, but the short dimensions will increase in an amount that is often too small to measure.

The effect of moisture content on the dimensions of a molding can be predicted more accurately than the effect of annealing. The changes in dimension at various water contents for Zytel 101 and Zytel 31 are shown in Fig. 40. These data are for the annealed or stress-free condition. For most applications in air, the dimensions corresponding to those obtained in equilibrium with 50 percent relative humidity air are usually chosen as the average size expected when a moisture balance is established. As shown, Zytel 101 will increase 0.006 in./in. from the dry condition to equilibrium with 50 percent relative humidity. Zytel 31 under similar conditions will increase 0.0025 in./in. In many applications these changes are small enough and occur so slowly that they do not affect the operation of finished parts.

Stress-free objects of Zytel 101 and Zytel 31 will increase approximately 0.026 and 0.007 in./in., respectively, during the change from dry to completely saturated with water. In water service, if dimensions are critical, the molding can be moisture-conditioned before assembly, to allow for this change. Once a part is at or near saturation, little dimensional change can occur unless it is allowed to dry for long periods.

In a later section of this chapter, nylon is compared to some other thermoplastics in heat deflection (Fig. 56), tensile modulus (Fig. 58), tensile strength (Fig. 59), stress-strain data (Fig. 60), and dimensional changes due to absorbed moisture (Fig. 63).

Electrical properties Because of its combination of mechanical toughness, the ability to operate continuously at temperatures as high as 105°C, and good resistance to ambient atmospheres, nylon has become widely used in electrical applications. Examples are coil forms, plugs, strain-relief grommets, binding posts, switch parts, and tough overcoatings on wire insulations.

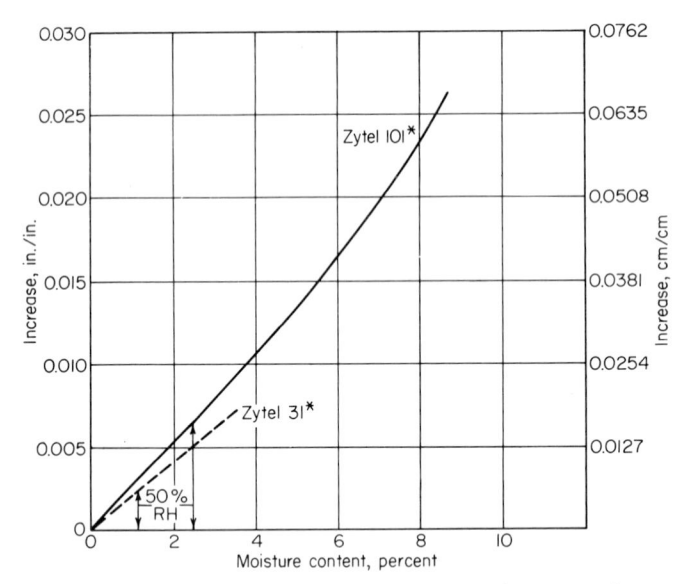

Fig. 40 Changes in dimensions with moisture content for two nylon materials in the stress-free, annealed conditions.[35] (* Trademark of E. I. du Pont de Nemours & Co., Inc.)

Electrically, nylon is not outstanding, but it is adequate for most 60-Hz power applications. Although the dissipation factor may be high under some conditions of temperature, this usually means little actual power loss in 110- to 220-volt applications. Mechanical requirements on insulation thickness will keep the voltage per mil value low. Power loss is proportional to (volts/mil)^2 × dielectric constant × dissipation factor × frequency.

Volume resistivity, dielectric strength, dielectric constant, and dissipation factor are all variable with moisture content and temperature of the nylon as well as with the specific kind of nylon.

The effect of frequency on the dielectric constant and dissipation factor of nylon is shown in Fig. 26. The effect of temperature on these two electrical properties is presented in Fig. 35. Resistivity decreases rapidly with moisture content, dropping as low as 10^{12} ohm-cm at 4 percent moisture, and as low as 10^9 ohm-cm at 8 percent moisture. Nylons do not compare favorably with the good electrical thermoplastics such as fluorocarbons or polystyrene in these two electrical parameters. Nylons are good general-purpose electrical materials, however, as noted in Table 14.

High-temperature nylon In addition to conventional nylon molding resins, a high-temperature grade of nylon exists. Known as Nomex, this high-temperature nylon retains about 60 percent of its strength at 475 to 500°F temperatures, which would melt conventional nylons. This high-temperature nylon has good dielectric strength (constant to 400°F and 95 percent relative humidity) and volume resistivity, and low dissipation factor, as shown in Fig. 41. Fabricated primarily in sheet, fiber, and paper form, this material is being used in wrapped electrical insulation constructions such as transformer coils and motor stators. It retains high tensile strength, resistance to wear, and electrical properties after prolonged exposure at temperatures up to 500°F.

Parylenes

Parylene is the generic name for members of a thermoplastic polymer series developed by Union Carbide, which are unique among plastics in that they are produced as thin films by vapor-phase polymerization. The basic member of the series, called parylene N, is poly-*p*-xylene, a completely linear, highly crystalline material. Parylene C, the second member of the series, is produced from the same monomer modified only by the substitution of a chlorine atom for one of the hydrogen atoms on the ring. Parylene C, or poly-(monochloro-*p*-xylene), is somewhat less crystalline than parylene N. Parylene N has the best dielectric properties, and parylene C has the best clarity and barrier properties. A third member of the series, the dichloroparylene D, has improved flame resistance. Many other members of this series are possible by substitution of chemical groups in the basic molecular structure.

Parylenes can also be deposited onto a cold condenser and then stripped off as a free film, or they can be deposited onto the surface of objects as a continuous, adherent coating in thicknesses ranging from 1,000 Å (about 0.004 mil) to 3 mils or more. Deposition rate is normally about 0.5 microns/min (about 0.02 mil). On cooled substrates, the deposition rate can be as high as 1.0 mil/min.

The material can be used at both elevated and cryogenic temperatures. The 1,000-hr service life for the N and the C members, respectively, is 200 to 240°F. Corresponding 10-year service in air is limited to 140 or 175°F. Parylenes are excellent in having low gas permeability, low moisture-vapor transmission, and low temperature ductibility. Dimensional stability is reported as better than that of polycarbonate, and overall barrier properties to most gases are reported superior to those of many other barrier films.

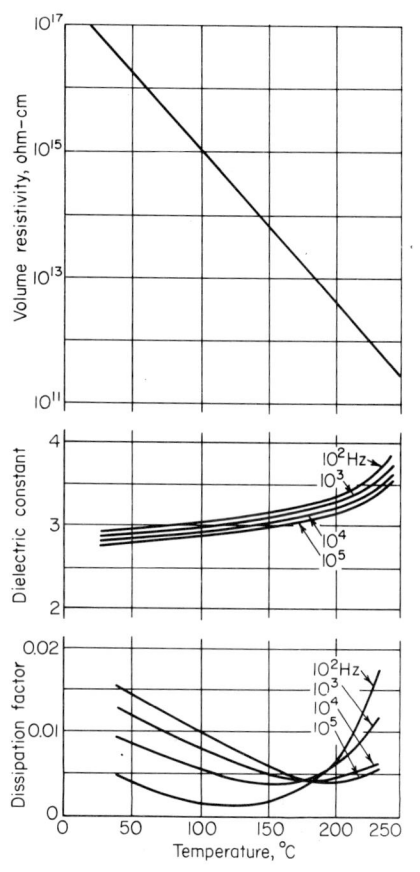

Fig. 41 Effect of temperature on electrical properties of 10-mil high-temperature nylon.[36]

Parylenes are good dielectric materials, with parylene N having a constant dielectric constant of 2.65 and a dissipation factor that increases from 0.0002 to 0.0006, over the range of 60 Hz to 1 MHz, at room temperature. The chemical modification of parylene C raises the dielectric constant to 2.9 to 3.1, and the dissipation factor to 0.012 to 0.020 over the above frequency range. Volume resistivity values approximate 10^{17} ohm-cm at room temperature, and short-time dielectric strength values are 5,600 volts/mil for parylene C, and 7,000 volts/mil for parylene N, in 1-mil test samples. Dielectric strength values decrease with increasing thickness, of course.

Phenoxies

The general properties of phenoxies are given in Tables 12 to 14. Phenoxy resins are transparent, high-molecular-weight thermoplastics that exhibit high rigidity, toughness, impact strength, tensile strength, elongation, and ductility, especially when compared to other rigid thermoplastics. Although not outstanding in any given property, phenoxies have an excellent combination of uniformly high properties.

Principal advantages of phenoxy resins are excellent processibility; low mold shrinkage, 0.003 to 0.004 in./(in.)(in.); and excellent dimensional stability and creep resistance. Regrind can be processed without loss of properties. The chief limitation is their heat-distortion temperature of 188°F. Recommended maximum continuous-use temperature is approximately 170°F. Phenoxies are generally good electrical insulating materials, but are not outstanding in any given characteristic.

Polyallomers, Polyethylenes, and Polypropylenes

This large group of polymers, basically divided into the three separate polymer groups listed under this heading, all belong to the broad chemical classification known as *polyolefins*. There are, of course, many categories or types within each of these three polymer groups. Although property variations exist among the polymer groups and among the subcategories within the polymer groups, there are also many similarities. The differences or unique features of each will be discussed later. The similarities are, broadly speaking, appearance, general chemical characteristics, and electrical properties. The differences are more notably in physical and thermal stability properties. Basically, polyolefins are all waxlike in appearance and extremely inert chemically, and they exhibit decreases in physical strength at somewhat lower temperatures than the higher-performance engineering thermoplastics. All these materials have nearly identical very low dissipation factors and dielectric constants, relatively independent of frequency and temperature over their operating range. These data for polyethylene are shown in Fig. 26. Resistivity values are very high among plastics. Polyethylenes were the first of these materials developed, and hence, for some of the original types, have the weakest mechanical properties. The later-developed polyethylenes, polypropylenes, and polyallomers offer improvements. The unique physical and mechanical features of each of these three polymer groups will be discussed at this point. The application information and physical and electrical properties of each are given in Tables 12 to 14.

Polyethylenes Polyethylenes are among the best-known plastics and come in three main classifications based on density: low, medium, and high. These density ranges are 0.910 to 0.925, 0.925 to 0.940, and 0.940 to 0.965, respectively. These three density grades are also sometimes known as Types I, II and III, respectively. All polyethylenes are relatively soft, and hardness increases as density increases. Generally, the higher the density, the better are the dimensional stability and physical properties, particularly as a function of temperature. Thermal stability of polyethylenes ranges from 190°F for the low-density material up to 250°F for the high-density material. Toughness is maintained to low negative temperatures.

The effect of temperature on the modulus of elasticity of polyethylenes is presented in Fig. 20. Temperature effect on the tensile yield strength is shown in Fig. 23, and 150°F deflection in Fig. 24. The stress-strain curve for high-density polyethylene is compared with that of some other thermoplastics later in this chapter (Fig. 60).

Polypropylenes Polypropylenes are chemically similar to polyethylenes, but have somewhat better physical strength at a lower density. The density of polypropylenes is among the lowest of all plastic materials, ranging from 0.900 to 0.915.

Polypropylenes offer more a balance of properties than a single unique property, with the exception of flex fatigue resistance. These materials have an almost infinite life under flexing, and they are thus often said to be "self-hinging." Use of this characteristic is widespread as plastic hinges. Polypropylenes are perhaps the only thermoplastics surpassing all others in combined heat resistance, rigidity, toughness, chemical resistance, dimensional stability, surface gloss, and melt flow, at a cost less than that of most others.

Because of their exceptional quality and versatility, polypropylenes offer outstanding potential in the manufacture of products through injection molding. Mold shrinkage is significantly less than that of other polyolefins; uniformity in and across the direction of flow is appreciably greater. Shrinkage is, therefore, more predictable; and there is less susceptibility to warpage in flat sections.

The general tensile yield strength of polypropylene, as a function of temperature, is shown in Fig. 23, and the tensile creep, under a variety of conditions, is given in Fig. 42. Elongation at yield for a general-purpose polypropylene is in the range of 5 to 7 percent, and yield stress is of the order of 5,000 psi at room temperature. The yield stress of high-impact polypropylenes can vary from about 2,500 to 4,000 psi, depending on the resin type.

The tensile moduli of elasticity of polypropylenes, over a temperature range of −50 to 250°F, may also be used for compressive modulus, since these moduli are approximately equal. However, the flexural moduli of elasticity of polypropylenes are different from the tensile strength values.

The ability of polypropylenes to absorb shock without fracture mainly depends on temperature. Above the glass transition temperature, impact strength is high and ductile failure occurs. However, at temperatures below the glass transition, where molecular movement has decreased, impact resistance drops markedly and brittle failure occurs.

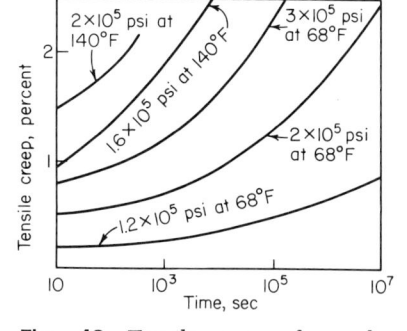

Fig. 42 Tensile creep for polypropylenes at several conditions.[20]

Polyallomers Polyallomers are also polyolefin-type thermoplastic polymers produced from two or more different monomers, such as propylene and ethylene, which would produce propylene-ethylene polyallomer. As can be seen, the monomers, or base chemical materials, are similar to those of polypropylene or polyethylene. Hence, as was mentioned above, and would be expected, many properties of polyallomers are similar to those of polyethylenes and polypropylenes. Having a density of about 0.9, they, like polypropylenes, are among the lightest plastics.

Polyallomers have a brittleness temperature as low as −40°F and a heat-distortion temperature as high as 210°F at 66 psi. Excellent impact strength plus exceptional flow properties of polyallomer provides wide latitude in product design. Notched Izod impact strengths run as high as 12 ft-lb/in. of notch.

Although the surface hardness of polyallomers is slightly less than that of polypropylenes, resistance to abrasion is greater. Polyallomer is superior to linear polyethylene in flow characteristics, moldability, softening point, hardness, stress-crack resistance, and mold shrinkage. The flexural fatigue-resistance properties of polyallomers are as good as or better than those of polypropylenes.

Crosslinked polyolefins While polyolefins have many outstanding characteristics, they, like all thermoplastics to some degree, tend to creep or cold-flow under the influence of temperature, load, and time. In order to improve this and some other properties considerable work has been done on developing crosslinked polyolefins, especially polyethylenes. The crosslinked polyethylenes offer thermal performance improvements of up to 25°C or more. Crosslinking has been achieved primarily by chemical means and by ionizing radiation. Products of both types are available. Radiation-crosslinked polyolefins have gained particular prominence in a heat-shrinkable form. This is achieved by crosslinking the extruded or molded polyolefin using high-energy electron-beam radiation, heating the irradiated material above its crystalline melting point to a rubbery state, mechanically stretching to an expanded form (up to four or five times the original size), and cooling the stretched material. Upon further heating, the material will return to its original size, tightly shrinking onto the object around which it has been placed. Heat-shrinkable boots, jackets, tubing, etc., are widely used. Also, irradiated polyolefins, sometimes known as irradiated polyalkenes, are important materials for certain wire and cable jacketing applications.

Polyamide-Imides and Polyimides

These two somewhat related groups of plastics have some outstanding properties for electronic applications. This is due to their combination of high-temperature stability up to and beyond 500°F, good electrical and mechanical properties which

1-62 Plastics for Electronics

are also relatively stable from low negative temperatures to high positive temperatures, dimensional stability (low cold flow) in most environments, excellent resistance to ionizing radiation, and very low outgassing in high vacuum. Polyamide-imides and polyimides have very good electrical properties, though not as good as those of TFE fluorocarbons, but they are much better than TFE fluorocarbons in mechanical and dimensional stability properties. This provides advantages in many high-temperature electronic applications. All these properties also make polyamide-imides and polyimides excellent material choices in extreme environments of space and temperature. These materials are available as solid (molded and machined) parts, films, laminates, and liquid varnishes and adhesives. Since data are relatively similar except for form factor, the data presented below are for solid polyimides, unless indicated otherwise. Films are quite similar to Mylar,* except for improved high-temperature capabilities, as shown in Table 21.

TABLE 21 Comparison of Properties of Kapton* Polyimide Film and Mylar* Polyester Film

Property	ASTM test	Temperature, °C	Value Kapton	Value Mylar
Dielectric strength at 60 Hz, 1 mil thick, volts/mil	D 149	25	7,000	7,000
		150	6,000	5,000
Dielectric constant at 1 kHz	D 150	25	3.5	3.1
		200	3.0	
Dissipation factor at 1 kHz	D 150	25	0.003	0.0047
		200	0.002	0.01
Volume resistivity, ohm-cm	D 257	25	10^{18}	10^{18}
		200	10^{14}	5×10^{11}
Surface resistivity at 1 kv, 50% RH, ohms	D 257	25	10^{16}	10^{16}
Tensile strength, psi	D 882	25	25,000	23,000
		200	17,000	7,000
Yield point, psi	......	25	14,000	12,000
		200	9,000	1,000
Stress to produce 5% elongation, psi	D 882	25	13,000	13,000
Ultimate elongation, %	D 882	25	70	100
		200	90	large
Tensile modulus, psi	D 882	25	430,000	550,000
		200	260,000	50,000
Bursting strength, psi	D 774	25	75	30
Density, g/cc	......	25	1.42	1.4
Coefficient of friction, kinetic, film to film	D 1505	25	0.42	0.45
Area factor, ft²/lb-mil	......	25	135	140

* Trademarks of E. I. du Pont de Nemours & Co., Inc., Wilmington, Del.

Mechanical wear and thermal properties Tensile and compressive stress-strain data in Fig. 43, show solid polyimide parts to be relatively strong, with an average tensile strength at 73°F of 10,500 psi, as indicated in Fig. 44. In tension, ultimate elongation at 73°F is about 5 to 6 percent. In compression, polyimides can be permanently distorted but will not break until a strain level of at least 20 percent is reached.

Polyimide parts are stiff, exhibiting a modulus of elasticity at 73°F of 450,000 psi (unfilled), as shown in Fig. 45. Graphite-filling (15 percent) increases stiffness about 40 percent.

The plot of apparent modulus versus time in Fig. 46 indicates the outstanding resistance to creep displayed by the polyimide parts, particularly those based on

* Trademark of E. I. du Pont de Nemours & Co., Wilmington, Del.

filled polyimide compositions. For example, the graphite-filled composition exhibits a 0.34 percent creep deflection plus an initial elastic deflection of 0.41 percent after 1,000 hr at 212°F and at a stress of 2,500 psi.

The unique feature of polyimide seals and bearings is their ability to perform without external lubrication at extremes of temperature. Operation at 700°F has been attained with graphite filler, and operation at −420°F with bronze filler. Poly-

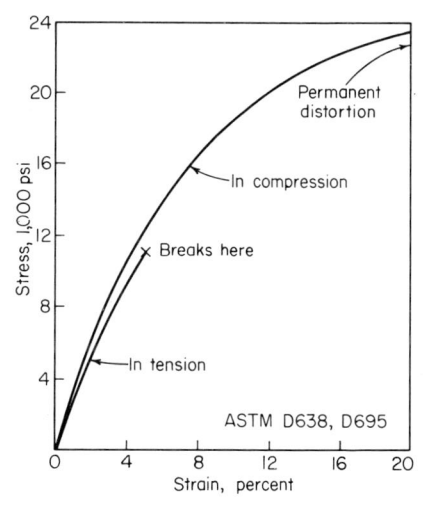

Fig. 43 Stress-strain curves for unfilled polyimide at 73°F.[37]

Fig. 44 Tensile strength versus temperature for unfilled and 15 percent graphite-filled polyimide.[37]

imide sleeve bearings and thrust washers operate at a higher PV than attainable with any other organic material. PV values as high as 100,000 psi × ft/min have been attained with the polyimide bearing surface operating against carbon steel without lubrication. PV values higher than 100,000 have been achieved in some applications involving intermittent operation and light lubrication. In an inert atmosphere (less than 10 ppm free oxygen), unlubricated wear rates over a broad range of PV values are extremely low compared to those other materials even at high PV values. In air the wear rate is accelerated. Excellent bearing performance in a hard vacuum is also attainable with MoS_2 filler.

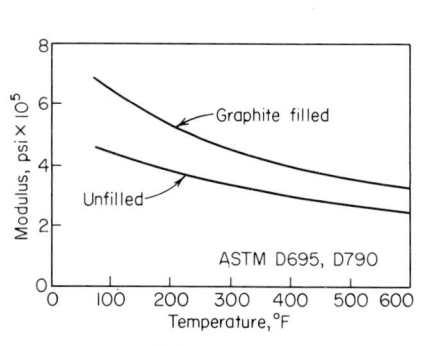

Fig. 45. Modulus of elasticity versus temperature for unfilled and 15 percent graphite-filled polyimide.[37]

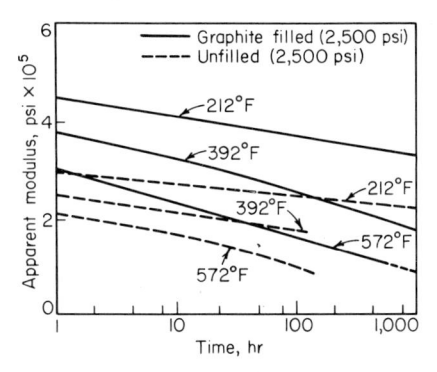

Fig. 46. Apparent modulus versus time for unfilled and 15 percent graphite-filled polyimide.[37]

1-64   Plastics for Electronics

The use of a lubricant improves the performance of polyimide seals and bearings by reducing both friction and wear. Specifically, under boundary lubrication conditions, 15 percent graphite-filled polyimide has significantly lower wear than babbitt and bronze; i.e., a lubricant would be more likely to fail to prevent "welding" in babbitt and bronze than in steel at boundary conditions.

Wear rate is not affected by varying the moisture level in polyimide parts. Without lubrication the dynamic coefficient of friction is high on start-up, but drops rapidly during the first minutes of operation. Figure 47 illlustrates the higher coefficient of friction at zero PV with decreasing friction at higher PV's.

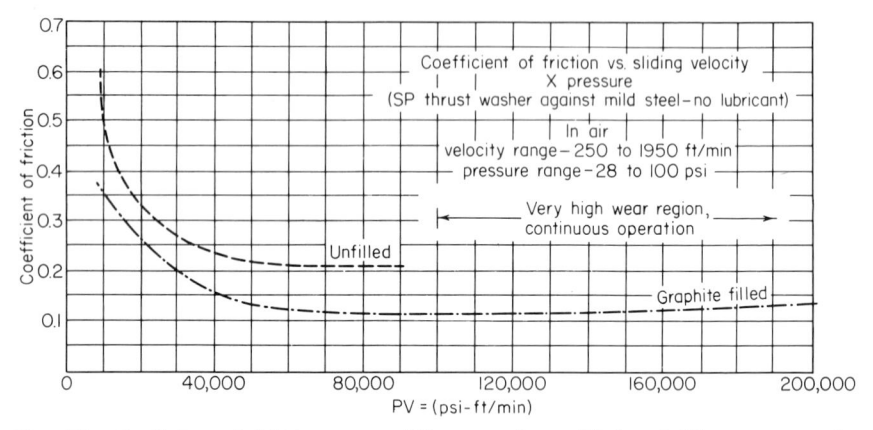

Fig. 47 Coefficient of friction versus PV curves for unfilled and 15 percent graphate-filled polyimide.[37]

Polyimide precision parts can be used continuously in air at temperatures up to 500°F. The continuous operating temperature can be raised to 600°F in an inert atmosphere or vacuum. For intermittent, short-term exposures, top temperatures appear to be in the 800 to 900°F range.

At elevated temperatures, polyimide parts retain both strength and stiffness, as shown in Figs. 44 and 45. Tensile strength at 600°F is 5,000 psi, and the flexural modulus is about 230,000 psi. Impact strength increases slowly with increasing temperature. Unnotched test bars have significantly higher impact strength than notched samples.

On prolonged exposure at 572°F, polyimide parts show an insignificant loss of tensile strength in nitrogen as compared to about a 40 percent decrease in air, as shown in Fig. 48.

As can be seen in Fig. 49, the weight loss of unfilled polyimide parts in air

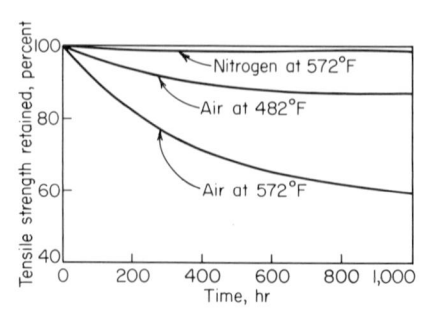

Fig. 48 Tensile strength versus time for unfilled polyimide at several conditions.[37]

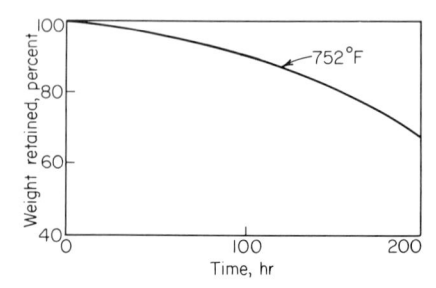

Fig. 49 Weight retention of unfilled polyimide resin aged at 752°F.[37]

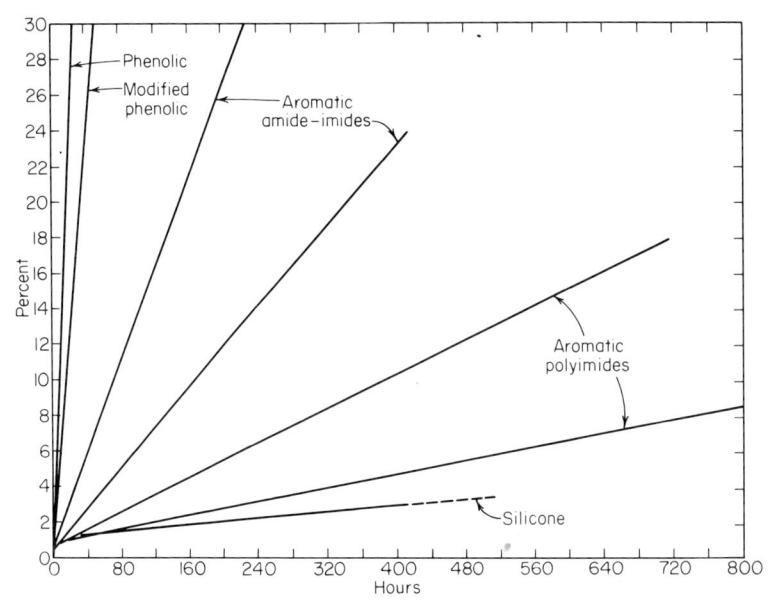

Fig. 50 Weight loss versus time at 320°C for thin films of several plastic materials.[38]

at 752°F is 10 percent after 100 hr exposure. The weight loss of a group of the more thermally stable plastics is compared in Fig. 50.

The thermal expansion of polyimide parts is between that of metals and conventional plastics. Use of certain fillers such as glass fibers reduces the expansion coefficient considerably. Like other materials, orientation during fabrication can have a pronounced effect on the thermal expansion of polyimide parts, as shown in Fig. 51.

Tests exposing polyimide parts in liquid nitrogen indicate possible applications in cryogenic systems at −320 to −420°F. Tensile strength at −320°F shows a 30 percent increase above the strength at room temperature. Shrinkage at −300°F, for example, is about 7 mils/in., as noted in Fig. 51.

Tests in high vacuum indicate that polyimide parts give exceptionally low outgassing at high temperatures. For example, for unfilled parts, the evaporation rate at 500°F and 10^{-7} mm Hg is less than 10^{-10} g/(cm^2)(sec). Other physical behavior in high vacuum environments is similar to that in inert atmospheres.

Fig. 51 Change in length versus temperature for unfilled and 15 percent graphite-filled polyimide.[37]

Electrical properties The combination of very good electrical properties, high strength, and excellent thermal and radiation resistance makes polyimide parts outstanding candidates for electrical applications in most severe environments. The dielectric constant, shown in Fig. 52, decreases gradually from 3.5 at room temperature to 3.0 at 500°F. At a given temperature, dielectric constant is essentially unchanged with frequency variations in the range of 100 Hz to 100 kHz. Dissipation factor, shown in Fig. 53, is influenced by both temperature and frequency. Up

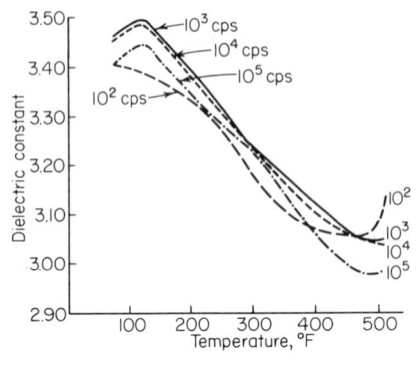

Fig. 52 Dielectric constant versus temperature and frequency for unfilled polyimide.[37]

Fig. 53 Dissipation factor versus temperature and frequency for unfilled polyimide.[37]

to about 212°F, dissipation factor increases with increasing frequency. From 212 to 400°F, frequency has essentially no effect, while above 400°F, dissipation factor decreases with increasing frequency.

Both dielectric constant and dissipation factor increase with increasing moisture content. For example, at 1 kHz and room temperature, the dielectric constant of a dry test bar of unfilled polyimide is 3.1, and the dissipation factor is 0.001. With a moisture content of 2.4 percent (obtained after 300 hr immersion in water at room temperature), these values are 4.0 and 0.002, respectively. Drying will restore the original values.

The volume resistivity of samples molded from unfilled polyimide is 10^{17} ohm-cm at room temperature, as shown in Fig. 54. This value decreases linearly to 10^{11} ohm-cm at 572°F. Surface resistivity is 10^{16} ohms/sq at room temperature, and decreases linearly to 5×10^{10} at 572°F.

Fig. 54 Volume resistivity versus temperature for unfilled polyimide.[37]

Polyimide parts as fabricated from unfilled resin have a very high arc resistance of 230 sec. However, when exposed to arcing, a carbon track is formed. The corona resistance is superior to that attainable with fluorocarbons and polyethylenes. For example, at 200 volts/mil (60 Hz and room temperature), corona life is 2,200 hr.

A potentially weak point of amide-imides and polyimides is moisture absorption in high humidities, especially in film forms, and attendant effects on physical and electrical properties. This varies with the material and condition, and should be investigated for a given application. Some amide-imide films are chemically etchable, especially in caustic solutions, and this presents design opportunities using such etchable amide-imide films.

Polycarbonates

This group of plastics is among those which are broadly classified as engineering thermoplastics, due to their high performance characteristics in engineering designs. Polycarbonates are tough, rigid, and dimensionally stable and are available as transparent or colored parts. They are easily fabricated with reproducible results, using molding or machining techniques. An important molding characteristic is the low and predictable mold shrinkage, which sometimes gives polycarbonates an advantage over nylons and acetals for close tolerance parts. As with most other plastics containing aromatic groups, radiation stability is high. The six most commonly useful properties of polycarbonates are creep resistance, high heat resistance, dimensional stability, good electrical properties, self-extinguishing properties, and the exceptional impact strength which compares favorably with that of some metals and exceeds that for essentially all competitive plastics. In fact, polycarbonate is sometimes considered to be competitive to zinc and aluminum casting. Although such comparisons have limits, the fact that the comparisons are sometimes made in material selection for product design indicates the strong performance characteristics possible in polycarbonates.

In addition to the performance of polycarbonates as engineering materials, as discussed in the following paragraphs, polycarbonates are also alloyed with other plastics, to increase the strength and rigidity of these other plastics. Notable among the plastics with which polycarbonates have been alloyed are the ABS plastics. Information and data on such alloys are presented in the preceding section entitled ABS Plastics. The application information on polycarbonates is given in Table 12, the physical and mechanical data appear in Table 13, and the electrical data in Table 14. More detailed mechanical and electrical data are given below.

Mechanical properties Polycarbonates are mechanically outstanding, as generally described above. One of the most oustanding advantages of polycarbonates over the other engineering-quality thermoplastics is high impact resistance. Izod impact strength on ⅛-in. ASTM standard bars is 12 to 16 ft-lb/in. of notch. On ¼-in. and ½-in. bars the value is between 2 and 3 ft-lb/in. of notch. Unnotched bars show impact resistance greater than 60 ft-lb/in. Polycarbonate's strength compares very favorably with that of some metals, and is up to four times better than that of nylon and acetal plastics.

On ⅛-in.-thick bars the break is ductile. On thicker bars it is the brittle type. With polycarbonates there is a limiting thickness above which brittle breaks result on notched samples. The transition region lies between 0.140 and 0.160 in. at room temperature. Lowering the temperature shifts the region. At approximately −18°F, brittle breaks result on ⅛ in. bars. Impact values drop to 1.62 and 1.52 per in. of notch at −35 and −65°F, respectively.

Heat aging at elevated temperatures progressively reduces the thickness at which ductile breaks on notched bars result. Immersion in water at elevated temperatures similarly reduces the thickness of the ductile break region.

The lowered impact required to break a notched bar suggests that care be exercised in designing molded parts. Sharp inside or stress-concentrating corners should be avoided. The impact strength of several thermoplastics is compared in Fig. 55.

Fig. 55 Impact strength versus temperature for several thermoplastics.[39,40]

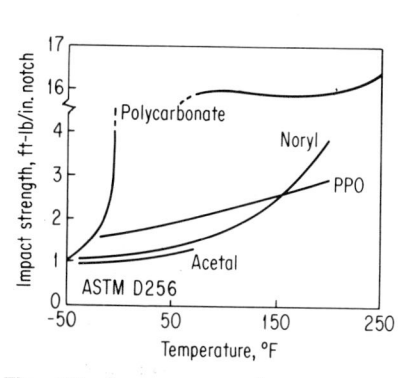

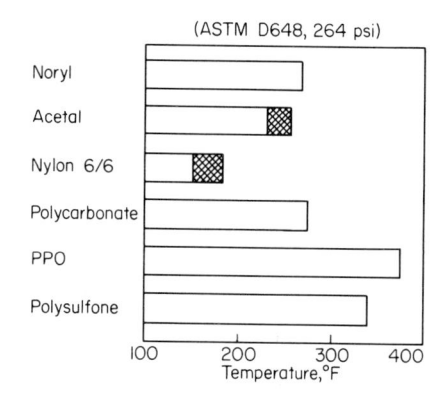

Fig. 56 Heat-deflection temperature of several thermoplastics at 264 psi.[39,40]

1-68 Plastics for Electronics

The polycarbonate's heat-deflection temperature of 270°F at 264 psi fiber stress is higher than that of nearly all other thermoplastics and is one of its outstanding characteristics. This suggests its use in applications formerly limited to the opaque thermosets, contributing the possibility of transparency in addition to lower processing costs.

Unloaded polycarbonate parts have outstanding dimensional stability up to 250°F, as shown in Fig. 56. Extended aging at this temperature will effect a progressive

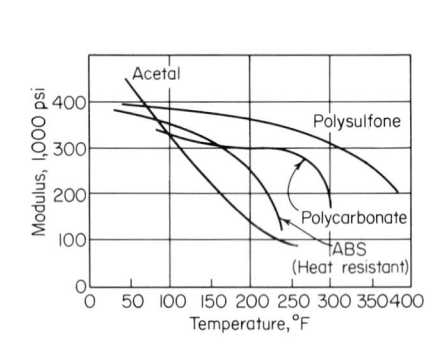

Fig. 57 Flexural modulus versus temperature for several thermoplastics.[41]

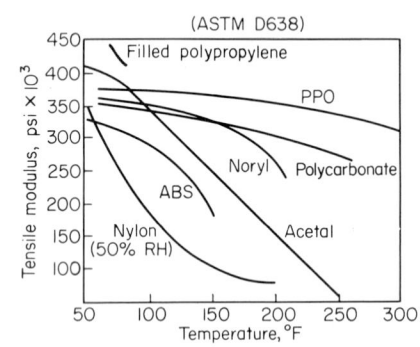

Fig. 58 Tensile modulus versus temperature for several thermoplastics.[39,40]

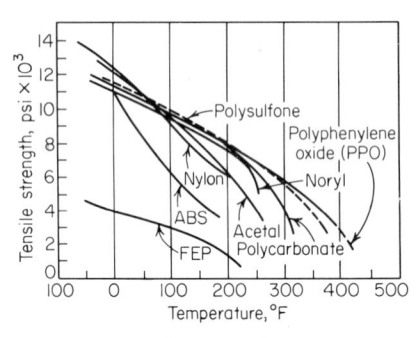

Fig. 59 Tensile strength versus temperature for several thermoplastics.[39,40]

increase in the tensile modulus, with an accompanying reduction in elongation and notched Izod impact strength.

Oxidative stability on heating in air is good, and immersion in water, and exposure to high humidity at temperatures up to 212°F have little effect on dimensions. Steam sterilization is another advantage that is attributable to the resin's high heat stability. However, if the application requires continuous exposure in water, the temperature should be limited to 140°F.

Other curves illustrating the relative thermal stability of polycarbonates are Fig. 57, which shows flexural modulus versus temperature; Fig. 58, which shows tensile modulus versus temperature; and Fig. 59, which shows tensile strength versus temperature. Each of these curves gives similar data for other engineering thermoplastics. The stress-strain curves for polycarbonate are shown, along with that for some other thermoplastics, in Fig. 60.

Exhibiting among the best creep resistance over a broad temperature range of any thermoplastic, polycarbonates can be utilized in plastics applications previously open only to thermosets. Loads up to the yield strength of approximately 8,000 psi can be tolerated for short periods at room temperature. Room-temperature creep data for several thermoplastics are shown in Figs. 39 and 61.

As with all materials, the resistances of polycarbonates to creep—permanent deformation under continued stress—changes with environmental and load variables. For long-term service at room temperatures, loadings should not exceed 1,500 to 2,000 psi. Applications in which bolts, rivets, or other fastening devices are used are quite feasible.

Polycarbonates are among the most stable plastics in a wet environment, as shown in Figs. 62 and 63.

Thermoplastics 1-69

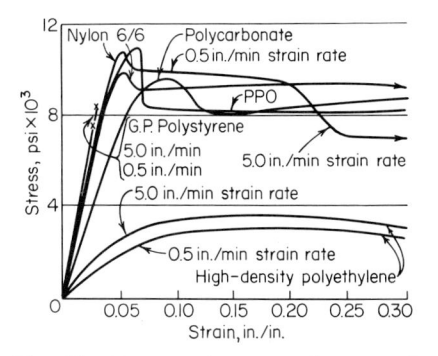

Fig. 60 Stress-strain curves for several thermoplastics. × indicates test bar broke. (*From Product Engineering, April 25, 1966. Copyright 1966 by McGraw-Hill, Inc., used by permission of Product Engineering.*)

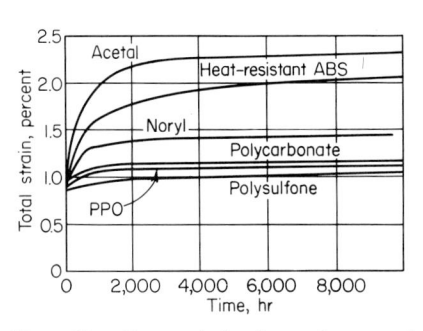

Fig. 61 Creep behavior of several thermoplastics at 72°F in air, and under 3,000 psi tensile stress.[41]

Electrical properties The basic electrical properties of polycarbonates are very good. The six most important electrical characteristics of polycarbonates are shown in Figs. 64 to 69, inclusive. The dissipation factor and dielectric constant, as a function of temperature, are shown in Figs. 65 and 67. It can be observed that there is no major change up to about 150°C. Both begin to increase beyond the 150 to 175°C temperature area, however. The dielectric constant value remains at about 3 up to 100 MHz or higher, whereas the power factor or loss factor does increase somewhat in this frequency range. Electrical properties remain relatively stable in high-humidity environments, which also helps in many design areas.

Polyesters

Thermoplastic polyesters are broadly inclusive, and there are many similar films; but one polyester film that has gained extremely widespread acceptance, and is more commonly referred to simply as Mylar, is *polyethylene terephthalate*, the polymer formed by the condensation reaction of ethylene glycol and terephthalic acid. This plastic is available only in film form. Its unusual balance of properties makes it especially useful in the electrical and electronic fields as a replacement for heavier, conventional insulating materials.

Mylar polyester film has a dielectric strength of 7,000 volts/mil for 1-mil film, a tensile strength of 23,000 psi excellent resistance to most chemicals and moisture, and can withstand temperature extremes from −60 to +150°C. It remains flexible and does not become brittle with age.

Mylar can be laminated, embossed, metalized, punched, dyed, or coated. Adhesives are available for laminating this material to itself and practically any other material. The film can also be coated with heat-sealable, friction, or abrasive materials. Heavier gages can be shaped in different ways such as stamping or vacuum forming. Lighter gages can be wound into spiral tubes.

There are eight types of Mylar polyester film. Each is manufactured in roll

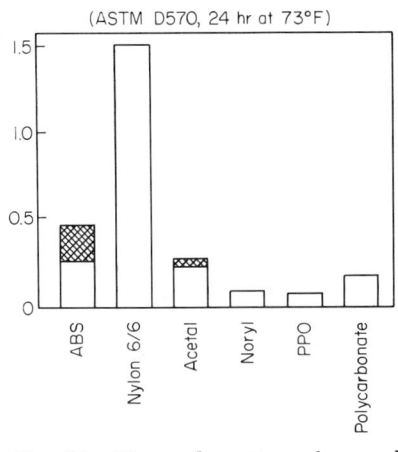

Fig. 62 Water absorption of several thermoplastics.[39,40]

1-70   Plastics for Electronics

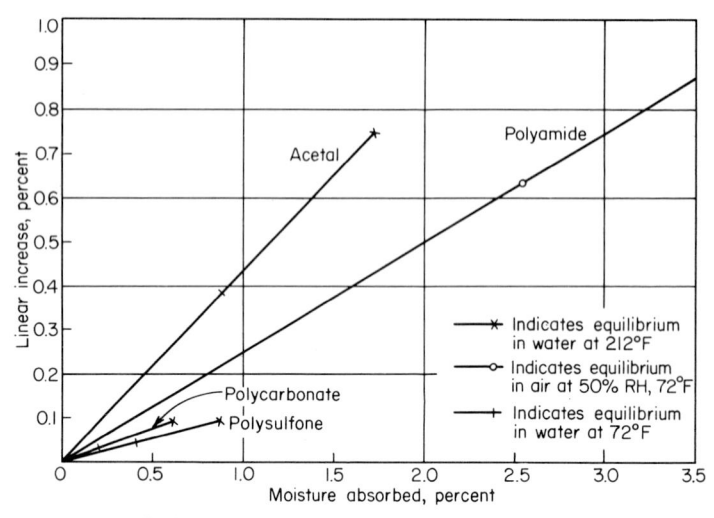

Fig. 63 Dimensional changes of several thermoplastics due to absorbed moisture.[39,40]

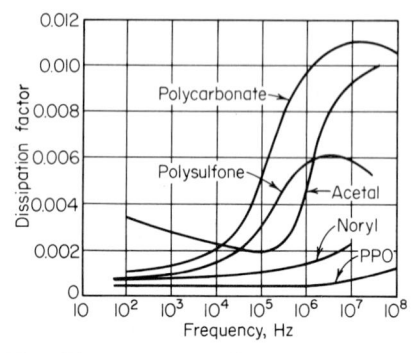

Fig. 64 Dissipation factor versus frequency for several thermoplastics.[43]

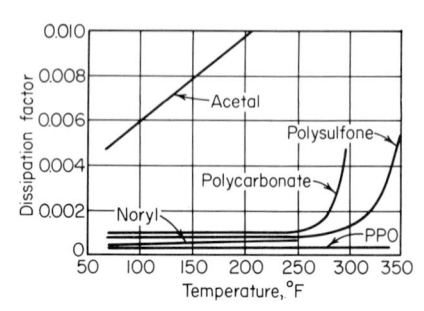

Fig. 65 Dissipation factor at 60 Hz versus temperature for several thermoplastics.[43]

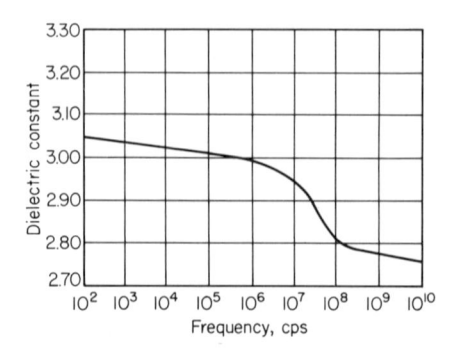

Fig. 66 Dielectric constant versus frequency for polycarbonates at 23°C.[39]

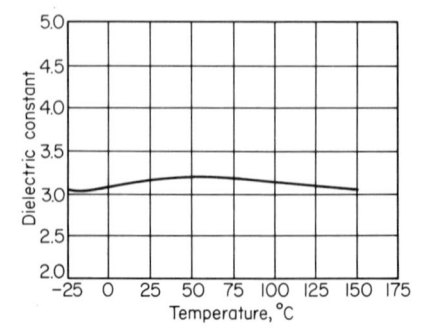

Fig. 67 Dielectric constant versus temperature for polycarbonates at 60 Hz.[39]

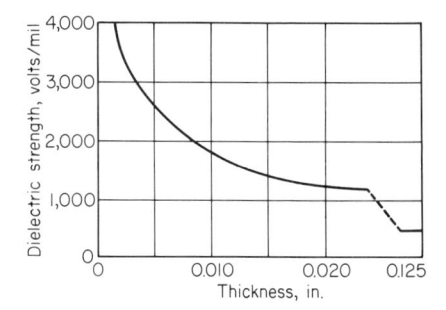

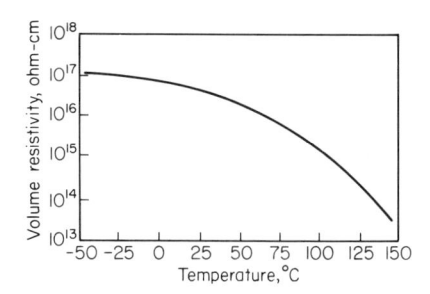

Fig. 68 Dielectric strength versus thickness for polycarbonates.[39]

Fig. 69 Volume resistivity versus temperature for polycarbonates.[39]

or sheet form in a range of thicknesses from ¼ mil (0.00025 in.) to 14 mils (0.0140 in.), and in widths from ½ in. to 60 to 72 in., depending on gage and type. It is important to know the correct type for properly specifying. The individual types are as follows:

1. *Type A.* General-purpose type, having exceptional high-temperature durability properties. It is available up to 14 mils thick. This type is normally used for motors, wire and cable, and laminations.

2. *Type C.* Has a higher level of insulation resistance than Type A films; it also has more clarity and sparkle than Type A film. This type has been especially designed for capacitors, and it is also normally used for metallizing and metallic yarn.

3. *Type D.* Has high transparency in the heavier gages. This type is used in cartography and drafting and for graphic arts applications where appearance and dimensional stability are important.

4. *Type HS.* Is used for the jackets of capacitors. It shrinks 30 percent in both directions when heated to 100°C. The film remains strong and tough after shrinking.

5. *Type M.* Has a heat-sealable polymer coating and is highly moistureproof. Like all types of Mylar polyester film, it does not get brittle with age and has a long shelf life.

6. *Type S.* For stationery supplies, book jackets, and sheet protectors. This is a very transparent film with a brilliant surface.

7. *Type T.* Has an unusually high tensile strength in the longitudinal direction. This type has more longitudinal stiffness than equivalent gages of the other films. It is used for magnetic tape and pressure-sensitive tapes.

8. *Type W.* Is resistant to the sun's ultraviolet radiation. It was specifically developed for exterior glazing applications. Greenhouses made with it last four to seven years in sun or snow.

The important physical, thermal, and electrical properties of Mylar are given in Table 22.

Polyphenylene Oxides

These materials are another of the high-performance engineering thermoplastics which are very useful for electronic and electrical applications. There are two groups in this family of plastics, namely, PPO* and Noryl.* Noryl is the lower-cost member of this family, at a sacrifice in some properties. Polyphenylene oxides have good all-around mechanical and electrical properties, with some of the PPO materials being especially good in their combination of mechanical properties and high-frequency electrical properties. For application purposes in electronics, polyphenylene oxides might be compared with acetals, polycarbonates, polysulfones, and nylons (polyamides). The major points of strength for PPO thermoplastics are:

* Trademarks of General Electric Company, Schenectady, N.Y.

1-72　Plastics for Electronics

TABLE 22 Important Properties of Mylar Polyester Film[44]

Properties	Typical values		Test condition*
Physical	Type A	Type T	
Ultimate tensile strength (MD), psi	25,000	40,000	25
Yield point (MD), psi	12,000 at 4%	Indeterminate	25
Stress to produce 5% elongation (MD), psi	>14,000	22,000	25
Ultimate elongation (MD), %	120	50	25
Tensile modulus (MD), psi	550,000	800,000	25
Impact strength, kg-cm/mil	6.0	6.0	25
Folding endurance (MIT), Hz	14,000	....	25
Tear strength:			
Propagating (Elmendorf), g/mil	15	12	25
Initial (Graves), g/mil	900	590	25
lb/in	2,000	1,300	
Bursting strength (Mullen), psi	66	55	25
Density	1.395	1.377	25
Coefficient of friction, kinetic, film to film	.45	.38	
Refractive index (Abbé)	1.64_nD 25	....	25
Area factor, $in.^2/(lb)(mil)$	20,500	20,700	25
Thermal			
Melting point, °F	480–510		
°C	250–265		
Service temperature, °F	−75 to +300		
°C	−60 to +150		
Coefficient of linear expansion, in./(in.) (°F) × 10^5	1.5		70–120°F
Coefficient of thermal conductivity:			
$(Btu)(in.)/(ft^2)(hr)(°F)$	1.035		25
$(cal)(cm)/(cm^2)(sec)(°C) \times 10^4$	3.63		25
Heat sealability	No		
Specific heat, cal/(g)(°C)	0.315		
Shrinkage, %	2–3		30 min at 150
Electrical			
Dielectric strength (1 mil) at 60 Hz, volts/mil	7,000		25
	5,000		150
Dielectric constant:			
At 60 Hz	3.25		25
At 1 kHz	3.1		25
At 1 MHz	3.0		25
At 1,000 MHz	2.8		25
At 60 Hz	3.7		150
Dissipation factor:			
At 60 Hz	0.0021		25
At 1 kHz	0.0047		25
At 1 MHz	0.016		25
At 1,000 MHz	0.003		25
At 60 Hz	0.0064		150
Volume resistivity, ohm-cm	1×10^{18}		25
	1×10^{13}		150
Surface resistivity, ohms:			
At 0% RH	1×10^{12}		25
At 100% RH	5×10^{11}		25
Insulation resistance, megohm-mfd	5,000		100
	400		130
	100		150

TABLE 22 Important Properties of Mylar Polyester Film[44] **(Continued)**

Chemical resistance to	Tensile strength	Elongation	Tear strength	Days immersed at room temperature
	Percent retained			
Acetic acid, glacial...................	75	107	83	33
Hydrochloric acid, 18%..............	82	99	113	18
Sodium hydroxide, 10%..............	83	30	33	33
Ammonium hydroxide, concentrated....	0	0	0	3
Acetone...........................	63	151	67	33
Hydrocarbon oil.....................	92	88	87	500 hr at 100°C
Phenolic resin, GE1678..............	92	3	73	Baked 168 hr at 150°C
Moisture absorption.................	Less than 0.8%			Immersion for 24 hr at 23°C

* In degrees centigrade unless otherwise specified.

1. Wide temperature range, brittle point approximately 275°F, and heat-distortion temperature of 375°F (at 264 psi)
2. Good mechanical properties relatively constant over a wide temperature range
3. Excellent electrical properties over a wide range of temperature and frequency
4. Unusual resistance to aqueous environments, acids, bases, and steam
5. Processibility on conventional extrusion and injection molding equipment

The most outstanding differences between Noryl and PPO are in low- and high-temperature resistance. Brittle point for Noryl is below −40°F, and heat deflection at 264 psi is 265°F, compared with −275 and +375°F, respectively, for PPO.

Mechanical properties The heat-deflection temperature of the polyphenylene oxides ranks high among the more rigid thermoplastics, as indicated in Fig. 56. Polyphenylene oxides also rank well in tensile strength versus temperature. Tensile modulus is shown in Fig. 58, and tensile strength in Fig. 59. Stress-strain characteristics appear in Figs. 60 and 70. Dimensional stability also rates well among thermoplastics, as shown by tensile creep data (Fig. 61). Water absorption is low among thermoplastics, (Fig. 62), as are the attendant changes in dimensions and weight associated with water absorption. Impact strength is not as good as that of polycarbonates at elevated temperatures, however, as is shown in Fig. 55.

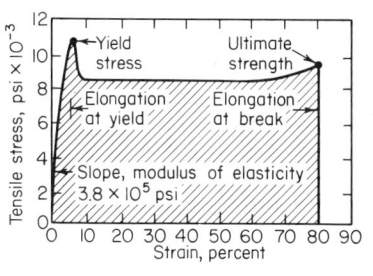

Fig. 70 Important tensile stress-strain characteristics for PPO.[40]

Electrical properties Polyphenylene oxides have excellent electrical properties which remain relatively constant with frequency and temperature over their safe operating temperature range. The effect of frequency on dissipation factor is given in Fig. 64, and the effect of temperature on 60-Hz frequency in Fig. 65. Like PPO, Noryl also has a relatively flat dissipation factor with temperature, with typical 60-Hz values being 0.0004 at 73°F, 0.0006 at 140°F, and 0.008 at 220°F, as indicated in Fig. 65. The dielectric constant of PPO, with frequency and temperature, is given in Table 23. The dielectric constant of Noryl is also low and relatively stable up to the 220 to 225°F range. Dielectric strength of both polyphenylene oxide classes is 500 to 550 volts/mil in ⅛-in. sections. Electrical properties of both materials are also relatively unaffected by humidity.

TABLE 23 Dielectric Constant of PPO at Various Temperatures and Frequencies[40]

Frequency, cps	Dielectric constant
60	2.56
10^3	2.55
10^6	2.55
10^6	2.55
10^9	2.59

Temperature, °F	Dielectric constant at 60 cps
360	2.52
190	2.54
73	2.56
20	2.55
−320	2.57

Polystyrenes

Polystyrenes represent an important class of thermoplastic materials in the electronics industry because of very low electrical losses. Mechanical properties are adequate within operating temperature limits, but polysytrenes are temperature-limited with normal temperature capabilities below 200°F. Polystyrenes can, however, be crosslinked to produce a higher-temperature material, as noted in Table 12. Crosslinked polystyrenes are actually thermosetting materials, and hence do not remelt, even though they may soften. The improved thermal properties, coupled with the outstanding electrical properties, hardness, and associated dimensional stability, make crosslinked polystyrenes the leading choice of dielectric for many high-frequency, radar-band applications. The radar bands are defined in Table 24.

TABLE 24 Identification of Radar Band Frequencies

Band	Frequency, GHz*
P	0.225–0.390
L	0.390–1.550
S	1.550–5.200
C	3.900–6.200
X	5.200–10.90
K	10.90–36.00
Q	36.00–46.00
V	46.00–56.00

* G (giga) is 1,000 mega; or 1,000 million cycles per second.

Also, polystyrenes are foamed to produce the very widely used low-density foam products. This, too, is discussed further below.

Conventional polystyrenes are essentially polymerized styrene monomer alone. By varying manufacturing conditions or by adding small amounts of internal and external lubrication, it is possible to vary such properties as ease of flow, speed of setup, physical strength, and heat resistance. Conventional polystyrenes are frequently referred to as normal, regular, or standard polystyrenes.

Since conventional polystyrenes are somewhat hard and brittle, having low impact strength, many modified polystyrenes are available. Modified polystyrenes are materials in which the properties of elongation and resistance to shock have been increased by incorporating into their composition varying percentages of elastomers. Hence, these types are frequently referred to as high-impact, high-elongation, or rubber-modified polystyrenes. The so-called superhigh-impact types can be quite rubbery.

Mechanical properties The effect of stress and time on the deflection of polystyrenes is compared with that of some of the other thermoplastics in Fig. 24. The strength properties of polystyrene drops off rapidly at 100°C or lower. The stress-strain curve for general-purpose polystyrene is shown, along with curves for several other thermoplastics, in Fig. 60. As was mentioned above, general-purpose polystyrenes are somewhat hard and brittle, but many modifications are available for improved impact strength, improved elongation, and improved thermal resistance. Other sacrifices often result, however, especially in electrical properties.

Polystyrenes are subject to stresses in fabrication and forming operations, and often require annealing to minimize such stresses for optimized final-product properties. Parts can usually be annealed by exposing them to an elevated temperature approximately 5 to 10°F lower than the temperature at which the greatest tolerable distortion occurs.

Polystyrenes have generally good dimensional stability and low mold shrinkage, and are easily processed at low costs. They have poor weatherability and are chemical attacked by oils and organic solvents. Resistance is good, however, to water, inorganic chemicals, and alcohols.

Electrical properties The exceptionally low dissipation factor and dielectric constant for polystyrene as a function of frequency is shown in Fig. 26. The dielectric constant and dissipation factor of some polystyrenes increase rapidly above 1,000 to 10,000 MHz. Hence, the specific material and application should be checked above this frequency area. The low dissipation factor coupled with the relative rigidity of polystyrene compared to polyethylene and TFE gives polystyrene advantages in many electronic applications requiring material hardness and extremely low electrical losses, particularly at high frequencies. As was mentioned above, the crosslinked polystyrenes are especially useful here.

The dielectric strength of polystyrenes is excellent, and the resistivity properties of polystyrenes are outstanding. The excellence of the resistivity characteristic can be seen in Fig. 71. These properties coupled with the other above-mentioned

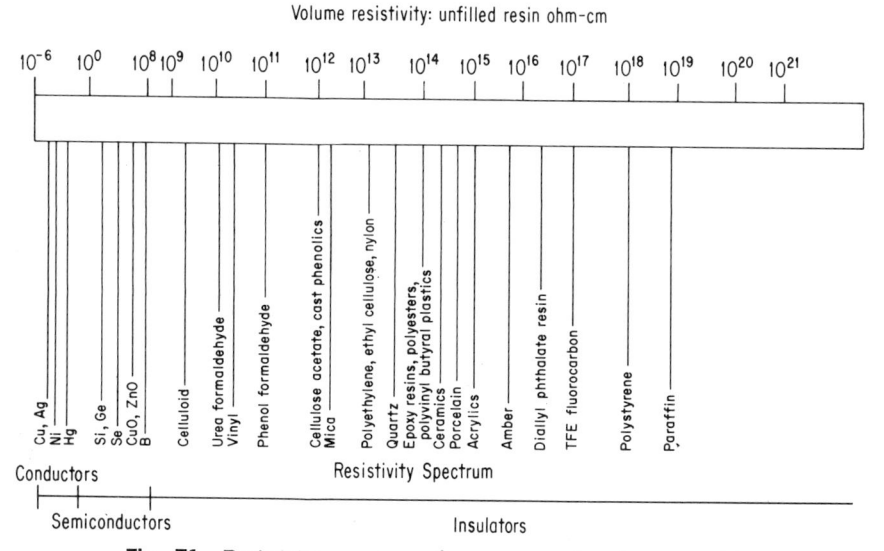

Fig. 71 Resistivity spectrum for some engineering materials.

excellent properties of polystyrenes make these materials most useful in high-frequency electronic applications, with the other application limitations.

Expanded polystyrene Expanded or foamed polystyrene is another useful form of polystyrene. While most plastics can be made into some low-density forms, perhaps none are made on a large scale at densities as low as expanded polystyrene. Densities in the 2-pcf (pound per cubic foot) area and lower are common. This low density presents design advantages in applications requiring thermal insulation, flotation, fragile component handling and cushioning and in other areas where low weight, low density, and soft (but not flexible) cushioning are important. The temperature limitations of conventional polystyrenes, as mentioned above, also apply to expanded polystyrene. Likewise, strength limitations exist, as for any soft, low-density structure. In thermal insulation properties, expanded polystyrene has a

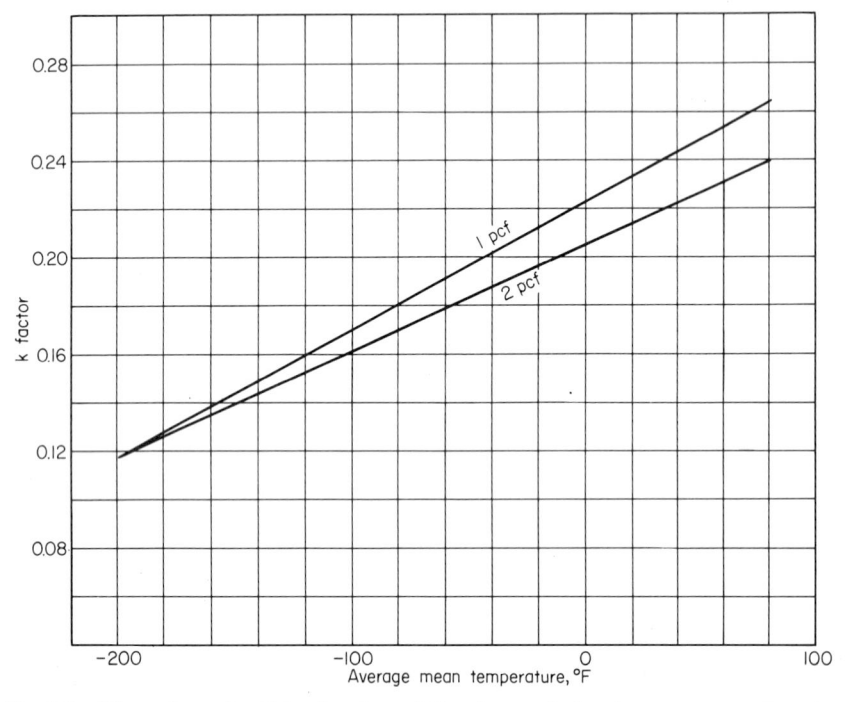

Fig. 72 Thermal conductivity for 1- and 2-pcf expanded polystyrenes as a function of temperature.[45]

slightly lower K factor than fibrous glass and corkboard, and approximately half that of cellular glass, expanded silica, and vermiculite. The thermal conductivity for 1- and 2-pcf expanded polystyrene, as a function of temperature, is shown in Fig. 72. In acoustical insulation properties, expanded polystyrene is not especially efficient. It is similar to wood, plaster, and brick for low-frequency sound, but better than these for higher-frequency sound. It is not as good as commercial sound-absorbing materials.

Polysulfones

Polysulfones are very useful thermoplastics for electronic design, having excellent strength versus temperature properties, good electrical properties (though not outstanding for high frequency), and outstanding strength retention over long periods of aging up to 300°F or over. Application guideline information is given in Table

12, and quantitative physical, mechanical, and electrical data are presented in Tables 13 and 14. Important properties will be detailed below.

Mechanical properties The 264-psi heat-deflection temperature of polysulfones rates very high among thermoplastics and is compared with that of several other thermoplastics in Fig. 56. Flexural modulus versus temperature is higher than for many other thermoplastics as indicated in Fig. 57. The tensile strength properties of polysulfones are generally similar to those of polycarbonates and polyphenylene oxides, and these are shown in Fig. 59. The dimensional changes of polysulfones due to absorbed moisture are low, being similar to those of polycarbonates, as indicated in Fig. 63. Although moisture absorption does increase slightly at 212°F, the linear dimensional change is little different from that at 72°F. Creep versus time is low, as is shown in Fig. 61. Creep is only slightly over 2 percent after 10,000 hr at 212°F and 4,000 psi, compared to about 4 percent after 1,500 hr at 210°F for polycarbonates. The retention of strength properties upon prolonged heat aging, mentioned above as perhaps the outstanding feature of polysulfones, is given in Fig. 73. Like many thermoplastics, stress relieving is sometimes required for polysulfones.

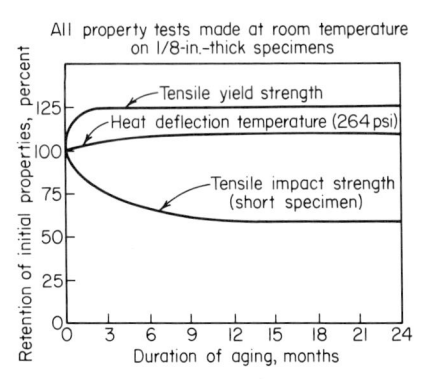

Fig. 73 Effect of heat aging at 300°F on three properties of polysulfones.[46]

Fig. 74 Effect of temperature on volume resistivity of polysulfones.[46]

Two methods of annealing are recommended. The most rapid of these is to immerse the part in hot liquid at 330°F for a short period. A pure refined mineral oil or glycerin may be used in the bath. Glycerin has the advantage of being water-soluble and, therefore, is easily removed from the annealed part by rinsing in water. The annealing time will vary, depending on thickness, from 1 to 5 min; 2 min is required for an 0.125-in. thickness, and 5 min for an 0.400-in. thickness. With thick parts it is advisable to lessen the thermal shock by immersing the part in boiling water for 5 min before and after the high-temperature annealing step.

A second recommended method is to anneal in air, again at 330°F. The time will vary from 2 to 4 hr, again depending on thickness. An 0.060-in. thickness requires 1 hr, an 0.125-in. thickness 3 hr, and an 0.400-in. thickness 4½ hr.

Electrical properties The dissipation factor versus frequency is shown in Fig. 64, and the dissipation factor versus temperature at 60 Hz in Fig. 65. The dielectric constant of polysulfones is approximately 3.1 up to 1 MHz, and decreases slightly at 10 MHz. Volume resistivity, as a function of temperature is shown in Fig. 74. The other important electrical properties of polysulfones are also good and are satisfactory for most electronic applications. The electrical properties of polysulfones are maintained to approximately 90 percent of their initial values after one year or more exposure at 300°F. Also, the basic electrical properties are generally stable up to about 350°F, and under exposure to water or high humidity.

Vinyls

Vinyls are good general-purpose electrical insulating materials, but are not outstanding from the electronics application viewpoint. The wide range of basic formulations, and the wider range of modifications, makes a detailed study of specific materials desirable for a given use. General guidelines are given in Tables 12 to 14. Perhaps their widest applications in electronics are as hookup wire jacketing, and as sleeving and tubing. These applications are discussed in a separate chapter on wires and cables. There are many grades and types of vinyls, among which are some special electrical grades, which should be considered for any electronic applications. Owing to the many variations available, consultation with leading suppliers is especially important. In addition to various basic vinyl classifications, vinyls may be rigid, flexible, or foamed. Further, they may be filled in many ways, alloyed with other plastics, plasticized with various plasticizers. Some vinyls are particularly outstanding in their resistance to corrosive chemicals.

Basically, the vinyl family is comprised of seven major types. These are polyvinyl acetals, polyvinyl acetate, polyvinyl alcohol, polyvinyl carbazole, polyvinyl chloride, polyvinyl chloride-acetate, and polyvinylidene chloride.

Polyvinyl acetals consist of three groups, namely, polyvinyl formal, polyvinyl acetal, and polyvinyl butyral. These materials are available as molding powders, sheet, rod, and tube. Fabrication methods include molding, extruding, casting, and calendering.

Polyvinyl chloride (PVC) is perhaps the most widely used and highest-volume type of the vinyl family. PVC and polyvinyl chloride-acetate are the most commonly used vinyls for electronic and electrical applications.

Vinyls are basically tough and strong, resist water and abrasion, and are excellent electrical insulators. Special tougher types provide high wear resistance. Excluding some nonrigid types, vinyls are not degraded by prolonged contact with water, oils, foods, common chemicals, or cleaning fluids such as gasoline or naphtha. Vinyls are affected by chlorinated solvents.

Generally, vinyls will withstand continuous exposure to temperatures ranging up to 130°F; flexible types, filaments, and some rigids are unaffected by even higher temperatures. These materials also are unaffected by even higher temperatures. These materials also are slow burning, and certain types are self-extinguishing; but direct contact with an open flame or extreme heat must be avoided. Temperature effects on modulus of elasticity are shown in Fig. 20, and effects on tensile yield strength are shown in Fig. 23.

PVC is a material with a wide range of rigidity or flexiblity. One of its basic advantages is the way it accepts compounding ingredients. For instance, PVC can be plasticized with a variety of plasticizers to produce soft, yielding materials to almost any desired degree of flexibility. Without plasticizers, it is a strong, rigid material that can be machined, heat-formed, or welded by solvents or heat. It is tough, with high resistance to acids, alcohol, alkalies, oils, and many other hydrocarbons. It is available in a wide range of colors. Typical uses include profile extrusions, wire and cable insulation, and foam applications. It also is made into film and sheets.

Polyvinyl chloride raw materials are available as resins, latexes, organosols, plastisols, and compounds. Fabrication methods include injection, compression, blow or slush molding, extruding, calendering, coating, laminating, rotational and solution casting, and vacuum forming.

A convenient-to-use form of vinyls is the polyvinyl chloride dispersion, which is available in liquid form, and solidifies upon application of heat. Hence, these dispersions can be used for casting, potting, and dip coating types of applications—somewhat similar to RTV silicones—except with considerably different properties, of course. Properties favor silicones, and costs favor the vinyl dispersions where properties are acceptable.

GLASS-FIBER-REINFORCED THERMOPLASTICS

Basically, thermoplastic molding materials are developed and can be used without fillers, as opposed to thermosetting molding materials, which are more commonly used with fillers being incorporated into the compound. This is primarily because shrinkage, hardness, brittleness, and other important processing and use properties necessitate the use of fillers in thermosets. Thermoplastics, on the other hand, do not suffer from the same shortcomings as thermosets, and hence, can be used as molded products, without fillers. However, thermoplastics do suffer from creep and dimensional stability problems, especially under elevated-temperature and load conditions. Because of this shortcoming, most designers find difficulty in matching the techniques of classical stress-strain analysis with the nonlinear, time-dependent strength-modulus properties of thermoplastics. Glass-fiber-reinforced thermoplastics,

TABLE 25 Comparison of Some Important Properties of Raw and Glass-filled Thermoplastics[47]

Material	Percent glass loading	Tensile strength at 73°F, psi × 10^{-3} (D 638)*	Elongation at 73°F, % (D 638)	Flexural strength at 73°F, psi × 10^{-3} (D 790)	Impact strength at 73°F, notched Izod (D 256)	Heat-distortion temperature at 264 psi, °F	Rockwell hardness	Specific gravity
Nylon 6/10:								
Raw	...	8.5	85–300	....	1.2	....	R111	1.09
Short fiber	30	17–19	3.0	22–26	1.4–2.2	400	E35-45, R118	1.30
Long fiber	30	19.0	1.9	23	3.4	420	E70-75	1.30
Nylon 6/6:								
Raw	...	9.0	60–300	12.5	1.0–2.0	150–186	R108-R118	1.13–1.15
Short fiber	30	18.5–23	3.0	26.5–32	1.2–2.0	400–470	E50-55-R120	1.37
Long fiber	30	20	1.5	28	2.5	498	E60-70	1.37
Nylon 6:								
Raw	...	7.0	25–320	8	1.0–3.6	152–158	R103-R118	1.12–1.14
Short fiber	30	17–24	3	22.5–32	1.3–2.0	400–420	E45-50, M90	1.37
Long fiber	30	21.0	2.0	27	3.0	420	E55-60	1.37
Polycarbonate:								
Raw	...	9.5	60–110	13.5	2.5	265–280	M70-R118	1.2
Short fiber	20	12–18.5	2.5–3	17–25	1.5–2.5	285–295	M92-R118	1.35
Long fiber	20	14–18.5	2.2–5	18.5	2.5–3.0	295	H80-90	1.35
Polypropylene:								
Raw	...	4.3	200–700	6	....	135–145	R85-110	0.90–0.91
Short fiber	20	6.0	3.0	7.5	1.0	230	M40	1.05
Long fiber	20	8	2.2	10	3.5	283	M50	1.05
Polyacetal:								
Raw	...	10.0	15	14	1.4	255	M94-R120	1.425
Short fiber	20	10–13.5	2–3	14–15	0.8–1.4	315–325	M70-75-95	1.55
Long fiber	20	10.5	2.3	15	2.2	325	M75-80 (Shore)	1.55
Polyethylene:								
Raw	...	1.2	50–600	4.8	0.5–16	90–105	D50-60	0.92–0.94
Short fiber	20	6	3.0	7	1.1	225	R60	1.10
Long fiber	20	6.5	3.0	8	2.1	260	R60	1.10
Polysulfone:								
Raw	...	10.2	50–100	....	1.3	345	M69-R120	1.24
Short-fiber	30	16	2	21	1.8	360	....	1.41
Long fiber	30	18.5	2.0	24	2.5	333	E45-55	1.37
PPO:								
Raw	...	11.6	20–40	16.5	1.2	345	M78	1.06
Short fiber	30	18.0	4.0–6.0	24.4	1.7	360	M94	1.27

* ASTM Test Method.
NOTE: Most favorable figures for short-fiber performance are based upon results with nominal ¼-in. fibers. Not included in the table are glass-reinforced styrene, SAN, ABS, and polyurethane, for which comparable data between short and long fibers are not available.
SOURCES: Data for raw-resin and long-glass properties for all resins—Fiberfil Inc.; short-glass polysulfone, polyethylene, polypropylene, all three nylons—Fiberfil and Liquid Nitrogen Processing Corp.; polycarbonate—Fiberfil, LNP, and General Electric Co.; polyacetal and PPO—LNP, Fiberfil, Celanese Corp., and Du Pont Co.; nylon 6/6—Fiberfil, LNP, and Polymer Corp.

1-80 Plastics for Electronics

or FRTP's, help to simplify these problems. For instance, 40 percent glass-fiber-reinforced nylon outperforms its unreinforced version by exhibiting two and a half times greater tensile and Izod impact strength, four times greater flexural modulus, and only one-fifth of the tensile creep.

Thus, glass-fiber-reinforced thermoplastics fill a major materials gap in providing plastic materials which can be reliably used for strength purposes, and which in fact can compete with metal die castings. Strength is increased with glass-fiber reinforcement, as is stiffness and dimensional stability. The thermal expansion of

TABLE 26 Effect of Glass Content on Some Important Properties of Polysulfone[48]

Properties	ASTM Method	Unfilled	20	30	40
Physical					
Specific gravity at 73°F	D 792	1.24	1.38	1.45	1.55
Specific volume, in.³/lb	....	22.40	20.1	19.0	17.9
Water absorption in 24 hr at 73°F, %	D 570	0.20	0.20	0.20	0.18
Equilibrium content, immersion at 73°F	....	0.60	0.60	0.58	0.58
Mold shrinkage, in./in	D 955	0.007–0.008	0.003–0.004	0.002–0.003	0.0015–0.002
Thermal					
Heat deflection temperature, °F:	D 648				
At 264 psi	....	345	360	365	370
At 66 psi	....	358	370	375	380
Thermal conductivity, Btu/(hr)(ft²)(°F)(in.)	C 177	1.8	2.0	2.2	2.5
Coefficient of linear thermal expansion, in./(in.)(°F) × 10^5	D 696	3.1	1.7	1.4	1.2
Flammability, in./min	D 635	Self-Ext.	Self-Ext.	Self-Ext.	Self-Ext.
	U.L. Subj. 94		S.E. Grp. 1	S.E. Grp. 1	S.E. Grp. 1
Mechanical					
Tensile strength, psi:	D 638				
At −40°F	....	12,200	19,000	22,000	24,000
At 73°F	....	10,200	15,000	18,000	20,000
At 210°F	....	7,500	10,000	13,000	16,000
At 300°F	....	5,800	7,000	9,500	12,000
Elongation at 73°F, %	D 638	50–100	3.0	3.0	2.0
Flexural strength, psi:	D 790				
At −40°F	....	....	25,000	27,000	32,000
At 73°F	....	15,400	21,000	24,000	27,000
At 210°F	....	12,300	17,000	19,000	23,000
At 300°F	....	10,000	13,000	15,500	16,800
Flexural modulus, psi:	D 790				
At 73°F	....	390,000	850,000	1,200,000	1,600,000
At 210°F	....	365,000	780,000	1,125,000	1,500,000
At 300°F	....	315,000	740,000	1,000,000	1,400,000
Compressive strength, psi	D 695	13,900	21,000	24,000	25,500
Shear strength, psi	D 732	9,000	9,000	9,500	10,000
Deformation under load at 122°F, %:	D 621				
2,000 psi	....	0.10	0.08	0.07	0.05
4,000 psi	....	0.20	0.12	0.10	0.06
Izod impact, ¼-in. bar, ft-lb/in.:	D 256				
Notched:					
At −40°F	....	1.2	1.2	1.6	2.0
At 73°F	....	1.2	1.3	1.8	2.0
Unnotched	D 256	>60	12	14	16
Tensile impact, S type, ft-lb/in.²	D 1822	250	48	63	82
Rockwell hardness	D 785	M69, R120	M92, L107	M92, L108	M92, L109
Taber abrasion (1,000-g load, CS-17 wheel), mg/kHz	D 1044	20	35–40	35–40	35–40
Fatigue endurance limit at 2 MHz, psi	D 671	1250	....	5400	

the FRTP's is reduced. Creep is substantially reduced, and molding precision is much greater.

Dimensional stability of glass-reinforced polymers is invariably better than that of the nonreinforced materials. Mold shrinkages of only a few mils per inch are characteristic of these products. Low moisture absorption of reinforced plastics ensures that parts will not suffer dimensional increases under high-humidity conditions. Also, the characteristic low coefficient of thermal expansion is close enough to that of such metals as zinc, aluminum, and magnesium that it is possible to design composite assemblies without fear that they will warp or buckle when cycled over temperature extremes. In applications where part geometry limits maximum wall thickness, reinforced plastics almost always afford economies for similar strength or stiffness over their unreinforced equivalents. A comparison of some important properties for unfilled and glass-filled (20 and 30 percent) thermoplastics is shown in Table 25. Property variations possible with changes in loading from 20 to 40 percent are shown for polysulfone in Table 26.

TABLE 27 Electrical Properties of Several Glass-reinforced Thermoplastics[2]

Property	Nylon, 20–40% filled	Polycarbonate, 20–40% filled	Polypropylene, 20–40% filled	Polyphenylene oxides, 30% filled PPO	Polyphenylene oxides, 30% filled Noryl	Polystyrene, 20–30% filled
Volume resistivity at 50% RH, ohm-cm	1.53–5.5×10^{15}	1.4–1.52×10^{15}	1.7×10^{16}	2.5×10^{17}	2.1×10^{17}	3.2–3.7×10^{16}
Dielectric strength, ⅛ in., volts/mil:						
Short time	408–503	475	475	500	550	350–425
Step by step	375–450	475	375			350–430
Dielectric constant:						
At 60 Hz	4.0–4.6	3.7	2.37	2.9	2.9	2.8–3.1
At 1 kHz	3.9–4.4	3.7	2.36	2.9	2.9	2.8–3.0
1 MHz	3.4–3.9	3.2–3.5	2.38	2.9	2.9	2.8–3.0
Dissipation factor:						
60 Hz	0.018–0.025	0.003–0.005	0.0022	0.0009	0.0009	0.004–0.014
1 kHz	0.020–0.025	0.002–0.004	0.0017	0.0009	0.0009	0.001–0.004
1 MHz	0.017–0.022	0.009	0.0035	0.0016	0.0015	0.001–0.003
Arc resistance, sec	92–148	5–120	74	120	75	25–40
Water absorption, ⅛-in. after 24 hr, %	0.2–2.0	0.07–0.10	0.01–0.05	0.07	0.06	0.05–0.10
Maximum continuous-use temperature, °F	300–400	275–300	300–320	315–365	225–250	180–200

One significant property to electrical-electronic engineers is temperature resistance. Glass-reinforced nylons, for example, have a deflection temperature up to 500°F, compared to 150°F for the same material in unreinforced form. This makes possible use of these materials in electrical or electronic products where the material might be adjacent to soldering where heat might accumulate. Electrical properties of several glass-reinforced thermoplastics are shown in Table 27. In addition to the property improvements mentioned above, other properties of the base resin are also either enhanced or undergo no significant change. In most cases the electrical properties are parallel to or better than those of the base resin. In general, they have excellent electrical insulating properties, suiting them for connectors, insulators, and structural portions of electrical components. Electrical properties of polymers such as nylon are generally improved by glass reinforcement, particularly at high humidities. Volume resistivity of type 6/6 nylon, for example, is 10^{14} while the value for glass-filled nylon is 10^{15}. Improvement here is probably due to the reduction in moisture absorption with addition of glass fibers, as shown in Fig. 75. Chemical resistance is essentially unchanged, except that environmental

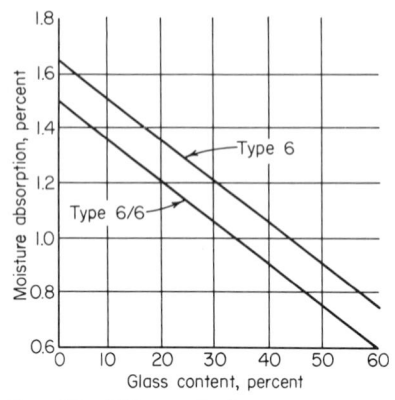

Fig. 75 Effect of glass content on moisture absorption of two nylon types.[49]

stress-crack resistance of such polymers as polycarbonate and polyethylene is markedly increased by glass reinforcement.

PLASTIC PROCESSING METHODS AND DESIGN GUIDES

Although most electronic industry users of plastics buy plastic parts from plastic processors, they should still have some knowledge of plastic processing, as such information can often be helpful in optimizing product design. Also, an increasing number of electronic companies are doing some in-house processing. For these reasons, some guideline information on plastic processing and some guidelines on design of plastic parts are presented here. It should be mentioned that this information applies broadly to all classes of plastics and types of processing. Most plastic suppliers will provide very specific data and guidelines for their individual products. This invaluable source of guidance is too often unused. A strong recommendation is made to utilize plastic suppliers more fully for product design guidance. The information presented at this point will be valuable for making initial design and process decisions, however.

Table 28 explains the major ways that plastic materials can be formed into parts, and the advantages, limitations, and rough cost of each processing method. In general, a plastic part is produced by a combination of cooling, heating, flowing, deformation, and chemical reaction. As previously noted, the processes differ, depending on whether the material is a thermoplastic or thermoset.

The usual sequence of processing a thermoplastic is to heat the material so that it softens and flows, force the material in the desired shape through a die or in a mold, and chill the melt into its final shape. By comparison, a thermoset is typically processed by starting out with partially polymerized material which is softened and activated by heating (either in or out of the mold), forcing it into the desired shape by pressure, and holding it at the curing temperature until final polymerization reaches the point where the part hardens and stiffens sufficiently to keep its impressed shape.

The cost of the finished part basically depends on the material and the process used. A very rough estimate of the finished cost of a part can be obtained by multiplying the material cost by a factor ranging from 1.5 to 10. The cost factors shown in Table 28 are based on general industry experience in pricing.

Table 29 gives guidelines on part design for the various plastic processing methods shown in Table 28. The design of a plastic part frequently depends on the processing method selected to make the part. Also, of course, selection of the best processing method frequently is a function of the part design. Listed in Table 29 are the major plastic processing methods and their respective design capabilities, such as minimum section thicknesses and radii and overall dimensional tolerances.

The basic purpose of this guide is to show the fundamental design limits of the many plastic processing methods. As mentioned above, it is always an advantage to review the design with the materials supplier and/or experienced plastics processor. Their suggestions on design and configuration, at an early stage, may avoid costly production and performance problems.

EMBEDDING MATERIALS AND PROCESSES

Embedding materials, in the electronics field, are those materials used to encase or embed an electronic part or assembly, so that the embedded assembly is protected

TABLE 28 Descriptions and Guidelines for Plastic Processing Methods[50]

Process	Description	Key advantages	Notable limitations	Cost factor*
Blow molding	An extruded tube (parison) of heated thermoplastic is placed between two halves of an open split mold and expanded against the sides of the closed mold by air pressure. The mold is opened, and the part is ejected.	Low tool and die costs; rapid production rates; ability to mold relatively complex hollow shapes in one piece.	Limited to hollow or tubular parts; wall thickness and tolerances often hard to control.	1.5–5, 2–3
Calendering	Doughlike thermoplastic mass is worked into a sheet of uniform thickness by passing it through and over a series of heated or cooled rolls. Calenders also are used to apply plastic covering to the back of other materials.	Low cost; sheet materials are virtually free of molded-in stresses; i.e., they are isotropic.	Limited to sheet materials; very thin films not possible.	1.5–3, 2–5.5
Casting	Liquid plastic (usually thermoset except for acrylics) is poured into a mold (without pressure), cured, and removed from the mold. Cast thermoplastic films are made by depositing the material, either in solution or in hot-melt form, against a highly polished supporting surface.	Low mold cost; ability to produce large parts with thick cross sections; good surface finish; suitable to low-volume production.	Limited to relatively simple shapes; except for cast films, becomes uneconomical at high-volume production levels; most thermoplastics not suitable.	1.5–3, 2–2.5
Compression molding	A thermoplastic or partially polymerized thermosetting resin compound, usually preformed, is placed in a heated mold cavity; the mold is closed, heat and pressure are applied, and the material flows and fills the mold cavity. Heat completes polymerization, and the mold is opened to remove the part. The process is sometimes used for thermoplastics, e.g., vinyl phonograph records.	Little waste of material and low finishing costs; large, bulky parts are possible.	Extremely intricate parts involving undercuts, side draws, small holes, delicate inserts, etc., not practical; very close tolerances difficult to produce.	2–10, 1.5–3
Cold forming	Similar to compression molding in that material is charged into split mold; it differs in that it uses no heat—only pressure. Part is cured in an oven in a separate operation. Some thermoplastic sheet material and billets are cold-formed in process similar to drop hammer-die forming of metals. Shotgun shells are made in this manner from polyethylene billets.	Ability to form heavy or tough-to-mold materials; simple; inexpensive; often has rapid production rate.	Limited to relatively simple shapes; few materials can be processed in this manner.	

*Material cost × factor = purchase price of a part: top figure is overall range, bottom is probable average cost.

TABLE 28 Descriptions and Guidelines for Plastic Processing Methods[50] (Continued)

Process	Description	Key advantages	Notable limitations	Cost factor*
Extrusion	Thermoplastic or thermoset molding compound is fed from a hopper to a screw and barrel where it is heated to plasticity and then forwarded, usually by a rotating screw, through a nozzle having the desired cross section configuration.	Low tool cost; great many complex profile shapes possible; very rapid production rates; can apply coatings or jacketing to core materials, such as wire.	Limited to sections of uniform cross section.	2–5, 3–4
Filament winding	Continuous filaments, usually glass, in form of rovings are saturated with resin and machine-wound onto mandrels having shape of desired finished part. Once winding is completed, part and mandrel are placed in oven for curing. Mandrel is then removed through porthole at end of wound part.	High-strength reinforcements are oriented precisely in direction where strength is needed; exceptional strength-to-weight ratio; good uniformity of resin distribution in finished part.	Limited to shapes of positive curvature; openings and holes reduce strength.	5–10, 6–8
Injection molding	Thermoplastic or thermoset molding compound is heated to plasticity in cylinder at controlled temperature; then forced under pressure through a nozzle into sprues, runners, gates, and cavities of mold. The resin solidifies rapidly, the mold is opened and the part(s) ejected. In modified version of process—runnerless molding—the runners are part of mold cavity.	Extremely rapid production rates, hence low cost per part; little finishing required; good dimensional accuracy; ability to produce relatively large, complex shapes; very good surface finish.	High initial tool and die costs; not practical for small runs.	1.5–5, 2–3
Laminating, high pressure	Material, usually in form of reinforcing cloth, paper, foil, etc, preimpregnated or coated with thermoset resin (sometimes a thermoplastic), is molded under pressure greater than 1,000 psi into sheet, rod, tube, or other simple shape.	Excellent dimensional stability of finished product; very economical in large production of parts.	High tool and die costs; limited to simple shapes and cross-section profiles.	2–5, 3–4
Matched-die molding	A variation of conventional compression molding, this process uses two metal molds having a close-fitting, telescoping area to seal in the plastic compound being molded and to trim the reinforcement. The reinforcement, usually mat or preform, is positioned in the mold, and the mold is closed and heated (pressures generally vary between 150 and 400 psi). Mold is then opened and part lifted out.	Rapid production rates; good quality and reproducibility of parts.	High mold and equipment costs; parts often require extensive surface finishing, e.g., sanding.	2–5, 3–4

1-84

Rotational molding	A predetermined amount of powdered or liquid thermoplastic or thermoset material is poured into mold. Mold is closed, heated, and rotated in the axis of two planes until contents have fused to inner walls of mold. The mold is opened and part removed.	Low mold cost; large hollow parts in one piece can be produced; molded parts are essentially isotropic in nature.	Limited to hollow parts; in general, production rates are slow.	1.5–5, 2–3
Slush molding	Powdered or liquid thermoplastic material is poured into a mold to capacity. Mold is closed and heated for a predetermined time to achieve a specified build up of partially cured material on mold walls. Mold is opened, and unpolymerized material is poured out. Semifused part is removed from mold and fully polymerized in oven.	Very low mold costs; very economical for small-production runs.	Limited to hollow parts; production rates are very slow; limited choice of materials that can be processed.	1.5–4, 2–3
Thermoforming	Heat-softened thermoplastic sheet is placed over male or female mold. Air is evacuated from between sheet and mold, causing sheet to conform to contour of mold. There are many variations, including vacuum snapback, plug assist, drape forming, etc.	Tooling costs generally are low; produces large parts with thin sections; often economical for limited production of parts.	In general, limited to parts of simple configuration; limited number of materials to choose from; high scrap.	2–10, 3–5
Transfer molding	Thermoset molding compound is fed from hopper into a transfer chamber where it is heated to plasticity. It is then fed by means of a plunger through sprues, runners, and gates of closed mold into mold cavity. Mold is opened and the part ejected.	Good dimensional accuracy; rapid production rate; very intricate parts can be produced.	Molds are expensive; high material loss in sprues and runners; size of parts is somewhat limited.	1.5–5, 2–3
Wet lay-up or contact molding	Number of layers, consisting of a mixture of reinforcement (usually glass cloth) and resin (thermosetting), are placed in mold and contoured by roller to mold's shape. Assembly is allowed to cure (usually in an oven) without application of pressure. In modification of process, called spray molding, resin systems and chopped fibers are sprayed simultaneously from spray gun against mold surface; roller assist also is used. Wet lay-up parts sometimes are cured under pressure, using vacuum bag, pressure bag, or autoclave.	Very low cost; large parts can be produced; suitable for low-volume production of parts.	Not economical for large-volume production; uniformity of resin distribution very difficult to control; mainly limited to simple shapes.	1.5–4, 2–3

* Material cost × factor = purchase price of a part: top figure is overall range, bottom is probable average cost.

TABLE 29 Guidelines on Part Design for Plastic Processing Methods[50]

Design rules	Blow molding	Casting	Compression molding	Extrusion	Injection molding	Wet lay-up (contact molding)	Matched-die molding	Filament winding	Rotational molding	Thermoforming	Transfer molding
Major shape characteristics	Hollow bodies	Simple configurations	Moldable in one plane	Constant cross-section profile	Few limitations	Moldable in one plane	Moldable in one plane	Structure with surfaces of revolution	Hollow bodies	Moldable in one plane	Simple configurations
Limiting size factor	M	M	ME	M	ME	MS	ME	WE	M	M	ME
Minimum inside radius, in.	0.125	0.01–0.125	0.125	0.01–0.125	0.01–0.125	0.25	0.06	0.125	0.01–0.125	0.125	0.01–0.125
Undercuts	Yes	Yes[a]	NR[b]	Yes	Yes[a]	Yes	NR	NR	Yes[c]	Yes[a]	NR
Minimum draft, degrees	0	0–1	>1	NA[b]	<1	0	1	2–3	1	1	1
Minimum thickness, in.	0.01	0.01–0.125	0.01–0.125	0.001	0.015	0.06	0.03	0.015	0.02	0.002	0.01–0.125
Maximum thickness, in.	>0.25	None	0.5	6	1	0.5	1	3	0.5	3	1
Maximum thickness buildup, in.	NA	2–1	2–1	NA	2–1	2–1	2–1[d]	NR	NA	NA	2–1
Inserts	Yes	Yes	Yes	Yes	Yes	Yes	Yes	Yes	Yes	NR	Yes
Built-in cores	Yes	Yes	No	Yes	Yes	Yes	Yes	Yes	Yes	No	Yes
Molded-in holes	Yes	Yes	Yes	Yes[e]	Yes	Yes	Yes	No	Yes	Yes	Yes
Bosses	Yes	Yes	Yes	Yes	Yes	Yes	No[f]	No[g]	Yes	Yes	Yes
Fins or ribs	Yes	Yes	Yes	No	Yes	Yes	Yes	No	Yes	Yes	Yes
Molded in designs and nos.	Yes	Yes	Yes	No	Yes	Yes	Yes	No	Yes	Yes	Yes
Overall dimensional tolerance, in./in.	±0.01	±0.001	±0.001	±0.005	±0.001	±0.02	±0.005	±0.005	±0.01	±0.01	±0.001
Surface finish[h]	1–2	2	1–2	1–2	1	4–5	4–5	5	2–3	1–3	1–2
Threads	Yes	Yes	Yes	No	Yes	No	No	No	Yes	No	Yes

Notes:
M — Material. ME — Molding equipment. MS — Mold size. WE — Winding equipment.
[a] Special molds required.
[b] NR—not recommended; NA—not applicable.
[c] Only with flexible materials.
[d] Using premix: as desired.
[e] Only in direction of extrusion.
[f] Using premix: yes.
[g] Possible using special techniques.
[h] Rated 1 to 5: 1 = very smooth, 5 = rough.

TABLE 30 Basic Considerations of the Various Embedding Processes[51]

Method	Advantages	Limitations	Material requirements	Applications
Casting consists of pouring a catalyzed or hardenable liquid into a mold. The hardened cast part takes the shape of the mold, and the mold is removed for reuse.	Requires a minimum of equipment and facilities; is ideal for short runs.	For large-volume runs, molds, mold handling, and maintenance can be expensive; assemblies must be positioned so they do not touch the mold during casting; patching or surface defects can be difficult.	Viscosity must be controlled so that the embedding material completely flows around all parts in the assembly at the processing temperature and pressure.	Most mechanical or electromechanical assemblies within certain size limitations can be cast.
Potting is similar to casting except that the catalyzed or hardenable liquid is poured into a shell or housing which remains as an integral part of the unit.	Excellent for large-volume runs; tooling is minimal; presence of a shell or housing assures no exposed components, as can occur in casting.	Some materials do not adhere to shell or housing; electrical short-circuiting to the housing can occur if the housing is metal.	Same material requirements as for casting except that materials which will bond to the shells or housings are required.	Most mechanical or electromechanical assemblies, subject to certain size limitations and housing complexity limitations.
Impregnating consists of completely immersing a part in a liquid so that the interstices are thoroughly soaked and wetted; usually accomplished by vacuum and/or pressure.	The most positive method for obtaining total embedding in deep or dense assembly sections such as transformer coils.	Requires vacuum or pressure equipment which can be costly; in curing, the impregnating material tends to run out of the assembly creating internal voids unless an encapsulating coating has first been applied to the outside of the assembly.	Low-viscosity materials are required for the most efficient and most thorough impregnation.	Dense assemblies which must be thoroughly soaked; electric coils are primary examples.
Encapsulating consists of coating (usually by dipping) a part with a curable or hardenable coating; coatings are relatively thick compared with varnish coatings.	Requires a minimum of equipment and facilities.	Obtaining a uniform, drip-free coating is difficult; specialized equipment for applying encapsulating coatings by spray techniques overcomes this problem, however.	Must be both high viscosity and thixotropic; that is, material must not run off the part during the cure.	Parts requiring a thick outer coating, such as transformers.
Transfer molding is the process of transferring a catalyzed or hardenable material, under pressure, from a pot or container into the mold which contains the part to be embedded.	Economical for large-volume operations.	Initial facility and mold costs are high; requires care so that parts of assemblies are not exposed: some pressure is required, and processing temperatures are often higher than for other embedding operations.	Should be moldable at the lowest possible pressure and temperature, and should cure in the shortest possible time for lowest processing cost.	For embedding small electronic assemblies in large-volume operations.

1-87

Plastics for Electronics

against environments to which the nonembedded part or assembly would be susceptible. Although any material achieving these objectives might be broadly considered, a particular group of liquid and low-pressure molding materials offers the greatest advantages for embedding electronic and electrical assemblies. It is this group of organic and silicone materials and the processes for using them which will be discussed in this section. All these materials and processes are extensively covered in Ref. 51, listed at the end of this chapter.

Embedding Processes

Since the same basic processes are used for all the organic and silicone embedding materials discussed in this chapter, a summary of these processes will be useful before proceeding to the discussion on materials. There are, of course, some differences dictated by the specific material formulation. However, the common point among all these processes is that they consist of either (1) converting a liquid resin into a solid mass at atmospheric pressure, usually by the addition of curing agents and application of heat, or (2) converting a partially reacted powdered resin into a solid mass, by application of low molding pressures and heat. The liquid resin embedding processes are casting, potting, impregnating, and encapsulating. The common powdered resin embedding process is transfer molding. A summary of the basic points of each of these processes is given in Table 30.

Casting and potting As noted in Table 30, the casting process utilizes molds, and the potting process utilizes shells or housings. The choice between these two processes is usually one of economics, that is, cost of tooling to produce shells or housings versus costs of molds for casting and the associated operating costs. Costs of tooling for shells and housings can be high unless amortized over a sufficient volume of parts produced. Manufacturing costs can be high in the casting process, owing to mold handling and assembly costs. A crossover point should be calculated, based on all cost factors and the volume of parts to be made, in order to determine whether to use casting or potting to embed a given product. Some further considerations for selecting between casting and potting are given in Table 31. In addition, some important considerations for selection of mold type for casting are given

TABLE 31 Considerations in Selection of Casting or Potting Processes[51]

Characteristic	Casting	Potting
Skin thickness	Difficult to control; components can become exposed in high-component-density packages.	Controlled minimum wall or skin thickness, due to thickness of shell or housing.
Surface appearance	Cavities and surface blemishes often require reworking.	Established by surface appearance of shell or housing, though problems can arise if resin spillage not controlled.
Repairability	Resin exposed for easy access.	Shell or housing must be removed and replaced.
Handling	Handling and transfer of unhoused assembly can reduce yield.	Most handling of unembedded unit can be in housing.
Assembly	If molds are not well maintained, or if unit fits tightly into mold, handling can cause breakage of components.	Assembly is simplified since new shells or housings are always used, and wall thickness is controlled.
Manufacturing-cycle efficiency	Production rate usually limited by quantity of molds.	Output not limited by tools.
Tool preparation and maintenance	Relatively expensive.	Costs are minimal.

in Table 32. Some considerations for selection of shell or housing for potting are given in Table 33.

Transfer molding Transfer molding offers advantages in economy and increased production rates for those assemblies that adapt to this technique and that are produced in large quantities. These advantages consist of a large reduction in processing steps (over casting or potting), and a shorter curing time for the embedding compound. Transfer molding materials cure in minutes; liquid casting and potting resins require hours in most instances.

Major limitations of the transfer molding process are: (1) the assembly must be able to withstand pressures of 50 to 500 psi; (2) the assembly must be able

TABLE 32 Considerations for Selection of Molds for Casting Process[51]

Mold material and fabrication	Advantages	Disadvantages
Machined steel	Good dimensional control; can be made for complex shapes and insert patterns. Good heat transfer; surfaces can be polished.	Assembly sometimes difficult; can corrode; usually requires mold release.
Machined aluminum	Same as machined steel except more easily machined.	Same as machined steel, except for corrosion; easily damaged, because of softness of metal.
Cast aluminum	None over machined aluminum, except lower mold costs for high volume operations.	Same as machined aluminum; surface finish and tolerances usually not as good as for machined aluminum; complex molds not as accurate as for machined metal.
Sprayed metal*	None over machined metal. Good surface possible.	Use usually limited to simple forms; not always easy to control mold quality; number of quality parts per mold limited; requires mold release.
Dip molded* (slush casting)	Same as sprayed-metal molds.	Same as sprayed-metal molds.
Cast epoxy	Good dimensional control; surface can be polished; can be made for inserts and multiple-part molds; long life and low maintenance.	Dimensional control not quite as good as in machined-metal molds; requires mold release and cleaning; low thermal conductivity compared with that of metals.
Cast plastisols	Parts easily removed from molds; molds are easy to make.	Short useful life; poor dimensional control.
Cast RTV silicone rubber	Same as for plastisols; better life than plastisols.	Poor dimensional control, though better than plastisols.
Machined TFE fluorocarbon	No mold release required; convenient to make for short runs and simple shapes; withstands high-temperature cures.	Poor dimensional control.
Machined polyethylene and polypropylene	Same as listed for TFE fluorocarbon except temperature capability and lower cost.	Poor dimensional control.
Molded polyethylene and polypropylene	Same as listed for TFE fluorocarbon except temperature capability and lower cost.	Poor dimensional control.

* Although sprayed-metal molds and dip-molded molds are similar, differences in method of making these two types may give one an advantage over the other in specific instances.

TABLE 33 Considerations for Selection of Shell or Housing for Potting Process[51]

Housing or container	Advantages	Disadvantages
Steel..............	Many standard sizes available; easily plated for solderability; good thermal conductivity; easily cleaned by vapor degreasing; good adhesive bond formed with most resins; easily painted; flame-resistant.	Can corrode in salt spray and humidity; fitting of lids sometimes a problem; cutoff of resin-filled can is sometimes difficult; possibility of electric short-circuiting.
Aluminum.........	Same as for steel except plating ease; lightweight; corrosion-resistant.	Same as for steel except aluminum is more corrosion-resistant; not easily soldered.
Molded thermosets (epoxy, alkyd, phenolic, diallyl phthalate, etc.)	Many standard sizes available; good insulator; corrosion-resistant; color or identification can be molded in; terminals can sometimes be molded in; cutoff of resin-filled shell easier than for metal cans; same type of material can be used for shell and filling resin, resulting in good compatibility.	Does not always adhere to resin, especially if silicone mold releases used to make shell; sealing of leakage joints can be difficult; physically weaker than steel, especially in thin sections; molding flash can cause fitting problems; cleaning of resin spillage can break shells.
Molded thermoplastics (nylon, polyethylene, polystyrene etc.)	Same as listed for thermosets except last two items; often less prone to cracking than thermosetting shells although this depends on resiliency of material.	Same as first three items listed in thermosets; adhesion can be poor owing to excellent release characteristics of most thermoplastics. Shell can distort from heat. Cutoff can be a problem due to melting or softening of thermoplastics under mechanically generated heat.

to take the curing temperatures of 250 to 350°F; (3) production volume must be large enough to justify equipment expenditures.

In transfer molding, a dry, solid molding compound—usually in powder or pellet form—is heated in a molding press to the point of becoming flowable or liquid, at which time it flows (is transferred) under pressure into a mold cavity containing the assembly to be embedded, as shown in Fig. 76. The plastic remains in the heated mold for a short time until curing is completed.

The transfer mold in Fig. 76 embeds two similar parts, although a larger number of parts can be molded simultaneously. Multicavity molds are common in transfer molding, and represent one of the major points of economy for large-volume runs. Many cavities can be filled nearly as rapidly as a single cavity, thereby reducing the cost per part. Although mold cost increases as the number of cavities increases, cost does not increase proportionately. Overall mold cost per part produced can be further reduced by incorporating cavities of different shapes into

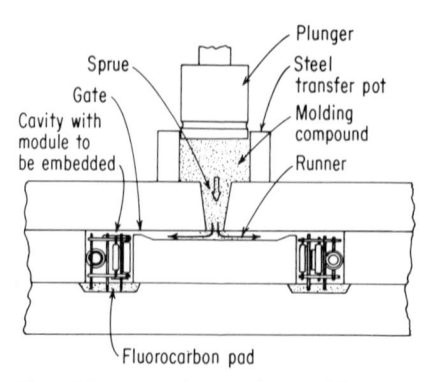

Fig. 76 Typical transfer molding assembly, showing molding compound flow.

the same mold in proportion to the production volumes required, or by using mold inserts to vary the cavity configuration as required by changing production needs.

A gate scar remains on the finished part at the point where the molding compound goes out of the runners from the transfer cylinder into the cavity. The small gate scar is usually unobjectionable.

Processing Characteristics of Embedding Resins

Nearly all embedding resins have important basic processing characteristics of viscosity and exotherm. Each resin-curing agent-curing cycle system provides its own unique set of viscosity and exotherm curves, however. The nature of these two important processing characteristics will be explained below. It should be mentioned that even transfer molding materials, while solid, as received, go through a liquid phase in their cure, and have viscosity and exothermic characteristics equally important to those of liquid resin systems. Although there are numerous chemical types of resins, each having different end properties, most resins cure by heat and/or curing-agent influences, give off heat during the curing process, and are thermosetting. Resin viscosity or fluidity, and the time-temperature curve (exotherm) for the exothermic reaction vary with each resin system, and are key properties describing the nature of the individual resin through the curing cycle.

Viscosity Flow properties of resins are important because of the need for flow and penetration at atmospheric or low pressures. When the viscosity is too high, the formulation is difficult to pour and does not flow properly around inserts or components, thus allowing internal cavities to form. A high-viscosity resin is usually too thick to allow evacuation of entrapped air, which also causes cavity formations. High viscosity also makes mixing difficult. On the other hand, a resin whose viscosity is too low may cause problems of leakage through openings in the mold or container.

For most embedding applications, there is an optimum range of viscosity. For impregnating operations, extremely low viscosities—100 centipoises or less—are desirable because complete impregnation of the parts under vacuum is required. In practice, however, impregnation is often achieved with viscosities considerably higher—up to 1,000 centipoises. However, the higher the viscosity, the longer the cycling time and/or the higher the vacuum required for complete impregnation.

For embedment operations such as casting or potting, there is no limit as to how low the viscosity can be, provided that the mold or container is tight enough to prevent leakage. Usually, however, if impregnation is not required and if the components are not packed tightly, viscosities in the range of 1,000 to 5,000 centipoises are satisfactory for casting and potting operations.

An encapsulation coating requires a thixotropic (nonflowing) material with an extremely high viscosity because the part is dipped into the compound and the coated part cured without the use of a mold or container. The coating must now flow off during the curing operation.

Viscosity usually can be lowered by heating the resin (Fig. 77) or by adding diluents, and it can be raised by addition of fillers. All resins do not exhibit as great a viscosity dependence on temperature as that shown in Fig. 77; silcones, for example, have a relatively flat viscosity curve.

Exothermic properties Most polymeric resins used for embedment have an exothermic reaction; that is, heat is produced as the reaction progresses. It is essential that the exothermic properties of a particular system be known and that they be controlled. Too much heat can cause resin cracking during cure, and the heat generated can also affect heat-sensitive components.

Three specific values are commonly used for control measurements of these exothermic properties: gel time, peak exothermic temperature, and time to peak exothermic temperature. These characteristics are measured from a single graphic plot of exothermic temperature versus time for a given resin–catalyst–curing-agent system. A typical exothermic curve for a polyester resin is shown in Fig. 78. Although shapes of these curves vary widely from system to system, the curve for

1-92   Plastics for Electronics

a given system should be closely reproducible. Exothermic curves also vary with the mass of resin, as indicated in Fig. 79.

Gel time is the interval from the time when the exothermic reaction temperature reaches 150°F to the time when it is 10° above the bath temperature (see Fig. 78). The reason for starting the timing from a 150°F exothermic temperature

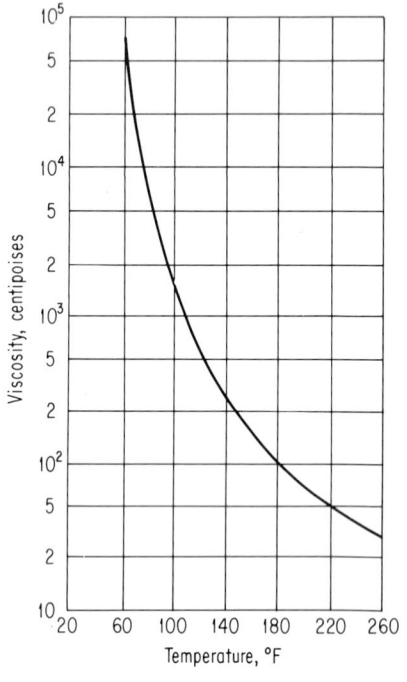

Fig. 77 Viscosity-temperature relationship for a bisphenol epoxy resin.

Fig. 78 Exothermic reaction curve for a typical polyester resin, using benzoyl peroxide catalyst and cured at 180°F.

rather than from the time at which the catalyst and resin are initially mixed is that it is not always practical to have the temperature of the ingredients precisely the same when the reaction starts, that is, during and immediately after the initial

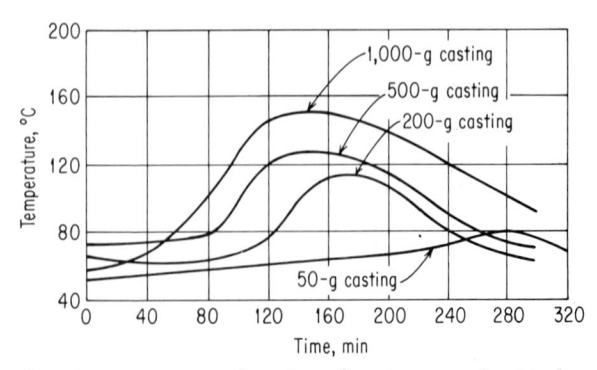

Fig. 79 Exothermic curves, as a function of resin mass, for bisphenol epoxy and 5 percent piperidene curing agent, cured at 60°C.

mixing. The common base point is used to assure better reproducibility. Gelation usually occurs by the time the exothermic temperature has slightly exceeded the bath temperature for this type of resin system.

The exothermic temperature rise is much greater after gelation has occurred than it is before gelation. This tendency is common for these materials. All resin-curing agent-curing cycle changes vary this curve, which can be nearly flat, or very steep.

Embedding Resins

Many chemical groups of embedding resins are available, and there are many variations in each group. The most important of these are discussed in the following sections. Typical mechanical, physical, thermal, and electrical properties of several of these classes are shown in Tables 34 and 35.

TABLE 34 Typical Mechanical and Physical Properties of Several Common Embedding Resins[53]

Material	Tensile strength, psi	Elongation, percent	Compression strength, psi	Impact strength, Izod, ft-lb/in. of notch	Hardness	Linear shrinkage during cure, percent	Water absorption, percent by weight
Epoxy:							
Rigid, unfilled	9,000	3	20,000	0.5	Rockwell M 100	0.3	0.12
Rigid, filled	10,000	2	25,000	0.4	Rockwell M 110	0.1	0.07
Flexible, unfilled	5,000	50	8,000	3.0	Shore D 50	0.9	0.38
Flexible, filled	4,000	40	10,000	2.0	Shore D 65	0.6	0.32
Polyester:							
Rigid, unfilled	10,000	3	25,000	0.3	Rockwell M 100	2.2	0.35
Flexible, unfilled	1,500	100	....	7.0	Shore A 90	3.0	1.5
Silicone:							
Flexible, unfilled	500	175	....	No break	Shore A 40	0.4	0.12
Urethane:							
Flexible, unfilled	500	300	20,000	No break	Shore A 70	2.0	0.65

Epoxies The most used of the embedding resins are the epoxies, in many types and modifications. All classes of epoxies have certain outstanding characteristics which are important in electronic assemblies. Chief among these properties are low shrinkage, excellent adhesion, excellent resistance to most environmental extremes, and ease of application for casting, potting, or encapsulation.

Bisphenol Epoxies. The original class of epoxies—the bisphenols—are the workhorses of the electronics industry. They are available as liquids over a wide viscosity range, and as solids. The resins are syrupy, commonly having viscosities in the range of 10,000 to 20,000 centipoises at room temperature. Their viscosity can be reduced, of course, by heating, as shown in Fig. 77, or by addition of reactive or nonreactive diluents.

End properties of bisphenol epoxies, as well as of other epoxies discussed here, are controlled by the type of curing agent used with the resin. Major types of curing agents are aliphatic amines, aromatic amines, catalytic curing agents, and acid anhydrides, as summarized in Table 36.

Aliphatic amine curing agents produce a resin–curing-agent mixture which has a relatively short working life but which cures at room temperature or at low

TABLE 35 Typical Thermal and Electrical Properties of Several Common Embedding Resins[53]

Material	Heat distortion temp., °C	Thermal shock per MIL-I-16923	Coefficient of thermal expansion, ppm/°C	Thermal conductivity, (cal)(cm)/(cm²)(sec)(°C)	Dissipation factor[°]	Dielectric constant[°]	Volume resistivity,[°] ohm-cm	Dielectric strength,[°] volts/mil	Arc resistance, sec
Epoxy:									
Rigid, unfilled	140	Fails	55	4×10^{-4}	0.006	4.2	10^{15}	450	85
Rigid, filled	140	Marginal	30	15×10^{-4}	0.02	4.7	10^{15}	450	150
Flexible, unfilled	<RT	Passes	100	4×10^{-4}	0.03	3.9	10^{15}	350	120
Flexible, filled	<RT	Passes	70	12×10^{-4}	0.05	4.1	3×10^{15}	130	360
Polyester:									
Rigid, unfilled	120	Fails	75	4×10^{-4}	0.017	3.7	10^{14}	440	125
Flexible, unfilled	<RT	Passes	130	4×10^{-4}	0.10	6.0	5×10^{12}	325	135
Silicone:									
Flexible, unfilled	<RT	Passes	400	5×10^{-4}	0.001	4.0	2×10^{15}	550	120
Urethane:									
Flexible, unfilled	<RT	Passes	150	5×10^{-4}	0.016	5.2	2×10^{12}	400	180

[°] Dissipation factor and dielectric constant are at 60 cps and room temperature; volume resistivity is at 500 volts d-c; and dielectric strength is short time.

baking temperatures, in a relatively short time. Aliphatic-amine-cured resins usually develop the highest exothermic temperatures during the curing reaction. Thus, the mass of material which can be cured at one time is limited because of possible cracking, crazing, or even charring of the resin system if too large a mass is mixed and cured. Also, physical and electrical properties of epoxy resins cured with aliphatic amines tend to degrade as operating temperature increases. Aliphatic-amine-cured epoxies find their greatest usefulness where small masses can be used, where room-temperature curing is desirable, and where the operating temperature is below 100°C.

Aromatic-amine-cured epoxies have a considerably longer working life than aliphatic-amine-cured epoxies, but they commonly require curing at 100°C or higher. These cured resin systems can operate at temperatures considerably above those used for aliphatic-amine-cured resin systems. However, aromatic amine curing-agent systems are not as easy to work with because of the solid nature of the curing agents and the fact that some (such as MPDA) sublime when heated, causing stains and residue deposition.

Catalytic curing agents also have longer working lives than the aliphatic amine materials; and, like the aromatic amines, catalytic curing agents normally require curing of the epoxy system at 100°C or above. Resins cured with these systems have good high-temperature properties compared with those of aliphatic-amine-cured epoxies. With some of the catalytic curing agents, the exothermic reaction becomes high as the mass of resin mixture increases, as shown in Fig. 79.

Acid anhydride curing agents are particularly important curing agents for epoxy resins, especially the liquid anhydrides. High-temperature properties are better than those of aromatic-amine-cured resin systems. Some anhydride-cured epoxy systems retain most electrical properties up to 150°C and higher, and are little affected physically, even after prolonged heating aging up to 200°C. In addition, the liquid anhydrides are extremely easy to work with—they blend easily with the resins and reduce the viscosity of the resin system. Also, working life of the liquid acid anhydride systems is long compared with that of aliphatic-amine-resin mixtures, and odors are slight. Amine promoters such as benzyldimethylamine (BDMA) or DMP-30 are used to promote the curing of acid anhydride–epoxy resin mixtures. Thermal stability of epoxies is improved by anhydride curing agents, as shown in Fig. 80.

TABLE 36 Curing Agents for Epoxy Resins[51]

Curing-agent type	Characteristics	Typical materials
Aliphatic amines	Aliphatic amines allow curing of epoxy resins at room temperature, and thus are widely used. Resins cured with aliphatic amines, however, usually develop the highest exothermic temperatures during the curing reaction, and therefore the mass of material which can be cured is limited. Epoxy resins cured with aliphatic amines have the greatest tendency toward degradation of electrical and physical properties at elevated temperatures.	Diethylene triamine (DETA) Triethylene tetramine (TETA)
Aromatic amines	Epoxies cured with aromatic amines usually have a longer working life than do epoxies cured with aliphatic amines. Aromatic amines usually require an elevated-temperature cure. Many of these curing agents are solid and must be melted into the epoxy, which makes them relatively difficult to use. The cured resin systems, however, can be used at temperatures considerably above those which are safe for resin systems cured with aliphatic amines.	Metaphenylene diamine (MPDA) Methylene dianiline (MDA) Diamino diphenyl sulfone (DDS or DADS)
Catalytic curing agents	Catalytic curing agents also have a working life better than that of aliphatic amine curing agents, and, like the aromatic amines, normally require curing of the resin system at a temperature of 200°F or above. In some cases, the exothermic reaction is critically affected by the mass of the resin mixture.	Piperidene Boron trifluoride-ethylamine complex Benzyl dimethylamine (BDMA)
Acid anhydrides	The development of liquid acid anhydrides provides curing agents which are easy to work with, have minimum toxicity problems compared with amines, and offer optimum high-temperature properties of the cured resins. These curing agents are becoming more and more widely used.	Nadic methyl anhydride (NMA) Dodecenyl succinic anhydride (DDSA) Hexahydrophthalic anhydride (HHPA) Alkendic anhydride

Novolak Epoxies. Excellent high-temperature properties are characteristic of novolak epoxy systems, especially using high-temperature curing agents. These resins contain more of the benzene ring or phenolic-type structure in the molecule and thus combine the excellent thermal stability of the phenolics with the reactivity (with curing agents and catalysts) and versatility of the epoxies. Because their average epoxide functionality* is greater than 3 or 4, tightly crosslinked structures are readily formed, producing cured masses that exhibit higher heat-distortion temperature, better chemical resistance, and better adhesion than do bisphenol epoxies having an epoxide functionality of about 2. The functionality of novolak epoxy

---

* Average epoxide functionality is the average number of epoxide or epoxy groups per molecule. Since crosslinking is through the epoxy group, a higher functionality means more crosslinking, more rigidity, and more resistance to thermal degradation. Unfortunately, it also often means more brittleness and tendency to crack.

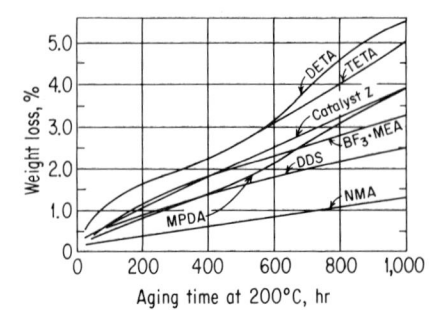

Fig. 80 Weight loss at 200°C for Epon 828 cured with various curing agents. DETA and TETA are aliphatic amines, MPDA and DDS are aromatic amines, and NMA is a liquid acid anhydride.[54]

ERR-0100, produced by Union Carbide, averages about 5 epoxy groups per molecule, and of D.E.N. 438, produced by Dow, 3.3 epoxy groups per molecule, compared with a functionality of 2.10 and below for most conventional epoxy resins. The close-knit crosslinking of the cured novolak epoxy assures excellent retention of mechanical properties at high temperatures and thereby enlarges those application areas in which properties of epoxies are needed.

Viscosity of many novolak epoxies is originally high, but can be reduced by conventional ways: with solvents, with lower-viscosity resins, with diluents, with heat, or with low-viscosity hardeners. Hardeners or curing agents used with conventional bisphenol epoxy resins are also used with novolak epoxies because curing reaction is through the epoxy groups of the molecule. Postcuring is usually necessary to develop maximum properties of the novolak epoxy resins.

Cycloaliphatic Diepoxides. These materials, also known as peracetic epoxies, do not contain phenolic rings which are associated with most epoxy resins. They offer unusual combinations of low viscosity, low vapor pressure, and high reactivity toward acidic curing agents such as polycarboxylic acids and anhydrides. A wide range of cured resin properties are possible, depending on selection of epoxide and hardener. Either rigid or flexibilized cured resins are possible. These formulations, which are free of aromatic structures, show outstanding resistance to the formation of carbon tracks under an electric arc, and are widely used for applications where surface arcing and tracking must be minimized. They have excellent electrical properties. They also resist discoloration when exposed to ultraviolet light.

Flexibilized and Modified Epoxies. Four major flexibilizers used with epoxy resin systems are polyamides, polysulfides, polycarboxylic acids, and polyurethanes. Although flexibilizers improve thermal shock resistance and reduce internal stresses, some degradation in electrical properties results. Dielectric constant, dissipation factor, and resistivity can be seriously degraded. However, it is often possible to achieve a compromise between electrical and physical properties desired. Properties of several flexibilized epoxy systems are shown in Tables 37 through 39. A large variety of flexibilizers and flexibilized resins are available from resin formulators. These suppliers should be consulted for final selection of a flexibilized system.

Polysulfides also can be used alone for potting and coating assemblies. They provide a flexible end product with low moisture permeability, but with relatively low thermal endurance. Polyurethanes can also be used alone for some applications. Both are discussed separately, below.

TABLE 37 Properties of Epoxy-polyamide Systems[55]

Epoxy-polyamide weight ratio	80:20	70:30	60:40	50:50	40:60	30:70	25:75
Heat-distortion temperature, °F	220	215	136	100	65		
Hardness	B70°	B66	B58	B50	B30	A70†	A40
Specific resistivity, ohm-cm	10^{15}	10^{15}	10^{14}	10^{12}	10^{10}	10^9	10^8
Moisture absorption, %	0.15	....	0.20	....	0.50	1.0	2.0

° Barcol M.
† Shore A.

TABLE 38 Properties of Epoxy-polysulfide Systems[56]

Epoxide/polysulfide weight ratio	1:0	2:1	1:1	1:2	1:3	0:1
Elongation, %	1	10	30	50	100	200
Hardness, Shore A	100	98	95	70	50	35
Specific resistivity, ohm-cm	10^{14}	10^{13}	10^{12}	10^{11}	10^{10}	10^{9}
Dielectric constant at 10^6 cps	3.5	3.7	4.0	5.5	6.5	7.5
Loss tangent at 10^6 cps	0.005	0.01	0.03			

* Average epoxide functionality is the average number of epoxide or epoxy groups per molecule. Since crosslinking is through the epoxy group, a higher functionality means more crosslinking, more rigidity, and more resistance to thermal degradation. Unfortunately, it often also means more brittleness and tendency to crack.

Another flexibilizing system for epoxy resins—linear polyazelaic polyanhydride (PAPA)—is reported to offer several advantages. Epoxies flexibilized with PAPA are said to not increase in hardness, and to maintain their toughness with under 1 percent weight loss after aging for periods of up to eight weeks at 300°F.[58]

Fire-retardant grades of epoxies are obtained by using chlorinated basic epoxy resins, or chlorinated curing agents or modifiers such as chlorendic anhydride or triphenyl phosphate, or flame-retardant fillers such as antimony trioxide. While fire retardance is often a key requirement, resins modified for flame or fire retardance require property trade-offs, and all these should be considered for a given application. Particularly affected are electrical properties of the resin system, with dielectric constant and dissipation factor often being seriously degraded.

Silicones Silicones, based on the silicon-oxygen linkage rather than the carbon-carbon linkage of organic resins, are unique among embedding resins in many ways. First, they are available in a wide variety of forms, such as room-temperature-vulcanizing (RTV) one-component and two-component elastomeric formulations, as clear or pigmented resilient resins or gels of varying consistencies, as rigid resins, and as transfer molding compounds. Although a wide range of variations are possible in such parameters as viscosity, curing temperature, and time, etc., all silicones offer an important combination of advantages over organic resins, namely, the combination of thermal stability and excellent electrical properties, especially the low values of dielectric constant and dissipation factor, which are relatively stable with temperature and frequency. This is illustrated in Fig. 81. Not only are the excellent electrical properties stable with elevated temperature, but also mechanical properties are relatively stable with reduced temperature. Electrical properties of many silicones are stable up to 400 to 500°F. Flexibility of some silicones is maintained to quite low temperatures, with stiffening temperature of some silicones, especially the resilient resins such as Sylgard,* being −130°F or lower, as measured by the Gehman flexure test (ASTM 1053). Furthermore, silicones are very stable for space applications.[60] (See also Fig. 18.) The viscosity

* Trademark, Dow Corning Corporation, Midland, Mich.

TABLE 39 Properties of Epoxy-polyurethane Systems[57]

Epoxy/polyurethane weight ratio	0:100	25:75	50:50	75:25	100:0
Ultimate tensile strength, psi	5,000	2,050	6,000	10,000	10,000
Ultimate elongation, %	420	350	10	10	10
Hardness, Shore A	90	60			
Hardness, Shore D	40	15	80	85	90
Heat-distortion temperature, °F	*	*	100	176	260

* Too low to measure.

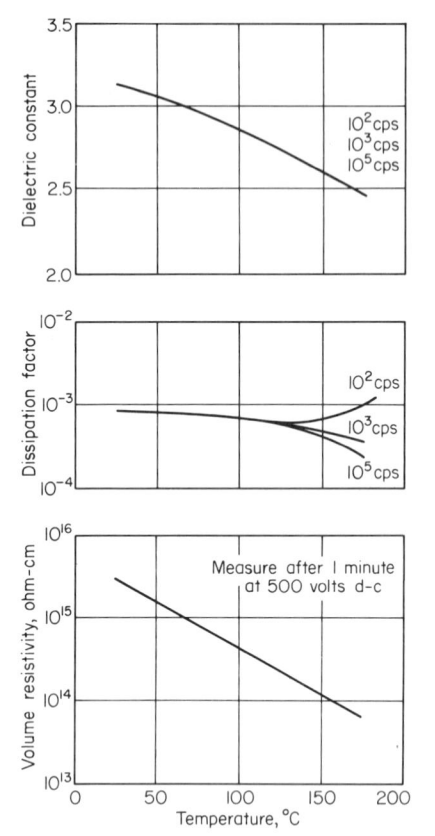

Fig. 81 Effect of temperature on three important properties of a clear, flexible silicone resin.[59]

of silicones is also relatively stable with temperature, as compared to that of organic resins. Silicones usually have low exotherms during cure. The major disadvantages of silicones are cost and poor adhesion of many to most substrates.

There are several basic types and forms of silicones, as mentioned above. RTV silicones and flexible resins are widely used in the electronics industry. These flexible materials have excellent thermal shock resistance and low internal curing stresses. RTV silicone rubbers are generally supplied as two components that must be mixed before use, which introduces the problem of pot life. The gel time of RTV silicone rubbers can be varied from minutes to more than 8 hr by controlling the catalyst concentration. Catalysts are generally soluble organic-metallic compounds such as tin octoate. Viscosity of RTV silicone rubber compounds can be varied from readily pourable liquids to stiff pastes. Low-viscosity compounds are most suitable for plotting and impregnation; the high-viscosity compounds for encapsulation. Complete curing of RTV silicone rubbers generally requires 24 hr or more at room temperature. The cure can be accelerated by using elevated-temperature cures.

One-component compounds are also available that include a volatile inhibitor with precatalyzed RTV. On exposure to air, the volatile inhibitor is slowly lost, and the catalyst effects the cure. Since some one-component RTV silicones contain inhibitors which are activated by moisture, they should not be used for potting sealed devices without a few minutes of open time under humid conditions to allow the inhibitor to escape and the surface to absorb water. Thick sections may also prevent inhibitor volatilization, and thus hinder resin cure.

Although RTV compounds are usually pigmented or colored, several clear, flexible resins are available. These materials are increasing in usage since they have most of the good properties of the pigmented materials in addition to their optical clarity. The combination of flexibility and clarity facilitates cutting and repairing when needed. The cure of some clear, flexible resins is inhibited when these resins contact certain materials. Notable inhibiting materials are sulfur-vulcanized rubber and certain RTV silicone rubbers. This problem can usually be solved by applying a coating of a noninhibiting material to the inhibiting surface.

Silicone gels, as the name implies, exist in a gel state after being cured. Although these materials are very tough, they are usually used in a can or case. One interesting use of a silicone gel is for electric checking of circuits and components. After the test probes are withdrawn, the memory of the gel is sufficient to heal the portion that has been broken by the probes. Cure-inhibiting surfaces must also be considered with silicone gels.

Rigid solventless silicones are not used as widely as the other groups of silicones, because their resistance to thermal shock and cracking is not as good as that of the flexible materials, and because the rigid solventless resins are not as convenient to work with as the room-temperature-curing materials. However, when the general properties of silicones are desired and rigidity is preferred to flexibility, the rigid solventless silicone resins should be considered.

Viscosity of these rigid resins can be controlled by blending high- and low-viscosity resins. Curing of the resin is usually accomplished with organic peroxides, such as dicumyl peroxide, ditertiary butyl peroxide, and tertiary butyl perbenzoate. Curing with organic peroxide or free-radical catalysts is subject to inhibition by certain impurities and oxygen. Tacky surfaces will result if the surface is not protected from air during cure. High-temperature final cures, i.e., at 200°C, will harden such tacky surfaces if they form. Dicumyl peroxide and ditertiary butyl peroxide are the most useful catalysts for solvent-free silicone resins. They are less sensitive to inhibition by dissolved impurities, and dicumyl peroxide is least sensitive to air inhibition. The curing temperature used will depend upon the catalyst.

Urethanes Cured urethane, or polyurethane, resins are generally very tough and have outstanding abrasion resistance. Their maximum tear strength is high compared with that of most other resilient or flexible materials. Polyurethanes have very good thermal shock resistance. Their combination of toughness and resistance to cracking in thermal shock offers advantages in many applications. Urethanes can be considered for applications from cryogenic temperatures to continuous service at 130°C, with short-term capabilities to 155°C. One of their unique advantages is the low stress produced on fine electronic components at low temperatures. These stresses, at −40°C, are considerably lower than the best attainable with flexible epoxies. Cure is effected at room temperature or at low baking temperatures. Electrical properties are excellent. Their adhesion to most materials is excellent—an advantage over silicones in some instances. Their thermal resistance does not compare with that of silicones, however. Because some liquid polyurethanes react with moisture, careful drying of parts to be embedded may be required to avoid bubbles. In fact, some cured urethanes will revert to a liquid under the influence of high humidity and heat. This reversion caused a major industry problem in 1968 and 1969, owing to failure of aircraft connectors potted in urethane. This factor should be reviewed with the material supplier for applications experiencing this environment.

Many liquid polyurethane resin systems are available. Two general resin systems are used to produce the final products: namely, one-shot and prepolymer. In the prepolymer system, diisocyanate and other reactants are reacted partially before use, and final curing is effected by adding a catalyst. The one-shot system requires mixing the raw materials and effecting the cure in one stage. In most applications, the prepolymer system is preferable since it avoids handling the raw diisocyanates, which are lung irritants. However, the one-shot systems usually react more rapidly and produce a material that has better mechanical and electrical properties. The one-shot system produces a much higher exotherm than the prepolymer system, and this must be considered.

Polyesters These materials were among the first of the liquid resins to be used for embedding electronic devices, and they are available in all degrees of flexibility and a wide range of viscosities. Most general-purpose polyesters are copolymers of a basic polyester resin and styrene monomer. However, other monomers are also used, and particularly good high-temperature stability is obtained in polyesters that are copolymerized with triallyl cyanurate.

Most polyesters have higher shrinkage, less thermal shock resistance, less humidity resistance, and lower adhesion than the epoxies. Lower-viscosity polyesters, containing higher percentages of monomer, are usually most prone to cracking and highest shrinkage. Exotherms are often quite high in the curing of polyester resins. Electrical properties of polyesters, especially loss properties, are generally superior to those of epoxies. Polyesters materials are widely used in commercial applications where cost is most important, and in applications where minimum electrical losses are required.

Polybutadienes Another class of resins with excellent electrical loss properties, which are relatively stable with temperature and frequency, is the polybutadiene

TABLE 40 Properties of One Class of Polybutadiene Casting Resins[61]

Properties	ASTM test method	Value
Mechanical		
Tensile ultimate strength, psi	D 638	8,200
Tensile yield strength, psi	D 638	8,000
Elongation, %	D 638	4
Compressive strength, psi	D 695	36,000
Compressive modulus, psi	D 695	212,000
Compression to break, %	......	25
Specific gravity	......	1.00
Shrinkage from B to C stage, %	......	5
Hardness:		
Barcol	......	45
Rockwell M	......	100
Impact strength, ft-lb/in. notch	......	0.66
Electrical		
Dielectric strength short time, 0.020-in. thickness, volts/mil	D 149	1,800
Dielectric constant at 10^5–10^7 Hz	D 150	2.4 to 2.7
Dissipation factor at 10^5–10^7 Hz	D 150	0.0048 to 0.0057

class of materials. The properties of one interesting group of polybutadiene embedding resins are given in Table 40.

This group of polybutadienes, known as Hystl,* contain high concentrations of vinyl groups.[61] In addition, certain resins of this group have chemically functional terminal groups. The resins offered contain over 90 percent vinyl unsaturation. These high-vinyl-content resins provide rapid curing properties with peroxide initiation and yield thermally stable plastic products which are hard and tough. Product properties indicate that the plastic may be characterized as a crosslinked ladder structure stemming from condensed cyclohexane groups interspersed with some residual unsaturation.

In addition to the advantages accruing from the high vinyl content of this group of resins, additional significant advantages stem from the chemically functional terminal groups. Carboxy- and hydroxypolybutadiene resins can be significantly advanced in molecular weight, before curing, by chemical reaction with preferably polyfunctional epoxides and isocyanates, respectively, employing temperatures appreciably lower than the curing temperature. The rate of the reaction allows convenient workable pot life. Thus, without affecting the curing reactivity of the resins, castable resin systems may be formulated for preparing B-stage materials of the consistency desired for processing. These resins with terminal functional groups, therefore, provide the means whereby filler in high concentration may be readily mixed into the resin, and the mixture cast without the aid of solvents. The casting can then be chemically advanced to yield, if desired, very viscous or rubbery material. Unique, highly filled rubbery sheets, thereby, may be formed as the B stage. The sheets drape and conform to match die molds, vacuum forming etc., and subsequently yield (upon curing) hard, tough, useful products.

Chemical properties of Hystl products are outstanding. The cured resins, for example, are stable in liquid N_2O_4, fuming nitric acid, strong bases, glacial acetic acid, and phosphoric acid.

The hydrophobic nature of these resins, due to the high hydrocarbon content of the resin matrices, imparts resistance to water absorption. Thus, electrical properties of the cured resins are relatively insensitive to the relative humidity of the atmosphere. Polybutadiene resins have excellent electrical loss properties.

* Trademark of Hystl Development Company, Redondo Beach, Calif.

Cured products exhibit mechanical and thermal properties equivalent to those of products derived from high-performance acid-anhydride-cured epoxies, and electrical properties better than those of epoxies. In contrast to those epoxy products, however, these products are fabricated in relatively short time spans. And in addition, the B-stage materials may be readily formulated to obtain a wide range of desired B-stage consistencies. These B-stage materials do not require refrigeration; they are shelf-stable at room temperature without apparent loss of curing properties for over six months, and at 130°F for over 750 hr.

Polysulfides These materials are available as liquids for use in potting of electrical connectors. The cured polysulfide rubber is flexible and has excellent resistance to solvents, oxidation, ozone, and weathering.[56] Gas permeability is low, and electrical insulation properties are good at temperatures between —65 and +250°F.

Polysulfide rubber resins are of the same chemical class as the polysulfide rubber resins used in modifying epoxy resins. Chemically, polysulfide rubbers are organic compounds containing sulfur, the sulfur groups in the polymer chain being known as mercaptan groups.

Low-density Foams Plastic foams are cellular materials made from a variety of plastics by chemically or mechanically expanding the resins. This expansion can be part of the chemical reaction, such as liberation of carbon dioxide which causes expansion in urethane foams; or it can be caused by gassing of a chemical blowing or gassing agent added to the resin mix. Also, expansion can occur by mechanical mixing of some type. The type of expansion used depends on the plastic and the desired end product, but the expansion technique must be controlled. Properly controlled, foams can be made from many plastics, with either open- or closed-cell structures, and in a variety of densities. For most foams, a fairly narrow density range is common, however, usually at low densities of 1 to 5 pcf. For some foams, notably rigid urethanes, densities up to the area of 15 to 20 pcf are practical. Another type of low-density material, also often classified as syntactic foams, is produced by adding hollow fillers such as glass or phenolic microballoons to the resin mix. Here, considerably higher-density materials can be produced.

Foams are used for design purposes such as thermal insulation, buoyancy, cushioning, low-weight packaging, acoustical insulation, and electrical insulation. They are available in rigid and flexible forms. Most foams can be obtained from suppliers in various forms such as blocks, sheets, rods, and molded shapes. Some of them can be foamed in place, and hence are especially useful for embedding electronic packages. Most common here are urethanes, epoxies, and silicones, especially the rigid foam-in-place materials.

Since urethanes can be made as either flexible or rigid foams, and can be foamed in place, they are very practical for any laboratory or shop to use. High-temperature varieties, up to 400°F or so, as well as flame-retardant urethanes are available. Epoxy foams are usually closed-cell, all rigid or semirigid foams, which can be obtained in various forms and shapes, or in formulations that can be processed in-house. They have the usual good chemical properties of epoxies, tempered in the areas where density is a controlling factor. Silicone foams are available in rigid or flexible formulations, and like epoxies, either in forms and shapes or as formulations that can be processed in-house. As with high-density silicones, silicone foams are outstanding in high-temperature and low-temperature stability and in electrical properties. Phenolic foams are rigid, open-cell materials, having good heat resistance, among foams, as an outstanding property. This also gives them good dimensional stability among foam materials.

Many basic properties of foams, such as thermal stability, chemical resistance, etc., are not dependent on density, and hence are similar to the properties previously discussed for the specific high-density plastic-base material. Other properties, such as strength, electrical, and acoustical properties are at least partially controlled by density. The air-filled cellular structure of foams gives low dielectric constants and dissipation factors. The values for these two characteristics increase with increasing foam density, often almost linearly to the values of the specific high-density plastic involved. These low values for foams can sometimes be misleading, however, in that the cell walls are still composed of the high-density plastic; and in some instances, especially as high-frequency power levels increase, breakdown of the cell walls occurs, even though the averaged characteristic values for the foam are low. The breakdown mechanism is usually dielectric heating of the cell walls, which ultimately leads to melting, burning, or carbonization of the plastic wall.

Mechanical properties of foams are inherently low, owing to the weak structure. Tensile and compressive strengths of 25 to 100 psi are common, and deformation under load can be high, especially as a function of increasing temperature, especially with thermoplastics. Abrasion resistance of thermosetting foams and of some thermoplastic foams is quite low. Many of these properties depend heavily on both the material and the density, and hence they must be evaluated almost on an individual basis.

Fillers for Embedding Resins

Fillers play a most important role in the application of resins for embedding applications. They are additives, usually inert, which are used to modify nearly

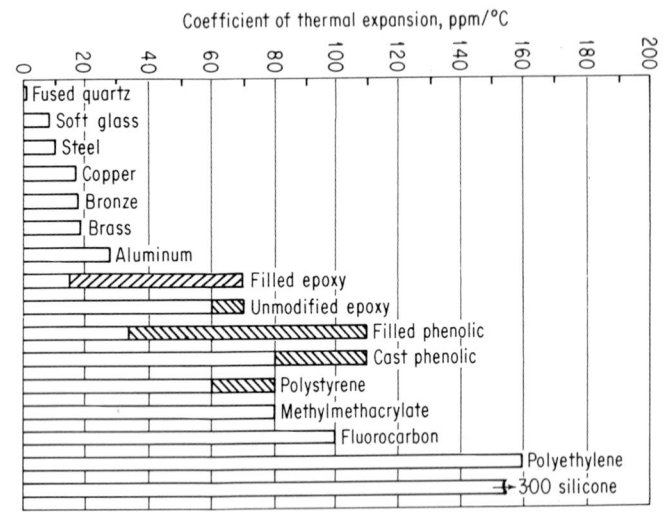

Fig. 82 Effect of filler content on coefficient of thermal expansion of epoxy resins compared with coefficients for various other materials. Ranges are shown cross-hatched.[62]

all basic resin properties in the direction desired. Fillers overcome many of the limitations of the basic resins. Through proper use of fillers, major improvements can be made in basic resin properties such as thermal conductivity, coefficient of thermal expansion, shrinkage, thermal shock resistance, density, exotherm, viscosity, and cost.

A major problem with most resin systems is their tendency to crack because of the difference in thermal expansion between embedded parts and embedment material. Figure 82 compares the thermal expansion of various materials with that of filled and unfilled epoxy resins. Addition of sufficient filler can reduce the thermal expansion coefficient of epoxies to the range of coefficients for metals. This is also true, to some extent, of other embedding resins. Although the trend is the same for most fillers, effects of specific fillers vary to some degree.

Viscosity of resin compounds is also increased by addition of fillers—especially by thixotroping fillers, such as finely dispersed silica, as shown in Fig. 83. This type of filler is often used to produce a thixotropic resin for encapsulation applications.

Effects of fillers on thermal properties of resins In addition to the beneficial effects of reducing the thermal expansion characteristics of a resin, fillers also increase thermal conductivity and reduce the weight loss (during heat aging) of a resin. The more filler incorporated into a resin, the lower will be the weight loss of that system during heat aging. Although the resin portion of the system will degrade upon heat aging, overall performance is almost always improved by the incorporation of fillers. This results not only from reduced weight loss, but also from shrinkage reduction and thermal conductivity increase produced by incorporation of filler into the system. Examples of possible improvements in thermal conductivity are shown in Fig. 84.

Another beneficial effect of fillers on properties of a resin system is reduction in the exothermic heat of the system during the curing cycle. This effect, plus reduced shrinkage and the decreased thermal expansion, gives many resin systems minimal resin-cracking characteristics. Thus, the addition of filler can change a normally unsatisfactory system into a very usable system with respect to cracking. Fillers

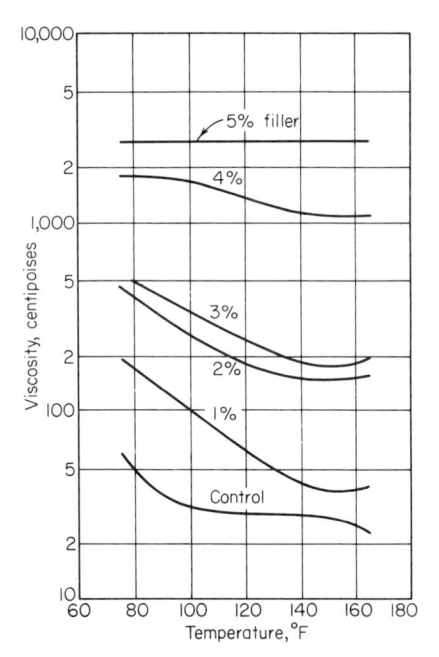

Fig. 83 Effect of high-surface-area silica filler on viscosity of a polyester resin.[63]

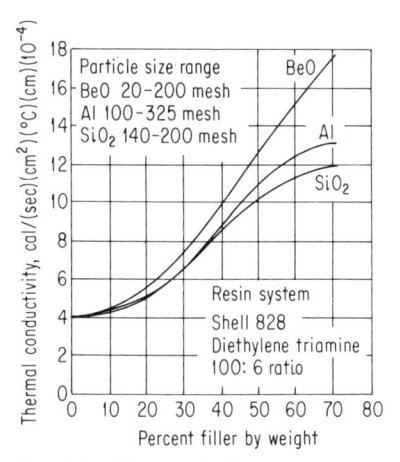

Fig. 84 Effect of filler content on thermal conductivity for three filler materials.[64]

can also increase pot life of resin system. The extension of pot life is related to the control of exothermic heat of the system.

Still another thermal property that can be improved by the use of fillers is fire resistance or burning rate. The burning rate is reduced considerably through the addition of a filler such as antimony oxide. Certain phosphates can also be used to reduce flammability of embedding resins, as was mentioned earlier.

Effects of fillers on mechanical properties of resins Fillers can have a major effect on mechanical properties of resins; however, the type of filler is also important. With respect to their effects on mechanical properties, fillers are classed as non-reinforcing (bulk, or nonfibrous) and reinforcing, with fillers such as silica being bulk, and fillers such as glass fibers being reinforcing.

Hardness and machinability characteristics depend on specific fillers; but, generally, hardness is increased and machinability decreased by the use of fillers. Abrasive fillers such as silica and sand can produce particularly difficult machining problems.

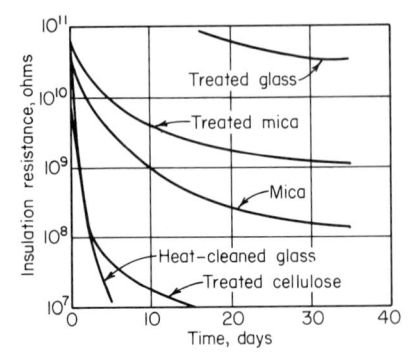

Fig. 85 Insulation resistance of epoxy castings at 140°F and 95 percent RH, using various fillers and filler treatments. Filler treatment indicated is Du Pont Volan chromic chloride sizing.[5]

Impact strength and tensile strength can be increased by the use of reinforcing fillers and are normally decreased by the use of bulk fillers. Milled or chopped glass fibers are especially good reinforcing fillers.

Effects of fillers on electrical properties of resins Although certain electrical characteristics of a resin system can be improved by the incorporation of fillers, these effects are minor compared with improvements in mechanical and thermal properties of the system. Dielectric strength may even be decreased if the filler has absorbed moisture or a contaminant. Dissipation factor and dielectric constant usually can be controlled, however, by the use of low-density fillers and other selective fillers such as barium titanate, etc., depending on the change desired.

An important factor in the selection of a treated filler for dielectric applications is the type of binder used on the filler, because it affects the insulating properties of the filled compound. Binders or sizings are commonly applied to some fillers, especially glass fibers, to improve adhesion of the fillers to the resin and hence the strength properties of the system. The binders or sizings can be a starch, silane, chromic chloride, or other types of treatment. There are pros and cons concerning which sizing is best for specific applications, but in general, a starch sizing allows moisture penetration along the fibers if it is subjected to humidity conditions, and thus lowers insulation resistance of the compound. This can be improved through use of other treatments, as shown in Fig. 85.

Stresses in resins Among the mechanical effects that embedding resins have on components and critical circuits are the internal stresses created either by the shrinkage of the embedding resin during cure, or in the unit during thermal excursions, caused either by operating or storage conditions or testing to some set of specifications. Basically, embedding resins have a considerably higher coefficient

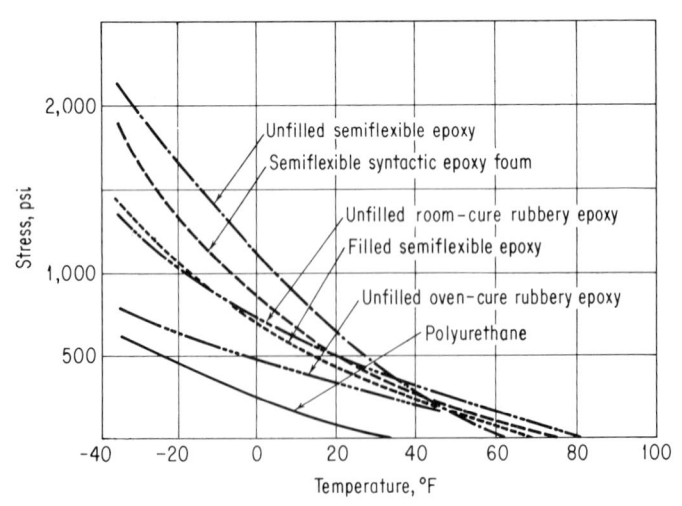

Fig. 86 Comparison of stress curves for several resin systems.[66]

of thermal expansion than glasses and metals. This is shown graphically in Fig. 82, which compares thermal coefficients for a filled and an unfilled epoxy resin with those for several other materials, as was discussed in more detail earlier. The object, of course, in designing with these data, is to attempt to match the coefficient of expansion of the resin system with that of the construction material of the critical component to be packaged.

Much can be and has been written on stress effects. Each situation is an individual one, and must be evaluated accordingly. Many components are unaffected by the highest stresses, whereas others are extremely sensitive. One generalization that can be stated, however, is that rigid, unfilled resins usually exhibit higher stresses than soft resins such as silicone rubbers, urethanes, and flexibilized epoxies. Silicones and urethanes are better at lower temperatures since they maintain better resiliency or flexibility at these low temperatures. A general set of curves appears in Fig. 86. Although there are many points of debate on protecting critical components by a resilient coating, such a coating quite often gives improved yield and performance. One example of this is given in Fig. 87.

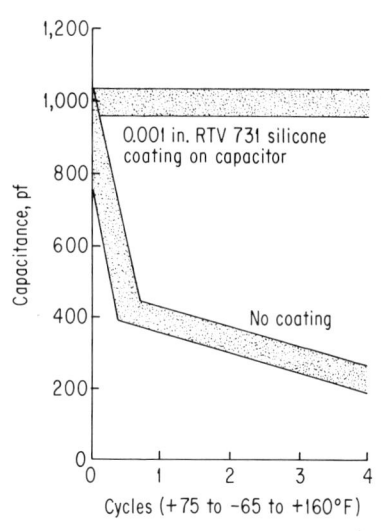

Fig. 87 Effect of silicone coating on behavior of capacitors potted in rigid, silica-filled epoxy.[67]

REFERENCES

1. Harper, Charles A. (ed.): "Handbook of Electronic Packaging," McGraw-Hill Book Company, New York, 1969.
2. Harper, Charles A.: Electrical Insulating Materials, *Mach. Des.*, Sept. 28, 1967.
3. Allied Chemical Corporation, Plastics Division: technical bulletin entitled Plaskon Plastics and Resins.
4. Carpenter, R. E.: Insulating and Structural Properties of Alkyd Molding Compounds, *Electro-Technol.*, June, 1965.
5. Parry, H. L., et al.: High Humidity Insulation Resistance of Resin Systems, *Soc. Plast. Eng. J.*, October, 1957.
6. O'Toole, J. L., et al.: Thermosets vs. New Thermoplastics: Elevated Temperature Behavior, *Soc. Plast. Eng. Reg. Tech. Conf.* Chicago, December, 1966.
7. Sunderland, G. B., and A. Nufer: Aminos, *Mach. Des.*, Plastics Reference Issue, December, 1968.
8. Beacham, H. H., and J. C. Thomas: Allylics, *Mach. Des.*, Plastics Reference Issue, December, 1968.
9. Jessup, J. N., and H. H. Beacham: Effects of Temperature and Humidity on Electrical Properties of Thermosets, *Mod. Plast.*, December, 1968.
10. Chottiner, J.: Dimensional Stability of Thermosetting Plastics, *Mat. in Des. Eng.*, February, 1962.
11. Chapman, J. J., and L. J. Frisco: A Practical Interpretation of Dielectric Measurements up to 100 Mc, Dielectrics Laboratory Report, The Johns Hopkins University, Baltimore, December, 1963.
12. Hauck, J. E.: Heat Resistance of Plastics, *Mat. in Des. Eng.*, April, 1963.
13. Buchoff, L. S.: Effect of Humidity on Surface Resistance of Filled Epoxy Resins, *Soc. Plast. Eng. Ann. Nat. Tech. Conf.*, Chicago, January, 1960.
14. Bainbridge, R. W.: Phenolics, *Mach. Des.*, Plastics Reference Issue, December, 1968.
15. Martino, C. F.: Phenolics Too Good to Be Outmoded, *Soc. Plast. Eng. Reg. Tech. Conf.*, Chicago, December, 1966.
16. Dow Corning Corporation: technical bulletins on Silicone Molding Materials.

17. Kookootsedes, G. J., and F. J. Lockhart: Silicone Molding Compounds for Semiconductor Devices, 153 ACS Meeting, Division of Organic Coatings and Plastics Chemistry, vol. 27, no. 1, pp. 56–72, 1967.
18. Madorsky, S. L., and S. Straus: Stability of Thermoset Plastics at High Temperatures, *Mod. Plast.*, February, 1961.
19. Staff Report: Contemporary Thermoplastics Materials Properties Chart, *Plast. World*, October, 1966.
20. Hauck, J. E.: Long-term Performance of Plastics, *Mat. Eng.*, November, 1965.
21. Patten, G. A.: Heat Resistance of Thermoplastics, *Mat. in Des. Eng.*, May, 1962.
22. Hauck, J. E.: Alloy Plastic to Improve Properties, *Mat. Eng.*, July, 1967.
23. Barken, H. E., and A. E. Javitz, Plastics Molding Materials for Structural and Mechanical Applications, *Elec. Mfg.*, May, 1960.
24. E. I. du Pont de Nemours & Co., Inc.: technical booklet entitled Delrin Acetal Resins.
25. E. I. du Pont de Nemours & Co., Inc.: technical data bulletin entitled Delrin AF.
26. Ragolia, A. J.: Dimensional Stability of Acrylic Resins, *Mod. Plast.*, July, 1966.
27. Riley, M. W.: Selection and Design of Plastics, *Mat. & Methods*, June, 1957.
28. Koo, G. P., et al.: Engineering Properties of a New Polytetrafluorethylene, *Soc. Plast. Eng. J.*, September, 1965.
29. Staff Report: Creep Properties of Plastics, *Mod. Plast.*, Encyclopedia Issue, October, 1968.
30. Staff Report: Fluorocarbon Plastics, *Mat. in Des. Eng.*, February, 1964.
31. E. I. du Pont de Nemours & Co., Inc.: technical booklet entitled Mechanical Design Data for Teflon Resins.
32. E. I. du Pont de Nemours & Co., Inc.: technical booklet entitled Teflon for Electrical and Electronic Systems.
33. Jolly, C. E., and J. C. Reed: The Effects of Space Environments on TFE and FEP Resins, *11th Ann. Signal Corps Wire and Cable Symp.*, Ashbury Park, N.J., November, 1962; and Radiation Tolerance of Teflon Resins, *Du Pont J. of Teflon*, February, 1969.
34. Ehner, W. J.: Thermoplastic Parts, *Mach. Des.*, August, 1963.
35. E. I. du Pont de Nemours & Co., Inc.: technical booklet entitled Zytel Nylon Resins.
36. E. I. du Pont de Nemours & Co., Inc.: technical bulletin entitled Nomex High Temperature Insulation.
37. E. I. du Pont de Nemours & Co., Inc.: technical bulletin entitled Vespel Precision Parts.
38. Freeman, J. H., et al.: New Organic Resins for High Temperature Reinforced Laminates, *Soc. Plast. Eng. Trans.*, July, 1962.
39. General Electric Company: technical booklet entitled Lexan Polycarbonate Resin.
40. General Electric Company: technical bulletins on Noryl and PPO Resins.
41. Union Carbide Corp., Plastics Division: technical booklet entitled Polysulfone.
42. Staff Report: Why Plastics Don't Live Up to Design Data, *Prod. Eng.*, Apr. 25, 1966.
43. Gowan, A., and P. Shenian, Properties and Applications for PPO Plastic, *Insul.*, September, 1965.
44. E. I. du Pont de Nemours & Co., Inc.: technical bulletin entitled Mylar Polyester Film.
45. Sinclair-Koppers Co.: technical manual entitled Dylite Expandable Polystyrene.
46. Union Carbide Corp.: technical manual entitled Bakelite Polysulfone.
47. Staff Report: Can RTP Meet Your Product Requirements, *Mod. Plast.*, March, 1966.
48. Liquid Nitrogen Processing Corp.: technical data bulletin entitled LNP Glass Fortified Polysulfone.
49. Metz, E. A.: Fiber Glass Reinforced Nylon, *Mach. Des.*, Feb. 17, 1966.
50. Hauck, J. E.: Engineer's Guide to Plastics, *Mat. in Des. Eng.*, February, 1967.
51. Harper, Charles A.: "Electronic Packaging with Resins," McGraw-Hill Book Company, New York, 1961.
52. Harper, Charles A.: Embedding Processes and Materials, *Mach. Des.*, June 9, 1966.
53. Hinkley, J. R.: Resins for Packaging Electronic Assemblies, *Electro-Technol.*, June, 1965.
54. Lee, M. M., and R. D. Hodges: Heat Resistant Encapsulating Resins, *Proc. 15th Ann. Tech. Conf. Soc. Plast. Eng.*, New York, January, 1959.

55. General Mills, Inc.: technical data bulletins entitled Versamids.
56. Thiokol Chemical Corp.: technical data bulletins entitled Thiokols.
57. Du Pont, Inc., Elastomer Chemicals Dept.: technical data bulletins entitled Adiprenes.
58. Black, R. G.: Linear Polyazelaic Polyanhydride as a Converter for Epoxy Resin Systems, *Proc. 20th Ann. Tech. Conf. Soc. Plast. Eng.*, Atlantic City, N.J., January, 1964.
59. Dow Corning Corp.: Technical data bulletin entitled Sylgard 184 Resin.
60. Thorne, J. A., C. L. Whipple, and A. B. Boehm: Space Environmental Effects on Silicone Insulating Materials, *Proc. 6th Elec. Insul. Conf.*, New York, September, 1965.
61. Hystl Development Co.: technical data bulletin entitled Hystl 110 Polybutadiene Resins.
62. Lee, H., and K. Neville: "Handbook of Epoxy Resins," McGraw-Hill Book Company, New York, 1966.
63. Cabot Corp.: technical data bulletin entitled Cab-O-Sil.
64. Wolf, D. C.: Trends in the Selection of Liquid Resins for Electronic Packaging, *Proc. Nat. Electron. Packag. Conf.*, New York, June, 1964.
65. Parry, H. L., et al.: Effect of Fillers on High Humidity Insulation Resistance of Epoxy Resin Systems, *Proc. Soc. Plast. Eng. Reg. Tech. Conf.*, Lowell, Mass., June, 1957.
66. Minnesota Mining and Manufacturing Co.: technical data bulletins entitled Scotchcast Electrical Insulating Resins.
67. Johnson, L. I., and R. J. Ryan: Encapsulated Component Stress Testing, *Proc. 6th Elec. Insul. Conf.*, New York, September, 1965.

Chapter 2

Laminates, Reinforced Plastics, and Composite Structures

RONALD N. SAMPSON

Research and Development Center, Westinghouse
Electric Corporation, Pittsburgh, Pennsylvania

Basic Properties and Testing	2–2
Laminates	2–11
Definition	2–11
Design Criteria	2–12
Reinforcing Fibers	2–13
Laminated Materials	2–17
NEMA Industrial Laminates	2–48
NEMA Grades	2–48
Properties of Industrial Laminates	2–62
Thermal Aging of NEMA Grade Laminates	2–63
Tolerances of NEMA Grade Laminates	2–63
Vulcanized Fiber	2–63
Prepreg Materials	2–66
Prototypes	2–70
Filament Winding	2–83
Space-age Laminates	2–84
Machining Plastics	2–90
Joining	2–93
Adhesive Bonding and Adhesives	2–96
Printed-circuit Laminates	2–105

Testing Copper-clad	2–107
Copper Foil	2–109
Dimensions and Tolerances	2–111
Multilayer Printed Circuits	2–111
Flexible Printed Circuits	2–113
Microwave Strip Line	2–114
Films and Flexible Insulations	2–115
Films—Definition	2–115
Film Properties	2–115
Flexible Insulations—Definition	2–127
References	2–131

BASIC PROPERTIES AND TESTING

Composite materials differ from other materials in that they are anisotropic. This directionality of properties is related to the nature of the materials and to the way in which they are manufactured. It has been found necessary to develop specific test methods for plastic materials. Those test methods have come from a wide variety of sources including governmental bodies, industry laboratories, industry associations, and professional societies.

Testing standards are prepared by both the American Society for Testing and Materials (ASTM) and the Federal government. A selected tabulation of these standards is shown in Table 1.[1,2] Many of the test methods are identical in the two sources, but not always.

Dielectric strength is the voltage gradient at which electric failure results. The failure is characterized by an excessive flow of current (arc) and by partial destruction of the material. Dielectric strength is measured through the thickness of the

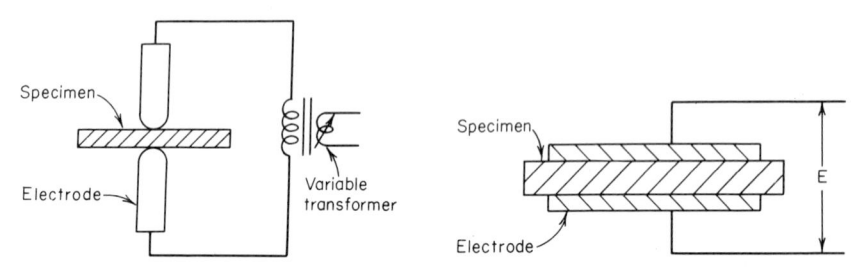

Fig. 1 Dielectric breakdown test.

Fig. 2 Dielectric constant test.

material, as shown in Fig. 1, and is expressed in volts per unit of thickness. Two principal methods of measuring this value exist. The short-time test is performed by increasing the voltage from zero at a predetermined speed (100 to 3,000 volts/sec) to breakdown. The rate is specified by the investigator. The step-by-step test differs in that an initial voltage of 50 percent of the short-time voltage is applied to the material, and then the voltage is increased in equal increments and held for periods of time as determined by the investigator. Dielectric strength is influenced by temperature, humidity, voids or foreign materials in the specimen, electrode configuration, frequency, and specimen geometry. Because of this it is often difficult to compare breakdown data from different sources unless all test conditions are known.

Dielectric constant is the ratio of the capacitance formed by two plates with a material between them and the capacitance of those same plates with a vacuum between them (Fig. 2). The value is of particular interest to designers of microwave equipment. For example, in radar the thickness of the part is dictated by

Basic Properties and Testing 2-3

TABLE 1 Standard Test Methods[1,2]

Test method	Federal Standard Method No.	ASTM No.
Abrasion wear (loss in weight)	1091	
Accelerated service tests (temperature and humidity extremes)	6011	D 756-56
Acetone extraction test for degree of cure of phenolics	7021	D 494-46
Arc resistance	4011	D 495-61
Bearing strength	1051	D 953-54, Method A
Bonding strength	1111	D 229-63T, pars. 40–43
Brittleness temperature of plastics by impact	2051	D 746-64T
Compressive properties of rigid plastics	1021	D 695-63T
Constant-strain flexural fatigue strength	1061	
Constant-stress flexural fatigue strength	1062	
Deflection temperature under load	2011	D 648-56, Procedure 6(a)
Deformation under load	1101	D 621-64, Method A
Dielectric breakdown voltage and dielectric strength	4031	D 149-64
Dissipation factor and dielectric constant	4021	D 150-64T
Drying test (for weight loss)	7041	
Effect of hot hydrocarbons on surface stability	6062	
Electrical insulation resistance of plastic films and sheets	4052	
Electrical resistance (insulation, volume, surface)	4041	D 257-61
Falling-ball impact	1074	
Flame resistance	2023	
Flammability of plastics 0.050 in. and under in thickness	2022	D 568-61
Flammability of plastics over 0.050 in. in thickness	2021	D 635-63
Flexural properties of plastics	1031	D 790-63
Indentation hardness of rigid plastics by means of a durometer	1083	D 1706-61
Interlaminar and secondary bond shear strength of structural plastic laminates	1042	
Internal stress in plastic sheets	6052	
Izod impact strength	1071	D 256-56, Method A
Linear thermal expansion (fused-quartz tube method)	2031	D 696-44
Machinability	5041	
Mar resistance	1093	D 673-44
Mildew resistance of plastics, mixed culture method, agar medium	6091	
Porosity	5021	
Punching quality of phenolic laminated sheets	5031	D 617-44
Resistance of plastics to artificial weathering using fluorescent sunlamp and fog chamber	6024	D 1501-57T
Resistance of plastics to chemical reagents	7011	D 543-60T
Rockwell indentation hardness test	1081	D 785-62, Method A
Salt-spray test	6071	
Shear strength (double shear)	1041	
Shockproofness	1072	
Specific gravity by displacement of water	5011	
Specific gravity from weight and volume measurements	5012	
Surface abrasion	1092	D 1044-56
Tear resistance of film and sheeting	1121	D 1004-61
Tensile properties of plastics	1011	D 638-64T
Tensile properties of thin plastic sheets and films	1013	D 882-64T
Tensile strength of molded electrical insulating materials	1012	D 651-48
Tensile time—fracture and creep	1063	
Thermal expansion test (strip method)	2032	
Warpage of sheet plastics	6054	D 1181-56
Water absorption of plastics	7031	D 570-63

the frequency, physical loads, and dielectric constant. Most plastics have dielectric constants between 2 and 10. Dielectric constant is related to the temperature and to frequency. For many plastics dielectric constant decreases with frequency and increases with temperature.

Dissipation factor is the ratio of the parallel reactance to parallel resistance, i.e., the tangent of the loss angle or the cotangent of the phase angle. Figure 3 shows this relationship. It should be noted that dissipation factor (tan δ) is not identical to power factor (cos θ). However, at low values of tan δ (less than 0.10) the values for power factors are nearly identical. The dissipation factor is related to the energy dissipated and hence to the efficiency of the insulation material. Some designers use another term, *loss factor*, which is the product of the dielectric constant and the dissipation factor. This term is related to the total loss of power (in watts) occurring in the insulation material. Dissipation factors for most plastics tend to decrease with increasing frequency and increase with increasing temperature. Since this property may vary widely through any temperature or frequency range, care should be used with reported values that do not include the test conditions. Many sound data on plastics are available at 60 cycles and room temperature.[3]

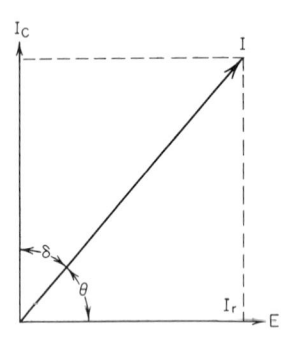

Fig. 3 Vector diagram of an equivalent parallel circuit. E = voltage; I = current; I_r = resistance component of current = E/R_r; I_c = capacitive component of current = $2\pi f E C_p$; R_r = parallel resistance; f = frequency; C_p = parallel capacitance; θ = phase angle, and δ = loss angle.

Insulation resistance is the ratio of voltage applied to a sample to the current flowing in the sample. It is made up of two components: surface resistance and volume resistance. Measuring surface and volume resistivities (the specific resistances) involves separating these two components from the insulation resistance. This is accomplished by using three electrodes as shown in Fig. 4. The outermost (guard) electrode is at the same voltage as the measuring electrode.

Surface resistivity is the resistance to current leakage along the surface of an insulator. The units of measurements are in ohms per square. Surface resistivity

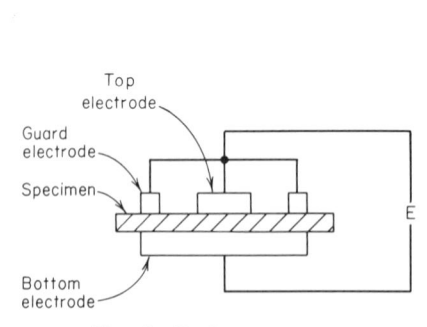

Fig. 4 Resistance test.

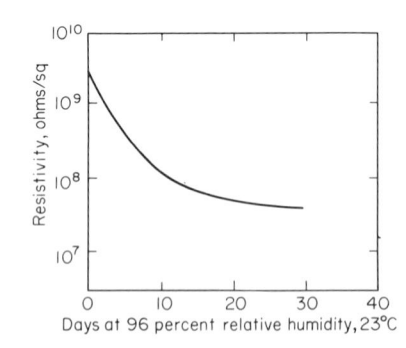

Fig. 5 Surface resistivity of a paper-phenolic laminate as a function of time at 96 percent relative humidity.[4]

is very sensitive to humidity, surface cleanliness, and surface contour. Figure 5 shows the effect of moisture on the surface resistivity of a typical phenolic laminate.

Volume resistivity is the resistance to current leakage through the body of an insulator. The units of measurement are ohm-centimeters or ohm-inches. Volume resistivity is related to the nature of the insulator, moisture in the material, and temperature. Figure 6 shows the effect of temperature on a typical polyester–glass-cloth laminate.

Arc resistance is a measure of the breakdown of the surface of an insulator caused by an arc which tends to form a conducting path. Many testing methods have been developed and are useful for specific problems. However, only four have been agreed upon by ASTM. The earliest of these is ASTM D 495-61 which is a high-voltage low-current test under clean conditions. Figure 7 schematically shows the electrode arrangement. Other tests are specified to relate more closely with contamination and surface conditions found in practice. These tests all rely on the introduction of some contaminant in the arcing area. The dust-fog test is specified by ASTM D 2132-66T. This test is performed at 1.5 kv on a sample in a fog chamber and with a standardized dust applied to the sample surfaces. Failure is characterized by erosion of the sample or tracking. ASTM D 2302-65T describes the differential wet-track test. This test makes use of a 3-kv arc at several power levels. The sample is inclined and partially immersed in a water solution of ammonium chloride and a wetting agent. Failure is by tracking. The inclined-plane test is defined by ASTM D 2303-64T. In this test

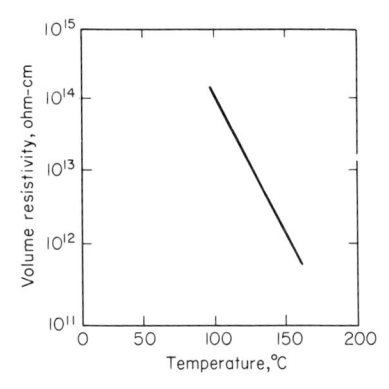

Fig. 6 Volume resistivity of a glass-cloth–polyester laminate as a function of temperature.[5]

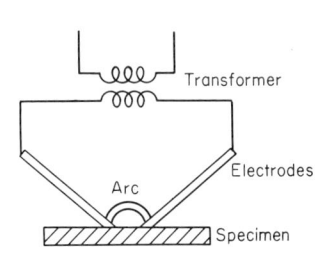

Fig. 7 Arc resistance test.

a specimen is inclined at 45°, and electrodes are placed on the underneath side. An electrolyte is fed onto the surface at a controlled rate, and voltage is increased simultaneously. Failure is by erosion and tracking.

Certain plastic materials track easily. For example, polymers based on aromatic rings such as phenolics or polystyrene track readily. Certain other plastics which are based on long straight-chain polymers are more resistant. In general the addition of mineral fillers improves arc and track resistance. Certain hydrated fillers exemplified by hydrated alumina[6] dramatically improve the tracking properties. Fibers generally decrease wet-tracking resistance and improve dry arc resistance. Typical properties for selected plastic materials are given in Table 2.

Tensile strength is the force per unit area required to break a plastic in tension. The test is shown schematically in Fig. 8. A typical stress-strain curve for a glass-reinforced plastic is shown in Fig. 9. Other tensile parameters which are of use are also shown. Young's modulus E is the slope of the curve between 0 and point a. The secant modulus is the slope of a line between any point such as b and the origin. Unlike metals, reinforced plastics generally do not have a clearly defined yield point. One can be defined by measuring a distance representing a certain percentage of total strain (usually 0.2 percent) as point c and then drawing a line through c and parallel to the modulus line. The intercept

2-6   Laminates, Reinforced Plastics, and Composite Structures

TABLE 2 Arc and Track Resistance of Plastics[7]

Material tested	D 495-61 Equiv. sec/10	D 2132-62T Hours at 1.5 kv	D 2302-64T Watt-minutes	D 2303-64T Voltage kv
Polyvinyl chloride.............	0.5 Tr	0.5 Tr		
Phenolic laminate, paper base...	0.5 Tr	0.5 Tr	0.2* Tr	0.5 Tr
Epoxy resin, unfilled...........	1.7 Tr	0.5 Tr	1.6 Tr	
Polyamide resin................	58+ Er	0.5 Tr	1.3− Int	
Silicone resin, glass cloth.......	54 Tr	1.0 Tr	1.8 Tr	1.5 Tr
Melamine resin, glass cloth.....	47 Tr	3.5 Tr	2.3 Tr	2.3 Tr
Polyethylene..................	13 Tr	27 Tr		
Polymethyl methacrylate.......	100+ Er	90 Er	8.1+ Er	6 F
Polypropylene.................	310 Er	180 Er	8.1+ Er	3.8 Tr
Epoxy resin, Hmf..............	100+ Er	200 Er		
Polyester, glass mat, Hmf......	51 Tr	350 Tr	6.4 Tr	3 Tr
Butyl rubber, Hmf.............	100+ Er	450 Er	8.1+ Er	6 F
Silicone rubber, Nmf...........	5 Tr	750 Er	....	3.7 Tr
Polytetrafluorethylene.........	310+ Er	2,700 Er	8.1+ Er	7 F

Hmf   Hydrated mineral filled.   Tr   Tracked.   Int   Internal.
Nmf   Nonhydrated mineral filled.   Er   Eroded.   F   Flame.
* Failed at 1.3 watts in 1 sec.

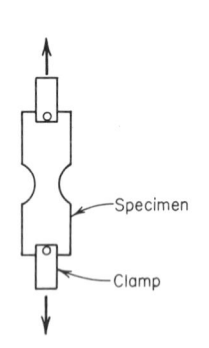

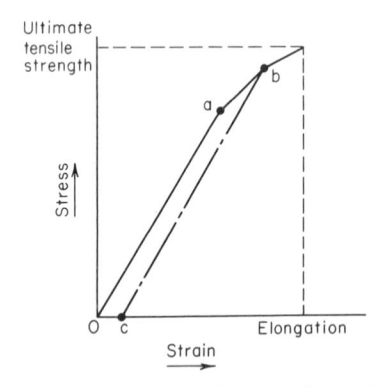

Fig. 8 Tensile test.

Fig. 9 Stress-strain diagram for reinforced plastics.

with the curve *b* is then defined as the 0.2 percent offset yield. Tensile properties of plastics are influenced by environmental conditions and the nature of the materials. The ultimate properties are of little use to a designer since he must be more concerned with deformations and elongations which occur at lower loads. Tensile strengths are useful in comparing and selecting materials. Tensile properties depend to a great extent upon the nature and amount of reinforcements present in the plastics. For example, an unreinforced polyethylene may have a tensile strength of 1,000 psi, a laminated phenolic a strength of 45,000, and a filament-wound cylinder a strength of 100,000 psi. Tensile modulus values are of help in comparing the stiffness of materials and the ability of the materials to resist deformation under short-time loads.

Poisson's ratio is the ratio of lateral strain to axial strain in axial-loaded specimens; for example, for a rod in tension it is the ratio of the decrease in diameter to the increase in length. The method for determining Poisson's ratio is given in ASTM E 1321. This ratio does not vary greatly for plastic materials. Thermoplastic materials have a ratio of about 0.5, and thermosetting laminates have ratios between 0.4 and 0.5.

Compressive strength is the force per unit area required to break a plastic in compression. Many of the comments in the tensile portion of this section also apply to compressive strength. Compressive strengths are also influenced by temperature and environmental conditions but not to the same extent as tensile properties.

Flexural strength is the force per unit area required to break a plastic in bending. The loads are applied as shown in Fig. 10. Flexural modulus is the ratio of

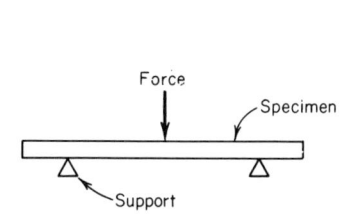

Fig. 10 Flexural strength test.

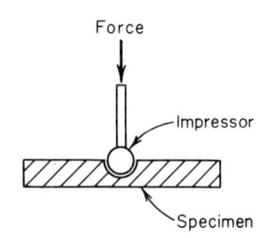

Fig. 11 Hardness test.

the stresses to the unit deformations. Modulus values vary with influences such as temperature as do the tensile properties. Flexural modulus values are of particular interest since they relate to the stiffness of a material under load. Modulus values vary from 0.3×10^6 psi for thermoplastics to 5×10^6 psi for filament-wound products.

Hardness is the ability of a plastic to resist penetration, as shown in Fig. 11. Several methods are used such as Brinell, Vickers, and Rockwell. Figure 12 shows the relationship between different hardness methods. One hardness method used by some workers is based on the Barcol Impressor sold by the Barber-Coleman Company. This expresses hardness values in arbitrary units, is easy to use, and has good repeatability. A similar test is specified in ASTM D 1706-64T and is based on a durometer instrument with a pointed impressor. The units of this test are also arbitrary, and the test is most useful on materials that creep easily. Hardness values stay relatively constant to the glass transition temperature, at which point they decrease.

Creep is nonrecoverable deformation which occurs before the yield point. This stretching is caused by uncoiling and slipping of the molecules of a plastic. Creep may be studied with any of the types of loading, i.e., tensile, flexural, or compressive, or combinations of these. When a sample is loaded, there is an immediate elastic deformation followed by a slow deformation at a decreasing rate called primary creep. This deformation rate soon stabilizes and holds constant for some time which is called secondary creep. Failure often occurs during this period, but some materials exhibit tertiary creep during which time the deformation rate suddenly increases to failure. Creep rate is very sensitive to total load and temperature. Other environmental factors such as moisture content also influence creep. In general, creep is similar to tensile elongation. The greater the elongation, the more a material will exhibit creep. Reinforcing a plastic radically improves the creep resistance, and so fiber-filled—particularly glass fiber—plastics creep much

2-8   Laminates, Reinforced Plastics, and Composite Structures

less than unfilled materials. Thermosetting plastics have less creep than thermoplastics. Fabric-reinforced laminates creep less than paper- or mat-reinforced materials. Figure 13 shows the creep behavior of a glass-cloth-polyester laminate.

Impact strength is the resistance of a material to a sudden loading applied as an impact. Two tests are described in ASTM D 256-56: the Izod and the Charpy. An Izod test is made on a notched sample held as a cantilever beam. A

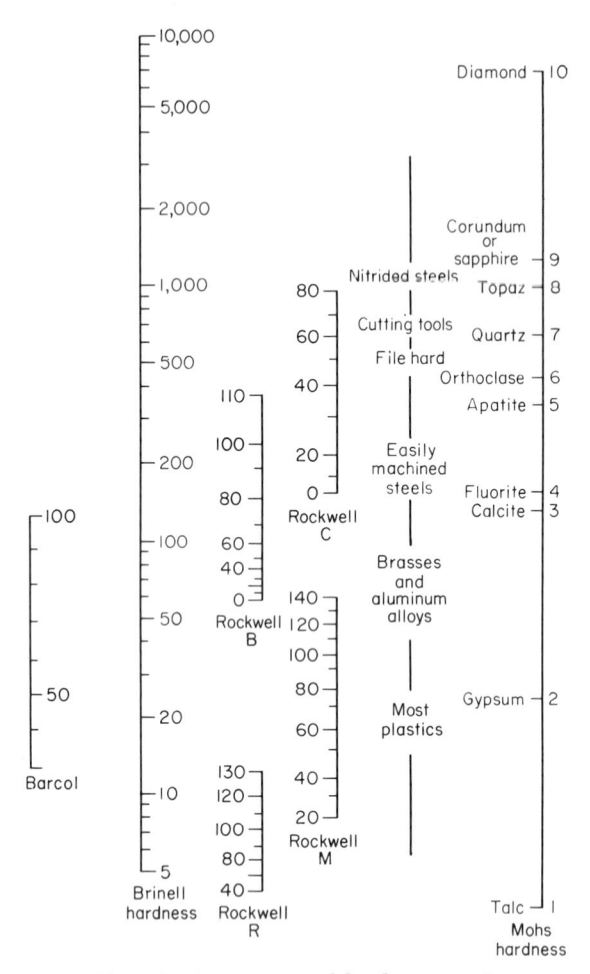

Fig. 12   Comparison of hardness tests.[8]

Charpy test is made on either a notched or an unnotched sample which is supported as a simple beam at each end. The sample arrangement for an Izod test is shown in Fig. 14. The Charpy test is more useful for materials at other than room temperature. Impact tests are useful in comparing the energy-absorbing characteristics of materials, and impact values are often used in standards. Impact tests are often used to specify the amount of fiber in a reinforced plastic. Care must be taken, however, when using the test to fully appreciate the notch sensitivity that occurs in some materials—particularly thermoplastics like nylon. A tensile impact test is defined by ASTM D 1822-61T and is being used with success.

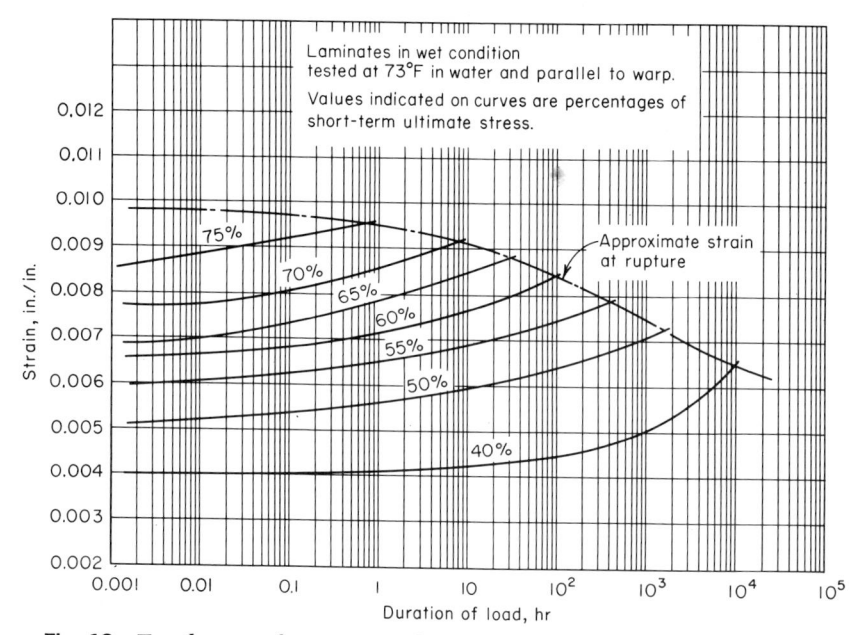

Fig. 13 Tensile creep for continuously loaded 10-oz cloth-polyester laminates.[9]

Deflection temperature is a test devised to specify the heat resistance of plastic materials. Formerly called the *heat-distortion test*, the procedure is specified in ASTM D 648-56. A specimen 5 in. long by ½ in. wide by any thickness from ⅛ to ½ in. is placed on two supports 4 in. apart, and a load equal to 66 or 264 psi is applied to the center while the temperature is increased by 2°C per min. The temperature reached when the sample deflects 0.010 in. is reported as deflection temperature. The test is useful in comparing materials and for quality control purposes, but it does not describe differences in thermal capabilities of plastics.

Thermoplastic and thermosetting resins behave differently according to temperature. Thermoplastics soften as temperature increases. Thermosetting materials may soften or harden depending on many factors. If further curing or crosslinking takes place, the material will grow stiffer. Some polymers, fully cured, will soften with increasing temperature. Finally, at high enough temperatures polymers oxidize and undergo pyrolysis (decomposition), and these chemical processes relate to the stiffness and the deflection temperature. Hence the deflection temperature for any plastic can vary widely depending on what conditions have occurred. Table 3 shows typical deflection temperatures for some plastics.

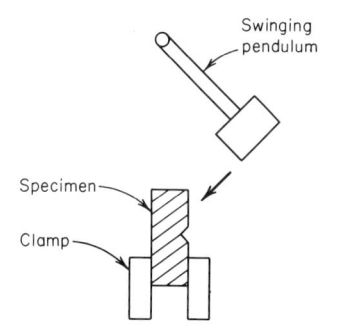

Fig. 14 Izod impact test.

Flame resistance is the ability of a plastic to withstand burning. Many test methods are in use. Table 4 lists the standard flammability tests used for plastic materials. ASTM D 635-63 is widely used throughout the plastics industry. The test employs a bunsen burner, and burning time and rate are measured. Simplicity and ease of testing are characteristic of this method. The electrical industry uses extensively Federal Method 2023. This test consists of a heating coil which sur-

2-10 Laminates, Reinforced Plastics, and Composite Structures

rounds the specimen. A spark plug suspended over the sample is used to ignite the gases which are evolved from the sample. Time to ignition and burning time are reported. This method is less variable but more costly.

Water absorption is the amount of water absorbed into the mass of a sample. Usually a high water absorption is accompanied by some change in dimension

TABLE 3 Deflection Temperatures

Material	Filler	°F at 264 psi
Silicone	Glass cloth	840
Polyimide	....	650
Phenolic	Glass cloth	600
Epoxy	Glass cloth	500
Melamine	Glass cloth	400
Polyester	Glass cloth	400
Polyphenylene oxide	....	375
Polysulfone	....	345
Polytetrafluoroethylene	....	250
Polyethylene	....	105

TABLE 4 Flame Resistance Tests

Test*	Sample size, in.	Test title
D 568-61	1 × 18 × <0.050	Flammability of Plastics 0.050 in. and Under in Thickness
D 635-63	½ × 5 × >0.050	Flammability of Rigid Plastics over 0.050 in. in Thickness
D 757-49	½ × 5 × ⅛	Flammability of Plastics—Self-extinguishing Type
D 1433-58	3 × 9 × t†	Flammability of Flexible Thin Plastic Sheeting
D 1692-59T	2 × 6 × t†	Flammability of Plastics Foams and Sheeting
E 84-67	10 × 25 ft × t†	Surface-burning Characteristics of Building Materials
E 136-65	1½ × 1½ × 2	Determining Noncombustibility of Elementary Materials
E 162-67	5¼ × 17½ × t†	Surface Flammability of Materials Using a Radiant Heat Energy Source
E 286-65T	13¾ × 8 ft × t†	Method of Test for Surface Flammability of Building Materials Using an 8-foot Tunnel
Federal Standard 406, Method 2023	½ × 5 × ½	Flame Resistance

* ASTM unless otherwise designated.
† *t* represents thickness.

and in electrical and physical properties. Water absorption is a function of the nature of the plastic in question, but is also related to the form of the plastic. Some laminated materials, if not well coupled to the reinforcing fibers, will absorb large quantities of water. Cellulose-based materials have high water absorption, as do nylon materials.

The coefficient of thermal expansion is a measure of the dimensional changes which occur when a sample is heated or cooled. The value reported is generally the linear value, and the volumetric coefficient is considered to be three times the linear. The thermal expansion of plastics is substantially higher than that of most metals. Table 5 lists typical values for plastics and metals. Because

TABLE 5 Typical Thermal Expansions and Thermal Conductivities of Materials

Material	Linear coefficient of thermal expansion, in./(in.)(°F) × 10⁻⁵	Thermal conductivity, Btu/(hr)(ft²)(°F/in.)
Aluminum	1.30	1390
Carbon steel	0.67	360
Copper	0.93	310
Stainless steel	0.96	113
Plastic laminates	2.0	2.0
Cast resin and filler	1–3	6.0
Premix molding material	1.5	4.0
Thermoplastic	2–17	0.3–2

of the difference in expansion, care must be used in molding metal inserts into plastics.

When a material is molded, a certain amount of shrinkage occurs in the mold, varying from ½ to 1½ percent. The shrinkage is made up of thermal contraction from the molding temperature and the shrinkages associated with the phase change of the plastic from a liquid to a solid. These changes result in parts that are smaller than the molding cavity. Cooling jigs are often used to preserve critical dimensions. Other dimensional changes which occur on some materials with heating are stress relief of molded-in strains and expulsion of adsorbed moisture.

Thermal conductivity is a basic property of materials relating to the flow of heat through the material. With few exceptions those materials which are good electrical insulators are poor thermal conductors. The thermal conductivities of the basic resins used in plastics are very similar, the chief differences in the conductivities of the final plastics being related to the fillers and other modifying ingredients. The conductivities of various plastics and metals are shown in Table 5. Conductivities can be increased with increasing additions of fillers, but the relationship is not linear. Fillers such as alumina, silica, and calcium carbonate are commonly used.

Environmental conditioning standards for conditioning samples in various environments are defined in ASTM D 709-67:

Condition A—as received, no special conditioning
Condition C—humidity conditioning
Condition D—immersion conditioning in distilled water
Condition E—temperature conditioning
Condition T—test condition

Note that some differences exist between this ASTM specification which covers laminates and a more general specification ASTM D 618 which covers plastics. The condition designation is followed by a number defining the duration of conditioning in hours. After a slant a second number indicates the conditioning temperature in degrees centigrade, and after another slant the third number defines the humidity in those cases where humidity is controlled. For example, Condition C-24/23/50 reads humidity condition of the sample for 24 hr at 23°C and 50 percent relative humidity.

LAMINATES

Definition Laminates are plastic materials made by bonding together two or more sheets of reinforcing fibers. The fibrous sheets may be fabric, paper, or mat made of cellulose, asbestos, glass fibers, or synthetic plastics. The binders—called resins—can be almost any polymeric material but in general are based on thermo-

2-12 Laminates, Reinforced Plastics, and Composite Structures

setting resins such as a phenolic, melamine, polyester, epoxy, or silicone. These resins are dissolved in suitable solvents and used to impregnate the reinforcing sheets in a treating tower, as shown in Fig. 15. In the treater oven the solvents are removed, and in some cases the resins are advanced to a semicured condition called the *B stage* in which the resins are dry and hard but yet capable of further flow. The sheets are then placed into a molding press and pressed at high temperature (250 to 400°F) and high pressure (200 to 3,000 psi) until they are cured. In the curing (thermosetting) process the resins become permanently infusible and insoluble. The finished laminates are usually cooled under pressure, removed from the press, and trimmed to size.

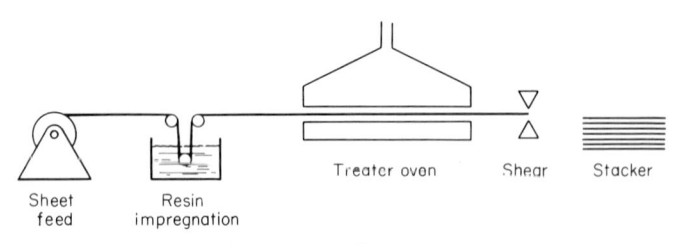

Fig. 15 Horizontal treater tower.

Design criteria The presence of the reinforcing media (paper, cloth, etc.) differentiates laminates from other plastics and results in material with substantially improved physical properties. These high mechanical strengths make it possible to use laminates in structural applications where regular plastics could not be used. The high strength-to-weight ratio of certain of these laminates rivals even alloy steels, and one therefore can find these reinforced-plastic materials being used in applications such as rocket motor casings. However, when comparing laminates with metals, one should remember that laminates differ in behavior quite dramatically, and the designer must allow for these differences. The biggest difference between metals and plastics is that plastics are not ductile. This means that whereas metals can be designed to be used up to their yield point, plastics (lacking a well-defined yield) can be used almost to their ultimate strength. The lack of ductility also means that reinforced plastics will not change greatly in shape when loaded, and stress risers are not easily formed as is the case with metals.

The first step in design selection of a laminate is to define carefully the service that the laminate is expected to meet. Are the requirements physical or electrical? What specific properties are important, and what values are to be assigned to these properties?

The second step is selection of the laminate. The choice of the laminate should be made—if possible—from standard available materials to avoid costly and time-consuming development programs. Remember when consulting NEMA Standard property tables that the values listed are usually conservative, and perhaps a lower-cost grade should be considered. At this point one may also find that several laminates are candidate, and—particularly where cost is a vital factor—compromises may have to be made in one's requirements to obtain a material with the best balance of properties.

Many properties of laminates are conflicting. For example: materials with high resin content usually have improved electrical properties and reduced physical properties. Cloth-based materials are more easily punched than paper-based materials, but they have poorer electrical properties. Cloth-based materials are higher cost than paper materials, but finished parts may be less expensive because punching costs less than machining. Flame-retardant materials are often lower in physical properties than similar nonflame-retardant grades.

The next step in the design process is to ensure that the material selected is available in the form and size that are required. Not all materials are available in all forms (for example, many materials are not available in tubes), and each laminate has certain thickness limits. If you have selected incorrectly and the form or size is critical, you may have to go back and select a laminate which compromises the properties required.

The next step is to determine whether the tolerances you require are available. If you have selected nonstandard tolerances, you can expect the cost of your laminate to be substantially increased. The use of nonstandard tolerances should be carefully considered. Can some element of the design be modified so that standard tolerances can be used?

The last design step is to check your results. Have you under- or overspecified? For example, have you specified a glass-based laminate to obtain water resistance when sufficient resistance is available in a paper-based material? Have you carefully considered machining or processing cost in relation to the cost of the laminate?

Have you specified accurately in relation to the following features—all of which are cost adders?

    Fabric base
    Flame resistance
    Nonstandard sizes or tolerances
    Sanded surfaces
    Thermal stability
    Punchability
    High mechanical strength
    Machinability
    Improved electrical properties
    Arc and track resistance

Reinforcing fibers Many reinforcing fibers are used in laminates. The selection of the reinforcing media is related to the properties desired and, to a certain extent, to the resin system desired. The reinforcements include fiber glass, asbestos, cellulose, and synthetic fibers. Table 6 lists common reinforcements and the special properties which they contribute to a laminate.

TABLE 6 Laminate Properties in Relation to Reinforcing Media

Reinforcing medium	Mechanical strength	Electrical properties	Impact resistance	Chemical resistance	Machinability	Heat resistance	Moisture resistance	Abrasion resistance	Low cost	Punchability	Appearance
Glass fabric	X	X	X	X	...	X	X	...	...	...	...
Glass mat	...	...	X	X	...	X	X	...	X	...	...
Woven roving	...	...	X	X	...	X	...	...	X	...	...
Asbestos paper	...	X	...	...	...	X	...	...	X	...	...
Asbestos fabric	X	X	X	...	...	X	...	...	...	...	...
Kraft paper	X	...	...	...	X	...	...	...	X	X	...
α-Cellulose, paper	...	...	...	...	X	...	...	...	...	X	X
Rag paper	...	X	...	...	X	...	...	...	...	X	X
Cotton fabric	X	X	X	...	X	...	...	...	...	...	...
Nylon	...	X	X	X	...	...	...	X	...	...	...
Polyester	...	X	...	X	...	...	X	X	...	...	...
Acrylic	...	...	...	X	...	...	X	...	...	...	...

2-14 Laminates, Reinforced Plastics, and Composite Structures

Glass fabric reinforcements are cloths made in a variety of woven patterns. The weaves commonly used are plain, leno, satin, crowfoot satin, and twill. With over one hundred weaves to choose from, the fabrics are available in thicknesses from 0.001 to 0.067 in. Fabrics and their use are defined in MIL-C-9084B and MIL-Y-1140C. All fabrics (and most glass fiber products) to be used with polymers are coated with a coupling agent called a *finish*. The purpose of these finishes is to increase the adhesion of the plastics to the glass surface. Hence the type of finish is highly dependent on the type of polymer to be used. Table 7 describes common glass finishes and their uses. Glass fiber finishes are defined in MIL-F-9118A.

TABLE 7 Glass Fiber Finishes

Finish	Use
111	For high-pressure melamine laminates.
516	Similar to 111, but with improved electrical properties.
210	With silicone rubber.
112	Designates fabric that has been heated to high temperature to remove all organic constituents. Also known as "heat-cleaned." Used with silicone laminates.
Neutral pH	Fabric for silicones which is washed in deionized water to remove alkali.
Volan A*	Based on methacrylate chromic chloride. Widely used with polyester, epoxy, and phenolic resins.
Garan†	Used with polyesters where white color and high wet strength is required.
A-172	A vinylsilane material used with polyesters.
A-1100	Used with phenolic, epoxy, melamine, phenolic, and polyimide resins, this material is an aminosilane.
A-1100 Soft	Similar to A-1100, but treated fabrics are much more drapable.

* Trademark of E. I. du Pont de Nemours & Co., Wilmington, Del.
† Trademark of Johns-Manville Corp., New York, N.Y.

Glass mat reinforcement is composed of randomly deposited strands of glass from ¼ to 1 in. long. The strands are held together with a light coating of some polymer called a *binder*. Mats are available in a range of thicknesses and with the same glass treatments described in the previous section. Mat laminates do not possess physical properties comparable to fabric laminates, but they are less costly and can be molded to more complex shapes. For molded shapes a mat preform is sometimes made. This is a formed-in-place mat made in a special machine and also consisting of randomly deposited fibers with a binder. Another special case of a mat laminate is the process used primarily by boat manufacturers called spray-up. The spray-up process consists of chopping fibers of glass into small lengths, mixing them with a resin, and then spraying them onto a mold surface. In this fashion a mat with pre-added resin is assembled at the mold. All these methods produce products with similar properties.

Woven roving laminates are a compromise in physical properties between mat and cloth. This material is made from roving—continuous strands of fiber glass—which is very simply woven in an over-under pattern. The mat is easily handled and can be molded in rather complex shapes. The cost is intermediate between those of mat and cloth. Such constructions are available in 15 styles comprising a variety of strengths in warp and fill directions. A special example of this material is unidirectional roving. In this material the roving strands are laid parallel and held in position with a light fabric or a crosstie thread. The plies are highly directional in properties, and laminates can be made with a wide range of properties by varying the direction of each ply.

Asbestos paper is used in higher-temperature laminates. Asbestos is a naturally occurring mineral consisting of $3MgO \cdot 2SiO_2 \cdot 2H_2O$ with varying quantities of silicates of iron and calcium. Another asbestos form, crocidolite, is also used. The asbestos fibers are separated from the rock strata, purified, and made into a paper in a conventional papermaking machine. Various resins are used as binders. The papers are fairly weak compared to cellulose.

Asbestos fabric is cloth made from asbestos thread. The fibers in the thread are staple fibers—that is—discontinuous. The fabrics are available in several woven patterns and thicknesses.

Cellulose paper is widely used for a reinforcing sheet. The three most common papers are kraft, α-cellulose, and rag paper. Kraft paper is the cheapest and is made by the sulfate process involving hydrolysis of the lignin in cellulose. Kraft paper laminates have good physical properties. α-Cellulose papers are made from purified pulp blended with waste printed stock such as magazine or letterhead. Laminates from this paper have improved appearance, machinability, and electrical properties. The paper is more expensive than kraft. Rag paper is made from pure cotton fibers obtained from scrap textiles. The fibers are purified by being cooked in caustic solutions which remove all waxes, dirt, and polymers. Rag paper laminates have the best electrical properties and machinability and are not badly affected by water. This paper is the most costly of the three grades.

Cellulose fabrics are often used in laminates and are available in weights from 2 to 10 oz/yd^2 and in many weave patterns. The fabrics are used in laminates for mechanical rather than electrical purposes.

Synthetic fibers are used in certain laminates for special purposes. These fibers most often are used in textile form, but mats and nonwoven batts are sometimes useful. Nylon (polyamide) fabric-based laminates are high in electrical properties, impact strength, and chemical resistance. Not all resin binders may be used with nylon because of the difficulty in obtaining a good resin-to-fiber bond. Polyester-based fabrics such as Dacron* result in laminates with excellent water and chemical resistance. Abrasion resistance is also improved. Dynel† and Orlon* are trade names for acrylic fibers. Laminates containing these fabrics are characterized by mildew resistance, acid resistance, and dimensional stability. Some other synthetic fibers such as Nomex* (aromatic polyamide) and polypropylene are used for special applications, but the cost of such laminates is high, and their use is limited.

* Trademarks of E. I. du Pont de Nemours & Co., Wilmington, Del.
† Trademark of Union Carbide Corporation, New York.

TABLE 8 Laminate Selector Chart

Laminate	Cost	Electrical properties	Chemical resistance	Thermal stability	Arc resistance	Flame resistance	Humidity resistance	Dimensional stability	Mechanical properties
Phenolic	X	...	...	X	...	...	...	X	X
Melamine	...	X	...	...	X	X	...	...	...
Epoxy	...	...	X	...	X	...	X	X	X
Polyester	X	...	X	...	X	X	...	...	...
Phenylsilane	...	...	...	X	...	...	...	...	X
Silicone	...	...	...	X	X	X	...	...	...
Diallyl phthalate	...	X	X	...	...	...	X	X	...
Teflon*	...	...	X	X	...	X	X	...	...
Polyimide	...	X	...	X	...	...	...	...	...
Polybenzimidazole	...	...	...	X	...	...	...	...	X

* Trademark of E. I. du Pont de Nemours & Co., Wilmington, Del.

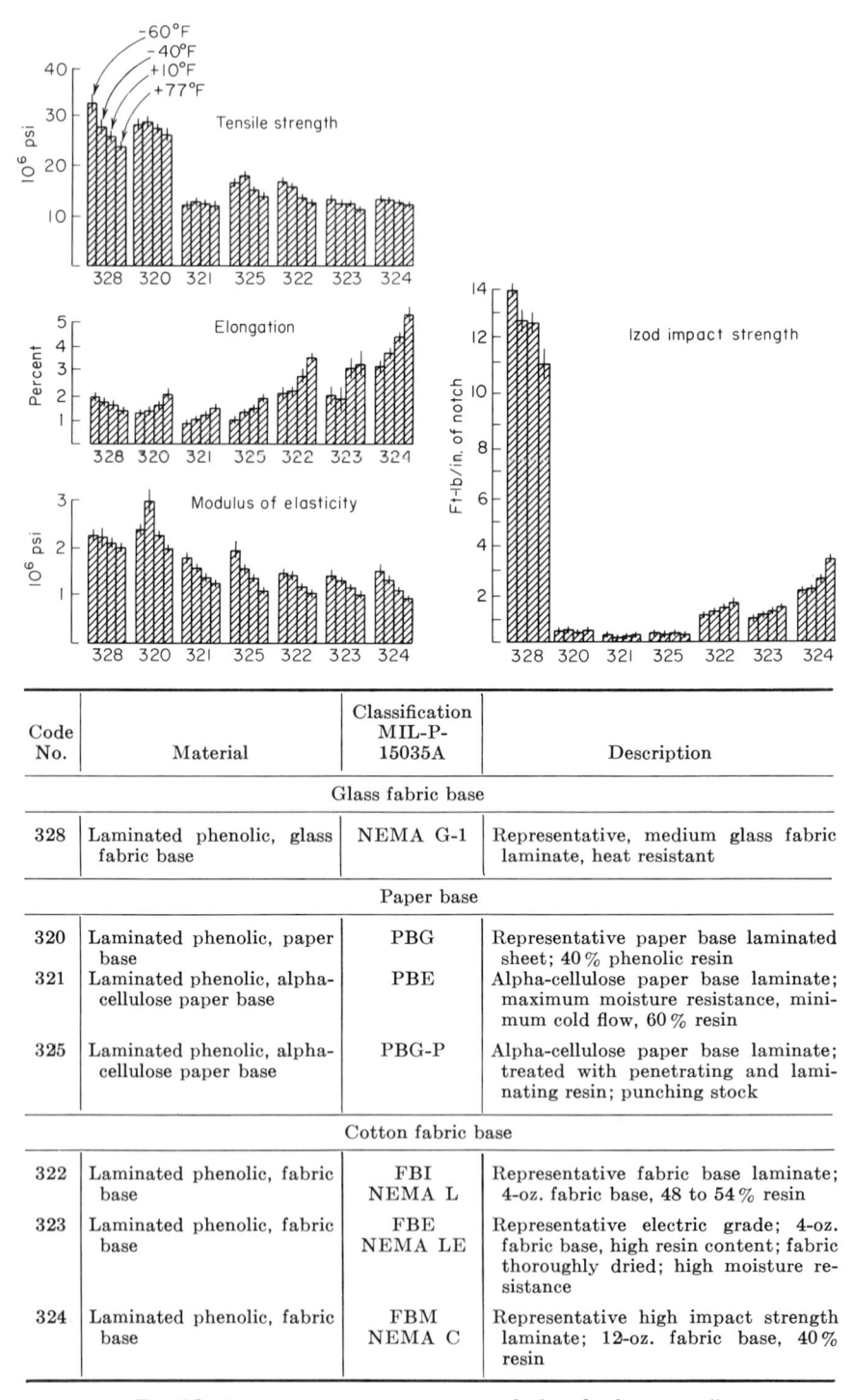

Code No.	Material	Classification MIL-P-15035A	Description
		Glass fabric base	
328	Laminated phenolic, glass fabric base	NEMA G-1	Representative, medium glass fabric laminate, heat resistant
		Paper base	
320	Laminated phenolic, paper base	PBG	Representative paper base laminated sheet; 40% phenolic resin
321	Laminated phenolic, alpha-cellulose paper base	PBE	Alpha-cellulose paper base laminate; maximum moisture resistance, minimum cold flow, 60% resin
325	Laminated phenolic, alpha-cellulose paper base	PBG-P	Alpha-cellulose paper base laminate; treated with penetrating and laminating resin; punching stock
		Cotton fabric base	
322	Laminated phenolic, fabric base	FBI NEMA L	Representative fabric base laminate; 4-oz. fabric base, 48 to 54% resin
323	Laminated phenolic, fabric base	FBE NEMA LE	Representative electric grade; 4-oz. fabric base, high resin content; fabric thoroughly dried; high moisture resistance
324	Laminated phenolic, fabric base	FBM NEMA C	Representative high impact strength laminate; 12-oz. fabric base, 40% resin

Fig. 16 Low-temperature properties of phenolic laminates.[11]

Laminated materials The large number of reinforcing materials combined with an equally large selection of resin binders could theoretically result in an infinite number of laminated materials. The next sections will present the properties and uses of the most commonly used materials. Table 8 lists the most important laminate properties to aid in proper selection.

Phenolic Laminates. Phenolic resins are among the oldest plastics. They are available in an infinite number of variations depending on the nature of the starting materials, the ratios used, the catalysts, plasticizers, lubricants, fillers, and pigments employed. The properties of phenolic laminates based on papers, fabrics, and impregnated wood are shown in Table 9. Military Specifications MIL-R-9299 and MIL-P-25515B describe phenolics. These laminates are excellent general-purpose materials. They have a wide range of mechanical and electrical properties, and by selecting the correct reinforcement, products with almost any property can be obtained. Phenolics are used widely for both high- and low-temperature service. Figure 16 presents tensile strength, elongation, modulus, and Izod impact at -60, -40, $+10$, and $+77°F$ for phenolic laminates based on glass cloth, cotton cloth, and paper. Figure 17 shows the room-temperature flexural, tensile, and compressive strength of a high-temperature glass-cloth-reinforced phenolic laminate after 200 hr at various temperatures.

Phenolics are often used because their cost is lower than that of laminates with other resins. They have excellent water resistance and are often chosen for marine applications. Their wear and abrasion resistance make them well suited for gears, wheels, and pulleys. Their stability under a variety of environmental conditions makes them useful in printed circuits and terminal blocks. Two disadvantages limit their use. Because of the nature of the resin the laminates are available only in dark colors, usually brown or black. The second disadvantage is the rather poor resistance to arcs. High filler loading can improve the low-power arc resistance as measured by ASTM D 495 but the presence of moisture, dirt, or higher voltages usually results in complete arcing breakdown.

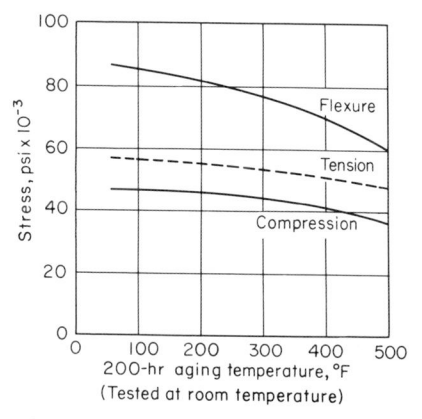

Fig. 17 Physical properties of high-temperature phenolic laminate after aging.[12]

Melamine Laminates. Melamine-formaldehyde polymers are widely used in the home in the form of high-quality molded dinnerware. This application takes advantage of their heat resistance and water-white color. Also used in the home are melamine-surfaced decorative laminates. Typical properties of melamine laminates are shown in Table 9. Melamine resin can be combined with a variety of reinforcing fibers, but properties are best when glass cloth is the reinforcement. Table 10 shows that electrical properties are maximized when the filler is glass. Melamines retain their properties at low temperatures. Table 11 shows the physical properties of melamine with several reinforcements down to $-65°F$. For many years melamines were exclusively used where arc resistance was desired. Recently, however, filled polyester and epoxy materials have begun to compete with melamine on both a cost and an arc-resistance basis. Melamines are often used where resistance to caustic is required. Other uses include switchgear, terminal blocks, circuit breaker parts, slot wedges, and bus bar supports. Melamine laminates have poor dimensional stability, particularly when the part is exposed to alternating cycles of high and low humidity.

Epoxy Laminates. Epoxies were first synthesized in 1931, but they did not

TABLE 9 Laminates Property Chart[10]

Property*	Diallyl phthalate laminates — Glass fabric base	Diallyl phthalate laminates — Cotton fabric base	Polyester laminates — Glass fabric base	Polyester laminates — Glass mat base	Polyester laminates — Paper base	Polyester laminates — Cotton base
Physical						
1. Laminating temperature, °F	200–350	200–300	R.T.-300	R.T.-300	R.T.-300	R.T.-300
2. Laminating pressure, psi	10–1,500	10–1,500	0–50	0–500	0–10	0–10
3. Specific gravity (D 792)	1.65–1.8	1.38	1.5–2.1	1.5–2.1	1.2–1.5	1.2–1.4
4. Specific volume, in.³/lb (D 792)	16.8	20.1	18.5–13.2	18.5–14.4	23.0–18.5	23.0–19.7
5. Tensile strength, psi (D 638)	21,000–55,000	8,500–15,000	18,000–60,000	10,000–20,000	6,100–14,300	7,000–9,000
6. Modulus of elasticity in tension, psi × 10^{-5} (D 638)			10–28	10–19	8–12	
7. Compressive strength, psi (D 695)	45,000	40,000	25,000–60,000	25,000–60,000	19,700–25,000	23,000–24,000
8. Modulus of elasticity in compression, psi × 10^{-5} (D 695)			30–40			
9. Flexural strength, psi (D 790)	42,000–70,000	12,000–21,000	18,000–80,000	15,000–40,000	13,000–28,000	13,000–18,000
10. Modulus of elasticity in flexure, psi × 10^{-5} (D 790)			10–30	10–18	8–13	5–12
11. Shear strength, psi (D 732)	17,800–20,000	14,000–14,700	12,000–18,000	10,000–20,000		
12. Modulus of elasticity in shear, psi × 10^{-5}			5–7			
13. Bearing strength, psi (D 953)			40,000–50,000			
14. Izod impact strength, ft-lb/in. of notch (D 256)	4.3–20	0.8–1.2	19–35	8–18		1.5–2.3
15. Bond strength, lb (D 952)	1,300	1,100–1,500	2,000–3,000	850–1,700	1,000–2,000	1,000–1,500
16. Hardness, Rockwell (D 785)	M120	M104–M110	M100–M110	M70–M110		M50–M90
17. Water absorption in 24 hr, ⅛-in. thickness, % (D 570)	0.12	0.60	0.2–2.5	0.1–0.8	0.10–5.0	1.3–3.5
18. Effect of sunlight	Nil	Nil	Approx. nil	Approx. nil	Slight darkening	
19. Machining qualities	Fair	Good	Fair	Good	Excellent	Excellent
Thermal						
20. Thermal conductivity, cal/(sec)(cm²)(°C/cm) × 10^4 (C 177)		6.3–6.6		1.7–2.0		
21. Specific heat, cal/(°C)(g)			0.25–0.26			
22. Thermal expansion, per °C × 10^5 (D 696)				2.0–2.2	3.1	
23. Resistance to heat, continuous† °F	300		200–350	250–300	220–250	250
24. Heat distortion temperature, °F (D 648)					240	

			Slow to nil Depending on resin	Slow to nil	Moderate to self-extinguishing
25. Burning rate (D 635)	......	......	......	......	......
Electrical					
26. Insulation resistance (96 hr at 90% relative humidity and 35°C, megohms (D 257)	150	......	......	100–50,000	......
27. Volume resistivity at 50% relative humidity and 25°C, ohm-cm (D 257)	......	......	......	10^6–10^{12}	10^{13}
Dielectric strength:					
28. Short-time, ⅛-in. thickness, volts/mil (D 149)	530	390	250–700	300–500	600–800	300–500
29. Step by step, ⅛-in. thickness, volts/mil (D 149)	75	......	220–600	200–400	400–600	270–400
30. Step by step, parallel to lamination, kv/in.	......	40–70	45–50	40–85	40–60	......
Dielectric constant (D 150):						
31. At 60 Hz	4.3	5.1	4.0–6.0	4.4–6.0	5.1	......
32. At 1 kHz	......	......	......	4.2–5.0	......	......
33. At 1 MHz	4.1	4.8	3.0–4.0	3.0–6.0	3.0–4.2	2.9–3.6
Dissipation factor (D 150):						
34. At 60 Hz	0.013	0.015	0.02–0.04	0.005–0.05	0.1	......
35. At 1 kHz	......	......	......	0.004–0.04	......	......
36. At 1 MHz	0.015	0.018–0.040	0.007–0.03	0.007–0.04	0.02–0.03	0.02–0.04
37. Arc resistance, sec (D 495)	150	130	80–140	100–185	28–75	70–85
Chemical						
38. Effect of weak acids (D 543)	None	None	None	None	None	None
39. Effect of strong acids (D 543)	Slight	Decomposes	Some attack	Some attack	Some attack	Some attack
40. Effect of weak alkalies (D 543)	None	None	Slight to none	Slight to none	Slight	Slight
41. Effect of strong alkalies (D 543)	Slight	Attacked	Some to severe	Some to severe	Attacked	Attacked
42. Effect of organic solvents (D 543)	None	None	Generally nil	Generally nil	Generally nil	Generally nil

* D numbers refer to ASTM test methods.

† The values reported for this property depend on the size and shape of the testpiece and the molding conditions. The maximum permissible service temperatures will vary with the formulation of the material, design of the part, and service conditions such as amount of stressing, humidity, etc. Requirements for resistance to heat should be considered a special problem.

2-19

TABLE 9 Laminates Property Chart[10] (Continued)

Property*	Teflon laminates — Glass nonwoven fiber base	Teflon laminates — Ceramic nonwoven fiber base	Teflon laminates — Glass fabric base	Silicone laminates — Glass fabric base	Silicone laminates — Asbestos fabric base
Physical					
1. Laminating temperature, °F	....	....	....	325–480	350–400
2. Laminating pressure, psi	....	....	....	30–2,000	30–2,000
3. Specific gravity (D 792)	2.15	1.9–2.2	2.2	1.6–1.9	1.75
4. Specific volume, in.³/lg (D 792)	13.1	14.6–12.8	12.8	17.3–14.5	15.8
5. Tensile strength, psi (D 638)	5,500–7,500	1,500–4,000	12,000–20,000	10,000–37,600	....
6. Modulus of elasticity in tension, psi × 10⁻⁵ (D 638)	....	....	....	15–20	....
7. Compressive strength, psi (D 695)	7,000	....	20,000	25,000–46,000	40,000–50,000
8. Modulus of elasticity in compression, psi × 10⁻⁵ (D 695)	....	....	....	....	....
9. Flexural strength, psi (D 790)	10,000–15,000	....	11,000–17,000	10,000–38,000	12,000–16,000
10. Modulus of elasticity in flexure, psi × 10⁻⁵ (D 790)	2.4–3.7	....	....	10–35	....
11. Shear strength, psi (D 732)	....	....	11,000	16,500–20,000	....
12. Modulus of elasticity in shear, psi × 10⁻⁵	....	....	....	....	....
13. Bearing strength, psi (D 953)	....	....	....	....	....
14. Izod impact strength, ft-lb/in. of notch (D 256)	1.3–2.5	1.1–2.5	4–6	5–13	6–9
15. Bond strength, lb (D 952)	1.50	....	600–900	600–1,300	....
16. Hardness, Rockwell (D 785)	....	....	M60	M100	....
17. Water absorption in 24 hr, ⅛-in. thickness, % (D 570)	0.005–0.25	0.5–2.0	0.02	0.070–0.65	1.0–1.5
18. Effect of sunlight	Nil	Nil	Nil	Nil	Nil
19. Machining qualities	Excellent	Fair to good	Good	Fair	Fair
Thermal					
20. Thermal conductivity, cal/(sec)(cm²)(°C/cm) × 10⁴ (C 177)	1.8	....	1–2	7	....
21. Specific heat, cal/(°C)(g)	0.23	0.25	0.2	0.25–0.27	....
22. Thermal expansion, per °C × 10⁵ (D 696)	....	3.7–3.8	....	0.5–1.0	....
23. Resistance to heat, continuous† °F	500	500	480	400–700	>480
24. Heat distortion temperature, °F (D 648)	7,500	>500	....	500	>400
25. Burning rate (D 635)	Nil	Nil	Nil	Nil	Nil

Electrical

26. Insulation resistance (96 hr at 90% relative humidity and 35°C, megohms (D 257))	400,000	30	5,000 to 10⁶	2,500–3,000,000
27. Volume resistivity at 50% relative humidity and 25°C, ohm-cm (D 257)	>10⁷	>10⁷	>10⁶	....
Dielectric strength:				
28. Short-time, ⅛-in. thickness, volts/mil (D 149)	300	150–250	300–750	180–480	50–150
29. Step by step, ⅛-in. thickness, volts/mil (D 149)	250	....	250–700	150–350	50–100
30. Step by step, parallel to lamination, kv/in.	>55	....	50–70	25–60	....
Dielectric constant (D 150):				
31. At 60 Hz	2.2	....	2.6–2.8	3.7–4.3
32. At 1 kHz	2.2	2.5–2.7	2.6–2.8	3.7–4.3
33. At 1 MHz	2.2	2.4–2.5	2.6–2.8	3.7–4.3
Dissipation factor (D 150):				
34. At 60 Hz	0.004	0.002–0.003	0.0005	0.0005–0.0055
35. At 1 kHz	0.001	....	0.0004	0.0006–0.0035
36. At 1 MHz	0.0005	0.001–0.003	0.0006	0.0015–0.0050
37. Arc resistance, sec (D 495)	180	190	180+	150–250
Chemical				
38. Effect of weak acids (D 543)	Nil	None	None	None	Very slight
39. Effect of strong acids (D 543)	....	Attacked by HF only	Some attack	Some attack
40. Effect of weak alkalies (D 543)	None	None	None	Very slight	Very slight
41. Effect of strong alkalies (D 543)	Slight	Slight	Slight	Attacked	Attacked
42. Effect of organic solvents (D 543)	None	None	None	Attacked by some

* D numbers refer to ASTM test methods.

† The values reported for this property depend on the size and shape of the testpiece and the molding conditions. The maximum permissible service temperatures will vary with the formulation of the material, design of the part, and service conditions such as amount of stressing, humidity, etc. Requirements for resistance to heat should be considered a special problem.

2-21

TABLE 9 Laminates Property Chart[10] (Continued)

Property*	Epoxy laminates — Nonwoven continuous glass filament base	Epoxy laminates — Glass fabric base	Epoxy laminates — Cellulose paper base	Melamine-formaldehyde laminates — Cellulose paper base	Melamine-formaldehyde laminates — Cotton fabric base
Physical					
1. Laminating temperature, °F	250–330	R.T.–370	250–300	270–320	260–320
2. Laminating pressure, psi	10–100	10–1,800	600–1,000	500–1,800	1,000–1,500
3. Specific gravity (D 792)	1.75–2.00	1.7–2.0	1.52	1.40–1.55	1.35–1.52
4. Specific volume, in.³/lb (D 792)	15.8–14.9	16.3–14.5	18–19.5	19.1–17.8	20.5–18.5
5. Tensile strength, psi (D 638)	110,000	35,000–100,000	10,000–19,000	10,000–25,000	7,000–17,000
6. Modulus of elasticity in tension, psi $\times 10^{-5}$ (D 638)	55	20–35	7–12		10–19
7. Compressive strength, psi (D 695)	100,000	35,000–90,000	28,000–30,000	30,000–48,000	33,000–50,000
8. Modulus of elasticity in compression, psi $\times 10^{-5}$ (D 695)					
9. Flexural strength, psi (D 790)	120,000	40,000–100,000	16,600–24,000	14,000–20,000	14,000–28,000
10. Modulus of elasticity in flexure, psi $\times 10^{-5}$ (D 790)	53	20–50	10–13		16–20
11. Shear strength, psi (D 732)		17,000–30,000	12,200		16,500–18,000
12. Modulus of elasticity in shear, psi $\times 10^{-5}$					
13. Bearing strength, psi (D 953)					
14. Izod impact strength, ft-lb/in. of notch (D 256)	36.0	5.5–25	0.6–0.8	0.3–1.5	0.5–2.4
15. Bond strength, lb (D 952)		1,600–3,200	1,200	1,000	1,500–2,500
16. Hardness, Rockwell (D 785)	M100–M108	M105–M120	M100–M110	M110–M125	M110–M120
17. Water absorption in 24 hr, ⅛-in. thickness, % (D 570)	0.04–0.30	0.04–0.30	0.15–0.50	1.0–2.0	0.9–3.5
18. Effect of sunlight	Slight	Slight color change	Slight color change	Slight color change	
19. Machining qualities	Excellent	Good	Excellent	Fair	Fair
Thermal					
20. Thermal conductivity, cal/(sec)(cm²)(°C/cm) $\times 10^4$ (C 177)		7.1–7.4	7.0		
21. Specific heat, cal/(°C)(g)	0.217	0.35–0.40	0.35–0.40		
22. Thermal expansion, per °C $\times 10^5$ (D 696)	0.9–2.5	1.0–1.2	2.0		0.7–2.5
23. Resistance to heat, continuous† °F	325	250–300	300	210–260	210–260
24. Heat distortion temperature, °F (D 648)	350–400	355–375	275		
25. Burning rate (D 635)	0.1 in./min.	Slow to nil	Slow	Approximately nil	

Electrical

26. Insulation resistance (96 hr at 90% relative humidity and 35°C, megohms (D 257))	52,000	50,000–1,000,000	100,000–500,000	100–400,000
27. Volume resistivity at 50% relative humidity and 25°C, ohm-cm (D 257)	3.3×10^{13}	1×10^{15}	7×10^{15}	....
Dielectric strength:				
28. Short-time, ⅛-in. thickness, volts/mil (D 149)	450	400–750	475–700	200–450
29. Step by step, ⅛-in. thickness, volts/mil (D 149)	....	300–600	325–540	150–350
30. Step by step, parallel to lamination, kv/in.	36	35–70	60–80	12–75
Dielectric constant (D 150):				
31. At 60 Hz	5.9	4.2–5.3	4.0–6.0	7.5–8.6
32. At 1 kHz	5.6	....	....	7.3–8.3
33. At 1 MHz	5.4	4.5–5.3	4.0–4.6	6.2–8.0
Dissipation factor (D 150):				
34. At 60 Hz	0.005	0.003–0.015	0.010–0.025	0.06–0.15
35. At 1 kHz	0.007	....	....	0.03–0.09
36. At 1 MHz	0.018	0.010–0.025	0.30–0.035	0.03–0.07
37. Arc resistance, sec (D 495)	60	15–180	30–100	120–186
Chemical				
38. Effect of weak acids (D 543)	None	None		Slight to marked
39. Effect of **strong** acids (D 543)	Slight	Slight		Decomposes
40. Effect of weak alkalies (D 543)	Slight	Slight		None
41. Effect of **strong** alkalies (D 543)	Slight	Attacked		Attacked
42. Effect of organic solvents (D 543)	Slight	Slight to none	Slight	None on bleed-proof materials

* D numbers refer to ASTM test methods.
† The values reported for this property depend on the size and shape of the testpiece and the molding conditions. The maximum permissible service temperatures will vary with the formulation of the material, design of the part, and service conditions such as amount of stressing, humidity, etc. Requirements for resistance to heat should be considered a special problem.

2-23

TABLE 9 Laminates Property Chart[10] (Continued)

Property*	Melamine-formaldehyde laminates — Asbestos paper or fabric base	Melamine-formaldehyde laminates — Glass fabric base	Melamine-formaldehyde laminates — Glass mat	Phenol-formaldehyde laminates — Cellulose paper base	Phenol-formaldehyde laminates — Cotton fabric base
Physical					
1. Laminating temperature, °F	270–320	270–300	270–320	275–350	275–350
2. Laminating pressure, psi	1,000–1,800	1,000–1,800	1,000–1,800	1,000–1,800	1,000–1,800
3. Specific gravity (D 792)	1.75–1.95	1.82–1.98	1.75–1.85	1.28–1.40	1.30–1.40
4. Specific volume, in.³/lb (D 792)	15.8–14.9	15.2–13.9	16.9–15.0	21.8–19.7	21.3–20.4
5. Tensile strength, psi (D 638)	6,000–12,000	25,000–40,000	16,000–25,000	8,000–20,000	7,000–16,000
6. Modulus of elasticity in tension, psi × 10⁻⁵ (D 638)	16–22	20–25		8–20	5–15
7. Compressive strength, psi (D 695)	27,000–50,000	25,000–85,000	60,000	20,000–40,000	30,000–44,000
8. Modulus of elasticity in compression, psi × 10⁻⁵ (D 695)		47		8	6
9. Flexural strength, psi (D 790)	12,000–24,000	35,000–85,000	21,000–30,000	10,500–30,000	14,000–30,000
10. Modulus of elasticity in flexure, psi × 10⁻⁵ (D 790)		15–30	13	8–15	8–10
11. Shear strength, psi (D 732)	13,000–17,000	19,000–35,000		6,000–14,000	7,000–18,000
12. Modulus of elasticity in shear, psi × 10⁻⁵					
13. Bearing strength, psi (D 953)	0.75–4	5–15	5–6.5	0.3–1.10	0.8–3.0
14. Izod impact strength, ft-lb/in. of notch (D 256)	800–1,700	1,500–2,300	1,400–2,000	700–1,400	1,000–2,500
15. Bond strength, lb (D 952)	M110–M118	M115–M125	M115–M125	M70–M120	M90–M115
16. Hardness, Rockwell (D 785)	1.0–5.0	0.20–2.5	1.25–2.8	0.2–6.0	0.7–4.4
17. Water absorption in 24 hr, ⅛-in. thickness, % (D 570)	Slight color change			Lowers surface resistance and general darkening	
18. Effect of sunlight					
19. Machining qualities	Fair	Fair	Fair	Fair to excellent	
Thermal					
20. Thermal conductivity, cal/(sec)(cm²)(°C/cm) × 10⁴ (C 177)		7.12		1.8–7.0	2.1–7.0
21. Specific heat, cal/(°C)(g) (D 696)		0.23–0.40		0.38–0.41	0.35–0.40
22. Thermal expansion, per °C × 10⁵ (D 696)		0.7–12.0		1.4–3.0	1.8–3.2
23. Resistance to heat, continuous† °F	225–275	300	290	225–250	225–250
24. Heat distortion temperature, °F (D 648)				250–>320	>320
25. Burning rate (D 635)	Approximately nil	Nil	Nil	Very low	Very low

Electrical					
26. Insulation resistance (96 hr at 90% relative humidity and 35°C, megohms (D 257)	5	30–10,000	10–500	35–1,000,000	10–5,000
27. Volume resistivity at 50% relative humidity and 25°C), ohm-cm (D 257)	......	......	......	10^{10}–10^{13}	10^{10}–10^{12}
Dielectric strength:					
28. Short-time, ⅛-in. thickness, volts/mil (D 149)	50–150	200–500	300–450	300–1,000	150–600
29. Step by step, ⅛-in. thickness, volts/mil (D 149)	60–120	150–350	250–350	260–800	120–440
30. Step by step, parallel to lamination, kv/in.	7	23–75	35–55	15–90	10–60
Dielectric constant (D 150):					
31. At 60 Hz	......	6.5–10.0	4.5–6.5	4.5–7.5	5.0–10
32. At 1 kHz	......	6.1–9.5	4.2–6.0	4.2–6.0	4.2–6.5
33. At 1 MHz	8.0–9.6	6.0–9.0	4.2–7.5	3.6–6.0	5.0–7.0
Dissipation factor (D 150):					
34. At 60 Hz	......	0.04–0.10	0.004–0.05	0.02–0.10	0.04–0.50
35. At 1 kHz	......	0.012–0.03	0.004–0.05	0.03–0.07	0.04–0.09
36. At 1 MHz	0.12–0.22	0.011–0.025	0.006–0.08	0.02–0.08	0.05–0.10
37. Arc resistance, sec (D 495)	......	175–200	170–200	4–75	4
Chemical					
38. Effect of weak acids (D 543)	None	None	None	None	None
39. Effect of strong acids (D 543)	None	Decomposes	None	Slight	Slight
40. Effect of weak alkalies (D 543)	None	None	None	Slight	Slight
41. Effect of strong alkalies (D 543)	None	Attacked	Attacked	Attacked	Attacked
42. Effect of organic solvents (D 543)	None on bleed-proof materials		None	None	None

* D numbers refer to ASTM test methods.

† The values reported for this property depend on the size and shape of the testpiece and the molding conditions. The maximum permissible service temperatures will vary with the formulation of the material, design of the part, and service conditions such as amount of stressing, humidity, etc. Requirements for resistance to heat should be considered a special problem.

2-25

TABLE 9 Laminates Property Chart[10] (Continued)

Phenol-formaldehyde laminates

Property*	Cotton web base	Nylon fabric base	Glass fabric base	Asbestos fabric base	Asbestos paper base	Wood base
Physical						
1. Laminating temperature, °F	275–350	275–325	275–350	300–350	300–350	300–320
2. Laminating pressure, psi	1,000–1,800	15–1,800	15–2,000	300–1,800	1,000–1,800	1,000–2,000
3. Specific gravity (D 792)	1.31–1.37	1.15–1.19	1.4–1.95	1.55–1.92	1.65–1.83	1.3
4. Specific volume, in.³/lb (D 792)	21.1–20.2	24.0–23.2	19.7–15.4	17.8–15.4	16.8–15.1	21.3
5. Tensile strength, psi (D 638)	14,000–16,000	5,000–9,500	9,000–50,000	10,000–13,000	5,000–15,000	16,000–32,000
6. Modulus of elasticity in tension, psi × 10⁻⁵ (D 638)	6–11	3.5–5.0	12–25	10–17	16–25	37
7. Compressive strength, psi (D 695)	32,000–47,000	28,000–36,000	34,000–75,000	30,000–55,000	40,000	12,000–21,000
8. Modulus of elasticity in compression, psi × 10⁻⁵ (D 695)	4–8			7	5	
9. Flexural strength, psi (D 790)	18,000–25,000	9,000–22,000	16,000–80,000	10,000–35,000	11,000–30,000	25,000–40,000
10. Modulus of elasticity in flexure, psi × 10⁻⁵ (D 790)	6–10	5–10	10–40	10–15	14–23	
11. Shear strength, psi (D 732)	12,600–14,000	10,000–13,700	17,200–24,000	12,000–24,000	7,500–13,500	
12. Modulus of elasticity in shear, psi × 10⁻⁵						
13. Bearing strength, psi (D 953)						
14. Izod impact strength, ft-lb/in. of notch (D 256)	1.4–2.5	2.0–4.0	4.0–18.0	3–5	0.6–1.5	4–8
15. Bond strength, lb (D 952)	1,000–1,600	1,000–1,500	800–2,000	1,500–2,200	600–2,000	
16. Hardness, Rockwell (D 785)	M100–M110	M100–M110	M100–M110	M70–M115	M90–M111	M90–M105
17. Water absorption in 24 hr, ⅛-in. thickness, % (D 570)	1.2–1.6	0.15–0.60	0.12–2.70	0.3–2.5	0.4–2.0	2.5–11
18. Effect of sunlight		Lowers surface resistance and general darkening				
19. Machining qualities	Fair to excellent			Fair to good		Good
Thermal						
20. Thermal conductivity, cal/(sec)(cm²)(°C/cm) × 10⁴ (C 177)			1.4–1.8	4.6		0.2–0.4
21. Specific heat, cal/(°C)(g)	1.3–1.6	0.35–0.40	0.23–0.30	0.30	0.30	0.6‖, 6.5‖
22. Thermal expansion, per °C × 10⁵ (D 696)		4	1.5–2.5	1.5–2.5	1.0–1.6	150–200
23. Resistance to heat, continuous†°F	225	165	250–500	275	275	>320
24. Heat distortion temperature, °F (D 648)	>320	165	>320	>320	>320	Very low
25. Burning rate (D 635)	Very low	Slow	Nil	Approx. nil	Approx. nil	

Electrical

26. Insulation resistance (96 hr at 90% relative humidity and 35°C, megohms (D 257)	5–200	30,000–1,000,000	25–5,000	....	0.25–335
27. Volume resistivity at 50% relative humidity and 25°C), ohm-cm (D 257)	....	7×10^{13}	2.5×10^{10}	....	....
Dielectric strength:					
28. Short-time, ⅛-in. thickness, volts/mil (D 149)	200	360–600	300–700	160–250	75–500
29. Step by step, ⅛-in. thickness, volts/mil (D 149)	150	300–450	250–650	95–200	....
30. Step by step, parallel to lamination, kv/in.	10–40	40–100	30–50	4–25	....
Dielectric constant (D 150):					
31. At 60 Hz	....	3.7–6.0	4.0–10.0	....	....
32. At 1 kHz	....	3.6–4.1	4.8–6.3	7.0	....
33. At 1 MHz	5.5–7.0	3.3–4.5	3.7–6.6	5.2	5
Dissipation factor (D 150):					
34. At 60 Hz	0.10–0.5	0.02–0.06	0.01–0.10	....	0.20
35. At 1 kHz	....	0.01–0.02	0.015–0.042	0.15–0.20	....
36. At 1 MHz	0.06–0.50	0.015–0.040	0.005–0.05	0.08–0.14	0.05
37. Arc resistance, sec (D 495)	Tracks	10	Tracks	4	Tracks

Chemical

38. Effect of weak acids (D 543)	None to slight depending on acid
39. Effect of strong acids (D 543)	Decomposed by oxidizing acids; reducing and organic acids none to slight effect
40. Effect of weak alkalies (D 543)	Slight to marked depending on alkalinity and grade
41. Effect of strong alkalies (D 543)	Attacked by strong alkalies unless a special alkali-resistant resin is used
42. Effect of organic solvents (D 543)	None on bleed-proof materials

* D numbers refer to ASTM test methods.

† The values reported for this property depend on the size and shape of the testpiece and the molding conditions. The maximum permissible service temperatures will vary with the formulation of the material, design of the part, and service conditions such as amount of stressing, humidity, etc. Requirements for resistance to heat should be considered a special problem.

TABLE 10 Electrical Properties of Melamine Laminates Related to Reinforcements[13]

Property	Reinforcement		
	Asbestos	Glass	Cellulose
Insulation resistance, ohms...............	10^8	10^9	10^7
Arc track resistance, volts (D 2303).......	1,500	1,500	1,000
Dielectric strength, volts/mil.............	50	300	250

TABLE 11 Mechanical Properties of Laminated Melamine at Varying Temperatures[11]

Property*	Temperature, °F	Melamine† glass	Melamine‡ asbestos	Electrical melamine-formaldehyde§	α-Melamine-formaldehyde¶
Tensile strength, psi × 10^{-3}	77	32.5	13.5	5.4	7.8
	10	32.9	14.5	6.7	6.9
	−40	38.1	15.4	5.7	6.9
	−65	37.2	14.1	5.6	6.7
Modulus of elasticity, psi × 10^{-6}	77	2.130	2.146	1.060	1.270
	10	2.290	3.040	1.610	1.640
	−40	1.430	3.390	1.540	1.730
	−65	1.580	3.060	1.390	1.880
Elongation at break, %	77	2.15	0.82	0.54	0.62
	10	2.17	0.64	0.50	0.44
	−40	2.36	0.60	0.40	0.39
	−65	2.75	0.55	0.38	0.37
Work to produce failure, ft-lb/in.³	77	31.0	......	1.32	2.08
	10	36.7	......	1.40	1.33
	−40	44.3	......	1.02	0.86
	−65	52.2	......	0.97	1.13
Proportional limit, psi × 10^{-3}	77	15.8	14.5	33.2	25.9
	10	7.1	9.5	10.5	9.5
	−40	5.4	6.7	5.7	5.6
	−65	7.8	6.9	6.9	6.7
Izod impact strength, ft-lb/in. notch (D 256)	77	11.12	0.98	0.33	0.31
	10	12.64	0.90	0.37	0.28
	−40	13.43	0.96	0.32	0.28
	−65	14.68	0.93	0.28	0.29

* Average values.
† Laminated melamine, glass fabric base.
‡ Laminated melamine, asbestos paper base.
§ Melamine-formaldehyde, cellulose filler, electrical grade.
¶ Melamine-formaldehyde, α-cellulose filler.

become popular until about 1955. The family consists of many resins based on a large variety of chemical modifications. Typical properties are given in Table 9. Note the extremely high physical-strength values of laminates made from unidirectional fiber mat as shown in the first column. This laminate approaches filament-wound materials in properties. Epoxies are frequently used in cryogenic applications where resistance to resin crazing is desired. Figure 18 shows the tensile properties of laminates to −423°F. Military Specifications MIL-R-9300A and MIL-P-25421 describe epoxy laminates. Epoxies are frequently used in electronic applications where their excellent dielectric properties are of value. In general, however, these properties increase with temperature, limiting the electrical application of epoxies to about 250°F. With the correct choice of hardener, laminates can be made with good retention of physical properties up to 500°F. One such glass-cloth-epoxy

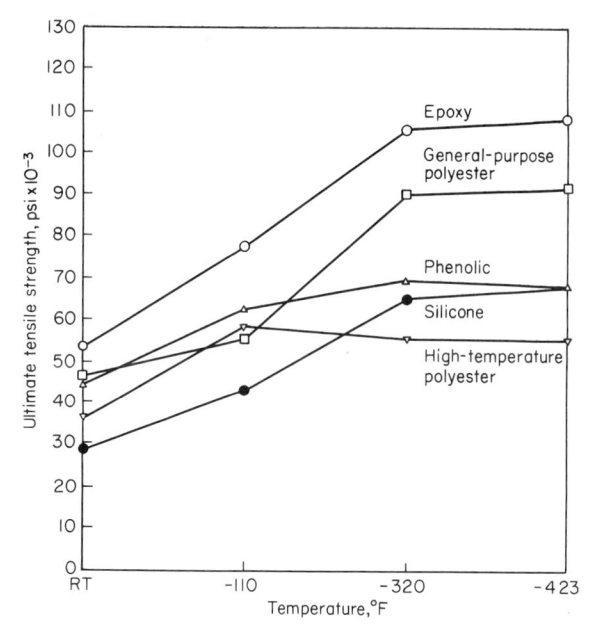

Fig. 18 Cryogenic tensile strength of various laminates.[14]

laminate is shown in Figs. 19 to 22 along with the changes in flexural strength, water absorption, and electric strength from aging at 329, 392, 436, and 482°F. The changes in volume resistivity, dielectric constant, and power factor versus temperature and at several frequencies for the same material are given in Figs. 23 to 25. Epoxy laminates are used in printed circuits and chemical-resistant applications. The cost of these materials is higher than that of phenolic, melamine, or polyester composites.

Polyester Laminates. Polyester laminates are described in MIL-R-7575B and MIL-P-8013C. The resins were first used during World War II for a variety of structural and insulating applications. Commercial polyesters usually consist of the polyester polymer dissolved in a monomer like styrene. Typical properties are shown in Table 9. The resins are usually combined with glass fibers since many cellulosic reinforcements inhibit their cure. Table 12 shows the effect of reinforcement media on several properties. One of the chief virtues of these resins is the excellent chemical resistance of the laminates. These materials resist acids and bases for long times even at 200°F. They are attacked by most chlorinated solvents. Table 13 gives the results of exposure by four different polyester resin laminates to chemicals. Polyesters are useful over a fairly wide temperature range

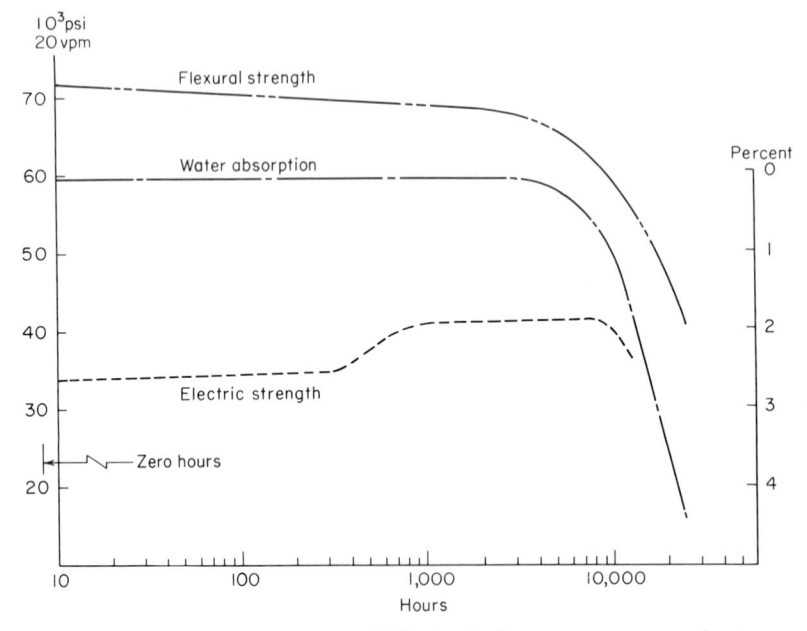

Fig. 19 Properties versus time at 329°F for high-temperature epoxy laminate.

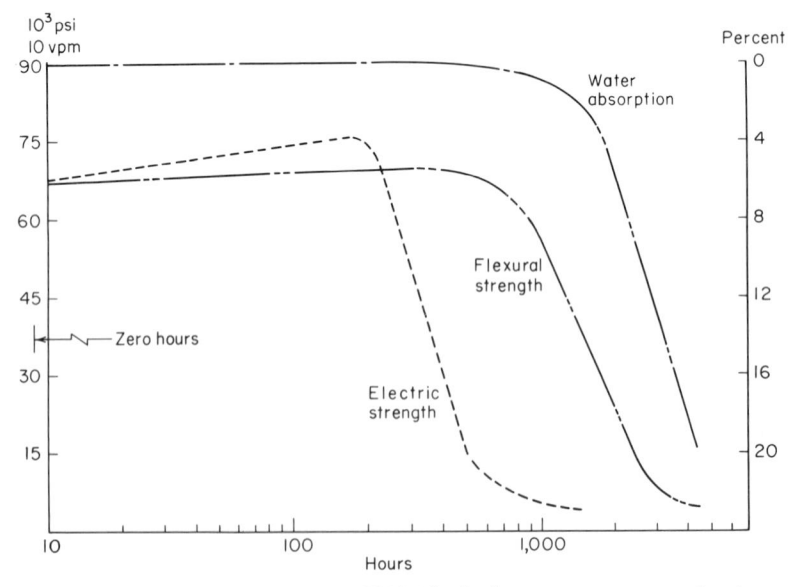

Fig. 20 Properties versus time at 392°F for high-temperature epoxy laminate.

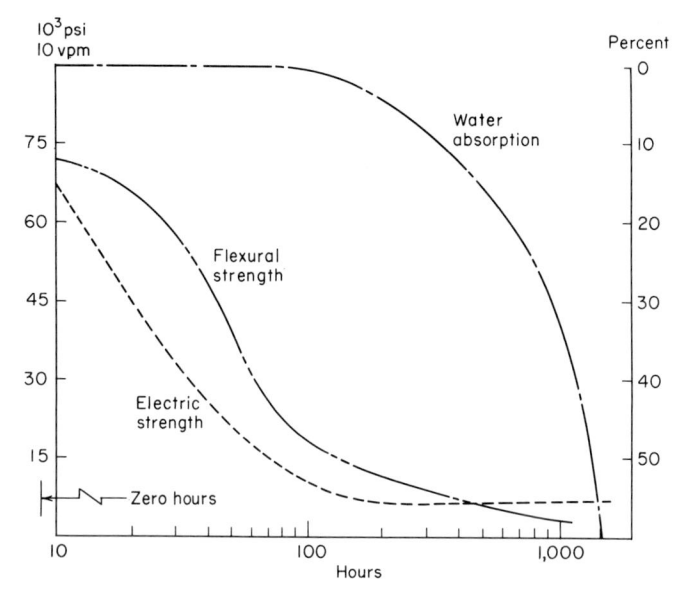

Fig. 21 Properties versus time at 436°F for high-temperature epoxy laminate.

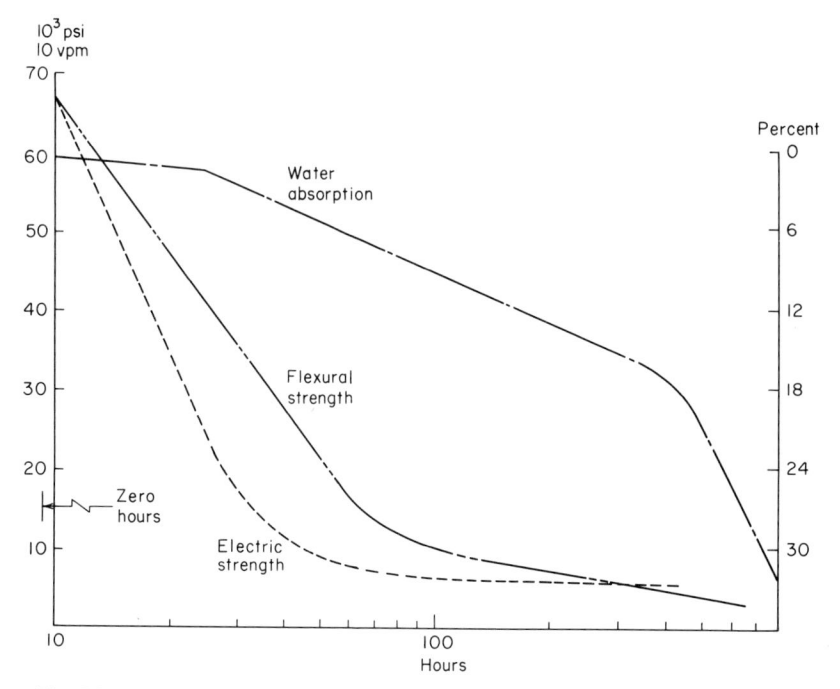

Fig. 22 Properties versus time at 482°F for high-temperature epoxy laminate.

2-32 Laminates, Reinforced Plastics, and Composite Structures

but are not good at temperatures above 250°F. Table 14 shows physical properties for a typical polyester laminate to −60°F. Polyester laminates are widely used in the switchgear industry. They can be made highly arc-, track-, and flame-resistant. The polyester resins themselves possess good arc and track resistance, which is further enhanced by the addition of mineral fillers, particularly aluminum oxide trihydrate. Flame resistance is obtained by modifying the basic resin, usually by the incorporation of chlorine. Polyesters are moderate in cost. They are also used in applications requiring dimensional stability and water resistance. They are not in extensive use by the electronics industry since epoxies can often supply better dielectric properties.

Diallyl Phthalate Laminates. Properties of diallyl phthalate laminates are given in Table 9. This resin has only two principal forms; diallyl orthophthalate (DAP) and diallyl isophthalate (DAIP). The isophthalate is more thermally stable, being useful up to 350°, whereas the orthophthalate is useful only to about 300°F. In most other respects the resins are similar. Diallyl phthalate laminates are used almost exclusively as electronic insulations. Their low-loss characteristics are attractive over a range of frequencies and up to their maximum temperature limits. Radar and microwave applications make use of DAP materials because of these low losses. Figures 26 and 27 show the influence of frequency on dielectric constant and loss factors of DAIP and DAP laminates in comparison with various laminates.

In these curves the epoxy MPDA and CL are epoxy-cured with m-phenylenediamine, epoxy DICY is epoxy-cured with dicyandiamide, and TAC polyester refers to a high-temperature polyester resin incorporating triallyl cyanurate monomer. All materials were laminated with style 181 glass cloth.

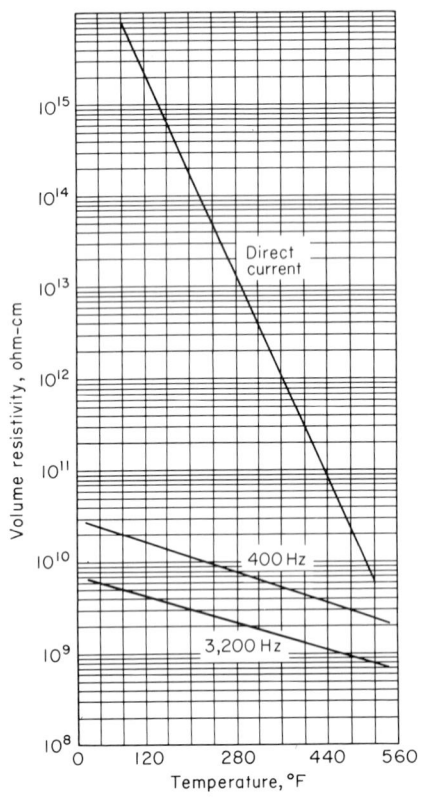

Fig. 23 Volume resistivity versus temperature at 400 and 3,200 Hz and dielectric constant for high-temperature epoxy laminate.[15]

DAIP has the lower dielectric constant and retains it virtually unchanged up to 10^{10} Hz. DAIP does not show an inflection point in the loss-factor curve as is the case with many other laminates. Figures 28 and 29 show the effect of temperature. DAIP again demonstrates flat curves with only the more expensive silicones rivaling its performance. However, as shown in Fig. 30, when moisture is present, DAIP is better than the silicone. These data are for laminates exposed to 100 percent relative humidity for 24 hr.

Diallyl phthalate laminates are used for microwave and radar applications, where they appear as radomes, terminal boards, feedomes, and connectors. The cost of these materials is fairly high. They are respected for their dimensional stability, being little affected by humidity changes. In hermetic and high-vacuum applications, care must be taken to ensure that these materials are well cured and postbaked because undercured laminates outgas DAP monomer which can be detected as white crystals.

Laminates 2-33

Silicone Laminates. Table 9 gives typical values of silicone laminates, and these are further defined by Military Specification MIL-R-25506A and MIL-P-25518A. The silicone resins were the first of the true high-temperature plastics. Silicone polymers are similar to organic polymers except that certain carbon atoms are replaced by silicon and oxygen atoms. Silicone polymers analogous to the corresponding organic polymers are available in the form of liquids, gels, elastomers,

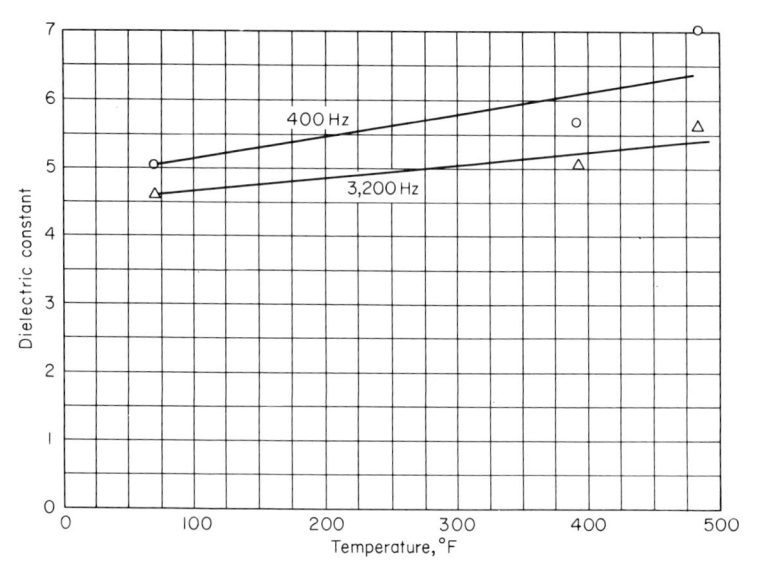

Fig. 24 Dielectric constant versus temperature at 400 and 3,200 Hz for high-temperature epoxy laminate.[15]

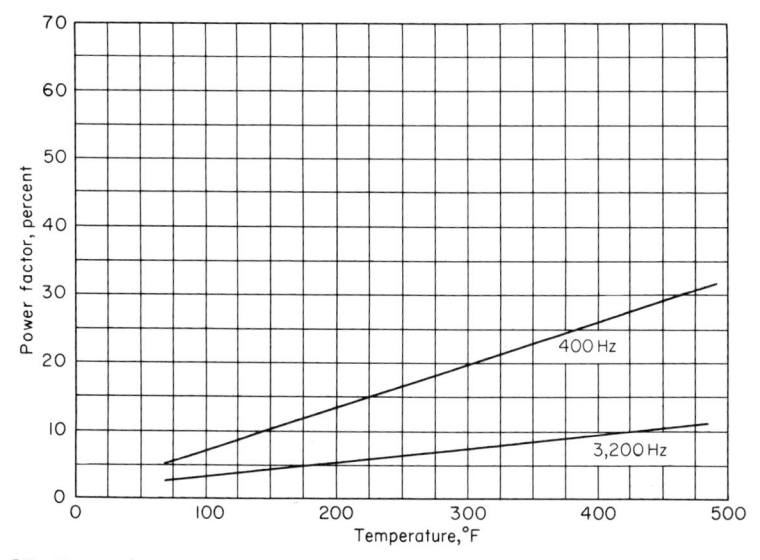

Fig. 25 Power factor versus temperature at 400 and 3,200 Hz for high-temperature epoxy laminate.[15]

and brittle solids. These latter polymers when combined with glass fabrics or asbestos constitute the silicone laminates known to the electronics and electrical industries. The materials are universally acceptable for Class 180°C (Class H) operation. As might be expected, these materials are useful from cryogenic temperatures to about 500°F. Cryogenic properties of the silicone laminates are given in Table 15 where their physical properties are compared with those of other laminating resins. Dielectric properties of silicones are particularly useful. The dissipation factor is low at room temperature and stays relatively constant to 300°F.

TABLE 12 Relationship of Properties to Reinforcement for Polyester Laminates

Reinforcement material	Moisture absorption, %	Ultimate tensile strength, psi × 10^{-3}				Ultimate flexural strength, psi × 10^{-3}			
		At room temperature		At −65°F (frozen)		At room temperature		At −65°F (frozen)	
		Dry	Wet	Dry	Wet	Dry	Wet	Dry	Wet
Glass cloth......	0.25	52.9	48.7	64.7	62.8	64.8	47.7	72.9	69.4
Glass mat.......	0.28	13.0	11.8	16.2	13.3	31.1	23.1	31.9	29.4
Cotton cloth....	2.6	5.5	6.0	5.9	7.2	13.2	....	12.8	

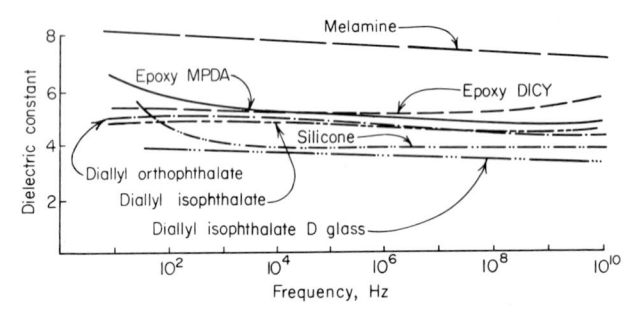

Fig. 26 Dielectric constant of various laminates versus frequency.[18]

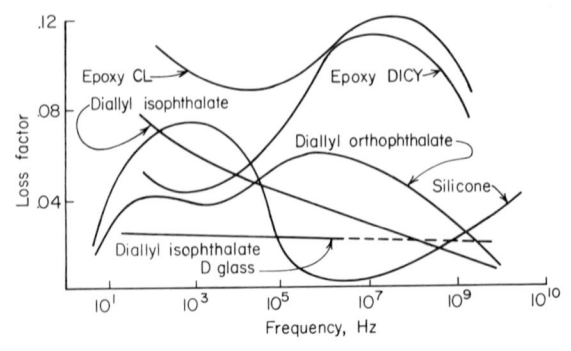

Fig. 27 Loss factor of various laminates versus frequency.[18]

Figure 31 compares glass-cloth-reinforced silicones with epoxy laminates. In this curve and the others following two silicones are shown, one designed for low-pressure laminating and the other for high-pressure, and they are compared with NEMA G-10 and G-11 epoxy-glass laminates. Notice the flat slope of the silicones. The epoxies characteristically show an inflection point near their glass transition temperature, and their losses get progressively worse. The response of dissipation factor to frequency is shown in Fig. 32. Thermal aging does little to influence these properties. Table 16 shows the effect of aging at 572°F on the dielectric constant,

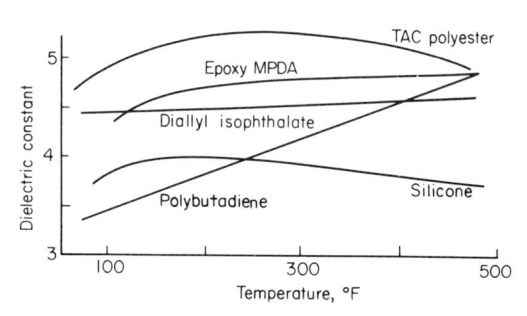

Fig. 28 Dielectric constant at 9,375 MHz of various laminates versus temperature.[18]

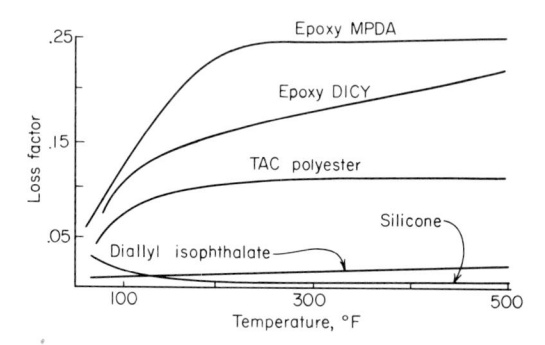

Fig. 29 Loss factor at 9,375 MHz of various laminates versus temperature.[18]

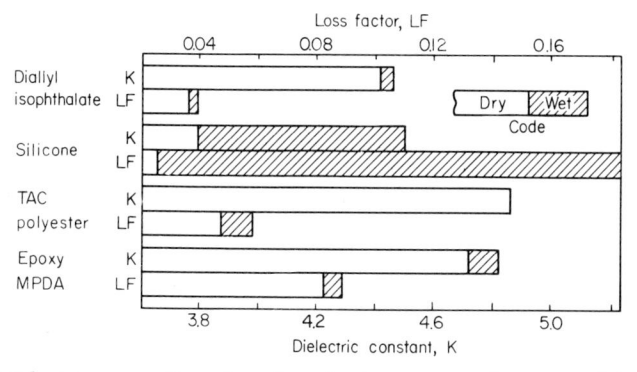

Fig. 30 Dielectric properties of various laminates as a function of humidity at 9,375 MHz.[18]

TABLE 13 Chemical Exposure of Polyester Laminates[16]

Resin (1)	Reagent (2)	Time* on test (3)	Flexural strength, psi Initial (4)	Final (5)	% change (6)
A	Distilled water	36	17,000	15,400	− 9.5
B		36	17,000	17,500	+ 3.0
C		36	17,700	16,400	− 7.5
H		36	15,300	15,600	+ 2.0
A	Ethanol	36	17,000	12,800	−25.0
B		36	17,000	15,200	−11.0
C		36	17,700	12,100	−32
H		24	Very bad internal delamination—failure		
A	Methyl ethyl ketone	12	Severe delamination—failure	. . .	
B		12	Severe delamination—failure	. . .	
C		24	Severe delamination—failure	. . .	
H		12 days	Severe delamination—failure	. . .	
A	Monochlorobenzene	12	Severe delamination—failure	. . .	
B		12	Severe delamination—failure	. . .	
C		24	Severe delamination—failure	. . .	
H		12 days	Severe delamination—failure	. . .	
A	Perchloroethylene	36	Severe delamination—failure	. . .	
B		Void	. . .	. . .	. . .
C		36	17,700	11,200	−37.0
H		12	Severe delamination—failure	. . .	
A	Xylene	36	17,000	15,600	− 8.0
B		36	17,000	8,000	−53.0
C		24	Severe delamination—failure	. . .	
H		12	Severe delamination—failure	. . .	
A	Acetic acid, 25%	36	17,000	16,400	− 4.0
B		Void			
C		36	17,700	13,000	. . . . . .
H		36	15,300	16,900	+11.0
A	Chromic acid, 40%	36	17,000	15,800	− 7.0
B		36	17,000	10,200	−40.0
C		2	Severe delamination—failure	. . .	
H		10	Severe delamination—failure	. . .	
A	Hydrochloric acid, 10%	36	17,000	16,700	− 2.0
B		36	17,000	. . . . . .	. . . . . .
C		36	17,700	15,000	−15.0
H		36	15,300	14,700	− 4.0
A	Nitric acid, 5%	36	17,000	15,000	−12.0
B		36	17,000	18,000	+ 6.0
C		36	17,700	15,600	+12.0
H		36	15,300	15,300	0
A	Sulfuric acid, 35%	36	17,000	17,300	+ 2.0
B		36	17,000	13,000	−24.0
C		36	17,700	15,700	−14.0
H		36	15,300	14,800	− 3.0
A	Sodium hydroxide, 10%	36	17,000	12,800	−25.0
B		36	17,000	5,000	−70.0
C		36	17,700	14,900	+16.0
H		36	15,300	13,900	− 9.0

* Months, unless otherwise specified.
A   2:1:1—propylene glycol/isophthalic acid/maleic anhydride.
B   3:2:1—propylene glycol/isophthalic acid/maleic anhydride.
C   1:1—propylene oxide adduct of bis-phenol A/maleic anhydride.
H   2:1:1—2,2,4-Trimethyl-1,3-pentanediol/isophthalic acid/maleic anhydride.

TABLE 13 Chemical Exposure of Polyester Laminates[16] (Continued)

Flexural modulus, psi × 10⁻⁵			Barcol			Appearance at end of test (13)
Initial (7)	Final (8)	% change (9)	Initial (10)	Final (11)	% change (12)	
5.6	7.64	+36.0	45	56	+25.0	No visible damage
5.5	7.80	+42.0	45	58	+29.0	No visible damage
6.4	7.30	+14.0	38	50	+32.0	Slight fiber bloom
5.0	6.5	+30.0	45	43	− 4.5	Slight fiber bloom
5.6	6.2	+11.0	45	16	−65.0	Severe internal delamination
5.5	6.0	+10.0	45	22	−50.0	Some internal delamination
6.4	9.5	+49.0	38	20	−48.0	Some internal delamination
6.4	5.2	−19.0	38	0	−100	Surface cracks—some internal delamination
5.6	6.3	+13.0	45	23	− 50	Some internal delamination
5.5	2.0	−64.0	45	0	−100	Some internal delamination
5.6	7.8	−39.0	45	53	−18.0	No visible damage
6.4	8.4	......	38	45	......	Some surface cracks and delamination—slight fiber bloom
5.0	7.8	......	45	46	+ 2.0	
5.6	6.3	+13.0	45	27	−11.0	Severe surface cracks and some delamination
5.5	2.9	−48.0	45	8	−83.0	Some internal delamination —surface etched
5.6	7.3	+30.0	45	50	+11.0	Some small surface cracks
5.5	....	......	45	50	......	No visible damage
6.4	14.5	+127.0	38	...	......	Some surface bloom
5.5	6.5	+30.0	45	40	0	Some internal delamination
5.6	7.2	+28.0	45	58	+29	No visible damage
5.5	7.5	+36.0	45	55	+11.0	No visible damage
6.4	7.3	......	38	40	......	Some fiber bloom
5.5	6.4	+25.0	45	59	+31.0	No visible damage
5.6	7.7	+38.0	45	53	+18.0	Some fiber bloom
5.5	6.0	+ 9.0	45	45	+16.0	No visible damage
6.4	6.9	+ 8.0	38	50	+24.0	No visible damage
5.5	7.0	+56.0	45	46	− 2.0	Panels became opaque
5.6	5.5	0	45	43	− 7.0	Very bad surface cracks and delamination
5.5	4.8	−13.0	45	28	−50.0	Very bad surface cracks and delamination
6.4	8.5	+33.0	38	52	+37.0	Some internal delamination
5.5	6.2	−23.0	45	40	−11.0	Some surface cracks

2-38 Laminates, Reinforced Plastics, and Composite Structures

TABLE 14 Low-temperature Properties of Polyester Laminates[17]

Property*	−60	−40	−20	0	23	50	80
Modulus of elasticity, primary:							
High rate, psi × 10^3	4,770	4,480	4,540	4,460	5,545	3,500†	3,360†
Static, psi × 10^3	3,760	2,910	2,690	2,820	2,825	2,500†	2,240†
Increase, %	27	54	69	58	96	40	50
Modulus of elasticity, secondary:							
High rate, psi × 10^3	1,930	2,090	1,870	1,990	2,352		
Static, psi × 10^3	1,840	1,890	1,610	1,890	2,086		
Increase %	5	11	16	5	13		
Tensile strength:							
High rate, psi × 10^{-3}	73.6	72.5	68.8	66.47	62.34	58.00	54.2
Static, psi × 10^{-3}	51.58	44.84	41.4	39.0	36.74	34.13	29.4
Increase, %	43	62	66	70	70	70	84
Elongation, %:							
High rate	3.20	3.24	2.91	2.78	2.41	2.42	2.10
Static	2.46	2.21	2.01	1.83	1.66	1.65	1.48
Increase	30	47	45	52	45	47	42
Work to produce failure:							
High rate, ft-lb/in.^{3}	111.5	111.2	97.2	90.1	74.6	70.4	52.4
Static, ft-lb/in.^{3}	60.7	45.3	38.6	32.9	27.4	25.4	19.6
Increase, %	84	145	152	174	172	177	167
Time to failure:							
High rate, msec	10.7	10.7	10.1	10.1	9.9	9.8	9.0
Static, msec × 10^4	7.8	6.8	6.5	6.2	5.9	5.8	5.8

* ASTM D 638, type I specimen: ⅛-in.-thick hand lay-up panels formulated with Stypol 25 resin, vacuum pressure >12 psi; gelled under heat lamps; postcured 3 hr at 250°F.
† Value at 0.5% strain.

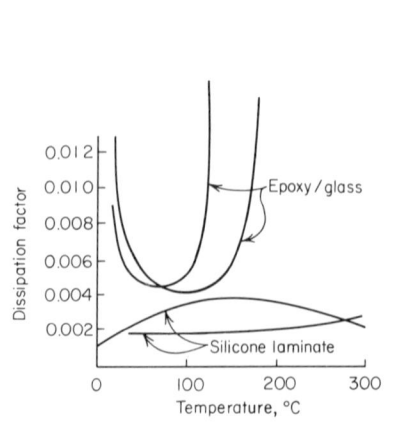

Fig. 31 Dissipation factor versus temperature for silicone and epoxy laminates.[19]

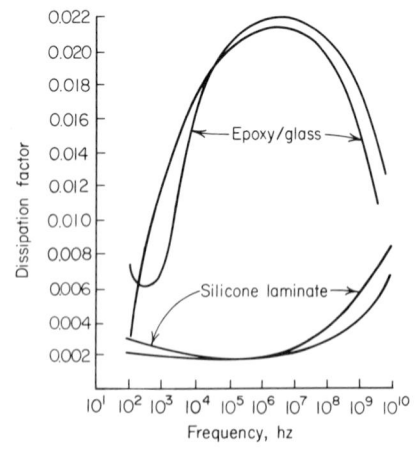

Fig. 32 Dissipation factor versus frequency for silicone and epoxy laminates at room temperature.[19]

TABLE 15 Strength of Glass-reinforced Plastic Laminates Laminated with 181 Glass Cloth[14] (In psi × 10⁻³)

Resin system	Ultimate tensile strength					Ultimate flexural strength					Ultimate compressive strength				
	RT*	−110°F	−320°F	−424°F		RT*	−110°F	−320°F	−424°F		RT*	−110°F	−320°F	−424°F
Epoxy														
Epon[a] 828	40.9	62.3	91.6	97.9		77.9	99.3	150.4	145.0		43.5	85.4	109.9	108.8
Epon 1001	49.0	71.4	94.9	100.7		83.2	105.5	154.0	122.0		58.6	86.2	89.4	78.9
Epon 826 NMA	58.5	82.0	127.2	121.1		86.0	116.5	166.7	171.2		50.9	84.4	113.9	129.1
Epon 826 DDS	61.0	98.1	115.2	118.0		94.9	129.6	167.5	172.4		71.7	107.4	138.1	135.2
Narmco 522 (prepreg)	54.0	75.1	104.4	108.6		75.6	99.1	152.6	163.7		55.8	93.2	122.9	135.5
Phenolic														
Narmco 506	38.1	58.3	79.4	72.2		63.4	76.2	84.5	83.0		46.5	55.1	81.8	76.7
CTL 91 LD	51.7	65.7	65.1	64.9		77.4	100.1	109.0	110.0		70.4	88.1	99.1	91.6
Polyester (general purpose)														
Hetron 92[b]	39.2	63.2	71.7	84.9		71.4	82.1	87.2	81.9		22.9	35.8	43.6	41.4
Paraplex P43[c]	47.9	58.5	81.3	78.2		68.8	79.4	84.2	84.3		31.5	43.2	51.3	47.9
Narmco 527 (prepreg)	49.5	66.9	102.1	104.1		75.2	80.9	93.8	106.3		36.6	55.1	66.6	65.8
Hetron 31	50.0	68.1	105.3	100.1		71.2	104.1	126.7	129.0		19.9	41.3	68.5	66.4
Polyester (high-temperature)														
Laminac[d] 4232	41.0	60.7	56.9	55.5		63.3	74.7	70.0	65.4		28.4	42.0	47.6	44.1
Vibrin[e]	32.7	57.6	52.0	54.6		52.0	76.2	78.1	75.1		26.3	33.5	38.1	36.1
Silicone														
Narmco 513	29.6	47.8	76.3	80.5		34.1	40.7	52.1	52.8		25.3	28.7	39.1	43.3
Trevarno F 130[f]	25.7	46.7	67.9	74.6		36.1	43.5	63.7	72.8		17.7	23.6	34.9	30.8
Trevarno IF 131[f]	29.4	47.4	70.4	76.2		37.9	45.8	66.9	65.0		21.0	39.0	42.7	46.1
Dow-Corning 7146	31.1	32.3	50.7	42.4		23.8	29.4	18.9	13.4		22.9	35.9	19.0	No test
Phenylsilane														
Narmco 534 (prepreg)	34.4	55.3	63.4	61.2		69.0	102.1	105.4	110.4		45.7	59.6	59.5	55.8
Nylon-epoxy														
Metlbond 406[g]	29.5	57.2	78.8	76.4		33.8	76.0	96.0	81.7		21.6	76.1	82.4	82.2
Polybenzimidazole	37.4	50.9	77.1	77.6		78.9	103.7	129.8	144.0		65.4	45.2	71.1	69.1
Flexible polyurethane														
Adiprene L-100[b]	11.6	76.1	73.8	70.2		Too flexible to test 19.4	51.8	58.7	48.5		4.9	48.6	90.1	91.1
Teflon														
FEP	24.5	43.0	65.7	81.3			37.4	67.3	66.3		7.3	11.4	22.6	33.3

* RT Room temperature.
[a] Trademark of Shell Chemical Co., New York.
[b] Trademark of Hooker Chemical Corporation, Niagara Falls, N.Y.
[c] Trademark of Rohm & Haas Co., Philadelphia, Pa.
[d] Trademark of American Cyanamid Co., Wayne, N.J.
[e] Trademark of United States Rubber Co., Naugatuck, Conn.
[f] Trademark of Coast Manufacturing and Supply Co., Livermore, Calif.
[g] Trademark of Whittaker Corp., Costa Mesa, Calif.
[h] Trademark of E. I. du Pont de Nemours & Co., Wilmington, Del.

TABLE 16 Effect of Accelerated Heat Aging on Silicone-Glass Laminates[19]

Hours aging at 572°F	Condition A	Condition D 48/50
Dielectric constant at 1 MHz		
0	3.78	3.82
24	3.72	3.76
48	3.77	3.80
96	3.75	3.79
192	3.71	3.90
Dissipation factor at 1 MHz		
0	0.0009	0.0027
24	0.0013	0.0024
48	0.0012	0.0030
96	0.0012	0.0033
192	0.0012	0.0230
Surface resistivity, ohms		
0	3.4×10^{13}	8.5×10^{13}
24	4.2×10^{14}	4.6×10^{13}
48	7.9×10^{14}	2.6×10^{14}
96	3.3×10^{14}	1.5×10^{14}
192	3.6×10^{14}	5.5×10^{13}
Perpendicular electric strength in air, volts/mil		
0	>328	>337
24	>379	>361
48	>375	349
96	351	315
192	106	100

dissipation factor, surface resistivity, and electric strength of a silicone laminate ⅛ in. thick formed from 116 glass cloth and Dow Corning 2105A resin. Figure 33 shows the effect of aging at 300°C on the insulation resistance. Because of the presence of the silicon atom, silicone laminates have good arc and track resistance. Table 17 shows typical tracking properties. Physical properties of silicones are not greatly influenced by aging, but, compared with laminates based on other resins, the flexural and tensile strengths are not unusually high. Figures 34 and 35 show the influence of heat aging at 250°C on flexural and compressive strength.

TABLE 17 Track Resistance of Silicone-Glass Laminates[20]

IEC CTI,* volts........................... 320
Dust fog (ASTM D 2132-62T) hr............ 1.0
Inclined plane, liquid contaminated, kv........ 1.5
Differential wet track:
   Discharge power, watts.................. 1.3
   Time to track, sec....................... 8.0

*International Electrotechnical Commission—Critical Tracking Index.

Laminates 2-41

Silicone laminates are used as radomes, structures in electronic heaters, slot wedges, ablation shields, terminal boards, and printed circuits. As printed-circuit boards they have bond strengths to copper much less than those of epoxies (epoxy bond strength about 6 lb/in.). Physical properties at elevated temperature do not compare favorably with those of other high-temperature laminates. The cost of silicones is high but not as high as some of the other "space-age" materials described in later sections. The use of silicones should be carefully considered in applications where they may be adjacent to sliding electrical surfaces (such as carbon brushes) or moving contacts (such as switches). Certain silicones give off volatile fragments when exposed to elevated temperatures, which fragments condense on the surfaces in question, insulating these surfaces with a minute layer of silicon dioxide.

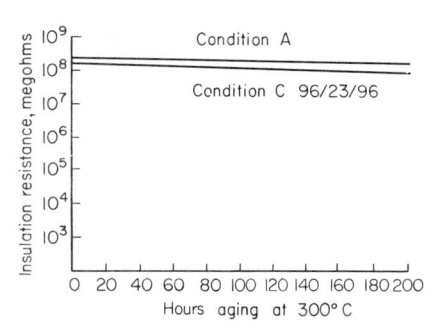

Fig. 33 Effect of aging on insulation resistance of silicone laminates.[19]

Phenylsilane Laminates. The phenylsilane resins are a hybrid version of phenolics and silicones. Phenylsilane laminates are used primarilly for structural purposes at elevated temperatures. Electrical properties are relatively poor compared to those of the silicones or DAP, for example, but they change very little with changing temperature. Table 18 presents physical properties of a ⅛ in. style 181 glass-cloth-phenylsilane laminate based on Monsanto's SC 1013 resin and compares these properties with the requirements of MIL-R-9299A. Table 19 shows the effect of seven days' immersion of the same laminate in various fluids and indicates the excellent chemical resistance of these laminates. Table 20 shows the influence of thermal exposure on dielectric constant and loss tangent. The high cost of phenylsilanes limits their use of aerospace and specialty applications. Long postcure schedules must be followed in order to obtain maximum properties.

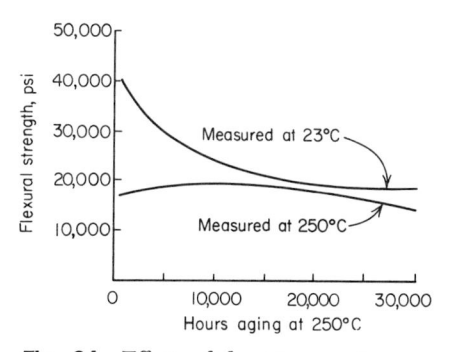

Fig. 34 Effect of long-term aging on flexural properties of silicone laminates.[19]

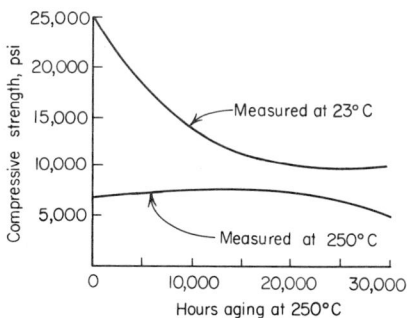

Fig. 35 Effect of long-term aging on compressive properties of silicone-glass laminate.[19]

Teflon Laminates. The excellent properties of Teflon resins were adequately described in the first chapter of this handbook. Teflon laminates were developed to improve the creep, cold flow, and other physical properties of unmodified Teflon without disturbing the electrical, chemical, and thermal properties. Physical properties for Teflon laminates are shown in Table 9. The physical properties are much improved at only slight sacrifice in dielectric properties. These laminates are used almost exclusively in electrical insulation where they meet the requirements of

2-42   Laminates, Reinforced Plastics, and Composite Structures

TABLE 18   Physical Properties of Phenylsilane Laminates[21]

Property	Phenylsilane, press grade — At room temperature — Dry	Wet	At 500°F after: ½ hr	200 hr	MIL-R-9299A, Class 2 — At room temperature — Dry	Wet	At 500°F after ½ hr
Flexural strength, psi × 10⁻³....	64	62.6	53.8	49	50	45	40
Flexural modulus, psi × 10⁻⁶....	3	2.5	3.01	2.6	3	2.5	3
Compressive strength, psi × 10⁻³	59	59	49.9	48.9	35	30	30
Tensile strength, psi × 10⁻³.....	42	39.5	33.1	33.6	40	38	30
Laminate resin content, %......	29.8						
Barcol hardness...............	76			55 min.			
Specific gravity...............	1.8						
Laminate thickness, in.........	0.12						
Flammability, in./min..........	Self-extinguishing			1.0 (max.)			
Water absorption after 24 hr immersion, % change in weight..	0.15			+1¼ (max.)			

TABLE 19   Chemical Resistance of Phenylsilane Laminates[21]

Fluid	% change in weight	% change in thickness	Ultimate flexural strength, psi
Hydraulic aircraft oil (MIL-H-5606).......	+0.011	0.049	72,200
Isopropyl alcohol (MIL-F-5566)..........	−0.056	0.000	75,000
Ethylene glycol (MIL-E-5559)............	−0.020	0.001	74,400
Hydrocarbon fluid (MIL-S-3136)..........	−0.005	0.000	73,200

TABLE 20   Effect of Thermal Aging on Electrical Properties of Phenylsilane Laminates[22]

Temperature, °F	No previous exposure — Dielectric constant	Loss tangent	After 24 hr at 550°F — Dielectric constant	Loss tangent
75	5.10	0.0234	5.03	0.0214
140	5.16	0.0248	5.10	0.0227
212	5.25	0.0267	5.17	0.0248
302	5.34	0.0282	5.25	0.0240
401	5.34	0.0254	5.34	0.0227
500	5.48	0.0212	5.35	˙0.0199

Class 180°C. They are used as slot wedges, coil separators, terminal blocks, and printed circuits.

Polyimide Laminates. Polyimide resins first appeared in the early 1960s, and they represent a new concept in thermally stable polymers. Also available as copolymer amide-imides, the resins represent a new polymeric family with many resins potentially available. The materials do not oxidize quickly, and the rate of pyrolysis is slow to about 650°F. The polyimides (also called polyaromatics) have certain similarities to nylon. They are available in the form of films, molding materials, wire enamels, adhesives, and laminates. The polyimides are really thermoplastic resins, their molecules being linear and not crosslinked. However, like Teflon, their melting point is virtually identical with the decomposition point, and the

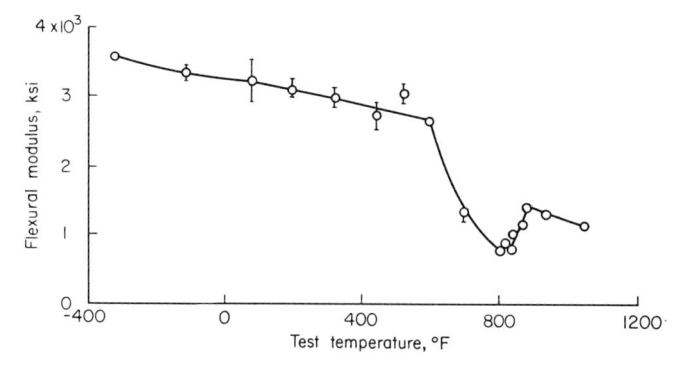

Fig. 36 Variation of flexural modulus with temperatures for polyimide-glass laminate.[23]

resin is thermally destroyed before it melts. This means that in the molding process laminates do not flow and that the various plies become bonded together by a cold flow or creep process. Therefore the molding process calls for pressures of at least 200 psi and temperatures reaching 700°F.

The physical properties of polyimides are good to about 600°F. Flexural strength and modulus for a duPont polyimide–glass-cloth laminate with A-1100 finish are shown in Figs. 36 to 39. Polyamide-imide laminates exhibit a similar response

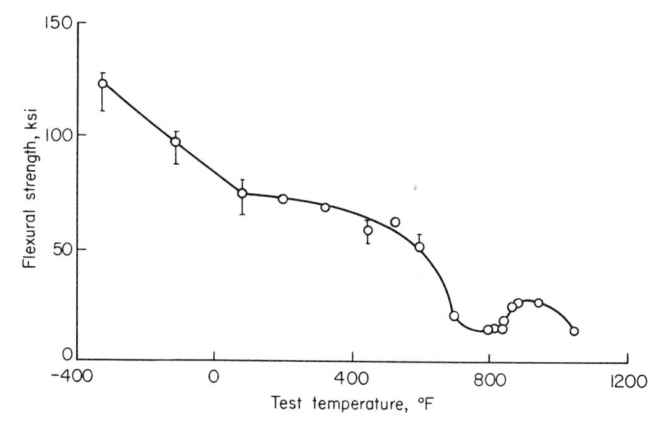

Fig. 37 Variation of flexural strength with temperature for polyimide-glass laminate.[23]

of flexural properties with aging, as shown in Fig. 40. This curve compares the aging at 300°C of several different polyamide-imides and silicone and phenolic laminates. The polyimides show little change in physical properties at low temperatures. Figures 41 and 42 show the flexural, tensile, and compressive strengths and moduli for Westinghouse I-8 laminates on 181-A-1100 glass cloth. Electrical

2-44   Laminates, Reinforced Plastics, and Composite Structures

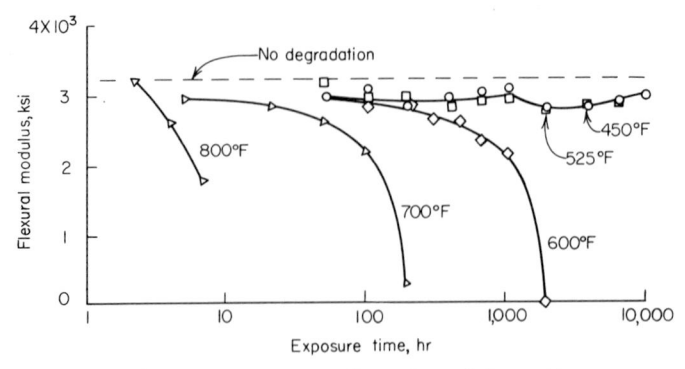

Fig. 38 Variation of room-temperature flexural modulus with exposure time for polyimide laminate.[23]

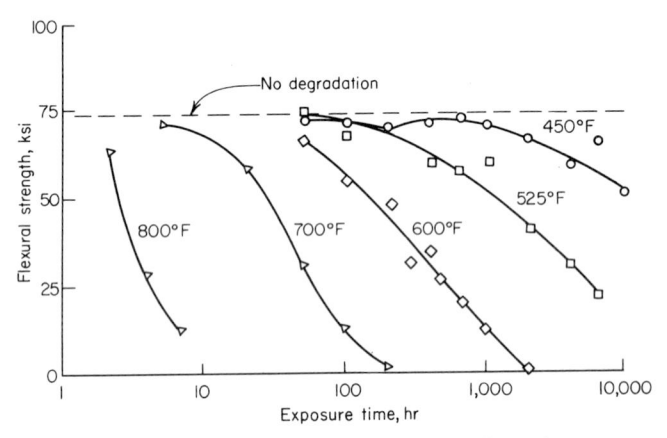

Fig. 39 Variation of room-temperature flexural strength with exposure time for polyimide laminate.[23]

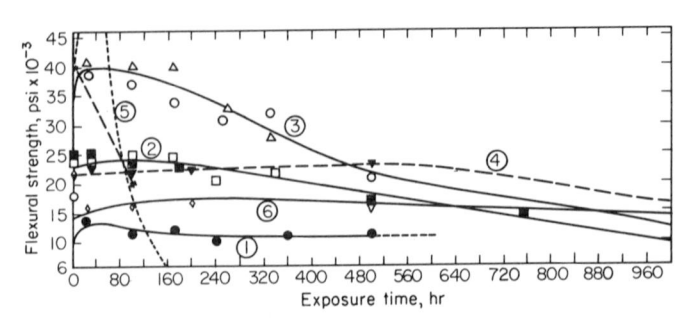

Fig. 40 Flexural strength of laminates aged and tested at 300°C in air (181 glass cloth): 1,6-silicones; 2,3,4-aromatic amide-imides; 5-phenolic, MIL-R-9299 for 260°C.[24]

Laminates 2-45

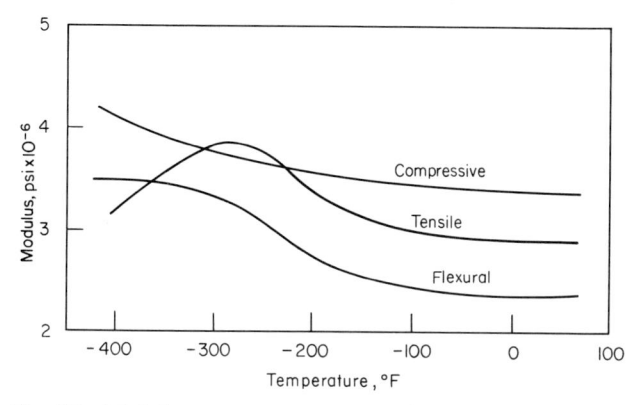

Fig. 41 Modulus versus temperature for polyimide laminate.[25]

properties are equally insensitive to temperature up to 650°F. The effect of temperature on dielectric constant, power factor, and volume resistivity for I-8 polyimide are shown in Figs. 43 to 45. These curves are given for direct current, 400 and 3,200 Hz. Complete electrical values for Skybond 700* polyimide laminates are given in Table 21.

TABLE 21 Electrical Properties of Skybond 700 Polyimide Laminates[26]

Property	As is	D 24/23	D 48/50	C 96/35/90
Dielectric strength:				
Short time parallel to laminate, volts	55,000	....	32,000	
Step by step parallel to laminate, volts	38,000	....	16,000	
Short time, volts/mil	179	....		
Stepwise, volts/mil	140	....		
Dielectric constant at 1 MHz	4.10	4.30	4.81	
Dissipation factor at 1 MHz	0.00445	0.00639	0.01650	
Insulation resistance, megohms	1.9×10^7	....	....	1.4×10^2
Volume resistivity, ohm-cm	2.47×10^{15}	....	....	1.16×10^{11}
Surface resistivity, ohms	3.35×10^{14}	....	....	2.90×10^{10}

Crossband data at 8.5 kMc

Temperature, °C	Dielectric constant	Dissipation factor
Room temperature	3.74	0.016
50	3.74	0.015
100	3.74	0.014
150	3.74	0.018
200	3.74	0.013
250	3.74	0.010
300	3.70	0.015

Polyimides are too new to be in widespread use, but they are being considered for the radome and certain structural applications in the Super Sonic Transport. The cost of these materials is somewhat higher than that of silicones.

* Trademark of Monsanto Chemical Co., St. Louis, Mo.

2-46　Laminates, Reinforced Plastics, and Composite Structures

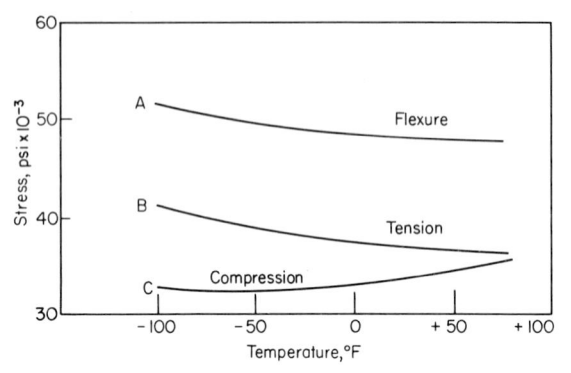

Fig. 42 Stress versus temperature for polyimide laminate.[25]

Polybenzimidazole Laminates. This new polyaromatic space-age material is being marketed by Narmco, trade-named Imidite 1850.* Available as both an adhesive and a laminate, polybenzimidazoles (PBI) are recommended for service from -423 to $700°F$ and for very short term service to $1200°F$. The material is characterized by chemical inertness, being little affected by acids, bases, and oils. The electrical properties are comparable to those of phenylsilanes. Table 22 gives typical physical

TABLE 22 Mechanical Properties of Polybenzimidazole 1581/994S Glass Laminates[27]
(Laminate: 80 wt. % Glass)

Property	Temperature, °F		
	−65	75	250
Tension			
Yield, psi × 10^{-3}	53.9	53.5	44.9
Ultimate, psi × 10^{-3}	100.8	87.0	77.1
Retention, %	115	100	88
Modulus, psi × 10^{-6}	5.08	4.74	4.64
Modulus retention, %	107	100	92
Compression			
Ultimate, psi × 10^{-3}	70.9	70.1	68.9
Retention, %	101	100	97
Modulus, psi × 10^{-6}	5.40	5.40	4.88
Modulus retention, %	100	100	90
Flexure			
Ultimate, psi × 10^{-3}	121.0	111.8	103.2
Retention, %	108	100	93
Modulus, psi × 10^{-6}	4.66	4.82	4.60
Modulus retention, %	97	100	96
Interlaminar shear			
Ultimate, psi	6,810	6,275	6,290

properties of PBI from -65 to $250°F$ for a laminate made with high-strength S994 glass cloth. Figures 46 and 47 show the effect of temperature on flexural and tensile properties. Figure 48 gives the effect of temperature on dielectric constant and loss tangent. As is the case for all high-temperature materials, the

* Trademark of Whittaker Corp., Costa Mesa, Calf.

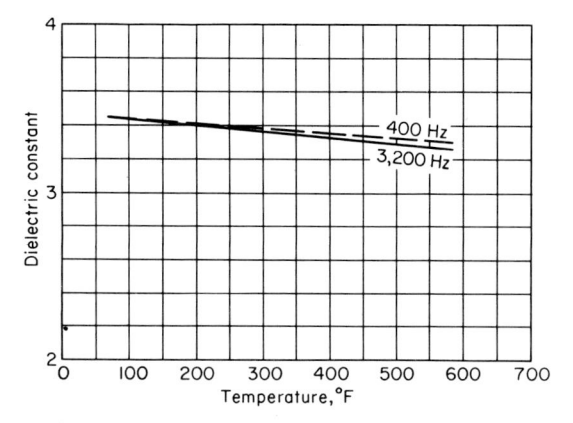

Fig. 43 Dielectric constant versus temperature for polyimide laminate.[15]

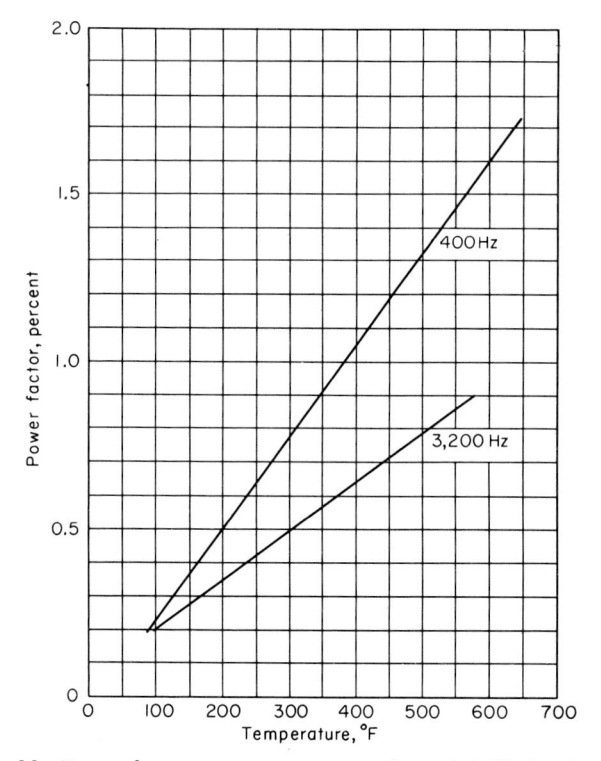

Fig. 44 Power factor versus temperature for polyimide laminate.[15]

PBI requires a postbake to develop maximum properties. Suggested applications for PBI laminates include radomes, turbine blades, aircraft leading edges, and ducts.

Diphenyl Oxide Laminates. Another new high-temperature laminate is based on diphenyl oxide (DO) resins. These materials have good chemical resistance and retain their properties well up to 500°F. The changes in flexural strength

2-48   Laminates, Reinforced Plastics, and Composite Structures

and dielectric strength of diphenyl oxide–resin laminate are shown in Fig. 49, while dielectric constant, power factor, and volume resistivity as a function of temperature and frequency are given in Figs. 50 to 52. DO laminates have been used as terminal boards, slot wedges, and coil forms.

Thermoplastic Laminates. Laminates with a reinforcing base of glass or asbestos and containing a thermoplastic resin binder are just becoming available. Mostly aimed at the automotive markets, these laminates can be easily shaped on simple tools by heating and pressing. Some of the materials can be stamped on metal-forming equipment. Tables 23 and 24 present physical properties for several kinds of thermoplastic laminates. Table 25 presents relative cost and the maximum longtime service temperature for all the laminating materials described in the preceding sections.

NEMA industrial laminates Industrial laminates are those materials in sheets, tube, and rod form used chiefly by the electrical and electronic industries for primary insulation and for dielectric purposes where structural requirements must be met. Standards for the laminates belonging to this category have been developed by the National Electrical Manufacturers Association (NEMA).[32]

NEMA grades The NEMA laminates are described below. All resins are phenolic except where specified differently.

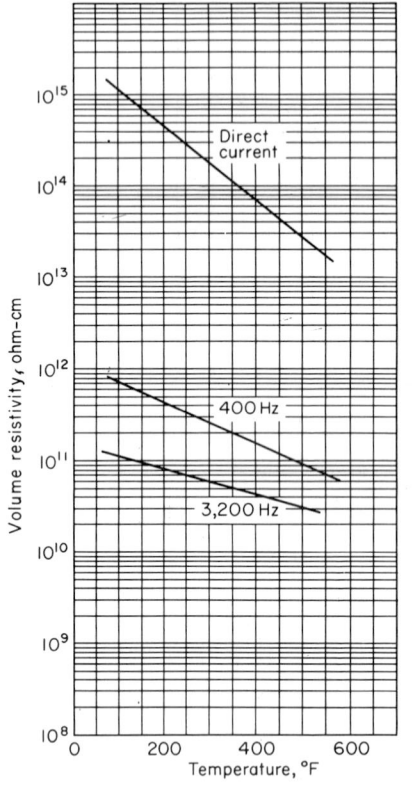

Fig. 45 Volume resistivity versus temperature for polyimide laminate.[15]

Table 26 lists the NEMA laminates and shows the outstanding properties of each grade. In addition to NEMA, many specifications relating to laminates have been developed by the military. A comparison of NEMA and military grades is given in Table 27.

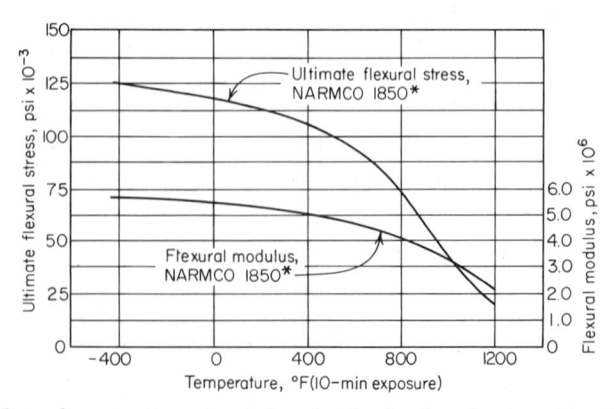

Fig. 46 Flexural properties of polybenzimidazole-glass laminate.[28] (*Trademark of Whittaker Corp.)

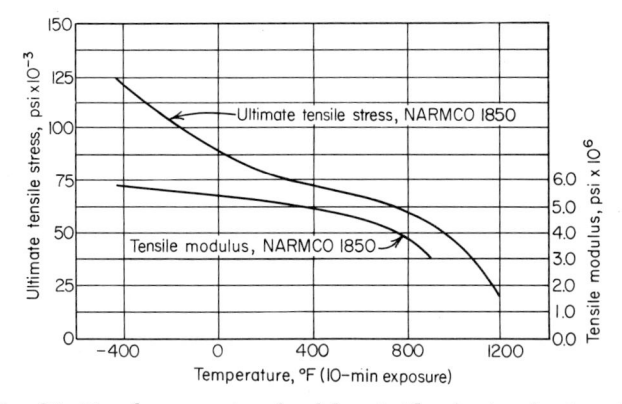

Fig. 47 Tensile properties of polybenzimidazole-glass laminate.[28]

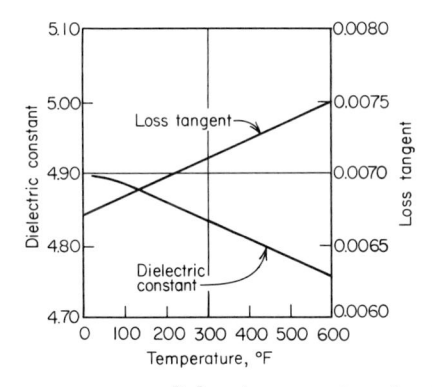

Fig. 48 Effect of temperature on dielectric properties of polybenzimidazole-glass laminate.[29]

TABLE 23 Properties of Thermoplastic Laminates[30]

Resin used	Reinforcement	Tensile strength, psi	Flexural strength, psi	Flexural modulus, psi $\times 10^{-6}$	Flatwise impact strength, ft-lb/in.
Polymethylmethacrylate	Nil	10,000	15,000	0.45	0.5
	Glass mat	15,000	26,000	1.1	11
Polystyrene........	Nil	7,000	13,000	0.5	0.5
	Glass mat	15,000	25,000	1.2	10
	Glass woven roving	34,000	18,000	....	....
	Asbestos felt	20,000	28,000	3.0	7.0
Polyvinylchloride...	Nil	7,000	14,000	0.5	2.0
	Glass mat	17,000	24,000	1.1	14
	Glass woven roving	30,000	19,000	....	30
	Asbestos felt	18,000	21,000	2.5	8.0

2-50   Laminates, Reinforced Plastics, and Composite Structures

Grade X. Paper-based sheet and tubes intended for mechanical applications where electrical properties are of little importance. These materials are humidity-sensitive and are not equal to fabric-based materials in impact strength. The rolled tubes have good power factor and electric strength when dry, and they can be punched well.

Grade XP. Paper-based sheet used for hot punching. More flexible and not as

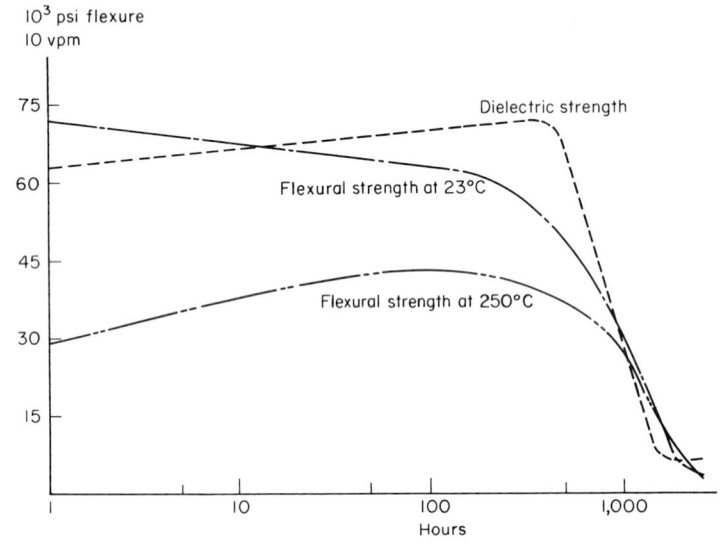

Fig. 49 Effect of temperature on flexural strength and dielectric strength for diphenyl oxide–glass laminate.

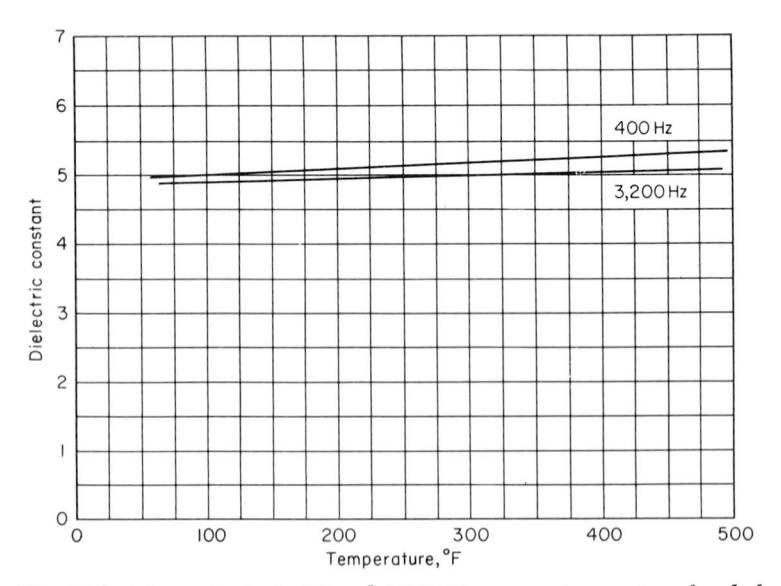

Fig. 50 Dielectric constant at 400 and 3,200 Hz versus temperature for diphenyl oxide–glass laminate.[15]

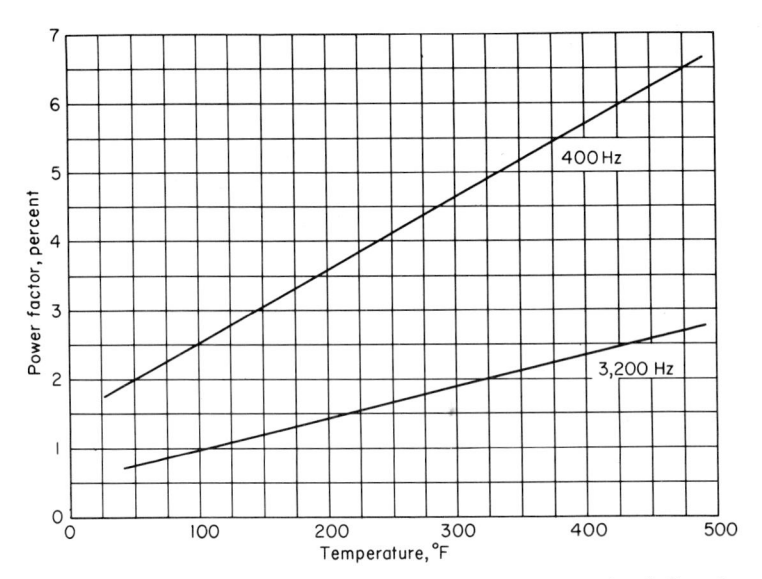

Fig. 51 Power factor at 400 and 3,200 Hz versus temperature for diphenyl oxide–glass laminate.[15]

TABLE 24 Properties of Glass-mat-reinforced Thermoplastics[31]

Property	Polypropylene	Styrene-acrylonitrile copolymer	Polyvinyl chloride
Flexural strength at RT, psi.	26,900	32,400	31,300
Flexural modulus at RT, psi.	1,030,000	1,300,000	1,260,000
Tensile strength at RT, psi.	16,500	18,400	19,100
Tensile modulus at RT, psi.	860,000	1,100,000	950,000
Flexural strength at 180°F, psi.	14,000	22,000	
Flexural modulus at 180°F, psi.	720,000	1,200,000	
Heat distortion temperature at 264 psi, °F.	327	255	221
Maximum short-term-use temperature, °F.	300	200	200
Notched Izod impact, ft-lb/in.	11.2	15.7	27
Falling-dart impact, in.	16	36	
Wet flexural strength, psi.	24,600	26,900	
Wet flexural modulus, psi.	1,000,000	1,300,000	
Coefficient of thermal expansion, in./(in.)(°F) $\times 10^{-5}$	1.5	1.3	1.61
Glass fiber content, %.	44	42	36
Lb/ft² at 100 in.	0.65	0.70	0.85
Weight for equivalent stiffness.	0.7	0.7	
Specific gravity.	1.26	1.36	

strong as Grade X. The moisture resistance and electrical properties are better than those of Grade X and less than those of Grade XX.

Grade XPC. Paper-based sheet similar to Grade XP but can be punched cold. Slightly more flexible but with lower flexural strength and higher cold flow than Grade XP.

Grade XX. Paper-based sheet, tube, and rods with good machinability and good electrical properties. Not as strong as Grade X but better in moisture resistance. Electric strength of thin-walled tubes may be low at seam areas.

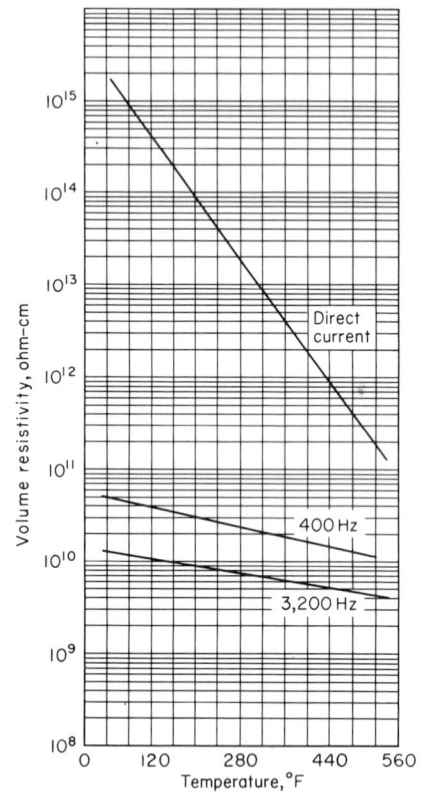

Fig. 52 Volume resistivity versus temperature for diphenyl oxide–glass laminate.[15]

Grade. XXP. Paper-based sheet with better electrical and moisture resistance than Grade XX. Punching and cold-flow qualities are intermediate between those of Grades XP and XX.

Grade XXX. Paper-based sheet, tube and rods for radio-frequency applications. Has good resistance to humidity and possesses low cold flow. Rolled tubes are superior to all other paper-based grades when exposed to high humidity.

Grade XXXP. Paper-based sheet intermediate between grades XXP and XX in punching. Has high insulation resistance and low losses at high humidity.

Grade XXPC. Paper-based sheet similar to XXXP but with lower punching temperatures.

Grade FR-2. Paper-based modified chemically so as to be self-extinguishing after ignition source is removed. Similar in all other properties to Grade XXXPC.

Grade FR-3. Paper-based epoxy resin sheet. The epoxy resin contributes excellent stability of electrical properties at high humidity. The laminate is self-extinguishing, has high flexural strength, and is punchable.

Grades ES-1, ES-2, and ES-3. Paper-based sheets used for nameplate applications.

Grade C. Cotton-fabric-based sheet, tube, and rod made from a cloth weighing more than 4 oz/yd². Mechanical grade; not recommended for electrical applications. Used for gears and other applications requiring high impact strength.

Grade CE. Cotton-fabric-based sheet, tube, and rod. Similar to Grade C. Used for electrical applications requiring greater toughness than that possessed by Grade XX or for mechanical applications requiring better moisture resistance than that possessed by Grade C. Not recommended for primary insulation at voltages exceeding 600 volts.

Grade L. Cotton-fabric-based sheet, tube, and rod made from a cloth weighing less than 4 oz/yd². This grade machines better than Grades C or CE but is not as tough. A mechanical material and not recommended for electrical applications. Has good moisture resistance.

Grade LE. Cotton-fabric-based sheet, tube, and rod similar to Grade L. Not recommended for primary insulation at voltages over 600 volts. Used for applications requiring greater toughness than Grade XX or better machining than Grade CE. Available in thin sheets.

Grade CF. Cotton-fabric-based sheet made from a cloth weighing more than 4 oz/yd². The laminates are postforming or capable of being shaped by the application of heat and pressure.

Grade A. Asbestos-paper-based sheet and tube. More resistant to flame and heat than cellulose-based materials. Limited to service at voltages less than 250 volts. Moisture causes small dimensional changes.

TABLE 25 Cost and Thermal Rating of Laminates

Resin	Reinforcement	Relative cost (paper phenolic = 1)	Maximum use temperature, °F
Phenolic	Paper	1	220
Phenolic	Glass cloth	3.5	500
Phenolic	Asbestos cloth	3.2	350
Phenolic	Asbestos paper	1.2	350
Phenolic	Cotton cloth	1.5	250
Phenolic	Nylon cloth	4.0	250
Melamine	Glass cloth	2.8	300
Melamine	Cotton cloth	3.0	300
Epoxy	Glass cloth	3.3	250
Polyester	Glass cloth	3.0	250
Polyester	Glass mat	1.1	250
Diallyl phthalate	Glass cloth	5.0	300
Silicone	Glass cloth	6.5	500
Phenylsilane	Glass cloth	6.5	500
Teflon	Glass cloth	7.0	400
Polyimide	Glass cloth	25.0	600
Polybenzimidazole	Glass cloth	100.0	600
Diphenyl oxide	Glass cloth	5.0	450
Thermoplastic	Glass mat	0.75	180

Grade AA. Asbestos-fabric-based sheet and tube. Similar to Grade A but with better physical properties and thermal resistance. Not recommended for primary insulations.

Grade G-2. Glass-cloth-based sheet. Cloth is made of a staple (noncontinuous) fiber. Good electrical properties under moist conditions. Weakest of all glass-based grades. Lower in dielectric losses than all other glass-based grades except the silicone.

Grade G-3. Glass-cloth-based sheet, tube, and rod. Cloth is made of a continuous fiber. High mechanical properties except for bond strength which is low. Good electrical properties under dry conditions.

Grade G-5. Glass-cloth-based sheet, rod, and tube with a melamine resin binder. Cloth is made of continuous fiber. Maximum physical properties. Hardest of all grades. Good flame resistance, arc resistance, dimensional stability. Tubing has excellent burst strength.

Grade G-7. Glass-cloth-based sheet and tube with a silicone resin binder. Cloth is made of continuous fiber. Extremely good dielectric loss and insulation resistance when dry. Good electrical properties when wet. Excellent heat and arc resistance. Second only to Grade G-5 in flame resistance. Meets IEEE requirements for Class 180°C insulation.

Grade G-9. Glass-cloth-based sheet, tube, and rod with a heat-resistant melamine resin binder. Cloth is made of continuous filaments. Similar to Grade G-5 except more thermally stable. Second only to Grade G-7 in arc and heat resistance.

Grade G-10. Glass-cloth-based sheet, tube, and rod with an epoxy resin binder. Cloth is made of continuous fiber. Extremely high mechanical strength at room temperature. Good dielectric loss and electric strength properties both wet and dry. Insulation resistance at high humidity is better than that of Grade G-7.

Grade G-11. Glass-cloth-based sheet, tube, and rod with an epoxy resin binder. Cloth is made of continuous fibers. Similar to Grade G-10 except that it retains at least 50 percent of its room-temperature flexural strength at 150°C after 1 hr at 150°C.

Grade FR-4. Glass-cloth-based sheet, rod, and tube with an epoxy resin binder. Similar to Grade G-10 but is self-extinguishing when ignition source is removed.

TABLE 26 NEMA Laminate Selector Chart

Grade	X	XP	XPC	XX	XXP	XXX	XXXP	XXXPC	FR-2	FR-3	C	CE	L	LE	CF	A	AA	G-2	G-3	G-5	G-7	G-9	G-10	G-11	FR-4	FR-5	GPO-1	GPO-2	N-1
Mechanical application	X	X	X								X	X	X	X	X	X	X		X	X		X	X	X	X	X	X	X	
Electrical application		X		X	X	X	X	X	X	X		X			X			X			X	X	X	X	X	X	X	X	X
Punching qualities	X																												
Tubes																													
Rods																													
Flame resistance									X	X																			
Impact resistance				X	X	X	X	X			X	X		X	X	X	X		X	X	X	X	X	X	X	X	X	X	X
Flexural strength											X	X		X	X		X		X	X	X	X	X	X	X	X			
Thermal stability																					X	X	X	X	X	X	X	X	X
Cold flow																													
Humidity resistance						X	X	X																					X
Electronic application			X			X	X	X	X	X	X	X	X	X		X		X		X	X	X	X	X	X	X	X	X	X
Machinability													X																X
Arc resistance				X									X	X						X	X	X			X	X	X	X	X

2-54

TABLE 27 Cross Reference of NEMA and Military Designations for Plastic Laminates[33]

Resin: reinforcement	NEMA grade	MIL type	Military (MIL) Specification
Sheets			
Epoxy:			
Glass cloth	G-10	GEE	
Glass cloth	G-11	GEB	
Paper	FR-3	PEE	P-3115C
Glass (copper-clad)	G-10, G-11	GE, GB	P-13949B
Glass (copper-clad)		GF	P-13949B
Melamine:			
Paper	ES-1, ES-3	NDP	P-78A
Glass cloth	G-5	GMG	P-15037C
Glass (copper-clad)	G-5	GM	P-13949B
Phenolic:			
Paper	X, P		
Paper	XP	HSP	P-78A
Paper	XPC		
Paper	XX	PBG	P-3115C
Paper	XXX, XXP	PBE, PBE-P	P-3115C
Paper	XXXP, XXXPC	PBE-P	P-3115C
Paper	ES-2	NDP	P-78A
Paper	FR-1	PBG	P-3115C
Paper	FR-2	PBE-P	P-3115C
Cotton fabric	C	FBM, FBG	P15035C
Cotton fabric	CF		P-8655A
Cotton fabric	L	FBI	P-15035C
Cotton fabric	LE	FBE	P-15035C
Asbestos paper	A, AA	PBA, FBA	P-8059A
Glass fiber	G-2		
Glass fiber		II-2	P-25515A
Glass cloth	G-3		
Nylon cloth	N-1	NPG	P-15047B
Glass (copper-clad)	XXXP, XXXPC	PP	P-13949B
Polyester:			
Glass fiber	GPO-1		
Glass fiber	GPO-1	1, 2, 3	P-8013C
PTFE:			
Glass cloth	GTE	GTE	P-19161A
Silicone:			
Glass cloth	G-6, G-7	GSG	P-997C
Glass (copper-clad)	G-6	GS	P-13949B
Rods and tubes			
Phenolic:			
Paper	X	PBM	P-79C
Paper	XX	PBG	P-79C
Paper	XXX	PBE	P-79C
Cotton fabric	C	FBM	P-79C
Cotton fabric	CE	FBG	P-79C
Cotton fabric	L	FBI	
Cotton fabric	LE	FBE	P-79C
Glass cloth	G-3		
Melamine:			
Glass cloth	G-5	GMG	P-79C

TABLE 28 Typical Values for NEMA Laminates (Condition A)[32]

NEMA and ASTM grades	X	XP	XPC	XX	XXP	XXX	XXXP	ES-1	ES-2	ES-3
Equivalent MIL-P Specification No.				3115		3115	3115			
MIL type				PBG		PBE	PBE-P			
Tensile strength, psi × 10⁻³:										
Lengthwise	20	12	10.5	16	11	15	12.4	12	13	15
Crosswise	16	9	8.5	13	8.5	12	9.5	8.5	9	12
Modulus of elasticity in tension, psi × 10⁻⁶:										
Lengthwise	1.9	1.2	1	1.5	0.9	1.3	1			
Crosswise	1.4	0.9	0.8	1.2	0.7	1	0.7			
Modulus of elasticity in flexure, psi × 10⁻⁶:										
Lengthwise	1.8	1.2	1	1.4	0.9	1.3	1			
Crosswise	1.3	0.9	0.8	1.1	0.7	1	0.7			
Compressive strength, psi × 10⁻³:										
Flatwise	36	25	22	34	25	32	25			
Edgewise	19			23		25.5				
Rockwell hardness, M	110	95	75	105	100	110	105	118	118	120
Deformation and shrinkage (cold flow at 4,000 psi for ⅛-in. thickness), % change				0.90		0.80				
Electric strength perpendicular to laminations, volts/mil:										
Short time:										
³⁄₃₂ in. thickness	950	900	850	950	950	900	900	750		
¹⁄₁₆ in. thickness	700	650	600	700	700	650	650			
⅛ in. thickness	500	470	425	500	500	470	470			
Step by step:										
³⁄₃₂ in. thickness	700	650	625	700	700	650	650	550	550	
¹⁄₁₆ in. thickness	500	450	425	500	500	450	450			
⅛ in. thickness	360	320	290	360	360	320	320		400	
Insulation resistance, megohms (Condition C-96/35/90)				60	500	1,000	20,000			
Density, g/cc	1.36	1.33	1.34	1.34	1.32	1.32	1.30	1.45	1.40	1.38
Specific volume, in.³/lb	20.4	20.8	20.6	20.6	21.0	21.0	21.3	19.1	19.8	20.1
Thermal expansion, cm/(cm)(°C)	←――――――――――― 2.0 × 10⁻⁵ ――――――――――→									
Thermal conductivity, cal/(sec)(cm²)(°C/cm)	←――――――――――― 7.0 × 10⁻⁴ ――――――――――→									
IEEE insulation class	←――――――――――― 105 ―――――――――――→									
Specific heat	←――――――――――― 0.35–0.40 ――――――――――→									
Effect of acids	All grades except Grades G-5 and G-9 are resistant to dilute solutions of most acids.									
Effect of alkalies	Not recommended for use in alkaline solutions except for melamine Grades G-5 and G-9 which are resistant to dilute alkaline solutions.									
Effect of solvents	Unaffected by most organic solvents except acetone which may soften the punching grade stocks. Aromatic hydrocarbons and chlorinated aliphatics may affect silicone Grade G-7.									

TABLE 28 Typical Values for NEMA Laminates (Condition A)[32] (Continued)

NEMA and ASTM grades	C	CE	L	LE	A	AA	G-2	G-3	G-5	G-7
Equivalent MIL-P Specification No.	15035 FBM	15035 FBG	15035 FBI	15035 FBE					15037 GMG	997 GSG
MIL type										
Tensile strength, psi $\times 10^{-3}$:										
Lengthwise	10	9	13	12	10	12	16	23	37	23
Crosswise	8	7	9	8.5	8	10	11	20	30	18.5
Modulus of elasticity in tension, psi $\times 10^{-6}$:										
Lengthwise	1	0.9	1.2	1	2.5	1.7	1.8	2	2.3	1.8
Crosswise	0.9	0.8	0.9	0.85	1.6	1.5	1.2	1.7	2	1.8
Modulus of elasticity in flexure, psi $\times 10^{-6}$:										
Lengthwise	1	0.9	1.1	1	2.3	1.6	1.3	1.5	1.7	1.4
Crosswise	0.9	0.8	0.85	0.85	1.4	1.4	1	1.2	1.5	1.2
Compressive strength, psi $\times 10^{-3}$:										
Flatwise	37	39	35	37	40	38	38	50	70	45
Edgewise	23.5	24.5	23.5	25	17	21	15	17.5	25	14
Rockwell hardness, M	103	105	105	105	111	103	105	100	120	100
Deformation and shrinkage (cold flow at 4,000 psi for ⅛-in. thickness), % change									0.30	0.30
Electric strength perpendicular to laminations, volts/mil:										
Short time:						No values recommended				
¹⁄₃₂ in. thickness	No values recommended	500	No values recommended	700	225		500	750	.....	450
¹⁄₁₆ in. thickness		360		500	160		425	700	350	400
⅛-in. thickness				360				600	260	350
Step by step:										
¹⁄₃₂ in. thickness		300		450	135		360	550	.....	400
¹⁄₁₆ in. thickness		220		300	95		300	500	220	350
⅛ in. thickness				220				450	160	250
Insulation resistance, megohms (Condition C-96/35/90)							5,000		100	2,500
Density, g/cc	1.36	1.33	1.35	1.33	1.72	1.70	1.50	1.65	1.90	1.68
Specific volume, in.³/lb	20.4	20.8	20.5	20.8	16.1	16.3	18.5	16.8	14.6	16.5
Thermal expansion, cm/(cm)(°C)	2.0 × 10⁻⁵				1.5 × 10⁻⁵	1.8 × 10⁻⁵		1.0 × 10⁻⁵	1.0 × 10⁻⁵	
Thermal conductivity, cal/(sec)(cm²)(°C/cm)	7.0 × 10⁻⁴								12.0 × 10⁻⁴	7.0 × 10⁻⁴
IEEE insulation class	105					130				180
Specific heat	0.35–0.40				0.30				0.26	0.25
Effect of acids	All grades except Grades G-5 and G-9 are resistant to dilute solutions of most acids.									
Effect of alkalies	Not recommended for use in alkaline solutions except for melamine Grades G-5 and G-9 which are resistant to dilute alkaline solutions.									
Effect of solvents	Unaffected by most organic solvents except acetone which may soften the punching grade stocks. Aromatic hydrocarbons and chlorinated aliphatics may affect silicone Grade G-7.									

TABLE 28 Typical Values for NEMA Laminates (Condition A)[32] (Continued)

NEMA and ASTM grades	N-1	G-9	G-10	G-11	FR-2	FR-3	FR-4	FR-5
Equivalent MIL-P Specification No.	15047	15037	18177	18177	3115	22324	18177	18177
MIL type	NPG	GME	GEE	GEB	PBE-P	PEE	GEE	GEB
Tensile strength, psi × 10^{-3}:								
Lengthwise	8.5	37	40	40	12.5	14	40	40
Crosswise	8	30	35	35	9.5	12	35	35
Modulus of elasticity in tension, psi × 10^{-6}:								
Lengthwise	0.4	2.3	2.5	2.5	1	1.2	2.5	2.5
Crosswise	0.4	2	2	2	0.8	1	2	2
Modulus of elasticity in flexure, psi × 10^{-6}:								
Lengthwise	0.6	2.5	2.7	2.8	1.1	1.3	2.7	2.8
Crosswise	0.5	2	2.2	2.3	0.9	1	2.2	2.3
Compressive strength, psi × 10^{-3}:								
Flatwise		70	60	60	25	29	60	60
Edgewise		25	35	35			35	35
Rockwell hardness, M	105	120	111	112	97	100	111	114
Deformation and shrinkage (cold flow at 4,000 psi for $\frac{3}{16}$-in. thickness), % change			0.25	0.20	0.85	0.50	0.25	0.20
Electric strength perpendicular to laminations, volts/mil:								
Short time:								
$\frac{1}{32}$ in. thickness	850	450	750	750	800	650	750	750
$\frac{1}{16}$ in. thickness	600	400	700	700	650	600	700	700
$\frac{1}{8}$ in. thickness	450	350	550	550	470	475	550	550
Step by step:								
$\frac{1}{32}$ in. thickness	650	400	500	500	550	500	500	500
$\frac{1}{16}$ in. thickness	450	350	450	450	450	450	450	450
$\frac{1}{8}$ in. thickness	300	275	350	350	320	325	350	350
Insulation resistance, megohms, (Condition 3-96/35/90)	50,000	10,000	200,000	200,000	20,000	100,000	200,000	200,000
Density, g/cc	1.15	1.90	1.80	1.80	1.33	1.42	1.85	1.85
Specific volume, in.³/lb	24.1	14.6	15.3	15.3	21.0	19.5	14.9	14.9
Thermal expansion, cm/(cm)(°C)		1.0×10^{-5}	0.9×10^{-5}	7.0×10^{-5}	2.0×10^{-4}	1.0×10^{-5}	1.0×10^{-5}	0.9×10^{-5}
Thermal conductivity, cal/(sec)(cm²)(°C/cm)					105			
IEEE insulation class					$0.35{-}0.40$			
Specific heat								
Effect of acids	All grades except Grades G-5 and G-9 are resistant to dilute solutions of most acids.							
Effect of alkalies	Not recommended for use in alkaline solutions except for melamine Grades G-5 and G-9 which are resistant to dilute alkaline solutions.							
Effect of solvents	Unaffected by most organic solvents except acetone which may soften the punching grade stocks. Aromatic hydrocarbons and chlorinated aliphatics may affect silicone Grade G-7.							

TABLE 29 Properties of Molded Tubing[34]

Grade	Filler	Resin	Colors available	Specific gravity	Compressive strength, axial, psi	Tensile strength, psi	Power factor at 1 MHz	Dielectric constant at 1 MHz	Water absorption, in 24 hr, % — Wall thickness, in. $\frac{1}{32}$	$\frac{1}{16}$	$\frac{3}{32}$	$\frac{1}{8}$	Dielectric strength, volts/mil (short time) perpendicular to laminations — Wall thickness, in. Over $\frac{1}{16}$ to $\frac{1}{8}$, incl.	Over $\frac{1}{8}$ to $\frac{1}{4}$, incl.	Over $\frac{1}{4}$ to $\frac{1}{2}$, incl.
X	Paper	Phenolic	Natural and black	1.25	18,000	12,000	...	...	6.0	4.5	4.0	3.6	270	180	100
XX	Paper	Phenolic	Natural and black	1.25	18,000	11,000	0.040	5.5	...	2.0	1.8	1.6	220	150	110
XXX	Paper	Phenolic	Natural and black	1.22	20,000	9,000	0.040	5.5	...	1.4	1.2	1.1	220	150	110
C	Coarse-weave cotton fabric	Phenolic	Natural and black	1.2	19,000	9,000	...	...	...	3.5	2.6	2.2	160	110	75
CE	Coarse-weave cotton fabric	Phenolic	Natural and black	1.25	19,000	8,500	...	...	...	3.0	2.2	2.0	175	125	90
C-M	Coarse-weave cotton fabric	Melamine	Grayish brown	1.30	19,000	8,500	...	...	...	6.2	4.7	3.9	150	100	80
L	Fine-weave cotton fabric	Phenolic	Natural and black	1.25	18,000	9,000	...	...	6.5	3.5	2.2	1.8	160	110	75
LE	Fine-weave cotton fabric	Phenolic	Natural and black	1.25	19,000	8,500	...	...	4.5	2.2	1.8	1.5	175	125	90
A	Asbestos paper	Phenolic	Natural	1.55	18,000	10,000	...	...	...	2.0	1.8	1.6	85	70	55
AA	Asbestos fabric	Phenolic	Natural	1.55	18,000	9,000	...	...	...	2.0	1.8	1.6	90	75	60
G-3	Continuous-filament glass fabric	Phenolic	Grayish brown	1.55	20,000	30,000	...	...	6.5	3.7	3.2	3.0	170	120	85
G-5	Continuous-filament glass fabric	Melamine	Grayish brown	1.80	20,000	30,000	...	...	5.0	3.9	3.7	3.0	155	105	75
N-1	Nylon fabric	Phenolic	Natural	1.18	6,500	6,000	0.040	5.0	...	1.0	0.8	0.6	200	160	80

2-59

TABLE 30 Properties of Rolled Tubing[35]

Grade	Filler	Resin	Colors available	Specific gravity	Compressive strength, axial, psi	Tensile strength, psi	Power factor at 1 MHz	Dielectric constant at 1 MHz	Water absorption, in 24 hr, % Wall thickness, in. 1/32	1/16	3/32	1/8	Dielectric strength, volts/mil (short time) perpendicular to laminations Wall thickness, in. Over 1/16 to 1/8, incl.	Over 1/8 to 1/4, incl.	Over 1/4 to 1/2, incl.
X	Paper	Phenolic	Natural and black	1.10	10,000	8,500	.....	.....	8.0	5.0	4.3	4.0	325	200	145
XX	Paper	Phenolic	Natural and black	1.10	13,000	8,000	0.040	5.0	6.0	3.0	2.5	2.0	290	200	145
XXX	Paper	Phenolic	Natural and black	1.12	13,000	7,000	0.040	5.3	...	1.5	1.3	1.0			
C	Coarse weave cotton fabric	Phenolic	Natural and black	1.12	12,000	6,000	.....	.....	...	5.0	3.6	3.0	120	110	75
CE	Coarse weave cotton fabric	Phenolic	Natural and black	1.12	13,000	6,000	.....	.....	...	4.7	3.3	2.7	140	120	85
C-M	Coarse weave cotton fabric	Melamine	Grayish brown	1.12	13,000	6,000	.....	.....	...	7.0	5.0	4.2	150	100	75
L	Fine weave cotton fabric	Phenolic	Natural and black	1.12	12,000	6,000	.....	.....	8.0	5.0	3.3	2.8	120	110	75
LE	Fine weave cotton fabric	Phenolic	Natural and black	1.12	13,000	7,000	.....	.....	7.5	5.0	3.0	2.5	140	120	85
A	Asbestos paper	Phenolic	Natural	1.25	10,000	7,000	.....	.....	...	2.2	2.0	1.8	90	75	60
G-3	Continuous filament glass fabric	Phenolic	Grayish brown	1.50	13,000	25,000	.....	.....	7.0	4.0	3.5	3.2	175	125	90
G-5	Continuous filament glass fabric	Melamine	Grayish brown	1.70	13,000	25,000	0.012	7.0	5.0	3.9	3.7	3.0	160	110	85
G-7	Continuous filament glass fabric	Silicone	Grayish brown	1.55	6,000	.....	.....	.....	1.0	1.0	0.8	0.8	125	140	140
G-10	Continuous filament glass fabric	Epoxy	Grayish brown	1.55	25,000	15,000	0.005	7.0	1.0	0.6	0.4	0.3	300	250	200
N-1	Nylon fabric	Phenolic	Natural	1.15	5,500	5,000	0.040	5.0	...	1.5	1.2	1.0	220	180	120

2-60

TABLE 31 Properties of Laminated Rods[36]

Grade	Filler	Resin	Range of diameters available, in., incl.	Colors available	Specific gravity (min.)	Flexural strength, psi (min.)	Compressive strength, psi (axial)	Tensile strength, psi (average)	Water absorption, in 24 hr, % Diameter, in. 1/8	1/4	1/2	1
X	Paper	Phenolic	1/8–1 3/4	Natural and black	1.30	16,000	20,000	10,000	3.0	2.0	1.5	1.5
XX	Paper	Phenolic	1/8–1 3/4	Natural and black	1.30	15,000	20,000	10,000	2.5	1.5	1.0	1.0
XXX	Paper	Phenolic	1/8–1 3/4	Natural and black	1.25	13,000	20,000	9,000	1.5	1.0	0.75	0.75
C	Coarse weave cotton fabric	Phenolic	1/4–4	Natural and black	1.28	16,000	19,000	9,000	...	2.5	2.0	2.0
CE	Coarse weave cotton fabric	Phenolic	1/4–4	Natural and black	1.26	13,000	20,000	8,000	...	1.7	1.3	1.0
L	Fine weave cotton fabric	Phenolic	3/16–4	Natural and black	1.28	16,000	19,000	11,000	2.5	2.0	1.5	1.2
LE	Fine weave cotton fabric	Phenolic	3/16–4	Natural and black	1.26	12,000	20,000	10,000	2.2	1.4	1.1	1.0
A	Asbestos paper	Phenolic	3/8–2	Natural	1.70	15,000	18,000	6,000	2.5	1.5	1.0	1.0
AA	Asbestos fabric	Phenolic	3/8–4	Natural	1.60	17,000	20,000	8,000	...	1.5	1.0	1.0
G-3	Continuous filament glass fabric	Phenolic	1/4–4	Grayish brown	1.65	30,000	22,000	30,000	...	4.2	2.5	2.5
G-5	Continuous filament glass fabric	Melamine	1/4–4	Grayish brown	1.80	30,000	22,000	30,000	...	4.5	3.0	3.0
N-1	Nylon fabric	Phenolic	1/4–4	Natural	1.15	10,000	20,000	6,000	0.8	0.6	0.5	0.5

2-62 Laminates, Reinforced Plastics, and Composite Structures

Grade FR-5. Glass-cloth-based sheet, rod, and tube with an epoxy resin binder. Similar to Grade G-11 but is self-extinguishing when ignition source is removed.

Grade GPO-1. Glass-mat-based sheet with a polyester resin binder. General-purpose mechanical and electrical grade.

Grade GPO-2. Glass-mat-based sheet with a polyester resin binder. Similar to Grade GPO-1 except has low flammability and is self-extinguishing.

Grade N-1. Nylon-cloth-based sheet. Excellent electrical properties at high humidity. Good impact strength but subject to creep and cold flow.

Properties of industrial laminates The properties of each grade standardized by NEMA are fully described in Ref. 32. Table 28 gives typical values for these sheets. These properties should not be used as standards but can be used as guides for the selection of materials. NEMA Standards also describe tubing and rods. Tubing is made by rolling impregnated sheet on a mandrel between heated pressure rollers and then either curing the tubes in an oven (rolled tubing) or pressing them in a mold (molded tubing). The properties of the finished tube depend to some extent on the curing method used. Table 29 gives properties of molded tubing, while Table 30 gives properties of rolled tubing. Table 31 shows the properties of laminated rods.

Polyester–glass-mat laminates Grades GPO-1 and GPO-2 have recently been standardized by NEMA and the Society of the Plastics Industry, Inc. These materials are made of random fiber glass mat and polyester resins and are used both mechani-

TABLE 32 Properties of NEMA Grades GPO-1 and GPO-2

Property	GPO-1	GPO-2
Flexural strength, psi $\times 10^{-3}$	18*	18*
Flexural modulus, psi $\times 10^{-6}$:		
Lengthwise	1.2	1.2
Crosswise	1.0	1.0
Tensile strength, psi $\times 10^{-3}$:		
Lengthwise	12	10
Crosswise	10	9
Tensile modulus, psi $\times 10^{-6}$:		
Lengthwise	1.0	1.0
Crosswise	0.9	0.9
Compressive strength, psi $\times 10^{-3}$:		
Flatwise	30	
Edgewise	20	
Shear strength, psi $\times 10^{-3}$	14	
Impact strength, ft-lb/in. notch	8*	8*
Bond strength (Condition A), lb	850*	850*
Water absorption, %, 1/8 in. thickness	0.7*	0.6*
Thermal expansion per °C	2.0×10^{-5}	
Flame resistance, sec:		
Ignition		75*
Burning		85*
Electric strength, volts/mil	400	400
Dielectric constant:		
At 1 MHz	4.3	
At 60 Hz	4.5	4.5
Dissipation factor:		
At 1 MHz	0.03	
At 60 Hz	0.05	0.05
Arc resistance, sec	100*	100*

* Values shown with an asterisk are guaranteed maximum or minimum values. Other properties are not to be considered as standards but rather as design guides. All values quoted are Condition A—as received.

cally and electrically, particularly in switchgear applications. Table 32 illustrates the properties of these two laminates.

Thermal aging of NEMA grade laminates Materials used in electrical devices often are exposed to elevated temperatures. Both electrical and physical properties of plastic materials degrade as a function of time and temperature. The IEEE has classified insulation materials in various classes related to temperature, as shown on Table 33. In designing electrical devices the effect of temperature on electric

TABLE 33 Definitions of Insulating Materials[37]

Class	Definition
90°C (Class O)	Materials or combinations of materials such as cotton, silk, and paper without impregnation. Other materials or combinations of materials may be included in this class if, by experience or accepted tests, they can be shown to be capable of operation at 90°C.
105°C (Class A)	Materials or combinations of materials such as cotton, silk, and paper when suitably impregnated or coated or when immersed in a dielectric liquid such as oil. Other materials or combinations of materials may be included in this class if, by experience or accepted tests, they can be shown to be capable of operation at 105°C.
130°C (Class B)	Materials or combinations of materials such as mica, glass fiber, and asbestos with suitable bonding substances. Other materials or combinations of materials, not necessarily inorganic, may be included in this class if, by experience or accepted tests, they can be shown to be capable of operation at 130°C.
155°C (Class F)	Materials or combinations of materials such as mica, glass fiber, and asbestos with suitable bonding substances. Other materials or combinations of materials, not necessarily inorganic, may be included in this class if, by experience or accepted tests, they can be shown to be capable of operation at 155°C.
180°C (Class H)	Materials or combinations of materials such as silicone elastomer, mica, glass fiber and asbestos with suitable bonding substances such as appropriate silicone resins. Other materials or combinations of materials may be included in this class if, by experience or accepted tests, they can be shown to be capable of operation at 180°C.
220°C	Materials or combinations of materials which, by experience or accepted tests, can be shown to be capable of operation at 220°C.
Over 220°C (Class C)	Insulation which consists entirely of mica, porcelain, glass, quartz, and similar inorganic materials. Other materials or combinations of materials may be included in this class if, by experience or accepted tests, they can be shown to be capable of operation at temperatures over 220°C.

strength and flexural strength are often of importance. The decrease in these properties after 1, 168, and 1,000 hr at a variety of temperatures is shown in Figs. 53 to 60 for several NEMA laminates.

Impact strength, insulation resistance, and power factor for NEMA grade laminates from −100 to 200°F are shown in Figs. 61 to 63. Other conditions influence the thermal properties of laminates. For example, storage of laminates at high relative humidity often decreases the physical properties. This is caused by the plasticizing effect of water on the polymers. Figures 64 to 66 illustrate the effect of high (90% RH) humidity exposure on the hot flexural properties and hardness of a typical G-10 laminate.

Tolerances of NEMA grade laminates NEMA grade laminates are available in a wide variety of sheet sizes from different manufacturers, depending on the molder's available presses. In most cases sheets can be obtained 48 × 96 in. The range of thickness available in each of the grades is shown in Table 34. Thickness tolerances for sheets are given in Table 35.

Vulcanized fiber Vulcanized fiber is made by soaking rag paper in a zinc chloride bath. Zinc chloride causes the cellulose to gelatinize, thereby making it formable.

2-64   Laminates, Reinforced Plastics, and Composite Structures

The term *vulcanize* is meant to describe this softening process and not the thermal vulcanization normally associated with rubber. The soft cellulose is then soaked in water to leach out the zinc chloride, and the sheet is dried and calendered, resulting in a hard, dense, regenerated cellulose.

Vulcanized fiber is made in many grades, the grade and properties thereof being determined by the quality of the rag paper and by the processing. Four grades

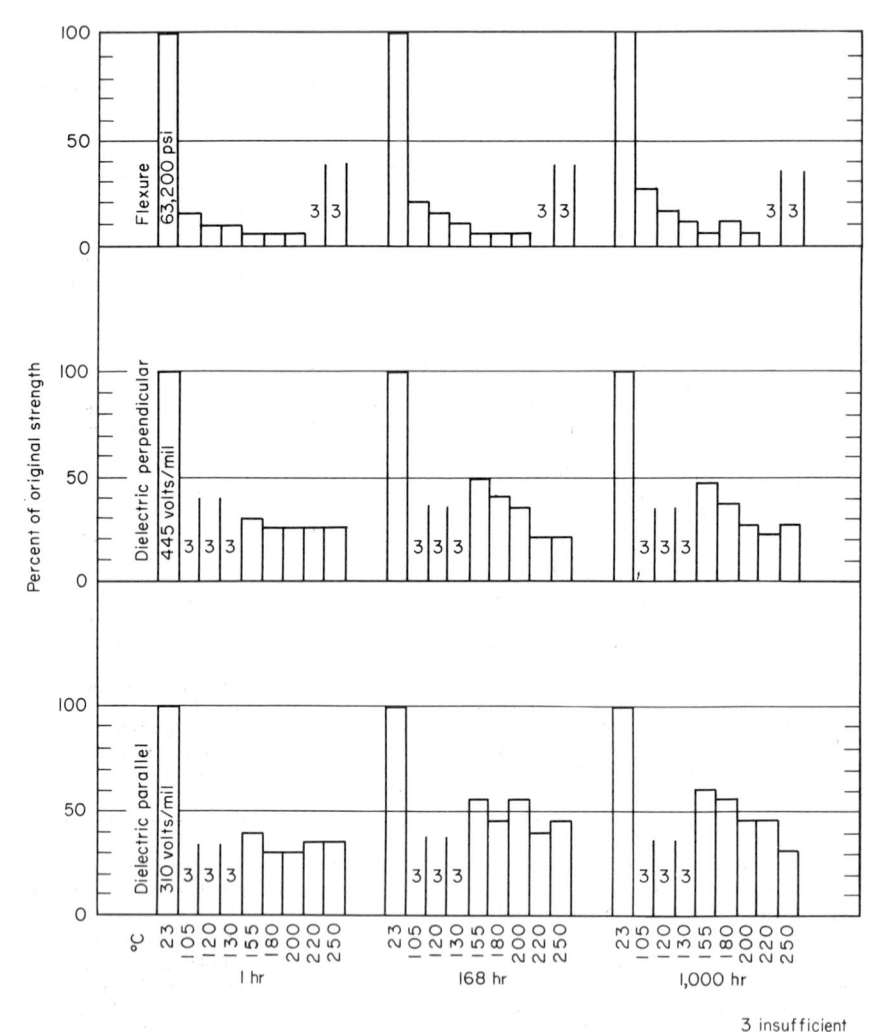

3 insufficient material

Fig. 53   Elevated-temperature properties of NEMA grade G-10.[38]

have been standardized by NEMA: commercial, bone, electrical insulation (fish paper), and white. These materials and applicable specifications are shown in Table 36. Other grades for nonelectrical uses can be obtained from suppliers. Available in rod, sheet, and tube form, vulcanized fiber is easily formed into shaped articles. Moisture plasticizes fiber and permits deep drawing. For example, if a sheet of fiber is soaked in hot water, it can be drawn in conventional shaping equipment in thickness up to 3/32 in. Wet fiber can be stretched up to 25 percent of the original dimension in tension and 50 percent in compression. Dry fiber can also be formed after being heated to 350°F.

Fiber is noted for being hard, and tough with good electrical strength. It is low in cost and possesses unusual arc resistance. When it is subjected to an arc, large quantities of carbon dioxide and water vapor are generated. These

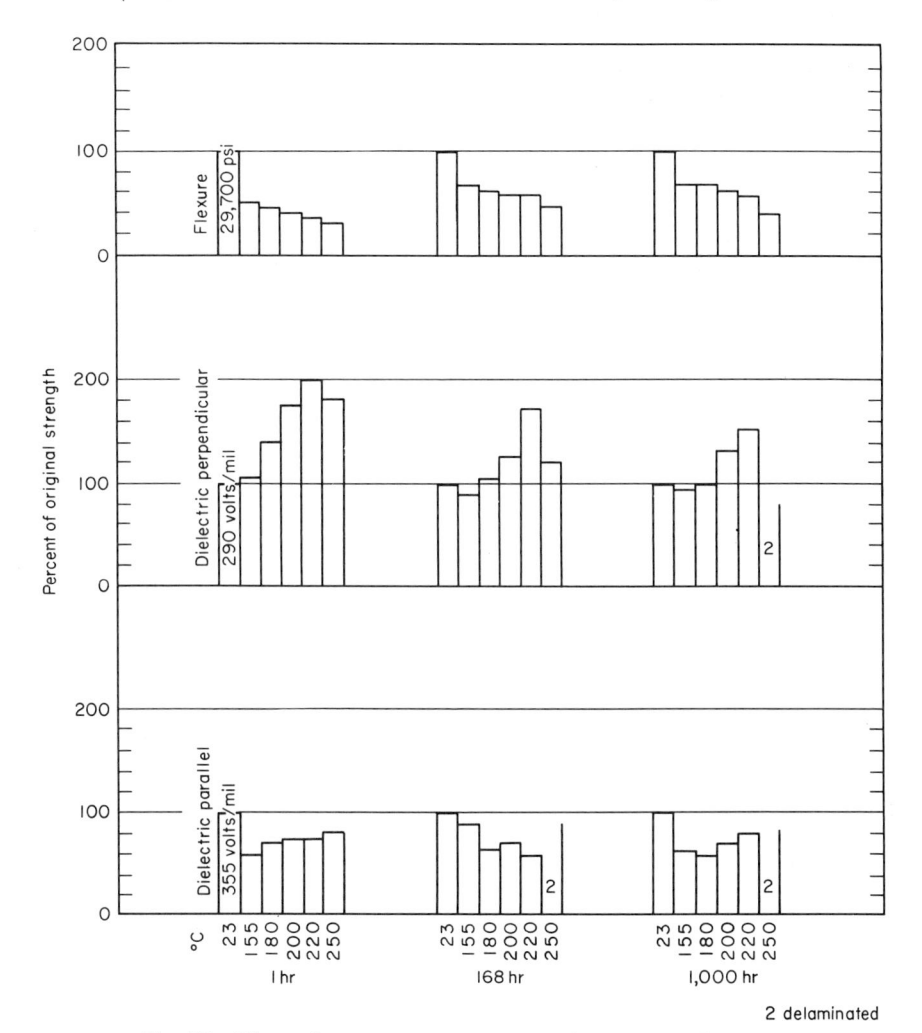

Fig. 54 Elevated-temperature properties of NEMA grade G-7.[38]

gases cool the arc, suppress combustion, and deionize the air. Hence the fiber does not form a conductive path.

Fiber is resistant to abrasion, gasoline, oil, and many solvents and machines similarly to wood. Table 37 presents some of the properties of vulcanized fiber. Testing standards for fiber are given in ASTM D 619-63.

Typical uses for fiber include spacers, slot liners, washers, relay parts, lightning arresters, fuses, and railroad track insulation. One disadvantage of fiber exists

2-66   Laminates, Reinforced Plastics, and Composite Structures

in applications where precise dimensional stability is required. In such cases the swelling of fiber accompanying water absorption would exclude its use. Figure 67 shows the absorption of water by vulcanized fiber after being exposed to various humidities.

Prepreg materials   Prepregs can be defined to be any resin treated on a reinforcing web and sold to the purchaser in an uncured form. The user of the prepreg can then mold his own laminated structure. The resin on a prepreg may be

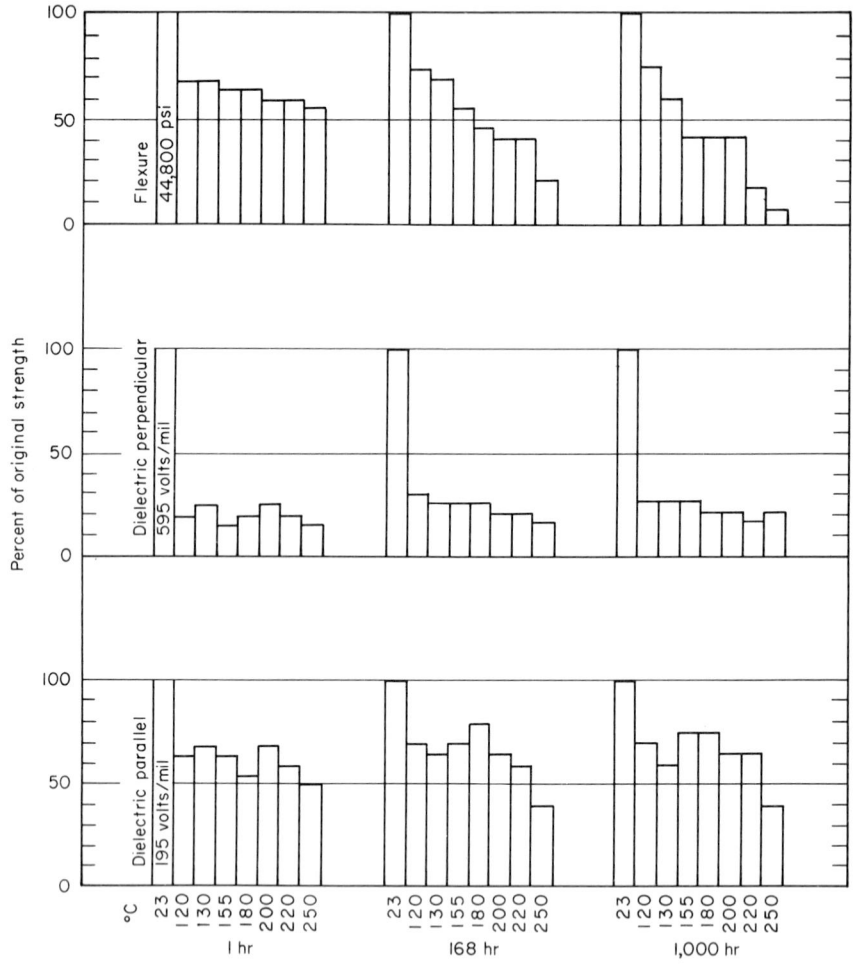

Fig. 55   Elevated-temperature properties of NEMA grade G-5.[38]

"B-staged," which means partly cured to a non- or semitacky state, or, in certain resin systems, the resin may normally be a solid at room temperature.

Prepregs have several advantages over wet lay-up molding.

1. For making small runs and prototypes this form of the molding is most convenient—particularly for smaller molders.

2. By pretreating the reinforcement the prepreg supplier removes much of the cost of resin treatment from the user. Better housekeeping results for the molder

by the elimination of the sticky, wet lay-up resins. The prepreg supplier can often be more efficient, and treating costs are lower as a result.

3. The molder is relieved of the necessity to maintain extensive inventories of many items.

4. A prepreg is often of higher quality than wet molding materials. Resin contents accurate to 2 percent are standard.

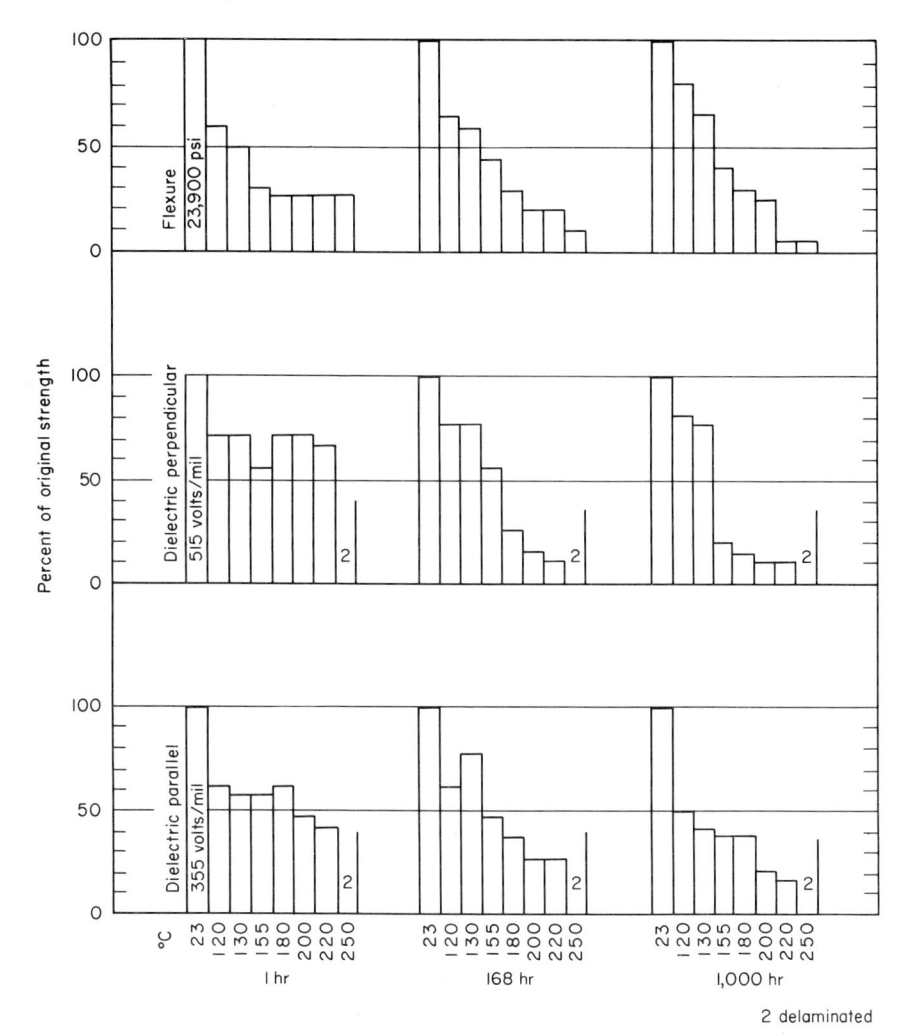

Fig. 56 Elevated-temperature properties of NEMA grade GPO-1.[38]

5. Prepreg supplies will supply die-cut material, thereby saving the molder labor expense.

6. Prepreg sometimes permits more efficient mass-production runs because of its uniformity and the ability to handle it with automated equipment.

7. Molding lay-up can often be made external from the press, thereby saving valuable press time. Various layers of prepreg can be tacked together with a hot iron to form complicated lay-ups.

2-68 Laminates, Reinforced Plastics, and Composite Structures

8. In the special case of filament-winding prepregs, better tension, pattern accuracy, and quality are possible.

These advantages must be weighed against the following disadvantages.

1. Room-temperature cures are impossible, which may limit the material and applications possible.

2. The prepregs are much more expensive than the cost of the raw materials.

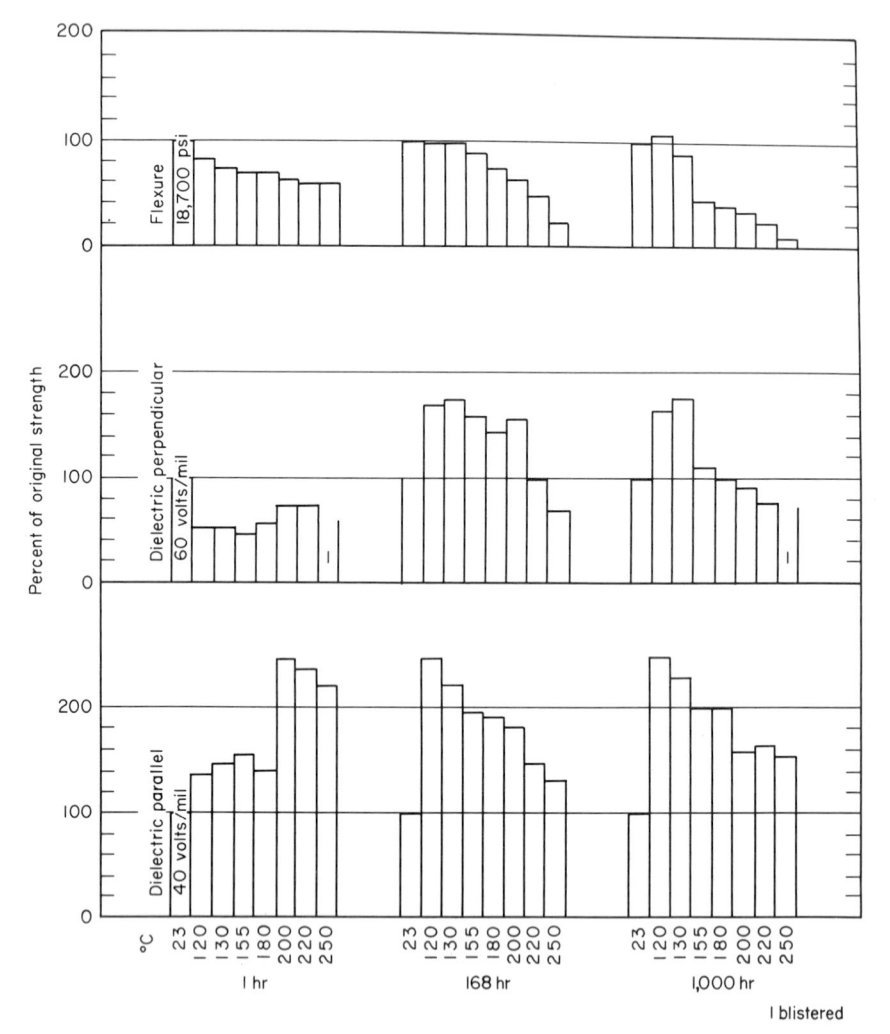

Fig. 57 Elevated-temperature properties of NEMA grade AA.[38]

Depending on the molding volume and other factors, the molder may find that economics dictate making his own resin treatment.

3. These prepregs are often most useful to vacuum-bag molders to whom the efficiencies of prepregs are not important because of the slow nature of this molding process.

4. Prepregs are best kept under refrigeration which necessitates expensive storage

equipment. Sometimes under adverse conditions or if the molder grows careless, the prepreg may spoil, resulting in poor-quality parts and extra costs.

Prepreg materials are available with most of the reinforcing webs described in the previous section dealing with reinforcements. Most common resins are phenolic, epoxy, diallyl phthalate, polyester, phenylsilane, silicone, and polyimide. Other resins can sometimes be obtained on special order. Over 30 suppliers of prepreg

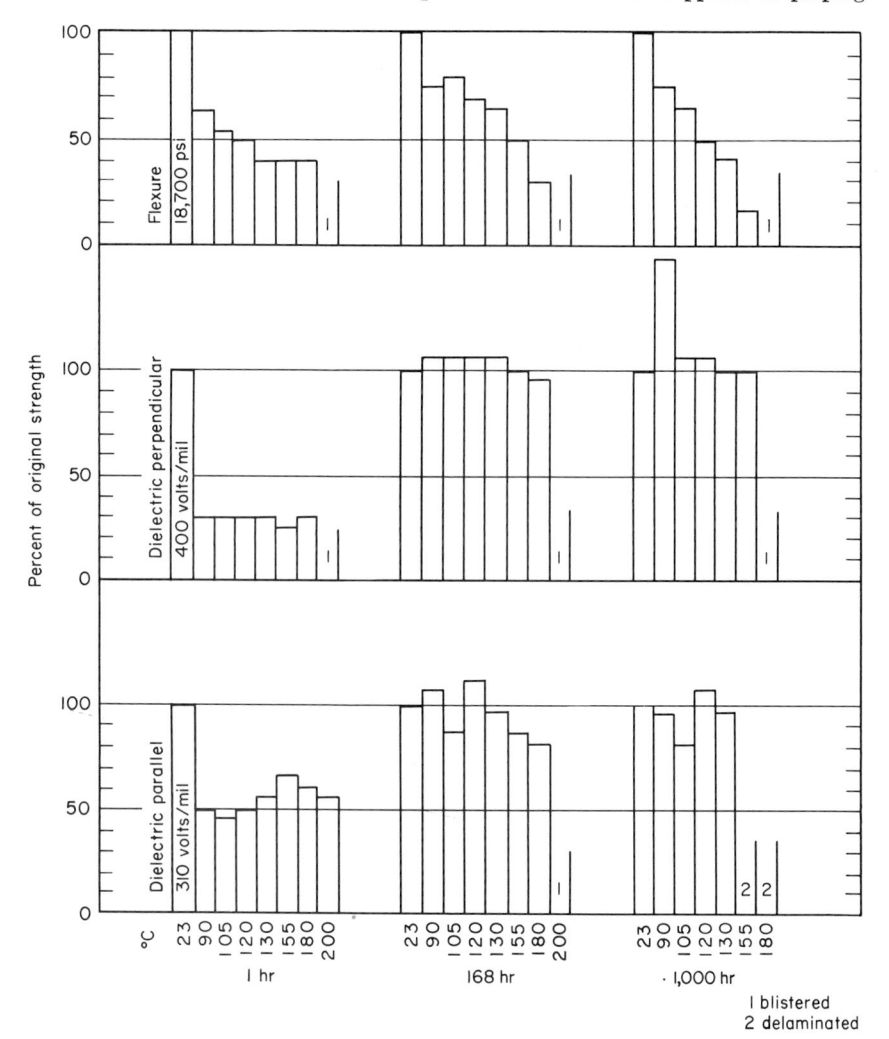

Fig. 58 Elevated-temperature properties of NEMA grade LE.[38]

exist, but fewer than 10 supply a full range of prepregs—the remainder supplying special materials. The prepregs are sold in 38-, 44-, 50-, 54-, and 60-in. widths on a wide range of reinforcing bases. Style 181 glass cloth is popular because of its formability. The prepregs consist of 30 to 55 weight percent resin with carefully controlled flow and gel time. All the prepregs have some volatiles, with the epoxies and silicones averaging less than 2 percent and the other materials less than 10 percent.

2-70 Laminates, Reinforced Plastics, and Composite Structures

Many of the prepregs are available in two forms—a vacuum-bagging grade and a press-curing grade. Typical vacuum-bagging cure cycles consist of molding for 240 min at 300 to 375°F. Press cycles average 90 min at 350°F and 200 to 1,000 psi. Moldings destined for high temperature must be postcured, for which a typical schedule is: preheat to 200°F for a fairly long time, and then gradually increase the temperatures to 400°F. The exact time and temperature cycle has

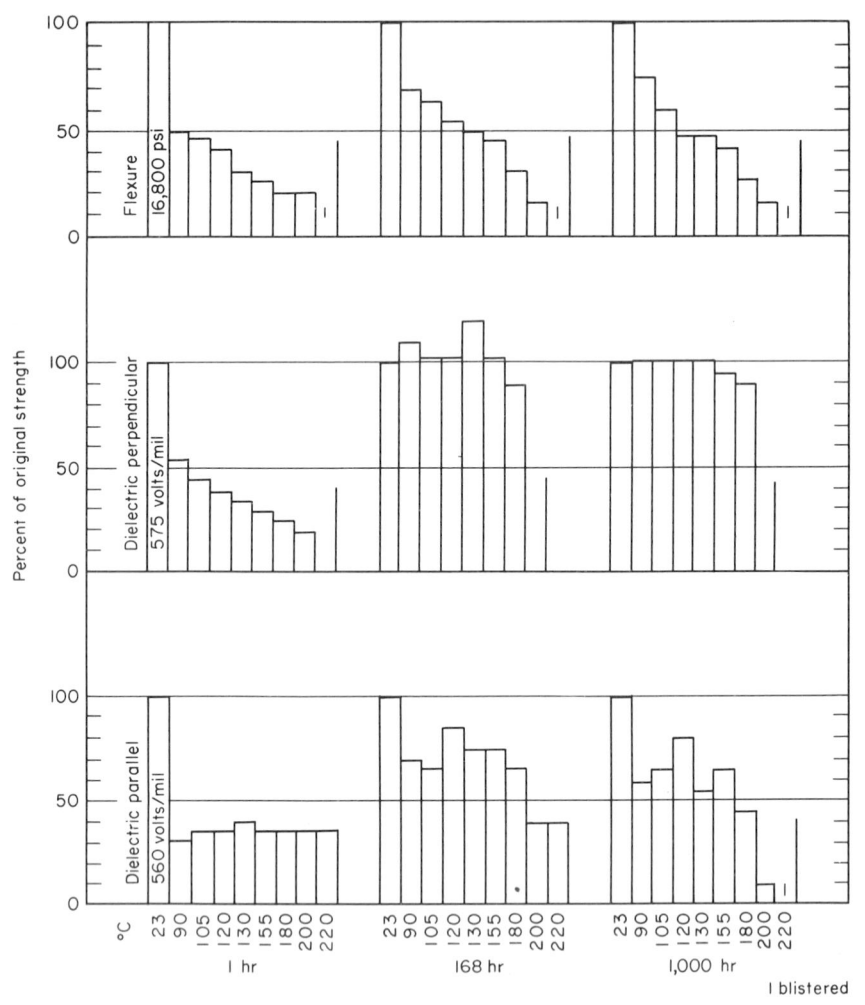

Fig. 59 Elevated-temperature properties of NEMA grade XXXP.[38]

been established by the manufacturer for each prepreg requiring postcure and is related to the nature of the resin and the thickness of the molding.

Prototypes Often during the design process it is desirable to produce parts to check dimensions or obtain more design information. Several methods of making such prototypes are available, and these methods—with others—are also used in production. Table 38 compares the features for each molding method to aid in process selection.

Laminates 2-71

Hand Lay-up. The simplest method of making prototypes is the hand lay-up or contact lay-up method. It is best used to make large parts and is frequently employed by boat manufacturers. Hand lay-up can be used on either male or female molds. The molds can be wood, metal, plastic, or plaster. When wood or plaster is used, the pores must be sealed with varnishes or lacquers. After the sealing compound is dry, the mold must be coated or covered with a release

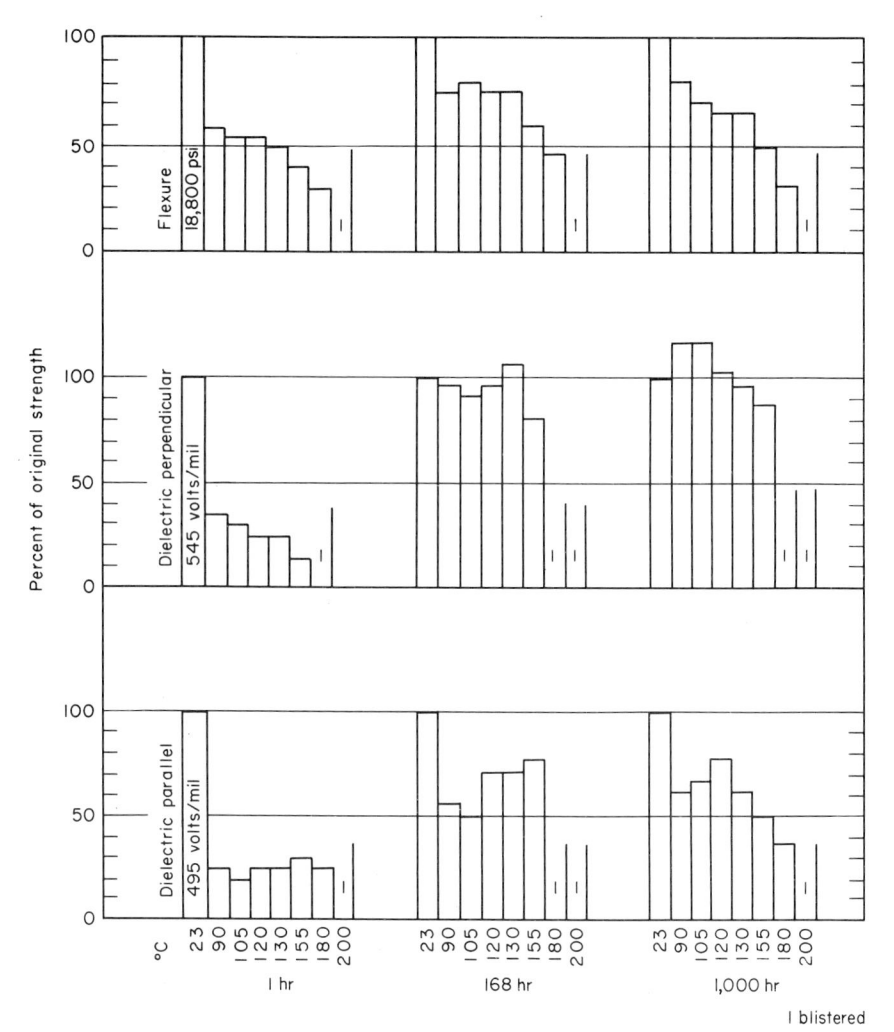

Fig. 60 Elevated-temperature properties of NEMA grade XX.[38]

agent such as floor wax or a film forming polymer like polyvinyl alcohol (PVA). At this point the reinforcing web (glass cloth, mat, or woven roving) is placed on the mold, and a low-viscosity resin is painted onto the reinforcement, the resin generally being a room-temperature-curing polyester or epoxy. The air from the reinforcement is worked out carefully with a paint roller or a rubber squeegee. Additional layers of reinforcement and resin are added until the correct thickness is obtained. The exposed topmost layer is sometimes covered with cellophane or Mylar* to make a smooth surface. Some polyesters are slightly inhibited by air, and thus film overlays help to complete the cure. When the resin has hardened, the part is removed and given a postbake. Although this process is cheap and easy to perform, the moldings are not of the highest quality and may not be completely representative of the properties expected. The laminates contain more resin than is obtained from other methods, and more voids may be present. Little control can be exercised over the uniformity of wall thickness.

Spray-up. A special case of hand lay-up is the rather new spray-up process. This is accomplished with a special type of spray gun. Molds similar to those

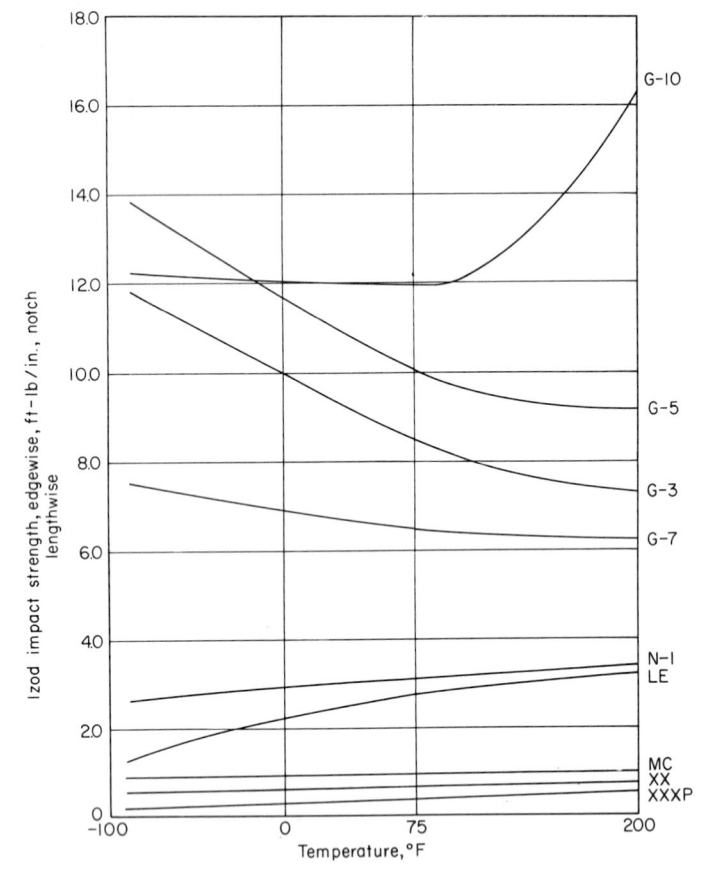

Fig. 61 Impact strength of NEMA grade laminates versus temperature.[39]

used in hand lay-up are used. The gun chops fiber glass into predetermined lengths, mixes them with resin, and deposits them onto the mold surface. The mold surface is often pretreated with a mineral-filled resin which has been permitted to partially cure. This surface is called a *gel coat*. When enough glass has been deposited, an impregnated glass mat has been formed. From this point on, the process is identical to the hand lay-up process. The process is attractive because it makes use of fiber glass in its least expensive form. Parts with complicated shapes can

---

* Trademarks of E. I. du Pont de Nemours & Co., Wilmington, Del.

be made, and size is not critical. Skilled operators are required if control over wall thickness is required.

Vacuum-bag Molding. Vacuum-bag molding is one of the most popular methods of making prototype or short-run parts. Molds similar to those just discussed are used. Either wet resins or prepregs can be employed. Figure 68 shows the molding process. Over the mold surface is placed a release sheet such as PVA or release-treated Mylar. The reinforcement is placed onto the release film and impregnated

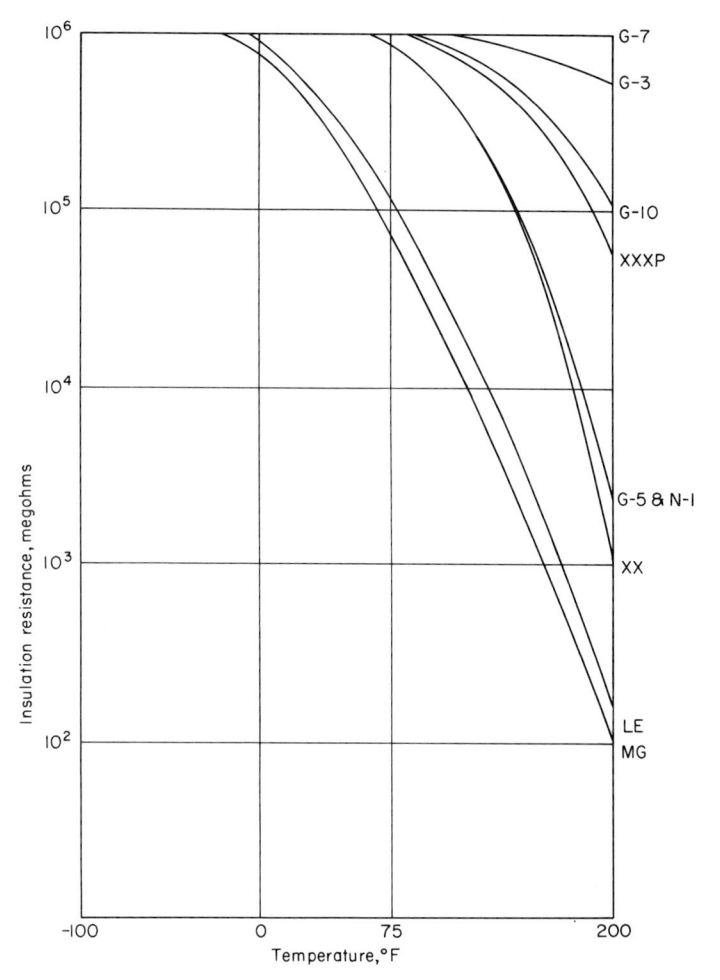

Fig. 62 Insulation resistance of NEMA grade laminates versus temperature.[39]

by brushing or spraying if prepregs are not being used. Next bleeders are installed along the top of the mold. These are made of burlap, muslin, osnaburg cloth, or felt, and are intended to provide a space from which the air in the mold can be evacuated. The impregnated reinforcement is sometimes covered with a perforated release film: PVA, Teflon, or treated Mylar. Special release fabrics are also used. Over this is often placed dry cloth to assist in air evacuation. In other cases the bag can contact the lay-up directly, depending on the shape and size of the part. Finally the vacuum bag is clamped into place.

2-74   Laminates, Reinforced Plastics, and Composite Structures

Vacuum bags are made from many materials. Cellophane is cheap and easy to install, its chief limitations being tear resistance and temperature. Polyvinyl alcohol (PVA) can be used up to about 150°F, and it is tougher and more tear-resistant than cellophane. PVA is highly sensitive to humidity, being water-soluble. Polyethylene has been used where temperatures permit (room temperatures only). Polyvinyl chloride (PVC) bags are useful up to 150°F and are less expensive than PVA. Certain PVC polymers inhibit polyester resins and prevent a tack-free

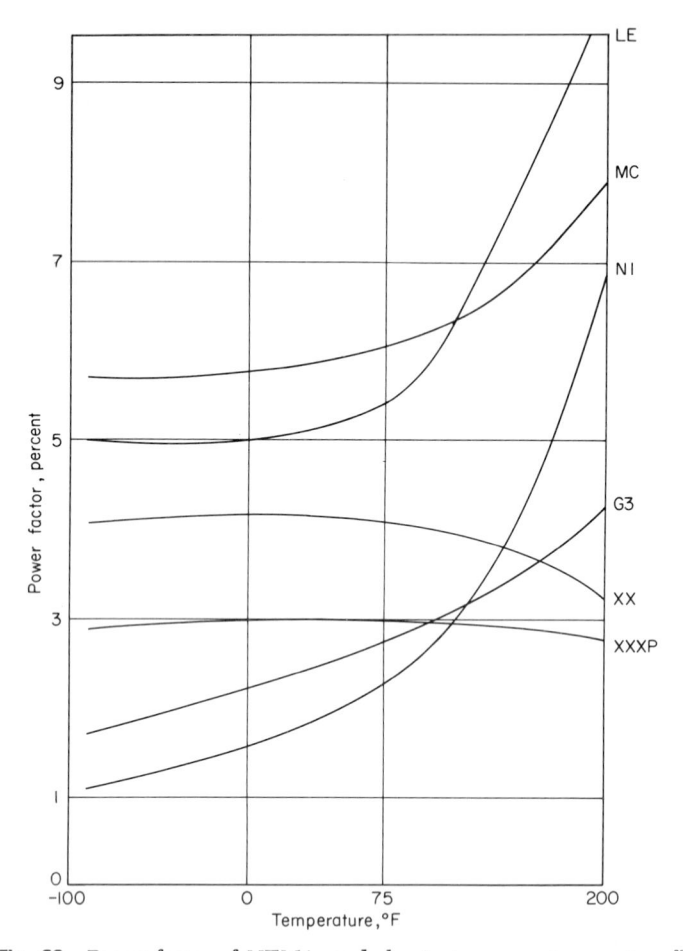

Fig. 63   Power factor of NEMA grade laminates versus temperature.[39]

cure. When high curing temperatures or steam autoclave curing are required, the bag should be made of neoprene rubber or some similar elastomer. Neoprene and some other rubbers also inhibit polyester resins, and if polyesters are to be cured, the bag should have a film such as Mylar or Kapton* between the lay-up and the bag.

The edge of the bag must be well sealed to prevent loss of vacuum and introduction of air into the lay-up. A common method is to use zinc chromate paste which is supplied in ribbons or beads. The paste is pressed into intimate contact

* Trademark of E. I. du Pont de Nemours & Co., Wilmington, Del.

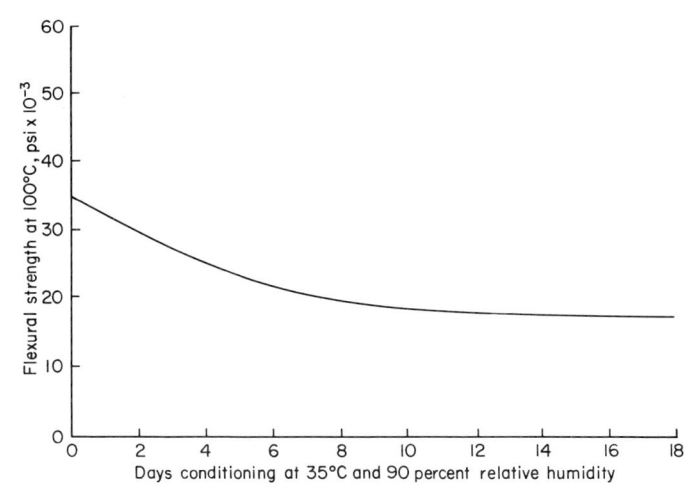

Fig. 64 Effect of high-humidity storage on elevated-temperature flexural strength for G-10 laminate.

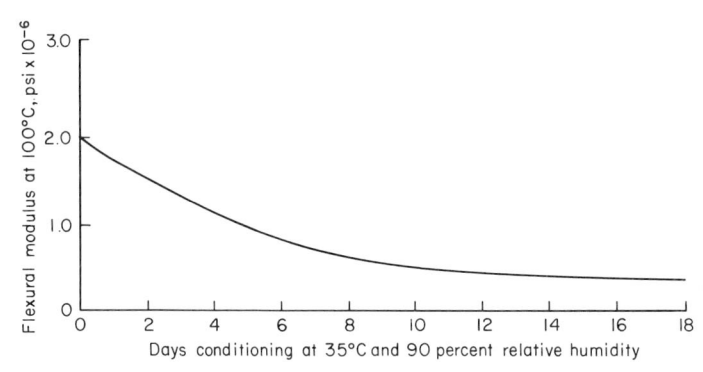

Fig. 65 Effect of high-humidity storage on elevated-temperature flexural modulus for G-10 laminate.

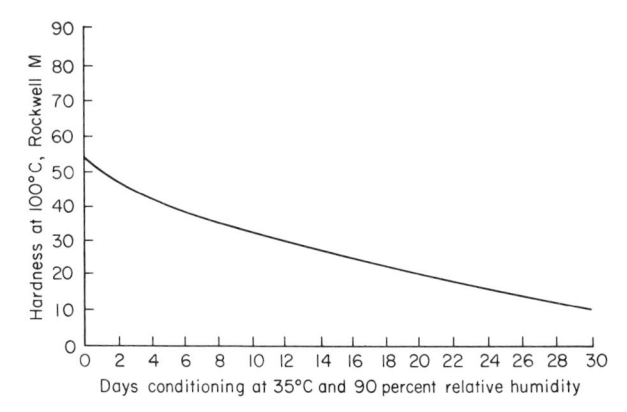

Fig. 66 Effect of high-humidity storage on elevated-temperature hardness for G-10 laminate.

TABLE 34 Thickness of NEMA Grade Laminates

Grade No.	Minimum	Maximum
X	0.010	2
XP	0.010	¼
XPC	¹⁄₃₂	¼
XX	0.010	2
XXP	0.015	¼
XXX	0.015	2
XXXP	0.015	¼
XXXPC	¹⁄₃₂	¼
C	¹⁄₃₂	10
CE	¹⁄₃₂	2
L	0.010	2
LE	0.015	2
A	0.025	2
AA	¹⁄₁₆	2
G-2	¹⁄₃₂	2
G-3, G-7	0.010	2
G-5, G-9	0.010	3½
G-10, G-11, N-1	0.010	1
FR-2, FR-3	¹⁄₃₂	¼
FR-4, FR-5	0.010	1

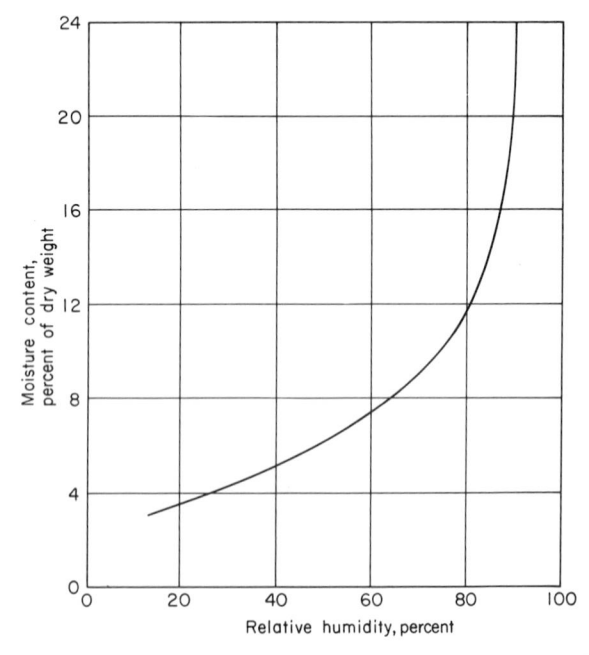

Fig. 67 Equilibrium moisture content of vulcanized fiber.[40]

Laminates 2-77

with the mold, and the bag is pressed into the paste. This material holds a vacuum well and is easily removed from the assembly after molding. Other rubber gaskets can be used along with hold-down or clamp rings, either to replace or to supplement the paste. Getting a good seal is critical to the success of the vacuum-bag process, and care exercised at this step will avoid much difficulty. Vacuum is then applied, bringing air pressure to bear on all surfaces. The use of a resin trap in the vacuum line as shown in Fig. 69 is essential. The trap plays the same role that an oil separator plays in an air line. As the resin softens and flows during the molding, it will enter the vacuum line. If not removed, it will decrease the vacuum, clog and ruin the lines, and, if permitted to enter the pump, will cause pump failure. Mechanical working of the bag with a roller or squeegee aids in air removal when wet resins are used.

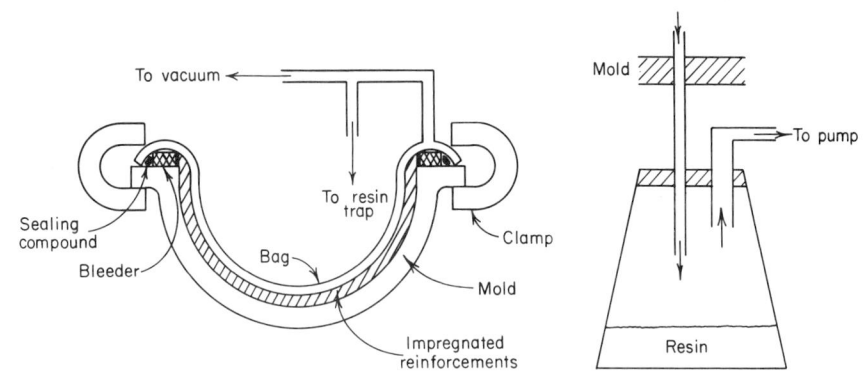

Fig. 68 Vacuum-bag molding. **Fig. 69** Resin trap.

If prepregs are used, the whole assembly is placed in an oven for cure. If prepregs requiring temperatures greater than 300°F are employed, all release sheets should be made of Mylar, Teflon, or nylon. A typical cure schedule consists of increasing the temperature to 250°F in 30 min, to 300°F in 15 min, and to 350°F in 20 min. When the part is cured, temperatures should be reduced to about 150°F before the vacuum is released.

Advantages of the vacuum-bag process are:
1. The molds are relatively inexpensive.
2. The process is faster than hand lay-up.
3. Prepregs can be used.
4. The product is not as rich in resin as in hand lay-up, and properties are improved.
5. The process does not require highly skilled personnel.

Disadvantages are:
1. Only the surface adjacent to the mold surface is smooth. The other surface may be duller and show folds and marks from the bag.
2. The part still contains more resin than those made by high-pressure techniques, and hence properties are not optimized.
3. Certain shapes with complicated surfaces or sharp corners are difficult to form.
4. Dry spots and voids are common.

Pressure-bag Molding. A vacuum-bag molding process augmented with an externally supplied pressure is known as *pressure-bag molding*. A pressure plate is added to the top of the mold, as shown in Fig. 70. The bleeder strips may need to be repositioned to accommodate the top plate. The molding is the same as in the previous case except that pressures up to 50 psi can be introduced into the mold. Care must be taken to ensure that the mold is capable of withstand-

TABLE 35 Thicknesses and Thickness Tolerances of NEMA Grade Laminates (In inches)

Nominal thickness	X, XP, XPC, XX, XXP, XXX, XXXP, XXXPC, FR-2, FR-3 Plus or minus	ES-1, ES-2, ES-3, CE, A Plus or minus	C Plus or minus	L Plus or minus	LE Plus or minus	AA Plus or minus	G-2 Plus or minus	G-3, G-5, G-7, G-9, G-10, G-11, FR-4, FR-5 Plus or minus	N-1 Plus or minus
0.010	0.002	.....	.....	0.003	.....	.....	.....	0.002	0.003
0.015	0.0025	.....	.....	0.0035	0.0035	.....	.....	0.003	0.0035
0.020	0.003	.....	.....	0.004	0.004	.....	.....	0.004	0.004
0.025	0.0035	0.005	.....	0.0045	0.0045	.....	.....	0.005	0.0045
1/32	0.0035	0.0065	0.0065	0.005	0.005	.....	0.008	0.0065	0.0065
3/64	0.0045	0.0075	0.0075	0.0055	0.0055	.....	0.010	0.0075	0.0075
1/16	0.005	0.0075	0.0075	0.006	0.006	0.018	0.010	0.0075	0.0075
3/32	0.007	0.009	0.009	0.007	0.007	0.018	0.012	0.009	0.009
1/8	0.008	0.010	0.010	0.008	0.008	0.020	0.012	0.012	0.010
5/32	0.009	0.011	0.011	0.009	0.009	.....	0.015	0.015	0.011
3/16	0.010	0.0125	0.0125	0.010	0.010	0.024	0.019	0.019	0.0125
7/32	0.011	0.014	0.014	0.011	0.011	.....	0.021	0.021	0.014
			Plus only	Plus only					
1/4	0.012	0.015	0.030	0.024	0.012	0.028	0.022	0.022	0.015
5/16	0.0145	0.0175	0.035	0.029	0.0145	0.034	0.026	0.026	0.024
3/8	0.017	0.020	0.040	0.034	0.017	0.038	0.030	0.030	0.032
7/16	0.019	0.022	0.044	0.038	0.019	0.044	0.033	0.033	0.040
1/2	0.021	0.024	0.048	0.042	0.021	0.048	0.036	0.036	0.048
5/8	0.024	0.027	0.053	0.048	0.024	0.058	0.040	0.040	0.054
3/4	0.027	0.029	0.058	0.054	0.027	0.068	0.043	0.043	0.058
7/8	0.030	0.031	0.062	0.060	0.030	0.076	0.046	0.046	0.062
1	0.033	0.033	0.065	0.065	0.033	0.086	0.049	0.049	0.066

Nominal Thickness							G-5 and G-9 only
1⅛	0.035	0.035	0.069	0.035	....	0.053	0.053
1¼	0.037	0.037	0.073	0.037	....	0.055	0.055
1⅜	0.039	0.039	0.077	0.039	0.106	0.058	0.058
1½	0.041	0.041	0.081	0.041	0.124	0.061	0.061
1⅝	0.043	0.043	0.085	0.043	....	0.064	0.064
1¾	0.045	0.045	0.089	0.045	0.144	0.067	0.067
1⅞	0.047	0.047	0.093	0.047	....	0.070	0.070
2	0.049	0.049	0.097	0.049	0.160	0.073	0.073
2¼	....	....	0.105	....	....	....	0.079
2½	....	....	0.113	....	....	....	0.085
2¾	....	....	0.121	....	....	....	0.090
3	....	....	0.130	....	....	....	0.097
3½	....	....	0.146	....	....	....	0.110
4	....	....	0.163	....	....	....	
4½	....	....	0.179	....	....	....	
5	....	....	0.190	....	....	....	
5½	....	....	0.210	....	....	....	
6	....	....	0.230	....	....	....	
6½	....	....	0.240	....	....	....	
7	....	....	0.260	....	....	....	
7½	....	....	0.280	....	....	....	
8	....	....	0.290	....	....	....	
8½	....	....	0.310	....	....	....	
9	....	....	0.320	....	....	....	
9½	....	....	0.340	....	....	....	
10	....	....	0.360	....	....	....	

NOTES:
1. For a sheet having a nominal thickness not listed in the table, the tolerances shall be the same as for the next larger nominal thickness.
2. The minimum and maximum thickness for each grade shall be in accordance with LI 1-4.05.

TABLE 36 Specifications for Vulcanized Fiber

NEMA grade	Description	Applicable specifications	Sheets Color	Sheets Thickness range, in.	Max. width, in.	Approx. width, full rolls, in.
Commercial	An electrical and mechanical grade having good bending, punching, stamping, and forming characteristics	ASTM: D-170 commercial NEMA: VU 1-1963 commercial MILITARY: MIL-F-1148 type CH MILITARY: MIL-F-13526-Red ASA: C-59.20-1949 commercial FEDERAL: HH-P-91 type CH AMS: 3564 grade CH	Gray Red Black	0.004–⅛ Also ⅛–3 built-up layers of fiber ranging from 0.030 to 0.090 plies bonded by adhesive	56 48	56
Bone	An electrical and mechanical grade tougher and denser than commercial grade, therefore providing finer machined surfaces	ASTM: D-710 bone NEMA: VU 1-1963 bone MILITARY: MIL-F-1148 type BH FEDERAL: HH-P-91 type BH RAILROAD: AAR-13-58 type bone	Gray	¹⁄₃₂–⅛ Also ⅛–3 built-up layers of fiber ranging from 0.030 to 0.090 plies bonded by adhesive	52 48	52
Electrical insulation	Primarily intended for electrical applications and others involving difficult bending or forming operations. Thin sections are often referred to as fish paper	ASTM: D-710 electrical insulation NEMA: VU 1-1963 electrical insulation MILITARY: MIL-I-695A type F ASA: C-59.20-1949 electrical insulation	Gray	0.004–⅛	56	56
White	A grade having good finish, a minimum of impurities, and a good surface for printing	NEMA: VU 1-1963 white	White	0.010–⅛	54	54

2-80

TABLE 37 Properties of Vulcanized Fiber

Property	Value
Dielectric strength, volts/mil:	
Short time	150–300
Step by step	125–225
Dielectric constant at 60 Hz	4.7
Dissipation factor at 60 Hz	0.03–0.08
Arc resistance, sec	100
Tensile strength, psi $\times 10^{-3}$:	
Crosswise	6.5–8
Lengthwise	13–18
Tensile modulus, psi $\times 10^{-6}$:	
Crosswise	0.8
Lengthwise	1.2
Flexural strength, psi $\times 10^{-3}$:	
Crosswise	12–14
Lengthwise	14–18
Flexural modulus, psi $\times 10^{-6}$:	
Crosswise	0.7
Lengthwise	1.0
Izod impact strength, ft-lb/in. notch:	
Crosswise	1.1–1.2
Lengthwise	1.4–1.6
Water absorption, % in 2 hr	30–52
Specific gravity	0.9–1.55
Hardness, Rockwell R	50–80
Bond strength, lb	1,000
Thermal conductivity, Btu/(hr)(ft^2)(°F/in.)	0.168
Thermal expansion per °F $\times 10^{-5}$:	
Crosswise	1.1
Lengthwise	1.7
Specific heat, Btu/(lb)(°F)	0.4
Burning rate	Slow

ing this extra load. The advantage obtained by pressure-bag molding is the extra consolidation of the lay-up and the decrease in resin content with attendant increase in properties.

Autoclave Molding. Autoclave molding is similar to the preceeding methods except the pressure plate is removed and the entire mold and the lay-up is placed into an autoclave. Steam or hot pressurized air up to 100 psi is admitted supplying both pressure and heat. If steam is used, the covering bag must have no holes

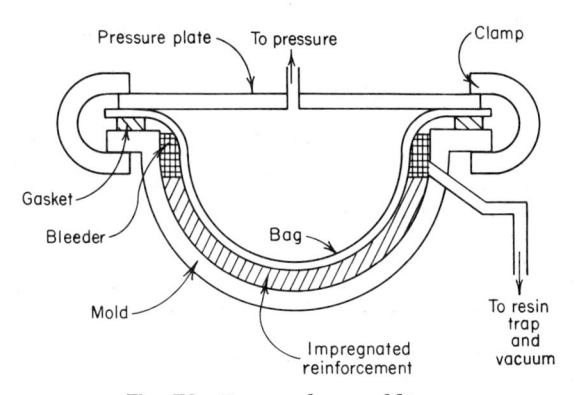

Fig. 70 Pressure-bag molding.

Laminates, Reinforced Plastics, and Composite Structures

TABLE 38 Molding Methods for Prototypes

Feature	Hand lay-up	Spray-up	Vacuum-bag molding	Pressure-bag molding	Auto-clave molding	Vacuum-injection molding	Matched metal die
Mold cost	Low	Low	Medium	Medium	Medium	High	Very high
Facility cost	Low	Medium	Low	Low	High	Low	Very high
Production rate	Low	High	Low	Low	Low	Medium	Very high
Physical properties	Very low	Medium	Low	Low	Medium	Low	Very high
Quality of molding	Very low	Medium	Low	Medium	Medium	Medium	High
Surface appearance	Poor	Good	Poor (1 side)	Poor (1 side)	Poor (1 side)	Good	Excellent
Resin content	High	Medium	High	High	Medium	Very high	Low
Finishing required	Much	Little	Much	Much	Much	Little	Very little
Operator skill required	Low	High	Medium	Medium	Medium	Medium	Low

or leaks because steam will enter the lay-up and ruin the molding. The chief advantages of this method are faster cycles and increased properties.

Vacuum-injection Molding. This method is very useful in making prototype moldings because of its simplicity and low cost. Figure 71 shows the process. Both male and female molds are required, although the female mold may be a bag. The mold surfaces must be very smooth and very well coated with release agents. Dry reinforcements (mat or cloth) are arranged over the male mold to the thickness desired. The female section is carefully—to avoid moving the reinforcements—

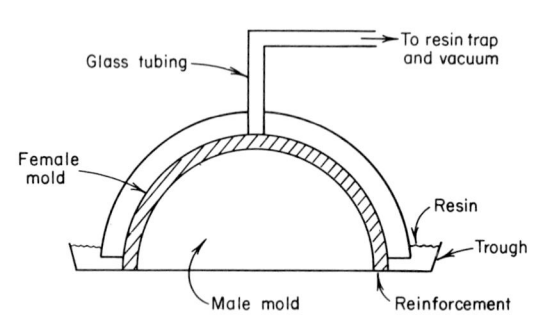

Fig. 71 Vacuum-injection molding process.

lowered into place over the male. Catalyzed room-temperature-curing resin (epoxy or polyester) is poured into the trough surrounding the bottom of the male mold, and vacuum is applied. As the air is pumped from the molds, the resin will rise through and saturate the reinforcements. When the resin is seen to reach the glass tubing (which must be at the highest place in the mold) and become bubble-free, the vacuum line is clamped and the pump turned off. After cure the part is usually stuck to the male mold and may be removed by blowing compressed air between the mold and the part. The advantages for the vacuum-injection process are:
1. Both surfaces are glossy and good in appearance.
2. Little finishing is required.
3. The process is neat and clean.

Disadvantages include:
1. The parts are high in resin content.
2. It is more costly to manufacture two molds.
3. Dry spots can occur and will not be detected until the part is finished.
4. Poor selection and application of mold release may cause sticking.

Matched Metal Die Molding. This method is not usually a candidate for prototype manufacture. Instead it represents the ultimate in molding speed and part quality. The cost of the molded part is directly related to the number of parts desired since expensive molds must be amortized and high-cost molding facilities are used. Molds are generally made of high-quality alloy steels and are machined to exacting tolerances. Aluminum-, cast-iron-, and beryllium-based alloys are sometimes used for special purposes. The molds usually contain hardened cutoff edges which automatically trim the parts, eliminating costly finishing operations. The mold surface is polished to a mirror finish and is usually chrome-plated to aid in part ejection.

Filament winding Filament winding refers to the process of wrapping resin-impregnated continuous fibers around a mandrel forming a surface of revolution. When enough layers have been wound, the part is cured and the mandrel is removed. The continuous filaments are generally tapes or rovings of fiber glass, but other reinforcements are also used. Epoxies are popular impregnating resins although others are used. Typical properties of glass-epoxy filament-wound parts are shown in Table 39. Of special notice are the high tensile and modulus values,

TABLE 39 Properties of Filament-wound Products

Property	Value
Specific gravity	2.0
Density, psi	0.072
Thermal conductivity, Btu/(hr)(ft^2)(°F/in.)	2.2
Thermal expansion, per °F $\times 10^{-6}$	7.0
Specific heat, Btu/(lb)(°F)	0.227
Maximum use temperature, °F	<400
Hoop tension, psi $\times 10^{-3}$:	
Unidirectional windings	230
Helical windings	135
Compressive strength, psi $\times 10^{-3}$	70
Flexural strength, psi $\times 10^{-3}$	100
Bearing strength, psi $\times 10^{-3}$	35
Shear strength, psi $\times 10^{-3}$:	
Interlaminar	6
Cross	18
Modulus of elasticity, tension, psi $\times 10^{-6}$	6.0
Modulus of rigidity, torsion, psi $\times 10^{-6}$	2
Dielectric strength, step-by-step, volts/mil	400

Filament-wound parts contain a high percentage of reinforcing fiber and approach the properties of the fiber. The role of the resin is to bond adjacent fibers together and to transmit stress from fiber to fiber. The resin content is determined by the amount of resin on the glass and the winding tension. Table 40 shows the effect of tension on the properties of some wound parts, and Table 41 compares filament-wound pressure vessels with metal ones.

In designing filament-wound parts one attempts to orient the fibers in the direction of stress. If, because of the geometry of the pair, it is impossible to so align the fibers, then a balanced structure is used wherein all fibers regardless of direction have equal stress. Careful design consideration must be given to holes and flanges to avoid interrupting the continuous nature of the fibers.

2-84 Laminates, Reinforced Plastics, and Composite Structures

The resin used can be applied at the time of the winding process, or prepreg rovings can be used. Such prepreg rovings are available from D-glass, E-glass, and S-glass in 12, 20, 30, and 60 ends. The resin constitutes 17 to 23 percent by weight of the virgin roving. In most cases the materials are shipped frozen and must be kept refrigerated. Resins available include epoxy, phenolic, diallyl phthalate, and polyester.

TABLE 40 Hoop Strength Data—Biaxial Test Specimen[41]

Spec. No.	Tension at mandrel, lb/end	Bandwidth per 30 ends, in.	Density, lb/in.³	Glass content by wt. %	Voids volume, %	Burst pressure, psi	Ultimate hoop stress, psi	Ultimate stress per single glass end, psi × 10⁻⁵	Strength-weight ratio, in. × 10⁻⁶
1A	0.12	0.180	0.0694	73.87	3.25	4,100	126,000	3.09	1.82
1B	0.12	0.180	0.0694	73.87	3.25	4,600	143,000	3.48	2.06
1C	0.12	0.180	0.0694	73.87	3.25	3,800			
Avg.						4,350	134,000	3.28	1.92
2A	0.20	0.180	0.0740	79.70	1.92	3,900			
2B	0.20	0.180	0.0740	79.70	1.92	5,200	163,000	3.80	2.21
2C	0.20	0.180	0.0740	79.70	1.92	5,150	161,600	3.76	2.18
Avg.						5,175	162,250	3.78	2.19
6A	0.5	0.180	0.0756	80.752	3.31	4,600	155,000	3.24	2.05
6B	0.5	0.180	0.0756	80.752	3.31	4,300	145,000	3.05	1.92
6C	0.5	0.180	0.0756	80.752	3.31	4,200	142,000	2.96	1.88
Avg.						4,360	147,000	3.08	1.95

Filament-wound parts are used in the electrical industry as fuse tubes, lightning arrester tubes, fuses, bushings, and bands on motors and generators.

Space-age laminates The need for lighter-weight, higher-temperature resistance and greater stiffness has compelled the aerospace industry to develop new exotic fibrous reinforcements. These reinforcements take two forms: continuous and discrete whiskers. Table 42 lists new filament reinforcements, and Table 43 shows whiskers. The continuous materials are often used in filament winding, but some tapes and fabrics have been made. As might be expected, the cost of these materials is very high, thus limiting their use to applications requiring their unique properties.

Boron Fiber Laminates. Boron filaments are made by depositing elemental boron

TABLE 41 Properties of Internal-pressure Vessel Materials[42]

Material	Ultimate hoop (tensile) strength, psi × 10⁻³	Density, lb/in.³	Specific strength,* in. × 10⁻⁶	Thermal conductivity†	Tensile modulus of elasticity, in. × 10⁻⁶	Compressive strength, psi × 10⁻³
Glass-resin‡ (unidirectional)	130–170	0.077	1.6–2.1	2.0–5.0	6.0–9.0	70–175
Glass-resin (bidirectional)	60	0.072	0.7	2.0–5.0	2.0–3.5	40–60
Steel wire–resin (bidirectional)	150	0.166	1.0		12	
Titanium (homogeneous)	50	0.163	0.9		16.5	135
Steel (homogeneous)	280	0.280	0.9	314	30.0	
Aluminum (homogeneous)	80	0.097	0.8	1,416	10.0	40
Magnesium (homogeneous)	32	0.064	0.5			

* Specific strength equals ultimate tensile strength/density (approx.).
† Btu/(hr)(ft²)(°F)/in.
‡ Interlaminar shear strength = 6,000 to 8,000 psi (parallel to laminations); 20,000 psi (perpendicular to laminations).
    Axial bearing strength = 20,000 to 40,000 psi
    Compression strength = 50,000 to 70,000 psi

TABLE 42 Properties of Fibrous and Wire Reinforcements[43]

Fiber/wire	Density, lb/in.^{3}	Specific gravity	Melting point, °F	Tensile strength Ultimate, psi × 10^{-3}	Ratio to density, × 10^{-6}	Modulus of elasticity psi × 10^{-6}	Ratio to density, × 10^{-6}
Aluminum	0.097	2.70	1220	90	0.9	10.6	110
Aluminum oxide	0.144	3.97	3780	100	0.7	76	530
Aluminum silica	0.140	3.90	3300	600	4.3	15	110
Asbestos	0.090	2.50	2770	200	2.2	25	280
Beryllium	0.067	1.84	2343	250	3.7	44	660
Beryllium carbide	0.088	2.44	3800	150	1.7	45	510
Beryllium oxide	0.109	3.03	4650	75	0.7	51	470
Boron	0.093	2.59	3812	500	5.4	60	650
Carbon	0.051	1.40	6700	250	4.9	27	530
Glass:							
E-glass	0.092	2.55	2400	500	5.4	10.5	110
S-glass	0.090	2.49	3000	700	7.8	12.5	140
Graphite	0.051	1.40	6600	250	4.9	37	720
Molybdenum	0.367	10.20	4730	200	0.5	52	140
Polyamide	0.041	1.14	480	120	2.9	0.4	10
Polyester	0.050	1.40	480	100	2.0	0.6	10
Quartz (fused silica)	0.079	2.20	3500	1000	12.7	10	130
Steel	0.282	7.87	2920	600	2.1	30	110
Tantalum	0.598	16.60	5425	90	0.2	28	50
Titanium	0.170	4.72	3035	280	1.6	16.7	100
Tungsten	0.695	19.30	6170	620	0.9	58	80
Tungsten monocarbide	0.565	15.70	2500	106	0.2	104	200

onto a thin (0.0005 in.) tungsten wire core. The final diameter of the fiber is about 0.004 in. These filaments have a modulus of 60×10^6 psi, making them five times stiffer than glass and only slightly heavier. Obtaining good adhesion of the resins to the boron surfaces has been a problem that is being solved with various acid-etching processes. Physical properties of epoxy-boron laminates related to etching with nitric acid are shown in Table 44. Most of the laminates of boron which are available contain epoxy resins, but some work has been done recently with polyimides (Table 45) to gain the extra thermal stability of this resin. Obtaining a minimum void content is important to maximize properties, and much development work is now under way to decrease voids.

TABLE 43 Physical Properties of Whiskers[43]

Material	Density, g/cc	Melting point, °F	Tensile strength, psi × 10^{-6}	Young's modulus, psi × 10^{-6}	Strength/density, in. × 10^{-6}
Aluminum oxide	3.9	3780	2–4	100–350	14–28
Aluminum nitride	3.3	3990*	2–3	50	13–21
Beryllium oxide	1.8	4620	2.0–2.8	100	31–43
Boron carbide	2.5	4440	1	65	11
Graphite	2.25	6500	3	142	37
Magnesium oxide	3.6	5070	3.5	45	27
Silicon carbide (alpha)	3.15	4200*	3–5	70	26–44
Silicon carbide (beta)	3.15	4200*	1–3	100–150	0.9–26
Silicon nitride	3.2	3450	0.5–1.5	55	4.2–13

* Dissociates.

TABLE 44 Properties of Epoxy-Boron Laminates as a Function of Etching Process[44] (In psi $\times$ 10^{-3})

Property	Not etched	Cold-etched	Hot-etched
Fiber content, volume %	65.2	60.1	67.3
Tensile strength	50.5	37.5	71.7
Tensile standard deviation	4.4	5.3	3.8
Flexural strength	73.8	65.7	107.0
Flexural modulus	24.6	22.8	23.9
Flexural standard deviation	8.4	9.0	16.8
Compressive strength	140.0	124.5	148.2
Compressive modulus	....	21.9	22.2
Compressive standard deviation	25.3	11.6	7.0

NOTES:
1. ERLA 0400/CL resin system used throughout.
2. Data reported are averages of eight replicates.
3. Compressive results based on edgewise compression of 0.5 $\times$ 2.0-in. specimens; flexural tests based on four-point loading; 32:1 span-depth ratio on 0.5 $\times$ 4.0-in. specimens.

Carbon Fiber Laminates. Carbon and graphite fabrics and yarns have been available for some years based on carbonized rayon. Recently, however, changes in fiber orientation have resulted in novel graphite fibers with drastically improved properties. In the United States the Union Carbide Corporation has introduced Thornel 25, 40, and 50—the number indicating the tensile modulus—and other firms are offering similar fibers. The process of producing these fibers is continuous, and the rayon fibers are stretched and oriented while they are hot. A British fiber RAE is stronger and has a higher modulus but, being made in batch

TABLE 45 Mechanical Properties of Boron-Polyimide Laminates[45]
(In pounds per square inch)

Property	Average value	Maximum value obtained
Flexural strength	192,330	224,000
Flexural modulus	23.4 $\times$ 10^{-6}	35.8 $\times$ 10^{-6}
Compressive strength	62,464	72,500
Compressive modulus	28.4 $\times$ 10^{-6}	28.4 $\times$ 10^{-6}
Tensile strength	123,750	136,000
Tensile modulus	37.4 $\times$ 10^{-6}	40.5 $\times$ 10^{-6}
Horizontal shear strength	5,350	6,540
Transverse flexural strength	6,210	6,420
Tensile strength at RT after 100 hr at 600°F	114,000	114,000
Tensile modulus at RT after 100 hr at 600°F	34.4 $\times$ 10^{-6}	34.4 $\times$ 10^{-6}
Average boron content	58% by vol.	
Average resin content	13% by wt.	
Average void content	15% by vol.	
Molding pressure	100 psi	

fashion, it is only about 40 in. long. These fibers are stretched and oriented before the graphitization process. The physical strengths of these graphite fibers are compared with others in Table 46. Coupling agents to encourage bonding of the resin to the fiber are not yet optimized for graphite, and various coupling

TABLE 46 Properties of Graphite Fibers

Property	Density, lb/in.^{3}	Tensile strength, psi $\times 10^{-3}$	Tensile strength/density, in. $\times 10^{-6}$	Modulus, psi $\times 10^{-6}$	Modulus/density, in. $\times 10^{-6}$
Thornel 25	0.054	200	3.70	25.0	464
Thornel 40	0.054	250	4.64	40.0	742
Thornel 50	0.054	285	5.26	50.0	910
RAE	0.072	300	4.16	60.0	834
Boron	0.090	500	5.55	60.0	666
S-glass	0.072	650	9.00	12.5	174
E-glass	0.072	250	3.47	9.0	125
Maraging steel	0.260	319	1.10	27.0	93
Titanium	0.160	113	1.81	16.2	101

materials and etches are being investigated. The fibers contribute exceptional stiffness to laminates and also make possible the use of high temperature resins. Epoxy resins have been used as the resin binder. Figures 72 to 74 show physical properties of graphite-reinforced epoxy laminates in comparison with other fibers. Table 47 gives physical properties of polyimide-graphite laminates at room temperature and 600°F. In this table the polyimide was Du Pont's PI-3301, and PVA on and off refer to the presence or absence of a polyvinyl alcohol size on the graphite fibers.

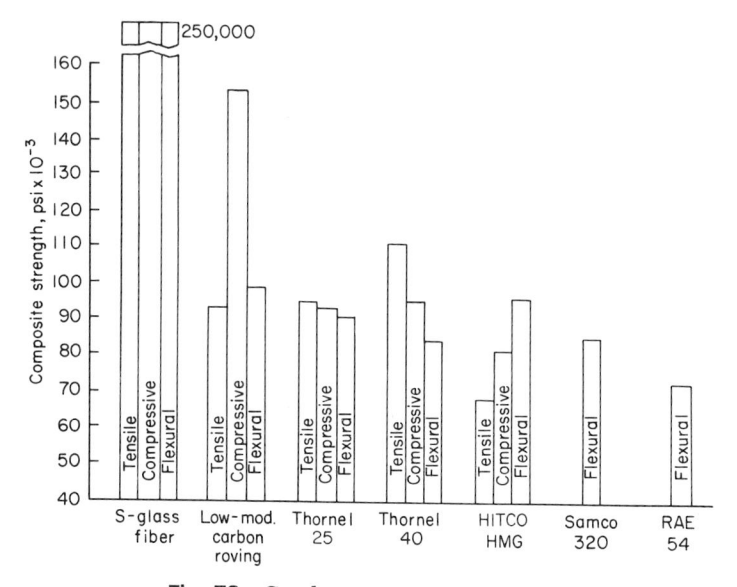

Fig. 72 Graphite composite strengths.[46]

Beryllium Fiber Laminates. Beryllium fibers are also being used as reinforcements. Beryllium has a high modulus-to-density ratio and satisfactory tensile properties. These laminates demonstrate a yielding-type stress-strain diagram with the proportional limit 40 percent of ultimate and a yield at 96 percent of ultimate. As with the other exotic fibers, coupling the resin to the fiber surface is difficult, and acid etches seem to be best. Properties of a beryllium laminate are compared with other fibers in Table 48 with an epoxy resin used as a binder.

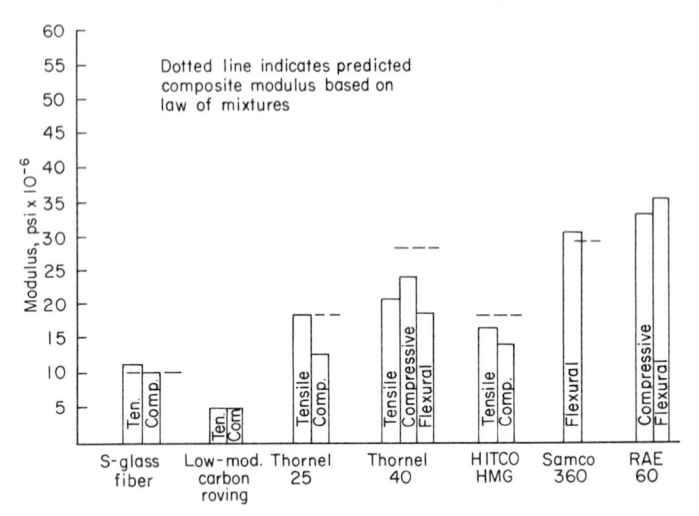

2-88 Laminates, Reinforced Plastics, and Composite Structures

Fig. 73 Graphite composite moduli.[46]

Whisker-Reinforced Laminates. Whiskers are short, single-crystal fibers with extremely high physical properties. Since they are in reality grown crystals, the cost is very high. Most common are aluminum oxide (sapphire), boron carbide, silicon carbide, and silicon nitride. The fibers are available as loose fibers or as papers. Modulus values as high as 50×10^6 psi have been obtained for epoxy-sapphire whiskers laminates. One use of whiskers is to add small amounts to a resin matrix being used to impregnate some other fiber. The interlaminar shear strength is greatly improved thereby, making possible more efficient structures.

TABLE 47 Properties of Polyimide-Graphite Laminates[45]

Material or process variable	Thornel 25 (250 psi, PVA on)	Thornel 40 (250 psi, PVA on)	Thornel 40 (250 psi, PVA off)	Thornel 40 (PVA off, vacuum bag)	Thornel 40 (PVA off, 500 psi)	Thornel 40 (PVA off, 1,000 psi)
Horizontal shear strength, psi:						
At RT	1,775	1,350	1,700	1,850	1,875	1,600
At 600°F	2,775	1,300	1,950			
At 600°F after 100 hr at 600°F	1,225	1,050	1,425			
At 600°F after 250 hr at 600°F	1,175	575	475			
At 700°F		1,175	1,650			
At RT after 2-hr H$_2$O boil		1,250	1,700			
Flexural strength (psi $\times 10^{-3}$)/modulus (psi $\times 10^{-6}$):						
At RT	35.3/8.5	56.3/18.1	59.0/16.4	63.5/18.7	62.5/16.0	59.8/16.5
At 600°F after 100 hr at 600°F	19.0/6.5	31.0/16.8	30.2/12.7	26.6/13.3	30.5/13.9	27.0/13.6
At 600°F after 250 hr at 600°F	10.9/5.1		17.6/10.1			
Tensile modulus at RT, psi $\times 10^{-6}$			20.5	22.0	19.4	17.9
Physical properties:						
Resin content, % vol	38	36	32	27	33	33
Fiber content, % vol	52	59	57	63	57	58
Void content, % vol	10	6	11	10	10	9
Specific gravity	1.24	1.42	1.34	1.35	1.33	1.36

TABLE 48 Properties of Beryllium-Wire Laminate* Compared with Other Fibers[43] (Epoxy Resin Used as Binder†)

| Property | Composite fiber/wire |||||| None |
	Beryllium	Boron	Graphite	Carbon	S-glass‡	
Density, lb/in.^{3}	0.058	0.074	0.067	0.054	0.074§	0.043
Specific gravity	1.61	2.05	1.87	1.50	2.05	1.20
% volume matrix	35	35	35	35	35	100
Ultimate tensile strength, psi $\times 10^{-6}$	97	320	160	80	260	12
Ratio to density, psi $\times 10^{-6}$	1.67	4.32	2.39	1.48	3.53	20.8
Tensile modulus of elasticity, psi $\times 10^{-6}$	28.0	36.0	20.0	4.3	7.6	0.5
Ratio to density, psi $\times 10^{-6}$	482	486	299	80	103	11

* Unidirectional orientation.
† Epoxy resin: tensile strength 12,000; tensile modulus of elasticity 0.5×10^6 psi; elongation 4.47%; compressive strength 18,000 psi; compressive modulus of elasticity 0.5×10^6 psi; deformation 6%.
‡ S-glass epoxy composite: tensile elongation 3.2%, compressive strength 175,000 psi; compressive modulus of elasticity 7.5×10^6 psi; interlaminar shear 11,000 psi; flexural strength 190,000 psi; Poisson's ratio 0.2.
§ 20 wt. %.

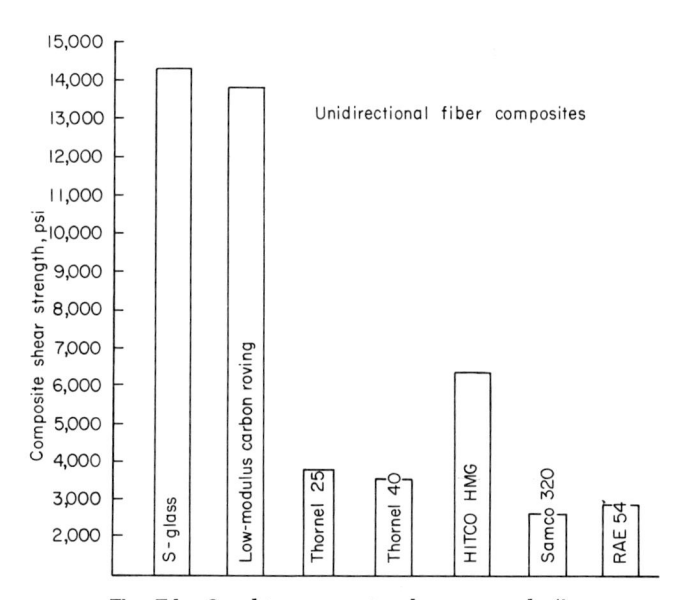

Fig. 74 Graphite composite shear strengths.[46]

Table 49 shows the influence of adding silicon carbide to epoxy-boron and other laminates. It is also possible to grow these crystals onto the surface of other reinforcing fibers, the only limit being that the parent fiber be able to withstand processing temperatures. Figure 75 shows the improvement in shear stress of graphite-fiber-based laminates with grown-in-place whiskers of silicon carbide.

TABLE 49 Addition of SiC to Resin Matrices[47]

Reinforcing fibers	Matrix	Whisker content of matrix, vol. %	Interlaminar shear strength, psi $\times 10^{-3}$	As % of control	Flexural modulus of matrix, psi $\times 10^{-6}$
4-mil boron....	Epoxy resin	0	9.27		0.41
4-mil boron....	Epoxy resin	4½	11.97	129	0.72
6-mil steel.....	Epoxy resin	0	6.5		0.41
		4½	10.4	160	0.72
6-mil steel.....	Phenolic resin	0	4.10		0.75
		2½	4.56	111	1.1
		4½	5.23	128	1.3

Machining plastics Plastic materials present unique machining requirements because of their basic property differences from wood and metals. The resin systems are very similar in their hardness and machining properties, the chief differences in machinability being caused by the type of reinforcement. Reinforcements can be listed in order of increasing difficulty of machining as:
  Paper
  Fabrics
  Asbestos
  Glass
The cellulose-paper- and fabric-based materials machine much like wood. Glass- and asbestos-reinforced laminates should be treated somewhat like hard brass. The glass-based materials, and laminates with certain mineral fillers, tend to wear tools rapidly because of the abrasiveness of the fibers. Thus, if glass-based laminates are to be machined, hardened tools should be used.

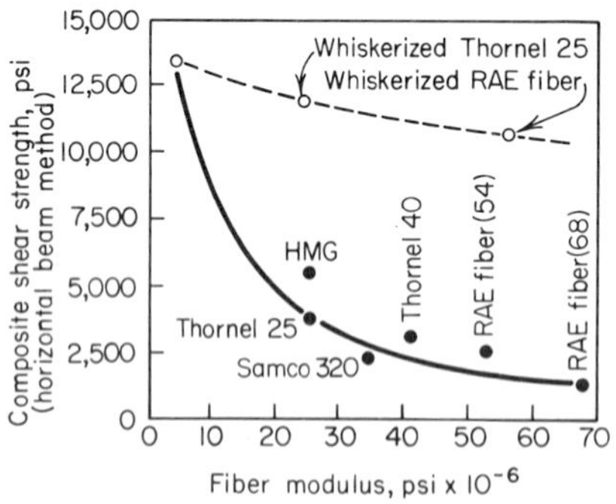

The differences in machinability of laminates are caused by several basic property peculiarities:

1. The materials are in laminar form. Laminates are in reality sheets of reinforcement glued together. The bond strength of the glue (resin) is much lower than the strength of the reinforcements. Hence machining forces tend to separate (delaminate) the plies of the reinforcements. For example, holes should be drilled perpendicular to the plane of the reinforcements and not parallel. If a hole is to be drilled parallel, clamps and slow speeds are required.

2. The brittleness of the resins is severe. All resins have similar moduli. They are considered to be brittle materials with low elongations. Hence machining with dull tools causes chipping and gouging. This condition is at its worst in glass-reinforced materials since the glass fibers are also brittle. Glass laminates require hard, sharp tools and carefully controlled cutting speeds.

Fig. 75 Whiskerized fibers.[48]

3. Resilience of laminates is much greater than that of metals. As plastics are machined, they yield away from the tool only to spring back when the deforming forces are removed. Dull tools emphasize the condition. Tools require greater clearance and less rake to minimize the condition. All holes and machined areas will end out smaller than the tool which formed them. Hence tools must be oversized to obtain correct dimensions.

4. The thermal conductivity of plastics is much lower than that of metals. If forced hard during cutting, both laminate and tools may become hot with decomposition of the material and dulling of the tool as a result. Carbide tips are superior because of their heat resistance. Coolants—either air or liquids—should be used wherever possible. If coolants are not available, the cutting should be interrupted periodically to permit natural cooling.

5. The densities and hence the weights of plastics are lower than those of metals. This means that larger parts can be machined on the machines and higher speeds are often possible.

6. Tolerances are different. In general, the tolerance obtainable in metals cannot be achieved in plastics. Mold stresses, water absorption, and thermal effects prevent the preciseness obtained with metal parts.

The following sections describe particular machining processes and the best methods of machining.

Shearing. All laminates may be sheared, the limit being—as in metals—the thickness of the part. Paper-based laminates up to $\frac{1}{16}$ in. and fabric-based materials up to $\frac{1}{8}$ in. may be sheared on conventional guillotine shears. For thicknesses up to $\frac{1}{4}$ in., rotary shears should be used. Some grades must be heated to 250°F to be sheared, but dimensional stability may then introduce tolerance problems. For glass based materials shearing is not possible over $\frac{1}{16}$ in., and thicker laminates should be sawed.

Sawing. Paper- and fabric-based grades can be cut with ease on conventional band saws having teeth with 4 to 10 points per in. The speeds can reach 9,000 ft/min. Circular saws should have hollow-ground no-set teeth and can be operated up to 13,000 surface ft per min.

Glass-based laminates require hardened blades. A regular high-speed steel saw blade will be useless after 3 min use. Carboloy*-tipped teeth should be used on circular saws, and similar band saw blades are available. These should have 10 to 12 teeth per in., and speeds should be limited to about 2,000 ft/min. Cutoff wheels are useful but should be equipped with diamond blades. Air or water coolants should be used. If air is used, provisions should be made to collect the dust which may cause dermatitis.

Drilling. Holes should be drilled perpendicular to laminations. If parallel drilling is required, then work should be clamped securely between plates, and feeds should be slow. Frequent back-out of the drill will keep the hole clean and avoid chipping. Air cooling should be used. Special drills for plastic laminates are available from all suppliers. These have 70° to 90° points, "slow spirals," and a high helix angle. In all cases holes will decrease in size after the tool is removed. Drills should be 0.002 to 0.004 in. oversized, and tolerances of ±0.001 in. can be obtained.

Paper and fabric can be drilled with high-speed steel drills. Surface speeds of 350 ft/min at 10,000 rpm can be maintained.

Glass laminates should be drilled with polished and chrome-plated tungsten carbide drills at speeds up to 250 ft/min. Feed rates should approach to 0.003 in. per revolution. Diamond drills can be run at high speeds up to 35,000 rpm.

All laminates can be countersunk or counterbored. Speeds one-half those of drilling should be used. Tolerances can be held to ±0.005 in.

Punching. Laminates can be punched with standard presses and dies. Because of the resilience of the materials, all holes should be punched 4 percent oversize. The dies must be equipped with a spring-loaded stripper plates to avoid lifting the punched part on the pins and to minimize the delamination forces. The stripper

---

* Trademark of General Electric Co., Schenectady, N.Y.

plate should have 0.001 in. clearance from the die. All hole spacing should be at least 1.5 times the laminate thickness. Die clearances should be one-half those for steel. Best results have been found with Carpenter No. 610 steel hardened to Rockwell C-60. Minimum hole size should be 75 percent of laminate thickness. Laminates up to $3/32$ in. can be punched with sharp, clean holes. Holes in thicker laminates get rough and chipped. The number of holes capable of being punched in a laminate with a given press is a function of the shear area and the stripper-plate spring force. Holes should be no smaller in diameter than the laminate thickness and no larger than three times the laminate thickness. All grades can be cold-punched, but maximum punchable thickness is related to both resin type and reinforcements. In general, the following thicknesses are the maximum for cold punching:

Paper-based laminates............ $1/16$ in.
Fabric-based laminates........... $1/8$ in.
Glass-polyester laminates......... $5/32$ in.
Glass-silicone laminates.......... $1/4$ in.

Heating the laminates makes punching easier. Most laminates should be heated to 150°F, but some grades require 250°F. Glass-based grades should be hot-punched because of the dimensional changes that occur when the laminate cools. Best punching of glass-based grades results when carbide tools are used. Quality criteria for punched phenolic laminates are presented in ASTM D 617-44.

One is often faced with the decision as to whether to punch or drill a laminate, particularly in printed-circuit work. The correct economic answer is a function of the number of units to be punched, the dimensions of laminate and holes, the shape of the part, and the facilities available. Table 50 lists the factors to be considered in making a correct decision.

TABLE 50 Punching and Drilling Laminates

Factor	Punching	Drilling
Machine tool cost...................	High	Low
Die or jig cost.....................	High	Low
Labor cost.........................	Low	High
Production rate....................	High	Low
Maintenance needed................	Little	Much
Tool life...........................	Long	Short
Size of hole.......................	Limited	Not limited
Ability to work curved surfaces......	No	Yes
Irregular-shaped holes..............	Yes	No

Tapping and Threading. Maximum accuracy for threads is obtained with a 75 percent thread and a Class 2 fit. The tools should be made oversized to allow for the resilience of the laminates. Tap holes need to be deep enough for at least three threads, and the edges should be chamfered to avoid chipping of the walls. Tools should have a 5° to 10° negative rake on front of lands.

The taps for paper- and fabric-based grades can be made of high-speed plated steel with four flutes and should be about 0.002 in. oversized. Taps for glass-based laminates should be nitrided and plated with two or three flutes. The same rake should be used, and they should be 0.002 to 0.008 in. oversized depending on the resin. Speeds of no more than 80 ft/min are used.

Dies for paper- and fabric-based materials should have a chamfer of about 33° on the lead and a negative rake of 10°. Cutting speeds can approach 400 ft/min. External threads on glass-based laminates are not recommended.

Lathe Work. Laminates cut much like brass. Critical to success are highly sharpened tools and air cooling. When cutting laminates, never stop the tool, or a mark will remain on the surface of the work. Excessive pressure or dull tools may cause delaminations to occur.

Paper- and fabric-based laminates can be machined with high-speed steel tools with a 10° negative rake and about a 35° clearance. Rough cuts ⅛ in. deep can be made at surface speeds up to 500 ft/min.

Glass-based laminates should be cut with tools made of carbide or diamond. The rake should be 10° negative, and a 30° clearance is needed. Surface speeds should be less than 400 ft/min, and feeds of 0.010 in. per revolution can be maintained. Dust must be collected to avoid dermatitis problems.

Milling. Laminates must be carefully handled on a milling machine since milling tool forces act to delaminate the work. Tools should be of carbide, should be operated at surface speeds of less than 1,000 ft/min, and should have a negative rake. Feed speeds of about 30 in./min are common. The materials cut like brass.

Joining Laminates can be joined with any of the conventional joining methods such as rivets, bolts, and screws. One must always remember the tendency of the materials to delaminate and the anisotropic nature of the laminates.

Rivets. Riveting practices should be those normally used. Riveted joints should be designed to act in shear rather than tension if possible. If tension loads must be used, they should amount to no more than one-half the shear loads defined for the rivets. Rivet holes should be large enough to permit the rivets to be slid in easily. Remember in drilling the holes to allow for the resiliency of the plastic materials. Too large holes will permit the rivets to buckle or bend, thereby reducing the rivet strength. Too small holes will cause damage to the hole walls as the rivet is driven. In general, the rivet hole should be about ¹⁄₆₄ in. larger than the rivet diameter. Rivet holes should be at least 4½ rivet diameters from

TABLE 51 Holding Forces of Screws in Glass-reinforced Plastics[41]

Fastener size threads	Plastic material polyester—181 cloth, axial holding force				Type of fastener machine screw, lateral holding force			
	Minimum		Maximum		Minimum		Maximum	
	DoP, 16th in.	Force, lb	DoP, 16th in.	Force, lb	DoP, 16th in.	Force, lb	DoP, 16th in.	Force, lb
4–40	2	130	5	500	1	210	2	350
6–32	2	170	6	820	1	250	3	700
8–32	3	320	7	1,140	2	440	3	810
10–32	3	420	8	1,500	2	850	4	1,200
¼–20	4	700	10	2,800	2	1,200	5	1,900
⁵⁄₁₆–18	4	820	12	4,200	3	2,100	6	2,800
⅜–16	5	1,160	14	6,000	4	3,300	7	3,900
⁷⁄₁₆–14	5	1,300	16	8,000	4	3,800	8	5,200
½–13	6	1,660	17	9,800	5	5,300	10	7,500
⁹⁄₁₆–12	6	1,820	18	12,000	5	5,800	11	9,000
⅝–11	7	2,200	20	15,300	6	7,300	13	11,900
¾–10	7	2,550	22	20,600	7	8,900	15	15,000

Plastic mechanical properties; psi:
  Tensile strength: 30,000–45,000.
  Edge compressive strength: 18,000–27,000.
  Shear strength (Johnson): 14,000–17,000.

2-94 Laminates, Reinforced Plastics, and Composite Structures

the edge of the laminate. The rivet diameter should be larger than the laminate thickness. Best results are found with type 56S or 2S aluminum rivets with truss heads. These larger heads distribute the loads on the plastic surface better than the normal button-headed rivets. The length of the rivets should be equal to the thickness of the parts being joined plus one rivet diameter. It is better to have a rivet too long than too short.

In addition to the standard truss-headed rivets described, other special types are useful in joining plastic laminates. These include blind rivets, "Pop" rivets, drive rivets, and plastic rivets.

Bolts. Care must be exercised in bolting laminates so that the laminate is not crushed or placed under such high loads that creep will cause failure. As in riveting, the bolt hole should be 1/64 in. larger than the bolt diameter, and the bolt diameter should be larger than the laminate thickness. Always use washers or load strips for the head and nut to bear against. All plastic materials are notch-sensitive. Glass mat laminates, in particular, exhibit stress sensitivity at holes, and hence bolts should be located at least 5 times the bolt diameter in from the side of the laminates. Fabric and paper laminates can have bolts located only 4½ times the bolt diameter in from the side.

Screws. Screws work about as well in plastic laminates as in wood. Tapping of holes for screws was described in a previous section. Screwed joints possess excellent resistance to axial loads (loads in the direction of the screw axis) and lateral loads (loads perpendicular to the direction of the screw axis). These forces are related to both the type of resin and the type of reinforcement. Tables 51 to 53 give the holding forces for polyester laminates based on several types of glass reinforcements.

Self-tapping screws are becoming increasingly popular. Their strengths approach those of screws in tapped holes, and they eliminate one machining step. Frequently

TABLE 52 Holding Forces of Screws in Glass-reinforced Plastics[41]

Fastener size threads	Plastic material polyester—1000 cloth, axial holding force				Type of fastener machine screw, lateral holding force			
	Minimum		Maximum		Minimum		Maximum	
	DoP, 16th in.	Force, lb	DoP, 16th in.	Force, lb	DoP, 16th in.	Force, lb	DoP, 16th in.	Force, lb
4–40	1	90	3	500	1	180	2	350
6–32	1	110	4	700	1	220	2	450
8–32	2	230	5	1,000	1	280	3	880
10–32	2	260	6	1,500	2	750	3	1,100
¼–20	2	320	8	2,700	2	1,000	4	1,800
5/16–18	3	520	10	4,000	2	1,350	6	2,000
3/8–16	3	620	11	4,850	3	2,600	7	3,300
7/16–14	4	950	13	6,800	3	3,000	9	4,500
½–13	4	1,100	15	7,720	4	4,500	10	6,000
9/16–12	5	1,600	16	9,000	5	6,000	12	9,000
5/8–11	5	1,750	18	11,600	5	6,500	13	10,000
¾–10	6	2,500	22	24,000	6	8,000	16	15,000

Plastic mechanical properties, psi:
  Tensile strength: 30,000–45,000.
  Edge compressive strength: 18,000–27,000.
  Shear strength (Johnson): 14,000–17,000.

TABLE 53 Holding Forces of Screws in Glass-reinforced Plastics[41]

Fastener size threads	Plastic material polyester— 1½-oz mat, axial holding force				Type of fastener machine screw, lateral holding force			
	Minimum		Maximum		Minimum		Maximum	
	DoP, 16th in.	Force, lb	DoP, 16th in.	Force, lb	DoP, 16th in.	Force, lb	DoP, 16th in.	Force, lb
4–40	2	40	5	450	1	150	2	290
6–32	2	60	6	600	1	180	2	380
8–32	2	100	7	1,150	1	220	3	750
10–32	2	150	8	1,500	2	560	4	1,350
¼–20	3	300	10	2,300	3	1,300	5	1,900
5⁄16–18	3	400	12	3,600	3	1,600	7	2,900
3⁄8–16	4	530	14	5,000	4	2,600	10	4,000
7⁄16–14	4	580	16	6,500	5	3,800	12	5,000
½–13	4	620	18	8,300	6	5,500	14	6,000
9⁄16–12	4	650	20	10,000	7	6,500	15	8,000
5⁄8–11	4	680	22	12,000	7	6,800	16	11,000
¾–10	4	700	24	13,500	7	7,000	17	17,000

Plastic mechanical properties, psi:
  Tensile strength: 6,000–25,000.
  Edge compressive strength: 10,000–22,000.
  Shear strength (Johnson): 10,000–13,000.

used with adhesives, self-tapping screws are used to join either two laminates together or one laminate to some other material. Literally dozens of sheet-metal designs can be chosen for use with laminates, and some screws have been developed specifically for laminates. These screws are of two types: thread forming (in which the material is forced aside), and thread cutting (in which the material is cut or machined). Thread-forming screws are available in hardened alloy steel and stainless. They can be inserted into punched or drilled holes and resist tension and shear forces well. Tables 54 and 55 present pullout strengths for Type Z screws in cloth and mat laminates.

Thread-cutting screws cut a thread that is similar to a Class 2 tolerance. They are available in coarse and fine threads, both of which have similar holding power,

TABLE 54 Pullout Strength of Type Z Thread-forming Screws[41]
(Style 1000 cloth—66.5 wt. % Glass)

Property	Fastener size threads			
	4–24	6–20	8–18	14–14	3⁄8–12
Initial penetration, 16th in.................	2	4	4	5	4
Pullout strength at initial penetration, lb.....	70	200	120	300	100
Pullout strength per 1⁄16 in., lb.............	150	165	170	300	410
Range, 16th in........................	2–6	4–9	4–15	5–18	2–16
Approximate break load of fastener, lb.......	650	1,100	2,000	4,300	

TABLE 55 Pullout Strength of Type Z Thread-forming Screws[41]
(1½-oz Mat—37 wt. % Glass)

Property	\multicolumn{5}{c}{Fastener size threads}
	4–24	6–20	8–18	14–14	⅜–12
Initial penetration, 16th in..................	2	4	4	5	4
Pullout strength at initial penetration, lb.....	110	110	180	100	200
Pullout strength per 1/16 in., lb..............	115	85	145	265	250
Range, 16th in...........................	4–6	4–10	4–14	5–20	4–8

but the fine thread causes less delamination. Tables 56 and 57 present pullout strengths for Type F screws in cloth and mat laminates.

Since the strength of a screw is proportional to the shear area and hence the screw diameter, fasteners of increased diameters represented by inserts are increasingly popular. They are invaluable when a screw is to be removed, frequently

TABLE 56 Pullout Strength of Type F Thread-cutting Screws[41] (1½-oz Mat—37 wt. % Glass)[41]

Property	\multicolumn{4}{c}{Fastener size threads}
	4–40	6–32	8–32	¼–20
Initial penetration, 16th in...............	2	4	4	5
Pullout strength at initial penetration, lb..	10	130	360	400
Pullout strength per 1/16 in., lb..........	150	90	230	300
Range, 16th in........................	2–6	4–12	4–12	5–17

because threads in a laminate do not wear nearly so well as threads in a metal. These inserts can be placed in a molded laminate or, where design permits, molded in place. The inserts are self-tapping or can be pushed into a properly sized hole. Bonding of inserts with appropriate adhesives is also possible.

Adhesive bonding and adhesives One widely accepted method of fastening laminates together (as well as other substrates) is adhesive bonding. An adhesive

TABLE 57 Pullout Strength of Type F Thread-cutting Screws[41] (Style 1000 Cloth—66.5 wt. % Glass)

Property	\multicolumn{4}{c}{Fastener size threads}
	4–40	6–32	8–32	¼–20
Initial penetration, 16th in...............	2	4	4	5
Pullout strength at initial penetration, lb..	70	250	400	350
Pullout strength per 1/16 in., lb..........	130	185	195	365
Range, 16th in........................	2–6	4–9	4–12	5–12
Approximate break load of fastener, lb...	650	1,300	2,200	4,000

has been defined as any substance capable of holding two materials together by surface attachments. The mechanism of this attachment is not yet well defined, and many theories exist that explore the phenomenon of adhesion. These theories are shown below:

Adsorption. Adhesion results from molecular contact between two materials and is related to the free energies of the surfaces.

Mechanical. The adhesive flows into and fills cavities in the substrate and is held mechanically.

Diffusion. The two contacting surfaces will interdiffuse in relation to their chemical similarity.

Chemical. A chemical reaction occurs between adhesive and substrate, forming permanent bonds.

Electrical. Electrostatic charges form attractive forces between adhesive and substrate.

Probably many or all of these factors play a part in adhesion, but the adsorption theory seems most accurate.

Adhesive bonding presents several distinct advantages to the designer. One of the most important is distribution of stresses. Plastic laminates are notch-sensitive, and therefore conventional mechanical fasteners cause stress risers. Adhesive bonding provides a large area over which the stresses are distributed. Other advantages and disadvantages are given in Table 58.

TABLE 58 Advantages and Disadvantages of Adhesive Bonding of Laminates

Advantages	Disadvantages
1. Large stress-bearing area.	1. Surfaces must be cleaned.
2. Excellent fatigue strength.	2. Inspection of joint is difficult.
3. Adhesive tends to damp vibration.	3. Jigs and fixtures may be needed.
4. Adhesive may absorb shock.	4. Long cure times may be needed.
5. Eliminates galvanic corrosion when metals used.	5. Special facilities required.
6. Joins all thicknesses.	
7. Appearance improved—smooth contours.	
8. Seals joint.	
9. May be less expensive.	

Adhesives Specifications. A variety of Federal specifications describing adhesives and test methods have been prepared and are described in Table 59. Adhesives are also defined in a variety of ASTM standards, and selected standards are given in Table 60.

Joint Design. The veracity of an adhesive joint is a function of the adhesive, the quality of the joint, joint design, and the environment. Before an adhesive is selected, the joint should be carefully designed. Any joint design should take into consideration the following factors:

1. Joints can be loaded in shear, tension, or compression, but most loads are combinations of these.
2. The joint should keep the stresses on the bond line to a minimum.
3. Use rigid adhesives for shear loads and flexible adhesives in peel.
4. Minimize tensile loads.
5. Use as large a bond area as possible.
6. Stress the adherends in the direction of greatest strength. In a laminate this is parallel to the laminations.

The simplest joints to make are butt joints, as shown in Fig. 76. Similar butt joints for angles are shown in Fig. 77. Butt joints cannot withstand bending loads, and thick bond lines induce internal stresses which result in poor performance. The most popular modification of joint design is the lap joint. Tensile loading causes the lap joint to be stressed in shear. This induces an unbalance of strain in

2-98   Laminates, Reinforced Plastics, and Composite Structures

TABLE 59   Government Adhesive Specifications

Specification number	Title	Comment
Government Federal Test Method, Standard 175	Adhesives: Method of Testing	Sampling, inspection, and testing.
Military MIL-A-397	Adhesives: Room Temperature and Intermediate Temperature, Resin	Thermosetting adhesives for wood to wood and wood to plastic.
MIL-A-5090	Adhesive: Airframe Structural, Metal-to-Metal	
MIL-A-25457	Adhesives: Air-drying, Silicone Rubber	Two-component silicone adhesive for bonding silicone to itself and to aluminum.
MIL-A-8623	Adhesives: Epoxy Resin, Metal-to-Metal Structural Bonding	Epoxy adhesives for use up to 200°F. Three classes based on curing temperature.
MIL-A-25463	Adhesives: Metallic Sandwich Construction	For bonding metal facings to metal cores for exposure to 500°F
MIL-A-9067	Adhesives: Bonding Process and Inspection Requirements	Processing policies and surface preparation recommendations.
Aerospace Military AMS 3690	Adhesives: Compound-Epoxy, Room-temperature Curing	For bonding electronic components operating not higher than 250°F.
AMS 3691	Adhesives: Compound-Epoxy, Medium-temperature Application	For bonding electronic components operating not higher than 250°F.
AMS 3692	Adhesives: Compound-Epoxy: High-temperature Application	For bonding electronic components operating not higher than 500°F.

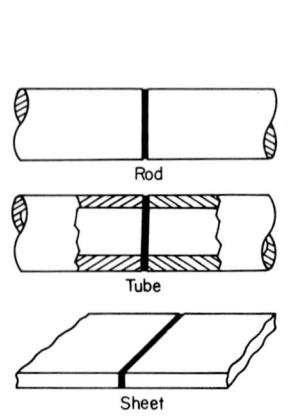

Fig. 76   Simple butt joints.[49]

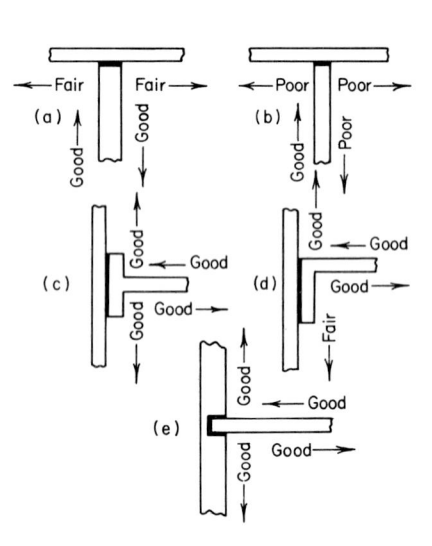

Fig. 77   Angle joints.[49]

TABLE 60 ASTM Adhesive Standards*

Adhesion

Adhesion of Bondable Silicone Rubber Tapes Used for Electrical Insulation, Testing (D 2148) 29
Bond of Oil and Resin-base Caulking Compounds, Test for (D 2450) 21
Bond Strength of Plastics and Electrical Insulating Materials, Test for (D 952) 27
Peel or Stripping Strength of Adhesive Bonds, Test for (D 903) 16
Preparation of Bar and Rod Specimens for Adhesion Tests, Recommended Practice for (D 2094) 16
Pressure-sensitive Adhesive Coated Tapes Used for Electrical Insulation, Testing (D 1000) 29
Rubber Insulating Tape, Specification for (D 119) 28

Adhesives

Adhesives, Definitions of Terms Relating to (including Tentative Revision) (D 907) 16
Adhesives for Structural Laminated Wood Products for Use under Exterior (Wet Use) Exposure Conditions, Specification (D 2559) 16
Adhesives Relative to Their Use as Electrical Insulation, Testing (D 1304) 16
Apparent Viscosity of Adhesives Having Shear-rate-dependent Flow Properties, Test for (D 2556) 16
Applied Weight per Unit Area of Liquid Adhesive, Test for (D 899) 16
Applied Weight per Unit Area of Dried Adhesive Solids, Test for (D 898) 16
Atmospheric Exposure of Adhesive Bonded Joints and Structures, Recommended Practice for (D 1828) 16
Blocking Point of Potentially Adhesive Labers, Test for (D 1146) 16
Climbing Drum Peel Test for Adhesives, (D 1781) 16
Consistency of Adhesives, Tests for (D 1084) 16
Cross-lap Specimens for Tensile Properties of Adhesives, Testing (D 1344) (16
Density of Adhesives in Fluid Form, Test for (D 1875) 16
Exposure of Adhesive Specimens to High-energy Radiation, Recommended Practice for (D 1879) 16
Flow Properties of Adhesives, Test for (D 2183) 16
Hydrogen Ion Concentration of Dry Adhesive Films, Test for (D 1583) 16
Integrity of Glue Joints in Structural Laminated Wood Products for Exterior Use, Test for (D 1101) 16
Peel Resistance of Adhesives (T-peel Test), Test for (D 1876) 16
Penetration of Adhesives, Test for (D 1916) 16
Preparation of Metal Surfaces for Adhesive Bonding, Recommended Practice for (D 2651) 16
Preparation of Surfaces of Plastics Prior to Adhesive Bonding, Recommended Practice for (D 2093) 16
Pressure-sensitive Tack of Adhesives, Test for (D 1878) 16

Adhesives (continued)

Resistance of Adhesives to Cyclic Laboratory Aging Conditions, Tests for (D 1183) 16
Shear Strength and Shear Modulus of Structural Adhesives, Test for (E 229) 30

Adhesives, Bonding Strength

Cleavage Strength of Metal-to-Metal Adhesive Bonds Test for (D 1062) 16
Determining Strength Developments of Adhesive Bonds, Recommended Practice for (D 1144) 16
Effect of Bacterial Contamination on Permanence of Adhesive Preparations and Adhesive Bonds, Test for (D 1174) 16
Effect of Moisture and Temperature on Adhesive Bonds, Test for (D 1151) 16
Effect of Mold Contamination on Permanence of Adhesive Preparations and Adhesive Bonds, Test for (D 1286) 16
Impact Strength of Adhesive Bonds, Test for (D 950) 16
Peel or Stripping Strength of Adhesive Bonds, Test for (D 903) 16
Resistance of Adhesive Bonds to Chemical Reagents, Test for (D 896) 16
Shear-block Test for Quality Control of Glue Bonds in Scarf Joints, Conducting (D 1759) 16
Storage Life of Adhesives by Consistency and Bond Strength, Test for (D 1337) 16
Strength of Adhesive Bonds on Flexural Loading, Test for (D 1184) 16
Tensile Properties of Adhesive Bonds, Test for (D 897) 16
Working Life of Liquid or Paste Adhesives by Consistency and Bond Strength, Test for (D 1338) 16

Adhesives, Creep Properties

Conducting Creep Tests of Metal-to-Metal Adhesives, Recommended Practice for (D 1780) 16
Creep Properties of Adhesives in Shear by Compression Loading (Metal-to-Metal), Test for (D 2293) 16
Creep Properties of Adhesives in Shear by Tension Loading (Metal-to-Metal), Test for (D 2294) 16

Adhesives, Strength Properties

Strength Properties of Adhesives in Shear by Tension Loading in the Temperature Range of -267.8 to -55 C (-450 to -57 F), Test for (D 2557) 16
Strength Properties of Adhesives in Shear by Tension Loading at Elevated Temperatures (Metal-to-Metal), Test for (D 2295) 16
Strength Properties of Adhesive Bonds in Shear by Compression Loading, Test for (D 905) 16
Strength Properties of Adhesives in Shear by Tension Loading (Metal-to-Metal), Test for (D 1002) 16
Strength Properties of Metal-to-Metal Adhesive by Compressive Loading (Disk Shear), Test for (D 2182) 16
Tensile Strength of Adhesives by Means of Bar and Rod Specimens, Test for (D 2095) 16

* Number following test method number is volume number of ASTM book.

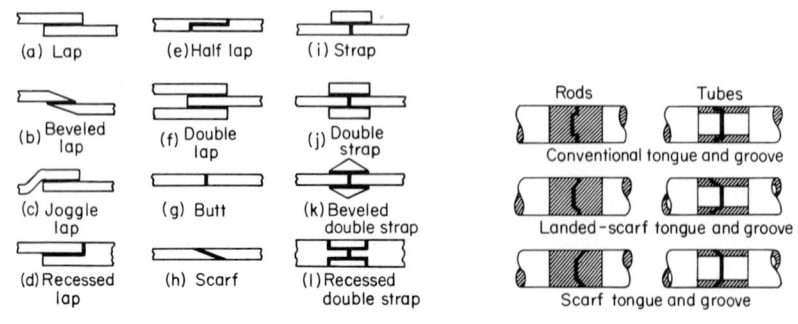

Fig. 78 Joints for sheets and bars.[49]

(a) Lap
(b) Beveled lap
(c) Joggle lap
(d) Recessed lap
(e) Half lap
(f) Double lap
(g) Butt
(h) Scarf
(i) Strap
(j) Double strap
(k) Beveled double strap
(l) Recessed double strap

Fig. 79 Rod and tube joints.[49]

Rods — Tubes
Conventional tongue and groove
Landed-scarf tongue and groove
Scarf tongue and groove

the adhesive and adherends, and hence a bending of the joint. These factors result in stresses at the ends of the lap joint substantially greater than the stresses elsewhere. Breaking strength of lap joints increases linearly with overlap length up to some limiting value. Strengths of lap joints are proportional to overlap area only at constant lap length. Modifications of joint design which result in stronger joints include:
1. Reducing joint bending by stiffening the adherends.
2. Making the adherends thinner.
3. Using a tougher adhesive.
4. Using a thinner bond line.
5. Decreasing the overlap length.

Joints for sheets and bars with properties improved over the lap joint are shown in Fig. 78. The scarf joint has no bending forces but is expensive to make. A joggle lap is useful because the jog can be molded into the plastic. This joint is not as good as a scarf or bevel joint. Strap joints are designed to limit bending. Best is the beveled double strap, followed in decreasing order by recessed double strap, double strap, and single strap. Methods of applying these same joint principles to rods and tubes are shown in Fig. 79. Here the tongue-and-groove joint is the cheapest. Careful design methods will increase reliability of a joint. Figure 80 shows what can be done to decrease peeling tendencies in a simple joint, and Fig. 81 shows four methods of increasing the stiffness of an assembly.

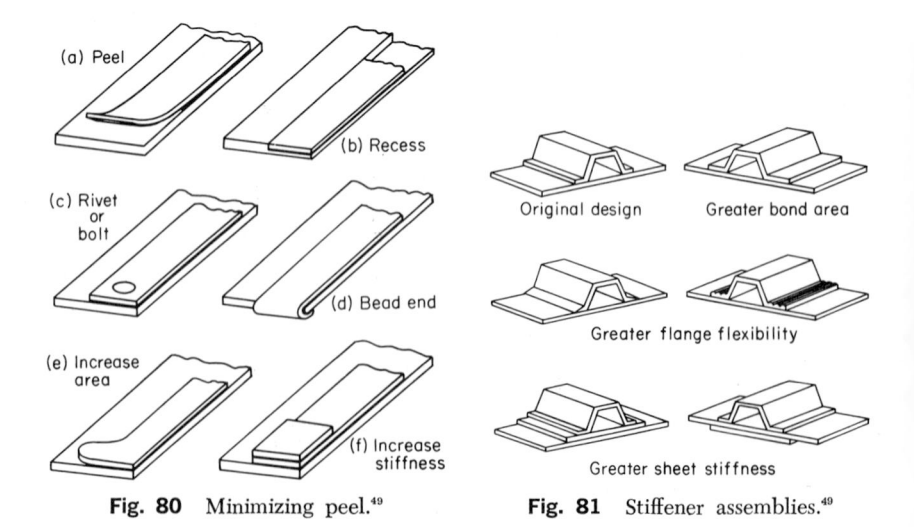

(a) Peel
(b) Recess
(c) Rivet or bolt
(d) Bead end
(e) Increase area
(f) Increase stiffness

Fig. 80 Minimizing peel.[49]

Original design — Greater bond area
Greater flange flexibility
Greater sheet stiffness

Fig. 81 Stiffener assemblies.[49]

Adhesives. Adhesives are made from almost every polymer yet developed. The adhesives described in this section are limited to those structural adhesives which are useful for bonding laminates and moldings. Table 61 describes structural adhesives and presents some of their properties. Thermal properties of adhesives are often better than those of similar polymers in other forms because the bond lines are little exposed to air and oxidation.

TABLE 61 Properties of Adhesives[50]

Adhesive	Temperature, °F Max.	Temperature, °F Min.	Lap shear	Peel	Type of bond
Epoxy-amine	150	−50	Medium to high	Low	Rigid
Epoxy-polyamide	150	−60	Low to medium	Low to medium	Tough and moderately flexible initially but becomes brittle when exposed to heat
Epoxy polysulfide	150	−100	Low to medium	Medium	Flexible
Epoxy-anhydride	500	−60	Medium	Poor	Rigid
Epoxy-phenolic	500	−425	Low	Medium	Tough and moderately flexible
Epoxy-silicone	500	−60	Low	Low	Rigid
Nylon-epoxy	180	−425	High	High	Tough
Silicone	400	−60	Low	Low	Flexible
RTV silicone	500	−300	Low	Medium	Flexible
Phenolic-nitrile	300	−100	Medium to high	Medium to high	Tough and moderately flexible to rigid
Vinyl-phenolic	225	−60	High	Medium	Tough and moderately flexible to rigid
Polyurethane	150	−300	Low to medium	Medium to high	Tough and flexible
Phenoxy	180	−70	Low to medium	Good	Tough and moderately flexible
Cyanoacrylate	180	−60	Very low	Very low	Rigid
Acrylic acid diester	200	−60	....	....	Rigid
Polyimide	600	....	Medium	Low	Rigid
PBI	500	....	Medium	Low	Rigid

Epoxy adhesives may be cured with amines, polyamides, and anhydrides. Amine cures are often at room temperature and have excellent shear strength at room temperature. Formulations are available that cure in a few minutes at room temperature. Polyamide curing agents lend flexibility to the adhesive and improve low-temperature properties. The flexibility is often lost as the temperature is increased. High-temperature properties are obtained from anhydride curing agents. These resins require elevated-temperature cures. Properties of some of these epoxy adhesives are given in Table 62.

A variety of polymers can be blended to and reacted with epoxy resins to contribute certain properties. Polysulfide resins combined with epoxies provide adhesives with excellent flexibility and chemical resistance.

Phenolic resins coreact with epoxies to produce adhesives with good thermal stability. These resins produce water during the cure. They have low peel strengths and limited thermal shock resistance but retain their properties over a wide temperature range, as shown in Fig. 82. These are often used on honeycomb applications.

Epoxy silicones are also designed for elevated-temperature use. Room-temperature bond strengths are very low. Figure 83 shows the effect of temperature on shear strength of an epoxy-silicone.

Epoxy-nylon adhesives are used for cryogenic applications. Figure 84 shows their properties down to −400°. These adhesives are supplied in film form and are moisture-sensitive.

TABLE 62 Influence of Curing Agent on Bond Strength Obtained with Various Base Materials[51]

Curing agent	Amount*	Cure cycle, hr at °F	Polyester glass mat laminate	Polyester glass cloth laminate	Cold-rolled steel	Aluminum	Brass	Copper
Triethylamine	6	24 at 75, 4 at 150	1,850	2,100	2,456	1,810	1,765	655
Trimethylamine	6	24 at 75, 4 at 150	1,054	1,453	1,385	1,543	1,524	1,745
Triethylenetetramine	12	24 at 75, 4 at 150	1,150	1,632	1,423	1,675	1,625	1,325
Pyrrolidine	5	24 at 75, 4 at 150	1,250	1,694	1,295	1,733	1,632	1,420
Polyamid amine equivalent 210–230	35–65	24 at 75, 4 at 150	1,200	1,450	2,340	3,120	2,005	1,876
m-Phenylenediamine	12.5	4 at 350	780	640	2,150	2,258	2,150	1,650
Diethylenetriamine	11	24 at 75, 4 at 150	1,010	1,126	1,350	1,420	1,135	1,236
Boron trifluoride, monoethylamine	3	3 at 375	.....	.....	1,732	1,876	1,525	1,635
Dicyandiamide	.....	4 at 350	530	432	2,680	2,785	2,635	2,550
Methyl nadic anhydride	85	6 at 350	600	756	2,280	2,165	1,955	1,835

Epoxy resin was derived from bisphenol A and epichlorohydrin and had an epoxide equivalent of 180–195; the adhesives contained no filler.
* Per hundred parts by weight of resin.

Laminates   2-103

Room-temperature vulcanizing (RTV) silicones can be used up to 500°F. The resins have low strengths and are frequently employed as sealants. They polymerize with water in air and release acetic acid in the process.

Phenolic-nitrile adhesives are useful up to about 300°F. They are characterized by excellent strengths and toughness and can be obtained as solutions and supported and unsupported films. Table 62 shows the effect of temperature on phenolic nitriles in comparison with other resins.

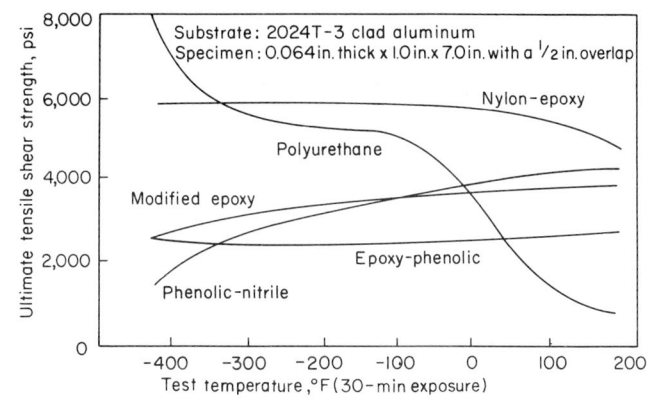

Fig. 82 Effect of temperature on shear stress of adhesives.[52]

Vinyl-phenolic adhesives have excellent shear and peel strengths. Their heat resistance is good—about 300°F; and they are chemically resistant. They are copolymers of a phenolic resin with polyvinyl butyral or polyvinyl formal. Properties are shown in Tables 63 and 64.

Polyurethanes form tough bonds with high peel strength. They are excellent adhesives for cryogenic applications. Generally used as a two-part system, the

TABLE 63   Shear Strength of High-temperature Adhesives[53]

Property	Nitrile-phenolic	Vinyl-phenolic	Epoxy-phenolic	Modified phenolic	Poly-benzimi-dazole	Epoxy-novolak	Epoxy-silicone
Initial room-temperature shear strength, psi	3,200–4,100	3,800–4,500	3,240	2,160	2,920*	3,800*	1,700*
% of original shear strength							
After 10-min soak:							
At 180°F	35–70	35–106	94	101	99	83	
At 250°F	31–42	12–79	81	121	99	74	
At 350°F	22–25	6–11	66	111	98		
At 500°F			54	81	83	28	80
At 650°F			43	52	74		
At 800°F			15	21			80
After 1,000-hr soak:							
At 180°F	62–78	60–122	96	107			
At 250°F	52–66	33–82	81	112			
At 350°F	27–56	13–30	43	94			
At 500°F					6–7		31

* Stainless steel used instead of aluminum as adherend metal.

TABLE 64 Tensile Shear Strengths for Vinyl-Phenolics[53] (In pounds per square inch)

Supplier	At room temperature	At 180°F	At 250°F	At 350°F
A	5,168	3,799	1,068	242
B	4,220	3,137	1,462	562
C	5,075	3,028	736	78
D	2,800	1,100	500	315
E	4,500	4,700–4,800	2,900–3,500	400–500
F	4,000	1,200	480	250
G	3,800	1,350	750	350

adhesives are adversely affected by water. They can be used to bond films and elastomers, but their maximum temperature is limited to about 200°F. The cryogenic shear strength of polyurethane is shown in Fig. 85 along with other adhesives.

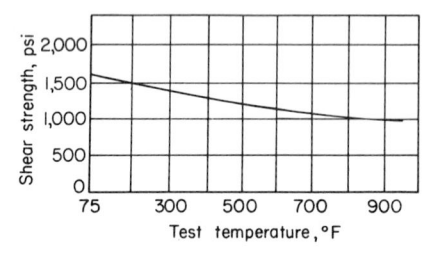

Phenoxy* adhesives are new epoxylike polymers with good shear strength and creep resistance. They are thermoplastic and soften at 180°F. Below this temperature their bonding properties are excellent.

Acrylate acid diester resins are anaerobic resins (cures when air is excluded) with good bonding properties. A family of one-part systems is sold under the trade name Loctite† for liquid lock nuts and bonding gears, shafts, and bushings. They are quick setting and useful from −65 to 300°F, and are available in a wide range of viscosities.

Fig. 83 Effect of temperature on epoxy-silicone adhesives.[53]

Cyanoacrylate adhesives are single-component adhesives which polymerize with the aid of moisture on the surfaces of the adherends. Bonds result after only a few seconds. Adhesive properties are rather low, and temperatures are limited to about 150°F. The cost of these adhesives is very high.

* Trademark of Union Carbide Corp., New York, N.Y.
† Trademark of Loctite Corporation, Newington, Conn.

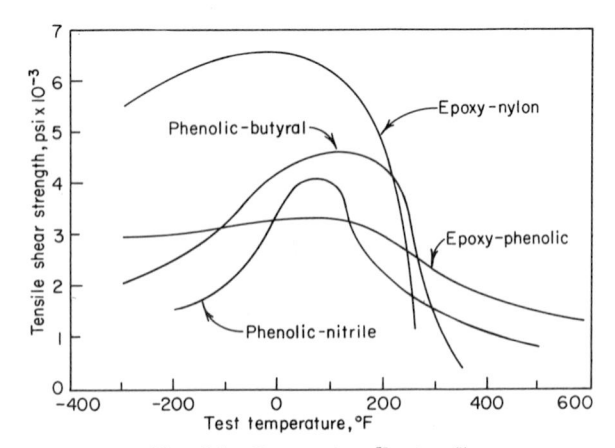

Fig. 84 Cryogenic adhesives.[51]

Polyimide and polybenzimidizole (polyaromatic or PBI) adhesives are useful up to 600°F. They are available coated onto glass cloth and can be exposed to 1000°F for short times. Cure temperatures are from 600 to 850°F. The cost of the adhesives is high.

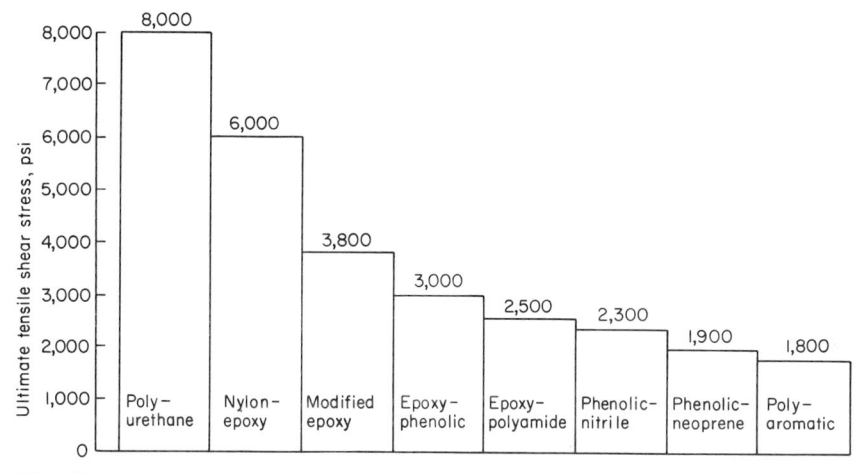

Fig. 85 Shear strength of adhesives at liquid hydrogen temperature (−423°F).[52]

PRINTED-CIRCUIT LAMINATES

The many advantages of printed circuits have resulted in their ever-increasing use in electronic applications. A *printed circuit* is defined to be an electric conducting path reproduced onto an insulation medium. In the most common case the insulating medium consists of plastic laminate with copper foil bonded to the surface. The desired electric circuit is printed onto the copper foil by any one of many reproduction techniques such as photography, offset printing, or silk screening, the object being to deposit in a precise pattern some polymer (resist) which is insoluble in a copper-etching media. The printed board is then passed through the etchant (often hot ferric chloride) where all unprinted copper foil is dissolved, leaving behind the desired circuit. The board is then passed through another bath where the resist is removed. Washing, drying, and in some cases plating conclude the process. Printed circuits offer to the designer weight reductions, decreased cost, and increased reliability, and are admirably suited for mass-production methods.

A vital part of a printed circuit is the plastic (copper-clad) laminate, for it is this material that dictates the board's physical and insulation properties. Theoretically any plastic or ceramic material can serve as the base for a printed circuit, and special applications have required the use of many unusual combinations. However, certain materials are more useful and have been standardized by NEMA (ASTM Standards are identical) and the Federal Government. A list of the NEMA grades of copper-clad is given in Table 65. Many copper-clad materials are also specified in Military Specifications. Frequently the Military Specification is not identical to commercial standards but requires a higher level of performance. However, most commercial copper-clad laminates can qualify under both commercial and military standards. Table 66 presents military standards for copper-clad laminates in comparison with similar NEMA Specifications. Military Specification MIL-P 13949D referenced in the table defines the copper-clad laminates and their properties, and is similar in many respects to NEMA LI 1-1965, part 10, and ASTM D 1807-64T. Design standards are defined in MIL-275B, and quality control and reliability standards are given in MIL-P-55110A.

2-106 Laminates, Reinforced Plastics, and Composite Structures

Phenolic paper copper-clad is often used for commercial applications where great strength and extreme environments are not encountered. The NEMA XXXPC is similar to NEMA XXXP but can be punched at about 140°F. FR-2 laminates have the properties of the XXXPC but are flame-retardant as well. All materials are medium brown in color and opaque, with an identifying code molded into the surface by the manufacturer.

TABLE 65 NEMA Copper-clad Laminates

NEMA grade	Resin	Reinforcement	Description
XXXP	Phenolic	Paper	Hot-punching grade for general use
XXXPC	Phenolic	Paper	Room-temperature punching for boards with small, close holes
G-10	Epoxy	Glass cloth	General-purpose glass base. Excellent electrical and physical properties. Good moisture resistance
G-11	Epoxy	Glass cloth	Similar to G-10 but more thermally stable
FR-2	Phenolic	Paper	Similar to XXXPC but flame-retardant
FR-3	Epoxy	Paper	Room-temperature punching, flame-retardant, better physical properties than XXXPC
FR-4	Epoxy	Glass cloth	Similar to G-10 but flame-retardant
FR-5	Epoxy	Glass cloth	Similar to G-11 but flame-retardant

TABLE 66 Specifications for Military Copper-clad

Military Specification	Similar NEMA standard	Description
MIL-P-13949D, Type PH	.....	Paper base, epoxy resin, hot punch, flame-retardant
MIL-P-13949D, Type PX	FR-3	Paper base, epoxy resin, flame-retardant
MIL-P-13949D, Type GB	G-11	Glass fabric base, epoxy resin, temperature-resistant
MIL-P-13949D, Type GC	.....	Glass fabric and nonwoven fabric base, polyester resin, flame-retardant
MIL-P-13949D, Type GE	G-10	Glass fabric base, epoxy resin, general purpose
MIL-P-13949D, Type GF	FR-4	Glass fabric base, epoxy resin, flame-retardant
MIL-P-13949D, Type GH	FR-5	Glass fabric base, epoxy resin, temperature resistant and flame-retardant
MIL-P-13949D, Type GP	.....	Glass fabric base, polytetrafluoroethylene resin
MIL-P-13949D, Type GT	.....	Glass fabric base, polytetrafluoroethylene resin
MIL-P-22324A, Type PEE	FR-3	Paper base, epoxy resin, flame-retardant
MIL-P-19161A, Type GTE	.....	Glass fabric base, polytetrafluoroethylene resin
MIL-P-997, Type GS	.....	Glass fabric base, silicone resin
MIL-P-15037, Type GM	.....	Glass fabric base, melamine resin
MIL-P-8013C, Types 1, 2, 3	.....	Glass fabric base, polyester resin

Epoxy paper copper-clad is NEMA grade FR-3. This material has physical properties superior to those of NEMA XXXPC and is flame-retardant. Type PH in MIL P-13949D is similar in properties to NEMA XXXP but is flame-retardant and, like NEMA XXXP, must be punched hot. These materials are light yellow to white in color and partly translucent.

Epoxy glass NEMA G-10 and G-11 are made with glass cloth and epoxy resin and are light green and semitransparent. They are used when optimum physical properties are required. Moisture resistance and dielectric properties are also excellent. Both grades behave similarly except that NEMA G-11 retains more than 50 percent of its room-temperature flexural strength when measured at 150°C after 1 hr aging at 150°C. Thus this grade has better thermal stability and should be employed when physical properties at elevated temperatures are critical. NEMA grades FR-4 and FR-5 are flame-retardant versions of NEMA G-10 and G-11, respectively. The NEMA FR-4 is yellow-green while NEMA FR-5 varies in color from green to brown depending on the manufacturer. Table 67 lists the properties of the NEMA copper-clad laminates.

TABLE 67 Properties of NEMA Copper-clad[54]

Property	Conditioning procedure by ASTM Methods D 618*	XXXP	XXXPC	FR-2	FR-3	FR-4	FR-5	G-10	G-11
Peel strength, min., lb/in. width									
1-oz copper:									
After solder dip	A	6	6	6	7	7	7	7	7
After elevated temperature	E-1/140†	6	6	6	7	7	7	7	7
2-oz copper:									
After solder dip	A	7	7	7	9	9	9	9	9
After elevated temperature	E-1/140†	7	7	7	9	9	9	9	9
Volume resistivity (min.), megohm-cm	C-96/35/90	10,000	10,000	10,000	100,000	100,000	100,000	100,000	100,000
Surface resistance (min.), megohms	C-96/35/90	1,000	1,000	1,000	1,000	1,000	1,000	5,000	5,000
Dielectric breakdown parallel to laminations, min., kv, step by step	D-48/50	15	15	15	30	30	30	30	30
Dielectric constant (avg. max.) at 1 MHz.‡	D-48/50	5.3	5.3	5.3	5.0	5.8	5.8	5.8	5.8
Dissipation factor (avg. max.) at 1 MHz.‡	D-48/50	0.05	0.05	0.05	0.045	0.045	0.045	0.045	0.045
Flexural strength (avg. min.), psi									
Lengthwise§	A	12,000	12,000	12,000	20,000	55,000	55,000	55,000	55,000
Crosswise§	A	10,500	10,500	10,500	16,000	45,000	45,000	45,000	45,000
Lengthwise¶		......	......	......	......	50,000	50,000	50,000	50,000
Crosswise¶		......	......	......	......	40,000	40,000	40,000	40,000
At elevated temperature, % of Condition A value retained, lengthwise:	E-1/150 T-150								
1/32-in. thickness		......	......	......	......	......	50	......	50
1/16- and 3/32-in. thicknesses		......	......	......	......	......	40	......	40
1/8- and 1/4-in. thicknesses		......	......	......	......	......	50	......	50
Flammability (avg. max.), seconds to extinguish	A	......	......	15	15	15	15		
Water absorption (avg. max.), %									
1/32-in. thickness		1.30	1.30	1.30	......	0.80	0.80	0.80	0.80
1/16-in. thickness		1.00	0.75	0.75	0.65	0.35	0.35	0.35	0.35
3/32-in. thickness		0.85	0.65	0.65	......	0.25	0.25	0.25	0.25
1/8-in. thickness		0.75	0.55	0.55	0.50	0.20	0.20	0.20	0.20

* Methods of Conditioning Plastics and Electrical Insulating Materials for Testing (ASTM Designation D 618).
† For grades XXXP, XXXPC, FR-2, and FR-3, use condition E-1/120.
‡ Applies only to 3/32- and 1/8-in. thicknesses.
§ Applies only to 1/32- to 3/32-in. thicknesses.
¶ Applies only to 1/8- and 1/4-in. thicknesses.

Testing copper-clad Certain properties relating to interactions of laminate and copper foil are unique to copper-clad, and specific test methods have been devised to measure them.

Oven Blister Test. This NEMA test is to evaluate the delamination of copper foil from the surface of the laminate. It consists of exposing the laminate to

2-108   Laminates, Reinforced Plastics, and Composite Structures

elevated temperature for various times and inspecting for blisters in the foil. No blistering should occur in the foil under the following conditions:

Grade	Temperature, °C	Time, min
XXXP, XXXPC, FR-2	120 ± 2	30
FR-3	120 ± 2	60
FR-4, FR-5, G-10, G-11	140 ± 2	60

Solder Float Test. This test is made on the copper-clad both before and after a pattern is etched. The test identifies the resistance of the copper bond to soldering conditions. In the case of unetched samples the laminate is floated copper side down in a solder bath for the times and temperature given in Table 68. At

TABLE 68   Solder Float Test[54]

	Composite thickness, in.				
	\$1/32\$ up to \$1/16\$			\$1/16\$ and over	
Grades	Time, sec	Temperature, °F	Time, sec	Temperature, °F
XXXP, XXXPC and FR-2	10	475 ± 3.6 (246.1 ± 2°C)	5	500 ± 3.6 (260 ± 2°C)
FR-3	10	500 ± 3.6 (260 ± 2°C)	10	500 ± 3.6 (260 ± 2°C)
FR-4, FR-5, G-10, and G-11	20	500 ± 3.6 (260 ± 2°C)	20	500 ± 3.6 (260 ± 2°C)

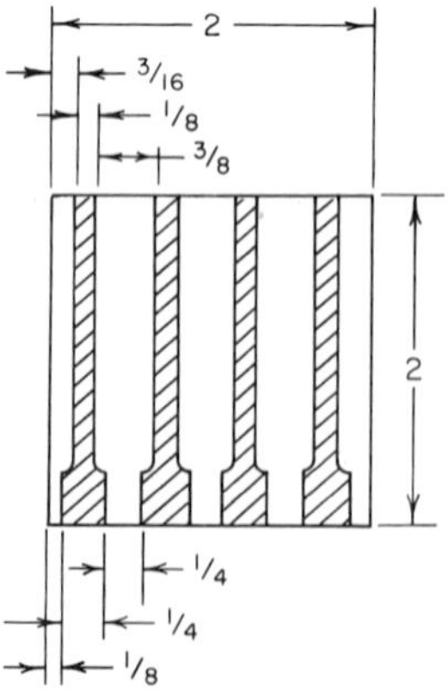

the end of the test no blistering or delamination should be seen in either foil or laminate. For etched samples a pattern like that shown in Fig. 86 is made. Etching processes are detailed in ASTM D 1825-65T. The process just described is performed, and a similar inspection is made. If no blistering has occurred the etched specimens are tested for peel strength as described next.

After exposure to solder the laminates should be inspected for color change and for "measling." Measling, a phenomenon sometimes observed in glass-fabric-based laminates, is caused by separation of individual glass fiber bundles at their crossover points and is observed as white spots. The Institute of Printed Circuits in the publication "Acceptability of Printed Circuit Boards" defines measling and provides rejection criteria. It should be noted that measling is purely cosmetic, and no impairment of any property has ever been attributed to this condition. Table 67 lists the properties of the NEMA copper-clad laminates.

Peel Strength Test. The sample from the solder float test is clamped horizontally, and about 1 in. is hand-peeled. A suitable machine then peels the foil from the surface at about 2 in./min. Peel strength are also often measured after oven aging.

Fig. 36   Solder float test sample.[54]

Solvent Resistance. To remove the resist from the completed wiring the circuit boards are generally exposed to trichloroethylene. Thus NEMA and the military suggest evaluating the boards for delaminations and blisters after exposure for 2 min over boiling trichloroethylene. Table 69 is a chart designed to aid in the selection of the proper copper-clad material. The value judgments offered are comparative only and not absolute.

TABLE 69 Copper-clad Selector Chart

Property	XXXP	XXXPC	G-10	G-11	FR-2	FR-3	FR-4	FR-5
Cost index	1.2	1.0	2.6	2.6	1.4	2.1	2.6	2.6
Copper bond strength	Good	Good	Excellent	Excellent	Good	Excellent	Excellent	Excellent
Flexural strength	Low	Low	High	High	Low	Medium	High	High
Water absorption	High	High	Low	Low	High	High	Low	Low
Flammability	Burns	Burns	Burns	Burns	Self-extinguishing	Self-extinguishing	Self-extinguishing	Self-extinguishing
Thermal stability	Low	Low	Medium	High	Low	Low	Medium	High
Punchability (room temperature)	Poor	Excellent	Good	Good	Excellent	Good	Good	Good
Dip solder	Good	Good	Excellent	Excellent	Good	Good	Excellent	Excellent
Electric strength	Good	Good	Excellent	Excellent	Good	Excellent	Excellent	Excellent
Insulation, resistance	Low	Low	High	High	Low	Low	Low	Low
Impact, strength	Low	Low	High	High	Low	High	High	High

Copper foil Copper foil used in printed circuits may be either rolled or electroformed, but most of it is electroformed. Most manufacturers treat the copper to oxidize the surface, which results in higher bond strength. Adhesives are used on the foil on phenolic copper-clad but not on epoxy copper-clad. The copper foil is supplied in thicknesses of 0.0014 in. (1 oz/ft^2) and 0.0028 in. (2 oz/ft^2). The standard tolerances for foil are as follows:

Thickness, in.	Tolerance, in. Plus	Minus
0.0014	0.0004	0.0002
0.0028	0.0007	0.0003

Other foil thicknesses are available in special orders. The copper surface in relation to pits and dents is described in MIL-P-13949D which assigns point values to the dimensions of the pit. A score exceeding 30 points per sq ft is sufficient cause for rejection. Another method is by comparison with photographic overlays which provide a standard with which to measure each defect in terms of area. The areas are summed, and the board is rejected at some critical value. The user should remember that in the typical printed circuit 95 percent of the copper is etched away; therefore the probability of a defect occurring in the circuit is small. Even when such a defect does occur within the pattern, it may not be of significance unless it is in a critical area. The purchaser must carefully examine his needs to avoid paying premium price for specially prepared boards free from defects.

Other metal foils are available such as aluminum, nickel, and steel but these materials are special-order items with attendant premiums.

TABLE 70 Thicknesses and Plus or Minus Thickness Tolerances of Copper-clad Sheet[55] (In inches)

Nominal overall thickness, including copper	Class I tolerances			Class II tolerances	
	XXXP, XXXPC, FR-2, and FR-3	FR-4, FR-5, G10, and G-11	XXXP, XXXPC, FR-2, and FR-3	FR-4, FR-5, G-10, and G-11		
	1-oz copper, 1 side	1-oz copper, 2 sides; and 2- to 5-oz copper, 1 and 2 sides	All weights of copper, 1 and 2 sides	
1/32 (0.031)	0.004	0.0045	0.0065	0.003	0.004
3/64 (0.046)	0.005	0.0055	0.0075	0.0035	0.005
1/16 (0.062)	0.0055	0.006	0.0075	0.004	0.005
3/32 (0.093)	0.007	0.0075	0.009	0.005	0.007
1/8 (0.125)	0.0085	0.009	0.012	0.006	0.009
5/32 (0.156)	0.0095	0.010	0.015	0.007	0.010
3/16 (0.187)	0.010	0.011	0.019	0.008	0.011
7/32 (0.218)*	0.011	0.012	0.021	0.009	0.012
1/4 (0.250)*	0.012	0.012	0.022	0.009	0.012

* These values do not apply to Grades XXXPC and FR-2.

TABLE 71 Thicknesses and Plus or Minus Thickness Tolerances in Length and Width of Fabricated Panels Cut by Sawing[55]

Nominal thickness, in............. 1/32–1/4
Tolerance:
  6 in. and under................. 0.010
  Over 6 to 18 in................. 0.015
  18 in. and over................. 1/32

TABLE 72 Thicknesses and Plus or Minus Thickness Tolerances in Length and Width of Fabricated Panels Cut by Shearing, in Lengths Less than 48 in.[55]

Nominal thickness, in............ 1/32–1/16
Tolerance
Squaring sheared:
  2 in. and under................. 0.010
  Over 2 to 6 in.................. 0.015
  Over 6 to 24 in................. 1/32
  Over 24 in...................... 3/64
Rotary sheared:
  3/64 to 3/4 in................... 0.005
  Over 3/4 to 3 in................ 0.005
  Over 3 to 6 in.................. 0.010

Dimensions and tolerances Since a copper-clad is a composite made of layers of insulation and of foil, the total thickness tolerance is determined by the tolerances of the individual layers and the care that the laminator exercises during the molding process. Table 70 presents standard tolerances prepared by NEMA.

Copper-clad laminates are cut to shape either by sawing on a band or circular saw or by shearing. The area tolerances depend on which method is used and are presented in Tables 71 and 72.

Copper-clad laminates received from the manufacturer invariably demonstrate a certain warp or twist. This is caused by strains of molding and differences in thermal expansion between copper and laminate. NEMA identifies the maximum twist of copper-clad as shown in Table 73. The change in dimension of a copper-

TABLE 73 Maximum Warp or Twist of Copper-clad Sheet[55]
(Based on 36-in. dimension,* percent)

Range of thickness, in.	Class A			Class B†		
	1- and 2-oz copper, 1 side	1- and 2-oz copper, 2 sides		1- and 2-oz copper, 1 side	1- and 2-oz copper, 2 sides	
		Glass‡	Paper§		Glass‡	Paper§
$\frac{1}{32}$ to $\frac{3}{64}$	12	5	6	10	2	5
Over $\frac{3}{64}$ to $\frac{1}{16}$	10	5	6	5	1	2½
Over $\frac{1}{16}$ to $\frac{1}{8}$	8	3	3	5	1	2½
Over $\frac{1}{8}$ to $\frac{1}{4}$	5	1.5	1.5	5	1	1½

* These values apply only to sheet sizes as manufactured and to cut pieces having either dimension not less than 18 in. In the case of warp, this percentage is stated in terms of the lateral dimension (length or width); in the case of twist, the percentage is stated in terms of the dimension from one corner to the diagonally opposite corner.

† Where the intended use requires Class B warp or twist, the purchaser should so specify. (This note has been approved as Authorized Engineering Information.)

‡ Glass grades: G-10, G-11, FR-4, and FR-5.

§ Paper grades: XXXP, XXXPC, FR-2, and FR-3.

clad is always a problem to the designer. When the copper is etched from the surface of a laminate, locked-in strains are removed, and certain dimensional changes occur. These changes may vary from 0.001 to 0.010 in./in. laminate and are rarely predictable. It is always wise to investigate new patterns for distortion. It is equally wise to carefully identify every batch of copper-clad throughout the process so that manufacturing problems can be referred to the copper-clad manufacturer when appropriate.

Multilayer Printed Circuits

Development of Multilayer Circuits. Multilayer printed circuits are individual circuit layers which have been bonded together and interconnected. First used experimentally in the early 1960s, they are frequently employed with integrated circuits. Multilayer circuits are extremely precise because of the complicated miniaturized patterns required. Circuits one foot square with 10 layers and a total thickness of 0.100 in. are common. Such a circuit may contain many plated-through holes 0.020 in. in diameter on 0.050 in. centers. Circuits have been made with as many as 40 layers.

Manufacturing Process. As many manufacturing techniques for multilayer circuits exist as there are manufacturers. In general, however, the following basic process is used. The thin copper-clad sheets have the desired circuit impressed onto the

2-112 Laminates, Reinforced Plastics, and Composite Structures

copper—often by photographic methods. The copper is etched using conventional etching practices. The individual layers are laminated together using B-stage prepreg. Holes are then drilled at the appropriate places and to the correct depth. The internal connections from layer to layer are then made by plating copper through the holes, resulting in the completed multilayer circuit.

Multilayer Copper-clad. The copper-clad laminates used in multilayer circuits are the same materials employed in conventional printed circuits except they are much thinner. These laminates are available in 0.002 in. increments from 0.002 to $\frac{1}{32}$ in. in thickness. Thickness tolerances for the base copper-clad as specified by NEMA are shown in Table 74, and tolerances for cut sheets of copper-clad

TABLE 74 Thicknesses and Plus or Minus Thickness Tolerances of Multilayer Base Laminates after Etching[56]

Nominal thickness, excluding copper, in.	Tolerance, in.
0.002 to 0.0045	0.001
Over 0.0045 to 0.006	0.0015
Over 0.006 to 0.012	0.002
Over 0.012 up to but not including $\frac{1}{32}$	0.003

less than 18 in. on a side are given in Table 75. Copper tolerances are the same as those for conventional printed circuits presented in the previous section.

The condition of the copper surface is of more concern in multilayer laminates than in the regular printed circuits because of the need for exact alignment of circuit lines. NEMA has provided a variety of copper quality guides including pinholes, roughness, scratches, pits, and indentations.

Although any polymer could be used as the base in multilayer circuits, the excellent electrical properties and bond strength of epoxy resins make the choice of epoxies particularly attractive. Two grades have been standardized by NEMA: G-10-UT and FR-4-UT. These materials—except for their thicknesses—are similar

TABLE 75 Thicknesses and Plus or Minus Tolerances in Length and Width of Cut Sheets of Multilayer Copper-clad[56]

Thickness, in.	2 in. and under	Over 2 to 6 in.	Over 6 to 18 in.
Under 0.010	0.015	0.031	0.046
0.011 to 0.015	0.015	0.020	0.031
0.016 to under 0.031	0.010	0.020	0.031

to the previously discussed NEMA grades G-10 and FR-4. Table 76 shows the properties of these two materials.

Multilayer Prepreg Materials. Prepregs as used in multilayer circuits consist of a glass fiber fabric saturated with a resin—usually epoxy—and partly reacted to a dry tack-free condition. With heat and pressure the resin softens, flows, and acts as an adhesive to bond adjacent circuit layers together. Many epoxy prepreg materials are available from the same suppliers who make prepreg materials for laminates and also from suppliers of the copper-clad laminates. In use the prepreg is cut to the correct shape and inserted between the various circuits layers. Locating pins in a molding frame hold the various layers in register. The frame and lay-up are placed into a molding press with platens aligned accurately to

Printed-circuit Laminates

TABLE 76 Property Values for NEMA Grades G-10-UT and FR-4-UT Laminates[56]

Property to be tested	Conditioning procedure	Grade G-10-UT	Grade FR-4-UT	Test methods
Peel strength, min., lb/in. width:				
1-oz copper................	A	6	6	LI 1-13.08
2-oz copper................	A	8	8	LI 1-13.08
Volume resistivity (min.), megohm-cm....................	C-96/35/90	10^6	10^6	LI 1-13.09
Surface resistance (min.), megohms.....................	C-96/35/90	10^4	10^4	LI 1-13.09
Dielectric constant (max. avg.) at 1 MHz (1/32-in. laminate)..	D-24/23	5.4	5.4	LI 1-13.10
Dissipation factor (max. avg.) at 1 MHz (1/32-in. laminate)....	D-24/23	0.035	0.035	LI 1-13.10
Flammability................	A	......	Self-extinguishing	LI 1-13.11

within 0.001 in. in parallel. The lay-up is then heated at a rate and to a temperature as specified by the prepreg supplier. Careful control of pressure is needed. Pressing temperatures vary from 300 to 450°F, and molding pressures average 500 psi. Molding times may reach 1 hr. The molded circuit must be cooled under pressure to 120°F, and then it can be removed from the press.

Most of the rejections of multilayer circuits result from an improper molding process. Causes of rejects include blisters, voids, delamination, measling, displaced layers, and board distortions. The last-named condition is very common and results from stresses induced in the molding process. Warped circuits can often be straightened by clamping them between flat plates and heating them at 250°F for 1 hr.

Flexible Printed Circuits

A special case of printed circuits is the flexible printed circuit in which the reinforced laminate base is replaced by a very thin, flexible insulation member such as a fiber, film, or paper. Advantages of flexible printed circuits include

TABLE 77 Properties of Flexible Circuit Materials

Base insulation	Polyester film	Varnished glass cloth	Epoxy glass cloth	Resin-impregnated paper
Insulation thickness, mils.......	1–5	3–12	1–3	1–3
180° bond strength, 1 oz copper, lb/in. width...............	8	8	8	
Tensile strength, lb/in., width, machine direction..........	125	125	125	30
Dissipation factor at 1 MHz....	0.016	0.026	0.020	0.032
Dielectric constant at 1 MHz...	3.0	3.2	3.0	2.5
Dielectric strength, volts/mil, short time.................	5,000	2,000	2,000	1,300
Surface resistivity, ohms/sq.....	10^{12}	10^{12}	10^{12}	10^{11}
Fold endurance at 1 kg, machine direction, cycles.............	14,000	10,000	150	1,300

2-114   Laminates, Reinforced Plastics, and Composite Structures

space saving; ability to form the circuit into a complicated, irregular area; automated manufacturing processes; and multilayer capability. Flexible circuits are finding increasing use in computers, communications devices, printed flat cables, and electronic applications wherein it is desired to move one end of the circuit with respect to the other as in drawer applications. Table 77 shows typical properties of certain flexible circuit materials. Copper-foil weights available include 1, 2, and 3 oz/ft^2.

Microwave Strip Line

Microwave strip line is a unique application of a specialized printed circuit. Microwave energy necessitates insulation materials with low loss factors. Although some traditional polymeric materials possessed these qualities, they frequently had too much variability to permit their use. Recently, however, quality advances have been made, and several resin systems can be used for strip line.

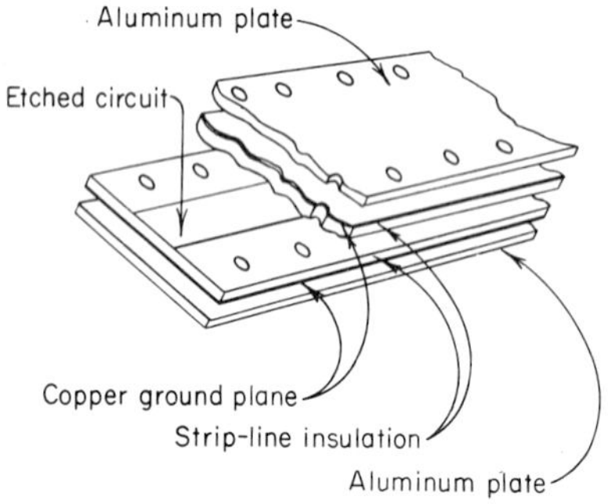

Strip line is a type of coaxial cable but in the form of a printed circuit. To make a strip line a sheet of double copper-clad (copper on both sides) is etched on one side to produce the necessary circuit. A second copper-clad board with copper on one side only is placed over top of the etched board with the insulation side adjacent to the circuit. These two boards are then sandwiched between aluminum cover plates held together with screws or rivets. The screws not only hold the assembly together but also provide a grounding plane on the sides. This arrangement is shown in Fig. 87.

Fig. 87   Strip-line circuit.[57]

Three materials are customarily used for strip line: polytetrafluoroethylene, polystyrene, and radiation crosslinked polyethylene. By virtue of cost and stability, the last of these is greatly popular. Table 78 presents data on the electrical properties of strip-line insulations.

TABLE 78   Electrical Properties of Strip-line Insulations

Property	Polyethylene	Polytetrafluoro-ethylene-glass	Polystyrene
Dielectric constant at 10MH$_3$–10GH$_3$	2.32	2.30	2.53
Dissipation factor:			
At 1 MHz	0.00004	0.0006	0.00012
At 10 MHz	0.00006	0.0006	0.00025
At 10 GHz	0.00018	0.0006	0.00006
Volume resistivity, ohm-cm	>10^{16}	>10^{13}	>10^{16}
Surface resistivity, ohms/sq	>10^{12}	>10^{14}	>10^{14}
Dielectric strength, step by step, volts/mil	500	300	500
Specific gravity	0.94	2.15	1.05
Tensile strength, psi	3,000	7,500	7,000
Flexural strength, psi	8,000		11,500
Impact strength, ft-lb/in	3.0	2.0	0.3
Coefficient of thermal expansion, in./(in.)(°C) × 10^{-5}	10	3	7
Thermal conductivity, Cal/(sec)(cm^2)(°C/cm) × 10^{-4}	12.7	2	3.5
Operating temperature range, °C	−80 to 100	−80 to 250	−60 to 100

FILMS AND FLEXIBLE INSULATIONS

Films-definition *Films* are thin sections of the same polymers described in the first chapter of this handbook. Most films are thermoplastic in nature because of the great flexibility of this class of resins. Although films can be made from all thermoplastics, many of them are used in applications other than electrical and hence are of no interest here.

Films are made by extrusion, casting, calendering, and skiving. Certain of the materials are also available in foam form. The films are sold in thicknesses from ½ to 10 mils. Thicknesses in excess of 10 mils are properly called *sheets*.

Tapes are films slit to some acceptable size and are frequently coated with adhesives. The adhesives are either thermosetting or thermoplastic. The thermoset adhesives consist of rubber, acrylic, silicones, and epoxies whereas the thermoplastic adhesives are generally acrylic or rubber. Tackifying resins are generally added to increase the adhesion. The adhesives all deteriorate with storage. The deterioration is marked by loss of tack and bond strength and can be inhibited by storage at low temperature.

Film properties Films differ from similar polymers in other forms in several key properties but are identical in all others. Since the first chapter in this handbook described in detail most of the thermoplastic resins, this section will be limited to film properties only. Table 79 presents the properties of common electrical films. To aid in selection of the proper films the most important features are summarized in Table 80.

Films differ from other polymers chiefly in improved electric strength and flexibility. Both of these properties vary inversely with film thickness. Electric strength is also related to the method of manufacture. Cast and extruded films have higher electric strength than skived films. This is caused by the greater incidence of holes in the latter films. Some films can be oriented, which substantially improves their physical properties. Orientation is a process of selectively stretching the films, thereby reducing the thickness and causing changes in the crystallinity of the polymer. This process is usually accomplished under conditions of elevated temperature, and the benefits are lost if the processing temperatures are exceeded during service.

Most films can be bonded to other substrates with a variety of adhesives. Those films that do not readily accept adhesives can be surface-treated for bonding by chemical and electrical etching. Films can also be combined to obtain bondable surfaces. Examples of these combined films are polyolefins laminated to polyester films and fluorocarbons laminated to polyimide films.

Cellulose films are low-cost films with a wide range of properties. Based on wood, they take various chemical forms such as cellophane, cellulose acetate, cellulose acetate butyrate, and cellulose triacetate. They are used as transformer insulation, phase separators, and coil ground sheets. Their temperature limit is about 80°C.

Fluorocarbon films are versatile films which are widely used in electronic applications. These films include polytetrafluoroethylene (PTFE), fluoronated ethylene polymer (FEP), polytrifluorochloroethylene, and polyvinyl fluoride (PVF). With the exception of the last-named film, these polymers can be used at 180° and in some applications to 220°C. The FEP and PTFE materials (trade names Teflon 100-X and Teflon, respectively) have very low dielectric properties which are steady with increasing frequency and temperature. The films do not absorb water and are unaffected by most chemicals. Creep and cold cut-through are the chief weaknesses of the materials. They are used as microwave insulation and in coaxial cables and cable and wire wrap.

Polyamide films (nylons) and polymethyl methacrylate (acrylic) are specialty films little used for insulation in the United States. The nylons are water-sensitive, and both nylons and acrylics have marginal electrical properties. However, they are low in cost and have excellent physical properties. Polyethylene films are in wide use as electronic insulations, coaxial cables, and coil insulations. With a low density and resin cost, the polyethylenes are among the most economic

TABLE 79 Properties of Common Electrical Films[58]

Property	ASTM test No.	Cellulose acetate	Cellulose triacetate	Cellulose acetate butyrate	FEP fluorocarbon	Polyamide Nylon 6
Type of plastic base						
1. Method of processing	.....	Casting, extrusion	Casting	Casting, extrusion	Extrusion	Extrusion
2. Forms available	.....	Sheets, rolls, tapes	Sheets, rolls	Sheets, rolls	Rolls, sheets, tubes	Rolls, tubes
3. Thickness range, in.	.....	0.0005–0.250	0.0008–0.020	0.0011–0.150	0.0005–0.020	0.0005–0.030
4. Maximum width, in.	.....	40–60	40–46¼	40 cast / 42 extruded	46–48	54
5. Area factor, in.²/(lb)(mil)	.....	21,000–22,000	21,000–22,000	23,000–23,300 (1-mil)	12,900	24,500
6. Specific gravity	D 1505-63T	1.28–1.31	1.28–1.31	1.19–1.20	2.15	1.13
7. Tensile strength, psi	D 882-61T	8,000–16,400	9,000–16,000	5,000–9,000	2,500–3,000	9,000–18,000
8. Elongation, %	D 882-61T	15–70	10–40	50–100	300	250–550
9. Bursting strength at 1-mil thickness, Mullens points	D 774-63T	30–60	50–70	40–70	11	Elongates
10. Tearing strength, g[a]	D 1922-61T	4–10 (1-mil)	4–10	5–10 (1-mil)	125	50–90
11. Tearing strength, lb/mil[b]	D 1004-61	1–2	55–395	80–105	600	1,000–1,210
12. Folding endurance	D 2167-63T	500–2,000 (1-mil)	1,000–4,000 (1-mil)	800–1,200 (1-mil)	4,000	>250,000
13. Water absorption after 24 hr, %	D 570-63	5–9 (5-mil)	2–4.5	1–2	<0.01	9.5
14. Water-vapor permeability: g/(100 in.²)(24 hr)(mil) at 25°C g/(m²)(24 hr)(mm thickness) at 25°C	E-96-635(E)	3,000 / 475–1,200	2,000 / 790	2,000 / 790	0.40	5.4–20 at 38°C.
15. Permeability to gases, cc/(100 in.²)(mil thickness)(24 hr)(atm) at 25°C:	D 1434-63					
CO_2		860–1,000	880	6,000	1,670	9.7–45
H_2		835	.....	.....	2,200	90–250
N_2		30–40	30	250	320	0.9–6
O_2		117–150	150	950	750	2.6 (Dry)–25

Resistance rating[c]					
16. Strong acids	D 543-63T	F-P	F	F-G	E	P
17. Strong alkalies	D 543-63T	P	P	P	E	E
18. Grease and oils	D 722-45	G	E	G	E	E
19. Organic solvents	D 543-63T	P	F to P	P	E	E
20. Water	E 96-63T	G	G	G	E	E to P
21. High relative humidity	D 756-56	G	G	G	E	E to P
22. Sunlight	D 1435-58	E	E	E	E	F to G
23. Resistance to heat, °F	D 759-48	150–200	300–400	120–180	400–525	200–400
24. Resistance to cold, °F	D 759-48	−15			−425	−100
25. Change in linear dimensions at 100°C for 30 min	D 1204-54	+0.2 to −3.0	0 to −0.8	+0.1 to −3.0	<1	+2 to −2
26. Flammability	D 1433-58	Slow to self-extinguishing	Self-extinguishing	Slow burning	Nonflammable	Self-extinguishing
27. Burning rate, in./sec		0.2–2.2 (Based on 1-mil)	0.2–0.4 (Based on 1-mil)	0.2–1.2 (Based on 1-mil)		
Electrical						
Dielectric constant:						
28. At 1 kHz	D 150-64T	3.6	3.2	2.9	2.0	3.7 (dry)
29. At 1 MHz	D 150-64T	3.2	3.3	2.5	2.0	3.4 (dry)
30. At 10³ MHz	D 150-64T	3.2	3.2	2.8		
Dissipation factor						
31. At 1 kHz	D 150-64T	0.013	0.016	0.013	0.0002	0.016 (dry)
32. At 1 MHz	D 150-64T	0.038	0.033	0.030	0.0007	0.025 (dry)
33. At 10³ MHz	D 150-64T					
34. Dielectric strength, volts/mil		3,200–5,000	3,700	3,100	5,000	1,300–1,500 (2-ml)
35. Volume resistivity, ohm-cm		10¹⁰–10¹⁴	10¹³–10¹⁵	10¹¹–10¹⁵	10¹⁹	
36. Heat-sealing temperature range, °F		350–450			540–700	380–450

[a] Propagating. [b] Initial. [c] Code: E = excellent, G = good, M = moderate, F = fair, P = poor. [d] Commercial standard CS192-53.

2-117

TABLE 79 Properties of Common Electrical Films[58] (Continued)

Property	ASTM test No.	Polyamide Nylon 66	Polytrifluoro-chloro-ethylene copolymers	Polymethyl methacrylate	Polyethylene Low density	Polyethylene Medium density	Polyethylene High density
Type of plastic base							
1. Method of processing		Extrusion	Extrusion	Extrusion, two-way stretch, orientation	Extrusion	Extrusion	Extrusion
2. Forms available		Rolls	Rolls	Rolls, sheets	Rolls, sheets, tapes, tubes	Rolls, sheets, tapes, tubes	Rolls, sheets, tapes, tubes
3. Thickness range, in.		0.005–0.020	0.0005–0.030	0.005–0.010	0.0003 & up	0.0003 & up	0.0004 & up
4. Maximum width, in.		20	54	43	480	240	60
5. Area factor, in.^{2}/(lb)(mil)		24,200	13,000	23,400	30,000	29,500	29,000
6. Specific gravity	D 1505-63T	1.14	2.08–2.15	1.18–1.19	0.910–0.925	0.926–0.940	0.941–0.965
7. Tensile strength, psi	D 882-61T	9,000–12,000	5,000–8,000	8,200–8,800	1,500–3,000	2,000–3,500	2,400–6,100
8. Elongation, %	D 882-61T	200	50–150	4–12	100–700	50–650	10–650
9. Bursting strength at 1-mil thickness, Mullens points	D 774-63T		23–31		10–12		
10. Tearing strength, g[a]	D 1922-61T		8–29		100–500	50–300	15–300
11. Tearing strength, lb/mil[b]	D 1004-61		330–900	340–380	65–575		
12. Folding endurance	D 2167-63T		Good		Excellent	Very high	Very high
13. Water absorption after 24 hr, %	D 570-63	8.9	Nil	0.3–0.4	<0.01	<0.01	Nil
14. Water-vapor permeability: g/(100 in.^{2})(24 hr)(mil) at 25°C g/(m^2)(24 hr)(mm thickness) at 25°C	E-96-635(E)	3–6	0.025–0.055	0.5	1.0–1.5 0.512	0.7	0.3
15. Permeability to gases, cc/(100 in.^{2})(mil thickness)(24 hr)(atm) at 25°C:	D 1434-63						
CO_2		9.1	16–40		2,700	2,500	580
H_2			220–330			1,950	
N_2		0.35	2.5		180	315	42
O_2		5.0	7–15		500	535	185

2-118

Resistance rating[c]						
16. Strong acids	D 543-63T	P	E	E	E	E
17. Strong alkalies	D 543-63T	E	E	E	E	E
18. Grease and oils	D 722-45	E	E	P	G	G to E
19. Organic solvents	D 543-63T	E	E	E	G (60°C)	G (80°C)
20. Water	E 96-63T		E		E	E
21. High relative humidity	D 756-56		E	G	E	E
22. Sunlight	D 1435-58	P	E	G	F to G	F to G
23. Resistance to heat, °F	D 759-48		300–390	155–190	180–200	220	250
24. Resistance to cold, °F	D 759-48		−320	Good	−70	−70	−50
25. Change in linear dimensions at 100°C for 30 min	D 1204-54		+2 to −2		Depends on orientation		Nil
26. Flammability	D 1433-58	Self-extinguishing	Non-flammable	Slow burning	Slow burning	Slow burning	Slow burning
27. Burning rate, in./sec[d]				0.1	<1.2		
Electrical							
Dielectric constant:							
28. At 1 kHz	D 150-64T	4.0	2.5–2.7	3.5–4.0	2.2	2.2	2.3
29. At 1 MHz	D 150-64T	3.4	2.3–2.4	3.0–3.5	2.2	2.2	2.3
30. At 10³ MHz	D 150-64T		2.3	2.58	2.2	2.2	2.3
Dissipation factor							
31. At 1 kHz	D 150-64T		0.022–0.024	0.040	0.0003	0.0003	0.0005
32. At 1 MHz	D 150-64T		0.009–0.017	0.030	0.0003	0.0003	0.0005
33. At 10³ MHz	D 150-64T		0.004	0.009	0.0003	0.0003	0.0005
34. Dielectric strength, volts/mil		850	1,000–3,700	400	4,700	500	500 (Varies with thickness)
35. Volume resistivity, ohm-cm		4.5 × 10¹³	10¹⁸	10¹⁵	10¹⁶	10¹⁶	10¹⁶
36. Heat-sealing temperature range, °F			350–400		250–400	250–400	275–400

[a] Propagating. [b] Initial. [c] Code: E = excellent, G = good, M = moderate, F = fair, P = poor. [d] Commercial standard CS192-53.

TABLE 79 Properties of Common Electrical Films[58] (Continued)

Property	ASTM test No.	Polystyrene	Regenerated cellulose (cellophane)	Polycarbonate	Polyimide	Polyvinyl fluoride
Type of plastic base						
1. Method of processing		Extrusion oriented	Extrusion into bath	Casting, extrusion		Extrusion
2. Forms available		Rolls, sheets	Sheets, rolls, ribbons	Rolls (cast) Rolls (extr.)	Rolls	Rolls
3. Thickness range, in.		0.001	0.0008–0.0019	0.0005–0.020	0.0005–0.005	0.0005–0.002
4. Maximum width, in.		40	60	45 cast, 36 ext.	18	138
5. Area factor, in.²/(lb)(mil)	D 1505-63T	26,100	11,600–25,000	23,100	19,400	17,200
6. Specific gravity	D 882-64T	1.05–1.07	1.40–1.50	1.20	1.42	1.38
7. Tensile strength, psi	D 882-64T	7,000–12,000	7,000–18,000	8,400–8,800	25,000	7,000–18,000
8. Elongation, %	D 774-63T	10	10–50	85–105	70	115–250
9. Bursting strength at 1-mil thickness, Mullens points		30	55–65	No break 4 mil, 25–35	75	19–70
10. Tearing strength, g[a]	D 1922-61T	2–8	2–20	20–25	8	12–40
11. Tearing strength, lb/mil[b]	D 1004-61		110–515	1,150–1,570	510 gm./mil	997–1,400
12. Folding endurance	D 2167-63T	Low	45–115	250–400	10,000	5,000–47,000
13. Water absorption after 24 hr, %	D 570-63	0.04		0.35	2.9	<0.5
14. Water-vapor permeability:						
g/(100 in.²)(24 hr)(mil) at 25°C	E-96-635 (E)	6.2	0.4–134	11.0	5.4	3.24
g/(m²)(24 hr)(mm thickness) at 25°C				6.7		
15. Permeability to gases, cc/(100 in.²)(mil thickness)(24 hr)(atm) at 25°C:						
CO₂	D 1434-63	926	0.4–6.0	1,075	45	15
H₂			1.2–2.2	1,600	250	58
N₂		42	0.5–1.6	50	6	0.25
O₂		213	0.2–5.0	300	25	3

2-120

Resistance rating[c]						
16. Strong acids	D	543-63T	G	P	G	E
17. Strong alkalies	D	543-63T	E	P	P	E
18. Grease and oils	D	722-45	G	E	E	E
19. Organic solvents	D	543-63T	P	E to P	E	E
20. Water	E	96-63T	E	G	E	E
21. High relative humidity	D	756-56	E	G	E	E
22. Sunlight	D	1435-58	F	M	E	
23. Resistance to heat, °F	D	759-48	160	270	750	220–250
24. Resistance to cold, °F	D	759-48	. . .	−150	−450	−100
25. Change in linear dimensions at 100°C for 30 min	D	1204-54	Nil	0	30 min at 250°C, 0.3%	−1
26. Flammability	D	1433-58	Slow burning	Flammable	Self-extinguishing	Slow to self-extinguishing
27. Burning rate, in./sec[d]			. . .	0.7–2.3	. . .	. . .
Electrical						
Dielectric constant:						
28. At 1 kHz	D	150-64T	2.4	3.2 (uncoated)	3.5	8.5
29. At 1 MHz	D	150-64T	2.4	. . .	3.4	7.4
30. At 10³ MHz	D	150-64T	. . .	. . .	. . .	. . .
Dissipation factor						
31. At 1 kHz	D	150-64T	0.005	0.015 (uncoated)	0.003	1.6
32. At 1 MHz	D	150-64T	0.005	. . .	0.010	. . .
33. At 10³ MHz	D	150-64T	. . .	. . .	. . .	. . .
34. Dielectric strength, volts/mil			500	2,000–2,500 (uncoated)	7,000	3,500
35. Volume resistivity, ohm-cm			. . .	10¹¹	10¹⁸	3 × 10¹³
36. Heat-sealing temperature range, °F			250–350	200–350	. . .	400–425

[a] Propagating.   [b] Initial.   [c] Code: E = excellent, G = good, M = moderate, F = fair, P = poor.   [d] Commercial standard CS192-53.

2-121

TABLE 79 Properties of Common Electrical Films[58] (Continued)

Property	ASTM test No.	Polypropylene Extrusion (cast)	Polypropylene Biaxially oriented	Polyester (PE terephthalate)	Polytetra-fluoroethylene
Type of plastic base					
1. Method of processing	....	Extrusion (cast)	Orientation	Casting, extrusion	Skiving, casting, extrusion
2. Forms available	....	Sheets, rolls, tapes	Rolls	Sheets, rolls, tapes, tubes	Sheets, tapes, tubing
3. Thickness range, in	....	0.0005 & up	0.0005–0.00125	0.00015–0.014	Up to 0.010
4. Maximum width, in	....	60	48–72	60–120	38
5. Area factor, in.^{2}/(lb)(mil)	....	30,900–31,300	30,600	19,800–22,600	12,800
6. Specific gravity	D 1505-63T	0.885–0.9	0.902–0.907	1.380–1.399	2.1–2.2
7. Tensile strength, psi	D 882-64T	4,500–10,000	12,000–33,000	20,000–30,000	1,500–4,000
8. Elongation, %	D 882-64T	550–1,000	50–200	40,000 (Type T) 70–120; 50 (Type T)	100–350
9. Bursting strength at 1-mil thickness, Mullens points	....	....	....	55–80	....
10. Tearing strength, g^a	D 774-63T	600-TD 25-MD	7–20	12–27	....
11. Tearing strength, lb/mil^b	D 1922-61T	....	....	1,000–1,300	10–100
12. Folding endurance	D 1004-61	Very high	Excellent	>100,000	....
13. Water absorption after 24 hr, %	D 2167-63T	0.005 or less	<0.005	<0.8	0.00
14. Water-vapor permeability:	D 570-63				
g/(100 in.^{2})(24 hr)(mil) at 25°C	E 96-63T(E)	0.4–1.0	0.35–0.45	1.7–1.8	
g/(m^2)(24 hr)(mm thickness) at 25°C	....	0.1575	0.14		
15. Permeability to gases, cc/(100 in.^{2})(mil thickness)(24 hr)(atm) at 25°C:					
CO_2	D 1434-63	800	370	16	
H_2	....	1,700	....	100	
N_2	....	48	....	0.7–1.0	
O_2	....	240	120	3.5–6.0	

Resistance rating[c]					
16. Strong acids.	D	543-63T	E	E	E
17. Strong alkalies.	D	543-63T	E	E	E
18. Grease and oils.	D	722-45	G	G to E	E
19. Organic solvents.	D	543-63T	G-E	G	E
20. Water.	E	96-63T	E	E	E
21. High relative humidity.	D	756-56	E	E	E
22. Sunlight.	D	1435-58	F to E	F to G	E
23. Resistance to heat, °F.	D	759-48	270–300	285	M; E (Type W) 300
24. Resistance to cold, °F.	D	759-48	0	−60	−75
25. Change in linear dimensions at 100°C for 30 min	D	1204-54		Under 2% if head = set	<0.5
26. Flammability.	D	1433-58	Slow burning	Slow burning	Slow burning to self-extinguishing
27. Burning rate, in./sec[d]					Nonflammable
Electrical					
Dielectric constant:					
28. At 1 kHz.	D	150-64T	2.0–2.1	2.2	3.2	2.0–2.1
29. At 1 MHz.	D	150-64T	2.0–2.1	2.2	3.0	2.0–2.1
30. At 10³ MHz.	D	150-64T	2.0–2.1		2.8	2.0–2.1
Dissipation factor					
31. At 1 kHz.	D	150-64T	0.0003		0.005	0.0002
32. At 1 MHz.	D	150-64T	0.0003		0.016	0.0002
33. At 10³ MHz.	D	150-64T	0.0003		0.008	0.0002
34. Dielectric strength, volts/mil.			3,000–4,500		7,500 (1-mil)	430
35. Volume resistivity, ohm-cm.			3 × 10¹⁵		10¹⁸	10¹⁸
36. Heat-sealing temperature range, °F.			285–400	320–340		

[a] Propagating. [b] Initial. [c] Code: E = excellent, G = good, M = moderate, F = fair, P = poor. [d] Commercial standard CS192-53.

TABLE 80 Film Selector Chart

Film	Cost	Thermal stability	Dielectric constant	Dissipation factor	Strength	Electric strength	Water absorption	Folding endurance
Cellulose	Low	Low	Medium	Medium	High	Medium	High	Low
FEP fluorocarbon	High	High	Low	Low	Low	High	Very low	Medium
Polyamide	Medium	Medium	Medium	Medium	High	Low	High	Very high
PTFE polytetrafluoroethylene	High	High	Low	Low	Low	Low	Very low	Medium
Acrylic	Medium	Low	Medium	Medium	Medium	Low	Medium	Medium
Polyethylene	Low	Low	Low	Low	Low	Low	Low	High
Polypropylene	Low	Low	Medium	Low	Low	Low	Medium	Low	High
Polyvinyl fluoride	High	High	High	High	High	Medium	Low	High
Polyester	Medium	Medium	Medium	Low	High	High	Low	Very high
Polytrifluorochloroethylene	High	High	Low	Low	Medium	Medium	Very low	Medium
Polycarbonate	Medium	Medium	Medium	Medium	Medium	Low	Medium	Low
Polyimide	Very high	High	Medium	Low	High	High	High	Medium

of materials. Low losses up to 10^{11} Hz and 60°C make the resins useful in microwave applications. The electric strength of polyethylene is not spectacular and is influenced by temperature and humidity. Figure 88 shows electrical properties of polyethylene films as a function of frequency. The films are also affected by corona, as shown in Fig. 89.

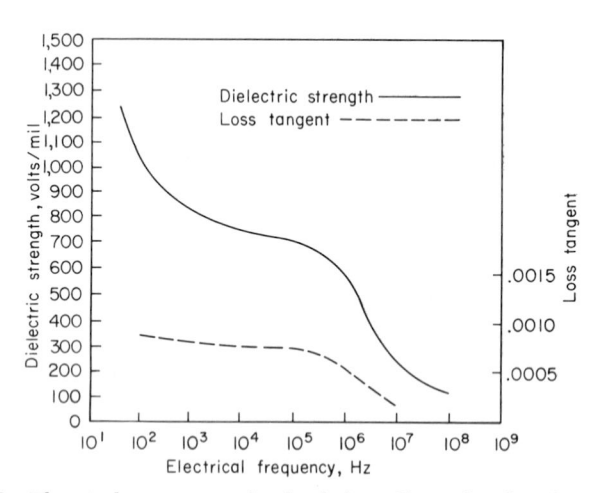

Fig. 88 Electrical properties of polyethylene films related to frequency.[59]

Polystyrene film is used electrically primarily as a capacitor insulation where its low loss characteristics are of value. It can also be worked at high-voltage levels. Figure 90 shows the voltage life of polystyrene as compared to nylon.

Polyester films are the most popular of film insulations. These films are based on polyethylene terephthalate and are exemplified by Du Pont's Mylar. This film can be used up to 150°C and possesses excellent physical properties even at cryogenic temperatures, as shown in Table 81. This film is oriented and highly crystalline.

Films and Flexible Insulations   2-125

Mylar is often used where its corona resistance is of value, as seen in Table 82. Electrical properties are also excellent with temperature, as shown in Fig. 91.

Polyester films are thermoplastic and begin to soften at about 180°C. Being polyester, they can hydrolyze under humid alkaline conditions. They are used in motors, transformers, coils, and capacitors.

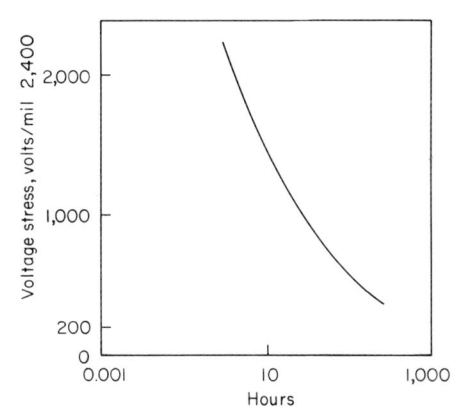

Fig. 89 Dielectric strength versus time under corona conditions for polyethylene.[60]

Polyimide films are the most thermally stable film materials available. They have a temperature limit of over 200°C and are one of the strongest films. Table 83 shows the effect of temperature on the tensile strength of H-film (now called Kapton by Du Pont) as compared to other films. Electrical properties are excellent. The resistivity is extremely high and decreases within acceptable limits up to 300°C, as shown in Fig. 92 for 1-mil Kapton.

In like fashion dielectric properties are little responsive to temperature and frequency. Figures 93 and 94 show the dissipation factor and dielectric constant

TABLE 81 Tensile Properties of 3-mil Mylar Film of Two Different Crystallinities at Cryogenic Temperatures[62]

		Crystallinity	
Property	Temperature, °F	15%	55%
Tensile yield strength, psi × 10⁻³......	−423 −320 −297	42.5 39.0 37.0	33.5 33.5 32.5
Ultimate tensile strength, psi × 10⁻³...	−423 −320 −297	42.5 39.0 37.0	33.5 33.5 32.5
Tensile modulus, psi × 10⁻⁶..........	−423 −320 −297	1.78 1.70 1.63	1.44 1.42 1.36
Elongation, %.....................	−423 −320 −297	8 31 35	20 21

Laminates, Reinforced Plastics, and Composite Structures

TABLE 82 Voltage Endurance of Mylar Film[63]

Sample thickness, mils	No. of films in sample thickness	Average voltage stress, volts/mil	Average breakdown time, hr	Breakdown time range, hr	No. of tests
2.0	1 film 2 mils thick	3,000	0.2	0.1–0.3	6
		2,000	1.0	0.3–2.9	18
		1,500	3.1	1.7–5.5	12
		1,000	16.4	5.5–24.8	23
		750	34.4	17.1–54.6	11
		500	106.8	45.6–160.6	5
10.0	1 film 10 mils thick	900	17.2	13.2–26.9	6
		800	23.1	18.0–30.2	20
		700	49.0	24.1–89.3	17
		400	152.4	125.2–193.4	6
4.0	2 films 2 mils thick	1,375	5.2	1.8–10.1	10
		1,000	15.1	12.0–17.5	6
6.0	2 films 3 mils thick	920	21.0	13.6–31.8	5
		670	50.6	43.9–59.5	3
10.0	5 films 2 mils thick	870	21.8	16.2–24.4	6
		770	34.8	32.5–38.7	6
		660	46.9	39.0–54.0	6
10.0	2 films 5 mils thick	790	33.7	28.5–39.8	6
12.0	6 films 2 mils thick	580	74.5	61.4–83.4	5
		500	111.9	90.1–124.4	5
		420	149.6	89.3–205.6	7
		330	273.9	182.7–306.1	5
		250	450.4	354.8–540.1	5

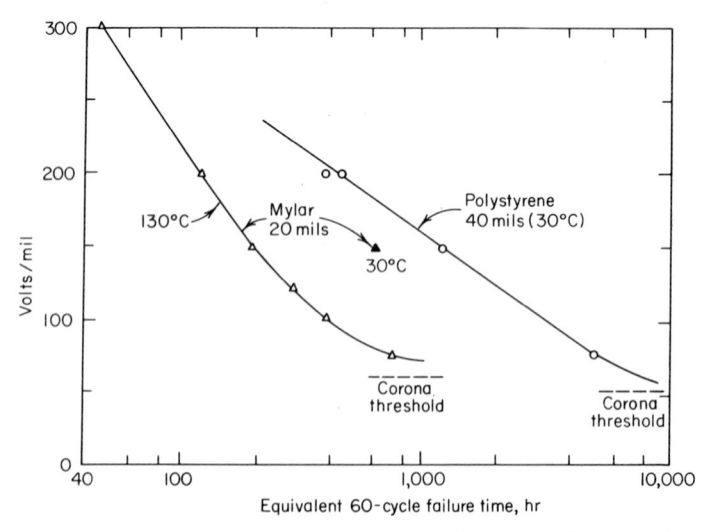

Fig. 90 Electric strength versus time for Mylar and polystyrene.[61]

TABLE 83 Effect of Temperature on Ultimate Tensile Strength of Plastic Films[62] (In pounds per square inch)

Film material	Temperature, °F			
	390	200	77	−320
Mylar	5,000	.....	23,000	39,200
H-film	12,000	.....	21,300	25,000
Kel-F* (low-crystallinity)	......	2,000	5,500	25,000
Kynar† (quenched)	......	5,100	7,200	29,700

* Trademark of 3M Company, St. Paul, Minn.
† Trademark of Pennsalt Chemical Corporation, Philadelphia, Pa.

from −40 to 250°C and from 10^2 to 10^5 Hz. Figure 95 shows the effect of temperature on the ac and dc dielectric strength of Kapton, using ¼-in. electrodes. As is generally true, the dc strength is most affected by temperature but in Kapton this effect is small. The corona resistance of Kapton is similar to that of Mylar. Table 84 gives the voltage endurance data for Kapton as a single film and for several layers stacked together.

Polyimide films are used for flexible printed circuits, cable warp, coil insulation, motor and generator insulation, and high-temperature capacitors.

Flexible insulations—definition Flexible insulation materials are those electrical barriers which are not rigid and are capable of being deformed through severe bends without losing their physical or electrical continuity. The film materials just discussed fit this definition but differ from other flexible insulation systems in that the films are continuous, homogeneous materials while other flexible insulations are often combinations of two or more materials.

These insulation systems are usually better insulations than air but this is not always true. For example, certain electric systems depend on insulations like glass cloth tape which has a breakdown strength about the same as air (see Fig. 96). This is due to the fact that this tape is porous and air-filled. The purpose of the tape in such an application

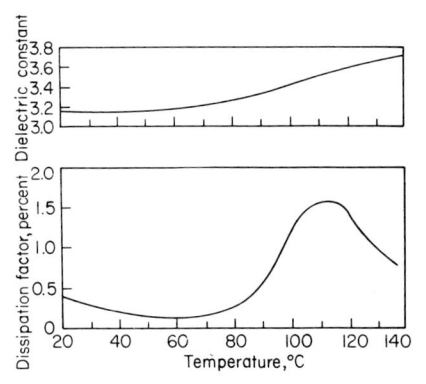

Fig. 91 Electrical properties of polyethylene terephthalate.[64]

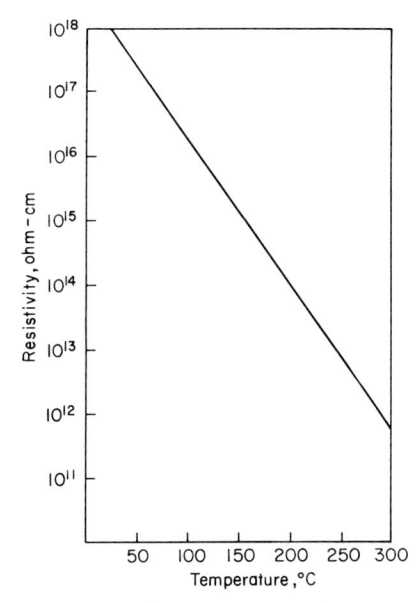

Fig. 92 Volume resistivity of Kapton film.[65]

is as a spacer only. Of course, high humidity conditions and contaminants of many kinds can reduce the electric strength of air, and so such insulation is of specialized use only and limited to low voltages.

The properties of most importance in a flexible insulation are electric strength, breaking strength, dissipation factor, and water absorption. The decrease of these

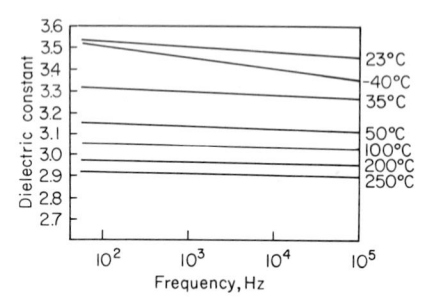

Fig. 93 Dielectric constant versus frequency for Kapton.[65]

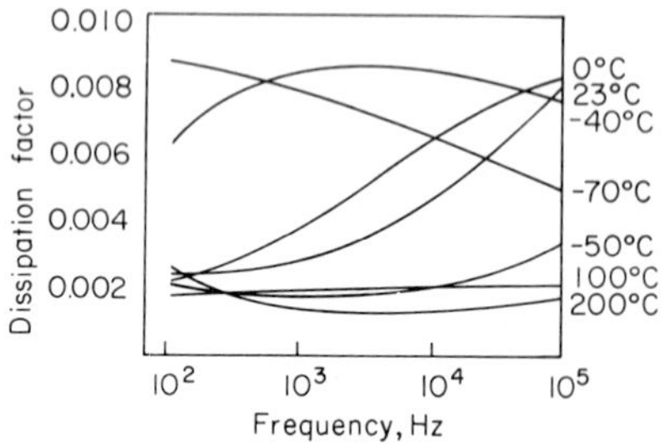

Fig. 94 Dissipation factor versus frequency for Kapton.[65]

properties determine the thermal life of the system. Of most concern is the change of dielectric strength with temperature and time. ASTM D 1830 describes the most commonly used test for evaluating thermal life.

Varnished Cloth. Perhaps the oldest insulation systems are the varnished fabrics. Although any fabric may be chosen, cotton and glass are most common. These materials, because of the fabric base, are physically strong and have good abrasion resistance. The thermal behavior is most related to the resin (varnish) treatment, which includes such resin as oleoresins, phenolic, alkyd, epoxy, and silicone. Fabrics are available in any of the temperature classifications. The fabric-based insulations have electric strength of about 1,500 to 2,000 volts/min. Some of these systems are available with B-stage resins which can be further cured to form rigid insulation. Other tapes can be obtained with pressure-sensitive adhesives.

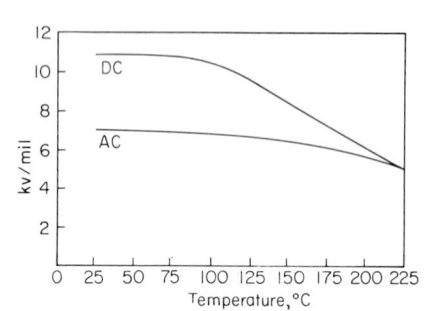

Fig. 95 Dielectric strength of Kapton versus temperature.[65]

Bias-weave (diagonal) tapes are also available. These tapes, because of the bias, have great conformity and elongation and result in snug wraps on coils, cables, and bus bars.

Inorganic Paper Insulations. Many insulation systems are based on asbestos papers which have the trade names Quinterra,° Quinorgo,° and Novabestos.† These papers are described in MIL-I-3503A. The Class A papers consist of asbestos and wood pulp with about 20 percent starch or rubber binder. Class B papers contain a polyvinyl acetate, Class F has epoxy, and Class H silicone. The physical properties of these papers are low and are related to the amount of binder used. Electric strengths average 1,000 volts/min. The papers are often laminated to films, which increases the electric strength to 7,000 volts/min, or to glass cloth for physical property improvement.

° Trademarks of Johns-Manville Corp., New York, N.Y.
† Tradename of Raybestos—Manhattan, Inc., Manheim, Pa.

TABLE 84 Voltage Endurance and Corona Resistance of Kapton[63]

Sample thickness, mils	No. of films in sample	Average voltage stress, volts/mil	Average breakdown time, hr	Breakdown time range, hr	No. of tests
1.0	1 film 1 mil thick	1,590	10.1	6.0–12.9	6
		1,040	34.7	16.9–54.6	5
		560	115.8	114.4–117.1	2
4.0	4 films 1 mil thick	490	113.5	72.3–163.1	5
6.0	6 films 1 mil thick	470	150.6	109.4–216.1	9

Some insulation systems are based on glass flake paper. Glass flake results from glass drawn thin at high speeds and broken into random-sized flakes. The thickness of each flake is about 0.0002 in. These papers are composed of the glass flakes with a resin binder and are available in 2- to 5-mil thicknesses.

Cellulose Paper-Film Composites. These insulations are available as two-ply and three-ply combinations where the film can be polyester, acetate, or even polyimide. For general-purpose applications the composites are bonded with rubber adhesives. For somewhat greater thermal stability the binders can be polyester. Special hermetic grades are also available.

Mat-Film Composites. These constructions consist of a single ply of a film such as polyester bonded between two layers of a mat. The mats may be based on acrylic or polyester fibers and may be of chopped fibers or of continuous strands held together with a binder. With the proper adhesive these constructions can be used at 130°C. These composites are very strong and have high cut-through resistance. The mat permits resin or oil saturation for excellent electrical properties.

Nylon-paper Composites. Nylon paper was developed by Du Pont and trade named Nomex. This material is an aromatic nylon with excellent thermal stability. The thermal stability of Nomex is shown in Fig. 97. Nomex is often combined with many other fibrous materials, cloths, or films where specified properties are

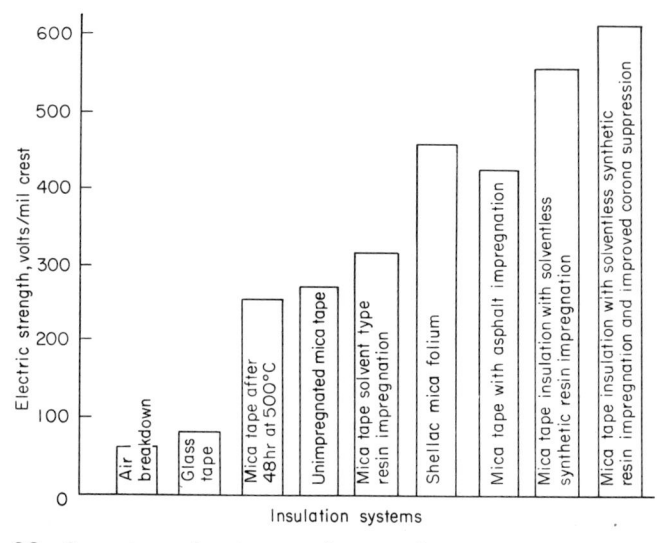

Fig. 96 One-minute electric strength on test bars of various insulations.[66]

2-130 Laminates, Reinforced Plastics, and Composite Structures

desired. For example, Nomex has been laminated to asbestos mat, glass cloth, and Kapton film.

Mica Composites. Mica flexible insulations are available in a wide variety of products. Large mica splittings were first used with shellac or asphalt as shown in Fig. 96. Later many synthetic resins were used as binders. Mica is also pro-

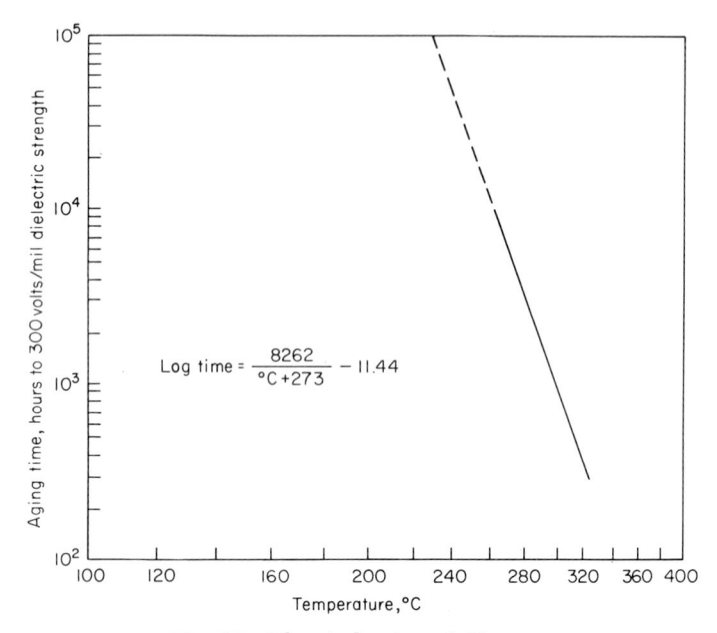

$$\text{Log time} = \frac{8262}{°C + 273} - 11.44$$

Fig. 97 Electrical aging of Nomex.

duced in paper form. Mica products have excellent electric strength and thermal stability, and the corona resistance is better than that of any other insulation system.

Rubber Tapes. Elastomeric insulations are either vulcanized or unvulcanized. Most common are the tapes based on silicone rubber. Useful up to 200°C, silicones are waterproof and resistant to chemicals and insulating fluids. Tapes are available

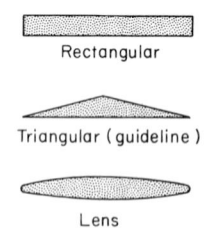

Fig. 98 Tape cross sections.

from 0.005 to 0.060 in. and in a variety of cross sections, as shown in Fig. 98. Rectangular tapes are either half-lapped or butt-wrapped, and lens tapes are half-lapped. These two tapes are usually applied in two layers to ensure that the system is void-free. Triangular tape has a white stripe at the apex which permits ease and accuracy of wrapping and usually makes two layers unnecessary.

REFERENCES

1. ASTM Standards, parts 27 and 20, 1967.
2. Plastics: Methods of Testing, *Federal Test Method Standard* 406, (1961).
3. Von Hippel, A.: "Dielectric Materials and Applications," John Wiley & Sons, Inc., New York, 1954.
4. Baer, E.: "Engineering Design for Plastics," p. 510, Reinhold Book Corporation, New York, 1964.
5. *Ibid.*, p. 507.
6. Norman, R. S., and A. A. Kessel: Internal Oxidation Mechanism for Non-tracking Organic Insulations, *AIEE Trans.*, vol. 77, part III, p. 632, 1958.
7. Mandelcorn, L., and G. M. L. Sommerman: Tracking and Arc Resistance of Materials, *Proc. 5th Elec. Insul. Conf.*, p. 91, 1963.
8. Kinney, G. F.: "Engineering Properties and Applications of Plastics," p. 202, John Wiley & Sons, Inc., New York, 1957.
9. Engineers of Gibbs & Cox Inc.: "Marine Design Manual for Fiberglass Reinforced Plastics," pp. 5–29, McGraw-Hill Book Company, New York, 1960.
10. "Modern Plastics Encyclopedia," McGraw-Hill, Inc., New York, 1968.
11. Tish, H. A.: Mechanical Properties of Rigid Plastics at Low Temperature, *Mod. Plast.*, vol. 33, no. 11, 1955.
12. Brenner W., et al.: "High Temperature Plastics," p. 60, Reinhold Book Corporation, New York, 1962.
13. Laurie, W. A.: New Concepts for Thermosets, *Reg. Tech. Conf.-Plast. in Elec. Insul.*, Society of Plastics Engineers, 1964.
14. Landrock, A. H.: Properties of Plastics and Related Materials at Cryogenic Temperatures, *Plastec Rep.* 20, Picatinny Arsenal, 1965.
15. Kueser, P. E., et al.: Electrical Conduction and Electrical Insulation Materials Topical Report, *NASA Rep.* CR-54092, October, 1964.
16. Wigington, G. D.: Chemical Resistance of Polyester Resins Cured under Field Conditions, *23d Ann. Tech. Conf.*, Reinforced Plastics/Composites Division, Society of Plastics Industry, 1968.
17. McAbee, E., and M. Chmura: Effect of Rate and Temperature on the Tensile Properties of 181 Glass Cloth Reinforced Polyester, *21st Ann. Tech. Conf.*, Society of Plastics Industry, 1965.
18. Raech, H., and J. M. Kreinik: Prepreg Materials for High Performance Dielectric Applications, *20th Ann. Tech. Conf.*, Reinforced Plastics Division, Society of Plastics Industry, 1965.
19. Herberg, W. F., and E. C. Elliott: Reinforced Silicone Plastics Retain Physical and Electrical Properties with Little Regard to Temperature or Frequency, *Plast. Design Process.*, vol. 6, no. 1, 1966.
20. Korb, L. L.: Arc and Track Resistance Tests, *Reg. Tech. Conf.—Plast. in Elec. Ind.*, Society of Plastics Engineers, 1964.
21. Coast Manufacturing & Supply Co.: *Data Bull.* 13b, 1966.
22. Miglarese, J.: Heat Resistant Reinforced Laminates, *Soc. Plast. Eng. J.*, vol. 16, no. 5, 1960.
23. Pride, R. A., et al.: Mechanical Properties of Polyimide Resin/Glass Fiber Laminates for Various Time, Temperature, and Pressure Exposures, *23d Ann. Tech. Conf.*, Reinforced Plastics/Composites Division, Society of Plastics Industry, 1968.
24. Freeman, J. H., et al.: New Organic Resins for High Temperature Reinforced Laminates, *Soc. Plast. Eng. Trans.*, vol. 2, no. 3, 1962.
25. Milek, J. T.: Polyimide Plastics: a State of the Art Report, Electronic Properties Information Center, October, 1965.
26. Monsanto Chemical Co.: *Skybond* 700 *Tech. Bull.* 5042B.
27. Mackay, H. A.: Evaluation of Polybenzimidazole Glass Fabric Laminates, *Mod. Plast.*, vol. 43, no. 5, 1966.
28. Whittaker Corp.: Imidite 1850, Data Sheet.
29. New Polymer Looks Good for 1000°F, *Mater. Design Eng.*, vol. 57, no. 5, p. 92, 1963.
30. Wicker, G. L.: A New Concept in Reinforced Thermoplastics Sheet, *20th Ann. Tech. Conf.*, Society of Plastics Industry, 1965.
31. GRTL Company: Azdel Product Sheet.
32. Industrial Laminated Thermosetting Products, *NEMA Standard* LI-1-1965, 1965.
33. Harper, C. A.: "Plastics for Electronics," Kiver Publications, Chicago, 1964.
34. "Modern Plastics Encyclopedia," p. 546, McGraw-Hill, Inc., New York, 1967.

35. *Ibid.*, p. 547.
36. *Ibid.*, p. 548.
37. IEEE Publication 1, Institute of Electrical and Electronics Engineers, 1957.
38. High Temperature Properties of Industrial Thermosetting Laminates, *NEMA Standard* LI-3-1961, 1961.
39. Skow, N. A.: Performance Properties of Industrial Laminates at High Temperature, *1st Nat. Conf. Appl. Elec. Insul.*, Institute of Electrical and Electronics Engineers, 1958.
40. Miller, G. J.: Humidity Effects on Vulcanized Fiber and Its Relation to Storage, *1st Nat. Conf. Appl. Elec. Insul.*, Institute of Electrical and Electronics Engineers, 1958.
41. Oleesky, S. S., and J. G. Mohr: "Handbook of Reinforced Plastics of the SPI," Reinhold Book Corporation, New York, 1964.
42. Rosato, D. V., and C. S. Grove, Jr.: Filament Winding, Interscience. Publishers, a division of John Wiley & Sons, Inc., New York, 1964.
43. "Modern Plastics Encyclopedia," McGraw-Hill, Inc., New York, 1968.
44. Jaffe, E. H.: Boron Fibers in Composites, *21st Ann. Tech. Conf.*, Reinforced Plastics Division, Society of Plastics Industry, 1966.
45. Copeland, R. I., et al.: Polyimides—Reinforced and Unreinforced, *23d Ann. Tech. Conf.*, Reinforced Plastics/Composites Division, Society of Plastics Industry, 1958.
46. Simon, R. A., and S. P. Prosen: Graphite Fiber Composites: Sheer Strength and Other Properties, *23d Ann. Tech. Conf.*, Reinforced Plastics/Composites Division, Society of Plastics Industry, 1968.
47. Economy, J., and L. C. Woher: The Potential of Whisker Reinforced Plastics, *Plast. Design Process.*, vol. 7, no. 7, p. 31, 1967.
48. Busche, M. G.: High Modulus Carbon Fiber: New Entry in FRP Race, *Mater. Eng.*, vol. 67, no. 2, p. 46, 1968.
49. Sharpe, L. H.: Assembling with Adhesives, *Mach. Design*, Aug. 18, 1966.
50. Burgman, H. A.: Adhesive Bonding of Materials and Plastics, *Electron. Packag. & Prod.*, vol. 5, no. 9, 1965.
51. Burgman, H. A.: Selecting Structural Adhesive Materials, *Electrotechnol.*, vol. 75, no. 6, p. 69, 1965.
52. Kuno, J. K.: Adhesive Bonding of Insulation for Temperature Extremes: Cryogenic to Re-entry, *Proc. 7th Nat. Symp. on Adhes. & Elastomers for Environ. Extremes*, Society of Aerospace Materials & Process Engineers, 1964.
53. Licari J. J.: High Temperature Adhesives, reprinted from *Prod. Eng.*, Dec. 7, 1964, Copyright, 1964, by McGraw-Hill, Inc.
54. Copper-clad Thermosetting Laminate for Printed Wiring, *ASTM Standard* D-1867-64T.
55. Industrial Laminated Thermosetting Products, *NEMA Standard* LI-1-1965, part 10, Copper-clad Laminates, 1965.
56. *Ibid.*, part 13, Thin Copper-clad Laminates, 1965.
57. Rausch, J. M.: Microwave Printed Circuitry for High Performance and Reliability, *6th Nat. Conf. Appl. Elec. Insul.*, Institute of Electrical and Electronics Engineers, 1965.
58. "Modern Plastics Encyclopedia," p. 532, McGraw-Hill, Inc., 1967.
59. Castagna, E. G.: High Energy Joining Methods for Plastics, *Reg. Tech. Conf.—Joining Plast.*, Society of Plastics Engineers, 1963.
60. O'Toole, J. L.: Testing and Evaluation of Plastics, *Symp. Design Criteria for Plast.*, Picatinny Arsenal, 1963.
61. Dakin, T. W.: Corona Discharges and Their Effects on Insulation, *4th Nat. Conf. Appl. Elec. Insul.*, Institute of Electrical and Electronics Engineers, 1962.
62. Mowers, R. E.: Program of Testing Non Metallic Materials at Cryogenic Temperatures, AD 294,772, Defense Documentation Center, 1962.
63. Bowman, R. L.: The Effect of Oriented Mica Flakes on Voltage Endurance and Corona Resistance on Electrical Insulation, *5th Nat. Conf. Appl. Elec. Insul.*, Institute of Electrical and Electronics Engineers, 1963.
64. Watson, M. T.: Properties of a New Polyester Film, *17th Ann. Tech. Conf.*, Society of Plastics Industry, 1961.
65. Tatum, W. E., et al.: H Film: Du Pont's New Polyimide Film, *5th Nat. Conf. Appl. Elec. Insul.*, Institute of Electrical and Electronics Engineers, 1963.
66. Botts, J. C.: High Voltage Generator Insulation, *4th Nat. Conf. Appl. Elec. Insul.*, Institute of Electrical and Electronic Engineers, 1962.
67. Petrie, E. M.: Reinforced Polymers for High Temperature Microwave Applications, *IEEE Trans. Electrical Insulation*, March, 1970.

Chapter 3

Elastomers

JOSEPH G. CERYAN

Aeroquip Corporation, Republic Rubber Division,
Youngstown, Ohio

Introduction	3–2
Commercially Available Elastomers	3–6
Natural (NR)	3–6
Isoprene (IR)	3–7
Styrene-Butadiene Copolymer (SBR)	3–7
Neoprene (CR)	3–8
Nitrile (NBR)	3–8
Butyl (IIR)	3–8
Ethylene Propylene EPT (EPM) or (EPDM)	3–8
Hypalon (CSM)	3–9
Acrylic (ABR)	3–9
Polysulfide (T)	3–9
Silicones (FSI) (PSI) (VSI) (PVSI) (SI)	3–9
Urethane (U)	3–10
Fluoroelastomers (FPM)	3–10
Butadiene (BR)	3–10
Carboxylic (COX)	3–10
Hydrin (CO) (ECO)	3–10
Polyblend (PVC + NBR)	3–11
Relative Elastomer Costs	3–11
Physical Properties	3–12
Chemical Properties	3–15
Electrical Properties	3–18
Specifications	3–28
Material Specifications	3–29

Product Specifications	3-29
Rubber Manufacturers Association	3-29
Test Methods	3-30
Specification Sources	3-30
Manufacturing Methods of Rubber Products	3-31
Millable Gums	3-31
Latex	3-32
Solutions	3-32
Castables	3-32
Compounding	3-33
Vulcanizers	3-33
Reinforcers	3-33
Softeners	3-33
Protectants	3-34
Colorants	3-34
Diluents	3-34
Processing Aids	3-35
Economics	3-35
Effects on Electrical Properties	3-36
Cellular Sponge and Foam Rubber	3-36
Sponge Rubber	3-38
Closed-cell versus Open-cell Sponge	3-38
Uses of Sponge	3-38
Foam	3-38
Specifications	3-39
References	3-39

INTRODUCTION

The organic chemist of yesterday unlocked the secrets of polymerizing large macromolecules leading to the synthesis and development of completely new families of engineering materials; and, thus, he himself became a specialist of his own—a polymer chemist. Many of these new materials had elastic properties similar to the familiar natural rubber so that some groupings of the macromolecules acquired a special name of their own also—elastomers.

Today, therefore, we find a synonymous usage of the terms rubber, synthetic rubber, and elastomer in the manufacture of rubber products. However, in truly encompassing all such materials, the use of *elastomer* is now more fitting.

ASTM Designation D 1566-66T[1] defines elastomers as "macromolecular material that returns rapidly to approximately the initial dimensions and shape after substantial deformation by a weak stress and release of the stress." It also defines a rubber as:

... material that is capable of recovering from large deformations quickly and forcibly, and can be, or already is, modified to a state in which it is essentially insoluble (but can swell) in boiling solvent, such as benzene, methyl ethyl ketone, and ethanol toluene azeotrope.

A rubber in its modified state, free of diluents, retracts within one minute to less than 1.5 times its original length after being stretched at room temperature (20° to 27°C) to twice its length and held for one minute before release.

More specifically,[2]

Introduction 3-3

An elastomer is a rubber-like material that can be or already is modified to a state exhibiting little plastic flow and quick and nearly complete recovery from an extending force. Such material before modification is called, in most instances, a raw or crude rubber or a basic high polymer and by appropriate processes may be converted into a finished product.

When the basic high polymer is converted (without the addition of plasticizers or other diluents) by appropriate means to an essentially nonplastic state, it must meet the following requirements when tested at room temperature (60°–90°F; 15°–32°C):

A. Is capable of being stretched 100%.
B. After being stretched 100%, held for 5 minutes and then released, it is capable of retracting to within 10% of its original length within 5 minutes after release (ASTM D 412-51T).

The rubber definition with its swelling test certainly limits it to only the natural latex tree source, whereas the elastomer definition is more in line with our modern new synthetics. Definitions are somewhat confusing and probably always will be since they are arrived at after the fact. Because of this, we most likely will never have an explicit term, since the rate of development of new polymers is difficult to keep up with, even for those in the field.

The fantastic rate of development must be recognized when we realize that our commercially available synthetics started in the late 1930s. Polymer chemistry had its start earlier, of course, but still in most of our lifetimes. The important discoveries leading to our polymer science can possibly be credited as beginning with Stardinger's work in the later 1920s on the glucose unit and the start of cellulose. The 1930s saw work by Corothers and Florey on polycondensates such as esters and polyamides (nylon). Also, the 1930s saw the important start of the addition polymers, resulting in high molecular weights of vinyl and diene polymers. The 1940s showed rapid advance of the addition polymers: copolymerization, control of molecular weights for plasticity, emulsion polymerization, development of physics in polymers, solution properties of polymers, intrinsic viscosities, effect of *cis* or *trans* isomerism, and mechanical behavior of polymers. In the 1950s we learned the detailed structure of polymers and by means of special catalysts (Ziegler) progressed stereospecific polymers and finally a true synthetic natural rubber.

Many carbon-hydrogen chain elements, it was found, tend to gather and behave in basic groups. The hydrogen in a basic group could be substituted by some other element or group to form a new group. These basic groups could react with each other in a head-to-tail fashion to make large molecules and larger macromolecules. Also, a group can react with a different group to form a new larger group which can then react with itself in a head-to-tail fashion to produce the desired new macromolecule. These basic groups are called *monomers;* and when they head-to-tail or polymerize into long chains, they then form *polymers.* The reaction between different basic groups to form a new basic recurring group is called *copolymerization.* These copolymers can also reoccur in a head-to-tail fashion to form long-chain molecules.

The basic monomers are usually gases or very light liquids; and, as polymerization continues, the molecular weight and viscosity both increase until solidification and the formation of the gum or solid products results. The degree of polymerization can be controlled so that the end product has the desired processing properties for the rubber industry but lacks the final properties for engineering use. The long chains in this state have a great deal of mobility between them which also varies greatly with slight changes in temperatures. Some such materials are usable as thermoplastic material; but generally, to obtain the properties of elasticity, this mobility must be arrested. These long chains can react further with themselves or with other chemicals. The type of reaction desired for elastomers is not a continuation of chain length at this point, but a tying or linking of the chains together at certain points to form a network and thus reduce plastic mobility. As the chains lie close to each other, especially when the reactive centers are near each other as double bonds between carbon atoms, they can be made to join

Elastomers

TABLE 1 Nomenclature for Elastomers and General Properties

ASTM D 1418 designation	Trade names or common names	Chemical description	General properties
NR	Natural rubber Hevea, pure gum	Polyisoprene plus resins. Isoprene natural	Low hysteresis, high resilience, low water absorption.
IR	Isoprene rubber, polyisoprene, Coral, rubber, Natsyn, Ameripol SN, synthetic natural	Polyisoprene, isoprene rubber synthetic	Same as NR but fewer impurities, better water resistance.
SBR	GR-S, buna-S	Styrene-butadiene rubber	General-purpose use, better heat resistance than natural, but less rubbery properties. Cheap filler insulator.
CR	Neoprene, Denka chloroprene, Perbunan C, Petro-Tex Chloroprene	Polychloroprene	General-purpose oil-resistant and weather-resistant rubber. Widely used in cover insulation.
NBR	Buna-N nitrile rubber, Hycar, Chemigum, Paracril, FR-N, Polysar Krynac	Butadiene-acrylonitrile copolymer	General-purpose oil-resistant rubber where better oil resistance is needed than with neoprene. Needs special compounding for weather resistance. Good general-purpose chemical resistance. High power factor.
IIR	Butyl	Isobutylene-isoprene copolymer	Low rebound, high damping, low air permeation, good weathering, good electrical properties.
EPM	EPR, ethylene rubber, Epsyn, Royalene, Vistalon, Nordel	Ethylene-propylene copolymer	Generally same as IIR but better heat resistance and higher air permeation.
EPDM	EPT, Epsyn, Royalene, Vistalon, Nordel	Terpolymer of ethylene, propylene, and a diene side chain	Generally same as EPM but faster curing. Should become the standard general-purpose rubber for all non-oil or solvent use. Good electrical properties and inexpensive. Resists hot water, steam, dry heat, ozone.
CSM	Hypalon	Chlorosulfonated polyethylene	Very similar to neoprene but good for color stability, colored covers, spark plug boots. Good heat resistance, chemical resistance.
ABR	Acrylic, Cyanacril, Hycar, Krynac, Thiacryl, Acrylon, Paracril OHT	Acrylate butadiene	Similar to NBR but better hot oil resistance.
T	Polysulfide Thiokol	Organic polysulfide	Good solvent resistance, odorous, good weather resistance, tender.

TABLE 1 Nomenclature for Elastomers and General Properties (Continued)

ASTM C 1418 designation	Trade names or common names	Chemical description	General properties
SI, FSI, PSI, VSI, PVSI	Silicones, SE Rubber, Si-O-Flex, Silastic	Polydemethyl siloxane variations	Outstanding at high and low temperatures, nonconductive decomposition products. Special grades for oil resistance. Good ozone and weather resistance, tender.
U	Urethane, isocynate rubbers, Cyanaprene, Vibrathene, Elastothane, Adiprene, Estane, Genthane, Texin, Roylar, Castomer, Conothane	Reaction products of diisocynates and polyalkylene ether glycols	High tensile strength, high abrasion resistance, high load capacity, wide hardness range, tough. Morplastic than rubber in wide property range.
FPM	Fluoroelastomer Fluorel, Viton	Copolymer of vinylidene fluoride and hexafluoropropylene	Outstanding chemical solvent resistance and good to about 450°F.
	Kel-F	Copolymer of chlorotrifluoroethylene and vinylidene fluoride	
BR	Butadiene, Cis-4, Diene Rubber, Ameripol CB, Budene, Cisdene, Taktene	Polybutadiene	Improves compression set, wear, low-temperature rebound when added to NR or SBR. Mostly used to improve wear of SBR tires.
COX	Carboxylic Hycar 1072	Butadiene-acrylonitrile modified with carboxylic groups	Improves wear and cold-temperature properties over NBR.
CO, ECO	Hydrin 100, Hydrin 200	Homopolymer of epichlorohydrin, copolymer of epichlorohydrin with ethylene oxide	Fuel and oil resistance, chemical resistance to acids and bases, good weathering, a blend of NBR and CR properties.
	Polyblend Paracril OZO, Hycar, Chemivic	A blend of polyvinyl chloride and NBR	Adds weather and ozone resistance to NBR. Allows light color covers.

to each other directly or through some element such as sulfur which will hold the two chains together. These joiners can be loosely called linking agents, and more loosely, vulcanization agents; and again the crosslinked material in the rubber industry can now be called a *vulcanized* material.

One of our basic monomers is gaseous ethylene which is polymerized to the well-known polymer, polyethylene. Ethylene is the basis for many other monomers such as styrene, acrylonitrile, isobutylene, vinyl alcohol, vinyl chloride; which in turn can be polymerized to form polystyrene, polyacrilonitrile, polyisobutylene, polyvinyl alcohol, and polyvinyl chloride. Another basic monomer is butadiene which is the basis for other monomers such as isoprene and chloroprene. The reactions of these two groups of monomers yield the familiar copolymers of butadiene-styrene (GR-S), butadiene-acrilonitrile (buna-N), isoprene with isobutylene (butyl), etc.

The mechanical properties of solid insulators are often more important than their electrical properties. Most organic materials are good insulators if they can be formed into continuous films that exclude moisture. The general principles relating mechanical properties of polymers to structures have been known for many years. Rubbers, plastics, and fibers, for example, are not intrinsically different materials.

3-6 Elastomers

Their differences are a matter of degree rather than kind. If the forces of attraction between the molecular chains are small and the chains do not fit readily into a geometric pattern, lattice, or network, the normal thermal motion of the atoms tends to cause the chains to assume a random, more or less coiled arrangement. These conditions lead to a rubberlike character. In practical rubbers, a few cross-links are added to prevent slippage of the molecular chains and permanent deformation under tension. With such polymers, when the stress is released, the normal thermal motion of the atoms causes them to return to a random coiled arrangement. If the forces between the chains are strong and the chains fit easily into a regular geometric pattern, the material is a typical fiber. In cases where the forces are moderate and the tendency to form a regular lattice is also moderate, the result is a typical plastic. Some polymers are made and used as three different materials: rubber, plastic, and fiber. Polyethylene, for example, is used as a substitute for natural rubber in wire covering, as a plastic in low-loss standoff insulators and insulating films, and as a fiber in acid-resistant filter cloths where high fiber strength is not as important as chemical resistance.[3]

The rubber industry as such was long occupied in processing "gumlike" natural rubber. The user expected a material from this industry to be elastic in nature. Therefore, the chemical plants, in their designing of polymers for the rubber industry, kept two requirements in mind: substitutes for natural rubber with its elasticity, and also polymers that could be processed with the conventional equipment. Plastics, as such, behave quite differently from rubber in processing, and this fact led to the establishment of a new industry. Therefore the incorporation of plastic materials into the rubber inventory was not accomplished. This separation of the two industries has resulted in separate technical groups and societies.

Elastomers, when compared with other engineering materials, are characterized by large deformability; low shape, rigidity; large energy-storage capacity; nonlinear stress-strain curves; high hysteresis; large variations with stiffness, temperature, and rate of loading; and compressibility of the same oder of magnitude as most liquids. Certain of the elastomeric materials possess additional useful characteristics to a relative degree, such as corrosive chemical resistance, oil resistance, ozone resistance, temperature resistance, and other environmental conditions.

The future of polymer chemistry is certainly unlimited, and thousands of polymers have been tested. The goal is in the requirements of the rubber industry for chemical and environmental properties in addition to simply elastic properties, and to this end much has been accomplished in a very short time.

COMMERCIALLY AVAILABLE ELASTOMERS

Natural rubber (ASTM designation NR),[4] **a polyisoprene** Source: natural, and, as the name implies, a truly natural-occurring product mostly derived from a *Hevea brasiliensis* tree under cultivation in large plantations around the Malay peninsula area. There are, however, other sources such as the wild rubbers of the same tree growing in Central America; guayule rubber, coming from shrubs grown mostly in Mexico; and balata. Balata is a resinous material of *cis-trans* isomerism and cannot be tapped as the Hevea tree. It is actually cut down and *boiled* to extract balata, which cures to a hard and tough product as golf ball covers.

Contrary to popular opinion, there are many grades of natural rubber which must be selected for specific uses. In producing natural rubber, the tree is tapped and the free-flowing latex is then coagulated with an acid and coagulant, then milled or creped to reduce moisture, and further smoked or air-dried for further drying. Smoking of rubber also adds mildew protection by the introduction of creosote.

Most natural rubber is from the Hevea tree, but the grading which affects properties is based generally on the cleanliness of the latex before coagulation. Grade 1 smoked sheets are plantation-made and generally free of any dirt. Grades 2 to 5 are from small holders, and qualities are graded on the amount of bark, sand, air inclusion, or moisture. Blankets are graded according to color only, and may contain any amount of dirt. Flat bark crepe is the scrap coagulant picked off the bark and any other pickings of small holders, and so it may contain a very high amount of bark and dirt and also may be resinous from excessive heat in smoking. Pale crepe is of the finest quality and is used for light-colored or odor-free products. River water is used for washing and at times is dirtier than the rubber. Guayule has 20 percent resin and needs to be washed to bring it up to smoked sheet quality. Wild rubber gives long-fingered cements; otherwise, it has the same physical properties as grade 1 smoked sheets, and also needs washing to rid it of about 20 to 30 percent residual moisture.

Another rubber that needs mentioning is plantation leaf gutta-percha.[5] As the name implies, this material is produced from the leaves of trees grown in bush formation or on plantations. These leaves are plucked, and the rubber is boiled out as with the balata. Gutta-percha has been used successfully for submarine cable insulation for better than 40 years, a lasting tribute to the pioneers who made this selection. The dielectric constant of gutta-percha was reported to be 2.6, while reports during the same period for Hevea rubber were 3.90 to 4.31.

Needless to say, natural rubber, properly selected, processed, and compounded, produces very high quality rubber properties and was the original basis of electrical insulation. It is still a good basis in comparison with the new synthetic elastomers. It possesses high tensile strength, tear, resilience, wear, and cut growth. It has good electrical insulation properties, but by special compounding can be made conductive to a specific resistance as low as 2,000 ohm-cm. It has very low compression set and good resistance to cold flow, and it can be compounded to function as low as $-70°F$. Resilience is the main superior quality over the synthetics. On the negative side, it is not as good for sunlight, oxygen, and ozone resistance as some of the synthetics, but special compounding offsets this to a degree. Its aging properties are not as good as those of the synthetics, but synthetics will harden with time while natural rubber will show a reversal and start softening. Also, natural rubber is not satisfactory in resistance to oils from petroleum, vegetable, or animal origin. On the other hand, it has good resistance to acids and bases with the exception of oxidizing agents.

Isoprene rubber[6] **(IR) a polyisoprene** Trade names: Shell Isoprene (Shell Chemical), Coral Rubber (Firestone), Ameripol SN (Goodrich-Gulf), Natsyn (Goodyear Chemical). This was the ultimate dream of all polymer chemists—to truly synthesize natural rubber—and it has been accomplished.

As a synthetic, isoprene offers the rubber industry the purity of all synthetics and the freedom from worry of a small supply, as was experienced during World War II. No longer is it necessary to sort, wash, screen, dry, blend, and masticate for processing qualities. Synthesizing to controlled molecular weights gives the industry uniform processing qualities and a bonus on the side—odorlessness.

As of this writing, it can fairly safely be used in almost all natural rubber applications.

SBR (SBR), a styrene-butadiene copolymer Trade names: None as such. There are many manufacturers of SBR, but generally it is known as GR-S or buna-S.

This was used for the synthetic rubber tire that was developed because of the natural rubber shortage during World War II; and since the government undertook the development, it became known as government rubber styrene type (GR-S). There are now many variations of SBR, generally basic in the ratio of butadiene to styrene, temperature of polymerization, and type of chemicals used during polymerization. This last aspect is important in its initial choice for insulation since the type of salts and moisture content affect the quality of insulation.

The original purpose of GR-S, of course, was to substitute for the then hard-to-obtain imported materials. In this role it became and still is the workhorse of the industry, even though it does not match the superior physical properties of natural rubber. It lacks in tensile strength, elongation, resilience, hot tear, hysteresis; but it more than makes up for itself in low cost, cleanliness, slightly better heat aging, slightly better wear than natural rubber for passenger tires, and availability at a stable price. Generally speaking, natural rubber was not employed with its ultimate properties but in a cheapened and compounded-down product to meet the desired engineering needs. In this range, SBR is certainly not a substitute any longer but the prime choice.

Neoprene (CR), polychloroprene Trade Names: Neoprene (E. I. du Pont), Denka Chloroprene (Denki Kagaku Kogyo), Perbunan C (Farben-Fabriken Bayer), Petro-Tex Chloroprene (Distugil-France).

Neoprene was actually our first commercial synthetic rubber. Discovered in the laboratories of Notre Dame University and developed by E. I. du Pont de Nemours & Company, it was first used in the 1930s as an oil-resisting rubber. Today it is classified as a moderately oil resisting rubber with very good weather- and ozone-resisting properties and other properties closer to those of natural rubber than SBR. Because of its crystalizing nature, it has inherent high tensile strength, elongation, and wear properties at pure gum (not extended or hardened) levels. In this respect, it compares with natural rubber, whereas SBR needs reinforcing materials for good tensile strength and thus has no pure gum properties. Electrically, Neoprene ranks below natural rubber or SBR because of its polar chlorine group. It can be compounded for temperatures as low as $-67°F$ but crystallizes rapidly, and it is not as good in this respect as either natural rubber or SBR. It has excellent flame resistance and is, in fact, self-extinguishing. Because of this property, it is a must in coal mining operations and other areas where fire is extrahazardous. Also, it has good resistance to oxidative chemicals.

Nitrile rubber (NBR), a butadiene-acrilonitrile copolymer Common name: Buna-N. Trade names: Hycar (B. F. Goodrich), Chemigum (Goodyear Chemical), Paracril (Uniroyal), FR-N (Firestone), Polysar Krynac (Polymer Corp., Canada).

With NBR, we have now covered the four most generally used elastomers. NBR, with its nitrile group is above Neoprene in polarity and thus poorer in electrical insulation. Its big advantage is that it is considerably more resistant to oils, fuel, and solvents than Neoprene. Its other properties fall in line with those of the butadiene group, and thus it closely matches the physical properties of SBR except that it has much better heat resistance. Also, considering cost (slightly above Neoprene), it has the broadest resistance to chemicals with a balance of properties than any other polymer. It has a gain in oil resistance over that of Neoprene but at a sacrifice to weathering and ozone resistance.

Butyl rubber (IIR), isobutylene-isoprene copolymer Trade names: Enjay Butyl (Enjay Chemical), Petro-Tex Butyl (Petro-Tex Chemical), Polysar Butyl (Polysar), Bucar Butyl (Columbian Carbon Co.)

Butyl rubber's outstanding physical properties are low air permeability (about one-fifth that of natural rubber) and high energy-absorbing qualities. It has excellent weathering and ozone resistance, excellent flexing properties, excellent heat resistance, good flexibility at low temperature, tear resistance about that of natural rubber, tensile strength about in the range of SBR, and very good insulation properties. It has very poor resistance to petroleum oils and gasoline, but excellent resistance to corrosive chemicals, dilute mineral acids, vegetable oils, phosphate ester oils, acetone, ethylene, glycol, and water. It is also very nonpolar.

EPT (EPM), a copolymer of ethylene and propylene; or (EPDM), a terpolymer of ethylene, propylene, and a diene Trade names: Nordel (E. I. du Pont), Epsyn (Copolymer Chem.), Royalene (Uniroyal), Vistalon (Enjay Chem.)

This is a new elastomer which may replace SBR as the general-purpose rubber. In fact, it most certainly will become the standard for wire insulation because it has so much to offer: inexpensive, excellent weathering; and excellent ozone, heat, hot water, and steam resistance. In general, it is very similar to butyl rubber with the extra pluses just mentioned.

Because of its saturated nature and its slow air permeability, butyl rubber has met and still meets considerable resistance from fabricators who otherwise might use it. Slight contamination affects its cure and physical properties, and the air retention causes high defects due to air entrapment. The EPT family, with its terpolymer and air-retention properties of natural rubber overcomes the two large objections against the use of butyl rubber. Therefore, it is foreseen that compounders may become less reluctant to use it and recommend its use. If its processing characteristics eventually match those of SBR, there will be no doubt as to the future success of EPT—in fact, it is inevitable. *Caution:* Like butyl rubber, it is not oil-resistant; in fact, it is completely miscible.

Hypalon (CSM), chlorosulfonated polyethylene. (E. I. du Pont) A very close match to Hypalon is neoprene, and in many instances, substitutions can be made with hardly any difference in properties. However, Hypalon does offer some exceptional added properties over neoprene such as improved heat and ozone resistance, improved electrical properties, improved color stability, and improved chemical resistance. Light-colored neoprene will darken with age, and Hypalon will not; so if neoprene properties are desired along with color stability, Hypalon is the choice. Its very good electrical properties plus added heat resistance, color stability, and oil resistance are the reasons why it is preferred for spark plug boots over the costlier silicone rubber and for dust covers on high-voltage ground cables for TV picture tubes, etc.

Acrylic rubber (ABR), an acrylic ester copolymer Trade names: Cyanacril (American Cyanamid), Hycar 4021, 4031, 4032 (B. F. Goodrich), Krynac 882 (Polymer Corp.), Thiacryl (Thiokol Chem.), Acrylon (Borden Chem. Co.), Paracril OHT (Uniroyal).

Oils for hot applications are often fortified with sulfur-bearing chemicals; and this sulfur, in a system such as automotive transmissions, will continue to react and harden a nitrile-base rubber to a point of excesssive hardening. Acrylic rubber is not affected by hot-sulfur-modified oils; and for this reason, it is used in applications of this sort. Where test temperatures are normally run at 250°F for nitrile base, 302°F is used for the acrylic rubber in conjunction with transmission oils. Acrylics, therefore, can be said to have better oil, heat, and ozone resistance than nitrile, and good sunlight resistance; but they are not as good as nitrile for low-temperature work. They are usually used in hot-oil applications.

Polysulfide rubber (T), organic polysufide Trade name: Thiokol (Thiokol Chem. Co.). Thiokol is especially resistant to hydrocarbon solvents, aliphatic liquids, or blends of aliphatic with aromatic. The common alcohols, ketones, and esters used in paints, varnishes, and inks have little effect on it. It is also resistant to some chlorinated solvents, but preliminary tests should be made before using it for this purpose. Nitrile-base rubbers will swell considerably in such solvents, and Thiokol has a special use in applications using such solvents. Compared to nitrile, it has poor tensile strength, pungent odor, poor rebound, high creep under strain, and poor abrasion resistance. Thiokol has most of its applications in solvent-carrying hose, printers rolls, and newspaper blankets; and because of its excellent weathering properties, it is used heavily for calking purposes.

Silicones (FSI) (PSI) (VSI) (PVSI) (SI), polydimethylsiloxane and variations Trade names: SE Rubber (General Electric Co.), Silastic (Dow Corning Corp.), Silicones (Union Carbide Corp.), Si-O-Flex (Stauffer Chem.)

Silicones are an elastomer family of their own since their polymer backbone consists of silicone and oxygen atoms rather than the carbon structure of all other elastomers. Silicone is in the same chemical group as carbon; and since it is a more stable element, chemists predicted more stable compounds from it if it could be substituted for carbon in the chain. Its outstanding property is its heat resistance (500°F$^+$). It is also very flexible at below -100°F, and it has very good electrical properties. The outstanding property of silicone elastomers for insulation purposes is that their decomposition product on extreme heat is still an insulating silicon dioxide, whereas the decomposition product for organic compounds is conductive carbon black, which can sublime and thus leave nothing for insulation. The first user of silicone during World War II took advantage of this property in purposely undersizing motors where weight was a factor and letting them run hot. Silicone has resistance to oxidation, ozone, and weathering corona; and special modifications by inclusion of halogen groups can make it also oil-resistant.

Silicones generally do not have high tensile strength, but their overall properties of good compression set, improved tear resistance, and stability over a wide temperature range certainly make them suitable for many engineering uses. Silicones are the most heat-resisting elastomers available today and the most flexible at low temperatures. The basic structure is modified with groups such as vinyl or fluoride which enhance properties of tear and oil resistance, etc., so that we now have a family of silicone rubbers covering a wide range of physical and environmental needs.

Urethane (U), reaction products of diisocynates and polyalkylene ether glycols Common names: polyurethane, isocynate rubbers. Trade names: Cyanaprene (American Cyanamid Co.), Vibrathene (U.S. Rubber), Elastothane (Thiokol Chem.), Adiprene (E. I. du Pont), Estane (B. F. Goodrich), Genthane (General Tire and Rubber), Texin Mobay Chem.) Roylar (Uniroyal), Castomer (Isocynate Products Inc.), Conothane (Conap, Inc.)

Urethanes crosslink as well as undergo chain extension to produce a wide variety of compounds. They are available as "castable," or liquids, and as "solids," or gums, to be used on conventional rubber processing equipment; and as thermoplastic resins to be processed similarily as polyvinyl chloride. They can be cured to tough elastic solid rubbers with outstanding load-bearing capacity, resilience, abrasion resistance, and tensile strength. The outstanding properties are exceptionally high wear and two to three times the tensile strength of natural rubber. Urethanes exhibit very good resistance to oils, solvents, oxidation, and ozone. They have poor resistance to hot water and are not recommended for temperatures above 175°F, and also are quite stiff at low temperatures. They are useful where load-bearing and wear properties are desirable, and can be employed on insulation as a protective coating over an underlay.

Fluoroelastomers (FPM), copolymer of vinylidiene fluoride and hexafluoropropylene, copolymer of chlorotrifluoroethylene and vinylidene fluoride Trade names: Fluorel (Minnesota Mining & Mfg.), Viton (E. I. du Pont), Kel-F (Minnesota Mining & Mfg.)

Again the halogen groups indicate stability, and these do. The fluorinated rubbers are exceptionally good for high-temperature service but are below silicone in this respect. They resist most of the lubricants, fuels, and hydraulic fluids encountered in aircraft missiles; a wide variety of chemicals, especially the corrosive variety; and also, most chlorinated solvents. They have good physical properties, somewhere near those of SBR at the higher hardness levels. It is valuable in automotive use for its extreme heat and oil resistance and is on a much higher level in this respect than the acrilic elastomers. It has weathering properties superior to those of neoprene, but is very expensive.

Butadiene (BR), polybutadiene Trade names: Cis-4 (Phillips Chem.), Diene Rubber (Firestone), Ameripol CB (Goodrich-Gulf), Budene (Goodyear), Cisdene (American Synthetic Rubber), Taktene (Polymer Corp.)

A stereospecific controlled structure like isoprene rubber, its outstanding properties are excellent resilience and hysteresis (almost equivalent to those of natural rubber) and superior abrasion resistance compared to SBR. Butadiene is most similar to SBR and finds wide use as an admixture with SBR in tire treads to improve wear qualities. It is somewhat difficult to process, and for this reason it is hardly ever used in a larger amount than 75 percent of the total polymer in a compound.

Carboxilic (COX) butadiene-acrilonitrile modified with carboxilic groups Trade name: Hycar 1072 (B. F. Goodrich). Similar in properties to nitrile rubber, the carboxilic groups results in vulcanization with outstanding abrasion resistance (approaching that of urethane), improved ozone resistance, and superior cold-temperature properties compared to nitrile.

Hydrin rubber (CO) (ECO), homopolymer of epichlorophydrin, copolymer of epichlorohydrin, with ethylene oxide Trade names: Hydrin 100 and Hydrin 200 (B. F. Goodrich Chem.).

Hydrin rubbers are recognized as having excellent resistance to swelling in oils and fuels, in addition to good chemical resistance to acids, bases, and water. They also have very good aging properties, including ozone resistance. The high chlorine content imparts good-to-fair flame retardance. Hydrin rubber appears to possess a blend of the good properties of neoprene and nitrile. However, since the rubber is somewhat new in the industry, it is too early yet to be able to determine its acceptance and use.

Polyblend (PVC + NBR), a blend of butadiene-acrilonitrile and polyvinyl chloride Trade names: Paracril Ozo (Uniroyal), Hycar 1203 and 1205 (B. F. Goodrich), Chemivic (Goodyear Chem.).

This is a nitrile rubber that has been modified by vinyl resins. The resulting elastomer gives all the rubberlike properties of nitrile rubber, in addition to excellent ozone and weather resistance. Blends are available to give nearly the same oil-resisting properties and near ozone resistance of neoprene. Color stability is excellent; and in this respect, it is superior to neoprene and nearer to Hypalon. It extrudes quite readily and matches in properties previously mentioned polymers of similar price. It is highly recommended for use in wire and cable jackets, especially where color is desired.

TABLE 2 Relative Elastomer Costs

Polymer	Specific gravity	Cost/lb	Lb-vol.-cost ratio (SBR = 1.0)
NR	0.92	0.24	1.0
SBR	0.94	0.23	1.0
CR	1.23	0.41	2.3
NBR	1.00	0.51	2.3
IIR	0.92	0.25	1.1
EPT	0.86	0.30	1.2
CSM	1.1	0.45	2.3
ABR	1.09	1.35	6.8
T	1.34	1.15	7.1
SI	1.10	3.60	18.2
FSI	1.38	11.98	76.3
U	1.07	2.50	11.9
FPM (hexafluor)	1.85	10.25	87.0
FPM (chlorotrifluor)	1.85	16.00	136.8
IR	0.91	0.23	1.0
BR	0.91	0.25	1.0
COX	0.98	0.66	3.0
(CO)(ECO)	1.36	1.25	7.8
Polyblend	1.07	0.44	2.2

Relative elastomers costs Table 2 indicates the relative costs of the gum polymers, but it must be remembered that these represent only the raw costs to the rubber companies. Quantity purchases, freight, and credit terms will also modify these costs.

Parameters affecting cost are many and varied:
1. Mastication to bring the rubber to proper processing plasticities
2. Compounding ingredients necessary for reinforcement, curing, processing aids, protectants, diluents, colorants, etc., many of which are costlier than the gum
3. Mixing of the batch which depends and varies as to size and type rubber
4. Quality control assurance testing, depending upon extent and severity of specification
5. Processing variables of the various elastomers
6. Cure times and/or special curing techniques such as post cures
7. Defective percentages which are known to be higher for certain elastomers in various operations
8. Type of product being made which may entail special processing for certain degrees of physical properties

9. Possible need of clean room or refrigerated room because of contamination or pot life of mixed compounds, etc.

Engineering products are made to specific dimensions instead of weight, and for this reason a pound-volume-cost ratio of cost is given since it is the only true basis of comparison.

PHYSICAL PROPERTIES

The proper selection and application of elastomers is difficult for design engineers in many instances because engineering terms in conventional usage have different meanings when applied to rubber properties. Rubber is an organic material and reacts in a completely different manner from metals. For this reason definitions* need to be explained.[7]

Tensile strength in rubber refers to the force per unit of original cross section on elongating to rupture. As such, it is not really an important property in itself since rubber is rarely required to possess tensile strength. However, it is an indication of other qualities which correlate with tensile strength. Properties that improve with tensile strength are wear and tear, resistance, resilience, cut resistance, stress relaxation, creep, flex fatigue, and in some polymers such as neoprene better ozone resistance. In insulation, it should indicate better cut, abrasion resistance, and generally a tougher and more durable shield.

Elongation is the maximum extension of a rubber at the moment of rupture. As a rule of thumb, a rubber with less than 100 percent elongation will usually break if doubled over on itself.

Hardness is an index of the resistance of rubber to deformation and is measured by pressing a ball or blunt point into the surface of the rubber. The most commonly used instrument is the durometer made by the Shore Instrument Company. There are several Shore instruments such as A, B, C, D, and O, designed to give different readings covering soft sponge up through ebonite-type materials. Shore A duro readings are the most common, and these are the readings appearing in most specifications. On the Shore A scale, 0 would be soft and 100 hard; as a comparison, a rubber band would be about 35 and rubber tire tread about 70. Hard rubber (ebonite) is much more solid and is read on the Shore D scale; as a comparison, a hard rubber pipe stem may read about 60 Shore D, and a bowling ball about 90 Shore D. Most products will fall between 40 and 90 duro (unless otherwise specified, Shore A is assumed).

Modulus, in rubber, refers to the force per unit of original cross section to a specific extension. Most modulus readings are taken at 300 percent elongation, but lower extensions can be used. It is a ratio of the stress of rubber to the tensile strain but differs from that of metal as it is not a Young's modulus stress-strain-type curve: stress-strain values are extremely low for slight extensions but increase logarithmically with increased extension. Rubber has the useful property, for some products, of being extendible to 10 times its original length. Modulus can vary from the same hardness and does affect the stiffness of insulation.

Hyteresis is energy lost per loading cycle. This mechanical loss of energy is converted into heating of the rubber product and could reach destructive temperatures.

Heat buildup is a term used to express temperature rise in a rubber product resulting from hysteresis. However, the term also can mean use of high frequencies on rubber where the power factor is too high.

Permanent set is the deformation of a rubber that remains after a given period of stress followed by release of stress.

Stress relaxation is the loss in stress that remains after the rubber has been held at constant strain over a period of time.

Creep of rubber refers to change in strain when the stress is held constant.

Compression set of rubber is the permanent creep that remains after the rubber has been held at either constant strain or stress and in compression for a given time. Constant strain is most generally employed, and is reported as a percentage

---

* Reprinted from *Design News,* Mar. 31, 1965; a Cahners publication.

of the permanent creep divided by the amount of original strain. A strain of 25 percent is most common.

Abrasion resistance is one of the remarkable properties of rubber and one most difficult to measure. The term refers to the resistance of a rubber composition to wear and is usually measured by the loss of material when a rubber part is brought into contact with a moving abrasive surface. It is specified as percent of volume loss of sample as compared with a standard rubber composition, and it is almost impossible to correlate these relative values of life expectancy. Also, since many formulations contain wax-type substances which exude to the surface, test results can be erroneous because of the lubricating effect of the waxes.

Flex fatigue is the result of rubber fracturing after being subjected to fluctuating stresses.

Impact resistance is the resistance of a rubber to abrading or cutting when hit by a sharp object.

Tear resistance is a measure of stress needed to continue rupturing a sheet of rubber, usually after an initiating cut.

Flame resistance is the relative flammability of a rubber. Some rubbers will burn profusely when ignited, whereas others are self-extinguishing when the igniting source is removed.

Low-temperature properties indicate a stiffening range and brittle point of rubber. Stiffening range is the more useful of these two but is difficult to measure. Brittle point also has little meaning unless the deforming force and rate are known. Time is also an important consideration since some characteristics change as the temperature is lowered and held: hardness, stress-strain rate, and modulus, for example. With many materials crystallization occurs, at which time the rubber is brittle and will fracture easily. There are many tests to measure cold-temperature properties; but for a general comparison it is simplest to bend the specimens by hand at the test temperature for a difference in "feel" of stiffness or actual breakage. All tests must be made in the cold box and all precautions taken so that the rubber is not warmed during the test. The specimens (0.075 standard ASTM slabs) warm very quickly and will produce false results unless extreme care is taken in the testing.

Heat resistance: No rubber is completely heat-resistant; time and temperature have their aging effects. Heat resistance is measured usually as change in tensile strength, elongation, and durometer readings from the original values, usually after a 72-hr period.

Aging: Elastomeric properties can only be destroyed by further chain growth and linkage, which would result in a hard, rigid material, or a chain rupture, which would result in a plastic or resinous mass. The deteriorative agents mostly considered in this category are sunlight, heat, oxygen, stress with atmospheric ozone, atmospheric moisture, and atmospheric nitrous oxide. Chain growth or crosslinkage will usually decrease elongation and increase hardness and tensile strength whereas chain rupture will have the opposite effect. Some elastomers will continue to harden and some to soften; and some will show an initial hardening followed by softening. All are irreversible responses.

Radiation: Deteriorating effects of radiation are similar and complementary to those of aging. It has been fairly well demonstrated that damage is dependent only on dosage, or amount of radiation energy absorbed, irrespective of the form of radiation within a factor or two. Dosage rate is an important factor only where significant degradation from other agents, such as oxygen, has access to the system. Surprisingly enough, the least heat resistant of the elastomers displays the most radiation resistance.

Table 3 presents the physical properties of various rubber compounds. The values listed are only relative ones, in an attempt to differentiate among the rubbers; and so extreme caution must be exercised in the use of the table.

Cold-temperature stiffening of some rubbers is shown in Fig. 1. Again, all these curves can be altered with compounding and, in fact, can easily be made to interchange their respective position.

3-14 Elastomers

TABLE 3 General Physical Properties of Rubber Compounds

Compound	Tensile strength, psi (upper limit)	Durometer Shore A	Percent elongation	Service temperature, °F	Tear resistance	Abrasion resistance	Compression set	Flame resistance
Natural	4,500	25–95	800	−80 to 175	Very good	Excellent	Good	Poor
SBR	3,000	40–95	450	−67 to 200	Good	Good to excellent	Good	Poor
Chloroprene	3,200	30–90	700	−67 to 250	Good	Excellent	Fair to good	Good
NBR	3,000	40–90	650	−67 to 275	Good	Good to excellent	Good	Poor to fair
Butyl	2,500	35–90	500	−67 to 250	Very good	Good	Poor to fair	Poor
EPT	2,800	35–90	350	−80 to 300	Good	Excellent	Fair	Poor
Chlorosulfonated polyethylene	3,300	40–90	500	−67 to 275	Good	Excellent	Fair to good	Good
Acrylic	1,800	40–95	350	−20 to 350	Fair	Good	Fair	Poor to fair
Polysulfide	1,250	20–80	400	−40 to 225	Good	Poor	Poor	Poor
Silicone	1,500	20–90	750	−140 to 550	Poor	Poor	Good to excellent	Fair to good
Urethane	8,000	60–95	700	−40 to 250	Excellent	Superior	Excellent	Poor to fair
Fluorocarbon	2,700	60–90	300	−40 to 450	Fair	Good	Good to excellent	Excellent
Isoprene	4,000	25–95	600	−80 to 175	Very good	Excellent	Good	Poor
Butadiene	3,500	35–95	550	−100 to 175	Very good	Excellent	Good	Poor
Carboxilic	2,200	50–85	400	−90 to 275	Very good	Very excellent	Fair	Poor
Epichlorohydrin	2,500	50–85	450	−40 to 325	Good	Good	Good	Fair to good
Polyblend	2,800	45–90	400	−30 to 250	Good	Excellent	Poor	Fair

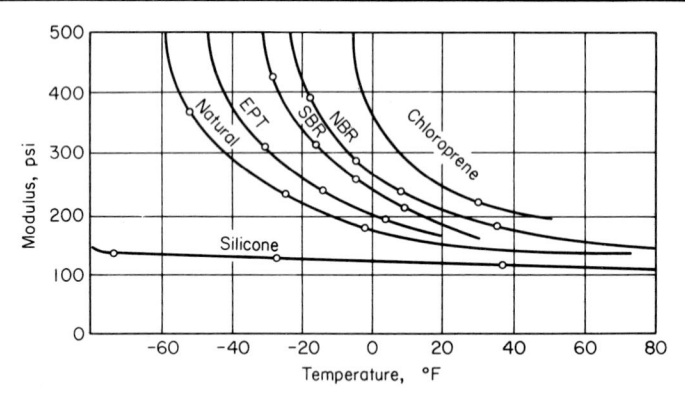

Fig. 1 Low-temperature stiffness of elastomers.[8]

One of the most difficult problems is the recommendation of a service temperature, and/or expectant service life at specific temperatures. Without data on a specific compound for a specific application, almost all recommendations must be based on previous experience coupled with data gathered from correlating laboratory results. Many new elastomers simply have not existed long enough for the time-span expectancy demands of some designs to be known. Most comparisons are made on a percentage of property changes from original values. The general rule of twice the speed of reaction for approximately every 15°F change in temperature follows for elastomers.

Table 4 compares thermal and radiation stabilities of various elastomers.

A good illustration of the effects of time and heat on elongation is shown in Fig. 2.[2]

Chemical Properties 3-15

TABLE 4 Comparison of Thermal and Radiation Stabilities of Various Elastomers[2]

Elastomer	Temperature at which 25% of original tensile strength is lost after 8-hr aging, °F	Order	Dosage at which 25% change in original properties occurs, ergs/g	Order
Silicone:				
Aliphatic side group	480	1	4.2×10^8	7
Aromatic side group		1	12.0×10^8	4
Fluoroelastomer	450	2	4×10^8	8
Acrylic	425	3	3.3×10^8	9
Urethane	350	4	4.2×10^9	1
Chlorosulfonated polyethylene	350	4	4.2×10^8	7
NBR	340	5	7×10^8	5
Isobutyliene	335	6	4×10^8	8
Chloroprene	325	7	5.5×10^8	6
Polysulfide	285	8	1.5×10^8	10
SBR	275	9	1.5×10^9	3
Natural	210	10	2.5×10^9	2

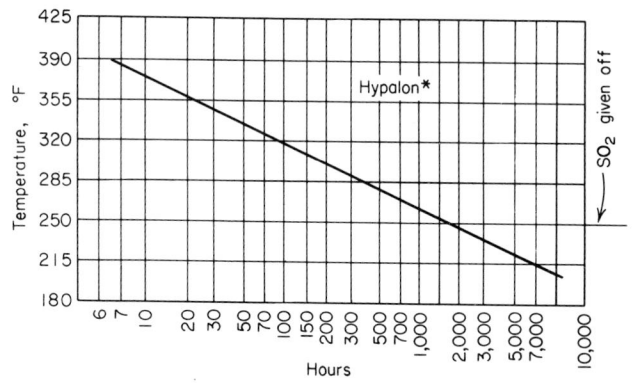

Fig. 2 Heat exposure time in which Hypalon elongation dropped to 100 percent.[9] (*Trademark of E. I. du Pont de Nemours & Co.)

Silicone is the best commercially available rubber for heat resistance, and Fig. 3[2] indicates its life expectancy at elevated temperatures.

A common misconception is that original room-temperature tensile strengths and elongations will prevail at various temperatures. Figure 4[2] clearly dramatizes the varying tensile value with varying temperatures. All polymers are similarly affected, but not to the same degree, so that if the mechanical properties are of design importance at elevated temperatures, then supporting data on a particular compound are needed.

CHEMICAL PROPERTIES

Rubber is a desired end product of chemical reactions, and as such is not thought of as an intermediate chemical for further reactions. However, as desirable as it would be to preserve the end properties unchanged, this is not possible. As with most other materials, time and environment have their effects. All rubbers are organic materials and as such are relatively less stable than many other engineering materials. In fact, in products employing rubber as a component material, it is

3-16 Elastomers

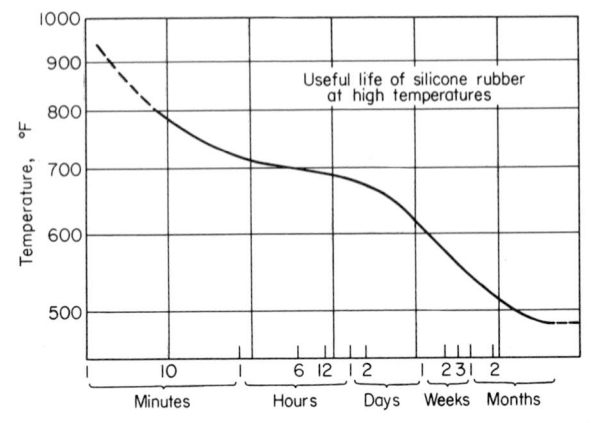

Fig. 3 Aging of silicone rubber at elevated temperatures.[9]

the rubber that is expected to deteriorate first (as insulation on copper wire, for example). The expectant life and time interval of useful life are often measured only in hours; but for most practical applications, a 5- to 10-year life would suffice, and for long-term aging, as in transformers, up to 20 years. All rubbers gradually and continually undergo slow changes in measured physical properties until the total changes result in a product no longer suitable for the original intended use.

The chemical interaction, possibly aging would be a more suitable term, is dependent on the external and internal factors. External factors would be the environmental exposure to oxygen, heat, ozone, light and weather, radiation, chemicals, and oils. The internal factors are the type of rubber, compounding materials, and mixing. Two relationships of primary importance in the interaction of the two factors are solution and reaction. Another way of saying the same is "swell" and "decomposition."

Swell: The theories of solution chemistry apply to rubbers and their media in that the general "like likes like" applies. The various elastomers range from relatively nonpolar rubbers, such as butyl, to relatively high-polar rubbers, such as nitrile; and most media also show this range in polarity. Therefore, a nonpolar butyl rubber with a nonpolar petroleum oil is readily soluble and results in the

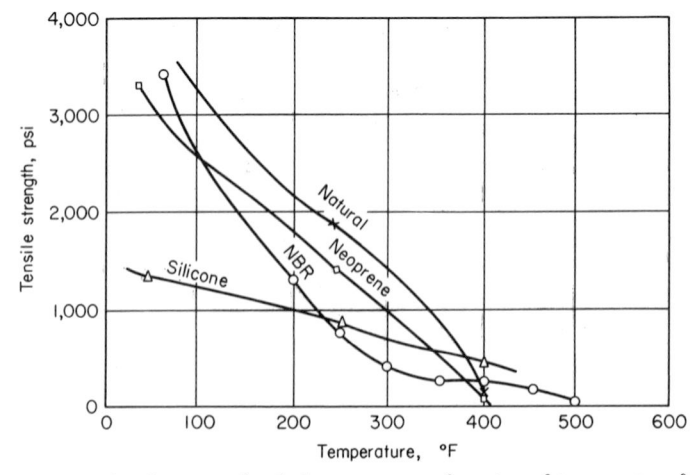

Fig. 4 Tensile strength of elastomers as a function of temperature.[9]

swelling of the rubber, whereas butyl, with a polar ester fluid, is not affected. The swelling media are separated into two groups: aliphatic and aromatic. Aliphatics are straight-chain hydrocarbons and their derivatives such as low-octane gasolines, hexane, butane, heptane, propane, etc. Aromatics are benzene and its derivatives such as toluene, xylene, etc. Aromatics swell rubbers more than aliphatics, and so that it is useful to have a measure of this property in oils to estimate its effects. Measurement of aromaticity is called aniline point (lowest temperature at which the hydrocarbon is miscible with an equal volume of aniline). Figures 5 and 6 illustrate these swelling effects on several rubbers.

The hydrocarbon group is followed by other basic groups such as:
1. Nitrogen compounds such as aniline and nitrobenzene.
2. Oxygenated compounds such as alcohols, ketones, esters, and ethers.
3. Halogenated hydrocarbons such as methyl chloride, carbon tetrachloride, ethylene dichloride. These have greater solvency power than the parent hydrocarbon.

Swelling of rubber products will enlarge the specimen, lower the hardness, and decrease tensile strength, modulus, and tear and wear resistance. The degree of swelling in relation to its design parameters determines the usefulness. All rubbers

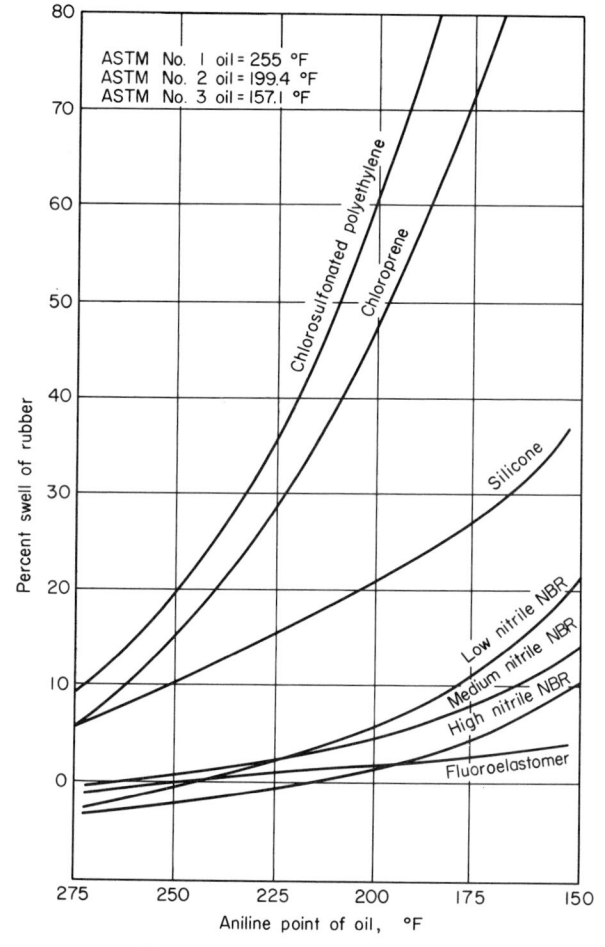

Fig. 5 Effect of aniline point on swell of elastomers.[10]

3-18 Elastomers

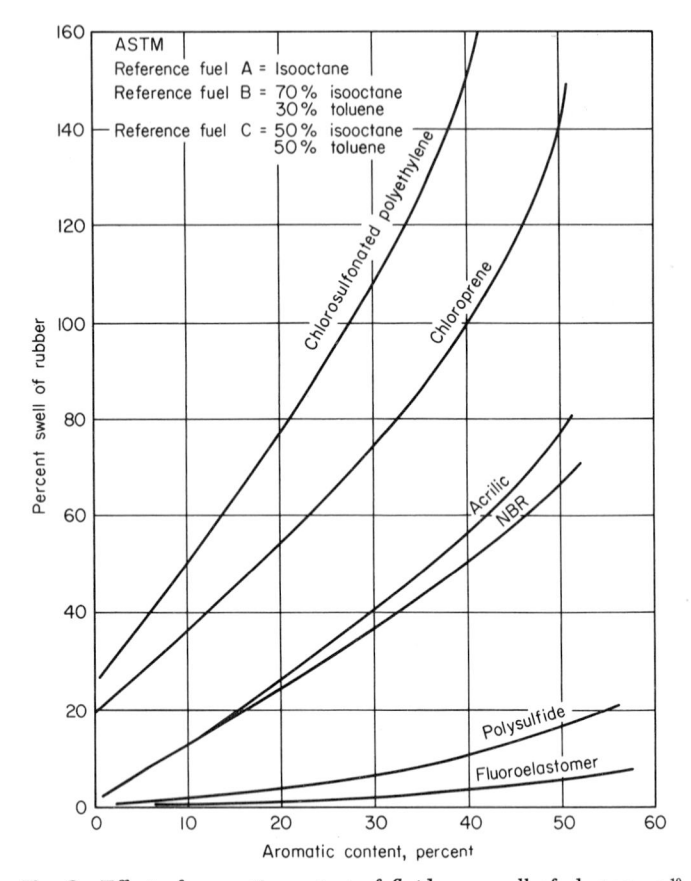

Fig. 6 Effect of aromatic content of fluids on swell of elastomers.[10]

in most instances will swell somewhat. Rubber in swelling solutions will have leaching or extracting effects or both. Oils and other compounding materials as well as the rubber having an affinity for a certain fraction of the media can and do affect physical properties of the rubber, especially if low-temperature plasticizers are extracted out of the rubber. Figure 7[8] indicates the relative swell and cold-temperature properties of several elastomers.

Decomposition: Swelling is only a physical condition; and if the agent is removed, most properties (less extraction) will restore themselves to their original condition. Decomposition is used here to describe a condition in which the second factor chemically reacts with the polymer and generally renders it unfit for use. There are many such degradants such as acids, alkalies, and radiation effects, in addition to the environmental factors mentioned above.

Table 4 indicates the relative resistance of the various elastomers. Various laboratories have accumulated data on specific media for rubber, showing concentration, time, and effect, but these are far too voluminous to include in this writing. Table 5 lists the effects of specific chemicals on specific elastomers.

ELECTRICAL PROPERTIES

A good conductor such as copper has a resistivity of 1.7×10^{-6} ohm-cm, whereas a good insulating material has a resistivity of about 10^{18} ohm-cm. The vast difference in resistivity between conductors and insulators makes possible the construction

TABLE 5 Effect of Chemicals on Elastomers

Elastomer	Ozone	Aliphatic hydro-carbons	Aromatic hydro-carbons	Halo-genated hydro-carbons	Ketones	Animal or vegetable oils	Acids dilute	Acids concen-trated	Alkalies	Water	Weather resis-tance	Oil resis-tance	Fuel resis-tance
Natural	P	P	P	P	G	P to G	F to G	F	F	VG	F	VP	VP
SBR	P	P	P	P	G	P to G	F to G	F	F	G	F	VP	VP
Chloroprene	VG	G	F	P	P	G	G	F	G	P to G	VG	G	F
NBR	P	VG	F	P	P	E	G	F	F	G	P	VG	G
Butyl	E	P	P	P	G	E	VG	G	G	VG	VG	VP	VP
EPT	E	P	P	P	G	E	VG	G	G	VG	E	VP	VP
Chlorosulfonated polyethylene	E	G	F	P to F	P	G	E	G	E	E	E	G	F
Acrylic	E	VG	P	P	P	E	F	F	P	VP	VG	VG	F
Polysulfide	E	E	VG	F to G	G	E	G	F	G	VG	E	E	E
Silicone	E	P	F	P to G	F	F	G	F	G	E	E	F	E
Urethane	E	G	G	F to G	F	G	F	F	P	F	E	E	P
Fluorocarbon	E	VG	G	G	F	E	E	E	E	VG	E	E	VG
Isoprene	P	P	P	P	G	P to G	F to G	F	E	VG	F	VP	E
Butadiene	P	P	P	P	G to F	P to G	F to G	F	F	VG	F	VP	VP
Carboxilic	G	G	F	P	...	...	...	...	...	P	E	G	G
Epichlorohydrin	E	VG	F	P to F	P	G	G	F	F	G	E	G	F
Polyblend	VG	VG	F	P	P	G	...	F	F	G	E	G	F

E   Excellent.
VG Very good.
G   Good.
F   Fair.
P   Poor.
VP Very poor.

3-20 Elastomers

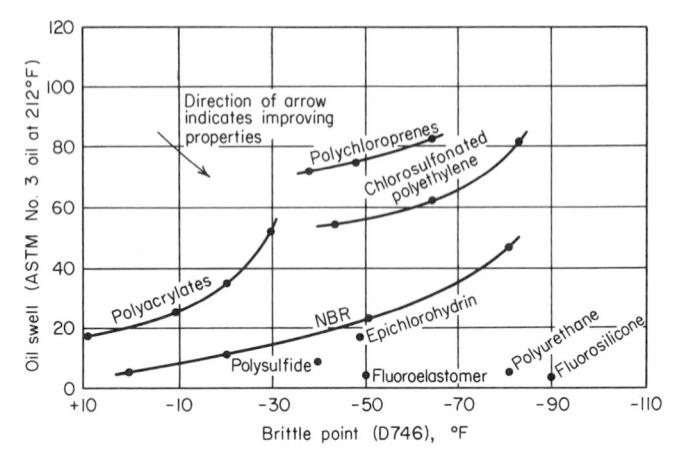

Fig. 7 Oil resistance versus brittle point of elastomers.[8]

of modern electric apparatus with different potentials.[3] The resistivity of various materials is illustrated in Fig. 8,[11] especially the spread of rubber values from conductive rubber through nonconductive rubber (ebonite).

The principal function of an insulating material is to direct current flow through conductors; and usually the mechanical, thermal, and cost considerations determine the choice of an insulation. Electrical losses in many applications are secondary; however, they do become important in high-frequency applications, such as large ac capacitors, where it is desirable to minimize heat buildup. Cost considerations dictate the type and minimum insulation thickness consistent with reliability of use; in this respect thermosetting organic insulators are a proved choice in many applications.

Figure 8 indicates the usage of rubber for conductive purposes as well as for insulation, and these applications are certainly numerous. Normally, conductive rubber is used for antistatic applications such as sandblast hose, machine mountings, bearing mountings, carpet backing, containers, chutes, shoe soles, conveyor belts, trolleybus tires, and castor wheels. It is also employed for heating purposes as drying, airplane deicing equipment, and antifogging. It also finds application as conductive sheaths between conductive cables and insulators to prevent corona formation, as coatings over metals to reduce electrolyses, as electrostatic shielding, as suppressors of electromagnetic radiation, as attenuators in the very high frequency field, and as conductive gaskets to provide complete shielding. Conductive rubbers are usually defined as having resistivities up to 10^6 ohm-cm, and by special compounding can readily be reduced to 30 to 500 ohm-cm. Resistivities up to 10^6 ohm-cm are effective for reduction of static charges, and conducting flooring is satisfactory up to 10^5 ohm-cm.[11]

The insulation properties, rather than conductive properties, are of major consideration. The properties that most concern the rubber compounder are dielectric strength, insulation resistance, dielectric constant, and power factor. Following is a description of the various properties.[12]

Dielectric strength: The dielectric strength of a material is the stress required to puncture a sample of known thickness. This property is expressed in terms of volts per mil thickness and is influenced by the rate of rise of the applied voltage and the total length of time that the voltage is applied. A slow rate of rise usually will cause the material to puncture at a lower voltage than a rapid rate of rise. Similarly, a material may withstand a relatively high voltage for a short time but will be punctured by prolonged exposure to a considerably lower voltage.

The test for dielectric strength is made by determining the minimum voltage that will puncture a sample of known thickness placed between electrodes of specified size and shape. Most frequently the rate of increase is, starting from zero,

Electrical Properties 3-21

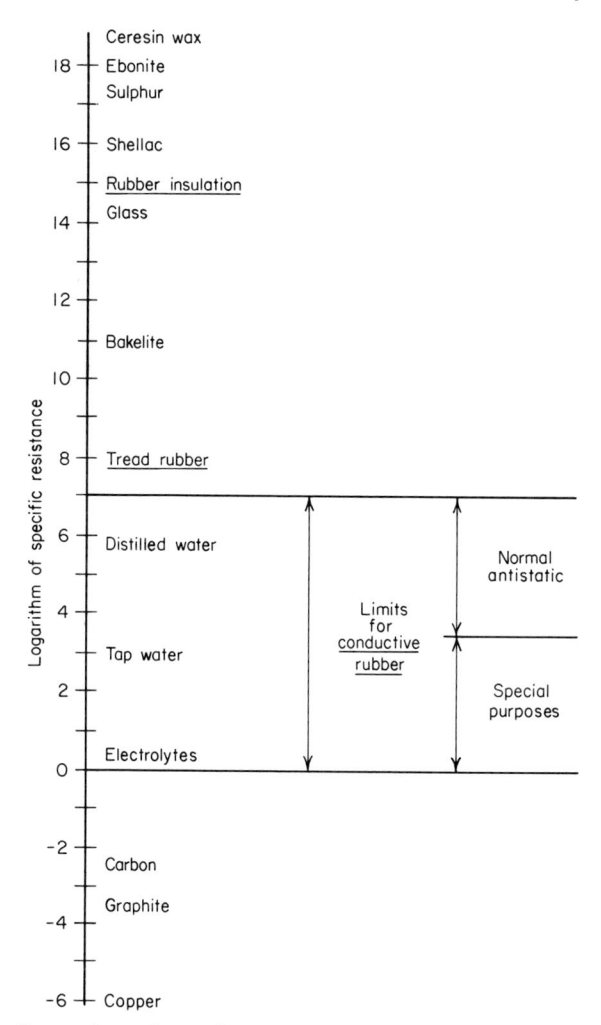

Fig. 8 Comparison of specific resistance of rubber with other materials.[11]

500 volts/min until breakdown occurs. It should be noted that values obtained by this test are not absolute because the dielectric strength in volts per mil varies somewhat, depending on the thickness of the sample tested. Ordinarily, thin test specimens show a higher dielectric constant than thick ones.

The dielectric strength of a rubber compound is involved whenever the compound is used as an insulating material. Mechanical design factors require dimensions of the insulation such that the electric stress (volts per mil) is far below the breakdown point. A liberal safety factor should be provided because of the dependency of dielectric strength on time and voltage.

Insulation resistance: The resistivity or specific resistance of a material is the electrical resistance offered by an element of the material having unit length and unit cross-sectional area. The ohm is the basic unit of resistance. The measured resistance of the insulation of a device or product is called the insulation resistance. This measurement is taken along the path over which the insulation is intended to be effective.

The resistivity of a material is determined by measuring the resistance of a

3-22 Elastomers

sample of known dimensions. This is accomplished by placing the sample between two electrodes, impressing a voltage between these, and measuring the current flowing through the sample. The insulation resistance test is essentially the same as for resistivity measurements. To test insulated wire, tanks must be provided for immersing the coils or reels in water and providing a well-insulated test lead for making connections to the conductors of the wire under test. The insulation resistance of a molded product with current-carrying metal inserts is usually measured between these inserts. The insulation resistance of insulated wire is the resistance between the conductor and the outside of the insulation. When the insulation is covered by a metallic sheath or braid, the measurement is made between the conductor and the sheath. Insulator wire with no sheath is usually immersed in water and the resistance measured between the conductor and the water.

The results of the insulation resistance tests are expressed in ohms. For insulated wire, the insulation is expressed in ohms per unit length. The unit length usually chosen is either 1,000 ft or 1 mile. Resistance measurements are seriously affected by temperature, humidity, and surface conditions of the test specimen.

Insulation resistance or resistivity is important for a rubber compound or an insulating materials since there is a minimum value of resistivity below which the material will not function satisfactorily.

Dielectric constant or specific inductive capacity (SIC): Dielectric constant is a specific property of an insulating material that is both defined and measured by the ratio of the electric capacitance of a capacitor having that material as the dielectric to the capacitance of the same capacitor having air as the dielectric. (The dielectric of a vacuum is unity; but for most practical purposes, air is considered as unity even though its constant is slightly higher.)

Two parallel plates of conducting material separated by an insulating material, called the *dielectric*, constitutes an electric capacitor. These two plates may be electrically charged by connecting them to a source of dc potential. The amount of electric energy that can be stored in this manner is called the *capacitance* of the capacitor, and is a function of the voltage, area of the plates, thickness of the dielectric, and the characteristic property of the dielectric material called dielectric constant. The capacitance of this capacitor is measured by connecting it in a bridge circuit so that the capacitance may be compared with that of a calibrated variable standard capacitor. This measurement is made using alternating current in the bridge circuit. The choice of frequency of the alternating current is governed by the end use of the insulating material; but for most rubber insulation, 1,000 Hz is used.

The ac phenomena associated with electric capacitance are of interest chiefly in the wire and cable application field. In actual use, almost every wire will have a conducting sheath, will be run in a conduit, or will be adjacent to other wires, so that a capacitance is set up between the conductor and these surrounding or adjacent metal objects. At most commercial voltages and frequencies used in power distribution, the capacitance effects are negligible. At relatively high voltages, the current due to capacitance may reach sufficient value to affect the circuit; and insulation for such an application is designed for moderately low dielectric constant. In communications where higher frequencies are used, the capacitance effect is pronounced and is of great importance. Among the other effects, the capacitance of the wire determines the maximum length of line over which the signal may be transmitted without excessive distortion. The lower the dielectric constant, the greater the distance over which signals may be transmitted satisfactorily.

Power factor: The power factor (or loss angle) of an insulating material is determined by the ratio of the energy loss expressed in watts to the volt-ampere product of the alternating current flowing in a capacitor using the material as a dielectric.

This property is more easily understood by considering the following hypothetical direct-current explanation. A capacitor is connected to a dc source, and the amount of energy required to charge the capacitor is measured. The capacitor is then disconnected from the source and discharged. The amount of energy given up by the capacitor during discharge is measured. The difference between these two measured values will be the energy loss of the capacitor. The power factor is the ratio of this loss to the energy required to charge the capacitor, and may be expressed as a decimal or as a percent of the charging energy.

The dielectric losses of an insulating material are caused partly by insulation resistance, and partly by an electrical stress hysteresis which is characteristic for each material. In both instances the loss results in heating of the dielectric.

In measuring the power factor of a material, a capacitor is made, using the material as the dielectric. This test capacitor is connected in one arm of an ac bridge circuit. The corresponding arm of the bridge contains a calibrated variable capacitor in parallel with a calibrated variable resistor. When the bridge is balanced, the capacitance of the calibrated capacitor is equivalent to the capacitance of the test capacitor, and the calibrated resistor will be equivalent to the apparent parallel resistance of the test capacitor. From these values, the power factor of the material may be calculated.

Power factor, like dielectric constant, is a property of interest primarily in the wire and cable insulation field. This property represents a power loss that takes place when a wire insulation becomes the dielectric of a capacitor because of a surrounding sheath or other conducting medium. In commercial power wiring this loss does not become serious if the power factor of the insulation is kept below 10 percent. Above 10 percent, the power factor losses under certain conditions will cause a temperature rise in the insulation that may result in failure or reduced life of insulation. In communications wiring the power factor of the insulation plays a more important role. Here the actual power loss can represent an appreciable portion of the total energy in the circuit. In addition, this loss will disturb the circuit characteristics of the equipment at both ends of the line.

Electrical properties for various applications of rubber are varied and many. The products range from the large-volume usage of insulated wires to electrical equipment parts, electrical sockets, plugs, spacers, aprons, mats, gloves, etc. Wires and cables must be designed for practically 0 to higher than 25 kv, and frequencies from direct current to 10,000 MHz. In actual service, rubber insulators are often subjected to temperature extremes, chemicals, moisture, oils, crushing abrasion, impacts, sunlight, ozone, and high operating stress. All these effects must be recognized and provided for by both proper selection of material and adequate design.

The most suitable materials for electrical purposes would be those possessing high dielectric strength and resistivity, with dielectric constant and power factor at their lowest. A combination of this sort is rarely attained since other physical and chemical properties are also required; and so compromise values are the rule rather than exception.

In the field of wire and cable, emphasis on properties varies with the type of circuit. Table 6 will aid in visualizing this.

TABLE 6 Properties for Types of Service[11]

| Application | Relative value of property needed ||||
	Dielectric strength	Dielectric constant	Power factor	Resistivity
Power:				
Low voltage	M to L	M to H	M to H	M to L
High voltage	H	M to L	M to L	M to L
Signal	M to L	M to L	M to L	M to H
Communication:				
Low frequency	M to L	L	M to L	M to H
High frequency	M to L	L	L	L

H High.
M Medium.
L Low.

3-24 Elastomers

TABLE 7 Summary of Electrical Insulating Qualities of Selected Elastomers[11]

	Examples of insulation values		
	Volume resistivity, ohm-cm	Power factor, %	Dielectric strength, volts/mil
Excellent:			
Silicone	10^{14}	1	600
Fluorocarbon	10^{13}	4	500
Good:			
Butyl	10^{16}	2	600
Natural rubber	10^{16}	5	500
Hypalon	10^{14}	3	600
SBR	10^{13}	7	500
Fair:			
Urethane	10^{12}	4	500
Polysulfide	10^{12}	30	250
Neoprene	10^{10}	20	350
Acrylic	10^{9}	20	350
Poor:			
Buna-N insulating	10^{9}	30	350
Buna-N conductive	10^{2}		

Electrical values of elastomers: Table 7 is a summary of electrical insulating qualities of selected elastomers. The ratings in order of merit were based on factors such as degradation product of a silicone being a nonconductive silica, flame resistance, etc.; and the general elastomer evaluation study was directed for use in aerospace systems. Additional values are given in Table 8.

Table 9[1] presents more comparative ratings of elastomers for electrical insulation.

TABLE 8 Electrical Values of Elastomers[13]

Elastomer	Resistivity, ohm-cm	Dielectric strength (short time), volts/mil	Dielectric constant at 1 kHz	Dissipation factor at 1 kHz
NBR	10^{10}	....	13.0	0.055
SBR	10^{15}	....	2.9	0.0032
IIR (butyl)	10^{17}	600	2.1–2.4	0.0030
CSM (chlorosulfonated polyethylene)	10^{14}	500	7–10	0.03–0.07
EPR (EPM)	10^{15}–10^{17}	900	3.17–3.34	0.0066–0.0079
EPT (EPDM)	10^{15}–10^{17}	900–1,050	3.0–3.5	0.004 at 60 Hz
FPM (hexafluor)	10^{14}	613	5.9	0.053
FPM (chlorotrifluor)	10^{13}	250–750	....	0.03–0.04
FSI (fluorosilicone)	10^{13}–10^{14}	340–350	6.9–7.4 at 100 Hz	0.03–0.07 at 100 Hz
CR (neoprene)	10^{11}	150–600	9.0	0.030
NR or IR (natural or synthetic)	10^{15}–10^{17}	....	2.3–3.0	0.0023–0.0030
Polysulfide	10^{12}	250–600	7.0–9.5	0.001–0.005
Urethane	10^{11}–10^{14}	350–525	5–8	0.015–0.09
SI (silicone)	10^{11}–10^{17}	100–655	3.0–3.5	0.001–0.010

TABLE 9 Properties of Polymers Used as Electrical Insulation[11] (Comparative ratings based on best attainable values)

Property	Natural rubber	GR-S	Butyl	Neo-prene	Nitrile rubbers	Thiokol	Silicone	Poly-vinyl chloride	Poly-ethylene	Nylon	Teflon	Fluoro-thene	Mylar
Dielectric strength	H	H	MH	M	M	L	M	VH	VH	MH	VH	M	VH
Power factor	M	M	M	MH	H	VH	M	MH	VL	MH	VL	L	L
Resistivity	H	M	VH	L	L	VL	M	M	VH	L	VH	MH	MH
Dielectric constant	M	M	M	MH	M	VH	L	MH	VL	M	VL	L	L
Abrasion resistance	H	H	MH	VH	MH	M	L	H	MH	H	..	H	H
Flexibility	H	H	H	MH	MH	H	H	M	L	L	L	L	VH
Resistance to compression cutting	MH	MH	M	M	MH	M	L	H	H	VH	VH	H	VH
Cold flow at room temperature	L	L	L	L	L	M	MH	L	L	VL	VL	VH	VH
Cold cracking temperature	VL	M	L	M	MH	MH	VL	L	VL	M	VL	M	VL
Sunlight resistance	MH	MH	H	H	M	H	H	H	H*	MH*	VH	VH	VH
Ozone resistance	L	M	H	H	MH	H	H	VH	VH	H	VH	VH	H
Corona resistance	M	M	M	MH	M	..	..	H	M	..	..	..	H
Heat resistance	MH	H	VH	MH	MH	H	VH	H	M	H	VH	VH	M
Flame resistance	VL	VL	VL	MH	VL	L	M	H	VL	MH	VH	VH	M
Water resistance	H	H	H	M	M	M	H	VH	VH	L	VH	VH	VH
Oil resistance	VL	VL	VL	M	MH	H	M	H	M	H	VH	VH	H

* Only if suitably pigmented
VH Very high. MH Medium high. L Low.
H High. M Medium. VL Very low.

Water-absorbing effect: The dielectric constant and power factor of a rubber insulation are increased considerably by small amounts of absorbed water; hence, wire insulation for communications use usually must have a dielectric constant as stable as possible in the presence of water or moisture. Aged rubber will also change the subsequent immersion values since the inclusion of polar oxygen in the aging process enhances the affinity for water absorption and thus varies the dielectric and power properties. For this reason, the earlier mention of EPT (or EPR) rubber as a substitute for butyl, since it has a built-in stability on aging and exposure to moisture which makes it highly suitable for building, service entrance and weatherproof insulation, portable cord insulation, and power cable (600 volts to 15 kv). Tables 10 to 17 follow illustrating the properties of butyl rubber compared to those of filled crosslinked polyethylene and EPT.[14]

TABLE 10 Physical Properties of Butyl, Polyethylene, and EPR[14]

Property	Butyl	Filled crosslinked polyethylene	EPR
Tensile strength, psi	800	2,300	900
Elongation, %	500	350	500
IPCEA set, 64th in	24	60	20
Deformation at 121°C, %	14	10	15
Tear, lb/in. thickness	35	75	35
Hardness, Shore A	60	95	60
Compression cut, lb/0.10 in	450	1,200	725

TABLE 11 Cold Flexibility of Polyethylene, Butyl, and EPR[14]

Elastomer	Stiffness in flexure, psi (ASTM D 747)	Brittle temperature, °C (ASTM D 746)	IPCEA cold bend, °C
Filled crosslinked polyethylene	6,600	−80	−65
Butyl	670	−47	−65
EPR	640	−75	−80

TABLE 12 Rate of Oxidation and Percent Retention of Unaged Properties of Butyl, Filled Crosslinked Polyethylene, and EPR

Property	Butyl	Filled crosslinked polyethylene	EPR
Tensile strength, psi:			
Unaged	800	2,300	900
After 7 days in 121°C air oven	77	98	98
After oxygen pressure test, 168 hr at 80°C	75	98	98
After air-pressure heat test, 168 hr, 80 psi, at 127°C	75	87	98
Elongation, %:			
Unaged	500	350	500
After 7 days in 121°C air oven	75	92	92
After oxygen pressure test, 168 hr at 80°C	90	93	98
After air-pressure heat test, 168 hr at 127°C	80	95	98

TABLE 13 Dielectric Strength—Average RMS AC Stress at Breakdown with Conductor at 25°C[14] (In volts per mil)

Property	Butyl	Filled crosslinked polyethylene	EPR
Impulse:			
1/0 AWG, 0.297-in. wall	730	1,000*	860
Rising ac voltage:			
14 AWG, 0.047-in. wall	513	778	510
1/0 AWG 0.156-in. wall	490	646*	500
1/0 AWG, 0.297-in. wall	323	.....	350
IPCEA short-time ac voltage:			
1/0 AWG, 0.297-in. wall	245	375*	330

* 4/0 AWG, 0.110-in. wall.

TABLE 14 Dielectric Loss Comparison of Butyl, Filled Crosslinked Polyethylene, and EPR[14] (80 volts/mil at 60 cycles)

Temperature, °C	Butyl K	Butyl 100 tan δ	Butyl $K \times$ 100 tan δ	Polyethylene K	Polyethylene 100 tan δ	Polyethylene $K \times$ 100 tan δ	EPR K	EPR 100 tan δ	EPR $K \times$ 100 tan δ
\multicolumn{10}{c}{14 AWG, 0.047-in. wall in water}									
20	3.79	0.75	2.83	2.62	0.38	0.99	2.88	0.44	1.26
50	3.79	0.96	3.64	2.60	0.46	1.20	2.88	0.50	1.44
75	3.83	2.21	8.46	2.56	0.65	1.66	2.88	1.00	2.88
90	3.90	4.98	19.42	2.51	0.84	2.10	2.88	1.79	5.15
\multicolumn{10}{c}{Shielded cable in air}									
20	3.3	0.8	2.64	2.29	0.25	0.57	2.45	0.36	0.88
75	....	....	.....	2.13	0.59	1.28	2.45	0.96	2.35
100	....	....	.....	2.00	0.83	1.66	2.45	1.30	3.18
121	3.3	1.5	4.95	1.98	1.15	2.27	2.45	1.63	4.00
140	....	....	.....	1.96	2.05	4.01	2.45	2.31	5.66

Butyl cable: 4 AWG, 0.156-in. wall.
Polyethylene cable: 4/0 AWG, 0.110-in. wall.
EPR cable: 1/0 AWG, 0.256-in. wall.

TABLE 15 Insulation Resistance of Butyl, Filled Crosslinked Polyethylene, and EPR[14]
(In megohms at 1,000 ft)

Butyl	200,000
Polyethylene	50,000
EPR	200,000

3-28 Elastomers

TABLE 16 Electrical Stability in 75°C Water of Butyl, Filled Crosslinked Polyethylene, and EPR[14]

Immersion time, days	Butyl	Polyethylene	EPR
0	513	778	510
7	551	647	541
14	466	725	656
28	399	699	561

TABLE 17 Corona Resistance of Butyl, Filled Crosslinked Polyethylene, and EPR[14]
(IPCEA Vertical U-bend—hours to failure)

Elastomer	Average stress, volts/mil 100	150
Butyl*	>100	168
Polyethylene†	52	35
EPR*	>100	168

* 1/0 AWG, 0.297-in. wall.
† 6 AWG, 0.110-in. wall.

The data in these tables were based on a peroxide-cured ethylene-propylene copolymer EPR (EPM). The terpolymer EPT (EPDM) should give similar electrical properties if also in peroxide cure except that the polymer may cure tighter with resultant slightly higher tensile strength.

Earlier it was stated that the principal function of an insulator was to direct current flow through conductors; and usually the mechanical, thermal, and cost considerations determine the choice of an insulation. When one now considers special electrical properties, coupled with long-term aging effects, the selection of the most suitable insulation becomes difficult. A choice and recommendation will depend on the experience at hand, coupled with availability of commercially ready materials. Choice of any new materials must be based on, first, assurance of a continuous supply; followed by the property comparisons on standard laboratory testing parameters, coupled with processibility and comparable finished product costs; then field testing; and finally recommendations for commercial use. Years often pass before a new material eventually finds its proper place; and if the past 30 years of elastomer development can be used as a barometer, then surely in the near future years insulation recommendations can be expected to become even more complex.

SPECIFICATIONS

A standard specification for the selection of elastomers for the electrical industry is not available. There are several popular specifications which are written regarding properties of certain materials or elastomers, but these specifications are based on "black-loaded" compounds intended for mechanical property needs rather than electrical needs and are thus nearly worthless. A material is selected because of its *performance* qualities related to a *product* need. Thus, a product need must be defined, for specific environmental conditions with an expectant life goal. Once a satisfactory product is developed, then a *product specification* can be written, along with special material component properties. A properly written product specification then becomes meaningful in assurance of quality and competitive procurements.

Rubber uses in the electrical industry are probably among the most varied of any industry needs. Products include grommets, gloves, lineman blankets, spark plug boots, wire insulations, etc.; with a myriad of uses such as underground, water, outdoor, missiles, electronics, chemical plants, shipboard, etc. This diversity results in a complexity of specifications, along with many sponsoring groups. As was mentioned earlier, material standard specifications for the electrical industry as a whole are not available. However, many special needs such as grommets, lid gaskets, dust shields, vibration isolators, etc., are covered by established material specifications. Therefore, popular material specification need to be accompanied with product specifications along with their sources if a complete tabulation of these is to be made. However, product specifications in their entirety are beyond the scope of this chapter; and so only the major available groups with information on how to obtain them will be given here.

Material specifications *ASTM D 2000 and/or SAE J-200.* This classification system was jointly sponsored by ASTM and SAE, and is the most up-to-date and the basis of all rubber specifications, both private and military. The group under this joint sponsorship is officially called the Technical Committee on Automotive Rubber (TCAR), and its membership is comprised equally of users and suppliers of automotive rubber. The basis of this specification is to classify elastomers by type and class: that is, heat resistance and oil-swell resistance. In this classification all elastomers fall within the range of tabulated commercial availabilities, and so the specification can be used as a material "line call-out" on procurement documents. (NOTE: This does not cover electrical properties.)

ASTM D 735 and/or SAE J-14. Both these specifications were superceded by D 2000 and J-200 above. The last revision on D 735 was made in 1961 and its publication in Part 28 of ASTM Standards was dropped after the April, 1967, edition. The "SAE Handbook" carries J-14 for information purposes only. All revisions since 1961 have been made only on the D 2000 and J-200 specifications.

MIL-R-3065, MIL-Std-417, AMS. These specifications follow somewhat the same format and values as D 735 and J-14, but not exactly. (NOTE: Again, no electrical values.)

Private Specifications. Industrial specification are numerous, and most major manufacturers use their own. Their basic format and properties follow very closely the D 735 and J-14 with additional special properties, as each industry feels that special values or tests are needed for their own uses. However, these special properties are allied with some product.

Product specifications Various technical societies have contributed greatly to specifications in conjunction with elastomeric or rubber goods for the electrical industry. ASTM specifications are basic; and, again, part 28 of ASTM Standards covers electrical insulating materials, rubber insulation for wire and cable, and rubber protective equipment. SAE covers the entire electrical requirements for the automotive industry. AMS (Aeronautical Material Specifications), a subdivision of the Aeronautics Committee of SAE, has jurisdiction over the specifications for aircraft parts and materials. Underwriters' Laboratories, along with the National Fire Protection Association, covers important specifications where any insurances are involved. This includes all commercial, industrial, and domestic applications. The Joint Industrial Council is noted for its contribution to the equipment needs; IEEE for contributions to the electronics field, and IPCEA and AEIC for contributions to the wire and cable industry. ANSI assumes the role of overall tie-in for all American specifications; and the rubber industry, through ANSI, has been constructive in international specifications under ISO (International Standards Organization).

Rubber Manufacturers Association RMA publishes several handbooks which may be helpful in selecting rubber materials: "Glossary of Terms Used by the Mechanical Rubber Goods Industry"; "Rubber Sheet Packing Handbook," which provides information on classification, types of surface finish, dimensional tolerances, materials

3-30   Elastomers

used, and storage; and "Specification for Rubber Products, Molded—Extruded—Lathe Cut—Open and Closed Cell Sponge," which provides information on dimensions, tolerances, finish, and type of flash. These publications should be complementary to the ASTM or SAE material specifications.

Test methods ASTM Standards, part 28, is a recommended must in a technical library where rubber usage on electrical components is important. This publication covers not only the material specification D 2000 but also all applicable physical testing methods and the following electrical materials tests: D 150, A-C Loss Characteristics and Dielectric Constant; D 257, D-C Resistance or Conductance; D 149, Dielectric Voltage and Dielectric Strength at Commercial Power Frequencies; D 470, Wire and Cable Testing; and D 991, Volume Resistivity of Electrical Conductive Rubber. Military Specifications such as ZZ-R-601 parallel the ASTM test methods. Almost all technical groups refer to the ASTM test methods; and in the rubber industry the call-out between supplier and user finds these specifications quite adequate. (Actually a specification is never completely adequate in a changing technology, and recognition is given to all sponsoring technical groups for their continuing work on specifications. It should be mentioned now that membership in particular specification groups is needed for the specialists in the field so that the specifications can be continually updated and rewritten for current industry needs so that they do not cease to be useful.)

Specification Sources

1. ASTM (American Society for Testing and Materials)
   1916 Race Street
   Philadelphia, Pennsylvania 19103
2. SAE (Society of Automotive Engineers, Inc.)
   485 Lexington Avenue
   New York, New York 10017
3. AMS (Aeronautical Material Specifications)
   See SAE
4. UL (Underwriters' Laboratories, Inc.)
   207 East Ohio Street
   Chicago, Illinois 60611
5. Federal Specifications
   General Service Administration
   7th and D streets, S.W.
   Washington, D.C. 20025
6. Army, Navy, Air Force, MIL, JAN, & DOD
   Naval Supply Depot
   5801 Tabor Avenue
   Philadelphia, Pennsylvania 19120
7. "Department of Defense Indexes of Specifications and Standards" (DODISS) (MIL, Air Force, Army, Navy) "Index to Federal Specifications, Standards, and Handbooks"
   Superintendent of Documents
   United States Government Printing Office
   Division of Public Documents
   Washington, D.C. 20025
8. "National Electric Code," NFPA, No. 70, ASACI
   National Fire Protection Association
   60 Batterymarch Street
   Boston, Massachusetts 02110
9. JIC (Joint Industrial Council)
   2139 Wisconsin Avenue
   Washington, D.C. 20007
10. IPCEA (Insulated Power Cable Engineers Association)
    283 Valley Road
    Montclair, New Jersey 07042
11. NEMA (National Electrical Manufacturers Association)
    155 East 44th Street
    New York, New York 10017

12. ASA (American Standards Association)
    Under new name: ANSI
13. ANSI (American National Standards Institute)
    10 East 40th Street
    New York, New York 10016
14. Association of Edison Illuminating Companies
    51 East 42d Street
    New York, New York 10017
15. RMA (Rubber Manufacturers Association, Inc.)
    444 Madison Avenue
    New York, New York 10022

MANUFACTURING METHODS OF RUBBER PRODUCTS

The methods of manufacturing rubber products can be categorized into four general processes: millable gums, latex, solutions, and castables. The use of millable gums is by far the largest process, such as tire production; followed secondly by the use of latex, as with foam cushions and latex paints. In simple form, the manufacturing descriptions are as follows:

Millable gums The simplest description of a gum would be the familiar "chewing gum," and the millable association is the use of a two-roll mill in mixing or masticating the gum. Raw natural rubber is a thermoplastic gum; and when it is "worked" between a set of rolls, the shearing action causes some chain rupture of the molecules which reduces its "nerve"; also the heat generated causes the gum to soften. The gum in this softened and plastic state can then be easily processed into finished products. As natural rubber was the only elastomer used before the 1930s, the processing machinery in the utilization of gums had already then been established; therefore, any synthetic had to be made to process on the same equipment. Therefore, almost all elastomers start as gums.

The introduction of compounding chemicals is performed when the gum is in this plastic state, and the whole mass is thus subjected to the shearing and mixing forces between the nip of the rolls. Two types of mixers are used: the simple two-roll mill is called open-roll mixing, and the second is called an internal mixer. The internal mixer is simply an encasement of the two rolls so that the shearing action is between the rolls and the walls of the encasement in addition to the small area between the nip. Most internal mixers use other than cylindrical rolls to enhance the shearing and mixing action. Economics favors the internal (Banbury is a common one) mixers because of the shorter mixing time and larger batches. Some compounds can only be mixed on an open mill, and small runs add to the versatility of products if a rubber plant has both mixers. The mixed rubber is then "sheeted" out or "pelletized" for further processing.

The sheeting of mixed rubber is done on a set of rolls called a *calender*, and its product is called a *calendered sheet*. This sheet can be interplied with fabric to make belting, or cured by itself for use as "packing" material, or rolled on a mandrel to make sleeves. The sleeves can be interplied with wire or fabric to be used as a hose. The raw sheet can also be cemented to storage tanks for special chemical resistance.

The mixed gum can also be put through an auger-type or piston-cylinder-type *extruder* to form extruded products. Extruded rubber products can be cured as is, or cut into slugs for molding operations. The extrusion process is the basic method of wire insulation coating in which the wire is passed through a die normal to the auger axis. As the wire is pulled through the die, it continues through heated alleys for a continuous extrusion and vulcanization process (CV cures). Should the wire require a protecting jacket as braided fabric or lead on the coated wire, it is usually put through the braiding operation or leading before curing. (Incidentally, remove the wire and you have a tubing or reinforced hose.)

The bulk of rubber goes into molded products, mostly for tire use. The simplest molding technique is to hollow out a matching cavity between two plates, insert a slug of raw rubber between the plates and in the cavity, and squeeze the two plates together, subjecting them to heat for a period of time. Then the release product is a compression-molded article. Placing the slug in a piston-cylinder (pot) machined part of a mold so that on compression the rubber is transferred from the cylinder to the cavity is called *transfer molding*. Injecting the raw slug from an external source into a mold cavity held under compression is called *transfer injection molding*. The use of pneumatic or hydraulic media for hollow molded articles is usually called *hollow molding*. In today's tire production a rubber bag is used to confine pressurized hot water and is a modification of hollow molding. This bag is inserted into the inner side of the raw tire, a clam-type mold is then closed, and hot water under pressure enters the bag which forces the tire to flow into the tread cavities. This is now commonly called *Bagomatic* type of tire presses.

The size, shape, dimensional tolerance, inserts as lead connectors, volume of production, and type of compound will dictate the type of mold for a particular product. Compression-type molds are the cheapest and fit most curing presses without special adaptations. Transfer molds offer better dimensional tolerances, and transfer injection types are most suitable for continuous molding on very small parts. Rubber companies can differ in their recommendations, depending on the basic equipment available and the cost analysis. However, a molding recommendation should not be considered as the only method since most parts can be made using different molding techniques, and reliance must be placed on the individual molder.

Latex This is a colloidal water dispersion of a rubber material and is the liquid product of tapping rubber trees. Like millable gums, some synthetics are also supplied in colloidal suspensions. Some products use these latices in their original form; but in the majority, the latex is compounded like the gums for the improved vulcanized physical properties. Latex is used mostly in the manufacture of foamed products, such as seat cushions and mattresses. However, many "dipped goods" products are also made. Dipping a form into a compounded latex results in a thin coagulant, which is then cured, and the article then stripped from the form. Articles such as toy balloons, electrical and surgical gloves, baby pants, coated metal parts, and blotters are made by this process. The predominant difference from molded rubber is that latex products are thin-walled, whereas molded products are usually above 45 mils in thickness.

Solutions These are mixed raw rubbers dissolved in appropriate solvents. Solutions can be made with various viscosities so that they can be used as free-flowing cements or thicker for spreading uses. Most commercial cements are elastomers of some sort and usually have the curatives left out of the compound. Some cements can be processed with the curatives, but the shelf life of the cement is much shorter. As a rule, a self-curing cement will show signs of setting up when it increases in viscosity and gets stringy and also starts to loose its tack or green strength. Coatings will almost always have either a curative or a hardening agent included so as to render the cured coating tackless.

Castables. A castable elastomer is an elastomer that is a liquid but does not contain water or solvent. The term *castable* suggests a product being made by casting a liquid into a form or mold rather than compressing a solid or gum. To fully utilize the wide range of urethane polymers, casting methods must be employed. Millable gum urethanes are available to the rubber industry, but their physical properties are in the range of about 65 to 90 hard, and tensile strengths are below 5,000 psi; whereas for the castables the range is from extreme softness on the Shore A scale through extreme hardness on the Shore D scale, and the full higher tensile strengths are up to 9,000 psi. The use of castable urethanes necessitates special handling equipment and in a sense, represents a continuation of a chemical polymer plant. For some urethane formulations only simple blending techniques need be employed; but to handle the complete range of compounds, very refined equipment and specially trained personnel are needed.

All methods and processing forms of elastomers are used in the electrical industry, and this should not cause any concern regarding the quality of product. The basic characteristics of the elastomers will, for the most part, be unchanged; the predominant difference will be in the additions to the compound. The exception most likely would be in cements since resins are commonly used for tack which could impair cold-temperature properties and hardness at elevated temperatures.

COMPOUNDING

A rubber compound consists of at least one elastomer and possibly a blend of two types, vulcanizing agents, reinforcing materials, softeners, protectants, colorants, diluents, and processing aids. As such, it is a highly technical mixture which has cured through the application of heat in an irreversible reaction to form a finished product. Contrary to many opinions, one cannot reheat the mixture to form a new part nor do hardness variations come about by cure-time variations. A basic mixture of about 13 components would be a normal type of compound in a rubber plant today.

Before today's elastomers when natural rubber was the only type available, compounding was a skilled art. Today, however, with the families of basic elastomers and their respective ingredients, coupled with the methods for their proper selection, the chemist is relying more and more on the reactions of chemicals and on systematic analysis in his choice of materials, so that one can truly, today, look at compounding as a science.

A company in business 30 years ago offering versatility of products could easily have settled on four grades of natural rubber with some 20-odd other ingredients. Today, to offer the same versatility, a company needs some 40 families and grades of elastomers along with another 120 other ingredients as a base. Also, a rubber chemist must know what is available for his use so that we have a complex mess of hundreds of ingredients which can be compounded into thousands of formulations. Add to "mother of invention" the growth in fabrics, metals, expansion of engineering uses, and specifications one finds that not only is highly trained competent staffing needed but also a well-equipped laboratory to supply factual support of theory. What a metallurgist may do in steel, his counterpart, the compounder, does in rubber; and on the basis of his judgment rests the product quality reputation of his company.

Eight basic parameters must be considered in the design of any compound. They may be described as follows.

Vulcanizers Crosslinking of the polymers in a transition from a plastic to an elastic state was mentioned earlier. Curing is another term commonly used; and along with vulcanizing agent, the word "curative" now has a synonymous meaning. Every compound must have a cure system designed for the intended end use of the rubber coupled with the process limitations. The selection of curatives will especially vary the resilience, compression set, heat resistance, water resistance; and to a degree, affect other properties such as odor, taste, toxicity, and surface appearance.

A cure system will usually consist of a basic vulcanizing agent, a metallic oxide, a fatty acid, an activator, and an accelerator. The combination and types of each are varied for processing and end-use conditions.

Reinforcers Some synthetic rubbers have low gum physical properties (elastomer plus curatives), and had it not been for the development of special additives, these synthetics would never have reached commercial use. SBR is a good example: the basis of our automobile tires. Most reinforcing materials are micro size and high-structure carbon blacks which enhance the pure gum tensile strength of SBR, for example, from 400 to over 3,000 psi.

Softeners A dry gum and reinforced compound would be high in durometer hardness. Therefore, special oils are added to soften or lower the durometer hardness. Petroleum-based oils are mostly used with the nonpolar polymers; and synthetic esters, phosphates, etc., are commonly used with the polar polymers. Cold-temperature and heat-resistant properties are affected by the choice and use of these. Generally speaking, an improvement in cold-temperature properties of a particular compound takes place at the sacrifice of heat resistance, and conversely so. At times a compound can be saturated with an oil to minimize its later swelling upon oil exposure.

Protectants Often cheaper and poorer aging elastomers can be used instead of costlier and better aging elastomers if proper protection is designed into the compounds. In fact, all elastomers contain a protectant of some sort, and the goal is to extend the useful life or aging properties of the compounds. Antioxidants are always added; and for various applications waxes, lacquers, fungicides, flame resistors, etc., are also added.

Colorants Most, but not all, rubbers are black. Reinforcing blacks will improve the tensile and wear properties, but this toughness is sacrificed if colors are desired. In electrical applications blacks may not be used at all, and so nonblack reinforcers must be used. Should black be added, it would possibly be a very small percentage to uniformly hold a black color for economic reasons. Should color be needed, it must be remembered that high gloss and pastel shades (as with plastics) are impossible in rubber colors; other than this, a wide variety of colors are available. Inorganic colors such as lead oxide, titanium dioxide, chromic oxide green, Mapico yellow, and Prussian blue are somewhat dull but fast on heat aging, while the organic lakes which tend to give brighter hues are sensitive to heat and light.

Diluents Once all the basic needs are designed into a compound, it could very well be overdesigned; and for economic reasons, relatively inexpensive materials are included. Natural occurring clays, whiting, soapstone, coal dust, petroleum, and coke oven residues are commonly used.

TABLE 18 Effect of Various Mineral Fillers on Wet Electrical Properties[15]

Filler	Dielectric loss factor* after 7 days in 75°C water, %†
Nonwettable:	
Translink 37 clay	2
Calcene TM[a]	6
Mistron vapor[b]	9
Kaophobe 2 clay	12
Kaophobe 5, 7, 45, 90 and C-12 clays	17–45
Super-Multiflex[c]	180
Translink clay	>200
Wettable:	
Paragon clay[d]	42
McNamee clay[e]	72
Atomite whiting	96
Bentonite clay	110
Blanc fixe	>200
Crown clay	>200
Dixie clay[e]	>200
Iceberg clay	>200
Suprex clay[d]	>200
ASP-106 clay[f]	>200
Whitex clay	>200
Catalpo clay	>200
Buca clay	>200

* 100 tan δ × dielectric constant K (specific inductive capacity).
† Nordel[g] 1070, 100; zinc oxide. 5; filler. 90; tetramethylthiuram monosulfide, 1.5; 2-mercaptobenzothiazole, 0.5; sulfur, 2. Cure: 15 min at 320°F.
[a] Trademark of Pittsburgh Plate Glass Co. Chemical Division, Pittsburgh, Pa.
[b] Trademark of Sierra Talc & Chemical Corporation, South Pasadena, Calif.
[c] Trademark of Diamond Alkali Co., Cleveland, Ohio.
[d] Trademarks of J. M. Huber Corporation, New York.
[e] Trademarks of R. T. Vanderbilt Co., Inc., New York.
[f] Trademark of Minerals & Chemicals Philipp Corporation, Menlo Park, N.J.
[g] Trademark of E. I. du Pont de Nemours & Co., Wilmington, Del.

TABLE 19 Effects of Silicone Compounds on Wet Electrical Properties[15]

Formula	Parts
EP terpolymer*	100
Silicone (as shown below)	3
Zinc oxide	5
Calcined clay	120
Petroleum oil†	40
Sulfur	3
Tetramethylthiuram monosulfide	1.5
2-Mercaptobenzothiazole	0.5

Properties	No silicone added	Silicone rubber‡	Silicone fluid§
Physical properties (cure: 15 min at 320°F)			
300% modulus, psi	415	280	315
Tensile strength, psi	1,380	1,440	1,470
Elongation, %	740	780	740
Resilience, %	70	70	72
Compression set, method B, %, 22 hr at 70°C	32	28	28
Electrical properties			
Original:			
100 tan δ, %	0.5	0.4	0.4
Dielectric constant K	3.2	3.2	3.2
100 tan $\delta \times K$, %	1.6	1.3	1.3
After 1 day in water at 75°C:			
100 tan δ, %	>20	1.5	2.6
Dielectric constant K	7.8	3.3	3.4
100 tan $\delta \times K$, %	190	5.0	8.8
After 7 days in water at 75°C:			
100 tan δ, %	>20	1.2	2.0
Dielectric constant K	9.6	3.3	3.5
100 tan $\delta \times K$, %	>200	4.0	7.0
After 14 days in water at 75°C:			
100 tan δ, %	....	0.9	1.5
Dielectric constant K	....	3.3	3.5
100 tan $\delta \times K$, %	....	3.0	5.2

* Nordel 1070.
† Flexon 765 (Humble Oil & Refining Co.).
‡ Silastic 440 (Dow Corning Corp.).
§ Dow Corning 550 fluid.

Processing aids Hardly ever will a compound selected from the literature or a compound from another plant work properly without some modifications. Raw rubber compounds will vary in degree of plasticity and nerve, depending on the history of the polymer itself and also its mixing history. Rubber in the raw is not completely plastic as such, and this varying degree of opposing elasticity must be adjusted in some manner so that the compound will process in a given plant. A compound is worthless if it meets a specification but cannot be processed. Small amounts of special organic blends are usually added, and as a rule have little or should have no deleterious effects on the final rubber.

Economics The end use coupled with processibility coupled with choice of materials determines the worth of a compound. Many new elastomers show ideal end properties but may be costly or comparatively difficult to process; costly; compounding experience may be lacking; and future availability and life expectancy may be uncertain. Also, each new elastomer added creates a warehousing problem, especially since almost always other new chemicals need to be used along with the new polymer. This also increases the demands on the technical staff. The net general effect is that rubber companies will be hesitant to incorporate new additions since in toto the cost level of its materials increases and affects the cost of all the compounds. Many elastomers have overlapping properties, and special needs can be enhanced by proper compounding. For these reasons, it is very important for the compounder to know the end needs; and, depending on his experience and warehouse situation, he can select different elastomer from different rubber companies.

Compounding effects on electrical properties It is most important to remember that once an elastomer has been chosen, it must also be properly compounded; and this is where an experienced compounder is needed. Tables 18 to 21 illustrate the effects of compound ingredients very effectively.[15] The comparisons were made on the same base elastomer EPT (EPDM).

TABLE 20 Effect of Curing Systems on Wet Electrical Properties[15]

Formula and properties	Compound 1	Compound 2	Compound 3*
EP terpolymer†	100	100	100
Zinc oxide	5	5	5
Calcined clay	120	120	90
Petroleum oil‡	20	20	
Lead dioxide			10
Quinone dioxime			3
Sulfur	2.5	1	1
Zinc dimethyldithiocarbamate	2.5	1	
Tellurium diethyldithiocarbamate		1	
Tetramethylthiuram disulfide		1	
2-Mercaptobenzothiazole	0.5	1	
Mooney scorch at 250°F, minimum	21	20	77
Minutes to 10-point rise	20	14	1
Physical properties			
Cure: min/°F	6/350	6/350	5/356
300% modulus, psi	400	300	690
Tensile strength, psi	975	850	1,540
Elongation, %	740	770	720
Electrical properties			
Original:			
100 tan δ, %	0.4	0.4	0.5
Dielectric constant K	3.3	3.2	3.4
100 tan δ × K, %	1.3	1.3	1.7
After 1 day in 75°C water:			
100 tan δ, %	>20	12.2	1.6
Dielectric constant K	>10	5.7	3.5
100 tan δ × K, %	>200	69.5	5.6

* Data on compound 3, courtesy of Dr. J. R. Albin.
† Compounds 1 and 2, Nordel 1040; compound 3, Nordel 1070.
‡ Circosol 2XH (Sun Oil Co.).

CELLULAR SPONGE AND FOAM RUBBER

Cellular rubber products is the more fitting term describing a rubber or elastomeric material containing many small hollow cells. These cells may be interconnecting and open which would allow the passage of air or liquid from cell to cell, or closed and not interconnecting.

Cellular rubbers have many uses, but the outstanding properties requiring their

TABLE 21 Effect of Combining a Hydrophobic Filler, Heat Treatment, a Metal Stearate, and a Preferred Curing System[15]

Formula and properties	1 (inferior system)	2 (preferred system)
EP terpolymer*	100	100
Calcined clay	120	
Translink 37 clay		120
Zinc stearate		1
Petroleum oil	20	20
Quinone dioxime	1	0.75
Above ingredients hot-mixed 5 min at 365–380°F in a Banbury, cooled, and added on a mill to the following materials:		
Zinc oxide	5	20
Sulfur	2.5	1
Zinc dimethyldithiocarbamate	2.5	1
Tellurium diethyldithiocarbamate		1
Tetramethylthiuram disulfide		1
2-Mercaptobenzothiazole	1	1
Electrical properties†		
After 16 hr in 60°F water:		
Specific insulation resistance megohms/1,000 ft	14,800	23,600
100 tan δ, %	2.5	0.1
Dielectric constant K	3.1	3.1
100 tan δ × K, %	7.8	0.3
After 1 day in 75°C water:		
100 tan δ, %	17.9	1.3
Dielectric constant K	7.1	3.1
100 tan δ × K, %	127	4.0
After 7 days in 75°C water:		
100 tan δ, %	17.3	1.4
Dielectric constant K	11.0	3.3
100 tan δ × K, %	190	4.6
After 6 months in 75°C water:		
100 tan δ, %		2.4
Dielectric constant K		4.1
100 tan δ × K, %		9.8

* Nordel 1040.
† Steam cure: 60 sec at 225 psi. Electrical measurements made on the insulated wire while it was immersed in the hot water.

use would be good resilience, low compression, low thermal conductivity, and low density. Sponge rubber can be compressed to half its height or thickness and cycled a half million times with a loss of perhaps 5 percent of its original height.[16] Sponge can be compressed to 50 percent for 48 hr at room temperature and only have a set averaging 2 to 3 percent.

Several crude rubber sponges have been tested at 50 percent compression for 6 months and had sets averaging just over 9 percent, the highest being 11.4 percent. This means that some sponge can be used as gasket material for a 6-month uninterrupted period and still recover to 90 percent of its original height. A comparison of coefficients of heat conductivity for some materials is as follows:[9] hard sponge, 0.30; cork, 0.25; glass wool, 0.29; soft rubber sponge, 0.41 to 0.65; rock wool, 0.26. The low density of cellular products lends itself to wide uses for flotation products, especially the hard-cellular materials.

The manufacturing methods of cellular materials vary widely in both the techniques and materials used. *Sponge rubber* is the oldest-known term defining a cellular material, followed by *latex film*. As a vast array of materials have lent themselves to cell structure, the broader term of *expanded cellular products* now finds wider acceptance in technical usage.

Sponge rubber The manufacture of this product is analogous to a baker baking bread in that both depend largely on gas producing sodium bicarbonate to form the cells, and both depend on a heating or baking process to "set" or "fix" the cell structures into the familiar products.

Sponge has been made from all the elastomers mentioned in this chapter. The basic difference in the processing of the gum stocks or milled sponge is that a chemical "blowing agent" such as sodium bicarbonate is added to the raw mix which decomposes at curing temperatures and the resultant gas evolution expands to raise and form the cell structure.

During the compounding, extra oils and lower-viscosity rubbers are used so that the expansion pressures are reduced; also, to aid in the decomposition of the bicarbonate, additional stearic or other fatty acids are employed. Because of the extra oil and the lower-viscosity rubbers, the final mix tends to flow and stick to almost everything, and so small batches are transported throughout the plant and dusted very heavily with antistick materials such as soapstone.

Special handling techniques, coupled with the contaminating nature of soapstone, necessitate a separation of manufacturing facilities; and for this reason, sponge plants rarely offer solid rubber products along with their sponge products.

Most sponge is cured in molds with approximately a 35 percent solid loading, which then produces a sponge consisting of about 65 percent air cells. As the charge or "slug" of raw rubber is heated in the mold, the first reaction must be the gas evolution to expand the rubber; and the second and final reaction is the crosslinking or vulcanizing of the surrounding rubber to form the permanent product. Such a sponge will have a naturally cured "skin" on the outside, approximately 1 to 2 mils thick. Heavier skins are sometimes needed for wear resistance and moisture resistance, and these are acquired either by the application of a sheet of nonblowing rubber before molding or by solution coating by cement or latex dipping.

A soft red sponge without a skin is still on the market and is used as a washing sponge. This type of sponge is cured in a steam autoclave and is allowed to expand freely. After curing, the sponge is simply cut into the desired shape and, of course, the skin is removed.

Closed-cell versus open-cell sponge Should the gas cells rupture during the "blowing" or "forming" stage, the cells will then be interconnecting and are called *open cells*. If they do not open, they are then called *closed cells*. Open-cell sponge will allow the passage of gas and liquids, while closed cell will not. Also, closed cell is harder to compress than open cell since the gas has no way of escape. The use of sodium bicarbonate as a blowing agent will produce open-celled sponge, while the use of nitrogen-producing blowing agents produces closed-celled sponge.

There are several blowing agents which release nitrogen in the same way that carbon dioxide is released from sodium bicarbonate, and they are used in exactly the same manner of compounding. The only difference in manufacturing is that normally a two-stage blowing operation is used. The first is a precure in a tight mold, followed by a cure in a larger mold which then uniformly allows the cells to form and finally cure to the required shape.

Uses of sponge Open-cell sponge is widely used in products such as lamp gaskets, vibration mountings, cushioning material, sealing agents, sound absorbents, armrests, car door and trunk lid gaskets, kneeling pads, cushions, burners, stamp pads, bushings, heel pads, mounts for electronic instrument parts, earphone pads, ink applications, bath sponges, toys, etc.

Closed-celled sponge with its inpermeability feature is widely used for divers' suits, football clothing, and various other athletic padding; many thermal applications; and vibration arrestors for small electronic uses. The hard closed-cell sponges are used for flotation gear, insulation for airborne-type electronic gear, plying for sheet-metal plying structures, etc.

Foam Latex foam, along with urethane foam, today comprises the largest user of elastomers in cellular materials. A completely different manufacturing method is used in making foam rubber than in making sponge. Foam uses liquids while sponge uses the millable gums. Latex is rubber held in a water emulsion, and in the liquid state all the necessary rubber compounding materials are added as in normal latex. At this point, it is whipped creamlike with air included mechanically, and then this cream is poured into molds for expansion and curing. The resulting sponge is extremely soft and resilient and contains about 90 percent air with about 250,000 air cells per in.[3] Foam is approximately three times lighter than milled sponge rubber: 0.19 to 0.34 against 0.05 to 0.10 oz/in.[3] The cells are interconnecting; and since a purer rubber is employed, it is much stronger and better wearing. Its uses cover the complete transportation seating material, domestic seat cushioning material, mattresses, pillows, and the large upholstering field.

Urethane foam is cast with a water inclusion in its monomer blending operation which during the reaction process releases carbon dioxide, or else gas is included during the blending. This blend is then cast, as latex, and allowed to rise and cure. Urethane foam lacks the resilient properties of latex foam for seating purposes; however, its thermal insulating qualities, coupled with "on-the-spot" foaming, have resulted in major changes in several industries. Refrigeration insulation is now an assembly-line operation; truck insulation is incomparably improved in insulating value with a truly closed-celled and water-resistant insulation. Home construction on-the-spot insulating is feasible, and its use is expected to increase in many other fields.

Specifications The two most up-to-date specifications on sponge are ASTM D 1056, Testing Sponge and Expanded Cellular Rubber Products, and ASTM D 1055, Standard Specification and Methods of Test for Latex Foam Rubbers.

D 1056 first separates sponge in Class O for open-cell sponge and Class E for closed-cell or expanded sponge. Each class is then graded into Type R for non-oil resistance, Type SB for low oil swell, and Type SC for medium oil swell. They then contain some basic properties as compression deflection forces at a 25 percent deflection, oil aging, and heat aging. Other suffix call-outs such as low temperature and compression set are also provided for. The basic importance of this specification is the allowable call-out of the stiffness of sponge ranging from 1 through 20½ psi needed to deflect the test specimen 25 percent.

ASTM D 1055 is based only on natural rubber or SBR type latex, and oil requirements are therefore omitted. In the casting of foam, cores are often employed, which then leave large void areas in the bottom side of the sponge. This results in different compressive forces needed to deflect the product to a given percentage, and the specification is thus divided into two grades: RC for cored, and RU for uncored. The basic table lists the froce and limits to compress a 50-in.^{2} area 25 percent; for cored, the values are 5 through 90 lb, and for uncored, 11 through 150 lb. Other basic call-outs are specified such as hot-air aging, compression set, flex test, and low-temperature tests.

Both specifications also list tolerances on dimensions for general applications.

REFERENCES

1. ASTM Standards, part 28, April, 1967.
2. Pickett, A. G., and M. M. Lemcoe, Southwest Research Institute. "Handbook of Design Data on Elastomeric Materials Used in Aerospace Systems," A SDRT61-234, Aeronautical Systems Division, Air Force Systems Command, United States Air Force, Wright-Patterson Air Force Base, Dayton, Ohio, June, 1961.
3. Swiss, Jack: Insulation for Electronics, "Encyclopedia of Chemical Technology," 2d ed., vol. 2, John Wiley & Sons, Inc., New York, 1966.
4. ASTM Standards part 28, D 1418-67, April, 1968.
5. Davis and Blake: "Chemistry and Technology of Rubber," ACS Monograph, Reinhold Publishing Corporation, New York, 1937.
6. "1968 Rubber Redbook," Palmerton Publishing Co., Inc., New York, 1968.
7. Elastomers: Tailor Made Design Materials, *Design News*, Mar. 31, 1965. Cahners Publishing Co., Inc., Rogers Publishing Division, Englewood, Colo.

8. Hauck, Jack E.: *Mater. Design Eng.*, April, 1966.
9. Burton, Walter E.: "Engineering with Rubber," McGraw-Hill Book Company, New York, 1949.
10. Duke, N. G.: Compounding for Special Applications Including Oil and High Temperature, University of Akron Rubber Product Compounding Lectures.
11. McPherson and Klemin: "Engineering Uses of Rubber," Reinhold Book Corporation, New York.
12. Winspear, George G.: in William Arnott (ed.), "The Vanderbuilt Rubber Handbook," R. T. Vanderbuilt Co., New York.
13. "Insulation Directory Encyclopedia," no. 7, June–July, 1967, Lake Publishing Co., Libertyville, Ill.
14. Blodgett and Fisher: A New Corona and Heat Resistant Cable Insulation, (*IEEE Paper* 63-162), New York, 1963.
15. "The Rubber World Handbook of New Compounding and Processing Technology," Bill Brothers Publishing Corp., New York, 1965.
16. Sereque, Arthur F., "Technical Bulletin on Sponge Rubber," The Sponge Rubber Products Co., Shelton, Conn.

Chapter 4

Wires and Cables

ARTHUR G. SCHUH

Materials Engineering Laboratories, Martin Marietta Corporation,
Orlando, Florida

Introduction	4–2
Conductors	4–2
Materials	4–2
Coatings	4–9
Construction	4–12
Wire and Cable Insulation	4–15
Materials	4–15
Construction and Application Methods	4–21
Shielding	4–22
Materials and Construction	4–22
Shield Jackets	4–24
Materials and Construction	4–24
Design Considerations	4–28
Wire-gage Selection	4–28
Insulation-material Selection	4–31
Interconnection and Hookup Wire	4–31
Definitions	4–31
Interconnection Wire	4–33
Hookup Wire	4–33
Outer Space Applications	4–41
Coaxial Cables	4–46
Design Considerations	4–46
Cable Selection	4–49
Multiconductor Cables	4–49
Airborne Cables	4–49
Ground-electronics Cables	4–54

Ground-support Cables	4–54
Flat Flexible Cable	4–55
Miscellaneous Cables	4–66
Design Considerations	4–66
Magnet Wire	4–68
Conductors	4–68
Insulations	4–70
Design Considerations	4–74
Wire and Cable Terminations	4–76
Terminating Hardware	4–76
Identification	4–79
Associated Hardware	4–80
References	4–86

INTRODUCTION

The selection of wires or cables is often neglected within the total design concept. Hookup and interconnection wiring are frequently the last areas of consideration. More and more engineers are realizing the importance of selecting conductors, insulation, shielding, jacketing, and cabling that offer the best combination of size, weight, environmental protection, and handling resistance, coupled with lowest cost and ease of availability and maintenance. If the intended environment is abnormal, proper wire and cable selection should be verified by adequate environmental testing for complete success in application. In addition, the designer must adequately evaluate the environment and mechanical stress and consider the compatibility of his materials with all possible encapsulants, fuels, chemicals, explosives, and the new and varied types of associated hardware and fabrication techniques.

The purpose of this chapter is to provide the designer with sufficient background information and guideposts for proper wire and cable selection. Definitions of various terms that will be used appear in Table 1.

CONDUCTORS

Materials Following is a brief discussion of conductor materials most widely used in the aerospace and electronics industry. Table 2 presents a summary of conductor materials and coatings.

Copper. ELECTROLYTIC TOUGH PITCH (ETP): ETP copper makes up the majority of conductor strands used in industry. ETP copper is controlled in composition, conductivity, and purity by ASTM B3. Properties of individual strands, such as tensile strength, elongation, and dimensions, are covered by ASTM B33 or QQ-W-343.

OXYGEN-FREE HIGH CONDUCTIVITY (OFHC): OFHC differs from ETP copper in its fabrication. Electrolytic slabs are melted into bars for extrusion or rolling in an inert atmosphere excluding oxygen. OFHC copper, covered by ASTM B170, has improved properties at temperatures above 1000°F and is resistant to hydrogen embrittlement.

Copper Alloys. Pure copper has poor mechanical characteristics; breakage will occur from tensile pull, flexing, and vibration. As a result of progressive miniaturization, various high-strength copper alloys have been developed. Figure 1 presents a comparison of flex endurance between high-strength copper alloys and copper. Since drawing and annealing techniques used by different suppliers play a significant part in the final physical and electrical properties of high-strength alloys, the recommended practice is to specify performance requirements of the conductor instead of requiring a specific alloy. One exception to this practice would be the use of zirconium alloy for high-temperature applications.

TABLE 1 Terms and Definitions

Abrasion resistance. Ability to resist surface wear.
Aging. The change in properties of a material with time under given conditions.
Ambient temperature. The temperature of the surrounding atmosphere.
Attenuation. The power or signal loss in a circuit, expressed in decibels (db).
AWG (American Wire Gauge). A standard for copper wire sizes. The diameters of successive sizes vary in geometric progression.
Braid. A woven outer covering. It may be composed of any filamentary material such as fiber, plastic, or metal.
Braid angle. The angle between the axis of cable and the axis of any one member or strand of the braid.
Breakdown. A disruptive discharge through insulation.
Bunch lay. The twisting of strand members in the same direction without regard to geometrical arrangement.
Capacity. That property of a system of conductors and dielectrics which permits the storage of electricity when potential differences exist between conductors. Its value is expressed as the ratio of a quantity of electricity to a potential difference.
Coat. To cover with a finishing, protecting, or enclosing layer of any compound.
Coax. Abbreviation for coaxial cable.
Concentric lay cable. A concentric lay cable is composed of a central core surrounded by one or more layers of helically wound insulated conductors.
Concentric lay stranding. Composed of a central core surrounded by one or more layers of helically laid wires.
Conductor. A slender rod or filament of metal or group of such rods or filaments not insulated from one another, suitable for transmitting an electric current.
Corona. A luminous discharge caused by the ionization of the gas surrounding a conductor around which exists a voltage gradient exceeding a certain critical value.
Corona resistance. The time for which insulation will withstand a specified level field—intensified ionization that does not result in the immediate complete breakdown of the insulation.
Crosslinking. The setting up of chemical bonds between molecular chains.
Crosstalk. Undesirable electromagnetic coupling between adjacent signal-carrying conductor pairs.
Cut-through resistance. Resistance of a solid material to penetration by an object under conditions of pressure, temperature, etc.
Decibel (db). Unit employed to express the ratio between two amounts of power, voltage, or current between two points.
Density. Weight per unit volume of a substance.
Dielectric. A nonconducting material or a medium having the property that energy required to establish an electric field is recoverable, in whole or in part, as electric energy.
Dielectric constant. That property of a dielectric which determines the electrostatic energy stored per unit volume for unit potential gradient.
Dielectric loss. The time rate at which electric energy is transformed into heat in a dielectric when it is subjected to a charging electric field.
Dielectric loss factor. The product of its dielectric constant and the tangent of its dielectric loss angle.
Dielectric strength. The voltage that an insulating material can withstand before breakdown occurs, usually expressed as a voltage gradient (such as volts per mil).

TABLE 1   Terms and Definitions (Continued)

Direction of lay. The direction in which individual members of a multiconductor cable or stranded conductor spiral over the top of the cable in a direction going away from an observer who is standing behind the twisting device.
Drain wire. An insulated, stranded, or solid conductor that is located directly under and in intimate contact with a shield.
Elastomer. A material which at room temperature stretches under low stress to at least twice its length and snaps back to its original length upon release of stress.
Elongation. The fractional increase in length of a material stressed in tension.
Extrusion. Compacting a natural or synthetic material and forcing it through an orifice in a continuous fashion.
Farad. Unit of capacitance. The capacitance of a capacitor which, when charged with a coulomb, gives a difference of potential of one volt.
Filler (cable). Fillers are used in multiconductor cable to occupy space or interstices formed by the assembled conductors. Fillers are employed to obtain circularity.
Flame resistance. Ability of a material to extinguish flame, once the source of heat is removed.
Flammability. Measure of the ability of a material to support combustion.
Flex life. The life of a material when subjected to continuous bending.
Flexural strength. The strength of a material in bending.
Foamed insulation. Resins in flexible or rigid sponge form with cells closed or interconnected.
Hard-drawn. Refers to the temper of conductors that are drawn without annealing or that may harden in the drawing process.
Hydrolysis. Chemical decomposition of a substance involving the addition of water.
Hygroscopic. Tending to absorb moisture.
Impact resistance. Relative susceptibility of material to fracture by shock.
Impedance. The apparent resistance in a circuit to the flow of an alternating current, analogous to the actual resistance to a direct current.
Insulation resistance. The insulation resistance of an insulated conductor is the electrical resistance offered by its insulation to an impressed current tending to produce a leakage current. Normally measured in megohms per 1,000 ft for insulated wire.
Insulator. A material of such low electric conductivity that the flow of current through it can usually be neglected.
Interstice. A minute space between one thing and another, especially between things closely set or between the parts of a body.
Jacket. A protective sheath or outer covering applied over an insulated conductor or cable.
Layer. Consecutive turns of a coil lying in a single plane.
Lay length. The lay length of any helically wound strand or insulated conductor is the axial length of one turn of the helix.
Magnet wire. Insulated wire intended for use in windings on motor and transformer coils.
Migration of plasticizer. Loss of plasticizer from an elastomeric plastic compound, with subsequent absorption by an adjacent medium of lower plasticizer concentration.
Moisture resistance. The ability of a material to resist absorbing moisture from the air when immersed in water.

TABLE 1 Terms and Definitions (Continued)

Nylon. The generic name for synthetic fiber-forming polyamides.
Ohm. Unit of electrical resistance. Resistance of a circuit in which a potential of one volt produces a current of one ampere.
Ozone. A faintly blue, gaseous, allotropic form of oxygen, obtained by the silent discharge of electricity in ordinary oxygen or in air.
Plastic. High-polymeric substances, including both natural and synthetic products, but excluding the rubbers, that are capable of flowing under heat and pressure at one time or another.
Plasticizer. A chemical agent added to plastics to make soft and more flexible.
Polyamide. A polymer in which the structural units are linked by amide or thiamide groupings.
Polychloroprene. The chemical name for neoprene, a polymer of chloroprene, a combination of vinyl acetylene and hydrogen chloride.
Polyethylene (PE). A thermoplastic material composed of the polymers of ethylene.
Polypropylene. A plastic made by the polymerization of high-purity propylene gas in the presence of an organometallic catalyst at relatively low pressures and temperatures.
Polyurethane. A copolymer of urethane similar in properties to neoprene.
Polyvinyl chloride (PVC). A family of thermoplastic insulating compounds composed of polymers of polyvinyl chloride or its copolymer, vinyl acetate, in combination with certain plasticizers, stabilizers, fillers, and pigments.
Polyvinylidene fluoride. A thermoplastic-resin crystalline high-molecular-weight polymer of vinylidene fluoride.
Primary insulation. A nonconductive material placed directly over a current-carrying conductor whose prime function is to act as an electrical barrier for the applied potential.
Quad. A four-conductor cable.
Relative humidity. Ratio of the quantity of water vapor present in the air to the quantity that would saturate it at any given temperature.
Resistance. Property of a conductor that determines the current produced by a given potential difference.
Resistivity. The ability of a material to resist passage of electric current either through its cross section or on the surface. The unit of volume resistivity is the ohm-centimeter; of surface resistivity, the ohm.
rf. Abbreviation for *radio frequency.* Usually considered the frequency spectrum above 10,000 Hz.
rms. Abbreviation for *root mean square.* When the term is applied to voltages and currents, it means the effective value, i.e., that it produces the same heating effect as a direct current or voltage of the same magnitude.
Rope lay. In a rope-lay conductor or cable, stranded members are twisted together with a concentric lay; the stranded members themselves may have a bunched, concentric, or rope lay.
Serve. The spiral application of a material, as opposed to braid.
Shield. A metallic sheath placed around an insulated conductor or group of conductors to protect against extraneous currents and fields.
Shielded conductor. An insulated conductor that has been shielded to reject extraneous electric fields.
Silicone. Polymeric materials in which the recurring chemical group contains silicone and oxygen atoms as links in the main chain.

TABLE 1   Terms and Definitions (Continued)

Sintering. Forming articles from fusible powders at a temperature below melting point.
Solvent. A liquid substance that dissolves other substances.
Specific gravity. The density (mass per unit volume) of any material divided by that of water at a standard temperature.
Stabilizer. An ingredient used in some plastics to maintain physical and chemical properties throughout processing and service life.
Strand. A single metallic conductor.
Tape. A relatively narrow woven or cut strip of fabric, paper, or film material.
Tear strength. Force required to initiate or continue a tear in a material under specified conditions.
Tensile strength. The pulling stress required to break a given specimen.
Thermoplastic. A classification of resin that can be readily softened and resoftened by repeated heating.
Volt. Unit of electromotive force. It is the difference of potential required to make a current of one ampere flow through a resistance of one ohm.
Volume resistivity. The electrical resistance between opposite faces of a 1-cm cube of insulating material, commonly expressed in ohm-centimeters.
VSWR. Abbreviation for voltage standing-wave ratio, the ratio of voltage maximum to voltage minimum in a transmission line.
Water absorption. Ratio of the weight of water absorbed by a material to the weight of a dry material.
Wire. A conductor of round, square, or rectangular section, either bare or insulated.
Working voltage. The recommended maximum voltage of operation for an insulated conductor.

Because of high annealing temperatures during processing, high-strength alloys are available only with high-temperature coatings such as silver or nickel. Tin-coated high-strength-alloy conductors are still under development. A description of the more popular high-strength alloys follows.

CADMIUM-COPPER ALLOY (CADMIUM BRONZE): This is used in fine-wire applications for increased strength; ASTM B105 covers hard-drawn temper. No specification is available for temper applicable to high-strength-alloy conductors of higher elongation. This alloy is not recommended for temperature applications exceeding 400°F.

ZIRCONIUM-COPPER ALLOY: The required physical and electrical characteristics of high-strength-alloy conductors as defined are difficult to meet with zirconium alloy; however, zirconium alloy shows best retention of physical and electrical properties at temperatures up to 400°C.

CHROMIUM-COPPER ALLOY: This provides the best conductivity of the high-strength-alloy conductors; however, it is the most difficult alloy to process consistently. Chrome-copper alloy is a precipitation-hardened alloy; if improperly heat-treated, the chromium is not retained in solution, causing inconsistency in physical properties.

CADMIUM-CHROMIUM-COPPER ALLOY: This most recent addition to high-strength alloys appears to present the most consistent and best physical properties, although minimum conductivity is slightly less than in chrome alloy.

Copper-covered Steel. This conductor is available in two consistencies: hard drawn for high strength (125,000 to 150,000 psi), or annealed (50,000 to 80,000 psi) for greater ductility; and two conductivities: 40 and 30 percent. At high frequencies, either 30 or 40 percent copper-covered steel has a conductance almost equivalent to that of copper; therefore, it is commonly used in coaxial cables. At low frequencies or with direct current, the conductance is 30 or 40 percent of an equivalent size

TABLE 2 Bare and Coated Conductor Properties

Conductor material and coating	Tensile strength, psi	Elong, %	Flex. life	Conductivity min., %	Cont. oper. temp. max., °C	Oxidation resist.	Galv. corrosion resist.	Solderability	Availability	Cost	Specifications
Bare copper (ann.)	34,000	10	Fair	100	150	Poor	Good	Fair	Good	Low(1)	QQ-W-343 ASTM B3
TC copper (ann.)	34,000	10	Fair	100	150	Good	Good	Good	Good	Low(2)	QQ-W-343 ASTM B33
SC copper	34,000	10	Fair	102	200	Good	Poor	Good	Good	Medium	ASTM B298
NC copper	34,000	10	Fair	96	260	Good	Good	Poor	Fair	Medium	ASTM B355
Aluminum	20,000–32,000	15	Poor	61	150	Poor	Good	Poor	Good	Low	MIL-W-7071 (insulated)
SC alum.	20,000–32,000	15	Poor	63	150	Good	Poor	Good	Fair	Medium	
Silver	25,000	15	Fair	104	260	Fair	Good	Good	Poor	High	
SC copper-clad steel (HD)	110,000–125,000	1	Poor	30–40	200	Good	Medium	Good	Fair	Medium	QQ-W-345
SC copper-clad steel (ann.)	55,000–65,000	8	Good	30–40	200	Good	Medium	Good	Fair	Medium	
SC cadmium-copper (HD)	90,000	1	Poor	80	200	Good	Poor	Good	Fair	Medium	
SC cadmium-copper (ann.)	55,000	5	Good	80	200	Good	Poor	Good	Fair	Medium	ASTM B268
SC zirconium	55,000	5	Good	85	200	Good	Poor	Good	Fair	Medium	
SC chrome (ann.)	55,000	6	Good	85	200	Good	Poor	Good	Fair	Medium	
SC cadmium-chrome	60,000	5	Good	84	200	Good	Poor	Good	Fair	Medium	

TC = tin-coated. SC = silver-coated. NC = nickel-coated. HD = hard-drawn. ann. = annealed.

Source: Martin Marietta Corporation.

4-8 Wires and Cables

of copper conductor. Table 3 presents the physical properties of the small sizes of copper-covered steel wire. Specifications for copper-covered steel wire are contained in QQ-W-345.

Aluminum. The major use of aluminum conductors has been in large-gage power conductors. Aluminum offers a weight advantage of approximately 3:1, possessing 62 percent conductance of an equivalent copper conductor. Compared with copper, weight savings of approximately 2:1 can be attained for equivalent current-carrying

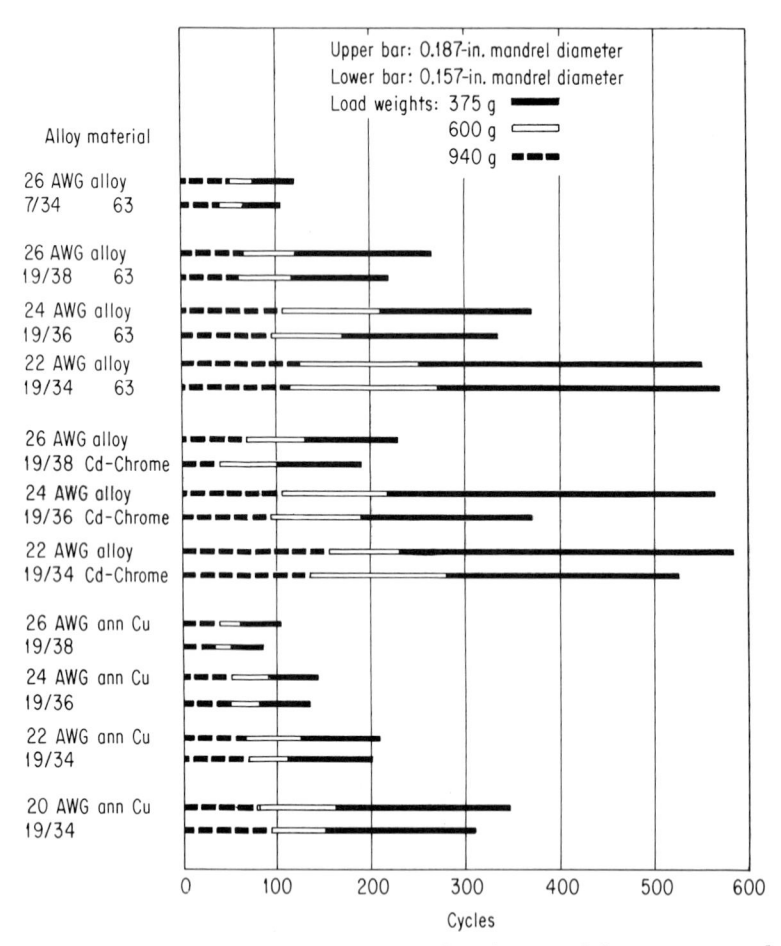

Fig. 1 Flex endurance tests (concentric standings)—annealed copper versus high-strength alloys.[1]

capacity. The major disadvantages of aluminum conductors versus copper are shorter fatigue life, poor corrosion resistance, and decreased solderability.

The Military Specification for insulated aluminum wire is MIL-W-5088.

Stainless Steel. Although very seldom utilized by itself as a conductor because of low conductivity, stainless steel (in strands) has been employed to reinforce stranded copper conductors. This reinforcement is normally used in small hookup wire (No. 24 AWG and smaller) and in ground-support cable (No. 18 AWG and smaller).

Silver. Recent observations on the corrosion problems of silver-plated copper conductors have resulted in investigations into the use of pure silver and silver-alloy conductors. The physical properties of silver materials available to date are covered in Table 2.

Coatings Bare copper conductor is rarely used in the aerospace and military electronics industry, since copper will oxidize from exposure to the atmosphere. Corrosion will impair the conductors and reduce solderability. Some insulating materials also tend to corrode bare copper. Following is a list of the most widely used conductor coatings.

Tin. For temperatures below 150°C, the least expensive protective coating is

TABLE 3 Physical Properties of Copper-clad Steel Wire, Small Sizes[2]

Size, AWG	Nominal dia., in.	Weight lb/1,000 ft	ft/lb	Average dc resistance, ohms/1,000 ft at 68°F 40%	30%	Nominal strength, lb Annealed wire (soft), avg. 40%	30%	Hard-drawn wire, min. 40%	30%	Size, AWG
13	0.072	14.4	70	5.1	6.8	200	280	390	530	13
14	0.064	11.4	88	6.4	8.6	160	230	310	440	14
15	0.057	9.04	110	8.1	10.8	130	180	240	330	15
16	0.051	7.17	140	10.2	13.7	100	140	190	270	16
17	0.045	5.68	180	12.9	17.2	80	110	150	200	17
18	0.040	4.51	220	16.3	21.7	64	89	120	170	18
19	0.036	3.58	280	20.5	27.4	51	71	96	130	19
20	0.032	2.84	350	25.9	34.5	40	56	76	110	20
21	0.028	2.25	440	32.6	43.5	32	45	60	81	21
22	0.025	1.78	560	41.2	54.9	25	35	48	64	22
23	0.023	1.41	710	51.9	69.2	20	28	38	51	23
24	0.020	1.12	890	65.5	87.3	16	22	30	40	24
25	0.018	0.89	1,100	82.5	110	13	18	24	32	25
26	0.016	0.71	1,400	104	139	10	14	19	25	26
27	0.014	0.56	1,800	131	175	7.9	11	15	20	27
28	0.013	0.44	2,300	166	221	6.3	8.8	12	16	28
29	0.011	0.35	2,800	209	278	5.0	7.0	9.5	13	29
30	0.010	0.28	3,600	263	351	3.9	5.5	7.5	10	30
31	0.0089	0.22	4,500	332	442	3.1	4.4	5.9	7.9	31
32	0.0080	0.18	5,700	418	558	2.5	3.5	4.7	6.3	32
33	0.0071	0.14	7,200	528	703	2.0	2.8	3.7	5.0	33
34	0.0063	0.11	9,100	665	887	1.6	2.2	3.0	4.0	34
35	0.0056	0.87	11,000	839	1,120	1.2	1.7	2.4	3.1	35
36	0.0050	0.069	14,000	1,060	1,410	1.0	1.4	1.9	2.5	36
37	0.0045	0.055	18,000	1,330	1,780	0.8	1.1	1.5	2.0	37
38	0.0040	0.044	23,000	1,680	2,240	0.6	0.9	1.2	1.6	38
39	0.0035	0.035	29,000	2,120	2,830	0.5	0.7	0.9	1.2	39
40	0.0031	0.027	36,000	2,670	3,570	0.4	0.5	0.7	1.0	40

tin. In addition to protecting copper from oxidation, it aids soldering. Tin should not be used as a protective coating above 150°C, as it will rapidly oxidize.

TIN DIP: Protective tin coating can be applied to a copper conductor by two methods: tin dip or electroplating. Tin dip is the process of passing strands through a molten tin bath. The major disadvantage of tin dip is poor control of coating thickness; thickness can vary from 20 to 200 μin. Specifications for tin-coated wire are covered in ASTM B33 and QQ-W-343.

Several variations of the standard tin-coated copper conductor have been utilized as a fabrication aid to eliminate strand twisting and pretinning. Following are some of the variations available.

TABLE 4 Properties of Solid Conductors[4]

Gauge, AWG or B. & S.	Nominal diameter, in.	Area, cir mils	Weight, lb/ 1,000 ft	Length, ft/lb	ohms/ 1,000 ft	Resistance at 68°F ft/ohms	ohms/lb
0000	0.4600	211,600	640.5	1.561	0.04901	20,400	0.00007652
000	0.4096	167,800	507.9	1.968	0.06180	16,180	0.0001217
00	0.3648	133,100	402.8	2.482	0.07793	12,830	0.0001935
0	0.3249	105,500	319.5	3.130	0.09827	10,180	0.0003076
1	0.2893	83,690	253.3	3.947	0.1239	8,070	0.0004891
2	0.2576	66,370	200.9	4.977	0.1563	6,400	0.0007778
3	0.250	62,500	189.1	5.286	0.1659	6,025	0.000877
4	0.2294	52,640	159.3	6.276	0.1970	5,075	0.001237
	0.2043	41,740	126.4	7.914	0.2485	4,025	0.001966
5	0.188	35,344	106.98	9.425	0.2934	3,407	0.00276
6	0.1819	33,100	100.2	9.980	0.3133	3,192	0.003127
7	0.1620	26,250	79.46	12.58	0.3851	2,531	0.004972
8	0.1443	20,820	63.02	15.87	0.4982	2,007	0.007905
	0.1285	16,510	49.98	20.01	0.6282	1,592	0.01257
9	0.1144	13,090	39.63	25.23	0.7921	1,262	0.01999
10	0.1019	10,380	31.43	31.82	0.9989	1,001	0.03178
11	0.09074	8,234	24.92	40.12	1.260	794	0.05053
12	0.08081	6,530	19.77	50.59	1.588	629.6	0.08035
13	0.07196	5,178	15.68	63.80	2.003	499.3	0.1278
14	0.06408	4,107	12.43	80.44	2.525	396.0	0.2032
15	0.05707	3,257	9.858	101.4	3.184	314.0	0.3230
16	0.05082	2,583	7.818	127.9	4.016	249.0	0.5136
17	0.04526	2,048	6.200	161.3	5.064	197.5	0.8167
18	0.04030	1,624	4.917	203.4	6.385	156.5	1.299
19	0.03589	1,288	3.899	256.5	8.051	124.2	2.065
20	0.03196	1,022	3.092	323.4	10.15	98.5	3.283

21	0.02846	810.1	2.452	407.8	12.80	78.11	5.221
22	0.02535	642.4	1.945	514.2	16.14	61.95	8.301
23	0.02257	509.5	1.542	648.4	20.36	49.13	13.20
24	0.02010	404.0	1.223	817.7	25.67	38.96	20.99
25	0.01790	320.4	0.9699	1,031	32.37	30.90	33.37
26	0.01594	254.1	0.7692	1,300	40.81	24.50	53.06
27	0.01420	201.5	0.6100	1,639	51.47	19.43	84.37
28	0.01264	159.8	0.4837	2,067	64.90	15.41	134.2
29	0.01126	126.7	0.3836	2,607	81.83	12.22	213.3
30	0.01003	100.5	0.3042	3,287	103.2	9.691	339.2
31	0.008928	79.7	0.2413	4,145	130.1	7.685	539.3
32	0.007950	63.21	0.1913	5,327	164.1	6.095	857.6
33	0.007080	50.13	0.1517	6,591	206.9	4.833	1,364
34	0.006305	39.75	0.1203	8,310	260.9	3.833	2,168
35	0.005615	31.52	0.09542	10,480	329.0	3.040	3,448
36	0.005000	25.00	0.07568	13,210	414.8	2.411	5,482
37	0.004453	19.83	0.06001	16,660	523.1	1.912	8,717
38	0.003965	15.72	0.04759	21,010	659.6	1.516	13,860
39	0.003531	12.47	0.03774	26,500	831.8	1.202	22,040
40	0.003145	9.888	0.02993	33,410	1,049	0.9534	35,040
41	0.00280	7.8400	0.02373	42,140	1,323	0.7559	55,750
42	0.00249	6.2001	0.01877	53,270	1,673	0.5977	89,120
43	0.00222	4.9284	0.01492	67,020	2,104	0.4753	141,000
44	0.00197	3.8809	0.01175	85,100	2,672	0.3743	227,380
45	0.00176	3.0976	0.00938	106,600	3,348	0.2987	356,890
46	0.00157	2.4649	0.00746	134,040	4,207	0.2377	563,900

4-11

Heavy tin is the application of a minimum thickness of 100 μin. to No. 31 AWG and smaller, and of 150 μin. to No. 30 AWG and larger. Utilized with the aid of high-frequency induction heaters, tin is melted and flowed, tacking the strands together in the area where the conductor is to be stripped.

Prefused conductor utilizes heavy tin-coated strands that are melted and fused together for the entire length of the insulated conductor. Although fabrication costs may be reduced, conductor flexibility is impaired. This process defeats the purpose of stranded conductors for increased flex life.

Overcoated conductor consists of individually tinned copper strands with an overcoating of tin over the conductor, bonding all strands together. Conductor flexibility is reduced significantly.

Top-coated conductor has bare, untinned copper strands with an overcoating of tin. The finished conductor is bonded along its entire length; it is less expensive than the overcoated conductor but has less corrosion resistance.

ELECTROPLATED TIN: The electroplating process applies to the copper strand a tin coating of controlled thickness, and is becoming increasingly important for automated stripping and termination equipment. There is no individual specification covering the application of tin coating by electroplating.

Silver. This coating is utilized for continuous service of conductors in excess of 150°C to a maximum of 200°C and as required for the application of certain high-temperature insulation materials. Silver-coated conductors are widely used with fluorocarbon, polyimide, and silicone rubber insulation. Silver is very readily soldered; this at times is considered a disadvantage in that solder flows under the insulation and potentially reduces flex life at the conductor termination.

Silver-coated copper is susceptible to corrosion caused by the galvanic interaction between copper and silver. Microscopic breaks in the silver coating, caused by stranding or coating porosity, allow galvanic reaction to take place between bare copper and silver in the presence of moisture or other electrolytes. Considerable work is under way to solve this problem. Protective coatings applied over the silver, increased silver thickness, and intermediate barrier platings are the subjects of current research. These approaches have been successful in reducing but not eliminating this potential problem.

ASTM B298 contains the specifications for silver-coated copper conductors.

Nickel. This is a good high-temperature protective coating, suitable for continuous use up to 300°C, and not susceptible to the corrosion potential of silver. The main disadvantage of nickel is its poor solderability. Even with activated fluxes and high soldering temperatures, good solder terminations are extremely difficult to achieve. Nickel coating can be applied to copper conductors by two methods: cladding and electroplating.

Cladding is the application of a relatively thick nickel outer coating under heat and pressure over the copper billet. The process is similar to the fabrication of copper-covered steel wire. Cladding offers superior protection at extremely high-temperature applications (1000°F maximum, continuous). Cladding is not advantageous for lower temperatures because of the higher resistance of nickel.

Electroplating is applied by standard electroplating techniques, normally with 50 μin. minimum thickness.

Construction Conductor construction plays a significant role in the proper functioning and reliability of any selected conductor. Solid wire, while low in cost and weight, will not withstand even mild flexure without breaking because of work hardening and fatigue. Bunch stranding offers extreme flexibility. Although it offers low cost and fatigue resistance, it does not provide a consistent circular cross section; therefore, it is not recommended for thin-wall extruded insulated-wire construction. Major conductor constructions are discussed below.

Solid. Table 4 contains the characteristics of solid conductors. Solid-conductor usage is limited for reasons given above. The major application of solid wire, magnet wire excepted, is for short jumper and bus wires not subjected to vibration or flexing.

TABLE 5 Details of Stranded Conductors[5]

Size designation, AWG	Nominal conductor area, cir mils	Number of strands	Allowable no. of missing strands	Nominal diameter of individual strands, in.	Max diameter of stranded conductor, in.	Tin-coated copper	Silver-plated copper	Nickel-plated copper	Silver-plated high-strength copper alloy
						\multicolumn{4}{c}{Max resistance of finished wire, ohms/1,000 ft at 20°C}
30	112	7	0	0.0040	0.013	107.0	101.0	109.0	116.0
28	175	7	0	0.0050	0.016	67.6	62.9	68.3	72.2
26	304	19	0	0.0040	0.021	39.3	36.2	40.1	41.5
24	475	19	0	0.0050	0.026	24.9	23.2	25.1	26.6
22	754	19	0	0.0063	0.033	15.5	14.6	15.5	16.8
20	1,216	19	0	0.0080	0.041	9.70	9.05	9.79	10.4
18	1,900	19	0	0.0100	0.052	6.08	5.80	6.08	6.65
16	2,426	19	0	0.0113	0.060	4.76	4.54	4.76	5.23
14	3,831	19	0	0.0142	0.074	2.99	2.87	3.00	3.30
12	7,474	37	0	0.0142	0.102	1.58	1.48	1.59	1.70
10	9,361	37	0	0.0159	0.118	1.27	1.20	1.27	1.38
8	16,983	133	0	0.0113	0.176	0.700	0.661	0.680	0.760
6	26,818	133	0	0.0142	0.218	0.436	0.419	0.428	0.483
4	42,615	133	0	0.0179	0.272	0.274	0.263	0.269	0.302
2	66,500	665	2	0.0100	0.345	0.179	0.169	0.174	0.194
0	104,500	1,045	3	0.0100	0.432	0.114	0.105	0.109	0.123

4-14 Wires and Cables

Stranded. Besides increased flexibilty, construction characteristics of stranded conductors may limit the application of insulating materials. Table 5 contains detailed characteristics of stranded conductors that meet military specifications. Following is a brief description of the various stranding constructions.

TRUE CONCENTRIC: This construction presents the most consistent circular cross section of all available strandings. Alternate layers, applied in opposite directions with increasing lay, hold strands in place and prevent "strand popping" and high strands. Insulated wire manufacturers prefer this stranding for conductors with

Concentric lay strands

| 3 wire | 7 wire | 12 wire | 19 wire |

| 37 wire | 61 wire | 91 wire | 127 wire |

Concentric rope-lay strands

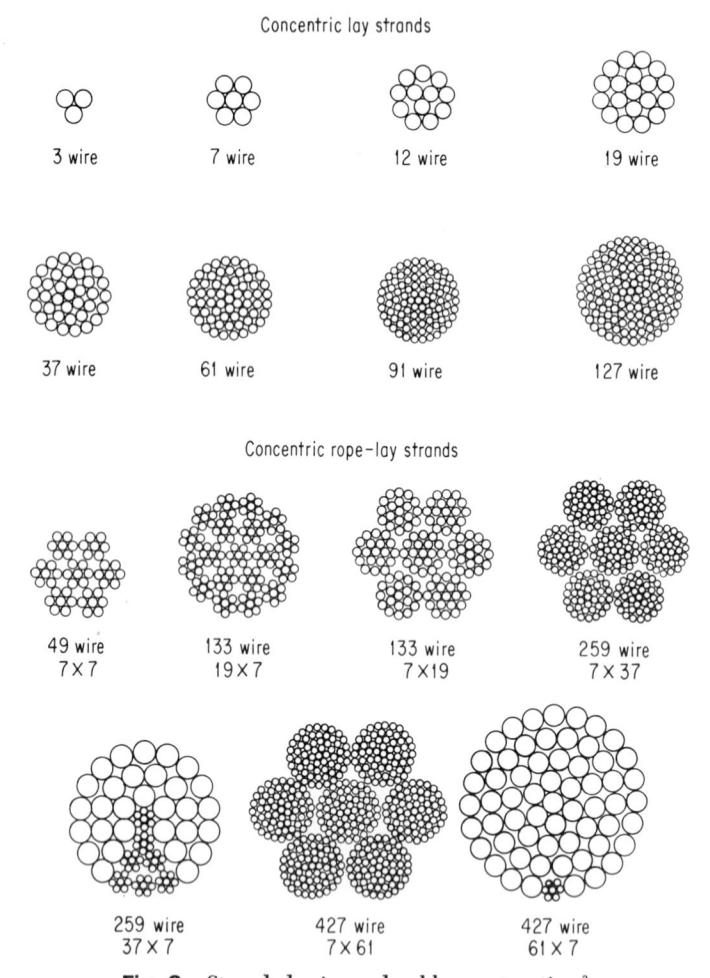

| 49 wire 7×7 | 133 wire 19×7 | 133 wire 7×19 | 259 wire 7×37 |

| 259 wire 37×7 | 427 wire 7×61 | 427 wire 61×7 |

Fig. 2 Stranded wire and cable construction.[a]

extruded thin-wall insulation. Construction of concentric stranded conductors is shown in Fig. 2.

UNIDIRECTIONAL CONCENTRIC: This construction differs from true concentric in that the lay of successive layers is applied over a core in the same direction; it does not alternate directions. Unidirectional lay is more flexible and has greater flex endurance than true concentric. A comparison of flex endurance among true concentric, unidirectional concentric, bunch, and unilay conductor constructions is shown in Fig. 3.

BUNCH: Since bunch stranding consists in twisting a group of strands with the same lay length in the same direction, without regard to geometric arrangement, the strands are susceptible to movement, and circular cross section is not ensured. If conductor circularity is important, as demanded by extrusion of thin-wall insulating materials, bunch conductor is not recommended.

UNILAY: The advantages of unilay are a smaller diameter, flexibility approaching bunch construction, and superior flex endurance (Fig. 3).

EQUILAY: This is a variation of concentric construction that has reversed lay of layers but equal length of lay and a stiffer construction than unidirectional concentric lay.

ROPE LAY: Rope lay is basically a large-gage (No. 10 AWG and larger) conductor construction that consists of a central-core stranded member with one or more layers of stranded members surrounding the core. The stranded members may

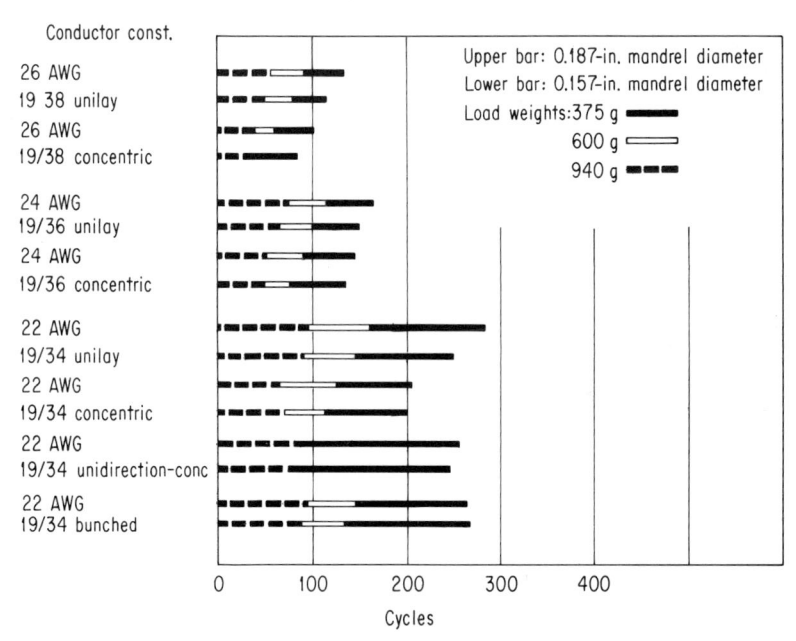

Fig. 3 Flex endurance tests on annealed conductors (copper) in different stranding configurations.[1]

be bunch-stranded (ASTM B172) or concentric (ASTM B173). Rope construction shown in Fig. 2 provides a uniform circular cross section with good flexibility.

WIRE AND CABLE INSULATION

Materials The properties of insulating and jacketing materials used with wire and cable can be divided into three main categories: mechanical, electrical, and chemical.

The mechanical and physical properties to consider are tensile strength, elongation, specific gravity, abrasion resistance, cut-through resistance, and mechanical temperature resistance (cold bend and deformation under heat).

The electrical properties are dielectric strength, dielectric constant, loss factor, and insulation resistance.

The chemical properties are fluid resistance, flammability, temperature resistance, and radiation resistance.

4-16 Wires and Cables

Tables 6 through 8 give these major properties for the most commonly employed wire insulating and jacketing materials. Following is a brief description of the insulating materials.

Polyvinyl Chloride (PVC). This is a low-cost primary insulating and jacketing material available in many formulations tailored to meet applications. Two varieties of PVC, most widely used in aerospace and electronics, are: (1) plasticized PVC, 105°C operating temperature applicable under most military specifications that re-

TABLE 6 Mechanical and Physical Properties of Insulating Materials

Insulation	Common designation	Tensile strength, psi	Elongation, %	Specific-gravity	Abrasion resist.	Cut-through resist.	Temperature resistance (mechanical)
Polyvinyl chloride	PVC	2,400	260	1.2–1.5	Poor	Poor	Fair
Polyethylene	PE	1,400	300	0.92	Poor	Poor	Good
Polypropylene	....	6,000	25	1.4	Good	Good	Poor
Crosslinked polyethylene	IMP	3,000	120	1.2	Fair	Fair	Good
Polytetrafluoroethylene	TFE	3,000	150	2.15	Fair	Fair	Excellent
Fluorinated ethylene propylene	FEP	3,000	150	2.15	Poor	Poor	Excellent
Monochlorotrifluoroethylene	Kel-F	5,000	120	2.13	Good	Good	Good
Polyvinylidine fluoride	Kynar	7,100	300	1.76	Good	Good	Fair
Silicone rubber	Silicone	800–1,800	100–800	1.15–1.38	Fair	Poor	Good
Polychloroprene rubber	Neoprene	150–4,000	60–700	1.23	Good	Good	Fair
Butyl rubber	Butyl	700–1,500	500–700	0.92	Fair	Fair	Fair-good
Fluorocarbon rubber	Viton	2,400	350	1.4–1.95	Fair	Fair	Fair-good
Polyurethane	Urethane	5,000–8,000	100–600	1.24–1.26	Good	Good	Fair-good
Polyamide	Nylon	4,000–7,000	300–600	1.10	Good	Good	Poor
Polyimide film	Kapton	18,000	707	1.42	Excellent	Excellent	Good
Polyester film	Mylar	13,000	185	1.39	Excellent	Excellent	Good
Polyalkene	....	2,000/7,000	200/3,001.2	1.76	Good	Good	Fair-good
Polysulfone	....	10,000	50–100	1.24	Good	Good	Good
Polyimide-coated TFE	TFE/ML	3,000	150	2.2	Good	Good	Good
Polyimide-coated FEP	FEP/ML	3,000	150	2.2	Good	Good	Good

Source: Martin Marietta Corporation.

quire PVC insulation; and (2) semirigid PVC, a less flexible but more cut-through-resistant version used primarily with automated termination devices (wire wrap, Termi-point,° etc.).

Polyethylene (PE). This possesses excellent properties for wire and cable insulation. Electrically, because of its dielectric constant, loss factor, and insulation resis-

° Trademark of AMP, Inc.

tance, PE can be matched only by some fluorocarbons. Solvent resistance, moisture resistance, and low-temperature performance are superior. PE has three distinct disadvantages: it is flammable, its maximum operating temperature is only 80°C, and it is stiff. Flame retardants can be added to render PE self-extinguishing. Crosslinking can be achieved either by chemical methods or by irradiation methods, increasing the continuous operating temperature to the 135 to 150°C range, and reducing its susceptibility to crack under thermal and environmental stresses.

Polypropylene. A member of the polyolefin family, like polyethylene, polypropylene exhibits good electrical and chemical properties. Mechanically polypropylene has properties superior to those of polyethylene; its abrasion and cut-through resistance is better and is comparable to that of nylon. Until recent improvements, polypropylene could cause copper poisoning (embrittlement of conductor). Poly-

TABLE 7 Electrical Properties of Insulation Materials

Insulation	Common designation	Dielectric strength, volts/mil	Dielectric constant, 10^3 Hz	Loss factor, 10^3 Hz	Volume resistivity, ohm-cm
Polyvinyl chloride	PVC	400	5–7	0.02	2×10^{14}
Polyethylene	PE	480	2.3	0.005	10^{16}
Polypropylene	....	750	2.54	0.006	10^{16}
Crosslinked polyethylene	IMP	700	2.3	0.005	10^{16}
Polytetrafluoroethylene	TFE	480	2.1	0.0003	10^{18}
Fluorinated ethylene propylene	FEP	500	2.1	0.0003	10^{18}
Monochlorotrifluoroethylene	Kel-F	431	2.45	0.025	2.5×10^{16}
Polyvinylidine fluoride	Kynar	1,280 (8 mils)	7.7	0.02	2×10^{14}
Silicone rubber	Silicone	575–700	3–3.6	0.003	2×10^{15}
Polychloroprene rubber	Neoprene	113	9.0	0.030	10^{11}
Butyl rubber	Butyl	600	2.3	0.003	10^{17}
Fluorocarbon rubber	Viton	500	4.2	0.14	2×10^{13}
Polyurethane	Urethane	450–500	6.7–7.5	0.055	2×10^{11}
Polyamide	Nylon	385	4–10	0.02	4.5×10^{13}
Polyimide film	Kapton	5,400 (2 mils)	3.5	0.003	10^{18}
Polyester film	Mylar	2,600	3.1	0.15	6×10^{16}
Polyalkene	....	1,870	3.5	0.028	6×10^{13}
Polysulfone	....	425	3.13	0.0011	5×10^{16}
Polyimide-coated TFE	TFE/ML	480	2.2	0.0003	10^{18}
Polyimide-coated FEP	FEP/ML	480	2.2	0.0003	10^{18}

Source: Martin Marietta Corporation.

propylene has poor low-temperature (cold-bend) characteristics; it is not recommended for low-temperature applications. Its main usage to date has been in telephone and communication wire and cable insulation.

Polytetrafluoroethylene (TFE). This is undoubtedly the best-known and most widely used material in the fluorocarbon family. It has excellent electrical, chemical, and thermal properties. TFE is not a particularly tough material; its abrasion and cut-through resistance are rated only fair. However, it maintains mechanical resistance at temperatures exceeding 200°C. TFE can withstand short-time contact with a hot soldering iron without damage. TFE-insulated wire is in wide use because of its solder-resistance characteristics: it reduces the problem of insulation overheating at solder terminations and allows soldering of high-density wiring where soldering without contact of nearby insulation is unavoidable. TFE insulating and jacketing applications are high-temperature interconnection and hookup wires and cables (also multiconductor) and coaxial cable. TFE is suitable neither for high-voltage applications, because of poor corona resistance, nor for radiation environments, because of mechanical degradation at levels in excess of 10^6 rads. TFE's radiation resistance improves in oxygen-free atmosphere.

Fluorinated Ethylene Propylene (FEP). A melt-extrudable counterpart of TFE, FEP has almost identical electrical and chemical properties; but it melts to a fluid at 290°C, and thus its solder-iron resistance and temperature resistance are

TABLE 8 Chemical Properties of Insulation Materials

Insulation	Common designation	Fluid resistance	Flammability	Radiation resistance, rads gamma exposure	Temperature resistance, °C	Comments
Polyvinyl chloride	PVC	Good	Slow to self-extinguishing	10^6–10^7	−55–105	
Polyethylene	PE	Excellent	Flammable	10^8	−65–80	
Polypropylene	....	Good	Self-extinguishing	10^8	−20–125	
Crosslinked polyethylene	IMP	Good	Self-extinguishing	10^8	−65–150	
Polytetrafluoroethylene	TFE	Excellent	Non-flammable	10^6	−80–260	
Fluorinated ethylene propylene	FEP	Excellent	Non-flammable	10^6	−80–200	
Monochlorotrifluoroethylene	Kel-F	Good	Non-flammable	10^6	−80–200	Fluids tend to permeate at high temperature
Polyvinylidine fluoride	Kynar	Good	Self-extinguishing	10^8	−65–130	
Silicone rubber	Silicone	Poor	Flammable	10^8	−65–200	
Polychoroprene rubber	Neoprene	Good oil resistance	Self-extinguishing	10^7	−55–80	
Butyl rubber	Butyl	Good	Flammable	10^6	−55–85	Poor resistance to hydrocarbons
Fluorocarbon rubber	Viton	Excellent	Self-extinguishing	10^6–10^7	−40–200	Poor resistance to oxygenated alcohols
Polyurethane	Urethane	Good	Flammable	10^7–10^8	−55–85	
Polyamide	Nylon	Good	Self-extinguishing	10^7	−55–105	Soluble in alcohol
Polyimide film	Kapton	Excellent	Non-flammable	10^9	−80–260	
Polyester film	Mylar	Good	Flammable	10^8	−65–120	
Polyalkene	....	Good	Self-extinguishing	10^8	−65–135	
Polysulfone	Polysulfone	Fair	Self-extinguishing	....	−65–150	Soluble in chlorinated hydrocarbon
Polyimide-coated TFE	TFE/ML	Excellent	Non-flammable	10^6	−80–260	
Polyimide-coated FEP	FEP/ML	Excellent	Non-flammable	10^6	−80–260	

SOURCE: Martin Marietta Corporation.

inferior. FEP is widely used as jacketing material, since it is melt-extrudable by thermoplastic techniques (TFE cannot be melt-extruded). Used as primary insulation, FEP offers the following advantages: lower cost (in quantities), longer continuous lengths, and easier identification (hot stamp).

Monochlorotrifluorethylene (Kel-F*). As part of the fluorocarbon family, this is less temperature-resistant but substantially tougher than TFE or FEP. It has good electrical, chemical, and mechanical properties, but is susceptible to crystalliza-

* Trademark of 3M Company, St. Paul, Minn.

tion manifested in poor shelf aging. Kel-F has seen limited use as wire insulation in the military, aerospace, and electronics industries.

Vinylidene Fluoride (Kynar*). Kynar is a member of the fluorocarbon family of wire and cable insulation materials. It is extremely tough and has excellent abrasion and cut-through resistance. Its electrical, thermal, and chemical properties are inferior to those of TFE and FEP; however, Kynar has superior radiation resistance. Kynar has been utilized by itself as insulation, in combination with polyethylene (polyalkene), and as jacketing. Some problems have been encountered on insulation shrinkage, and flaring while soldering, although the problems can be corrected by controlled extrusion techniques.

Silicone Rubber (Silicone). Silicone is a good, flexible, high-temperature elastomeric insulator; however, it is not noted for its mechanical toughness or fluid resistance. Its electrical properties are good, and its radiation resistance is superior for an elastomer but generally inferior to that of most plastic insulations. Excellent corona resistance makes it useful as a high-voltage insulation. Applications include shipboard wire and cable, nuclear cable, high-voltage cable, and aircraft wire and cable. Silicone rubber is flammable but has the unique feature of leaving a white, nonconductive ash which, if contained, forms good insulation.

Neoprene Rubber (Neoprene). Neoprene finds major usage in the aerospace and electronic industry as jacketing material. Its electrical insulating properties are inferior: it is suitable only for low-voltage, low-frequency applications. Its resistance to abrasion and to mechanical bending and impact in the -55 to $+70°C$ range are excellent. Neoprene is extruded as a jacket for ground-support cable applications. Neoprene tubing may be used as harness jacketing where extremely heavy use is anticipated.

Butyl Rubber (Butyl). Butyl rubber has better moisture resistance and electrical properties than neoprene. It is used as insulation for larger ground-support cable sizes. More flexible than polyethylene, butyl is used in thick-wall insulation. As a jacket material butyl does not possess the mechanical toughness of neoprene but has the advantage of compatibility with some of the more exotic missile fuels.

Fluorocarbon Rubber (Viton†). This has some outstanding characteristics: high temperature, oil, and fuel resistance. Its mechanical properties are inferior to those of neoprene, and its low-temperature cold-bend resistance is poor. Electrically it is not recommended as a primary insulation except for low-voltage, low-frequency application.

Polyurethane Elastomers (Urethane). Two basic urethane elastomers are utilized in the wire and cable industry: polyether-based and polyester-based. The polyether-based material possesses superior low-temperature characteristics and humidity resistance. The polyester-based material exhibits superior high-temperature resistance but is susceptible to hydrolysis under high-humidity and high-temperature conditions. Major application of urethanes has been as jacketing material, presenting an extremely tough abrasion- and tear-resistant covering.

Polyamide (Nylon). This is used almost exclusively as jacketing because of its high moisture absorption. Nylon is a tough, abrasion-resistant material. Because of its stiffness and susceptibility to cracking it is not recommended for extruded jacketing over cores greater than 0.210 in. in diameter. Nylon provides a good protective jacket over thin-wall thermoplastic insulations compatible with its temperature characteristics. However, nylon cracking has also been observed when the material is used in a low-humidity environment or if stored for a length of time in a formed condition.

Polyimide. Two forms of polyimide resin are at present in use in the wire and cable industry: (1) a coating material, applied as an insulation on magnet wire and as a mechanical barrier on TFE and FEP, enhancing their mechanical strength; (2) a film which, when coated with FEP, becomes heat-sealable. The film can be applied over a conductor by tape wrapping where the final sintering provides a fused homogeneous insulation. Polyimide possesses exceptionally high

* Trademark of Pennsalt Chemicals Corporation, Philadelphia, Pa.
† Trademark of E. I. du Pont de Nemours & Co., Wilmington, Del.

heat resistance, excellent mechanical properties, high radiation resistance, and very good chemical resistance. As a supplementary coating applied to TFE or FEP in 0.0005 to 0.001 in. thickness, polyimide more than doubles abrasion and cut-through resistance. Standard film construction provides a thin wall of approximately 0.007 in., which saves space and weight without sacrifice of mechanical strength or abrasion and cut-through resistance, compared with many larger and heavier insulations.

Polyester (Mylar). Mylar is available to date only in yarn or film, although several extrudable polyester-based resins have been applied to conductors experimentally. Polyester film is widely used in cable fabrication as a separator or binder wrap. A heat-sealable construction is available with either PE or PVC coating. The tape is applied by wrapping, like the FEP-coated polyimide film previously mentioned. When applied and heat-sealed, the coated polyester film exhibits me-

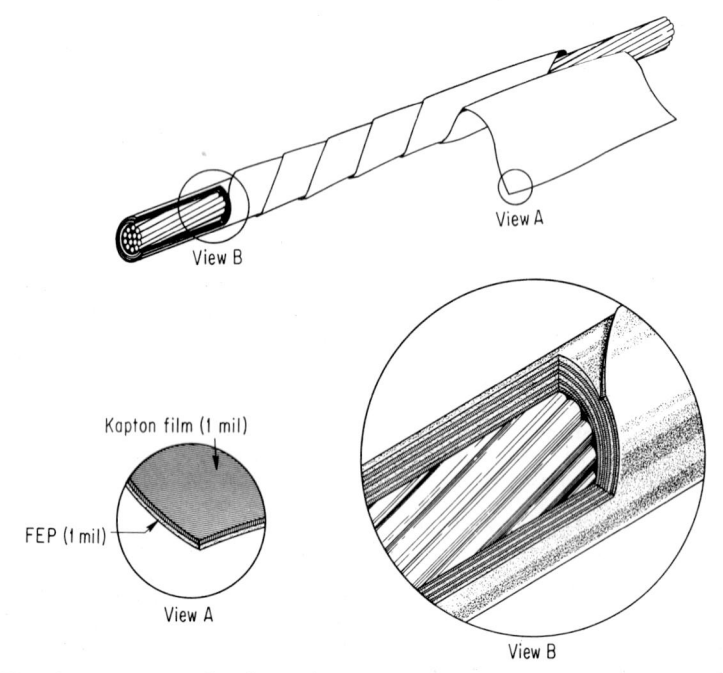

Fig. 4 Tape-wrap detail, single-wrap construction: two-thirds overlap.[6]

chanical properties similar to those of the coated polyimide film. Heat resistance of polyester-coated film is inferior to that of polyimide. Major application of the coated polyester film insulation has been to date in the computer industry for tough, thin-walled, insulated wire used in conjunction with wire wrap and other automated wiring methods.

Polyalkene. Polyalkene is a dual extrusion of polyolefin and polyvinylidene fluoride. Both these materials are crosslinked for increased heat resistance and greater mechanical strength. Combined use of the two insulating compounds mutually offsets their individual disadvantages. Polyvinylidene fluoride provides mechanical toughness not inherent in polyolefin, while its main disadvantage, high dielectric constant, is tempered by the excellent electrical properties of polyolefin. Polyalkene exhibits good properties for thinner-walled, lighter-weight wire construction.

Polysulfone. This is manufactured by Union Carbide and is an extrudable resin used mostly for thin-walled wires and cables in computer applications. Polysulfone is tough, suitable for high-density computer wiring; its electrical properties are adequate; and it is self-extinguishing.

Construction and application methods Reference has been made to extrusion, tape wrapping, and coating as methods of applying specific materials to a conductor. Certain materials are available only in a tape or solution form and cannot be obtained in extruded form. Following is a brief description of methods of application:

Extrusion. Many materials covered in the preceding section are thermoplastic; i.e., they are heated to softness, formed into shape, then cooled to become solid again. Thermoplastics are shaped around a conductor by the extrusion process. This consists in forcing the plastic material under pressure and heat through an orifice. The suitability for extrusion presents certain advantages to both the manufacturer and the user: longer continuous lengths of insulated conductor, smooth outer surfaces for easier connector sealing, homogeneous insulation, fast processing, quick delivery, and usually, lower cost due to simpler processing.

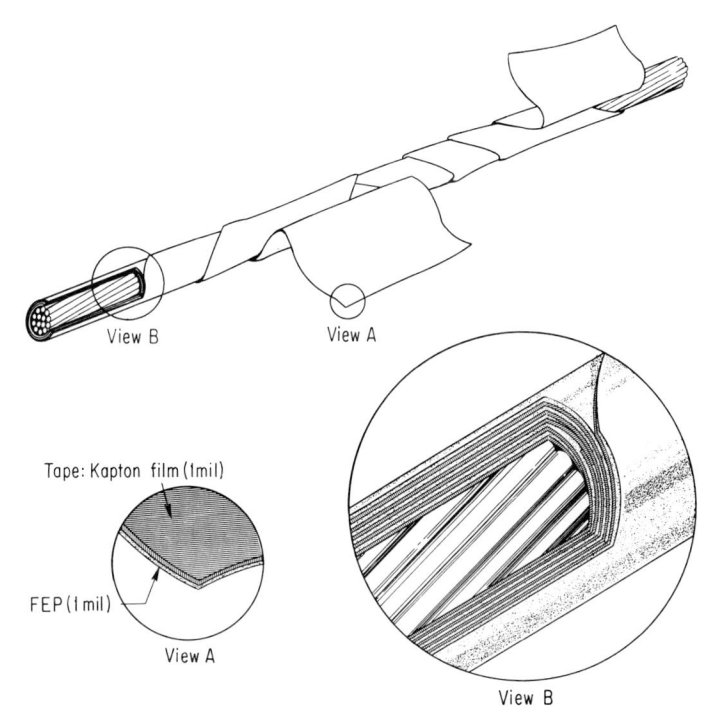

Fig. 5 Tape-wrap detail, double-wrap construction: one-half overlap.[6]

Tape Wrap. Tape wrapping is the application of insulation in the form of a thin film or tape. The tape is normally applied to the conductor with minimum overlap to ensure wire flexibility without baring the conductor. Layers of tape can be built up to achieve the desired insulation-wall thickness. Successive layers are wrapped in opposite directions, ensuring coverage of tape overlaps. Finally, after the tape is applied, the wire is sintered, cured, or fused with sufficient heat to seal the layers of tape into a homogeneous mass. Tape wrap provides good control of wall thickness; there is no concentricity problem with tape construction as opposed to extruded insulation. Tape wrap has the disadvantage of slower fabrication. Problems of poor adhesion between tape layers have also been encountered. Details of representative tape-wrapped Kapton* film insulated construction are shown in Figs. 4 and 5.

* Trademark of E. I. du Pont de Nemours & Co., Wilmington, Del.

Coating. Insulation by dip coating is limited almost exclusively to magnet wire. However, for small-diameter Teflon° TFE and FEP applications, a polyimide overcoating of less than 0.001 in. significantly enhances the mechanical strength, abrasion protection, and solder resistance of the insulation. The coating of fluorocarbon base is a difficult process; it requires complete surface preparation to promote good adhesion of the polyimide coating material.

SHIELDING

Materials and construction Although there is no real substitute for braided copper as a general-purpose shield, certain applications may allow the use of a lighter, less bulky shield construction. The purpose of a shield is (1) to prevent external fields from adversely affecting signals transmitted over the center conductor, (2) to prevent undesirable radiation of a signal into nearby or adjacent conductors, or (3) to act as a second conductor in matched or tuned lines. Following is a brief description of various available shield types. Table 9 presents the shielding effectiveness at tested frequency ranges of the systems described.

Braided Round Shield. Braided round copper shielding is the most commonly utilized by industry today as flexible coaxial or shielded cable. The braided-shield technique is highly effective in the frequency range below 0.5 Hz. Above this frequency range the gaps in the interstices of the shield braid cause loss in effectiveness. Braiding consists in interweaving groups (carriers) of strands (ends) of metal over an insulated conductor. The braid angle at which the carriers are applied with reference to the axis of the core, the number of ends, the number of carriers, and the number of carrier crossovers per inch (picks) determine the percent coverage, which is a measure for the shielding gap. As reference, a solid metal tube is equivalent to 100 percent coverage.

Braided-shield strands normally consist of coated copper. The coating must be compatible with the insulation material employed: tin coating is used for low-temperature insulations such as PVC, PE, and polyalkene; silver or nickel coating are used with high-temperature insulations.

Silver-plated aluminum strands are employed for lighter weights, although a significant galvanic corrosion potential exists. Stainless steel or other metals can be employed for added strength if conductivity is not the prime requisite.

Flat-braided Shield. The application and considerations are the same as with round shielding, except for shape: the individual strands are not round but flat. Flat braid combines the advantages of reduced shield buildup and weight with good coverage. Termination is a problem solved satisfactorily with a shrinkable solder sleeve.

Metal Tape. Metal tapes of either copper or aluminum with a minimum overlap of 10 percent can be employed as an effective shield. Because of their stiffness, metal tapes are normally employed only in large single-conductor or multiconductor cable construction.

Solid Shield. Solid or tubular shields may be applied by tube swaging or by forming interlocked, seam-welded, or soldered tape around the dielectric. These methods offer 100 percent shielding and couple as protective armor. Corrugated tapes facilitate cable bending, but their flexibilty is poor. Solid shield is usually applied to buried cables or where 100 percent shielding is mandatory.

Served Shield. A served, or spiral, shield consists of a number of metal strands wrapped flat as a ribbon over a dielectric in one direction. Weight and size of shielding are approximately one-half those of braided shields, as there is no strand crossover or overlap. A served shield is known to be effective in the audio-frequency range. Above audio frequencies the possibility of inductive effects caused by strands spiraling in one direction should be examined.

Foil Shield. Foil tape materials such as copper or aluminum-coated Mylar° have been employed in such techniques as normal overlap taping and longitudinal wrapping with various interlocking approaches. Shielding effectivity at low frequen-

° Trademark of E. I. du Pont de Nemours & Co., Wilmington, Del.

TABLE 9 Shield Effectivity in Volts Peak to Peak[7]

Sample description	30 Hz	100 Hz	300 Hz	1 kHz	3 kHz	10 kHz	30 kHz	100 kHz	300 kHz	1 MHz
Control unshielded	<0.600	<1.6	4.0	10.0	12.0	12.5	13.5	12.5	12.5	14.0
Braid, No. 36 AWG, tinned copper, 90% coverage	<0.001	<0.001	0.0025	0.005	0.006	0.00625	0.0075	0.0075	0.007	0.008
Aluminum tape, No. 22 AWG drain wire	<0.001	<0.001	0.001	0.002	0.002	0.002	0.005	0.0085	0.012	0.014
Semiconductive PVC, No. 26 AWG drain wire	<0.001	<0.001	<0.00325	0.020	0.060	0.120	0.240	0.450	0.540	1.85
Serve—No. 36 AWG tinned copper, 90% coverage, 4 ends reversed	<0.001	<0.001	0.001	0.002	0.0025	0.0025	0.004	0.006	0.0065	0.013
Braid—8 carriers, No. 36 AWG tinned copper, 8 carriers, conductive glass yarn	<0.001	<0.001	<0.001	0.001	0.00125	0.00125	0.002	0.003	0.005	0.012
Braid—No. 36 AWG flat ribbon silver-plated copper	0.00125	0.003	0.008	0.018	0.024	0.024	0.028	0.028	0.027	0.029
Serve—No. 36 AWG tinned copper	<0.001	0.003	0.008	0.018	0.026	0.026	0.029	0.029	0.030	0.031
Solid—cadmium-plated copper	<0.001	<0.001	<0.001	<0.001	<0.001	<0.001	<0.001	<0.001	<0.001	<0.001

cies is poor, and termination is difficult. The shield is terminated with a single strand of solid wire (drain wire) in contact with the metal coating under the tape.

Conductive Plastics. PVC and PE compounds have been formulated with conductive additives for the purpose of shielding. Effectiveness has been poor, normally limited to the low audio-frequency range. As with foil, termination is a problem; a drain wire can be employed for terminating.

Conductive Yarns. Conductive yarns, such as impregnated glass, provide weight reduction, but lack of conductivity and difficulty in termination make them suited only for special applications.

Metal Shield and Conductive Yarn. An effective marriage between metal strands and conductive yarns has been developed and tested which provides effective shielding when braided and which reduces shield weight. Terminations can be accomplished by normal techniques.

SHIELD JACKETS

Materials and construction It is the basic function of a shield jacket to insulate a shield from ground (structure), other shields, or conductors. Secondarily, a shield jacket may serve as a lubricant in multiconductor cables, enabling free movement during bending; as an insulator, to allow varying potentials between adjacent shields; and as a moisture and abrasion barrier. A few of the materials described in the section on Wire and Cable Insulation are appropriately used as jacketing materials, because of either superior mechanical strength or compatibility with the chosen primary insulation system.

Extrusion, braiding, and taping techniques are employed in the application of shield jackets. Fused tapes and extruded materials are the only reliable methods for assuring a moistureproof jacket. Consistent wall thickness is best maintained with tapes; however, reliable fusion across interstices of multiple shielded conductors is frequently a problem. Lacquered fiber braids allow a large conductor or multiple conductors more flexibility than extruded or taped jackets but offer limited moisture and humidity resistance.

Table 10 lists the characteristics of extruded and taped jackets suited for low- and medium-temperature applications; Fig. 6 provides shield-jacket abrasion-resistance data. In comparing abrasion-resistance values, differences in wall thickness should be considered. Following is a discussion of available shield-jacket materials.

Polyamide (Nylon). Nylon has excellent mechanical properties within its temperature capabilities (120°C maximum). Electrical properties, relatively unimportant in jacket materials, are poor owing to rather high moisture absorption. Nylon jacket is likely to crack when heated and bent at less than a 10× mandrel. Jacketing is applied by extrusion or in the form of braided nylon yarn. Extrusion is not recommended over cores greater than 0.210 in. because of stiffness and cracking.

Polyvinyl Chloride (PVC). This provides good electrical isolation but requires substantial wall thickness to achieve adequate mechanical protection. PVC has seen much use as a coaxial cable jacket formulated with a noncontaminating compound containing a nonbleeding plasticizer. It is applied by extrusion with practically no limitation to core diameter.

Polyethylene (PE). Polyethylene is used rarely as a shield jacket because of poor high-temperature characteristics, stiffness, and inadequate mechanical resistance in thin walls. Crosslinked PE with improved temperature and mechanical properties is used with compatible primary insulation materials. PE jackets are applied by extrusion with no minimum limitation on core diameter.

Teflon (TFE). This is supplied as a shield jacket by extrusion or taping. Tape jackets are the only solution for larger shields owing to practical extrusion limitations. Close control must be exercised by the manufacturer in the curing of TFE to prevent the flow of inner primary conductor insulation.

Teflon (FEP). As outlined in the section on Wire and Cable Insulation, FEP is much easier to fabricate as a jacket material than TFE. However, the lower

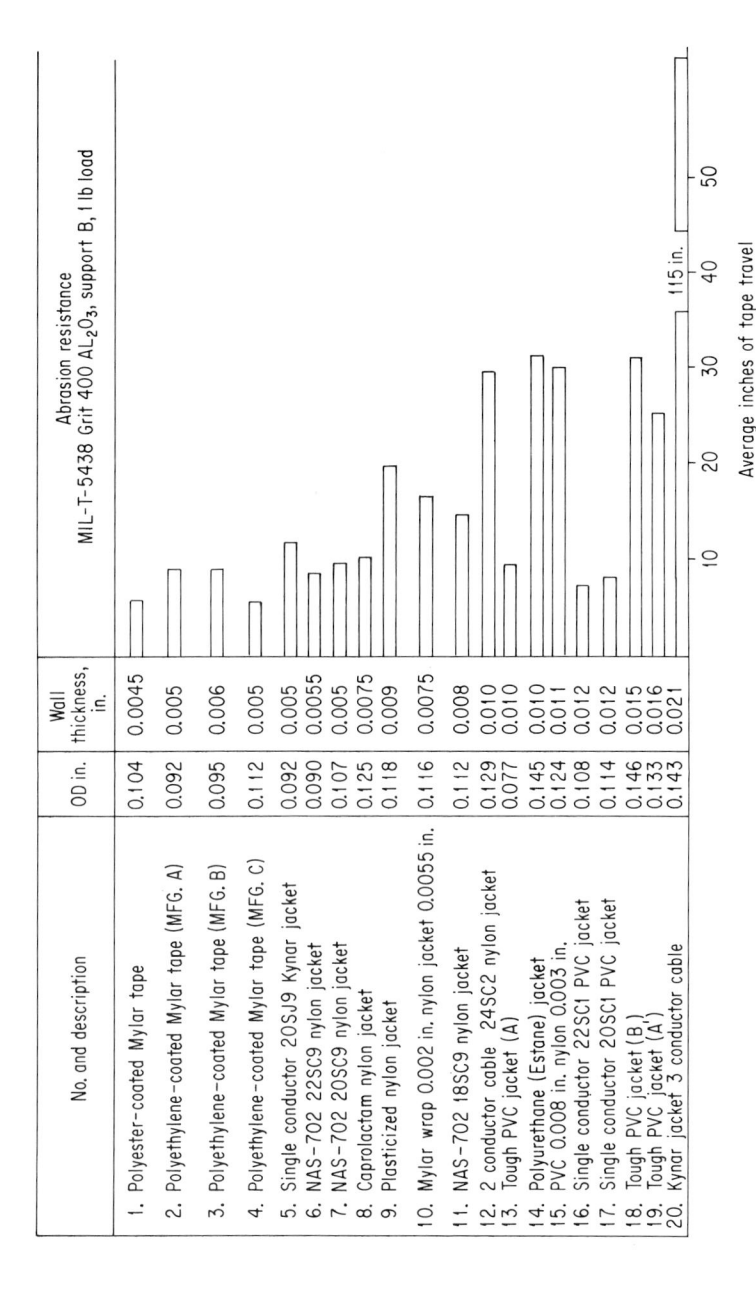

No. and description	OD in.	Wall thickness, in.	Abrasion resistance MIL-T-5438 Grit 400 AL_2O_3, support B, 1 lb load
1. Polyester-coated Mylar tape	0.104	0.0045	
2. Polyethylene-coated Mylar tape (MFG. A)	0.092	0.005	
3. Polyethylene-coated Mylar tape (MFG. B)	0.095	0.006	
4. Polyethylene-coated Mylar tape (MFG. C)	0.112	0.005	
5. Single conductor 20SJ9 Kynar jacket	0.092	0.005	
6. NAS-702 22SC9 nylon jacket	0.090	0.0055	
7. NAS-702 20SC9 nylon jacket	0.107	0.005	
8. Caprolactam nylon jacket	0.125	0.0075	
9. Plasticized nylon jacket	0.118	0.009	
10. Mylar wrap 0.002 in. nylon jacket 0.0055 in.	0.116	0.0075	
11. NAS-702 18SC9 nylon jacket	0.112	0.008	
12. 2 conductor cable 24SC2 nylon jacket	0.129	0.010	
13. Tough PVC jacket (A)	0.077	0.010	
14. Polyurethane (Estane) jacket	0.145	0.010	
15. PVC 0.008 in. nylon 0.003 in.	0.124	0.011	
16. Single conductor 22SC1 PVC jacket	0.108	0.012	
17. Single conductor 20SC1 PVC jacket	0.114	0.012	
18. Tough PVC jacket (B)	0.146	0.015	
19. Tough PVC jacket (A')	0.133	0.016	
20. Kynar jacket 3 conductor cable	0.143	0.021	115 in.

Fig. 6 Shield-jacket abrasion resistance. (*Martin Marietta Corporation.*)

TABLE 10 Shield-jacket Properties

No.	Sample description	OD, in.	Wall, in.	Thermal[a] shock	Aging stability[b] Visual	Aging stability[b] Dielectric, kv	Heat resistance[c] Visual	Heat resistance[c] Dielectric, kv	Cold[d] bend	Cut through[e] 120°C	Cracking test[f] Visual	Cracking test[f] Dielectric, kv
1	Polyester-coated Mylar tape	0.104	0.0045	Not applicable	Not applicable	....	Cracked	No test	Passed	120 hr	Passed 3X	3–4
2	Polyethylene-coated Mylar tape	0.092	0.005	Not applicable	Not applicable	....	Passed	4–5	Passed	120 hr	Passed 3X	4–5
3	Polyethylene-coated Mylar tape	0.095	0.006	Not applicable	Not applicable	....	Passed	6.5–7.5	Passed	120 hr	Passed 3X	3.5–4
4	Polyethylene-coated Mylar tape	0.112	0.005	Not applicable	Not applicable	....	Cracked	No test	Passed	120 hr	Passed 3X	Failed up to 9X
5	Kynar jacket	0.092	0.005	Not applicable	Not applicable	....	Passed	9.5	Passed	120 hr	Passed 5X	4
6	NAS-702-22SC9 nylon jacket	0.090	0.0055	Not applicable	Not applicable	....	Cracked	No test	Passed	120 hr	Failed up to 9X	No test
7	NAS-702-20SC9 nylon jacket	0.107	0.005	Not applicable	Not applicable	....	Cracked	No test	Passed	120 hr	Failed up to 9X	No test
8	Caprolactam nylon jacket	0.125	0.0075	Not applicable	Not applicable	....	Passed	4.5–5	Failed	6 min	Failed up to 9X	No test
9	Plasticized nylon jacket	0.118	0.009	Not applicable	Not applicable	....	Cracked	No test	Passed	120 hr	Passed 9X	5.5–6
10	Mylar wrap 0.002 nylon jacket	0.116	0.0075	Not applicable	Not applicable	....	Cracked	No test	Passed	120 hr	Failed up to 9X	No test
11	NAS-702 18SC7 nylon jacket	0.112	0.008	Not applicable	Not applicable	....	Cracked	No test	Passed	120 hr	Failed up to 9X	No test
12	2 cond. cable nylon jacket	0.129	0.010	Not applicable	Not applicable	....	Cracked	No test	Passed	120 hr	Failed up to 9X	No test

4-26

#	Description											
13	Tough PVC jacket	0.077	0.010	Not applicable	Passed	11.5–14	Not applicable	......	Passed	5 min	Passed	8–8.5
14	Polyurethane (Estane) jacket[a]	0.145	0.010	Jkt. softened and flowed	Passed	5.5–10	Not applicable	......	Passed	5 min	Jkt. softened flowed	No test
15	PVC 0.008 in. Nylon 0.003 in.	0.124	0.011	Not applicable	Not applicable	......	Passed	11–12.5	Passed	120 hr	Passed 3X	10–11.5
16	22SCI PVC jacket	0.108	0.012	Failed	Passed	2–15	Not applicable	......	Passed	5 min	Passed 3X	5–6.5
17	20SCI PVC jacket	0.114	0.012	Passed	Passed	16–17.5	Not applicable	......	Passed	5 min	Passed 3X	5–6
18	Tough PVC(B)	0.146	0.015	Passed	Passed	14	Not applicable	......	Passed	5 min	Passed 3X	6–7.5
19	Tough PVC (A')	0.153	0.016	Passed	Passed	12.5–13.5	Not applicable	......	Passed	5 min	Passed 3X	11.5–12
20	Kynar jacket 3 cond.	0.143	0.021	Not applicable	......	......	Passed	......	Passed	120 hr	Passed 3X	15–20

[a] Conducted in accordance with MIL-C-27500.
[b] Conducted in accordance with MIL-C-27500, with the addition of 1,000 volts (rms) for 1 min, then raised to breakdown.
[c] Conducted in accordance with MIL-C-27500–dielectric breakdown noted.
[d] Conducted in accordance with MIL-C-27500.
[e] Specimen subjected to 120°C oven temperature, 800-g load, over a 0.052-in. mandrel, 1,000 volts, rms applied continuously–time of breakdown noted. Test discontinued at 120 hr.
[f] Specimen conditioned 24 hr at 150°C wrapped around mandrel 3, 5, 7, and 9 times cable diameter, cooled in a desiccator, cable straightened and visually inspected, then subjected to 1,000 volts rms for 1 min, raised to breakdown after 1 min.
[g] Estane is a trademark of B. F. Goodrich Chemical Co., Columbus, Ohio.

SOURCE: Martin Marietta Corporation.

4-28 Wires and Cables

temperature rating of FEP must be kept in mind; the material should not be used above 200°C if subject to any appreciable stress.

Kynar. This is a good, tough jacket material with mechanical resistance and superior crack resistance comparable to those of nylon.

Polyethylene-coated Mylar. Excellent for shield jacketing and applied by taping, this causes minimum size buildup. Some problems are encountered in obtaining a good seal over irregular multiconductor cores.

Dacron° Braid. This has good temperature resistance (150°C) and good abrasion resistance. Braid construction is flammable and prone to moisture absorption.

Glass Braids. Glass braids are utilized for high-temperature, nonflammable applications. Glass braids have poor abrasion resistance and are prone to fraying because satisfactory lacquering is difficult to achieve. Among glass-braid constructions, Teflon-coated glass braids appear most satisfactory where individual fiber strands are coated with a Teflon dispersion, and then braided and cured to form a more homogeneous nonfraying construction.

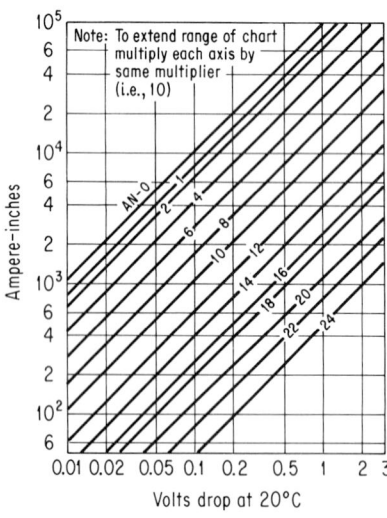

Fig. 7 Dc voltage drop of copper wire.[8]

DESIGN CONSIDERATIONS

Wire-gage selection The following factors must be considered for proper wire-gage selection:
1. Voltage drop
2. Current-carrying capacity
3. Circuit-protector characteristics

Voltage drop is a major factor in low-voltage systems, except where leads are very short. At high ambient temperatures or high-voltage installations current-carrying capacity controls selection. If several loads are supplied by a single protector, the circuit protector becomes the significant factor. If voltage drop is the major consideration, a single wire should be used to save weight; but if current-carrying capacity is significant, two or more parallel wires will generally weigh less than a single wire of the same total current-carrying capacity. Parallel wires should be of the same gage and length for even current split.

Voltage Drop. Voltage-drop calculations should be based on anticipated load current at nominal system voltage. Voltage drop through an aluminum structural ground return can be considered zero for all practical purposes. Normal voltage-drop limits do not apply to starting currents of equipment such as motors. The voltage at load during start-up should be considered to ensure proper operation of equipment. Figure 7 may be used to select wire gage. Ampere-inches are the product of wire length between terminations in inches and wire current in amperes.

Current-carrying Capacity. The following factors must be considered in determining the current-carrying capacity of a wire:
1. Continuous-duty rating
2. Short-time rating
3. Effect of ambient temperature
4. Effect of wire grouping
5. Effect of altitude

CONTINUOUS-DUTY RATING: This applies if a wire is to carry current for 1,000 sec or more. The continuous current-carrying capacity of copper and aluminum wire in amperes for aerospace applications is shown in Table 11. The following

° Trademark of E. I. du Pont de Nemours & Co., Wilmington, Del.

TABLE 11 Maximum Current Capacity, Amperes

Size, AWG	MIL-W-5088 Copper Single wire	MIL-W-5088 Copper Wire bundled	MIL-W-5088 Aluminum Single wire	MIL-W-5088 Aluminum Wire bundled	National Electrical Code	Underwriters Laboratory +60°C	Underwriters Laboratory +80°C	American Insurance Association	500 cir mils/amp
30	...	...	...	...	...	0.2	0.4	...	0.20
28	...	...	...	...	...	0.4	0.6	...	0.32
26	...	...	...	...	...	0.6	1.0	...	0.51
24	...	5	...	...	...	1.0	1.6	...	0.81
22	9	7.5	...	...	...	1.6	2.5	...	1.28
20	11	10	...	...	...	2.5	4.0	3	2.04
18	16	13	...	...	6	4.0	6.0	5	3.24
16	22	17	...	...	10	6.0	10.0	7	5.16
14	32	23	...	...	20	10.0	16.0	15	8.22
12	41	33	...	...	30	16.0	26.0	20	13.05
10	55	46	58	36	35	...	...	25	20.8
8	73	60	86	51	50	...	...	35	33.0
6	101	80	108	64	70	...	...	50	52.6
4	135	100	149	82	90	...	...	70	83.4
2	181	125	177	105	125	...	...	90	132.8
1	211	150	204	125	150	...	...	100	167.5
0	245	175	237	146	200	...	...	125	212.0
00	283	200	...	...	225	...	...	150	266.0
000	328	200	...	...	275	...	...	175	336.0
0000	380	225	...	...	325	...	...	225	424.0

SOURCE: Martin Marietta Corporation.

4-30 Wires and Cables

criteria apply to the ratings in the table:
1. Abient temperature is
   57.2°C (135°F) for 105°C insulated wire
   92°C (197.6°F) for 135°C insulated wire
   107°C (225°F) for 150°C insulated wire
   157°C (315°F) for 200°C insulated wire
2. "Wire bundled" indicates 15 or more wires in a group.
3. The sum of all currents in a bundle is not more than 20 percent of the theoretical capacity of the bundle, which is calculated by adding up the bundle ratings of the individual wires.

SHORT-TIME CURRENT RATINGS: These apply when a wire is to carry a current for less than 1,000 sec. The short-time rating is generally applicable to starter lead applications. Figure 8 presents curves for various wire gages in harness (bundle).

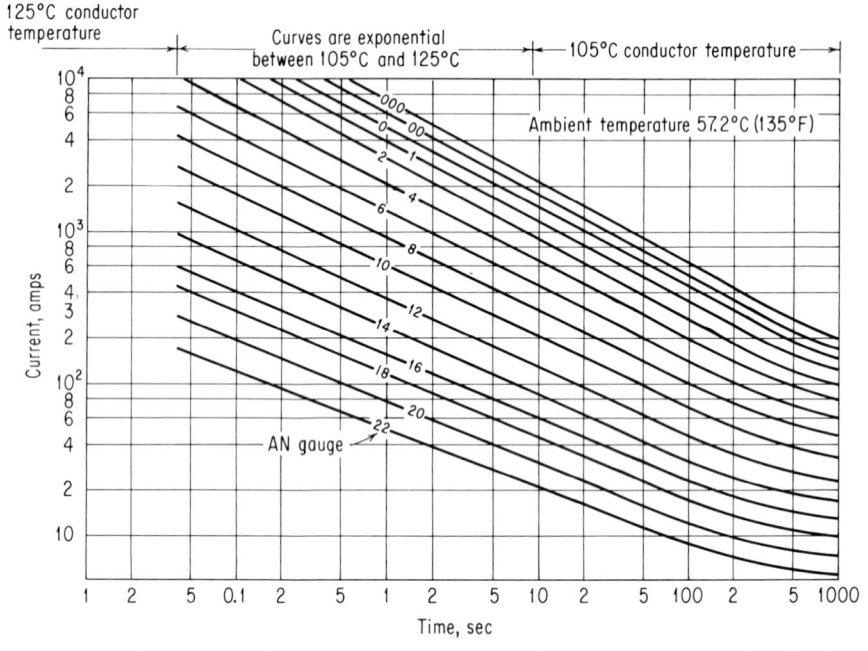

Fig. 8 Short-time working curves for 105°C insulated copper wire in bundles.[8]

TEMPERATURE-CURRENT RELATIONSHIP: The following equation provides a means of rerating the current-carrying capacity of wire and cable at any anticipated ambient temperature:

$$I = I_r \sqrt{\frac{t_c - t}{t_c - t_r}}$$

where I = current rating in ambient temperature t
  I_r = current rating in rated ambient temperature t_r (Table 11)
  t = required ambient temperature
  t_r = rated ambient temperature (continuous-duty rating, above)
  t_c = temperature rating of insulated wire or cable

Figures 9 and 10 present curves showing the effect of temperature on the current-carrying capacity of copper wire with 10- and 15-mil insulation, respectively.

EFFECTS OF WIRE GROUPING: When wires are grouped (bundled, harnessed) together, their current ratings must be reduced because of restricted heat loss. Table 11 and Fig. 6 take into account reduced ratings based on grouping 15 or more wires. If a harness of wire bundle contains fewer than 15 wires, the allowable capacity may be increased toward the rating of a single wire in free air. In grouping wires, it is good practice to use wires of the same gage in one bundle. Grouping wires of widely differing gages sacrifices the current capacity of smaller gages.

EFFECTS OF ALTITUDE: Air density decreases with increasing altitude. Since lower density reduces the dielectric properties of air, trapped air between conductors and insulators presents a problem. In addition, air is retained in voids that occur primarily in stranded conductors, although voids cannot be eliminated from solid insulated wire either. A direct effect of increased altitude on insulated wire is lower corona threshold, resulting in lower peak operating voltage. Wire insulated with organic materials should always be operated below the corona extinction voltage, as corona has a degrading effect on all organic materials. Operating below corona threshold but above corona extinction is risky, for a surge may start corona, which will continue until the voltage is lowered to the extinction level.

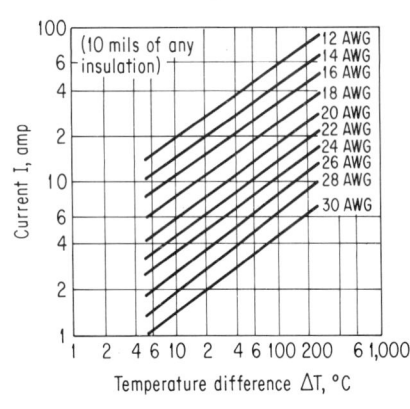

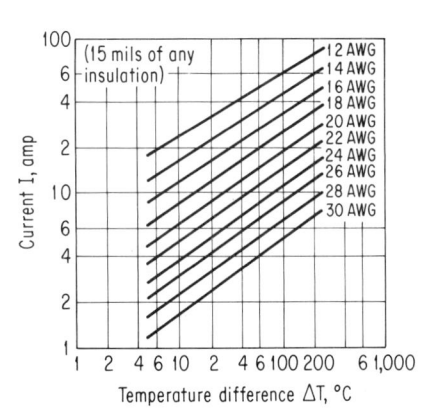

Fig. 9 Effects of temperature on current-carrying capacity of copper conductors (10-mil insulation).[9]

Fig. 10 Effects of temperature on current-carrying capacity of copper conductors (15-mil insulation).[9]

Figure 11 presents curves on insulation breakdown voltage of air as a function of altitude. In addition to proper derating of operating voltage on insulated wires and cables, derating of termination spacing in air must be considered (see Table 28).

Circuit-protector Characteristics. The circuit protector must be selected carefully. The circuit-protector rating must be low enough to protect the smallest gage wire connected to it against damage from overheating, smoke, or fire from short circuits.

Insulation-material selection Factors governing selection of an optimum insulation material cover a wide range and are less theoretical in nature than those for wire-gage selection. In addition to meeting specific system electrical and environmental requirements, related fabrication techniques must be considered. The insulation must withstand mechanical abuse and the heat from soldering or application of associated shrinkable devices. Furthermore, the insulation must be compatible with encapsulants, potting compounds, conformal coating, and adhesives.

Following is a designer's checklist for insulation selection.

INTERCONNECTION AND HOOKUP WIRE

Definitions For purposes of this discussion, the difference between interconnection and hookup wire is determined by the amount of mechanical stress applied

4-32 Wires and Cables

Insulation Selection Checklist

Requirement	Considerations
1. Environment:	
a. Temperature extremes:	
(1) Continuous operating	Refer to Tables 5 and 7.
(2) Short-term operating	May require test that simulates specific application.
(3) Fabrication temperature	Check for soldering iron resistance in high-density packaging; cure temperatures of encapsulant; compatibility with shrinkable devices, if employed.
(4) Storage	Check for embrittlement, long-term storage, low-humidity conditions.
b. Altitude:	
(1) Outgassing	Weight loss, smoke, condensation.
(2) Corona	Maintain voltage below corona level, especially with insulations susceptive of erosion.
c. Radiation	Refer to Table 7.
d. Weather	Moisture resistance, aging, ultraviolet radiation.
e. Flame	Refer to Table 7.
f. Fluids	Refer to Table 7.
2. Electrical:	
a. Capacitance	$C = \dfrac{7.36K}{\log(10D/d)}$ where C = capacitance, pf/ft K = dielectric constant (Table 6) D = insulated wire diameter, in. d = conductor diameter, in.
b. Dielectric strength	Refer to Table 6.
c. Volume resistivity	Refer to Table 6.
d. Loss factor	Refer to Table 6.
3. Mechanical:	
a. Installation and handling	Check for minimum bend radius, special tooling, clamping, stresses, chaffing. Refer to Table 5 for abrasion, cut-through, and mechanical resistance.
b. Operating	Refer to Table 5.
4. Size	Refer to applicable specification for outside dimensions.
5. Weight	Refer to applicable specification for maximum weight. If not listed, use the following equation for insulation weight. $W = \dfrac{D \times d^2}{2} \times K \times G$ (lb/1,000 ft) where D = diameter over insulation, in. d = diameter over conductor K = 680 G = specific gravity of insulation (Table 5)

to the wire. Interconnection wire is used to connect electric circuits between pieces of equipment and must withstand rough handling, the abrasion of pulling through conduit, and potential accidental damage during installation that results from the slip impact of hand tools. Hookup wire is used to connect electric circuits within a unit of equipment or "black box." This type of equipment may range from miniaturized airborne equipment to a massive ground installation. With today's emphasis on miniaturization, many applications require the use of hookup wire where interconnection wire was used in the past.

Interconnection and Hookup Wire 4-33

Interconnection wire As stated, interconnection wire is by nature bulky; it must withstand severe mechanical abuse; and, since it has major application in the aircraft industry, it must have minimum weight and size to allow greater payload. Insulation suppliers and wire manufacturers maintain a continuous quest to develop an interconnection wire insulation achieving the primary requisite of mechanical resistance with minimum bulk and weight.

In addition to the development of higher-strength insulation materials, innovations such as the addition of reinforcing mineral fillers to existing resin systems have proved feasible (MS-17411, MS-17412, MS-1800, and MS-18001). The addition of filler materials enhances the abrasion resistance of the basic resin system without wire diameter increase and with minimum penalty to weight. Table 12 presents a listing of interconnection wires most widely used in industry. The tabulation is made on the basis of 22 AWG insulated conductor so that weight and size can be analyzed and compared for the different types. Figure 12 presents thermal life curves for the five Military Specification wires and one nonmilitarized wire

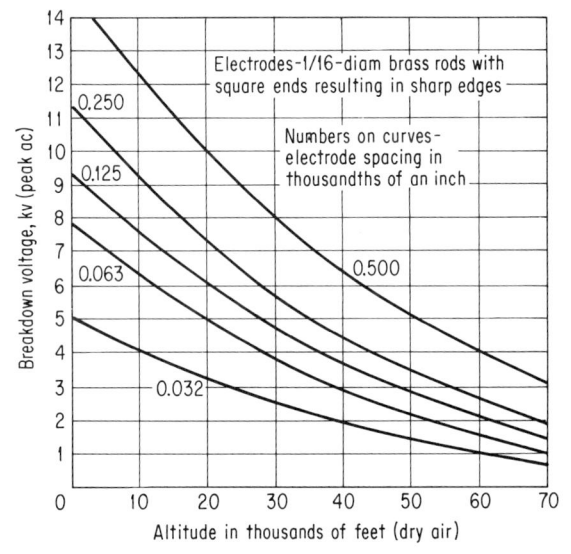

Fig. 11 Average breakdown voltage versus altitude. (*Martin Marietta Corporation.*)

construction with 0.018-in. wall thickness of extruded polyvinylidene fluoride (Kynar). Figures 13 and 14 present thermal life curves on polyalkene insulated wire in accordance with MIL-W-81044/4, using different sets of mandrels and weights.

Hookup wire Although the name hookup wire is likely to become a catchall for all wire constructions, it represents the area where by far the highest footage of military and electronic wire is consumed. Under the demands of miniaturization, many aerospace and missile designs call for hookup wire rather than the heavier, bulkier interconnection wire.

The hookup wire field is the most dynamic in the wire and cable industry, as pointed out by A. H. Lybeck, "Hookup Wires Are the Proving Grounds for Any New Plastic or Elastameric Compound."[27] The hookup wire manufacturer must be cognizant of all newly developed resin systems, fabrication techniques, and their applicability as a wire insulation. In addition, the manufacturer must be adaptable to the many and varied user requirements ranging from a conventional back-panel application to the complicated environment of space flight. Working

TABLE 12 Interconnection Wire Data—Copper Conductor

Basic specification	Class type or MS No.	Size range, AWG	Conductor coating	Primary insulation	Jacket	Voltage rating, rms	Temperature rating, °C	Diameter rating, max.	Weight rating, max.
MIL-W-5086	MS-25190 Ty 1	22–12	Tin	PVC	Nylon	600	−55 to 105	1.0	1.0
	MS-25190 Ty 2	22–4/0	Tin	PVC	Glass nylon	600	−55 to 105	1.1	1.07
	MS-25190 Ty 3	22–4/0	Tin	PVC	Glass PVC nylon	600	−55 to 105	1.25	1.3
	MS-25190 Ty 4	22–16	Tin	PVC	Nylon	3,000	−55 to 105	1.46	1.64
MIL-W-8777	MS-25471	22–2/0	Silver	Silicone rubber	Dacron braid	600	−55 to 200	1.25	1.32
	MS-27110	22–4	Silver	Silicone rubber	FEP	600	−55 to 200	1.25	1.61
MIL-W-22759	MS-17331	22–8	Silver	TFE	Asbestos	600	−55 to 260	1.1	1.25
	MS-17332	22–8	Nickel	TFE	Asbestos	600	−55 to 260	1.1	1.25
	MS-17411	24–4	Silver	Reinforced TFE		600	−55 to 200	1.25	1.82
	MS-17412	24–4	Nickel	Reinforced TFE		600	−55 to 200	1.25	1.82
	MS-18000	24–4	Silver	Reinforced TFE		600	−55 to 200	1.04	1.36
	MS-18001	24–4	Nickel	Reinforced TFE		600	−55 to 200	1.04	1.36
	MS-90294	22–2/0	Silver	TFE and glass tape	Glass braid FEP	600	−55 to 200	1.07	1.3
MIL-W-81044	/1	24–4	Silver	Crosslinked polyalkene	Crosslinked Kynar	600	−55 to 135	1.0	0.98
	/2	24–4	Tin	Crosslinked polyalkene	Crosslinked Kynar	600	−55 to 135	1.0	0.98
	/5	24–0	Silver	Crosslinked polyalkene	Crosslinked Kynar	600	−55 to 150	1.0	0.925
	/6	24–0	Tin	Crosslinked polyalkene	Crosslinked Kynar	600	−55 to 150	1.0	0.925
	/7	26–20	Silver-high strength alloy	Crosslinked polyalkene	Crosslinked Kynar	600	−55 to 150	1.0	0.925
	/8	24–0	Silver	Crosslinked polyalkene	Crosslinked Kynar	600	−55 to 150	0.90	
	/9	24–0	Tin	Crosslinked polyalkene	Crosslinked Kynar	600	−55 to 150	0.90	
	/10	26–20	Silver-high strength alloy	Crosslinked polyalkene	Crosslinked Kynar	600	−55 to 150	0.90	
MIL-W-25038	MS-27125	22–4/0	Nickel clad	Asbestos glass TFE tape	Glass braid	600	−55 to 288	1.62	2.26
MIL-W-7139	Class 1	22–4/0	Silver	TFE and glass	Glass braid	600	−55 to 200	1.25	1.82
	Class 2	22–4/0	Nickel	TFE and glass	Glass braid	600	−55 to 260	1.25	1.82
MIL-W-81381	/3	26–2	Silver	Polyimide FEP film	FEP dispersion	600	−55 to 200	0.84	0.89
MIL-W-81381	/4	26–2	Nickel	Polyimide FEP film	FEP dispersion	600	−55 to 260	0.84	0.89

TABLE 12 Interconnection Wire Data—Copper Conductor (Continued)

Basic specification	Class type or MS No.	Duty rating	Cost rating	Avail-ability	Mechanical properties	Electrical properties	Chemical resistance	Solder	Bond-ability	Strip-pability	Marking
MIL-W-5086	MS-25190 Ty 1	M	1.0	RA	Fair	Fair	Good	Poor	Good	Good	Good
	MS-25190 Ty 2	M	1.5	RA	Fair	Fair	Good	Poor	Good	Fair	Good
	MS-25190 Ty 3	H	1.7	RA	Good	Fair	Good	Poor	Good	Fair	Good
	MS-25190 Ty 4	M	1.3	RA	Fair	Fair	Good	Poor	Good	Good	Poor
MIL-W-8777	MS-25471	H	9.6	LS	Good	Good	Fair	Fair	Good	Fair	Good
	MS-27110	H	9.1	LS	Good	Good	Good	Fair	Poor	Fair	Poor
MIL-W-22759	MS-17331	M	10.5	LS	Good	Good	Good	Good	Fair	Fair	Poor
	MS-17332	M	12.0	LS	Good	Good	Good	Good	Fair	Good	Poor
	MS-17411	H	10.0	RA	Excellent	Good	Excellent	Good	Poor	Good	Poor
	MS-17412	H	11.0	RA	Excellent	Good	Excellent	Good	Poor	Good	Poor
	MS-18000	M	7.8	RA	Good	Good	Excellent	Good	Poor	Good	Poor
	MS-18001	M	8.6	RA	Good	Good	Excellent	Good	Poor	Good	Good
	MS-90294	M	9.3	LS	Good	Good	Good	Good	Poor	Fair	Good
MIL-W-81044	/1	M	5.6	LS	Good	Good	Good	Fair	Good	Good	Good
	/2	M	4.8	LS	Good	Good	Good	Fair	Good	Good	Good
	/5	M	5.6	LS	Good	Good	Good	Fair	Good	Good	Good
	/6	M	4.8	LS	Good	Good	Good	Fair	Good	Good	Good
	/7	M	7.0	LS	Good	Good	Good	Fair	Good	Good	Good
	/8	L	5.4	LS	Good	Good	Good	Fair	Good	Good	Good
	/9	L	4.6	LS	Good	Good	Good	Fair	Good	Good	Good
	/10	L	6.8	LS	Good	Good	Good	Fair	Good	Good	Good
MIL-W-25038	MS-27125	H	15.0	RA	Excellent	Fair	Good	Good	Poor	Fair	Poor
MIL-W-7139	Class 1	H	12.2	RA	Excellent	Good	Good	Good	Fair	Fair	Fair
	Class 2	H	12.3	RA	Good	Good	Good	Good	Fair	Fair	Fair
MIL-W-81381	/3	M	6.0	RA	Good	Good	Excellent	Good	Fair	Fair	Fair
MIL-W-81381	/4	M	6.7	RA	Good	Good	Excellent	Good	Fair	Fair	Fair

L = light.   M = medium.   H = heavy.   RA = readily available.   LS = limited sources (fewer than 4).
SOURCE: Martin Marietta Corporation.

4-35

4-36 Wires and Cables

in such a fluid and nebulous environment imposes substantial burden on both user and manufacturer. Many requirements are necessarily established without the ability to thoroughly analyze either the requirement or the capability of a selected insulation to meet the required environment. The cost and availability of test equipment to simulate many user environments is for all practical purposes prohibitive to a wire manufacturer, and he must depend on user evaluation to determine the worthiness of his product.

Table 13 presents a summary of hookup wire types, including thin-walled insulations that have been covered in specifications.

In addition to the consideration of documented thin-walled insulations, certain applications may be conducive to the use of ultra-thin-walled insulation. Kapton Type F film, used in the construction of MIL-W-81381/1 through 4 wire, is composed of 1-mil polyimide film coated on one or two sides, as applicable with

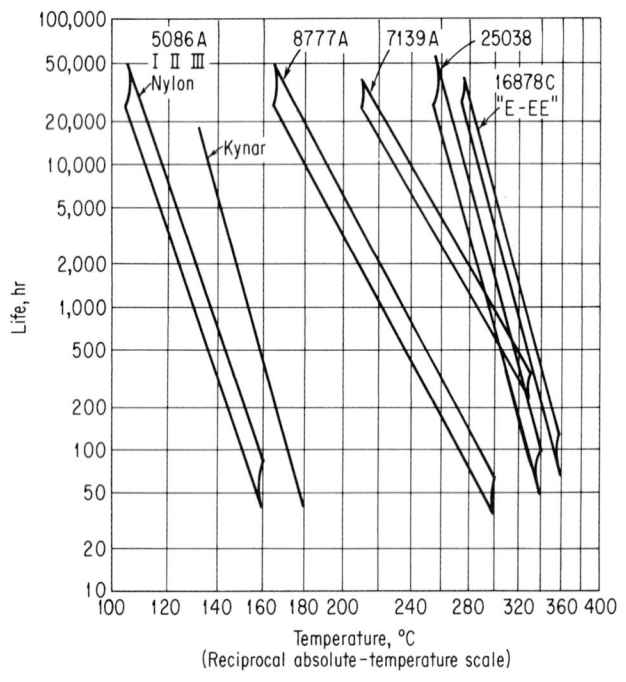

Fig. 12 Summary graph—ranges of life-temperature curves of MIL Specification wires.[10]

½ mil of FEP Teflon. The thin-walled wire construction using this Kapton film results in a nominal insulation thickness of 7.5 mils. Kapton Type XF film reduces the thickness to 5.0 mils. Figure 15 and Table 14 compare wire cross sections and weights of Kapton Type F and Type XF insulated wire with various other commonly used insulations of varying wall thickness. Figure 16 and Table 15 compare cut-through resistance and abrasion resistance of the previous insulation systems.

Another Kapton film variation has become available: Kapton 120XHD utilizing 0.1 mil of FEP coating on both sides of 1-mil polyimide film, used in a 5-mil nominal wall construction. This configuration offers greater cut-through and abrasion resistance than Type XF film because of the increased thickness of the stronger polymide film.

Interconnection and Hookup Wire 4-37

Kapton 120XHD film also offers the potential of an even thinner wall construction (3-mil nominal wall thickness) with a single-layer wrap construction. Use of the FEP coating on both sides of the polyimide film makes a single-layer construction with a good seal between overlaps feasible; and the potential disadvantage of poor strippability, caused by placing the FEP next to the conductor, is greatly reduced owing to the small amount of FEP used for sealing. Table 16 presents

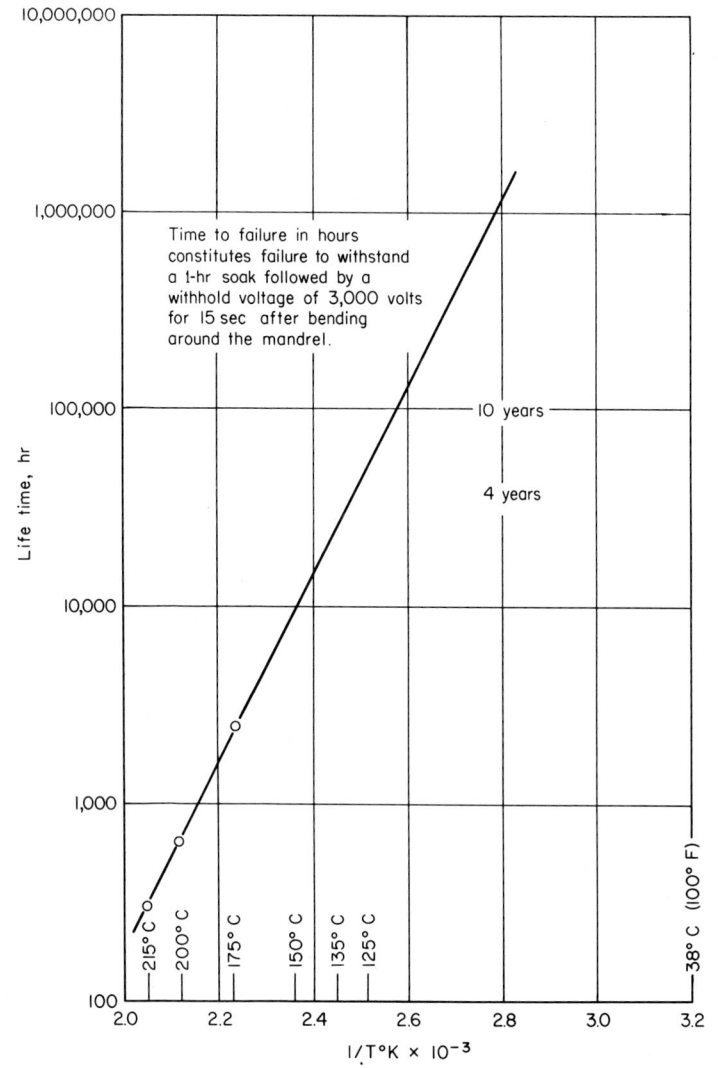

Fig. 13 Arrhenius plot of MIL-W-81044/4 wire (50D mandrel).[11]

a comparison of various significant properties of Kapton Types XF, 120 XD, and F constructions. The comparison is based on a two-tape, cross-wrapped, 50 percent minimum overlap construction.

Table 17 presents the properties and characteristics of a specialized hookup wire used with automated termination techniques, such as wire wrap and Termi-point.

4-38 Wires and Cables

These termination techniques require a special set of criteria. The major conductor size is 24 AWG with a nominal 10-mil wall of insulation. A 30 AWG conductor with nominal 5-mil wall of insulation is gaining increasing acceptance. Important considerations for automated termination wire are:

1. *Stiffness:* It is undesirable to have wire "take a set" and "pop up" on wiring panels.
2. *Cut-through:* To achieve a satisfactory wire wrap, termination pins have very sharp edges.
3. *Long lengths:* Long uninterrupted wire lengths are desirable to increase efficiency of operation.
4. *Strippability:* Machine automatically strips insulated wire. Wall thickness and concentricity must be controlled to close tolerances.

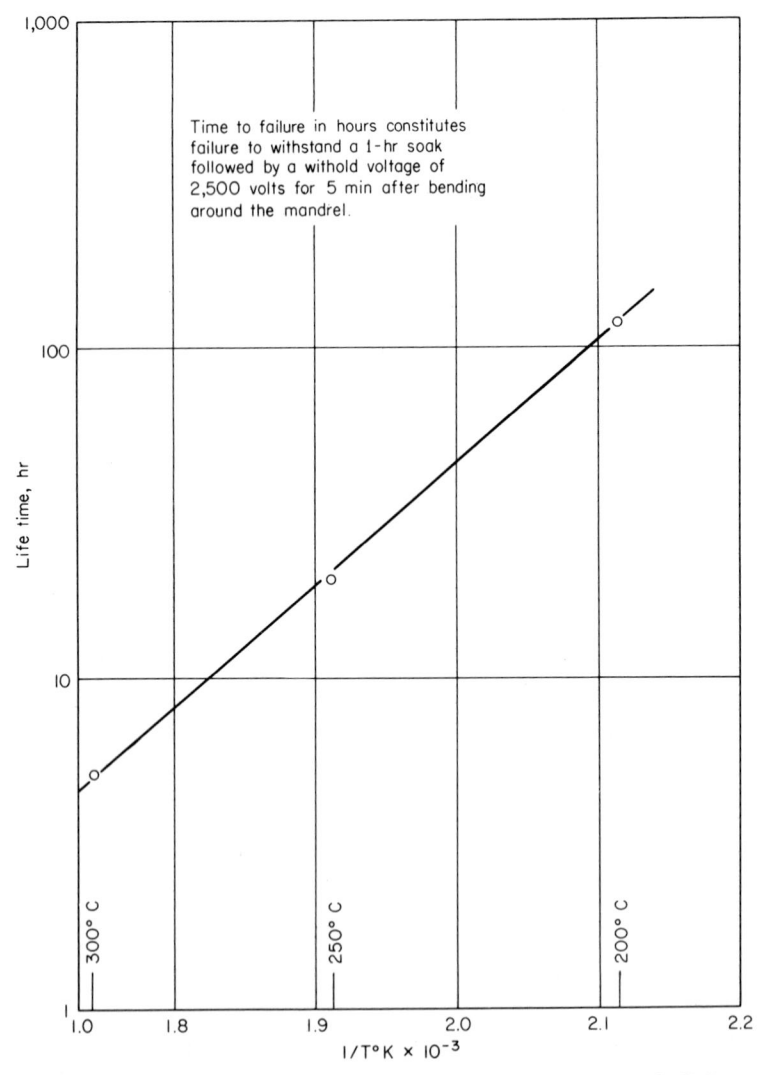

Fig. 14 Arrhenius plot of MIL-W-81044/4 wire (7D mandrel).[11]

TABLE 13 Hookup Wire Data—Copper Conductor

Basic specification	Type class or MS No.	Size range, AWG	Conductor coating	Primary insulation	Jacket material	Voltage rating, rms	Temperature rating, °C	Diameter rating	Weight rating	Duty rating
MIL-W-16878/1	Type B	32–14	Tin	PVC	....	600	−55 to 105	1.0	1.0	M
MIL-W-16878/1	Type B/N	32–14	Tin	PVC	Nylon	600	−55 to 105	1.15	1.15	H
MIL-W-16878/2	Type C	26–12	Tin	PVC	....	1,000	−55 to 105	1.28	1.22	M
MIL-W-16878/3	Type D	24–1/0	Tin	PVC	....	3,000	−55 to 105	1.81	1.88	M
MIL-W-16878/4A	Type E	32–10	Silver or nickel	TFE	....	600	−55 to 200 or 260	1.02	1.22	M
MIL-W-16878/5A	Type EE	32–8	Silver or nickel	TFE	....	1,000	−55 to 200 or 260	1.21	1.48	M
MIL-W-16878/6A	Type ET	32–20	Silver or nickel	TFE	....	250	−55 to 200 or 260	0.86	0.97	L
MIL-W-16878/7	Type F	24–4/0	Tin, silver, or nickel	Silicone rubber	....	600	−55 to 200	1.21	1.1	M
MIL-W-16878/8	Type FF	24–4/0	Tin, silver, or nickel	Silicone rubber	....	1,000	−55 to 200	1.83	1.75	M
MIL-W-16878/10A	Type J	24–4/0	Tin	Polyethylene	....	600	−55 to 75	1.13	0.98	M
MIL-W-16878/11	Type K	32–10	Silver	FEP	....	600	−55 to 200	1.02	1.22	M
MIL-W-16878/12	Type KK	32–8	Silver	FEP	....	1,000	−55 to 200	1.21	1.48	M
MIL-W-16878/13	Type KT	32–20	Silver	FEP	....	250	−55 to 200	0.86	0.97	L
MS-18104	MS-18104	28–12	Silver	TFE	Polyimide dip-coated	600	−55 to 200	0.98	1.15	H
MS-18105	MS-18105	28–12	Nickel	TFE	Polyimide dip-coated	600	−55 to 260	0.98	1.15	H
MIL-W-22759	MS-21985	28–12	Silver	TFE	....	600	−55 to 200	0.98	1.24	M
MIL-W-22759	MS-21986	28–12	Nickel	TFE	....	600	−55 to 260	0.98	1.24	M
MIL-W-22759	MS-18113	28–8	Silver	TFE	....	1,000	−55 to 200	1.17	1.48	M
MIL-W-22759	MS-18114	28–8	Nickel	TFE	....	1,000	−55 to 260	1.17	1.48	M
MIL-W-81044	/3	30–12	Silver	Crosslinked polyalkene	Crosslinked Kynar	600	−55 to 135	0.92	0.99	H
MIL-W-81044	/4	30–12	Tin	Crosslinked polyalkene	Crosslinked Kynar	600	−55 to 135	0.92	0.99	H
MIL-W-81044	/11	30–12	Silver	Crosslinked polyalkene	Crosslinked Kynar	600	−55 to 150	0.92	0.99	H
MIL-W-81044	/12	30–12	Tin	Crosslinked polyalkene	Crosslinked Kynar	600	−55 to 150	0.92	0.99	H
MIL-W-81044	/13	30–20	Silver	Crosslinked polyalkene	Crosslinked Kynar	600	−55 to 150	0.92	0.99	H
MIL-W-81381	/1	26–10	Silver	Polyimide/FEP film	FEP dispersion	600	−55 to 200	0.95	1.0	H
MIL-W-81381	/2	26–10	Nickel	Polyimide/FEP film	TFE dispersion	600	−55 to 260	0.95	1.0	H

4-39

TABLE 13 Hookup Wire Data—Copper Conductor (Continued)

Basic specification	Type class or MS No.	Cost rating	Avail- ability	Mechanical properties	Electrical properties	Chemical properties	Solder iron resistance	Solder- ability	Strip- pability	Marking
MIL-W-16878/1	Type B	1.0	RA	Poor	Fair	Fair	Poor	Good	Good	Good
MIL-W-16878/1	Type B/N	1.2	RA	Good	Fair	Good	Poor	Good	Good	Good
MIL-W-16878/2	Type C	1.1	RA	Fair	Fair	Fair	Poor	Good	Good	Good
MIL-W-16878/3	Type D	1.5	RA	Fair	Fair	Fair	Poor	Good	Good	Good
MIL-W-16878/4A	Type E	6.4	RA	Fair	Excellent	Excellent	Excellent	Poor	Fair	Poor
MIL-W-16878/5A	Type EE	8.8	RA	Fair	Excellent	Excellent	Excellent	Poor	Fair	Poor
MIL-W-16878/6A	Type ET	6.4	RA	Poor	Excellent	Excellent	Excellent	Poor	Fair	Poor
MIL-W-16878/7	Type F	6.4	LS	Fair	Good	Poor	Fair	Fair	Good	Fair
MIL-W-16878/8	Type FF	9.0	LS	Fair	Good	Poor	Fair	Fair	Good	Fair
MIL-W-16878/10A	Type J	1.1	RA	Poor	Excellent	Good	Poor	Poor	Good	Good
MIL-W-16878/11	Type K	5.5	RA	Poor	Excellent	Excellent	Poor	Poor	Good	Fair
MIL-W-16878/12	Type KK	7.8	RA	Poor	Excellent	Excellent	Poor	Poor	Good	Fair
MIL-W-16878/13	Type KT	5.3	RA	Poor	Excellent	Excellent	Poor	Poor	Good	Fair
MIL-W-22759	MS-18104	16.1	LS	Good	Excellent	Excellent	Excellent	Fair	Fair	Poor
MIL-W-22759	MS-18105	18.1	LS	Good	Excellent	Excellent	Excellent	Fair	Fair	Poor
MIL-W-22759	MS-21985	6.2	RA	Fair	Excellent	Excellent	Excellent	Poor	Fair	Poor
MIL-W-22759	MS-21986	6.2	RA	Fair	Excellent	Excellent	Excellent	Poor	Fair	Poor
MIL-W-22759	MS-18113	8.9	RA	Fair	Excellent	Excellent	Excellent	Poor	Fair	Poor
MIL-W-22759	MS-18114	9.2	RA	Fair	Excellent	Excellent	Excellent	Poor	Fair	Poor
MIL-W-81044	/3	5.1	LS	Good	Good	Good	Fair	Good	Good	Good
MIL-W-81044	/4	4.2	LS	Good	Good	Good	Fair	Good	Good	Good
	/11	5.1	LS	Good	Good	Good	Fair	Good	Good	Good
	/12	4.2	LS	Good	Good	Good	Fair	Good	Good	Good
	/13	5.6	LS	Good	Good	Good	Fair	Good	Good	Good
MIL-W-81381	/1	8.3	RA	Good +	Good	Excellent	Good	Fair	Fair	Fair
MIL-W-81381	/2	9.0	RA	Good +	Good	Excellent	Good	Fair	Fair	Fair

L = light. M = medium. H = heavy. RA = readily available. LS = limited sources (fewer than 4).
SOURCE: Martin Marietta Corporation.

Interconnection and Hookup Wire 4-41

Outer-space applications Hookup wire configurations lend themselves to outer-space usage because of the prime importance of weight, which directly affects the useful load that can be carried into outer space by a vehicle. The requirements for an outer-space wire are directly related to the performance of the insulation, being the element most susceptible to the applicable environmental extremes. Proper selection of insulated wire requires that the environment be defined for a specific vehicle, mission, and trajectory.

Insulation	Wall thick, mils	Space-wire cross section 10^{-4} in.^{2}
Conductor AWG #22	—	
Kapton type XF	5.0	
Kapton type F	7.5	
MIL-W-81044/3	9.2	
Crosslinked polyalkene	10.0	
Crosslinked polyalkene	13.5	
Silicone rubber	17.0	
MIL-W-81044/1	18.5	

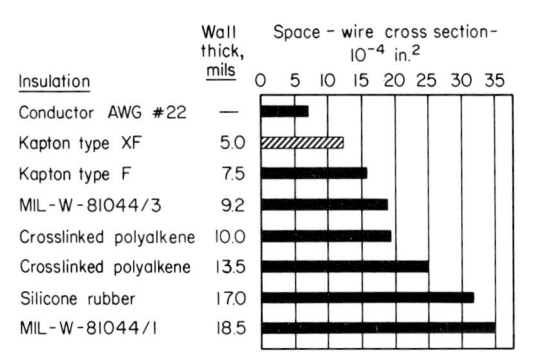

Fig. 15 Wire cross-section comparison.[12]

Environment. The environment discussed here will be limited to those factors peculiar to space flight. The ability of an insulated wire to meet previously discussed criteria such as abrasion, handling, vibration, and shock are considerations common to all applications and not limited to space usage. The principal space-environmental conditions that may have damaging effects in wire insulations are temperature, pressure, and radiation. Figure 17 presents a composite picture of the space environment. The extent of environmental extremes that wiring may be subject to will depend on the specific design of the vehicle and the location of material with the vehicle.

TABLE 14 Wire Weight Comparison[12]

Insulation	Wall thickness, mils	Weight, lb/1,000 ft
Kapton Type XF	5.0	2.75
MIL-W-81044/3	9.2	2.95
Kapton Type F	7.5	2.98
Crosslinked polyalkene	10.0	3.27
Crosslinked polyalkene	13.5	3.64
Silicone rubber	17.0	3.81
MIL-W-81044/1	18.5	4.10

Design Considerations. TEMPERATURE ENVIRONMENT: If insulated wire is to be utilized in non-environmental-controlled areas Teflon TFE, FEP, and Kapton insulations offer the broadest range of temperature resistance. These insulations will withstand 180° bending at −184°C without insulation fracture, and maximum continuous temperatures of 260°C for TFE and 200°C for Kapton and FEP. Crosslinked polyolefin insulation can be utilized at −184°C; however, it will not take as severe bending and its temperature rating is 135°C.

PRESSURE ENVIRONMENT: The effect of extreme high vacuum on wire insulations manifests itself primarily as an initial weight loss due to loss of water and absorbed gases. A weight-loss rate that does not rapidly approach zero is an important

4-42 Wires and Cables

indication of long-term continuous weight loss and slow volatization of potential, undesirable materials that may eventually degrade both the physical and electrical properties of the wire insulation or seriously effect other equipment in the vehicle. This secondary effect of insulation volatization of condensable materials is extremely significant to any vehicle with optical systems that may fog up as a result of

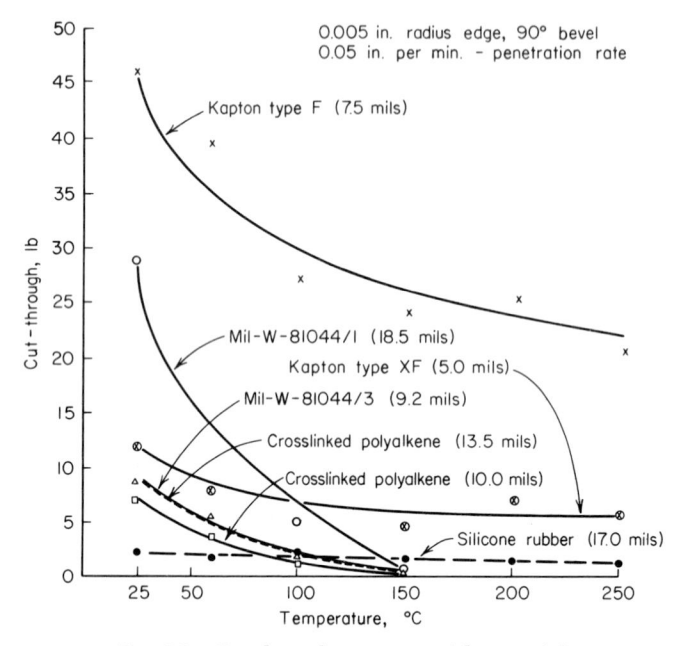

Fig. 16 Cut-through resistance (dynamic).[12]

TABLE 15 Abrasion Resistance Comparison—G. E. Scrape Abrader[12]

Insulation	Wall thickness, mils	Weight, lb, to produce failure in 3–4 scrapes, 0.016-in. OD needle
Kapton Type F............	7.5	5.75
MIL-W-81044/1...........	18.5	5.75
Crosslinked polyalkene......	13.5	3.50
Kapton Type XF..........	5.0	3.00
MIL-W-81044/3...........	9.2	3.00
Crosslinked polyalkene......	10.0	3.00
Silicone rubber............	17.0	0.75

the release of condensable contaminants into the atmosphere. Table 18 presents a comparison of percent weight loss for various insulation materials under varying temperatures and times. For comparative purposes this table has been compiled from the results published in Refs. 16 and 17, and additional data obtained from E. I. du Pont on Kapton. Novathene is a specially formulated radiation crosslinked polyalkene system designed for space applications. The data indicate low weight loss experienced in the Teflon and Kapton; Novathene exhibits somewhat higher weight loss due to greater absorbed water and gases.

Interconnection and Hookup Wire 4-43

RADIATION ENVIRONMENT: All space vehicles are subjected to some form of radiation, solar radiation, Van Allen belt, and potential use of a nuclear power source. The effects of solar radiation appear to be negligible, and degradation of insulation material is a secondary thermal effect arising from the absorption of electromagnetic energy.

The radiation resistance of materials varies widely; Fig. 18 presents a general broad spectrum of the effects of radiation on man, electronic components, organics, and inorganics. Of all the organic materials used as a wire insulation, Teflon TFE shows the lowest resistance to radiation. The threshold dose of TFE resin in air is approximately 7×10^4 rads. At 1×10^6 rads, the tensile strength is about 50 percent of original and the elongation less than 5 percent of original. In the

TABLE 16 Kapton Film Comparison—AWG No. 22 (19/34) Unilay Conductor[13]

Property	2 tapes, Type XF	Cross-wrapped, 120 XHD	+50% lap, Type F
Finished wire, OD, in.	0.040	0.040	0.044
Wall thickness, in.	0.005	0.005	0.007
Weight, lb/1,000 ft.	2.75	2.67	2.90
Tank test at 2.5 kv.	Pass	Pass	Pass
Dielectric strength, kv.	16	15	23
Shrinkage at 250°C in 6 hr, in.	0	0.04	0.05
Thermal shock, in.	0.02	0.04	0.06
Cold bend at −65°C.	Pass	Pass	Pass
Insulation resistance, ohms/1,000 ft.	3×10^{10}	5×10^{10}	1×10^{11}
Humidity resistance, ohms/1,000 ft.	2×10^8	1×10^9	8×10^9
Thermal aging for 500 hr at 250°C.	Pass	Pass	Pass
Flammability:			
After flame.	None	None	None
Length burned, in.	1.75	1.75	1.75
Static cut-through, °C (2,000 g, 0.005-in. radiused edge).	100	550	550
Dynamic cut-through, lb			
At 25°C.	12	48	46
At 1000°C.	5	42	26
At 150°C.	4	32	24
At 200°C.	7	26	26
Scrape abrasion, cycles (0.016-in. OD mandrel)			
3.5 lb.	......	23	37
3.0 lb.	......	39	57
2.5 lb.	15	71	141
2.0 lb.	46	107	363
Tensile strength of insulated wire, lb (conductor, 23 lb)	28	36	36
Printing.	O.K.	O.K.	O.K.
Stripping.		——— Under study ———	

absence of oxygen TFE shows significant improvement in retention of tensile strength and elongation. Figure 19 shows that in a vacuum the tensile strength of Teflon is reduced about 20 percent, and the elongation approximately 25 percent, at a dose of 10^6 rads. FEP Teflon reacts somewhat differently from TFE; crosslinking can occur, and when it is irradiated in the absence of oxygen with doses greater than 2.6×10^6 rads, some improvement of physical properties; yield stress, and deformation resistance can occur. In terms of elongation, an important characteristic for a flexible wire insulation, the radiation tolerance of FEP is approximately 10 times that of TFE.[18]

The effects of radiation on the electrical properties of TFE and FEP Teflon can be summarized as follows:

4-44 Wires and Cables

1. *Volume resistivity:* After threshold dose rate is reached, resistivity decreases rapidly until a dose of 5.5×10^5 rads, where equilibrium is reached. Equilibrium value is greater than 1×10^{12} ohm-cm for a 20-mil specimen.[18]

2. *Dielectric strength:* The dielectric strength of 3-, 5-, and 11-mil specimens showed no change at a dose of 5.7×10^7 rads.[18]

3. *Dielectric constant and dissipation factor:* The dielectric constant and dissipation factor of FEP Teflon are unaffected when irradiated to a dose of 8×10^6 rads by x-ray, in the absence of oxygen at measured frequencies[16] of 100 Hz to 100 kHz.

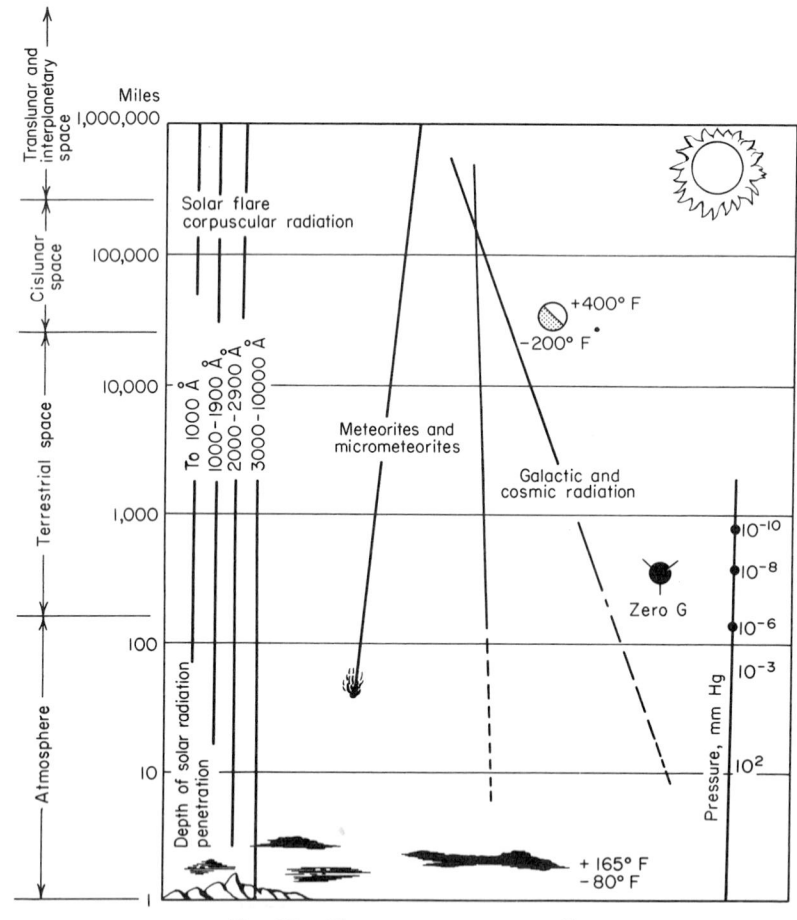

Fig. 17 The space environment.[14]

Figures 20 and 21 summarize the effects of x-ray irradiation on dielectric constant and dissipation factor of TFE Teflon.

The foregoing data indicate that Teflon insulation may be satisfactorily used in limited radiation environments and that the limiting level of usage may be increased if the material is kept in an oxygen-free environment. As denoted in Table 8, there are several insulation materials which exhibit resistance to radiation superior to that of Teflon; these insulations would be preferable in a radiation environment provided all other constraints, such as resistance to temperature ex-

TABLE 17 Automated-termination Wire Data—No. 30 AWG Conductor (Nominal 0.005 in. wall thickness)

Insulation material	TFE Teflon	TFE/ML	FEP/ML	Vinylidene fluoride	Polyethylene-coated Mylar	Polysulfone	Polyalkene + Kynar	Kapton
Conductor coating	Silver or nickel	Silver or nickel	Silver	Tin	Tin	Tin	Tin or silver	Silver or nickel
Temperature rating, °C	200 or 260	200 or 260	200	135	125	125	135	200 or 260
Cut-through resistance	Poor	Fair	Fair	Good	Good	Excellent	Good	Excellent
Abrasion resistance	Poor	Fair	Fair	Good	Good	Good	Good	Excellent
Dielectric constant	2.1	2.1	2.1	7.7	2.8	3.2	3.4	3.2
Flexibility (stiffness)	Good	Good	Good	Good	Good	Good	Good	Good
Chemical resistance	Good	Good	Good	Fair	Fair	Fair	Fair	Fair
Cost	Excellent	Excellent	Excellent	Fair	Good	Poor	Fair	Excellent
	Medium	Medium to high	Medium	Low	Low	Low	Low to medium	High
Availability	RA	LS	LS	RA	LS	LS	LS	DR
Long lengths	Poor	Poor	Fair	Good	Fair	Good	Good	Fair

RA = readily available. LS = limited sources (fewer than 3 manufacturers). DR = development required.
SOURCE: Martin Marietta Corporation.

4-46   Wires and Cables

tremes, minimum outgassing, and electrical and mechanical properties are satisfactory for the application.

TABLE 18   A Comparison of Gross Weight Loss in Ultrahigh Vacuum   (at 10^{-6} mm Hg)

Sample	Temperature, °F	Time in vacuum, hr	Weight loss, %*
Irradiated modified polyolefin	77	240	0.053
Polytetrafluoroethylene	77	240	0.006
Novathene	78	96	0.09
Irradiated modified polyolefin	122	138	0.30
Polytetrafluoroethylene	212	100	0.012
Novathene	212	200	0.66
Novathene	232	42	0.78
Polytetrafluoroethylene	250	120	0.018
Novathene	250	200	0.99
Irradiated modified polyolefin	300	240	1.30
Kapton	392	½	0.036
Kapton	392	30	0.036

* All data based upon insulation weight only.

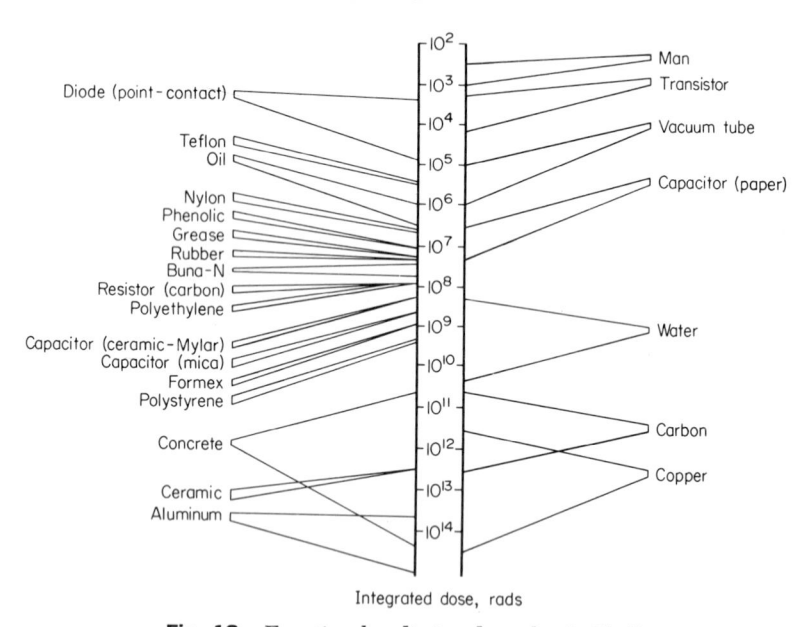

Fig. 18   Functional radiation-dose thresholds.[15]

COAXIAL CABLES

Design considerations   Coaxial cable consists of a center conductor, an insulation shield, and, usually, an outer jacket. It is essentially a shielded and jacketed wire. The term coaxial not only implies construction but also connotes usage at radio frequencies.

Background.   The purpose of a coaxial cable is to transmit radio-frequency energy from one point to another with minimum loss (attenuation). Loss of radio-fre-

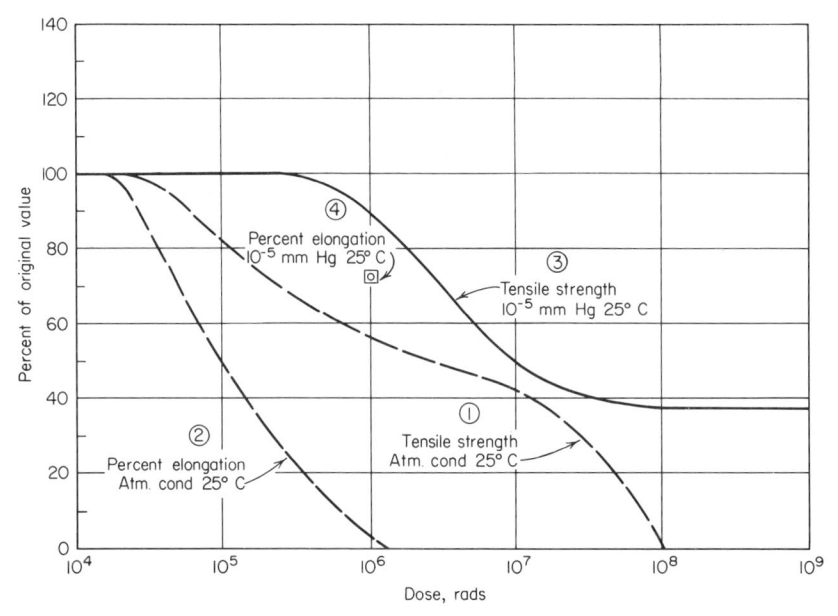

Fig. 19 Tensile strength and elongation—irradiated TFE resins.[16]

quency energy in a coaxial cable can occur (1) in the conductor, which is a power loss due to heating caused by currents passing through a finite resistance; and (2) in the dielectric, caused by the use of materials with high power factor. Unlike a shielded and jacketed wire used at low frequencies where the major loss is incurred in resistance of the conductor, high-frequency transmission invokes

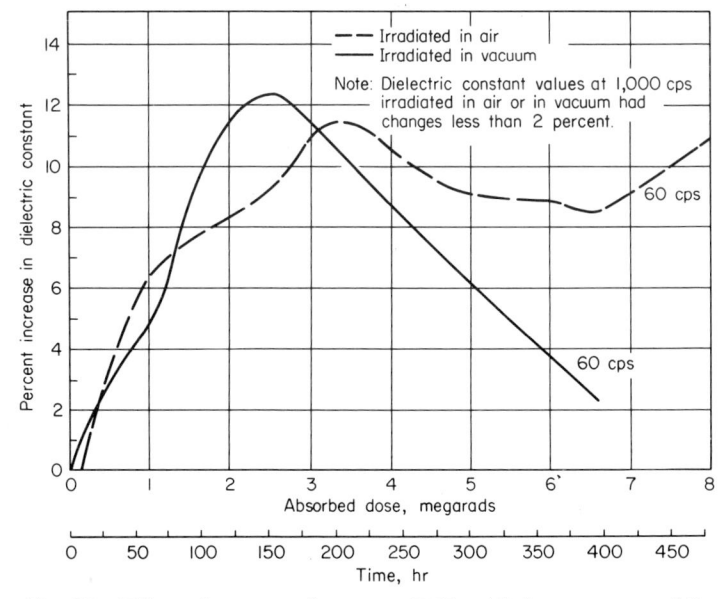

Fig. 20 Effect of x-ray irradiation on TFE-6 (dielectric constant).[16]

4-48 Wires and Cables

a phenomenon called *skin effect,* where currents travel on the outer surface (skin) of a conductor and partly through the adjacent insulating material; hence, loss in the insulation itself becomes more significant.

Electrical. In addition to loss, other important electrical characteristics of coaxial cables are velocity of propagation, impedance, capacitance, and corona extinction point. These are discussed in detail below.

VELOCITY OF PROPAGATION: Velocity is a function of the insulation dielectric constant, $V = 1/K$, and is expressed in percentage of the speed of light.

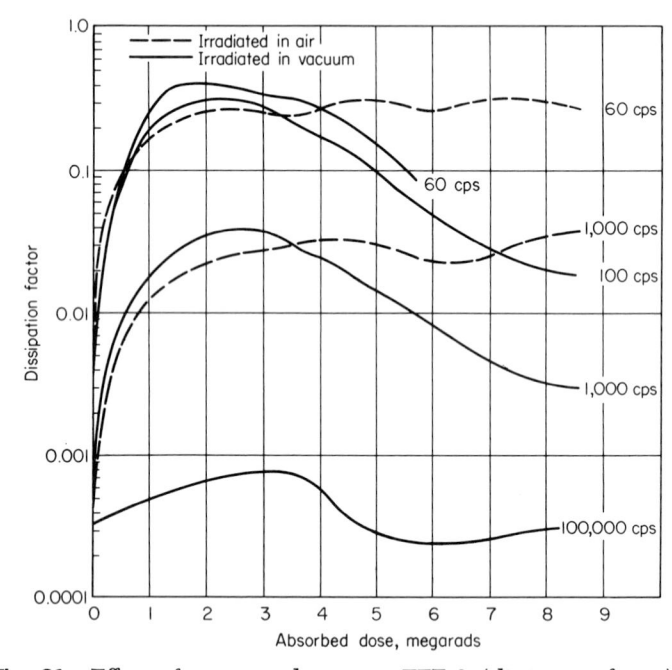

Fig. 21 Effect of x-ray irradiation on TFE-6 (dissipation factor).[16]

IMPEDANCE. The three common impedance values for coaxial cables are 50, 75, and 95 ohms. Impedance can be determined by the following formula:

$$Z_0 = \frac{138}{K} \log \frac{D}{d}$$

where Z_0 = characteristic impedance, ohms
D = diameter over the insulation, in.
d = diameter over the conductor, in.
K = dielectric constant of the insulation

CAPACITANCE: It is usually desirable to have minimum capacitance for minimum coupling and crosstalk. Capacitance, like impedance, is a logarithmic function of dimensions and is also dependent on the dielectric constant. The equation for calculating capacitance is

$$C = \frac{7.36K}{\log_{10}(D/d)}$$

where C = capacitance, pf/ft
D = diameter over the insulation
d = diameter over the conductor
K = dielectric constant of the insulation

CORONA EXTINCTION POINT: This determines the maximum voltage at which a coaxial cable may be operated. The corona extinction point of a cable is determined experimentally by gradually raising the voltage on a sample of cable until corona is detected, and then lowering the voltage until no further ionization is present. If the cable is operated consistently below this level, corona will not occur within the cable. Corona can cause noise at higher frequencies and eventual degradation of certain insulating materials.

Mechanical. The dependence of significant electrical properties, such as attenuation, capacitance, and impedance, on the relative size of conductor and insulation has ramifications in respect to the mechanical strength of coaxial cables. Attenuation can be reduced by increasing conductor size, which in turn forces an increase in insulation wall thickness if capacitance and impedance are to be maintained. An additional means of maintaining low capacitance with increased conductor size is through foamed or air dielectrics. Introduction of air bubbles in an insulation material or the use of air itself reduces the dielectric constant. The introduction of air into a solid insulation material such as polyethylene or FEP Teflon can reduce the dielectric constant to as low as 1.4.

Other coaxial constructions make use of a spiral thread or a web of insulation to center the conductor, which is essentially surrounded by air as the dielectric. From a mechanical standpoint, skin effect aids the coaxial-cable user in that he may use a stronger conductor core coated with a thin layer of copper without significant attenuation but with significant increase in mechanical strength. Copper-coated steel conductors are widely used in coaxial-cable applications. For applications where the cable is to be flexed, an annealed conductor is recommended for greater flex life.

Coaxial cables must be treated with care in handling and installation, since important electrical properties are dependent on conductor-to-insulation dimensions. Any flow or movement of the conductor can seriously affect electrical properties. For installation, the minimum allowable bend radius should be at least 10 times the cable diameter, in order to minimize stresses and preclude any cable deformation.

Environmental. Coaxial cables are normally fabricated with low-loss low-dielectric-constant insulation materials covered in the section on Wire and Cable Insulation. Teflon (TFE and FEP), PE, and irradiated PE, including foamed versions, are the most common, dielectric materials. For coaxial-cable application, one additional requirement is imposed on shield-jacket materials: they must be noncontaminating. Shield jackets must withstand the required environment without allowing any contamination of the dielectric which might affect its loss characteristics. Contamination is usually associated with PVC compounds, which contain plasticizers that could migrate into the dielectric material. In coaxial-cable applications, moisture resistance of the cable jacket is important. If water, which has a relatively high dielectric constant, penetrates the core, cable performance can be seriously affected.

Cable selection A guide to military coaxial-cable selection is MIL-HDBK-216. Specific cable types are documented in MIL-C-17, which covers requirements for approximately 150 different cable configurations. Power ratings of MIL-C-17 cable types are covered in Table 19. Table 20 presents nominal attenuation figures for MIL-C-17 cables at specific frequencies ranging from 1 to 10,000 MHz. MIL-C-23806 and MIL-C-22931 are specifications covering semiflexible cables with foamed dielectric and air-spaced dielectric.

MULTICONDUCTOR CABLES

Multiconductor cables fall into four major categories: airborne, ground electronics, ground support, and miscellaneous.

Airborne cables The primary design considerations for airborne multiconductor cables are size and weight. Airborne cables are often fabricated by a user who selects the appropriate interconnection or hookup wires, lays the insulated wires in a bundle or harness, then laces, spot-ties, and applies insulating tubing over

TABLE 19 Power Ratings of Coaxial Cable[4]

| RG/U cable | \multicolumn{9}{c}{Maximum input power rating, watts at frequencies, MHz} |||||||||
	1.0	10	50	100	200	400	1,000	3,000	5,000	10,000
5, 5A, 5B, 6, 6A, 212	4,000	1,500	800	550	360	250	150	65	50	25
7	4,100	1,550	810	540	370	250	140	70	50	30
8, 8A, 10, 10A, 213, 215	11,000	3,500	1,500	975	685	450	230	115	70	
9, 9A, 9B, 214	9,000	2,700	1,120	780	550	360	200	100	65	40
11, 11A, 12, 12A, 13, 13A, 216	8,000	2,500	1,000	690	490	340	200	100	60	
14, 14A, 74, 74A, 217, 224	20,000	6,000	2,400	1,600	1,000	680	380	170	110	40
17, 17A, 18, 18A, 177, 218, 219	50,000	14,000	5,400	3,600	2,300	1,400	780	360	230	
19, 19A, 20, 20A, 220, 221	110,000	28,000	10,500	6,800	4,200	2,600	1,300	620	410	
21, 21A, 222	1,000	340	160	115	83	60	35	15		
22, 22B, 111, 111A	7,000	1,700	650	430	290	190	110	50		
29	3,500	1,150	510	340	230	150	95	50	35	
34, 34A, 34B	19,000	7,200	2,700	1,650	1,100	700	390	140	80	
35, 35A, 35B, 164	40,000	13,500	5,500	3,800	2,500	1,650	925	370	210	
54, 54A	4,400	1,580	675	450	310	210	120	60	40	
55, 55A, 55B, 223	5,600	1,700	2,700	480	320	215	120	60	40	
57, 57A, 130, 131	10,000	3,000	1,250	830	570	370	205	95		
58, 58B	3,500	1,000	450	300	200	135	80	40	20	
58A, 58C	3,200	1,000	425	290	190	105	60	25	20	
59, 59A, 59B	3,900	1,200	540	380	270	185	110	50	30	
62, 62A, 71, 71A, 71B	4,500	1,400	630	440	320	230	140	65	40	15
62B	3,800	1,350	600	410	285	195	110	50	31	15
63, 63B, 79, 79B	8,200	3,000	1,300	1,000	685	455	270	130	75	35
87A, 116, 165, 166, 226, 227	42,000	15,000	6,250	4,300	3,000	2,050	1,200	620	480	250
94	62,000	15,500	5,900	4,300	2,900	1,900	1,400	650	480	200
94A, 226	64,000	18,000	9,600	6,800	4,600	3,300	1,750	775	540	250

4-50

Type										
108, 108A	1,300	360		145	100	70	45	30	15	5	15
114, 114A	5,300	1,350		475	345	230	150	85	40	25	170
115, 115A, 235	33,000	9,900		4,200	2,900	2,000	1,380	830	600	450	490
117, 118, 211, 228	200,000	66,000		25,000	19,000	12,800	8,500	4,800	2,200	1,400	490
119, 120	100,000	31,000		13,000	9,000	6,100	4,100	2,400	1,100	770	250
122	1,000	240		100	65	45	30	15	10	5	
125	8,500	2,300		910	620	435	285	165	75	45	
140, 141, 141A	19,000	6,300		2,700	1,700	1,200	830	450	220	140	65
142, 142A, 142B	19,000	5,700		2,600	1,800	1,300	900	530	265	175	100
143, 143A	26,000	8,700		3,750	2,600	1,800	1,250	750	390	275	160
144	51,000	17,000		7,500	5,400	3,700	2,500	1,400	700	440	20
149, 150	7,100	1,900		740	485	315	200	105	45	25	
161, 174	1,000	350		160	110	80	60	35	15	10	
178, 178A, 196	1,300	640		330	240	180	120	75	40		
179, 179A, 187	3,000	1,400		750	480	420	320	190	100	73	
180, 180A, 195	4,500	2,000		1,100	800	570	400	240	130	90	50
188, 188A	1,500	770		480	400	325	275	150	80	55	
209	180,000	55,000		22,000	15,000	8,500	6,000	3,400	1,600	1,000	310
281	150,000	47,000		19,000	13,500	8,800	6,000	3,300	1,650	1,150	625

Power-rating conditions: ambient temperature 104°F.
Center-conductor temperature 175°F with polyethylene dielectric.
Center-conductor temperature 400°F with Teflon dielectric.
Altitude—sea level.

TABLE 20 Attenuation Ratings of Coaxial Cables[4]

RG/U cable	\multicolumn{8}{c}{Nominal attenuation, db/100 ft at frequencies, MHz}
	1.0	10	50	100	200	400	1,000	2,000	5,000	10,000
5, 5A, 5B, 6, 6A, 212	0.26	0.83	1.9	2.7	4.1	5.9	9.6	23.0	32.0	56.0
7	0.18	0.64	1.6	2.4	3.5	5.2	9.0	18.0	25.0	43.0
8, 8A, 10, 10A, 213, 215	0.15	0.55	1.3	1.9	2.7	4.1	8.0	16.0	27.0	>100.0
9, 9A, 9B, 214	0.21	0.66	1.5	2.3	3.3	5.0	8.8	18.0	27.0	45.0
11, 11A, 12, 12A, 13, 13A, 216	0.19	0.66	1.6	2.3	3.3	4.8	7.8	16.5	26.5	>100.0
14, 14A, 74, 74A, 217, 224	0.12	0.41	1.0	1.4	2.0	3.1	5.5	12.4	19.0	50.0
17, 17A, 18, 18A, 177, 218, 219	0.06	0.24	0.62	0.95	1.5	2.4	4.4	9.5	15.3	>100.0
19, 19A, 20, 20A, 220, 221	0.04	0.17	0.45	0.69	1.12	1.85	3.6	7.7	11.5	>100.0
21, 21A, 222	1.5	4.4	9.3	13.0	18.0	26.0	43.0	85.0	>100.0	>100.0
22, 22B, 111, 111A	0.24	0.80	2.0	3.0	4.5	6.8	12.0	25.0	>100.0	>100.0
29	0.32	1.20	2.95	4.4	6.5	9.6	16.2	30.0	44.0	>100.0
34, 34A, 34B	0.08	0.32	0.85	1.4	2.1	3.3	5.8	16.0	28.0	>100.0
35, 35A, 35B, 164	0.06	0.24	0.58	0.85	1.27	1.95	3.50	8.6	15.5	>100.0
54, 54A	0.33	0.92	2.15	3.2	4.7	6.8	13.0	25.0	37.0	>100.0
55, 55A, 55B, 223	0.30	1.2	3.2	4.8	7.0	10.0	16.5	30.5	46.0	>100.0
57, 57A, 130, 131	0.18	0.65	1.6	2.4	3.5	5.4	9.8	21.0	>100.0	>100.0
58, 58B	0.33	1.25	3.15	4.6	6.9	10.5	17.5	37.5	60.0	>100.0
58A, 58C	0.44	1.4	3.3	4.9	7.4	12.0	24.0	54.0	83.0	>100.0
59, 59A, 59B	0.33	1.1	2.4	3.4	4.9	7.0	12.0	26.5	42.0	>100.0
62, 62A, 71, 71A, 71B	0.25	0.85	1.9	2.7	3.8	5.3	8.7	18.5	30.0	83.0
62B	0.31	0.90	2.0	2.9	4.2	6.2	11.0	24.0	38.0	92.0
63, 63B, 79, 79B	0.19	0.52	1.1	1.5	2.3	3.4	5.8	12.0	20.5	44.0
87A, 116, 165, 166, 225, 227	0.18	0.60	1.4	2.1	3.0	4.5	7.6	15.0	21.5	36.5
94	0.15	0.60	1.6	2.2	3.3	5.0	7.0	16.0	25.0	60.0
94A, 226	0.15	0.55	1.2	1.7	2.5	3.5	6.6	15.0	23.0	50.0

108, 108A	0.70	2.3	5.2	7.5	11.0	16.0	26.0	54.0	86.0	>100.0
114, 114A	0.95	1.3	2.1	2.9	4.4	6.7	11.6	26.0	40.0	65.0
115, 115A, 235	0.17	0.60	1.4	2.0	2.9	4.2	7.0	13.0	20.0	33.0
117, 118, 211, 228	0.09	0.24	0.60	0.90	1.35	2.0	3.5	7.5	12.0	37.0
119, 120	0.12	0.43	1.0	1.5	2.2	3.3	5.5	12.0	17.5	54.0
122	0.40	1.7	4.5	7.0	11.0	16.5	29.0	57.0	87.0	>100.0
125	0.17	0.50	1.1	1.6	2.3	3.5	6.0	13.5	23.0	>100.0
140, 141, 141A	0.30	0.90	2.1	3.3	4.7	6.9	13.0	26.0	40.0	90.0
142, 142A, 142B	0.34	1.1	2.7	3.9	5.6	8.0	13.5	27.0	39.0	70.0
143, 143A	0.25	0.85	1.9	2.8	4.0	5.8	9.5	18.0	25.5	52.0
144	0.19	0.60	1.3	1.8	2.6	3.9	7.0	14.0	22.0	50.0
149, 150	0.24	0.88	2.3	3.5	5.4	8.5	16.0	38.0	65.0	>100.0
161, 174	2.3	3.9	6.6	8.9	12.0	17.5	30.0	64.0	99.0	>100.0
178, 178A, 196	2.6	5.6	10.5	14.0	19.0	28.0	46.0	85.0	>100.0	>100.0
179, 179A, 187	3.0	5.3	8.5	10.0	12.5	16.0	24.0	44.0	64.0	>100.0
180, 180A, 195	2.4	3.3	4.6	5.7	7.6	10.8	17.0	35.0	50.0	88.0
188, 188A	3.1	6.0	9.6	11.4	14.2	16.7	31.0	60.0	82.0	>100.0
209	0.06	0.27	0.68	1.0	1.6	2.5	4.4	9.5	15.0	48.0
281	0.09	0.32	0.78	1.1	1.7	2.5	4.5	9.0	13.0	24.0

the wiring assembly. Several variations to this type of harness construction have been utilized in an effort to reduce size and weight and increase mechanical protection. In one variation, small-diameter hookup wire is encapsulated in a hard, physically tough epoxy compound. This approach offers greater wire density, reduced overall weight, and effective mechanical protection. Its disadvantages are limited flexibility, repairability, and wire interchangeability. In a second variation, a braided fibrous covering or jacket impregnated with resin for greater abrasion and fray resistance is applied over the cabled wires.

Ribbon cable consisting of conventional wires arranged in single or multiple parallel layers offers particular geometric advantages in many applications. Tape cable, consisting of copper-foil conductors laminated between with a dielectric, offers even greater advantages of space savings in one plane. However, the designer must remember that by the use of tape cabling, versatility (as offered by shielding, twisted pair, shielded pair, termination, etc.) is sacrificed in favor of space savings. Table 21 contains a comparative evaluation of the mechanical, electrical, and application features of conventional harness, ribbon-cabling, and tape-cabling concepts in airborne or ground-electronic applications.

Specifications MIL-C-7078 and MIL-C-27500 cover multiconductor cables utilizing interconnection and hookup wires in a round configuration. These specifications include single shielded and up to seven multiconductor cables with or without an overall shield and with or without an overall jacket. All conductors must be of the same gage; no individually shielded conductors are permissible. Various shield and jacket options are available offering compatibility with the chosen primary wire. In the interest of minimum weight and size, fillers are not used. Table 22 presents recommended MIL-C-27500 options, including construction details and a mechanical-usage rating.

Ground-electronics cables The term ground-electronics cabling embraces rack and panel interconnection, equipment cabling installed in conduit (as used with fixed computer and data-processing installations), or cabling placed beneath flooring that is not subjected to extreme mechanical abuse or environment.

Military Specification coverage for multiconductor cables in this area of usage is extremely poor; MIL-C-7078 and MIL-C-27500 are frequently used for lack of adequate specification coverage. MIL-C-27072 is broad enough in scope to cover heavy-duty usage in this area but has had little industrial application to date. The most commonly used heavy-duty construction for this category of cabling contains PVC-insulated nylon-jacketed primary conductors. These are cabled and jacketed with a PVC sheath. Variations from this general construction are unjacketed PVC-insulated conductors (not recommended for individually shielded members because of potential shield-end puncture); shielded or unshielded single conductors, pairs, triplets, quads, etc.; shield-jacketed with PVC or nylon material; equipped with overall shield, if applicable; PVC-sheathed.

Ground-support cables Ground-support cables for tactical systems should receive early attention from the designer. These cables are tailored to system needs, require considerable lead time for delivery, and can amount to a considerable system cost if no effort at standardization is made. A major manufacturing cost in the production of this type of multiconductor cable is cabling-machine setup. Many cable manufacturers require a minimum order of 500 to 1,000 ft of cable for a given configuration. After setup additional cable footage can be produced more economically. Following is a brief discussion of ground-support multiconductor cable applications.

Permanent Installation. Cables that are buried or placed in conduits, open ducts, troughs, or tunnels are considered permanent cables. These cables are not handled, flexed, reeled, or dereeled except at the time of installation.

For permanent installation either neoprene or polyethylene cable sheaths are preferred. Polyethylene offers greater moisture and water protection and has a lower coefficient of friction for easier installation in conduit where the cable must be pulled for substantial distances. Neoprene offers good mechanical protection and greater flexibility, affording easier termination. Cable-pulling compounds can be applied to neoprene-sheathed cable for easier installation in conduits.

Portable Installation. Portable heavy-duty multiconductor ground-support cables are designed to withstand installation and use under the following conditions:
1. Rocky, uneven, or sandy terrain
2. Mechanical abuse such as frequent reeling and dereeling; heavy vehicle traffic; twisting, kinking, jerking, and impact by heavy objects
3. Operating temperature −65 to +165°F, sand, dust, water immersion, high humidity, coastal (salt-water) atmosphere, and ultraviolet radiation

MIL-C-13777 is used as the basis for design of heavy-duty portable cables. The basic construction of this cable is as follows:
1. Conductor: tin-coated, annealed copper, stainless-steel-reinforced, No. 18 AWG and smaller
2. Insulation: polyethylene
3. Jacket: extruded nylon or nylon braid for No. 10 AWG and larger
4. Binder: Mylar wrap
5. Separator: braided cotton
6. Sheath: double-layer reinforced neoprene
7. Component shielding: No. 36 AWG tin-coated copper braid
8. Shield jacket: extruded or braided nylon, or fused Mylar
9. Overall shield: No. 34 AWG tin-coated copper braid.

A working example of utilization of recent techniques in thin-wall insulation, high-strength conductors, and high-strength sheath materials is the lightweight small-diameter portable cable developed for the Pershing Weapons System. Conventional heavy-duty cable was not compatible with the Pershing concept of a quick-reaction weapon system, not only from the standpoint of ease of handling but also from that of the size and weight of associated hardware and components such as cable reels and storage facilities. Cable size and weight reduction have been accomplished by incorporation of higher-strength conductors, tougher insulation, shield innovations, and tougher sheath materials discussed in the previous sections. For example, a conventional 60-conductor No. 20 AWG MIL-C-13777 cable would have a nominal diameter of 1.121 in. and a cable weight of 890 lb per 1,000 ft, as compared with the lightweight cable having a diameter of 0.518 in. and a weight of 370 lb per 1,000 ft.

Flat flexible cable There are two basic types of flat flexible cable: flat-conductor cables, often referred to as "tape cable," and flexible printed wiring. The major differences between the two types of cable is their manner of construction and general application. Tape cable is used primarily to interconnect individual electronic units; flexible printed wiring is used to provide interconnection within a unit. Advantages and disadvantages of the use of tape cable compared with other more conventional cable constructions are covered in Table 21. Following is detailed information for each flat-flexible-cable type.

Flat-conductor Cable. CONSTRUCTION: Flat-conductor cable is constructed by the encapsulation of flat rectangular conductors between layers of dielectric film. Figure 22 depicts the manufacturing technique for laminating flat-conductor cable. Conductor ribbons are parallel-prepositioned with a uniform spacing. Although different conductor widths and spacings are possible within a single cable, in the interest of cost and availability it is best to use uniform constructions as specified in National Aerospace Standard NAS 729 (see Figs. 23 and 24). Cable can be stacked in multiple layers to achieve desired circuit density, and copper tapes may be interleaved between layers to provide shielding.

MATERIALS: Conductors are usually rolled copper, annealed in accordance with QQ-C-576. Other conductor materials and protective coatings may be obtained to meet unusual application requirements. For selecting proper conductor size to satisfy current and voltage requirements refer to Table 23 and Figs. 25 and 26.

Table 24 presents a comparison of flat-cable insulation materials. Selection of a specific insulation material for flat-cable application is subject to an equivalent type of analysis to what one would perform for a standard wire insulation. For instance, if soldering is to be employed as a termination technique, and repairability is required, high-temperature materials such as Teflon TFE, supported or unsup-

TABLE 21 Cabling-concept Comparison[19]

Conventional cable. Separate wires laced or tied into bundles or jacketed	Ribbon cable. Conventional wires bonded together in ribbon form	Tape cable. Copper-foil conductors sandwiched between thin sheets of dielectric
\multicolumn{3}{c}{Mechanical Features}		
Greatest variety of conductor configurations: Single wires—all gauges Shielded wires—all gauges Twisted pair, triples—all gauges Twisted and shielded—all except very large gauges Coaxial cables—all sizes Jacketed overall for heavy duty, if needed	Wide variety of conductor configurations: Single wires—limited to smaller gauges Shielded wires—limited to smaller gauges Twisted pair, triples—limited to smaller gauges Twisted and shielded—limited to smaller gauges Coaxial cables—limited to smaller gauges Not suited for jacketed cables	Limited variety of conductor configurations—some are compromises: Single conductor—smaller gauges only Shielded conductor—smaller gauges only Twisted pair—simulated by zigzag crossovers in sandwich Twisted and shielded—simulated in sandwich construction Coaxial cable not practical—twin-line construction must be used Not suited for jacketed cables
Round-bundle configuration is relatively self-supporting	Flat-bundle configuration requires moderate support	Tape configuration requires continuous support
Stranded wire usually required for increased mechanical life	Stranded wire is desirable. On larger quantities of conductors per cable, solid wire can be used owing to the mutual mechanical effects of adjacent wires	Foil is ordinarily used. Some modifications have used stranded wire configured to a rectangular cross section
Termination preparation—simple mechanical or hot-wire stripper	Termination preparation—simple mechanical or hot-wire stripper after separating wires from each other	Termination preparation—minimum practical production process exists for removal of insulation to expose conductors. Certain types of insulation respond to new welding techniques which do not require insulation removal

Electrical Features

Rugged insulation achieved by selective combination of dielectric materials in layers	Rugged insulation achieved by selective choice of primary dielectric plus an overall protective ribbon skin. Cables without this skin permit undesired separation of conductors even with careful handling	Edges of tape insulation or complete laminate should be reinforced because, once nicked, the cable can be torn across conductors with relatively small force. A deep scratch or cut in the outer layer of the cable insulation will allow conductors to fracture with sharp bend at the scratch
Standard voltage-drop characteristics; use standard current ratings. Bundled circuits require derating because of heat rise if circuits are operated simultaneously	Standard voltage-drop characteristics; use standard current ratings. Derating not normally necessary	Voltage-drop characteristics will vary more widely than standard. Broad, tinsellike conductors permit much higher current limits compared with the equivalent cross section of circular conductors. In general this permits use of much less copper for a comparative current requirement if the increased voltage drop can be tolerated
Requires no heat sink in normal application	Requires no heat sink in normal application	Required *continuous* heat sink in normal application to utilize the high current-carrying capability
Interconductor capacitance is high; spacing generally random; capacitance values are unpredictable; capacitance to structure relatively low	Interconductor capacitance is high, spacing fixed, capacitance predictable. Circuits can be selectively spaced to minimize effects. Capacitance from cable to metal support or shielding is high	Interconductor capacitance is low, spacing fixed, capacitance predictable. Circuits can be selectively spaced to minimize effects. Shielding needed in stacked cables. Capacitance from cable to metal support or shielding foil is high
Crosstalk is uncontrolled because of random spacing; individual shielding usually required	Crosstalk can be controlled by conductor placement; shielded conductor not usually required. CAUTION: In stacking cables, overall interlayer shielding may be needed	Crosstalk can be controlled by circuit placement. Overall interlayer shielding may be necessary when cables are stacked

Wires and Cables

TABLE 22 Multiconductor Cable Options—Recommended MIL-C-27500

Basic primary wire specifications	Specification symbol	Size range, AWG	Shielded Shield style	Shielded Jacket style	Jacketed Shield style	Jacketed Jacket style	Shield-jacketed Shield style	Shield-jacketed Jacket style	Voltage rating, rms	Temperature rating, °C
MIL-W-5086: MS-25190, TY I	A	22–12	T	O	U	1 or 3	T	1, 2, or 3	600	−55 to 105
MS-25190, TY II	B	22–4/0	T	O	U	3	T	1 or 3	600	−55 to 105
MS-25190, TY III	C	22–4/0	T	O	U	3	T	1 or 3	600	−55 to 105
MS-25190, TY IV	P	22–16	T	O	U	3	T	1 or 3	3,000	−55 to 105
MIL-W-7139, CL 1	D	22–4/0	S	O	U	7	S	6 or 7	600	−55 to 200
MIL-W-7139, CL 2	E	22–4/0	N	O	U	7	N	6 or 7	600	−55 to 200
MIL-W-8777: MS-25471	H	22–2/0	S	O	U	4	S	4	600	−55 to 150
MS-27110	F	22–4	S	O	U	5	S	5	600	−55 to 200
MIL-W-22759: MS-17411	V	24–4	S	O	U	6 or 7	S	6 or 7	600	−55 to 200
MS-17412	W	24–4	N	O	U	6 or 7	N	6 or 7	600	−55 to 260
MS-18000	S	24–4	S	O	U	6 or 7	S	6 or 7	600	−55 to 200
MS-18001	T	24–4	N	O	U	6 or 7	N	6 or 7	600	−55 to 260
MS-18113	LA	28–8	S	O	U	6	S	6	1,000	−55 to 200
MS-18114	LB	28–8	N	O	U	6	N	6	1,000	−55 to 200
MS-21985	R	28–12	S	O	U	6	S	6	600	−55 to 200
MS-21986	L	28–12	N	O	U	6	N	6	600	−55 to 260
MS-90294	N	22–2/0	S	O	U	6 or 7	S	6 or 7	600	−55 to 200
MIL-W-25038: MS-27125	J	22–4/0	F	O	U	7	F	7	600	−55 to 750
MIL-W-81044/1	M	24–4	S	O	U	4° or 5	S	4° or 5	600	−55 to 135
MIL-W-81044/2	MA	24–4	T	O	U	4 or 5	T	4 or 5	600	−55 to 135
MIL-W-81044/3	MB	30–12	S	O	U	4 or 5	S	4 or 5	600	−55 to 135
MIL-W-81044/4	MC	30–12	T	O	U	4 or 5	T	4 or 5	600	−55 to 135

° A jacket compatible with primary insulation system not available to date. Kynar, polyethylene-coated Mylar, and crosslinked Kynar are proposed additions to specification.

SOURCE: Martin Marietta Corporation.

Multiconductor Cables 4-59

TABLE 22 Multiconductor Cable Options—Recommended MIL-C-27500 (Continued)

Basic primary wire specifications	Specification symbol	Mechanical duty rating	Conductor	Primary insulation	Shield	Jacket
MIL-W-5086:						
MS-25190, TY I	A	Medium	Tinned copper	PVC/nylon	Tinned copper	(1) PVC
MS-25190, TY II	B	Medium; fire-resistant	Tinned copper	PVC/glass nylon	Tinned copper	(2) Extruded nylon
MS-25190, TY III	C	Heavy	Tinned copper	PVC/glass PVC/nylon	Tinned copper	(3) Nylon braid
MS-25190, TY IV	P	Medium	Tinned copper	PVC/nylon	Tinned copper	
MIL-W-7139, CL 1	D	Heavy	Silver-coated copper	Teflon TFE	Silver-coated copper	(6) Taped TFE Teflon
MIL-W-7139, CL 2	E	Heavy	Nickel-coated copper	Tapes and glass braid	Nickel-coated copper	(7) Glass braid
MIL-W-8777:						
MS-25471	H	Heavy	Silver-coated copper	Silicone rubber	Silver-coated copper	(4) Dacron braid
MS-27110	F	Medium	....	....	....	(5) Extruded FED Teflon
MIL-W-22759:						
MS-17411	V	Heavy	Silver-copper	Mineral-filled Teflon (TFE)	Silver-copper	(6) Taped TFE Teflon
MS-17412	W	Heavy	Nickel-copper		Nickel-copper	(7) Glass braid
MS-18000	S	Medium	Silver-copper	....	Silver-copper	
MS-18001	T	Medium	Nickel-copper	....	Nickel-copper	
MS-18113	LA	Light	Silver-copper	Extruded TFE Teflon	Silver-copper	(6) Taped TFE Teflon
MS-18114	LB	Light	Nickel-copper	....	Nickel-copper	
MS-21985	R	Light	Silver-copper	....	Silver-copper	
MS-21986	L	Light	Nickel copper	....	Nickel-copper	
MS-90294	N	Medium	Silver-copper	TFE-glass-FED	Silver-copper	(6) Taped TFE Teflon (7) Glass braid
MIL-W-25038:						
MS-27125	J	Heavy; fire-resistant	Nickel-clad copper	TFE tapes and glass braid	Stainless steel	(7) Glass braid
MIL-W-81044/1	M	Medium	Silver-copper	Polyalkene and Kynar (cross-linked)	Silver-copper	(4) Dacron braid°
MIL-W-81044/2	MA	Medium	Tinned copper	....	Tinned copper	(5) Extruded FED Teflon
MIL-W-81044/3	MB	Light	Silver-copper	....	Silver-copper	
MIL-W-81044/4	MC	Light	Tinned copper	....	Tinned copper	

ported, and polyimide are required to satisfy the application. Polyester, polyimide, polytetrafluoroethylene, and fluorinated ethylene propylene are the most commonly used dielectric films.

DESIGN CONSIDERATIONS: The following important criteria should be considered by the designer and used to formulate optimum cable construction, materials selection, terminations, installation, fabrication, and handling:

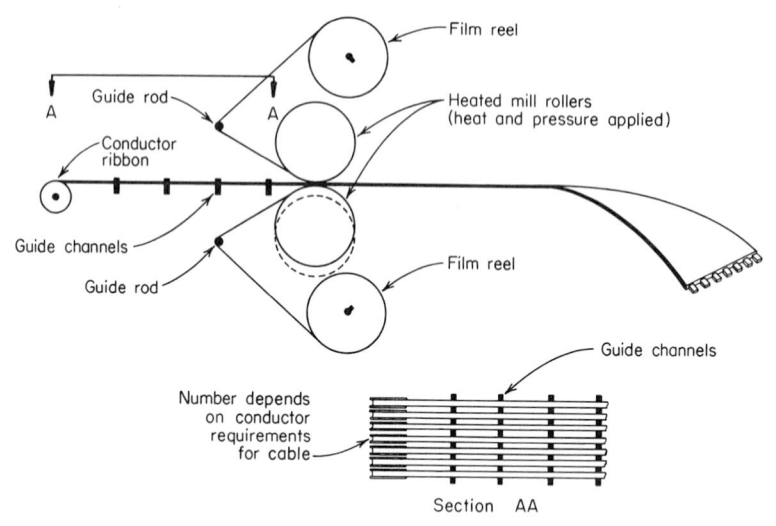

Fig. 22 Flat-conductor laminating process.[20]

1. Electrical requirements:
    a. Current-carrying capacity
    b. Voltage drop
    c. Overload rating
    d. Impedance
    e. Capacitance
    f. Shielding requirements
    g. Derating due to stacking of cables
2. Mechanical requirements:
    a. Conductor width
    b. Conductor thickness
    c. Conductor spacing
    d. Insulation tear resistance
    e. Insulation puncture or cut-through resistance
    f. Insulation and conductor flexure resistance
3. Environmental requirements:
    a. Temperature extremes
    b. Flame resistance
    c. Vibration and shock requirements
    d. Temperature cycling
    e. Humidty requirements
    f. Altitude requirements
    g. Material compatibility
    h. Fungus resistance
    i. Radiation resistance
    j. Aging resistance
4. Termination requirements:
    a. Spacing compatible with connector
    b. Type of termination-eyelet, rivet, welded, soldered, braze, crimp, etc.
    c. Termination type compatible with material
5. Fabrication requirements:
    a. Insulation strippability technique- mechanical abrasion, chemical, thermal, or piercing.

Multiconductor Cables   4-61

    b. Conductor support
6. Installation requirements:
    a. Cable support
    b. Cable routing—avoid interference with high heat sources, component installation, and sharp edges

Flexible Printed Wiring. CONSTRUCTION: The majority of flexible printed-wiring cables are manufactured by standard printed-circuit etching techniques; however, cables have been fabricated by die-stamping or die-premolding processes. Briefly, the standard etching process entails bonding or laminating conductor foil to a dielectric substrate, application of an etchant resist, which presents the proper conductor pattern to the conductor foil, and finally etching removal of the unwanted foil. An insulating cover layer can be laminated over the conductor pattern, or insulating coatings may be applied. The insulating cover layer may be prepunched before lamination, to achieve access to terminations by mechanical or chemical

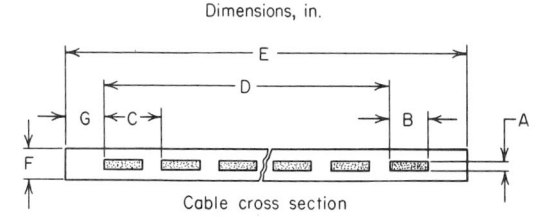

Dimensions, in.

Cable cross section

Dash-number Table (Types A and B Cables)

Dash No.	Equiv. AWG No.	No. of conductors	Conductor Thickness A ±0.0005	Width B ±0.002	Spacing C ±0.005	D ±0.005	Cable Width E ±0.005	Thickness F (a)	Margin G ±0.008
1	32	17	0.0020	0.025	0.050	0.800	1.000	0.010	0.087
2	32	37	0.0020	0.025	0.050	1.800	2.000	0.010	0.087
3	32	57	0.0020	0.025	0.050	2.800	3.000	0.010	0.087
4	30	17	0.0030	0.026	0.050	0.800	1.000	0.010	0.087
5	30	37	0.0030	0.026	0.050	1.800	2.000	0.010	0.087
6	30	57	0.0030	0.026	0.050	2.800	3.000	0.010	0.087
7	32	12	0.0020	0.025	0.075	0.825	1.000	0.010	0.075
8	32	25	0.0020	0.025	0.075	1.800	2.000	0.010	0.075
9	32	38	0.0020	0.025	0.075	2.725	3.000	0.010	0.075
10	30	12	0.0030	0.026	0.075	0.825	1.000	0.010	0.075
11	30	25	0.0030	0.026	0.075	1.800	2.000	0.010	0.075
12	30	38	0.0030	0.026	0.075	2.775	3.000	0.010	0.075
13	28	12	0.0030	0.042	0.075	0.825	1.000	0.010	0.075
14	28	25	0.0030	0.042	0.075	1.800	2.000	0.010	0.075
15	28	38	0.0030	0.042	0.075	2.775	3.000	0.010	0.075
16	26	9	0.0030	0.062	0.100	0.800	1.000	0.010	0.069
17	26	19	0.0030	0.062	0.100	1.800	2.000	0.010	0.069
18	26	29	0.0030	0.062	0.100	2.800	3.000	0.010	0.069

(a) Cable thickness tolerance—Type A: ±0.003; Type B: +0.005, −0.003.

Fig. 23 Standard flexible flat copper conductor 300-volt cables per NAS-729 (Type A and B cables).[21]

Wires and Cables

Dash-number Table (Types A and B Cables) (Continued)

			Conductor		Spacing		Cable		
Dash No.	Equiv. AWG No.	No. of conductors	Thickness A ±0.0005	Width B ±0.002	C ±0.008	D ±0.008	Width E ±0.010	Thickness F (a)	Margin G ±0.008
19	24	9	0.0050	0.063	0.100	0.800	1.000	0.012	0.069
20	24	19	0.0050	0.063	0.100	1.800	2.000	0.012	0.069
21	24	29	0.0050	0.063	0.100	2.800	3.000	0.012	0.069
22	22	6	0.0050	0.100	0.150	0.750	1.000	0.012	0.100
23	22	12	0.0050	0.100	0.150	1.650	2.000	0.012	0.100
24	22	19	0.0050	0.100	0.150	2.700	3.000	0.012	0.100

(a) Cable thickness tolerance—Type A: ±0.003; Type B: +0.005, −0.003.

Dash-number Table (Type C Cable)

			Conductor		Spacing		Cable		
Dash No.	Equiv. AWG No.	No. of conductors	Thickness A ±0.0005	Width B ±0.002	C ±0.010	D ±0.010	Width E ±0.010	Thickness F ±0.005	Margin G ±0.008
1	32	12	0.0020	0.025	0.075	0.825	1.000	0.017	0.075
2	32	25	0.0020	0.025	0.075	1.800	2.000	0.017	0.075
3	32	38	0.0020	0.025	0.075	2.775	3.000	0.017	0.075
4	30	12	0.0030	0.026	0.075	0.825	1.000	0.017	0.075
5	30	25	0.0030	0.026	0.075	1.800	2.000	0.017	0.075
6	30	38	0.0030	0.026	0.075	2.775	3.000	0.017	0.075
7	28	12	0.0030	0.042	0.075	0.825	1.000	0.017	0.075
8	28	25	0.0030	0.042	0.075	1.800	2.000	0.017	0.075
9	28	38	0.0030	0.042	0.075	2.775	3.000	0.017	0.075
10	26	9	0.0030	0.063	0.100	0.800	1.000	0.017	0.069
11	26	19	0.0030	0.063	0.100	1.800	2.000	0.017	0.069
12	26	29	0.0030	0.063	0.100	2.800	3.000	0.017	0.069
13	24	9	0.0050	0.063	0.100	0.800	1.000	0.017	0.069
14	24	19	0.0050	0.063	0.100	1.800	2.000	0.017	0.069
15	24	29	0.0050	0.063	0.100	2.800	3.000	0.017	0.069
16	22	6	0.0050	0.100	0.150	0.750	1.000	0.017	0.100
17	22	12	0.0050	0.100	0.150	1.650	2.000	0.017	0.100
18	22	19	0.0050	0.100	0.150	2.700	3.000	0.017	0.100

Fig. 24 Standard flexible flat copper conductor 300-volt cables per NAS-729 (Type C cable).[21]

means. Layers of foil insulation may be built up to construct a multilayer flexible cable.

MATERIALS: As was true for flat-conductor cable, copper is the most widely used conductor material. Two basic types of copper are available: rolled copper foil and electrodeposited. Rolled copper foil exhibits superior flex life but is more difficult to achieve adherence with. Chemical oxide treatments are used to improve

TABLE 23 Copper Conductor Characteristics—Flat Conductor[23]

Flat-conductor dimensions		Cross section		Nearest equivalent AWG wire size based on equivalent:		Resistance, milliohms/ft at 20°C	Current, amp for 30°C rise
Thickness in.	Width, in.	sq mils	cir mils	Cross section	Current rating		
0.0027	0.030	81	102	30	28	100	3.4
	0.045	122	154	28	27	67	3.8
	0.060	162	204	27	25	50	5.1
	0.075	202	254	26	24	40	5.8
	0.090	243	306	25	23	34	6.5
	0.125	338	425	24	22	24	8.2
	0.155	418	527	23	21	19.5	9.2
	0.185	500	630	22	20	16.2	10.7
	0.250	675	850	21	18	12	13.5
0.004	0.030	120	151	28	27	67	4.0
	0.045	180	227	26	25	45	5.2
	0.060	240	302	25	24	34	6.0
	0.075	300	378	24	23	27	7.0
	0.090	360	454	23	22	22.5	7.8
	0.125	500	630	22	20	16.2	10.0
	0.155	620	780	21	19	13	11.8
	0.185	740	930	20	18	11	13.5
	0.250	1,000	1,260	19	17	8	17.0
0.0055	0.045	248	312	25	24	33	6.0
	0.060	330	415	24	23	25	7.2
	0.075	412	520	23	22	20	8.2
	0.090	495	624	22	21	16.5	9.5
	0.125	687	865	21	20	12	12.2
	0.155	852	1,075	20	19	9.5	14.8
	0.185	1,020	1,285	19	17	8	17
	0.250	1,375	1,730	18	16	6	21
0.008	0.045	360	454	23	22	23	7.8
	0.060	480	605	22	21	17	9.8
	0.075	600	755	21	20	13.5	11.5
	0.090	720	905	20	19	11.2	13.2
	0.125	1,000	1,260	19	17	8	17
	0.155	1,240	1,560	18	16	6.5	20
	0.185	1,480	1,860	17	14	5.5	23
	0.250	2,000	2,520	16	13	4.1	26

the adherence of the copper foil to the insulation. The use of electrodeposited copper should be carefully evaluated with respect to flex-life requirements; improved flex-life electrodeposited copper is available, but this material still does not equal the flex resistance obtainable with rolled copper.

Flexible printed wiring is available fabricated with the film materials specified in Table 24. Since few of the films used for fabrication are thermoplastic, the films must be bonded to the conductor foil by means of an adhesive; this is normally accomplished under heat and pressure in a press operation, but may be automated for continuous production by roll laminating similar to the procedure outlined in Fig. 22. The flexible printed-cable system is only as good as its weakest link, and the adhesive represents the weak link. The properties listed for the dielectric films in Table 24 are for the pure unlaminated film; processing and the addition of adhesives where required will produce some changes in the end properties of the system.

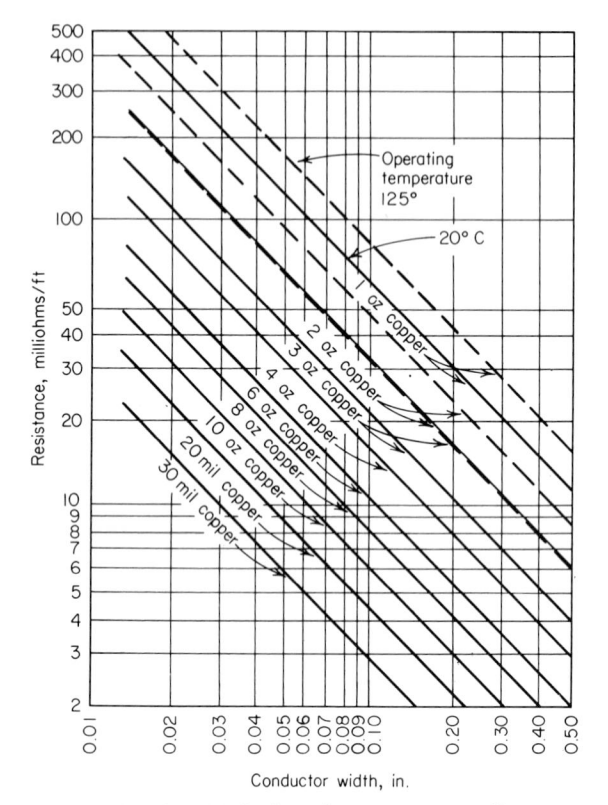

Fig. 25 Etched-conductor resistance.[22]

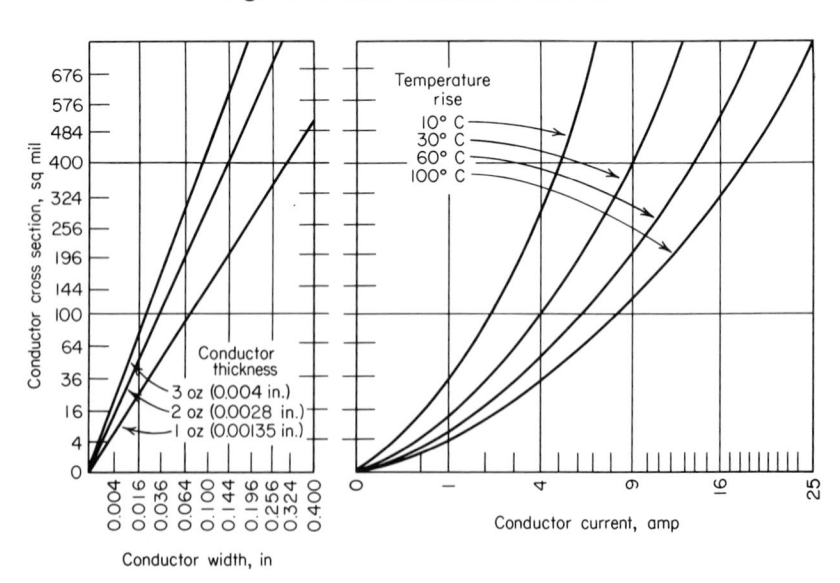

Fig. 26 Etched-conductor current capacity.[22]

TABLE 24 Flat-flexible-cable Insulating Materials[23]

Property	TFE fluoro-carbon	TFE glass cloth	FEP fluoro-carbon	FEP glass cloth	Poly-mide	Poly-chloro-trifluoro-ethylene	Poly-vinyl fluoride	Poly-propylene	Poly-ester	Poly-vinyl chloride	Poly-ethylene
Specific gravity	2.15	2.2	2.15	2.2	1.42	2.10	1.38	0.905	1.395	1.25	0.93
Square inches of 1-mil film per lb	12,800	13,000	12,900	13,000	19,450	12,000	20,000	31,000	21,500	22,000	30,100
Service temperature, °C											
Minimum	−70	−70	−225	−70	−250	−70	−70	−55	−60	−40	−20
Maximum	250	250	200	250	+250	150	105	125	150	85	60
Flammability	Nil	Nil	Nil	Nil	Nil	Nil	Yes	Yes	Yes	Slight	Yes
Appearance	Translucent	Tan	Clearbluish	Tan	Amber	Clear	Clear	Clear	Clear	Translucent	Clear
Thermal expansion, in./(in.)(°F) × 10[6]	70	Low*	50	Low*	11	45	28	61	15		
Bondability with adhesives	Good†	Good†	Good†	Good†	Good	Good†	Good†	Poor	Good	Good	Poor
Bondability to itself	Good	Poor	Good	Good	Poor	Good	Good	Good	Poor	Good	Good
Tensile strength at 77°F, psi	3,000	20,000	3,000	20,000	20,000	4,500	8,000	5,700	20,000	3,000	2,000
Modulus of elasticity, psi	80,000	3	70,000	3	430,000	200,000	280,000	170,000	550,000		50,000
Volume resistivity, ohm-cm	2 × 10[16]	1 × 10[16]	1 × 10[17]	1 × 10[16]	1 × 10[16]	1 × 10[16]	3 × 10[13]	1 × 10[16]	1 × 10[18]	1 × 10[19]	1 × 10[16]
Dielectric constant at 10[2]–10[8] Hz	2.2	2.5/5[3]	2.1	2.5/5[3]	3.5	2.5	7.0	2.0	2.8–3.7	3–4	2.2
Dissipation factor at 10[2]–10[8] Hz	0.0002	0.0007/0.001*	0.0002	0.0001/0.001*	0.002/0.014	0.015	0.009–0.041	0.0002/0.0003	0.002–0.016	0.14	0.0006
Dielectric strength at 5 mils thickness, volts/mil	800	650/1,600	3,000	650/1,600	3,500	2,000	2,000	750†	3,500	800	1,500
Chemical resistance	Excellent	Excellent	Excellent	Excellent	Excellent	Excellent	Good	Excellent	Excellent	Good	Excellent
Water absorption, %	0	0.10/68	0	0.18/30	3	0	15	0.01	0.5	0.10	0
Sunlight resistance	Excellent	Excellent	Excellent	Excellent	Excellent	Excellent	Excellent	Low	Fair	Fair	Low

* Depends on percent glass cloth.
† Must be treated.
‡ At 0.125 in. thickness.

4-66   Wires and Cables

DESIGN CONSIDERATIONS: The design considerations applicable to printed flexible cable are analogs to those previously established for flat-conductor cable. Since the fabrication techniques involved are essentially identical to those used for conventional rigid printed wiring, the design standards for conductor spacing, line width, etc., as specified in MIL-STD-275 apply. Artwork for flexible printed cable should be prepared with the same careful consideration as for standard printed wiring, and should be precise, clear original artwork with free lines, clean edges, and gentle contours. Size is normally four to five times the desired final size, but may require greater magnification to meet desired layout-size tolerances. Figures 25 and 26 present design criteria for selection of optimum conductor size. Figure 25 shows the etched-conductor resistance for varing conductor cross section, and Fig. 26 gives current-capacity data.

Miscellaneous cables *Shipboard Cables.* Specifications MIL-C-915 and MIL-C-21984 cover electric cables for installation in fixed wireways on combat ships. The available configurations are covered by referenced specifications. In general, shipboard cable design has the following features:

1. Conductors are uncoated copper made watertight by filling the strands with a flexible material that is compatible with the insulation.
2. Insulation is either extruded silicone rubber or silicone-rubber-treated glass tape.
3. Fillers are used to obtain a well-rounded cable core.
4. A qualified binder may be used at the manufacturer's option.
5. A separator is employed only with specific approval.
6. An impervious sheath, normally neoprene rubber, is applied over the cable core.
7. A braided metal armor is applied over the sheath.
8. Finally, the metal armor braid is painted for corrosion protection.

Shipboard cables are designed for watertightness, to prevent water transmission between sealed bulkheads, and for heat and flame resistance, to prevent flame travel if exposed to fire or current overload.

Rubber-insulated Cables. IPCEA Publication S-19-81 was combined with NEMA Publication WC3 for the IPCEA-NEMA Standards Publication entitled Rubber-insulated Wire and Cable for the Transmission and Distribution of Electrical Energy. These standards apply to materials, construction, and testing of rubber-insulated wires and cables for installation and service for indoor, aerial, underground, portable, or submarine applications.

Construction and materials covered in the standards vary with application. General features are as follows:

1. Conductors are annealed, coated or uncoated copper or aluminum; sizes AWG 8 and smaller are normally solid; but stranded for specific applications.
2. Insulations are vulcanized natural or synthetic rubber compounds.
3. Insulation shielding is metallic, nonmagnetic, and made of tape, braid, serve, or tubular.
4. Coverings—single-conductor wires or cables, AWG 8 and smaller, rated at 600 volts or less have at least one covering applied over the insulation.

All other constructions have at least two coverings applied: i.e., two braids, a braid and a serve, or a tape and two servings. Coverings shall be compound-filled tape, cotton yarn, rubber, or thermoplastic jackets.

Building Wires and Cables. Building wire and cable applicable to the wiring of buildings or apparatus installed in buildings is described in the National Electrical Code as published by Underwriters' Laboratories, Inc.

There are 26 different types of building wire and cable as listed in the 1965 edition of the National Electrical Code. Table 25 presents the various constructions available as listed in Table 310-2(b) of the National Electrical Code. Federal Specifications J-C-580 and J-C-129 partly cover these types.

Design considerations Design of the cabling installation is an integral part of the mechanical design of any equipment or system requiring electrical interconnection. Design planning for electrical installation should be concurrent with the layout of the mechanical design. Quality installation design requires that each of the following major considerations be thoroughly evaluated and a positive design approach determined.

Environment. Vibration, acceleration, and shock are dynamic environments which are controlling factors from a design viewpoint. Acceleration places a load on cables, supports, brackets, connectors, and mounting points for black boxes. Wherever practicable, these items must be so designed as to be in compression against a structural member. Connectors should be oriented so that the possibility of inadvertent disconnection is minimized. Consideration shall also be given to deceleration forces.

Vibration sets up varying stresses in cables and supports, proportional to the mass being supported. Shock, including acceleration and deceleration, also contributes to an overstressed condition.

Other environmental conditions include high dynamic pressure, elevated temperatures, and lowered atmospheric pressure. High temperature has the most severe effect on network installations. Insulation against high temperature and selection of materials capable of withstanding high temperature are the basic approaches for controlling problems caused by temperature extremes. Low atmospheric pressure is insignificant except in outer-space applications. Here the effects can be significant: outgassing and deterioration of plastics with time.

Several different phenomena are generated by a nuclear burst: radiation, heat pulse, shock wave, and electromagnetic pulses (EMP). The initial radiation may be of such type and intensity and over a sufficient time interval so that it may seriously degrade the quality of materials, including metals. The heat pulse and shock wave generate conditions similar to those already discussed above. The EMP can produce voltages and currents through the structure. Thus, installation hardware can become electrically energized with electric stresses far higher than those normally encountered.

Ground-environmental factors include high and low temperatures, humidity, and dynamic parameters due to handling and transportation. The levels of these ground-environmental factors are usually far less severe than those of flight environment. However, despite the lower level, the duration of these stresses is far in excess of normal flight time. The accumulated stress or degradation under these conditions may be quite appreciable.

Routing and Grouping. Interconnecting cables and networks should be designed and installed to minimize the adverse effects of electromagnetic interference and to control crosstalk between circuits. To eliminate the adverse effects, special grouping, separation, and shielding practices should be followed, for which the following general guidelines are recommended:

1. Direct-current supply lines. Use twisted pair. Separate from ac power and control lines.

2. Alternating-current power lines. Use twisted lines. Separate from susceptible lines. Shield ac circuits in which switching transients occur, and ground the shield at both ends.

3. Low-level signals. Use shielded twisted pair, and ground the shield at one end.

4. High-level signals. Use shielded twisted pair, and ground the shield at both ends.

5. Provide adequate filtering to prevent conducted-noise problems.

6. Follow a single-point ground concept where possible. Analyze flow of parasitic chassis currents, and design ground conductor for worst case.

7. Plan the separation of signal and power circuits with maximum distance between runs.

8. Hold wire and cable length to minimum.

9. Locate high-heat-generating wires on the outside.

10. Plan the cable routing in coordination with the structural-design effort. Plan cable runs and tie-down points in the early design phase for incorporation into the structure. Plan for minimum length. Attempt to optimize cable installation; compromise only in the solution of installation and maintenance problems.

11. Hold bend radius of coaxial and multiconductor cables to at least 10 times the cable outside diameter.

TABLE 25 Building-wire Insulations*

Trade name	Type letter	Insulation	Outer covering
Code	R	Code rubber	Moisture-resistant, flame-retardant, nonmetallic covering†
Heat-resistant	RH RHH	Heat-resistant rubber	Moisture-resistant, flame-retardant, nonmetallic covering†
Moisture-resistant	RW	Moisture-resistant rubber	Moisture-resistant, flame-retardant, nonmetallic covering†
Moisture and heat-resistant	RH-RW	Moisture and heat-resistant rubber	Moisture-resistant, flame-retardant, nonmetallic covering†
Moisture and heat-resistant	RHW	Moisture and heat-resistant rubber	Moisture-resistant, flame-retardant, nonmetallic covering†
Heat-resistant latex rubber	RUH	90% unmilled, grainless rubber	Moisture-resistant, flame-retardant, nonmetallic covering
Moisture-resistant latex rubber	RUW	90% unmilled, grainless rubber	Moisture-resistant, flame-retardant, nonmetallic covering
Thermoplastic	T	Flame-retardant, thermoplastic compound	None
Moisture-resistant thermoplastic	TW	Flame-retardant moisture-resistant thermoplastic	None
Heat-resistant thermoplastic	THHN	Flame-retardant, heat-resistant thermoplastic	Nylon jacket
Moisture and heat-resistant thermoplastic	THW	Flame-retardant, moisture- and heat-resistant thermoplastic	None

MAGNET WIRE

The field of film-insulated wire is so vast and, in general, unrelated to insulation materials and to practices connected with hookup and interconnection wires that some descriptions of conductors and insulation must be restated in this section in a form applicable to magnet wire.

Conductors *Materials and Construction.* COPPER: The most common conductor is bare, round, solid annealed copper wire in accordance with USA Standards Specification C7.1. Square and rectangular copper wire is available as described in C7.9. Copper strip can be obtained in accordance with QQ-C-576. Rounded edges should be specified to preclude any roughness or sharp projections.

TABLE 25 Building-wire Insulations* (Continued)

Trade name	Type letter	Insulation	Outer covering
Moisture- and heat-resistant thermoplastic	THWN	Flame-retardant, moisture- and heat-resistant thermoplastic	Nylon jacket
Thermoplastic and asbestos	TA	Thermoplastic and asbestos	Flame-retardant, nonmetallic covering
Thermoplastic and fibrous braid	TBS	Thermoplastic	Flame-retardant, nonmetallic covering
Synthetic heat-resistant	SIS	Heat-resistant rubber	None
Mineral-insulated metal-sheathed	MI	Magnesium oxide	Copper
Silicone-asbestos..	SA	Silicone rubber	Asbestos or glass
Fluorinated ethylene propylene	FEP	Fluorinated ethylene propylene	None
	FEPB	Fluorinated ethylene propylene	Glass braid Asbestos braid
Varnished cambric	V	Varnished cambric	Nonmetallic covering or lead sheath
Asbestos and varnished cambric	AVA and AVL	Impregnated asbestos and varnished cambric	AVA-asbestos braid or glass AVL-lead sheath
Asbestos and varnished cambric	AVB	Impregnated asbestos and varnished cambric	Flame-retardant cotton braid (switchboard wiring) Flame-retardant cotton braid
Asbestos	A	Asbestos	Without asbestos braid
Asbestos	AA	Asbestos	With asbestos braid or glass
Asbestos	AI	Impregnated asbestos	Without asbestos braid
Asbestos	AIA	Impregnated asbestos	With asbestos braid or glass
Paper	.....	Paper	Lead sheath

* Excerpted from table 310-2(b), National Electrical Code 1965, USAS C1-1965.
† Outer covering is not required over rubber insulations which have been specifically approved for the purpose.

Bare copper will oxidize rapidly at temperatures approaching 200°C; at these temperatures copper should be protected with silver or nickel coating. Nickel plating can safely be used for oxidation protection of copper up to 260°C. Above 260 and up to 400°C nickel cladding can be used on a continuous basis with permissible time exposures up to 700°C. Hollow copper conductors through which a coolant is forced are also available. Stainless-steel-clad copper has been used up to 650°C.

4-70 Wires and Cables

ALUMINUM: Although copper is the most widely used conductor material, there is an increasing application of aluminum conductors. Unfortunately, in most applications, direct substitution of copper is impossible without significant change in design because aluminum conductors have lower conductivity and increased brittleness. A significant area for the application of aluminum is in insulated strip conductors for distribution and power transformers. Anodization (surface oxidation) of aluminum conductors presents a unique approach to both mechanical protection and electrical insulation. Figure 27 is a chart for the dielectric strength of various anodized-aluminum surface thicknesses. Aluminum oxide is inorganic and therefore possesses many desirable electrical insulation properties, such as resistance to radiation, to aging at high temperatures (melting point 3600°F), and to chemical attack. The film is somewhat porous, but sealing treatments for protection against moisture are available.

Conductors for High-temperature Applications. The usable temperature range of copper conductors (bare and with protective coatings) is evaluated on the basis of oxidation, melting point, grain growth, and solid-state diffusion. Temperature ranges have been established in the form of a spectrum as follows:

1. Very low temperature (VLT) 70 to 90°C (160 to 195°F)
2. Low temperature (LT) 90 to 120°C (195 to 250°F)
3. Medium temperature (MT) 120 to 170°C (250 to 340°F)
4. High temperature (HT) 170 to 250°C (340 to 480°F)
5. Very high temperature (VHT) 250 to 400°C (480 to 750°F)
6. Ultrahigh temperature (UHT) 400 to 650°C (750 to 1200°F)
7. Superhigh temperature (SHT) 650 to 1000°C (1200 to 1830°F)
8. Extremely high temperature (EHT) 1000 to 1500°C (1830 to 2730°F)

Table 26 presents bare conductors, coated conductors, and magnetic materials suitable for use at each of the eight temperature ranges.

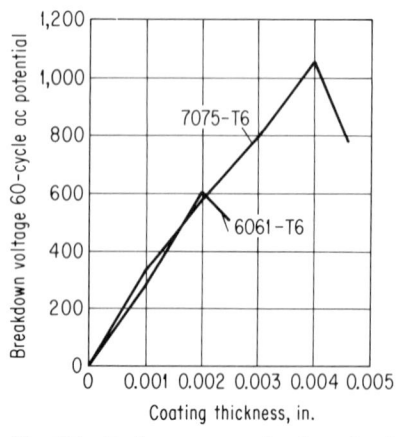

Fig. 27 Dielectric strength of anodized aluminum. (*Martin Marietta Corporation.*)

Insulations *Film Insulation.* ACRYLIC: Rated for 103°C temperature operation, acrylic coating is available with modifiers to produce an enamel with solderable characteristics. The basic unmodified resin is resistant to refrigerants and solvents. Major usage of the insulation has been in hermetically sealed motors. The nonsolderable version is covered by NEMA Standard MW4.

CERAMIC: Ceramic insulation may be used for temperatures as high as 650°C depending on the conductor utilized. Ceramic insulation exhibits good resistance to radiation but is difficult to handle and has poor moisture resistance. At present there is no specification covering ceramic insulation.

CERAMIC WITH OVERCOAT: Ceramic insulation with overcoats of polyimide, silicone, or polytetrafluoroethylene for moisture barrier and crack protection is available. Since it decomposes at temperatures below that of the ceramic, the overcoat material is the temperature-limiting factor. None of the overcoats is specifically covered in MIL-W-583. NEMA Standards MW8 and MW7 provide coverage for the Teflon- and silicone-overcoated constructions, respectively.

EPOXY: Epoxy enamels may be rated as high as 130°C. They exhibit good moisture, chemical, and corona resistance. A cement-coated epoxy magnet wire is available offering a self-bonding coating that can be used at elevated temperatures. This type of construction is suitable for self-supported coils. The cement-coated epoxy can be bonded by oven heating, resistance heating, or solvent activa-

TABLE 26 Thermal Spectrum for Electrical Insulation, Conductors, and Magnetic-circuit Elements from 70 to 150°C[24]

Class of materials	VLT (very low temperature) 70°C 90°C / 160°F 195°F	LT (low temperature) 90°C 120°C / 195°F 250°F	MT (medium temperature) 120°C 170°C / 250°F 340°F	HT (high temperature) 170°C 250°C / 340°F 480°F	VHT (very high temperature) 250°C 400°C / 480°F 750°F	UHT (ultrahigh temperature) 400°C 650°C / 750°F 1200°F	SHT (superhigh temperature) 650°C 1000°C / 1200°F 1830°F	EHT (extremely high temperature) 1000°C 1500°C / 1830°F 2730°F
Electrical insulation	Untreated cotton, paper, silk, etc.	Oil-filled or varnished cotton, paper, silk, enamels such as Formvar, nylon, etc.	Varnished glass and mica, Mylar and other polyester films, enamels such as polyurethane, polyester, epoxy, and combinations	Polyimide, silicone, TFE-fluorocarbon, silicone-varnished glass fibers, mica, etc., resins plus ceramic, polyimide plus glass fiber	Ceramic-coated wires, glass-bonded fiber glass, glass enamel, glass-bonded mica and asbestos, etc.	Glass-bonded fibers, glass plus ceramic, glass-bonded synthetic mica, glass enamel, etc.	Glass plus refractories, crystallized glass, quartz, ceramic fibers plus glass	Pure refractory oxides, sapphire, beryllia, magnesia
Conductors	Copper, aluminum	Copper, aluminum	Copper, aluminum	Nickel-plated copper, aluminum-copper at 180°C	Nickel-plated copper, nickel-clad copper, aluminum	Nickel-clad copper, stainless-steel-clad copper, Cufenic (nickel-iron-clad copper), nickel-clad silver	Inconel plus barrier over Inconel-clad silver*	Platinum
Magnetic materials	Iron	Iron	Iron	Iron	Iron	Iron (to 500°C), cobalt alloys	Cobalt alloys, cobalt	None available

* Trademark of Huntington Alloy Products Division, The International Nickel Co., Inc., Hungtington, W. Va.

4-72 Wires and Cables

tion. Epoxy-insulated magnet wire meets the requirements of MIL-W-583, Class 130, Types B, B2, B3, and B4. NEMA Standards MW14 and MW9 cover rectangular (square wires) and round wires, respectively. To date, there is no specification coverage of the cement-coated epoxy.

POLYAMIDE (NYLON): Nylon is used mainly for overcoating of magnet wire. It provides a tough, smooth surface for better windability, solvent resistance, abrasion resistance, and varnish compatibility. It can be soldered through with resin-alcohol flux and tin-lead solder. Its moisture resistance is poor, which may preclude its use where high insulation resistance is required.

Nylon is covered by MIL-W-583, Class 105, Types T1, T2, T3, and T4, and NEMA Standard MW6.

OLEORESINOUS: This film, consisting of a cured varnish made with a natural resin and a drying oil, is the oldest of the enameled magnet wires. During recent years natural resins have been replaced with synthetic materials. Oleoresinous coating is preferred for paper-filled coils and where low initial cost is important.

Oleoresinous enamels are covered by MIL-W-583, Class 105, Types E and E2. The NEMA Standard Classification is MW1.

POLYAMIDE-POLYIMIDE (AMIDE-IMIDE): A high-temperature insulation, amide-imide presents a host of desirable properties such as (1) toughness, smoothness, and the abrasion resistance of nylon, (2) good dielectric strength in a humid environment, (3) resistance to deformation at high temperatures, (4) operating temperature range up to 220°C, (5) resistance to solvents, (6) compatibility with insulating varnishes and encapsulants, and (7) good radiation resistance, withstanding gamma exposure up to 3×10^9 rads. Amide-imide insulation will meet requirements of MIL-W-583, Class 220, Types M1, M2, M3, and M4. No NEMA Standard has been prepared to date.

POLYESTER: Polyester is one of the oldest insulating-film materials. Significant improvements in the heat resistance of the coating have been made: temperature ratings range from 130 to 200°C. Polyester insulation is generally provided with an overcoat for improved physical performance. The insulation is susceptible to hydrolysis; encapsulation is recommended. Polyester-polyimide, a basic polyester material that has been modified with polyurethane and polyimide, was recently introduced; it is tougher and more heat-resistant than its predecessors. Polyester materials offer good solvent resistance and compatibility with insulating varnishes and encapsulating materials. Specifications MIL-W-583, Classes 155, 180, and 200, depending on formulation and overcoating, are applicable for Types L1, L2, L3, and L4; H1, H2, H3, and H4; and K, K2, K3, and K4. NEMA Standards are MW5 for Class 155, Round Wire; MW13 for Class 155, Rectangular and Square Wire; and MW25 for Class 180, Round Wire.

POLYTETRAFLUOROETHYLENE (TEFLON): Teflon is a high temperature insulation that may be used up to 260°C maximum. Silver- or nickel-coated conductors are recommended for use above 200°C. The insulation is applied as a dispersion coating, and then cured to remove the carrier. Teflon exhibits good electrical properties, flexibility, fair abrasion resistance and plasticity, but poor adhesion properties unless treated. Teflon magnet wire meets the requirements of MIL-W-583, Class 200, Types K1, K2, K3, and K4. NEMA Standard MW10 is applicable to coated round copper wire.

POLYURETHANE: Polyurethane polymers, a family of magnet-wire insulation materials, have captured a significant portion of the market. Polyurethane-coated wire can be soldered without prior removal of insulation. The film presents a tough coating that has good chemical, moisture, and corona resistance.

Polyurethane-coated, round copper magnet wire is covered by NEMA Standard MW2, thermal classification 105°C. (MIL-W-583 does not specifically cover polyurethane-coated wire.)

Polyurethane-insulated magnet wire is available with various overcoats: a friction surface overcoat is applied containing inorganic materials, which allows winding of basket as universal weave coils without the use of adhesives; polyvinyl butyral resin overcoat is applied to achieve a bondable wire with coating that may be

Magnet Wire 4-73

activated by heat or solvent; a combination overcoat or nylon and polyvinyl butyral is available for increased cut-through resistance in a bondable, 130°C version; nylon overcoat is available for improved windability and varnish compatibility. The construction is solderable without prior removal of the insulation. NEMA Standard MW28 is applicable to polyurethane-nylon coated round copper in thermal class 130°C. The construction is not specifically covered by MIL-W-853 but will meet the requirements of Class 130, Types B, B2, B3, and B4.

POLYVINYL FORMAL(FORMVAR*): This resin is the "old reliable" of film-insulated wire. Though challenged by many newly developed resins, Formvar is still the most widely used. Formvar insulation is suitable for use in Class A electric equipment. It has a high dielectric strength, good abrasion resistance, and good windability characteristics, and is compatible with most electrical insulations and varnishes. The thermal classification for polyvinyl formal is 105°C. NEMA Standards MW15 and MW18 cover the insulation for round and rectangular (square) wires, respectively. MIL-W-583, Class 105, Types T, T2, T3, and T4 is applicable.

Polyvinyl formal has been modified with isocyanates for resistance fluorinated refrigerants in hermetic use. NEMA Standard MW27 (proposed) covers this specific construction for round wire. Polyvinyl formal is also available with a nylon or a polyvinyl butyral overcoat for improved windability of self-bonding. NEMA Standards are MW17 for nylon-coated polyvinyl formal, and MW19 for self-bonding overcoat round wire. Nylon-overcoated construction meets the requirements of MIL-W-583, Class 105, Types T, T2, T3, and T4. Self-bonding overcoat construction is not covered by Military Specifications.

POLYIMIDE (PYRE-M.L.†): The development of Pyre-M.L. coating is an outstanding advance for magnet-wire insulation. Pyre-M.L. is the only enameled wire which has a 220 thermal classification. It is chemically inert and therefore is compatible with all varnishes and encapsulating compounds. Polyimide coating provides an extremely tough, abrasion-resistant film which exhibits high nuclear-radiation resistance, excellent thermal resistance, and good windability. NEMA Standard MW16 covers polyimide-coated round wire. MIL-W-583, Class 220, Types M, M2, M3, and M4, is applicable to polyimide-coated round, square, and rectangular wire.

Textile and Composite Insulation. ASBESTOS FIBER: Asbestos-fiber-covered magnet wire is available with phenolic or asphaltic impregnation for 130°C usage and with silicone for 180°C applications. There are no NEMA Standards for these constructions. MIL-W-583, Class 130, Type AV, covers asbestos varnish construction.

CELLULOSE ACETATE FIBER: Insulation is applied by one or more servings of fiber. There is no NEMA Standard. Construction meets the requirements of MIL-W-583, Class 90,‡ Type F and F2.

COTTON: Cotton-insulated wire is available with a single or double serve; impregnated or unimpregnated, it is the oldest such construction. Primary insulation is achieved through space separation. NEMA Standards are MW11, Class 90 or 105 for (impregnated) round copper wire, and MW12 for rectangular wire (MW12 has been omitted from the latest proposed standards). MIL-W-583, Class 90, Types C and C2 covers cotton insulation.

GLASS FIBER: Glass-fiber insulation is available with a wide variety of impregnating and bonding agents and in combination with polyester fibers. These constructions are employed when long service at high temperatures and high cut-through strength are required. Glass insulation provides insulation by spacing where extreme temperatures may drive out binders. Glass, used in combination with polyester fibers which are subsequently fused, provides a smooth surface and prevents fraying. Glass-fiber combinations are available in the following forms:

1. Glass-fiber-covered and impregnated round copper wire, Class 155, NEMA

---

* Trademark of Monsanto Co., St. Louis, Mo.
† Trademark of E. I. du Pont de Nemours & Co., Wilmington, Del.
‡ Obsoleted by Revision C. Not to be used for new design.

4-74 Wires and Cables

Standard MW41. Square or rectangular wire, MW42. Construction is covered by MIL-W-583, Class 130, Types GV and G2V.

2. Glass-fiber-covered, silicone-treated, round copper wire, Class 180, NEMA Standard MW44. Rectangular and square wires are covered by MW43. Construction is in accordance with MIL-W-583, Class 200, Types GH and G2H.

3. Polyester-glass-fiber-covered round copper wire, Class 155, NEMA Standard MW45. Rectangular and square wires are covered by MW46. MIL-W-583. Class 130, Types DG and DG2, covers the construction.

4. Polyamide (nylon)-fiber insulation is used in small rotating machines, in instruments, and where more spacing of conductors is needed than can be realized with enameled insulation. Nylon-fiber insulation is covered by NEMA Standard MW22 and MIL-W-583. Class 90, Types F and F2.

5. Paper insulation is used primarily in oil-filled transformers. NEMA Standards for round and rectangular conductors are MW31 and MW33, respectively. MIL-W-583 Class 90. Types P and P_2, is applicable to paper-insulated round wire.

6. Silk insulation is applied with one or two serves of yarn. NEMA Standard MW21 and MIL-W-583, Class 90, Types S and S2, are applicable.

Design considerations Table 27 contains the characteristics of most commonly used magnet wires. Following is a brief discussion of environmental, electrical, and mechanical considerations.

Environmental Requirements. In contrast to interconnection and hookup wire, the environmental requirements for magnet wire are largely self-imposed. High-temperature extremes are imposed by demands for increased efficiency, higher operating temperatures, and overload protection. Insulation compatibility is required to cope with a multitude of varnishes, encapsulants, and system fluids.

In recent years, radiation resistance has become a significant environmental consideration in industrial use and in military applications.

In addition to radiation, moisture is a significant external factor in the environment. High humidity can cause insulation degradation with the use of materials that are susceptible to hydrolysis or high moisture absorptions.

Electrical Requirements. The electrical requirements of magnet wire are not normally as stringent as those of exposed wire systems. A continuous insulation film is certainly desirable; however, the chance of breakdown caused by two discontinuities lined up in adjacent windings and facing each other is extremely small. The likelihood of this mode of failure is further decreased by coil impregnation, encapsulation, or potting.

Mechanical Requirements. Mechanical property considerations for magnet-wire insulation and conductors are as follows:

Conductor properties:
  1. Malleability
  2. Dimensional uniformity
  3. Solderability

Insulation properties:
  1. Surface condition of insulation—windability
  2. Abrasion resistance
  3. Flexibility
  4. Resistance to flow
  5. Solderability

Conductor malleability is an important quality in magnet wire. A conductor that has been hardened by excessive tension will not form an evenly wound package and will tend to spring out in unreeling and coiling winding. In addition, if a conductor has been hardened, reduced cross section and higher resistance are indicated.

Dimensional uniformity results in proper winding buildup and minimum "hot spots." The surface condition of the conductor, such as the lack of oxidation prior to the application of a conductive or insulating coating, has a significant effect on solderability.

TABLE 27 Film-insulated Wire Characteristics[25]

Insulation type	Class	Snap	Maximum flexibility	Abrasion Repeated, strokes	Abrasion Single, g	Dielectric strength, volts/mil	Solubility Naphtha	Solubility Toluol	Solubility Alcohol	Solubility Mild acid	Solubility Mild alkali	Completeness of cure	Cut-through, °C	Solderability, °F	Remarks
Black enamel (oleo resins)	A	OK	15% + 1 time	1–2	900	1,200–1,500	OK	OK	OK	OK	OK	Fails	160	No	Minimum abrasion, no solder
Formvar	A	OK	Snap + 1 time	60	2,700	2,000–2,500	OK	OK	OK	OK	OK	OK	225	1000	Excellent windability, excellent moisture resistance
Nyform	A	OK	Snap + 1 time	40	2,400	2,000–2,500	OK	OK	OK	OK	OK	OK	220	1000	Same as Formvar, slight moisture absorption
Thermoplastic-overcoated Formvar	A	OK	Snap + 1 time	30	No test	2,000–2,500	OK	OK	Softens for self-bond	OK	OK	No test	200	1000	Self-bonding when alcohol is applied
Solderable Acrilac	A	OK	Snap + 1 time	50	2,000	2,000–2,500	OK	OK	OK	OK	OK	OK	200	850	Good solderability, slight moisture absorption
Polyurethane	A	OK	Snap + 3 times	40	2,300	2,000–2,500	OK	OK	OK	OK	OK	OK	220	680	Excellent solderability
Thermoplastic-overcoated polyurethane	A	OK	Snap + 1 time	30	No test	2,000–2,500	OK	OK	Softens for self-bond	OK	OK	No test	220	680	Excellent solderability, self-bonding
Epoxy	B	OK	Snap + 3 times	25	1,400	2,000–2,500	OK	OK	OK	OK	OK	Fails	175	900	Chemically inert
Nylon over polyurethane	B	OK	Snap + 1 time	30	2,300	2,000–2,500	OK	OK	OK	OK	OK	OK	240	680	Excellent solderability and windability
Thermoplastic-overcoated epoxy	B	OK	Snap + 3 times	25	1,500	2,000–2,500	OK	OK	Softens for self-bond	OK	OK	Fails	175	900	Self-bonding
Polyester	F	OK	Snap + 1 time	50	2,000	2,000–2,500	OK	OK	OK	OK	OK	OK	300	No	No solder
Linear polyester over polyester	F	OK	Snap + 1 time	60	1,700	2,000–2,800	OK	OK	OK	OK	OK	OK	300	No	No solder, excellent abrasion resistance
Polythermalax	H	OK	Snap + 2 times	60	1,700	2,000–2,800	OK	OK	OK	OK	OK	OK	300	No	No solder, high abrasion resistance and thermal properties
Polyamide	250C	OK	Snap + 3 times	25	1,500	2,000–3,000	OK	OK	OK	OK	OK	OK	Over 500	No	No solder, excellent thermal properties

4-75

WIRE AND CABLE TERMINATIONS

Terminating hardware Wire and cable terminations are of major importance to design reliability. Careful selection of proper terminating hardward and the reduction of the number of terminations to a minimum should be primary design goals.

The following conditions must be evaluated in the selection and use of terminations: (1) termination life; (2) connection density; (3) comptability; (4) environment; (5) preparations; (6) mass production; (7) process control; (8) inspectability; (9) current and voltage and resistance limits; (10) maintenance tools; (11) repairability, including time and skill required; (12) contractual constraints.

A number of wire-attachment methods are used in wire terminations such as (1) crimping, (2) soldering, (3) clamping, (4) welding, (5) wrapping, and (6) friction.

Terminating devices normally used in electric installations are (1) studs, (2) lugs (crimp), (3) terminal posts (solder or wrap), (4) connectors (with solder or crimp-type contact terminal), (5) splices (crimp or solder), (6) compression

TABLE 28 Allowable Voltage between Terminals, Volts

Min. air space, in.	Creepage distance, in.	At sea level			At 50,000 ft			At 70,000 ft		
		Flash-over, rms	Working dc	Working ac	Flash-over, rms	Working dc	Working ac	Flash-over, rms	Working dc	Working ac
*	3/64	800	280	200	300	100	75	200	70	50
1/32*	1/16	1,400	490	350	500	190	125	375	125	90
3/64	5/64	2,000	700	500	700	210	175	500	175	125
1/16	7/64	2,500	840	600	900	315	225	600	210	150
5/64	1/8	3,000	1,050	750	1,050	360	260	675	230	165
3/32	5/32	3,600	1,260	900	1,200	420	300	750	260	185
1/8	3/16	4,500	1,550	1,100	1,400	490	350	900	310	225
3/16	1/4	6,100	2,000	1,500	1,800	630	450	1,100	375	275
1/4	5/16	7,300	2,500	1,800	2,000	700	500	1,300	455	325
5/16	3/8	8,500	2,900	2,100	2,300	810	575	1,420	500	355

NOTE: The allowable voltage is determined by the actual creepage distance or the minimum air space, whichever provides a lower rating. At 70,000 ft visible corona has been recorded by voltages as low as 350 volts rms. Consequently, at these elevations corona may be the limiting factor rather than flashover.

* Continuous insulation should be provided between electrical connections of 1/32 in. or less.

SOURCE: Martin Marietta Corporation.

screw lugs, (7) screw terminals (usually limited to use on barrier strips), (8) ferrules, (9) taper pins, and (10) pads and eyelets.

Terminals. TERMINAL LUGS: These are designed to establish electric connection between a wire and a connection point such as a stud.

TERMINAL POSTS: Terminal posts are used on terminal boards in the assembly type of wiring and on many components, such as electric connectors (solder-type) relays, transformers, lampholders, switches, etc.

DESIGN GUIDES FOR TERMINALS: Following are the "dos" and "don'ts" of terminals and wire terminations:

DO:
Use special, prebussed connector terminals where required.
Apply supplementary insulation sleeving over axial terminations where continuous insulation is not provided between adjacent terminations.
Ensure that electrical spacings between terminals conform to Table 28.

DON'T:
Use solder cap adapters to accommodate additional connectors or larger gages in connector terminals. Connect more than three leads to one terminal.

Wire and Cable Terminations 4-77

Twist multiple wires or leads to effect terminations. Terminate more than one wire in a connector terminal.

Terminal Boards. These are used for junctions or terminations of wire or cable assemblies as an aid to installation and maintenance.

STUD TERMINAL BOARD: The stud terminal board is generally a threaded post with the axial portion of its body firmly anchored into a mounting panel; it requires the use of tools for attachment of wire lugs. The features of the stud terminal board are listed below:
1. Provisions for four wires per stud
2. Greater mechanical strength than the barrier type
3. Adaptability to bus connections
4. Suitability for larger wire gages
5. Vibration resistance by proper choice of locking nut or vibration-proof washer

BARRIER TERMINAL BOARD: The barrier terminal board is molded of insulating material; it has integral raised barriers between pairs of screw terminals. Its features are listed below:
1. Longer leakage path than the study type between adjacent terminals
2. More connections in a given length of board
3. Limited current-carrying capacity
4. Poor adaptability to applications with high levels of dynamic stress

For maximum wire sizes that are applicable to terminal board termination refer to Table 29.

TABLE 29 Maximum Wire Gage—Terminal Board Termination

	Stud-type terminal board			Barrier-type terminal board	
	Stud size			Screw size	
	6	10	1/4–3/8	5/40	6/32
Copper wire (AWG)	12	6	2/0	18	18
Aluminum wire (AWG)	None	6	1/0	None	None

SOURCE: Martin Marietta Corporation.

TAPER-PIN TERMINAL BLOCK: This is composed of molded insulating material containing metal inserts designed to hold taper pins. Its features are listed below:
1. High-density construction
2. Provisions for mutual connection of four wires (in dual-insert type)
3. Limited range of wire gages
4. Special tooling requirement for pin insertion and pin-to-wire crimping

SPECIFICATIONS: Terminal boards should be installed in accordance with MIL-E-7080 for aircraft and MIL-E-25366 for missiles unless other specific requirements are established.

Splices. Permanent splices, available for both shielded and unshielded cables, should be used only when absolutely required. Conductor splices in interconnecting wiring should be grouped and located in designated areas selected for ready access. Where leads from electric equipment are spliced into a cable assembly, the splice area should be located as near to the equipment as practical. Nonpermanent splices should be avoided; however, certain very special applications may require their use.

Shield-wiring Terminations. There are two basic shield terminations: terminated to a shield common, and floating.

Shield termination merits careful consideration from the design phase through production. A judicious grouping of wires and careful examination of the need for shielding will alleviate termination problems. Following are generalized design recommendations:
1. Minimize the use of shielded wiring.
2. Avoid shielding of leads less than 4 in. long.

Wires and Cables

3. Provide lead segregation instead of shielding where this is practical.
4. Become familiar with all facets of shield termination techniques (see subsequent paragraphs here and on page 4-79) to ensure that the techniques fully satisfy the design environment.

Some of the above is, of course, not applicable to ac, pulsed, or rf leads and cables with significant radiation potential. The following shield terminations are in general practice:
1. Direct shield termination (pigtail)
2. Ferrule termination (crimp attachment)
3. Solder sleeve termination (solder attachment)

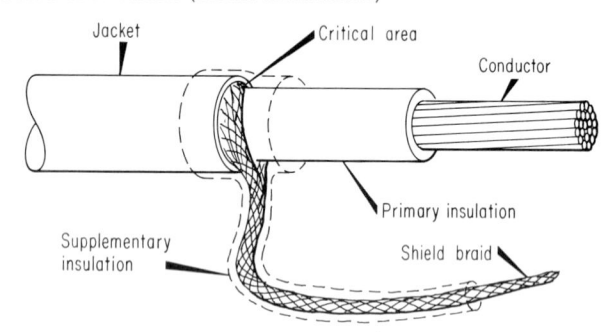

Fig. 28 Shield braid in pigtail. (*Martin Marietta Corporation.*)

Direct Shield Termination. An established practice is to form the shield braid into a pigtail as in Fig. 28. No special tooling is required, but the pigtail must be the braid at the breakout point or the shield braid strands. No external pressure should be applied to the breakout point by clamp, tie, or flexure. Supplementary insulation (sleeving) over the breakout or the braid and the braid itself is required.

Shield Ferrule Termination. Shield termination ferrules are available in two basic types: two-piece preinsulated, and single-piece uninsulated. The preinsulated

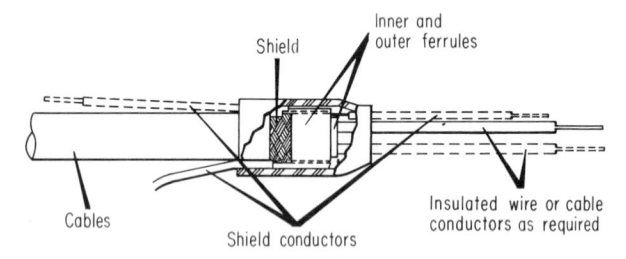

Fig. 29 Insulated ferrule shield termination. (*Martin Marietta Corporation.*)

type satisfies applications where the shield braid must be electrically isolated from other shields. The uninsulated ferrule would be unsuitable for this application. Although ferrules are lightweight, they add to the bulk of the harness or cable trunk. The bulk problem can be remedied by the staggering of ferrule positions back along the harness or cable trunk. However, this practice leads to degradation of overall shielding effectiveness. A typical application of two-piece insulated ferrule is shown in Fig. 29.

Solder Sleeve Termination. Solder sleeve is another shield termination device. Before it is selected, compatibility between solder melting temperature, insulating-sleeve shrink temperature, and the temperature resistance of the primary-wire insulation, as well as that of the jacket over the shield, must be determined. Grouping of shield conductors for solder sleeve applications is shown in Fig. 30. Solder sleeves permit the use of center-strip shield terminations to minimize the bulk of shield terminations at connector back shells and to allow continuation of shields closer to the point of termination for the shielded conductor.

Further design considerations involve shield grouping and collection. With direct shield termination the shield conductor (pigtail) may be formed by the twisted braid terminated in a crimped lug for connection to stud or screw. In certain applications the pigtail may be directly soldered to a suitable terminal post.

Multiple shield braids or multiple shield conductors extending from shield ferrules or solder sleeves may be collected into a single shield conductor as shown in Fig. 30, which also gives typical shield-conductor grouping schemes.

In floating terminations, a preferred practice, a short length of the shield is folded back over the outer sheath, and a short length of close-fitting insulating sleeving is applied over the fold-back. The heat-shrinkage tubing will provide snugly fitting insulation that is safely retained in place. This method should be used for floating shield terminations previously discussed.

RF Cable (Coaxial) Termination. Terminations for rf cables may be selected from MIL-HDBK-216. Straight rf connectors of the TNC (threaded coupling) type are desirable. The right-angle type of rf connectors and adapters should be avoided because of the inherent mechanical weakness of many designs which use brazed metal housings.

Identification Identification of wiring and cabling includes the marking or coding of individual wire leads, harnesses, cables, and termination devices. Wire identification facilities design control and traceability (wiring diagram to hardware, etc.), manufacturing efficiency, and maintenance (troubleshooting). Identifying markings on harness and cable assemblies usually provide usage information, interconnection instructions, and part numbers. The marking may also include serial number, source, assembly date, lot number, etc. Marking is also useful for inventory control and supply stock records.

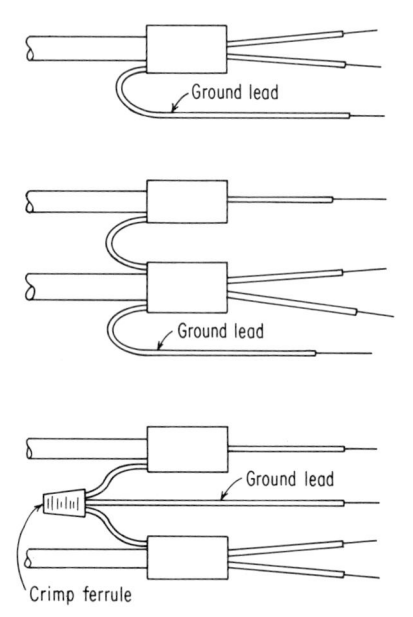

Fig. 30 Typical shield terminating grouping practice. (*Martin Marietta Corporation.*)

Wire Marking. MIL-STD-681 is applicable to various wire-marking methods. Color stripes, bands, and numbers are acceptable. Numbering can be a relatively simple, sequential matter, beginning with number 1 and progressing consecutively to the highest number required for an assembly.

Certain military requirements specify a coded marking for individual wires, which includes (1) unit number, (2) equipment identity or circuit function, (3) wire number, (4) wire segment letter, (5) wire gage, and (6) ground, phase, or thermocouple letter. Hot impression stamping or color banding of the required wire identification is a practical user production marking method. Some wire-insulation materials are difficult to mark by this method. In these cases, a short length of close-fitting insulating sleeving or shrinkable tubing with the marking is slipped over the wire, adjacent to its point of termination.

Harness Marking. Wire-harness assembly marking should be as simple as possible to convey the information required. A simple method is to use a short length of close-fitting insulating sleeving over the harness trunk, adjacent to each termina-

4-80 Wires and Cables

tion. Identification can be applied by hot impression stamping of the thermoplastic sleeve.

Cable Marking. Sheathed cables are identified in a manner similar to wiring-harness assemblies. The most significant difference between the two involves the materials used in the actual marking. A cable-marking method for production cables uses a reflective label with pressure-sensitive, adhesive backing. Other types of cable marking used in military electric applications have specific disadvantages:

1. Embossed metal bands or straps are difficult to read under poor lighting conditions. Retention is difficult. The band may constitute a hazard to cable sheath and personnel.

2. Hot-impression-stamped thermoplastic sleeves, the type used on wiring harnesses, are difficult to retain in position. From experience, this type of marking on sheathed cables may discolor under field conditions, and the marking information may thus be obliterated.

Associated hardware *Insulation Sleeving and Tubing.* Insulation sleeving and tubing serve a multiple purpose in electrical assembly and harness fabrication. They are used for insulating, protection from chafing or abrasion, jacketing, strain relief, thermal or chemical protection, and identification. For best selection of sleeving a thorough knowledge of the physical and chemical properties of the materials employed and of the available construction and configurations is required. Following is a review of sleeving and tubing characteristics.

MATERIALS: Extruded tubings are made from all the plastic and rubber materials discussed in the section on Wire and Cable Insulation. Material characteristics described there apply to insulating tubing as well and will not be reviewed here.

A major category not previously covered is braided insulating sleeving. Braided sleeving is made from basic uncoated yarns, lightly treated yarns, or yarns heavily coated with various insulating varnishes or resins. Although practically any yarn can be used for braiding, the materials most frequently used are fiber glass, cotton, rayon, and asbestos. Untreated braided sleevings have excellent flexibility, but the only space advantage they offer is with respect to electrical insulation. Sleevings may be coated with various resin compounds such as PVC, oleoresinous varnish, acrylic varnish, polyester, polyurethane, epoxy, polyimide, polytetrafluoroethylene, silicone varnish, and rubber. Such coating reduces flexibility; however, elasticity of the finished sleeving will thus largely depend on the choice of coating material. Elastomeric materials, such as silicone rubber, yield the highest flexibility.

Military Specifications. MIL-I-631, MIL-I-22076, and MIL-I-7444 are the specifications for extruded polyvinyl chloride tubing. Refer to Table 30 to determine the characteristics for proper selection.

Extruded polytetrafluoroethane (TFE) tubing is covered by MIL-I-22129 for thin-wall electric applications and by MIL-P-22296 for heavier-wall, larger-size-range tubing designed mainly for mechanical applications.

Neoprene tubing for harnessing applications that require severe abrasion resistance is covered by MIL-R-6855.

Federal Specification ZZ-R-765 covers silicone rubber tubing for extraflexible heat-resistant harness applications.

Specifications for braided sleeving with optional coatings are available in MIL-I-3190, MIL-I-18057, and MIL-I-21557. Refer to Table 31 to determine the proper selection for a specific application.

INDUSTRIAL SPECIFICATIONS: ASTM D 922, Grades A, B, and C, covers general-purpose low-temperature and high-temperature extruded PVC tubing. NEMA VS-1 and ASTM D 372 cover coated braided sleeving.

Shrinkable Tubing. Shrinkable tubing is based on the theory of elastic memory. Under specific thermal and mechanical conditions, molecules of certain materials may be overexpanded, and then frozen in place in a strained condition; finally, when heated, the material tends to return to its original shape and size as strains are released.

MATERIALS: Heat-shrinkable tubing is available in many of the plastics and elastomers listed in the section on Wire and Cable Insulation. The properties of

TABLE 30 Extruded Tubing

Tubing type	Size range	Wall thickness nominal, in.	Temperature range, °C	Colors	Flammability	Applicable specifications	Characteristics
Polyvinyl chloride (PVC)	AWG 20 to 2 in.	0.016–0.060	−20 to +105	Clear and colors	Self-extinguishing	MIL-I-631 Gr. C; Cl. I, Cat. I	Fungus- and flame-resistant
High-temperature	AWG 24 to 2 in.	0.012–0.060	−10 to +105	Clear and colors	Self-extinguishing	ASTM D922, Gr. C, UL–105	Flame-resistant
Polyvinyl chloride, general-purpose	AWG 20 to 2 in.	0.016–0.060	−30 to +80	Clear and colors	Self-extinguishing	MIL-I-631, Gr. a, Cl. I, Cat. I	Fungus- and flame-resistant
	AWG 24 to 2 in.	0.012–0.060	−30 to +60	Colors only	Self-extinguishing	ASTM D922 Grade A	Good printability
Polyvinyl chloride (PVC) low-temperature	AWG 20 to 2 in.	0.016–0.060	−46 to +80	Clear and colors	Self-extinguishing	MIL-I-631, Gr. 6, Cl. I, Cat. I	Fungus- and flame-resistant, good dielectric
	AWG 24 to 2 in.	0.012–0.060	−55 to +80	Clear and colors	Self-extinguishing	MIL-I-22076	Good low and high temp, noncorrosive, flame- and fungus-resistant
	AWG 24 to 2 1/2 in.	0.012–0.070	Cl I: −68 to +80 Cl. II: −55 to +80	Clear and colors	Self-extinguishing	MIL-I-7444	Corrosion-resistant, good low temperature
Polytetrafluoro-ethylene (TFE)	AWG 30 to 0	0.009–0.020	−70 to 250	Natural and colors	Non-burning	MIL-I-22129	Excellent dielectric, chemically inert
	1/16 to 3 in.	As specified	−70 to 250	Natural	Non-burning	MIL-P-22296	Mechanical applications, heavy wall, abrasion-resistant
Polychloroprene (neoprene)	1/8 to 1/8 to	Light, 3/64–1/16 Heavy, 3/32–1/8 Extra heavy, 3/16–1/4	−40 to 75	Black	Self-extinguishing	MIL-R-6855, Class II	Mechanical applications, oil-resistant
Silicone rubber	AWG 24 to 2 in.	0.012–0.060	−75 to 200	White	Self-extinguishing	ZZ-R-765, Class III, Grade 60	Tear-resistant, good flexibility

SOURCE: Martin Marietta Corporation.

TABLE 31 Braided Sleeving

Tubing type	Size range	Temperature range, °C	Dielectric strength, volts	Wall thickness, in.	Color	Specifications	Characteristics
Varnished cotton, rayon, or nylon	AWG 24 to 1 in.	−10 to 105	To 7,000	……	Natural, yellow, and black std. for classes A and B	MIL-I-3190, Cl. A NEMA vs 1, Ty 1 ASTM D372	Low moisture absorption; oil- and acid-resistant; high tensile strength; good flexibility
Varnished glass	AWG 24 to 1 in.	−10 to 130	To 7,000	……	Natural, yellow, and black std.	MIL-I-3190, Cl. B, NEMA vs 1, Ty 2 ASTM D372	Chemical-resistant flexible; tear and moisture-resistant
Silicone-varnished glass	AWG 24 to 1 in.	−60 to 200	To 7,000	……	Natural	MIL-I-3190, Cl. H, NEMA vs 1, Ty 4	Heat-resistant, compatible with magnet wire coatings
Vinyl-coated glass	AWG 24 to 1 in.	−10 to 130	To 8,000	……	……	MIL-I-21557, NEMA vs 1, Ty 3	Oil- and solvent-resistant, hot-spot temperature 130°C
Acrylic-coated glass	AWG 24 to 1 in.	−10 to 155	To 7,000	……	Natural, yellow, and black std.	NEMA vs 1, Ty 6 ASTM D372	Tough; abrasion-resistant; chemical-resistant
Silicone-rubber-coated glass	AWG 24 to 1 in.	−70 to 200	To 8,000	0.030–0.075	Natural	MIL-I-19057, NEMA vs 1, Ty 5	Extra-flexible; resists bending and flexing; radiation resistant
Polytetrafluoroethylene-coated glass	AWG 24 to 1 in.	−80 to 250	……	……	Natural	Not covered to date—tentative NEMA Ty 4	Excellent resistance to fluids; high heat resistance; nonflammable
Polyimide-coated glass	AWG 24 to 1 in.	−70 to 250	……	……	Natural	Not covered to date—tentative NEMA Ty 4	Tough; high heat resistance; nonflammable

Source: Martin Marietta Corporation

these heat-shrinkable materials are comparable with those of conventional materials. Heat-shrinkable tubing is available in:
1. Polyvinyl chloride (PVC)
2. Polyolefin (irradiated)
3. Polyvinylidene fluoride (Kynar)
4. Polychloroprene (neoprene)
5. Silicone rubber
6. Butyl rubber
7. Polytetrafluoroethylene (TFE)
8. Fluorinated ethylene propylene (FEP)

Table 32 presents typical properties of heat-shrinkable tubings.

SPECIFICATIONS: Military Specification MIL-I-23053 covers heat-shrinkable polyvinyl chloride, polyolefin, and polytetrafluoroethylene materials. The specification has six classes which are given in Table 33.

Industrial SAE Specifications AMA-3636, 3637, 3638, and 3639, respectively, cover the following heat-shrinkable polyolefin tubing: pigmented, flexible; clear, flexible; pigmented, semirigid; and clear, semirigid.

Shrinkable Devices. MELTABLE LINER: A variation of the standard shrinkable tubing is of dual-wall construction. The inner wall is composed of noncrosslinked material. With the proper amount of heat the inner wall melts, and the outer shrinks, forcing the melted material into voids. The meltable-liner construction is applicable to tubing and end-cap devices, offering the additional advantage of moisture sealing.

MOLDED SHAPES: A great variety of shrinkable molded shapes (boots, etc.) that offer bend and strain relief are available in polyolefin, neoprene, silicone, and butyl materials. In addition to connector boots, shrinkable molded breakouts are available for harnessing applications. Adhesives are used to attach the molded shapes to connector shell and cable jacket. The interior of the boot or breakout may be potted to ensure complete encapsulation and moisture resistance.

SOLDER SLEEVE: This shield termination device consists of a crosslinked, shrinkable plastic tube containing a flux-cored, preformed solder ring at the center, with thermoplastic rings at either end. The device is placed over a cable shield and ground wire; then heat is applied to melt the solder and the thermoplastic rings and to shrink the outer sleeve, all in one operation. Softening of the primary thermoplastic insulation and flow of the conductor must be prevented by careful heating.

Tubular Zipper Tubing. Tubular jacketing with zipper closure is available in polyvinyl chloride, polyethylene (plain or irradiated), polyvinyl chloride-impregnated nylon cloth, or fiber glass and polytetrafluoroethylene-impregnated glass cloth. Combinations of the above materials are available in conjunction with shielding materials of aluminum, conetic, or netic foil, for shielding from magnetic or low-frequency interference.

Zipper tubing offers the unique advantage of applying a harness jacket without regard to size, shape, and quantity of conductor or size of conductors. Reworking or repair of cable harness can be accomplished with minimum complexity. The zipper track may be permanently sealed with an adhesive for optimum moisture and water protection. Special configurations such as breakouts, multiple channels, or boots are available for specific applications.

Wire and Cable Mounting and Spacing Hardware. Good installation-design practice requires adequate space not only for wiring and cabling but also for the supporting hardware (clamps, sleeving, grommets, guides, etc.). Space is also needed for manipulating tools during initial installation as well as during maintenance and replacement.

HARDWARE: Cable mounting with the MS type of cable clamp is a proved method. Many variations of this clamp are available from specialty suppliers. The principal advantages of MS clamps are low cost, light weight, high strength, ready adaptability, and ease of installation and servicing. Clamps can be installed on any structure or skin of adequate strength that can be drilled. If the structure or skin cannot be drilled, bonding is recommended. One technique is to bond a cable-supporting

TABLE 32 Typical Heat-shrinkable Tubing Properties[26]

Properties	Irradiated polyolefin Flexible opaque	Flexible clear	Semirigid opaque	Semirigid clear	Dual wall	Irradiated PVF$_2$	Flexible PVC	Flexible irradiated PVC	Semirigid irradiated PVC	Neoprene rubber	Silicone rubber	Butyl rubber	PTFE
Tensile strength, psi	2,500	2,500	3,000	3,000	2,000	7,000	3,000	3,000	5,000	1,900	900	1,600	4,500
Ultimate elongation, %	400	400	400	400	400	300	300	300	250	220	300	350	250
Brittleness temperature, °C	−60	−85	−60	−90	....	−73	−20	−20	−20	−40	−75	....	−90
Hardness	98A	90A	....	....	....	....	85A	....	....	85A	70A	80A	....
Specific gravity	1.3	0.93	1.3	0.95	0.94	1.76	1.4	1.35	1.4	1.4	1.2	1.2	2.2
Water absorption, %	0.05	0.01	0.05	0.01	0.1	0.1	....	0.6	0.6	0.5	0.5	0.1	0.01
Dielectric strength, volts/min	1,300	1,300	1,300	1,300	1,100	1,500	750	750	900	300	300	130	1,200
Volume resistance, ohm-cm	10^{15}	10^{17}	10^{15}	10^{17}	10^{16}	....	10^{12}	10^{12}	$>10^{13}$	10^{11}	10^{15}	10^{12}	10^{18}
Dielectric constant	2.7	2.3	2.7	2.4	2.4	....	5.4	....	....	....	3.3	....	2.1
Power factor	0.003	0.0003	0.003	0.0003	0.0005	....	0.12	....	....	....	....	....	0.0002
Fungus resistance	Inert	Inert	Inert	Inert	Inert	Inert	Inert	Inert	Inert	Inert	Inert	Inert	Inert
Fuel and oil resistance	Excellent	Excellent	Excellent	Excellent	Excellent	Excellent	....	R exc.	Excellent	Good	Fair	Fair	Excellent
Hydraulic fluid resistance	Excellent	Excellent	Excellent	Excellent	Excellent	Excellent	....	Excellent	Excellent	Fair	Poor	Good	Excellent
Solvent resistance	Good	Good	Good	Good	Good	Excellent	....	Excellent	Excellent	Fair	Fair	Fair	Excellent
Acid and alkali resistance	Excellent	Excellent	Excellent	Excellent	Excellent	Excellent	....	Excellent	Excellent	Good	Good	Good	Excellent
Flammability	Self-extinguishing	Burns slowly	Self-extinguishing	Burns slowly	Excellent	Non-burning	Self-extinguishing	Self-extinguishing	Self-extinguishing	Self-extinguishing	Self-extinguishing	Burns slowly	Non-burning

device or pad to the supporting area, and then strap, tie, or clamp the cable to it. An alternative technique is to bond the entire cable to the supporting area for a very secure installation. Since this technique inhibits design flexibility, it is best used with firm design or under more flexible conditions where spare wires can be installed. In areas where large spaces must be spanned without available support, either an integral cable support or an auxiliary structure must be installed. Integral cable supports can be tubes or geometric extruded shapes in the center of the cable. A variation of this method is to mold the entire cable into a rigid mass, which, however, greatly reduces design flexibility. If bonding is the only possible means of attachment, MS nylon, reinforced nylon, or Kynar harness straps, mounting plates, and a compatible bonding material can be used. Specification MIL-S-23190 covers adjustable plastic cable straps in military use.

TABLE 33 Classification of Heat-shrinkable Tubing Materials in Accordance with MIL-I-23053

Classification	Material	Consistency	Flame resistance	Pigmentation	Temp. rating, °C	Colors
Class 1 ...	Polyolefin (irradiated)	Flexible	Flame-retarded	Yes	135	Black, red, yellow, blue, white
Class 2 ...	Polyolefin (irradiated)	Flexible	Non-flame-retarded	No	135	Clear
Class 3 ...	Polyolefin (irradiated)	Semirigid	Flame-retarded	Yes	135	Black, red, yellow, blue, green, slate, brown, white
Class 4 ...	Polyolefin (irradiated)	Semirigid	Non-flame-retarded	No	....	Clear
Class 5 ...	Polyvinyl chloride (irradiated)	Flexible	......	Yes	105	Black
Class 6 ...	Polytetrafluoro-ethylene	....	......	No	250	Clear

SOURCE: Martin Marietta Corporation.

SUPPORT SPACING: The spacing of clamps and other cable-support tie-down devices can be determined from experience and development mock-ups. Applicable electric system specifications generally establish bundle tie-down spacing by stating a maximum distance between supports (MIL-W-8160 maximum spacing 24 in.). For adequate design of electric cable installations in missiles and space vehicles, the spacing must be resolved analytically and tested for verification. The spacing will be determined by the dynamic environment in which the cabling must reliably perform. Therefore, the installation designer must coordinate his cable tie-down spacing with the dynamics specialist for mechanical stability under extreme dynamic conditions. Complete design coordination must exist between the structures, dynamics, and electrical installations engineers in order to meet system requirements. Dynamic tests on development hardware are recommended early in a program for verifying the installation.

Drastic changes in cable stiffness or section size caused by the ending or branching of wires may lead to points of dynamic weakness. Firm support is recommended

4-86 Wires and Cables

on both sides immediately adjacent to these points, regardless of the spacing of other tie-downs or clamps.

The spacing of supporting devices on a high-acceleration missile system can be determined by the formula

$$F = \Sigma LANG$$

where F = design load of attachment device, lb
L = unsupported length, in.
A = unit weight per length, lb/in., for each wire size
N = number of each wire size in bundle
G = maximum dynamic environmental load in g's

A sample calculation to illustrate a design example is given below.

Given harness parameters are:
  10 unshielded wires size 20 AWG
  10 shielded and jacketed cables size 20 AWG
  10 twisted, shielded, and jacketed pairs size 26 AWG

$$G \text{ load} = 150g$$
$$F = 50 \text{ lb}$$
$$A_{20u} = 4.02 \times 10^{-4} \text{ lb/in.}$$
$$A_{20} = 6.36 \times 10^{-4} \text{ lb/in.}$$
$$A_{20w3} = 4.98 \times 10^{-4} \text{ lb/in.}$$

F is indicated as 50 lb. However, a safety factor of 2:1 changes this value to

$$F = \frac{50}{2} = 25 \text{ lb}$$

From this,

$$L = \frac{25}{150} \frac{1}{10(4.02 \times 10^{-4}) + 10(6.36 \times 10^{-4}) + 10(4.98 \times 10^{-4})}$$

$$L = \frac{1}{6} \frac{1}{1.536 \times 10^{-2}}$$

$$L = 10.8 \text{ in.}$$

The sample harness must be clamped or attached every 10.8 in. to satisfy the given conditions.

The above details indicate a technique evolved under a specific set of requirements and are presented as a guide only.

SUPPORT AND CLAMPING OF CABLES TO CONNECTORS: The cable clamp associated with a multipin connector is used primarily to support the wire(s) or the cable terminating at the connector and also to relieve strain from the terminations. Soft telescoping bushings (in accordance with Specification AN3420) are available for cables smaller than the cable clamp opening. The bushings permit the cable to be centered and anchored securely without excessive padding. Selection of the proper size of bushing permits easy assembly of the clamp. There should be adequate clamping pressure without bottoming the two halves of the clamp. Clamp screw-thread engagement should be equal to ⅔ to ½ times the major nominal screw diameter.

Major differences in size between cable and connector can be corrected with step-up or step-down telescoping extension sleeves instead of bushings.

REFERENCES

1. Schuh, A. G., and J. Penkacik: High Strength Alloy Conductors, Symposium on Communication Wires and Cables, December, 1964.
2. Copperweld Steel Company: Copperweld Wire for Electronic Applications.
3. Rome Cable Division: "The Rome Cable Manual of Technical Information," copyright 1957, Rome Cable Corp.

4. Amphenol Corporation: Cable Products Catalog ACD-5.
5. *Military Specification* MIL-W-81044.
6. Bigelow, N. R.: Development and Evaluation of a Lightweight Airframe and Hookup Wire for Aerospace Applications, Bureau of Naval Weapons Symposium, Oct. 13–14, 1964.
7. Martin Marietta Corp.: Extra Flexible Tactical Cable Report no. 3, December, 1964.
8. The Martin Company: "Electrical Design," copyright, 1958.
9. Reed, J. C.: Save Space by Hookup Wire Insulated with Teflon, *J. Teflon*, 1964.
10. Campbell, F. J., C. L. Baggett, R. J. Flaherty, and J. A. Kimball: Wire Insulation Thermal Life Studies, Bureau of Naval Weapons Symposium, Oct. 12–13, 1965.
11. Heslop, W. R.: Evaluation of MIL-W-81044 Wire, October, 1966.
12. Lewis, L. L.: Ultra Thin-wall Wire Insulation from "Kapton" Polyimide Film, Type XF, Naval Air Systems Command Symposium, October, 1966.
13. E. I. du Pont de Nemours & Co.: Technical data.
14. Schwartz, S., and D. L. Wells: Processing of Plastics in Space, *SPE J.*, August, 1962.
15. Prise, W. J.: When the Gamma Heat Is On. Insulators, *Electronic Design*, May 23, 1968.
16. Jolley, E. E., and J. C. Reed: The Effects of Space Environments on Insulation of Teflon TFE and FEP Resins, Signal Corps Symposium, November, 1962.
17. Lanza, V. L., and R. M. Halperin: The Design and Development of Wire Insulators for Use in the Environment of Outer Space, Signal Corps Symposium, December, 1963.
18. Reed, J. C., and J. T. Walbert: "Teflon" Fluorocarbon Resins in Space Environments.
19. French, E. M.: Electrical Interconnections and Cabling, *TOS 521 Study Rep.*, December, 1964.
20. Byram, K. C.: Flat Conductor Cabling and Connectors.
21. *National Aerospace Standard* NAS-729.
22. "Flexible Flat Cable Handbook," Institute of Printed Circuits, copyright, 1965.
23. "Flexprint Circuit Design Handbook," *Sanders Associates Bull.* FT-169, 1965.
24. Pendleton, W. W.: Advanced Magnet Wire Systems, *Electro-technol.*, October, 1963.
25. Martin, G. C.: Insulated Magnet WIRE Characteristics, EDN August, 1963.
26. *Insulation*, Directory Encyclopedia Issue, May-June, 1966.
27. Lybeck, A. H.: Hookup Wires: State of Art Review, *Insulation*, May, 1966.

Chapter **5**

Coatings for Electronics[*]

J. J. LICARI
and
E. R. BRANDS

Autonetics, Division of North American Rockwell
Corporation, Anaheim, California

Functions of Coatings	5–2
Nature of Coatings	5–2
End-use Coatings	5–3
Properties of Electronic Coatings	5–25
Electrical Protection	5–25
Environmental Protection	5–36
Mechanical Protection	5–48
Application Technology	5–51
Cleaning and Cleaning Methods	5–51
Application Methods	5–55
Specifications	5–72
Material Specifications	5–72
Acceptance Specifications	5–72
Process Specifications	5–74
References	5–74

[*] Adapted from J. J. Licari, "Plastic Coatings for Electronics," McGraw-Hill Book Company, New York, 1970.

FUNCTIONS OF COATINGS

Nature of Coatings

Most of the electrical properties data available for plastic molding and casting compounds apply also to coatings, because the majority of coatings are based on the same polymer resin systems. Other basic properties, such as resistance to chemicals, solvents, and environmental conditions, will also be similar for coatings as for the bulk plastics. Basic information on the various classes of polymeric (plastic) materials can be found in Chap. 1, and will therefore not be repeated here for coatings. Likewise, many basic properties will not be covered where bulk plastic properties suffice for coatings. Many electrical and electronics-related properties are defined and discussed in Chap. 2, to which the reader is also referred. However, additional formulating is necessary to obtain properties such as adhesion, low viscosity, good application properties in thin films, cohesive strength, flexibility, low moisture-vapor transmission, and high dielectric properties. Other resins or functional chemicals are often added in order to modify polymer basic properties. For instance, more active curing agents may be necessary in thermosetting coatings than are used in bulkier sections of resin (such as potting compounds) wherein exothermic heat can accumulate to speed the reaction.

Other problems, directly related to thinness, are the need for low porosity, and freedom from pinholes and foreign inclusions. Most coatings are applied as solutions of the resins in suitable solvents. This enables the application of uniformly thin coatings (and often limits the possible thickness), but causes other problems. Freeing the films of the last vestiges of these solvents is sometimes difficult, but necessary in order to develop the full polymer properties.

The following is a discussion of function and end use of coatings for electronics, specific test methods, properties for coatings, and application technology, including surface preparation and applicable specifications.

Property—Function—End Use. The following definitions of property, function, and end use are those used in the discussion of coatings.

PROPERTY: A property is that single characteristic which can be separated and measured in terms of established units. Examples are dielectic breakdown voltage, hardness, and moisture-vapor transmission.

The significant properties of coatings are, of course, directly related to the desired end use, or functions, of the coating. As mentioned, a given end-use coating must fulfill a number of different functions and, hence, often possess a myriad of different properties. There presently exists no perfect coating, since there are no combinations of perfect properties. All coatings, therefore, will exhibit advantages and disadvantages for a given end use. Presumably, those coatings having a preponderance of advantages are being or have been employed for the given end use. A listing of end uses and recommended coatings is to be found in Table 1.

FUNCTION: A function is defined as a general area of performance such as electrical insulation or environmental protection and consists of a collection of "properties." For instance, electrical insulation depends upon properties such as dielectric strength, volume resistivity, surface resistivity, and dielectric constant. The different conditions present during performance of the function will determine which of the required properties are most important. For example, in a high-humidity environment, properties such as moisture absorption and freedom from electrolytes (purity) would be major factors. In a high-temperature environment, the softening temperature and chemical stability would be influential. Environmental protection would depend upon some integrations of properties such as moisture absorption, moisture transmission, adhesion, hardness, and toughness.

END USE: End use is the actual service to which a coating is put. Typically, a given end-use coating must fulfill a number of different functions. An example might be conformal coatings for circuitry. Here, a minimum listing of required functions would include electrical insulation, environmental resistance, filleting and bonding of smaller components, ruggedization to handling abuse, and vibration damping.

In order to choose a coating for a given end use, one must analyze and decide which of the many available coatings will best satisfy the functional requirements for the desired end use. Data are provided here as an aid to making such selections.

End-use Coatings

Printed circuit-board coatings *Coating Types.* The most commonly used coatings for circuit boards are the polyurethanes, epoxies, silicones, polystyrenes, and polyvinyl fluoride. Their choice depends largely on the operating and storage requirements of the circuit boards. For high-temperature applications, for example, only silicones should be used; whereas for resolderability, polyurethanes are outstanding; and for adhesion and moisture resistance, epoxies are preferred. The most widely used circuit-board coatings are covered by MIL-I-46058, according to which three classes of coatings are defined as follows:

    Class A. General-purpose coating:
        Type ER—epoxy
        Type PUR—polyurethane
    Class B. Heat-resistant coating:
        Type SR—silicone
    Class C. Low-loss dielectric coating:
        Type PO—Polystyrene

Reliability. Printed-circuit boards are usually coated as the last step in their fabrication to improve the reliability of the entire assembly. This is particularly important for boards used in military electronics and astrionics, where the value of a completed board may easily be in the $3,000 to $5,000 range and where storage and operational time may be as long as eight years. It has been demonstrated numerous times and by many firms that without a protective coating, these assemblies will fail owing to such mechanisms as moisture degradation, large decreases in electrical insulation resistance, electric short circuiting and corrosion. Even for commercial applications such as television and radio, organically coated circuit boards prolong service life.[1]

After six months in a 95 percent relative-humidity environment, a board coated with 1.5 mils (0.0015 in.) of polyurethane showed no signs of corrosion and passed all electrical functional tests. A similar unprotected board showed excessive corrosion on components, solder, and conductor lines after only two days of the same exposure. To ensure maximum reliability of circuit boards and other electronic assemblies, defense contracts require hardware to meet stringent environmental tests called out in specifications such as MIL-E-5272, MIL-E-5400, MIL-E-16400E, and MIL-STD-202. These specifications call for resistance to humidity and temperature exposure (Fig. 1), salt spray, abrasion, impact, fungus, and other tests. Table 2 gives a list of tests defined in MIL-STD-202. Many of these are applicable to coatings either per se or in combination with a component. Thus, to meet these requirements, the need for a coating or some type of encapsulant protection becomes mandatory. Figure 2 shows the corrosive effects of 50-hr salt spray on an uncoated circuit board and the protection afforded by a 1-mil polyurethane coating under the same conditions. An insulation resistance drop of 10^6 to 10^7 ohms is often experienced with printed circuitry on uncoated epoxy or phenolic laminates when exposed to warm, humid environments, compared to 10^2- to 10^3-ohm drops for coated boards (Fig. 3 and Table 3).

In addition to minimizing electrical degradation of laminates, coatings also minimize or prevent some discoloration effects caused by moisture and processing conditions. With some laminates, including epoxy-glass, whitening (also referred to as "measling") occurs when the board, which has gone through various processing steps, is subjected to a warm, humid environment. Although structurally not serious, this effect has been the cause for rejection of many completed assembled boards because of the poor aesthetics. A coating possessing good humidity-resistance prop-

TABLE 1 Properties of Coatings by Polymer Type

	Electrical properties				Physical characteristics			
Coating type	Volume resistivity, ohm-cm (ASTM D 257)	Dielectric strength, volts/mil	Dielectric constant	Dissipation factor	Maximum continuous service temperature, °F	Adhesion to metals	Flexibility	Approximate Sward hardness (higher number is harder)	Abrasion resistance
Acrylic	10^{14}–10^{15}	450–550	2.7–3.5	0.02–0.06	180	Good	Good	12–24	Fair
Alkyd	10^{14}	300–350	4.5–5.0	0.003–0.06	200 250 T.S.	Excellent	Fair to good Low temperature—poor	3–13 (air dry) 10–24 (bake)	Fair
Cellulosic (nitrate butyrate)		250–400	3.2–6.2		180	Good	Good Low temperature—poor	10–15	
Chlorinated poly-ether (Penton*)	10^{15}	400	3.0	0.01	250	Excellent	Good		
Epoxy-amine cure	10^{14} at 30°C 10^{10} at 105°C	400–550	3.5–5.0	0.02–0.03 at 30°C	350	Excellent	Fair to good Low temperature—poor	26–36	Good to excellent
Epoxy-anhydride, Dicy		650–730	3.4–3.8	0.01–0.03	400	Excellent	Good to excellent Low temperature—poor	20	Good to excellent
Epoxy-polyamide	10^{14} at 30°C 10^{10} at 105°C	400–500	2.5–3.0	0.008–0.02	350	Excellent	Good to excellent Low temperature—poor	20	Fair to good
Epoxy-phenolic	10^{12}–10^{13}	300–450			400	Excellent	Good Low temperature—fair		Good to excellent
Fluorocarbon TFE	10^{18}	430	2.0–2.1	0.0002	500	Can be excellent; primers required.	Excellent		
FEP CTFE	10^{18} 10^{18}	480 500–600	2.1 2.3–2.8	0.0003–0.0007 0.003–0.004	400 400	Can be excellent; primers required	Excellent		
Parylene (polyxylylenes)	10^{16}–10^{17}	700	2.6–3.1	0.0002–0.02	240°F (air) 510°F (inert atm.)	Good	Good		

Phenolics	10^9–10^{12}	100–300	4–8	0.005–0.5	350	Excellent	Poor to good Low temperature—poor	30–38	Fair
Phenolic-oil varnish					250	Excellent	Good Low temperature—fair		Poor to fair
Phenoxy	10^{13}–10^{14}	500	3.7–4.0	0.001	180	Excellent	Excellent		
Polyamide (nylon)	10^{13}–10^{15}	400–500	2.8–3.6	0.01–0.1	225–250	Excellent			
Polyester	10^{12}–10^{14}	500	3.3–8.1	0.008–0.04	200	Good on rough surfaces; poor to polished metals.	Fair to excellent	25–30	Good
Chlorosulfonated (polyethylene (Hypalon)†			6–10	0.03–0.07	250	Good	Elastomeric	Less than 10	
Polyimide	10^{16}–10^{18}	400 3,000 (10 mil)	3.4–3.8	0.003	500	Good	Fair to excellent		Good
Polystyrene	10^{10}–10^{19}	500–700	2.4–2.6	0.0001–0.0005	140–180		Poor to fair		
Polyurethane	10^{12}–10^{13}	450–500 3,800 (1 mil)	6.8 (1 kHz) 4.4 (1 MHz)	0.02–0.08	250	Often poor to metals. (Excellent to most non-metals.)	Good to excellent. Low temperature—poor	10–17 (castor oil) 50–60 (polyester)	
Silicone	10^{14}–10^{16}	550	3.0–4.2	0.001–0.008	500	Varies, but usually needs primer for good adhesion.	Excellent Low temperature—excellent	12–16	Fair to excellent
Vinyl chloride (poly-)	10^{11}–10^{15}	300–800	3–9	0.04–0.14	150	Excellent, if so formulated.	Excellent Low temperature—fair to good	5–10	
Vinyl chloride (plastisol, organisol)	10^9–10^{16}	400	2.3–9	0.10–0.15	150	Requires adhesive primer.	Excellent Low temperature—fair to good	3–6	
Vinyl fluoride	10^{13}–10^{14}	260 1,200 (8 mil)	6.4–8.4	0.05–0.15	300	Excellent, if fused on surface.	Excellent Low temperature—excellent		
Vinyl formal (Formvar‡)	10^{13}–10^{15}	850–1,000	3.7	0.007–0.2	200	Excellent			

* Trademark of Hercules Powder Co., Inc., Wilmington, Del.
† Trademark of E. I. du Pont de Nemours & Co., Wilmington, Del.
‡ Trademark of Monsanto Co., St. Louis, Mo.

TABLE 1 Properties of Coatings by Polymer Type (Continued)

Coating type	Resistance to environmental effects					Film formation				Typical uses
	Chemical and solvent resistance	Moisture and humidity resistance	Weather-ability	Resistance to micro-organisms	Flamma-bility	Repairability	Method of cure	Cure schedule	Application method	
Acrylic	Solvents—poor Alkalies—poor Dilute acids—poor to fair	Good	Excellent resistance to UV and weather	Good	Medium	Remove with solvent.	Solvent evaporation	Air dry or low-temperature bake	Spray, brush, dip	Coatings for circuit boards quick dry protection for markings and color coding.
Alkyd		Poor	Good to excellent	Poor	Medium	Poor	Oxidation or heat	Air dry or baking types	Most common methods	Painting of metal parts and hardware.
Cellulosic (nitrate butyrate)	Solvents—good Alkalies—good Acids—good	Fair		Poor to good	High	Remove with solvents.	Solvent evaporation	Air dry or low-temperature bake	Spray, dip	Lacquers for decoration and protection. Hot-melt coatings.
Chlorinate polyether (Penton*)		Good			Low		Powder or dispersion fuses	High temperature fusion	Spray, dip, fluid bed	Chemically resistant coatings.
Epoxy-amine cure	Solvents—good to excellent Alkalies—good Dilute acids—fair	Good	Pigmented—fair; clear—poor (chalks)	Good	Medium	No	Cured by catalyst reaction	Air dry to medium bake	Spray, dip, fluid bed	Coatings for circuit boards. Corrosion-protective coatings for metals.
Epoxy-anhydride, Dicy	Solvents—good Alkalies— Dilute acids—	Good		Good	Medium	No	Cured by chemical reaction	High bakes 300 to 400°F	Spray, dip, fluid bed, impreg.	High-bake, high-temperature-resistant dielectric and corrosion coatings.
Epoxy-polyamide	Solvents—fair Alkalies—good Dilute acids—poor	Good		Good	Medium	No	Cured by coreactant	Air dry or medium bake	Spray, dip	Coatings for circuit boards. Filleting coating.
Epoxy-phenolic	Solvents—excellent Alkalies—fair Dilute acids—good	Excellent	Pigmented—fair; clear—poor	Good	Medium	No	Cured by coreactant	High bakes 300 to 400°F	Spray, dip	High-bake solvent and chemical resistant coating.
Fluocarbon TFE	Solvents—excellent Alkalies—good Dilute acids—excellent	Excellent		Good	None	No	Fusion from water or solvent dispersion	Approx. 750°F	Spray, dip	High-temperature-resistant insulation for wire.
FEP CTFE		Excellent		Good	None	No	Fusion from water or solvent dispersion	500–600°F	Spray, dip	High-temperature-resistant-insulation. Extrudable.
Parylene (polyxylylenes)		Excellent			None		Vapor phase deposition and polymerization requiring special license from Union Carbide.			Very thin, pinhole-free coatings, possible semi-conductable coating.

5-6

Phenolics	Solvents—good to excellent Alkalies—poor Dilute acids—good	Excellent	Fair	Poor to good	Medium	No	Cured by heat	Bake 300–500°F	Spray, dip	High-bake chemical and solvent-resistant icoatings.
Phenolic-oil varnish	Solvents—poor Alkalies—poor Dilute acids—good to excellent	Good	Good	Poor, unless toxic—additive	Medium	Poor	Oxidation or heat		Spray, brush, dip-impregnate	Impregnation of electronic modules, quick protective coating.
Phenoxy		Good		Good			Cured by heat			Chemical resistant coating.
Polyamide (nylon)		Fair				Fairly solderable				Wire coating.
Polyester	Solvents—poor Alkalies—poor to fair Dilute acids—good	Fair	Very good	Good	Medium	Poor	Cured by heat or catalyst	Air dry or bake 100–250°F	Spray, brush, dip	
Chlorosulfonated polyethylene (Hypalon†)	Solvents—poor Alkalies—good (dilute) poor (concentrated) Dilute acids—good	Good		Good	Low		Solvent evaporation	Air dry or low temperature bake	Spray, brush	Moisture and fungus proofing of materials.
Polyimide	Solvents—excellent Alkalies— Dilute acids—	Good		Good	Low	Poor	Cured by heat	High bake	Dip, impregnate, wire coater	Very high temperature resistant wire insulation.
Polystyrene		Good		Good	High	Dissolve with solvents.	Solvent evaporation	Air dry or low bake	Spray, dip	Coil coating, low dielectric constant, low loss in radar uses.
Polyurethane	Solvents—good Dilute alkalies—fair Dilute acids—good	Good		Poor to good	Medium	Excellent; melts, solder-through properties	Coreactant or moisture cure	Air dry to medium bake	Spray, brush, dip	Conformal coating of circuitry, solderable wire insulation.
Silicone	Solvents—good Alkalies—good (dilute) poor (concentrated) Dilute acids—good	Excellent	Excellent	Good	Very low (except in O$_2$ atm.)	Fair to excellent. Cut and peel.	Cured by heat or catalyst	Air dry (RTV) to high bakes	Spray, brush, dip	Heat-resistant coating for electronic circuitry. Good moisture resistance.
Vinyl chloride (poly-)	Solvents—alcohol, good Alkalies—good	Good	Pigmented—fair to good Clear—poor	Poor to good (depends on plasticizer)	Very low	Dissolve with solvents.	Solvent evaporation	Air dry or elevated temperature for speed	Spray, dip, roller coat	Wire insulation. Metal protection (especially magnesium, aluminum).
Vinyl chloride (plastisol, organisol)		Good		Poor to good (depends on plasticizer)	Low	Poor	Fusion of liquid to gel	Bake 250–350°F	Spray, dip, reverse roll	Soft-to-hard thick coatings, electroplating racks, equipment.
Vinyl fluoride		Good	Excellent	Good	Very low	Poor	Fusion from water or solvent dispersion	Bake 400–500°F	Spray, roller coat	Coatings for circuitry. Long-life exterior finish.
Vinyl formal (Formvar‡)		Good		Good	Medium	Poor	Cured by heat	Bake 350–500°F	Roller coat, wire coater	Wire insulation (thin coatings) coil impregnation.

* Trademark of Hercules Powder Co., Inc., Wilmington, Del.
† Trademark of E. I. du Pont de Nemours & Co., Wilmington, Del.
‡ Trademark of Monsanto Co., St. Louis, Mo.

5-8 Coatings for Electronics

erties can minimize or eliminate this defect. The curing schedule for the coating also has a significant effect on its moisture-barrier properties, with an extended cure at an elevated temperature enhancing the performance of the coating. Improvement is probably due to elimination of residual solvent in the coating film and further crosslinking.

The use of conformal coatings for printed circuits also allows design engineers greater freedom in achieving narrower conductor lines and closer spacings. Without coatings, impurities, moisture, and other contaminants will easily bridge the conductors, resulting in decreases in insulation resistance and even arcing between conductors. The dielectric breakdown voltage across a clean, dry surface is very good, but maintaining the surface in this condition is extremely difficult. The increased insulation protection afforded by an epoxy coating on a G-10 laminate board with $1/16$-in. conductor spacings was shown by McGuiness.[3] The flashover voltage for

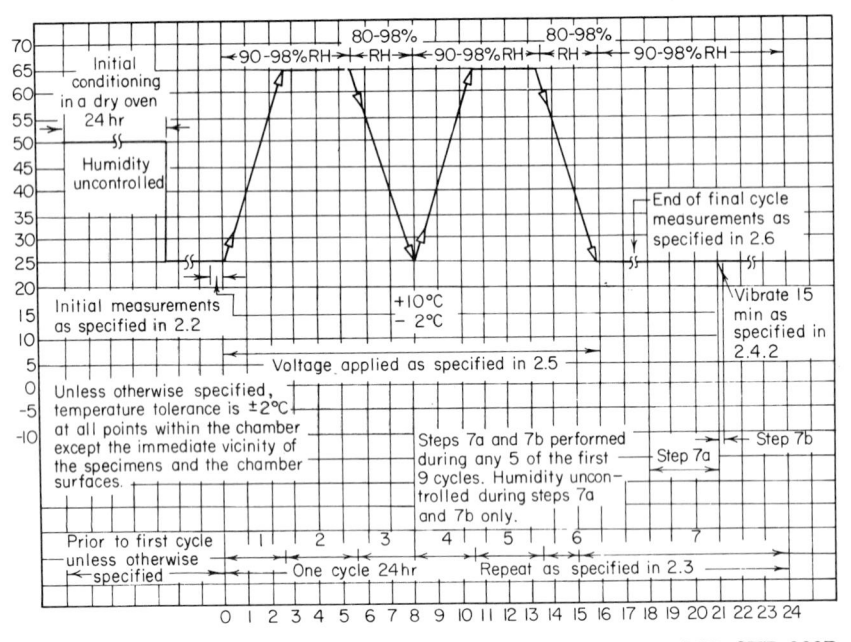

Fig. 1 Graphical representation of moisture-resistance test per MIL-STD-202B, Method 106.

the coated board was 5,000 volts as compared with 1,600 volts for an uncoated board when run at 80°F in 50 percent relative humidity. Allowable spacings between conductors for various applied voltages are given in Table 4. The reduction in spacing made possible by the use of conformal coatings is also apparent from this table. These data are specified in MIL-STD-275B, Printed Wiring for Electronic Equipment.

Coatings also serve other functions which contribute to the overall assembly reliability. These include the rigidity imparted to thin leads, solder joints, and components, thus preventing breakage or lifting during normal handling and especially during vibration.

Coating Thickness. Theoretically, the thicker the coating, the better the humidity barrier since moisture permeability is inversely related to thickness. This holds true fairly well for the 100 percent solids coatings. However, it is not always true for solvent-based coatings since the probability of entrapment of solvent volatiles in the cured coating is greater, the thicker the coating. With trapped solvent molecules, a more porous structure results in which water can be more readily absorbed and transmitted, thereby causing blistering, corrosion, and large decreases in insulation-resistance values. Solvent-based coatings, however, may be formulated that contain various additives, permitting rapid release of solvents on curing; and cure cycles can be optimized also to achieve the same results. Thus, with proper techniques, both 100 percent solids and solvent-based coatings will afford better humidity protection by increasing their thickness.

Most circuit-board manufacturers use 0.3- to 3-mil-thick coatings. Excessive thicknesses have been reported to cause cracking of components such as glass

TABLE 2 MIL-STD-202 Test Methods and Numbers

Test method	Method No.
Environment tests (100 class):	
Salt spray (corrosion)	101C
Temperature cycling	102A
Humidity (steady state)	103B
Immersion	104A
Barometric pressure	105C
Moisture resistance	106B
Thermal shock	107B
Life (at elevated ambient temperature)	108A
Explosion	109A
Sand and dust	110
Flammability (external flame)	111
Seal	112A
Physical-characteristics tests (200 class):	
Vibration	201A
Shock (specimens weighing not more than 4 lb)	202B
Shock (specified pulse)	213
Random drop	203A
Vibration, high frequency	204A
Vibration, random	214
Shock, medium impact	205C
Life (rotational)	206
High-impact shock	207A
Solderability	208B
Resistance to solvents	215
Electrical-characteristics tests (300 class):	
Dielectric withstanding voltage	301
Insulation resistance	302
Dc resistance	303
Resistance-temperature characteristic	304
Capacitance	305
Q factor	306
Contact resistance	307
Current-noise test for fixed resistors	308
Voltage coefficient of resistance	309
Contact-chatter monitoring	310

diodes or glass-sealed resistors. Cracking is variously attributed to (1) stresses resulting from shrinkage of the plastic coating when the solvent evaporates; (2) high shrinkage from polymerization during curing; or (3) large differences in coefficients of expansion between the glass and the plastic coating. The cracking may be manifest soon after drying, after curing, or much later during testing or rework operations where additional stresses are imposed. For stress-sensitive components, it is therefore important to avoid very thick coatings; and if filleting is used, to avoid bridging between components.

Inspection. Many manufacturers are providing dyed or fluorescent coatings to allow visual or ultraviolet-light inspection, assuring that all areas have been coated

5-10 Coatings for Electronics

adequately. Bare spots, pinholes, and other discontinuities are easily detected, thus facilitating quality control. Fluorescent indicators are also being used to determine whether particles of coating are migrating and contaminating other portions of a system, because particulate matter is especially critical to the functioning of parts in the immediate vicinity of the circuit board, such as gyros, accelerometers, bearings, and rotating memory disks.

Reworkability of Coated Assemblies. Many commercial electronic modules are designed as throw-away items, so that no coating or, at most, a relatively inexpensive varnish or polyester is employed. For more complex and expensive electronic gear, speciality coatings are needed to achieve long-term reliability. Because of the

Fig. 2 Effects of salt spray on coated and uncoated circuit boards (board on right was coated).

probability that one or more defective components or solder joints will have to be repaired, it is imperative that the module be reworkable. Hence, the coating selected, in addition to meeting the numerous engineering and manufacturing requirements, must be amenable to being removed so that defective components may be separated and replaced. The removal technique must be one that does little or no damage to good adjacent components, surfaces, and markings. Combining all these desirable features in one material is, of course, a very difficult problem. Where service temperatures are lower than 275°F, polyurethanes have been popular because, being thermoplastic, they can be melted in localized areas by a hot solder-iron tip (Fig. 4), and indeed behave as a partial solder flux. Epoxy-polyamides and epoxy-amines also may be softened and removed by a hot solder iron; however, with much more difficulty. Heat may decompose and carbonize the epoxy polymer, resulting in poor solderability and darkened areas.

Silicone elastomeric coatings are easily removed by cutting into them with a knife blade because of their softness. Being thermosetting, they will not soften or melt with heat, but will eventually decompose. Silicones and other thermosetting plastics may also be swelled by prolonged contact with chlorinated or fluorinated solvents such as methylene chloride or trichloroethylene, and then removed mechanically.[4] However, there are four problems encountered with solvent-removal techniques: (1) the process is slow, requiring immersion for one or two days; (2) solvents do not really dissolve the plastic, but only swell it; (3) the solvent cannot be localized, and so there is the probability of the other areas being damaged; (4) swelling of the plastic can generate large stresses which then induce failure modes in other devices.

Addition of 3 to 10 percent of an additive, such as silica aerogel, will convert the solvent to a paste which will not flow under normal conditions. This method is also extremely slow and subject to solvent evaporation.

Solder resists and maskants In the assembly and processing of electronic components, resists or maskants may be needed to "stop off" certain areas from conformal coating or solder. For example, in the conformal coating of printed-circuit boards, the edge connector strip should be left uncoated to allow electric contact with an edge connector.

In soldering, of course, solder splatter is undesirable, especially where bridging and short-circuiting of closely spaced conductors can occur. To prevent solder wetting of

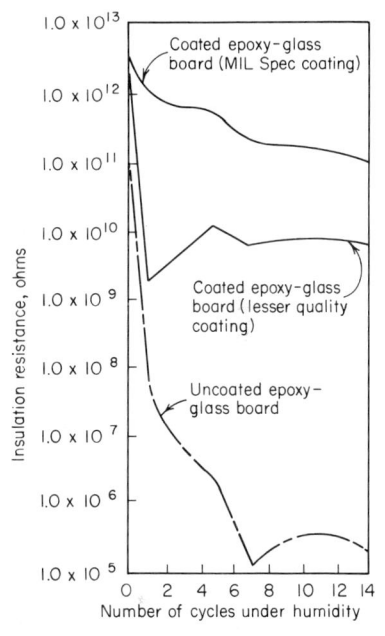

Fig. 3 Insulation resistance of coated and uncoated boards as a function of humidity cycling (MIL-I-46058).[2]

TABLE 3 Humidity Effects on Insulation Resistance of Circuit Boards[2]

Sample	Insulation resistance, ohms		
	Initial	5th cycle	10th cycle
A Uncoated..........	1.5×10^{12}	3×10^7	5×10^4
Coated............	1×10^{12}	3×10^9	2×10^9
B Uncoated..........	3×10^{12}	5×10^7	5×10^3
Coated............	3.1×10^{12}	5×10^{11}	4×10^{11}
C Uncoated..........	3×10^{12}	5.5×10^7	1.5×10^6
Coated............	3×10^{12}	3×10^{11}	4.1×10^{10}

such areas, solder maskants, also referred to as solder stop-offs, are employed. They are usually epoxy formulations because of the need to withstand exposure to solder temperatures of 475°F and higher without degrading. Most nonepoxy types are reported to blister, lift, outgas, or decompose in some other way. Solder maskants are normally applied by screening, although spraying through a stencil and brushing have also been used. The main limitation in the use of epoxy solder maskants is

5-12 Coatings for Electronics

that once they have been applied and cured, repair is virtually impossible. A fully cured epoxy coating is strongly adherent, and all methods required to remove it will invariable result in damage to the substrate or the circuitry.[4]

TABLE 4 Minimum Allowable Spacings between Conductors on Printed-circuit Boards for Conformance to MIL-STD-275B

Voltage between conductors (dc or ac peak)	Minimum spacing, in.
Uncoated boards:	
Sea level to 10,000 ft:	
0–150	0.025
151–300	0.050
301–500	0.100
Greater than 500	0.002*
Over 10,000 ft:	
0–50	0.025
51–100	0.060
101–170	0.125
171–250	0.250
251–500	0.500
Greater than 500	0.001*
Coated boards:	
All altitudes:	
0–30	0.010
31–50	0.015
51–150	0.020
151–300	0.030
311–500	0.060
Greater than 500	0.00012*

* Inch per volt.

Fig. 4 Soldering through a polyurethane conformal coating on a ceramic printed-circuit assembly.

The maskants described above as useful for stopping off either conformal coatings or solder may take a number of forms, as follows:
1. Screenable coatings
2. Strippable dip or spray coatings
3. Tapes, adhesive-backed
4. Molded rubber pressed into close contact by means of a clamp fixture

A list of representative examples is given in Table 5.

Coatings for thin-film circuits Coatings, either alone or in conjunction with encapsulants, are being used more and more as the method of packaging of both thin- and thick-film circuits. Large cost savings can be achieved over the normal packaging methods involving hermetically sealed metal cans. However, because of the close value tolerances to which thin- or thick-film elements such as resistors must be held, and because of the sensitivity of these elements to changes, hermetically sealed packages with an inert gas must still be used in many cases. Where organic coatings are to be used, they should be checked for compatibility with the various circuit elements (resistors, capacitors, conductors, etc.) before use. Compatibility is commonly determined by (1) assessing changes in electrical values before and after coating and after environmental testing such as thermal cycling

TABLE 5 Typical Resist Coatings and Maskants Used in Assembly of Printed Circuits

Type	Name	Manufacturer	Characteristic feature
Strippable coatings.....	Spraylat 1071	Spraylat	Vinyl solution, brush, dip, or spray. Air dry, good toughness. Heat resistance 200–250°F.
Adhesive tapes........	No. 35	Minnesota Mining & Mfg. Co.	Vinyl tape, acrylic adhesive.
	No. 63		Teflon tape, acrylic adhesive.
	EE 6379	Permacel	Polyimide tape, silicone adhesive.
	No. 470	Minnesota Mining & Mfg. Co.	Vinyl tape, synthetic rubber adhesive.
Screenable solder resists	DSM 727	Dynachem Corp.	A thermosetting, screenable, one-component coating, available in a number of colors. An elevated-temperature cure is recommended. Stable for 1 min at 525°F.
	184 Series	Wornow Process Paint Co.	Reinforced alkyd-based transparent green. Heat curing from 200 to 300°F required for permanent resist. Stable at 500°F for 15 sec.
	PRH-403	Permacel	Two-component epoxy.
	Lonco PC33R	London Chemical	Melamine-polyester.
	PC 535		Silicone-polyester.

and humidity exposure; (2) determining degree of adhesion of the coating to the various substrates and surfaces comprising the circuit; and (3) checking for corrosivity to metallic surfaces by long-term or accelerated testing.

Because many thin-film resistors must maintain close tolerances (± 0.1 percent) and most coatings effect changes greater than this, it is a common practice to passivate the resistors using special composition glasses or silicon monoxide. After such glassification, the circuit chip can be plastic-packaged by coating, transfer molding, or other methods.

Coatings for thick-film circuits So-called thick-film circuits are formed by the deposition of conductive, resistive, dielectric, and insulative coatings on ceramic or plastic substrates. The coatings, often in the form of screenable inks or pastes, are applied as circuitry lines, resistors, or capacitors.

Most of the criteria previously described as applicable to the selection of conformal

5-14 Coatings for Electronics

coatings for etched printed-circuit boards also apply to thick-film circuit protection. Hence, moisture resistance, stability of electrical insulation properties, adhesion, solder-through properties, and transparency are usually required. Many coatings used for the protection of epoxy circuit boards cannot be used for ceramic circuits because of differences in adhesion to the substrate. Adhesion of organic coatings to ceramics is generally poorer than to an epoxy substrate. Primers may have to be used to achieve both adhesion and other desirable properties.

TABLE 6 NEMA Standards and Manufacturers' Trade Names for Wire Insulation

Manufacturer	Plain enamel	Polyvinyl formal	Polyvinyl formal modified	Polyvinyl formal with nylon overcoat	Polyvinyl formal with butyral overcoat	Poly-amide	Acrylic	Epoxy
Thermal class	105°C	105°C	105°C	105°C	105°C	105°C	105°C	130°C
NEMA Standard[6]	MW 1	MW 15	MW 27	MW 17	MW 19	MW 6	MW 4	MW 9
Anaconda Wire & Cable Co.	Plain enamel	Formvar	Hermetic Formvar	Nyform	Cement coated Formvar	....	....	Epoxy epoxy-cement coated
Asco Wire & Cable Co.	Enamel	Formvar	....	Nyform	Formbond	Nylon	Acrylic	Epoxy
Belden Manufacturing Co.	Beld-enamel	Formvar	....	Nyclad	....	....	....	Epoxy
Bridgeport Insulated Wire Co.	....	Formvar	....	....	Quickbond	Quick-Sol	....	....
Chicago Magnet Wire Corp.	Plain enamel	Formvar	....	Nyform	Bondable Formvar	Nylon	Acrylic	Epoxy
Essex Wire Corp.	Plain enamel	Formvar	Formetex	Nyform	Bondex	....	Ensolex/ ESX	Epoxy
General Cable Corp.	Plain enamel	Formvar	Formetic	Formlon	Formeze	....	Solderable acrylic	Epoxy
General Electric Co.	....	Formex	....	....	....	Nylon	....	....
Haveg-Super Temp Div. Hitemp Wires Co. Division Simplex Wire & Cable Co.	....	....	....	....	....	....	....	....
Hudson Wire Co.	Plain enamel	Formvar	....	Nyform	Formvar AVC	Ezsol	....	....
New Haven Wire & Cable, Inc.	Plain enamel	....	....	....	....	....	....	....
Phelps Dodge Magnet Wire Corp.	Enamel	Formvar	Hermeteze	Nyform	Bondeze	....	....	....
Rea Magnet Wire Co., Inc.	Plain enamel	Formvar	Hermetic Formvar special	Nyform	Koilset	Nylon	....	Epoxy
Viking Wire Co., Inc.	Enamel	Formvar	....	Nyform	F-Bondall	Nylon	....	....

Courtesy of Rea Magnet Wire Co., Inc.

Polymer coatings or encapsulants used over thick-film circuits should be carefully chosen for purity and stability. Outgassing or contaminants of a reducing nature such as hydrogen gas evolving from the polymer will degrade resistance values, especially of palladium-palladium oxide-silver composition inks. This is due to the reduction of the oxides in such compositions to free metal which increases conductivity. Hence, plastic overcoatings may have a very damaging effect rather than the intended beneficial effect. It is reported that the platinum composition resistors are unaffected by hydrogen and are less sensitive to impurities.[5]

Wire and coil coatings and varnishes In general, wire coatings consist of high-molecular-weight polymers, either of the thermoplastic (linear) or thermosetting (crosslinked) types. The thermoplastic types have definite softening or melting points, above which the materials are not usable especially if a stress is applied in a localized area. Examples include Teflon* and nylon. The thermosetting coatings are more resistant to *cut-through*—required in all wire-wound devices—and have improved solvent resistance. Examples include silicones, polyester, and poly-

TABLE 6   NEMA Standards and Manufacturer's Trade Names for Wire Insulation (Continued)

Teflon	Poly-urethane	Poly-urethane with friction surface	Poly-urethane with nylon overcoat	Poly-urethane with butyral overcoat	Poly-urethane with nylon and butyral overcoat	Polyester	Polyester with overcoat	Poly-imide	Polyester polyimide	Ceramic, ceramic-Teflon, ceramic-silicon
200°C MW 10	105°C MW 2	105°C	130°C MW 28	105°C MW 3	130°C MW 29	155°C MW 5	155°C MW 5	220°C MW 16	180°C	180°C+ MW 7
....	Analac	....	Nylac	Cement-coated analac	(PROP) Cement coated nylac	....	Anatherm D, Anatherm 200	Al 220 M.L.	Anatherm N Anamid M (amide-imide)	
....	Poly	....	Nypol	Asco bond-P	Asco bond	Ascotherm	Isotherm 200	M.L.	Ascomid	
....	Beldure	....	Beldsol	....	....	Isonel	Polyther-maleze	M.L.		
....	Polyure-thane	Uniwind	Poly-nylon	Polybond	....	Isonel 200				
....	Soderbrite	....	Nysod	Bondable polyure-thane	....	Polyester 155				
....	Soderex	....	Soderon	Soder-bond	Soder-bond N	Thermalex F	Polyther-malex/ PTX 200	Allex		
....	Enamel "G"	....	Genlon	....	....	Gentherm	Polyther-maleze 200			
Teflon Temprite	....	....	....	....	....	Alkanex Isonel				
....	Hudsol	Gripon	Nypoly AVC	Hudsol AVC	Nypoly AVC	Isonel 200	Isonel 200-A	M.L.	Isomid	
....	Impsol	....	Impsolon	....	....	....	Imp-200			
....	Sodereze	Gripeze	Nyleze	....	S-Y Bondeze	....	Polyther-maleze 200 II	M.L.		
....	Solvar	....	Nylon solvar	Solvar koilset	....	Isonel 200	Polyther-maleze 200	Pyre M.L.	Isomid	Ceroc
....	Polyure-thane	....	Poly-nylon	P-Bondall	....	Isonel 200	Iso-poly	M.L.	Isomid Isomid-P	

vinyl formal. A list of some wire coating types, trade names, and manufacturers is given in Table 6.

As described in Table 6, many wire coatings consist of a combination of polymers. Polyvinyl formal may be coated with nylon to attain the lubricating properties of nylon which improves winding properties. Polyvinyl formal is frequently overcoated with polyvinyl butyral to form a coating that is self-bonding by either heat or solvent action to make formed coils.

* Trademark of E. I. du Pont de Nemours & Co., Wilmington, Del.

5-16 Coatings for Electronics

A description of wire coating types and their properties follows. A summary of properties is found in Table 7.

Coil Coatings. Wire that is to be wound in coils around forms, cores, or the like to form inductors, transformers, fields and armatures, or solenoids, of course, requires primary insulation on the wire. In addition, these coils are usually further treated to encapsulate or fix the wires in place. Furthermore, such inductive devices may have to withstand severe environments, as in military or aerospace equipment, or even in household washing machines. These are usually encapsulated, potted in plastic cups, or molded or cast in synthetic resins designed to meet the severe conditions. These treatments are largely described in Chap. 1.

TABLE 7 Summary of Properties of Various Coated Wires

Coating	Thermal rating, °C	Advantages	Limitations
Formvar.............	105	Tough, high dielectric strength, compatibility, heat shock	Crazes in polar solvents*
Polyesters...........	155	Toughness, dielectric strength, chemical resistance, cut-through	Hydrolyzes in moist, sealed atmosphere
Polyurethane........	105	Dielectric strength, chemical resistance, compatibility, solderable without stripping, moisture and corona resistance	Reduced abrasion resistance
Nylon...............	105	Dielectric strength, solvent resistance, solderability, toughness, windability	Moisture absorption, high electrical loss at all frequencies
Polytetrafluoroethylene (Teflon)	180	Thermal stability, chemical stability, dielectric strength, low dielectric constant, low friction	Gas permeability, cold flow, adhesion
Formvar-nylon......	105	Solvent resistance, toughness, windability, heat shock	Nylon portion subject to same limitation as nylon above
Formvar-Butvar†....	105	Bondable, dielectric strength, heat shock	Vibration, high mechanical stress
Polyurethane-nylon	105	Solderability, solvent resistance	Moisture absorption, high-frequency losses
Polyimide...........	220	Thermal resistance, chemical stability, radiation resistance	Stripping difficulty, solvent crazes*

*A prebake at 125°C for 2 to 4 hr after winding relieves film stress so that crazing does not subsequently occur during varnish or resin encapsulation.
† Trade name of Shawinigan Resins Co., division of Monsanto Chemical Co.

Wire Coating Types. OLEORESINOUS COATINGS: These coatings are essentially varnishes derived from natural drying oils such as linseed or tung oil containing numerous carbon-to-carbon double bonds and polymerizing by oxidative reaction of the double bonds. The varnishes may be "beefed up" by "cooking in" fossil or synthetic resins to improve their hardness and solvent resistance. Because of the dark color of the finished wire, they are commonly referred to as *black enamels.*

POLYURETHANE COATINGS: Polyurethane coatings used in wire enamels are generally the reaction products of hydroxylated polyesters, and hydroxylated polyesters which have been reacted with an excess of tolylene diisocyanate.

If two such materials are mixed, the reaction proceeds too rapidly for application. For this reason, for wire coating, the polyester containing the excess isocyanate groups is reacted with phenol, which forms a moderately stable compound and gives the total mixture sufficient stability for application. After application of the material to the wire, the phenol is released during the bake; and the resulting isocyanate can then react with the hydroxyl groups of the other polyester forming the urethane linkages necessary for polymerization.

POLYAMIDES: Polyamides are the result of the reaction of a dicarboxylic acid and an aliphatic diamine. The best-known example is nylon.

POLYVINYL FORMAL: Coatings based on polyvinyl formal, commonly referred to as Formvar, are probably the most widely used of all wire coatings. The base polymer is prepared by partial hydrolysis of polyvinyl acetate followed by condensation with formaldehyde. Polyvinyl formal used alone is deficient in solvent resistance and hardness; but these properties are greatly improved by the addition of reactants such as phenol-formaldehyde and cresol-formaldehyde resins. These resins condense with the hydroxyl groups of the polyvinyl formal, producing a coating that is exceedingly tough and flexible, yet possessing good resistance to abrasion and solvents. Strong solvents are used in formulating these.

POLYESTER COATINGS: Polyester resins are produced by the condensation of a dibasic acid, such as fumaric, phthalic, or maleic; a glycol; and perhaps an unsaturated monomer such as styrene. When mixed and cooked in the proper proportions, partial polymerization occurs, leaving the product soluble in solvents. After application to the wire and during the cure, condensation occurs with release of water to produce a crosslinked network. Polyester wire coatings derive their excellent thermal properties from the stability of the ester linkages and the symmetry of the para substitution on the benzene ring.

POLYIMIDE COATING: Varnish suitable for application as a wire coating consists of polyamide acid resins soluble in strong solvents such as dimethyl formamide. Removal of water from the polyamide-acid during the coating cure results in formation of the polyimide.

POLYFLUORO COATING: These coatings differ from previously discussed coatings in that polymerization does not occur during application of the coatings. The coatings are applied as water dispersions of the polymer and then heat-fused or sintered during application. Examples would be Teflons TFE and FEP, polyvinyl fluoride, and CTFE.

Testing of Coated Wire. Coated wire testing is conducted most often according to procedures described in NEMA (National Electrical Manufacturers Association) Publication MW 1000, "Magnet Wire"[6]; or in Military Specification MIL-W-583C, Wire, Magnet, Electrical,[7] in the case of government procurement. Considerable duplication exists between these two standards since many tests are identical. In the following discussion, some of the more commonly used tests are described.

FILM THICKNESS: Film thickness is measured by micrometer readings of the coated wire, removal of the coating in a manner nondestructive to the wire, and determining the wire diameter after removal of the coating. Half the difference between these diameters gives the thickness of the coating. One method of removing the coating from the copper wire is to burn off the coating by heating, followed by quick immersion of the hot wire in ethyl alcohol to reduce any copper oxide formed.

ADHESION: Adhesion is measured by elongating a 10-in. length of wire at 12 in./min for sizes 13 AWG (American Wire Gauge) and larger or by sudden pull for sizes 14 and finer. Examination of the wire for separation or cracking at specified magnification for different wire sizes determines failure.

FLEXIBILITY: Flexibility is determined by winding the wire around a mandrel of specified size followed by examination at magnification dependent on the size of the wire. Flexibility is frequently expressed as $1\times$, $2\times$, $3\times$, etc., signifying that the particular wire coating will withstand winding around a mandrel 1, 2, or 3 times the AWG size of the wire involved.

Adhesion and flexibility tests are frequently followed by elongation and by a mandrel-wind test similar to the above.

HEAT SHOCK: Heat is determined by elongation or stretching, winding around a mandrel of specified size, dependent on wire size, followed by exposure to oven heat. The specimen is then examined for cracking or separation from the metal under specified magnification.

SCRAPE RESISTANCE: Scrape resistance is measured in two ways: repeated scrape and unidirectional scrape. The repeated-scrape test is conducted using a machine that rubs a 0.016-in.-diameter needle or wire, oriented at 90° to the wire, but rubbed in a direction parallel to the axis of the wire under test with provision for a specified load at a rate of 60 strokes/min. Each stroke consists of a 360° rotation of an eccentric drive. The machine is equipped with an electric circuit which stops the action, and a counter which records the strokes at the time of break-through of the film.

The unidirectional-scrape resistance test consists of a scraping action in one direction only. The machine employed provides for a 0.009 in.-diameter needle rubbing in one direction on the coated wire at 16 in./min at right angles to the wire with constantly increasing load. The machine is equipped with an electric circuit for detecting the time to failure. Failure is recorded as "grams to fail."

THERMOPLASTIC FLOW: This property is frequently referred to as *cut-through*. Two pieces of 18 AWG wire are placed at right angles to each other and pressed together by a load of 2,000 g. The conductors of the wires are connected to a 110-volt power supply with an indicating device in the circuit such as an argon lamp. The unit is placed in an oven and the temperature raised at a rate of less than 5°C per min. When the indicating device shows short circuiting, the temperature of the plate supporting the wires is measured and recorded.

DIELECTRIC STRENGTH: Dielectric strength of most wire coatings used in electronic equipment is determined by means of "twisted pairs." The specimens used consist of two wires twisted under a specified tension and total number of twists, dependent on wire size, for a length of approximately 4.75 in. The conductors of each pair are connected to a 60 Hz ac power source, and the voltage is increased at a rate of 500 volts/sec until breakdown occurs.

CONTINUITY: Continuity is measured by passage of 100 ft of wire through a mercury bath at 100 ft/min. A discontinuity-indicating circuit operates at voltages from 20 to 75 volts direct current dependent on wire size. The discontinuity device operates when the resistance between the bath and conductor is less than 5,000 ohms but not more than 10,000 ohms. The test circuit is arranged so as to record the number of dielectric breaks. Results are reported as number of breaks per 100 ft of wire.

High-voltage continuity is measured by passage of 100 ft of wire between grooved metal rollers with the wire grounded to the take-off drum. The test circuit is arranged so as to impose a 60-Hz voltage between the take-off drum and the roller, and has a sensitivity of 1 megohm at the test voltage to indicate dielectric breaks during test. The test voltages to be used are dependent on the thickness of the coating and the size of wire.

SOLVENT RESISTANCE AND COMPLETENESS OF CURE: Degradation of wire coatings due to solvent exposure may be determined by an accelerated test in which the wire specimens are exposed to various solvents and their scrape resistance is measured. The scrape-resistance value may then be compared with that of a control unexposed wire. Weight loss of insulating coating after extraction with solvents is another method of assessing solvent resistance as well as the degree of cure of the coating. Solvent extractables are usually determined by Soxhlet methods using toluene followed by methanol; or, in the case of polyvinyl formal–urethane, using boiling methanol. Extractables are calculated from the original weight of coated wire, the weight of the extracted solution after stripping off the solvent, or the weight of wire after all the coating has been removed. For most wires, the extractable content provides a good index of degree of cure, particularly where the wire is later to be exposed to the action of various liquids such as in oil- or fluid-filled transformers.

Still a third method involves microscopic examination of the coated wire after it has been exposed to solvent. In the case of polyvinyl formal wire, the wire is exposed for 5 min to a mixture of 30 percent toluene, 70 percent denatured alcohol (by volume). Removal of self-bonding resin from the polyvinyl formal or swelling within 0.5 in. from the end is considered a failure. For other insulations such as polyimide, the completeness of cure may be determined by dissipation factor measurements.

SOLDERABILITY: The solderability of wire (principally nylon- and polyurethane-insulated) is tested by immersion in rosin-alcohol flux followed by molten 50:50 tin-lead solder at temperatures dependent on size of wire followed by visual examination. The sample is prepared by twisting the ends of a 12-in. length of wire for a distance of ¾ to 1 in. and cutting off the end of the twist, leaving 5 to 10 turns.

BOND STRENGTH OF SELF-BONDING WIRE: Twisted pairs are prepared similar to those used for dielectric strength tests and bonded by heating for ½ hr at 125°C. The bond strength is then measured by using a tensile testing machine with one end of one wire clamped in one jaw of the machine and the opposite end clamped in the other jaw. The rate of pull is 12 in./min, and the load causing one wire to slide along the other is recorded.

THERMAL RATING AND THERMAL STABILITY: Thermal ratings are based on heat-aging tests of twisted-pair specimens of 18 AWG wire with a heavy film coating in accordance with ASTM D 2307. According to this rather complex test, sets of 10 or more twisted-pair specimens are exposed to temperatures at least 20°C apart for cycle periods specified, followed by proof-voltage tests dependent on coating thickness. The proof voltages (60 Hz alternating current) are selected to produce a voltage stress of approximately 300 volts/mil. Using statistical procedures outlined in ASTM D 2307, the results are extrapolated for a predicted life of 20,000 hr. This method is the basis for the thermal ratings presented in Tables 6 and 7.

The thermal stability of wire coatings may also be determined by differential thermal analysis (DTA)[8] and by thermal gravimetric analysis (TGA)[9]. According to the DTA procedure, all endothermic and exothermic reactions that the material undergoes in being exposed to a programmed temperature sweep are recorded. These heat changes correspond to phase changes occurring at the indicated temperatures; for example, they may consist of glass transitions, softening or melting, oxidation, sublimation, or decomposition. Differential thermal analysis, however, does not indicate weight changes since phase changes may occur without weight loss or gain. For exact weight changes, thermal gravimetric analysis may be used. According to TGA, weight loss is plotted dynamically (again during a programmed temperature rise), using a very sensitive electrobalance. By running both a DTA curve (known as a thermogram) and a TGA curve (known as a pyrogram), an almost complete thermal profile of the material may be obtained. Both methods are being used to study the thermal and oxidative stability of other polymeric coatings and for plastics in general. The reader is referred to other books and articles for further details and applications.[8-13] With reference to wire coatings, the thermograms for polyethylene, polyvinyl chloride, Kel F, Teflon, Formex, Nylon-Formex, Alkenex, polyurethane, and polyimide coatings have been reported by Murphy and Hill.[8]

Coatings used in packaging electronic devices Organic coatings, used alone or in combination with transfer-molded or cast resins, provide an extremely important and economical approach for packaging and protecting electronic components. Plastic encapsulated devices are rapidly replacing hermetically sealed canned devices in many applications. Resistors, capacitors, transistors, diodes, and integrated circuits are all now available in plastic packages. So great has been the impact of plastics that production of encapsulated transistors alone now numbers in the tens of millions of devices per month. Indeed, the sealing of devices in plastics appears to be the dominant trend, especially in the semiconductor industry.

The popularity of using plastics stems from the simplicity and rapidity with which they can be processed and applied, resulting in substantial cost savings.[14] Dramatic cost reductions have been achieved, especially in the case of transistors and integrated circuits. Devices that previously sold for as high as $3 to $5 in metal cans have been reduced to 50 cents to $1 per unit by transfer-molding them in either epoxy or silicone resins. Plastics may also be molded in an almost

5-20 Coatings for Electronics

infinite variety of configurations; however, most of them are designed to be interchangeable with their metal-can counterparts (Fig. 5).

Functions of Plastic Packaging Materials. Both coatings and transfer-molding compounds are generally employed as a complete system.

The combination of a coating and a molding plastic provides greater rigidity, dimensional stability, impact resistance, and moisture protection over the use of a coating alone. The main functions of the coating, also referred to as the "inner coating," "barrier coating," or "junction coating," may be listed as follows:

1. To hold delicate components and fine interconnection wires together, allowing them to be handled, dropped, vibrated, etc., without electrical or mechanical damage. This function is sometimes referred to as "ruggedization."

2. To impart stress relief. Stresses arising from the transfer-molding operation, resin shrinkage, differences in coefficients of expansion between the plastic and substrates, thermal cycling, and thermal shock may be equalized and dissipated by using the proper barrier coating.

3. To provide some degree of moisture, gas, and dust protection, preventing corrosion and electrical malfunctioning.

4. To provide an insulation barrier, thus preventing electric short circuiting.

Fig. 5 Plastic-encapsulated semiconductor devices.

Among the functions of the transfer-molding or casting compounds are the following:

1. To provide a well-defined structure with exact dimensional tolerances comparable to those for hermetic cans. This permits standardization and interchangeability of parts.

2. To provide high impact and mechanical shock resistance.

3. To provide a longer path for moisture, thus reducing moisture penetration. The use of the coating alone does not afford a sufficient moisture seal.

Application of Coatings Used in Plastic Packaging of Electronic Devices. Barrier coatings may be applied to components by dipping, spraying, fluidized bed, or eye dropping. These methods are applicable to resistors and capacitors, but the coating of uncased semiconductor devices and microcircuits is limited to the eye-dropping method. Barrier or stress-relief coatings applied to semiconductor devices must be localized so that they cover only the active surface, metallization, and a portion of the internal flying leads. The coating should not be allowed to spread to the edges of the device or onto the external leads where it could affect the adhesion and moisture permeability of the subsequently applied molding compound or the solderability of the leads. Because of these restrictions, the only practical method of applying the coating is by dispensing it in small drops from a microsyringe or from an eyedropper.

The structural plastic is subsequently applied by transfer molding, casting, or dipping, and is successfully being used for the encapsulation of thin-film resistors, wire-wound resistors, diodes, bipolar and MOS-FET transistors, integrated circuits, coils, and many other components. The relative position of the barrier coating with respect to the molding compound is shown in Fig. 6 for a commonly used 14-lead integrated-circuit package.

Plastic Parameters Affecting Device Reliability. The main plastic parameters that can affect the reliability of a device are moisture penetration, purity, corrosivity, stresses, and adhesion. Some of these parameters will be treated separately in the following discussion. However, it should be emphasized that they are all closely interrelated. For example, lack of adhesion will increase moisture penetration; impurities will cause corrosion; and corrosion in turn can affect both adhesion and moisture penetration.

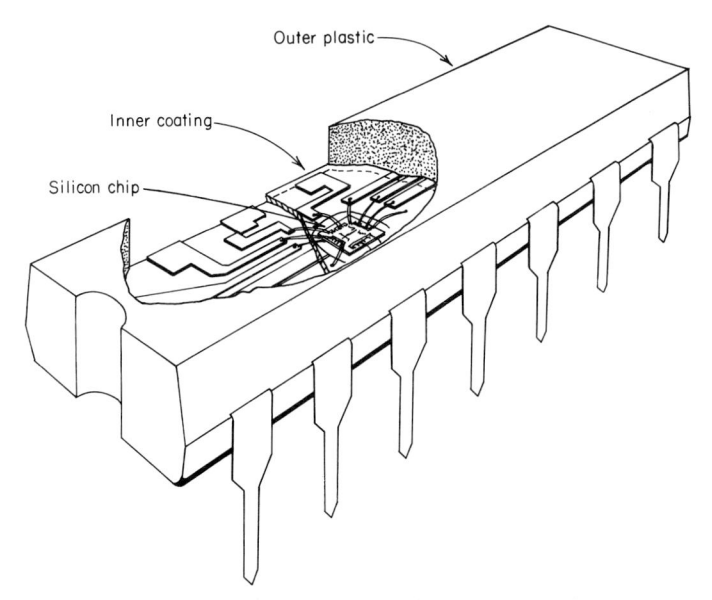

Fig. 6 Cross section of plastic-packaged integrated circuit.

MOISTURE RESISTANCE: All plastics absorb and transmit moisture to a finite extent. When used to package devices, they should not therefore be considered hermetic in the same sense that metal cans or metal-to-glass seals are; nor do the same hermeticity and leakage tests apply. In spite of this, plastic-packaged devices have been found to pass the moisture requirements for many military specifications, and devices have been shown to be operational after limited exposures to high humidity and elevated temperatures. Their long-term resistance to humidity and warm, humid environments is, however, questionable. In many cases, the key to the success of plastic-packaged transistors and integrated circuits has been not the plastic itself but the reliability and hermeticity of the passivation layer used over the device surface. Significant improvements in passivation layers such as the development of silicon nitride and silicon nitride–silicon dioxide composites and the deposition of certain low-melting glasses have greatly reduced the criticality of plastic properties.

PURITY: The degree of purity of coatings and plastics is an extremely critical factor when these materials are used on any electronic assembly, but more so when used directly on active microelectronic devices. However, if the active devices are well passivated, plastic purity is not as critical a factor. Indeed, the success of many plastic-packaged devices has been due primarily to improvements in the reliability of inorganic passivation layers such as glass and silicon nitride.[15,16] Silicon nitride, in particular, has been shown to be one of the best barriers to sodium ions even at elevated temperatures and with an applied bias.[17] Sodium, potassium, and lithium ions cause inversion layers, high leakage currents, and low breakdown voltages in diodes and transistors.[18-20] These effects and the degradation of electrical insulating properties of coatings are enhanced by an increase in mobility of the charged carriers, brought about by moisture, electrical stress, or elevated temperature. Ionic impurities are present in most commercially available coatings.

CORROSIVITY: The immediate or long-term effect of moisture may be the corrosion of metal portions of a device. Most susceptible to corrosion are the external leads, internal flying leads, interconnection bonds, and metallized conductors. Besides moisture, ionic and other impurities on the device surface or within the plastic play a key role in initiating and accelerating corrosion. Chloride ion contamination, for example, chemically corrodes aluminum through the formation of aluminum chloride. Other ions, both cationic and anionic, contribute to electrolytic corrosion between external leads or between closely spaced internal conductors. Again, this type of corrosion can be minimized or avoided by employing extremely pure plastics.

In many instances, catalysts used for curing resins, have been known to induce corrosion of metal surfaces.

STRESSES AND STRESS RELIEVING: Barrier coatings can be used to minimize or eliminate stresses imposed by subsequently used transfer-molding or casting compounds and to protect electronic components from vibration and shock. The stress-relief coatings selected must, however, be flexible and possess a low modulus of elasticity. Silicones, both the Silastic* and RTV† types, meet these requirements and have found many applications in component packaging. For example, in the packaging of transistors, one firm coats components with 10 to 20 mils of silicone, and then encapsulates them with a m-phenylenediamine-cured epoxy. Other semiconductor manufacturers employ small amounts of silicone barrier coatings, and then transfer-mold with silicone or epoxy.

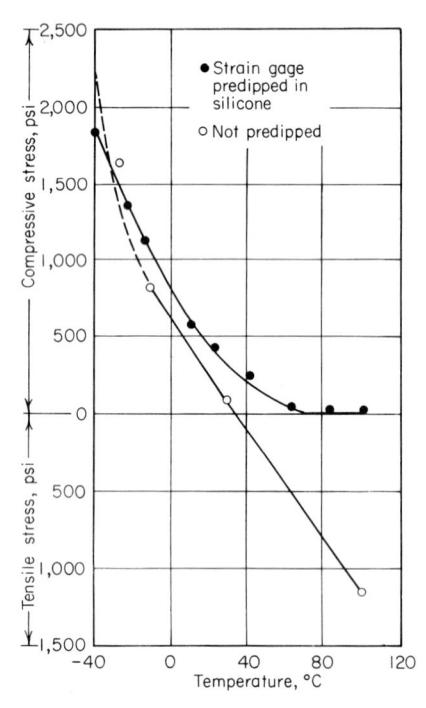

Fig. 7 Effect of flexible precoat on stress of DGEBA-polyamide cured system.[27]

The validity of using flexible dip coatings to relieve stresses is apparent from data presented in Fig. 7. In other work, total stresses of 5,000 psi at −40°C on components embedded in a silica-filled epoxy resin have been reported; but when the components were precoated with RTV elastomeric silicone, the total stress imposed and stress concentrations were reduced. A 1-mil silicone coating at −40°C reduced the total pressure to 3,500 psi, and an 8-mil-thick coating further reduced this pressure to below 400 psi.[21,22]

The use of plastic systems involving more than one coating or plastic may present

---

* Trademark of Dow Corning Corporation, Midland, Mich.
† Trademark of General Electric Co., Schenectady, N.Y.

problems if the inner coating has a higher coefficient of expansion than the outer plastic. An increase in temperature will obviously cause the inner coating to exert pressure against the confining outer plastic. An example of this is the use of an inner unfilled silicone coating and an outer epoxy shell in some resistor and semiconductor devices. Stress problems can be avoided by either minimizing the amount of inner silicone coating or by providing an empty volume between the inner coating and the outer shell into which the silicone can expand.

ADHESION: Good adhesion of plastics to components is essential for two basic reasons: (1) to effect a satisfactory seal against moisture and contaminants; and (2) to allow the encapsulating plastic to dissipate mechanical stresses to the leads during handling, installation, and service. Adhesion can be improved by coatings.

Several factors affecting adhesion may be itemized as follows:
1. Type of substrate material (metal or nonmetal)
2. Surface treatment prior to encapsulation
3. Surface roughness or porosity
4. Intrinsic adhesive properties of the polymer coating
5. Presence of internal release agents in transfer-molding plastics

The nature and preparation of the substrate are the strongest factors affecting adhesion. For example, the ease of bonding (bondability) varies greatly among metals, with the noble metals being more difficult to bond to. Examples of metals known to be difficult to bond to are gold, Kovar,* nickel alloys, solder, solder plating, and cadmium.

A clean surface is essential for good adhesion. Methods for cleaning vary from solvent cleaning to etching of surfaces. For difficult-to-bond metals, primers are sometimes used. For example, epoxy primers are used over gold before coating with polyurethane. Special primers are frequently used to enhance the adhesion of silicones, particularly the RTV types.

The intrinsic bondability of polymers varies from type to type. Epoxies and phenolics, for example, are known for their inherently good adhesive properties. Polyurethanes exhibit fair to excellent bonds, while silicones exhibit adhesive properties that vary from very poor to good.

With reactive metals, other critical adhesion factors are the type and mechanical strength of the oxide which forms rapidly on the surface under normal ambient conditions. This oxide layer may actually assist in improving adhesion.

Coatings for Semiconductors. Early experimental work on coatings for the protection of semiconductor junctions was performed between 1955 and 1960. Many of these early compositions consisted of waxes combined with plastics.[23,24] Later, silicone resins filled with lead tetroxide or mercuric oxide were used to coat transistors and germanium rectifying diodes.[25] In 1962, the first epoxy-packaged transistors were mass-produced for the consumer and automotive markets.[26] Finally, in 1964 plastic-packaged devices were introduced for industrial and computer applications and in 1966 and 1967 for some military applications.

Presently, several types of coatings and molding compounds are employed in plastic-packaged semiconductor devices and other electronic components. These types are predominantly epoxies and silicones, with phenolics and polyesters being used to a lesser extent. A list of some commonly used barrier and junction coatings and their characteristics is given in Table 8.

Commercial versus Military Uses of Plastic Packages. For the most part, plastic-packaged devices are used for consumer and industrial applications such as for television, radios, computers, and in-house test equipment. There is no question about their cost savings and performance for the applications. Greater precautions and proof of long-term reliability are, however, necessary for military systems because of the often more severe environments that these systems will encounter. The main drawback to the more extensive use of plastics for military electronics has been their inability to act as sufficient moisture barriers to meet stringent requirements such as those defined in MIL-E-5272, MIL-E-5400, MIL-E-16400 E, MIL-STD-202,

* Trademark of Westinghouse Electric Corporation, Pittsburgh, Pa.

Coatings for Electronics

TABLE 8 Barrier and Junction Coatings

Commercial designation	Supplier	Type	Characteristics
DC 643	Dow Corning	One-part unfilled silicone solution coating	A high-purity coating compatible with transistors and n-p-n junctions. Cures in 12 hr at 200°C, or 4 hr at 250°C. Does not cure at 150°C.
DC 644	Dow Corning	One-part unfilled silicone solution coating	Used for diode, transistor, and rectifier junctions. Cures in 12 hr at 200°C, or 4 hr at 250°C.
DC 645	Dow Corning	One-part silicone solution coating	Used for diode, transistor, and rectifier junctions. Cures in 12 hr at 200°C, or 4 hr at 250°C.
DC 646	Dow Corning	One-part unfilled silicone solution coating	A high-purity coating compatible with diodes, controlled rectifiers, p-n-p, and p-n-p+ type junctions. Cures in 12 hr at 200°C, or 4 hr at 250°C.
DC 51	Dow Corning	A silicone dielectric gel	A very soft, high-purity material used to coat devices in hermetically sealed metal cases. Stabilizes life character of devices by reducing junction temperatures and cushioning the junction against shock.
R-60087	Dow Corning	Two-part clear, unfilled, solventless silicone	Provides physical protection and support of small interconnecting wires plus protection against contamination of junction surfaces. Cures in 2 hr at 200°C.
R-60093	Dow Corning	Two-part black, filled, solventless silicone	High-purity silicone resin designed for use on semiconductor junction surfaces. Cures in 2 hr at 200°C.
R-62044	Dow Corning	One-part clear, unfilled silicone solution coating	Contaminant-free silicone resin which contains no catalysts or driers. Cures in 24 hr at 200°C.
XR-60092	Dow Corning	Two-part clear silicone	Same as R-60093 except does not contain filler.
R-90703	Dow Corning	Black-filled two-part solventless silicone	A high-purity coating for semiconductor junction protection. Cures in 4 hr at 65°C or 2 hr at 150°C.
XR-60095	Dow Corning	Two-part solventless silicone	A high-purity coating designed for protecting active semiconductor surfaces and to support delicate interconnecting wires. Cures in 4 hr at 65°C.
SR-98	General Electric	One-part silicone varnish	Very hard film over wide temperature range with excellent stability at high temperatures. Cures in 8 hr at 200°C.
R-620	Union Carbide	One-part silicone varnish	A high-temperature coating which retains electrical properties under adverse conditions. Cures in about 12 hr at 200°C.

and more recently, MIL-STD-883. In attempting to meet these requirements, some device manufacturers have resorted either to the use of multiple coats of silicones or epoxies, with high-temperature bakes after each application, or to alternating several different coating types, e.g., a silicone followed by an epoxy. In active devices and integrated circuits, improved passivation layers such as silicon nitride or special glasses can serve as the primary barriers for moisture and ionic contaminants. Less reliance therefore needs to be placed on the plastic to achieve these functions. Some firms report a very high level of reliability with plastic-packaged devices primarily because of the improved passivation layers which they have developed.

PROPERTIES OF ELECTRONIC COATINGS

Coatings for electronics may serve numerous functions; but, in general, the two most important ones are (1) to provide high electrical insulation between conductors, interconnections, components, and other electrical parts, and (2) to provide protection from environmental damage. Protection from one or more of a number of severe environments may be necessary. Examples include moisture, heat, salt spray, oxygen, radiation, microorganisms, chemicals, and solvents. In addition, for rather fragile microelectronic assemblies, coatings are used to protect from abrasion, handling, shock, and vibration damage. Specially synthesized or formulated coatings may also serve a number of functional requirements such as to provide electric conductivity for radio-frequency interference (RFI) and for ohmic contacts, or to provide either thermal insulation or thermal conductance. Hence, coatings play a very vital part in the successful operation of electronic assemblies; and indeed, if it had not been for the development of superior electrical coatings, many of the highly dense and intricate windings, coils, and microelectronic assemblies would not have been possible.

Electrical Protection

One of the most important functions of organic coatings is to provide high electrical insulation and dielectric isolation for active or passive electronic components. Their effectiveness in this respect is expressed in terms of insulation resistance, volume resistivity, surface resistivity, and dielectric strength. Other functions include the storage of electric current and the conductance of current. These are expressed in terms of dielectric constant, capacitance, dissipation factor, and conductivity. A knowledge of exact electrical values and how these vary with changes in composition, purity, structure, or environment is important in selecting the most reliable coating for electronic equipment. The following will define these electrical parameters; show how they change with temperature, frequency, and other variables; and present tables of values for the more commonly used coatings. More theoretical discussions of electrical parameters may be found in other books.[28-30]

Insulation resistance and resistivity Insulation resistance is simply the ohmic relationship of the ratio of applied voltage to the total current between two electrodes in contact with a specific material. It is directly proportional to the length of a specimen and inversely proportional to the cross-sectional area according to the equation

$$R = \frac{\rho l}{A}$$

where R is insulation resistance in ohms, l is length, A is area, and ρ is a proportionality constant called the specific resistance or resistivity. When dimensions are in centimeters, the units are ohm-centimeters; when in inches, the units are ohm-inches.

Resistivity and other properties of plastics in general are treated in Chap. 3.

Resistivity is an important value because it reduces resistance measurements to a common denominator and permits comparisons between different materials. Volume resistivity, for example, is the ohmic resistance of a cube of bulk dielectric

Coatings for Electronics

TABLE 9 Volume Resistivities of Polymer Coatings

Material	Volume resistivity at 25°C, ohm-cm	Source of information
Acrylics	10^{14}–10^{15}	a
	$>10^{14}$	b
	7.6×10^{14}–1.0×10^{15}	Columbia Technical Corp., Humiseal
Alkyds	10^{14}	a
Chlorinated polyether	10^{15}	b
Chlorosulfonated polyethylene	10^{14}	b
Depolymerized rubber	1.3×10^{13}	H. V. Hardman, DPR Subsidiary
Diallyl phthalate	10^{8}–2.5×10^{10}	a
Epoxy (cured with DETA)	2×10^{16}	c,d
Epoxy polyamide	1.1–1.5×10^{14}	
Phenolics	6×10^{12}–10^{13}	a
Polyamides	10^{13}	
Polyamide-imide	7.7×10^{16}	e
Polyethylene	$>10^{16}$	
Polyimide	10^{16}–10^{18}	f
Polypropylene	10^{10}–$>10^{16}$	a
Polystyrene	$>10^{16}$	b
Polysulfide	2.4×10^{11}	g
Polyurethane (single component)	5.5×10^{12}	h
Polyurethane (single component)	2.0×10^{12}	h
Polyurethane (single component)	4×10^{13}	Columbia Technical Corp.
Polyurethane (two components)	1×10^{13}	Products Research & Chemical Corp. (PR-1538)
	5×10^{9} (300°F)	
Polyvinyl chloride	10^{11}–10^{15}	b
Polyvinylidene chloride	10^{14}–10^{16}	a
Polyvinylidene fluoride	2×10^{14}	h
Polyxylylenes (parylenes)	10^{16}–10^{17}	Union Carbide Corp.
Silicone (RTV)	6×10^{14}–3×10^{15}	Stauffer Chemical Co., Si-O-Flex SS 831, 832, & 833
Silicone, flexible dielectric gel	1×10^{15}	Dow Corning Corp.
Silicone, flexible, clear	2×10^{15}	Dow Corning Corp.
Silicone	3.3×10^{14}	Columbia Technical Corp. Humiseal 1H34
Teflon TFE	$>10^{18}$	b
Teflon FEP	$>2 \times 10^{18}$	b

[a] *Mater. Eng.*, Materials Selector Issue, vol. 66, no. 5, Chapman-Reinhold Publication, mid-October, 1967–1968.

[b] *Insulation*, Directory Encyclopedia Issue, no. 7, June–July, 1968.

[c] Lee, H., and K. Neville: "Epoxy Resins," McGraw-Hill Book Company, New York, 1966.

[d] Tucker, Cooperman, and Franklin: Dielectric Properties of Casting Resins, *Electron. Equip.*, July, 1956.

[e] Freeman, J. H.: A New Concept in Flat Cable Systems, *5th Ann. Symp. on Advan. Tech. for Aircraft Elec. Syst.*, Washington, D.C., October, 1964.

[f] Milek, J. T.: Polyimide Plastics: A State of the Art Report *Hughes Aircraft Rep.* S-8, October, 1965.

[g] Hockenberger, L.: *Chem.-Ing. Tech.*, vol. 36, 1964.

[h] Hughson Chemical Co. Tech. Bull. 7030A; Pennsalt Chemicals Corp. Prod. Sheet KI-66a, Kynar Vinylidene Fluoride Resin, 1967.

Properties of Electronic Coatings 5-27

material, one unit on a side, and is expressed as ohm-centimeters or ohm-inches. Surface resistivity is the resistance between two electrodes on the surface of an insulating material in ohms per square centimeter. All materials may be roughly classified according to their ability to conduct or impede the flow of electricity. They range from metals, which are extremely good conductors, to plastics, which are very good insulators. A rough classification of types may be made as follows:

Class	Volume resistivity, ohm-cm
Good conductors	$10^{-6}-10^0$
Semiconductors	10^0-10^8
Good insulators (poor conductors)	10^8-10^{12}

Organic coatings are, in general, good electrical insulators and as such are useful for electronic hardware (see Table 9). Although the resistivity of plastics is quite high (usually greater than 10^{12} ohm-cm), and the current which they conduct is considered negligible for most power equipment, still for some electronic assemblies even this small current flow can be a critical factor. Therefore, a knowledge of the electrical tolerances that a circuit or device must meet is important before a coating selection is made. A comparison of volume resistivities of some of the more familiar materials is presented in Fig. 8. The wide difference (up to 24 orders of magnitude) between the most conductive materials such as copper or silver and the least conductive such as polytetrafluoroethylene or polystyrene should be noted.

Effect of Variables on Resistivity: As was pointed out above, resistivity is a constant characteristic of a material. There are, however, many variables which can affect it. First of all, changes in the chemical composition or formulation may have pronounced effects. Besides gross compositional changes, minor amounts of impurities will affect resistivity values. Ionic impurities in plastics coupled with the presence of moisture are known to degrade resistivity values by as much as 6 to 11 orders of magnitude. The degree of cure or advancement in the state of polymerization will affect resistivity values, which increase as the cure of the resin advances.

Variation of resistivity of coatings with temperature is also important. Large decreases in values can occur at elevated temperatures. Hence a coating that is an adequate insulator at room temperature may be completely unsatisfactory when the equipment is operating at higher temperatures. Examples of temperature-resistivity curves for some polymer coating types are given in Fig. 9. Finally, volume- and surface-resistivity values also depend on frequency, voltage, pressure, light (photoconductivity of some organic crystals and polymers), and on the conditioning of the sample.

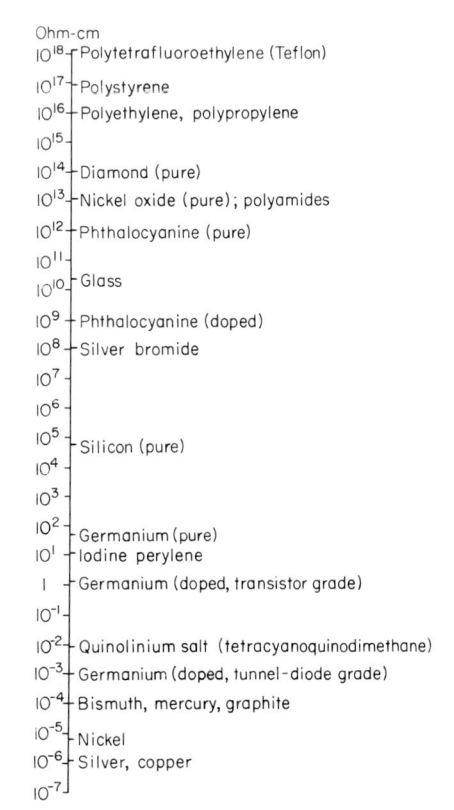

Ohm-cm
10^{18} — Polytetrafluoroethylene (Teflon)
10^{17} — Polystyrene
10^{16} — Polyethylene, polypropylene
10^{15} —
10^{14} — Diamond (pure)
10^{13} — Nickel oxide (pure); polyamides
10^{12} — Phthalocyanine (pure)
10^{11} —
10^{10} — Glass
10^9 — Phthalocyanine (doped)
10^8 — Silver bromide
10^7 —
10^6 —
10^5 — Silicon (pure)
10^4 —
10^3 —
10^2 — Germanium (pure)
10^1 — Iodine perylene
1 — Germanium (doped, transistor grade)
10^{-1} —
10^{-2} — Quinolinium salt (tetracyanoquinodimethane)
10^{-3} — Germanium (doped, tunnel-diode grade)
10^{-4} — Bismuth, mercury, graphite
10^{-5} — Nickel
10^{-6} — Silver, copper
10^{-7} —

Fig. 8 Comparative electrical resistivities of materials.

5-28 Coatings for Electronics

Surface Resistivity: Moisture and contaminants have a more pronounced effect on surface resistivity than on volume resistivity. Whereas it may take several days or weeks for volume resistivity to change when subjected to humid or contaminating

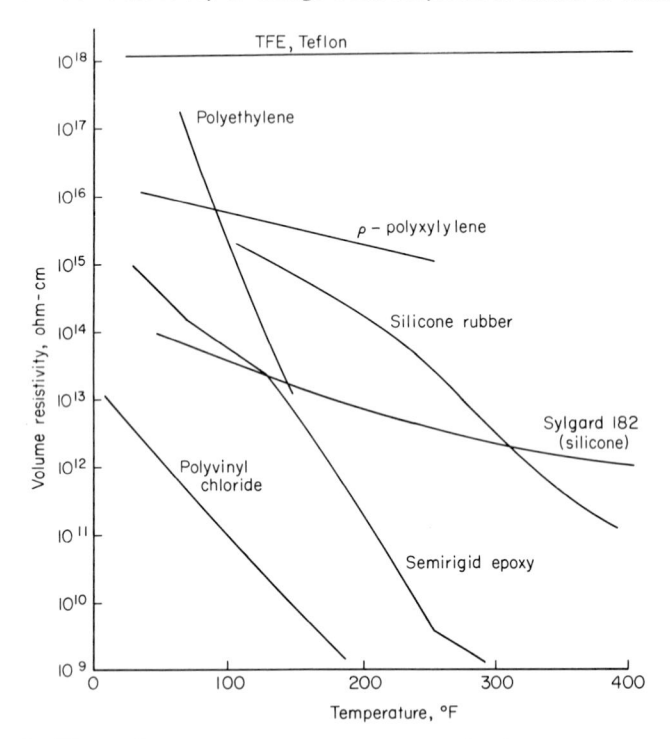

Fig. 9 Electrical resistivity–temperature curves of polymer coating types.

TABLE 10 Effects of Contaminants on Surface Resistivity[31]

Material	Surface resistivity, ohms/sq At 50% RH	At 96% RH
Polyethylene (as received)	$>10^{18}$	2.4×10^9
Polyethylene (contaminated with fingers)	.....	4.6×10^7
Polyethylene (freshly shaved surface)	.....	2.2×10^{11}
Polystyrene (as received)	$\sim 10^{18}$	1.8×10^{11}
Polystyrene (contaminated with fingers)	.....	5.5×10^9
Polymethyl methacrylate (as received)	$\sim 10^{18}$	2.0×10^{14}
Polymethyl methacrylate (contaminated with fingers)	.....	1.2×10^{12}
Clean silica glass	10^{16}	5.0×10^8

environments, surface effects are almost immediate. Drastic reductions in surface resistivity occur through contamination, even for coatings that have very high initial values. Table 10 lists some examples.

Conductance and conductivity The reciprocal of resistance is called conductance (expressed in $ohms^{-1}$ or mhos), and the reciprocal of resistivity is the specific conductance (expressed in ohms^{-1}cm^{-1} or mhos/cm). Coatings may be formulated with metal or conductive fillers to render them electrically semiconducting or conducting. These coatings are useful in applications such as providing ohmic contact for circuits and bridging of conductors, providing radio-frequency interference protection, and for the dissipation or bleeding off of electrostatic charges. Silver, gold, copper, and carbon blacks are the most commonly used fillers; whereas epoxies, polyurethanes, silicones, vinyls, and acrylics are typical resin binders. Compositions are available or may be formulated in all viscosities ranging from thick pastes to sprayable types.[32]

TABLE 11 Conductivity Data for Metal-filled Plastics
(mhos/cm)

Composition	Cure	Initial	After 40 days humidity cycling	After 50 hr 20% salt spray
Silver-filled epoxy....	6 hr at 350°F	3,000	1,250	1,400
Silver-filled epoxy....	Air dry at 36 hr	100	700	
Gold-filled epoxy.....	6 hr at 120°F	200–700		

Conductivity values vary widely, depending on the conductivity of the filler chosen, the cleaning and processing of the filler particles, and the percentage (loading) of the filler. Except for the noble metals (gold, platinum, etc.), meticulous cleaning of the filler metal is necessary to remove high-resistance surface oxide layers. Even after removal, metal oxides may again slowly form; and over long periods of time or under accelerated conditions, conductivity will decrease. The degree of cure of the resin binder also affects conductivity values and, in general, better values are obtained with the baking-type formulations than with the air-drying varieties (see Table 11).

TABLE 12 Conductive Compositions Analysis

Metal filler	Advantages	Disadvantages
Silver...................	Very high conductivity	High cost; silver migration can occur under certain conditions; tarnishing and corrosion.
Gold....................	Very high conductivity, very inert and stable	Cost higher than that of silver. Subject to government controls and audit.
Copper..................	Low cost, high conductivity	Requires extra steps for cleaning to remove oxides; conductivity values usually decrease on aging.
Carbon blacks (lampblack, acetylene black).........	Low cost	Very low conductivity.

For very high conductance, silver or gold fillers are employed; and with optimum formulation and cure, conductivity values of the order of 10^3 mhos/cm may be obtained. A comparison of several conductive compositions is shown in Table 12.

Dielectric constant, dissipation factor, power factor *Dielectric Constant of Materials.* The significance of dielectric constant is described in Chap. 3. As can be seen from the tables, the dielectric constants of gases are low and only slightly greater than one, but values up to 100 are common for many organic compounds, especially the very polar types, and values up to 1,000 are known for some inorganic materials such as barium titanate.

TABLE 13 Dielectric Constants of Polymer Coatings (at 25°C)

Coating	60–100 Hz	10^6 Hz	$>10^6$ Hz	Reference source
Acrylic		2.7–3.2		a
Alkyd			3.8 (10^{10} Hz)	b
Asphalt and tars			3.5(10^{10} Hz)	b
Cellulose acetate butyrate		3.2–6.2		a
Cellulose nitrate		6.4		a
Chlorinated polyether	3.1	2.92		a
Chlorosulfonated polyethylene (Hypalon)	6.19 7–10(10^3 Hz)	~5		E. I. du Pont de Nemours & Co.
Depolymerized rubber (DPR)	4.1–4.2	3.9–4.0		H. V. Hardman, DPR Subsidiary
Diallyl isophthalate	3.5	3.2	3(10^8 Hz)	a,c
Diallyl phthalate	3–3.6	3.3–4.5		a,c
Epoxy-anhydride—castor oil adduct	3.4	3.1	2.9(10^7 Hz)	Autonetics, Division of North American Rockwell
Epoxy (one component)	3.8	3.7		Conap Inc.
Epoxy (two components)	3.7			Conap Inc.
Epoxy cured with methylnadic anhydride (100:84 pbw)	3.31			d
Epoxy cured with dodecenyl-succinic anhydride (100:132 pbw)	2.82			d
Epoxy cured with DETA	4.1	4.2	4.1	e
Epoxy cured with m-phenylenediamine	4.6	3.8	3.25(10^{10} Hz)	e
Epoxy dip coating (two components)	3.3	3.1		Conap Inc.
Epoxy (one component)	3.8	3.5		Conap Inc.
Epoxy-polyamide (40 % Versamid* 125, 60 % epoxy)	3.37	3.08		e
Epoxy-polyamide (50 % Versamid 125, 50 % epoxy)	3.20	3.01		e
Fluorocarbon (TFE, Teflon)	2.0–2.08	2.0–2.08		E. I. du Pont de Nemours & Co.
Phenolic		4–11		a
Phenolic	5–6.5	4.5–5.0		c
Polyamide	2.8–3.9	2.7–2.96		
Polyamide-imide	3.09	3.07		
Polyesters	3.3–8.1	3.2–5.9		c
Polyethylene		2.3		a
Polyethylenes	2.3	2.3		e
Polyimide-Pyre-M.L.† enamel	3.8	3.8		f
Polyimide–Du Pont RK-692 varnish	3.8			g
Polyimide–Du Pont RC-B-24951	3.0(10^3 Hz)			g
Polyimide–Du Pont RC-5060	2.8(10^3 Hz)			g
Polypropylene		2.1		a
Polypropylene	2.22–2.28	2.22–2.28		c
Polystyrene	2.45–2.65	2.4–2.65	2.5(10^{10} Hz)	b
Polysulfides	6.9			h
Polyurethane (one component)	4.10	3.8		Conap Inc.
Polyurethane (two components—castor oil cured)		2.98–3.28		i
Polyurethane (two components)	6.8(10^3 Hz)	4.4		Products Research & Chemical Corp. (PR-1538)
Polyvinyl butyral	3.6	3.33		a
Polyvinyl chloride	3.3–6.7	2.3–3.5		a
Polyvinyl chloride–vinyl acetate copolymer	3–10			a
Polyvinyl formal	3.7	3.0		a
Polyvinylidene chloride		3–5		a
Polyvinylidene fluoride	8.1	6.6		j
Polyvinylidene fluoride	8.4	6.43	2.98(10^9 Hz)	k

TABLE 13 Dielectric Constants of Polymer Coatings (Continued)

Coating	60–100 Hz	10⁶ Hz	>10⁶ Hz	Reference source
p-Polyxylylene:				
Parylene N	2.65	2.65	....	Union Carbide Corp.
Parylene C	3.10	2.90	....	Union Carbide Corp.
Parylene D	2.84	2.80	....	Union Carbide Corp.
Shellac (natural, dewaxed)	3.6	3.3	2.75(10⁹ Hz)	l
Silicone (RTV types)	3.3–4.2	3.1–4.0	....	General Electric and Stauffer Chemical Cos.
Silicone (Sylgard ‡ type)	2.88	2.88	....	Dow Corning Corp.
Silicone, flexible dielectric gel	3.0	....	....	Dow Corning Corp.
FEP dispersion coating	2 1(10³ Hz)	....	....	E. I. du Pont de Nemours & Co.
TFE dispersion coating	2.0–2.2(10³ Hz)	....	....	E. I. du Pont de Nemours & Co.
Wax (paraffinic)	2.25	2.25	2.22(10¹⁰ Hz)	l

* Trademark of General Mills, Inc., Kankakee, Ill.
† Trademark of E. I. du Pont de Nemours & Co., Wilmington, Del.
‡ Trademark of Dow Corning Corporation, Midland, Mich.
[a] *Mater. Eng.*, Materials Selector Issue, vol. 66, no. 5, Chapman-Reinhold Publication, mid-October 1967–1968.
[b] Volk, M. C., J. W. Lefforge, and R. Stetson: "Electrical Encapsulation," Reinhold Publishing Corporation, New York, 1962.
[c] *Insulation*, Directory Encyclopedia Issue, no. 7, June–July, 1968.
[d] Coombs, C. F. (ed.): "Printed Circuits Handbook," McGraw-Hill Book Company, New York, 1967.
[e] Lee, H., and K. Neville: "Handbook of Epoxy Resins," McGraw-Hill Book Company, New York, 1967.
[f] Learn, J. R., and M. P. Seegers: Teflon-Pyre-M.L. Wire Insulation System, *13th Symp. of Tech. Progr. in Commun. Wire and Cable*, Atlantic City, N.J., Dec. 2–4, 1964.
[g] *Du Pont Bull.* H65-4, Experimental Polyimide Insulating Varnishes, RC-B-24951 and RC-5060, January, 1965.
[h] Hockenberger, L.: *Chem.-Ing. Tech.* vol. 36, 1964.
[i] *Spencer-Kellogg* (division of Textron, Inc.) *Bull.* TS-6593.
[j] Barnhart, W. S., R. A. Ferren, and H. Iserson: *17th ANTEC of SPE*, January, 1961.
[k] Pennsalt Chemicals Corp. Prod. Sheet KI-66a, Kynar Vinylidene Fluoride Resin, 1967.
[l] Von Hippel, A. R. (ed.): "Dielectric Materials and Applications," Technology Press of MIT and John Wiley & Sons, Inc., New York, 1961.

Dielectric constants for polymers used for insulating and protective coatings range from 2 to 8. Those coatings that have low dielectric constants and low dissipation factors and are able to maintain these low values over a wide temperature and humidity range are preferred as electrical insulating materials. On the other hand, those combining high dielectric constants and low dissipation factors are useful as capacitors, since they can store and hold large amounts of electric energy.

Dielectric constants of 4.5 maximum at 1,000 Hz and 77°F are normally considered satisfactory for insulating materials used for electronic and electrical assemblies. This is the minimum requirement specified in MIL-I-16923. However, for microelectronic and miniaturized circuitry operating at high frequencies, capacitance effects must be minimized, and materials with lower dielectric constants are necessary. A knowledge of dielectric constants at high frequencies is therefore important to the engineer designing such circuits. Values of dielectric constants at 60 Hz and at 10⁶ (or 1 MHz) are to be found in Table 13.

Dissipation Factor and Power Factor. The dissipation factor D is the ratio of the resistive (loss) component of current I_r to capacitive component of current I_c and equals the tan of the dielectric loss angle δ as follows:

$$D = \frac{I_r}{I_c} = \tan \delta$$

5-32 Coatings for Electronics

For dielectric materials, the power factor and dissipation factor are essentially equal; and these terms are used interchangeably in the literature. Military specifications call for values no greater than 0.020 at 1,000 Hz and 77°F (MIL-I-16923).

Loss factor is the product of the power factor and dielectric constant and is a measure of signal absorption according to the following equation:

$$\text{Loss factor} \cong \text{watts loss} \cong K \tan \delta \cong KD$$

For all these parameters, low values are desirable, especially for high-speed high-frequency circuitry operating in the 10^7- to 10^{10}-Hz range.

A low value indicates low conversion of electric energy to heat energy and reduced power loss for the network. The dissipation factor as well as other electrical parameters is dependent upon frequency, temperature, humidity, and purity of the sample. A compilation of dissipation factors is given in Table 14.

DIELECTRIC CONSTANT AND DISSIPATION FACTOR VARIATION WITH TEMPERATURE: At constant frequency, both dielectric constant and dissipation factor for insulating coatings will, in general, increase with increasing temperature. Because coating formulations are not homogeneous but consist of many ingredients that will volatilize or change on heating, electrical values may decrease initially and then increase, or with some materials, may be quite erratic so that no simple straight-line curve can be drawn.

Dielectric strength and breakdown voltage Still another important electrical property is dielectric strength or the ability of a coating to withstand applied voltage without breakdown. It is the maximum voltage (usually reported per mil thickness) at or below which no breakdown occurs. Some prefer to speak about dielectric breakdown, that is, the voltage threshold at or above which actual failure occurs. This is also reported in volts per mil thickness. The two terms are often used interchangeably. The physical breakdown of coatings at a critical voltage is believed to be due to a large increase in electrons (electron avalanche effect) within the insulation.[33] In even the best insulation, a small number of ions and free electrons exist owing to thermal agitation or molecular imperfections. Here a very high voltage will be necessary to cause breakdown. Other less perfect insulation, containing impurities, voids, and moisture, will have lower breakdown voltages.[34]

Accurate dielectric strength data are important in designing reliable electrical parts of high circuit or component density intended for operating at high voltages. For both low- and high-voltage operating parts, organic coatings are widely used since they have sufficiently high dielectric strength values and are used in sufficient thicknesses so that a very high margin of safety exists. In fact, dielectric strengths for organic coatings are, in general, much higher than for inorganic or ceramic materials.

Dielectric strength values must be carefully obtained under specified conditions to obtain reproducible results and to allow reliable comparisons. A large number of test conditions are known to affect the dielectric strength values. As an example, it is always important to standardize on the electrode configuration and on the thickness of the specimen. Although values are all ultimately reduced to volts per mil thickness, differences in values are obtained, depending on the thickness of the sample used (Figure 10). In general, thin specimens give higher per unit thickness values than thicker ones. In most cases, the thickness specified in the ASTM procedure of 17 mils is too thick to be applicable to coatings. Therefore, this procedure is followed but at lower thicknesses of 2 to 5 mils. Other variables that affect readings are the manner in which the voltage is applied, whether continuously or stepwise; the rate of voltage increase; the frequency of the applied power; the integrity of the coating film (lack of pinholes and air voids); and the purity of the sample. With respect to the last named, the dielectric strength will decrease with an increase in sample imperfections or impurities.

Properties of Electronic Coatings 5-33

TABLE 14 Dissipation Factors of Polymer Coatings (at 25° C)

Coating	60–100 Hz	10^6 Hz	$>10^6$ Hz	Reference source
Acrylics	0.04–0.06	0.02–0.03		a
Alkyds	0.003–0.06			a
Chlorinated polyether	0.01	0.01		a
Chlorosulfonated polyethylene	0.03	0.07(10^3 Hz)		b
Depolymerized rubber (DPR)	0.007–0.013	0.0073–0.016		H. V. Hardman, DPR Subsidiary
Diallyl phthalate	0.010	0.011	0.011	b
Diallyl isophthalate	0.008	0.009	0.014(10^{10} Hz)	b
Epoxy dip coating (two components)	0.027	0.018		Conap Inc.
Epoxy (one component)	0.011	0.004		Conap Inc.
Epoxy (one component)	0.008	0.006		Conap Inc.
Epoxy polyamide (40% Versamid 125, 60% epoxy)	0.0085	0.0213		c
Epoxy polyamide (50% Versamid 115, 50% epoxy)	0.009	0.0170		c
Epoxy cured with anhydride–castor oil adduct	0.0084	0.0165	0.0240	Autonetics, North American Rockwell
Phenolics	0.005–0.5	0.022		a
Polyamide	0.015	0.022–0.097		
Polyesters	0.008–0.041			a
Polyethylene (linear)	0.00015	0.00015	0.0004(10^{10} Hz)	d
Polymethyl methacrylate	0.06	0.02	0.009(10^{10} Hz)	d
Polystyrene	0.0001–0.0005	0.0001–0.0004		
Polyurethane (two component, castor oil cure)		0.016–0.036		e
Polyurethane (one component)	0.038–0.039	0.068–0.074		Conap Inc.
Polyurethane (one component)	0.02			Conap Inc.
Polyvinyl butyral	0.007	0.0065		
Polyvinyl chloride	0.08–0.15	0.04–0.14		a
Polyvinyl chloride, plasticized	0.10	0.15	0.01(10^{10} Hz)	d
Polyvinyl chloride–vinyl acetate copolymer	0.6–0.10			
Polyvinyl formal	0.007	0.02		
Polyvinylidene fluoride	0.049	0.17		f
	0.049	0.159	0.110	g
Polyxylylenes:				
Parylene N	0.0002	0.0006		Union Carbide Corp.
Parylene C	0.02	0.0128		Union Carbide Corp.
Parylene D	0.004	0.0020		Union Carbide Corp.
Silicone (Sylgard 182)	0.001	0.001		Dow Corning Corp.
Silicone, flexible dielectric gel	0.0005			Dow Corning Corp.
Silicone, flexible, clear	0.001			Dow Corning Corp.
Silicone (RTV types)	0.011–0.02	0.003–0.006		General Electric
Teflon FEP dispersion coating	0.0002–0.0007			E. I. du Pont de Nemours & Co.
Teflon FEP	<0.0003	<0.0003		a
Teflon TFE	<0.0003			a
Teflon TFE	0.00012	0.00005		Union Carbide Corp.
Other materials:				
Alumina (99.5%)		0.0001		h
Beryllia (99.5%)		0.0003		h
Glass silica	0.0006	0.0001	0.00017(10^{10} Hz)	d
Glass, borosilicate		0.013–0.016		Corning Glass Works
Glass, 96% silica		0.0015–0.0019		Corning Glass Works

[a] *Mach. Des.*, Plastics Reference Issue, vol. 38, no. 14, Penton Publishing Co., 1966.
[b] *Insulation*, Directory, Encyclopedia Issue, no. 7, June-July, 1968.
[c] Lee, H. and K. Neville: "Handbook of Epoxy Resins," McGraw-Hill Book Company, New York, 1967.
[d] Mathes, K.: Electrical Insulation Conference, 1967.
[e] *Spencer-Kellogg* (Division of Textron, Inc.) Tech. Bull. TS-6593.
[f] Barnhart, W. S., R. A. Ferren, and H. Iserson: *17th ANTEC of SPE*, January, 1961.
[g] Pennsalt Chemicals Corp. Prod. Sheet KI-66a, Kynar Vinylidene Fluoride Resin, 1967.
[h] *Mach. Des.*, Design Guide, Sept. 28, 1967.

5-34 Coatings for Electronics

Atmospheric conditions, such as temperature and humidity, under which the test is carried out are also important variables. A plot of the variation of dielectric strength with temperature shows the same general shaped curve for all polymers. There is fairly little change at low temperature, but a rapid decline occurs above a critical temperature, characteristic of the coating being tested[35] (Fig. 11).

Although temperature and moisture exposure have pronounced degrading effects in some cases, final values may still be sufficiently high to be of little concern. As an example, the dielectric breakdown voltage of an 8-mil silicone coating DC630 with an initial 1,000-volt/mil value, dropped to 400 volts/mil after 24 hr immersion in distilled water and to 500 volts/mil after 24 hr exposure at 500°F.[36]

Several standardized procedures for measuring dielectric strength are available, such as ASTM D 149. A list of dielectric strength values for the commonly used organic coatings and for some ceramic and nonorganic materials is given in Table 15.

Arc resistance Arc resistance is the time in seconds that an arc may play across the surface of a material without rendering it conductive or the time before electrical breakdown occurs along the surface. An arc of high voltage and low current is usually used to simulate service conditions such as those existing in ac circuits operating at high voltages with currrents limited to the order of milliamperes. The ASTM D 495 test method,

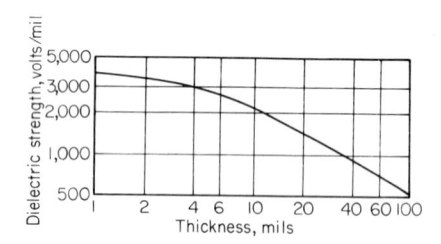

Fig. 10 Effect of coating thickness on dielectric strength of Teflon TFE.

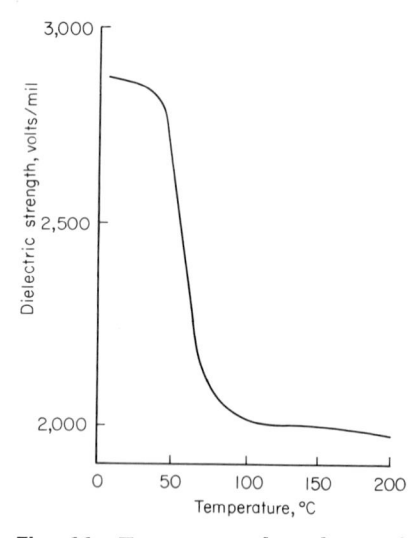

Fig. 11 Temperature dependence of the dielectric strength of polytetrafluoroethylene[35] (5-mil film, ¼-in. electrode).

normally used, calls for exposure of 300 sec or more without failure. There are three modes by which failures due to arcing can occur in organic coatings:

1. By tracking, that is, by the formation of a thin wiry line between the electrodes.
2. By carbonization of the surface due to heating. A conductive path is formed because of the formation of carbonaceous material which has less resistance to the flow of current.
3. The coating may burst into flame without the formation of any visible conducting path.

Not all coatings may be defined in terms of precise arc-resistance values. Some may melt, gum up, or mechanically degrade by cracking. Where precise values are obtained, they will be a function of the inherent stability of the material, the purity of the material, and its surface cleanliness. With regard to the last mentioned, arc resistance may be improved by maintaining the surface dry and free from contaminants. Simply handling the surface with the fingers will reduce the arc resistance because of moisture, salts, and grease contamination transferred from the fingers. Typical arc-resistance values of some polymeric compounds are found in Table 16.

Properties of Electronic Coatings 5-35

TABLE 15 Dielectric Strengths of Polymer Coatings

Material	Dielectric strength, volts/mil	Comments*	Source of information
Polymer coatings:			
Acrylics............	450–550	Short-time method	a
	350–400	Step-by-step method	a
	400–530	....	b
	1,700–2,500	2-mil-thick samples	Columbia Technical Corp., Humiseal Coatings
Alkyds.............	300–350	....	b
Chlorinated polyether	400	Short-time method	a
Chlorosulfonated polyethylene..........	500	Short-time method	a
Diallyl phthalate.....	275–450	....	b
	450	Step-by-step method	a
Diallyl isophthalate...	422	Step-by-step method	a
Depolymerized rubber (DPR)	360–380	....	H. V. Hardman Co., DPR Subsidiary
Epoxy..............	650–730	Cured with anhydride–castor oil adduct	Autonetics, Div. of North American Rockwell
Epoxy..............	1,300	10-mil-thick dip coating	
Epoxies, modified.....	1,200–2,000	2-mil-thick sample	Columbia Technical Corp., Humiseal Coatings
Neoprene...........	150–600	Short-time method	a
Phenolic............	300–450	....	b
Polyamide..........	780	106 mils thick	
Polyamide-imide......	2,700		
Polyesters..........	250–400	Short-time method	a
	170	Step-by-step method	a
Polyethylene........	480	....	b
	300	60-mil-thick sample	c
	500	Short-time method	a
Polyimide..........	3,000	Pyre-M.L., 10 mils thick	d
	4,500–5,000	Pyre-M.L. (RC-675)	e
	560	Short-time method, 80 mils thick	a
Polypropylene.......	750–800	Short-time method	a
Polystyrene.........	500–700	Short-time method	a
	400–600	Step-by-step method	a
	450	60-mil-thick sample	c
Polysulfide..........	250–600	Short-time method	a
Polyurethane (single component)........	3,800	1-mil-thick sample	f
Polyurethane (two components)/castor oil cured..........	530–1,010	....	g
Polyurethane (two components, 100% solids)	275	125-mil-thick sample	Products Research & Chem. Corp. (PR-1538)
	750	25 mils thick sample	
Polyurethane (single component)	2,500	2-mil-thick sample	Columbia Technical Corp., Humiseal 1A27
Polyvinyl butyral.....	400		
Polyvinyl chloride....	300–1,000	Short-time method	
	275–900	Step-by-step method	
Polyvinyl formal......	860–1,000	....	b
Polyvinylidene fluoride	260	Short-time, 500-volt/sec, ⅛-in. sample	h
	1,280	Short-time, 500-volt/sec, 8-mil sample	h
	950	Step by step (1-kv steps)	h

TABLE 15 Dielectric Strengths of Polymer Coatings (Continued)

Material	Dielectric strength, volts/mil	Comments*	Source of information
Polyxylylenes:			
Parylene N	6,000	Step by step	Union Carbide Corp.
	6,500	Short time	Union Carbide Corp.
Parylene C	3,700	Short time	Union Carbide Corp.
	1,200	Step by step	Union Carbide Corp.
Parylene D	5,500	Short time	Union Carbide Corp.
	4,500	Step by step	Union Carbide Corp.
Silicone	500	Sylgard 182	Dow Corning Corp.
Silicone	550–650	RTV types	General Electric & Stauffer Chemical Co. bulletins
Silicone	800	Flexible dielectric gel	Dow Corning Corp.
Silicone	1,500	2-mil-thick sample	Columbia Technical Corp., Humiseal 1H34
TFE fluorocarbons	400	60-mil-thick sample	c
	480	Short-time method	a
	430	Step by step	a
Teflon TFE dispersion coating	3,000–4,500	1–4-mil-thick sample	E. I. du Pont de Nemours & Co.
Teflon FEP dispersion coating	4,000	1.5-mil-thick sample	E. I. du Pont de Nemours & Co.
Other materials used in electronic assemblies:			
Alumina ceramics	200–300		b
Boron nitride	900–1,400		b
Electrical ceramics	55–300		b
Forsterite	250		b
Glass, borosilicate	4,500	40-mil sample	c
Steatite	145–280		b

* All samples are standard 125 mils thick unless otherwise specified.
a Insulation, Directory Encyclopedia Issue, no. 7, June–July, 1968.
b Mater. Eng., Materials Selector Issue, vol. 66, no. 5, Chapman-Reinhold Publication, mid-October, 1967–1968.
c Kohl, W. H.,: "Handbook of Materials and Techniques for Vacuum Devices," p. 586, Reinhold Publishing Corporation, New York, 1967.
d Learn, J. R., and M. P. Seegers: Teflon-Pyre-M. L. Wire Insulation System, *13th Symp. on Tech. Progr. in Commun. Wire and Cables*, Atlantic City, N.J., Dec. 2–4, 1964.
e Milek, J. T.: Polyimide Plastics: A State of the Art Report, *Hughes Aircraft Rep.* S-8, October, 1965.
f Hughson Chemical Co. Bull. 7030A.
g Spencer-Kellogg (Division of Textron, Inc.) *Bull.* TS-6593.
h Pennsalt Chemicals Corp. Prod. Sheet KI-66a, Kynar Vinylidene Fluoride Resin, 1967.

Environmental Protection

Moisture protection Probably the most important function of coatings for printed circuits and indeed for all electronic assemblies is to act as moisture and gas barriers preventing corrosion and electrical insulation breakdown.

It should be understood that all present state-of-the-art coatings absorb water and are permeable to moisture and other gases to a greater or lesser extent and cannot be considered true hermetic seals in the sense that welded metal packages can. However, it is possible to select plastic coatings that will reduce the rate of moisture transmittance such that assemblies can pass long-term storage and operational requirements.

Failures due to penetration of moisture may be of two types. Moisture mobilizes ionic contaminants in the plastic or on the surface of the device. This results in the normally observed deterioration of electrical insulation or in the formation of inversion layers in semiconductor devices. Secondly, moisture, again in the presence of impurity ions, acts as a vehicle permitting electrolytic corrosion to take place. It is for these reasons that both active and passive devices such as

TABLE 16 Arc Resistances of Polymer Coatings

Material	Arc resistance, sec	Information source
Acrylic	35–55	a
Diallyl phthalate	118	b
Diallyl isophthalate	123–128	b
Epoxy-polyamide (60% epoxy, 40% Versamid 125)	82	c
Epoxy-polyamide (50% epoxy, 50% Versamid 115)	76	c
Epoxies, modified	45–55	a
Epoxies	45–120	b
Epoxy (Epocast 202/D40) = A	64	d
A + 100% silica	124	d
A + 100% gypsum	102	d
A + 75% mica	79	d
A + 100% alumina	78	d
Phenolic	Tracks	b
Polyamide (nylon 6/6)	140	b
Polyamide (nylon 6/10)	130	b
Polyimide	183	e
Polystyrene	60–80	b
Polyurethanes (two components)—castor oil cured	88–140	f
Polyurethane (one component—air drying)	40	a
Polyvinylidene fluoride	50–70	g
Silicones	120–200	h
Silicones (RTV types)	90–130	h
Silicone	60	a
Teflon TFE	>300	i
Teflon FEP	>300	i

[a] Columbia Technical Corp.: Humiseal Protective Coatings, technical data sheet.
[b] *Insulation*, Directory Encyclopedia Issue, no. 7, June–July, 1968.
[c] Lee, H. and K. Neville: "Handbook of Epoxy Resins," p. 10–8, McGraw-Hill Book Company, New York, 1967.
[d] Delmonte, J.: *Plast. Technol.*, March, 1958.
[e] Milek, J. T.: Polyimide Plastics, A State of the Art Report, *Hughes Aircraft Rep.* S-8, October, 1965.
[f] *Spencer-Kellogg* (Division of Textron, Inc.) *Bull.* TS-6593.
[g] *Pennsalt Chemicals Corp. Prod. Sheet* KI-66a, Kynar Vinylidene Fluoride Resin, 1967.
[h] Dow Corning Corp.: Technical bulletins.
[i] E. I. du Pont de Nemours & Co.: Teflon for Electrical and Electronic Systems, technical bulletin, 1964.

transistors, diodes, and resistors are hermetically sealed in metal packages and why entire electronic assemblies are sealed in an inert gas atmosphere. For most commercial applications, hermetic sealing is expensive and impractical. In these cases, organic coatings or plastic encapsulants may provide sufficient protection.

Since all organic coatings are permeable to moisture in varying degrees, quantitative values for their resistance to water are important. Water resistance may be defined in terms of three factors: (1) the percent absorption of water by the coating

5-38 Coatings for Electronics

at constant temperature over a period of time, (2) the rate of water-vapor transmission through the coating, and (3) survival of coated electronic assembly after simulated or accelerated humidity cycling tests.

Water Absorption. The percentage of water absorption eventually becomes constant at constant temperature for each material. In many cases, however, equilibrium is only achieved after several months of immersion. Unfortunately, most of the reported data are for a 24-hr water immersion and do not give the true

TABLE 17 Water Absorption Data for Polymer Coatings

Material	% absorption in 24 hr*	Source of information
Chlorinated polyether (Penton)	Negligible	Hercules Powder Co.
Depolymerized rubber	0.5	H. V. Hardman, DPR Subsidiary
Diallyl phthalate	0.09	a
Epoxy	0.04	a
Epoxies	0.11–1.84	b
Epoxy cured with DETA	0.11–0.12	c
Fluorocarbon (Teflon)	0.01	a
Fluorocarbon (Teflon TFE dispersion coating)	<0.1	Du Pont
Fluorocarbon (Teflon FEP dispersion coating)	<0.01 0.3 –0.4	Du Pont
Phenolic	0.9	e
Phenoxy	0.13	d
Polyesters (copolymerized with styrene)	0.17–0.28	e
Polyethylene	0.01	a
Polyimide	0.7	Union Carbide
Polyimide-amide	0.5 –1.9 (in boiling water)	
Polystyrene	0.04	a
Polyvinyl fluoride	0.04	Product Techniques, Inc., PT-207 Coating
Polyvinyl formal (Formvar)	0.75–1.1	
Polyvinylidene chloride–polyvinyl chloride (Saran)	0.05	
Polyvinylidene fluoride	0.04	f
Polyxylylene (Parylene C)	0.06	Union Carbide
Polyxylylene (Parylene N)	0.01	Union Carbide
Silicone	0.15	a

* At room temperature, unless otherwise specified.

[a] Piser, J.: *Electron. Des.* vol. 17, Aug. 16, 1967.

[b] *Celanese Corp. Tech. Bull.* 1164.

[c] Lee, H., and K. Neville: "Epoxy Resins," McGraw-Hill Book Company, New York, 1966.

[d] Lee, H., D. Stoffey, and K. Neville: "New Linear Polymers," McGraw-Hill Book Company, New York, 1967.

[e] Parker, E. E., and E. W. Moffett: *Ind. Eng. Chem.*, vol. 46, p. 1615, 1954.

[f] Barnhart, W. S., R. A. Ferren, and H. Iserson: *17th ANTEC of SPE*, January, 1961.

[g] "Modern Plastics Encyclopedia" (annual), McGraw-Hill, Inc., New York.

equilibrium values, which are often considerably higher. Twenty-four-hour data will be meaningful and can serve as a means of comparison for applications where they are equal to or more severe than the part will see in actual operation. For the most part, the percent of water absorption is not in itself a very meaningful value. For some coatings, electrical properties are adversely affected, even though only a small amount of water has been absorbed, and conversely, others with very large water absorptions may be quite stable electrically.[37] In general, however, surface resistivity and volume resistivity decrease and dissipation factor increases with increase in percent of water absorbed, and therefore water absorbed or permeating a coating is undesirable and should be avoided.[38] It is in this light that comparative water-absorption values should be studied and used (Table 17).

Water-vapor Permeability. Moisture may seep through a coating by one or both of two mechanisms. One involves passage through microcracks, channels, or pinholes arising from imperfections in the coating. Pinholes and pores can be reduced considerably by applying a number of consecutive coats, each one further reducing the pinhole probability. The second mechanism is one inherent with the molecular structure of polymer molecules and within polymers.

This water-vapor permeability or moisture-vapor transmission rate (MVTR) is more significant than water absorption in defining water resistance. It is expressed according to the equation

$$\text{MVTR} = \frac{Ql}{at\,\Delta p}$$

where MVTR = $g/(hr)(cm^2)(cm/cm\ Hg)$
Q = water vapor permeating the film, g
l = film thickness, cm
a = film area, cm^2
t = time, hr
Δp = vapor-pressure difference, cm Hg

Organic coating polymers differ widely in permeability rates because of differences in their molecular structures. An example of this is polyvinylidene chloride (Saran),* a material which because of its polar chloro groups, is an excellent moisture barrier.

Permeability is also a function of the nature and amount of pigment or filler used. Coatings have decreasingly lower permeabilities as pigment is increased up to a critical pigment-volume ratio, after which the permeability rises sharply. Pigments in flake form such as aluminum will considerably improve resistance to moisture penetration. Several other factors that affect permeability are as follows:

Factor	Usual effect on permeability
1. Solvent entrapment	Increase
2. Plasticizers	Increase
3. Structure of film:	
Polar	Increase
Nonpolar	Decrease
4. Degree of crosslinking	Decrease

A list of permeabilities may thus serve as a preliminary selection guide (Table 18). The list should not be used as the sole basis for selection because these values are usually obtained by using carefully fabricated films for the test. In actual usage, a sprayed or dipped coating may contain many imperfections such as pinholes, solvent entrapment, and contaminants, all of which increase its porosity.

Humidity Testing. Both water-absorption and water-permeability values may serve as preliminary data for selecting coatings for further evaluation. As was pointed out above, to ensure long-term reliability, electronic assemblies should be subjected to long-term or accelerated environmental testing. A humidity test such as that called out in MIL-E-5272 or MIL-STD-202B is one of the most severe and critical. Generally, the test requires a 10-day exposure to moisture in both the vapor and condensate forms with temperature cycling from 25 to 65°C. A 100-volt dc applied stress is also required according to MIL-STD-202B. The test is an accelerated one but does give an indication of the long-term reliability of the assembly under outdoor and operating conditions. Because of the variety of

* Trademark of Dow Chemical Company, Midland, Mich.

5-40 Coatings for Electronics

substrates, metal combinations, and sensitive components, very few pieces of uncoated hardware will pass this test without serious physical or electrical degradation. Even when a protective coating is employed, care should be taken in selecting one that will not only shield the assembly but also will not itself degrade by blistering, lifting, whitening, softening, or absorbing excessive amounts of water. Unfortunately, since coating formulations vary widely, it is impossible to give a list of good coatings and poor ones from which to select. It is imperative that humidity testing of some sort be performed in each case before recommendation and use.

TABLE 18 Moisture-vapor Transmission Rates (MVTR) of Plastic Coatings and Film (in $g/(mil)(in.^2)(24\ hr)$)

Coating or film	MVTR	Information source
Epoxy-anhydride	2.38	Autonetics data (25°C)
Epoxy–aromatic amine	1.79	Autonetics data (25°C)
Neoprene	15.5	Baer[29] (39°C)
Polyurethane (Magna X-500)	2.4	Autonetics data (25°C)
Polyurethane (isocyanate-polyester)	8.72	Autonetics data (25°C)
Olefane,* polypropylene	0.70	Avisun data
Cellophane (type PVD uncoated film)	134	Du Pont
Cellulose acetate (film)	219	Du Pont
Polycarbonate	10	FMC data
Mylar†	1.9	Baer[29] (39°C)
	1.8	Du Pont data
Polystyrene	8.6	Baer[29] (39°C)
	9.0	Dow data
Polyethylene film	0.97	Dow data (1-mil film)
Saran resin (F120)	0.097 to 0.45	Baer[29] (39°C)
Polyvinylidene chloride	0.15	Baer[29] (2-mil sample, 40°C)
Polytetrafluoroethylene (PTFE)	0.32	Baer[29] (2-mil sample 40°C)
PTFE, dispersion cast	0.2	Du Pont data
Fluorinated ethylene propylene (FEP)	0.46	Baer[29] (40°C)
Polyvinyl fluoride	2.97	Baer[29] (40°C)
Teslar	2.7	Du Pont data
Parylene N	14	Union Carbide data (2-mil sample)
Parylene C	1	Union Carbide data (2-mil sample)
Silicone (RTV 521)	120.78	Autonetics data
Methyl phenyl silicone	38.31	Autonetics data
Polyurethane (AB0130-002)	4.33	Autonetics data
Phenoxy	3.5	Lee, Stoffey, and Neville[40]
Alkyd-silicone (DC-1377)	6.47	Autonetics data
Alkyd-silicone (DC-1400)	4.45	Autonetics data
Alkyd-silicone	6.16–7.9	Autonetics data
Polyvinyl fluoride (PT-207)	0.7	Product Techniques Incorp.

* Trademark of Avisun Corporation, Philadelphia, Pa.
† Trademark of E. I. du Pont de Nemours & Co., Wilmington, Del.

Corrosion protection Coatings are applied to metal surfaces where protection from the corrosive effects of the surrounding medium is necessary. The most commonly encountered corrosive environments are listed in Table 19. In most cases, elevated temperature will accelerate the degradation effects due to these environments.

Metals corrode in different ways depending on their nature and the environment to which they are exposed. A partial list of the various modes of failure for the metals and alloys commonly employed in electronic assemblies is given in Table 20.

In general, corrosion occurs by either a chemical or an electrochemical process. In either process, moisture is invariably necessary. Hence the moisture barrier and

Properties of Electronic Coatings 5-41

TABLE 19 Environments and Their Corrosive Constituents

Environment	Corrosive constituents
1. Normal air ambient	Moisture, oxygen, sulfur dioxide, carbon dioxide
2. Water immersion	Calcium salts and other metal salts
3. Salt-water immersion or salt spray	Sodium and chloride ions, marine organisms
4. Chemicals and solvents	Numerous acids, bases, oxidizing or reducing agents, solvents strippers, etc. which may come into contact with the part during processing or operation.
5. Soil	Moisture, fungus, other microorganisms

resistance of a coating become a key factor in corrosion prevention. For many metals, oxide inorganic films will normally form on storage in air. These films are quite passive and will arrest the further degradation of the metal. In other cases, artificially produced inorganic coatings may be formed; and these will also provide

TABLE 20 Corrosion Modes for Metals and Alloys Commonly Used in Electronic Assemblies

Metal/Alloy	Type of corrosion
Copper	Thermal-air exposure results in black oxidation product. Forms green copper carbonate (verdigris) which inhibits further corrosion except as accelerated by galvanic coupling. Exposure to salt environment produces high electrical resistance due to formation of copper salts.
Copper-nickel alloys	Tarnishes in sulfur-containing ambients.
Copper-beryllium alloys	Tarnishing in sulfur-containing ambients and in moist air.
Aluminum (pure or 5052, 6061, 1100, 3003)	White oxide formations, normally superficial with no structural damage.
Aluminum high-strength alloys: 2024, 2014	Susceptible to tunneling, exfoliation, and stress corrosion.
7075, 7079	Susceptible to stress corrosion.
Aluminum alloys, general	Galvanic corrosion from dissimilar metal couples.
Magnesium AZ31B	Rapid local dissolving occurs at breaks in metallic coatings or at points of coupling with nobler metals. Oxidizes slowly in moist air.
Magnesium-lithium alloy	Interaction of moisture with the lithium portion of the alloy results in rapid evolution of hydrogen and formation of white lithium hydroxide. Organic protective coatings used are permeable to moisture and become blistered and lift owing to hydrogen generation. This alloy also reacts with carbon dioxide from the ambient to give white lithium carbonate.
Magnesium-thorium alloy	Very susceptible to moist ambient.
Beryllium	Stable in air. On heating in air, forms a metal oxide protective layer. Will react rapidly with methyl alcohol. Attacked by alkalies with evolution of hydrogen.
Nickel	Little corrosion problem except for galvanic couples. Very stable in air and water owing to presence of nickel oxide layer.
Solder (lead-tin)	Little corrosion problem except for galvanic couple, such as with copper. Tin salts produced.
Silver	Black silver sulfide "tarnish" on exposure to ambient due to interaction with SO_2 and H_2S contaminants in air.
Gold	Extremely inert, no corrosion products.

the same protection. These films, called chemical conversion coatings, are formed by the reaction of the metal surface with oxygen, phosphates, chromates, or other chemicals. Other passive films are electrochemically formed, such as the anodized films. For very harsh environments or for very long-term exposure, organic coatings are applied over these conversion coatings and provide a very effective barrier system. Lacquers, baking enamels, and highly resistant epoxies, vinyls, polyurethanes, or silicone coatings are often used.

Wide differences in the effectiveness of these coatings are possible because of differences in surface preparation and cleanliness of the substrate and in thickness of the coating. In general, the thicker the coating, the better corrosion protection is afforded. Thickness affords protection by reducing the probability of imperfections in the coating, by increasing protection from abrasion and mechanical damage during use, and by increasing the moisture resistance of the film.

Magnesium. Because of its high strength-to-weight ratio, good fatigue strength, and good dimensional stability, magnesium has become a very popular metal for electronic applications, especially those intended for flight hardware. Heat sinks, housings, coolant tubes, and frames are but a few examples of the use of magnesium. For benign environments such as indoor exposure or a black box containing inert gas, a protective finishing system is usually not necessary. Even for normal weathering or short-term intermittent outdoor exposure, an electrochemical conversion coating alone or a single coating of clear lacquer may provide sufficient protection.

Other chemical treatments, useful as protective coatings in themselves or as bases for primers and topcoats, are Dow Chemical Treatments No. 1 (chrome pickle); No. 7 (dichromate); and No. 19 (dilute chromic acid).[41] The treatments most commonly used for electronic hardware are Dow 17, HAE, or fluoride anodizing, which are excellent paint bases for long-term exposure to humid and salt environments; Dow 7, used for a corrosion-protective paint base; and Dow 19, used as a touch-up applied by brush, spray, or swab for rework on Dow 7 or Dow 17 surfaces. The various treatments are compared in Table 21.

These surface treatments are all acid types in order to either neutralize or acidify the normally alkaline magnesium surface. Suppression of alkali formation is essential in improving the durability and adhesion of paints and coatings. To be effective, primers for magnesium should be based on alkali-resistant vehicles such as polyvinyl butyral (as contained in wash primers), acrylic, polyurethane, vinyl, epoxy, or baked phenolic. Zinc chromate, strontium chromate, and basic zinc chromate are often employed as further corrosion inhibitors.[41,42]

For severe or extensive exposure as encountered in ground-support, ground-operating equipment, or marine equipment, the best finish system will usually consist of three parts: a conversion coating, a primer, and finally an organic topcoat.

Magnesium-Lithium Alloys. Magnesium-lithium alloys such as LA141 are attractive to designers of airborne electronic equipment because of their stiffness and high strength-to-weight ratio. By using these alloys, a 20 to 25 percent weight savings over magnesium or other magnesium alloys may be achieved. However, the high reactivity of magnesium-lithium with moisture and carbon dioxide limits their use to hermetically sealed boxes with inert gas ambients. Often, even in an enclosed box, an incompatible situation may arise owing to the evolution of moisture from plastics or from electronic components contained in the box. Moisture reacts with the lithium portion of the alloy, forming hydrogen gas and lithium hydroxide. In the presence of carbon dioxide, white lithium carbonate is formed on the alloy surface. In addition to the formation of these chemical corrosion products and deterioration of the alloy, a secondary, perhaps even more harmful effect may result by the evolution and reaction of hydrogen gas with active devices, chemically sensitive thick- or thin-film resistors and other components. Although there are reported to be several coatings which will protect magnesium-lithium in normal environments, no organic coating or coating system is currently known to provide adequate protection to pass military environmental specifications such as MIL-E-5272, temperature and humidity cycling, or MIL-STD-151, salt-spray test.[43]

TABLE 21 Chemical and Anodic Treatments Used in Finishing of Magnesium[11]

Name	Type	Military specification	Appearance	Alloys used on	Uses	Remarks
Dow 1	Chrome pickle	MIL-M-3171A —type 1	Matte gray to yellow-red	All	For protection during shipment and storage.	Simple dip treatment, slight dimensional loss, good paint base. Does not materially affect dimensions.
Dow 7	Dichromate	MIL-M-3171A —Type 3	Brassy to dark brown	All except Th-containing alloys	Good combination of paint base and protective qualities.	
Dow 9	Galvanic anodize	MIL-M-3171A —Type 4	Dark brown to black	All	Used on alloys that will not take No. 7 or where a black coat is needed.	Requires galvanic couple between work and steel cathode. No external current required.
Dow 10	Sealed chrome pickle	MIL-M-3171A —Type 2	Matte gray to yellow-red-brown	All	Corrosion resistance equal to No. 7.	Chrome pickle plus dichromate boil. Improved protection over No. 1. Slight dimensional loss.
Dow 17	Anodize thin	MIL-M-45202	Light gray to light green	All	Excellent protective and decorative finish.	Thin: 0.0001 to 0.0003 in. Full: 0.001 to 0.0015 in.
Dow 17	Heavy	MIL-M-45202	Dark green	All	Excellent protective and decorative finish.	Coating is stable above melting point of metal. Best abrasion resistance, best paint base, and most consistent of all Dow treatments for magnesium. License required.
Dow 23	Stannate immersion	None	Medium to dark gray	All	Good paint base and protective finish. Used with dissimilar metal inserts (except aluminum or in assemblies).	Must be neutralized if painted. Retards galvanic corrosion. License required.
HAE	Anodize	MIL-M-45202 MIL-C-13335 AMA 2476	Tan to brown	All	Excellent protective and decorative finish. Good paint base.	H. A. Evangelides, at Pitman-Dunn Lab, Frankford Arsenal.
Fluoride	Anodize	Proprietary	Light gray	All	Good base for resin impregnation or painting. Very clean surface.	Dr. W. F. Higgins, Magnesium Elektron, Ltd.

5-44 Coatings for Electronics

Tin-Lead. For maximum protection of solder joints and solder-plated surfaces, conformal coatings should be used.[44] However, as with other metals, corrosion can be prevented or minimized by employing certain design principles. For example, bimetallic combinations of tin-lead solder with either copper or aluminum should be avoided in high-humidity or salt-water environments where corrosion is known to occur.[45] The coating of such bimetallic combinations will only provide temporary protection. Other difficulties associated with corrosion of solder joints or solder

TABLE 22 Resistance of Coating Resins to Microorganisms[46]

Coating type	Resistance to attack
Acrylic	G
Alkyd	P
Cellulosics:	
Cellulose acetate	G to P
Cellulose acetate butyrate	G
Cellulose nitrate	P
Ethylcellulose	G
Chlorinated rubber	G
Chlorosulfonated polyethylene	G
Epoxy-amine	G
Epoxy-polyamide	G
Epoxy-phenol-formaldehyde	G
Epoxy-urea-formaldehyde	G
Melamine-formaldehyde	G to P
Phenol-formaldehyde	G to P
Phenoxy	G
Polyester	G
Polyethylene:	
Low molecular weight	P
High molecular weight	G
Polychlorofluoroethylene	G
Polyisobutylene	G
Polystyrene	G
Polyurethane	P
Polyvinyl acetate	P
Polyvinyl butyrate	P
Polyvinyl chloride	G*
Polyvinyl chloride–acetate	G
Polyvinylidene chloride	G to P
Polyvinylidene fluoride	G
Silicone	G
Silicone-alkyds	G to P
TFE fluorocarbon	G
Urea formaldehyde	G

* Except when plasticized.
G Good
P Poor

plate are caused by the incomplete removal of corrosive fluxes during cleaning operations. Conformal coatings will not provide protection unless these contaminants are adequately removed. A discussion of cleaning procedures for flux residues is given in another section.

Aluminum. Like magnesium, aluminum is also widely used because of its low weight. It, too, must be protected from corrosion, although the normal oxide film which forms on its surface on exposure to air is protective and prevents deterioration of the bulk of the metal. A number of oxide coatings (anodize) may be purposely formed, as are chemical surface treatments such as chromates, to function as protective coatings or to improve the adhesion of subsequent organic coatings.

Microorganism Protection. As with moisture and salt spray, microorganisms also cause millions of dollars of loss per year due to material deterioration.

Microbial growth may cause corrosion of metals and bimetallic combinations, may deteriorate a coating or substrate mechanically or electrically, or may roughen a surface, causing an increase in friction and drag; the latter property being especially critical for underwater vehicles. The resistance of coatings to biological attack from microorganisms such as fungi, bacteria, etc., is particularly important for electronic hardware that will be stored or operated in humid, tropical, or semitropical environments, or contact of moist soil. Table 22 gives some coating resins and the resistance to attack.

As a rule, the newer synthetic coatings such as epoxies, silicones, and polyurethanes possess excellent fungus-resistance properties. Some formulations may not be as good as others because of plasticizers or other nutrient additives used in the formulation.

Thermal protection One of the more destructive influences on most organic materials is heat and high temperature. The heat transfer (conductivity) of organic materials compared to that of air spaces is excellent, but compared to metals, or even some ceramics, it is very poor.

Heat Resistance. Heat degradation is an effect of time and temperature. Hence, one usually refers to temperatures that can be resisted for short, intermittent times (minutes or hours) or to temperatures for the long term or "service life" (months or years). A rough value for representative type coatings can be found in Table 1.

Thermal Conductivity. The total heat conducted through a material is directly proportional to the surface area, time of contact, and temperature gradient, and inversely proportional to the thickness of the sample according to Eq. (1),

$$Q \propto \frac{(T_2 - T_1)At}{d} \qquad (1)$$

where Q = total heat flow
A = surface area, cm^2
d = thickness of sample, cm
T_2 = temperature of the hot surface, °C
T_1 = temperature of the cold surface, °C
t = time, in seconds

When a proportionality constant k is introduced, the equation becomes

$$Q = \frac{k(T_2 - T_1)At}{d} \qquad (2)$$

The constant k is a material constant called the *coefficient of thermal conductivity*. It is the time rate of heat flow under steady-state conditions, through a unit area and unit thickness, per unit temperature gradient.

Organic coatings, as a rule, are seldom used for the prime purpose of conducting heat away. Unfilled coatings and plastics have rather low k values and are very good thermal barriers or insulators. Furthermore, there is very little difference in thermal conductivity among the different polymer types; the order of magnitude is about 10^{-4} cal/(sec)(cm^2)(°C/cm). This value is to be compared with those for metals or metal oxides which are thousands of times higher. Table 23 gives some representative thermal conductivities for coatings and other electronic materials.

From a design standpoint, the low thermal conductivity of organic coatings may be desirable as in the case of ablative and thermal protective materials, or undesirable as in the case of high heat-dissipating electronic assemblies.

For the most part, plastics used in intimate contact with electronics such as spacers, filleting compounds, substrates, heavy coatings, and encapsulants require high thermal conductance. Assemblies or modules having a high density of electronic components require the dissipation of large amounts of heat. Properly insulated metal heat sinks or beryllium oxide spacers are commonly employed. No plastic yet exists

TABLE 23 Thermal Conductivity Data for Polymer Coatings

Material	k value,* cal/(sec)(cm²)(°C/cm) × 10⁴	Source of information
Unfilled plastics:		
Acrylic	4–5	a
Alkyd	8.3	a
Depolymerized rubber	3.2	H. V. Hardman, DPR Subsidiary
Epoxy	3–6	b
Epoxy (electrostatic spray coating)	6.6	Hysol Corp., DK-4
Epoxy (electrostatic spray coating)	2.9	Minnesota Mining & Mfg., No. 5133
Epoxy (Epon† 828, 71.4% DEA, 10.7%)	5.2	
Epoxy (cured with diethylenetriamine)	4.8	c
Fluorocarbon (Teflon TFE)	7.0	Du Pont
Fluorocarbon (Teflon FEP)	5.8	Du Pont
Nylon	10	d
Polyester	4–5	a
Polyethylenes	8	a
Polyimide (Pyre-M.L. enamel)	3.5	e
Polyimide (Pyre-M.L. varnish)	7.2	f
Polystyrene	1.73–2.76	g
Polystyrene	2.5–3.3	a
Polyurethane	4–5	n
Polyvinyl chloride	3–4	a
Polyvinyl formal	3.7	a
Polyvinylidene chloride	2.0	a
Polyvinylidene fluoride	3.6	h
Polyxylylene (Parylene N)	3	Union Carbide
Silicones (RTV types)	5–7.5	Dow Corning Corp.
Silicones (Sylgard types)	3.5–7.5	Dow Corning Corp.
Silicones (Sylgard varnishes and coatings)	3.5–3.6	Dow Corning Corp.
Silicone (gel coating)	3.7	Dow Corning Corp.
Silicone (gel coating)	7 (150°C)	Dow Corning Corp.
Filled plastics:		
Epon 828/diethylenetriamine = A	4	b
A + 50% silica	10	b
A + 50% alumina	11	b
A + 50% beryllium oxide	12.5	b
A + 70% silica	12	b
A + 70% alumina	13	b
A + 70% beryllium oxide	17.8	b
Epoxy, flexibilized = B	5.4	i
B + 66% by weight tabular alumina	18.0	i
B + 64% by volume tabular alumina	50.0	i
Epoxy, filled	20.2	Emerson & Cuming, 2651 ft
Epoxy (highly filled)	15–20	Wakefield Engineering Co.
Polyurethane (highly filled)	8–11	International Electronic Research Co.
Other materials used in electronic assemblies:		
Alumina ceramic	256–442 (20–212°F)	a
Aluminum	2767–5575	a
Aluminum oxide (alumina), 96%	840	j
Beryllium oxide, 99%	5500	j
Copper	8095–9334	a
Glass (Borosill, 7052)	28	k
Glass (pot-soda-lead, 0120)	18	k

TABLE 23 Thermal Conductivity Data for Polymer Coatings (Continued)

Material	k value,* cal/(sec)(cm²) (°C/cm) × 10^h	Source of information
Glass (silica, 99.8% SiO₂)	40	l
Gold	7104 (20–212°F)	a
Kovar	395	m
Mica	8.3–16.5	a
Nichrome‡	325	m
Silica	40	k
Silicon nitride	359	m
Silver	9995 (20–212°F)	a
Zircon	120–149	a

* All values are at room temperature unless otherwise specified.
† Trademark of Shell Chemical Co., New York, N.Y.
‡ Trademark of Driver-Harris Co., Harrison, N.J.
^{a} *Mater. Eng.*, Materials Selector Issue, vol. 66, no. 5, Chapman-Reinhold Publication, mid-October, 1967.
^{b} Wolf, D. C., *Proc. Nat. Electron. and Packag. Symp.*, New York, June, 1964.
^{c} Lee, H., and K. Neville: "Handbook of Epoxy Resins," McGraw-Hill Book Company, New York, 1966.
^{d} Davis, R.: *Reinf. Plast.*, October, 1962.
^{e} *Du Pont Tech. Bull.* 19, Pyre-M.L. Wire Enamel, August, 1967.
^{f} *Du Pont Tech. Bull.* 1, Pyre-M.L. Varnish RK-692, April, 1966.
^{g} Teach, W. C., and G. C. Kiessling: Polystyrene, Reinhold Publishing Corporation, New York, 1960.
^{h} Barnhart, W. S., R. A. Ferren, and H. Iserson: 17th *ANTEC of SPE*, January, 1961.
^{i} Gershman, A. J., and J. R. Andreotti: *Insulation*, September, 1967.
^{j} American Lava Corp. Chart 651.
^{k} Shand, E. B., "Glass Engineering Handbook," McGraw-Hill Book Company, 1958.
^{l} Kingery, W. D.: Oxides for High Temperature Applications, *Proc. Int. Symp.*, Asilomar, Calif., October, 1959, McGraw-Hill Book Company, New York, 1960.
^{m} Kohl, W. H.: "Handbook of Materials and Techniques for Vacuum Devices," Reinhold Publishing Company, New York, 1967.
^{n} "Modern Plastics Encyclopedia," McGraw-Hill, Inc., New York, 1968.

combining the high thermal conductivity and low electric conductivity of beryllia, which is unique among materials in this respect.

Where some degree of thermal conduction in organic coatings is required, one or both of two approaches may be used: (1) employing a coating that is formulated with a high, thermally conductive filler and/or (2) employing as thin a coating layer as possible, since thermal conduction is inversely related to thickness. In practice, it is found that what is gained in filling a coating is lost in that the coating may be more difficult to apply in thin layers. Except for those cases where electrical conductivity is not a problem, fillers used should be good electrical insulators. Commonly used fillers meeting these requirements are aluminum oxide (Al_2O_3) also known as alumina, silicon dioxide (SiO_2) also known as silica, mica, and beryllium oxide (BeO), known as beryllia. The improvement in the thermal conductivity of an Epon 828-diethylenetriamine (DEA) formulation by filling it with silica, aluminum, or beryllium, oxide is about three-to-fourfold but still far below the thermal conductivity of the filler alone. There is a sparcity of thermal conductivity data for coatings, probably due to the difficulty in preparing thick enough samples to run the thermal conductance. Many of these materials reported are molding or casting types where thick enough samples can be formed (see Table 23).

Test Methods. Several methods for measuring thermal conductivity are available, of which the Cenco-Fitch and the Guarded Hot Plate methods are the most widely

5-48 Coatings for Electronics

used. The most accurate method for plastic samples having thermal conductivities of 3.4×10^{-3} or less (most filled or unfilled plastics will fall in this category) is the guarded hot-plate method described in ASTM C 177.[47] A discussion of this and other methods is given by Anderson.[48]

The Cenco-Fitch method is based on the calorimeter principle, in which the sample to be measured is placed between a constant-temperature heat source and a heat drain (an insulated copper plug), and the heat transfer over a period of time is accurately measured. The procedure is given in ASTM D 1674 and MIL-I-16923.

For fairly accurate results, this method requires a uniform pinhole-free, air-free sample of 0.125 in. thickness.

TABLE 24 Abrasion Resistance of Organic Coatings

Coating	Taber wear index, mg/kHz
Polyurethane Type 1	55–67
Polyurethane Type 2 (clear)	8–24
Polyurethane Type 2 (pigmented)	31–35
Polyurethane Type 5	60
Urethane oil varnish	155
Alkyd	147
Vinyl	85–106
Epoxy-amine-cured varnish	38
Epoxy-polyamide enamel	95
Epoxy-ester enamel	196
Epoxy-polyamide coating (1:1)	50
Phenolic spar varnish	172
Clear nitrocellulose lacquer	96
Chlorinated rubber	200–220
Silicone, white enamel	113
Catalyzed epoxy, air-cured (PT-401)	208
Catalyzed epoxy, Teflon-filled (PT-401)	122
Catalyzed epoxy, bake-Teflon-filled (PT-201)	136
Parylene N	9.7
Parylene C	44
Parylene D	305
Polyamide	290–310
Polyethylene	360
Alkyd TT-E-508 enamel (cured for 45 min at 250°F)	51
Alkyd TT-E-508 (cured for 24 hr at room temperature)	70

The buildup of such thick coatings presents a problem with many solvent-based coatings but can be achieved by multiple coating and baking operations. As with the guarded hot-plate method, thinner coatings or coatings on a metal substrate may be used but with less accuracy.

Mechanical Protection

Abrasion and frictional resistance   For applications involving excessive handling, rubbing, or other contact with objects, a high degree of abrasion resistance will be required. For other applications where repeated sliding or insertions must take place as in some electronic drawers or consoles, surfaces with a low coefficient of friction will render the insertions easier and will minimize surface abrasion and deterioration of the coating.

Abrasion Resistance.   Abrasion resistance may be determined by periodic examinations performed while the coating is in actual use. However, several accelerated and quantitative test methods are available that may be correlated with actual service performance or used to compare the relative abrasion resistances of coatings. These include projection of a controlled stream of abrasive such as Carborundum*

---

* Trademark of The Carborundum Co., Niagara Falls, N.Y.

(ASTM D 658-44) or sand (ASTM D 968-51) against the sample and recording the amount of abrasive required to wear through a unit thickness of coating, use of abrasive tape (Armstrong test, ASTM D 1242), controlled scraping with a metal tool until the coating is penetrated,[49] or use of rotating abrasive wheels. The last mentioned (referred to as the Taber Abrasion Test) is the most widely used for plastic coatings and is described in Federal Test Methods 141 No. 6192 and ASTM Methods D 412 and D 1044. According to the last-mentioned test, abrasion is effected through the action of two standardized abrading wheels made to rotate against the coating sample. The sample consists of a 4- by 4-in.-square coated metal panel having a center spindle hole. The panel is preweighed to the nearest milligram and placed on a turntable operating at a constant speed while two abrad-

TABLE 25 Coefficients of Friction of Typical Coatings

Coating	Coefficient of friction, μ	Information source
Polyvinyl chloride	0.4–0.5	a
Polystyrene	0.4–0.5	a
Polymethyl methacrylate	0.4–0.5	a
Nylon	0.3	a
Polyethylene	0.6–0.8	a
Polytetrafluoroethylene (Teflon)	0.05–0.1	a
Catalyzed epoxy air-dry coating with Teflon filler	0.15	b
Parylene N	0.25	c
Parylene C	0.29	c
Parylene D	0.31–0.33	c
Polyimide (Pyre-M.L.)	0.17	d
Graphite	0.18	d
Graphite–molybdenum sulfide:		
Dry-film lubricant	0.02–0.06	e
Steel on steel	0.45–0.60	e
Brass on steel	0.44	e
Babbitt on mild steel	0.33	e
Glass on glass	0.4	e
Steel on steel with SAE no. 20 oil	0.044	e
Polymethyl methacrylate to self	0.8 (static)	e
Polymethyl methacrylate to steel	0.4–0.5 (static)	e

[a] Bowder, F. P., *Endeavor*, vol. 16, no. 61, p. 5, 1957.
[b] Product Techniques Incorporated: Bulletin on PT-401 TE, Oct. 17, 1961.
[c] Union Carbide data.
[d] Du Pont Tech. Bull. 19, Pyre-M.L. Wire Enamel, August, 1967.
[e] Electrofilm, Inc. data.

ing wheels each ½-in. wide are fixed on each side of the spindle and in direct contact with the coating.

A load up to 1,000 g may be placed on the wheels in addition to their own weight. A built-in tachometer records the number of turntable revolutions, and a vacuum attachment removes plastic particles as they are abraded from the surface. The instrument is operated for a specific number of cycles or until a portion of the substrate becomes exposed. Results are reported as weight loss in milligrams for a specified number of cycles ranging from 100 to 1,000, referred to as Taber Wear Index, or may be reported as the number of cycles to failure per mil thickness of coating. Quantitative data for a number of coatings may be found in Table 24.

Coefficient of Friction. The coefficients of friction of plastics, with exceptions such as some fluorocarbons, are of the same order of magnitude as metals. With plastics, however, values are more dependent on test conditions such as temperature, humidity, loading, operating speed, and on material conditions such as degree of cure and moisture content. As an example, the coefficient of friction of nylon will vary from 0.91 to 1.19, with a variation of moisture content from 0.2 to 10 percent, respectively.[50] The surface against which the sample is tested is also an important variable. A highly polished steel surface is often used, but polished glass or wood have also been employed. In any event, the nature of this surface should be specified.

Teflon appears to be unique among both plastics and metals in possessing one of the lowest friction coefficients. The coefficients of friction of nylon and of some specially formulated epoxies and polyurethanes containing Teflon fillers, although not as low as for Teflon, are still considered satisfactory for many sliding or insertion applications. The use of Teflon as a filler for epoxy coatings can reduce the coefficient of friction and the Taber Wear Index (Tables 24 and 25).

TABLE 26 Sward Hardness Values for Coatings (Polished Glass = 100)

Coating	Value	Information source
Two-component polyurethane (polyester cure)	68	a
Two-component polyurethane (castor oil cure)	46	a
Chlorinated rubber	12	a
Epoxy	22	a
Polyvinyl	6	a
Polyurethane clear coating (TDI–castor oil)	10	b
Polyurethane clear coating (MDI–castor oil)	17	b
Polyurethane (moisture cure) ASTM, Type 2	48	c
Acrylics	~20	d
Epoxy-amine paint (PT-401)*	18–22	e
TT-E-489B, Class A*	13	e
TT-E-489, Class A*	5–8	e
MIL-E-7729, Type I*	8	e
TT-E-529, Class A*	7	e
Kemacryl M49WC15 (Sherwin-Williams), acrylic*	12	e
Aero Cati-Coat F55W-P-20 (Sherwin-Williams), epoxy*	26	e
TT-E-527, alkyd*	3–4	e
MIL-E-15090, Type III, alkyd, Class 1*	5–7	e
MIL-E-7729, Type III, alkyd (baked 15 min at 320°F)	8–9	e
MIL-L-7178, nitrocellulose*	11–12	e

* Cure 7 days at room temperature.
[a] *Chem. Eng.*, Feb. 9, 1959, pp. 144–148.
[b] Dombrow, B. A.: "Polyurethanes," Reinhold Publishing Corporation, New York, 1965.
[c] Patton, T. C.: *J. Paint Technol.*, vol. 39, no. 512, September, 1967.
[d] Teach, W. C., and G. C. Kiessling, "Polystyrene," Reinhold Publishing Corporation, New York, 1960.
[e] Autonetics Materials and Processes Laboratories data.

Hardness Hardness values are also valuable in establishing how resistant a coating may be to erosion, scratching, or other mechanical abuse. One of the more popular methods for determining hardness makes use of the Sward hardness tester. Values obtained are relative to polished plate glass which is taken as 100. Although the Sward hardness test is nondestructive, it often cannot be used on an actual part because the tester will accommodate only a flat (about 0.032-in.) surface from 3 by 6 in. to about 4 by 12 in. Sward hardness values for some coatings and paints are presented in Table 26.

The pencil hardness test employs the use of a series of pencils of varying hardness to scratch through the coating to the substrate. It is simple but highly subjective and destructive in nature. Numerous other hardness and scratch tests have been developed.[51]

APPLICATION TECHNOLOGY

Cleaning and Cleaning Methods

The efficient cleaning of an assembly prior to coating is critical to the performance of the coating and the assembly. Surface contaminants cause poor adhesion, resulting in peeling, or leave water-soluble materials that cause blistering and corrosion, especially after humidity exposure. Therefore, proved cleaning procedures should always be specified. A discussion of cleaners, methods, and equipment follows.

Potential contaminants Contamination, always a serious problem, has become critical for high-density microelectronics, where the active portions of the device may be of the same size as the contaminant particle. Ultraclean areas, such as Class 100 clean rooms,[52] laminar-flow work stations, spray booths, and ovens, should always be employed for these highly sensitive operations. Some of the commonest contaminants to be found on electronic hardware are categorized in Table 27.

TABLE 27 Typical Contaminants and Their Sources

Contaminant	Possible source
Fibers (nylon, cellulose, etc.)	Clothes of personnel in assembly areas, cleaning papers
Silicates, fly ash	Rocks, sand, soil
Oxides and scale	Oxidation products from some metals
Oils and greases	From machining, fingerprints, handling, hair sprays, tonics, and ointments
Silicones	Hair sprays, shaving cream, aftershave lotions, hand lotions, and soap
Metals	Grinding, machining, and fabricating metal parts, storage cans, metal containers
Ionic residues	Perspiration, fingerprints (sodium chloride), residues from cleaning solutions containing ionic detergents, certain fluxes such as the glutamic acid hydrochloride types, residues from previous chemical steps such as etching or plating
Nonionic residues	Rosin fluxes, nonionic detergents, organic processing material
Solvent residues	Cleaning solvents

The incomplete removal and entrapment of these contaminants under a coating may result in electric short circuits, corrosion, deterioration of the coating, and increased penetration of moisture and contaminants through the degraded coating. Several dramatic examples of the seriousness of leaving fingerprint contaminants may be cited. In thin-film circuits the entire disappearance of Nichrome resistors has been observed when tested with applied voltage in a humid environment. In another instance fingerprints left under the coating become visible after humidity exposure through blistering of the coating over the fingerprints.

To establish a scientific basis for the selection of a cleaning method, the coatings specialist should answer the following questions:
1. What contaminants are on the surface?
2. How damaging are they?
3. How can they be removed?
4. What tests can be used to assure complete removal of contaminants?

These are all difficult questions to answer.

An empirical approach is often used to establish the cleaning method. A number of cleaning solutions and methods may be tried, and the one found to be most convenient or least expensive is selected. If no trouble is encountered, no further work is performed.

A number of cleaning techniques which have been employed are described below. To evaluate the method, the contaminants must first be detected and monitored in some fashion to assure the effectiveness of the cleaning procedures. Examples are by magnification, by rendering the particles fluorescent or radioactive, or by determining the wettability or the water break-free surface properties.

Solvents and Solutions Used for Cleaning. Cleaning solvents and solutions may be classified in three groups, according to Table 28. The hydrophobic types are those that will effectively remove water-insoluble contaminants such as greases, oil, or organic material. The hydrophilic types will easily remove water-soluble contaminants such as ionic salts. Since both types of contaminants are normally present on electronic assemblies, a safe practice is to use both types of cleaners in a two-step process or to use a combination cleaner of the hydrophobic-hydrophilic type. A dramatic example of what can happen if only the hydrophobic solvent is used is shown in Fig. 12. In example *a*, where only the hydrophobic solvent was used, considerable blistering of the coating occurred after humidity whereas in example *b* where both hydrophobic and hydrophilic (aqueous solution) solvents were used, no blistering occurred.

TABLE 28 Cleaning Solution Types

I. Hydrophobic
   Organic solvents—methyl ethyl ketone
   Freons*
   Chlorinated solvents, such as trichloroethylene
II. Hydrophilic
   A. Alkaline, acid, detergent-water solutions
   B. Nonionic, detergent-water
   C. Water deionized or distilled
   D. Alcohols such as isopropanol or ethanol
III. Hydrophobic-hydrophilic
   A. Alcohol-naphtha (50:50)
   B. TWD-602 (a Freon-water emulsion-containing surfactant)

* Trademark of E. I. du Pont de Nemours & Co., Wilmington, Del.

Solvent Purity. Solvents used in cleaning may contain varying amounts of particulate contaminants and nonvolatile residues. Though amounts may be extremely small, such residues may be critical to the performance of the electronics. The purity of the cleaning media may be determined by residue from evaporation, by conductivity of water extract for ionizable matter, and particle-count analysis after filtering through standard-size membranes.[53]

Nature of Particles. In interpreting particle-analysis results, not only the amount but also the nature of the particles is important. The identification of particles after they have been separated requires considerable skill and experience. A widely used procedure involves microscopic examination and comparison with standards. Descriptions and photographs of hundreds of particles may be found in "The Particle Atlas."[54] Metal particles, even in small quantities, are serious. For example, even a single metal particle, wedged between two closely spaced conductors can be catastrophic. An electric short circuit could occur immediately; or, in time, ionic corrosion could be produced. Metal particles have been detected in several commercial-grade cleaning solvents and have probably been introduced from storage cans or during the transfer of solvent from one container to another. Other particles such as fibers, greases, and soil minerals have been isolated and identified and could be detrimental, depending on the application. For space applications, some firms are specifying that deionized water and other solvents must contain no particles larger than 20 microns in size. The purest commercial solvents seldom meet this requirement; and if they do initially, they will quickly become contaminated through storage conditions and handling. It is a safe practice for all microelectronic fabrication processes to use solvents and cleaning solutions that have been filtered through membrane filters just before use.[55] The two major suppliers of membrane filters in the United States are Millipore Corporation and Gelman Industries.

Many Federal and Military Specifications are available for the procurement of solvents and chemicals, but few are tight enough to provide control over the very small amounts of particulate contamination that has just been described. Hence, many firms have found it necessary to supplement these specifications with their own, calling out the tighter requirements.

(a)

(b)

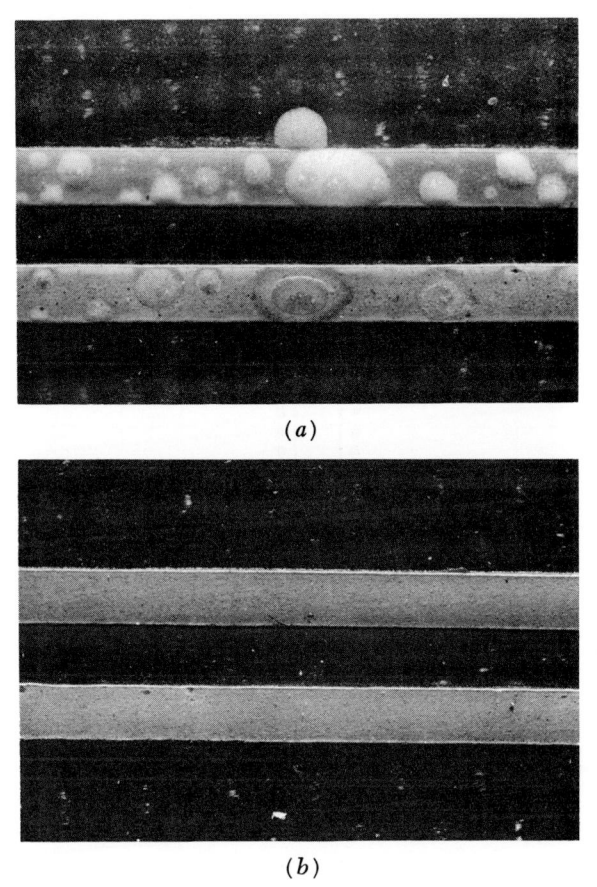

Fig. 12 Effect of aqueous cleaning in preparing circuit boards for coating. (a) Test board D-D prepared without aqueous cleaning. (b) Test board Y-Y prepared with aqueous cleaning.

Cleaning methods There are essentially three mechanisms by which contaminants may be removed from a surface. These are:
1. By dissolving in the solvent or cleaning solution used. An example is the dissolving of salt in water.
2. By reacting chemically to give products that can be dissolved. Examples include the removal of metal oxides and scale by acid or alkaline treatments.
3. By being dislodged mechanically and then removed with liquids. Examples include metal and fiber particles removed by jet spray or ultrasonic energy. In practice, cleaning methods employ combinations of two or all three of these mechanisms to achieve most efficient cleaning.

5-54 Coatings for Electronics

A limited number of basic cleaning methods exist, but numerous modifications and combinations of these basic types are possible and are being used. The basic methods are:
1. Solvent spray
2. Solvent dip
3. Brush
4. Vapor degrease
5. Ultrasonic immersion
6. Pulsating spray

Vapor Degreasing. Vapor degreasing is a very widely used and efficient method for cleaning electronic parts. According to this method, the part is suspended in a specially designed chamber (commerically available) so that the vapors of a heated solvent condense on and flush the surface clean. The part is thus being washed repeatedly with fresh solvent as opposed to hand scrubbing or other manual cleaning methods where the solvent is stagnant and soon becomes contaminated. Combinations of vapor condensation and clean liquid spray are also used (Fig. 13). Detailed instructions for vapor degreasing are given in the literature.[56] The two most widely used vapor-degreasing solvents for removing greases and

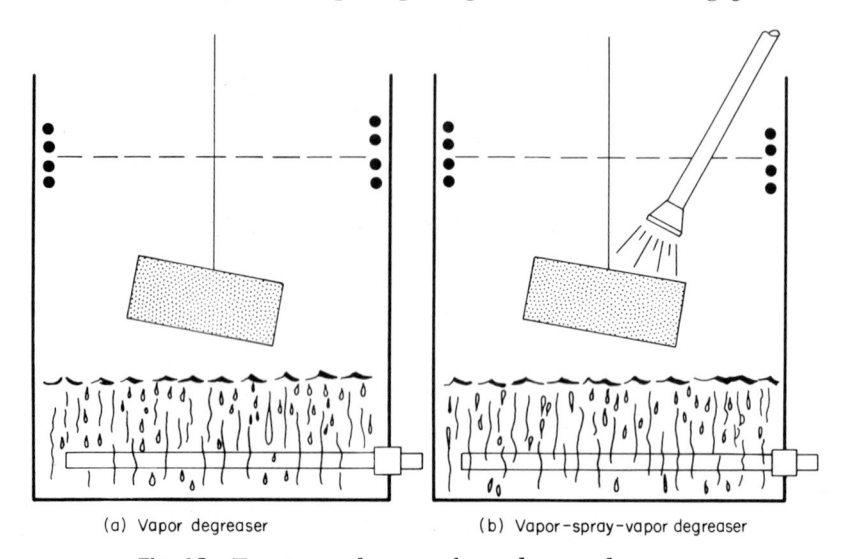

(a) Vapor degreaser (b) Vapor-spray-vapor degreaser

Fig. 13 Two types of commonly used vapor degreasers.

other hydrophobic contaminants are trichloroethylene and perchloroethylene. They may be purchased from a number of manufacturers under different trade names. Both are relatively nontoxic and stable. Trichloroethylene, will contain stabilizers to prevent hydrolytic breakdown, whereas perchoroethylene, because of its unreactive structure, requires little or no stabilization. As with other cleaning methods, the compatibility of these solvents with the substrates should be verified before they are specified for production parts. For example, it is reported that with some circuit-board laminates, cleaning with trichloroethylene causes leaching-out of resin, delamination, and discoloration effects.

Ultrasonic Cleaning. One of the best and most effective methods of cleaning employs an ultrasonically agitated solvent bath into which the part to be cleaned is immersed. Because of the high energy imparted to the solvent, contaminants are rapidly and easily removed, even from difficultly accessible areas. According to the method, sound vibrations at frequencies higher than 20,000 Hz (usually 20 to 40 kHz) are imparted to a liquid solvent or cleaning solution. This results in what is called *cavitation* or the rapid formation and sudden collapse of thousands of microscopic bubbles in the liquid. The transferal of such energy to the part to be cleaned thus results in a very strong scrubbing action. Ultrasonic generators, and indeed complete ultrasonic cleaning systems, are available from a number of equipment manufacturers.

Ultrasonic cleaning is a very rapid process requiring only a few minutes per part and lends itself well to automated production lines. A large number and variety of off-the-shelf solvents and cleaning solutions are available, and many others have been formulated to achieve removal of very specific contaminants. The solvents used are limited only to the corrosion which they may impart to the tank, to their toxicity or flammability if any, and to their vibration dampening characteristics. Concentrated detergent solutions and acetone have high dampening effects and are not as effective as other solvents such as the Freons. The solvent is normally kept at room temperature, however. Water detergent or alkaline solutions are more effective at slightly elevated temperatures of 100 to 140°F. Ultrasonic agitation will itself heat the solution in time.[57]

Although it is one of the most effective methods, ultrasonic cleaning has several limitations that should be recognized. For example, as the cleaning bath becomes contaminated, it will require frequent filtering or purification of the solvent or periodic replenishment with fresh solvent.

Secondly, mechanically fragile components such as glass diodes and other glass-sealed semiconductor devices have been known to fail catastrophically owing to the ultrasonic cracking of the glass or by lead separation. Where such failures do not immediately occur, there is always the question of the effect that ultrasonic stresses can have on the reliability and life expectancy of the device. Therefore, ultrasonic cleaning has found its greatest usefulness in the cleaning of bare substrates or of circuits prior to the attachment of sensitive devices.

Pulsating-spray Cleaning. The pulsating-spray method is unique in that the fresh solvent or cleaning solution used is pulsed at frequencies of 20,000 to 30,000 Hz at the time it is sprayed. The spray impinges upon the part at a pressure of about 500 psi. The main advantages of this high-energy method over ultrasonic cleaning are the avoidance of cavitation erosion of metal surfaces and the elimination of recontamination from drag-out or static dip-tank rinsing. A commercially available pulsating-spray unit is the Heinicke parts washer manufactured by the Heinicke Corporation. The equipment has been used successfully for cleaning printed-circuit boards, wiring, wire harnesses, and many other electronic and electrical parts. A completely automated cleaning cycle for such a washer consists of (1) a high-pressure wash combining pulsating jet stream of cleaning solution, (2) tap water rinse, and (3) final rinse using distilled or deionized water.

Pulsating-spray cleaning has been shown to be one of the most efficient methods for removing particulate contamination 5 microns or larger in size. Most of the work has employed aqueous detergent cleaners followed by hot tap water and filtered deionized water rinses. However, the equipment also lends itself to the use of organic solvents. Each operation is automatically timed to a predetermined cycle with a total cleaning time seldom exceeding 5 min. The number of parts that can be cleaned simultaneously is limited only by their size and orientation within the washer.

Application Methods

Because of the wide variety and complexity of electrical and electronic parts to be coated, there are presently in use numerous application methods. The coating of tiny semiconductor junctions on the one hand and thousands of feet of wire on the other, impregnation of all voids in huge motor rotors and stators, and applying ultrathin continuous dielectric layers on the thin-foil wrap of a capacitor require various and often ingenious deposition and polymerization techniques.

About a dozen major application methods exist. Some are amenable to rapid and even mechanized production, whereas others are very specialized, being suitable to the treatment of but one part at a time. A discussion of the coating processes

TABLE 29 Application Methods for Coatings

Method	Advantages	Limitations	Typical applications
Spray	Fast, adaptable to varied shapes and sizes. Equipment cost is low.	Difficult to completely coat complex parts and to obtain uniform thickness and reproducible coverage.	Motor frames and housings, electronic enclosures, circuit boards, electronic modules.
Dip	Provides thorough coverage, even on complex parts such as tubes and high-density electronic modules.	Viscosity and pot life of dip must be monitored. Speed of withdrawal must be regulated for consistent coating thickness.	Small- and medium-sized parts, castings, moisture and fungus proofing of modules, temporary protection of finished machined parts.
Brush	Brushing action provids good "wetting" of surface, resulting in good adhesion. Cost of equipment is lowest.	Poor thickness control; not for precise applications. High labor cost.	Coating of individual components, spot repairs, or maintenance.
Roller	High-speed continuous process; provides excellent control on thickness.	Large runs of flat sheets or coil stock required to justify equipment cost and setup time. Equipment cost is high.	Metal decorating of sheet to be used to fabricate cans, boxes.
Impregnation	Results in complete coverage of intricate and closely spaced parts. Seals fine leaks or pores.	Requires vacuum or pressure cycling or both. Special equipment usually required.	Coils, transformers, field and armature windings, metal castings, and sealing of porous structures.
Fluidized bed	Thick coatings can be applied in one dip. Uniform coating thickness on exposed surfaces. Dry materials are used, saving cost of solvents.	Requires preheating of part to above fusion temperature of coating. This temperature may be too high for some parts.	Motor stators; heavy-duty electrical insulation on castings, metal substrates for circuit boards, heat sinks.
Screen-on	Deposits coating in selected areas through a mask. Provides good pattern deposition and controlled thickness.	Requires flat or smoothly curved surface. Preparation of screens is time-consuming.	Circuit boards, artwork, labels, masking against etching solution, spot insulation between circuitry layers or under heat sinks or components.
Electrocoating	Provides good control of thickness and uniformity. Parts wet from cleaning need not be dried before coating.	Limited number of coating types can be used; compounds must be specially formulated ionic polymers. Often porous, sometimes nonadherent.	Primers for frames and bodies, complex castings such as open work, motor end bells.
Vacuum deposition	Ultrathin, pinhole-free films possible. Selective deposition can be made through masks.	Thermal instability of most plastics: decomposition occurs on products. Vacuum control needed.	Experimental at present. Potential use is in microelectronics, capacitor dielectrics.
Electrostatic spray	Highly efficient coverage and use of paint on complex parts. Successfully automated.	High equipment cost. Requires specially formulated coatings.	Heat dissipators, electronic enclosures, open-work grills and complex parts.

most commonly used in the electronics industry is given in the following sections. Ten of these processes, their advantages, disadvantages, and typical applications, are summarized in Table 29.

Spray coating The most widely used application method for electrical coatings is by spraying. Spraying in this case includes not only the common compressed-air vaporization, but also airless pressure spray, hot spray, hot vapor-impelled spray, electrostatic spray, dry-powder resin spray, electrostatic powder spray, and even the ubiquitous aerosol spray can.

Pressure Liquid Spray. Most liquid-spray methods employ the basic principle of atomizing the paint into a fine mist which is directed upon the surface to be coated. Atomization is accomplished by the use of compressed air, pressurized volatile solvents, or high-velocity jet stream. With the proper technique, a thin but continuous wet layer of lacquer or enamel may be deposited without rundown of drips or "sags" and without excessive "orange peel" or the unsightly "sandiness" of dry overspray. A skilled spray operator can apply a very uniform coating and can usually touch up defective areas during the process or later.

In applying coating to objects such as circuit boards having discrete components, the operator should build up the total desired thickness by spraying from four different directions so as to cover the top and all sides of components and to force coating underneath the components as well. Freedom from pinholes, skips, or voids should be a major consideration for such environmental protective coatings. Turntables, either hand- or power-rotated, are useful in orienting the work for directional spraying. Parts conveyers can be utilized to move parts past a spray painter or past automatic spray heads, as in the case of fully mechanized painting lines. Parts platforms or holders can be mounted on rotating spindles to make all surfaces accessible to the paint spray.

All paint spraying should, of course, be done in a ventilated booth which exhausts the overspray and solvent vapors to the outside or preferably to a fume and solvent trap system. Not only are the fumes and paint aerosol obnoxious, but also they may be a flammability hazard, or may be toxic, especially when inhaled from repeated exposures.

Vapor Spray (Pressurized Volatile Solvents). For solventless, very high vicosity coatings which are difficult or impossible to apply by the conventional techniques of spraying or dipping, the Zicon vapor carrier or spray method[*] may be employed. Zicon equipment is successfully being used by many companies for applying viscous coatings to circuit boards and other electronic assemblies. This technique utilizes a preheated, dry, inert vapor such as that of a hydrocarbon, in place of compressed air, to atomize and transport the coating to the part. Besides its ability to apply viscous coatings, other advantages of this method include precision control of thickness, surface uniformity, and coating integrity. The probability of pinholes and other imperfections due to moisture or contaminants is virtually eliminated because the hydrocarbon gas used is dry, inert, and of high purity, and because the console such as the series 9000 (Zicon) employs an ultraclean horizontal laminar-flow chamber.

A second advantage, deriving from the first, is that dry film coating thicknesses from 3 to 30 mils may be easily applied, simply by continuous spraying. Thus, heavy conformal coatings may be applied not only for protection, but also for vibration damping and ruggedization. Zicon sprayers are used for applying conformal coatings to printed-circuit boards, photoresists for integrated circuit and semiconductor device manufacture, magnetic oxide coatings to memory disks, and numerous types of protective coatings to metal substrates. Commonly used coating types are epoxies, polyurethanes, and fluorocarbons.

Airless Spray (High-velocity Jet Stream). One disadvantage of pressure air spraying is the tendency for the blown cloud of atomized paint to "swirl" out of recesses, holes, and corners owing to the velocity of the accompanying airstream, leaving these areas lightly coated or bare. Airless spray equipment has been developed

---

[*]Marketed and trademarked by the Zicon Corporation, Mount Vernon, N.Y.

5-58   Coatings for Electronics

(available from firms such as Nordson Corporation or the Gray Company) which depends, not on atomization by high-velocity air, but on producing the spray pattern by forcing the material under high pressure through tiny orifice nozzles. The coating material, which is circulated continuously in the storage system, is normally also heated to reduce its viscosity and to volatilize solvents more rapidly. This system is efficient, especially for coating parts with many recesses or for coating the inside of hollow objects.

Hot-melt Flame Spraying. According to the hot-melt flame-spray method, a cloud of powdered resin suspended in air is interjected into a special spray gun and melted on being blown through the flame-shrouded nozzle of the gun. The melted plastic solidifies immediately on impinging on the object. This technique is limited to plastics that are stable above their melting temperatures. Powders used are normally thermoplastics such as polyethylenes, polypropylenes, other polyolefins, and polytetrafluoroethylenes. Techniques have also been worked out for spraying even some unstable plastics providing they have a sufficiently long induction period before decomposition occurs.[58]

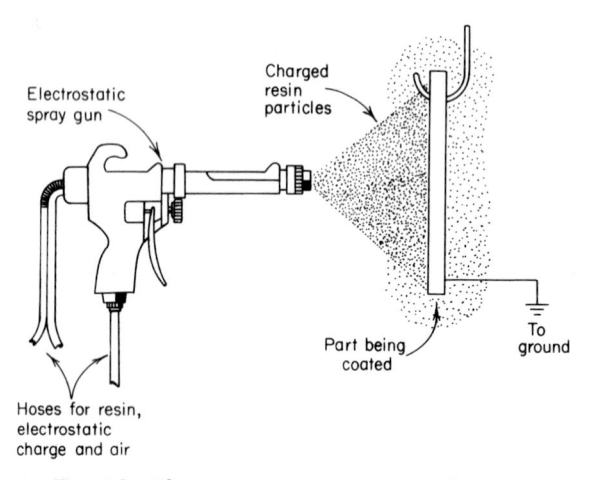

Fig. 14 Electrostatic spray-coating technique.

Other Spray Equipment. Coatings, such as epoxies or polyurethanes, provide a high degree of chemical and solvent resistance but have been difficult to apply in production because they are two-component systems and have relatively short pot lives. Equipment is now available which facilitates the application of these two-component coatings. This equipment continuously proportions and mixes the two components and feeds the mixture to the spray head or extrusion nozzle for application. After loss of solvent, if any, from the film, the coating readily polymerizes under the proper curing conditions of temperature and time.

Electrostatic spray coating   Several modifications of the spray method have been developed and are in use. One of these, the electrostatic spray process, developed by the Ransburg Electro-Coating Company, is extensively used for coating objects having intricate shapes or a high proportion of open areas. The technique consists in the deposition of negatively charged spray-paint droplets or plastic powders onto a metal part. The particles are rendered electrostatic as they are propelled through a diffuser on the front of the spray gun and are then rapidly attracted to the part (Fig. 14). Subsequent drying or curing of paint is done by usual methods, whereas powdered-resin coatings are fused in as little as 20 to 60 sec at 400°F, or at lower temperatures for a longer period of time.

Rules governing other forms of liquid spraying apply to electrostatic spraying. In powdered-resin spraying, coatings may be applied in thicknesses from 2 to over

20 mils, depending on the plastic powder selected. The method is very amenable to thermoplastic coatings, but thermosetting types such as epoxies are being used extensively.

Particular advantages of the electrostatic spray process over other spray processes include the high deposition efficiency whereby very little material is lost as overspray and the uniform buildup that can be obtained in one application, saving time and cost. Paint savings of 25 to 50 percent over conventional spray coatings are not uncommon. The process is also less susceptible to human errors, lending itself very well to automated production and to the coating of a wide variety of metal articles. In the electronics industry, for example, this method is valuable for the electrical insulation and corrosion-protective coating of heat dissipators (Fig. 15), metal brackets, "black box" enclosures, circuit-board housings, and motor armatures and stators before winding.

Dip coating As the name implies, dip coating involves the immersion of a part into a liquid composition (or dry powder, in the case of fluidized-bed coating), withdrawal, and drying or curing. It is a highly desirable method from the stand-

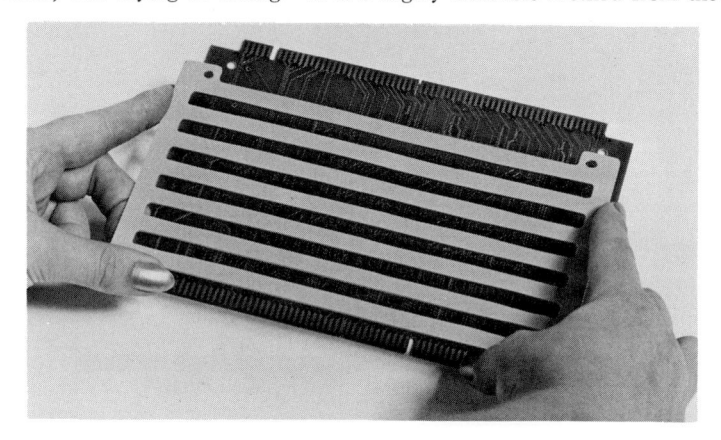

Fig. 15 Heat-sink rail electrostatically insulated with epoxy coating.

point of achieving a continuous automated manufacturing operation and has the advantage over spray coating of assuring coverage of all areas and crevices. Dip coating is most suitable for one-component plastic systems having sufficiently long pot lives. Since many of the high-reliability coatings used in electronics production are two-component systems of limited pot life, it therefore has limited usage. Techniques that have been used to maintain a stable dip bath include:

1. Replenishing solvents (lost by evaporation) to maintain constant viscosity
2. Refrigerating or maintaining the dip bath at a low temperature to extend its pot life

The final thickness of a dip-coated part (which may range from a fraction of a mil to several hundred mils) is dependent upon the viscosity of the bath, the rates of immersion and withdrawal, the termperature of the bath, and the number of dips. Adequate drainage must be provided to avoid trapping liquid coating in pockets or recesses in the part. Continued paint flow after withdrawal of the part may result in "tears" along the edge of the part, which may be removed by "touching off" or by electrostatic removal. Adequate air drying to remove solvents is recommended since entrapped volatiles may result in blistering of the coating upon subsequent oven curing. Hot-dip coating may be used to ensure greater wettability and minimize pinholes. This variation consists of preheating the part (where allowable) to an elevated temperature, followed by dipping into a bath maintained at room temperature, and results in faster drying or even fusion of the coating, as in plastisol dipping.

Here preheated parts are dipped in vinyl plastisol, which becomes gelled by the heat, leaving a solid layer on the surface. The cooler outer layer which is picked up on withdrawal remains liquid and drains off. An alternative technique involves dipping parts at ambient temperature into a molten resin such as asphalt, in which solidification of the coating occurs on removal from the bath and cooling. These coating resins will normally be of a thermoplastic type.

With high-viscosity coatings, such as the 100 percent solid types, numerous air bubbles may be entrapped after mixing of the two components. In these instances a technique commonly employed with encapsulants—vacuum degassing—may be used. Evacuation of the mixture, at 25 to 20 in. mercury (about water aspirator pressure) for several minutes, is recommended.

Fluidized bed A unique variation of dip coating and one that is finding ever-increasing uses is the fluidized-bed process, also known as the aerated-bed process. The basic process and modifications of it are the subject of numerous patents,

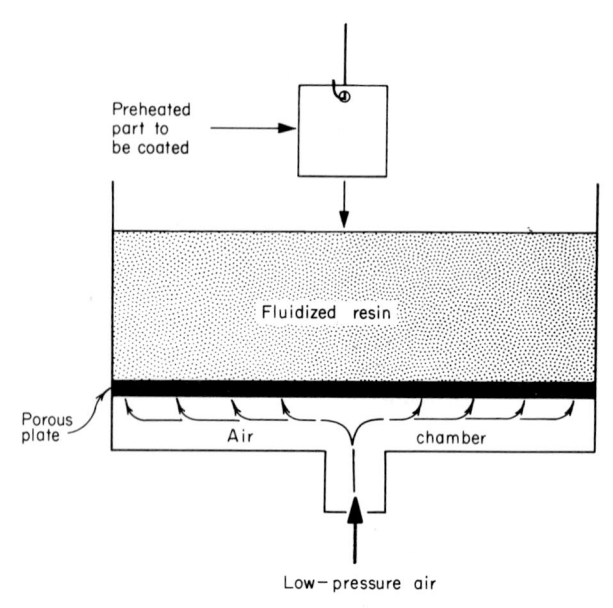

Fig. 16 Fluidized-bed-coating technique.

both in the United States and in other countries.[59,60] According to the method, dry powdered resins or partially reacted powdered coatings are placed in a container in which they are suspended and set in motion by controlled velocity air introduced from the bottom of the container through a screen and porous membrane (see Fig. 16). The resultant low-density powder is in constant motion and behaves like a low-viscosity, low-surface-tension fluid. The part to be coated is then preconditioned by one of several methods to promote adhesion of the particles to the surface and then is dipped into the fluidized powdered resin. The principal pretreatment methods are as follows:

1. Preheating the part (usually metal) to a temperature above that required to melt and fuse the plastic powder.

2. Applying a thin adhesive primer. This patented proprietary process has an advantage over preheating in that areas of a part may be selectively coated.

3. Applying an electrostatic charge to the part, to attract oppositely charged coating particles.

In all three cases, the part is normally heated in an oven after removal from the fluidized bed. For epoxy and other thermosetting coatings, this postbaking step results in further fusing and curing the coating, thus improving its physical and electrical properties. For thermoplastic coatings, the postbaking operating is important in smoothing out the surface and in obtaining a less porous coating.

Process Parameters. The number of process parameters for fluidized-bed coatings is very large, with a list of at least 20 being reported.[61] However, once these parameters have been optimized, fixed, and controlled, the process can be automated and gives highly reproducible results.[62,63]

The thickness of the coating deposited depends on a number of factors, including time of immersion (Fig. 17), heat retention of the part, mass of the part, and characteristics of the powder. Probably the most important parameter is the ability of the part to retain heat, which, in turn, is a function of the specific heat, thermal conductivity, thermal stability, and mass of both the part and the coating material.

Coating Materials. Both thermoplastic and thermosetting coatings can be applied by fluidized bed. Some thermoplastic materials, notably vinyls and polyethylenes, can often be completely fused by the heat retained in the preheated part; thermosetting coatings usually require postcuring after the dipping step. Epoxies, chlorinated polyethers, polyethylene, and silicones are other examples of coatings commonly applied by fluidized bed. Although TFE and FEP fluorocarbon powders may be applied by fluidized bed, very little practical use has been found here, probably because of the very high preheat temperature required and adhesion problems. Poor adhesion is a problem common to other coatings also, and in these cases primers should always be used.

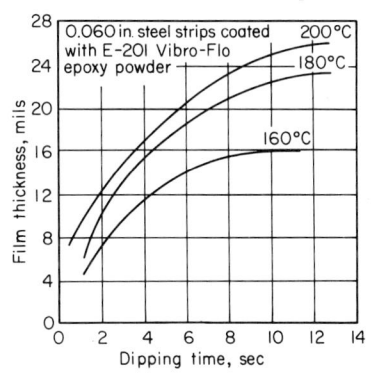

Fig. 17 Effect of preheat temperature and immersion time on coating thickness when fluidizing a steel bar with epoxy fluidizing material. (*Armstrong Products Corp.*)

Examples include cellulose acetate butyrate, nylons, and vinyls. Other coating types such as epoxies, polyethylenes, and chlorinated polyether normally do not require the use of primers. A list of some powder coating types and some important deposition parameters is given in Table 30. A list of some proprietary formulations and manufacturers is given in Table 31.

Sprayed-powder Process. The sprayed-powder process involves the spraying of powdered coatings onto preheated metal parts using flock-coating spray guns. As in the fluidized-bed process, powder particles contacting a hot surface will fuse, forming an integral coating. Flocking spray guns may be purchased from DeVilbis (no. GB), Binks (no. 171), Paasche (FF ¾ in.), and others. The same powders sold for fluidized-bed coating are normally used for spray-powder coating (see Table 31).

Electrostatic Fluidized-bed Process. As the name implies, this process combines elements of both the fluidized-bed and the electrostatic spray processes. The normal fluidized bed is used, but, in addition, the plastic particles are given a negative charge by applying a high-voltage (~90,00-volt) direct current to the container. The part to be coated is then given the opposite electric charge and suspended in the bed, so that the charged particles can be attracted toward the part.

Advantages and Limitations. The main advantage of the fluidized-bed process is that it allows the application of coatings without use of solvents, and in rather thick films (5 to 25 mils) in only one or two dips. These thicknesses are not readily achieved by other coating methods. Rather complex parts having sharp edges and corners are easily and uniformly coated. Extra-thick coatings thus produced solve many chemical, solvent, erosion, and corrosion problems. Other advantages over the conventional methods of liquid dipping and spraying include: (1) lower cost because of the elimination of waste material; (2) greater ease of handling

TABLE 30 Summary of Plastics for Fluidized Bed Coating

Resin	Fluidizing conditions			Fluidized-bed powder		
	Preheat temperature, °F	Cure or fusion		Maximum operating temperature, °F	Adhesion	Weather resistance
		Temperature, °F	Time, min			
Epoxy...............	250–450	250–450	1–60	200–400	Excellent	Good
Vinyl................	450–550	400–600	1–3	225	Poor	Good
Cellulose acetate butyrate.............	500–600	400–550	1–3	225	Poor	Good
Nylon...............	550–800	650–700	1	300	Poor	Fair
Polyethylene........	500–600	400–600	1–5	225	Fair	Good
Polypropylene.......	500–700	400–600	1–3	260	Poor	Good
Penton..............	500–650	450–600	1–10	350	Poor	Good
Teflon...............	800–1000	800–900	1–3	500	Poor	Good

since the powder resins are one component; (3) time savings since no weighing, mixing, or deaerating are necessary, as for the two-component coatings; (4) no cost for solvents and reduction of fire hazard, because no solvents are employed. Some disadvantages should also be recognized. Among these are: (1) parts that can be coated are limited to metals or other materials having a high heat retention; (2) heat-sensitive electronic components and assemblies cannot be coated; parts must be capable of withstanding preheat temperatures of 400 to 650°F; (3) thin, pinhole-free films less than 5 mils thick are difficult to achieve; (4) some coating powders are degraded by the high processing temperatures used; and (5) the process is economical only for large production runs.

TABLE 31 Typical Examples of Fluidized Bed Coatings*

Chemical type	Proprietary designation	Manufacturer or supplier
Epoxy...............	Scotchcast 260, 265, XR-5133, and XR-5162	Minnesota Mining & Manufacturing Co.
Epoxy...............	PK Series Nos. 60 (gray) and 61 (blue)	Hysol, Division of Dexter Corp.
Epoxy...............	Dri-Kote Series: DK-1, DK-2-02, DK-4, DK-4-01	Hysol, Division of Dexter Corp.
Epoxy...............	E Series Nos. 200, 300, 600, 7000, and 8000	Armstrong Products Co.
Epoxy...............	Epocast Series: 2613, 2612, 268	Furane Plastics
Epoxy...............	Corvel Epoxies Series	The Polymer Corp.
Epoxy-silicone.......	DC-3100	Dow Corning Corp.
Silicone.............	R-5061, R-5071	Dow Corning Corp.
Vinyl................	Miccron 400 Series	Michigan Chrome and Chem.
Vinyl................	Corvel 1289 and 1325	The Polymer Corp.
Nylon...............	Corvel Nylons Series	The Polymer Corp.
Cellulosic............	Corvel Cellulosics Series	The Polymer Corp.
Penton chlorinated polyether..........	Corvel Penton Series	The Polymer Corp.
Polyolefin...........	Corvel Polyethylene	The Polymer Corp.
Phenolformaldehyde	Durez 175	Hook Chemical Co.

* Most of these are also suitable for powder spray and electrostatic spray coating.

Because of the high speed and automation possible with the fluidized-bed process, it is being used for many diverse electronic and electrical applications. Among these uses are the coating of electronic enclosures, housings, covers, brackets, wire harness guides, heat sinks and dissipators; and the insulation of transformers, motor rotors, and stators. Fluidized-bed coatings are also being used for the moisture sealing and packaging of resistors and capacitors.

Vacuum impregnation Still another variation of dip coating, known as vacuum impregnation, employs vacuum or vacuum-pressure cycling during the dipping operation. This technique is particularly useful for the thorough coating (impregnation) of intricate parts or assemblies containing numerous small cavities, such as coils, windings, porous metal, and ceramic surfaces. Associated hardware such as porous metal castings may be sealed to maintain vacuum or hold pressures or to contain certain gaseous environments. Thorough impregnation is important, especially for high voltage, because any small void may act as a moisture trap or a corona region, resulting in insulation breakdown.

Many porous specimens may be impregnated by simply soaking in the resin or coating; however, vacuum evacuation techniques are helpful in reducing the amount of entrapped air and moisture. Air is first removed, and then replaced with impregnation material. In order to do this more effectively, it is desirable to evacuate the specimen before immersion in the resin. After immersion, vacuum is again applied, and then interrupted periodically to force the liquid into the evacuated pores. A combination of vacuum followed by pressure is even more effective for impregnation of very closely wound coils or porous substances. The amount of vacuum and time of immersion will be determined by the temperature, viscosity, volatility, and pot life of the resin.

Screening Coating formulated with suitable thixotropic and wetting agents may be applied by squeegee through a silk, nylon, or stainless steel screen. To apply coatings by the screening process, surfaces must be fairly flat. The thickness of the applied coating may be varied from a fraction of a mil to above 6 mils, depending on the screen-mesh size and on the number of coats applied. Once all parameters have been established, very reproducible thicknesses are obtained. In order to minimize pinholes or other imperfections, it is advisable, as with other coating methods, to apply and superimpose two or three layers of coating, each layer being partially or fully cured before application of the next one. By using screens with a photoresist pattern, insulation coatings may be applied to substrates, such as ceramics or epoxy laminates, in selected areas only. This is often necessary in isolating a thin- or thick-film conductor or resistor from an active device which is to be placed above it. A number of screenable epoxies, polyurethanes, and other coating types are available. The screening process for applying conductor, resistor, and capacitor units in selected areas for fabricating thick-film circuits is a highly developed art. Automated equipment for high-production schedules is available. Photoemulsion compositions such as the gelatin-chromate or photoresist types are formed into masks to yield the desired pattern of open screen. Line thickness of 10 mils with 10-mil spacings are easily attainable, but advances are being made in improving definition to 2-mil lines and 2-mil spacings.[64]

Rollercoating Rollercoating is a highly mechanized and closely controlled method for coating flat stock. It provides a means for applying a fixed amount of liquid coating to one or both surfaces of any sheet or strip requiring an even coating. Thus, it is finding use for the application of photoresist coating to copper laminate stock, upon which etched circuitry is to be developed, and for applying thin films on metal foil for use in capacitors, as well as for coating sheet metal for fabrication into cans, boxes, or covers for housing or shielding electronics. In this process, the liquid coating material is transferred by means of a series of metering and transfer rolls from a reservoir pan to the flat surface. The printing roll may be cut out to lay down a pattern, or the metering roll may be lithographed to print a functional or decorative design. After the coating operation, the material passes through a zoned drying and curing oven, and the sheets are stacked or the strip is rerolled. Besides its usage for applying photoresist, this is the method universally employed for metal decorating, the process whereby stock for such end products as cans, bottle caps, toys, TV trays, wastebaskets, automobile instrument panels, and lighting fixtures is coated.

Brush Brushing is not usually a primary method of applying electronic coatings. However, when small areas on a part are to be coated, or when rather inaccessible areas require coating, a skillfully handled brush is often the answer. Thickness will not be carefully controlled, and outlines will not be strictly observed (unless masking is used), but the job is quickly done with a minimum of investment but a high labor cost.

Brushing is extensively used for touch-up and repair work. Examples would be to repair skips or damaged areas of the coating, to cover spots where components have been removed and replaced or pinholes left from electrical test probes.

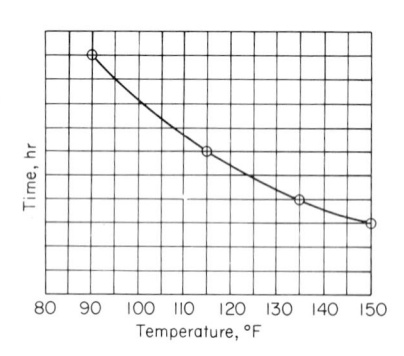

Fig. 18 Time-temperature cure schedule for a polyurethane conformal coating.

Thin-film deposition methods Thin polymeric films will be defined for purposes of this discussion as solid organic polymer coatings which are less than 3 microns thick (30,000 Å or about ⅛ mil). The practical applications of such films have been confined largely to capacitor dielectrics; however, they are recently being used as microcircuit insulation, corrosion-protective coatings, and as barrier coatings and passivation layers for semiconductor devices. One of the main problems in the fabrication of microelectronic circuits is the difficulty in obtaining continuous ultrathin, pinhole-free dielectric coatings deposited in layers or in specific geometric patterns. To date, the inorganic materials have dominated the semiconductor device and microcircuits industry; but as new high-temperature-stable organic materials are developed and as techniques are refined for the deposition of thin films, more uses will be made of organic thin films. Among the advantages of ultrathin organic coatings are their high dielectric breakdown strengths, ability to form pinhole-free layers, and in some cases, high purity of the coating, thus avoiding tunneling effects and leakage currents. Among their disadvantages are thermal and oxidative instability inherent in so many of the presently available polymers.

Thin polymer films normally cannot be formed by the usual methods of deposition such as spraying and dipping, or by the use of two-component systems such as epoxies. Thus only a limited number of coating types and specific deposition techniques lend themselves to thin-film formation. The five most important techniques are:

1. Vacuum or vapor deposition
2. Ultraviolet-light polymerization
3. Electron-beam polymerization
4. Glow-discharge polymerization
5. Gamma irradiation

Thermoplastic linear polymers are most amenable to thin-film deposition, especially those polymerizing by a free-radical or addition mechanism. Hence, monomers such as styrene, divinylbenzene, and other vinyl and acrylic compounds are very suitable and are widely used.

Vapor Deposition. The vapor or vacuum deposition of coatings, although a highly desirable method, can be employed for only a few polymer types. Unlike metals, most plastics cannot be heated to vaporization without gross decomposition or changes in physical and electrical properties. The evaporation method is most successful with the low-molecular-weight polymers, such as polyethylene which recombines with little or no interaction and hence gives rather soft films.[65] There have been some reports of Teflon vapor deposition. Teflon was depolymerized at 1000°F in a vacuum of 10^{-5} to 10^{-6} torr and reported to repolymerize as an adherent film on a cold substrate.[66] However, by far the greatest success in vapor deposition has been achieved with the polyxylylenes. Solid dimers of p-xylene are pyrolyzed under moderate vacuum, forming gaseous diradicals which in turn combine on a substrate into long-chain linear polymers.[*] The equipment for deposition consists essentially of a pyrolysis tube and a vacuum chamber.

Ultraviolet-light Polymerization. Gaseous monomers adsorbed on a surface may be polymerized on exposure to ultraviolet light in either the presence or absence of air. Monomers most suitable for this type of polymerization are those containing double bonds (unsaturation). Styrene, butadiene, methyl methacrylate, acrolein, and divinylbenzene are a few examples of monomers that have been successfully polymerized.

TABLE 32 Source of Specifications

Abbreviation	Full name	Source of document
AMS	Aerospace Material Specifications	Society of Automotive Engineers 485 Lexington Ave., New York, N.Y.
ASTM	American Society for Testing and Materials	American Society for Testing and Materials 1916 Race St., Philadelphia, Pa.
ERF	Epoxy Resin Formulators	Society of Plastics Industry 250 Park Ave., New York, N.Y.
Fed. Std.	Federal Standard	General Services Administration 30 Church St., New York, N.Y. 300 N. Los Angeles St., Los Angeles, Calif. or 7th and D Sts., S.W., Washington, D.C., and Naval Supply Depot, 5801 Tabor Ave., Philadelphia, Pa.
MIL or MIL-STD	Military or Military Standard	Naval Supply Depot, 5801 Tabor Ave., Philadelphia, Pa.
CS	Commercial Standard	Superintendent of Documents, U.S. Government Printing Office, Washington, D.C.
AS	American Standard	American National Standards Institute 10 East 40th St., New York, N.Y.
AATCC	American Association of Textile Chemists and Colorists	American Association of Textile Chemists and Colorists, Lowell Technical Institute, Box 28, Lowell, Mass.
(Letter)-(Letter)-(No.)	....	Same as for Fed. Std.
IEC	International Electrotechnical Commission	American National Standards Institute 10 East 40th St., New York, N.Y.

Films, a few millionths of an inch thick, which are pinhole-free and highly adherent, have been prepared by A. N. Wright of General Electric.[67] By employing a mask in front of the substrate during the ultraviolet irradiation, very fine, well-defined insulation layers useful in microcircuits may be deposited.

Ultraviolet light has a distinct advantage over both electron-beam and glow-discharge methods in that the equipment necessary is fairly simple and inexpensive, requiring a mercury arc or some other ultraviolet source, a chamber consisting of a quartz tube or a bell jar with a quartz window to allow entrance of the ultraviolet light, and a means for evacuating, purging, and introducing the gaseous monomer into the chamber. The high-vacuum systems necessary for electron and glow-discharge bombardment are therefore not required. Besides the polymerization of gaseous monomers just described, ultraviolet polymerization may occur in the solid and semisolid state. A good example of this is the curing of photoresists as in the fabrication of printed-circuit boards or microelectronic circuits.

[*] Parylenes, a trademark of Union Carbide Corporation, New York, N.Y.

TABLE 33 Electrical and Electronic Requirements

Document	Title	Description
ANA Bulletin 400	Electronic Equipment: Aircraft and Guided Missiles, Applicable Documents	A comprehensive listing of specifications and other documents pertinent to aircraft and aerospace electronic systems.
MIL-STD-202	Test Methods for Electronic and Electrical Component Parts	Establishes standard uniform methods for testing electronic and electrical component parts including environmental, physical, and electrical tests.
MIL-STD-275B	Printed Wiring for Electronic Equipment	Establishes design principles governing the fabrication of printed wiring and the mounting of parts and assemblies.
MIL-STD-446A	Environment Requirements for Electronic Parts, Tubes, and Solid State Devices	Establishes the research and development objectives, environmental design requirements for use in current and future electronic planning.
MIL-STD-454	Standard General Requirements for Electronic Equipment	The basic Department of Defense document standardizing the more common requirements for electronic equipment.
MIL-STD-750	Test Methods for Semiconductor Devices	Specification describing tests and environmental conditions for semiconductor devices.
MIL-STD-883	Test Methods and Procedures for Microelectronics	Establishes uniform methods and procedures for testing microelectronic devices, including monolithic, multichip, film, hybrid microcircuits, and microcircuit arrays for Military-NASA uses.
MIL-W-583C	Wire, Magnet, Electrical	Covers electric magnet wire of various classes and types used for fabricating coils and transformer coils.
MIL-T-945	Test Equipment, for Use with Electronic Equipment, General Specification	General requirements for design and manufacture of test equipment used in testing of electronic parts.
MIL-I-983	Interior Communication Equipment, Naval Shipboard, Basic Design Requirements for	Covers the basic design requirements, test, and operating conditions for interior communication equipment used on Naval ships.
MIL-E-4158	Electronic Equipment, Ground, General Requirements for	Requirements for design and fabrication of electronic ground equipment.
MIL-E-5400	Electronic Equipment, Aircraft, General Specification for	Gives general requirements for electronic gear, primarily in piloted aircraft by classes defined as altitude and temperature range.
MIL-E-8189	Electronic Equipment, Guided Missiles, General Specification for	Covers the philosophy of design and general requirements for the design and manufacture of electronic systems for guided missiles.
MIL-P-11268	Parts, Materials, and Processes Used in Communication Equipment	Covers the selection, application, and use of materials and processes in the construction of electronic communication equipment.
MIL-E-11991	Electrical-Electronic Equipment, Surface Guided, Missile Weapon Systems, General Specification for	Covers requirements which are common to guided-missile-systems electronic, electrical, and electromechanical equipment.
MIL-E-16400	Electronic Equipment, Naval Ship and Shore: General Specification	

TABLE 33 Electrical and Electronic Requirements (Continued)

Document	Title	Description
MIL-T-17296	Test, Checkout and Evaluation Equipment Guided Missiles (Fixed Installations) General Specification	Covers general requirements for guided-missile test, checkout for fixed installations on board ship or missile depots and schools.
MIL-F-18870	Fire Control Equipment, Naval Ship and Shore: General Specification	Covers general requirements for design and manufacture of surface, shore, antiaircraft and underwater fire control equipment.
MIL-S-19500	Semiconductor Devices, General Specification for	
MIL-T-21200	Test Equipment for Use with Electronic and Fire Control Systems, General Specification for	Covers general requirements for design and construction of equipment used in testing electronic and fire control equipment.
NEMA, MW-1000	Magnet Wire	Gives standards for magnet wire and wire coatings and acceptance tests.
NEMA	Temperature Classifications for Electrical Insulation	

Electron-bombardment Polymerization. In 1954 it was suggested that electron-beam polymerization could be useful as a new way of depositing ultrathin films;[68] and later, in 1958, the significance of this technique in fabricating microelectronic circuits and devices was reported.[69]

Gases or liquids that have been polymerized on electronic devices by electronic bombardment are normally those containing double bonds (ethylenic groups), such as monomeric tetrafluoroethylene, vinylidene fluoride, acrylic, methacrylic, acrylonitrile, styrene, and organopolysiloxanes.

The deposition of thin films may be selective by exposing only certain areas of a substrate. Hence, by employing a mask or by programming and directing the beam, very fine lines and areas may be defined.[70]

Germanium diodes coated with the liquid monomers, cured with high-energy electrons for periods up to 3 min at a total dose of 10^6, showed little effect on their reverse-current characteristics. The advantages of applying coatings in this manner is the rapidity with which they can be cured and the fact that heat is not required, thus obviating the possible deterioration of heat-sensitive devices.[71]

Glow-discharge Polymerization. The glow-discharge method for depositing and polymerizing thin films (also known as the electric or silent discharge method) involves striking an electrical discharge of several hundred to several thousand volts between two electrodes in the presence of a monomer gas.

Under direct-current conditions, the substrate to be coated is either made the anode or attached to the anode in some way. In ac discharge, the substrate is placed between the anode and the cathode. Still a third variation of glow discharge makes use of radio-frequency (Rf) excited molecules and does not require the use of electrodes at all.

The physical characteristics of these films depend on the conditions of glow discharge (voltage, amperage, time) and also on the nature of the surface on which the coating is deposited. Some surfaces lend themselves to the formation of strong chemical bonds, whereas others are rather passive and only weak adhesion results.

The key advantage of the glow-discharge method is the complete coating coverage which is provided to the part, since the polymers form within and behind all crevices and cover all surface imperfections resulting in an integral coating.[72] Other advantages include: (1) the temperature of the substrate may be kept low, even at room temperature; (2) the process if fairly simple and rapid—a very high vacuum system is not necessary as in other methods since vacuum of 10^{-3} and 10^{-4} torr only is required; and (3) the choice of the monomer starting material is wide and not limited to the conventional monomer types.

A large number of monomers have been polymerized by glow discharge, yielding

TABLE 34 Finishing Systems

Document	Title	Description
MIL-HDBK-132	Military Handbook: Protective Finishes	Details inorganic and organic protective finish systems which have proved satisfactory at various ordnance corps installations.
MIL-STD-171	Systems for Preparation, Painting, and Finishing Metal and Wood Surfaces	A comprehensive index of specifications and guide to the preparation and finishing of many types of surfaces.
MIL-STD-186	Painting and Finishing Systems for Rocket and Guided Missile Material	Establishes minimum requirements for procedures, materials, and systems for cleaning, plating, and painting of electronic components for rockets and guided missiles.
MIL-STD-194	Painting and Finishing Systems for Fire Control Instruments	Establishes minimum requirements for surface preparation, surface treatment, painting, and finishing of fire control instruments.
MIL-T-704	Treatment and Painting of Material	Covers materials to be used and procedures to be followed in cleaning, treating, and painting surfaces and equipment to provide protection from environment and corrosion (including sources for color call-out).
MIL-F-7179	Finishes and Coatings; General Specification for Protection of Aerospace Weapons Systems, Structures, and Parts	Specifies coating systems, mainly for the protection on nonelectronic surfaces.
MIL-F-14072	Finishes for Ground Signal Equipment	Design guide for selection of compatible materials and specifications for individual finishing system by substrate and severity of exposure to prevent deterioration in the anticipated environments.
MIL-F-18264	Finishes: Organic, Weapons System, Application and Control of	Gives recommended materials, their characteristics, application equipment and procedures, and troubleshooting painting problems.
MIL-C-22751	Coating System, Epoxy-Polyamide, Chemical and Solvent Resistant, Process for Application of	Complete coating system for protection of aluminum and magnesium. Consists of MIL-C-8514 pretreatment etch primer, MIL-P-23377 primer, and MIL-P-22750 enamel.

films from several hundred to 10,000 Å thick.[73,74] Of these, the best films in terms of electrical and physical properties were those derived from dimethylpolysiloxane, triethylsilane, diethylvinylsilane, vinyltrimethylsilane, heptene-2, cyclohexene, styrene, valeronitrile, and 2,5-dimethyl-2,4-hexadiene.

The glow-discharge technique has also been used to produce thin polymeric films from halogenated olefins including fluorinated derivatives.[75] These films, found suitable as dielectric and electronic device coatings, were formed by the application of several hundred volts potential difference between electrodes.[13]

Application Technology 5-69

TABLE 35 Coatings, Paints, and Primers

Document	Title	Description
TT-L-20	Lacquer, Camouflage	Camouflage colors for metal; 18 colors.
TT-L-32	Lacquer, Cellulose Nitrate, Gloss, for Aircraft Use	Durable cellulose nitrate gloss lacquer. Clear and 16 colors.
TT-V-119	Varnish, Spar, Phenolic Resin	Clear, air-drying spar varnish of the phenolic resin type.
TT-E-485	Enamel, Semi-gloss, Rust-inhibiting	One grade of a semigloss rust-inhibiting enamel for use on metal; four colors.
TT-E-489	Enamel, Alkyd, Gloss (for Exterior and Interior Surfaces)	Alkyd gloss enamel for use on primed exterior and interior wood and metal surfaces; 43 colors.
TT-E-527	Lustreless Enamel	Alkyd-type air-drying or baking lustreless enamel finish; 25 colors.
TT-E-529	Semigloss Enamel	Alkyd semigloss enamel for exterior and interior wood and metal surfaces for military and special equipment; 47 colors.
MIL-V-173	Varnish, Moisture- and Fungus-resistant (for Treatment of Communications, Electronic and Associated Electrical Equipment)	Describes a commonly used moisture- and fungus-resistant varnish consisting of p-phenyl phenol formaldehyde resin combined with tung oil and solvents.
MIL-V-1137	Varnish, Electrical-insulating (for Electromotive Equipment)	An impregnation varnish used for wire insulation. Also used to impregnant glass fiber.
MIL-I-2707	Insulating Varnish, Electrical, Impregnating, High Temperature	Used for wire insulation and coil impregnation.
MIL-E-5557	Enamel, Heat Resisting Glyceryl Phthalate Black	Air-drying and baking type, black heat-resistant enamel.
MIL-C-7439	Rain Erosion Coating	Thermally reflective rain-erosion-resistant and antistatic coating for glass-fabric-reinforced plastic laminates.
MIL-E-7729	Enamel, Gloss	Air-dry and baking types; 16 colors.
MIL-P-7962	Primer Coating, Cellulose-nitrate Modified Alkyd Type, Corrosion-inhibiting, Fast-drying (for Spray Application over Pretreatment Coating)	Lacquer primer for use only over MIL C-8514 and under nitrocellulose or acrylic-nitrocellulose lacquer.
MIL-C-8514	Coating Compound, Metal Pretreatment, Resin-Acid	Pretreatment wash primer coating for metals to be applied before complete coating system. Has adhesion-promoting and anticorrosive properties. Air Force and Navy-Weps version.
MIL-S-8516	Coating, Sealing, and Reinforcing of Electrical Connectors	Covers accelerated organic polysulfide liquid polymer for sealing and reinforcing electrical apparatus.
MIL-P-8585	Primer Coating, Zinc Chromate Low Moisture Sensitivity	Anticorrosive primer for aircraft aluminum (alkyd). (Replaced MIL-P-6889.) (Not for use on magnesium.)
MIL-E-14002	Enamel, Corrosion Inhibiting	Quick-drying lustreless synthetic enamel for steel, magnesium, and aluminum.
MIL-L-14486	Lusterless Lacquer, Vinyl	For use as finish coat.
MIL-E-15090	Alkyd Light Gray Equipment Enamel	Air-drying and baking enamel suitable for use on equipment, furniture, and switchboard installation; one color.
MIL-C-15328	Primer, Pretreatment (Formula No. 117 for Metals)	Wash primer similar to MIL-C-8514. Navy-Ships and DOD version.

TABLE 35 Coatings, Paints, and Primers (Continued)

Document	Title	Description
MIL-P-15929	Primer, Vinyl, Red Lead	Used for ferrous alloys, only over MIL-C-8514, MIL-C-15328.
MIL-P-15930A	Primer, Vinyl, Zinc Chromate	Used only over acid-etch primers MIL-C-8514, etc.
MIL-P-15932-6	Enamel, Exterior (Vinyl Alkyd)	A vinyl-alkyd enamel for use on pretreated coating, black, white, grays.
MIL-I-16923D	Insulating Compound, Electrical, Embedding	Requirements for several types of compounds for embedding or encasing electrical components.
MIL-I-17384	Insulating Compound, Electrical, Quick-drying	Covers lacquer of several types for use on electrical panels or cases and for sealing the edges of phenolic laminates.
MIL-L-19537	Lacquer, Acrylic-nitrocellulose Gloss (for Aircraft use)	General-purpose exterior coating for metals, resistant to diester lubricating oils. Clear and 13 colors.
MIL-L-19538	Lacquer, Acrylic-nitrocellulose, Camouflage (for Aircraft use)	Same as MIL-L-19537 except in lusterless camouflage colors. White, black, and nine colors.
MIL-P-22750	Coating, Epoxy-Polyamide, Chemical- and Solvent-resistant, for Weapons Systems	Protective coating for magnesium and aluminum, over MIL-C-8514, MIL-C-8514 plus MIL-P-23377, or direct on metal. Clear, black, white, red, and four grays.
MIL-P-22808	Paint, Epoxy, Hydraulic, Fluid Resistant	An epoxy-polyamide enamel for protection of metals at temperatures up to 500°F. White and four colors.
MIL-P-23377	Primer Coating, Epoxy-Polyamide, Chem. & Solv. Resist.	Useful over Mg and Al alloys and for separation of dissimilar metals.
MIL-I-24092	Insulating Varnish, Electrical, Impregnating	Characterizes insulating varnishes, baking type for electric coils and structures.
MIL-C-27227	Coating, Polyurethane, Thermal Resistant for Aircraft Application	Gloss finish coat. May be used for magnesium.
MIL-I-46058	Insulation Compound, Electrical (for Coating Printed Circuit Assemblies)	Covers various types (epoxy, polyurethane, silicones, polystyrene low-dielectric) conformal coatings for printed-circuit assemblies.
MIL-L-52043	Semigloss Lacquer	May be used for magnesium.
MIL-P-52192	Primer Coating, Epoxy	Used for magnesium.
MIL-C-52210	Coating, Epoxy, Spray Type, for Printed Circuitry	Epoxy-polysulfide-type coating having fair flexibility for use over epoxy laminate printed-circuit boards.

General properties of thin-film polymer coatings Many of the properties of polymer thin films are independent of their chemical structure or composition. The most important of these is the very high dielectric breakdown strengths which all thin films possess. The increase of dielectric strength with decrease in film thickness is a well-known phenomenon previously described.[39]

Many of the processes for depositing thin films result in films that are pinhole-free and continuous. The lack of pinholes and integrity of the film are largely dependent on the method of deposition, surface treatment, and cleanliness of the surface and ambient in which the coating is applied.

Still another property common to most thin films is their high degree of purity. With adequate control, the polymerization reaction may be directed so that no side reactions or by-products are formed.

Application Technology 5-71

TABLE 36 Solvents, Resins, and Other Coating Ingredients

Document	Title	Description
TT-R-271	Resin, Phenol-Formaldehyde, Paraphenyl	
TT-T-548	Toluol (Toluene)	Solvent for use in organic coatings.
TT-T-775	Tung Oil, Raw (China Wood)	Used as a varnish ingredient.
TT-X-916	Xylene	Solvent for use in organic coatings.
MIL-M-261	Methyl Ethyl Ketone	For use in organic coatings.
TT-E-463	Ethyl Alcohol	
MIL-A-6091	Alcohol, Ethyl, Specially Denatured	
MIL-I-10428	Isopropyl Alcohol, Technical	
MIL-T-27602	Trichloroethylene (commercial)	Used as a cleaning solvent.

TABLE 37 Cleaning and Surface Treatments

Document	Title	Description
TT-C-490	Cleaning Methods and Pretreatment of Ferrous Surfaces for Organic Coatings	Cleaning and pretreatment of ferrous metals, and ferrous metals having zinc or aluminum surfaces.
MIL-M-3171	Magnesium Alloy, Processes for Pretreatment and Protection of Corrosion	Covers Dow 1, 7, 9, and 19 treatments.
MIL-S-5002	Surface Treatments and Metallic Coatings for Metal Surfaces of Weapons Systems	Preparation, surface treatments, and metallic coatings for metal surfaces of weapons systems prior to application of organic finishes.
MIL-C-5541	Chemical Films and Chemical Film Materials for Aluminum and Aluminum Alloys	A chromate conversion coating for aluminum, covers Alodine,* Iridite,‡ and other proprietary treatments.
MIL-A-8625	Anodic Coatings, for Aluminum and Aluminum Alloys	Covers chromic and sulfuric acid anodization processes.
MIL-T-12879	Treatments, Chemicals, Prepaint and Corrosion Inhibitive, for Zinc Surfaces	Prepaint and corrosion-inhibitive chemical treatments for plated, hot-dipped, and solid zinc surfaces.
MIL-C-13335	Coating for Magnesium and Magnesium Alloys	Covers HAE anodize treatment.
MIL-C-38334	Corrosion Removing Compound, Prepaint, for Aircraft Aluminum Surfaces	Covers use of nonflammable phosphoric acid base liquid.
MIL-M-45202	Magnesium Alloys, Anodic Treatment of	Covers Dow 17, HAE, and CR-22 anodic treatments.
MIL-M-46080	Magnesium Castings; Process for Anodic Cleaning and Surface Sealing of	Covers fluoride-anodize treatment plus Dow 7, plus MIL-C-46070, epoxy baking coating.
AMS-2476	Electrolytic Treatment for Magnesium Base Alloys	Covers HAE anodize treatment.
AMS-2478	Anodic Treatment of Magnesium Base Alloys	Covers Dow 17 anodize treatment.

*Trademark of Amchem Products, Inc.
† Trademark, of Allied Research Products, Inc.

SPECIFICATIONS

Material Specifications

A coating material specification is written to control the engineering requirements, to establish receiving tests on batches, or even to establish explicit formulas for materials to be furnished. When written around engineering requirements, there may be many or perhaps no qualified materials on the market; whereas when written as a formulation or narrow performance specification, it may be presumed that at least one already tested and qualified material exists.

TABLE 38 Test Procedures—General Specifications

Document	Title	Description
Fed. Std. 141 (TT-P-141)	Paint, Varnish, Lacquer, and Related Materials; Methods of Inspection, Sampling and Testing	Describes methods to determine physical and chemical properties of paint, varnish, and lacquer for conformance to Federal and Military Specifications. Covers 278 methods.
Fed. Std. 406	Plastics: Methods of Testing	A compilation of 76 methods for testing plastics.
MIL-STD-202	Test Methods for Electronic and Electrical Component Parts	Establishes uniform methods for testing electronic and electrical component parts such as capacitors, resistors, switches, relays, transformers, and jacks. Applies only to small parts such as transformers and inductors weighing up to 300 lb.
MIL-STD-810A	Military Standard Environmental Test Methods for Aerospace and Ground Equipment	A compilation of methods previously described in a variety of other specifications, as, for example, MIL-E-5272, MIL-T-5422, and MIL-E-4970.
MIL-E-4970 (USAF)	Environmental Testing, Ground Support Equipment (inactive for new design)	Procedures for testing ground-support equipment under simulated and accelerated conditions.
MIL-E-5272 (ASG)	Environmental Testing, Aeronautical and Associated Equipment	Procedure for subjecting equipment to simulated and accelerated environmental conditions to ensure satisfactory operation and freedom from deterioration in use.
MIL-T-5422 (ASG)	Environmental Testing for Aircraft Electronic Equipment	Testing procedures under simulated and accelerated environmental conditions to meet requirements of MIL-E-5400.
MIL-I-16923	Insulating Compound, Electrical Embedding	Describes or references many electrical and other tests applicable to plastics and coatings.
ASTM D 115-55	Methods of Testing Varnishes Used for Electrical Insulation	Covers tests for electrical properties, solids percent, viscosity, dry time, flexibility.

Acceptance Specifications

For electronic parts or assemblies that are coated by vendors, an acceptance specification detailing requirements and quality control tests to be performed on the coated parts may be useful. Destructive or nondestructive tests may be called out to assure that the part is adequately coated to perform the required function. Examples of tests that are commonly specified are adhesion, dielectric strength, insulation resistance, humidity, and salt-spray exposures. The fact that a material meets a military, federal, or trade association specification does not necessarily assure its suitability for a given application. Additional or tighter requirements

TABLE 39 Specific Test Methods for Coatings

Test	ASTM	Fed. STD. 141a, method	MIL-STD-202, method	Fed. STD. 406, method	Others
Abrasion	D 968	6191 (Falling Sand) 6192 (Taber)	....	1091	Fed. Std. 601, 14111
Adhesion	D 2197	6301.1 (Tape Test, Wet) 6302.1 (Microknife) 6303.1 (Scratch Adhesion) 6304.1 (Knife Test)	....	1111	Fed. Std. 601, 8031
Arc resistance	D 495	....	303	4011	
Dielectric constant	D 150	....	301	4021	Fed. Std. 101, 303
Dielectric strength (breakdown voltage)	D 149 D 115	....	....	4031	Fed. Std. 601, 13311
Dissipation factor	D 150	....	....	4021	
Drying time	D 1640 D 115	4061.1			
Electrical insulation resistance	D 229 D 257	....	302	4041	MIL-W-81044, 4.7.5.2
Exposure (exterior)	D 1014	6160 (On Metals) 6161.1 (Outdoor Rack)			
Flash point	D 56, D 92 D 1310 (Tag Open Cup)	4291 (Tag Closed Cup) 4294 (Cleveland Open Cup)	....	....	Fed. Std. 810, 509
Flexibility	....	6221 (Mandrel) 6222 (Conical Mandrel)	....	1031	Fed. Std. 601, 11041
Fungus resistance	D 1924	....	....	....	MIL-E-5272, 4.8 MIL-STD-810, 508.1 MIL-T-5422, 4.8
Hardness	D 1474	6211 (Print Hardness) 6212 (Indentation)			
Heat resistance	D 115 D 1932	6051			
Humidity	D 2247	6071 (100% RH) 6201 (Continuous Condensation)	103 106A	....	MIL-E-5272, Proc. 1 Fed. Std. 810, 507
Impact resistance	....	6226 (G.E. Impact)	1074		
Moisture-vapor permeability	E 96 D 1653	6171	7032		
Nonvolatile content	....	4044			
Salt spray (fog)	B 117	6061	101C	6071	MIL-STD-810, 509.1 MIL-E-5272, 4.6 Fed. Std. 151, 811.1 Fed. Std. 810, 509
Temperature-altitude	....	....	....	....	MIL-E-5272, 4.14 MIL-T-5422, 4.1 MIL-STD-810, 504.1
Thermal conductivity	D 1674 (Cenco Fitch) C 177 (Guarded Hot Plate)	....	....	....	MIL-I-16923, 4.6.9
Thermal shock	....	....	107	....	MIL-E-5272, 4.3 MIL-STD-810, 503.1
Thickness (dry film)	D 1005 D 1186	6181 (Magnetic Gage) 6183 (Mechanical Gage)	....	2111, 2121, 2131, 2141, 2151	Fed. Std. 151, 520, 521.1
Viscosity	D 1545 D 562 D 1200 D 88	4271 (Gardner Tubes) 4281 (Krebs-Stormer) 4282 (Ford Cup) 4285 (Saybolt) 4287 (Brookfield)			
Weathering (accelerated)	D 822	6151 (Open Arc) 6152 (Enclosed Arc)	....	6024	

A more complete compilation of test methods is found in J. J. Licari, "Plastic Coatings for Electronics," McGraw-Hill Book Company, New York, 1970.
The major collection of complete test methods for coatings is "Physical and Chemical Examination of Paints, Varnishes, Lacquers, and Colors," by Gardner and Sward, Gardner Laboratory, Bethesda, Md. This has gone through many editions.

5-74 Coatings for Electronics

may be needed and can be incorporated as extra provisions of a company internal specification.

Process Specifications

Process specifications are detailed procedures to be used for a manufacturing operation, but should allow manufacturing as much freedom as possible and still meet the engineering requirements. As an example, if there are a number of different cure schedules which have been proved equivalent, the choice of any one of them should ordinarily be allowed, provided none of the temperatures is so high as to damage the assembly. An example of such a time-temperature cure schedule is plotted in Fig. 18. Sources of specifications are listed in Table 32. Tables 33 to 37 provide lists of many specifications with brief descriptions. Test specifications and specific test procedures are listed in Tables 38 and 39.

REFERENCES

1. Etchason, P. T., and J. J. Licari: High Reliability Polyurethane Insulation Coating for Circuit Board Protection, *Insulation*, June, 1962.
2. *Motorola Rep.* 2, DA 36-039-sc-89136.
3. McGuiness, E. W.: Improved Reliability for Printed Circuits with Protective Coatings, *Soc. Plast. Eng. Tech. Papers*, vol. 7, January, 1961.
4. Manko, H. H.: "Solders and Soldering," p. 223, McGraw-Hill Book Company, New York, 1964.
5. Saito, Y., and T. Hino: *Proc. AIEE* (1960).
6. *NEMA Standard Publ.* MW 1000, Magnet Wire.
7. *Military Specification* MIL-W-583C, Wire, Magnet, Electrical, November, 1965.
8. Murphy, C. B., and J. A. Hill: Thermal Stability and Characterization of Wire Coatings by Differential Thermal Analysis, *Insulation*, August, 1962.
9. Duval, C.: "Inorganic Thermogravimetric Analysis," 2d ed., American Elsevier Publishing Company, Inc., New York, 1963; Hunt, C. F, and A. H. Markhart: *Insulation*, vol. 6, November, 1960.
10. Anderson, D. A., and E. S. Freeman: *Anal. Chem.*, vol. 31, p. 1697, 1959.
11. Ke, B., and A. W. Sisko: *J. Polym. Sci.*, vol. 50, p. 87, 1961.
12. Murphy, C. B.: Differential Thermal Analysis, *Anal. Chem.*, vol. 30, pp. 867–872, 1958; vol. 32, pp. 168R–171R, 1960; vol. 34, pp. 298R–300R, 1962.
13. U.S. Patent 3,068,510; U.S. Patent 3,069,283.
14. Robinson, W. M., and H. R. Lee: Pitfalls and Progress in Plastic Encapsulation of Semiconductors, *Insulation*, December, 1967.
15. Hamill, A. T.: Westinghouse Goldilox Integrated Circuits. *μ-Notes, Inform. on Microelectron. for Navy Equip.*, no. 25, U.S. Naval Ammunition Depot, Crane, Ind., August, 1968.
16. Flood, J. L.: Reliability of Plastic Integrated Circuits, *ibid.*
17. Dalton, J. V.: Sodium Drift and Diffusion in Silicon Nitride Films, Electrochemical Society Meeting, Cleveland, May, 1966.
18. Snow, E. H., A. S. Grove, B. E. Deal, and C. T. Sah: Ion Transport Phenomena in Insulating Films, *J. Appl. Phys.*, vol. 36, p. 1664, 1965.
19. Yon, E., W. H. Ko, and A. B. Kuper: Sodium Distribution in Thermal Oxide on Silicon by Radiochemical and MOS Analysis, *IEEE Trans. Electron. Devices*, vol. ED-13, 1966.
20. Schneer, G. H., W. Van Gelder, V. E. Hauser, and P. F. Schmidt: A Metal Insulator–Silicon Junction Seal, *IEEE Trans. Electron. Devices*, vol. ED-15, 1968.
21. MacLeod, Ross W.: Final Protection of Printed Wiring Modulus by Organic Materials, Institute of Metal Finishing, Printed Circuit Group, Annual Symposium, Bristol, England, September, 1968.
22. Schwartz, S. (ed.): "Integrated Circuit Technology," p. 144, McGraw-Hill Book Company, New York, 1966.
23. Pankove, J. I.: Method of Sealing a Semiconductor Device, U.S. Patent 2,807,558, Apr. 12, 1954.
24. Rulison, R.: Surface Treatment of Germanium Circuit Elements, U.S. Patent 2,768,100, September, 1956.
25. John, H. F.: Protective Treatment for Semiconductor Devices, U.S. Patent 2,937,110, May 17, 1960.

26. Herr, E.: "Reliability Assessment of Epoxy Transistors," μ-*Notes, Inform. on Microelectron. for Navy Equip.*, no. 25, U.S. Naval Ammunition Depot, Crane, Ind., August, 1968.
27. Wahlgren, E. B.: Use of Epoxy Resins in High Reliability Transformers, *Electron. Equip. Eng.*, January, 1956.
28. Von Hippel, A. R. (ed.): "Dielectric Materials and Applications," Technology Press of MIT and John Wiley & Sons, Inc., New York, 1961.
29. Baer, E. (ed.): "Engineering Design for Plastics," Reinhold Publishing Corporation, New York, 1964.
30. Harper, C. A.: "Electronic Packaging with Resins," McGraw-Hill Book Company, New York, 1961.
31. Mathes, K. N.: Electrical Properties of Insulating Materials, *Proc. 7th Elec. Insul. Conf.*, Chicago, Oct. 15–19, 1967.
32. Delmonte, J.: "Metal Filled Plastics," Reinhold Publishing Corporation, New York, 1961.
33. O'Dwyer, J. J.: An Examination of Avalanche Theories of Dielectric Breakdown in Solids, *Elec. Insul. Conf. Publ.* 1356, National Academy of Sciences, 1966.
34. Philofsky, H. M., and E. M. Fort: Degree of Correlation between Tan δ and Electric Strength of Insulation, *Proc. 7th Elec. Insul. Conf.*, Chicago, October, 1967.
35. Oakes, W. J.: *Proc. IEE*, vol. 96, no. 1, p. 37, 1949.
36. *Dow Corning Bull.* 07-062, May, 1963.
37. Mathes, K. N.: Selection of an Insulation System in Product Design, *Insulation*, October, 1967.
38. Killam, D. L.: Effect of Humidity on the Dielectric Properties of Some Polymers, *Elec. Insul. Conf. Publ.* 1238, National Academy of Sciences, 1965.
39. Whitehead, S.: "Dielectric Breakdown of Solids," Clarendon Press, Oxford, 1951; O'Dwyer, J. J.: "Theory of Dielectric Breakdown of Solids," Clarendon Press, Oxford, 1964.
40. Lee, Henry, Donald G. Stoffey, and Kris Neville: "New Linear Polymers," McGraw-Hill Book Company, New York, 1967.
41. Dow Chemical Company: Magnesium Finishing, Midland, Mich., 1963.
42. Stevens, J. A.: "Designing for Long Service: a Primer of Magnesium Finishing," *Magnesium Topics*, October, December, 1958.
43. Brands, E. R., and L. C. Kilgore: Finishing Systems for Magnesium-Lithium Alloy LA141, presented at Air Force Material Laboratory, Corrosion of Military and Aerospace Equipment Symposium, Denver, Colo., May, 1967.
44. Bailey, J. C., and J. A. Hirschfield: *Research*, vol. 7, p. 320, 1954.
45. LaQue, F. L., and H. R. Copson (eds.): "Corrosion Resistance of Metals and Alloys," Reinhold Publishing Corporation, New York, 1963.
46. Licari, J. J., and E. R. Brands: Organic Coatings for Metal and Plastic Surfaces, *Mach. Des.*, May 25, 1967.
47. *ASTM Standard* C117-63, part 14, pp. 15–26, 1965.
48. Anderson, D. R.: "Thermal Conductivity of Polymers," *Physical and Chemical Reviews*, June, 1965.
49. Smits, P.: *Metal Progr.*, vol. 80, no. 3, p. 75, 1961.
50. Shooter, K. V.: *Proc. Roy. Soc.*, vol. A212, p. 490, 1952; *Plastics*, vol. 26, pp. 281, 117, 1961.
51. Gardner, H. A., and G. G. Sward: "Paint Testing Manual: Paints, Varnishes, Lacquers, and Colors," Gardner Laboratory, Bethesda, Md., 1962.
52. *Federal Standard* 209.
53. Austin, P. R., and S. W. Timmerman: "Design and Operation of Clean Rooms," chap. 9, Business News Publishing Co., Detroit, Mich., 1965.
54. McCrone, W. C., R. G. Draftz, and J. J. Delly: "The Particle Atlas," Ann Arbor Science Publishers, Ann Arbor, Mich.
55. Heuring, H. F.: Cleaning Electronic Components and Subassemblies, *Electron. Packag. and Prod.*, June, 1967.
56. Werner, M.: Solvent Vapor Degreasing and Recovery, in "Metal Finishing Guidebook," pp. 162–172, Metals and Plastics Publications, Westwood, N.J., 1967.
57. Jeffrey, L. R.: Ultrasonic Cleaning, in "Metal Finishing Guidebook," pp. 192–196, Metals and Plastics Publications, Westwood, N.J., 1967.
58. Von Fischer, W., and E. G. Bobalek: "Organic Protective Coatings," Reinhold Publishing Corporation, New York, 1953.
59. Davis, W. J.: Methods and Apparatus for Applying Protective Coatings, U.S. Patent 3,004,861, October, 1961; Dettling, C. J.: Fluidized Bed Coating Method, U.S. Patent 2,974,060, March, 1961; Dettling, C. J., and R. E. Hartline: Process

and Apparatus for Producing Continuous Coatings, U.S. Patent 2,987,413, June, 1961; Gemmer, E.: Fluidized Bed Coating Process, U.S. Patent 2,974,059, March, 1961; Gemmer, E.: Fluidized Bed Coating Process for Coating with Thermosetting Materials, U.S. Patent 3,090,696, May, 1963.
60. Gemmer, E.: Process and Apparatus for the Preparation of Protective Coatings from Pulverulent Synthetic Thermoplastic Materials, German Patent 933,019, September, 1955.
61. Landrock, A. H.: The Coating of Aluminum with Plastics by the Fluidized Bed and Electrostatic Powder Techniques, *Plastec Note* 18, Plastics Technical Evaluation Center, Picatinny Arsenal, N.J., February, 1968.
62. Pettigrew, C. K.: Fluidized Bed Coatings. *Mod. Plast.*, part 1, vol. 43, no. 12, August, 1966 and part 2, vol. 44, no. 1, September, 1966.
63. Gaynor, J., A. H. Robinson, M. Allen, and E. E. Stone: Variables in Fluidized Bed Coatings, *Mod. Plast.*, vol. 43, no. 5, January, 1966.
64. Short, O. A.: Conductor Compositions for Fine Line Printing, *Electron. Packag. and Prod.*, February, 1968.
65. White, M.: Vacuum Evaporation of Polythene, *Vacuum*, vol. 15, p. 449, 1965.
66. *Proc. 7th Conf. Soc. Vacuum Coaters*, pp. 78–81, 1964.
67. General Electric Co.: Thin Film Polymers, G. E. Public Information Unit, Schenectady, N.Y., July, 1967.
68. Ennos, A. E.: *Brit. J. Appl. Phys.*, vol. 5, p. 27, 1954.
69. Buck, D. A., and K. R. Shoulders: *Proc. East. Joint Comput. Conf.*, Philadelphia, 1958.
70. White, P.: Preparation and Properties of Dielectric Layers Formed by Surface Irradiation Techniques, *Insulation*, May, 1967.
71. Schmitz, J. V., and E. J. Lawton: Process for Applying Protective Coatings by Means of High Energy Electrons, U.S. Patent 2,900,277, August, 1959.
72. Caswell, H. L.: Fabrication of Thin Film Insulation, *Elec. Insul. Conf. Publ.* 1238, National Academy of Sciences, 1965.
73. Bradley, A., and J. Hammes: Electrical Properties of Thin Organic Films, *J. Electrochem. Soc.*, January, 1963.
74. Smolinsky, G., and J. H. Heiss: "Thin Organic Films Formed by a Glow Discharge Technique," ACS 155th Meeting, April, 1968, *Proc. ACS*, vol. 28, no. 1.
75. Goodman, J.: *J. Polym. Sci.*, vol. 44, p. 551, 1960; U.S. Patent 2,676,145; Bamford, C. H., et al: Uses of Electric Discharge in Polymer Science, *Soc. Chem. Ind. Monogr.* 17, 1963; Otozai, K: *Bull. Chem. Soc. Japan*, vol. 27, p. 476, 1954.

Chapter **6**

Ceramics, Glasses, and Micas

DON E. HARRISON
and
CHRISTY J. MORATIS[*]
Westinghouse Electric Corporation,
Research and Development Center,
Pittsburgh, Pennsylvania

Introduction	6–2
Electrical Tests and Measurements	6–2
Resistivity	6–2
Dielectric Constant	6–2
Microwave Properties	6–3
Dielectric Strength	6–3
Magnetic Properties	6–3
Piezoelectric Properties	6–3
Insulators	6–3
Ceramics	6–3
Glass	6–27
Mica	6–35
Ceramic and Glass Substrates	6–38
Glass-Ceramics	6–42
Seals	6–42
Thick-film Components	6–54
Dielectrics	6–56
Ceramic Capacitors	6–56
Glass Capacitors	6–75
Mica Capacitors	6–77

[*] Mr. Moratis is now with Allis-Chalmers Manufacturing Company, Pittsburgh, Pa.

Piezoelectricity 6–79
Magnetic Ceramics 6–87
References 6–113

INTRODUCTION

This chapter was written from the viewpoint of the materials technologist. Its aim is to provide the electrical or electronics design engineer with a concise description of the more important electrical and magnetic characteristics of ceramics, glasses, and micas. The contents are divided into four principal sections, namely, insulators, dielectrics, piezoelectrics, and ceramic magnets. For our purposes, we classify materials with permittivities under 12 as *insulators* and those with values over 12 as *dielectrics*. Since the most important application of high-dielectric-constant materials is for capacitive devices, the section on dielectrics is devoted entirely to capacitors. The sections on piezoelectrics and ceramic magnets give descriptions of the phenomena and the extent to which they are affected by the crystal chemistry and the microstructure of the ceramic. In addition to these four topics, reference to some of the more common measuring and testing techniques is given as well. The symbols used are listed in Table 1.

We are indebted to the literature generously supplied by the companies cited in the text. In particular, we wish to acknowledge Mr. T. I. Procopowicz of the Sprague Electric Company for supplying us with much unpublished data on ceramic capacitors and for reviewing the section on capacitors, Dr. H. W. Stetson of the Western Electric Company for giving us information on ceramic substrates, and finally from the Westinghouse Research Laboratories thanks are due Dr. A. I. Braginski for reviewing the section on magnetic ceramics, and Dr. G. Mott and Mr. J. H. Thompson for reviewing the section on piezoelectricity.

ELECTRICAL TESTS AND MEASUREMENTS

The properties of ceramics, glasses, and micas that are of most interest to the electrical or electronics design engineer include resistivity, dielectric constant, dielectric loss, dielectric strength, and magnetic permeability. The manner in which these properties vary as a function of temperature, frequency, and electric or magnetic field determines the suitability of materials for specific applications. A description of the more common measuring and testing methods is given in the following references.

Resistivity

Guard electrode: ASTM D 257.[2]
Electrode materials: ASTM D 257,[2] Sauer and Flaschen,[3] Landis.[4]
Measuring circuits:
Voltmeter-Ammeter (V-I) Method: ASTM D 257.[2]
Comparison Method: ASTM D 257.[2]
Wheatstone Bridge: Sauer and Shirk.[5]

Dielectric Constant

Dielectric measurements:
Electrodes: ASTM D 150.[6]
Two-terminal System: ASTM D 150.[6]
Three-terminal System: ASTM D 150.[6]
Micrometer-Electrode System: ASTM D 150.[6]
Dielectric constant test methods:
Frequency Range (10^0 to 10^7 Hz): ASTM D 150,[6] Von Hippel.[7]
Frequency Range (10^4 to 10^8 Hz): ASTM D 150,[6] Von Hippel,[7] ASTM C 525-63T.[8]

Insulators   6-3

Microwave Properties

    Saturation magnetization: ASTM C 527-63T, Foner.[1]
    Spin-wave line width: Ollom, Von Aulock.[149]
    Complex dielectric constant: ASTM C 525-63T.[8]
    Line width and gyromagnetic ratio: ASTM C 524-63T.

Dielectric Strength

    Short-time test; slow-rate-of-rise test; step-by-step test: ASTM D 149-64.
    Effect of specimen thickness: Comeforo.[71]
    Testing frequency: Frisco.[9]

Magnetic Properties

    Magnetizing force H: ASTM Publication 371, Magnetic Testing.
    Magnetic flux density B: ASTM Publication 371, Magnetic Testing.
    Hysteresis curves: Storm.[10]

Piezoelectric Properties

    Piezoelectric constants: IRE Standards on Piezoelectric Crystals.[11]

INSULATORS

The properties of inorganic materials with dielectric constants less than 12 are described in this section under the headings of Ceramics, Glasses, Micas, Substrates, Glass-Ceramics, and Thick Films. Since micas and glasses are also used as dielectrics, the properties associated with capacitor applications are discussed in the section on Dielectrics.

Ceramics

Ceramic materials offer mechanical and electrical properties that make them especially suitable for the electrical and electronics industries. The bulk of electrical ceramics are used either as insulators or as dielectrics. Electrical grade ceramics have more exacting property requirements than the ceramics used for refractory or structural purposes. Properties such as dielectric strength, dielectric constant, dissipation factor, and thermal and electric conductivity are closely related to microstructure as well as to composition and processing.

In common with other insulating materials, ceramic insulators must have low dielectric constants so as to avoid capacitance effects, adequate dielectric strength to withstand the applied voltage without breakdown, low dissipation in order to avoid excessive electrical losses, and mechanical strength sufficient to withstand service conditions.

Oxide ceramics are characterized by their chemical inertness, oxidation resistance, moderately high refractoriness (see Fig. 1) and resistivity (see Fig. 2), and by their low thermal expansions (see Fig. 3), thermal conductivities (see Fig. 4), and densities (see Fig. 5). Both the electrical resistivity and the thermal conductivity of oxide ceramics decreases with increasing temperature.

Ceramics are subject to failure by thermal shock due to the disruptive stresses that result from the differential dilation between the surface and core of a body. With the relatively slow cooling rates usually encountered in most electrical and electronic applications and with the exception of materials with very low thermal expansion coefficients, the thermal conductivity usually determines the resistance of a ceramic to failure by thermal shock. Listed in Table 2 (on page 6-9) are the thermal stresses or shock resistances of several ceramics as derived from the relation $R'' = C'\sigma/\alpha E$, where C' is the thermal conductivity, σ is the tensile strength, α is the coefficient of thermal expansion, and E is Young's modulus.

    Electrical porcelain   A typical electrical porcelain body consists of approximately 50 percent clay [$Al_2Si_2O_5(OH)_4$] and 25 percent each of flint (SiO_2) and of feldspar ($KAlSi_3O_8$). The high clay content gives the green body plasticity, which facilitates easy fabrication. Feldspar reacts with the clay at high temperatures (1200 to

TABLE 1 Symbols Used in Test

Symbol	Meaning	mks	cgs	English
ac	Alternating current			
Å	Angstrom			
AV	Anion vacancies			
B	Magnetic induction	webers/meter^2	gauss	lines/in.^{2}
B_{max}	Maximum magnetic induction	webers/meter^2	gauss	lines/in.^{2}
B_r	Residual induction	webers/meter^2	gauss	lines/in.^{2}
$B_d H_d$	Peak energy product	(webers)(amp-turns)/meters^3	gauss-oersteds	(lines)(amp-turns)/in.^{3}
c_{ij}	Elastic stiffness constant	newtons/meter^2	dynes/cm^2	lb weight/in.^{2}
C'	Thermal conductivity	cal/(sec)(meter)(°C)	cal/(sec)(cm)(°C)	Btu/(sec)(in.)(°F)
C	Capacitance	farads	farads	farads
CV	Cation vacancies			
d, t	Thickness	meter	cm	in.
D_{ij}	Piezoelectric strain constants	meters/volt = coul/newton	10^7 cm/volt = coul/dyne	in./volt = 0.113 coul/lb weight
d_p	Planar piezoelectric strain constant	meters/volt = coul/newton	10^7 cm/volt = coul/dyne	in./volt = 0.113 coul/lb weight
d_h	Hydrostatic piezoelectric strain constant	meters/volt = coul/newton	10^7 cm/volt = coul/dyne	in./volt = 0.113 coul/lb weight
dc	Direct current			
D	(superscript) At constant electric displacement			
D, D_1 D_3	Electric displacement and its components	coul/meter^2	coul/cm^2	coul/in.^{2}
$D =$ tan δ	Dissipation factor			
e	Charge on electron	coul	coul	coul
E	(superscript) At constant electric field			
ϵ_r''	Relative loss factor			
E, E_1 E_3	Electric field intensity and its components	volts/meter	volts/cm	volts/in.
E, Y_{ij}	Young's modulus	kg/meter^2	g/cm^2	lb weight/in.^{2}
E	Energy	joules	ergs	ft-lb weight
f, Hz, ω	Frequency	Hz	Hz	Hz

g	Landé g factor			
g_{ij}	Piezoelectric "voltage" constant; electric field/stress at constant charge or strain/charge density at constant stress	volt-meters/newton = meters^2/coul	volt-cm/dyne = 10^7 cm^2/coul	volt-in./lb weight = 8.85 in.^{2}/coul
g_p	Planar piezoelectric voltage constant	volt-meters/newton = meters^2/coul	volt-cm/dyne = 10^7 cm^2/coul	volt-in./lb weight = 8.85 in.^{2}/coul
g_h	Hydrostatic piezoelectric voltage constant	volt-meters/newton = meters^2/coul	volt-cm/dyne = 10^7 cm^2/coul	volt-in./lb weight = 8.85 in.^{2}/coul
ΔH	Resonance absorption line width	amp-turns/meter	oersteds	amp-turns/in.
H	Magnetic field strength	amp-turns/meter	oersteds	amp-turns/in.
H_c	Magnetic coercive force	amp-turns/meter	oersteds	amp-turns/in.
H_i	Internal magnetic field	amp-turns/meter	oersteds	amp-turns/in.
$H\theta^a$	Magnetic anisotropy field	amp-turns/meter	oersteds	amp-turns/in.
H_a	Applied magnetic field	amp-turns/meter	oersteds	amp-turns/in.
J	Magnetic moment	amp/meter^2	amp/cm^2	amp/in.^{2}
k_{31}	Transverse coupling factor			
k_{33}	Longitudinal coupling factor			
k_{15}	Shear coupling factor			
k_t	Thickness coupling factor			
k_p	Planar coupling factor			
K_1, K_2	Magnetic anisotropy constants			
L	Angular momentum	(kg)(meter^2)/sec	(g)(cm^2)/sec	(lb weight)(in.^{2})/sec
L	Inductance	henrys	henrys	henrys
m	Mass of electron	kg	grams	lb weight
M	Magnetization	webers/meter^2	gauss	lines/in.^{2}
M_s	Magnetic saturation	webers/meter^2	gauss	lines/in.^{2}
$N_{x,y,z}$	Demagnetization coefficients			
P	Electric polarization	coul/meter^2	coul/cm^2	coul/in.^{2}
P.F.	Power factor			
Q	Mechanical quality factor			
R''	Thermal stress or shock resistance	cal/(sec)(meter)(°C)^2	cal/(sec)(cm)(°C)^2	Btu/(sec)(in.)(°F)^2
R	Resistance of sample	ohms	ohms	ohms
R_s	Magnetic squareness ratio			
rf	Radio frequency			
s_{ij}	Elastic compliance constants (superscript) At constant strain	meters^2/newton	cm^2/dyne	in.^{2}/ft-lb weight
S				
S	Spin momentum	(kg)(meter^2)/sec	(gram)(cm^2)/sec	(lb weight)(in.^{2})/sec
T	(superscript) At constant stress			
T_c	Curie temperature	°C	°C	°F

6-5

TABLE 1 Symbols Used in Test (Continued)

Symbol	Meaning	Practical units mks	cgs	English
VHF	Very high frequency	cycles/sec	cycles/sec	cycles/sec
Z	Impedance	ohms	ohms	ohms
α	Linear thermal expansion coefficient	per °C	per °C	per °F
γ	Gyromagnetic ratio			
γ	Surface energy	joules/meter^2	ergs/cm^2	lb-weight/in.^{2}
δ	Loss angle			
ε:	Permittivity of free space	farads/meter	farads/cm	farads/in.
ε	Permittivity of medium	farads/meter	farads/cm	farads/in.
ϵ_r, K	Relative permittivity or dielectric constant	farads/meter	farads/cm	farads/in.
$\epsilon_r{}^*$	Complex dielectric constant ($\epsilon_r{}^* = \epsilon_r' - j\epsilon_r''$)	farads/meter	farads/cm	farads/in.
ϵ, ϵ_{ij}	Strain			
θ	Phase angle			
λ_{ijk}	Magnetostrictive linear dialation	meters	cm	in.
μ_0	Permeability of free space	henrys/meter	gauss/oersted	lines/(amp-turns)(in.)
μ	Permeability of medium	henrys/meter	gauss/oersted	lines/(amp-turns)(in.)
μ_r	Relative permeability			
$\mu_r{}^*$	Complex relative permeability ($\mu_r{}^* = \mu_r' - j\mu_r''$)			
tan δ_m	Magnetic dissipation factor			
ρ	Density	kg/meter^3	g/cm^3	lb/in.^{3}
ρ	Electrical resistivity	ohm-meters	ohm-cm	ohm-in.
σ_{ij}	Stress	newtons/meter^2	dynes/cm^2	lb weight/in.^{2}
χ	Susceptibility	henrys/meter	gauss/oersted	lines/(amp-turns)(in.)

1300°C) to give mullite and a viscous liquid phase. Solution of flint (1300 to 1400°C) increases the viscosity of the liquid phase and helps to maintain the shape of the body during firing. These compensating changes give porcelain an unusually long firing range and a great tolerance for compositional variations.[16]

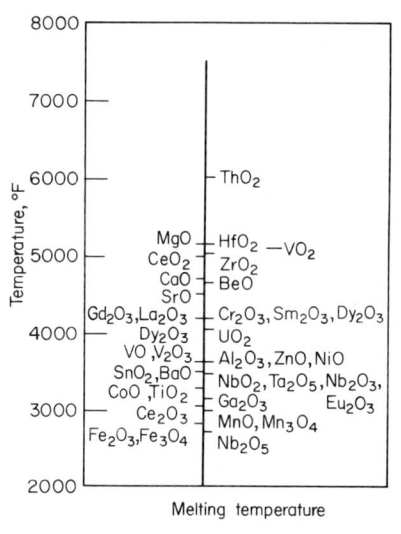

Fig. 1 Melting temperatures of typical oxide ceramics.[65]

Fig. 2 Relationship between resistivity and temperature of typical oxide ceramics.[66]

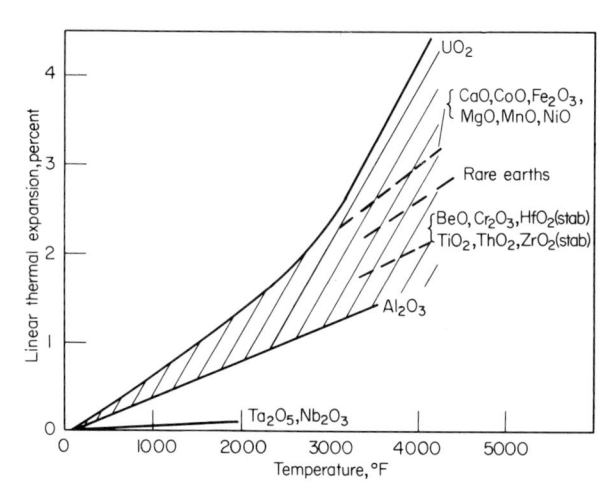

Fig. 3 Linear thermal expansion of typical oxide ceramics.[65]

The high-loss factor of porcelain is due to the large glass content and to the high mobility of the alkali ions. These ions also cause porcelain to have a relatively low resistivity (see Fig. 6.) The electrical properties can be improved by replacing the alkali ions with larger and less mobile alkaline-earth ions (Ca^{2+}, Mg^{2+}, Ba^{2+}) and by lowering the glass content. High-alumina porcelains have a dielectric constant nearly constant through the temperature range of most interest (−50 to +250°C) in electronic device application (see Fig. 7).

6-8 Ceramics, Glasses, and Micas

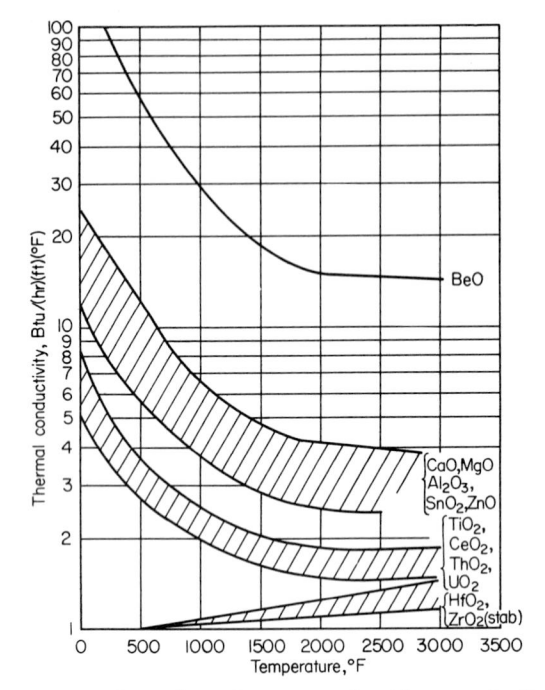

Fig. 4 Thermal conductivities of oxide ceramics.[65]

Steatite Steatite porcelains are low-loss materials that are widely used as components for variable capacitors, coil forms, electron tube sockets and general structural insulation (bushings, spacers, support bars, etc.). These bodies can be manufactured to close dimensional tolerances using automatic dry pressing and extrusion

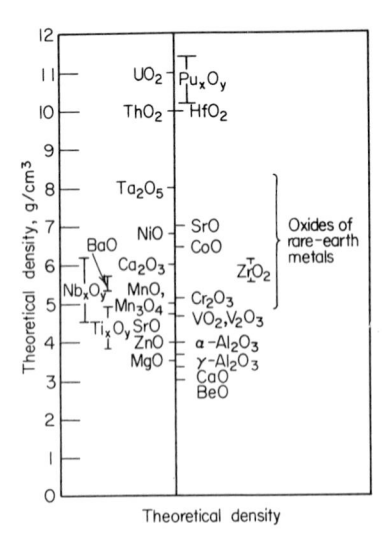

Fig. 5 Density of oxide ceramics.[65]

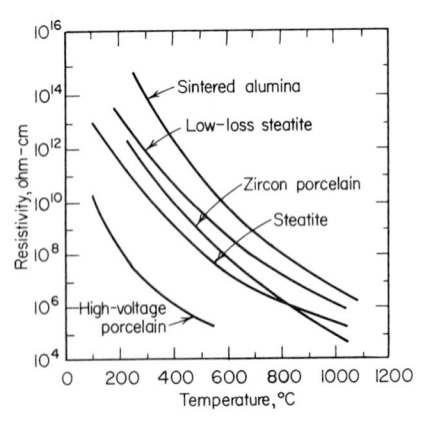

Fig. 6 Decrease in resistivity with temperature for some typical oxide ceramics.[68]

TABLE 2 Thermal Stress Resistance Factor R'' for Various Materials, Calculated from Published Values of C', α, σ, and E[67]

Material	C', cgs	$\alpha \times 10^{-6}$	σ, psi	E, psi $\times 10^{-6}$	$R'' = C'\sigma/\alpha E$
Beryllium oxide	0.53	9.0	10,000	40.0	14.7
Fused silica	0.004	0.5	15,500	10.9	11.4
Pyroceram 9605†	0.01	1.4	20,000*	20.0	7.0*
Sapphire (Linde)	0.065	6.7	35,000	50.0	6.8
Coors AI200	0.05	6.7	26,000	40.2	4.8
Zircon (Coors Z14)	0.009	3.36	11,500	19.4	1.6
Pyroceram 9606†	0.007	5.7	20,000*	18.0	1.36*
Aluminosilicate glass (Corning 1723)	0.005*	4.6	10,000	6–13	1.1*
Steatite (AlSiMag 228‡)	0.006	6.4	10,000	10*	0.94*
Forsterite (AlSiMag 243‡)	0.008	9.1	10,000	10*	0.88*
Soda-lime-silica glass	0.004	9.0	10,000	9.5	0.47
Fireclay	0.0027	5.5	750	2.3	0.16

* Estimated values.
† Trademark of Corning Glass Works, Corning, N.Y.
‡ Trademark of American Lava Corp, Ridgefield, N.J

methods. Unlike clay-flint-feldspar porcelains, steatite bodies require close control of the firing temperature, since the firing range is short to obtain vitrified materials. Commercial steatite compositions are based on 90 percent talc [$Mg_3Si_4O_{10}(OH)_2$] plus 10 percent clay. Feldspar additions greatly extend the firing range, but they degrade the electrical properties because of the introduction of alkali ions. Low-loss steatite compositions use additional magnesia to combine with the excess silica, and barium oxide as the fluxing agent. The fired body consists of enstatite ($MgSiO_3$) crystals bonded together by a glassy matrix.[16]

Steatite is characterized by a dielectric constant and a dissipation factor that increase with temperature at low frequencies, but are relatively independent of

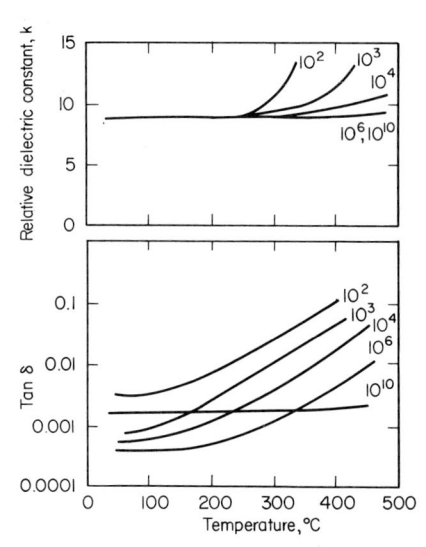

Fig. 7 Dielectric constant and tan δ for an alumina porcelain over a range of temperatures and frequencies.[69]

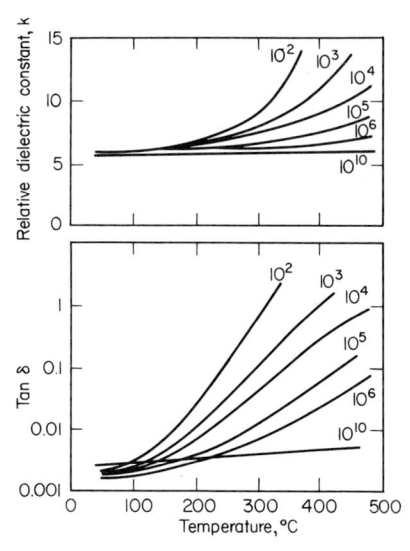

Fig. 8 Dielectric constant and tan δ for a steatite ceramic over a range of temperatures and frequencies.[69]

temperature at microwave frequencies (see Fig. 8). The physical properties of steatite are listed in Table 3.

Cordierite The low thermal expansion coefficient, and consequently the high thermal shock resistance, makes cordierite ($Mg_2Al_4Si_5O_{18}$) bodies useful for electric heater plates, resistor cores, thermocouple insulators, and burner nozzles. Like steatite, vitrified bodies are difficult to make because of the short firing range. When the intended use is for other than electrical applications, feldspar is added as the fluxing agent in order to increase the firing range. Typical physical properties are listed in Table 3.

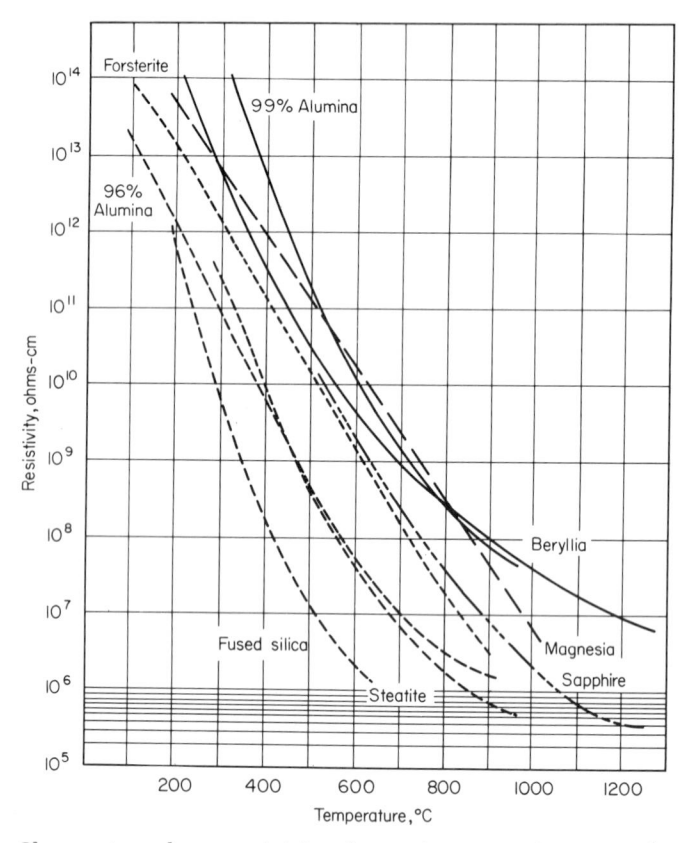

Fig. 9 Change in volume resistivity for various ceramics as a function of temperature.[71]

Forsterite In contrast to either steatite or cordierite, forsterite (Mg_2SiO_4) bodies present few firing problems. The absence of alkali ions in the vitreous phase give forsterite insulators a higher resistivity (see Fig. 9) and a lower electric loss with increasing temperature than steatite bodies. Since these low-loss dielectric properties persist at high frequencies, forsterite is used for small microwave tubes such as nuvistors.[17]

The high thermal expansion coefficient makes forsterite suitable for ceramic-to-metal seals, but it also causes the material to have poor thermal shock resistance. Table 3 lists the physical properties of forsterite.

Alumina The desirable electrical and mechanical properties possessed by alumina (Al_2O_3) make it suitable for a wide variety of applications. The high strength and refractoriness of alumina permit it to be used for spark-plug insulators, power resistor cores, and missile nose cones. Because of its high dielectric strength

(200 volts/mil) and resistivity (>10^{14} ohm-cm), alumina is used for bushings, insulators, and circuit-breaker components. The relatively low dielectric constant (see Fig. 7) and the low-loss factor over a wide frequency range make alumina suitable for radome housings, microwave windows, and electron-tube spacers and envelopes.

The dielectric constant and the loss tangent of alumina are affected by impurities, e.g., Si, Ti, Mg, and Ca.[18] Substitution of either Si^{4+} or Ti^{4+} for Al^{3+} in alumina creates donor levels at the impurity sites and acceptor levels at the compensating cation vacancies. Conversely, the substitution of either Mg^{2+} or Ca^{2+} for Al^{3+} creates acceptor levels at the impurity sites and donor levels at the compensating interstitials. These donors and acceptors contribute charge carriers which affect the dielectric and loss characteristics of alumina. Since the solubility of Si^{4+} in Al_2O_3 is limited, excess SiO_2 leads to the formation of a glassy phase at the grain boundaries, and to ionic conduction.

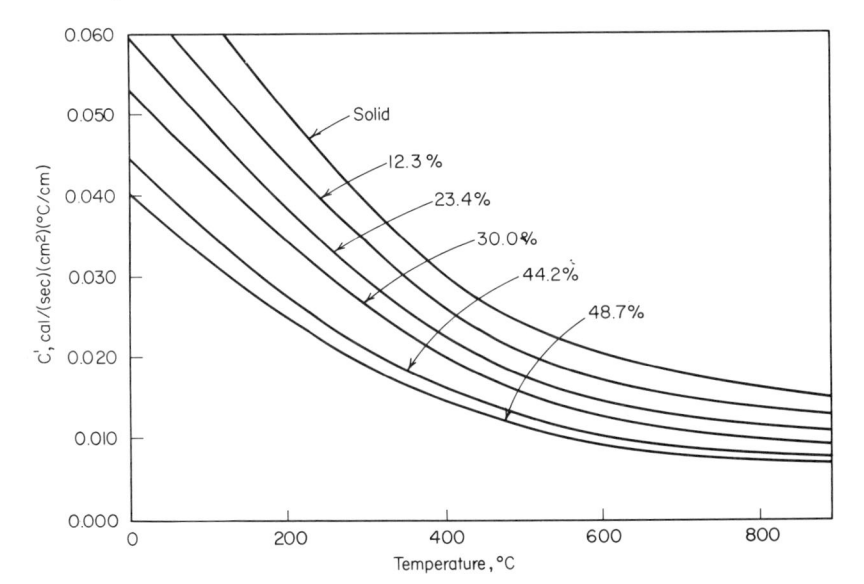

Fig. 10 Thermal conductivity of Al_2O_3 with various amounts of porosity.[19]

Ultra-low-loss alumina can be made by eliminating the glassy phase. This is accomplished by sintering pure, fine-grained powder at high temperatures (1800 to 1900°C) to produce a low-porosity body having very small grain size. The problem of residual porosity in Al_2O_3 ceramics affects both the thermal and the optical properties. Below a red heat, thermal conductivity decreases with increasing porosity, since the pores act as a thermal impedance (see Fig. 10).[19] The effect of even a small amount of porosity on the optical transmission of Al_2O_3 is drastic (see Fig. 11), and emphasizes the difficulty involved in making a translucent ceramic. High-density polycrystalline alumina bodies have been made with sufficient translucency to be used as envelopes for high-intensity lamps.[20]

Alumina ceramics have high elastic moduli (~50×10^6 psi) and high strengths (bend strength 20 to 40×10^3 psi), giving them the highest fracture strength among the refractory oxides. In common with other oxide ceramics, the strength begins to decrease rapidly above 1000°C.

Alumina is stable in air, vacuum, water vapor, hydrogen, carbon monoxide, nitrogen, and argon at temperatures up to 1700°C. Hydrogen fluoride will react with alumina. At high temperatures (~1700°C) alumina will vaporize as Al_2O in the presence of either water vapor or reducing atmospheres.[21] A compilation of the electrical and physical properties of alumina is listed in Table 4.

Beryllia BeO offers the unusual combination of high thermal and high electric

TABLE 3 Typical Physical Properties of Ceramic Dielectrics[70]

	Vitrified products						
	1	2	3	4	5	6	7
Material	High-voltage porcelain	Alumina porcelain	Steatite	Forsterite	Zircon porcelain	Lithia porcelain	Titania, titanate ceramics
Typical applications	Power-line insulation	Spark-plug cores, thermocouple insulation, protection tubes	High-frequency insulation, electrical appliance insulation	High-frequency insulation, ceramic-to-metal seals	Spark-plug cores, high-voltage high-temperature insulation	Temperature-stable inductances, heat-resistant insulation	Ceramic capacitors, piezoelectric ceramics
Specific gravity, g/cm³	2.3–2.5	3.1–3.9	2.5–2.7	2.7–2.9	3.5–3.8	2.34	3.5–5.5
Water absorption, %	0.0	0.0	0.0	0.0	0.0	0.0	0.0
Coefficient of linear thermal expansion, at 20–700°C, 10^{-6} in./(in.)(°C)	5.0–6.8	5.5–8.1	8.6–10.5	11	3.5–5.5	1	7.0–10.0
Safe operating temperature, °C	1,000	1,350–1,500	1,000–1,100	1,000–1,100	1,000–1,200	1,000	
Thermal conductivity, (cal/cm²)/(cm)(sec)(°C)	0.002–0.005	0.007–0.05	0.005–0.006	0.005–0.010	0.010–0.015		0.008–0.01
Tensile strength (psi)	3,000–8,000	8,000–30,000	8,000–10,000	8,000–10,000	10,000–15,000		4,000–10,000
Compressive strength, psi	25,000–50,000	80,000–250,000	65,000–130,000	60,000–150,000	80,000–150,000	60,000	40,000–120,000
Flexural strength, psi	9,000–15,000	20,000–45,000	16,000–24,000	18,000–20,000	20,000–35,000	8,000	10,000–22,000
Impact strength (½-in. rod), ft-lb	0.2–0.3	0.5–0.7	0.3–0.4	0.03–0.04	0.4–0.5	0.3	0.3–0.5
Modulus of elasticity, psi × 10^{-6}	7–14	15–52	13–15	13–15	20–30		10–15
Thermal shock resistance	Moderately good	Excellent	Moderate	Poor	Good	Excellent	Poor
Dielectric strength (¼-in.-thick specimen), volts/mil	250–400	250–400	200–350	200–300	250–350	200–300	50–300
Resistivity at room temperature, ohm/cm³	10^{12}–10^{14}	10^{14}–10^{15}	10^{13}–10^{15}	10^{13}–10^{15}	10^{13}–10^{15}		10^{8}–10^{15}
Te value, °C	200–500	500–800	450–1,000	above 1,000	700–900		200–400
Power factor at 1 MHz	0.006–0.010	0.001–0.002	0.0008–0.0035	0.0003	0.0006–0.0020	0.05	0.0002–0.050
Dielectric constant	6.0–7.0	8–9	5.5–7.5	6.2	8.0–9.0	5.6	15–10,000
L grade (JAN Spec. T-10)	L-2	L-2–L-5	L-3–L-5	L-6	L-4	L-3	

Semivitreous and refractory products

Material	8	9	10	11
	Low-voltage porcelain	Cordierite refractories	Alumina, aluminum silicate refractories	Massive fired talc, pyrophyllite
Typical applications	Switch bases, low-voltage wire holders, Light receptacles	Resistor supports, burner tips, heat insulation, arc chambers	Vacuum spacers, high-temperature insulation	High-frequency insulation, vacuum-tube spacers, ceramic models
Specific gravity, g/cm³	2.2–2.4	1.6–2.1	2.2–2.4	2.3–2.8
Water absorption (%)	0.5–2.0	5.0–15.0	10.0–20.0	1.0–3.0
Coefficient of linear thermal expansion, at 20–700°C, 10⁻⁶ in./(in.)(°C)	5.0–6.5	2.5–3.0	5.0–7.0	11.5
Safe operating temperature, (°C)	900	1,250	1,300–1,700	1,200
Thermal conductivity, (cal/cm²)(cm)(sec)(°C)	0.004–0.005	0.003–0.004	0.004–0.005	0.003–0.005
Tensile strength, psi	1,500–2,500	1,000–3,000	700–3,000	2,500
Compressive strength (psi)	25,000–50,000	20,000–45,000	15,000–60,000	20,000–30,000
Flexural strength, psi	3,500–6,000	1,500–7,000	1,500–6,000	7,000–9,000
Impact strength (½-in. rod), ft-lb	0.2–0.3	0.2–0.25	0.17–0.25	0.2–0.3
Modulus of elasticity, psi × 10⁻⁶	7–10	2–5	2–5	4–5
Thermal shock resistance	Moderate	Excellent	Excellent	Good
Dielectric strength (¼-in.-thick specimen), volts/mil	40–100	40–100	40–100	80–100
Resistivity at room temperature, ohm/cm³	10¹²–10¹⁴	10¹²–10¹⁴	10¹²–10¹⁴	10¹²–10¹⁵
Te value, °C	300–400	400–700	400–700	600–900
Power factor at 1 MHz	0.010–0.020	0.004–0.010	0.0002–0.010	0.0008–0.016
Dielectric constant	6.0–7.0	4.5–5.5	4.5–6.5	5.0–6.0
L grade (JAN Spec. T-10)				

TABLE 4 Mechanical, Thermal, and Electrical Properties of Coors Alumina and Beryllia Ceramics

Property*	Test	(1) AD-85		(2) AD-90		(3) AD-94		(4) AD-96	
Specific gravity:									
Typical	ASTM C 20-46	3.42		3.58		3.62		3.72	
Minimum		3.37		3.53		3.57		3.67	
Hardness (typical), Rockwell 45N	ASTM E 1867	75		79		78		78	
Surface finish, μin. (arithmetic avg.):									
Typical, as fired	Profilometer	65		65		65		65	
Typical, ground	(0.030-in. cutoff)	45		40		50		50	
Ultimate, lapped		13		3		10		10	
Crystal size, microns:									
Range		2–12		2–10		2–25		2–20	
Average		7		4		12		11	
Water absorption	ASTM C 373-56	None		None		None		None	
Gas permeability†		None		None		None		None	
Color		White		White		White		White	
Compressive strength (typical), psi × 10⁻³:									
At 25°C	ASTM C 528-63T	280		360		305		300	
At 1000°C		—		75		50		—	
Flexural strength, psi × 10⁻³:									
At 25°C:									
Typical	ASTM C 369-56	43		49		51		52	
Minimum†	(¼-in.-diam.	39		44		46		47	
At 1000°C:	rods)								
Typical		25		—		20		25	
Minimum‡		20		—		17		20	
Tensile strength (typical), psi × 10⁻³:	Brazil test	22		32		28		28	
At 25°C									
At 1000°C									
Modulus of elasticity, psi × 10⁻⁶		33		15		15		14	
Shear modulus, psi × 10⁻⁶	Sonic method	13		39		41		44	
Bulk modulus, psi × 10⁻⁶		20		16		17		18	
Sonic velocity, 10³ m/sec		8.2		23		24		25	
Poisson's ratio		0.22		8.7		8.9		9.1	
Maximum use temperature, °C(°F)		1400 (2550)		0.22		0.21		0.21	
				1500 (2725)		1700 (3100)		1700 (3100)	

		°C	°F	°C	°F	°C	°F	°C	°F
Thermal coefficient of linear expansion, 10⁻⁶ cm/(cm)/(°C):	ASTM C 372-56								
−200 to 25°C		3.4	1.3	3.4	1.9	3.4	1.9	3.4	1.9
25 to 200°C		5.3	3.0	6.1	3.4	6.3	3.5	6.0	3.3
25 to 500°C		6.2	3.5	7.0	3.9	7.1	4.2	7.4	4.1
25 to 800°C		6.9	3.8	7.7	4.3	7.6	4.2	8.0	4.4
25 to 1000°C		7.2	4.0	8.1	4.5	7.9	4.4	8.2	4.6
25 to 1200°C		7.5	4.2	8.4	4.7	8.1	4.5	8.4	4.7

Property	ASTM	Mat. 1	Mat. 2	Mat. 3	Mat. 4
Thermal conductivity, cal/(sec)(cm²) (°C/cm):	C 408-58				
20°C		0.035	0.040	0.043	0.043
100°C		0.029	0.032	0.035	0.035
400°C		0.016	0.017	0.017	0.017
800°C		0.010	0.010	0.010	0.010
Specific heat at 100°C, cal/(g)(°C)	C 351-61	0.22	0.22	0.21	0.21
Dielectric strength, volts/mil (avg. rms values). Specimen thickness, in.:	D 116-69	AC / DC	AC / DC	AC / DC	AC / DC
0.250		240 / —	235 / —	220 / —	210 / —
0.125		340 / 940	320 / 920	300 / 1050	275 / 740
0.050		440 / 1250	450 / 1100	425 / 1100	370 / 840
0.025		550 / 1550	580 / 1300	550 / 1140	450 / 900
0.010		720 / 1750	760 / 1600	720 / 1150	580 / 980
Dielectric constant:	D 150-65T	25°C / 500°C	25°C	25°C / 500°C / 800°C	25°C / 500°C / 800°C
1 kHz		8.2 / 13.9	8.8	8.9 / 11.8 / —	9.0 / 10.8 / —
1 MHz		8.2 / 8.9	8.8	8.9 / 9.7 / —	9.0 / 9.6 / —
100 MHz		8.2 / —	8.8	8.9 / — / —	9.0 / — / —
1 GHz		8.2 / 8.3	8.7	8.9 / 9.1 / 9.4	8.9 / 9.4 / 9.9
10 GHz		— / —	—	8.7 / — / —	8.7 / — / —
50 GHz		—	—	—	—
Dissipation factor:	D 150-65T				
1 kHz		0.0014 / 0.580	0.0006	0.0002 / 0.215 / —	0.0011 / 0.200 / —
1 MHz		0.0009 / 0.024	0.0004	0.0001 / 0.008 / —	0.0001 / 0.0039 / —
100 MHz		0.0009 / —	0.0004	0.0005 / — / —	0.0002 / — / —
1 GHz		0.0014 / —	—	0.0008 / 0.002 / 0.004	0.0001 / 0.0009 / —
10 GHz		0.0019 / 0.003	0.0009	0.0010 / 0.018 / 0.038	0.0006 / — / 0.0028
50 GHz		—	—	0.0021 / — / —	0.0068 / — / —
Loss index:	ASTM #??				
1 kHz		0.011 / 8.06	0.005	0.002 / 2.54 / —	0.010 / 2.16 / —
1 MHz		0.007 / 0.214	0.004	0.001 / 0.078 / —	0.001 / 0.037 / —
100 MHz		0.007 / —	0.004	0.004 / — / —	0.002 / — / —
1 GHz		0.011 / —	—	0.007 / 0.018 / 0.028	0.001 / 0.008 / —
10 GHz		0.016 / 0.025	0.008	0.009 / — / 0.038	0.005 / — / 0.028
50 GHz		—	—	0.018 / — / —	0.059 / — / —
Volume resistivity, ohm/cm²/cm:	D 1829-66				
25°C		>10^{14}	>10^{14}	>10^{14}	>10^{14}
300°C		4.6×10^{10}	1.4×10^{11}	9.0×10^{11}	3.1×10^{11}
500°C		4.0×10^{8}	2.8×10^{8}	2.5×10^{9}	4.0×10^{9}
700°C		7.0×10^{6}	7.0×10^{6}	5.0×10^{7}	1.0×10^{8}
1000°C		—	8.6×10^{6}	5.0×10^{5}	1.0×10^{6}
Te value, °C		850	960	950	1000

* Footnotes are on page 6-20.

TABLE 4 Mechanical, Thermal, and Electrical Properties of Coors Alumina and Beryllia Ceramics *(Continued)*

Property*	Test	(5) AD-99		(6) AD-995		(7) AD-998	
Specific gravity:							
Typical	ASTM C 20-46	3.83		3.84		3.82	
Minimum		3.78		3.80		3.78	
Hardness, typical Rockwell 45N	ASTM E 1867	80		81		79	
Surface finish, μin. (arithmetic avg.):	Profilometer (0.030-in. cutoff)						
Typical, as fired		55		55		—	
Typical, ground		35		35		—	
Ultimate, lapped		3		3		—	
Crystal size, microns:							
Range		5–50		10–50		10–35	
Average		22		20		12	
Water absorption	ASTM C 373-56	None		None		None	
Gas permeability†		None		None		None	
Color		White		Pink		Ivory	
Compressive strength (typical), psi × 10⁻³:							
At 25°C	ASTM C 528-63T	345		330		320	
At 1000°C		130		140		—	
Flexural strength, psi × 10⁻³:							
At 25°C:							
Typical	ASTM C 369-56 (½-in.-diam. rods)	48		45		48	
Minimum‡		45		40		43	
At 1000°C:							
Typical		30		33		28	
Minimum‡		25		28		—	
Tensile strength (typical), psi × 10⁻³:	Brazil test						
At 25°C		31		28		—	
At 1000°C		12		15		—	
Modulus of elasticity, psi × 10⁻⁶	Sonic method	50		52		50	
Shear modulus, psi × 10⁻⁶		21		22		21	
Bulk modulus, psi × 10⁻⁶		29		30		—	
Sonic velocity, 10³ m/sec		9.4		9.7		9.4	
Poisson's ratio		0.21		0.21		0.21	
Maximum use temperature, °C(°F)		1725 (3140)		1750 (3180)		1950 (3540)	
		°C	°F	°C	°F	°C	°F
Thermal coefficient of linear expansion, 10⁻⁶ cm/(cm)(°C):	ASTM C 372-56						
−200 to 25°C		3.4	1.9	3.4	1.9	3.4	1.9
25 to 200°C		6.3	3.5	6.3	3.5	6.7	3.7
25 to 500°C		7.3	4.0	7.3	4.1	7.3	4.1
25 to 800°C		7.9	4.4	7.8	4.4	7.8	4.4
25 to 1000°C		8.2	4.6	8.1	4.5	8.0	4.5
25 to 1200°C		8.4	4.7	8.3	4.6	8.3	4.6

Property	ASTM	Material 1						Material 2						Material 3
		25°C AC	25°C DC	500°C AC	500°C DC	800°C AC	800°C DC	25°C AC	25°C DC	500°C AC	500°C DC	800°C AC	800°C DC	
Thermal conductivity, cal/(sec)(cm²)(°C/cm):														
20°C	ASTM C 408-58	0.070						0.075						0.070
100°C		0.055						0.065						0.055
400°C		0.025						0.028						0.025
800°C		0.015						0.017						0.015
Specific heat at 100°C, cal/(g)(°C)	ASTM C 351-61	0.21						0.21						0.21
Dielectric strength, volts/mil (avg. rms values): Specimen thickness, in.:														
0.250	ASTM D 116-69	215	—					225	—					
0.125		290	800					310	840					
0.050		390	900					450	1050					
0.025		480	980					550	1150					
0.010		600	1100					625	1300					
Dielectric constant:														
1 kHz	ASTM D 150-65T	9.4		11.3	—	—	—	9.4		10.3	—	—	—	
1 MHz		9.4		10.0	—	—	—	9.4		—	—	—	—	
100 MHz		9.4		10.0	10.5	—	—	9.4		10.4	11.0	—	—	
1 GHz		9.4		10.0	10.4	—	—	9.4		9.8	10.2	—	—	
10 GHz		—		—	—	—	—	—		—	—	—	—	
50 GHz		—		—	—	—	—	—		—	—	—	—	
Dissipation factor:														
1 kHz		0.0042		0.1500	—	—	—	0.0004		—	—	—	—	
1 MHz		0.0002		0.0047	—	—	—	0.0001		0.0023	—	—	—	
100 MHz	ASTM D 150-65T	0.0002		0.0003	0.0005	—	—	0.0001		0.0002	0.0003	—	—	
1 GHz		0.0002		0.0002	0.0006	—	—	0.0001		0.0003	0.0006	—	—	
10 GHz		—		—	—	—	—	—		—	—	—	—	
50 GHz		—		—	—	—	—	—		—	—	—	—	
Loss index:														
1 kHz		0.040		1.70	—	—	—	0.004		0.024	—	—	—	
1 MHz	ASTM #??	0.002		0.047	—	—	—	0.001		—	—	—	—	
100 MHz		0.002		0.003	0.005	—	—	0.001		0.002	0.003	—	—	
1 GHz		0.002		0.002	0.006	—	—	0.001		0.003	0.006	—	—	
10 GHz		—		—	—	—	—	—		—	—	—	—	
50 GHz		—		—	—	—	—	—		—	—	—	—	
Volume resistivity, ohm/cm²/cm:														
25°C	ASTM D 1829-66	$>10^{14}$						$>10^{14}$						$>10^{14}$
300°C		1.0×10^{13}						1.5×10^{11}						$>10^{13}$
500°C		6.3×10^{10}						1.4×10^{7}						6.3×10^{11}
700°C		5.0×10^{8}						4.0×10^{7}						3.4×10^{9}
1000°C		2.0×10^{6}						8.0×10^{5}						7.8×10^{6}
Te value, °C		1050						980						1140

* Footnotes are on page 6-20.

6-17

TABLE 4 Mechanical Thermal, and Electrical Properties of Coors Alumina and Beryllia Ceramics (Continued)

Property*	Test	(8) AD-999	(9) Vistal	(10) BD-995-2
Specific gravity:				
Typical	} ASTM C 20-46	3.96	3.99	2.90
Minimum		3.94	3.98	2.86
Hardness (typical), Rockwell 45N	ASTM E 1867	90	85	67
Surface finish, μin. (arithmetic avg.):				
Typical, as fired	} Profilometer	20	25	22
Typical, ground	} (0.030-in. cutoff)	35	35	20
Ultimate, lapped		<1	<1	—
Crystal size, microns:				
Range		1–6	50–45	10–40
Average		3	20	24
Water absorption	} ASTM C 373-56	None	None	None
Gas permeability†		None	None	None
Color		Ivory	Translucent white	White
Compressive strength (typical), psi × 10⁻³:				
At 25°C	} ASTM C 528-63T	550	370	310
At 1000°C		280	70	40
Flexural strength, psi × 10⁻³:				
At 25°C:				
Typical	} ASTM C 369-56	95	41	40
Minimum‡	} (½-in.-diam.	89	—	35
At 1000°C:	} rods)			
Typical		70	25	—
Minimum‡		65		—
Tensile strength (typical), psi × 10⁻³:				
At 25°C	} Brazil test	48	30	20
At 1000°C		32	15	5
Modulus of elasticity, psi × 10⁻⁶		56	57	51
Shear modulus, psi × 10⁻⁶		23	23.5	20
Bulk modulus, psi × 10⁻⁶				—
Sonic velocity, 10³ m/sec	} Sonic method	9.9	9.9	11.1
Poisson's ratio		0.22	0.22	0.26
Maximum use temperature, °C(°F)		1900 (3450)	1900 (3450)	1850 (3360)

Thermal coefficient of linear expansion, 10⁻⁶ cm/(cm)(°C):		°C	°F	°C	°F	°C	°F
−200 to 25°C	} ASTM C 372-56	3.6	2.0	3.4	1.9	2.4	1.3
25 to 200°C		6.5	3.6	6.5	3.6	6.4	3.6
25 to 500°C		7.4	4.1	7.4	4.1	7.7	4.3
25 to 800°C		7.8	4.3	7.8	4.3	8.5	4.7
25 to 1000°C		8.0	4.5	8.0	4.5	8.8	4.9
25 to 1200°C		8.3	4.6	8.3	4.6	9.4	5.2

		25°C	500°C	800°C		25°C			25°C
							AC	DC	
Thermal conductivity, cal/(sec)(cm²)(°C/cm):									
20°C	⎫	0.074				0.095		0.67	
100°C	⎬ ASTM C 408-58	0.055				0.070		0.48	
400°C	⎭	0.031				0.030		0.20	
800°C						—		0.07	
Specific heat at 100°C, cal/(g)(°C)	ASTM C 351-61	0.21				0.21		0.31	
Dielectric strength, volts/mil (avg. rms values):		AC	DC			AC		AC	DC
Specimen thickness, in.:									
0.250	⎫	240	—			230		260	—
0.125	⎪	325	920			340		340	830
0.050	⎬ ASTM D 116-69	460	1050			510		490	—
0.025	⎪	590	1200			650		610	—
0.010	⎭	800	1450			—		800	—
Dielectric constant:		25°C	500°C	800°C		25°C		25°C	
1 kHz	⎫	9.9	—	—		10.1		6.8	
1 MHz	⎪	9.8	—	—		10.1		6.8	
100 MHz	⎬ ASTM D 150-65T	—	—	—		10.1		6.8	
1 GHz	⎪	—	—	—		—		6.8	
10 GHz	⎪	9.8	10.3	10.7		10.1		6.7	
50 GHz	⎭	—	—	—		—		—	
Dissipation factor:									
1 kHz	⎫	0.002	—	—		0.0005		0.001	
1 MHz	⎪	0.0002	—	—		0.00004		0.0003	
100 MHz	⎬ ASTM D 150-65T	—	—	—		0.00006		0.0006	
1 GHz	⎪	—	—	—		—		0.0006	
10 GHz	⎪	0.00006	0.0003	0.0008		0.00009		0.0003	
50 GHz	⎭	—	—	—		—		—	
Loss index:									
1 kHz	⎫	0.020	—	—		0.005		0.007	
1 MHz	⎪	0.002	—	—		0.0004		0.002	
100 MHz	⎬ ASTM #??	—	—	—		0.0006		0.004	
1 GHz	⎪	—	—	—		—		0.004	
10 GHz	⎪	0.0006	0.003	0.009		0.001		0.002	
50 GHz	⎭	—	—	—		—		—	
Volume resistivity, ohm/cm²/cm:									
25°C	⎫	$> 10^{16}$				—		$> 10^{17}$	
300°C	⎪	1.0×10^{15}				—		$> 10^{15}$	
500°C	⎬ ASTM D 1829-66	3.3×10^{12}				—		5.0×10^{13}	
700°C	⎪	9.0×10^{9}				—		1.5×10^{10}	
1000°C	⎭	1.1×10^{7}				—		7.0×10^{7}	
Te value, °C		1170				—		1240	

* Footnotes are on page 6-20.

Footnotes to Table 4

SOURCE: Coors Porcelain Company.

* Ceramic property values vary somewhat according to method of manufacture, size, and shape of part. Closer control of values is possible.

† No helium leak through a 1-in.-diameter plate, 0.001 in. thick, measured at 3×10^{-7} torr vacuum versus approximately 1 atm helium pressure for 15 sec at room temperature.

‡ Minimum flexural strength is a minimum mean for a sample of 10 specimens.

(1) AD-85: Nominally 85% Al_2O_3. A good all-around high-alumina ceramic for both electrical and mechanical applications.
(2) AD-90: Nominally 90% Al_2O_3. A tough, fine-grained alumina ceramic especially well suited for demanding mechanical applications.
(3) AD-94: Nominally 94% Al_2O_3. A very good alumina ceramic for metallizing; ideal for all but the most critical electrical and mechanical applications.
(4) AD-96: Nominally 96% Al_2O_3. An excellent alumina ceramic for special electronic applications and many mechanical applications.
(5) AD-99: Nominally 99% Al_2O_3. A very strong, impervious alumina ceramic designed for virtually all kinds of critical electrical and mechanical applications.
(6) AD-995: Nominally 99.5% Al_2O_3. An extremely low-loss alumina ceramic used widely in many electronic applications and some mechanical applications.
(7) AD-999: Nominally 99.9% Al_2O_3. The hardest, strongest, purest alumina ceramic available; recommended for use in ultrasevere mechanical applications and/or highly hostile environments.
(8) Vistal: Registered trademark. Nominally 99.9% Al_2O_3. A translucent, high-purity alumina ceramic for highly critical electrical and electronic applications; strong, excellent resistance to chemical attack.
(9) BD-995-2: Nominally 99.5% BeO. A much-improved beryllia ceramic possessing low dielectric loss, high electrical resistivity, good dielectric strength; meant for use where high thermal conductivity is required.

GENERAL NOTES:

All measurements are typical for the materials shown. For some specific applications it may be necessary to measure values for use in design formulas. Dielectric constant specifically can be controlled.

Dashes (—) and blank spaces indicate values not measured at this time. All values shown are measured at room temperature unless otherwise specified. All data are typical unless otherwise specified.

Composition Control: Alumina and beryllia contents of Coors ceramics are controlled using chemical, spectrographic, and x-ray fluorescent methods for quantitative determination of minor ingredients.

Chemical Resistance: Coors sintered alumina and beryllia ceramics are highly resistant to chemical attack and corrosion. For optimum material selection, it is recommended that specific data on chemical resistance be obtained for particular applications.

conductivity. At room temperature, the thermal conductivity is about one-half that of copper, but with increasing temperature the conductivity of beryllia decreases much more rapidly than that of copper (see Fig. 12). The high strength and the high thermal conductivity gives BeO good thermal shock resistance.

Beryllia is stable in air, vacuum, hydrogen, carbon monoxide, argon, and nitrogen at temperatures up to 1700°C.[22] Of all the oxides in contact with graphite at high temperatures, BeO is the most stable, and it is resistant to corrosion by liquid

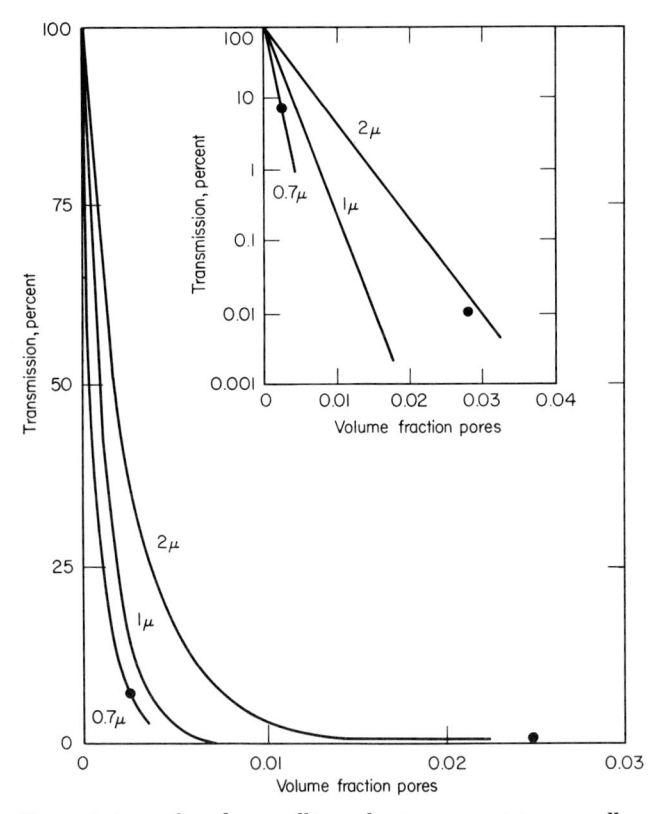

Fig. 11 Transmission of polycrystalline alumina containing small amounts of residual porosity.[72]

alkali metals. Beryllia will react with water vapor above 1650°C to form volatile Be(OH)$_2$, and it will decompose in atmospheres containing halogens or sulfur.

BeO has a lower dielectric constant than alumina, but one with about the same temperature dependence (see Fig. 13). The temperature dependence of the resistivity of BeO is again similar to that of Al$_2$O$_3$ (see Fig. 9).

The high cost of BeO has limited its use to critical applications. It is employed for structural and insulating members in electron and traveling-wave tubes, microwave windows in high-power klystrons, and heat sinks for transistors.[23]

In massive form, BeO is not dangerous; however, in the form of powder or dust, beryllia is hazardous to health.[24]

Magnesia MgO is a better electrical insulator than Al$_2$O$_3$, particularly at high temperatures. Because of its high thermal expansion (see Fig. 3) and low strength, it has poor thermal shock resistance. Although MgO shows little tendency to hydrate in large masses, it will hydrate in powdered form. The combination of high thermal and low electric conductivity makes MgO suitable for insulating thermocouple leads and for heating-core elements.

Zirconia Dense, crack-free zirconia (ZrO_2) ceramics are difficult to produce because of the disruptive volume change (~ 10 percent) that occurs during the tetragonal to monoclinic phase transformation at 1000°C. This disruptive phase change is eliminated by stabilizing the cubic form of zirconia with solid solutions of CaO, Y_2O_3, Yb_2O_3, Nd_2O_3, or Sc_2O_3. Stabilized ZrO_2 has relatively poor thermal resistance because of its high thermal expansion coefficient [$\sim 11 \times 10^{-6}$ cm/(cm)(°C), about 1½ times that of alumina] and its relatively low thermal conductivity [0.004 cal/(sec)(cm^2)(°C/cm), about one-fourth that of alumina].

The electrical resistivity of yttria- and calcia-stabilized ZrO_2 decreases with temperature (see Fig. 14), primarily owing to the increase in the oxygen ion diffusion rate.[25] This property has led to use of stabilized ZrO_2 for the electrolyte in high-tem-

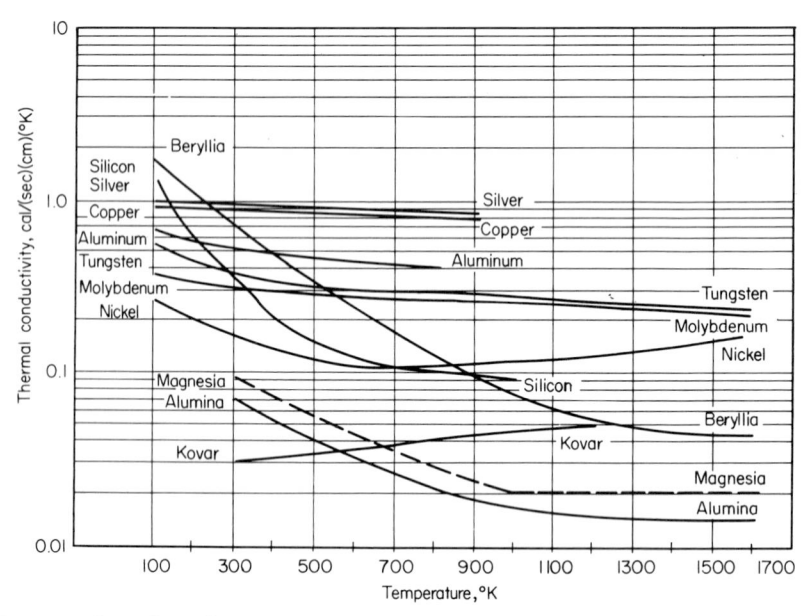

Fig. 12 Thermal conductivity as a function of temperature for various materials.[71]

perature fuel cells.[26] At temperatures between 400 and 1000°C, oxygen ions diffuse through the ceramic electrolyte, whereas gases such as hydrogen, nitrogen, carbon dioxide, carbon monoxide, and water vapor cannot penetrate the zirconia electrolyte. Electronic conduction is negligible under these conditions. A sensitivie oxygen detector system operating in the range of 10^{-4} to 10^{+3} torrs and using a stabilized zirconia electrolyte is shown in Fig. 15. When the electrolyte is maintained at 850°C, the electric output as a function of pressure follows the relation, emf = 0.056 log (0.2/p), where p in torrs is the oxygen pressure to be measured. Sintered zirconia ceramic heating elements containing up to 15 mole percent MgO have been operated as high as 2000°C. The resistivity of this ceramic decreases from 10^4 ohm-cm at 600°C to 6 ohm-cm at 1200°C.[27]

Carbides Compounds in this group are among the most refractory materials known (see Fig. 16). Most carbides are easily oxidized; therefore, they require protective atmospheres at high temperatures (>1000°C). Silicon carbide, on the contrary, has excellent oxidation resistance because of the formation of a protective surface layer of silica. Maximum reported working temperatures for SiC are approximately 1700°C in oxidizing atmospheres and 2200°C in neutral atmospheres.[28]

Silicon Carbide. SiC heating elements are made by pressing rods of granular

Insulators   6-23

SiC with a temporary binder and firing at 2000 to 2500°C. After sintering, the bulk resistance of the body is essentially the same as that of the individual grains. A typical resistivity of these rods at room temperature is 0.2 ohm-cm; this decreases to 0.1 ohm-cm at 1000°C (see Fig. 17). In the low-temperature region (0 to 750°C), the number of conducting electrons increases progressively with increasing temperature as more and more electrons are released by thermal excitation from impurity centers. Between 750 and 1500°C, thermal vibration decreases the

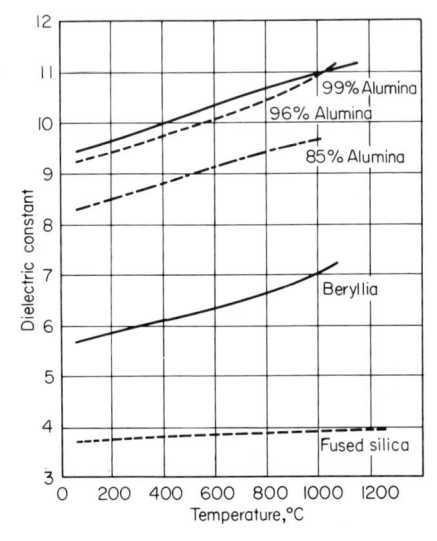

Fig. 13 Dielectric constant as a function of temperature measured at a test frequency of 4 GHz.[71]

Fig. 14 Temperature-dependent resistivities of zirconia-calcia solutions.[73]

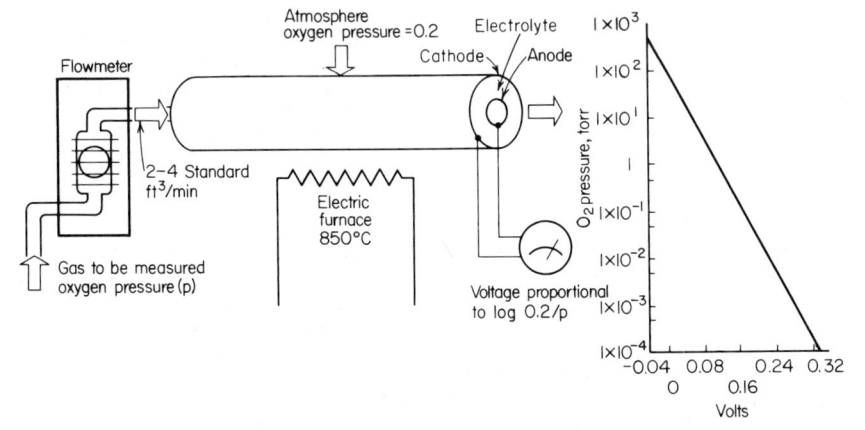

Fig. 15 Oxygen detector using a stabilized zirconia electrolyte.[74] To detect oxygen, gas to be measured is passed through heated fuel cell. Voltage generated by cell is displayed on the voltmeter.

6-24 Ceramics, Glasses, and Micas

mobility of the conduction electrons, which in turn causes higher resistance. Above 1500°C, intrinsic conduction becomes significant, and the resistivity falls abruptly with increasing temperature.

Nonlinear Silicon Carbide. Nonlinear electrical grade SiC is used for lightning arrestors and in voltage-limiting resistor applications. Practically the entire voltage drop in SiC occurs at the interface between the grains, and it is described by the relation. $V = KI^n$, where K is a constant depending on the geometry of the body, and n is a constant determined by the manufacturing process. The exponent n generally varies from 0.1 to 0.35. The high resistance of nonlinear SiC is attributed to a grain-boundary barrier, which has been given several explanations. One is that the barrier is a layer of silica or silaceous material; another one is that the barrier results from the distribution of space charges in the SiC lattice.[29] The typical relationship between voltage and current in SiC (see Fig. 18) shows that an ohmic or linear relationship exists at both very low and very high voltages. In the intermediate voltage range, the V-I relationship is nonlinear.

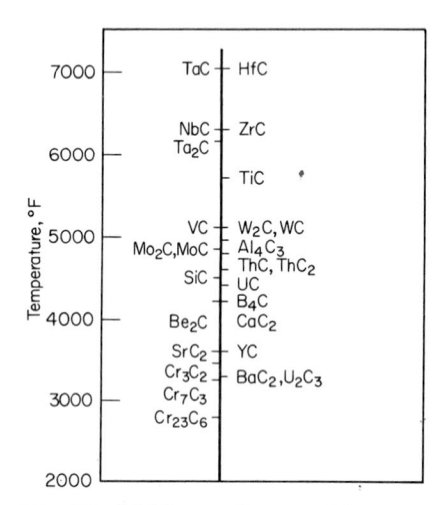

Fig. 16 Melting or decomposition temperature of various carbides.[65]

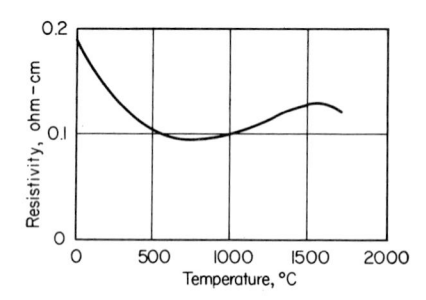

Fig. 17 Resistivity as a function of temperature for a typical SiC heating element.[29]

Nitrides Nitride ceramics have both low thermal expansions and low thermal conductivites (see Figs. 19 and 20). Several nitride ceramics have high melting points (see Fig. 21); however, they are not oxidation-resistant and are not recommended for use above 1100°C.

Boron Nitride. Massive boron nitride (BN) is made by hot pressing at 2000°C and 1,000 psi. It can be easily machined by conventional edge-cutting methods. Unreacted B_2O_3 will hydrate during exposure to moist air, and it may cause BN to exhibit erratic electrical behavior and spalling during rapid heating. These difficulties can be avoided by thoroughly dehydrating BN at 350°C.

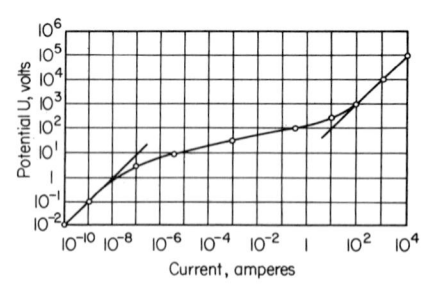

Fig. 18 Voltage-current characteristic of nonlinear SiC.[75]

The dielectric constant and loss tangent of hot-pressed BN are relatively low at mircowave frequencies (10^9 Hz) and fairly independent of temperature between 25°C and 500°C (see Figs. 22 and 23). By careful control of the microstructure, the anistropy in the dielectric constant can be as little as 2 percent. BN has good arc-tracking characteristics,[30] although its electrical resistance is two decades lower than that of alumina (see Fig. 24). Hot-

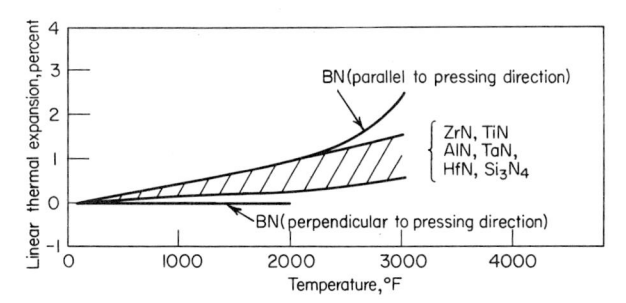

Fig. 19 Linear thermal expansion of various nitrides.[65]

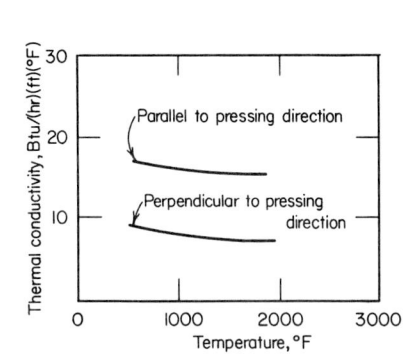

Fig. 20 Thermal conductivity of boron nitride.[65]

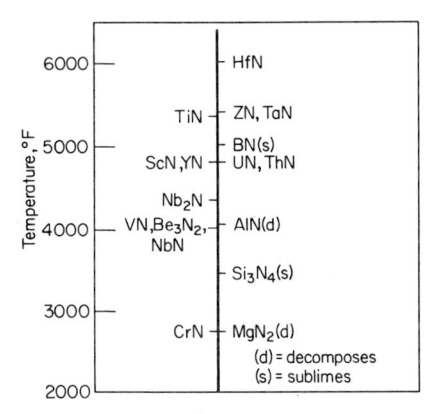

Fig. 21 Melting or decomposition temperature of various nitrides.[65]

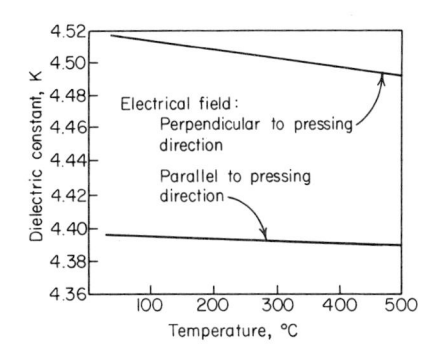

Fig. 22 Dielectric constant of hot-pressed BN over a range of temperatures—test frequency 8.5×10^9 Hz.[76]

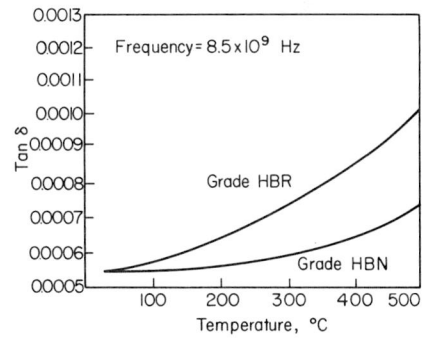

Fig. 23 Loss tangent of hot-pressed BN over a range of temperatures—test frequency 8.5×10^9 Hz.[76]

pressed BN is used for high-temperature insulation in plasma-arc heaters, refractories for magnetohydrodynamic generators and electric furnaces, heat sinks, and high-temperature capacitors.

Pyrolytic BN in the form of a pure, dense, anistropic film can be deposited at high temperatures from a mixture of ammonia and boron halide. It has a higher electrical resistivity than hot-pressed BN (see Fig. 24), but its dielectric

6-26   Ceramics, Glasses, and Micas

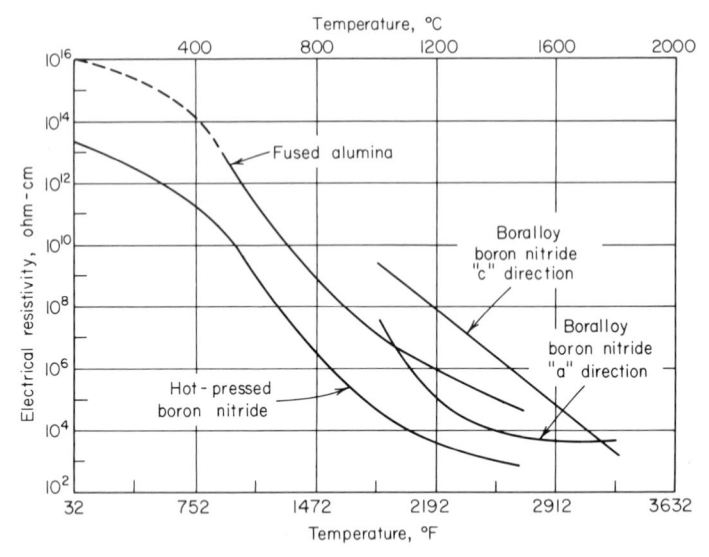

Fig. 24 Electrical resistivity of hot-pressed and pyrolytic (Boralloy) BN.[77] (Boralloy is trademark of Union Carbide Corp.)

constant and loss tangent are about the same. The dielectric strength of BN is rated at 4,000 volts/mil at room temperature, which makes it an excellent insulator. As is illustrated in Fig. 25, the oxidation rate of pyrolytic BN is significantly less than that of pyrolytic graphite below 2000°C.

Silicon Nitride. Si_3N_4 is characterized by high resistivity ($>10^{12}$ ohm-cm at 200°C), high strength ($>25,000$ psi at 1200°C), excellent thermal shock resistance, and chemical inertness. High-purity Si_3N_4 powder cannot be sintered or hot pressed. However, silicon powder can be nitrided in situ by heating it in nitrogen at between 1250 and 1450°C.[31] Partially nitrided mate-

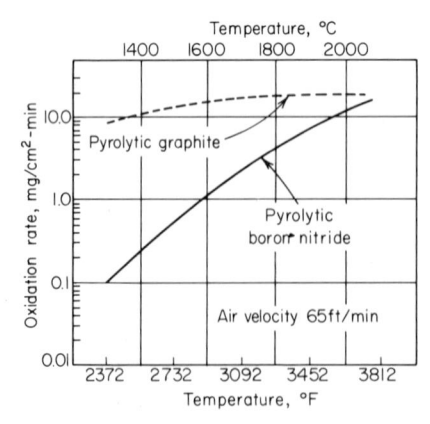

Fig. 25 Oxidation rate of pyrolytic forms of BN and graphite.[77]

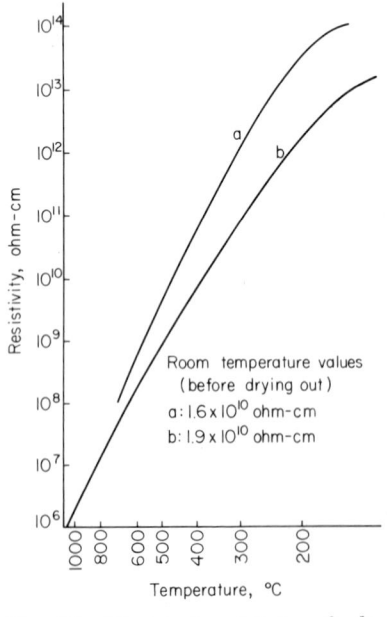

Fig. 26 Electrical resistivity of silicon nitride.[31]

rial can be worked by conventional edge-cutting tools. Since little volume change (~1 percent) occurs on nitriding, intricate shapes having close dimensional tolerances can be produced. The porosity of these reaction-sintered bodies depends on the density of the compacted silicon powder, and it is rarely less than 20 percent because of the absence of sintering. Silicon nitride can be hot pressed to theoretical density by using small additions of MgO. As shown in Fig. 26, Si_3N_4 has high electrical resistivity; however, unreacted silicon degrades the electrical properties severely.

Pure silicon nitride in the form of a thin, dense electrically insulating layer can be pyrolytically deposited on molybdenum or graphite by heating the substrate to 1000°C in ammonia and passing over it silicon tetrachloride vapors in a nitrogen carrier gas.

Glass

Reproducible electrical properties, ease of fabrication into complex shapes, and low cost are the principal reasons why glass is used in many electrical and structural applications in the electrical industry, e.g., tube envelopes, capacitor dielectrics, and substrates. On the order of 500 to 600 varieties of glass are commercially available, covering a wide range of electrical properties. Silicate glasses exhibit

TABLE 5 Chemical Compositions (percent by weight) of Some Typical Commercial Glasses[12]

Constituent oxide	Soda-lime-silica glasses			Borosilicate glasses			Lead glass	Alkali-free aluminosilicate glasses	
	Soft soda glass for lamp and valve envelopes	Sheet glass		Chemical and heat resisting glass	A glass for sealing to Kovar	A glass for sealing to tungsten	Glass for lamp and valve pinches	Glass for high-pressure mercury vapor lamps	Electrical fiber glass
SiO_2	70.5	72.8		80.0	65.0	74.6	56.5	54.5	53.5
Al_2O_3	1.8	1.4		2.1	2.3	1.0	1.0	21.1	14.5
B_2O_3	....	....		13.2	24.0	18.0	....	7.4	10.0
CaO	6.7	8.1		....	....	0.3	....	13.3	17.5
MgO	3.4	3.8		....	....	....	....	....	4.5
BaO	....	....		....	....	....	....	3.5	
PbO	....	....		....	....	....	30.0	....	
Na_2O	16.7	12.8		4.1	4.0	4.2	5.1	....	
K_2O	0.8	0.7		0.2	4.2	1.7	7.2	....	

dielectric constants ranging from 3.8 for pure silica to 10 for high-lead glasses and to as much as 25 for certain tellurium- and barium-oxide glasses. Loss factors vary from a tan δ of approximately 0.0009 for vitreous silica to about 0.009 at 60 Hz for soda-borosilicate glasses. At room temperature, most glasses are good electrical insulators (resistivity $>10^{14}$ ohm-cm); however, certain ones are poor insulators with resistivities between 10^2 and 10^9 ohm-cm, and others are semiconductors.

Glass is made up of randomly crosslinked networks of polyhedra, e.g., SiO_4^{4-}, BO_3^{3-}, PO_4^{3-}. In order to lower the viscosity and softening range of pure oxide glasses, alkalies and alkaline-earth oxides are added to the melt. As a result of making the glass structure more open, these ions are able to migrate more freely through the polyhedra and produce ionic conductivity. Thus, if an alkali metal is used as the fluxing agent, potassium is preferred to sodium since it has a lower mobility because of its larger size. The properties of a glass, therefore, can be altered substantially by making the appropriate additions, e.g., Al_2O_3 to improve the strength, PbO to improve the optical quality, and B_2O_3 to increase the resistance to thermal shock.

Commercial glasses Typical compositions are listed in Table 5. The general types of commercial glasses and some of their applications include: (1) soda-lime glasses: inexpensive; used for lamp envelopes, bottles, and window glazing, (2) borosilicate glasses: inexpensive; used for sealing to Kovar* and tungsten, and for ultraviolet transmission and chemical glassware, (3) high-silica glasses: expensive; used for ultrasonic devices, antenna shields, missile nose cones, and high-temperature-service chemical glassware, (4) high-lead glasses: moderately expensive; used for electric-light bulb stems, neon sign tubing, optical devices, quality glassware, and radiation shields for absorbing x-rays and gamma rays, and (5) alkali-free aluminosilicate glasses: moderately expensive; possessing properties similar to those of high-silica glasses; used in high-performance, high-power electron tubes; and in optical-quality aluminosilicate glass employed for electron-tube faceplates and for space-vehicle windows.

Electrical resistivity Because of its high resistivity, glass is widely used for electrical insulation (see Fig. 27). Resistivity is largely determined by the composi-

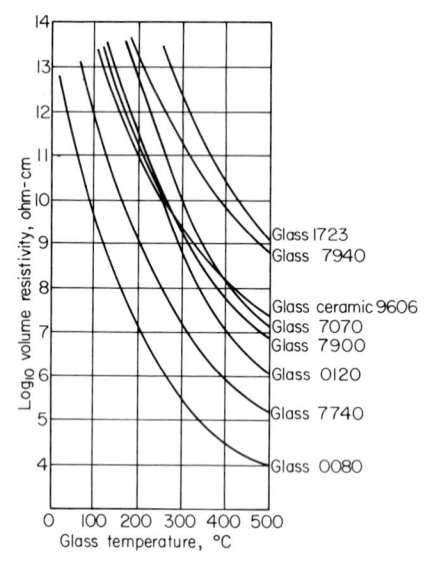

Fig. 27 Volume resistivity as a function of temperature for several Corning glasses.[78]

Fig. 28 Effect of increasing the amount of Na$_2$O in a soda-silica glass on volume resistivity.[79]

tion of the glass; however, it is affected by temperature, moisture, and structural defects. Since the factors that affect volume and surface resistivity are somewhat different, these properties are treated separately.

Volume resistivity Electric conductivity in most oxide glasses occurs by ion transport under the influence of an electric field. Because ions such as Li^{1+} and Na^{1+} are relatively small, they can easily move through the glass structure. The K^{1+}, Mg^{2+}, Ca^{2+}, Pb^{2+}, and Ba^{2+} ions, however, are larger and less mobile and, therefore, less able to contribute to the electric conductivity. With increasing alkali content, the degree of crosslinking in the network polyhedra is lowered; consequently, the volume resistivity is decreased in proportion to the alkali content (see Fig. 28).

Semiconducting glasses exhibit electronic rather than ionic conductivity. Glasses with resistivities in the range of 10^2 to 10^9 ohm-cm have been reported in the Ge-P-V oxide system[32,33] and in the elemental glasses of either S, Se, or Te combined with one or more of the elements Si, Ge, P, As, Th, and Pb.[34] Electronic devices are described that use semiconducting glasses as the solid-state switch or memory component.[35,36]

* Trademark of Westinghouse Electric Corporation, Pittsburgh, Pa.

Insulators   6-29

Temperature. With increasing temperature, the volume resistivity of oxide glasses decreases (see Fig. 27) because of the thermal weakening or breakdown of the bonds between the individual polyhedra. This causes the volume of the interstices to increase, which in turn permits higher mobility of the modifying cations. At the melting temperature, the resistivity of the glass decreases to as low as 10^1 to 10^2 ohm-cm, thereby making it possible to use the glass itself as the electric conductor for melting and sealing operations.

Surface resistivity The relatively low surface resistivity exhibited by glass under certain conditions is usually caused by either adsorbed moisture or electrolytic substances in the surface.

Moisture. The effect of atmospheric moisture becomes pronounced above 50 percent relative humidity (see Fig. 29). Below 50 percent, the surface water does not form a coherent monolayer; therefore, it does not provide an electrical path for conduction. Treating the glass with water-repellent materials such as silicones will lessen the effect of humidity on surface resistivity.

Electrolytic Effects. The chemical resistance of a glass determines its stability in the presence of moisture. Glasses with loosely bound ions, e.g., Na^{1+}, Li^{1+}, will give up these ions to surface moisture and in that way contribute to the surface conductivity. This condition can be avoided by selecting glasses with low alkali contents, e.g., high-silica or aluminosilicate glasses, and by rinsing the surface with a mineral acid to remove surface impurities.[37]

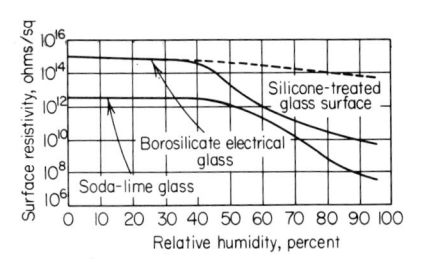

Fig. 29 Effect of relative humidity on surface resistivity of several glasses.[78]

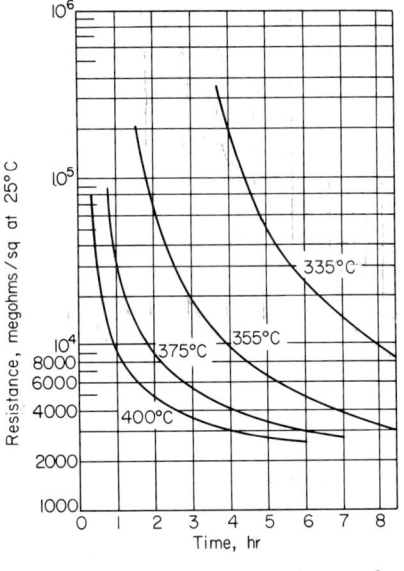

Fig. 30 Surface resistivity of a Ba-Pb-Si oxide glass at room temperature after heating at 335 to 400°C in hydrogen for various periods of time.[38]

Techniques for Lowering the Surface Resistance. By applying a thin film of either a semiconductor or a metal to the glass, the surface resistivity can be deliberately lowered so as to either eliminate a static charge or to give the desired resistance to the surface. For example, SnO can be applied on glass as an optically transparent conductive coating by hydrolysis of stannic chloride at elevated temperatures. Defrostable aircraft windshields are made by this process.

The surface resistance of high-lead glass can be lowered by a thermal treatment in hydrogen (see Fig. 30). This process forms a surface layer of metallic lead approximately 100 Å thick having a resistivity of 800 ohms/sq or greater.[38]

Dielectric strength A practical requirement of a dielectric is that it be able to withstand the applied voltage stress. Dielectric breakdown can occur by thermal and intrinsic processes. Thermal breakdown is due to localized heating caused by inhomogeneities in the electric field and in the dielectric itself. With increasing temperature, dielectric strength of typical alumina ceramics decreases by about 40 to 50 volts/ml per 100°C (See Table 6). If thermal heating is permitted

6-30   Ceramics, Glasses, and Micas

to increase without limit, chemical breakdown of the dielectric will result in the passage of even higher currents, which in turn leads to fusion, vaporization, and finally puncture of the dielectric.

TABLE 6   Dielectric Strength of Alumina at Several Temperatures[71]

	Volts/mil at 60 Hz			
Ceramic	25°C	400°C	800°C	1000°C
Alumina porcelain (0.125 in. thick)	400	200	50	30–40
Alumina, 99% (electrode spacing 0.038 in.)	450 (250)*	180 (70)*	55 (35)*	35 (17)*

* Safe maximum values; sample withstood 2,000 hr at these voltages.

Intrinsic or electronic breakdown occurs when conduction electrons are accelerated to sufficiently high energies by local field gradients to liberate valence electrons by collision. This avalanche effect continues at an accelerating rate until finally dielectric breakdown results. The dielectric strength of various materials at several frequencies is given in Table 7.

TABLE 7   Dielectric Strengths of Various Insulating Materials at Frequencies from 60 Hz to 100 MHz   (rms volts/mil)[80]

Material	Thickness, mils	60 Hz	1 kHz	38 kHz	180 kHz	2 MHz	18 MHz	100 MHz
Polystyrene (unpigmented)	30	3174	2400	1250	977	725	335	220
Polyethylene (unpigmented)	30	1091	965	500	460	343	180	132
Polytetrafluoroethylene (Teflon*)	30	850	808	540	500	375	210	143
Monochlorotrifluoroethylene (Kel-F)‡	20	2007	1478	1054	600	354	129	29†
Glass-bonded mica	32	712	643	—	360	207	121	76
Soda-lime glass	32	1532	1158	—	230	90	55	20†
Dry-process porcelain	32	232	226	—	90	83	71	60†
Steatite	32	523	427	—	300	80	58	56†
Forsterite, (AlSiMag-243)	65	499	461	455	365	210	112	74
Alumina, 85% (AlSiMag-576§)	55	298	298	253	253	178	112	69

* Trademark of E. I. du Pont de Nemours & Co., Inc., Wilmington, Del.
† Puncture with attendant volume heating effect.
‡ Trademark of Minnesota Mining and Manufacturing Co., St. Paul, Minn.
§ Trademark of American Lava Corp., Ridgefield, N.J.

The effect of the testing medium on the dielectric strength is shown in Fig. 31. The intrinsic breakdown strengths of borosilicate glass A and soda-lime-silicate glass B are both very high and linearly dependent on the specimen thickness. When these glasses are tested in various mediums, e.g., insulating and semiconducting oils, the breakdown strengths are much lower and nonlinearly related to specimen thickness. A carefully designed glass insulator can have an impulse-voltage strength of 1.7×10^5 volts/cm in air.[39]

Temperature and Time.   The dielectric strength of glass decreases with increasing temperature because of the greater ionic conductivity at the higher temperatures (see Fig. 32). Regions of instability due to the channeling of the current and the formation of hot spots result in dielectric breakdown by the puncture of the dielectric material.

The combined effects of temperature and duration of applied voltage stress on Pyrex* glass are shown in Fig. 33. Below −50°C, the breakdown strength is quite high and independent of both temperature and duration of applied voltage stress.

* Trademark of Corning Glass Works, Corning, N.Y.

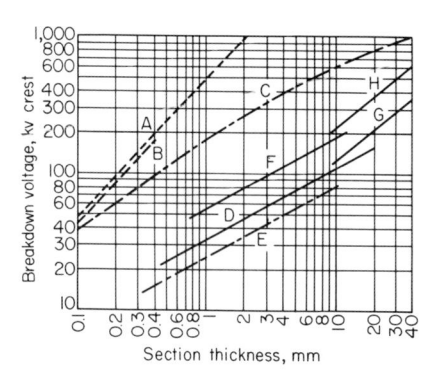

Fig. 31 Breakdown voltage versus thickness of glass for different conditions at room temperature—60-Hz voltage raised continuously. A, Intrinsic dielectric strength of borosilicate glass. B, Intrinsic dielectric strength of soda-lime glass. C, Highest test values available for borosilicate glass. D, Borosilicate glass plate immersed in insulating oil. E, Soda-lime glass plate immersed in insulating oil. F, Borosilicate glass plate immersed in semiconducting oil. G, Borosilicate glass power-line insulator immersed in insulating oil. H, Borosilicate glass power-line insulator immersed in semiconducting oil.[81]

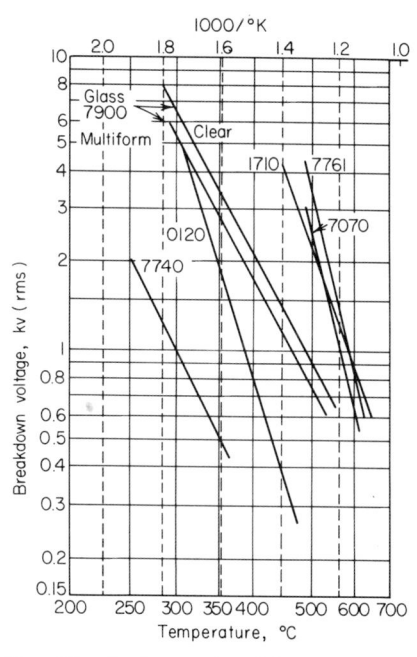

Fig. 32 Dielectric breakdown of several Corning glasses at higher temperature—1-min breakdown for thickness of 2 mm at 60 Hz.[81]

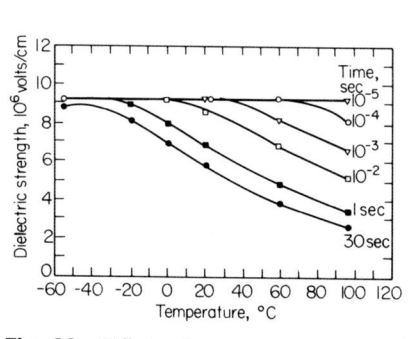

Fig. 33 Effect of test temperature and test duration on the breakdown strength of a Pyrex glass.[82]

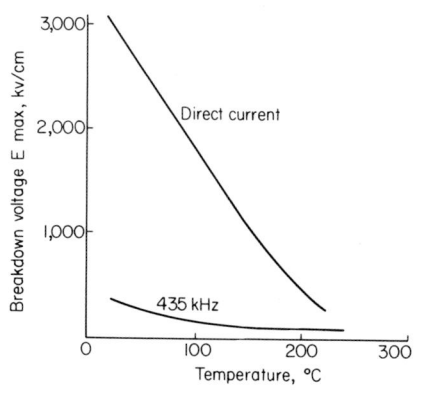

Fig. 34 Breakdown voltage in air for lime glass as a function of temperature for direct current and for 435-KHz alternating current.[83]

Above room temperature, however, the combined effects of higher temperatures and longer testing times do cause a decrease in the dielectric strength.

Frequency. The dielectric strength of glass is lowered at higher frequencies by dipole relaxation processes (see Fig. 34). Data listed in Table 8 indicate that the dielectric strength of ordinary glass at 100 MHz is 2 to 5 percent of the corresponding value measured at 60 Hz.

TABLE 8 Dielectric Breakdown of Ordinary Glass as a Function of Frequency[84]

Frequency	Breakdown voltage, rms kv	
	0.030 in. thick	0.125 in. thick
60 Hz	38.2	38.4
1 kHz	28.3	
14 kHz	21.	
150 kHz	7.5	
2 MHz	2.65	9.25
4.2 MHz	2.0	
9.8 MHz	1.7	
18 MHz	1.7	3.4
100 MHz	0.63	1.9

Dielectric constant The dielectric constants of the more common types of glass (silicates, phosphates, borosilicates, and aluminosilicates) vary from a low of 3.8 for fused silica to 8 to 10 for high-lead glasses (see Fig. 35). The high polarizability of the Pb^{2+} ion gives rise to the high dielectric constant of high-lead glasses, whereas the absence of such ions in fused quartz accounts for the low permittivity. Tellurium- and barium-oxide glasses with dielectric constants of 25 are reported.[40]

High-dielectric-constant glass-ceramic capacitors are made by crystallizing a ferroelectric phase from the glass.[41] This process requires careful control of the nucleation and growth steps in order to obtain the required electrical properties.

Temperature. The dielectric constant of glass increases with rising temperature in proportion to the polarizability of the

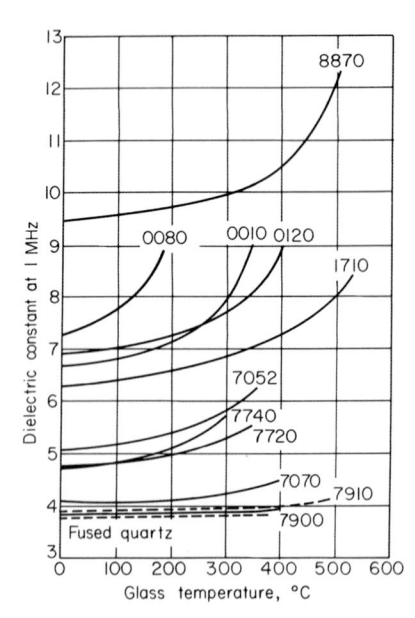

Fig. 35 Dielectric constants of several Corning glasses at 1 MHz as a function of temperature.[81]

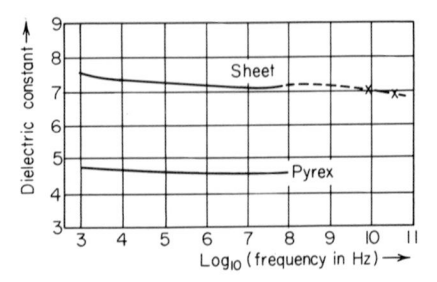

Fig. 36 Dielectric constant as a function of frequency for sheet and Pyrex glasses.[85]

TABLE 9 Dielectric Properties of Several Commercial Glasses[86]

Glass	Composition	Temperature, °C	Dielectric constant						Dissipation factor					
			60 Hz	1 kHz	1 MHz	100 MHz	3 GHz*	25 GHz*	60 Hz	1 kHz	1 MHz	100 MHz	3 GHz*	25 GHz*
Corning 0010	Soda-potash-lead silicate ~20% lead oxide	24	6.70	6.63	6.43	6.33	6.10	5.87	0.0084	0.00535	0.00165	0.0023	0.0060	0.0110
Corning 0120	Soda-potash-lead silicate	23	6.76	6.70	6.65	6.65	6.64	6.51	0.0050	0.0030	0.0012	0.0018	0.0041	0.0127
Corning 1990	Iron-sealing glass	24	8.41	8.38	8.30	8.20	7.99	7.84	....	0.0004	0.0005	0.0009	0.00199	0.0112
Corning 1991		24	8.10	8.10	8.08	8.00	7.92	....	0.0027	0.0009	0.0005	0.0012	0.0038	
Corning 7040	Soda-potash-borosilicate	25	4.85	4.82	4.73	4.68	4.67	4.52	0.0055	0.0034	0.0019	0.0027	0.0044	0.0073
Corning 7050	Soda-borosilicate	25	4.90	4.84	4.78	4.75	4.74	4.64	0.0093	0.0056	0.0027	0.0035	0.0052	0.0083
Corning 7060 (Pyrex)	Soda-borosilicate	25	....	4.97	4.84	4.84	4.82	4.65	....	0.0055	0.0036	0.0030	0.0054	0.0090
Corning 7070	Low-alkali potash-lithia-borosilicate	23	4.00	4.00	4.00	4.00	4.00	3.9	0.0006	0.0005	0.0008	0.0012	0.0012	0.0031
Corning 7720	Soda-lead borosilicate	24	4.75	4.70	4.62	....	4.60	....	0.0093	0.0042	0.0020	....	0.0043	
Corning 7750	Soda-borosilicate ~80% silicon dioxide	25	....	4.42	4.38	4.38	4.38	....	....	0.0033	0.0018	....		
Corning 7900	96% silicon dioxide	20	3.85	3.85	3.85	3.85	3.84	3.82	0.0006	0.0006	0.0006	0.0006	0.00068	0.0013
Fused silica 915c	Silicon dioxide	25	....	3.78	3.78	3.78	3.78	....	....	0.00026	0.00001	0.00003	0.0001	
Quartz (fused)	100% silicon dioxide	25	3.78	3.78	3.78	3.78	3.78	3.78	0.0009	0.00075	0.0001	0.0002	0.00006	0.00025

* GHz = gigaHertz = 10^9 Hz.

medium. When stability of the dielectric properties is required, glasses with rigid structures, e.g., fused silica and high-silica glasses, are used.

Frequency. The dielectric constant of glass decreases very slightly with increasing frequency (see Fig. 36). At frequencies above 50 Hz, the relatively long relaxation times typical of silicate glass structures prevents an increase in the dielectric constant. Table 9 lists the dielectric constants at several frequencies for a number of commercial glasses.

Dielectric loss The total dielectric loss in glass is the sum of four different loss mechanisms, namely, conduction, dipole relaxation, vibration, and deformation losses. As shown in Fig. 37, the magnitude of each of these loss mechanisms varies with frequency and temperature. Conduction and dipole relaxation losses (1 and 2) are the predominant loss mechanisms at low frequencies; whereas vibration and deformation losses (3 and 4) are the controlling mechanisms at high frequencies. These losses generally increase with rising temperature.

Conduction Losses. This loss mechanism is important only at frequencies less than about 50 Hz, where it is described by the relation, $\tan \delta = (2\pi f \rho \epsilon' \epsilon_0)^{-1/2}$. Since the conductivity $1/\rho$ of glass increases with rising temperature, so also does the conduction loss. Likewise, alkali-rich glasses exhibit greater conduction losses than high-silica glasses.

Dipole Relaxation Losses. Up to about 10^6 Hz, dipole relaxation losses are important. They result from the energy that is given up by mobile ions as they jump over small distances in the network. Dipole losses are greatest when the frequency of the applied field is equal to the relaxation time of the dipole. At frequencies either far above or far below the resonant frequency, the losses are the least. Because of its random structure, glass has more than one relaxation time. This results in a broad distribution of dipole relaxation losses with frequency, rather than one discrete loss peak (see curve 2 in Fig. 37).

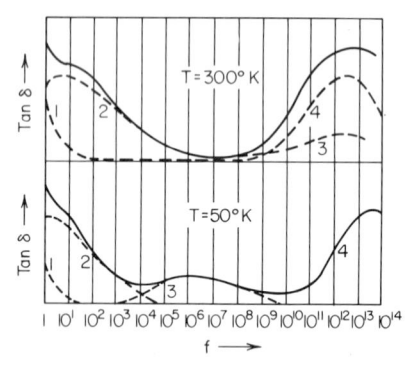

Fig. 37 The general shape of tan δ as a function of the frequency at 300 and 50°K. The fully drawn curves give the total losses. The sum of four different contributions: (1) the conduction losses, (2) the dipole relaxation losses (sometimes called relaxation losses), (3) the vibration losses, (4) the deformation losses.[87]

The dissipation factor at 1 MHz and 250°C for various commercial glasses (see Fig. 38) reveals a correlation between high resistivity and low dielectric losses. This is further illustrated in Fig. 39 by glasses with low alkali content, having not only lower losses but also showing greater temperature stability. Above 200°C losses become excessive in practically all glasses because of increased ionic conduction.

Deformation Losses. Relaxation and deformation losses are similar in that they are associated with small displacements of ions in the glass network. Deformation losses involve smaller ion movements than do either conduction or dipole relaxation loss mechanisms, and they occur in the region of 10^9 Hz at room temperature (see curve 4 in Fig. 37).

Vibration Losses. These losses occur by a resonance phenomena involving both the network-forming ions and the network-modifying ions. As shown in Fig. 37 (curve 3), vibration losses take place over a broad frequency range because of the variation in both the mass and the location of the different ions in the glass network. The resonant frequency of this loss mechanism is given by $f_{res} = (A/M)^{1/2}$, where A is a constant relating the displacement and the restoring force, and M is the mass of the ion.

In general, heavy ions, e.g., Pb^{2+} or Ba^{2+}, will decrease the resonant frequency of the network, whereas lighter ions will shift it to higher frequencies. This is shown in Fig. 40 where glass 7910, a high-silica glass, has a slowly increasing dissipation factor at 10^{10} Hz whereas glass 8870, a high-lead glass, has a rapidly increasing loss factor at this frequency.

Mica

Mica is used in the electrical industry as a dielectric in capacitors and as an electrical insulator in motors, electron tubes, transistors, and appliances such as toasters and flatirons. Commercially important mica is a naturally occurring form of a potassium-alumina-silicate mineral. Muscovite or ruby mica [$KAl_3Si_3O_{10}(OH)_2$] is the most desirable form for capacitor use because of its low dissipation factor and high dielectric strength. Phlogopite or amber mica [$KMg_3AlSi_3O_{10}(OH)_2$] has better resistance at high temperature than muscovite, but its electrical properties

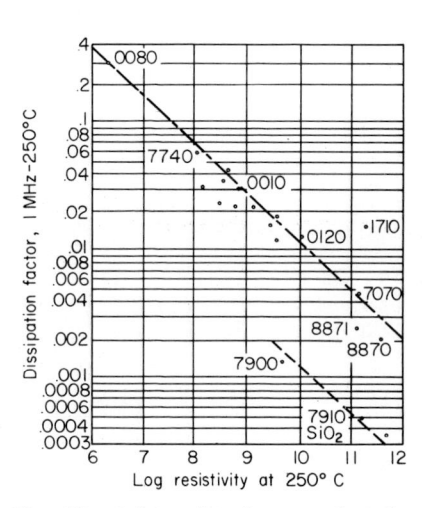

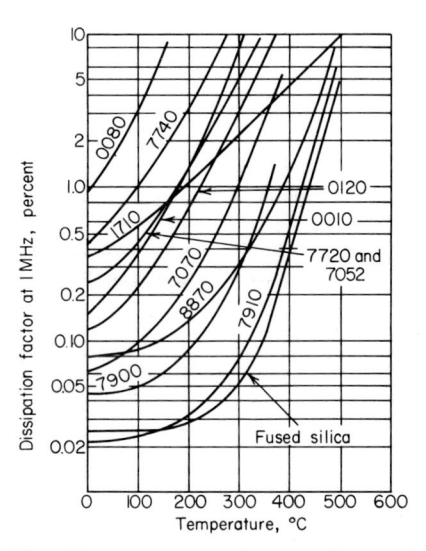

Fig. 38 Relationship between log dissipation factor and log resistivity of several Corning glasses at 250°C.[81]

Fig. 39 Dissipation factors of several Corning glasses at 1 MHz as a function of temperature.[81]

are not so well suited for use as capacitor dielectrics. Phlogopite mica is used, however, to insulate motor and generator commutator segments, as well as flatiron and toaster heater elements. Fluorophlogopite is a synthetic form of mica ($KMg_3AlSi_3O_{10}F_2$), where the hydroxyl ions have been replaced by fluorine ions. The absence of water in the synthetic composition has resulted in a form of mica that is more stable at elevated temperatures. Other types of commercial mica include glass-bonded mica and ceramoplastic and reconstituted mica.

Mica has a unique combination of electrical properties not available in other materials. The relatively stable dielectric constant, the low-loss tangent, and the high dielectric strength of mica are used to form stable and reliable capacitors for critical circuit applications. Disadvantages of mica are its small capacitance-to-volume ratio and its high cost in the form of large natural sheets. A listing of ASTM specifications and methods of testing mica and mica products is given in Table 10.

Muscovite—ruby mica Muscovite is chemically inert when in contact with the chemicals typically used in electrical insulation. It is attacked by HF and H_2SO_4, but it is resistant to other acids, solvents, and alkalies. Unless precautions are taken, liquids such as water or oil can work their way in between the laminae

6-36 Ceramics, Glasses, and Micas

TABLE 10 ASTM Specifications and Methods of Test for Mica and Mica Products

D 748-59........	Natural block mica and mica films suitable for use in fixed mica-dielectric capacitors
D 351-62........	Natural muscovite mica based on visual quality
D 2131-65T.....	Natural muscovite mica splittings
D 352-63........	Pasted mica used in electrical insulation
D 1082-54.......	Power factor and dielectric constant of natural mica
D 1677-62.......	Untreated mica paper used for electrical insulation sampling and testing
D 374-57T......	Thickness of solid electrical insulation
D 1039-65.......	Testing glass-bonded mica used as electrical insulation
F 12-64T.......	Mica bridges for electron tubes
F 48-64T.......	Dimensioning mica bridges
F 652-61T......	Measuring mica stampings or substitutes used in electron devices and lamps

of mica by capillary action. Water lowers the electrostatic force of attraction between the layers of mica by an exchange mechanism of hydroxyl for potassium ions.

Outgassing becomes intense above 600°C (Fig. 41), and prolonged exposure at higher temperatures results in further release of water and finally in decomposition of the mica.[42] For this reason, mica spacers and insulators should be outgassed before being used in vacuum tubes or other devices where outgassing is a problem.

Muscovite is easily split into thin uniform sheets because of its planar cleavage habit. Efforts to mechanize the splitting process have not been successful, and splitting is still a hand operation.

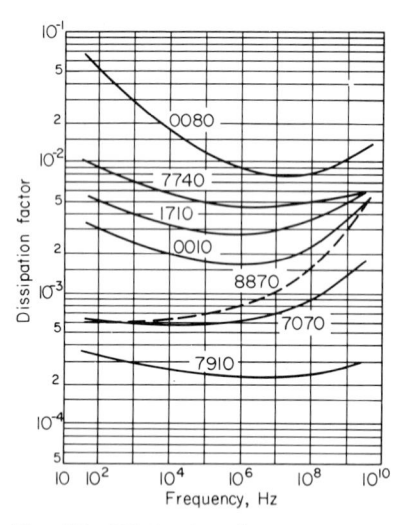

Fig. 40 Dissipation factors at room temperature over a range of frequencies for several Corning glasses.[81]

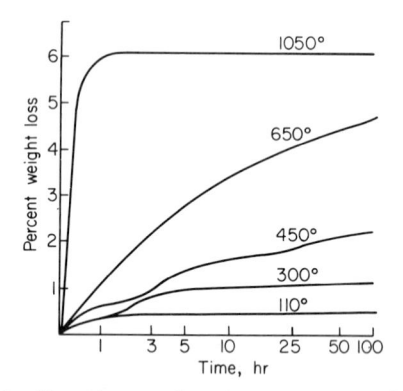

Fig. 41 Change of total percentage weight loss as a function of time for pulverized muscovite held at various temperatures in air.[42]

The electrical resistivity of muscovite decreases with increasing temperature, as shown in Fig. 42. Because of the removal of moisture from the surface as well as from between the exfoliated layers, the electrical resistivity of the degassed mica is higher than that of the untreated material.

Muscovite has a high dielectric strength that averages between 3,000 and 6,000 volts/mil at 60 Hz for specimens 1 to 3 mils thick. It has a relatively stable dielectric constant over a wide frequency range, as is shown in Fig. 43. Muscovite also has a dissipation factor (tan δ) that decreases with frequency. This makes ruby mica an especially useful dielectric for high-frequency applications.

Phlogopite—amber mica Phlogopite is more stable at higher temperatures (850 to 1000°C) than muscovite. However, the higher dissipation factor and lower dielectric strength of phlogopite prevents it from being used in capacitor applications. Phlogopite is used mainly for insulation in transformers, motors, and soldering

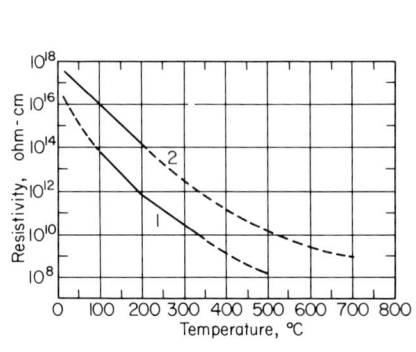

Fig. 42 Electrical resistivity of muscovite mica as a function of temperature.[88]

Fig. 43 Dielectric constant and tan δ for muscovite mica over a range of frequencies.[89]

irons, and as insulating washers or disks. The thermal conductivity of phlogopite is about the same as that of muscovite.

Synthetic mica—fluorophlogopite A synthetic form of mica is made by heating a mixture of silica, alumina, magnesia, and fluoride compounds. Fluorophlogopite

TABLE 11 Properties of Mica[13]

	Natural mica		Synthetic mica
Property	Muscovite	Phlogopite	Fluorophlogopite
Volume resistivity, ohm-cm:			
At 20°C	10^{16}–10^{14}	Lower than muscovite at low temperature	10^{16}–10^{17}
At 500°C	10^8–10^{10}	Higher at high temperatures	
Dielectric strength (1–3-mil specimen), volts/mil	3,000–6,000	3,000–4,000	2,000–3,000
Dielectric constant:			
At 100 Hz	5.4		
At 1 MHz	5.4	6	5–6
At 100 MHz	5.4		
Dissipation factor:			
At 100 Hz	0.0025		
At 1 MHz	0.0003	0.001–0.01	0.0004–0.0007
At 100 MHz	0.0002		
Safe operating temperature, in vacuo, °C	350–450	....	1,000
Water absorption	Practically zero	Practically zero	Practically zero
Compressive strength, psi	25,000	15,000	

mica has better temperature-resistance properties than muscovite mica (Table 11) because the structure contains no water.

The cleavage of synthetic mica is similar to that of natural mica. Synthetic fluorophlogopite has a lower dielectric strength than natural muscovite (Fig. 44). However, it can be used at a much higher operating temperature than muscovite. A method of obtaining single-crystal sheets of fluorophlogopite up to 3 in. in diameter and 3 to 10 mils thick has been developed.[43]

Reconstituted mica Reconstituted mica is a sheet material made from thin flakes of either natural or synthetic mica that have been ground up, pressed into sheet form, and heated under pressure to bond the tiny mica flakes together. Reconstituted natural mica evolves more gas than natural mica, up to 500°C.[44] At 800°C, however, natural mica becomes more gassy than reconstituted mica. Reconstituted mica is available in tape and sheet form, and it is used as insulation for motors, generators, and transformers. A dielectric strength of approximately 450 to 900 volts/mil for the synthetic material is somewhat lower than that of natural mica. The properties of reconstituted mica are highly reproducible in sheet form.

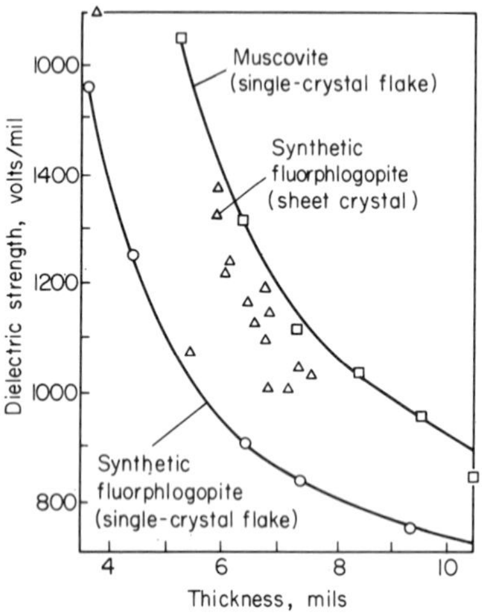

Glass-bonded mica (ceramoplastics)
Glass-bonded mica consists of fine particles of either natural or synthetic mica mixed with low-melting glass powder. The glass must become plastic at temperatures below the decomposition temperature of the mica. An advantage of this material is that it can be formed into various shapes by either compression or injection molding. It can be molded or machined to close tolerances, which makes it useful for components such as bases and sockets for electronic tubes, coil supports, brush holders, switchboard panels, washers, and spacers. Since the thermal expansion rate of glass-bonded mica is identical to that of mild steel, metal inserts can be attached directly to it. Glass-bonded mica is easily outgassed, thereby making it especially useful for structural members in vacuum tubes,[45] as well as for high-temperature vacuum-tight windows.[46] Techniques for attaching metal gasket seals to glass-bonded mica are available.[47]

Electrical properties of a commercially available glass-bonded mica with respect to temperature frequency and humidity are shown in Figs. 45 to 48. The low dissipation factor and relatively stable dielectric constant at high frequencies persist to fairly high temperatures. Engineering properties of glass-bonded mica are listed in Table 12.

Fig. 44 Relationship between dielectric strength and thickness of mica specimens.[43]

Ceramic and Glass Substrates

The demand for miniaturization and for increased reliability of electronic equipment is responsible for the increased use of ceramic and glass substrates. These materials serve as the base upon which electrical components such as resistors, capacitors, and conductors are deposited by silk screening, vacuum depositing, sputtering, and painting processes. Discrete components can be attached to the substrate by either soldering or compression welding.

A substrate must be physically, electrically, and chemically compatible with the material deposited on it, and in particular it should be selected on the basis of the following properties:

Surface smoothness The surface finish requirements for thick-film deposition ($\sim 10^{-3}$ cm) are not as stringent as those for thin-film deposition ($\sim 10^{-6}$ cm). Con-

Insulators 6-39

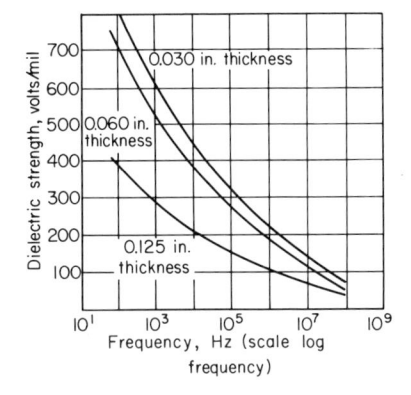

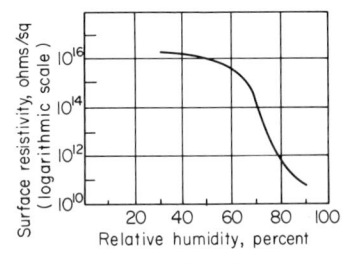

Fig. 45 Relationship between dielectric strength and frequency for three thicknesses of Mycalex 400. (*Mycalex Corp. of America.*)

Fig. 46 Variation of surface resistivity of Mycalex 400 with relative humidity. (*Mycalex Corp. of America.*)

sequently, either a glass or a glazed ceramic is required for thin-film deposition, whereas an "as-fired" ceramic substrate is satisfactory for thick-film deposition.

The smoothness of a ceramic depends upon its microstructure and density. A high-density body composed of small grains will form a smoother substrate than

Fig. 47 Variation of volume resistivity of Mycalex 400 with (*a*) frequency and (*b*) temperature. (*Mycalex Corp. of America.*)

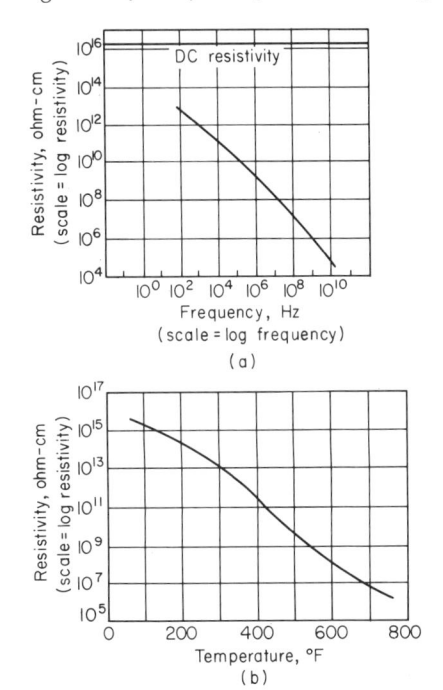

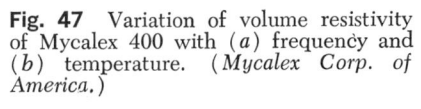

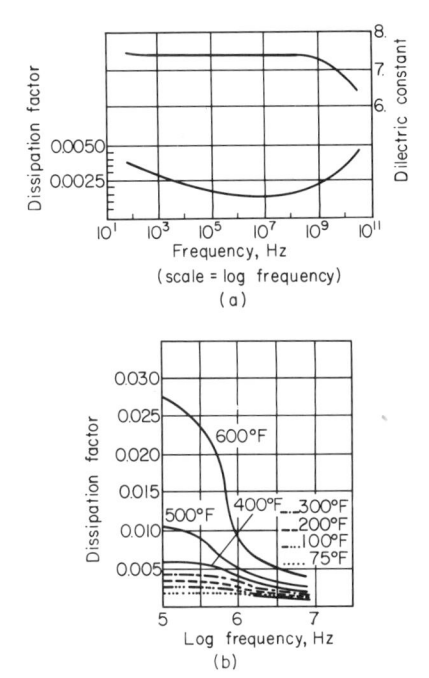

Fig. 48 Variation of dissipation factor and dielectric constant of Mycalex 400 with (*a*) frequency and (*b*) frequency at several temperatures. (*Mycalex Corp. of America.*)

TABLE 12 Properties of Glass-bonded Mica[13]

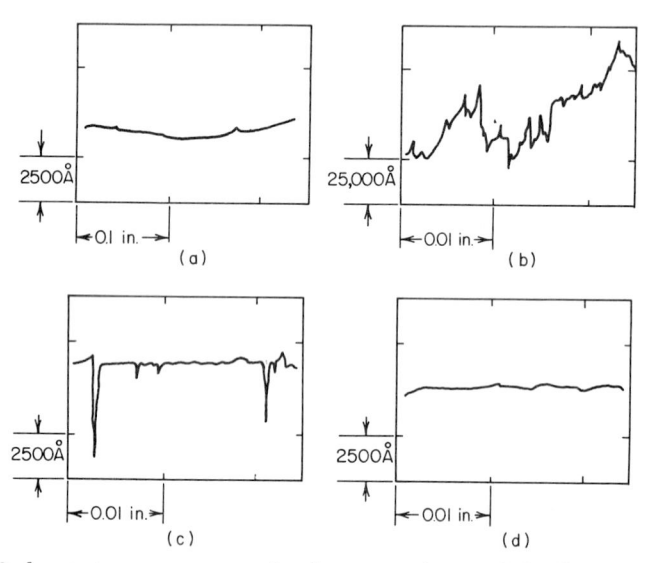

Property	Moldable grade (injection-molded)	Machinable grade (compression-molded)
Volume resistivity, ohm-cm:		
At 20°C	10^{14}–10^{16}	10^{17}–10^{18}
At 500°C	10^8	10^7
Dielectric strength, volts/mil	270–400	250–500
Dielectric constant at 1 MHz	6.7–8.8	6.8–6.9
Dissipation factor at 1 MHz	0.0014–0.0023	0.0013–0.0020
Maximum temperature endurance (unstressed), °C	345–650	370–845
Thermal expansion coefficient at 20–300°C, 10^{-7} in./(in.)(°C)	94–114	112–120
Compression strength, psi	25,000–40,000	35,000–40,000
Impact strength (Charpy), ft-lb/in.^{2}	1.2–2.0	1.2–2.4
Vacuum tightness (He leakage rate), cm^3/sec	Less than 2×10^{-10}	

Fig. 49 Stylus instrument traces of substrate surfaces. (a) Glazes and drawn glasses, (b) as-fired, large-grain polycrystalline ceramic, (c) polished, large-grain polycrystalline ceramic, and (d) polished, small-grain polycrystalline ceramic.[49]

a ceramic consisting of larger grains (see Fig. 49). As-fired alumina substrates with a 2 μin. surface finish are reported,[48] and glass suubstrates with a surface smoothness of ¼ μin. (60 Å) are commercially available (Corning 7059).

Glazed ceramic substrates have somewhat lower thermal conductivities than as-fired ones, but compared to glass substrates they have good thermal conductivities. Glazed ceramic substrates, therefore, combine the properties of surface smoothness and good thermal conductivity. Ceramic substrates having a glaze several mils thick and a surface finish of approximately one microinch are reported.[49]

Electric conductivity Both the volume and the surface resistance of a substrate must be high in order that leakage currents are minimized either through the bulk of the substrate or along the surface between the components. Alkali-metal ions, especially Na⁺, should be avoided in the substrate as well as in the deposited material.[50,51] This ionic species migrates relatively easily under the influence of an electric field. Furthermore, it has a high thermal diffusion coefficient, which can lead to device degradation.

The aging effect of tantalum nitride resistors deposited on substrates containing differing amounts of sodium is shown in Fig. 50. Although the glazed alumina with 3 percent Na₂O contains more alkali than the 7059 glass, the glazed alumina body is still more stable than the glass substrate. This is because the thermal conductivity of the alumina is greater than that of the glass; therefore, both the temperature of the alumina substrate and the Na⁺ ion diffusion rate are lower than in the 7059 glass.

Thermal conductivity Substrates with high thermal conductivites permit greater component packing density since surface temperatures can be lowered and the danger of hot spots minimized through the use of heat sinks. Since ceramics have higher thermal conductivities than glasses, glazing ceramic substrates to improve their surface finish will decrease their thermal conductivities. For example, the thermal resistance of 1 ml of glaze is equivalent to that of 30 mils of Al₂O₃ or 190 mils of BeO.

A high alumina content is required to achieve a high thermal conductivity in alumina-based substrate. This effect is illustrated by the data in Table 13, which

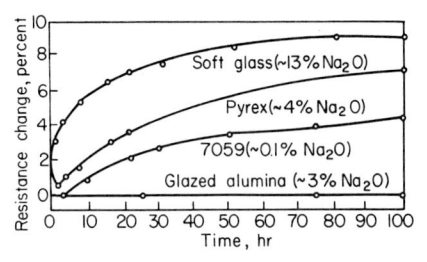

Fig. 50 Effect of substrate alkali content on overload testing of tantalum nitride resistors.[90]

TABLE 13 Effect of Alumina Content on the Thermal Conductivity of Alumina Substrates

% Alumina	Thermal conductivity cal/(sec)(cm²)(°C/cm)	% change in thermal conductivity
99	0.070	
98	0.061	−13
96	0.043	−39
85	0.035	−50

shows that the thermal conductivity decreases by 13 percent when the alumina content is lowered from 99 to 98 percent, and by 50 percent when the alumina content is lowered to 85 percent.

BeO has a higher thermal conductivity than Al₂O₃; however, the conductivity of Al₂O₃ is adequate for most applications, and alumina is much less expensive than beryllia.

Thermal shock resistance This property is important particularly as it affects the ability of a substrate to withstand soldering and joining operations. For the heat transfer rates usually encountered in electronic packaging operations, thermal conductivity, thermal expansion, and strength largely determine the ability of a substrate to withstand thermal shock. The high thermal shock resistance of BeO is due to its high thermal conductivity, whereas that of fused silica is due to its low rate of thermal expansion (see also Table 2).

Differential thermal expansion To help assure long device life, the thermal expansion mismatch between substrate and components, interconnections, and seals should be as small as possible so as to minimize residual stresses.

Dielectric constant At high frequencies, the dielectric constant of the substrate is especially important. A substrate with a high dielectric constant may have a sufficiently high capacitive reactance to be no problem at low frequencies, whereas at higher frequencies low reactance paths may develop. Ceramic and glass substrates are available with dielectric constants in the range from 4 to 9 (see Tables 14 and 15). In substrate form, $BaTiO_3$ can have a dielectric constant of 6,500; however, it also has a higher dissipation factor and a lower volume resistivity than other substrate materials.

TABLE 14 Properties of Ceramic Substrates and Glazes[91]

Material	Tensile strength, psi	Expansion coefficient, 10^{-6} in./(in.)(°C)	Coefficient of heat transfer, (watts)(in.)/(in.²)(°C)	Relative dielectric constant	% dissipation factor	Volume resistivity at 150°C, ohm-cm $\times 10^{-12}$
Alumina	25,000	6.4	~0.89	9.2	0.03	>100
Beryllia	15,000	6.0	5.8	6.4	0.01	>100
Corning 7059 glass	~10,000	4.6	~0.03	5.8	0.1	>100
Modified $BaTiO_3$	4,000	9.1	0.007	6,500	1.8	0.2
Modified TiO_2	7,500	8.3	0.017	80	0.03	0.5
Glaze for alumina:						
2.5 % sodium oxide	~10,000	5.5	~0.03	6.3	0.16	>100
Alkali-metal-free	~10,000	5.3	~0.03	7	0.2	>100

Fabrication The higher thermal conductivity and strength of ceramic over glass substrates permits them to be processed by automatic machinery. Holes, perforations, and scribed lines (to facilitate breaking) can be easily made while the ceramic is in the "green" or unfired state. In contrast to this, it is difficult to put holes in a glass substrate during the drawing operation when the smooth surface is formed. Critical applications requiring a substrate with both high thermal conductivity and surface smoothness use single-crystal sapphire.[52]

Glass-Ceramics

The advantage that a glass-ceramic offers over either glass or ceramic materials is the combination of easy fabrication and outstanding mechanical properties. Before the crystallization process, the glassy body is worked by conventional glass-forming methods, i.e., blowing, pressing, drawing, and rolling. After the forming operation, the glass is cooled to approximately 100°C above the annealing temperature for a time sufficiently long to allow stable nuclei to separate from the vitreous phase. After nucleation, the body is heated to approximately 100°C below the softening point of the glass for rapid crystallization. Nucleating agents commonly used include TiO_2, Au, Ag, ZrO_2, Cr_2O_3, and NiO.[53] Several glass compositions and the precipitating crystalline phases are listed in Table 16. Since glass-ceramic bodies have zero porosity and submicron grains, they are mechanically stronger than many conventional glass and ceramic bodies. The mechanical properties of several glass-ceramic, glass, and ceramic materials are listed in Table 17.

Certain glass-ceramic compositions are used in microwave applications because of their low dissipation factors (tan δ = 0.0003) at high frequencies (10 MHz), as shown in Fig. 51.

Seals

Glass-to-metal seals Incandescent and vapor lamps and electron tubes are some of the many applications that use glass-to-metal seals for the combination of either optical or insulating qualities and metallic conductivity. The requirements for bonding glass and metal are: (1) The glass must be wet and adhere to the metal, and (2) the thermal expansions of the glass and metal should closely match over the

TABLE 15 Properties of Glass Substrate Materials[147]

Glass code number	0080	0211	1715	1723	7059	7740	7900	7940
Glass type	Soda lime	Alkali zinc borosilicate	Lime alumino-silicate, alkali-free	Lime alumino-silicate, alkali-free	Barium alumino-silicate, alkali-free	Alkali borosilicate	96% silica	Fused silica
Dielectric constant:								
At 100 Hz:								
25°C	8.3	6.8	6.0	6.4	5.9	4.9	3.9	3.9
200°C	....	....	6.1	6.6	6.1	....	4.1	3.9
400°C	....	....	7.2	7.3	7.4	....	....	....
At 1 MHz:								
25°C	6.9	6.6	5.9	6.4	5.8	4.6	3.9	3.9
200°C	9.3	7.4	6.1	6.5	5.9	5.1	3.9	3.9
400°C	....	....	6.3	6.7	6.1	....	3.9	3.9
Loss tangent:								
At 100 Hz:								
25°C	0.078	0.01	0.0018	0.0008	0.0011	0.027	0.0006	0.00006
200°C	....	....	0.0047	0.0024	0.0062	....	0.07	0.0003
400°C	....	....	0.2	0.09	0.19	....	....	....
At 1 MHz:								
25°C	0.01	0.0047	0.0024	0.0013	0.0011	0.0062	0.0006	0.00002
200°C	0.17	0.032	0.0028	0.0014	0.0018	0.03	0.001	0.00002
400°C	....	....	0.0061	0.0034	0.0071	....	0.026	0.0003
Viscosity data, °C:								
Annealing point	512	542	866	710	650	565	910	1050
Softening point	696	720	1060	910	872	820	1500	1580
Linear coefficient of thermal expansion at 0–300°C, 10⁻⁷ in./(in.)(°C)	92	72	35	46	45	32.5	8	5.6
Density, g/cm³	2.47	2.57	2.48	2.63	2.76	2.23	2.18	2.20
Volume resistivity (log₁₀) ρ, ohm-cm:								
At 250°C	6.4	8.3	13.6	14.1	13.5	8.1	9.7	11.8
At 350°C	5.1	6.7	11.3	11.8	11.3	6.6	8.1	10.2
Thermal conductivity, cal/(sec)(°C)(cm)								
At 25°C	0.0025	0.002	0.002	0.0032	0.0030	....	....	0.0035
At 300°C	0.0032	....	....	....	0.0036	....	....	0.0042

6-43

temperature range of intended use. Glasses suitable for sealing to various metals are given in Table 18.

Bonding In order to form a strong glass-to-metal bond, an oxide layer is needed on the surface of the metal. Ideally, the metal oxide will diffuse into the molten glass to form the bond. Bonding, therefore, depends on how well the molten glass wets the metal (see Fig. 52). The degree of wetting is described

TABLE 16 Some Representative Glass-Ceramic Composition Fields[53]

Glass	Crystal phases	Catalysts
MgO-Al_2O_3-SiO_2	$2MgO$-$2Al_2O_3$-$5SiO_2$	TiO_2
Li_2O-Al_2O_3-SiO_2	Li_2O-Al_2O_3-$2SiO_2$	TiO_2
	Li_2O-Al_2O_3-$4SiO_2$	TiO_2
	LiO-Al_2O_3-$6SiO_2$	TiO_2
	Li_2O-SiO_2	Au, Ag, Cu, Pt
	Li_2O-$2SiO_2$	Au, Ag, Cu, Pt
Na_2O-BaO-SiO_2	BaO-$2SiO_2$	Au, Ag, Cu, Pt

by the wetting or contact angle θ, which is related to the energy between the solid-liquid, the solid-vapor, and the liquid-vapor interfaces by the relation $\cos \theta = (\gamma_{sv} - \gamma_{sl})/\gamma_{lv}$, where γ is the surface energy. When $\theta < 90°$, the liquid partially wets the surface; when $\theta > 90°$, the liquid is nonwetting; and when $\theta = 0°$, the liquid completely wets the surface.

The surface energy of a liquid or solid arises from the unsatisfied bonds of the surface atoms. Since these atoms are more strongly bound by internal forces,

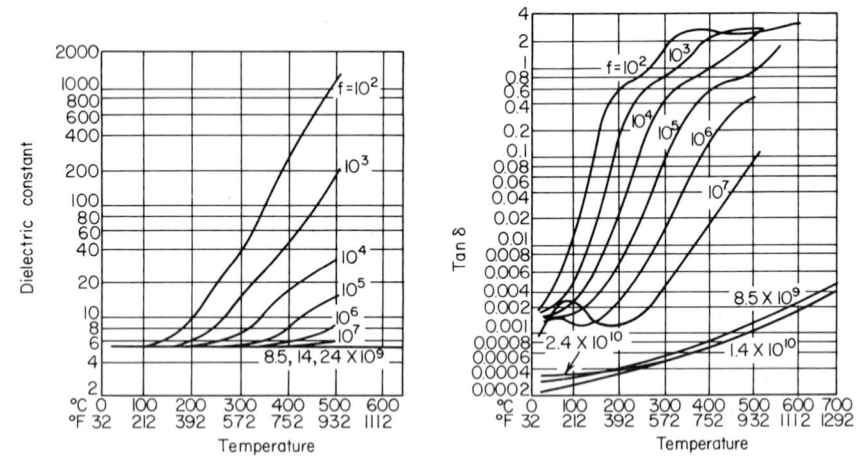

Fig. 51 Dielectric constant and loss tangent of Pyroceram 9606 glass as a function of temperature at various frequencies. (*Corning Glass Works.*)

work (surface free energy) must be done against the internal forces to bring additional atoms to the surface and create new surface area.

The strongest glass-to-metal bond results when the interfacial energy is lowered. This occurs when the metal ions from the metal oxide interdiffuse with the oxygen ions of the glass. If the metal oxide layer is dissolved or is very thin, a weaker (van der Waals) bond is produced.[54]

Stresses The stresses that develop in a glass-to-metal seal are proportional to the differential contraction between the two materials. When molten glass is first joined to a metal, no stresses develop until the glass becomes rigid. Glass is considered rigid at its setting point, which is approximately 20°C below its annealing temperature. Below the setting point, the differential contraction rate of the metal and the glass will determine the stresses in the seal (see Table 19). Tangential stresses in annealed tubular joints can be calculated from the relation $\sigma_t = \bar{E} \Delta L/2L$, where σ_t is the tangential stress, $\bar{E}$ is the average of the elastic moduli for the two materials, and $\Delta L/L$ is the differential contraction between the two materials at the setting point of the glass, or if two glasses are joined at the lower setting point. Stresses from 0.5 to 1.5 kg/mm² can be tolerated in tubular butt seals, depending on the size and quality of the seal. A method for measuring residual stresses in cylindrical metal-to-glass seals is given in ASTM Standard F 218-50.

A useful guide for evaluating combinations of sealing materials is given in Tables 20 and 21 in terms of the differential expansion between glasses, ceramics, or metals. The following general rules apply: (1) Differential expansions <100 ppm provide very good sealing conditions, (2) differential expansions between 100 and

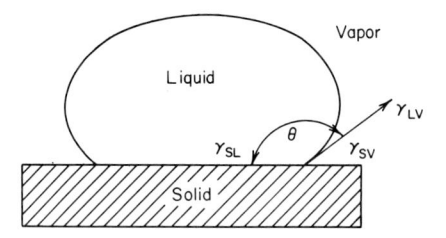

Fig. 52 Relationship between interfacial energy and contact angle of a sessile drop.

500 ppm satisfy conditions for medium-size seals, and (3) differential expansions between 500 and 1,000 ppm are tolerable for either small seals of short length or seals in which the critical stresses are compressive.

Housekeeper Seals. Seals of this type are used to join relatively soft metals (e.g., copper) that have high rates of thermal expansion to glasses that have low rates of thermal expansion. The seal is made by forming a taper or feather edge on the metal, and then sealing the glass to the taper (see Fig. 53). Mismatch between the expansion rates of the metal and glass is equalized by the flexibility of the thin taper. This seal was originally developed for high-power water-cooled radio tubes.

Kovar Seals. Kovar is a Fe-Ni-Co alloy with a thermal expansion rate which matches that of several hard or low-expansion glasses such as Corning 7052 and 7040. It is available as a powder as well as in tube, sheet, and rod stock form. An advantage of Kovar seals is that they can be fired in air since the oxidation rate of Kovar is low compared to the softening rate of hard glass (see Fig. 54). The problem of power losses at high frequencies and the resultant heating of the glass-metal joint can be overcome by plating the Kovar with silver, gold, or chromium, to lower its effective resistance.[55] Kovar seals are used extensively in vacuum technology and in packaging of solid-state components.

Fused Silica-to-Metal Seals. The high viscosity of fused silica makes it particularly suitable for glass-to-metal seals for high-temperature applications, e.g., seals for mercury-arc lamps. Since fused silica has a very low thermal expansion rate, special attention is given to minimizing the strain between the metal and glass. These techniques include using an intermediate glass between the fused silica and metal and making the metal leads very thin. The intermediate glass method can be used to seal tungsten rods as large as 3 mm in diameter to fused silica.[56]

TABLE 17 Comparison of Properties of Pyroceram, Glass and Ceramic[13]

Property	Pyroceram* 9606	Pyroceram* 9608	Glass Fused silica 7940	Glass Vycor* 7900	Glass Pyrex* 7740	Glass Lime glass 0080	High-purity alumina (93%+)	Ceramic Steatites $MgO-SiO_2$	Ceramic Forsterite $2MgO-SiO_2$
Specific gravity at 25°C	2.61	2.50	2.20	2.18	2.23	2.47	3.6	2.65–2.92	2.9
Water absorption, %	0.00	0.00	0.00	0.00	0.00	0.00	0.00	0–0.03	0–0.01
Gas permeability	0	0	0	0	0	0	0	......	0
Thermal:									
Softening temperature, °C†	......	......	1584	1500	820	696	1700	1349	1349
Specific heat (25°C)	0.185	0.190	0.176	0.178	0.186	0.200	0.181		
Specific heat mean (25–400°C)	0.230	0.235	0.223	0.224	0.233	0.235	0.241		
Thermal conductivity, cgs, at 25°C mean temperature	0.0087	......	0.0032	0.0036	0.0026	0.0025	0.042–0.086¶ 73(20–500°C)	0.0062–0.0065 81.5–99. (20–500°C)	0.010 99. (20–500°C)
Linear coefficient of thermal expansion at 25–300°C, 10^{-7} in./(in.)(°C)	57	4–20‡	5.5	8	32	92			
Mechanical:									
Modulus of elasticity, psi × 10^{-6}	17.1	12.5	10.5	10.0	9.1	10.2	40–50	15	20
Poisson's ratio	0.25	0.25	0.17	0.19	0.20	0.22	0.21		
Modulus of rupture (abraded), psi × 10^{-3}	18–20	12–14	5–9	5–9	6–10	6–10	50§	20§	19§
Knoop hardness:									
100 g	657	593	......	463	418	......	1850		
500 g									
Electrical:									
Dielectric constant:									
At 1 MHz:									
25°C	5.58	6.78	3.78	3.8	4.6	7.2	8.81	5.9	6.3
300°C	5.60	......	......	3.9	5.9		9.03		
500°C	8.80								
At 10 GHz:									
25°C	5.45	6.54	3.78	3.8	4.5	6.71	8.79	5.8	5.8
300°C	5.51	6.65	3.78				9.03		
500°C	5.53	6.78	3.78						

Dissipation factor:								
At 1 MHz:								
25°C	0.0015	0.0030	.....	0.0005	0.0046	0.009	0.00035	0.0013	0.0003
300°C	0.0154	.....	.....	0.0042	0.0130	.....	0.012		
500°C	.....	.....	.....	.....	.....	.....			
At 10 GNz:								
25°C	0.00033	0.0068	.....	0.0009	0.0085	0.017	0.0015	0.0014	0.0010
300°C	0.00075	0.0115	.....	.....	.....	.....	0.0021		
Lo 500°C	0.00152	0.040	.....	.....	.....	.....			
ss factor:								
At 1 MHz:								
25°C	0.008	0.02	.....	0.0019	0.0212	0.065	0.0031	0.0077	0.0019
300°C	0.086	.....	.....	0.0164	0.0566	.....	0.108		
500°C	.....	.....	.....	.....	.....	.....			
At 10 GHz:								
25°C	0.002	0.045	.....	0.0036	0.0382	0.114	0.0132	0.0082	0.0058
300°C	0.004	0.077	.....	.....	.....	.....	0.019		
500°C	0.008	0.27	.....	.....	.....	.....			
Volume resistivity ($\log_{10}$), ohm-cm:								
At 250°C	10	8.1	12.0	9.7	8.1	6.4	14.0(100°C)	14(20°C)	14(20°C)
At 350°C	8.6	6.8	9.7	8.1	6.6	5.1	12.95(300°C)		

* Trademarks of Corning Glass Works, Corning, N.Y.
† Softening temperature; method of evaluation: (a) glass: ASTM C 338-54T; (b) ceramics: ASTM C 24-56.
‡ Expansion coefficients depend on heat treatment.
§ Unabraded values.
¶ 99.5 % pure; 98.0 % dense.

6-47

TABLE 18 Glass-Metal Combinations for Seals[92]

Metal or alloy	Trade name or type	Thermal conductivity, cal/(sec)(cm)(°C)	Electrical resistivity at 20°C, ohm-cm × 10⁻⁶	Matching glass Corning	Matching glass Other	Remarks
Cold-rolled steel, SAE 1010, AlSi C1010	Other grades of soft steel or iron also used	0.108	18	1990 1991	5643[a]	Tends to oxidize excessively. Plating with Cu, Cr, or Ag often used to prevent this. External rings of iron are frequently sealed to Pt-sealing glasses.
17% chrome iron, Al51 430A	Allegheny Telemet[c]	0.06	72	0129 9019	3720[a]	Used in metal TV picture tube as a ring seal outside a glass plate of lower expansion, putting glass in compression.
28% chrome iron	Allegheny Sealment-1[c] Ascaloy 446[d] Carpenter-27	0.06	72	9012	K-51[b]	Used in TV tubes. Pt-sealing glasses can also be used under certain conditions and types of seals. Alloy should be preoxidized in wet H₂ gas.
Platinum		0.165	10.6	0010	R-5[b] R-6[b]	Now used mainly for scientific apparatus. Preoxidizing unnecessary.
Composite material: core, 42 Ni, 58 Fe Sheath, Cu	Dumet[d]	0.04	4–6	0080 0120 7570	KG-12[b]	Copper sheath bonded to core. Surface usually coated with borax to reduce oxidation. Expansion: Radial, 90 × 10⁻⁷ per °C. Axial, 63 × 10⁻⁷ per °C. Wire size usually limited to 0.020 in. D.
Nickel-chrome iron: 42 Ni, 6 Cr, 52 Fe	Sylvania 4 Allegheny Sealmet HC-4 Carpenter 426	0.032	34	8160 9010		Matches Pt-sealing glasses. Relatively large seals can be made between this alloy and suitable glasses. Pretreatment in H₂ atmosphere furnace essential.
Nickel-cobalt iron: 28 Ni, 18 Co, 53 Fe	Kovar[d] Fernico[e] Rodar[f]	0.046	47	7040 7050 7052	K-650[b] K-704[b] K-705[b]	Low-expansion sealing alloy. Should be annealed after cold working and pretreated in H₂ atmosphere furnace.
Molybdenum		0.35	5.7	7040 7052	7055 8830	Metal rods usually ground and sometimes polished. Surfaces should be cleaned in fused nitrite and not overoxidized.
Tungsten		0.38	5.5	3320 7720	1720	Same as for molybdenum.
Copper	OFHC grade (oxygen-free high-conductivity)	0.92	1.75	Most glasses	7050 8830 K-772[b]	Copper-glass seals are made with Housekeeper technique with thin metal sections. Care must be taken to prevent overoxidation of copper.

Physical properties of metals and alloys from Monack and Partridge.
[a] Trademarks of Pittsburgh Plate Glass Company, Pittsburgh, Pa.
[b] Trademarks of Kimble Glass Division, Ohens-Illinois Glass Company, Montclair, N.J.
[c] Trademark of Allegheny Ludlum Steel Corp., Windsor, Conn.
[d] Trademark of Westinghouse Electric Corporation, Pittsburgh, Pa.
[e] Trademark of General Electric Company, Schenectady, N.Y.
[f] Trademark of Wilbur B. Driver Co., Newark, N.J.

Because of its low dielectric losses at microwave frequencies, fused silica is used for microwave windows. By first metallizing the silica with TiH_2 and a solder containing In, Sn, and Pb, which are needed to form an adherent alloy, the silica can be sealed to metallic waveguides. In this process, the metallized silica is clamped to the metal part, which must be coated with either Ag or Au, and then the combination is heat treated to produce a diffusion joint.[57]

TABLE 19 Stresses in Glass Parts of Seals—Metal Rod or Wire Surrounded by Glass

Type of stress	Glass contraction greater than metal	Glass contraction less than metal
Axial	Tensile	Compressive
Tangential	Tensile	Compressive
Radial	Compressive	Tensile

Ceramic-to-metal seals These seals are commonly used in vacuum tubes, semiconductor devices, feed-through insulators, bushings, and transformer and capacitor terminals. They are particularly useful in operation either at high temperatures or in corrosive atmospheres. Properties that often dictate the use of ceramic-to-metal rather than glass-to-metal seals are greater strength, better dielectric ratings, greater refractoriness, and better themal shock resistance. The principal disadvantages are opacity to most radiations and high cost when compared to glass-to-metal seals.

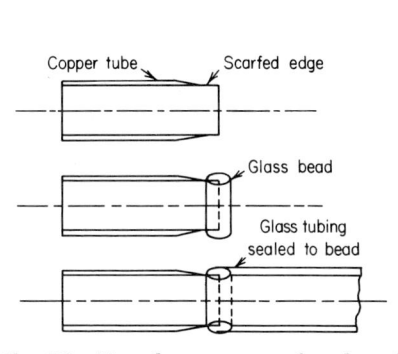

Fig. 53 Housekeeper type of seal, and procedure for making it.[92]

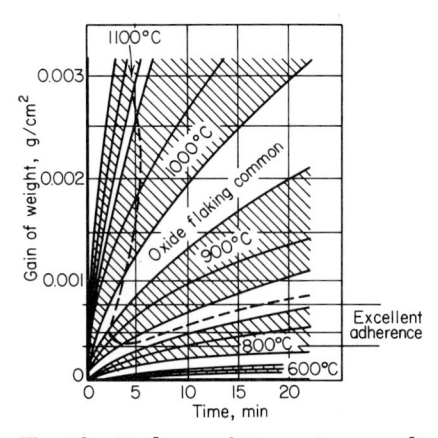

Fig. 54 Oxidation of Kovar: Areas inside V-shaped dotted curve indicate conditions under which greatest tendency for oxide flaking exists.[54]

Bonding The mechanism of ceramic-to-metal bonding depends on the sealing technique that is used, and it may be due to chemical reaction between the seal members, mechanical interlocking of microscopically rough surfaces, interfacial diffusion, or interpenetration of glassy phases.

Molybdenum-Manganese Process. The sintered-metal-powder process is used to form very high strength seals for klystrons, traveling-wave tubes, and large feed-through insulators. A slurry, which can be deposited by spraying, painting, or

TABLE 20 Differential Expansion between Glasses, Ceramics or Metals[148]—
Low Expansion Range (differential in ppm)

Metal	Glass code no.	7761	7070	7740	7760	1715	7720	7251	5420	9700	7780	9741	3320	7750	1720	7331	1723	7052	7050	7040	1826	8830	7510
	7761	0	380	50	210	230	400	630	210	180	470	860	400	770	640	610	740	+	890	+	+	+	980
	7070	380	0	290	160	130	20	290	150	150	50	680	60	520	300	250	350	730	500	700	+	+	600
	7740	50	290	0	180	90	470	550	100	50	400	690	440	630	550	690	680	880	690	940	+	650	970
	7760	210	160	180	0	30	230	400	310	50	230	140	180	550	420	390	650	850	630	820	+	990	800
	1715	230	130	90	30	0	310	330	90	470	400	50	50	550	890	510	980	790	800	820	830	850	810
	7720	400	20	470	230	310	0	120	270	550	310	330	190	480	230	570	240	720	660	770	830	830	560
Tungsten	7251	630	290	550	400	330	120	0	360	250	0	550	90	450	150	140	680	660	520	690	810	540	390
	5420	210	150	220	0	80	270	360	0	50	180	410	310	80	350	120	520	320	680	800	610	690	910
	9700	180	100	310	50	190	250	0	80	180	250	570	190	510	80	440	610	480	250	690	910	850	610
	7780	470	50	400	260	230	90	180	250	200	270	200	150	470	290	330	390	760	630	800	750	630	520
	9741	860	680	690	630	550	180	570	550	580	550	0	100	500	450	550	480	710	650	870	870	840	840
	3320	400	60	440	180	310	90	180	190	150	310	0	0	430	190	150	290	750	630	800	800	790	620
	7750	770	520	630	550	480	450	80	510	470	500	430	320	0	220	240	560	660	690	+	790	740	390
	1720	640	300	580	420	390	570	140	120	350	380	290	450	190	0	180	180	450	450	+	860	840	790
	7331	610	250	690	390	570	140	120	390	390	550	150	170	500	450	0	20	110	450	170	590	640	520
	1723	740	390	680	510	980	180	310	550	560	480	560	290	550	190	170	0	20	110	80	40	60	60
	7052	+	730	880	790	720	680	320	440	250	560	0	80	430	150	150	80	0	0	50	50	30	20
	7050	+	730	+	720	720	680	320	440	250	+	180	140	190	150	240	140	120	80	110	10	20	10
	7052	890	500	820	690	660	520	250	680	550	+	480	260	320	180	260	310	0	0	20	0	10	0
	7050	+	820	+	+	×	×	×	×	×	×	110	560	480	250	350	×	0	×	180	×	×	280
Molybdenum	7040	+	690	+	820	+	550	+	0	0	180	40	0	10	0	0	40	20	270	×	70	75	15
	1826	+	700	940	820	770	690	370	690	610	+	110	560	480	250	350	40	270	380	180	80	10	10
	8830	+	+	+	+	+	+	990	610	+	+	+	40	640	180	540	240	890	510	90	80	10	120
	7510	980	600	970	800	810	560	390	740	420	670	420	180	790	370	360	70	0	250	380	300	10	120

(columns at right, top-to-bottom labels):
7550, 46 Ni-Fe, 42 Ni-Fe, 7330, Alumina, 7340, 7280, 7520, 8800, 7056, 7055, Kovar, 7510, 8830, 1826, 7040, Molybdenum, 7050, 7052, 1723, 7331, 1720, 7750, 3320, 9741, 7780, 9700, 5420, 7251, 7720, 1715, 7760, 7740, 7070, 7761

6-50

Kovar		+	770	+	+	+	860	×	+	+	230	+	890	+	20	270	×	70	320	140	280	0	130	130	30	30	260	260	570	×
	7055	+	690	+	+	910	700	900	860	940	710	240	480	510	380	0	180	80	10	300	30	120	130	0	0	320	320	520	630	750	860	840	—
	7056	+	690	+	+	910	920	910	850	940	720	240	480	510	390	0	180	80	10	300	30	120	130	0	0	320	320	520	630	740	870	850	—
	8800	+	+	+	+	+	980	+	+	+	+	620	+	890	+	390	530	710	380	80	370	450	30	320	320	0	0	10	290	270	610	810	—
	7520	+	+	+	+	+	980	+	+	970	470	620	+	890	+	390	530	710	380	80	370	450	30	320	320	0	0	10	290	270	610	810	—
	7280	+	+	+	+	+	+	+	+	970	470	+	+	+	+	590	730	950	580	280	570	630	260	520	520	10	10	0	330	160	550	—	—
	7340	+	+	+	+	+	+	+	+	+	650	+	810	+	+	690	840	+	680	380	680	760	260	630	630	290	290	330	0	80	310	530	800
Alumina		+	+	+	+	+	+	+	+	+	750	+	910	+	+	800	940	+	800	480	800	870	×	750	740	270	270	160	80	0	300	×	×	730
	7530	+	+	+	+	+	+	+	+	+	830	+	+	+	+	900	+	+	900	580	910	990	570	860	870	610	610	550	310	300	0	210	450	530
42 Ni-Fe		+	+	+	+	+	+	+	+	+	940	+	+	+	+	+	+	×	720	270	840	+	×	840	850	810	810	+	530	×	×	210	0	×	330
46 Ni-Fe		+	+	+	+	+	+	+	+	+	490	+	+	+	+	+	+	×	+	680	+	+	×	+	+	+	+	+	+	750	×	450	0	80
	7550	+	+	+	+	+	+	+	+	+	960	+	+	+	+	+	+	+	+	920	+	+	+	+	+	+	+	+	990	800	730	530	330	0
	0280	+	+	+	+	—	—	+	+	+	+	+	+	+	+	+	+	+	+	+	+	+	+	+	+	+	+	+	+	950	750	570	320	250
Platinum		+	+	+	+	—	—	+	+	+	+	+	+	+	+	+	+	×	+	+	+	×	+	+	+	+	+	+	+	+	×	780	×	320
	7570	+	+	+	+	—	—	+	+	+	+	+	+	+	+	+	+	+	+	+	+	+	+	+	+	990	990	840	770	720	620	+	560	360

NOTE: Expansion differentials between Corning glasses and metals or ceramics in parts per million are given at the setting point of the glass for glass-to-metal (or ceramic) seals, and at the setting point of the softer glass for glass-to-metal seals. The setting point is arbitrarily chosen as lying 5°C above the strain point of the glass. A minus sign above a number indicates that for this combination of materials, the one on the ordinate has a lower effective expansion than the one on the abscissa. So too, a large plus or minus sign indicates that the effective expansion is in excess of 1,000 ppm.

6-51

TABLE 21 Differential Expansion between Glasses, Ceramics or Metals[148]—High Expansion Range (differential in ppm)

Metal	Glass code no.	7340	Alumina	7530	42 Ni Fe	46 Ni Fe	7550	0280	Platinum	7570	0041	7560	0281	28% Cr steel	0122	9010	Sylvania no. 4	0120	8161	9361	8870	0050	8160	0081	0010	52 Ni Fe	0080	0088	9823	9821	8361	0128	430 Ti steel	9019	0129	8871	7290	1010 steel	1990
	8800	$\overline{290}$	$\overline{270}$	$\overline{610}$	$\overline{810}$	—	—	—	—	$\overline{990}$	—	—	—	—	—	—	—	—	—	—	—	—	—	—	—	—	—	—	—	—	—	—	—	—	—	—	—	—	—
	7520	$\overline{290}$	$\overline{270}$	$\overline{610}$	$\overline{810}$	—	—	—	—	$\overline{990}$	—	—	—	—	—	—	—	—	—	—	—	—	—	—	—	—	—	—	—	—	—	—	—	—	—	—	—	—	—
	7280	$\overline{330}$	$\overline{160}$	$\overline{550}$	—	—	$\overline{990}$	—	—	$\overline{840}$	—	—	—	—	—	—	—	—	—	—	—	—	—	—	—	—	—	—	—	—	—	—	—	—	—	—	—	—	—
	7340	0	80	$\overline{310}$	$\overline{530}$	$\overline{750}$	$\overline{800}$	—	—	$\overline{770}$	$\overline{970}$	—	—	—	—	—	—	—	—	—	—	—	—	—	—	—	—	—	—	—	—	—	—	—	—	—	—	—	—
Alumina		$\overline{80}$	0	$\overline{300}$	×	×	$\overline{730}$	$\overline{950}$	×	$\overline{720}$	$\overline{880}$	—	—	×	$\overline{970}$	—	×	—	$\overline{950}$	—	—	—	—	—	—	×	—	—	—	—	—	—	×	—	—	—	—	×	—
	7530	310	300	0	$\overline{210}$	$\overline{450}$	$\overline{530}$	$\overline{750}$	$\overline{780}$	$\overline{620}$	$\overline{780}$	—	—	—	$\overline{870}$	$\overline{900}$	—	$\overline{890}$	$\overline{840}$	$\overline{980}$	$\overline{930}$	$\overline{950}$	$\overline{960}$	—	$\overline{970}$	—	—	—	—	—	—	—	—	—	—	—	—	—	—
42 Ni Fe		530	×	210	0	×	$\overline{330}$	$\overline{570}$	×	—	—	$\overline{860}$	$\overline{970}$	×	—	—	×	—	—	—	—	—	—	—	—	×	—	—	—	—	—	—	×	—	—	—	—	×	—
46 Ni Fe		750	×	450	×	0	$\overline{80}$	$\overline{320}$	×	$\overline{560}$	$\overline{720}$	$\overline{600}$	$\overline{680}$	×	$\overline{830}$	$\overline{880}$	×	$\overline{860}$	$\overline{820}$	$\overline{890}$	$\overline{900}$	$\overline{910}$	$\overline{930}$	$\overline{870}$	$\overline{940}$	×	$\overline{950}$	$\overline{940}$	×	$\overline{860}$	$\overline{980}$	—	×	—	—	—	—	×	—
	7550	800	730	530	330	80	0	$\overline{250}$	$\overline{320}$	$\overline{360}$	$\overline{450}$	$\overline{540}$	$\overline{610}$	—	$\overline{550}$	$\overline{590}$	—	$\overline{590}$	$\overline{540}$	$\overline{640}$	$\overline{640}$	$\overline{630}$	$\overline{660}$	$\overline{780}$	$\overline{660}$	$\overline{570}$	$\overline{850}$	$\overline{850}$	$\overline{890}$	$\overline{790}$	$\overline{920}$	—	—	—	—	—	—	—	—
	0280	+	950	750	570	320	250	0	$\overline{110}$	$\overline{240}$	$\overline{280}$	$\overline{300}$	$\overline{380}$	$\overline{860}$	$\overline{400}$	$\overline{440}$	—	$\overline{440}$	$\overline{390}$	$\overline{470}$	$\overline{490}$	$\overline{480}$	$\overline{510}$	$\overline{560}$	$\overline{520}$	$\overline{350}$	$\overline{630}$	$\overline{630}$	$\overline{700}$	$\overline{560}$	$\overline{680}$	$\overline{820}$	—	$\overline{920}$	$\overline{980}$	$\overline{920}$	—	—	—
Platinum		+	×	780	×	×	320	110	0	60	30	$\overline{130}$	$\overline{160}$	×	$\overline{90}$	$\overline{120}$	×	$\overline{130}$	$\overline{80}$	$\overline{160}$	$\overline{180}$	$\overline{170}$	$\overline{200}$	$\overline{300}$	$\overline{210}$	×	$\overline{370}$	$\overline{380}$	$\overline{410}$	$\overline{340}$	$\overline{510}$	$\overline{520}$	×	$\overline{620}$	$\overline{680}$	$\overline{610}$	—	×	—
	7570	770	720	620	+	560	360	240	$\overline{60}$	0	120	120	50	480	40	$\overline{10}$	50	$\overline{70}$	$\overline{10}$	$\overline{80}$	$\overline{130}$	$\overline{70}$	$\overline{100}$	$\overline{60}$	$\overline{110}$	$\overline{260}$	$\overline{110}$	$\overline{110}$	$\overline{170}$	$\overline{80}$	$\overline{210}$	$\overline{210}$	$\overline{610}$	$\overline{300}$	$\overline{320}$	$\overline{620}$	—	—	—
	0041	970	880	780	+	720	450	280	$\overline{30}$	$\overline{120}$	0	110	30	$\overline{630}$	$\overline{210}$	$\overline{230}$	$\overline{380}$	$\overline{270}$	$\overline{210}$	$\overline{180}$	$\overline{320}$	$\overline{290}$	$\overline{340}$	$\overline{110}$	$\overline{350}$	$\overline{210}$	$\overline{170}$	$\overline{190}$	$\overline{290}$	$\overline{160}$	$\overline{290}$	$\overline{410}$	$\overline{800}$	$\overline{510}$	$\overline{550}$	$\overline{780}$	—	—	—
	7560	+	+	+	860	600	540	300	130	$\overline{120}$	$\overline{110}$	0	$\overline{120}$	$\overline{620}$	$\overline{250}$	$\overline{270}$	$\overline{740}$	$\overline{290}$	$\overline{240}$	$\overline{300}$	$\overline{340}$	$\overline{330}$	$\overline{360}$	$\overline{330}$	$\overline{370}$	$\overline{70}$	$\overline{410}$	$\overline{390}$	$\overline{500}$	$\overline{300}$	$\overline{380}$	$\overline{640}$	$\overline{860}$	$\overline{730}$	$\overline{790}$	$\overline{780}$	—	—	—
	0281	+	+	+	970	680	610	380	160	$\overline{50}$	$\overline{30}$	120	0	$\overline{570}$	$\overline{170}$	$\overline{200}$	$\overline{620}$	$\overline{220}$	$\overline{160}$	$\overline{220}$	$\overline{270}$	$\overline{250}$	$\overline{290}$	$\overline{220}$	$\overline{300}$	0	$\overline{320}$	$\overline{290}$	$\overline{400}$	$\overline{180}$	$\overline{290}$	$\overline{550}$	$\overline{800}$	$\overline{630}$	$\overline{700}$	$\overline{710}$	—	—	—
28% Cr. steel		+	×	+	×	×	+	860	×	480	630	620	570	0	470	450	×	410	470	450	350	390	340	410	330	×	330	330	270	390	240	130	×	40	$\overline{20}$	$\overline{170}$	$\overline{880}$	×	—
	0122	+	970	870	+	830	550	400	90	$\overline{40}$	210	250	170	$\overline{470}$	0	$\overline{10}$	$\overline{150}$	$\overline{90}$	$\overline{30}$	$\overline{20}$	$\overline{180}$	$\overline{90}$	$\overline{160}$	40	$\overline{170}$	$\overline{100}$	$\overline{20}$	$\overline{30}$	$\overline{120}$	0	$\overline{150}$	$\overline{200}$	$\overline{630}$	$\overline{310}$	$\overline{350}$	$\overline{700}$	—	—	—
	9010	+	+	900	+	880	590	440	120	10	230	270	200	$\overline{450}$	10	0	$\overline{130}$	$\overline{60}$	0	0	$\overline{150}$	$\overline{70}$	$\overline{140}$	60	$\overline{150}$	$\overline{70}$	0	$\overline{10}$	$\overline{100}$	20	$\overline{120}$	$\overline{190}$	$\overline{600}$	$\overline{300}$	$\overline{330}$	$\overline{640}$	—	—	—
Sylvania no. 4		+	×	+	×	×	+	+	×	$\overline{50}$	380	740	620	×	150	130	0	60	130	220	$\overline{30}$	60	$\overline{20}$	410	$\overline{30}$	×	310	350	180	460	360	$\overline{30}$	×	90	$\overline{160}$	$\overline{670}$	—	×	—
	0120	+	+	890	+	860	590	440	130	70	270	290	220	$\overline{410}$	90	60	$\overline{60}$	0	60	30	$\overline{70}$	$\overline{10}$	$\overline{70}$	80	$\overline{80}$	$\overline{60}$	30	20	$\overline{60}$	50	$\overline{100}$	$\overline{140}$	$\overline{570}$	$\overline{250}$	$\overline{280}$	$\overline{580}$	—	—	—
	8161	+	950	840	+	820	540	390	80	10	210	240	160	$\overline{470}$	30	0	$\overline{130}$	$\overline{60}$	0	$\overline{20}$	$\overline{140}$	$\overline{70}$	$\overline{140}$	30	$\overline{150}$	$\overline{120}$	$\overline{30}$	$\overline{40}$	$\overline{120}$	0	$\overline{150}$	$\overline{200}$	$\overline{620}$	$\overline{320}$	$\overline{340}$	$\overline{650}$	—	—	—
	9361	+	+	980	+	890	640	470	160	80	180	300	220	$\overline{450}$	20	0	$\overline{220}$	$\overline{30}$	20	0	$\overline{90}$	$\overline{60}$	$\overline{110}$	80	$\overline{120}$	$\overline{20}$	0	0	$\overline{110}$	30	$\overline{110}$	$\overline{240}$	$\overline{620}$	$\overline{340}$	$\overline{390}$	$\overline{580}$	—	—	—
	8870	+	+	930	+	900	640	490	180	130	320	340	270	$\overline{350}$	180	150	30	70	140	90	0	70	10	140	0	$\overline{20}$	80	80	0	110	$\overline{40}$	$\overline{70}$	130	$\overline{500}$	$\overline{180}$	$\overline{220}$	$\overline{530}$	—	—
	0050	+	+	950	+	910	630	480	170	70	290	330	250	$\overline{390}$	90	70	$\overline{60}$	10	70	60	$\overline{70}$	0	$\overline{70}$	120	$\overline{80}$	$\overline{30}$	60	40	$\overline{40}$	80	$\overline{70}$	$\overline{120}$	$\overline{540}$	$\overline{240}$	$\overline{270}$	$\overline{590}$	—	—	—

8160	+	930	660	200	100	340	360	290	370	160	140	20	70	140	110	1̄0	70	0	160	1̄0	0	100	9̄0	10	120	3̄0	6̄0	500	180	220	550	—	—	—	—	—	—	—
0081	+	960	+	870	780	560	300	60	110	330	220	4̄0	6̄0	3̄0	80	3̄0	120	1̄0	0	160	180	1̄00	60	230	30	90	390	60	620	480	540	600	—	—	—	—	—
0010	+	+	970	940	660	520	210	110	350	370	300	330	170	50	30	150	120	20	80	10	160	0	10	110	100	20	130	5̄0	2̄0	170	480	200	540	—	—	—
52 Ni Fe	+	×	+	×	570	350	×	260	210	70	0	×	100	70	×	30	20	20	30	0	180	1̄0	0	230	270	190	310	1̄0	350	×	460	520	410	—	×	—
0080	+	+	+	+	950	850	630	370	110	170	410	320	3̄0	0	3̄10	30	0	8̄0	60	1̄00	100	110	1̄00	0	50	140	110	1̄0	326	550	400	460	550	—		
0088	+	+	+	+	940	850	630	380	110	190	390	290	330	10	350	2̄0	0	0	8̄0	4̄0	90	60	100	230	0	5̄0	100	2̄0	310	5̄0	440	460	550	—		
9823	+	+	+	+	+	890	700	410	170	290	500	400	270	120	100	180	30	40	1̄0	40	1̄0	230	2̄0	270	140	1̄50	0	210	70	1̄50	470	270	330	490		
9821	+	+	+	+	+	860	790	560	340	80	160	300	180	390	0	2̄0	460	5̄0	0	3̄0	1̄10	8̄0	120	3̄0	190	1̄10	1̄00	210	0	110	350	620	440	500	570	
8361	+	+	+	+	+	980	920	680	510	210	290	380	290	240	150	120	360	100	150	110	40	70	30	90	20	310	10	20	7̄0	110	0	210	480	310	370	440
0128	+	+	+	+	+	+	820	520	210	410	640	550	130	200	190	30	140	260	240	70	120	60	390	50	350	320	310	190	350	210	0	310	7̄0	130	440	
430 Ti steel	+	×	+	×	+	×	610	800	860	800	×	570	620	620	130	540	500	620	480	×	550	550	470	620	480	310	0	230	170	3̄0	720	×				
9019	+	+	+	+	+	+	920	620	300	510	730	630	4̄0	310	300	90	250	340	500	240	180	480	170	460	400	270	440	310	70	230	0	7̄0	340	—		
0129	+	+	+	+	+	+	+	980	680	320	550	790	700	20	350	330	160	280	340	390	180	270	220	540	200	520	460	330	500	370	130	170	70	0	320	—
8871	+	+	+	+	+	+	+	920	610	620	780	780	710	170	700	640	670	580	650	580	220	590	550	600	540	410	550	490	330	370	440	30	340	320	0	500	790	810
7290	+	+	+	+	+	+	+	+	+	+	+	+	880	+	+	+	+	+	+	+	530	+	+	+	+	+	+	+	+	+	720	+	+	500	0	220	330
1010 steel	+	×	+	×	+	+	+	+	+	+	+	+	+	×	+	+	+	+	+	+	+	+	+	+	+	+	×	+	×	+	×	770	+	+	790	220	0	5̄0
1990	+	+	+	+	+	+	+	+	0	+	+	+	+	+	+	+	+	+	+	+	+	+	+	+	+	+	+	+	+	+	+	0	+	+	810	330	30	0

NOTE: Expansion differentials between Corning glasses and metals or ceramics in parts per million are given at the setting point of the glass for glass-to-metal (or ceramic) seals, and at the setting point of the softer glass for glass-to-metal seals. The setting point is arbitrarily chosen as lying 5°C above the strain point of the glass. A minus sign above a number indicates that for this combination of materials, the one on the ordinate has a lower effective expansion than the one on the abscissa. So too, a large plus or minus sign indicates that the effective expansion is in excess of 1,000 ppm.

6-53

silk screening, is made from powdered molybdenum and manganese, with small additions of Fe, Si, TiO_2, Al_2O_3, and CaO. The bond is formed between the ceramic and metal parts by heating them to between 1300 and 1500°C in a wet hydrogen atmosphere. At these temperatures, the oxides form a eutectic melt that reacts with the ceramic and the oxide on the metal and with the sintered molybdenum and manganese to form the bond. Low-purity alumina (94 to 96 percent Al_2O_3) will usually yield a stronger bond than high-purity alumina (99 percent Al_2O_3) because of the larger amount of eutectic melt formed by the impurities in the former. Evidence indicates that the bond strength of alumina-to-metal seals increases with the average grain size of the alumina ceramic (see Fig. 55).[58]

Active-metal Process. This process is based on the ability of Ti and Zr metals to adhere to the oxide surfaces of the parts to be joined without the assistance of a fluxing phase. It is accomplished at a lower temperature than the moly-manganese process; however, the bond strength is slightly lower as well. With this technique, seals can be made between different ceramics, or ceramics and metals, or different metals. For example, the ceramic and metal parts can be coated with a slurry of TiH_2 or ZrH_2 in nitrocellulose and bonded together by heating to 900°C in vacuum to decompose the hydrate and release the free metal. In

TABLE 22 Comparison of Printed-film Resistor Systems[91]

Property	Carbon	Metal	Cermet
Resistivity range (sheet)	100 ohms–20 megohms	1 ohm–15 kilohms	2–50 kilohms
Tolerance, %	±10	±0.5	±0.5
Temperature coefficient, ppm per °C	−2,000	±500	±150
Voltage coefficient, ppm/volt	500	30	20
Noise, μv/volt per frequency decade	2	0.1	0.5
Typical load life*	−7	+0.25 (+0.025% on alkali-metal free underglaze)	+0.25

* 1,000 hr at 85°C with 0.2 watt on a 0.050- by 0.150-in. resistor.

metal-to-metal joining, the seal is formed by the alloying effect of the Ti or Zr, whereas in either metal-to-ceramic or ceramic-to-ceramic joining, it is formed by chemical reaction.[59]

Thick-film Components

Circuit networks can be built up by depositing resistors, capacitors, and interconnectors in the form of films on insulating substrates. Thick films are usually made by silk screening, brushing, dipping, or spraying, whereas thin films are made by vacuum deposition, sputtering, or chemical deposition. The temperature of the substrate depends on the number of resistors, the power dissipation per unit area, and the cooling method. Techniques for optimizing the thermal conditions by the correct placement of the resistors and various cooling techniques are available.[60] This section describes several of the materials used in making thick-film components.

Resistors Thick-film resistor technology can be divided into three areas, namely, carbon, cermet, and precious metal films. A general comparison of their electrical properties is given in Table 22.

Carbon. One of the materials first used for thick-film resistors was carbon. Early carbon-film resistors had a negative coefficient of resistance (NTC) from 0.01 to 0.05 percent per °C, whereas later resistors had an NTC from 0.005 to 0.02 percent per °C.[61] This improvement in performance was achieved by adding boron to help prevent oxidation of the carbon. A metallic dispersion in the carbon film also improves the temperature coefficient; the NTC of the carbon is balanced by the PTC of the metallic dispersion. Carbon and boron-carbon films have the highest electrical noise level of the film-type resistors. Noise is generated by the small fluctuations in the resistance caused by imperfections, variations in the particle-to-particle contact, and hot spots.

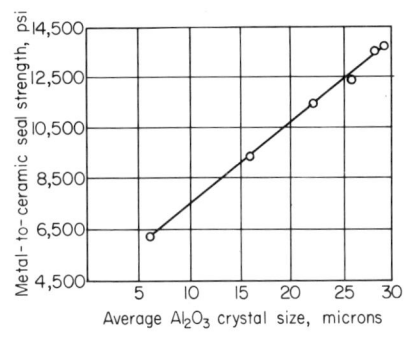

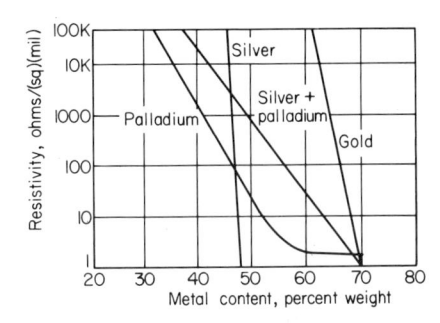

Fig. 55 Effect of alumina crystal size on metal-to-ceramic bond strength when metallized at 1550°C.[58]

Fig. 56 Relationship between resistivity and concentrations of metal powders in $PbO-B_2O_3-SiO_2$ frit.[93]

Cermets. Resistor films of this type are usually either mixtures of precious metals (e.g., Ag or Au) and glass frit or mixtures of PdO and glass frit. Cermet resistive inks are made by mixing a cermet powder with an organic vehicle of suitable viscosity. This mixture can be screened onto a substrate and cured by firing. Resistor films are typically 0.25 to 2.0 mils thick, and usually they do not require an overglaze for protection. Cermet films are sufficiently hard and abrasion resistant to be used as slide resistors in potentiometers. Since thick cermet films are not very sensitive to substrate surface roughness, special surface preparation usually are not necessary.

TABLE 23 Effect on Temperature Coefficient of Resistance and Noise of Silver Content in Palladium-Lead Borosilicate Compositions[93]

Silver, %	Resistivity, ohms/mil²	TCR at 25–105°C, ppm per °C	Noise, db/decade
0	200,000	+1720	+22
10	7,300	+ 440	+12
20	400	+ 380	+ 4
30	20	+ 175	− 8
40	0.5	+ 320	− 4

The resistance of a cermet film is determined by the metal-to-glass ratio. For example, the resistance of lead-borosilicate glass and either Ag or Au changes by five orders of magnitude with a few percent change in metal content (see Fig. 56), whereas either Pd or Pd + Ag glass mixtures are much less sensitive; therefore, they are more easily reproduced on a production basis. The addition of silver to palladium improves the temperature coefficient of resistance and decreases the electrical noise and resistivity of the films (see Table 23).

A casting method of forming Pd-Ag-glass resistor films was developed that is less wasteful of material than either the screening or the brushing process.[62] In the casting method, cermet powder is made into a slurry and cast into a continuous flexible sheet, which can be stored indefinitely before being applied to the substrate.

6-56　Ceramics, Glasses, and Micas

With this technique, any resistance value between 50 ohms and 1 megohm can be made by fusing together a maximum of four films, each having a different resistivity.

The conductivity mechanism in cermet films is not well understood; however, the available evidence indicates that conduction occurs by chains or networks of touching conductive particles embedded in an insulating matrix. In these films, the temperature coefficient of resistance varies from negative at low temperatures to positive at high temperatures.

Metal Films. Resistors of this type are made by depositing a layer of metal on a glazed substrate and then overglazing the film for protection (see Fig. 57). The glaze should be alkali-free in order to prevent ionic conduction and degradation. Both the underglaze, which provides a smooth surface for deposition, and the overglaze, which encapsulates and protects the resistor, should be in slight compression after firing, to lessen the effects of thermal and mechanical shock. The temperature coefficient of resistance changes from negative and nonlinear to zero, and finally to positive and linear as the film thickness increases and the resistivity decreases (see Fig. 58).

Dielectrics Techniques for forming dielectric films are similar to those used to form resistive ones. Films used for capacitor applications should have high dielectric constants, whereas those used for either insulator or crossover functions should have low dielectric constants in order to minimize both capacitance effects and crosstalk or coupling between

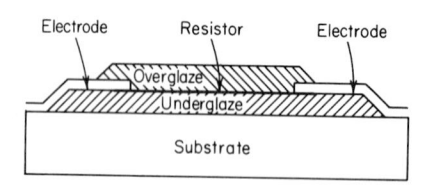

Fig. 57　Precious metal film-resistor structure.[94]

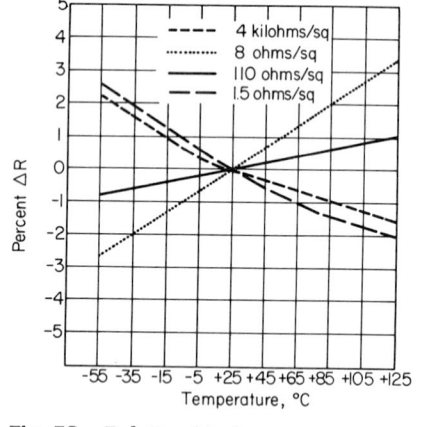

Fig. 58　Relationship between resistance and temperature of films having different resistivities.[94]

components. By using various materials, e.g., glasses, titanates, niobates, and stannates, capacitors can be made with values from 6,500 to 150,000 pf/in.^{2} with a dielectric strength of 500 dc volts/mil.[63] Dielectric constants between 20 and 400 can be obtained by varying the degree of crystallization in a ferroelectric glass-ceramic. Higher-dielectric-constant films (400 to 1,200) can be obtained by crystallizing $BaTiO_3$ from the glassy phase.[64]

DIELECTRICS

The most important application of high-dielectric-constant materials is for capacitive devices. This section describes the properties of ceramic, glass, and mica dielectrics, and it gives the characteristics of capacitors made from these materials. For more information on ceramic, glass, and mica capacitors the reader is referred to the books by Henny and Walsh[95] and by Dummer and Nordenberg.[96] The state-of-the-art review by Gruver, Buessem, and Dickey gives much useful information on the properties of ferroelectric ceramic materials.[97]

Ceramic Capacitors

Ceramic dielectric materials are conveniently divided into two classes. Class I materials are linear dielectrics that have a predictable change in capacitance with temperature. They are especially suited for resonator or oscillator circuit applications requiring good capacitance stability and high Q but not maximum capacitance per unit volume. Class II dielectrics are nonlinear materials having dielectric constants as high as 16,000. They are used in general-purpose capacitors to meet the demand for miniaturization in coupling, filtering, and bypass circuit applications where large changes in capacitance and high losses are not critical. A summary chart of ceramic capacitors is given in Table 24, and the range of capacitance values available is given in Table 25.

Construction Ceramic capacitors are customarily made in three styles: (1) disks and hollow cylinders, (2) multilayers of many dielectric layers in parallel, and (3) junction or barrier layers.

Disk and Hollow-cylinder Types. Disk- or tubular-shaped ceramic dielectrics are metallized (usually with silver) on opposing faces to form the electrodes. The dielectric thickness may vary from 0.01 to about 1.0 in. Although ceramic capacitors can be made in a variety of shapes, disk, tubular, and multilayer styles are the most suitable for critical temperature-compensating applications because the linearity of the temperature coefficient is affected by the metallic mass in contact with the capacitor. Disk styles offer greater capacitance per unit volume and less inductance per unit area than tubular types; consequently, they have higher resonant frequencies than the tubular styles.

Disk capacitors are available in sizes ranging from 3/16 to 1 in. diameter with ratings up to 6 kv dc for continuous operation, and up to 40 kv dc for pulsed voltage applications. Tubular styles, however, are generally rated at 1 kv dc or less. Axial-lead tubular types are useful in high-voltage circuits since this configuration affords maximum separation of the electrodes.

For disk styles, the volume of the capacitor varies roughly as the square of the voltage rating since the dielectric strength is proportional to the thickness of the active element (Fig. 59). The requirement of sufficient mechanical strength for automated manufacturing methods limits the minimum thickness of the dielectric to about 0.005 in. Voltage ratings below about 100 volts dc therefore

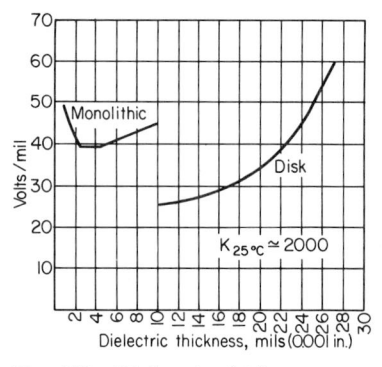

Fig. 59 Dielectric thickness as a function of the voltage rating of typical Monolythic and disk capacitors. (Monolythic is trademark of Sprague Electric Co.)

are not accompanied by a decrease in the volume of the unit.

Multilayer Types. In this style of ceramic capacitor, thin layers (ca. 0.001 to 0.005 in.) of unfired metallized ceramic are stacked and fired to form multilayer units consisting of many dielectric layers arranged in parallel circuit configuration. The inherent advantage of high capacitance per unit volume afforded by the multilayer construction comes at a high cost since current technology dictates the use of platinum, palladium, or gold internal electrodes instead of lower-cost silver ones. The absence of self-inductance in these capacitors permits operation at VHF. For 200-volt ratings, the sizes range from 0.4×0.2 in. at 100 pf to 0.5×0.3 in. for 5,000-pf units. Both Class I and II multilayer capacitors are available.

Barrier-layer Types. After a $BaTiO_3$ disk has been made conducting to 0.02 to 2 $(ohm-cm)^{-1}$ by firing it in a reducing atmosphere, an insulating barrier is formed by oxidizing about a 0.5-mil-thick layer on the surface of the disk.[98,99] Electrodes may be applied to the disk either before or after the insulating barrier is formed. Since the dissipation factors are high (5 to 10 percent), barrier-layer capacitors are usually rated at 25 volts dc or less. At ratings of 25 volts dc, disk range in diameter from 0.25 to 0.875 in. for capacitances of 0.005 to 0.1 μf, respectively.

TABLE 24 Ceramic-Capacitor Summary Chart[110]

Capacitor type	Use	Outstanding characteristic	Dc working voltage	Specifications Commercial	Specifications MIL	Type of ceramic	Insulation resistance (minimum), megohms	Dissipation factor (maximum)	Operating temperature, °C
Disk									
Temperature compensating	Compensate for drift in tuned circuits	High-Q and stable capacitance characteristics	Up to 500 volts 1-6 kv	RS198 Class I RS165 Class I	MIL-C-20D —	NPO-N750	50,000	0.1 % at 1 MHz*	−55 to 125
Extended range temperature compensating	Circuits requiring high-Q and stability	High-Q and stable capacitance characteristics	Up to 500 volts 1-6 kv	RS198 Class I RS168 Class I	— —	N1000-N5250	50,000	0.6 % at 1 MHz*	−55 to 85
Hi-K ceramic dielectric	Bypass and coupling	High capacitance in small package	Up to 500 volts 1-6 kv	RS198 Class II RS165 Class II	MIL-C-11015D MIL-C-11015D Some	Hi-K	20,000	1.5 % at 1 kHz	−55 to 200
Ac rated	Line bypass and antenna isolation	High capacitance in small package	150 volts rms	UL 492	—	NPO-N750 N1000-N5250 Hi-K	50,000 50,000 20,000	0.2 % at 1 MHz 0.6 % at 1 MHz 0.15 % at 1 kHz	−55 to 85
Low voltage (Ultra-Kap†)	Low-voltage circuits	Very high capacitance in small package	3-25 volts dc	—	—	Special	2,000 ohm–10 megohm	5-15 % at 1 kHz depending on voltage	−55 to 85
Tubular									
Temperature compensating	Compensate for drift in tuned circuits	High-Q and stable capacitance characteristics	Up to 1 kv	RS198 Class II	MIL-C-20D	NPO-N750	50,000	0.1 % at 1 MHz*	−55 to 125
Extended-range temperature compensating	Circuits requiring high-Q and stability	High-Q and stable capacitance characteristics	Up to 1 kv	RS198 Class I	—	N1000-N5250	50,000	0.6 % at 1 MHz	−55 to 85
Hi-K ceramic dielectric	Bypass and coupling	High capacitance in small package	Up to 1 kv	RS198 Class II	—	Hi-K	20,000	1.5 % at 1 MHz	−55 to 85
Feedthrough	Tuner and other feedthrough applications	Dual function, low cost	Up to 60 kv	—	—	NPO-N750 N1000-N5250 Hi-K	10,000 10,000 10,000	0.2 % at 1 MHz* 0.6 % at 1 MHz* 2.5 % at 1 kHz	−55 to 125 −55 to 85 −55 to 125

			Special			
Transmitting	Transmitters, induction heaters, electronic welding, x-ray, diathermy	High-Q and stable capacitance characteristics	Up to 60 kv	—	NPO-N750	10,000	0.1% at 1 MHz*	−55 to 85
					N1000-N5250	10,000	0.2% at 1 MHz*	−55 to 85
					Hi-K	10,000	1.5% at 1 kHz	−55 to 125

Miscellaneous

Glass dielectric capacitors	Circuits requiring high-Q and stability	High-Q and stable capacitance characteristics	Up to 500 volts	—	MIL-C-11272B	P140	100,000	0.1% at 1 kHz	−55 to 125
Porcelain dielectric capacitors	Tuned circuits, high-frequency circuits		500 volts	—	MIL-C-11272B	P105	100,000	0.1% at 1 kHz	−55 to 125
Multilayer ceramic	These capacitors are extensions of the standard disk capacitor. Very high capacitance is obtained through the use of multiple plates.		25–200 volts	—	MIL-C-11015D	TC	100,000	0.2% at 1 MHz	−55 to 150
						Hi-K		1.5% at 1 kHz	

* Limits are for capacitance values above 30 pf. See RS198 or MIL-C-20 for limits below 30 pf.
† Trademark of Centralab, Division of Globe Union Corp.

6-59

TABLE 25 Ceramic Capacitor Value Ranges[110]

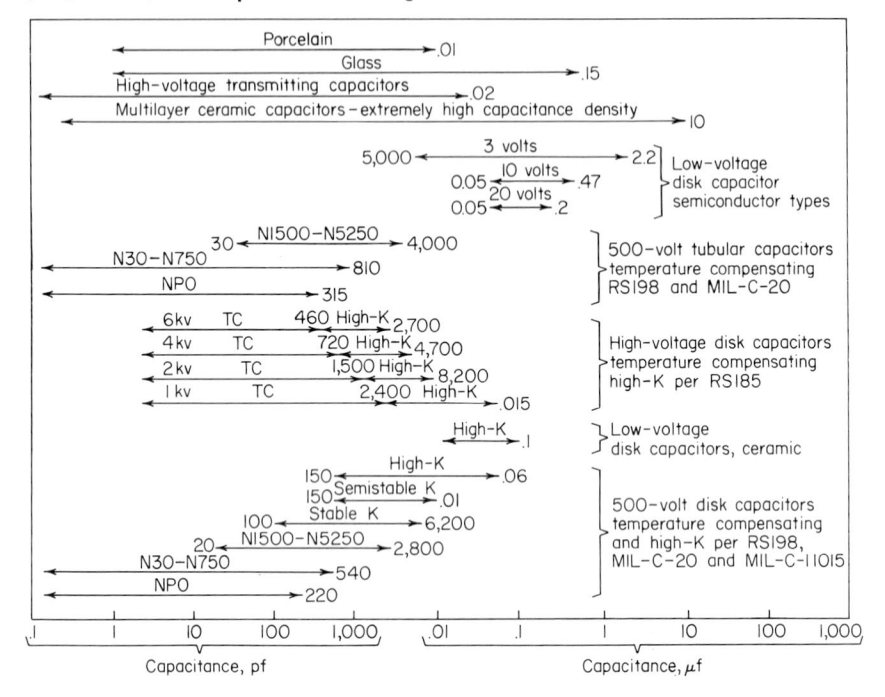

[Table showing capacitance ranges for various ceramic capacitor types, with two scales: Capacitance in pf (0.1 to 1,000) and Capacitance in μf (.01 to 1,000)]

Range (pf)	Type
Porcelain	.01
Glass	.15
High-voltage transmitting capacitors	.02
Multilayer ceramic capacitors – extremely high capacitance density	10
5,000 ← 3 volts → 2.2	Low-voltage disk capacitor semiconductor types
0.05 ← 10 volts → .47	
0.05 ← 20 volts → .2	
30 ← N1500–N5250 → 4,000	500-volt tubular capacitors temperature compensating RS198 and MIL-C-20
N30–N750 → 810	
NPO → 315	
6kv TC, 460 High-K → 2,700	High-voltage disk capacitors temperature compensating high-K per RS185
4kv TC, 720 High-K → 4,700	
2kv TC, 1,500 High-K → 8,200	
1kv TC, 2,400 High-K → .015	
High-K → .1	Low-voltage disk capacitors, ceramic
High-K → .06	
150 Semistable K → .01	500-volt disk capacitors temperature compensating and high-K per RS198, MIL-C-20 and MIL-C-11015
100 ← Stable K → 6,200	
20 ← N1500–N5250 → 2,800	
N30–N750 → 540	
NPO → 220	

Temperature-compensating ceramic capacitors Class I capacitors have predictable temperature coefficients of capacitance because the dielectric materials used are free from ferroelectric effects over the temperature interval of interest. Since the change in capacitance with temperature is not exactly linear, the meaning of the nominal temperature coefficient of capacitance should be appreciated.

Temperature Coefficient of Capacitance Definition.[100] The change in capacitance with temperature is expressed in parts per million per degree centigrade as defined by the following method: Curve F in Fig. 60 is the measured capacitance in the temperature range from -55 to $85°C$. Line A in this figure is drawn through the 25 and $85°C$ points, and line B through the -55 and $25°C$ points. At all temperatures from -55 to $85°C$, curve F lies between these two limiting straight lines. The difference between the slopes of lines A and B is a measure of the departure from linearity of curve F. The slope of line A is the basis for the nominal temperature coefficient of capacitance (TCC):

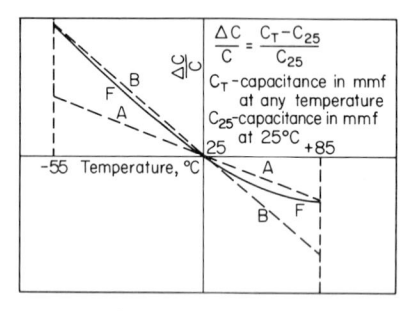

$$\frac{\Delta C}{C} = \frac{C_T - C_{25}}{C_{25}}$$

C_T - capacitance in mmf at any temperature
C_{25} - capacitance in mmf at $25°C$

Fig. 60 Method of obtaining temperature coefficient of capacitance.[95]

$$TCC = \frac{C_T \times 10^6}{C_{25} \times \Delta T}$$

where TCC is in parts per million (ppm) per degree centigrade; C_{25} and C_T are the capacitances at $25°C$ and the test temperature, respectively; and ΔT is the difference in temperature between 25°C and the test temperature. Since the above equation is valid only between 25 and 85°C, TCC is based on a straight line drawn between these temperature points. In a similar manner the TCC over the range −55 to 25°C is based on line B. Actually, of course, the coefficient varies over every small part of curve F, with greater variations occurring in capacitors having larger changes in capacitance with temperature.

Because of the spread of capacitance values from the straight-line nominal value, the temperature coefficients must also have a tolerance (TTC). This tolerance of the temperature coefficient is in addition to the tolerance on the nominal capacitance at 25°C. Thus, a given capacitor may be labeled as 100 pf + 10 percent $N330 \pm 60$ ppm. At +55°C, its capacitance may fall anywhere between 88.95 and 109.1 pf as determined from the following relation:

$$C_T = \frac{(\text{TCC} + \text{TTC})(C_{25})(\Delta T)}{10^6} + C_{25}$$

Therefore,

$$C_{55} = \frac{(-330 + 60)(100 + 10)(30)}{10^6} + 100 \pm 10$$

Standard coefficients and tolerances are described in EIA, Class I, RS-165, RS-168, and RS-198 Specifications, and in Military Specification MIL-C-20D. TCC for commercial applications is specified from −55 to 25°C and from 25 to 85°C, and in some cases to 125°C, or the maximum operating temperature. Military specifications can be obtained free of charge from the Commanding Officer, U.S. Naval Supply Depot, 5801 Tabor Avenue, Philadelphia, Pennsylvania 19120.

TABLE 26 Representative Formulations of Class I and Class II Dielectric Materials

Capacitor	Formulation
Class I (capacitance change, ppm per °C) Disk type:	
−1,500	$BaTiO_3$ + high % $CaZrO_3$ or $BaTiO_3$ + $CaTiO_3$ + $BaZrO_3$
−750	TiO_2 + low % ZrO or $CaTiO_3$ and rare-earth titanates*
0	$BaTiO_3$ + TiO_2 + low % ZrO_2 or rare-earth titanates†
+100	Mg_2TiO_4 or $MgTiO_3$ or mixture‡
Multilayer:	
−30 ($K = 36$)	$BaTiO_3$ + TiO_2 + low % ZrO_2†
−750 ($K = 110$)	TiO_2 + $CaTiO_3$ + rare earths*
Class II Disk type:	
$K = 325$	$BaTiO_3$ + $CaTiO_3$ + low % $Bi_2Sn_3O_9$
$K = 700$	$BaTiO_3$ + $CaTiO_3$ + low % $Bi_2Sn_3O_9$
$K = 1,150$	$BaTiO_3$ + low % $Bi_2Sn_3O_9$§
$K = 2,100$	$BaTiO_3$ + low % $CaZrO_3$ and Nb_2O_5
$K = 2,850$	$BaTiO_3$ + low % TiO_2¶
$K = 6,500$	$BaTiO_3$ + low % $CaZrO_3$ or $CaTiO_3$ + $BaZrO_3$
Multilayer:	
$K = 2,100$	$BaTiO_3$ + low % $CaZrO_3$ and Nb_2O_5
$K = 8,000$	$BaTiO_3$ + low % $CaZrO_3$

SOURCE: Sprague Electric Co.
* U.S. Patent 2,398,088.
† U.S. Patent 2,429,588.
‡ U.S. Patent 2,305,327.
§ U.S. Patent 2,658,833.
¶ U.S. Patent 2,966,420.

Ceramics, Glasses, and Micas

TC Capacitor Characteristics. Class I capacitors are produced with TCC ranging from +120 (P120) ppm capacitance change per °C to −5,600 (N5600)ppm per °C. Materials with TC's of +150 to −750 and dielectric constants from 5 to 110 have nearly linear temperature coefficients of capacitance while the extended TCC series with TC's of −1,000 to −5,600 ppm per °C and dielectric constants from 100 to 570 are more nonlinear. Temperature control is achieved by combining materials having different temperature coefficients of capacitance. For example, $MgTiO_3$ with a TCC of +100 ppm per °C combined with $CaTiO_3$ with a TCC of −1,500 ppm per °C yields dielectrics with TC's ranging from +100 to −1,500 ppm per °C. Since $MgTiO_3$ and $CaTiO_3$ have dielectric constants of 16 and 150, respectively, materials with less negative TC's have lower dielectric constants. Thus, a $(Mg,Ca)TiO_3$ composition with a TCC of −750 ppm per °C has a dielectric constant of 40. Low-loss steatite ($MgSiO_3$) capacitors have TC's from +100 to +140 ppm per °C and dielectric constants from 5 to 7 at 1 MHz. Temperature

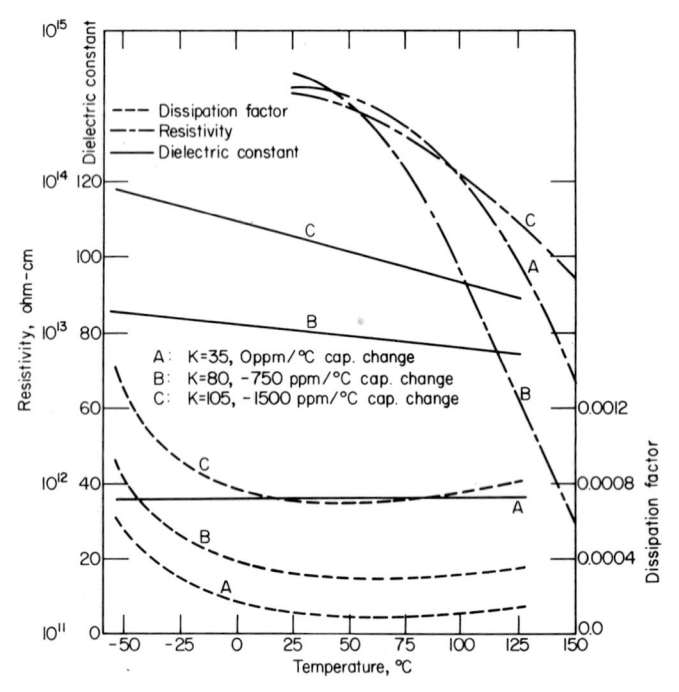

Fig. 61 The variation with temperature of the dielectric constant, dissipation factor, and resistivity of Class I disk capacitors with TCC of 0, −750 and −1,500 ppm per °C. (*Sprague Electric Co.*)

control of capacitance is generally achieved using compositions based largely on $BaTiO_3$, TiO_2, $CaTiO_3$, $MgTiO_3$, $SrTiO_3$, and rare-earth titanates. Some representative formulations are listed in Table 26.

The dielectric constant, dissipation factor, and resistivity as a function of temperature of several commercial disk capacitors having temperature coefficients of 0 ppm per °C (0NP), −750 ppm per °C, and −1,500 ppm per °C are shown in Fig. 61. The same properties are shown in Fig. 62 for commercial multilayer capacitors having temperature coefficients of −30 and −750 ppm per °C. Class I materials have negligible changes in capacitance and dissipation factor with either frequency (from 60 Hz to 100 MHz) or applied voltage. The color code used to mark TC capacitors is shown in Table 27.

General-purpose ceramic capacitors Class II capacitors are nonlinear materials (i.e., capacitance is not a linear function of applied voltage) that are characterized by high dielectric constants, polarization saturation, and ferroelectric hysteresis. They have neither the stability nor the high Q of temperature-compensated capacitors. These dielectrics are usually based on $BaTiO_3$, which exhibits a maximum dielectric constant at 120°C. A variety of other compounds are combined with $BaTiO_3$ to broaden the temperature interval of maximum K and shift it to 25°C. The compounds that are frequently combined with $BaTiO_3$ include $SrTiO_3$, $CaZrO_3$, $BaZrO_3$, $CaTiO_3$, $Bi_2Sn_3O_9$, $MgTiO_3$, Nb_2O_5, and Ta_2O_5.[97,101] Representative formula-

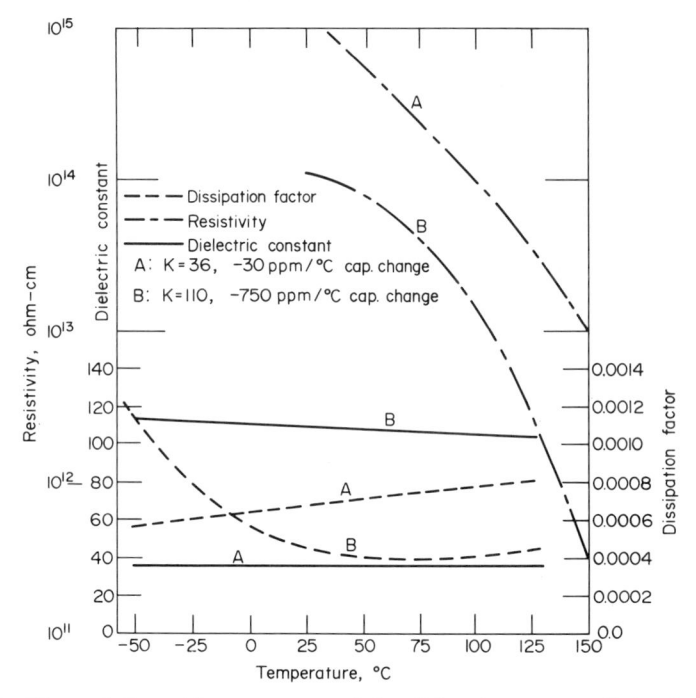

Fig. 62 The variation with temperature of the dielectric constant, dissipation factor, and resistivity of Class I Monolythic capacitors with TCC of -30 and -750 ppm per °C. (*Sprague Electric Co.*)

tions are listed in Table 26. Military specifications for Class II ceramic fixed capacitors are given by MIL-C-39014A and MIL-C-11015D.

The important properties of ferroelectric Class II capacitors are (1) the variation of the dielectric constant with temperature, voltage, and frequency, (2) change in the dissipation factor with temperature, (3) dielectric strength, and (4) effects of time (aging and degradation). In general, the temperature stability of the dielectric constant is inversely proportional to the applied dc field, but no explicit function can be assigned to the field dependence. This effect is shown in Fig. 63 for the disk capacitors with dielectric constants ranging from 325 to 6,500.

Disk and Multilayer Capacitors. As a consequence of the nonlinearity of ferroelectric materials, the dielectric constant is a function of the applied ac measuring field and bias field.[102] The results of this nonlinear behavior on the change in the dielectric constant with small ac measuring fields for several disk capacitors are shown in Fig. 64 and for several multilayer capacitors in Fig. 65. These curves show that the nonlinearity is most pronounced in materials having the larger dielectric constant.

TABLE 27　Ceramic-capacitor Color Code[110]

Color	Temperature coefficient			Temperature range, °C (D)	Capacitance change over temperature range, % (E)	Significant figure		Decimal multiplier (H)	Capacitance tolerance, % (I)	Tolerance of capacitance	
	ppm per °C (A)	Significant figure (B)	Multiplier (C)			1st (F)	2d (G)			<10 pf (J)	>10 pf (J)
Black	0	0.0	−1		±2.2	0	0	1	±20	±2 pf	±20%
Brown	−33	.....	−10	+10 to +85	±3.3	1	1	10	.....	±0.1 pf	±1%
Red	−75	1.0	−100		±4.7	2	2	100	.....	.....	±2%
Orange	−150	1.5	−1,000		±7.5	3	3	1,000	.....	.....	±3%
Yellow	−220	2.2	−10,000		±10	4	4	10,000	+100 to 0	.....	.....
Green	−330	3.3	+1		±15	5	5	.....	±5%	±0.5 pf	±5%
Blue	−470	4.7	−10		±22	6	6				
Violet	−750	7.5	+100		+22 to −33	7	7				
Gray	+100 to −1500	−1000 to −5200*	+1,000		+22 to −56	8	8	0.01	+80 to −20	±2.5 pf	
White	+100 to −750	.....	−10,000	−55 to +85	+22 to −82	9	9	0.1	±10	±1 pf	±10%
Gold	.....	.....	.....	−30 to −85	±1.5						
Silver	.....	.....	.....								

* Multiplier dot is black.

Note: Color-code bands used on axial units
Color-code dots used on radial units

6-64

The effect of temperature on the dielectric constant at several levels of applied dc bias field is displayed in Figs. 66 to 69 for disk capacitors having dielectric constants ranging from 1,150 to 6,500. The most abrupt variation in the dielectric constant as a function of applied dc field occurs in the higher dielectric-constant materials. Similar behavior is observed in multilayer capacitors (see Figs. 70 and 71).

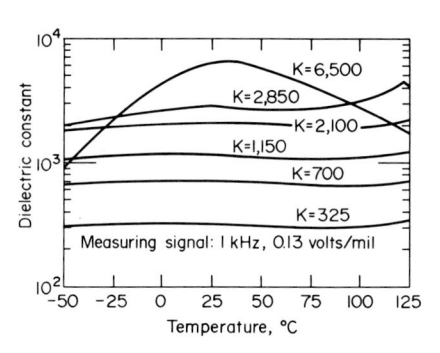

Fig. 63 Temperature stability of dielectric constants of Class II disk capacitors (measuring signal 0.13 ac volt/mil at 1 KHz). (*Sprague Electric Co.*)

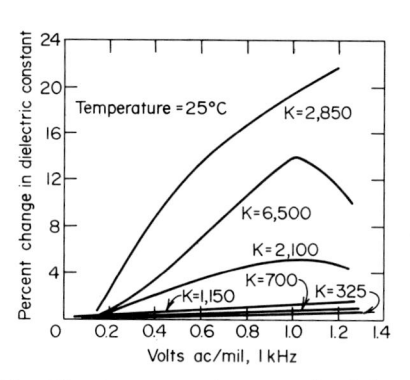

Fig. 64 Percentage change in the dielectric constant with applied field at 1 KHz for Class II disk capacitors at 25°C. (*Sprague Electric Co.*)

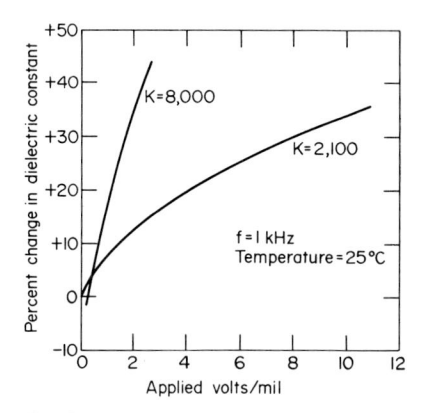

Fig. 65 Percentage change in the dielectric constant with applied field at 1 KHz for Class II Monolythic capacitors at 25°C (dielectric thickness 0.0025 in.) (*Sprague Electric Co.*)

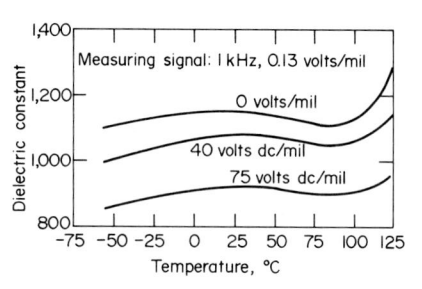

Fig. 66 Effect of dc bias field on the temperature dependence of the dielectric constant of a Class II disk capacitor with $K = 1,150$ (measuring signal 0.13 volt/mil at 1 KHz). (*Sprague Electric Co.*)

The sensitivity of the dielectric constant to frequency depends largely on the additives used to achieve temperature stability. For example, bismuth is sometimes used to improve the temperature stability of the dielectric constant, but this additive does so only at the expense of causing a greater change in capacitance (see Fig. 72) and a greater increase in the dissipation factor (see Fig. 73) with increasing frequency compared to bismuth-free compositions.

As shown in Fig. 74, the sensitivity of the dissipation factor at 25°C to the applied field correlates with the dielectric constant of the disk capacitors. With

6-66 Ceramics, Glasses, and Micas

K less than 1,150, the voltage dependence of the dissipation factor is negligible, but with higher-dielectric-constant materials the dissipation factor becomes increasingly sensitive to the applied field. Multilayer capacitors behave in a similar manner, as can be seen in Fig. 75. In general, the dissipation factor decreases with increase in temperature at a rate that is fairly independent of the dielectric constant (see Figs. 76 to 78).

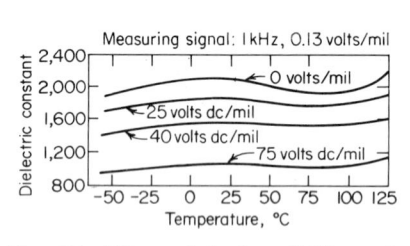

Fig. 67 Effect of dc bias field on the dependence of the dielectric constant of a Class II disk capacitor with $K = 2,100$ (measuring signal 0.13 volt/mil at 1 KHz). (*Sprague Electric Co.*)

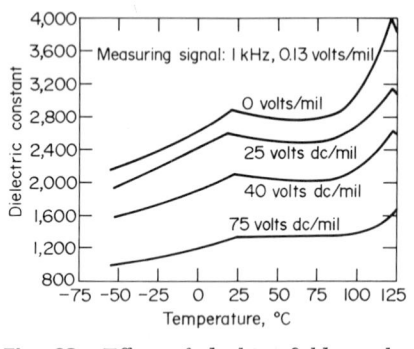

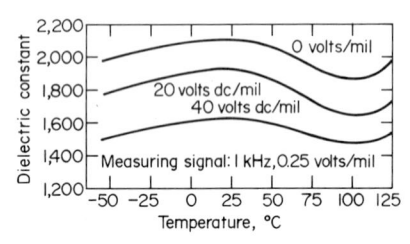

Fig. 68 Effect of dc bias field on the temperature dependence of the dielectric constant of a Class II disk capacitor with $K = 2,850$ (measuring signal 0.13 volt/mil at 1 KHz). (*Sprague Electric Co.*)

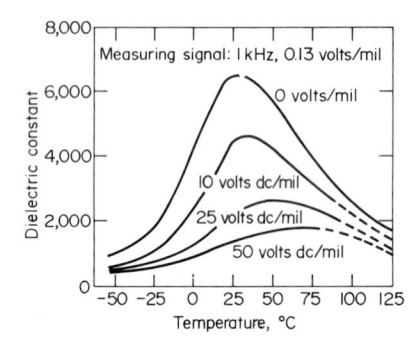

Fig. 69 Effect of dc bias field on the temperature dependence of the dielectric constant of a Class II disk capacitor with $K = 6,500$ (measuring signal 0.13 volt/mil at 1 KHz). (*Sprague Electric Co.*)

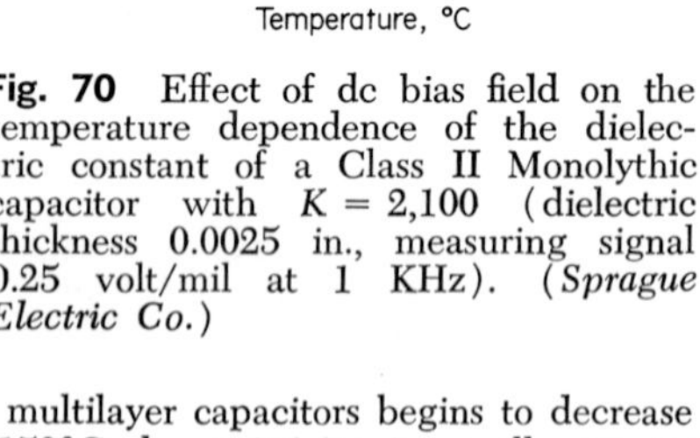

Fig. 70 Effect of dc bias field on the temperature dependence of the dielectric constant of a Class II Monolythic capacitor with $K = 2,100$ (dielectric thickness 0.0025 in., measuring signal 0.25 volt/mil at 1 KHz). (*Sprague Electric Co.*)

The insulation resistance of both disk and multilayer capacitors begins to decrease rather abruptly above 85°C, but even at 150°C the resistivity is usually greater than 10^{11} ohm-cm (see Figs. 79 and 80).

Barrier-layer Capacitors. Relatively high capacitance per unit volume is achieved with capacitors of this type, but the insulation resistance is lower and the dissipation factor is higher than is found either in disk or multilayer types. Barrier-layer capacitors are used extensively in transistor circuits for bypass and coupling applications. Since the semiconducting interior of the barrier-layer capacitor presents a finite series resisitance, the dissipation factor increases from 3 to 6 percent in the audio range. These capacitors have less insulating resistance and greater decrease

Dielectrics

in resistance with applied voltage than either disk or multilayer types. For these reasons such units are best suited for low impedance, low voltage, and low frequency (<30 kHz applications).

Barrier-layer capacitors are normally cataloged by their voltage ratings. Table 28 lists some industry rating and the specific capacitances of typical capacitors. These units are normally rated to withstand life tests of 250 to 1,000 hr at 85°C

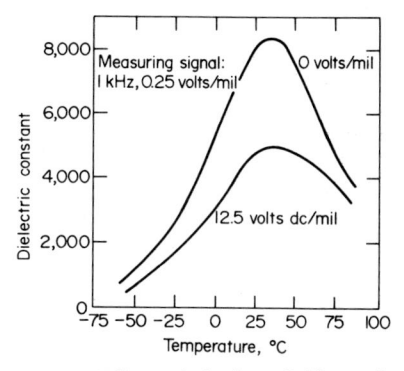

Fig. 71 Effect of dc bias field on the temperature dependence of the dielectric constant of a Class II Monolythic capacitor with $K = 8,000$ (dielectric thickness 0.002 in., measuring signal 0.25 volt/mil at 1 KHz). (*Sprague Electric Co.*)

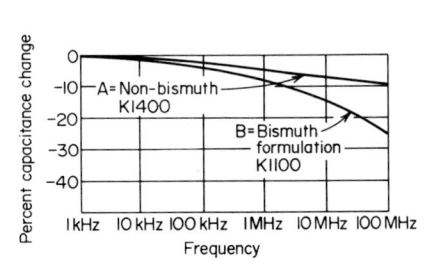

Fig. 72 Percentage capacitance change with frequency of bismuth and nonbismuth formulations (parallel-plate capacitor 0.001 in. thick). (*Sprague Electric Co.*)

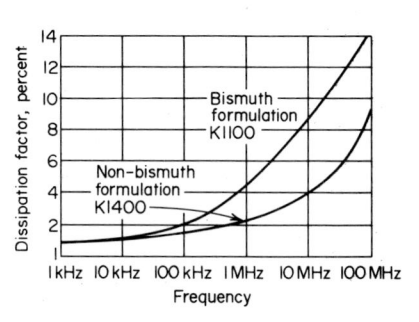

Fig. 73 Dissipation factor (%) dependence on frequency of bismuth and nonbismuth formulations. (*Sprague Electric Co.*)

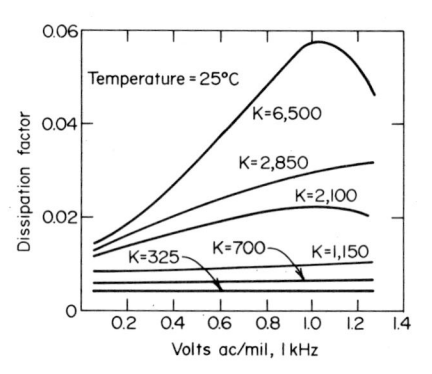

Fig. 74 Dependence of dissipation factor on applied field for Class II disk capacitors at 25°C. (*Sprague Electric Co.*)

and voltages up to twice the rated voltage. The leakage resistance at the end of the life tests generally goes up with higher voltage ratings and with greater capacitance values. It varies from 5,000 ohm for 2.2-μf 3-volt units to 500 megohm for 0.01-μf 25-volt units. As shown in Figs. 81 to 84, the insulation resistance goes up logarithmically as the applied voltage decreases.

Barrier-layer capacitors are made in a wide range of temperature coefficients. The capacitance change with temperature at several ambient temperatures for units with voltage ratings of 3 to 30 volts is listed in Table 29. In Fig. 85 are shown the shapes of these curves for 3-, 10-, 16-, and 25-volt units. The percent change in the insulation resistance with temperature for typical 3-, 10-, 16-, and 25-volt units is shown in Figs. 86 and 87. Figures 88 to 90 give the variation of capacitance and dissipation factor with frequency.

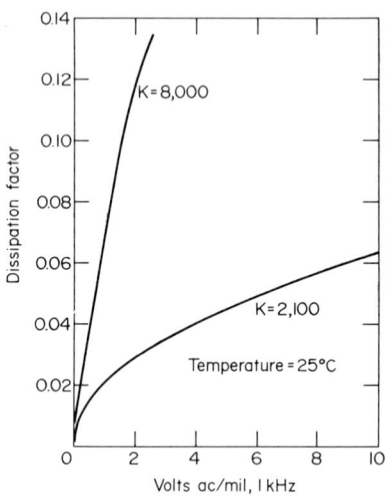

Stability Two factors can affect the stability of capacitors: (1) aging, which is the spontaneous decrease in capacitance or dissipation factor of Class II materials with time, and (2) degradation, which is the change in these properties with time under the influence of direct or alternating current, mechanical stress, or alterations in the conditions of atmosphere, humidity, radiation, etc. Aging occurs only in the

Fig. 75 Dependence of dissipation factor on applied field of class II Monolythic capacitors at 25°C (dielectric thickness 0.0025 in.). (*Sprague Electric Co.*)

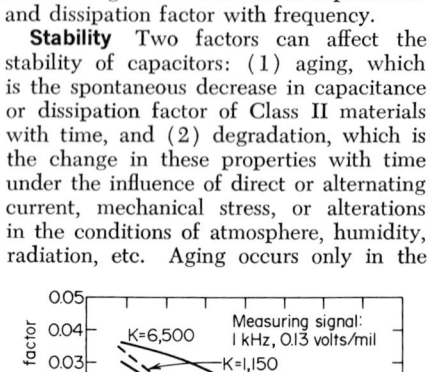

Fig. 76 Dependence of dissipation factor on temperature for Class II disk capacitors (measuring signal 0.13 ac volt/mil at 1 KHz). (*Sprague Electric Co.*)

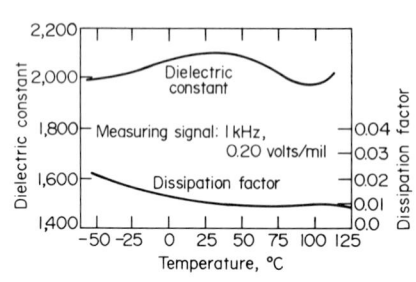

Fig. 77 Dependence of dissipation factor and dielectric constant on temperature for Class II Monolythic capacitor with K = 2,100 (dielectric thickness 0.0025 in., measuring signal, 0.20 volt/mil at 1 KHz). (*Sprague Electric Co.*)

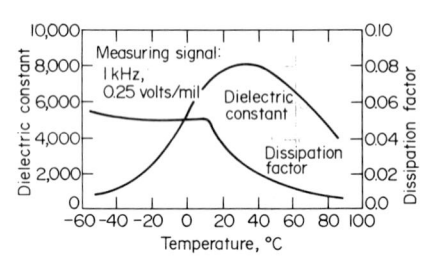

Fig. 78 Dependence of dissipation factor and dielectric constant on temperature for Class II Monolythic capacitor with K = 8,000 (dielectric thickness 0.002 in., measuring signal 0.25 volt/mil at 1 KHz). (*Sprague Electric Co.*)

TABLE 28 Specific Capacitances of Barrier-layer Capacitors with 3- to 30-volt Ratings

Rating, volts	Specific capacitance of uncompleted capacitors, $\mu f/in.^2$
3	6–7
10, 12	1–1.5
16	0.5–0.8
25, 30	0.3–0.6

Source: Sprague Electric Co.

Dielectrics 6-69

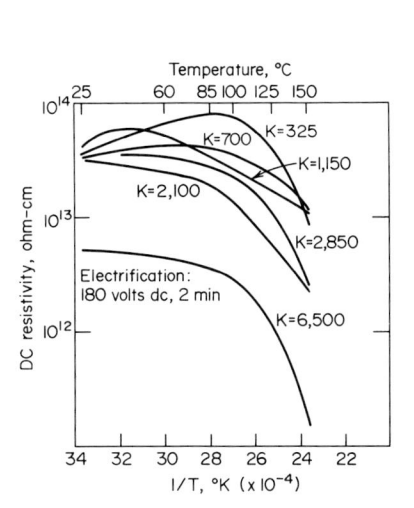

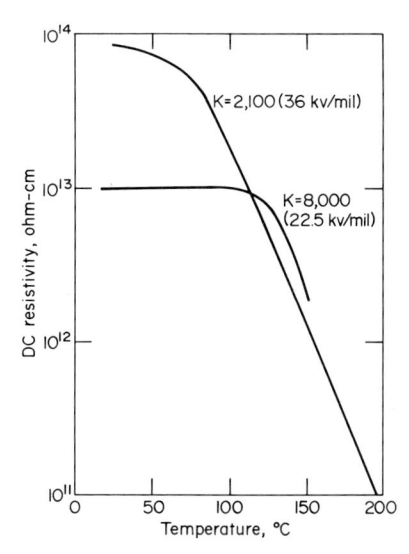

Fig. 79 Temperature dependence of resistivity of Class II disk capacitors (2 min electrification at 180 volts dc). (*Sprague Electric Co.*)

Fig. 80 Temperature dependence of resistivity of Class II Monolythic capacitors. (*Sprague Electric Co.*)

ferroelectric state whereas degradation can occur in either the ferroelectric or the paraelectric state; therefore, it includes Class I and Class II capacitor materials.

Aging.[103] As a consequence of aging, Class II capacitors lose their capacitance with time. The aging rate is usually expressed as a percentage x per decade of time t in the logarithmic time expression.

$$K_t = K_1 \left(1 - \frac{x}{100} \log t\right)$$

Fig. 81 Effect of applied voltage on insulation resistance of 3-volt barrier-layer capacitors. (*Sprague Electric Co.*)

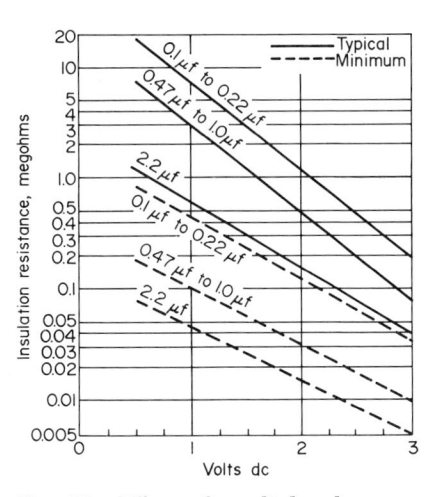

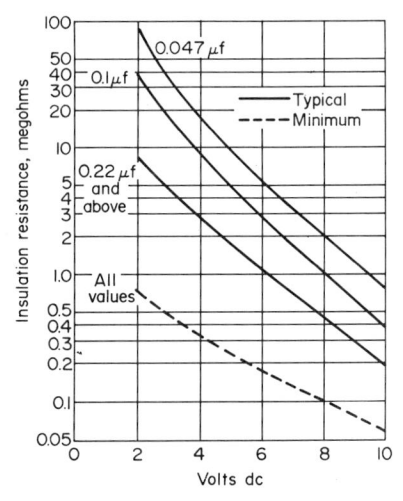

Fig. 82 Effect of applied voltage on insulation resistance of 10-volt barrier-layer capacitors. (*Sprague Electric Co.*)

6-70 Ceramics, Glasses, and Micas

This empirical relation holds true for only several decades of time after one minute; nevertheless, it illustrates the aging phenomenon. For example, a material with an aging rate of 3 percent per decade and the initial measurement made after 1 minute will have aged 15 percent after 10^5 min. Over the first year the aging rate will be about 2 percent per year; over the next 10 years, the rate will be only ½ percent per year. The usual industrial practice, however, is to make the initial measurement after 24 hr.

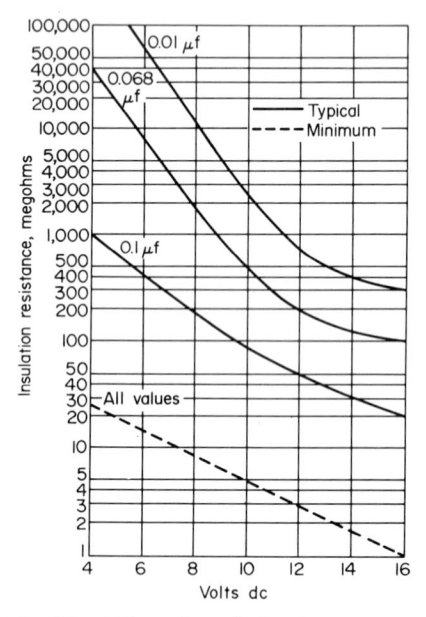

Fig. 83 Effect of applied voltage on insulation resistance of 16-volt barrier-layer capacitors. (*Sprague Electric Co.*)

Fig. 84 Effect of applied voltage on insulation resistance of 25-volt barrier-layer capacitors. (*Sprague Electric Co.*)

TABLE 29 Percentage Capacitance Change of Barrier-layer Capacitors at Several Temperatures

Voltage rating, volts	Manufacturer	Temperature, °C		
		-55	$+10$	$+85$
3	S	-10	-3	$+12$
10	S	-23	-5	$+17$
10	C	-50	-7	-14
10	D	-85	-8	-60
12	SI	-71	-17	-50
25	S	-22	-4	$+4$
25	R	-45	-7	-25
25	E	-75	-8	-30
30	SI	-29	-11	$+0.5$

S   Sprague   R   RMC
C   Centralab   E   Erie
Si   Siemens   D   Dielectron

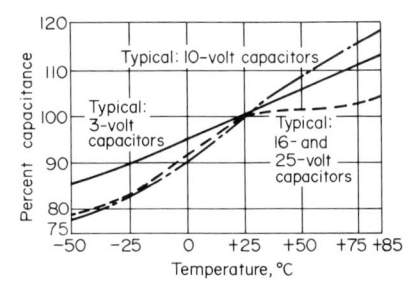

Fig. 85 Capacitance change (%) with temperature of typical 3-, 10-, 16-, and 25-volt barrier-layer capacitors. (*Sprague Electric Co.*)

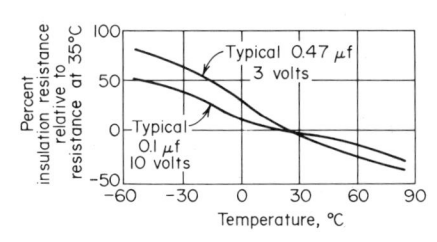

Fig. 86 Effect of temperature on the relative insulation resistance of typical 3- and 10-volt barrier-layer capacitors. (*Sprague Electric Co.*)

Reheating the capacitor above the Curie temperature, i.e., the temperature at which the material changes from a ferroelectric to a paraelectric state, will completely restore the initial capacitance. Partial de-aging will occur as a result of any temperature change or from mechanical or electrical stresses. Therefore, the extent to which pre-aging can be used to stabilize the properties of a capacitor will depend on its sensitivity to de-aging. For these reasons capacitors requiring close tolerances should be made from either Class I materials or low-aging-rate formulations. Some typical aging rates for Class II capacitors are listed in Table 30. The logarithmic dependence of the aging rate at 25°C is shown in Fig. 91 for typical Class II disk and monolithic capacitors with K ranging from 325 to 6,500. Capacitance aging rates usually change as a function of temperature.

Degradation.[104] Three different failure modes are recognized: (1) dielectric breakdown, which is an instantaneous electronic process occurring at high fields ($\simeq 10^9$ volts/mil); (2) thermal breakdown, which is a thermal and electrical instability resulting from excessive heating within the dielectric; and (3) degradation, which is a slow process involving a steady increase in conductivity of the dielectric until ohmic heating leads to thermal breakdown. The relation among these three modes of failure is illustrated diagrammatically in Fig. 92. This diagram of allowed temperature-field combinations illus-

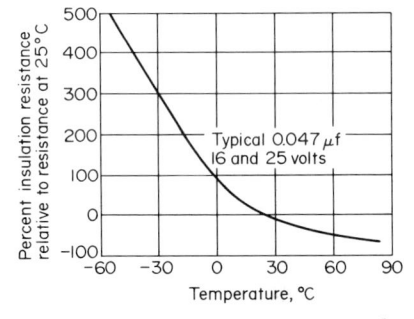

Fig. 87 Effect of temperature on the relative insulation resistance of typical 16- and 25-volt barrier-layer capacitors. (*Sprague Electric Co.*)

TABLE 30 Typical Aging Rates of Class II Capacitors at 25°C

Dielectric constant	Aging rate, % per decade
700	1.50
1,200	2.00
2,000	2.00
2,500	3.25
8,000	4.00

From *Sprague Tech. Paper* TP-67-15A.

6-72 Ceramics, Glasses, and Micas

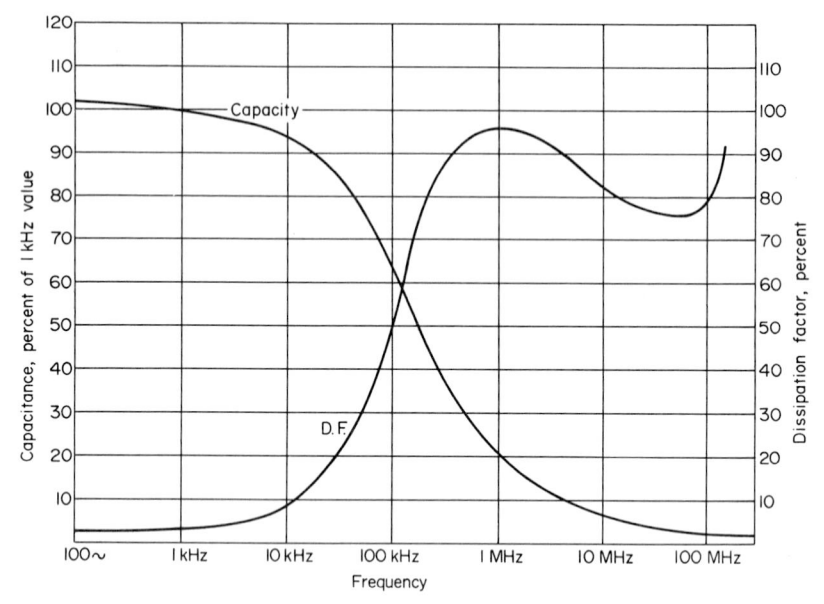

Fig. 88 Dependence of the capacitance and dissipation factor on frequency of typical 3-volt barrier-layer capacitors. (*Sprague Electric Co.*)

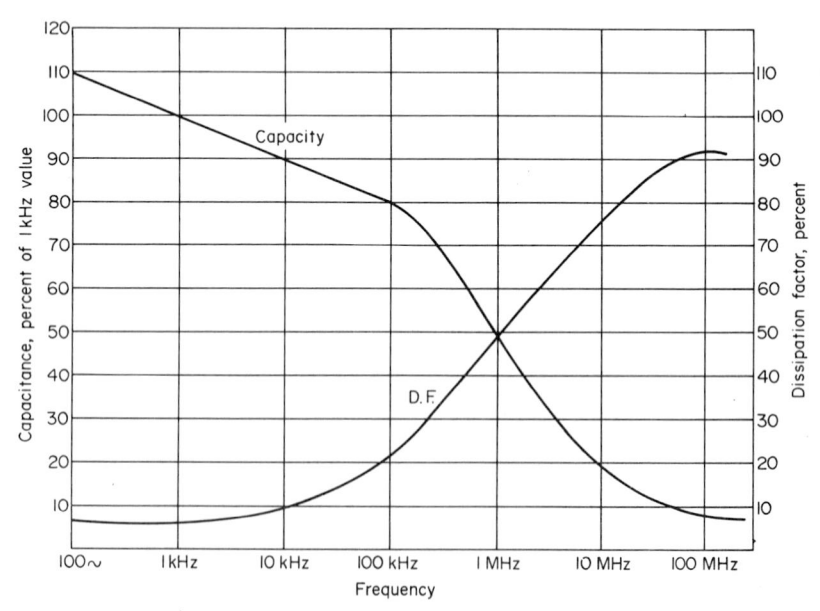

Fig. 89 Dependence of the capacitance and dissipation factor on frequency of typical 10-volt barrier-layer capacitors. (*Sprague Electric Co.*)

trates that the thermal breakdown boundary is shifted to lower temperatures as the time is increased.

Ionic processes occurring in the dielectric are the principal reason for degradation. These processes cause alterations in the concentration and mobility of ions and

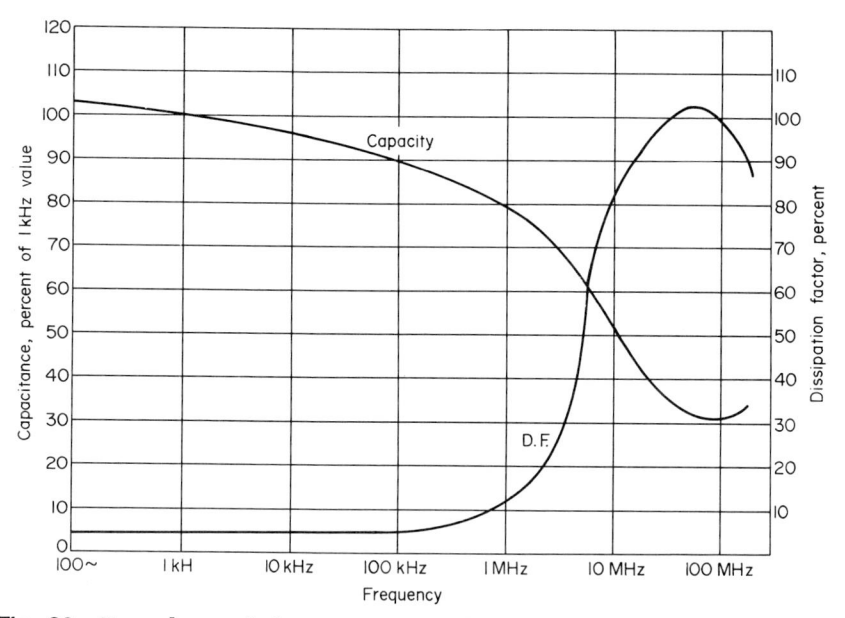

Fig. 90 Dependence of the capacitance and dissipation factor on frequency of typical 25-volt barrier-layer capacitors. (*Sprague Electric Co.*)

vacancies, which produce changes in the electric conductivity. During firing at $\simeq 1300°C$ in air, titanate ceramics are reduced appreciably. Anion vacancies (AV) left by the missing oxygen ions carry an effective charge of 2^+. This charge is neutralized by the electrons left by the missing oxygen atoms in localized energy levels of nearly Ti^{4+} ions, which as a result become Ti^{3+} ions. At low temperatures a small fraction of these $AV-Ti^{3+}$ ion complexes are dissociated, which enables them to contribute to the conduction process. The undissociated $AV-Ti^{3+}$ ion complexes cannot participate in the conduction process, but they do experience a torque that leads to a dielectric polarization and a dielectric absorption current.

As a consequence of the manufacturing history, there is a much higher concentration of AV in the interior than at the surface of a ceramic grain. The higher concentration of charge carriers in the interior of the grains gives the bulk of the titanate a much higher conductivity than the grain boundaries. Under an applied dc field most of the voltage drop takes place across these insulating barriers. The field at the grain boundaries is further increased as more AV move out of this region toward the cathode until Zener breakdown occurs. In this process, two electrons from each contributing O^{2-} ion at the boundary are transferred to the conduc-

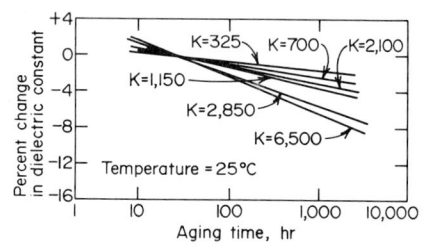

Fig. 91 Aging of typical Class II disk and barrier-layer capacitors at 25°C. (*Sprague Electric Co.*)

6-74   Ceramics, Glasses, and Micas

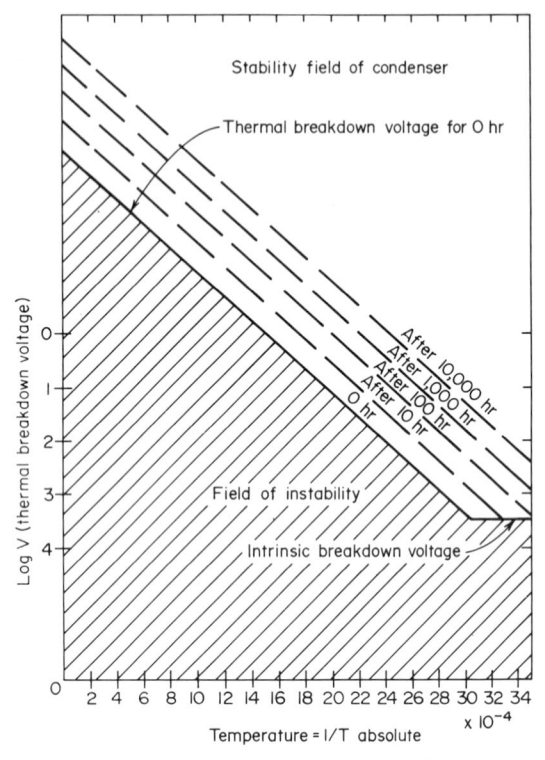

Fig. 92 Schematic diagram of breakdown voltage as a function of $(1/T)$ per °K.[97]

tion band; atomic oxygen is subsequently freed at the anode, and the newly formed AV is swept toward the cathode. On an atomic scale the degradation process has thus begun. How far it will go on a macroscopic scale will depend on the fate of the oxygen atoms emerging at the anode. If they cannot escape, the influx of AV

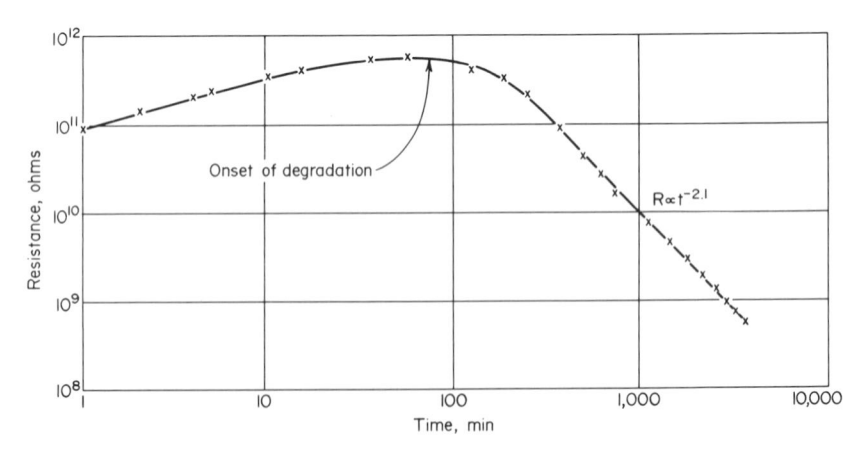

Fig. 93 Change of resistance with time for a C67 case size I Monolythic capacitor (8,000 pf). (*Sprague Electric Co.*)

will stop. One method of arresting this process is to seal the anode with a coating containing either U^{6+}, W^{6+}, Nb^{5+}, or Ta^{5+}, La^{3+}, Th^{4+}. Another method is to stabilize the titanate ceramic with these ions, which is the easier industrial alternative. They have the effect of decreasing the concentration of AV in both the bulk and surface of the titanate. Thus, the difference in the conductivity between the grain boundaries and the bulk is minimized, and the principal cause of degradation is diminished.

An accelerated test has been developed to determine the onset of degradation in Monolythic* capacitors. The onset of degradation depends on temperature and applied direct voltage according to the following relation:[105]

$$t_1 = t_2 \left(\frac{E_2}{E_1}\right)^n \exp\left(\frac{W}{kT_1} - \frac{W}{kT_2}\right)$$

where t_1 = calculated time to onset of degradation
t_2 = measured time to onset of degradation
E_2 = applied dc field for accelerated degradation test
E_1 = rated dc field
n = 2.7 (specified to 067 Sprague Monolythic capacitor)
W = 0.9 ev (specific to 067 Sprague Monolythic capacitor)
k = Boltzman constant
T_2 = measured temperature for accelerated degradation test, °K
T_1 = rated temperature, °K

The resistance of a barium titanate Monolythic capacitor as a function of time with a field gradient of 80 dc volts/mil applied continuously at 150°C is shown in Fig. 93. As the data illustrate, the onset of degradation does not indicate failure since the capacitor could operate for at least two more decades before ohmic heating would be great enough to initiate thermal breakdown.

Glass Capacitors

Glass capacitors were developed by Corning Glass Works during World War II as a direct replacement for mica capacitors. Unlike natural mica, the properties of the glass dielectric can be closely controlled and reproduced. This control over the properties plus the ability to encase the capacitor stack in the same dielectric material has given glass capacitors a reputation for reliability, stability, and performance predictability.

The distinctive characteristics of glass capacitors include high Q at high frequencies and temperatures, negligible hysteresis, zero voltage coefficient of capacitance, and very low capacitance drift with load life. They are used for timing, blocking, coupling, and bypassing functions in circuits requiring high Q and stability. Military specifications are described in MIL-C-11272B, which can be obtained free of charge from the Commanding Officer, U.S. Naval Supply Depot, 5801 Tabor Avenue, Philadelphia, Pennsylvania 19120.

Construction The glass dielectric is made from a lead-potash glass having dielectric properties similar to those of mica. Thin ribbons of glass (0.5 to 1.0 mil thick) are stacked with alternate layers of aluminum foil and fused into a multilayer block.[106] Leads are welded to the electrodes, and the stack is embedded between two slabs of glass having the same composition as the dielectric. This ensures that the electrical properties are entirely those of the glass dielectric. During encapsulation, hermetic glass-to-metal seals are formed between the lead wires and the glass case. Instead of using a glass case, the stacks can be potted in plastic shells suitable for upright mounting on circuit boards. Since the cost of the plastic-encased capacitors is less, they can compete with TC-ceramic capacitors. The insulation resistance of the plastic-potted units, however, is less than that of the glass-encased capacitors.

Low-voltage (300- to 500-volt dc) glass units are made by connecting the capacitor stacks in parallel while medium- and high-voltage (6,000-volt dc) units are made by connecting the stacks in series-parallel or series-series combinations, respec-

---
* Trademark, Sprague Electric Co.

6-76 Ceramics, Glasses, and Micas

tively. The maximum voltage of low-capacitance units is determined by the onset of corona.

Operating characteristics Capacitance values in glass capacitors are available from 1 to 150,000 pf. With 1 kHz as the reference frequency, the percentage capacitance change at 1 MHz is less than 0.5 percent for values up to 10,000 pf, and −10 percent for values up to 100,000 pf (see Figs. 94 and 95). A

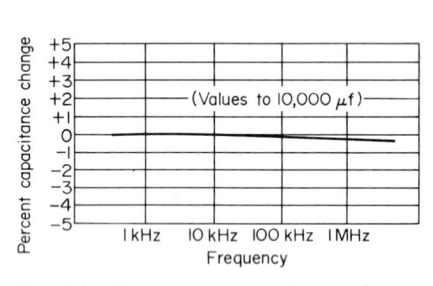

Fig. 94 Percentage capacitance change with frequency of 10,000-µf glass capacitor.[110]

Fig. 95 Percentage capacitance change with frequency of 20,000-, 39,000-, and 100,000-pf glass capacitors. (*Corning Glass Works.*)

similar sensitivity of the high-value capacitors is shown by the capacitance change with temperature. With 25°C as the reference temperature, the capacitance change for values up to 10,000 pf is about −2 percent at 125°C, and for values up to 100,000 pf, it is about −20 percent (see Figs. 96 and 97).

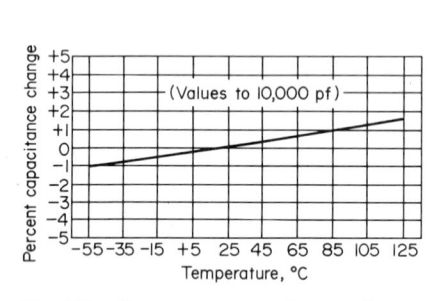

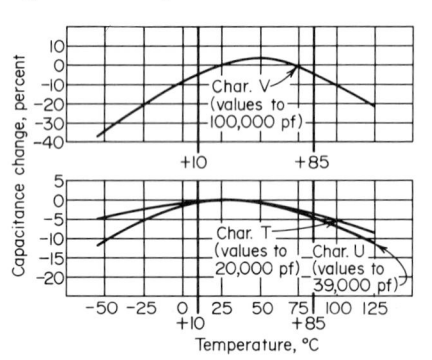

Fig. 96 Percentage capacitance change with temperature of 10,000-pf glass capacitor.[110]

Fig. 97 Percentage capacitance change with temperature of 20,000-, 39,000-, and 100,000-pf glass capacitors. (*Corning Glass Works.*)

In common with other inorganic insulators, the resistivity of glass capacitors decreases exponentially with temperature. A typical glass capacitor rated at 300 volts dc and 10,000 pf has at 25°C an insulation resistance of greater than 50,000 megohms; at 175°C, it has decreased to about 1,150 megohms. Q at 1 MHz and 25°C is typically greater than 2,000 for values up to 10,000 pf (see Fig. 98).

Regardless of the capacitance value or the size, the TCC of glass capacitors is 140 ± 25 ppm per °C at 100 kHz. Since the retrace characteristic of these units is excellent, the TCC will not deviate from the curve by more than 5 ppm at any temperature.

As a consequence of the very low self-inductance of glass capacitors, the capacitance at the resonance frequency is much higher than in mica units. For example, at a self-resonance frequency of 100 MHz, the capacitance of a glass unit was 800 pf compared to 100 pf for the comparable mica capacitor.[107]

Mica Capacitors

The combination of perfect cleavage and outstanding electrical properties prompted the early use of mica as a dielectric for capacitors. Of the various varieties of mica, ruby mica has the best dielectric properties. Since it is a mineral, the chemical composition and purity vary. But the better grades of mica typically have the following properties: dielectric strengths of 3,000 to 6,000 volts/mil, volume resistivities $>2 \times 10^{13}$ ohm-cm, dielectric constants from 6.5 to 8.7, dissipation factors of ~0.0001 at radio frequency and ~0.001 at 60 Hz, and chemical stability to 400°C.[108,109]

There are two main classes of mica capacitors: the transmitting and the receiving types. Transmitting capacitors have capacitances ranging from 0.00001 to 1 μf at voltages from 200 to 50,000 volts dc, and up to 100 amp at radio frequencies. Receiving capacitors are used in lower-power circuits at milliampere current levels. They are rated from 0.000001 to 0.1 μf with operating voltages of 50 to 5,000 volts dc. The inherent electrical and dimensional stability of mica results in capacitors that have low dielectric losses at VHF; good temperature, frequency, and aging characteristics; low power factors; and low equivalent resistances at VHF.

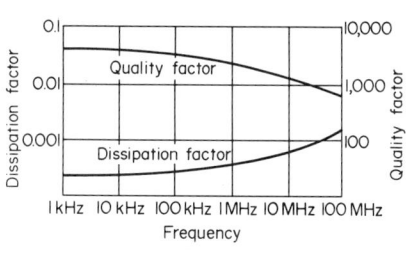

Fig. 98 Dependence of dissipation and quality factors of 10,000-pf glass capacitor on frequency.[110]

Stacked mica-foil capacitors Sheets of mica as thin as 0.1 mil are interlayered with tin-lead foil conductors in a "sandwich" fashion. Each set of foil conductors, which extends alternately from opposite sides of the mica stack, is joined together and connected to one of the capacitor terminals by either pressure clamping or soldering. These individual stacks may be connected either in series or in parallel, or in a combination of both to achieve the desired capacitance-voltage rating. Since the dielectric breakdown strength per mil is less in thick sections than in thin ones, the stacks in high-voltage transmitting capacitors are connected in series to give a voltage-dividing effect and to decrease the electrical stress on the mica sheets. This scheme requires greater plate area per unit of capacitance and, therefore, results in a much larger volume per unit of capacitance. Thus, molded receiving-type capacitors are generally 0.1 to 1.0 in.³ in volume compared to 3 to 500 in.³ for the potted transmitting types.

The danger of flashover and corona between conductors is minimized by leaving a margin between the electrode foil and the edge of the mica sheet. This danger is further reduced by potting the mica stacks in order to exclude moisture and gaseous elements.

Silvered-mica capacitors The exacting demands of precise frequency-selective and timing circuits could not be met by the foil-mica construction because of the relative motion of the foil-mica lamellae. These difficulties were overcome by bonding an accurately positioned layer of silver on the mica dielectric by a screening and firing process. Extremely stable capacitors having closely repeatable temperature coefficients are possible with silvered mica by elimination of the possibility for any relative motion between the conductor and the dielectric. Since the conductor is bonded to the mica, thermal instabilities due to the expansion and contraction of air pockets and other foreign matter are eliminated. Errors due to the positioning or wrinkling of the foil are eliminated by the silvered-mica process. One difficulty with silvered-mica capacitors, however, is silver ion migration, which lowers the dielectric strength, the insulation resistance, and the capacitance.

Ceramics, Glasses, and Micas

This effect is accentuated by high humidity, high temperature, and constant dc potential.

Mica-paper capacitors Even in 0.1-mil thickness, mica sheets are still too brittle to be rolled without fracture. Greater flexibility is introduced by first shredding mica into flakes and then reconstituting it into paper. Integrated mica sheets make use of the natural adhesive forces that are present in thin plates of freshly cleaved material. After the mica is disintegrated by high-velocity water jets, the flakes are deposited onto a belt which passes over a vacuum plate to keep the flakes in place during drying. The inherent fragility of the paper is overcome by treating it with silicone resins. Compared to sheet mica, the electrical properties of integrated mica are somewhat degraded (see Table 31). Two types of construction are used to make mica-paper capacitors: (1) the stacked-plate type similar to that used for ordinary mica capacitors and (2) the rolled type similar to that used for paper or plastic capacitors.

TABLE 31 Mica Capacitor Characteristics[110]

Characteristic	Mica receiving	Mica transmitting	Mica reconstituted
Capacitance range, µf	0.000001–0.1	0.00001–1	0.1–4
Tolerance, %	To ±0.25	To ±1	To ±5
Dc operating volts	50–5,000	200–50,000	200–15,000
Ac operating volts	Seldom used	Rf voltage varies with current and frequency	100–7,500 at 60 Hz
Dissipation factor, %:			
At 60 Hz	0.001–0.01	Seldom used	Seldom used
At 1 kHz	0.02–0.5	0.04–0.07	0.5
At 1 MHz	0.01–0.1	0.03–0.06	0.7–0.9
Insulation resistance:			
At 25°C, megohms/µf	100,000 megohms, or 1,000 megohms/µf	15,000	10,000
At 85°C compared with 25°C	1:5	1:7	1:8
Temperature range, °C	−55 to +150	−55 to +70	−55 to −315
Temperature coefficient, ppm per °C	+50 to −200 (depends on capacitance)	−20 to +100	−350 to −500
Stability (ΔC) with temperature aging	Very small	Very small	Good

Encapsulation Mica capacitors are enclosed in various materials to exclude moisture and to provide mechanical protection. Low-power receiving capacitors are usually molded or dipped, whereas high-power transmitting types are potted in either molded plastic cases or ceramic tubes fitted with metal terminations. Before potting, the mica stacks are clamped under pressure to maintain dimensional integrity. Wax, phenolic, and epoxy resins are frequently used potting materials.

Molded Types. Phenolic resin is still a widely used molding material despite its poor properties at high temperatures. Phenolics have limited flow during molding (which minimizes contamination of the capacitance field), good moisture resistance, and hardness. At low temperatures, the electrical properties are excellent. Since phenolics will not bond to the pigtail leads, moisture seals are made by impregnating the capacitor with wax.

Softer plastics are gaining in importance because of their better molding characteristics and superior high-temperature properties. The better bonding to the lead wires exhibited by these softer materials has in some instances eliminated the need for impregnation. In general, however, the softer plastics have higher dissipation factors than phenolic resins and poorer resin moisture resistance. The principal disadvantage of encapsulation by molding is that the heat and pressure decrease the reliability of the capacitor and lower its life expectancy.

Dipped Types. Stresses related to molding are largely eliminated by the dip-coat method of encapsulation. In this process, the necessary dielectric insulation and moisture barrier are build up by a series of resin coatings. Dip coating improves the reliability of the capacitor and gives more stable capacitor operation. Excellent moisture resistance, satisfactory electrical properties to 150°C, and lower fabrication costs than with molding methods are also attributed to the dip-coat process.

Operating characteristics Transmitting capacitors are primarily used in rf coupling and bypassing applications because of their low inductances and low power factors. There is no intrinsic upper-frequency limit for stacked mica capacitors; however, the inductive reactance due to the leads, the lead assembly, and the metallic internal structure of the active capacitive section set practical limits. Since the mica stacks are clamped under pressure, the stability of the capacitor will depend almost entirely on the response of the supporting assembly to thermal dilation. For this reason, potted capacitors can exhibit a complex and nonlinear variation of capacitance with temperature. Properly designed high-power mica capacitors, nevertheless, will exhibit minimum changes in capacitance with temperature and can withstand high electrical stresses and pass heavy direct currents without noticeable deterioration for many years.

Receiving capacitors made using foil-mica construction are inherently less stable than the silvered-mica types; they have, therefore, been largely replaced by the silvered micas and TC-ceramic capacitors. Compared to the TC ceramics the silvered micas have better long-term performance characteristics, less sensitivity to frequency and voltage variations, and lower dissipation factors. A compendium of receiving, transmitting, and reconstituted mica capacitor characteristics is given in Table 31. Published standards for military applications are given by MIL-C-5C and for the Electronic Industries Association Standard by RS-153-A. A mica-capacitor color code is shown in Table 32. Military specifications can be obtained free of charge from the Commanding Officer, U.S. Naval Supply Depot, 5801 Tabor Avenue, Philadelphia, Pennsylvania 19120.

PIEZOELECTRICITY

In preparing this section, the authors relied heavily on two sources: "Physical Properties of Crystals," by J. F. Nye,[111] and "Physical Acoustics" edited by W. P. Mason.[112] The book by Nye should be consulted for a clear derivation of the effects of crystal symmetry on piezoelectricity. In Mason's book, the chapter by D. A. Berlincourt, D. R. Curran, and H. Jaffe gives a full account of the properties of piezoelectric materials. Two useful compendiums of piezoelectric properties are given in government reports AMRA MS 64-05 and AFML-TR-66-164. For references on applications and equivalent circuits of piezoelectric transducers the books of Hueter and Bolt,[113] Mason,[114] and Albers[115,116] should be consulted. Current publications on piezoelectrics may be found in the *Journal of the Acoustical Society of America*.

When a stress is applied to certain crystals, an electric polarization develops whose magnitude is proportional to the applied stress. This is known as the direct piezoelectric effect. Each component of the polarization P_i is linearly related to all the components of the general stress σ_{ij}. Since polarization is a vector with three components, while stress is a tensor with nine components, the piezoelectricity is described by 27 moduli, d_{ijk}. Each component of P_i can be written as

$$P_1 = d_{111}\sigma_{11} + d_{112}\sigma_{12} + d_{113}\sigma_{13} + d_{121}\sigma_{21} + d_{122}\sigma_{22} + d_{123}\sigma_{23} + d_{131}\sigma_{31} + d_{132}\sigma_{32} \\ + d_{133}\sigma_{33} \quad (3)$$

The P_2 and P_3 components have similar equations. In condensed form, the relation between P_i and σ_{ij} is, in dummy suffix convention,

$$P_i = d_{ijk}\sigma_{jk} \quad i, j, k = 1, 2, 3$$

Since the 27 coefficients (d_{ijk}) form a third-rank tensor, the array is in the shape of a cube rather than a square as with a second-rank tensor. This diffi-

TABLE 32 Mica Capacitor Color Code[110]

The electrical value of a mica capacitor can be determined from an examination of the six-dot color code imprinted on the body of the capacitor.

When the capacitor is being read, the direction indicator (an arrow on the body in some form) should point to the right of the reader. The dots are read from left to right across the top row and from right to left across the bottom row, in that order.

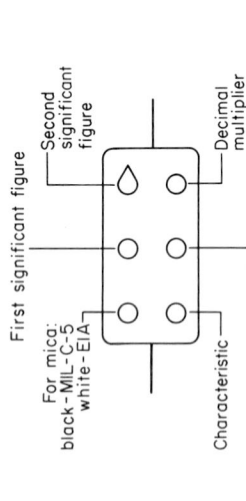

First significant figure — Second significant figure — Decimal multiplier

For mica:
black – MIL-C-5
white – EIA

Characteristic — Tolerance

Color	1st and 2d significant figure	Decimal multiplier	Tolerance, %	Characteristic
Black	0	1	±20	B
Brown	1	10	±1	C
Red	2	100	±2	D
Orange	3	1,000	±3	E
Yellow	4			F
Green	5		±5	
Blue	6			
Purple	7			
Gray	8			
White	9			
Gold	…	0.1	±5	
Silver	…	0.01	±10	

The only exception that may be made occurs when three significant figures are to be found in the characteristic. In this case the upper three dots represent the three significant figures, and the identification color is eliminated.

Characteristic Code

Letter	Temperature coefficient, ppm per °C	Capacitance drift
B	MIL Not specified EIA ±500	MIL Not specified EIA ±(3% + 1 pf)
C	−200 to +200	MIL ±(0.5% + 0.1 pf) EIA ±(0.5% + 0.5 pf)
D	−100 to +100	±(0.3% + 0.1 pf)
E	−20 to +100	±(0.1% + 0.1 pf)
F	0 to 70	±(0.05% + 0.1 pf)

If the color code is to be used to identify a MIL molded or dipped mica capacitor, a nine-dot sequence is used. The six dots on one face of the capacitor remain and have the same identification as the six-dot code previously discussed. There are three additional dots on the reverse side of the capacitor, and these are identified as follows:

Voltage rating — Temperature range — Vibration grade

MIL 9-dot Code

Color	Voltage rating, volts dc	Temperature range, °C	Vibration grade, cps
Black	…	−55 to +70	10–55
Brown			
Red		−55 to +85	
Orange	300		
Yellow		−55 to +125	10–2,000
Green	500		
Blue	…	−55 to +150	

6-80

culty of concisely presenting a three-dimensional array is overcome by neglecting the body torques and putting for convenience $d_{ijk} = d_{ikj}$, thereby lowering the number of independent components to 18 and making possible a two-dimensional matrix notation. This is done by abbreviating the second and third suffix of d_{ijk} and both suffixes of σ_{ij} with a single suffix running from 1 to 6 as follows:

Tensor notation............ 11 22 33 23, 32 31, 13 12, 21
Matrix notation............ 1   2   3   4   5   6

and by introducing the factor 2, where

$$2d_{ijk} = d_{in} \quad n = 4, 5, 6$$

Accordingly, the tensor notation of Eq. (3) is transformed into the matrix notion as follows:

$$P_1 = d_{11}\sigma_1 + \tfrac{1}{2}d_{16}\sigma_6 + \tfrac{1}{2}d_{15}\sigma_5 + \tfrac{1}{2}d_{16}\sigma_6 + d_{12}\sigma_2 + \tfrac{1}{2}d_{14}\sigma_4 + \tfrac{1}{2}d_{15}\sigma_5 + \tfrac{1}{2}d_{14}\sigma_4 + d_{13}\sigma_3$$

or

$$P_1 = d_{11}\sigma_1 + d_{12}\sigma_2 + d_{13}\sigma_3 + d_{14}\sigma_4 + d_{15}\sigma_5 + d_{16}\sigma_6$$

or in dummy suffix convention the direct piezoelectric effect is

$$P_i = d_{ij}\sigma_j \quad i = 1, 2, 3 \quad j = 1, \ldots, 6$$

The converse piezoelectric effect, i.e., the dilation of certain crystals upon application of an electric field, has a one-to-one correspondence with the direct effect. In matrix notion and using the dummy suffix convention, the converse piezoelectric effect is written as

$$\epsilon_j = d_{ij}E_i \quad i = 1, 2, 3 \quad j = 1, \ldots, 6$$

Thus a piezoelectric material can be described by a matrix of piezoelectric coefficients, d_{ij}, which are summarized as follows:

```
           → ε₁  ε₂  ε₃  ε₄  ε₅  ε₆
           → σ₁  σ₂  σ₃  σ₄  σ₅  σ₆

  ↓  ↓
  E₁ P₁     d₁₁ d₁₂ d₁₃ d₁₄ d₁₅ d₁₆
  E₂ P₂     d₂₁ d₂₂ d₂₃ d₂₄ d₂₅ d₂₆
  E₃ P₃     d₃₁ d₃₂ d₃₃ d₃₄ d₃₅ d₃₆
```

Read horizontally by rows, it gives the direct effect, e.g., $P_1 = d_{11}\sigma_1$; read vertically my columns, it gives the converse effect, e.g., $\epsilon_1 = d_{11}E_1$.

The number of independent coefficients that actually exist in a material depends on the crystal symmetry. As an extreme example, crystals with a center of symmetry cannot be piezoelectric because the polarization is zero. Thus of the 32 crystal classes only 20 are piezoelectric. Whether or not a crystal belonging to one of these noncentrosymmetric classes is piezoelectric has to be established experimentally. In Table 33 are shown the d_{ij} matrices for each of the 20 piezoelectric classes written in the matrix form summarized above but with the E_i, P_i, ϵ_j, and σ_j omitted.

The physical significance of the matrix is illustrated by considering class 32 for right-handed quartz.

```
           → ε₁  ε₂  ε₃  ε₄  ε₅  ε₆
           → σ₁  σ₂  σ₃  σ₄  σ₅  σ₆

  ↓  ↓
  E₁ P₁    ⎛ -2.3  2.3   0   -0.67   0     0  ⎞
  E₂ P₂    ⎜   0    0    0      0   0.65  4.6 ⎟  × 10⁻¹² coul/newton
  E₃ P₃    ⎝   0    0    0      0    0     0  ⎠
```

Ceramics, Glasses, and Micas

TABLE 33 Elastopiezodielectric Matrices of the Various Crystal Classes*[11]

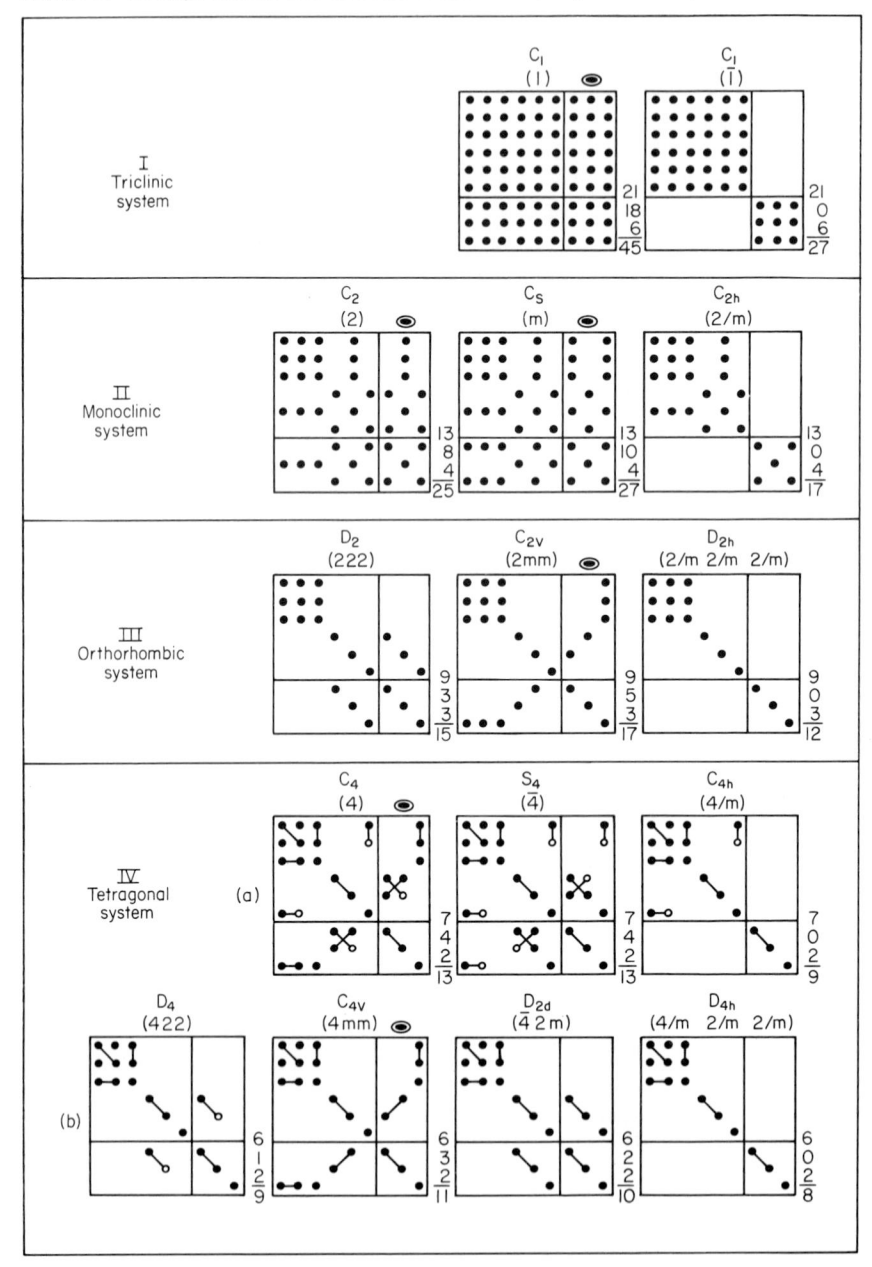

Piezoelectricity

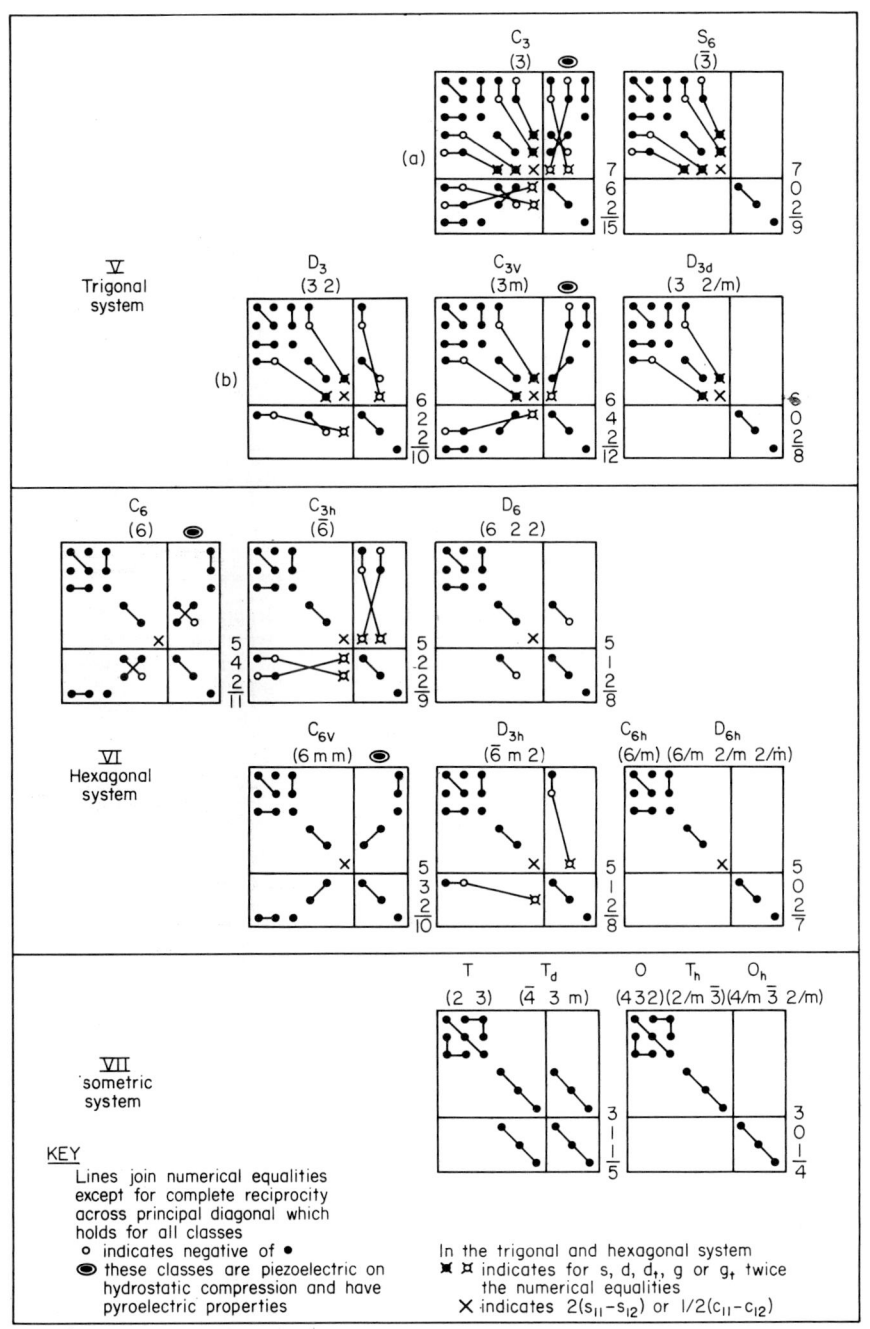

V Trigonal system

(a) C_3 (3); S_6 ($\bar{3}$)

(b) D_3 (3 2); C_{3v} (3m); D_{3d} (3 2/m)

VI Hexagonal system

C_6 (6); C_{3h} ($\bar{6}$); D_6 (6 2 2)

C_{6v} (6mm); D_{3h} ($\bar{6}$ m 2); C_{6h} (6/m); D_{6h} (6/m 2/m 2/m)

VII Isometric system

T (2 3); T_d ($\bar{4}$ 3 m); O (4 3 2); T_h (2/m $\bar{3}$); O_h (4/m $\bar{3}$ 2/m)

KEY

Lines join numerical equalities except for complete reciprocity across principal diagonal which holds for all classes
- ○ indicates negative of ●
- ◉ these classes are piezoelectric on hydrostatic compression and have pyroelectric properties

In the trigonal and hexagonal system
- ✖ ¤ indicates for s, d, d_t, g or g_t twice the numerical equalities
- ✕ indicates $2(s_{11} - s_{12})$ or $1/2(c_{11} - c_{12})$

*The numbers on the right side of each scheme indicate, from top to bottom, the number of the independent elastic, piezoelectric, and dielectric constants.

6-84   Ceramics, Glasses, and Micas

If a tensile stress σ_1 of 1 kg/cm² (9.81×10^4 newton/meter²) is applied parallel to the diad axis x_1 of quartz, the polarization along the x_1 axis is given by

$$P_1 = d_{11}\sigma_1 = (-2.3 \times 10^{-12})(9.81 \times 10^4) = -2.3 \times 10^{-7} \text{ coul/meter}^2$$
$$P_2 = 0 \qquad P_3 = 0$$

On the other hand, a compressive stress σ_2 along x_2 produces no polarization parallel to x_2, but it does produce a polarization along x_1, since

$$P_1 = d_{12}\sigma_2 = (2.3 \times 10^{-12})(-9.81 \times 10^4) = -2.3 \times 10^{-7} \text{ coul/meter}^2$$
$$P_2 = 0 \qquad P_3 = 0$$

Therefore, a given polarization along x_1 can be produced either by a tensile stress along x_1 or by a compressive stress along x_2.

A polarization can also be produced by a shear stress σ_4 about x_1 since

$$P_1 = d_{14}\sigma_4 \qquad P_2 = 0 \qquad P_3 = 0$$

Conversely, the strain along the diad axis x_1 produced by an electric field E_1 of 10^4 volts/meter is given by

$$\epsilon_1 = d_{11}E_1 = (-2.3 \times 10^{-12})(10^4) = -2.3 \times 10^{-8}$$

This field also produces an expansion along x_2 (of 2.3×10^{-8}) and a shear strain (of -6.7×10^{-8}) about x_1, since

$$\epsilon_2 = d_{12}E_1 = (2.3 \times 10^{-12})(10^4) = 2.3 \times 10^{-8}$$
and
$$\epsilon_4 = d_{14}E_1 = (-0.67 \times 10^{-12})(10^4) = -6.7 \times 10^{-8}$$

Relation among the piezoelectric, dielectric, and elastic properties A description of piezoelectricity must consider both the elastic and the dielectric properties of a material because an applied stress produces mechanical strain as well as dielectric polarization. Providing that the stress is below the elastic limit, the relation between stress and strain is given by Hooks' law:

$$\epsilon_{ij} = s_{ijkl}\sigma_{kl}$$

where s_{ijkl} are the compliances of the material. There are 81 s_{ijkl} coefficients in this forth-rank tensor; however, only 36 of them are independent, since $s_{ijkl} = s_{jikl}$. Similarly, $\sigma_{ij} = c_{ijkl}\epsilon_{kl}$, where c_{ijkl} are the stiffness constants. The dielectric constants form a second-rank tensor by relating two vectors: the electric field E and the electric displacement D. Since Gauss' law, $D = \epsilon E + P$, relates P to D and E, there are five variables P, D, E, ϵ, σ, and Gauss' equation. Two equations can be written, therefore, that relate two dependent variables to two independent variables, e.g.:

$$\epsilon = f(\sigma, E)$$
$$D = f'(\sigma, E)$$

In matrix form this set of linear equations has 45 independent coefficients for crystals of lowest symmetry, but it has only 4 independent coefficients for crystals of highest symmetry. A convenient graphical presentation of the elastopiezodielectric matrices for the various crystal classes is given in Table 33, reprinted from the IRE Standards on Piezoelectric Crystals for 1958. From the four sets of equations between D, E, σ, and ϵ, the definitions of the piezoelectric constants are derived:

$$d = \left(\frac{\partial \epsilon}{\partial E}\right)_\sigma = \left(\frac{\partial D}{\partial \sigma}\right)_E$$

$$g = \left(-\frac{\partial E}{\partial \sigma}\right)_D = \left(\frac{\partial \epsilon}{\partial D}\right)_\sigma$$

$$e = \left(-\frac{\partial \sigma}{\partial E}\right)_\epsilon = \left(\frac{\partial D}{\partial \epsilon}\right)_E$$

$$h = \left(-\frac{\partial \sigma}{\partial D}\right)_\epsilon = \left(-\frac{\partial E}{\partial \epsilon}\right)_D$$

These four piezoelectric constants are interrelated through the dielectric constant K_{nm} by the following equations:

$$d_{mi} = K_{nm}{}^\sigma g_{ni} = e_{mj}s_{ji}{}^E$$
$$g_{mi} = \beta_{nm}{}^\sigma d_{ni} = h_{mj}s_{ji}{}^D$$
$$e_{mi} = K_{nm}{}^\epsilon h_{ni} = d_{mj}c_{ji}{}^E \quad m, n = 1, 2, 3$$
$$h_{mi} = \beta_{nm}{}^\epsilon e_{ni} = g_{mj}c_{ji}{}^D \quad i, j = 1, \ldots, 6$$

where β is the dielectric impermeability which is equal to $1/K$ except for the triclinic and monoclinic crystal classes.

Piezoelectric coupling factor The amount of energy of one kind stored in a material as energy of the second kind is called the *piezoelectric coupling factor* k_{ij} defined by the equation

$$k_{ij} = \left(\frac{U_m{}^2}{U_e U_d}\right)^{1/2}$$

where U_m = the mutual elastic and dielectric energies, U_e = the elastic energy, and U_d = the dielectric energy. The energy expressions, which are extremely complicated in the general case, simplify for simple geometries when most of the stresses are zero. For example, the coupling factor for the fundamental axial mode of an axially poled rod having electrodes on either end is given by

$$k_{33} = \frac{d_{33}}{(K_{33}{}^\sigma s_{33}{}^E)^{1/2}} \quad \text{when } \sigma_1 = \sigma_2 = 0, \sigma_3 \neq 0$$

$$\epsilon_1 = \epsilon_2 \neq 0, \epsilon_3 \neq 0$$

Typical values of k_{33} are 0.1 for quartz and tourmaline, 0.5 for $BaTiO_3$, 0.7 for $Pb(Zr,Ti)O_3$, and 0.9 for rochelle salt at its most favorable temperature.

Pyroelectricity and ferroelectricity The polarization observed in quartz arises from an unbalancing of the permanent dipole moments by the applied stress. In certain materials, a polar axis exists in the absence of an applied stress. When such materials are heated or cooled, thermal dilation changes the magnitude of the polarization, and consequently a charge appears across the polar axis. This phenomenon is called pyroelectricity. Wurtzite (hexagonal ZnS), for example, has a polar axis and, as a result, it is both pyro- and piezoelectric. Sphalerite (cubic ZnS), in contrast, does not have a polar axis in the absence of an applied stress; therefore, it is only piezoelectric.

Ferroelectric materials have a polar axis similar to that of pyroelectric crystals. But in this case the polarization arises from a spontaneous alignment of the dipole moments. Unlike the polar axis of a nonferroelectric material, which is fixed by crystal symmetry and is not reversible, the polar axis of a ferroelectric material is free to rotate in response to a sufficiently strong electric field. Thus, only in ferroelectric materials can a domain structure appear that is analogous to ferromagnetic domains. In ferroelectric single crystals biased with a strong electric field, the relation between the stress and strain components on one hand and the electric field and displacement on the other is linear only at small input signals; larger signals cause hysteresis and gross deviations from linearity.

Piezoelectric ceramics The piezoelectric ceramics of greatest technological importance have slightly distorted perovskite structures. These ceramics include $BaTiO_3$, $Pb(Ti,Zr)O_3$, and $NaNbO_3$. Antiferroelectric $NaNbO_3$ becomes ferroelectric and piezoelectric when small quantities of K^{1+}, Pb^{2+}, or Cd^{2+} are substituted for Na^{1+}. The wide range of properties available in $Pb(Ti,Zr)O_3$ makes this material the most important piezoelectric ceramic. Lead metaniobate is the only widely used piezoelectric ceramic that does not have the perovskite structure.

The ideal perovskite structure is cubic and, therefore, it is neither ferroelectric nor piezoelectric. In $BaTiO_3$, for example, each Ti^{4+} ion is surrounded by six O^{2+} ions and eight Ba^{2+} ions (see Fig. 99). Below a critical temperature, called the *Curie temperature*, certain titanium ions spontaneously move closer to one of their six surrounding oxygen ions. The resulting change in dipole configuration causes the neighboring titanium ions to move in the same direction. As a result of this spontaneous polarization, the center of symmetry is destroyed, and the crystal acquires both ferroelectric and piezoelectric properties.

Before poling, a ferroelectric ceramic has no remanent polarization, and consequently no piezoelectric effect. When such a material is poled by a strong dc electric field, a torque is exerted on the polar axes of the individual crystals that are inclined to the field direction. Since there are six alternate directions for movement of the titanium ions, each crystallite can change its domain configuration in response to the direction of the electric field. There are two types of domain switching. The first type is by 180°, and the second type is by an angle less than 180° (90° in tetragonal perovskites, 71 to 109° in rhombohedral perovskites, and 90° or 60 to 120° in orthorhombic perovskites). Electric fields may cause switching of either type in virgin material, but only 180° domain switching occurs in poled material. Domain switching of the first type is virtually complete, whereas switching of the second type is only partially complete. In $BaTiO_3$ the 90° domain (mechanical) switching is only about 13 percent complete, whereas in $Pb(Ti,Zr)O_3$ it is about 50 percent complete, which accounts for the higher coupling factors exhibited by $Pb(Ti,Zr)O_3$ materials.

The stresses that develop from misorientations between mechanical domains gradually relax with time by domain reorientations, which affect the permanent dipole moments. This "aging" effect progresses linearly with the logarithm of time, and it is usually expressed as the percentage change per decade of time. Certain substitutions such as $(Nb,Ta)^{5+}$ for $(Zr,Ti)^{4+}$, or $(La,Nd)^{3+}$ for Pb^{2+} in $Pb(Ti,Zr)O_3$, appear to facilitate domain wall motion. This permits most of the stresses to be relaxed immediately upon removal of the poling field, and as a result, little further aging occurs. Reversible domain wall motion contributes to higher permittivities, and elastic compliances as well as to higher dielectric and mechanical losses since there is an activation energy associated with domain wall movement.

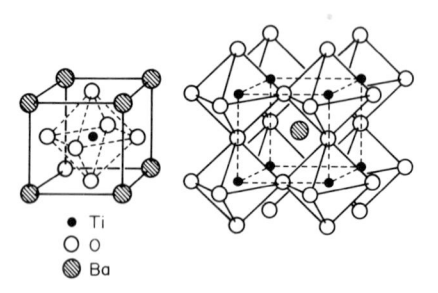

Fig. 99 Perovskite structure ($BaTiO_3$).

When the input signal is small compared to the dc polarization, the remanent polarization is sufficient to keep the device operating on the reversible-strain versus electric-field curve, just as was the case with piezoelectric single crystals. With larger input signals, however, hysteresis effects are pronounced, and the response between strain and electric field is nonlinear.

Applications Piezoelectric ceramics are used in applications that can be divided into three categories: (1) low-stress devices, (2) high-stress devices, and (3) energy-storage devices (see Table 34). Low-stress devices, e.g., microphones, demand a strong electric output from a weak driving force. Stresses may vary from <1 to >10 psi, and the corresponding voltages from<1 to >100 volts/in. In this range, the response between stress and strain is essentially linear. High-stress devices, e.g., ultrasonic generators and spark pumps, operate at stress levels of 100 to >1,000 psi and voltages to >1,000 volts/in. Since materials used for these applications are operated at saturation, the relation between stress and strain is nonlinear. Certain energy-storage devices use dielectrics made from piezoelectric ceramics as capacitors. In fact, at low frequencies many piezoelectric devices exhibit capacitor characteristics.

The properties of experimental piezoelectric materials are available in Ref. 97, and the properties of commercially available piezoelectric ceramics are listed in Table 35. By remembering that the first suffix refers to the electrical direction while the second suffix refers to the mechanical direction, the meaning of the matrix notation is easily recalled. Thus, with reference to an orthogonal set of axes x_1, x_2, and x_3, d_{33} is a measure of the electrical effect along the polar axis x_3 resulting from a mechanical deflection in the same direction. The coefficient d_{31} measures the electrical effect along the x_3 axis when the force is along the x_1 axis, and finally d_{15} measures the shear deflection around the x_2 axis caused by a voltage along the x_1 axis.

MAGNETIC CERAMICS

The purpose of this section is to give a brief review of the behavior of magnetic ceramic materials and to list some examples of commercially available ferrites. Two guides to the current literature on magnetism are particularly noteworthy:
  1. "Magnetism and Magnetic Materials Digest," Academic Press, New York, gives an annual survey of the technical literature of the preceding year.
  2. "Index to the Literature of Magnetism," published by the Bell Telephone Laboratories, Inc., gives a semiannual comprehensive listing of the titles of new papers in the field of magnetism.

TABLE 34 Piezoelectric Ceramic Applications[146]

Types of ceramic	Requirements	Applications
Low stress BaTiO$_3$ (Ba,Ca)TiO$_3$ Pb(Zr,Ti,Nb)O$_3$ (Na,K)NbO$_3$	High strain coefficient; high compliance; stability with temperature, voltage, and/or time; high dielectric constant	Microphones, phonograph pickups, accelerometers, IF filters, remote-control units, delay lines, flaw detectors, etc.
High stress (Ba,Pb)TiO$_3$ (Pb,Sr)(Zr,Ti)O$_3$	High strain coefficient; stability with temperature and voltage; low electrical and mechanical losses; high dielectric constant	Sonar generators, spark pumps, emulsifiers, pigment dispersers, degassers, ultrasonic cleaners, and machiner, tools, etc.
Energy storage Pb(Zr,Ti,Bi)O$_3$ Pb(Zr,Ti,Sr,La)O$_3$	(A) High remanent polarization; low coercive force; minimum of flaws; high reliability; high dielectric constant (B) Reproducibility of charge versus field; stability with temperature, voltage, and time	(A) Detonation fuses for shells and other projectiles, and high-energy storage capacitors (B) Memory devices

In preparing this review, the authors relied largely on two sources: (1) "Handbook of Microwave Ferrite Materials," edited by W. H. von Aulock,[117] and (2) "Ferrites" by J. Smit and H. P. J. Wijn.[118] The available literature on commercial ferrites was useful, especially that supplied by Airtron Division of Litton Industries, Allen-Bradley Company, Ceramic Magnetics, Inc., Ferroxcube Corporation, Huges Aircraft Company, Indiana General Corporation, Magnetics, Inc., Trans-Tech, Inc., and Sperry Rand Corporation Univac Division.

Magnetism Certain oxides containing iron ions and ions from either the first-transition or the rare-earth series have both high resistivities and high permeabilities. The magnetic behavior of these materials is similar to that of ferromagnetic metals in that below the Curie temperature the permeability is very much higher than it is above this temperature. In both instances the high permeability originates from the spontaneous alignment of the individual atomic ion magnetic moments. The feature that distinguishes ceramic magnetic materials from metallic ones is that the magnetic ions in ceramic magnets are distributed over two or more interpenetrating sublattices.[119] All the ion moments on each sublattice are aligned, but the moments of the sublattices are antiparallel to each other. The term *ferrimagnetic* denotes the case when a net magnetic moment results, and *antiferrimagnetic* the case when the sublattice vectors cancel.

TABLE 35 Properties of Commercial Piezoelectric Materials

Company		Clevite						Gulton				
Trade name	Ceramic-B	PZT-2	PZT-4	PZT-5A	PZT-5H	HD-11	HS-21	G-53	HDT-31	G-1408	G-1500
Type of material*	A	C	C	C	C	A	A	C	C	C	C
Temperature coefficient T_c, °C	115	370	328	365	193	>135	>125	>330	>330	>300	>360
Aging, % k_p/decade	−1.8	−1.8	−2.3	−0.2	−0.3	—	—	—	—	—	—
Density ρ, g/cm³	5.55	7.6	7.5	7.75	7.5	>5.6	>5.6	>7.6	>7.6	>7.5	>7.6
Elastic constants:											
c_{11}^E	15.8	13.5	13.9	12.1	12.6	13.7	11.5	8.1	8.1	9.0	6.3
c_{33}^E	15.0	11.3	11.5	11.1	11.7	13.0	11.1	6.5	6.7	8.2	4.9
Dielectric constants:											
K_{33}^T	1,200	450	1,200	1,700	3,400	600	1,150	720	1,300	1,000	1,700
K_{33}^S	910	260	635	830	1,470	—	—	—	—	—	—
tan δ	0.006	0.005	0.004	0.02	0.02	0.006	0.015	0.022	0.006	0.003	0.015
Coupling coefficients:											
k_{33}	0.48	0.626	0.70	0.705	0.752	0.45	0.51	0.60	0.66	0.60	0.69
k_{31}	0.194	0.28	0.334	0.344	0.388	0.16	0.18	0.29	0.35	0.26	0.34
k_{15}	0.48	0.701	0.71	0.685	0.675	0.42	0.48	0.64	0.59	0.60	0.65
k_p	0.33	0.47	0.58	0.60	0.65	0.26	0.30	0.50	0.57	0.50	0.58
Frequency constants:											
N^{33}	104	82	79	74	79	105	103	77	80	84	73
N^{31}	90	66	65	55	56	100	89	65	63	69	58
N_p	—	—	—	—	—	134	123	88	88	93	78
Piezoelectric constants:											
d_{33}	149	152	289	374	593	86	148	190	280	200	370
d_{31}	−58	−60.2	−123	−171	−274	−30	−50	−84	−120	80	−166
d_{15}	242	440	496	584	741	125	225	300	360	315	540
g_{33}	14.1	38.1	26.1	24.8	19.7	16	16	30	23	22	25
g_{31}	−5.5	−15.1	−11.1	−11.4	−9.11	−5.5	−5.2	−13	−11	−9	−11
g_{15}	21.0	50.3	39.4	38.2	26.8	25	25	36	29	29	36
Q_m	400	680	500	75	65	>800	300	140	>500	>1,200	80

* Type = composition of base material:
 A Barium titanate.
 C Lead zirconate titanate.
 E Lead metaniobate.

TABLE 35 Properties of Commercial Piezoelectric Materials (Continued)

Company	Gulton								Electro-Ceramics			
Trade name	HST-41	G-1512	G-2000	EC-31	EC-51	EC-55	EC-57	EC-64	EC-65	EC-69	EC-70
Type of material*	C	C	E	A	A	A	A	C	C	C	C
Temperature coefficient											
T_c, °C	>270	>240	>400	>115	>130	>115	>140	>320	>350	>300	220
Aging, % k_p/decade	—	—	—	−2.0	−4.0	−1.4	−5.0	−2.1	−0.3	−1.5	—
Density ρ, g/cm³	>7.6	>7.4	>5.8	5.55	5.60	5.55	5.30	7.60	7.65	7.60	7.45
Elastic constants:											
c_{11}^E	7.0	6.3	4.0	—	—	—	—	—	—	—	—
c_{33}^E	5.9	5.4	4.7	11.6	11.0	11.9	11.6	8.05	6.6	8.42	6.42
Dielectric constants:											
K_{33}^T	1,800	2,600	250	1,170	1,250	1,220	600	1,300	1,725	1,050	2,600
tan δ	0.022	0.018	0.006	0.007	0.005	0.005	0.006	0.005	<0.02	0.005	0.16
Coupling coefficients:											
k_{33}	0.66	0.72	0.38	0.48	0.47	0.46	0.38	0.65	0.70	0.62	0.74
k_{31}	0.35	0.37	0.04	0.19	0.19	0.19	0.15	0.33	0.34	0.31	0.37
k_{15}	0.69	0.78	0.33	—	—	—	—	—	—	—	—
k_p	0.59	0.63	0.066	0.32	0.32	0.31	0.25	0.55	0.59	0.52	0.63
Frequency constants:											
N_{33}	77	74	59	112	105	112	112	82	70	84	68
N_{31}	59	59	55	91	88	92	91	66	53	68	52
N_p	81	79	70	124	117	125	124	87	78	89	76
Piezoelectric constants:											
d_{33}	325	500	80	152	51	150	87	280	355	220	480
d_{31}	−157	−232	−10	−59	−56	−58	−30	−120	−160	−95	−225
d_{15}	625	680	115	—	—	—	—	—	—	—	—
g_{33}	22	20	36	14.8	14.3	14.3	16.2	24.5	25.0	24.5	20.9
g_{31}	−11	−9.3	−4.5	−5.8	−5.3	−5.6	−5.5	−10.5	−11.5	−10.5	−9.8
g_{15}	37	35	50	—	—	—	—	—	—	—	—
Q_m	70	70	10	400	400	550	600	400	100	960	75

* Type = composition of base material:
  A Barium titanate.
  C Lead zirconate titanate.
  E Lead metaniobate.

TABLE 35 Properties of Commercial Piezoelectric Materials (Continued)

Company		Channel						Erie				
Trade name	300	1300	700	5400	5800	5500	1005	1006	1008	1009	1011
Type of material*	A	A	A	C	C	C	A	A	A	A	A
Temperature coefficient T_c, °C	>115	>115	>145	>300	>300	>345	120	120	115	135	145
Aging, % k_p/decade	—	—	—	—	—	—	−2.0	−2.0	−1.6	−2.5	−1.6
Density ρ, g/cm³	>5.5	>5.5	>5.4	>7.6	>7.6	>7.7	5.6	5.5	5.4	5.6	5.35
Elastic constants:											
c_{11}^E	11.2	11.3	12.6	8.45	9.10	6.40	—	—	—	—	—
c_{33}^E	10.6	10.7	12.1	6.75	7.30	5.55	1.02	11.5	9.4	11.0	11.2
Dielectric constants:											
K_{33}^T	1,250	1,350	550	1,300	980	1,750	1,700	1,200	1,200	1,400	700
K_{33}^S	980	1,070	470	715	600	895	—	—	—	—	—
tan δ	—	—	—	—	—	—	0.008	0.009	0.005	0.008	0.006
Coupling coefficients:											
k_{33}	0.46	0.45	0.38	0.65	0.62	0.69	0.47	0.48	0.55	0.52	0.51
k_{31}	0.19	0.18	0.16	0.31	0.29	0.32	0.20	0.18	0.19	0.19	0.16
k_p	0.32	0.30	0.26	0.56	0.50	0.58	0.33	0.31	0.32	0.32	0.27
Frequency constants:											
N_{33}	106	106	104	77	78	71	97	103	101	101	98
N_{31}	91	91	91	65	66	60	85	98	86	88	89
N_p	124	124	124	86	89	82	116	123	118	118	121
Piezoelectric constants:											
d_{33}	160	150	105	270	222	360	188	146	154	160	107
d_{31}	−60	−55	−30	−125	−95	−160	−71	−56	−60	−62	−38
g_{33}	15	15	18	25.3	25.1	23.5	11.9	13.6	16.4	13.1	17.2
g_{31}	−6.08	−5.95	−5.30	−10.6	−10.8	−10.7	−4.9	−5.2	−6.4	−5.1	−6.1
Q_m	450	600	1,200	450	600	1,200	380	360	600	360	350

* Type = composition of base material:
  A Barium titanate.
  C Lead zirconate titanate.
  E Lead metaniobate.

Magnetic Ceramics 6-91

Magnetic characteristics At temperatures much below the Curie point, all the elementary magnetic moments in a sufficiently small particle (<1-micron diameter for ferrites) will become spontaneously aligned and create an external magnetic field. The energy associated with this field can be decreased if a fraction of the elementary dipoles become aligned in the opposite sense. This can happen providing that the decrease in field strength is more than the energy required to form a boundary separating two domains having opposite magnetic moments. Since larger particles can better accommodate domain-wall strain energy, growth will take place with increasing particle size until the external magnetic field vanishes.

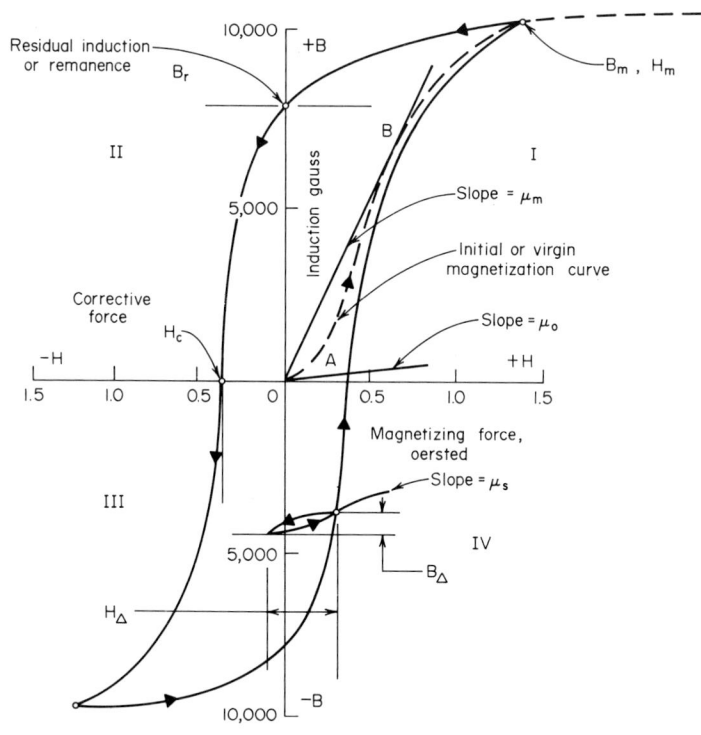

Fig. 100 Typical hysteresis curve of a ferrimagnetic material.[1]

Magnetic hysteresis Measurable external magnetization can be induced into such a magnetic material by applying an external magnetic field that will align the individual domain moments. This induced magnetization M is described by the magnetization curve (Fig. 100) derived by plotting either M or the magnetic induction B as a function of field strength H, where $B = H + 4\pi M$. The ratios B/H and M/H are called the permeability μ and the magnetic susceptibility χ, respectively. Initially, the domains that are most nearly aligned with the magnetic field grow at the expense of the misaligned domains by the movement of domain walls. As the field strength is increased, easy domain-wall movement leads to maximum permeability, after which the permeability decreases. Eventually most of the domain walls disappear, with the remainder reaching stable positions. Further increase in the magnetization occurs by rotation of the elementary ion moments in the direction of the applied field. When all these moments are parallel with the applied field, magnetic saturation is reached.

Upon decreasing H, the magnetic induction does not retrace the original curve, but instead it lags behind the applied field to give the hysteresis curve. The trace of this curve is determined by the ease with which domain walls nucleate, move through the material, and disappear. At zero H, a residual induction B_r remains; in order to decrease it to zero, a reverse field H_c called the *coercive force*, must be applied. Larger fields produce saturation in the reverse direction. The areas enclosed by the hysteresis loop is proportional to the energy loss per cycle.

Initial Permeability. For small magnetization, $B = \mu_0 H$, where the initial permeability μ_0 is a complex number:

$$\mu_0^* = \mu_0' + j\mu_0''$$

μ_0' is the real part of the permeability, and μ_0'' is the imaginary part taken to account for the losses. Energy dissipation is usually expressed as $\tan \delta_m = \mu_0''/\mu_0'$. Both μ_0' and μ_0'' vary as a function of frequency, as shown in Fig. 101 for a Mg-spinel ferrite. The resonance peak at about half $(\mu_0' - 1)$ maximum is due to domain wall motion, and the one at the higher frequency is due to magnetic dipole rotational effects.[118]

Anisotropy Energy. In general, magnetic crystals exhibit an easy or preferred direction of magnetization because the free energy of the system is dependent on the orientation of the magnetization with respect to the crystallographic axes. The energy required to change the magnetization from the easy direction is known as the *anisotropy energy* E_a. For cubic crystal symmetry, this energy is given to the second order by

$$E_a = K_1(\alpha_1^2\alpha_2^2 + \alpha_2^2\alpha_3^2 + \alpha_3^2\alpha_1^2) + K_2\alpha_1^2\alpha_2^2\alpha_3^2$$

where K_1 and K_2 are the first- and second-order anisotropy constants, and α_1, α_2, and α_3 are the direction cosines of the magnetic moment relative to the edges of a cube. Usually K_2 is negligible compared to K_1. Negative values of K_1 indicate that the easy direction of magnetization is parallel to the body diagonal of a

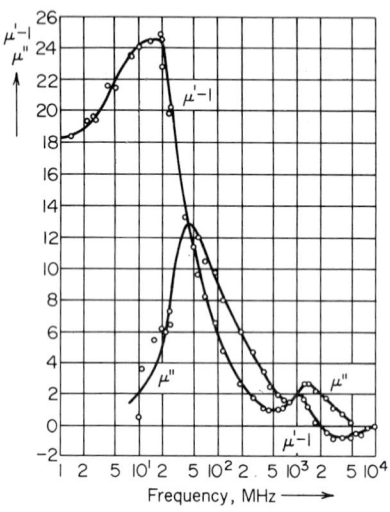

Fig. 101 Magnetic spectrum of a sintered Mg-ferrite showing two natural resonances.[127]

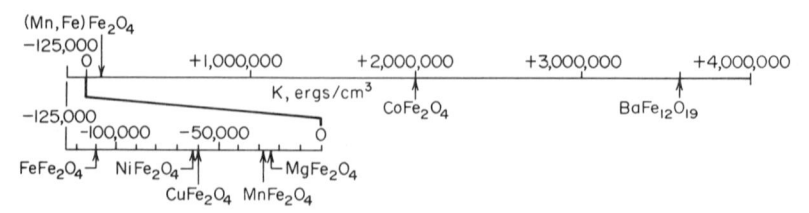

Fig. 102 First-order anisotropy constants of simple spinel ferrites and $BaFe_{12}O_{19}$.[123]

cube, whereas positive values signify that the preferred direction is parallel to a cube edge. A comparison of the K_1 constants for some spinel and hexagonal ferrites is given in Fig. 102.

Magnetostriction. The second contribution to the anisotropy is magnetostriction, which generally includes any changes in the dimensions of a material during magnetization. Usually the fractional change in length per unit length is measured. For a polycrystalline material, the average lambda is

$$\lambda_s = \tfrac{2}{5}\lambda_{100} + \tfrac{3}{5}\lambda_{111}$$

where λ_{100} and λ_{111} are the dilations in the crystallographic directions [100] and [111], respectively. Lambda is given a positive or negative sign depending on whether the material expands (+) or contracts (−) in the direction of magnetization (see Fig. 103).

The energy E_{stress} generated by magnetostrictive deformation against an applied stress can be of magnitude comparable to the energy of cubic crystal anisotropy, since $E_{stress} = \tfrac{3}{2}\lambda\sigma$. For example, an energy of 1.65×10^5 ergs/cm^2, which is about 8 percent of the anisotropy energy for CoFe$_2$O$_4$, is developed during magnetization when a stress $\sigma = 1,000$ kg/cm^2 (10^9 dynes/cm^2) is applied to a ceramic having a $\lambda_s = 110 \times 10^{-6}$ cm.

Demagnetization Energy. The third contribution to anisotropy depends only on the physical shape of a magnetic material. As a result of the free magnetic poles on the surface of a body, a demagnetizing field is generated that is proportional to the magnetization. This demagnetizing energy modifies the applied field so that the internal static field H_i is related to the applied field H_a by Kittel's equation:

$$H_i = \{[H_a - (N_x - N_z)4\pi M_s][H_a + (N_y - N_z)4\pi M_s]\}^{1/2}$$

where N_x, N_y, and N_z are the demagnetizing coefficients. For an infinite medium or a sphere, $H_i = H_a$; for an infinite plane parallel to the applied field, $H_i = (H_a^2 + 4\pi M_s H_a)^{1/2}$; whereas for the perpendicular direction, $H_i = H_a - 4\pi M_s$. Thus, the appropriate anisotropy energy, the demagnetizing field energy, and the

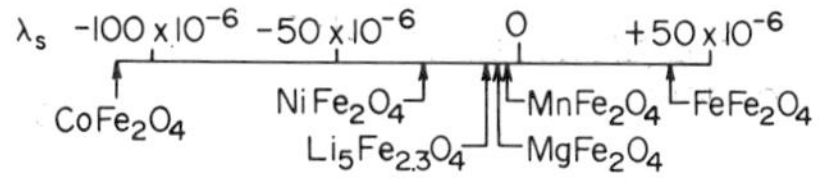

Fig. 103 Average magnetostriction of polycrystalline ferrites.[123]

applied field combine to give the static internal magnetic field energy. In ceramic magnetic materials, internal demagnetization will occur at pores, at nonmagnetic inclusions, and at grain boundaries.

Spinel ferrites[120] The largest category of ferrimagnetic oxides is the spinel ferrite with the general formula AB$_2$O$_4$. This structure, which consists of a nearly close-packed cubic array of oxygen ions possess two crystallographically distinguishable interstical sites (see Fig. 104). One-half of the available octahedral *b* sites and one-eighth of the tetrahedral *a* sites are filled by cations. When the *b* sublattice consists of magnetic ions and the *a* sublattice has nonmagnetic ones, the superexchange interaction between the oxygen ions and the magnetic ions in the *b* sublattice causes the individual magnetic moments to align themselves in an antiparallel order, thereby rendering the structure nonmagnetic. The same situation is true if the magnetic ions only occupy the *a* sites. This antiparallel *bb* or *aa* arrangement is overcome by the much stronger superexchange forces that develop when the magnetic ions are distributed between the *a* and *b* sites. In this case, all the magnetic dipoles in either the *a* sublattice or the *b* sublattice are aligned essentially parallel, but now the magnetic moment of the *a* lattice is aligned antiparallel to the moment of the *b* lattice. At the Curie temperature, ferrimagnetic materials transform from a ferromagnetic to a paramagnetic state, as thermal energy overcomes the exchange energy that holds the magnetic dipoles in alignment. Since the magnetic behavior of ferrimagnetic materials is so similar to that exhibited by metallic magnets, the term *ferromagnetism* is often applied to both materials.

There are two idealized spinel structures called *normal* and *inverse*. The normal spinels have all the divalent cations on the *a* sites and all the trivalent ions on the *b* sites, whereas the "inverse" spinels have one-half the trivalent ions on the *a* sites, while the other half together with all the divalent cations occupy the

6-94   Ceramics, Glasses, and Micas

b sites. The distribution of cations between the a and b sites depends on the preference of the divalent cation. For example, Zn^{2+} ions preferentially enter the a sites. As a result, zinc ferrite is nonmagnetic since the magnetic moment of the ferric ions is canceled out by bb antiparallel interaction, and the zinc ions are nonmagnetic. On the other hand, nickel ferrite is magnetic. Contrary to the

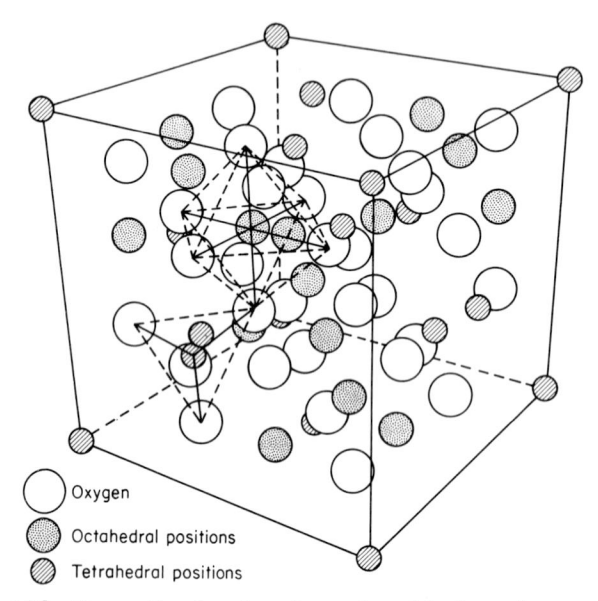

Fig. 104   Perspective drawing of crystal model of spinel unit cell.

situation with zinc ferrite, the nickel ions preferentially occupy the b sites, thereby causing the ferric ions to be distributed between the a and b sites. The magnetic moments of the ferric ions are again canceled, but this time by the antiparallel ab interaction. As a consequence of this ab interaction, the moments of the nickel ions on the b sublattice are parallel and produce, therefore, a net magnetization.

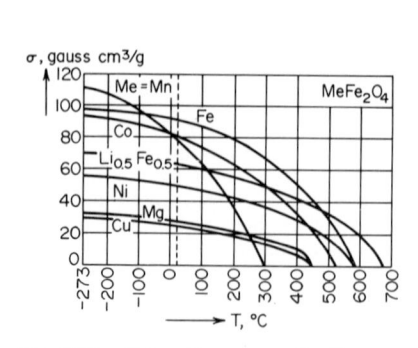

Fig. 105   Saturation magnetization per gram σ for simple spinel ferrites as a function of temperature.[128]

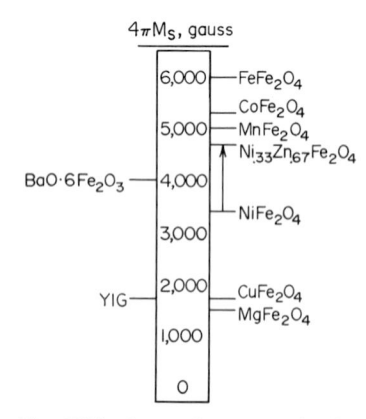

Fig. 106   Saturation magnetization of simple ferrites at room temperature.[123]

The saturation magnetization per gram of several spinel ferrites as a function of temperature is shown in Fig. 105. A comparison of the saturation magnetization at room temperature of several ferrites, $Y_3Fe_5O_{12}$ and $BaFe_{12}O_{19}$, is given in Fig. 106.

The saturation magnetization can be changed not only by solid solution but also by thermal treatment. At low temperatures there is a slight preference by the magnesium ions in magnesium ferrite for the b sites; consequently, the nonmagnetic inverse spinel structure is formed. Whereas at sufficiently high temperatures, there is a random distribution of Mg^{2+} ions between the a and the b lattice sites. As the quenching temperature is raised, a greater number of magnesium ions can be frozen in the a sites, thereby increasing the net magnetic moment of the ferric ions.

Zinc and cadmium form normal spinels, whereas Fe^{2+}, Co^{2+}, Cu^{2+}, Mg^{2+}, and Ni^{2+} form inverse spinels. Manganese spinel is about 80 percent normal spinel structure. Monovalent lithium in combination with Fe^{3+}, which is necessary for charge balance,

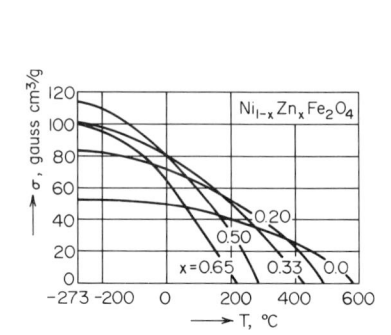

Fig. 107 Saturation magnetization per gram σ as a function of temperature for some ferrites of the $Ni_{1-x}Zn_xFe_2O_4$ series.[129]

Fig. 108 Saturation magnetization of polycrystalline specimens of $NiAl_xFe_{2-x}O_4$ as functions of temperature.[130]

enters the octahedral b sites to form an inverse spinel. In general, the substitution of nonmagnetic ions (Zn^{2+}, Cd^{2+}, In^{2+}), which enter the a sites, initially increase the saturation magnetization (see Fig. 107); whereas nonmagnetic ions (Al^{3+}, Sc^{3+}, Ti^{3+}, Cr^{3+}, Ga^{3+}) which occupy the b sites, decrease the saturation moment (see Fig. 108). Substitution of either type lowers the Curie temperature by weakening the ab sublattice interaction (see Fig. 109). The Curie temperatures of several spinel ferrites, $BaFe_{12}O_{19}$, and $Y_3Fe_5O_{12}$, are given in Fig. 110.

The spinel structure can readily accommodate ion pairs, such as Ni^{2+}-Ni^{3+} or Fe^{2+}-Fe^{3+} in $NiFe_2O_4$, since only a fraction of the cation sites are occupied. Ni^{3+} can form when there is an excess of oxygen, and Fe^{2+} can occur when there is an oxygen deficiency. The couple that results provides an easy conduction path for electron transfer. One solution to this low-resistivity problem is to introduce a third ion with a lower valence state that has a greater affinity for oxygen than Ni^{2+}, and a higher valency state with a lesser electron affinity for oxygen than Fe^{2+}. Manganese and cobalt meet this requirement, and although present in multiple-valency states, they do not provide easy conduction paths (see Fig. 111).

Cobalt ferrite is unique among the spinels in that it exhibits a large positive anisotropy constant (see Fig. 102). The high coercive force ($\simeq 500$ oersteds) associated with the anisotropy energy places $CoFe_2O_4$ intermediate in properties between magnetically soft and hard ferrites. Materials having nearly zero anisotropy energy can be prepared by forming solid solutions of spinels with positive and negative anisotropies.

Hexagonal ferrites[121] Of the hexagonal ferrites, the magnetoplumbite materials are of greatest commercial importance. Strontium ferrite has in recent years re-

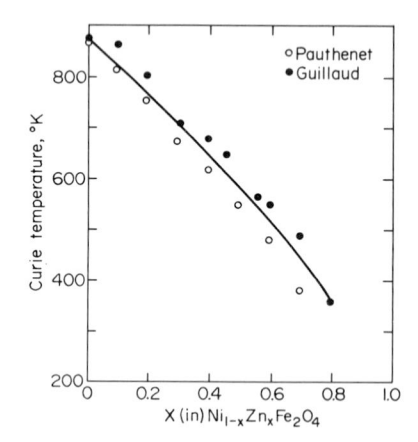

Fig. 109 Curie temperature of $Ni_{1-x}Zn_xFe_2O_4$ ferrites as a function of x.[131]

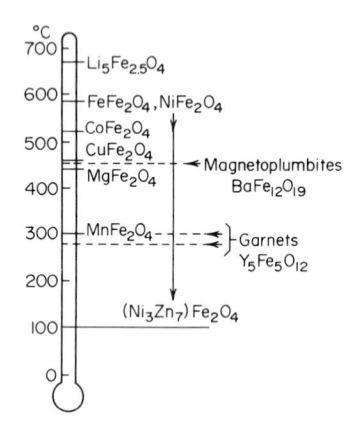

Fig. 110 Curie temperature of simple spinel ferrites, $BaFe_{12}O_{19}$ and $Y_5Fe_5O_{12}$.[123]

placed barium ferrite because it can be made with a higher energy product $(B_dH_d)_{max}$, and a higher coercive force H_c. Magnetoplumbites or M-type hexagonal ferrites have the formula unit $MeFe_{12}O_{19}$, where Me is usually Ba^{2+}, Sr^{2+}, or Pb^{2+}. The structure consists of a hexagonal packing of oxygen ions containing five crystallographically distinguishable ferric ion sites. In a unit cell (see Fig. 112), which contains two formula weights, there are tetrahedral and trigonal lattice sites and three different types of octahedral sites. Seven octahedral ions and one trigonal ion contribute to the majority α sublattice, and two octahedral and two tetrahedral ions contribute to the minority β sublattice. A unit cell, therefore, has a net majority moment equivalent to eight ferric moments equal to 40 Bohr magnetons, or on the basis of the unit formula 20 Bohr magnetons.

Replacement of Al^{3+}, Cr^{3+}, or Ga^{3+} for Fe^{3+} lowers the saturation magnetization since these ions preferentially enter the majority sublattice (see Fig. 113). As a result of weakening the α-β sublattice interaction, substitution of these ions lowers the Curie temperature (see Fig. 114). Substitution of divalent transition-metal ions, e.g., Co^{2+}, Ni^{2+}, or Zn^{2+}, for Fe^{3+} rapidly decreases the magnetic anisotropy

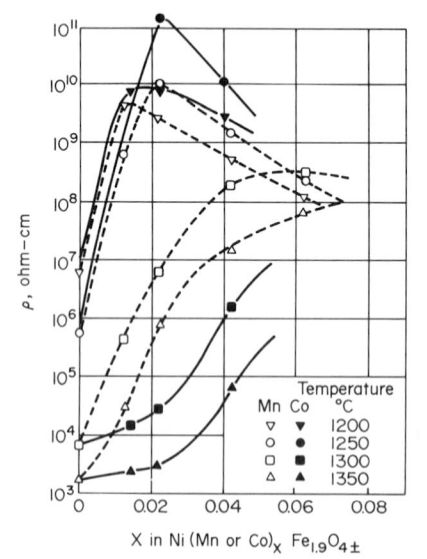

Fig. 111 Dependence of the resistivity of nickel ferrite on manganese and cobalt additions for various firing temperatures.[132]

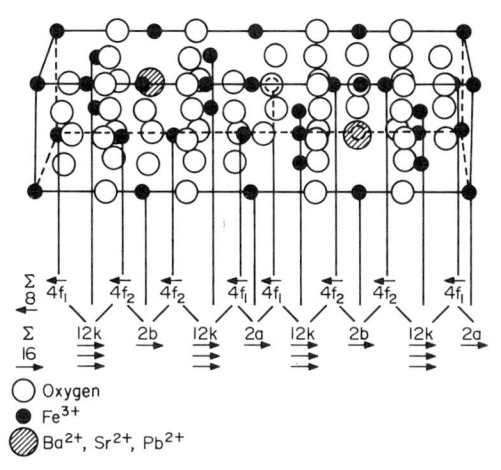

Fig. 112 Perspective drawing of crystal model of magnetoplumbite unit cell.

(see Fig. 115) as well as lowering the saturation moment and the Curie temperature. Charge balance is maintained by the replacement of Fe^{3+} ions with divalent plus tetravalent ions, for example, $Co^{2+} + Ti^{4+}$, $Ni^{2+} + Ti^{4+}$, or $Zn^{2+} + Ti^{4+}$. The saturation magnetization M_s, the crystal anisotropy K_1, and the anisotropy field $H_\theta{}^a$ of $BaFe_{12}O_{19}$ as a function of temperature are shown in Fig. 116.

Planar ferrites[121] At room temperature, the easy direction of magnetization in Y-type ferrites is in the basal plane rather than parallel to the crystallographic c axis,

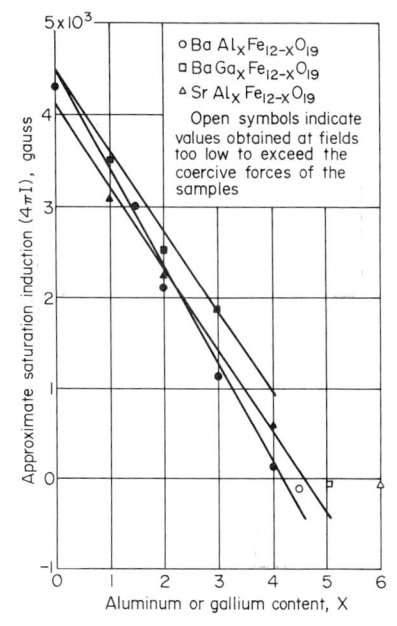

Fig. 113 Saturation magnetization of aluminum- and gallium-substituted M-type ferrites as a function of composition.[133]

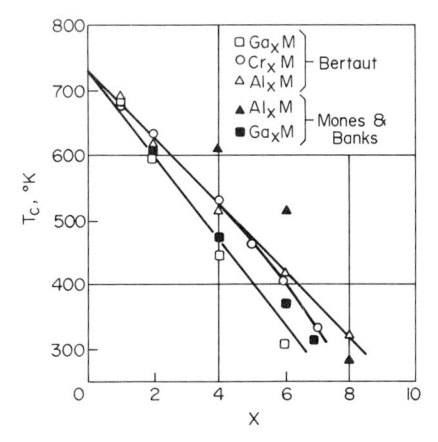

Fig. 114 Curie temperature of aluminum-, chromium-, and gallium-substituted $BaFe_{12}O_{19}(M)$ as a function of composition.[134]

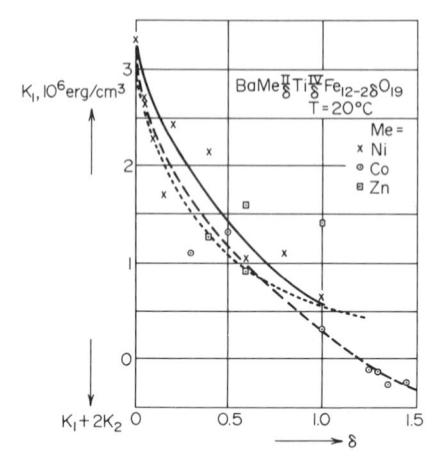

Fig. 115 Room temperature anisotropy constants of Co-Ti, Ni-Ti, and Zn-Ti substitutions in $BaFe_{12}O_{19}$ as functions of composition.[135]

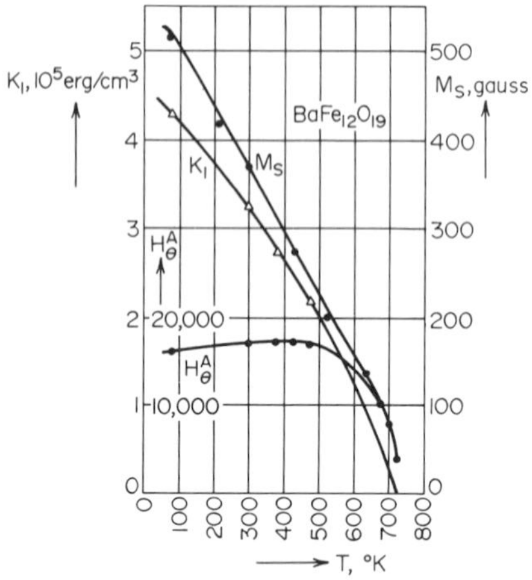

Fig. 116 Saturation magnetization M_s, crystal anisotropy K_1, and anisotropy field H_θ^A of $BaFe_{12}O_{19}$ as a function of temperature.[136]

as found in the other hexagonal ferrites. Y-type ferrites have the general formula $Ba_2Me_2Fe_{12}O_{22}$, where Me_2 is either Mg_2, Ni_2, Zn_2, $Fe_{1.5} + Zn_{0.5}$, Co_2, or Mn_2. Since cobalt tends to induce planar anisotropy in all hexagonal ferrites, the largest negative value of the crystal anisotropy $(K_1 + 2K_2)$ is found in the cobalt compound, as would be expected. The saturation magnetization, crystal anisotropy, and anisotropy field of $Ba_2Co_2Fe_{12}O_{22}$ as a function of temperature are shown in Fig. 117. In planar ferrites, two effective internal fields can be defined: one corresponds to the motion of the magnetization vector in the easy plane, and the other to the vector perpendicular to this plane. Since the effective field for resonance is the

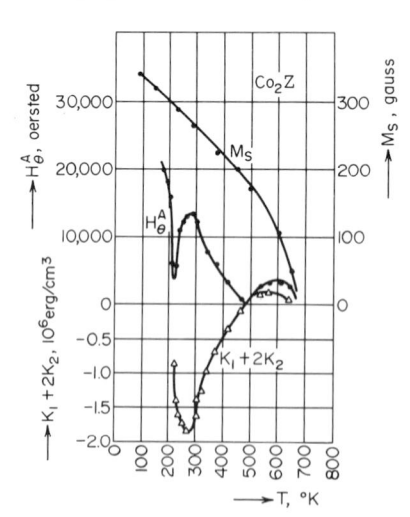

Fig. 117 Saturation magnetization M_s, crystal anisotropy $(K_1 + 2K_2)$, and anisotropy field H_θ^A of Co_2Y as a function of temperature.[137]

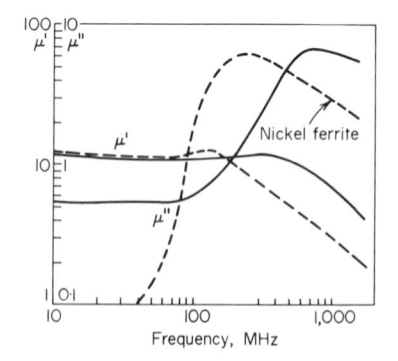

Fig. 118 Magnetic spectrum of polycrystalline Mg_2Y compared with that of a nickel ferrite.[138]

geometric mean of these fields, planar ferrites exhibit much larger anisotropy fields than spinel ferrites while still possessing soft magnetic characteristics. As a result, the useful frequency range is extended in planar ferrites beyond that available in spinel ferrites (see Fig. 118).

Garnets[124] Among the ferrimagnetic oxides, garnets have one of the most complicated crystal structures. The unit cell contains eight formula units of $N_3Fe_5O_{12}$, where N represents trivalent rare-earth ions including yttrium and lanthanum. There are three types of interstitial sites formed by the cubic packing of oxygen ions. Ferric ions occupy the octahedral a and tetrahedral d sites, and the larger trivalent rare-earth ions reside in the dodecahedral c sites. The array of cations in two octants of the unit cell is illustrated in Fig. 119. As a result of the very strong interaction between the ferric ions on the a and d sublattices, the magnetic moments of these two lattices are aligned antiparallel. The net contribution of the a-d sublattices to the magnetic moment is equivalent, therefore, to one ferric

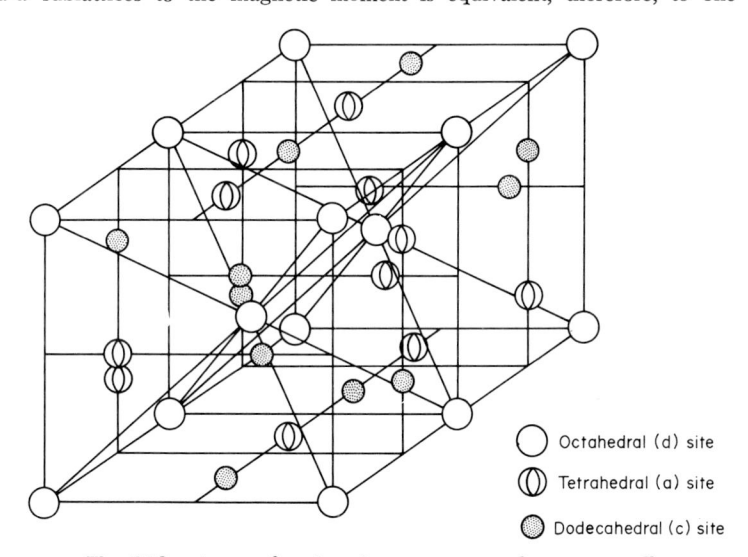

Fig. 119 Array of cations in two octants of a garnet cell.

ion per formula unit since they contain three tetrahedral and two octahedral ferric ions. In the case of yttrium-, lanthanum-, and lutecium-garnets, there is no further contribution to the magnetic moment.

Partial substitution of other trivalent ions for the ferric ions affects the saturation magnetization and the Curie temperature in a predictable way. Since Al^{3+} and Ga^{3+} ions are smaller than Fe^{3+}, they occupy tetrahedral positions, thereby decreasing the majority lattice contribution to the magnetic moment and consequently the saturation magnetization. Cr^{3+}, Sc^{3+}, and In^{3+} ions, however, prefer octahedral positions, which decrease the minority lattice contribution; therefore, they serve to increase the saturation magnetization. In either case, the a-d sublattice interaction is weakened; as a result, the Curie temperature is lowered by these substitutions. Because all the metal ion sites in garnet are filled with trivalent cations, electron transfer is limited, and consequently the electrical resistivity is high.

An additional contribution to the magnetic moment is introduced when the c sites are occupied by magnetic rare-earth ions. With the exception of Pr^{3+} and Nd^{3+}, the magnetic rare-earth ion moments are aligned antiparallel to the net moment of the ferric sublattices by weak exchange forces. As a result, the weaker c-sublattice alignment is more readily overcome by thermal vibrations than is the stronger a-d sublattice interaction. Consequently, the magnetic rare-earth ions have their major effect on the saturation moment at low temperatures. Because the moment of the c sublattice is antiparallel to the net ferric ion moment, at some temperature these two vectors will just cancel each other. This point is called the *compensation temperature*, and it is marked by a nearly zero magnetization. At higher temperatures, only the ferric ions contribute to the magnetization; consequently, all the rare-earth garnets have about the same Curie temperature with the differences observed due to the change in lattice parameters caused by the rare-earth ions (see Fig. 120).

Unlike the $3d$ band electrons of the first-transition-series ions, which are perturbed by the crystal field forces, the better-screened $4f$ band electrons of the rare-earth ions have an orbital as well as a spin-moment contribution to the magnetization. When the $4f$ band is less than half filled, the spin and orbital moments of the electrons are oppositely directed. As a result of this, the magnetic moments of the rare-earth ions are aligned parallel with the net ferric ion moment. Thus,

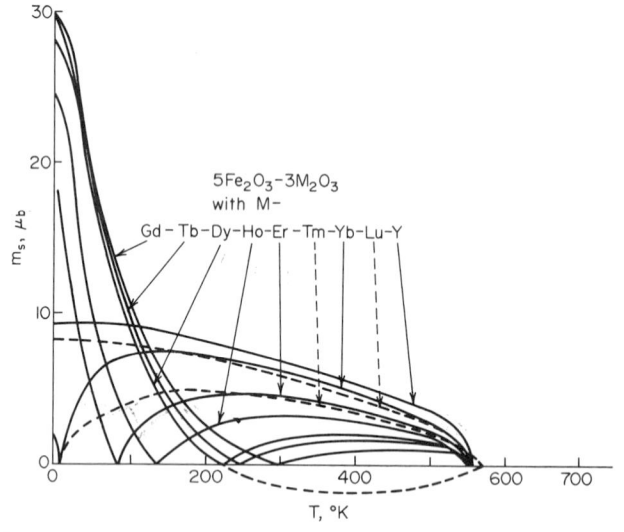

Fig. 120 Spontaneous magnetization Bohr magnetons per gram molecule as a function of temperature for the family of ferrimagnetic garnets.[139]

the three-sublattice system with the a-d lattice taken as a unit is in ferromagnetic alignment; consequently, there is no compensation temperature.

The Pr^{3+} and Nd^{3+} ions are too large to form pure garnets. Since these ions have large orbital moments, partial substitution of them for yttrium in YIG increases the saturation magnetization at low temperatures. Sm^{3+} and Eu^{3+} substitutions produce only small changes in the saturation magnetization of YIG since the spin and orbital moments nearly cancel. The Eu^{3+} ion moment should couple ferromagnetically with the net ferric ion moment because the $4f$ band is less than half filled. Actually, however, it couples antiparallel as do the heavier rare-earth ions. Beginning with samarium, all the heavier rare-earth ions are small enough to form pure garnets.

When the $4f$ band is half or more filled, the orbital and spin moments are in the same direction; as a result, the heavy rare-earth ions couple antiparallel with the net ferric ion moment. Gadolinium has a large spin moment, but unlike the other magnetic rare-earth ions it has no orbital angular momentum. Consequently, Gd^{3+} ions do not interact strongly with the crystal field of the lattice and cause resonance damping. Gadolinium garnets have attracted considerable commercial interest because they exhibit a low resonance damping, a high saturation magnetization at low temperatures, and a compensation point near room temperature

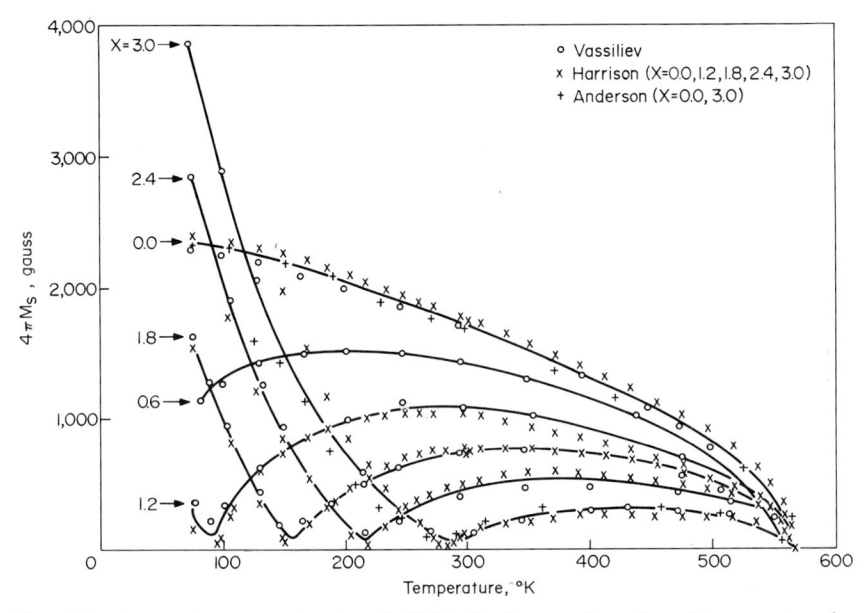

Fig. 121 Saturation magnetization of $Gd_xY_{3-x}Fe_5O_{12}$ as a function of temperature for $x = 3.0$, 2.4, 1.8, 1.2, 0.6, and 0.0.[140]

that can be lowered to absolute zero by solid solution with YIG (see Fig. 121). The magnetization curves of YIG and Gd through Lu rare-earth-ion garnets as a function of temperature are shown in Fig. 120.

Classification of magnetic materials On a functional basis, ceramic magnetic materials can be divided into four groups: (1) soft ferrites, (2) hard ferrites, (3) square-loop ferrites, and (4) microwave ferrites.

Soft Ferrites. Magnetically soft ferrites are easily magnetized, but they lose their magnetization when the applied field is removed. Typically a soft ferrite has a high initial permeability and a narrow, rounded S-shaped hysteresis loop indicative of a low coercive force. Since the energy dissipated in a magnetic core under ac conditons is proportional to the area of the hysteresis loop, soft ferrites are also called *low-energy materials*.

Pores or nonmagnetic inclusions raise the coercive force by two mechanisms: (1) the applied field is decreased by the demagnetizing field associated with the nonmagnetic inclusions and (2) additional work must be supplied to free domain walls that become pinned by these inclusions. Magnetostriction and crystal anisotropy add to the coercive force required for demagnetization since work is done during deformation. Materials exhibiting a minimum coercive force and, therefore, a minimum hysteresis loss have a minimum in porosity, magnetostriction, and crystal anisotropy;

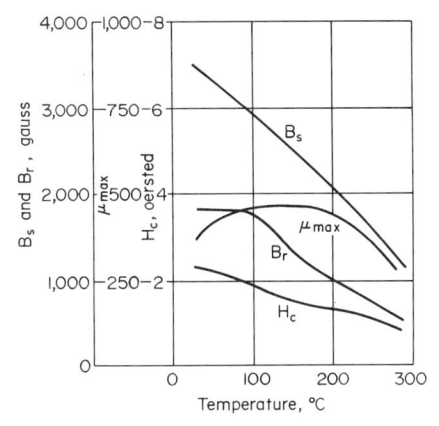

Fig. 122 Dependence of maximum permeability μ_{max}, saturation magnetic induction B_s, coercive force H_c, and residual induction B_r on temperature. (*Indiana General.*)

TABLE 36 Typical Ferrite Applications[144]

Application	Ferrite property	Typical ferrite*
Filter inductor	High μ, low loss	Mn-Zn
Transformer	Low loss, temperature stability, high μ	Ni-Zn; Mn-Zn
Antenna rod	High μ, low loss	
Adjustable inductors	High μ, low loss	Mn-Zn; Ni-Zn
Flyback transformers	Low loss, good temperature coefficient	Mn-Zn; Ni-Zn
Deflection yoke	Low loss, high skin resistance	
Suppression beads	High loss at specific frequency	Mn-Zn; Ni-Zn
Recording heads	Low loss, mechanically strong	Ni-Zn; Mn-Zn
Memory cores	Hysteresis-loop rectangularity, moderately low H_c	Mg-Mn
Switching cores	Hysteresis-loop rectangularity, moderately low H_c	Mg-Mn
Multiaperture cores	Hysteresis-loop rectangularity, moderately low H_c	Mg-Mn
Magnetic amplifiers	Hysteresis-loop rectangularity, moderately low H_c	Mg-Mn
Isolators, attenuators, modulators, switches	Faraday rotation, ferromagnetic resonance	Mg-Mn
Delay lines	High magnetostriction	Ni-Zn
Filters and oscillators		Ni-Zn
Temperature controls	Low Curie temperature, steep temperature coefficient	Ni-Zn

* This refers to major constituents only.

and a maximum in saturation magnetization. These same factors also give maximum initial permeability.[123]

Nickel-zinc and manganese-zinc ferrites have high initial permeabilities, low coercive forces, and low losses. They are used extensively as core materials for transformers, inductors, and relays. The variation of μ_{max}, B_s, and H_c with temperature for a typical Ni-Zn spinel ferrite is shown in Fig. 122. Mn-Zn ferrites are used in the lower-frequency range from 1 to 1,500 kHz, whereas Ni-Zn ferrite materials are used at higher frequencies of 0.1 to 200 MHz. Typical soft ferrite applications

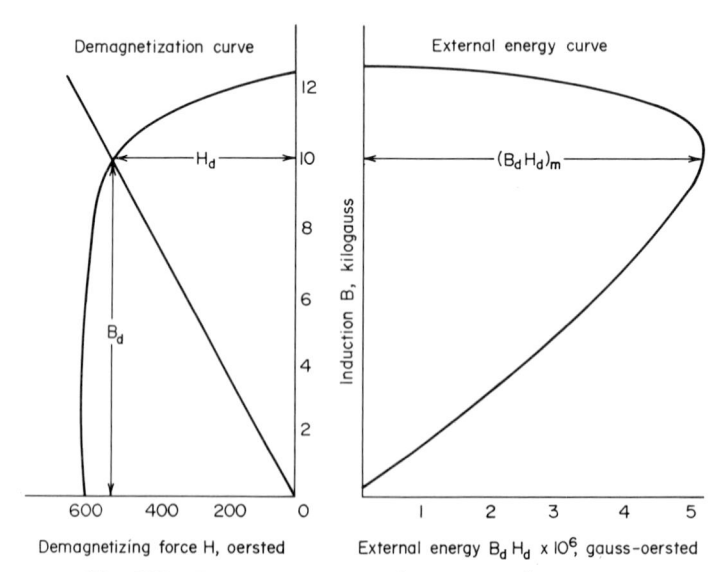

Fig. 123 Demagnetization and energy product curves.

TABLE 37 Wide-tolerance* General-purpose Core Material

Company	Ferroxcube					Ceramic Magnetics, Inc.	Indiana General				Ceramic Magnetics, Inc.	Indiana General	Ceramic Magnetics, Inc.
Trade designation	3E	3E2A	3E3	3C5	3B7	MN-60	O-6	O-5	T-1	O-3	CM-2002	TC-3	CN-20
Initial permeability μ_0:													
At 5 kHz	2,700	5,000	≥10,000	—	2,300	6,000	—	—	—	—	—	—	—
At 100 kHz	—	—	—	800	—	—	4,700	3,000	2,000	1,500	1,200	1,000	800
Maximum permeability μ_{max}	—	—	—	—	—	10,000	6,000	4,800	3,600	4,000	6,000	2,000	—
Maximum flux density B_{max}, gauss:													
10 oersteds at 23°C	3,500	4,100	4,000	4,200	3,400	—	—	—	—	—	3,500	—	3,680
16 oersteds at 25°C	—	—	—	—	—	4,220	—	—	—	—	—	—	—
20 oersteds at 25°C	—	—	—	—	—	—	4,700	4,700	4,400	4,500	—	3,600	—
25 oersteds at 25°C	—	—	—	—	—	1,200	1,400	1,000	1,000	1,600	2,500	700	2,565
Retentivity B_r, oersteds	0.1	0.08	0.1	0.2	0.2	0.1	0.10	0.12	0.18	0.15	0.22	0.25	0.19
Coercive force H_c, oersteds													
Temperature coefficient T_c of μ_0, % per °C:													
-25 to $+75°C$	—	—	—	—	-0.14 to $+0.14$	—	—	—	—	—	—	—	—
0 to $+60°C$	—	—	—	—	—	—	—	—	—	—	0.9	0.08–0.35	1
0 to $+100°C$	—	—	—	—	—	0.020	—	—	—	—	—	—	—
+23 to +55°C	0 to 1.1	1.0	—	—	—	—	—	—	—	—	—	—	—
Curie temperature, °C	≥125	≥170	≥125	≥200	≥170	>180	210	215	180	190	170	165	150
Resistivity, ohm-cm	≥30	—	≥5	—	≥100	80	Low	Low	Low	Low	10^3	Low	10^4
Loss factor (×10^6), $1/\mu_0 Q$:													
At 4 kHz	≤2.5	—	≤2.5	—	≤1	2.6	—	—	—	—	—	—	—
At 100 kHz	≤14	≤10	≤50	30	≤5	6.7	6	7	25	33	70	3.3	75
At 500 kHz	≤90	—	—	—	—	—	—	—	—	—	95	—	95
At 1 MHz	—	—	—	—	—	—	—	—	—	—	150	—	150
Frequency range, maximum	500 kHz	300 kHz	200 kHz	500 kHz	300 kHz	100 kHz	400 kHz	400 kHz	400 kHz		2 MHz	1.5 MHz	2 MHz

* Tolerance on magnetic properties ±20% or better.

are listed in Table 36, and some examples of commercially available ferrite materials are listed in Tables 37 to 39.

Hard Ferrites. Magnetically hard ferrites retain their magnetization after the applied field is removed. They are characterized by high coercive forces and high remanences. The high coercive force arises from the very high strength of the anisotropy field found in hexagonal ferrites and the small grain size of ceramic magnets (see Fig. 116). That portion of the hysteresis curve between B_r and H_c is called the *demagnetizing curve* (see Fig. 123). At any point on this curve,

TABLE 38   Narrow-tolerance* High-stability Core Materials

Company	Ferroxcube			Indiana General			
Trade designation	3B7	3B9	3D3	TC-8	TC-7	TC-3	H
Initial permeability μ_0:							
At 100 kHz	2,300	1,800	750	1,500	1,400	1,000	—
At 1 MHz	—	—	—	—	—	—	850
Maximum permeability μ_{max}	—	—	—	2,400	1,850	2,000	4,000
Maximum flux density B_{max}, gauss:							
10 oersteds at 25°C	3,500	3,300	3,500	—	—	—	—
25 oersteds at 25°C	—	—	—	4,000	3,500	3,600	3,400
Remanence B_r, oersteds	—	—	—	1,600	550	700	1,470
Coercive force H_c, oersteds	—	—	—	0.42	0.25	0.25	0.18
Temperature coefficient T_c of μ_0, % per °C:							
+25 to +80°C	—	—	—	—	—	—	0.66
0 to +60°C	—	—	—	0.15 to −0.10	0.11 to 0.31	0.08 to 0.35	—
+20 to +70°C	−0.14 to +0.14	—	—	—	—	—	—
−30 to +70°C	—	0.25 to 0.40	0.08 to 0.23	—	—	—	—
Disaccommodation factor (decay of μ, 10–100 min) $\times 10^6$, $\Delta\mu/\mu_0^2$	<3.5	<2.0	<12.7	≤10	<4.5	<40	—
Curie temperature, °C	≥170	≥145	≥150	180	140	165	150
Resistivity, ohm-cm	≥100	≥50	≥150	Low	Low	Low	Medium
Loss factor ($\times 10^6$), $1/\mu_0 Q$:							
At 4 kHz	≤1	≤2	2	—	—	—	—
At 100 kHz	≤5	≤6	5	6.7	5.0	3.3	—
At 500 kHz	25	25	≤14	—	—	—	—
At 1 MHz	120	120	≤30	60.0	—	—	300
Frequency range, maximum	300 kHz	300 kHz	2.5 MHz	500 kHz	500 kHz	1.5 MHz	1 MHz

* Tolerance on magnetic properties ±50% or better.

the product $B_d H_d$ indicates the magnetic energy per unit volume that can be stored by the external magnetic field of the magnet. The peak energy product $(B_d H_d)_{max}$ is used as a criterion for comparing different permanent magnet materials (see Fig. 123).

Improved polycrystalline permanent magnets are made by orientating all the single-domain particles in the same direction. When this is accomplished, the remanent magnetization is only slightly less than the saturation value. Randomly orientated materials, in contrast, have remanent magnetizations that are about half the saturation moment (see Fig. 124). Randomly oriented materials typically have an energy product of 1×10^6 gauss-oersteds, whereas uniaxially orientated magnets have a value of over 3×10^6. By comparison, $(B_d H_d)_{max}$ for metallic magnets can be as high as 10×10^6 gauss-oersteds.

Since sintered ceramic magnets are hard, brittle, and machinable only by grinding, plastic- or rubber-bonded permanent magnets have been developed.[124] These flexible

TABLE 39 High-frequency* Core Materials

Company	Indiana General				Ferroxcube				
Trade designation	Q-1	TC-4	Q-2	Q-3	4-B	4-C	4C4	4-D	4-E	1Z2
Initial permeability μ_0:										
At 100 kHz	—	—	—	—	250	125	125	50	15	—
At 1 MHz	125	95	40	16	—	—	—	—	—	15
At 100 MHz	—	—	—	—	—	—	—	—	—	—
Maximum permeability μ_{max}	400	420	115	42	—	—	—	—	—	—
Maximum flux density B_{max}, gauss:										
20 oersteds at 23°C	—	—	—	—	3,300	2,800	3,300	—	—	—
25 oersteds at 25°C	3,300	3,300	2,400	2,600	—	—	—	2,400	1,800	—
40 oersteds at 25°C	—	—	—	—	—	—	—	—	—	—
60 oersteds at 25°C	—	—	—	—	—	—	—	—	—	—
Remanence B_r, oersteds	1,800	2,300	750	1,470	—	—	—	—	—	—
Coercive force H_c, oersteds	2.1	3.2	4.7	21.0	—	—	—	—	—	—
Temperature coefficient T_c of μ_0, % per °C:										
+25 to +80°C	0.1 max	—	0.1 max	—	0 to 0.2	0 to 0.15	−15 to 0	0 to 0.08	0 to 0.03	0.12
+23 to +55°C	—	0.03–0.06	—	—	—	—	—	—	—	—
0 to +60°C	—	—	—	—	—	—	—	—	—	—
Curie temperature, °C	350	425	450	500	250	350	≥350	400	500	—
Resistivity, ohm-cm	High	High	High	High	≥10⁵	≥10⁵	≥10⁵	≥10⁵	≥10⁵	10⁶
Loss factor (×10⁶), $1/\mu_0 Q$:										
At 1 MHz	20	47	—	—	≤90	≤120	≤40	—	—	—
At 2 MHz	—	—	—	—	—	≤300	≤60	≤210	≤300	—
At 5 MHz	—	—	—	—	—	—	≤100	≤300	≤360	—
At 10 MHz	60	—	85	—	—	—	—	—	—	—
At 50 MHz	—	—	170	—	—	—	—	—	—	—
At 150 MHz	—	—	—	420	—	—	—	—	—	1,000
Frequency range, maximum	10 MHz	20 MHz	50 MHz	225 MHz	2 MHz	5 MHz	20 MHz	20 MHz	50 MHz	6,000 200 MHz

* Tolerance on magnetic properties ±20% or better.

magnets can be readily fabricated into intricate shapes by usual edge-cutting tool methods. The effect of dilution of the magnetic material by the resin bond on the energy product can be partially overcome by developing a highly orientated ferrite particle texture. Orientation raises the energy product from about 0.4×10^6 to 1.0×10^6 gauss-oersteds, a value comparable to that of isotropic-sintered-ceramic magnets.

The main advantages of ceramic permanent magnets compared to metallic ones are: (1) much higher ohmic resistance (10^8 ohm-cm), (2) higher coercive force (1,500 to 2,000 versus 50 to 1,500 oersteds for metallic magnets), (3) lower cost and weight, and (4) lesser chemical reactivity. The high resistivity simplifies the insulation of line-operated appliances, and the low eddy-current losses permit applications involving superposed high-frequency fields. As a result of the high coercive force, magnet shapes are possible that otherwise might substantially weaken the magnetic field by self-demagnetizing effects. For the same reason, ceramic magnets are not easily demagnetized by stray magnetic fields. In microwave applications, the large anisotropy energy of $2K_1/M_s$ creates a "built-in" biasing field about 17,000 oersteds, which makes possible the construction of resonance insolators for operation in the 50- to 70-GHz range. Examples of some commercially available ceramic magnets are listed in Table 40.

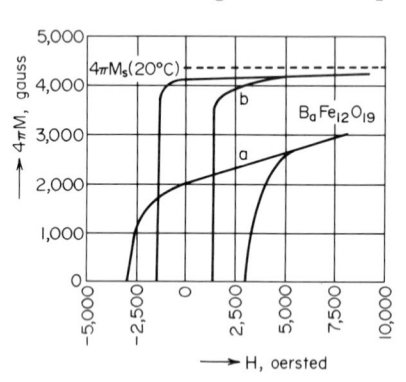

Fig. 124 Hysteresis curves of $BaFe_{12}O_{19}$ for (a) isotropic and (b) crystal-oriented magnets of $BaFe_{12}O_{19}$.[141]

Square-loop Ferrites. Certain materials have high remanences and low coercive forces. Instead of the rounded S-shaped hysteresis loop found in most soft ferrites, materials such as Mn-Mg, Mn-Cu, and Li-Ni spinel ferrites, and doped yttrium-iron-

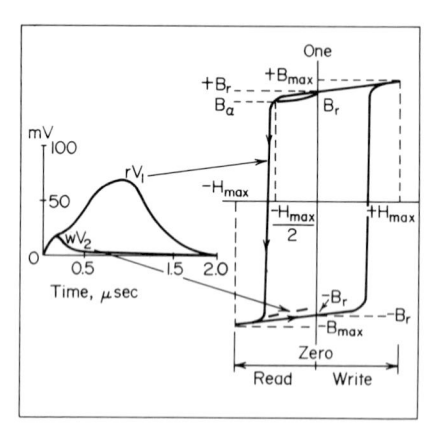

Fig. 125 Operation mode of memory core.[142]

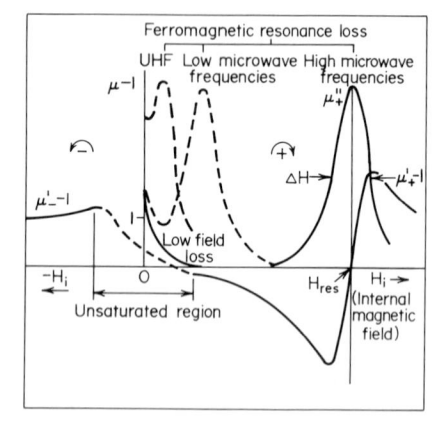

Fig. 126 Microwave susceptibility dispersion.[143]

garnets have rectangularly shaped loops. Use is made of the abrupt change in magnetic flux with applied field in memory cores for high-speed digital computers. The principle of operation of these devices is illustrated in Fig. 125. A magnetic core is first cycled so that it is in remanence state $+B_r$ or $-B_r$, corresponding by convention to the "one" or "zero" storage states, respectively. To find the sense of the stored

TABLE 40 Permanent Ceramic Magnets

Company	Allen Bradley						Ceramic Magnetics, Inc.	Indiana General				
Trade designation	M-01	M-06C	M-05C	M-058	M7	M8	PM-4	Inoox. I	Inoox. II	Inoox. VI-A	Inoox. 7	Inoox. V
Peak energy product $(B_dH_d)_{max}$ $(\times 10^{-6})$	1.0	2.5	2.6	3.3	2.7	3.5	1.1	1.0	1.65	2.6	2.8	3.5
Residual induction B_r, gauss	2,200	3,300	3,300	3,800	3,400	3,850	2,210	2,200	2,700	3,300	3,450	3,840
Coercive force H_c, oersteds	1,800	2,800	2,300	2,050	3,250	2,950	1,930	1,825	2,250	3,000	3,200	2,200
Curie temperature, °C	450	450	450	450	460	460	450	450	450	450	450	450
Reversible temperature coefficient (magnetic), % per °C	−0.20	−0.20	−0.20	−0.20	−0.20	−0.20	−0.20	−0.19	−0.19	−0.19	−0.19	−0.19
Density, g/cm³	—	4.9	4.85	5	4.75	4.90	5.1	4.95	4.55	4.5	4.6	5.0
Incremental permeability $\Delta B/\Delta H$	1.2	1.09	1.09	1.07	1.06	1.06	—	1.1	1.15	1.06	1.06	1.05
Resistivity, ohm-cm	10⁹	—	—	—	—	—	10¹⁰	10⁶	—	—	—	10⁴
Peak magnetizing force required, oersteds × 10⁻³	10	10	10	10	10	10	10	10	10	10	10	10
Permeance coefficient B/H at $(B_dH_d)_{max}$	1.2	1.1	1.2	1.1	—	—	1.1	1.2	1.1	1.06	—	1.05
Peak intrinsic energy product $(B_iH_i)_{max}$, gauss-oersteds × 10⁻⁶	4.2	8.4	6.6	6.7	12.0	10.5	—	—	—	—	—	—

TABLE 41 Typical Ferrite Memory Core Data* at 25°C   (Sperry Rand Corporation Univac Division)

Univac core type	10181	20180	20291	20361	20370	20284	30370	30470	30450	30460	30551	30660	30560	31030	50941	50943	51050	51330	51341	82940
						WT		ET				ET								
Outside diameter, mils	18	20	23	23	23	23	30	30	31	31	30	30	31	30	50	50	50	50	50	80
Test conditions at 25°C:																				
Full drive current, ma	850	840	875	640	700	825	725	700	520	660	475	575	575	280	450	475	530	350	400	366
Partial drive current, ma	510	504	525	385	420	495	435	420	315	400	285	345	345	165	270	285	320	210	240	220
Pulse rise time, μsec	0.05	0.05	0.05	0.05	0.05	0.05	0.10	0.10	0.10	0.10	0.10	0.1	0.10	0.50	0.20	0.20	0.20	0.20	0.20	0.50
Pulse duration, (μsec)	0.4	0.4	0.4	0.4	0.4	0.4	1.0	1.00	0.6	0.6	0.5	1.0	0.5	2.0	2.0	2.0	2.0	3.0	3.0	5.0
Performance characteristics:																				
Undisturbed one response, mv	36	39	43	37	41	41	50	51	45	44	46	54	45	21	62	70	64	48	53	63
Disturbed zero response, mv	9.0	9.0	10.0	10.0	7.0	8.5	9.0	12	8.5	8.0	10.5	11	8.5	1.7	12.0	13.0	11.0	10.5	9.0	9.5
Peaking time, μsec	0.085	0.097	0.095	0.110	0.120	0.125	0.170	0.190	0.210	0.200	0.210	0.21	0.210	0.750	0.450	0.420	0.450	0.630	0.615	1.45
Switching time, μsec	0.155	0.175	0.185	0.240	0.250	0.285	0.310	0.360	0.410	0.400	0.420	0.41	0.440	1.255	0.930	0.895	0.920	1.375	1.250	2.70

* These core types are indicative of those being produced in the industry.

information, the core is pulsed with a magnetic field of strength $-H_{max}$. If on one hand the core was in the $+B_r$ or "one" state, it is driven through $-B_{max}$ to $-B_r$ or "zero" state. The emf due to the flux reversal has the time response shown to the left of the figure. If on the other hand the core was in the $-B_r$ or "zero" state, the emf due to the flux reversal is much smaller, as shown in Fig. 125. Thus, a core threaded with two conductors, each carrying a pulse of strength $-H_{max}/2$, will give a "one" or "zero" (emf versus time) response that can then be detected by means of a third wire that also threads the core. Other cores threaded by only one of the conductors will be merely perturbed by the $-H_{max}/2$ pulse and return to their original one or zero positions. Typical ferrite memory-core data are listed in Table 41. Standard terms and definitions relating to ferrite memory cores can be found in ASTM Document C 526-63T.

Microwave Ferrites.[125,126] A variety of nonreciprocal microwave devices are designed around the gyromagnetic behavior of the elementary magnetic dipoles or uncompensated electron spins of ferrimagnetic materials. When a ferrite is placed in a magnetic field, it will have a resonance frequency given by $f = \gamma H_i$, where f is the mid-band resonance frequency, γ is the gyromagnetic ratio ($\simeq 2.8$ MHz/oersted), and H_i is the magnetic field inside the material. At a given frequency, the varia-

TABLE 42 Room-temperature Properties of Polycrystalline Ferrites[145]

Material	$4\pi M_s$, gauss	T_c, °C	ΔH, oersteds	g factor	$2\lvert K_1 \rvert/M_s$ oersteds	ϵ	tan δ_ϵ	ρ, ohm-cm	Distinctive feature
MgFe$_2$O$_4$	1,450	352	900	2.06	650	8.5	0.016	10^8	Low $4\pi M_s$ High resistivity Characteristic sensitive to heat treatment
MnFe$_2$O$_4$	4,800	302	600	2.2	180	9.3	0.051	10^4	High $4\pi M_s$ Low resistivity
NiFe$_2$O$_4$	3,200	587	500	2.4	460	8.9	0.017	10^4	High T_c Low resistivity
BaFe$_{12}$O$_{19}$	4,600	450	50	2.05	17,000	....	.....	10^8	High anisotropy field
Y$_3$Fe$_5$O$_{12}$ (YIG)	1,740	277	50	2.00	80	12.0	0.002	10^{10}	High resistivity Narrow line width

tion of the susceptibility ($\mu - 1$) with the internal static magnetic field H_i is shown in Fig. 126. If a plane electromagnetic wave moves through a ceramic magnet parallel to H_i, it is transformed into two counterrotating circularly polarized components. The permeability μ'_- of the negatively polarized component goes asymptotically to unity, whereas the permeability μ'_+ of the positively polarized component experiences resonance at a particular value of H_i, H_{res}. At this resonance point, the loss component of the permeability μ''_+ has an absorption peak. The width of this peak at half amplitude ΔH is used as a measure of the resonance absorption losses. Since the negatively polarized wave does not undergo resonance, there is no peak absorption loss for this component. Static field strengths below H_{res} are accompanied by lower losses until the magnet becomes unsaturated and normal hystersis losses are important.

For values of the internal static field above saturation and about $5\Delta H$ below resonance, the losses are small, and the value of μ'_+ is also small. Because the corresponding value of μ'_- is much larger than that of μ'_+, the negatively polarized wave component moves more slowly through the material than the positively polarized component. As a result of this effect, a plane-polarized wave moving through such a medium experiences a rotation of its axis of polarization. Since the rotation depends only on the direction of the internal static field and not on the direction of wave propagation, the initial plane wave and the first reflected wave will be rotated in the same direction by the same angle, which with appropriate device geometry, eliminates interference between these waves. Faraday rotation devices such as phase shifters, isolators, and gyrators operate on this principle.

TABLE 43 Properties of Commercial Microwave Ferrites

Supplier	Trans-Tech	Hughes		Airtron		Trans-Tech		Airtron		Trans-Tech		
Material code	G-113	H-101	H-SC-1	—	—	G-1600	G-1400	—	—	G-1210	G-1200	G-1021
Composition	YIG	YIG	YIG Single crystal	YIG Single crystal	YGalG Single crystal	YGadG	YGadAlG	YGalG Single crystal	YGalG Single crystal	YAlG	YGadAlG	YGadAlG
Saturation magnetization $4\pi M_s$, gauss	1,780	1,750	1,785	1,780	1,600	1,600	1,400	1,400	1,200	1,200	1,200	1,100
Resonance line width ΔH at 25°C, oersteds:												
At 9.4 GHz	45	35	0.50	—	≤0.45	50	50	≤0.50	≤0.55	40	50	90
At 5 GHz	—	—	—	0.35	—	—	—	—	—	—	—	—
Landé g factor at 9.4 GHz	1.97	2.010	2.005	275	265	1.98	1.98	255	243	1.98	1.98	1.99
Curie temperature T_c, °C	280	280	290	—	—	270	265	—	—	220	260	280
Dielectric constant K	15.0	16.3	—	—	—	15.1	15.1	—	—	14.8	15.1	15.2
Dielectric loss factor, tan δK	<0.0002	0.002	<0.0002	—	—	<0.0002	<0.0002	—	—	<0.0002	<0.0002	<0.0002
Density, g/cm³	—	5.08	5.13	—	—	—	—	—	—	—	—	—

Supplier	Trans-Tech			Airtron		Trans-Tech				Airtron
Material code	G-1000	G-1002	G-1010	—	—	G-1003	G-810	G-1004	G-800	G-1005	—
Composition	YGadAlG	YGadG	YAlG	YGalG Single crystal	YGalG Single crystal	YGadG	YAlG	YGadAlG	YGadAlG	YGadG	YGalG Single crystal
Saturation magnetization $4\pi M_s$, gauss	1,000	1,000	1,000	1,000	800	870	800	800	800	725	700
Resonance line width ΔH at 25°C, oersteds:											
At 9.4 GHz	55	100	40	≤0.60	≤0.65	140	40	75	55	200	≤0.70
At 5 GHz	—	—	—	—	—	—	—	—	—	—	—
Landé g factor at 9.4 GHz	1.99	1.99	1.99	230	215	2.00	1.99	2.00	2.00	2.02	205
Curie temperature T_c, °C	250	280	210	—	—	280	200	240	230	280	—
Dielectric constant K	14.7	15.4	14.7	—	—	15.4	14.6	14.8	14.7	15.4	—
Dielectric loss factor, tan δK	<0.0002	<0.0002	<0.0002	—	—	<0.0002	<0.0002	<0.0002	<0.0002	<0.0002	—

TABLE 43 Properties of Commercial Microwave Ferrites (Continued)

Supplier		Trans-Tech				Airtron	Trans-Tech				Airtron
Material code	G-600	G-610	G-510	G-500			G-400	G-1006	G-350	G-300	
Composition	YGadAlG	YAlG	YAlG	YGadAlG		YGalG Single crystal	YAlG	YGadAlG	YAlG	YAlG	YGalG Single crystal
Saturation magnetization $4\pi M_s$, gauss	680	680	550	550		500	400	400	350	300	300
Resonance line width ΔH at 25°C, oersteds:											
At 9.4 GHz	60	40	40	65		≤0.95	<45	65	<45	<45	≤2.2
At 5 GHz	—	—	—	—		—	—	—	—	—	—
Landé g factor at 9.4 GHz	2.02	2.00	2.00	2.00		—	2.01	2.01	2.01	2.02	—
Curie temperature T_c, °C	105	185	155	180		180	135	150	130	120	150
Dielectric constant K	13.8	14.5	14.3	14.4		—	14.1	14.2	14.0	14.0	—
Dielectric loss factor, tan $\delta\kappa$	<0.0002	<0.0002	<0.0002	<0.0002		—	<0.0002	<0.0002	<0.0002	<0.0002	—

Supplier		Trans-Tech				Hughes	Trans-Tech				
Material code	G-250	TT2-111	TT2-101	TT2-102	TT2-125	H-300	TT2-120	TT2-118	TT2-115	TT2-116	TT2-130
Composition	YAlG	NiZn	NiCo	Ni	NiAl	NiAlTi	NiCoAl	NiCoAl	NiCoAl	NiCoAl	NiAl
Saturation magnetization $4\pi M_s$, gauss	250	5,000	3,000	2,500	2,100	2,200	1,900	1,800	1,600	1,400	1,000
Resonance line width ΔH at 25°C, oersteds:											
At 9.4 GHz	<45	160	350	490	460	500	880	775	370	260	320
Landé g factor at 9.4 GHz	2.02	2.11	2.19	2.25	2.30	2.45	2.60	2.45	2.40	2.40	2.60
Curie temperature T_c, °C	105	375	585	570	560	550	545	500	450	425	400
Dielectric constant K	13.8	12.5	12.8	12.7	12.6	12.0	9.0	9.0	12.3	12.3	12.0
Dielectric loss factor, tan $\delta\kappa$	<0.0002	<0.0010	<0.0025	<0.0020	<0.0010	0.002	<0.0020	<0.0015	<0.0010	<0.0010	<0.0010
Density, g/cm³	—	—	—	—	—	4.80	—	—	—	—	—

TABLE 43 Properties of Commercial Microwave Ferrites (Continued)

Supplier	Trans-Tech	Trans-Tech	Indiana General	Indiana General	Trans-Tech	Trans-Tech	Trans-Tech	Trans-Tech	Indiana General	Trans-Tech	Indiana General
Material code	TT2-113	TT1-390	R-4	R-1	TT1-105	TT1-1500	TT1-109	TT1-1000	R-5	TT1-414	R-6
Composition	NiAl	MgMn	MgMn	MgMnAl	MgMnAl	MgMnAl	MgMnAl	MgMnAl	MgMnAl	MgMnAl	MgMnAl
Saturation magnetization $4\pi M_s$, gauss	500	2,150	1,780	1,760	1,750	1,500	1,300	1,000	1,110	750	730
Resonance line width ΔH at 25°C, oersteds:											
At 9.4 GHz	150	540	380	490	225	180	135	100	220	120	120
Landé g factor at 9.4 GHz	1.54	2.04	2.06	2.11	1.98	1.98	1.98	1.98	2.05	1.98	2.00
Coercive force H_c, oersteds	—	—	1.4	2.1	—	—	—	—	0.8	—	0.7
Curie temperature T_c, °C	120	320	320	290	225	180	140	100	140	100	100
Dielectric constant K	9.0	12.7	12	12	12.2	12.0	11.8	11.6	12	11.3	12
Dielectric loss factor, tan δ_K	<0.0008	<0.00025	0.0005	0.0005	<0.00025	<0.00025	<0.00025	<0.00025	0.0003	<0.00025	0.0004
Density, g/cm³	—	—	4.15	4.10	—	—	—	—	3.95	—	3.90

At resonance, the microwave material can absorb energy from the positively polarized wave while the negatively polarized component undergoes little if any absorption. This phenomenon is used to achieve nonreciprocal attenuation in devices such as resonance isolators.

The properties and distinctive features of spinel ferrites, $BaFe_{12}O_{19}$ and yttrium-iron-garnet are listed in Table 42. In Table 43 are given the properties of some commercial microwave materials.

REFERENCES

1. Foner, S.: Versatile and Sensitive Vibrating-sample Magnetometer, *Rev. Sci. Instrum.*, vol. 30, pp. 548–557, 1959.
2. *ASTM Designation* D 257–66, D-C Resistance or Conductance of Insulating Materials.
3. Sauer, H. A., and S. S. Flaschen: Positive Temperature Coefficient of Resistance Thermistor Materials for Electronic Applications, *Proc. 1956 Elec. Comput. Conf.*, pp. 41–46, 1956.
4. Landis, H. M.: Electrodes for Ceramic $BaTiO_3$ Type Semiconductors, *J. Appl. Phys.*, vol. 36, no. 6, p. 2000, June, 1965.
5. Sauer, H. A., and W. H. Shirk: A D-C Wheatstone Bridge for Multi-terohm Measurements with High Accuracy Capability, *IEEE Commun. and Electron.*, March, 1964.
6. *ASTM Designation* D 150, Tests for Dielectric Constant of Electrical Insulation.
7. Von Hippel, A. R.: "Dielectric Materials and Applications," Technology Press of MIT and John Wiley & Sons, Inc., New York, 1954.
8. *ASTM Rep.* C 525-63T, Test for Complex Dielectric Constant of Nonmetallic Magnetic Materials at Microwave Frequencies.
9. Frisco, L. J.: Frequency Dependence of Electric Strength, *Electro-Technol.*, vol. 68, pp. 110–116, August, 1961.
10. Storm, H. F.: "Magnetic Amplifiers," pp. 53–58, John Wiley & Sons, Inc., New York, 1955.
11. IRE Standards on Piezoelectric Crystals: Measurements of Piezoelectric Ceramics, *1961 Proc. IRE*, vol. 49, no. 7, pp. 1162–1169, July, 1961; vol. 46, no. 4, pp. 764–778, April, 1958.
12. Birks, J. B., Modern Dielectric Materials, p. 19, Heywood and Co., London, England, 1960.
13. Kohl, W. H.: "Handbook of Materials and Techniques for Vacuum Devices," p. 586, Reinhold Publishing Corp., New York, 1967. Copyright 1967 by Reinhold Publishing Corp., by permission of Van Nostrand Reinhold Co.
14. Mason, W. P., and H. Jaffe: Methods for Measuring Piezoelectric, Elastic, and Dielectric Coefficients of Crystals and Ceramics, *Proc. IRE*, vol. 42, pp. 921–930, June, 1954.
15. Von Hippel, A. R.: "Dielectrics and Waves," p. 1, John Wiley & Sons, Inc., New York, 1954.
16. Kingery, W. D.: "Introduction to Ceramics," pp. 420–426, John Wiley & Sons, Inc., New York, 1960.
17. Kohl, *op. cit.* p. 77.
18. Atlas, N. M., and H. H. Nakamura: Control of Dielectric Constant and Loss in Alumina Ceramics, *J. Amer. Ceram. Soc.*, vol. 45, no. 10, pp. 467–471, October, 1962.
19. Francl, J., and W. D. Kingery: Thermal Conductivity: IX, Experimental Investigation of Effect of Porosity on Thermal Conductivity, *J. Amer. Ceram. Soc.*, vol. 37, no. 2, pp. 99–107, February, 1954.
20. Coble, R. L.: Transparent Alumina and Method of Preparation, U.S. Patent 3,026,210, March, 1962.
21. Kingery, W. D.: "Oxides for Higher Temperature Applications," pp. 76–89, McGraw-Hill Book Company, New York, 1960.
22. Smith, R., and J. P. Howe: Beryllium Oxide, *Proc. 1st Int. Conf. on BeO*, North-Holland Publishing Company, Amsterdam, 1964.
23. McPhee, K. H.: Cooling Transistors with Beryllia Heat Sinks, *Electron.*, vol. 34, pp. 76–78, May 5, 1961.
24. Kohl, *op. cit.* p. 89.
25. Strickler, D. W., and W. G. Carlson: Ionic Conductivity of Cubic Solid Solu-

tions in the System $CaO-Y_2O_3-ZrO_2$, *J. Amer. Ceram. Soc.*, vol. 47, no. 3, pp. 122–127, March, 1964.
26. Weissbart, J., and R. Ruka: A Solid Electrolyte Fuel Cell, *J. Electrochem. Soc.*, vol. 109, no. 8, pp. 723–726, 1962.
27. Keler, E. K., and E. N. Nitikiv: High Temperature Ceramic Heaters, *J. Appl. Chem. (USSR)* (English Transl.), vol. 32, no. 9, pp. 2033–2036, 1959; *Ceram. Abstr.* p. 140f, June, 1962.
28. Pearl, H. A., J. M. Nowak, and H. G. Deban: Mechanical Properties of Selected Alloys at Elevated Temperatures: II, Design Criteria of SiC, *WADC-TR-702 Tech. Rep.*, Contract AF-33(616)-5760, March, 1960.
29. Fetterly, G. H.: Electrical Conduction in SiC, *J. Electrochem. Soc.*, vol. 104, no. 5, pp. 322–327, May, 1957.
30. Shepard, C. E., V. R. Watson, and H. A. Stine: Evaluation of a Constricted-arc Supersonic Jet, *NASA Tech. Note* D-2066, January, 1964.
31. Popper, P., and S. N. Ruddlesden: The Preparation, Properties and Structure of Silicon Nitride, *Trans. Brit. Ceram. Soc.*, vol. 60, pp. 603–626, 1961.
32. Janakirama-Rao, B. H. V.: Structure and Mechanism of Semiconductor Glasses, *J. Amer. Ceram. Soc.*, vol. 48, no. 6, pp. 311–319, June, 1965.
33. Mackenzie, J. D.: Semiconducting Oxide Glasses: General Principles of Preparation, *J. Amer. Ceram. Soc.*, vol. 47, no. 5, pp. 211–214, May, 1964.
34. Owen, A. E.: Electronic Conduction Mechanisms in Glasses, *The Glass Industry*, part I, pp. 637–642, November, 1967; part II, pp. 695–699, December, 1967.
35. Sideris, G.: Transistors Face an Invisible Foe, *Electronics*, vol. 39, pp. 191–195, Sept. 19, 1966.
36. Perschy, J. A.: On the Threshold of Success: Glass Semiconductor Circuits, *Electron.*, vol. 40, pp. 74–84, July 24, 1967.
37. Kohl, *op. cit.*, pp. 34–35.
38. Blodgett, K. B.: Surface Conductivity of Lead Silicate Glass after Hydrogen Treatment, *J. Amer. Ceram. Soc.*, vol. 34, no. 1, pp. 14–27, January, 1951.
39. Birks, *loc. cit.*, p. 218.
40. Stanworth, J. E.: Telluride Glasses, *J. Soc. Glass Technol.*, vol. 38, no. 183, pp. 425–435, August, 1954.
41. Corning Glass Works: Glass-K Capacitors, Ref. file CE-1.02, December, 1966.
42. Roy, R.: Decomposition and Resynthesis of the Micas, *J. Amer. Ceram. Soc.*, vol. 32, no. 6, pp. 202–209, June, 1949.
43. Levine, A. K., and S. Nathansohn: Growth of Sheet Crystals of Fluoropholgopite, *Amer. Ceram. Soc. Bull.*, vol. 45, no. 3, pp. 307–311, March, 1966.
44. Crawford, S. C., and J. C. Hickle: Electrical Quality Classification of the Suitability of Reconstituted Mica Paper for Use as Receiving Tube Spacers, in D. Slater (ed.), "Advances in Electron Tube Techniques," *Proc. 5th Nat. Conf.*, 1960, Pergamon Press, New York, 1961.
45. Hatch, R. A.: Synthetic Mica Investigation: IX, Review of Progress from 1947–1955, *Bur. Mines, Rep. of Invest.* 5337, June, 1957.
46. Gould, L.: Improved Keep-alive Design for TR Tubes, *Proc. IRE*, vol. 45, no. 4, pp. 530–533, April, 1957.
47. Lindsay, W. C.: Improved Demountable High Temperature Mica Vacuum Window, *Rev. Sci. Instrum.*, vol. 32, no. 6, pp. 748–749, June, 1961.
48. Stetson, H. W., and W. J. Gyurk: Development of Two Micro-inch (CLA) As-fired Alumina Substrates, presented at American Ceramic Society Meeting, New York, May, 1967.
49. Brown, R.: Substrates for Tantalum Thin-film Circuits, *Amer. Ceram. Soc. Bull.*, vol. 45, no. 8, pp. 720–726, August, 1966.
50. Cox, S. M.: Ion Migration in Glass Substrates for Electronic Components, *Phys. Chem. Glasses*, vol. 5, no. 6, pp. 161–165, December, 1964.
51. Mackenzie, J. D.: Fine Structure in Glass from Ionic and Volumetric Considerations, *Proc. VII Int. Congr. Glass*, paper No. 22, Brussels, June, 1965.
52. Cramer, H. A., W. F. Johnston, and D. W. Callaway: Industrial Sapphire: a Key to Reliable Microelectronics, *IEEE Trans. Component Parts*, vol. CP11, no. 2, pp. 120–128, June, 1964.
53. Stookey, S. D., and R. D. Maurer: Catalyzed Crystallization of Glass: Theory and Practice, *Progr. Ceram. Sci.* vol. 2, pp. 77–102, Pergamon Press, New York, 1962.
54. Pask, J. A.: New Techniques in Glass-to-Metal Sealing, *Proc. IRE.*, vol. 36, pp. 286–289, February, 1948.
55. Dusing, W.: Seal Metals with Precious Metal Coatings for Microwave Tubes, *Glastech. Ber.*, vol. 31, pp. 137–142, April, 1958.

56. Dalton, R. H.: How to Design Glass-to-Metal Joints, *Prod. Eng.*, vol. 36, pp. 62–69, April, 1965.
57. Bondley, R. J.: Quartz-Metal Seals, *Proc. Int. Congr. on Microwave Tubes*, pp. 598–601, Academic Press, Inc., New York, 1961.
58. Floyd, J. R.: Effect of Composition and Crystal Size of Alumina Ceramics on Metal-to-Ceramic Bond Strength, *Amer. Ceram. Soc. Bull.*, vol. 42, no. 2, pp. 65–70, February, 1963.
59. Wisser, G. R., and M. W. Hagadorn: An Improved Nickel-Titanium Ceramic-to-Metal Seal, *Gen. Tel. & Elec. Res. & Develop. J.*, vol. 1, pp. 43–46, January, 1961.
60. Hatzipangos, D., and J. H. Power: Thermal Characteristics of Film Resistor Modulus, *Proc. IEEE Electron. Components Conf.*, pp. 69–78, Washington, D.C., 1968.
61. "Reference Data for Radio Engineers," 4th ed., pp. 79–83, International Telephone & Telegraph Corp., American Book-Stratford Press, Inc., New York, 1961.
62. Buzard, J. B.: Cermet Resistor Process for Commercial Applications, *Proc. IEEE Electron. Components Conf.*, pp. 79–83, Washington, D.C., 1968.
63. Asher, J. W., and C. P. Pratt: Screen-printed Ferro-electric Glass-Ceramic Capacitors, *Proc. IEEE Electron. Components Conf.*, pp. 239–245, Washington, D.C., 1968.
64. Herczog, A.: Microcrystalline $BaTiO_3$ by Crystallization from Glass, *J. Amer. Ceram. Soc.*, vol. 47, no. 3, pp. 107–115, March, 1964.
65. Hague, J. R., J. F. Lynch, A. R. Rudnick, F. C. Holden, and W. H. Duckworth: "Refractory Ceramics for Aerospace," published by American Ceramic Society, 1964.
66. Ceramic Parts, *Mater. Des. Eng.*, Manual 239, November, 1966.
67. Priest, D. H., and R. Talcott: Thermal Stresses in Ceramic Cylinders Used in Vacuum Tubes, *Amer. Ceram. Soc. Bull.*, vol. 38, no. 3, pp. 99–105, March, 1959.
68. Kingery, W. D.: "Introduction to Ceramics," John Wiley & Sons, Inc., New York, p. 682, 1960.
69. *Ibid.*, pp. 723–724.
70. Von Hippel, ref. 7, p. 181.
71. Comeforo, J. E., Properties of Ceramics for Electronic Applications, *Electron. Eng.*, April, 1967.
72. Kingery, W. D.: "Introduction to Ceramics," p. 534, John Wiley & Sons, Inc., New York, 1960.
73. Dixon, J. M., L. D. LaGrange, U. Merten, C. F. Miller, and J. T. Porter: Electrical Resistivity of Stabilized Zirconia at Elevated Temperatures, *J. Electrochem. Soc.*, vol. 110, no. 4, pp. 276–280, April, 1963.
74. Hickman, W. M., and J. F. Zamaria: Furnace Control by Fuel Cell, *Instrum. and Contr. Syst.*, vol. 40, p. 87, August, 1967.
75. Goffaux, R.: Electrical Properties of SiC Varistors, *Proc. Symp. on SiC: A High Temperature Semicond.*, pp. 462–481, Pergamon Press, New York, 1960.
76. Fredrickson, J. E., and W. H. Redanz: Boron Nitride for Aerospace Applications, Carbon Products Div., Union Carbide Corp.
77. *Union Carbide Bull.* 713-204 EF.
78. "Engineering with Glass," Corning Glass Works, Library of Congress Catalog Card No. 62-19567.
79. Sedden, E., E. J. Tippett, and W. E. S. Turner: The Electrical Conductivity of Sodium Metasilicate–Silicate Glasses, *J. Soc. Glass Technol.*, vol. 16, pp. 450T–477T, 1932. Kohl, *op. cit.*, p. 30.
80. Frisco, L. J.: Frequency Dependence of Electric Strength *Electro-Technol.*, vol. 68, pp. 110–116, August, 1961. Kohl, *op. cit.*, p. 587.
81. Shand, E. B.: "Glass Engineering Handbook," 2 ed., pp. 67–90, McGraw-Hill Book Company, New York, 1958.
82. Vermeer, J.: The Impulse Breakdown Strength of Pyrex Glass, *Physica*, vol. 20, pp. 313–325, 1954.
83. Guyer, E. M.: Electrical Glass, *Proc. IRE* vol. 32, pp. 743–750, December, 1944.
84. Chapman, J. J., and L. J. Frisco: *Elec. Mfg.*, vol. 53, p. 136, May, 1959; Shand, *op. cit.*, p. 79.
85. Birks, *op. cit.*, p. 214.
86. "Reference Data for Radio Engineers," 4th ed., p. 64, International Telephone & Telegraph Corp., American Book–Stratford Press, Inc., New York, 1961.

87. Stevels, J. M.: "Electrical Properties of Glass," *Handb. Phys.*, vol. 20, pp. 350–391, 1957.
88. Kohl, *op. cit.*, p. 121.
89. Von Hippel, ref. 7, p. 313.
90. McLean, D. A., N. Schwartz, and E. D. Tidd: Tantalum Film Technology, *Proc. IEEE* (Special Issue on Integrated Electronics), vol. 52, no. 12, pp. 1450–1462, December, 1964.
91. Martin, J. H.: The Manufacture of Ceramic-based Microcircuits, *Sprague Tech. Paper* TP-66-10, Sprague Electric Co., North Adams, Mass.
92. Shand, *op. cit.*, pp. 123–126.
93. Hoffman, L. C.: Precision Glaze Resistors, *Amer. Ceram. Soc. Bull.*, vol. 42, no. 9, pp. 490–493, September, 1963.
94. Burks, D. P., and B. Greenstein: Screened Thick Film Resistors, *Sprague Tech. Paper* TP-67-2.
95. Henney, K., and C. Walsh: "Electronic Components Handbook," vol. 1, chap. 4, McGraw-Hill Book Company, New York, 1957.
96. Dummer, G. W. A., and H. M. Nordenberg: "Fixed and Variable Capacitors," pp. 280, McGraw-Hill Book Company, New York, 1960.
97. Gruver, R. M., W. R. Buessem, and C. W. Dickey: "State-of-the-Art Review on Ferroelectric Ceramic Materials," *Air Force Mater. Lab. Tech. Refs.* 66-164, Air Force Systems Command, May, 1966.
98. Birks, *op. cit.*, p. 191.
99. Hamer, D. W.: Reduced Titanate Chips for Thick Film Hybrid IC's, *Proc. IEEE Electron. Components Conf.*, pp. 256–264, Washington, D.C., 1968.
100. Henney and Walsh, *op. cit.*, p. 104.
101. Jonkers, G. H.: Capacitor Materials with High Dielectric Constant, *Philips Tech. Rev.*, vol. 17, no. 5, pp. 129–137, November, 1955.
102. Brown, F.: High Dielectric Constant Ceramics, *IRE Trans. Component Parts*, vol. CP6, no. 4, pp. 238–251, December, 1959.
103. Gruver et. al., *op. cit.*, p. 102.
104. *Ibid*, p. 118.
105. Prokopowicz, T. I., and P. M. Kennedy: "Research and Development Program Intrinsic Reliability Subminiature Ceramic Capacitor," DA-36-039-SC-90705, U.S. Army Signal R&D, Sprague Electric Co., AD-468839, July, 1961.
106. Henney and Walsh, *op. cit.*, pp. 125–127.
107. Dummer and Nordenberg, *op. cit.*, pp. 132–140.
108. *Ibid.*, pp. 100–114.
109. Henney and Walsh, *op. cit.*, pp. 127–132.
110. *Electron. World*, pp. 37–60, July, 1965.
111. Nye, J. F.: "Physical Properties of Crystals," Clarendon Press, Oxford, 1967.
112. Mason, W. P. (ed.): "Physical Acoustics," vol. I, part A, Academic Press, Inc., New York, 1964.
113. Hueter, T. F., and R. H. Bolt: "Sonics," John Wiley & Sons, Inc., New York, 1955.
114. Mason, W. P.: "Physical Acoustics and the Properties of Solids," D. Van Nostrand, Company, Inc., Princeton, N.J., 1958.
115. Albers, V. M.: "Underwater Acoustics Handbook," Pennsylvania State University Press, University Park, 1960.
116. Albers, V. M.: "Underwater Acoustics," Pennsylvania State University Press, University Park, 1963.
117. von Aulock, W. H. (ed.): "Handbook of Microwave Ferrite Materials," Academic Press, Inc., New York, 1965.
118. Smit, J., and H. P. J. Wijn: "Ferrites," John Wiley & Sons, Inc., New York, 1959.
119. Néel, L.: Propriétés Magnétiques des Ferrites; Ferrimagnétisme et Antiferromagnétisme, *Ann. Phys.*, vol. 3, ser. 12, pp. 137–198, March-April, 1948.
120. von Aulock, *op. cit.*, sec. 3.
121. *Ibid.*, sec. 4.
122. *Ibid.*, sec. 2.
123. Brockman, F. G.: Magnetic Ceramics: a Review and Status Report, *Amer. Ceram. Soc. Bull.*, vol. 47, no. 2, pp. 186–194, February, 1968.
124. Blume, W. S.: U.S. Patents 2,999,275; 3,141,050; 2,964,793; and 3,246,060. R. E. Schernstheimer: U.S. Patent 3,070,841.
125. Snelling, E. C.: The properties of Ferrites in Relation to Their Application, *Proc. Brit. Ceram. Soc.*, no. 2, pp. 151–174, December, 1964.

References 6-117

126. *Tech. Briefs* 651-6512, Trans-Tech, Inc., Gaithersberg, Md., 1967.
127. Rado, G. T., V. J. Folen, and W. H. Emerson: Effect of Magnetocrystalline Anisotropy on the Magnetic Spectra of Mg-Fe Ferrites, *Proc. IEE (London)*, vol. 104, part B, pp. 198–212, 1957.
128. Smit and Wijn, *op. cit.*, p. 156.
129. *Ibid.*, p. 158.
130. Gorter, E. W.: Saturation Magnetization and Crystal Chemistry of Ferrimagnetic Oxide, thesis, University of Leyden, 1954. Reprinted in *Philips Res. Rev.*, vol. 9, pp. 295–320, 321–365, and 403–443, 1954. L. R. Maxwell and S. J. Pickart: Magnetization in Nickel Ferrite–Aluminates and Nickel Ferrite–Gallates, *Phys. Rev.*, vol. 92, p. 120, 1953. von Aulock, *op. cit.*, p. 397.
131. Guillaud, C.: Propriétés Magnétiques des Ferrite, *J. Phys. Radium*, vol. 12, p. 239, 1951. R. Pauthenet: Aimantation Sportanée des Ferrites, *Ann. Phys.*, vol. 7, p. 710, 1952. von Aulock, *op. cit.*, p. 382.
132. Van Uitert, L. G.: High-resistivity Nickel Ferrites: the Effect of Minor Additions of Manganese or Cobalt, *J. Chem. Phys.*, vol. 24, pp. 306–310, 1956.
133. Van Uitert, L. G.: Magnetic Induction and Coercive Force Data and Members of the Series $BaAl_xFe_{12-x}O_{19}$ and related Oxides, *J. Appl. Phys.*, vol. 28, no. 3, pp. 317–319, March, 1957.
134. Bertaut, F., et al.: Substitution dans les Hexaferrites de l'Ion Fe^{3+} par Al^{3+}, Ga^{3+}, Cr^{3+}, *J. Phys. Radium*, vol. 20, p. 404, 1959. A. H. Mones and E. Banks: Cation Substitutions in $BaFe_{12}O_{19}$, *J. Phys. Chem. Solids*, vol. 4, p. 217, 1958. von Aulock, *op. cit.*, p. 478.
135. Smit and Wijn, *op. cit.*, p. 208.
136. *Ibid.*, p. 205.
137. *Ibid.*, p. 207.
138. Jonker, G. H., H. P. J. Wijn, and P. B. Braun: A New Class of Oxidic Ferromagnetic Materials with Hexagonal Crystal Structures, *Proc. IEE (London)*, vol. 104, part B, pp. 249–254, 1957.
139. Bertaut, F., and R. Pauthenet: *Proc. IEE (London)*, vol. 104, part B, pp. 261–264, 1957.
140. Anderson, E. E., et al.: Magnetic Effects of Indium and Gallium Substitutions in Yttrium Iron Garnet, *J. Phys. Soc. Japan*, vol. 17, Suppl. B–I, p. 365, 1962. G. R. Harrison and L. R. Hodges, Jr.: Temperature Stable Microwave Hybrid Garnets, *J. Appl. Phys.*, Suppl. to vol. 33, p. 1375, 1962. A. Vassilieu et al.: Sur les Propriétiés des Grenatas Mixtes d'Yttrium-Gadolinium, *C. R. Acad. Sci.*, vol. 252, p. 2529, 1961. von Aulock, *op. cit.*, p. 179.
141. Smit and Wijn, *op. cit.*, p. 325.
142. Snelling, *op. cit.*, p. 160.
143. *Ibid.*, p. 165.
144. Economos, E.: Ferrites, "Encyclopedia of Chemical Technology," vol. 8, pp. 881–901, John Wiley & Sons, Inc., New York, 1965.
145. von Aulock, *op. cit.*, p. xxiii.
146. "Piezoelectric Ceramics," *Ceramics Div. ITT Research Inst. Rep.* 16, July–August, 1968.
147. *Corning Glass Works Bull.* CEP 2/5M/9-62.
148. Hagy, H. E.: "Thermal Expansion Differential Tables, 1962 Revision," Research and Development Division, Corning Glass Works, Corning, N.Y.
149. Ollom, J. E., and W. H. von Aulock: Measurements of Microwave Ferrites at High Signal Levels, *IRE Trans. Instrum.*, vol. 9, pp. 187–193, September, 1960.

Chapter 7

Semiconductor Materials

WALTER R. RUNYAN
and
STACY B. WATELSKI

Texas Instruments Incorporated, Dallas, Texas

Properties of Semiconductors 7–3
   General Properties 7–3
   Crystal Structure 7–4
   Plane Indices 7–8
   Effect of Crystal Symmetry on Physical Properties 7–11
   Crystallographic Defects 7–11
   Individual Properties 7–17
   Index of Properties 7–23
Measurement Techniques 7–24
   Orientation Determination 7–24
   Defect Observation 7–28
   Conductivity Type 7–33
   Resistivity 7–34
   Lifetime 7–60
   Measurements on Finished Devices 7–67
   Layer Thickness 7–68
Crystal Growth 7–75
   Crystal-growing Environment 7–75
   Czochralski 7–77
   Float Zoning 7–78
   Zone Leveling 7–79
   Verneuil or Flame Fusion 7–79
   Bridgeman, Stockbarger, or Bridgeman-Stockbarger 7–79

Ribbon Growth	7–79
Flux Growth (including Solution Epitaxy)	7–80
Peritectic Growth	7–81
Behavior of Impurities during Melt Growth	7–81
Purification during Growth	7–82
Uniform Distribution	7–83
Facet Effect	7–84
Impurity Striations	7–84
Dynamic Vacuum Evaporation	7–84
Closed-tube Transport via Intermediary	7–85
Open-tube Transport	7–86
Shaping	7–90
Slicing	7–90
Smoothing	7–91
Mechanical Damage	7–93
Polishing	7–93
Ultrasonic Cutting	7–94
Sandblasting	7–94
Abrasives	7–94
Effect of Orientation on Grinding	7–96
Cleaving (Scribe and Break)	7–97
Chemical Shaping and Polishing	7–98
Diffusion	7–99
Diffusion from an Infinite Source	7–102
Bilateral Diffusion	7–102
Diffusion from a Limited Source	7–103
Diffusion from a Concentration Step	7–104
Effect of Time-varying Temperature	7–104
Three-dimensional Solution	7–105
Error Function Algebra	7–105
Transfer of Impurities to the Semiconductor	7–105
References	7–109

This chapter is arranged in three sections. Number 1 considers the properties of semiconductors, first in broad general terms, next with capsule descriptions of some of the more familiar materials, and finally with tabulations of various properties. The tables follow the general format of presenting comparative data on a given property and then, when appropriate, detailed information on the three most widely used semiconductors: silicon, germanium, and gallium arsenide.

The second section is devoted to methods of making the common measurements required by the semiconductor industry. Here the format is a tabular summary of methods, followed by more detailed discussion as appropriate.

Finally, the third section gives a brief review of the various material technologies required for the conversion from raw semiconductor to finished device. These include crystal growth, cutting processes, and diffusion and other methods of producing junctions. Photolithography and metallization, though both required in the production of devices, are excluded from the discussion since they do not represent *semiconductor* material technology.

PROPERTIES OF SEMICONDUCTORS

General Properties[1,2]

Semiconductors, as indicated in Fig. 1, may be loosely characterized as having resistivities intermediate between those of metals and good insulators, i.e., in the 10^{-3} to 10^{+8} ohm-cm range. Semiconductors and insulators do not have overlapping bands as do the metals, and so their electric conductivity in general will increase with increasing temperature (often many orders of magnitude in the span of a few hundred degrees). Either ionic or electronic conduction mechanisms can provide the resistivity range and the temperature dependency, but semiconductors all have electronic conduction. The main contributor to heat conduction is not electrons as it is in metals, but rather phonons. Thus the Wiedeman-Franz law, which for metals relates thermal conductivity to electric conductivity, is no longer valid, and indeed for a given semiconductor material, increasing the electric conduc-

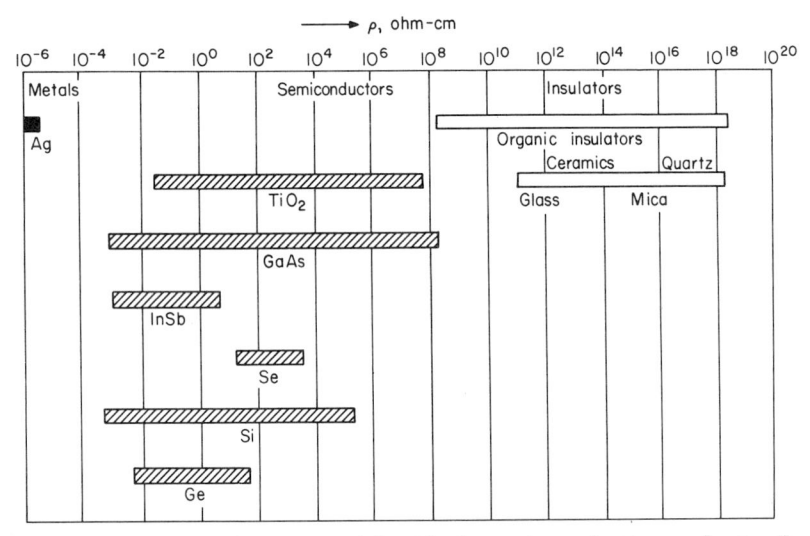

Fig. 1 Resistivity of some materials with electronic conduction mechanism.[95]

tivity usually slightly decreases thermal conductivity because of various scattering processes. The materials may be glassy or crystalline, organic or inorganic, though the most common are inorganic and crystalline. Further, they may be elemental, compound, or alloy. The last-mentioned is single-phase material composed of two or more semiconductors in solid solution. The individual semiconductors of the alloy may be elemental, binary compounds, ternary compounds, or yet more complex structures. In most cases elemental semiconductors are refined in a normal metallurgical fashion, and compound semiconductors are usually prepared by pyrosynthesis from carefully purified elements. Further processing is then provided in order to obtain the required purity level for semiconductor devices. This level varies with the material and device to be made, but ordinarily must be less than 10^{17} atoms/cm^3, and often must be less than 10^{14} atoms/cm^3 (parts per million to parts per billion).

The conventional semiconductors are characterized by some degree of covalent bonding, and will lie on the covalent-ionic leg of the Grimm tetrahedron of Fig. 2. Table 1 shows in a general way how various physical properties vary with bonding and indicates that semiconductors will normally be brittle, have good cleavage, be quite hard, and have reasonable strength.

7-4   Semiconductor Materials

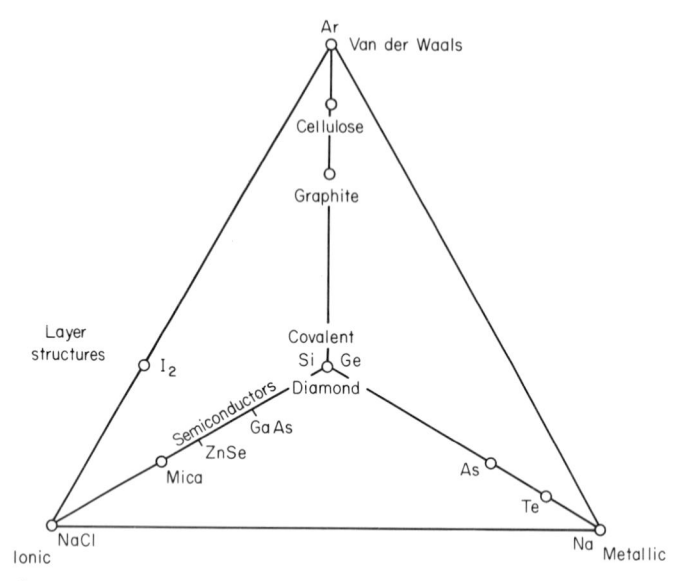

Fig. 2 The Grimm tetrahedron showing graphically the four principal types of bonding at the corners; the mixed bondings are shown along the edges joining the corners.[2]

Crystal Structure

The majority of semiconductors are either cubic or hexagonal, though some, e.g., boron, have lower symmetry structures, and occasionally more than one modification. The shape of the various crystallographic unit cells is shown in Fig. 3,

TABLE 1   Variation of Physical Properties with Bond Type[2]

Property	Covalent	Ionic	Metallic	van der Waals
Electrical conductivity	Low to moderate	Low	High	Low
Color	Colorless to gray, or metallic	Wide variety of colors	Generally gray, with few exceptions (Au, Cu)	Color derived from molecule, not crystal
Cleavage	Fairly good on high-atomic-density crystallographic planes	Very good on crystal planes of high atomic density	Difficult, except at low temperature, because of high ductility	Easy and uncontrolled
Compressibilities	Lowest	Moderate	Low	High
Density	Intermediate	Intermediate	High	Low
Melting point	High (if any)	Intermediate	Intermediate	Low
Hardness	Highest	High	High	Low
Tensile strength	High	Low	High	Low
Ductility	Low	Low	High	Moderate
Coefficient of thermal expansion	Lowest	Moderate	Low	High
Thermal conductivity	Low to high	Low	High	Low

along with the minimum symmetry required for each shape.[3][*] It should be remembered that besides the atoms at the corners of the cells, there may be additional ones in the interior so that the structure may be quite complicated. Again, for boron, the β-rhombohedral unit cell contains over 100 atoms.

The elemental semiconductors, diamond, silicon, germanium and gray tin, have the cubic diamond lattice which is shown in various aspects in Fig. 4. Such a lattice can be considered as two face-centered interpenetrating cubes with their origin displaced $\frac{1}{4},\frac{1}{4},\frac{1}{4}$ with respect to each other. This is illustrated in Fig. 4b, where the same atoms as Fig. 4a are shown, as well as enough additional

[*] These six systems are subdivided into 32 crystal classes which are distinguished from one another according to the various combinations of symmetry elements which they possess, and they represent the total number of different combinations possible in a repetitive structure.

Shape	Description	Minimum Symmetry
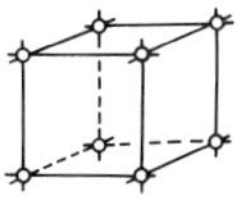 Cubic	Three equal-length mutually perpendicular axes	Four 3-fold axes
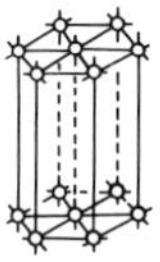 Hexagonal	Three coplanar axes of equal length, making a 120° angle with one another, and a fourth axis of different length perpendicular to the others; or two equal-length axes making a 120° angle with each other and perpendicular to a third axis of different length	One 6-fold axis or one 3-fold axis
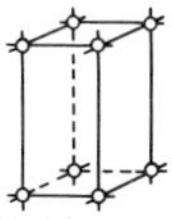 Tetragonal	Three mutually perpendicular axes with two of equal length	One 4-fold axis
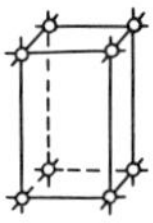 Orthorhombic	Three mutually perpendicular unequal-length axes	Three mutually perpendicular 2-fold axes
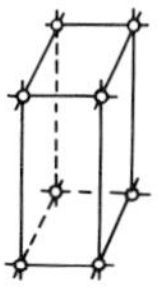 Monoclinic	Three unequal-length axes, two of which are perpendicular	One 2-fold axis
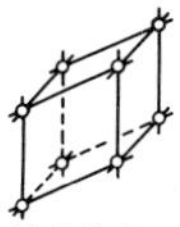 Triclinic	Three unequal-length axes not at right angles to one another	One 1-fold axis

Fig. 3 Unit cell shapes.

7-6 Semiconductor Materials

ones to allow the two cubes to be outlined. The III-V compounds have the zinc blende structure, and differ from diamond only in that one of the interpenetrating face-centered cubes is exclusively component one and the other cube exclusively component two.

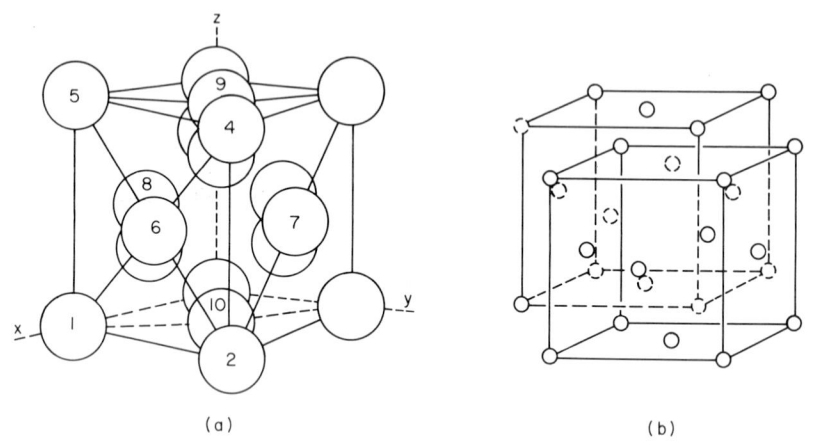

Fig. 4 Diamond lattice.

As an alternate method of visualization, consider first a single layer of close-packed atoms which will appear as in Fig. 5. Next, a second layer is added by placing atoms in the darkened depressions of the first layer, and then a third layer is added by placing atoms over the undarkened depressions of the first layer. The next layer will then repeat the first, etc. Thus, a stacking sequence $ABCABC$ is initiated. This construction is just that needed to give a face-centered cube, and if each atom is replaced by a pair separated by some distance and lying one above the other, the diamond lattice is developed. Figure 6 presents a view of the latter structure designed to delineate the double-layer effect, and at the same time show the bond directions drawn in. As further indicated on this figure, the [111] direction* is normal to the layers just described.

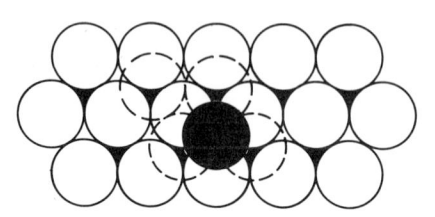

Fig. 5 A plan view of cubic close packing. The first layer is indicated by the solid line, unfilled circles; the second layer by the dotted circles positioned over the depressions which point to the bottom of the page (darkened); and the third layer by the solid circle(s) which lies above the second layer and centered over the unshaded depression of the first layer. The fourth layer is a repeat of the first one.

In the zinc blende structure, the couples used to build up the lattice are made up of component one on one end and component two on the other. It might be noted that the atoms of the two close-spaced planes are held to one another by three bonds while the next double layer is attached to the first with only one bond per atom. Thus, cleavage between the single-bonded planes is to be expected and is observed in the diamond lattice. However, in the zinc blende structure, the additional ionic bonding arising from the oppositely charged atoms of the A-α bond of Fig. 6 is often sufficient to prevent separation, so many III-V compounds cleave between some other planes

* Direction terminology is described in the following section.

(usually (110)s where the ionic bonding averages out).[4] Chemical etching usually is slowest when proceeding in a [111] direction, but again, because of the alternate single-triple bonding, the α-B, β-C, or γ-A double layer will be the one predominately exposed at any given time. If a diamond structure is being so etched, since there is no difference between an α layer and a B layer, whether etching is from the top down (in Fig. 6) or from the bottom up, the result is the same. However, if the material were, for example, GaAs, the α layer might be arsenic and the B layer gallium. In this case, depending on whether etching were from the top of a (111) slice or from the bottom, either arsenic or gallium planes would be exposed, and the surface behavior and appearance will be different. By convention, when the Group III layer is exposed, it is called a ($\bar{1}\bar{1}\bar{1}$) face; and the Group V layer, a (111) face.[5]

If the atomic pairs are of different constituents and stacked in the sequence ABABAB rather than ABCABC, the hexagonal wurtzite structure is formed and is typical of several of the compound semiconductors. If a section is taken through Fig. 5 in the [110] direction, the atoms exposed are arranged as in Fig. 7a. The wurtzite sequence ABABAB just described is shown in Fig. 7b. Some materials, notably silicon carbide, have not only ABAB and ABCABC sequences but also more complicated ones such as are shown in Fig. 7c. These are called *poly types*, and maintain hexagonal symmetry. For a given material they all have almost

Fig. 6 View of diamond lattice showing bond direction and emphasizing the double layer character of the structure.

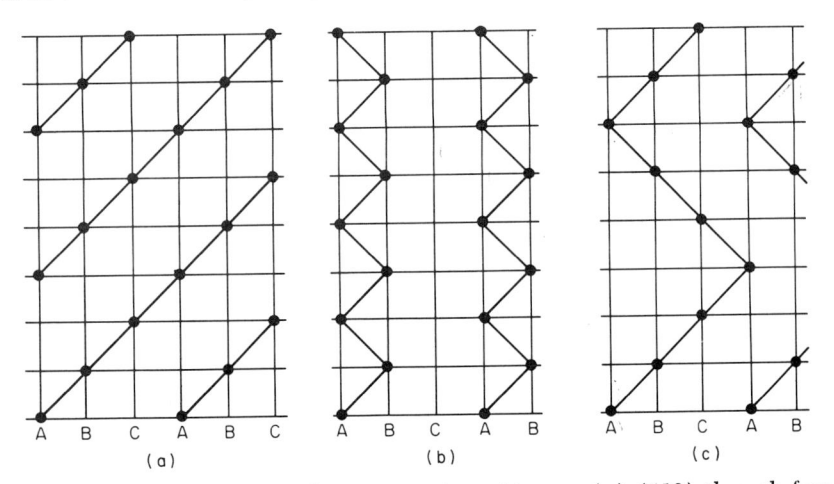

Fig. 7 Crystal-lattice section showing atomic positions. (a) (110) through face-centered cubic; (b) (11$\bar{2}$0) through wurtzite; (c) (11$\bar{2}$0) through a SiC polytype, Si atoms only.[7]

identical a spacings, but the c dimension (vertical in the drawing) varies in integral multiples of some minimum value.[6,7] Some poly types have c directions which are exceedingly large, and it is not presently clear what the orienting forces are that can act over such distance and enable a growing crystal to continually repeat the sequence.

7-8　Semiconductor Materials

Plane Indices[a]

The various planes that pass thorough a crystal may be described in terms of the reciprocal of the intercepts of that plane with the crystallographic axes. These reciprocals are usually expressed as the smallest possible integer having the same ratio and are written as (hkl). For the hexagonal crystal system, in which there are three coplanar axes, as well as one perpendicular to the plane of the first three, the indices are $hkil$. h, k, and i are the reciprocals of the intercepts of the plane in question with the three coplanar axes, and $h + k = -i$. Since h and k together unequivocalably determine i, it is not necessary to write all four indices, but some method of indicating that they are hexagonal is required. This can be accomplished by $hk:l$. Figure 8 shows examples of some of the more

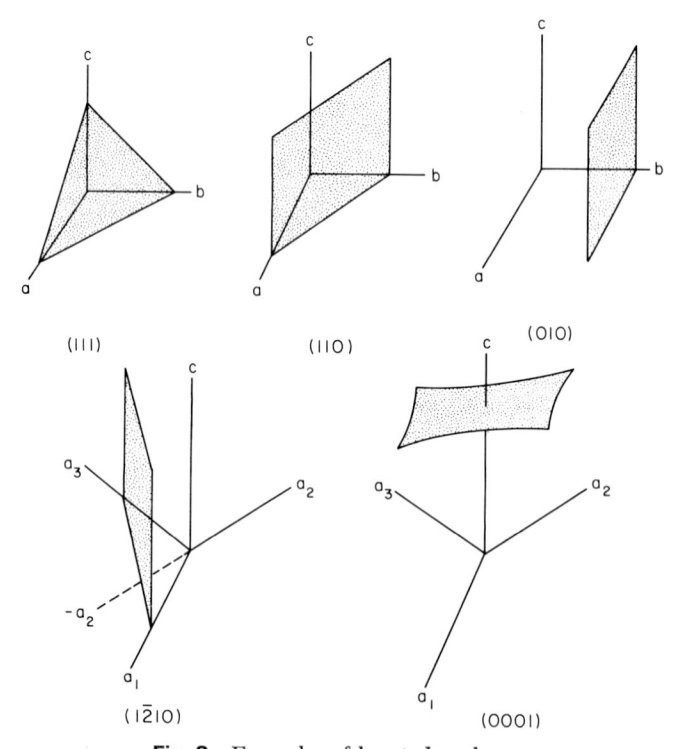

Fig. 8　Examples of low-index planes.

common low-index planes. A complete family, composed of all possible planes resulting from permuting a given set of indices (including negative values) is denoted by braces, e.g., {111}.

The indices of a direction through the crystal are determined by:

1. Considering that travel starts at the origin and is to go through some point p in the lattice.
2. Reaching that point by going a distance u along the x axis, a distance v parallel to the y axis, and a distance w parallel to the z axis.
3. Expressing the ratio of $u:v:w$ as the smallest possible set of integers $h:k:l$. These are the direction indices and are written as $[hkl]$. A complete set of equivalent directions is written as $\langle hkl \rangle$.

For a cubic system, a direction will always be perpendicular to a plane with the same indices, but in other systems this is not generally true.

Properties of Semiconductors 7-9

TABLE 2   Formulas for Calculating Spacings between hkl Planes and Angles between $h_1k_1l_1$ and $h_2k_2l_2$ Planes[3]

Triclinic

$$\frac{1}{d^2} = \frac{1}{V^2}(S_{11}h^2 + S_{22}k^2 + S_{33}l^2 + 2S_{12}hk + 2S_{23}kl + 2S_{13}hl)$$

where $S_{11} = b^2c^2 \sin^2 \alpha$ $S_{12} = abc^2(\cos \alpha \cos \beta - \cos \gamma)$
$S_{22} = a^2c^2 \sin^2 \beta$ $S_{23} = a^2bc(\cos \beta \cos \gamma - \cos \alpha)$
$S_{33} = a^2b^2 \sin^2 \gamma$ $S_{13} = ab^2c(\cos \gamma \cos \alpha - \cos \beta)$
$V = abc\sqrt{1 - \cos^2 \alpha - \cos^2 \beta - \cos^2 \gamma + 2 \cos \alpha \cos \beta \cos \gamma}$

and α = angle between b and c axes
β = angle between c and a axes
γ = angle between a and b axes

Monoclinic

$$\frac{1}{d^2} = \frac{h^2}{a^2 \sin^2 \beta} + \frac{k^2}{b^2} + \frac{l^2}{c^2 \sin^2 \beta} - \frac{2hl \cos \beta}{ac \sin^2 \beta}$$

Orthorhombic

$$\frac{1}{d^2} = \left(\frac{h}{a}\right)^2 + \left(\frac{k}{b}\right)^2 + \left(\frac{l}{c}\right)^2$$

Tetragonal

$$\frac{1}{d^2} = \frac{h^2 + k^2}{a^2} + \frac{l^2}{c^2}.$$

Hexagonal

$$\frac{1}{d^2} = \frac{4}{3} \cdot \frac{h^2 + hk + k^2}{a^2} + \left(\frac{l}{c}\right)^2$$

Cubic

$$\frac{1}{d^2} = \frac{h^2 + k^2 + l^2}{a^2}$$

Cubic

$$\cos \phi = \frac{h_1h_2 + k_1k_2 + l_1l_2}{[(h_1^2 + k_1^2 + l_1^2)(h_2^2 + k_2^2 + l_2^2)]^{1/2}}$$

Tetragonal

$$\cos \phi = \frac{\dfrac{h_1h_2 + k_1k_2}{a^2} + \dfrac{l_1l_2}{c^2}}{\left[\left(\dfrac{h_1^2 + k_1^2}{a^2} + \dfrac{l_1^2}{c^2}\right)\left(\dfrac{h_2^2 + k_2^2}{a^2} + \dfrac{l_2^2}{c^2}\right)\right]^{1/2}}$$

Hexagonal

$$\cos \phi = \frac{h_1h_2 + k_1k_2 + \dfrac{1}{2}(h_1k_2 + h_2k_1) + \dfrac{3\,a^2}{4\,c^2}l_1l_2}{\left[\left(h_1^2 + k_1^2 + h_1k_1 + \dfrac{3\,a^2}{4\,c^2}l_1^2\right)\left(h_2^2 + k_2^2 + h_2k_2 + \dfrac{3\,a^2}{4\,c^2}l_2^2\right)\right]^{1/2}}$$

Orthorhombic

$$\cos \phi = \frac{\dfrac{h_1h_2}{a^2} + \dfrac{k_1k_2}{b^2} + \dfrac{l_1l_2}{c^2}}{\left[\left(\dfrac{h_1^2}{a^2} + \dfrac{k_1^2}{b^2} + \dfrac{l_1^2}{c^2}\right)\left(\dfrac{h_2^2}{a^2} + \dfrac{k_2^2}{b^2} + \dfrac{l_2^2}{c^2}\right)\right]^{1/2}}$$

7-10 Semiconductor Materials

The spacing d between parallel planes in a crystal may be calculated from the indices of the planes by the appropriate equations of Table 2. For nonparallel planes, the angle between them may be computed by equations which are also given in Table 2. In addition, angles between various low-index planes of the cubic system are tabulated in Table 3. However, since the expressions for the other crystal systems involve various ratios of the cell dimensions, and since these will differ for different materials, the angles between them cannot be uniquely determined.

TABLE 3 Angles between Crystallographic Planes (and between Crystallographic Directions) in Crystals of the Cubic System[*]

{HKL}	{hkl}	Values of angles between HKL and hkl planes (or directions)						
100	100	0.00	90.00					
	110	45.00	90.00					
	111	54.74						
	210	26.56	63.43	90.00				
	211	35.26	65.90					
	221	48.19	70.53					
	310	18.43	71.56	90.00				
	311	25.24	72.45					
110	110	0.00	60.00	90.00				
	111	35.26	90.00					
	210	18.43	50.77	71.56				
	211	30.00	54.74	73.22	90.00			
	221	19.47	45.00	76.37	90.00			
	310	26.56	47.87	63.43	77.08			
	311	31.48	64.76	90.00				
111	111	0.00	70.53					
	210	39.23	75.04					
	211	19.47	61.87	90.00				
	221	15.79	54.74	78.90				
	310	43.09	68.58					
	311	29.50	58.52	79.98				
210	210	0.00	36.87	53.13	66.42	78.46	90.00	
	211	24.09	43.09	56.79	79.48	90.00		
	221	26.56	41.81	53.40	63.43	72.65	90.00	
	310	8.13	31.95	45.00	64.90	73.57	81.87	
	311	19.29	47.61	66.14	82.25			
211	211	0.00	33.56	48.19	60.00	70.53	80.40	
	221	17.72	35.26	47.12	65.90	74.21	82.18	
	310	25.35	40.21	58.91	75.04	82.58		
	311	10.02	42.39	60.50	75.75	90.00		
221	221	0.00	27.27	38.94	63.61	83.62	90.00	
	310	32.51	42.45	58.19	65.06	83.95		
	311	25.24	45.29	59.83	72.45	84.23		
310	310	0.00	25.84	36.87	53.13	72.54	84.26	
	311	17.55	40.29	55.10	67.58	79.01	90.00	
311	311	0.00	35.10	50.48	62.96	84.78		

[*] A more complete listing may be found in Peavler and Lenusky, Angles between Planes in Cubic Crystals, *IMD Spec. Rep.* 8, American Institute of Mining, Metallurgical, and Petroleum Engineers.

Properties of Semiconductors

A plane belongs to a zone $[uvw]$ if it is parallel to the line $[uvw]$. The condition for this is that

$$hu + kv + lw = 0$$

One requirement for zone determination sometimes arises in etching studies where it is found that various etchants expose not just a single slow-etching face, but a number of them, all belonging to the same zone.

If the trace of one plane intersecting another is desired, its equation may be determined from the general equation of the line of intersection of two planes. Given that the Miller indices of the planes are (hkl) and $(h'k'l')$, the direction $[uvw]$ of the line is

$$u = \begin{vmatrix} k & l \\ k' & l' \end{vmatrix} \qquad v = \begin{vmatrix} l & h \\ l' & h' \end{vmatrix} \qquad w = \begin{vmatrix} h & k \\ h' & k' \end{vmatrix}$$

Markings (small facets) usually occur on crystals being grown from the melt where the slow-growing planes are tangent to the melt-crystal interface. Their location is thus determined from the traces of the slow-growing planes on the growth plane.

Effect of Crystal Symmetry on Physical Properties

By making use of the symmetry of the crystal and the tensor rank of a given property, the crystal classes for which the property will be zero, those in which it will be isotropic, and those crystallographic directions in which it will have the same value can be predicted. This ability is based on *Neumann's principle* which requires that the symmetry elements of any physical property of a single crystal must include the symmetry elements of the point group of the crystals.[8,9] Table 4 summarizes some of these results and shows, for example, that since the diffusion coefficient is a second-rank tensor, its value will be the same in any direction in a cubic crystal. The piezoelectric coefficients form a third-rank tensor and are zero in crystals with a center of symmetry. Thus crystals with the diamond lattice (e.g., silicon) do not exhibit the piezoelectric effect, whereas those with the zinc blende structure (like GaAs) do. The elastic modulii comprise a fourth-rank tensor so that the elastic properties are not the same in all directions, even in a cubic crystal. However, though not indicated in the table, for the stress in any direction in a (111) plane, Young's modulus is isotropic.[10] In very thin sections such as inversion layers, the crystal may appear to be two-dimensional so that its symmetry is different from the bulk. For these cases, the tensor element must be recalculated and show, for example, that the electronic mobility becomes anisotropic in silicon.[11]

Crystallographic Defects

In general, the most perfect crystal lattice possible is desired in finished semiconductor devices, but unfortunately there are many crystallographic defects which occur during growth and subsequent processing. Table 5 is a partial listing of these and illustrates the wide range of mistakes possible. Some, such as vacancies, are thermodynamically stable in small quantities and so cannot be completely eliminated. Others, such as impurity atoms, are required for device operation, but most of those listed are undesirable. Table 6 summarizes deleterious effects which may arise from these various errors. Of these, dislocations, twins, and grain boundaries are the most troublesome in the melt growth. Vapor growth usually yields the same defects as well as stacking faults, the various shape defects, micropolycrystalline regions, and hillocks.

The compound semiconductors may be troubled by all the defects listed.

Placement errors In compound semiconductors, an occasional error in placement (antistructure) may occur; i.e., an indium atom might appear in an antimony site of an InSb crystal or vice versa. Furthermore, impurities might occupy either site and perhaps have different properties. For example, silicon (a Group

IV element) replacing an indium atom behaves as an n-type dopant, whereas if it replaces antimony, it is a p-dopant.

Foreign atoms Foreign atoms widely dispersed (as in normal doping) cause local lattice strain, but othewise little difficulty. Because of differences in ionic radii of the impurity and the host, local introduction of high concentrations of impurities, as in planar diffusions, can cause stresses great enough to produce plastic flow at the diffusion temperatures.[12]

Dislocations[13-15] Edge dislocations may be thought of as occurring when extra layers of atoms are introduced in some areas of the crystal. The perfect array of Fig. 9a would then be modified to that of Fig. 9b. Actually, the "inserted" atoms may not be a single plane, but rather a double layer, as shown in Fig. 9c.

TABLE 4 Tensor Properties of Crystals[8]

Tensor rank	Relates:	Properties	Relating equations (symbols not defined)	Isotropic in:	Zero-valued in:
0	A scalar to a scalar	Density Heat capacity	$m = \delta V$ $Q = C_p \Delta T$	All classes	None
1	A vector to a scalar	Pyroelectric coefficient	$\Delta P = p_i \Delta T$	None	Those with center of symmetry (10 classes)
2	A vector to a vector	Electrical conductivity Electrical mobility Thermal conductivity Diffusion coefficient	$j_i = \sigma_{ik} E_k$ $j_i = \mu_{ik} qn E_k$ $Q_i = k_{ij} \dfrac{\partial T}{\partial X_j}$ $j_i = D_{ij} \dfrac{\partial C}{\partial X_j}$	All cubic	None
	A scalar to a second-rank tensor	Thermal expansion	$\epsilon_{ij} = \alpha_{ij} \Delta T$		
3	A vector to a second-rank tensor	Piezoelectric coefficient	$p_i = d_{ijk} S_{jk}$	None	Those with center of symmetry
4	Two second-rank tensors	Elastic constants Piezoresistance	$S_{ij} = c_{ijkl} \epsilon_{kl}$ $\dfrac{\Delta \rho_{ij}}{\rho} = \pi_{ijkl} S_{kl}$	None	None

This arises from the fact that for the only defect introduced to be an edge dislocation, the stacking sequence of planes must remain the same except in the immediate vicinity of the dislocation. In the diamond lattice, dislocations usually lie in {111} planes and in ⟨110⟩ directions; but if jogs occur, other directions will be involved. During growth, dislocations can be initiated by large thermal gradients along the growth interface and by the chance inclusion of foreign matter. They can also be propogated into the growing crystal from those already present in the seed.[16] These in turn occur either from having been grown into the seed or from a mechanically damaged surface. In epitaxial growth onto foreign substrates, there is an additional possibility for dislocations to occur periodically at the interface in order to accommodate the differences in lattice spacing between the substrate and the new growth.[17] Dislocations can also be generated after growth

Properties of Semiconductors 7-13

TABLE 5 Summary of Lattice Defects

Local defects (minimum of long-range disorder)
Vacancies
Interstitials
Antistructure
Foreign atoms (singly)
Edge dislocations
Stacking faults
Twins
Single grain boundary

Aggregate defects
Polycrystalline regions
Void
Cracks
Inclusions (separate phase)
Inclusions (compositional variation)

Shape defects (no atomic misplacement)
Faceting
Habit change
Variation in surface contour
Miscellaneous
Lattice strain

TABLE 6 Device Problems Associated with Crystal Defects

Defect	Problem	Reference
Divacancy	Recombination centers—could cause reduction in lifetime.	
Edge dislocation (moving)	Enhanced diffusion—could cause emitter-collector short circuits.	a
Edge dislocation (stationary)	Enhanced metal precipitation—causes emitter-to-base short circuits, excess p-n junction leakage current, microplasmas.	a, b, c
Stair-rod dislocations	Microplasmas, probably due to metal precipitates.	d
Twins	Produces regions of differing orientation.	
Stacking faults	Source of stair-rod dislocations, enhanced metal precipitation.	
Grain boundaries	Enhanced diffusion, reduced mobility across boundary, enhanced metal precipitation.	e, f
Inclusions	Regions of different mobility, melting points, and resistivity. Produces high fields.	

[a] J. E. Lawrence, Behavior of Dislocations in Silicon Semiconductor Devices: Diffusion, Electrical, *J. Electrochem. Soc.*, vol. 115, pp. 860–865, 1968.
[b] G. H. Schwuttke, Semiconductor Junction Properties as Influenced by Crystallographic Imperfections, *Sci. Rep.* 1 and 2, AF Contract AF 19 (628) 5059.
[c] A. Goetzberger and W. Shockley, *J. Appl. Phys.*, vol. 31, p. 1891, 1960.
[d] H. J. Queisser and A. Goetzberger, *Phil. Mag.*, vol. 8, p. 1063, 1963.
[e] H. S. Queisser, K. Hubner, and W. Shockley, Diffusion along Small Angle Grain Boundaries in Silicon, *Phys. Rev.*, vol. 123, pp. 1245–1254, 1961.
[f] K. E. Bean, H. P. Hentzschel, and D. Coleman, Thermal and Electrical Anisotropy of Polycrystalline Silicon, *J. Appl. Phys.*, vol. 40, pp. 2358–2359, 1969.

7-14 Semiconductor Materials

by plastic flow and are a common occurrence during the cooling of crystals grown from the melt[18] and during diffusion heating and cooling cycles.[19]

Twinning[14,20] A twin plane separates two regions of crystalline material which, though having different orientations, can share the twin plane. Under such circumstances, nearest-neighbor positions are satisfied, and lattice disruption is thus minimized. A pictorial representation of twinning is show in Fig. 10 and illustrates the idea of a common plane. For some materials the twin plane may be a sheet of atoms, but for diamond and zinc blende the plane falls between atoms. The orientation of twin boundaries is restricted by both symmetry and energy considera-

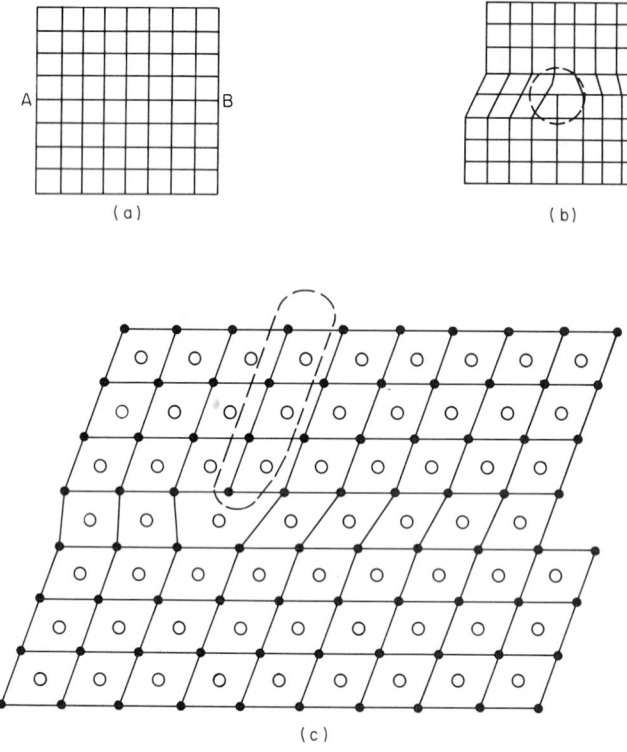

Fig. 9 Dislocation formed by insertion of an extra plane (b) or by an extra double layer (c) into the "perfect" lattice of (a).

tions. For example, based on symmetry, the diamond lattice should twin in both (111) and (112) planes. Experimentally, only (111) twins have been observed.

For the diamond lattice, since the atoms are alike, all twinning is equivalent, but for materials with the zinc blende structure this is no longer true, since the two sublattices are nonequivalent. In the top portion of Fig. 10b, a twin is shown which has nearest neighbors across the twin in the correct position, and in addition, has them of the proper type. The lower portions show another configuration which results in the correct position, but with wrong bonds. The upper type would appear more likely to occur and indeed is the one reported.

Large areas of twin boundary are often found in crystals grown from the melt, but during vapor-phase growth twinned regions are usually very local in nature. This is primarily due to the relatively small thickness of most vapor-grown crystals. After twinning, the new growth direction can be determined by the use of a suitable twinning matrix, as is indicated in Fig. 11.

Stacking faults[14] Stacking faults occur when complete layers are omitted or added, thus producing a change in the stacking sequence. Instead of $ABCABC$ it might be $ABCBCABC$.

In the diamond and zinc blende structure, the faulted areas occur between (111) planes so that on the surface, a stacking fault will have a shape defined by the traces of (111) planes intersecting the surface. These are shown in Fig. 12 for various surface orientations.[21] Stacking faults seldom occur in crystals grown from the melt, but are widespread in vapor-phase-grown material (i.e., epitaxy). By properly etching the seed surface before growth is initiated, they may be greatly minimized but are seldom completely eliminated. Because of the well-defined geometrical relation that exists between the surface outline of the faulted region and its

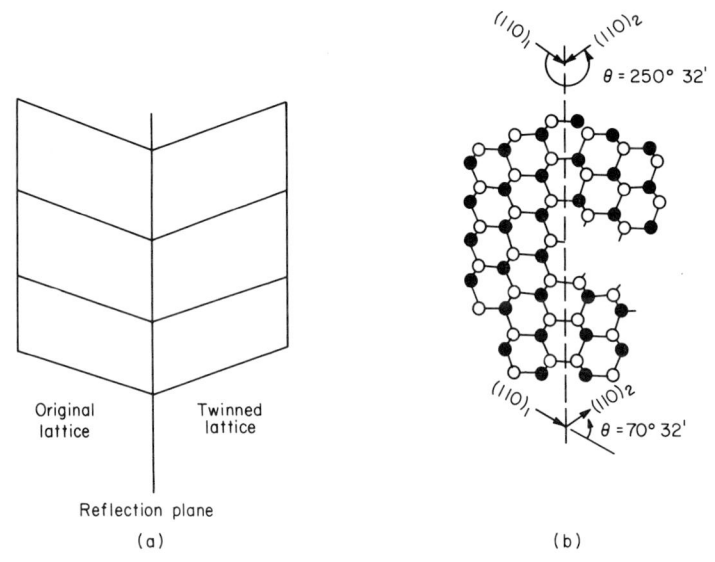

Fig. 10 Twin structures. The upper portion of (b) shows a zinc blende twin with nearest neighbors across the twin of correct component, while the lower portion shows one with the nearest neighbors in the correct position, but the positions of the two constituents have been interchanged.[20]

point of origin, such measurements are sometimes used to determine growth rate or the thickness of epitaxial layers since most faults originate at the original growth interface.

Grain boundaries Grain boundaries occur when regions with gross differences in orientation grow together. Various orientations can arise when randomly nucleated regions grow together, or when regions again come in contact after several intervening twinning steps. Because of these large differences, there will be a high density of dislocations at the interface. Thus, if the boundary is subjected to a dislocation etch, the spacing of dislocations will be so close together that individual etch pits will not be separable, and a groove will result.

Lineage Lineage is used to describe mosaic regions with small-angle deviations from one another. The intersection of two such blocks gives a line of dislocations closely spaced but individually distinguishable. Such a definition is admittedly subjective and may vary from observer to observer. In order to promote clarity, ASTM has established the specifications for lineage in silicon as being a linear

7-16 Semiconductor Materials

density of dislocations greater than 25 per mm along a line of 0.5 mm minimum length.[22]

Separate phases (same composition) When growing materials which may have either the zinc blende or wurtzite structure, both phases sometimes occur in the same crystal. Unequivocable determination of such regions is complicated because of the usual small differences between diffraction patterns of the two modifications.

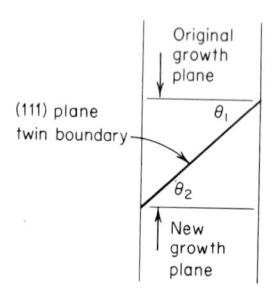

Schematic diagram of twinning geometry

The indices of the new plane with respect to the axes of the old one may be calculated from the following three simultaneous equations:*

$$ht + ku + lv = -hp - kq - lr$$
$$kt - hu = kp - hq$$
$$lt - hv = lp - hr$$

where (hkl) = indices of the twin plane
(pqr) = indices of the original plane
(tuv) = indices of the new plane

or from the appropriate twinning matrix:†

$$\begin{bmatrix} t \\ u \\ v \end{bmatrix} = \begin{bmatrix} m_{11} & m_{12} & m_{13} \\ m_{21} & m_{22} & m_{23} \\ m_{31} & m_{32} & m_{33} \end{bmatrix} \begin{bmatrix} p \\ q \\ r \end{bmatrix}$$

where $[M]$ for the various (111) twin planes is given below:

$$[111] \text{ or } [\bar{1}\bar{1}\bar{1}] \quad \begin{bmatrix} -1 & 2 & 2 \\ 2 & -1 & 2 \\ 2 & 2 & -1 \end{bmatrix} \qquad [\bar{1}11] \text{ or } [1\bar{1}\bar{1}] \quad \begin{bmatrix} -1 & -2 & -2 \\ -2 & -1 & 2 \\ -2 & 2 & -1 \end{bmatrix}$$

$$[1\bar{1}1] \text{ or } [\bar{1}1\bar{1}] \quad \begin{bmatrix} -1 & -2 & 2 \\ -2 & -1 & -2 \\ 2 & -2 & -1 \end{bmatrix} \qquad [11\bar{1}] \text{ or } [\bar{1}\bar{1}1] \quad \begin{bmatrix} -1 & 2 & -2 \\ 2 & -1 & -2 \\ -2 & -2 & -1 \end{bmatrix}$$

* Chester B. Slawson, Twinning in the Diamond, Am. Mineral., vol. 35, pp. 193–206, 1950.

† K. F. Hulme and J. B. Mullin, Indium Antimonide, Solid State Electron., vol. 5, pp. 211–274, 1962.

Fig. 11 Twinning geometry.

Crystals such as silicon carbide which have a wide variety of hexagonal stacking sequences often have included regions of different stacking, though they remain hexagonal.

Inclusions Inclusions of foreign materials usually but not always give rise to gross polycrystallinity. The exceptions appear to be in those cases in which the included material does not itself promote nucleation, and is not suddenly introduced in a manner which leads to excessive supercooling or supersaturation. Gas bubbles of rather large size have been observed in melt-grown silicon crystals. Germanium crystals several inches in diameter have been grown (from the melt) around graphite plugs, and crystals of water-soluble salts grow very well around support rods or strings. Vapor-phase-grown gallium arsenide will cover substantial areas of SiO_2 without becoming polycrystalline. Silicon phosphide crystallites have been observed in heavily phosphorus-diffused regions of silicon.

Individual Properties

Bismuth telluride[23] Bismuth telluride has the formula Bi_2Te_3, a complex rhombohedral structure, and very anisotropic thermal and electrical properties. Fine-grained

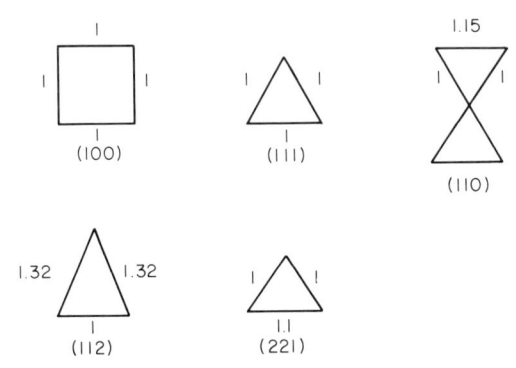

Fig. 12 Stacking-fault outlines on various planes [for (111) fault system]. The numbers by the sides of the outlines give the ratios of the lengths.

polycrystalline material is produced either by sealing the constituent elements in an evacuated fused silica capsule, heating above the melting point, and quenching, or by sintering a mixture of component powders under pressure. Single crystals can be grown by zone leveling or Czochralski pulling.

Lattice constants:
  Rhombohedral.......................... $a = 10.473$ Å, $\alpha = 24°9'32''$
Density................................ 7.8587 g/cc
Melting point.......................... 585°C
Latent heat of fusion.................. 29.0 kcal/mole
Thermal conductivity at room temperature:
  Perpendicular to c axis................ 0.015 watt/(cm)(°C)
  Parallel to c axis..................... 0.007 watt/(cm)(°C)
Band gap at room temperature........... 0.20 ev
μ_n at room temperature................... 1,200 cm²/volt-sec
μ_p at room temperature................... 500 cm²/volt-sec
Room temperature resistivity........... 0.007 ohm-cm

Bismuth telluride is used primarily as a thermoelectric material.

Boron[24,25] Boron is a very hard material with a dark, nonmetallic luster. It is a semiconductor with a band gap in the 1.5- to 2.0-ev range. At least three crystallographic modifications have been reported: low-temperature α-rhombohedral, tetragonal, and high-temperature β-rhombohedral. If crystal growth is below 1200°, the alpha modification usually occurs; between 1100° and 1300°, the tetragonal; and above 1500°, the beta. The details of crystal growth are still not well developed, and large crystals are quite scarce. Vapor deposition of a polycrystalline rod from a halide followed by vertical zone leveling has thus far been the most successful method. Purification is most easily achieved before deposition by distillation of the halide.

7-18   Semiconductor Materials

Lattice constants:
  α rhombohedral............ $a = 5.06$ Å, $\alpha = 58°4'$
  β rhombohedral............ $a = 10.12$ Å, $\alpha = 65°28'$
  Tetragonal................ $a = 8.75$ Å, $c = 5.06$ Å
Density....................... 4.82 g/cc
Melting point................. 2000°C
Hardness...................... 2,400–2,500 Knoop
Breaking strength (bending)... 76,000 psi
Band gap...................... 1.5–2 ev
Etchants...................... Water and red fuming nitric acid (1:1), 20% KOH + 20% $K_3Fe(CN)_6$

Because of the large activation energy, boron appears suitable for use as a thermistor material.

Cadmium sulfide[26,27]   Cadmium sulfide is a brittle material which may vary in color from nearly transparent to orange to black, depending on the kind and amount of impurities present. Single crystals are usually grown by sublimitation from CdS powder. A temperature gradient is required, but the approximate temperature is 1200°C. Under a 100-atm pressure, they have been grown from the melt, but this method is seldom used. Both cubic and hexagonal modifications occur, though the cubic form is reported to be unstable.

Lattice constants:
  Zinc blende................ $a = 5.82$ Å
  Wurtzite................... $a = 4.1368$ Å, $c = 6.7163$ Å
Density....................... 4.82 g/cc
Melting point................. Sublimes under 1 atm
μ_e at room temperature... 210 cm²/volt-sec

Cadmium sulfide is most used as a thin-film photoconductor. Applications include light meters and various other optical sensors designed to operate in the visible spectrum. Thin-film CdS polycrystalline solar cells with efficiencies of a few percent have also been made, though surface stability is a problem.

Diamond[28]   Two types of diamond have now been reported. Type I behaves as an insulator, whereas the considerably rare Types II has semiconductor properties. The crystallographic perfection of both types is very poor when compared to that of the more conventional semiconductors such as silicon and germanium, and appears to have a mosaic structure. Furthermore, all diamonds appear to be highly strained. The room temperature resistivity of Type I is of the order of 10^8 ohm-cm. Type II diamonds are p-type and have resistivities in the 10- to 100-ohm-cm range. The impurity level(s) appear to be about 0.34 ev above the valence band.

Lattice constant:
  Diamond.................... $a = 3.5668$–3.5672
Density....................... 3.52 g/cc
Melting point................. Sublimes at atmospheric pressure
Hardness...................... 10,000–50,000 Vickers
Band gap...................... 5.6 ev
μ_n at room temperature... 1,250 cm²/volt-sec
Etchant....................... Molten potassium nitrate

The semiconducting form of diamond is quite rare and has not been used commercially. Because of a high thermal conductivity coupled with good electrical insulating properties, ordinary diamonds are occasionally used as mounting blocks for high-power semiconductor devices.

Gallium arsenide[29,30]   GaAs is a dark gray material with the zinc blende structure. It is made by melting two constituents together in a fused silica boat. Because of the relatively high vapor pressure of arsenic over GaAs at the melting point, crystal growing from the melt must be in some variety of sealed system. Both Czochrolski and zone leveling work satisfactorily. In addition, layers are grown both from the vapor using various disproportionation reactions and from low-temperature gallium melts.

Lattice constant:
 Zinc blende .... $a = 5.654$ Å
Density .... 5.316 g/cc
Melting point .... 1238°C
Band gap at room temperature .... 1.47 ev
μ_n at room temperature (cal. max.) .... 11,000 cm²/volt-sec
μ_p at room temperature (cal. max) .... 450 cm²/volt-sec
n_i at room temperature .... ~10⁶ cm⁻³

Gallium arsenide was originally heralded as a high-temperature transistor material. Instead, its primary uses have arisen because of its band structure and high mobility. Applications include varactor diodes, Schottky barrier diodes, light-emitting diodes and injection lasers, and Gunn and LSA mode oscillators.

Germanium Germanium is a brittle silvery gray metallic element with an atomic number of 32 and atomic weight of 72.6[31]. It is in Group IV of the periodic table, along with carbon, silicon, tin, and lead. It is prepared by hydrolyzing $GeCl_4$ with high-purity water to GeO_2 and then reducing the oxide with hydrogen. Crystal growth from the melt is by zone leveling or Czochralski, and from the vapor either by the thermal decomposition of germane or by the hydrogen reduction of $GeCl_4$. At room temperature there is a little indication of plastic flow, and along with a hardness of 6 on the Mohs scale it behaves as a brittle material.

Lattice constant at 20°C:
 Diamond .... $a = 5.257$ Å
Density .... 5.323 g/cc
Melting point .... 937.4°C
Hardness .... 6 Mohs
Band gap at 300°K .... 0.665 ev
n_i^2 at 300°K .... 5.6×10^{26} cm⁻⁶

During World War II, germanium was investigated for possible use as a microwave detector. The transistor, invented in 1948, originally used germanium. At the present time, germanium usage in transistors and diodes is steadily declining in favor of silicon. It is, however, expected to have continued usage in those devices which require very low forward-voltage drop, e.g., low-voltage inverter applications, photodetectors for the 1.1- to 2-micron range, and some high-frequency applications. It might be noted that although the tabulated mobilities for electrons and holes in germanium are greater than those of silicon (and this should favor germanium for high-frequency applications), most very high-frequency devices operate in regions of high fields, where the mobility values are quite close together.

Gray tin[26,32] Tin has two modifications. A gray semiconductor cubic (α) form stable below 13.2°C and a metallic white tetragonal structure (β) stable from 13.2°C to the melting point (231.9°C). It may be purified by zone refining. Crystals of white tin may be grown from the melt by either the Czochralski or the Bridgeman method. Cubic crystals, however, must be grown below the transformation temperature of 13°C. The most successful way of doing this has been to grow them from a tin-saturated mercury solution. Growth rates are in the inches-per-month range. It may be doped n-type or p-type by using Group III or Group V elements. Antimony and aluminum have been used most often.

Lattice constants:
 Diamond α .... $a = 6.5041$ Å
 Tetragonal β .... $a = 5.8313$ Å, $c = 3.1812$ Å
Atomic number 50
Atomic weight 118.70
Density:
 Cubic .... 5.75 g/cc
 Tetragonal .... 7.28 g/cc
Band gap .... 0.0 ev
Room temperature resistivity .... 2×10^{-4} ohm-cm
$\mu_e = 2,500(T/300)^{-1.65}$ cm²/volt-sec
$\mu_n = 2,400(T/300)^{-2}$ cm²/volt-sec
Etchant .... Potassium chlorate, HCl

Indium antimonide (InSb)[30,33,34]   Indium antimonide is a bright, shiny material with a relatively low melting point of 523°C. It is compounded by the direct fusion of stoichiometric quantities of indium and antimony, both of which will usually have been previously zone-refined. Further zone refining after compounding is required, after which single crystals may be grown in the same boat by zone leveling. Alternatively, the material can be transferred to either another zone leveler or to a Czochralski puller. H_2 is the usual atmosphere. For doping, Group VI elements are used for n-type, Group II elements for p-type.

Lattice constant:
   Zinc blende.......................... $a = 6.48$ Å
Density................................ 5.789 g/cc
Melting point.......................... 523°C
Hardness.............................. 240 Vickers
Band gap............................. 0.17 ev
μ_n at room temperature............... 270,000 cm²/volt-sec
μ_n at liquid nitrogen temperature....... 7×10^5–10^6 cm²/volt-sec
μ_p.................................... ~750 cm²/volt-sec
n_i at room temperature................ ~1.6×10^{16} cm^{-3}

The primary use of indium antimonide is in the construction of photovoltaic infrared detectors. However, it can also be used in either the photoconductive or the PEM mode. It has a very large magnetoresistive effect which has been utilized in the construction of various displacement gages and variable resistors. Because of the high mobility, the Hall voltage is also quite large. For this reason InSb Hall bars have been used for such applications of flux meters and analog multipliers.

Lead sulfide (PbS)[34]   Lead sulfide is a dark metallic-appearing material which can occur naturally as large crystals (galena). Single crystals may be grown by the Bridgeman method, and polycrystalline layers are formed by vacuum evaporation. Thin single-crystal epitaxial layers can be formed by evaporation onto a suitable substrate such as rock salt.

Lattice constant:
   FCC............................. $a = 5.9$ Å
Density............................... 7.5 g/cc
Mobility at room temperature:
   μ_n................................ 600 cm²/volt-sec
   μ_p................................ 700 cm²/volt-sec
Band gap............................. 0.34–0.37 ev
n_i at room temperature................ 10^{15} cm^{-3}
Carrier lifetime....................... 10^{-10}–10^{-5} sec

A point-contact diode made from a wire whisker and naturally occurring galena was used as an rf detector for many years, but more recently the primary use has been in infrared detectors.

Selenium[34]   Selenium is found in Group VI, subgroup oxygen, in the same series as sulfur and tellurium. The atomic number is 34 and the atomic weight 78.96. Selenium exists in monoclinic, hexagonal, and amorphous forms. The monoclinic form in turn has both an α and β modification. Hexagonal selenium is gray or metallic-looking and consists of zigzag chains running parallel to the z axis. The amorphous variety begins to soften at 50 to 60°C and to flow at 100°C.

Density:
   Hexagonal form.......... 4.81 g/cc
Melting point:
   Hexagonal form.......... 217°C
Boiling point............... 688°C
Hardness.................. 2 mohs
Young's modulus.......... 5,550 kg/cm²

Selenium is used in rectifiers and in photoconductive applications. The output of selenium is relatively small. For example, in 1958 the United States production was estimated at only 300 tons.

Silicon[35] Silicon, from the Latin *silex* for flint, is the 14th element of the periodic series and was first reported by Berzelius in 1817. Elemental silicon has never been found naturally, but in various minerals such as silica and the silicates, it accounts for approximately 25 percent of the earth's crust. It has a gray metallic luster, and an atomic weight of 28.083. Low-purity silicon is made by the reduction of quartz sand with coke. It is then converted to a halide; usually either $SiCl_4$ or $SiHCl_3$, purified by repeated distillation, and then reduced with hydrogen. Crystal growth from the melt is either by Czochralski pulling or float zoning. Vapor growth is normally by the hydrogen reduction of $SiHCl_3$ or $SiCl_4$ or the thermal decomposition of SiH_4.

Silicon has a hardness intermediate between that of germanium and quartz and is normally considered a brittle material. There is a ductile region above 800°C, but work hardening occurs and temperatures near 1300°C are required for appreciable deformation.

Lattice constant:
  Diamond.............................. $a = 5.42$ Å
Atomic radius............................ 1.17 Å
Density................................. 2.3290 g/cc
Melting point........................... 1412°C
Boiling point........................... 3145°C
Heat of fusion.......................... 12.1 kcal/mole
Surface tension at freezing point....... 720 dynes/cm
Band gap at 300°K....................... 1.107 ev
n_i at room temperature............... 1.37×10^{10} cm^{-3}

Silicon is used in the production of integrated circuits; rectifiers, diodes, transistors, SCR's and Triacs;* positive-temperature-coefficient resistors, solar cells, and near-infrared detectors.

Silicon carbide[36] Silicon carbide, SiC, is primarily a man-made compound and is known to occur naturally only in meteor specimens. At atmosphere pressure, it decomposes at 2830°C without melting. Depending on the impurities present, its color may vary from black to green to orange to pale blue, to virtually clear. Manufacturing of the low-purity industrial grade is usually by the direct reaction of silicon and carbon in an electric furnace. Single crystals may be grown from a carbon-rich silicon melt, by sublimitation of a polycrystalline charge, or by various vapor-phase reactions between silicon and carbon-bearing compounds, or by the thermal decomposition of some volatile material containing both carbon and silicon.

The crystal structure of silicon carbide is close-packed, but because of variations in stacking sequences, a wide variety of hexagonal polytypes as well as a zinc blende form occur.

Lattice constants:
  Cubic form........ $a = 4.3596$ Å
  Hexagonal form... $a = 3.0806$ Å
                    $c =$ multiples of 2.52
Hardness............ 9 Mohs
Band gap:
  Cubic............. 2.2 ev
  2H................ >3 ev
  4H................ 3.1 ev
  6H................ 2.9 ev
  21H............... 2.8 ev
Etchants............ Molten salts such as NaOH and Na_2CO_3
                     and gaseous chlorine at 1000°C

Uses: The oldest and still by far the largest use of silicon carbide is as an abrasive, but it has also been utilized as a semiconductor for many years. Earliest of these was the bonded varistor, but single-crystal diodes and rectifiers have recently been marketed. Because of its high thermal conductivity it also finds some application as a heat sink for high-power semiconductor devices.

* Trademark of General Electric Company, Schenectady, N.Y.

Semiconductor Materials

Tellurium[37] Tellurium is a white, metallic-looking element with atomic number 52 and atomic weight 127.61. Single crystals can be grown rather easily by either the Czochralski or the Bridgeman method. It may be purified by distillation. The thermal, elastic, and electrical properties are all highly anisotropic.

Lattice constants:
  Hexagonal form .... $a = 4.44$ Å, $c = 5.912$ Å
Density .... 6.25 g/cc
Melting point .... 452°C
Boiling point .... 1390°C
Hardness .... 2–3 Mohs
Band gap .... 0.32–0.38 ev
Intrinsic resistivity at room temperature:
  Parallel to c axis .... 0.29 ohm-cm
  Perpendicular to c axis .... 0.59 ohm-cm
n_i (average of 3 parallel-perpendicular axes) at room temperature 9.3×10^{15} cm^{-3}

Elemental tellurium has been little used as a semiconductor, though thin-film tellurium transistors have been reported. It finds more application as a constituent of various semiconducting compounds, e.g., the thermoelectric material Bi$_2$Te$_3$, and photoconductors such as PbTe.

TABLE 7 Band Gaps of Various Materials (in electron volts)

Alphabetical				Numerical			
AlAs	2.16	HgTe	0.025	HgTe	0.025	InP	1.27
AlN	3.8	InAs	0.36	PtSb$_2$	0.07	Bi$_2$S$_3$	1.3
AlP	3.0	InN	2.5	Cd$_3$As$_2$	0.13	P	1.45
AlSb	1.6	InP	1.27	InSb	0.17	GaAs	1.47
As$_2$Se$_3$	1.6	InSb	0.17	Bi$_2$Te$_3$	0.20	CdTe	1.50
As$_2$Te$_3$	1.0	In$_2$S$_3$	2.28	PbSe	0.26	B	2.0
B	2.0	In$_2$Se$_3$	1.25	Bi$_2$Se$_3$	0.28		1.53
	1.53	In$_2$Te$_3$	1.0	PbTe	0.29	AlSb	1.6
BAs	5	Mg$_2$Ge	0.74	Sb$_2$Te$_3$	0.30	As$_2$Se$_3$	1.6
Bi$_2$S$_3$	1.3	Mg$_2$Si	0.77	Te	0.32	CdSe	1.7
Bi$_2$Se$_3$	0.28	Mg$_2$Sn	0.36	InAs	0.36	Sb$_2$S$_3$	1.7
Bi$_2$Te$_3$	0.20	P	1.45	Mg$_2$Sn	0.36	Se	1.8
BN	4.6	PbS	0.37	PbS	0.37	Ca$_2$Si	1.9
BP	6.0	PbSe	0.26	Ca$_2$Pb	0.46	HgS	2.0
Ca$_2$Pb	0.46	PbTe	0.29	CdSb	0.46	AlAs	2.16
Ca$_2$Si	1.9	PtSb$_2$	0.07	ZnSb	0.53	GaP	2.25
Ca$_2$Sn	0.9	S	2.65	HgSe	0.6	ZnTe	2.26
CdAs$_2$	1.0	Sb$_2$S$_3$	1.7	Ca$_2$Sn	0.9	In$_2$S$_3$	2.28
Cd$_3$As$_2$	0.13	Sb$_2$Se$_3$	1.2	Ge	0.66	InN	2.5
CdS	2.59	Sb$_2$Te$_3$	0.30	GaSb	0.68	CdS	2.59
CdSb	0.46	Se	1.8	Mg$_2$Ge	0.74	S	2.65
CdSe	1.7	Si	1.107	Mg$_2$Si	0.77	ZnSe	2.67
CdTe	1.50	SiC	2.8	ZnAs$_2$	0.90	SiC	2.8
GaAs	1.47	Te	0.32	Zn$_3$As$_2$	0.93	AlP	3.0
GaN	3.3	ZnAs$_2$	0.90	In$_2$Te$_3$	1.0	GaN	3.3
GaP	2.25	Zn$_3$As$_2$	0.93	As$_2$Te$_3$	1.0	ZnS	3.58
GaSb	0.68	ZnS	3.58	CdAs$_2$	1.0	AlN	3.8
Ge	0.66	ZnSb	5.3	Si	1.107	BN	4.6
HgS	2.0	ZnSe	2.67	Sb$_2$Se$_3$	1.2	BAs	~5
HgSe	0.6	ZnTe	2.26	In$_2$Se$_3$	1.25	BP	6.0

Properties of Semiconductors 7-23

Index of Properties

		Page
Absorption coefficients:		
GaAs	Fig. 13	7-24
Ge	Fig. 13	7-24
Si	Fig. 13	7-24
Band gap at room temperature:		
Tabulation	Table 7	7-22
Variation with temperature:		
Ge	Fig. 14	7-25
Si	Fig. 14	7-25
Breakdown voltage	Fig. 14	7-26
Density	Table 8	7-26
Diffusion coefficients:		
Impurities in Ge	Table 9, Fig. 16	7-27, 7-28
Impurities in Si	Table 9, Fig. 17	7-27, 7-28
Impurities in III-V's	Table 10, Fig. 18, Fig. 19	7-30, 7-31, 7-32
Elastic constants	Table 11	7-31
Hardness	Table 12	7-32
Impurity levels	Fig. 20	7-33
Intrinsic carrier concentration:		
Room temperature tabulation	Table 13	7-34
As a function of temperature:		
Ge	Fig. 21	7-34
GaAs	Fig. 21	7-34
InSb	Fig. 21	7-34
Si	Fig. 21	7-34
Latent heat of fusion	Table 14	7-35
Lattice spacing	Table 15, 16	7-35, 7-37
Lifetime:		
Versus impurity concentration (silicon)	Fig. 22	7-36
Versus dislocation density (silicon)	Fig. 23	7-37
Versus gold concentration (silicon)	Fig. 24	7-38
Melting points	Table 17	7-38
Mobility of carriers:		
Tabulation	Table 18	7-39
GaAs	Fig. 25	7-38
Ge	Fig. 25	7-38
Si	Fig. 25	7-38
Piezoresistance coefficients	Table 19	7-40
Refractive index tabulation	Table 20	7-41
Refractive index versus wavelength:		
Si	Fig. 26	7-40
Ge	Fig. 26	7-46
Resistivity versus impurity concentration:		
GaAs	Fig. 27	7-42
Ge	Fig. 28	7-45
Si	Fig. 29	7-46
Segregation coefficients	Table 21, 22	7-43, 7-44
Solid solubility limits:		
Elements in Ge	Fig. 30	7-47
Elements in Si	Fig. 30	7-47
Surface energy	Table 23	7-45
Surface tension at melting point	Table 24	7-46
Thermal conductivity:		
Room temperature tabulation	Table 25	7-48
As a function of temperature	Fig. 32	7-49
As a function of composition	Fig. 33	7-50
Thermal expansion coefficients:		
Tabulation at room temperature	Table 26	7-48
GaAs	Fig. 34	7-50
Ge	Fig. 34	7-50
InSb	Fig. 34	7-50
Si	Fig. 34	7-50

Semiconductor Materials

MEASURING TECHNIQUES

Orientation Determination

Semiconductor processing steps are almost always designed for use on a specific crystallographic plane. As examples, alloy devices are usually fabricated on the (111) plane so that the alloy front will remain parallel to the wafer surface. Selec-

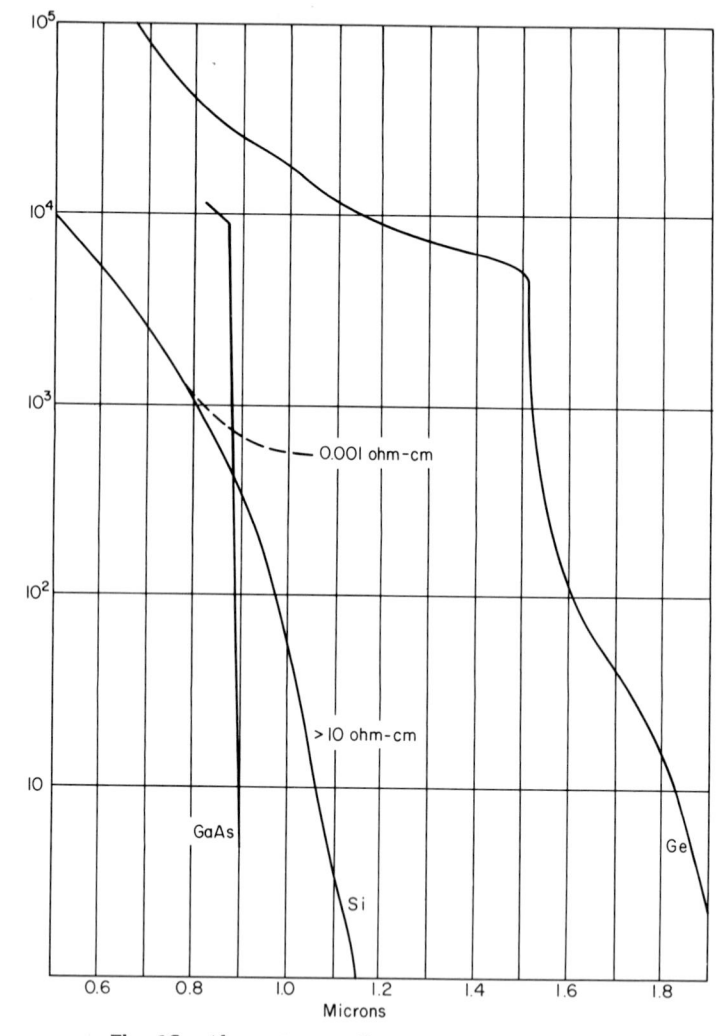

Fig. 13 Absorption coefficient versus wavelength.

tive etchants are sometimes used which perform better on specific orientations, and epitaxial growth is more controllable when grown in preferred directions.

Before being sliced, a crystal may be roughly oriented by referring to markings which may have developed during growth. This method is particularly applicable to Si, Ge, and most III-V's grown in or near ⟨111⟩ or ⟨100⟩ directions since ordinarily only {111} faces occur, and it is virtually impossible to grow crystals without some trace of them being visible. These may be in the form of large, well-developed planes, or as ridges where a series of tiny facets have formed along the edge of a crystal. Figure 35 shows sketches of crystals grown in ⟨111⟩ and ⟨100⟩ directions and may be used as a guide in orienting.

For more accurate orientation, etch-pit or stacking-fault observations or x-ray diffraction may be used. Difficulty in interpretation restricts the use of etch pits and stacking faults to the simple planes such as (111), (100), and (110), but x-rays may be used to determine any random orientation. The use of etch pits depends on the availability of a very selective etch which preferentially exposes a given orientation. For diamond and zinc blende structures, etchants specifically for (111) planes are most readily available and commonly used. For this class of etchant, if the surface were originally near a (111) plane, there would be little three-sided inverted pyramids etched in the surface. For a (100) plane, four-sided pyramids will result. In the case of (110)s a diamond-shaped aperture is formed. It should be noted that overetching may expose planes other than (111) and in some cases can lead to erroneous interpretation.

The etch-pit shape can be observed in a microscope and used to make an approximate determination of plane:[38] i.e., equilateral triangle near (111), square near (100), diamond near (110); but the most useful aspect is that it allows directions in the plane of a thin slice to be rather easily determined (a thing not readily done otherwise). These directions are shown in Fig. 36. More accurate optical orientation can be accomplished by observing the pattern reflected back from a beam of collimated light impinging normal to the surface.[39] Typical patterns are shown in Fig. 37 along with directions in the plane of the crystal. If the actual surface is slightly misoriented from the crystallographic plane, the image will be asymmetrical. By measuring the amount of tilt necessary to bring the pattern back to symmetry, the amount of misorientation is directly determined.

If stacking faults are visible, they too can be used for orientation purposes. In the diamond and zinc blende structures such faults occur along (111) planes, and hence their intersection with the surface produces the same pattern as that of an etch pit. They do have the advantage, however, of not rounding off as etch pits often do. To obtain the sharpest definition, use the shortest possible etch time required to clearly define the fault outline.

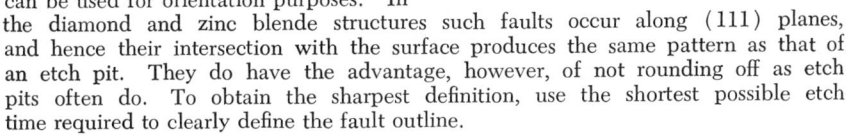

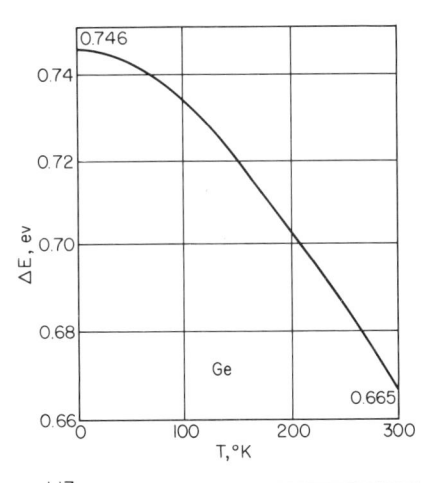

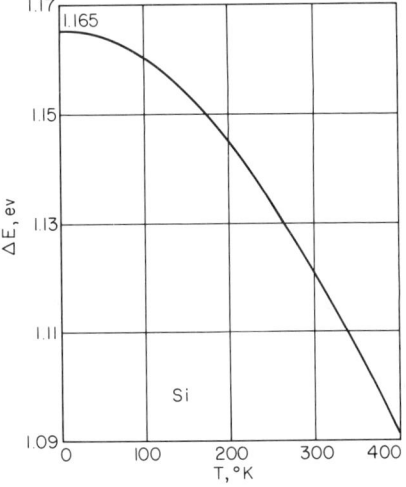

Fig. 14 Variation of band gap with temperature. (*From R. A. Smith, "Semiconductors," Cambridge University Press, London, England, 1961, pp. 351 and 352.*)

7-26 Semiconductor Materials

Stacking faults coming to a (111) silicon surface may be revealed by etching in Sirtl etch, dislocation copper etch, or 1-3-10. Usual Sirtl etch times are 2 to 30 sec; dislocation copper etch and 1-3-10 etch times are 20 to 120 min. Stacking faults in (110) and (100) silicon are delineated by use of a Sirtl etch or Dash etch. Stacking faults in (111), (110), and (100) germanium are delineated by etching 2 to 30 min in ferricyanide etch at 80°C. Table 27 gives etch compositions.

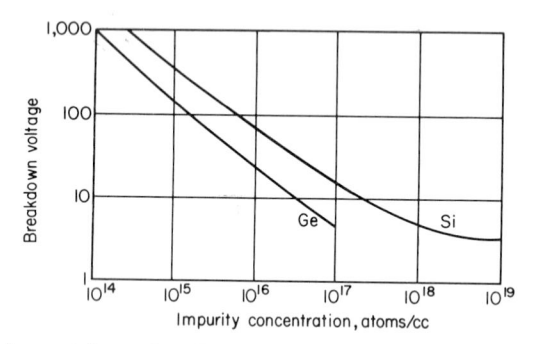

Fig. 15 Breakdown voltage for abrupt junction diodes. From 10^{15} to 10^{18}, GaAs values are very close to those of Si.[97]

For the diamond lattice ($\bar{1}\bar{1}\bar{1}$) and (111) faces are equivalent, but for the zinc blende case they are not, and thus additional provisions must be made for distinguishing between them. This is usually done by visually observing the surface after an appropriate etch.

TABLE 8 Densities of Various Materials

Density	g/cc	Density	g/cc
AlAs	3.81	HgSe	8.25
AlP	2.85	HgTe	8.42
AlSb	4.218	InAs	5.7
CdO	8.238	InP	4.787
CdS	4.820	InSb	5.7751
CdSe	5.81	Si	2.329
CdTe	6.06	SiC	3.21
Diamond	3.51	ZnO	5.664
GaAs	5.307	ZnS	4.1
GaSb	5.619	ZnSe	5.42
Ge	5.327	ZnTe	5.72

X-ray orientation is usually done either by Laue patterns or by the use of an x-ray goniometer.[40] The former is most useful when the orientation is completely unknown, and an accuracy no greater than ±1° or 2° is desired. Interpretation is somewhat involved and ordinarily restricts usage to x-ray laboratories. Goniometers are used when the orientation is approximately known and a precise determination is required. These instruments are direct-reading and may be used for production control. The principle is as follows: If the wavelength of the x-rays used and the spacing between the desired plane are known, the angle at which coherent

TABLE 9 Diffusion Coefficients for Elements in Germanium and Silicon*

Group	Impurity		D	Group	Impurity		D
			Germanium				*Silicon*
I	Li	n	$0.0025 \exp(-0.51/kT)$	I	Li	n	$0.0023 \exp(-0.66/kT)$
	Cu	p	$3 \cdot 10^{-5}$ ($T1173$)		Cu†		$4.7 \times 10^{-3} \exp(-0.43/kT)$ interstitial
	Ag	p	$0.044 \exp(-1.0/kT)$		Au‡		$2.4 \times 10^{-4} \exp(-0.387/kT)$ interstitial
	Au	n	$12.6 \exp(-2.25/kT)$				$2.75 \times 10^{-3} \exp(-2.04/kT)$ substitutional
II	Be	p	$<10^{-11}$ ($T1123$)	II	Zn	p	$1 \cdot 10^{-6} - 1 \cdot 10^{-7}$ ($1273 < T < 1573$)
	Zn	p	$0.65 \exp(-2.5/kT)$				
	Cd	p	$1.75 \cdot 10^9 \exp(-4.4/kT)$	III	B	p	$10.5 \exp[(-3.6 \pm 40\%)/kT]$
			—		Al	p	$8.0 \exp[(-3.4 \pm 40\%)/kT]$
					Ga	p	$3.6 \exp[(-3.5 \pm 40\%)/kT]$
					In	p	$16.5 \exp[(-4.0 \pm 40\%)/kT]$
					Tl	p	$16.5 \exp[(-4.0 \pm 40\%)/kT]$
III	B	p	$1.8 \cdot 10^9 \exp(-4.55/kT)$	IV	Ge		$6.26 \cdot 10^5 \exp(-5.3/kT)$
	Al	p	—				
	Ga	p	$3.5 \cdot 10^{-13}$ ($T1123$)	V	P	n	$10.5 \exp[(-3.6 \pm 40\%)/kT]$
	Tl	p	$7.0 \cdot 10^{-14}$ ($T1123$)		As	n	$0.32 \exp[(-3.6 \pm 40\%)/kT]$
	In	p	$20 \exp(-3.0/kT)$		Sb	n	$5.6 \exp[(-3.9 \pm 40\%)/kT]$
					Bi	n	$10.30 \exp[(-4.7 \pm 40\%)/kT]$
IV	Ge		$(7.8 \pm 3.4) \exp(-2.98/kT)$	VI	S	n	$0.92 \exp(-2.2/kT)$
V	P	n	$2.4 \cdot 10^{-11}$ ($T1123$)	VII	Mn	n	$\gg 2 \cdot 10^{-7}$ ($T1473$)
	As	n	$2.12 \exp(-2.4/kT)$				
	Sb	n	$(1.3 \pm 1.0) \exp[(-2.26 \pm 0.07)/kT]$	VIII	Fe	n	$0.0062 \exp(-0.87/kT)$
	Bi	p	$4.7 \exp(-2.4/kT)$				
VI	O	n	$\sim 10^{-8}$				
	S	n	$1 \cdot 10^{-9}$ ($T1193$)				
	Se	n	$\sim 10^{-10}$ ($T1193$)				
	Te	n	$\sim 10^{-11}$ ($T1193$)				
VII	Fe	p	$0.13 \exp(-1.1/kT)$				
	Ni	p	$0.8 \exp(-0.91/kT)$				

There is a range of D values reported for most impurities. The ones presented here are representative, but for a more complete listing of the variations, see, for example, A. Seeger and K. P. Chik, Diffusion Mechanisms and Point Defects in Silicon and Germanium, *Phys. Stat. Solids*, vol. 29, p. 455, 1968.

* C. Benoit a la Guillaume et al., "Selected Constants Relative to Semiconductors," Pergamon Press, New York, 1961.
† R. N. Hall and J. H. Racette, *J. Appl. Phys.*, vol. 35, p. 379, 1964.
‡ W. R. Wilcox and T. J. LaChapelle, *J. Appl. Phys.*, vol. 35, p. 240, 1964.

scattering or "reflection" is expected can be calculated from

$$\sin \theta = \frac{\lambda}{2d}$$

when the geometry is as in Fig. 38, λ is the x-ray wavelength, and d the crystal lattice spacing. Therefore, if the detector is preset at an angle 2θ appropriate for receiving x-rays from a given plane, and the crystal rotated independently until a maximum does occur, the difference in angle between the crystal face and θ represents the amount of misorientation of the surface from the plane. Goniometers made specifically for orientation purposes have provisions for separately varying the crystal angle and the detector angle, but those designed primarily for spacing studies of powder samples will have the two movements mechanically coupled together so that when the crystal holder moves to any given angle θ, the detector is automatically positioned to 2θ. The latter equipment cannot be used for orientation work unless the decoupling can be arranged. The wavelength to be used is determined by the x-ray tube, and d values for any given orientation can be calculated from the lattice constants of the material being used (see section on properties), and the expressions of Table 2.

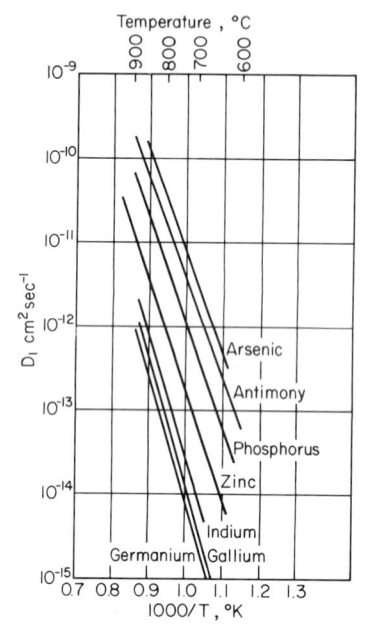

Fig. 16 Diffusion coefficients for impurities in germanium.[14]

Defect Observation

The region near a dislocation line usually etches more rapidly than the rest of the crystal, and thus affords a simple method of determining the number of dislocations which intersect the surface. The etchant must be rather carefully chosen in order to prevent etch pits developing from other causes, or at least to produce a characteristic pattern easily recognized under the microscope. Surface damage may produce etch pits, but they will usually become flat-bottomed as etching proceeds and the damage is removed. Dislocation pits, however, continue to etch rapidly for the length of the dislocation and thus usually maintain pointed bottoms. Because of its shape, a true dislocation etch pit will appear dark, whereas a portion of a surface-damage-induced etch pit will appear bright. Etch pit delineation is usually orientation-sensitive, and indeed many etches will work only on specific planes. Furthermore, for silicon and germanium, heavy doping reduces the effectiveness of the etches currently in use. For more highly doped samples as well as samples having non-surface-intersecting dislocations, x-ray topography is more suitable, as well as being nondestructive.[41,42] Table 27 lists a wide variety of etch formulations reported for dislocation delineation.

Twin lines can often be directly observed because of slightly different rates of growth of the two orientations. Sandblasting will usually produce an easily discernible difference because as the material fractures, the cleavage planes make different angles with the surface as the crystallographic orientation changes. This in turn produces a sharp change in the reflectivity as the twin plane is crossed. A similar effect is produced by a selective etch which will provide etch pits. Any etch suitable for crystal orientation is useful for the optical differentiation of twinned regions. Since the shape of etch pits depends on the orientation of the material, a microscopic examination of such pits may reveal small local twinned regions not otherwise visible.

Most solutions used for chemical-polishing a given semiconductor will also reveal polycrystalline areas, but the delineation of these areas can be enhanced by a proper etchant. In any case, the surface should be chemically polished first. Etching silicon in Sirtl, 1-3-10, or 1-3-6 for 0.5, 60.0, and 6.0 min, respectively, is satisfactory. The first two will also reveal dislocations; the latter one will not. Germanium polycrystalline areas may be made visible by etching in ferricyanide at 80°C, Wag, 1-1-1 or CP4 etch. Etch times are 20, 5, 1.5, and 2 min, respectively.

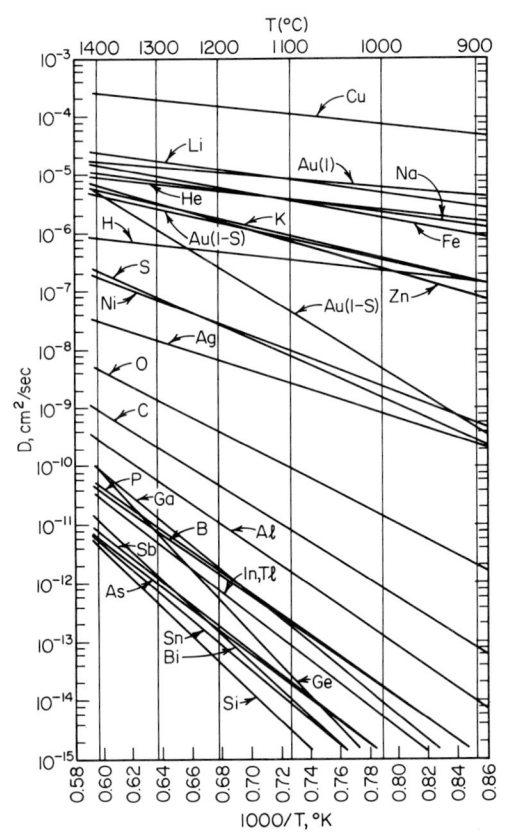

Fig. 17 Diffusion coefficients for impurities in silicon.[111]

Thin polycrystalline layers of silicon or germanium may be revealed by using the etches previously mentioned but decreasing the times to a minimum of one-fourth of those indicated.

Polycrystalline areas may also be delineated by lapping the surface in an aqueous slurry of number 1800 abrasive, or by sandblasting the surface with an abrasive whose particle size is less than 5 μm. Either preparation will present a surface whose polycrystalline areas appear as different shades of gray. It may be necessary to vary the position of the sample with respect to the incident light in order to see the polycrystalline areas.

Semiconductor Materials

TABLE 10 Reported Impurity D_0 and Q Values in III-V Compounds* [111]

Compound and diffusant	D_0, cm²/sec	Q, electron volts	Comments
AlSb:			
Zn	3.3×10^{-1}	1.93	Apparently independent of Zn conc., T
Cu	3.5×10^{-3}	0.36	T
GaP:			
Zn	1.0×10^{0}	2.1	T, +
GaAs:			
S	4.0×10^{3}	4.04	Large vaporization losses, T
S	2.6×10^{-5}	1.86	Al₂S₃ source with powdered GaAs, pn
S	1.2×10^{-4}	1.8	2 atm As pressure, pn
S	*1.6×10^{-5}*	*1.63*	Minimized vapor loss, ISR, T
Se	3.0×10^{3}	4.16	Ga₂Se₃ layer formed, T
Sn	6.0×10^{-4}	2.5	T, +
	3.8×10^{-2}	*2.7*	Average $\bar{D}$, T
Zn	1.5×10^{1}	2.49	T, +
	2.5×10^{-1}	*3.0*	Estimate for intrinsic GaAs, no excess As, T
	6.0×10^{-7}	0.6	Estimated $\bar{D}$ at 3×10^{19} cm⁻³, T
	3.0×10^{-7}	1.0	Average $\bar{D}$, ISR
Cd	5.0×10^{-2}	2.43	T, ++
	5.0×10^{-2}	*2.8*	Average $\bar{D}$, ISR
Mg	1.4×10^{-4}	1.89	ISR
	2.1×10^{-2}	2.5	ISR
	2.6×10^{-2}	*2.7*	Purest Mg, SR
Mn	*6.5×10^{-1}*	*2.49*	8.5×10^{-3} atm As
	8.5×10^{-3}	1.7	Average $\bar{D}$, no excess As, ISR
Cu	*3.0×10^{-2}*	*0.53*	Interstitial D, T
	1.0×10^{-3}	*0.53*	D estimate for int-sub mechanism
Tm	2.3×10^{-16}	(−)1.0	Retrograde with temperature, T
Ag	3.9×10^{-11}	0.33	Q, D_0 seem low, T
	2.5×10^{-3}	1.5	Artifact due to vapor reaction, T
	4.0×10^{-4}	*0.80*	T
Au	*1.0×10^{-3}*	*1.0*	T
Li	5.3×10^{-1}	1.0	Interstitial-substitutional pairing
GaSb:			
In	1.2×10^{-7}	0.53	T
Sn	2.4×10^{-5}	0.80	T
Te	3.8×10^{-4}	1.20	T
InAs:			
Mg	1.98×10^{-6}	1.17	pn
Zn	3.11×10^{-3}	1.17	pn, +
Cd	4.25×10^{-4}	1.17	pn
Ge	3.74×10^{-6}	1.17	pn
Sn	1.49×10^{-6}	1.17	pn
S	6.78×10^{0}	2.20	pn
Se	1.26×10^{1}	2.20	pn
Te	3.43×10^{-5}	1.28	pn, surface erosion
Cu		*0.52*	Interstitial D, T

* Values in italics are typical of diffusion at the low concentration limit.
T   Tracer.
pn p-n junction depth measurements.
+   Concentration dependence not taken into account.
ISR Incremental sheet resistance.
SR Sheet resistance and assumed distribution.
++ Average $\bar{D}$ typical of high concentration.

TABLE 10 Reported Impurity D_0 and Q Values in III-V Compounds*[111] (Continued)

Compound and diffusant	D_0, cm²/sec	Q, electron volts	Comments
InSb:			
Te	1.7×10^{-7}	0.57	T
Sn	5.5×10^{-8}	0.75	T
Zn	5.0×10^{-1}	1.35	T, ++, +
	1.6×10^{6}	2.3	Vapor diffusion, pn, +
	1.4×10^{-7}	0.86	$\bar{D}$ at lower conc., T
	8.7×10^{-10}	0.7	$\bar{D}$ at low conc., T
	9.0×10^{-10}	0	$\bar{D}$ at 2×10^{20} cm^{-3} Zn, T
	6.3×10^{8}	2.61	High Zn conc., T
	5.3×10^{7}	2.61	Lower Zn conc., T
	5.5×10^{0}	1.60	Vapor diffusion, pn, +
Cd	1.3×10^{-4}	1.2	T, ++, +
	1.23×10^{-9}	0.52	Average $\bar{D}$, T, +
	1.0×10^{-5}	1.10	T, +
	1.26×10^{0}	1.75	T, +
Hg	4.0×10^{-6}	1.17	T
Cu	9.0×10^{-4}	1.08	Surface vacancy-controlled, T, ISR
	3.0×10^{-5}	0.37	Dislocation-controlled, T
Au	7.0×10^{-4}	0.32	T
Co	1.0×10^{-7}	0.25	T
	2.7×10^{-11}	0.39	T, low D_0
Ag	1.0×10^{-7}	0.25	T
Fe	1.0×10^{-7}	0.25	T

* Values in italics are typical of diffusion at the low concentration limit.
T   Tracer.
pn p-n junction depth measurements.
+ Concentration dependence not taken into account.
ISR Incremental sheet resistance.
SR Sheet resistance and assumed distribution.
++ Average $\bar{D}$ typical of high concentration.

TABLE 11 Elastic Constants

(in dynes/cm²)

Material	C_{11} $\times 10^{11}$	C_{12} $\times 10^{11}$	C_{44} $\times 10^{11}$
Ge*	12.92	4.79	6.70
Si*	16.740	6.523	7.959
GaAs†	1.192	0.5986	0.538
GaSb†	8.85	4.04	4.33
InSb†	6.72	3.67	3.02

* E. M. Conwell, Properties of Silicon and Germanium, *Proc. IRE*, vol. 40, pp. 1327–1337, 1952.

† J. H. Westbrook, Mechanical Properties of Intermetallic Compounds: a Review of the Literature, in J. H. Westbrook (ed.), "Properties of Intermetallic Compounds," John Wiley & Sons, New York, 1960.

Semiconductor Materials

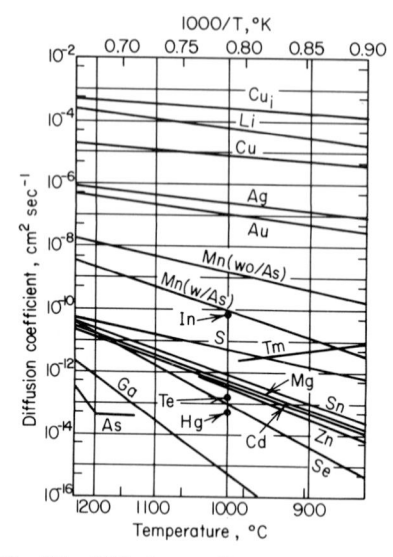

Fig. 18 Diffusion coefficients in GaAs at low concentration limit.[111]

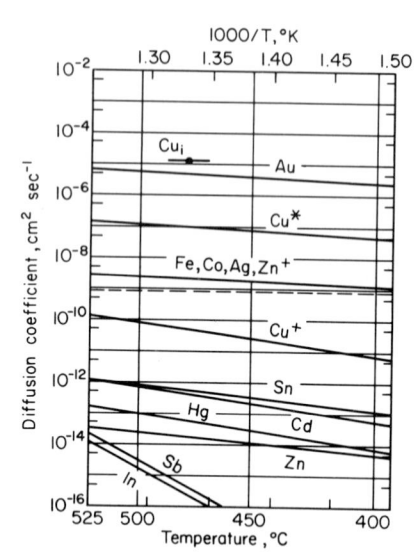

Fig. 19 Diffusion coefficients in InSb at low concentration limit except as noted. * 10^4 dislocations cm^{-2}; † At 2×10^{20} cm^{-3}; ‡ Dislocation free.[111]

TABLE 12 Knoop Hardness

Material	Moh	Value,* kg/mm²	Indenter load, grams
AlSb†	4.0	360	25
BP†	....	3,200	100
CdS‡	....	55	25
Diamond‡	10	8,800	100
GaAs†	4.5	750	25
GaP†	5	950	25
GaSb†	4.5	450	25
Ge‡	6	750	25
InAs†	....	380	25
InP†	....	535	25
InSb†	....	220	25
Si§	6.5	950	100
SiC‡	9	2,880	25
ZnS‡	....	180	25

* Orientation-dependent.

† N. A. Goriunova, A. S. Borschchovskii, and D. N. Tretiakov, Hardness, in R. K. Willardson and Albert C. Beer (eds.), "Semiconductors and Semimetals," vol. 4, Academic Press, Inc., New York, 1968.

‡ G. A. Wolff, L. Tolman, N. J. Field, and J. C. Clark, *Proc. Int. Coloq. Semicond. and Phosphors*, Interscience Publishers, Inc., New York, 1958.

§ A. A. Giardini, A Study of Directional Hardness in Silicon, *Am. Mineral.*, vol. 43, pp. 957–969, 1958.

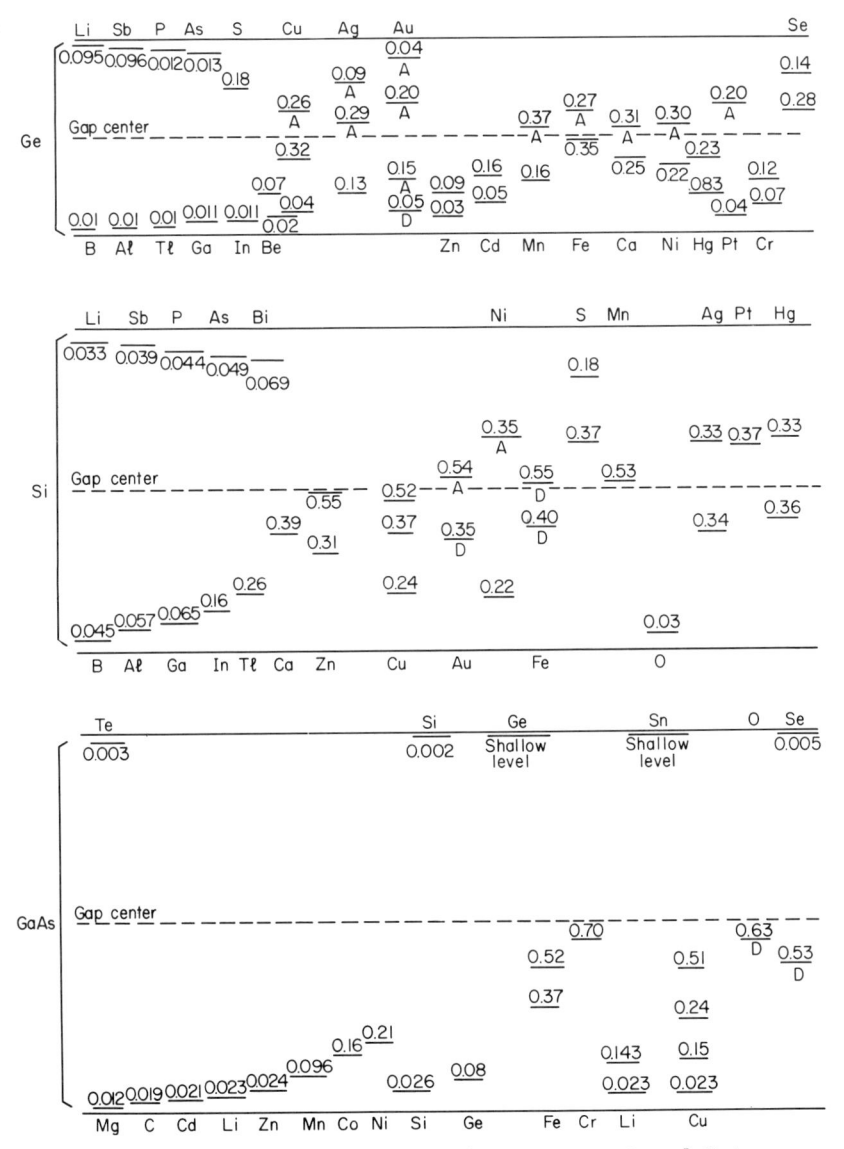

Fig. 20 Impurity energy levels for various elements in Ge, Si and GaAs at room temperature.[98]

Conductivity Type

Generally, the more useful methods of determining conductivity type are the polarity of thermoelectric voltage and the direction of current flow through a point contact to the semiconductor. However, various types of selective chemical stains and the polarity of the Hall voltage are also used. In addition, the problem of locating the position of p-n junctions often arises. This may be done by mapping the confines of one type or the other, by detecting the very high resistivity that

Semiconductor Materials

TABLE 13 Intrinsic Carrier Concentration n_i at Room Temperature

Material	n_i, carriers/cm³
Si*	1.4×10^{10}
Ge*	2.3×10^{13}
InSb*	1.35×10^{16}
Mg₂Sn*	2.8×10^{17}
PbS*	3×10^{15}
PbSe*	$<10^{17}$
PbTe*	$<7 \times 10^{16}$
GaAs†	1.4×10^{6}

* R. A. Smith, *Semiconductors*, Cambridge University Press, London, 1961.
† David Richman, Some Special Features Make the Difference, *Electron.*, Nov. 13, 1967, pp. 108–110.

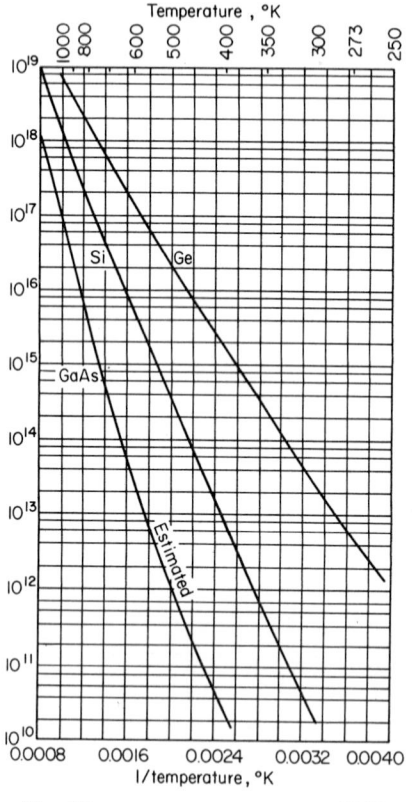

Fig. 21 n_i versus temperature.[113-115]

occurs at the transition, or by detecting the high electric field that can be developed between n and p regions.

Thermoelectric method A hot probe touching the surface of an n-type semiconductor becomes positive with respect to an ambient temperature contact placed elsewhere on the material, and negative when the material is p-type. The hot probe may be a soldering iron operated at reduced voltage (e.g., 30 volts). The galvanometer should have high sensitivity and a center zero. Either a moving coil or an electronic type is satisfactory, though the electronic variety are less likely to be damaged by accidental overloads. If the semiconductor resistivity is very high, the additional local heating may move the material into the intrinsic conductivity region. When this occurs, most materials indicate n-type because of the greater mobility of n-type carriers. By the use of a cold probe rather than a hot one, this difficulty is prevented.

Ac rectification If a dc ammeter is placed in series with an ac source and the circuit is completed by a (point-contact)–(semiconductor large-area contact) combination, the deflection of the ammeter will indicate the material type. A simple microammeter will suffice for low resistivity, but a much more sensitive instrument is required for resistivities in the 100- to 1,000-ohm-cm range.

Hall voltage The sign of the Hall constant is negative for n-type and positive for p-type material.

Resistivity

The basic resistivity measurements are made by affixing contacts to the ends of a bar of uniform cross section A. Current I is passed through the bar, and

TABLE 14 Latent Heat of Fusion

Material	Value, kcal/mole
GaSb*	8.0
Ge†	7.6
InAs*	9.2
InSb*	5.6
Se‡	1.2
Si†	12.1

* R. H. Cox and M. J. Pool, *J. Chem. Eng. Data*, vol. 12, p. 247, 1967.
† R. E. Honig, *RCA Rev.*, vol. 23, p. 569, 1962.
‡ C. A. Hogarth (ed.), "Materials Used in Semiconductors," Interscience Publishers, N.Y. (1965).

the potential drop V between two points a distance l apart on the bar is measured. The resistivity ρ is then given by

$$\rho = \frac{V}{I}\frac{A}{l} \qquad (1)$$

For V in volts, I in amperes, and the dimensions in centimeters, the resistivity is expressed in ohm-centimeters.

Various other methods of measuring resistivity are summarized in Table 28. It

TABLE 15 Cubic Lattice Spacing*

Alphabetical		Numerical	
Material	a, angstroms	Material	a, angstroms
AlSb	6.1355	C (diamond)	3.5668
C (diamond)	3.5668	SiC	4.3596
CdS	5.820	ZnS	5.4093
CdSe	6.05	Si	5.4307
CdTe	6.481	GaP	5.4504
GaAs	5.6533	GaAs	5.6533
GaP	5.4504	Ge	5.6575
GaSb	6.0961	ZnSe	5.6687
Ge	5.6575	CdS	5.820
InAs	6.0584	HgS	5.851
InP	5.8687	InP	5.8687
InSb	6.4788	PbS	5.935
HgS	5.851	CdSe	6.05
HgSe	6.084	InAs	6.0584
HgTe	6.460	HgSe	6.084
PbS	5.935	GaSb	6.0961
PbSe	6.122	ZnTe	6.1037
PbTe	6.460	PbSe	6.122
Si	5.4307	AlSb	6.1355
SiC	4.3596	HgTe	6.460
Sn	6.5041	PbTe	6.460
ZnS	5.4093	InSb	6.4788
ZnSe	5.6687	CdTe	6.481
ZnTe	6.1037	Sn	6.5041

* W. B. Pearson, "Handbook of Lattice Spacings and Structures of Metals and Alloys," vol. 2, Pergamon Press, New York, 1967.

should be remembered, however, that the resistivity, carrier concentration, and impurity concentration are all interrelated, and for most impurities in most semiconductors, the relations are well known. Thus a determination of any of the three will usually suffice. Table 29 summarizes some of the carrier-concentration methods

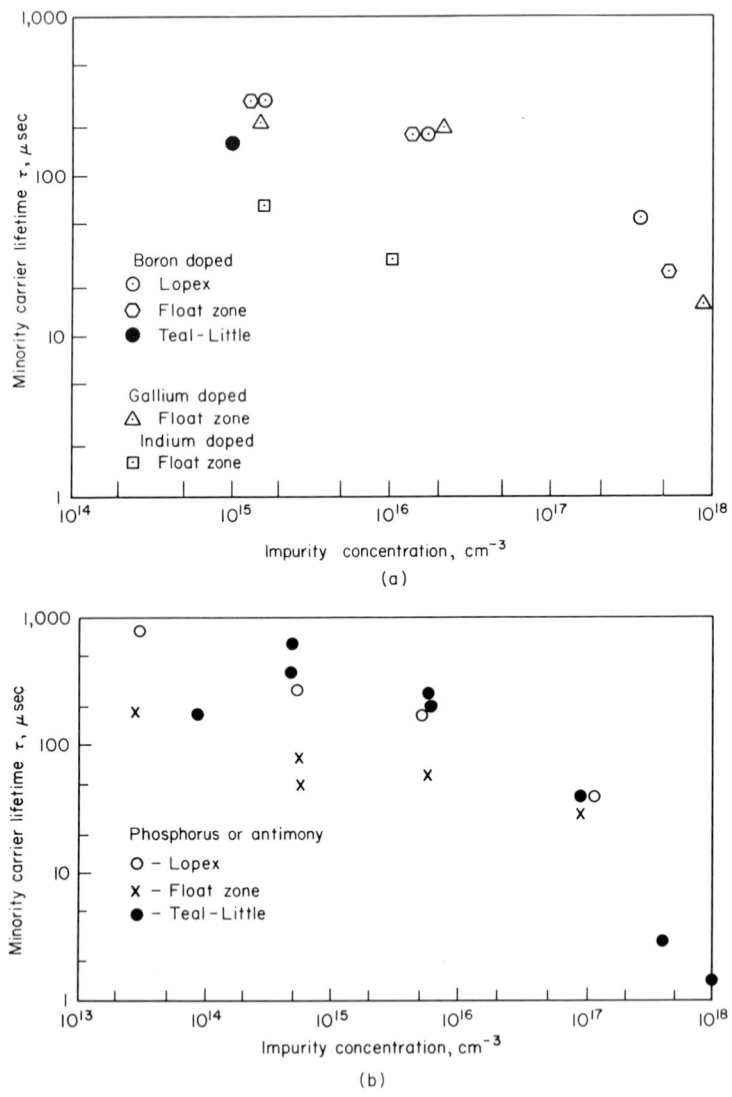

Fig. 22 Dependence of lifetime on carrier concentration in (*a*) p-type silicon and (*b*) n-type silicon.[116]

(many of which are calibrated directly in terms of resistivity). Table 30 lists the commonly used analytical methods.

Direct method The simplest way of finding resistivity is to measure the sample dimensions, I and V, and use Eq. (1). A major disadvantage is that the calculated

TABLE 16 Hexagonal Lattice Spacing*

		Angstroms	
Material	Polytype	a	c
CdSe		4.2985	7.0150
CdS		4.1368	6.7163
SiC	4H	3.073	10.053
	6H	3.073	15.079
	15R	3.073	37.70
	21R	3.073	52.78
	33R	3.073	82.94
	51R	3.073	128.18
ZnS	2H	3.819	6.246
	4H	3.814	12.46
	6H	3.821	18.73
	8H	3.82	24.96
	10H	3.824	31.20
	9R	3.82	28.08
	12R	3.82	37.44
	15R	3.83	46.88
	21R	3.82	65.52

* Leonid V. Azaroff, "Introduction to Solids," McGraw-Hill Book Company, New York, 1960.

resistivity contains a contact-resistance term, which for semiconductors can be appreciable.

Two-point probe The effect of contact resistance can be eliminated by use of a two-point probe. Current contacts are affixed to the ends of a specimen whose cross section is relatively uniform. Two voltage probes are used to measure the potential drop at specific positions along the length of the crystal, and Eq. (1) is used to compute the resistivity. Measurement restrictions are that the current

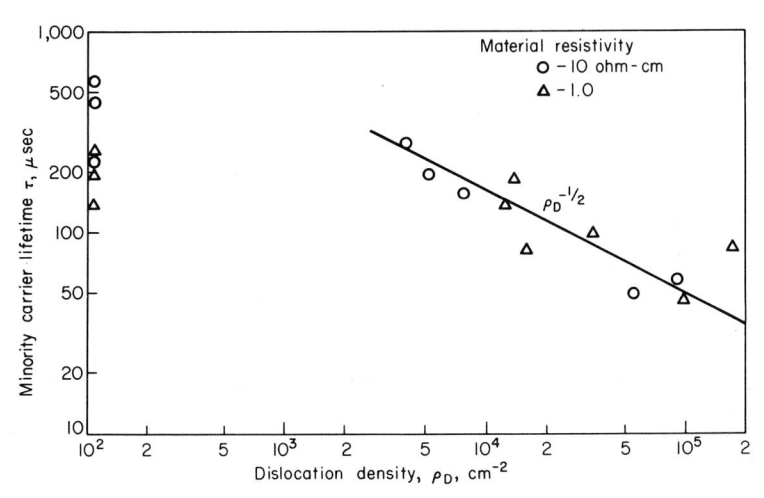

Fig. 23 Dependence of lifetime on dislocation density in n-type silicon.[116]

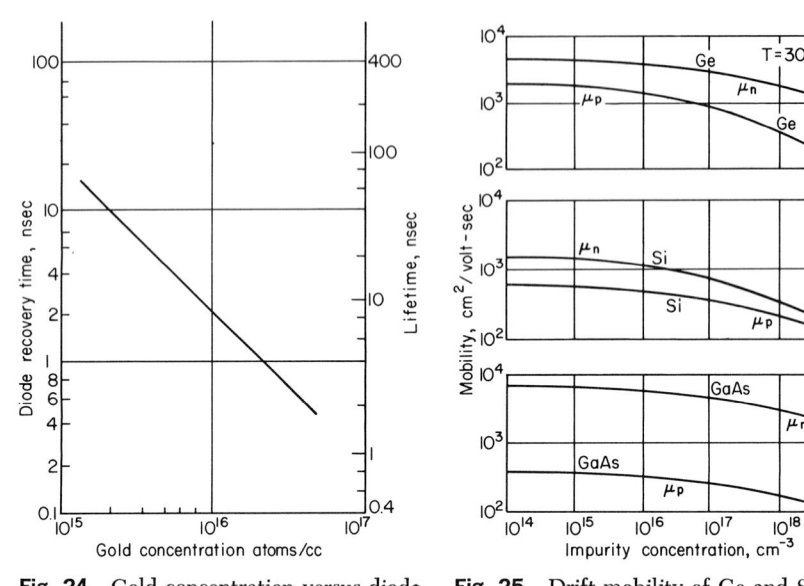

Fig. 24 Gold concentration versus diode recovery time.[117]

Fig. 25 Drift mobility of Ge and Si and Hall mobility of GaAs at 300°K versus impurity concentration.[98]

TABLE 17 Melting Points of Various Materials*

Material	Temperature, °C	Material	Temperature, °C
AlAs	>1600	InP	1070
AlP	>1500	In_2S_3	1050
AlSb	1050	InSb	525
As_2Se_3	608	In_2Se_3	890
B	2075	In_2Te_3	667
Bi_2S_3	850	Mg_2Ge	1115
Bi_2Se_3	706	Mg_2Sn	778
Bi_2Te_3	580	Mg_2Si	1102
Ca_2Pb	1110	PbS	1077
Ca_2Si	920	PbSe	1062
Ca_2Sn	1122	PbTe	904
$CdAs_2$	621	$PtSb_2$	1240
Cd_3As_2	721	Sb_2S_3	546
CdSe	1350	Sb_2Se_3	612
CdSb	456	Sb_2Te_3	620
GaAs	1237	Se	217
GaP	1465	Si	1417
GaSb	712	Te	450
Ge	937	$ZnAs_2$	768
HgS	1450	Zn_3As_2	1015
HgSe	800	ZnSb	546
HgTe	670	ZnSe	1515
InAs	942	ZnTe	1238

* From E. L. Kern and E. Earleywine, New Developments in Semiconductor Materials, *Solid State Technol.*, October, 1965.

TABLE 18 Mobilities of Various Materials*

Group	Material	μ_n, cm²/volt-sec	μ_p, cm²/volt-sec
III	B	6,000	4,000
IV	Ge Si SiC	3,640 1,400 4,690	1,900 380
VI	S Se Te	100 2 2,200	17 1,000
III-V	BP AlSb GaP GaAs GaSb InP InAs InSb	900 300 7,200 5,000 4,600 33,000 80,000	300 400 150 200 1,000 100 450 450
II-VI	ZnS ZnSe ZnTe CdS CdSe CdTe HgSe HgTe	120 530 530 340 600 700 18,500 22,000	16 900 18 65 160
Lead salts	PbS PbSe PbTe	600 1,400 6,000	200 1,400 4,000
II-IV	Mg₂Si Mg₂Ge Mg₂Sn	370 530 210	65 110 150
II-V	ZnSb ZnAs₂ Zn₃As₂ CdSb CdAs₂ Cd₃As₂	10 300	350 50 10 1,000 100 15,000
V-VI	Bi₂S₃ Bi₂Se₃ Bi₂Te₃ Sb₂Se₃ Sb₂Te₃ As₂Se₃ As₂Te₃	200 600 10,000 15 15 170	400 45 270 45 80
VIII-V	PtSb₂	200	1,400
III-VI	In₂Se₃ In₂Te₃	30 340	

* From E. L. Kern and E. Earleywine, New Developments in Semiconductor Materials, *Solid State Technol.*, October, 1965.

Semiconductor Materials

TABLE 19   Piezoresistance Coefficients at Room Temperature*

Material	ρ, ohm-cm	Concentration 10^{18} cm^{-3}	π_{11} × 10^{-12}	π_{12} × 10^{-12}	π_{44} × 10^{-12}	π_p × 10^{-12}	π_A × 10^{-12}
n-Ge	9.9	.....	−4.7	−5.0	−137.9	+14.7	+138.2
Ge							
p-Ge	1.1	.....	−3.7	+3.2	+96.7	−2.7	−103.6
n-Si	11.7	.....	−102.2	+53.4	−13.6	−4.6	−142.0
Si							
p-Si	7.8	.....	+6.6	−1.1	+138.1	−4.4	−130.4
n-PbTe	.....	1 to 3	+20.0	+25.0	−107.0	−70.0	+102.0
PbTe							
p-PbTe	.....	∼3	+24.0	+15.0	+215.0	−54.0	−206.0
p-PbTe	.....	1 to 3	+35.0	+40.0	+185.0	−115.0	−190.0
p-PbSe	.....	1 to 3	+24.0	+19.0	+57.0	−62.0	−52.0
n-PbS	.....	1 to 3	+11.6	+6.6	−11.2	−24.7	+16.2
n-PbS	.....	6 to 9	+6.9	+2.9	−1.2	−12.8	+5.2
PbS							
InSb	.....	.....	−81.6	−114.2	+33.0	+310.0	−0.4
InSb							
p-InSb	0.54	0.003	−70.0	−115.0	−10.0	+300.0	+55.0

$$\pi_p \equiv -(\pi_{11} + 2\pi_{12}) \qquad \pi_A \equiv \pi_{11} - \pi_{12} - \pi_{44}$$

* Warren P. Mason, "Physical Acoustics," vol. 1, part B, Academic Press, Inc., New York, 1964.

must be kept low enough to prevent heating of the sample, and the voltmeter must have a high input impedance.

A Hall bar with two arms on each side can be used as a two-point resistivity probe and can be quite accurate. It is particularly applicable to high-resistivity materials.

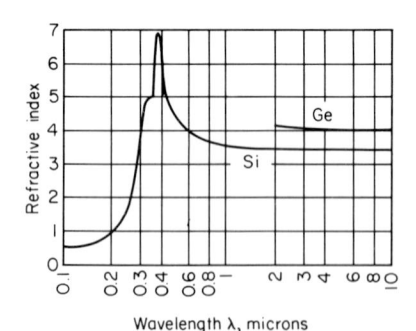

Fig. 26   Index of refraction.[118,119]

Linear four-point probe   In the semiconductor industry, the most generally used technique for the measurement of resistivity is the four-point probe. It may be used directly for measuring large-volume crystals, but with proper correction for thickness and/or area it is applicable to thin slices or to epitaxial layers where the layer is electrically isolated from the substrate by being of the opposite conductiv-

TABLE 20 Refractive Indices

λ, microns	Te $E \parallel C$ axis	Te $E \perp C$ axis	InSb	B	CdS	ZnS	PbS	PbSe	PbTe	Si	Ge	GaAs	C (diamond)	SiC
0.1														
0.2														
0.3					2.83									2.9
0.4					2.58	2.5							2.46	2.7
0.5				2	2.67	2.43								
0.6					2.5	2.35				4.0		3.6	2.41	2.6
0.8					2.4	2.3								
1.0				3.2	2.35	2.28				3.5		3.6		
2.0				3.2	2.3	2.25				3.45		3.6		
3.0							4.1	4.59	5.35	3.43				
4.0	6.3	4.95								3.42				
6.0	6.2	4.95								3.42				
8.0	6.2	4.8	4.0							3.42				
10.0			3.96											
12			3.93											
14	6.25	4.8	3.91											
16			3.88											
18			3.86											
20														

ity type. The method is normally nondestructive; however, the probe points may damage certain semiconductor materials when excessive probe pressure is applied.

The usual geometry is to place the probes in a line and use equal probe spacing. Current is passed through the outer two probes (current probes), and the potential developed across the inner two probes (voltage probes) is measured. For probes resting on a semi-infinite medium, the resistivity ρ_0 is given by[43]

$$\rho_0 = \frac{2\pi(V/I)}{\dfrac{1}{S_1} + \dfrac{1}{S_3} - \dfrac{1}{S_1 + S_2} - \dfrac{1}{S_2 + S_3}} \qquad (2)$$

where ρ_0 = resistivity, ohm-centimeters
 V = floating potential across inner probes, volts
 I = current through outer pair of probes, amperes
 S = probe spacing, centimeters

7-42 Semiconductor Materials

When the probes are equally spaced, $S_1 = S_2 = S_3$, and Eq. (2) becomes

$$\rho_0 = \frac{V}{I} 2\pi S \tag{3}$$

When the sample is of limited size, various corrections to Eq. (3) are required. These corrections usually take one of two forms. Either they are designed to be applied to readings from meters already calibrated according to Eq. (3) to read resistivity directly, i.e.,

$$\rho = F\rho_0 \tag{4}$$

where ρ_0 is the meter reading; or they may be designed to give true resistivity upon multiplication by the measured V/I ratio.

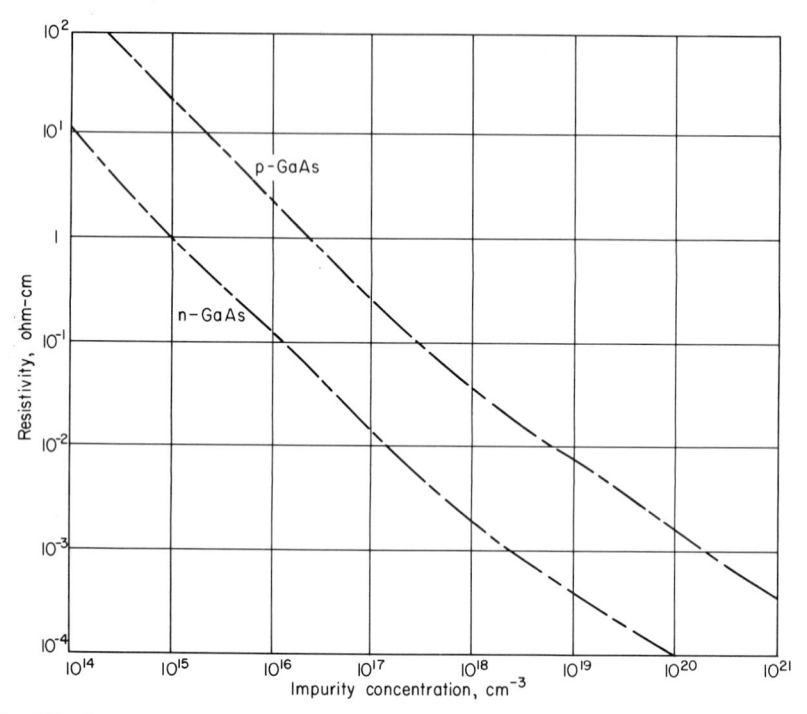

Fig. 27 Resistivity versus impurity concentration for GaAs at room temperature.[98]

Probes close to an edge[43] Figure 39 gives resistivity corrections for meters already calibrated in terms of ρ_0 for probes close to an edge of an otherwise infinite sample. The specific probe locations are shown beside the individual curves.

Thin slices of infinite lateral expanse[43,44] These corrections are shown in Fig. 40. For an electrically isolated slice, F approaches $0.72W/S$ (where W = the slice thickness) as W/S becomes less than 1. If the back of the slice is covered with a conducting layer, e.g., a metal layer, or a very low resistivity substrate as in the case of epitaxial layers, dependable results are possible only if W/S is greater than about 0.5. Thus, in order to accurately measure layers a few microns thick, very close probe spacing is required.

Thin slices of finite extent For a W/S ratio of less than 0.5, the resistivity of circular slices measured with the probes centered is given by

$$\rho = \bar{F}W\frac{V}{I} \tag{5}$$

where $\bar{F}$ is given in Table 31.[45] For a rectangular bar with the probes symmetrically placed,

$$\rho = \bar{\bar{F}} W \frac{V}{I} \qquad (6)$$

where $\bar{\bar{F}}$ is given in Fig. 41 for the probe row in the long direction of the bar.[46]

TABLE 21 Segregation Coefficients of Various Elements in Germanium and Silicon*

Group	Element	Germanium	Silicon
IA	Lithium	0.002	0.01
IB	Copper Silver Gold	1.5×10^{-5} 4×10^{-7} 1.3×10^{-5}	4×10^{-4} 2.5×10^{-5}
IIB	Zinc Cadmium	4×10^{-4} $>1 \times 10^{-5}$	$\sim 1 \times 10^{-5}$
IIIA	Boron Aluminum Gallium Indium Thallium	17 0.073 0.037 0.001 4×10^{-5}	0.80 0.0020 0.0080 4×10^{-4}
IVA	Silicon Germanium Tin Lead	5.5 1 0.020 1.7×10^{-4}	1 0.33 0.016
VA	Nitrogen Phosphorus Arsenic Antimony Bismuth	.... 0.080 0.02 0.0030 4.5×10^{-5}	$<10^{-7}(?)$ 0.35 0.3 0.023 7×10^{-4}
VIA	Oxygen Sulfur Tellurium	.... $\sim 10^{-6}$	0.5 10^{-5}
Transition elements	Vanadium Manganese Iron Cobalt Nickel Tantalum Platinum	$<3 \times 10^{-7}$ $\sim 10^{-6}$ $\sim 3 \times 10^{-5}$ $\sim 10^{-6}$ 3×10^{-6} $\sim 5 \times 10^{-6}$	 $\sim 10^{-5}$ 8×10^{-6} 8×10^{-6} 10^{-7}

* F. A. Trumbore, Solid Solubilities of Impurity Elements in Germanium and Silicon, *BSTJ*, vol. 39, pp. 205–233, 1960.

Square four-point array[47] For the probes arranged in a square rather than a linear array,

$$\rho = 10.7 \, S \frac{V}{I} \qquad (7)$$

Semiconductor Materials

TABLE 22 Segregation Coefficients for Various Elements in Group III-V Compound Semiconductors

Group	Impurity element	AlSb	GaSb	GaAs	InAs	InP	InSb
IB	Copper Silver Gold	0.01	....	<0.002 0.1	....	....	6.6×10^{-4} 4.9×10^{-5} 1.9×10^{-6}
IIA	Magnesium Calcium	....	....	0.3 <0.02	0.7 		
IIB	Zinc Cadmium	0.02 0.002	0.2–0.3 0.02	0.1–0.9 <0.02	0.77 0.13	....	4 0.26
IIIA	Boron Aluminum Gallium Indium	0.01–0.02 .... ~1	....	 3 .... 0.1	....		 2.4
IVA	Carbon Silicon Germanium Tin Lead	0.6 0.4 0.3 2×10^{-4} 8×10^{-4} 0.01	.... ~1 0.3 0.01	0.8 0.1 0.2–0.3 0.03 <0.02	 0.4 0.07 0.09 <0.01	 0.05 0.03 	 0.045 0.057
VA	Phosphorus Arsenic Antimony	....	.... 2–4	2 .... <0.02	....	....	0.16 5.4
VIA	Sulfur Selenium Tellurium	0.003 0.003 0.01	0.06 0.2–0.4 0.4	0.3 0.44–0.55 0.3	1.0 0.93 0.44	~0.8 ~0.6	0.1 0.2–0.5 0.5–3.5
Transition elements	Vanadium Manganese Iron Cobalt Nickel	0.01 0.01 0.02 0.002 0.01	....	 0.05 0.003 <0.02	....	....	 6×10^{-5}

Sheet resistance Quite often it is desirable to characterize a thin layer by its sheet resistance, that is, its resistance per unit area. Sheet resistance is given in units of ohms per square and is related to resistivity by the factor of thickness:

$$\rho = W\rho_s \quad (8)$$

where ρ_s = sheet resistance, ohms per square
W = thickness, centimeters

For equally spaced in-line probes on an infinite sheet, the sheet resistance is given by[46]

$$\rho_s = \frac{V}{I} \frac{\pi}{\ln 2} = 4.532 \frac{V}{I} \quad (9)$$

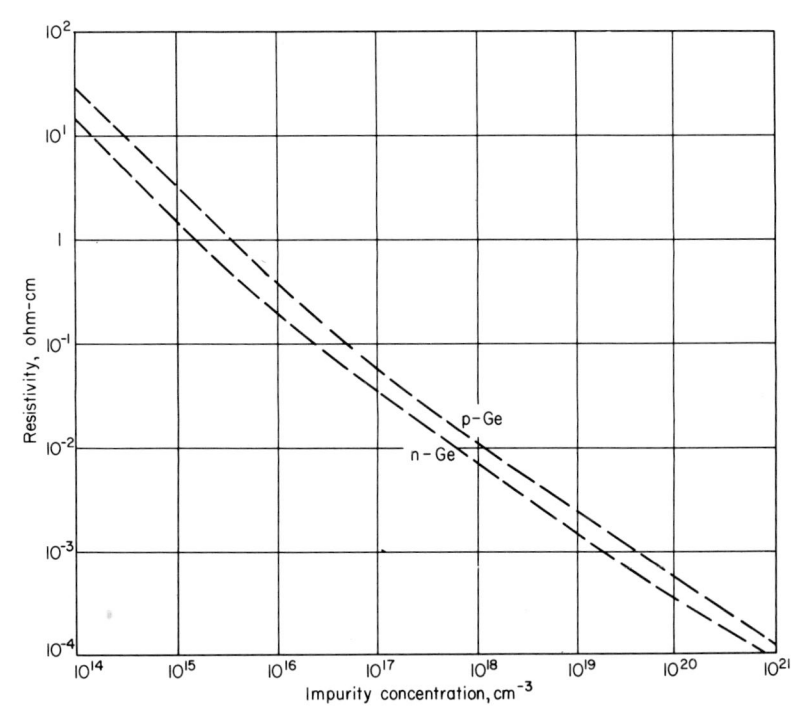

Fig. 28 Resistivity versus impurity concentration for Ge at room temperature.[98]

TABLE 23 Surface Energies

Material	{100}	{110}	{111}	Reference
AlAs	2,600	1,800	1,500	(calc)*
AlP	3,400	2,400	2,000	(calc)*
AlSb	1,900	1,300	1,100	(calc)*
Diamond	9,820	....	5,650	(calc)†
GaAs	2,200	1,500	1,300	(calc)*
GaP	2,900	2,000	1,700	(calc)*
GaSb	1,600	1,100	910	(calc)*
InAs	1,400	1,000	840	(calc)*
InP	1,900	1,300	1,100	(calc)*
InSb	1,100	750	600	(calc)*
Si	2,130	1,510	1,000, 1,240	(exp)‡§

Face, ergs/cm²

* J. W. Cahn and P. E. Hanneman, (111) Surface Tensions of III-V Compounds and Their Relationship to Spontaneous Bending of Thin Crystal, *Surface Sci.*, vol. 1, pp. 387–398, 1964.

† William D. Harkins, Energy Relations of the Surface of Solids, *J. Chem. Phys.*, vol. 10, pp. 268–272, 1942.

‡ R. J. Jaccodine, Surface Energy of Germanium and Silicon, *J. Electrochem. Soc.*, vol. 110, pp. 524–527, 1963.

§ J. J. Gilman, Direct Measurements of the Surface Energies of Crystals, *J. Appl. Phys.*, vol. 31, pp. 2208–2218, 1960.

7-46 Semiconductor Materials

For a square-probe array, the sheet resistance is given by

$$\rho_s = \frac{2\pi}{\ln 2} \frac{V}{I} = 9.06 \frac{V}{I} \qquad (10)$$

To obtain the sheet resistance of a finite sample, various corrections must be applied and may be obtained by eliminating W from the bulk correction for thin slices, that is,

$$\rho_s = \bar{F} \frac{V}{I} \qquad (11)$$

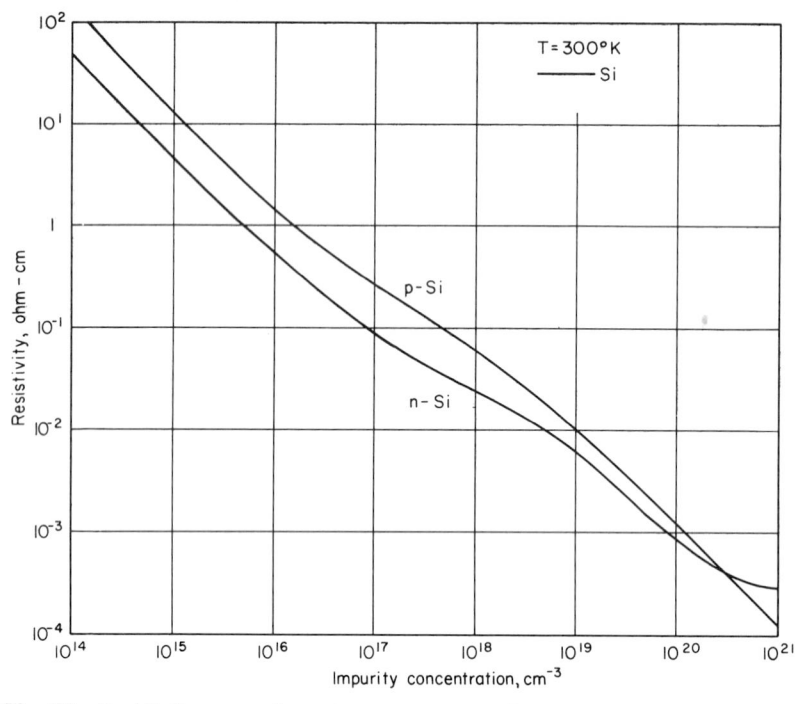

Fig. 29 Resistivity versus impurity concentration for Si at room temperature.[98]

TABLE 24 Surface Tensions near Melting Point

Material	Dynes/cm
Germanium*	600
Silicon*	720
Selenium†	92.5

* P. H. Keck and W. Van Horn, The Surface Tension of Liquid Silicon and Germanium, *Phys. Rev.*, vol. 91, pp. 512–513, 1953.

† C. A. Hogart (ed.), "Materials Used in Semiconductor Devices," Interscience, N.Y. (1965). Publishers, a division of John Wiley & Sons, Inc., New York, 1965.

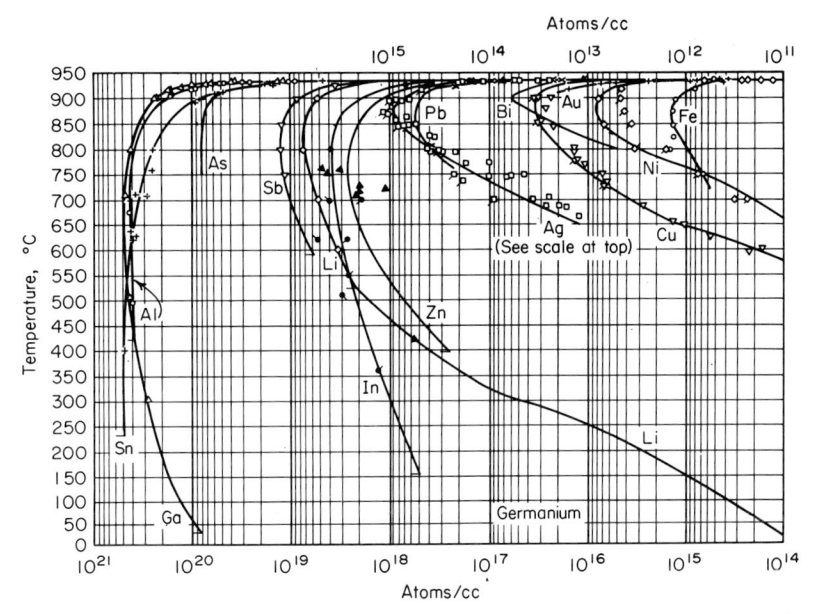

Fig. 30 Solid solubility of various impurities in Ge.[120,123] (*From F. A. Trumbore, Solid Solubilities of Impurity Elements in Germanium and Silicon, BSTJ, vol. 39, Fig. 1 on p. 208, 1960. Copyright 1960, The American Telephone and Telegraph Co., reprinted by permission.*)

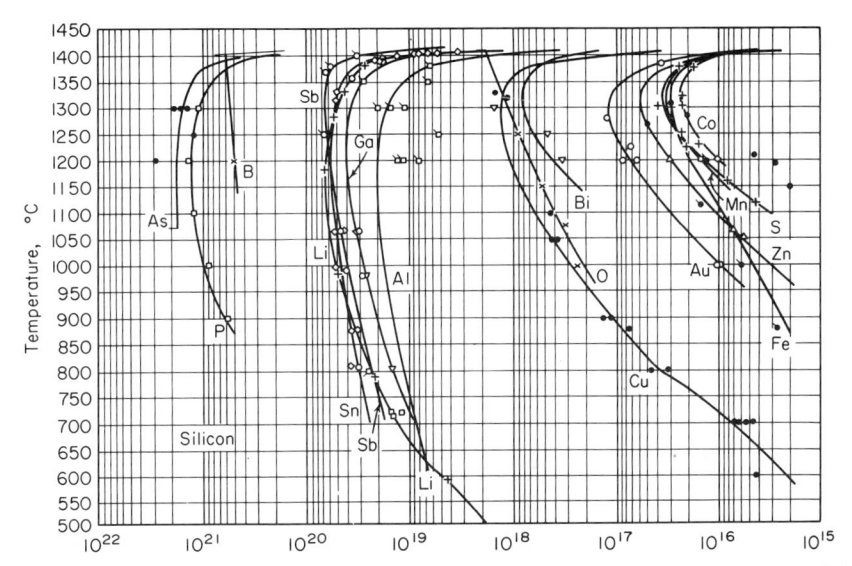

Fig. 31 Solid solubility of various impurities in Si. (*From F. A. Trumbore, Solid Solubilities of Impurity Elements in Germanium and Silicon, BSTJ, vol. 39, Fig. 2 on p. 210, 1960. Copyright 1960, The American Telephone and Telegraph Co., reprinted by permission.*)

TABLE 25 Thermal Conductivities at Room Temperature

Material	cal/(sec)(cm)(°C)*	Material	cal/(sec)(cm)(°C)*
AlSb[a]	0.11	Ge[d]	0.14
Bi$_2$Te$_3$[b]	0.0044 ∥ to C axis	InAs[d]	0.055
	0.0019 ⊥ to C axis	InP[d]	0.16
Diamond[c]	5	InSb[d]	0.038
GaAs[d]	0.09	Se[e]	0.003–0.007
GaP[d]	0.19	Si[d]	0.35
GaSb[d]	0.08	SiC[c]	1.2

* To convert from cal/(sec)(cm)(°C) to watts/(cm)(°C), multiply by 4.185.

[a] G. L. Pearson and F. L. Vogel, Plastic Deformation of Semiconductors, in Alan F. Gibson (ed.), "Progress in Semiconductors," vol. 6, John Wiley & Sons, Inc., New York, 1962.

[b] R. A. Smith, "Semiconductors," Cambridge University Press, London, 1961.

[c] Glen A. Slack, Thermal Conductivity of Pure and Impure Silicon, Silicon Carbide, and Diamond, *J. Appl. Phys.*, vol. 35, pp. 3460–3466, 1964.

[d] P. D. Maycock, Thermal Conductivity of Silicon, Germanium, III-V Compounds, and III-V Alloys, *Solid State Electron.*, vol. 10, pp. 161–168, 1967.

[e] C. A. Hogarth (ed.), "Materials Used in Semiconductor Devices," Interscience Publishers, a division of John Wiley & Sons, Inc., New York, 1965.

Sheet resistance, van der Pauw method[48] Rather than depend on miscellaneous corrections for finite sheets, it is possible, by placing four contacts A, B, C, and D sequentially spaced about the periphery of the sample, to determine ρ_s directly:

$$\rho_s = \frac{\pi}{2 \ln 2}(R' + R'')f\left(\frac{R'}{R''}\right) \qquad (12)$$

where $f(R'/R'')$ is van der Pauw's function and is shown in Fig. 42. R' is

TABLE 26 Thermal Expansion Coefficients at Room Temperature*

Material	Per °C	Material	Per °C
Diamond	1.2 × 10^{-6}	InAs	5.3
GaAs	5.7	InP	4.5
GaSb	6.9	InSb	5.0
GaP	5.3	Si	2.3
Ge	5.7	SiC	4.5

* G. L. Pearson and F. L. Vogel, Plastic Deformation of Semiconductors, in Alan F. Gibson (ed.), "Progress in Semiconductors," vol. 6, John Wiley & Sons, Inc., New York, 1962.

the potential difference between the contacts D and C per unit current through the contacts A and B, and R'' is the potential difference between the contacts A and D per unit current through the contacts B and C.

If the probes are placed symmetrically about a line through any pair of nonadjacent probes of the four-point array so that $R'/R'' = 1$, van der Pauw's function also becomes 1. In addition to requiring contacts on the periphery, the method also must have very small contacts, a uniform thickness sample, and no isolated holes in the interior of the sample.

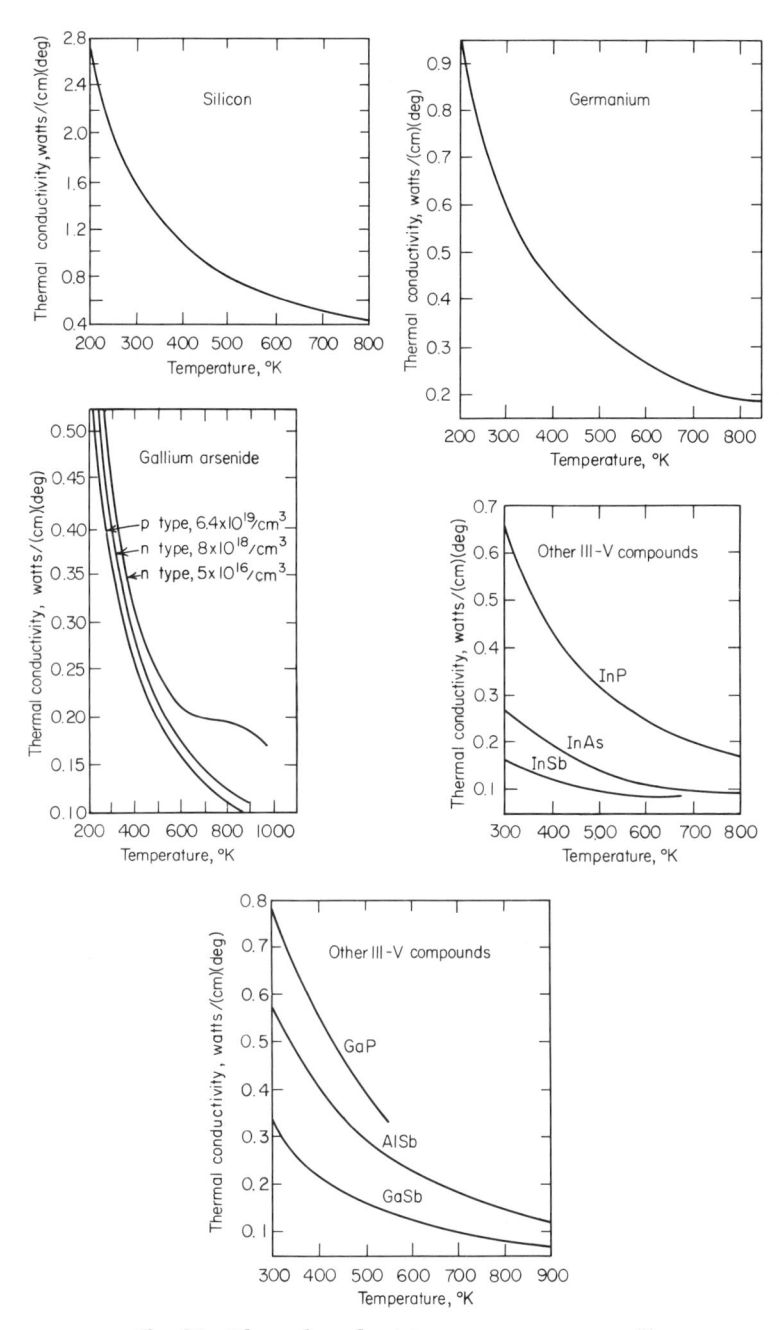

Fig. 32 Thermal conductivity versus temperature.[121]

7-50   Semiconductor Materials

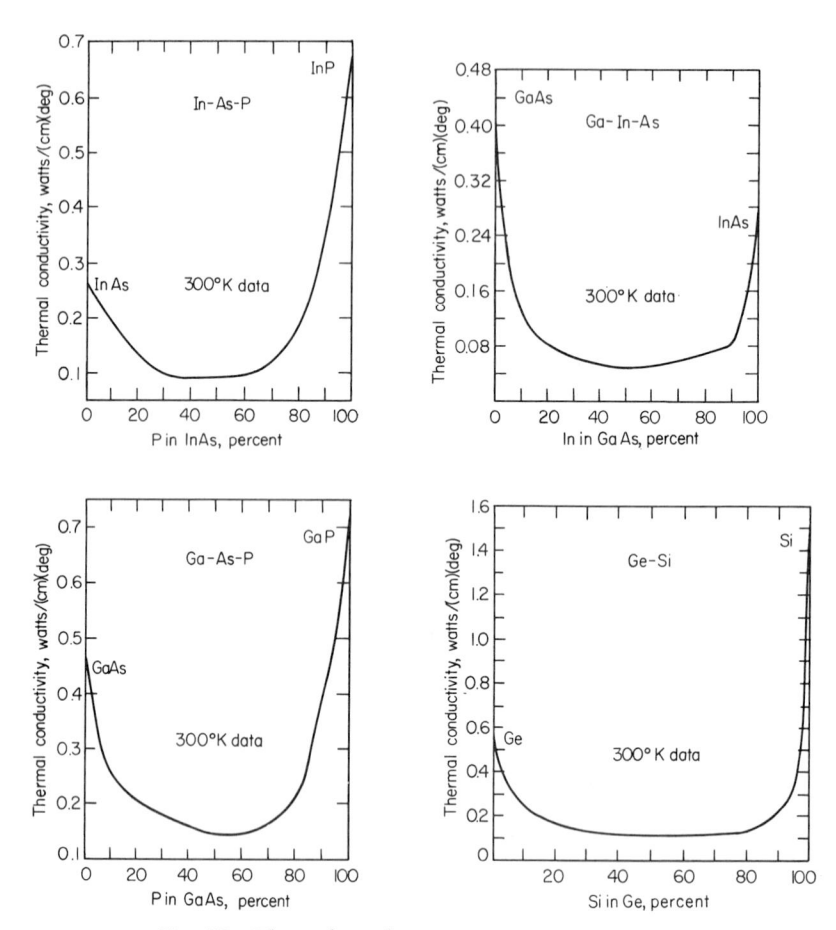

Fig. 33   Thermal conductivity versus composition.[121]

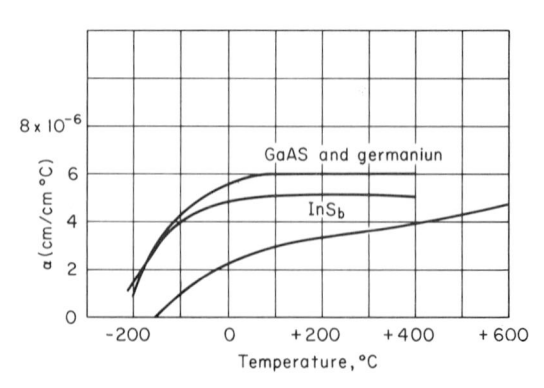

Fig. 34   Expansion coefficients versus temperature for the more common semiconductors.[35, 127, 128]

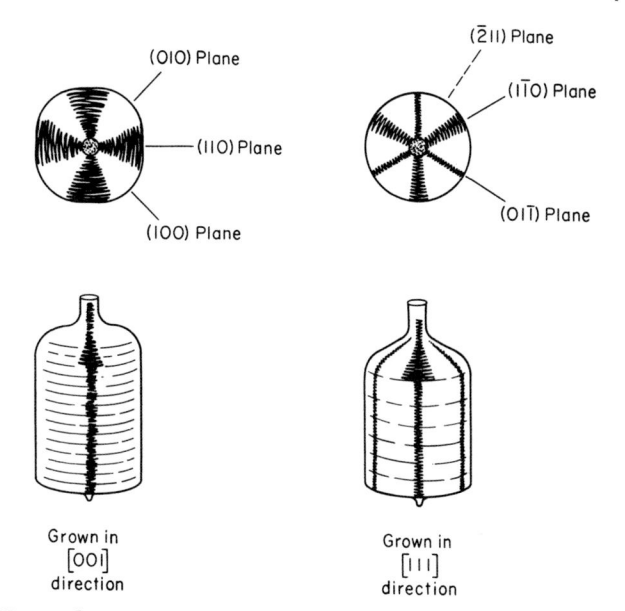

Fig. 35 Markings on silicon and germanium crystals grown from the melt.

Spreading resistance[49] In order to minimize the corrections required, the spreading resistance of a point contact may be used to determine resistivity. If the probe is assumed to be hemispherically tipped and the full hemisphere embedded, the spreading resistance is

$$\rho_{sp} = \frac{\rho}{2\pi r} \qquad (13)$$

where r is the tip radius. If the tip is assumed to be flat rather than hemispherical,

$$\rho_{sp} = \frac{\rho}{2r} \qquad (14)$$

Because of difficulties in determining tip radius, the kind of indent, and effect of contact resistance, calibration curves for a particular tip are usually made.

Capacitance-voltage measurements[50] As the voltage bias is changed on a p-n junction, the space-charge width and consequently the junction capacitance change.

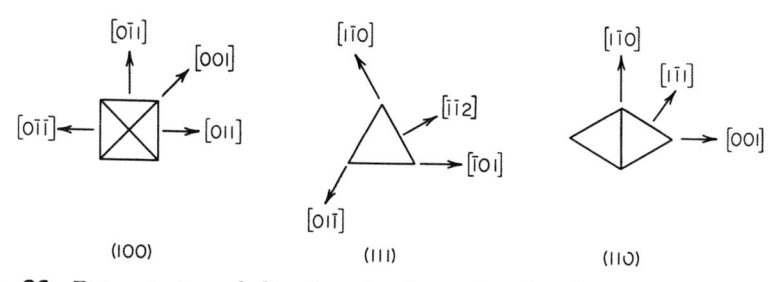

Fig. 36 Determination of directions in plane of a slice by use of etch pits or stacking faults. Method applicable to cubic crystals and a (111) selective etch or a (111) stacking-fault system.

TABLE 27 Dislocation Etches*

Material	Orientation	Etch name and composition	Etch time	Remarks
B_2Te_3		0.5 M $K_2Cr_2O_7$	10 min at 95°C	Etch pits on A surfaces.
Cds		6 fuming HNO_3 6 acetic 1 H_2O	2 min	Sharply defined hexagonal pits on B surfaces.
CdS		0.5 M $K_2Cr_2O_7$ in 16 N H_2SO_4	5–10 min at 95°C	Dislocation etch on B surface.
CdSe		30 HNO_3 0.1 HCl 20 18N H_2SO_4	8 sec at 40°C rinsing in concentrated H_2SO_4 to dissolve Se film	A surface hexagonal pits.
CdSe		1 HNO_3 3 HCl		Sharply beveled pits on A surfaces.
CdTe		3 HF 2 H_2O_2 1 H_2O	2 min	Etch pits on the B surfaces.
GaAs	(111)	HCl—HNO_3 2 ml HCl 1 ml HNO_3 2 ml deionized H_2O	10 min	Mix fresh— shows pits on Ga side.
GaA	(100) Ga(111), (110) As(111)	2 ml H_2O 8 mg $AgNO_3$ 1 g CrO_3 1 ml HF	10 min	65°C
GaSb		1 ml bromine 10 ml methanol	20 sec	Shallow pits on A surface.
GaSb		1 ml H_2O_2 1 ml HCl 2 ml H_2O	1 min	Develops etch figures.
Ge	(111)	CP4: 15 ml HF 25 ml HNO_3 0.3 ml bromine 15 ml acetic acid	1–3 min	
Ge	(111)	WAg 4 ml HF 2 ml HNO_3 4 ml H_2O 0.2 g $AgNO_3$		
Ge	(111) (110)	Ferricyanide 100 ml deionized H_2O 9.7 g $K_3Fe(CN)_6$ 13.7 g KOH	1–10 min	80°C

* Extracted from H. C. Gatos and M. C. Lavine, Chemical Behavior of Semiconductors Etching Characteristics, *Lincoln Lab. Tech. Rep.* 293, January, 1963.

TABLE 27 Dislocation Etches* (Continued)

Material	Orientation	Etch name and composition	Etch time	Remarks
Ge		Superoxal 1 ml HF 1 ml H_2O_2 4 ml H_2O		
Ge	(100)	(100) etch 75 ml HF 25 ml HNO_3 50 ml H_2O 3 g $Cu(NO_3)_2$		
Ge	(100)	Peroxide: 1 ml HF 1 ml H_2O_2 (30%) 1 ml acetic acid	1 min	
HgSe		6 HCl 2 HNO_3 3 H_2O	Start with chemical polishing surface 5 min, films removed by immersing briefly in 50 HNO_3 10 acetic 1 HCl 20 18N H_2SO_4 then brushing under H_2O. Repeat process.	B surfaces develop triangular figures.
HgTe		1 HCl 1 HNO_3	Start with chemical polishing surface in 6HNO_3 1 HCl 10-15 min 1 H_2O 3-1 min etching with H_2O rinsing in between.	Pits on A surface with background figures.
InAs	(111)	Bromine-methanol 5-10% bromine in methanol by volume	13-30 sec	
InAs		15 ml HF 75 ml HNO_3 15 ml acetic 0.06 Br_2	5 sec	Etch pit on A surface.
InAs		HCl		Develops etch figures.

TABLE 27 Dislocation Etches* (Continued)

Material	Orientation	Etch name and composition	Etch time	Remarks
InAs GaAs	(111) (110) (100)	Silver nitrate–chromic acid 2 ml deionized H_2O 8 mg $AgNO_3$ 1 g CrO_3 1 ml HF	10 min	
InP		0.4 N ferric ion in concentrated HCl	1.5 min	Develops etch figures.
InSb		1 HF 1 H_2O_2 8 H_2O 0.4 % n-butylthiobutane		Develops β-dislocation etch pits on A and B surfaces.
PbS		1 HCl 3 10% thiourea	1–10 min at 60°C	Reveals dislocation.
PbTe		2 15% NaOH 1 saturated $Na_2S_2O_8$	10 min	Reveals dislocation.
Si		Copper etch: 4 ml HF 2 ml HNO_3 4 ml H_2O 0.2 g $Cu(NO_3)_2 \cdot 3H_2O$		
Si	(111)	Copper etch: 24 g copper nitrate 2.4 g bromine 600 ml HF 300 ml HNO_3 Mix 1 part to 10 parts deionized water to use.	2 hr, ultrasonic	>0.02 ohm-cm
Si	(110) (100) (112)	Dash etch: 1 ml HF 3 ml HNO_3 10 ml acetic	4 hr	
Si	(111)	Sirtl: 1 g CrO_3 2 ml H_2O Mix with 1 ml HF just before using.	1–7 min	
SiC	(111)	NaOH	900°C	Dislocation.
SiC		Na_2O_2	350°C	Dislocation.
ZnS		0.5 M $K_2Cr_2O_7$	10 min at 95°C	Etch pits on A surfaces.

TABLE 28 Resistivity Measurement Methods

Method	Geometry	Advantages	Disadvantages
Direct		Simple in concept, requires no special equipment.	Requires sample with known regular dimensions; includes effect of contact resistance.
Two-point probe		Accuracy very good.	Requires sample of known regular dimensions.
Four-point probe		Quick to use, needs minimum of sample preparation.	Requires special probe head. If sample is not very large, must use various correction factors.
Spreading resistance		Can be used in very small areas.	Accuracy poor, dependent on probe pressure.
Contactless (inductive)		Requires no direct mechanical contact.	Accuracy poor, geometry-dependent.
Contactless (capacitance)		Requires no direct mechanical contact.	Accuracy poor, most applicable only to high-resistivity materials.

The relation between voltage and width is dependent on the concentration of ionized impurity (N_A or N_D) in the vicinity of the junction and may be calculated from Poisson's equation. Thus, if the capacitance is measured as a function of applied voltage, $N_{A,D}$ may be determined.

It can be shown that for an abrupt junction,

$$N(x - x_j) = \frac{1}{q\epsilon\epsilon_0 A^2} \frac{C^3}{dC/dV}$$

$$x = \frac{\epsilon\epsilon_0 A}{C} \tag{15}$$

where q is the electronic charge, V the applied voltage, C the capacitance at voltage V, A the area of the diode, ϵ the dielectric constant, and ϵ_0 the permittivity of free space.

7-56 Semiconductor Materials

There are a number of limitations to this procedure which must be considered. One of the more obvious ones is that it requires diodes to be fabricated. The high-temperature operation associated with such processing may change the impurity concentration to be measured, and in any event, unless the diodes are required anyway, the method is destructive.

Three-point probe[51] The breakdown voltage of a diode formed between a point-contact probe and a semiconductor is a function of the carrier concentration and can be calibrated directly in resistivity. Besides the point-contact diode probe, one contact to the surface is required for a voltage pickoff, and another is required for the return current path (hence "three-point probe"). A sketch of the equipment is shown in Fig. 43. Because of damage due to local heating, pulsed operation is the normal mode. As long as the point is slightly further away from a conducting or high recombination surface boundary than the width of the depletion layer at breakdown, the reading is independent of boundary effects. When it is closer, the breakdown voltage is proportional to the separation between the diode probe and the surface and not to carrier concentration.

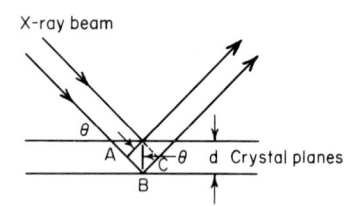

Maxima in intensity occur when the distance $\overline{ABC}$ equals an integral number of wavelengths, i.e.
$$\sin\theta = n\lambda/2d$$

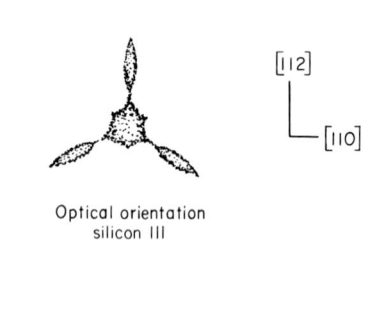

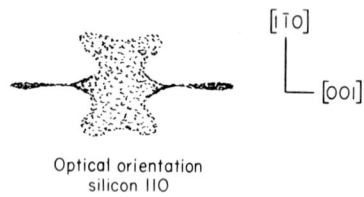

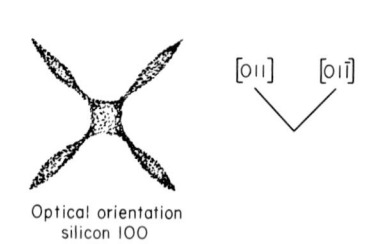

Fig. 37 Optical orientation patterns.

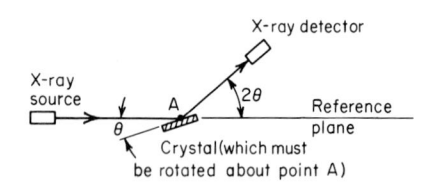

Fig. 38 X-ray orientation geometry.

Profiling It is often desirable to determine the impurity profile normal to the surface of a slice after some operation (e.g., epitaxial deposition or diffusion) or to measure resistivity fluctuations laterally across the surface. Each of these requirements usually necessitates a finer resolution than is normally employed and thus requires special procedures. For example, a diffused layer may be only a few microns thick, and even with a fine point on a spreading resistance probe, the active volume could easily encompass the whole layer depth. Various direct methods are summarized in Fig. 44, and in addition chemical or electrochemical methods can be used. These latter depend on differences of etching or plating speed with impurity concentration and are most useful in looking at variations across a large surface area. It is difficult to determine the actual magnitude of resistivity in

TABLE 29 Carrier Concentration Measurement

Method	Geometry	Advantages	Disadvantages
p-n junction Voltage—capacitance		Applicable to finished devices.	Requires p-n junction. Limited range for given structure.
Hall voltage		Applicable to wide range of concentrations.	Requires magnetic field, comparatively large sample.
Metal-insulation Semiconductor voltage—capacitance		Applicable to thin layers not electrically isolated from substrate.	Difficult to interpret.
Three-point probe		Useful for high-resistivity layer on low-resistivity substrate.	Accuracy poor, surface-dependent, must depend on standards for calibration.
Thermoelectric emf			Accuracy poor, requires hot probe, depends on standard for calibration.
Thermal rebalance		Can be used on thin high-resistivity layers on low-resistivity substrate.	Requires ohmic alloy contact, destructive.

this manner, but spatial variations become clearly visible. Sirtl,* copper,* or anodic[52] etching have all been used for delineation.

For profiling normal to the surface, use incremental lapping techniques and measure sheet-resistance change.† If the sheet resistance R_1 is measured, a thin layer of thickness Δx removed, and the sheet resistance again measured (R_2), the average

---

* See Tables 27 and 35 for various basic etch formulations. Individual variations of these various etches often prove more effective than the specific ones mentioned.

† This presupposes that the whole region to be profiled is thin compared to four-point probe spacing. If it is a diffused layer which is being examined, there will ordinarily be a p-n junction separating it from the bulk of the semiconductor. For profiling diffusions into material of the same type, a different interpretation is required, and the reader is referred to Ref. 1.

Semiconductor Materials

resistivity of the layer removed is given by

$$\rho = R_1 R_2 \frac{\Delta x}{R_2 - R_1} \tag{16}$$

It is often convenient to convert sheet resistance versus thickness to bulk resistivity versus thickness by a method which minimizes the effect of experimental error in the sheet-resistance value and does not depend upon the value of Δx used. This is done by relating the logarithm of the sheet-conductance values to depth, from which the bulk resistivity ρ for any thickness x can then be found. If $G(x)$

TABLE 30 Impurity Concentration Determination

Method	Advantage	Disadvantage
Neutron activation	Great sensitivity.	Special equipment, not applicable to routine evaluation.
Radio tracer	Great sensitivity.	Special equipment, impurity to be studied must be available as a radioactive isotope. Not applicable to routine evaluation.
Solid mass spectroscopy	Great sensitivity.	Very susceptible to surface contamination. Not applicable to routine evaluation.
Emission spectroscopy	More readily available equipment.	Limited sensitivity. Not applicable to routine evaluation.
Optical absorption	Equipment widely available.	Limited sensitivity, limited applicability, requires special sample preparation.
X-ray fluorescence	Nondestructive.	Special equipment, limited accuracy. Not applicable to routine evaluation.
Wet analysis	Usually available.	Very low sensitivity. Not applicable to routine evaluation.

is the sheet conductance for a thickness of material x, then the conductivity $\sigma(x)$ is given by

$$\sigma(x) = \frac{dG}{dx} \tag{17}$$

The logarithm of sheet conductance has some functional dependance on depth, which may be defined as $F(x)$: i.e.,

$$\ln G(x) = F(x) \tag{18}$$

Differentiation with respect to x gives

$$\frac{1}{G}\frac{dG}{dx} = \frac{dF}{dx} \tag{19}$$

So that by combining Eqs. (17) and (19)

$$\rho(x) = \frac{1}{G\, dF/dx} \tag{20}$$

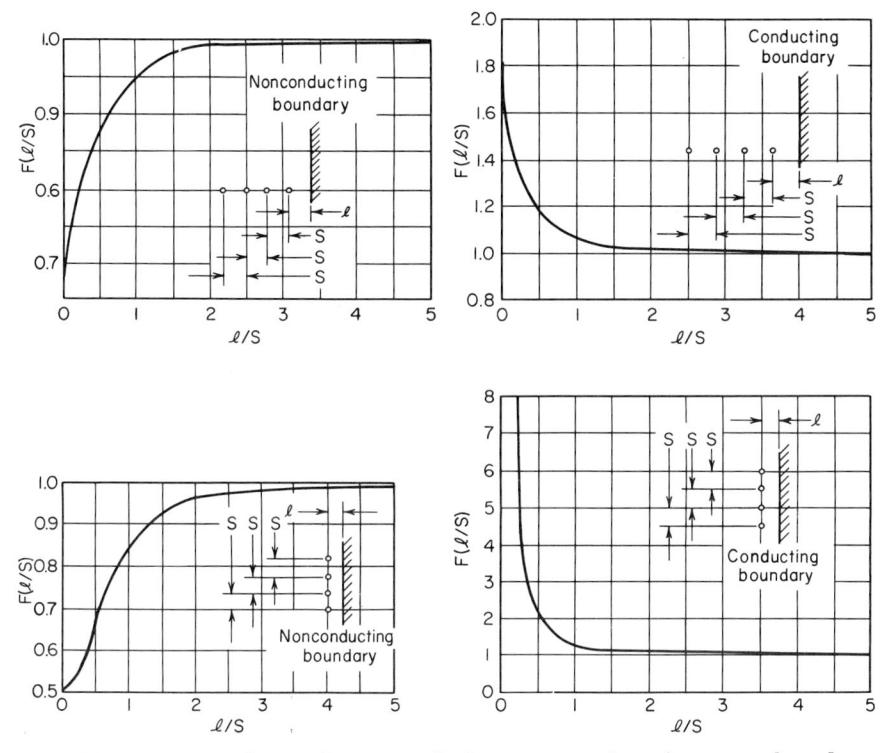

Fig. 39 Correction factors for use with four-point probes close to a boundary $\rho = 2\pi S(V/I)F$.[43]

If it is desired to measure a concentration profile in which the concentration increases with increasing depth, conventional incremental lapping from the front surface is not appropriate since the removal of the first layers may hardly change the resistance of the remaining sample. In such cases, beveling or parallel-layer removal from the opposite side is used.

TABLE 31 Correction Factor $\bar{F}$ as a Function of the Ratio of Probe Separation $\bar{S}$ to Slice radius r[45]

$\bar{S}/r$	$\bar{F}$	$\bar{S}/r$	$\bar{F}$
0	4.532	0.11	4.417
0.01	4.531	0.12	4.395
0.02	4.528	0.13	4.372
0.03	4.524	0.14	4.348
0.04	4.517	0.15	4.322
0.05	4.508	0.16	4.294
0.06	4.497	0.17	4.265
0.07	4.485	0.18	4.235
0.08	4.470	0.19	4.204
0.09	4.454	0.20	4.171
0.10	4.436		

Lifetime

Lifetime of a carrier (average time between generation and recombination) may be measured either by temporarily injecting an excess, or nonequilibrium, number of carriers and measuring the time required to return to equilibrium, or by maintaining an excess and relating it to the recombination (or generation) rate through an appropriate function of lifetime. The latter measurement almost always involves an intermediate determination of the diffusion length, from which the lifetime may be obtained if the diffusion coefficient D_n or D_p is known. A great number of variations of these two basic approaches have been proposed and are summarized in Table 32.

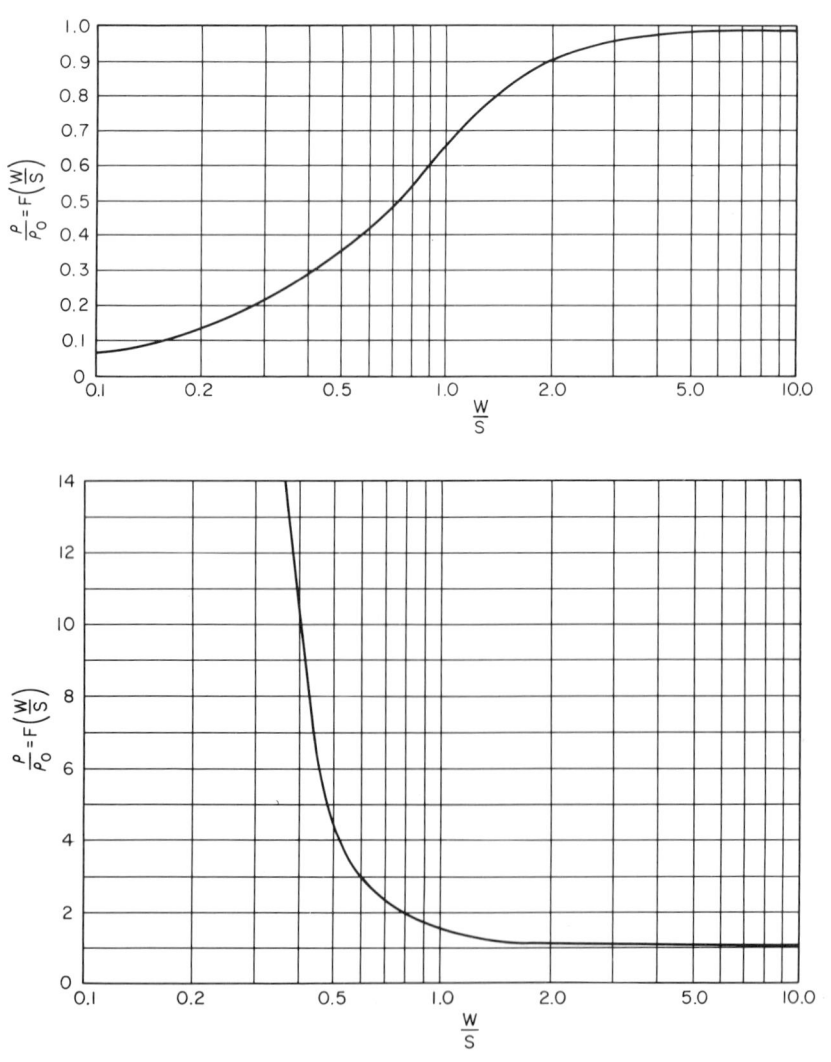

Fig. 40 Correction factors for use with four-point probe and thin layers. The top graph is for slices with nonconducting surfaces; the bottom graph is for slices with a conducting lower surface.[44]

Measurement Techniques 7-61

The choice of a particular method or variation depends primarily on the magnitude of the lifetime and the sample size, and whether the sample can only be a piece of bulk material or if a complete device is available. The usual working range for several of these methods is shown in Table 33. It should be remembered,

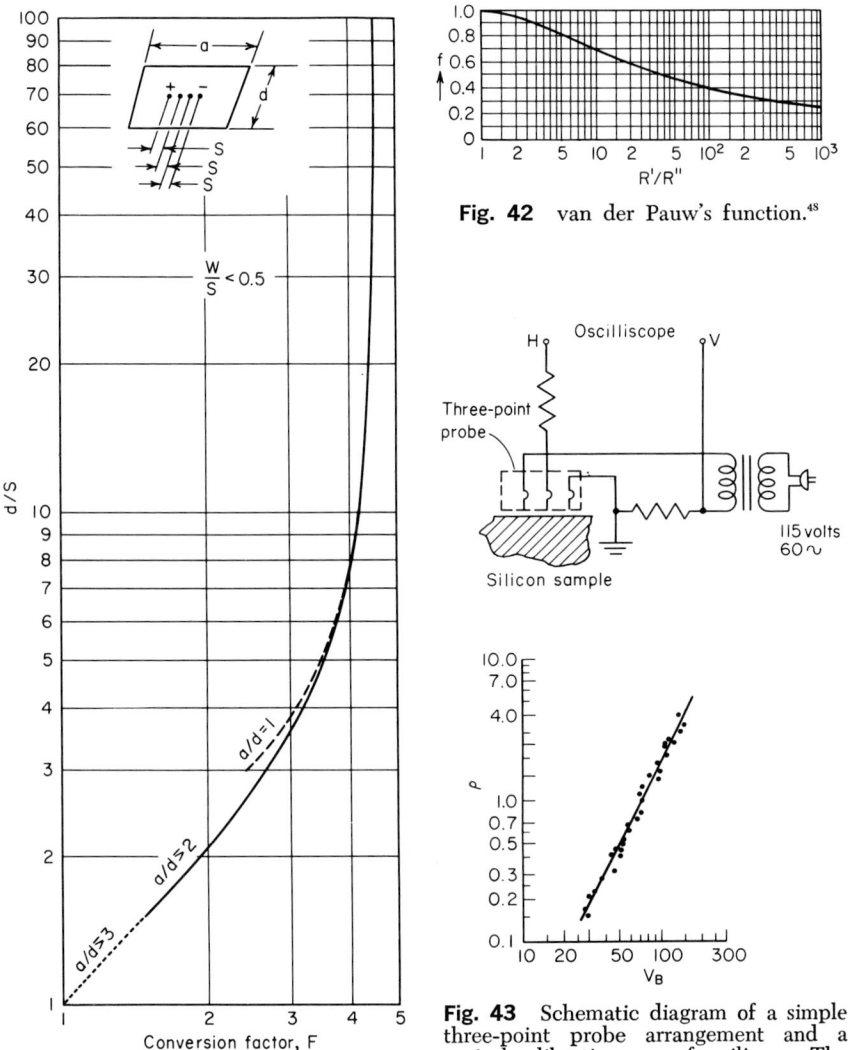

Fig. 41 Correction factor for use with four-point probe and thin layers.[46]

Fig. 42 van der Pauw's function.[48]

Fig. 43 Schematic diagram of a simple three-point probe arrangement and a typical calibration curve for silicon. The instruments must be individually calibrated.[51]

however, that with special emphasis and attention to a particular case, nearly any method can be extended in either direction. However, the photodecay (PC) method is so simple relatively, that it is used whenever possible. Thus there has been little inclination to extend other methods into the longer-lifetime region. The one exception has been that as longer and longer lifetimes have been achieved after

7-62 Semiconductor Materials

various processing stages, the PC method sometimes becomes size-limited (for example, with slices). In such circumstances surface photovoltage, which allows approximately a 4-to-1 reduction in thickness for a given lifetime measurement, becomes very attractive, even though it is quite tedious.

Method	Description	Value measured	Limitations
4-point probe resistivity	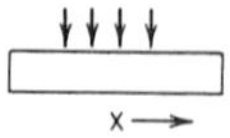	$\rho(x)$	Poor resolution
4-point probe or spreading resistance combined with beveling	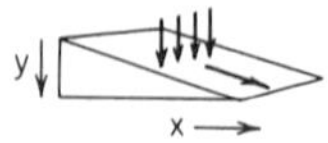	$\rho(y)$	Destructive of sample—assumes uniformity in x direction. Requires large x dimensions, $Y <$ probe spacing.
2-point probe	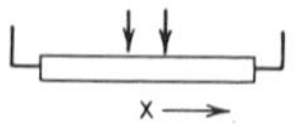	$\rho(x)$	Requires careful contacting to ends. Sample must have large length to area ratio.
1-point probe (spreading resistance)	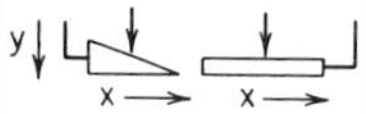	$\rho(x)$	
Voltage-capacitance	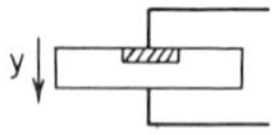	$N(y)$	Requires an abrupt p-n junction; area of junction must be accurately known; values of y which can be examined are restricted to the difference between width of space-charge region caused by built-in potential and width at avalanche breakdown.
4-point probe combined with successive layer removal	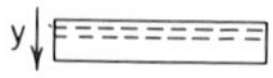	$\rho(y)$	Requires accurate method of removal, ρ must remain constant or increase with y.
Neutron activation combined with successive layer removal		$N(y)$	Impurity must be amenable to neutron activation.
Tracer analysis combined with layer removal		$N(y)$	Radioactive impurity.
X-ray probe	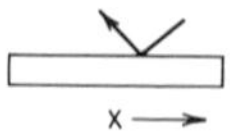	$N(x)$	Accuracy subject to surface contamination.

Fig. 44 Profiling methods.

Photoconductive decay In this method the excess carriers are generated by irradiating the sample with light of short enough wavelength to produce hole-electron pairs. The conductivity of the sample is directly proportional to the number of carriers, and so monitoring the conductivity after the light is removed allows the

TABLE 32 Methods of Measuring Carrier Lifetime*

Conductivity decay methods
1. Photoconductive decay:
   Direct observation of resistivity
   Q changes
   Microwave reflection
   Microwave absorption
   Spreading resistance
   Eddy-current losses
2. Pulse decay:
   Direct observation of resistivity
   Microwave absorption
3. Bombardment decay

Conductivity modulation methods
4. Photoconductivity:
   Steady state
   Modulated source
   Infrared detection, steady state
   Infrared detection, modulated source
   Q changes
   Microwave absorption
   Eddy-current losses
   Spreading resistance, modulated source
5. Pulse injection—spreading resistance

Magnetic field methods
6. Suhl effect (and related effects)
7. Photomagnetoelectric effect:
   Steady state
   Modulated source
   Transient decay

Diffusion length methods
8. Traveling spot:
   Steady state
   Modulated source
9. Flying spot
10. Dark spot

Diffusion length methods (Continued)
11. Sweep-out effects:
    Pulse injection
    Photoinjection
12. Drift field
13. Pulse delay
14. Emitter point efficiency

Junction methods
15. Open-circuit voltage decay
16. Reverse recovery
17. Reverse-current decay
18. Diffusion capacitance
19. Junction photocurrent:
    Steady state
    Decay
20. Junction photovoltage
21. Stored charge
22. Current distortion effects
23. Current-voltage characteristics

Transistor methods
24. Base transport
25. Collector response
26. Alpha cutoff frequency
27. Beta cutoff frequency

Other methods
28. MOS capacitance
29. Charge collection efficiency
30. Noise
31. Surface photovoltage:
    Steady state
    Decay
32. Bulk photovoltage:
    Steady state
    Modulated source
33. Electroluminescence
34. Photoluminescence
35. Cathodoluminescence

* From W. Murray Bullis, Measurement of Carrier Lifetime in Semiconductors: an Annotated Bibliography Covering the Period 1949–1967, *Nat. Bur. Stand. Tech. Rep.* AFML-TR-68-108, June, 1968.

time to return to equilibrium to be determined. In the simplest case,

$$\Delta n = Ae^{-t/\tau}$$
$$\Delta p = Be^{-t/\tau}$$
(21)

where τ = lifetime
$\Delta n, \Delta p$ = excess number of mobile carriers
t = time
A, B = constants

Since the conductivity change ΔG is proportional to the number of extra carriers, it too will decay with a time constant τ.

Surface recombination can seriously affect both the form and interpretation of photodecay curves, and so it is recommended practice to either calculate the surface contribution or else use very large samples.[53] In any event, light with wavelengths near the band edge (and thus with relatively low absorption coefficients) should

7-64 Semiconductor Materials

be used to ensure that the carriers are generated in the body of the semiconductor and not all at the surface.

The use of large samples and sandblasted surfaces is the procedure recommended by the IRE Standards.[54] The exact size of sample required is of course somewhat subjective, and varies with the diffusion coefficient of the minority carriers. Figure 45, based on Ref. 53, gives a suggested minimum sample dimension as a function of lifetime to be measured for various diffusion coefficients. For convenience, typical high-resistivity D values are also tabulated.

If the dimensions are not as large as indicated in Fig. 45, the bulk lifetime can be obtained by applying a suitable correction factor to the measured value. If

TABLE 33 Range and Limitations of Lifetime Measuring Methods*

Method	Lower limitation	Lifetime in seconds and the scale 10^{-9} 10^{-6} 10^{-3}	Upper limitation	Miscellaneous limitations
Diode recovery...	Pulsing instrumentation		Device geometry	
PME-PC........	Adequate signal to noise		....	Complex equipment very sensitive to surface effects
Surface photovoltage	Inadequate signal to noise		....	Time-consuming measurement
Photodecay......	Inadequate signal to noise, lack of rapid cutoff source	†	Sample size	
MOS capacitance	Doping level to n_i ratio		Surface recombination	

* Based on Fig. 2 in Joseph Horak, Minority Carrier Lifetime Measurements on Silicon Material for Use in Electron Irradiation Studies, *AF Contract* F19628-67-0043 *Sci. Rep.*, May, 1968.

† May be extended into this range by using high-energy electron excitation.

the sample is rectangular, with length A and cross section BC, and is sandblasted as before so that the surface recombination term is very large,

$$\frac{1}{\tau_{meas}} = \frac{1}{\tau_{bulk}} + \pi^2 D \left(\frac{1}{A^2} + \frac{1}{B^2} + \frac{1}{C^2} \right) \quad (22)$$

where D is an ambipolar diffusion coefficient given by

$$D = \frac{n + p}{n/D_p + p/D_n} \quad (23)$$

For reasonably heavily doped material, D reduces to D_n for p-type, and D_p for n-type. It can be shown analytically and can also be seen from the decay curve that the initial portion of the curve is not exponential because of surface-bulk interactions. IRE Standards thus recommend that no measurement of the decay slope be made until the signal has decayed to less than 60 percent of its peak value if half or less of the width of the sample is exposed to illumination. If more than half of the width is illuminated, no measurement should be made until the signal is less than 25 percent of the peak value. If trapping is present, a long tail on the decay curve will be observed. Shining a continuous background light on the sample or heating it to about 70°C should remove the effect and allow useful measurements to be made.

The simplest method of observing the carrier decay is to observe the voltage drop across a resistor in series with the sample. If current is supplied by a constant-current source, then it can be shown[54,55] that the time constant for the voltage decay is related to the carrier time constant by

$$\tau_{\text{sample}} = \tau_{\text{volt}} 1 - \frac{\Delta V}{V_0} \qquad (24)$$

if ΔV is small compared to V_0, the dark IR drop developed by the constant current across the portion of the sample illuminated by the chopped light. Ordinarily, conditions are chosen so that $\Delta V/V_0$ is very small, and the correction can be neglected. Actual measurement of the time constant is by means of an oscilloscopic display and by using the variable sweep to match a curve drawn on the face; using a fixed sweep and measuring the time for an amplitude reduction to $1/\epsilon$ of the value at the start of the time measurement; or generating a second curve from a variable but known RC time constant and matching it to the photodecay curve.[56]

Noncontacting detecting schemes which have been occasionally suggested but have not received widespread acceptance include inserting the sample in a microwave cavity[57,58] or in the coil of a radio-frequency bridge, and capacitance-coupling a high-frequency constant current to the sample and detecting the carrier modulation.[58,59]

Drift method[60,61] This is one of the oldest methods of measuring lifetime (and mobility) and is direct in its approach. A pulse of current (carriers) is injected into a relatively long, narrow rod of material, swept along by an applied field, and collected some distance away. By varying the external field, the time of arrival of the pulse can be varied.

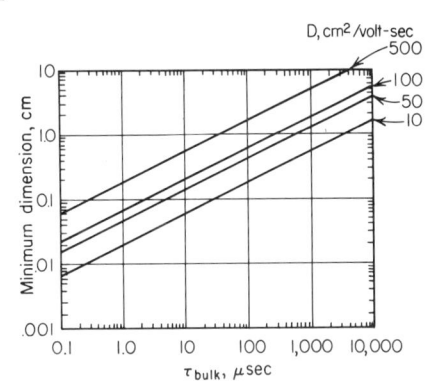

Fig. 45 Minimum sample thickness required for lifetime measurement.[53]

Mat	D, cm2/volt-sec	
	n	p
Ge	100	50
Si	35	13

The received amplitude may then be plotted as a function of time, and the lifetime determined. This procedure is not suited to very short lifetimes since the spacing between emitter and collector would become exceedingly small.

MOS capacitor transient response[62,63] If an appropriate polarity voltage (negative to the metal for n-type, positive for p-type) is suddenly applied to the plates of an MOS capacitor, a depletion layer will quickly form in the semiconductor. As time progresses, the layer thickness will decrease as carriers are generated within the space-charge region, and an inversion layer forms at the surface. The measured capacitance will thus suddenly decrease as V is applied, and then increase in value with a characteristic time constant T_0. By making various assumptions such as that the minority- and majority-carrier lifetimes are equal, and that the recombination centers are located mid-gap, T_0 can be related to τ through Eq. (25).

$$T_0 = \frac{2\tau N_{A,D}}{n_i} \qquad (25)$$

where $N_{A,D}$ is the doping level of the semiconductor, and n_i is the intrinsic carrier concentration. The constant T_0 may in turn be found from the experimental curve by

$$T_0 = C^2 \frac{1 - (C/C_f)}{dC/dt} \qquad (26)$$

where dC/dt may be measured at any point on the curve (though for the first few seconds surface recombination may contribute), and C_f is the final value of capacitance.

Diffusion length measurements Instead of directly measuring the recombination, as the previous methods have done, it is possible to deduce lifetime from a measurement of diffusion length ($\sqrt{D\tau}$). For example, suppose that carriers are generated by shining light on the surface and collected some distance away by a suitable biased point-contact electrode (or built in p-n junction if one is available). The carrier motion can be described by the continuity equation,

$$\nabla^2 n - \frac{n}{L_n^2} = 0$$
$$\nabla^2 p - \frac{p}{L_p^2} = 0 \tag{27}$$

where L is the diffusion length.

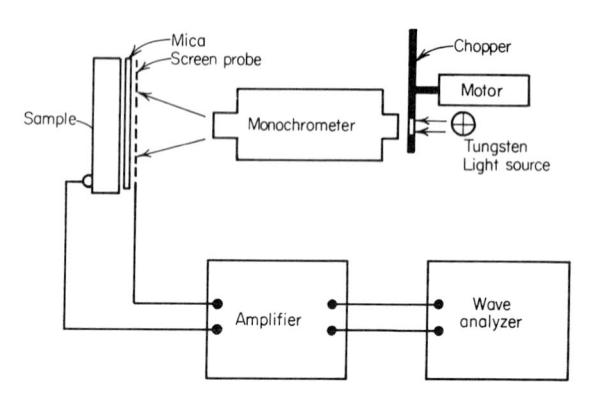

Fig. 46 Diagram of surface photovoltage technique.

For the particular geometry used, Eq. (27) can be solved for n or p collected as a function of generation and collector separation.[64] Because of various difficulties in evaluating constants, experimental measurements are usually made for several separations, and a curve of collector current versus distance is plotted. The relation between distance and current depends on the geometry, but usually there will be a region which can be approximated by

$$I = I_0 \exp \frac{X}{L} \tag{28}$$

where X is the distance from spot to collection, and L is the diffusion length. Therefore, if response versus distance is plotted on semilog paper, the diffusion length is given by the slope of the line.

If a junction is available, but diffusion lengths are very short so that response is obtained only if the spot is exceedingly close to the junction, some relief occurs if a beveled configuration is used. The spot can be moved rather large distances laterally while remaining quite close to the junction.[65,66]

As an alternative to moving the spot, the short-circuit current of an irradiated p-n junction may be measured, and if the generation rate and distance from the junction to each electrode is known, it is sometimes possible to calculate the diffusion length. In general,

$$i_{sc} = qG(L_1 + L_2) \tag{29}$$

where i_{sc} is the short-circuit current density, q the electronic charge, G the generation rate, and L_1 and L_2 are functions of L_n and L_p. If the electrodes are many diffusion lengths away from the junction, Eq. (29) reduces to

$$i_{sc} = qG(L_n + L_p) \tag{30}$$

Then, if the diffusion length of one carrier is much greater than that of the other (e.g., if one mobility is much higher, or if one lifetime is much greater than the other), one L will dominate. Alternatively, if one electrode is very close to the junction and other several diffusion lengths away, the current contribution will be primarily from the wide region, and discrimination of one L can be effected.

Surface photovoltage[67,68] The minority-carrier diffusion length can be determined from the spectral dependence of the surface photovoltage V_s. This voltage, which is developed at the illuminated surface of a semiconductor, is proportional to the excess minority-carrier density, and may be detected with an experimental arrangement such as shown in Fig. 46; thus

$$V_g = f(\Delta p) \quad \text{for n-type material} \tag{31}$$

where
$$\Delta p = \frac{\beta I_0 (I - R)}{S + D/L} \frac{\alpha L}{1 + \alpha L} \tag{32}$$

and Δp = excess holes
β = quantum yield
I_0 = light intensity
R = reflection coefficient
α = optical absorption coefficient for wavelength of light being used
S = surface recombination velocity
d = sample thickness
d_1 = thickness of surface space-charge layer

if $\alpha d \gg 1$, $\alpha d_1 \ll 1$, $d/L \gg 1$, $d_1/L \ll 1$, and Δ_p is much less than the equilibrium electron concentration.

It has been shown that Eqs. (31) and (32) can be combined and rewritten as

$$I_0 = f(V_s) M \, 1 + \frac{1}{\alpha L} \tag{33}$$

where
$$M = \frac{S + D/L}{\beta(1 - R)}$$

from which
$$I_0 = 0 \quad \text{when } L = -\frac{1}{\alpha}$$

If the quantum yield β and the reflection coefficient R are independent of the wavelength of incident light, then M is a constant for a given material and surface. Since the absorption coefficient of most semiconductors is a strong function of wavelength in the wavelength region where hole-electron pairs are produced, α can be changed rather easily by changing the incident-light wavelength. In addition, if the intensity I_0 is changed simultaneously so that V_s and, therefore, $f(V_s)$ remain constant, a plot of I_0 versus $1/\alpha$ can be extrapolated to $I_0 = 0$ to give L. This is illustrated in Fig. 47. This method is more tolerant of geometry than photoconductivity decay and is thus preferred when only small samples are available. Figure 48 gives a comparison of adequate sample thickness for each of the two procedures.

Measurements on Finished Devices[69-72]

Since processing steps either by intent or otherwise can radically alter the lifetime, it is often desirable to determine lifetime on a finished device. As was shown

Semiconductor Materials

in Table 32, there are a number of ways to do this, all quite dependent on device geometry and mode of operation. Figure 49 shows two of the more common methods. In the first, a p-n junction is first forward-biased and then disconnected from the current source, and the open-circuit voltage is observed. If contacts or other junctions are a few diffusion lengths away from the junction being biased, the voltage will appear as in Fig. 49a, and the lifetime is given by

$$\tau = \frac{kT}{q}\frac{\Delta t}{\Delta V} \tag{34}$$

where $\Delta V/\Delta t$ is the slope of the linear portion of the decay curve, T the temperature, k Boltzmann's constant, and q the electronic charge. Instead of observing open-circuit voltage the diode may be switched from forward to reverse bias and the

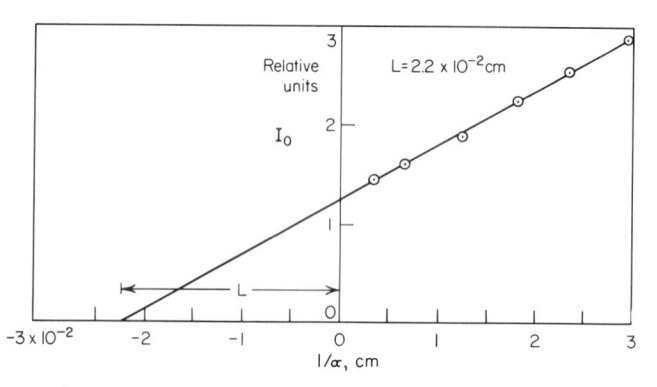

Fig. 47 Use of surface photovoltage to determine the diffusion length L from I_0 and α.

current transient observed. It will look approximately as shown in Fig. 49b and with the same restrictions as before:

$$\tau = \frac{t_r}{0.23} \tag{35}$$

Layer Thickness

In semiconductor work, layers to be measured can be broken down into two categories: (1) thicknesses of the variously doped regions within the semiconductor itself; (2) thickness of external layers of foreign material, e.g., oxides or nitrides. If the properties of the layer and its substrates are such that some region of the layer can be etched away to leave a step, the height may be measured either by mechanical profiling or by interferometry. When this is not possible or desirable, optical differences between the layer and substrate are most often used for differentiation. Optical viewing may be either in plan or in section. The second case is almost always visual, and one usually stains to improve contrast and uses mechanical beveling to increase magnification. The use of beveling is shown in Fig. 50. By sectioning at some small angle to the surface, a magnification of 1/sin θ is achieved; angles of 1° and 3° are common, although 7° is sometimes used. If the lateral dimensions of the sample are very small, a low-angle lap may require removal of all the top surface before the bottom of the layer is reached (i.e., S of Fig. 50b reduces to zero). Since both the top surface and the intersection of the beveled surface with the interface are required for measurement, a larger lap angle must then be used. Actual measurement of thickness may be either by measuring with a bifilar eyepiece or traveling stage, or by laying a partially silvered piece of cover glass on the reference surface S and counting the number of fringes between X and Y. The latter procedure removes the requirement for accurate knowledge of the angle θ. Directions for delineating various combinations of layers are given in Table 34.

Stacking faults[73] Stacking faults often originate during the initiation of epitaxial overgrowth, and because such faults always have a predictable geometry, they can be used to measure the thickness of a layer which grew after their inception. Specifically, after the trace of the fault system on the surface has been delineated, and this will be a polygon (see Crystallographic Defects), the distance from the surface to the point of origin of the fault is given by the length of a side of the polygon multiplied by the appropriate number in Table 35. It should be remembered that except for the specific cases of (111) and (100) planes, the sides will not be all equal in length. Thus, more than one multiplication factor is given in those instances. Note, however, that any side multiplied by its own multiplication factor will give the same answer as any other.

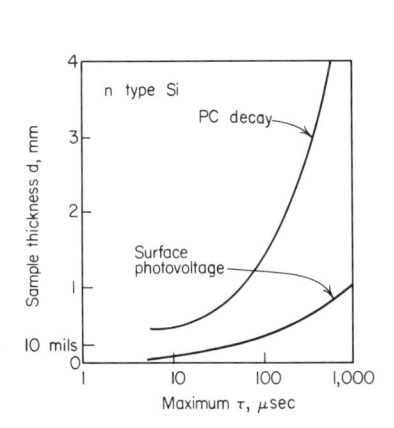

Fig. 48 Maximum lifetime determinable for a given thickness.[116]

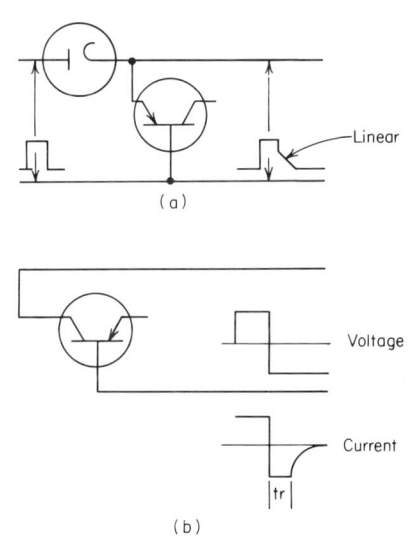

Fig. 49 Methods for measuring lifetime in finished devices.

Interferometry The general principle of interferometric thickness measurement is illustrated in Fig. 51. The incident ray I impinges on the front surface and is partially reflected (1), and partially transmitted. If there is sufficient discontinuity in the index of refraction at the layer boundary, a portion of the transmitted ray will be reflected and emerge as ray 2. Phase change may occur during each reflection unless the optical paths between rays 1 and 2 differ by an integral number of wavelengths, and destructive interference will be observed. By assuming that the phase change during reflection is independent of wavelength, a determination of the wavelengths at which maxima or minima in intensity occur can be used to calculate the layer thickness.

The thickness is given by[74]

$$T = \frac{x(\lambda_m \lambda_{m+x})}{2n \cos \phi (\lambda_m - \lambda_{m+x})} \qquad (36)$$

where T = layer thickness
x = integral number of fringes from λ_m to λ_{m+x}
n = index of refraction of layer
ϕ = angle of incidence of beam on slice
λ_m, λ_{m+1}, etc., = wavelengths at which successive intensity maxima occur

TABLE 34 Etches for Delineating Layers in Semiconductors

Sample	Delineant	Method of application*	Illumination	Sectioning angle, deg	Visible results†	Layer delineated	Delineation time, sec	Resistivity range
Silicon									
Diffused									
n⁺/p	1-3-10 50-6	1, 2, 4	Yes	1, 3, 7	1	p	1-15	
p⁺/n	1-3-10 50-6	1, 2, 4	Yes	1, 3, 7	1	p⁺	1-15	$\leq 0.00x/\leq 0.0x$
n⁺/n	1-3-10 50-6	1, 2, 4	Yes	1, 3, 7	1	n⁺	1-15	
p⁺/p	1-3-10 50-6	1, 2, 4	Yes	1, 3, 7	1	p⁺	1-15	
Alloyed									
p⁺/n	1-3-6 Sirtl A	1, 2, 4 1	No	7	3	...	10-60 1-30	$\leq 0.00x/\leq 0.0x$
n⁺/p	1-3-6 Sirtl A	1, 2, 4 1	No	7	3	...	10-60 1-30	
Combination (diffused and epitaxial)									
	50-6 followed by 1-3-6	1	Yes	1, 3, 7	1, 3	...	1-15 10-60	$\leq 0.0x/\geq 0.00x$
	1-3-10	1	Yes	1, 3, 7	2	...	10-60	
Epitaxial									
n/n⁺	1-3-10 1-3-10-Cu 1-3-10-Cu-Mo 50-6-Cu 1-3-5 1-3-6	1, 2, 4 2, 3, 4 4 2, 3, 4 1, 2, 4 3	Yes Yes Yes Yes Yes No	1, 3, 7	1 2 2 2 1 1 and/or 3	n⁺	1-15 1-15 1-15 1-15 1-15 10-60	$\geq 0 \cdot x/\leq 0.0x$

p/p⁺	1-3-10	1, 2, 4	Yes	1, 3, 7	1	p⁺	1–15
	1-3-10-Cu	2, 3, 4	Yes		2		1–15
	1-3-10-Cu-Mo	4	Yes		2		1–15
	50-6-Cu	2, 3, 4	Yes	1, 3, 7	2		1–15
	1-3-5	1, 2, 4	No		1		10–60
	1-3-6	3			1 and/or 3		
n/p	1-3-10	1, 2, 4	Yes		1	p	1–15
p/n	1-3-10-Cu	2, 3, 4	Yes		2		1–15
	1-3-10-Cu-Mo	4	Yes	1, 3, 7	2		1–15
	50-6-Cu	2, 3, 4	No		2		10–60
	1-3-6	3	Yes		1 and/or 3		1–15
	50-6	1, 2, 4			1		
p/p	1-3-6	1, 3	Yes	1, 3, 7	3, 1	...	5–60
n/n	1-3-10						
Miscellaneous							
Si-SiO₂	None	......	...	......	4	...	
Si-SiC	None	......	...	......	4	...	
Si-polySi	Sirtl A	3	No	......	3	...	10–30
Si-ceramic	Sirtl A	1	No	1, 3, 7	3	...	1–30

$\geq 0.0x / \leq 0.00x$

All

$\leq 0 \cdot x / \leq 0 \cdot x$

* Methods of application:
1. Swab with cotton swab dipped in delineate.
2. Squirt delineate on sample, using an acid-resistant squirt bottle.
3. Dip in delineate.
4. Drip delineate on sample, using squirt bottle or cotton swab.
5. Dip swab in 5-1 and then in Cu solution, and allow to drip on sample.

† Visible results are:
1. Stain will appear.
2. Stain and/or variable-darkness copper plate.
3. Etched step.
4. Color difference.

TABLE 34 Etches for Delineating Layers in Semiconductors (Continued)

Sample	Delineant	Method of application*	Illumination	Sectioning angle, deg	Visible results†	Layer delineated	Delineation time, sec	Resistivity range
Germanium								
Epitaxial								
n/n^+	50-6 50-6-Cu 1-3-10-Cu	4	Yes No Yes	3, 7	2	n^+	<5	
p/p^+	1-3-10 1-3-10-Cu 50-6-Cu 5-1 5-1-Cu	1, 2, 3 4 4 4 5	Yes	3, 7	4 2 2 1 4	p^+	10-60 <5 <5 <20 <20	$\leq 0 \cdot x / \leq 0.00x$
n/p	50-6-Cu $HF-H_2O_2$ 1-3-10-Cu 1-3-10	4	Yes	3, 7	2 2 2 3	p n	<5 <30 <60 <60	
p/n	50-6-Cu	4	Yes	3, 7	2	n	<5	
Ge/GaAs	1-3-10-Cu 1-3-10 None	1, 2, 3	Yes ...	3, 7 3, 7	2, 3 3 4	Ge Ge	<30 <30	All
Diffused								
n^+/p p^+/n n^+/n p^+/p	50-6-Cu 1-3-10-Cu 5-1-Cu	4 4 5, 4	Yes Yes Yes	3, 7 3, 7 3, 7	2 and 3 2 and 3 2	Plus Plus n^+	<30 <30 <20	
Alloyed								
p^+/n n^+/p	1-3-6-Cu	4	Yes	3, 7	3	....	<30	
Combination (diffused and epitaxial)								
$n^+/(p/p^+)$	50-6-Cu 1-3-10-Cu 5-1-Cu	4 5	Yes	3, 7	2 3	n^+ $p-p^+$	<30	
$p^+/(n/n^+)$	50-6 1-3-10-Cu 5-1-Cu	4	Yes	3, 7	2	n^+	<30	

7-72

Epitaxial InAs/GaAs	1:1 Clorox:water	1	Yes	1, 3, 7	1	GaAs	1	All
InAs/InAs	2 ml H$_2$O 8 mg AgNO$_3$ 1 g CrO$_3$ 1 ml HF	1	Yes	1, 3, 7	1	InAs	15	
GaAs/GaAs	2 ml H$_2$O 8 mg AgNO$_3$ 1 g CrO$_3$ 1 ml HF	1	Yes	1, 3, 7	1	GaAs	600	
GaAs/GaAs (n/n$^+$)	200 g KOH 10 g KAu(CN)$_4$ H$_2$O to make 1 liter	1	Yes	1, 3, 7	Gold plate	n$^+$	2	
GaAs/GaAs (n/n$^+$)	1 ml HNO$_3$ 5 ml HCl Allow to stand 1 hr, 3 parts with 2 parts H$_2$O	1	Yes	1, 3, 7	1	n$^+$	900–1,200	
GaAs/GaAs	1 ml HNO$_3$ 9 ml H$_2$O 0.8 g Fe‡	1	Yes	1, 3, 7	1	p	5	
Ge/GaAs	50 ml HF 5 ml HNO$_3$ (5 ml)		Yes	1, 3, 7	2	Ge	10	
	20 g CuSO$_4$ 80 ml H$_2$O 1 ml HF (5 drops)	1						
	MoO$_3$ saturate							

InAs, GaAs

* Methods of application:
  1. Swab with cotton swab dipped in delineate.
  2. Squirt delineate on sample, using an acid-resistant squirt bottle.
  3. Dip in delineate.
  4. Drip delineate on sample, using squirt bottle or cotton swab.
  5. Dip swab in 5-1 and then in Cu solution, and allow to drip on sample.

† Visible results are:
  1. Stain will appear.
  2. Stain and/or variable-darkness copper plate.
  3. Etched step.
  4. Color difference.
‡ Sixpenny finishing nail.

Semiconductor Materials

For measuring the thickness of high-resistivity epitaxial layers on a low-resistivity substrate, modifications of commercial double-beam spectrometers are ordinarily used. A reflectance attachment brings the beam out to a slice holder and back into the instrument, and sometimes, because of the very low reflection coefficient at n-n⁺ or p-p⁺ interfaces, the reference beam intensity has to be changed in order to obtain reasonable sensitivity. The ratio of resistivities required for good readings is at least 10:1 (layer ρ:substrate ρ), but the substrate must have a resistivity of about 0.1 ohm-cm or less (for higher-purity materials there is much less dependence of refractive index on resistivity).

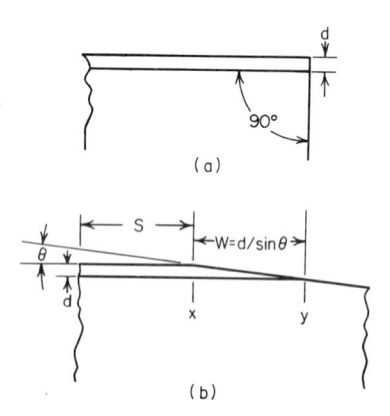

Fig. 50 Mechanical beveling magnification.

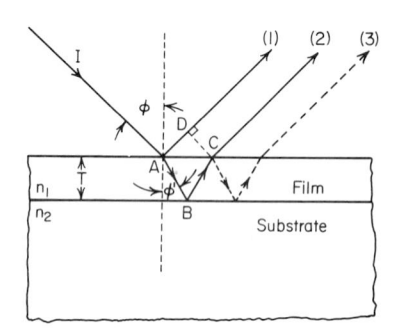

Fig. 51 Ray diagram for reflection from a film deposited on a thick substrate. The amplitudes of rays 1, 2, 3, etc., add vectorially, resulting in maxima and minima in the reflectance spectrum.

Visual determination When transparent layers are viewed under white light, interference colors characteristic of the thickness and refractive index may be seen. As is true with such interferometric methods, the colors repeat as the thickness increases, so that not only the color but also the order of the interference must be known. Tables are sometimes used to relate color to thickness, but color descriptions (e.g., pale blue) mean different things to different observers and are, therefore, quite inaccurate. Printed color charts are also occasionally used but, because of the difficulty of obtaining good color rendition, are not particularly desirable. Actual

TABLE 35 Factors for Converting from the Length of Side of a Stacking Fault to Depth at Which Fault Originated [(111) Fault System][73]

Orientation	Multiplication factors		
1. {110}	0.5	0.577	
2. {221}	0.707	0.785	0.236
3. {111}	0.816		
4. {334}	0.85	0.776	0.142
5. {112}	0.866	0.655	0.288
6. {114}	0.5	0.833	0.575
7. {100}	0.707		

samples independently calibrated and mounted are most satisfactory and have the advantage of allowing observation at angles other than normal. This is of particular importance when trying to determine the order of a particular color (the thicker the sample, the smaller the change in angle required to appreciably alter the color).

Ellipsometry[75] As an alternative to ordinary interferometry, the thickness of a thin transparent film on a reflecting substrate (e.g., SiO_2 on Si) can be determined by ellipsometry. This method depends on the amplitude and phase differences between incident and reflected beams of polarized light. Interpretation requires either fairly complex calculation or graphical methods but can give both thickness and index of refraction. Furthermore, the minimum measurable thickness is only a few angstroms, rather than the few hundred commonly associated with light interferometry. By using infrared,[76] ellipsometric measurements can also be employed to determine the thickness of epitaxial layers of Si on Si or Ge on Ge if the layer resistivity is sufficiently higher than that of the substrate.

VAMFO (variable-angle monochromenter)[77] In this system, instead of varying the wavelength to obtain maxima and minima of intensity, the path length through the film is varied by changing the viewing angle. This allows a much simpler and less expensive system to be constructed. One observes bright (maxima) and dark (minima) fringes on the sample layer as the stage and sample are rotated about an axis perpendicular to the beam. The angular positions (angle of incidence or reflection) of the sample associated with the observed maxima and minima are read from a dial attached to the shaft of the rotating stage. The film thickness is given by

$$d = \frac{\Delta N \lambda}{2n(\cos \nu_2 - \cos \nu_1)} = \frac{\lambda}{2n(\Delta \cos \nu)} \tag{37}$$

where λ = wavelength of filtered light
n = refractive index of the layer
ν = angle of refraction at that fringe for which angle of incidence is θ. Thus, $\sin \nu = (\sin \theta)/n$.
ΔN = number of fringes observed between θ_1 and θ_2
$\Delta \cos \nu = (\cos \nu_2 - \cos \nu_1)/\Delta N$

CRYSTAL GROWTH[78-86]

Because most devices require single-crystal semiconductors for best performance, the growth of single crystals has been studied extensively, and depending on the desired properties, a number of processes are available.

Crystal-growing Environment

Crystal growth can be divided into three broad categories, each of which has several subdivisions.
1. Crystallization from a one-component system. (Note that small amounts of impurity may shift growth from this type to growth from a multicomponent system.)
    a. Crystallization from a liquid of the same composition.
    b. Crystallization from a vapor of the same composition. Evaporation and subsequent regrowth in a vacuum is an example of this.
    c. Crystallization in the solid state. This could include polymorphic transitions and the growth of larger crystallites from small ones during heat treatment.
2. Crystallization from a multicomponent system, with the same three subdivisions as the one-component system. An example would be a silicon crystal grown from a silicon-gold mixture.
3. Crystallization from a multicomponent system coupled with chemical reaction(s), e.g., the growth of germanium on a hot substrate from the hydrogen reduction of germanium tetrachloride.

7-76   Semiconductor Materials

Semiconductor crystal-growth processes have representation in each of these categories, but they are more often classified by whether growth is from the melt or the vapor, and by the particular mechanical arrangement used. A number of these different methods are shown schematically in Figs. 52 and 53. It is much more difficult to obtain the same degree of crystal perfection with vapor-phase growth than is possible in growth from the melt, and, in addition, the growth

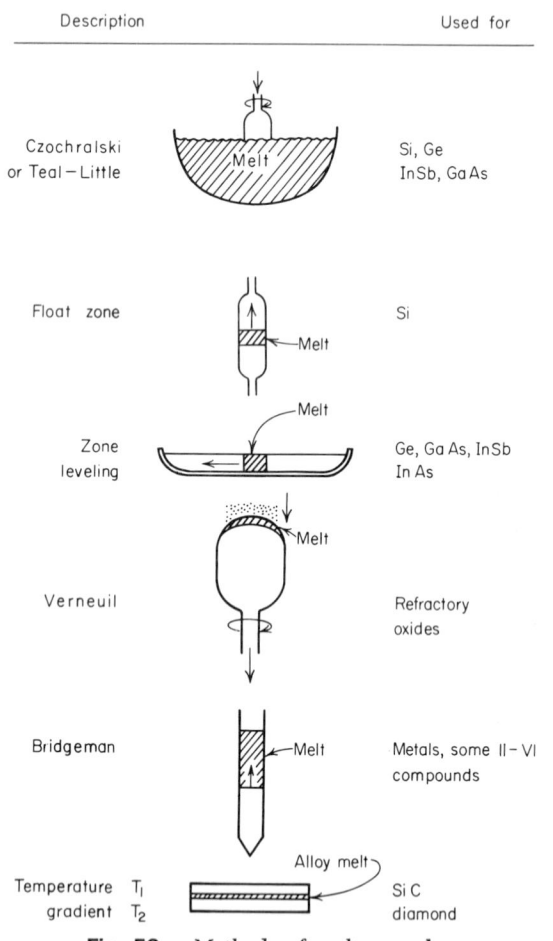

Description		Used for
Czochralski or Teal–Little		Si, Ge InSb, GaAs
Float zone		Si
Zone leveling		Ge, GaAs, InSb InAs
Verneuil		Refractory oxides
Bridgeman		Metals, some II–VI compounds
Temperature gradient T_1 T_2		SiC diamond

Fig. 52   Methods of melt growth.

rates are much slower. The latter occurs not only because of transport limitation, but also because the maximum rate possible for single-crystal growth decreases with decreasing temperature. For most semiconductors and a given deposition rate, as the temperature is reduced, a region is first reached in which only polycrystals can be grown; then finally there is a transition to a very high resistivity amorphous material.

Because of these limitations, growth from the vapor is ordinarily used only if it is difficult to grow from the melt, or to allow doping impurities to be changed so that layers of quite different resistivities can be grown sequentially without compensation. The more widely used semiconductors (e.g., Si, Ge, GaAs) are required to meet rather rigid specification with regard to defect densities and impurity control. Consequently, a great deal of crystal-growth technology has been developed with these specific goals in mind. Conversely, for many of the newer and less-studied materials emphasis has been primarily on defining applicable methods.

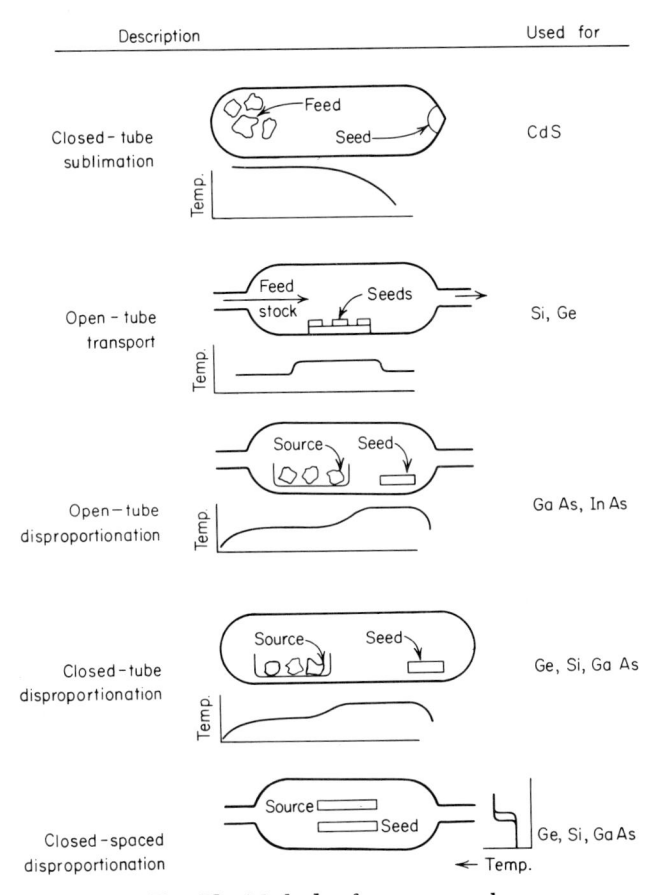

Fig. 53 Methods of vapor growth.

Czochralski

The most common and least expensive one (used for Si, Ge, GaAs, and InSb) is the Czochralski (named after J. Czochralski who first reported the process in 1918, but sometimes referred to as Teal-Little in honor of G. K. Teal and J. B. Little who first applied the process to semiconductor materials). It proceeds by allowing the melt to slowly freeze onto a single-crystal seed which may be simultaneously rotated and withdrawn from the melt. The freezing (growth) rate is controlled by a combination of the temperature of the melt and the amount of heat lost from the crystal by conduction up the seed and radiation from its surface. The relation among these variables may be expressed by the following heat-balance equation:

Latent heat + heat transferred to crystal from melt
= heat conducted away from the interface by the crystal

$$L\frac{dm}{dt} + k_1 \frac{dT}{dx} A_1 = k_s \frac{dT'}{dx_2} A_2 \qquad (38)$$

where
L = latent heat of fusion
dm/dt = amount freezing per unit time
T = temperature
k_1 = thermal conductivity of the liquid
dT/dx_1 = thermal gradient in the liquid at some point x_1 close to the interface
A_1 = area of the isotherm which goes through x_1
k_s = thermal conductivity of the solid
dT/dx_2 = thermal gradient in the solid near the interface
A_2 = area of the isotherm through x_2, which will be approximately the area of the crystal (exactly, if the isotherm is planar and perpendicular to the growth direction)

$$\frac{dm}{dt} = vA_c\sigma = (V_p + V_d)A_c \qquad (39)$$

where v is the growth velocity and is equal to the pull rate V_p plus the rate of drop of the liquid surface V_d, and A_c is the area of the crystal at the liquid-solid interface.

$$k_s \frac{dT}{dx_2} A_2 = Q_C + Q_R \qquad (40)$$

where Q_C is the amount of heat conducted up the seed, and Q_R is the amount radiated. Solutions to these equations show that the crystal diameter varies inversely with growth (or pull) rate, that the maximum growth rate occurs when the thermal gradients in the melt are reduced until $Q_S + Q_R$ must account only for the latent heat generated as the crystal freezes. Stirring via crystal rotation minimizes unsymmetrical crystal growth owing to uneven heating of the melt, and in general reduces dT/dx_1. Thus increasing crystal rotational velocity will normally produce a larger-diameter crystal if all other external variables are kept constant. Crystals as small as ⅛ in. in diameter and as large as 6 in. in diameter have been grown by this process. Growth rates vary considerably but are ordinarily in the inches-per-hour range.

The primary advantage of this method is that it puts few physical constraints on the growing crystal. This is particularly important since most semiconductors expand several percent when they freeze, and in addition, many bond to container walls so that differential expansion can cause fracture of both crystal and container.

Float Zoning

In order to minimize the purity problems, float zoning (invented independently by Paul Keck, Henry Theuerer, and R. Emeis)[87] is often used. Here a polycrystalline rod is held vertically, and a molten zone is caused to traverse the length of the rod. A seed crystal is attached to one end and the zone initiated partially in it and partially in the polycrystalline material. Both ends of the rod are usually rotated, and provisions are made to vary the vertical spacing between the end jaws so that the crystal may be "stretched" if desired. The maximum-diameter zone which can be supported depends on its length and the surface tension of the material. Heating is usually by radio-frequency induction, and so some additional zone support can be obtained by properly designing the coil to maximize interaction of coil currents with induced currents in the melt (levitation). The ambient may be either an inert gas or a vacuum.

Since only the ends of the semiconductor rod touch any foreign material, and they can be kept cool, the highest-purity crystals are produced by the process. The major disadvantage is a relatively small diameter, e.g., 1 to 2 in. for silicon.

Zone Leveling[88]

The concept of a molten zone traversing a horizontal charge confined in a boat (zone leveling) preceded float zoning by several years and was introduced by Pfann. The equipment is simpler than that required for float zoning, and works quite well for materials such as germanium, which is relatively easy to grow and which does not stick to the boat. As will be shown later, a molten zone moving through a long rod, be it horizontal or vertical, provides a very effective means of metallurgically purifying materials. When used for that purpose the process is referred to as *zone refining* rather than zone leveling.

Verneuil or Flame Fusion

In this system, the top of the seed crystal is melted, and powdered feedstock is slowly added. Simultaneously with this addition the crystal is slowly lowered so that freezing occurs on the bottom of the molten zone. Thus it is not unlike float zoning except that the zone is replenished, not by melting the upstream part of a bar, but by melting a fine stream of powder. Its original use (1890s) was for growing sapphire and other refractory oxides. For this purpose heating could be by flame (hence, flame fusion). Modern-day experiments have used alternatives such as electron bombardment and arc imaging.

Bridgeman, Stockbarger, or Bridgeman-Stockbarger

If the material to be grown is not particularly reactive and contracts upon freezing, a vertical tubular container may be used to hold a melt, which is allowed to freeze from the bottom up by passing the container through a rather abrupt temperature gradient. Either a single-crystal seed or a sharp-pointed container can be used to promote a single-crystal growth. This method is very simple and is widely used for growing metal crystals.

Ribbon Growth[89]

Most semiconductor applications require thin, flat single-crystal slices with damage-free surfaces for subsequent processing. All the methods discussed thus far generate long rods from which slices must be sawed, lapped, and polished. In attempts to circumvent these steps, several modifications have been proposed in order to directly grow single-crystal ribbon. The simplest one uses a series of vertical spacers running the length of a zone-leveling boat. Engineering problems make it difficult to maintain spacing, and the surfaces still need finishing. Another process grows the crystal through a heated orifice, i.e., uses thermal gradients to force a rectangular cross section. The crystal orientation is chosen so that the sides of the crystal are bounded by the slowest-growing planes—(111) for Si and Ge. Difficulties occur in keeping a uniform thickness and a wide enough ribbon to be useful.

If thermal gradients in the melt are arranged so that dT/dx_1 of Eq. (38) is negative, i.e., a supercooled melt, the transmission of the latent heat up the growing crystal is no longer the growth-limiting step, and high-velocity dendritic growth occurs. The orientation required is again such that atomically flat surfaces are obtained—growth in [112] direction with (111) exposed faces. Besides being very difficult to grow, these dendrites are very narrow, and all have several twin planes running the length of the ribbon. To extend the width, two dendrites originating from the same seed but separated by up to ¾ in. can be simultaneously pulled from the melt. As these two dendrites are withdrawn, they drag up a thin web of molten semiconductor between them. This web freezes and produces a good-quality very thin single-crystal ribbon. Both germanium and silicon have been grown in this way, but the process is not in widespread use.

Flux Growth (including Solution Epitaxy*)

Growth from fluxes is often used in order to obtain lower growth temperature. The lower temperatures may be desirable because of convenience or because of some higher-temperature transition. Flux growth may also be advantageous for materials which sublime under ordinary pressures. If the semiconductor-flux forms a system such as shown in Fig. 54 and the flux is saturated with the semiconductor at some temperature, T_1, then as the temperature is reduced, growth can occur on a suitable seed. The crystal will in turn be saturated with the flux so that

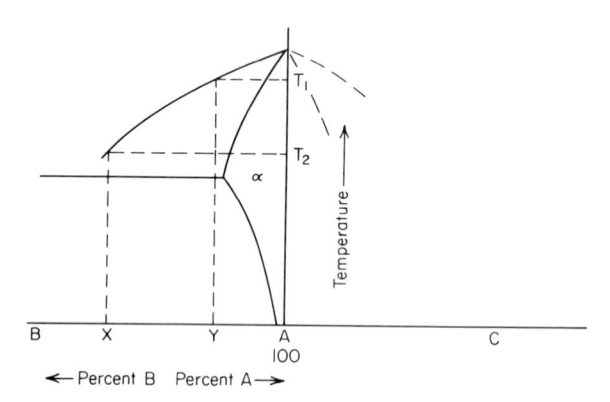

Fig. 54 Growth of component A from flux B. As the temperature decreases from T_1 to T_2, the equilibrium melt composition changes from Y to X and thus $(Y-X)$ of component A is available for growth. For most semiconductor systems, the α region is very narrow.

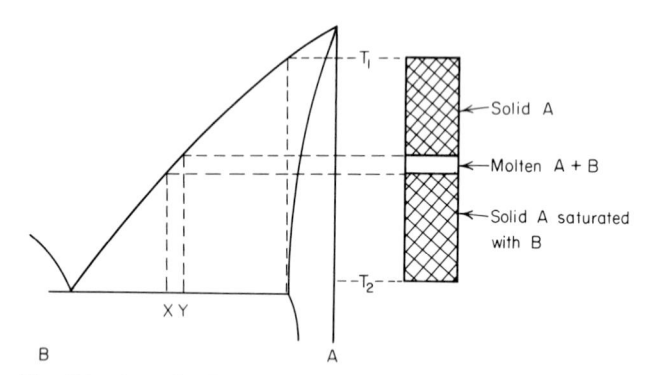

Fig. 55 Growth of component A from a flux of component B.

if high purity is desired, a system must be chosen so that the α region is very narrow. Occasionally, however, very heavy doping is desired, and such growth can afford a means of accomplishing it. If component A is a compound, the phase diagram will be extended to the right as shown by the dotted lines. This allows the possibility of producing either B- or C-rich material.

If instead of gradually lowering the temperature of the whole melt, a thin layer of component B is placed on one end of a block of A held at temperature T_2 (Fig. 55), B will slowly move through the block if the other end is held at temperature T_1. At the cool molten interface, the concentration of impurity in

* For further discussion of epitaxy see page 7-89.

the molten region is equal to x, whereas at the warm interface the impurity concentration in the molten region is equal to y. Since x is greater than y, there is an impurity concentration gradient across the molten zone. The upper solid is continuously dissolved in an attempt to eliminate the concentration gradient, and solidification occurs continuously at the lower interface to maintain the saturation conditions and the material balance. This process has been used for growing crystals of many semiconductors and is referred to as *temperature gradient growth*.

Peritectic Growth

The growth of a peritectic compound is quite similar to growth from a flux and is illustrated in Fig. 56. The compound β is to be grown, and the initial melt composition must lie between X and Y if only β is to be grown. If the crystal is to be produced by zone melting, the starting charge must have a composition β, but the zone itself must be in the XY region. Then as the zone moves at a constant temperature somewhere between T' and T'', the β phase is merely transported through the liquid from the melting to the freezing interface, and the zone composition remains invariant. It should be remembered that if the β phase is not exceedingly sharp, it may not be close to stoichiometry, and furthermore, it will always be C-rich.

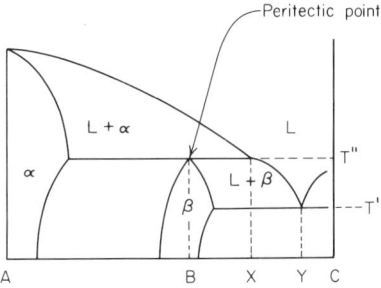

Fig. 56 Peritectic phase diagram.

Behavior of Impurities during Melt Growth

As a material freezes, the concentration of impurities incorporated into the solid is usually different from the concentration in the liquid at the interface. The ratio of these two concentrations is defined as the equilibrium segregation coefficient k_0, i.e.,

$$N_s = k_0 N_l \qquad (41)$$

where N_s is the concentration in the solid, and N_l the concentration in the liquid. It is possible to determine k_0 directly from a phase diagram, as shown in Fig. 57.

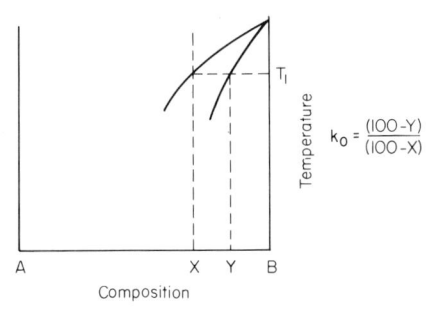

Fig. 57 Determination of the segregation coefficient k_0 from phase diagram.

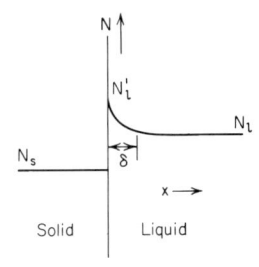

Fig. 58 Impurity pile-up at crystal-melt interface.

If the initial melt composition is x, it will remain at that value as the temperature is dropped until the liquidus line is reached at T_1. At that time composition y will begin freezing, and k_0 is given by $(100-y)/(100-x)$. Often the terminal β phase is very narrow so that the solidus line overlays the 100 percent B line on an ordinary phase diagram. In such cases, either an expanded-scale solidus curve may be used, or k_0 may be determined by separate experiment. Data shown in Tables 21 and 22 give k_0 values for several major semiconductors and their common impurities.

If growth is not extremely slow, there can be either an impurity-concentration depletion or enhancement at the melt-solid interface, depending on whether k_0 is less than or greater than 1 (it ordinarily is less than 1). This arises when the rejection (or acceptance) rate is higher than that which can be transported by diffusion and is illustrated in Fig. 58. If an attempt is made to calculate N_s on the basis of an equilibrium k_0 as previously defined and the concentration N_l in the main body of the liquid, the increased concentration at the interface can cause serious error. This is normally corrected by considering an "effective" k dependent on growth rate, stirring, geometry, and impurity species, which is defined as the ratio of N_s to N_l away from the crystal. k_{eff} is then given by N/N_l (of Fig. 58) times the actual k_0, where N_i is the concentration in the liquid at the interface. Stirring, whether by crystal rotation, convection, or melt rotation, tends to reduce the width of the artificial layer δ through which the impurity must diffuse. Because of these various effects, k_{eff} is very much geometry-dependent, but is usually less than four times k_0, and must be experimentally determined for each specific case.

If growth conditions are maintained so that k_{eff} does not vary during the growth, and if no additional impurities are added, the impurity concentration along a crystal grown from a melt (all material initially melted) is given by

$$N_s(l) = N_0 k_{eff} (1-l)^{k_{eff}-1} \tag{42}$$

where l is the fraction of the original which has solidified. If k_{eff} is close to 1, $N_s \sim N_0$ independent of the amount grown. When k_{eff} is very much less than 1,

$$N_s = \frac{N_0 k_{eff}}{(1-l)} \tag{43}$$

and the functional dependence of N_s on the amount grown is independent of k_{eff}, though the absolute value is directly proportional to k_{eff}.

For the case when only a zone is molten at any given time (float zone, zone leveling, Verneuil),

$$N_s(x) = 1 - (1 - k_{eff}) \exp\left(-k_{eff} \frac{X}{L}\right) \tag{44}$$

where X is the total length of the bar, and L is the zone length.

Both the normal freezing distribution and the zone distribution are shown plotted for comparison in Fig. 59.

Purification during Growth

The preceding figure shows that for k's different from one, purification accompanies freezing, and in principle is quite independent of a single-crystal growth.* Since one of the most important aspects of semiconductor materials is the high purity required, use is often made of this mode of purification.

When using normal freezing, if additional impurity removal is required over that shown in Fig. 59a, the last portion of the melt frozen (for $k < 1$) can be discarded and the process repeated. For the second melt-freeze sequence, the initial concentration is given by

$$N_{0,2} = [1 - (1-l')^{k_{eff}}] \frac{N_0 k_{eff}}{l'} \tag{45}$$

where l' is the fraction of the original volume used in the second operation. By continued cropping and remelting, considerable purification can be accomplished.

* In practice, impurities often segregate at grain boundaries, and so it is almost always better to grow single crystals, even when the goal is only purification.

Any given zone-refining pass is not as effective as a single normal freeze (compare curves of comparable k in Fig. 59), but it is possible to make multiple-zone sweeps without loosing material and without undue mechanical difficulty. Calculation of the impurity distribution along a bar after some arbitrary number of passes

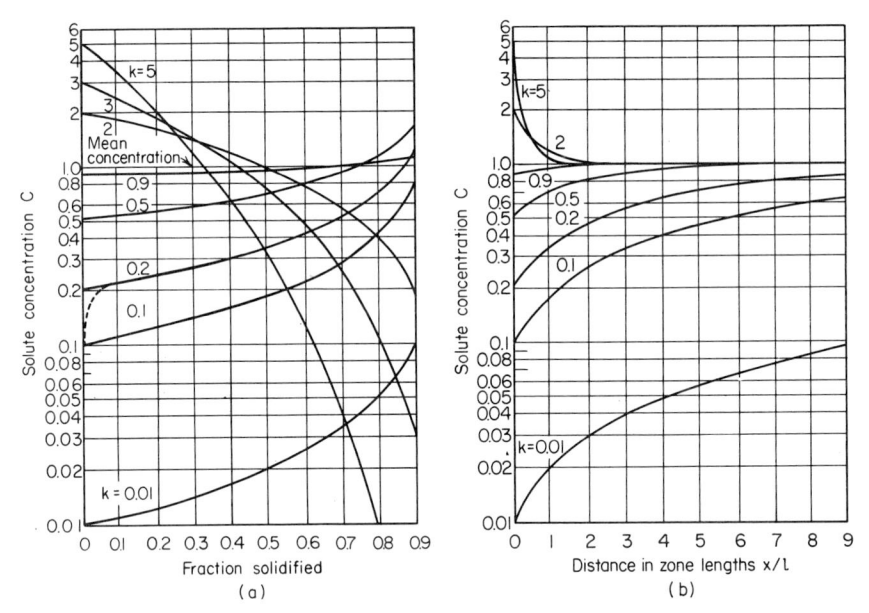

Fig. 59 Comparison of impurity distributions along a crystal for a normal freeze (a) and one pass of a molten zone (b). (Adapted from William G. Pfann, "Zone Melting," 2d ed., John Wiley & Sons, Inc., New York, 1966, Fig. 2.3 on p. 11 and Fig. 3.3 on p. 25. Used by permission of the author.)

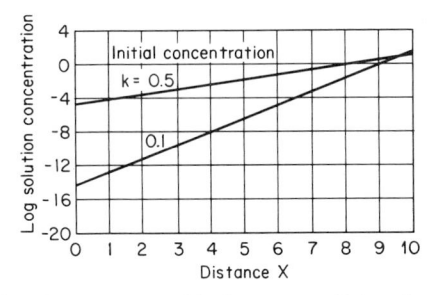

Fig. 60 Ultimate distribution attainable by zone refining for several values of k, for an ingot 10 zones long. (Adapted from William G. Pfann, "Zone Melting," 2d ed., John Wiley & Sons, Inc., New York, 1966, Fig. 3.10 on p. 42. Used by permission of the author.)

is quite involved, but the ultimate distribution after a large number of passes can be determined in a straight-forward fashion and is plotted in Fig. 60.

Uniform Distribution

Ordinarily, one of the goals during crystal growth is a *uniform* distribution of impurities at some predetermined level rather than a variation as predicted by Eqs. (42) and (44) or simple purification as described in the previous paragraph.

For pulled crystals, if k is not near one, the choices for uniformity are few. Either k_{eff} can be varied during growth by changing growth rate or stirring (see next section), or an impurity can be chosen which will evaporate during growth. It is sometimes possible to keep the net* impurity level $|N_A - N_D|$ relatively constant by compensating during growth, but that is usually undesirable from a device standpoint as well as being hard to control. A floating crucible in which a small portion of the melt from which the crystal is being grown is isolated from the main reservoir by a narrow passage can be used to approximate zone leveling. As will be shown in the next paragraph, the latter offers uniformity opportunities not available in a normal freezing environment.

If a doping impurity is added in a zone at one end of the bar,

$$N_s = k_{eff} N_d \exp \frac{-k_{eff} X}{L} \tag{46}$$

when the exponential term can be made small, either by a small k_{eff} or a large L/X ratio, N_s will be quite constant with distance. Thus, for uniformity, if the segregation coefficient is close to 1, pulling is best, but for k very small, zone leveling is superior. Unfortunately, many k's are neither very large nor very small, but if k is less than 0.5, alternate sweeps of the zone from end to end coupled with an initial dope injection at each end will produce a distribution over a major portion of the bar in a reasonable number of passes. Alternatively, a series of notches can be cut in the bar before the zone is swept through, and dope placed in each of them. This procedure has an additional advantage of allowing the bar to be profiled initially and varying amounts added as required to produce the desired profile.

Facet Effect

With crystals grown in or near the [111] direction, often a well-defined (111) face is formed on the bottom of the crystal. In many semiconductors the segregation coefficient is much higher in the faceted region and causes impurity "coring." The explanation for this phenomenon probably lies in the fact that nucleation is more difficult on (111) planes than on the rest of the growing interface, and that considerable supercooling of the melt contacting the facet is required before nucleation occurs. However, once a layer does nucleate, latent heat can be transferred very rapidly to the supercooled liquid, and growth proceeds laterally with a high velocity. Because of this high velocity, an excessive number of impurities are trapped during growth.

Impurity Striations

Cyclic impurity gradients (appearing as spirals in a plane normal to the growth axis of a cylindrical crystal) can usually be found in crystals grown from the melt. Some of these are due to nonuniform thermal gradients causing a varying growth rate as the crystal rotates; others occur because of vibration of the melt due to the growing machinery; and finally, some are caused by convection currents within the melt.

Dynamic Vacuum Evaporation[90-91]

If cleanliness can be maintained, vacuum evaporation onto a heated substrate is a reasonably simple means of transporting material, but unfortunately, many semiconductor compounds dissociate when heated in vacuum and the higher-vapor-pressure constituents evaporate first. Original crystal-growth applications involved oriented overgrowth studies of single-component metal films and thus presented no problems with regard to noncongruent evaporation.

In order to minimize such difficulties, several approaches have evolved. One is to flash-evaporate. That is, small particles are allowed to fall slowly onto a very hot heater. Another method merely sequentially evaporates the components

---

* Difference between the acceptor concentration N_A and donor concentration N_D.

in very thin layers, and then by subsequent diffusion forms the compound. Alternatively, the separate components may be evaporated separately, but simultaneously; then, if the evaporation rate at the source is accurately controlled, layers approaching stoichiometry can be produced. To simplify the rate-control problem, evaporation may sometimes be controlled at the substrate surface rather than at the source. For example, if the properties of the components are such that a temperature can be found in which the sticking coefficient of one of them is much less than the other, then the compound formation will be determined by the slower rate, and the excess of the other component will reevaporate.

The relatively poor vacuum systems sometimes encountered can produce severe contamination from pump oil or residual air. For some materials this can be tolerated, and even for the more sensitive ones an increased growth rate will sometimes allow reasonably defect-free single crystals to be grown. For example, both device-quality silicon and germanium have been grown by using growth rates of a micron per minute. In any event, the use of ultrahigh-vacuum systems alleviates the problem.

Closed-tube Transport via Intermediary[86]

In this process, the source and seed are sealed in an evacuated tube, after which the assembly is placed in a temperature gradient. If the source temperature is higher than that of the seed, there will be a net transport of material to the seed. Several possible ways of arranging the temperature gradients present themselves. The simplest would be as shown in Fig. 61. Such a system would be appropriate for growing masses of material which might contain large crystals, but it does not favor growth from one point only. This deficiency can be corrected by appropriately varying the temperature gradient with time. Such a change might be accomplished by gradually cooling one end of the tube (Fig. 61b) or by moving the tube and holding the temperature gradient fixed, as shown in Fig. 61c and d.

Closed-tube Transport via Intermediary[86]

Instead of depending on sublimination to provide the gaseous source of material to be transported, it is feasible to provide additional components in the tube which will react with the feedstock to give some transportable specie at one temperature, and yet upon a temperature change of tens of degrees, decompose to yield the material to be grown. In this manner, materials can be grown well below temperatures where their vapor pressures are high enough for closed-tube sublimination. There are a wide variety of reactions that can be used, the following showing a few examples of each of three common methods.

1. Disproportionation:

$$2SiI_2 \rightleftharpoons SiI_4 + Si$$
$$4GaAs + 2GaI_3 \rightleftharpoons As_4 + 6GaI$$

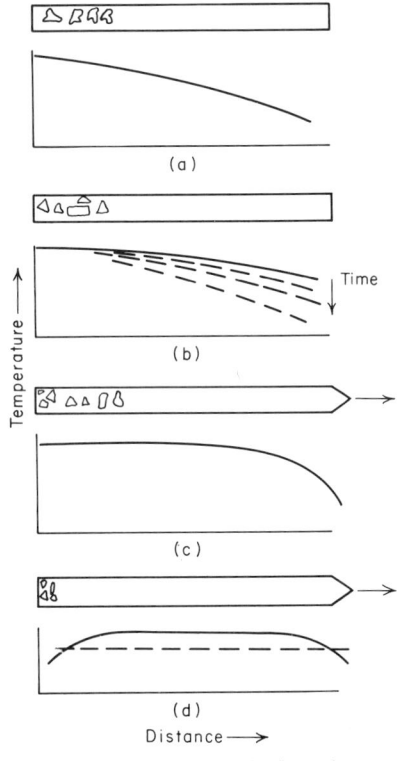

Fig. 61 Various methods of providing temperature gradients for closed-tube vapor growth.

2. Transport with halides:

$$Ge + 2HI \rightleftharpoons GeI_2 + H_2$$
$$Si + AlCl_3 \rightleftharpoons SiCl_2 + AlCl$$

3. Transport with water vapor (via suboxide):

$$Ga + XH_2O \rightleftharpoons GaO_x + XH_2$$

Table 36 summarizes the agents which have been used.

TABLE 36 Agents for Closed-tube Transport

Boron	Boron subhalides
CdS	H_2, NH_4Cl, I_2
CdSe	I_2
GaAs	$AlCl_3$, $CaCl_2$, $CdCl_2$, $CuCl_2$, GaI_3, $H_2O + H_2$, HCl, I_2, $MgCl_2$, $SnCl_2$, $ZnCl_2$
GaP	$H_2O + H_2$, HCl, I_2, $SnCl_2$, $ZnCl_2$
GaSb	I_2, $SbCl_3$
Ge	$GeCl_4$, I_2, Se, Te
InAs	$InCl_3$, InI_3
InP	I_2, InI_3
Si	$AlCl_3$, I_2, $SiBr_4$, $SiCl_4$, SiF_4, Se, Te
ZnS	H_2, I_2, NH_4Cl
ZnSe	I_2
ZnTe	I_2

Open-tube Transport[92]

The closed-tube system just described has the disadvantage of usually requiring a sealed ampul, and since it is a closed system, changes during growth, e.g., changes in doping, are difficult to accomplish. Open-tube configurations can be devised for the disproportionation reactions commonly used in closed tubes, and in addition, reduction and decomposition reactions can be accommodated. Table 37 is a summary of methods, both open and closed, for Si, Ge, and GaAs.

Disproportionation reactors have two temperature zones as in the closed system, but a continuous supply of carrier gas is used to transport material between the two zones. Decomposition and reduction reactors will usually have only one temperature zone. If large volumes of crystal are to be grown, rod-shaped seeds are often used, and are heated by current flow through them. When thin layers are to be overgrown on slices, a flat external heater is more convenient, though individual direct radio-frequency heating of slices has been used.

By the proper choice of a transport agent and the presence of a suitable thermal gradient, material can be transported between the adjacent surfaces of two close-spaced semiconductor wafers.[93] Close spacing of the source and substrate has a number of advantages. If the spacing is much less than the diameter of the source and substrate, the chemical transport conditions are largely independent of the conditions elsewhere in the system. Reuse of the transporting agent means that deposition conditions are not affected by the source and substrate size (provided their diameter is much greater than the spacing). Close spacing also means direct transport of each component of material. This is particularly advantageous for compounds comprised of elements with widely different physical properties. Furthermore, in such a system the dopant is transported in approximately the same concentration as it was in the seed, so that excessive autodoping often seen in wide-space flow systems will not be observed.[122]

The early work used an externally heated tube for deposition, but later designs have provided for selective heating of the substrate. Such designs have the twofold advantage of minimizing contamination from hot container walls and of preventing deposition on all walls.

A popular reactor uses a long horizontal tube, either circular or rectangular in cross section, as illustrated in Fig. 62a. Heating may be by radio frequency, motor generator, internal-resistance element, or external radiant heater. Usually layers from a horizontal reactor will have a taper, with the front end being thicker.

TABLE 37 Vapor-phase Crystal Growth

Semi-conductor	Transport configuration	Reaction	Normal growth temperature, °C	Comments
Silicon	Open tube	H_2 red. of $SiCl_4$	1050–1250	Most widely used.
	Open tube	H_2 red. of $SHiCl_3$	1050–1250	Faster growth rates than for $SiCl_4$ system.
	Open tube	H_2 red. of $SiBr_4$	....	Expensive, seldom used.
	Open tube	H_2 red. of SiI_4	....	Expensive, seldom used.
	Open tube	Thermal decomposition of SiH_4	1050–1200	Very high deposition rate at reduced pressures, hydrogen only by-product.
	Closed-tube disproportionation	$2SiI_2 \rightarrow Si + SiI_4$	450–650	Has been used for very low temperature growth, closed tubes inconvenient to use.
	Open-tube disproportionation	$Si + HCl \rightarrow SiCl_4 + \cdots$ $SiCl_4 + H_2 \rightarrow Si + \cdots$	1050–1250	Seldom used intentionally, often occurs on back of slices.
	Vacuum evaporation	....	1050–1300	Crystal quality usually poor.
	Sputtering	....	1050–1300	Crystal quality usually poor.
Germanium	Open tube	H_2 red. of $GeCl_4$	600–800	Most widely used.
	Open tube	Thermal decomp. of GeH_4	500–800	
	Closed-tube disproportionation	$2GeI_2 \rightarrow Ge + GeI_4$	350–450	
	Vacuum evaporation			
Gallium arsenide	Open-tube disproportionation	HCl transport		
	Combination	HCl transport of gallium, reduction of $AsCl_3$	....	Very commonly used.
	Close-spaced disproportionation	Water vapor		

However, in the very first slices, this is sometimes reversed because as the cold gases become warmer, the efficiency of the reaction increases. The downstream taper occurs because of the changing composition of the stream as it flows across the slices. In order to minimize this effect, some designs incline the slices slightly with respect to direction of gas flow. Also, since deposition rate (over some tem-

7-88 Semiconductor Materials

perature ranges) is temperature-sensitive, it is possible to establish a thermal gradient along the heater length in order to compensate for changes in growth rate.

To promote uniform flow across slices, various diffusers and deflectors have been investigated, but one of the better ones is simply to direct the incoming gas to the rear of the tube. As an example of the effect of improper gas direction, if the inlet is directed axially, the stream follows a spiral path down the tube. Thus slices which are not directly in the flow path of the fresh gas receive little deposition.

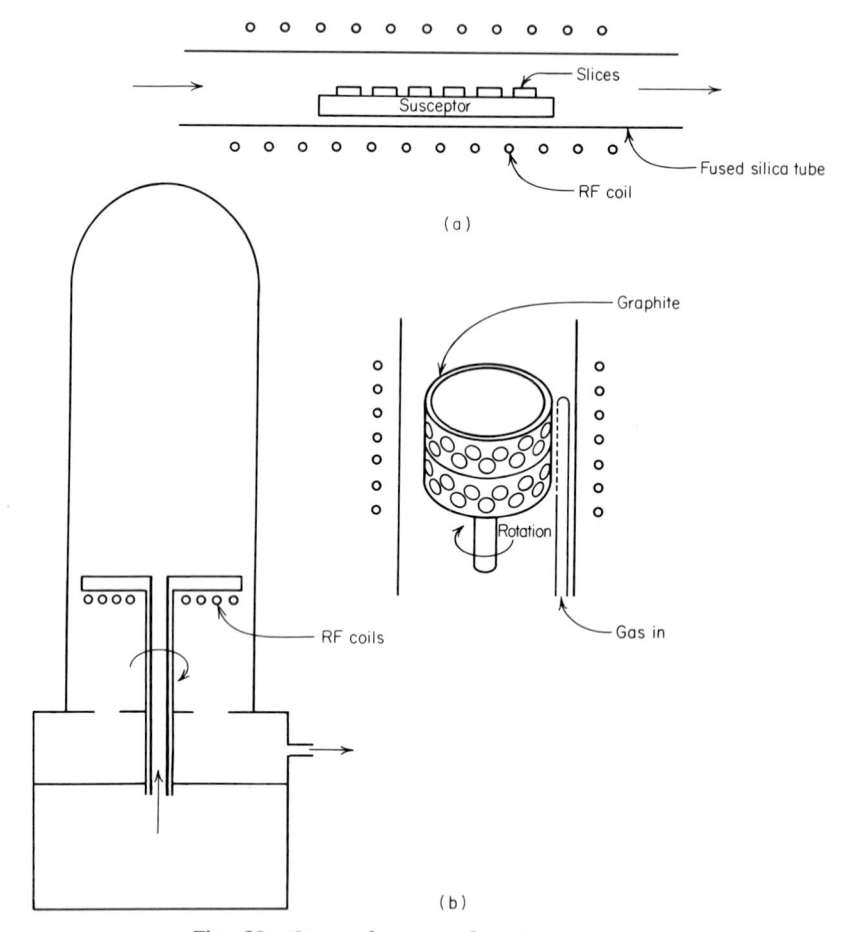

Fig. 62 Open-tube vapor-deposition systems.

As a means of eliminating the downstream taper of the horizontal reactor, radially symmetrical reactors are also in use. One of the simplest of the multiple-slice vertical reactors is shown in Fig. 62b. Gas admitted coaxially as shown is deflected from the top, passes over the slices, and out at the bottom. The diameter of the heater may vary from 4 or 5 in., hold one row of slices up to 24 in., and have a capacity of 50 slices. Thickness uniformity from such reactors is generally better than from the horizontal ones, even though total flow rates are much less.

By combining the idea of a long horizontal reactor and radial symmetry, a drumlike holder has been evolved.[94] Such a machine is shown in Fig. 62c, and has the advantage of a very large capacity.

Flow systems Flow systems can be conveniently broken up into flow controllers, flow monitors, pressure regulators, and interconnections. The simplest sort of system would be one in which flow is controlled from full-off to full-on by a single valve, and read on a rotameter or similar gage. If a wide range of flows is desired, several flowmeters may be paralleled. Furthermore, because smooth flow control usually requires a valve with a low-angle taper, and because clamping it tight enough to close off the flow completely may damage resettability, a second valve in series with the first is often used. Flowmeters are sensitive to the gas pressure, and are usually calibrated for exit at essentially atmospheric pressure. Hence, they should be installed downstream from throttling valves.

When many branches are required, e.g., a hydrogen line, a H_2-$SiCl_4$ line, and a dope line, in order to minimize the effects of each on the others it appears best to design a high-input impedance circuit. That is, each source gas is maintained at a pressure well above that which might be expected in the reactor. The reactor pressure itself will usually be low, but will vary somewhat with flow, exit line plugging, etc. Alternatively, differential regulators and associated bleed lines may be used to maintain a proper balance of pressure, but generally are more troublesome than helpful.

Doping Most vapor-deposition systems to be used for the growth of semiconductor materials require suitable dopants. Because of the limited amounts needed, special metering methods are normally used. In any event, the dopant compound used must be amenable to reduction to elemental form under the same conditions which produce the desired semiconductor.

For the specific cases of silicon and germanium growth by the hydrogen reduction of the halide, the required amount of dope halide (liquid) can be added to the germanium or silicon halide. If hydrogen is bubbled through the mixture, because of differences in the vapor pressures of the two components the relative concentrations will gradually change. To prevent this, the mixture may be metered as a liquid and then flash-vaporized. Alternatively, if the carrier gas sweeps over the surface of the mixture, instead of bubbling through it, uniform composition can be maintained.

Just as it is most convenient to transport and meter the feedstock as a gas, so it is for dopants as well. If they are available as liquids, a carrier gas may be saturated by bubbling through a container of the compound, but in order to eliminate the need for very low temperatures (to reduce the vapor pressure), a double dilution system may be required. Occasionally, compounds are used which have very low vapor pressures. It is then necessary to heat them and their lines in order to transfer material. It may, however, be possible by a rather short line to introduce the dope into a high-flow carrier gas line and dilute it sufficiently to reduce its partial pressure to less than its vapor pressure at room temperature. If this can be done, line heating is not required beyond that point.

If the dopant is a gas at room temperature, then it can be directly metered. Again, however, the total flows may be so low that accurate measurement becomes difficult. To circumvent this, a double-dilution system is occasionally used, but more often the doping gas (such as phosphine) is premixed with a diluent and stored in high-pressure cylinders. For the specific cases of hydrides, they may also be made in controlled quantities by using a spark discharge between electrodes of the desired material while in a hydrogen atmosphere.

Epitaxy[78,79,89,92] The word *epitaxy* was coined in the late 1920s to describe oriented overgrowth of one crystalline material onto another. Studies of these orientation effects have been underway for over a hundred years and started with observations of the orientation of water-soluble salts grown from droplets placed on various mineralogical samples. Present usage has broadened the term to include growth not only onto foreign substrates but also onto substrates of the same material. In order to differentiate between them, heteroepitaxy and homoepitaxy, respectively, are now accepted terminology. Furthermore, with the exception of *liquid epitaxy* used to describe, for example, the growth of layers of GaAs from a gallium melt, epitaxy in the semiconductor field implies vapor-phase growth of relatively thin layers on broad-area seeds. Thus, the previous sections relating to vapor-phase growth could have equally as well been under the heading of Epitaxy, and indeed most present-day literature on vapor-phase growth of semiconductors will be indexed under epitaxy. Early work assumed that orientation on foreign surfaces occurred only if the two lattice spacings were rather closely matched; but with better surface cleanup procedures, that no longer appears to be a restriction. If the spacing is not perfect, there will, of course, be dislocations at the interface, and as the original individual nuclei grow together, there will be dislocations at their boundaries. Thus, the crystalline perfection of heteroepitaxy seldom approaches that of homoepitaxy. Also, because most crystal growth requires high temperatures, thermal mismatches often cause cracking. Silicon has been grown on a wide variety of substrates, and reasonably good quality has been obtained on sapphire and spinel. Germanium and GaAs can be grown on each other without difficulty.

SHAPING

The conventional semiconductors are all characterized as brittle materials, and can be shaped by the same processes used for glass* and ceramics. There are, however, two additional factors to be considered: (1) that the single-crystal semiconductors are anisotropic in their mechanical properties, and (2) that the amount of crystal-lattice damage which occurs during processing must be minimized.

Slicing

After the semiconductor crystal is grown, the first step is to convert it into slices. The cross section of the slice may be circular and range from ½ to 3 in. in diameter (typical of silicon). It may also be trapezoidal (typical of germanium grown by zone leveling) or roughly semicircular (GaAs, InSb). Ordinarily such slicing is done normal to the long axis of the crystal, but occasionally it is desired to grow in one direction and cut slices in some other orientation. This requires an oblique cut with the attendant chance of the blade sliding out of alignment before the cut actually starts.

The main criteria for the choice of a slicing method are (not necessarily in order):
1. Width of cut (kerf loss)
2. Depth of damage
3. Smoothness of cut
4. Economy of operation

The width of cut is particularly important since material removed during cutting is lost. It is usually desirable to make the slices very thin in order to get more slices per unit length of crystal. The device design almost always specifies some minimum thickness after all saw damage has been removed so that the as-sawed slice thickness must allow for saw damage. The final surface after mechanical and chemical processing must be quite flat and have a high finish. It is, therefore, important to minimize undulations in the cut since their removal again represents wasted material. Finally, the ease of machine operation, rapidity of cut, initial cost and upkeep of equipment must all be considered.

For small quantities of very large diameter thick slices such as might be used for optical purposes, it is feasible to use a band saw with a diamond-loaded cutting surface. Such blades are quite thick, and in addition, considerable blade wobble occurs. Wire saws are sometimes used for slicing. They are inexpensive, and by using a small wire, will cut with a very low kerf loss. Wires as small as 5 mils in diameter have been used in them. Sawing speed is slow, and wire wear is troublesome. If attempts are made to increase cutting speed by increased pressure, the cut will wobble. If soft, easily loaded wire such as copper is used, cutting speed increases, but at the expense of wire life. When cutting is done with low loading at slow speeds, the depth of damage is small. Large numbers of thin blades are sometimes used to cut slices. Again, as in the case of wires,

---

* As a general reference for the cutting and polishing of glass, F. Twyman, "Prism and Lens Making," Hilger and Watts, Ltd., London, 1957, is suggested.

cutting speed is slow, but large numbers may be ganged and used simultaneously.[39] The blades may be thin so that kerf loss can be minimized and stretched very tightly in order to provide rigidity. Such blades are 4 to 6 mils thick and about 1/8 in. wide. Actual cutting may be either from diamonds plated on the blade or by abrasive slurry fed into the cut. By far the most common slicing method is the use of rotating diamond-impregnated blades. The diamond may be on the outer periphery (O/D), as is common in most saws used for cutting brittle materials, or the blade may be supported by its outer edge and have a large hole in it with the cutting being done by the inside edge of the hole. The latter is referred to as I/D cutting.

If conventional cutting blades are used, they must be reasonably thick to prevent excessive wobble. In order to use thinner blades, either they must be rotated at higher speeds so that centrifugal force will prevent whipping, or spacers must be used so that the blade is effectively much thicker in the middle. Typically, a 5-in.-diameter blade capable of cutting a 2-in.-diameter crystal must be at least 15 mils thick and will make a saw cut approximately 20 mils wide. If a 6-mil-thick blade is desired, then all but the outer 1/4 to 1/2 in. must be supported so that only a very shallow cut can be made. Attempts have been made to support blades by high-velocity-cutting fluid jets impinging on either side, but have never been very successful.

An I/D saw allows thin blades to be used, and yet by attaching them on a heavy flange and stretching in the manner of a drumhead, flexing can be prevented. Cuts only 6 to 8 mils wide through 3-in.-diameter crystals can be made, using this arrangement. Coolant introduction during I/D sawing is more difficult than in conventional sawing because instead of feeding the liquid onto the blade near the hub and letting centrifugal force sweep it to the edge and into the work, it must be sprayed directly into the cut. This occurs because the effect of the rotating blade is now to throw the coolant away from the cutting edge. The jets must thus be rather close to the work, and considerable care need be taken to make sure that as the crystal travels into the blade, it does not deflect the flow away from the work.

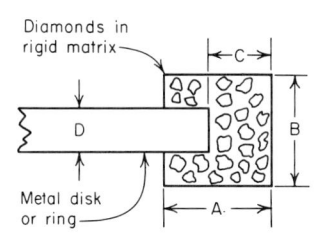

Fig. 63 Section of diamond saw blade.

A diamond blade consists of a thin metal disk, or an annular ring in the case of an I/D blade, with a diamond-impregnated rim as shown in Fig. 63. Each of the dimensions shown must be specified as well as blade diameter, the type of matrix bond used to hold the diamonds, the concentration of diamonds, and their size. Table 38 summarizes some of the choices available. Furthermore, though not ordinarily specified, resin-bonded wheels perform better with needle-shaped diamond chips, and blocky chunks are more suited to metal bonding.

No comprehensive study has been made of the cutting characteristics versus grit size, peripheral speed, feed speed, etc., for semiconductors; but some general observations can be outlined. The blade peripheral speed is not particularly critical but should be in the range of 3,000 to 4,500 ft/min for silicon and germanium. Excessive feed speeds cause the slices to be cut dish-shaped on an I/D saw and wedge-shaped on a normal saw. Increasing the diamond concentration will increase the unit blade cost, but over a wide range of values will also increase blade life and in many cases reduce overall cost.

Smoothing

After the semiconductor has been sliced, it is then smoothed by some process, and, in general, both sides are made flat and parallel.

Generally the smoothing will involve a lapping operation, of which there are two distinct variations. In one the slices are allowed to free-float between two

7-92 Semiconductor Materials

lapping plates so that material is removed uniformly from both sides. This results in slices which can be plane and parallel to within 0.0001 in. or better. In the other the slices are glued to blocks and then lapped from one side only. This method may produce warped slices for at least two separate reasons. The first is that if the slice was bowed before lapping, it would probably be flattened while being attached to the block. After lapping and subsequent release from the block, it will spring back to its original contour. Secondly, if the amount of damage is uneven on the two sides, bowing will occur. This effect is most noticeable where one side is polished and the other rough-lapped, but is almost always present. The bowing is probably due to lapping debris being forced into cracks in the material in the manner of a wedge, producing tension in the vicinity

TABLE 38 Diamond-saw and Grinding-wheel Data

Type of bond	Characteristics
Metal....................	Most wear resistant, almost always used for semiconductor cutting.
Vitrified..................	Compromise between metal and resin in both wear resistance and free-cutting capability.
Resin....................	Free cutting, but least wear resistant.

Diamond concentration	Carats/cm^3 of bonding material
50	2.2
75	3.3
100	4.4
125	5.5
150	6.6
175	7.7
200	8.8

Use	Mesh Number	Size, microns
O/D saw, ceramics...........	50	300
O/D saw, ceramics...........	80	177
I/D saw, Si and Ge..........	180	80
O/D saw, Si and Ge.........	400	37
Blanchard grinder, Si and Ge	1,800	8

of the fracture tip. That the lapped surface is left under tension can be verified by observing the direction of bending after lapping one side of a polished slice (or polishing one side of a lapped slice) and by birefringence measurements.

While the fixed-to-block lapping is at a disadvantage with regard to flatness, it can, when used with diamond stops, allow very close and positive control of final thickness.

Surface grinding with diamond-loaded cutters also works very well, cuts rapidly, and generates less damage than comparable grit used in a lapping operation.[100] Blanchard grinders, widely used in the optical industry, are directly applicable to working most semiconductors. Diamond disk grinding can also be used for semiconductor work. In this method the abrasive (diamond grit) is preembedded in a metal disk and used without further addition of diamonds. Grinding is very rapid, but careful attention must be given to the work pressures used. In particular, pressures must be considerably less than that normally used in lapping; otherwise disk life will be seriously shortened.

Mechanical Damage[101-102]

Regardless of which of the mechanical working methods just described is used, some damage to the semiconductor occurs. The damaged layer appears to be contained in two sublayers. The one immediately below the surface is highly distorted and fragmented. Underneath this layer is a region containing an excessive density of dislocations, but no cracks and fissures. This depth is roughly proportional to the diameter of the abrading particle, but it also depends on the material, the coolant, and the specific operation being used, i.e., lapping, blanchard grinding, or sawing. In particular, depth of damage depends on whether the abrasive particle can roll or is held fixed. A somewhat different fracturing process appears to be involved, and rolling particles give the most damage. Figure 64 shows the effect of particle size and semiconductor material for lapping on glass flats with Al_2O_3. There are many ways of estimating the damage depth, but the two most common are: (1) determining the thickness of material that must be removed by etching before the etch rate becomes constant, and (2) sequentially removing thin layers by either chemical or mechanical polishing and checking for dislocations until the dislocation density shows no change. The latter method always indicates a thicker damaged layer than the first, and so when comparing the results from different investigations, the exact method used must be known.

Surface damage such as scratches and probe marks and areas of improper or incomplete mechanical or chemical polishing can be delineated by mounting the sample on a glass slide with wax and etching, with agitation, for a few minutes. A proper choice of etchant will exaggerate fine scratches and probe damage and produce deep moats which are easily observed by optical microscopy. Incomplete polished surfaces can still appear bright and specular, whereas in reality there are many regions which have semiconductor and abrasive particles packed in cracks and polished over so that microscopically the surface looks smooth and continuous. A 10 percent aqueous sodium tetraborate solution at 80°C will remove the packed-in particles; or alternately, a light etch can be used.

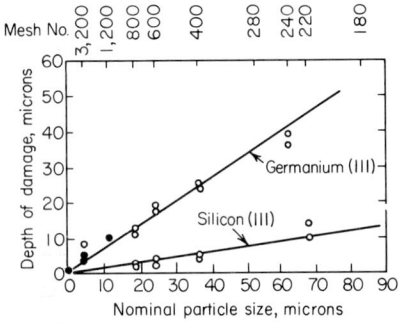

○ Etching rate method
● Etching rate method and photomagnetoelectric method

Fig. 64 Depth of damage on germanium and silicon as a function of abrasive particle size.[126]

Polishing[103,104]

The requirements for semiconductor polishing are considerably more stringent in some respects than those for optical element preparation. In the latter, emphasis is on following a specified surface contour and on having a smooth surface. Semiconductor polishing, on the other hand, is tolerant of surface undulations, but there must be no smearing of semiconductor across grooves and burying of polishing compound.

In optically polishing silicon and germanium their high thermal conductivity requires higher-than-normal tool pressures in order to maintain proper surface temperature. The added pressure makes it more difficult to keep the shape of the polishing surface the same as the piece being polished. This in turn makes the polishing more time-consuming so that, for example, to polish silicon to a good optical finish takes 30 to 50 percent longer than for an equivalent fused-silica surface. Polishing pitches such as coal tar and burgundy used with cerox on alumina are appropriate for the harder semiconductors.

Semiconductor finishing ordinarily uses cloth, rather than the pitches common to the optical industry, for holding the polish compound. The polishing compounds themselves are the same, though some are more preferred for particular semiconductors. For example, diamond can be used on all of them, but the incidence of scratches is higher than for some other choices. Cubic Al_2O_3 is quite good for all semiconductors, but cubic zirconia is most favored for silicon. Chromic oxide is reported to be well suited for GaAs.

The performance of each of these also greatly depends on the type of cloth used. For example, harder cloths are more prone to scratching but produce flatter surfaces, and nonwoven cloth must be "broken in" by polishing scrap slices before the removal rate becomes constant. Attention must also be given to the mixing of the polishing compound with its liquid carrier (dry polishing produces much more damage than wet polishing). Diamond is used with oil, and most others with water, though occasionally the pH must be altered by the addition of HCl. In any event, the solids content affects performance, and is dependent on cloth type.

Ultrasonic Cutting

In order to cut special shapes such as small disks or cylinders and Hall bars, ultrasonic cutting is often used. This machine utilizes a cutting head which vibrates at ultrasonic frequency (20 to 40 kHz) normal to the surface of the work. An abrasive slurry is fed into the cut around the tool and does the actual cutting. Tool pressure is light, and its excursion is small so that damage is usually not very deep. For shallow cuts (a few mils deep) it works very well, but as the depth of cut increases, the walls of the cut begin to flare at the top because of continued abrasion of the slurry, and the cutting speed decreases.

Sandblasting

Sandblasting, either wet or dry, is occasionally used for working semiconductors. Broad-area jets are useful for abrading crystals before looking for twining, or for slice cleaning. Fine jets can be used for cutting intricate shapes from flat slices, but tolerances are hard to hold, and such a process is not presently considered a precision method of cutting.

Abrasives

All the mechanical processes described require an abrasive as part of the operation. Parameters considered in the choice of a particular one are:[105]

  Hardness
  Particle size
  Fracture characteristics
  Melting point
  Chemical reactivity
  Cost

In semiconductor operations, cost is ordinarily secondary to performance and seldom plays a very important role. Most abrasive operations are performed in either a water or an oil environment so that virtually any abrasive is acceptable from a reactivity standpoint. Hardness is a most important parameter because the abrasive must be harder than the workpiece. Figure 65 gives a comparative listing of several abrasives and semiconductors. From this it can be seen that based on hardness alone, most common abrasives are acceptable for use with semiconductors. Because of the high temperatures involved, polishing compounds should be of higher melting point than the material to be polished, but again, this requirement can be easily satisfied for semiconductors.

The particle size, size distribution, and shape are very important with regard to cutting speed, depth of damage, and type of damage. There are several methods of describing the size, but the most common is based on standard sieves. Table 39 gives a comparison of sieve designation and mesh size over the range used

Shaping 7-95

in semiconductor work.[105] The actual distribution about a nominal value will depend on the method of separation and the standards used. As an example, Fig. 66 shows the distribution of a particular commercial product and indicates that for 1,200-grit aluminum oxide (1,200 grit = 12 microns), 14 percent has a diameter greater than 20 microns, and 6 percent has a diameter less than 7.5 microns. Since larger grits will produce more damage, tight distributions about the chosen size are desired except for rough work. Reproducibility of distribution from lot to lot of abrasive is also important since the distribution will affect cutting speed, and the latter combined with time is often used to estimate the amount of material removed. Figure 67 indicates in an approximate way the particle sizes used for lapping and polishing.

Fracture characteristics describe the form of the resultant particles as the abrasive breaks apart in use. If needle-like chips occur, scratching is likely. Silicon carbide, in particular, fractures in this manner, and so though it has great hardness and high cutting speed, it should be used only for rough grinding. If the edges wear away and round off instead of continuously fracturing and exposing new sharp cutting edges, the abrasive will have a short life. If it easily fractures into a multitude of small particles, cutting speed will quickly decrease since the material will then have a finer grit. Most of the naturally occurring abrasives suffer from this defect because of the presence of weaker cementing materials between grains.

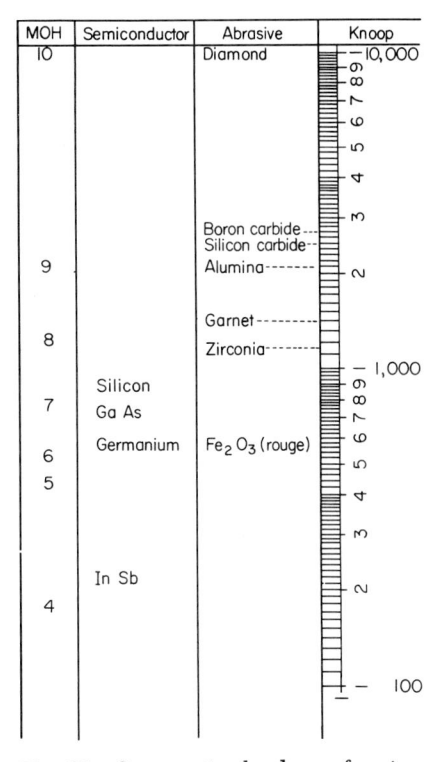

Fig. 65 Comparative hardness of various semiconductors and abrasives.[105]

TABLE 39 Sieve Sizes

Sieve	Particle size, microns
60	250*
80	177*
100	149*
200	74*
220	68
240	62
325	44*
400	37*
600	25
800	19
1,200	12
3,200	5

* Part of the U.S. Standard Screen Series. The Tyler Standard Series is slightly different in size, but ordinarily can be used interchangeably. The unasterisked sizes are in the proper ratio with the others. However, the set of standard sieves does not include those particular values.

Effect of Orientation on Grinding

With the exception of scribing, all the mechanical shaping operations were described as being independent of the orientation of the worked face. There are, however, directional properties which affect the grinding behavior. The effect is very pronounced in diamond and not very noticeable in silicon, but few data are available for other semiconductors.

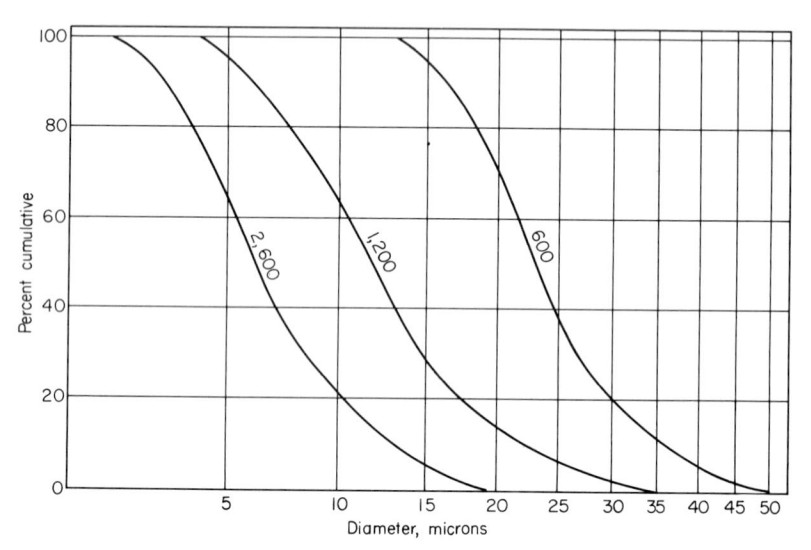

Fig. 66 Cumulative distribution of particle sizes for one supplier's natural corundum.

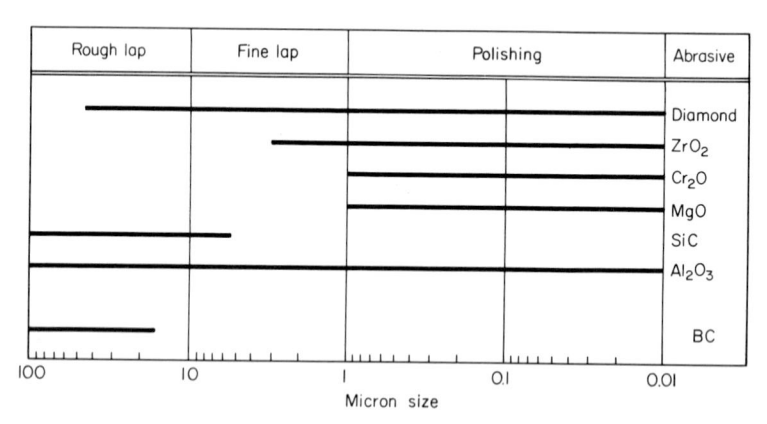

Fig. 67 Particle sizes versus usage.[124]

Because of its difficulty of working, diamond has been extensively studied as a matter of necessity, whereas the more common semiconductors are relatively soft (6 to 7 mohs) and are thus readily cut by most abrasives. In addition, the use of relatively harder abrasives makes the effect much less noticeable.

For diamond, the easiest plane from which to remove material is the (100), but only if the direction of grinding is also in a [100] direction. If grinding is done in a [110] direction, the material removal rate is decreased nearly a thousand-

Shaping 7-97

fold. Apparently the rate is a smooth function of grinding direction. Similar effects have been observed on the octahedral or (111) faces, and on (110) faces. The (111) is quite hard in all directions, and grinding it is not recommended. Silicon behaves in a similar fashion, but without the large variation observed in diamond. Table 40 summarizes some data for silicon.

TABLE 40 Effect of Grinding Direction on Comparative Hardness

Grinding on plane	In the direction toward:	Comparative hardness
(101)	(001)	2
(001)	(101)	1.95
(112)	(101)	1.55
(101)	(112)	1.50
(001)	(111)	1.50
(111)	(001)	1.5
(101)	(111)	1.2
(111)	(101)	1.05

At this point, it might be well to comment that much of the literature describes the grinding direction, not in terms of its indices, but rather according to which crystallographic face it points. For example, consider Fig. 68 which is a cube with vestige (111) planes exposed. The arrows shown point in a [011] direction, but also toward the (111) plane; and this might be described as pointing in the octahedral or [111] direction. Examples for (111) and (110) planes are also shown in Fig. 68.

Cleaving (Scribe and Break)

It is possible to scribe a line across a thin slice of a semiconductor, and since it is brittle, break it in the manner of glass. However, while glass is isotropic and scribes well in any direction, most semiconductors are anisotropic in their physical properties. Silicon, germanium, and diamond cleave most easily between (111) planes. Therefore it is advantageous to scribe so that breaking can occur in that manner. This requires that the scribe lines be drawn along traces of (111) planes intersecting the slice in question. If the slice is (100), then the traces delineate squares. Since square or rectangular dice are generally required, this works well, but if the surface is (111), {111} traces form diamond-shaped figures. Such scribing is occasionally used but is wasteful of material. The second-easiest cleavage plane is the (110), and so scribing along {110} is preferable to random orientation. By combining scribing along alternate {111} and {110} traces it is possible to scribe squares or rectangles in (111) slices. It is this

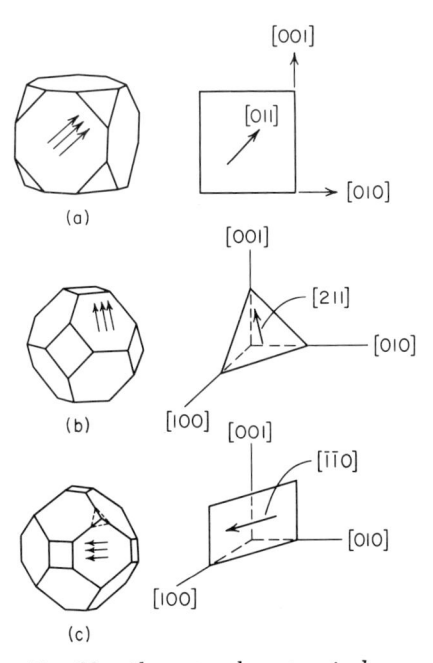

Fig. 68 Alternate plane terminology.

Semiconductor Materials

method that is usually used. In order to provide a reference for the scribing operation, slices may have a preoriented flat along one side.

Chemical Shaping and Polishing

The removal of material by chemical rather than mechanical means affords the advantage of not propagating any additional damage into the semiconductor but has other limitations. These are the likelihood of leaving some unwanted and difficult-to-remove contaminant on the surface and the problem of producing or maintaining the large flat areas which are often required in device manufacture.

TABLE 41 Chemical Polishing Methods

Semi-conductor	Method	Etchant	Typical rate, microns/min	Temperature, °C	Comments
Silicon	Basket etching	Planar etch	4.5	Room	Many slices at a time can be polished but edge feathering and orange peel constitute a problem.
	Electrochemical	....	0.1–1.0	Room	Works reasonably well for very low resistivities. Otherwise an intense light is required to produce carriers at the etching surface.
	Vapor	HCl	0.5–2.0	1100–1250	Widely used just before epitaxial deposition.
	Vapor	HI	....	1100–1200	Expensive to use because of iodine. Is reported to be very nonselective.
Germanium	Basket etching	CP4	25	Room	
	Electrochemical	KOH	0.4–1.25	Room	
	Chemimechanical	NaOCl	0.6	Room	Works well on (111) (100).
	Vapor	HCl	8.0	900	Used to clean substrate surface just before epitaxial deposition.
Gallium arsenide	Basket etching				
	Chemimechanical	NaOCl			
	Vapor	HCl	0.1–5.0	600–800	Tends to produce faceting, but is almost always used before epitaxial deposition.

The contaminant problem arises from several sources. Most etchants are masked by organics, and so if the surface to be etched is not free of all such contaminants the organic, as well as some material directly underneath will remain after etching. Occasionally there is abrasive powder or other small particulate matter initially on or embedded in the surface. Many times these are not attacked, or if so, only slowly; and because of strong chemical binding to the surface, they will remain attached even as etching proceeds. This phenomenon is particularly troublesome in silicon. One of the more dependable ways of removing such particulate matter is by mechanical swabbing after initial etching. If there are small amounts of contaminating elements in the etch which are lower in the electromotive series than the semiconductor, they will continuously replate on the surface, and no amount of etching will remove them. Thus it is particularly difficult to keep elements such as copper and gold from an etched surface.

Some of these problems can be eliminated by using a vapor etch instead of the more common water-based ones. In principle, if all reaction products are gaseous, they can be easily carried away. However, such etching is usually at high temperatures, and fast-diffusing impurities may enter the semiconductor more rapidly than the surface can be removed. Additionally, most vapor-phase reactions are rather easily reversed and thus may also redeposit etch products. Another potential disadvantage of high-temperature etching is that if the semiconductor is heated before the more serious surface damage is removed, residual stresses due to the damage can cause severe slip when the temperature is raised enough for plastic flow to occur. Vapor etching to date has found widest application when used in conjunction with some other process which has already removed most damage and contaminants. In particular, it is very effective when used to remove a slight amount of seed material just before starting vapor-phase crystal growth.

Occasionally etching is done just to remove damage, i.e., for stress relief only; but more often a polish is desired. For this purpose an etch which is insensitive to structure and damage is required. In addition, for device use, relatively large-area slices free of ripples, "orange peel," and feathering at the edges are required. Feathering is particularly difficult to prevent by etching alone, but a combination of etching and mechanical rubbing can often produce surfaces with the flatness normally associated with a mechanical polish and the high quality given by chemically etching. Table 41 is a summary of etchants and processes applicable to silicon, germanium, and GaAs.

DIFFUSION[35,107-111]

Diffusion studies can be broken into three rather distinct areas. One concerns itself with the mechanism(s) by which a given impurity moves, and the means of calculating and predicting values for diffusion coefficients. Another is primarily a study in boundary values. Given diffusion coefficients and concentrations at some initial time t_0, what is the position of the concentration profiles at some later time t? The third area deals with the methods of establishing those initial conditions. The following discussion will be confined to the more commonly used solutions to the diffusion equation and to methods for setting concentration values.

The simple theory of diffusion assumes that one species of impurity will diffuse independent of any other and that the diffusion rate is independent of concentration. Although neither of these assumptions is rigorously true, each is accurate enough for many calculations. It can then be shown that

$$\frac{\partial N}{\partial T} = D \nabla^2 N \quad \text{(Fick's law)} \tag{47}$$

and if D is not constant,

$$\frac{\partial N}{\partial T} = \nabla \cdot (D \nabla N) \tag{48}$$

Most diffusions are made normal to a large surface area and do not extend very far into the parent material so that the one-dimensional solution

$$\frac{\partial N}{\partial T} = \frac{D \, \partial^2 N}{\partial x^2} \tag{49}$$

is usually adequate. D is a strong increasing function of temperature and is of

TABLE 42 Table of the Error Function*

z	erf z	z	erf z	z	erf z
0.00	0.000 000	0.50	0.520 500	1.00	0.842 701
0.01	0.011 283	0.51	0.529 244	1.01	0.846 810
0.02	0.022 565	0.52	0.537 899	1.02	0.850 838
0.03	0.033 841	0.53	0.546 464	1.03	0.854 784
0.04	0.045 111	0.54	0.554 939	1.04	0.858 650
0.05	0.056 372	0.55	0.563 323	1.05	0.862 436
0.06	0.067 622	0.56	0.571 616	1.06	0.866 144
0.07	0.078 858	0.57	0.579 816	1.07	0.869 773
0.08	0.090 078	0.58	0.587 923	1.08	0.873 326
0.09	0.101 281	0.59	0.595 936	1.09	0.876 803
0.10	0.112 463	0.60	0.603 856	1.10	0.880 205
0.11	0.123 623	0.61	0.611 681	1.11	0.883 533
0.12	0.134 758	0.62	0.619 411	1.12	0.886 788
0.13	0.145 867	0.63	0.627 046	1.13	0.889 971
0.14	0.156 947	0.64	0.634 586	1.14	0.893 082
0.15	0.167 996	0.65	0.642 029	1.15	0.896 124
0.16	0.179 012	0.66	0.649 377	1.16	0.899 096
0.17	0.189 992	0.67	0.656 628	1.17	0.902 000
0.18	0.200 936	0.68	0.663 782	1.18	0.904 837
0.19	0.211 840	0.69	0.670 840	1.19	0.907 608
0.20	0.222 703	0.70	0.677 801	1.20	0.910 314
0.21	0.233 522	0.71	0.684 666	1.21	0.912 956
0.22	0.244 296	0.72	0.691 433	1.22	0.915 534
0.23	0.255 023	0.73	0.698 104	1.23	0.918 050
0.24	0.265 700	0.74	0.704 678	1.24	0.920 505
0.25	0.276 326	0.75	0.711 156	1.25	0.922 900
0.26	0.286 900	0.76	0.717 537	1.26	0.925 236
0.27	0.297 418	0.77	0.723 822	1.27	0.927 514
0.28	0.307 880	0.78	0.730 010	1.28	0.929 734
0.29	0.318 283	0.79	0.736 103	1.29	0.931 899
0.30	0.328 627	0.80	0.742 101	1.30	0.934 008
0.31	0.338 908	0.81	0.748 003	1.31	0.936 063
0.32	0.349 126	0.82	0.753 811	1.32	0.938 065
0.33	0.359 279	0.83	0.759 524	1.33	0.940 015
0.34	0.369 365	0.84	0.765 143	1.34	0.941 914
0.35	0.379 382	0.85	0.770 668	1.35	0.943 762
0.36	0.389 330	0.86	0.776 100	1.36	0.945 561
0.37	0.399 206	0.87	0.781 440	1.37	0.947 312
0.38	0.409 009	0.88	0.786 687	1.38	0.949 016
0.39	0.418 739	0.89	0.791 843	1.39	0.950 673
0.40	0.428 392	0.90	0.796 908	1.40	0.952 285
0.41	0.437 969	0.91	0.801 883	1.41	0.953 852
0.42	0.447 468	0.92	0.806 768	1.42	0.955 376
0.43	0.456 887	0.93	0.811 564	1.43	0.956 857
0.44	0.466 225	0.94	0.816 271	1.44	0.958 297
0.45	0.475 482	0.95	0.820 891	1.45	0.959 695
0.46	0.484 655	0.96	0.825 424	1.46	0.961 054
0.47	0.493 745	0.97	0.829 870	1.47	0.962 373
0.48	0.502 750	0.98	0.834 232	1.48	0.963 654
0.49	0.511 668	0.99	0.838 508	1.49	0.964 898

*For a more complete table, see L. J. Comrie, "Chambers Six Figure Mathematical and Its Derivative, *Nat. Bur. Stand. Appl. Math. Ser.* 41, Oct. 22, 1954.

z	erf z	z	erf z	z	erf z
1.50	0.966 105	2.00	0.995 322	2.50	0.999 593
1.51	0.967 277	2.01	0.995 525	2.51	0.999 614
1.52	0.968 413	2.02	0.995 719	2.52	0.999 634
1.53	0.969 516	2.03	0.995 906	2.53	0.999 654
1.54	0.970 586	2.04	0.996 086	2.54	0.999 672
1.55	0.971 623	2.05	0.996 258	2.55	0.999 689
1.56	0.972 628	2.06	0.996 423	2.56	0.999 706
1.57	0.973 603	2.07	0.996 582	2.57	0.999 722
1.58	0.974 547	2.08	0.996 734	2.58	0.999 736
1.59	0.975 462	2.09	0.996 880	2.59	0.999 751
1.60	0.976 348	2.10	0.997 021	2.60	0.999 764
1.61	0.977 207	2.11	0.997 155	2.61	0.999 777
1.62	0.978 038	2.12	0.997 284	2.62	0.999 789
1.63	0.978 843	2.13	0.997 407	2.63	0.999 800
1.64	0.979 622	2.14	0.997 525	2.64	0.999 811
1.65	0.980 376	2.15	0.997 639	2.65	0.999 822
1.66	0.981 105	2.16	0.997 747	2.66	0.999 831
1.67	0.981 810	2.17	0.997 851	2.67	0.999 841
1.68	0.982 493	2.18	0.997 951	2.68	0.999 849
1.69	0.983 153	2.19	0.998 046	2.69	0.999 858
1.70	0.983 790	2.20	0.998 137	2.70	0.999 866
1.71	0.984 407	2.21	0.998 224	2.71	0.999 873
1.72	0.985 003	2.22	0.998 308	2.72	0.999 880
1.73	0.985 578	2.23	0.998 388	2.73	0.999 887
1.74	0.986 135	2.24	0.998 464	2.74	0.999 893
1.75	0.986 672	2.25	0.998 537	2.75	0.999 899
1.76	0.987 190	2.26	0.998 607	2.76	0.999 905
1.77	0.987 691	2.27	0.998 674	2.77	0.999 910
1.78	0.988 174	2.28	0.998 738	2.78	0.999 916
1.79	0.988 641	2.29	0.998 799	2.79	0.999 920
1.80	0.989 091	2.30	0.998 857	2.80	0.999 925
1.81	0.989 525	2.31	0.998 912	2.81	0.999 929
1.82	0.989 943	2.32	0.998 966	2.82	0.999 933
1.83	0.990 347	2.33	0.999 016	2.83	0.999 937
1.84	0.990 736	2.34	0.999 065	2.84	0.999 941
1.85	0.991 111	2.35	0.999 111	2.85	0.999 944
1.86	0.991 472	2.36	0.999 155	2.86	0.999 948
1.87	0.991 821	2.37	0.999 197	2.87	0.999 951
1.88	0.992 156	2.38	0.999 237	2.88	0.999 954
1.89	0.992 479	2.39	0.999 275	2.89	0.999 956
1090	00992 790	2.40	0.999 311	2.90	0.999 959
1.91	0.993 090	2.41	0.999 346	2.91	0.999 961
1.92	0.993 378	2.42	0.999 379	2.92	0.999 964
1.93	0.993 656	2.43	0.999 411	2.93	0.999 966
1.94	0.993 923	2.44	0.999 441	2.94	0.999 968
1.95	0.994 179	2.45	0.999 469	2.95	0.999 970
1.96	0.994 426	2.46	0.999 497	2.96	0.999 972
1.97	0.994 664	2.47	0.999 523	2.97	0.999 973
1.98	0.994 892	2.48	0.999 547	2.98	0.999 975
1.99	0.995 111	2.49	0.999 571	2.99	0.999 976

Tables," vol. 2, W. & R. Chambers, Ltd., Edinburgh, 1949, or Tables of the Error Function

Semiconductor Materials

the form $D = D_0 \exp(-\Delta E/kT)$ where the preexponential D_0 and the activation energy ΔE are constants, T is the temperature and, k is Boltzmann's constant.[*]

There are two general categories of solutions of equations to be considered. One is the steady-state condition in which the concentration at any point is invariant with time. Such a set of circumstances could arise from a source on one side of a plate diffusing through the plate and evaporating from the other side, or from diffusion into a surface which is slowly evaporating. The other solutions (and the ones to be discussed in this section) are for transient conditions and are applicable over the time range before the concentration becomes time-independent.

Diffusion from an Infinite Source

If the initial concentration in the bulk of the solid is negligible, and the surface concentration is N_0 impurity atoms/cm³ and remains constant, then

$$N(x,t) = N_0 \left(1 - \operatorname{erf} \frac{x}{2\sqrt{Dt}}\right) \text{[†]} \tag{50}$$

When the concentration N_1 in the bulk is uniform, but not negligible, and is the same element as N_0,

$$N(x,t) = N_1 + (N_0 - N_1)\left(1 - \operatorname{erf} \frac{x}{2\sqrt{Dt}}\right) \tag{51}$$

If an n-type diffusion is being made into p-type material of concentration N', or vice versa, then a junction will occur where $N(x,t) = N'$, or when

$$N_0 \left(1 - \operatorname{erf} \frac{x_j}{2\sqrt{Dt}}\right) = N' \tag{52}$$

If a second diffusion, with impurity type opposite to that of the first, is now made, a second junction will occur when

$$N_0'\left(1 - \operatorname{erf} \frac{x}{2\sqrt{D't'}}\right) + N' = N_0\left(1 - \operatorname{erf} \frac{x}{2\sqrt{Dt}}\right) \tag{53}$$

where N_0' is surface concentration of the second diffusant, and $D't'$ is its diffusion-time product (Fig. 69).

Sometimes the two diffusions are done simultaneously, but more often they will be done in sequence, and quite likely at different temperatures. The Dt product required in the equation for each diffusion is then the sum of Dt's for that step and all following ones. Although tedious calculations are required to get precise junction locations and profiles for the various distributions considered, it is often quite satisfactory to use simple graphical analysis.

Bilateral Diffusion

If simultaneous diffusions from an infinite source are made from each side of a slab of finite thickness a, the distribution is given by

$$N(x,t) = N_0 \left[1 - \frac{4}{\pi}\left(\epsilon^{-y}\sin\frac{\pi x}{a} + \frac{1}{3}\epsilon^{-9y}\sin\frac{3\pi x}{a} + \frac{1}{5}\epsilon^{-25y}\sin\frac{5\pi x}{a} + \cdots\right)\right] \tag{54}$$

where $y = \pi^2 Dt/a^2$.

It should be observed that as y becomes smaller, progressively more terms are required, and that for very small y, corresponding to a semi-infinite medium, the Case I solution is easier to use.

[*] Values for D_0, E, and $D(T)$ may be found in Tables 9 and 10 and Figs. 16 through 19.

[†] erf is an abbreviation for "error function." Some of its properties are given on page 7-105. Its values are given in Table 42.

In the event that diffusion is only from one side of the thin slice of thickness a, as, for example, if an oxide mask were used on one side, then the solution would be the same as for the bilateral case for a thickness of $2a$. That is,

$$N(x,t) = N_0 \left[1 - \frac{4}{\pi}\left(\epsilon^{-y'} \sin\frac{\pi x}{2a} + \frac{1}{3}\epsilon^{-9y'} \sin\frac{3\pi x}{2a} + \cdots \right)\right] \tag{55}$$

where $y' = \pi^2 Dt/4a^2$.

Diffusion from a Limited Source

If instead of having an infinite source of concentration N_1 at the surface, only a limited supply of S impurities per unit area is available, the solution is

$$N(x,t) = \frac{S}{\sqrt{\pi Dt}} \epsilon^{-x^2/4Dt} \tag{56}$$

In this case, since the total amount of impurity is fixed, continued diffusion reduces the surface concentration, as shown in Fig. 70.

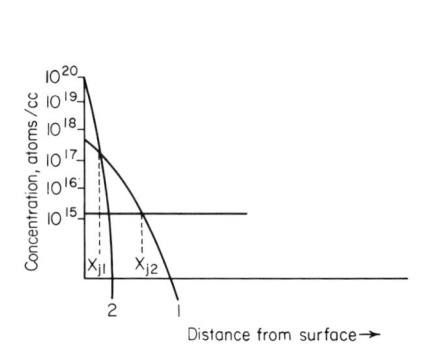

Fig. 69 Distribution of impurities after two diffusions. If the background were n-type and diffusion 1 were p-type, then x_{j2} could be a collector-base junction. If diffusion 2 were n-type, then an emitter-base junction would occur at x_{j1}.

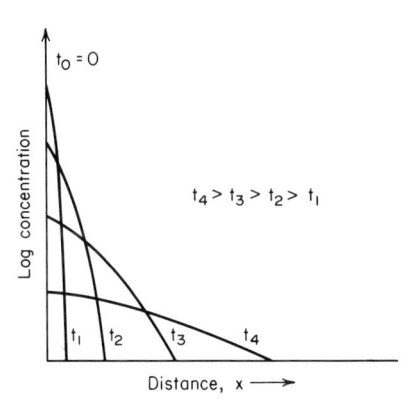

Fig. 70 Change of concentration profile with time when diffusing from a limited source.

Often, in order to get controlled low-surface concentration, a two-step diffusion is used. A very short normal diffusion of the type described by Eq. (50) is first made. For this first diffusion, N_0 will be high and determined probably by either the solid-solubility limit of the impurity or the concentration in a glassy layer that sometimes forms on the surface. After the initial short diffusion is made, the slices are removed from the furnace, any glassy layer source is removed, and the slices are put into a clean furnace for further diffusion. For this second step, the thin layer from the first diffusion will act as a limited source for the second step. In the event that the Dt product of the second diffusion is long compared to the first one, the initial layer will appear as a surface source, so that Eq. (56) is applicable. The value for S is the total number of impurities per square centimeter which entered the surface during the first diffusion and is given by

$$S = \int_0^\infty N(x)\, dx = \int_0^\infty N_0 \left(1 - \text{erf}\frac{x}{2\sqrt{Dt}}\right) dx = \frac{2N_0}{\sqrt{\pi}}\sqrt{Dt} \tag{57}$$

If this value for S is substituted into Eq. (56),

$$N(x,t) = \frac{2N_0}{\pi} \sqrt{\frac{Dt}{D'T'}} \epsilon^{-x^2/4D't'} \tag{58}$$

The primes indicate the second diffusion. The possibility of D' being different from D arises because the two steps of the diffusion may be made at different temperatures. Sometimes the second diffusion will not be long compared to the first one, and so the assumption that the first diffusion will behave as a thin source is not valid. This case may be solved analytically (see, for example, R. C. T. Smith[112]); but more often experimental curves are constructed and used for prediction purposes.

Diffusion from a Concentration Step

If at $t = 0$, $N(-x)$ equals a constant N_1 and $N(+x)$ equals a constant N_2,

$$N(x,t) = \frac{N_1}{2}\left(1 - \text{erf}\frac{x}{2\sqrt{Dt}}\right) + \frac{N_2}{2}\left(1 + \text{erf}\frac{x}{2\sqrt{Dt}}\right) \tag{59}$$

so that the diffusion from each side can be considered independent of the other, and the actual distribution is given by the sum of the individual solutions, regardless of whether N_1 and N_2 are of the same impurity. In the event the step is not at $x = 0$, but rather at $x = a$, it then becomes necessary to substitute $x - a$ for x in the above equation. This kind of distribution arises most often from the growth of a layer on a layer of differing impurity concentration. For such cases it is common to assume that at time zero, i.e., at the instant when layer growth started, the distribution was indeed a step; it is then only necessary to calculate any diffusion which might be expected to occur owing to subsequent processing steps (including completion of the growth cycle).

Effect of Time-varying Temperature

Often it is desirable to calculate the total amount of diffusion that occurs during a series of temperature cycles. For example, in order to know the final impurity profile in the vicinity of an epitaxial layer-substrate interface after a device has been constructed, it is necessary to consider the diffusion that occurred not only during the deposition cycle but also during subsequent diffusion, oxide growth, and contact addition steps. Similarly, if a slow cool after diffusion is used, it is necessary to consider the amount of diffusion which took place during the cool-down cycle. In order to compute the total diffusion, an effective Dt product must be found. If a series of discrete times at fixed temperatures is used, then the solution is

$$Dt_{\text{eff}} = \Sigma \; D_1 t_1 + D_2 t_2 + D_3 t_3 + \cdots \tag{60}$$

If the temperature varies continuously with time, then $\int_0^t D(T)\,dt$ is required. Since $D = D_0 \exp(-A/T)$, and T is now a function of time,

$$Dt_{\text{eff}} = D_0 \int \exp\left(\frac{-A}{T}\right) dt \tag{61}$$

This equation has been explicity solved for the special case of a linear decrease of temperature with time and is given by

$$Dt_{\text{eff}} = \frac{D(T_0)T_0^2}{AR}\left[1 - \frac{2T_0}{A} + \left(\frac{6T_0}{A}\right)^2 + \cdots\right] \tag{62}$$

where $T = T_0 - Rt$ defines both T_0 and R. In general, however, it appears more appropriate to plot $D(t)$ versus t and use numerical integration to obtain an approximate Dt_{eff}.

Three-dimensional Solution

If the diffusion source cannot supply diffusant to all parts of the surface, i.e., if selective diffusion is done via surface masking, a one-dimensional solution is not applicable near the mask edge. To calculate the amount of diffusion under the mask, either a two or a three-dimensional solution to Eq. (47) is required. Results for a long mask edge are shown in Fig. 71.

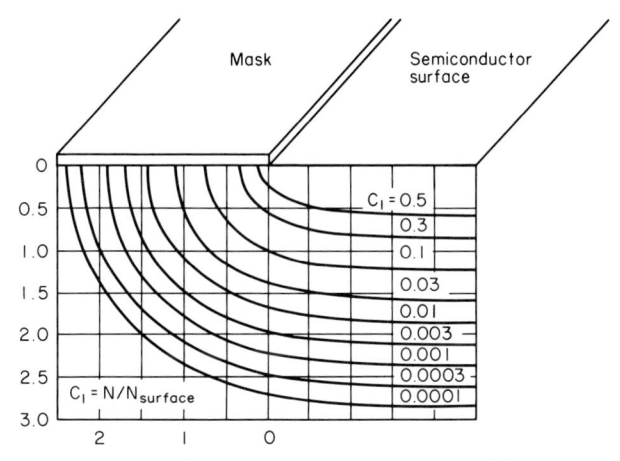

Fig. 71 Impurity profile under the edge of a mask after diffusion from an infinite source.[125]

Error-function Algebra

$$\text{erf } z = \frac{2}{\sqrt{\pi}} \int_0^z \epsilon^{-\alpha^2} d\alpha = \frac{2}{\sqrt{\pi}} \left[z - \frac{z^3}{3 \cdot 1!} + \frac{z^5}{5 \cdot 2!} + \cdots \right]$$

$$\text{erfc } z = (1 - \text{erf } z) = \frac{2}{\sqrt{\pi}} \int_z^\infty \epsilon^{-\alpha^2} d\alpha$$

$$\text{erf } (0) = 0$$
$$\text{erf } (\infty) = 1$$
$$\text{erf } (-z) = -\text{erf } z$$

$$\frac{d}{dz} (\text{erf } z) = \frac{2}{\sqrt{\pi}} \epsilon^{-z^2}$$

$$\frac{d^2}{dz^2} (\text{erf } z) = \frac{-4}{\sqrt{\pi}} z \epsilon^{-z^2}$$

$$\int_0^z \text{erfc } \alpha \, d\alpha = z \text{ erfc } z + \frac{1}{\sqrt{\pi}} (1 - e^{-z^2})$$

$$\int_0^\infty \text{erfc } z \, dz = \frac{1}{\sqrt{\pi}}$$

Transfer of Impurities to the Semiconductor

Impurities may be transported to the semiconductor surface coincidental with diffusion by some vapor-phase mechanism at high temperature, or they may have been previously added to the surface by painting on, electroplating, evaporating, or ion bombardment. The actual diffusion may be done in either a closed or an open tube. In the first, the semiconductor is placed in an evacuated, sealed tube, along with a suitable dopant. The dopant may be in direct contact with the semiconductor, e.g., plated on the surface, but more likely would be contained in a bulb connected by a small orifice to the main chambers in order to prevent surface attack or alloying. During open-tube diffusion, either a continuous flow of carrier gas plus an impurity-bearing gas is maintained across the surface, or else it is exposed to a stagnant ambient. Figure 72 illustrates both open- and closed-tube configurations and shows several methods of gas-phase transport.

Obtaining a uniform, reproducible, and predetermined surface concentration is ordinarily the most difficult part of a diffusion process. The simplest approach is to quickly supply an excess of impurities at the surface and raise the impurity

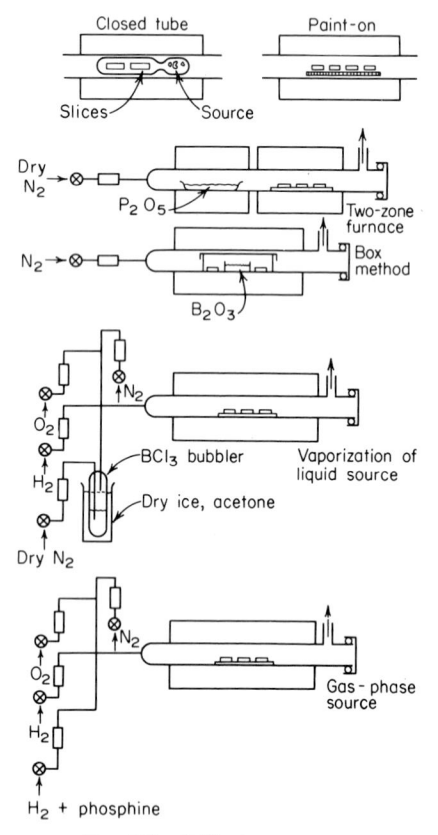

Fig. 72 Diffusion systems.

concentration to the solid-solubility limit at that temperature. If a lower final concentration is desired, a two-step diffusion can be done (see Diffusion from a Limited Source in a previous section). If the dopant is prone to form compounds with the semiconductor in question, then some method of limiting the rate of introduction such as control of the vapor pressure can be used. For many semiconductors such control can be exercised in either an open or a closed tube, but for others, such as silicon, the native oxide on the surface acts as a rate-limiting barrier and leads to great variability. Indeed, for silicon, the more successful methods all provide a glassy source on the surface which simultaneously acts as an infinite source and dissolves any oxide.

For any given impurity, there are obviously a variety of diffusion methods avail-

TABLE 43 Impurity Sources for Open-tube Diffusion of Boron*†

Original impurity source	Room temperature state	Temperature range of source during diffusion, °C	Impurity concentration range	Advantages	Disadvantages
Boric acid	Solid	600–1200	High and low	Readily available. Proved source.	Source contaminates tube. Control is difficult.
Boron tribromide	Liquid	10–30	High and low	Clean system. Good control over wide range of impurity concentration. Can use in nonoxidizing diffusion.	Geometry of system is important.
Methyl borate	Liquid	10–30	High	Simple to prepare and operate.	
Boron trichloride	Gas	Room temperature	High and low	Same as boron tribromide. Accurate control by gas metering equipment.	Restricted to high surface concentration.
Diborane	Gas	Room temperature	High and low	Easy installation and operation. Same as boron trichloride.	Highly toxic.

* From A. M. Smith and R. P. Donovan, Impurity Diffusion in Silicon, *3d Annu. Microelectron. Symp.*, St. Louis, Mo., April, 13, 1964.

TABLE 44 Impurity Sources for Open-tube Diffusion of Phosphorus*

Original impurity source	Room temperature state	Temperature range of source during diffusion, °C	Impurity concentration range	Advantages	Disadvantages
Red phosphorus	Solid	200–300	Low ($<10^{20}$ cm^{-3})	Low surface concentration.	Variable composition and vapor pressure.
Phosphorus pentoxide	Solid	200–300	High ($>10^{20}$ cm^{-3})	Proved source for high surface concentration.	Sensitive to water vapor. Requires frequent tube cleaning.
Ammonium phosphate	Solid	450–1200	High and low	Avoids water vapor dependence.	Purification is marginal.
Phosphorus oxychloride	Liquid	2–40	High and low	Clean system.	Geometry of system is important.
Phosphorus tribromide	Liquid	170	High and low	Good control over wide range of impurity concentrations. Same as phosphorus oxychloride. Can be used in nonoxidizing diffusion.	
Phosphine	Gas	Room temperature	High and low	Same as phosphorus tribromide. Accurate control by gas metering. Easy installation and operation.	Highly toxic.

* From A. M. Smith and R. P. Donovan, Impurity Diffusion in Silicon, *3d Annu. Microelectron. Symp.*, St. Louis, Mo., April, 13, 1964.

able. The final choice must comprehend numerous compromises, but in general should provide:
1. A surface concentration reproducible from run to run
2. A means of changing surface concentration (may require a two-step process)
3. No compounds difficult to remove from the surface
4. No undue stress in the semiconductor due to layer formation
5. A method for preventing decomposition of high-vapor-pressure semiconductors
6. Means of preventing spurious doping from contaminants
7. The possibility of simultaneously processing large numbers of slices

Tables 43 and 44 list variations of procedures available for the two most commonly used dopants, boron and phosphorus.

REFERENCES

1. Hannay, N. B. (ed.): "Semiconductors," Reinhold Book Corporation, New York, 1959.
2. Mason, Donald P.: Lecture Notes, University of Michigan Short Course on Semiconductors, 1965.
3. Barrett, Charles S., and T. B. Massalski: "Structure of Metals," 3d. ed., McGraw-Hill Book Company, New York, 1966.
4. Pfister, Von H.: Spaltbarkeit Der A^{III} B^{V}—Verbindungen InSb and AlSb, Z. Naturforsch, vol. 10A, p. 79, 1955.
5. Dewald, J. F.: J. Electrochem Soc., vol. 104, p. 244, 1957.
6. Verma, Ajit Ram, and P. Krishna: "Polymorphism and Polytypism in Crystals," John Wiley & Sons, Inc., New York, 1966.
7. Knippenberg, W. F.: Growth Phenomena in Silicon Carbide, Philips Res. Rep., vol. 18, pp. 161–274, 1963.
8. Nye, J. F.: "Physical Properties of Crystals," Oxford University Press, Fairlawn, N.J., 1960.
9. Mason, Warren P.: "Piezoelectric Crystals and Their Application to Ultrasonics," D. Van Nostrand Company, Inc., Princeton, N.J., 1950.
10. Riney, T. D.: Residual Thermoelastic Stresses in Bonded Silicon Wafers, J. Appl. Phys., vol. 32, pp. 454–460, 1961.
11. Coleman, D., R. T. Bate, and J. P. Mize: Mobility, Anisotropy and Piezoresistance in Silicon p-type Inversion Layers, J. Appl. Phys., vol. 39, pp. 1923–1931, 1968.
12. Queisser, H. J.: J. Appl. Phys., vol. 32, p. 1776, 1961.
13. Reed, W. T.: "Dislocations in Crystals," McGraw-Hill Book Company, New York, 1953.
14. Rhodes, R. G.: "Imperfections and Active Centres in Semiconductors," Pergamon Press, New York, 1964.
15. Newkirk, J. B., and J. H. Wernick (eds.): "Direct Observations of Imperfections in Crystals," Interscience Publishers, a division of John Wiley & Sons, Inc., New York, 1962.
16. Light, T. B.: Imperfections in Germanium and Silicon Epitaxial Films, in John B. Schroeder (ed.), "Metallurgy of Semiconductor Materials," Interscience Publishers, a division of John Wiley & Sons, Inc., New York, 1962.
17. Holt, D. B.: Misfit Dislocations in Semiconductors, J. Phys. Chem. Solids, vol. 27, pp. 1053–1067, 1966.
18. Billig, E.: Some Defects in Crystals Grown from the Melt: I, Defects caused by Thermal Stresses, Proc. Roy. Soc. London Ser. A, vol. 235, pp. 37–55, 1956.
19. Morizani, Kenji, and Paul S. Glein: Thermal Stress and Plastic Deformation of Thin Silicon Slices, J. Appl. Phys., vol. 40, pp. 4104–4107, 1969.
20. Holt, D. B.: Defects in Epitaxial Films of Semiconductor Compounds with the Sphalerite Structure, J. Mater. Sci., pp. 280–295, 1966.
21. Mendelson, S.: Growth and Imperfections in Epitaxially Grown Silicon on Variously Oriented Silicon Substrates, in M. H. Francombe and H. Sato (eds.), "Single Crystal Films," The Macmillan Company, New York, 1964.
22. ASTM Specification F47-64T.
23. Goldsmid, H. J.: Bismuth Telluride, in C. A. Hogarth (ed.), "Materials Used in Semiconductor Devices," Interscience Publishers, a division of John Wiley & Sons, Inc., New York, 1965.

24. Kohn, J. A., W. F. Nye, and G. K. Gaule (eds.): "Boron: Synthesis, Structure, and Properties," Plenum Press, Plenum Publishing Corporation, 1960.
25. Gaule, Gerhart K. (ed.): "Boron: Preparation, Properties, and Applications," Plenum Press, Plenum Publishing Corporation, New York, 1965.
26. Reynolds, P. C.: Sulfides, in J. J. Gilman (ed.), "Art and Science of Growing Crystals," John Wiley & Sons, Inc., New York, 1963.
27. Lambe, John, and Clifford C. Klick: Electronic Processes in Cadmium Sulphide, in Alan F. Gibson (ed.), "Progress in Semiconductors," vol. 3, John Wiley & Sons, New York, 1958.
28. Proc. 1st Int. Congr. on Diamonds in Ind., Industrial Diamond Information Bureau, London, 1963.
29. Scrupski, Stephen E. (ed.): The Many Facets of Gallium Arsenide, Electron., Nov. 13, 1967, pp. 106–136.
30. Willardson, Robert K, and Harvey L. Goering (eds.): "Compound Semiconductors," Reinhold Book Corporation, New York, 1962.
31. Davydov, V. I.: "Germanium," Gordon and Breach, Science Publishers, Inc., New York, 1966.
32. Groves, S., and W. Paul: Band Structure of Grey Tin, in "Physics of Semiconductors," Proc. 7th Int. Conf. Semicond. Phys., Paris, 1964.
33. Hume, K. F., and J. B. Mullin: Indium Antimonide: a Review of Its Preparation, Properties, and Device Applications, Solid-State Electron., vol. 5, pp. 211–247, 1962.
34. Hogarth, C. A. (ed.): "Materials Used in Semiconductor Devices," Interscience Publishers, a division of John Wiley & Sons, Inc., New York, 1965.
35. Runyan, W. R.: "Silicon Semiconductor Technology," McGraw-Hill Book Company, New York, 1965.
36. O'Connor, J. R. (ed.): "Conference on Silicon Carbide," Pergamon Press, New York, 1960.
37. Blakemore, J. S., D. Long, K. C. Nomura, and A. Nussbaum: Tellurium, in Alan F. Gibson (ed.), "Progress in Semiconductors," vol. 16, John Wiley & Sons, Inc., New York, 1962.
38. Honess, Arthur P.: "The Nature, Origin, and Interpretation of Etch Figures on Crystals," John Wiley & Sons, Inc., New York, 1927.
39. Schwuttke, G. H.: Determination of Crystal Orientation by High Intensity Reflectograms, J. Electrochem. Soc., vol. 106, pp. 315–317, 1959.
40. Wood, Elizabeth A.: "Crystal Orientation Manual," Columbia University Press, New York, 1963.
41. Lang, A. R.: Direct Observation of Individual Dislocations by x-ray Diffraction, J. Appl. Phys., vol. 29, pp. 597–598, 1958.
42. Schwuttke, G. H.: New x-ray Diffraction Microscopy Technique for the Study of Imperfections in Semiconductor Crystals, J. Appl. Phys., vol. 36, pp. 2712–2721, 1965.
43. Valdes, L. B.: Resistivity Measurements on Germanium for Transistors, Proc. IRE, vol. 42, pp. 420–427, 1954.
44. Knight, G.: Measurement of Semiconductor Parameters, in Lloyd P. Hunter (ed.), "Handbook of Semiconductor Electronics," 2d ed., McGraw-Hill Book Company, New York, 1962.
45. ASTM Tentative Method of Test for Resistivity of Silicon Using Four-point Probes, F84–68T.
46. Smits, F. M.: Measurement of Sheet Resistivities with the Four-point Probe, BSTJ, vol. 37, pp. 711–718, 1958.
47. Uhlir, A., Jr.: The Potentials of Infinite Systems of Sources and Numerical Solutions of Problems in Semiconductor Engineering, BSTJ, vol. 34, p. 105, 1955.
48. Van der Pauw, L. J.: A Method of Measuring Specific Resistivity and Hall Effect of Discs of Arbitrary Shape, Philips Res. Rep., vol. 13, pp. 1–9, 1958.
49. Mazur, R. G., and D. H. Dickey: A Spreading Resistance Technique for Resistivity Measurements on Silicon, J. Electrochem. Soc., vol. 113, p. 255, 1966.
50. Hilibrand, J., and R. D. Gold: Determination of the Impurity Distribution in Junction Diodes from Capacitance Voltage Measurements, RCA Rev., vol. 21, p. 245, 1960.
51. Brownson, John: A Three-point Probe Method for Electrical Characterization of Epitaxial Films, J. Electrochem. Soc., vol. 111, pp. 919–924, 1964.
52. Inskeep, C. N., and Earl Riggs: Rev. Sci. Instrum., vol. 34, p. 310, 1963.

53. Stevenson, Donald T., and Robert J. Keyes: Measurement of Carrier Lifetimes in Germanium and Silicon, *J. Appl. Phys.*, vol. 26, pp. 190–195, 1955.
54. IRE Standards on Solid State Devices: Measurement of Minority Carrier Lifetime in Germanium and Silicon by the Method of Photoconductive Decay, *Proc. IRE*, vol. 49, pp. 1291–1299, 1961.
55. Mattis, R. L., W. E. Phillips, and W. M. Bullis: Measurement and Interpretation of Carrier Lifetime in Silicon and Germanium, *Nat. Bur. Stand. Tech. Rep.* AFML-TR-68-81, July, 1968.
56. Armstrong, H. L.: Comparator Method for Optical Lifetime Measurements on Semiconductors, *Rev. Sci. Instrum.*, vol. 28, p. 202, 1957.
57. Jacobs, H., A. P. Ramsa, and F. A. Brand: Further Consideration of Bulk Lifetime Measurement with a Microwave Electrodeless Technique, *Proc. IRE*, vol. 48, pp. 229–233, 1960.
58. Atwater, H. A.: Microwave Determination of Semiconductor Carrier Lifetimes, *Proc. IRE*, vol. 49, pp. 1440–1441(L), 1961.
59. Weingarten, I. R., and M. Rothberg: Radio-frequency Carrier and Capacitive Coupling Procedures for Resistivity and Lifetime Measurements on Silicon, *J. Electrochem. Soc.*, vol. 108, pp. 167–171, 1961.
60. Arthur, J. B., W. Bardsley, A. F. Gibson, and C. A. Hogarth: On the Measurement of Minority Carrier Lifetime in n-Type Silicon, *Proc. Phys. Soc.*, vol. 68, pp. 121–129, 1955.
61. Haynes, J. R., and W. Shockley: The Mobility and Life of Injected Holes and Electrons in Germanium, *Phys. Rev.*, vol. 81, pp. 835–843, 1951.
62. Heiman, Frederic P.: On the Determination of Minority Carrier Lifetime from the Transient Response of an MOS Capacitor, *Trans. Electron. Devices*, vol. ED-14, pp. 781–784, 1967.
63. Jund, C., and R. Poirer: Carrier Concentration and Minority Carrier Lifetime Measurement in Semiconductor Epitaxial Layers by the MOS Capacitance Method, *Solid State Electron.*, vol. 9, pp. 315–319, 1966.
64. Valdes, L. B.: Measurement of Minority Carrier Lifetime in Germanium, *Proc. IRE*, vol. 40, pp. 1421–1423, 1952.
65. Jungk, G., and H. Menninger: *Phys. State Solids*, vol. 5, p. 169, 1964.
66. Norwood, M. H., and W. G. Hutchinson: Diffusion Lengths in Epitaxial GaAs by Angle Lapped Junction Method, *Solid-State Electron.*, vol. 8, pp. 807–811, 1965.
67. Bergmann, F., C. Fritzsche, and H. D. Riccius: *Tele. Leist.*, vol. 37, p. 186, 1964.
68. Goodman, A. M.: *J. Appl. Phys.*, vol. 32, p. 2550, 1961.
69. Kumari, K. Santha, and B. A. P. Tantry: Measurement of Lifetime of Minority Carriers in Junction Transistors, *Indian J. Pure Appl. Phys.*, vol. 3, pp. 380–384, 1965.
70. Lederhandler, S. R., and L. J. Giacoletto: Measurement of Minority Carrier Lifetime and Surface Effects in Junction Devices, *Proc. IRE*, vol. 43, pp. 477–482, 1955.
71. Kingston, Robert H.: Switching Time in Junction Diodes and Junction Transistors, *Proc. IRE*, vol. 42, pp. 829–834, 1954.
72. Byczkowski, M., and J. R. Madigan: Minority Carrier Lifetime in p-n Junction Devices, *J. Appl. Phys.*, vol. 28, pp. 878–881, 1957.
73. Mendelson, S.: Stacking Fault Nucleation in Epitaxial Silicon on Variously Oriented Silicon Substrates, *J. Appl. Phys.*, vol. 35, p. 1540, 1964.
74. Albert, M. P., and J. F. Combs: Thickness Measurement of Epitaxial Films by the Infrared Interference Method, *J. Electrochem. Soc.*, vol. 109, p. 709, 1962.
75. Archer, R. J.: Determination of the Properties of Films on Silicon by the Method of Ellipsometry, *J. Opt. Soc. Amer.*, vol. 52, pp. 970–977, 1962.
76. Hilton, A. R., and C. E. Jones: Measurement of Epitaxial Film Thickness Using an Infrared Ellipsometer, *J. Electrochem. Soc.*, vol. 113, pp. 472–478, 1966.
77. Pliskin, W. A., and E. E. Conrad: Non-destructive Determination of Thickness and Refractive Index of Transparent Films, *IBM J. Res. Develop.*, vol. 8, p. 43, 1964.
78. Buckley, H. E.: "Crystal Growth," John Wiley & Sons, New York, 1951.
79. Crystal Growth, *Discuss. Faraday Soc.* 5, 1949, Butterworth's Scientific Publications, London, 1959.
80. Mullin, J. W.: "Crystallization," Butterworth & Co. (Publishers), Ltd., London, 1961.

81. Lawson, W. D., and S. Nielsen: "Preparation of Single Crystals," Butterworth Scientific Publications, London, 1958.
82. Brice, J. C.: "The Growth of Crystals from the Melt," North-Holland Publishing Company, Amsterdam, 1965.
83. Strickland-Constable, R. F.: "Kinetics and Mechanism of Crystallization," Academic Press, Inc., New York, 1968.
84. Peiser, H. Steffen (ed.): "Crystal Growth," Pergamon Press, New York, 1967.
85. Gilman, J. J. (ed.): "The Art and Science of Growing Crystals," John Wiley & Sons, Inc., New York, 1963.
86. Schafer, Harald: "Chemical Transport Reactions," Academic Press, Inc., New York, 1964.
87. Petritz, Richard L.: Contributions of Materials Technology to Semiconductor Devices, *Proc. IRE,* vol. 50, pp. 1025–1038, 1962.
88. Pfann, W. G.: "Zone Melting," 2d ed., John Wiley & Sons, Inc., New York, 1966.
89. Grubel, Ralph O. (ed.): "Metallurgy of Elemental and Compound Semiconductors," Interscience Publishers, a division of John Wiley & Sons, Inc., New York, 1961.
90. Hirth, J. P., and G. M. Pound: "Condensation and Evaporation," The Macmillan Company, New York, 1963.
91. Francombe, Maurice H., and Hiroshi Sato: "Single Crystal Films," The Macmillan Company, New York, 1964.
92. Powel, Carroll F., Joseph H. Oxley, and John M. Blocher, Jr.: "Vapor Deposition," John Wiley & Sons, Inc., New York, 1966.
93. Sirtl, Erhard: "Sandwich-Methode" Einneuserfahren zur Herstellung Epitaktisch Gewachsener Halbleiterschichten, *J. Phys. Chem. Solids,* vol. 24, pp. 1285–1289, 1963.
94. Ernst, E. O., D. J. Hurd, G. Feeley, and P. J. Olshefski: High-capacity Epitaxial Machine, *J. Electrochem. Soc.,* vol. 112, 1965.
95. Strutt, J. O.: "Semiconductor Devices," Academic Press, Inc., New York, 1966.
96. Smith, R. A.: "Semiconductors," Cambridge University Press, London, England, 1961.
97. Sze, S. M., and G. Gibbons: Avalanche Breakdown Voltages of Abrupt and Linearly Graded p-n Junctions in Ge, Si, GaAs, and GaP, *Appl. Phys. Lett.* no. 8, pp. 111–113, 1966.
98. Sze, S. M., and J. C. Irvin: Impurity Levels in GaAs, Ge, and Silicon at 300°K, *Solid State Electron.,* vol. 11, no. 6, pp. 599–602, June, 1968.
99. Briody, Thomas F., Joseph Santangini, and Karl F. Kipp: Slicing Semiconductor Materials, *West. Elec. Eng.,* vol. 7, pp. 27–32, 1963.
100. Jensen, E. W.: Diamond Disc Grinding Techniques, *Geosci. Tech. Rep.* 13A, May, 1965.
101. Faust, J. W., Jr.: Factors That Influence the Damaged Layer Caused by Abrasion on Si and Ge, *Electrochem. Tech.,* vol. 2, pp. 339–346, 1964.
102. Stickler, R., and J. W. Faust, Jr.: Comparison of Two Different Techniques to Determine the Depth of Damage, *Electrochem. Technol.,* vol. 4, pp. 399–401, 1966.
103. Jensen, E. W.: Polishing Silicon Wafers, *Solid-State Technol.,* pp. 19–24, February, 1967.
104. Mendel, Eric: Polishing of Silicon, *Solid-State Technol.,* pp. 27–39, 1967.
105. Mendel, E., and E. W. Jensen: Survey of Lapping Abrasives, *Geosci. Rep.* TR-10A, February, 1965.
106. Dalla Valla, J. M.: "Micromeritics," Pitman Publishing Corporation, New York, 1943.
107. Jost, W.: "Diffusion in Solids, Liquids, Gases," Academic Press, Inc., New York, 1952.
108. Shewmon, Paul G.: "Diffusion in Solids," McGraw-Hill Book Company, New York, 1963.
109. Boltaks, B. I.: "Diffusion in Semiconductors," Academic Press, Inc., New York, 1963.
110. Burger, R. M., and R. P. Donovan (eds.): "Fundamentals of Silicon Integrated Device Technology," Prentice-Hall, Inc., Englewood Cliffs, N.J., 1967.
111. Kendall, Don L.: Diffusion, in R. K. Willardson and Albert C. Beer (eds.), "Semiconductors and Semimetals," vol. 4, Academic Press, Inc., New York, 1968.
112. Smith, R. C. T.: Conduction of Heat in the Semi-infinite Solid with a Short Table of an Important Integral, *Aust. J. Phys.,* vol. 6, pp. 127–130, 1953.

113. Morin, F. J., and J. P. Maita: Conductivity and Hall Effect in the Intrinsic Range of Germanium. *Phys. Rev.*, vol. 94, pp. 1525–1529, 1954.
114. Morin, F. J., and J. P. Maita: Electrical Properties of Silicon Containing Arsenic and Boron, *Phys. Rev.*, vol. 96, pp. 28–35, 1954.
115. Hall, R. N., and J. H. Racette, Diffusion and Solubility of Copper in Extrinsic and Intrinsic Germanium, Silicon, and Gallium Arsenide, *J. Appl. Phys.*, vol. 35, pp. 379–397, 1964.
116. Horak, J. B.: Unpublished data, Texas Instruments.
117. Bullis, W. M.: Properties of Gold in Silicon, *Solid State Electron.*, vol. 9, pp. 143–168, 1966.
118. Salzberg, C. D., and J. J. Villa: Infrared Refractive Indexes of Silicon, Germanium, and Selenium Glass, *J. Opt. Soc. Amer.*, vol. 47, p. 244, 1957.
119. Philipp. H. R., and E. A. Taft: Optical Constants of Silicon in the Region 1 to 10 ev, *Phys. Rev.*, vol. 120, pp. 37–38, 1960.
120. Trumbore, F. A.: Solid Solubilities of Impurity Elements in Germanium and Silicon, *BSTJ*, vol. 39, pp. 205–233, 1960.
121. Maycock, P. D.: Thermal Conductivity of Silicon, Germanium, III-V Compounds, and III-V Alloys, *Solid State Electron.*, vol. 10, pp. 161–168, 1967.
122. Kendall, Don L., and Dale B. DeVries: Diffusion in Silicon, in Rolf Habernecht and E. L. Kern (eds.), "Semiconductor Silicon," The Electrochemical Society, Inc., New York, 1969.
123. Trumbore, F. A., W. G. Spitzer, R. A. Logan, and C. L. Luke: Solid Solubilities of Antimony, Arsenic, and Bismuth in Germanium from a Saturation Diffusion Experiment, *J. Electrochem. Soc.*, vol. 109, pp. 734–738, 1962.
124. *A. B. Metal Digest*, vol. 10, no. 1, 1964.
125. Kennedy, D. P., and R. R. O'Brien: Analysis of the Impurity Atom Distribution Near the Diffusion Mask for a Planar p-n Junction, *IBM J.*, vol. 9, pp. 179–186, 1965.
126. Faust, J. W., Jr.: Damage Caused by Abrasion on Si and Ge, *Electrochem. Tech.*, vol. 2, pp. 339–346, 1964.
127. Dutta, B. N.: Lattice Constants and Thermal Expansion of Silicon up to 900°C by x-Ray Method, *Phys. Stat. Sol.*, vol. 2, pp. 984–987, 1962.
128. Willardson, R. K., and Albert C. Beer (eds.): "Semiconductors and Semimetals," vol. 2, Pergamon Press, New York, 1966.

Chapter 8

Ferrous Metals

WARREN S. EBERLY
Carpenter Technology Corporation, Reading, Pennsylvania

Introduction	8–1
Magnetic Properties	8–2
High-permeability Magnetic Alloys	8–2
Permanent Magnetic Alloys	8–17
Thermal Expansion Properties of Alloys	8–23
Electrical-resistance Alloys and Electrical-resistivity Properties of Some Ferrous Alloys	8–33
References	8–33

INTRODUCTION

An alloy containing 50 percent or more of iron is considered a ferrous alloy. This chapter, however, also includes information on alloys that contain from nil to 50 percent iron because their specific desired properties are similar to those of the ferrous alloys. This is the only chapter presenting these specific properties.

Ferrous alloys are divided into several categories such as typical compositions, specific properties, and standard trade designations. In this chapter all three classes are employed; the class designations selected are those most commonly used in the industry. Many types of properties are sought from these alloys; i.e., magnetic properties, thermal expansion, electrical resistivity, thermal conductivity, corrosion resistance, and physical properties. Alloys discussed herein are classified by the following methods:

1. Typical composition when produced by two or more sources and each having a different brand name

8-2 Ferrous Metals

2. Industry-designated brand names utilized by all sources
3. Brand name if only available from one source

Various properties for these alloys are explained in terms commonly used by many in the industry. If the desired properties are not expressed in the units desired, conversion factors, in all probability, can be found. Properties not reviewed here may be available from producers of the alloys.

MAGNETIC PROPERTIES

A wide range of magnetic characteristics are available from the many alloy combinations known to date. These magnetic characteristics are expressed in many specific parameters that have a significant meaning. To understand the information presented, one should be familiar with the general terms. A knowledge can be obtained by reviewing the references provided.

A general classification of magnetic alloys described herein is:
1. High-permeability magnetic alloys
2. Permanent magnetic alloys
3. Stainless steel alloys
4. Temperature-compensator alloys

Magnetic parameters discussed here are typical for the respective alloy as determined by test methods having approval of ASTM Committee A-6, Magnetic Properties.[1] An introduction to magnetic testing can be acquired by referring to *ASTM Special Technical Publication* 371, Magnetic Testing: Theory and Nomenclature. In order to discuss specific magnetic parameters of a given alloy, the method of test must be a standard test procedure that has been qualified by round-robin magnetic tests using an identical magnetic-core configuration. In those cases where a standard test does not exist, test methods must be identical and the test equipment must be of identical circuitry. Reproducibility between test equipment must be verified via the round-robin tests. This approach is employed, in many instances, between two or several laboratories because the ASTM standard methods of test[2] do not provide all property information to predict performance in specific applications. Many nonstandard tests are employed today to express given properties in terms mutually agreed upon to express performance. These nonstandard test methods and procedures cover evaluating at all frequency levels, shapes of hysteresis loops, residual magnetism, types of magnetic field excitation, and using a selected magnetic-core configuration. If a magnetic core or magnetic-core section cannot be evaluated properly because of shape, then a magnetic-core shape capable of being evaluated by standard methods of test can be subjected to the identical treatments. Standard magnetic tests will then provide property data in general terms.

Many magnetic materials, both high-permeability and permanent magnetic alloys, require thermal heat treatments or processes to achieve their designated optimum magnetic parameters. Each given magnetic material is capable of exhibiting a variety of magnetic properties as a result of heat treating and manufacturing processes; hence, it is necessary to specify details of heat treatment and test procedures. The manufacturing source must also control processing procedures so as to have minimum variation within a lot and from lot to lot.

High-permeability Magnetic Alloys

High-permeability magnetic alloys, also known as soft magnetic alloys, are required to have low magnetic hysteresis loss resulting from variations in the magnetic flux density produced within the alloy and a low residual magnetism after being subjected to a high magnetic field strength. This class of alloys does not retain permanent magnetic poles of any significant degree after being highly magnetized. There are several alloy families that produce magnetic properties of interest to the industry. Within each family, specific alloy compositions have been selected by the industry as standard grades because of best magnetic properties for a large number of applications.

Nickel-iron alloys This family of alloys contains 30 to 80 Ni balance Fe and exhibits a variety of magnetic and expansion properties. In some cases, small percentages of other elements are added to obtain more desirable magnetic properties for certain applications. Figure 1 shows the effect of nickel content on initial permeability and Curie range and Curie points. All the nickel-iron alloys used for magnetic properties contain low carbon and cannot be hardened or have their hardness increased by thermal heat treatment. Hardness can only be increased

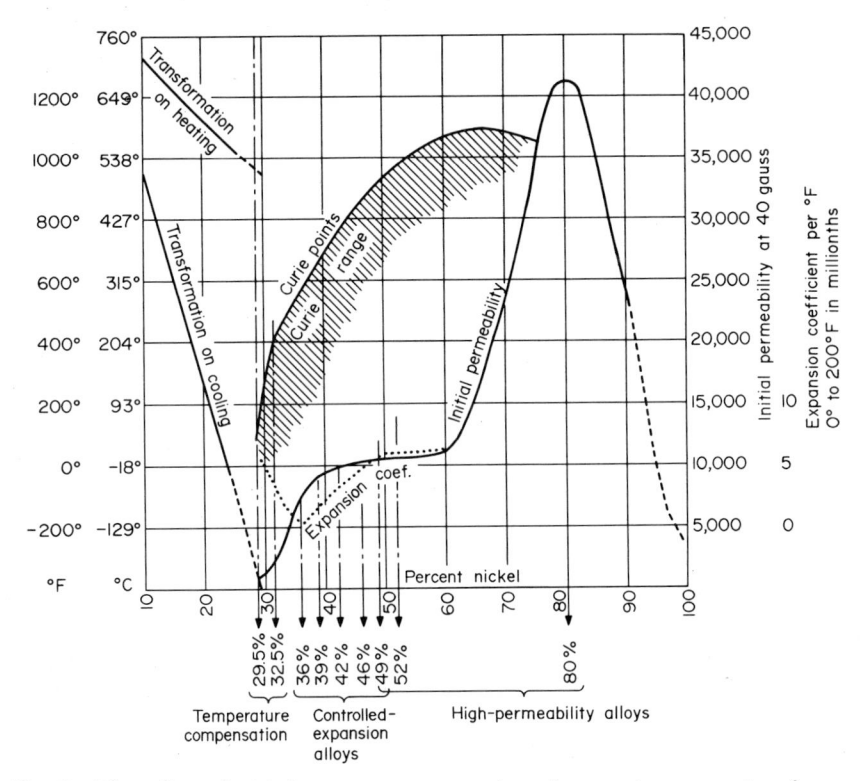

Fig. 1 The effect of nickel content on magnetic and expansion properties of some high nickel-iron alloys.

by cold-working the alloy, and the higher the hardness, the lower the high-permeability characteristics. Thermal heat treatment will reduce the hardness, and the lower the hardness, the higher the high-permeability characteristics will be. Hardness, however, in the softest condition, is no reflection on the degree of optimum high-permeability melt to melt or lot to lot from the same melt.

Producers of steel-mill products do not supply the alloys in the high-permeability condition in forms of bar and strip because subsequent fabrication operations such as machining, bending, forming, and deep drawing reduce the high-permeability properties by the introduction of cold-work stresses. Parts must be subjected to the thermal heat treatments to achieve most uniform desired magnetic properties.

High-permeability materials can be purchased in several types of physical conditions for most ease of fabrication, depending upon the part to be made. Suggested

8-4 Ferrous Metals

physical conditions are as follows:

Blanking of flat parts.................	Rockwell B-90 minimum
Blanking and forming................	Rockwell B-75/85
Best forming or deep drawing..........	Rockwell B-75 maximum

Each alloy is capable of exhibiting a variety of magnetic properties. Lowest magnetic properties are observed in the cold-work or high-hardness condition; hence, any thermal heat treatment relieving cold-work stresses will improve the magnetic characteristics over that originally shown. Heat treatments should be conducted in a nonoxidizing, noncarburizing, noncontaminating atmosphere.

Atmospheres generally employed are dry hydrogen, dissociated ammonia, argon, dry nitrogen, and vacuum. It is recognized that the best atmosphere is dry hydrogen. The heat treatment is conducted in a sealed retort or equivalent. A continuous flow of atmosphere in the retort assures removal of undesirable gas compounds given off by the alloy, and this helps to improve the magnetic quality. Elements

TABLE 1 Typical DC Magnetic Properties of Some Nickel-Iron Alloys

Property	45 Ni bal. Fe	49 Ni bal. Fe	49 Ni 0.15 Se bal. Fe	47 Ni 3 Mo	78.5 Ni bal. Fe	77 Ni 1.5 Cr 5 Cu	80 Ni 4 Mo	80 Ni 5 Mo
Heat treatment, H_2......	←————1140/4/1————→				1040/2/ 600 Q	1140/4/1	←——1140/4/5——→	
B_{40} permeability..........	5,000	7,000	3,200	4,000	8,000	20,000	50,000	70,000
Maximum permeability.....	60,000	75,000	10,000	70,000	120,000	150,000	200,000	250,000
Approximate B at maximum, permeability............	8,000	7,000	6,000	5,000	3,000	3,000	3,500	3,500
Coercive force, oersted.....	0.07	0.04	0.25	0.08	0.02	0.02	0.01	0.009
From B, gauss............	10,000	10,000	10,000	10,000	10,500	5,000	7,000	7,000
Saturation induction, gauss.	15,000	15,500	15,000	13,500	13,000	6,500	8,700	8,500
Curie temperature, °C.....	460	500	500	460	600	400	460	450
Electrical resistivity, microhm-cm............	45	48	48	80	20	60	58	63
Density, g/cc.............	8.17	8.25	8.25	8.27	8.45	8.50	8.74	8.75

Q Quench from temperature.
Heat treatment: 1140°C for 4 hr at temperature; 1°C per min cool rate (1140/4/1).

removed from the alloy are carbon, oxygen, and sulfur. It is necessary to study the magnetic properties of the parts heat-treated by varying temperature, time at temperature, and cool rate for establishing the treating procedure for a given facility.

Recommended heat treating for highest permeability in the nickel-iron family of alloys is as follows:

1. Retort should be capable of being gas-tight or vacuum-tight: i.e., a sealed retort.
2. Purge or flush retort—to free it of ambient atmosphere.
3. Introduce protective or purifying atmosphere.
4. Heat retort to 1120 to 1170°C.
5. Hold 2 to 4 hr at temperature.
6. Cool at a rate recommended for the alloy.
7. Use protective or purifying atmosphere until the chamber is cooled to 200°C or less.
8. Introduce an inert atmosphere until cooled below 100°C.
9. Open retort.

Typical dc magnetic properties of the most common high-permeability alloys are given in Table 1. Recommended heat treatment for obtaining these magnetic

Magnetic Properties 8-5

properties are shown for each alloy. These dc data apply to cold-rolled strip items of 0.030 in. or thicker and bar items. Figure 2 provides the dc normal induction curves for the various alloys.

Alternating-current magnetic properties vary considerably, depending upon lamination thickness, degree of interlaminar resistance, lamination shape, and frequency of the magnetizing current. Since each alloy exhibits a wide variety of ac magnetic characteristics, each commercial grade will be reviewed separately.

80 Ni 4 Mo balance Fe. This alloy is recognized as one of the highest initial permeability alloys and is used for its response to very low magnetizing forces. Well-known brand names are:

4-79 Permalloy (30)
HyMu 80 (29)
Hipernom (31)

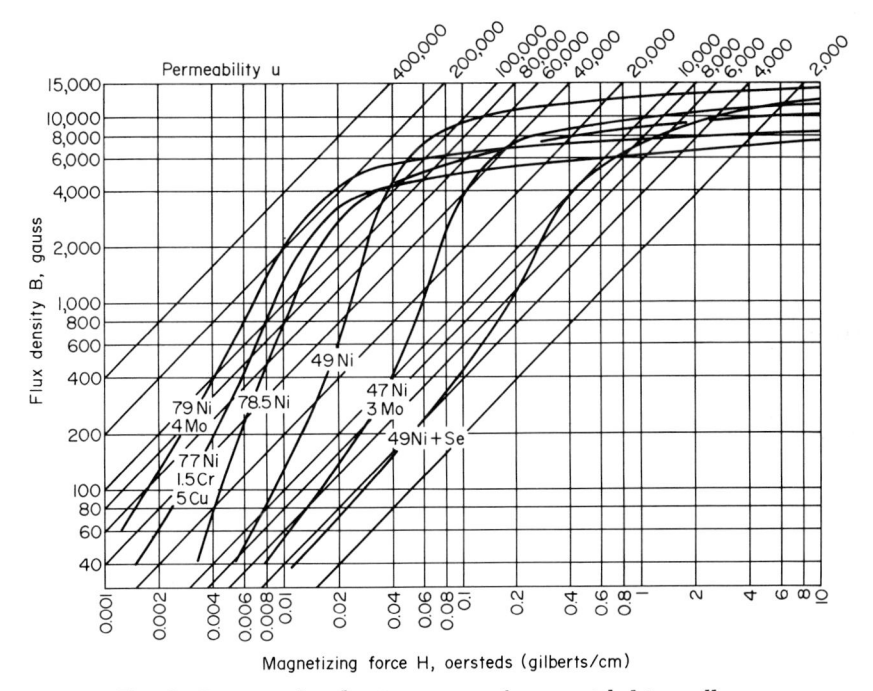

Fig. 2 Dc normal induction curves of some nickel-iron alloys.

Magnetic-core and lamination manufacturers have applied various other brand names to the alloy. Most common application for this alloy is in the form of laminations 0.006 and 0.014 in. thick; hence, 60- and 400-Hz data are readily available. See Figs. 3 and 4 for typical data. Core-loss data are given in Fig. 5. The alloy is available in strip form as thin as ⅛ mil (0.000125 in.), which is used to manufacture small bobbin cores having -specific square-hysteresis-loop properties. Thicknesses in the range of ½ to 4 mils are used to manufacture numerous types of magnetic cores such as tape toroids for high initial permeability, tape toroids having square-hysteresis-loop characteristics, and analog- and digital-computer laminated pickup heads. Typical magnetic properties of strip at thicknesses of 1, 2, and 4 mils are given in Table 2.

As stated previously, a wide variety of magnetic characteristics are obtainable by various heat treatments, as shown by 0.006 in. lamination strip in Fig. 6. The

8-6 Ferrous Metals

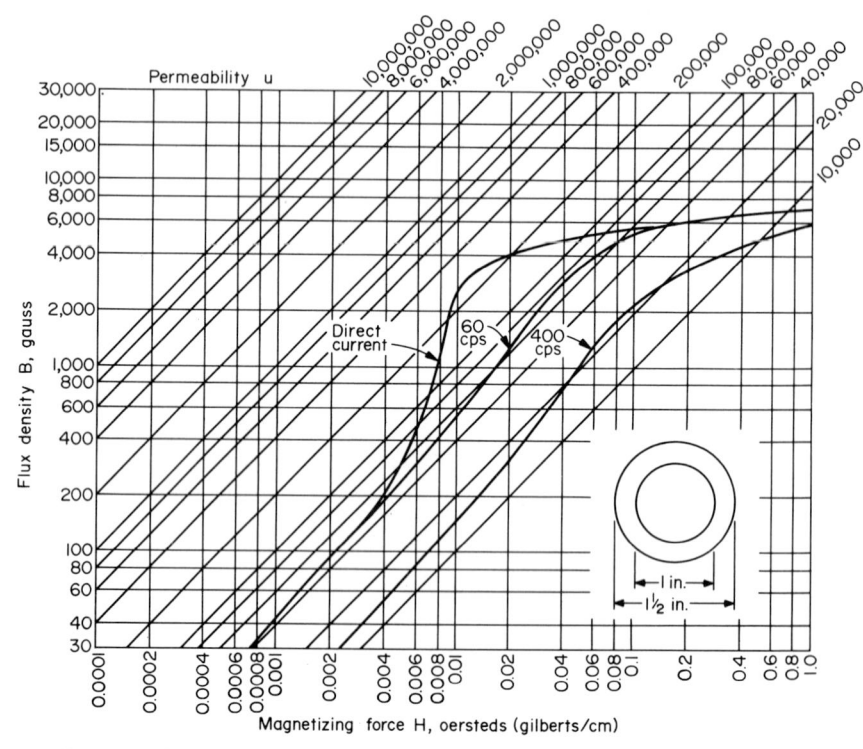

Fig. 3 Dc and ac magnetic properties of 80 Ni 4 Mo balance Fe lamination strip 0.014 in. thick. Heat treatment: 1140°C for 4 hr; 5°C per min cool rate.

magnetic properties will also be altered by time at temperature and cool rate; i.e., in some cases, magnetic-core parts are heat-treated through the continuous furnace and produce acceptable magnetic properties for the given application.

Since the alloy shows a very low hardness value (Rockwell B-50/55) after heat treatment, parts cannot be stressed or deformed without a loss and change in magnetic properties. Extreme care must be exercised during handling. Magnetic-

TABLE 2 Typical Magnetic Properties of 80 Ni 4 Mo Balance Fe

Thickness, in.	B_{40} at 60 Hz†	B_m, gauss	$B_m - B_r$, gauss	H_1, oersteds	ΔH	μ at B_{40} and 15 kHz†
0.001	70,000	7,000	1,000	0.030	0.006	20,000
0.002	80,000	7,000	1,200	0.030	0.006	12,000
0.004	80,000	7,000	1,500	0.035	0.009	

1-, 2-, and 4-mil square-loop properties*

* Constant-current flux-reset method of test, per AIEE No. 432. Heat treatment: 1160°C for ½ hr at temperature; 1–5°C per min cool rate.

† Heat treatment: 1170°C for 1–4 hr at temperature; 5–15°C per min cool rate. Tape toroid ID/OD 0.80 ratio.

Magnetic Properties   8-7

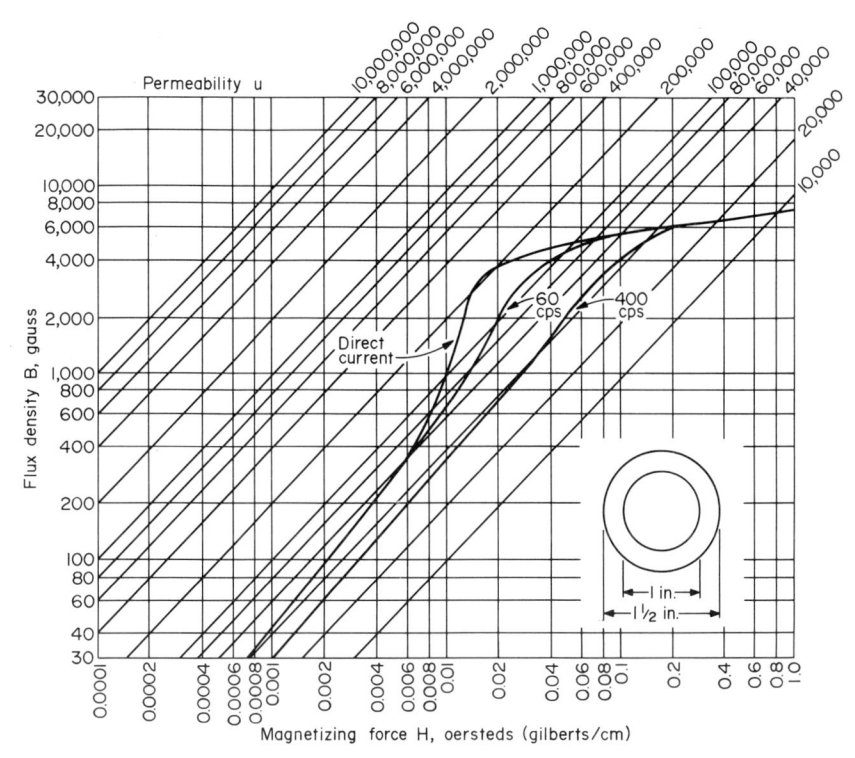

Fig. 4 Dc and ac magnetic properties of 80 Ni 4 Mo balance Fe lamination strip 0.006 in. thick. Heat treatment: 1140°C for 4 hr; 5°C per min cool rate.

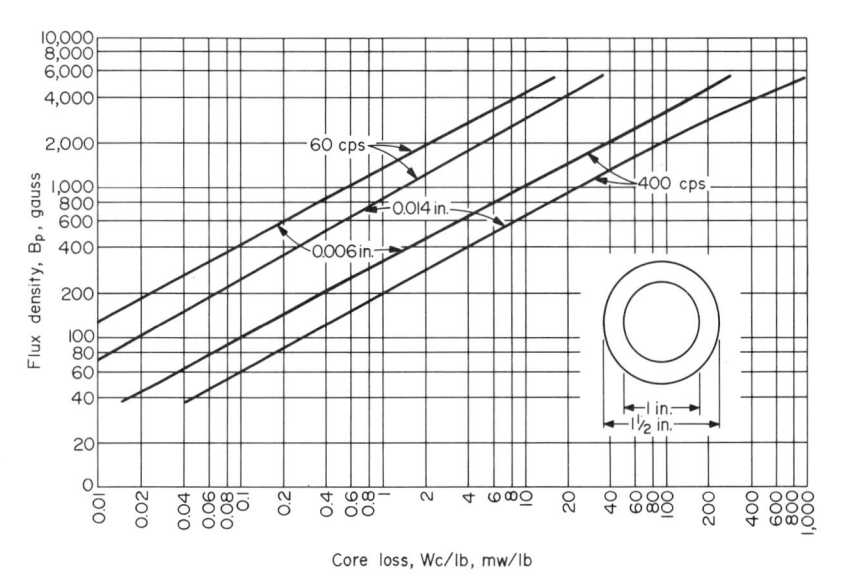

Fig. 5 Core loss of 80 Ni 4 Mo balance Fe lamination strip. Heat treatment: 1140°C for 4 hr; 5°C per min cool rate.

8-8 Ferrous Metals

core parts heat-treated at lower temperatures are less strain-sensitive. Reference 6 provides information regarding permeability-stress relationship.

General literature on other basic magnetic properties is available from the alloy producers in the United States:

Carpenter Technology Corp., Reading, Pa.
Allegheny-Ludlum Steel Corp., Pittsburgh, Pa.
Westinghouse Electric Corp., Blairsville, Pa.

77 Ni 1.5 Cr 5 Cu balance Fe. This alloy has high initial permeability properties and is well known as Mumetal.[30] Its handling characteristics are similar to those of the 80 Ni 4 Mo balance Fe alloy. General magnetic characteristics for laminations, magnetic cores, and shields are available from the producer.[7]

49 Ni balance Fe. The alloy composition of 49 Ni balance Fe is capable of producing a fairly high initial permeability with a moderately high magnetic satura-

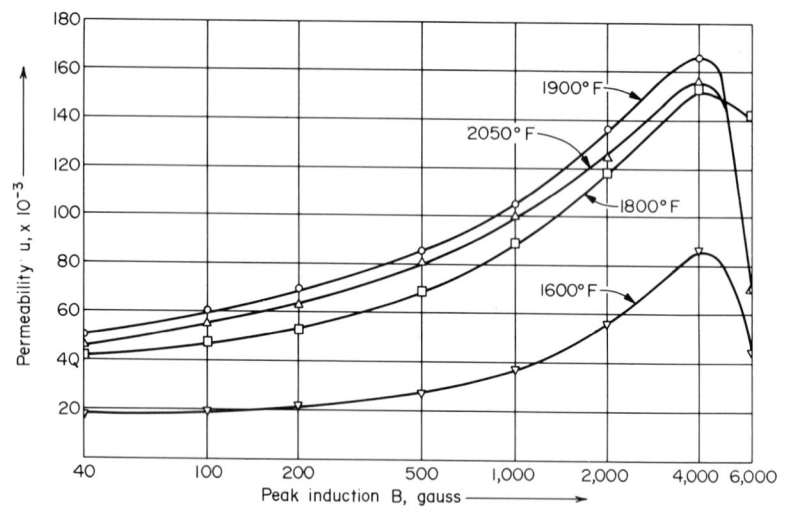

Fig. 6 Ac permeabilities versus induction at 60 Hz for 80 Ni 4 Mo balance Fe 0.006-in. stamped rings for various 4-hr anneals.

tion. Magnetic-core components are used where response is necessary in a fairly weak magnetic field; they have minimum residual magnetism where very small air gaps are in the magnetic circuit, and show fairly good properties in ac magnetic fields. General usage is in the form of lamination strip, tape toroids, fabricated relay parts, and flux-carrying members and precision castings. A free-machining grade is available; however, there is a substantial sacrifice in the magnetic properties.

Through controlled melting practice and strip manufacturing procedures, several types of strip products are available in the industry to achieve specific types of magnetic properties that can be utilized in specific types of magnetic cores. This applies to strip products 0.020 in. and less. The strip products are referred to as *highly oriented*, *semioriented*, and *random-oriented*. In each case, the strip product must be fabricated and heat-treated as recommended to achieve the magnetic properties desired.

The semioriented grade is preferred for transformer lamination application because of having the highest magnetic properties when heat-treated above 1000°C. Lamination strip ac properties are given in Figs. 7 and 8. Core-loss values appear in Table 3.

Magnetic Properties 8-9

The random-oriented grade is recommended for rotor and stator laminations that must have minimum directional magnetism. Most suitable method of detecting this grade is a very fine grain size when heat-treated at 1100°C (mean grain diagonal of 0.040 in.). The semioriented grade will show a very coarse grain when heat-treated under similar conditions. Basic magnetic properties of the random-oriented grade are slightly lower than those of the semioriented grade; however, they are superior to those of the semioriented grade when heat-treated below 1000°C. Effect of temperature on magnetic properties is shown in Fig. 9.

Highly oriented 49 Ni balance Fe is marketed in strip form from ½ to 14 mils and is used for its square-hysteresis-loop properties after heat-treating the

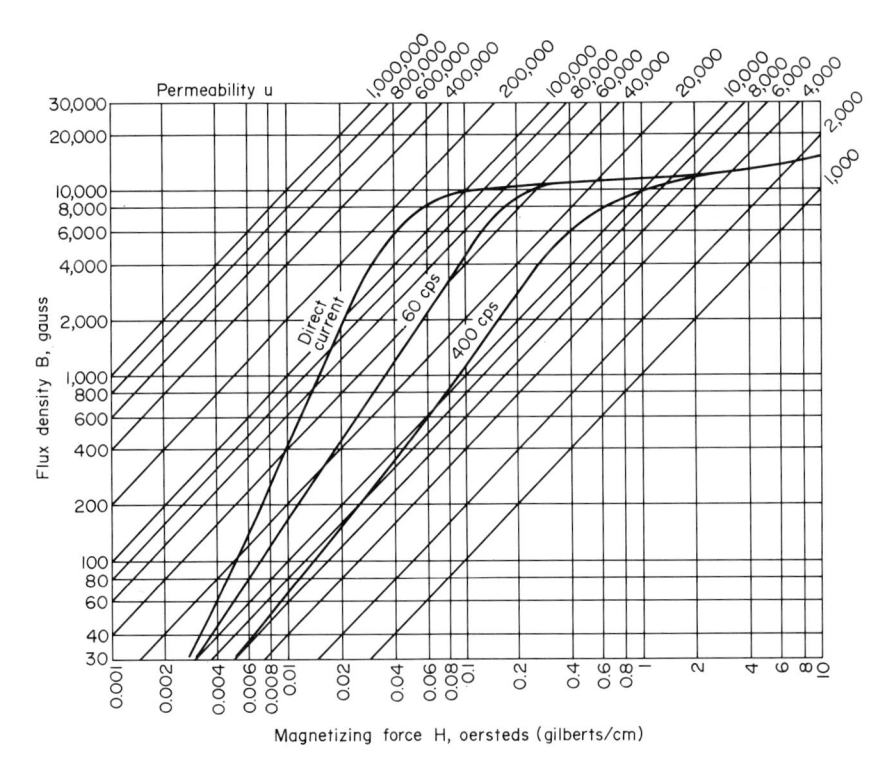

Fig. 7 Dc and ac magnetic properties of 49 Ni balance Fe semioriented lamination strip 0.014 in. thick. Heat treatment: 1140°C for 4 hr; 1°C per min cool rate.

fabricated cores. Table 4 lists the basic magnetic parameters. Laminations 0.006 and 0.014 in. thick are available for certain types of magnetic cores.

Commercial brand names marketed are as follows:
Allegheny-Ludlum 4750 (Transformer and Motor) and Deltamax
Carpenter Technology Corp. High Permeability 49 (Transformer and Rotor) HyRa 49
Westinghouse Electric Corp. Hipernik and Hipernik V
Universal-Cyclops
Simonds Saw & Steel Co.
Armco Steel Corp.

Soft magnetic alloys are employed to shield magnetic components and electron beams from undesirable external magnetic fields that have an effect upon their function. This is accomplished by utilizing the soft magnetic alloy to deflect or

8-10 Ferrous Metals

carry around the undesirable magnetic field. Shields can also be used to contain undesirable magnetic fields by surrounding the undesirable source. Shielding efficiency db, a function of the magnetic alloy thickness, magnetic permeability, and shape of the shield can be calculated for long, thin cylinders, using the equation

$$\text{S.E.} = 20 \log \frac{1}{2} \frac{\mu t}{R_0}$$

where μ = permeability
t = thickness
R_0 = outer radius of shield

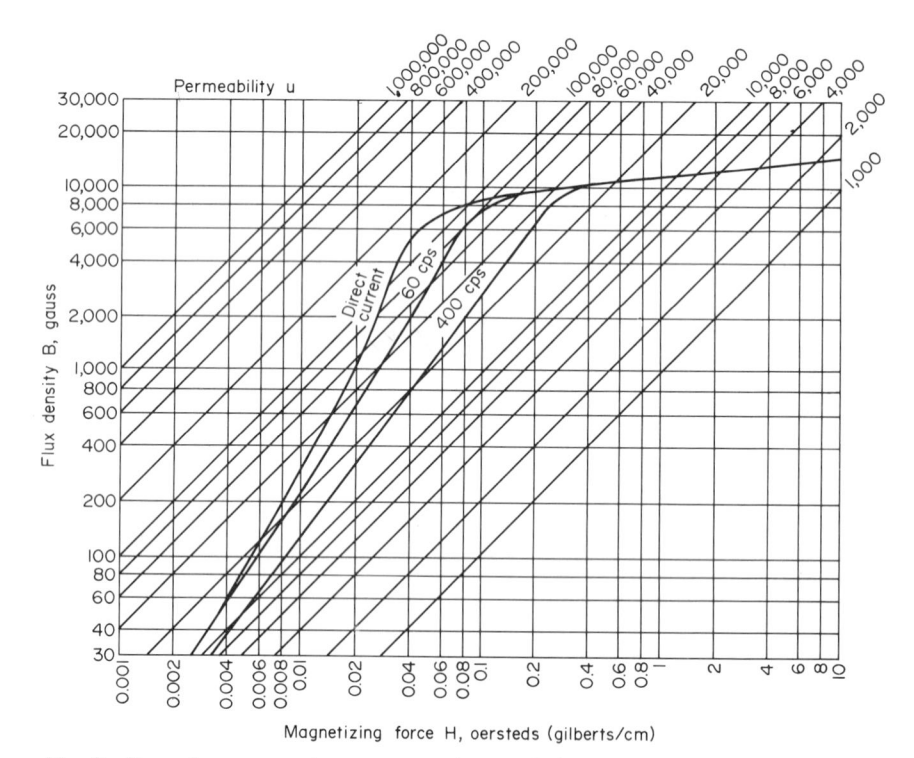

Fig. 8 Dc and ac magnetic properties of 49 Ni balance Fe semioriented lamination strip 0.006 in. thick. Heat treatment: 1140°C for 4 hr; 1°C per min cool rate.

This equation can be useful in determining how much the shield dimensions should be modified to compensate for a change in shield material permeability. Magnetic shields should be tested in accordance with ASTM A346 to provide an understanding of the db value for the given shield.

Laminated shields provide greater efficiency. To protect from very high frequency fields, copper and soft magnetic alloy laminants are employed. Most common alloys employed are:
  80 Ni 4 Mo balance Fe
  77 Ni 1.5 Cr 5 Cu balance Fe
  49 Ni balance Fe
  Si–Fe
  Commercial Fe

More specific information regarding magnetic shielding is presented in Ref. 8.

Silicon-iron alloys The family of silicon-iron alloys has a very important role in the field of magnetics because of their low cost and because of having been engineered into systems where an appreciable amount of electric power is involved Even though their response to weak magnetic fields is inferior to the nickel-iron alloys, their magnetic characteristics surpass those of other alloys in the field of electronics under certain conditions. Considerable amount of property evaluation has been conducted for the electrical industry and electric core-loss requirements have been established for the electrical sheet industry. Basic magnetic requirements of commercial grades are given in Ref. 9. Dc magnetic induction curves are given in Fig. 10.

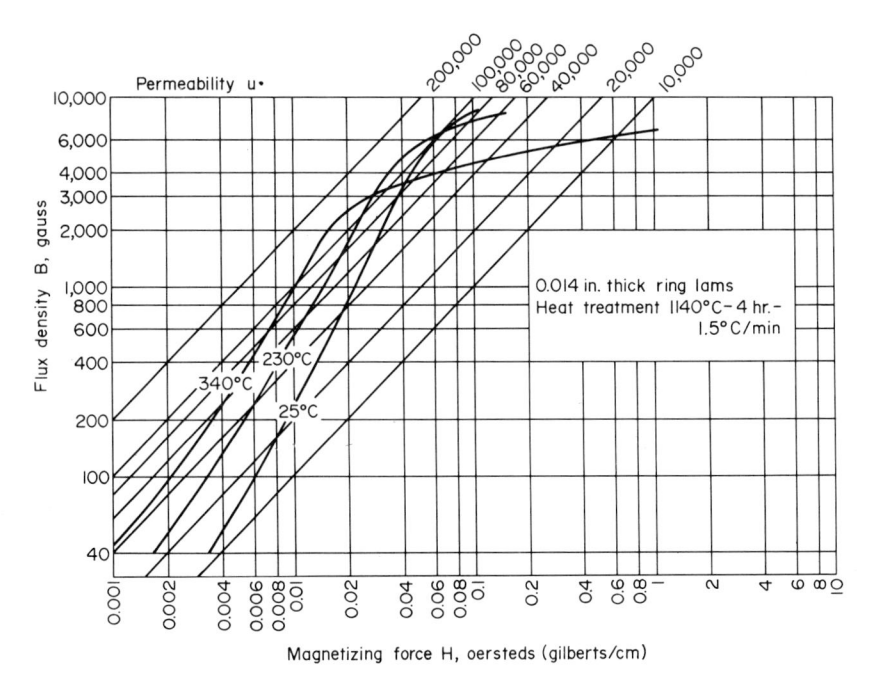

Fig. 9 60-Hz permeability versus temperature for 49 Ni balance Fe random-oriented alloy.

Three types of silicon-iron alloys in thick strip and bar forms are being fabricated into many shapes of magnetic cores and magnetic flux-carrying members. Machining operations, forgings, and precision castings make all shapes and sizes available. Commercial available grades have nominal silicon contents of 1, 2.5, and 4 percent. For improved machining properties, the silicon contents of 1 and 2.5 percent are available in free-machining grades. As with all other high-permeability alloys, a thermal heat treatment is required to obtain uniform magnetic properties from part to part. Magnetic properties improve with increasing heat-treating temperature as illustrated in Table 6 and Figs. 11, 12, and 13.

Cobalt-iron alloys The family of cobalt-iron alloys and the various elements added to this family have been thoroughly investigated for basic magnetic properties.[5,10] Five basic alloys have found commercial application; however, only three are being manufactured in production quantities in various forms. These alloys are used primarily for their high magnetic saturation and relative low coercive force in ac and dc applications. Typical dc magnetic properties are given in Fig. 14.

8-12 Ferrous Metals

TABLE 3 Core Loss of 49 Ni Balance Fe "Semioriented" and "Random-oriented" Strip (60-Hz core loss, milliwatts per pound)

°C for 4 hr	\multicolumn{6}{c	}{B, gauss}
	1,000	2,000	4,000	6,000	8,000	10,000

0.006-in. rings semioriented grade

°C for 4 hr	1,000	2,000	4,000	6,000	8,000	10,000
920	9.60	26.3	68.5	120.0	173.0	240.0
982	2.66	8.45	26.6	52.0	85.5	135.0
1038	1.83	6.03	19.7	39.8	66.6	107.0
1093	1.51	5.20	17.8	36.4	60.0	101.0
1180	1.58	5.43	18.2	37.4	66.1	103.0

0.006-in. rings random-oriented grade

°C for 4 hr	1,000	2,000	4,000	6,000	8,000	10,000
870	6.21	18.5	67.0	85.7	131.3	193.0
920	4.42	13.44	51.5	66.8	104.1	154.5
982	3.26	10.2	41.8	54.8	86.5	131.6
1180	2.07	6.98	31.6	42.7	72.4	112.3

0.014-in. rings semioriented grade

°C for 4 hr	1,000	2,000	4,000	6,000	8,000	10,000
920	9.95	28.0	74.3	132.0	195.0	275.0
982	3.02	9.85	31.1	62.5	107.0	174.0
1038	2.69	8.91	28.8	58.0	99.9	164.0
1093	2.47	8.50	27.9	57.0	99.0	162.0
1180	2.64	8.90	28.6	57.6	99.5	164.0

0.014-in. rings random-oriented

°C for 4 hr	1,000	2,000	4,000	6,000	8,000	10,000
920	5.34	16.2	62.2	82.1	130.9	199.2
982	4.2	13.2	53.9	72.4	119.2	185.0
1180	2.61	8.97	40.2	55.0	93.3	151.3

TABLE 4 Highly Oriented 49 Ni Balance Fe Alloy (400 Hz CCFR Data)

Thickness, in.	B_m, gauss	$B_m - B_1$, gauss	H_1, oersteds	ΔH
0.001	14,600	400	0.20	0.025
0.002	14,600	425	0.21	0.030
0.004	14,500	550	0.26	0.035

Constant-current flux-reset method at 400 Hz, per AIEE No. 432.
Core ID/OD 0.80 ratio.

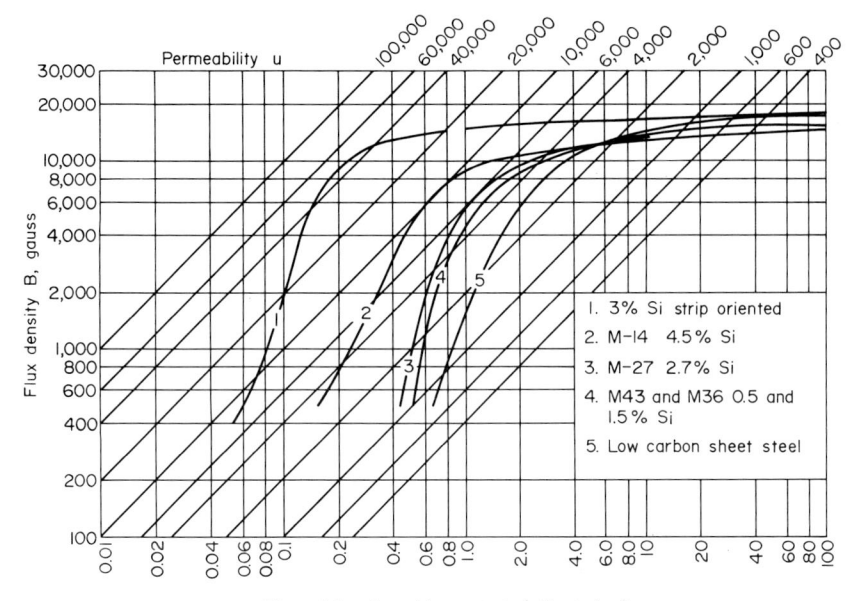

Magnetic Properties 8-13

Fig. 10 Dc normal induction curves of Si-Fe alloys.

Curves:
1. 3% Si strip oriented
2. M-14 4.5% Si
3. M-27 2.7% Si
4. M43 and M36 0.5 and 1.5% Si
5. Low carbon sheet steel

Axes: Flux density B, gauss vs. Magnetizing force H, oersteds (gilberts/cm); Permeability u.

TABLE 5 Mechanical Properties of Some Magnetic Alloys

Alloy	80 Ni 4 Mo bal Fe	77 Ni 1.5 Cr 5 Cu bal Fe	49 Ni bal Fe	49 Co 2 V bal Fe	27 Co bal Fe
Stamping quality, Rb90 min.:					
Tensile strength ksi	125–160	....	100–160	170–215	165–200
Yield strength ksi	115–150	....	95–155	164–205	157–196
Elongation in 2 in., %	3–4	....	2–5	1–3	1–3
Elastic modulus, ksi × 10^3	33.7	....	24.0		
Mill-annealed:					
Tensile strength, ksi	90	85	75		
Yield strength, ksi	40	32.5	26		
Elongation in 2 in., %	35	40	40		
Hardness, Rockwell B	75	....	65		
1170°C hydrogen-annealed:					
Tensile strength, ksi	79	64	64	70–87*	
Yield strength, ksi	22	18.5	22	46–57	
Elongation in 2 in., %	64	27	29	5–6	
Hardness, Rockwell B	50	....	47	....	75–80†
Hardness, Rb:					
1090°C anneal	51	....	50	....	40–45
982°C anneal	56	....	55	....	15–20
870°C anneal	63	....	59	....	85

* 790°C anneal 2 hr in hydrogen.
† 800–900°C anneal 2–4 hr, fast cool.

8-14　Ferrous Metals

TABLE 6　DC Magnetic Properties for Silicon-Iron Alloys

Annealing temperature, °F for 2 hr	Silicon content, %	DC maximum permeability	Flux at 50 oersteds	From 10,000 gauss B_r	H_c
1350	2.5, FM	3,280	16,900	6,600	0.913
	2.5	3,790	17,500	6,200	0.865
	1.0	4,280	17,780	8,100	0.960
	1.0, FM	3,380	17,020	7,800	1.250
1550	2.5, FM	4,990	16,660	6,000	0.600
	4.0	4,750	16,500	5,300	0.600
	2.5	4,630	16,900	5,700	0.611
	1.0	4,545	16,940	7,400	0.792
	1.0, FM	4,240	17,100	7,400	0.864
1750	2.5, FM	7,120	16,500	7,000	0.480
	4.0	5,200	16,600	5,400	0.540
	2.5	6,280	17,000	6,200	0.515
	1.0	7,580	16,700	8,300	0.574
	1.0, FM	7,050	16,900	6,800	0.475
1950	2.5, FM	16,250	17,180	8,600	0.310
	4.0	17,700	16,600	8,400	0.506
	2.5	14,600	17,000	8,200	0.330
	1.0	8,060	17,280	7,800	0.480
	1.0, FM	8,030	17,120	8,030	0.444

FM　Free machining.

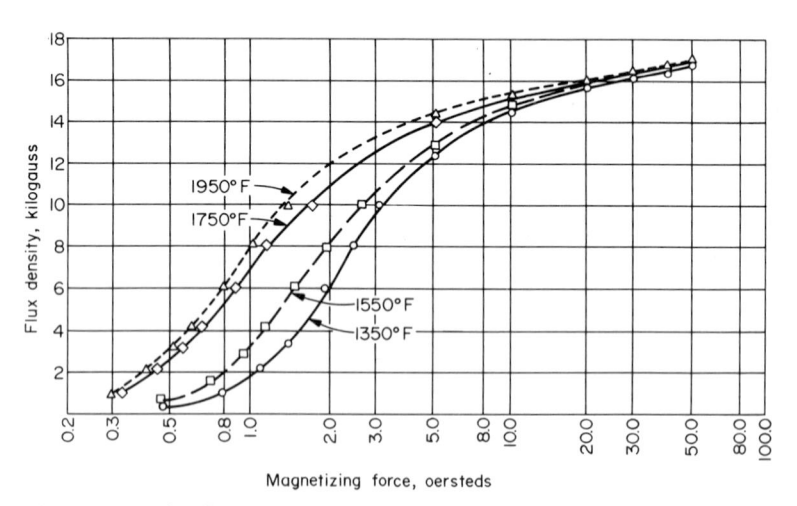

Fig. 11　Dc normal induction curves for 1 Si and 1 Si FM alloys versus heat-treating temperature.

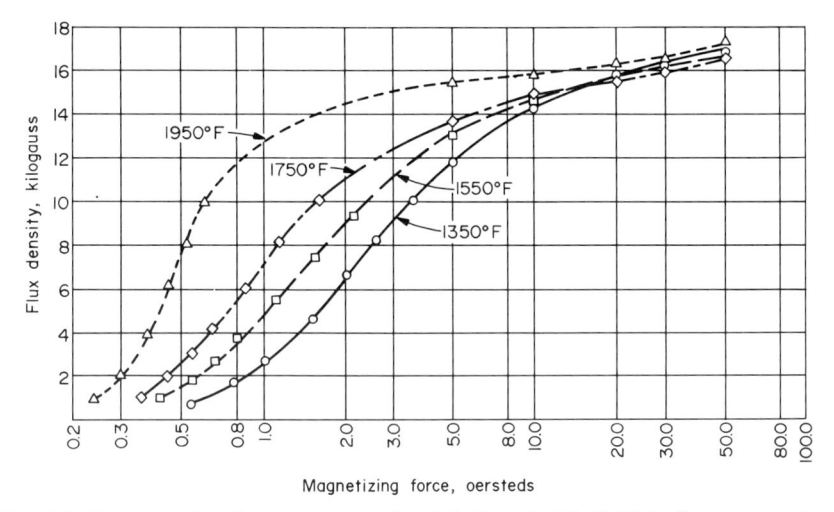

Fig. 12 Dc normal induction curves for 2.5 Si and 2.5 Si FM alloys versus heat treatment.

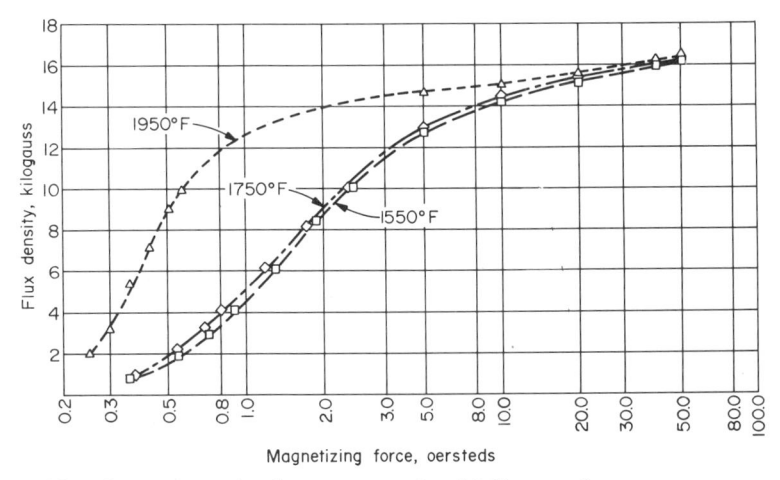

Fig. 13 Dc normal induction curves for 4.0 Si versus heat treatment.

Alloys being produced on a commercial basis in 1970 were:
  Supermendur—thin strip 0.006 in. and less
  50 Co 50 Fe—bar form only—Permendur, Hy-Sat 50[29]
  49 Co 49 Fe 2 V—strip, bar, forgings—Hiperco 50,[31] Hy-Sat 48,[29] 2-Vanadium Permendur[30]
  49 Co 49 Fe 2 V FM—bar form only—Hy-Sat 48 FM[29]
  35 Co 1 Cr balance Fe—strip, bar, forgings—Hiperco 35[31]
  27 Co 0.6 Cr balance Fe—strip, bar, forgings—Hiperco 27,[31] Hy-Sat 27[29]
Items available in strip form have been investigated for ac magnetic applications.[11]

Commercial irons and commercially pure irons Many commercial irons are used for their magnetic properties because of their relative low cost in comparison to other more expensive nickel-iron, silicon-iron, and cobalt-iron alloys The magnetic properties obtainable vary considerably, depending upon heat treatment applied to the fabricated parts. Magnetic data on these alloys are very limited, as the properties are not too critical. Armco Ingot Iron, a well-known commercially pure iron, has been surveyed extensively, and detailed literature is available from Armco Steel Corp., Middletown, Ohio.[12]

Applications requiring a minimum degree of nonmetallics and freedom of internal discontinuities demand a product manufactured by other precedures such as consumable electrode melting and/or vacuum-induction melting. These methods of manufacture also assure a very low carbon content, lowest gas content, and minimum

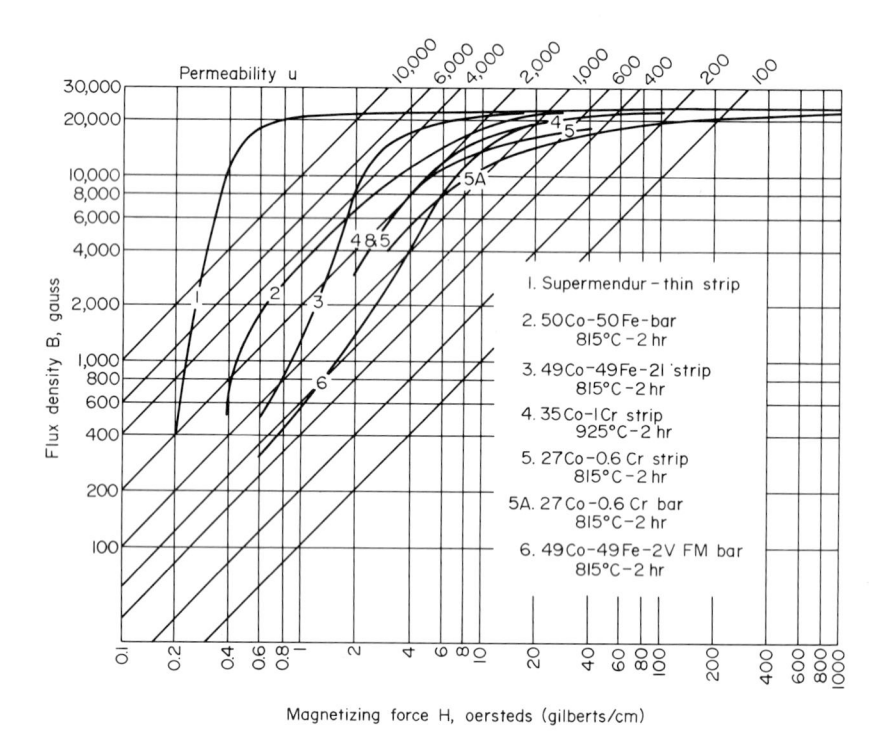

Fig. 14 Dc normal induction curves for Fe-Co alloys.

1. Supermendur – thin strip
2. 50Co-50Fe-bar 815°C-2 hr
3. 49Co-49Fe-2I strip 815°C-2 hr
4. 35Co-1Cr strip 925°C-2 hr
5. 27Co-0.6 Cr strip 815°C-2 hr
5A. 27Co-0.6 Cr bar 815°C-2 hr
6. 49Co-49Fe-2V FM bar 815°C-2 hr

residual elements. Their characteristics assure best performance in vacuum envelopes and assure minimum contamination over long periods of time. Brand names of these special irons are: Consumet Core Iron,[29] Vacumet Core Iron,[29] Vacumet Consumet Core Iron[29] and Ferrovac E.[35] Typical magnetic properties of these irons are given in Table 7. Physical properties of Armco Magnetic Ingot Iron are given in Table 8. These can be employed as references for the remainder of the commercial irons and commercial pure irons; however, slight variations can be anticipated.

Mechanical properties and suggested machining information for magnetic alloys General information has been compiled for some of the alloys reviewed. Thermal expansion properties are given in Table 9, and suggested machining references are given in Table 10.

Additional information regarding the properties of several alloys is available from Engineering Alloys Digest, Inc., Upper Montclair, N.J. (see Table 11).

Magnetic Properties

TABLE 7 Typical Magnetic Properties of Some Commercial Irons and Commercially Pure Irons

Property	Ingot iron	B1113	Consumet* core iron	Vacumet* core iron	Ferrovac E†
850°C for 2 hr, fast cooling:‡					
Dc permeability μ at B_{1000}	1,200	800	1,200	6,000	
Dc permeability μ_{max}	4,500	2,200	3,000	14,000	
H_c from 10^4 gauss	1.1	1.4	1.3	0.30	
1000°C for 2 hr, fast cooling:‡					
Dc permeability at B_{1000}	1,500	.....	1,500	6,000	6,700 §
Dc permeability μ_{max}	5,200	.....	4,800	19,000	37,200
H_c from 10^3 gauss	0.90	.....	0.90	0.26	0.14

* Carpenter Technology Corp.
† Crucible Steel Co.
‡ Dry hydrogen atmosphere.
§ Annealed at 854°C in wet hydrogen for 35 hr.

TABLE 8 Physical Properties and Mechanical Properties of Armco Ingot Iron[12]

Physical constants at 68°F	
Specific gravity	7.86
Density, lb/in.^{3}	0.284
Melting temperature, °F	2790
Thermal conductivity, Btu/(hr)(ft^2)(°F/in.)	508
Specific heat at 20°C, cal/(g)(°C)	0.1075
Thermal coefficient of expansion at 68–750°F, in./(in.)(°F)	0.0000076
Mass resistivity, ohms/mile × lb/mile (mile-ohms at 20°C)	4,800
Magnetic saturation, gauss (ferric induction)	21,550
Volume electrical resistivity at 20°C, microhms/cm^3	10.7
Modulus of elasticity, psi	30,000,000

	Sheet and Strip		Bar and Rod		Annealed at:	
Property	Hot rolled	CR Surface B55 max.	Hot rolled	Cold drawn	1400°F	1700°F
Tensile strength, psi	45,000	43,000	44,000	65,000	42,000	40,000
Yield strength, psi	30,000	28,000	30,000	60,000	27,000	20,000
Elongation in 2 in., %	38	38	40	20	38	40
Reduction of area, %	......	......	......	......	73	78
Hardness, Rb*	50	44	50	80	40–50	30–40
Fatigue strength, psi	25,000	24,000	25,000	32,000	23,000	22,000

* Rockwell B.

Permanent Magnetic Alloys

Permanent magnetic alloys are generally referred to as *hard magnetic materials*. These materials retain magnetic poles (north and south) after being subjected to a strong magnetic field. Magnetic property tests are generally conducted on specific bar shapes, heat-treated as prescribed, and evaluated in accordance with ASTM A341. Most common magnetic parameters employed to compare and rate permanent magnetic materials are:

8-18   Ferrous Metals

TABLE 9   Thermal Expansion Properties of Some Magnetic Alloys*

Temperature, °C	80 Ni 4 Mo bal. Fe	77 Ni 1.5 Cr 5 Cu bal. Fe	49 Co 2 V bal. Fe	27 Co bal. Fe	1 Si bal. Fe	2.5 Si bal. Fe	4 Si bal. Fe	Commercial iron
20–100	11.51	12.5	....	.....	11.2	11.3	11.8	12.7
200	12.62	14.4	....	10.10	12.0	12.2	12.4	13.8
300	13.20	....	9.5	10.35	12.4	12.7	12.9	14.6
400	13.67	....	....	.....	12.8	13.2	13.5	15.8
500	13.95	....	9.8	10.90	....	....	....	16.7

* Coefficient of expansion: in./in. $\times 10^{-6}$ per °C.
NOTE: Information for 36 Ni balance Fe and 49 Ni balance Fe can be found in the text.

TABLE 10   Suggested References for Machining Magnetic Alloys

Alloy	Hardness, Rockwell B	Turning Speed, sfm	Turning Feed, mpr	Drilling Speed, sfm	Drilling Feed, mpr	Milling Speed, sfm	Milling Feed, mpr	Tapping speed, sfm
80 Ni 4 Mo...	90	50	0.7–2	35	1–4	40	2–5	10–15
49 Ni bal. Fe	90–100	60	0.7–2	45	1–5	45	2–5	15–20
49 Ni, FM....	90–100	100	1–3	75	1–4	85	3–5	20–25
1 Si bal. Fe...	90–100	85	1–2.5	60	1–5	65	2–5	15–20
1 Si, FM.....	90–100	115	1–2.5	80	2.8	90	3–6	25–30
2.5 Si bal. Fe	90–100	80	1–2	55	1–5	60	2–5	15–20
2.5 Si, FM...	90–100	120	1–3	85	2–8	95	3–6	25–30
4 Si.........	90–100	75	1–3	50	1–5	50	2–5	10–15
Commercial Fe........	85–95	100	1–3	80	1–5	65	2–5	20

sfm   Surface feet per minute.
mpr   Mils (0.001 in.) per revolution.
FM   Free machining.

TABLE 11   High-permeability Alloys*

Code no.	Composition	Brand name
Fe-4	49 Ni bal. Fe	Allegheny 4750[30]
Fe-10	Commercial iron	Armco Ingot Iron[32]
Fe-13	49 Co 2 V bal. Fe	Hiperco[31]
Fe-19	4 Si bal. Fe	Hipersil[31]
Fe-25	49 Ni bal. Fe	Carpenter High Permeability 49[29]
Fe-26	27 Co bal. Fe	Hiperco 27[31]
Fe-36	1, 2.5, 4 Si bal. Fe	Silicon Core Iron[29]
F-18	49 Ni bal. Fe	Hipernik V[31]
Ni-5	80 Ni 4 Mo	HyMu 80[29]
Ni-25	77 Ni 1.5 Cr 5 Cu	MuMetal[30]
Ni-26	49 Ni bal. Fe	Hipernik[31]
Ni-62	49 Ni bal. Fe	Conpernik[31]
Co-52	49 Co 2 V, FM, bal. Fe	Covandur[29]

* Engineering Alloys Digest, Inc., Upper Montclair, N.J.
FM   Free machining.

1. *Coercive force* H_c: The demagnetizing force that must be applied to a magnet to reduce the magnetic induction to zero.
2. *Demagnetization curve:* The second-quadrant portion of the hysteresis loop relating induction in a magnet to its magnetizing force.
3. *Energy product* BH_{max}: The product of $B \times H$ in the second quadrant of the hysteresis loop. A figure of merit for a magnet. It is proportional to the amount of external energy available from a magnet of given dimensions.
4. *Residual induction* B_r: The magnetic induction that is retained by a magnet after removal of a saturating magnetizing force.

TABLE 12 Magnetic Properties and Chemical Compositions[13]

Magnet material	Nominal chemical composition	Residual flux density B_r, gauss (nominal)*	Coercive force H_c, oersteds (nominal)*	Maximum energy product BH_{max}, megagauss-oersteds (nominal)*
3½% Cr steel	3.5 Cr, 1 C, bal. Fe	10,300	60	0.30
3% Co steel	3.25 Co, 4 Cr, 1 C, bal. Fe	9,700	80	0.38
17% Co steel	18.5 Co, 3.75 Cr, 5 W, 0.75 C, bal. Fe	10,700	160	0.69
36% Co steel	38 Co, 3.8 Cr, 5 W, 0.75 C, bal. Fe	10,400	230	0.98
Alnico 1	12 Al, 21 Ni, 5 Co, 3 Cu, bal. Fe	7,200	470	1.40
Alnico 2	10 Al, 19 Ni, 13 Co, 3 Cu, bal. Fe	7,500	560	1.70
Alnico 3	12 Al, 25 Ni, 3 Cu, bal. Fe	7,000	480	1.35
Alnico 4	12 Al, 27 Ni, 5 Co, bal. Fe	5,600	720	1.35
Alnico 5	8 Al, 14 Ni, 24 Co, 3 Cu, bal. Fe	12,800	640	5.50
Alnico 5 DG	8 Al, 14 Ni, 24 Co, 3 Cu, bal. Fe	13,300	670	6.50
Alnico 5 Col	8 Al, 14 Ni, 24 Co, 3 Cu, bal. Fe	13,500	740	7.55
Alnico 6	8 Al, 16 Ni, 24 Co, 3 Cu, 1 Ti, bal. Fe	10,500	780	3.90
Alnico 7	8.5 Al, 18 Ni, 24 Co, 3.25 Cu, 5 Ti, bal. Fe	7,700	1,050	2.85
Alnico 8	35 Co, 15 Ni, 4 Cu, 5 Ti, 7 Al, bal. Fe	8,200	1,650	5.3
Alnico 9	35 Co, 15 Ni, 4 Cu, 5 Ti, 7 Al, bal. Fe	10,500†	1,500†	9.0†
Sintered Alnico 2	10 Al, 19 Ni, 13 Co, 3 Cu, bal. Fe	7,100	550	1.50
Sintered Alnico 5	8 Al, 14 Ni, 24 Co, 3 Cu, bal. Fe	10,900	620	3.95
Sintered Alnico 6	8 Al, 16 Ni, 24 Co, 3 Cu, 1 Ti, bal. Fe	9,400	790	2.95
Sintered Alnico 8	35 Co, 15 Ni, 4 Cu, 5 Ti, 7 Al, bal. Fe	7,400	1,500	4.00
Ceramic 1	MO·6 Fe$_2$O$_3$	2,300	1,860–3,250‡	1.05
Ceramic 2	MO·6 Fe$_2$O$_3$ M represents one or more	2,900	2,400–3,000‡	1.8
Ceramic 3	MO·6 Fe$_2$O$_3$ of the metals chosen from	3,300	2,200–2,400‡	2.6
Ceramic 4	MO·6 Fe$_2$O$_3$ the group barium, stron-	2,500	2,300–3,800‡	1.45
Ceramic 5	MO·6 Fe$_2$O$_3$ tium, lead.	3,800	2,400	3.40
Ceramic 6	MO·6 Fe$_2$O$_3$	3,200	2,820–3,300‡	2.45
E.S.D. 31	20.7 Fe, 11.6 Co, 67.7 Pb	5,000	1,000	2.30
E.S.D. 32	18.3 Fe, 10.3 Co, 72.4 Pb	6,800	960	3.00
E.S.D. 41	20.7 Fe, 11.6 Co, 67.7 Pb	3,600	970	1.10
E.S.D. 42	18.3 Fe, 10.3 Co, 72.4 Pb	4,800	830	1.25
Cunife 1	60 Cu, 20 Ni, 20 Fe	5,500	530	1.40
Vicalloy 1	10 V, 52 Co, bal. Fe	7,500	250	0.80
Remalloy	12 Co, 15 Mo, bal. Fe	9,700	250	1.00

* Values derived from major hysteresis loop.
† Tentative properties.
‡ Intrinsic coercive force.

5. *Intrinsic curve* $(B - H)$: A plot of intrinsic induction in the second and third quadrants of the hysteresis loop. Intrinsic induction is not zero at the H_c point. This means that a field intensity H which is large enough to hold B at zero does not completely demagnetize the magnet, owing to recoil. Complete demagnetization occurs only when $B - H = 0$.

Many alloy compositions have been evaluated for their permanent magnetic characteristics; however, the industry has standardized on certain compositions to meet the general requirements. Magnetic properties and chemical compositions standardized by the permanent magnetic industry are given in Table 12.[3] MMPA Standard 0100-66 defines the physical and magnetic characteristics, dimensions, and tolerances of commercial grades of permanent-magnet materials. Cobalt and chromium steels shown in Table 12 are available in wrought (rolled) and cast shapes. Data on several other commercial grades are shown in Table 13. Physical properties of

TABLE 13 Other Commercial Permanent-magnet Grades*

Magnetic material	Composition	Residual flux density B_r, gauss (nominal)	Coercive force H_c, oersteds (nominal)	Maximum energy product BH_{max}, megagauss-oersteds (nominal)
0.65 % C steel...	0.65 C 0.85 Mn	10,000	42	0.18
1 % C steel.....	1.0 C 0.50 Mn	9,000	51	0.20
Remalloy.......	20 Mo 12 Co bal. Fe	8,550	355	1.25
P-6†...........	45 Co 45 Fe 6 Ni 4 V	14,200	60	0.41

* Engineering Alloys Digest, Inc., Code C-18.
† General Electric Co. (strip and wire form).

TABLE 14 Nominal Physical Properties of Permanent-magnet Alloys

Material	Resistivity at 25°C, microhm-cm	Mean coefficient of expansion at 20–300°C, in./in. $\times 10^{-6}$	Hardness, Rockwell C	Tensile strength, ksi	Density, lb/in.^{3}
0.65 % C steel........	....	13.3	60–65	300	0.283
1 % C Steel..........	....	12.4	60–65	300	0.282
3½ % Cr steel........	29	12.6	60–65	300	0.281
17 % Co steel........	27	15.9	60–63	300	0.302
36 % Co steel........	26	17.2	60–65	300	0.296
Remalloy 17 % Co....	45	....	60–63	....	0.303
Remalloy 20 % Co....	45	....	60–63	....	0.303
P-6.................	30	....	....	....	0.285
Alnico 1, cast........	75	12.6	45	4.1	0.249
Alnico 2, cast........	65	12.4	45	3.0	0.259
Alnico 3, cast........	60	13.0	45	1.2	0.249
Alnico 4, cast........	75	13.1	45	9.1	0.253
Alnico 5, cast........	47	11.5	50	5.4	0.264
Alnico 5 DG, cast.....	47	11.5	50	5.4	0.264
Alnico 5 Col., cast....	47	11.5	50	....	0.264
Alnico 6, cast........	50	11.4	50	23.0	0.265
Alnico 7, cast........	58	11.4	60	....	0.265
Alnico 8, cast........	50	11.0	56	....	0.262
Alnico 2, sintered.....	68	12.4	43	65	0.247
Alnico 5, sintered.....	50	11.3	44	50	0.253
Alnico 6, sintered.....	53	11.2	44	....	0.254
Vicalloy 1...........	60–67	12.0	60–63	....	0.295
Cunife 1............	18	....	....	100	0.311

the grades are given in Table 14. When permanent magnet units are very small (less than several ounces), the sintering manufacturing procedure may be preferred for economical reasons and the obtaining of higher physical properties. Since permanent-magnet alloys are very hard in the heat-treated condition, the unit cannot be machined or formed. Metal removal can only be achieved by grinding operations, and these must be conducted very carefully so as not to fracture the unit.

Ferrite magnets are also known as ceramic magnets. General composition is of a $M \cdot 6Fe_2O_3$, where the M represents barium, strontium, lead or combinations thereof. This type of magnet is very hard and brittle, is a good electrical insulator and a poor conductor of heat, and is chemically inert. Careful grinding procedures are recommended for removing material. Considerable development work is in progress on this type of permanent-magnet material, and it is necessary to maintain constant contact with the producers to ascertain new products commercially available and details of forms available. Table 15 lists literature that can be procured from Engineering Alloys Digest, Inc.

Stainless steels Stainless steels are divided into two groups: namely, ferritic stainless steels (400 series) and austenitic stainless steels (300 series.) Ferritic stainless steels are magnetic and can be further subdivided into martensitic (harden-

TABLE 15 Permanent-magnet Alloys*

Code No.	Title
Fe-17	Alnico V
Co-18	General Electric P-6
Co-39	Alnico VIII
Co-42	Vicolloy 1
SA-44	Cobalt Magnet Steel
Cu-101	Cunico
Cu-105	Cunife 1

* Engineering Alloys Digest, Inc., Upper Montclair, N.J.

TABLE 16 Typical Magnetic Properties of Hardened Martensitic Stainless Steels

Types	Heat treatment	Hardness, Rockwell C	μ_{max}	H_c, oersteds	Saturation, 10^3 gauss
410 and 416	940°C, Q	41	1,000	36	14
420	1025°C, Q	50	400	45	14
440	1025°C, Q	55	62	64	12

Q Oil quench plus 1 hr at 370°C.

able) and ferritic (nonhardenable) groups. Types 410, 416, 420, and 440 can be hardened by heat treatment, and in this condition the alloy acts as a weak permanent magnet, as illustrated in Table 16 by the H_c coercive values. In the full-annealed (low-hardness) condition, the magnetic values are considered as weak, soft magnetic properties. Table 17 lists the general magnetic properties and includes the ferritic grades such as Types 430, 430F, and 446.

Obviously, the Type 400 series of stainless steels are employed for their magnetic properties when other soft magnetic alloys lack corrosion resistance to a given media and plated cores cannot be employed. Type 400 series of stainless steels provide corrosion resistance to fresh water, mine water, steam, gasoline, crude oil, perspiration, alcohol, ammonia, soap, sugar solutions, and most foodstuffs.

Austenitic stainless steels (Type 300 series) are generally considered as nonmagnetic alloys and as having good corrosion resistance to ambient conditions. All Type 300 series stainless steels have maximum permeabilities of 1.005 or less in the annealed (lowest hardness condition). Cold work introduced through processing

TABLE 17 Typical DC Soft Magnetic Properties of Type 400 Stainless Steels

Types	Hardness, Rockwell B	\multicolumn{4}{c}{μ at B gauss}	μ_{max}	H_c, oersteds from 10^3 gauss
		500	1,000	5,000	10,000		
410 and 416*..	80	200	380	900	1,020	1,080	5.0
430 and 430 F*	75	400	550	1,600	1,600	1,840	1.7
446†.........	85	350	500	1,100	700	1,100	4.0

* Heat-treated at 815°C for 2 hr, furnace-cooled.
† Heat-treated at 900°C for 2 hr, furnace-cooled.

and fabrication will increase the maximum permeability. The rate of increase depends upon the chemical composition and the stability of the austenite.[14]

Effect of cold work on several austenitic stainless steels is shown in Fig. 15. Magnetic properties of many ferrous and nonferrous materials have been reported in Ref. 15.

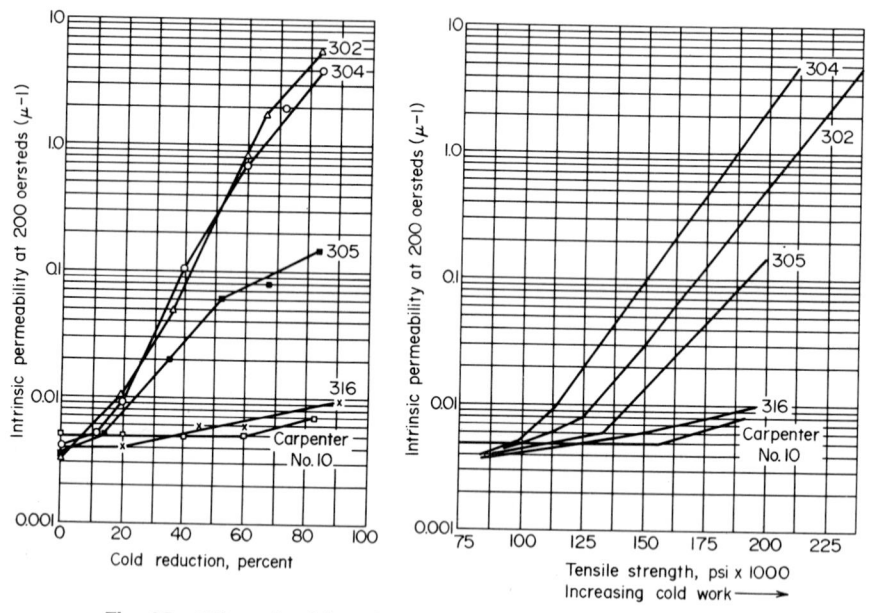

Fig. 15 Effect of cold work on several austenitic stainless steels.

Temperature-compensator alloys As the nickel content in nickel-iron alloys increases above 29 percent, the Curie temperature increases (Fig. 1). Commercial alloys containing 29 to 39 percent nickel are utilized as temperature-compensator alloys because their magnetic saturation decreases with increasing temperature. This characteristic provides a means of controlling a dc magnetic flux density produced by a permenent-magnet or an electromagnet field which decreases with increasing temperature. To achieve this performance from the temperature-compensator alloy, the magnetic source must produce a magnetic flux density in the alloy above the knee of the dc $(B - H)$ curve. Most data available for these alloys have been determined at 46 oersteds, which is in the saturation range.

Generally, the magnetic pole strength of a permanent magnet decreases with increasing temperature. To achieve a uniform functional magnetic field, the temperature-compensating alloy is employed to shunt magnetic lines of force from the poles and away from the functional magnetic area. As the temperature decreases and the magnet pole strength increases, the compensating alloy shunts or robs more flux from the functional area. With increasing temperature and decreasing magnet pole strength, the compensating alloy shunts less flux. By proper selection of compensating alloys, thickness of compensator cross section, and design of shunt, the functional magnetic field can be controlled.

The temperature-compensating characteristics of these alloys can be varied by the amount of cold-work stresses introduced into the alloy. Cold work decreases

TABLE 18 Magnetic Flux Densities versus Temperature of Temperature-compensating Alloys at $H = 46$ Oersteds

Ni content, %...	29.0	29.8	29.8	32.5	36.0	38
Type*...........	5	4	2	1		
Temperature, °C	\multicolumn{6}{c}{Flux density, gauss}
−60	.....	.....	.....	10,450	14,900	
−40	4,000	5,450	5,650	9,900	14,400	
−20	2,980	4,620	4,900	9,350		
0	1,350	3,480	3,920	8,700	13,400	14,200
+15	370	2,500	3,000			
+25	148	1,760	2,240	7,800	12,700	13,400
+40	40	750	1,120	7,250		
+50	.....	340	550	.....	11,870	13,200
+80	.....	.....	.....	5,450	10,700	
+90	.....	.....	.....	4,800	10,500	
+100	.....	.....	.....	.....	10,000	12,000
+150	.....	.....	.....	.....	7,000	9,200
+200	.....	.....	.....	.....	.....	8,800

* Carpenter Technology Corp. type numbers.

the flux-carrying capacity and reduces the change in flux density per degree temperature from that shown by the alloy in the annealed condition (condition generally supplied by the producer). Thermal treatments up to 480°C for several hours help to make the properties more stable with time and increase the flux-carrying capacity slightly. If the degree of cold work exceeds 10 percent reduction, it is recommended that the part be heat-treated as recommended by the producer.

Temperature-compensator alloys are used to compensate permanent magnets in instruments, switches, watthour meters, tachometers, speedometers, microwave tubes, etc. These alloys can also be used for thermal switches and temperature regulators.

Nominal flux density changes with temperature of commercial grades are given in Table 18.

Thermal Expansion Properties of Alloys

Many ferrous alloys are used in various applications for their thermal expansion characteristics. All metals expand when the temperature is increased above 25°C (77°F), and contract when the temperature is decreased. The rate of expansion and contraction depends upon chemical composition and physical condition; hence, a wide variety of thermal expansion characteristics is available from ferrous alloys. Typical thermal expansion properties of various alloys employed in the industry for this particular characteristic will be reviewed. Test procedures have been established by ASTM to determine the expansion characteristics of materials.[16] Thermal expansion properties are expressed in several ways:

1. Linear thermal expansion is the change in length per unit length resulting from a change in temperature of the material.

2. Mean coefficient of linear expansion α_m between temperatures T_1 and T_2 is defined as

$$\alpha_m = \frac{L_2 - L_1}{L_0(T_2 - T_1)} = \frac{\Delta L/L_0}{\Delta T}$$

3. Instantaneous coefficient of linear thermal expansion α_t at temperature is defined by the expression

$$\alpha T = T_1 \lim \rightarrow T_2 \frac{L_2 - L_1}{L_0(T_2 - T_1)} = \frac{dL/dT}{L_0}$$

where ΔL = observed change in length
L_0 = length of specimen at reference temperature T_0
L_1 and L_2 = specimen lengths at temperature T_1 and T_2, respectively
ΔT = change of temperature

Data presented herein are expressed in the form of mean (average) coefficient of linear expansion in units per degree Fahrenheit. These data can be employed to calculate and plot the linear thermal expansion. From the linear thermal expansion, the instantaneous coefficient of linear thermal expansion can be determined.

Nickel-iron alloys and related alloys This family of alloys has been thoroughly investigated for thermal expansion properties. The alloy containing 36 Ni balance Fe (Invar) exhibits the lowest expansion properties of all metals (see Fig. 1) in the temperature range of −459 to approximately +350°F; hence, the 36 percent nickel alloy is used predominantly in applications where minimum size change is necessary. The dimensional stability with time and thermal heat treatment to achieve same has been thoroughly investigated.[17] Investigations are in progress to reduce the expansion properties further by controlling the residual elements commonly found in commercial alloys. A free-machining grade, having slightly higher expansion characteristics, is available to realize lower machining costs. Invar itself is difficult to machine because of gumminess of the alloy.

Figure 1 illustrates that as the nickel content decreases below 36 percent and increases above 36 percent, the 25°C (77°F) coefficient of expansion increases; however, with increasing nickel content up to 65 percent, the Curie temperature increases. Below the Curie temperature, the nickel-iron alloys show fairly uniform coefficients of expansion; however, in the Curie range, the expansion properties increase. Above the Curie temperature, the alloys expand at a very rapid rate, and similar to a true austenitic alloy having an instantaneous coefficient in the range of 9 to 10 × 10⁻⁶ per °F. In order to achieve specific expansion properties, other elements can be added, such as chromium. The addition of other elements increases the coefficient of expansion and lowers the Curie temperature. Expansion properties and physical properties of standard commercial grades are given in Table 19.

Expansion data, except for 36 Ni balance Fe, are given for the annealed condition (free of cold-work stresses by heat treating at 1925°F for 2 hr in a nonoxidizing atmosphere). Expansion characteristics in conditions other than annealed are not reproducible, and change as cold-work stresses are relieved.

Thermal expansion properties below 77°F of several alloys are given in Table 20. For the effect of elevated temperature on the physical properties of a nickel-iron alloy, Fig. 16 provides nominal data observed on 42 Ni balance Fe.

Stainless steels (ferritic and austenitic) Thermal expansion properties of ferritic stainless steels are presented in Table 21, and these properties are greater than those of the nickel-iron family having 49 percent nickel and less. Lower cost and other specific properties favor these alloys in some applications. Exposure

Magnetic Properties 8-25

to high elevated temperatures can alter the expansion properties owing to metallurgical phase changes.[18] The expansion properties are altered when austenite is formed in the alloy and the deletion of the austenitic phase is sluggish on cooling after being formed. As the chromium content increases, the ferrite is more stable with increasing temperature; titanium increases the ferrite stability.

TABLE 19 Thermal Expansion and Physical Properties of Nickel-Iron Alloys and Related Alloys

Type analysis and properties	36 Ni	36 Ni, FM	39 Ni	42 Ni	42 Ni 6 Cr
Carbon	0.03	0.05	0.05	0.03	0.06
Manganese	0.35	0.90	0.40	0.50	0.50
Silicon	0.30	0.35	0.25	0.25	0.25
Nickel	36.0	36.0	39.0	42.0	42.5
Other elements		Se 0.20			Cr 5.75
Iron	bal.	bal.	bal.	bal.	bal.
Physical constants:					
Specific gravity	8.05	8.05	8.08	8.12	8.12
Density, lb/in.³	0.291	0.291	0.292	0.293	0.294
Thermal conductivity, Btu/(hr)(ft²)(°F/in.)	72.6	72.6	73.5	74.5	87.0
Electrical resistivity, ohms/cir mil-ft	495	495	440	430	570
Curie temperature, °F	536	536	644	716	563
Specific heat, Btu/(lb)(°F)	0.123	0.123	0.121	0.120	0.120
Coefficient of thermal expansion, in./in. $\times 10^{-6}$ per °F (annealed):					
At 77–212°F	0.655†	0.89†	1.22	2.57	3.64
392	0.956	1.62	1.48	2.54	3.94
572	2.73	3.33	1.88	2.71	4.59
662	3.67	4.20	2.60	2.78	5.02
752	4.34	4.93	3.34	3.14	5.56
842	4.90	5.45	4.01	3.83	5.89
932	5.40	5.92	4.54	4.32	6.39
1112	6.31	6.67	5.33	5.50	6.99
1292	7.06	7.17	6.11	6.12	7.45
1472	7.48	7.56	6.64	6.66	7.87
1652	7.70	8.12	7.10	7.10	8.17
1832			7.45		
Mechanical properties (as mill-annealed):					
Tensile strength, ksi	65	65	75	82	80
Yield strength, ksi	40	40	42	40	40
Elongation in 2 in., %	35	35	30	30	30
Hardness, Rockwell B	70	70	76	76	80
Elastic modulus, ksi $\times 10^3$	20.5	20.5	21.0	21.0	23

* Also available at carbon 0.01% (max).
† Unannealed.
‡ Fully aged.

Properties of several austenitic stainless steels are given in Table 22. This series of alloys have very high coefficients of expansions; the values are slightly lower than the 22 Ni 3 Cr balance Fe alloy given in Table 19.

Glass-to-metal sealing alloys A glass-to-metal seal is a vacuum-tight bond between a glass and a metal. The seal, depending upon design, is used to conduct a form of electricity into a chamber or to provide a structural advantage such as a transparent window or a support. In order to achieve a vacuum-tight and strong glass-to-metal seal, the glass and metal must have practically identically thermal

8-26 Ferrous Metals

expansion characteristics below the setting point of the glass so as to have minimum stress at the glass-to-metal interface and within the glass.[19-21] Metal parts, fabricated to the final shape employed in making the glass-to-metal seal, must have a thoroughly clean surface. Strongest seals are achieved when a tight metal oxide film is produced on the surface of the metal before the glass comes in contact with the sealing surface. This oxide can be generated by a prior thermal treatment or

TABLE 19 Thermal Expansion and Physical Properties of Nickel-Iron Alloys and Related Alloys (Continued)

45 Ni 6 Cr	46 Ni	48.5 Ni	50.5 Ni*	22 Ni 3 Cr	29 Ni 17 Co	NiSpan C
0.05	0.03	0.05	0.03	0.10	0.01	0.03
0.30	0.50	0.50	0.50	0.50	0.30	0.40
0.30	0.25	0.40	0.25	0.25	0.20	0.40
45.0	46.0	48.5	50.5	22.0	29.0	42.0
Cr 6.0	Ti 0.40	....	....	Cr 3.0	Co 17.0	Cr 5.5 Ti 2.5 Al 0.40
bal.	bal.	bal.	bal.	bal.	bal.	bal.
8.14	8.17	8.25	8.30	8.18	8.36	8.18
0.295	0.295	0.290	0.30	0.294	0.302	0.294
87.0	85.0	90.0	97.0	....	91.0	90
570	320	290	258	462	294	480‡
563	960	932	986	None	815	380‡
0.12	0.12	0.12	0.12	....	....	0.12
4.22	3.95	4.80	6.53	10.55	3.25	4.5
4.54	4.09	5.20	6.14	10.94	2.89	5.3
4.86	4.16	5.17	5.79	11.06	2.85	6.8
5.00	4.13	5.14	5.66	11.08	2.72	
5.56	4.12	5.07	5.67	11.00	2.81	
5.84	4.40	5.36	5.68	11.05	2.92	
6.24	4.82	5.40	5.74	11.06	3.41	
6.80	5.58	6.00	6.28	11.11	4.34	
7.24	6.10	6.51	6.70	11.14	5.06	
7.62	6.57	7.06	7.09	....	5.73	
....	....	7.38	7.47	....	6.25	
80	82	85	80	70	75	200‡
40	40	40	40	40	50	180‡
30	27	35	35	35	30	7‡
80	76	70	70	74	68	42 Rc‡
23	23	24	24	27	20	27

by heating the metal for sealing. The oxide roughens the surface of the metal and is soluble in the molten glass, thus increasing the area of contact. Small bubbles within the glass-to-metal interface are undesirable and are an indication of a weak bond. The bubbles can be eliminated by a prior treatment of the metal in a wet hydrogen atmosphere to clean and/or decarburize the metal surface. After the sealing operation, an annealing treatment should be employed so that the nonequilibrium stresses within the glass are removed.

Chromium-Iron Alloys. Chromium-irons such as Type 430, Type 430Ti, Type 446, and 28 Cr balance Fe are employed. Selection depends upon the sealing temperature employed. Type 430 is not recommended for temperatures above

TABLE 20 Coefficients of Thermal Expansion of Nickel-Iron Alloys Annealed at or below Room Temperature

Temperature, °F	In./in. × 10⁻⁶ per °F		
	36 Ni bal. Fe	42 Ni bal. Fe	49 Ni bal. Fe
0–77	0.8	3.1	5.1
−100−	0.9	3.4	5.2
−200−	0.9	3.5	5.1
−300−	1.0	3.5	4.8

1700°F so as not to encounter the formation of austenite in the alloy. The presence of austenite increases the degree of contraction upon cooling and can result in cracked or overstressed seals. Type 446 will withstand higher sealing temperatures up to 2000°F, depending upon chromium content and other elements in the analysis.[18] Type 430 Ti and 28 Cr balance Fe have the most stable ferrite of the commercial grades. To obtain the strongest glass-to-metal seal, the parts should be heated in a wet hydrogen atmosphere to produce a greenish black oxide which is very

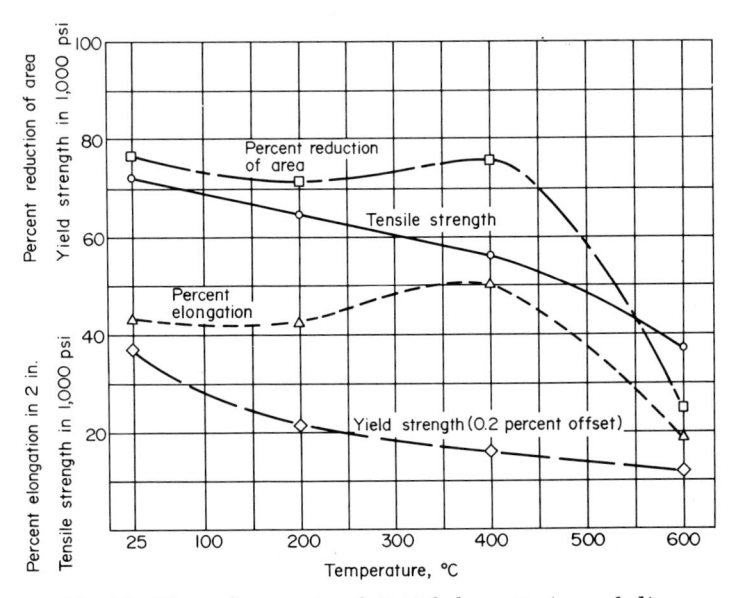

Fig. 16 Physical properties of 42 Ni balance Fe (annealed).

tightly adhering to the base metal. Heating in air or gas-firing flame for sealing to produce a sealing oxide is being used successfully. In both cases, the oxidizing procedure must be carefully controlled. Some of the common glasses employed with these alloys are 001, 008, 012, 024, and 8160. The oxide produced with wet hydrogen atmosphere is more difficult to remove. Recommended procedure is:

1. Immerse in an aqueous solution of 10 percent sodium hydroxide, 10 percent potassium permanganate. Recommended solution temperature is 200°F.

8-28 Ferrous Metals

2. Wash thoroughly in water.
3. Immerse in an aqueous solution of 5 percent ammonium citrate at a temperature of 200°F or a 25 to 50 percent hydrochloric acid solution at a temperature of 180°F.
4. Wash thoroughly in water.
5. Immersion times in items 1 and 3 have to be varied depending upon the nature of the oxide.

TABLE 21 Thermal Expansion and Physical Properties of Ferritic Stainless Steels: Type 400 Series

Property	Type 410, Type 416*	Type 430, Type 430 F*	Type 430 Ti	Type 446	28% Cr alloy
Chromium content, %...	13.0	17.5	18.2	24.0	28.0
Specific gravity.........	7.7	7.7	7.7	7.7	7.60
Density, lb/in.³.........	0.28	0.28	0.28	0.28	0.27
Thermal conductivity at 32–212°F, Btu/(hr)(ft²)(°F/ft)	14.4	15.1	15.1	12.1	13.2
Electrical resistivity, ohms/cir mil-ft	343	361	361	385	385
Specific heat, Btu/(lb)(°F).............	0.11	0.11	0.11	0.12	0.14
Structure.............	Martensitic	Ferritic	Ferritic	Ferritic	Ferritic
Coefficient of thermal expansion, in./(in.)(°F) (annealed):					
At 77–212°F...........	5.5	5.6	5.10	5.2	5.25
392..............	6.1	5.8	5.50	5.6	5.60
572..............	6.3	6.1	5.80	6.0	5.86
752..............	6.5	6.3	6.05	6.0	5.98
932..............	6.7	6.5	6.30	6.2	6.19
1112..............	6.8	6.6	6.50	6.3	6.25
1292..............	7.0	6.8	......	6.4	6.46
1472..............	....	7.0	......	6.5	6.48
1652..............	....	7.1	......	7.0	7.09
Mechanical properties (as annealed):					
Tensile strength, ksi.....	70	75	75	85	85
Yield strength, ksi......	45	45	45	55	55
Elongation in 2 in., %..	25	30	30	25	25
Hardness, Rockwell B...	80	75	75	80	85
Elastic modulus, ksi × 10³.................	29	29	29	30	30

* S 0.30 for better machinability.

Some of the commercial brand names are:
  Type 430Ti—Glass Sealing 18[29]
  28 Cr balance Fe—Glass Sealing 27[29]
  Sealmet #1[30]

Nickel-Iron Alloys.[24] Most common grades contain the nominal nickel contents of 42, 46, 48, and 50.5 percent. Most grades must be degassified in wet hydrogen before sealing. A typical treatment is to heat to 1750°F in wet hydrogen for 30 min. Preoxidizing treatments or flame oxidizing can be used. The 42 percent nickel alloy is generally used for ring (housekeeper type) seals with hard glasses whereby a feathered edge is inserted in the glass. This thin wall absorbs the expansion difference between the metal and the glass. The other alloys can be used for various internal, external, butt, and other types of seals,[19] depending upon the match with the various types of soft glasses—having a coefficient greater than 2.8×10^{-6} in./(in.)(°F).

Gas-free types are available in several of these grades, and these have low carbon contents or other additives.

TABLE 22 Thermal Expansion and Physical Properties of Austenitic Stainless Steels: Type 300 Series

Property	Type 301	Types 303, 302, 304	Type 316	Type 347	Type 310	Type 330
Typical chromium content, %	17.0	18.0	17.0	18.0	25.0	18.0
Typical nickel content, %	7.0	10.0	12.0	11.0	21.0	35.0
Specific gravity	8.0	8.0	8.0	8.0	8.0	8.0
Density, lb/in.³	0.29	0.29	0.29	0.29	0.29	0.29
Thermal conductivity at 32–212°F, Btu/(hr)(ft²)(°F/ft)	9.4	9.4	9.4	9.3	8.0	10.8
Electrical resistivity, ohms/cir mil-ft	435	433	445	438	469	600
Specific heat, Btu/(lb)(°F)	0.12	0.12	0.12	0.12	0.12	0.11
Coefficient of thermal expansion, in./in. $\times 10^{-6}$ per °F (annealed):						
At −300 to +70°F	7.6	7.4	7.1	7.5	7.0	5.8
−200	7.8	7.7	7.4	8.1	7.5	6.5
−100	8.2	8.2	7.8	8.5	7.8	7.2
0	8.7	8.7	8.2	8.7	8.0	7.6
+70 to +200	9.2	8.8	8.9	9.0	8.4	8.1
300	9.4	9.0	9.2	9.2	8.6	8.3
400	9.5	9.2	9.3	9.4	8.8	8.5
500	9.6	9.4	9.4	9.5	8.9	8.7
600	9.7	9.5	9.6	9.7	9.0	8.9
700	9.8	9.7	9.7	9.8	9.2	9.0
800	9.9	9.8	9.8	9.9	9.3	9.1
900	10.0	9.9	9.9	10.0	9.4	9.3
1000	10.1	10.0	10.0	10.1	9.5	9.4
Mechanical properties (as annealed):						
Tensile strength, ksi	110	85	84	95	95	70
Yield strength, ksi	40	42	42	40	45	35
Elongation in 2 in., %	60	55	50	45	45	30
Hardness, Rockwell B	85	80	80	85	85	85
Elastic modulus, ksi $\times 10^3$	28	28	29	28	29	

Some of the commercial grades are:
 Glass Sealing 42[29]
 Glass Sealing 42 Gas Free[29]
 Glass Sealing 46 Gas Free[29]
 Glass Sealing 49[29]
 Glass Sealing 52[29]
 Glass Sealing 52 Gas Free[29]
 Allegheny 42[30]
 4750[30]
 142 Alloy[33]
 146 Alloy[33]
 52 Alloy[33]
 Uniseal 42[36]
 Uniseal 52[36]

8-30 Ferrous Metals

42 Ni 6 Cr balance Fe.[25] Because of the chromium content in this alloy, a very tight greenish black oxide can be formed on the alloy in a wet hydrogen atmosphere. Treatments to form the oxide range from 1950 to 2350°F with varying time until the thin tight oxide is formed. This varies from facility to facility, each of which must be evaluated. To remove the oxide after cleaning, refer to comments on Chromium-Iron Alloys. Suggested glasses for sealing are 001, 012, and 8160. Commercial grades are:
  Glass Sealing 42-6[29]
  Sealmet #4[30]
  Sylvania #4[37]

TABLE 23 Properties of Ceramvar[33]

Density, lb/in.³	0.295
Specific gravity	8.17
Thermal conductivity at 20°C, cal/(cm²)(sec)(cm)	0.40
Electrical resistivity at 20°C, ohms/cir mil-ft	220
Average coefficient of thermal expansion, in./in. × 10⁻⁶ per °C:	
At 20 to 100°C	8.80
200	8.70
300	8.39
400	7.96
500	7.65
600	8.32
700	9.46
800	10.47
900	11.34
1000	12.12

TABLE 24 Expansion Alloys*

Code no.	Trade name
Fe-21	Ceramvar[33]
Fe-24	Invar
Fe-3	Kovar[31]
Fe-16	Nilo K
Fe-8	Nilvar[34]
Fe-2	NiSpan C
Fe-6	Rodar[33]
Fe-11	Therlo[34]

* Engineering Alloys Digest, Inc., Upper Montclair, N.J.

28 Ni 17 Co balance Fe.[26] This alloy is most commonly used to make glass-to-metal seals with the hard glasses of the Pyrex[11] type; hard glasses are defined as having a coefficient of expansion less than 2.8×10^{-6} in./(in.)(°F) up to 572°F. Because of the high sealing temperature, these alloys should be degassified as prescribed previously to produce good glass-to-metal seals. Preoxidizing or flame oxidizing can be employed before making the seal. If glass-to-metal seals are to be exposed to temperatures below −100°F, the alloy should be checked for phase transformation, which alters the expansion properties that could result in cracked seals or overstressed seals.[26]

Commercial grades available are:
  Nicoseal[29]
  Kovar[31]
  Rodar[33]
  Therlo[34]
  Uniseal 29-17[36]

TABLE 25 Electrical Resistivities of Some Alloys and Factors for Calculating Values at Elevated Temperatures

Composition	Resistivity at 68°F ohms/cir mil-ft	200°F	400°F	600°F	800°F	1000°F	1200°F	1400°F	1600°F	1800°F	2000°F	2100°F
36 Ni bal. Fe..............	484	1.09	1.185	1.25	1.3	1.35	1.386	1.42	1.45	1.48		
NiSpan C.................	625	1.03	1.06	1.11	1.16	1.23						
80 Ni 20 Cr 1.5 Si.........	650	1.016	1.037	1.054	1.066	1.070	1.064	1.064	1.066	1.072	1.078	1.084
60 Ni 15 Cr bal. Fe........	675	1.019	1.044	1.070	1.092	1.108	1.112	1.118	1.130	1.145		
35 Ni 15 Cr bal. Fe........	600	1.029	1.067	1.105	1.137	1.167	1.187	1.206	1.223			
23 Cr 6.2 Al 1.9 Co bal. Fe.	836	1.002	1.005	1.009	1.014	1.020	1.024	1.028	1.037	1.045	1.050	1.055
22.6 Cr 4.5 Al 2 Co bal. Fe.	812	1.002	1.005	1.009	1.014	1.020	1.024	1.028	1.037	1.045	1.055	1.053
16 Cr 5 Al bal. Fe.........	800	1.007	1.018	1.032	1.053	1.084	1.116	1.137				
Types 302 and 304.........	435	1.08	1.18	1.28	1.37	1.44	1.50	1.65	1.75	1.82		
Type 316.................	445	1.07	1.17	1.26	1.34	1.41	1.59	1.68				
Type 321.................	435	1.10	1.25	1.36	1.45	1.55	1.62	1.58	1.64			
Type 347.................	440	1.09	1.20	1.25	1.35	1.46	1.52	1.58	1.58			
Type 309.................	470	1.07	1.17	1.26	1.35	1.42	1.48	1.53	1.58	1.62		
13 Cr 4 Al................	672	1.005	1.02	1.04	1.06	1.09	1.14	1.18				
Type 410.................	345	1.11	1.26	1.43	1.58	1.73	1.90	2.02				
Type 430.................	360	1.11	1.27	1.42	1.58	1.72	1.86	1.96				
Type 446.................	405	1.1	1.24	1.37	1.50	1.60	1.70	1.76				

8-32 Ferrous Metals

Mild Steel. SAE 1010 can be used to make certain types of internal and external glass-to-metal seals. The expansion coefficient of expansion over the range of 77 to 572°F is 6.95×10^{-6} in./(in.)(°F). This alloy is sometimes plated before being subjected to a degassification treatment. Preoxidizing or flame oxidizing can be employed.

General comments on metal fabrication are as follows:

Deep Drawing. Best ductility is exhibited by the nickel-iron alloys: 42 Ni 6 Cr balance Fe and the 28 Ni 17 Co balance Fe alloy. First draws as much as 50 percent are achievable; second redraws up to 25 percent. Less ductility is exhibited by the 28 Cr balance Fe alloy with which first draws should not exceed 20 percent.

Welding. With the exception of the chromium-irons, all grades can be welded by all techniques with precautions. Chromium-iron alloys can be spot-welded with precautions.

Photochemical Machining. All grades can be photochemically machined with precautions. The vendor should be advised if this processing will be employed so as to achieve best performance.

TABLE 26 Some Commercially Available Brands of Electrical Resistance Alloys

Composition	Brand name	Producer	Code no.*
13 Cr 4 Al bal. Fe............	1-JR	Carpenter Technology	SS-144
	Ohmaloy	Allegheny-Ludlum	
22.6 Cr 4.5 Al 2 Co bal. Fe........	Kanthal D	Kanthal Corp.	SS-34
23 Cr 6.2 Al 1.9 Co bal. Fe........	Kanthal A	Kanthal Corp.	
35 Ni 15 Cr bal. Fe............	Hoskins 502	Hoskins Mfg. Co.	SS-89
	Chromax	Driver-Harris Co.	SS-6
60 Ni 15 Cr bal. Fe............	Nichrome	Driver-Harris Co.	Ni-41
	Tophet C	W. B. Driver Co.	Ni-106
80 Ni 20 Cr 1.5 Si............	Tophet A	W. B. Driver Co.	Ni-39
	Nichrome V	Driver-Harris Co.	Ni-30

* Engineering Alloys Digest, Inc., Upper Montclair, N.J.

Ceramic-to-Metal Sealing Alloys. To obtain the best ceramic-to-metal seal, both ceramic and metal must have matching expansion characteristics. Generally the bond is made with a brazing alloy. Precautions must be exercised during this treatment, as overheating may tend to cause excessive penetration of the brazing alloy into the metal. Intergranular penetration of the brazing alloy can be retarded by copper-, silver-, or nickel-plating the metal in the area to be sealed. Common sealing alloys are the 28 Ni 17 Co balance Fe alloy and 27 Ni 25 Co 47 Fe alloy known as Ceramvar[33] (refer to Table 23).

Table 24 lists expansion alloys covered by the Engineering Alloys Digest.

Constant Modulus of Elasticity Alloys. The most commonly used constant modulus of elasticity alloys are known as NiSpan C[38] and NiSpan C Alloy 902.[39] The iron-nickel-chromium-titanium alloy is age-hardenable and exhibits a constant modulus in the temperature range of −50 to +150°F. The alloy is available in the forms of wire, rod, strip, round tube, and forgings. To achieve the lowest thermoelastic coefficient in the range of -10 to $+10 \times 10^{-6}$ per °F, the parts must be thermally aged as a final heat treatment. Through utilization of an annealing solution treatment of approximately 1850°F and rapid quench and/or percentages of cold reduction (if possible on the cross section), variations in high tensile strength, hardness, and near-zero thermoelastic coefficients are obtainable. Slight differences in chemical analysis between melts result in minor variations; hence, for very precise

Electrical-resistance Alloys and Electrical-resistivity Properties of Some Ferrous Alloys

Electrical-resistance alloys are generally used to produce heat or to control electric current. In some cases, the electrical resistance or resistivity is a minor property for the selection of a given alloy. Standard procedures have been established to determine these electrical characteristics.[27,28] Table 25 provides the electrical resistivity of some alloys and the effect of temperature upon this property. Commercially available brands are listed in Table 26. The electrical resistivity of most alloys increases with increasing temperature, and properties are generally expressed as follows:

$$R = \sigma \frac{l}{A}$$

Temperature coefficient of resistivity =

$$\frac{\sigma_1 - \sigma_0}{\sigma_0(T_1 - T_0)} \text{ ohms/(ohm) (deg temperature)}$$

where R = electrical resistance, ohms
l = length, ft
A = area, cir mils
σ = ohms/cir mil–ft
σ_0 = resistivity at temperature T_0
σ_1 = resistivity at temperature T_1

Ferrous alloys and pure metals are not recommended for precision resistors because of having high-temperature coefficients. Several nonferrous alloy series[28] are available for manufacturing wire-wound resistors to meet a variety of requirements.

REFERENCES

1. ASTM Standards Relating to Magnetic Properties.
2. ASTM Standards:
   A 34-68, Magnetic Materials, Testing.
   A 219-58, Local Thickness of Electrodeposited Coatings, Test for.
   A 340-65, Magnetic Testing, Definitions of Terms, Symbols, and Conversion Factors Relating to.
   A 341-64, Normal Induction and Hysteresis of Magnetic Materials, Test for.
   A 342-64, Permeability of Feebly Magnetic Materials, Test for.
   A 343-68, Alternating Current Magnetic Properties of Materials at Power Frequencies Using Wattmeter-Ammeter-Voltmeter Method and 25-cm Epstein Test Frame, Test for.
   A 344-68, Electrical and Mechanical Properties of Magnetic Materials, Test for.
   A 345-55(1964), Flat-Rolled Electrical Steel, Specification for.
   A 346-64, Alternating Current Magnetic Properties of Laminated Core Specimens, Test for.
   A 347-68, Alternating Current Magnetic Properties of Materials Using the Modified Hay Bridge Method with 25-cm Epstein Frame, Test for.
   A 348-68, Alternating Current Magnetic Properties of Materials Using the Wattmeter-Ammeter-Voltmeter Method, 100 to 10,000 Hz and 25-cm Epstein Frame, Test for.
   A 349-68, Alternating Current Magnetic Properties of Materials Using the Wattmeter-Ammeter-Voltmeter Method, 50 to 60 Hz and 50-cm Epstein Frame, Test for.
   A 566-68, Alternating Current Magnetic Properties of Materials Using an Alternating-Current Potentiometer and 25-cm Epstein Frame, Test for.

8-34 Ferrous Metals

3. Iron-Nickel and Related Alloys of the Invar and Elinvar Types (30% to 60% Nickel), section 8, Data Sheet A, The International Nickel Co., Inc., New York.
4. Iron-Nickel Alloys for Magnetic Purposes (20% to 90% Nickel), Section 8, Data Sheet B, The International Nickel Co., Inc., New York.
5. Bozorth, Richard M.: "Ferromagnetism," D. Van Nostrand Company, Inc. Princeton, N.J.
6. Mendelsohn, L. I., E. D. Orth, and P. A. Robbins: Experimental Determination of Permeability-Stress Relationships, *J. Appl. Phys.*, vol. 35, no. 3, March, 1964.
7. "Electrical Materials Handbook," Allegheny-Ludlum Steel Corp.
8. *Proc. Nat. Electron. Conf.*, vol. 9, February, 1954, pp. 580–590.
9. Flat Rolled Electrical Steel, American Iron and Steel Institute, New York.
10. Barranger, John: Investigation of Magnetically Soft, High Temperature Cobalt-Iron Alloy, *NASA Rep.* TND-3693, October, 1966.
11. Pavlovic, D. M., and J. J. Clark: Magnetic Behavior of High-saturation Core Materials at High Temperatures and Frequencies up to 3200 CPS, *IEEE Paper* 31, pp. 66–476.
12. Easton, Robert W.: The Use of Armco Magnetic Ingot Iron in Magnetic Components. Armco Steel Corp., Middletown, Ohio.
13. Standard Specifications for Permanent Magnetic Materials, *MMPA Standard* 0100-66, Magnetic Materials Producers Association, Chicago.
14. Post, C. B., and W. S. Eberly: Stability of Austenite in Stainless Steels, *ASM Trans.*, 1947, pp. 868–890.
15. The Measurement of Magnetic Characteristics of "Non-magnetic Materials," *NAVORD Rep.* 2415, May 1, 1952.
16. ASTM Standards:
    B 95-39, Test for Linear Expansion of Metals.
    E 228-66A, Linear Thermal Expansion of Rigid Solids with a Vitreous Silica Dilatometer.
    E 289-65T, Linear Thermal Expansion of Rigid Solids with Interferometry.
17. Lement, B. S., B. L. Averbach, and Morris Cohen: The Dimensional Behavior of Invar, *ASM Trans.*, vol. 43, pp. 1072–1096, 1950.
18. Post, C. B., and W. S. Eberly: Formation of Austenite in High-chromium Stainless Steels, *ASM Trans.*, vol. 43, 1950.
19. Pask, Joseph A.: Glass-to-Metal Seals, *Prod. Eng.*, January, 1950.
20. Monach, A. J.: Glass-to-Metal Seals in Electronic Components and Applications, *Elec. Mfg.*, February, 1947.
21. Comer, Jacque: Glass-to-Metal Seals, *Elec. Mfg.*, March, 1958.
22. Specification for 17% Chromium-Iron Alloy for Sealing to Glass, *ASTM Standard* F 256-53.
23. Specification for 28% Chromium-Iron Alloy for Sealing to Glass, *ASTM Standard* F 257-53.
24. Specification for Iron-Nickel Sealing Alloys, *ASTM Standard* F 30-68.
25. Specification for 42% Nickel 6% Chromium Sealing Alloy, *ASTM Standard* F 31–68.
26. Specification for Iron-Nickel-Cobalt Sealing Alloys, *ASTM Standard* F 15-68.
27. Recommended Practice for Determining Temperature–Electrical Resistance Characteristics (EMF) of Metallic Materials, *ASTM Standard* F 189-63.
28. Standard Specifications for Wire for Use in Wire-wound Resistors, *ASTM Standard* B 267-68.
29. Carpenter Technology Corporation.
30. Allegheny-Ludlum Steel Corporation.
31. Westinghouse Electric Corporation.
32. Armco Steel Corporation.
33. Wilbur B. Driver Company.
34. Driver-Harris Company.
35. Crucible Steel Company of America.
36. Cyclops Corporation.
37. Sylvania Electric Products, Inc.
38. H. A. Wilson Company.
39. International Nickel Company.

Chapter **9**

Nonferrous Metals

EUGENE W. BROACHE
Westinghouse Electric Corporation, Aerospace and
Electronics Systems Division, Baltimore, Maryland

Introduction	9–3
Considerations in Alloy Selection	9–3
Composition	9–3
Material Manufacturing Variables	9–3
Strengthening Mechanisms	9–5
Mechanical Properties	9–6
Physical Properties	9–7
Manufacturing Considerations	9–7
Comparative Data on Nonferrous Metals	9–11
Yield Strength Comparisons	9–11
Modulus of Elasticity and Stiffness	9–11
Strength-Density Ratios	9–11
Thermal Conductivity	9–12
Electrical Resistivity	9–12
Heat Absorption	9–12
Aluminum and Aluminum Alloys	9–12
General Properties	9–12
Aluminum Alloy Groups	9–13
Aluminum Alloy Temper Designations	9–14
Composition—Wrought Aluminum Alloys	9–14
Mechanical Properties—Wrought Aluminum Alloys	9–14
Effect of Temperature on Mechanical Properties of Aluminum Alloys	9–17
Physical Properties of Aluminum	9–17

Fabrication Properties of Aluminum	9–17
Cast Aluminum Alloys	9–37
Beryllium and Beryllium Alloys	9–39
Availability	9–39
Chemical Composition	9–40
Mechanical Properties	9–41
Physical Properties	9–41
Corrosion Resistance and Finishing	9–41
Fabrication of Beryllium Alloys	9–42
Cost	9–43
Copper and Copper Alloys	9–43
General Characteristics	9–43
High-electric-conductivity Alloys	9–43
Brasses	9–45
Phosphor Bronzes	9–46
Beryllium-Copper	9–47
Magnesium Alloys	9–50
Forms Available	9–51
Alloy Designations	9–52
Alloy Availability	9–52
Magnesium Alloy Castings	9–53
Fabrication of Magnesium Alloys	9–58
Nickel and Nickel Alloys	9–59
Forms Available	9–59
Fabrication	9–59
Chemical Composition of Nickel Alloys	9–62
Physical Properties of Nickel Alloys	9–62
Mechanical Properties of Nickel Alloys	9–62
Titanium and Titanium Alloys	9–62
Titanium Alloys	9–63
Corrosion Resistance of Titanium Alloys	9–65
Contamination	9–66
Forms Available	9–67
Fabrication of Titanium	9–67
High-density Materials	9–68
Lead and Lead Alloys	9–68
Sintered Tungsten	9–68
Depleted Uranium	9–68
Refractory Metals	9–68
Mechanical Properties	9–69
Modulus of Elasticity	9–69
Density	9–69
Thermal Expansion	9–69
Electrical Properties	9–70
Solders	9–70
Precious Metals	9–72
References	9–74

INTRODUCTION

The nonferrous metals and alloys offer a wide variety of physical and mechanical properties for use in the electronics and electrical industries. However, because of the demanding requirements on materials in electronic equipment, there are many trade-offs that have to be made in selecting metals. Electric conductivity must often be sacrificed in order to gain strength, while strength is sometimes given up in order to get a suitable thermal expansion. These are only two simple trade-offs made in electronic design. Often the selection involves even more mechanical and/or physical properties along with decisions on economics and manufacturability.

CONSIDERATIONS IN ALLOY SELECTION

Intelligent selection of material entails consideration of the functional requirements with respect to physical and mechanical properties along with the feasibility and economics of manufacturing. It does little good to select a material with optimum mechanical properties if the material cannot be fabricated into the desired shape at the required cost. The total design job requires consideration of all the factors of functional material properties and manufacturing processes.

Some of the detailed factors to be considered are discussed below.

Composition

Chemical composition is the most commonly specified criterion for metals. The composition determines the basic mechanical and physical properties of the metal or alloy. Other factors, such as the fabrication method and heat treatment, will influence these properties, but the composition provides fundamental characteristics. Many metals are used in essentially their pure state with minor impurity elements. Addition of other elements to a basis metal is common, to provide alloys with altered physical or mechanical properties. Alloying elements are controlled to specified limits in order that the desired properties may be controlled to given limits. Standardized alloys are available with specified composition limits so that the designer may select the desired properties.

Material Manufacturing Variables

As mentioned above, the basic chemical composition of a metal or alloy provides the fundamental physical and mechanical properties. However, these properties may be greatly influenced by the processing history during manufacture to attain the final shape. Material bought from a metals producer may have been subjected to a number of operations which will affect its final properties. It is important in many cases that the designer be aware of these variables in selecting a material.

Metals received into an electronics plant for use in manufacturing can be classified as either *cast* or *wrought* forms. These are discussed separately below.

Cast metals Cast metal forms are made by casting molten metal into a mold of the desired shape. No metal deformation occurs, and therefore the properties of the metal are dependent upon the casting process, soundness of the casting, chemical composition, and heat treatment. Cast billets for machining into a final shape are sometimes purchased, but in most cases cast parts are molded to final shape for most dimensions so that a minimum of machining is required to provide a finished part. Casting of complex shapes can often eliminate costly joining, machining, and forming operations, resulting in cost savings in manufacture of such parts.

A major factor in the properties of cast metals is their soundness. Nearly all castings have some defects such as porosity, gas holes, and shrinkage voids. These defects are dependent upon foundry practice, mold design, pouring practice, and the design of the part. It is important that casting designers consult with foundries during design to arrive at agreement on defect limits, tolerances, etc., since the allowable defects and dimensional tolerances will have a larger effect on the cost of the part.

The specified mechanical properties for castings are often based on separately cast test bars. These test-bar values may vary widely from the actual strength of a casting owing to defects existing in the casting. Where strength is critical, many foundries will guarantee properties in the actual casting in lieu of separately cast test-bar values. In any case, the properties expected from the casting itself should be made clear to the foundry.

The cooling rate of metal has a major effect on the properties of the casting. The grain size of the metal decreases as the cooling rate increases, and small grain size generally means increased strength.

Cooling rate is affected by the thickness and mass of the metal shape being cast. Thus a small, thin-wall casting will be of higher strength than a thicker-wall, massive casting of equal soundness.

The mold material will also affect cooling rate and thus strength. A metal mold (known as a permanent mold) will cool a part faster than a sand mold. Mass of the mold, preheating practice, etc., will also affect cooling rate.

Castings have a wide use in manufacture of many parts with the least amount of machining, joining, and forming. However, they require special care in specifying defect limits and property limits in order to assure that required functions are met.

It should be kept in mind that not all metals can be satisfactorily cast into shapes of any great complexity, although practically all metals are cast into ingots for working into shapes by deformation methods. In the copper, aluminum, and magnesium alloy families, specific alloys have been developed for castings and for wrought alloys. There are exceptions where an alloy may be available as both a casting and as a wrought alloy, but the general separation remains.

Wrought metals The majority of metals and alloys are furnished in a wrought form; that is, the metal has been subjected to some shaping operation by deformation in the solid state. Among these operations are rolling, drawing, extruding, and forging.

Such operations will have major effects on the final properties of the material. These operations affect grain size, crystallographic orientation, homogeneity, soundness, size and shape of inclusions, size and shape of metallurgical phases, etc. These factors in turn affect strength and ductility, formability, directional physical and mechanical properties, as well as reaction to subsequent heat treatment.

The temperature at which the metal is deformed or worked is important in its effect on properties. *Hot working* is defined as deformation done at temperatures above the recrystallization temperature of the metal. The recrystallization temperature is that temperature above which a metal will form new, strain-free grains as rapidly as the old grains are strained by the working process. In hot working the metal is not hardened by the working process; thus large amounts of deformation may be carried out without fracturing the metal.

Cold working takes place below the recrystallization temperature and thus results in strain hardening of the material since new grains cannot be formed to relieve the strain set up by deformation.

Wrought metals are formed to shape by either hot working alone or by combinations of hot working and cold working. Simplified descriptions of some of the wrought product procedures follow.

Extrusion. This process generally uses a *billet* as raw material. In most cases the billet is a casting. The billet is heated above the recrystallization temperature and then forced through a die of suitable shape to form the finished shape. As can be seen by the description, this is a hot-working process in essentially one direction. No appreciable work-hardening occurs in this process. In some metals the tensile strength is considerably higher than the compressive strength in the longitudinal direction.

Forging. This is generally a hot-working process which starts with a cast billet which is deformed by hammering or pressing to form a shape. In order to get optimum properties it is important that the cast structure be completely worked to form new grains and to close up and weld together any voids in the cast billet. The properties of forgings will depend to some extent on the flow direction of the metal during forging. For highest strength, forging suppliers should be consulted during design to assure that the correct flow pattern can be attained.

Rolling. Sheet, plate, strip, and many bars and shapes are formed by rolling. This process probably accounts for the longest tonnage of metal produced in wrought form.

Rolled parts usually start out with a large cast billet which is heated well above its recrystallization temperature and reduced to a slab or billet by rolling. This slab or billet is then hot-worked to some thin long shape such as sheet. Hot rolling aids in breaking up the large grained cast structure of the ingot and replacing it with finer, uniform, recrystallized grains. The rolling also tends to close in voids and weld the surfaces together, resulting in a more dense, defect-free metal. This is an essential step in getting a high-quality product.

Hot-rolled products, however, tend to have relatively rough surfaces, with some rolled-in scale, etc., which is not desirable for most electronic purposes. The mechanical properties of hot-rolled properties are generally close to the annealed (or lowest) properties for that particular composition.

Most rolled products used in the electronic industry are given final shaping to finished dimension by cold rolling after initial hot rolling. This rolling results in better surface finishes, closer dimensional tolerances, and under proper control can result in higher strength as a result of cold working. Many metals owe their high-strength properties to the strain hardening by cold working. Many aluminum, copper, magnesium, and nickel alloys and all pure metals cannot be strengthened beyond their annealed strength by any method other than strain hardening.

In metals and alloys that are hardened by strain hardening, the thickness of the section being used affects significantly the degree of strengthening. Thin sheet which can be thoroughly deformed through the entire thickness will have higher strength than thicker material which does not get worked thoroughly. The designer should be careful not to assume that strengths shown for one thickness apply to all thicknesses.

Strain hardening can give desirable increases in strength, but it also reduces ductility and formability. Reduction in electric and thermal conductivity can result from strain hardening. Stress corrosion can be another detrimental effect of strain-hardened materials.

Strengthening Mechanisms

In order to adequately specify metals and their processing, the designer should have a knowledge of the strengthening mechanisms and the interactions of these mechanisms with the manufacturing process. The major hardening mechanisms are discussed below, along with some of the limitations and advantages offered by these methods.

Strain hardening As discussed above, cold working results in distortion of the atomic structure of metals, thus causing an increase in strength and hardness. Practically all metals strain-harden to some extent, and it is the only way to strengthen some metals.

When strain-hardened metals are used in service, care must be taken that manufacturing processes do not exceed the recrystallization temperature of the metal. Such temperatures will result in loss of the strain hardening, and the metal will revert back to the annealed state. Such processes as welding and brazing are prime examples of materials of heat sources that will reduce work-hardened materials back to their annealed state.

Precipitation hardening (age hardening) Many alloy compositions are such that solid solubility of one metal into the other increases with temperature. By heating the alloy to a temperature at which the solubility is high, and then quenching quickly to room temperature, the solute is retained in solution. This process is known as *solution treating.* Subsequent aging at a suitable temperature precipitates out the solute uniformly throughout the crystallographic structure, causing lattice distortions which result in strengthening.

Materials of this type offer the advantage of allowing forming or other working in the annealed or soft condition with subsequent heat treatment to higher strength. Also, parts may be welded, brazed, or otherwise heated during fabrication, and then brought to high strength after these operations by solution treating and aging.

The sequence of solution treating and aging can be varied somewhat, depending upon circumstances. As mentioned above, both solution treating and aging may be carried out on the final assembly. One disadvantage of this procedure is that the fast cooling required to solution-treat may result in distortion. However, to get full strength from an assembly which has been heated above the aging temperature, this sequence is necessary.

Forming of material bought in the solution-treated condition is done in many cases. The solution-treated condition is usually relatively ductile, thus permitting reasonable formability. After forming, a relatively low-temperature aging will result in full strength and hardness. Another variation of this procedure is for the user to solution-treat the raw stock just before forming. This takes advantage of the fact that solution-treated material is softest immediately after solution treating.

Phase-transformation hardening Some nonferrous alloys, notably certain titanium alloys are hardenable by manipulation of *phase transformations.* This mechanism takes advantage of the fact that a normally two-phase alloy can be converted to a one-phase alloy by heating to elevated temperature. Quenching from this temperature results in this single phase being retained at room temperature. Subsequent aging at moderately elevated temperature will cause the alloy to revert to the two-phase structure, but with much finer particles of each phase than would occur under equilibrium conditions. This fine structure results in increased strength and hardness.

Many metals are supplied with combined strengthening mechanisms. The most common instance is that of a material which is solution-treated, and then cold-worked by rolling or stretching, after which it is aged. This combination results in higher strength than could be attained by heat treatment or cold working alone.

It is obvious that selection of a metal or alloy must take into consideration not only the final strength desired, but also the feasibility of maintaining or attaining that property in fabrication.

Mechanical Properties

After chemical composition, the most commonly specified properties are the tensile properties. Usually these consist of *ultimate strength, yield strength,* and *percentage elongation* as determined on a standard tensile test bar. These properties are often all that need to be known in order to make a material selection for mechanical design. However, it must be remembered that these values represent data taken on machined test bars at room temperature under one short-time loading. A number of factors which will reduce the design allowable well below the published tensile values are discussed below.

Repeated loads (fatigue) When loads are repeated, the allowable stress is reduced well below the tensile strength of the material. In fact, with most nonferrous materials subjected to 10 million cycles of stress, the breaking stress can be one-third of the tensile strength. In considering fatigue data, attention should be paid to the type of specimen, the type of loading, and the condition of the test specimen as regards notches. The presence of notches can further reduce fatigue stress to as low as one-fifth of the tensile strength. Improvement in fatigue has been brought about by improving surface finish, by working the surface by shot peening, and by various other methods.

The subject of fatigue is very complex and subject to some controversy. In designing for repeated loads special attention should be paid to the available data. Specific tests under service conditions are also recommended.

Sustained loads Failure under long-time sustained loads may occur as a result of creep or fracture at stresses below the yield strength, as determined by the standard tensile test. Although this type of failure is thought of as occurring at relatively high temperatures, such phenomena do occur at room temperature.

Temperature effects The strength of metals decreases as temperature increases. Ductility generally increases until temperatures near the melting point are reached. The reverse occurs on lowering temperature; that is, strength increases and ductility decreases. The decreasing ductility with decreasing temperature is not particularly severe with nonferrous alloys. No drastic embrittlement occurs even at cryogenic temperatures.

It can be seen that the room-temperature tensile test values can be useful in selection of materials, but the service conditions and type of loading can be important factors in selecting a safe design stress.

Physical Properties

Physical properties are a result of chemical composition, atomic arrangement, and temperature. In considering data on physical properties, the above factors should be considered. Practical situations in which significant changes in physical properties may be found are briefly discussed below.

Cold working Slight decreases in electric conductivity are experienced with cold working on strain hardening.

Heat-treatment condition Heat treatment has various effects on the structure and distribution of alloying elements and phases and thus will often have major effects on physical properties, including electric conductivity, thermal conductivity, density, thermal expansion, and sometimes elastic modulus. Many of these changes are insignificant for engineering purposes, but the electric conductivity differences can be quite dramatic.

Precipitation-hardenable alloys are especially prone to change in electric conductivity with change in heat-treat condition. The solution-treated condition usually has the lowest conductivity since the alloying elements are in solution, a condition which gives low conductivity. Annealing results in conglomeration of the alloying elements, thus giving the effect of a mixture where the two phases may average out. Aging of solution-treated materials will usually result in some increase in electric conductivity, but not to the extent that results from annealing.

Owing to atomic rearrangement, some metals change in density when heat-treated. Beryllium-copper is one such material. Beryllium-copper is also an example of a material that undergoes a change in elastic modulus due to heat treatment.

Temperature Physical properties change with temperature. Most of these changes are gradual. However, if the temperature changes result in a phase change, drastic changes in physical properties can occur. In magnetic materials, increasing temperature beyond the Curie temperature results in material becoming nonmagnetic.

It can be seen that in using physical property data attention should be paid to the details of service temperature and material condition.

Manufacturing Considerations

Selection of material must always take into consideration the manufacturing processes to be used in fabricating the part. Material must be selected which can be economically processed to the final requirements.

Metallurgical properties of the materials selected will greatly influence the machinability, formability, and joining ability. Many times there are compromises which can be made to enhance manufacturing of the item. Knowledge of available fabrication equipment is also useful.

Some of the basic material characteristics that can be used to select materials for fabrication are discussed below.

Formability The tensile test values for a material can be used as a reasonable guide for selection of formable material. The percentage elongation is a good indication of how much a material may be deformed. Also, a large difference between the ultimate tensile strength and the yield strength indicates good formability. Minimum bend radii are also published for most materials and are sometimes included in specifications.

9-8 Nonferrous Metals

For drastic forming operations such as deep drawing, more sophisticated specifications of material properties are required. Grain-size control is one of these. An optimum grain size along with a minimum elongation exists for most materials.

Machinability Most nonferrous alloys can be machined without great difficulty. Many alloys are made with specific alloy additions which enhance machinability. The additions usually have the advantage that small chips are formed during machining. Many of these alloys have the disadvantage of being less formable, more brittle, and less weldable because of these additions. Care should be taken that the free-machining additions do not interfere with other fabrication techniques.

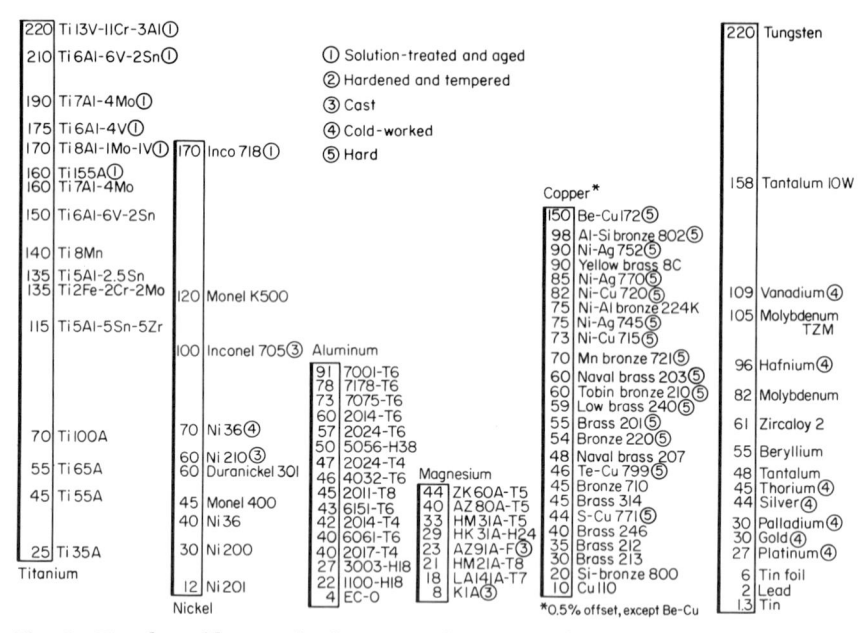

220	Ti 13V-11Cr-3Al ①		220	Tungsten
210	Ti 6Al-6V-2Sn ①	① Solution-treated and aged		
190	Ti 7Al-4Mo ①	② Hardened and tempered		
175	Ti 6Al-4V ①	③ Cast		
170	Ti 8Al-1Mo-1V ①	④ Cold-worked		
160	Ti 155A ①	⑤ Hard		
160	Ti 7Al-4Mo			

Titanium values (×1,000 psi):
- 220 Ti 13V-11Cr-3Al ①
- 210 Ti 6Al-6V-2Sn ①
- 190 Ti 7Al-4Mo ①
- 175 Ti 6Al-4V ①
- 170 Ti 8Al-1Mo-1V ①
- 160 Ti 155A ①
- 160 Ti 7Al-4Mo
- 150 Ti 6Al-6V-2Sn
- 140 Ti 8Mn
- 135 Ti 5Al-2.5Sn
- 135 Ti 2Fe-2Cr-2Mo
- 115 Ti 5Al-5Sn-5Zr
- 70 Ti 100A
- 55 Ti 65A
- 45 Ti 55A
- 25 Ti 35A

Nickel:
- 170 Inco 718 ①
- 120 Monel K500
- 100 Inconel 705 ③
- 70 Ni 36 ④
- 60 Ni 210 ③
- 60 Duranickel 301
- 45 Monel 400
- 40 Ni 36
- 30 Ni 200
- 12 Ni 201

Aluminum:
- 91 7001-T6
- 78 7178-T6
- 73 7075-T6
- 60 2014-T6
- 57 2024-T6
- 50 5056-H38
- 47 2024-T4
- 46 4032-T6
- 45 2011-T8
- 43 6151-T6
- 42 2014-T4
- 40 6061-T6
- 40 2017-T4
- 27 3003-H18
- 22 1100-H18
- 4 EC-0

Magnesium:
- 44 ZK60A-T5
- 40 AZ80A-T5
- 33 HM31A-T5
- 29 HK31A-H24
- 23 AZ91A-F ③
- 21 HM21A-T8
- 18 LA141A-T7
- 8 K1A ③

Copper* :
- 150 Be-Cu 172 ⑤
- 98 Al-Si bronze 802 ⑤
- 90 Ni-Ag 752 ⑤
- 90 Yellow brass 8C
- 85 Ni-Ag 770 ⑤
- 82 Ni-Cu 720 ⑤
- 75 Ni-Al bronze 224K
- 75 Ni-Ag 745 ⑤
- 73 Ni-Cu 715 ⑤
- 70 Mn bronze 721 ⑤
- 60 Naval brass 203 ⑤
- 60 Tobin bronze 210 ⑤
- 59 Low brass 240 ⑤
- 55 Brass 201 ⑤
- 54 Bronze 220 ⑤
- 48 Naval brass 207
- 46 Te-Cu 799 ⑤
- 45 Bronze 710
- 44 Brass 314
- 44 S-Cu 771 ⑤
- 40 Brass 246
- 35 Brass 212
- 30 Brass 213
- 20 Si-bronze 800
- 10 Cu 110

Other:
- 158 Tantalum 10W
- 109 Vanadium ④
- 105 Molybdenum TZM
- 96 Hafnium ④
- 82 Molybdenum
- 61 Zircaloy 2
- 55 Beryllium
- 48 Tantalum
- 45 Thorium ④
- 44 Silver ④
- 30 Palladium ④
- 30 Gold ④
- 27 Platinum ④
- 6 Tin foil
- 2 Lead
- 1.3 Tin

*0.5% offset, except Be-Cu

Fig. 1 Tensile yield strength of some nonferrous metals at room temperature (typical values × 1,000 psi).

Joining Alloy selection for joining is quite important in the nonferrous field. Often the alloy selection determines the joining method and vice versa. Bolting, riveting, and other mechanical methods offer no real problem for most metals, but the metallurgical processes of welding, brazing, and soldering require considerable thought.

Weldability. The ratings on weldability are qualitative and are generally based on ability to be welded with conventional techniques. A metal or alloy may get a poor weldability rating for a number of reasons, including (1) tendency to crack during welding, (2) poor corrosion resistance of the weld, (3) brittleness of the weld.

To a great extent weldability becomes a function of the process which is used for welding. Much of the success for welding of reactive metals such as aluminum, magnesium, and titanium is due to the use of the inert-gas shielded-arc method of welding, where the molten metal is shielded from the air atmosphere by an inert gas. The development of the electron beam with its narrow, deep-penetration weld has resulted in successful welding of alloys and combinations of alloys formerly considered nonweldable.

Brazing. Selection of an alloy to be brazed requires consideration of the melting point of the alloy to be brazed in comparison to melting point of the brazing

Considerations in Alloy Selection 9-9

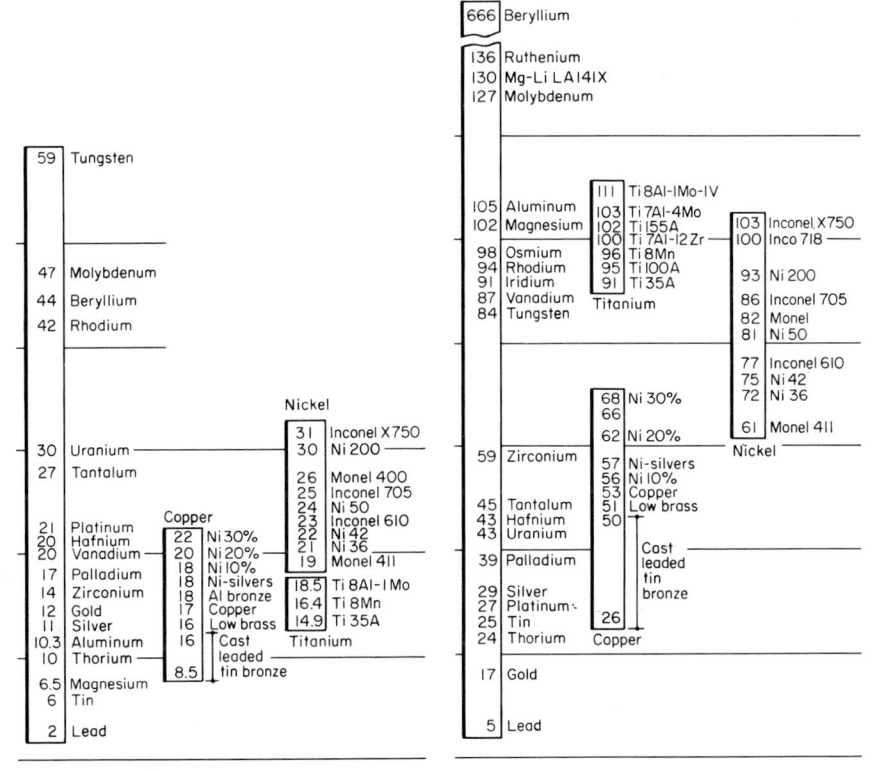

Fig. 2 Modulus of elasticity of some nonferrous metals at room temperature (psi × 10⁻⁶).

59	Tungsten					
47	Molybdenum					
44	Beryllium					
42	Rhodium					
				Nickel		
				31	Inconel X750	
30	Uranium			30	Ni 200	
27	Tantalum			26	Monel 400	
				25	Inconel 705	
				24	Ni 50	
21	Platinum	Copper		23	Inconel 610	
20	Hafnium	22	Ni 30%	22	Ni 42	
20	Vanadium	20	Ni 20%	21	Ni 36	
17	Palladium	18	Ni 10%	19	Monel 411	
14	Zirconium	18	Ni-silvers	18.5	Ti 8Al-1Mo	
12	Gold	18	Al bronze	16.4	Ti 8Mn	
11	Silver	17	Copper	14.9	Ti 35A	
10.3	Aluminum	16	Low brass		Titanium	
10	Thorium	16	Cast leaded tin bronze			
6.5	Magnesium	8.5				
6	Tin					
2	Lead					

Fig. 3 Specific stiffness of some metals at 70°F (modulus of elasticity/density × 10⁶ in.).

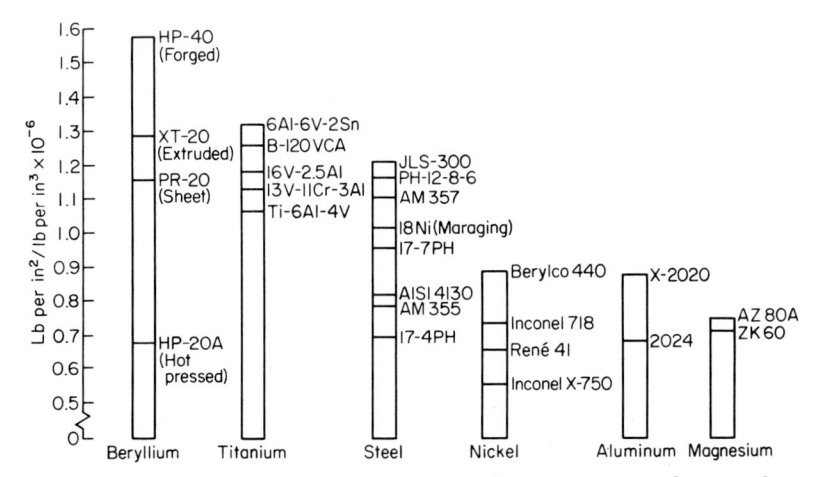

666	Beryllium					
136	Ruthenium					
130	Mg-Li LA141X					
127	Molybdenum					
				111	Ti 8Al-1Mo-1V	
105	Aluminum			103	Ti 7Al-4Mo	
102	Magnesium			102	Ti 155A	
				100	Ti 7Al-12Zr	103	Inconel X750
98	Osmium			96	Ti 8Mn	100	Inco 718
94	Rhodium			95	Ti 100A	
91	Iridium			91	Ti 35A	93	Ni 200
87	Vanadium				Titanium	86	Inconel 705
84	Tungsten					82	Monel
						81	Ni 50
						77	Inconel 610
						75	Ni 42
				68	Ni 30%	72	Ni 36
				66		
				62	Ni 20%	61	Monel 411
59	Zirconium						Nickel
				57	Ni-silvers	
				56	Ni 10%	
45	Tantalum			53	Copper	
43	Hafnium			51	Low brass	
43	Uranium			50		
					Cast leaded tin bronze	
39	Palladium					
29	Silver					
27	Platinum					
25	Tin			26		
24	Thorium				Copper	
17	Gold					
5	Lead					

Fig. 4 Ultimate tensile strength–density ratios of various structural materials at room temperature.

Lb per in²/lb per in³ × 10⁻⁶

Beryllium:
- HP-40 (Forged): 1.6
- XT-20 (Extruded): 1.25
- PR-20 (Sheet): 1.15
- HP-20A (Hot pressed): 0.65

Titanium:
- 6Al-6V-2Sn
- B-120VCA
- 16V-2.5Al
- 13V-11Cr-3Al
- Ti-6Al-4V

Steel:
- JLS-300
- PH-12-8-6
- AM 357
- 18Ni (Maraging)
- 17-7PH
- AISI 4130
- AM 355
- 17-4PH

Nickel:
- Berylco 440
- Inconel 718
- René 41
- Inconel X-750

Aluminum:
- X-2020
- 2024

Magnesium:
- AZ 80A
- ZK 60

9-10 Nonferrous Metals

alloy. The relationship of the brazing temperature to the heat treating to be performed on the part should also be considered. If it is desired to have the final properties of the part in the heat-treated condition, the brazing alloy must have a melting point above the heat-treating temperature.

The method to be used for brazing is also of importance. Torch brazing locally heats the joint and may result in distortion. Furnace brazing requires heating of the whole part, resulting in less distortion, but requiring fixturing in most cases. Induction brazing offers fast, localized heating. Dip brazing requires immersing the complete area to be brazed in a molten flux bath.

Soldering. Soldering is a low-temperature process that can be used on a wide variety of metals. Some nonferrous metals such as aluminum and magnesium alloys are soldered with great difficulty. Normal solders are generally galvanically different from these alloys, and joints with poor corrosion resistance in the presence of moisture will result. These alloys may be plated with solderable coatings and then soldered much as other metals.

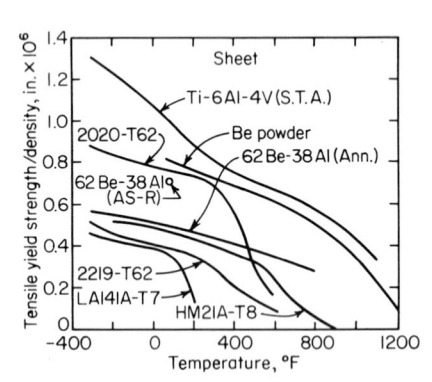

Fig. 5 Comparison of tensile yield–density ratios of some structural metals at various temperatures.

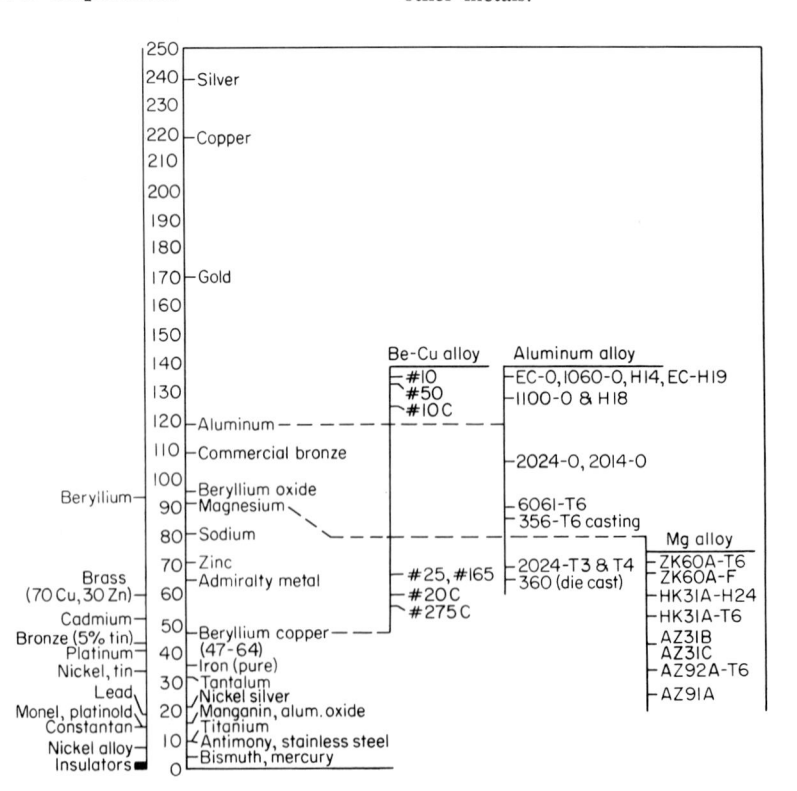

Fig. 6 Thermal conductivity at 212°F for some materials used in electronic equipment.

Comparative Data on Nonferrous Metals

The metallurgy of solders is discussed in a later section of this chapter (see pages 9-70 to 9-72).

COMPARATIVE DATA ON NONFERROUS METALS

It is obvious that with the thousands of metal and alloy compositions available and the varying degrees of strain hardening, the heat-treating conditions, and the varying service temperatures, it is impossible to list all the data for all conditions. Comparison of the range of properties available in an alloy group can usually narrow the selection greatly. Some general comparisons of a wide range of metals follow.

Yield Strength Comparisons

The yield strength of a metal is generally used as the maximum stress to which a part is subjected in service. Figure 1 gives a comparison of the yield strengths of a number of alloys. It can readily be seen that there is a wide overlap of yield strength among the commonly used alloys. At stress levels below 40,000

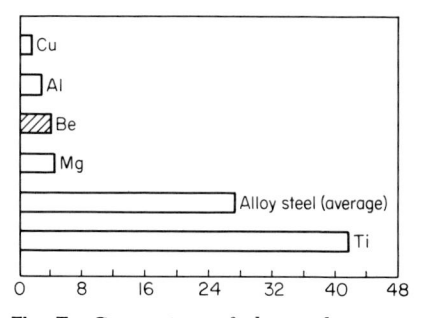

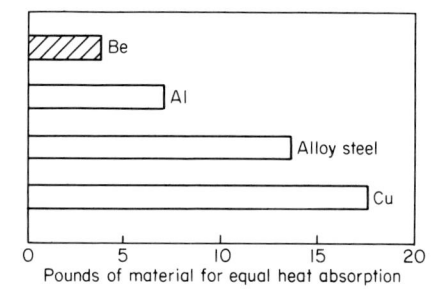

Fig. 7 Comparison of electrical resistivity of several pure metals.

Fig. 8 Comparison of heat-absorption ability of several pure metals.

psi most alloy families have alloys which are adequate, and selection would be made on the basis of some other property or on the basis of economic factors.

Modulus of Elasticity and Stiffness

Figure 2 compares the elastic moduli of a number of alloys. Since this is the property that indicates stiffness of a part, it plays an important part in design. Note there is little significant variation in elastic modulus for any one alloy family, even though yield strengths may vary as much as 2,000 percent in some alloy families.

In Fig. 3 the specific stiffnesses, calculated by dividing density into the elastic modulus of the various alloys, are compared. Note that the commonly used structural materials such as aluminum, magnesium, and titanium are about equal, whereas nickel and copper alloys have somewhat lower values. The only standout material in this property is beryllium with a value of 6 to 12 times that of common structural materials.

Strength-Density Ratios

In Fig. 4, the strength-weight ratios of a number of materials are compared. This figure shows that, when compared on a weight basis, some of the weaker, low-density materials are superior to the high-strength, high-density materials.

Figure 5 compares the effect of temperature on the yield strength–density of several alloys. This figure illustrates that room-temperature comparisons do not necessarily hold at elevated temperatures.

Nonferrous Metals

Thermal Conductivity

In Fig. 6 thermal conductivities are compared. It can be seen that the pure metals—silver, copper, gold, and aluminum—have the highest thermal conductivities. Alloying in general lowers the conductivity, sometimes drastically.

Electrical Resistivity

Figure 7 graphs the comparative electrical resistivities of certain metals. Again it should be pointed out that alloying will drastically affect electrical resistivity.

Heat Absorption

Figure 8 gives a comparison of the ability of metals to absorb heat. These values are for pure metals (with the exception of steel), and alloying can result in drastic changes.

In addition, Table 1 lists the thermal properties of a number of pure metals.

The above data were presented to give a broad brush look at comparisons among metals, in an effort to give the designer an idea of the range of properties available. Further detailed properties of specific alloys are presented in subsequent sections of this chapter.

TABLE 1 Thermal Properties of Several Pure Metals

Property	Be	Mg	Al	Ti	Fe	Cu
Melting temperature:						
°C	1277	650	660	1668	1536	1083
°F	2332	1202	1220	3035	2797	1981
Boiling temperature:						
°C	2770	1107	2450	3260	3000	2595
°F	5020	2025	4442	5900	5430	4703
Thermal conductivity, cal/(sec)(cm²)(°C/cm)	0.35	0.367	0.53	0.41	0.18	0.94
Linear thermal expansion:						
μin./(in.)(°C)	11.6	27.1	23.6	8.41	11.7	16.5
μin./(in.)(°F)	6.4	15.05	13.1	4.67	6.5	9.2
Specific heat, cal/(g)(°C)	0.45	0.25	0.215	0.124	0.11	0.092
Heat of fusion, cal/g	260	88	94.5	104*	65	50.6

* Estimated.

ALUMINUM AND ALUMINUM ALLOYS

General Properties

Aluminum and its alloys possess properties which find wide use in the electronics industry. Favorable physical properties, good strength-weight properties, good corrosion resistance, and low density, combined with economy in material cost and fabrication cost, make this alloy family a basic material of construction for electronic assemblies.

Electrical properties of aluminum Electric conductivity of the higher-purity grades is 62 percent of that of pure copper on a volume basis and exceeds copper by approximately twice on a per pound basis. Alloying to increase strength results in lower conductivity, but fairly strong alloys are available which still exceed the conductivity of copper on a weight basis.

Thermal conductivity of aluminum The high thermal conductivity of aluminum is an advantage in dissipating heat, often a requirement in electrical apparatus.

Density of aluminum The light weight of aluminum can often result in reduction in weight over materials which are stronger than aluminum on a volume basis. The fact that aluminum covers more area per pound when substituted gage for gage for copper alloys, nickel alloys, or steel results in more favorable material costs since the same weight of aluminum will cover a larger area than will the heavier alloys.

Strength-Weight of aluminum Although aluminum alloys cannot match high-strength steels, beryllium-copper, high-strength titanium alloys, or some nickel alloys in strength per unit area, its low density makes it competitive on a strength-weight basis at room temperature and at slightly elevated temperatures. At low temperatures even in the cryogenic range aluminum alloys retain good ductility, and thus have found use in low-temperature pressure vessels.

Corrosion of aluminum Corrosion resistance in aluminum is good to the extent that many applications require no protection. Corrosion resistance varies according to the alloying elements used, with the higher-purity alloys generally having best corrosion resistance. In severe environment, protection is needed, especially for the higher-strength alloys. This protection can be provided by anodizing or by conversion coating, either alone, or in combination with painting. The fact that aluminum is galvanically dissimilar to most common structural metals makes it necessary to pay special attention to applications where it comes into contact with other metals.

Availability Aluminum alloys are available in about every form, including sheet, plate, foil, pipe, tubing, forgings, and castings, as well as all types of extrusions and rolled shapes. Not all alloys are available in all shapes. This is especially true of castings which are furnished in an entirely different group of compositions than are wrought alloys.

Fabrication of aluminum The fabrication of aluminum can be carried out by all common means. Machining is very good, but the softer alloys and tempers tend to be gummy. Formability is excellent at room temperature.

Joining of aluminum Joining can be accomplished by mechanical means such as bolts, screws, and rivets, as well as by metallurgical processes of welding, brazing, and soldering. However, precautions must be taken in selection of alloys for the metallurgical processes. Certain alloys are difficult to weld. The brazing alloys melt at a temperature which exceeds the melting point of certain of the commonly used alloys, thus making selection of materials to be brazed subject to careful consideration. Solders are available for aluminum, but corrosion due to galvanic action between the aluminum and solder alloys may be a problem where moisture is present. Fluxes for brazing and soldering are highly active and will cause corrosion if left on the parts, thus making postjoining cleaning of paramount importance. Fluxless brazing has been done in vacuum furnaces and is currently under development. Successful fluxless soldering has been accomplished with ultrasonic devices. These processes eliminate the flux corrosion hazard.

Aluminum Alloy Groups

Table 2 shows the four-digit system for designating alloy groups on the basis of composition as devised by the Aluminum Association.

TABLE 2 Designations for Aluminum Alloy Groups

Group	Alloy No.
Aluminum—99.00% minimum and greater	1xxx
Aluminum alloys grouped by major alloying elements:	
Copper	2xxx
Manganese	3xxx
Silicon	4xxx
Magnesium	5xxx
Magnesium and silicon	6xxx
Zinc	7xxx
Other element	8xxx
Unused series	9xxx

Aluminum Alloy Temper Designations

Table 3 defines the temper designations which follows the alloy designation. Note that two types of designations are used. The –H designations represent cold-worked conditions and are used for non-heat-treatable alloys. The –T designations are used for heat-treatable alloys which are primarily strengthened by heat treatment. It will be noted that the –T designations also include conditions which result from cold working combined with heat treatment.

Composition—Wrought Aluminum Alloys

Composition limits for most of the available alloys are shown in Table 4. A number of composite or clad alloys are also available. The components of these products are shown in Table 5.

Mechanical Properties—Wrought Aluminum Alloys

Typical mechanical properties are shown in Table 6 for a number of alloys. In general these values are above the specified minimum except for the annealed (–O Temper) values which are slightly below the specified maximum. These values are the most probable figures to be found when testing a large number of specimens. The data presented may be used for comparison in selection of an alloy.

The properties will also depend to a significant extent on the form in which the alloy is used. Thin sheet will generally have higher strengths than thick plate of the same composition and heat treatment. Direction of loading will also influence the properties. In critical applications where safety margins are low, consultation with the producer for guaranteed properties is desirable. Standard specifications are available for most aluminum alloys in most forms. These specifications have minimum and/or maximum property limits which take into account section size, method of manufacture, and other process variables. Producers are able to guarantee compliance with these specifications. Table 7 gives a cross index of the Aluminum Association alloy designations and various government and standardizing agency specifications.

TABLE 3 Temper Designations for Aluminum Alloys

—F	As fabricated. Products acquire some temper from shaping processes, but there is no special control over strain hardening or thermal treatment. For wrought products, there are no mechanical property limits.
—O	Annealed, recrystallized. The softest temper of wrought products.
—H	Strain-hardened. Products have their strength increased by strain hardening, with or without supplementary heat treatments, to produce partial softening. The H is always followed by two or more digits. The first digit indicates the specific combination of basic operations as follows:
—H1	Strain hardening only. Products are strain-hardened without supplementary heat treatment. The number following the designation indicates the degree of strain hardening.
—H2	Strain-hardened and then partially annealed. Products are strain-hardened more than desired and then reduced in strength to the desired level by partial annealing. For alloys that age-soften at room temperature, —H2 tempers have approximately the same ultimate strength as corresponding —H3 tempers. For other alloys, —H2 tempers have approximately the same ultimate strength as corresponding —H1 tempers and slightly higher elongations. The number following the designation indicates the degree of strain hardening remaining after annealing.
—H3	Strain-hardened and then stabilized. Products are strain-hardened and then stabilized by low-temperature heating to lower their strength and increase ductility. The designation applies only to magnesium-containing alloys which, unless stabilized, gradually age-soften at room temperature. The number following the designation indicates degree of strain hardening remaining after hardening a specific amount and then stabilizing.

TABLE 3 Temper Designations for Aluminum Alloys (Continued)

Number 8 indicates tempers having a final degree of strain hardening equivalent to that resulting from approximately 75% reduction of area. Number 0 means annealed. Number 9 designates extra-hard tempers.

The third digit, when used, indicates that the degree of control of temper or that the mechanical properties are different from, but close to, those for the two-digit —H temper to which it is added. An exception is the —H112 temper which is defined as follows:

—H112 Products acquire some temper from shaping processes not having special control over strain hardening or heat treatment, but for which mechanical property limits or mechanical property testing are required.
—H311 Products strain-hardened less than H31.
—H321 Products strain-hardened less than H32.
—H323, —H343 Products specially fabricated to have acceptable resistance to stress-corrosion cracking.
—W Solution-heat-treated. An unstable temper—alloys age at room temperature after solution heat treatment. The designation is specific only when the period of natural aging is indicated (e.g., W½ hr).
—T Heat-treated. Stable tempers, other than —F, —O, or —H, are produced with or without supplementary strain hardening. The T is always followed by one or more digits. Numbers 1 through 10 indicate specific sequences of basic treatments, as follows:
—T1 A substantially stable condition is produced by cooling to room temperature from an elevated-temperature shaping process and naturally aging.
—T2 Annealed (cast products only). Annealing is used to improve ductility and increase dimensional stability of castings.
—T3 Solution-heat-treated and then cold-worked. Products are cold-worked to improve strength, or the effect of cold work in flattening or straightening is recognized in applicable specifications.
—T4 Solution-heat-treated and naturally aged to a substantially stable condition. Products are not cold-worked after solution heat treatment, or the effect of cold work in flattening or straightening may not be recognized in applicable specifications.
—T5 Artificially aged after cooling from an elevated-temperature shaping process to improve mechanical properties or dimensional stability or both.
—T6 Solution-heat-treated and then artificially aged. Products are not cold-worked after solution heat treatment, or the effect of cold work in flattening or straightening may not be recognized in applicable specifications.
—T7 Solution-heat-treated and then stabilized. Products are stabilized to carry them beyond the point of maximum strength to provide control of some special characteristic(s).
—T8 Solution-heat-treated, cold-worked, and then artificially aged. Products are cold-worked to improve strength, or the effect of cold work in flattening or straightening is recognized in applicable specifications.
—T9 Solution-heat-treated, artificially aged, and then cold-worked. Products are cold-worked to improve strength.
—T10 Artificially aged after cooling from an elevated-temperature shaping process, then cold-worked to improve strength.

The following additional digits have been assigned for stress-relieved tempers of wrought products:

—Tx51 Stress-relieved by stretching. Products are stretched the indicated amounts after solution heat treatment: Plate, 1 to 3% permanent set: rolled or cold-finished rod and bar, 1 to 3% permanent set. The products receive no further straightening after stretching.
—Tx51x Stress-relieved by stretching. Extruded rod, bar, shapes, and tube, and drawn tube are stretched the indicated amounts after solution heat treatment: Extruded rod, bar, shapes, and tube, 1 to 3% permanent set: drawn tube, ½ to 2½% permanent set.
—Tx510 Products receive no further straightening after stretching.
—Tx511 Products may receive minor straightening after stretching to comply with standard tolerances.
—Tx52 Stress-relieved by compressing. Products are stress-relieved by compressing after solution heat treatment to produce a nominal permanent set of 2½%.
—Tx53 Stress-relieved by heat treatment.

TABLE 4 Chemical Composition Limits of Wrought Aluminum Alloys[a,b,4]

AA designation	Silicon	Iron	Copper	Manganese	Magnesium	Chromium	Nickel	Zinc	Titanium	Others[c] Each	Others[c] Total	Aluminum, min.[d]
EC[e]	....	....	....	....	....	....	....	....	....	....	....	99.45
1060	0.25	0.35	0.05	0.03	0.03	....	....	0.05	0.03	0.03[f]	....	99.60
1100	1.0 Si + Fe		0.05–0.20	0.05	....	....	....	0.10	....	0.05[f]	0.15	99.00
1145[g]	0.55 Si + Fe		0.05	0.05	....	....	....	....	....	0.03	....	99.45
1175[h]	0.15 Si + Fe		0.10	....	....	....	....	....	....	0.02	....	99.75
1230[i]	0.7 Si + Fe		0.10	0.05	....	....	....	0.10	....	0.05	....	99.30
1235	0.65 Si + Fe		0.05	....	....	....	....	....	....	0.05	....	99.35
2011	0.40	0.7	5.0–6.0	....	....	....	....	0.30	....	0.05[j]	0.15	Bal.
2014	0.50–1.2	0.7	3.9–5.0	0.40–1.2	0.20–0.8	0.10	....	0.25	0.15	0.05[f]	0.15	Bal.
2017	0.8	0.7	3.5–4.5	0.40–1.0	0.20–0.8	0.10	....	0.25	....	0.05	0.15	Bal.
2018	0.9	1.0	3.5–4.5	0.20	0.45–0.9	0.10	1.7–2.3	0.25	....	0.05	0.15	Bal.
2024	0.50	0.50	3.8–4.9	0.30–0.9	1.2–1.8	0.10	....	0.25	....	0.05	0.15	Bal.
2025	0.50–1.2	1.0	3.9–5.0	0.40–1.2	0.05	0.10	....	0.25	0.15	0.05	0.15	Bal.
2117	0.8	0.7	2.2–3.0	0.20	0.20–0.50	0.10	....	0.25	....	0.05	0.15	Bal.
2218	0.9	1.0	3.5–4.5	0.20	1.2–1.8	0.10	1.7–2.3	0.25	....	0.05	0.15	Bal.
2219	0.20	0.30	5.8–6.8	0.20–0.40	0.02	....	....	0.10	0.02–0.10	0.05[k]	0.15	Bal.
2618	0.25	0.9–1.3	1.9–2.7	....	1.3–1.8	....	0.9–1.2	....	0.04–0.10	0.05	0.15	Bal.
3003	0.6	0.7	0.05–0.20	1.0–1.5	....	....	....	0.10	....	0.05[f]	0.15	Bal.
3004	0.30	0.7	0.25	1.0–1.5	0.8–1.3	....	....	0.25	....	0.05[f]	0.15	Bal.
4032	11.0–13.5	1.0	0.50–1.3	....	0.8–1.3	0.10	0.50–1.3	0.25	....	0.05	0.15	Bal.
4043	4.5–6.0	0.8	0.30	0.05	0.05	....	....	0.10	0.20	0.05[f]	0.15	Bal.
4045[l]	9.0–11.0	0.8	0.30	0.05	0.05	....	....	0.10	0.20	0.05	0.15	Bal.
4343[l]	6.8–8.2	0.8	0.25	0.10	....	....	....	0.20	....	0.05	0.15	Bal.
5005	0.40	0.7	0.20	0.20	0.50–1.1	0.10	....	0.25	....	0.05	0.15	Bal.
5050	0.40	0.7	0.20	0.10	1.1–1.8	0.10	....	0.25	....	0.05[f]	0.15	Bal.
5052	0.45 Si + Fe		0.10	0.10	2.2–2.8	0.15–0.35	....	0.10	....	0.05[f]	0.15	Bal.
5056	0.30	0.40	0.10	0.05–0.20	4.5–5.6	0.05–0.20	....	0.10	....	0.05[f]	0.15	Bal.
5083	0.40	0.40	0.10	0.30–1.0	4.0–4.9	0.05–0.25	....	0.25	0.15	0.05	0.15	Bal.
5086	0.40	0.50	0.10	0.20–0.7	3.5–4.5	0.05–0.25	....	0.25	0.15	0.05	0.15	Bal.
5154	0.45 Si + Fe		0.10	0.10	3.1–3.9	0.15–0.35	....	0.20	0.20	0.05[f]	0.15	Bal.
5252	0.08	0.10	0.10	0.10	2.2–2.8	....	....	....	....	0.03	0.10	Bal.
5257	0.08	0.10	0.10	0.03	0.20–0.6	....	....	0.03	....	0.02	0.05	Bal.
5356	0.50 Si + Fe		0.10	0.05–0.20	4.5–5.5	0.05–0.20	....	0.10	0.06–0.20	0.05[f]	0.15	Bal.
5454	0.40 Si + Fe		0.10	0.50–1.0	2.4–3.0	0.05–0.20	....	0.25	0.20	0.05	0.15	Bal.
5456	0.40 Si + Fe		0.10	0.50–1.0	4.7–5.5	0.05–0.20	....	0.25	0.20	0.05	0.15	Bal.
5457	0.08	0.10	0.20	0.15–0.45	0.8–1.2	....	....	....	....	0.03	0.10	Bal.
5557	0.10	0.12	0.15	0.10–0.40	0.40–0.8	....	....	....	....	0.03	0.10	Bal.
5657	0.08	0.10	0.10	0.03	0.6–1.0	....	....	0.03	....	0.02	0.05	Bal.
6003[m]	0.35–1.0	0.6	0.10	0.8	0.8–1.5	0.35	....	0.20	0.10	0.05	0.15	Bal.
6053	[n]	0.35	0.10	....	1.1–1.4	0.15–0.35	....	0.10	....	0.05	0.15	Bal.
6061	0.40–0.8	0.7	0.15–0.40	0.15	0.8–1.2	0.04–0.35	....	0.25	0.15	0.05	0.15	Bal.
6063	0.20–0.6	0.35	0.10	0.10	0.45–0.9	0.10	....	0.10	0.10	0.05	0.15	Bal.
6066	0.9–1.8	0.50	0.7–1.2	0.6–1.1	0.8–1.4	0.40	....	0.25	0.20	0.05	0.15	Bal.
6070	1.0–1.7	0.50	0.15–0.40	0.40–1.0	0.50–1.2	0.10	....	0.25	0.15	0.05	0.15	Bal.

[a] Composition in percent maximum unless shown as a range.

[b] For purposes of determining conformance to these limits, an observed value or a calculated value obtained from analysis is rounded off to the nearest unit in the last right-hand place of figures used in expressing the specified limit, in accordance with U.S.A. Standard Rules for Rounding Off Numerical Values (USAS Z25.1).

[c] Analysis is regularly made only for the elements for which specific limits are shown, except for unalloyed aluminum. If, however, the presence of other elements is suspected to be, or in the course of routine analysis is indicated to be, in excess of the specified limits, further analysis is made to determine that these other elements are not in excess of the amount specified.

[d] The aluminum content for unalloyed aluminum not made by a refining process is the difference between 100.00 % and the sum of all other metallic elements present in amounts of 0.010 % or more each, expressed to the second decimal.

[e] Electric conductor.

[f] Beryllium 0.0008 maximum for welding electrode and filler wire only.

[g] Foil.

[h] Cladding on Clad 1100 and Clad 3003 reflector sheet.

[i] Cladding on Alclad 2024.

[j] Also contains 0.20–0.6 % each of lead and bismuth.

[k] Vanadium 0.05–0.15; zirconium 0.10–0.25.

[l] Brazing alloy.

[m] Cladding on Alclad 2014.

[n] Silicon 45 % to 65 % magnesium content.

TABLE 4 Chemical Composition Limits of Wrought Aluminum Alloys[a,b,4] **(Continued)**

AA desig-nation	Silicon	Iron	Copper	Man-ganese	Mag-nesium	Chro-mium	Nickel	Zinc	Titanium	Others[c] Each	Others[c] Total	Alumi-num, min.[d]
6101[o]	0.30–0.7	0.50	0.10	0.03	0.35–0.8	0.03	....	0.10	....	0.03[p]	0.10	Bal.
6151	0.6–1.2	1.0	0.35	0.20	0.45–0.8	0.15 C.35	....	0.25	0.15	0.05	0.15	Bal.
6201	0.50–0.9	0.50	0.10	0.03	0.6–0.9	0.03	....	0.10	....	0.03[p]	0.10	Bal.
6253[q]	n	0.50	0.10	....	1.0–1.5	0.15–0.35	....	1.6–2.4	....	0.05	0.15	Bal.
6262	0.40–0.8	0.7	0.15–0.40	0.15	0.8–1.2	0.04–0.14	....	0.25	0.15	0.05[r]	0.15	Bal.
6463	0.20–0.6	0.15	0.20	0.05	0.45–0.9	....	....	....	....	0.05	0.15	Bal.
6951	0.20–0.50	0.8	0.15–0.40	0.10	0.40–0.8	....	....	0.20	....	0.05	0.15	Bal.
7001	0.35	0.40	1.6–2.6	0.20	2.6–3.4	0.18–0.35	....	6.8–8.0	0.20	0.05	0.15	Bal.
7039	0.30	0.40	0.10	0.10–0.40	2.3–3.3	0.15–0.25	....	3.5–4.5	0.10	0.05	0.15	Bal.
7072[s]	0.7 Si + Fe		0.10	0.10	0.10	....	....	0.8–1.3	....	0.05	0.15	Bal.
7075	0.40	0.50	1.2–2.0	0.30	2.1–2.9	0.18–0.35	....	5.1–6.1	0.20	0.05	0.15	Bal.
7079	0.30	0.40	0.40–0.8	0.10–0.30	2.9–3.7	0.10–0.25	....	3.8–4.8	0.10	0.05	0.15	Bal.
7178	0.40	0.50	1.6–2.4	0.30	2.4–3.1	0.18–0.35	....	6.3–7.3	0.20	0.05	0.15	Bal.

[o] Bus conductor.
[p] Boron, 0.06% maximum.
[q] Cladding on Alclad 5056.
[r] Also contains 0.40–0.7% each of lead and bismuth.
[s] Cladding on Alclad 2219, 3003, 3004, 6061, 7075, 7079, and 7178.

Effect of Temperature on Mechanical Properties of Aluminum Alloys

The effect of temperature on tensile properties of aluminum alloys is shown in Table 8. Note that these values were obtained from material which had been aged 10,000 hr at test temperature. Other properties such as fatigue strength and shear strength will also show similar changes. It is obvious that moderately elevated temperatures can have appreciable effects on mechanical properties of aluminum alloys. The time of heating can also have major effects. It will also be noted that the alloys that are strongest at room temperature are not necessarily the strongest at elevated temperature.

Physical Properties of Aluminum

Tables 9 and 10 show a compilation of electrical and physical properties at room temperature. These values will also change with change in temperature.

Fabrication Properties of Aluminum

Table 11 shows a comparison of relative fabrication characteristics of aluminum alloys along with relative corrosion resistance. These comparisons are highly qualitative and a D rating does not necessarily rule out the use of the alloy for that fabrication technique, nor does an A rating mean that complete freedom from precaution is implied.

Corrosion resistance The A-rated materials will perform satisfactorily in mild environment without protection; however, even these alloys may need protection in extreme environments such as salt spray, salt water or where chlorides of any kind exist. The C-rated materials can be given protection by anodizing or conversion coating. In some cases further protection is provided by painting where the environment is severe or for decorative purposes. A discussion of corrosion protection is contained in another chapter of this handbook.

Workability Most aluminum alloys form very well in the annealed condition since they are soft and ductile. A rough estimate of the formability of the metal may be gained from the percentage elongation as determined in the tensile test (see Table 6) and from the difference between the ultimate tensile strength and

TABLE 5 Components of Aluminum Alloy Clad Products[1]

Designation	Core	Cladding	Total thickness of composite product, inches	Sides clad	Cladding thickness per side, % of composite thickness Nominal	Min. average[†]
Alclad 2014‡ sheet and plate	2014	6003	Up thru 0.024	Both	10	8
			0.025–0.039	Both	7½	6
			0.040–0.099	Both	5	4
			0.100 and over	Both	2½	2
Alclad 2024 sheet and plate	2024	1230	Up thru 0.062	Both	5	4
			0.063–0.187	Both	2½	2
			0.188 and over	Both	1½, 2½ §¶	1.2, 2
Alclad 2219 sheet and plate	2219	7072	Up thru 0.039	Both	10	8
			0.040–0.099	Both	5	4
			0.100 and over	Both	2½	2
Alclad 3003 sheet and plate	3003	7072	All	One or both	5	4
Alclad 3003 tube	3003	7072	All	Inside	10	
			All	Outside	7	
Alclad 3004 sheet and plate	3004	7072	All	Both	5	4
Alclad 5056 rod and wire	5056	6253	All	Outside	20	16
					(of total cross-sectional area)	
Alclad 6061 sheet and plate	6061	7072	All	Both	5	4
Alclad 7075 sheet and plate	7075	7072	Up thru 0.062	Both	4	3.2
			0.063–0.187	Both	2½	2
			0.188 and over	Both	1½ ¶	1.2
Alclad one side 7075 sheet and plate	7075	7072	Up thru 0.062	One	4	3.2
			0.063–0.187	One	2½	2
			0.188 and over	One	1½ ¶	1.2
Alclad 7079 sheet	7079	7072	Up thru 0.062	Both	4	3.2
			0.063–0.187	Both	2½	2
			0.188 and over	Both	1½	1.2
Alclad 7178 sheet and plate	7178	7072	All	Both	4	3.2
No. 11 brazing sheet	3003	4343	Up thru 0.063	One	10	8
			0.064 and over	One	5	4
No. 12 brazing sheet	3003	4343	Up thru 0.063	Both	10	8
			0.064 and over	Both	5	4
No. 21 brazing sheet	6951	4343	Up thru 0.090	One	10	8
			0.091 and over	One	5	4
No. 22 brazing sheet	6951	4343	Up thru 0.090	Both	10	8
			0.091 and over	Both	5	4
No. 23 brazing sheet	6951	4045	Up thru 0.090	One	10	8
			0.091 and over	One	5	4
No. 24 brazing sheet	6951	4045	Up thru 0.090	Both	10	8
			0.091 and over	Both	5	4
Clad 1100 reflector sheet	1100	1175	Up thru 0.064	One or both	15	12
			0.065 and over		7½	6
Clad 3003 reflector sheet	3003	1175	Up thru 0.064	One or both	15	12
			0.065 and over		7½	6

* Cladding composition is applicable only to the aluminum or aluminum alloy bonded to the alloy ingot or slab preparatory to rolling to the specified composite product. The composition of the cladding may be subsequently altered by diffusion between the core and cladding due to thermal treatment.

† Average minimum thickness per side as determined by averaging cladding thickness measurements taken at a magnification of 100 diameters on the cross section of a transverse sample polished and etched for microscopic examination.

‡ Trademark of Aluminum Company of America.

§ Alclad 2024 sheet and plate in thicknesses of 0.188 in. and over are furnished with a nominal cladding thickness of 2½% unless the thinner cladding is specifically designated.

¶ For thicknesses of 0.500 in. and over with 1½% nominal cladding thickness, the average maximum thickness of cladding per side after rolling to the specified thickness of the plate will be 3% of the thickness of the plate as determined by averaging cladding thickness measurements taken at a magnification of 100 diameters on the cross section of a transverse sample polished and etched for microscopic examination.

Aluminum and Aluminum Alloys 9-19

TABLE 6 Typical Mechanical Properties[a] of Aluminum Alloys at Room Temperature[1]

Alloy and temper	Tension Strength, ksi (kg/mm²) Ultimate	Tension Strength, ksi (kg/mm²) Yield	Tension Elongation, % in 2 in. 1/16-in.-thick specimen	Tension Elongation, % in 2 in. 1/2-in.-diameter specimen	Hardness Brinell number, 500-kg load, 10-mm ball	Shear Ultimate shearing strength, ksi (kg/mm²)	Fatigue Endurance limit, ksi (kg/mm²)	Modulus Modulus of elasticity, ksi (kg/mm²) × 10³
EC-0[d]	12(8.4)	4(2.8)	...	...[e]	...	8(5.6)	....	10.0(7.0)
EC-H12[d]	14(9.8)	12(8.4)	...	...	...	9(6.3)	....	10.0(7.0)
EC-H14[d]	16(11.2)	14(9.8)	...	...	...	10(7.0)	....	10.0(7.0)
EC-H16[d]	18(12.7)	16(11.2)	...	...	...	11(7.7)	....	10.0(7.0)
EC-H19[d]	27(19.0)	24(16.9)	...	...[f]	...	15(10.5)	7(4.9)	10.0(7.0)
1060-0	10(7.0)	4(2.8)	43	...	19	7(4.9)	3(2.1)	10.0(7.0)
1060-H12	12(8.4)	11(7.7)	16	...	23	8(5.6)	4(2.8)	10.0(7.0)
1060-H14	14(9.8)	13(9.1)	12	...	26	9(6.3)	5(3.5)	10.0(7.0)
1060-H16	16(11.2)	15(10.5)	8	...	30	10(7.0)	6.5(4.6)	10.0(7.0)
1060-H18	19(13.4)	18(12.7)	6	...	35	11(7.7)	6.5(4.6)	10.0(7.0)
1100-0	13(9.1)	5(3.5)	35	45	23	9(6.3)	5(3.5)	10.0(7.0)
1100-H12	16(11.2)	15(10.5)	12	25	28	10(7.0)	6(4.2)	10.0(7.0)
1100-H14	18(12.7)	17(12.0)	9	20	32	11(7.7)	7(4.9)	10.0(7.0)
1100-H16	21(14.8)	20(14.1)	6	17	38	12(8.4)	9(6.3)	10.0(7.0)
1100-H18	24(16.9)	22(15.5)	5	15	44	13(9.1)	9(6.3)	10.0(7.0)
2011-T3	55(38.7)[g]	43(30.2)[g]	...	15	95	32(22.5)	18(12.7)	10.2(7.2)
2011-T8	59(41.5)	45(31.6)	...	12	100	35(24.6)	18(12.7)	10.2(7.2)
2014-0	27(19.0)	14(9.8)	...	18	45	18(12.7)	13(9.1)	10.6(7.5)
2014-T4, -T451	62(43.6)	42(29.5)	...	20	105	38(26.7)	20(14.1)	10.6(7.5)
2014-T6, -T651	70(49.2)[h]	60(42.2)[h]	...	13	135	42(29.5)	18(12.7)	10.6(7.5)
Alclad 2014-0	25(17.6)	10(7.0)	21	...	...	18(12.7)	....	10.5(7.4)
Alclad 2014-T3	63(44.3)[i]	40(28.1)[i]	20	...	...	37(26.0)	....	10.5(7.4)
Alclad 2014-T4, -T451	61(42.9)[i]	37(26.0)[i]	22	...	...	37(26.0)	....	10.5(7.4)
Alclad 2014-T6, -T651	68(47.8)[i]	60(42.2)[i]	10	...	...	41(28.8)	....	10.5(7.4)
2017-0	26(18.3)	10(7.0)	...	22	45	18(12.7)	13(9.1)	10.5(7.4)
2017-T4, -T451	62(43.6)	40(28.1)	...	22	105	38(26.7)	18(12.7)	10.5(7.4)
2018-T61	61(42.9)	46(32.3)	...	12	120	39(27.4)	17(12.0)	10.8(7.6)
2024-0	27(19.0)	11(7.7)	20	22	47	18(12.7)	13(9.1)	10.6(7.5)
2024-T3	70(49.2)	50(35.2)	18	...	120	41(28.8)	20(14.1)	10.6(7.5)
2024-T4, -T351	68(47.8)	47(33.0)[h]	20	19	120	41(28.8)	20(14.1)	10.6(7.5)
2024-T36	72(50.6)	57(40.1)	13	...	130	42(29.5)	18(12.7)	10.6(7.5)
Alclad 2024-0	26(18.3)	11(7.7)	20	...	...	18(12.7)	....	10.6(7.5)
Alclad 2024-T3	65(45.7)[i]	45(31.6)[i]	18	...	...	40(28.1)	....	10.6(7.5)
Alclad 2024-T4, -T351	64(45.0)[i]	42(29.5)[i]	19	...	...	40(28.1)	....	10.6(7.5)
Alclad 2024-T36	67(47.1)[i]	53(37.3)[i]	11	...	...	41(28.8)	....	10.6(7.5)
Alclad 2024-T81, -T851	65(45.7)[i]	60(42.2)[i]	6	...	...	40(28.1)	....	10.6(7.5)
Alclad 2024-T86	70(49.2)[i]	66(46.4)[i]	6	...	...	42(29.5)	....	10.6(7.5)
2025-T6	58(40.8)	37(26.0)	...	19	110	35(24.6)	18(12.7)	10.4(7.3)
2117-T4	43(30.2)	24(16.9)	...	27	70	28(19.7)	14(9.8)	10.3(7.2)
2218-T72	48(33.7)	37(26.0)	...	11	95	30(21.1)	....	10.8(7.6)
2219-0[k]	25(17.6)	11(7.7)	18	...	...	....	....	10.6(7.5)
2219-T42[k]	52(36.6)	27(19.0)	20	...	...	....	....	10.6(7.5)
2219-T31, -T351[k]	52(36.6)	36(25.3)	17	...	...	....	....	10.6(7.5)
2219-T37[k]	57(40.1)	46(32.3)	11	...	...	....	....	10.6(7.5)
2219-T62[k]	60(42.2)	42(29.5)	10	...	...	....	15(10.5)	10.6(7.5)
2219-T81, -T851[k]	66(46.4)	51(35.9)	10	...	...	....	15(10.5)	10.6(7.5)
2219-T87[k]	69(48.5)	57(40.1)	10	...	...	....	15(10.5)	10.6(7.5)
5456-0	45(31.6)	23(16.2)	...	24	...	....	....	10.3(7.2)
5456-H111	47(33.0)	33(23.2)	...	18	...	....	....	10.3(7.2)
5456-H112	45(31.6)	24(16.9)	...	22	...	....	....	10.3(7.2)
5456-H321	51(35.9)	37(26.0)	...	16	90	30(21.1)	....	10.3(7.2)

Nonferrous Metals

TABLE 6 Typical Mechanical Properties[a] of Aluminum Alloys at Room Temperature[1] (Continued)

Alloy and temper	Tension Ultimate Strength, ksi (kg/mm²)	Tension Yield Strength, ksi (kg/mm²)	Elongation, % in 2 in. 1/16-in.-thick specimen	Elongation, % in 2 in. 1/2-in.-diameter specimen	Hardness Brinell number, 500-kg load, 10-mm ball	Shear Ultimate shearing strength, ksi (kg/mm²)	Fatigue Endurance limit, ksi (kg/mm²)	Modulus Modulus of elasticity, ksi (kg/mm²) × 10³
6061-0	18(12.7)	8(5.6)	25	30	30	12(8.4)	9(6.3)	10.0(7.0)
6061-T4, −T451	35(24.6)	21(14.8)	22	25	65	24(16.9)	14(9.8)	10.0(7.0)
6061-T6, −T651	45(31.6)	40(28.1)	12	17	95	30(21.1)	14(9.8)	10.0(7.0)
Alclad 6061-0	17(12.0)	7(4.9)	25	...	...	11(7.7)	....	10.0(7.0)
Alclad 6061-T4, −T451	33(23.2)	19(13.4)	22	...	...	22(15.5)	....	10.0(7.0)
Alclad 6061-T6, −T651	42(29.5)	37(26.0)	12	...	...	27(19.0)	....	10.0(7.0)
6063-0	13(9.1)	7(4.9)	...	...	25	10(7.0)	8(5.6)	10.0(7.0)
6063-T1	22(15.5)	13(9.1)	20	...	42	14(9.8)	9(6.3)	10.0(7.0)
6063-T4	25(17.6)	13(9.1)	22	...	...	....	....	10.0(7.0)
6063-T5	27(19.0)	21(14.8)	12	...	60	17(12.0)	10(7.0)	10.0(7.0)
6063-T6	35(24.6)	31(21.8)	12	...	73	22(15.5)	10(7.0)	10.0(7.0)
6101-H111	14(9.8)	11(7.7)	...	...	...	....	....	10.0(7.0)
6101-T6	322(2.5)	28(19.7)	15	...	71	20(14.1)	....	10.0(7.0)
6262-T9	58(40.8)	55(38.7)	...	10	120	35(24.6)	13(9.1)	10.0(7.0)
6463-T1	22(15.5)	13(9.1)	20	...	42	14(9.8)	10(7.0)	10.0(7.0)
6463-T5	27(19.0)	21(14.8)	12	...	60	17(12.0)	10(7.0)	10.0(7.0)
6463-T6	35(24.6)	31(21.8)	12	...	74	22(15.5)	10(7.0)	10.0(7.0)
7001-0	37(26.0)	22(15.5)	...	14	60	....	....	10.3(7.2)
7001-T6, −T651	98(68.9)	91(64.0)	...	9	160	....	22(15.8)	10.3(7.2)
7039-T61, −T6151[k]	60(42.2)	50(35.2)	14	...	123	....	....	....
7075-0	33(23.2)	15(10.5)	17	16	60	22(15.5)	....	10.4(7.3)
7075-T6, −T651	83(58.4)[l]	73(51.3)[l]	11	11	150	48(33.7)	23(16.2)	10.4(7.3)
Alclad 7075-0	32(22.5)	14(9.8)	17	...	...	22(15.5)	....	10.4(7.3)
Alclad 7075-T6, −T651	76(53.4)	67(47.1)	11	...	...	46(32.3)	....	10.4(7.3)
7079-0	33(23.2)	15(10.5)	17	16	...	....	....	10.4(7.3)
7079-T6, −T651	78(54.8)	68(47.8)	...	14	145	45(31.6)	23(16.2)	10.4(7.3)
Alclad 7079-0	32(22.5)	14(9.8)	16	...	...	....	....	10.4(7.3)
Alclad 7079-T6	71(49.9)	62(43.6)	10	...	...	42(29.5)	....	10.4(7.3)
7178-0	33(23.2)	15(10.5)	15	16	...	....	....	10.4(7.3)
7178-T6, −T651	88(61.9)[l]	78(54.8)[l]	10	11	...	....	....	10.4(7.3)
Alclad 7788-0	32(22.5)	14(9.8)	16	...	...	....	....	10.4(7.3)
Alclad 7171-T6, −T651	81(56.9)	71(49.9)	10	...	...	....	....	10.4(7.3)

[a] These typical properties are average for various forms, sizes, and methods of manufacture, and may not exactly describe any one particular product.
[b] Based on 500,000,000 cycles of completely reversed stress using the R. R. Moore type of machine and specimen.
[c] Average of tension and compression moduli. Compression modulus is about 2% greater than tension modulus.
[d] Electric conductor grade, 99.45% minimum aluminum.
[e] EC-0 wire will have an elongation of approximately 23% in 10 in.
[f] EC-H19 wire will have an elongation of approximately 1½% in 10 in.
[g] Sizes greater than 1½ in. will have strengths slightly lower than these values.
[h] Extruded products more than ¾ in. thick will have strengths 15 to 20% higher than these values.
[i] Sheet less than 0.040 in. thick will have strengths slightly lower than these values.
[j] Sheet more than 0.062 in. thick will have strengths slightly higher than these values.
[k] Applicable to sheet and plate products only.
[l] Extruded products will have strengths approximately 10% higher than these values.

Aluminum and Aluminum Alloys 9-21

TABLE 7 Aluminum Mill Product Specifications[1]

Alloy	Product	Federal	Military	ASTM	U.S.A.
EC	ACSR.................................	....	....	B232-64T	C7.22-1965
	Wire; H19 temper.....................	....	....	B230-60	C7.20-1960
	Wire; H16 temper.....................	....	....	B262-61	C7.35-1961
	Wire; H14 temper.....................	....	....	B314-60	C7.40-1960
				B323-61	C7.42-1961
				B324-60	C7.43-1960
1060	Sheet and plate.......................	....	....	B209-67	H38.2-1967
	Wire, rod, and bar; rolled or cold-finished...	....	....	B211-67	H38.4-1967
	Rod, bar, shapes, and tube; extruded......	....	....	B221-67	H38.5-1967
	Tube; extruded, seamless...............	....	....	B241-67	H38.7-1967
	Tube; drawn..........................	....	....	B210-67	H38.3-1967
	Tube; condenser......................	....	....	B234-67	H38.6-1967
	Pipe; gas and oil transmission...........	....	....	B345-67	H38.13-1967
	Tube; condenser with integral fins........	....	....	B404-65T	
1100	Sheet and plate.......................	QQ-A-250/1d	....	B209-67	H38.2-1967
	Wire, rod, and bar; rolled or cold-finished...	QQ-A-225/1c	....	B211-67	H38.4-1967
	Rod, bar, shapes, and tube; extruded......	....	....	B221-67	H38.5-1967
	Tube; extruded, seamless...............	....	....	B241-67	H38.7-1967
	Tube; drawn..........................	WW-T-700/1d	....	B210-67	H38.3-1967
	Tube; welded.........................	....	....	B313-67	H38.11-1967
	Tube; coiled for special applications......	....	....	B307-67	H38.9-1967
	Tube; waveguide......................	....	MIL-W-85E		
			MIL-W-23068		
			MIL-W-23351		
	Rivet wire............................	QQ-A-430a	....	B316-67	H38.12-1967
	Spray-gun wire.......................	....	MIL-W-6712B		
	Forgings and forging stock..............	....	....	B247-67	H38.8-1967
	Welding rod; bare.....................	QQ-R-566a	MIL-E-16053K-1	B285-61T	
	Impact extrusions.....................	....	MIL-A-12545B		
	Foil.................................	....	MIL-A-148E		
1145	Foil.................................	....	....	B373-66	
1235	Foil.................................	....	MIL-A-148E	B373-66	
2011	Wire, rod, and bar; rolled or cold-finished...	QQ-A-225/3c	....	B211-67	H38.4-1967
2014	Sheet and plate.......................	....	....	B209-67	H38.2-1967
	Wire, rod, and bar; rolled or cold-finished...	QQ-A-225/4c	....	B211-67	H38.4-1967
	Rod, bar, shapes, and tube; extruded......	QQ-A-200/2c	....	B221-67	H38.5-1967
	Tube; extruded, seamless...............	QQ-A-200/2c	....	B241-67	H38.7-1967
	Tube; drawn..........................	....	....	B210-67	H38.3-1967
	Structural shapes; rolled or extruded......	....	MIL-A-25994		
	Forgings and forging stock..............	QQ-A-367g	MIL-A-22771B-1	B247-67	H38.8-1967
	Impact extrusions.....................	....	MIL-A-12545B		
Alclad 2014	Sheet and plate.......................	QQ-A-250/3d	....	B209-67	H38.2-1967
2017	Wire, rod, and bar; rolled or cold-finished...	QQ-A-225/5c	....	B211-67	H38.4-1967
	Rivet wire............................	QQ-A-430a	....	B316-67	H38.12-1967
2024	Sheet and plate.......................	QQ-A-250/4d	....	B209-67	H38.2-1967
	Wire, rod, and bar; rolled or cold-finished...	QQ-A-225/6c	....	B211-67	H38.4-1967
	Rod, bar, shapes, and tube; extruded......	QQ-A-200/3c	....	B221-67	H38.5-1967
	Tube; extruded, seamless...............	QQ-A-200/3c	....	B241-67	H38.7-1967
	Tube; drawn..........................	WW-T-700/3d	....	B210-67	H38.3-1967
Alclad 2219	Sheet and plate.......................	....	....	B209-67	H38.2-1967
2618	Forgings and forging stock..............	QQ-A-367g			
3003	Sheet and plate.......................	QQ-A-250/2c	....	B209-67	H38.2-1967
	Wire, rod, and bar; rolled or cold-finished...	QQ-A-225/2c	....	B211-67	H38.4-1967
	Rod, bar, shapes, and tube; extruded......	QQ-A-200/1b	....	B221-67	H38.5-1967
	Tube; extruded, seamless...............	QQ-A-200/1b	....	B241-67	H38.7-1967
	Tube; drawn..........................	WW-T-700/2d	....	B210-67	H38.3-1967

TABLE 7 Aluminum Mill Product Specifications[1] (Continued)

Alloy	Product	Federal	Military	ASTM	U.S.A.
5050	Sheet and plate.................	....	....	B209-67	H38.2-1967
	Tube; drawn...................	....	....	B210-67	H38.3-1967
	Tube; welded..................	....	....	B313-67	H38.11-1967
	Tube; coiled, for special applications......	....	....	B307-67	H38.9-1967
	Pipe; gas and oil transmission.....	....	....	B345-67	H38.13-1967
5052	Sheet and plate.................	QQ-A-250/8d	....	B209-67	H38.2-1967
	Wire, rod, and bar; rolled or cold-finished...	QQ-A-225/7b	....	B211-67	H38.4-1967
	Tube; drawn...................	WW-T-700/4d	....	B210-67	H38.3-1967
	Tube; hydraulic................	WW-T-700/4d	....		
	Tube; extruded.................	....	....	B221-67	H38.5-1967
	Tube; extruded, seamless.........	....	....	B241-67	H38.7-1967
	Tube; condenser................	....	....	B234-67	H38.6-1967
	Tube; condenser with integral fins......	....	....	B404-65T	
	Tube; welded..................	....	....	B313-67	H38.11-1967
	Tube; coiled, for special applications......	....	....	B307-67	H38.9-1967
	Pipe; gas and oil transmission.....	....	....	B345-67	H38.13-1967
	Rivet wire....................	QQ-A-430a	....	B316-67	H38.12-1967
5083	Sheet and plate.................	QQ-A-250/6e	....	B209-67	H38.2-1967
	Rod, bar, shapes, and tube; extruded......	QQ-A-200/4b	....	B221-67	H38.5-1967
	Tube; extruded, seamless.........	QQ-A-200/4b	....	B241-67	H38.7-1967
	Tube; drawn...................	....	....	B210-67	H38.3-1967
	Forgings and forging stock........	QQ-A-367g	....	B247-67	H38.8-1967
					H38.12-1967
5086	Sheet and plate.................	QQ-A-250/7d	....	B209-67	H38.2-1967
	Rod, bar, shapes, and tube; extruded......	QQ-A-200/5b	....	B221-67	H38.5-1967
5154	Sheet and plate.................	....	....	B209-67	H38.2-1967
	Wire, rod, and bar; rolled or cold-finished...	....	....	B211-67	H38.4-1967
	Rod, bar, shapes, and tube; extruded......	....	....	B221-67	H38.5-1967
	Tube; drawn...................	....	....	B210-67	H38.3-1967
5456	Sheet and plate.................	QQ-A-250/9e	....	B209-67	H38.2-1967
	Rod, bar, shapes, and tube; extruded..	QQ-A-200/7c	....	B221-67	H38-5.1967
	Tube; extruded, seamless.........	QQ-A-200/7c	....	B241-67	H38.7-1967
	Tube; drawn...................	....	....	B210-67	H38.3-1967
	Structural shapes; rolled or extruded......	....	MIL-A-25994		
	Pipe.........................	....	MIL-P-25995		
6053	Rivet wire....................	QQ-A-430a	....	B316-67	H38.12-1967
	Forgings and forging stock........	....	....	B247-67	H38.8-1967
	Rivets........................	....	MIL-R-1150A-1		
6061	Sheet and plate.................	QQ-A-250/11d	....	B209-67	H38.2-1967
	Tread plate....................	....	MIL-F-17132B		
	Wire, rod, and bar; rolled or cold-finished...	QQ-A-225/8c	....	B211-67	H38.4-1967
	Rod, bar, shapes, and tube; extruded......	QQ-A-200/8c	....	B221-67	H38.5-1967
	Structural shapes...............	....	MIL-A-25994	B308-67	H38.10-1967
	Tube; extruded, seamless.........	QQ-A-200/8c	....	B241-67	H38.7-1967
	Tube; drawn...................	WW-T-700/6d	....	B210-67	H38.3-1967
	Tube; hydraulic................	WW-T-700/6d	MIL-T-7081D-3		
	Tube; condenser................	....	....	B234-67	H38.6-1967
	Tube; condenser with integral fins......	....	....	B404-65T	
	Tube; welded..................	....	....	B313-67	H38.11-1967
Alclad 6061	Sheet and plate.................	....	....	B209-67	H38.2-1967
	Pipe; gas and oil transmission.....	....	....	B345-67	H38.13-1967
6063	Rod, bar, shapes, and tube; extruded..	QQ-A-200/9b	....	B221-67	H38.5-1967
	Tube; extruded, seamless.........	QQ-A-200/9b	....	B241-67	H38.7-1967
	Tube; drawn...................	....	....	B210-67	H38.3-1967
	Tube; waveguide................	....	MIL-W-23068		
			MIL-W-23351		
	Pipe.........................	....	MIL-P-25995	B241-67	H38.7-1967
	Pipe; gas and oil transmission.....	....	....	B345-67	H38.13-1967
	Structural pipe and tube; extruded........	....	....	B429-67	

Aluminum and Aluminum Alloys   9-23

TABLE 7 Aluminum Mill Product Specifications[1] (Continued)

Alloy	Product	Federal	Military	ASTM	U.S.A.
6066	Rod, bar, shapes, and tube; extruded	QQ-A-200/10c			
	Structural shapes; rolled or extruded		MIL-A-25994		
	Forgings and forging stock	QQ-A-367g			
6070	Rod, bar, shapes, and tube; extruded		MIL-A-46104		
	Impact extrusions		MIL-A-12545B		
6101	Bus conductor	QQ-B-825a		B317-66	
6151	Forgings and forging stock	QQ-A-367g		B247-67	H38.8-1967
6201	Wire; T81 temper			B398-63T	
				B399-63T	
7039	Sheet and plate			B209-67	H38.2-1967
	Armor plate		MIL-A-46063B-3		
	Extruded armor		MIL-A-46083-1		
	Forged armor		MIL-A-45225B		
7075	Sheet and plate	QQ-A-250/12d		B209-67	H38.2-1967
	Wire, rod, and bar; rolled or cold-finished	QQ-A-225/9c		B211-67	H38.4-1967
	Rod, bar, shapes, and tube; extruded	QQ-A-200/11c		B221-67	H38.5-1967
	Tube; extruded, seamless	QQ-A-200/11c		B241-67	H38.7-1967
	Tube; drawn			B210-67	H38.3-1967
	Forgings and forging stock	QQ-A-367g	MIL-A-22771B-1	B247-67	H38.8-1967
	Impact extrusions		MIL-A-12545B		
	Rivet wire	QQ-A-430a		B316-67	H38.12-1967
Alclad 7075	Sheet and plate	QQ-A-250/13d		B209-67	H38.2-1967
	Tapered sheet and plate			B209-67	H38.2-1967
Alclad one side 7075	Sheet and plate	QQ-A-250/18d		B209-67	H38.2-1967
7079	Sheet and plate	QQ-A-250/17d		B209-67	H38.2-1967
	Forgings and forging stock	QQ-A-367g	MIL-A-22771B-1	B247-67	H38.8-1967
	Rod, bar, shapes, and tube; extruded	QQ-A-200/12c		B221-67	H38.5-1967
	Tube; extruded, seamless	QQ-A-200/12c		B241-67	H38.7-1967
Alclad 7079	Sheet		MIL-A-8923	B209-67	H38.2-1967
7178	Sheet and plate	QQ-A-250/14d		B209-67	H38.2-1967
	Rod, bar, shapes, and tube; extruded	QQ-A-200/13a		B221-67	H38.5-1967
	Tube; extruded, seamless	QQ-A-200/13a		B241-67	H38.7-1967
Alclad 7178	Sheet and plate	QQ-A-250/15d		B209-67	H38.2-1967

the yield strength. The greater the elongation and the greater the difference between ultimate and yield, the better the formability.

It will be noted from Table 11 that heat-treated materials in the solution-treated condition or solution-treated and aged condition are less workable than –0 material. With such alloys the procedure is to form in the annealed condition, and then solution-treat after forming. Subsequent aging to –T6 condition can be carried out if desired. However, solution treatment which requires fast cooling in water after heating to 835 to 1000°F can cause significant distortion on complicated parts or on long slim parts. One remedy which is often used is to solution-treat the raw material just before forming. At this time the material is soft and formable. The forming operation will also tend to straighten any distortion resulting from the solution-treating operation. Subsequent low-temperature aging to –T6 condition can be accomplished with negligible distortion.

TABLE 8 Typical Tensile Properties of Aluminum Alloys at Various Temperatures*,1

Alloy and temper	Temperature, °F	Tensile strength, ksi Ultimate	Yield†	Elongation in 2 in., %	Alloy and temper	Temperature, °F	Tensile strength, ksi Ultimate	Yield†	Elongation in 2 in., %
1100-0	-320	25	6	50	2017-T4,	-320	80	53	28
	-112	15	5.5	43	-T451	-112	65	42	24
	-18	14	5	40		-18	64	41	23
	75	13	5	40		75	62	40	22
	212	10	4.6	45		212	57	39	18
	300	8	4.2	55		300	40	30	15
	400	6	3.5	65		400	16	13	35
	500	4	2.6	75		500	9	7.5	45
	600	2.9	2	80		600	6	5	65
	700	2.1	1.6	85		700	4.3	3.5	70
					2024-T3	-320	85	62	18
1100-H14	-320	30	20	45	(sheet)	-112	73	52	17
	-112	20	18	24		-18	72	51	17
	-18	19	17	20		75	70	50	17
	75	18	17	20		212	66	48	16
	212	16	15	20		300	55	45	11
	300	14	12	23		400	27	20	23
	400	10	7.5	26		500	11	9	55
	500	4	2.6	75		600	7.5	6	75
	600	2.9	2	80		700	5	4	100
	700	2.1	1.6	85	2024-T4,	-320	84	61	19
1100-H18	-320	34	26	30	-T351	-112	71	49	19
	-112	26	23	16	(plate)	-18	69	47	19
	-18	25	23	15		75	68	47	19
	75	24	22	15		212	63	45	19
	212	21	19	15		300	45	36	17
	300	18	14	20		400	26	19	27
	400	6	3.5	65		500	11	9	55
	500	4	2.6	75		600	7.5	6	75
	600	2.9	2	80		700	5	4	100
	700	2.1	1.6	85	2024-T6,	-320	84	68	11
2011-T3	75	55	43	15	-T651	-112	72	59	10
	212	47	34	16		-18	70	58	10
	300	28	19	25		75	69	57	10
	400	16	11	35		212	65	54	10
	500	6.5	3.8	45		300	45	36	17
	600	3.1	1.8	90		400	26	19	27
	700	2.3	1.4	125		500	11	9	55
						600	7.5	6	75
						700	5	4	100
2014-T6,	-320	84	72	14	2024-T81,	-320	85	78	8
-T651	-112	74	65	13	-T851	-112	74	69	7
	-18	72	62	13		-18	73	68	7
	75	70	60	13		75	70	65	7
	212	63	57	15		212	66	62	8
	300	40	35	20		300	55	49	11
	400	16	13	38		400	27	20	23
	500	9.5	7.5	52		500	11	9	55
	600	6.5	5	65		600	7.5	6	75
	700	4.3	3.5	72		700	5	4	100

TABLE 8 Typical Tensile Properties of Aluminum Alloys at Various Temperatures*,1 (Continued)

Alloy and temper	Temperature, °F	Tensile strength, ksi Ultimate	Tensile strength, ksi Yield†	Elongation in 2 in., %	Alloy and temper	Temperature, °F	Tensile strength, ksi Ultimate	Tensile strength, ksi Yield†	Elongation in 2 in., %
2024-T86	-320	92	85	5	3003-0	-320	33	8.5	46
	-112	81	77	5		-112	20	7	42
	-18	78	74	5		-18	17	6.5	41
	75	75	71	5		75	16	6	40
	212	70	67	6		212	13	5.5	43
	300	54	48	11		300	11	5	47
	400	21	17	28		400	8.5	4.3	60
	500	11	9	55		500	6	3.4	65
	600	7.5	6	75		600	4	2.4	70
	700	5	4	100		700	2.8	1.8	70
2117-T4	-320	56	33	30	3003-H14	-320	35	25	30
	-112	45	25	29		-112	24	22	18
	-18	44	24	28		-18	22	21	16
	75	43	24	27		75	22	21	16
	212	36	21	16		212	21	19	16
	300	30	17	20		300	18	16	16
	400	16	12	35		400	14	9	20
	500	7.5	5.5	55		500	7.5	4	60
	600	4.7	3.3	80		600	4	2.4	70
	700	2.9	2	110		700	2.8	1.8	70
2219-T62	-320	73	49	16	3003-H18	-320	41	33	23
	-112	63	44	13		-112	32	29	11
	-18	60	42	12		-18	30	28	10
	75	58	40	12		75	29	27	10
	212	54	37	14		212	26	21	10
	300	45	33	17		300	23	16	11
	400	34	25	20		400	14	9	18
	500	27	20	21		500	7.5	4	60
	600	10	8	40		600	4	2.4	70
	700	4.4	3.7	75		700	2.8	1.8	70
2219-T81, -T851	-320	83	61	15	3004-0	-320	42	13	38
	-112	71	54	13		-112	28	11	30
	-18	69	52	12		-18	26	10	26
	75	66	50	12		75	26	10	25
	212	60	47	15		212	26	10	25
	300	49	40	17		300	22	10	35
	400	36	29	20		400	14	9.5	55
	500	29	23	21		500	10	7.5	70
	600	7	6	55		600	7.5	5	80
	700	4.4	3.7	75		700	5	3	90
2618-T61	-320	78	61	12	3004-H34	-320	52	34	26
	-112	67	55	11		-112	38	30	16
	-18	64	54	10		-18	36	29	13
	75	64	54	10		75	35	29	12
	212	62	54	10		212	34	29	13
	300	50	44	14		300	28	25	22
	400	32	26	24		400	21	15	35
	500	13	9	50		500	14	7.5	55
	600	7.5	4.5	80		600	7.5	5	80
	700	5	3.5	120		700	5	3	90

TABLE 8 Typical Tensile Properties of Aluminum Alloys at Various Temperatures[*,1] **(Continued)**

Alloy and temper	Temperature, °F	Tensile strength, ksi Ultimate	Yield[†]	Elongation in 2 in., %	Alloy and temper	Temperature, °F	Tensile strength, ksi Ultimate	Yield[†]	Elongation in 2 in., %
3004–H38	−320	58	43	20	6063–T1	−320	34	16	44
	−112	44	38	10		−112	26	15	36
	− 18	42	36	7		− 18	24	14	34
	75	41	36	6		75	22	13	33
	212	40	36	7		212	22	14	18
	300	31	27	15		300	21	15	20
	400	22	15	30		400	9	6.5	40
	500	12	7.5	50		500	4.5	3.5	75
	600	7.5	5	80		600	3.2	2.5	80
	700	5	3	90		700	2.3	2	105
4032–T6	−320	66	48	11	6063–T5	−320	37	24	28
	−112	58	46	10		−112	29	22	24
	− 18	56	46	9		− 18	28	22	23
	75	55	46	9		75	27	21	22
	212	50	44	9		212	24	20	18
	300	37	33	9		300	20	18	20
	400	13	9	30		400	9	6.5	40
	500	8	5.5	50		500	4.5	3.5	75
	600	5	3.2	70		600	3.2	2.5	80
	700	3.4	2	90		700	2.3	2	105
5050–0	−320	37	10		6063–T6	−320	47	36	24
	−112	22	8.5			−112	38	33	20
	− 18	21	8			− 18	36	32	19
	75	21	8			75	35	31	18
	212	21	8			212	31	28	15
	300	19	8			300	21	20	20
	400	14	7.5			400	9	6.5	40
	500	9	6			500	4.5	3.5	75
	600	6	4.2			600	3.3	2.5	80
	700	3.9	2.6			700	2.3	2	105
5050–H34	−320	44	30		6101–T6	−320	43	33	24
	−112	30	25			−112	36	30	20
	− 18	28	24			− 18	34	29	19
	75	28	24			75	32	28	19
	212	28	24			212	28	25	20
	300	25	22			300	21	19	20
	400	14	7.5			400	10	7	40
	500	9	6			500	4.8	3.3	80
	600	6	4.2			600	3	2.3	100
	700	3.9	2.6			700	2.5	1.8	105
6061–T6, –T651	−320	60	47	22	6151–T6	−320	57	50	20
	−112	49	42	18		−112	50	46	17
	− 18	47	41	17		− 18	49	45	17
	75	45	40	17		75	48	43	17
	212	42	38	18		212	43	40	17
	300	34	31	20		300	28	27	20
	400	19	15	28		400	14	12	30
	500	7.5	5	60		500	6.5	5	50
	600	4.6	2.7	85		600	5	3.9	43
	700	3	1.8	95		700	4	3.2	35

TABLE 8 Typical Tensile Properties of Aluminum Alloys at Various Temperatures*,1 (Continued)

Alloy and temper	Temperature, °F	Ultimate Tensile strength, ksi	Yield†	Elongation in 2 in., %	Alloy and temper	Temperature, °F	Ultimate Tensile strength, ksi	Yield†	Elongation in 2 in., %
6262–T651	−320	60	47	22	7075–T73, –T7351	−320	92	72	14
	−112	49	42	18		−112	79	67	14
	−18	47	41	17		−18	76	65	13
	75	45	40	17		75	73	63	13
	212	42	38	18		212	63	58	15
	300	34	31	20		300	31	27	30
6262–T9	−320	74	67	14		400	16	13	55
	−112	62	58	10		500	11	9	65
	−18	60	56	10		600	8	6.5	70
	75	58	55	10		700	6	4.6	70
	212	53	52	10	7079–T6, –T651	−320	92	80	12
	300	38	37	14		−112	82	70	14
	400	15	13	34		−18	79	68	14
	500	8.5	6	48		75	78	68	14
	600	4.6	2.7	85		212	67	60	18
	700	3	1.8	95		300	33	28	37
7039–T61, –T6151	−320	83	59	15		400	16	13	60
	−112	65	53	14		500	11	8.5	100
	−18	63	52	14		600	7.5	6	175
	75	60	50	14		700	5.5	4.3	175
	212	53	46	16					
	300	45	39	20					
	400	32	29	29					
7075–T6, –T651	−320	102	92	9					
	−112	90	79	11					
	−18	86	75	11					
	75	83	73	11					
	212	70	65	14					
	300	31	27	30					
	400	16	13	55					
	500	11	9	65					
	600	8	6.5	70					
	700	6	4.6	70					

* Lowest strengths during 10,000 hr of exposure at testing temperature under no load; stress applied at 5,000 psi/min to yield strength and then at strain rate of 0.05 in./(in.)-(min) to failure. Under some conditions of temperature and time, the application of heat will adversely affect certain other properties of some alloys.

† Offset equals 0.2%.

Table 12 shows approximate bend radii for various alloys, temper, and thickness for sheet.

Machinability The major machinability problem of aluminum is its tendency to gall and smear owing to its relatively soft and ductile nature. For this reason, it is preferred to machine aluminum alloys in the heat-treated or cold-worked condition.

Brazeability of aluminum The brazing alloys used for aluminum have melting points above the solidus of many of the high-strength aluminum alloys. Such alloys as 2014, 2024, and other 2000 series alloys are not brazed for this reason. The same situation exists with most of the 7000 series alloys. However, 7005 alloy,

TABLE 9 Typical Thermal and Electrical Properties of Aluminum Alloys[1]

Alloy	Temper	Thermal conductivity at 25°C (77°F) Cgs units*	Thermal conductivity at 25°C (77°F) English units†	Electrical conductivity at 20°C (68°F), % of International Annealed Copper Standard — Equal volume	Electrical conductivity at 20°C (68°F), % of International Annealed Copper Standard — Equal weight	Electrical resistivity at 20°C (68°F) µohm/cm	Electrical resistivity at 20°C (68°F) ohms/cir mil-ft
EC‡	All	0.56	1625	62	204	2.8	17
1060	0	0.56	1625	62	204	2.8	17
	H18	0.53	1540	61	201	2.8	17
1100	0	0.53	1540	59	194	2.9	17
	H18	0.52	1510	57	187	3.0	18
2011	T3	0.36	1040	39	123	4.4	27
	T8	0.41	1190	45	142	3.8	23
2014	0	0.46	1340	50	159	3.4	21
	T4	0.32	930	34	108	5.1	31
	T6	0.37	1070	40	127	4.3	26
2017	0	0.46	1340	50	159	3.4	21
	T4	0.32	930	34	108	5.1	31
2018	T61	0.37	1070	40	127	4.3	26
2024	0	0.46	1340	50	160	3.4	21
	T3, T4, T36	0.29	840	30	96	5.7	35
	T6, T81, T86	0.36	1040	38	122	4.5	27
2025	T6	0.37	1070	40	130	4.3	26
2117	T4	0.37	1070	40	130	4.3	26
2218	T72	0.37	1070	40	126	4.3	26
2219	0	0.41	1190	44	138	3.9	24
	T31, T37	0.27	780	28	88	6.2	37
	T62, T81, T87	0.30	870	30	94	5.7	35
3003	0	0.46	1340	50	163	3.4	21
	H12	0.39	1130	42	137	4.1	25
	H14	0.38	1100	41	134	4.2	25
	H18	0.37	1070	40	130	4.3	26
3004	All	0.39	1130	42	137	4.1	25
4032	0	0.37	1070	40	132	4.3	26
	T6	0.33	960	35	116	4.9	30
4043	0	0.39	1130	42	137	4.1	25
5005	All	0.48	1390	52	172	3.3	20
5050	All	0.46	1340	50	165	3.4	21
5052	All	0.33	960	35	116	4.9	30
5056	0	0.28	810	29	98	5.9	36
	H38	0.26	750	27	91	6.4	38
5083	0	0.28	810	29	98	5.9	36
5086	All	0.30	870	31	104	5.5	34
5154	All	0.30	870	32	107	5.3	32
5252	All	0.33	960	35	116	4.9	30

TABLE 9 Typical Thermal and Electrical Properties of Aluminum Alloys[1] (Continued)

Alloy	Temper	Thermal conductivity at 25°C (77°F) Cgs units*	Thermal conductivity at 25°C (77°F) English units†	Electrical conductivity at 20°C (68°F), % of International Annealed Copper Standard Equal volume	Electrical conductivity at 20°C (68°F), % of International Annealed Copper Standard Equal weight	Electrical resistivity at 20°C (68°F) μohm/cm	Electrical resistivity at 20°C (68°F) Ohms/cir mil–ft
5454	0	0.32	930	34	113	5.1	31
	H38	0.32	930	34	113	5.1	31
5456	0	0.28	810	29	98	5.9	36
5557	All	0.45	1310	49	159	3.5	21
6053	0	0.41	1190	45	148	3.8	23
	T4	0.37	1070	40	132	4.3	26
	T6	0.39	1130	42	139	4.1	25
6061	0	0.43	1250	47	155	3.7	22
	T4	0.37	1070	40	132	4.3	26
	T6	0.40	1160	43	142	4.0	24
6063	0	0.52	1510	58	191	3.0	18
	T1‡	0.46	1340	50	165	3.4	21
	T5	0.50	1450	55	181	3.1	19
	T6, T83	0.48	1390	53	175	3.3	20
6066	0	0.37	1070	40	132	4.3	26
	T6	0.35	1010	37	122	4.7	28
6070	T6	0.41	1190	44	145	3.9	24
6101	T6	0.52	1510	57	188	3.0	18
	T61	0.53	1540	59	194	2.9	18
	T63§	0.52	1510	58	191	3.0	18
	T64	0.54	1570	60	198	2.9	17
	T65	0.52	1510	58	191	3.0	18
6151	0	0.49	1420	54	178	3.2	19
	T4	0.39	1130	42	138	4.1	25
	T6	0.41	1190	45	148	3.8	23
6262	T9	0.41	1190	44	145	3.9	24
6463	T1¶	0.46	1340	50	165	3.4	21
	T5	0.50	1450	55	181	3.1	19
	T6	0.48	1390	53	175	3.3	20
7001	T6	0.29	840	31	104	5.5	34
7039	T61	....	....	35	116	4.9	30
7072	0	0.53	1540	59	193	2.9	18
7075	T6	0.31	900	33	105	5.2	31
7079	T6	0.30	870	32	104	5.4	32
7178	T6	0.30	870	31	98	5.6	33

* Cgs units = cal/(sec)(cm²)(°C/cm).
† English units = Btu/(hr)(ft²)(°F/in.).
‡ Electric conductor grade: 99.45% minimum aluminum.
§ Formerly designated T42.
¶ Formerly T62.

TABLE 10 Typical Physical Properties of Aluminum Alloys: Density, Thermal Expansion, and Melting Point

Alloy	Density lb/in.³	Density lb/ft³	Specific gravity	Average coefficient* of thermal expansion per deg F (68–212°F)	Melting point (approx.), °F
EC†	0.098	169	2.70	13.2	1195–1215
1060	0.098	169	2.70	13.1	1195–1215
1100	0.098	169	2.71	13.1	1190–1215
2011	0.102	176	2.82	12.7	995–1190
2014	0.101	175	2.80	12.8	950–1180
2017	0.101	174	2.79	13.1	955–1185
2018	0.101	175	2.80	12.4	948–1180
2024	0.100	173	2.77	12.9	935–1180
2025	0.101	175	2.79	12.6	970–1185
2117	0.099	171	2.74	13.2	950–1200
2218	0.102	176	2.81	12.4	940–1175
2219	0.103	178	2.84	12.4	1010–1190
3003	0.099	170	2.73	12.9	1190–1210
3004	0.098	170	2.72	13.3	1165–1205
4032	0.097	167	2.69	10.8	990–1060
4043	0.097	167	2.69	....	1065–1170
5005	0.098	169	2.70	13.2	1170–1205
5050	0.097	168	2.69	13.2	1160–1205
5052	0.097	167	2.68	13.2	1100–1200
5056	0.095	165	2.64	13.4	1055–1180
5083	0.096	166	2.66	13.2	1060–1180
5086	0.096	166	2.66	13.2	1084–1184
5154	0.096	166	2.66	13.3	1100–1190
5252	0.097	167	2.68	13.2	1100–1200
5257	0.098	169	2.70	13.1	....
5454	0.097	167	2.68	13.1	1115–1195
5456	0.096	166	2.65	13.3	1060–1180
5557	0.098	169	2.70	13.1	1180–1215
5657	0.098	169	2.70	13.1	1180–1215
6053	0.097	167	2.69	12.8	1070–1205
6061	0.098	169	2.70	13.1	1080–1200
6063	0.098	169	2.70	13.0	1140–1205
6066	0.098	169	2.70	12.9	1050–1200
6070	0.098	169	2.71	....	1070–1200
6101	0.098	169	2.70	13.0	1140–1205
6151	0.098	169	2.70	12.9	1025–1200
6262	0.098	170	2.72	13.0	1100–1205
6463	0.098	169	2.70	13.0	1140–1210
6951	0.098	169	2.70	13.0	1140–1210
7001	0.102	176	2.81	13.0	890–1160
7039	0.099	171	2.74	....	....
7072	0.098	169	2.72	13.1	1195–1215
7075	0.101	175	2.80	13.1	890–1180
7079	0.099	171	2.74	13.1	900–1180
7178	0.102	175	2.81	13.0	890–1165

* Coefficient to be multiplied by 10^{-6}; Example: 12.2×10^{-6} = 0.0000122.

† Electric conductor grades: 99.45% minimum aluminum.

TABLE 11 Typical Fabrication Characteristics of Aluminum Alloys[*,1]

Alloy and temper	Resistance to corrosion	Workability (cold)	Machinability	Brazeability	Weldability Gas	Weldability Arc	Weldability Resistance (spot and seam)	Forgeability	Alloy and temper	Resistance to corrosion	Workability (cold)	Machinability	Brazeability	Weldability Gas	Weldability Arc	Weldability Resistance (spot and seam)	Forgeability
EC-O	A	A	D	A	A	A	B	—	5050-0	A	A	D	B	A	A	B	—
-H12	A	A	D	A	A	A	A	—	-H32	A	B	D	B	A	A	A	—
-H14	A	A	C	A	A	A	A	—	-H34	A	B	C	B	A	A	A	—
-H16	A	B	C	A	A	A	A	—	-H36	A	C	C	B	A	A	A	—
-H19	A	C	C	A	A	A	A	—	-H38	A	C	C	B	A	A	A	—
1100-O	A	A	D	A	A	A	B	A	5052-0	A	A	D	C	A	A	B	—
-H12	A	A	D	A	A	A	A	A	-H32	A	B	C	C	A	A	A	—
-H14	A	A	C	A	A	A	A	A	-H34	A	B	C	C	A	A	A	—
-H16	A	B	C	A	A	A	A	A	-H36	A	C	C	C	A	A	A	—
-H18	A	C	C	A	A	A	A	A	-H38	A	C	C	C	A	A	A	—
2011-T3	C	C	A	D	D	D	B	—	5056-0	A	A	D	D	C	A	B	—
-T8	C	D	A	D	D	D	B	—	-H38	C	C	C	D	C	A	A	—
2014-T4	C	C	B	D	D	B	B	C	5086-0	A	A	D	D	C	A	B	—
-T6	C	D	B	D	D	B	B	C	-H32	A	B	D	D	C	A	A	—
2017-T4	C	C	B	D	D	B	B	—	-H34	B	B	C	D	C	A	A	—
2018-T61	C	—	B	D	D	B	B	C	-H36	B	C	C	D	C	A	A	—
2024-T3	C	C	B	D	D	B	B	—	-H38	B	C	C	D	C	A	A	—
-T4	C	C	B	D	D	B	B	—	5154-0	A	A	D	D	C	A	B	—
-T36	C	D	B	D	D	B	B	—	-H32	A	B	D	D	C	A	A	—
2117-T4	C	B	C	D	D	B	B	—	-H34	A	B	C	D	C	A	A	—
2218-T72	C	—	B	D	—	B	—	D	-H36	A	C	C	D	C	A	A	—
									-H38	A	C	C	D	C	A	A	—
									-H112	A	B	D	D	C	A	A	—
3003-O	A	A	D	A	A	B	B	A									
-H12	A	A	D	A	A	A	A	A	6061-0	A	A	D	A	A	A	B	—
-H14	A	B	C	A	A	A	A	A	-T4	A	C	C	A	A	A	A	—
-H16	A	C	C	A	A	A	A	A	-T6	A	C	C	A	A	A	A	—
-H18	A	C	C	A	A	A	A	A	6063-0	A	A	D	A	A	A	B	—
3004-O	A	A	D	B	B	A	B	—	-T1†	A	B	C	A	A	A	A	—
-H32	A	B	D	B	B	A	A	—	-T4	A	B	C	A	A	A	A	—
-H34	A	B	C	B	B	A	A	—	-T5	A	B	C	A	A	A	A	—
-H36	A	C	C	B	B	A	A	—	-T6	A	C	C	A	A	A	A	—
-H38	A	C	C	B	B	A	A	—	-T83	A	C	C	A	A	A	A	—
5005-O	A	A	D	B	A	A	B	—	-T832	A	C	C	A	A	A	A	—
-H12	A	A	D	B	A	A	A	—	-T831	A	C	C	A	A	A	A	—
-H14	A	B	C	B	A	A	A	—	6066-0	B	B	D	A	A	A	B	—
-H16	A	C	C	B	A	A	A	—	-T4	B	C	B	A	A	A	A	—
-H18	A	C	C	B	A	A	A	—	-T6	B	C	B	A	A	A	A	—
-H32	A	A	D	B	A	A	A	—	7001-T6	C	D	B	D	D	D	B	D
-H34	A	B	C	B	A	A	A	—	7075-T6	C	D	B	D	D	D	B	—
-H36	A	C	C	B	A	A	A	—									
-H38	A	C	C	B	A	A	A	—									

* Resistance to corrosion, workability (cold), machinability and forgeability ratings A, B, C, and D are relative ratings in decreasing order of merit. Weldability and brazeability ratings A, B, C, and D are relative ratings defined as follows:
  A. Generally weldable by all commercial procedures and methods.
  B. Weldable with special technique or on specific applications which justify preliminary trials or testing to develop welding procedure and weld performance.
  C. Limited weldability because of crack sensitivity or loss in resistance to corrosion, and all mechanical properties.
  D. No commonly used welding methods have so far been developed.
† Formerly designated T42.

TABLE 12 Approximate Bend Radii* for 90-degree Cold Bend in Aluminum Alloys[1]

Alloy	Temper	Radii for various thicknesses expressed in terms of thickness *t*							
		1/64 in.	1/32 in.	1/16 in.	⅛ in.	3/16 in.	¼ in.	⅜ in.	½ in.
1100	–O	0	0	0	0	0	0	0	1t–2t
	–H12	0	0	0	0	0–1t	0–1t	0–1t	1t–3t
	–H14	0	0	0	0	0–1t	0–1t	0–1t	2t–3t
	–H16	0	0	0–1t	½t–1½t	1t–2t	1½t–3t	2½t–3½t	3t–4t
	–H18	0–1t	½t–1½t	1t–2t	1½t–3t	2t–4t	2t–4t	3t–5t	3t–6t
Alclad	–O	0	0	0	0	0–1t	0–1t	1½t–3t	3t–5t
2014	–T3	1t–2t	1½t–3t	2t–4t	3t–5t	4t–6t	4t–6t	5t–7t	5½t–8t
	–T4	1t–2t	1½t–3t	2t–4t	3t–5t	4t–6t	4t–6t	5t–7t	5½t–8t
	–T6	2t–4t	3t–5t	3t–5t	4t–6t	5t–7t	6t–10t	7t–10t	8t–11t
2024	–O†	0	0	0	0	0–1t	0–1t	1½t–3t	3t–5t
	–T3†‡	1½t–3t	2t–4t	3t–5t	4t–6t	4t–6t	5t–7t	6t–8t	6t–9t
	–T36†	2t–4t	3t–5t	4t–6t	5t–7t	5t–7t	6t–10t	7t–10t	8t–11t
	–T4†	1½t–3t	2t–4t	3t–5t	4t–6t	4t–6t	5t–7t	6t–8t	6t–9t
	–T81	3½t–5t	4½t–6t	5t–7t	6½t–8t	7t–9t	8t–10t	9t–11t	9t–12t
	–T86	4t–5½t	5t–7t	6t–8t	7t–10t	8t–11t	9t–11t	10t–13t	10t–13t
3003	–O	0	0	0	0	0	0	0	1t–2t
	–H12	0	0	0	0	0–1t	0–1t	0–1t	1t–3t
	–H14	0	0	0	0–1t	0–1t	½t–1½t	1t–2½t	1½t–3t
	–H16	0–1t	0–1t	½t–1½t	1t–2t	1½t–3t	2t–4t	2½t–4t	3t–5t
	–H18	½t–1½t	1t–2t	1½t–3t	2t–4t	3t–5t	4t–6t	4t–7t	5t–8t
3004	–O	0	0	0	0	0–1t	0–1t	½t–1½t	1t–2t
	–H32	0	0	0	0–1t	0–1t	½t–1½t	1t–2t	1½t–2½t
	–H34	0	0	0–1t	½t–1½	1t–2t	1½t–3t	2t–3t	2½t–3½t
	–H36	0–1t	½t–1½t	1t–2t	1½t–3t	2t–4t	2t–4t	2½t–5t	3t–5½t
	–H38	½t–1½t	1t–2t	1½t–3t	2t–4t	3t–5t	4t–6t	4t–7t	5t–8t
5005	–O	0	0	0	0	0	0	0	1t–2t
	–H12	0	0	0	0	0–1t	0–1t	0–1t	1t–3t
	–H14	0	0	0	0–1t	0–1t	½t–1½t	1t–2½t	1½t–3t
	–H16	0–1t	0–1t	½t–1½t	1t–2t	1½t–3t	2t–4t	2½t–4t	3t–5t
	–H18	½t–1½t	1t–2t	1½t–3t	2t–4t	3t–5t	4t–6t	4t–7t	5t–8t
	–H32	0	0	0	0	0–1t	0–1t	0–1t	1t–3t
	–H34	0	0	0	0–1t	0–1t	½t–1½t	1t–2½t	1½t–3t
	–H36	0–1t	0–1t	½t–1½t	1t–2t	1½t–3t	2t–4t	2½t–4t	3t–5t
	–H38	½t–1½t	1t–2t	1½t–3t	2t–4t	3t–5t	4t–6t	4t–7t	5t–8t
5050	–O	0	0	0	0	0	0	—	—
	–H32	0	0	0	0	0–1t	½t–1½t	2t–3½t	2½t–4t
	–H34	0	0	0	0–1t	½t–1½t	1t–2t	2t–4t	3t–4t
	–H36	0–1t	0–1t	½t–1½t	1t–2t	1½t–3t	2t–4t	2½t–4t	3t–5t
	–H38	½t–1½t	1t–2t	1½t–3t	2t–4t	3t–5t	4t–6t	4t–7t	5t–8t
5052	–O	0	0	0	0	0–1t	0–1t	½t–1½t	1t–2t
	–H32	0	0	0	0–1t	0–1t	½t–1½t	1t–2t	1½t–2½t
	–H34	0	0	0–1t	½t–1½t	1t–2t	1½t–3t	2t–3t	2½t–3½t
	–H36	0–1t	½t–1½t	1t–2t	1½t–3t	2t–4t	2t–4t	2½t–5t	3t–5½t
	–H38	½t–1½t	1t–2t	1½t–3t	2t–4t	3t–5t	4t–6t	4t–7t	5t–8t
5083	–O	—	—	0–½t	0–1t	0–1t	½t–1½t	1½t–2t	1½t–2½t
5086	–O	0	0	0	0–1t	0–1t	½t–1t	½t–1t	½t–1½t
	–H32	0–½t	0–1t	½t–1½t	1t–2t	1½t–2t	1½t–2½t	2t–2½t	2½t–3t
	–H34	0–1t	½t–1½t	1t–1½t	1½t–2½t	2t–3t	2t–3t	2½t–3½t	3t–4t
	–H36	—	—	—	2t–3½t	2½t–4t	3t–4½t	3t–5t	3½t–5½t
	–H112	0–½t	0–½t	½t–1t	½t–1½t	1t–1½t	1t–2t	1t–2t	1½t–2½t
5154	–O	0	0	0	0–1t	0–1t	½t–1½t	½t–1½t	1t–2t
	–H32	0	0	0–1t	½t–1½t	1t–2t	1½t–3t	2t–4t	2½t–5t
	–H34	0–1t	0–1t	½t–1½t	1t–2t	1½t–3t	2t–4t	2½t–4½t	3t–5t
	–H36	0–1t	½t–1½t	1t–2t	1½t–3t	2t–4t	2t–4t	2½t–5t	3t–6t
	–H38	1t–2t	1½t–3t	2t–4t	3t–5t	4t–6t	4t–6t	5t–8t	5t–8t
	–H112	—	—	—	—	—	1½t–3t	2½t–4t	3t–5t

TABLE 12 Approximate Bend Radii* for 90-degree Cold Bend in Aluminum Alloys[1] (Continued)

Alloy	Temper	1/64 in.	1/32 in.	1/16 in.	1/8 in.	3/16 in.	1/4 in.	3/8 in.	1/2 in.
5456	–O	—	—	—	0–1t	½t–1t	½t–1t	½t–1½t	½t–2t
	–H321	—	—	—	2t–3t	3t–4t	3t–4t	3t–4t	3t–4t
	–H323	—	—	1t–2t	1½t–3t	1½t–3½t	2t–4t	—	—
	–H343	—	—	1t–2t	1½t–4t	2t–4t	2½t–4½t	—	—
5457	–O	0	0	0	0	0	0	0	1t–2t
	–H38	½t–1½t	1t–2t	1½t–3t	2t–4t	3t–5t	4t–6t	4t–7t	5t–8t
6061	–O	0	0	0	0	0–1t	0–1t	½t–2t	1t–2½t
	–T4†	0–1t	0–1t	½t–1½t	1t–2t	1½t–3t	2t–4t	2½t–4t	3t–5t
	–T6†	0–1t	½t–1½t	1t–2t	1½t–3t	2t–4t	3t–4t	3½t–5½t	4t–6t
7075	–O	0	0	0–1t	½t–1½t	1t–2t	1½t–3t	2½t–4t	3t–5t
	–T6†	2t–4t	3t–5t	4t–6t	5t–7t	5t–7t	6t–10t	7t–11t	7t–12t
7079	–O	0	0	0–1t	½t–1½t	1t–2t	1½t–3t	2½t–4t	3t–5t
	–T6†	2t–4t	3t–5t	4t–6t	5t–7t	5t–7t	6t–10t	7t–11t	7t–12t
7178	–O	0	0	0–1t	½t–1½t	1t–2t	1½t–3t	2½t–4t	3t–5t
	–T6†	2t–4t	3t–5t	4t–6t	5t–7t	5t–7t	6t–10t	7t–11t	7t–12t

* Minimum permissible radius over which sheet or plate may be bent varies with nature of forming operation, type of forming equipment, and design and condition of tools. Minimum working radius for a given material or hardest alloy and temper for a given radius can be ascertained only by actual trial under contemplated conditions of fabrication.

† Alclad sheet can be bent over slightly smaller radii than the corresponding tempers of the uncoated alloy.

‡ Immediately after quenching, this alloy can be formed over appreciably smaller radii.

a relatively high strength alloy which can be brazed, has become available, and other similar alloys likely will follow. At present the 1000, 3000, and 6000 series account for the bulk of brazing done with aluminum. The 5000 series alloys can be brazed with the torch method, but limited success has been experienced in brazing these alloys by furnace or dip methods.

Temperature control is a must when brazing aluminum because of the small temperature difference between the melting point of the brazing alloy and the beginning of melting in the base alloy. This makes torch brazing a highly skilled art, while requiring no greater than ±10°F variation during furnace or dip brazing.

Dip Brazing of Aluminum. Dip brazing, where the heat source is a pot of molten brazing flux held at the brazing temperature, has probably been the most suitable brazing method for aluminum. This method provides (1) a means of uniform heating, (2) a copious amount of flux, (3) protection from the atmosphere, and (4) support for the metal by flotation. This support is important because the material is near its melting and will support little load in this condition. The biggest disadvantage of dip brazing is that the flux completely envelopes the part during brazing. The flux is corrosive and must be completely removed before parts are put into service. A part with difficult-to-clean configurations, such as closed passages, deep holes, and crevices, must be carefully processed in order to remove all flux.

Furnace Brazing of Aluminum. Use of a furnace as a heat source is satisfactory where adequate temperature control is maintained. Flux must be painted on the joint area in amounts sufficient to protect the joint and provide fluxing action. The amount of flux used is not as great as in dip brazing, but cleaning requirements are equally important.

Aluminum Brazing Alloys. Brazing alloys are basically aluminum-silicon alloys with minor additions of other alloys. They are available in wire, foil, and as a cladding on a brazeable base metal (see Table 5). A powdered brazing alloy mixed with flux and other ingredients is also available for use in dip brazing. This product can be mixed with water to form a paste which can be painted onto joints. This product is very widely used because of its ease of application.

Aluminum Brazing Fluxes. Fluxes used for aluminum brazing are mixtures of fluorides and chlorides which if left on the work will cause serious corrosion in the presence of moisture. All flux residues must be removed. Procedures for flux removal should be developed, based on recommendations of the aluminum suppliers. These procedures typically consist of soaking in hot water, cold-water rinses, and acid dips. Steam cleaning or ultrasonic agitation is often used where stubborn residues are present.

The corrosiveness of flux residues has led to investigation of procedure for brazing without flux. Vacuum brazing, at this time, has been developed as a usable process. Since no flux is involved, the flux problem is eliminated. There is also a process utilizing dry-air atmosphere furnace brazing which reduces the amount of flux required. This results in reduced cleaning requirements.

Effect of Brazing Temperature on Mechanical Properties. The brazing temperature will anneal alloys which are brazed. The 1000 and 3000 series alloys cannot be rehardened after brazing, but the 6000 series can be solution-treated and aged to −T6 condition if desired. Solution treatment may be performed by quenching parts from just below the brazing temperature. This is accomplished in dip brazing by removing the part from the salt and holding it until the flux solidifies, and then quenching it into water. Aging may follow with a resulting −T6 condition.

Weldability Gas welding can be performed on aluminum with satisfactory results on some alloys (see Table 11). However, a flux must be used, and a wide heat-affected zone is present. Very little gas welding is done because of these two factors.

The *tungsten inert-gas* (TIG) process is probably the most universally used method for welding aluminum alloys. The *metal inert-gas* (MIG) method is also satisfactory for welding aluminum, and is widely used for heavy sheet and for welding plate. Production speed is considerably greater than with TIG welding.

The *electron-beam welding* (EB) process has been developed for aluminum, and may overcome weldability problems of some of the alloys now considered difficult to weld. The narrow weld zone and narrow heat-affected zone given by EB welding hold promise as a means to minimize cracking, objectionable alloy segregation, and other detrimental metallurgical effects which now plague high-strength alloys during welding. Although this process now requires welding in vacuum and equipment cost is high, developments now under way may result in major breakthroughs.

As can be seen from Table 11, most of the 2000 and 7000 series are rated as being poor in weldability. However, procedures have been developed for welding these alloys in highly critical applications. Such welding must be done under very precise control and with attention to detail. Your aluminum supplier should be consulted before you attempt to weld these materials.

Weldable alloys have been developed in both the 2000 and the 7000 series. The most successful alloys have been 2219, 7005, and 7039. Consult your aluminum supplier for availability and procedures.

Strength of Welds. When welds are to be left as the welded condition, the strength across the weld will generally be the annealed strength of the base metal. Thus it is wise to select a base metal with a high strength in the annealed condition where joints are left in the as-welded condition. The 5000 series alloys excel in this characteristic, and thus are widely used for large structures which cannot be heat-treated after welding.

If the part can be heat-treated after welding, heat-treatable base materials can be brought back to the solution-treated and aged condition. Depending upon the filler metal selected, it may be possible to develop the original base-metal strength across the weld. Such alloys as 7005 and 7039 offer the advantage that

TABLE 13 Aluminum and Aluminum-alloy Welding Rods and Bare Electrodes*[2]

Alloys to be joined	43, 355, 356	214, A214, B214, F214	6061, 6062, 6063, 6151	5456	5454	5154, 5254	5086, 5356	5083	5052, 5652	5005, 5050	3004, Clad 3004	1100, 3003, Clad 3003	1060
1060	ER4043	ER4043	ER4043	ER5356	ER4043	ER4043	ER5356	ER5356	ER4043	ER1100	ER4043	ER1100	ER1260
1100, 3003, Clad 3003	ER4043	ER4043	ER4043	ER5356	ER4043	ER4043	ER5356	ER5356	ER4043	ER4043	ER4043	ER1100	
3004, Clad 3004	ER4043	ER4043	ER4043	ER5356	ER5356	ER5356	ER5356	ER5356	ER5356	ER4043	ER4043		
5005, 5050	ER4043	ER4043	ER4043	ER5356	ER4043	ER5356	ER5356	ER5356	ER4043	ER4043			
5052, 5652	ER4043	ER4043	ER4043	ER5356	ER5356	ER5356	ER5356	ER5356	ER5652				
5083	ER5356	ER5356	ER5356	ER5183	ER5356	ER5356	ER5356	ER5183					
5086, 5356	ER5356	ER5356	ER5356	ER5356	ER5356	ER5356	ER5356						
5154, 5254	ER5356	ER5356	ER5356	ER5356	ER5356	ER5254							
5454	ER5356	ER5356	ER5356	ER5556	ER5554								
5456	ER5356	ER5356	ER5356										
6061, 6062, 6063, 6151	ER4043	ER5356	ER4043										
214, A241, B214, F214	ER4043	ER5356											
43, 355, 336	ER4043												

Rods and electrodes

* Recommendations are for general-purpose welding and apply to arc-welding methods. For gas welding, only ER1100, ER1260, and ER4043 filler metals are ordinarily used.

9-35

9-36 Nonferrous Metals

aging alone results in increased strength of weldments. This eliminates the distortion which often results from solution treating.

Table 13 gives recommended filler metals for various alloys and combinations. Figure 9 gives typical values of weld strength. It can be seen that some latitude exists in selection of filler rod for the various alloys. Also, the selection of rod can depend upon the postweld heat treatment, if any, and ductility requirements.

Heat treatment of aluminum The heat-treatable alloys are strengthened by the processes of solution treating and aging. Aging may be carried out at room temperature or at elevation temperature, depending upon the alloy and condition desired.

Solution treating consists of heating the material to a temperature at which the alloying elements are in solid solution in the aluminum. This temperature varies with the alloy from 830 to about 1000°F. For any particular alloy the temperature must be controlled closely within the specified range to assure that complete alloy solution occurs but that melting of alloy phases does not occur. Furnace temperature uniformity is of prime importance. A maximum ±10°F variation

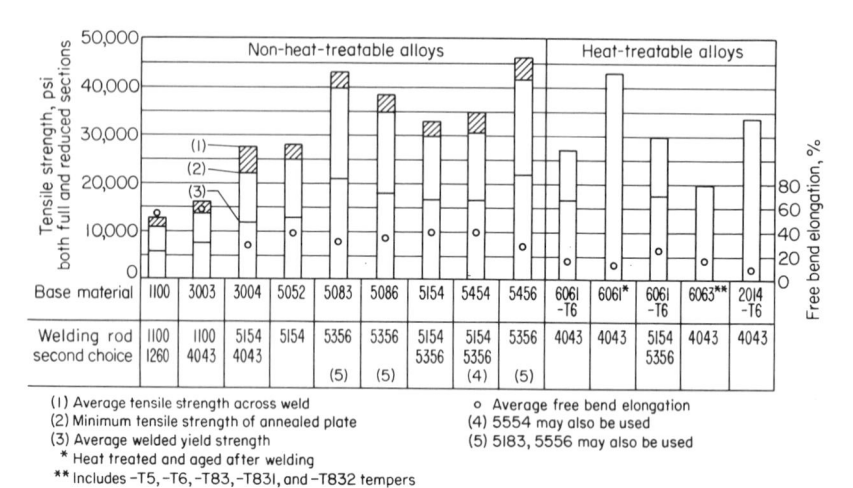

(1) Average tensile strength across weld
(2) Minimum tensile strength of annealed plate
(3) Average welded yield strength
* Heat treated and aged after welding
** Includes −T5, −T6, −T83, −T831, and −T832 tempers
○ Average free bend elongation
(4) 5554 may also be used
(5) 5183, 5556 may also be used

Fig. 9 Strength and ductility of butt welds in aluminum alloys (MIG process).

over the entire furnace is required. After the part has stabilized at the proper temperature, it is held for a suitable time, and then quenched into water. This fast cooling rate holds the alloying elements in a supersaturated solution. The material is in the T4 condition if left in this state.

Aging takes place after solution treating as the alloying elements come out of the solution to cause precipitation hardening. In the case of 2024 aluminum, maximum tensile strength and hardness will come about after aging at room temperature for about 4 days. Alloys such as 6061, 2014, 7075 require elevated-temperature aging to develop maximum strength. The aging temperature usually is about 350°F, with some alloys such as 7075 requiring only 250°F.

Standardized heat treatments are used to obtain the properties shown for the various tempers, as shown earlier in Table 6. Exact temperatures for each alloy and form are available from suppliers' literature or can be found in government specifications such as MIL-H-6088.

The major problem area in aluminum heat treatment is the distortion caused by the quench into water. Variations of quenching procedures have been used to reduce this distortion. Among these procedures are:

1. Quenching into boiling water

2. Quenching to mixtures of water and polyalkalene glycol Ucon A)*
3. Quenching into molten salt at slightly elevated temperature

The quenching into boiling water is an allowable procedure for certain alloys and is mandatory for some castings. The other two procedures have been used with success but to the author's knowledge have not been recognized as standard procedures by government or standardizing agencies.

As mentioned previously, it is sometimes advantageous to solution-treat materials and take advantage of the formability of the material immediately after solution treating to perform drastic forming operations. Material may then be aged to full strength with no appreciable distortion. After solution treating, the heat-treatable alloys will age to higher hardness and consequently less ductibility on standing at room temperature. Figure 10 shows the room-temperature aging characteristics of several alloys. These data can be used to determine the time available in which best formability exists. Lowering the temperature, such as by storing on Dry Ice can slow down the aging process significantly.

Cast Aluminum Alloys

Aluminum alloys can be cast by all available casting methods including sand, die, investment, permanent mold, and plaster mold.

Castings compositions are specifically designed for castings and differ significantly from those of the alloys used for wrought forms such as sheet, box, extrusions, and forgings. As a result, some of the highest electric and thermal conductivity values found in wrought alloys cannot be found in casting alloys. However, aluminum castings exhibit the favorable corrosion resistance and strength-weight ratio found in wrought alloys.

Table 14 gives typical mechanical properties of some of the most common aluminum casting alloys. It should be emphasized that these values are typical and are subject to considerable variation, depending upon casting soundness, section size of the casting, foundry practice, etc.

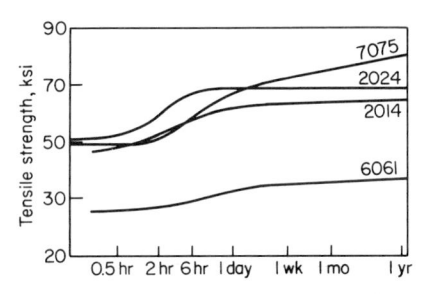

Fig. 10 Aging characteristics of aluminum sheet alloys at room temperature versus time after solution treating.

In design of castings it is important to work with the foundry as early in design as possible, since there is room for wide variation in strength and quality of castings, depending upon the factors mentioned above. Some major questions to be agreed upon between purchaser and vendor are:

1. How are mechanical properties to be determined? Are they to be determined on separately cast test bars or on test bars from the casting itself? Many specification properties are based on separately cast test bars, and properties of the casting can be considerably lower than those specified. On the other hand, foundries will agree to guaranteeing properties of the actual casting in some instances.

2. What level of defects is to be allowed, and what means of inspection is to be used?

A procedure covering means of specifying and controlling these factors is outlined in MIL-C-6021.

Selection of alloy Casting alloy selection is not so straightforward as wrought alloy selection. The casting selection is complicated by the foundry characteristics of the alloy and by the foundry practice used in production. It may often be the case that a sound, inherently low-strength material will exceed the strength of a high-strength material with poor quality.

Recommended aluminum casting alloys In electronics packaging, certain grades of castings may be recommended for specific purposes.

* Trademark of Union Carbide Corporation, New York, N.Y.

9-38   Nonferrous Metals

General-purpose Aluminum Alloy. Alloy 356 probably is the most widely used alloy in production. It has good foundry characteristics and very good mechanical properties. It is available in sand, permanent-mold, investment, and plaster-mold castings. A high-purity version, A356, offers somewhat higher strength.

Brazeable Compositions. Alloys A612, C612, and 40E are most commonly used for parts which must be brazed, since their melting points are high enough to withstand brazing temperatures.

TABLE 14 Typical Mechanical Properties of Aluminum Casting Alloys[3]

Alloy	ASTM No.	Tensile strength ksi	Yield strength ksi	Elongation in 2 in. %*	Shear strength ksi	Endurance limit (500 × 10⁶ cycles), ksi	Brinell hardness†
Sand casting:							
43–F	S5A	19	8	8.0	14	8.0	40
319–F	SC64C	27	18	2.0	22	10.0	70
355–T6	SC51A	35	25	3.0	28	9.0	80
356–T6	SG70A	33	24	3.5	26	8.5	70
D612	ZG61A	35‡	25‡	5.0‡	26‡	8.0	75‡
Permanent-mold casting:							
43–F	S5A	23	9	10.0	16	8.0	45
333–F	SC94A	34	19	2.0	27	14.5	90
333–T6	SC94A	42	30	1.5	33	15.0	105
355–T51	SC51A	30	24	2.0	24	....	75
355–T6	SC51A	42	27	4.0	34	10.0	90
C355–T61	SC51B	46	34	6.0	32	14.0	100
356–T6	SG70A	38	27	5.0	30	13.0	80
A356–T61	SG70B	41	30	10.0	28	13.0	90
Die casting:							
13	S12B	43	21	2.5	28	19	
43	S5C	33	16	9.0	21	17	
218	G8	45	27	8.0	29	20	
360	SG100B	47	25	3.0	30	19	
380	SC84B	48	24	3.0	31	21	
Premium-strength casting:							
220–T4	....	48	26	16			
355–T6	....	42	27	4			
356–T6	....	38	27	5			
C355–T62	....	49	36	6			
A356–T62	....	43	32	7			
354–T62	....	57	46	3			
A357–T62	....	53	42	3			
359–T62	....	50	42	5			
Tens 50–T6	....	50	44	6			

* Specimen ½-in. diam.
† Load, 500 kg; 10-mm ball.
‡ Tested 30 days after casting.

Low Thermal Expansion. The high-silicon-content (18 to 20 percent silicon) alloys have thermal expansion rates close to that of alloy steel. These alloys are difficult to cast and must be carefully poured with special procedures.

High Strength. Recently developed alloys have approached the strength of the higher-strength wrought aluminum alloys. Alloy KO-1* is claimed to give casting tensile properties approaching 60,000 psi. Tens 50 is an alloy with a tensile strength of 50,000 psi. Alloy 220 gives a tensile strength above 40,000 psi with good ductility but it is difficult to cast.

It cannot be emphasized too strongly the importance of close cooperation with foundries in early stages of casting design. Agreement should be reached on tolerance, quality standards, and strengths.

* Trademark of Electronic Specialties Co., Batavia, Ill.

BERYLLIUM AND BERYLLIUM ALLOYS

Beryllium has a combination of physical and mechanical properties which make it a natural material for use in electronic packaging as a structural material and as a material for thermal and electrical functions. The high cost of beryllium metal combined with difficulties in joining and forming have made beryllium structures very expensive. In addition, the lack of ductility of the metal has precluded its use in many structural applications. A combination of beryllium and aluminum called Lockalloy* shows promise of overcoming, to some extent, the forming, joining, and ductility problems, while retaining many of the desirable beryllium characteristics. However, material costs remain high. Despite its drawbacks, there are applications where the high stiffness-to-weight and strength-to-weight ratio of beryllium and Lockalloy make their use cost-effective because of weight reduction. Mirror blanks, gyro parts, airframe and missile body stiffeners, and control surfaces on airplane are examples of structural uses of these materials. The high thermal conductivity of pure beryllium along with its high heat capacity makes it a natural lightweight heat sink.

Availability

As mentioned above, beryllium is available in two basic compositions, these being essentially pure beryllium with various beryllium oxide contents and the material Lockalloy which is a combination of beryllium and aluminum. The production

TABLE 15   Impact Data for Beryllium and Lockalloy[4]

Material	Temper	Direction	Impact data
Prefatigue cracked Charpy specimens			*Impact strength W/A, in.-lb/in.^{2}*
62 Be 38 Al sheet....................	As rolled	L	60
		T	97
	Annealed	L	64
		T	103
Be powder sheet.....................	As received	L	12
		T	22
Unnotched specimens			*Energy absorbed (0.1 × 0.5 in.), in./lb*
62 Be 38 Al ext.....................	As extruded	ST	68
	Annealed	ST	44
Be powder sheet.....................	As received	ST	4

methods for both these alloys are advancing, with the result that products with improved properties are being developed, as well as capability for producing the materials in new forms. At this writing neither product is available in usable cast form. Close liaison with suppliers during the early stages of planning is recommended, in order that the most suitable product available is used.

A large amount of beryllium metal is produced in the form of hot-pressed block. This product is produced by first casting the metal into an ingot which is then reduced to a powder stage by mechanical means. The powder is then compacted into a block by heat and pressure. Many beryllium parts are produced by machining from the hot-pressed block.

Extruded rod, tube, and bar made by extruding hot-press block or billets is also available. Sheet which has been rolled from hot-pressed block is also available. Sheet is also being rolled directly from the cast ingot in some cases. Forged blocks are also available.

The working done by extrusion and rolling results in material of improved prop-

* Trademark of Lockheed Aircraft Corp., Burbank, Calif.

erties (both strength and ductility) over the hot-pressed block. However, beryllium is susceptible to developing directionality in properties because of working. Often, good ductility is present in one direction while being very poor in other directions. One method used in sheet production is to cross-roll. A designer should be aware of these directional properties and discuss his requirements with his supplier.

Lockalloy, at present, is produced primarily by extrusion in the form of bar and shapes. Dies for some simple structural shapes now exist. Since this material

TABLE 16 Chemical Compositions of Various Grades of Beryllium[5]

Grade*	Beryllium oxide content, % Maximum	Beryllium oxide content, % Typical	Beryllium assay, minimum
HP-20	2.0	1.8	98.0
HP-8	0.8	0.7	99.0
HP-12	1.2	1.0	98.5
HP-40	†	5.5	92.0
C-10	...	0.01	99.0

* Designations of Kawecki Berylco Inc.
† HP-40 material is produced to a minimum beryllium oxide content of 4.25%.

TABLE 17 Typical Mechanical Properties of Beryllium[5]

Grade	Ultimate tensile strength, ksi	Yield strength (0.2% offset), ksi	Elongation, %
HP-20	48	35	2
HP-8	37	27	3
HP-12	38	30	3
HP-40	65	60	1
C-10	20		
PR-20	77	55	9
IS-2	45	31	
IS-3	49	40	
XT-20	77	40	7
XT-40	90	52	5

is rapidly developing, consultation with the supplier on shape and size availability is recommended. Production of sheet and plate will probably be initiated in the near future.

Chemical Composition

Beryllium is produced with controlled beryllium oxide content. Typical compositions are shown in Table 16. The compositions represent the present practice of one manufacturer and are not presented as a standard.

The major Lockalloy composition is the 62 percent beryllium–38 percent aluminum composition. These are nominal values. It is expected that as usage becomes more general, standardized specifications, including limits on impurities, will be issued.

Mechanical Properties

Table 17 shows some typical properties of beryllium metal grades. These values give an indication of the mechanical properties now available. These values, however, do not cover all the conditions, such as forgings, extrusions, etc. For actual design purposes, the specific supplier should be consulted.

Table 18 covers some preliminary values for Lockalloy. They represent few tests and do not reflect any long-time production experience. These properties do not necessarily reflect the potential values which may be attained with further development.

TABLE 18 Preliminary Mechanical Properties for 62 Be 38 Al Alloy at 75°F*[4]

Property	Direction and plane[†]	Sheet, as rolled[‡]	Bar, as extruded[§]
F_{tu}, ksi	L	54.3	75.1
	T	56.2	71.3
F_{ty}, ksi	L	41.0	63.4
	T	42.1	59.7
T_{cy}, ksi	L	.....	64.5
	T	.....	59.0
F_{su}, ksi	yz-z	.....	46.8
	yz-y	30.4	39.7
	xz-z	.....	36.6
	xz-x	28.9	30.5
F_{bru}, ksi ($e/D = 2$)	L	102.0	122.0
$\frac{3}{16}$-in. pin	T	106.0	132.0
F_{bry}, ksi	L	74.0	115.0
	T	77.0	94.0
e, %	L	8.2	4.7
E, psi $\times 10^{-6}$	L	.....	0.5
	T	.....	.....

* Basis: limited data.
† Shear plane: xz = longitudinal, yz = transverse.
  Shear direction: x = longitudinal, y = long transverse, z = short transverse.
‡ Alloy 3A 21-30. Alloy 300A 21-34; extrusion reduction ratio 136:1.
§ Alloy 308 21-55; extrusion reduction ratio 13.6:1.

Physical Properties

Table 19 covers some physical properties of beryllium. Poisson's ratio is extremely low compared to that of other structural metals. This value should be considered as provisional and the supplier should be consulted as to actual values for his material since there seems to be some conflict in data.

Corrosion Resistance and Finishing

Beryllium is susceptible to corrosion by chlorides. This factor alone makes it necessary to protect the metal from corrosion since most equipment will be subjected to some chloride environment during service. Since field-performance data are practically nonexistent, it is recommended that testing under simulated service conditions be carried out.

Lockalloy possesses some of the favorable corrosion resistance of aluminum. However, since the material is essentially a mixture of beryllium and aluminum, there exists a possibility of some dissimilar metal-corrosion effects. At present, there exists few general-service data on corrosion, and therefore caution should be exercised. As with beryllium, corrosion testing under simulated service testing is recommended.

Nonferrous Metals

Protective coatings for beryllium and Lockalloy Both conversion coatings and anodic coatings are available for Lockalloy and beryllium. These coatings offer added corrosion resistance over that of the bare metals. Proprietary treatments are available from the metal producers.

Electroplating of beryllium is a production procedure. The common electroplated coatings may be applied. Generally, Lockalloy should accept platings by standard techniques used for aluminum.

In general, it appears that both beryllium and Lockalloy with suitable finish selected from conversion coatings, anodizing, or electroplating can meet the corrosion-resistance requirements for most electronic applications.

TABLE 19 Some Physical Properties of Beryllium[6]

Atomic diameter, A°	2.221	Latent heat of fusion, cal/g	250–275
Atomic number	4	Latent heat of vaporization	5390
Atomic weight	9.02	Magnetic susceptibility,	
Density, g/cc	1.85	gauss/oersted cgs	0.79
Density, psi	0.055	Thermal neutron absorp-	
Electrical conductivity, % IACS	38–40	tion cross section,	
Electrode potential, volts	−1.69	barn/atom	0.0090 ± 0.0005
Electrochemical equivalent,		Beryllium atoms per cm³	1.24×10^{23}
mg/coul	0.04674	Thermal-neutron scattering cross section,	
Photoelectric work function, ev	3.92	barn/atom	7.0
Hall coefficient, cgs electromagnetic units	0.0024 ± 0.001	Slowing-down length, cm (fission energy to thermal energy)	9.9
Mass susceptibility:		Velocity of sound:	
−180°C	−0.72 cgs units	ft/sec	41,300
+20°C	−1.00 cgs units	meters/sec	12,600
+300°C	−1.20 cgs units	Entropy, $S_{298.1}$, cal/(mole)(°K):	
Melting point, °C	1285	Solid	2.28 ± 0.02
Melting point, °F	2345	Liquid	32.56 ± 0.01
Boiling point, °C	2970	Reflectivity (white light)	50/55
Boiling point, °F	5378	Photoelectric work function, ev	3.92(9)
		Poisson's ratio	0.025 ± 0.005

Fabrication of Beryllium Alloys

Toxicity Breathing of beryllium dust particles is a health hazard, and therefore any operation which may possibly create fine particles must be done under conditions which prevent any possible breathing of these particles. The same precautions as taken with beryllium must be taken with Lockalloy. All machining operations must be done under conditions where the particles created are collected at the tool and isolated in a container which precludes their discharge into the atmosphere. Heating operations such as welding and brazing must be similarly controlled. Before processing beryllium or Lockalloy, consultation with plant safety engineer, local government medical authorities, vendor safety department, or other cognizant safety or health officials is mandatory.

Machining Beryllium can be machined by conventional machining methods. Tool wear will be high because of the abrasive nature of the material.

Most cutting-type operations leave a surface damage in the form of microcracks. These small cracks will act as failure initiation points. Removal of the surface layer by etching is required if structural integrity is to be maintained.

Lockalloy can be machined by conventional techniques. At present, it appears that the microcracks on the surface are not a problem with this material.

Forming The brittle nature of beryllium makes forming a problem. Most forming is done at elevated temperatures.

Lockalloy has greatly improved formability over beryllium. However, insufficient data are presently available to give concrete figures.

Developments in improving the ductility and thus formability of beryllium and

Lockalloy are under way. Contact with producers is recommended in order that material with best formability is obtained.

Joining Mechanical fastening of beryllium must be done with care since drilling of holes will cause microcracks. Riveting operations may cause cracking of the structure.

Lockalloy may be joined satisfactorily by riveting since its increased ductility will absorb impact, and surface damage from machining is not a major problem.

Welding of beryllium results in brittle welds which have not at present been considered satisfactory. Lockalloy appears to be weldable with either Lockalloy filler rod or with aluminum alloy filler rod.

Brazing of beryllium is practicable using either conventional aluminum brazing alloys or silver brazing alloys. Lockalloy may be brazed with aluminum brazing alloys.

Cost

Beryllium hot-pressed block sells for about $80 per pound. Sheet may run several hundred dollars per pound. Lockalloy is also priced in this region. It is obvious that these materials must offer major design advantages to be cost-effective over more conventional materials.

COPPER AND COPPER ALLOYS

General Characteristics

Copper and copper alloys to a great extent still dominate the electric-conductor applications in electronic equipment, even though aluminum and its alloys can show reduction of weight and a lower material cost. Among the characteristics that account for use of copper and its alloys are:

1. Pure copper has the highest electric conductivity of any metal except silver when considered on a volume basis. As a result, copper electric conductors require the lowest space for a given conductivity.
2. Forming characteristics of copper are exceptionally good and make economics of manufacture favorable.
3. Solderability of copper is good, although for most electric soldering a solderable finish is required.
4. Application of finishes by electroplating is easy and reliable, as is application of hot-dipped coatings.
5. Basic corrosion resistance of copper is good, and again finishes are easily and economically applied.
6. Availability of copper forms covers all sizes and shapes, including extremely fine wire, thin strip, castings, as well as sheet, plate, bar, forgings, and extrusions.
7. Copper is suitable for many plant forming operations and lends itself to all machining operations. Such methods as electroforming and chemical milling are especially useful in shaping copper alloys.
8. Basic costs of copper and its alloys are still reasonable.
9. Thermal conductivity is high, offering a good means of transferring heat from components to a heat sink.

The above characteristics apply in general to copper and copper alloys. However, it should be pointed out that alloying can drastically affect properties. For instance, some of the copper alloys have an electric conductivity as low as 5.5 percent. It is therefore necessary to examine carefully all properties when making a selection of alloy. Trade-offs will be required at times between such properties as strength and electric conductivity.

High-electric-conductivity Alloys

Hundreds of copper alloy compositions are available. These range from high-purity copper to multiple alloys. Table 20 shows selected compositions of the relatively pure coppers with electric conductivities of approximately 100 per-

TABLE 20 Nominal Compositions* of Wrought Copper with High Electric Conductivity[7]

Copper number	Previous commonly accepted trade name	Copper plus silver, % min.	Silver, oz/ton	Arsenic	Antimony	Phosphorus	Tellurium	Nickel	Bismuth	Lead	Other named elements
101	Oxygen-free, certified	99.96	. . . . . .	. . . . . .	. . . . . .	0.0003	. . . . . .	. . . . . .	. . . . . .	0.001	0.00040 sulfur 0.0003 zinc 0.0001 mercury
102	Oxygen-free	99.95	. . . . . .								
104	Oxygen-free with silver	99.95	8 min.								
105	Oxygen-free with silver	99.95	10 min.								
107	Oxygen-free with silver	99.95	25 min.								
110	Electrolytic tough pitch	99.90									
111	Electrolytic tough pitch, anneal-resistant	99.90									
113	Tough pitch with silver	99.90	8 min.								
114	Tough pitch with silver	99.90	10 min.								
116	Tough pitch with silver	99.90	25 min.								
120	Phosphorus deoxidized, low residual phosphorus	99.90	. . . . . .	. . . . . .	. . . . . .	0.004-0.012					
121	. . . . . .	99.90	4 min.	. . . . . .	. . . . . .	0.005-0.012					
122	Phosphorus deoxidized, high residual phosphorus	99.90	. . . . . .	. . . . . .	. . . . . .	0.015-0.040					
123	. . . . . .	99.90	4 min.	. . . . . .	. . . . . .	0.015-0.025					
125	Fire-refined tough pitch	99.88	. . . . . .	0.012	0.003		0.025	0.050	0.003	0.004	
127	Fire-refined tough pitch with silver	99.88	8 min.	0.012	0.003		0.025	0.050	0.003	0.004	
128	Fire-refined tough pitch with silver	99.88	10 min.	0.012	0.003		0.025	0.050	0.003	0.004	
130	Fire-refined tough pitch with silver	99.88	25 min.	0.012	0.003		0.025	0.050	0.003	0.004	
141	Arsenical tough pitch	99.40		0.15-0.50							
142	Phosphorus deoxidized, arsenical	99.40		0.15-0.50		0.015-0.040					
145	Phosphorus deoxidized, tellurium-bearing	99.90				0.004-0.012	0.40-0.6				
147	Sulfur-bearing	99.90									0.20-0.50 sulfur
150	Zirconium-copper	99.80									0.10-0.20 zirconium

* Percent maximum unless shown as a range or minimum.

cent. The compositions are Copper Development Association* designations. Other compositions are also available with minor variations. The alloys shown in this table are distinguished by several characteristics.

The tough-pitch grades are probably the most widely used for electric conductors. In the annealed condition their electric conductivities are approximately 100 percent IACS. The major drawback to these compositions compared to the other high-conductivity coppers is that they are subject to embrittlement by hydrogen at elevated temperature of 700°F and above. This is due to the presence of oxygen in the metal, which will combine with hydrogen, resulting in internal cracking of the metal. Since most brazing operations involve hydrogen either in the products of combustion of the heating gases or in furnace atmospheres, tough-pitch copper is usually not suitable where brazing is to be used in joining.

This problem can be overcome by use of copper with low oxygen content. Phosphorous deoxidized copper is one approach to solving this deficiency. Use of phosphorous as a deoxidizer results in a slight reduction of electric conductivity and somewhat reduced formability.

Oxygen-free copper which is produced without the use of metal deoxidizers results in a product which is free from susceptibility to hydrogen embrittlement and which retains 100 percent or better electric conductivity and has formability equivalent to tough-pitch copper.

Silver-bearing coppers High-conductivity coppers are also available with small amounts of silver present in the composition. These alloys retain the approximately 100 percent electric conductivity, with two additional advantages in improved creep resistance and resistance to softening at elevated temperatures. The latter advantage is useful where work-hardened material must be subjected to elevated temperature during processing it is desired to retain the work-hardened condition. The silver-bearing alloys will resist annealing at temperatures as high as 650°F, a temperature which will substantially anneal the pure coppers.

Other special-purpose, high-electric-conductivity copper alloys are available with special properties. Among these are:

1. *Zirconium-Copper.* This is an alloy containing about 0.15 percent zirconium which has improved creep strength at elevated temperature with a small sacrifice in electric conductivity.

2. *Tellurium-Copper.* The addition of small percentages of tellurium results in an alloy with major improvement in machinability over pure copper while retaining an electric conductivity of 90 percent IACS. Mechanical properties are moderately high compared to those of copper.

3. *Sulfur-bearing Copper.* The addition of sulfur to copper in small percentages results in substantial increase in machinability while retaining a relatively high electric conductivity of about 96 percent IACS.

It can be seen that a wide choice of compositions is available with electric conductivities in the 95 to 101 percent electric-conductivity range. Wide overlapping of properties, both physical and mechanical, can lead to many trade-offs in selection of materials. Very often the selection will be one of economics of fabrication weighed against material costs. Material suppliers can offer aid in selecting a specific material.

Brasses

Brass is by definition an alloy of copper and zinc, but in some cases trade names have been given to brasses which indicates they are bronze, which is by definition an alloy of copper and tin. Table 21 gives the composition of some of the copper-zinc alloys in use. It can be seen that there exists a wide range of compositions. Some generalizations may be made about the effect of zinc on copper. Among these are:

1. Tensile strength increases with increasing zinc content up to about 30 percent zinc.

* Copper Development Association, Inc., New York, N.Y.

9-46 Nonferrous Metals

2. Electric and thermal conductivity decreases as zinc content increases.
3. Ductility of soft annealed material increases as zinc content increases up to about 30 percent zinc. In the harder tempers ductility decreases in lean zinc alloys but begins to increase at around 10 percent zinc.

As indicated, these are generalizations and are subject to modifications by cold-working and annealing sequences which affect grain size. For critical forming or strength applications it is desirable to work with the material supplier to select the specific set of properties tailored to the application and processing procedures.

TABLE 21 Nominal Compositions* of Some Copper-Zinc Alloys (Brasses)[7]

Copper alloy number	Previous commonly accepted trade name	Copper	Lead	Iron	Zinc	Nickel	Aluminum	Phosphorus	Total other elements
205	....	97.0–98.0	0.02	0.05	Bal.	....	....	....	0.05
210	Gilding, 95%	94.0–96.0	0.05	0.05	Bal.	....	....	....	0.10
220	Commercial bronze, 90%	89.0–91.0	0.05	0.05	Bal.	....	....	....	0.10
226	Jewelry bronze, 87½%	86.0–89.0	0.05	0.05	Bal.	....	....	....	0.15
230	Red brass, 85%	84.0–86.0	0.05	0.05	Bal.	....	....	....	0.15
234	....	81.0–84.0	0.05	0.05	Bal.	....	....	....	0.15
240	Low brass, 80%	78.5–81.5	0.05	0.05	Bal.	....	....	....	0.15
250	....	74.0–76.0	0.05	0.05	Bal.	....	....	....	0.15
260	Cartridge brass, 70%	68.5–71.5	0.07	0.05	Bal.	....	....	....	0.15
261	....	68.5–71.5	0.05	0.05	Bal.	....	....	0.02–0.05	0.15
262	....	67.0–70.0	0.07	0.05	Bal.	....	....	....	0.15
268	Yellow brass, 66%	64.0–68.5	0.15	0.05	Bal.	....	....	....	0.15
270	Yellow brass, 65%	63.0–68.5	0.10	0.05	Bal.	....	....	....	0.15
274	Yellow brass, 63%	61.0–64.0	0.10	0.05	Bal.	....	....	....	0.20
280	Muntz metal, 60%	59.0–63.0	0.30	0.07	Bal.	....	....	....	0.20
298	Brazing alloy	49.0–52.0	0.50	0.10	Bal.	....	0.10	....	

* Percent maximum unless shown as a range or minimum.

Leaded brasses Most of the brass compositions are available with lead added as a means of improving machinability. The lead exists in the matrix of the alloy as free lead, not as an alloy. The presence of lead results in production of fine chips in machining, thus producing a free-cutting condition. Lead contents vary from a nominal 0.5 to as high as 4.5 percent, with the highest percentage giving the best machinability.

Lead has its detrimental qualities in that it reduces ductility. Problems may occur with the high-lead-content materials in forming operations. Terminals made from leaded alloys may crack in swaging, and in some cases fracture during soldering. The latter problem may occur especially when the terminals are swaged to thick plastic terminal boards. Selection of one of the lower-lead-content alloys is a solution for this problem, although machinability may be reduced.

Phosphor Bronzes

A series of bronzes containing 1 to 11 percent tin and 0.04 to 0.35 percent phosphorus are widely used for spring applications in electronics. Compositions of these alloys are shown in Table 22. These alloys are available in varying strengths up to a yield strength of 116,000 psi for the 10 percent tin alloy which has been cold-worked to the extra spring hard condition. However, this alloy has an electric conductivity in the range of 11 percent IACS. The 1.25 percent tin alloy, on the other hand, has an electric conductivity of about 48 percent, but its maximum attainable yield strength is 75,000 psi. This is an example of the penalty that sometimes must be paid in reduced electric conductivity in order to buy strength.

Beryllium-Copper

Copper when alloyed with beryllium forms an alloy which is capable of being strengthened by precipitation hardening. It is capable of attaining tensile strengths in the range of 200,000 psi by proper heat-treating and cold-working procedures. At this hardness level it still possesses good electric conductivity in excess of 22 percent IACS. As a result of this combination of high strength and good electric conductivity, beryllium-copper has become a widely used material for electrical spring applications.

TABLE 22 Nominal Compositions* of Phosphor Bronzes[7]

Copper alloy number	Previous commonly accepted trade name	Copper + tin + phosphorus, % min.	Lead	Iron	Tin	Zinc	Aluminum	Phosphorus
502	....	99.5	0.05	0.10	1.0–1.5	....	....	0.04
505	Phosphor bronze, 1.25% E	99.5	0.05	0.10	1.0–1.7	0.30	....	0.03–0.35
507	....	99.5	0.05	0.10	1.5–2.0	....	....	0.04
508	....	99.5	0.05	0.10	2.6–3.4	....	....	0.01–0.07
509	....	99.5	0.05	0.10	2.5–3.8	0.30	....	0.03–0.30
510	Phosphor bronze, 5% A	99.5	0.05	0.10	3.5–5.8	0.30	....	0.03–0.35
518	Phosphor bronze	99.5	0.02	....	4.0–6.0	....	0.01	0.10–0.35
521	Phosphor bronze, 8% C	99.5	0.05	0.10	7.0–9.0	0.20	....	0.03–0.35
524	Phosphor bronze, 10% D	99.5	0.05	0.10	9.0–11.0	0.20	....	0.03–0.35

* Percent maximum unless shown as a range or minimum.

Because it is a precipitation-hardening alloy, it has processing advantages in forming. The material is subjected to a solution-annealing treatment which leaves it in a relatively soft condition. Forming may be readily performed in this condition. A low-temperature aging treatment at about 600°F will result in increased strength with little distortion.

Table 23 shows the compositions available along with some properties and forms available. Table 24 shows a summary of properties available. Note that again the higher-strength alloys have the lower electric conductivity.

Figure 12 shows comparative strength and electric conductivity of two beryllium-copper alloys with some other copper alloys. Figure 13 also shows a comparison of several alloys in the effect of temperature on electric conductivity.

Figure 14 shows the effect of age-hardening time on electric conductivity of 2 percent beryllium-copper alloy. Similar effects on tensile strength and yield strengths are brought about by variations in aging temperatures, as shown in Fig. 15. The 600°F aging is usually specified since the time at temperature is not critical. Note that aging time at 600°F can vary several hours with little effect on tensile properties. At higher temperatures a few minutes of excess time may result in overaging with resulting drop in properties. However, it is practical to use shorter aging cycles at higher temperatures if proper controls of material condition and heat-treating time and temperature are exercised. The cold-working history of the material between solution treating and aging is important in the aging re-

9-48 Nonferrous Metals

TABLE 23 Beryllium-Copper Compositions and Availability[8]

Composition	Property evaluation	Tempers	Forms
Alloy 172 Beryllium 1.80–2.05% Cobalt 0.18–0.30% Copper balance	High strength Good conductivity Fatigue strength Wear resistance Nonmagnetic RWMA Class 4 material	Heat-treatable or heat-treated	Strip, bar, rod, wire, billets, plater bars, forgings, casting ingot
Alloy 170 Beryllium 1.60–1.80% Cobalt 0.18–0.30% Copper balance	Available in mill-hardened strip Good strength Good conductivity Nonmagnetic	Heat-treatable or mill-hardened	Strip, bar, billets, casting ingot
Alloy 175 Beryllium 0.40–0.70% Cobalt 2.35–2.70% Copper balance	High conductivity Good strength No heat treatment required Good wear resistance Higher operating temperatures RWMA Class 3 material	Heat-treatable or heat-treated	Strip, bar, rod, billets, plater bars, casting ingot

sponse. Determination of aging cycles by test is desirable when deviation is made from the 600°F aging temperature.

Figure 16 shows the effect of repeated stress on the cycles to failure in fatigue for a 2 percent beryllium-copper alloy. Note that at 10^8 cycles the stress is about 40,000 psi and that even at 10^5 cycles the tolerable stress level is less than 90,000 psi, even though the material has a yield strength of 150,000 psi. These data

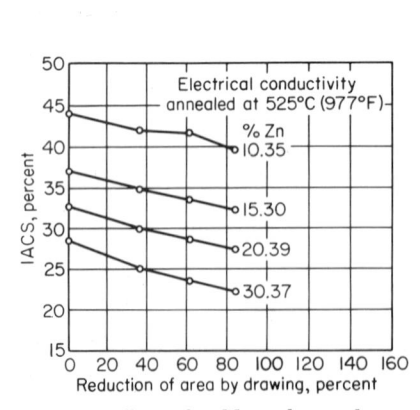

Fig. 11 Effect of cold work on electric conductivity of various copper-zinc alloys.

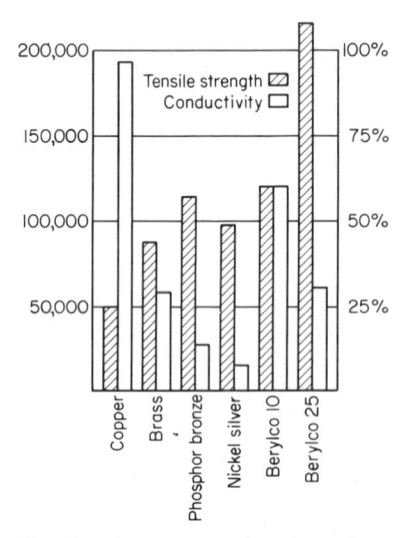

Fig. 12 Comparative electric conductivity and tensile strength of some representative copper alloys.

Copper and Copper Alloys   9-49

TABLE 24 Summary of Mechanical Properties of Beryllium-Copper Alloys[8]

Properties	Berylco 25 AT	Berylco 25 HT	Berylco 165 AT	Berylco 165 HT	Berylco 10 AT	Berylco 10 HT
Tensile strength (ultimate), ksi	165–190	190–215	150–180	180–200	100–120	110–130
Proportional limit, ksi	100–125	125–155	85–115	110–140	60–80	75–95
Yield strength at 0.20% offset, ksi	140–170	165–195	130–160	155–185	80–100	100–120
Rockwell C (nominal)	39	43	36	40	12	9
Electrical conductivity, IACS % (minimum)	22	22	22	22	45	48
Elongation (nominal)	6	3	7	3	12	8
Endurance strength at 10^8 load cycles:						
Strip, in reverse bending	36	45	35	44	......	35
Rod, rotating beam	......	51	......	......	40	......
Elastic modulus, at 20°C:						
In tension	18,500	......	18,500	......	19,000	......
In torsion	7,300	......	7,300	......	7,500	......
In compression	19,400	......	19,400	......		

Thermoelastic coefficient of Berylco 25 AT,
In tension.............. 0.00035
In torsion.............. 0.00033

Compressive properties of Berylco 165 AT, % contraction under load
0.01.............. 160
0.10.............. 180
0.20.............. 190

Fig. 13 Effect of temperature on electric conductivity and resistivity of representative copper alloys.

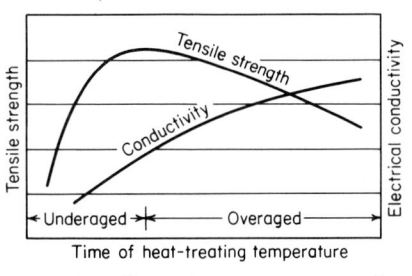

Fig. 14 Effect of aging time on electric conductivity and tensile strength for a typical age-hardening alloy. (Note: The conductivity is measured at room temperature after the aging.)

9-50   Nonferrous Metals

point out the importance of considering the number of load applications when designing for repeated loads.

MAGNESIUM ALLOYS

Magnesium alloys have found acceptance primarily because they have the lowest density of any structural metal available today. The commonly used alloys have

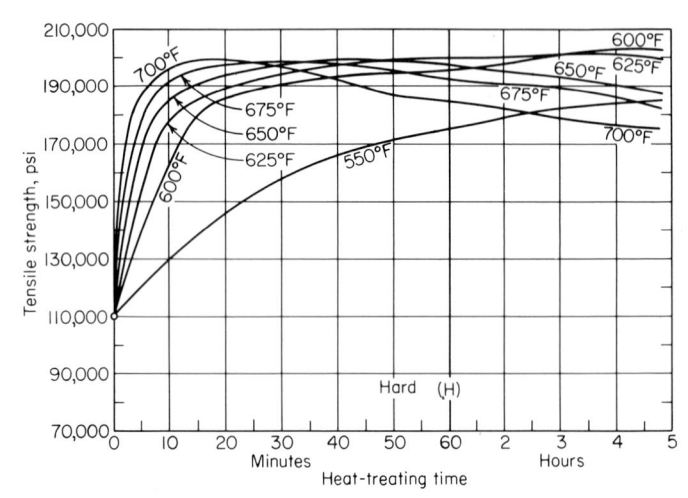

Fig. 15   Effect of various aging times and temperatures on tensile strength of 2 percent beryllium-copper alloy.

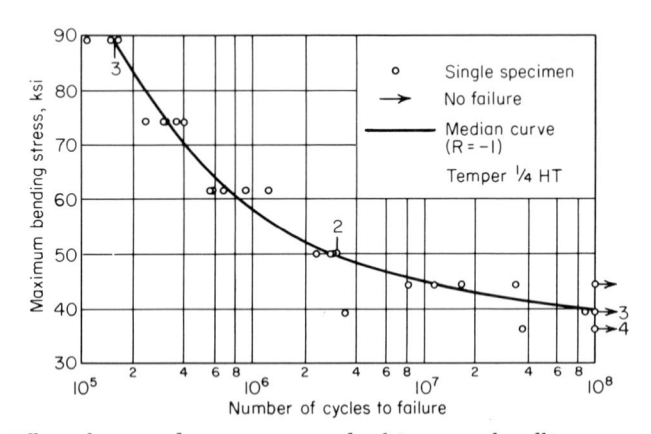

Fig. 16   Effect of repeated stress on strength of 2 percent beryllium-copper strip.

densities of about 0.065 lb/in.³ while the recently developed magnesium-lithium alloys have densities of less than 0.050 lb/in.³. These values may be compared to those of aluminum at approximately 0.100 and beryllium at approximately 0.066 lb/in.³. Table 25 gives specific gravities of typical magnesium alloys.

Poor corrosion resistance and poor formability have hindered the use of magnesium alloys in many applications. However, procedures are available for overcoming these problems and providing economical and functional units for a variety of applications.

Magnesium Alloys 9-51

TABLE 25 Specific Gravities of Various Magnesium Alloys[9]

Alloys	Specific gravity, g/cm³
Magnesium-lithium:	
LA141A	1.35
LA91A	1.45
LAZ933A	1.56
Magnesium:	
ZE10A	1.76
AZ31B	1.77
HM21A	1.78
AZ61A	1.80
AZ80A	1.80
HK31A	1.81
EK41A	1.81
AZ91A,B,C	1.81
AZ92A	1.82
HZ32A	1.83
AZ63A	1.84

Magnesium alloys have been used in many applications for reduction in weight over aluminum alloys and for less cost than beryllium structures.

Forms Available

Magnesium alloys are available in nearly all forms including sheet, plate, bar, forgings, and extruded shapes. Extrusion is generally used for shaping bar, tubing, and other shapes. Castings are also available, but in compositions specifically developed for casting. All types of castings including sand, die, permanent-mold, and investment castings are available. Table 26 shows a summary of availability of forms in terms of alloy numbers and government specifications. Investment castings are available in any of the casting compositions.

TABLE 26 Forms and Specifications of Magnesium Alloys[10]

Alloy	Sand cast	Permanent-semi-permanent cast	Die cast	Plate sheet	Bars, rods, wire, shapes, extruded	Tubing extruded	Forgings	Welding rod and wire
AZ31B	....	....	....	QQ-M-44 MIL-F-46048	QQ-M-31	WW-T-825		
AZ61A	....	....	....	....	QQ-M-31	WW-T-825	QQ-M-40	MIL-R-6944
AZ63A	QQ-M-56	QQ-M-55	....	....	....	....	....	MIL-R-6944
AZ80A	....	....	....	....	QQ-M-31	....	QQ-M-40	
AZ81A	QQ-M-56	QQ-M-55	....	....				
AZ91A	....	....	QQ-M-38					
AZ91C	QQ-M-56	QQ-M-55						
AZ92A	QQ-M-56	QQ-M-55						
EZ33A	QQ-M-56	QQ-M-55						
HK31A	QQ-M-56	....	....	MIL-M-26075	....	....	....	MIL-R-6944
HZ32A	QQ-M-56							
M1A	....	....	....	QQ-M-54	QQ-M-31	WW-T-825	QQ-M-40	MIL-R-6944
TA54A	....	....	....	....	....	....	QQ-M-40	
ZK10A	....	....	....	MIL-M-46037				
ZH62A	QQ-M-56							
ZK21A	....	....	....	....	MIL-M-46039			
ZK51A	QQ-M-56							
ZK60A	....	....	....	....	QQ-M-31 MIL-M-5354	WW-T-825	QQ-M-40	

Alloy Designations

Magnesium alloys are designated by a coding system which indicates approximate nominal alloy percents of the two major alloying elements. Designations for alloys consist of not more than two letters representing the two alloys present in the largest percentages, arranged in order of decreasing percentages. The letters are followed by numbers representing the percentage of the alloying element rounded off to the nearest whole number. The most frequently used letters are as follows:

A	Aluminum	L	Lithium
E	Rare earths	M	Manganese
H	Thorium	Z	Zinc
K	Zirconium		

An example designation is AZ31 which indicates an alloy of magnesium in which the two main alloying elements are aluminum and zinc as indicated by the AZ. The number 3 indicates 3 percent nominal aluminum while the 1 indicates 1 percent zinc.

Alloy Availability

Nominal compositions of conventional alloys available are shown in Table 27.

The sheet alloy most commonly used is the AZ31 (3 percent aluminum–1 percent zinc) composition. This is the general-purpose alloy for service near room temperature and is the most readily available alloy. The HK31 and HM21 alloys have been developed for improved strength at elevated temperature. Table 28 shows a comparison of the various alloys at room temperatures.

The LA141 alloy (14 percent lithium–1 percent aluminum) is an example of the lowest-density class of alloys. Note that the strength of this alloy is relatively low, but that the alloy possesses mechanical properties exceeding those of some commonly used metals such as annealed low-strength aluminum alloys. Other alloys based on the lithium-magnesium alloys are also available.

Extrusion alloys Considerably more alloys are available as extrusions than as sheet alloys. Table 29 summarizes the mechanical properties of these alloys. The ZK60 alloys show the highest room-temperature properties of any magnesium alloys and compare favorably on a strength-weight ratio with the high-strength aluminum alloys. Note however, that compressive yield strength of magnesium alloy extrusions is significantly lower than the tensile yield strength of that same alloy. This is an excellent example of the directional properties found in certain alloys, as discussed previously in this chapter. The magnitude of the difference between tensile and compressive yield strengths makes it mandatory that designers consider this difference in designs where stresses approach the yield strength.

Elevated-temperature properties of wrought magnesium alloys As with all metals and alloys, mechanical properties of magnesium are changed by temperature. Strength decreases as temperature increases, and ductility usually increases. Magnesium, being a relatively low melting point metal, can be drastically affected by moderately elevated temperature. For instance, the yield strength of AZ31B–H24 alloy decreases from 32,000 psi at room temperature to approximately 13,000 psi at 300°F. Alloys such as HK31A–H24 and HM21A–T8 are less affected by increases in temperature. HK31A–H24 has a yield strength of about 22,000 psi at 400°F while HM21A–T8 retains a yield strength of about 13,000 psi at 600°F. Thus it can be seen that useful properties exist in selected magnesium alloys for moderately high temperature service. Caution must be taken when using elevated-temperature properties, however, since the time at temperature, stress level, and the temperature itself also must be considered. Creep, that is, deformation under a steady load, may occur well below the stress levels mentioned above. Consultation with suppliers for detailed properties for specific loading conditions and environment is recommended when designing for elevated temperatures. In some cases tests simulating service conditions may be required.

Magnesium Alloy Castings

Magnesium alloy castings are poured in specific compositions primarily designed for casting. Usually these alloys contain higher percentages of alloying elements than are found in wrought compositions. Table 27 gives typical compositions of various casting alloys.

TABLE 27 Nominal Compositions of Magnesium Alloys[10]

Alloys	Li	Al	Mn, min	Zn	Th	Zr	Rare earths
Sand and permanent-mold casting:							
AM100A	....	10.0	0.1				
AZ63A	....	6.0	0.15	3.0			
AZ81A	....	7.6	0.13	0.7			
AZ91C	....	8.7	0.13	0.7			
AZ92A	....	9.0	0.10	2.0			
EK30A	....	....	....	...	...	0.3	3.3
EK41A	....	....	....	...	...	0.6	4.0
EZ33A	....	....	....	2.7	...	0.6	3.3
ZE41A	....	....	....	4.2	...	0.7	1.2
ZK51A	....	....	....	4.6	...	0.7	
ZK61A	....	....	....	6.0	...	0.7	
ZK61A	....	....	....	6.0	...	0.7	
HK31A	....	....	....	...	3.3	0.7	
HZ32A	....	....	....	2.1	3.3	0.7	
ZH42	....	....	....	4.0	2.0	0.7	
ZH62A	....	....	....	5.7	1.8	0.7	
Die casting:							
AZ91A[2]	....	9.0	0.13	0.7			
Extruded bars and shapes:							
M1A	....	....	1.2				
AZ31B	....	3.0	....	1.0			
AZ61A	....	6.5	....	1.0			
AZ80A	....	8.5	....	0.5			
HM31XZ	....	....	1.2	...	3.0		
ZK21A	....	....	....	2.3	...	0.45 min.	
ZK60A	....	....	....	5.5	...	0.45 min.	
ZK60B[3]	....	....	....	5.7	...	0.45 min.	
Sheet and plate:							
LA141A	14.0	1.0					
AZ31B	....	3.0	....	1.0			
HK31A	....	....	....	...	3.0	0.6	
HM31A	....	....	0.6	...	2.0		
ZE10A	....	....	....	1.3	...	....	0.17
PE	....	3.3	....	0.7			

Investment casting practice has been developed to the extent that wrought-type alloys can be cast for special applications. AZ31 alloy has been cast for applications where brazing was required. It seems logical to expect that other wrought compositions could be cast if the need arises.

K1 alloy, an alloy possessing high damping capacity, is another specialty alloy. This alloy has been used for parts where vibration damping is required. It is

TABLE 28 Mechanical Properties of Magnesium Sheet and Plate at Room Temperature: Typical Values (in ksi)[9]

Alloy	Thickness, in.	Ultimate tensile strength (Minimum)	Tensile yield strength (Minimum)	Compressive yield strength	Elongation, min. % in 2 in.	Shear	Bearing Ultimate tensile strength (Typical)	Bearing Yield
AZ31B-H24	0.050–0.249	39	29	24	6	29	77	47
	0.250–0.374	38	26	20	8	28	72	45
	0.375–0.500	37	24	16	8	27	70	40
	0.501–1.000	36	22	13	8	26	68	37
	1.001–2.000	34	20	10	8	26	66	35
	2.001–3.000	34	18	9	8	26	66	33
AZ31B-H26	0.250–0.374	39	27	22	6	28	72	46
	0.375–0.438	39	26	21	6	28	72	44
	0.439–0.500	38	26	18	6	28	72	44
	0.501–0.750	37	25	17	6	28	72	40
	0.751–1.000	37	23	16	6	28	70	39
	1.001–1.500	35	22	15	6	27	68	37
	1.501–2.000	35	21	14	6	26	68	36
AZ31B-0	0.050–0.249	32	15	12	12	26	66	37
	0.250–0.500	32	15	10	12	25	65	43
	0.501–2.000	30	15	10	10	25	65	33
	2.001–3.000	30	15	8	9	25	65	32
HK31A-H24	0.050–0.125	34	26	20	4	26	67	41
	0.126–0.250	34	24	22	4	26	56	41
	0.251–1.000	34	25	20	4	27	68	44
	1.001–3.000	33	25	17	4	26	65	40

9-54

HK31A-O	0.050–0.250	30	18	12	12	24	58	28
	0.251–0.500	30	16	10	10	24	58	27
	0.501–1.000	30	15	10	10	24	58	27
	1.001–3.000	29	14	10	10	24	57	27
HM21A-T8	0.050–0.250	33	18	15	6	19	63	37
	0.251–0.500	32	21	20	6	...	67	41
	0.501–1.000	30	21	17	6	...	63	39
	1.001–2.000	30	21	15	6	...	59	38
	2.001–3.000	30	21	14	6	...	59	36
HM21A-T81	0.050–0.250	33	25	...	4			
	0.251–0.312	34	25	...	4			
LA141A-T7	0.010–0.090	19	15	18	10	14	47	34
	0.091–0.250	19	14	17	10	14	42	27
	0.251–2.000	18	13	15	10	14	42	27

TABLE 29 Mechanical Properties of Magnesium Alloy Extrusions*,[11]

Properties	AZ31B–F or AZ31C–F				AZ61A–F				AZ80A–F				AZ80A–T5	
Least dimension, in.	Under 0.250	0.250–1.499	1.500–2.499	2.500–5.000	Under 0.250	0.250–2.499	2.500–5.000	Under 0.250	0.250–1.499	1.500–2.499	2.500–5.000	Under 0.250	0.250–1.499
Area, in.²	......	......	......	......	......	......	......	......	......	......	......	......	......
Tensile strength, ksi:													
Typical	38	38	38	38	46	45	45	49	49	49	48	55	55
Minimum	35	35	34	32	38	40	40	43	43	43	42	47	48
Tensile yield strength, ksi:													
Typical	28	29	28	28	33	33	31	36	36	35	36	38	40
Minimum	21	22	22	20	21	24	22	28	28	28	27	30	33
Elongation in 2 in., %:													
Typical	14	15	14	15	17	16	15	12	11	11	9	8	7
Minimum	7	7	7	7	8	9	7	9	8	6	4	4	4
Compressive yield strength, ksi:													
Typical	15	14	14	14	...	19	21	...	...	...	...	34	35
Minimum	...	12	12	10	...	14	14	...	17	17	17	...	28
Shear strength, ksi (typical)	19	19	19	19	23	22	22	22	22	22	22	24	24
Bearing strength, ksi (typical):													
Ultimate	56	56	56	56	65	68	68	68	68	68	68	60	60
Yield	34	33	33	33	38	40	42	48	48	48	48	57	58
Brinell hardness	...	49	...	...	...	60	...	60	60	60	60	82	82

TABLE 29 Mechanical Properties of Magnesium Alloy Extrusions*,[11] (Continued)

Properties	AZ80A-T5	AZ80A-T5	HM31A-T5	HM31A-T5	ZK21A-F	ZK60A-F	ZK60A-F	ZK60A-F	ZK60A-F	ZK60A-T5	ZK60A-T5	ZK60A-T5	(P)ZK60B-T5
Least dimension, in.	1.500–2.499	2.500–5.000	Under 1.000	1.000–3.999	Under 5.000	......	......	......	......	......	......	......	......
Area, in.[2]	......	......	......	......	......	Under 2.000	2.000–2.999	3.000–4.999	5.000–39.999	Under 2.000	2.000–2.999	3.000–4.999	Up to 2.999 Under 20.000
Tensile strength, ksi:													
Typical	53	50	44	43	42	49	49	49	48	53	52	51	49
Minimum	48	45	37	37	38	43	43	43	43	45	45	45	45
Tensile yield strength, ksi:													
Typical	39	38	38	34	33	38	37	36	37	44	43	42	38
Minimum	33	30	26	26	28	31	31	31	31	36	36	36	35
Elongation in 2 in., %:													
Typical	6	6	8	13	10	14	14	14	9	11	12	14	17
Minimum	4	2	4	4	4	5	5	5	4	4	4	4	4
Compressive yield strength, ksi:													
Typical	32	31	27	23	25	33	28	27	23	36	31	30	40
Minimum	27	26	19	15	20	27	26	25	20	30	28	25	35
Shear strength, ksi (typical)	24	24	22	21	21	24	24	24	24	26	26	25	25
Bearing strength, ksi (typical):													
Ultimate	66	68	62	62	60	76	76	76	75	79	78	77	76
Yield	54	53	49	44	46	56	50	49	44	59	53	52	65
Brinell hardness	82	82	...	...	...	75	75	75	75	82	82	82	...

* Temper:
 –F   As extruded.
 –T5 Artificially aged.

9-57

9-58   Nonferrous Metals

an alloy with about 1 percent zirconium. This alloy is available from most investment casting foundries. Mechanical properties are lower than those of most of the other magnesium casting alloys.

Table 30 lists some typical physical and mechanical properties of representative magnesium castings. Again as with other castings, the question of guaranteed mechanical properties should be agreed upon either by specification, drawing call-out, or on a purchase order. Most standardized specifications show mechanical properties of separately cast test bars and often allow the properties of the actual castings to be well below those specified. It is obviously important that the designer know whether the specified values are for test bars or for the casting itself. However, there is danger of unnecessarily raising the cost of castings if the maximum possible strength is specified for every section of the casting. A more reasonable procedure is to specify strength requirements in the critically stressed areas of the casting. This will allow the foundry to arrange its pouring practice and chilling practices to attain maximum soundness and strength in the critical area.

TABLE 30 Typical Physical and Mechanical Properties of Magnesium Alloy Castings

Properties	AZ91C-T4	AZ91C-T6	EZ33A-T5	HK31A-T6	QE22A-T6	AZ91A, B-F
Physical:						
Density, lb/in.^{3}	0.0653	0.0653	0.0660	0.0648	0.0655	0.0653
Thermal conductivity at 68°F, cal/(sec)(cm^2)(°C/cm)	0.11	0.13	0.24	0.22	0.25	0.12
Electrical resistivity at 68°F, μohm-cm	16.2	12.9	7.0	7.7	6.8	14.3
Tensile strength, ksi:						
Average	40	40	23	32	40	33
Minimum	34	34	20	27	35	
Mechanical (at room temperature):						
Tensile yield strength, ksi:						
Average	12	19	15	15	30	22
Minimum	11	16	14	13	25	
Elongation in 2 in., %:						
Average	14	5	3	8	4	3
Minimum	7	3	2	4	2	
Compressive yield strength, ksi:						
Average	12	19	15	15	30	22
Minimum						
Shear strength, ksi	17	20	22	22	......	20
Bearing strength, ksi:						
Ultimate	60	75	57	61		
Yield	44	52	40	40		

Elevated-temperature properties of magnesium castings Similarly to the wrought alloys, magnesium casting alloys suffer decrease in strength as temperature increases. The normal aluminum-zinc alloys such as AZ91 lose strength rapidly as temperature increases. EZ33 alloy maintains 50 percent of its room-temperature properties at 600°F, while HK31A-T6 still has 81 percent of its room-temperature strength at 500°F and 50 percent at 700°F. HK31A is another example of an alloy which is normally considered a wrought alloy but has found use in castings.

Fabrication of Magnesium Alloys

Forming Magnesium has a reputation of being difficult to form. However, under proper conditions magnesium alloys may be formed to complex shapes by many techniques, including deep drawing. Use of elevated temperatures during forming is often necessary to accomplish such forming. Facilities which have been set up with heated dies have been very successful in forming magnesium alloys.

It is desirable, of course, to form material at room temperature for reasons of economy and convenience. Developments in magnesium alloy AZ31 have resulted in a material capable of being formed as room temperature to a $2T$ radius (where T = metal thickness) in thicknesses of $\frac{1}{8}$ in. and less. This compares favorably with some of the aluminum alloys.

Joining Magnesium can be joined by most conventional joining methods, both mechanical and metallurgical, as discussed below.

Riveting. Riveting of magnesium is done by normal riveting practice. The riveting material recommended is aluminum alloy 5056. This alloy is sufficiently close to magnesium alloys in galvanic potential that no serious dissimilar metal problem exists. It is advisable that rivets be installed with a barrier such as zinc chromate primer which is applied wet to the rivet just before driving.

Fusion Welding. Fusion welding of magnesium alloys may be readily accomplished with the tungsten inert-gas process or with the metal inert-gas process. Filler rods must be selected in accordance with the alloy being welded. The aluminum-zinc alloys (AZ) are usually welded with AZ9Z alloy as filler. The high-temperature alloys are often welded with EZ33 rod but also are welded with rod matching the material composition. Selection of rod composition will depend upon the joint efficiency required, ductility required, etc. The individual magnesium supplier should be consulted for each specific case.

Stress Relieving of Magnesium Weldments. The alloys containing aluminum and zinc as alloying elements are subject to stress corrosion. Since welding nearly always results in some residual stress, stress-relieving is required after welding. Typical cycles are: (1) 500°F for 15 min AZ31 alloy annealed sheet and extrusions and as-extruded AZ61, ZK60, and AZ80; (2) 300°F for 60 min for AZ31 hard-rolled and ZK60 in the aged condition; (3) 500°F for 60 min for cast alloys.

Brazing. It is possible to braze magnesium by the dip method, utilizing brazing fluxes similar to aluminum brazing fluxes. Filler metal is usually a 12.5 percent aluminum alloy. The removal of flux is a must since the residues of flux are extremely corrosive to magnesium alloys. Only the higher-melting-point alloys such as AZ31 and M1 may be brazed.

There is little production experience with brazing of magnesium alloys, and the process should be specified only after careful consultation with magnesium suppliers.

NICKEL AND NICKEL ALLOYS

Nickel alloys are characterized by good strength, corrosion resistance, and high ductility. The metal nickel has relatively good electric conductivity, making it useful for electric conductors for specialized applications. The magnetostrictive properties of nickel and high-nickel alloys make them applicable to various devices such as ultrasonic transducers. Nickel and nickel alloys have Curie temperatures which vary with alloy content. This transition from magnetic to nonmagnetic behavior has been used in control devices.

The series of alloys based on nickel-copper are known as Monels.* These materials show exceptional corrosion resistance in salt-water immersion service. Some of these alloys are age-hardenable, notably the alloy K-500 and 500.

The Inconel* series of alloys is based primarily on nickel plus chromium plus iron with minor additions of other alloying elements. These alloys have exceptionally good elevated-temperature properties. Such alloys are useful at temperature approaching 2000°F.

A specialty alloy of beryllium and nickel is age-hardenable to strength in excess of 200,000 psi. This material retains good strength at 600°F. It has found use as a spring material.

Forms available Nickel and nickel alloys are available in most wrought forms. This includes all section sizes from ultrathin sheet and fine wire to heavy plate and forgings.

Fabrication Nickel and nickel alloys have exceptionally good fabrication properties, as discussed below.

* Trademark of Huntington Alloy Products Division, The International Nickel Co., Inc., Huntington, W. Va.

TABLE 31 Nominal Chemical Compositions of Some Nickel Alloys[12]

Designation	Ni	C	Mn	Fe	S	Si	Cu	Cr	Al	Ti	Others	Previous designation
Nickel 200	99.5	0.06	0.25	0.15	0.005	0.05	0.05	.....	.....	.....	.....	"A"* Nickel
Nickel 201	99.5	0.01	0.20	0.15	0.005	0.05	0.05	.....	.....	.....	.....	Low-carbon nickel
Nickel 204	95.2	0.06	0.20	0.05	0.005	0.02	0.02	.....	.....	.....	Co 4.50	Nickel 204
Nickel 205	99.5	0.06	0.20	0.10	0.005	0.05	0.05	.....	.....	0.02	Mg 0.04	"A" Nickel (electronic grade)
Nickel 211	95.0	0.10	4.75	0.05	0.005	0.05	0.03	.....	.....	.....	.....	"D"* Nickel
Nickel 220	99.5	0.06	0.12	0.05	0.005	0.03	0.03	.....	.....	0.02	Mg 0.04	"220"* Nickel
Nickel 230	99.5	0.09	0.10	0.05	0.005	0.03	0.01	.....	.....	0.003	Mg 0.06	"230"* Nickel
Nickel 233	99.5	0.09	0.18	0.05	0.005	0.03	0.03	.....	.....	0.003	Mg 0.07	"330"* Nickel
Nickel 270	99.98	0.01	<0.001	<0.001	<0.001	<0.001	<0.001	<0.001	.....	<0.001	Co <0.001 Mg <0.001	New product
Permanickel alloy 300	98.6	0.25	0.10	0.10	0.005	0.06	0.02	.....	.....	0.50	Mg 0.35	Permanickel* alloy
Duranickel alloy 301	94.0	0.15	0.25	0.15	0.005	0.55	0.05	.....	4.50	0.50	.....	Duranickel* alloy
Monel alloy 400	66.0	0.12	0.90	1.35	0.005	0.15	31.5	.....	.....	.....	.....	Monel* alloy
Monel alloy 401	44.5	0.03	1.70	0.20	0.005	0.01	53.0	.....	.....	.....	Co 0.50	Monel* "401" alloy
Monel alloy 402	58.0	0.12	0.90	1.20	0.005	0.10	40.0	.....	.....	.....	.....	Monel* "402" alloy
Monel alloy 403	57.5	0.12	1.80	0.50	0.005	0.25	40.0	.....	.....	.....	.....	Monel "403" alloy
Monel alloy 404	55.0	0.06	0.01	0.05	0.005	0.02	44.0	.....	0.02	.....	.....	New product
Monel alloy R-405	66.0	0.18	0.90	1.35	0.050	0.15	31.5	.....	.....	.....	.....	"R" Monel alloy
Monel alloy 406	84.0	0.12	0.90	1.35	0.005	0.15	13.0	.....	.....	.....	.....	LC Monel alloy
Monel alloy 474	54.0	0.01	Trace	0.01	0.001	<0.01	46.0	.....	Trace	.....	.....	New product
Monel alloy K-500	65.0	0.15	0.60	1.00	0.005	0.15	29.5	.....	2.80	0.50	.....	"K" Monel alloy
Monel alloy 501	65.0	0.23	0.60	1.00	0.005	0.15	29.5	.....	2.80	0.50	.....	"KR" Monel alloy

Alloy										
Inconel alloy 600	76.0	0.04	0.20	7.20	0.007	0.20	0.10	15.8			Inconel* alloy
Inconel alloy 604	74.0	0.04	0.20	7.20	0.007	0.20	0.10	15.8			Inconel "600" alloy
Inconel alloy 625	61.0	0.05	0.15	3.00	0.007	0.30	0.10	22.0		Cb 2.0	New product
										Cb 4.0 Mo	
										9.0	
Inconel alloy 700	46.0	0.12	0.10	0.70	0.007	0.30	0.05	15.0	3.00	Co 28.5 Mo 3.75	Inconel "700" alloy
										2.20	
Inconel alloy 702	79.5	0.04	0.05	0.35	0.007	0.20	0.10	15.6	3.40	0.70	Inconel "702" alloy
Inconel alloy 718	52.5	0.04	0.20	18.5	0.007	0.30	0.07	18.6	0.40	0.90 Cb 5.0 Mo	Inconel "718" alloy
										3.1	
Inconel alloy 721	71.0	0.04	2.25	7.20	0.007	0.12	0.10	16.0		3.00	Inconel "M" alloy
Inconel alloy 722	75.0	0.04	0.55	6.50	0.007	0.20	0.05	15.0	0.60	2.40	Inconel "W" alloy
Inconel alloy X-750	73.0	0.04	0.70	6.75	0.007	0.30	0.05	15.0	0.80	2.50 Cb 0.85	Inconel "X" alloy
Inconel alloy 751	72.5	0.04	0.70	6.75	0.007	0.30	0.05	15.0	1.20	2.50 Cb 1.00	Inconel "X-550" alloy
Incoloy alloy 800	32.0	0.04	0.75	46.0	0.007	0.35	0.30	20.5	0.30	0.30	Incoloy* alloy
Incoloy alloy 801	32.0	0.04	0.75	44.5	0.007	0.35	0.15	20.5		1.00	Incoloy "T" alloy
Incoloy alloy 804	42.6	0.06	0.85	25.4	0.007	0.50	0.40	29.3	0.25	0.40	Incoloy "804" alloy
Incoloy alloy 825	41.8	0.03	0.65	30.0	0.007	0.35	1.80	21.5	0.15	0.90 Mo 3.0	Ni-o-nel alloy
Ni-span-C alloy 902	42.0	0.02	0.40	48.5	0.008	0.50	0.05	5.4	0.65	2.40	Ni-span-C* alloy

* Trademark of International Nickel Co., Inc.

9-62 Nonferrous Metals

Forming. The ductility of nickel alloys makes forming by most conventional techniques extremely favorable.

Joining. Welding is practical for all nickel alloys by arc, gas, and resistance methods. Welding of nickel alloys to most steel as well as to some copper alloys is often done.

Brazing with the silver brazing alloys, copper, and nickel-base brazing alloys are all readily accomplished by furnace, torch, induction, or resistance brazing.

Chemical composition of nickel alloys Table 31 gives the compositions of a number of nickel alloys.

TABLE 32 Physical Properties of Nickel and Nickel Alloys[12]

Designation	Density lb/in.^{3}	Modulus of elasticity, psi × 10^{-6} Tension	Torsion	Melting range, °F	Specific heat at 70°F, Btu/(lb)(°F)	Thermal expansion, in./(in.)(°F) × 10^{-6} 70–200°F	70–500°F
Nickel 200	0.321	30.0	11.0	2615–2635	0.109	7.4	7.9
Nickel 201	0.321						
Nickel 204	0.321	29.2	....	....	.....	7.4	7.8
Nickel 205	0.321	....	....	....	.....	...	...
Nickel 211	0.315	....	....	....	.....	7.4	7.9
Nickel 220	0.321						
Nickel 230	0.321						
Nickel 233	0.321						
Nickel 270	0.321	30.0	....	2650	0.11	7.4	7.8
Permanickel alloy 300	0.316	30.0	11.0	....	0.106	6.8	7.9
Duranickel alloy 301	0.298	30.0	11.0	2550–2620	0.104	7.2	7.7
Monel alloy 400	0.319	26.0	9.5	2370–2460	0.102	7.7	8.7
Monel alloy K-500	0.306	26.0	9.5	2400–2460	0.100	7.6	8.2
Monel alloy 501	0.306	26.0	9.5	2400–2460	0.100	7.6	8.2
Inconel alloy 600	0.304	31.0	11.0	2500–2600	0.106	7.4	7.8
Inconel alloy 604	0.305	....	....	....	.....	6.8	7.6
Inconel alloy X-750	0.298	31.0	11.0	2540–2600	0.103	7.0	7.4
Inconel alloy 751	0.298	31.0	11.0	2540–2600			
Incoloy alloy 800	0.290	28.0	11.0	2475–2525	0.12	7.9	9.0
Incoloy alloy 801	0.287	29.0	....	....	.....	8.1	9.0
Incoloy alloy 804	0.286	28.0	....	....	.....	8.0	
Incoloy alloy 825	0.294	28.0	....	2500–2550	.....	7.8	8.4
Ni-span-C alloy 902	0.293	24–29	9–10	2650–2700	0.12	4.2	6.0

Physical properties of nickel alloys Table 32 shows values of a variety of properties of nickel alloys. The variety of magnetic and electrical properties obtainable by alloying makes nickel alloys extremely interesting to the electronics engineer.

Mechanical properties of nickel alloys Table 33 lists the range of properties available with selected nickel alloys. Note that all conditions have ranges for properties. As discussed before, these ranges will depend upon section size, manufacturing procedures, and heat-treating practice. Caution should be exercised when selecting materials of large section sizes since it is likely that the highest strength listed will not be available.

TITANIUM AND TITANIUM ALLOYS

The chief attributes of the titanium family are excellent corrosion resistance, moderate density, and good strength, especially when considered on a strength-weight basis. Titanium alloys have densities about 60 percent that of steel, therefore making the high-strength titanium alloys competitive with the high-strength steels when considered on a strength-weight basis. At temperatures around 600°F, titanium alloys excel practically all other metals in reducing weight.

The electric and thermal conductivities of titanium and its alloys are low compared to those of the aluminum, copper, and magnesium alloys. These low values are definite disadvantages in many applications while being an advantage in uses where thermal isolation is desired or where induced current from electric fields is to be minimized. Thermal expansion rates are slightly lower than for steels, but

Thermal conductivity, Btu/(hr)(ft²)(°F/in.)		Electrical resistivity, ohms/cir mil/ft		Curie temperature, °F		Permeability at 70°F (H = 200)			Poisson's ratio
70°F	500°F	70°F	500°F	Annealed	Age-hardened	Annealed	Age-hardened	
...	374	57	126	680	....	.....	......	0.31
...	...	65	...	770				
...	...	57	...					
306	296	102	195					
655	...	43	125	667				
400	...	94.5	...	600	563			
165	211	255	310	60–120	200	.....	....	0.31
151	204	307	342	20–50	....	.....	....	0.32
121	167	370	382	<−210	<−150	1.001	1.002	0.32
121	167	370	382	<−210	<−150	1.001	1.002	0.32
103	127	620	639	−192	....	1.010	....	0.29
101	126	666	687	<−320	...	1.004		
83	104	731	754	−225	−193	1.002	1.0035	
80	109	595	662	−175	....	1.009	....	0.30
...	...	638	...	<−320	....	1.003		
77	103	678	719	<−320	....	1.005		
84	114	611	...	340	380			

generally are close enough to cause no appreciable problems. Physical properties are shown in Table 34.

Titanium is available in commercially pure form (99 percent or more Ti) and in a number of alloyed compositions. A number of the alloys and forms available are shown in Table 35.

Commercially pure titanium is available in several purities. This material ranges from 60,000 to 80,000 psi in tensile strength depending upon purity. It has the best heat-corrosion resistance, weldability, and formability of any of the titanium alloys. The pure metal is generally lowest in cost also. See Table 35 for mechanical property ranges for titanium and alloys.

Titanium alloys A number of alloying elements are used to enhance the properties of titanium. These include aluminum, vanadium, iron, manganese, tin, chromium, and molybdenum. Alloys are most often designated by nomenclature which starts with Ti, the chemical symbol for titanium, followed by a number which is the nominal percentage of an alloying element followed by the chemical symbol

9-64 Nonferrous Metals

of the element. For example Ti 6 Al 4 V is titanium alloy containing 6 percent aluminum and 4 vanadium.

Three basic metallurgical types of alloys exist, namely, alpha, alpha-beta, and beta. The alpha type consists of alloys which retain the hexagonal crystal structure as found in pure titanium. On heating to 1625°F, unalloyed titanium transforms to a body-centered structure which is designated as the beta phase. By selection of alloying elements, the beta phase can be stabilized to the extent that it will exist at room temperature. Some alloys are alloyed to stabilize the entire structure in the beta phase at room temperature, and thus are obviously called beta alloys.

TABLE 33 Nominal Range of Mechanical Properties: Wrought Rods and Bars

Designation	Temper	Tensile strength, ksi	Yield strength at 0.2 % offset, ksi	Elongation in 2 in., %	Brinell hardness, 3,000 kg
Nickel 200	Hot finished	60–85	15–45	55–35	90–150
	Cold drawn	65–110	40–100	35–10	140–230
	Annealed	55–80	15–30	55–40	90–120
Nickel 201	Hot finished	50–60	10–25	60–40	75–100
	Cold drawn	60–100	35–90	35–10	125–200
	Annealed	50–60	10–25	60–40	75–100
Permanickel alloy 300	Hot finished	90–120	35–65	40–25	140–230
	Hot finished, age-hardened	160–200	120–150	20–10	285–360
	Cold drawn, as drawn	90–150	50–130	35–15	185–300
	Cold drawn, age-hardened	150–190	120–150	20–10	300–380
Duranickel alloy 301	Hot finished	90–130	35–90	55–30	140–240
	Hot finished, age-hardened	160–200	115–150	30–15	300–375
	Cold drawn, as drawn	110–150	60–130	35–15	185–300
	Cold drawn, age-hardened	170–210	125–175	25–15	300–380
	Annealed	90–120	30–60	55–35	135–185
	Annealed, age-hardened	150–190	110–140	30–20	285–360
Monel alloy 400	Hot finished	80–95	40–65	45–30	140–185
	Cold drawn, stress-relieved	84–120	55–100	35–22	160–225
	Annealed	70–85	25–40	50–35	110–140
Monel alloy 402 and Monel alloy 403	Hot finished	70–95	25–65	45–30	130–185
	Cold drawn, stress-relieved	75–120	45–100	35–15	150–225
	Annealed	65–85	23–40	50–35	100–140
Monel alloy R-405	Hot finished	75–90	35–60	45–30	130–170
	Cold drawn	85–115	50–100	35–15	160–240
	Annealed	70–85	25–40	50–35	110–140
Monel alloy K-500	Hot finished	90–135	40–110	45–20	140–260
	Hot finished, age-hardened	140–170	100–125	30–17	265–330
	Annealed	90–110	40–60	45–25	140–185
	Annealed, age-hardened	130–160	85–110	30–20	250–300
	Cold drawn	100–135	70–100	35–13	175–260
	Cold drawn, age-hardened	135–180	95–130	30–15	255–325
Inconel alloy 600	Hot finished	85–120	35–90	50–30	140–217
	Cold drawn	105–150	80–125	30–10	180–300
	Annealed	80–100	30–50	55–35	120–170
Incoloy alloy 800	Hot finished	85–120	35–90	50–25	140–217
	Cold drawn	100–150	75–125	30–10	180–300
	Annealed	75–100	30–50	50–30	120–170
Incoloy alloy 825	Annealed	85–105	35–65	50–30	120–180

By proper selection of alloying elements, alloys have been developed which consist of a combined alpha and beta two-phase structure at room temperature. These alloys retain some of the advantages of both the alpha and beta phases and also render such alloys capable of heat treatment to higher strengths. Heat treatment consists of heating to temperatures sufficiently high to transform the alpha phase to beta. Fast cooling suppresses the natural transformation of beta to alpha. On subsequent aging there is transformation of some of the beta to alpha. This alpha is distributed throughout the beta matrix to give a fine structure which results in increased strength.

Some beta alloys can be heat-treated by aging also. During aging, fine particles of alpha and intermetallic compounds are precipitated out to provide strengthening.

By alloying, the strength of titanium can be raised considerably above the strength of the unalloyed material. Yield strengths above 150,000 psi at room temperature are attainable with available alloys. Table 35 lists some of the available alloys, along with typical mechanical properties. Several factors other than mechanical properties must be considered when selecting an alloy. One of these is the availability of the alloy within the time required and in the quantity required. Ware-

TABLE 34 Typical Physical Properties of Titanium Alloys

Nominal composition, % (bal. Ti)	Density, lb/in.[3]	Thermal expansion mean coefficient per °F × 10⁻⁶ RT to 200°F	RT to 600°F	RT to 1000°F	Thermal conductivity, Btu/(hr)(ft)(°F) RT	800°F	Instantaneous specific heat, Btu/(lb)(°F) RT	800°F	Electrical resistivity, μohm-cm RT	800°F	Elastic moduli, psi × 10⁻⁶ B	G
99.5 Ti	0.163	4.8	5.2	5.5	9	....	0.124	....	57	....	14.9	6.5
99.2 Ti	0.163	4.8	5.2	5.5	9.5	....	0.125	....	56	....	14.9	6.5
99.0 Ti	0.163	4.8	5.2	5.5	9.5–11.5	10.5	0.125	0.151	48–57	117.7	15.0	6.5
99.0 Ti	0.164	4.8	5.2	5.5	9.8–10.1	10.0	0.129	0.155	55–60	122.3	15.1	6.5
98.9 Ti	0.164	4.8	5.2	5.5	9.8	....	0.129	....	58	....	15.5	6.5
0.15–0.20 Pd	0.163	4.8	5.1	5.4	9.5	....	0.125	....	56.7	....	14.9	6.5
Alpha alloys:												
5 Al 2.5 Sn	0.161–0.162	5.2	5.3	5.3	4.5	7.2	0.125	0.152	157	180	16.0	7.0
5 Al 2.5 Sn (low O)	0.161	5.2	5.3	5.4	4.5	....	0.125	....	157	....	16.0	
5 Al 5 Sn 5 Zr	0.166	...	...	...	....	....	....	....	....	....	16.0	
7 Al 12 Zr	0.165	...	...	...	....	....	....	....	....	....	16.5	
7 Al 2 Cb 1 Ta	0.159	...	...	...	....	....	....	....	....	....	17.7	
8 Al 1 Mo 1 V	0.158	4.7	5.0	5.6	....	....	....	....	199	203	18.5	
Alpha-beta Alloys:												
8 Mn	0.171	4.8	5.4	6.0	6.3	9.0	0.118	0.152	192	140	16.4	7.0
2 Fe 2 Cr 2 Mo	0.171	...	...	...	....	....	....	....	....	....	16.7	
2.5 Al 16 V	0.165										15.0	(Aged)
3 Al 2.5 V	0.162	...	...	...	....	....	....	....	....	....	15.5	
4 Al 4 Mn	0.163	4.9	5.1	5.4	4.2	7.4	0.126	0.159	153	172	16.4	7.3
4 Al 3 Mo 1 V	0.163	5.0	5.3	5.5	3.9	6.8	0.132	0.142	165	....	16.5	7.0
5 Al 1.25 Fe	0.162–0.163	5.2	5.3	5.5	....	....	....	....	....	....	16.8	
2.75 Cr											17.6	(Aged)
5 Al 1.5 Fe 1.4 Cr	0.162–0.163	5.2	5.5	5.7	4.7	7.0	....	....	163	180	16.5	6.3
1.2 Mo											17.0	(Aged)
6 Al 4 V	0.160	4.9	5.1	5.3	4.2	6.8	0.135	....	171	187	16.5	6.1
6 Al 4 V (low O)	0.160	5.3	5.3	5.3	....	....	0.135	....	171	....	16.5	6.1
6 Al 6 V 2 Sn	0.164	5.0	5.2	5.3	4.2	....	0.155	....	157	....	15.0	
1 (Fe, Cu)	....	...	...	...	....	....	....	....	157	170	16.5	(Aged)
7 Al 4 Mo	0.162	5.0	5.2	5.6	3.7	7.0	0.123	0.151	175	183	16.2	6.5
											16.9	(Aged)
Beta alloys:												
1 Al 8 V 5 Fe	0.168										14.2	6.2
8 Al 13 V 11 Cr	0.175–0.176	5.2	5.6	5.9	4.0	8.0	0.120	0.198	153	....	14.2	6.2
		...	...	...	....	....	....	....	142	....	14.8	(Aged)

house stocks are often sparse, and not all alloys are immediately available. The method of fabrication will also influence selection. Many of the alloys are not considered weldable since welds are brittle.

Corrosion resistance of titanium alloys One of the major attributes of titanium alloys is corrosion resistance. The major commercial applications of these materials have been in the chemical field where the superior resistance to chemical attack has resulted in major economies, due to longer life of titanium parts compared to that of other structural metals. Room-temperature atmospheric and salt-water environment are withstood well by titanium alloys. Stress corrosion in the presence of chlorides above 600°F is a problem, but such conditions are uncommon in electronic equipment, so that this is not a serious drawback.

Titanium is the noble metal in a galvanic couple with most structural materials except stainless steel and Monel. Protection to prevent a galvanic couple may be needed when titanium is attached to other metals, especially aluminum alloys, magnesium alloys, or beryllium.

TABLE 35 Typical Mechanical Properties of Some Titanium Alloys at Room Temperature

Typical composition, % (bal. Ti)	Forms*	Yield strength, ksi	Tensile strength, ksi	Elongation in 2 in., %
Unalloyed	S, B, E, W	55	65	18
Unalloyed	S, B, E, W	70	80	15
Unalloyed	S, B, E, W, T	40	50	20
8 Mn	S	110	120	10
6 Al 4 V	S, B, E, W	120	130	10
Unalloyed	B, F	70	80	15
2 Fe 2 Cr 2 Mo	S, B, E, W, F	120	130	
4 Al 4 Mn	B, W, F	130	140	10
5 Al 2.5 Sn	S, B, W	110	115	10
3 Al 5 Cr	B, F	135	145	10
6 Al 4 V	B, F	120	130	10
5 Al 1.5 Fe 1.5 Cr 1.2 Mo	B, F	135	145	10
Unalloyed	T (Welded)	40	50	20
Unalloyed	W	...	50	
5 Al 2.5 Sn	W	...	115	
5 Al 2.75 Cr 1.25 Fe	B	140	150	10
3 Mn 1.5 Al	S, B, W	100	110	12
3.25 Mn 2.25 Al	S, B, W	110	120	10
2.5 Al 16 V	S, B, W	55	90	12
4 Al 3 Mo 1 V	S	90	125	
6.5 Al 3 Mo 1 V	B	150	155	17
6.5 Al 3.75 Mo	B	152	162	
8 Al 2 Cb 1 Ta	S, B	120	127	16
8 Al 1 Mo 1 V	B	132	137	18
13 V 11 Cr 3 Al	S, B, W	120	125	10
8 Al 8 Zr 1(Cb + Ta)	S, B, F	125	135	16
7 Al 4 Mo	B, F	130	140	10

* S Rolled flat products: sheet, strip, plate.
  E Extrusions.
  T Tube.
  B Bar and billet.
  W Wire.
  F Forgings.

Contamination Titanium and its alloys are subject to pickup of hydrogen, oxygen, and nitrogen, which will generally result in decreased ductility. Such contamination is most likely to occur at elevated temperatures. Thus, caution must be used in selecting atmospheres for heat treatment, especially above 1200°F. Selection of heat-treating atmospheres which are rich in one or more of the deleterious elements should be avoided. Inert gases such as helium or argon are most satisfactory as protective atmospheres, while vacuum is good protection or in some cases may be used to remove hydrogen which has been picked up from other sources.

Plating and pickling are also sources of hydrogen pickup. Welding of titanium without full shielding is also a source of oxygen and nitrogen contamination.

Forms available There are titanium alloys available in about any wrought form. Casting developments are under way, but casting has not become a common technique. Sheet, plate, bar, wire, strip, tubing, and extrusions are available in one alloy or another. Warehouse stock is not reliable, although commercially pure sheet is generally available. Consultation with suppliers on availability of alloys and forms available in that alloy is recommended before specifying material for production. Table 35 indicates the forms in which the various alloys can be made.

Fabrication of titanium Conventional processes are used for fabricating titanium parts. However, special precautions and revised techniques are sometimes required.

Forming. Titanium alloys have reasonable formability similar to many other high-strength materials. Bend radii on commercially pure annealed, high-purity sheet may be as low as one times the sheet thickness (1T). The high-strength alloys in the solution-treated and aged condition may require radii as great as

TABLE 36 Typical Bend Radii for Titanium Alloys at Various Temperatures

Temperature, °F	Bend axis*	\\ Minimum bend radius T							
		Ti 8 Mn	Ti 8 Mn	Ti 8 Mn	Ti 5 Al 2.5 Sn	Ti 6 Al 4 V	Ti 6 Al 4 V	Ti 3 Mn 1.5 Al	Ti 3.25 Mn 2.25 Al	Ti 2.2 Fe 2.1 Cr 2 Mo
70	L	3½	3½	3	7	4½	6	4	4	3
	T	3½	3	3	6	7	5	4	2½	2½
400	L	2½	2½	2½	6	3½	4	3	2½	2
	T	2½	2½	2½	4½	6½	3½	3	2	2
600	L	2	2½	2½	6	3½	3½	3	2	2
	T	2½	2½	2½	4½	5½	3	2½	1½	2
800	L	2	2	2	5½	3	3½	2½	1½	1½
	T	2½	2	2	4	5	2½	2	1½	1½
1000	L	1½	1	1½	5	2½	3	2	1	1½
	T	1½	1½	2	3½	4	2½	1½	1	1
1200	L	...	...	...	3½	2	2½			
	T	...	...	...	3	3	2			
1400	L	...	...	...	2½	1½	1½			
	T	...	...	...	2½	2	1½			
1500	L	...	...	...	1½	1	1			
	T	...	...	...	2	1	1			

* L Bend axis perpendicular to rolling direction.
  T Bend axis parallel to rolling direction.

8T or more. Springback is particularly troublesome in titanium and its alloys because of their high yield strengths combined with relatively low elastic moduli. Hot forming is often used with titanium alloys to minimize springback and to decrease bend radii. See Table 36. Caution must be used in selecting temperatures for forming since temperatures above the aging temperature will overage the material, resulting in lower strength.

Welding. Tungsten inert-gas welding of the weldable alloys is practical. The major precaution required is that the heated weld area must be shielded from the atmosphere. Shielding gas should be supplied to both sides of the weld, and a trailing shield of gas is required to protect the weld during cooling. It may often be desirable to weld in an enclosure which is filled with inert gas. Glove box chambers for this purpose are commercially available.

Mechanical properties of welds are excellent. Most weldable alloys can be welded to 100 percent joint efficiency. The pure titanium grades and the alpha alloy grades are all weldable with good resulting mechanical properties and ductility. Certain of the alpha-beta alloys can be satisfactorily welded, but others are subject to formation of brittle welds. Caution should be used when specifying welding of these alloys and advice of materials suppliers should be sought on selection of welding rod, joint design, and welding procedures.

Resistance weldability of titanium is excellent.

High-density Materials

Although to a great extent electronics engineers are interested in producing lightweight equipment, there are occasions where a high-density material is required in order to place a concentrated mass in a small volume. A typical application is a counterweight.

Three categories of metals are available for these purposes. Of course, the moderately heavy metals such as steels, nickel alloys, and copper alloys could be used for such purposes if space is available. We will concentrate on the lead, tungsten, and depleted uranium metals since they have higher densities than the aforementioned metals, but still have reasonable economic feasibility when compared to high-density precious metals such as gold.

Lead and lead alloys Lead with a specific gravity of 11.35 g/cc (0.41 lb/in.^{3}) and a relatively low cost is often a choice for counterweights. The low melting point of lead results in its ability to be cast by low-cost techniques, resulting in further economies.

The major disadvantage of lead is its low strength. The yield strength of relatively pure lead is around 800 psi, thus making it unsuitable for load-carrying applications. Creep rate of lead at room temperature is appreciable at stresses of 300 psi.

Antimony when alloyed with lead in percentages of 1 to 9 percent can appreciably increase the strength of lead. These alloys are also precipitation-hardenable by heat treatment. For the 4 percent antimony alloy, values of above 11,000 psi tensile strength have been reported. As-cast values around 3,500 to 4,000 psi tensile strength have been found. Creep strengths do not appear to be drastically improved, however, indicating that sustained loads should be avoided.

Recently dispersion-hardened lead has been developed. This material is available in the dispersion-hardened condition, but will lose most of its favorable properties if melted.

Sintered tungsten Sintered tungsten material is available with specific gravities in the range of 17 g/cc. These materials are usually tungsten particles mixed with other materials such as copper or nickel, and then compacted and sintered. The metals other than tungsten melt and braze the tungsten particles together. A wide variety of proprietary materials are available. Density and mechanical properties will vary, depending upon the percentage of metals other than tungsten used.

The materials are available in bars and may be furnished in relatively complex shapes if quantities are sufficient to amortize dies.

Depleted uranium This material with a density in the range of 18.5 g/cc has become available. With mechanical properties similar to those of low-carbon steel, it offers good structural properties. It is also capable of being cast to shape and worked by conventional tools, making economies of manufacture possible.

Corrosion resistance of depleted uranium is poor, thus requiring corrosion protection by application of coatings.

REFRACTORY METALS

Refractory metals which are noted for their high metaling points and ability to be useful at high temperatures find some uses in electronic equipment at relatively low temperatures. These uses often are based on a combination of physical and mechanical properties. In addition, some of the metals have extremely good resistance to acids.

Refractory Metals

Table 37 summarizes the physical and mechanical properties of four important refractory alloys. It can be seen that a variety of properties are available among the metals tungsten, tantalum, molybdenum, and columbium. Alloys are available which modify these properties. Some of the properties are discussed below.

Mechanical properties The major importance of mechanical properties of these alloys is the useful strength which exists at elevated temperatures as high as 1000°C. This has been of major importance in consideration of these materials and structures.

TABLE 37 Typical Properties* of Refractory Alloys

Property	Tungsten	Tantalum	Molybdenum	Columbium
Atomic number	74	73	42	41
Atomic weight	183.92	180.88	95.95	92.91
Atomic volume	9.53	10.9	9.41	10.83
Mass:				
Density at 20°C:				
g/cc	19.3	16.6	10.2	8.57
lb/in.^{3}	0.697	0.600	0.368	0.31
Thermal properties:				
Melting point, °C	3400	2996	2625	2415
Boiling point, °C	5900–6700	7400	5560	3300
Linear coefficient of expansion per °C $\times 10^{-6}$	4.0	6.5	5.45	6.89
Thermal conductivity at 20°C, cal/(sec)(cm^2)(°C/cm)	0.399	0.130	0.349	0.13
Specific heat, cal/(g)(°C)	0.034 (100°C)	0.036 (0°C)	0.061 (20°C)	0.065 (0°C)
Electrical properties:				
Electrical conductivity, % IACS	31	13.9	36	13.3
Electrical resistivity, μohm-cm	5.48 (0°C)	12.4 (18°C)	5.17 (0°C)	14.2 (20°C)
Temperature coefficient of electrical resistivity per °C	0.00482 (20–100°C)	0.00382 (0–100°C)	0.0047 (20–100°C)	0.0395 (0.600°C)
Mechanical properties:				
Tensile strength, ksi:				
At room temperature	100–500	100–150	120–200	75–150
At 500°C	175–200	35–45	35–65	35
At 1000°C	50–75	15–20	20–30	
Young's modulus of elasticity, psi $\times 10^{-6}$:				
At room temperature	59	27	46	12–15
At 500°C	55	25	41	6.5
At 1000°C	50	22	39	
Working temperature, °C	1700 down	Room	1600	Room
Recrystallization temperature, °C	1300–1500	1050–1500	900–1200	900–1300
Stress-relieving temperature, °C	1200	900	800	800
Nuclear:				
Cross section, thermal neutrons, barns/atom	19.2	21.3	2.4	1.1

* Ranges only: data vary with type of sample and previous work history.

Modulus of elasticity The variation in modulus is wide, with tungsten possessing a value about twice that of steel, while columbium's modulus is little higher than that of aluminum. The retention of high-modulus values at elevated temperatures is another factor in the use of these materials at elevated temperature.

Density The high density of tungsten and tantalum has resulted in their use as counterweights, and as radiation shields. Molybdenum and columbium are slightly more dense than steel.

Thermal expansion The low-thermal-expansion rates of tungsten and molybdenum make good materials to be used in conjunction with some glasses and ceramics. Tantalum and columbium have expansion rates which are relatively low compared to those of steels.

9-70 Nonferrous Metals

Electrical properties The electrical properties of these metals, especially tungsten and molybdenum, make them applicable as electric conductors. On the other hand, these alloys are used as heating elements.

The properties shown in Table 37 gives a general view of the properties to be expected in alloys of this group. Alloys based on these metals will generally retain the properties of the base metal in density, modulus of elasticity, and high-temperature resistance. Attention should be paid, however, to the effect of alloying on electric conductivity and thermal conductivity as well as thermal expansion.

TABLE 38 Compositions and Melting Points* of Some Solder Alloys

\multicolumn{5}{	c	}{Nominal chemical composition, %}	\multicolumn{4}{	c	}{Melting range}
					\multicolumn{2}{	c	}{°C}	\multicolumn{2}{	c	}{°F}
Sn	Pb	Sb	Ag	Ni	In	Solidus	Liquidus	Solidus	Liquidus
99.75	....	...	...	....	.....	232	232	450	450
99	....	1	...	....	.....	232	234	450	453
95	....	5	...	....	.....	233	240	452	464
96	....	...	4	....	.....	221	221	430	430
69.75	30	...	...	0.25	.....	183	192	361	378
63	37	...	...	....	.....	183	183	361	361
60	40	...	...	....	.....	183	188	361	370
50	50	...	...	....	.....	183	212	361	414
50	47	3	...	....	.....	185	204	365	399
40	60	...	...	....	.....	183	238	361	460
40	58	2	...	....	.....	185	231	365	448
10	90	...	...	....	.....	268	300	514	573
10	88	...	2	....	.....	265	295	509	563
5	95	...	...	....	.....	300	313	572	596
	97.5	...	2.5	....	.....	305	305	581	581
1.0	97.5	...	1.5	....	.....	309	309	588	588
	94.5	...	5.5	....	.....	304	365	579	689
	95	...	...	....	5	293	314	560	597
	90	...	5	....	5	290	310	554	590
	....	...	...	....	99.99	156	156	313	313
	50	...	...	....	50	180	209	356	408
	75	...	...	....	25	227	264	441	507

*Estimated melting point.

SOLDERS

Solders are a widely used group of alloys in the electronic and electrical industries. The most common use of these alloys is in making electric connections to hook up wire and component leads. Another frequent use of solder is in sealing of cans and other types of containers for purposes of hermeticity, sealing in inert gases or insulating gases, maintaining pressure, or sealing out contaminants.

Most useful solders have melting points in the range of 370 to 600°F. Their compositions are based primarily on tin and lead with minor alloying elements such as silver and antimony. Table 38 shows some of the major compositions along with melting points.

Solders are low-strength materials at room temperature and decrease rather rapidly in strength, even at the moderately elevated temperatures experienced by electronic equipment. This is understandable since we are approaching to within 150°F of the melting point of 60 percent tin–40 percent lead solder when we reach the

Solders 9-71

TABLE 39 Effect of Temperature and Strain Rate on the Mechanical Properties of 63% Tin–37% Lead Solder

Test temperature, °C	Strain rate, in./min	Tensile strength, psi	0.2% yield strength, psi	Elongation, in 1.4 in., %	Reduction in area, %
−130	0.050	18,000	10,300	27.4	33.5 Q
	0.050	17,850	10,380	31.3	37.1 Q
+25	0.050	5,250	4,600	86.9+	77.0 Q
	0.050	5,180	4,550	66.3	78.4
+150	0.050	1,110	1,000	133.6	99.2 Q
	0.050	1,080	980	174.0	99.3 Q
−130	1.0	18,950	11,100	23.7	23.5 Q
	1.0	19,350	11,050	26.7	33.9
+25	1.0	6,980	5,870	51.1	68.0
	1.0	6,850	5,760	62.4	75.7
+150	1.0	2,445	2,430	118.6	98.7 Q
	1.0	2,240	2,090	276.0	99.2 Q
−130	20.0	20,300	19,450	20.9	23.1
	20.0	20,600	20,300	21.4	23.5
+25	20.0	8,050	8,000	48.3	66.0
	20.0	7,910	7,900	45.6	68.9
+150	20.0	3,590	3,570	87.3	92.1 Q
	20.0	3,670	3,660	63.7 F	91.8 Q

Q Quarter break.
+ Elongation in 1.5 in.
F Flaw in fracture.

TABLE 40 Effect of Temperature on the Modulus of Elasticity and Poisson's Ratio of 63% Tin–37% Lead Solder Alloy

Temperature, °C	Poisson's ratio	Modulus of elasticity, psi $\times 10^{-6}$
−130	0.28	3.51
	0.35	3.54
	(0.32 avg.)	(3.53 avg.)
−60	0.36	2.32
	0.39	2.30
	(0.38 avg.)	(2.31 avg.)
+25	0.45	2.24
	0.36	2.07
	(0.40 avg.)	(2.16 avg.)
+80	0.36	1.32
	0.44	1.18
	(0.40 avg.)	(1.25 avg.)
+120	0.56	1.08
	0.35	1.03
	(0.46 avg.)	(1.06 avg.)
+150	0.55	0.89
	0.48	0.87
	0.43	0.78
	(0.49 avg.)	(0.85 avg.)

9-72 Nonferrous Metals

boiling point of water. Even at room temperature we are at relatively high temperature when considered on the absolute temperature scale.

Table 39 shows the effect of three temperatures on the mechanical properties of 63 percent tin–37 percent lead solder. Table 40 shows the effect of the same temperatures on elastic modulus and Poisson's ratio of 63 percent tin–37 percent lead solder, while Table 41 shows shear strength of solders at various temperatures.

These tables give an indication of the major effects that a relatively small change in temperature can have on mechanical properties of solder. The striking mechanical property of solders is the elongation or ductility. This ability to undergo large

TABLE 41   Shear Strength of Some Solder Alloys at Various Temperatures

Solder alloy	Shear strength, psi		
	−130°C	RT	+150°C
Tin	14,200	2,740	1,020
	13,600	2,810	1,020
99 Sn 1 Sb	14,250	2,780	1,150
	16,150	3,050	1,080
95 Sn 5 Sb	18,300	4,610	1,860
	18,000	4,640	1,900
96 Sn 4 Ag	16,650	4,660	1,490
	17,600	4,640	1,550
69.75 Sn 30 Pb 0.25 Ni	12,650	3,980	1,190
	12,780	3,950	1,280
63 Sn 37 Pb	12,350	4,120	1,160
	13,200	4,140	1,170
50 Sn 50 Pb	11,400	3,520	1,110
	10,850	3,510	1,080
40 Sn 60 Pb	9,550	3,760	1,190
	9,570	3,810	1,340
40 Sn 58 Pb 2 Sb	11,650	3,970	1,080
	11,500	3,780	1,080
10 Sn 90 Pb	7,330	2,760	1,550
	7,240	2,830	1,460
97.5 Pb 2.5 Ag	5,410	2,600	1,510
	5,220	2,580	1,370
97.5 Pb 1.5 Ag 1.0 Sn	5,820	3,070	1,540
	5,970	3,020	1,500
95 Pb 5 In	4,750	2,590	1,440
	4,460	2,630	1,370
90 Pb 5 In 5 Ag	5,920	3,530	1,770
	6,420	3,420	1,740

deformation without fracturing is a major factor in the good performance of solder in electric connections. It will also be noted that the rate of straining had a major effect on the tensile and yield strength of the solder, especially at elevated temperatures, where a ratio of four to one is shown at 150°C. This is probably due to the fact that at these temperatures solder creeps very rapidly. Thus when loads are applied slowly, creep takes place to the extent the load is relieved as it is being applied. This factor must be taken into account when long-time loads are applied to solder. Under these circumstances appreciable deformation may occur well below the stresses indicated by short-time tests.

Precious Metals

The precious metals find rather frequent use in electronic equipment, generally for their physical and surface properties rather than for their mechanical properties. Physical properties such as electric conductivity account for the use of such alloys as gold and silver. The resistance to form surface compounds which interfere with electrical and/or optical properties of surfaces account for the use of such materials as gold, rhodium, and platinum.

Tables 42 and 43 give some of the properties of the most frequently used precious metals. These are typical figures for the pure metals. Most of these metals may be alloyed with other metals to enhance certain properties, but often with major

TABLE 42 Physical Properties of Precious Metals

Metal	Density lb/in.^{3}	Melting temperature, °F	Coefficient of linear expansion, per °F × 10^{-6}	Thermal conductivity at 212°F, Btu/(hr)(ft^2)(°F/ft)	Specific heat, Btu/(lb)(°F)	Electrical resistivity at 32°F, μohm-cm
Platinum....	0.775	3217	4.9	42	0.031	9.83
Palladium...	0.434	2826	6.5	41	0.058	10.0
Rhodium....	0.447	3560	4.6	50	0.059	4.51
Ruthenium	0.441	4190	5.1	...	0.057	7.6
Iridium.....	0.813	4447	3.8	34	0.031	5.3
Osmium.....	0.82	5432	3.6	...	0.031	9.5
Gold........	0.698	1945	7.9	172	0.031	2.19
Silver.......	0.379	1761	10.9	242	0.056	1.47

TABLE 43 Mechanical Properties of Precious Metals

Metal	Tensile strength, ksi	Yield strength, ksi	Elongation in 2 in., %	Vickers hardness	Modulus of elasticity psi × 10^{-6}
Platinum........	18–21* 28–30†	2–5.5* 27†	30–40* 2.5–3.5†	40* 100†	25
Palladium........	20–28* 47†	5* 30†	24–40* 1.5	40* 100†	18
Rhodium........	73* 300†	....	....	122* 300†	55
Ruthenium.......	55*	......	......	240* 750†	68
Iridium..........	80* 290†	......	......	230*	79
Osmium..........	......	......	......	350* 1000†	80
Gold............	19* 32†	Nil* 30†	45* 4†	.....	12
Silver...........	22* 54†	8* 44†	48* 2.5†	.....	11

* Annealed.
† Cold rolled 50%.

effects on other properties. The form of the metal must also be considered. This is especially true of the electroplated form, since many of these metals find their major use as electroplates. Many electroplates have additives which are present for purposes of aiding in throwing power, smoothness, luster, or for other reasons. These additives may have major effects on such properties of the deposited metal. In addition, many electroplates are actually alloys, which may have drastically different physical and mechanical properties from those quoted for pure wrought metals. The density of electroplates varies somewhat also, and will result in variation in properties.

Silver Silver has the highest electric conductivity and thermal conductivity of any of the metals. It has found use as electric conductors and as a contact material. Probably the largest use of silver is as an electroplate on various types of pins, terminals, and contacts. The tarnishing of silver by sulfur compounds which are present in most industrial atmospheres is a major problem in the use of silver as a contact alloy or as a finish for terminals which must be soldered. The sulfide of silver interferes with electric contact and makes soldering difficult. As a result, most silver surfaces are protected with an overplate of gold which will protect the surface from sulfide formation. Problems have been encountered with this system of protection since thin or porous electroplates do not offer sufficient protection to prevent sulfiding through the gold electroplate. At present, 30 millionths of an inch of gold is considered sufficient.

Silver is a low-strength material which is not suitable for most structural applications. Alloying with copper is a means of increasing the strength of silver. At 10 percent copper, which is known as coin silver, electrical resistivity increases to 2.2 ohm-cm as compared to about 1.6 ohm-cm for pure silver. The tensile strength of the 10 percent copper alloy in the annealed condition is almost twice that of silver in the annealed condition.

Gold Gold is used widely as an electroplate to give corrosion resistance, to provide a contact surface with low resistance, and in some special cases as a conductor material. The low strength of gold makes it of little use as a structural material or as a wear-resistant material. Alloys are available in both wrought and cast form and in the electroplated form. Alloys such as cobalt may increase the resistivity quite drastically. For instance, 1 percent cobalt increases the resistivity of gold by 710 percent.

Rhodium The high reflectivity and low electrical surface resistance of rhodium combined with its high hardness and the stability of the surface under corrosive conditions makes rhodium highly suitable for contact surfaces and for mirrors. Rhodium is primarily applied by electroplating.

REFERENCES

1. The Aluminum Association: Standards for Aluminum Products.
2. ASTM Standards: part 6, Light Metals and Alloys, 1964.
3. Kawecki Berylco Industries: *Technical Bulletin* 2200.
4. Kawecki Berylco Industries: *Technical Bulletin* 2100.
5. Copper Development Association: Standards Handbook—Wrought Mill Products.
6. Kawecki Berylco Industries: *Technical Bulletin* 304 03 PDI.
7. Brooks and Perkins: Technical Bulletin entitled "Magnesium-Lithium Products."
8. MIL HBK 693.
9. The Dow Metal Products Company: Technical Bulletin entitled "Magnesium Mill Products."
10. International Nickel Company: Handbook of Huntington Alloys.
11. Fansteel Metallurgical Corp.: Technical Bulletin entitled "Fansteel Metallurgy."
12. NASA Contract NAS8-21233, Final Report.
13. Brooks & Perkins: Technical Publication entitled "Metalscope."
14. Reynolds Metal Co.: Technical Bulletin entitled "Aluminum Structural Design."
15. Aluminum Company of America: Technical Bulletin entitled "Forming Alcoa Aluminum."
16. The Beryllium Corp.: Technical Bulletin entitled "Beryllium Copper Alloys."

Chapter **10**

Metallic and Chemical Finishes on Metals and Nonconductors

H. L. PINKERTON

*Aerospace and Electronic Systems Division,
Westinghouse Electric Corporation, Baltimore, Maryland*

Finishes and Substrates	10–2
Types of Finish	10–2
Types of Substrate	10–2
Reasons for Finishing	10–2
Electrodeposition	10–5
Basic Theory	10–5
Current and Metal Distribution	10–8
Bath Constituents and Their Functions	10–11
Bath Operation and Control	10–12
Design Considerations	10–15
The Substrate Metals	10–19
Electrodeposition of Metals and Alloys	10–26
Postplating Treatments	10–27
Properties of Electrodeposited Coatings	10–27
Brush Plating	10–30
Electroforming	10–32
Electromachining	10–35
Stripping of Electrodeposits	10–36
Nonelectrolytic Deposition of Metals	10–36
Deposition on Metallic Substrates	10–36
Deposition on Nonmetallic Substrates	10–39
Electroplating on Metallized Surfaces	10–44
Nonmetallic Coatings on Metals	10–45

Anodic Films	10–45
Conversion Coatings	10–48
Corrosion Resistance	10–49
Mechanism of Corrosion	10–49
Testing of Deposits and Coatings	10–58
Thickness Testing	10–58
Porosity Testing	10–59
Adhesion Testing	10–60
Stress Measurement	10–60
Solderability Testing	10–60
Mechanical Property Tests	10–60
Government Specifications for Finishes	10–61
References	10–61

FINISHES AND SUBSTRATES

Types of Finish

It is the intent of this chapter to describe the properties, advantages, and limitations of metallic and nonmetallic finishes on metals and on such nonconductors as plastics and ceramics. The nonmetallic finishes exclude organic coatings, but do include chromates, oxides, phosphates, and the like formed in situ. The entire approach, of course, is slanted to the electronics industry, which means that there will be an unusual interest in plating upon some exotic substrates, and often in depositing metals considered rare by the run-of-the-mill plater. Since the end use of much electronic gear is extremely sophisticated, performance is usually the overriding consideration rather than cost, and hence we will find design engineers specifying heavy deposits of such costly metals as gold or rhodium.

Designers of electronic equipment are not usually conversant with the art of metal finishing. The treatment of the subject matter of this chapter is therefore somewhat more basic than is usual in a handbook, in which much of the material can be easily comprehended by the knowledgeable reader when compressed and presented to him in graphic or tabular form. The purpose of the present approach is to give the designer not just a cold assembly of facts, but rather enough familiarity with plating technology so that he will be able to specify finishes intelligently. He should be informed sufficiently, for instance, to avoid geometries that are unnecessarily difficult and costly to plate, or even impossible. The parts he designs must also function with reliability and efficiency with respect to their physical, mechanical and electrical characteristics when finished as he specifies. They must also exhibit adequate corrosion resistance for the anticipated environment.

Types of Substrate

In air- and space-borne equipment, where weight is of utmost importance, the substrates that the designer turns to are the light metals—magnesium, beryllium, and aluminum—or the plastics which, besides their low density, have other properties functionally desirable for certain applications. Furthermore, the designer does not consider himself restricted to the most commonly plated plastics, such as ABS and a few others, but will feel free to use any plastic that has the particular properties he needs.

REASONS FOR FINISHING

The incentive to coat a nonmetal with a metal, or a metal with another metal or metallic compound may be simply to decorate it, or to protect it from some

TABLE 1 Definition of Terms

Activation. Elimination of a passive surface condition.
Addition agent. A material present in a plating bath (usually in small quantities) for modifying the character of the deposit.
Adhesion. (1) The attractive force between a deposit and its substrate, or, (2) the force required to separate them.
Anion. A negatively charged ion which migrates to the anode when direct current is passed through the solution.
Anode. The positive electrode in electrolysis, at which current enters, or electrons leave, the solution; where positive ions are formed and negative ions are discharged; where oxidizing reactions occur; the anode usually tends to dissolve.
Anodizing. Anodic treatment of metals (particularly aluminum).
Base (adj.). The opposite of noble (q.v.).
Basis (adj.). The substrate material upon which coatings are deposited.
Bipolar electrode. An electrode not directly connected to the power supply, but present in the solution, whereby the part nearest the anode becomes cathodic and the part nearest the cathode becomes anodic.
Brush plating. A method of plating in which a pad wetted with plating solution is connected as an anode and is moved over the cathode.
Buffer. Any substance in a solution which causes it to resist the normal change in pH when acid or alkali is added. Each buffer has a characteristic range of pH over which it is effective.
Burning. Refers to an unacceptable deposit formed at excessive current density.
Cathode. The negative electrode in electrolysis, at which current leaves, or electrons enter, the solution; where negative ions are formed and positive ions are discharged; where reducing actions occur; the electrode which receives the deposit of metal and/or hydrogen.
Cation. A positively charged ion.
Contact plating. Deposition without an outside source of current, by immersion of the work in solution while *contacting another selected metal.*
Conversion coating. A superficial layer of a compound of the basis metal, which is produced by chemical (or sometimes electrochemical) means.
Current density. (1) Average: total current divided by total area; (2) actual: the local value of current density at a selected point or area on the electrode.
Deionization. The removal of ions from a solution by ion exchange.
Drag-in. Water or solution adhering to an object entering a solution.
Drag-out. Solution removed from a bath which is clinging to an object removed from it.
Dummy. A cathode used for working a plating bath, which cathode is then discarded.
Efficiency (anodic or cathodic). The relation of the amount of desired electrode reaction per ampere hour to the theoretical maximum.
Electrobrightening. Treating briefly in an electropolishing bath.
Electrodeposition. The process of depositing a substance by electrolysis—includes electroplating, electroforming, electrorefining, and electrowinning.
Electroless deposition. Deposition without the use of an externally imposed current, which deposition continues indefinitely at a substantially constant rate. Also called catalytic deposition.
Electropolishing. Improvement in surface finish (rms) of a metal, effected by making it the anode in an appropriate solution.

TABLE 1 Definition of Terms (Continued)

Electrorefining. The electrometallurgical process of anodically dissolving metal from an impure anode and depositing it in a more pure state.

Electrowinning. The form of electrorefining in which insoluble anodes are used to deposit relatively pure metals from impure solutions derived from ores or scrap.

Embrittlement. See *Hydrogen embrittlement*.

Epitaxy. The assumption by a deposit of the crystal habit and orientation of the substrate.

Flash plate. A metal deposit of undefined (very thin) thickness.

Galvanic series. A list of metals and/or alloys arranged according to their relative potentials *in a given environment*.

Hard chromium. Chromium plated (usually to 0.1 mil or more) for engineering purposes—not actually harder than decorative chromium.

Hydrogen embrittlement. Loss of ductility of a metal (to the point of brittleness and incipient fracture) caused by absorption of hydrogen during cleaning, pickling, or plating.

Hydrogen overvoltage. The irreversible excess of potential required for liberation of hydrogen at an electrode.

Immersion deposit. A metallic deposit formed noncatalytically without an outside source of current, and therefore limited to (usually) about 20 millionths of an inch in thickness.

Inhibitor. A substance in a solution which reduces the rate of attack of the solution upon a metal surface, but not upon metal *compounds* on the surface.

Ion. An atom or group of atoms which has lost or gained one or more electrons, and thereby possesses a net electric charge.

Karat. A twenty-fourth part. Thus, 18-karat is $18/24$ pure.

Leveling action. The production of a deposit smoother than the basis metal.

Limiting current density. The maximum current density at which satisfactory electrode reactions will occur.

Matrix. The cathode form, mold, or mandrel used in electroforming.

Metallizing. The application of an electrically conductive layer upon a nonconductor, by nonelectrolytic means.

Noble. A noble metal does not readily tend to form ions, and therefore does not dissolve readily, nor easily enter into reactions, such as oxidation, dissolution, etc.

Passivity. A condition of a metal such that it assumes a more noble potential than usual.

Periodic reversal. A method of applying direct current in which the polarity is reversed at predetermined intervals.

Pitting. (1) Production of pores or depressions in an electrodeposit; (2) nonuniform corrosion causing isolated craters or holes.

Polarization. Potential change of an electrode during electrolysis such that the electrode voltage is always higher than would be the corresponding static potential.

Resist (noun). A material applied to a selected area of a cathode to render the surface nonconducting.

Rms (root mean square). A measure of surface roughness; actually the square root of the mean of the sum of the squares (in microinches) of the distances (of the trace of a surface) above or below a mean reference line.

Robber. See *Thief*.

TABLE 1 Definition of Terms (Continued)

Sacrificial protection. Corrosion protection provided by corrosion of the coating rather than of the basis metal, even though the latter be exposed to the corroding medium through pores, scratches, or even bared areas.

Shield (noun). A nonconductive medium interposed between anode and cathode for the purpose of altering current flow and distribution.

Stop-off. See *Resist*.

Thief. An auxilliary cathode of scrap metal which is positioned near an area of high current density in order to "steal" current from the workpiece and so prevent burning.

Substrate. The basis metal.

Throwing power. The improvement of metal distribution ratio over the primary current distribution ratio. The term is also useful for anodic processes such as the formation of oxides in anodizing.

Whisker. A microscopic metallic filament growing out from the surface of a metal (usually an electroplate), a phenomenon not completely understood. On electroplates, this usually occurs long after the deposit has been laid down.

---

anticipated corrosive environment, or to provide new functional properties which the original substrate did not possess, or any combination of these reasons.

In the electronics field, the decorative aspect is minimized, as compared with consumer markets. Protection, however, is of the utmost concern with respect to the reliability of the product, especially in air- or space-borne and naval equipment. In any case, there must be absolutely no deterioration in service, since even mild corrosion can often result in gross malfunction of the component.

Frequently in the manufacture of electronic gear, the primary reason for applying a finish is purely functional: to afford magnetic shielding, as by iron plating; to provide for rf shielding; to permit welding or soldering, or to preserve solderability over long storage periods in the manufacturing process; to improve the absorption or radiation of heat, or to yield a highly specular reflection of light.

Probably the most widely used procedure for attaining one or more of these objects is to electroplate the surface.

ELECTRODEPOSITION

Basic Theory

Faraday's laws In order to provide the designer with a background which will enable him to specify electroplated finishes that will serve his purposes adequately, it will be well to take a brief excursion into electroplating theory.

The basic laws of electrochemistry are, of course, Faraday's:

1. *In the process of electrolytic changes, equal quantities of electricity charge or discharge equivalent quantities of ions at each electrode.*

2. *One gram-equivalent weight of matter is chemically altered at each electrode for each faraday of electricity passed through the electrolyte.*

A faraday is $96,516 \pm 2$ coul on the physical scale (on the chemical scale, the atomic weight of the naturally occurring mixture of oxygen isotopes is taken as 16.0000, whereas on the physical scale this value is assigned to the isotope O_{16}) or $96,489 \pm 2$ coul on the chemical scale. One coulomb is a current of one ampere flowing for one second, or the quantity of electricity required to deposit 0.00111880 g of silver from a solution of silver nitrate.

Field theory Essentially, electroplating is the art of creating a field of force in a solution containing metallic ions, such that the ions are moved to the cathode and are there discharged (in the ideal situation) as a uniform layer of metal

10-6   Metallic and Chemical Finishes on Metals and Nonconductors

over it. The first application of the general theory of the potential field to electrolytes was made in 1855 by Riemann, but it was not until the work of Kasper in 1939 to 1942, and later of Kronsbein, that any serious application of the theory was made to electrodeposition. The subject is mathematically so complex that only the simplest geometric configurations can be rigorously handled, and even these require certain simplifying assumptions that will apply only approximately to a few real plating solutions. This serves to explain why plating is still an art. Even so, however, a general grasp of the fundamental theory of the potential field in electrolytes and its implications, in the form of a few generalizations, may prevent the designer from asking the impossible of the plater, and will certainly assist the plater in doing what the designer is within his right in asking for.

When a dc potential is imposed between two metallic conductors immersed in an electrolyte, *every* point in the electrolyte assumes some potential intermediate

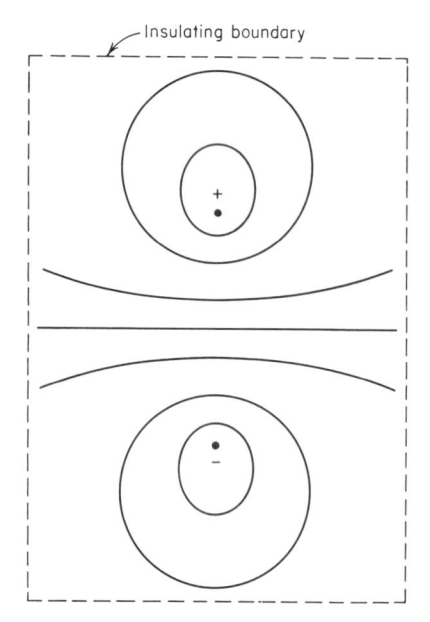

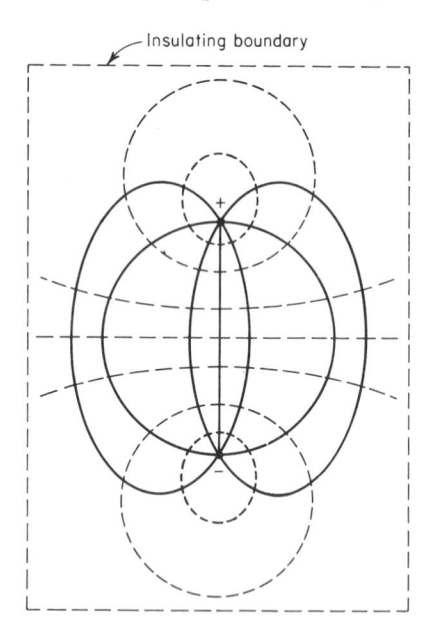

Fig. 1   Trace of equipotential surfaces in a plane of electrolyte with two line electrodes.

Fig. 2   Trace of surfaces of force in a plane of electrolyte with two line electrodes.

between those of the two electrodes. Since the electric conductivity of the metal electrodes is usually of the order of several million times that of the electrolyte, it may be said that every point on the surface of an electrode is at the same potential: i.e., it is an equipotential surface. Close to one electrode, this surface will resemble that electrode in shape; however, as it moves away from that electrode, the shape of the surface will change, until at last it conforms to that of the other electrode (Fig. 1).

It is a basic law in the potential field theory that the traces of the equipotential surfaces and of the surfaces of force form an orthogonal net—that is, they everywhere cross each other at right angles (Fig. 2). A surface of force is a path along which the current flows, and since the electrodes are themselves equipotential surfaces, it follows that the current enters or leaves *each electrode* (whatever its shape) perpendicularly to the electrode surface at every point. This is a most important consideration to keep in mind, when later we consider the metal distribution on the cathode.

Electrodeposition 10-7

If we replace the one-dimensional electrodes of Figs. 1 and 2 with a typical anode section and a more practical cathode (in this case an angle), the equipotential lines and the lines of force will assume something of the shape shown in Fig. 3. This figure is an approximation illustrating an important point, and is so drawn that between any two lines of force an equal number of amperes of current are flowing, and some smaller number between the outer line and the cell walls. Later, we will inquire more fully into the implications of this.

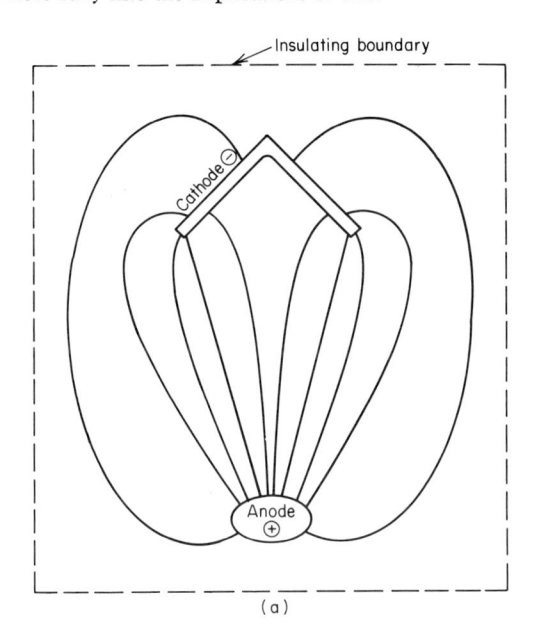

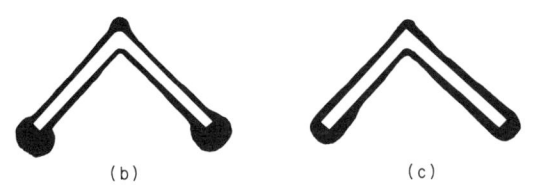

Fig. 3 (*a*) Lines of force in a hypothetical plating tank with insulating walls. (*b*) Metal distribution pattern on the cathode of Fig. 3*a* in a bath with 100% cathode efficiency. (*c*) Metal distribution pattern on the cathode of Fig. 3*a* in a bath with good throwing power.

It is also fundamental in field theory that any equipotential surface may be *totally* replaced by a (perfect) conductor without disturbing the field. Similarly, the *total* substitution of an insulating boundary for any surface of force is without influence. On the other hand, when anything, conductor or not, is placed in the solution crossing any of the lines of the net, the field is disturbed. As will be seen, these concepts can be put to good use.

It should be observed that these diagrams represent a horizontal planar section through an electrolyte contained within insulating walls. To picture the actual three-dimensional system, the diagrams must be moved perpendicularly to the paper. Practically, this implies that both electrodes extend from the surface of the solution to the bottom of the tank. When this is not so, the situation is, of course, altered.

Current and Metal Distribution

Primary current distribution The current distribution over an electrode in the absence of any disturbing effects occurring in the vicinity of the electrode (associated with dissolution or deposition) is called the primary current distribution and can be deduced at least qualitatively from field theory.

There are basically only three possible electrode configurations which give perfectly uniform primary current distribution: infinite parallel planes, infinite concentric cylinders, and concentric spheres. None of these arrangements has as yet been offered to any shop to be plated, but under the rules cited above we have certain freedoms: we may replace (1) any equipotential surface with an electrode or (2) any surface of force with an insulating boundary. Thus, we may box in our infinite parallel electrodes with *insulating* surfaces perpendicular to them: four vertical walls, a horizontal bottom, and an air surface. A plating rack filled with closely spaced flat parts, equidistant between two parallel rows of closely spaced anodes, is about as close to this ideal arrangement as the practical plater is ever likely to see.

In the same way, we may place our infinite cylinders absolutely vertically, cut them off with a horizontal insulating bottom and a parallel air surface at the top, and obtain perfect primary current distribution, if the cylinders are absolutely concentric. This arrangement is approached in the plating of gun barrels and is complicated only by the presence of lands in the barrel and by a few assorted engineering difficulties. We are also free to use radial sections of these cylinders if we need to go to this trouble. One interesting property of this concentric cylinder arrangement is that, for a given current density (current per unit area of electrode) the voltage required is a function only of the *ratio* of their radii, not of their absolute value, which may be mils or miles.

The case of concentric spheres is understandably rare in the real world and deserves no further discussion.

Secondary current distribution While the primary current distribution is almost always the major controlling factor in determining metal distribution, it is modified by other disturbing influences in the cathode film which arise from the passage of current and produce what is called the secondary current distribution. This is what actually does the work of deposition. The cathode film extends about 0.10 to 0.25 mm into the bath and is to be regarded as a region of altered composition and concentration rather than as a film in the usual sense.

Polarization. The potential across the cathode film while current is flowing varies from point to point over the cathode, depending upon various factors (and the local variations in current density). The difference between the dynamic and static potentials is known as the *cathode polarization.* The major factors causing this potential difference are:

1. A reduction in the concentration of metallic ions present in the cathode film. This is called *concentration* polarization.
2. The voltage required to move metal ions through the field and to deposit them. This is known as *activation or deposition* polarization, and varies with the depositing metal and with the substrate.
3. The reduction of the effective cross section of the cathode film caused by evolution of hydrogen at the cathode, called *gas* polarization.

Somewhat similar effects occur at the anode, but because they are usually more uniform, anode polarization* affects only the total bath voltage required, but not the current distribution.

---

* Platers sometimes speak of a "polarized anode"—this refers, not to what we are discussing, but to an anode which is not dissolving properly and which is causing a more than normal voltage drop.

With reference to the cathode, the electrochemical properties which determine the secondary current distribution are the cathode polarization and the conductivity of the solution. The *primary* current distribution between two points 1 and 2 on a cathode (Fig. 4) is given by the equation

$$\frac{i_1}{i_2} = \frac{d_2}{d_1} \tag{1}$$

where i_1 and i_2 are the current densities at points 1 and 2, respectively, at distances d_1 and d_2 from the anode *measured along the lines of force traversed by the current*. The secondary current distribution is generally accepted to be given by the equation

$$\frac{i_1'}{i_2'} = \frac{d_2 + k(dE/dI)}{d_1 + k(dE/dI)} \tag{2}$$

where k is the solution conductivity and dE/dI is the slope of the cathode polarization curve between the current densities i_1 and i_2. The second terms in the

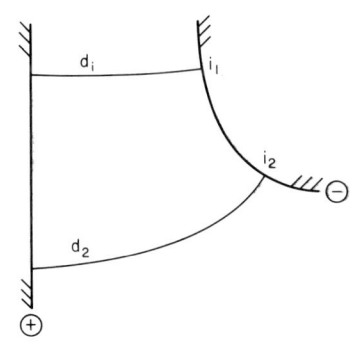

Fig. 4 Primary current distribution between points on a cathode.

numerator and denominator are identical and have the dimensions of a length. The effect of polarization is therefore equivalent to adding an equal solution length to the path lengths d_1 and d_2. Also, therefore, the secondary current distribution is always more or less of an improvement over the primary distribution.

Metal distribution The rate of metal deposition at any point on a cathode is determined by the current density existing at that point in the secondary current distribution and by the cathode efficiency of the bath *at that current density*. Thus, referring to Fig. 3, we may observe that the current density will be highest at the ends of the angle shanks and on the outer corner of the angle. If the cathode efficiency remains constant over that range of current density, the metal distribution will be of a shape shown in Fig. 3b.

Since the actual secondary current distribution is not known, the metal distribution cannot be predicted accurately. With a sufficient knowledge of a given system, however, the *directional* effect of a certain change in variables can often be stated, and an improvement in metal distribution can thereby be effected. It should still be remembered that the primary current distribution, which can be controlled in an empirical sort of way, is more often than not the controlling factor.

Control of the secondary current distribution is complicated by the fact that the governing variables are completely interrelated: e.g., a temperature change affects the resistivity and viscosity of the cathode film, the ionization equilibrium of complex compounds, the hydration and mobility of ions, and the cathode efficiency. These may have conflicting effects on metal distribution and the result of their interaction is unpredictable.

Since the variables that control the secondary current distribution lie almost entirely within the plater's province, they are not the designer's concern.

Throwing Power. Long ago, platers invented the term *throwing power* to describe the ability of certain plating baths to "throw" more metal into areas of low current density than other baths would under the same circumstances. Thus, referring to Fig. 3, if the bath electrochemistry is such that the cathode efficiency falls off as the current density increases, less metal will be deposited on high-current-density areas than would be expected, and the metal distribution will be more uniform than the current distribution (Fig. 3c). This bath is said to have good throwing power.

In 1923, Haring and Blum[3] described their well-known throwing-power box (the Haring cell) and defined throwing power as a number derived from the deviation of metal distribution from the primary current distribution. Their number is not a property of the solution, and no useful calculations can be made from it, but it does serve to arrange various plating baths qualitatively on their ability to throw. Thus, chromium plating baths have miserable throw; other acid plating baths can be expected to throw considerably better (usually almost what would be expected from the primary current distribution); most cyanide baths throw quite well (their efficiency–current-density curves would indicate this); and stannate tin baths are about the best.

In 1949 Graham[4] and coworkers made the surprising observation that in micropores such as are present in sintered powder metal compacts, the usual order of the throwing power of baths is completely reversed. Kardos[5] showed theoretically that when the dimensions of a cavity approach the order of magnitude of the cathode film thickness, microthrowing power is exhibited (just as Graham had found), and he offered an elegant explanation.

Deposition Potential. As we have seen, it requires a certain voltage to deposit the ions of a given metal on a particular metallic substrate (we speak here only of the voltage across the cathode film), and this voltage is different for the various valences of a single metal as it exists in the given solution (e.g., monovalent copper in a cyanide solution, and divalent copper in an acid solution).

Hydrogen Overvoltage. Hydrogen deposition requires a similar potential, or overvoltage, and this also varies with the composition of the solution, the cathode metal and its surface condition, the current density, the temperature, and also with impurities in the solution, even in extremely minute traces.

Effect of Cathode Efficiency. Whether we deposit only metal, only hydrogen, or a mixture of the two depends upon their relative deposition potentials under the operating conditions, and in any case Faraday's laws will hold. If the potentials are close, both metal and hydrogen will be deposited, but when they are widely separated (e.g., 0.1 volt or more) only one will be deposited. The cathode efficiency is the percentage of the current useful in depositing metal. For instance, in acid zinc sulfate solution, hydrogen discharge requires at least -1.04 volt, whereas the metal reduction potential is only -0.76 volt. Hence, zinc can be deposited at deposition potentials (remember this is not the total bath voltage) of -0.76 to -0.9 volt at substantially 100 percent efficiency. In cyanide baths, the hydrogen overvoltage is considerably lower and is close to that of zinc or copper; consequently in cyanide zinc baths efficiency of metal deposition is usually much less than 100 percent. (In silver cyanide baths and in gold cyanide baths, the cathode efficiency is 100 percent. The modern type of acid gold bath—still a cyanide type, but containing no *free* cyanide—often gives efficiencies as low as 40 percent.) To illustrate, in a zinc cyanide bath operated at a cathode current density of 10 asf (amperes per square foot) the cathode efficiency will be about 90 percent; at 20 asf it may be 75 percent, at 25 asf 60 percent, and only 45 percent at 30 asf. The useful current density effective in depositing metal, thus, would be, respectively, 9, 15, 15, and 13.5 asf. There would be no advantage, then, in operating this bath at current densities much over 20 asf. But it can be seen that this bath will have excellent throwing power. If the current density on the high points were, for example, 30 asf, and 10 asf in recesses, the metal thicknesses on these points would be in the ratio of 1.5:1 instead of 3:1. This bath obviously would have good throwing power, as illustrated in Fig. 3b.

Bath Constituents and Their Functions

Major constituents The plating bath must contain a soluble metal cation and an anion which usually serves to dissolve the anode. The bath may also contain buffer salts which are designed to maintain it in the proper pH range; salts to improve the conductivity; wetting agents to prevent pitting; and additives for brightness, leveling, or other purposes.

Cations. The soluble metal salt is usually present in a fairly high concentration, because this permits the use of higher current densities and hence faster plating. In the precious-metal baths, such as gold and rhodium, low concentrations are usually

TABLE 2 Anions Commonly Used for Plating and Some of Their Properties

Anion	Used for plating:	Anode corrosion	Bath stability	Bath corrosivity to equipment
Chloride	Fe, Co, Ni	Excellent	Good	High
Chromate	Cr	*†	Excellent	High
Cyanide	Cd, Cu, Au, In,* Ag, Zn	Restricted to low-anode current density	Needs careful control	Very low
Fluoborate	Cd, Co, Cu, In, Fe, Pb, Ni, Sn	Excellent	Excellent	Very high
Phosphate	Rh	*	Good	High
Pyrophosphate	Cu	Good	Needs careful control	Low
Stannate	Sn	‡	Needs careful control	Very low
Sulfamate	Ni,§ Co	Good	Good	Moderate
Sulfate	Co, Cu, In, Fe, Ni,§ Rh, Zn	Good§	Excellent	Moderate (except Fe baths)

NOTES: Some metals are also plated from mixed anion baths, as nickel from sulfate-chloride (Watts' type), tin from chloride-fluoride (Halogen bath) and silver from cyanide-nitrate (Wood's bath). Alloy baths also are often mixed-anion formulations.
  * Insoluble anodes are used.
  † Anion is metal source, and needs catalyst—control ticklish.
  ‡ Requires special attention.
  § Limited by very poor anode corrosion for nickel when no chloride is present.

used because of the high metal cost; plating speeds are therefore much slower in these baths.

Anions. The anion is chosen on the basis of a number of properties: solubility of the particular metal salt; the degree to which the metal is complexed (held in a complex radical, such as the cyanide in contrast to a simple, readily disassociated molecule such as a chloride); the corrosivity of the anion for the anode metal; and its chemical stability. The most commonly used anions are listed in Table 2. For various reasons, nitrates, nitrites, chlorates, perchlorates, iodides, bromides, borates, carbonates and organic acid radicals are not used as the metal salts, although some of these may be used for other purposes.

Minor constituents *Conductive salts* are sometimes added for the sole purpose of improving bath conductivity, which is important for two reasons: a highly conductive bath materially reduces the operating cost; and also in a highly conductive bath the factors that control the secondary current distribution will have more effect, as may be seen from Eqs. (1) and (2).

Wetting Agents. Surfactants are added to certain baths to avoid pitting. Pitting of a deposit is disastrous from every point of view. It occurs most often in the acid baths and can be traced to several causes, the most common of which are: suspended insoluble matter in the bath, organic contamination of the solution, and (sometimes associated with this) gas pitting. We will discuss contamination later; gas pitting occurs most often on nonvertical surfaces in baths with less than 100 percent cathode efficiency. The evolved hydrogen may cling to the surface (sometimes even a vertical one) and the bubble sits there preventing metal deposition, while a small round depression is formed in the plate, often with a streak above it caused by gas release as the bubble grows larger. Surface tension holds the bubble on the metal; consequently, by reducing the surface tension of the bath, we can prevent it from clinging and causing an observable pit. The surface tension of a plating bath is usually close to that of water—about 72 dynes/cm; by reducing this to 30 to 40 dynes/cm, pitting can often be prevented. For low levels of organic contamination, pitting from this source can also be relieved.

TABLE 3 Control Tests Used to Monitor Bath Contamination

Test	Method of operation	Notes
Hull cell.............	Sample is plated under controlled conditions of cell geometry to give a defined range of current density variations over the panel. Various refinements and variations of cell type are available.	Defects are shown at a lower impurity level than that which would cause rejected work. Experience is required to evaluate test panels.
Bent cathode........	Sample is plated on a specially formed cathode (U-, J- or pigtail). More limited than Hull cell, but widely used for certain baths.	Same.

Both tests can detect the presence of impurities that would defy any analytical method.

Buffers. Buffers are compounds used in a number of (usually acidic) baths to help maintain pH at the desired level. These are simply compounds which (over a specific narrow pH range for each buffer) resist normal pH changes that would otherwise occur owing to local depletion of metal content and other chemical imbalances due to electrolysis.

Additives. Specially selected compounds are added in minor amounts (sometimes in the range of parts per million) to achieve special objectives, of which the most common are brightness and leveling. These compounds are discovered purely empirically, often by evaluating hundreds until one is found which works. Nothing more need be said, except that the best of them produce bright plates, and some will produce a plate that is smoother than the surface on which it is deposited; whereas in the normal course of events, the plate is always rougher than the substrate.

Bath Operation and Control

Solution control All solutions in a plating line must be analytically controlled on a scheduled basis. The most critical solution is the plating or processing bath, which should be analyzed daily and maintained within specified limits.

Impurities. In a plating bath that is otherwise within control limits impurities can cause every defect known to the art. Modern practice, therefore, is to use specially designed tools (Table 3) for their detection and control.

Metal Concentration. The total metal concentration in the bath is important in that the upper limit of the current density range is raised with increasing metal content. The voltage required for a given current density decreases with increasing metal content, and hence operating cost is lower. On the other hand, the increased metal losses due to drag-out of solution are directly proportional to the metal concentration of the solution. The drag-out volume depends on the orientation of the surface (vertical or horizontal) and on the care or sloppiness with which the transfer is made:

	Drag-out, gal/1,000 ft^2
Absolute minimum for vertical sheet	0.4
Vertical surface, typically drained	2.0
Vertical surface, poorly drained	4.0
Horizontal surface, well drained (minimum)	0.8
Horizontal surface, poorly drained	10.0
Rough average for production parts	3.0–7.0

The final choice of metal concentration is based upon a proper balance between these two factors, since the initial cost of the bath is a nonrecurring capital expense.

Operating Temperature. The bath temperature chosen is also a balance between several cost factors, but in general, it is most economical to operate a bath at an elevated temperature when this is compatible with the bath chemistry.

Agitation. Some form of agitation is always desirable, but the choice of method is important. Slow movement of the part itself provides the best agitation and is to be preferred. Solution agitation (by stirrer or pump) can lead to undesirable impingement effects; air agitation is less effective, and can cause pitting, and also roughness by stirring up sediment.

Anode performance Besides being the positive electrode, the anode usually also serves as the source of metal replenishment to the solution, and in the ideal case, it dissolves at a current efficiency equal to the cathode current efficiency, whereby no metal salts need ever be added to the solution after the initial bath makeup. Practically, the anode efficiency is almost always slightly higher than the cathode efficiency, and since the plated parts drag out solution (and its metal content), this often results in a balance, with the same desirable outcome. In fact, very few baths require metal replenishment, and in some (notably the cyanide baths) some insoluble anodes may have to be used to correct for the higher anode efficiency as compared to the efficiency of deposition.

Insoluble Anodes. In some situations, all anodes in the bath may be insoluble. Then metal salts must continually be added to the solution under analytical control. Chromium baths are always operated in this way, while for many years (and for obvious reasons) soluble gold anodes have been almost universally replaced with insoluble anodes. When insoluble anodes are used, oxygen is evolved at the anode in considerable quantities, which causes some mechanical loss of solution, and frequently represents a health hazard. Exhaust ventilation is usually necessary.

Auxiliary Anodes. A problem occurs when the geometry of the part is such that it is difficult to obtain the desired metal thickness in a recess without gross overplating of the high-current-density areas, sometimes to an extent that is beyond dimensional tolerances.

The plater's solution is to use either auxiliary or bipolar anodes. An auxiliary anode is electrically connected to the anode circuit, and it is preferably positioned on, but well-insulated from, the plating rack, and its function is obvious. A bipolar anode, on the other hand, is not electrically connected. It is merely hung (again, preferably positioned on the rack, from which it must be insulated) in a strategic position; because it is electrically so much more conductive than the solution, it acts as a funnel, and carries much larger currents to the starved areas. The bipolar anode may be of carbon or a metal insoluble in the solution, or it may be made of anode metal. In the former case replacement is not needed, but metal will deposit on the end nearest the anode, while at the other end hydrogen will be evolved. Provision must be made to allow this gas to escape freely: if it is trapped, there will be areas that are bare or will be only thinly plated. The soluble metal anode will build up on one end while the other end will dissolve. At some point this type of bipolar anode must be replaced (sometimes it is sufficient merely to turn it around).

Current density limitations Each bath formulation will have a practical range of current density over which no untoward effects will be expected at either anode or cathode. Since usually the anode and cathode current density ranges will differ, we frequently find that some anode-to-cathode area ratio is specified, which should be adhered to for extended periods of bath operation. For short periods (an hour or so) there will be only slight disturbance of the bath chemistry if the recommended anode current density is exceeded. The limiting current density is a function of the bath composition and of the operating conditions of temperature and agitation.

Excessive Anode Current Density. This results in a lowered anode efficiency and evolution of oxygen from the anode surface. This sometimes causes a tenacious oxide film on the anode which cannot yield metal ions even when normal current density is restored. The plater speaks of this as a "polarized" anode, and he must remove the film mechanically or chemically before the anode can again function normally.

Excessive Cathode Current Density. This causes dark, rough corners and edges, and in worse situations, a "burnt" deposit—spongy and nonadherent. There are available a number of expedients for correcting such a situation other than by simply lowering the current density:

1. Increase the metal content of the bath, or use a different type of bath.
2. Raise the bath temperature.
3. Change the position of the cathode.
4. Use increased agitation, particularly of the cathode itself.
5. Use shields or thieves.

A *thief* is simply a piece of wire or scrap metal connected to the cathode rod or to the cathode itself to "steal" some of the excess current. A thief is cheap to use, but it may be expensive to throw away; no one would consider using a thief in a rhodium bath where the metal costs $2,200 a pound.

Far better, and much more elegant, is the use of a shield, but proper application of this requires at least a smattering of field theory. A *shield* is a nonconducting barrier placed in the current path (frequently attached to the plating rack) in such a position relative to the areas of excessive current density that the current path is so significantly lengthened that a more acceptable current density will be imposed on the areas where burning has been occurring. The use of shields yields a dual benefit: not only is an acceptable part produced, but also the improved metal distribution means that a specified deposit thickness can be obtained in low-current-density areas with a reduction in the total amount of metal deposited. For expensive metals, this can represent a real saving.

Current Form Variations. The usual direct current obtained from a rectifier has a ripple (varying from 3 to 5 percent for three-phase to over 20 percent for single-phase rectification) which has virtually no effect on the character of plate deposited. In special cases, an alternating current has been superimposed purposely for some claimed advantages, the substantiation of which has not yet clearly been proved. Periodic reversal of direct current (PR plating) has well-substantiated advantages, and has been particularly useful in certain baths, leading to smoother deposits, and also better metal distribution. The pioneer work in the field was done by Jernstedt[6] and his associates. The penalty paid, obviously, is a loss of current efficiency.

Chromium: the exception There is as yet no completely acceptable theory to explain chromium plating, and all the generalities we have advanced for the other metals operate exactly in reverse for chromium. Thus, the cathode efficiency *increases* with increasing current density, and *decreases* with increased temperature and with the metal concentration. Whole books have been written on the subject, which in sum, add up to the opening sentence.

Nonelectrolytic chromium is deposited by the Electrolizing Corporation[7] by a secret process which is capable of depositing uniform coatings of a 96 to 97 percent chromium alloy. Especially on complicated configurations this is particularly useful, owing to the miserable throwing power of electrolytic chromium.

Power requirements While the cost of metal is usually higher than that of power, still the metal cost for a specified deposit is fixed, whereas the current density used affects two other factors. Power costs increase as a direct function of current density, but the tank time is decreased directly in the same way. Since tank time involves the productivity of labor as well as of equipment, it is most usually the case that operating at higher current densities represents a savings in overall costs.

Metal costs Typical metal costs, based on 1969 prices, are shown in Table 4.

Design Considerations

The designer is in a position to ease the plater's problems of arranging for uniform metal distribution by recognizing certain design limitations. The designer must, however, be aware of certain basic design principles if the part is to remain within tolerance after plating. Figure 5 shows typical thickness distributions found in one design study in which the ratio of average to minimum thickness was determined. Figure 6 illustrates some other principles in a general way.

Corners and edges (Figs. 5a and 6a). Corners and edges should be given a modest radius (of 1/64 in. or more) whenever possible. This not only reduces the actual thickness buildup at these points, but also the final finished surface will more closely approach the original design concept before the radius was added, and is less likely to be out of tolerance.

Angles Theory shows that when a perfect zero-fillet angle is plated, the primary current distribution is such that on the sharp external corner the current density approaches infinity, while on the internal corner it is zero. Thus it is theoretically impossible to deposit any plate in the interior angle (even one of 179°), while the deposit on the external corner must inevitably be burnt, statements which any practical plater will find easily believable. Kronsbein[8] made quantitative measurements on the effect of fillet radius in improving the metal distribution in the interior angle and concluded that, as a general rule, no further improvement was afforded by increasing the fillet radius beyond 0.01 in./in. of angle shank (measured at right angles to the intersection of the two surfaces forming the angle). No similar rule is available for exterior angles, but in this case the only upper limit of desirable radius is that set by other functional considerations.

Obtuse angle sections plate more uniformly than acute angles (Fig. 5b). Convex contours are preferable to concave (Fig. 5c).

Holes When there must be a hole in the part, a through hole is preferable to a blind hole, and in any case the ratio of hole diameter to depth should be as large as possible, and preferably not less than 1:1. Threaded holes should be counterbored where possible (Fig. 5d).

Special shapes Wires, rods, and cylinders receive a surprisingly even plating thickness when the anodes are arranged as shown in Fig. 5e. (This is theoretically predictable from field theory.) When parallel ribs or valleys appear in the design, a curving contour is preferred (Fig. 6b), with the valleys or slots as wide as permissible, and with tapering walls, rounded corners, and cambered surfaces. On a flat surface (Fig. 6c) which is bright plated, a slight surface waviness can cause a distorted image, which is less annoying if the surface is slightly crowned. This is most particularly applicable to the plating of plastic surfaces. The latter are usually smoother and intrinsically more specularly reflective than polished metal, but are more apt to be slightly wavy or undulating.

Joints When overlaps are to be joined by welding or brazing, avoid if possible spot welds or intermittent beads, and use a smoothly contoured *continuous* weld

TABLE 4 Typical Costs of Anode* Metals and Substrate Metals

Metal	Form	Cost/lb	Cost/ft^2-mil†
Rhodium	....	$2,450.00	$158.00
Platinum	....	1,550.00	223.00
Gold	....	612.00	62.00
Palladium	....	490.00	30.00
Beryllium	....	78.00	
Columbium	....	40.00	
Tantalum	....	35.00	
Indium	....	27.00	1.00
Silver	....	23.00	1.20
Molybdenum	....	8.00	
Haynes 25‡	....	5.80	
Waspalloy	....	5.15	
Zirconium	....	5.00	
Hastelloy C‡	....	4.00	
Hastelloy B‡	....	3.90	
Vanadium	....	3.45	
Hastelloy F‡	....	3.33	
Hastelloy X‡	....	3.12	
Cadmium	Anode	3.00	0.14
Tungsten	Powder	2.50	
Bismuth	....	2.25	
Beryllium copper	....	1.98	
Cobalt	....	1.65	0.08
Titanium	....	1.32	
Chromium	....	1.31	0.06
Tin	Anode	1.28	0.05
Tin	....	1.18	
Nickel	Anode	1.05	0.05
Nickel	....	0.87	
Phosphorus bronze	....	0.78	
Tungsten	Billet	0.62	
Manganese bronze	....	0.58	
Bronze	....	0.55	
Naval brass	....	0.54	
Antimony	Lump	0.42	
Copper	Anode	0.43	0.02
Stainless Steel, ferritic	....	0.36	
Manganese	....	0.33	
Magnesium	....	0.30	
Aluminum	Alloy	0.28	
Stainless Steel, martensitic	....	0.32	
Aluminum	Primary	0.25	
Zinc	Anode	0.20	0.007
Zinc	Alloy	0.16	
Zinc	Primary	0.14	
Lead	....	0.13	
Steel	Alloy	0.04	
Steel	Carbon	0.035	
Iron	Anode	0.032	0.004
Iron	Pig	0.03	

* Always higher than current metal prices for primary metal.
† Cost of metal in plated coating, dollars per square foot for one mil thickness.
‡ Trademark of Union Carbide Corporation, New York, N.Y.

Electrodeposition 10-17

or braze. We are concerned here not so much with uniformity of metal distribution as with corrosion protection. A spot-welded overlap, for instance (and any other similar configuration), is a virtually unrinsable solution trap. The implications of this are most serious and will be discussed later at length. (See Corrosion Resistance.)

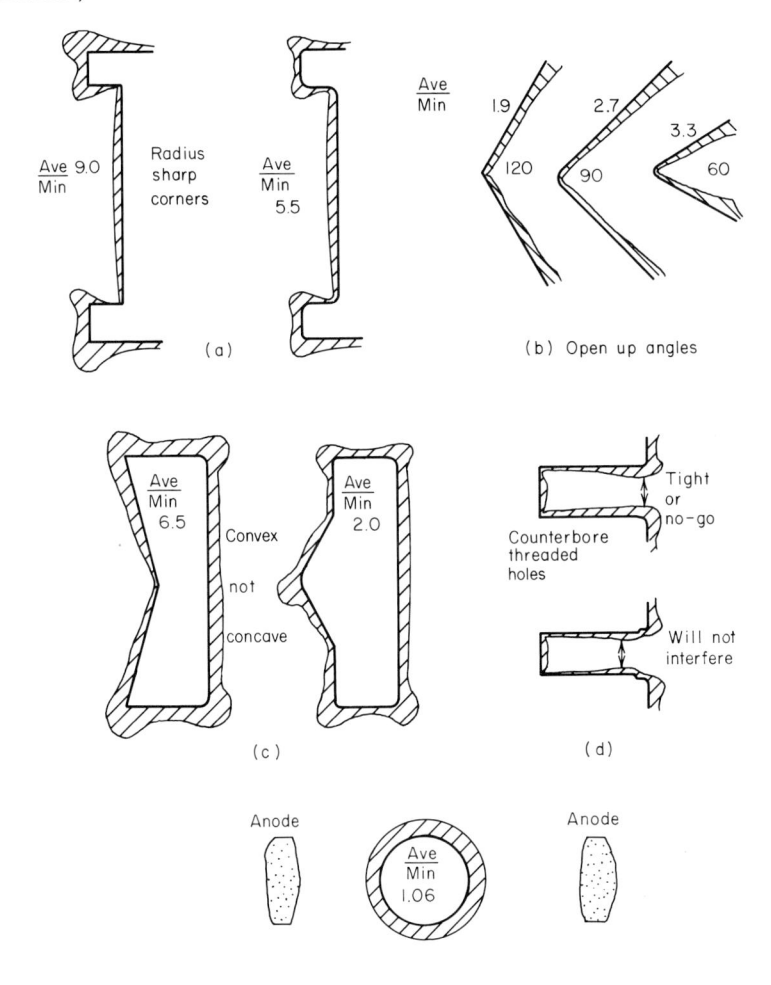

Fig. 5 Actual thickness distribution in a design study of the plating of selected shapes in which the ratio of average to minimum thickness was determined.

Gas traps Parts must be designed for the escape of gases. Suppose we have a box that is a 4 in. cube with an open top, except that we want a continuous narrow flange around the opening for later assembly with some other unit (Fig. 7). When this part is immersed in a solution, air must inevitably be trapped in one corner or another, unless the part is tilted and juggled to let the trapped air escape. In the same way, hydrogen formed during plating will similarly be trapped and will prevent the deposition of metal in those areas unless very elaborate precautions are taken during plating to rotate the piece in a controlled manner.

10-18 Metallic and Chemical Finishes on Metals and Nonconductors

Even this will not guarantee a completely satisfactory job. Such designs *must* be reworked so that:
 1. The corners are not solidly closed, which permits gases to escape (Fig. 7a),
 2. The top or one side of the box is left completely open, and the other components are assembled to it in a different manner for sealing after plating (Fig. 7b).

Remember that *all* solutions, preparatory as well as plating, must reach all surfaces to be plated, and these surfaces must also be easily and freely accessible to the

(a)

(b)

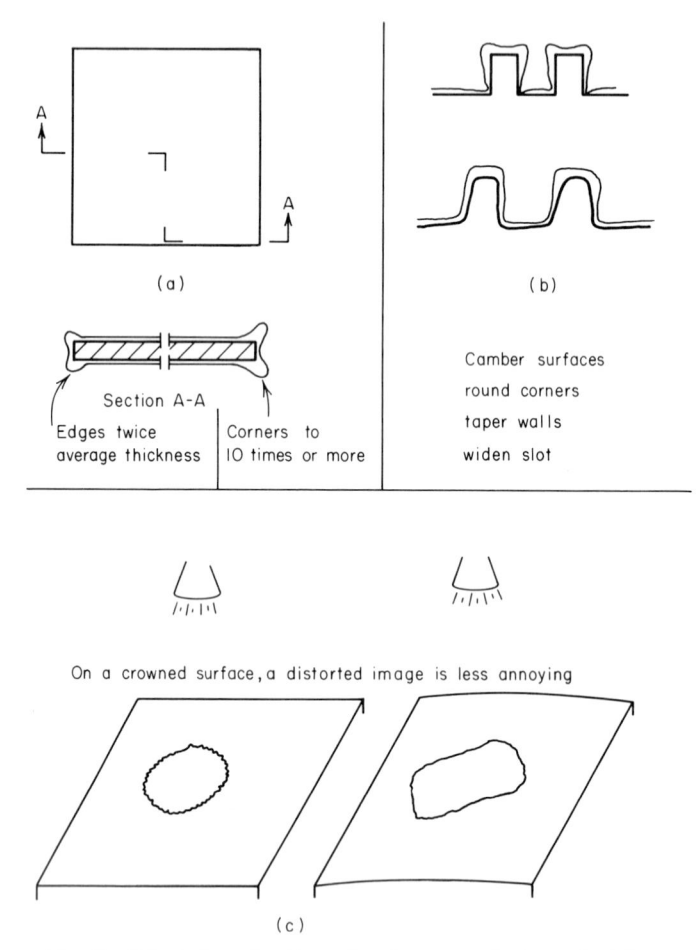

Section A-A
Edges twice average thickness | Corners to 10 times or more

Camber surfaces
round corners
taper walls
widen slot

On a crowned surface, a distorted image is less annoying

(c)

Fig. 6 Selected applications showing some design principles.

rinsing operations. The box we have just been considering (with the solid corners) presents another very great problem to the plater: drag-out of solution, which is both expensive and deleterious to plating quality.

Contacts Provision must also be made in the design for electric contacts to be made. Usually, firm contacts are necessary, and often the contact will leave a mark, sometimes unsightly, and sometimes creating a small area where corrosion protection is reduced or lacking. Sometimes contact can be made on an area that is to be removed later from the part. This is quite common in printed-circuit plating. But for parts of complicated shape, the designer will do well to consult a plating engineer with regard to contact points to be provided. This problem

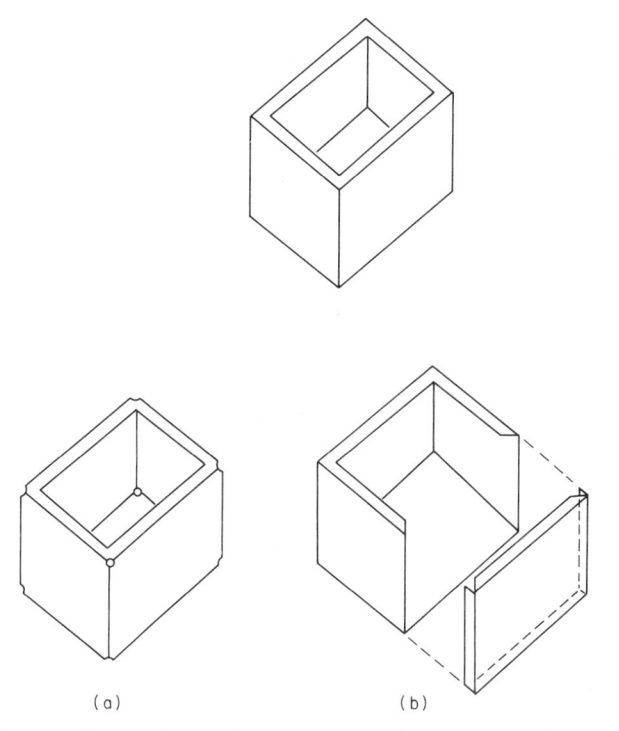

Fig. 7 A housing design destined to become a plating reject. (*a*) One possible solution to the housing of Fig. 7: open the corners. (*b*) A second solution to the housing problem: plate two simpler pieces, and assemble later.

can be very complicated, and much grief can be avoided by this type of design consultation.

The Substrate Metals

Many metals, for example mild steel, are easy to prepare for plating and will withstand almost any cleaning cycle necessary to remove surface dirt, corrosion products, and the like. Some metals form natural oxides very swiftly on the freshly cleaned surface, as, for example, on active surfaces such as aluminum, or on passive surfaces such as the stainless steels. Others are so active or so passive that special procedures are required to plate them, and often some basis metals can be plated only in certain bath formulations. A very rough appraisal of the difficulty of plating upon various metals is indicated in the listing below:

Easy to plate upon	Difficult	Very difficult
Iron	Stainless steel	Tantalum
	Powdered metal compacts	
Copper	Aluminum	Tungsten
Zinc	Beryllium	Niobium (columbium)
Cobalt	Magnesium	Thorium
Nickel	Chromium (electrodeposited)	Uranium
Silver	Lead	Silicon
Cadmium	Titanium	Molybdenum
Gold	Zirconium	Germanium

10-20   Metallic and Chemical Finishes on Metals and Nonconductors

Special circumstances such as alloying elements may make a metal easier or more difficult to plate upon (but usually the latter, as for example beryllium-copper or leaded brass).

The designer need not be concerned with details of procedures preparatory to plating a metal, except to realize one vital point relative to *combinations* of metals to be plated:

When attempting to plate an assembly of two or more different metals, the plater may be faced with an impossible task, or at best a difficult and very expensive one. Always consult an expert in such cases.

Preparing the metal substrate for electroplating It is not the purpose of this section to do more than indicate the kind of preparatory cycle that may be used. For each substrate to be plated, the cycle chosen will depend not only on the basis metal but also on the type of bath selected for the metal deposition. Always, however, there must be a cleaning step or steps, followed by an acid treatment (sometimes both are repeated at least once). For some passive metals, there must also be an activation step—for instance, stainless steels are often given a strike in a very acid nickel bath formulated for low efficiency, where the codeposited hydrogen serves to activate the surface. On the other hand, aluminum is so active that it always has a very thin oxide film which must be removed before adherent deposits can be obtained. Once this film is removed, the surface must be coated immediately *and without exposure to air*. For aluminum and similar metals, the alkaline zincate baths serve both purposes—the oxide film, as well as the active metal surface beneath it, is soluble in the alkali, and zinc deposits on it by immersion (replacement of the dissolved aluminum by zinc), resulting in a zinc surface which, although still quite active, can be plated adherently in certain types of bath that are not too aggressive to the zinc.

The light substrate metals Because weight is of such importance in aircraft and aerospace equipment, the basis metals most used are aluminum, magnesium, and beryllium, in that order. The cost of the metal and also the relative difficulty of plating upon it determines the order of choice.

Aluminum. Aluminum may be plated upon in several ways (Table 5). Certainly the most widely used is the zincate method. In this procedure, the cleaned aluminum is dipped briefly into an alkaline zincate bath, whereby the normal oxide coating on aluminum is dissolved and replaced with a less reactive coating of zinc about one-millionth of an inch thick. After a copper strike plate, further electroplates can be applied in conventional fashion.

With the proper preparatory steps, it is also possible to deposit nickel electrolessly directly upon the aluminum. When nickel can be used, this process is frequently very advantageous, because a protective coating of nickel can be uniformly applied over the most complicated geometries. Any further plating, such as silver or gold that may be required for ease of soldering or other special purposes, may then be quite thin, since the uniform undercoating of nickel can be as thick as is required to afford corrosion resistance without violating dimensional tolerances.

Magnesium. Wherever weight is a consideration in design, magnesium comes to mind as the basis metal; yet here the finishing is of utmost importance, because magnesium is such a reactive metal. Most of the development work on magnesium finishing has been done by Dow Chemical (Tables 6 and 7), and many of the processes are subject to licensing arrangements (now available through MacDermid of Waterbury). Table 6 lists the Dow treatments used for commercial finishing of wrought alloys. Table 7 shows the relative costs of these treatments and those for organic finishing. Where the latter can be used, they are highly recommended, since the greater protection afforded by organic coating over a chemically protected magnesium surface is well worth the added cost.

When magnesium is to be electroplated, it should be cleaned, pickled in chromic or phosphoric acid formulations, activated in a phosphoric acid-fluoride mixture, then zinc-coated in a special zinc immersion-coating bath, usually a zinc-pyrophosphate composition containing fluorides at a moderate pH of about 10.3 (compared to the highly alkaline zincate bath used for aluminum). A copper strike is then

TABLE 5 Coatings for Aluminum

Coating	Application	Finish and thickness	Usual purpose	Used on:
Sulfuric anodizing	Electrolytic at room temperature for 30–60 min	Hard, dense; 300-hr salt spray; 0.2–0.8 mil usually	Corrosion resistance; dyed finishes; paint base	All alloys—but not recommended for assemblies
Chromic anodizing	Electrolytic at room temperature for 20–45 min	Very hard, dense; 300-hr salt spray; not over 0.02–0.06 mil	Corrosion resistance; dyed finishes; paint base	Not for alloys high in Cu, Si—OK on assemblies
Chromate conversion	Room-temperature dip, brush, spray	Clear to yellow or brown; 300-hr salt spray; 0.02 mil or less; electrically conductive	Corrosion resistance; paint base; decorative effect	All alloys
Phosphate coating	Spray or dip at 130°F	Crystalline; 0.01–0.02 mil	Paint base	Wrought alloys
Zincate	15- to 120-sec immersion at room temperature	About 1 millionth; valueless until electroplated	Preparation for electroplating	All alloys
Electroless nickel	Immersion at 160–220°F	Deposits about 1 mil/hr—no practical limitation, evenly distributed; costly	Final finish or base for further electroplating	Directly on all alloys, with suitable preparatory steps
Electroplating	Conventional	Any desired metal or thickness	Functional or decorative	Over zincate or electroless nickel

10-21

TABLE 6 Chemical Treatments for Magnesium*

Dow No.	Name	Military Specification	Appearance	Alloys used on	Uses	Remarks
1	Chrome pickle	MIL-M-3171A, Type 1	Matte gray to yellow-red	All	For protection during shipment and storage.	Simple dip treatment, slight dimensional loss, good paint base. Does not materially affect dimensions.
7	Dichromate	MIL-M-3171A, Type 3	Brassy to dark brown	All except Th-containing alloys	Good combination of paint base and protective qualities.	Requires galvanic couple between work and steel cathode. No external current required.
9	Galvanic anodize	MIL-M-3171A, Type 4	Dark brown to black	All	Used on alloys which will not take No. 7 or where a black coat is necessary.	Chrome pickle plus dichromate boil. Improved protection over No. 1. Slight dimensional loss.
10	Sealed chrome pickle	MIL-M-3171A, Type 2	Matte gray to yellow-red-brown	All	Corrosion resistance equal to No. 7.	Neutralizing required if parts are to be painted. Hard, abrasion-resistant coating. Special-purpose treatment. Superseded by Dow No. 17.
12	Caustic anodize	MIL-M-3171A, Type 5	Light gray to light tan	All	Good abrasion resistance and protective value.	
17	Anodize: Thin Heavy	MIL-M-54202 MIL-M-45202	Light gray to light green Dark green	All	Excellent protective and decorative finish.	Thin: 0.0001 to 0.0003 in. Full: 0.001 to 0.0015 in. Coating is stable above melting point of metal. Best abrasion resistance, best paint base, and most consistent of all Dow treatments for magnesium. License required.
23	Stannate immersion	None	Medium to dark gray	All	Good paint base and protective finish. Used with dissimilar metal inserts (except aluminum or in assemblies).	Must be neutralized if painted. Retards galvanic corrosion. License required.

* Dow Chemical Company.

applied from a bath containing either fluorides or Rochelle salts and in either case with low free cyanide. From the latter bath, heavy deposits can be obtained if periodic reversal of current is used after the initial coverage in the strike. Otherwise, conventional plating procedures may be used over either strike plate.

Beryllium. This quite expensive basis metal is used only where its special properties (low neutron cross section, lightness, stiffness, and high-temperature properties) warrant it. It is best plated directly in electroless nickel, although a zincate treatment similar to aluminum can also be used when nickel is not desirable. Provided the proper pretreatment is used, beryllium is not difficult to plate, but because massive beryllium is formed by a sintering process, the plate is likely to possess the inherent porosity of the basis metal.

Special treatments *Electropolishing.* This is a special form of electrolytic acid treatment in which metal is removed from the surface by anodic treatment in

TABLE 7 Estimated Costs* of Chemical Treatments for Magnesium (In cents per square foot)

Process	Chemical cost	Labor and overhead cost	Range of total cost
Dow No. 1.........	0.9–1.2	1.7–3.3	2.6–4.5
Dow No. 7†........	0.9–1.3	2.5–4.2	3.4–5.5‡
Dow No. 9.........	0.6–0.9	2.0–4.0	2.6–4.9‡
Dow No. 17:			
Thin	0.2–0.3	3.3–5.0	3.5–5.3‡,§
Heavy...........	1.2–1.5	5.0–6.7	6.2–8.2‡,§
Dow No. 23........	0.9–1.3	2.5–4.2	3.4–5.5
Organic finishes, cents/mil/ft^2.....	....	....	1.5–3.0

* Based on $10 per hr for labor and overhead and 100 ft^2 of surface being treated.
† Based on two dichromate boil tanks. Double the time of labor and overhead cost for one tank.
‡ Complicated racking could increase cost.
§ Not including amortization of electric equipment costs which depend on size and volume processed (estimate 5–15 cents/ft^2).

a suitable strongly acidic solution in which the water content may be critical, since etching rather than polishing occurs when the total content of acid is too low. It is applicable to many metals and alloys, including steels, stainless steels, aluminum, brass, bronze, copper, silver, cobalt, zinc, lead, tin, lead-tin, and other alloys. Compared to a mechanically polished surface, a properly electropolished surface is usually superior in respect to total reflectivity and the absence of scratches, strains, metal debris, and embedded abrasive particles. The surface is usually faintly wavy with tiny crystalline protuberances, and the definition of a reflected image is frequently poorer. Electropolished surfaces tend to be passive, and often require activation to obtain good plate adhesion.

Strike Plates. In many cases, and for a variety of reasons, a preliminary, or "strike" plate is necessary, but the basic purpose is to obtain an adherent deposit. When clean steel is immersed in an acid copper solution, for instance, it is covered at once with an immersion deposit of copper, which has little or no adhesion, and the subsequent electrodeposit will then peel. If it is desired to plate on steel with copper from an acid bath, it is necessary to coat it first with copper from a cyanide bath. A strike of only several millionths of an inch is sufficient.

Silver deposits by immersion on most substrates. Hence a silver strike, consisting of a bath with low metal and high cyanide content, is used to prevent roughness

10-24   Metallic and Chemical Finishes on Metals and Nonconductors

due to too high a rate of deposition from the standard bath, as well as to eliminate the tendency for immersion deposits to form, which cause lack of adhesion.

The nickel strike previously mentioned is used to deposit a relatively active metal on a normally passive one. The atomic hydrogen evolved in this strike operation serves to activate the surface, and prevents the poor adhesion which would result from attempting to plate in the normal fashion.

Hydrogen Embrittlement. Embrittlement of a basis metal is often caused by hydrogen absorption during pickling, cleaning, and plating operations. Certain metals and alloys are particularly susceptible, notably the spring steels, high-strength steels, nickel-base alloys, and titanium. The mechanism is not entirely clear, but the metals become quite brittle and fracture well below their normal limits of strength, especially under alternating flexural stresses.

It is well to specify in such cases that the piece be given a hydrogen embrittlement relief treatment both *before* the metal is processed and also *immediately after* the plating is completed. The reason for the foregoing statement is that the absorption of hydrogen is cumulative, and embrittlement is a threshold phenomenon. Thus, a piece with tolerable hydrogen content when plated in an embrittling type of bath may well absorb enough more hydrogen during plating to exceed the threshold. If the part is in a stressed condition from forming operations, cracking can even take place during plating.

Embrittlement relief procedures vary with the alloy and its temper, and a metallurgist should specify the treatment. Usually it involves heating the part long enough (1 to 24 hr) at a temperature sufficiently high (usually 275 to 400°F) to remove the hydrogen.

In the plating operations, any step that involves the release of hydrogen at the surface of the sensitive metal can lead to embrittlement. Cathodic cleaning should thus be avoided; acid pickling time must be held to a minimum, and inhibitors should be present in the solution. Inhibitors materially reduce acid attack on the metal itself and restrict the bulk of the acid action to the dissolution of surface oxides and similar compounds, thus restricting the quantity of hydrogen evolved.

Embrittlement can occur in any plating bath operating below 100 percent efficiency, but it is most serious in the cyanide zinc and especially so in cadmium baths. Where zinc is used, an acid bath is to be preferred, where the efficiency is close to 100 percent and little hydrogen is generated; in the case of cadmium, no completely satisfactory substitute for the cyanide bath is as yet available, although a fluoborate type of bath has found some acceptance. Since both zinc and cadmium are relatively impermeable to hydrogen, they tend to "plaster" the hydrogen firmly into the basis metal. If the metal is one that is metallurgically susceptible to embrittlement, this will occur. Hence such metals can rarely be finished with cadmium; zinc is preferred, particularly if deposited from an acid bath.

Etching. Etching of the basis metal (as a purposeful step, and not merely as an undesirable degradation of surface finish in one of the preplating steps) is not ordinarily considered as a usual preparatory step. In the printed-circuit industry, however, there comes a point where etching is required. Just when in the manufacturing process etching is specified depends on the process being used to make the circuit, but in any case an etching resist of one sort or another is used in a desired pattern; then the unprotected areas of the copper laminate are etched away by some corrosive solution such as ferric chloride, sodium chlorite, cupric chloride, ammonium persulfate, or the like. Unfortunately, the solution does not just eat straight down like a trench digger. As soon as sidewalls form under the resist, they too are eaten back, undercutting the resist, and the result is a trench with a slanting wall. The amount of undercutting will be proportionate to the thickness of the copper cladding. In microcircuitry this can be of considerable concern, since the desired line width to be left is approaching the order of magnitude of the thickness.

Fortunately, another industry faced this problem long ago. The photoengravers have developed etching solutions and techniques that should be invaluable to the etcher of printed circuits.

Rinsing It is quite safe to say that a major portion of a day's rejected plating work can be laid to poor rinsing, which not only leaves unwanted fouling residues on a part; worse, it carries this undesirable freight into the next processing bath to contaminate it as well, until the poor plater wonders what's wrong with which bath. Admittedly, this is not the designer's problem, but I thought it would be comforting to point out that other people have their troubles, too.

TABLE 8 Major Bath Types for Plating Common Metals from Aqueous Solutions

Metal	Chief salt(s)	pH	C.D., asf*	Operating temperature, °F	Cathode efficiency, %
Cadmium	Fluoborate	0.8–1.7	125–350	Room–170	95–100
	Cyanide	>9	5–50	Room	90–95
Chromium	Chromic acid	<1	100–350	110–130	About 13
Cobalt	Sulfate	3–5	20–50	Room	95–100
	Chloride	2.3–4.0	45–60	125–160	95–100
	Fluoborate	3.5	60	120	95–100
Copper	Cyanide	12–12.6	10–40	Room–160	30–50
	Pyrophosphate	8.2– 8.8	10–75	110–140	95–100
	High-metal cyanide	>13	30–60	170–180	100
	Fluoborate	0.8– 1.7	75–350	80–170	95–100
	Sulfate	1.2– 1.7	20–100	70–120	95–100
Gold	Cyanide	10–11.5	1–10	Room–160	95–100
	"Acid" cyanide	3–7	1–12	Room–130	25–50
Indium	Cyanide	>13	15–30	Room	70–95
	Fluoborate	1.5–2.0	5–25	Room	80–90
	Sulfate	2–2.7	5–50	Room–140	30–80
Iron	Sulfate	4.5–6.0	50–100	100	95–100
	Chloride	1.0–1.5	60	195	95–100
	Fluoborate	3.0–3.7	20–90	130–190	95–100
Lead	Fluoborate	<1.5	5–300	75–160	100
Nickel	Sulfate-chloride	2.0–5.2	10–60	130	95–100
	Chloride	0.9–1.1	30–100	130	90–100
	Fluoborate	3.0–4.5	30–300	130	95–100
	Sulfamate	3–5	50–300	120	98–100
Rhodium	Sulfate	N.D.	10–100	100–120	80
	Phosphate	N.D.	10–100	105–115	80
Silver	Cyanide	N.D.	5–80	Room–130	100
	Cyanide-nitrate	N.D.	5–150	Room–125	100
Tin	Stannate	N.D.	10–65	140–180	60–100
	Fluoborate	N.D.	25–125	75–100	100
Zinc	Cyanide	>13	10–50	Room–125	50–95
	Sulfate	3.5–4.2	10–30	Room	95–100

* Current density, amperes per square foot.
N.D. Not determined for control purposes.

Poor rinsing leads to contaminated and short-lived baths, and trouble inevitably follows. But also old cleaning and acid baths simply get tired, and must be dumped and made afresh, even though the control analyses claim they are sound enough. These baths do a lot of work removing gunk (there is no more descriptive term, and no analytical procedures are available for gunk concentration). Their average life varies with the amount of work they do, but for most active plating shops, cleaners and acids last only a week or two at the most.

Plating baths, on the other hand, are a different (and much more expensive) breed. Providing there is adequate rinsing preceding a plating bath, its life is (apparently) infinite, because of drag-out. In the usual busy shop, the entire chemical content of a plating bath will be dragged out by the work leaving it and will have been replaced by the maintainance chemicals added to it in the matter of a few months, or at most a year, as can be seen by applying the previously cited table of drag-out rates to your own shop, and calculating the life of a given bath. As a consequence, there is *never* any reason (other than grossly poor operation) for dumping a plating tank and renewing it.

Electrodeposition of Metals and Alloys

Only 13 of the common metals are readily electrodeposited from aqueous solution. As will be seen from Table 8, some can be deposited only from a single bath type and others from a choice of two or three types. The operating conditions shown are typical. Many other bath types are available, some of which serve special purposes. Speaking generally, the best baths are proprietary and are offered and serviced by a few specialized supply houses.

A number of other metals can be deposited (some with difficulty) from aqueous solutions and are of varying interest and commercial importance:

1. Platinum, palladium, iridium, and ruthenium are of some commercial interest.
2. Molybdenum and tungsten have not been deposited in the pure state, but alloys with other metals (chiefly iron, cobalt, nickel) can be deposited.
3. Antimony and manganese are deposited chiefly for electrowinning purposes.
4. Mercury, gallium, thallium, arsenic, bismuth, selenium, tellurium, polonium, technetium, and rhenium can be deposited, but are presently of no interest.
5. The deposition of osmium has received no attention, owing to scarcity, cost, and lack of demand.

Aluminum and germanium can be deposited from nonaqueous organic solutions, but not from aqueous solutions. The technical difficulties are many, including the problem of maintaining a water-free bath.

The following metals have not been deposited from either aqueous or nonaqueous baths (but most can be deposited from fused salt baths): beryllium, magnesium, calcium, strontium, barium, radium, scandium, yttrium, the lanthanide and actinide series, titanium, zirconium, hafnium, vanadium, niobium, tantalum, and the alkali metals.

Plating from fused salt baths is of marginal use only. For many applications, the high temperatures involved are too limiting, as for plating on plastics or metals of low melting point.

Alloy plating The number of alloys that have been deposited is legion, comprising combinations of almost all depositable metals, but most are of academic interest only. Commercial alloy deposition is chiefly restricted to those shown below:

Alloy	Bath type	Usual composition range
Copper-zinc	Cyanide	70–80% copper*
Copper-tin	Cyanide-stannate	Any
Lead-tin	Fluoborate	Any
Tin-zinc	Cyanide-stannate	10–85% zinc
Tin-nickel	Chloride-fluoborate	65% tin

* These alloys with higher copper content have a bronze color, and have been used for decorative plating; with lower copper content, the color is very light: for example, the "white brass" used on some automobile bumpers and trim during World War II as a rather unsuccessful substitute for nickel.

Ternary alloys have also been plated, but the problems of controlling the composition of the deposit from a three-metal system are extremely difficult.

Alloys of cobalt or nickel with phosphorus have also been electrodeposited, but these have not yet been received with much favor, in view of the fact that similar alloys can be deposited without the use of current, as will be discussed later under electroless plating.

Postplating Treatments

Bright dipping Cyanide zinc and cadmium deposits are frequently bright-dipped after plating. The solutions are oxidizing in nature, such as nitric acid or a chromate-sulfuric mixture and are quite weak (¼ vol % nitric acid, for example). These give brighter surfaces which fingerprint less readily.

Water-dip lacquers Special water-dispersed organic formulations may be used on zinc and cadmium, with or without prior bright dipping. These lacquers are used in the final rinse. They are clear and extremely thin, and are used to avoid fingerprinting.

Electropolishing Electrolytic polishing has also been used to improve the brightness of nickel, copper, and brass plate before chromium plating. Since the amount of metal removed must be very restricted, the processing times are short—this operation has been called "electrobrightening" to distinguish it from true electropolishing.

Conversion coatings Both phosphate and chromate types of conversion coatings have been used on suitable deposits—for instance, zinc or cadmium. Discussion of these will be deferred to a later section, since there is intrinsically no difference between these coatings on a massive or on a plated metal.

Properties of Electrodeposited Coatings in General

The general properties desirable in electrodeposits are soundness, adhesion, purity, and a level of internal stress that is acceptable. Corrosion resistance is also desirable, of course, and is so important that it will be discussed later in a separate section.

Soundness Soundness of the deposit refers to lack of such defects as pits, blisters, cracks, inclusions, roughness, burnt plate, and skip plate (isolated areas of no or very little plating scattered over a surface). These defects are caused by poor mechanical surface condition of the basis metal, poor preparatory steps, a dirty or contaminated plating bath, or one that is being operated at too high a current density.

When the basis metal is pitted, it is rare that the deposit fills the pits. Usually, either the pit is bridged over with plate, or else no metal deposits at all if the pit is filled with nonconductive debris. In the former case, gases trapped under the bridged plate may expand (e.g., in the hot rinse) and cause blisters. There are also many other causes of pits and blisters, including the mechanical inclusion of solid particles from the bath, and it serves no purpose to discuss these in detail.

Cracking is typical of all chromium plating, and is caused by the high internal stress existing in the deposit, a subject on which more will be said later. Cracking of other metal deposits may arise from flexure of a brittle coating, such as bright nickel, or from a combination of stress and poor adhesion, in which case the deposit often exfoliates and curls up from the basis metal, or pops off. General roughness is caused by solid particles not removed by the cleaning cycle, or from solid matter floating in the plating bath itself, which particles are then plated over. Too high a plating current density causes burnt plate, while skip plate is caused by local areas that are either not clean or are passive.

Adhesion The adhesion of an electrodeposit to the basis metal, *when conditions are proper*, is usually determined by the tensile strength of the weaker metal, which will tear when an attempt is made to separate the two. For certain combinations of plate and basis metal, an intermediate alloy layer forms at the interface, and sometimes, when this is weaker than either metal, rupture occurs at the interface, even when the adhesion, per se, is perfect. But this degree of adhesion is almost always satisfactory, in contrast to the situation where preparation of the basis metal is poor. When metallographic examination of the cross section reveals some continuity of crystal structure (even if there may be an area of distortion due to differing crystal structure or of lattice parameters), adhesion will be perfect.

Purity The purity of an electrodeposit depends on the absence of codepositable metals in the plating bath. In some cases, there are no ill effects—most nickel baths contain small amounts of cobalt, for instance, which codeposit to form a low-cobalt alloy. But lead, for example, which will also codeposit with nickel, will have a profound effect on the appearance and corrosion resistance of the plate, often causing peeled and brittle deposits.

Brightness Brightness is not a normal attribute of a plated metal, except in very thin deposits on a bright or buffed surface. A good example is the chromium plating on an automobile, which is normally only 10 millionths of an inch thick, and has been deposited on a bright nickel. Heavier deposits from the same bath will be dull.

The nickel is usually a mil or more in thickness, and its brightness must be obtained by adding compounds (usually organic) to the dull nickel bath in amounts ranging from only a few to several thousand parts per million. The crystal structure of the deposit is changed, as are many of its physical properties. Ductility is sharply reduced (substantially to zero); tensile strength may be greatly increased; the internal stress, using modern brighteners, is usually reduced to low tensile values or even changed to a low contractile stress. All now-known nickel brightener systems contain at least one organic sulfur compound which codeposits to some extent with the metal, and the sulfur causes the phenomenon of hot-shortness. Thus bright nickel deposits crack and frequently crumble to powder, if heated above 500°F.

Crystal structure The structure of electrodeposits is often different from the usual structure of the massive metal. Also changes in the deposition variables may cause a change from the normal crystal habit of the deposit. Nickel, for example, usually deposits in the face-centered cubic arrangement, but under certain conditions it can be plated in a hexagonal close-packed structure.

Some electrodeposited alloys are laid down from baths operated at, say, 140°F in a condition only obtainable in thermal alloys at temperatures much higher, such as 1000°F. An example is the cobalt-tungsten alloy described by Brenner et al.[10]

It is important that the designer avoid specifying (and that the plater avoid using) any procedures involving temperatures which would degrade the effects of any prior heat treatment. Operations which involve high temperatures, such as polishing and buffing, or special heat treatments for the purpose of hydrogen embrittlement relief and the like, are suspect, and a competent metallurgist should be consulted.

Internal stress All deposited metals (electroless as well as electrolytic) are laid down in a condition of more or less internal stress, which varies with the bath composition as the controlling variable and is further influenced by bath temperature, part geometry, and (for electrodeposits) by current density. These stresses are most often tensile, tending to crack the deposit, but are sometimes compressive, whereby the deposit tends to expand. If the adhesion is poor, a deposit with tensile stress may exfoliate and curl up, while a compressively stressed deposit will blister (this is only one of several causes of blistering, however).

The range of stress values for an electrodeposited metal varies widely, as shown in Table 9, and the table does not pretend to show the limits possible. For a given basic bath formulation, very minor quantities of certain compounds may increase or decrease stress markedly. Also in any one bath the stress varies with temperature, current density, and the form of current (direct, or modified with superimposed alternating, or by use of periodic reversal). For instance, the basic Watts nickel bath (40 oz/gal nickel sulfate, 6 oz/gal nickel chloride, and 0.05 oz/gal sodium lauryl sulfate) at 140°F and a current density of 40 asf will normally give a tensile stress of about 20,000 psi in the deposit. If the wetting agent is replaced with hydrogen peroxide (to give 5 to 10 ppm of free oxygen), as is frequently done in engineering plating, the stress rises dangerously close to the ultimate tensile strength and can cause rupture of the deposit. For reasons not now known, the addition of the same amount of peroxide to a fluoborate nickel bath has no observable effect on the stress in the deposit. In any nickel bath, additions of saccharin or p-toluenesulfonamide brighteners in the amount of 0.1 to 0.2 oz/gal can reduce the stress to zero and even change it to compressive. It must be recognized that a chosen stress value (of zero for instance) will not be maintained over the entire surface of a piece, because of current density variations. From what has previously been said, it follows that impurities in the bath may sometimes have disastrous results on the stress and mechanical properties of the deposit.

Stress Relief. Many engineers and metallurgists are accustomed to thinking of relieving stress in a component after mechanical deformation by means of some

TABLE 9 Typical Range of Stress in Electrodeposits from Selected Baths*

Metal	Bath	Stress,† psi $\times 10^{-3}$
Chromium	Standard bath	-17 to over 50‡
Copper	Cyanide	5 to 12
	Sulfate	0.4 to 2
	Fluoborate	0.7 to 3
Gold	Alkaline cyanide	Low compressive
Iron	Chloride	12 to 24
Nickel	Watts, no additions	16 to 30
	Watts, with peroxide	40 to over 50
	Watts, with brighteners	-20 to 20
	All chloride	30 to 45
	Fluoborate	14 to 25
	Fluoborate, with peroxide	14 to 25
	Sulfamate, no chloride	0.5 to 8
	Sulfamate, with chloride	8 to 16
	All sulfate	16 to 20
Silver	Cyanide	1 to 5

* Brighteners and addition agents can cause values lower or higher than those shown.
† Positive values are in tension.
‡ Stress in chromium depends on extent of cracking.

elevated temperature treatment. In this process, the high temperature in time allows a relaxation of the strained crystal structure into what may be thought of as a more comfortable arrangement. By contrast, stress relief in a deposit is much more limited, and is usually confined to *equalizing* the stresses within the deposit, which originally will vary from place to place, owing to local current density variations. The average stress remains, and true relief occurs only through cracking of the deposit, as in most chromium plates, or deformation of the piece, which latter is the basis for most stress-measurement procedures.

Mechanical properties The mechanical properties of electrodeposits depend chiefly on bath composition and to a lesser extent on the conditions of operation, within the operable range. Table 10 shows the range of mechanical properties for various metals obtained from different bath types.

Low-temperature mechanical properties of electrodeposits have particular importance for aerospace designers but, unfortunately, few data have been published in this area. Figure 8 has been compiled from data abstracted from *ASTM Special Technical Bulletin* 318 (1962) and shows the tensile properties of nickel deposited from two bath types over a wide range of testing temperatures.

Physical properties In contrast to mechanical properties, physical properties of electrodeposits are characteristic of the metal itself and are independent of bath and deposition variables. Table 11 lists a number of the physical properties of the important electrodepositable metals.

Brush Plating

This is a technique developed over the last decade or so; it has developed from a strictly manual process to a mechanized production operation. In its simplest form, a small anode (usually insoluble) is wrapped with a suitable absorbent cloth, dipped into a proprietary[9] plating solution and swabbed over the area to be plated

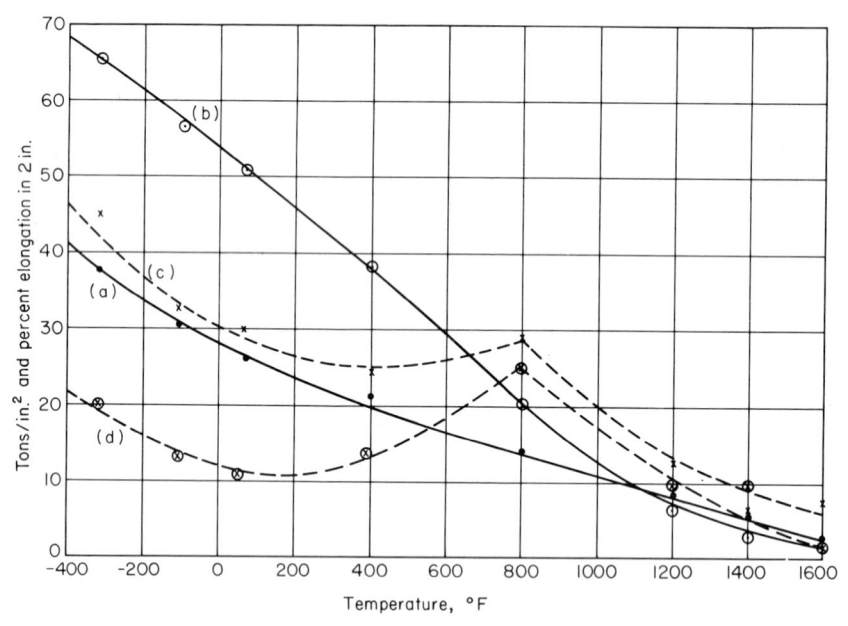

Fig. 8 Tensile properties of electroformed nickel from two typical baths, tested over a wide range of temperature (deposits 0.02 to 0.1 in. thick). (*a*) Tensile strength of Watts' nickel deposits, tons per square inch. (*b*) Tensile strength of sulfamate nickel deposits, tons per square inch. (*c*) Watts' deposits, elongation in 2 in., percent. (*d*) Sulfamate deposits, elongation in 2 in., percent.

or repaired, which is connected as the cathode. More recently, means have been developed for mechanically moving the stylus and/or the work to cover the desired area, while in some cases the cloth has been dispensed with by maintaining a fixed distance (usually about 1/64 in.) between the anode and work, while a feed of fresh solution through openings in the stylus is used to maintain the electric circuit.

All the common metals can be brush-plated, as well as precious metals such as gold, indium, palladium, platinum, and silver, and such unusual metals as antimony, arsenic, bismuth, gallium, mercury, rhenium, and ruthenium. By mixing the solutions, many alloys can also be plated, including the binary alloys of iron, nickel, cobalt, and tungsten for hard, magnetic coatings, as well as gold-antimony for electronic purposes.

Present applications of brush plating in the electronic field include:
    Plating through stencils onto tiny well-defined areas
    Plating adherently on semiconductors

TABLE 10 Range of Mechanical Properties of Some Electrodeposits

Metal	Bath	Ultimate tensile strength, psi × 10⁻³	Elongation in 2 in., %	Vickers hardness
Cadmium	Cyanide	....	......	(12–22)*
Chromium†	Conventional	10–30	0–0.1	300–1000
Copper	Sulfate	20–68	15–40	40–85
	Sulfate, with additions	69–90	1–20	75–180
	Fluoborate	17–50	6–25	40–75
	High-speed cyanide	....	30–50	100–160
	Above, with PR	100	6–9	150–220
Gold	Cyanide (conventional)	....	......	20–65
	Proprietary	....	......	110–130
Iron	Chloride	50–110	3–15	125–220
	Fluoborate, annealed	35–47	25–40	50–90
Nickel‡	Watts, no additions	50–88	20–25	330
	Watts, organic additions	170–230	0–5	350–820
	Cobalt-type baths	200	4	250–500
	All-chloride	95–140	2–20	200–400
	Fluoborate	55–120	5–30	125–300
	Sulfamate, no Cl	60–130	6–30	200–550
	Sulfamate with Cl	94–155	3–5	200–625
Palladium	Various	....	......	190–435
Platinum	Amminonitrite	....	......	600–640
Rhodium	Acid sulfate	....	......	590–635
Silver	Cyanide	35–50	10–22	50–125
Tin	Stannate	....	......	5–9
Zinc	Cyanide	....	......	35–45

* Brinell hardness.
† For usual thin deposits used as topcoat.
‡ See also Fig. 8.

TABLE 11 Physical Properties of Electrodeposits

Electrodeposit	Melting point, °C	Electrical resistivity, microhm/cm	Thermal conductivity, percent Ag	Coefficient of linear expansion, 10⁶ per °C	Modulus of elasticity, × 10⁻⁶
Cadmium	321	6.83 (0°)	22.2	29.8	8
Copper	1083	1.692 (20°)	91.8	16.6	17
Gold	1063	2.44 (20°)	70.0	14.2	10.8
Indium	157	8.8 (22°)	6	24.8	1.57
Nickel	1452	7.8 (20°)	14.2	13.5	30
Nickel (electroless)*	890	60	14	13	30
Platinum	1769	10.58 (20°)	16.6	8.9	22
Rhodium	1960	4.7 (0°)	21.0	8.5	41.2
Silver	961	1.59 (20°)	100	19.68	11
Tin	232	11.5 (20°)	15.5	20	5.9

* Nickel-phosphorous alloy—typical values.

Gold and indium plating of printed-circuit boards before or after assembly
Plating metal-plastic combinations or wired assemblies that cannot be immersed in a plating bath
Deposition of very pure metals for special purposes

Electroforming

Electroforming may be defined as the art of *producing or reproducing metallic objects by electrodeposition upon a master form or matrix which is then removed, in whole or in part.*

When a design engineer should consider electroforming is best illustrated by citing its advantages and disadvantages:

Advantages of Electroforming

1. By proper selection of a suitable metal and bath, and operating under controlled conditions, the mechanical and physical properties of the electroform can be predetermined.

2. With a proper choice of matrix material, parts can be produced in quantity with a very high order of dimensional accuracy, and all parts produced from this matrix will be dimensionally *identical*.

3. The reproduction of fine detail can be matched by no other mass-production method. For example, in a microgroove phonograph record, the lateral excursion of the recording needle is under half a thousandth of an inch, while the dimension of the smallest modulation considered necessary for the high-fidelity reproduction of sound must be held to about a half a millionth of an inch. Yet this fidelity of reproduction can be held through three generations of electroform: from the original mother to master negative, to several matrix positives, and finally to a number of printing negatives.

4. There is virtually no limit to the size of object that can be electroformed. Articles that have been so produced range from hypodermic needles, foil 0.0001 in. thick, and 400-mesh screen, to bulky objects of complicated geometry weighing several tons.

5. Shapes can be made that are not amenable to other fabrication methods, such as radar waveguides of complicated shape having internal surfaces with close dimensional tolerances and high surface finish, and frequently composed of two or more layers of metal.

Some articles that have been electroformed are: electrotypes, phonograph-record masters, embossing and graining plates, foil and sheet, screen, seamless tubing, various molds and dies, pitot tubes and venturi meters, roughness gages, computer cams, heat sinks for missile nose cones, musical-instrument components such as tuba bells, kettle drums; and hosts of others.

Disadvantges of Electroforming

1. The per piece cost is relatively high, especially for short runs.
2. The production rate is slow, often measured in days per piece.
3. There are some limitations in design, with respect to sharp angles, recesses, great and/or sudden changes in cross section or in wall thickness (unless the latter can be obtained by a final machining operation).
4. Just because of the perfect reproducibility, any scratches and imperfections in the master will appear in all the pieces produced from it.

Electroforms have been chiefly produced in nickel, copper, and iron. Silver and gold (and gold alloys) have been used where the cost is warranted, and cobalt can also be employed if its special properties as compared to nickel are of importance.

Internal Stress. Although a high order of dimensional accuracy can be achieved, due attention must be paid to considerations of internal stress. Especially in thin sections, after the part is separated from the master, internal stress may cause warpage, especially if the piece is heated to elevated temperatures. Excessive stress may also cause the electroform to grip the matrix so firmly that parting becomes impossible.

Choice of matrix material The matrix material determines the method used to prepare it for plating and the parting procedure that will be used. Both permanent and expendable matrices are used (see also Table 12):

Permanent matrices	Expendable matrices
Aluminum	Aluminum, zinc
Stainless steel	Low-melting alloy, e.g., Wood's metal
Invar, Kovar	Wax, graphited
Brass (chromium-plated)	Plastics
Nickel	Wood, plaster of paris, etc., waxed and graphited
Glass	
Plastics	

TABLE 12 Properties of Matrix Materials for Electroforming

Matrix material	Machinability	Finish attainable	Damage resistance	Cost of matrix	Metallize?	Parting difficulties	Hold tolerances
Permanent:							
Aluminum*	Easy	Good	Poor	High	No[a]	Varies	Close[b]
Invar,† Kovar‡	Difficult	Good	Good	Very high	No	Easy[c]	Close
Stainless steel	Difficult	Good	Good	High	No	Easy[d]	Close
Brass, plated with Ni-Cr	Easy	Good	Fair[e]	Low	No	Easy	Close
Nickel	Difficult	Good	Good	High	No	Easy[d]	Close
Glass§	Very difficult	Good	Fragile	High	Required	Easy	Close
Plastics	Easy	Varies	Poor	Moderate	Required	Easy	2–3 mils
Expendable:							
Aluminum, zinc*	Easy	Good	Poor	Moderate	No[a]	Slow[f]	Close
Low-melting alloy	Cast[g]	Fair	Poor	Low	No	Easy[h]	Fair
Wax	Cast[g]	Good	Very poor	Low	Required	Easy	Poor[h]
Plastics	Easy	Varies	Poor	Moderate	Required	Slow[f]	2–3 mils

* Light, easily corroded.
† Trademark of Carpenter Steel Co.
‡ Trademark of Westinghouse Electric Corporation, Pittsburgh, Pa.
§ Lost, if damaged.
[a] Zincate treatment required for aluminum.
[b] When aluminum is cleaned in a non-etch cleaner.
[c] By taking advantage of differential expansion on heating or cooling.
[d] When passivated; impossible otherwise.
[e] Not readily repairable if damaged.
[f] Matrix dissolved out.
[g] Rarely if ever machined.
[h] Except in special areas, e.g., phonograph records.

Permanent matrices are preferred for accuracy and especially for high production runs, in which their initial high cost can be spread over a number of pieces. Expendable matrices may be used (often they are castings) for short runs of parts, usually with loosely controlled tolerances. Expendable molds are sometimes required when the matrix cannot be removed afterward because of the part geometry. Sometimes, as in the case of venturi nozzles, the mold is made in two or more pieces firmly held together. The mated surfaces are carefully lapped so that no discernible blemish appears on the electroform.

Preparation of the matrix The surface preparation used for plating depends upon the material, and upon the dimensional accuracy required in the finished electroform. Machined permanent matrices usually require only conventional cleaning and acid treatment; some (e.g., stainless steel) require passivation, which must be carefully controlled to avoid either too much or too little adhesion of the electrodeposit. Chromium plate, for example, is so passive naturally that in some cases it may be found necessary to graphite the surface to *improve* the adhesion to the point where exfoliation does not occur. Aluminum usually requires a zinc immersion coating to obtain a sound deposit. This zinc coating is never over 3 μin. thick, and, since it replaces aluminum dissolved by the zincate solution, causes almost no dimensional changes.

Nonconducting matrices must be impervious to the processing solutions or must be made so by impregnating them with a suitable wax, lacquer, or synthetic resin, which must then be rendered conductive by one of the means described below.

Parting Expendable molds are melted or dissolved out:

Aluminum alloys are readily soluble in a strong hot sodium hydroxide solution.

Zinc alloys are dissolved out by a hydrochloric acid solution.

Low-melting alloys are melted and shaken out, and the alloys can be used over and over. Many fusible alloys have a tendency to "tin" the surface of the electroform when they are melted out. The lead-bismuth eutectic usually does not cause tinning, but this must be melted out in a hot-oil bath at 250°F or higher, depending on alloy composition. Graphiting the matrix sometimes overcomes the tendency for tinning, but may be otherwise objectionable, and not all metals can be successfully graphited. Another method that has been used is to plate the fusible matrix first with a mil or so of copper, which is desirably electropolished before deposition of the electroforming metal. Dimensional allowances must be made for the copper (since this will be dissolved), and there will also be some loss of dimensional accuracy due to variations in the thickness of the copper deposit. After melting out the matrix, the copper, when dissolved in nitric acid or other suitable reagent, carries with it any particles of the fusible alloy that may have clung to it.

Thermoplastic matrices, when simple mechanical separation is impossible, may be softened by heat and withdrawn, after which the electroform is cleaned with a suitable solvent. A costly alternative is to dissolve the entire matrix. Plastics are more often suitable for permanent than for expendable molds.

Waxes are melted out, and the electroform is then suitably cleaned. All the nonconductive matrix materials are, of course, metallized before plating. The metallized coating may:

1. Be used as the parting surface, in which case it must be passivated suitably, or
2. Be removed with the electroform and later dissolved from it, or
3. Remain as a permanent part of the electroform.

The last two methods are preferred, since it is rare that a metallized coating can adhere sufficiently to a plastic to withstand the stresses of parting.

Electroforms are removed from permanent matrices mechanically, by using one or a combination of several of the following techniques:

1. Heating with a torch or hot-oil bath can be used, either to melt or soften a parting compound, or to take advantage of a favorable differential in the coefficients of expansion of matrix and deposit.

2. Cooling, as in a mixture of dry ice and acetone, may be required. The choice between heating and cooling is dictated by the relative coefficients of expansion and by the geometry. Certain configurations, of course, are not separable by either heating or cooling.

3. Impact, as by a sudden hammerblow or sharp jerking, sometimes assists in separation.

4. Gradual force is often applied, as by a hydraulic ram, jackscrew, or wheel puller, to draw or push the pieces apart.

5. Prying is used to separate relatively flat pieces such as phonograph-record stampers or engraving plates from their matrices.

Machining of the electroform is frequently done before parting, to provide added support (this is rarely possible on matrices which are surfaces of revolution, however, since the electroform may slip during machining). Reference marks should be provided on the matrix, from which the finished dimensions of the electroform can be measured, and also to permit proper registry in later machining and assembly.

Whenever possible, permanent matrices should be tapered at least 0.001 in./ft to facilitate parting. Where a taper is not permissible, a material with a low coefficient of expansion such as Invar may be used, so that upon heating, the greater expansion of the electroform will permit withdrawal of the matrix. Alternatively, a parting medium such as wax-graphite or a low-melting electrodeposit (tin or lead) may be used with some sacrifice in dimensional accuracy.

Even when it is not functionally necessary, a fine surface finish is a *must* on permanent matrices, in order that there be no hang-up in parting—a finish of 2 μin. rms is frequently specified, and is achieved by lapping or (where tolerances permit) by electropolishing.

Electromachining

Electromachining may be best described as electroforming in reverse. It is a very specialized technique involving the use of costly high-precision equipment and requiring the application of sound fundamental theory.

Here the workpiece is made the anode in a suitable electrolyte, and a cathode precisely machined to dimensions slightly smaller than the cavity, and in the form of a negative, is brought very close to it while the electrolyte is pumped at high velocity between them. As the anode dissolves, the cathode is advanced at a predetermined rather high rate averaging about $\frac{1}{10}$ in./min until the desired conforming shape is achieved. Very high current densities are employed, and the circulating stream must be cooled and filtered.

The electrolyte must be one in which the anode metal dissolves freely to a high concentration, preferably with little or no formation of insoluble matter. Obviously it must also be a solution from which the dissolving metal does not form adherent deposits, and which forms no objectionable films of any other sort upon the cathode. Simple electrolytes such as sodium chloride, nitrate, and the like are used. The cathode itself, of course, must not be degraded in any way by the solution, even at the very high temperatures that will obtain in the thin film between anode and cathode at high current density.

Setting up the parameters so that the cavity does conform precisely to the shape of the cathode is somewhat of an art, and the extent of trial-and-error experimentation required depends on the experience of the engineer involved. For a suitable shape, the parameters of the process are:
1. Solution composition
2. Current density
3. Temperature to be maintained at the interface
4. Rate of tool advance
5. Interfacial distance to be maintained
6. Velocity of solution across interface
7. Cathode material

Obviously the parameters are almost all interrelated. Control must be excercised during electromachining to maintain the desired range of current density. The *total* current will therefore be continually changing as the machining progresses and the cavity expands.

Electromachining may be considered when conventional machining is difficult, time-consuming, and therefore expensive, and when the number of replicates to be produced is sufficient to justify the high cost of the equipment and of the necessary setup and experimental time. Because of the complexity of the many interrelated variables, no general estimates can be given for the time required for electromachining, except to say that it will be faster than conventional machining. The machined finish may vary from 4 to 100 rhr [this unit, rhr, or roughness height rating in microinches, is superceding the root mean square (rms) designation for surface finish], depending on the machining parameters used.

Understandably, certain configurations involving overhangs, undercuts, and the like will not usually lend themselves to electromachining techniques.

Stripping of Electrodeposits

Stripping of defective deposits is probably the last thing the designer should have to worry about, except to realize that chemically inert metals such as rhodium and platinum are difficult if not impossible to strip from the more active substrates such as aluminum, zinc, or steel, unless there is an intermediate deposit more reactive than rhodium or platinum, such as nickel or (better) copper in the finishing system. Then the plater can arrange to attack and dissolve the more soluble undercoating so that the noble metal falls off mechanically.

NONELECTROLYTIC DEPOSITION OF METALS

Deposition on Metallic Substrates

The title of this section is somewhat misleading. Nonelectrolytic deposition is achieved without the *purposeful* imposition of an outside potential. Of course in the total absence of any voltage, metal ions will not move in a preferred direction, and no deposit can form.

The source of such a potential in a system can arise in several ways, and will be described rather loosely, since the actual theory involves electrochemical concepts that are rather abstruse.

Noncatalytic deposition *Immersion Deposits.* Almost any metal can be satisfactorily deposited (coherently and reasonably adherently) from a properly formulated electrolyte *on a more electropositive substrate* without the use of an imposed electric potential. It must be emphasized that "electropositive," as used in this sense, is not a specific property of the metal itself. The polarity is determined by the relative potentials of the substrate and of the depositing metal in the *particular solution* in question. The familiar emf series of the metals (Table 13), for instance, is only an arrangement of the metals in the order of their potentials against a *standard specified concentration* of simple solutions of their *own salts* at a definite temperature.

A "simple" solution is one in which the metal ionizes freely. For example, in a copper sulfate solution, the dissolved copper is almost entirely present as free ions. In a cyanide copper solution, on the other hand, the dissolved copper is almost entirely tied up in a complex cuprocyanide $[Cu(CN)_2]^-$ ion and only a *very* minute fraction of the copper is present as free metal ions. If a piece of steel, for instance, is immersed in each of these solutions, we will find that nothing happens in the cyanide solution, but in the copper sulfate solution a deposit will form. This can range from a nonadherent smut to a sound deposit, depending upon factors too complicated to discuss here, which determine the potential between the copper ions in solution and the substrate metal. The greater this potential, the faster the deposition, and the more likely it is that the deposit will be mechanically unsound or nonadherent.

In the case of a sound adherent deposit, the driving potential disappears when the substrate is coated, and the thickness of most immersion deposits is limited to about 20 μin., although in rare cases up to 100 μin. or somewhat more can be obtained. The adhesion of any immersion deposit is always much less than is obtainable by electrodeposition.

Contact plating is similar to immersion plating in that no outside current source is required. In this case the basis metal does not dissolve, but instead, a still more electropositive metal *which is in contact with* the basis metal dissolves, while both it and the basis metal are coated with the depositing metal. For instance, if a piece of type metal is placed on a zinc tray and is covered with copper sulfate solution, both the tray and the type metal will be quickly coppered, whereas in a glass tray the action is very slow or does not take place at all. Similarly, aluminum can be tinned quite heavily (200 millionths or more) when in contact with magnesium in a stannate bath.

Zinc deposits by immersion on aluminum, magnesium, and beryllium from an alkaline zincate solution, and the resulting zinc film is used as the basis metal for subsequent electrodeposits, as will be more fully discussed later.

Catalytic or "electroless" deposits *Catalytic Nickel Deposition.* In a properly prepared solution containing nickel and hypophosphite ions (with some other additives for controlling the reaction and otherwise improving the operation), nickel can be deposited catalytically. Although this type of reaction was discovered early in the last century, it was not until 1946 that Brenner[10] and his coworkers made practical use of it to produce useful nickel and cobalt deposits by what they

TABLE 13 Emf Series: Standard Electrode Potentials of Metals versus Solutions of Their Own Ions (E_0 at Unit Molal Activity and 25°C)

Metal/metal ion	E_0, potential, volts
Base (or anodic) end of series:	
Magnesium/Mg^{++}	−2.37
Beryllium/Be^{++}	−1.85
Uranium/U^{3+}	−1.80
Aluminum/Al^{3+}	−1.66
Titanium/Ti^{++}	−1.63
Zirconium/Zr^{4+}	−1.53
Manganese/Mn^{++}	−1.18
Vanadium/V^{++}	ca. −1.18
Columbium/Cb^{++}	ca. −1.1
Zinc/Zn^{++}	−0.763
Chromium/Cr^{3+}	−0.74
Iron/Fe^{++}	−0.440
Cadmium/Cd^{++}	−0.403
Indium/In^{3+}	−0.342
Cobalt/Co^{++}	−0.277
Nickel/Ni^{++}	−0.250
Tin/Sn^{++}	−0.136
Lead/Pb^{++}	−0.126
Iron/Fe^{3+}	−0.036
Hydrogen $H_2/2H^+$	0
Antimony/Sb^{3+}	0.2
Copper/Cu^{++}	0.337
Silver/Ag^+	0.799
Palladium/Pd^{++}	0.987
Plutonium/Pu^{++}	ca. 1.2
Gold/Au^{3+}	1.50
Gold/Au^+	ca. 1.68

termed the "electroless" method. (Unfortunately, today a number of so-called "electroless" processes for other metals are offered which are simply, or at least have the same limitations as, noncatalytic methods.) The nickel already deposited catalyzes the reaction so that deposition continues indefinitely. The reaction velocity is temperature-controlled, and this type of bath is usually operated at 190°F or more. Recently, a differently formulated bath based on a borane complex of nickel has been announced and is commercially available on a limited basis. The deposition speed of this bath is also temperature-dependent, but useful rates are said to be obtained at much lower temperatures than from the hypophosphite baths, which are not useful below about 150°F. The character of the deposits from the two baths also varies. A nickel-phosphorus alloy with phosphorus contents varying from 6 to 14 percent is obtained from the hypophosphite bath, while deposits of 99 percent nickel are claimed from the borane system (presumably the remainder is a nickel boride). A newer hypophosphite bath is available, operating at lower temperatures and producing deposits with 2 percent phosphorous or less.

The hypophosphite nickels are more pore-free than electrodeposited nickel of equivalent thickness, and have remarkable chemical resistance. (Stripping of a defective deposit is much more difficult than of an equivalent thickness of electrodeposited nickel.) The coating is relatively brittle, but this can be reduced markedly by annealing at 1150 to 1550°F. The hardness of the deposit is about 500 DPH (diamond pyramid hardness) as deposited. This can be increased to 950 to 1100 DPH by heat treating at temperatures up to 752°F, followed by slow cooling in an inert atmosphere. At higher temperatures, increased ductility is obtained, as has been noted, but the hardness decreases with both temperature and heat-treatment time, returning to about the original hardness in 2 hr at 1300°F, for instance.

The thickness of electroless nickel obtainable is not limited, so long as the proper bath composition is maintained during the deposition. Deposition rates, however, are usually less than 1 mil/hr, whereas electrolytic nickel can easily be deposited at 2 to 3 mils/hr, and in special cases at rates many times higher.

Certain metals, such as aluminum, magnesium, beryllium, titanium, iron, cobalt, nickel, rhodium, palladium, and gold, are catalysts for the reaction, and deposition on them is spontaneous. Some bath-composition variations are necessary for magnesium and for some aluminum alloys. Noncatalytic metals, which are more noble than nickel, such as copper, silver, gold, manganese, high-strength ferrous alloys, can be plated if some nickel is first deposited by application of a brief external voltage, as by contacting (or hanging) the piece with an aluminum wire or actually making it cathodic for a few seconds in a dc circuit.

Finally, a number of metals are anticatalysts, especially lead, tin, zinc, cadmium, arsenic, bismuth, molybdenum, vanadium, tungsten, and ziconium. Not only do these metals not receive a deposit; but also many, such as zinc and lead, cause spontaneous decomposition of the bath, amounting to a slow explosion. In this reaction, hydrogen is evolved so fast that the bath literally boils out of the tank.

Catalytic Deposition of Other Metals. Electroless deposition of only a few other metals is possible at present. Cobalt, which is very similar chemically to nickel, can be so deposited from similar systems, either along with or associated with nickel. Copper can be deposited catalytically from proprietary baths, and these are widely used in the printed-circuit industry but find little application elsewhere. Deposition of palladium and gold catalytically is possible; catalytic deposition of chromium has also been claimed, but it has not yet been demonstrated that the latter process is truly autocatalytic, that is, that deposition will take place when the basis metal is the same as that depositing.

Advantages of Catalytic Deposits. Since electroless deposits are much more expensive (by a factor of 10 for nickel, for example) to produce than electrodeposits, their use must be predicated on some special advantage.

The major advantage offered by an electroless deposit—and sometimes this is a very compelling one—is the uniformity of deposition over even very complex shapes. Earlier, we discussed at some length the difficulty of obtaining uniform thickness of an electrodeposit even over a flat plane, and spoke of thickness variations of 10:1, or much more, over complex shapes, and of the difficulties presented by angles, corners, and threads.

By contrast, an electroless deposit will coat the surface of the most complex part with a deposit of uniform thickness, to within no more than 5 or 10 percent variation from the average. There is absolutely no tendency for edge buildup; nor are threads, sharp corners, or angles any problem.

There naturally must be some reservations attached to such sweeping statements. In the deposition reaction, a large amount of hydrogen is evolved, and this must be free to escape. If it is trapped, as in a blind downward-facing cavity, for instance, it will prevent the solution from reaching the surface of the upper part of the cavity, and metal cannot deposit there. Such parts must be suspended differently, or rotated slowly during deposition. In very small blind holes which are also deep, the small volume of solution within the hole may become depleted of metal before the desired thickness is obtained, although the escaping hydrogen will assist in removing the (now much less dense) solution and replacing it with undepleted bath. More serious for such small holes is the chance that they may entrap air when first immersed, and hold it by capillary action. In such cases it may be necessary to probe into such holes with a fine wire when the part is first immersed.

In spite of the large amount of hydrogen evolved, electroless deposits do not have the tendency to pit (so often associated with electrolytic nickel, for example). They are thus intrinsically less porous than electrodeposits, and so a thinner coating can be used for equivalent corrosion resistance, usually by a factor of about 0.75 in the thickness range of 1 to 2 mils.

Deposition on Nonmetallic Substrates

There are a great many methods for coating a nonmetallic surface with a metal (Table 14). Some are purely mechanical; others are chemical. In all cases, however, the surface must be prepared so that the deposit will have reasonable adherence. Whatever metallizing method is used, the adhesion will be less, and usually far less, than is obtained by electrodeposition on a metal.

TABLE 14 Comparison of Metallizing Methods for Plastics

Method	Conductivity after metallizing	Dimensional reliability	Readily electroplated	Cost of equipment	Cost of application
Mechanical:					
Wax-graphite........	Low	Poor	With care	Very low	
Conductive rubber...	Variable–low	Good	If properly formulated		
Conductive lacquers..	Variable–low	Poor	If properly formulated	Very low	Low
Vacuum deposition...	Not a factor	Not a factor	Usually cannot be plated	Very high	Very low
Metal spraying.......	Good	Poor	Yes, but rough	Moderate	Moderate
Chemical:					
Silvering (spray).....	Fair	Excellent	Yes	Moderate	Moderate
Immersion deposits...	Fair	Excellent	Yes	Low to moderate	Moderate
Autocatalytic (electroless (deposits)	Fair to excellent	Excellent	Yes	Moderate to high	Moderate

A clean surface is required, as would be expected, and if the substrate is porous (wood, plaster of paris, woven or felted fabrics, leather, paper, and the like) it must be sealed whenever the metallizing method involves immersion in a liquid. Otherwise, liquid will be trapped beneath the coating and will later cause blistering, corrosion, and other difficulties.

Mechanical coating of plastics This is typified by vacuum deposition, flame spraying, paste silvering (at low firing temperatures), the wax-graphite process, the use of conductive lacquers, or the formulation of a conductive plastic substrate.

Of these avenues, vacuum deposition is the most promising, but the adhesion of the very thin metal coating thus deposited is usually of a very low order (Table 15). Any attempt to electroplate over such a deposit usually results in a rupture of the weak plastic-metal bond. Flame spraying of low-melting metals and alloys on a plastic surface is surprisingly successful, but the resultant surface is rough, and its variable thickness can present a problem with dimensional tolerances. Paste silvering or gilding with conventional formulations requires firing temperatures far in excess of those tolerable for most plastics. Lower-temperature formulations are available, but their adhesion is of a lesser order and their reliability remains to be proved. The remaining processes all leave much to be desired.

Mechanical coating of ceramics This is less limited than metallic film formation on plastics, because much higher temperatures can be used. Thus, we can employ paste silvering or gilding (as is used for gilding glassware and china) where firing temperatures of the order of 1000°F are used. More sophisticated coatings are also used when certain physical or electrical properties are needed, as in the case of molymanganese films, which after firing, leave a layer of molybdenum or molybdenum alloy, on which electrodeposited films can be built up. As would be expected, the adhesion of such metallized coatings to the ceramics is of a very high order, and subsequent electroplating results in a strong structure. These processes are finding use in the field of microcircuitry.

Metallized Ceramics. Integral metal-ceramic bonding was developed primarily for electronic use. Before this, assembly was by mechanical means only, by cementing, force fit, and the like.

TABLE 15 Characteristics of Chemical- and Vacuum-deposited Metals

Characteristic	Type of deposit			
	Immersion	Catalytic (electroless)	Vacuum-metallized	Chemical, (e.g., silvering)
Use...............	Wide	Wide	Limited	Chiefly in record manufacture
Metals deposited.......	Almost any	Au, Ni, Co, Cu, (Cr?)	Almost any	Silver, rarely copper or others
Preparation cycle.......	Complex	Complex	Simple	Clean surface
Adhesion to plastic.....	Poor to fair	Fair to good	Almost none	Poor
Thickness range, mil....	to about 0.05 or 0.1	Not limited	Usually to about 0.02	Less than 0.02
Uniformity of thickness	Excellent	Excellent	Varies	Good
Can be electroplated....	Yes	Yes	Rarely	Yes
Equipment and maintenance cost.........	Low	Moderate	High	Varies
Direct labor cost.......	Moderate	Moderate	Low	Varies

In the electronics industry, it is becoming increasingly important to achieve reliable metallizing on ceramic substrates. One of the earliest methods was the fired-on silver paste which can be copper-plated to form an excellent bond and can then be successfully soldered using soft solders. This system is limited to soldering temperatures below 800°F.

More recently, these low-temperature metallizing compounds have been modified with additives such as glass frit, specifically for use on ceramic materials for better bonding than the simpler silver pastes. For higher temperatures, such as are encountered in silver soldering and the like, either the so-called *active metal* or *refractory metal process* can be used.

The *active metal* process[11] is based on the use of an active metal such as titanium. When brought to high temperature in a vacuum atmosphere, titanium will enter into a molten solution with silver, nickel, or other transition metals, and react with most oxide ceramic materials to form a strong chemical bond. Unlike the silver-glass frit or the refractory-metal–powder processes, the active metal process requires that flanges, pins, or other hardware be assembled and jigged in position during the one-shot vacuum-firing operation. In the vacuum furnace the braze material becomes molten and quite fluid. Since commercial gases are not pure enough to provide a sufficiently protective atmosphere for this operation, it is performed in vacuum of 10^{-4} torr, or better. Vacuum brazing has other advantages. It is essentially inert and thus causes little change in the properties of any of the components except for an annealing of the attached metal hardware.

A principal advantage of this process is that assemblies are made in a single-vacuum firing operation. The process is suited to quantity production of small assemblies. Since all assemblies must be held in place during bonding, the necessary jigs and fixtures must be carefully designed. The tendency of the active metal to adhere to fixture materials on contact must also be considered. When feasible, designs that result in self-jigging should be used.

The *refrectory-metal-powder*[11] process, sometimes referred to as the molymanganese process, uses powdered molybdenum, or its oxide, alone or combined with catalytic materials such as manganese, iron, nickel, titanium, chromium, silica, calcia, or glass frit. A wide variety of mixes, when properly selected, will bond well with the appropriate high alumina or beryllia ceramics to provide parts with good mechanical and electrical properties at elevated temperatures.

The process steps in the refractory-metal–powder process are quite different from those in the active metal process. The refractory-metal–powder mix is suspended in liquid organic materials which serve both as a temporary binder and as a vehicle for the mix. This paste is then applied to the ceramic by one of several methods. The coating is dried and then sintered in a protective atmosphere of hydrogen, dissociated ammonia ($75H_2$, $25N_2$), or similar gas. The sintered coating, now firmly bonded to the base ceramic, is at this point ready for electroplating.

The electroplating is usually nickel. This is then sintered in protective atmosphere at a lower temperature to improve adherence and reduce any tendency to oxidize. This sintered plating facilitates bonding with solders or brazes. Other metals may be electroplated to the base coat, or they may be applied over the nickel as an addition.

A variety of materials can be used for soldering or brazing. Soldering or low-temperature installations do not require protective atmosphere, but high-temperature brazing should be done in a dry protective atmosphere. This atmosphere can be hydrogen, dissociated ammonia ($75H_2$ $25N_2$), or forming gas ($90N_2$ $10H_2$). The selected brazing temperature should be approximately 75°C (167°F) above the flow temperature of the braze material.

Hermetic seals can be made with unusual bond strength for special service requirements. Low-expansion alloys, such as nickel-cobalt-iron, are often used to match the expansion properties of available ceramics. If high corrosion resistance is required, nickel, copper, or stainless steel may be used. The seal is strong to the point of ceramic fracture, and average tensile strength values on alumina ceramics exceed 15,000 psi when tested according to American Society for Testing and Materials Method F19-61T.

Refractory-metal–powder coatings are applied only to selected unglazed areas of the ceramic part. A high-temperature glaze may be applied to other areas, so long as the glaze is capable of withstanding the temperatures and atmospheres necessary for metallizing and assembly operations, and so long as the glaze and metallizing coats are separated one from the other by a gap of at least $\frac{1}{16}$ in. A glaze reduces flashover and corona tendencies in service and provides a surface which may be readily cleaned from time to time.

The refractory-metal–powder coating is nonmagnetic and thus has proved excellent for applications in the electronic industry. The process is also suited to the precision metallizing of many microminiature designs. Table 16 shows some practical tolerances applicable to these processes. Further electroplating tolerances must also be applied.

Thermal Expansion of Metals and Ceramics. One problem that has plagued designers of metallized ceramics is the often quite large difference in the coefficients of thermal expansion of the two types of material (Table 17). For good results the metallizing must be chosen so that the differential expansion is as small as possible. For this reason, chromium metallizing has been very popular by reason of the very close coefficients of expansion of chromium and ceramics. Thereafter,

10-42   Metallic and Chemical Finishes on Metals and Nonconductors

further plated coatings can be applied as needed, and the strains between ceramic and metal with changes in temperature are minimized.

Chemical metallizing of nonmetals   This method is at present much more reliable than any of the mechanical methods in terms of the adhesion of the metallized coating. The same formulations that have been described for nonelectrolytic deposition on a metal are also used for plastics, but now an additional step is necessary. The plastic must be sensitized by a black-magic kind of operation called sensitizing.

TABLE 16   Practical Metallizing Tolerances for Plating on Ceramics

Plating method	Pattern tolerance	Base metal thickness	Plate thickness	Solder coat thickness	Total metallized thickness
Spraying......	±0.010	0.0004–0.0015	0.0002–0.001	0.0002–0.0015	0.0008–0.0035
Screening......	±0.005 or 1 %, whichever is larger	0.0004–0.0015	0.0002–0.001	0.0002–0.0015	0.0008–0.0035
Brushing......	±0.020	0.0004–0.0015	0.0002–0.001	0.0002–0.0015	0.0008–0.0035
Banding by machine.....	±0.015	0.0004–0.0015	0.0002–0.001	0.0002–0.0015	0.0008–0.0035
Photoetching..	±0.001	0.0004–0.0015	0.0002–0.001	0.0002–0.0015	0.0008–0.0035

NOTES:
1. Tolerances apply to all low-temperature processes along with high-temperature refractory metal powder process.
2. Closer tolerances may be held on specific designs at increased cost.
3. Platings over this base metallizing affect these tolerances.

TABLE 17   Linear Coefficient of Thermal Expansion of Metals and Ceramics

Material	Linear expansion coefficient, in./in. $\times 10^{-6}$	Temperature range, °C
Ceramics..........	6.0–6.9	25–300
Ceramics..........	7.5–7.8	25–700
Ceramics..........	7.2–7.9	25–900
Chromium.........	6.2	20–200
Iridium...........	6.8	20–200
Rhodium..........	8.3	20–200
Titanium..........	8.4	20–200
Platinum..........	8.9	20–200
Beryllium.........	11.6	20–200
Palladium.........	11.8	20–200
Nickel............	13.3	20–200
Cobalt............	13.8	20–200
Gold..............	14.2	20–200
Copper............	16.5–17.7	20–200
Silver.............	19.7	20–200
Aluminum	19.4–24.1	20–200
Magnesium........	25.2–27.1	20–200
Cadmium.........	29.8	20–200
Zinc..............	39.7	20–200

Sensitizing.   A nonconductive surface must be conditioned to receive a nonelectrolytic metallic deposit, if a subsequent deposit is to be sound and cohesive. Probably the most familiar of such sensitizing procedures is that used for silvering glass. This produces the conventional mirror, and this can be plated upon successfully. In this case, the sensitizing solution is stannous chloride, and stannous ions are absorbed on the glass surface. Even after an extremely thorough rinsing, enough isolated stannous ions remain to ensure that when the glass is immersed in the silvering solution, the silver film formed will be sound and continuous. In this particular case the silvering solution is unstable, and the chief function of the sensitizer is to make sure that the desired surface is completely covered. Unsensitized surfaces will also be silvered, but probably not completely.

More stable metallizing solutions are often used, which will deposit metal only on sensitized nonconducting surfaces. The coppering solutions used for metallizing the epoxy-glass surfaces of printed-circuit boards in the plated-through-hole process are one example. These solutions contain palladous (or sometimes other) catalytic metal ions which act in a manner precisely similar to the stannous ions used for the silvering process.

TABLE 18 Comparison of Platable Plastics

Plastic	Cost/lb, dollars	Adhesion, lb/in.	Uses
Widely used:			
ABS	0.40	5–10	Many
Polypropylene	0.19–0.28	20–30	Expected to compete with ABS
Polysulfone	1.00	Good thermal cycling	Housewares Auto trim
Epoxy, filled (TS)	1.60 or less depending on filler	5 and up—good thermal cycling. (on vapor-honed surface)	Printed circuits (copper plate) Aircraft potential (heavy nickel plate)
Not in general use, but with potential:			
Polyesters (TS)	0.16	20 reported	Have been used
Polyphenylene oxide (PPO)	0.75	N.A.	Not widely used
Polyvinyl chloride (PVC)	0.24–0.29	N.A.	None reported
Polystyrene	0.15	3–5 reported	Good potential low cost
Nylon	0.90	N.A.	Not yet commercial
Polyacetal	0.65	N.A.	Still experimental
Polyethylene	0.18–0.20	Reported good	None as yet
Phenolics (TS)	0.20	Less than 5	Have been used
Polytetrafluoroethylene (PTFE)	4.50 and up	Under development	Outstanding electrical properties

N.A. Not available.
TS Thermosetting.

The process of silvering a piece of glass for the purpose of making a mirror consists of first cleansing the glass very rigorously, and finally using a solution of chromic acid in concentrated sulfuric acid. Then, after a very thorough rinsing, a stannous chloride solution is poured over it, and the glass is again rinsed as thoroughly as possible with distilled or deionized water. Now this very clean glass with a few stannous ions still clinging to it will silver nicely. In the same way, other nonconductors are sensitized, sometimes with stannous ions, sometimes with gold or palladium, after which they may be metallized, in the electronics field, usually with gold, copper, or nickel.

Adhesion of metallized coatings on plastics (Table 18). In no sense can the adhesion of a metallized coating on plastic approach that of an electrodeposited metal on a metal substrate. In the latter case the adhesion approaches and usually exceeds the tensile strength of the weaker of the two metals involved. In the former case it is *always* possible to separate the metal coating from the plastic, and the measure of adhesion is only a measure of the force required to separate the two. Instead of the usual pounds per square inch measure, which describes the

10-44　Metallic and Chemical Finishes on Metals and Nonconductors

force necessary to effect separation between two areas supposed to be adherent, it is customary to measure adhesion to a plastic in terms of a peeling mechanism, and here the measure of adhesion is the pounds of pull per *linear* inch of the parting surfaces required to separate them. (Of course, this measure is in no way comparable to a true adhesion test performed in a tensile testing machine.)

Until fairly recently, a linear pull of 5 lb/in. was considered acceptable for coatings plated on plastics. On certain formulations such as ABS, however, we are now looking for 10 to 15 lb/in. as standard, while on others 25 to 35 is considered attainable. These adhesion values vary with the plastic, and such high adhesion values are achieved only with a few, such as polypropylene and some polyesters.

TABLE 19 Specific Gravity and Relative Volume Costs of Thermoplastics and Metals

Material	Specific gravity (water = 1.0)	Cost/lb, dollars	Approximate relative cost/volume (low-density polyethylene = 1.0)
Plastics:			
Polyethylene	0.93–0.97	0.20	1.0
ABS	1.01–1.15	0.39–0.47	2.2
Polysulfone	1.21	1.00	6.4
Polypropylene	0.98	0.19–0.28	1.0–1.4
Polystyrene	1.04–1.07	0.15–0.32	0.8–1.3
Polycarbonate	1.2	1.05	6.3
Polyacetal	1.42	0.65	4.55
Polytetrafluoroethylene (PTFE)	2.13–2.22	4.50 and up	5.0–8.5
Metals:			
Magnesium	1.8	0.30–0.40	3.0
Aluminum	2.7	0.24–0.28	4.2
Zinc (anodes)	7.14	0.12–0.20	4.8–9.5 (fluctuates)
Tin (anodes)	7.3	1.18	48
Cadmium (anodes)	8.65	0.38–0.50	145
Nickel (anodes)	8.90	1.05	53*
Copper (anodes)	8.93	0.38–0.48	16
Silver (anodes)	10.5	23.00	1,320
Lead (anodes)	11.3	0.13–0.15	8.
Chromium	7.1	†	53
Gold‡	19.3	612.00	63,500

* Electroless nickel costs are relatively 690 per unit volume.
† Metallic chromium is not used as anode, but metal costs $1.31 per lb.
‡ The cost of gold plated from proprietary baths is about 30% higher.

Comparison of metals and plastics　For applications where weight is of concern, the use of a metallized plastic is advantageous (Table 19). Also, since plastics are often more expensive (per pound) than common metals, the cost per unit volume is reduced to a very competitive basis, as the table shows.

Table 18 also shows the current status of plastics considered commercially platable, their cost, and some of their properties and uses.

Electroplating on Metallized Surfaces

The only intrinsic difficulty in electroplating over a metallized coating on plastic is that in most cases the metallizing is very thin. Therefore the plating bath must be chosen with due regard to the possibility that it may dissolve the metallizing before a thick enough electrodeposit can be formed to protect it. This is usually the plater's problem rather than the designer's.

NONMETALLIC COATINGS ON METALS

Anodic Films

Anodizing of aluminum This process is very simple, compared to any electroplating operation. If a piece of aluminum is made the anode in certain solutions, an oxide coating forms. This coating can also be formed in the same solution when it is only *sometimes* the anode, as when an alternating current is used for anodizing, although the best coatings are customarily formed with the use of direct current.

The comparatively good corrosion resistance of aluminum in the atmosphere, despite its highly anodic position in any of the various galvanic series, is due to a naturally formed thin oxide film. The thickness of this film, normally only 0.2 to 0.6 μin. is yet sufficient to protect this very active metal surprisingly well. An artificial oxide film, formed purposely by anodic treatment, serves even better. Such films are thicker, mechanically superior, and more corrosion resistant. They can be formed in a variety of electrolytes under different conditions. The two solutions

TABLE 20 Formation and Thickness of Anodized Coatings on Aluminum

Electrolyte	Temperature, °F	C.D., asf[a]	Volts	Usual film thickness, mils[b]	Notes
Conventional CrO_3....	70–80	<1–5	40	0.04–0.12	
Conventional H_2SO_4...	60–75	8–16[c]	15	0.3–0.4	
H_2SO_4.............	32–35	20–36	20–100	1.0–4.0	Martin Hard Coat
H_2SO_4 plus oxalic acid.	47–50	20	To 60	Varies	Alumilite 225 and 226 processes
6% oxalic acid........	40	12–15	60[d]	1–3	British Hardas process
10% boric acid.......	160–210	Very low	50–500	0.0002–0.04	[e]

[a] Current density, amperes per square foot.
[b] The approximate dimensional *increase* of the piece is about half this.
[c] Varies with aluminum alloy.
[d] Superimposed ac on dc.
[e] Barrier film used to form capacitors.

most commonly used in the United States are sulfuric acid and chromic acid formulations, and each has its preferred area of application. An oxalic acid solution is widely used in Germany and Japan, while a boric acid bath can be used to produce thin dielectric films for use in electrolytic capacitors.

The as-formed oxide film is relatively brittle and porous. It is commonly sealed in a boiling solution, which hydrates the film and swells it to a dense, compact surface. If desired, certain dyes may be used for decorative effects, and these are used to treat the porous oxide film before sealing.

The anodized film is a nonconductor, but it is easily penetrated by a sharp point (except films formed in the so-called "hard anodizing" process) to make contact with the metal beneath. It is thus not safe from the design standpoint to allow bare soldered wire terminals, for instance, to make a rubbing type of contact with the anodized film and to depend on its insulating properties to prevent electrical short-circuiting.

Mechanism of Anodizing Aluminum. Since aluminum oxide is nonconductive, the distribution of the film thickness over a part is very uniform, because the current will be automatically directed to those areas of a part not yet fully coated with the insulating film, until a coating of the limiting thickness is everywhere present. This thickness is distributed approximately half above and half below the original surface of the aluminum (see Table 20).

Because the film is nonconductive, and because of its distribution with respect to the original surface of the piece, there are certain limitations imposed on the design of the part, and on the refinishing of rejects.

The part must be capable of being held firmly, so that the point of contact cannot shift and cause a loss of voltage or a complete breaking of the electric contact. Also, the points of contact will themselves not be anodized and should preferably be in an inconspicuous and noncritical area. If a part cannot be so designed (consider a solid cube where all faces may be critical, for instance), the designer must settle for two tiny (0.01 in. in diameter or less) untreated areas which may be touched up if necessary, by a manual application of a chromate conversion coating after anodizing. MIL-A-8625C recognizes this fact and allows for such spots.

A rejected anodized coating may be easily stripped and refinished, but there will be dimensional changes owing to the above-described distribution of the coating above and below the original surface. The stripped part will be smaller by about half the coating thickness and after reanodizing will return to approximately the original dimensions as machined. In most cases, as will be seen, this is not dimensionally too significant, unless the tolerance of the finished piece is less than about 0.3 mil and unless the anodizing process used is sulfuric acid. When an anodized part must be remachined locally (to add missing holes, for instance), it is possible to reanodize the part directly, after thorough degreasing, with minimal dimensional change in the original anodized surface. Contact is made either on the machined surface or on the old surface by piercing it with a sharp contact.

The first-formed oxide layer, of course, is located at the extreme outer surface. As formed, the outer layer is relatively porous, and the action of the current is able to penetrate it and oxidize the metal beneath, until finally the electrical resistance becomes too high and there is no further action. While this has been going on, however, the upper layers of oxide, which have been in long contact with the electrolyte, do dissolve somewhat and become more porous. The last-formed layer is located at the metal-oxide interface and is known as the barrier layer. It is nonporous and extremely thin, but it does have a significant effect on the electrical resistivity of the coating and on its corrosion resistance. The thickness of the barrier layer is theoretically 10 to 14 Å volt used in its formation.

Sulfuric Acid Anodizing of Aluminum. In conventional room-temperature anodizing in sulfuric acid solutions, the "growth" of the surface is about 0.3 to 0.35 mil. Martin Hard Coat (MHC) and similar very low temperature processes produce films of the order of 2 mils or more. Anodized films produced in chromic, oxalic, or boric acid are very much thinner than even the conventionally formed sulfuric acid coatings.

Because the solution for sulfuric acid anodizing is intrinsically so corrosive, this process cannot be used where faying surfaces such as spot-welds, overlaps, etc., are present in the design. Such crevices trap electrolytes which can later seep out to ruin the finish and cause damaging corrosion. In such cases, chromic acid anodizing is used, because if chromic acid is trapped and later exudes from the crevice, the only effect is to form a chromate film, which is in itself protective.

Sulfuric acid anodizing is carried out in a 15 volume percent solution of the acid at room temperature or somewhat below. The usual treatment time is about an hour, and the voltage used depends upon the alloy being treated, but is from 10 to 18 volts. A most important consideration is that an assembly of different alloys may be impossible to anodize satisfactorily, because one alloy may anodize faster than the other, and then while the second component is coming to completion of the process, the first may be deteriorating, as has been described. This can involve both dimensional changes and loss of protective value in the first component.

Hard coating usually refers to sulfuric anodizing carried out at lower temperatures (to 35°F and below) and at higher voltages and current densities. Under these conditions, films as thick as 2 to 4 mils may be formed, using longer times. These processes (of which there are several) are usually not recommended for alloys high in copper (over 4 percent) or silicon (over 7.5 percent).

Chromic Acid Anodizing of Aluminum. This process produces films of the order of 0.02 to 0.05 mil thick. This is advantageous dimensionally, since the machined dimensions are held fairly closely (the surface "grows" only by about half this amount); there is no harmful effect on fatigue strength; and the coatings are highly protective, thin as they are. The process is almost universally preferred for parts subject to stress and is particularly to be specified for assemblies or parts of intricate design where electrolyte may be trapped between mating surfaces or in small crevice. The inhibitive effect of hexavalent chromium enhances the corrosion resistance for such parts. Specification MIL-A-8625C prohibits the use of sulfuric acid anodizing for parts where solution may be trapped, and calls for chromic anodizing of such parts, except where the alloy contains over 5 percent copper or over 7 percent silicon. The film produced by chromic anodizing is typically colored, whereas sulfuric anodizing produces gray or colorless finishes depending on the alloy, but sealing in a dichromate bath (which adds to the corrosion resistance) is often used. A sulfuric acid coating so sealed will also have a reddish yellow tinge.

Sealing. Because of the soft and porous nature of the oxide film, sealing is necessary after either anodizing process. When this surface is immersed in boiling water, the porous oxide takes up water chemically to form a compound similar to boehmite (Al·OH). The resultant volume increase swells the oxide and seals the pores. If a solution other than water is used, dissolved materials such as dyes can be mechanically sealed in to give decorative effects, or there may be a chemical reaction, as when a dichromate seal is employed. When dyes are used, these may be inorganic compounds (which in chromic acid anodizing are limited to darker shades and give rather "muddy" effects) or organic compounds, which can be chosen and regulated to give shades of color from pastel to brilliant or very deep hues, including an intense black. These dyes vary in their lightfastness and in their resistance to bleeding out and fading or weathering under various conditions. Properly selected dyes can have a high order of permanence.

Other Anodizing Solutions. Other solutions have been developed but have limited use in the United States. Oxalic and (especially) boric acids do not have the solubilizing effect on the oxide film exhibited by sulfuric and chromic acids; and hence film formation is interfered with less as the anodizing proceeds, and the resultant coating is thin, with relatively high dielectric properties.

Properties of anodized aluminum (Table 21) *Film Thickness.* The increase in film thickness is not linear with anodizing time, but often an equilibrium may be reached between the rate of film growth and the rate of film dissolution in the electrolyte. In the case of some alloys (e.g., Al-Mg and Al-Mg-Zn) there even can be an actual decrease in film thickness after a maximum has been reached.

Anodizing at first increases the dimensions of the work, and then after the maximum is reached, reduces them, because even if the film thickness *does* remain constant, the basis metal continues to be anodized while the film dissolves at an equivalent rate (or even faster for some alloys, as was mentioned above).

Adhesion. Because the film is formed from the parent metal, its adhesion to the remaining unanodized basis metal is normally better than that of most electrodeposits. The film, however, tends to be weak vertically to the surface. When it is bent, the coating cracks in parallel lines but will not strip, as brittle electrodeposits may, especially on aluminum.

Hardness. Although aluminum oxide is very hard of itself, the film is too thin to increase the effective hardness of the piece. It will not protect the metal from strong pressure, but it will resist surface scratching and will protect the appearance and reflectivity of the coating on a polished surface. Anodized aluminum mirrors have good retention of reflectivity in outdoor environment, for instance.

Harder coatings are obtained in the less aggressive electrolytes, e.g., oxalic or boric acid, and at lower anodizing temperatures and higher current densities. Coatings produced by dc electrolysis are harder than those produced under ac conditions. The hardness of the film decreases more or less sharply from the barrier layer out to the surface.

Abrasion Resistance. Abrasion resistance (one criterion of that nonproperty, hardness) is decreased by sealing, because hydrated aluminum oxide is softer than unhydrated. In aging, an unsealed film slowly hydrates, with a consequent loss of abrasion resistance.

Flexibility. In general, the brittleness of anodic coatings limits any forming operations after anodizing to those in which elongation of the film is not over 0.5 percent. More elastic films are obtained when anodizing is done at higher temperatures, and also with alternating current.

Anodic films on magnesium Maximum protection and good paint-base properties for magnesium alloys are provided by anodic treatments. Light gray to a full medium green shade are obtained in an acid solution (Dow No. 17), based on a combination of fluorides, phosphates, and chromates. The full coating is obtained in 10 to 25 min at direct voltages up to 110, and the film will be 0.9 to 1.2 mil thick.

TABLE 21 Properties of Films Produced on Aluminum by Chromate Treatment

Thickness	0.04–0.1 mil (rarely 0.2 mil).
Dimensional change	Usually negligible.
Color	Depends on process; may be very pale to orange, or (for chromate-phosphate processes) iridescent bluish green.
Corrosion resistance	Surprisingly good salt-spray resistance (200 or more hr), especially when painted (500–1,000 hr).
Weldability	Good, with some increase in required voltage.
Paint adhesion	Improved.
Electrical resistivity	Depends on contact pressure and on the chromating process used.*

* In one series of tests, results were as follows:

Contact pressure, psi	Contact resistivity, microhms/in.²	
	Untreated	Chromated
10	1,200	8,000
100	470	1,900

An olive-drab to tan or brown color is obtained in the "HAE" process in a highly alkaline solution containing aluminum salts with fluoride, phosphate, and manganate or permanganate ions, operated up to an hour or more at alternating voltages ranging from 8 to over 90. The high-voltage coating obtained by this process is currently the hardest available for magnesium. Magnesium is usually painted after anodizing for best corrosion protection, and when this is done, the HAE coating must be pretreated before painting to neutralize the alkali.

Conversion Coatings

Conversion coatings are applied by a simple dip into a solution (this may or may not be heated) which by chemical action transforms the surface into the desired compound of the metal or of the alloy components. The films are extremely thin, and the surface appearance and properties are altered. There is some attack, of course, on the metal surface, which loss normally is limited to 10 to 100 μin.

Chromate coatings These are widely used on a number of metals and are today usually formed in proprietary solutions. The finish ranges from colorless to a typical yellow to an olive drab. On a given piece, there sometimes may be variations in the color (depending on the process used and variations in the structure of the metal), but usually a pleasing uniform color is obtained. On different alloys of a given metal, wide color variations must often be expected. The chemical composition of the film is normally a chromate (accounting for the usual orange color), but chromates of alloy constituents may have a different color, or the constituent may be converted to an oxide, which darkens the finish.

The films do provide some additional corrosion resistance, but it is frequently less than that supplied by anodizing. They are also useful for improving the bonding and corrosion resistance of a subsequent paint coating. Clear films provide an excellent surface for soldering with rosin fluxes, but the heavier films make soldering very difficult, except when use of chloride fluxes is permissible. The properties of chromated films are shown in Table 21.

Phosphate coatings An adherent surface layer of insoluble metallic phosphates is formed in proprietary aqueous solutions at moderately elevated temperatures (200°F). The layer may be composed chiefly of basis metal converted to phosphate; but usually it contains, in addition, phosphates of zinc, iron, manganese,

TABLE 22 Phosphate Coatings

Basis metal	Type of coating	Applied by	Usual coating time, min	Coating weight, mg/ft^2	Used for
Steel.......	Zinc phosphate	Dip, spray, or brush	1-5	100-600	Paint adhesion
		Dipping, tumbling	20-30	1,000-4,000	Oiled or waxed for guns, hardware
	Manganese phosphate	Dipping, tumbling	15-30	1,000-4,000	With or without oil or wax for hardware
	Iron phosphate	Power spray	2-3	30-100	Paint adhesion
Zinc and cadmium	Zinc phosphate	Hand brush or spray	2-15	200-600	On galvanized steel, for paint adhesion
	Zinc phosphate	Power spray	0.5-2	75-500	On Zn or Cd electroplate for paint adhesion
Aluminum	Zinc phosphate	Power spray, dip	1-3	100-300	Paint adhesion
	Al-Cr phosphate	Power or manual brush or spray	0.3-3	100-250	Strip, cast, or wrought alloys, to be painted or not

calcium, or mixtures of these that were present in the treatment solutions. The chief applications of a phosphate coating are:

1. Paint-bonding to iron, zinc, cadmium, and aluminum
2. Improving corrosion resistance of unpainted iron or aluminum, with or without the aid of oils or waxes
3. For steel, improvement in cold-working properties and reduction in surface friction

Table 22 shows typical properties and uses of phosphate coatings.

Other Coatings Numerous colored coatings (oxides, sulfides, etc.) are applied to many metals, usually by proprietary processes. Their application is chiefly decorative or, for black coatings, to reduce light reflectivity, or to increase heat absorption.

CORROSION RESISTANCE

Mechanism of Corrosion

Except in cases where a metal is readily soluble in the corroding medium, most corrosion is due to galvanic action. Broadly, galvanic corrosion occurs when, for any reason, an electric potential exists between two areas of a part or of an assembly, while at the same time these areas are covered with or immersed in an electrolyte. This electrolyte may be only moisture condensed from the ambient atmosphere which, although it is intrinsically not a very conductive electrolyte, may quickly become one by dissolving carbon dioxide, chlorides, sulfates, nitrates, and the like from the air. Worse, if because of poor rinsing, residues of soldering fluxes or of plating solutions are left on the part, such condensation will *at once* form highly conductive electrolytes, which may be corrosive in themselves, and are also capable of carrying very appreciable currents at low voltages. Without the imposition of any external potential, a battery cell is formed, and drastic corrosion can occur. If, in addition, there is an externally applied direct voltage (which is often the case in an electronic assembly), the corrosion can be catastrophic.

Complete inhibition of corrosion Only in a hermetically sealed space (e.g., with a metal-to-metal or metal-to-glass seal), which is completely free of both moisture and/or materials that may give off moisture or corrodents, can freedom from corrosion be assured. Other "seals" that may be used, such as rubber or elastomeric gaskets, leather, tapes, and shrinkable tubing, as well as liquid or paste, which solidifies in place to form seals, do not constitute true hermetic seals. All will "breathe" to a greater or lesser extent with changes in atmospheric pressure, and so draw moisture into the enclosure.

In every instance, the effect of moisture is damaging. Since a molecule of water has a diameter of only 3.4 Å, it can penetrate a very small opening, and so will diffuse through seals of rubber, paint, and many other materials.

Galvanic corrosion This arises from the use of dissimilar metals in contact and is probably the greatest single source of corrosion difficulties in sophisticated electronic gear, in which it is almost a mandatory feature of the design that many different metals must be used.

Numerous tables have been prepared for assisting the designer to avoid the most damaging combinations of metals. These tables, however, are not universally applicable, because each is based on one selected set of conditions. The familiar electromotive force (emf) series (Table 13) is probably the least useful, since it is based on the *single* electrode potential of one metal against a rather concentrated solution of its *own* ions, nor does it consider the other metal, which in practical cases forms the second electrode of the battery. The *galvanic series*, on the other hand, can be used to estimate the voltage generated between two immersed metals, but *it is necessarily limited to one electrolyte*, such as seawater, for example. Thus, we find one galvanic series for seawater (Table 23), another for the atmosphere applicable to guided missiles (Table 24), and so on. Table 25 is another listing taken from MIL-STD-454, and there are many more. All have been compiled for a specific set of circumstances, and the best design procedure is to select or find a table most closely describing the environmental conditions to be met. Unfortunately, no clearer guidelines can now be drawn.

Undesirable dissimilar metal combinations on different pieces can be made tolerable by plating each part. It is *extremely dangerous*, however, to extrapolate this freedom to the design of a *single* component which is constructed of several widely dissimilar metals joined together for plating. Such an agglomeration may quite often be impossible to plate satisfactorily, because the *required* pretreatment for one metal may cause catastrophic destruction of another, or at the very least may be responsible for the plate blistering on one of the other metals. When only two metals are involved, a compromise may often be worked out. Proposed untested combinations should first be cleared with a fully experienced plating engineer. The farther apart the metals are in a galvanic series, the greater is the likelihood of serious trouble.

Crevice Corrosion. Under spot-welded joints, in crevices at threaded fasteners, or under washers there will be a lack of available oxygen. In the presence of moisture, an electric cell is formed, and there will be metal attack. The attack can occur either where the two metals meet, or wholly underneath the joint. The latter is the most dangerous, since the progress of the corrosion is not visible until it is so far advanced that the joint may be in danger of failure.

Sealing the joint to exclude moisture will prevent this attack. (Sealing is necessary even when nonmetallic washers are used, since these keep out oxygen just as a metal would.) An epoxy, polyurethane, or silicone rubber sealant is preferred. For "stainless" steels in marine atmospheres, effective seals are 50:50 lead-tin solder or 10 percent copper powder in petrolatum.

Surface Contamination. Corrosion contributable to surface contamination can be wholly unrelated to galvanic corrosion (e.g., caused by a fingerprint), but when the opportunity for galvanic corrosion also is present, the effect of the contamination is enhanced. Even mild corrosion of contact surfaces can be ruinous to electronic components, causing noise in potentiometers, poor solderability, and other defects.

To avoid this type of attack, the designer should:
1. Select inherently resistant metals.
2. Use protective coatings where practical.
3. Improve resistance by passivating.

TABLE 23 Galvanic Series in Seawater

Noble (cathodic) Platinum Gold Graphite	Naval brass Manganese bronze Muntz metal
Silver	Tin Lead
Passive 316 stainless (18-8-3) Passive 304 stainless (18-8)	Active 316 stainless (18-8-3) Active 316 stainless (18-8)
Titanium Passive 410 Cr stainless (13%)	Active 410 Cr stainless
67 Ni 33 Cu	Cast iron Wrought iron Mild steel
Passive 76 Ni 16 Cr 7 Fe Passive nickel Silver solder M-B bronze G-bronze 70-30 cupronickel	Aluminum 2024
	Cadmium
Silicon bronze Copper Red brass Aluminum brass Admiralty brass Yellow brass	Alclad Aluminum 6053
	Galvanized steel Zinc Magnesium alloys Magnesium
Active 76 Ni 16 Cr 7 Fe Active nickel	*Base or active (anodic)*

NOTE: Within a group, galvanic effects will usually not be appreciable. With other combinations, intensity of galvanic effects will vary with the distance apart in the series and with the relative areas of the materials (small anodic area worst).

Localized Differences in Metal Surface (or Surfaces). Commercial metals and alloys are not homogeneous. In aluminum alloys, the grain boundaries are anodic to the grain centers and are thus subject to corrosion. To avoid this:
1. Use the 5000 or 6000 series rather than the 2000 or 7000 series.
2. In heat treating, use a rapid quench.
3. Use clad aluminum rather than bare.
4. Anodize where practical.

Another type of local cell is formed when a small surface of one metal (e.g., a rivet head) is surrounded by a large area of another metal. If the smaller area is anodic, the total corrosion current results in a high current density, and rapid corrosion at the anode (e.g., an aluminum rivet in a steel sheet). The reverse situation is not unduly troublesome, unless a mild diffused corrosion of the large area cannot be tolerated.

Galvanic corrosion as applied to aerospace components Special problems are also introduced when aircraft or aerospace conditions are considered.

Cycling Temperature Changes. These changes can be a major problem, since with a drop in temperature, relative humidity rises, and water condenses rapidly. A return to high temperature accelerates the corrosion as long as the water remains. There can also be "breathing" in closed but not hermetically sealed compartments.

TABLE 24 Galvanic Couples as Listed in MIL-STD-186B for Guided Missiles

Group	Metallurgical category	EMF (volt)
1	Gold, gold-platinum, platinum	0.15
2	Rhodium, graphite	0.05
3	Silver	0
4	Nickel, Monel*, high nickel-copper alloys, titanium	−0.15
5	Copper, low brass or bronze, silver solder, German silver, high copper-nickel alloys, nickel chromium, austentitic (type 300) stainless steels	−0.20
6	Commercial yellow brass and bronze	−0.25
7	High brass and bronze; naval brass, Muntz metal	−0.03
8	18% chromium steels	−0.35
9	Chromium, tin, 12% chromium steels	−0.45
10	Tin-plate, tin-lead solders, terneplate	−0.50
11	Lead, high-lead alloys	−0.55
12	Aluminum, 2000 series wrought	−0.60
13	Iron, low-alloy steels, Armco iron	−0.70
14	Aluminum, 3000, 6000, and 7000 series; aluminum-silicon castings	−0.75
15	Aluminum castings (other than silicon alloys); cadmium	−0.80
16	Hot-dip zinc and galvanized steel	−1.05
17	Zinc	−1.10
18	Magnesium	−1.60

○ = cathodic   ● = anodic

NOTE: Groups joined with arrows are permissible. Other combinations (usually over 0.1 volt potential difference) can result in harmful corrosion. Magnesium, for instance, *must* be isolated from any other metal, preferably by a combination of plating and painting.

† Trademark of Huntington Alloy Products Division, The International Nickel Co., Inc., Huntington, W. Va.

Cycling Pressure Conditions. Cycling pressure changes (as in aircraft alternately aloft and aground) cause massive breathing effects, with similar results.

Plastics. Plastics, normally thought of as inert and noncorrosive (especially when incompletely cured), give off corrosive vapors by outgassing at high altitudes or in space, which vapors attack metals in the vicinity, or metals plated on them (see Table 26). Cadmium is especially susceptible to this form of attack.

Exaggerated galvanic corrosion Exaggerated galvanic corrosion is likely to occur in an electronic assembly, where an externally applied direct voltage, sometimes of considerable magnitude, may be imposed between parts subject to wetting with an electrolyte. Currents may pass through the electrolyte and influence the corrosion type and rate (as well as having other undesirable effects; for example, causing silver migration and a subsequent short circuit). These currents may act to reinforce a galvanically generated current and so accelerate the corrosion, or they may oppose it, actually reversing the current direction, to cause corrosion of the normally more noble member of the couple. Intermittent passage of such a current could conceivably cause corrosion of both the base and noble metals comprising the couple.

TABLE 25 Groups of Compatible Metals in MIL-STD-454.

Group I	Group II
Magnesium and alloys	Aluminum and alloys
Aluminum 5052, 5056, 5356, 6061, 6063	Zinc
Tin	Tin
Group III	Cadmium
Zinc	Stainless steel
Cadmium	Tin-lead
Steel	Solder
Tin	*Group IV*
Lead	Copper and alloys
Stainless steel	Nickel and alloys
Nickel and alloys	Chromium
Tin-lead	Stainless steel
Solder	Silver
	Gold

All metals in one group are considered *similar* and unlikely to corrode from galvanic action, but are *dissimilar* to metals in another group.
This listing is similar to that in MIL-E-5400.

TABLE 26 Organics as Source of Vapor Corrosive to Metals

Material	Severely corrosive	Somewhat corrosive	Not corrosive
Adhesive............	Urea-formaldehyde	Phenol-formaldehyde	Epoxy
Gasket.............	Neoprene*-asbestos	Nitrile-asbestos	
	Resin-cork	Glue-cellulose	
Insulation (wire).....	Vinyl	Teflon*	Polyurethane
	Polyvinyl chloride	Nylon	Polycarbonate
	Vinylidene fluoride	Polyimide	
Sealer.............	Polysufide	Epoxy	Silicone
Sleeving...........	Vinyl	Silicone	
	Polyvinyl chloride		
Tubing............	Neoprene,* shrinkable		
Plastics............	Melamine	Polyester	Silicone
	ABS	Diallyl phthalate	Epoxy
	Phenolic		Polyurethane
Varnish...........	Vinyl	Alkyd	

NOTE: Extent of attack is worse when plastic is incompletely cured; at high temperatures of operation, it may vary with the plastic used.
* Trademark of E. I. du Pont de Nemours & Co., Wilmington, Del.

When other factors, such as vibration may be involved, entirely similar considerations apply.

Other causes of failure At least two other types of equipment failure may be caused by metal-finishing operations: (1) hydrogen embrittlement and (2) metal whiskers. When problems of this nature arise, they are best dealt with by metallurgical experts, but the designer should at least be aware of the conditions that usually give rise to each.

Hydrogen Embrittlement. Some metals and alloys containing absorbed hydrogen become quite brittle, particularly when subjected to alternating stresses. This is primarily a problem associated with spring steels, high-strength iron-base and nickel-base alloys, and titanium; particularly when electrocleaned cathodically; or pickled in acids; or plated with chromium or cyanide copper, zinc or cadmium. Since hydrogen may be evolved in the corroding process, embrittlement may also occur along with the corrosion.

The relief of hydrogen embrittlement involves a baking stress relief before and *immediately* after plating, usually for 3 or preferably 24 hr at 180°C (375°F), but this operation must be conducted with due regard for the metallurgical history of the workpiece. Embrittlement is most likely to occur in heat-treated parts, and every precaution must be taken to avoid baking times and temperatures that may deteriorate prior heat treatment.

Metal Whiskers. Metal whiskers may grow in storage (even in hermetically sealed units). Electroplated tin, zinc, cadmium, copper, and iron exhibit this phenomenon which, though fairly rare and not well understood, can cause serious short-circuiting of electronic circuits, especially when the physical size of the circuits is reduced as in microcircuitry elements. Whisker growth is not prevented but is retarded under conditions of low moisture and low temperature. If tin plating is used (the phenomenon seems to be most common with tin), it should be hot-dipped when practical, or reflowed by fusing after electroplating.

Designing for corrosion-resistant finishes Table 27 has been prepared to assist the designer in preventing galvanic corrosion. When doubts or uncertainties arise, some tests described below may assist in deciding whether certain combinations in the design are feasible or not.

Salt-spray Test. This test does not correlate with service life, but it can be a valuable tool for locating major weaknesses in the selection of materials. It should be noted that salt deposits may bridge insulating barriers and cause failure in any subsequent electrical testing, especially under humid conditions. Unless the part is specifically designed to function under conditions of salt plus humidity, it should be well rinsed and dried after salt spray before any electrical testing is done.

Repeated salt-spray tests (with drying in between) are frequently more severe than a single test. In any case, testing beyond 50 hr rarely causes more corrosion spots; usually each existing spot merely becomes larger.

Tests with Imposed Current and Other Factors. Although the ultimate answer to a design problem must be an environmental performance test of the complete assembled unit under simulated service conditions, it may sometimes be possible to impress the operating voltage on a subassembly still in the experimental design stage, while this unit is in a controlled test environment, such as cycling temperatures and humidity. When this can be done in advance of the final performance test (when both time and cost will be matters of more pressing concern) considerable savings can be achieved.

It is to be anticipated that such a test will be more significant than a simple environmental exposure without an imposed current, and the sooner the combined effect of galvanic and imposed currents can be established, the less need there will be for hasty, last-minute design changes and their associated cost and inefficiency.

Plating thickness Assuming that the designer has selected the optimum combination of coatings for minimizing galvanic effects, it remains for him to specify an adequate *thickness* for the plated coating. Two general problems must be considered.

The coating is anodic to the substrate (e.g., zinc on aluminum, or cadmium on steel). Corrosion of the basis metal is prevented by the sacrificial protection of the coating. Even if the coating is porous initially or (as it will) corrodes away to form pits or pores, protection is afforded until the coating is too severely eaten away. Exposure to service conditions can therefore be expected to result at first in worsening the appearance of the part. The duration of protection of the substrate is a direct function of coating thickness. In electronic parts, however, *no* corrosion products of either substrate or coating can be tolerated, as a rule; and this combination is thus generally to be avoided.

The coating is cathodic to (*i.e., nobler than*) the *substrate* (e.g., lead-tin on aluminum; silver on nickel, or on bronze; brass or nickel on steel). The protection offered by the coating in this case depends solely on its integrity. Once perforated, the coating then causes *accelerated* attack on the basis metal. Since it can only protect the basis metal so long as it can exclude the corrosive environment from contact with the substrate, a minimum deposit thickness is required to avoid the formation of pinholes.

TABLE 27 Designing for a Corrosion-resistant Finish

Problem	Solution	Example
Dissimilar metals	1. Select metals from appropriate table of permissible couples (Tables 23 to 25).	Use nickel or rhodium, not brass or bronze, next to silver.
	2. Plate with compatible metal to reduce potential difference.	Tinplate aluminum and bronze used together.
	3. Keep affected area of less-noble metal as large as possible.	Stainless steel hardware in sheet aluminum *may* be satisfactory because of large area of aluminum (but *not* the reverse!).
Contact.........	4. Apply corrosion inhibitors such as zinc chromate paste.	Assemble dissimilar hardware with zinc chromate paste.
	5. Interpose inert barrier or gaskets to prevent contact (extend ¼ in. beyond joint).	Vinyl tape, rubber gasket (and sometimes a plated washer).
	6. Paint both metals (or cathode at least) with *alkali-resistant* organic coating.	MIL-P-52192 or MIL-P-15930.
Electrolyte......	7. Avoid designs where moisture can be trapped.	Use sealant bead on crimped, spot-welded, and threaded joints.
	8. Use desiccant.	Useful only in hermetically sealed compartment.
	9. Seal joint with organic insulation.	MIL-S-7124.
	10. Seal metal faces against contact with electrolyte.	Primer, paint, or sealant.
General.........	11. Where possible, avoid use of magnesium.	Protection of magnesium requires very special attention.
	12. Do not zinc-plate on aluminum; use cadmium.	
	13. Avoid using cadmium in high vacuum.	

Because the porosity of deposits varies widely, and because the corrosivity of environments cannot be strictly defined, it is not possible to make a general statement as to the minimum deposit thickness required in a given case. However, less than 0.0010 in. of any commercially electrodeposited metal is almost certain to show some porosity, and hence to be of very doubtful value in an aggressive atmosphere (see Table 28).

Effects of Thickness. Whenever, for structural, electrical, or other reasons, dissimilar metals must be used in parts having electric contact with one another, whereby an unacceptable corrosion voltage could be set up, the effects of this voltage can be minimized or negated by plating the two parts with a sufficient thickness of the same metal. The metal to be plated will be chosen with a view to its functional properties of conductivity, magnetism, etc., and will, as has been stated, usually be cathodic to the substrate. Whether the effect of the corrosion potential is wholly or only partly suppressed will depend on the porosity of the

10-56 Metallic and Chemical Finishes on Metals and Nonconductors

deposit, which in turn is a direct function of the plating thickness. A thickness of 0.5 mil of most commercial electrodeposits is only relatively pore-free, and 1.0 mil is far superior. In severely aggressive environments and especially when a superimposed exterior potential exists, 3 to 5 mils may be required.

TABLE 28 Guide for Plating Thickness Selection

Plated metal	Thickness, mils	Probable protection rating*	Notes
Cadmium	0.1–0.3 0.5 1.0 2.0	0 1 2 3	
	0.2 0.3–0.5	0 1	Cd on aluminum
	0.2 0.3 0.5	0 1 2	Chromated Cd
Gold	Corrosion resistance depends on undercoat rather than on gold (up to 0.1 mil)		
Nickel	0.1–0.3 0.5–0.1 1.5 2.0	0 1 1+ 2	Electroplated
	0.5 1.0–1.5 2.0	1 2 3	Electroless
	1.0 2.0	2 3	Electroless on Al
Silver	0.1–0.2 0.3–0.5 0.7–1.0 2.0–5.0	0 1 2 2–3	
	0.1–0.2 0.3–0.5 0.7–1.0 2.0–3.0	0 1 2 2–3	On aluminum with 0.4-mil Cu strike

* Ratings:
0 Minimal protection.
1 Marginal protection.
2 Good protection.
3 Excellent protection.

When considerations of weight and closeness of dimensional tolerance are paramount, plate thicknesses of marginal corrosion resistance value may have to be considered, and the probable consequences of plating failure must be weighed against the weight or tolerance penalty.

Variations in Thickness. In design, due allowance must be made for the variations in plating thickness that must be expected. Unless special precautions are taken (which are usually not practiced in production tanks), the edge-to-center plating-thickness ratio of a flat plate will be of the order of 2:1. For parts with only reasonably complex geometry (e.g., a topless cubical box) a ratio of 10 or more to 1 will be likely without special fixturing. By using suitable fixtures, which must be designed and tested *in advance* of production (just as any other machine tool or fixture must be), this ratio can be made to approach more closely the flat plate ratio of 2:1.

Geometric factors having a strong influence on plate distribution have been discussed previously under Metal Distribution. For complicated shapes, electroless nickel (where nickel is applicable) is strongly recommended.

Except when electroless nickel is being used, another factor to consider when specifying a thicker plate is that the variation in plating thickness over a part is increased at least in proportion to the increase in thickness, and thus dimensional tolerances may be exceeded, even though the designer has made due allowance for the *average* thicknesses called out.

Specifying plate thickness *General Considerations.* Whenever a single metal can be deposited directly on a substrate and is able to perform its function over a useful service life without additional protection, there is no need for a multilayer plate. For various reasons, however, multilayer plates are frequently used and are called out in finish specifications as required. For estimating the probable relative corrosion resistance of such plates, neglect individual metal thicknesses of 0.0002 in. or less, and consider that corrosion resistance will be afforded by the remaining major constituent(s), of which the thickest layer is most significant.

Service Conditions. The expected service conditions under which a component must operate elude precise definition, but obviously the opportunity for massive "breathing" exists, and for at least part of the time the atmosphere thus drawn in will be highly corrosive to electronic gear.

Consequence of Malfunction Due to Corrosion. The designer is in the position to decide whether corrosion will cause a malfunction that will be *critical* (immediate and total loss of function, as by a dead short circuit), *major* (tolerable loss of function), or *minor,* or none at all.

For purposes of ensuring reliability, it is better for the designer to err on the side of too thick rather than too thin a plate, and the cost differential is minimal compared to the incalculable cost of malfunction of a critical component in military or space equipment. Naturally, as the criticality of the component diminishes, the cost of a heavier plating assumes a greater importance.

Selecting Thickness. Table 28 may be used as a rough guide in selecting thicknesses of deposits, but the final choice may have to be based on other factors, and the final responsibility is the designer's.

It should be emphasized that this table is necessarily only the roughest of guides. It is based on the usual porosity level of deposits and is related to the anticipated corrosion, as this is considered by the designer to affect the reliability of performance:

Rating	Probable protective value	Likelihood and effects of corrosion
0	Minimal	Corrosion may very likely occur, but it has no functional relationship. Any deterioration will relate to esthetic values only.
1	Marginal	If corrosion occurs, no appreciable loss of functional reliability is expected.
2	Good	Some corrosion may occur in time. When it does, there may be a tolerable partial loss of function.
3	Excellent	Corrosion will require an extended period of time under severe conditions. If it occurs, corrosion may cause complete component failure. Periodic inspection can be expected to detect onset of corrosion before massive failure of the component.

TESTING OF DEPOSITS AND COATINGS

Testing methods are of little real interest to the designer, other than to inform him what properties are measurable and with what accuracy, and also whether the finish or even the piece itself will be damaged or destroyed by the test. No attempt will be made, therefore, to describe the testing instruments or the method of making the test. Instead, the expected order of accuracy of the test measurement will be estimated, noting the degree of skill required, and whether the finish or the piece itself will survive the testing.

Thickness Testing

Destructive tests *Microscopic Determination.* Microscopic determination of thickness is less accurate than might be expected. Not only is the accuracy very dubious at thicknesses of 0.5 mil or less, but also considerable skill is required, and the procedure is costly. Especially for soft metals, such as silver, gold, tin, and especially lead, an overplate of a heavy layer of a harder metal (such as nickel) is absolutely necessary; and even with an overplate, these softer metals tend to smear in polishing and obscure the interface. Cutting the specimen before polishing must be done at exactly 90° to the surface (or some other exactly known angle for which suitable correction can be made), or the results will be grossly in error. Determining thickness on a curved surface is thus very difficult; and measuring the coating in a small hole (as in a printed-circuit board, where the hole diameter may be of the order of $\frac{1}{16}$ in. or less) is further complicated by the uncertainty as to whether the plane of measurement, after cutting and polishing, is *exactly* on the diameter of the hole. In any case, even for thicknesses of the order of 1 mil, an accuracy of no better than ±5 percent is to be expected.

Stripping the Deposit. The deposit may be stripped from an accurately known area, and analysis of the stripping solution for the metal or metals in the original coating is the most accurate way of determining the *average* thickness of the deposit. If the basis metal is immune to attack by the stripping solution, weight loss is a simple and accurate measure, or sometimes the basis metal may be dissolved, and the unattacked coating weight may be very accurately determined. This method is frequently used for deposits on aluminum, which is readily soluble in caustic soda which does not attack the noble metals.

LOCALIZED STRIPPING: Stripping of a defined area by impinging a suitable corrosive solution, either as a jet or in discrete drops, on a very small area, is also used for a number of combinations of topcoat and basis metal. The indication that the coating has been penetrated is visual, and the test is empirically calibrated in terms of thickness versus number of drops used. The accuracy is of a very low order (±10 to 30 percent), but for decorative thicknesses of chromium, which may be only 10 or 20 millionths of an inch, there is presently no better method available to the average shop.

Electrochemical stripping is also sometimes used, in which the instrument is arranged to give a response on a meter when penetration occurs, but there is little improvement in accuracy over the jet or dropping methods.

OTHER METHODS: These include many various ways of cutting mechanically through the coating in a controlled fashion to expose the basis metal, but they are not very accurate, and are applicable only to relatively heavy deposits of one mil or more.

Nondestructive tests *Magnetic Methods.* Magnetic methods are limited, of course, to nonmagnetic coatings on a magnetic substrate, and are based usually on measuring the pull necessary to detach a small calibrated magnet from the plated surface. Geometric errors will be introduced if the surface is not perfectly smooth and flat. Also any disturbance or joggling of the instrument, the piece, or its support will result in premature disengagement of the magnet and thus give a falsely high reading. There are several variations of this type of test.

X-ray and Interference Microscopy Methods. The technique for these methods is very specialized, and the equipment is costly and unsuitable for operation by the average technician.

Eddy-current Gages. These devices are very reliable for nondestructive thickness measurements and can be operated by trained technicians. These gages employ a probe that can be quite small and can reach into cavities and rather small bores. The probe contains a coil carrying a high-frequency current. The electromagnetic field of the coil introduces eddy currents, which should be such that they penetrate the coating and extend a short distance into the basis metal. By using probes of different frequencies, quite a range of thickness can be covered. The gage depends upon the electric conductivity of the coating being different from that of the substrate. It can be used, therefore, for measuring the thickness of nonconductive coatings such as anodizing, paint films, and the like on any metal; but it is rather insensitive for metal combinations of very similar conductivity, such as silver on copper. It can, however, be used to measure (silver plus copper) on steel, for example. One such widely used and very versatile instrument is the Dermitron.

Beta-ray Backscatter Gages. These instruments consist of a probe containing both a radioactive isotope (such as strontium-90 or cobalt-60, and others) and a detector for measuring the intensity of the rays backscattered by the test specimen. Some of the beta rays entering the specimen collide with the metal atoms and are reflected back toward the source, where the detector, which is a sensitive type of Geiger counter, measures them. The intensity of these backscattered beta rays is a function of the coating thickness, provided the atomic numbers of coating and basis metals are sufficiently different. By varying the intensity (Mev) of the source, different penetrations may be achieved to cover different thicknesses of coating.

These gages are most suited for measuring thin coatings, and can be used, when properly calibrated, for many combinations of coating and basis metals, with the limitation noted just above, and they are also useful for organic coatings on any metal. There is an upper limit of coating thickness for accurate measurement (3 mil of lead on copper, for instance); at the other extreme, gold over nickel can be measured at less than 10 μin. For coatings of the order of 100 μin. the accuracy of measurement is about 5 percent.

Porosity Testing

Determination of the porosity of a coating is not usually a routine control operation, but is a rather specialized laboratory procedure.[12,13] Most commercial electrodeposits are inherently porous, and the pores may not all be closed even at a thickness of one mil. Electrodeposits specially prepared under controlled laboratory conditions have been demonstrated to be pore-free at 0.05 mil, but this is not practical commercially. These pores are due to inherent defects in the basis metal, to pitting of the plate not related to basis-metal pores, to occluded particles in the plate, and to other causes. In the thickness range of 0.5 to 1.5 mil, the number of pores per square foot in a commercially prepared deposit may vary from several hundred to a few isolated occurrences. The familiar salt-spray test, which has long been generally discredited as a quality control test, is (for coatings, such as nickel, which are not themselves corroded by the salt) simply a porosity test, and bears little or no relation to the coating thickness.

Porosity testing methods are based on contacting the plated surface with a piece of specially prepared filter paper, which is wetted with a solution designed to corrode the basis metal but not the plate, and to register by a color reaction where the solution has found a pore through to the basis metal. A more sophisticated method is to arrange for a known dc potential to be exerted for a definite time in such an arrangement, whereby the basis metal, if exposed through a pore, is dissolved and driven to the recording paper. The latter method has been commercialized, and instruments are available.

Adhesion Testing

Other than a few research techniques, adhesion tests are crude and only qualitative, involving various means of attempting to separate the plate from the basis metal. Semiquantitative tests are used to measure the force necessary to separate a plate from a nonmetallic substrate, and the results are reported on an empirical basis.

Stress Measurement

Until rather recently, stress determinations have been confined to research investigations, but two instruments have been marketed for routine measurement of stress in a deposit. These do require some skill and understanding in their operation.

Stress is of particular significance in plating on plastics, since on a metal substrate, stress in the deposit is very rarely enough to overcome the adhesion forces and rupture the coating. But on plastics, even a relatively low value of stress is sufficient to cause the deposit to blister or exfoliate. For this reason, when plastics are to be plated, baths known to produce highly stressed deposits (Table 6) should be avoided, and even acceptable baths will need close monitoring.

Solderability Testing

Solderability is not an inherent property of a metal. It depends upon the surface cleanliness of the metal, and upon all the variables of the soldering technique. Numerous tests have been devised for special needs, such as the capillary-rise test used by the canning industry, but for electronic design purposes, the question is simply which coating on a given substrate will offer the best performance for a particular application. The coatings most suitable are tin and its alloys, silver, and gold; and the usual problem is how thick the chosen plating must be.

All three of the above classes of coatings are readily soldered, but the solderability of silver deteriorates with time, since a film of sulfide forms under the usual storage conditions, and frequently a thin (10 to 20 millionths) overcoat of gold is used to preserve the solderability. (Too heavy a gold coat can sometimes cause embrittlement of the solder joint, due to alloy formation.)

The part geometry will influence solderability to the extent that the time and temperature relationship will depend on how fast the soldered area heats up and on how rapidly the heat is conducted away from the joint. The solder spread test on a mock-up of the part can be used to optimize the variables. In this test, a solder button of controlled size and weight is heated under various precisely controlled temperatures and times, and the spread of solder is measured (either as area of spread, or height of the bead) on the mock-up.

Mechanical Property Tests

Hardness Hardness is not a measure of any intrinsic property of a metal; yet it is widely used. Any given hardness test actually measures a complex mixture of properties, and there are many varieties of tests. For electrodeposits, or any similar thin coating, a special form of the indent method is used, because it is necessary to avoid allowing the basis metal to exert any influence on the test results.

The Knoop indenter, developed at the Bureau of Standards, is used in the Vickers type of test equipment under loads measured in grams, and gives very reliable results when the deposit is sufficiently thick (usually several mils) so that the depth of the indent is not over 7 percent of the plate thickness. This indenter is an unsymmetrical diamond point, and the length of the indent, which is measured, is many times the actual penetration. Measurements can even be made on the cross section of a deposit if it is 5 to 10 mils thick.

Strength and ductility Strength and ductility of electrodeposits are extremely difficult to measure in conventional tensile equipment, because of the thin sections. Ductility measurements must be made on specimens of *exactly* the thickness that will be actually used. Tensile strength measurements are quite constant over the thickness range of 0.3 to 3 mil.

These measurements are best made on the hydraulic bulge tester, as modified by Prater and Read.[14,15] The specimen is deposited on polished and passivated stainless steel, and the foil is carefully separated, avoiding wrinkling or other mechanical damage, and its thickness is accurately determined by weighing or by other means. The specimen is clamped between two circular platens, and it is ruptured by oil pressure from beneath. From the bulge height at rupture and the oil pressure required, both strain and tensile strength are readily calculated. The

TABLE 29 Government Specifications for Finishes

Finish	Specification
Anodized aluminum (chromic acid)	MIL-A-8625*
Anodized aluminum (sulfuric acid)	MIL-A-8625†
Cadmium plating	QQ-P-416*
Cadmium plating, chromate finish	QQ-P-416†
Chromium plating	QQ-C-320‡
Chromate film on aluminum	MIL-C-5541
Chromate film on magnesium	MIL-M-3171
Copper plate	MIL-C-14550
Gold plate	MIL-G-45204
Nickel electroplating	QQ-N-290
Nickel, electroless coating	MIL-C-26074
Oxidized copper (black)	MIL-F-495
Oxidized stainless steel (black)	MIL-C-13924
Passivation of stainless steel	QQ-P-35 and MIL-S-5002
Rhodium plating	MIL-R-46085
Silver plating	QQ-S-365
Tin plating	MIL-T-10727
Tin-lead plating	QQ-S-571
Zinc phosphate coating on steel	TT-C-490
Zinc plating	QQ-Z-325

NOTES:
1. Specifications may include a letter (e.g., MIL-A-8625C) indicating a revision.
2. Specifications are frequently divided into several subclassifications (e.g., type II, grade B, class 2, etc.). The connotations of these divisions are not standard, but vary from one specification to another, and may relate to further treatment, type of process used, or other variants, including thickness.
* Type I.
† Type II.
‡ Class 1 buffed; Class 2 as plated.

bulge tester is especially suited to brittle materials, and hence to electrodeposits, because the sensitivity is greatest when the strain (bulge height) is small. No special machining of the specimen is necessary, and there is no alignment problem as is the case in the conventional tensile test.

GOVERNMENT SPECIFICATIONS FOR FINISHES

Table 29 lists a number of government specifications frequently used for finishes.

REFERENCES

1. Kasper, C.: *Monthly Rev. Amer. Electroplaters' Soc.*, vol. 26, pp. 11–26, 91–109, 1939; *Trans. Electrochem. Soc.*, vol. 77, pp. 353–383, 1940; vol. 78, pp. 131–161, 1941; vol. 82, pp. 153–185, 1942.
2. Kronsbein, J.: *J. London Math. Soc.*, vol. 17, pp. 152–157, 1942; *Proc. London Math. Soc.*, vol. 49(2), pp. 260–281, 1947; *Plating*, vol. 36, pp. 851–854, 1950; vol. 39, pp. 165–170, 1952.

3. Haring, H. E., and W. Blum: *Trans. Electrochem. Soc.*, vol. 44, pp. 313–345, 1923.
4. Graham, A. K., E. A. Anderson, H. L. Pinkerton, and C. E. Reinhard: *Plating*, vol. 36, pp. 702–709, 1949.
5. Kardos, O.: *Proc. Amer. Electroplater's Soc.*, vol. 43, pp. 181–194, 1956.
6. Jernstedt, G. W.: Numerous patents assigned to Westinghouse Corp.
7. The Electrolyzing Corp., 148 W. River Street, Providence, R.I.
8. Kronsbein, J.: *Plating*, vol. 40, pp. 898–901, 1953.
9. Selectrons, Ltd., 116 E. 16th Street, New York.
10. Brenner, A., et. al., *Proc. 3d Int. Electrodeposition Conf. (London)*, pp. 147–155, 1948.
11. *American Lava Corporation Tech. Data Bull.* 672, Chattanooga, Tenn.
12. Macnaughton, D. J., *Trans. Faraday Soc.*, vol. 27, no. 111, part 8, pp. 465–480, 1930; *J. Electrodepositors' Tech. Soc.*, vol. 5, pp. 135–151, 1930.
13. Electrographic Porosity Tester by the Meaker Co., subsidiary of Sel-Rex Corp., Nutley, N.J.
14. Prater, T. A., and H. J. Read: *Plating*, vol. 36, p. 1221, 1949.
15. Read, H. J., and T. U. Whalen: *Proc. Amer. Electroplaters' Soc.*, vol. 34, p. 74, 1947.

General References

Graham, A. K., and H. L. Pinkerton: "Electroplating Engineering Handbook," 2d ed., Reinhold Publishing Corporation, New York, 1962.
Wernick and Pinner: "Surface Treatment of Aluminum," 3d ed., Robert Draper, Ltd., Teddington, England, 1964.
Foulke and Crane: "Electroplaters' Process Control Handbook," Reinhold Publishing Corporation, New York, 1963.
Lowenheim (ed.): "Modern Electroplating," 2d ed., John Wiley & Sons, Inc., New York, 1963.
LaQue and Copson: "Corrosion Resistance of Metals and Alloys," 2d ed., ACS Monograph Series, Reinhold Publishing Corporation, New York, 1963.

Chapter 11

Thin Films

ROSEMARY BEATTY
NASA Electronics Research Center,
Cambridge, Massachusetts

Introduction	11-2
Thick Films versus Thin Films	11-2
Thick-film Materials and Techniques	11-3
Thin Films	11-6
Fabrication Techniques	11-8
Thin-film Materials	11-18
Thin-film Capacitors	11-19
Thin-film Resistors	11-23
Interconnections	11-31
Reliability	11-36
Instrumentation for Microelectronics	11-38
References	11-59

INTRODUCTION

Intense interest in thin-film technology has been spurred by the growing importance of microelectronics. By definition, microelectronics encompasses the entire body of the electronic art which is connected with or applied to the realization of electronic circuits, subsystems, or the entire system, from extremely small electronic parts (devices). To avoid confusion the EIA and IEEE have defined the following:

1. *Integrated circuits:* The physical realization of a number of circuit elements, inseparably associated on or within a continuous body to perform the function of a circuit.

2. *Thin-film integrated circuits:* The physical realization of a number of circuit elements entirely in the form of thin films deposited in a patterned relationship on a structure-supporting material.

3. *Semiconductor integrated circuit (monolithic integrated circuit):* The physical realization of a number of circuit elements inseparably associated on or within a continuous body of semiconductor material to perform the function of a circuit.

4. *Hybrid integrated circuit:* An arrangement consisting of one or more integrated circuits in combination with one or more discrete devices. Alternatively, the combination of more than one type of integrated circuit into a single integrated component.

This chapter on thin films will be mainly concerned with their importance and applications in microelectronics. The field of microelectronics evolved from the need for microminiaturization in complex systems having limitations on weight, space, and power. The squeeze on weight and space generally ruled out the use of discrete components fitted into welded modules. In the thin-film approach to integrated circuits, the components lose their identity, and the overall circuit performance is of prime concern.

Thin-film technology is important in the fabrication of all types of integrated circuits—as interconnections, insulators, resistors, and capacitors. The thin-film elements and their interconnections must be compatible with the rest of the integrated circuit in which they appear, with the various chemical and thermal processing and sealing steps, and they must withstand accelerated life-test procedures.

THICK FILMS VERSUS THIN FILMS

Thin films are defined as coatings up to a thickness of a few microns (1 micron = 10^{-4} Å). On the low end, a thin-film layer thickness ranges from one monolayer (about 5 Å) to 1 micron. There are many pros and cons as to the relative merits and disadvantages of thick films and thin films. The hybrid circuit may contain two or more semiconductor chips on a substrate with either thick- or thin-film resistors and capacitors. In the design of a hybrid circuit the engineer must choose among thick films, alloy thin films, and single-metal thin films. Though thick films are not the most sophisticated approach, economically they can be very attractive. The stencil-screening technique for printing interconnecting conductive patterns for resistors, capacitors, and connection pads for attachment of additional external elements has the important advantages of capability for complete automation of the assembly of components, reduction in size and weight, absence of interactions between elements and good high-frequency performance. The screened-circuit approach offers the following advantages permitting utilization in commercial applications with rapidity and ease of implementation:

1. The design cycle is extremely short: a set of screening masters can be designed and obtained in a matter of a few days. In addition, tailoring of resistor values within a range of approximately 20 percent from an established nominal value can be implemented without modifying the screening master set.

2. New module circuit designs, based upon an existing physical configuration, can be made available at low cost within a few weeks.

3. Reliability is improved over discrete component assemblies owing to the large decrease in the number of soldered interconnections and improved drift characteristics of the screened resistive components.[13]

The overall advantages to be gained by using thick films may be listed as:
1. Initial investment to buy equipment is low. Masking costs are relatively inexpensive.
2. The process lends itself to the techniques of mass production.
3. Many resistivities are available for flexibility in resistor values (1 ohm/sq to 20 kilohms/sq).
4. Trimming is easily automatable by the use of sandblasting or other means of mechanical trimming.
5. Power ratings on resistors are high.
6. Capacitors are possible.

On the other hand there are disadvantages:
1. Thick films tend to be unstable at high temperatures (350°C).
2. End-of-life data are questionable.
3. Thick films are noisy at high resistivities.
4. Line definitions are gross compared to those of thin films.
5. Each resistance change greater than a factor of 10 above the basic resistivity selection requires an additional screening.
6. In mass-production runs, more than two screenings seriously affect the yield.

A thumbnail comparison of thick- and thin-film applications may be seen in Table 1. Tables 2 to 5 have been devised to present an all-inclusive pictorial format for classifying integrated circuit (IC) and hybrid integrated circuit (HIC) technologies.

TABLE 1 Thumbnail Comparisons of Thick- and Thin-film Applications*

Circuit requirements	Thick film	Thin film
1. Mass-producible	Lower cost	Higher cost
2. Close initial tolerance, %	±10 absolute ±1 match	±5 absolute ±0.1 match
3. Close trimming tolerance, %	±1 easily mass-producible	±1 mass-producible (using tantalum)
4. Temperature coefficient, ppm per °C	±200	0
5. Power handling, watts/in.²	50	15
6. Line width, mils	10 ± 2	0.25 ± 0.02
7. Resistivities, ohms/sq	1–20,000	0.1–1,000
8. End-of-line tolerance, %	±2	±0.1
9. Capacitance available, maximum	10,000 pf	0.01 µf

* *Electron. Des.*, March, 1967.

THICK-FILM MATERIALS AND TECHNIQUES

Screen deposition is a technique used to fabricate all ceramic, thick-film microelectronic circuits. Alumina oxide (96 Al_2O_3) is the usual substrate material for the following reasons:
1. Has a planar surface.
2. Has minimum camber for ¾-in. substrate (0.002 in./in.).
3. Has no surface imperfections that prevent contiguous deposition.
4. Thermal conductivity most closely matches that of deposited materials.
5. Has typical requirements for surface resistivity (10^{10} ohms/sq).
6. Is chemically and physically inert.

Resistor pastes consist of mixtures of organics, a relatively low melting glass, a noble metal, and an oxide. It may be noted here that manufacturers currently in thick films have developed their own formulations to improve the range of resistivity and reduce the temperature coefficient of resistance (TCR). For example,

11-4   Thin Films

TABLE 2   Classification of Microcircuit and Devices Categories *

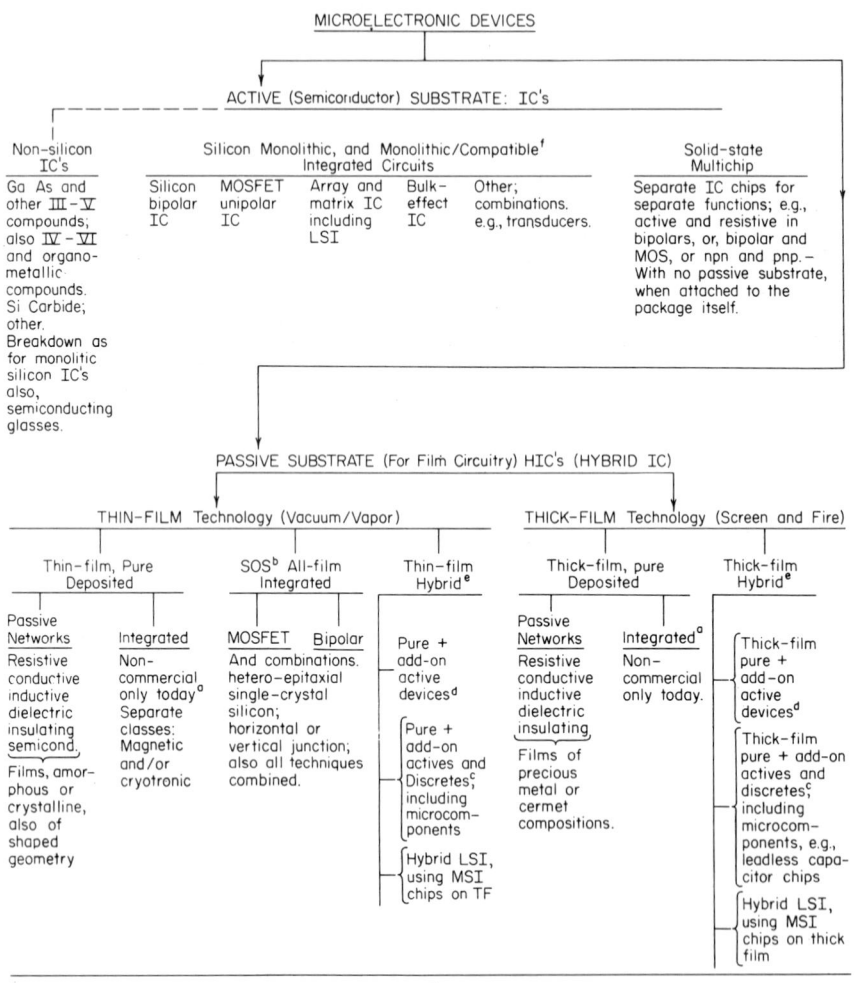

MICROELECTRONIC DEVICES

ACTIVE (Semiconductor) SUBSTRATE: IC's

Non-silicon IC's	Silicon Monolithic, and Monolithic/Compatible[f] Integrated Circuits					Solid-state Multichip
Ga As and other III-V compounds; also IV-VI and organo-metallic compounds. Si Carbide; other. Breakdown as for monolitic silicon IC's also, semiconducting glasses.	Silicon bipolar IC	MOSFET unipolar IC	Array and matrix IC including LSI	Bulk-effect IC	Other; combinations. e.g., transducers.	Separate IC chips for separate functions; e.g., active and resistive in bipolars, or, bipolar and MOS, or npn and pnp.– With no passive substrate, when attached to the package itself.

PASSIVE SUBSTRATE (For Film Circuitry) HIC's (HYBRID IC)

THIN-FILM Technology (Vacuum/Vapor)			THICK-FILM Technology (Screen and Fire)	
Thin-film, Pure Deposited	SOS[b] All-film Integrated	Thin-film Hybrid[e]	Thick-film, pure Deposited	Thick-film Hybrid[e]
Passive Networks — Resistive conductive inductive dielectric insulating semicond. Films, amorphous or crystalline, also of shaped geometry	Integrated Non-commercial only today[a] Separate classes: Magnetic and/or cryotronic	MOSFET Bipolar And combinations. hetero-epitaxial single-crystal silicon; horizontal or vertical junction; also all techniques combined.	Pure + add-on active devices[d] Pure + add-on actives and Discretes[c] including microcomponents Hybrid LSI, using MSI chips on TF	Passive Networks — Resistive conductive inductive dielectric insulating. Films of precious metal or cermet compositions.	Integrated[a] Non-commercial only today.	Thick-film pure + add-on active devices[d] Thick-film pure + add-on actives and discretes[c] including microcomponents, e.g., leadless capacitor chips Hybrid LSI, using MSI chips on thick film

[a] Incorporating film active elements.
[b] Silicon-on-sapphire.
[c] Compatible inductors, toroids, transformers, high-value capacitors, crystals, etc.
[d] Transistors, diodes, other semiconductors, integrated circuits; prepackaged, or as chips.
[e] Hybrid = hybrid-integrated, or HIC.
[f] IC's with optimized passive circuit elements deposited over the oxide passivation surface.

* *Electron. Commun.*, vol. 2, 1968.

mixtures of ruthenium oxide, iridium oxide, palladium oxide, and inorganic vehicles such as lead borosilicate glass are commonly used. A screen stencil is printed and mounted. The screening paste is pressed through the pattern of openings onto the substrate. The binder solvent is evaporated at temperatures from 150 to 200°F. Removal of the binder solvent before high-temperature firing is important; otherwise spots and voids will form because of outgassing. The firing cycle is divided into three phases of preheating, firing or vitrifying, and postheating. During the preheating portion of the cycle, the temperature is gradually increased to complete the outgassing of the binder solvent and reduce thermal shock. The vitrifying phase establishes the characteristic value of the material. The characteris-

Thick-film Materials and Techniques 11-5

TABLE 3 Classification of Microelectronic Devices by Functional-Circuitry Generic Categories*

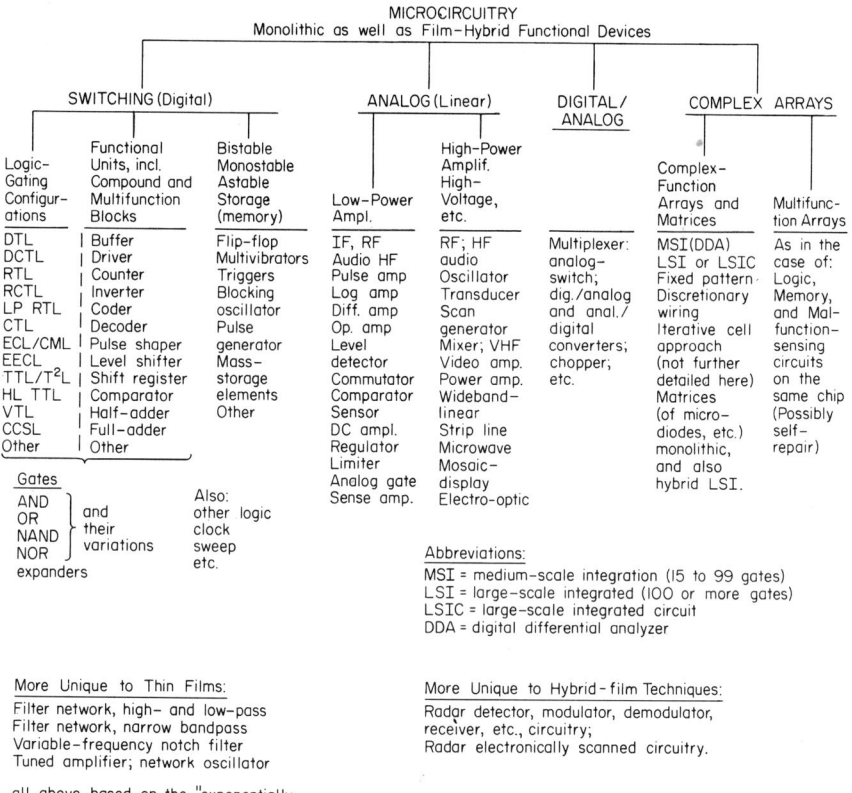

MICROCIRCUITRY
Monolithic as well as Film-Hybrid Functional Devices

SWITCHING (Digital)			ANALOG (Linear)		DIGITAL/ ANALOG	COMPLEX ARRAYS	
Logic-Gating Configurations	Functional Units, incl. Compound and Multifunction Blocks	Bistable Monostable Astable Storage (memory)	Low-Power Ampl.	High-Power Amplif. High-Voltage, etc.	Multiplexer: analog-switch; dig./analog and anal./digital converters; chopper; etc.	Complex-Function Arrays and Matrices	Multifunction Arrays
DTL DCTL RTL RCTL LP RTL CTL ECL/CML EECL TTL/T²L HL TTL VTL CCSL Other	Buffer Driver Counter Inverter Coder Decoder Pulse shaper Level shifter Shift register Comparator Half-adder Full-adder Other	Flip-flop Multivibrators Triggers Blocking oscillator Pulse generator Mass-storage elements Other	IF, RF Audio HF Pulse amp Log amp Diff. amp Op. amp Level detector Commutator Comparator Sensor DC ampl. Regulator Limiter Analog gate Sense amp.	RF; HF audio Oscillator Transducer Scan generator Mixer; VHF Video amp. Power amp. Wideband-linear Strip line Microwave Mosaic-display Electro-optic		MSI(DDA) LSI or LSIC Fixed pattern Logic, Discretionary wiring Iterative cell approach (not further detailed here) Matrices (of micro-diodes, etc.) monolithic, and also hybrid LSI.	As in the case of: Logic, Memory, and Malfunction-sensing circuits on the same chip (Possibly self-repair)

Gates
AND ⎤
OR ⎥ and
NAND⎬ their
NOR ⎦ variations
expanders

Also:
other logic
clock
sweep
etc.

Abbreviations:
MSI = medium-scale integration (15 to 99 gates)
LSI = large-scale integrated (100 or more gates)
LSIC = large-scale integrated circuit
DDA = digital differential analyzer

More Unique to Thin Films:
Filter network, high- and low-pass
Filter network, narrow bandpass
Variable-frequency notch filter
Tuned amplifier; network oscillator

all above based on the "exponentially tapered, distributed-parameter, thin-film network concepts". Also:
Functional (Ta) film component, substituting for a whole RC network, by being shaped.

More Unique to Hybrid-film Techniques:
Radar detector, modulator, demodulator, receiver, etc., circuitry;
Radar electronically scanned circuitry.

CIRCUIT MODIFIERS:
Micropower; ultra-fast switching; radiation-hardened; thermally compensated; sterilization-processed; high-reliability; majority-voting; redundant (at what level?); adaptive; cryo-electronic; pattern-recognition; magnetic; bionic; implantable (into animals and humans); etc.

*Courtesy of George Szekely, Aerojet General Corp.

tic resistivity of each resistive paste is established by controlling the peak temperature during the vitrifying, oxidizing, and sintering phases of the firing cycle, as may be seen in Fig. 1. Resistors can be adjusted after firing to achieve a resistive tolerance of 1 percent.

Capacitors are fabricated by screening glass frits on a conducting plate previously screened on a substrate. The glass frit is fired, forming a layer of high dielectric constant. Capacitances in the order of 30,000 pf/cm^2 can be obtained in this manner, depending upon the dielectric constant of the glass frit. Dielectric compositions for film capacitors include titanium oxide, barium titanate, glasses, and mixtures of such materials. Perhaps the most widely used materials are based on mixtures of glasses and barium titanate with the addition of modifying oxides. During the liquid-phase sintering the glass tends to become continuous surrounding the titanate particles.

11-6   Thin Films

TABLE 4   Modifiers for Monolithic Integrated-circuit Fabrication Technology*

Classification of IC-Device Design and Fabrication Modifiers (Semiconductor-Based).

Wafer Processing Planar	Isolation Technique	MOS(MIS[g])FET Technology	Intraconnections (internal)
Excellence of artwork[a] Single diffused Double and triple diffused Diffused epitaxial[b] Multiple epitaxial other combinations, using microphotolithographic mask/etch cycles and diffusion, passivated by oxide, or nitride, or both; extras as "buried layer," guard ring; also compatibly npn and pnp on same chip.	Diffused junction isol. Air isolation Ceramic isolation Dielectric isolation Epitaxial isolation Glass/Quartz isolation SiO_2 isolation, with polycrystalline Si substr. PIN isolation	n-channel, or p-channel; Junction-gate, or surface-gate, i.e., insulation-gate; enhancement- or depletion-mode; dual gate, and complementary (n- and p-channel FET's on the same chip). MOS[h] arrays; also MNS[g] VTT (very thin tech)	Au fine-wire leads Al fine-wire leads Metallization: Al, or Au Metallization Al with underlayer Two-layer intraconnect Three-layer intraconnect Multi-layer intraconnect Discretionary wiring Crossover (film) Crossunder (diffused) Beam lead Electrooptical Opto-electronic

Other process for junction forming:
Ion implantation.

Circuit – Complexity Levels* (Functional)
(Relating to Number of Contact Cuts)

Gate-level complexity
Flip-flop-level complexity
Shift register-level complexity
Approx. 500 circuit-elements-level
MSI-level (up to 99 gate-equivalents)
LSI-1 level (>100 gate-equivalents)
LSI-2 level (>1,000 gate-equivalents)
LSI-3 level (?   gate-equivalents)

Internal Joints and Techniques	Package and Lead, and Chip Attachment	
Au to Au Au to Al Al to Al TC ball bond TC wedge bond TC stitch bond TC bird-beak bond Ultrasonic bond Mask and evaporate[c] Electroforming[d] Plating[e] other	Eutectic preform solder Eutectic preform braze Soft-glass frit, scrubbed Uncased, face-up, for hybrids Uncased, glassivated, hybrids Uncased, flip-chip, for hybrids Flip-chip in dual in-line Dual in-line (DIP) packages TO-5 modified can packages Std. flat pack, metal/glass[f] Std. flat pack, ceramic[f] Oversized flat packs, for LSI[f] Flat pack, with heatsink Beam lead, sealed-junction	High-density matrix level *Aside from circuit complexity, there is also another kind, referred to properly as: process complexity. –It contains constituent elements such as: how many process steps; how many bonds or layers of intraconnects; how much surface area used on the wafer; and what the resolution of the masking is. – Process complexity may very well have a greater effect on device/circuit reliability than circuit complexity, i.e., functional complexity.

[a] Lines, spacing, registration, resolution, etc.
[b] Homo-epitaxial.
[c] Multilayer intraconnect(s).
[d] Beamleads.
[e] Bumps: tinned silver.
[f] Square or round; multilead (lead frames).
[g] Metal-Insulator-Semiconductor, as MNS, made with silicon nitride (not oxide).
[h] Or, MTOS, with "thick" oxide.

* Courtesy of George Szekely, Aerojet General Corp.

It clearly becomes the designer's choice whether his particular application can be best accomplished by thick-film techniques.

THIN FILMS

Thin films are not new; but the techniques for depositing thin films and thin-film applications have advanced considerably in the last few years. Conductive, resistive, and insulating materials may be deposited as thin films (100 to several thousand angstroms) on passive substrates (glass or ceramic) or on active substrates such as silicon. The physical characteristics of thin films are not usually the same as those of bulk material, and can be dependent upon deposition procedures and methods. For example, in Fig. 2 is a plot of sheet resistance versus thickness

TABLE 5 Modifiers for Thick- and Thin-film Pure and Hybrid Integrated-circuit Fabrication Technology*

MODIFIERS for THIN-/THICK-FILM PURE and HYBRID-INTEGRATED (HIC) CIRCUITS FABRICATION TECHNOLOGY.

THIN FILMS (Vapor/Vacuum) PURE and HYBRID

Substrates	Deposition, etc. Process	Film Materials	Leads and Connections[a]
Glazed alumina	Vacuum Evaporation:	Resistive:[b]	Au; Al; Ni
Glazed beryllia	Source resist. Heating	Nichrome(s)	Au-plated Kovar
Glass: microsheet	flash evaporation	Ta, gold-doped; nitride	Cu tinned
Glass: Corning 7059	RF induction heating	Chromium and CrAu	Dumet
Glass: Corning Pyrex	Electron beam vaporiz.	Cr-SiO Cermet	Multilayer paths and
Quartz	Sputtering:	Cr-Ti, alloy, nitride	laminates and stacks
Glass-ceramics:	DC, and reactive;	Ti; Tin-oxide, doped	other
Pyro- and Fotoceram	RF-added co-sputtering	Moly; Hafnium; Rhenium	
Sapphire, polished	Gas or Vapor Plating,	Capacitive (dielectric)[d]	Lead Attach. Methods:[a]
SiO_2 on silicon IC	incl. Epitaxial Growth:	Al oxide (Al_2O_3)	Solder (hand; dip; wave; etc.)
For Thick Films:	Thermal decomposition	Ba Titanate ($BaTiO_3$)	Weld (resistance, and
Alumina; Zirkon	Hydrogen reduction	TiOxides: TiO and TiO_2	percussive arc weld)
Beryllia	Conversion and Forming:	TaPentoxide (Ta_2O_5)	Parallel-gap weld
Ba Titanate	Anodizing, wet, or plasma	Si oxides, SiO and SiO_2	Parallel-gap solder
Steatite; MgO; etc.	Electro- and Electroless	Conductive:	Solder reflow
Substrate Prepar:	Plating; other process.	Aluminum	TC-bond (4 alternatives)
Drill; feedthroughs	Etching and Electron Beam:	Au, or Ag w. Cr underlayer	Ultrasonic bond
eyelets; curing; other,	Selective, sequential;	Cu-Au with Cr underlayer	Thermal pulse bond
e. g., thermoswaging	subtractive etch cycles[f]	Ni-Au, co-evaporated	Laser beam
contact pins, for	Electron beam scribing	Mo; Pt; Sn; Cu-Be; others.	Electron beam
thick films.	Trimming; Stabilizing; etc.	Semiconductors (Si; CdS)	Infrared energy
	Post-treat, stress-remove	Organic semiconductor	Conductive adhesives

THICK FILMS (Screen Printed and Fired-on, or "PAF") HYBRID (HIC)

Deposition and Process:	Add-on Actives[a]	Add-on Attachment[a]	Package and Enclosure[a]
Metal-screen printing;	and Passives,	of Actives and of	Flat pack types, hermetic,
or spraying; sequential	Discrete or IC:	Discrete Passives	standard, 14-leads (pins);
or co-firing; trimming,	Transistors, and	Direct-bonded:	Flat pack "jumbo", multilead
or not; underglaze, or	Diodes, capac's,	Eutectic solder or braze;	peripheral, of glass, metal,
under- and overglaze.	IC's arrays,	Face-down bonded:	or ceramic construction.
Variables:	packaged compat-	Flip-chip, bumps on	Closure techniques: solder,
Air dry, oven dry;	ibly, or as chips	the chip, or on the	braze, weld, koldweld
firing temperatures	uncased. Also	passive substrate;	(oxidized Kovar with glass).
line width, mask prep.	with tabs, or as	ultrasonic or solder-	TO-5 can, modified;
Compos. Inks, Pastes:[c]	LID[e] or channel-	reflow-bonded; also, beam	Dual in-line packages;
Resistive:	pack, etc.-axial,	leads, and conductive	Epoxy- or plastic-encapsulated
Pd; Pd-Ag cermets;	or coplanar	adhesives.- automated	Precast plastic shell
Indium- and Tin-oxides	leads.	"Batch-attachments".	Epoxy transfer molded
High ohm/sq/mil			Glassivated, and glass-encaps'd
others			
Conductive:			
Pt; Pt-Au; Au; Ag; Pd; Ni			
Dielectric:			
$BaTiO_3$; other.			
For Active Elements:			
Experimental only today			

[a] Common modifiers for thin and/or for thick film circuits/devices.
[b] Important ones listed only, out of an estimated 60 materials tried.
[c] Screenable compositions are a mixture of pigment (metal or dielectric), flux or glass frit binder, filler, reducing agent, all in an organic vehicle or solvent (dispersed), usually with a thixotropic agent added.
[d] Also organic dielectrics.
[e] Leadless inverted devices; also, compatible leadless pellet microcomp't.
[f] "Wet process" patterning.

* Courtesy of George Szekely, Aerojet General Corp.

of a typical alkali metal. This curve is divided into four areas. In area *a* the film is not continuous. It consists of isolated islands of one or more atoms of one atomic layer, and free electrons cannot traverse the film as in bulk metals. In area *b* the clusters of atoms are several atomic layers thick. The islands begin to touch, and metallic conduction does occur, governed by surface roughness. In area *c* the surface becomes smoother, and the film is continuous, bound by plane and parallel boundaries. In the fourth area, *d*, the film sheet resistance approaches bulk properties. The thin-film circuit elements of major importance at the present time are resistors and capacitors. These are depicted schematically in Fig. 3, which illustrates the insulating substrate, metal-film contacts, the resistor, and the dielectric film. The performance advantages of thin-film resistors and capacitors are:

1. Reduction of parasitic capacitance and series resistance
2. Larger range of component values
3. Lower temperature coefficients
4. Tighter process control

Fabrication Techniques

Thin-film component geometries can be tailored to a particular substrate and spatial arrangement. Long, straight geometry for resistors (resistivity greater than 100 ohms) is preferable. Three basic techniques are used for patterning thin films on passive substrates: photolithography, metal masks, and screening. As mentioned previously, screening finds greater application in thick films.

Photolithography as a masking procedure is usually a subtractive process. The photoresist, which is photosensitive polymeric material, is applied as a film over the entire substrate. A patterned mask is placed over the resist film and exposed to ultraviolet light. The portion of the resist exposed by the mask to the ultraviolet will polymerize. The thin film is then deposited onto the substrate, and immersion in a solvent will lift the metal film and photoresist in predetermined areas where the resist was not polymerized. The reverse process can also be used, in which the metal is deposited, then photoresist is applied, and the metal plus resist are selectively removed by etching.

Fig. 1 Typical variations in electrical properties, as a function of peak firing temperature, for resistive thick-film pastes.[13]

Metal masks are commonly used to selectively deposit thin-film circuit elements. The mask is placed in contact with the substrate, and films are deposited through openings in the mask. Using a series of masks, the resistive films are deposited first, followed by the interconnection pattern and/or the lower electrode of the capacitors. Next, the insulating films are deposited. These are used as the capacitor dielectric and as surface protection. Then the conductive films for the upper electrodes of capacitors are deposited. The advantages of metal masks are the simplicity and lack of the need for elaborate cleaning procedures after masking. However, metal masks do not provide fine-line resolution and sharpness of definition.

The various thin-film deposition techniques will be discussed in detail before the individual devices of thin-film circuitry are introduced. An attempt will be made to refer to the specific deposition techniques most commonly employed for resistors, capacitors, etc. The selection of a particular method of thin-film deposition should be governed by the characteristics desired in the film.

Vacuum evaporation Vacuum evaporation is by far the most widely used fabrication technique. This is particularly true for thin films whose composition and purity must be precisely controlled. With the evaporation process, the temperature of the material to be deposited is elevated in high-vacuum environment with the substrate positioned a predetermined distance from the evaporant. As the temperature rises, a point is reached at which the vapor pressure of the source material exceeds that of the environment. Evaporation or sublimation of the mate-

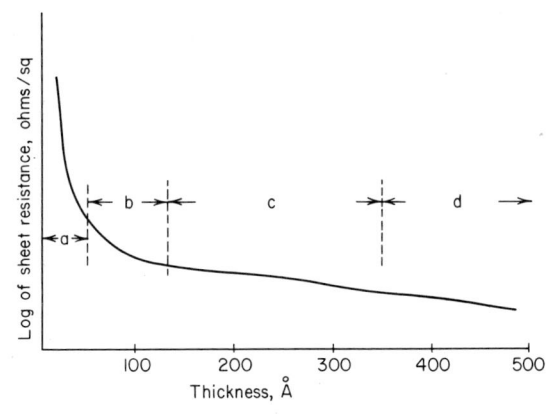

Fig. 2 Sheet resistance versus film thickness of typical alkali metal.[14]

rial takes place. It is common practice to evacuate the system to 10^{-6} to 10^{-7} torr before initiating the evaporation. (During the course of the evaporation the pressure may increase to 10^{-5} owing to the heat generated.) At these pressures the mean free path of the typical vapor molecule is about 40 cm. The *mean free path* may be defined as the average distance that a vapor molecule can be expected to travel without collision with another vapor molecule or residual gas molecule. An expanded explanation of the mean-free-path theory can be found in Ref. 8. If the dis-

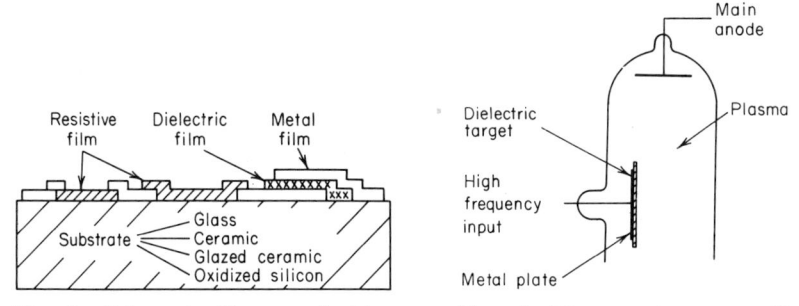

Fig. 3 Schematic diagram of thin-film resistor and capacitor.

Fig. 4 Rf sputtering system.[15]

tance from the evaporation source is equal to the mean free path, approximately 50 percent of the evaporated molecules will arrive at the substrate. Gas molecules may react with vapor molecules and contaminate the film. One method for prevention of this type of contamination is to keep the source-to-substrate distance as much shorter than the mean free path as is practical.

Many kinds of heat sources may be used: resistance heating, induction heating, radiant heating, and electron-beam bombardment. Again the choice is dependent

TABLE 6 "How to" Suggestions for Thin-film Deposition*

Name	Symbol	Melting point	Density	Vapor temperature at: 10^{-8} torr	10^{-6} torr	10^{-4} torr	Resistance sources: Filament	Type	Boats	Electron beam/bomb	Other	Remarks
Aluminum	Al	660	2.7	950	1085	1280	W, Ta	Coil	Graphite	Problematical	Sputter	Oxide forms in beam; causes bursting.
Aluminum fluoride	AlF	1040	...	410	490	700	...	...	W, Ta	Poor	...	Loses fluorine. Disproportionates.
Aluminum fluoride	AlF₃	...	...	Sublimes	...	...	...	...	...	...	...	...
Aluminum	Al₂O₃	2020	3.6	1320	1480	1600	W, Ta	Basket	W, Ta	Excellent	...	Rate control important. Some reduction in beams.
Antimony	Sb	630	6.7	552	618	700	Chromel Ta	Basket	Ta, C	Bursts	Toxic	Wets chromel, no carbides.
Arsenic	As	Subl	5.7	380	423	480	Graphite	Slotted strip	Al₂O₃, BeO, C	Poor	...	Always sublimes, toxic.
Barium	Ba	725	3.5	545	627	735	W, Ta, Mo, Cb, N	Basket	Ta, Mo	O.K.	Sputter	Violently reacts with ceramics, Wets refractory metals. Always yields some free Ba.
Barium titanate	BaTiO₃	...	...	Compound composition controller from 2 sources only practical technique								
Beryllium	Be	1284	1.9	980	1150	1270	Ta, W, Mo	Basket	Carbons	Excellent	...	Extreme caution: vapors, powders and oxides toxic.
Bismuth	Bi	271	9.8	600	682	790	Chromel, W, Ta, Mo	Basket	Al₂O₃, C	Poor	...	Vapors are toxic.
Bismuth titanate	BiTiO₃	...	...	Decomposes	...	...	Compound composition controller from 2 sources only practical technique					
Boron	B	2550	2.3	2100	2220	2400	No	...	No	Excellent	Sputter	Material explodes on rapid cooling.
Boron carbide	B₄C	2450	...	2500	2580	2650	No	...	No	Excellent	Sputter	Thermal shock requires slow heat-up.
Cadmium	Cd	321	8.6	347	390	450	Chromel, Ta, Mo, W, Ni, Fe	Basket	Mo, Ta, fused quartz	...	...	Wets Cb and chromel.
Cadmium sulfide	CdS	1750	4.8	760	840	920	Graphite	Strip	Al₂O₃	Fair	...	Sticking coefficient strongly affected by substrate temperature.
Calcium	Ca	842	1.5	555	630	732	W	Basket	Al₂O₃	...	...	Deposit corrodes in air.
Calcium fluoride	CaF₂	1360	...	1490	1600	1690	Ta, W	Basket	Ta, W	Poor	Sputter	Rate control important. Disproportionates except in sputtering.
Calcium titanate	CaTiO₃	...	...	...	...	...	No	...	...	...	...	
Carbon	C	3700	1 to 2	1930	2140	2410	Point contact carbon rods. High Rat point	...	Graphite	Excellent	Sputter	Solid "sparks" from point source, density variable.
Cerium	Ce	804	6.9	1246	1420	1650	W	Helix, coil	Al₂O₃, C	O.K.	Sputter	Attacks boat. Rate control important.
Cerium oxide	CeO₂	2600	6.9	1890	2000	2310	...	...	W	Good	Electrodeposited on W wire	Can be sublimed.
Chromium	Cr	1890	6.9	1110	1250	1430	W	Basket	...	...	...	
Nichrome	Ni/Cr	1350	8.2	1120	1260	1490	W, Ta	Coil	Ta, W, Mo, Al₂O₃	Excellent	Sputter	Smooth deposit from resistance sources.
Cobalt	Co	1495	8.9	1200	1340	1530	Cb	...	Al₂O₃, BeO	Excellent	Sputter	Forms low-melting eutectic with refractories.
Columbium	Cb(Nb)	2500	8.5	2030	2260	2550	W	Coil	W	Excellent	Sputter	Controlled high rates required as superconductors.
Copper	Cu	1083	8.9	1095	1110	1230	Pt, or Pt/Ir Mo, Ta	Coil	Mo, Ta, Al₂O₃	O.K.	Sputter	Does not wet sources.
Gallium	Ga	30	5.9	892	1015	1180	No	Fused quartz	Al₂O₃, BeO	...	...	Alloys with metals.

11-10

Material	Symbol	col3	col4	col5	col6	col7	Source material	Support	Evaporation technique	Quality	Remarks
Gallium phosphide	GaP	1350					Ta, W			Excellent	Does not decompose. Rate control important.
Gallium arsenide	GaAs						Two sources with alloy control only		Sputter		Compound decomposes.
Germanium	Ge	959	5.3	1085	1120	1210	Ta, Mo, W	Basket	Sputter	Excellent	Wets Ta and Mo.
Germanium oxide	GeO₂	1115	19.3	1080	1220	1405	Ta, Mo, W	Basket		Fair	Similar to SiO
Gold	Au	1063			870	1015	W, Mo	Helix or Basket		Good	Wets Ta well and reacts.
Indium	In	157	7.4	760			W, Fe	Basket			Wets W.
Indium arsenide	InAs	940					Two sources with alloy control		Sputter, flash		Compound decomposes.
Indium antimonide	InSb								Sputter, flash		Compound can be reconstituted by deposition on substrate at 900°C.
Indium phosphide	InP						Two sources with alloy control		Sputter, flash		Deposits P rich.
Iridium	Ir	2454	22.4	1850	2080	2380	Ta, W	Basket	Graphite Al₂O₃, BeO	Excellent	
Iron	Fe	1535	7.9		700	770	No W	Basket	Al₂O₃, BeO W, Ta	Excellent	Some carbon from boat. Attacks W if exceed ½ total weight.
Lead	Pb	328	11.3	615			Fe, Ni Chromel	Basket	Pt, Ir	Fair	Toxic. Very high controlled rates required for superconductors.
Lead stannate	PbSnO	1115	8.1	670	780	905	Pt, Ir Alloy control required	Basket	Al₂O₃		Disproportionates.
Lead telluride	PbTe	917	8.1	780	910	1050				Poor	Toxic vapors, Te-rich deposits.
Lithium	Li	186	0.5	500	580	680			Fused quartz steel	Good	Oxide destroys fused quartz.
Lithium fluoride	LiF	870	2.6	875	1020	1180			Carbon	Good	Rate control important for optical films.
Magnesium	Mg	651	1.7	458	520	600	W, Fe, Ta, Mo	Basket	Carbon	Excellent	Sublimes at high rate.
Magnesium fluoride	MgF₂			1090	1220	1540	W, Ta	Basket			Rate control important for optical films.
Manganese	Mn	1260	7.2	780	845	920	W, Ta, Mo	Basket	Al₂O₃	Good	Wets W resistance sources.
Molybdenum	Mo	2622	10.2	1865	2095	2390	No	Coil	No	Excellent	Very volatile, oxide.
Nickel	Ni	1455	8.9	1200	1345	1535	Heavy W		Al₂O₃, BeO	Excellent	Forms low-melting eutectic with metals.
Nichrome	Ni/Cr	1350	8.2	1120	1260	1490	W, Ta	Coil	Ta, W, Mo	Excellent	Smooth deposit from resistance sources.
Permalloy	Ni/Fe	1395	8.7	1220	1360	1580			Al₂O₃	Fair	Ni content low in film. Use 84% Ni source.
Supermalloy	Ni/Fe/ Mo	1410	8.9				Requires permalloy and Mo sources with alloy control				Permalloy from boat, Mo from beam.
Palladium	Pd	1550	12.0	1115	1265	1465	W	Coil	Al₂O₃, BeO	Excellent	Resistance sources including impurities.
Platinum	Pt	1774	21.5	1565	1765	2020	W	Coil		Excellent	Alloys with metals.
Rhodium	Rh	2149	12.1	1550	1745	1980				Good	Very low pressure for W source.
Selenium	Se	234	4.3	190	240	270	Chromel,	Basket	Fused quartz	Poor	Oxide reacts with quartz poisons, vacuum system.
Silicon	Si	1420	2.4	1265	1420	1610	SiC	Rod	BeO, C	Excellent	Oxygen inclusions cause low mobility.
Silicon monoxide	SiO	Softens no distinct melting point	2.1	870	990	1250	W, Ta, Mo	Basket	W, Ta	Excellent	Sputter Furnace. Pinholes reduced by low rate.

11-11

TABLE 6 "How to" Suggestions for Thin-film Deposition* (Continued)

Name	Symbol	Melting point	Density	Vapor temperature at: 10^{-8} torr	10^{-6} torr	10^{-4} torr	Filament	Type	Boats	Electron beam/bomb	Other	Remarks
Silicon dioxide	(Glass)	Softens	2.1	Influenced by composition			Decomposes			Good	Sputter	Glass films cannot be deposited from metal boats.
Silver	Ag	961	10.5	847	958	1105	Ta, Mo, Fe, Ni	Coil or basket	Good on Mo	Electro-deposited		Does not wet W.
Sodium chloride	NaCl	801	2.2	675	435	1014			Ta, W, C	Poor	Furnace	Hygroscopic film.
Sodium fluoride	NaF	980	2.8	945	1080	1200	No		Ta, W	Poor	Furnace	Soluble film.
Tantalum	Ta	2996	16.6	2230	2510	2860	Ta, W		No	Excellent	Sputter	Getters oxygen.
Tantalum oxide	Ta₂O₅	1470	8.7	1550	1780	1920		Basket	Ta, W	Poor		Forms smooth spectral film, high dielectric constant.
Tellurium	Te	452	6.2	430	480	550	W, Ta, Mo, Fe	Basket	Fused quartz	Poor		Poisons system, toxic vapors.
Thallium	Tl	302	11.9	556	630	740	Ni, Fe, Cb	Basket	Quartz, Al₂O₃	Poor		Very toxic vapor.
Tin	Sn	232	5.7	955	1080	1270	Chromel Ta, Mo	Coil	Al₂O₃, C, Te	Fair		Wets Mo and Ta.
Tin oxide	SnO₂	Decomposes	6.4						W	Poor	Sputter	Sublimes.
Titanium	Ti	1800	4.5	1335	1500	1715	W, Ta	Coil, basket	Graphite	Excellent		Getters gas, oxide film on melt, Reacts with W.
Titanium oxide	TiO₂	Decomposes	4.3				Ta, W	Basket		Poor		Suboxide form, must be reoxidized.
Tungsten	W	3370	19.3	2390	2680	3030	No		No	Good	Sputter	Forms volatile oxides, film oxidize.
Tungsten oxide	WO₃	1473	12.1	1120	1290	1460	W	Basket	Ta, W	Fair	Sputter	Forms smooth semiconductive film.
Tungsten carbide	W₂C	2860	15.7				No		Graphite	Excellent		Smooth conductive hard film.
Uranium	U	1132	18.7	1405	1600	1855	W	Basket		Good		Films oxidize.
Vanadium	V	1710	5.9	1435	1605	1820	W, Mo	Basket	Al₂O₃, Fe, C	Excellent		Wets Mo, reacts with W.
Zinc	Zn	419	7.1	296	350	420	W, Ta, Mo	Basket		Poor control		Wets refractory metals, no reaction.
Zinc oxide	ZnO	1800	5.6								Furnace	
Zinc fluoride	ZnF₂	872	2.9	790	905	1035			Ta, Mo, C		Furnace	Partially decomposes, sticking coefficient varies greatly with substrate temperature.
Zinc sulfide	ZnS	1900		950	1085	1265	TaMo	Basket				
Zirconium	Zr	1857	6.4	1750	1975	2260	W	Basket		Excellent		Film oxidizes readily.
Zirconium oxide	ZrO₂	2950	5.7				No	No		Excellent		Smooth clear film, no reduction.

* Sloan Instruments Corp.

on the material to be evaporated. Tungsten or molybdenum filaments have been used as resistance heaters. Crucibles of refractory metals, alumina or zirconia, may be resistance-heated by filaments or may serve as containers for the evaporants. The purity of the heating source is most important to prevent contamination of the deposited film. At high temperatures the filament is likely to evaporate along with the source, forming another contaminate means. Electron-beam bombardment techniques are especially advantageous for high-temperature materials. The beam of electrons is emitted from a thermionic cathode and directed at the source material. The beam can be regulated to a pinpoint of energy lending excellent process control and deposition rate control. Table 6 gives "how to" suggestions for thin-film deposition compiled by Sloan Instruments.

RF sputtering An rf sputtering system is shown in Fig 4. The rf electrode is immersed in the plasma of an auxiliary low-pressure discharge. A third electrode (not shown) can either be a mercury-pool-type cathode or a thermionic cathode. A self-sustained rf glow discharge can be maintained at pressures down to 2×10^{-3} torr. The use of a self-sustained glow discharge for depositing films by sputtering has the advantage that extremely uniform film-thickness distributions can be obtained

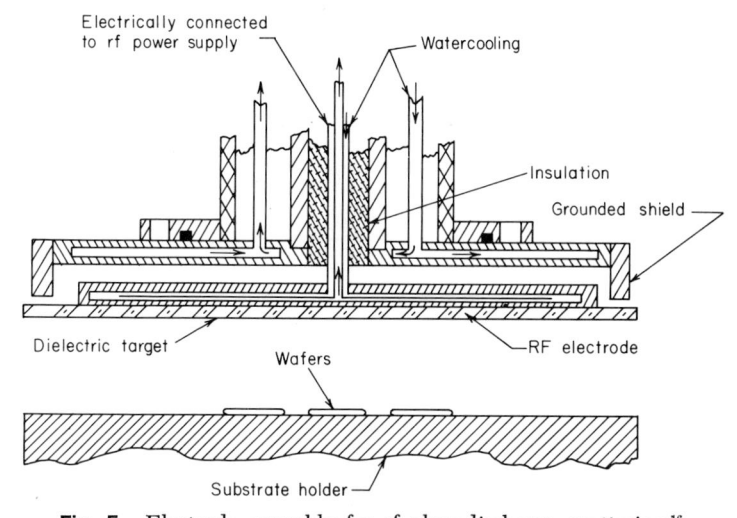

Fig. 5 Electrode assembly for rf glow-discharge sputtering.[16]

readily over large areas. In addition, the system is inherently simple. The low pressure at which an rf glow discharge operates combined with capacitive coupling allows the deposition of pure metal films by rf sputtering. Also, insulator films can be formed by rf reactive sputtering at deposition rates considerably higher than previously possible with dc methods. These new developments make rf glow-discharge sputtering an easily instituted and simple technique for thin-film deposition. Figure 5 shows an outline drawing of the electrodes for an rf glow-discharge sputtering system. The back of the target is metallized, for example with an evaporated chromium-copper film, and soldered directly to the electrode. If sputtering is done in the "up" direction, the target can simply be placed on the rf electrode. A grounded metal shield is placed close to the other side of the electrode to extinguish the glow on that side and to prevent sputtering of the metal electrode. Because the working pressure is considerably higher than in evaporation, there are numerous collisions between the sputtered atom and residual gas, and propagation across the cathode-anode is a diffusion process. It is usual to observe a nonluminous region surrounding the cathode. As the temperature in this region is high, the substrate is placed beyond the edge of the nonluminous or dark region.

Thin Films

The frequency seems to have few or second-order effects on the deposition rate. The effect of rf potential on deposition rate may be seen in Fig. 6. Figure 7 illustrates the effect of magnetic field intensity. The magnetic field causes the electrons to spiral around the lines of force, thus increasing their path length and the amount of ionization. Also, electrons that would otherwise be lost to the walls of the system are confined inside the glow. At high field intensities saturation effects occur. Figure 8 illustrates the effect of substrate temperature on the deposition rate. For this evaluation, the target was fused quartz, and the electrode diameter was 15 cm. For good thermal control, the substrates were provided with a gallium back to the heater block. Because of the temperature dependence of the deposition rate, it is obvious that a uniform substrate temperature is needed if a uniform film-thickness distribution is to be obtained.

Addition of oxygen to the sputtering gas results in a reduction of deposition rate of more than 50 percent. This is shown in Fig. 9. The large concentration of negative oxygen ions, which act as electron traps, is the most commonly accepted explanation of this effect.

A quick and accurate way to compare sputtering rates has been reported.[17] Sputtering yields for various dielectric materials were measured by placing a thin sample of each material on a large quartz target which was resting on an rf electrode. By applying rf power to the electrode, all materials were subjected to the same sputtering conditions. Direct calculation of the sputtering yields is not possible because of the unknown and varying ion

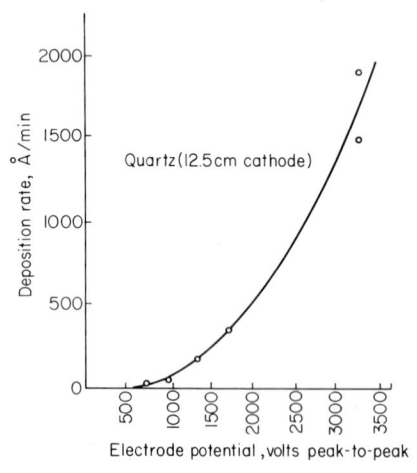

Fig. 6 Deposition rate versus electrode potential.[15]

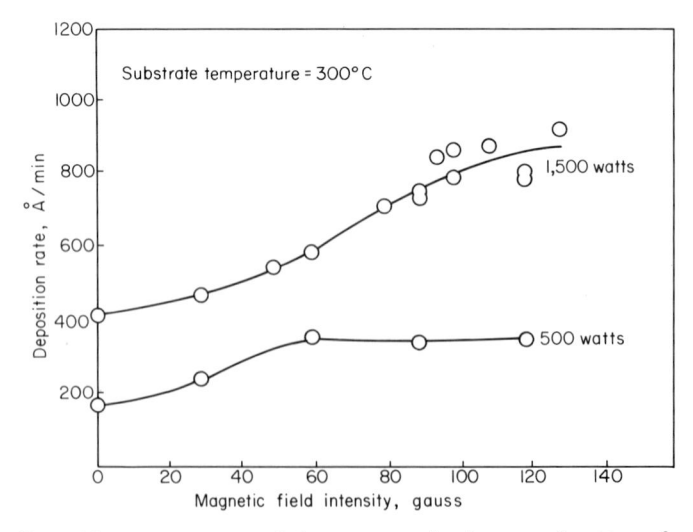

Fig. 7 Deposition rate at two rf input-power levels as a function of magnetic field intensity.[16]

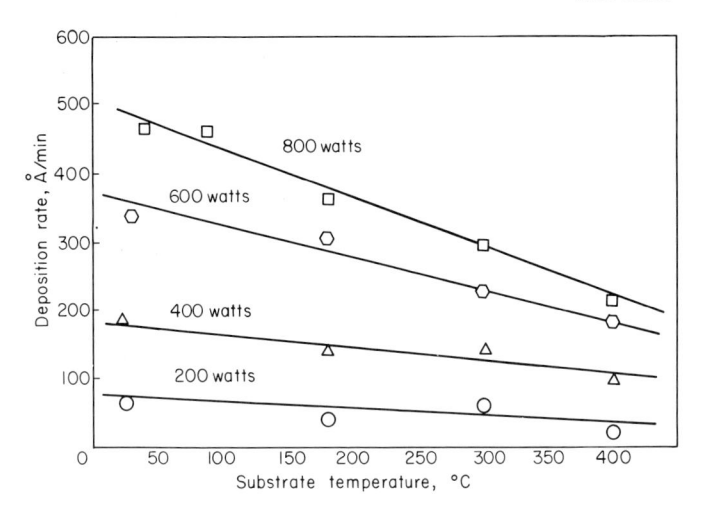

Fig. 8 Deposition rate versus substrate temperature for various rf input-power levels (magnetic field: 100 gauss).[16]

current density during rf sputtering. By sputtering simultaneously with the dielectric samples, materials with known dc sputtering yields, it is possible to calculate "equivalent dc sputtering yields" (see Fig. 10). In general, rf-sputtered films are found to be of higher quality than films made by low-temperature deposition processes.

Low-energy sputtering is recently receiving considerable attention as a technique for making thin-film circuits. Schematically, the configuration of the various electrodes is shown in Fig. 11. The filament in this system is the source of electrons. It is heated to a temperature that induces electron emission of approximately 2 amps/cm^2. A potential is applied to the anode, and a pressure of one micron is established in the chamber by backfilling with argon. At a pressure of one micron a filament emission current of 3 to 5 amp is realized. With an emission current of this magnitude very high plasma densities are obtained. The material to be sputtered and deposited as a

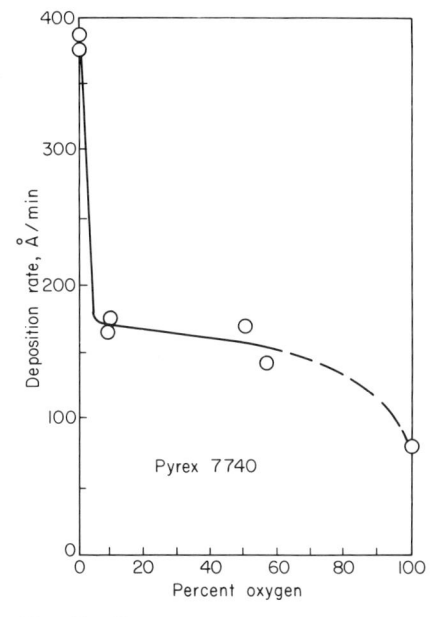

Fig. 9 Deposition rate versus oxygen percentage in sputtering atmosphere (magnetic field: 110 gauss).[16]

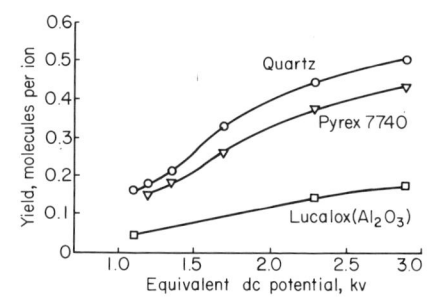

Fig. 10 Sputtering yields for various insulators.[17]

11-16   Thin Films

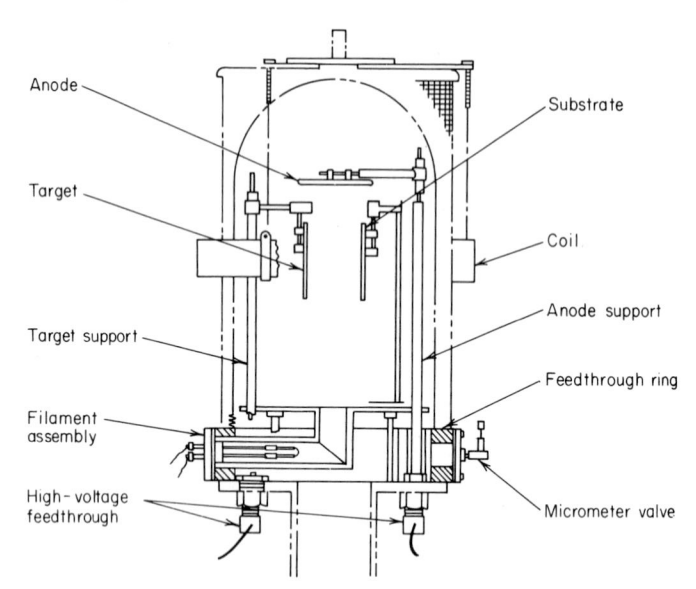

Fig. 11 Electrode configuration.[18]

thin film on the substrate is called the target. With negative bias applied to the target, ions are generated in the plasma which strike the target material and remove atoms from its surface. There is no direct heat involved, and although the target temperature may rise, it never is heated to the point at which the vapor pressure results in evaporation. A magnetic coil serves to confine the majority of the electrons that would otherwise be lost to the bell jar wall where they would be unavailable for performing their ionizing function. Use of a magnetic field makes it possible to increase the plasma density and consequently the sputtering rates. Figure 12 shows the relationship between target current and magnet current at three different target voltages.

With low-energy sputtering the distance between the source material and the substrate is generally between 2 and 4 cm. Operating at a pressure of one micron with a mean free path of 6 cm ensures that the sputtered atoms removed from the target will encounter very few collisions with residual gas atoms in traversing the distance from target to substrate. The result is an increase in deposition rates as well as improved purity of the deposited films. The thickness uniformity of a thin film is dependent upon many factors. Thickness control determines the resistance value of the film. Many report that they have not had to use thickness or rate monitors in connection

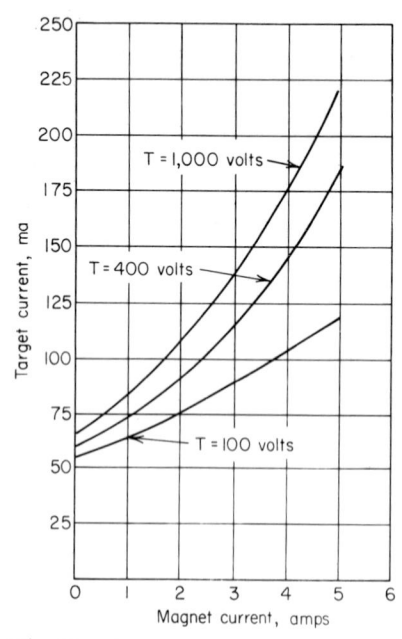

Fig. 12 Target current versus magnet current for various target voltages.[18]

with their deposition work with low-energy sputtering systems. They have been able to merely reproduce a given target voltage for a given material and maintain that target-voltage-current relationship for a predetermined time.

Getter sputtering Getter sputtering is a method that greatly simplifies the preparation of thin films where composition control is critical. The major difficulty in sputtering and vacuum-deposition techniques is the contamination of the film caused by residual gases. Even when the pressure is 10^{-6} torr with a slow deposition rate (10 Å/sec), there are as many contaminating gas collisions with the substrate as there are with film-forming metallic atoms. With highly reactive metals such as tantalum, the resulting films have vastly different properties from those of the starting material. A technique reported by Theuerer and Hauser[19] used a conventional vacuum system with residual pressures in the 10^{-6}-torr range. With this method, sputtering is confined within an anode can; and by geometric design, sputtering itself is used to purify the argon at the lower and upper parts of the system by the gettering action of the reactive metal. The central region of the system where coating occurs is shielded from gaseous contamination. The sputtering conditions used are given in the Table 7. Also included is the coating efficiency as

TABLE 7 Sputtering Conditions for Various Metals[19]

Metal	Volts	Current, ma	Argon,* 10^{-3} torr	Deposition rate, Å per min	Coating efficiency, %
Pb	1,000	2	31	372	3.2
Al	1,500	2	27	56	4.3
Al	1,500	10	73	65	1.9
Cu	1,500	9.5	59	357	2.7
Pt	1,000	2	30	88	4.4

* Uncorrected pressures as measured with a Pirani gage calibrated for air.

determined from the cathode weight loss and weight gain per square centimeter of coating area. The properties of the individual metallic films are summarized in Table 8.

Diffusion Diffusion techniques for the fabrication of thin-film resistors and capacitors are used in monolithic or functional electronic blocks for integrated circuits. Both active and passive circuit elements and their interconnections are fabricated within or on the surface of silicon. This type of circuit is difficult to design and produce. The diffused resistors are commonly formed at the same time as the transistor base diffusion, resulting in the formation of a thin layer of p-type silicon on n-type silicon substrate. The resistance values depend on the diffusion profile, diffusion depth, and length-to-width ratio of the diffused area. The p-n junction formed during transistor diffusion cycles is used for the formation of diffused capacitors. There are numerous problems and disadvantages involved with both diffused resistors and diffused capacitors, which will be discussed in a later section.

Chemical deposition Chemical deposition is not a widely used technique in present-day thin-film microelectronics. When it is used to deposit a metal oxide film, the film is formed by the hydrolysis of a mixture of the chlorides at the hot surface of the substrate.[20,21] The mixture of chlorides in solution is sprayed onto the hot substrate where the oxide is formed at the surface. The chlorides are mixed in the right proportion to give the required mixture of oxides, and the amount of water is controlled to ensure that the hydrolysis occurs at the surface of the substrate and not in the bulk of the spray. The temperature of the substrate is very high—in the case of glass near the softening point. The film thickness used can vary between 10^2 and 10^4 Å and is controlled by the concentration of the spray and the time duration of the deposition.

TABLE 8 Properties of Various Metallic Films Prepared by Getter Sputtering[19]

Metal	Thickness, Å	Table temperature, °C	T_C	T_B	ρ, ohm-cm
Pb	75	−195	....	...	∞
	85	−195	7.2	...	1.2×10^{-5}*
	185	−195	7.3	7.2	
	3,720	−195	7.3	...	1.4×10^{-5}*
Al	190	−195	1.9		
	8,400	−195	1.52	1.2	2.4×10^{-5}
	650	300	1.25	...	6.6×10^{-6}
Cu	600	−195	....	...	18.5×10^{-5}
	1,000	−195	....	...	2.1×10^{-5}
	10,070	−195	....	...	1.1×10^{-5}
Pt	65	−195	....	...	6.4×10^{-5}
	175	−195			
	2,620	−195			

* Measured with magnetic field of 17.4 kg to destroy superconductivity.

Chemically deposited films are very hard, crystalline in nature, and firmly adherent to the substrate. They are chemically inert and highly resistive to moisture. These films are for the most part used as resistors. For resistor fabrication, the nature of the substrate is important. It must be free from contaminating alkalies. The chemically deposited oxide film can be operated at high temperatures, and resistors fabricated by such techniques on ceramic substrates have been reported to operate up to 500°C. Another metal film that is chemically deposited and has very good resistive properties is nickel-cobalt.[22] The films of nickel-cobalt alloys are formed by a catalytic reduction process at a surface in contact with a solution of metal salts. The choice of a suitable substrate and the preparation and treatment of the surface are most important. It is claimed that the process can be rigidly controlled to produce films which meet the requirements of high-quality resistors. The film is susceptible to the action of moisture and to electrolysis. Therefore, to be corrosion-free the film must be protected.

Thin-film Materials

  Substrate Certain ceramics and glass are employed as substrates for thin films. The substrate must be physically inactive and should be carefully chosen for its compatibility with device-processing sequences, and for its physical strength, surface condition, and thermal characteristics. The substrate material must remain chemically inert during processing, withstand temperature and forces of processing and handling, and provide a surface that will promote adhesion and minimize contamination of the deposited films. It should closely approximate the thermal coefficient of expansion of the deposited films or other devices mounted on it, and provide good thermal conductivity for extraction of heat. It should also be electrically inert.

  Typical substrate materials are glasses and ceramics. Glasses are generally inexpensive, chemically inert, and high-temperature-tolerant. Unfortunately, even though glass provides an optimum surface, it is brittle and is a poor thermal conductor. Ceramics are usually more expensive and more porous than glasses, but are superior in thermal conductivity. The selection of a particular material is usually a compromise, depending on the importance attached to various properties in any given application. Ceramics are frequently glazed to provide a glasslike surface without severely compromising other desired properties.

Some of the commonly used substrate materials are listed below with the typical values of thermal constants.

Substrate material	Thermal conductivity, watts/(in.)(°C)	Coefficient of linear expansion, 10^{-6} per °C
Alumina (96 Al_2O_3)	0.6	4–7
Beryllia (99 BeO)	4.1	6–9
Borosilicate glass	0.03	5
Pyroceram	0.54	8
Quartz (SiO_2)	0.3	0.3

Typical cleaning procedures for ceramic wafers are as follows:
1. 10 min hot trichloroethylene ultrasonically
2. 5 min demineralized water ultrasonically
3. 10 min 1:1 solution of HCl and HNO_3
4. 10 min demineralized water ultrasonically
5. 5 min demineralized water ultrasonically
6. 5 min methanol ultrasonically

After the cleaning process the wafers are air-dried in a dust-free atmosphere.

To compensate for the porosity of ceramics, glazes are often applied. A typical application is described as follows: Most commercially available glazed ceramic wafers contain a doping element either as a major constituent or as an impurity. This fact necessitates the formulation of an acceptable glass from reagent-grade material in the laboratory. By compounding raw materials to conform to the limitations of coefficient of expansion and softening point, the following glazes are made:
1. Na_2O 15.4, MgO 10.0, SiO_2 74.6 percent
2. Na_2O 15.75, SiO_2 74.4, MgO 8.6, A_2O_3 1.75 percent
3. K_2O 19.2, SiO_2 72.6, MgO 8.2 percent
4. Na_2O 18.0, MgO 3.9, CaO 5.4, SiO_2 72.7 percent
5. Na_2O 10.0, SiO_2 70.0, CaO 20.0 percent

The oxides, carbonates, or hydrates necessary to supply the proper composition are thoroughly ground by porcelain mortar and pestle to a particle size less than 300 mesh. Ball milling using stainless steel balls cannot be used because the nickel impurities in steel add to the glaze and will alloy with the deposited films. The glass mixtures contain either carbonate or water of hydration, which is driven off with an initial firing at 800°C. The friable mixture is reground to less than 300-mesh size and fired in a platinum crucible for fusion at about 1400°C. The glass is then ground by mortar and pestle to less than 300-mesh size and mixed in a 1:1 ratio by weight with an organic binder, such as oil, which serves as a vehicle for the powdered glass and is driven off in subsequent baking. The silk-screen method is used to apply the glass and oil mixture to the ceramic wafer. The wafer is held on a jig below the silk screen. The mixture is spread through the screen across the wafer to form a uniform layer. The water is placed on a hot plate to bake off the organic vehicle and allow the glass particles to settle uniformly on the surface of the wafer. The final step is an empirical determination of the best temperature and time for fusing the glass to the wafer to form a smooth glaze. A typical temperature range is 1300 to 1400°C for 3 to 4 hr.

Thin-film Capacitors

Fabrication of thin-film capacitors requires three deposition steps: the lower plate area, the dielectric film, and the upper plate area. The mathematical expression[23] for the capacitance of a double-plate capacitor is

$$C = \frac{0.225K(N-1)A}{t}$$

where C = capacitance, pf
 K = dielectric constant
 N = number of plates
 A = area, in.^{2}
 t = dielectric thickness, in.

In choosing capacitor materials for microelectronics the dielectric must withstand temperatures of a few hundred degrees centigrade. Thus, we are limited to evaporated inorganic films or oxidized metal films. In considering evaporated films the standard specifications are:

1. Must not decompose during the evaporation-deposition process.
2. Should reach a vapor pressure of about 10 between 1000 and 1800°C.
3. Should not be hygroscopic or water-soluble.
4. Must adhere well to dielectric substrates and metals, and should be hard and not crack when temperature-cycled.

SiO, CaF$_2$, MgF$_2$, ZnS and Al$_2$O$_3$ meet these requirements. For a complete list see L. Holland.[8]

Silicon monoxide is the most commonly used thin-film capacitor. It has a dielectric constant of 6.0 (dependent on the deposition conditions) and a breakdown voltage of 2×10^6 volts/cm. SiO is easily oxidized and is hygroscopic; thus normal practice involves the deposition of a protective overlaying dielectric. For comparison, MgF has a dielectric constant of 6.5 and a breakdown of 2×10^6. In addition, MgF has low dc leakage and good dielectric properties. Unlike SiO$_2$, control and cleanliness of deposition conditions of MgF are most stringent. ZnS has a dielectric constant of 8.2 and a breakdown of 2×10^5. Good reproducibility of the process is possible, and ZnS has low losses at low frequencies.

Dielectrics of oxide layers of silicon, aluminum, and titanium may be grown from deposits of the base metal employing either thermal or electrochemical oxidation. By comparison, TiO$_2$ capacitors show a typical temperature coefficient of capacitance of $+100$ to $+450$ ppm per °C measured at -50 to $+80$°C, dielectric constants of 20 to 30, dielectric strength of 0.5 to 2 mv/cm, and leakage resistance of 10^{13} to 10^{15} ohms.[24] SiO$_2$ capacitors exhibit similar capacitance per unit area. Thermally grown SiO$_2$ layers are not as hygroscopic as evaporated SiO$_2$. Results of reactively sputtered oxides of silicon, tantalum, niobium, zirconium, and titanium, using a sputtering atmosphere of oxygen and a cathode of pure base metal, are shown in Table 9. Table 10 lists the properties of various dielectric films.

A further advancement in the development of tantalum capacitors is obtained by the thermal or electrolytic oxidation of sputtered tantalum films. An example is a tantalum oxide–silicon oxide duplex dielectric thin-film capacitor. A sketch of the duplex configuration is shown in Fig. 13.

TABLE 9   Reactively Sputtered Oxides of Silicon, Tantalum, Niobium, Zirconium, and Titanium[25]

Cathode	Power factor, %	Resistance, ohms/farad	Dielectric constant
Tantalum	3.0	200	14
Niobium	7.0	95	39
Zirconium	4.5	450	25
Titanium	6.0	190	62

TABLE 10 Properties of Various Dielectric Films[26]

Material	Method of deposition	Dielectric constant, ϵ	tan δ	Breakdown stress, Mv/cm	Capacitance, μf/cm^2
Silicon dioxide (SiO$_2$)......	Reactive sputtering	4	0.001	3.0	0.015
Magnesium fluoride (MgF$_2$)	Evaporation	5	0.016	1.0	0.01
Silicon monoxide (SiOx)...	Evaporation	5–7	0.010	1.2	0.01
Aluminum oxide (Al$_2$O$_3$)...	Plasma oxidation	8	0.005	2.0	0.10
Aluminum oxide (Al$_2$O$_3$)...	Anodic oxidation	8	0.005	4.0	0.20
Tantalum oxide (Ta$_2$O$_5$)...	Reactive sputtering	20	0.003	1.0	0.10
Tantalum oxide (Ta$_2$O$_5$)...	Anodic oxidation	27	0.005	3.0	0.15
Titanium oxide (TiOx)....	Anodic oxidation	30–40	0.030	1.0	0.30
Lead titanate (PbTiO$_3$)....	Reactive sputtering	80	0.040	0.6	0.20

A brief recapitulation of the advantages of tantalum oxide as a dielectric for thin-film capacitors: (1) a relatively high dielectric strength, (2) a high capacitance-to-area ratio, and (3) good mechanical and electrical stability. While SiO possesses the fundamental limitation of the incidence of pinhole defects, by deposition SiO on Ta$_2$O$_5$ to form a duplex dielectric structure, low-value capacitors (0.01 μf/cm^2) and large-area distributed RC networks can be realized and incorporated in tantalum integrated circuitry. Duplex dielectric capacitors have been or are commonly prepared on glass substrates by anodization of sputtered tantalum electrodes and vacuum sublimation of bulk silicon monoxide. These capacitors perform reliably at 50 volts and 85°C. Capacitors with gold counterelectrodes (as seen in Fig. 13) and the thicker films of silicon oxide exhibited a lower failure rate at the higher life-test voltages and temperatures.

Thin-film capacitors are of several types, and consist of metal top and bottom electrodes with a dielectric between. The sketch shown in Fig. 14 is illustrative

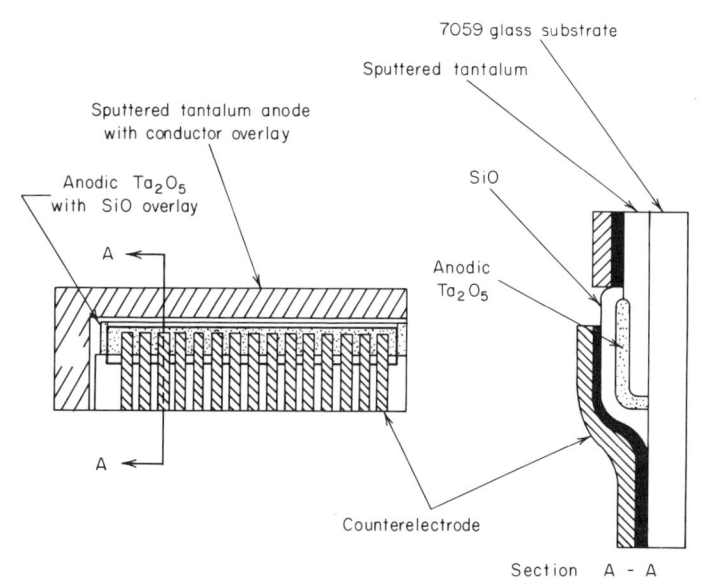

Fig. 13 Sketch of tantalum oxide–silicon oxide duplex dielectric thin-film capacitor.

11-22 Thin Films

of a capacitor deposited on a ceramic or silicon substrate. Metal-oxide-silica (MOS) capacitors may be formed, using either compatible thin-film techniques or all-silicon composition. In MOS capacitors the silicon substrate may form the lower electrode. The emitter diffusion may be used to provide a low sheet resistance under the capacitor. Thermally grown SiO_2 or a thin film of deposited glass may be used as the dielectric. The upper electrode is usually the same material as the interconnection metallization, and is deposited at the same time. For the sake of comparison a diffused capacitor is a reverse-biased p-n junction, usually the base-collector junction. There are disadvantages to diffused capacitors, such as:

1. Low capacitance per unit area.
2. High parasitic capacitance.
3. Capacitance is voltage-dependent.
4. Series resistance is high.

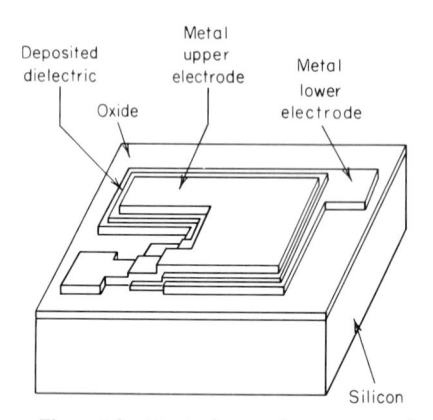

Fig. 14 Typical configuration of thin-film capacitor.

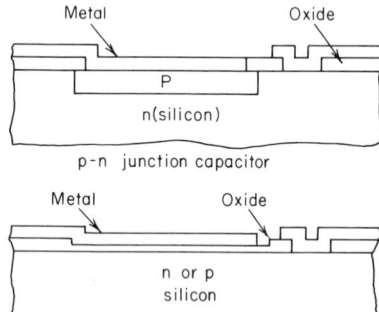

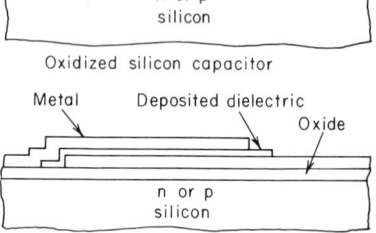

Fig. 15 Cross section of p-n junction thin-film capacitor. (*Courtesy of A. Lesk, Motorola Semiconductor Products, Inc.*)

For example: for a 0.5-ohm-cm n-type silicon:

Voltage, volts	Capacitance, pf/mil²
−1	0.2
−6	0.085
−10	0.06

To obtain large values of capacitance requires large areas: i.e., 100 pf with 6 volts takes 1,200 mil². A cross section of a p-n junction thin-film capacitor is shown in Fig. 15. The p-n junction is essentially a p-n junction diode, the capacitance of which is utilized to form the capacitor. The p-n junction must be maintained at reverse bias for realization of capacitance. Metal contacts to the electrodes of the capacitive elements continue over the passivating oxide to the other parts of the circuit. The center sketch in Fig. 15 is an oxidized silicon capacitor which uses the silicon for one electrode, a metal film for the other electrode, and silicon oxide for the dielectric. Therefore the terminology MOS. The oxide is normally much thinner than for average capacitor fabrication. The bottom sketch in Fig. 15 is a thin-film capacitor formed entirely on top of a passivating oxide layer. The bottom and top electrodes are metal films, the dielectric is a glass film deposited on the lower electrode, and the upper electrode is deposited last. One figure of merit for a capacitor is the capacitance per unit area, as was mentioned previously. The thinner the dielectric, the higher the capacitance per unit area. For instance, silicate dielectrics are preferred, as can be seen by the following data:

Material	Capacitance per unit area	Breakdown voltage	Usable range of values in typical integrated circuits, pf
Silicate	0.25 1.0	50 20	Less than 100 Less than 400

In a MOS capacitor the silicon substrate is the lower electrode. An n-type epitaxial layer with a shallow n$^+$ diffusion may be used. The n$^+$ region may be isolated by either channel diffusion around the n region or a p-type base diffusion under the n$^+$. The thermally grown SiO_2 dielectric is 500 to 1000 Å in thickness. The capacitance per unit area varies from 0.2 to 0.4 pf/mil^2, with a breakdown of 20 to 40 volts. The parasitic capacitance which is caused by the p-n junction formed between the lower silicon electrode and the isolation region can be appreciable. Unlike the diffused capacitor, the MOS capacitor is not voltage-dependent.

In summary, the critical features of a capacitive element are: a pinhole-free continuous layer for both electrodes and dielectric, high dielectric constant, low loss factor at the desired frequency, ease of making a low-loss compatible contact, temperature of formation (especially if active elements or easily alloyed metals are already in the circuit), and the ability to resist cracking from thermal stressing. Table 11 has been prepared for reference, but again one must bear in mind that the electrical characteristics of thin-film elements are highly dependent on the particular deposition techniques employed.

Of the three most commonly used dielectric materials, silicon dioxide, alumina, and tantalum oxide, the following characteristics may be expected:

Characteristic	SiO_2	Al_2O_3	Ta_2O_5
Maximum capacitance per unit area, pf/mil	0.25–0.4	0.3–0.5	2.5
Maximum value, pf	500	1,000	5,000
Maximum voltage	50	20–50	20
Q at 10 MHz	10–100	10–100	Good
Voltage coefficient	0	0.5	0
Dissipation factor at 1 MHz	0.7	0.5	5
Tolerance, %	±20	±20	±20

Thin-film Resistors

Thin-film resistors are finding increasingly wide application in compatible integrated circuits to satisfy performance requirements difficult or unachievable with diffused resistors. It is not within the scope of this survey to discuss diffused resistors, but for the sake of comparison a few points will be mentioned. Thin resistive films several times higher than the typical 200 ohms/sq base-diffusion sheet resistivity are readily available. Table 12 shows a few comparative values. A cross-sectional view comparing diffused and thin-film resistors is shown in Fig. 16.

Parasitic capacitance is a very important consideration in the design of any circuit. In thin-film devices the parasitic capacitances are directly proportional to the area, whereas the parasitic capacitance connected with diffused resistors are both area- and voltage-dependent. The primary effect of parasitic capacitance on resistors is to limit their frequency response. For example, a diffused resistor of 200 ohms/sq

TABLE 11 Characteristics of Thin-film Capacitor Materials[28]

Material	How formed	Dielectric constant at 25°C	Breakdown voltage above 1,000 Å, kv/cm	Dissipation factor, Hz at 25°C
SiO	Evaporated from powder with Al electrodes	6.0	1.6	0.025×10^3
Ta_2O_5	Anodize sputtered or evaporated, Ta with Ta, Au, or Al electrodes	....	....	0.008×10^3
SiO_2	Thermal oxidation			0.7×10^6
Al_2O_3	Anodize evaporated Al with evaporated counterelectrode or decomposition of alumino-organic into Ni-covered Al	....	8(9)	0.5×10^6 0.01×10^3
$BaTiO_3$	Evaporated	Approx 1,000		
ZnS	Evaporated	8.2	0.2	
MgF_2	Evaporated	6.5	2	
TiO_2	Oxidized evaporated Ti	....	....	0.004×10^3

fabricated from 0.5-ohm-cm silicon has a capacitance of 0.66 pf for 1 mil² geometry and 2.7 pf for 2 mil² geometry at a reverse bias of 1 volt. A 1-mil-wide resistor has a typical cutoff frequency of 19.3 MHz, and similarly the 2-mil side resistor will give a frequency cutoff approximately 75 percent lower. More heavily doped substrates (0.1 ohm-cm) yielding 200 ohms/sq have a capacity range of 1.35 and 4.7 pf for 1- and 2-mil² geometry, respectively, at 1 volt reverse bias. The same

TABLE 12 Comparison of Thin-film and Diffused Resistors*

Property	Thin film	Diffused
Sheet resistance, ohms/sq............	Nichrome 400 up Tantalum 400 up Cermets 1,000 up	200
Temperature coefficient of resistance, ohms/sq........................	50–250	200
Parasitic capacitance, ppm/°C........	Directly proportional to area	Area- and voltage-dependent
Geometry.......................	Less line width, parasitic capacitance reduced as square of line width	
Fabrication tolerance...............	15% untrimmed	20%

* Cross-sectional view comparing diffused and thin-film resistors is shown in Fig. 16.

resistors read 0.64 and 2.3 pf for a reverse bias of 10 volts. In the same fashion, the frequency response of the more heavily doped substrates falls to one-half value for the same area. For thin-film resistors in the range of 0.01 pf/mil², 200-ohm resistors of 1- and 2-mil widths have a capacitance of 0.1 and 0.3 pf, respectively. Going to 400 ohm/sq, the values drop to 0.06 and 0.2 pf. The cutoff frequency for a 200-ohm/sq and 1-mil-width-geometry thin-film resistor is roughly 100 MHz, but increasing the geometry to 2 mils drops the frequency 75 percent. It can be seen from the equation of cutoff frequency

$$f_{co} = \frac{2RC}{2}$$

that only one-half the total parasitic capacitance is considered. The tolerance of thin-film resistors is process-dependent and is largely determined by the degree of accuracy in the control of sheet resistance. A good deposition system is normally capable of providing sheet resistivities that vary less than 1 percent across the substrate and to within a couple of percent from substrate to substrate. Sheet resistance must be kept within tolerances during any further processing. A critical step is lead attachment which subjects the device to elevated temperatures. Under certain conditions the changes in sheet resistance due to further processing at elevated temperatures can be prevented by depositing a silicate glass passivating layer over the resistors before the heat cycles. Dimensional tolerances and aspect ratios are also critical. In a one-mil-wide resistor maintaining a tolerance of 5 percent, the width must not vary more than 0.05 mil or approximately one micron, to hold a tolerance of 1 percent in absolute resistance. Five percent tolerances are

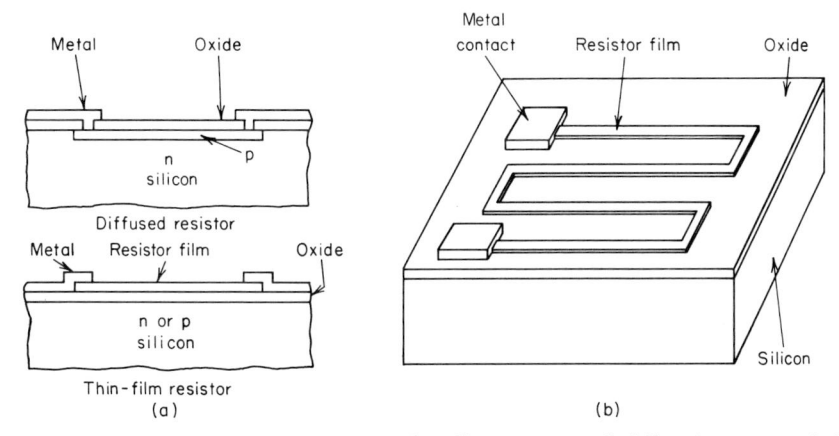

Fig. 16 Cross-section view comparing thin-film resistor and diffused resistor. (a) Cross-sectional view comparing diffused and thin-film resistors. (b) Typical configuration of a thin-film resistor. (*Courtesy of A. Lesk, Motorola Semiconductor Products, Inc.*)

not practically attainable; 10 percent is more realistic. Some rules to keep in mind in designing resistors are:
1. As large a tolerance as practical should be allowed.
2. The resistor should be designed for the lowest values possible.
3. Large contact areas should be used.
4. Spacing between contacts should not be less than one mil.
5. Ends and corners in the geometry should be avoided.
6. Subsequent process temperatures should be held to a minimum.

To meet the demand for deposited thin-film resistors which are compatible with the other circuit elements in the method of deposition and suitable electrical characteristics means that resistors having low temperature coefficients and high stabilities must be deposited on flat substrates to a close tolerance and covering a wide range of values. Many graphs and charts, such as those in Fig. 17 have been developed to eliminate tedious calculations for thin-film resistor design.[29] Table 13 shows the resistive lengths for given constants (ohms per square and milliwatts per square inch).

The resistance of thin-film resistors is determined by the dimensions of the resistive material (approximately 200 Å) deposited on a substrate. The resistive constant is the measured resistance of a geometric square (length equals width), regardless of the actual dimensions, deposited at a constant thickness, and is expressed as ohms per square. Since thickness is constant, resistance is length times width. The

11-26 Thin Films

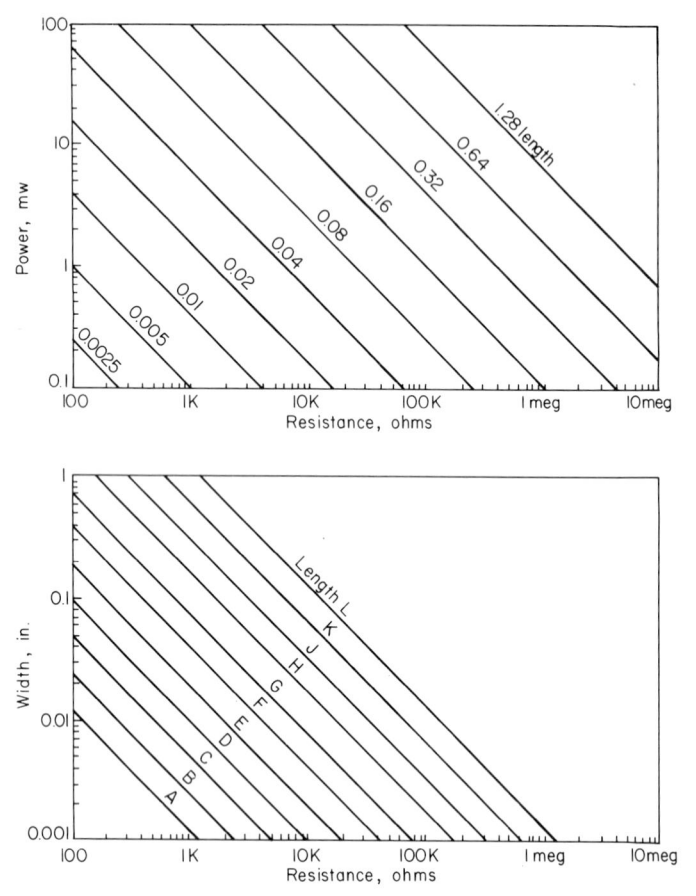

Fig. 17 Resistance is shown as a function of power in upper graph and as a function of resistor width in lower graph.[29]

power dissipation is determined by the resistor area and is an experimentally determined constant. At the present state of the art, resistance and power-dissipation calculations can be tedious. Referring again to the work of Cook,[29] the length

TABLE 13 Resistive Lengths for Given Constants

Constants		Length, in.									
ohms/sq	mw/in.²	A	B	C	D	E	F	G	H	J	K
125	32,000	0.01	0.02	0.04	0.8	0.16	0.32	0.64	1.28		
250	16,000	0.005	0.01	0.02	0.04	0.08	0.16	0.32	0.64	1.28	
500	8,000	0.0025	0.005	0.01	0.02	0.04	0.08	0.16	0.32	0.64	1.28
1,000	4,000	0.00125	0.0025	0.005	0.01	0.02	0.04	0.08	0.16	0.32	0.64
2,000	2,000	......	0.00125	0.0025	0.005	0.01	0.02	0.04	0.08	0.16	0.32
4,000	1,000	......	......	0.00125	0.0025	0.005	0.01	0.02	0.04	0.08	0.16

of a thin-film resistor required to obtain a desired resistance may be determined from

$$L = \left(\frac{RP}{K_r K_p}\right)^{1/2}$$

where L is the length in inches, R is the required resistance in ohms, P is the required power dissipation in milliwatts, K_r is the resistive constant in ohms per square, and K_p is a power constant in milliwatts per square inch. Length is determined by the equation, so that the resistor will be capable of the required heat dissipation. Because of the direct relationship between length and width in establishing the resistance of a thin-film resistor, width can be found from $W = LK_r/R$, where W is the width in inches. Table 14 and graphs 1 and 2 of Fig. 17 were constructed from the two equations.

TABLE 14 Resistance of Films One Unit in Length for Various Resistivities[30]

Resistivity, ohms/sq	25	50	100	150	200	500	1,000	1,400	2,000
Line width, units					Resistance, ohms				
0.006	4,166	8,333	16,666	25,000	33,332	83,333	166,666	233,324	333,333
0.007	3,560	7,120	14,240	21,360	28,480	71,200	142,400	200,360	284,800
0.008	3,140	6,280	12,560	18,840	25,120	62,800	125,600	175,840	251,200
0.009	2,790	5,580	11,160	16,740	22,320	55,800	111,600	156,240	223,200
0.010	2,500	5,000	10,000	15,000	20,000	50,000	100,000	140,000	200,000
0.011	2,275	4,550	9,100	13,650	18,200	45,500	91,000	127,400	182,000
0.012	2,080	4,160	8,320	12,480	16,640	41,600	83,200	116,480	166,400
0.013	1,920	3,840	7,680	11,420	15,360	38,400	76,800	107,520	153,600
0.014	1,780	3,560	7,120	10,680	14,240	33,600	67,200	99,680	134,400
0.015	1,665	3,330	6,660	10,000	13,320	33,300	66,600	93,240	133,200
0.016	1,565	3,130	6,260	9,390	12,520	31,300	62,600	87,640	125,200
0.017	1,470	2,940	5,880	8,820	11,760	29,400	58,800	82,320	117,600
0.018	1,385	2,770	5,540	8,310	11,080	27,700	55,400	77,560	110,800
0.019	1,315	2,630	5,260	7,890	10,520	26,300	52,600	73,640	105,200
0.020	1,250	2,500	5,000	7,500	10,000	25,000	50,000	70,000	100,000
0.021	1,170	2,340	4,680	7,020	9,360	23,400	46,800	65,520	93,600
0.022	1,135	2,270	4,540	6,810	9,080	22,700	45,400	63,560	90,800
0.023	1,085	2,170	4,340	6,510	8,680	21,700	43,400	60,760	86,800
0.024	1,040	2,080	4,160	6,240	8,320	20,800	41,600	58,240	83,200
0.025	1,000	2,000	4,000	6,000	8,000	20,000	40,000	56,000	80,000
0.050	500	1,000	2,000	3,000	4,000	10,000	20,000	28,000	40,000
0.100	250	500	1,000	1,500	2,000	5,000	10,000	14,000	20,000

Film resistivity is measured in ohms per square and is the dc resistance measured between opposite faces of a square. The advantage in using ohms per square to specify resistance is that the measurement is independent of the size of the square. A uniform film of given thickness will have the same resistance per square centimeter as per square inch. Ohms per square resistivity is controlled by the thickness of the film. Table 14 shows unit resistance values of line widths for 25 to 2,000 ohm/sq films.

Temperature coefficient of resistance of ideal films is zero from −55 to 150°C. Metals have high TCR but can be made usable by alloying or metal insulator mixes. The voltage coefficient of resistance is defined as fractional changes in resistance versus voltage. This is usually of negligible concern. Of great importance in resistors is stability. Resistors are rated as percentage change in resistance over a period of time at elevated temperatures (70 to 100°C) under electric load. Any change can be due to:

1. Mismatch in linear expansion between the film and the substrate
2. Contamination of the film from the substrate
3. Reaction of the film and atmosphere
4. Reaction of the film with surrounding material
5. Structural changes in the film

Nickel chromium resistive films The most commonly used thin-film resistor material is evaporated nickel-chromium alloys, mainly because of low TCR. Through the years people have developed highly stable alloys for NiCr over normal working temperatures. NiCr may be deposited on glass or ceramic substrates. The NiCr alloy may be evaporated from a boat, but there is reason to believe that this technique can result in film contamination. For greater purity the alloy is sublimed from heated coils in a vacuum of about 10^{-5} torr. The substrate is heated to 300°C during deposition. For even greater stability, the NiCr film is heated at 300°C in air for a short period of time (15 to 30 min). The sublimation technique for evaporating NiCr-resistive films is described in detail by Manfield.[31] Briefly, a solid source, usually a wire, is heated to just below its melting point. The source tends to deplete; the first films are chromium-rich. As an alternative, a flash-evaporation technique can be used. The constituents of the alloy are mixed in correct proportions, and the mixture is fed into the evaporator at specific intervals. The temperature is such that the material evaporates instantly, and as the vapor is a constant, composition of the films are homogeneous. Techniques for depositing thin-film Nichrome* resistors are a matter of choice. Of great importance, however, are substrate temperature, source temperature, vacuum, deposition rate, and type and size of the source. The typical characteristics are:

Sheet resistivity.......................... 100–300 ohms/sq
Temperature coefficient of resistance........ 5–100 ppm per °C
Stability............................... <0.5% per 1,000 hr at full load and 70°C

NiCr resistors must be protected against high humidity. It has been reported that very high sheet resistivity NiCr resistors can be produced by depositing nickel-chromium on alumina substrates.[32] After deposition, the film surface is roughened to remove high spots and to produce a scratched surface. There are not many substantiating data on the process.

Tantalum and tantalum oxide films Of the refractory metals, tantalum is one of the most suitable materials. To increase the resistance values, tantalum resistors can be trimmed by anodic oxidation. Sputtering is one of the most popular techniques for depositing Ta films.[33–35] Vacuum pressures of 10^{-5} torr are reached, and argon is admitted to 10^{-2} torr. The tantalum source is the cathode, and the substrate holder is the anode. When a few kilovolts are applied, the argon ionizes. The argon ions hit the tantalum cathode with sufficient energy to remove Ta atoms which diffuse across to the substrate. The resulting films are homogeneous in thickness and resistivity. The Ta films are stabilized by oxidation at 250 to 300°C. In addition to stabilizing the films, the oxidation process also increases sheet resistance by decreasing the thickness of Ta by conversion to TaO.

It has been reported that there is no direct correlation among a uniform film thickness, decrease of conduction layer, and oxidation of the film.[36] A substantial

* Trademark of Driver-Harris Co., Harrison, N.J.

improvement in sheet resistance and TCR was reported for Ta films deposited in air at 10^{-5} torr, then heat-treated 8 hr at 200°C, and anodized as shown in the following data:

Resistor	Before heat treatment	After heat treatment
	Resistance change	
1	611.0	706.0
2	434.0	494.3
3	471.0	520.6
	Temperature coefficient of resistance, ppm per °C	
1	−390	−430
2	−410	−460

Heat treating in air, however, tends to produce unstable films. Attempts to improve this situation have been made by incorporating gold doping[38,39] and by nitriding during deposition.[40,41] Unfortunately the usable sheet resistance of Ta is relatively limited.

Chromium-resistive films Chromium films have been vacuum-deposited on glass substrates and operated at 300°C. The TCR varied from large negative values for thin films to large positive values for thick films. Typical values are ±100 ppm per °C for 200 ohms/sq. A comparison of resistivities of films evaporated at pressures of 10^{-5} to 10^{-9} torr show about a 50 percent higher resistivity for the films deposited at the higher pressures, while increasing substrate temperature from 200 to 300°C reduced resistivity to one-third.[42] This was explained as film formation of grains of relatively pure Cr and impurities at their boundaries, and the grain size and impurity content were dependent on substrate temperature and deposition pressure. Subliming chromium from a chromium-plated tungsten wire onto a glass or ceramic substrate results in widely varied TCR for different deposition rates. For example, for a film of 300 ohms/sq at a TCR of −300 ppm per °C at a deposition rate of 2 Å/sec the temperature coefficient was zero; whereas when the deposition rate was increased to 18 Å/sec, there was a 10 percent decrease in resistivity at 25 MHz. Grain size increases with increasing substrate temperature. The results of the above-referenced study concluded that the best deposition conditions for chromium films are a substrate temperature of 270°C and a deposition rate of 210 Å/sec.

Low-sheet-resistivity films have been produced by alloying a small amount of nickel with the chromium.[44] Films were deposited by evaporating a mixture of Cr and Ni powders plus a small amount of chromium oxide. Typical results were 5 to 50 ohms/sq and a TCR less than 20 ppm per °C at 30 ohms/sq. Chromium resistors ranging from several hundred to several thousand ohms were made by ion-beam deposition.[45] In this technique the chromium is vaporized, and the vapor is ionized by electron beam. A collimated beam of Cr ions is extracted and directed toward the substrate. Reported values of 0.007- by 0.030-in. resistors were 5,000 ohms and had a TCR of 25 ppm per °C.

Titanium and titanium oxide films Recently, progress has been made in depositing titanium and titanium oxide thin-film resistors.[46,47] Such films are evaporated at 10^{-5} torr, but must be deposited at high rates to minimize contamination. By heating a Ti strip source and a Ta supporting boat in parallel, a 4000 Å film can be deposited in 15 sec. The film is anodized chemically. Depending on the degree of anodization, sheet resistivity of $1-10^9$ ohms/sq is possible with a TCR from −1,000 to +1,000 ppm per °C. Table 15 shows the structure and electrical properties of reactively sputtered Ti and TiO films prepared in a mixed gas of argon and oxygen at room temperature. Because of the difficulty in controlling the Ti and TiO films and their sensitivity to oxygen, the TiN films proved more advantageous. A high deposition rate is obtained when nitrogen gas pressure ranges from 10 to 50 percent of the argon gas pressure.

TABLE 15 Structure and Electrical Properties of Reactively Sputtered Titanium and Titanium Oxide Films[46,47]

Sample No.	Partial nitrogen pressure, torr*	Deposition rate, Å/min	Film thickness, Å	Specific resistivity, microhm-cm†	Temperature coefficient of resistance, ppm per °C†	Structure
1	$<1 \times 10^{-6}$	120	1500	200	+300	Crystalline Ti
2	7×10^{-5}	65	1500	300	+90	
3	1×10^{-4}	60	1500	300	−20	Crystalline TiN
4	3×10^{-4}	40	1000	290	−120	

* Total pressure (argon + nitrogen): 6×10^{-4} torr.
† Bulk value of Ti cathode: 53 microhm-cm + 4,150 ppm per °C.

Cermet films Cermet films may be defined as a mixture of a metal–silicon monoxide. These films may be coevaporated by flash evaporation techniques. A common type of cermet is chromium–silicon monoxide. The composition of the film is determined by the proportion of the ingredients, control-fed to a heated tantalum strip. The tantalum strip is held at temperatures well above the evaporating temperature of chromium and silicon monoxide so that the mixture evaporates instantly and the vapor has the same composition. Figures 18 and 19 show the

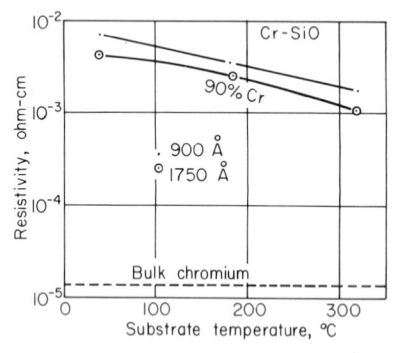

Fig. 18 Chromium–silicon monoxide resistivity versus substrate temperature.[50]

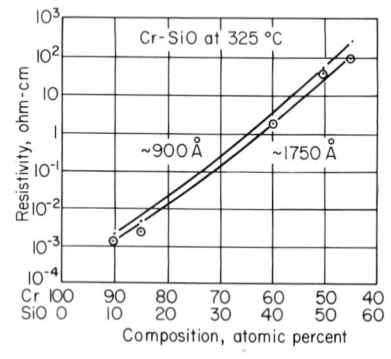

Fig. 19 Chromium–silicon monoxide resistivity versus composition.[50]

resistivity of this particular cermet versus substrate temperature and composition, respectively.

Temperature coefficients are highly negative for low chromium content. This type of cermet film also has the properties of hardness, good adhesion, uniformity, and stability, as reported by Beckemen and Thun.[49]

As a comparison of the various resistive films discussed, a reference chart has been compiled and is shown on page 11-31.

Material	Method of deposition	Resistivity, ohms/sq	Temperature coefficient, ppm per °C	Stability, % per 1,000 hr
Ni-Cr	Sublimation	100–300	+100	0.2
Ta	Sputtering	50–600	±150	1.0
TaN	Sputtering	10–100	−60	1.0
SiO-Cr	Flash evaporation	600	−100	0.5
Re	Electron bombardment evaporation	100–800	+150	0.1

In using the above values, it must be considered that the electrical properties of thin films are very deposition-, technique-, and parameter-sensitive. A very complete list was compiled by Jones[50] and appears in Table 16.

TABLE 16 Electrical Properties of Thin Films[50]

Films	Substrate	Resistivity, ohms/sq	Resistance range, ohms	Maximum temperature coefficient, ppm per °C	Maximum operating temperature, °C	Stability (loaded), %/hr/°C
Vacuum-deposited:						
Nickel-chromium (80/20)	Glass	100–300	10–15 × 10³	100	100	0.2/2,000/70
	Ceramic	100–1,000	10–1 × 10⁶	100	150	1.0/7,000/70
	Ceramic	1–10 × 10³	....	100	...	0.4/2,000/70
Tantalum	Glass	50–600	10–1.5 × 10⁶	±150	...	1.0/1,000/100
	Glass	4,000	....	−150	...	
Tantalum-gold	Glass	50	....	±50	...	0.25/1,700/150
Tantalum-nitride	Glass	10–100	10–300 × 10³	−60	...	1.0/1,000/150
Chromium	Glass	200–1,000	....	±100	300	
Chromium-nickel (95/5)	Glass	5–50	....	±20	...	
Titanium	Glass	50	....	Near zero	...	0.5/1,000/65
Stainless steel	Glass	100	....	200	100	0.2/1,000/100
Cermet (Cr-SiO)	Silicon monoxide	250	....	−50	300	0.2/1,000/200 2.0/1,000/300
	Ceramic	2–20 × 10³	....	±100	...	0.2/2,000/125
Silicon	Quartz	200	....	−200	600	
Chromium-silicon	Glass	100–10 × 10³	....	±500	250	1.0/1,000/250
Germanium	Glass	5,000	....	2,000	300	1.0/yr/70
Tungsten	Glass	200–600	....	Near zero	125	10 ppm/hr/125
	Glass	2,000	....	−200		
Platinum	Quartz	2	....	200		
Ruthenium	....	30	....	600		
Osmium	....	50	....	250		
Niobium	Quartz	50	....	150–300		
Vanadium	....	150–300	....	−450 to −780		
Zirconium	....	300	....	1,000		
Molybdenum	Quartz	100	....	350		
Rhenium	....	100–800	....	50–150		
Tungsten	....	100–300	....	50–200		
Chemically deposited:						
Metal oxide (SnO-SbO)	Glass	10–400	10–2 × 10⁶	±250	150	0.5/2,000/70
	Ceramic	....	10–500 × 10³	±250	150	1.0/2,000/70
Conductive glaze (Pd-PdO-Ag)	Ceramic	10–100 × 10³	50–500 × 10³	±250	500	5/1,000/70
Vishalloy (Cr-Ni)	Glass Plastic	0.6	15–500 × 10³	±0.5	125	0.02/yr/70
Nickel-cobalt	Ceramic	1–10 × 10³	40–360 × 10³	±30	70	0.2/2,000/70
Gold-platinum	Glass	1–100	10–3 × 10⁶	350	200	0.1/long term
Cracked carbon	Ceramic	10–10 × 10³	10–22 × 10⁶	−500	100	0.2–2/1,000/70

Interconnections

Interconnections of circuit elements produced on silicon wafers in combination with compatible thin-film elements over an oxide dielectric may be accomplished by a metal evaporation. The evaporation may be patterned by the use of a metal

mask or by photolithography. Of primary concern is that the interconnection material:
1. Make ohmic contact to the active devices.
2. Be easy to apply.
3. Be available in high-purity form.
4. Lend itself to production techniques.
5. Be nonreactive with all materials with which it is in contact.
6. Not creep or disappear, resulting in hot spots and open circuits.
7. Not change electrical characteristics owing to storage at high temperatures.
8. Minimize electrolysis due to alkali ions or humidity.

Properties of metal film conductors or interconnections are shown in Table 17.

TABLE 17 Properties of Metal Film Conductors for Interconnects

Metal	Melting point, °C	Volume conductivity at 20°C	Specific heat, cal/(g)(°C)	Adherence quality to SiO$_2$	Etching quality	TC bonds with Au wire	Other pertinent information
Silver	961	108.5	−2.59	Poor	Good	Feasible	Forms solid solutions with gold
Copper	1083	100	−34.9	Fair	Good	Feasible	Forms solid solutions with gold
Gold	1063	77.7	+39	Poor	Excellent	Feasible	Eutectic with Si at 377°C
Aluminum	660	61.2	−376(Al$_2$O$_3$)	Excellent	Excellent	Feasible	Brittle AuAl$_2$ with Au
Magnesium	650	38.7	−136.1	Very good	Good	Feasible	Chemically reactive
Rhodium	1906	38.3		Poor	Good	Feasible	High cost
Tungsten	3410	32.5	−182.5(WO$_3$)	Very good	Good	Difficult	Alternative for Mo
Molybdenum	2625	30.9	−162(MoO$_3$)	Very good	Very good	Difficult	Slight mutual solubility with Au
Cobalt	1495	27.6	−51(CoO)	Good	Good	Difficult	Difficult to deposit
Nickel	1455	25.2	−51.7	Good	Good	Difficult	Forms solid solution with Au
Platinum	3224	16.3		Poor	Good	Feasible	Expensive
Palladium	1554	20.0	−52.2	Poor	Good	Feasible	Expensive
Chromium	1890	13.8	−250(Cr$_2$O$_3$)	Excellent	Good	Difficult	Au solubility ∼20 wt % at 900°C
Tantalum	2850	13.2	−471(Ta$_2$O$_5$)	Excellent	Excellent	Difficult	High resistance
Lead	621	11.1	−45.3	Poor	Fair	Difficult	
Vanadium	1860	6.5	−271(V$_2$O$_3$)	Excellent	Fair	Difficult	Solution in Au 14 wt % at 500°C
Zirconium	1750	4.2	−244(ZrO$_2$)	Excellent	Good	Difficult	
Titanium	1820	3.2	−204(TiO$_2$)	Excellent	Excellent	Difficult	Au solution 7.8 wt % at 700° C

Table 18 lists three metallization systems currently used for integrated circuits and indicates typical thickness ranges.

Aluminum is by far the most commonly used material for the metallization of silicon integrated circuits. Typically a layer on the order of 1 micron in thickness is deposited by vacuum evaporation. Aluminum adheres well to the SiO$_2$ surface of metal-overlay devices and integrated circuits, in addition to affording good ohmic contact with both degenerate p-type and n-type silicon. It is easily patterned

TABLE 18 Commonly Used Metallization Systems

Metal	Thickness, microns	Metal-silicon contact
Al	0.5–1.5	Al-Si
Mo-Au	0.3 Mo, 0.7 Au	
Ti-Pt-Au	0.15 Ti, 0.35 Pt, 12.5 Au	Mo-Si, Al-Si, Pt-Si, Pt$_5$Si$_2$-Si

and contacted with gold or aluminum wires. Aluminum does, however, have certain limitation, some of which are listed below:
1. At elevated temperatures, aluminum and gold interact to form various intermetallic compounds which increase the contact resistance and can result in open bonds.

2. When a high-density direct current is passed through an aluminum interconnection strip, mass migration of the aluminum toward the positive terminal may be observed.
3. Above 500°C the aluminum attacks the underlying dielectric.
4. Aluminum is difficult to use with MOS devices.
5. Although aluminum is inherently about as conductive as gold, it is very sensitive to oxygen or water vapor in the vacuum system during deposition and so, generally, has a higher sheet resistance.

It might be pertinent at this time to mention that it is impossible to discuss interconnection techniques and materials without also considering packaging. The two must be compatible and therefore will be discussed in these terms. As aluminum is the most natural choice of interconnection material, it seems advantageous to discuss in greater detail the limitations that are listed above. At elevated temperatures for extended periods of time, aluminum interconnections and gold bonding wires interact to form various intermetallic compounds which increase the contact resistance and can result in open bonds. This effect occurs rapidly at 450°C and can be seen in a few hours at 300°C. This reaction can even occur at room temperature over an extended period of time. This interaction may be eliminated at the bonding pad by using aluminum wire bonding. The terminals of the package can be aluminum-clad Kovar,* thus eliminating gold-aluminum interaction throughout the device. When a high current density (above 10^6 amp/cm^2) is passed through an aluminum interconnection, mass migration of the aluminum occurs at a slow rate. If the temperature is elevated, the migration rate is greatly accelerated, to the point where open circuits can occur in a few hours. This appears to be a fundamental limitation to the application of aluminum to high-current-density applications. Above 500°C the aluminum attacks the underlying SiO$_2$ rapidly and will short-circuit to the silicon substrate of the integrated circuits if held at these high currents for any length of time. Even at temperatures of about 300°C over long periods of time aluminum films can nodularize, giving some very high crystalline spots and depleting others. This again results in openings in fine-line metallization and burnouts in thin-film capacitors or multilayer metallization structures. Although satisfactory MOS devices can be made with aluminum interconnections, it is virtually impossible to obtain, in combination, all the most desirable MOS properties. It is necessary to compromise among adhesion, threshold potential, oxide bulk stability, oxide surface stability, and cost. Where thick metallization is required, patterning aluminum into fine lines is difficult; whereas gold may be patterned as a thinner film and subsequently plated to increase the thickness and conductivity.

Chromium-gold metallization has been used as a substitute for aluminum in the metallization of silicon integrated circuits. Chromium adheres well to glass layers and will provide ohmic contacts to both p- and n-type silicon if the surface regions of the contact areas are degenerately doped by appropriate diffusions. Two deposition methods are used: thermal evaporation and low-energy inert-gas sputtering. In each method it is possible to blend the two metals to form a continuous interface between the two metals. The substrate is heated to 250°C before the deposition of chromium from a basket filament containing a chromium charge. The gold evaporation follows immediately with no appreciable cooling. The pressure is maintained at less than 10^{-5} torr, and the substrate is cooled to less than 100°C before exposure to atmosphere. Typical values of resistivity at various thicknesses of chromium and gold are shown in Table 19.

One of the principal degradation mechanisms in the Au-Cr system is diffusion of chromium into the gold, which decreases the electric conductivity. There can also be a problem of chromium diffusing through pinholes in the gold film.[51-53] Molybdenum-gold systems are more stable at elevated temperatures than the chromium-gold systems; but the molybdenum is sensitive to combinations of elevated temperatures and high humidities. This is not necessarily a limitation where a glass overlying passivation is employed. A standard vacuum-deposition cycle is used to

---

* Trademark of Westinghouse Electric Corporation, Pittsburgh, Pa.

TABLE 19 Resistivity of Evaporated Chromium-Gold Metallizations

Thickness of chromium, Å	Thickness of gold, Å	Total thickness, Å	Resistivity, microhm-cm
<100	4292	4292	3.6
<100	6300	6300	2.7
350	3770	4120	4.7
625	4850	5475	5.9
665	3535	4200	3.7
665	5400	6065	5.1
700	4330	5030	7.2
860	4330	5290	5.4
900	5040	5940	6.3
1900	5160	7060	10.2
2700	3500	6200	10.4
3000	3650	6650	13.2

deposit molybdenum-gold. The substrate is heated to 260°C, and about 1000 Å of molybdenum is deposited, followed by about 300 Å of gold. The substrate is cooled to less than 100°C, and approximately 3,000 Å of additional gold is deposited. The pressure is maintained at less than 10^{-5} torr, and substrates are cooled to less than 40°C before the system is opened to atmosphere. Typical thickness-versus-resistivity values are listed in Table 20.

TABLE 20 Resistivity of Evaporated Molybdenum-Gold Metallizations

Mo thickness, Å	Total thickness, Mo + Au, Å	Resistivity as deposited, microhm-cm vs. molybdenum thickness
<100	2800	2.7
360	2661	2.9
330	6880	3.9
390	2668	4.0
750	3000	4.1
840	2800	4.3 (320°C substrate)
1200	3810	4.1 (320°C substrate)
1400	4100	4.1
2080	3380	4.9

Films of gold and nickel-chromium alloy are prepared by evaporating Ni-Cr alloy to a thickness of about 0.1 micron, followed by the evaporation of an overlay of 0.1 micron of gold. Aging the films in dry nitrogen at temperatures of 200 to 400°C shows a large increase in resistance in 24 hr. There is no indication in the literature that a compound is formed in the gold-chromium system, and the solubility of either gold in chromium or chromium in gold is not indicated by the literature sources. It has been noted that some alloying occurs at 150°C in evaporated nickel films over gold.

A 0.3-micron layer of lead evaporated onto a substrate, followed by a 0.3-micron layer of gold, and aged at 275°C, indicates no compound formation. A significant increase in resistance occurs, however, because at about 275°C a small percentage of lead is soluble in gold.

The gold-titanium metallization system consists of a conductive upper layer of gold and a reactive underlying layer of titanium, which adheres well to most types of substrates. There are several compound formations at relatively low temperatures; those compounds lack the desirable properties of the original metals. In the application of gold-titanium for beam lead techniques, a barrier of platinum is placed between the gold and the titanium. This is due to the large changes in resistivity in the gold-titanium system on silicon and silicon dioxide aged at 250 and 350°C.

In the aluminum-tantalum system there is definite compound formation. Electron-diffraction studies have shown a small amount of the mixing of aluminum and tantalum at 100 to 150°C, but gross mixing at 250 and 350°C. The formation of compounds results in increased sheet resistivity.

The need for a technique for connecting a number of circuit elements to one another and to other circuits and components is obvious. The aluminum metallization system with thermocompression gold bonds or ultrasonically bonded aluminum wire is still the mainstay of the industry. Other methods have been suggested. Attempts to make a planar array by filling the gaps between chips with glass frit and firing were not very encouraging, but they do permit the use of evaporated and etched metallization. The flip-chip technique which is now reasonably well developed involves turning the circuit upside down and attaching it to mating pads on an interconnection substrate. Joining is accomplished by heating until the joining alloy melts and wets the attachment pad. The method has the disadvantage that joints cannot be readily inspected and its heat-transfer characteristics are inferior to those of a method in which the whole circuit is attached to the substrate.

Any method of making interconnections between elements must meet a large number of requirements. Some are:

1. Method must be adaptable for batch processing.
2. Temperature and chemical processes are limited.
3. Critical test points must be available.
4. Final product must have high resistance to the effects of stress environments such as heat, vibration, and moisture.
5. Heat transfer should be good, and the mounting substrate itself should be a good heat conductor.
6. Process should be inexpensive and lead to high yields.
7. Crossing capacitances and stray capacitance should be low.
8. Resistance in conductors should be minimized.

There are three general geometrical approaches to the problem:

1. Provide a surface level with the surface of the elements on which the conductors can be fabricated.
2. Interconnection array can then be fabricated and elements attached, such as beam lead or flip-chip.
3. A prefabricated substrate with recesses to hold the attached elements can be used.

The approach to providing a level surface upon which to deposit conductors in insulation, if necessary, has the advantage that the conductor deposition and patterning can be done in a batch manner by standard thin films and semiconductor techniques. Precision positioning of the elements would be required. Methods to do so are the deposition of a thick insulating layer over the surface, placing elements facedown against a flat surface and filling gaps between by plastic or molten glass. Any of these choices presents difficulties.

The concept of flip-chip is not new. In the early days of thin-film circuits, many experimenters tried to place individual uncased transistors. facedown against a matching thin-film wiring pattern. In this technique multiple thermocompression bonds were required. At the time, surface protection of semiconductor devices was inadequate or nonexistent and thin-film components suffered from stability problems. Recently techniques have been improved, and metallization and patterning for both the interconnection substrate and the active substrate are done by vacuum-evaporating a metal, followed by the application of photoresist patterns and etching. In the preparation of the active substrate, the planar technique is used. After the devices have been formed, the conductive pattern and thin-film resistors can be formed by vacuum evaporation and etching. At this point, the circuits have become vulnerable to temperatures above 400°C, and all subsequent processing must be done below this temperature. Connection pads can be defined by photoresist and plated up to the desired thickness. A thick coating of insulation is deposited and etched to expose the contact pads. The wafer is then immersed in a dip-solder pot to form the contact solder bumps at each interconnect pad.

The interconnection substrate should be economical, readily available, reasonably smooth, resistant to thermal shock, and have good thermal conductivity, high mechanical strength, and compatibility with metallizing systems. Glazed alumina meets these requirements. Photoetching is employed to define the interconnections and resistor patterns. The photoetching process consists of sequentially vacuum-depositing the resistive (cermet) and conductive (permalloy, copper, gold) films on the substrate and selectively etching the conductor and resistor patterns. The masked circuit-fabrication process consists of sequentially depositing the resistive and conductive films through different metal masks.

The addition of uncased active and passive devices to thin-film circuits to produce thin-film hybrid circuits is a popular concept. Compatible joining procedures require reliable diffusion (thermocompression) bonds to be made not only to device pads but to thin-film circuitry as well. Present-day technology utilizing photoetch techniques can achieve thin-film conductors as small as a few microns.

There is no straightforward solution to the interconnection material problem. Much work is currently being done to remedy the situation.

RELIABILITY

The following thoughts have been advanced to support claims for higher reliability in microelectronic circuits:[23]

1. Smaller size of microelectronic circuits allows better protection within constant weight and volume against hazards. The strength-to-weight ratio is increased.

2. Fewer connections of dissimilar materials have to be made in integrated circuits; therefore, there are less connection failures.

3. There is lower cost for microelectronic circuits, which may be traded for higher reliability by product improvement.

4. Coupling between components is stronger in microelectronic circuits—thermally, mechanically, chemically, and electrically because of their close proximity to each other. Therefore devices tend to exhibit identical changes in characteristics.

5. There is less handling of components in fabrication of integrated circuits; therefore, more uniform product.

6. Use of redundancy is less costly in integrated-circuit than in conventional device assemblies.

In addition, Keonjian points out a number of factors that are detrimental to reliability.[23]

1. Small components may be more difficult to fabricate than larger ones and, therefore, have more flaws.

2. Small production volume means less perfection in fabrication, and thus poor reliability because sufficient engineering may not be possible within the given cost limits.

3. Strong coupling between elements, though favorable in some respects, is unfavorable in others. A catastrophic failure would spread to higher circuit levels.

4. Small size means small inertia: thermal, mechanical, and electrical, and therefore sensitivity to dynamic overload (thermal, mechanical, or electrical).

5. New constructions may mean poor reliability initially until sufficient engineering has been done. This is temporary but is the main reason why microelectronics in its most advanced form has not yet found its way into space applications on any appreciable scale.

6. Excessive packing density may lead to higher operating temperature and, therefore, shortened life of components. Figure 20 and Table 21 represent this temperature problem.

Of concern now is to define the state of the art in reliability of integrated circuits to determine their present limitations as building blocks of complex but reliable electronic equipment, and to predict and explore possible future reliability improvements. Many test data are available, and it is still not possible to assign "a failure rate" to an integrated circuit. The data summarize results of tests performed under a great variety of conditions on a great many different devices. There is general agreement that temperature is a prime failure-accelerating factor, and yet many manufacturers do not apply an acceleration factor to their testing procedures. They determine failure rates at one or two temperatures and leave it to the customer to interpret these data. Where current or power is stress criterion, as is the case with resistors or transistors, derating results in longer life through a lower operating temperature as a result of less dissipated power. The established stress criterion for capacitors is the applied voltage or electric field, and derating is accomplished by reducing this voltage and field, resulting in a larger margin for transients and slower ionization effects. In an integrated circuit, lowering the supply voltage below a specific value may have little effect on the life of a certain capacitor, but it may cause the circuit to cease operating. If it is not possible to provide a simple stress definition for the circuit as a whole, we must ensure safe and conservative application by specifying parameters and test conditions.

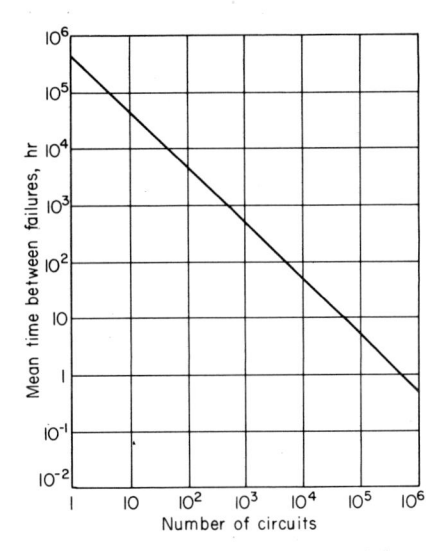

Fig. 20 Mean time between failures versus number of flip-flop circuits in system.[26]

It has been contended that failures are device-dependent rather than random, and that a certain operator producing good devices will probably produce nothing

TABLE 21 Typical Failure Rates of Flip-flop Circuit Devices[54]

Components	Number in circuit	Failure rate, % per 1,000 hr
Transistors, silicon...............	2	0.07
Diodes, silicon....................	2	0.02
Capacitors, paper.................	4	0.00125
Resistors, carbon composition......	8	0.0043

but good devices, and an operator not properly trained will probably produce unreliable devices. In other words, the human element is of utmost importance. Although most failures can be classified as to type and mechanism (chemical, physical, or mechanical), the primary underlying cause of failure is believed to be the human element. Metallization and interconnections today are the primary source of reliability problems. Many defects are scratches or smears caused by tweezer contact in handling. Occasionally there are shadowing effects during metal-

11-38 Thin Films

lization which cause differences in the thickness of the metallization across the circuit. Excessive differences in thickness will result in localized heating and eventually cause open or short circuits.

The overall reliability of integrated circuits is good. There are many people concerned with proving and improving integrated-circuit processing techniques and testing procedures and establishing comprehensive reliability data. To the circuit designer it is mandatory to know the best components to specify, the most meaningful parameters to designate, and what sort of reliability to demand.

INSTRUMENTATION FOR MICROELECTRONICS

There is a vast amount of instrumentation currently available for the chemical, physical, and metallurgical analysis of thin films. Tables 22 and 23 show the property measured by each type of instrument and how well the property is measured; some of the sensitivities were obtained from instrument makers' literature

TABLE 22 Sensitivities of Instruments[13]

Instrument	Property measured	Sensitivity, resolution or power
Optical microscope	Size and shape	3 to 2000X, 0.5 micron
Phase contrast	Stacking faults	Up to 1000X, 0.5 micron
Electron microscope	Size and shape	<10 Å or better
	Dislocation and stacking faults	100 Å apart
Interferometers	Film thickness or surface roughness	
Single beam		300 Å
Multiple beam		25–10 Å
Profilometer	Same	25–100 Å
Contour analyzer	Surface contour along a line shadow	1 micron
Ellipsometer	Film thickness, dielectric	2 Å
	Film thickness, silicon	10 microns or less
Electron diffraction	Crystal identification	
Reflection		(Needs 1-mm sample)
Transmission		<1000-Å sample size
Low-energy electron diffraction (<400 volts)	Surface structure	1 to 2 atomic layers deep
Charged-particle spectroscopy	Surface-contaminant identification	Atomic number difference of 1
Electron microprobe		Less than 1 monolayer, 10 ppm in
X-ray mode	Detection of chemical elements	bulk (about 1,000 ppm for lightest elements)
Specimen current mode	Device topography	0.5–1 micron
Scanning electron microscope	Device topography	0.05–0.5 micron
Radiography unit	Inner topography	0.1 micron
Oscilloscopes	Current and voltage	1 nanoamp
Ammeters and electrometers	Current	10^{-15} amp
Voltmeters	Voltage	10^{-9} volt
Capacitance bridges	Capacitance	10^{-17} farad static
Ohmmeters	Resistance	10^{-5} ohm
Absorption spectroscopy	Detection of chemical elements	
Atomic		<1 ppm of Al, Fe, Co, or Ni
		1 ppm, combined with extraction
IR, attenuated total reflection	Surface composition to a depth of about 1 wavelength (depends on the angle of reflection)	Depth about 1 wavelength and depends on the angle of reflection
X-ray	Chemical analysis. Oxidation state	10–1,000 ppm

TABLE 22 Sensitivities of Instruments[13] (Continued)

Instrument	Property measured	Sensitivity, resolution or power

Emission spectroscopy
  Visible
  X-ray
  Neutron activation
	Chemical analysis
	Same
	Same
		1 ppm of most elements
		10 ppm of most elements
		1 ppb of most elements, 1 ppm of oxygen, 0.01 ppb of a few elements

Mass spectroscopy
  Gas
  Spark
  Sputter
	Chemical analysis
	
	Analysis of surface layer (10–100 monolayers removed per sec)
		
		0.02–200 ppm, 10^{-13} torr
		1 ppb of many elements, 20–200 ppb of H, N, C, O
		1–10 ppm. Area 0.1 mm across may be analyzed.

Gas chromatography	Chemical analysis	1 ppb hydrocarbons, 1 ppm H, 10 ppm A, 500 ppm water
EPR (ESR)	Chemical analysis	10^{11} unpaired electron spins/cm^3, Equivalent to 10^{12} atoms/cm^3 of P in Si, and 10^{14} atoms/cm^3 of Sb in Si, and 10^{-6} moles/liter
NMR	Chemical analysis. Mainly suitable for structural characterization	2,000 ppm
X-ray diffraction	Phase analysis Crystallite size Dislocation mapping Residual stress	0.5–10 % 0.05- to 0.3- and 10- to 1,000-micron ranges Over 5 microns apart 5,000 psi or 500 ppm strain over area 0.010 in. across
Strain gages	Strain, thus stress	1 ppm over 0.015 in. with good accuracy, 100 ppb over 0,010 in. less accurately
Tensile testers	Bond strength	0.1 g
Thermal plotter	Surface temperature	0.5°C over 0.0015 in. or 2°C over 0.00035 in.
Refractometer	Index of refraction, thus some chemical analysis	0.0001 unit

Wet chemistry
  Colorimetric
  Fluorimetric
  Ion exchange
  Carbon analyzer
  Vacuum fusion
  Etching
	Chemical analysis
	
	None
	Carbon determination
	Chemical analysis
	Dislocations or stacking faults
		
		10^{-11} g
		10^{-12} g of many elements
		Concentrates 100 times
		10 ppm
		50 ppb H, 200 ppb of O or N
		10 microns apart

and are the most sensitive values quoted. Table 23 was compiled primarily as an aid to the integrated-circuit engineer but should be equally applicable to all thin-film applications.

For topographic examination the binocular microscope reveals gross mechanical damage, chemical corrosion, discoloration, and overall poor-quality films. For more detailed observations the metallurgical microscope and metallograph are required. A typical example of a research metallograph is the Bausch & Lomb pictured in Fig. 21. In the metallograph, higher useful magnification up to a nominal, 1,000× with a resolution as fine as 0.5 micron is available, but at the sacrifice of the use of natural illumination, depth of field, working distance, and binocular depth perception. Research metallographs are available that make accurate alignment of the polished face of the sample unnecessary; as the specimen is placed on the stage, it is automatically aligned perpendicular to the optical axis of the microscope. Twin light sources are available: a low-voltage lamp and a superpressure

TABLE 23 Use of Instruments in the Analysis of Thin Films[13]

Instrument	Abnormality observed	Relation to failure modes	Failure mechanisms deduced or likely	Relevant process step and recommendation
Unaided eye.	Stains or discoloration.	The chemical reaction, heat or contamination causing the abnormality may also cause faulty bonds or seals, electrical inversion, etc.	Poor wetting at braze or other bonds may lead to electrical or mechanical failure.	Review cleaning and handling techniques.
	Opened bonds.	Opens.	Contaminated land surface weakens bond, high temperature forms brittle intermetallics, etc.	Review and optimize bonding process.
	Broken dice.	Changes resistance.	Pressure during brazing or bonding causes cracks which fracture later.	Review and optimize braze wetting.
Binocular microscope, 3–120 magnification.	All above.			
	Surface voids in fillet, poor solder fillet.	Indicative of failure-prone or high-resistance joint.	Dirt leads to poor wetting.	Review cleaning. Optimize gas ambient and temperature.
	Corrosion products.	May cause electrical leakage, open circuits, or weakened connection.	Incomplete rinsing, contaminated ambient, or leak.	Cleaning. Package sealing.
	Cracks in seals.	May indicate leakers or weakened leads.	Rough handling, misalignment, thermal mismatch.	Redesign or reinforce package.
	Pits and pyramids on dice.	High leakage or hot spots at thinner base areas.	Poor epitaxial growth control allows local thin-base areas (after diffusion) at which hot spots occur.	Review and optimize epitaxial deposition.
	Poor registry or masking.	May allow high leakage, or short or open circuits.	Narrow insulating path may short-circuit, or inversion may cause high leakage current. Meager connection may open circuit.	Review and optimize registry, masking, and cleaning.
	Intermetallics on bonds.	Bond may be brittle or on way to further interaction and failure.	Intermetallics form by diffusion, may be accompanied by voids, and break to form open circuits.	Use barrier metal layer between reactive metals. Change materials.
	Scratches on dice or intraconnects.	Intraconnects scratched or over scratches are open or have thin areas prone to opening.	Burnouts.	Improve inspection, or automate.

	Rub marks.	As with scratches. Also may be contaminated and will cause inversion.	Improve assembler training, or automate.
	Microplasma in operating device.	Soft junctions.	Use better silicon, or inspect finished wafers by x-ray diffraction topography.
		As with scratches; or charge migration and inversion.	
		Current concentrates at stacking faults, dislocations, or thin spots in base.	
Metallograph. Photomicrography to 1000X.	All above. Poorly adherent interconnects or bonds.	Opens likely.	Monitor cleaning and joining steps.
	Voids in potting.		Change potting process or materials.
		Dirt or improper deposition allows contact breakage.	
		Missing strength from potting compound leaves weaker bonds.	
	Dislocation etch pits.	Mechanically weak, may leak in moisture. If dislocation at junction, may short-circuit.	Use better starting silicon, or inspect finished wafers by x-ray diffraction topography.
		Microplasma forms hot spot, then short circuit, at a stacking fault or dislocation.	
	Abnormal junction depth (by angle lapping).	May correlate with improper I–V characteristic.*	Review cleaning, etching, and photoresist removal.
		Dirt, oxide, residual photoresist affect diffusion and, thus, device parameters.	
	Diffusion of metal into silicon.	Hot spot was there.	Review bonding step for possibility of cracking dice.
		Metal causes short circuit.	
	Voids under die or in TC bonds.	Bond has high electrical resistance and may be mechanically weak.	Review cleaning and brazing procedures.
		May cause hot spots, poor electrical characteristics or may open-circuit.	
	Incomplete welds.	Bonds may break. Cans may leak.	Review and optimize welding cycle. Change welding equipment.
		Open circuit as a result of poor weld. Inversion from leakage through poor welds.	
	Thick or thin brazements.	Unusual stresses set up may cause breakage. Resistance improper.	Monitor braze preforms.
		Open circuit results from fracture of overstrained braze.	
	Unflat junction.	Hot spots affect resistance and thereby I–V characteristics.*	Review and optimize silicon polishing. Review and optimize cleaning of silicon.
		Rough junction has variable base thickness and therefore has hot spots.	
	Flaked areas.	May lead to opens.	Review cleaning of oxide layer.
		Poor adhesion causes intraconnect breakage and opens.	

* Current-voltage.

TABLE 23 Use of Instruments in the Analysis of Thin Films[13] (Continued)

Instrument	Abnormality observed	Relation to failure modes	Failure mechanisms deduced or likely	Relevant process step and recommendation
Phase-contrast microscope. Detects slight changes in surface level.	Stacking faults.	Soft junction.	Soft junction is due to microplasmas at stacking faults.	Use better silicon, or inspect by x-ray diffraction topography.
	Contaminants, transparent.	Surface inversion.	Ionic contaminants migrate and result in reverse leakage.	Cleaning.
	Improper oxide topography.	Electrical leakage and short or open circuits.	Open or short circuits due to misalignment. Narrow insulating path may short-circuit or may leak as result of inversion. Meager contacts may melt and open circuit.	Review mask making and registry.
Interferometer. Measures thickness of transparent films to about 25 Å.	Oxide and surface topography off design.	Same.	Same.	Same.
	Oxide thickness off design.	Stray capacitance, leakage, or inversion.	Metal may diffuse through and cause leakage.	Review oxidation step.
	Metal and resistor thickness off design.	If thickness nonuniform, may burn open.	Open-circuits due to burnout.	Review vacuum deposition.
Dark-field microscope. Emphasizes asperities.	Photoresist residues, dust.	Thermal runaway, inversion.	Inversion due to diffusion of charged contaminants. Uneven diffusion causes thin base areas which allow thermal runaway.	Photoresist removal. Ambient control.
Dye penetrant. Detects fine cracks.	Cracks. Pores. Residues.	See "Unaided eye." May falsely indicate gas leakage. See "Binocular and phase contrast microscopes."	Inversion due to diffused contaminants or adsorbed ambients from the pores.	Change sealant processing or materials.
Water condensate-wetting test. Detects some very thin films.	Photoresist residues or contaminants cause uneven wetting.	Surface inversion.	Ionic migration.	Improve cleaning.

Hot-stage metallograph. Views samples at 400X while heated in controlled ambient.	Contaminant melting point and reactivity (for identification).	Various.	Deduce contaminant source from identity and location; then correct process involved.
	Wettability by brazes and solders.	Joints may have improper resistance or be weak if wettability is poor.	Process control.
	Interdiffusion of metals, e.g., Al-Au, Mo-Au, Ti-Au, Kovar-Au.	Parameter drift.	See "Binocular microscope."
	Whisker growth.	Short and open circuits.	See "Binocular microscope, intermetallics."
			Change materials.
	Grain growth.	Parameter drift.	Age bonds or change alloys.
	Surface diffusion of metal films.	Short or open circuits, or parameter drift.	Investigate effect of ambient; change materials.
Electron microscope, transmission. Views samples less than 0.5 micron thick at 10 Å resolution.	Dislocation and stacking-fault distribution, crystallinity of thin-film deposits, contaminant and corrosion location.	See "Phase-contrast microscope." Affects stability of electrical properties. See "Binocular microscope."	Not applicable—process investigations. Metal whiskers grow and make short circuits. Removal of metal leaves open circuits. I–V off because braze, bond, or interconnect resistance drifts.* I–V off because films cause short circuits or inversion. Opens from reaction and fracture of conductor stripes.* Recrystallization causes drift of electrical properties. Dislocations and stacking faults allow uneven diffusion.
			Deposit at higher temperature.
Views shadowed replica of surface at 20 Å resolution. Cast may include loose particles.	Etch pits, scratches, dust and deposit roughness.	Open and short circuits, drift.	Surface roughnesses cause thin spots in deposits which may open-circuit.
	Rub marks.	Open circuits, inversion.	Same also migration of charge can cause inversion.
	Pattern alignment and topology errors.	Short or open circuits.	Narrow contacts burn open. Narrow isolation areas short-circuit across.
	Undercut etched edges.	Open circuits, inversion.	Contaminants may cause corrosion and open circuits, or migrate and cause inversion.
	Weld porosity	May hold contaminants, q.v.	Same.
			Review and optimize weld cycle.

* Current-voltage.

11-43

TABLE 23 Use of Instruments in the Analysis of Thin Films[13] (Continued)

Instrument	Abnormality observed	Relation to failure modes	Failure mechanisms deduced or likely	Relevant process step and recommendation
	Diffusion couple porosity.	Same. Weak bonds.	Same. Kirkendall effect causes voids.	Change materials or bonding cycle.
	Scribe feather cracks.	Open circuits if cracks propagate.	As with roughness. Thermal or mechanical stress may cause crack to propagate and cause an open circuit.	
Electron diffraction. Identifies crystalline particles less than 1 micron thick.	Crystalline contaminant or corrosion product identity.	See "Electron microscope."		
	Identity of intermetallic compounds.	$AuAl_2$ and Ni_3Si_2 are associated with voids and open circuits.	Intermetallics formed by diffusion are brittle and may crack. Volume changes involved form weak spots.	Change materials or process.
	Deposit crystallinity.	Affects stability of electrical properties.	Recrystallization causes drift of electrical properties.	Process temperature control.
Low-energy electron diffraction. Examines surface structure to depth of about 10 Å.	Adsorbates.	Inversion.	Migration of charged adsorbates may cause inversion.	Cleaning.
	Surface contaminants. Surface structure.	Same.	Same.	Same.
Electron microprobe, elemental chemical analysis. Analyzes 2-micron diameter volumes. Scans 400-micron-diameter area.	Contaminant or rub mark residue identity.	May short-circuit invert, or cause corrosion and open circuits. Relate to process control.	Migration of charged contaminants may cause inversion.	Review and optimize cleaning steps.
	Dopant and dopant concentration. Intermetallic analysis.	As above for "Electron diffraction." See "Electron microscope, transmission."		
	Corrosion product identity. Deposit topography map by chemical element.	See "Electron microscope, replica."		

11-44

Specimen current operation. See "Electron backscatter operation" below.	Deposit thickness map by chemical element.	Open circuits.	Review metal deposition and optimize.
	Crack, pit, and pinhole maps.	Open circuits, thin areas, short circuits.	Monitor cleanliness. Double deposit oxide.
	Crack, pit, and pinhole maps.	Same.	Thin areas may melt and open circuit.
	Junction misalignment or movement.	See "Phase-contrast microscope." Inversion.	Thin or narrow areas on interconnects may open circuit; pinholes may allow short circuits.
	Deposit topography.	Same.	Same.
	Open or short circuit, and current and/or voltage nonuniformity map.	See "Unaided eye" and "Phase-contrast microscope."	Moving junction is evidence of inversion.
	Surface-contaminant map.	Same.	Location may indicate cause.
	Oxide or resist residue map.	See above "Elemental chemical analysis" mode of operation.	Review indicated process steps.
	Scratch, asperity, and chip map.		
Electron backscatter operation. (Scanning electron microscope). Examines surface to resolution of 0.5 micron showing surface potential and average atomic number.	Same as for specimen current operation. Also: poor capacitors, improper junction diffusion.		Nonapplicable—process investigations. Review indicated process.
Electron backscatter thickness meter. Gages thickness using backscattered electrons from area 20 mils in diameter.	Thickness of oxide, metal, or photoresist.	As above.	As above.

11-45

TABLE 23 Use of Instruments in the Analysis of Thin Films[13] (Continued)

Instrument	Abnormality observed	Relation to failure modes	Failure mechanisms deduced or likely	Relevant process step and recommendation
Curve tracer. Measures static and dynamic current-voltage relationships.	Short circuits.		Sagging wire, punch-through, surface creep of metal, etc.	Indeterminate. Use supplemental tests.
	Open circuits, intermittent contacts.		Mishandling, intermetallics, etc.	Same.
	Soft junctions.		Surface leakage, microplasmas.	Check cleanliness, silicon material.
	Abnormal resistance.		Mechanical damage, recrystallization, corrosion.	Indeterminate.
	Leakage currents and inversion.		Surface charge migration, surface contamination.	Same.
	As an accessory to other instruments: Poorly diffused resistors.		Excessive heating.	Review and optimize diffusion.
	Poor capacitors.		Various.	Process control.
	Inversion cure activation energy.		Relates to identity of migrating charged species.	Review and optimize cleanliness, reagent purity, passivation.
	Inversion cure by ambient change.		Same.	Review and optimize packaging and ambient.
	Inversion by ambient change.		Same.	Same.
	Abnormal response to light.			Indeterminate.
Capacitance probe. (Zisman, Kelvin, etc.). Measures work function of surface.	Inversion layer movement.		Surface-contaminant charge migration.	Cleaning.
Infrared-absorption spectrograph. Analyzes for chemical bonds (compounds) present	Abnormal film thickness.	Abnormal electrical parameters.		Processing error.
	Abnormal oxygen content in Si.	Same.		Material deficiency.

11-46

As interferometer, measures film thickness down to about 1000 Å.	Abnormal epitaxy thickness.	Same.	Processing error.
	Impure photoresist.	Relates to processing.	Processing deficiency.
	Water.	Surface current leakage.	Leaky package or passivation layer may allow water entry.
	Depth of sharp dopant concentration change.	Relates to processing.	Processing variable.
Visible and ultraviolet absorption spectrograph. Analyzes for chemical elements, measures film thicknesses down to 300 Å.	Abnormal transparent film thickness.	Abnormal electrical parameters.	Processing error.
	Nonuniform transparent film thickness.	Same.	Contamination or processing error.
	Impurities in reagent solutions.	Relates to processing.	Processing deficiency.
Emission ultraviolet and visible spectrograph. Analyzes for chemical elements to 1 ppm.	Abnormal dopant concentration.	Low concentration eases inversion.	Processing error.
	Analysis of contaminants.	Parameter drift.	Contaminants diffuse and change electrical properties.
	Impurities in materials.	Parameter drift.	Same.
	Impurities in reagents.	May become surface contaminants.	Same.
	Identity of materials.	Wrong material may corrode or break.	Processing error.
Neutron activation analyzer. Analyzes for chemical elements to 0.001 ppm.	Same as emission spectograph but at lower concentrations.	Same.	Same.
Radiotracer analyzer. Analyzes for chemical elements to 1 ppm.	Incomplete cleaning.	Parameter drift.	Inversion by charge migration. Open circuits by corrosion.
	Extent of diffusion of dopants.		Processing control.
Monitors migration of chemical elements.	Incomplete photoresist removal.	Hot spots, inversion, open circuits.	Diffusion leads to thin base areas. Surface migration causes inversion.

11-47

TABLE 23 Use of Instruments in the Analysis of Thin Films[13] (Continued)

Instrument	Abnormality observed	Relation to failure modes	Failure mechanisms deduced or likely	Relevant process step and recommendation
	Extent of movement of ions through SiO_2	Inversion.	Process study.	
	Extent of movement of ions over surface.	Same.	Same.	
	Diffusion between films.	Inversion, poor adhesion, intermetallic formation.	Same.	
Gas chromatograph. Analyzes volatiles, such as in enclosed ambient of IC package, for compounds present.	Abnormal enclosed ambients.	Inversion.	Absorption, then charge migration causes inversion.	Review and optimize cleaning and drying.
	Leakers, by presence of air or test fluid.	Same.	Same.	Reject leakers or improve sealing.
	Desorbed volatiles.	Same.	Same.	Review and optimize cleaning and drying.
	Volatile impurities, in reagents or materials.	Same.	Same.	Change materials.
Mass spectrograph. Analyzes same for chemical compounds or elements.	As above. Also: reaction of device to ambient change in the spectrograph.	May allow inversion.	Same.	
Gas chromatograph plus mass spectrograph.	As above.			
Bubble tester. Detects gas emitted from heated package.	Gross leaks in packages.	Channeling or inversion. Open or short circuits.	Same. Corrosion may cause open or short circuits.	
Electron spin and nuclear magnetic-resonance spectrometers.	Hydrogen atoms in glass or oxide films. With etching, migration of hydrogen in oxide.	May cause inversion.	Migration causes inversion.	
	Dopants and interactions between dopants.	Indeterminate.		

11-48

X-ray diffraction. Analyzes for crystalline chemical compounds. Shows crystal defects.	Dislocation types and distribution. Stacking-fault distribution. Cracks.	See "Electron microscope, transmission." Same.	
		Open circuits. Inversion.	Crack formation causes open circuits. Contaminants in crack lead to inversion. Enhanced diffusion gives thin base which fails.	Review and optimize silicon wafer inspection.
	Scratches.	Low-voltage avalanching. Open circuits.		
	Chipping.	Same.		
	Identity of contaminants.	See "Unaided eye."		
	Identity of corrosion products.	See "Binocular microscope."		
	Identity of materials.	See "Emission ultraviolet spectrograph."		
	Abnormal film thickness.	See "Interferometer."		
Etching and microscopy. Detects crystal defects.	Dislocation distribution. Pits, cracks, and chips.	See "Electron microscope, transmission." Assesses semiprocessed silicon.	Processing control.	
Detects flaws in resistant layers.	Oxide and resist residues. Pinholes and cracks in oxide insulating layer.	Same. See "Electron microprobe."	Same. Also a processing control.	
Controlled etching and radiotracers.	Depth of inversion charge.	Inversion.	Failure mechanism investigation.	
Radiographic equipment. Renders all materials semitransparent with opacity dependent on atomic number and on specific gravity.	Voids in brazes.	Overheating, thermal runaway, also broken die or bond.	Poor heat conduction causes hot spots and thermal runaway. Stress concentration breaks bond.	
	Metal whiskers.	Short and open circuits.	Short circuit forms along whisker. Departed metal leaves weak bonds.	Change materials.
	Contaminant particles.	Short circuits.	Metallic contaminants move when package is jarred and may cause a short circuit.	
	Cracks.	Open circuits, overheating.	Cracks propagate and open connections. Cracks are poor thermal path.	Review and optimize cleanliness and mechanical processing.

11-49

TABLE 23 Use of Instruments in the Analysis of Thin Films[13] (Continued)

Instrument	Abnormality observed	Relation to failure modes	Failure mechanisms deduced or likely	Relevant process step and recommendation
	Overly long wires.	Short circuits.	Wires sag under acceleration and may cause short circuits.	Shorten leads.
	Metal migration.	Short circuits. Broken bonds.	Same as with whiskers.	Change materials.
	Solder balling.	Short circuits.	Same as with contaminant particles.	Change solder to higher-melting variety.
	Misalignment of metal parts.	Leakage. Open and short circuits.	Misalignment may cause weak bonds which cause open circuits or leaky packages or may leave too narrow an insulating gap which allows a short circuit.	Tighten inspection.
Strain gages. Measure strain or distortion in solids.	Loose headers or dice.	Thermal runaway.	Nonadhesion of braze shows in thermal expansion of can. Poor thermal path leads to overheating.	Change braze procedure or materials brazed.
	Poorly brazed headers or dice.	Same.	A poor braze fatigues and breaks.	Same.
	Braze flow.	Same.	Leads to fatigue and failure of braze.	Same.
Centrifuge. Stresses structures.	Short circuits.		Overlong wires sag and short circuit.	Shorten leads.
	Open circuits.		Weak bonds break.	See "Unaided eye-opened bonds."
Vibration tester. Stresses structures.	Same as "Centrifuge."			
Tensile tester. Measures material or bond strength.	Weak bonds.	Open circuits.	Process monitoring.	See "Unaided eye-opened bonds."
	Weak brazes.	Thermal runaway.	Same.	See "Strain gages."
	Weak welds.	Leakers.	Same.	Review and optimize welding procedure.

11-50

Helium leak tester or radioactive gas tester. Detects He or radioactive gas leaking out of or leaked into device packages.	Small leaks.	Uncontrolled ambient allows inversion and short or open circuits.	Charged adsorbates from ambient migrate and cause inversion. Corrosion by ambient causes open or short circuits.	Review and optimize package sealing.
	Porosity outside package.	None.	May cause unnecessary rejection.	Same.
Thermal plotter. Measures surface temperature of 0.3-mil-diameter areas.	Hot spots due to voids.	See "Radiographic equipment."		
	Hot spots due to thin-base areas.	Thermal runaway.	Contaminants or roughness cause uneven diffusion and thin-base areas.	Review and optimize silicon cleaning and polishing.
	Hot spots due to dislocations or stacking faults.	Same.	Faster diffusion along fault causes thin-base area.	Use better silicon, or inspect wafers by x-ray diffraction.

11-51

11-52   Thin Films

mercury-vapor lamp. The main feature of the mercury-vapor lamp is its high luminous density, and it proves particularly useful in all microscopic techniques requiring an extremely intense light source (e.g., dark-ground illumination, incident-light phase contrast) as well as in microprojection. For black-and-white photomicrography the use of the mercury-vapor lamp results in an appreciable reduction in exposure time. It also permits cinemicrography of changing phenomena, as for example, transformations in metals and alloys during heat treatment. Attachments may be added for grain-size determination, interference-surface testing, incident-light phase contrast, incident-light interference contrast, and microdrawing and microprojection. Polarized light permits the rapid recognition and examination of anisotropic materials and is frequently used to achieve increased contrast of certain structural details. Interference contrast is the latest method for increasing optical contrast and is used on specimens where the structural details are not visible sufficiently clearly in bright or dark ground illumination. Phase contrast also with incident light has already become one of the classical methods for examining unetched polished specimens and sintered metals. The addition of a microhardness tester allows one to establish the so-called single-crystal hardness of each separate

Fig. 21   Bausch & Lomb research metallograph.   (*Bausch and Lomb.*)

structural constituent. The small indentation loads and the high magnifications, moreover, permit the hardness testing of thin-walled or of small specimens and the determination of the relative hardness of even small specimen zones.

The hot-stage metallograph pictured in Fig. 22, though presently limited in magnification, enables one to follow changes in materials and devices in controlled atmospheres at temperatures up to 1600°C. Observation of materials at room temperature is limited to structures that either are stable under these conditions or can be obtained in the metastable state by quenching from higher temperatures. But static observations can give only a very limited picture of the actual dynamics of metal formation. Using high-temperature metallography, changes in metal structure can be observed at the actual temperatures at which they occur and photographed to provide a continuous record of the transition process. There is hardly an aspect of metals study which cannot profit from the direct observation made possible by the high-temperature microscope. Some of the more important studies in which it has proved of value are:

1. Phase transformations during heating and cooling
2. The growth of crystal grains at high temperature
3. Grain-boundary studies

4. Surface-diffusion phenomena
5. Plastic deformation and fracture at high temperatures
6. Brittleness at high temperature
7. The sintering of metals and nonmetals
8. The fusion and solidification of metals with consequent supercooling and dendrite formation
9. Oxidation and corrosion processes

To effectively study the structure of thin films it is sometimes necessary to know the crystalline structure. The atomic structure of crystals is on much too fine a scale to be seen with the optical or electron microscope. The most effective way of observing the atomic structure of crystals is through the study of the diffraction effects that occur when a beam of x-rays is passed through a crystal. A crystal acts as a three-dimensional diffraction grating for x-rays. The resulting pattern is used to identify crystals, determine the presence of impurities, and show stress or deformation in the crystalline structure. The powder method of x-ray diffraction is used to measure the lattice parameters of cubic crystals, and by observing the presence or absence of x-ray reflections from certain crystal planes,

Fig. 22 Hot-stage metallograph. (*Unitron Instrument Co.*)

the type for crystal lattice can be determined. X-ray diffraction and x-ray fluorescence instrumentation is pictured in Figs. 23 and 24. Two applications of x-ray diffraction are of particular concern: (1) x-ray diffraction microscopy or topography, and (2) x-ray diffraction analysis of secondary phases. The first technique has become established as a sensitive means for revealing the distribution of crystalline imperfections in depth and for establishing their geometry relative to the host crystal. It depends for its sensitivity on the marked relative enhancement produced in the effective diffracting power of crystal lattice regions which have suffered minute departures from ideal atomic order. Since microdefects such as dislocations are surrounded by large strain fields, minute imperfections are readily resolved, and their appearance or transformation during thin-film processing can be monitored nondestructively.

Diffraction analyses of foreign phases cannot be applied directly to the identification of microdefects since their low concentration places them far below the limit of detection. However, x-ray diffraction is suitable for identifying the relatively large volume of secondary phases such as intermetallic compounds which are formed by interdiffusion at metal-semiconductor or metal-metal interfaces. To analyze successfully the early stages of diffusion, maximum sensitivity is needed; this is achieved

11-54   Thin Films

Fig. 23 X-ray diffraction instrumentation. (*Norelco.*)

Fig. 24 X-ray fluorescence instrumentation. (*Norelco.*)

only by using high-resolution, well-monochromated diffraction systems, together with low-noise recording equipment. The use of such systems in combination with high-temperature x-ray cameras should also permit the growth of reaction products to be monitored as a function of time and temperature.

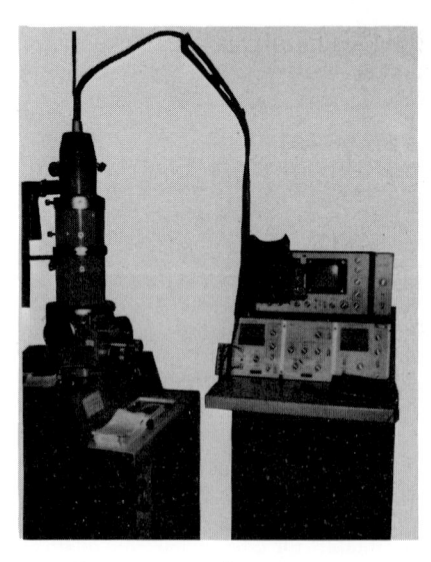

Fig. 25 Scanning electron microscope. (*Japan Electron Optics Laboratory Co., Ltd.*)

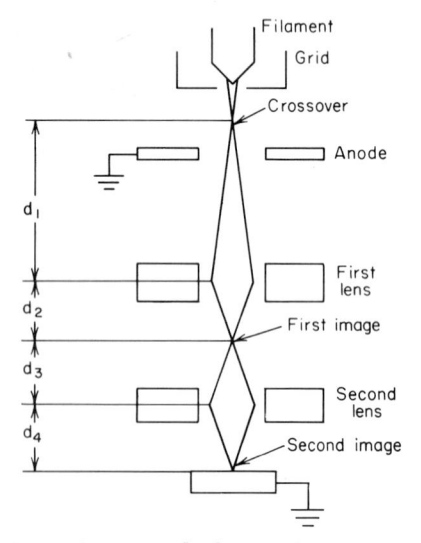

Fig. 26 Focused electron beam using two magnetic lenses.[13]

High-energy electron diffraction (50 to 100 kev) can be used for evaluating the structure of surfaces to a depth of 100 to 200 Å or of films of thickness below about 1,000 Å. Electron diffraction is used to best advantage in studying selectively interfaces such as those exposed after fracture, or in transmission studies of reaction processes in thin-film diffusion couples whose total thickness is but a few hundred angstroms.

The scanning electron microscope pictured in Fig. 25 is a relatively new and powerful instrument for scientific applications. The electron beam is focused by

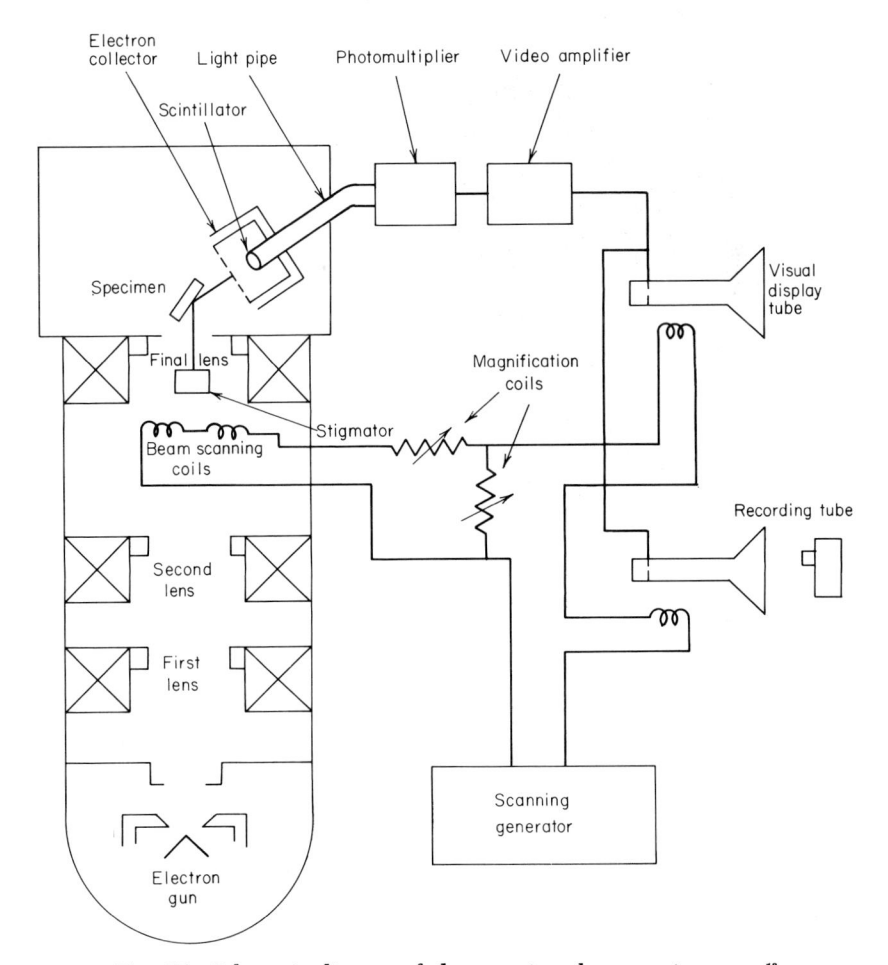

Fig. 27 Schematic diagram of the scanning electron microscope.[13]

a series of magnetic lenses, as shown in Fig. 26.[13] Present-day instruments have an electron spot size of 100 Å diameter, requiring three magnetic lenses, as shown in Figure 27.[13] The beam current in a scanning electron microscope is usually of the order of 10^{-11}. A higher beam current would cause enlargement of the spot size due to space-charge effects of the electrons interacting with one another. The magnification of the instrument is primarily controlled by the ratio of currents in the deflector coils of the microscope column and the cathode-ray tube, respectively. Various attenuation techniques have been worked out to vary the magnification in a known manner. Low magnification of about 30 can be obtained, while the high magnification is limited by the resolution of about 100 Å, giving a useful magnification of about 25,000×. The ability of the scanning electron microscope to detect variations in surface electric fields as well as in materials has made it a powerful tool for the investigation of semiconductor devices and microcircuits.

Image formation in the scanning electron microscope differs from that of the conventional electron microscope and optical microscope, whose images are formed directly by lenses; in the scanning microscope the image is formed on a cathode-ray tube after first converting the information from the specimen surface into a train of electric signals. In a scanning electron microscope a finely focused electron probe scans the specimen under study in a somewhat similar manner to a television raster; and by bombarding the specimen, a wealth of information can be obtained through secondary electrons, backscattered electrons, absorbed electrons, transmitted electrons, cathodoluminescence, and electromotive force. Bulky specimens can be observed directly without the necessity of slicing and replication, as in the case of conventional microscopy. The scanning electron microscope is perhaps best suited for observing various kinds of fractures and defects in thin films. Previously, in order to make such observations it was necessary to use the replica method in conjunction with a conventional electron microscope. It is now possible to obtain three-dimensional, clear images even if the specimen surface is very rough. In addition to the observation of specimen surface, the elucidation of physical and chemical properties can also be performed: namely, the backscattered and absorbed electron images clarify the distribution of the specimen components. In the case of microelectronics, it is important to study their electrical characteristics in the working stage. Electromotive-force images and secondary-electron images which are capable of exhibiting a voltage contrast will no doubt promote these studies, not to mention the fact that the observation of specimens in their working state has been appreciably facilitated.

Fig. 28 Electron microscope. (*Japan Electron Optics Laboratory Co., Ltd.*)

The conventional electron microscope as seen in Fig. 28, although not as versatile or sophisticated as the scanning electron microscope, permits reliable and convenient observation of most materials and furnishes the investigator with a base from which he can, with a little ingenuity, proceed to develop new methods suited to his special problems. Replica methods are available for the study of all manner of surfaces, thus circumventing the difficulties of observation by reflection. The cutting of materials in thicknesses down to about $\frac{1}{20}$ micron has become a reliable operation. Almost every noteworthy published account of an application in electron microscopy contains some innovation in techniques of sample preparation. The competent investigator will always attempt to improve or modify methods as he goes along, guided by an understanding of the electron microscope and by the special knowledge of his major field. The electron microscope is a valuable tool in the study of crystallographic defects, such as misplaced atoms, grain boundaries, or gross strain resulting from defects. A dislocation is one form of a collection of misplaced atoms and contains a linear array of atoms with an environment different from that in a perfect crystal. Associated with dislocations are variations in interatomic distances and strained interatomic bonds. In the immediate vicinity of a dislocation, the resultant difference in interplanar spacings results in a difference of diffracted intensity recorded on the x-ray diffraction topograph[55,56] or electron micrograph.[57]

Infrared radiometry has become increasingly important as miniaturization and packing density of integrated circuits has been emphasized. Infrared radiometry provides a nondestructive thermal distribution profile of a working circuit. The

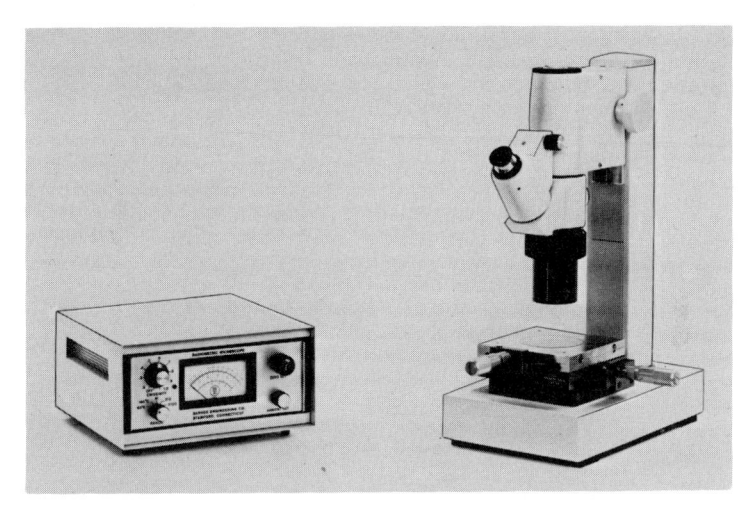

Fig. 29 Radiometric microscope. (*Barnes Engineering Co.*)

performance and reliability of microelectronic devices are dependent on their thermal behavior. Even though power is kept to milliwatt levels, the small physical size of junctions and the use of thin films cause power densities to reach astonishing values (10^6 amp/cm^2). Infrared microscopes make it possible to measure temperatures of a target as small as 0.001 in. They respond to infrared radiation that is naturally emitted by all devices under power. No physical contact is needed; the radiant energy is gathered by an optical system and focused onto a thermal detector, which in turn converts the radiant energy into an electric signal. Temperature-sensitive electrical parameters, such as V_{BE} can be used as internal thermometers to determine junction temperature. The parameter can be measured at a known temperature in an unpowered condition and related to a similar measurement under operating conditions. A typical instrument of this kind is shown in Fig. 29. Figure 30[13] demonstrates the type of thermal plot that may be generated using a radiometric microscope. This plot was taken of a deposited resistor on fused alumina substrate.

The electron-probe microanalyzer has the significant capability of analyzing minute volumes at the surface of solids, both nondestructively and quantitatively, with accuracies unattainable by any other method. Volumes of a few cubic microns

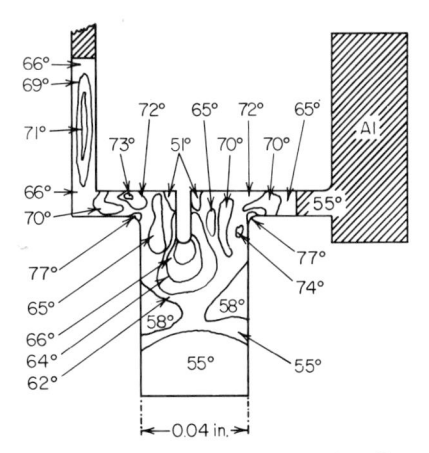

Fig. 30 Thermal plot of thin-film resistor.[13]

can be quantitatively analyzed to accuracies of about 1 percent. In some cases, concentrations of one part in 10^5 can be detected. The electron-probe x-ray microanalyzer contains an electron optical column which generates a very small diameter beam of high-energy electrons. The electron beam is focused on the surface of a specimen, which is viewed by an optical microscope. Exact registration of the electron beam on the point of interest is assured by aligning the microscope reticle and electron beam on the spot to be analyzed. As a result of the electron bombardment, the specimen gives off characteristic x-rays corresponding to the elements present. The x-rays are analyzed by diffracting them with a crystal and measuring them with an appropriate detector. The output of the detector, consisting of electric pulses of varying amplitude, is analyzed by means of several types of electronic signal-processing equipment for both qualitative and quantitative measurements. Concentrations of a specific element may be determined by comparing the relative intensities generated by the element in the solid and a standard. Semiquantitative and qualitative measurements may be made over small areas of the specimen (10×10 to 450×450 microns) by scanning the area with the electron beam. The

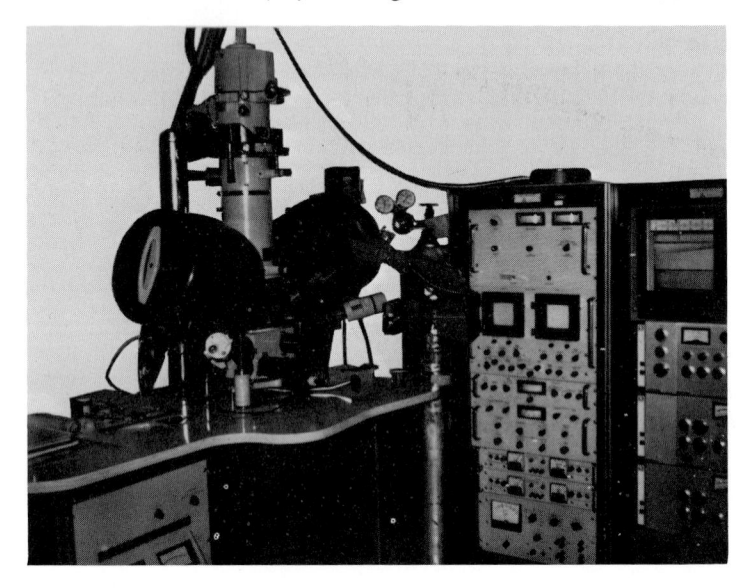

Fig. 31 Electron-probe x-ray microanalyzer. (*Acton Laboratories, Inc.*)

output of the spectrometer, tuned for a specific spectral line, is an indication of the presence and relative concentration for the corresponding element. An approximate indication as to the average of atomic number of elements present on the surface being scanned may be determined by measuring the specimen current or the backscattered electrons. Elements with a high atomic number scatter electrons more effectively; therefore, variations in specimen current are sensitive to the atomic number of the elements present in the scanned area.

In the past few years electron-beam microanalysis has been applied very effectively in fields of geology, mineralogy, biology, and microelectronics. A miniaturization of electronic components proceeds, such as integrated circuits, epitaxial structures, and thin-film devices; analysis, testing, and fabrication must be done on a microscale. To name a few areas of applicability:[13]

1. Metallurgical studies:
    a. Diffusion zones with steep gradients
    b. Localized impurities
    c. Peak-shift studies to determine the extent of chemical bonding

2. Scanning x-ray and electron microscopy:
    a. Specimen-current images showing bonding integrity, masking misregistration, pinholes, etc.
    b. Secondary-electron images showing potential distribution around activated components
    c. X-ray images showing impurities encountered in the component fabrication
    d. Observation of depletion-layer broadening with increased reverse-bias, n-type inversion layers on passivated high-resistivity p-type silicon surfaces
3. Microdiffraction Kossel patterns:
    a. Measurement of lattice-spacing variations
    b. Crystallographic orientation studies of thin films

An electron probe x-ray microanalyzer is shown in Fig. 31.

REFERENCES

1. Mansbridge, G. F.: British Patent 19,451, October, 1900.
2. Alexander, P., and E. L. Cranstone: British Patent 551,757, September, 1941.
3. Godley, P.: *Iron Age*, vol. 161, p. 90, 1949.
4. Wehe, H. G.: *Bell Lab. Rec.*, vol. 27, p. 317, 1949.
5. McLean, D. A., and G. G. Wehe: *Proc. IRE*, vol. 42, p. 1779, 1952.
6. National Defense Research Committee (US), division 14: Report, 360, 521, 534.
7. Schulze, A., and H. Eicke: *Deut. Elektrotech.*, vol. 6, p. 616, 1952.
8. Holland, L.: Vacuum Deposition of Thin Films, *Vacuum*, vol. 1, p. 223, 1951.
9. Libbey-Owens-Ford Glass Co., and H. A. McMaster: British Patent 632,256, May, 1950.
10. Gomer, R.: *Rev. Sci. Instrum.*, vol. 24, p. 993, 1953.
11. Dill, H. G.: *Semicond. Prod.*, vol. 5, p. 30, 1962.
12. Motorola Series in Solid-State Electronics, the Engineering Staff, Motorola, Inc., Semiconductor Products Division, 1965.
13. Schwartz, Seymour (ed.): "Integrated Circuit Technology," McGraw-Hill Book Company, New York, 1966.
14. Halaby, S. A.: *Electrotechnol.*, September, 1963.
15. Davidse, P. D., and L. I. Maissel: *J. Appl. Phys.*, vol. 37, p. 574, 1966.
16. Davidse, P. D., and L. I. Maissel: paper presented at *3d Int. Vacuum Congr.*, Stuttgart, June 28–July 2, 1965.
17. Davidse, P. D., and L. I. Maissel: Paper presented at *12th Nat. Symp. Amer. Vacuum Soc.*, New York, Sept. 12–Oct. 1, 1965.
18. Nickerson, John: Some Practical Considerations in the Use of Low-energy Sputtering, *Solid State Technol.*, December, 1965.
19. Theuerer, H. D., and J. J. Hauser: Getter Sputtering for the Preparation of Thin Film Interfaces, *Trans. Met. Soc. AIME*, vol. 233, March, 1965.
20. Phillips, L. S.: Transparent Conductive Metal Oxide Films, *Res. & Develop. for Ind.*, no. 19, p. 46, March, 1963.
21. Burkett, R. H.: *J. Brit. Inst. Radio Eng.*, vol. 21, p. 301, 1961.
22. Kirby, P. L.: *Electron. Components*, vol. 6, p. 521, 1961.
23. Keonjian, E. (ed.): "Microelectronics: Theory, Design, and Fabrication," McGraw-Hill Book Company, New York, 1963.
24. Rudenberg, H. G., J. R. Johnson, and L. C. White: *Proc. Electron Components Conf.*, May, 1962.
25. Smith, E. E., and S. G. Ayling: *Proc. Electron. Components Conf.*, May, 1962.
26. Lloyd, P.: Review of Thin Film Techniques for Microelectronics, *Microelectron. & Rel.*, vol. 6, pp. 177–187, 1967.
27. Keller, H., C. T. Kennerer, and C. Naegele: Tantalum Oxide–Silicon Oxide Duplex Dielectric Thin Film Capacitors, *IEEE Trans. on Pts., Mater., & Packag.*, vol. PMP-3, no. 3, September, 1967.
28. Kern, E. L., and L. A. Teichthesen: Passive and Process Materials for Semiconductor Device Fabrication, *Solid State Technol.*, October, 1966.
29. Cook, H. L.: New Effects Speed Thin Film Resistor Design, in "Microcircuits and Applications," McGraw-Hill, Book Company, New York, 1965.
30. Brandt, I.: Fabricating Thin Film Resistors, *Electronics*, CBS Electronics Information Services, 1957.
31. Manfield, H. G.: *Microelectron. & Rel.*, vol. 3, p. 13, 1964.

32. Fulmer Research Institute Ltd: Provisional patent, application no. 8919, 1964.
33. Lloyd, P.: *Ind. Electron.*, p. 22, October, 1962.
34. Philco Staff: *Electron. Rel. & Micromin.*, vol. 2, pp. 99, 1963.
35. Halaby, S. A.: *Electro-Technol.*, p. 97, September, 1963.
36. Basseches, H.: The Oxidation of Sputtered Tantalum Films and Its Relationship to the Stability of the Electrical Resistances of These Films, *IRE Trans. Component Pts.*, vol. CP-8, pp. 51–56, June, 1961.
37. Weber, R. J.: Some Structural Dependent Electrical Properties of Tantalum–Tantalum Oxide Thin Film Resistors, *IEEE Trans. on Pts., Mater., & Packag.*, vol. PMP-3, no. 1, March, 1967.
38. Schnable, P. M., and L. I. Maissel: Grain Boundary Diffusion in Sputtered Tantalum Films, *Vacuum Technol. Trans.*, Pergamon Press, New York, 1962.
39. Maissel, L. I.: Electrical Properties of Sputtered Tantalum Films, *Vacuum Tech. Trans.*, Pergamon Press, New York, 1962.
40. Layer, E. H., C. M. Chapman, and E. R. Olson: *Proc. 2d Nat. Conf. Mil. Electron.*, 1958.
41. Layer, E. H.: *Vacuum Tech. Trans.*, Pergamon Press, New York, 1959.
42. Gould, P. A.: The Resistance and Structure of Chromium Films, *Mullard Res. Labs. Rep.* 64, October, 1963.
43. Chapman, R. M.: *Vacuum*, vol. 13, p. 213, 1963.
44. Development of Metal Film Conical Resistors, G. V. Planer, Ltd., Rep. Ref. D 328/F, November, 1964.
45. Wolter A. R.: *Microelectron. & Rel.*, vol. 4, p. 101, 1965.
46. Gerstemberg, D.: *Ann. Physik*, vol. 11, p. 354, 1963.
47. Huber, F.: *Microelectron. & Rel.*, vol. 4, 283, 1965.
48. Lakstmanan, T. K., C. A. Wysocki, and J. Slegensky: *IEEE Trans. Component Pts.*, vol. CP-11(2), p. 14, 1964.
49. Berkerman, M., and R. E. Thus: *Vacuum Technol. Trans.*, Pergamon Press, New York, 1961.
50. Jones, D. E. H.: Electrical Properties of Vacuum and Chemically Deposited Thin and Thick Films, *Microelectron. & Rel.*, vol. 5, no. 4, November, 1966.
51. Schnable, G. L., and R. S. Kenn: Study of Contact Failures in Semiconductors. *RADC Final Rep.* TR-66-165, April, 1966.
52. Schnable, G. L., R. S. Keen, and L. R. Loewenster: Study of Contact Failures in Integrated Circuits, *RADC Final Rep.* TR-67-331.
53. Gianelle, W. H.: Analysis of Seven Semiconductor Metallurgy System Used on Silicon Planar Transistors, in M. E. Goldberg and J. Vaccaro (eds.), "Physics of Failure in Electronics," vol. 4, pp. 46–57, RADC, USAF, 1966.
54. Earles, D. R.: Reliability Growth Predictions during the Initial Design Analysis, *Proc. 7th Nat. Symp. Rel. & Qual. Contr.*, Philadelphia, January, 1961.
55. Schwuttke, G. H.: New X-ray Diffraction Microscopy Technique for the Study of Imperfections in Semiconductor Crystals, *J. Appl. Phys.*, vol. 36, pp. 2712–2721, 1965.
56. Blech, I. A., et al.: X-ray Surface Topography of Diffusion-generated Dislocations in Silicon, *Appl. Phys. Lett.*, vol. 7, pp. 176–178, 1965.
57. Thomas, G.: "Electron Microscopy and Strength of Crystals," Interscience Publishers, a division of John Wiley & Sons, Inc., New York, 1963.

Chapter **12**

Thick Films

ALBERT E. LINDEN
Advanced Microelectronics Division, Optimax, Inc.,
Colmar, Pennsylvania

Introduction	12–3
Thin Films	12–3
Thick Films	12–3
Comparisons	12–5
Terms and Definitions	12–5
Layout Criteria	12–5
Power Dissipation	12–8
Layout Suggestions	12–8
Artwork	12–11
Tape Method	12–12
Cut-and-strip Method	12–12
Coordinatographs	12–13
Automatic Artwork Generation	12–14
Photoreduction	12–15
Trapezoidal Effect	12–16
Artwork-film Flatness	12–17
Trapezium Distortion	12–17
Extensive Distortion	12–17
Optical Distortion	12–17
Processing Problems	12–18
Lenses	12–18
Reduction Cameras	12–19
Inspection Systems	12–19
Thick-film Screens	12–21
Mesh Sizes Available	12–21

Screen Tension	12–21
Screen Emulsion Application	12–23
Exposure	12–25
Metal Mask Screens	12–26
Substrate Materials	12–27
Substrate Properties	12–28
Substrate Machinability	12–28
Flatness	12–30
Thick-film Cermet Materials	12–31
Thick-film Resistors	12–31
Thick-film Conductors	12–37
Thick-film Insulators	12–40
Active Thick-film Materials	12–45
Dry Transfer Tapes	12–46
Thick-film Printing Processes	12–47
Printed-film-thickness Measurement	12–50
The Drying Process	12–52
Firing Processes	12–54
The Furnace	12–54
Resistor Trimming	12–56
Air Abrasion	12–57
Pulse Trimming	12–58
Scribe Trimming	12–59
Laser Trimming	12–60
Solder Assembly	12–61
Burnishing	12–61
Cleaning	12–61
Fluxing	12–62
Tinning	12–62
Reflow Assembly	12–63
Chip and Wire Assembly	12–65
Eutectic Die Bonding	12–66
Alloy Soldering	12–66
Paste Bonding	12–67
Film Bonding	12–67
Flip-chip Bonding	12–67
Thermal Compression Wire Bonding	12–69
Ultrasonic Wire Bonding	12–71
Probing Equipment	12–73
Hybrid Circuit Packages	12–74
Encapsulation	12–76
Hermetic Sealing	12–78
Leak Detection	12–80
Cleanliness Requirements	12–80
Ultrasonic Cleaning	12–81
Vapor Degreasing	12–82
Deionized Water Rinsing	12–82
Vacuuming or Dry Nitrogen Blowoff	12–82

INTRODUCTION

Before delving into the intricacies of thick-film materials and processes, it is necessary to have a thorough understanding of the nature of the thick-film hybrid circuit and the numerous technologies associated with this unique form of microelectronic circuitry. There exists a lack of understanding in the minds of electronic engineers who are not directly associated with the processes, concerning the differences between the products derived from the thick-film and the thin-film technologies. The major differences are not so much in the final product (although definite differences do exist) as in the processes utilized to arrive at the final product.

Thin films The process utilized for the fabrication of thin-film circuits is almost always associated with vacuum technology. This includes vapor deposition, sputtering, and other vacuum techniques for depositing metals and insulating materials on substrates. The thickness of the materials deposited on the substrates by thin-film techniques is usually expressed in angstroms, and seldom exceeds 5000 Å.

There are advantages and disadvantages to the thin-film process, but it is not the intent of this chapter to explore them in detail. It is sufficient to say that thin films can be deposited or etched to tolerances in the order of 0.0001 in. Since a major portion of the processing must be done in a vacuum, the process tends

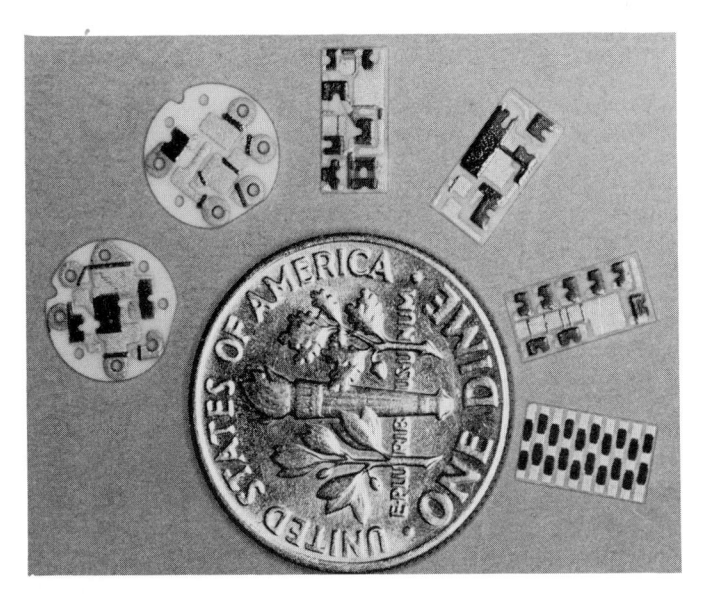

Fig. 1 Examples of thick-film substrates before attachment of parts. (*Dictograph Products, Inc.*)

to be expensive. The tooling required is usually extensive, precise, and custom made for the particular process or circuit.

Thick films The process utilized for the fabrication of thick-film circuits is almost always associated with a stainless steel screening and cermet firing technology. Screenable materials for conductors, resistors, and insulators are readily available from numerous vendors, as is processing equipment. Most of the tooling requirements involve photolithographic techniques and are, as a result, fairly simple and inexpensive. The thickness of the materials deposited on a substrate by the thick-film process is seldom less than 0.0005 or more than 0.002 in. in thickness.

The advantages of the thick-film process include the lack of extensive or expensive tooling for each circuit, a quick-response process, and skills easily mastered by inexperienced help. Limitations exist in the maximum attainable working frequen-

12-4 Thick Films

TABLE 1 Comparison of Thick and Thin Films

Criterion	Thick film	Thin film
Initial resistor tolerances, %	±10 to 20 as fired	±5 as deposited
Trimming tolerance, %	±0.5	±0.1
Power handling capabilities, watts/in^2	~50	~15
Resistor temperature coefficients, ppm per °C	±100	0
Resistance, ohms/sq	1–50,000	0.1–1,000
Line-width capabilities, mils	5 ± 1	0.2 ± 0.02
Capacitance limitations	10,000 pf practical	0.01 µf practical
10,000-hr drift, %	±1	±0.1
Cost each per 100 circuits	$30 each, plus $1,000 tooling	$55 each, plus $2,500 tooling

TABLE 2 Terms and Definitions

Integrated circuit. The physical realization of a number of circuit elements inseparably associated on or within a continuous body for performing the function of a circuit.

Semiconductor integrated circuit. The physical realization of a number of electrical elements inseparably associated on or within a continuous body of semiconductor material for performing the function of a circuit.

Thin-film integrated circuit. The physical realization of a number of electrical elements entirely in the form of thin films deposited in a patterned relationship on a structural supporting material.

Thick-film integrated circuit. The physical realization of a number of electrical elements entirely in the form of thick films, printed in a patterned relationship on a structural supporting material.

Hybrid integrated circuit. An arrangement consisting of one or more integrated circuits in combination with one or more discrete devices. Alternatively, the combination of more than one type of integrated circuit into a single integrated component.

Substrate. The physical material upon which a circuit is fabricated. Used primarily for mechanical support, but may serve a useful thermal or electrical function.

Active substrate. A substrate for an integrated component in which parts of the substrate display transistance. Examples of active substrates are single crystals of semiconductor materials within which transistors and diodes are formed, and ferrite substrates within which flux is steered to perform logical, gating, or memory functions. It can also serve as a passive substrate.

Passive substrate. A substrate for an integrated component which may serve as physical support and thermal sink to a thin- or thick-film integrated circuit, but which exhibits no transistance. Examples of passive substrates are glass, ceramic, and similar materials.

Glaze. A substance used as a protective cover or insulator which is placed over other components (resistors, conductors, etc.) on a film circuit.

Rheology. The science of the flow of materials. In thick films, those criteria dealing with the film flow after printing and during drying and firing.

Binder. The material used to hold the resistive or conductive powders together so that they may be used in a practical manner.

cies (about 2,000 MHz) and the density of the lines and spaces that can be achieved. Three-mil lines with three-mil spaces seem, at present, to be a practical limitation for common thick-film circuitry. Special techniques may be used to improve upon this limitation, but only with great difficulty.

Comparisons A comparison of the technical capabilities of thick-film versus thin-film components is listed in Table 1. The table contains many approximations, as it would be impossible to produce accurate data that would fit the processes and labor rates of each particular producer.

TERMS AND DEFINITIONS

In order to better comprehend the language of the industry, it is well to have a commom understanding of the terms used. A list of terms is supplied in Table 2. Some of these definitions have been selected from published documents while others have been defined by the author in order to give as complete coverage as possible.

LAYOUT CRITERIA

Operating rules for layout criteria must be established that are compatible with the equipment and processes in place at a given installation. Where a line-width minimum of 0.003 in. may be satisfactory for one facility, a minimum line width of 0.010 in. may be the least acceptable for another facility. This logic extends to resistors, capacitors, and even the insulation, since all process criteria are "layout-dependent" if reasonable yields are to be expected. The question of yield is most important to the success of any operation, and it is especially important in the field of thick-film hybrid circuits. One major factor affecting the process yield can cause many repercussions in the cost and the delivery of the end item. It is for these reasons that layout criteria should be carefully established with extreme consideration for each and every process step.

Some typical guidelines for layout criteria are listed in Table 3.

TABLE 3 Typical Guidelines for Layout Criteria

Criterion	Design	Minimum
Conductor width, in.	0.010–0.020	0.005
Conductor spacing, in.	0.010 or larger	0.005
Resistor length, in.	0.040	0.020
Resistor width, in.	0.040	0.010
Resistor spacings, in.	. . . .	0.025 (dependent on abrasion technique and clearance required)
Resistor dissipation, watts/in.^{2}		25–50 (dependent on heat-sinking methods)
Solder land, in.		0.050 × 0.050 (dependent on component lead size)
Wire-bond land, in.	. . . .	0.010 × 0.010 (0.020 × 0.020 preferred)
Die-bond land, in.	. . . .	0.005 on each side, larger than the chip
Edge of substrate clearance (for resistors), in.	. . . .	0.020
Resistor-conductor overlap, in.	. . . .	0.010

It is important to know the available area that a given substrate or package allows for the necessary active or passive circuitry. On small packages (up to ⅜- by ⅜-in. flat packs) this is best measured in square mils. On larger packages, however, the area would be expressed in fractions of a square inch, or square

12-6 Thick Films

inches. Table 4 lists the single substrate area availability for some typical package and substrate sizes.

Obviously, the area available for resistors, capacitors, and add-on parts is completely dependent on the total substrate area that can be used. Input-output pad areas must be subtracted from this total area available.

TABLE 4 Typical Package Areas Available

Size	Available area
To 5 (round)	26,200 mils²
To 8 (round)	136,900 mils²
⅜ × ⅜ in.	61,600 mils²
½ × ½ in.	0.25 in.²
⅝ × ⅝ in.	0.4 in.²
¾ × ¾ in.	0.56 in.²
1 × 1 in.	1 in.²

The following guidelines have been prepared to assist the circuit designer in making the transition from the breadboard to the hybrid thick film. These guidelines set forth the proper design of conductor patterns, cermet resistors, and semiconductor mounting pads, and outline the proper techniques for attaching die to the substrate, and for wire bonding where necessary. These are practical guides, and are not meant to describe current state-of-the-art techniques but rather to establish a set of rules which, if followed, will allow the resultant layout to be manufactured with good yield at a reasonable cost.

Thick-film Resistor Value Calculations. The basic equation for resistance is

$$R = \frac{pL}{A} \tag{1}$$

where R = resistance, ohms
 p = bulk resistivity
 L = length of resistor
 A = area of resistor cross section

Expanding gives

$$R = \frac{pL}{A} = \frac{pL}{tW}$$

where t = thickness of resistor
 W = width of resistor

Also, the resistance is inversely proportional to the thickness, or

$$p_s = \frac{p}{t} \quad \text{or} \quad t = \frac{p}{p_s}$$

where p_s = effective sheet resistivity, ohms/sq

Substitution of (3) in (2) provides

$$R \text{ ohms} = p_s \frac{L}{W}$$

$$= p_s \times \text{number of squares } N$$

and

$$N \text{ squares} = \frac{R}{p_s} = \frac{L}{W}$$

EXAMPLE: If an ink with sheet resistivity p_s = 12.5 kilohms/sq was screened with dimensions of L = 0.090 and W = 0.030 in., the resistance would be

$$R = 12.5 \text{ kilohms} \times \frac{0.090}{0.030}$$

$$= 12.5 \text{ kilohms} \times 3 \text{ sq or } 37.5 \text{ kilohms}$$

EXAMPLE: If an ink with sheet resistivity $p_s = 12.5$ kilohms/sq is to be used to screen a resistor of 37.5 kilohms, the number of squares will be

$$N = \frac{R}{p_s} = \frac{37.5}{12.5} = 3 \text{ sq}$$

If the maximum resistor width can be 0.050 in., then the length would be

$$L = W \times N = 0.050 \times 3 = 0.150 \text{ in.}$$

As a general practice, resistor geometry should be made as large as the available substrate will allow.

Allowance for Trimming. Thick-film resistors can be screened and fired to an accuracy of approximately 10 to 20 percent. Subsequently, if tighter tolerances are required, it is necessary to trim the resistor, usually by an air-abrasive technique. It is standard practice to trim all resistors, and for this reason the area of the resistor should be chosen such that its value is approximately 85 percent of the desired circuit value. The resistor will then be brought to the correct value by trimming.

EXAMPLE: If the final circuit value of a resistor is to be 25 kilohms, the screened and untrimmed resistor should be

$$R = 25K \times 85\% = 21.3 \text{ kilohms}$$

If it is assumed that the sheet resistivity of the ink with which the resistor is to be printed is 12.5 kilohms/sq, the number of squares N should be

$$N = \frac{R}{p_s} = \frac{21.3}{12.5} = 1.7 \text{ sq}$$

If the resistor is 0.050 in. wide, the length will be

$$L = W \times N = 0.050 \times 1.7 = 0.085 \text{ in. long}$$

The final value will be attained by trimming a portion of the resistance material away from the substrate, as shown in Fig. 2. This has the effect of increasing the number of squares, and thus the resistance.

Resistor Aspect Ratio. The aspect ratio of a film resistor is the ratio of its length to its width, or L/W. This is also equal to the number of squares N in the resistor. For the condition where the length of the resistor is greater than the width, the maximum aspect ratio should not exceed 10:1, and for good design, it should be 5:1 or less.

For resistors where the width is greater than the length, the aspect ratio should be 1:2 or greater, and never more than 1:3.

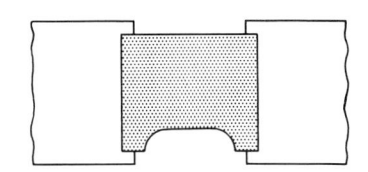

Fig. 2 Trimmed resistor.

Number of Resistor Inks per Substrate. The number of resistor inks per substrate should be kept to a minimum, and should never exceed 3. Each resistor ink requires separate artwork, its own screen, and a separate screening operation. For this reason, inks should be selected so that their sheet resistivities p_s cover the largest range of required values within the proper range of aspect ratios.

EXAMPLE: Resistors in a circuit include:

$$R_1 = 100 \text{ kilohms}$$
$$R_2 = 50 \text{ kilohms}$$
$$R_3 = 10 \text{ kilohms}$$
$$R_4 = 500 \text{ ohms}$$

The ink combinations could be:
Combination 1: 37.5 kilohms/sq and 375 ohms/sq

12-8   Thick Films

Combination 2: 125 kilohms/sq, 37.5 kilohms/sq, and 375 ohms/sq
Combination 3: 125 kilohms/sq, 1.25 kilohms/sq, and 375 ohms/sq

The best combination would be the first since it has one less screening operation.

On many occasions several possible ink combinations will appear equally advantageous. Under such a condition, the total area required for each combination should be calculated, after which a more meaningful decision can be made, based on considerations of substrate size and available resistor area.

Power dissipation   *Resistors.* A safe value for heat dissipation in thick-film resistors is 35 watts/in.^{2} of resistor at 125°C. The power to be dissipated by a resistor in a circuit may be calculated from

$$P = I^2 R = EI = \frac{E^2}{R}$$

The area of a resistor required to dissipate its generated heat may be calculated from the relation that

$$A = \frac{P}{P_r}$$

where A = resistor area, in.^{2}
P = power dissipated by resistor, watts
P_r = rated resistor power, watts/in.^{2}

or if the area is known, the power that a resistor can dissipate can be readily calculated from

$$P = A \times P_r$$

EXAMPLE: Area of resistor is 0.060 by 0.030 in., and rated power is 35 watts/in.^{2} The maximum power that the resistor can safely dissipate is

$$P = A \times P_r = (0.060 \times 0.030) \times 35 = 0.063 \text{ watt}$$

If the voltage drop across the resistor is 12 volts, and the current is 4 ma, the power to be dissipated is

$$P = EI = (12)(0.04) = 0.048 \text{ watt}$$

The resistor can dissipate the heat since its capacity is 0.063 watt.

Substrates. Similar calculations must be made for the completed substrate. A good design factor is 5 watts/in.^{2} The total heat dissipated by each component must be calculated, totaled, and compared to the total dissipation capability for that particular size substrate.

Layout Suggestions

As mentioned before, layout criteria are highly dependent upon the in-house or vendor's process, and therefore should be closely coordinated with the process engineering details. A knowledge of the limitations of the process can easily effect reductions in costs and increased yields if those process limitations are factored into the layouts. The following comments and illustrations show many of the dos and don'ts of thick-film layout:

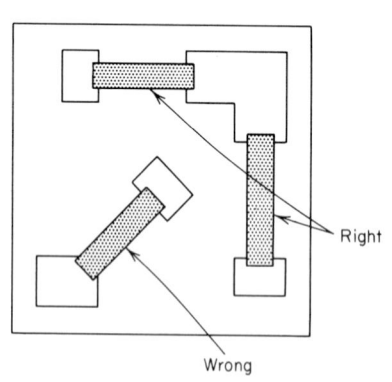

Fig. 3 Resistor orientation.

1. All final layout drawings should be to a minimum scale of 10:1.
2. Every effort should be made to have all resistors on the same side of the substrate. Double-sided substrate screenings are more expensive.

3. All lines should run parallel to the two major axes. This is particularly important for resistors because resistor trimming stations are normally, and more easily, operated along the X and Y axes, as shown in Fig. 3.

4. Avoid resistor loops since closed-path resistors cannot be measured and trimmed. This is shown in Fig. 4. The proper way to lay out a closed-loop resistor path is to break the path until the circuit is assembled. After the resistors have been trimmed, the loop is closed with a wire bond or suitable jumper, as shown in Fig. 5.

5. Crossovers should be avoided whenever possible. Two types of crossovers are available: a conductor crossover and a wire crossover.

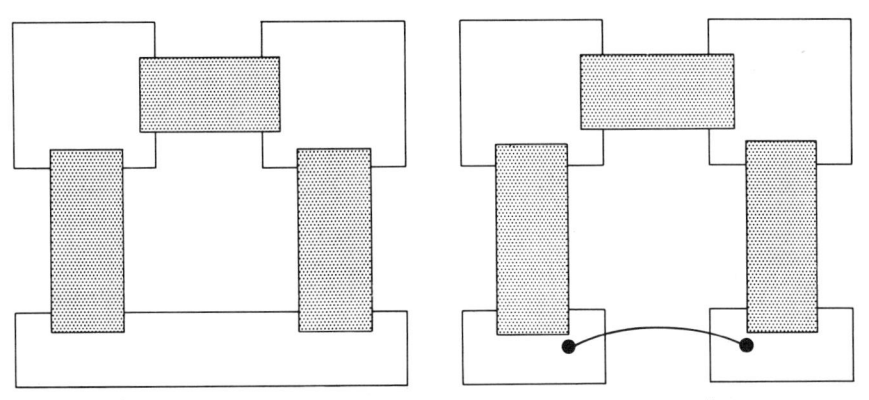

Fig. 4 Closed loop. **Fig. 5** Open loop.

Conductor crossovers require two additional screenings: one for the crossover insulator and one for the conductor.

Wire crossovers sometimes require an additional insulator screening to cover that portion of the circuit under the wire. This is not always necessary if there is little danger of the crossover wire short-circuiting a conductor or resistor under it.

When making wire crossovers, a wire should never be allowed to cross another wire or a semiconductor die (see Fig. 6).

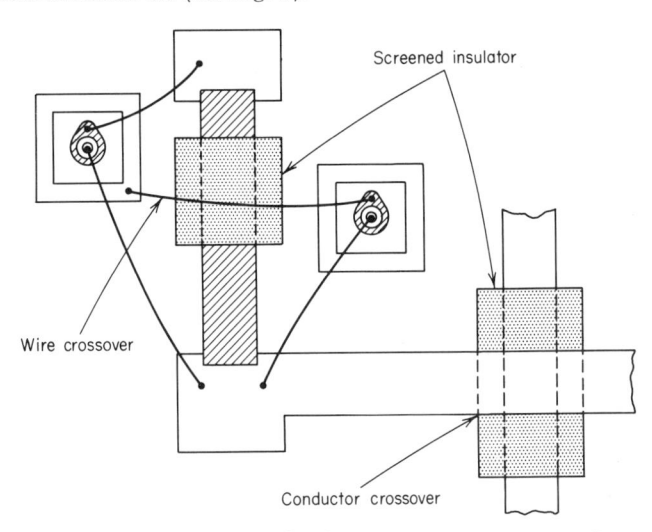

Fig. 6 Crossovers.

12-10   Thick Films

6. Resistor patterns should be rectangular and not zigzag (see Fig. 7). Zigzag patterns are difficult to trim and may develop hot spots at the corners.

7. For high-aspect-ratio situations, it is recommended that a "hat"-shaped geometry be used. This eliminates sharp corners and greatly reduces the probability of hot spots. This shape has particular application where aspect ratios are between 5:1 and 10:1 (see Fig. 8).

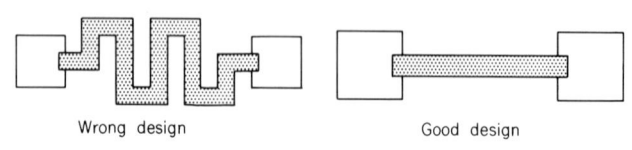

Fig. 7   Resistor patterns.

(a) "Hat" shape resistor before trimming

(b) "Hat" shape resistor after trimming

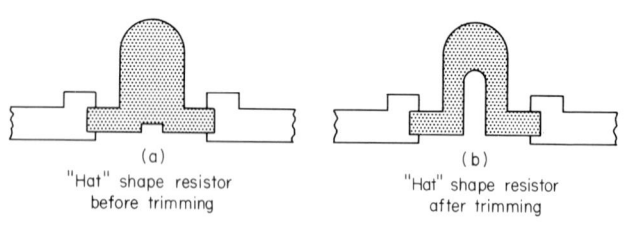

Fig. 8   High-aspect ratio.

8. Conductor paths should be as short as possible to minimize resistance. If a long path is required, it should be as wide as possible. This is of particular importance in ground and collector paths.

9. The minimum clearance between a resistor and the edge of a substrate should be 0.020 in.

10. The minimum resistor size should be 0.020 by 0.020 in., and this should be used only when absolutely necessary. Resistors should be as large as practical.

11. The conductor should overlap the resistor by 0.010 in. around the ends and sides of the resistor, as shown in Fig. 9.

12. Conductor width and spacing should not be less than 0.010 and 0.020 in. Greater width and spacing are recommended in order to hold conductive path resistance to a minimum.

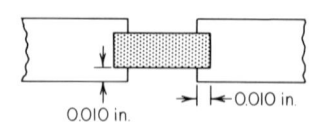

Fig. 9   Conductor-resistor overlap.

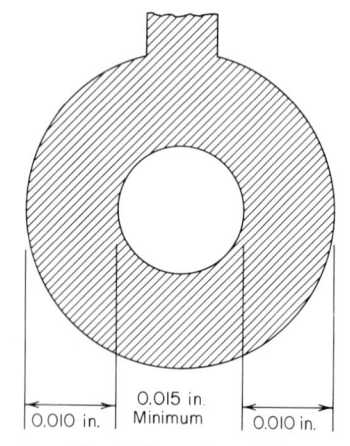

Fig. 10   Hole and pad dimensions.

13. Spacing between pads, resistors, and conductors should be a minimum of 0.020 in.; and a minimum trimming space of 0.025 in. should be maintained between resistors, to allow sufficient working room for the air-abrading tool used for trimming.

14. Holes in ceramic substrates should be avoided whenever possible because they are costly, and through-hole connections are very difficult to effect. The hole diameter should be a minimum of one-half the substrate thickness, and should

not be less than 0.015 in. Hole clearance should be 0.005 to 0.008 in. greater than the lead wire size.

For feed-through purposes, it is necessary that the hole have a pad screened around it on each side of the substrate. The diameter of this pad should be a minimum of 0.020 in. wider than the hole diameter, and should encircle the hole 360° if possible. In no case should the pad be less than 180°. For TO-5 substrates that are to be mounted in TO-5 headers, the diameter of the screened pad can be 0.015 in. It is necessary to screen the pad on only a single side if its function is to provide a mounting pad to the header pin or a bonding pad (see Fig. 10).

15. Pad areas for mounting semiconductor dice should allow a minimum 0.005-in. margin, in each direction, greater than the die's dimension. These areas should be made larger by at least 0.015 in. if ball or ultrasonic bonds are also to be connected to the pad, as shown in Fig. 11.

16. Pad areas for ultrasonic, wedge, or stitch bonding should be a minimum of 0.010 by 0.010 in. If solder is to be used during the assembly process, then solder pads should not be smaller than 0.050 by 0.050 in.

17. Wire length for connections and crossovers should be kept to a minimum. Maximum wire length should be 0.100 in.

18. Direct bonds should not be made between semiconductor dies. Instead, an intermediate bonding pad should be provided between them.

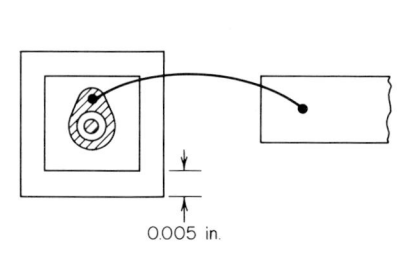

Fig. 11 Semiconductor pad area.

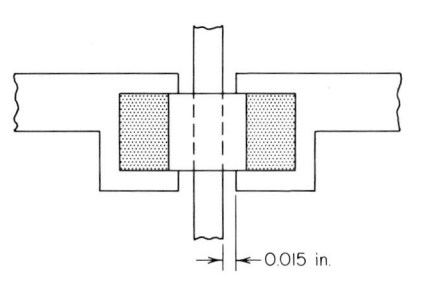

Fig. 12 Chip capacitor mounted over conductor.

19. Avoid having dice and long conductive paths close to the edge of the substrate. If they are too close to the edge, they are more easily damaged in handling.

20. External lead-attach pad areas should be at least two to three times the lead diameter in width, and not less than 0.050 in. in length.

21. Normally, discrete components should not be mounted directly across a conductor or resistor. However, multilayer ceramic chip capacitors may be mounted this way since they are well insulated. Care must be taken to assure that adequate clearance exists between the underlying conductor or resistor and the end terminations of the capacitor chip. A minimum of 0.015-in. clearance should be provided on each side, as shown in Fig. 12.

Figures 3 through 12, and selected text in Layout Suggestions, are by courtesy of the Varadyne Corporation. Following the general rules laid out herein will tend to keep the novice designer out of serious trouble. Again it must be stressed that the layout criteria are highly process dependent.

ARTWORK

Once the rough layout drawing has been made, it must be transformed into an accurate piece of artwork. This is usually done to a scale of at least 10:1, and on many occasions much larger than the final size requirements for the finished substrate; 20:1 is the most commonly selected size for the artwork. By working with artwork that is many times the size of the finished hardware, tolerances can be held quite easily; and registration, from one piece of artwork to another, can easily be accomplished. There are two basic methods for producing artwork, and both are in common usage. They are: (1) the tape method and (2) the cut-and-strip method. The tape method will suit many requirements, but where accuracy and registration are a factor, the cut-and-strip method is a great improvement over the tape method. In addition to these two basic methods, there are numerous automatic drafting machines and numerically controlled artwork generators that will assist in obtaining the required accuracies. These mechanical aids must seriously be considered for a company engaged in large numbers of constantly changing designs.

Tape method As initially mentioned, although taped artwork can be a satisfactory method for fabricating the necessary artwork, the degree of accuracy obtainable may be a problem, depending upon the criticality of the circuits involved. The tapes that are available have adhesive backings and can be placed rather neatly on Mylar[*] or other drafting media which have reasonable temperature stability. The most common available tapes are the black and red. The black tape has the consistency of masking tape, and the red (rubylith) has the consistency of Mylar. Each kind of tape comes in a variety of sizes to suit the many multiple-size scale factors that may be selected. Below are listed the typical tape sizes available, in inches:

$1/64$ by 648
$1/32$ by 648
$1/16$ by 648
$3/32$ by 648
$1/8$ by 324
$3/16$ by 324
$1/4$ by 324
$1/2$ by 324
$3/4$ by 324
1 by 324
2 by 324

A large variety of colors are available to suit any purpose of coding. In addition, registration targets may also be purchased in tape form. "Prestape" type T473 is considered suitable for registration marks where three or more items of artwork are required. Simpler targets, carefully used, are also suitable. Figure 13 shows a suitable target.

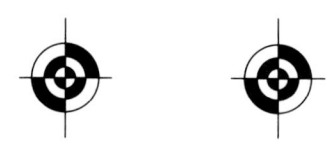

Fig. 13 Registration targets. (*Prestape*.)

Cut-and-strip method While the tape method essentially restricts the designer to the available dimensions in the "artwork scale multiples," the cut-and-strip method offers no such restrictions. Any width of line can be formed in order to accommodate any design requirement. The two basic requirements for the cut-and-strip method are:

  A dimensionally stable film, covered with an actinically opaque but a visually transparent, peelable surface
  A set of special tools, capable of accurately cutting the peelable surface with the necessary accuracy to achieve the required dimensional tolerances

The dimensionally stable materials are available from a number of manufacturers and several different types of material are obtainable. The ones listed below are

[*] Trademark of E. I. du Pont de Nemours & Co., Wilmington, Del.

typical of the kind used for thick-film artwork preparation:

M3 Rubylith Ulano—a hand-cut stripping film on stable Mylar. This is available on clear or milky Mylar base material in thicknesses of 0.0075, 0.005, and 0.003 in. Standard sheet sizes are 20 by 25, 22 by 28, and 24 by 30 in. Rolls are available in 40- and 44-in. widths up to 300 in. in length.

K and E also makes Stabilene stable-base peelable materials in a wide variety of sizes, colors, and thicknesses.

Sensitized stable cut-and-strip films are also available.

Utilizing these films, a cut-and-strip piece of artwork can be made from line drawings. To accomplish this, a sensitized film is exposed to an accurate line drawing, and after development of the film, the appropriate areas can be stripped away, leaving either positive or negative artwork.

Fig. 14 Linear coordinatograph. (*Development Associates Controls.*)

Coordinatographs In order to easily implement the cut-and-strip method of fabricating artwork, mechanical aids are necessary. The most common is the coordinatograph. This is essentially a very sophisticated drafting machine which guides a very sharp knife blade in an x,y axis. Attachments are available for circular cuts (compass), but it is difficult to cut angles on the (x,y) type of coordinatograph.

Tolerances on these machines are very good, and repeatability of positioning or measurement, on even a medium-priced machine, is held to 0.001 in. or better.

Figure 14 shows a typical linear (x,y) coordinatograph. It should be noted that these machines are available with a multitude of attachments which will increase the utility and automation possibilities. The coordinatograph shown in Fig. 14 has in the left background a digital readout for both x and y coordinates. It can be manually operated or set up to be driven by digital controls.

There are times when angular lines and large circles are required in the artwork, and for these special purposes polar or rotary coordinatographs have been developed. The basic difference between this machine and the linear, or (x,y), coordinatograph is that the worktable can be rotated so that precise angular cuts can be made. All other modes of operation of this machine are similar to those of the linear (x,y) coordinatograph. Figure 15 shows an example of a rotary coordinatograph.

Automatic artwork generation There are numerous methods of automating the artwork generation required for hybrid thick-film microcircuits, and these machine-aided methods go by various names, such as "digitizers," "numerically controlled drafting," "(x,y) plotters," etc. These devices should not be confused with computer-aided design systems.

In the automatic artwork generation, the design must have been created and reduced to numerical values. These numerical values are then put into an automatic drafting machine which, from the numerical stimulus, creates the artwork. A machine that will do work of this nature is shown in Fig. 16. This machine has an x and y coordinate display, with the sign also continuously displayed on a

Fig. 15 Rotary coordinatograph. (*Development Associates Controls.*)

five- or six-digit visual display board. Readings are displayed and can be recorded to the nearest 0.0005 in. A sequence counter is also provided.

In the computer-aided design, the parameters, tolerances, substrate size, parts values, from-to interconnections list, and other pertinent data are fed into a computer. These data are digested by the computer and compared with previously stored rules. The computer output results in a set of information which can be used to drive an automatic artwork generator which, in turn, produces the artwork in accordance with the design rules.

While the automatic artwork generator shown in Fig. 16 uses a standard pen for delineating the artwork, these methods can obviously be applied to cut-and-peel techniques. In addition, optical heads are available which will operate directly on photosensitive films, thereby producing the artwork directly. Figure 17 shows such an optical head, and Fig. 18 the drafting system that it fits with.

A sample of the type of plots that can be done with these optical heads is shown in Fig. 19.

Fig. 16 Graphic coordinate digitizer. (*Auto-Trol Corporation.*)

PHOTOREDUCTION

Once the numerous pieces of artwork have been prepared for a single hybrid thick-film microcircuit, they must be reduced to 1:1 scale so that actual-size screens may be made from the negatives or masks. This photoreduction is not a simple matter, as it is in this step that inaccuracies are most likely to occur, and they will affect all subsequent processes. Inadequate photographic equipment can cause loss of registration between individual negatives in sets of masks, dimensional inaccuracies, trapezoidal effects, trapezium effects, and pincushion distortion. Each of these problems is caused by inaccuracies in either the mechanics of the camera or the lens, and it is for these and other reasons that cameras for the photoreduction of hybrid microcircuit artwork are massive and expensive.

It must be recognized that there are virtually no absolutes in photoreduction.

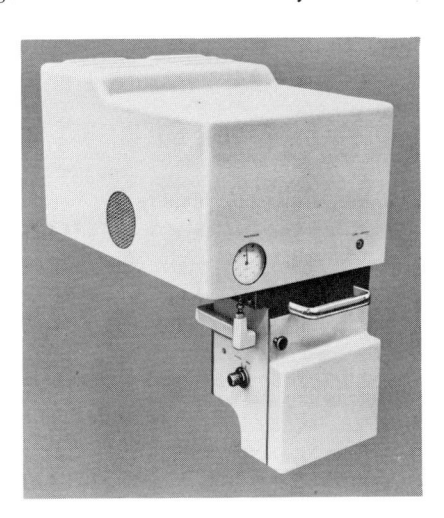

Fig. 17 Optical exposure head with single pen. (*Gerber Scientific Instrument Co.*)

12-16 Thick Films

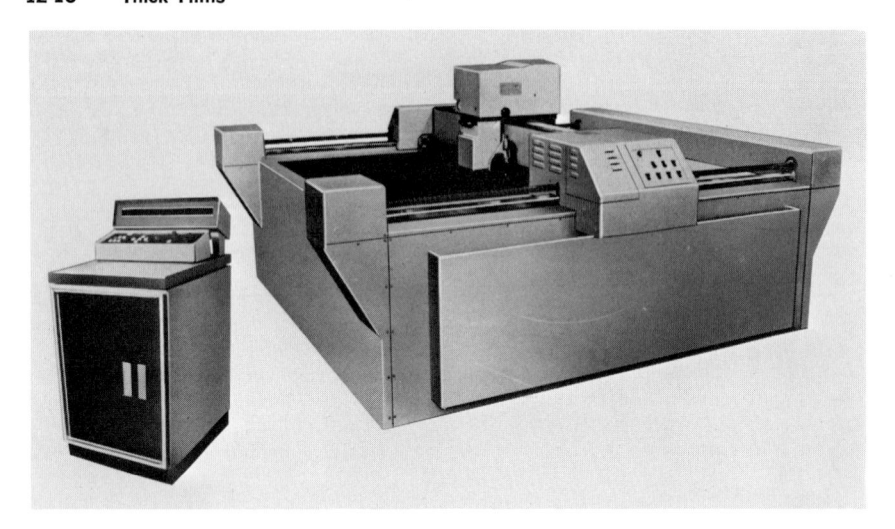

Fig. 18 Automatic drafting system with optical head. (*Gerber Scientific Instrument Co.*)

Rather, specific tolerances are required, and even these tolerances are limited by the capabilities of the available measurement equipment.

The problems most common in the photoreduction technology are discussed as follows.

Trapezoidal effect There are three principal planes associated with the reduction

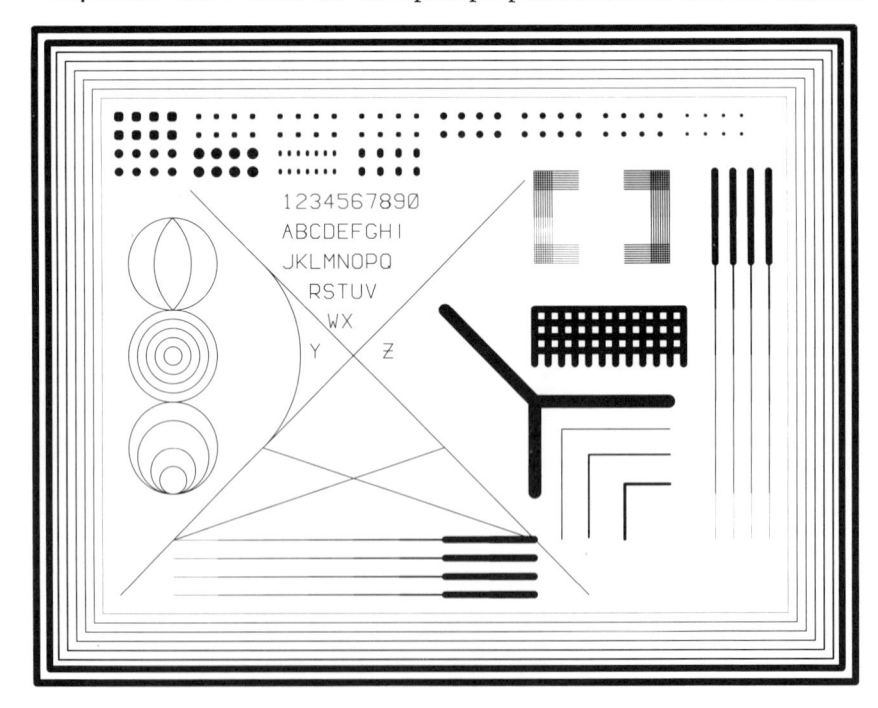

Fig. 19 Optical head plot. (*Gerber Scientific Instrument Co.*)

camera. These are the copyboard, the lens, and the film or plate holder. In order to reproduce a square or a rectangle faithfully, with the opposing sides parallel, all three planes must be parallel within very tight tolerances. Figure 20 depicts what can happen to the reproduction of a square if one of the planes is out of parallel. The resulting reproduction of the square has become trapezoid A. An autocollimation or trammel check will determine the plane that is out of parallel, and this must be corrected to eliminate the distortion.

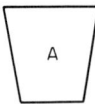

Fig. 20 Trapezoid resulting from one of three camera planes being out of parallel.

Artwork-film flatness If either the copy (artwork) or the film is not held flat, but the principal planes of the camera are parallel, a different type of distortion can occur. This is shown in Fig. 21, and manifests itself as the bulging shape shown as item B. To correct this problem, a vacuum system is needed at the copyboard and the film back, so that both materials are held flat.

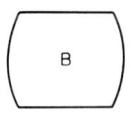

Fig. 21 Bulging sides caused by copy or film not being held flat.

This same type of distortion would be present if the glass photo plate had a curved surface.

Trapezium distortion Trapezium distortion can be defined as that condition which creates, from a square, a figure in which no two sides are equal. This will occur when two of the three planes are out of parallel. An example of this is shown in Fig. 22 with shape C.

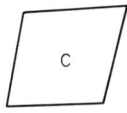

Fig. 22 A trapezium effect caused by two of the three camera planes being out of parallel.

Extensive distortion A combination of problems can easily occur and cause symptoms which may easily be confusing. Such a situation is shown in Fig. 23 with shape D. The situation depicted indicates a principal plane out of parallel, with an out-of-flatness condition at either the copyboard or film plane.

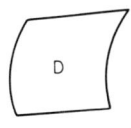

Fig. 23 Out of parallel and out of flatness.

Optical distortion Optical distortion manifests itself in basically two ways. The first is called *barrel distortion*. This is caused by a lens aberration which makes the magnification decrease toward the edge of the field. A typical example

12-18 Thick Films

of barrel distortion is depicted in Fig. 24 as shape E. This is the shape that would result from copying a square with a lens that had barrel distortion. *Pincushion distortion*, the second kind of optical distortion, causes the magnification to increase toward the edge of the field. Pincushion distortion is shown in Fig. 25 as shape F.

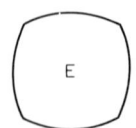

Fig. 24 Barrel distortion.

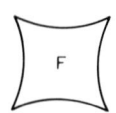

Fig. 25 Pincushion distortion.

Processing problems Assuming that all the mechanics and optics are aligned and performing properly, distortion can arise because of processing problems. The main processing steps are exposure and development. Improper control of either step can result in problems as serious as mechanical or optical distortion. Figure 26 presents three situations, shown by the fiducial target, which indicate proper

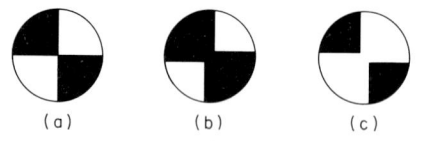

Fig. 26 (a) Correct exposure and development. (b) Overexposure or overdevelopment. (c) Underexposure or underdevelopment.

exposure and development, overexposure or overdevelopment, and underexposure or underdevelopment.

Lenses A variety of lenses are available for various reduction ratios. Most thick-film artwork will be produced with a scale factor of either 10:1 or 20:1, and the proper lens must be selected that will be compatible with the scale factor and therefore the reduction ratio. Table 5 lists a number of lenses that are available,

TABLE 5 Lenses Readily Available

Focal length, mm	Reduction range
45	20× to 50×
55	20× to 50×
70	15× to 38×
75	12× to 35×
105	10× to 25×
125	8× to 20×
210	4× to 11×
240	3× to 10×
250	3× to 10×

and their typical useful reduction ratios. Typical microphotography lenses are shown in Fig. 27.

Reduction cameras As indicated in the foregoing comments, the stability and accuracy of the reduction camera is of utmost importance. Inaccuracies that occur at this point in the hybrid thick-film circuit process are propagated throughout every step of the film printing and firing operations. It is for this reason that the selection and installation of the reduction camera should be done as carefully as possible.

Fig. 27 Microphotography lenses. (*R. W. Borrowdale Co.*)

Many very fine reduction cameras are available from different manufacturers, and it should be noted that many different types of cameras may be supplied by the same manufacturer. Instead of attempting to list the many variations available to the user, we present Figs. 28 to 32, which show examples of the many types of reduction cameras available.

Inspection systems After the plate or negative is processed, an inspection system is required so that processing inaccuracies may be measured. Although a single-axis measurement system, such as a microrule, will work well to establish

Fig. 28 Robertson Monitor III second-stage high-precision camera. (*Robertson Photo-Mechanix, Inc.*)

12-20 Thick Films

point-to-point dimensions, a two-axis unit will immediately detect coordinate errors or out-of-parallel conditions. Figure 33 shows a single-axis system being used to measure a processed photographic plate. Two-axis systems are available on the same basic frame.

Fig. 29 Robertson 31-in. multilayer camera. (*Robertson Photo-Mechanix, Inc.*)

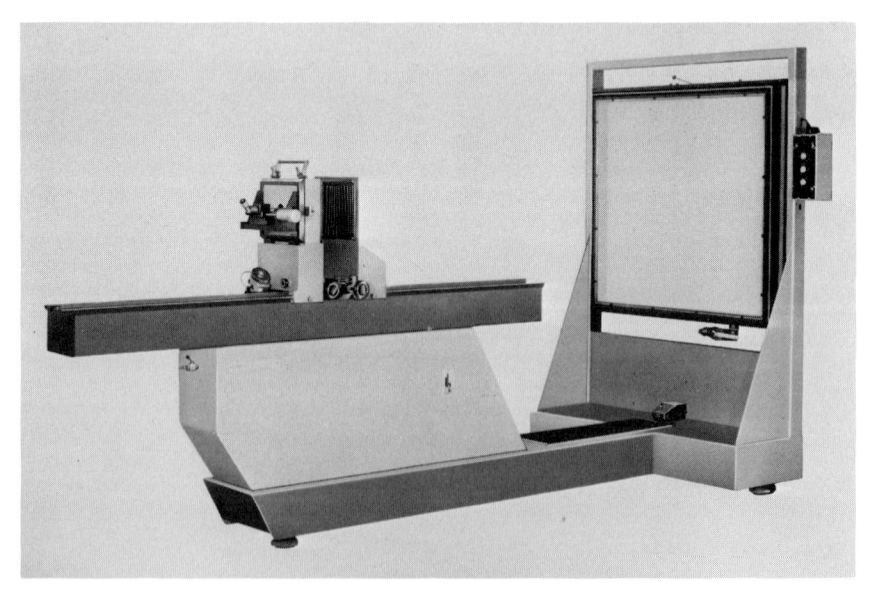

Fig. 30 Standard Dekacon III microphotography camera. (*HLC Manufacturing Co., Inc.*)

THICK-FILM SCREENS

The most common screens used for thick-film printing are made from a stainless steel mesh. The mesh is stretched taut across an aluminum frame (wood frames are sometimes used), and once the tension has been verified to be correct, a photoemulsion is applied to the mesh, and the screen is exposed to the negative and then developed to produce the finite pattern.

The choice of screen frame size is directly related to the size or area of the substrate to be printed. To obtain uniform and constant screening conditions, the screen frame should be at least large enough to permit a 2-in. border of unused screen around the substrate area. This will allow uniform deflection of the screen in the area directly over the substrate being printed.

In addition to the screens made from stainless steel mesh, there are suspended metal mask screens, solid metal etched screens and nylon or Nitex screens. Those

Fig. 31 "Pioneer" photoreduction camera. (*Borrowdale Photomechanical Systems, R. W. Borrowdale Co.*)

other than stainless steel mesh will be treated as special cases, and most of the information furnished here will relate to the stainless steel mesh, unless otherwise stated.

Mesh sizes available Table 6 is indicative of the large number of mesh sizes available. The information contained in this mesh dimension table is limited, but from it a great deal of other information may be gathered.

Screen tension In order to maintain printing repeatability and consistency, screen-tension tolerances must be given serious consideration. A tension deflection tolerance of ±0.010 in. is not considered too severe by most manufacturers of hybrid thick-film circuits. It is only by maintaining careful control of screen tension that resistor printing and firing yields can be maintained at an economically feasible value.

Table 7 gives recommended screen-tension deflections for stainless steel screens stretched across a 5- by 5-in. screen frame.

12-22 Thick Films

Fig. 32 Borrowdale camera back with coordinate plotter. (*Borrowdale Photomechanical Systems, R. W. Borrowdale Co.*)

TABLE 6 Mesh Sizes Available

Mesh size	Material	Wire (thread) diameter, in.	Mesh opening, in.
80	Stainless steel	0.0037	0.0070
105	Stainless steel	0.0030	0.0065
150	Stainless steel	0.0026	0.0041
165	Stainless steel	0.0020	0.0042
200	Stainless steel	0.0016	0.0034
200	Stainless steel	0.0021	0.0029
230	Stainless steel	0.0015	0.0029
250	Stainless steel	0.0016	0.0024
325	Stainless steel	0.0011	0.0020
400	Stainless steel	0.0010	0.0015
196	Nitex	0.0017	0.0033
260	Nitex	0.0015	0.0023
306	Nitex	0.0014	0.0019
330	Nitex	0.0014	0.0017
354	Perlon	0.00157	0.00122
125	12×× silk	0.0026	0.0045

NOTE: The above meshes are most widely used. Other types are available.
SOURCE: Industrial Reproductions, Inc. (IRI).

Fig. 33 Single-axis Optiplot measurement system. (*R. W. Borrowdale Co.*)

Figure 34 shows screens being tested for tension deflection. Although most manufacturers of screens run a test of this nature before shipping, it is well to repeat the test before use. A tension test gage is used which measures the tension through controlled deflection. The principal components of the test instrument are a force gage and a dial indicator.

Screen emulsion application There are essentially two methods in broad use for the application of photosensitive emulsion to the stainless steel screens. These methods are called *direct* and *indirect*.

In the direct emulsion method, the sensitized emulsion is poured onto the surface of the screen and spread over the entire surface by a squeegee or doctor blade. This forces the emulsion through the screen, and it is leveled by hand from both sides of the screen surface. Typically the thickness of the emulsion increases the thickness of the screen by from 0.0004 to 0.001 in. This, of course, directly relates to

TABLE 7 Recommended Screen-tension Deflection Values

Mesh size	Deflection, in.	Tolerance, in.
105	0.035	±0.010
150	0.035	±0.010
165	0.050	±0.010
200 (0.0016-in. wire)	0.055	±0.010
200 (0.0021-in. wire)	0.045	±0.010
230	0.055	±0.010
250	0.050	±0.010
325	0.065	±0.010

12-24 Thick Films

the thickness of the print applied. Figure 35 shows a cross section of a screen with direct emulsion applied.

An indirect emulsion screen is prepared by the application of a transfer film of a gelatin consistency backed up by a Mylar sheet. The transfer film is exposed and developed before application to the stainless steel screen. While the film is still wet from the developing process, it is applied to the screen. After it dries, the gelatin film adheres to the screen surface, and the Mylar backing film may

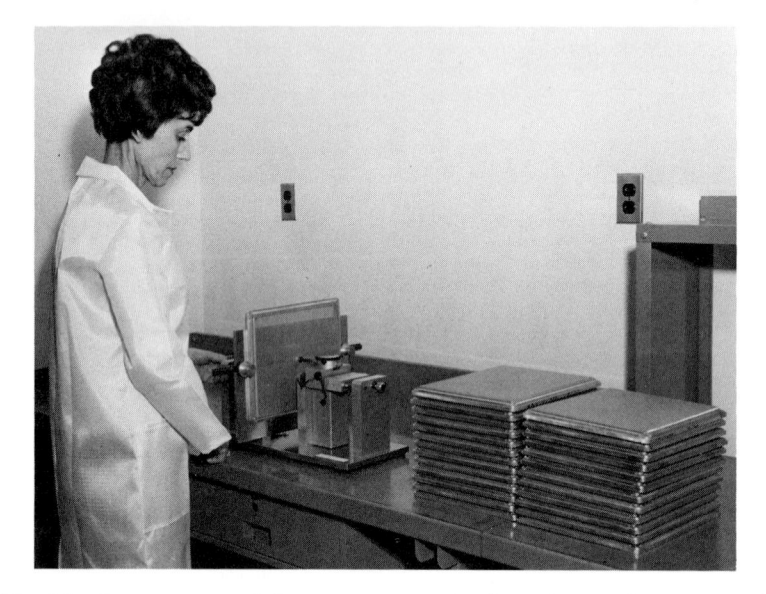

Fig. 34 Screen-tension deflection testing. (*Industrial Reproductions, Inc.*)

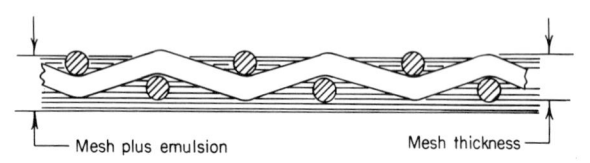

Fig. 35 Direct emulsion.

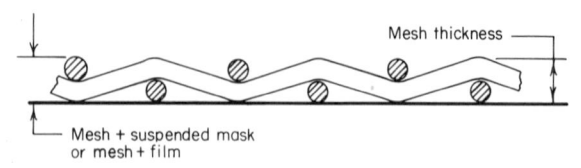

Fig. 36 Indirect emulsion.

be peeled away. Figure 36 shows pictorially how the indirect emulsion is suspended on the screen surface.

There are numerous pros and cons concerned with direct versus indirect emulsion screens. Table 8 gives some of them.

In addition to the thickness of the emulsion applied to the screen, the mesh size will also have a direct relationship to the thickness of the printed conductor, insulator, or resistor. Naturally as the resistor thickness varies, so does the resis-

TABLE 8 Direct versus Indirect Emulsion

Criterion	Direct	Indirect
Pattern detail	Good	Superior
Wear	10–25,000 prints	2–3,000 prints
Thickness of print	Controllable	Fixed

tance. A thick resistor will have a lower resistance value than a thin resistor, all other things being equal. Likewise, a small tolerance variation will have less effect on a thick resistor because the percentage change on the overall bulk of the item will be less. It is also interesting to note that as the resistor gets thinner, the TCR value becomes smaller.

A study by Du Pont, on four of their resistor formulations is shown in Table 9. A great deal of interesting information concerning screens may be derived from this table.

TABLE 9 Mesh Size and Film Thickness Comparisons

Du Pont composition no.	Mesh	Film thickness, mils Dry	Film thickness, mils Fired	Resistance, kilohms/sq	Hot (125°C) TCR
7826	105	1.64	0.99	0.153	662
	165	0.88	0.66	0.313	646
	200	0.75	0.45	0.532	594
	325	0.39	0.36	1.034	547
7827	105	1.55	0.94	1.245	404
	165	1.13	0.77	1.48	405
	200	0.82	0.55	2.051	319
	325	0.56	0.28	3.953	229
7828	105	1.48	0.98	2.65	148
	165	1.09	0.73	3.52	124
	200	0.84	0.58	4.72	81
	325	0.57	0.31	7.86	2.8
7832	105	1.59	1.10	5.56	65
	165	1.29	0.75	6.43	58
	200	0.99	0.53	8.23	31
	325	0.73	0.35	13.18	19

SOURCE: E. I. du Pont Data Sheet A-53826-3/67.

Test method was to screen-print a pattern of four squares 0.250 × 0.750 on a 1 × 1 × 0.025 AlSiMag 614 substrate with a Presco 100B, oil-check, and rigid-squeegee. Terminations were Du Pont 7553 Pd-Au. Terminations fired in box furnace, 1000°C, 10-min soak; resistors fired standard curve, 760°C peak in continuous-belt furnace. Dry-film thickness measured by calibrated optical focus. Fired film measured with mechanical gage.

Exposure While most facilities will expose the screen to the exposure light on an individual basis, the process may be accomplished in a multiple manner, using a deep-well vacuum frame.

The most important step associated with screen exposure is the alignment of the screen mask to the negative. This can be accomplished only if the artwork was prepared under the proper coordinate limitations. For instance, if a 200- by 200-mesh screen is to be used, the artwork pattern must be laid out on multiples of 0.005 in. This will permit the alignment of the pattern on the negative, to

12-26   Thick Films

the mesh of the screen. Any deviation from this will result in the negative-pattern edges falling between mesh lines, and will result in "stairstepping" or "sawtoothing" of the final print. Figure 37 shows the inspection of a screen for this condition. The tools needed are a light box and a microscope. For alignment of the negative pattern to the screen before exposure, a yellow light in the light box will be

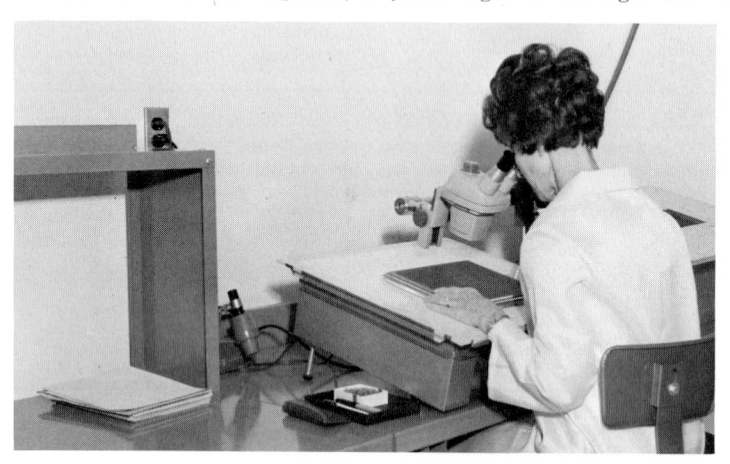

Fig. 37   Screen inspection techniques.   (*Industrial Reproductions, Inc.*)

Fig. 38   Multiple screen exposure.   (*Industrial Reproductions, Inc.*)

required. Figure 38 shows four screens held in a deep-well vacuum frame, being exposed to a carbon-arc light source. The negative outline can be noted. The negatives are held in position in relation to the screen mesh by the simple means of transparent adhesive tape.

Metal mask screens   As with regular mesh screens, there are two types of metal screens available: indirect and direct.

The indirect metal mask is formed by bonding a preetched foil to a stainless steel mesh which has been assembled to a screen frame. Most often, the foil thickness is 0.001 in., although some applications have required 0.002 in. The advantage of the metal mask over direct emulsion is close control of the paste deposited and resistance to pattern breakdown. It also is capable of reproducing intricate patterns with well-defined lines. Fine-line printing requires metal masks.

Tolerances with indirect metal masks can be held to the diameter of two wires of the mesh. This is required because the process is indirect, whereby contact of the mask to the mesh results in an immediate bond. Consequently, it is not possible to have the latitude in lining up the pattern under the microscope, as with a direct emulsion process.

Deposition is controlled because the etched foil is of a specific thickness and wire diameter can be easily predetermined. This means that more repeatable ink deposit is possible than with direct emulsion screens.

Pattern breakdown is forestalled in that an image formed by metal will withstand abrasion and the organic solvents of the pastes far better than one formed with emulsion. It should be remembered, however, that in applications where overstressing of the wire is the determining factor of screen life, this will not be improved with indirect metal masks.

Direct metal masks exhibit all the qualities of indirect masks and several more. The direct process involves etching the "mesh" and the pattern from opposite sides of a metal foil. These can be perfectly registered before etching. Mesh interference is far less of a problem than with an indirect approach. If a grid with a standard number of lines per inch is used, direct masks can be fabricated reasonably quickly, and at a moderate price. If, however, a pattern is so intricate, complex, or critical that a stock grid will offer intolerable interference, a special grid may be required. This has been done successfully, but obviously cannot be offered as inexpensively as a stock grid register.

An important advantage that the direct mask affords over all other types of printing screens is potentially longer screen life. This is because a solid piece of metal supports the image, and there is far less stretching than with conventional screening material. Since the largest single cause of deterioration is overstressing of the wire, the direct mask has a definite advantage. It should also be emphasized here that, if properly fastened and tensioned to the frame, direct metal masks can be successfully printed with conventional off-contact screening equipment and techniques. The mask will deflect and snap back if properly used. It is more common, however, to use contact printing when the process calls for metal masks.

SUBSTRATE MATERIALS

Ceramics are the most widely used materials for thick-film microcircuit substrates. One reason for this popular acceptance is the ability of the ceramic substrate to withstand temperatures far in excess of 1000°C. This is important since thick-film materials are fired at temperatures of about 1000°C and lower. There are many factors to be considered in selecting a substrate for a given thick-film system. These include, but are not necessarily limited to:

    Material
    Surface finish
    Flatness
    Camber
    Thermal coefficient of expansion
    Thermal conductivity
    Dielectric constant
    Water absorption
    Specific gravity
    Permeability
    Flexural strength
    Compressive strength

12-28   Thick Films

Volume resistivity
Dielectric strength
Loss tangent
Loss factor
Operating temperature capabilities
Compatibility to thick-film materials

Substrate properties   Considering the number of variables that this list could present to the design engineer, it is fortunate that the thick-film industry has pretty much standardized on 96 percent alumina (Al_2O_3). Although other ceramics, such as forsterite, beryllia, titanium dioxide, and steatite, can and are used, the bulk of applications are accomplished with 96 percent alumina. The primary reason is that this material (Al_2O_3) contains the necessary properties, both physical and electrical, that make it essentially compatible with the resistor, conductor, and insulating materials commonly used for the fabrication of hybrid thick-film microcircuits.

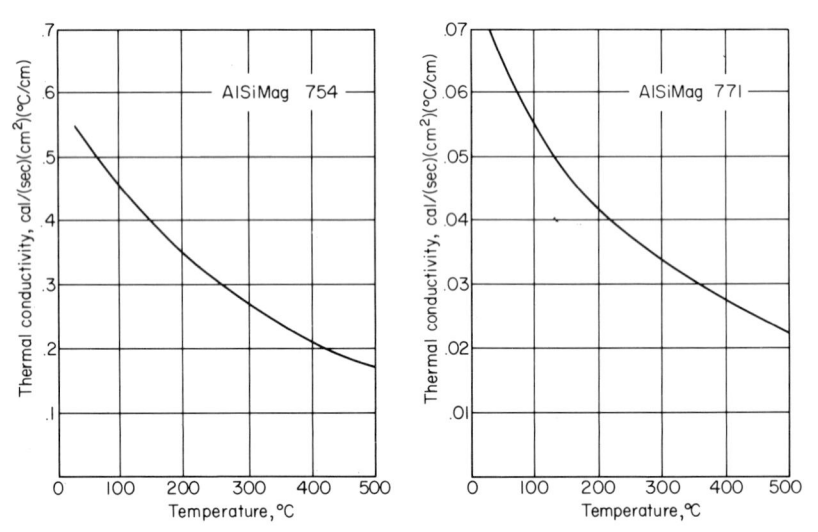

Fig. 39   Thermal conductivity. Note change in vertical scale. (*American Lava Corp.*)

The temperatures utilized in the manufacture of the ceramic substrates are approximately 1500 to 1900°C, and this high-temperature fabrication makes them naturally immune to the lower temperatures (1000°C and less) of thick-film processing.

Next to alumina, beryllia is the other most common substrate material, and beryllia is used only where high thermal conductivity is required. A comparison of the thermal conductivity of alumina and beryllia is shown in Fig. 39. The AlSiMag 754[*] shown on the left is a beryllia, while the AlSiMag 771[*] shown on the right is an alumina.

A third substrate material which is not commonly used is steatite. Steatite substrates were used during the early days of thick-film circuits primarily due to a cost advantage. The price reductions of alumina in large quantities, plus the many technical advantages, has relegated steatite to special uses where relatively thick bodies are required, such as discrete resistors and potentiometers. A comparison chart of the properties of steatite, alumina, and beryllia is shown in Table 10.

Substrate machinability   Cutting or drilling of cured ceramics is extremely difficult, and as a result, these operations are costly. In the case of beryllia, the

---

[*] Trademark of American Lava Corp., Ridgefield, N.J.

TABLE 10 Properties of Ceramics

Property[a]	Steatite MgO·SiO$_2$ AlSiMag 665[b] L 533	Alumina Al$_2$O$_3$ AlSiMag 576[c] L 524	Alumina Al$_2$O$_3$ AlSiMag 771[d] L 624	Alumina Al$_2$O$_3$ AlSiMag 614[e] L 624	Beryllia* BeO AlSiMag 754[f] L 623
Water absorption, %	0 Impervious	0 Impervious	0 Impervious	0 Impervious	0 Impervious
Specific gravity	2.7	3.40	3.62	3.70	2.88
Color[g]	White	White	White	White	White
Safe temperature at continuous heat: °C / °F	1000 / 1832	1100 / 2012	1500 / 2732	1550 / 2822	1500 / 2732
Hardness: Mohs' scale / Rockwell 45 N	7.5 / —	9 / 69	9 / 77	9 / 78	9 / 65
Linear coefficient of thermal expansion, 10^{-6} in./(in.)(°C): 25–300°C / 25–700°C / 25–900°C	6.9 / 7.8 / 8.0	6.5 / 7.5 / 7.2	6.0 / 7.2 / 7.4	6.4 / 7.5 / 7.9	6.0 / 7.8 / 8.5
Tensile strength, psi (kg/cm^2)	10,000 (700)	20,000 (1,400)	20,000 (1,400)	25,000 (1,760)	—
Compressive strength, psi (kg/cm^2)	90,000 (6,330)	275,000 (19,330)	315,000 (22,150)	375,000 (26,360)	>185,000 (>13,000)
Flexural strength, psi (kg/cm^2)	21,000 (1,470)	35,000 (2,460)	44,000 (3,090)	46,000 (3,230)	25,000 (1,760)
Resistance to impact, in.-lb (m-kg)	5.0 (0.058)	6.3 (0.073)	6.5 (0.075)	7.0 (0.081)	—
Modulus of elasticity, psi × 10^{-6} (kg/cm^2 × 10^{-6})	16 (1.12)	29 (2.04)	43 (3.02)	47 (3.30)	49 (3.45)
Shear modulus, psi × 10^{-6} (kg/cm^2 × 10^{-6})	6 (0.42)	12 (0.84)	17 (1.20)	19 (1.34)	19 (1.34)
Poisson's ratio	0.23	0.21	0.22	0.22	0.29
Thermal conductivity, (Btu)(in.)/(hr)(ft^2)(°F)[(cal)(cm)/(sec)(cm^2)(°C)]	41 [0.014] / 29 [0.010] / 26 [0.009] / —[—]	145 [0.050] / 70 [0.024] / 52 [0.018] / 35 [0.012]	192 [0.066] / 99 [0.034] / 70 [0.024] / 46 [0.016]	244 [0.084] / 119 [0.041] / 75 [0.026] / 58 [0.020]	1596 [0.55] / 813 [0.28] / 493 [0.17] / 232 [0.08]
Dielectric strength at 60 Hz ac, volts/mil (kv/mm) Test disks ¼ in. thick	230 (9.1)	200 (7.9)	210 (8.3)	210 (8.3)	220 (8.7)
Volume resistivity, ohm-cm: 25°C	>10^{14}	>10^{14}	>10^{14}	>10^{14}	>10^{14}
100°C	1.0 × 10^{14}	2.0 × 10^{13}	7.0 × 10^{13}	2.0 × 10^{13}	>10^{14}
300°C	1.4 × 10^{11}	5.0 × 10^{10}	4.4 × 10^{10}	1.1 × 10^{10}	>10^{14}
500°C	3.0 × 10^8	1.0 × 10^8	2.5 × 10^8	7.3 × 10^7	1.0 × 10^{13}
700°C	5.0 × 10^6	3.0 × 10^6	7.0 × 10^6	3.5 × 10^6	1.0 × 10^{11}
900°C	8.0 × 10^5	4.0 × 10^5	6.9 × 10^5	6.8 × 10^5	3.0 × 10^9
Te value[h]: °C / °F	870 / 1598	800 / 1472	840 / 1544	840 / 1544	1400 / 2552

* WARNING: In working with beryllia ceramics personnel should avoid exposure to dust- or fume-producing operations, such as sawing, grinding, drilling, or processing in moist atmospheres at high temperatures. Specialized equipment is necessary to prevent the dispersal of the dust and fumes into the air.

[a] Measurements shown are average values from unglazed test pieces. Production articles may vary slightly, depending on size, shape, and method of manufacture.
[b] Low-dielectric-loss steatite. An economical ceramic metallized primarily by the low-temperature process.
[c] An economical, high-strength alumina with good electrical and mechanical characteristics. 85% Al$_2$O$_3$.
[d] Good electrically and mechanically. Easily metallized by moly process. 94% Al$_2$O$_3$.
[e] High-strength, low-loss alumina for mechanical and electrical applications. 96% Al$_2$O$_3$.
[f] A dense 99.5% BeO with excellent electrical and thermal transfer properties for exceptionally demanding electronic and heat-sink applications. Possesses excellent nuclear properties.
[g] Standard glaze is white. Brown and other colors available on some bodies.
[h] Te value is the temperature at which a centimeter cube has a resistance of 1 megohm.

TABLE 10 Properties of Ceramics (Continued)

Property[a]	Steatite MgO·SiO$_2$ AlSiMag 665[b] L 533 25°C	Alumina Al$_2$O$_3$ AlSiMag 576[c] L 524 25°C	300°C	500°C	800°C
Dielectric constant:[i]					
1 MHz	6.3	8.3	8.4	9.8	15.3
1 GHz	—	8.1	—	—	—
10 GHz	6.0	8.0	8.1	8.3	—
25 GHz	—	—	—	—	—
Dissipation factor:[i]					
1 MHz	0.0008	0.0004	0.0023	0.0087	0.0594
1 GHz	—	0.0012	—	—	—
10 GHz	0.0020	0.0015	0.0021	0.0030	—
25 GHz	—	—	—	—	—
Loss factor:[i]					
1 MHz	0.0050	0.0033	0.0193	0.0853	0.9088
1 GHz	—	0.0097	—	—	—
10 GHz	0.012	0.0120	0.0170	0.0249	—
25 GHz	—	—	—	—	—

SOURCE: American Lava Corporation.
[i] AlSiMag 665 measured wet at 1 MHz, after immersion in water for 48 hr (MIL-I-10A).

dust caused by machining, drilling, grinding, or cutting can be hazardous to health. Even where thick-film resistors deposited on beryllia substrates are abraded to value, special precautions must be taken to preclude the inhalation of beryllia dust as a result of the abrasion.

The easiest way to obtain holes in substrates, or any other special sizing, shaping, or other mechanical detail, is to accomplish this in the "green" state. The "green" state refers to a condition of the ceramic, prior to curing, at which time it can be easily marked, by punching, drilling, molding, or pressing. Special requirements are most easily implemented by the ceramic vendor, rather than at the thick-film circuit manufacturer's plant after receipt of the cured substrates. An example of the complexity of hole patterns and sizes that can be obtained is shown in Fig. 40.

Machining can be accomplished on the cured ceramic substrate if it is absolutely necessary. Diamond-tipped drills can be used for holes, ultrasonic impact grinding (sonic drilling) can be used for holes or odd-shaped cuts, and a diamond scribe can be used to scribe and break the ceramic so as to reduce the size in one or both dimensions. As mentioned before, machining at this point in time is both difficult and costly. Figure 41 shows a machine especially designed to scribe and break ceramic substrates. It mechanizes the diamond scribe operation.

Flatness One last comment concerning substrates: Flatness can be very important in further processing. Most thick-film processes, including the assembly, use vacuum chucks and heat stages. If the substrate is badly warped, the vacuum chuck will draw air under the substrate, effecting a cooling action. In die- or wire-bonding

	Alumina Al$_2$O$_3$							Beryllia* BeO			
	AlSiMag 771^d L 624			AlSiMag 614^e L 624				AlSiMag 754^f L 623			
25°C	300°C	500°C	800°C	25°C	300°C	500°C	800°C	25°C	300°C	500°C	800°C
8.9	9.1	11.5	19.2	9.3	9.5	10.8	22.4	6.5	6.6	6.9	7.9
8.9	9.1	—	—	9.3	9.4	9.5	—	—	—	—	—
8.8	9.0	9.2	9.5	9.2	9.3	9.4	9.9	6.4	6.4	6.6	6.8
8.7	8.9	9.1	9.4	9.0	9.1	9.2	9.3	6.3	6.3	6.5	6.7
0.0003	0.0005	0.0078	0.0260	0.0003	0.0027	0.0131	0.0911	0.0001	0.0001	0.0004	0.0025
0.0010	0.0020	—	—	0.0003	0.0007	0.0015	—	—	—	—	—
0.0010	0.0010	0.0020	0.0030	0.0009	0.0010	0.0019	0.0062	0.0003	0.0003	0.0004	0.0005
0.0009	0.0009	0.0011	0.0043	0.0009	0.0009	0.0019	0.0057	0.0040	0.0040	0.0045	0.0063
0.0027	0.0046	0.0897	0.4992	0.0028	0.0257	0.1415	2.041	0.0007	0.0007	0.0028	0.0200
0.0089	0.0182	—	—	0.0028	0.0066	0.0143	—	—	—	—	—
0.0088	0.0090	0.0184	0.0285	0.0082	0.0093	0.0179	0.0614	0.0019	0.0019	0.0026	0.0034
0.0078	0.0080	0.0100	0.0404	0.0081	0.0082	0.0175	0.0530	0.0252	0.0252	0.0293	0.0422

* WARNING: In working with beryllia ceramics personnel should avoid exposure to dust- or fume-producing operations, such as sawing, grinding, drilling, or processing in moist atmospheres at high temperatures. Specialized equipment is necessary to prevent the dispersal of the dust and fumes into the air.

operations the leakage of air between the substrate and the heat stage can seriously affect the actual substrate temperature and thereby affect the bonding process. As a result of this, a flatness specification should be imposed on the substrate. As an example, a one-inch-square substrate should have a flatness of less than 0.005 in. from corner to corner.

THICK-FILM CERMET MATERIALS

Thick-film resistors The heart of the thick-film resistor industry is the palladium-silver resistor. Extensive laboratory and field testing has gone into this material system, and extensive testing has confirmed the high level of thermal and electrical stability that exists in the palladium-silver resistor material. Long-term life testing has been conducted on this basic resistor at 150°C in power densities of 50 watts/in.^{2}, and there are no data to indicate that there has ever been a catastrophic failure, even though test data accumulations from the major manufacturers indicate over 25 million component test hours.

Long-term drift and load-life stability are of major concern in any resistor material system, and load-life and drift curves are shown in Figs. 42 to 45 for 50, 500, 3,500, and 15,000 ohms/sq, respectively. It will be noticed that firing temperatures have a finite effect on the load-life characteristics.

Upon receipt of new lots of paste, the user should redisperse the solids in the

12-32   Thick Films

organic vehicle. This can usually be accomplished by hand stirring with a stainless steel spatula until no sediment remains at the bottom of the container and no lumps remain in the paste.

A laboratory stirrer may then be used to blend and completely disperse the solids in the vehicle system. If high stirring speeds are used or air is beaten

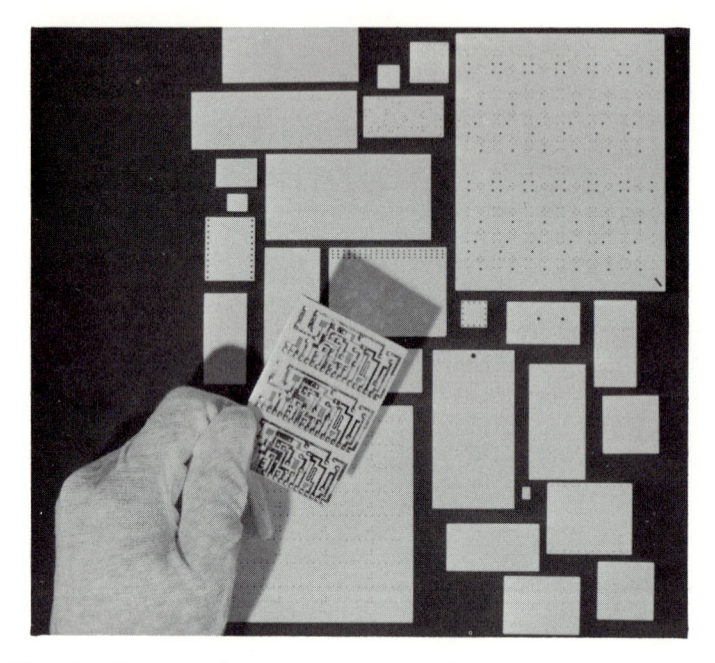

Fig. 40 Example of substrate type availability. (*Coors Ceramics.*)

Fig. 41 Substrate scribing machine. (*Mechanization Associates.*)

into the paste, the viscosity can be temporarily but drastically changed. Therefore, if viscosity measurements are to be made on vigorously stirred pastes, the paste must be stored on a roller storage device for 12 to 16 hr before accurate readings will be obtained.

If hand stirring only is used (or very mild mechanical stirring), the paste viscosity will not be significantly affected.

After initial dispersion, the pastes can be stored on a slowly rotating device so that the solids are prevented from settling out again. Such storage devices can be simply built, or commercial jar rolling mills can be used. The speed is not critical. Speeds of 3 to 10 rph are satisfactory. The basic consideration is that the solids be prevented from settling out. With this type of storage, the pastes remain homogeneous and can be quickly put into use on the production line.

Resistor compositions contain high-boiling, low-vapor-pressure solvents. During use or while jars are open, solvent will evaporate and should be replaced. This can simply be accomplished by hand stirring in the recommended solvent shown on the label until the original screening viscosity is achieved.

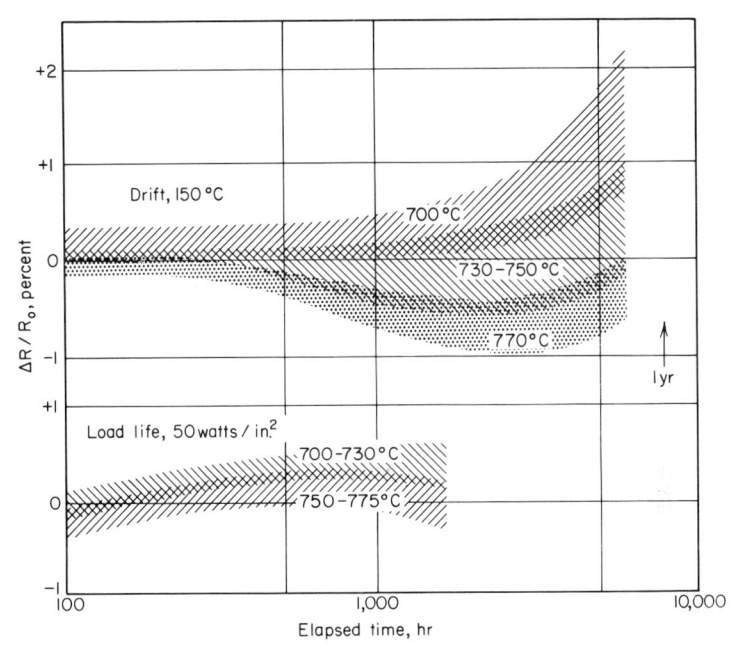

Fig. 42 Drift and load-life stability for 50-ohm/sq material. (*E. I. du Pont de Nemours & Co.*)

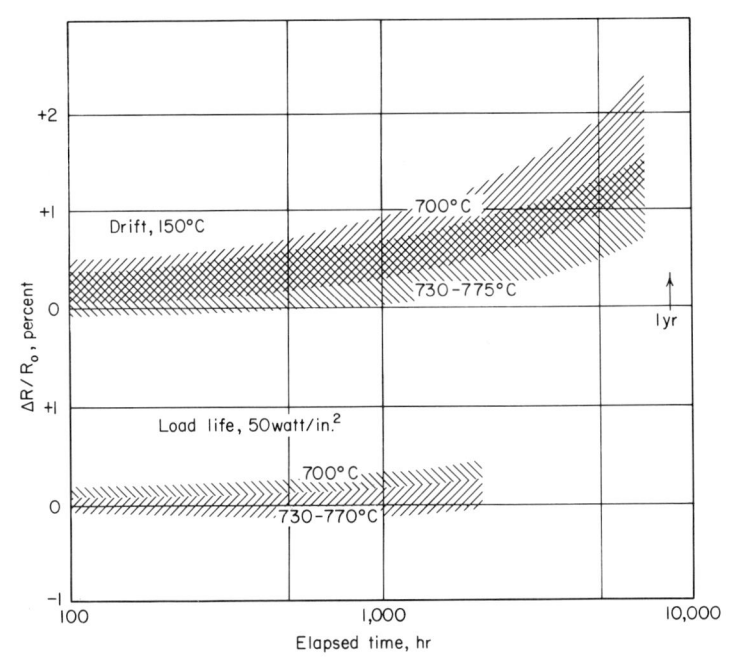

Fig. 43 Drift and load-life stability for 500-ohm/sq material. (*E. I. du Pont de Nemours & Co.*)

12-34 Thick Films

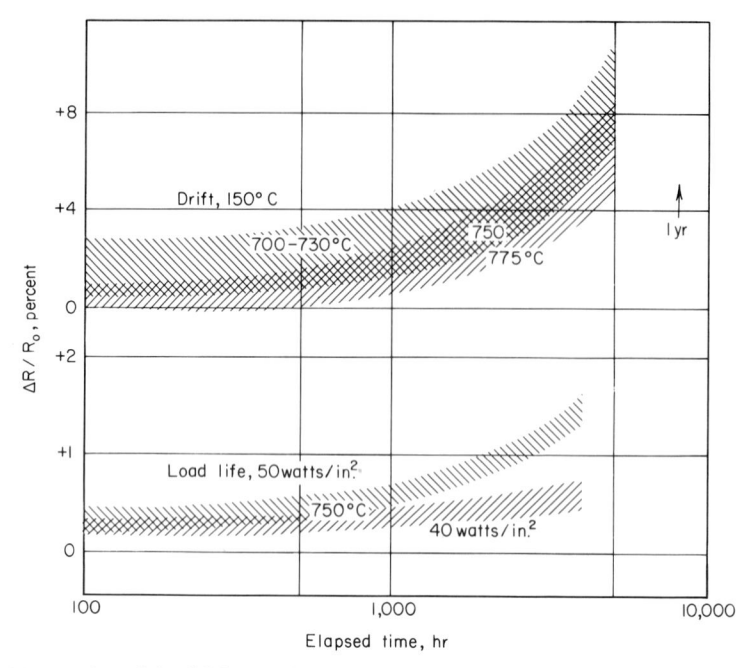

Fig. 44 Drift and load-life stability for 3,500-ohm/sq material. (*E. I. du Pont de Nemours & Co.*)

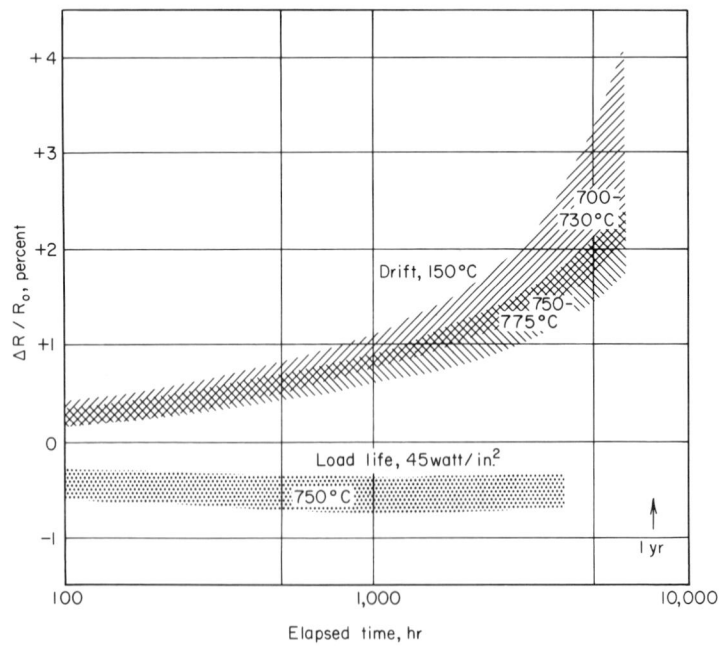

Fig. 45 Drift and load-life stability for 15,000-ohm/sq material. (*E. I. du Pont de Nemours & Co.*)

With resistor compositions, it is important that dilution be monitored because more than the original amount of solvent will change the density of the paste and result in a film-thickness variation that will change the resistor electrical properties. A viscometer is a good monitoring device for determination of proper solvent replacement. When preparing blends of pastes (to achieve intermediate resistance values, for example), each jar of paste should be stirred thoroughly before the required amount of composition is removed. When a small amount of one component is to be blended with a large amount of another, comparable volume of the larger component should be added to the smaller one, followed by good agitation. Subsequently, the larger component should be added in small increments and stirred to uniformity after each incremental addition until the blending has been completed.

Figure 46 is illustrative of the type of devices available for slowly rotating the thick-film pastes to prevent settling. The rolling speed should be kept slow enough so that the solids do not centrifuge to the sides of the container, but fast enough so that the solids do not settle. A speed of 3 to 10 rph is generally used.

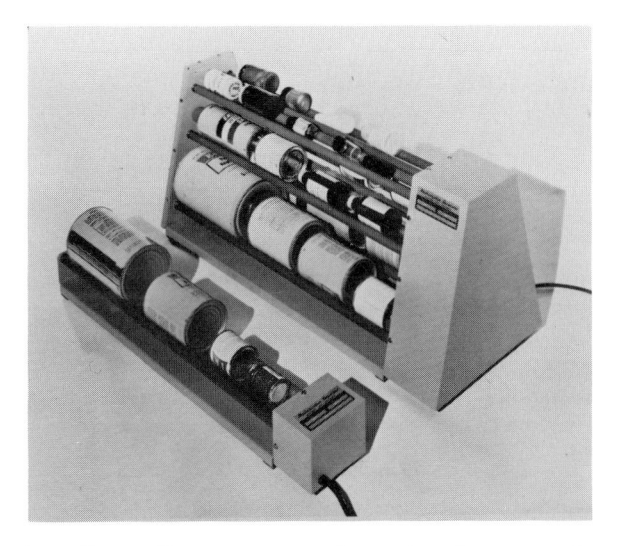

Fig. 46 Rolling mills to prevent settling. (*Mechanization Associates.*)

There are several variables which can affect the "as-fired" resistivity of any given value paste system. Awareness of these variables will permit higher yields and better repeatability of resistor characteristics. If we consider a series of resistor pastes designated as 7800 (Du Pont), it can be seen from Table 11 that the thickness of the screened and fired resistor will affect the TCR (temperature coefficient of resistance) to a large degree.

Taking this variation one step further, Figure 47 shows the effect of film thickness on the resistivity per square of a single paste.

Progressing another step further, we can observe how the TCR will vary with the resistivity of the thick film. This is illustrated in Fig. 48.

Noise is a characteristic that also varies with resistivity, and Fig. 49 shows the relationship between noise and resistivity on the same 7800 (Du Pont) paste system.

A look at Fig. 50 will indicate that encapsulation or glazing of the resistor materials, in addition to the actual trimming process, will vary the way that the load-life data occur. The load life is a measure of the drift rate of the resistor over a period of time, at an elevated temperature. The elevated temperature increases the rate at which any unoxidized palladium in the resistor combines with

TABLE 11 Effect of Film Thickness on Electrical Properties
(7800 Series Resistor Compositions)

Composition	Mesh	Film thickness, mils Dry	Film thickness, mils Fired	Resistance, ohms/sq	Hot (125°C) TCR
7800	105	1.23	0.66	0.15	133
	165	1.00	0.64	0.16	421
	200	0.59	0.45	0.21	643
	325	0.32	0.29	0.29	461
7826	105	1.64	0.99	153	662
	165	0.88	0.66	313	646
	200	0.75	0.45	532	594
	325	0.39	0.36	1,034	547
7827	105	1.55	0.94	1,245	404
	165	1.13	0.77	1,480	405
	200	0.82	0.55	2,051	319
	325	0.56	0.28	3,953	229
7828	105	1.48	0.98	2,650	148
	165	1.09	0.73	3,520	124
	200	0.84	0.58	4,720	81
	325	0.57	0.31	7,800	2.8
7832	105	1.59	1.10	5,560	65
	165	1.29	0.75	6,430	58
	200	0.99	0.53	8,230	31
	325	0.73	0.35	13,180	19
7860	105	1.68	0.98	9,830	403
	165	1.27	0.76	14,380	365
	200	0.89	0.50	32,900	262
	325	0.62	0.28	75,500	205

SOURCE: E. I. du Pont de Nemours & Co.

the oxygen in the ambient atmosphere, and this metallurgical combination causes a change in resistance, and thus the drift rate can be established.

As the hybrid thick-film industry matures, more companies are making available additional formulations of resistive, conductive, and insulating pastes. Yet, since there are so many variables which affect the "as-fired" resistance values of these pastes (thickness, time-temperature, batch variations, etc.) it is worthwhile investigating these possible variations. Table 12 shows two series of pastes, manufactured by Electro-Science Laboratories, Inc., with the possible batch variations that can be expected.

It is well to keep in mind that continued improvements are being made in resistive pastes which enhance the reproducibility of the obtainable resistors.

Resistor paste systems other than palladium-silver are appearing on the market. Some of the advantages of these new systems are increased stability, better repeatability from substrate to substrate and from batch to batch, and the lack of a protective film requirement. Since palladium-silver is susceptible to reduction by hydrogen atmosphere, it must be protected by an overglaze coat to assure resistor stability. Some of the new resistor systems claim the glaze coat to be unnecessary.

Any resistor system selected should be compatible with the conductor interface, and this resistor-conductor interface must be carefully considered before the materials are chosen. Parameters that can be measured are interface resistance, noise, and

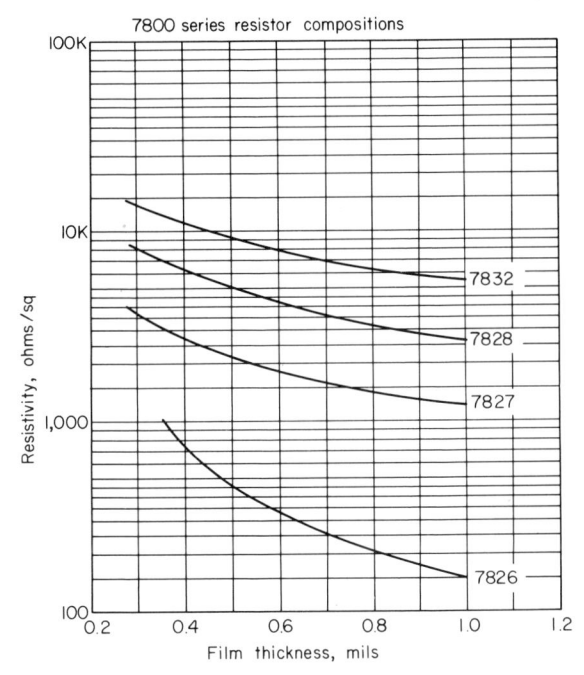

Fig. 47 Resistivity versus film thickness. (*E. I. du Pont de Nemours & Co.*)

metallic depletion from one material to the other. It is even important to consider the compatibility of any overglaze that might be used, with the conductor and resistor system.

Thick-film conductors Thick-film conductors essentially limit themselves to two categories of applications; these are soldering (lead-tin solders) and bonding (die

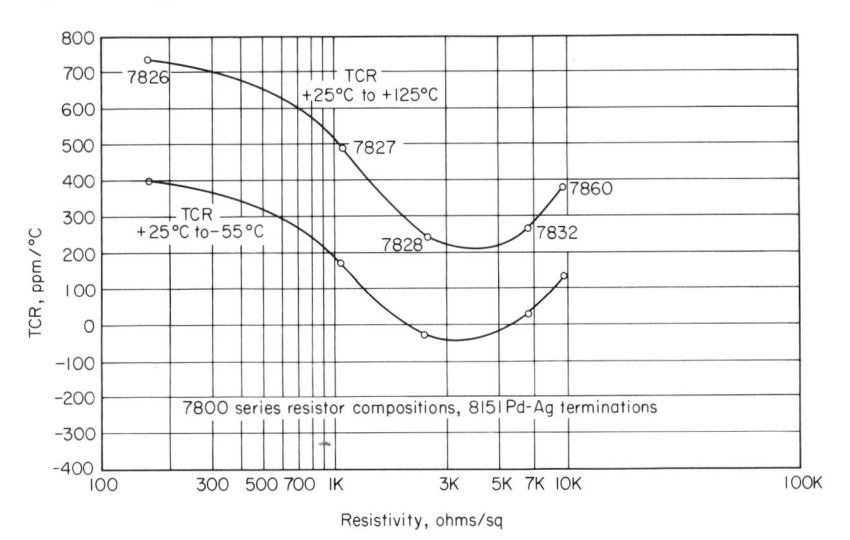

Fig. 48 TCR versus resistivity. (*E. I. du Pont de Nemours & Co.*)

12-38 Thick Films

and wire-thermal compression or ultrasonic). Of course, in pure resistive networks, where the conductors are merely used for interconnections, neither of these categories applies. The most popular conductor materials are listed in Table 13. An indication of their resistivity, which can be important in certain classes of circuits, is also given.

Each of the conductor materials listed in Table 13 has its own set of characteristics, advantages, and disadvantages. There are many manufacturers of these conductor materials, and each manufacturer's product differs from the others in some manner. It is, therefore, difficult to list all of the characteristics and parameters for each available material. A few, however, are given here for comparison. The characteristics of Du Pont gold conductors are listed in Table 14. It should be noted that, in general, gold conductor materials are not applicable to soft solder processes. The gold conductor will produce a highly dense film which is applicable to die and wire bonding, using either thermal compression or ultrasonic bonding techniques.

Platinum-gold conductors, on the other hand, are almost all compatible with soft solder processes. The characteristics of several platinum-gold conductors are listed in Table 15. A single palladium-gold is also listed. The excellent resistance

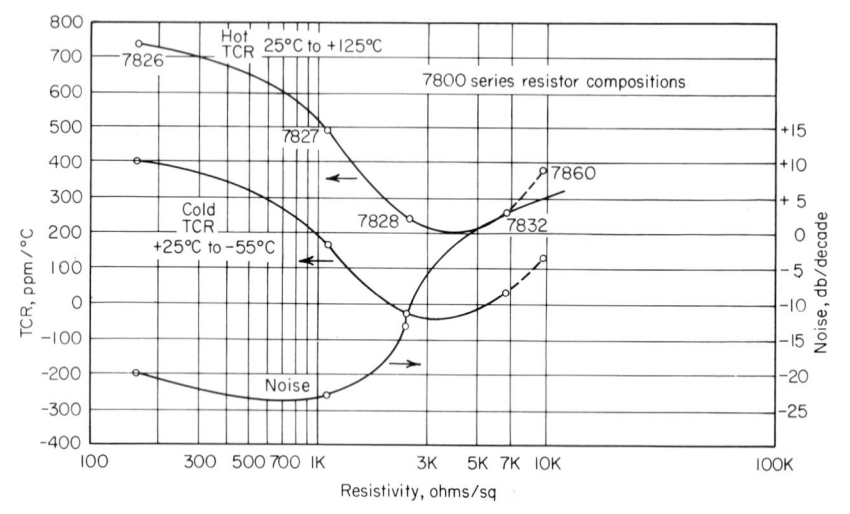

Fig. 49 Noise versus resistivity. (*E. I. du Pont de Nemours & Co.*)

of platinum-gold to solder leaching should be especially noted, as this is an important factor in any assembly that requires solder assembly.

Palladium-silver characteristics differ in many ways from the other metal formulations, and these characteristics are listed in Table 16. Primarily, palladium-silver was developed as a lower-cost solderable thick-film conductor.

Numerous combinations of thick-film conductor materials are under development by the suppliers of these products. The many criteria involved are always subject to improvement, and in any application involving high production rates, the primary criterion is cost. Economics involved in the final product, however, are not necessarily limited to material costs, and a less expensive material that would resist bonding or soldering, and thereby increase the labor content, would be of no advantage.

Table 17 summarizes the criteria associated with many of the conductor materials. The stock numbers of these materials are Du Pont's but the use area and bonding methods are similar, regardless of the supplier. It should be recognized, however, that even though a conductor has the same basic metallic content, the product may vary significantly from one supplier to another.

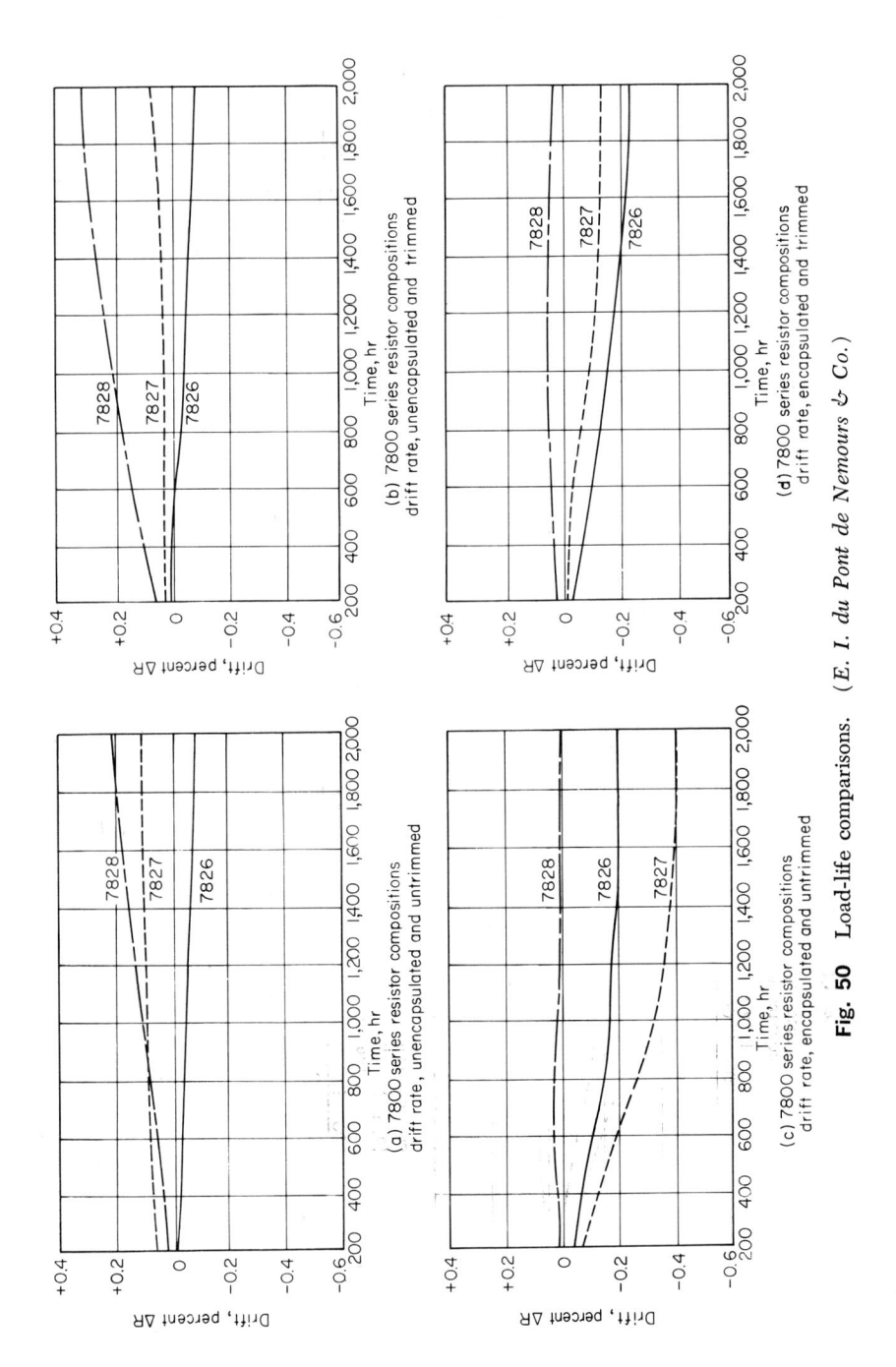

Fig. 50 Load-life comparisons. (*E. I. du Pont de Nemours & Co.*)

12-39

Thick Films

TABLE 12 Possible Batch Variations of Resistive Pastes

6900 series	7000 series	Nominal sheet resistivity, ohms/sq	Possible batch variation from nominal sheet resistivity, % 6900	7000	Typical TCR at −55 to +125°C, ppm per °C 6900	7000
6910		1	±50	—	+300–500	—
6950		5	±20	—	+300	—
6911		10	±15	—	+150	—
6951		50	±15	—	±150	—
6912	7012	100	±15	±20	±150	+150
6952	7052	500	±15	±15	±150	+150
6913	7013	1,000	±15	±15	±150	+150
6953	7053	5,000	±15	±15	±150	+150
6914	7014	10,000	±15	±15	±150	+150
6924	7024	20,000	±15	±15	−150	+150
6954 } Extended 6915 } ranges	7054 7015	50,000 100,000	±20 ±20	±15 ±20	−250 −350	±150 ±150
	7025 } 7045 } Extended 7016 } ranges	200,000 400,000 1,000,000	— — —	±25 ±25 ±30	— — —	−200–400 −300–600 −450–800

Higher and lower values of the 7000 series glazes are possible to obtain on an experimental basis.

Properties

Rheology............ Thixotropic pastes.
Brookfield viscosity.... (10 rpm; shortened No. 7 spindle; ¾ in. immersion) 150,000 ± 30,000 Hz at 26°C (after mild spatula stirring). Higher viscosities or gelled pastes can be made to special order. These are for printing thicker films so as to achieve tighter resistance distributions.
Shelf life............ 3–6 months; storage under mild vibration or slow rolling, may minimize settling of the dispersions. Sealing containers carefully and storing under refrigeration may be used to extend shelf life of pastes by minimizing the possibility of physical or chemical changes or interaction with the atmosphere.
Firing temperature.... 760–780°C peak.
Firing time.......... All pastes are calibrated in a 60-min cycle with 12–15 min at peak temperature. The higher the peak temperature, the less time is required at that temperature.

SOURCE: Electro-Science Laboratories, Inc.

Thick-film insulators Thick-film insulating materials (namely, glazes) have more than one purpose in a hybrid circuit. They provide insulation not only where two conductors may cross each other but also where a wire crosses a conductor. They hermetically seal resistors against the reducing effects of the atmosphere (namely, hydrogen), and they can also be used effectively to isolate pad areas so as to restrict solder flow or bonding to particular portions of the conductor. In

TABLE 13 Resistivity of Conductor Materials
(in ohms per square)

Gold	0.005
Platinum-gold	0.1
Palladium-gold	0.1
Palladium-silver	0.04
Silver	0.001

SOURCE: Varadyne, Inc.

TABLE 14 Gold Conductor Characteristics

Characteristic	Conductor			
	8115	DP 8237	DP 8233	DP 8260
Application	Screen print	Screen print	Screen print	Screen print
Viscosity, cp $\times 10^{-3}$	225–270	400–600	800–1,200	400–600
Brookfield instrument	HBF	HBF	HBF	HBF
Spindle No	4	6	6	6
Speed, rpm	5	10	5	10
Firing range, °C	850–1000	850–1000	850–1000	850–1000
Optimum firing, °C	950	950	95—	925
Solderability (60–40) at 215°C	No	No	No	Excellent
Refined solderability at 500–760°C	—	—	—	Excellent
Resistance to solder leaching, sec	—	—	—	20
Resistor termination	Poor	Fair	Poor	Good
Non-solder-bondable	Yes	Yes	Yes	Yes
Peel adhesion, lb/in. width	—	—	—	20
Sheet resistivity, milliohms	<10	<10	<10	20–30

SOURCE: E. I. du Pont de Nemours & Co.

addition to all these necessary uses, the insulating materials are also employed as the dielectric material for screened thick-film capacitors.

Glass pastes can be obtained that will fire at a wide range of temperatures, so that they may be integrated into the total screening and firing process. They may also be obtained in a variety of colors. As capacitors, they can be obtained over a wide range of sheet capacitances.

The glass materials generally consist of a suspension of vitreous particles suspended in a vehicle. At firing temperatures the glass will soften and coalesce into a continuous film. One of the important considerations to be aware of when using a thick-film glass as a crossover insulator is the dielectric constant of the glass. If the dielectric constant is too high, the crossover may act as a coupling capacitor. Another characteristic that must be considered is the leakage resistance of the glass. Like the capacitance problem, the resistance problem can also cause trouble. A high dc resistance will result in better dc isolation of the circuit, whereas a low dielectric constant will result in better ac isolation of the circuit.

Surface leakage is another parameter worth considering when dealing with critical circuits, and surface leakage will increase with humidity almost as soon as it exceeds 50 percent relative humidity.

To evaluate surface leakage, conductors can be printed on the crossover glass, and measurements may be taken at various voltage and humidity levels. Figure 51 shows an acceptable test pattern for measuring this phenomenon.

To evaluate the bulk leakage resistance, small conductors can be printed, one

TABLE 15 Platinum-Gold and Palladium-Gold Conductor Characteristics

	Conductor				
	Platinum-gold				Palladium-gold
Characteristic	7553	DP 8235	8036	8232	DP 8227
Application	Screen print	Screen print	Screen print	Screen print	Screen print
Viscosity, cp $\times 10^{-3}$	100–130	220–300	96–112	800–1,000	170–230
Brookfield instrument	HBF	RVT	HBF	HBF	RVT
Spindle No.	4	7	4	—	7
Speed, rpm	10	10	10	10	10
Firing range, °C	760–1000	760–1000	690–1000	760–1000	760–1000
Optimum firing, °C	1000	850	1000	1000	850
Solderability (60:40) at 215°C	Excellent	Excellent	Fair	Excellent	Excellent
Refined solderability at 500–760°C	Excellent	Excellent	Excellent	Excellent	Excellent
Resistance to solder leaching, sec	240	200	240	240	80
Resistor termination	Good	Fair	Good	Good	Very good
Non-solder-bondable	Yes	Yes	Yes	—	Yes
Peel adhesion, lb/in., width	5–10	26	15	21	28
Sheet resistivity, milliohms	80–100	80–100	80–100	80–100	80–100

SOURCE: E. I. du Pont de Nemours & Co.

TABLE 16 Palladium-Silver Conductor Characteristics

	Conductor				
Characteristics	8151	8153	8198	DP 8245	DP 8261
Application	Screen print	Dip	Brush	Dip band	Screen print
Viscosity, cp $\times 10^{-3}$	210–290	1.2–1.7	28–35	27–37	400–600
Brookfield instrument	HBF	RVT	RVT	HBF	HBF
Spindle No.	6	3	6	4	6
Speed, rpm	10	10	10	10	10
Firing range, °C	690–1000	690–1000	690–1000	690–1000	690–1000
Optimum firing, °C	850	850	850	850	850
Solderability (60:40) at 215°C	Excellent	Excellent	Excellent	Excellent	Excellent
Refined solderability at 500–760°C	Fair	Fair	Fair	Good	Very good
Resistance to solder leaching, sec	80	80	80	60	60
Resistor termination	Excellent	Excellent	Excellent	—	Very good
Non-solder-bondable	Yes	Yes	Yes	—	Yes
Peel adhesion, lb/in. width	30	30	30	28	30
Sheet resistivity, milliohms	40	40	40	—	50

SOURCE: E. I. du Pont de Nemours & Co.

underneath and the other on top of the crossover dielectric. A test pattern for measuring this parameter is shown in Fig. 52.

Probably the most important property that a crossover material must possess is a pinhole-free consistency after firing. Without this, there are bound to be short-circuit conditions occurring between the two conductor materials. In order to assure the minimum probability of short circuits it may be wise to consider two screenings and two firings of the glass materials.

A capacitor dielectric glass must be selected with several factors in mind. Since

TABLE 17 Conductor Criteria

Type	Number	Sheet resistivity, milliohms/(sq)(mil) (fired) film	Brookfield model	Spindle no.	Speed, rpm	Viscosity, cp × 10^{-3}	Adhesion (90° peel) lb	Firing temperature, °C	Bonding method	Use area
Pt-Au	7553	80–100	HBF	4	10	100–130	1–2	760–1000	Solder-TC bonding	Standard high reliability Pt-Au
Pt-Au	8208	80–100	HBF	6	10	1,200–1,600	1–2	760–1000	Solder-TC bonding	Fine-line printing Pt-Au
Au	448-6812R	80–100	HBF	6	10	1,200–1,600*	1–2	760–1000	Solder-TC bonding	Dense, fine line Pt-Au
Pt-Au	8036	80–100	HBF	4	10	88–120	1–2	760–1000	Solder-TC bonding	Acid plating Pt-Au
Pt-Pd-Au	448-7799R	80–100	RVT	7	10	170–230*	4–6	760–1000	Solder-TC bonding	High adhesion Pt-Pd-Au
Pd-Au	8206	80–100	HBF	4	10	170–230*	2–3	760–1000	Solder-TC bonding	Standard Pd-Au
Pd-Au	448-7708	80–100	RVT	7	10	170–230*	4–6	760–1000	Solder-TC bonding	High adhesion Pd-Au
Pd-Au	448-7559R	25	No specifications available yet				2–3	760–1000	Solder-TC bonding	Fine line Pd-Au nonsolder—more difficult to solder than 448-7781R
Pd-Au	448-7781R	120	No specifications available yet				2–3	760–1000	Solder-TC bonding	High R fine line Pd-Au for soldering
Au	8115	10	HBF	4	5	255–270	—	760–850	Eutectic die bonding	Low resistivity die bonding
Au	448-7674	10	No specifications available yet				—	760–850	Eutectic die bonding	Low-resistivity die bonding
Pd-Ag	8151	40	HBF	6	10	210–290	3–5	690–760	Solder-TC bonding	Standard screening Pd-Ag conductor
Pd-Ag	8183	40	RVT	3	10	1.2–1.7	3–5	690–760	Solder-TC bonding	Dip version 8151
Pd-Ag	8198	40	RVT	6	10	28–35	3–5	690–760	Solder-TC bonding	Brush version 8151

SOURCE: E. I. du Pont de Nemours & Co.
* Interim specifications.

Thick Films

a thick-film capacitor is really a system composed of a bottom conductor layer, a dielectric, and a top layer, the compatibility of materials and firing temperatures must be closely considered. Naturally the capacitance of a thick-film capacitor will depend on the geometry of the device (opposing electrode area), the dielectric

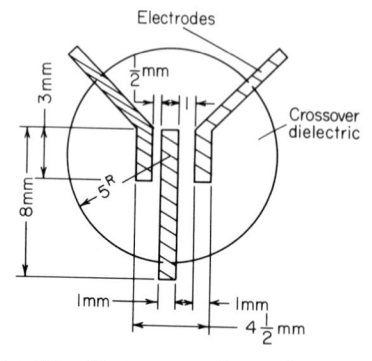

Fig. 51 Test pattern for surface conduction. (*E. I. du Pont de Nemours & Co.*)

Fig. 52 Test pattern for bulk conduction. (*E. I. du Pont de Nemours & Co.*)

constant of the glass material, and the thickness of the dielectric. Another factor that will affect the capacitance is the operating temperature. Figure 53 shows the capacitance of a system utilizing two Du Pont materials. The two curves represent the fact that electrodes and the dielectric were fired first at one temperature

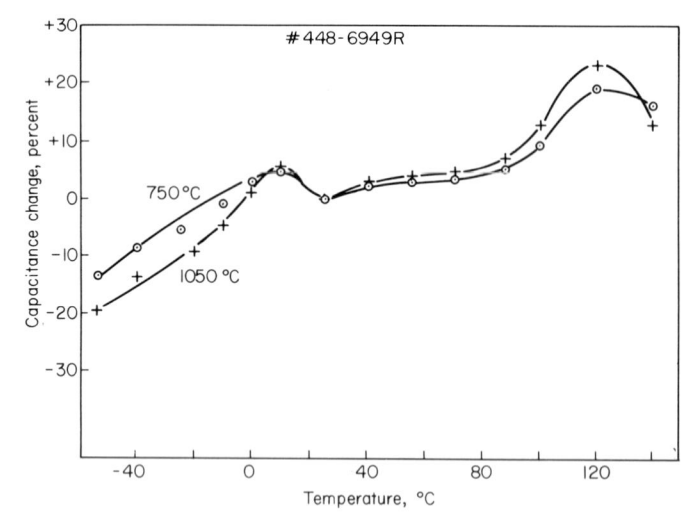

Fig. 53 Capacitance as a function of temperature. (*E. I. du Pont de Nemours & Co.*)

and then another. Dissipation factor can also be plotted as a function of temperature, and this is shown in Fig. 54.

The most important parameter to be considered when using a thick-film glass as an encapsulant is its resistance to moisture and gas. The firing cycle on glass encapsulants is designed so as to keep to a minimum the reaction between the encapsulant and the resistor that it is covering, so as to minimize any resistor change that occurs during the glass firing cycle. A measure of the effectivity of the glass encapsulant is to expose an encapsulated resistor to forming gas or hydrogen at high temperature. The resultant drift of the resistor under these conditions determines the effectiveness of the glass encapsulant in protecting the resistor.

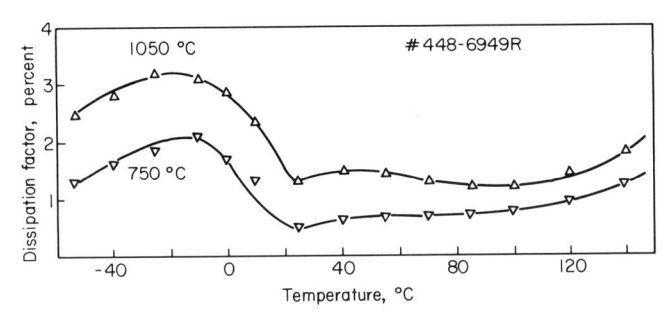

Fig. 54 Dissipation factor as a function of temperature. (*E. I. du Pont de Nemours & Co.*)

Once again, the importance of a pinhole-free surface cannot be overlooked. Also, the importance of protecting the abraded edge of a trimmed resistor should be considered. The process flow generally used calls for a coating of glass after the resistor has been fired. This first coat tends to stabilize the resistor value. Then, if the resistor is abraded to value through the glass, a second coating of glass can be applied to protect the abraded edge. This second coating of glass also has the effect of eliminating the pinholes that may have occurred in the first glass firing. Figure 55 shows a typical firing profile for an Electro-Science overglaze. Note that the maximum firing temperature required is about 500°C, and that the time through the furnace is only 30 min. The characteristics of the two Electro-Science glazes shown in Fig. 55 are listed in Table 18.

TABLE 18 Overglaze Characteristics

Property	4770	4771
Rheology	Thixotropic	Pastes
Viscosity (Brookfield)	>200,000 cp at 10 rpm, No. 7 spindle, ¾-in. immersion	
Drying	10–15 min at 120–140°C	
Firing, °C	500–525	525–540
Substrates	Matched to 96% alumina bodies	

SOURCE: Electro-Science Laboratories.

Active thick-film materials Studies have been conducted to pursue the feasibility of using thick-film screening and firing techniques for the fabrication of semiconductor devices. The advantages that a system of this sort could produce are obvious. If both the passive and active devices could be printed and fired, this could lead to the complete elimination of wire interconnections, the use of piece parts, a good portion of the assembly work, and most of the human factor. The high labor content in a thick-film hybrid circuit still lies in the assembly processes.

Cadmium sulfide (CdS) is an ingredient that has been used in experiments run by NASA (Langley)[*] to produce field-effect transistors. This, in combina-

[*] Applied Research in Thick Film Active Devices, *NASA Rep.* 66266, NASI-6285, Langley Research Center, Hampton, Va.

12-46 Thick Films

tion with silicon dioxide (SiO_2) as an insulating layer, was used to form thick-film field-effect transistors.[1] Figure 56 shows the construction of such a field-effect transistor.

Continued work is being carried on which some day may lead to the practical fabrication of the required variety of semiconductor devices needed, by the use

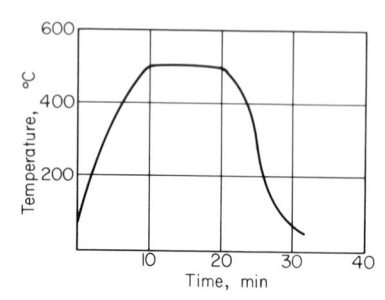

Fig. 55 Firing profile for overglaze. Note: Use 500 to 525° peak for No. 4770, 525 to 540° peak for No. 4771. (*Electro-Science Laboratories.*)

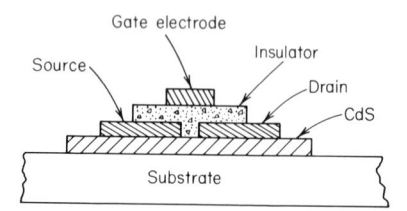

Fig. 56 Cross section of thick-film field-effect transistor. (*NASA Langley Research Center.*)

of thick-film processes. If and when this breakthrough does occur, it will succeed only if it offers economic advantages over the use of discrete parts utilizing advanced assembly procedures.

Dry-transfer tapes Departing from the common thick-film process of printing fluid pastes through screens, there appears the choice, in conductors, resistors, and insulators, of resorting to the use of dry-transfer processing. The transfer tape process allows the production of uniform layers of conductors and insulators. Thickness control over a range of 0.002 to 0.050 in. can be achieved.

The transfer tape is applied by pressure, which activates a special organic adhesive layer that transfers the material from the carrier tape, to the hybrid circuit surface.

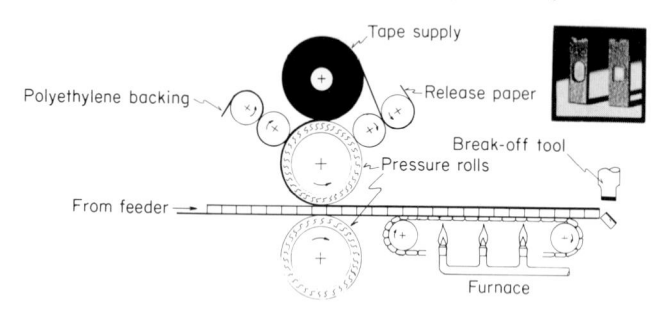

Fig. 57 Automated transfer tape process. (*Vitta Corp.*)

Figure 57 shows a flow diagram of a process for applying a low-temperature lead oxide sealing glaze to a ceramic surface. Resorting to this degree of automation can easily result in production speeds of 3,000 per hr.

The types of transfer tapes that are available are considerable, and a selection is listed as follows:
  Glass
  High-temperature ceramic metallizing
  Silver metallizing
  Gold

Platinum
Palladium
Gold-platinum
Other precious-metal combinations
Nickel
Resistors from 200 to 50,000 ohms/sq
Ceramics

The advantage of a dry-tape transfer process lies in the ability to exert thickness control on the materials before application to the substrate.

THICK-FILM PRINTING PROCESSES

The basic printing, or screening, process utilized for thick-film circuits is essentially the same as the silk-screening process developed by the Chinese centuries ago. The major differences revolve around the necessity for extreme accuracy of deposition, especially for resistors. In addition to deposition accuracy, registration accuracy is also of great importance.

Although it is necessary for a thick-film printing machine to be able to maintain very tight tolerances, it must also have extreme flexibility of adjustment. In addition to the ability to adjust screen height very accurately, it is also necessary to be able to adjust the substrate holder in x, y, and rotational positions. Only with a solidly built screen printer can reproducible results be achieved.

The basic elements of the screen printing machine are the platen, which holds the substrate, usually in a vacuum chuck, the frame for holding the screen, and a squeegee for forcing the paste through the selected openings of the screen.

There are many interrelations which affect the achievable reproducibility that can be obtained in the screen printer, and a portion of these have to do with the screen itself. Since the screen meters the paste through the mesh, the ratio of mesh openings to wire diameters assumes importance, as does the mesh count, the viscosity of the paste, the squeegee rate of travel and the weight on the squeegee. In addition, there are off-contact distances, contact breakaway, and squeegee drag to be considered.

The squeegee itself can be an item of intense study since 45, 60, and 90° angles are all used, as are a variety of materials. Items to be considered in squeegee selection are elasticity, durability, and coefficient of drag. The squeegee rate of travel must be maintained within tight limits, as an increase in the rate of squeegee travel will decrease the print thickness.

Table 19 shows the different parameters which must be considered in selecting a squeegee material and the associated squeegee hardness.

TABLE 19 Squeegee Parameters

Durometer	Classification	Life	Efficiency
30–50	Soft	Short	High
50–65	Medium	Satisfactory	Satisfactory
70–95	Hard	Long	Poor

Basic characteristics that should be sought when considering a screen printer are noted in the following list which has been abstracted from Du Pont Report A-53809:

1. Microadjustment and positive locking should achieve and maintain a screen position parallel to the substrate.

2. Microadjustment and positive locking should control the x,y axis to assure precise registration of the image with respect to the substrate.

3. Squeegee travel rate should be readily adjustable over a range of at least 2 to 10 in./sec.

12-48   Thick Films

4. Travel during the print stroke should be uniform and repeatable at any given preset speed.

5. If single-direction printing is to be employed, a flood bar, capable of repositioning a major portion of the paste to the printing area, should be provided.

6. Provision should be made for adjusting both squeegee pressure and limit of downward movement.

7. The platen carriage should move freely during substrate positioning, but must be constrained by positive stops during actual printing. Any movement of the carriage during printing will affect registration and may result in print smear.

8. There should be a ready means for adjusting the distance between the bottom of the screen and the top of the substrate (gap).

9. A rapid means of screen removal and installation of new screens is needed.

10. An efficient substrate feed, positioning, and takeoff mechanism is important.

Single-direction and double-direction printing techniques are used. Thickness control can be accomplished with slightly more precision in single-direction printing,

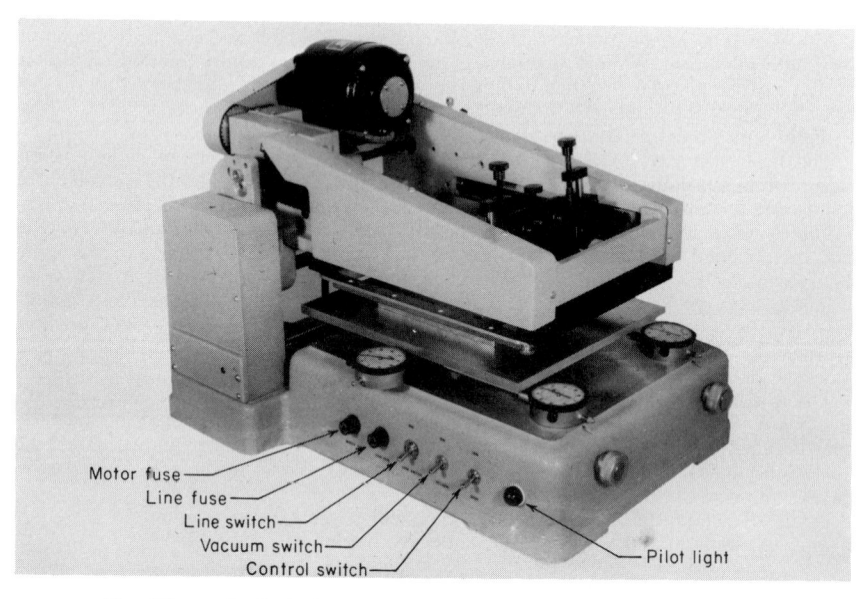

Fig. 58   Hybrid thick-film printer. (*Forslund Engineering Co.*)

but this requires the printing machine to have a flood bar recirculation device that will circulate the paste to the proper side of the squeegee bar for each single-direction pass.

Numerous manufacturers of screen printers exist, and as a result, the choice of a printer becomes difficult. A typical screen printer is shown in Fig. 58. The dial gages indicate the x, y, and rotational positioning of the screen. A machine of this nature is capable of imprinting in excess of 700 substrates per hr. This printer is unique, in that it is motor-driven rather than by hydraulics or air, and requires ac power for operation.

Another type of screen printing machine is shown in Fig. 59. This printer will accommodate an 8- by 10-in. screen, allowing the printing of substrate sizes up to 5 by 6 in. There is also the double-squeegee blade which allows single-direction or double-direction printing. Speeds of up to 800 substrates per hr are possible with a machine of this type.

Another hand-loaded type of hybrid screen printer is shown in Fig. 60. The screen size limitation on this machine is 5 by 5 in., with a work-area limitation of 2½ by 3 in. This illustration clearly shows all the essential parts of the printing

Fig. 59 AMI model RDS-65B FHT printer. (*Affiliated Manufacturers, Inc.*)

machine, i.e., the platen, the screen holder, and the squeegee. The one control required for operation is the knob at the lower right of the picture which controls the vacuum chuck.

Another type of hybrid thick-film screen printer uses a die set as the basic construction frame. This is shown in Fig. 61. As a result of this unique construc-

Fig. 60 Presco model 100 printer. (*Precision Systems Company, Inc.*)

12-50   Thick Films

tion, repeatable alignment between the substrate and the screen can be held as close as 0.003 in. Unlike most screen printers, the substrate on this machine remains stationary while the screen moves up and down. This permits both off-contact and contact printing, depending on how the machine is adjusted. Clearance between the screen and the platen, in excess of more than 7 in., permits easy loading and unloading of the substrates.

Printed-film-thickness measurement   There are several methods of measuring film thicknesses, and these include the use of an interferometer (for very thin films), a toolmaker's microscope with calibrated focus, a light-section microscope, or a profilometer. The most accurate of these methods is the interferometer, but unfortunately this can only be utilized for films that are measured in angstroms, since the measurement is limited by the wavelength of the light that is utilized.

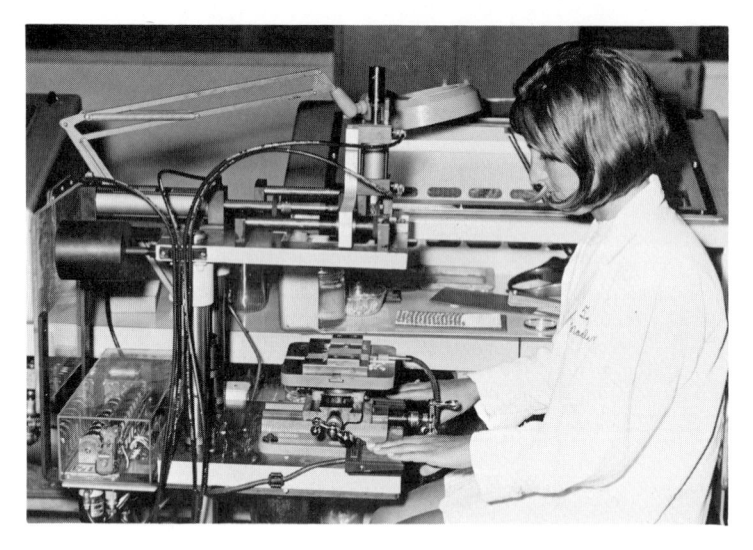

Fig. 61   Accu-Coat T.M. 3100 automatic screen printer.   (*Aremco Products, Inc.*)

The toolmaker's, or light-section microscope, is very handy for film-thickness measurement, but since this is a remote, off-contact type of measurement, which is somewhat operator-dependent, the accuracy of the measurement is limited. It is, however, very useful as a production-line inspection tool that will determine the continued repeatability of the printing machine. Since the ohmic value of the printed resistor is inversely proportional to the film thickness, a measurement of this thickness will be an important factor in maintaining high yields.

In order to virtually eliminate resistor thickness as a variable in checking the consistency of pastes, a profilometer can be used to measure the dried resistor film (before firing) to an accuracy of 0.03 mil (0.00003 in.). This is a nondestructive measurement and can be used in quality control sampling, or as a research tool. The instrument that will make this measurement is called a Surfanalyzer and is manufactured by Clevite Corporation.

Figure 62 shows a thick-film substrate being placed under the stylus of a Surfanalyzer for measurements. The weight of the stylus is in the order of 200 mg and the radius of the stylus is 0.001 in. The stylus can, therefore, be drawn across a printed and dried resistor film without scratching the surface of the resistor. It should be noted that the stylus is nondirectional so that the resistor may be moved in any direction under its point.

Not only does the Surfanalyzer measure the thickness of the resistor to a high degree of accuracy, but also since a graph is used as the display device, a complete

Thick-film Printing Processes 12-51

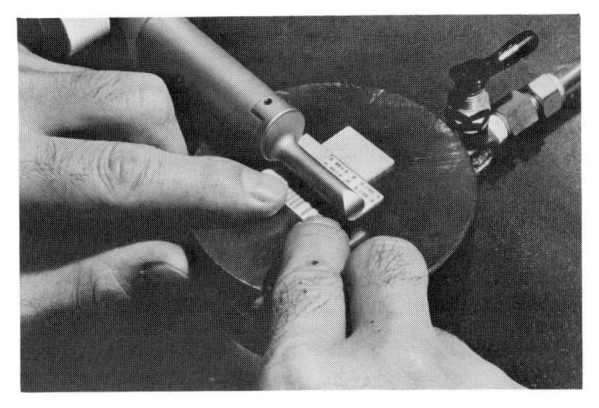

Fig. 62 Surfanalyzer stylus on thick-film substrate. (*Clevite Corporation.*)

profile of the resistor can be obtained. The profile measurement is displayed on a 10:1 ratio enlargement wherein the height is magnified 10 times more than the resistor width, on the resulting graph. Figure 63 shows a typical graph of resistor measurements taken with a Surfanalyzer.

Figure 64 shows the Surfanalyzer being used to measure resistor thicknesses. The instrument is bench-mounted, rather rugged, and does not require a clean room atmosphere to maintain its accuracy.

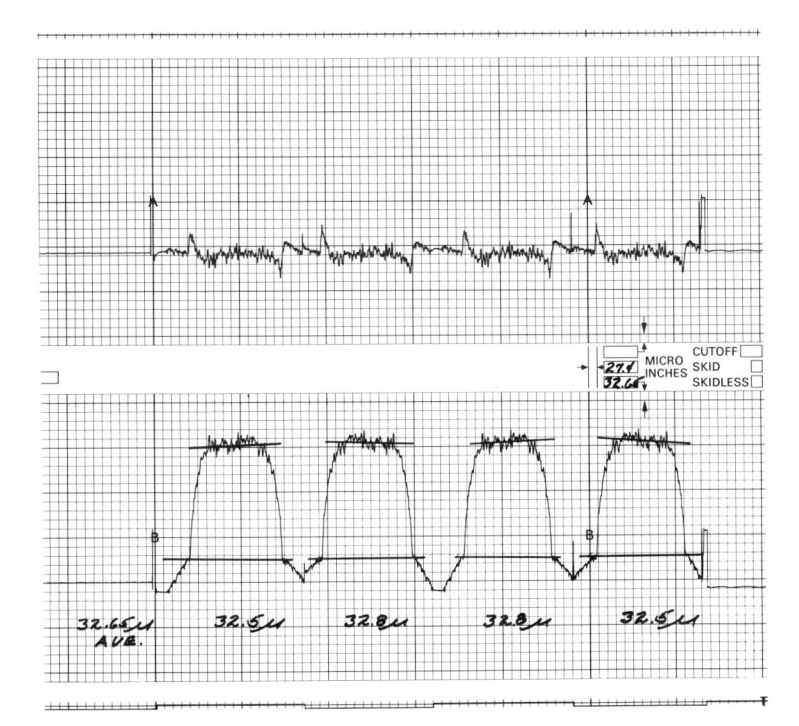

Fig. 63 Typical resistor measurement graph. (*Clevite Corporation.*)

Thick Films

It should be noted that the printed thickness of the resistor is not the only physical measurement that will affect the fired resistor value. Resistor "slump," that is, the edge runout as the screen leaves the surface of the substrate, will also affect the final value. The printing of repeatable resistor value is a precise process, and the only thing that is as important as a rugged, closely controlled screen printer is the possible variability in the pastes themselves. Careful attention to these points will produce higher yields.

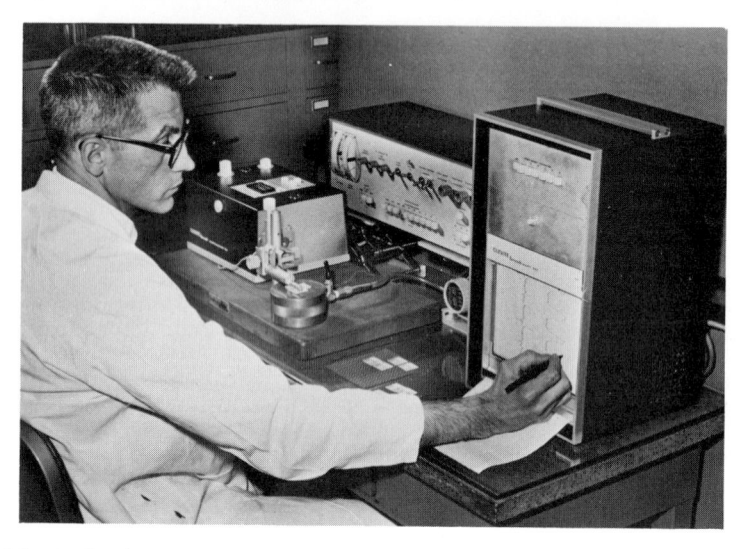

Fig. 64 Surfanalyzer in use. (*E. I. du Pont de Nemours & Co. and Clevite Corporation.*)

THE DRYING PROCESS

After printing, and before firing, the thick-film paste must be dried. In order to promote paste leveling, the printed substrate should be allowed to set for a few minutes after printing. The leveling of the freshly printed paste is visible to the naked eye. Dependent upon the viscosity of the paste, the leveling process will take from 1 to 3 min, but can take up to 20 min. The exact time devoted to leveling and drying can make a difference in the final film property, especially in resistors. At this point, the paste is ready for the drying process.

The prime purpose in drying the paste before firing is to evaporate the solvent portion of the vehicle out of the paste, leaving a dry film that may be handled without damage. There are a number of ways to accomplish the paste drying, and most are equally satisfactory.

Air drying may be accomplished, but this takes an excessively long period of time during which continued leveling may take place. Hot-air drying may be used to accelerate the air-drying process, but, as with other methods, this should be carefully controlled so as to avoid excessive drying speed which may leave cracks in the film. Microscopic inspection will be required to determine whether cracks are occurring during the drying process.

A drying oven may be used at temperature settings between 100 and 150°C. The temperature settings should be determined by the length of time it takes to dry the paste, which should not be under 15 min. Slower drying will reduce the risk of film cracking. Naturally, a belt furnace can be used as well as a box oven to accomplish the drying process.

Another suitable method of accomplishing the drying process is by the use of infrared heating. This has the advantage of causing the paste to heat more uniformly, rather than from the surface inward. In this manner crusting of the paste is reduced, and the solvent has a better chance to evaporate. Uniform heating, as accomplished by infrared techniques, will reduce the risk of film cracking during the drying process. A typical infrared drying oven is shown in Fig. 65. This oven is designed for use over a conveyer belt and contains fuzed-quartz heater elements. The unit is 36 in. long and can provide temperatures controllable from 100 to 250°C. The infrared wavelength is between 1.7 and 2.3 microns.

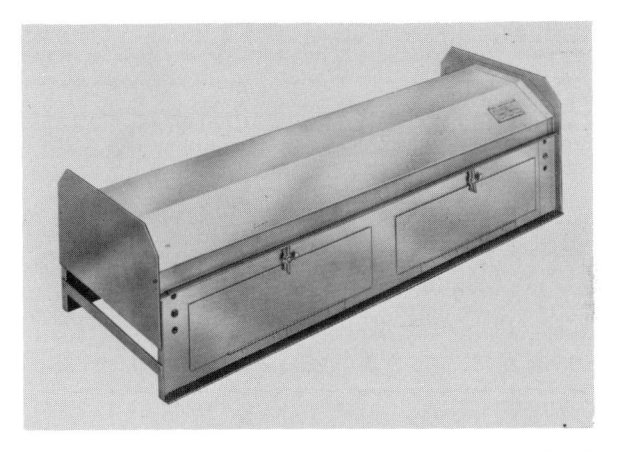

Fig. 65 Infrared drying oven. (*OAL Associates, Inc.*)

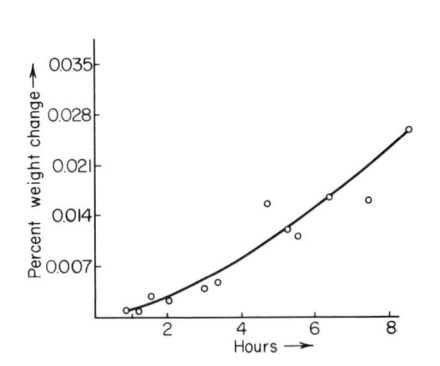

Fig. 66 Weight loss due to solvent evaporation. (*Electro Materials Corp. of America.*)

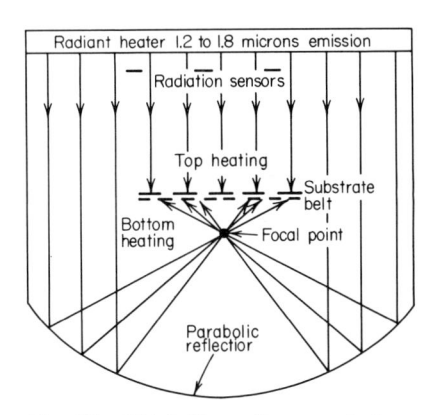

Fig. 67 Thick-film radiant top-bottom drying. (*BTU Engineering Corporation.*)

A determination of when the paste is fully dried can only be made by continued weight measurement. When the weight measurement no longer changes, then the volatiles have all left the paste. Rapid volatile loss is a hindrance during the printing process since pastes using highly volatile vehicles would show a rapid viscosity change while on the printer. Since volatiles are used which will evaporate slowly during the printing operation, it is necessary to use the higher temperatures of the drying process to eliminate all the volatiles before the firing process. Figure 66 shows a typically slow loss of weight due to solvent loss during an air-dry period of over 8 hr.

Generally for the best results, the vendor's recommendations should be followed concerning the leveling and drying times associated with each different paste.

A patented radiant heater drying system has been developed by BTU which uses a single heating source and a parabolic reflector to achieve top and bottom heating of the substrate and, therefore, more uniform drying. Figure 67 shows the principle behind this heating process.

FIRING PROCESSES

The firing of thick-film materials is a critical process. The resistor, conductor, or insulator that emerges from the printer and drying oven does not have the properties desired in the final product. These materials must be fired at temperatures between 450 and 1000°C to achieve these properties, and it should be well noted that minor changes in the furnace temperature, the belt speed (time-temperature integral), or for that matter, the ambient atmosphere can cause radical changes in the desired end-item properties. During the firing process, the following chemical changes must take place in the thick-film paste:

1. The organic binders must be burned off.
2. The metallic elements must be reduced or oxidized to develop the proper film parameters.
3. The colloidal glass within the paste materials must be sintered to develop the adhesion to the substrate.
4. The film must be cooled in an annealing manner to the room temperature.

The furnace The type of furnace most commonly used for the firing of thick-film hybrid circuits, is the moving-belt furnace. In this type of furnace there is an input opening and an exit opening, and therefore atmospheric pressure remains at ambient. Although artificial atmospheres can be used in the furnace to achieve special effects, most users supply a controlled flow of air through the furnace. In many cases, this airflow is the room air that exists. In many cases, the room air is temperature- and humidity-controlled so as to stabilize and improve the consistency of the fired films.

Figure 68 shows a simple belt furnace that contains most of the ingredients of the more expensive and more sophisticated types. The furnace incline can be adjusted from 0 to 3° from the horizontal plane so as to control the flow of air through the furnace, and dampers at each end can be further used to regulate the airflow. It should be noted that the higher the incline setting, the faster the air will flow through the furnace. In the burnout and sintering sections of the furnace, where the films emit fumes, a venturi jet system will distribute air over the surface of the substrates and direct the flow of fumes toward the entrance part of the furnace. To help eliminate contamination, a quartz muffle is used in the furnace. Metal muffles tend to oxidize and flake, causing contamination of the surface of the films.

In order to produce consistent results in the firing process, the time-temperature profile of the furnace must be kept constant within rather close limits, the atmosphere flow must be kept constant (as from exit to entrance as induced by an approximate 2° tilt), the belt speed must be closely controlled, and contamination must be limited.

Since it is very difficult and time-consuming to set these many variables, a single furnace is not easily usable for the production of thick-film hybrid circuits. The primary materials of a thick film are conductors, resistors, and insulators, and therefore a minimum of three firing furnaces would be needed to achieve a satisfactory production flow.

To understand what occurs in the firing furnace it is well to expand on the two primary phases of the firing process: i.e., burnout and sintering.

During burnout, which is the first phase of the firing process, the organic binders are burned off or expelled from the thick-film paste, and the fumes from this activity are expelled through the entrance part of the furnace. This burnout phase is accomplished at temperatures around 300 to 400°C, and it is important that

Firing Processes   12-55

the temperature rise to these burnout limits does not occur too rapidly, as it can cause blistering or lifting of the thick-film paste. The temperature rise should certainly not exceed a rate of 200°C per min. Once the burnout of the organic binders has occurred, the film is ready for the sintering phase.

To sinter the paste, it must be brought to temperatures high enough so that the glass frit particles begin to soften. Since this sintering phase will affect the final physical characteristics of the film, time and temperature are extremely important. For consistent results, the furnace temperature profile should be held within 2°C. When the film reaches the predetermined proper temperature and the oxidation-reduction reaction has occurred, the glass frit will melt and essentially seal the metallic particles from further atmospheric contact. The adhesion of the film to the substrate also occurs at this point in time.

Fig. 68   Explorer II firing kiln.   (*BTU Engineering Corporation.*)

Temperature control of the firing furnaces is quite critical—so much so, that dummy substrates are generally placed on the firing belt to simulate a fully loaded line, before a single substrate will be placed through the firing cycle. Feedback temperature controls are an absolutely essential element for controlling the firing temperatures. In addition, profiling with separate thermocouples and a chart recorder is necessary to determine and set the proper profile shape and tolerances.

Generally, profiling is accomplished by sending about three thermocouples through the furnace (distributed to measure the gradient across the belt) while recording the temperature-time product on a chart recorder. The thermocouples can be mounted on a large reel for convenience (no less than 2 ft in diameter), and the temperature-sensing end may be bonded to a substrate, or sandwiched between two substrates so as to stimulate actual conditions.

12-56 Thick Films

A typical firing profile for two types of resistor paste is shown in Fig. 69.

More complex firing ovens with multiple zones are available for thick-film firing processes. The advantages are usually in the many refinements that are included, such as more precise temperature and belt-speed control, longer dwell times in

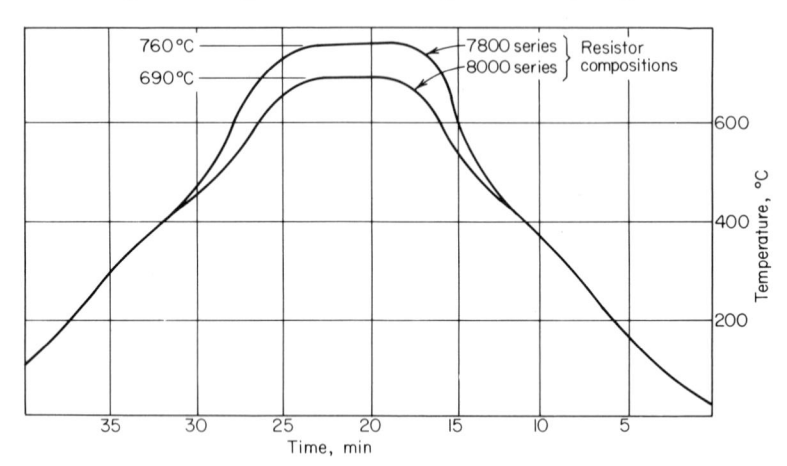

Fig. 69 Firing profile for two pastes. (*BTU Engineering Corporation.*)

the precisely controlled "flat" temperature zone, better atmosphere control, gas barriers between zones, overheat protection, and other refinements which, if properly utilized, will increase the product yield and reproducibility. Figure 70 shows a firing furnace of the type described.

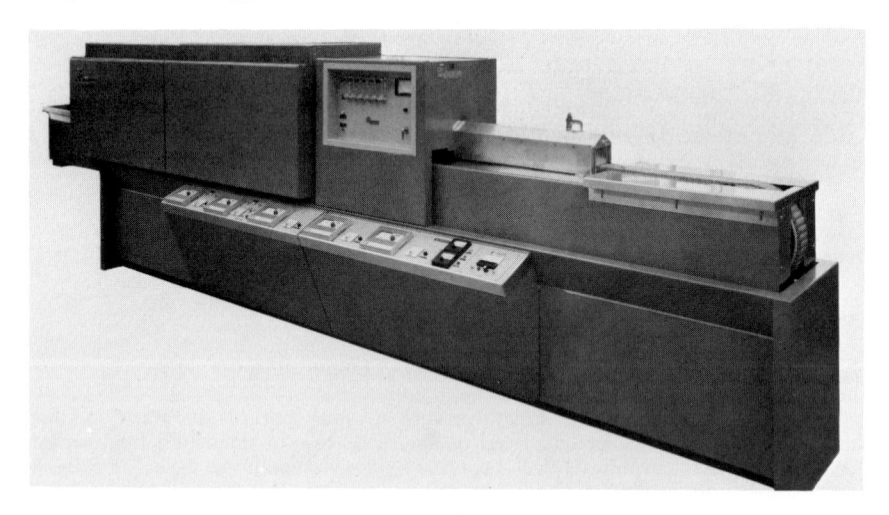

Fig. 70 Complex firing furnace. (*BTU Engineering Corporation.*)

RESISTOR TRIMMING

The many variables associated with thick-film resistor paste materials and thick-film resistor firing processes have led to a situation wherein the majority of the thick-film resistors are trimmed to their final values. It is practical, with good process control, to print and fire resistors to tolerances of plus or minus 15 percent of the desired ohmic value. Some manufacturers will claim that they can print and fire to 10 percent tolerances, but this can only be done with considerable care and reduced yields. It is, therefore, an accepted fact in the hybrid thick-film circuit industry that to achieve thick-film resistors with tolerances of better than 15 percent (and down to ½ percent) the resistors must, in some manner, be trimmed.

There are several methods for trimming thick-film resistors, and these include abrasive, pulse, scribe, and laser trimming. The most common method is abrasive trimming, primarily because of its high degree of development and simplicity of

TABLE 20 Comparison of Trimming Methods

Method	Technique	Upward	Downward	Comments
Abrasive...	Sandblast	Yes	No	Highly developed, inexpensive
Pulse......	High frequency, high voltage	Yes	Yes	Reacts differently with different materials
Scribe......	Vibrating diamond point	Yes	No	Inexpensive
Laser......	High-intensity light	Yes	Yes	Very reliable, very fast

operation. Table 20 lists the capabilities and characteristics of the various trimming methods.

It is possible to trim resistors downward by reduction, but this is not commonly done, owing to the difficulty of stopping the process at a precise resistor value.

Air abrasion This is the most common method of trimming resistors, but it has one disadvantage: that the resistor can be trimmed upward only. Nozzles of various sizes and shapes are available to suit the resistor geometry, and a choice of edge trimming or notch trimming is available. The abrasive generally used is alumina

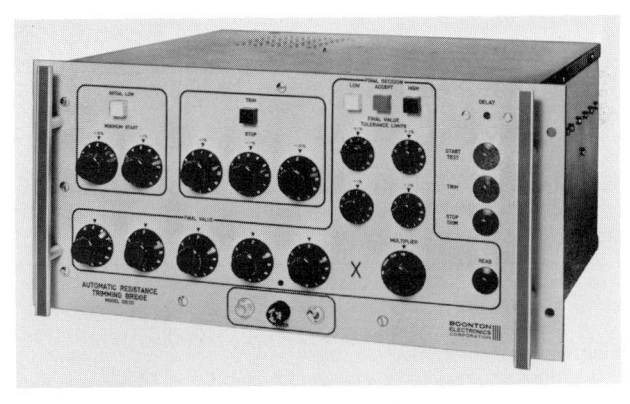

Fig. 71 Automatic resistance trimming bridge. (*Boonton Electronics Corporation.*)

grain (aluminum oxide) of about 100 mesh in size. The alumina must be dry and free flowing, and properly suspended in the airstream.

In order to accomplish the abrasive trimming process it is necessary to have, in addition to the trimming equipment, the essential measurement and control equipment. The measurement equipment in its simplest form contains a bridge circuit in which the resistor being abraided is compared with a standard. Sophisticated versions of these measurement bridges will control the entire abrasion process.

Figure 71 shows a resistance bridge which will first test the resistor to be trimmed to assure that it is within a practical trimming range. A resistor that is 50 percent

12-58 Thick Films

below the desired value, for instance, will require abrasion of 50 percent of its width to reach the required value. The bridge will continuously monitor the value of the resistor being trimmed, and then stop the abrasion action when the desired value is reached. After a slight delay, the instrument will then read out a high-low-accept indication on the finished resistor. The range of the instrument is from 10 ohms to 10 megohms.

For a greater degree of automation, bridges are available which may be programmed. On this type of instrument, external resistance standards may be used, and one trimmed resistor may be matched to another by using the first trimmed resistor as the external standard. An instrument with this feature is shown in Fig. 72.

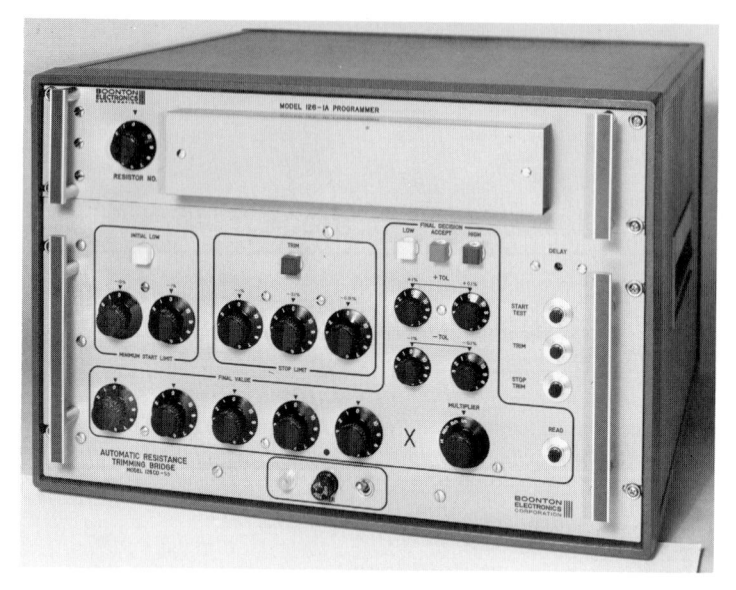

Fig. 72 A programmable resistor trimmer. (*Boonton Electronics Corporation.*)

An abrasive trimming station, in total, must consist of the electronic monitoring bridge; the electromechanical probes; the mechanical apparatus for accomplishing the x, y motion; and the abrasive section.

The monitoring bridge and its capability have been described. The electromechanical probes are usually specially designed to make good contact with the thick-film conductor. The probe points are made of a hard material such as tungsten carbide. Figure 73 shows a typical probe arrangement. Notice the x, y, and z adjustment knobs on the probe mechanism.

The (x,y) positioning arrangement may be hand-operated or automatic. Since the resistor bridge may be set to stop the abrasive action, a hand-operated table is quite satisfactory. In the automated version of the (x,y) trim table, there is still the necessity for hand-operated gross positioning adjustments. Figure 74 is illustrative of an (x,y) table.

The nozzle through which the abrasive grit is sprayed is the next essential to a trimming station. In addition, there must be a dust collection system which will collect and recirculate the abrasive grit. Figure 75 shows the nozzle positioned over the resistor to be trimmed. This picture does not clearly show the dust collection system.

The complete trim station is shown in Fig. 76. This includes all the equipment necessary to trim resistors at a rate of several hundred units per hour.

Pulse trimming Pulse trimming, also known as high-voltage discharge trimming, provides an initial resistor adjustment in the downward direction by fusing together the metallic particles within the resistor body. Dependent upon the time the pulse is applied to the resistor, this method will also cause upward movement of the resistor by burning out contact paths. A disadvantage associated with this method of resistor trimming is its dependence upon the rheology of the material involved.

Fig. 73 Resistor probe mechanism. (*S. S. White Industrial.*)

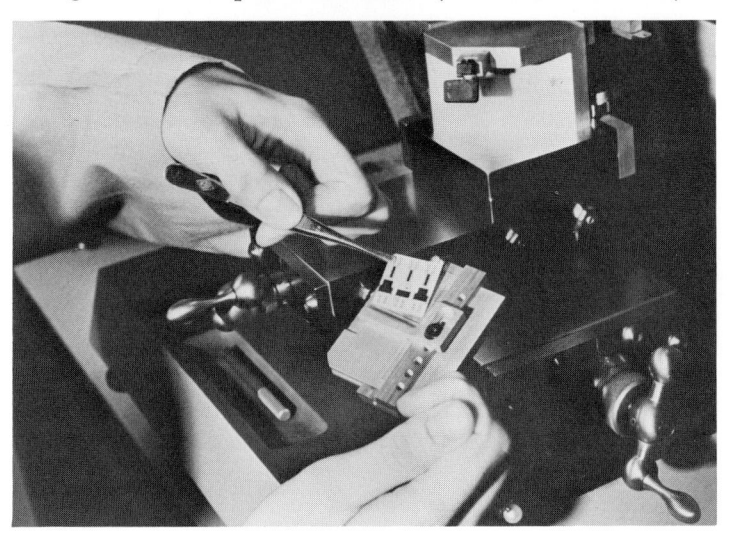

Fig. 74 (*x,y*) table for resistor trimming. (*S. S. White Industrial.*)

Also, resistors trimmed by this method appear to have a greater tendency to drift with time.

Scribe trimming Scribe trimming is a variation of abrasive trimming in that it is accomplished by removal of a portion of the resistive material. As a result, it can only adjust resistors in an upward direction. The equipment consists of

12-60   Thick Films

a diamond-pointed scribe which is driven in an oscillatory manner ultrasonically, and actually cuts away a portion of the resistor. The same contact arrangement and bridge measurement system can be used as is utilized with abrasive trimming. To accomplish the trimming action, the stage (table) on which the substrate is held is moved under the diamond scribe at a controlled rate. The advantage of this

Fig. 75   Abrasive nozzle in trim position.   (*S. S. White Industrial.*)

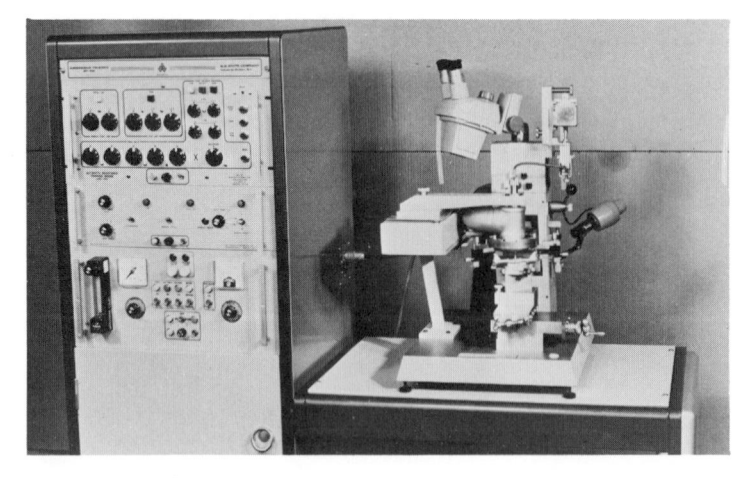

Fig. 76   Complete resistor trimming station.   (*S. S. White Industrial.*)

system is that it is relatively clean (does not use an abrasive blast), but the diamond scribes will wear more rapidly than an abrasive nozzle.

Laser trimming   Laser trimming is the newest technique to be developed for the adjustment of thick-film resistors. Since the laser beam can be used either to anneal or vaporize the resistor film, this technique can be adjusted to trim the resistor in either the upward or downward direction. An additional advantage lies in the fact that the laser beam re-fuses the resistor at the point of contact, and therefore, does not leave a vulnerable rough edge, as does abrasive or scribe trimming.

Since the laser trimming system has the capability of operating very rapidly, most available systems have the capability for fully automatic programmed operation.

One of the problems associated with laser trimming is the possibility of eye damage if the proper safety precautions are not observed. As a result of this, many laser trimming systems utilize a closed-ciruit television system for viewing and for alignment. Appropriate protective measures are used to prevent the laser light from affecting the television pickup tube. CO_2 lasers are safe for the eyes.

As with all other trimming systems, probes must be used so as to contact the resistor before, during, and after the trimming cycle.

Since the laser beam can accomplish upward trimming by cutting very fine lines (less than 0.020 in.), unique cuts can be made that will trim resistors over a wider range than other types of trimming. Figure 77 shows a method of trimming over an extended range.

SOLDER ASSEMBLY

Most of the available solder-assembly equipment was developed for use on printed circuits. Fortunately, a good deal of these techniques and hardware can be converted for use on thick-film microcircuits. It should be recognized that solder assembly of piece parts to thick-film substrates is just one method for attaching parts. Other methods, including thermal compression, eutectic, epoxy, and ultrasonic bonding, will be discussed later. When using solder assembly as a technique, it is generally used with packaged piece parts and not with bare semiconductor chips, as the two processes are incompatible. Transistors (in cases), diodes, capacitors, coils, resistors, and inductors can all be assembled to thick-film substrates by solder processes. Understanding a few of the problems and knowledge of many of the accepted techniques will help in the establishment of a satisfactory process.

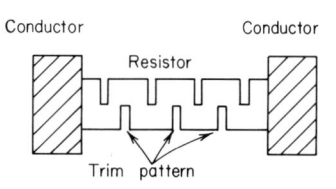

Fig. 77 Extended-range trim pattern.

Thick-film conductors may be soldered in the as-fired condition. Generally, however, some preparation is made before attaching the parts to the substrate. Improved tinning can be accomplished if the surfaces of the thick-film conductors are clear of finger oils or oxides. To achieve this, burnishing of the exposed conductor surfaces is easily performed.

Burnishing There are several ways that a finished substrate may be burnished. The act of burnishing will tend to compact the metallic surface of the thick-film conductor and make it more dense. In addition, burnishing will help to remove surface glass, oxidation, and other contaminants. Burnishing is most easily done with a fiber glass brush. This can be a hand operation with a straight brush or a machine operation with a rotary brush. Another accepted method is to employ a motorized erasing machine, as used in drafting operations, with an eraser of the texture of an ink eraser in the motorized device. Each of these methods leaves some residue on the substrate, and it therefore must be cleaned after burnishing.

Cleaning Substrate cleaning is best done ultrasonically before soldering. Naturally a large number of substrates can be cleaned at once by proper jigging and fixturing. The time in the cleaning fluids should not be shortened, and the substrates should be tinned as soon after cleaning and drying as possible. A usable cleaning cycle would be 5 min in each of two beakers of trichlorethylene with ultrasonics and mechanical agitation, followed by 5 min in methanol with ultrasonics and agitation. This may be followed by a nitrogen gas flow or an infrared heater,

12-62 Thick Films

preferably under a nitrogen cover gas. The effect of proper cleaning on conductor solderability should not be underestimated.

Fluxing Because of the difficulty of removing any flux completely, only the mildest of fluxes should be used. Strong fluxes may damage resistor, glass, or conductor materials. A good rule, therefore, is to use a mild flux and clean it off as soon after the tinning or soldering operation as is possible.

The most commonly used flux for thick-film conductor tinning is white resin. Mildly activated forms of resin fluxes containing small amounts of weak organic acids can also be used with care. This should be done only when necessary.

Fluxing of a thick-film substrate is most easily accomplished by dipping the entire substrate in a solution of flux and draining the excess off. This should be done immediately before the tinning operation.

If a flow- or wave-soldering machine is being used for tinning, then a "foam fluxer" may be used for the fluxing operation. This is usually a part of the wave soldering machine. Figure 78 shows a wave-soldering machine being utilized for

Fig. 78 Wave-soldering machine for tinning substrates. (*Electrovert, Inc.*)

substrate tinning. On the left can be seen the foam fluxer. Over the "hot solder" label, the substrates may be seen passing over the preheater and over the solder wave.

Tinning The substrate tinning operation can be done as simply as dipping the substrate in a solder pot. This method is quite satisfactory for small production runs or laboratory operations. Even with this simplified method, however, some technique should be provided to maintain the time period in the solder pot and the angle of withdrawal, so that consistent results are achieved. It may be found that with some conductor materials two fluxing and tinning operations are required to obtain an even and complete solder coating. It should be noted that the conductor firing profile through the firing furnace will affect the solderability of most conductor materials. It should be kept in mind also that surface cleanliness is the key to good solderability.

For high-production operations, wave- or flow-soldering machines can be used for the automatic tinning of thick-film substrates. Since most machines of this nature are designed for printed-circuit operation, some tooling will be required to adapt the process to thick-film substrate tinning. The fluxing operation, however, could be accomplished by the same machines, thereby saving an operation and the attendant hand labor. See Fig. 78, wave-soldering machine for tinning substrates. Figure 79 shows a substrate that has been tinned by the wave-soldering process. The miniscus and form of the solder pads should be noted, as should the complete solder coverage.

Common 60-40 solder is generally used for thick-film substrate tinning and solder assembly, but for special purposes other solder compositions may be used. Table 21 shows some properties of commonly used thick-film solders. Since leaching can be a problem, a small silver content is often used to inhibit the amount and rapidity of conductor leaching. This silver content rarely exceeds 2 percent of the total composition.

The lowest temperature, or eutectic combination, for melting of tin-lead solder is 63 percent tin and 37 percent lead. No treatise on solders would be complete without a phase diagram for tin-lead solder, and this is shown in Fig. 80.

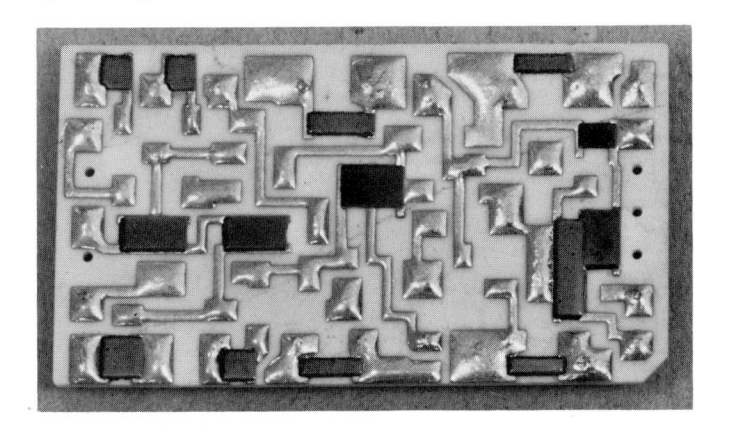

Fig. 79 Tinned thick-film substrate. (*Electrovert, Inc.*)

TABLE 21 Common Thick-film Solders

Type	Liquid temperature, °C	Tarnish resistance	Solderability
60-40 tin-lead.............	183	Good	Excellent
60-38-2 tin-lead-silver.......	185	Good	Excellent
90-10 lead-tin.............	280	Fair	Good
95-5 lead-indium...........	315	Fair	Excellent
100 tin...................	232	Good	Good

It is important to know the adhesion capabilities of the solder pads to which parts are to be attached. This can be accomplished by performing solder pull tests, which can take one of two forms. The most common method is to solder a wire to a 0.050- by 0.050-in. pad in a manner which places the wire vertical to the pad. A force is then applied to the wire until the solder pad is pulled loose from the cermet or the cermet is pulled loose from the substrate. Typically, a 0.050- by 0.050-in. pad should withstand several pounds of pull before breaking loose (about 4 lb).

Another method is to solder the wire parallel to the solder pad and then apply the force 90° to the pad. This will tend to peel either the solder from the cermet or the cermet from the substrate. The absolute value of the bond is less important than the repeatability of the tests, as an indication that the firing and tinning processes are under control.

Reflow assembly In almost all cases where parts are soldered to thick-film substrates, the reflow-soldering method is used. Essentially, this means that a part with a pretinned lead is soldered to a pretinned pad, without the application

12-64 Thick Films

of additional solder. Generally, flux is used to assist the reflow process. The heat required to cause the solder to reflow can be accomplished in a number of ways, and the parts attachment can be performed singly or all together, dependent upon production quantities required or the quality of work desired.

If part leads are being attached singly, it is wise to have the substrate on a preheated stage so that it is easier to bring an individual pad to the solder-reflow temperature. The temperature of the heated stage should be about 125°C. This is required to obtain a quality solder joint with good miniscus. If mass soldering is being utilized, a preheat stage should be used to reduce the thermal shock.

Numerous methods of applying heat to effect the solder reflow may be used. These can include hand-soldering irons, heated carbon tips, heated wire loops, hot gas jets, infrared lamps, hot plates, or parallel-gap heating. Timing devices and temperature controls will help to zero in on the proper solder reflow parameters. If working with the reflow of individual leads, optical assistance is necessary

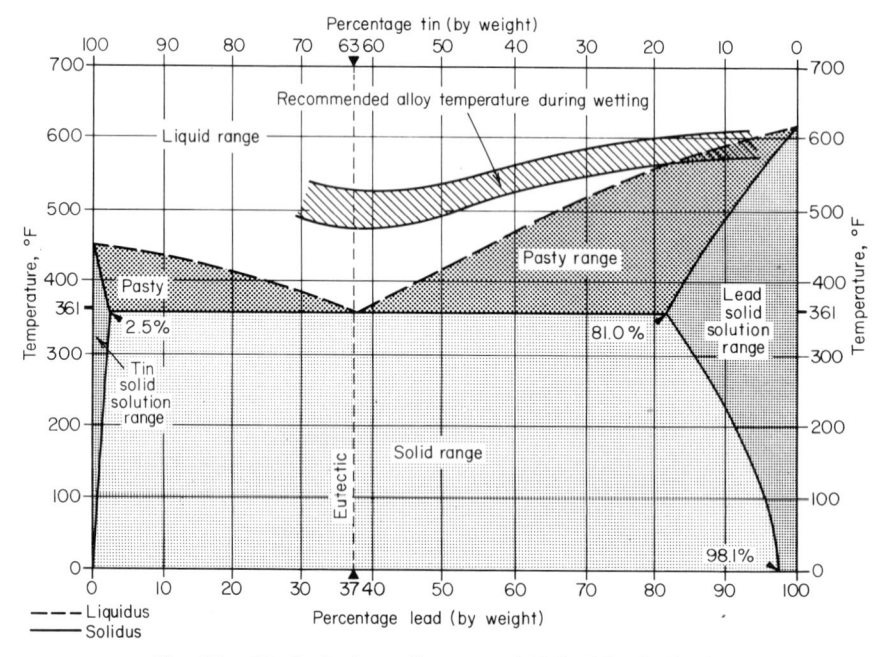

Fig. 80 Tin-lead phase diagram. (*Alpha Metals, Inc.*)

in the form of a microscope. A stereo zoom microscope is the most convenient type to use, but a fixed enlargement of 10 times may be satisfactory. Figure 81 shows a solder reflow station satisfactory for single-lead reflowing. For thick-film ceramic substrates the grease plate should be replaced by a heated stage. This reflow machine has built-in controls for time and temperature, and can be used in either an automatic or a manual mode. Temperatures can be controlled up to 1000°F, and the time cycle is variable from 100 msec to 3.5 sec.

A hand-soldering pencil, also with time and temperature controls, is shown in Fig. 82. Tools of this nature can be helpful in increasing the quality and repeatability of the workmanship.

When all parts are reflow-soldered to a substrate at one time, appropriate jigging and fixturing will be required to place the parts in the proper position, and then to hold them in place during the reflow operation. Figure 83 shows a substrate with capacitors, diodes, and transistors attached by the solder-reflow method.

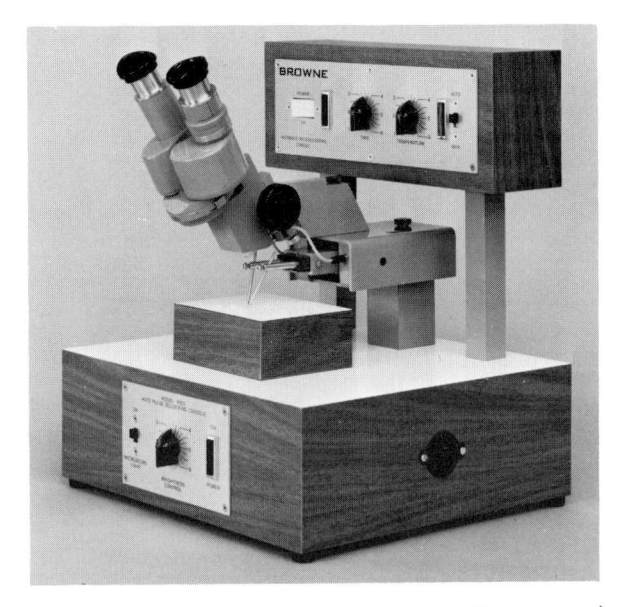

Fig. 81 Solder-reflow machine. (*Browne Engineering.*)

CHIP AND WIRE ASSEMBLY

When increased density is required, and for reasons such as economy, reliability, and circuit considerations, uncased semiconductor devices may be bonded and wired directly to thick-film conductor materials. This is done in approximately the same manner that the transistor manufacturers use to package the discrete devices. There are, of course, numerous techniques available for bonding and wiring the semiconductor devices to the substrate conductor materials, and these are:
1. Eutectic bonding
2. Alloy soldering
3. Epoxy paste bonding

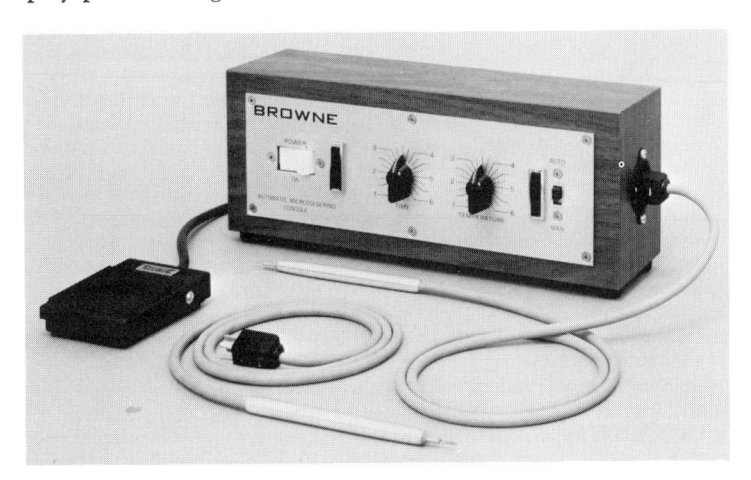

Fig. 82 Hand-soldering machine. (*Browne Engineering.*)

4. Plastic bonding (nonconductive)
5. Flip-chip bonding

The device wiring can also be accomplished in a manner similar to the techniques used by the semiconductor manufacturers, and these essentially include only:
1. Thermal compression bonding
2. Ultrasonic bonding

There are numerous variations on these methods for die and wire bonding; and any particular process selected must be compatible with all other processes associated with the fabrication and assembly of the hybrid thick-film circuit.

Eutectic die bonding Die bonding to a hybrid thick-film circuit presents some special problems that are different from those associated with discrete semiconductor packaging (in headers or flat packs). The most serious problem is the time-temperature integral and its effect on the large number of chip devices. When mounting a single silicon device to a header, the general practice is to bring the header up to the eutectic temperature of the silicon-gold combination (the silicon of the semiconductor and the gold of the header). By scrubbing the chip against the header at this eutectic temperature (above 385°C) a silicon-gold bimetallic compound is formed which bonds the silicon chip to the gold-plated header.

In order to bond large numbers of chips to a single substrate (or header) the eutectic temperature must be approached selectively. Generally, the method used here involves heating the substrate to about 25°C below the eutectic temperature, and then selectively raising the portion where the bond is to be made to the eutectic temperature. This can be done with resistance heating, hot gas heating, or by the use of a hot die collet. The object is to keep the other chips that have been bonded below the eutectic temperature while bonding successive chips. It should be noted that during all high-temperature processes associated with semiconductor bonding and wiring it is wise to have a cover gas of nitrogen flowing so as to prevent surface oxidation.

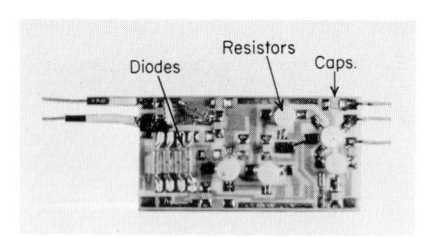

Fig. 83 Substrate assembled by solder-reflow technique. (*General Electric Company.*)

If the back surfaces of the silicon chips have become oxidized, it will be very difficult to obtain a good eutectic die bond since the silicon of the chip will have difficulty in coming in intimate contact with the gold conductor due to the oxide layer. To help prevent this problem, chips may be obtained with a gold backing. Assuming that the gold backing has been applied before oxide formation, a eutectic will be formed as soon as the chip reaches the proper temperature.

The scrubbing motion required to begin the eutectic formation during chip bonding may be implemented by hand movement of the manipulator, by ultrasonic action, or by vibratory or cam-actuated mechanical motion. No single method is recommended over any other, with the exception that five educated fingers can usually surpass any automated equipment.

Alloy soldering If lower temperatures than those required for silicon-gold eutectic bonding are desired, then preforms of lower-melting combinations may be utilized. Popular among these are gold-germanium and gold-tin combinations. It is even possible to use tin-lead solders to effect a satisfactory die bond. Circuit operating temperatures, the temperatures required by subsequent process steps, and the necessity to do rework will all affect the die-bond temperature chosen. For instance, thermal compression wire bonding requires temperatures in excess of 300°C. Therefore, a die bond made below that temperature would hardly be satisfactory.

The advantage of die bonding below the temperature of silicon-gold eutectic is that selective heating is no longer necessary, and mass bonding of several dies with hot-plate or furnace-heat application is possible.

Paste bonding There are several vendors of epoxy pastes that provide adhesives suitable for both conductive and nonconductive die bonding. In general, the conductive epoxy pastes are loaded with either silver or gold particles. Silver epoxy paste is in quite common usage for die bonding and for chip capacitor bonding. It does an adequate job in both applications. The problems associated with epoxy paste bonding are that the epoxy is difficult to apply and requires a high skill level; the pot life is short, leading to excessive waste; and the mixing and curing cycles are critical. Hydrogen outgassing can also occur.

Film bonding Included in the general category of paste bonding should be film bonding. This refers to bonding with nonconductive (for IC devices) films such as FEP Teflon* or similar adhesive films. The application of these films is usually accomplished at high temperatures (around 200°C), and they have the disadvantage of having marginal thermal conductivity properties.

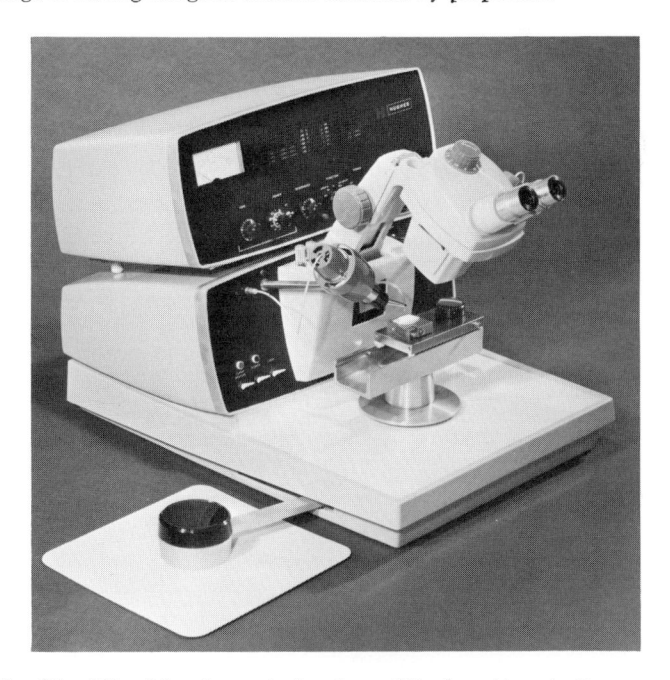

Fig. 84 Flip-chip ultrasonic bonder. (*Hughes Aircraft Company.*)

Flip-chip bonding Flip-chip bonding covers the case where the die itself is not really bonded, but the active interconnections to the chip device are bonded directly to the circuit paths. These circuit paths are utilized not only as the electric connections to the chip but also as the thermal paths. Flip-chip bonding can take essentially two forms: (1) where the protrusions are directly on the chip, and (2) where the protrusions are on the substrate. Bonding is performed by soldering, thermal compression, or ultrasonic bonding techniques. One of the problems associated with flip-chip bonding is the alignment of the chip to the substrate interconnections. Another problem is the planarity required on the substrate to assure that all points are bonded equally. The third problem associated with flip-chip bonding is that once it is accomplished, the bonded joints are uninspectable except by infrared techniques. Figure 84 is an example of a flip-chip bonder.

Most of the bonding machines associated with flip-chip bonding have unique

---

* Trademark of E. I. du Pont de Nemours & Co., Wilmington, Del.

12-68 Thick Films

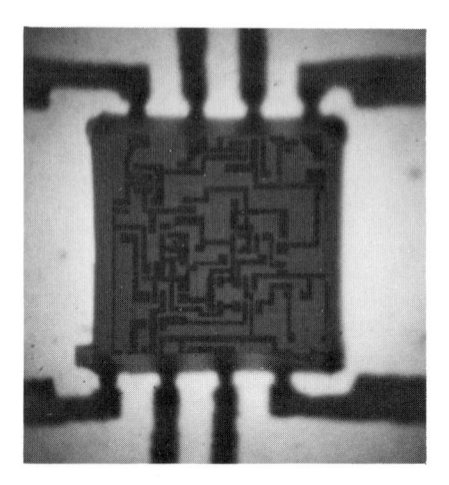

Fig. 85 Infrared view of flip-chip bond. (*Hugle Industries.*)

optical systems for the alignment of the chip pads to the substrate pads. The most advanced of these uses an infrared microscope to look through the substrate and the chip so as to produce a fluoroscopic view of the chip and substrate metallization. An infrared view of a chip being bonded to a substrate is shown in Fig. 85. This is the only method by which the bonding interface may be viewed during the actual bonding operation.

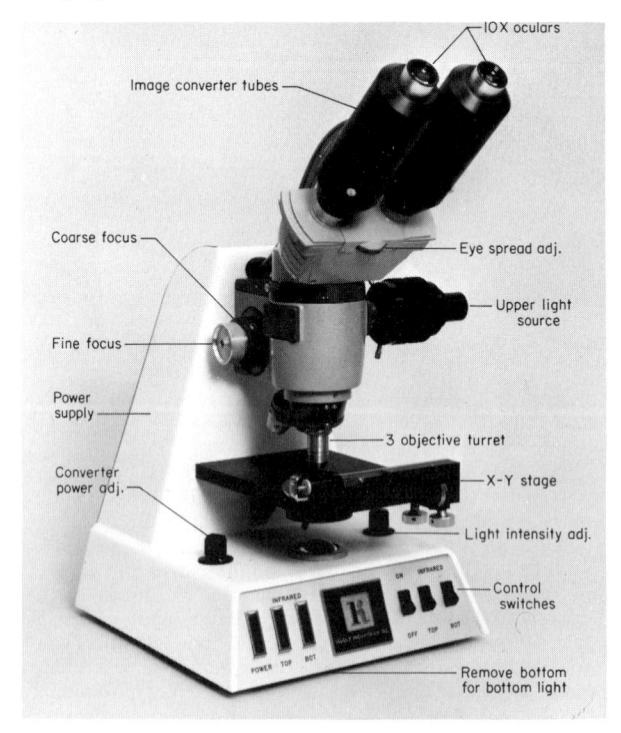

Fig. 86 Infrared microscope. (*Hugle Industries.*)

The type of optics that would be used for a flip-chip bonder with infrared capability is shown in Fig. 86. These optics can be adapted directly to the bonding machine and will see through silicon, gallium arsenide, and other infrared transparent materials.

Thermal compression wire bonding Thermal compression wire bonding is almost always accomplished using gold wire. The most common diameters are 0.0007, 0.001, and 0.002 in. Diameters as small as 0.0003 in. have been used for special purposes, and the wire is obtainable with a purity of 99.999 percent. Annealed wire is used since it will lay better on the spool and be less likely to shift or kink. The characteristics of the gold wire used for thermal compression bonding are listed in Table 22.

TABLE 22 Properties of Pure Gold Wire

Purity % of composition................................	99.999
Resistivity at 0°C, ohms/cm:	
Hard..	12.3
Annealed......................................	12.1
Temperature coefficient of resistance, at 0–100°C:	
Hard..	0.0039
Annealed......................................	0.0040
Tensile strength, psi $\times 10^{-3}$:	
Hard..	46
Annealed......................................	19
Elongation, %:	
Hard..	1.5
Annealed......................................	36
Melting point (solidus), °C............................	1063
Density, g/cm³..	19.30

SOURCE: Sigmund Cohn Manufacturing Co.

The actual thermal compression wire-bonding process is best accomplished on a burnished cermet gold conductor surface (for thick-film hybrid circuits). Both the capillary tip, which the wire runs through, and the substrate are usually maintained at high temperatures for the wire-bonding operation. Some of the factors that will influence the quality of the wire bond are as follows:

    The tip and substrate temperature
    The bonding pressure
    The dwell time
    The collet geometry
    The wire diameter
    The degree of wire stress relief
    The wire composition
    The cleanliness of the wire and bonding surfaces
    The composition of the bonding surface
    The firing temperature of the bonding surface

Two basic techniques are used for thermal compression wire bonding. They are nailhead (ball) and wedge bonding and stitch bonding. Nailhead (ball) and wedge bonding is accomplished by first cutting the gold wire with a precise flame, thereby forming a gold ball, or nailhead, on the end of the wire. The capillary is then brought down upon the target, and the ball is crushed, making the nailhead bond. The wire is then dragged to its second interconnection point, and the wedge bond is made by again crushing the wire against the target. The capillary is then raised, and the flame cuts the wire and readies it for the next bond. The wiretail that is left is then removed by breaking it at the wedge bond. Figure 87 graphically illustrates the requirements for nailhead (ball) and wedge bonding.

A photograph of a ball and wedge bonding capillary is shown in Fig. 88. To the right, the flame tip may be seen which is used for cutting the wire and forming the ball. The wire spool can be seen above the capillary.

12-70   Thick Films

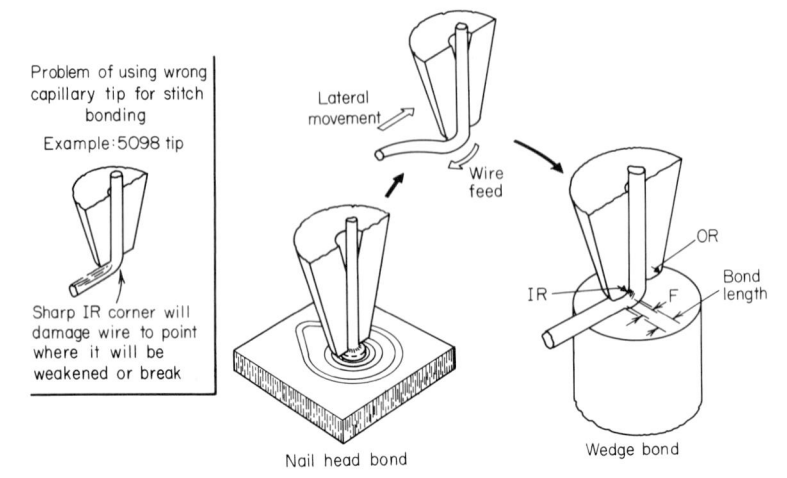

Fig. 87 Ball and wedge bonding geometry. (*Tempress Research Company.*)

IR—Must be large enough to allow smooth uninterrupted wire feed when capillary tip is moved laterally.

OR—The desired shape and dimensions of the bond to the post determines what the OR and F dimensions should be.

In the post and shear technique the OR dimension is usually 0.000, which gives a sharp edge for shearing the wire on the edge of the post.

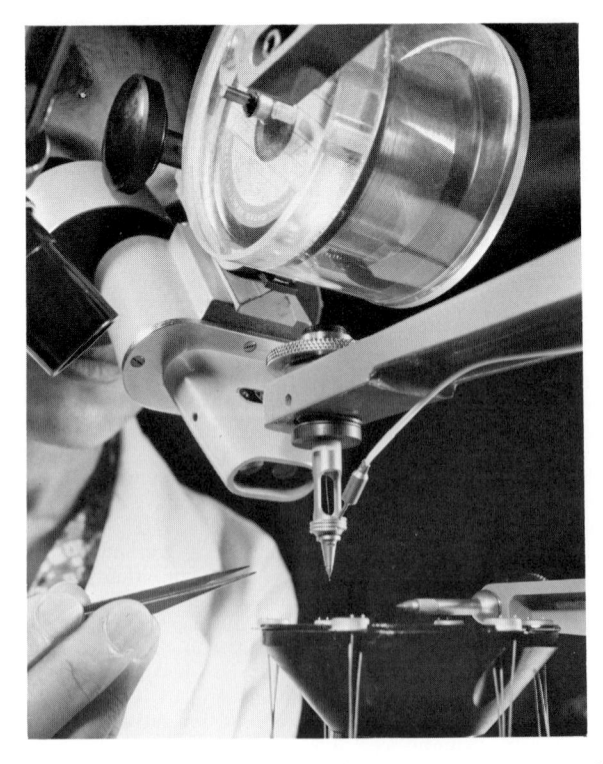

Fig. 88   Capillary and flame tip for ball bonding. (*Tempress Research Company.*)

Thermal compression stitch bonding is accomplished without the use of the ball or flame-off device, but essentially employs the same bonding parameters. A cutter technique is used to shear the wire in preparation for each bond. Figure 89 graphically illustrates the stitch-bonding technique.

Ultrasonic wire bonding Thermal compression wire bonding is dependent upon a hot stage (about 300°C) and pressure to form an intermetallic bond. Ultrasonic wire bonding accomplishes this intermetallic bond at room temperatures, utilizing ultrasonic energy and pressure. There are certain advantages in working at room temperature, especially when a large number of chips must be wired on a single

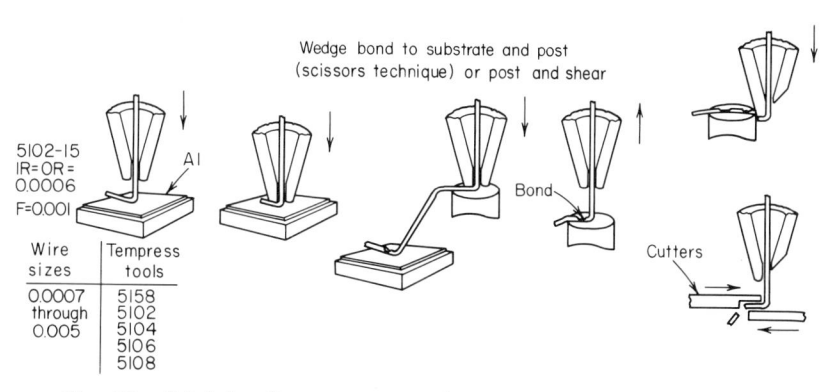

Fig. 89 Stitch-bonding geometry. (*Tempress Research Company.*)

substrate. High-temperature exposure for long periods of time will degrade a large variety of semiconductor devices and should therefore be avoided. It is for that reason that ultrasonic wire bonding has become popular. It should be remembered, however, that very small target metallizations may not be adaptable to ultrasonic bonding techniques since the ultrasonically bonded wire, approaching the target area parallel to the device surface, may easily short-circuit to adjacent metallization. A thermal compression ball bond approaches the target area vertically, and therefore runs less danger of short circuiting. Figure 90 shows the point in question.

This danger is eliminated, of course, in an integrated-circuit chip where all interconnection pads are on the periphery of the chip.

An ultrasonic wire bonder consists of an ultrasonic power supply which has as its output a high-frequency energy. The time and power level of this energy pulse must be able to be closely and repeatably controlled. In addition, there must be a transducer which converts the energy from the power supply into ultrasonic vibrations. The usual wire handling, clamping, and cutoff mechanism;

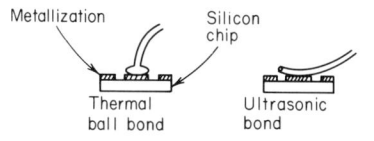

Fig. 90 Increased potential of short circuiting due to ultrasonic bond configuration.

the optics; and a manipulator comprise the rest of the machine. The tip, through which the wire is threaded, is a very important part of the machine, and the tip configuration will vary with the job being bonded. A typical ultrasonic bond sequence is shown in Fig. 91.

Because of the configuration of the ultrasonic bonding tip, the second bond on a given wire must be accomplished in a direction directly behind the first bond. This is unlike a ball and wedge bonding capillary, wherein the wire may be drawn at any angle away from the ball bond. As a result of this, the ultrasonic bonding stage must be rotatable over 360° to permit wire bonding from any point to any other point on the hybrid circuit. As a result of this requirement, many

12-72 Thick Films

ultrasonic wire bonders have a motorized work station which permits pushbutton rotation in either the clockwise or counterclockwise direction. A machine of this nature is shown in Fig. 92.

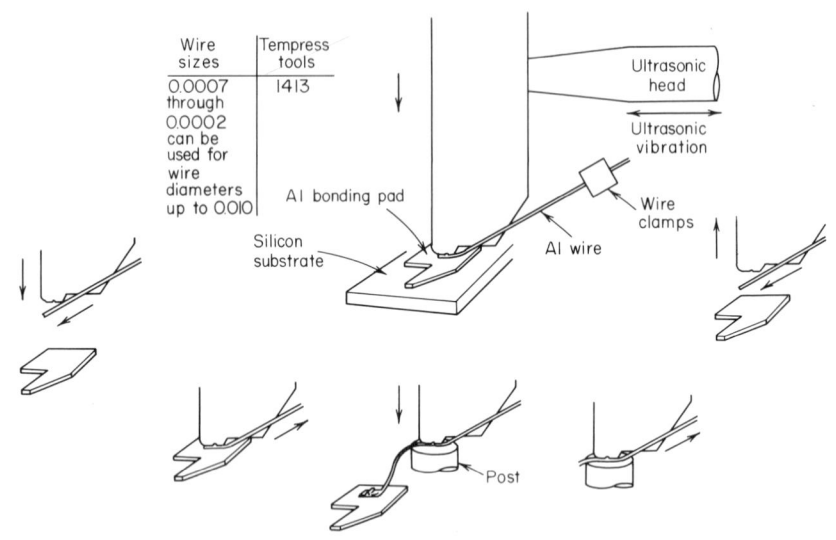

Wire sizes	Tempress tools
0.0007 through 0.0002 can be used for wire diameters up to 0.010	1413

Fig. 91 Ultrasonic bonding sequence. (*Tempress Research Company.*)

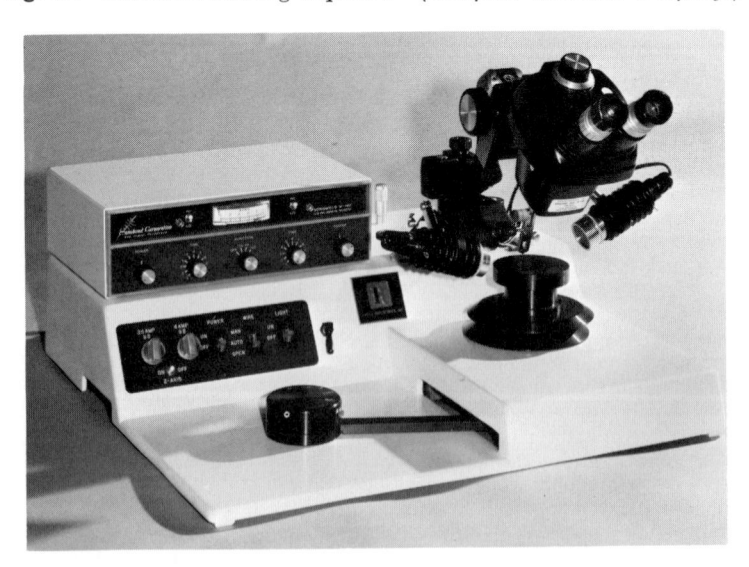

Fig. 92 Ultrasonic wire bonder. (*Hugle Industries.*)

Aluminum wire is most often used in ultrasonic wire bonding, and this may contain 1 percent silicon or 1 percent magnesium, added to the wire. Less frequently, gold wire is used for ultrasonic wire bonding, and this can have the same general specifications as the gold wire used for thermal compression wire bonding. The specifications for aluminum ultrasonic wire are listed in Table 23.

TABLE 23 Aluminum Wire for Ultrasonic Bonding

Suggested specification	1 % silicon aluminum		1 % magnesium aluminum	
Composition, %	Aluminum Silicon Calcium Manganese Iron Copper Magnesium	99.0 (nominal) 1.0 ± 0.15 <0.002 <0.002 <0.01 <0.01 <0.01	Aluminum Magnesium Calcium Manganese Iron Copper Silicon	99.0 (nominal) 1.0 ± 0.15 <0.002 <0.002 <0.01 <0.01 <0.01
Diameter, in., and temper	Breaking load range, g	Elongation range (nominal), %	Breaking load range, g	Elongation range (nominal), %
0.0007:				
Hard	6–9	1½–1	11–15	½–1
Ultrasonic bonding	8–12	—	8–12	—
Annealed	2–4	1–3	4–8	1–3
0.001:				
Hard	12–18	½–1	15–25	½–1½
Ultrasonic bonding	12–18	—	12–18	—
Stress relieved	—	—	11–15	1–3
Annealed	3–7	1–3	8–13	3–8
0.0015:				
Hard	25 40	½–1½	30–50	½–1½
Ultrasonic bonding	25–35	—	25–35	—
Stress relieved	—	—	22–35	1–3
Annealed	6–14	1½–5	20–27	3–8

SOURCE: Sigmund Cohn Corporation.

PROBING EQUIPMENT

Once a chip and wire hybrid thick-film circuit has been assembled, it is necessary to perform electrical tests so as to assure that the circuit is functioning according to an electrical specification. This will require a method of making contact to the input-output points on the hybrid circuit. There may also be in-process tests that require making contact to points within a partially completed substrate. In general, the areas to be contacted are small enough that both visual and mechanical aids are required. Many equipments have been produced which will make contact to transistors, diodes, and integrated circuits in wafer form. Some of these are adaptable to hybrid circuits, but most have limited-probe movement and therefore, are unusable. In the recent past, special probing equipments have been designed and made available specifically for hybrid circuits.

Kelvin contacts should be available on a probe used for hybrid circuits so as to accommodate the electric circuit that may require this refinement. The Kelvin probe system will eliminate contact resistance as a source of circuit test error by providing separate current and voltage paths to the circuit under test.

It should be noted that each probe head must have considerable latitude of adjustment in the x, y, and z motions. In order to achieve the Kelvin probe principle, each probe head must have two parallel insulated probe points that are spaced as closely as is reasonable (approximately 0.005 in.).

A hybrid circuit probe machine that utilizes these principles is shown in Fig. 93. It should be noted that a monocular microscope is supplied as part of the instrument, and this is required so as to adjust the probe positions relative to the circuit under test.

12-74 Thick Films

Some probe machines come supplied with binocular optics as part of the positioning and inspection system. A wafer probe that is adaptable to thick-film work is shown with this feature in Fig. 94.

The view shown in Fig. 95 is indicative of the large number of probe points that may be required so as to test a single hybrid circuit. It should be noted that 32 probes can be positioned around the periphery of a rather small substrate.

HYBRID CIRCUIT PACKAGES

There are several types of packages that may be used for hybrid circuits, and as yet, there has been little standardization in the industry, but in general, the basic types are:

    Round packages similar to transistor packages
    Flat packages similar to integrated-circuit flat packs
    Plastic packages of any convenient shape
    Molded packages conforming to the substrate outline

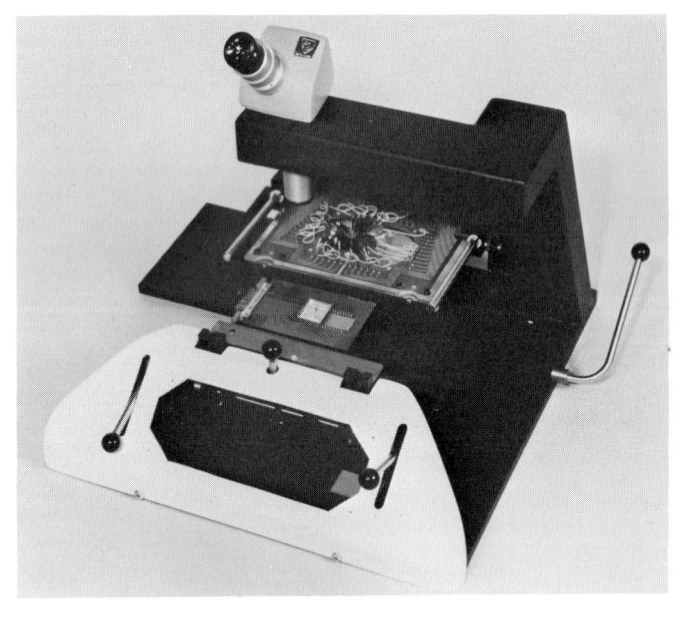

Fig. 93 Hybrid circuit probe instrument. (*Micro Dynamics Corporation.*)

For the most part, military or aerospace-type hybrid circuits use large flat packages of a size capable of accepting a ¼- by ⅜-in. or larger substrate, and it is generally required that these packages should be capable of being hermetically sealed. The common hermeticity requirement for a military hybrid circuit is in the order of 1×10^{-8} cm^3/sec He. This means, that after a helium bomb exposure, the package shall have a leak rate that is no greater than 1×10^{-8} cm^3 of helium per sec. As the hermetic package gets larger, this specification becomes more difficult to meet since the metal-to-glass and the metal-to-metal seal areas become greater. Another serious consideration is the distinct possibility of measuring surface adsorption due to the helium bomb exposure. To reduce the possibility of erroneous readings due to helium adsorption, many firms add a trace of helium into the inert atmosphere utilized in the hermetic sealing process. This eliminates the necessity of a helium bomb exposure and thereby greatly reduces the surface adsorption problem.

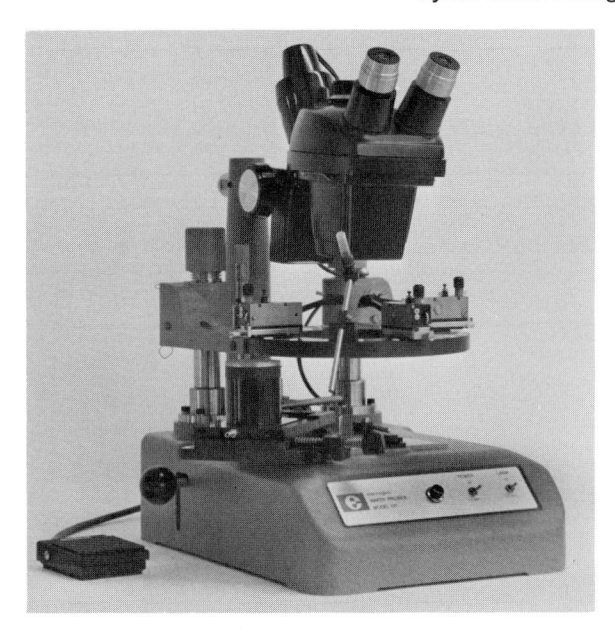

Fig. 94 Probe instrument with binocular optics. (*Electroglas Corporation.*)

MIL Standard 202C, Method 12, and more recently MIL Standard 883 present methods and specification requirements for hermetic seal testing. In MIL Standard 883 there is a formula for determining an acceptable leak rate which takes into consideration the size of the package and its cavity, and other factors which affect the hermetic capabilities of the package.

Fig. 95 Probe-point density. (*Wentworth Laboratories.*)

12-76   Thick Films

A typical hermetic package specification is listed in Table 24. This takes into consideration many of the initial requirements before sealing, such as flatness, and also references environmental requirements that are expected after sealing.

TABLE 24   Hermetic Package* Specification—Performance Requirements

Examination or test	MIL Std. reference	Specific conditions
Hermeticity	MIL Std. 202C Method 112 Condition C Procedure 1	$<1 \times 10^{-8}$ cm^3/sec
Lead strength (tension)	MIL Std. 750A Method 2036 Test condition A	1.5-lb load minimum
Lead strength (fatigue)	MIL Std. 750A Method 2036 Test condition E	3 bending cycles through 90° with 4-oz load
Thermal shock	MIL Std. 202C Method 107B Test condition C	-65 to $+200$°C at 5 Hz
Solderability	MIL Std. 202C Method 208	94% minimum coverage
Plating (24K gold)		425°C (air) for 5 min

* Metallized pad suitable for gold/tin eutectic substrate attachment. Sealing surface flat within 0.003 in.
SOURCE: Coors Porcelain Company.

The types of packages used where hermetically sealed devices are required are shown in Fig. 96. These differ from integrated-circuit flat packs and round packages only in the size and the numbers of available input-output leads. In addition to the packages shown in the figures, there is the solder preform, and the cover required so as to complete the enclosure.

ENCAPSULATION

There are two forms of encapsulation in common usage; they are, essentially, conformal coating and molding.

Conformal coating materials may range from wax through Durez° and, for special applications, epoxy. In cases where heat transfer is a requirement, an epoxy filled with alumina or other metallic powders might be used. In order to conformally coat a hybrid thick-film substrate it is only necessary to dip the substrate in the coating material, making sure that the input-output leads are in some way protected. A typical example of conformally coated hybrid thick-film substrates is shown in Fig. 97.

Transfer molding is another method of protecting the hybrid circuit from environments. For the transfer-molding process, thermosetting plastics are generally used. The cost reductions that can be realized in high-volume transfer-molding processes are significant. One of the advantages of transfer molding over conformal coating is that the resulting circuits are dimensionally uniform, dependent upon mold design. Of the problems associated with transfer molding, the temperature during the molding process stands out as the most important. The hybrid circuit in question

° Trademark of Hooker Chemical Corp.

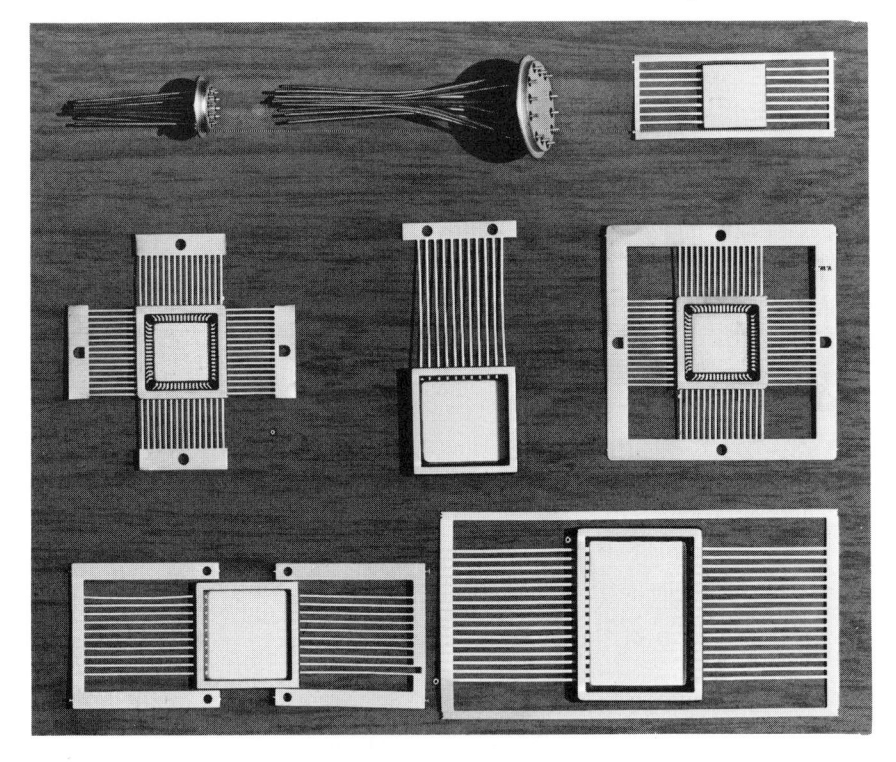

Fig. 96 Hybrid circuit hermetic packages. (*Alloys Unlimited.*)

must be able to withstand temperatures of up to 300°F for several minutes during the molding process. In addition, hydrostatic pressures of up to 200 psi are common in the transfer-molding process. It is important, of course, in the geometric and physical design of any hybrid thick-film circuit to take the final packaging into account.

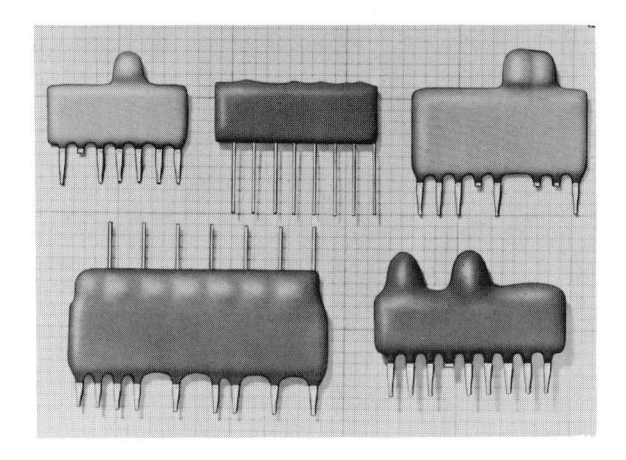

Fig. 97 Conformally coated thick-film circuits. (*Centralab Electronics Division.*)

Potting is the third method of protecting the hybrid circuit with a coating material. A potting shell, which remains as part of the final product, may be used, or a potting mold may be desired. When using a potting shell or a potting mold, the encapsulating material is usually first poured into the mold or the shell. If it is necessary to remove all air, a vacuum pump-out should follow the pouring. The curing of the encapsulant would be in accordance with the material used, and this cure cycle must be compatible with the ability of the hybrid thick-film circuit to withstand the curing temperatures.

Epoxy exothermic action can cause internal temperatures and stresses that will damage parts and cause shifts in resistor values. If it is found that a given encapsulating material displays these properties, the use of a buffer coating should be considered. Silicone rubber is often used as a buffer coating, and a conformal coating of 0.001 to 0.002 in. before encapsulation will usually solve most problems. The expansion coefficients of epoxy can be brought closer to that of the alumina substrate by the addition of metallic fillers. Some typical characteristics of molding resins are listed in Table 25.

TABLE 25 Typical Characteristics of Molding Resins

Property	Epoxy	Phenolic	Diallyl phthalate
Molding temperature, °C	150	150	150
Thermal expansion, in./(in.)(°C)	25–35	25–35	25
Water absorption in 24 hr, %	0.05	0.3	0.3
Dielectric constant	4	5	5
Dissipation factor, %	1	1	0.5
Maximum temperature use, °C	175	125	200

Hermetic sealing Hermetic sealing of hybrid microcircuits is usually required only when uncased semiconductor devices are used in the circuit fabrication and a hermetic specification is imposed. The hermetic packages, as previously described, can be sealed in essentially two ways, i.e., soldering or welding. A third method, used only with ceramic packages, involves "glassed sealing," and this will be discussed separately.

Soldering. Soldering is generally thought of as an operation requiring heat, solder, and flux. In the hermetic sealing of hybrid packages, however, flux is usually prohibited. It is therefore necessary that the surfaces to be soldered are scrupulously clean and free from oxides. As a result of these requirements, gold is generally used as the mating surfaces to be soldered. Compatible solders for fusing to the gold surfaces are gold-tin or gold-germanium.

To achieve the most reliable circuit condition, an inert atmosphere (nitrogen or helium) should be introduced into the package to replace the ambient atmosphere. Properly accomplished, the inert atmosphere precludes the possibility of harmful amounts of moisture being left in the enclosure.

Equipment is available that will evacuate the hermetic enclosure, backfill it with an inert atmosphere, and heat the solder preform so as to make the seal. The sealing process can also be performed in a furnace which is flushed with an inert atmosphere. The furnace necessary to accomplish hermetic package sealing would look identical to the firing furnaces previously described except that provisions for the introduction of special atmospheres would be included. Equipment designed for both solder sealing and glass sealing of hermetic packages is shown in Fig. 98.

A perimeter sealing type of machine, as shown, can seal in a wide variety of ranges and materials, including low-melting binder-filled epoxies, glass frits, and popular brazing materials. An operation sequence which purges the hermetic enclosure with inert gas, evacuates and backfills it, and then introduces the heat sequence at positive gas pressure can be adjusted to the needs of the selected sealing material.

Encapsulation   12-79

Welding. Welding the hermetic package has the advantage that the process does not subject the hybrid circuit to excessive temperatures. The welded hermetic seal is accomplished by making a large number of overlapping seam welds around the periphery of the enclosure. Since the welding heat is generated only at the small area of each weld, as it is made, the total heat rise of the package is very small.

Fig. 98 Perimeter-type hermetic sealing machine. (*GTI Corporation.*)

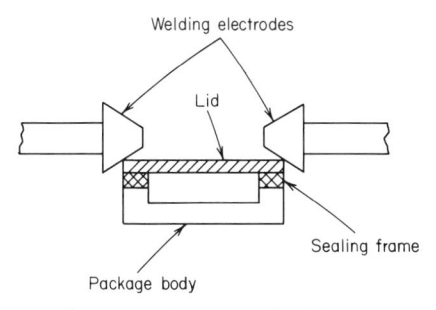

Fig. 99 Parallel-seam schematic diagram. (*Solid State Equipment Corporation.*)

The seam-welding system can be used to weld-seal or braze-seal the hermetic enclosure, dependent upon whether a solder preform is used. The weld seal is described schematically in Fig. 99. The welding heat is generated by the current flow through the high resistances at the small contact areas where the electrodes contact the lid.

12-80   Thick Films

Brazing or soldering with a seam welder can also be accomplished with certain advantages and some disadvantages when compared to a peripheral heater seal. Since a temperature at the welding electrodes of 1450°C is required to seal a welded cover, the impression of the electrode is readily visible. When using the welding principle for solder sealing, the temperature required at the interface of the seal frame, preform, and lid is only 280°C for gold-tin and 356°C for gold-germanium. As a result, the electrode marks are reduced. In addition, by adjusting the brazing current through the electrodes, a selectable portion of the preform can be melted, eliminating the possibility of solder balls within the enclosure. This is shown schematically in Fig. 100.

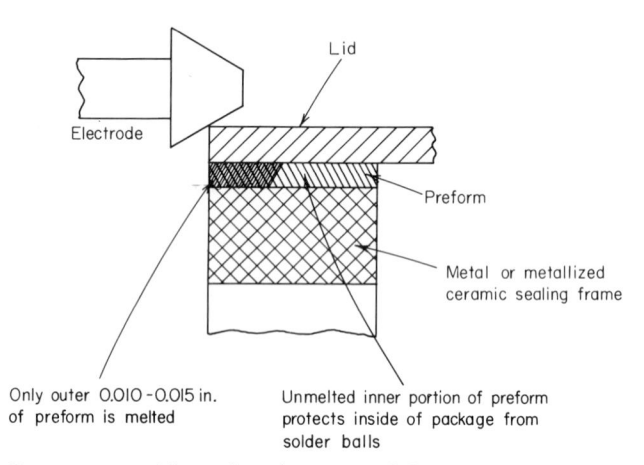

Fig. 100   Cross section of brazed package. (*Solid State Equipment Corporation.*)

Leak detection   Leak detection of hermetic package seals is a science unto itself. There are three methods of leak detection in common usage. Two methods are associated with fine leak detection, and they are helium leak detection by mass-spectrometer measurements and radioactive-tracer detection by Radi Flow techniques. Gross leak measurements are generally accomplished by immersing the package in a warm liquid and observing for the presence of bubbles.

Oils and glycerin fluids have been extensively used in the past for gross leak-detection fluids. Of late, however, fluorocarbons are becoming more popular because of their ability to detect (by bubble observation) a finer leak than either oil or glycerin, and because of their inert qualities. It is not even necessary to wash the package after immersion in fluorocarbon liquids since they will not combine with, or affect, any of the materials commonly used in microelectronic construction.

CLEANLINESS REQUIREMENTS

There are several points during the processing of a hybrid thick-film assembly where cleanliness of the parts and substrates and of the surrounding atmosphere is important. Cleaning processes should be incorporated, at the least, in the following process areas:

   Before substrate printing
   Before solder coating (if used)
   Before parts attachment
   Before encapsulation (if used)
   Before hermetic sealing

In addition to these general points in the process where cleaning is required, there are other cleaning precautions that may be taken, if needed by the atmospheric

Cleanliness Requirements 12-81

conditions in the assembly area. For instance, a mild vacuuming to remove any particulate matter, prior to any process requiring that the substrate be placed on a heated stage, will avoid certain contamination problems. The heated stage will melt many common airborne contaminants and render them impossible to remove. It is advisable, therefore, to remove all foreign particles before this becomes impossible.

Gloves, finger cots, tweezers, and triceps are all tools of the trade. Since finger oils are the most common source of contamination, every effort should be made to prevent their appearance, instead of relying on the probability of their removal during a cleaning process.

The cleaning processes used in hybrid thick-film microelectronics take four forms. They are:
   Ultrasonic cleaning
   Vapor degreasing
   Deionized water rinsing
   Vacuuming or dry nitrogen blowoff

TABLE 26 Ultrasonic System Selection Guide

Tank model	Working capacity, gal	Generator model	Average output power, watts	Watts/gal	Watts/in.^{2}
A1-D	1	AS-100 AS-200	100 200	100 200	2.05 4.11
A3-D	3	AS-200 AS-350	200 350	67 117	1.86 3.26
A5-D	5	AS-350 AS-700	350 700	70 140	2.55 5.11
A8-D	8	AS-700 AS-1000	700 1,000	88 125	4.16 5.95
A10-D	10	AS-1000 AS-1400	1,000 1,400	100 140	4.46 6.25
A15-W	15	AS-1400 AS-2000	1,400 2,000	93 133	4.60 6.57
A20-W	20	AS-2000 AS-3000	2,000 3,000	100 150	5.01 7.51
A30-W	30	AS-3000 AS-4000	3,000 4,000	100 133	5.76 7.69

SOURCE: American Process Equipment Corporation.

Ultrasonic cleaning Ultrasonic cleaners make use of acoustically introduced cavitation bubbles, in fluid, to dislodge dirt, grease, and particulate contamination which may cling to surfaces. The ultrasonic energy imparted to the cleaning fluids must be sufficient to create cavitation bubbles in order for the process to be of significant improvement over simple solvent cleaning. Ultrasonic cleaning machines are obtainable in a wide variety of sizes. Tanks are available from less than 1 gal to more than 1,000 gal capacity while the generators may be had with less than 25 watts or more than 30,000 watts of power. For microelectronic applications, the smaller sizes would be utilized. Table 26 shows fluid capacity with relation to ultrasonic power required, and can be used to estimate the requirements for an ultrasonic cleaning system.

12-82　Thick Films

Vapor degreasing　Vapor degreasing, after solvent immersion, assures that only pure solvent, free of soluble contaminants, oils, and particles, touches the microelectronic hybrid circuit. The solvent vapor is generated by raising the solvent above its boiling temperature. The height in the tank to which the vapor rises is controlled by a cooling coil, which condenses the vapor and permits it to return to the tank. As a result of this action, only pure distilled solvent can come in contact with the object being cleaned.

Deionized water rinsing　Certain types of contamination cannot be removed by solvents, and as a result, high-megohm water rinsing is used to remove ionic contamination. A continuous recirculation and repurification system should be used to keep the water free of organics, minute particles, and bacteria, and the water should have a resistivity of 15 to 18 megohms. Equipment for the continuous supply of high-resistance deionized water is available.

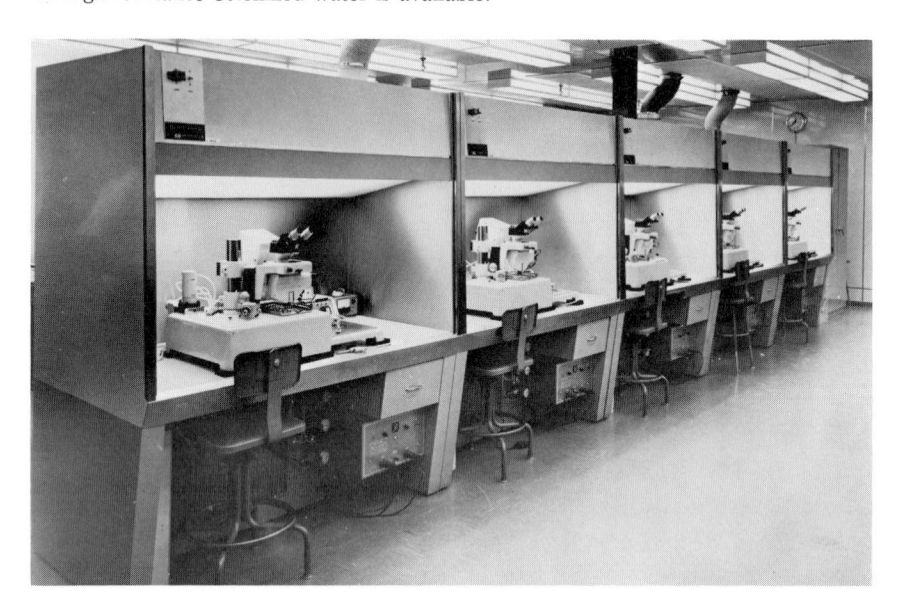

Fig. 101　Laminar flow work benches. (*Air Control, Inc.*)

Vacuuming or dry nitrogen blowoff　The only purpose of this type of cleaning is to remove particulate matter that may have settled on a partially assembled circuit. Certain types of particulate contamination will melt when the circuit goes through a heated assembly process (such as eutectic die bonding or thermal compression wire bonding), and the resulting foreign matter is generally impossible to remove. If all processing is accomplished in a Class 100 clean room atmosphere, or under a Class 100 laminar flow hood, this type of contamination will be less of a problem.

Laminar flow work stations may be used in lieu of a total clean room area so as to help prevent contamination during processing. An operation showing the use of laminar flow benches is presented in Fig. 101. Where a total clean environment is required, laminar flow benches are often used within a clean room area.

Chapter **13**

Metals Joining of Electronic Circuitry

JOHN E. McCORMICK
Rome Air Development Center, Griffiss Air Force Base, New York

Introduction	13–2
First-level Joining, Materials and Processes	13–8
Connecting Active Devices	13–8
Substrate (Inorganic) Internal Connections	13–55
Multilayer Printed-circuit-board Internal Connections	13–68
Second-level Joining, Materials and Processes	13–75
Solder Connections	13–75
Resistance-welded Connections	13–93
Other Welded Connections	13–103
Mechanical Connections	13–113
Comparison of Second-level Techniques	13–113
References	13–114

13-2　Metals Joining of Electronic Circuitry

INTRODUCTION

Figure 1 organizes and categorizes the processes, techniques, and materials discussed in this chapter. Table 1 presents a list of the terms and definitions appropriate to the chapter. Emphasis has been placed upon microelectronic connections and techniques. The methods that may be used to electrically intra- and interconnect unpackaged microelectronic parts and devices are categorized as first-level joining techniques. Substrates and multilayer printed boards that provide the complex matrices for electrically connecting the level-1 electronic assemblies have also been considered to be level-1 assemblies, but mechanical, rather than electronic

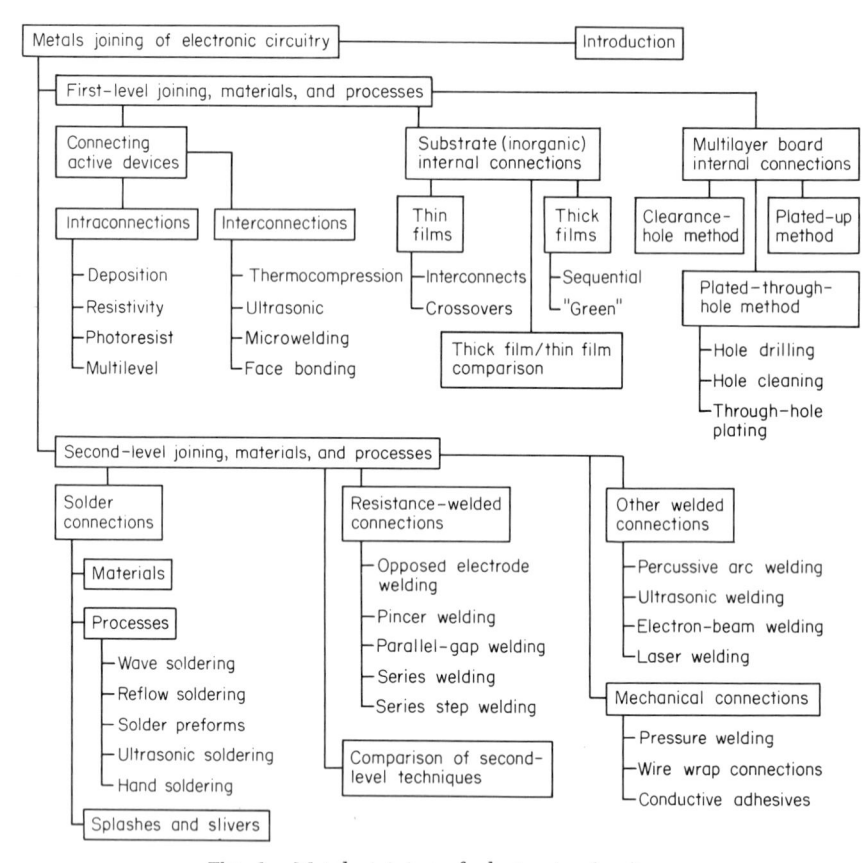

Fig. 1 Metals joining of electronic circuitry.

in nature. Therefore, the processes and techniques used to form the intraconnections within these matrices have also been categorized as first-level joining techniques.

The techniques that are used to attach the level-1 electronic assemblies to the level-1 mechanical assemblies, thereby forming a level-2 mechanical assembly, have been categorized as second-level joining techniques.

In the past electronic equipment has usually been described as being composed of a number of levels. The nomenclature of these levels differs somewhat, but the one fairly universal listing is system, set, group, unit, assembly, subassembly, and part. Figure 2 gives an example using this list of names. The system illustrated in Fig. 2 is made up of three functional sets. The radar set is made up

TABLE 1 Terms and Definitions[1-3]

Active element. An electronic-circuit element that displays power gain or control.
Beam leads. Electroformed thick-film extensions of the thin-film terminals of planar semiconductor devices or circuits. These "beams" extend beyond the edges of the device and may be bonded to appropriate contact points by a variety of techniques.
Bump. A means of providing connections to terminal areas of a device. A small amount is formed on the device (or substrate) pads and is utilized as a contact for facedown bonding.
Chip (Die, Dice). The individual circuit or component of a silicon wafer.
Circuit. The interconnected combination of a number of elements of electronic parts.
Clearance holes. These provide access to circuit pads on the different layers of clearance-hole multilayer printed wiring board.
Composite board. A completely laminated multilayer printed wiring board.
Conductive adhesives. Organic adhesive material that is heavily loaded with a conductive powder. Used to glue conductors together.
Conductor. A circuit element whose function is to conduct electric current.
Connector. A device used to make interconnections.
Cordwood. Assembling parts as close as possible and interconnecting them by welding or soldering the leads together.
Crossover. Deposition of a conductor over the insulated upper surface of another deposited conductor.
Deposition. The "laying down" of films of metals or insulators on a substrate.
1. *Evaporative deposition.* The technique of condensing a thin film of material on a substrate with the entire process taking place in a high vacuum. The source material may be radioactively heated by bombardment with electrons (electron beam) or may be heated by thermal conduction techniques.
2. *Sputtering.* The ejection of particles from the surface of a material resulting from bombardment by ions and atoms. The material may be used as a source for deposition.
3. *Plating.*
    a. *Electrolytic:* The process whereby an electric current is passed between electrodes submerged in an electrolyte. One electrode is the source material, and the other is the substrate upon which the material is to be deposited.
    b. *Vapor:* The deposition of a thin-film material as the result of a chemical reaction occuring in the vapor. Material is precipitated from the gas onto the substrate because of this reaction.

Device. An electronic part consisting of one or more active or passive elements.
Dielectric strength. The capability of an insulator to withstand an applied voltage without breaking down and conducting current.
Diffusion bonds. These processes are accomplished by bringing the two conductors to be joined into intimate contact and then inducing the atoms of one material to diffuse into the structure of the other.
1. *Thermocompression bonds.* These bonds are formed by the simultaneous application of heat and pressure.
    a. *Ball bond (Nailhead bond):* A wire is flame-cut to produce a ball-shaped end, which is then thermocompression-bonded to a metal pad.
    b. *Bird-beak bond:* The welding tool deforms the wire during the application of heat and pressure so that the resulting bond resembles the beak of a bird.

TABLE 1 Terms and Definitions[1-3] **(Continued)**

  c. *Wedge bond:* A type of wire joint that is bonded by a wedge-shaped tool (may also be a cold weld or an ultrasonic bond).
  2. *Ultrasonic bonds (welds).* In this process the wire is pressed against a bonding pad, and the pressing mechanism is ultrasonically vibrated at frequency above 10 kHz. These high-frequency vibration breakdowns disperse the oxide films that are present on the conductor surfaces. As these surface films are removed, diffusion of the conductor materials occur at the interface. The joints formed are metallurgically sound diffusion bonds.

Discrete element. An element or component manufactured in such a manner that it may be individually measured and transported.

Element. A constituent of a microcircuit or an integrated circuit that contributes directly to its operation.

Facedown bonding (flip-chip, registrative bonding). A method of attaching a component or circuit chip to a substrate by inverting the chip and bonding the chip contacts to the mirror-image contact points on the substrate. The pads or bumps that are used at the contact points may have originally been on either the chip or the substrate. The actual bonding process may be some type of thermocompression, ultrasonic, or solder technique.

Header. That portion of a device package to which the chip is attached and from which the external leads extend.

Heat sink. A mechanical device used to transfer heat away from an element or part.

Hybrid circuit. A circuit that utilizes two or more fabrication techniques to form the circuit, such as integrated-circuit chips attached to a substrate having thin-film devices and conductors.

Insulator. A material that does not conduct current.

Integrated circuit. A microcircuit consisting of interconnected elements inseparably associated and joined on or within a substrate.
  1. *Monolithic integrated circuit.* An integrated circuit consisting of elements formed on or within a semiconductor substrate with at least one of the elements formed within the substrate.
  2. *Film integrated circuit.* An integrated circuit consisting of elements that are films formed upon an insulating substrate.

Interconnection. The joining of one individual device with another.

Interlayer connection. A metal conductor that provides communication to or between layers.

Internal layer. A conductive pattern that is contained entirely within a composite board or substrate.

Intraconnection. The joining of elements within devices.

Lands (Pads). Metallized areas on the surface of a device or circuit chip to which bonds or interconnections and test probes may be applied.

Mechanical connections. A technique for holding two conductors in contact with each other, either by some external holding device, or by some deformation of the conductors themselves (and/or possibly deformation of the holding device).
  1. *Friction connections.* This technique is characterized by having a spring component somewhere in the joint, so that the spring supplies the necessary force on all components to hold the joint together.
  2. *Swaged (crimped) connections.* Some, or all, components of the connector are plastically deformed in such a manner as to establish the soundness of the joint.

TABLE 1 Terms and Definitions[1-3] **(Continued)**

   3. *Wire-wrapped connection.* This type of connection consists of a post with a specific number of turns of solid wire tightly wrapped around it.

Metallization. See *Deposition*.

Microelectronics. That area of electronic technology associated with or applied to the realization of electronic systems from extremely small electronic parts or elements.

Microcircuit. A small circuit having a high equivalent-circuit element density, which is considered as a single part composed of interconnected elements on or within a single substrate to perform an electronic circuit function.

Module. An assembly of microcircuits, or an assembly of microcircuits and discrete parts, designed to perform one or more electronic circuit functions, and constructed such that for the purposes of specification, testing, commerce, and maintenance it is considered indivisible.

Multilayer. A three-dimensional circuit board built of printed layers laminated together and interlayer-connected.

Ohmic contact. A contact between two materials across which the voltage drop is the same regardless of the direction of current flow.

Packaging system. The mechanical and electric devices that must be combined so that functional electronic circuits, capable of performing their function when exposed to specified environments, are attained.

Passive element. An electronic circuit element that displays no gain or control.

Pinhole. A very small imperfection that extends through a layer of material.

Pit. A very small imperfection that does not extend through a layer of material.

Plated post. A vertical, solid interlayer connection of the built-up process used to obtain electrical communication between individual layers of a multilayer printed wiring board.

Plated-through hole. A method for achieving electrical communication between individual layers of a multilayer printed wiring board.

Printed circuit. A pattern, comprising printed wiring and printed elements, all formed in a predetermined design on a nonconductive board or substrate.

Purple plague. A brittle gold aluminum intermetallic which sometimes forms at an interface of a gold-aluminum thermocompression bond. This intermetallic appears purple in the crystalline form.

Resistivity. The electrical resistance across the opposite faces of a cube of material.

Semipermanent connection. A connection capable of separation without destruction, but not specifically designed for separation.

Separable connection. A connection designed to withstand repeated separation and reconnection without deterioration.

Sheet resistivity. The electrical resistance measured across the opposite sides of a square of deposited thin-film material. Expressed in ohms per square.

Silicon dioxide. A dielectric material commonly used in the surface passivation of microelectronic circuits.

Silicon monoxide. A dielectric material often used in the fabrication of a microelectronic device to form an insulator.

Soldering. A process for joining two conductors together using a low-melting-point alloy. The process requires the use of a flux (cleaner) to promote the wetting of the conductors by the solder (alloy).

   1. *Dip soldering.* The conductors are dipped in molten solder. Thus, the solder and heat are applied simultaneously to the conductors. All the prefluxed metallic surfaces in contact with the molten solder are rapidly wetted.

TABLE 1 Terms and Definitions[1-3] **(Continued)**

2. *Hand soldering.* The solder and the prefluxed conductors are heated up to soldering temperature by a handheld soldering iron.
3. *Reflow soldering.* Solder is applied to the conductor previously by a number of techniques (dipping, plating, wiping, squeegee). Heat and pressure are applied to the solder-coated conductors when it is desired to effect bonding.
4. *Ultrasonic soldering.* This process eliminates fluxes and precleaning (in some cases) because of the cleaning action of the high-frequency vibrations.
5. *Wave soldering.* A soldering process whereby a mechanically induced wave of molten solder (in a solder bath) is brought into contact with the conductive surface of a printed-wiring board or substrate and results in the soldering of components to the board.

Substrate. A material upon which thick-film depositions or thin-film depositions are made.

Test coupon. A sample, or test pattern, usually made as an integral part of a multilayer printed wiring board, on which electronic and environmental tests may be made to evaluate the board design or process control, without destroying the basic board.

Thin-film circuit. A circuit fabricated by the deposition of material of several thousand angstroms in thickness (such as a circuit fabricated by vapor deposition).

Throw-away module. A module designed so that it is discarded at failure.

Undercut. The accentuation of etching at a material interface. Occurs during the etching of conductors and also during etching of windows in dielectric materials on a microelectronic semiconductor device.

Welding. Welded connections are made by fusing together two conductors using heat, pressure, or both.
1. *Cold welding.* The joining of two metals by the application of pressure only.
2. *Electron-beam welding.* The welding heat required is developed at the bond site by the impact of a beam of high-energy electrons. The electron beam can be focused into a spot approximately 10 microns in diameter. The heat penetrates much deeper into the material (in the direction of the beam) than with any other welding technique.
3. *Laser welding.* The heat required to fuse the conductors is transmitted by a beam of coherent light.
4. *Percussive arc welding.* This type of weld is accomplished by bringing the two conductors together during, or immediately after, the period of an electric arc between them. The arc melts the surfaces of both conductors. While these surfaces are still molten, the conductors are rapidly pressed together, and then allowed to cool.
5. *Resistance welding.* The welding heat in the parts to be welded is generated by the resistance of these parts to the passage of electric current.
   a. *Opposed-electrode Welding (Crossed-wire welding):* The conductors to be welded are clamped between two opposing electrodes. A current is passed through the joint, causing the metal to melt and the weld to result.
   b. *Parallel-cap welding:* Two parallel electrodes are placed on the top surface of the upper conductor to be welded. Current is passed from one electrode to the other through the conductors, causing the metal to be melted and the weld to result.
6. *Ultrasonic welding.* See *Diffusion bonds.*

of four functional groups. The various levels may be identified down to the part level.

Such level definitions are based on the electronic function, and do not necessarily have any significance with respect to joining and packaging. For example, the radar set of Fig. 2 consists of many large electronic equipment cabinets, large antennas, and associated hardware, whereas a radio set may very well be a single portable unit containing all circuitry, controls, and even an antenna in single handheld package. Thus, the physical characteristics of groups, units, assemblies, subassemblies, etc., are significantly different in the illustrated radar set and a handheld radio set.

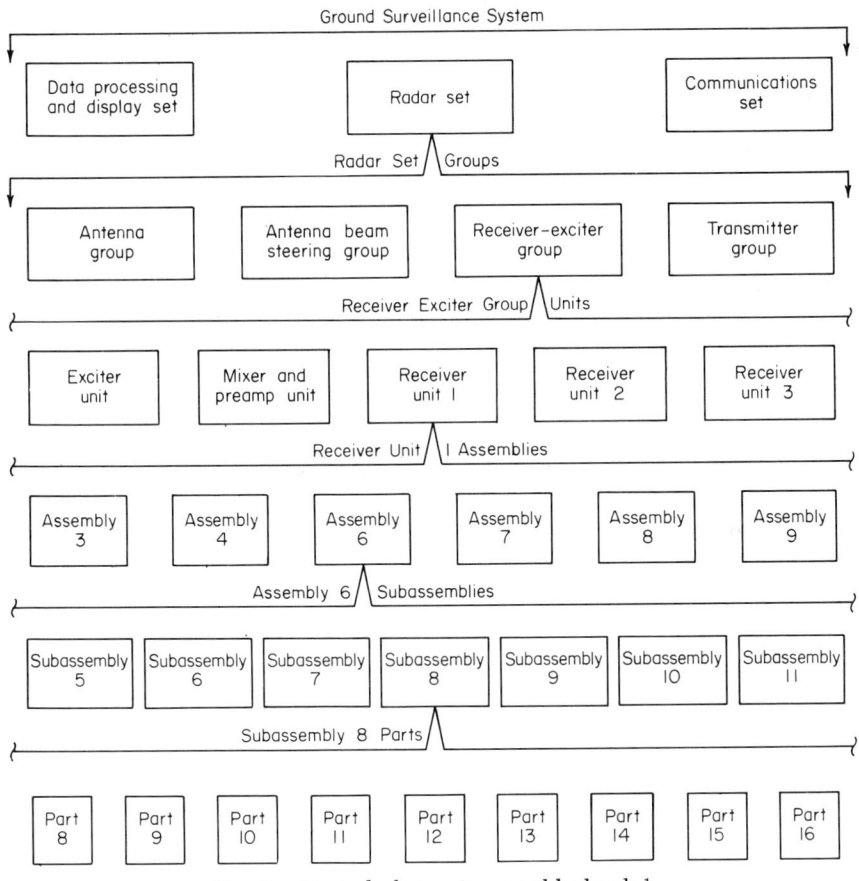

Fig. 2 Typical electronic assembly levels.[1]

The listing of levels as discussed above does not have enough significance to warrant use of it for a general categorization of packaging and joining levels. Therefore, it is necessary to establish designations for equipment levels that are based on the mechanical considerations used in joining electronic circuitry. Such a family of assembly levels has been defined in Table 2 and illustrated in Fig. 3.

The techniques that are used to electrically join or form connections within a level-1 package can be designated as first-level joining techniques. The techniques that are used to assemble a level-2 package may be designated as second-level joining techniques.

13-8   Metals Joining of Electronic Circuitry

Second-level joining techniques are also used to assemble certain portions of level-3, 4, and 5 packages (for example, back-plane soldering on a mother board). However, beginning with level 3, and continuing up through the last level, connections between packages are, in most cases, some form of nonmetallurgical electric contacts and therefore are not within the scope of this chapter.

TABLE 2   Mechanical Assembly-level Definitions[1]

Level	Definition
1	The lowest level of packaging an electronic device or circuit, commonly referred to as the "component level." *Example:* A discrete resistor or an integrated circuit in a TO 5 can.
2	The second level of assembly generally consisting of components grouped on a card, chassis, or module. *Example:* Circuit cards, cordwood modules, or flat-pack modules.
3	The third level of assembly consisting of groups of level-2 assemblies. *Example:* Card-mounted modules, card baskets, or chassis or assembly of subchassis on a chassis.
4	The fourth level of assembly consisting of groups of level-3 assemblies. *Example:* Chassis in drawers, hinged doors, or panels.
5	The fifth level of assembly consisting of groups of level-3 and/or level-4 assemblies. *Example:* Bays, cabinets, racks, or carrying cases.
6	The highest level of assembly at which the electronics, as packaged into individual self-supporting enclosures, are integrated into an installation with a particular arrangement and assembly methods.

FIRST-LEVEL JOINING, MATERIALS AND PROCESSES

Connecting Active Devices

Intraconnections This section of the chapter is concerned with the joining of the various electronic components that have been diffused into an integrated-circuit chip. Intraconnection of these diffused electronic components is made by vacuum-depositing metal on the surface of the insulating oxide layer that has been formed on the semiconductor. This metal layer is etched so that only the pattern of conductors needed to complete the circuit remains. The properties desired in this metal film are as follows:

1. It must be capable of making a good ohmic (nonrectifying) contact to the semiconductor at the contact points.

2. It must be an excellent conductor.

3. It must have metallurgical properties suitable for the lead-attachment procedures to be used.

Many metals have been used for the formation of ohmic contacts to semiconductor devices. The most common are gold, aluminum, nickel, lead, silver, and chromium. With silicon devices, aluminum has been found to be preferable. However, when using aluminum, one must be careful to avoid the formation of a p-type region that can result when aluminum is alloyed into silicon. The phenomenon of regrowth, which causes this problem, may be explained with the aid of Fig. 4, a phase diagram of the aluminum-silicon system. An alloy of composition Al 89 Si 11 has a lower melting or eutectic point than any other combination of the two metals. When the two metals are heated to the eutectic point, atoms of each metal begin diffusing into the other at the interface. When the eutectic point is reached, a very thin liquid layer forms at the interface, and this liquid phase very rapidly dissolves both metals in the appropriate proportion to form a large volume of eutectic alloy. If a limited amount of one of the two metals is available, this process will cease when the entire amount of the limited material has been consumed. The system is heated to a temperature higher than the eutectic, and more and more silicon is dissolved in the liquid phase, as seen on the phase diagram. If the system is now cooled to a point below the eutectic, this additional silicon is rejected from the liquid during cooling, forming a regrowth layer of silicon in the interface. This regrowth layer contains a small percentage of aluminum, as determined by the solid solubility of aluminum in silicon (approximately 0.001 percent). The liquid aluminum-silicon alloy will "freeze" and form an ohmic contact to the silicon, provided that the original silicon is either p-type or heavily doped n-type. If it is p-type to start with, there is no difficulty because the aluminum which has segregated out into the regrowth layer tends to make it more

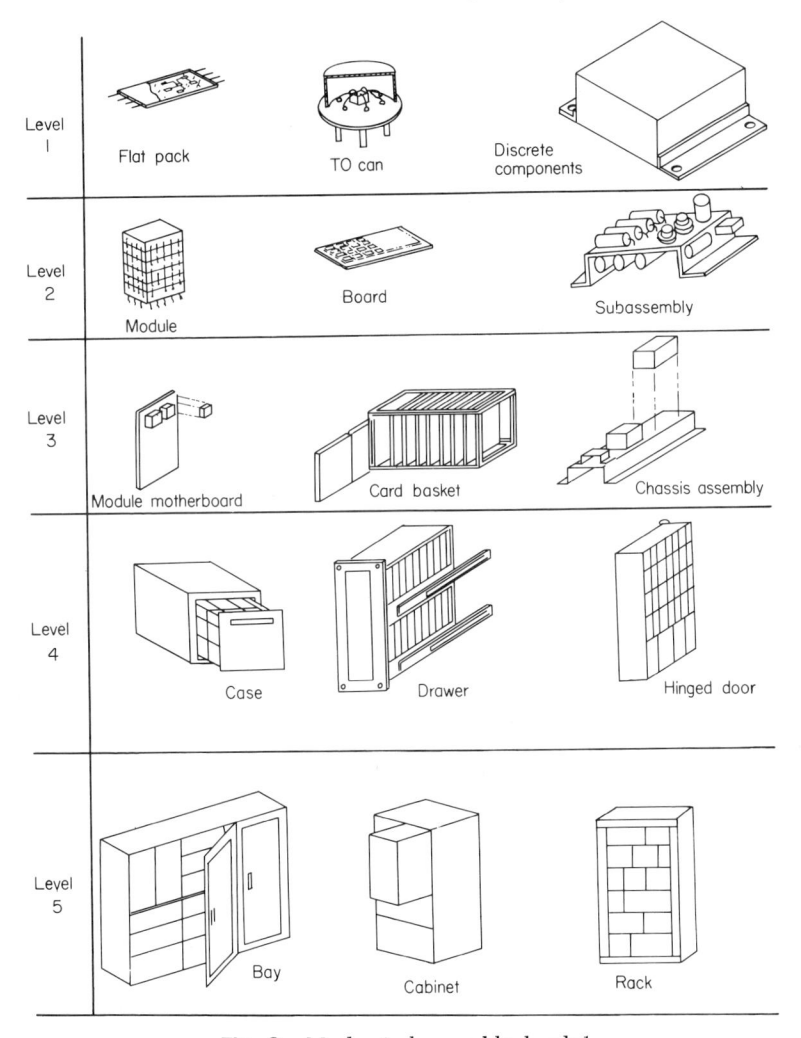

Fig. 3 Mechanical assembly levels.[1]

p-type. If it is n-type to start with, then there must be more n-type atoms than p-type aluminum atoms per unit volume in the regrown crystal; otherwise, a p-n junction will be formed at the interface between the regrowth layer and the original undisturbed silicon. The amount of aluminum present in the regrowth layer is approximately 5×10^{18} atoms/cm^3. Thus, unless there is an n-type species present in the regrowth layer in excess of this concentration, a rectifying contact will be formed. In practice, a typical metallizing sequence is as follows:

13-10 Metals Joining of Electronic Circuitry

After all diffusions have been completed, openings are cut in the oxide by photoresist techniques in the areas where ohmic contacts are to be formed. The wafers are then cleaned and placed in a vacuum-evaporation apparatus. The clean, etched wafers are placed under a tungsten filament in the bell jar, and the ohmic-contact metal desired (e.g., aluminum) is coiled around the tungsten filament. After the bell jar has been evacuated, the aluminum is first melted and then vaporized by

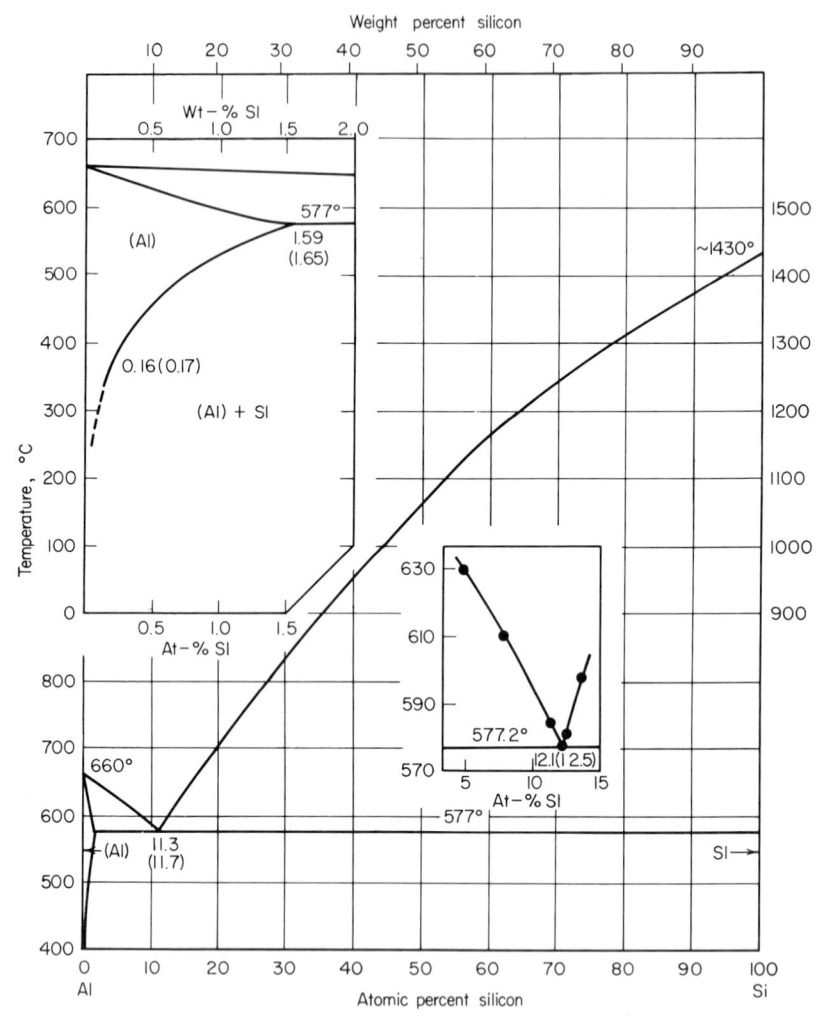

Fig. 4 Aluminum-silicon phase diagram.[6]

the heated filament. A thin film of aluminum metal is deposited on the wafers as well as on all interior parts of the system.

After all the metal has been evaporated, the bell jar is backfilled, and the wafers are removed. The metallized wafers are again coated with photoresist, exposed with a new mask which is essentially the inverse of the preceding mask, and developed. At this point, an appropriate etch such as sodium hydroxide is used to remove the aluminum in the unwanted areas. The metal is now alloyed into the surface of the silicon by heating the wafers to a temperature above the eutectic temperature.

A variation of the ordinary metallization procedure is a process known as overlay metallizing, which is useful in integrated circuits. In overlay metallizing the same basic principles apply as in ordinary metallizing. However, in addition to forming an ohmic contact to the silicon, the metal deposit forms extensions of the ohmic contact over the oxide. The metal deposit actually crosses the junctions but does not short-circuit them because of the intervening insulating oxide. The overlay contact adds about 0.03 pf/mil^2 of parasitic capacitance. With this technique, two new degrees of freedom are achieved. First, it is possible to make large ohmic-contact areas to extremely tiny junction areas. Second, it is possible to interconnect two or more otherwise isolated junction areas directly on a wafer without short-circuiting the junctions.

The choice of contact metal is much more restricted in the overlay process because the metal used must have the additional property of making a good mechanical bond to the SiO_2 without complete penetration of the oxide. Only two materials have been found satisfactory for this purpose: aluminum and chromium. Both aluminum and chromium are active reducing agents and can combine with the oxygen from the silicon dioxide layer. This reaction results in a disturbance of the surface of the layer and creates a good bond between the oxide and metallic film. Since it is difficult to use chromium because of its high melting point and certain other properties, aluminum is practically the only contact metal used for overlay metallizing. Pinholes formed during the photoresist process can seriously affect overlay metallized devices. Since integrated circuits employ large amounts of overlay metallization, the pinhole count must be kept extremely low to avoid loss of a significant number of devices.

Deposition Techniques.[7] Two methods of controlling the thickness of the aluminum films are discussed in this subsection. The first involved the use of a flash-deposition process which eliminated some of the variations in film thickness that occur from deposition to deposition. The second involved evaporation of the aluminum at a much slower but a more controlled rate. For both techniques resistance heating of the source material was used. The vacuum equipment used as a standard 4-in. oil-pumped system. The ultimate vacuum was 2×10^{-6} torr (with liquid nitrogen cooling). Power for the source heaters was controlled by manually operated rheostats. Substrates were heated by thermocouple-controlled infrared lamps. Film thickness was measured with a sensitive profilometer which has precise control of the vertical and horizontal movement of the stylus. It is possible to measure thickness of 100 ± 50 Å with this device.

In resistance heating, a current is passed through a filament which is in contact with the material to be evaporated. The heating of the filament above the vaporization temperature of the material causes the material to evaporate. For aluminum, either tungsten or tantalum may be used as the heater, which is generally in the shape of a solid rod or coil. The aluminum may be placed in direct contact with the heater, or it may be placed in a ceramic crucible or boat.

To deposit a film between 5,000 and 10,000 Å thick, the substrate was kept within 4 to 6 in. from the source. At this separation, a flash-deposition process was used to prevent discoloring of the deposited film. A solid tungsten heater, with a piece of high-purity aluminum wire placed directly on the heater, was used. Table 3 is a flow chart that indicates the sequence of operations involved in a flash-deposition process.

The aluminum deposited on land areas of silicon circuits and alloyed into the silicon to provide ohmic contact is 2,000 to 3,000 Å thick. To obtain this thickness, a single piece of aluminum wire (99.9999 percent pure) 0.062 in. in diameter and 0.375 in. long was used. After the land areas were alloyed, aluminum 8,000 10,000 Å thick was deposited for the interconnection pattern. To obtain this thickness, a single piece of 0.062-in.-diameter wire 0.500 in. long was used. The source-to-substrate separation was 3.75 in. for both the above cases. To control the film thickness by controlling the rate of evaporation, a heater coil of tungsten wire (three strands, each 30 mils in diameter) was tightly wound on a ¼ in. OD by ½-in.-long boron nitride crucible having an inside diameter of approximately ⅛ in. A 0.500-in. length of 0.062-in.-diameter aluminum wire (99.9999 percent pure) was standardized on as the source. Using the system previously described, evacuated to less than 4×10^{-5} torr, with an unheated substrate, the effects of

TABLE 3 Flash Evaporation: Sequence of Operations[7]

1. Place tungsten boat in vacuum system.
2. Place cleaned aluminum in boat.
3. Place wafers in substrate holder.
4. Place substrate holder in vacuum system.
5. Place shutter between wafers and boat.
6. Evacuate system to less than 3×10^{-5} torr.
7. Heat substrate to 200°C. Hold for 15 min.
8. Turn on source heater. Set manual rheostat at one-tenth of full scale for 1 min.
9. Set rheostat at one-fourth of full scale for 1 min. NOTE: Next three steps should be accomplished in 1 to 2 sec.
10. Turn rheostat to maximum.
11. Upon indication of aluminum outgassing (pressure rise), actuate shutter uncovering boat.
12. Turn rheostat to "off."
13. Cool wafers to less than 50°C.
14. Open system and remove substrate holder.

varying deposition time and source-to-substrate separation were determined. After the optimum conditions were found, a number of depositions were made at these conditions to determine whether repeatable results could be obtained. Table 4 lists the results of these variations. The spread, in thickness, 1,000 to 1,500 Å (in the last three entries), is considered satisfactory.

TABLE 4 Effects of Time and Spacing on the Thickness and Quality of Thin-film Aluminum[7]

Source-to-substrate separation, in.	Evaporation time, min	Heater current, amp	Thickness of deposited film, Å	Deposition rate, Å/sec	Remarks
7.5	1.50	87	750	8–9	Good deposition
7.5	1.75	87	1,000	9–10	
7.5	2.0	90	1,500	12–13	
7.0	3.0	87	1,700	9–10	
6.0	2.0	90	1,500	12–13	Discolored deposition
4.5	2.0	90	5,800	50	
7.5	3.0	62	1,000	5–6	Poor deposition
7.5	3.0	62	1,300	7–8	
7.5	3.0	65	1,500	8–9	

An x-ray diffraction scan of several specimens indicated that, in all cases, the aluminum films displayed a high degree of orientation in the 111 plane. The degree of preferred orientation or grain size (probably the latter) is dependent on the purity of the deposited film. Also, calculations indicate that the grain size is larger with thicker films, but is reduced if impurity content is high. A flash-evaporated sample with a thickness of 2,000 to 3,000 Å of aluminum showed a 111 orientation while being highly reflective. A sample deposited slowly to about 7,600 Å of aluminum had matte finish, but also 111 orientation. The latter was less reflectual because it was made up of larger grains.

Surface Resistivity.[8] Table 5 compares the surface resistivity of aluminum film, flash-evaporated by resistance heating, with chromium and chromium-gold films slowly evaporated by resistance heating.

Chromium films will peel if they are too thick. The maximum usable thickness for chromium films has been set at 2,000 Å. Although chromium films with thicknesses less than 1,300 Å have good adherence, their resistance is greater than 18 ohms/sq, and therefore they are not recommended for interconnections.

The gold film on chromium is deposited such that there are these regions: a chromium bonding layer, a phased-in chromium-gold layer, and the final gold layer. The two sources are side by side, and the gold source is brought up to about half power first, before the deposition is started. Failure to do this caused the chromium bonding film to be too thick.

TABLE 5 Surface Resistivity of Thin-film Conductors[8]

Property	Aluminum	Chromium	Gold*
Surface resistivity, ohms/sq	0.15–0.40	4–18	0.08–0.14
Film thickness, Å	3,000 ± 300	1,300—2,000	4,000 ± 400
Deposition rate, Å/sec	100–150	6–20	15
Substrate temperature, °C	100–120	100–120	100–120
Pressure during deposition, torr × 10^5	<3.0	<5.0	1.0–3.0

* On 100- to 400-Å chromium.

Photoresist Techniques.[9] A basic technique used in making thin-film conductors on solid silicon devices is masking. Masking is essentially a photographic process in which a photosensitive emulsion, known as photoresist, is exposed to ultraviolet light with a photomask used to delineate the pattern. When developed, the emulsion provides a chemically resistant pattern which conforms to that of the photomask. By this method it is possible to selectively etch thin-film aluminum conductor circuits on silicon dioxide–coated integrated circuits.

A photomask is a microphotograph that is a 1:1 scale pattern which is used to give the final geometric configuration within the accuracy specified in the master drawing. Dimensional stability of the photographic emulsion is of paramount importance and requires a flat stable base such as glass.

The master drawing is the starting point for making a photomask and is a scaled drawing showing the size, location, and dimensional limits of all conductors.

Artwork is prepared after the master drawing has been laid out and the total reduction determined. This artwork may be an ink drawing using reflected light, but actinically opaque Mylar-based material requiring transmitted light is preferred. This has a peelable coating on a clear Mylar* base. This coating may be cut with a sharp knife and stripped to any configuration desired, forming clear and opaque regions.

Microphotography is a process for making minute precision photographs of an object. Several methods are available to the microphotographer for making micronegatives. These techniques are combinations of single reductions, multiple reductions, step and repeats, or step and repeats with reduction.

Multiple reduction simply means reducing the artwork in a series of steps. For example, if a 200× reduction is required, the original might be reduced 10× and the resulting negative again reduced 20× for a total of 200×. This second reduction will now become a "positive" of the original. In other words, clear areas on the artwork will be clear on the second reduction. The original artwork can be reversed before the first reduction, and then the second reduction will become a negative of the artwork. This procedure is superior for making second-reduction contact reversals where both resolution and contrast are critical.

When multiple images of the artwork are required, two systems are generally available. The first is the use of a multiple-lens camera, and the second is a step-and-repeat process.

* Trademark of E. I. du Pont de Nemours & Co., Wilmington, Del.

13-14 Metals Joining of Electronic Circuitry

Process control is of extreme importance if consistent high-quality photomasks are to be produced. Items of particular concern are cleanliness of work areas, storage of film, use of fresh photographic solutions, and careful time-temperature developing.

The most widely used photoresists are negative working resists which polymerize when exposed to ultraviolet light. However, a positive working resist which depolymerizes when exposed to ultraviolet light is rapidly becoming popular for etching aluminum. Excellent results are consistently obtained when reasonable care is taken in cleaning the wafer and applying the photoresist. The positive resist has the following advantages:[10]

1. No pinholes result from dust particles.
2. Completely clean developing, with no remaining residues.

TABLE 6 Comparison of Photoresists for Aluminum[10]

	Interconnect pads				Lines	
	Small		Large			
Dimensions	Width, microns	% deviation (from photomask)	Width, microns	% deviation (from photomask)	Width, microns	% deviation from photomask
Design goals............	15		30		10	
Light-field photomask, dimensions..........	12.4		28.4		9.8	
Aluminum dimensions after completion of *negative* photoresist processing, masking, etching, and alloying	12.0	−3.3	28.8	+1.3	9.8	0
	12.0	−3.3	30.8	+8.0	9.1	−8
					8.7	−10
					8.12	−12
	13.1	+6			10.2	+4
			29.2	+3	10.0	+2
Dark-field photomask dimensions..........	13.5		31.4		10.6	
Aluminum dimensions after completion of *positive* photoresist processing, masking, etching, and alloying	11.2	−17	25.4	−19	8.3	−22
			25.4	−19	8.3	−22

3. Broad time limits for exposure and development, greatly reducing chance for error.
4. Complete removal within a few seconds.
5. Operation unaffected by humidity and temperature.
6. Aqueous developer and remover solutions.
7. Insensitive to fluorescent light for hours.
8. High solids content.
9. No softening or swelling of image during development results in ·better resolution.

Resist coatings are applied by dip, spraying, or spinning. Spin coating is the most advantageous for silicon work. Its main advantages are easy control of coating thickness, relatively even coating thickness, and rapid application.

Table 6 compares the two types of resists used with aluminum film. From the consistent negative error in the data, it was determined that the positive resist was being overexposed (causing reduction in the pad and line widths). In subsequent runs, correct exposure eliminated these undersize pads and lines, while retaining superior line definition and fewer pinholes.[10]

Multilevel Intraconnections.[11] Large-scale integrated-circuit array interconnection patterns are formed by successive levels of an insulation material and thin-film metal leads. The oxide that insulates the first-level metal leads from the silicon is thermally grown, whereas the insulation for the multilevel interconnection lead system must be deposited. Efforts to develop a suitable insulation-metallization system for LSI (large-scale, integration) have met with some success. However, there are still potential problems that have not been adequately investigated or evaluated. These include:
1. Dielectric breakdown, both bulk and interfacial
2. Ion migration (leakage) phenomena at interfaces
3. Metal over deposited insulator adhesion characteristics
4. Interlevel metal-to-metal continuity
5. Interlevel metal-to-metal short circuits through intervening insulation
6. Stresses in deposited metal and insulating films

The defect level for metal crossover short circuits is shown in Table 7. Other inorganic insulation includes insulation tests using anodized Al, thermally oxidized Mo and Ni, pyrolytically deposited SiO_2, SiO_2 formed from oxidized silane, and evaporated SiO_2 or SiO. The results of these techniques have not been encouraging, as the average defect level shows. One coat of 1-micron-thick, rf-sputtered SiO_2 has a defect level almost an order of magnitude greater than that for a double coat

TABLE 7 Crossover Defect Level[11]

Insulation type	Defects, short circuits per 100,000 crossovers
2 microns of rf-sputtered SiO_2	0.8–1
1 micron of rf-sputtered SiO_2	7–8
Other inorganic insulation	600–7,000
Organic insulation	5–6

of rf-sputtered SiO_2. The organic insulating systems were positive photoresists and polymerized hydrocarbons. However, evaporated aluminum or gold does not adhere to organic insulators. Another fault is that such insulators are not thermally stable. After a period of time at elevated temperatures, they evaporate.

Aluminum is the most commonly used metal for the construction of contacts, interconnections, and bonding pads on silicon monolithic integrated circuits. The metal is applied by thermal evaporation. Silicon dioxide is the most widely employed material for providing insulating coatings for construction of passivating films, interconnection crossovers, etc. Coating techniques include evaporation of quartz, oxidation or pyrolysis of substituted silanes, and various sputtering methods. However, the combination of these two materials (aluminum and silicon dioxide) does not yield a particularly good multilevel system.

There is an important phenomenon that occurs as high incidence of short or near short circuits. This is the inherent mechanical weakness of the system. Merely a light touch of a steel point to an $Al/SiO_2/Al$ crossover often causes a short circuit, a situation probably related to the extreme softness of the aluminum films and to the ease with which cracks can form in externally deposited silica films. Another more theoretical, undesirable feature of this system is its extreme thermodynamic instability. There is also the possibility that the aluminum will chemically react, etching its way through the SiO_2 insulating film. Therefore, molybdenum-gold-molybdenum was investigated for multilevel interconnections.

The sputtering of a molybdenum film in a triode system is performed as follows: The metal to be deposited is placed in the glow discharge created by the collision of electrons with argon, and a very strong negative voltage is applied. The metal (cathode) attracts the positively charged Ar⁺ ions. The Ar⁺ ions strike the surface of the metal with tremendous kinetic energy, there is a transfer of energy, and metal atoms are "sputtered" from the surface with considerable momentum. Sputtered metal atoms can be released from the surface of the cathode with much greater kinetic energy than is possible with resistance heating or electron-beam heating. Furthermore, sputtering is analogous to an elemental triode vacuum tube and offers a means for control by regulating the filament, anode, and cathode currents. Also, the glow discharge can be concentrated by an external magnetic field. By using a gold cathode in a similar triode arrangement located in the vacuum chamber, gold can be sputtered onto the molybdenum-coated silicon slices.

The etching of the various films is done as follows:

1. Buffered hydrofluoric acid is used to etch SiO_2 and rf-sputtered glasses. This generally leaves a beveled edge with one or more abrupt steps in the insulator.

2. The profile of the edge is dependent on the doping of the oxide and on the etching techniques.

3. Aluminum is etched in phosphoric acid, and generallly yields a very smooth, beveled edge. Lateral etching, particularly in thick aluminum, is quite extensive. Because of the porosity of evaporated aluminum, it is often difficult to properly rinse away the phosphoric acid after etching.

4. Molybdenum-gold-molybdenum is etched by immersion in successive solutions of buffered hydrofluoric acid and cyanide. The profile of the edge is dependent upon technique.

Interconnections The silicon integrated-circuit chips must be suitably mounted before they are useful. This mounting or bonding operation involves the placement of the chips onto a specified area of a substrate so that by some combination of heat, mechanical action, and intermediate material, a strong, continuous bond is formed between the silicon and the substrate. The substrate may be metal or ceramic and is most often part of the hermetic package. The intermediate bonding materials fall into two categories; hard solders, usually consisting of the gold-silicon eutectic, and solder glasses, usually of the devitrifying type. The gold-silicon eutectic is used for bonding the chip to a metal or metallized substrate. The eutectic point is at approximately 31 atomic percent silicon and 370°C. The eutectic bond is formed using one of several configurations: gold-plated silicon chip and gold-plated metal substrate in firm contact, unmetallized silicon die and gold-plated substrate with thin gold foil sandwiched between, or a gold-silicon eutectic preform between the silicon die and the gold-plated metal substrate. The substrate on which the silicon chip is to be bonded with gold silicon eutectic may be either the gold-plated Kovar* base of an integrated-circuit package, or the gold-plated metallized ceramic wafer or ceramic package base. This is usually a molymanganese layer fired to the ceramic at a high temperature (about 1500°C), which is plated with gold using the electroless immersion process. If the gold plate on the substrate is sufficiently thick, or if the silicon chip has been gold-plated on its bonding surface, then an intermediate preform may not be required. If, however, insufficient gold is available, a preform, usually of 0.001-in. gold foil, is required.

During mounting, the chip is held in position on the substrate (with the preform, if required) and is heated to above the gold-silicon eutectic temperature. This is done in either a continuous-belt furnace or on a special die-mounting apparatus. A special chip-mounting device with its own heat source is easily adapted to a variety of chip sizes, and may be closely controlled. The device consists of a heated station to hold the package or substrate, a vacuum pickup pencil for transferring the chip to the package and for holding it in position during bonding, a microscope for alignment and for observing the bonding cycle, and all the necessary controls. A silicon chip-to-substrate bond may also be made with solder glasses. Solder glasses have relatively low melting points compared to other glasses, contain a high percentage of lead oxide, and are designed for a variety of applications requiring the joining of higher-melting-point materials. Their density is higher

---

* Trademark of Westinghouse Electric Corporation, Pittsburgh, Pa.

and their chemical durability lower than those of hard glasses. The devitrifying types are opaque or translucent after firing owing to the crystalline phase. After firing the glass to obtain devitrification, the glasses do not melt, even on heating to temperatures above those of the initial cycle. There is, of course, a softening temperature above which they will flow. As heating is continued, crystallization increases at a rate which increases with temperature. The crystallization makes the glass stronger than vitreous glasses. A well-controlled thermal cycle is required to obtain reproducibility in properties. Both melting point and expansion coefficient depend on firing temperatures. Solder glasses are applied to the substrate as suspension in a vaporizable organic solution by silk screening or other convenient techniques. The mixture is applied to only one surface since the glass readily coats the silicon. The coated part is heated to drive off the organic vehicle before the chip is put in place. This minimizes voids in the bond caused by trapped vapor. After orientation of the chip in the glass-coated substrate, bonding is done in an oven. Heating rates, dwell time, and cooling rates should be closely controlled to that specified for the glass solder being used. A heating rate of 10°C per min, a firing temperature of 425°C, a dwell time of 2 hr, and a cooling rate of 5°C per min are approximate conditions.

A limitation in the use of solder glass is its thermal conductivity. Glass exhibits a thermal conductivity on the order of 10^{-2} watt/(cm)(°C) as compared to 1.5 watt/(cm)(°C) for gold. The thin layer of glass employed decreases the difference between eutectic bonds and glass bonds but the difference is significant at the higher power densities. Other factors of the package design must be considered before the effect of this difference in bonding materials can be assessed.

Solder-glass chip bonding is relatively easy since it eliminates the several metallizations required for eutectic bonding. Although it has been adequately demonstrated as capable of mounting reliable integrated circuits, there are problems regarding the effects of the mismatch of expansion coefficient.

After the silicon chip has been bonded to the base or substrate, the next step is to provide electric connectors with wire leads. The techniques used to attach gold or aluminum wires between the contact areas on the silicon chip and package leads or lands are thermocompression bonding, ultrasonic bonding, and resistance welding.

The lead-bonding processes depend on obtaining intimate contact between the two metallic surfaces so that the unsaturated atomic bonds on one surface interact with those of the second metallic surface and give a good bond. Atoms at the surface of any metal have atomic bonds that are not saturated by adjacent metal atoms of the same type. This is the definition of a surface or interface. If these unsaturated bonds become saturated with atoms of the same metal, then the surface or interface ceases to exist. If the bonds become saturated with metal atoms of a different type, then an intermetallic interface exists, and a true metallurgical bond is obtained. This type of bonding is, however, rare in normal practice. Even carefully ground metal surfaces have irregularities with peak-to-valley distances that average about 500 Å. More often the irregularities are two orders of magnitude greater than this. Thus, even if these metal surfaces were perfectly clean, contact between touching surfaces would be obtained on only a small fraction of the interface area.

In preventing intimate atomic contact between two metal surfaces, chemical contamination is just as important as surface irregularities. The surface atomic bonds attract and hold a variety of chemical species. Metal oxides are the most common form of this surface contamination, but many other types also exist. The nature of the oxide layers varies from one metal to another, but they are found even on gold. For a given metal in an oxidizing environment, the oxidation rate increases with temperature. Other forms of chemical contamination may either be substituted for surface oxides or form on top of oxides.

The contact areas on silicon devices, in general, have been of two types—direct contacts and expanded contacts. In the first of these, metallization is applied directly to the silicon, and the wire bond is formed directly on top of this metalliza-

13-18 Metals Joining of Electronic Circuitry

tion. The expanded contact is made with a metal film contact to the silicon and the wire bond located on the chip periphery. The metal film on top of the silicon dioxide passivation layer is extended to chip edge where a bonding pad is formed. The bonding pads are located on the edge of the chip in order to avoid contact between bare wires and metallization patterns. Only expanded contacts are used in integrated circuits. All contact points are located on the periphery of the chip, and a metallization pattern provides intraconnections between these and the active device structures. These contact structures are indicated schematically in Fig. 5. The oxide window providing a region for contact to the silicon typically is several tenths of a mil wide, while the diameter of the wire used in bonding is on the order of 1 mil. The dimensions in the figure are not to scale.

Thermocompression Bonding. Thermocompression bonding is a process in which the metals being joined are brought into intimate contact by pressure, using a shaped, smooth bonding tool at a temperature below that required for interface melting. In the region to which pressure is applied, plastic deformation and diffu-

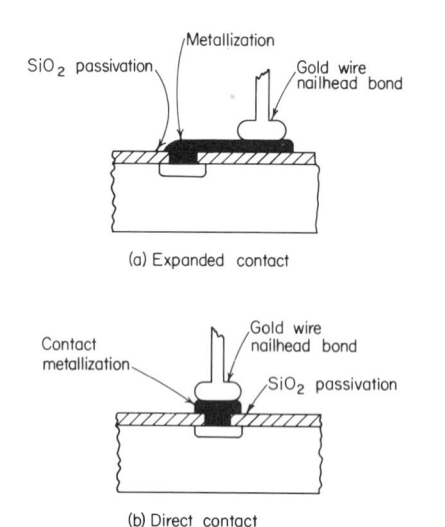

Fig. 5 Expanded and direct contacts.[12]

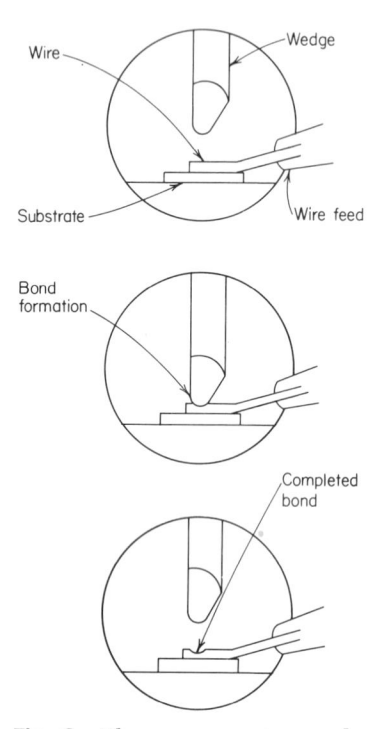

Fig. 6 Thermocompression wedge bonding.

sion occur at the interface during a controlled time, temperature, and pressure cycle. Plastic deformation at the interface is necessary not only to increase the contact area but also to destroy any interface films and bring the metal surfaces into intimate contact.

The amount of deformation that occurs in the wire being bonded to the silicon is dependent on the force applied through the bonding tool as well as on the bonding temperature. Consistency of the lead material and of the bonding configuration is a prerequisite for reproducible bonds. In most commercial equipments it is possible to set the bonding pressure over a considerable range.

The techniques for accomplishing thermocompression bonds are classified as follows:
1. Wedge bonding
2. Nailhead, or ball bonding

3. Stitch or scissors bonding
4. Bird-beak bonding

Wedge thermocompression bonds, as the name implies, are made with a wedge or chisel-shaped tool. The end of this tool is rounded with a radius one to four times that of the wire being bonded and is made of sapphire or similar hard material. It is used to apply pressure to the lead wire located on the bonding pad which has been heated to the bonding temperature. In the wedge-bonding procedure, the mounted silicon chip is positioned under a microscope. The wire is brought into the proper position on the silicon chip and held there by a glass capillary. After the wire is positioned, the wedge is brought into position and lowered to the wire. Thus, two precise positioning operations are required. Bonding is accomplished by an automatically controlled time sequence. In various equipments different methods are provided for precisely coaligning the bonding pad, wire, and wedge.

During the bonding operation, a gas curtain, either forming gas (a dilute mixture of hydrogen in nitrogen, 1 part hydrogen to 9 parts nitrogen) or dry nitrogen,

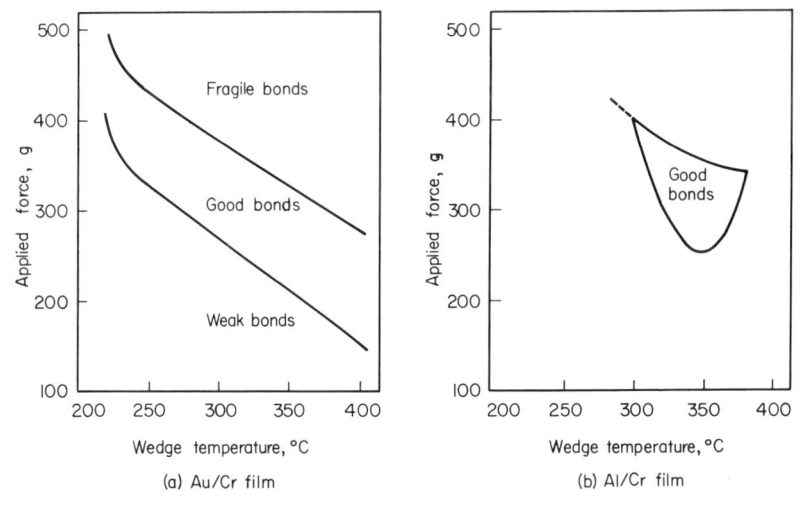

Fig. 7 Pressure for wedge bonds.[12] (4-mil gold wire, 10-mil wedge, to metal films on glass.)

surrounds the chip. This prevents oxidation of the aluminum or the gold from interfering with the bonding operation. Because the lead wire is very small and ductile, a special wire feed mechanism is required. Usually a spool of carefully specified wire is mounted in an enclosure which protects it from contamination. The wire from the spool is fed through the glass capillary for bonding. The bonding sequence is illustrated in Fig. 6.

Difficulties with wedge bonding may be traced to a number of sources. The most important of these are imprecise temperature control, poor wire, inadequately mounted silicon chips, or a poorly finished bonding tool. High temperatures are provided by means of a heat column which holds the package and sometimes by a heated wedge assembly. Uneven air flows and poor thermal contact between the package and the heat column are some of the difficulties that may be experienced in maintaining temperature control. The ductility of the wire is very important. Any bending or cold working of the wire is sufficient to cause poor bonds. The wire is annealed on the spools which fit the bonding apparatus, and if these spools are handled roughly or dropped, it may be sufficient reason to require that the wire be reannealed. Since it is very small, the wire is very weak. If it becomes exposed to a humid atmosphere or collects dust particles, it becomes difficult to use the wire without breaking. A common difficulty is that the wire sticks in the glass capillary through which it is fed to the work area. The wedge which is used for bonding must be polished to prevent sticking of the lead wire. At the bonding temperatures the wire can adhere to a tool with sufficient force to break the bond when the tool is raised. A 10- to 20-mμ finish on the tool surface is required.

Since in wedge bonding the cross-sectional area of the wire is reduced by the bonding operation, the wire is weakened. The shape of the wedge tool is an important factor in determining bond strength. The desired pressure is that which deforms the lead to one-half its original cross section. Excessive pressure may damage the silicon die or weaken the lead wire. An indication of the conditions required is shown in Fig. 7, which was obtained for wedge bonds on films on a glass substrate.

Ball or nailhead bonding is a technique for thermocompression bonding in which a small ball is formed on the end of the wire and deformed under pressure against the pad area on the silicon chip, giving a bond. It is used only with gold wire; the lead wire is perpendicular to the silicon chip as it leaves the bond area.

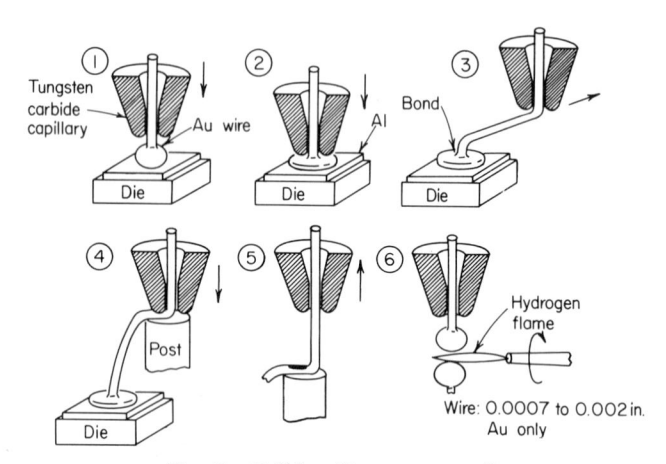

Fig. 8 Ball-bonding operation.[13]

Ball bonding requires that the small wire be fed through a quartz or tungsten carbide thick-walled capillary tube. The capillary tube, with one end tapered to a few mils in diameter on the outside, is mounted in a suitable mechanical fixture so that it can be moved both vertically and horizontally. The horizontal positioning must be accomplished by means of precision manipulators while being observed through a microscope. Positioning accuracies on the order of 20 millionths of an inch are required.

Before bonding, a small spherical ball is formed on the end of the gold wire by a hydrogen flame. The silicon chip to which the wire will be bonded is positioned on the work stage below the bonding capillary. The capillary is positioned over the bonding pad of the silicon chip and lowered until the ball is brought into contact with the capillary tip. The ball is then brought into contact with the bonding pad where a predetermined amount of force is applied. This deforms the ball and establishes intimate contact between the gold ball and the bonding pad. These operations are illustrated in Fig. 8. Wire lead attachment to the other terminal may be accomplished by bonding the wire with the edge of the capillary providing bonding pressure. The capillary is then raised and the hydrogen flame used to cut it off while forming another ball for the next bonding operation.

Ball bonding is used because of its relatively high speeds, economy, and strength.

The number of steps in this bonding operation are fewer than in alternative methods, and the strength of the bond obtained is superior to that of other methods. Aluminum wire cannot be used because of its inability to form a ball when severed with a flame. However, gold wire is an excellent electric conductor, is more ductile than aluminum, and is chemically inert. For ball bonding, hard gold wire may be used since the balling process determines the ductility of the gold to be deformed.

Among the disadvantages of ball bonding is the fact that a relatively large bonding pad is required. For a given device structure, the largest wire size practical should be used to give a higher bonding yield and greater mechanical strength. Two-mil gold wire typically requires a 5- by 5-mil aluminum pad; a 1½-mil gold wire requires a 4- by 4-mil bonding pad. Other bonding techniques allow smaller bonding-pad sizes. The gold-wire–aluminum system is used because of the excellent properties of aluminum as a contact material for silicon and of the gold as a bonding wire (its ductility, electric conductivity, and corrosion resistance). However, when gold and aluminum are placed in intimate contact and the combination is heated to modest temperatures in excess of 200°C, detrimental changes take place: gold-aluminum intermetallic compounds are formed. This intermetallic compound formation is further complicated by the presence of silicon and oxygen. A purplish material begins to appear at the junction between the two metals.

The Au-Al binary phase shows the formation of distinct compounds. These include $AuAl_2$, $AuAl$, Au_2Al, Au_5Al_2, and $AuAl_2$. All these compounds have been detected in gold-aluminum thermocompression bonds made over silicon dioxide. It has been found that the formation of gold-rich phases are responsible for the failing of Au-Al thermocompression bonds. The purple $AuAl_2$ has been shown to be a good electric conductor (resistivity 8×10^{-6} ohm-cm at 300°K) that does not weaken the mechanical strength of the bond, whereas the tan Au_2Al_1 is a poor electric conductor that can cause the bond to be brittle. The formation of all these compounds is the result of interdiffusion between gold and aluminum, the rate of which is temperature-dependent. One material diffusing faster than the other causes depletion of the faster diffuser in a region near the interface. In this case, the gold material is the faster-diffusing atom, and the result is a band of voids or pores in the gold-aluminum bond, weakening it mechanically. It eventually causes an electric open circuit if, as in the expanded contact, the aluminum is on silicon oxide. For gold-aluminum bonds to direct contacts the problem is not as severe. Since the depleted region is in a ring surrounding the wire, the open circuits are not formed.

Stitch or scissors bonding combines some of the advantages of both wedge and ball thermocompression bonding. The wire is fed through the bonding capillary, the bonding area is smaller than for ball bonds, and no hydrogen flame is required. The procedure is shown in Fig. 9. The wire is fed through the capillary and bent at a right angle by cutting with the scissors. As the capillary is lowered to the bonding pad, the bent wire is pulled against the capillary edge which performs the bonding operation much as a wedge. Either glass or tungsten carbide capillaries can be used. Capillary heating is possible. Either gold or aluminum wires can be bonded at a high rate.

The *bird-beak bond* is a form of thermocompression bond that is made with a split bonding tool. It feeds, bonds, and cuts the lead wire without requiring subsidiary equipment or leaving a pigtail. It is rapid, has a high yield, and requires small bond areas. A 1-mil wire gives a bond area of 1.3 by 1.0 mils. In the bonding tool, the lead wire is fed between a jeweled bonding tool and a jeweled holding tool. The underside of the bonding jewel is shaped so that it forms a rib, about 1 mil in diameter, on the wire as the bond is made. The rib adds mechanical strength and current-carrying capacity to the bond, while the flattened underside of the wire makes for excellent adhesion and electric contact to the underlying thin films. The bonding sequence is shown in Fig. 10. Once the bond is made, the vacuum clamp holds the wire. The wire breaks cleanly at the edge of the bond when the tool is moved horizontally away from the bond. This avoids short circuits due to tails or loose ends on the bonded wire.

13-22 Metals Joining of Electronic Circuitry

Bonding pressures vary from 25 to 125 g and bonding times from 1 to 6 sec, depending on the thickness, purity, and physical conditions of the aluminum film on the chip. Either gold or aluminum wire may be used. These are three methods by which heat is transmitted to the bond. Larsen[15] has described these:

METHOD 1: Heat is transferred to the workpieces through a heat column which is located under the parts being joined.

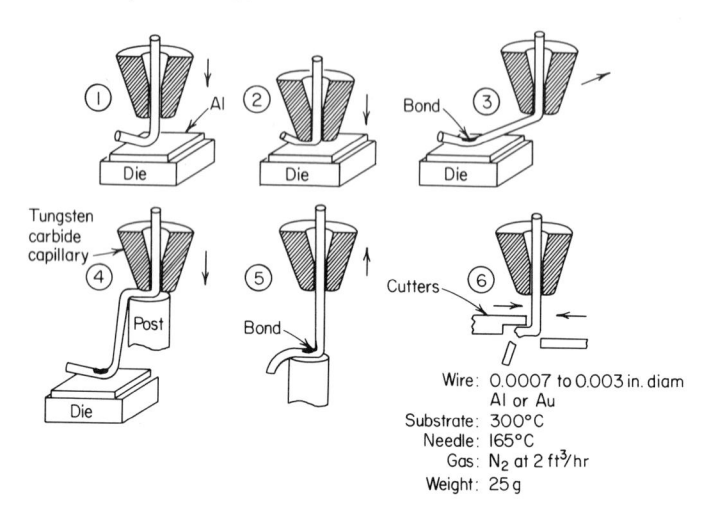

Fig. 9 Stitch bonding.[13]

METHOD 2: Heat is applied to the wire by means of a capillary bonding tip located over the parts being joined.

METHOD 3: Heat is applied to both the workpiece and wire by a combination of methods 1 and 2.

METHOD 1—HEAT COLUMN: With heat-column thermocompression bonding equipment, the microcircuit or workpiece to be bonded is placed on an electrically heated column where it is allowed to "soak up" enough heat to permit the lead to be bonded when a preset force is applied. Usually, the wire is fed to the workpiece through

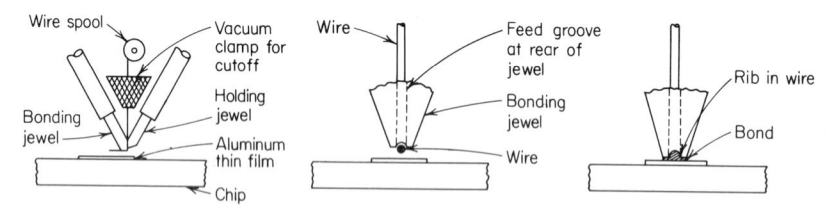

Fig. 10 Bird-beak bonding.[14]

a glass capillary tip. This same tip also is used to apply the pressure required for making the bond. This method has the following advantages and disadvantages:

Advantages:
  Since only a slight amount of heat is transferred to the capillary tip, oxidation at this point is minimized.
  The amount of heat transferred to the materials being bonded can be fairly well controlled.
  The temperature of the heat column can be automatically and continually monitored.

Disadvantages:
> The heat column is always hot while the equipment is in operation and, therefore, a burn hazard is present for the assembly operator.
> Because of the time required in bringing the heat column up to bonding temperature, the unit cannot be turned off and on readily.
> Some microcircuits such as thin films on glass do not transfer heat from the heat column to the bond area. The process will not work under these conditions.

METHOD 2—HEATED TIP: In the heated-tip method of thermocompression bonding, the heat for making the bond is applied directly through the capillary tip itself. The heat at the tip may be either continuous or intermittent. A continuously heated tip is, in a sense, like a heat column which is held at a constant temperature and transfers its heat to the materials being joined. The intermittently or pulse-heated tip, on the other hand, is heated only during that time in which the bond actually is being made. Heat is produced by passing an electric current either directly through the tip or through a heating element surrounding the tip. In the former case, the tip usually is made of high-resistance materials, and heat is built up rapidly during the current pulse. The conduction of heat from the tip to the materials being joined is sufficient to make the thermocompression bond.

Advantages:
> The bonding tip need not be kept hot all the time that the equipment is in operation.
> The equipment, generally, is less expensive than that which incorporates a heat column.
> Bonds can be made a little faster than with conventional heat-column bonding equipment.
> No warm-up time is required for pulse-heated tips.
> The formation of interface oxides is greatly reduced because the materials do not have to be held at elevated temperatures for extended periods of time.
> The process can be used successfully on some materials that cannot be bonded by other methods.

Disadvantages:
> Bonds cannot be made to base metals or other materials that are too thick and act as a "heat sink" to dissipate heat away from the joint interface.
> Metallic tips tend to corrode fast and to wear out sooner than the glass capillary tips which are used with conventional heat-column equipment.
> The heated-tip process, generally, is not as adaptable to high production as is the heat-column process.

METHOD 3—COMBINATION: The third method used in making thermocompression bonded joints is a combination of the two types just described. That is, the bonding equipment has both a heated tip and a heat column. This approach allows lower temperatures to be used at both the bonding tip and heat column. This condition helps to minimize the oxidation problem. Also, the lower temperatures make it possible to bond some delicate types of components that otherwise could not be joined. Reducing oxidation makes it possible to extend the useful life of tips before replacement or repair is needed.

Advantages:
> Tip life is increased.
> The equipment can be used to bond a wide variety of materials.
> Since the heat column is not kept at as high a temperature as when it is used alone, the development of nonconductive oxides at the bond interface is lessened.
> Normally, the equipment can be used selectively for the heated tip only, heated column only, or for a combination of the two methods.

Disadvantages:
> Setting up the correct bonding parameters normally takes more time than for other methods.

13-24 Metals Joining of Electronic Circuitry

The equipment required is a little more costly than that used in either method 1 or 2.

A warm-up delay is required for the heat column portion of the equipment.

The process parameters for thermocompression bonding will generally fall within the limits shown in Table 8.

TABLE 8 Thermocompression Process Parameters[15],*

Power supplies, ac	60 Hz, 0–1 kva
Pulse length	To 500 msec. Continuous for heat columns
Bonding tip force	10–250 g
Capillary tip size:	
Outside diameter	0.062 in.
Inside diameter	Approximately 0.0004 in. larger than wire being bonded
Wire sizes, diameter:	
Ball (nailhead)	0.0005–0.005 in.
Stitch (scissors)	0.0007–0.003 in.
Wedge	0.001–0.005 in.
Substrate metal, thickness:	
Minimum	250 Å
Actual	1,000–5,000 Å

* Reprinted from *Assembly Engineering*, November, 1966. Copyright, 1966, by the Hitchcock Publishing Company, Wheaton, Illinois.

The tensile strength of 1-mil gold wire bonded with a stitch bonder having a heated tip is shown in Table 9. The capillary tip was heated by two methods, continuous-resistive-heated, and pulsed-heated with direct current. Regardless of how the heat was applied to the bond area, or what land material was used,

TABLE 9 Thermocompression Stitch-bond Tensile Strengths[16],*,†

Thin-film conductor: material/thickness, Å	Tensile strength, grams Minimum	Tensile strength, grams Maximum	Average bond strength, grams	Tensile strength of wire, grams	Pulse time, sec
Pulse-heated:					
Gold/2,000	4.8	6.6	6.2	6.5	1.5
Copper/2,000	5.4	6.6	6.2	6.5	2
Aluminum/11,000	4.5	6.5	6.2	6.5	3
Resistive-heated:					
Gold/2,000	5.5	6.5	6.3	6.5	1.5
Copper/2,000	5.8	6.6	6.3	6.5	2
Aluminum/11,000	5.5	6.5	6.2	6.5	3

* 1-mil (0.001-in.) gold wire (99.999%) lead material.
For all cases:
  Pressure: 50 g
  Time, resistive-heated: 6 sec
  Capillary temperature, resistive-heated: 482°C
  Pulse heat: 1.99 volts
  Number of joints tested 45° to substrate: 25

† Reprinted from *Assembly Engineering*, November, 1966. Copyright, 1966, by the Hitchcock Publishing Company, Wheaton, Illinois.

the average bond tensile strength was quite constant and slightly less than the tensile strength of the gold wire.

In shear, 1-mil gold wire, ball-bonded to aluminum film (5-mil diameter bond), has an average strength of about 45 g. However, after 20 hr at 300°C, this average shear strength value dropped to about 10 g. Gold wire bonded to gold film had an average shear strength of about 60 g initially, with no change in strength after a 20-hr soak at 300°C.[17]

Ultrasonic Bonding. Ultrasonic bonding involves the same mechanism as thermocompression bonding except that the source of energy is mechanical rather than thermal. Ultrasonic bonding is very successful with aluminum because aluminum is soft and deformation occurs at relatively low pressures. Aluminum is coated with a refractory oxide which readily breaks down under ultrasonic stressing to provide an abrasive flux.

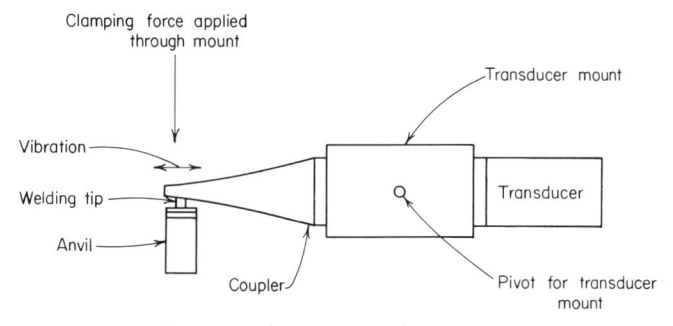

Fig. 11 Ultrasonic bonding head.[12]

In the ultrasonic bonding device, an elastic vibration is created by the rapid expansion and contraction of a magnetostrictive transducer driven by a source of high-frequency alternating current. These high-frequency stress waves travel through a coupler or mechanical transformer to the welding tip. The welding tip serves to transfer the vibrations to the materials to be welded. A portion of this coupler is tapered to provide a mechanical impedance match which increases the amplitude of vibration. The bonding tip undergoes excursions in a direction parallel to the interfaces of the weld, inducing a shear mode of vibration into the materials. A simplified schematic representation of a lateral-drive ultrasonic welder is shown in Fig. 11.

The materials to be bonded are clamped between the welding tip and the lower work stage called an *anvil*, so that the workpieces experience the stresses resulting from the clamping force and the superimposed vibration of the welding tip.

Since the ultrasonic welding process requires the conversion of electric energy into acoustical vibratory energy by means of a transducer element, this energy must be transmitted efficiently to the interface through a mechanical transformer and a welding tip. The energy delivered to the weld interface depends on the characteristics of this coupling system. The optimum value of rigidity and energy transmission is achieved by using a Fourier mechanical transformer. The Fourier horn and the standard exponential horn are shown in Fig. 12. The Fourier-shaped horn permits the use of lower welding power and pressure in bonding operations.

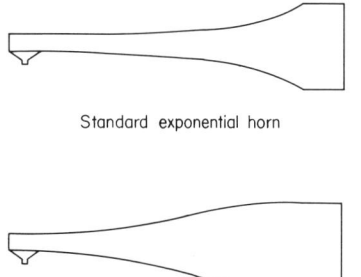

Fig. 12 Mechanical transformers.[12]

There are some supplementary phenomena which effect the welding process. The first is a temperature rise that results from the elastic hysteresis of the highly stressed portion of the weld zone during the welding sequence. The temperature rise has been measured by fine-wire thermocouple and fusible insert techniques and considered in the light of extensive electromicrography. These investigations revealed that the maximum transient temperature was between 30 and 50 percent of the absolute melting point of the metal in a similar metal joint. This temperature rise is highly localized in the weld metal and promotes plastic deformations which are associated with the process.

A second supplementary phenomenon derives from the fact that metallic crystal structures can be temporarily plasticized by high-frequency mechanical vibrations. This phenomenon has been observed in ultrasonically accelerated processes and in the nature and extent of the internal deformations in ultrasonic welds themselves. This plasticization occurs as a result of the acoustical excitation per se and is independent of temperature and the thermal plasticization previously described. However, it does accomplish the same purpose, that is, to facilitate the plastic deformation of the interface.

Figure 13 is a block diagram of the ultrasonic welding process. There are four parameters that may be varied to produce a specified weld: the welding tip, clamping force, power, and weld time.

The geometry of the welding tip is determined by the diameter of the wire. For an acceptable weld, the working surface of the tip should have a semicylindrical groove with a radius of curvature equal to the radius of the wire to be welded. This groove aids in the positioning of the wire and is necessary for the formation of a nugget weld. In addition, the length of this groove must be twice the diameter

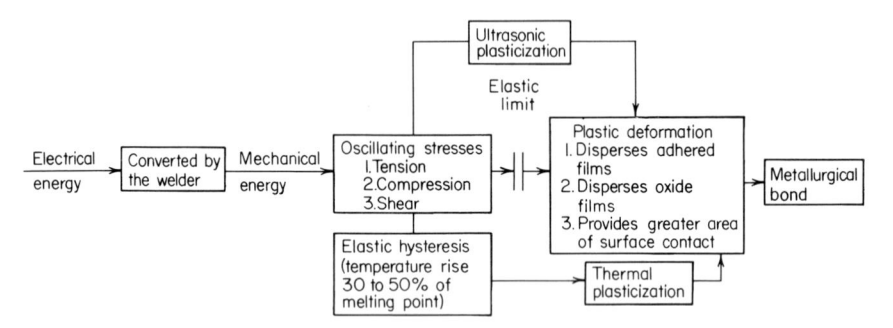

Fig. 13 Block diagram of the ultrasonic bonding process.[18]

of the wire. The tip must be positioned so that the long axis of the groove is parallel to the long axis of the coupler assembly. It is also very important to ensure that the working surface of the tip is parallel to the plane of the workpiece.

The clamping force used is of prime importance. This clamping force affords intimate contact among the tip, wire, and contact area of the device, thus providing good coupling. This force, which is applied perpendicularly to the plane of the weld, is varied by a continuously threaded rod fastened in cantilever fashion such that screw weights may be threaded to positions that afford a variety of clamping forces. These clamping forces may be measured at the welding tip by means of a gram gage. There is an optimum clamping force for each welding job.

The amplitude of tip vibration may be varied by means of power adjustment. The high power range delivers approximately 1 to 20 watts, and the low power range less than 1 watt. Each range is graduated into 20 steps: The vibration sets up a stress at the interface between the components being welded. There is a threshold for the minimum amount of stress required, and a power input lower than this minimum cannot produce an acceptable weld. It has been noted that there is also a maximum power input, which, if exceeded, will distort the materials being joined.

The time may be varied from zero to 1.5 sec (0.075-sec intervals). There is a special time interval of 32 sec that permits tuning of the instrument. The two parameters, time and power, determine the energy (watt-seconds) applied to the weld. The physical characteristics (hardness, thickness) of the materials being welded determine the appropriate time setting.

Because ultrasonic welding can create bonds between a very wide variety of dissimilar materials, it is an extremely flexible tool when applied to the problem of welding interconnections to microminiature circuit elements. Table 10 summarizes some of the combinations of films and wires that have been successfully bonded.

Void-free junctions are produced by ultrasonic bonding with relatively few foreign-material inclusions, making it a desirable means of creating high-quality, low-resistance electric junctions. Welding dissimilar metals at low temperature elimi-

TABLE 10 Electric Conductors That Have Been Ultrasonically Bonded to Metallized Surfaces[19]

Conductor material	Lead material	Lead diameter
On silicon substrate:		
Aluminum..........................	Aluminum wire	0.0005
		0.001
		0.003
Aluminum..........................	Gold wire	0.001
		0.001
		0.002
On ceramic substrate:		
Silver.............................	Aluminum wire	0.010
		0.004
On glass or glazed alumina substrates:		
Aluminum..........................	Aluminum wire	0.002
		0.004
Aluminum..........................	Gold wire	0.003
Nickel.............................	Aluminum wire	0.002
		0.004
Nickel.............................	Gold wire	0.002
		0.003
Copper............................	Aluminum wire	0.002
		0.004
Gold..............................	Aluminum wire	0.002
		0.004
Gold..............................	Gold wire	0.003
Tantalum..........................	Aluminum wire	0.002
		0.004
Chromel*.........................	Aluminum wire	0.002
		0.005
Chromel*.........................	Gold wire	0.003
Nichrome†........................	Aluminum wire	0.002

* Trademark of Hoskins Manufacturing Co., Detroit, Mich.
† Trademark of Driver-Harris Co., Harrison, N.J.

nates or greatly decreases the formation of intermetallic compounds and allows bonds to be made in the immediate vicinity of temperature-sensitive materials without adverse effects. A further advantage of ultrasonic welding is that it requires no preheating, thus eliminating temperature rises that can damage the assemblies or create problems of thermal distortion or warping.

Surface cleaning is not highly critical in preparing most materials for ultrasonic bonding. The vibratory displacements occurring during the welding operation disrupt normal oxide layers and other surface films at the interfaces.

In order to establish the optimum parameter settings for ultrasonic bonding, the weld-profile method was employed. The starting point was to set the machine power to zero, the time to 0.10 sec., and the bonding force to a value such that

13-28 Metals Joining of Electronic Circuitry

the wire was slightly deformed when the bonding tip was lowered, but no ultrasonic energy was applied.

The bonder consisted of a 10-watt 40-kHz vertical-action ultrasonic transducer, a zoom microscope, an X-Y and 360° rotational manipulator, and a vacuum-operated clamping chuck.

The machine variables are:

    Power: The setting is continuously variable from 0 to 10 in two ranges (the high range is 10 times the low range).

    Pulse time: Continuously variable from 0.10 to 0.35 sec.

    Bonding force: Approximate range from 25 to 600 g for 0.001- to 0.005-in.-diameter wire.

The machine parameters were gradually increased until the optimum parameters could be deduced from the weld-profile diagrams. Table 11 gives the weld sched-

TABLE 11 Weld Schedule for Microbonds Formed with Ultrasonic Bonder[20]

Substrate	Film	Film thickness, Å	Aluminum wire diameter, in.	Wire tensile strength, grams	Force, grams	Power	Range	Pulse width, sec
Sapphire	Au/Cr	24,000	0.001	14.5	30	0	Low	0.15
		24,000	0.002	54.2	90	2	High	0.15
		36,000	0.005	310	440	10	High	0.25
Beryllia	Au/Cr	43,000	0.001	14.5	35	0	Low	0.15
		36,000	0.002	54.2	95	5	High	0.15
Alumina	Au/Cr	23,000	0.001	14.5	25	0	High	0.15
Vycor*	Al	7,000	0.002	54.2	100	1	High	0.15
Oxidized silicon	Al	7,000	0.002	54.2	100	1	High	0.15
Sapphire	Al	7,000	0.002	54.2	100	9	Low	0.15

* Trademark of Corning Glass Works, Corning, N.Y.

ules employed to obtain the ultrasonically bonded microbonds. These schedules are given only as a guide for future work on similar bond systems.

Tables 12 shows the results of the pull tests. In general, the results show that the pull strength and thus the bond efficiency were greater for gold films than for aluminum films. This appears in a comparison of the B_2' bonds with the R, S', and T' bonds. The B_2' bonds are not included since the beryllia wafers had a much larger surface roughness. Also, the set of B' microbonds indicates that again bond efficiency decreases with increasing wire size. Although the bond resistance decreases with increasing wire size, this apparently is due to increased contact area. This occurs because the sheet resistance is the prime contributor to the bond resistance, as evidenced by the comparison of the B_2' microbonds and the R', S', and T' microbonds. In all these instances the wire is the same, but the aluminium film had a sheet resistance about four times greater than that of the gold film.

Microwelding. Microwelding is similar to the parallel-gap welding process; however, the series electrodes are smaller and closer together, and no fusion occurs.

First-level Joining, Materials and Processes 13-29

Such joints are produced by diffusion. In microwelding, a split electrode is brought into intimate contact with a small ductile lead (gold, platinum, copper), producing deformation under each electrode (Fig. 14). Direct current is pulsed through the electrodes which are short-circuited by the lead wire being bounded, producing heating of the lead and diffusion at the lead interface.

For microelectronic applications, the gap between the electrodes usually varies from ½ to 5 mils. A rule of thumb stipulates a 1:1 ratio between the gap size and the diameter of the bonding wire. Miniature split electrodes can be made

TABLE 12   Ultrasonic Bonds Pull-Strength Results[20]

Bond Type	Substrate	Film	Aluminum wire diameter, in.	Average wire strength, grams	Average bond strength, grams
K	Alumina	Au/Cr	0.001	14.5	6.8
R'	Vycor	Sintered Al	0.002	54.2	27.1
S'	Oxidized silicon	Sintered Al	0.002	54.2	21.5
T'	Sapphire	Sintered Al	0.002	54.2	21.5
B_1'	Sapphire	Au/Cr	0.001	14.5	9.2
B_2'	Sapphire	Au/Cr	0.002	54.2	31.6
B_5''	Sapphire	Au/Cr	0.005	310	172
B_1''	Beryllia	Au/Cr	0.001	14.5	5.8
B_2''	Beryllia	Au/Cr	0.002	54.2	17.4

100 samples each type.

of copper, tungsten, and molybdenum alloys similar to conventional miniature resistance-welding electrodes. However, for bonding small gold wires, tungsten, and molybdenum alloy, electrodes were found to be superior. Because of the relatively small size of the component leads, a microscope is used to aid in observing the joining process, and a micropositioner is required to locate the lead with respect to the split electrode and the conductor.

Bonding fine wires requires a controlled, repeatable pulse shape or envelope. Since the current through the weldment remains constant, the energy expended

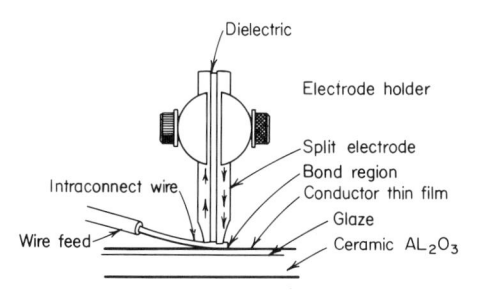

Fig. 14   Microwelder.[16]

in heating the area being welded depends on the resistance between the electrodes. This is a first-power dependence rather than a square power in the $H = I^2Rt$ equation. This permits a more nearly equal-power consumption from one weld to the next.

A three-weld pulse capability of the power supply provides an energy-time pulse output that prevents thermal shock to the fragile substrates and minimizes the mismatch in the film-substrate interface due to the difference in the thermal coefficient of expansion.

13-30　Metals Joining of Electronic Circuitry

A block diagram of the control circuit for the microbonder is shown in Fig. 15. The circuit is actuated by microswitches in the weld head through a foot pedal depressed by the machine operator. The path of the low-voltage high-amperage dc output pulse through the split-tip electrodes is shown in Fig. 16. Since both electrodes are on the same side (the lead material), the weld is not polarity-sensitive.

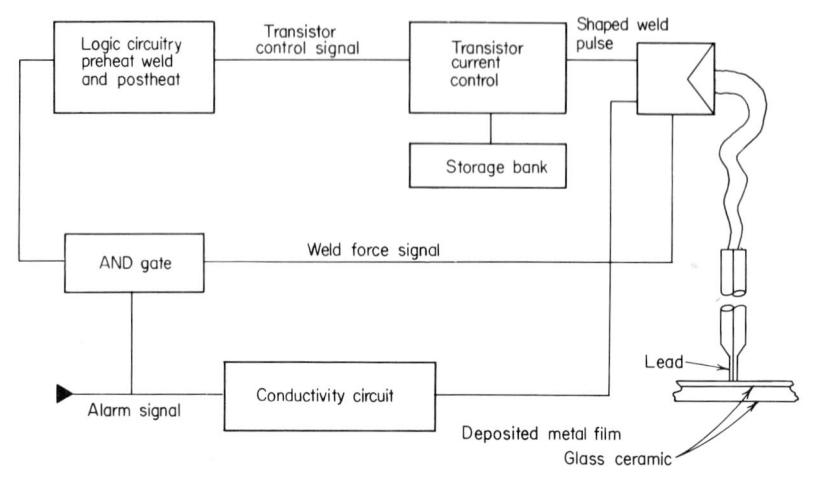

Fig. 15　Microwelder power-supply control circuit.[20]

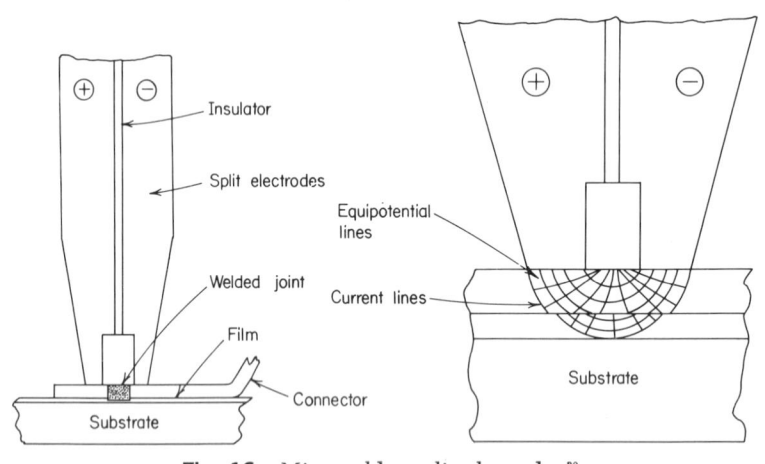

Fig. 16　Microwelder split electrodes.[20]

Weld schedules for substrates sensitive to thermal shock are readily developed by establishing the electrode force at a setting to produce the desired conductor deformation. Initial deformation varies with the conductor material hardness and shape. Welding energy is increased until bonding is obtained by using the middle or weld pulse controls. The preweld and postweld energy cycles are then increased, generally to approximately one-half the weld cycle, and final adjustments are made to shape the pulse as required.

TABLE 13 Weld Schedule for Microbonds Formed with a Split-tip Resistance Welder[20]

Substrate	Film	Film thickness, Å	Wire	Wire diameter, in.	Wire tensile strength, grams	Electrode	Force, grams	Pulse width, msec	Pulse height, amp
As-fired alumina	Au/Ni on fired Mo-Mn	190,000	Au	0.0007	2.84	EM1000	139	80	17.8
				0.002	26.7	EM1000	470	80	40
				0.005	182.5	EM1016	620	500	74
Sapphire	Au/Cr	10,000	Au	0.001	6.7	EM1000	220	80	16
		45,000		0.002	26.5	EM1000	139	80	29
		19,000		0.005	200.0	EM1002	370	150	39.5
Beryllia	Au/Cr	10,000	Au	0.001	6.7	EM1000	220	150	20
		45,000		0.002	26.5	EM1002	370	80	33
		24,000		0.005	200.0	EM102	520	300 –	43 –
								500 –	64 –
								300	43
Vycor	Au/Cr	40,000	Au	0.001	7.0	EM1000	220	80	12
		20,000		0.002	26.7	EM1000	370	150 –	25 –
								80 –	28.5 –
								150	25
		20,000		0.005	201.0	EM1002	710	300 –	35 –
								150 –	49 –
								300	35
As-fired alumina	Au/Cr	23,500	Au	0.001	6.7	EM1000	220	20	16
		25,000		0.002	27.47	EM1000	370	80	33.3
		25,000		0.005	182.5	EM1016	710	150	68
Glazed alumina	Au/Cr	32,000		0.001	7.0	EM1000	220	80	11
		25,000		0.002	27.47	EM1000	370	500 –	10 –
								300 –	18 –
								500	10
		25,000		0.005	182.5	EM1002	710	300	43
Oxidized silicon	Au/Cr	10,000	Au	0.001	6.7	EM1000	220	40	26
		30,000		0.002	26.7	EM1000	370	80	34
Oxidized silicon	Au/Cr	30,000	Au	0.005	182.5	EM1002	710	500 –	44 –
								1000 –	80 –
								500	44
Vycor	Au/Cr	25,000	Al	0.002	56.4	EM1000	220	300 –	10 –
								500 –	14 –
								300	10
As-fired alumina	Au/Cr	32,000	Al	0.002	56.4	EM1000	220	300 –	11 –
								500 –	22 –
								300	11
Glazed alumina	Au/Cr	30,000	Al	0.002	56.4	EM1000	220	300 –	10 –
								500 –	15.5 –
								300	10
Oxidized silicon	Au/Cr	30,000	Al	0.002	56.4	EM1000	220	680	25
Vycor	Al/Cr	20,000	Au	0.002	26.7	EM1000	370	300 –	25 –
								500 –	29 –
								300	25
		20,000		0.005	182.5	EM1002	710	300 –	35 –
								500 –	63 –
								300	35
Oxidized silicon	Al/Cr	25,000	Au	0.002	26.7	EM1000	370	500 –	27 –
								1000 –	36.5 –
								500	27
		25,000		0.005	182.5	EM1002	710	500 –	44 –
								1000 –	85 –
								500	44

13-32 Metals Joining of Electronic Circuitry

The electrodes used in the development of weld schedules for 0.001-, 0.002-, and 0.005-in.-diameter wires, respectively, were:
EM1000 (0.010 × 0.010 × 0.002 in.) molybdenum
EM1002 (0.020 × 0.020 × 0.004 in.) molybdenum
EM1016 (0.020 × 0.030 × 0.007 in.) molybdenum

Table 13 gives the weld schedules employed to form the split-tip welded micro-

TABLE 14 Summary of Results for Microbonds Formed with a Split-tip Resistance Welder[20]

Substrate	Film	Film thickness, Å	Wire	Wire diameter, in.	Wire tensile strength, grams	Pull test sample size	Average pull strength, grams	Pull strength, % of tensile strength	Wire resistance, ohms/in.	Microbond resistance, milliohms
As-fired alumina	Au/Ni on fired Mo-Mn	190,000	Au	0.0007	2.84	779	2.7	95	2.43	0.65
				0.002	26.7	106	22.81	85.5	0.273	0.39
				0.005	182.5	89	120.67	69.4		
Sapphire	Au/Cr	10,000	Au	0.001	6.7	779	6.5	97	1.09	0.73
		45,000		0.002	26.5	783	24.6	92.8	0.262	0.45
		19,000		0.005	200	782	159	79.5	0.04	0.16
Beryllia	Au/Cr	10,000	Au	0.001	6.7	785	6.5	97	1.09	1.5
		45,000		0.002	26.5	788	25	94.4	0.262	0.54
		24,000		0.005	200	782	141	70.5	0.04	0.14
Vycor	Au/Cr	40,000	Au	0.001	7.0	783	6.6	94.3	1.09	1.2
		20,000		0.002	26.7	869	21.37	80	0.272	0.62
		20,000		0.005	201	129	134.98	67.2	0.045	0.16
As-fired alumina	Au/Cr	23,500	Au	0.001	6.7	790	6.5	97	1.09	1.1
		25,000		0.002	27.47	772	22.7	82.6	0.273	0.53
		25,000		0.005	182.5	99	139.8	76.5	0.053	0.20
Glazed alumina	Au/Cr	32,000	Au	0.001	7.0	789	6.3	90	1.09	1.08
		25,000		0.002	27.47	786	20.61	75	0.272	0.57
		25,000		0.005	182.5	136	52.9	29	0.045	0.104
Oxidized silicon	Au/Cr	10,000	Au	0.001	6.7	776	6.6	98.5	1.09	1.3
		30,000		0.002	26.7	810	20.24	77	0.272	0.72
		30,000		0.005	182.5	79	102.8	56.3	0.045	0.22
Vycor	Au/Cr	25,000	Al	0.002	56.4	111	13.97	24.8	0.387	1.1
As-fired alumina	Au/Cr	32,000	Al	0.002	56.4	101	13.38	23.7	0.387	0.53
Glazed alumina	Au/Cr	30,000	Al	0.002	56.4	94	13.84	24.6	0.387	0.45
Oxidized silicon	Au/Cr	30,000	Au	0.002	56.4	106	16.87	30	0.387	3.05
Vycor	Al/Cr	20,000	Au	0.002	26.7	97	15.11	56.7	0.272	0.62
		20,000		0.005	182.5	104	90.92	49.7	0.045	0.22
Oxidized silicon	Al/Cr	25,000	Au	0.002	26.7	105	16.77	62.8	0.272	57
		25,000		0.005	182.5	101	84.77	46.5	0.045	0.18

bonds. Also included in this table are the wire strength, the average and range of bond pull strength, and the average resistance. Although these schedules yielded optimum results, deviations should be expected since other parameters such as surface roughness or electrode condition can also severely affect the bonds. However, the schedules can be used as a starting guide in selecting a weld schedule in a particular situation.

Table 14 shows the various substrate, wire, and film combinations studied. Included in the table are the results of pull tests of various sample sizes as well as the average electrical resistance of 25 microbonds of each type. The pull tests were performed at 90° to the weld surface, on an Instron Tensile tester while the resistance measurements were obtained by a four-probe technique which tended to eliminate the effects of wire leads and film resistivity.

Comparison of the results on the mixed material systems, gold-wire-to-aluminum films and aluminum-wire-to-gold films, to the gold-gold bonds, shows that much lower bond efficiencies can be expected with the mixed system. The probable cause of this is that the split-tip resistance-welding technique does not incorporate a method for removal of the oxide coating on the aluminum. Furthermore, the reliability of the mixed-systems bonds was also somewhat reduced.

Face Bonding. A rapidly growing joining technique for integrated circuits is the technique of inverting the chip and bonding it (at the electric-contact points only) to a substrate. The technique is called face bonding or flip chip because the device is flipped over for bonding.

Face-bonding techniques currently employed are beam leads, reflow soldering, and ultrasonic bonding. In all the methods, the interconnection pads on the chip are bonded to corresponding interconnection pads on the substrate, with the substrate and chip placed face to face. Two of these techniques have the added advantage that all bonds on a chip are made simultaneously, eliminating the need for painstakingly made individual bonds.

The use of face bonding for assembling monolithic circuits in individual packages, in place of conventional wire bonding, might lead to a small cost reduction, but the main potential of the technology is in the fabrication of hybrid arrays.

TABLE 15 Face-bonding Techniques

Face down	*Face up or down*
1. Ultrasonic:	1. Beam lead:
a. Bumps on chip	*a.* Thermocompression
b. Pillars on substrate	*b.* Ultrasonic
2. Reflow solder:	*c.* Solder
a. Pads	
b. Balls	

Table 15 summarizes the various techniques for implementing face bonding. The factors to consider in choosing a face-bonding technique are as follows:
1. Maximize reliability for the lowest cost.
2. Remove heat from the module.
3. Minimize the changes to conventional integrated circuits.
4. Optimize mechanical strength.
5. Protect the chips environmentally and mechanically.
6. Enable easy removal and replacement of a chip in a module.

With aluminum ultrasonic bonding and with solder pad bonding, there is negligible thermal resistance in the bond itself. In the case of solder balls with hard cores, a slight contribution to the thermal resistance will occur from the bond. The beam-lead approach to face bonding presents a method that has good mechanical strength but adds a significant amount to the thermal resistance of the structure (typically approximately 20°C per watt for a 14-lead device). Chips can be removed and replaced by cutting the leads and rebonding a new chip at a different position on the substrate interconnection.

Most presently available integrated circuits have aluminum metallizations in excess of 0.5 micron (usually between 1.2 and 1.8 microns) and should· be capable of being mounted to suitably chosen substrate bonding pads without difficulty. For use with alumina substrates, an absolute minimum total interconnection thickness of 4 microns should be used. This thickness is required in order to achieve bonds in which only small-area flatness need be considered. For practical applications employing chips of many different sizes, the thickness of metallization on the substrate must be sufficient to accommodate larger-area variations in flatness of both the substrate and the chip.

It was predicted theoretically that the best bonding results would be obtained with the bonding force applied centrally to the bonding pads. This was confirmed experimentally, and led to the conclusion that bonding pads should be arranged symmetrically about the center of the chip, and that the force should be applied to the center of the chip. For a given bond area, aluminum has substantially higher strength than solder, although the latter has adequate strength to pass standard military shock and acceleration tests.

The use of aluminum for the interconnection system would allow process temperatures of 300°C to be used without difficulty. Soft solders generally melt at around 200°C. Of the solders based on lead and tin with silver and gold, only lead-silver has any chance of providing a high-temperature capability, and this melts at approximately 300°C.

The substrates used for holding and interconnecting uncased chips of integrated circuitry are ceramic, glass, or epoxy-glass printed-circuit board material. The ceramics used are either beryllia or alumina. When a ceramic is chosen, it usually is alumina because of the higher cost and the health hazard in the working of beryllia. However, the thermal conductivity of beryllia is nearly seven times the value of high-purity alumina. This thermal conductivity advantage of beryllia obviously can be significant for a system with severe temperature requirements. Glass has been used as a substrate material to allow visual alignment of chips for bonding. Its thermal conductivity is very low.

Epoxy-glass printed-circuit boards have been developed by at least one company for face-bonding substrates. The successful use of epoxy-glass boards to hold face-bonded chips requires special processing of the bonds. An electroless tin plating over the copper conductor pattern is reported to be required to allow the chips to be reliably soldered to the boards. Once the tin plating is done, boards must be either used immediately or stored in an inert environment to prevent oxidation of the tin from degrading the final soldered assembly.

The advantages of epoxy-glass material for face-bonded substrates are low materials cost and, with the exception noted above, the use of standard printed-circuit-board processes. Since the thermal conductivity of epoxy-glass is very low, plain epoxy-glass substrates have power limitations comparable to those of glass substrates.

Registration of chips on a substrate is a problem, especially if the substrate is opaque. Therefore, methods of registering that do not rely on direct visual observation during assembly are necessary. In one method, the chips are placed on a mirror, registered with respect to a microscope eyepiece reticle, and then transferred to a substrate that also has been registered with respect to the reticle.

This type of system causes both a vertical and a horizontal offset of the reflected image of the chip and the actual image of the substrate due to the light traversing the glass region of the mirror. The horizontal image displacement may be corrected by interposing a piece of glass twice the thickness of the mirror between the microscope and the substrate during the registration of the substrate and the reticle; however, this would increase the vertical displacement.

The vertical displacement is important because it places a restriction on the type of vertical motion that can be used. A pivoted arm can cause horizontal displacement unless it is adjusted to make equal angular excursion on either side of the horizontal. The existence of an angle between the tip and the chip during chip bonding would result in a variation of bonding force from one lead to another on a circuit, as is shown in Fig. 17. The conclusions are:

1. During bonding, the tip must be coplanar with the substrate.
2. The vertical motion should be linear if there is a vertical offset, or may be rotational if the optical system gives no vertical offset.
3. The tip should be centrally located on the chip.

One technique for ultrasonic face bonding makes use of evaporated aluminum bumps or pillars on a glass substrate.[22]

A cross section through a bonded chip is shown in Fig. 18. The aluminum

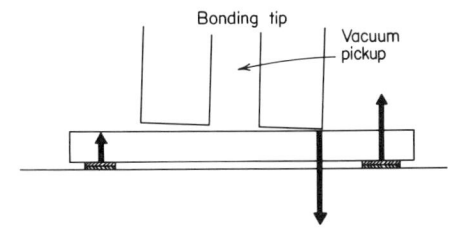

Fig. 17 Effect of angular misalignment of bonding forces.[21]

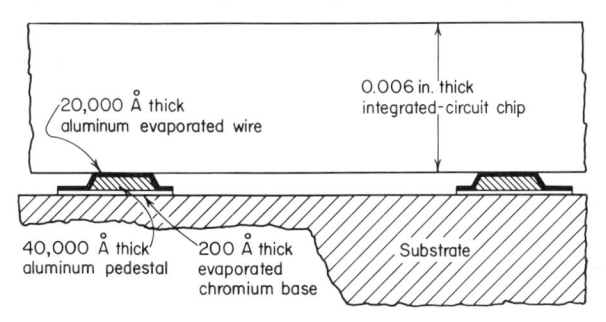

Fig. 18 Cross section through bonded chip and substrate.[22]

conductors are evaporated through masks onto the substrate. The conductors are raised on pedestals at the contact location to allow ease in bonding and to provide clearance under the chip for substrate wiring.

The bonding process is shown schematically in Fig. 19. The chip is lifted by means of a vacuum pickup and placed over the pedestals on the substrate. The chip is viewed through the substrate and is manipulated until the chip pads and substrate pedestals are aligned. The bonding tip in the transducer head is brought in contact with the back of the chip, and a downward clamping force is applied to the chip. When ultrasonic energy is applied, a lateral "scrubbing" between

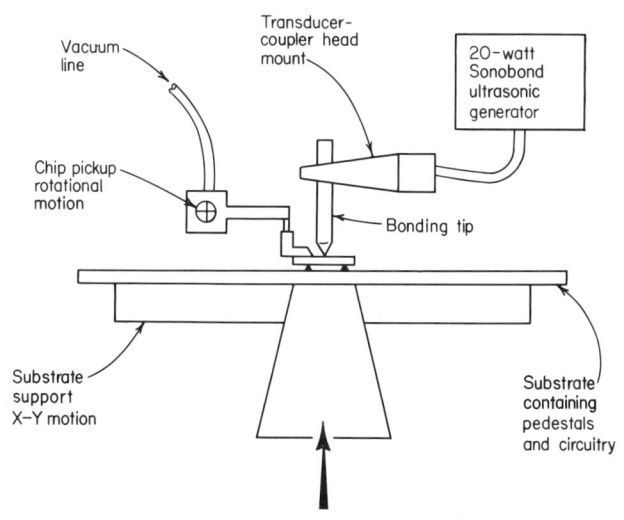

Fig. 19 Schematic diagram of chip-bonding process.[22]

the chip land area and the aluminum conductor over the pedestal takes place for a predetermined interval of time. The result is a molecular bond between each of the chip pads and the corresponding substrate wires.

The bonds made by this process are mechanically strong and have low electrical resistance. Chips can be removed and replaced, if the bonding parameters are properly chosen. To accomplish this, the clamping force is reduced until the bond becomes the weakest point in the structure; when a shear force is applied, the structure breaks cleanly at the bond. By proper choice of clamping force, bonds can be made sufficiently strong to pass vibration and shock tests, yet sufficiently weak to shear cleanly for replacement. The magnitude of clamping force will vary with pad and pedestal size and with the number of pedestals. Consequently, the appropriate magnitude has to be determined for each system.

Using Type 7059 Glass* as the substrate material, the aluminum pedestals and conductors were evaporated through electroformed nickel masks.

The evaporation process was carried out in the following steps: A 200 Å thick layer of chromium was evaporated to improve the adhesion between the subsequently deposited aluminum layers and the glass. Next the pedestals were evaporated, and then the interconnect conductors bring all the chip connections to convenient access points at the edge of the substrate. The pedestal diameter is nominally 0.0025 in. at the base. The diameter decreased during each evaporation because of the deposition of aluminum on the masks. The masks were cleaned after every fourth evaporation to limit the reduction in the pedestal diameter. The pedestals, made 40,000 Å high for the first three evaporations, were made 60,000 Å high on the fourth evaporation to compensate for the decrease in pedestal diameter.

Chips were facedown bonded to the evaporated aluminum pedestals on the glass substrates. The chip pads were usually aligned with the pedestals on the substrate and were then clamped to the pedestal by the bonding tool. The energy transfer from the ultrasonic transducer to the chip was made by means of the tip. Four tip designs are shown in Fig. 20. Good, reliable bonds are made when the truncated pyramid tip is used, provided the chips are square. Energy is transferred along the chip edges; consequently, chips that have damaged or have irregular edges do not bond well. Since this condition is common in chips that are obtained from scribed and broken wafers, and since this is the most general method of cutting wafers into chips, this tip has limited application. Good bonds are also made with the sharp-edged tip. Problems arise from incorporating the vacuum pickup into the tip; consequently, the usefulness of this design depends on the specific application. Poor bonds are made with the sharp needlepoint tip. Frequently, the chip is thrown out from under the bonding tool. Good bonds are made with the blunt needlepoint tip. This design was employed on the bonding equipment used in this study. Clamping forces of 0.5 and 1.0 lb were used. A spring-force gage was used to calibrate the clamping force at the point of application. Ultrasonic bonding energy was obtained from a 20-watt generator; the time setting on the generator was 2, and the power setting was 4. Chips bonded with

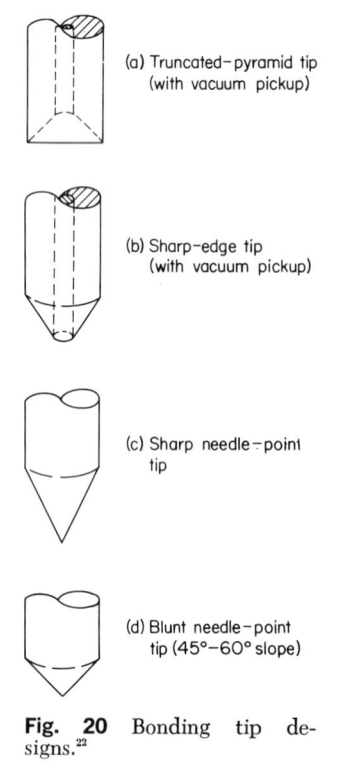

Fig. 20 Bonding tip designs.[22]

(a) Truncated-pyramid tip (with vacuum pickup)

(b) Sharp-edge tip (with vacuum pickup)

(c) Sharp needle-point tip

(d) Blunt needle-point tip (45°–60° slope)

* Trademark of Corning Glass Works, Corning, N.Y.

a 0.5-lb clamping force were replaceable. Those bonded with a 1-lb force were not.

To investigate the mechanical reliability of face bonded assembly designs, a matrix of input variables was established. The matrix was fully factorial and contained two levels on each of four independent variables. These independent variables were bonding pattern, planar circuit material, interconnection material, and interconnection (projection) size. The bonding pattern levels were defined in terms of the nominal strain state which they would induce during temperature excursions: uniaxial and biaxial.

Planar circuits were designed to produce the uniaxial and biaxial strain conditions. The planar circuit substrate for the uniaxial strain condition was designed to carry three face-bonded silicon chips and three (face-up) flying-lead bonded chips for temperature compensation. The planar circuit substrate for the biaxial strain condition was designed to carry two face-bonded chips and two temperature-compensating chips.

The planar circuit substrates measured 1.50 by 0.75 by 0.048 ± 0.004 in. Type 7059 Glass and microscope slides (soda-lime glass) were used. One side of each substrate was coated first with a vacuum-deposited layer of nickel-chromium, approximately 100 Å thick. Aluminum was subsequently vacuum-deposited on the nickel-chromium to a thickness of approximately 8,000 Å.

The vacuum-deposited films were tested for adhesion to the glass substrate and for bondability. An aluminum wire (1 percent silicon) measuring 0.001 in. in diameter was bonded ultrasonically to the film. A tensile load sufficient to cause fracture was then applied to the wire. Only when fracture occurred in the wire and outside the bond region was a film accepted for use in the program.

Acceptable metallized substrates were photomasked with negative resist. The aluminum was etched with a solution of methol alcohol and sodium hydroxide. The nickel-chromium was etched with a solution of cerric sulfate and nitric acid. The resist was then removed. The silicon chips were fabricated with five piezoresistive elements diffused into an 80- by 100-mil chip. The purpose of these elements was to serve as ancillary strain-measuring devices.

Conical projections were bonded ultrasonically to the metallized bond sites on glass planar circuit substrates. Gold and aluminum cones of two sizes were mounted. The gold was 99.99 percent pure. The aluminum was alloy 1100. The larger cone size has a base diameter of 0.005 in. and was 0.002 in. high. The smaller cone size had a base diameter of 0.003 in. and was 0.001 in. high.

The aluminum cones were mounted using a clamping force of 100 g. The larger cone size was bonded using 2.6 watts for 60 msec. The smaller cone size was bonded using 0.6 watts for 60 msec.

The gold cones were mounted using a clamping force of 130 g. The larger cone size was bonded using 1.5 watts for 60 msec. The smaller cone size was bonded using 0.95 watts for 60 msec. Table 16 summarizes the design of the face-bonded specimens.

After mounting projections on the planar circuits, silicon chips were face-bonded to the registered projections. This was accomplished using Ultrasonic Soldering Iron, modified to accept a variety of transducer tips. This unit provided a constant 10 watts at output frequencies between 32 and 42 kHz. The unit was equipped with a voltmeter across the output of the power supply to facilitate resonance adjustment.

The glass planar circuit substrates were clamped on the bonding stage. The chip was then placed facedown on the projections and was brought into registration. The transducer tip, preadjusted to be parallel to the back of the die, was then brought into contact with die at a predetermined load. Ultrasonic energy, 10 watts, was then applied for a predetermined time to form the bonds.

Several criteria were used to establish the face-bonding schedule. The schedule had to produce a minimum shear load (on the bar) to failure of 200 g, a face-bonded column height of 0.0002 to 0.0003 in. for small projections and of 0.0008 to 0.0010 in. for large projections, and electrical continuity in all resistor circuits.

13-38 Metals Joining of Electronic Circuitry

TABLE 16 Assembly Designs for Face-bonded Specimens[23]

Experimental cell number	Nominal strain condition	Planar circuit substrate material	Interconnection cone material	Projection cone height, in.
1	Biaxial	Soft glass	Gold	0.001
2	Biaxial	Soft glass	Gold	0.002
3	Biaxial	7059 glass	Gold	0.001
4	Biaxial	7059 glass	Gold	0.002
5	Biaxial	Soft glass	Aluminum	0.001
6	Biaxial	Soft glass	Aluminum	0.002
7	Biaxial	7059 glass	Aluminum	0.001
8	Biaxial	7059 glass	Aluminum	0.002
9	Uniaxial	Soft glass	Gold	0.001
10	Uniaxial	Soft glass	Gold	0.002
11	Uniaxial	7059 glass	Gold	0.001
12	Uniaxial	7059 glass	Gold	0.002
13	Uniaxial	Soft glass	Aluminum	0.001
14	Uniaxial	Soft glass	Aluminum	0.002
15	Uniaxial	7059 glass	Aluminum	0.001
16	Uniaxial	7059 glass	Aluminum	0.002

The chips, bonded according to Table 17, were inspected visually for mechanical damage and all circuits were inspected for electrical continuity. In each cell, several sacrificial dice were tested for shear strength to assure that the bonding schedule was producing the minimum 200-g shear strength. Typically, all schedules produced shear strengths in the range of 500 to 700 g.

TABLE 17 Face-bonding Schedule[23]

Cell numbers	Clamping force, grams	Power duration, sec	Power supply voltage
1, 3	700	0.6	7.5
2, 4	1,200	0.6	7.2
5, 7	1,200	0.6	8.0
6, 8	1,400	0.6	7.5
9, 11	1,100	0.6	8.7
10, 12	1,000	0.6	8.5
13, 15	1,200	0.6	8.7
14, 16	1,300	0.6	8.2

A primary output variable of this program was mechanical strain. The primary method of sensing mechanical strain was by means of foil-resistance strain gages. The foil gages were to be used to measure strains within their temperature capability and to provide strain calibration of the boron-diffused resistors. The diffused resistors initially believed to be necessary to measure strains at temperatures above 100°C were not needed since significant advances were effected in the temperature capability of the foil gages.

The small size of the silicon chips (0.100 × 0.080 in.) precluded the use of full-bridge strain-gage connections, and, considering the numbers of gage installations, it was not convenient to use half-bridge connections. Instead, the specimen measurements were made as quarter bridges, and a group of reference gages of each of the three types were also measured as quarter bridges. The readings of the reference gages were averaged by groups and the group averages used as corrections to the readings obtained with the three types of gages.

The reference gages were laid down on silicon wafers (111 out of a single crystal) that were not attached to anything. These devices were subjected to the temperatures used for the thermal cycling of the test specimens so that readings from them showed the effect of temperature on the resistance of the gage alone, and the strain effect created by the thermal incompatibility between the gage metal and the silicon wafer. The thermal coefficient of the gage metal is rated at 3.0×10^{-6} in./(in.)(°F), and the thermal coefficient for silicon (111 out) is 1.4×10^{-6} in./(in.)(°F) for the range of −65 to +350°F. This is an appreciable

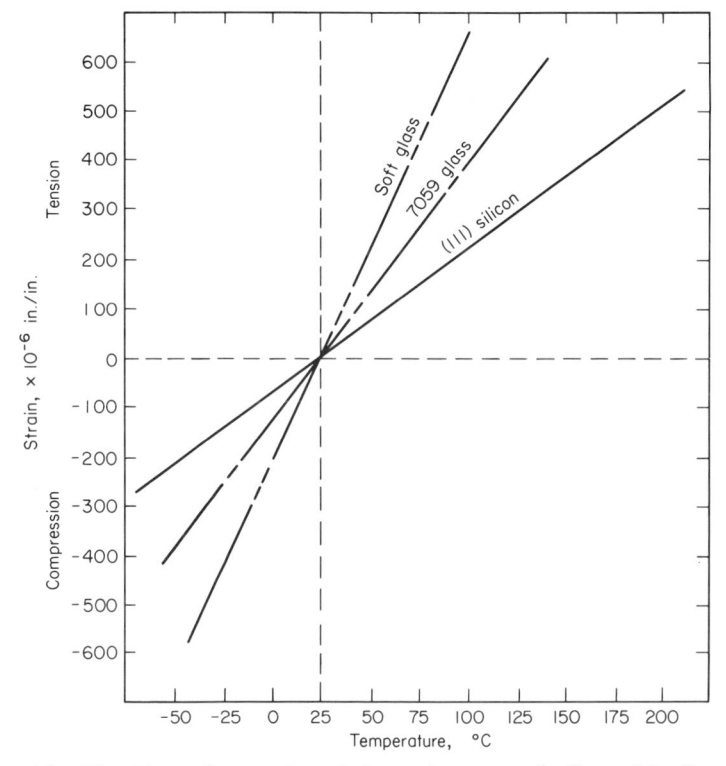

Fig. 21 Thermal expansion of glass substrates and silicon chips.[23]

degree of incompatibility, but unavoidable, because gages of lower thermal coefficient are not normally available.

The gages were bonded to the exposed surfaces of silicon chips with a specially prepared epoxy resin intended for the bonding of strain gages. The glass substrates to which the silicon chips were face-bonded were bonded into a 40-pin package with a polyester film adhesive. The strain gages were wired to the assigned pins on the package with gold wires attached to the strain-gage tabs by means of a modified form of thermocompression bonding. The leads from the package to the strain indicator were soldered to the package pins with high-temperature soft solder.

The three-wire system of leads was used to provide lead-resistance compensation, a very desirable feature for these tests because about half the length of the copper leads was heated or cooled within the oven during the temperature cycling tests.

13-40   Metals Joining of Electronic Circuitry

Otherwise the change of resistance of the lead wires would appear as an error in the results.

The strain-gage indications were manipulated and tested for validity in the following manner. Six readings were taken at each datum point, and the average of these, excluding obvious outliers, was accepted as the output value for the datum point. This produced six values per gage for each temperature cycle. Each cycle started at 24°C with zero strain being indicated. In order for the data from a cycle to be accepted, the cyclic return to zero strain at 24°C had to be less than 100 μin./in. Cyclic data not conforming to this requirement were rejected as invalid.

Another test of validity was that gages which indicated decreasing strain with increasing temperature were considered to be bonded unacceptably to the silicon chips. Cyclic data was also rejected unless all points were within the range of ±100 μin./in. The results from valid strain gages were examined for detectable strain between the glass substrate and the silicon chip. Figure 21 presents the experimentally determined thermal expansion behavior of the glass substrates and the silicon chips. Figure 22 presents the results of electrical continuity testing after three temperatures cycles. All circuits exhibited continuity in the fabricated condition.

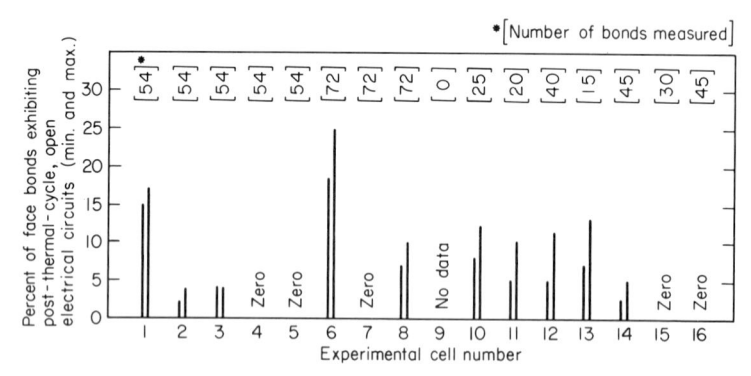

Fig. 22   Electric continuity test summary.[23]

Note that two levels of failure rate are indicated for each experimental cell in Fig. 22. The reason for this is that each circuit encompassed two face bonds. If a circuit was found electrically open, one or both bonds could have been damaged. In some cases each of the two could be isolated and checked. In other instances isolation was not possible. Thus, the two indicated levels are the possible extreme limits. The data in Fig. 22 were reduced and are summarized in Table 18. A number of circuits that exhibited electrical discontinuity after temperature cycling were analyzed metallographically.

A number of silicon chips were visually examined for bond-site damage after temperature cycling by observing through the glass substrate. This examination was performed primarily to detect differences related to bond material and was capable of detecting damaged metallization and damaged glass. Table 19 presents the results of this examination.

The measurements of thermal strain provided quantitative estimates of the strain levels which each design might exhibit. The strain data indicate that bond height is not particularly effective in changing the strain in the silicon bar. The metallographic information indicates no change in failure mechanism with bond height. The data in Table 18 indicates that the 2 to 3×10^{-4} in. bond height is the preferred level from the electrical continuity standpoint. Thus, it is judged that the 2 to 3×10^{-4} in. bond height is best.

Selection of substrate material is similarly clear-cut. The strain data and the data of Table 18 indicate selection of the Type 7059 Glass. The other output variables do not discriminate. Thus, it is judged that the 7059 glass is best.

Selection of bond material can be based on electrical continuity testing, metallographic analyses, and bond-site examination since the strain data do not discriminate. The data in Table 18 indicate that the lowest electrical failure rate was experienced with aluminum bonds. The data in Table 19 indicates much less bond-site damage

TABLE 18 Effect of Design Parameters on Electrical Continuity[23]

Assembly design parameter	Design parameter value	Lowest number of electrical failures
Strain condition	Uniaxial	×
	Biaxial	
Substrate material	Soft glass	
	7059 glass	×
Bond material	Gold	
	Aluminum	×
Bond height (fabricated)	2–3 × 10⁻⁴ in.	×
	8–10 × 10⁻⁴ in.	

for aluminum than for gold. The metallographic information shows clearly that the gold bonds are composed of gold plus a significant amount of gold-aluminum intermetallic compounds which are known to increase in amount with time and temperature. These compounds apparently activate a brittle fracture mechanism which was not observed in aluminum bonds. Furthermore, extensive damage to the glass under the bond pads was often observed with gold bonds, but seldom with aluminum bonds. Finally, the aluminum bonds were observed to assume

TABLE 19 Bond-site Damage[23]

Bond material	Number of sites examined	Percent of sites damaged
Gold	200	35
Aluminum	176	6.3

compliant shape, the lazy S, without a brittle fracture tendency. The lazy-S configuration bends slightly while temperature cycling. This affords substantial strain relief to the silicon bar. So long as the ends of this Euler column remain fastened to the bond pads, the partial fracture between bond and pad, necessary to form the lazy S, is a desirable feature. Furthermore, the intrinsic properties of aluminum are such that at the higher temperatures, where tensile strain on the silicon bar is greatest, the aluminum will flow plastically with minimal work hardening and embrittlement. Thus, it is judged that aluminum is the best bonding material.

13-42　Metals Joining of Electronic Circuitry

The most advantageous bonding pattern (nominal strain state) is, from the results, uniaxial. Table 18 indicates fewer electrical failures for the uniaxial pattern than for the biaxial pattern. However, utilization of the uniaxial pattern in real hardware would compromise seriously the packing density of devices.

This dilemma can be resolved by examining the results within the narrower scope of the preceding design parameter selections of aluminum bonds, 2 to 3 × 10^{-4} in. high, and the Type 7059 Glass substrate. The biaxial pattern used in cell 7 produced a very wide range of strains, as is depicted in Fig. 23. However,

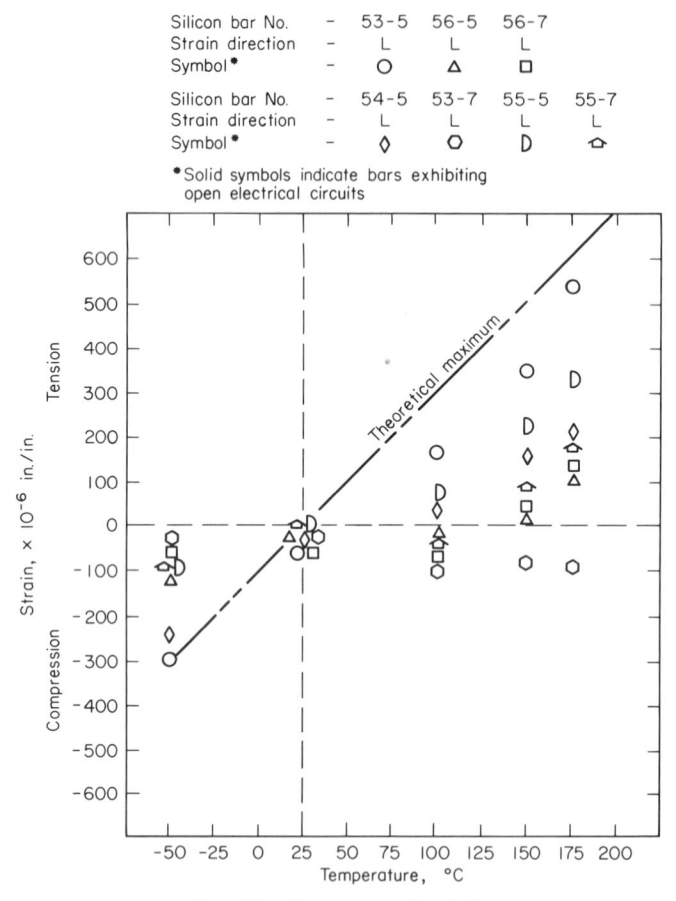

Fig. 23 Experimental cell.[23]

even where the strains approached the theoretical maximum, no electrical failures were detected. Thus, it might be reasoned that although high strains are not desirable, they can be tolerated in this case so as to avoid compromising the packing density. Thus, it is judged that the biaxial pattern should be selected. The more acceptable design parameters are, in summary, biaxial bonding pattern, aluminum bonds, bond height of 2 to 3 × 10^{-4} in., and Type 7059 Glass substrates.

Further work has been done with ultrasonic face bonding, using aluminum bumps on the chip. A standard series of experimental chips was designed for bond-strength evaluation. The form of the chips is shown in Fig. 24. The four-bond configuration was chosen since departures from flatness will affect the contact, and these effects

First-level Joining, Materials and Processes 13-43

are less likely to be masked by statistical variations in the strength of other bonds than they might be in multiple-contact chips. The chips used were of oxidized silicon with evaporated aluminum pads; the substrates were Type 7059 Glass with evaporated aluminum-contact areas.

A matrix approach was adopted. The matrix has bonding force, applied ultrasonic power level, and total aluminum-contact thickness (i.e., the sum of chip and substrate pad thicknesses) as the variables of prime importance. The effects of variations

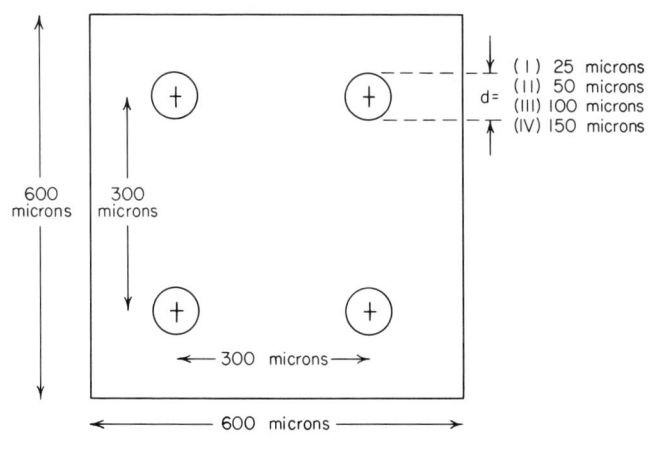

Fig. 24 Chip-pad configuration.[21]

of the chip aluminum-pad thickness for a constant total aluminum thickness and of mismatch in size between pads were also investigated.

Each point in the matrix was investigated by bonding 24 chips to a substrate and evaluating the bonds by measuring the force necessary to shear each chip from the substrate, and by visual inspection of the pad after the chip had been removed.

Preliminary experiments with a tubular bonding tool gave a considerable spread of shear strength. The distributions obtained proved to be unrepeatable from group to group, when nominally identical conditions were used, despite elaborate precautions to ensure that the tip and chip were coplanar. This was attributed to slight misalignment of tip and chip.

An experiment was performed in which the watt per second product of the ultrasonic bonder was kept constant while the duration of the bond pulse was reduced. It was found that the degree to which visible damage to the bonding pads occurred (deformation of the pads, smearing of the aluminum, etc.) decreased as the pulse duration decreased. The pulse length used for all ultrasonic bonding experiments was standardized at 0.06 sec.

A new form of bonding tool was designed in which a conical tip is used to transmit the ultrasonic energy. Such a tool makes a single central contact to

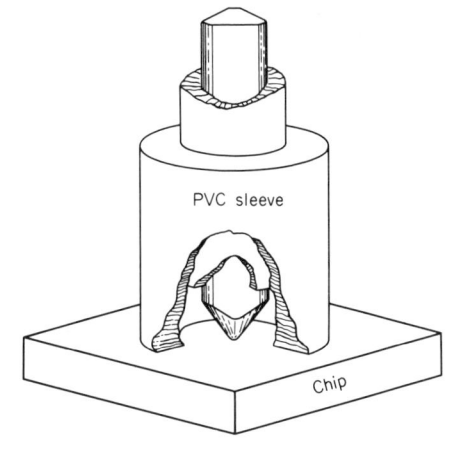

Fig. 25 Cutaway view of ultrasonic bonding tip.[21]

13-44 Metals Joining of Electronic Circuitry

the chip, ensuring an even distribution of the load between the bonding pads.

A detailed sketch of the new tip is shown in Fig. 25. It consists of a steel needle ground to a conical point with flats ground on the sides of the needle to give it a truncated equilateral triangular section. The shaped needle is forced into a circular cylinder with the flat sections on the needle providing vacuum channels for picking up the chip. A soft vinyl sleeve is placed around it to provide the vacuum connection to the chip without affecting the force applied to the chip during the bonding operation.

A comparison of sheer strength distributions for identical chips and substrates using both types of bonding tips is shown in Fig. 26. The power used in each case is the optimum value for that particular tip configuration.

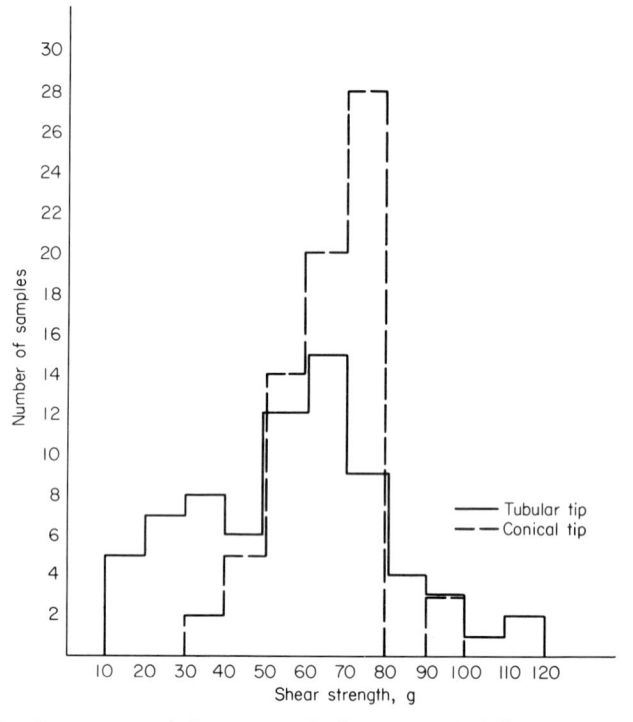

Fig. 26 Comparison of shear strength obtained with different tip designs.[21]

The observed improvement of shear strength distribution is typical of results that have been obtained with this tip on samples with different aluminum bonding-pad thicknesses and diameters. From these experiments it was concluded that only bonding tip designs that allow the chip to align itself to the substrate, without substantially altering the position at which the bonding force is applied, are capable of giving good results on thin films.

Using the new bonding tool design and the 0.06-sec pulse duration, the matrix experiment on aluminum bonding was performed.

Curves showing the effect of power on shear strength for 2-micron-thick pads are presented in Fig. 27. Further increase in power led to damage of the pads and a reduction in bond strength. The peak in strength occurred at different power levels for different thicknesses of aluminum pad, but was always in the range of 0.75 or 1.0 watt. The shear strengths observed were compared with the calculated values for bonds of this geometry. In a large number of the samples, the observed shear strength was significantly greater than the calculated value of 40 g for 50-micron-diameter bonds. The increase of bond strength with power beyond the calculated value of 40 g was taken to indicate that the aluminum pads, which are initially in an annealed state, are being work-hardened during the bonding process. A literature search revealed that the ultimate tensile strength of aluminum of 99.99 percent purity in a hardened state is 185 percent of that of the annealed form. This figure would lead to a value of approximately 75 g for the maximum expected shear strength. This is in reasonable accord with the observed distributions for well-formed bonds.

A comparison between 50- and 100-micron-diameter bonds is shown in Fig. 28 for 2-micron-thick films on both the substrate and the chip. The graph shows shear strength for the 100-micron bond on the left-hand V axis and shear strength scaled by a factor of 4 (the bond-area increase) on the right-hand axis. The power has been increased by approximately 3:1. The similarity in shape of these

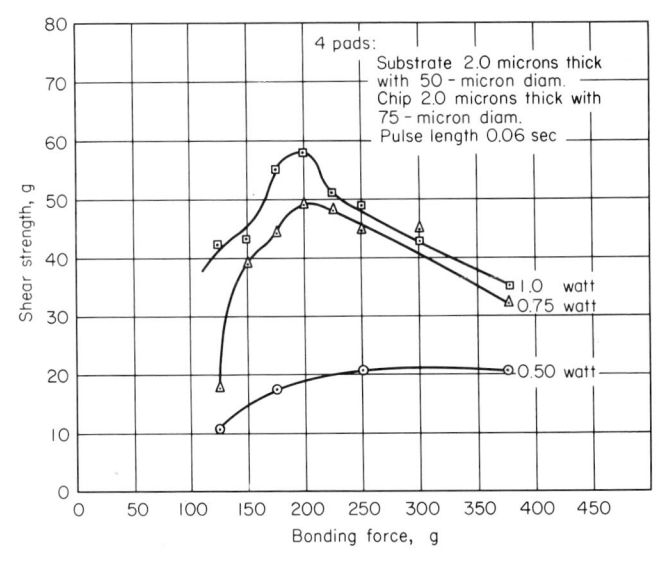

Fig. 27 Variation in shear strength with bonding force and power.[21]

scaled curves is striking. It is evident that bond strength is dependent on the power per unit area, but it appears that the effect of bonding force plays a greater role in the coupling of the ultrasonic energy to the chip than in providing additional force during the bonding operation once a threshold value of bonding force has been exceeded.

In both the 0.75- and 1.0-watt cases, the optimum bonding conditions were obtained with bonding forces between 200 and 300 g. depending on the thicknesses of the pads used. The use of 250 g as the bonding force would ensure considerable latitude in the choice of film thickness without substantial reduction of bond strength.

Table 20 gives the optimum bonding conditions for 0.5-, 1.0-, and 2.0-micron substrate film thicknesses with the total thickness kept constant at 4.0 microns for 0.75- and 1.0-watt power. The corresponding standard deviations and 2σ minimum limits are included.

It is concluded that: Best bonding uniformity is achieved with tips that allow the chip to align to the substrate while retaining the central point of application of the bonding force. A conical tip is a good example. Adequate bonding strength

13-46 Metals Joining of Electronic Circuitry

can be obtained over a relatively wide range of bonding force. The bonding force primarily ensures adequate coupling of the ultrasonic energy to the chip and plays a less important role in the actual bonding process. For a constant total aluminum thickness, the bond strength rises with a decreasing substrate film thickness in the range of 2.0 to 0.5 microns at the optimum bonding conditions.

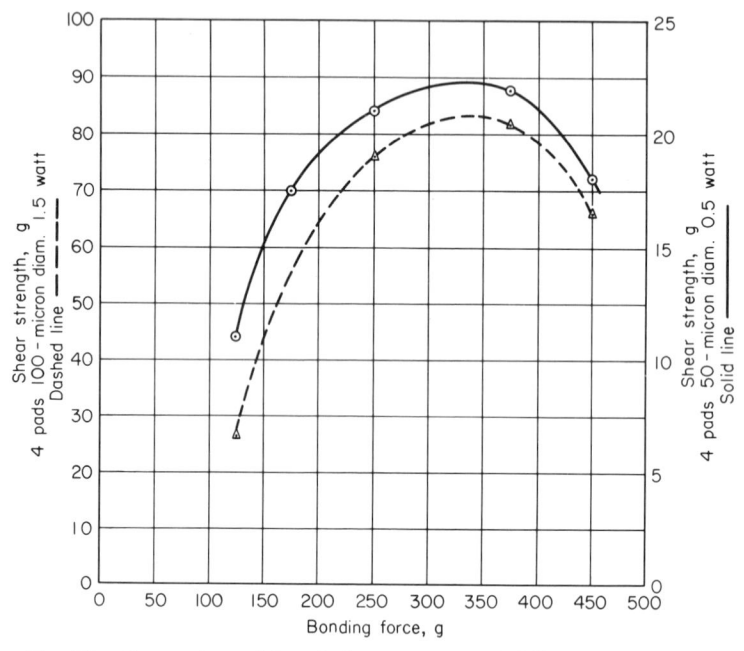

Fig. 28 Comparison of bond characteristics of different pad areas.[21]

Ultrasonic face bonding, utilizing solder bumps on silicon chips, was also studied. A number of methods of preparing solder pads on chips and substrates have been investigated in an attempt to obtain a wide range of solder thicknesses and good control. The methods tried were displacement plating, electrolytic plating, ultrasonic dip coating, and dip soldering. The last method was the most successful and was used in the preparation of all samples for experimental work.

TABLE 20 Optimum Bonding Conditions for 50-micron-diameter Bonds[21]

Substrate bonding film, microns	2.0	1.0	0.5
Chip bonding film, microns	2.0	3.0	3.5
For 0.75-watt power, grams:			
Bonding force	200.0	250.0	300.0
Mean shear strength	48.7	62.1	77.1
Standard deviation	11.9	14.1	18.1
2σ minimum	24.9	33.9	40.9
For 1.0-watt power, grams:			
Bonding force	200.0	200.0	325.0
Mean shear strength	57.9	57.5	72.9
Standard deviation	9.9	16.5	20.8
2σ minimum	37.7	24.5	31.3

Units with four pads, 100 microns in diameter, were ultrasonically bonded to glass substrates with similarly prepared pads. The conditions used were bonding force 450 g, bonding power 0.16 watt, and pulse duration 0.5 sec.

It was found that the distribution of bond strength was more tightly grouped than the thermal solder samples with no units showing less than 20 g shear strength and a few above 50 g. The theoretical shear strength for these samples is approximately 50 g. Some samples were heated above the melting point of the solder to reflow the bonds on the chip which had been attached by ultrasonic bonding. Shear strength tests on these showed no distinguishable change in the distributions from the sample ultrasonic bonds. Solder reflow face bonding of chips to Type 7059 Glass was also studied.

Methods aimed at improving the uniformity of wetting of solder pads to the metallization on both silicon chips and glass substrates were studied. A technique of building up the pads prior to solder dipping based on the electroless plating of nickel, and followed by gold onto Nichrome* and chromium metallized pads, offered no improvement in pad uniformity from substrate to substrate. Figure 29 shows the distribution of pad height within a substrate on which good solder wetting was obtained. Even on this sample, however, a fair number of metallized pads failed to wet during the dip-solder operation. Despite the less than optimum process for preparing solder pads, the chip quality was considered adequate for subsequent bonding experiments.

A bonding machine was designed for thermally attaching silicon chips onto glass substrates, both having solder pads. The basic requirements placed on the machine were:

1. A means of holding both chip and substrate in proper relation to each other, i.e., face to face.

2. A means of precise alignment of chip relative to substrate or vice versa.

3. A means of heating the chip to about 250°C.

4. A means of bringing chip and substrate together in proper alignment and advancing chip toward substrate by the proper amount (about 0.002 in.) as solder melts.

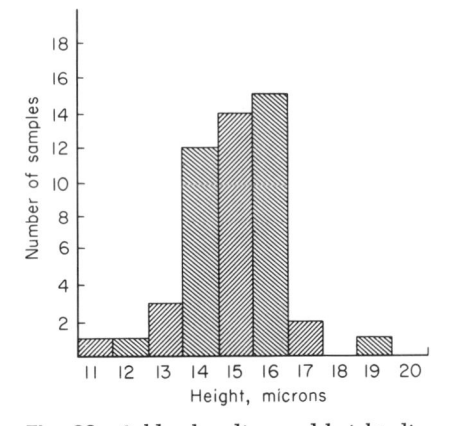

Fig. 29 Solder bonding pad-height distribution.[21]

A partially silvered mirror, held precisely in a horizontal plane, was introduced between chip and substrate. The chip was to be held facedown at a fixed distance above the mirror, and the substrate held horizontally the same distance below. Once registration was achieved, the chip was to be picked up by a vacuum chuck, the mirror removed, and the chip brought down to the mating pattern on the substrate.

A sketch of the mirror arrangement is shown in Fig. 30. Figure 31 shows how the registration is achieved using this mirror system. Based on the chosen method of operation, the machine was designed and built with the following features:

1. The registration jig, consisting of two disks of glass, held exactly parallel and the lower one partially silvered, is capable of moving freely in a fixed horizontal plane and provides visual access to the lower mirror from above.

2. The stylus which:

   a. Moves freely up and down, but does not deviate from a straight line.

   b. Clamps firmly at any point in its travel, with no residual motion up or down.

* Trademark of Driver-Harris Co., Harrison, N.J.

13-48   Metals Joining of Electronic Circuitry

  c. Holds a heatable vacuum chuck.
  3. The platform, vertically adjustable in precise increments with no residual lateral movement.

Bonding experiments were performed using the equipment described above. The chips used had four solder pads 100 microns in diameter prepared by dip

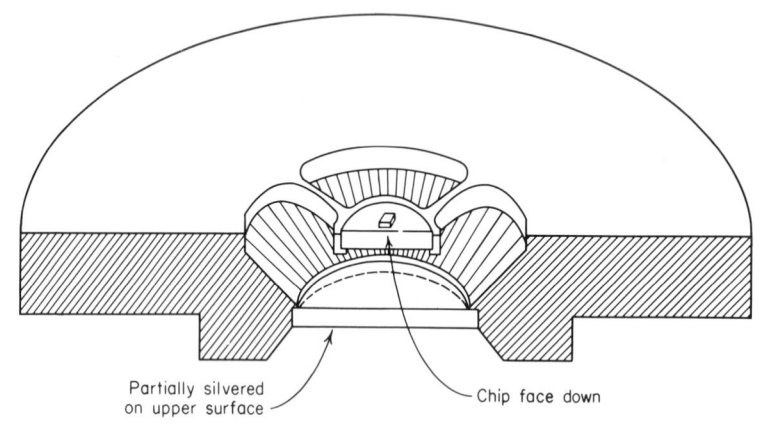

Fig. 30   Registration mirror.[21]

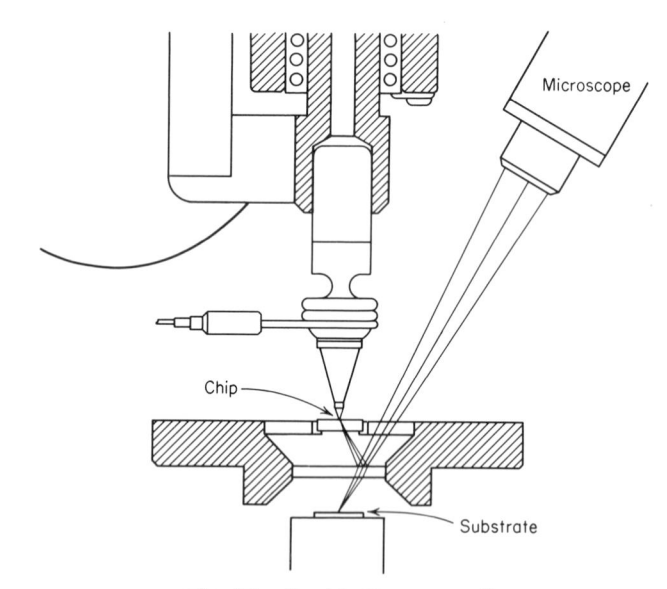

Fig. 31   Registration system.[21]

soldering. The pads on the substrates had geometries identical with those on the chip and were prepared in the same manner.

The chips were attached to the substrate by raising the tip temperature to approximately 230°C. Lowering the substrate during the bonding cycle alleviated but did not altogether eliminate the problem of "squashing" the bonding pads.

The bonds were evaluated by performing shear strength measurements and by microscopic evaluation of the sheared areas. Figure 32 shows a shear strength distribution obtained on 69 samples. The value expected on the basis of calculations reported in the first quarterly report is approximately 50 g. The spread of shear strength is typical of the results obtained and can be attributed, in most cases, to the jigging used in the assembly and the resultant lack of control of the bonded area. However, the results are distributed about the theoretically expected value and might be expected to group more tightly with equipment modification.

The incidence of shear strengths greater than the theoretical value is thought to be due to the spreading of the bonding pads, together with the formation of a "bond" between the oxidized solder and the glass. It should be noted that the distribution drops off sharply above the theoretically expected shear, and shows no indications of the anomalously high strengths obtained with aluminum ultrasonic welding.

Solder reflow facedown bonding of silicon chips to epoxy-glass printed-circuit boards has been accomplished.[24]

The microcircuit wafer was fabricated in the conventional manner through metallization. A layer of solderable metal was then applied to the bonding pads. Next, the wafer was vapor-plated with glass. The bonding pads were then exposed by etch-back techniques. The wafer fabrication was completed by dipping it into a solder bath to form solder pads. After completing the wafer processing necessary to provide tinned contact pads on the microcircuit in wafer form, the process continued with conventional die sorting, scribing and breaking, and visual inspection. The microcircuit chips are placed facedown on a mirrored glass surface which is subsequently placed into the assembly machine. With the aid of a microscope, the operator orients the chip while viewing its surface in the mirror, and picks up the chip using a vacuum probe. The printed-circuit board with its associated pad array and alignment marks are brought into the field of view beneath the chip. Since the operator is viewing the chip at approximately a 40° angle with the horizontal, the original scribed edge of the chip is clearly visible. As the chip is brought in near contact with the pad array on the printed-circuit board,

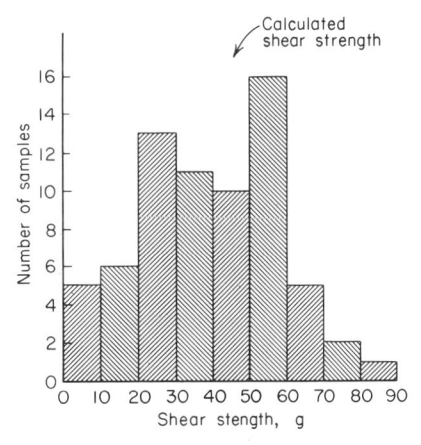

Fig. 32 Shear strength distribution for thermal soldered chips.[21]

the chip is positioned with the alignment marks. Rotation is achieved through the table supporting the board. Positioning of the chip in the x and y directions is accomplished through the micromanipulator to which the vacuum pickup is attached.

After the chip-to-board alignment is made and the chip put into contact with the substrate, the initial soldering cycle is activated. Heat is transmitted through the vacuum probe from a small resistance heater wound around the probe. A thermocouple embedded in the heater permits temperature control. Through judicious selection of time, temperature, and pressure, simultaneous solder joints are made between two parts. To restrict the solder flow to the pad area, a temperature gradient is provided by cooling the copper pattern beyond the chip with a flow of nitrogen. Selection of force is a function of the pad area; a force of 50 g was used for the units assembled on this program. The assembly operation does not require flux.

The second phase of the attachment process is accomplished by bringing the chip-and-board assembly beneath an infrared lamp source. A short heating cycle is provided to permit it to "float" on the solder joints, relieve the stress, and ensure a space between the chip and the substrate.

Beam-lead techniques are used as a means of face bonding. There are two accepted means of fabrication for beam-lead devices, the titanium-platinum technique and the molybdenum technique. Each technique may be used interchangeably for most applications. The most frequently used, but more complicated system, is the platinum-titanium–platinum-gold system. After diffusion of devices into the wafer, windows are etched through the oxide or other protective coating to the silicon device contacts. Then, platinum is sputtered approximately 500 Å thick and alloyed into the silicon contact area. The excess platinum is then removed by chemical etching with aqua regia. Titanium is then sputtered on a 500-Å thickness followed by 1,500 Å of platinum. The beam-lead pattern is defined in the platinum by using photoresist etching techniques. However, the titanium remains intact. Then, new photoresist is applied masking all areas except those to be raised by gold plating. After the gold plating, the photoresist is removed and the unprotected titanium chemically removed.

The back of the slice is lapped to 2 mils in thickness, and photoresist is applied. The areas of silicon between the devices are exposed and etched away to form overhanging beam leads. After etching, the components are ready for test and assembly.

The molybdenum technique substitutes a single molybdenum layer for the titanium and platinum layers. The major limitation of this technique is that the silicon surface impurity level must be in excess of 5×10^{18} atoms/cm^2 to ensure ohmic contact. Thermocompression bonding, heated substrate method, was capable of bonding beam-lead devices to gold traces on alumina substrates either facedown or faceup. However, the 360°C temperature required severely limited the substrate materials and restricted the devices. Ultrasonic bonding was also examined. However, ultrasonic bonding was not suitable for bonding beam leads to copper-clad epoxy circuit boards. Also, ultrasonic bonding proved difficult to handle on the small-size leads (1 mil wide by 2 mils long) without damaging the silicon chip. Parallel-gap welding could bond beam leads to gold-plated copper-clad epoxy boards or, if the substrate was tinned, the parallel-gap welder could be used to solder bond the beam leads. A difficulty is the distortion of the leads caused by the force required to penetrate the silicon dioxide layer on the surface of the beam lead. Furthermore, the 1- by 2-mil tungsten electrodes were difficult to make and difficult to keep dressed.

Soldering ranges were determined by bonding 0.004- by 0.016-in. solder-dipped Kovar* leads to 0.020 in. wide gold-plated traces on 2-oz copper printed-circuit board. It was found that the spaces between the electrodes, for optimum margins against blowout, should be at least equal to the width of a lead being bonded. Figure 33 presents the ranges found when using a 0.020-in. electrode spacing and 0.015- by 0.025-in. molybdenum electrodes. The coordinate values are the settings of the power supply. Some representative test measurement points are indicated in the figure. The bonding region shown at 40 msec remains constant as time is increased. Electrode force is not specified for these curves, as it is not a factor.

The soldering range shown in Fig. 33 has as its lower bound the lowest heat that will produce a satisfactory bond and as its upper bound the lowest heat that may produce a weld, no matter how slight. Any bond in the soldering range may be removed by application of a hot soldering iron. The welding range is not a true welding range for it includes all welds short of blowout. In the lower part of this range the weld may be just a slight tacking. This figure shows an important advantage of impulse soldering over parallel-gap welding: the soldering range is at least twice as great as the welding range and is remote from the blowout range. Even if the bond is made at a heat somewhat above the soldering range, the only result would be a weld. Unless this bond must later be broken, no damage would have been done.[26]

The thermal pulse-wedge technique was also investigated.[25] In this process a

* Trademark of Westinghouse Electric Corporation, Pittsburgh, Pa.

pulse of current (up to 500 A for 500 msec) passes from one clamp to the other through a tungsten-carbide wedge. No current passes through the weldments. The wedge can be brought to a bright cherry red (at about 600°C), but the bonding of beam leads does not require such high temperatures. The wedges were 1/16 in. in diameter and 5/32 in. long. It has been found feasible to make wedges with tips as small as 1 mil square. Because there is no inherent motion of the wedge and also because it can be made small, thermal pulse bonding was the only technique found feasible for bonding devices with leads only 1 mil wide by 2 mils long. Thermal pulse bonding proved capable of bonding beam leads to either ceramic or organic substrates. This method is capable of multilead bonding. It therefore met the need for a general-purpose beam-lead bonder. Figure 34 shows the bonding range determined using a 2.5- by 2.5-mil wedge to bond 0.5- by 1-mil beam leads to gold traces on alumina. All bonds in the bonding range withstood a stress of 75,000 g normal to the substrate. The tests were limited to this stress to avoid damaging the substrate or its attachment to the centrifuge. The bonds themselves have been shown to withstand 135,000 g, the

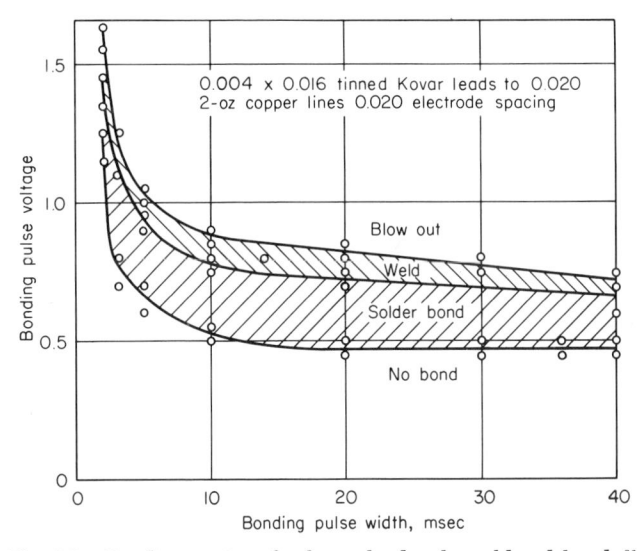

Fig. 33 Bonding regions for beam-lead pulse-soldered bonds.[26]

maximum of the centrifuge. The upper limit was determined by limiting the maximum lead deformation permitted to an increase in width of 50 percent. Actually considerably greater deformation can take place and still obtain a bond that will withstand 135,000 g, but the requirement of bonding 1-mil leads on 2.5-mil centers led to the limitation of a 50 percent increase maximum. Even wider tolerances can be obtained by using a large wedge, but the bonding range for a 2.5- by 2.5-mil wedge allows ±30 percent variation of bonding parameters, which is large compared to the tolerances of parameters of parallel-gap welding, for example. For this reason, thermal pulse bonding meets the requirement that special training of operators is not needed.

Thermal pulse bonding will make satisfactory bonds in the presence of considerable contamination. Tests were made by bonding through 0.2 mil of KMER.* The bonding range was reduced by about 50 percent, but the bonds withstood >5,000 g.

A serious problem with facedown-bonded (flip-chip) systems is the removal of heat from the chip. It has been conservatively assumed that conduction is the

* Trademark of Eastman Kodak Co., Rochester, N.Y.

13-52   Metals Joining of Electronic Circuitry

only major heat-transfer method by which heat will be removed from the active device. Estimates of thermal resistances of the various parts of the facedown-bonded system were made. These parts are the chip, the bonds, and the substrate. Values of thermal resistance of the chip are shown in Fig. 35, of various bonds in Fig. 36, and of various substrates in Fig. 37. Estimated conductive thermal resistances for face-bonded systems having chips with 4, 10, and 20 contacts are shown in Table 21.

For a module to be capable of operating over the full range of military environmental temperatures (−55 to +125°C) and keeping the maximum device junction temperature at or below 175°C, only a 50°C junction-to-ambient temperature rise is allowed. From the typical thermal resistance values for conduction alone, the maximum power dissipation per chip should be limited to approximately 1 watt for alumina, 2 watts for beryllia, and 65 mw for glass.

If forced air is used in addition to conduction, higher maximum chip dissipations could be allowed, but it is apparent that if glass is to be used as a substrate,

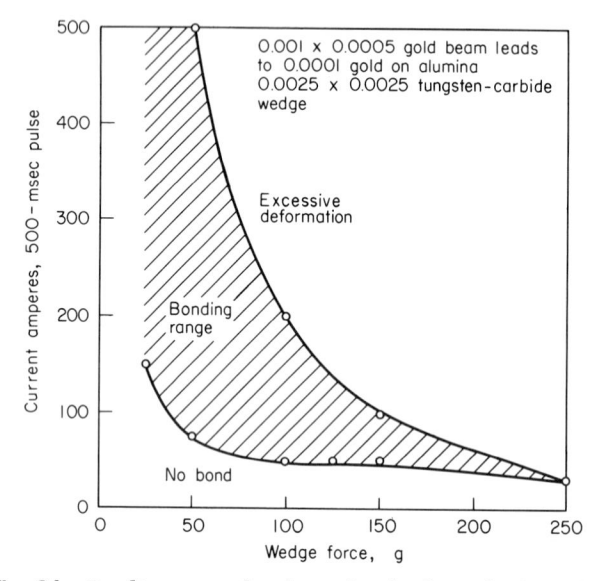

Fig. 34   Bonding range for thermal pulse beam-lead bonds.[25]

special thought will have to be given to the selection of chips or special provisions made for the removal of heat from the module.

It was concluded, therefore, that despite the advantages of the transparency of glass during module assembly, the majority of practical modules will be fabricated on opaque substrates of materials having good thermal conductivities. Therefore, it was concluded that: (1) Glass has inadequate thermal conductivity for use in face-bonded systems, except when moderately low power devices are used or in conjunction with an effective method of cooling such as high-velocity forced air; (2) alumina and beryllia can be used in modules where conduction is the only method of cooling for a wide range of power dissipations; and (3) thermal and mechanical considerations indicate that a large and constant number of pads should be used on a chip, whether or not they are actually used for electric connection.

The Electronic Industry Association has conducted a survey of face-bonded integrated-circuit users to determine preferred configurations for the present and future implementation of uncased integrated circuits. These configurations included flip chips with bumps, plain chips with bonding pads, and chips with beam loads. The

First-level Joining, Materials and Processes 13-53

makers were then surveyed as to the availability and physical characteristics of these chip forms. The responses formed the basis for the preliminary recommended standards for plain uncased chip, and beam-load uncased chip configurations.

Yield losses in face bonding must also be considered. Yield losses in bonding the chip are caused by registration errors, variation in pad heights or geometrics, contamination of pads, or broken interconnections.

Based on the assumption that the composite probability of success in assembly of a face-bonded chip is 95 percent, it can be shown, using yields from Fig. 38, that the effective module cost will increase very rapidly with complexity and

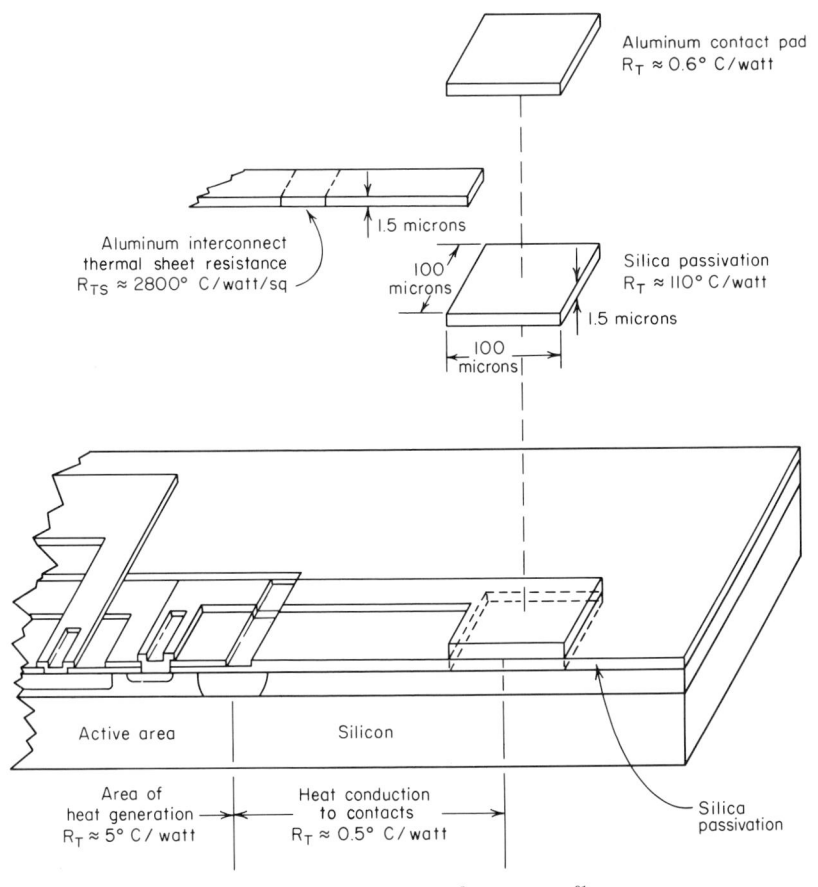

Fig. 35 Chip thermal resistance.[21]

will reach about five times the basic module construction cost at a complexity of only 32 chips. Thus, unless the yield problem can be overcome, face-bonded modules must either be of limited complexity or must have performance such that a premium price can be justified.

The most obvious way of improving the effective module yield is by electrically testing the chip after it has been mounted on the module and replacing defective units during the module manufacturing cycle.

Comparison of the costs of face-bonded modules and modules fabricated from discretely packaged integrated circuits has shown that the latter have cost advantages

13-54 **Metals Joining of Electronic Circuitry**

Silicon | 5 microns
Solder
Substrate 100 x 100 microns
$R_T \approx 12.5°\text{C/watt}$

Solder Pad

Silicon
Copper 100 micron diam. — Solder
Substrate
$R_T \approx 30°\text{C/watt}$

Solder Ball

100 microns
Silicon 100 microns
15 microns
Gold
Substrate
$R_T \approx 225°\text{C/watt}$

Beam Lead

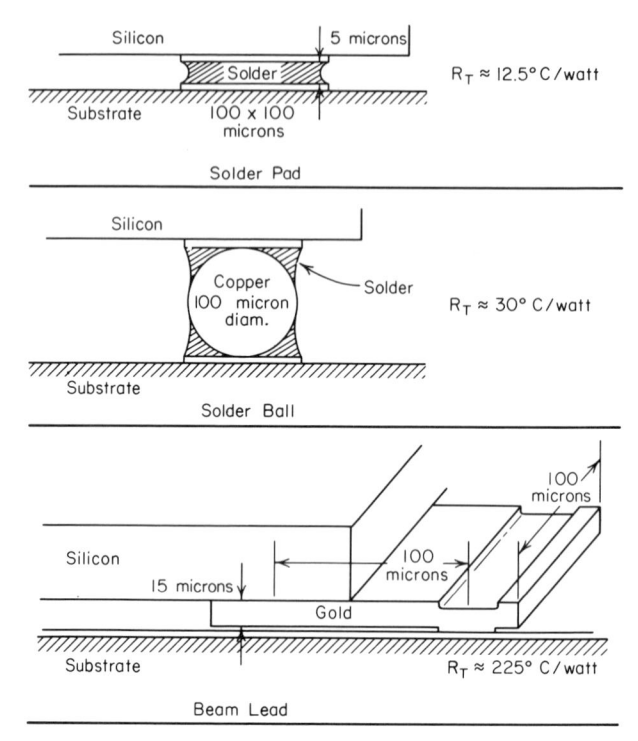

Fig. 36 Bond thermal resistance.[21]

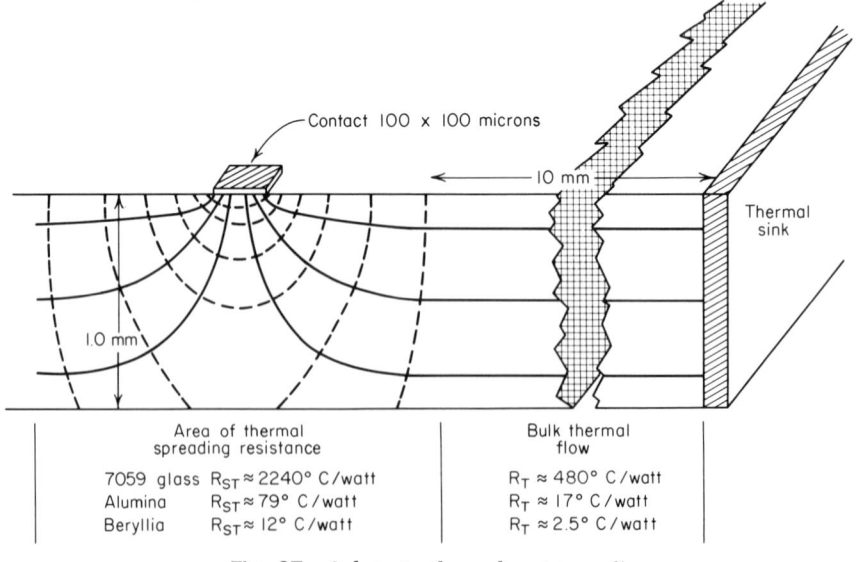

Contact 100 x 100 microns
10 mm
Thermal sink
1.0 mm

Area of thermal spreading resistance
7059 glass $R_{ST} \approx 2240°\text{C/watt}$
Alumina $R_{ST} \approx 79°\text{C/watt}$
Beryllia $R_{ST} \approx 12°\text{C/watt}$

Bulk thermal flow
$R_T \approx 480°\text{C/watt}$
$R_T \approx 17°\text{C/watt}$
$R_T \approx 2.5°\text{C/watt}$

Fig. 37 Substrate thermal resistance.[21]

TABLE 21 Estimated Conductive Thermal Resistance[21]

Thermal resistance components	Number of contacts								
	4			10			20		
	Degrees centigrade per watt								
Chip resistance:									
Active area.............	5			5			5		
Conductive area.........	11.1			4.5			4		
	($Q = 0.3$)			($Q = 0.7$)			($Q = 0.8$)		
Silica passivation.......	27.5			11.0			5.5		
Contact pad............	0.15			Negligible			Negligible		
Chip total............	43.75			20.5			14.5		
Bond system resistance:									
Ultrasonic weld.........									
Solder pad.............	3.1			1.25			0.63		
Solder ball.............	7.5			3.0			1.5		
Beam lead.............	56			22.5			11.2		
Substrate resistance	7,059	Al_2O_3	BeO	7,059	Al_2O_3	BeO	7059	Al_2O_3	BeO
Spreading resistance.....	560			224			112		
(Best case:		20			8			4	
parallel noninteracting									
combination).........			3.0			1.2			0.6
Bulk resistance.........	480			480			480		
		17			17			17	
			2.5			2.5			2.5
Substrate total........	1,040	37	5.5	704	25	3.7	592	21	3.1

over the flip-chip version at module complexities up to 4 to 6 chips. The reason is that the cost of the substrate and the cost of packaging the chips are greater than the cost of assembling conventional integrated circuits on an etched circuit board. If more than 10 chips per module are used, then a face-bonded hybrid assembly should prove to be substantially cheaper than its conventional integrated-circuit counterpart. However, if plastic encapsulated units are used as the basis of comparison, then the decision is not so clear-cut.

Another competing method of obtaining complex modules is by the use of large-scale integration techniques. There are two basic approaches to achieving large-scale integrated arrays: one uses discretionary wiring techniques, and the other relies on a single custom interconnection pattern and improved yield. The use of discretionary wiring would eliminate, to a large extent, the yield problems that will be encountered by the prewired approach, and will probably be somewhat cheaper than face bonding for large arrays, provided that the problems of automated interconnection-pattern generation can be solved on a real-time basis from test results and logical equations, and that the problems of final test can be solved.

Substrate (Inorganic) Internal Connections

Thin films There are two basic approaches to thin-film conductors. In the first, the thin-film conductor patterns are vacuum-deposited through a mask. In the second, the substrate or base surface is coated with a deposited layer of metal, and then the conductor pattern is achieved by an etching process combined with

13-56   Metals Joining of Electronic Circuitry

suitable photolithographic techniques. The first technique has some advantages over the second, when applied to multilayer conductors. The most important advantage is that in a majority of cases all sequential metal and dielectric deposition can be done in a single vacuum system, during one pump-down. Also, the dangers of contamination are eliminated because all the photoresist and etching processes are not needed.

Interconnections. Thin-film interfacial interconnections, electrically connecting two deposited films, are an integral part of thin-film, mono-, and multilayer conductor patterns. Such connections occur at the junction of two or more conductors, conductors and passive devices, and a conductor and a bonding pad. A failure of

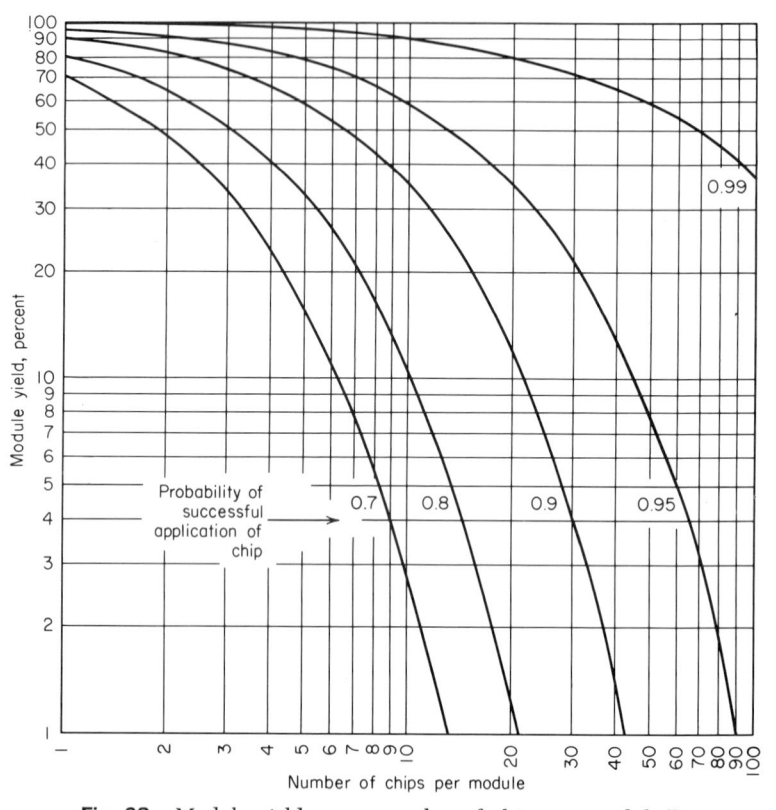

Fig. 38 Module yield versus number of chips per module.[28]

any such interconnecting joint will degrade the performance of the electronic device. Several interfacial-interconnection materials combinations have been evaluated.[8] A special interconnection pattern and several measuring and analytical techniques were developed to accomplish the evaluation. As a result of the evaluation, several criteria for interfacial interconnections have been established.

The electrical parameters studied were the dc interfacial resistance, the ac interfacial resistance, the rectification of the interconnection, and the current-carrying capability of the interconnection. The physical aspects studied were the adhesion of the film to its substrate and the compatability (or stability) of the film combinations. The above interconnection properties were studied under various environments. These environments were storage at −55, 125, and 200°C, for 1,000 hr;

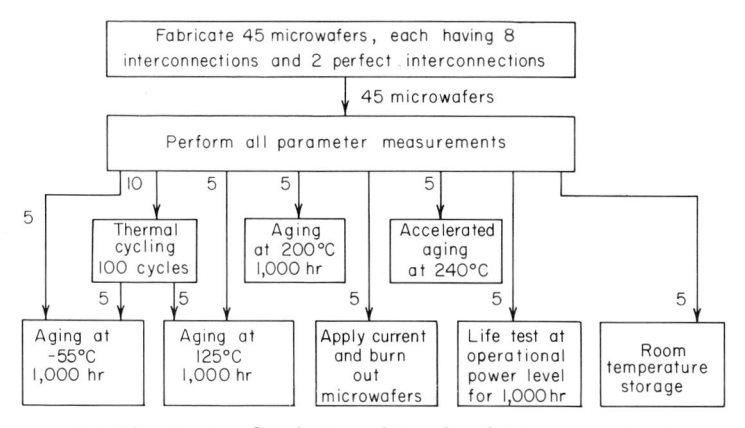

Fig. 39 Testing distribution of interfacial interconnections.[8]

100 thermal cycles from −65 to 125°C; accelerated aging at 240°C; and 1,000 hr operational life tests at a current level equal to 0.316 times the burnout value.

For all combinations of films, a sufficient number of interfacial interconnections were fabricated so that all the groups that were used for various testing were large enough to yield meaningful results. The minimum number of interfacial interconnections, on a statistical basis as well as on a scientific basis, is 305. There were eight interfacial interconnections and two monitors per microwafer; five microwafers were used for each test.

For each interconnection combination, 45 microwafers were selected and ultrasonically bonded to mother boards. The distribution of the 45 microwafers, for testing, is shown in Fig. 39.

The test pattern consisted of 10 butt-type in-line interfacial interconnections, each 0.010 in. square with 0.002-in. potential leads (see Fig. 40). The microwafer size is 0.290 in. long and 0.110 in. wide, allowing an 0.010-in. border around the pattern. One glass substrate yields 28 complete patterns and four resistivity monitor strips.

Except for sputtered tantalum, all conductor films were deposited in a 24-in. water-cooled stainless steel bell-jar system. An eight-position rotary-type mask changer was used.

Any one of eight substrates can be placed over any one of eight mechanical masks and the pair moved to any one of four evaporation positions. All motion is controlled from outside the system. The registration error of this changer is less than 0.001 in. The substrate holder can accommodate up to a 2- by 3¼ in. glass substrate. There are four substrate heaters, one over each source position. The complete system has an ultimate vacuum of 8×10^{-7} torr. The substrates used were an alkali zinc borosilicate Type 0211 Glass* with a thickness of 0.0075 to 0.0099 in.

Whenever possible, the thin-film interfacial interconnections were fabricated only

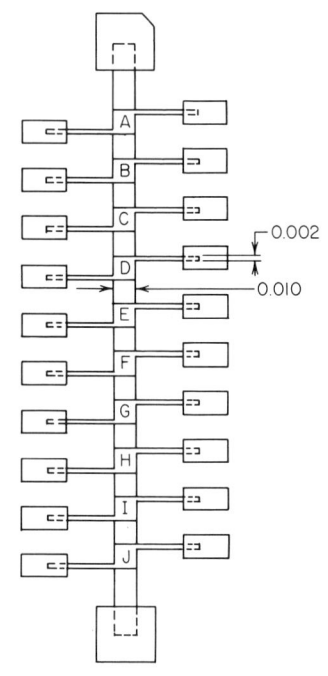

Fig. 40 Sketch of interfacial interconnection pattern.[8]

* Trademark of Corning Glass Works, Corning, N.Y.

13-58   Metals Joining of Electronic Circuitry

under vacuum conditions: i.e., the bottom conductor was only exposed to a vacuum environment before the deposition of the top conductor.

For the following interfacial interconnections:

    Type I-B. Chromium over chromium-gold
    Type I-C. Chromium-gold over chromium-gold
    Type I-D. Aluminum over chromium-gold
    Type I-G. Chromium-gold over chromium
    Type I-H. Aluminum over chromium

the following deposition sequence was used:
1. Deposition of the bottom conductor
2. Deposition of the top conductor
3. Deposition of the chromium-gold bonding pads

The fabrication of many interconnections in a single pump-down, having a small size and good registration, required the use of precision mechanical masks.

The starting material for the mask is a sheet of beryllium-copper 0.031 in. thick. Techniques employing photoresist, differential etching, and electroplating are employed to produce the completed mask. The wide lines in the pattern are 0.010 in. wide; the narrow lines are 0.002 in. wide.

The construction of the mask is such that the 0.031-in.-thick metal does not interfere with the vapor beam, but does form a structural support for the fine thin pattern on the other side of the mask.

The precision of these routinely fabricated masks has been established by many physical measurements using a measuring microscope having an accuracy of ±0.00005 in. In an area 3¼ by 5¾ in., all line widths down to and including 0.002 in. have an error of ±0.0002 in. More precise masks can be made. Masks having line widths of 0.001 in. have been fabricated with an error of ±0.001 in.

TABLE 22 Procedure Steps for the Fabrication of Tantalum-Aluminum Interfacial Interconnections[8]

Step No.	Type I-A, Ta-Al over Cr-Au	Type I-E, Cr-A over Ta-Al	Type I-F, Al over Ya-Al
1	Substrate cleaning.	Substrate cleaning.	Substrate cleaning.
2	Substrate cleaning.	Coat with resist.	Coat with resist.
3	Substrate cleaning.	Expose to ultraviolet.	Expose to ultraviolet.
4	Substrate cleaning.	Develop image.	Develop image.
5	Substrate cleaning.	Vacuum-bake.	Vacuum-bake.
6	Substrate cleaning.	Sputter tantalum.	Sputter tantalum.
7	Substrate cleaning.	Strip tantalum.	Strip tantalum.
8	Substrate cleaning.	Register to substrate holder.	Register to substrate holder.
9	Substrate cleaning.	Evaporate chromium-aluminum.	Evaporate chromium-aluminum.
10	Evaporate chromium-gold.	Evaporate chromium-gold.	Evaporate chromium-aluminum.
11	Evaporate chromium-gold.	Evaporate chromium-gold.	Evaporate aluminum.
12	Coat with resist.	Evaporate chromium-gold.	Evaporate aluminum.
13	Register pattern of top film.	Evaporate chromium gold.	Evaporate aluminum.
14	Expose to ultraviolet.	Evaporate chromium-gold.	Evaporate aluminum.
15	Develop image.	Evaporate chromium-gold.	Evaporate aluminum.
16	Vacuum-bake.	Evaporate chromium-gold.	Evaporate aluminum.
17	Sputter tantalum.	Evaporate chromium-gold.	Evaporate aluminum.
18	Evaporate chromium-aluminum.	Evaporate chromium-gold.	Evaporate aluminum.
19	Strip tantalum-aluminum.	Evaporate chromium-gold.	Evaporate aluminum.
20	Register lands.	Evaporate chromium-gold.	Evaporate aluminum.
21	Deposit chromium-gold lands.	Deposit chromium-gold lands.	Deposit chromium-gold lands.

Interfacial interconnections having tantalum-aluminum (actually tantalum-chromium-aluminum) were fabricated as shown in Table 22. A phased chromium-aluminum film is used instead of only aluminum because this latter film has poor adherence to tantalum once tantalum has been exposed to an etching and a photoresist stripping environment.

No protective coating was applied to any of the interconnections. The thickness of the thin-film conductors and dielectrics investigated are contained in Table 23.

TABLE 23 Thickness of Thin-film Materials[8]

Material	Thickness, Å
Chromium	1,300–2,000
Chromium over gold	100–400 Cr
	4,000 ± 500 Au
Aluminum	3,000 ± 300
Aluminum over tantalum	1,500 Al
	1,500 Ta
Silicon monoxide	5,000 ± 500

Five of the eight thin-film combinations are acceptable and are recommended for use. There is little difference between the overall merit rating of these five types. The five acceptable combinations are:

Chromium over chromium-gold
Chromium-gold over chromium-gold
Aluminum over tantalum-aluminum
Chromium-gold over chromium
Aluminum over chromium

The three unacceptable combinations are:

Tantalum-aluminum over chromium-gold
Aluminum over chromium-gold
Chromium-gold over tantalum-aluminum

These three are not recommended for use.

Crossovers. Thin-film insulating materials that prevent two thin-film conductors from making electric contact at a crossover are an important part of an overall thin-film connection system. A failure at a crossover will cause failure, or performance degradation, of the electronic device. Several thin-film crossover patterns and various measuring and analytical techniques have been devised. The quality of the crossover depends upon its electrical parameters and its physical characteristics. Based on this investigation, several criteria for the evaluation of any thin-film crossover have been established.

The electrical parameters studied were the dc insulation resistance, the 1,000-Hz capacitance, the 1,000-Hz dissipation factor, the dc breakdown voltage, and the withstanding voltage. The physical aspects studied were the adhesion of the film to its substrate and the compatability (or stability) of the film combinations. The above crossover properties were studied under various environments. The environments were storage at −55, 125, and 200°C for 1,000 hr; 100 thermal cycles from −65 to 125°C; accelerated aging at 300°C; and under high humidity with applied voltage.

Figure 41 shows the layout of a typical crossover pattern. Crossovers were deposited in four operations: lower-conductor film, dielectric film, upper-conductor film, and bonding pads. The pads that terminate the conductors were evaporated chromium-gold. Chromium-gold is used to facilitate bonding of the microwafer to the mother board.

The bottom conductors were 10 mils wide, deposited 20 mils apart, and connected together at each end. Top conductors were also 10 mils wide. The dielectric layer was evaporated over the lower conductors to the desired thickness. Then

13-60   Metals Joining of Electronic Circuitry

the upper conductors were deposited, spaced 15 mils apart, followed by the bonding pads.

For all the combinations a sufficient number of crossovers were fabricated so that all the groups (that are used for various tests) are large enough to yield meaningful results. The number of crossovers meeting this requirement on a statistical basis as well as on a scientific basis, were 40 microwafers. Since there are 18 crossovers per microwafer, 3 microwafers for each test are required.

For each crossover combination, 44 microwafers were selected and ultransonically bonded to their mother boards. The distribution of the 44 microwafers, for testing, is shown in Fig. 42. The deposition process for fabricating the various thin films was described in the preceding subsection.

Except in the case of crossovers having tantalum-aluminum or thick chromium-copper as one of the conductors, the crossover fabrication was done in a single pump-down. Because of this, the bottom-conductor–dielectric interface and the dielectric–upper conductor are only subjected to a vacuum environment.

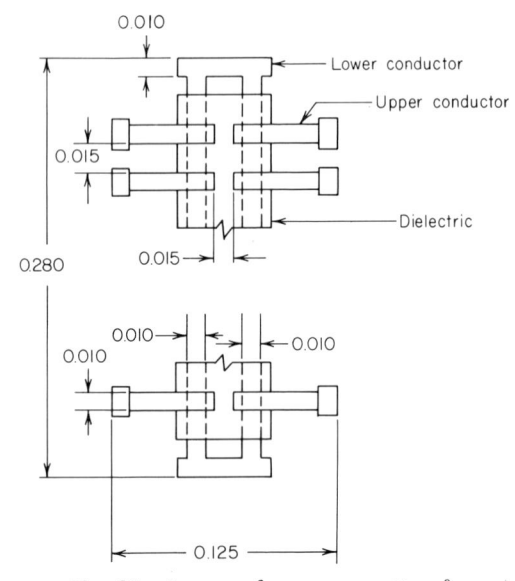

Fig. 41   Layout of crossover pattern.[8]

The evaporation rate and thickness of the dielectric and all the conductors (except tantalum) were controlled and recorded. The fabrication sequence was conventional, being:
1. The bottom conductor
2. The dielectric
3. The top conductor
4. The chromium-gold bonding pads

No protective coating was applied to any of the crossovers. Preliminary humidity and high-temperature experiments indicated that no such protective coating was necessary. The extensive environmental testing carried out definitely proves this to be true.

The first series of crossovers fabricated had dielectric films ranging in thickness from 5,000 to 16,400 Å. Table 24 lists these various crossover combinations and the associated dielectric thickness.

However, after initial tests, the dielectric thickness was chosen to be 5,000 Å for all crossover combinations. In the majority of cases in thin-film circuits where crossovers are used, the layout can be arranged so that dc power paths cross over other dc paths or grounds. In this case, the capacitance of a 0.010- by 0.010-in. crossover (~8 to 9 pf) with a dielectric thickness of 5,000 Å has a negligible effect on the circuit performance. In a situation where the crossover capacitance is important, an increase to twice the dielectric thickness (for the same area) reduces the capacitance only by one-half. This is usually not enough

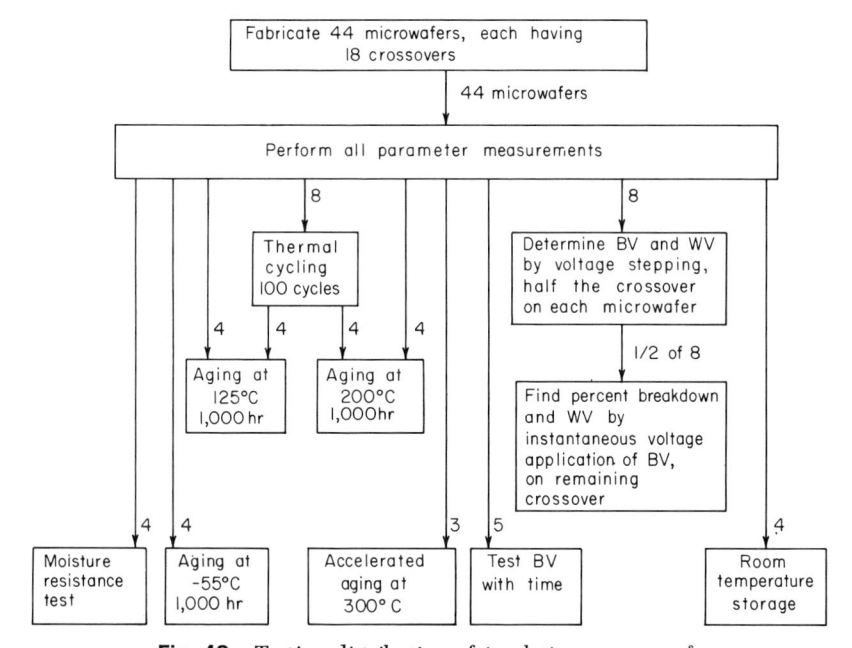

Fig. 42 Testing distribution of insulating crossovers.[8]

TABLE 24 Preliminary Crossover Types and Dielectric Thickness[8]

Crossover type*	Dielectric thickness, Å
II-A (Cr-Au/SiO/Cr-Cu)	5,000
II-B (Al/SiO/Cr-Cu)	5,000
II-C (Cr-Au/SiO/Ta-Al)	5,500
II-D (Al/SiO/Ta-Al)	16,400
II-E (Cr-Au/SiO/Cr)	5,000, 10,920
II-F (Al/SiO/Cr)	5,300
II-G (Cr-Au/SiO/Cr-Au)	5,400, 8,500
II-H (Al/SiO/Cr-Au)	9,600
II-I (Cr-Au/SiO/Al)	5,400
II-J (Al/SiO/Al)	5,400, 11,500

* Top conductor/dielectric/bottom conductor.

of a decrease to be effective. Crossovers of 5,000 Å have good dc voltage-breakdown values and reasonable fabrication yields, and process condition difficulties are at a minimum.

Crossovers manufactured with dielectric thicknesses less than 5,000 Å would have initial yield problems; process conditions would become more stringent; and the crossovers would more likely be damaged in the dicing and bonding process.

13-62　Metals Joining of Electronic Circuitry

The crossover combinations considered undesirable were eliminated, based on the results of preliminary environmental tests or on the physical degradation of the films.

Crossover types II-A (Cr-Au/SiO/Cr-Cu) and II-B (Al/SiO/Cr-Cu) were eliminated because of the extreme thickness of the lower chromium-copper electrode (50,000 Å). Samples were fabricated by initially evaporating a layer of phased-in chromium-copper (~5,000 Å) and then electroplating copper on the evaporated substrates to the desired thickness. The lower-conductor pattern was formed by etching, followed by an electropolishing to remove all sharp edges. The dielectric and top conductors were fabricated in the usual manner.

These crossover samples were unacceptable because the dielectric layer and the top conductor peeled immediately from the lower conductor.

Crossover types II-C (Cr-Au/SiO/Ta-Al) and II-D (Al/SiO/Ta-Al) were eliminated because of the poor adhesion of the lower aluminum to the tantalum. The adhesion values of these films were considerably lower than those of the other film combinations, and the adhesion values were also dependent upon minor varia-

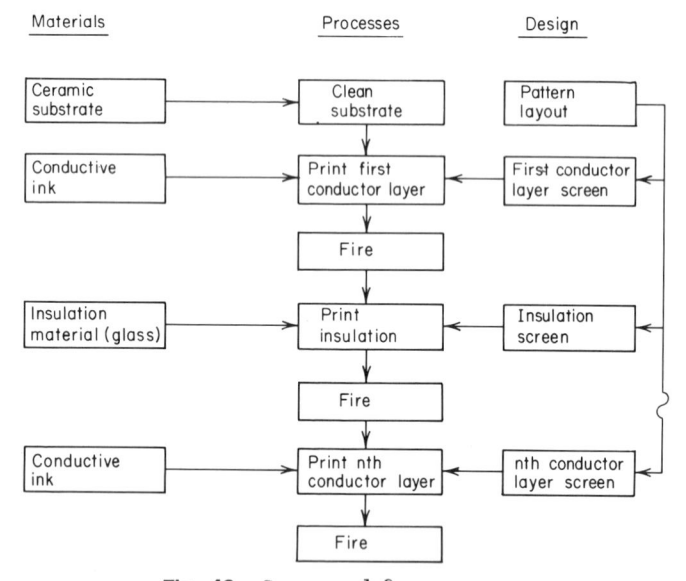

Fig. 43 Screen and fire process.

tions in the process conditions. Secondly, the use of a sputtered tantalum bonding film for the lower conductor involved an etching technique to form the pattern which resulted in alignment problems when the remainder of the film combinations were evaporated.

The last crossover combination, type II-F (Al/SiO/Cr) was rejected on the basis of its performance in the preliminary environmental tests. At that time, the results indicated that this combination was unstable. In addition, it was thought that chromium would not likely be used as a conductor material because of its high resistivity.

The remaining five crossover combinations were found to be acceptable after testing, and are listed below in their order of merit:

Al/SiO/Al
Cr-Au/SiO/Cr
Al/SiO/Cr-Au
Cr-Au/SiO/Al
Cr-Au/SiO/Cr-Au

First-level Joining, Materials and Processes   13-63

Thick films   There are two basic techniques for the fabrication of multilayer thick-film ceramic laminates. The first process involves the sequential layering of insulation and conductor, each operation being followed by a firing operation. Figure 43 is the flow diagram of such a process.

The second process involves the screening or printing of conductors onto layers of green ceramic (unfired ceramic in sheet form held together with binders), followed by the lamination of the printed green layers and a single firing operation. Figure 44 is the flow diagram for such a process.

The properties of the various ceramics used are contained in Table 25, and of the various glasses in Table 26.

Sequentially Fired Ceramics.   Ninety-six percent alumina is the most widely used substrate material. This material possesses a balance of physical and electrical properties which make it compatible with the resistor, capacitor, and conductive pastes commonly used. Other high-alumina bodies and other ceramics are available and also used, but in choosing the substrate, consideration must always be given to possible chemical effects. The conductor inks used in thick-film work are based

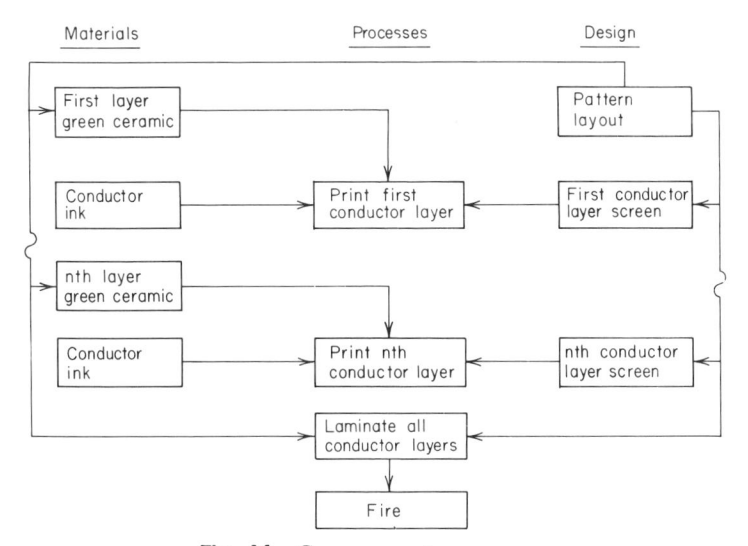

Fig. 44   Green ceramic process.

primarily on metal-glass mixtures. Compositions containing palladium-silver, gold-platinum, silver, or platinum are available. In order to meet the screening requirements, these ingredients are suspended in an organic carrier, such as butyl Cellosolve* acetate, butyl Carbitol* acetate, or terpineol. The conductor formulations consist of a finely divided suspension of the metal powders and glass in the organic vehicle (added to render the ink screenable). After proper drying and firing, the ink forms a matrix of metal particles in glass with the glass binding the agglomerate to the substrate. Table 27 lists the properties of some high-resolution inks.

The circuit pattern is generated on the screen by a photosubtraction method. A photosensitive emulsion is applied to the screen which, when exposed through a photographic negative to ultraviolet light, allows the circuit portion to be subtracted from emulsion by simple washing. The development process leaves clear areas on the screen, through which the ink can be deposited on the substrate.

Screens of 200 mesh are usually employed to deposit conductors. When narrow conductor paths are required, 5 mils or less, a finer mesh size of 270 to 325

---

* Trademarks of Union Carbide Corporation, New York, N.Y.

13-64   Metals Joining of Electronic Circuitry

TABLE 25   Properties of Ceramic Substrate Materials[29]

| | Dense alumina ||| Dense beryllia ||
Ceramic type	35 Al_2O_3	94 Al_2O_3 + CaO + SiO_2	96 Al_2O_3 + MgO + SiO_2	93 BeO	99.5 BeO
Code number*	576	719	614	735	754
Softening point, °C	1100	1500	1550	1600	1600
Thermal expansion coefficient, in./(in.) (°C) × 10^6	6.5	6.2	6.4	6.1	6.0
Thermal conductivity at 25°C, cal/(cm-sec)(°C)	0.060	0.073	0.084	0.50	0.55
Density, g/cc	3.40	3.58	3.70	2.90	2.88
Dielectric constant at 25°C, MHz	8.3	8.9	9.3	6.3	6.4
Loss tangent at 25°C, MHz	0.0058	0.0018	0.0028	0.0006	0.0006
Log volume resistivity at 300°C, ohm-cm	10.7	12.8	10.0	13.8	>14
Dielectric strength at 25°C and 60 Hz, volts/mil	230	230	230	255	260

* Code numbers are of AlSiMag Brand Ceramics, American Lava Corp., Chattanooga, Tenn.

may be required to obtain the necessary definition. After screening, the solvents are evaporated by heating to 100°C for about 15 min. Where continuous operation is desired, a bank of infrared heat lamps or a low-temperature belt furnace can be used to dry the ink just before high-temperature firing. This automated drying can, of course, be used with any printed deposit.

TABLE 26   Properties of Glass Substrate Materials[29]

Glass type	Soda lime	Alkali zinc boro-silicate	Lime aluminosilicate, alkali-free	Barium alumino-silicate, alkali-free	Alkali boro-silicate	96% silica	Fused silica
Code number*	0080	0211	1715	1723	7059	7740	7900	7940
Annealing point, °C	512	542	866	710	650	565	910	1050
Softening point, °C	696	720	1060	910	872	820	1500	1580
Thermal expansion coefficient, in./(in.)(°C) × 10^6	9.2	7.2	3.5	4.6	4.5	3.25	0.8	0.56
Thermal conductivity at 25°C, cal/(cm-sec)(°C)	0.0023			0.0032		0.0027	0.0038	0.0034
Density, g/cc	2.47	2.57	2.48	2.63	2.76	2.23	2.18	2.20
Dielectric constant at 25°C, MHz	6.9	6.6	5.9	6.4	5.8	4.6	3.9	3.9
Loss tangent at 25°C, MHz	0.01	0.0047	0.0024	0.0013	0.0011	0.0062	0.0006	0.00002
Log volume resistivity at 250°C, ohm-cm	6.4	8.3	13.6	14.1	13.5	8.1	9.7	11.8
Dielectric strength at 25°C, kv (rms)	0.35	2.	>10	>10	>10	2.	7.	>10
Weatherability, g/cm^2	>5.0	0.05–0.25	<0.01	<0.01	<0.01	0.05–0.25	<0.01	<0.01
Chemical durability, mg/cm^2:								
In 5% HCl for 24 hr	0.02	0.03	0.10	0.4	5.5	0.005	0.001	0.001
In 5% NaOH for 6 hr	0.5	2	1.2	0.3	3.7	1.1	1.1	0.7
In 0.02 N Na_2CO_3 for 6 hr	0.1	0.1	0.15	0.1	0.3	0.1	0.03	0.03

* Code numbers of Corning Glass Works, Corning, N.Y.

TABLE 27 Properties of Fine-line Printing Compositions[30]

Sample	Pd-Au	Pt-Ag	Pt-Pd-Au
Solids, %.	90	85	90
Viscosity (Brookfield) at 10 rpm, poises.	5,000	6,000	5,000
Spreading of 5-mil line, %.	<10	<10	<10
Conductivity, ohms/(in.)(5 mil).	5	4	10
Metal migration rate.	None	6–20 sec*	None
Adhesion,† lb.	2.5	4.0	2.0
Fissuring.	None	None	None
Soldering (Pb-Sn w/(rosin)).	OK	OK	OK

* Water-drop test: 600 volts/in. 5-mil gap at 3 volts dc—time to short circuit. The rate for silver is <0.5 sec.
† Peeling pull: 7,553 = 1.8 lb.

The firing cycle should be designed to optimize the conductivity, adhesion, and solderability of the conductor. Firing temperatures range from 500 to over 1,000°C, depending upon the ink used and the final application of the conductor.[31]

One of the problems with sequentially fired ceramic multilayers, is trying to fit a number of firings into the same thermal range. The metal-belt furnaces are limited to 1100°C. The best glass properties are obtained with glasses that fire from 500°C and up. This allows a total thermal range of 600°C. As glasses are fluids with a wide thermal softening range, one glass tends to intermix across the interface with another unless a 75 to 125°C temperature difference results. Therefore, if the total temperature range is from 500 to 1100°C, only five to six firing levels are available. Thus, for more than six layers some other method must be found. One method involves cofiring of several layers. This, however, causes problems in maintaining dimensional control of the small via windows through the dielectric required to make interlayer conductor connections. A second method involves the use of a devitrified glass which changes phase when heated above a peak temperature. The glass crystallizes and will not remelt until the temperature is 100°C or more higher than the original melting point.

"Green" Ceramic Multilayers. Most of the data and processes concerning this technique are proprietary. A recent paper,[32] however, contained an excellent description of the process, part of which is reproduced here.

Most ceramics have been fabricated into flexible green sheets. Other materials also fabricated were glasses and metals. The basic criteria for being processed appears to be whether the chemicals are available in a fine powder state and capable of being sintered to the desired physical condition.

The sheet-making process consists of preparing a paintlike suspension of the powder with suitable resins, plasticizers, wetting agents, and solvents. The chemical composition of the powder, its particle-size distribution, and its surface area are important variables. Specific binders used vary with investigators and applications. Examples are the polyvinyl chlorides, polystyrenes, methacrylates, polyvinyl alcohols, polyvinyl butyrals, etc. The plasticizers and wetting or deflocculating agents are only sometimes used, when necessary to meet specific processing conditions. In most cases organic solvents are used. Aqueous systems have not found wide acceptance.

The mixing or preparation of the suspension is generally done in a ball mill. Deairing of the suspension is important to minimize pinholes or porosity in the final product.

The most common method of forming the green sheets, with thicknesses of less than 1 to 10 mils, is doctor blading. This process is a controlled spreading of the suspension on a carrier. Other possible methods include centrifuging, extrusion, roll coating, and calendering.

Important process variables for doctor blading are the casting rate, rheology of the suspension, carrier, and release from this carrier. Typical carrier materials are glass, steel, Mylar,° Teflon,° etc. The bottom green sheet surface closely images that of the carrier. The upper surface of the doctor-bladed sheet is controlled by other factors, and can be detected by differences in glossiness. Drying can be accomplished with ovens, infrared heaters, air blowers, etc., but must be carefully controlled in any process. The green sheets have a relatively long shelf life. Accordingly, they offer a good time in the process to inventory stock. There are companies today who market green sheets or tapes. The green sheets can be made to vary from a paperlike to oilcloth or leathery condition. Their important feature is that shapes can be punched or stamped to very close tolerances by high-speed machines. After the cast sheets are inspected for defects and uniformity of green density, then working lengths, or squares, can be punched even with hand tools. Registration holes can also be made in the sheets which can provide for the positioning of these parts through all subsequent operations. The tooling should be standardized with suitable alignment pins. Different size and shape holes can also be used for making subsequent electric connections through the layers or for leaving cavities in the monolithic structure. The former are called *vias*.

The principle method of metallization of the green ceramic sheets is screening, similar to that used in the electronic industry for sintered ceramics. Special plate chucks can be used to hold the green sheets down. The registration holes position them in regard to the screening masks. Sometimes a vacuum chuck is used, which can also help the screened metal paste fill the via holes to make the electric connection through the sheet.

Reproducible results with screening require close control of the paste properties (e.g., viscosity, specific gravity, etc.), screens, and processes. Electric connections have been achieved in three-dimensional networks. This capability is important to future packaging and interconnection systems.

There are other methods for applying the metal powders. They can be made into green sheets and punched, similar to the method described for the ceramics. Some suspensions have been sprayed, and pure metals evaporated through masks onto the sheets. The metal patterns have also been first deposited on the carrier, and then the ceramic sheet has been formed over them. This sequence produces the metallization with a flush surface, embedded in the ceramic. In the other methods the metal patterns are slightly raised. The thermoplastic sheets conform to these patterns, however, during the subsequent lamination step.

The selection of the metal is vital, in that it must sinter compatibly with the ceramic. Palladium has been used widely with titanates, platinum with some ferrites, molybdenum or tungsten with aluminas or other porcelains. In each case, the primary metal may contain minor additives to optimize the wetting and sintering. In general, the refractory metals are most desirable for high electric conductivity and are required for hermetic sealing. The noble metals are used when an oxidizing atmosphere is required for sintering of the ceramics. If the thermal or electric conductivity of the metal phase is to be optimized, the metal phases should be near their theoretical density. With the conventional metallization techniques currently employed in the industry, whereby the internal conductors are created at the time of densification of the ceramic, the thermal expansion of the ceramics and metals should match as closely as possible in order to minimize internal strains. Designs with more ductile metals or very thin metal sections can tolerate greater mismatches. Heavy concentrations of metal in a multilayer ceramic structure during the sintering operation can alter the sintering cycle and shrinkage of the composite from that of either constituent. It is necessary, therefore, to make trial parts in determining the actual shrinkage of a complex design. Once the shrinkage of the laminate is established, however, it has been controlled to close tolerances, in the order of less than a percent. The lamination process is applying pressure to several stacked green sheets at a suitable temperature, to produce a monolithic

° Trademarks of E. I. du Pont de Nemours & Co., Wilmington, Del.

structure. The temperatures used have ranged from room temperature to several hundred degrees Fahrenheit. The pressures required vary inversely with the temperature and as a function of the type and amount of binder. A typical range is 200 to 20,000 psi. The interfaces between like materials are not detectable after proper lamination.

To ensure complete lamination, the entire surface of the part should receive equal pressure. Highly polished, parallel platens or conforming materials are sometimes used for this purpose. The surface of the laminate can be altered or embossed by the surface conditions of the laminating tools. Sufficient time must be allowed for all the layers to reach the process temperature. The parts are sometimes slightly densified during this process. This value must be taken into consideration if precision dimensional control is required.

Up to this point, many individual parts are usually processed in a batch on a large single laminated sheet, or on a continuous roll. The previous operations of punching, metallizing, and laminations may be repeated on the same material several times to obtain unique designs. For example, a cavity in the ceramic is produced by laminating a sheet with a hole to a solid sheet. Vias can go completely through a laminate or be "blind" within the structure. The individual parts are usually removed from the sheet in the step before sintering. They can be stamped as described before, which permits the fabrication of rather intricate shapes very cheaply. Straight-sided pieces have been sectioned with conventional cutoff saws.

The sintering operation involves heating the part to a specific temperature, time, and atmosphere program in order to remove the binders, complete any chemical reactions, densify the structure, complete the bonds between phases, control the grain and pore sizes, and establish the residual stresses. The thermodynamics of a given system should be thoroughly understood in order to control its manufacture. In this process, the materials shrink uniformly, as part of the densification, which is controlled in a manner similar to that of any other ceramic or powder metallurgical processing. The parameters include particle size, amount of binder, powder characterization, phase changes, heating cycles, etc. If the shrinkage of the metal does not approach that of the ceramic, the part may be weakened or deformed during sintering. In addition to the above-listed parameters, uniformity of heating, purity of materials, accuracy of controls, and the techniques of handling parts contribute to the success of the product. After this step, the laminate has the brittle characteristics of the ceramic used, but strengthened by the metals enclosed. It is now ready for further processing into the specific component being made.

The requirement that the internal-conductor metallurgy sinter compatibly with the ceramic can impose severe performance restrictions on the ceramic-metal composite. One must make a compromise between the properties of the ceramic and of the metals and choose a system that usually does not allow for maximum performance of either material. For example, if the application requires the strength and thermal conductivity of a 99 percent alumina ceramic, the metallurgies available are restricted to those that have melting points above 1500°C, the lower end of the sintering range for these types of ceramics. Thus, the need for desirable ceramic properties necessitated the selection of metals with less than optimum electric conductivity. Other compromises may be necessary with respect to maximum adhesion, hermeticity, metal migration, and the like.

Figure 45 shows comparisons of results actually obtained with screened palladium, screened molybdenum-manganese, and copper-filled molybdenum conductors. A reference curve is also plotted for the theoretical values of a solid copper line. A curve is also shown for the expected limiting values for the molybdenum-copper system. The copper-filled conductivities should not match that of pure copper, because of the volume required for the molybdenum and its interfacial regions. All copper lines in the laminated ceramic have been difficult to achieve. The important contribution of the capillary-fill technology to laminated ceramics at this time is the more than twofold improvement in conductivity over that of standard refractory and noble metals.

13-68 Metals Joining of Electronic Circuitry

Comparison between thick- and thin-film technologies In conclusion, Table 28 compares the differences that distinguish thick-film and thin-film techniques from each other.

Multilayer Printed-circuit-board Internal Connections

Three types of multilayer printed-circuit boards are commonly used. These are the clearance-hole type; the plated-up or sequential type; and the plated-through

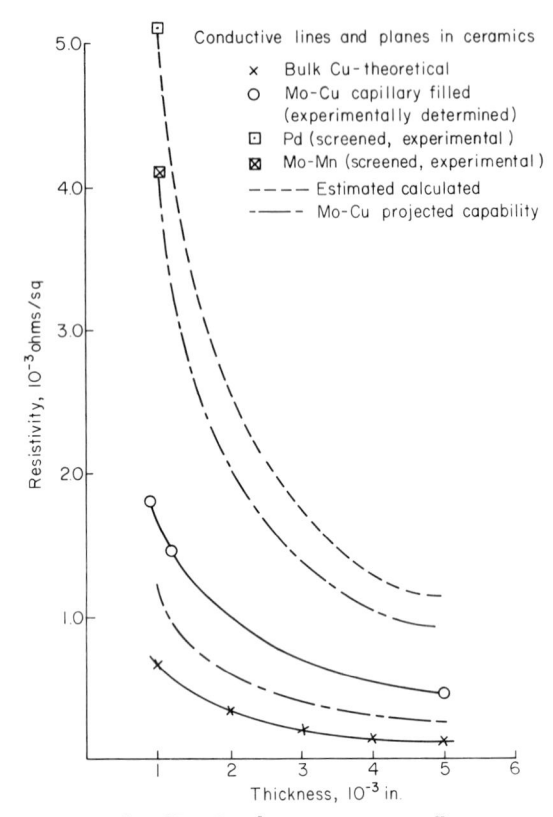

Conductive lines and planes in ceramics
- × Bulk Cu-theoretical
- ○ Mo-Cu capillary filled (experimentally determined)
- ▢ Pd (screened, experimental)
- ⊠ Mo-Mn (screened, experimental)
- ---- Estimated calculated
- —·— Mo-Cu projected capability

Fig. 45 Conductor resistivities.[32]

TABLE 28 Comparison of Thick and Thin Films

Technique	Thick film	Thin film
Substrate materials	Ceramics (fired and "green")	Glasses Ceramics
Conductor materials	Noble metals (Pd, Pt, Au, Ag)	(1) Aluminum (2) Gold
Insulating materials	(1) Glasses (2) Green ceramic	Oxides (SiO, SiO_2, Al_2O_3)
Pattern delineation	Screen	(1) Stencil (mask) (2) Etch
Deposition technique	Screening and firing	(1) Vacuum evaporation (2) Sputtering
Film thickness	0.5–10 mils	500–10,000 Å (2 to 40 μ in.)

TABLE 29 Interconnection Methods for Multilayer Printed-circuit Boards[36]

Characteristic	Clearance hole	Plated-through hole	Plated-up risers
Reliability of joint	Excellent	Fair	Excellent
Single- or double-sided circuits	Single	Both	N.A.
Soldered circuits	Easy	Easy	N.A.
Welded circuits	Easy	Hard	N.A.
Number of layers, maximum	8	20	8
Registration	Not critical	Not critical	Critical
Hole drilling	Not critical	Not critical	N.A.
Mechanical strength	Good	Good	Fair
Environmental resistance	Poor	Fair; good if conformal-coated	Excellent
Number of process steps	Few	Many	Many
Packaging density	Poor	Good	Excellent
Weight	Heavy	Light	Light
Thickness	Thick	Thin	Thin
Sockets or added terminals	Easy	Easy	Hard
Flexible circuits	Easy	Easy	Hard
Design changes	Easy	Hard	Hard
Cost	Low	Low	High
Production time	Low	Low	High
Automation possible	Yes	Yes	Yes
Equipment available	Yes	No	No
Visual inspection	Easy	Not possible	Easy
Repair	Easy	Hard	Hard
Proprietary process	No	Yes	Yes

N.A. Not available.

hole type. Multilayer boards normally use copper as conductors and epoxy-impregnated fiber glass as insulation. Table 29 gives a qualitative comparison of the three types.

Clearance-hole method This type of multilayer board (MLB) is made by laminating together single-sided printed-circuit boards which have clearance holes drilled for interconnections between layers. The clearance holes must be sufficiently large so that solder connections can be made directly to conductor termination pads at whatever layer a connection is required. A solder joint must be made down inside the clearance hole, since there are no plated interconnections. A small-tip soldering iron must be used, and each clearance hole made sufficiently large to accommodate the soldering-iron tip.

As each layer is added to the stack, clearance holes must be made for every termination beneath that layer, resulting in less and less board area available for conductor patterns as the number of layers is increased. Thus, each added layer is less effective than the previous one.

Figure 46 is a sketch of a cross section of the clearance-hole technique. Table 30 lists the design parameters of this technique.

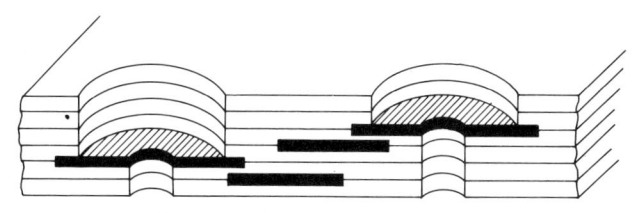

Fig. 46 Clearance-hole technique.

TABLE 30 Design Parameters for Clearance-hole Multilayer Boards[37]

Parameter	Description
Number of layers	By design option
Minimum distance between pad centers	0.150 in.
Minimum pad diameter	0.125 in. (0.100 in. hole in 8-layer board)
Minimum conductor width	0.015 in.
Minimum conductor spacing	0.015 in.
Minimum insulating material per layer	0.003 in.
Means of external interconnection	Pin or hardware
Ratio of hole diameter to depth	1:½
Thickness of wiring layer film	0.0025 in.
Thickness of copper foil	0.0028 in. (2 oz)

The plated-up MLB This type of multilayer board is constructed one layer at a time starting with a core board or from top to bottom. Since interlayer connections are made layer by layer, board area for interconnections is only on the layers where the interconnections are made. The top and bottom faces of the board can be reserved for component mounting only, and internal layers provide more area per layer for conductors. The total number of layers is considerably reduced. The cost per layer is relatively independent of the number of interconnections per layer, and so interconnection density can be high on each layer without cost penalty.

The core-board technique starts with a conventional two-sided board. The sequential layers are typically made of 1-oz copper foil over an epoxy-impregnated fiber glass insulation. After single top and bottom layers are laminated to the core board, interlayer connections are made by chemically etching blind holes through the new layers and then plating the holes to make the connections between layers. As each pair of layers is added, the laminate serves as the core board for the new pair of layers.

A variation on the plated-up MLB initiates the fabrication process by chemically drilling holes in the core board. These holes are copper-plated and filled flush with the board surface with conducting or nonconducting resin. The entire surface of the board including the surface of the resin is copper-plated. Then the core board is etched with the desired circuit pattern. This hole filling and plating procedure gives conductors uniformity of width over a hole location and thus maintains continuity in the circuitry on the board face. The next fabrication step is to laminate single layers to top and bottom and repeat the procedure: etch the holes, copper-plate the holes, fill the holes, plate the exposed surfaces of the new layers, etch the circuitry for the new layers, and repeat until all interconnections are made.

One advantage of this multilayer concept is that holes need not be completely through the board, and in such cases internal conductors may cross these sites. Also, connections from one layer to another can be made at any desired point internally. These features provide greater design freedom. A cross section of a typical board of this type is shown in Fig. 47.

Layers may be inspected as they are made. The process does not require holes, and because no space is used for holes, a greater wiring density is theoretically possible. The wiring is one continuous three-dimensional structure. The process is relatively expensive because of complex processing.

The plated-through-hole method The plated-through-hole (PTH) multilayer board is made by stacking preetched internal and solid external layers with uncured epoxy–fiber-glass sheets between layers for insulation. The layers are precisely registered, and the stack is placed in a laminating press. Carefully controlled heat, pressure, and cooling cycles are used to fuse the stack into a single solid board. Holes are drilled at all locations where interlayer connections are required, usually

First-level Joining, Materials and Processes 13-71

by numeric tape-controlled drill presses. Each drilled hole leaves an annular ring of copper wherever a copper pad was left for an interlayer connection. The holes and unetched copper top and bottom surfaces are cleaned, sensitized for conductivity, and plated. The required conductor patterns on top and bottom surfaces are then etched.

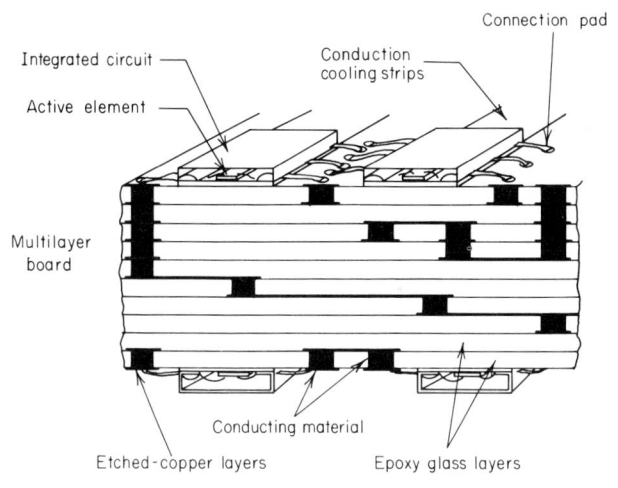

Fig. 47 Plated-up technique.[12]

The PTH MLB is the most widely used at the present time. It suffers from a similar decrease of available area per layer as the clearance-hole-type MLB, since each plated-through hole uses area on every layer. However, the plated-through holes can be considerably smaller than clearance holes. A cross section of a PTH MLB is shown in Fig. 48. The recommended dimensions and tolerances

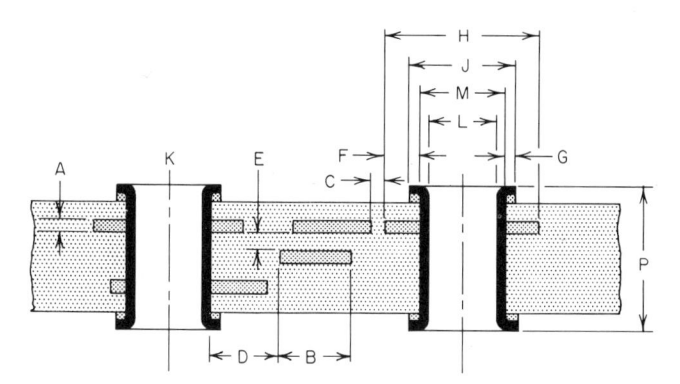

Fig. 48 Plated-through-hole MLB. (NOTE: Refer to Table 30 for dimensions.)

of the various parts of PTH MLB, three classes, have been prepared by the Institute of Printed Circuits (IPC) and are included in Table 31.

Eight variations of PTH and plated-up (sequential) are shown in Fig. 49. Table 32 lists the number of fabrication operators involved in each of these eight cases.

In the plated-through hole technique, hole drilling, hole cleaning, and hole plating are critical processes, all of which affect the reliability of the MLBs.

Metals Joining of Electronic Circuitry

TABLE 31 Dimensions and Tolerances for IPC-ML-910[38]

Symbol	Characteristic	Class I	Class II	Class III
A	Conductor thickness and tolerance	IPC-CF-150	IPC-CF-150	IPC-CF-150
B	Conductor width	0.020 min.	0.010 min.	0.005 min.
B_{tol}	Conductor-width tolerance	+0.004 / −0.005	+0.002 / −0.004	+0.001 / −0.002
C	Coplanar conductor spacing	0.015 min.	0.010 min.	0.005 min.
D	Conductor-to-hole spacing	0.010 min.	0.010 min.	0.010 min.
E	Layer-to-layer spacing	0.004 min.	0.003 min.	0.002 min.
F	Internal-layer annular ring	0.001 min.	0.001 min.	0.001 min.
G	External-layer annular ring	0.002 min.	0.002 min.	0.002 min.
H	Internal-layer terminal area	0.044 min.	0.034 min.	0.024 min.
J	External-layer terminal area	0.046 min.	0.036 min.	0.026 min.
K	Hole-location tolerance	0.014 diam.	0.010 diam.	0.006 diam.
		(For board dimensions up to 6 in.)		
L_{tol}	Unplated-hole diameter tolerance	0.004 / 0.008	0.002 / 0.006	0.002 / 0.004
M_{min}	Plated-hole diameter, min.	$\frac{1}{3}$ P_{max}	$\frac{1}{4}$ P_{max}	$\frac{1}{5}$ P_{max}
M_{tol}	Plated-hole diameter tolerance	L_{tol} + 2 × min. hole plating		
N	Layer registration	0.014 diam. TP	0.012 diam. TP	0.010 diam. TP
		(For distances within a 6-in. square)		
P_{tol}	Board thickness tolerance	±10% of P_{nom} and ±0.007 min.		

NOTE: All dimensions and tolerances are in inches unless otherwise specified. TP indicates time position.

Hole Drilling. The drilling of glass laminates is quite different from drilling homogeneous materials. In glass laminates, the fibers usually cross each other resulting in the intersections being harder than the rest of the board. When the drill hits the fiber or intersections, it tends to spot off-center. This will result in walkout unless the drill bit is closely held by the bushing until the drill has penetrated the fiber.

Uncured resin will also affect the drilling error: the gummy material tends to clog the flutes, resulting in the drill becoming chip-bound. Also, epoxy resin becomes smeared over the circuit connections, causing electric open circuits in the boards.

Hole Cleaning. To remove the epoxy and glass fibers that may be smeared over the copper in the drilled hole, an etch-back process may be used. Etch-back is the chemical process of removing glass fibers and epoxy between conductor layers for a given distance so that, after plating, a three-surface mechanical bond will be effected on the protruding portion of the conductor materials. The process is usually accomplished by a chemical immersion followed by a neutralizing process. It is recommended that the epoxy-glass be etched back a minimum of 0.0003 in. and not more than a maximum of 0.002 in.[40] A mixture of hydrofluoric acid, sulfuric acid, and trifluoroacetic acid has been used by one investigator as a back-etch.[41] By limiting the treatment to 2 min, the above constraints could be complied with. However, there are considerable hazards and dangers in handling

First-level Joining, Materials and Processes   13-73

Eight ways to make multilayer boards. Horizontal lines are conductor layers; rectangles are drilled holes. Solid blocks indicate layers and holes made by the sequential process. The lettering code for the holes is: A, intraconnections formed before lamination; B, C, and D, intraconnections formed after the first, second, and third laminations, respectively; X, layers that cannot be connected as single pairs in the laminate; Y, pair connections prohibited in subassembly laminates. Table 32 lists major fabrication steps for each kind of board.

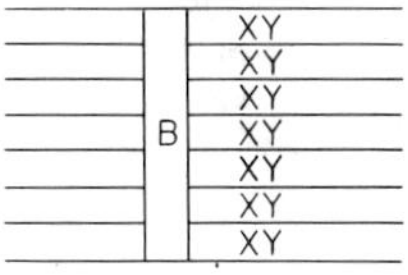

1. Simple through-hole board is lowest in interconnection density because each hole requires space in all layers. It is the simplest to design and make, but each hole location can be used for only one intraconnection, such as a top-to-bottom connection.

2. Through-hole boards with intermediate connections. Boards with single-sided external layers (left) are preferred when leads are to be joined in the through holes. Such boards are used to interconnect several other boards that are plugged into connectors on the through-hole board. The type with two-sided external layers (left) is preferred when leads are to be joined in the through pads on the board surface because with efficient layout the top two layers on either side can be used for as much as 50 percent of the interconnections.

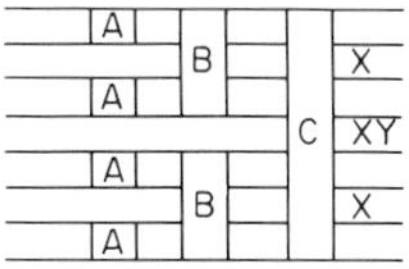

3. Through-hole board made by laminating subassemblies of through-hole boards and then drilling holes through the completed laminate. This allows two connections to the innermost planes in the same amount of space a through hole requires.

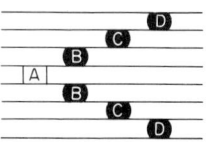

4. Sequential board gives the highest interconnection density and the greatest freedom in hole and conductor positioning. Intraconnections can be superimposed to provide the equivalent of through holes and through-hole subassemblies.

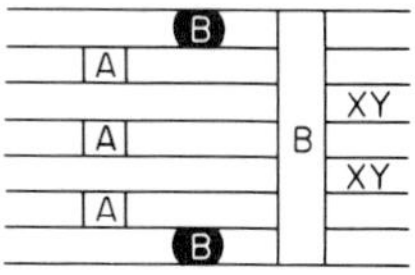

5. Combination board. The three two-sided core boards and the two single-clad outer layers are assembled in a single lamination. After lamination, both the outer connections and the through holes are made and plated.

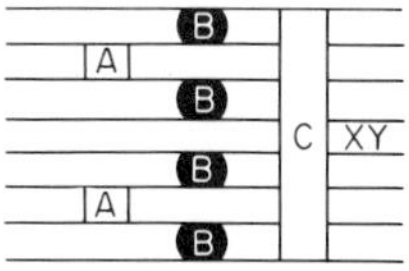

6. Eight-layer combination board is made by preparing two sequential subassemblies, laminating them, and then fabricating the through hole. This way of making an eight-layer board is more expensive than the previous design (5) because the fabrication of the subassemblies requires an additional plating step and two additional laminations, one for each sequential subassembly.

Fig. 49   Multilayer-board fabrication techniques.[39]

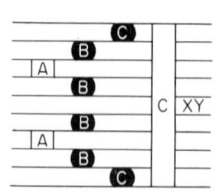

7. Ten-layer combination board. The process used to make this design is the same as that used for the single-lamination combination (5) except that two sequential subassemblies are used instead of three two-sided core boards.

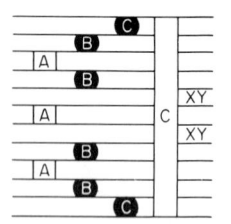

8. Twelve-layer board is also made like the combination board (5), except that two sequential subassemblies replace the outer core boards of the previous design.

Fig. 49 Multilayer-board fabrication techniques.[39] (*Continued*)

TABLE 32 Fabrication Operations for Multilayer Boards[39]

Figure number	1	2	3	4	5	6	7	8	PC*
Number of layers	8	8	8	8	8	8	10	12	2
Fabrication operations:									
Laminations	1	1	3	3	1	3	3	3	0
Drilling	1	4–5	7	7	5	5	6	7	1
Plating	1	4–5	7	7	4	5	5	6	1
Total operations	3	10	17	11	10	13	14	16	2
Fabrication sequences	2	2	3	4	2	3	3	3	1
Prohibited intraconnections	7	3–4	3	0	2	1	1	2	0

* Two-sided printed-circuit board.

a solution of hydrofluoric, sulfuric, and fluoroacetic acids, as it is very difficult to work with. Because of these difficulties, commercially available epoxy solvents were evaluated to see whether they could remove the epoxy smear. A phenol-based compound and a mixture of chlorinated solvents along with a solution of concentrated sulfuric acid were chosen. Samples were treated for 2 min in each bath, plated, tested, sectioned, mounted, and examined under a microscope.

The boards that were treated with the sulfuric acid and those treated with the chlorinated solvent had well-plated interfaces, thus indicating removal of the epoxy smear. The phenol-based compound did not remove the epoxy smear. The chlorinated solvent is much easier to handle, store, and use than the acid solutions.

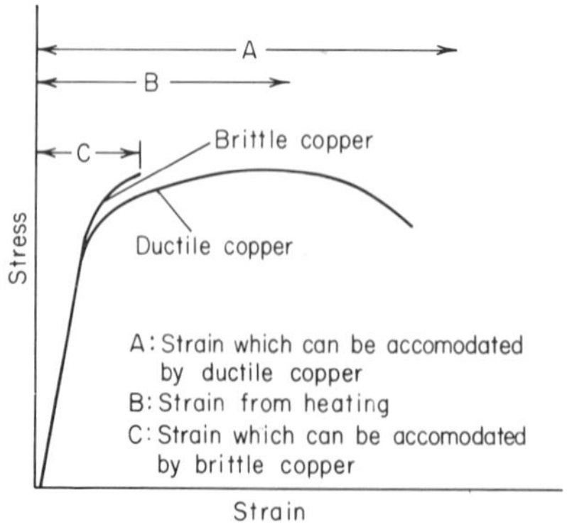

Through-hole Plating. The ductility of the copper that is plated or deposited on the wall of the drilled, etched-back hole is of primary importance. The thermal expansion of an epoxy-glass laminate is very high in the Z or thickness direction. Research in this area has shown that the coefficient of expansion is 80 μin./(in.)(°C) in the Z direction, as compared with 25 μin./(in.)(°C) in the X and Y directions. With such extreme differences in expansion coefficients, it is possible to exceed the fracture strength of brittle copper by the stress created by heating from room temperature to soldering temperature. A ductile copper, how-

Fig. 50 Strain on copper plate.[42]

A: Strain which can be accommodated by ductile copper
B: Strain from heating
C: Strain which can be accommodated by brittle copper

ever, will elongate to accommodate the increase in thickness of the board without fracturing. Figure 50 shows the effect of stress due to temperature on brittle and ductile copper plate.

Before the copper-deposition process, the surface of the epoxy-glass must be activated or sensitized so that the copper ions will attach themselves to the inside of the holes. This operation can involve one or more separate solutions which seed the hole surfaces with a microlayer of metal that catalyzes the deposition of copper from either an electroplating process or an electroless process.

Electroplating, or electrodeposition, is a method of depositing a metal on a conductive surface which is connected to form the cathode in an electrolytic bath. The metallic coating is produced by the reduction of the metal ions in solution through electrons supplied to the cathode from an external dc source.

Thickness of the electroplate obtained is proportional to the current density and the time of plating. Since variations in current density cause differences in plating thickness, minimum/maximum tolerances are not always economically feasible. A sensible solution is to specify a minimum plating thickness as in MIL-STD-275. The reason for this is the high cost of plating materials.

A main disadvantage of electroplating is that coatings of 100 μin. or less tend to be extremely porous. Metals commonly electroplated are copper, gold, tin, nickel, rhodium, and tin-lead alloys.

Electroless plating, or immersion plating, does not require the passage of electric current for deposition and is based on suitable oxidation-reduction reactions. Copper, gold, tin, and nickel are some of the metals generally deposited by this method. Electroless plating is easy to apply. The electroless plating process deposits uniform layers of fine-grain ductile copper into drilled or punched holes on the circuit board. The method requires fewer steps than the conventional way of electroplating.

A new electroless copper-plating process[43] may be used to prepare printed-circuit boards for conventional electroplating by depositing a preliminary coating on the surface within 4 to 10 min of immersion time. As the deposit is very pure and highly conductive, thinner coatings than usual may be used.

TABLE 33 Through-hole Plating Thickness[40]

Material	Through-hole plating, in.
Electroless copper	0.000025 min.
Electrodeposited copper	0.001 min.
Gold	0.000050 min.
	0.000100 max.
Nickel	0.0001 min.
Tin-lead	0.0003 min.
Tin-nickel	0.0003 min.

Specification MIL-P-55640 specifies that an electroless deposition system shall be used as a preliminary process for providing the conduction layer over nonconductive materials for subsequent electrodeposition.[40] After electroless coating, boards shall be copper-strike-plated with a minimum of 0.0002 in. copper in the hole if not immediately placed into the copper-electrodeposition process. The minimum thickness of the through-hole plate, specified in NSA 68-8 is contained in Table 33.

SECOND-LEVEL JOINING, MATERIALS AND PROCESSES

Solder Connections

Materials MIL-STD-454A, Standard General Requirements for Electronic Equipment, specifies that only noncorrosive and nonconductive rosin fluxes will be used for making electric connections. The standard specifies the 63 or 60 percent tin-lead alloy as solder metal for making electric connections.

13-76 Metals Joining of Electronic Circuitry

The fluxes that are recommended for use with various surface materials are listed in Table 34. These surfaces are the most likely to be encountered on printed-circuit boards or other types of connection matrices.

Water-white rosin is a nonsteam volatile fraction of pine sap and its specific composition varies with the source of the raw material. In general, rosin is a mixture of several isomeric diterpene acids. The three major components are sylvic acid, d-pimeric acid, and l-permeric acid. In an average analysis, abietic acid makes up 80 to 90 percent of the rosin, with the pimeric acids ranging between 10 and 15 percent. The commercial designation of "water-white rosin" refers to a grade of the material determined by colorimetric methods according to ASTM Designation D 509-55. Although it has a number of good qualities, it does not fulfill all the requirements for efficient fluxing. The abietic acid in the rosin is capable of reducing slight metallic tarnishes forming the appropriate metal abiets which, in themselves, are also nonconducting, nonhydroscopic, and noncorrosive. This will occur at a specific temperature range. Below or above this soldering temperature, the material rapidly becomes inactive.

TABLE 34 Types of Fluxes Recommended for Various Surface Materials[46]

Surface	Flux recommended: mildest possible — Freshly prepared	Flux recommended: mildest possible — Aged	Protective coating used to prevent aging	Restoring solderability	Contamination in solder — Dangerous levels	Contamination in solder — Effect on joints	General
Copper (bare)....	Water-white rosin	Activated rosin	Recommended	Chemical treatment	0.3–0.8	Grainy, with higher temperature required	Economy and best results
Gold, (over Cu, Ni, etc.)	Water-white rosin	Activated rosin	No	Remove plating	0.03–0.2	Dull and grainy	Avoid use if possible
Kovar...........	N.A.	Organic acid	No	Chemical treatment	None	N.A.	Glass matching and weldable leads
Silver...........	Water-white rosin	Activated rosin	Not normally	Chemical treatment	0.2–1.0	Grainy with higher temperature required	Avoid sulfur tarnishing
Tin (immersion electroless)	Water-white rosin	Loses solderability	Yes	Remove coating	N.A.	N.A.	Not recommended for storage
Tin-lead (plating)	Water-white rosin	Activated rosin	Not normally	Reflow plating	N.A.	N.A.	Excellent shelf life
Tin-lead (reflow or hot dip)	Water-white rosin	Activated rosin	Not normally	Reflow coating	N.A.	N.A.	Best shelf life
Tin-nickel.......	Activated rosin	Organic acid	Yes	Chemical treatment	None	N.A.	Hard to good for contact surfaces

N.A. Not available.

To improve the tarnish-removing capacity of rosin fluxes, a number organic and inorganic materials are added. These materials are labeled *activators*, and the flux is called *activated rosin*.

The activator serves as the tarnish remover, and the rosin fulfills the other fluxing functions such as surface activity and thermal stability. The nature of the activator and its quantity in the flux determine the corrosivity of the material. These fluxes are further divided into mildly activated fluxes and fully activated fluxes. The activator decomposes or evaporates at soldering temperatures, leaving a relatively pure rosin residue behind. When liquid rosin fluxes are applied, there is no way to assure that all the activator in the liquid flux is exposed to the soldering temperatures. Therefore, the activator in the material, a potential source of corrosion and electric current leakage, requires that flux residues and unused portions of the flux be removed from the surfaces. The total removal of activated-rosin flux can be accomplished only by total immersion of the assemblies in various processing solutions.

Second-level Joining, Materials and Processes 13-77

Various cleaning methods were developed, including ultrasonics, vapor degreasing, and other solvent systems where the major portion of the soldering flux was quickly removed from the surfaces, leaving them clean enough to perform their function adequately. These systems utilize various schemes for solvent recovery which lower the overall cost of the process. It was found that organic chlorinated solvent systems, widely used for safety reasons, were incapable of removing ionizable materials left behind by various sources of contamination, including soldering flux.

A second cleaning was used, based on aqueous solutions, to remove all ionizable materials contributed by various sources as those left behind by the rosin. A water-soluble flux system with the following advantages was developed:
 1. Good tarnish-removal capacity over a wide temperature range
 2. Fast and easy removal of the flux and its residues
 3. Nonspattering
 4. Suitable for a large variety of base metals, eliminating pretinning of the components.

Table 35 compares the various types of fluxes.

TABLE 35 Comparison of Various Types of Liquid Fluxes[46]

Assembly—tarnish	Cost	Joint reliability	Requires cleaning	Remarks
Water-white rosin:				
Clean—light............	Average	Good	No	Requires good process and material control.
Clean—heavy..........	N.A.	Poor	No	Not recommended.
Dirty—light...........	High	Medium	Optional*	Dirt might impair solderability
Dirty—heavy..........	N.A.	Poor	Optional*	Not recommended.
Activated rosin:				
Dirty—light...........	Average	Very good	Recommended*	Standard practice.
Dirty—heavy..........	Average	Good	Recommended*	Standard practice.
Organic water-base:				
Dirty—heavy..........	Low	Very good	Mandatory	Spatters excessively.
Water-soluble nonspitting:				
Dirty—heavy..........	Low	Very good	Mandatory	Requires least amount of solderability control.

N.A. Not available.
* Depends on the specific application required for critical cases.

Figure 51 is the tin-lead phase diagram. The Sn 63 Pb 37 composition is the eutectic alloy. Table 36 lists the recommended compositions of electric-connection solders as specified in Federal Specification QQ-S-571C. This specification indicates the type and maximum level of impurities in solder for electrical and electronic use.

The ASTM specifications for solder metal suitable for electric connection are shown in Table 37. Grade A refers to solder that has been alloyed from tin and lead extracted from ore.

Controls over the purity of tin and lead in these raw materials can be such that these levels of purity are not only maintained but also exceeded by many orders of magnitude. The Grade A levels of contamination as specified in Table 37 are maximum levels allowable for each particular contaminant. In addition, the specification imposes a maximum of 0.08 percent on all other elements not listed. Grade B is a specification similar to that of QQ-S-571C.

Processes There are two basic processes for effecting solder joints: dip soldering, and reflow soldering. The difference between these two processes is that in dip soldering, heat and solder are applied *simultaneously* to the prefluxed metal surfaces, while in reflow soldering, the solder has been applied *previously* by one of a

13-78   Metals Joining of Electronic Circuitry

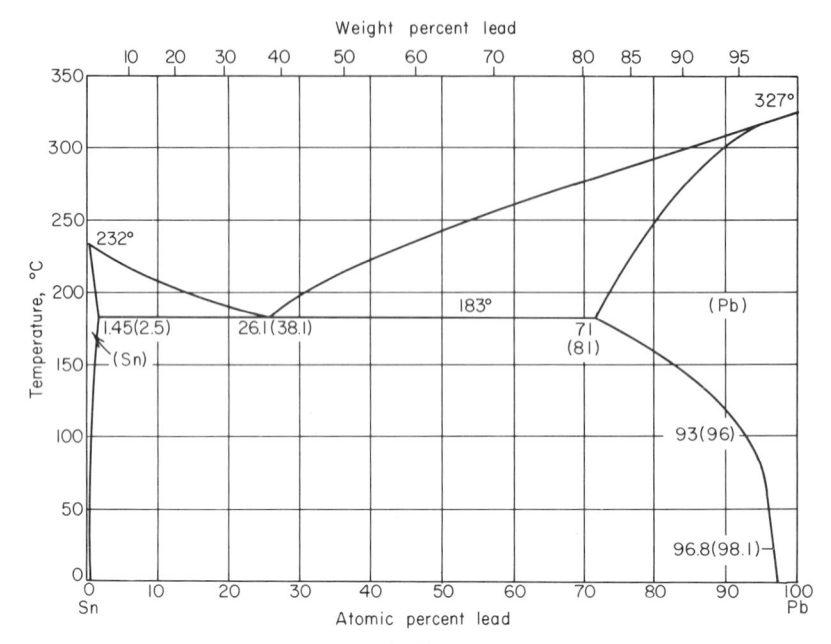

Fig. 51   Tin-lead phase diagram.[6]

variety of techniques (including dipping), and only heat and pressure are applied to cause bonding. Wave soldering is a type of dip soldering. Parallel-gap and resistance soldering are types of reflow soldering.

Wave Soldering.   Wave soldering is an automatic method of solder application where a continuous stream of solder is pumped up into an orifice, forming a standing wave of liquid solder through which the fluxed work can be passed. Figure 52 is a simplified illustration of wave soldering. The movement of the solder across the surface to be wetted decreases the soldering times appreciably. As a result, the heat distortion to the work as well as the overall temperature rise is smaller, and much more sensitive assemblies can be soldered. A printed-circuit board traveling through the solder wave receives some heat distortion in the immediate area of solder contact. As the board travels through the solder wave, the distortion travels through the card. By the time the printed-circuit board is out of the solder wave, no buckling or overall distortion is evident. This is the opposite case to regular solder dipping.

Fresh solder is pumped up into the solder head at all times, preventing the accumulation of dross on the solder surface from coming in contact with the work.

TABLE 36   Soft Solder-alloy Composition   (Federal Specification QQ-S-571C)

Code number	Tin	Lead	Antimony	Bismuth, max.	Copper, max.	Iron, max.	Zinc, max.	Aluminum, max.	Total all others, max.	Melting range* Solidus	Melting range* Liquidus
Sn 63	62.5–63.5	Bal.	0.10–0.25	0.10–0.25	0.08	0.02	0.005	0.005	0.080	360	360
Sn 62†	61.5–62.5	Bal.	0.20–0.50	0.25	0.08	0.02	0.005	0.005	0.080	350	372
Sn 60	59.5–61.5	Bal.	0.20–0.50	0.25	0.08	0.02	0.005	0.005	0.080	360	375
Sn 50	49.5–51.5	Bal.	0.20–0.50	0.25	0.08	0.02	0.005	0.005	0.080	360	420
Sn 40	39.5–41.5	Bal.	0.20–0.50	0.25	0.08	0.02	0.005	0.005	0.080	360	460

* In degrees Fahrenheit, approximately.
† Sn 62: silver 1.75–2.25.

TABLE 37 Chemical Composition of Soft Solders
(ASTM Specification B 32-58 T)

Alloy grade	Tin desired, %	Lead, nominal, %	Antimony, % Minimum	Antimony, % Desired	Antimony, % Maximum
60A	60	40	...	...	0.12
60B	60	40	...	...	0.50
50A	50	50	...	...	0.12
50B	50	50	...	...	0.50
45A	45	55	...	...	0.12
45B	45	55	...	...	0.50
40A	40	60	...	...	0.12
40B	40	60	...	...	0.50
40C	40	58	1.8	2.0	2.4

For elements other than those mentioned in the table, the maximum content in the alloy shall be as follows:

Alloy	Percent
Bismuth	0.25
Copper	0.08
Iron	0.02
Aluminum, zinc	Each shall not exceed 0.005

As a result, continuous skimming is not necessary. All flux and flux residues which are wiped off the work and would normally stay on top of the solder baths are carried down with the wave into a special reservoir where they do not come in contact with any future work.

The solder that is pumped out of the bottom of the solder container is always at the same temperature, and the solder reaching the head has no time to be cooled by air currents or other means. The temperature of the solder touching the work is always uniform and can easily be controlled and maintained. The combination of these properties makes the solder wave a useful tool for automation, suitable for large quantities of work with a high percentage of reproducible and reliable solder joints. Some automatic soldering systems use an inclined conveyer. The inclined conveyer, when combined with a wide solder wave pulls the board up and away from the solder wave as it moves along the conveyer. This gradual separation of the board from the wave (because of the incline) exerts an appreciable pulling action that aids the peel-back of the solder on the board as it passes over the wide wave. Further improvement in the performance of such an automatic soldering system can be obtained with the introduction of oil into the solder wave. This allows soldering at temperatures 10 to 15 percent lower than when oil is not used. The solder deposits are minimized and have a much shinier appearance. Also, the use of oil allows considerable variation in conveyer speed, solder temperature, and immersion depth. Any or all of these may be varied over a wide range, and still achieve satisfactory solder joints. The oil is usually a blown mineral oil, which reacts with the tin and lead oxides in the dross to form tin and lead soaps. It is a scavenger for dross, contained in the solder, and absorbs this dross without any undesirable side effects. It is, of course, necessary to solder at a high enough temperature so that the oil comes to the surface of the solder joint before the solder solidifies. The oil also prevents oxidation of a solder joint because it keeps air away from the joint until the solder has solidified.

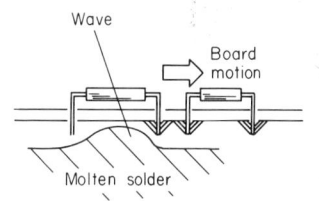

Fig. 52 Wave soldering.

13-80 Metals Joining of Electronic Circuitry

Compact, completely enclosed automatic-wave soldering systems are available that have individual stations for fluxing, flux drying, preheating, and soldering. Such systems have a built-in exhaust and a conveyor. Individual controls are positioned in front of each station, and meters are panel-mounted adjacent to each station.

Reflow Soldering. At some time before the actual joining process, solder has to be applied to the device leads and the printed-circuit boards that are to be reflow-soldered. Solder coatings may be applied by a number of methods. A method used to obtain extremely thin solder coatings, called *solder slinging*, has been developed. The bare boards without components are dipped in a solder pot that is covered with oil. As the board is removed, it is rotated at speeds up to 300 rpm in 1 sec, so that when it is withdrawn from the solder and travels through the oil on top of the solder, the centrifugal force removes all excess solder from the surface, leaving about 50 millionths of an inch of coating in a shiny easily wettable form (see Fig. 53).

Another method for solder coating printed-circuit boards is roll coating. Roll coating consists of passing the precleaned and fluxed printed-wiring board through

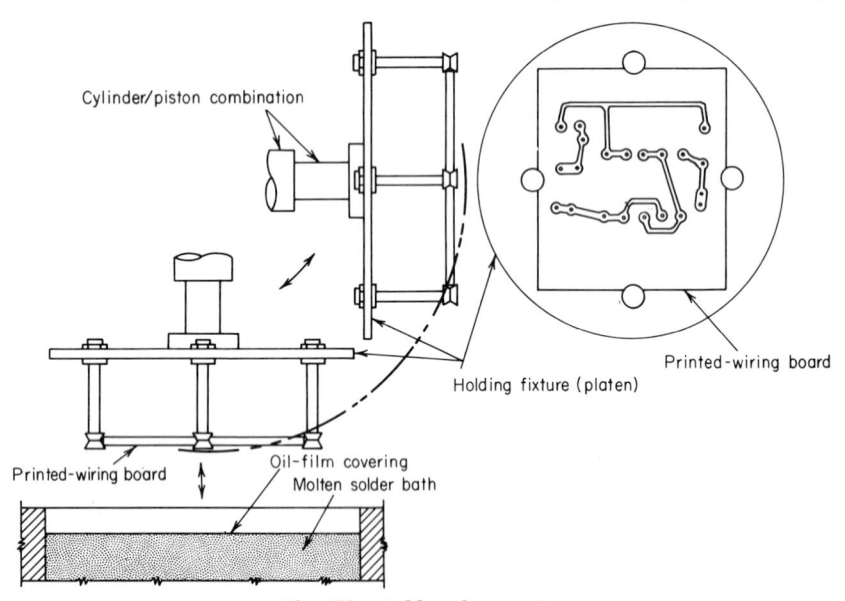

Fig. 53 Solder slinging.[48]

a pair of immersed pretinned high-carbon-steel rollers while the solder is in the molten state (450 to 550°F). As the printed-wiring board passes between the pressure rollers, the lower roller turns and deposits molten solder up from the bath and onto the conductive surface of the board. Because the solder is rolled on in the molten state, alloying occurs with the copper. This results in an excellent bond and a dense metallic coating. Coating thicknesses are determined by pressure control of the rollers. The main disadvantage to this coating system is the difficulty of obtaining a coating of uniform thickness. The minimum coating thickness per MIL-STD-275, Printed Wiring for Electronic Equipment, is 0.0003 in. Also, this coating should conform to composition Sn 60 or Sn 63 of Specification QQ-S-571.

Plating is another method for solder-coating printed-circuit boards. This type of plating is deposited from an acid fluoborate bath and is used primarily to provide a solderable finish on the printed-wiring board. As specified in MIL-STD-275, the plating must be a minimum of 0.0003 in. thick and must contain between 50 and 70 percent tin. The $^{60}/_{40}$ or $^{63}/_{37}$ tin-lead alloy, which is close to the eutectic composition, is preferred for solderable or etchant-resist platings.

The most important element in the tin-lead structure is the tin itself. Tin has the property of wetting and also of dissolving other metals, much in the manner that mercury produces an amalgam. In a truly wetted joint, the solder cannot be removed from the wetted surface without physical destruction of the bond. This ability of tin appears to be chemical in nature and has a rate of reaction directly proportional to temperature.

Tin-lead plating is particularly advantageous in the plated-through-hole process, as well as where fused eyelets are specified. To obtain a solder plating of the desired 60 or 63 percent tin alloy, plating conditions must be carefully controlled. Occasionally, improper plating conditions result in electrodeposits of the wrong composition. A small amount of antimony may be desirable in the deposit to prevent the formation of gray tin powder at low temperatures. With experience, it becomes possible to detect improper solder compositions during the etching process and to correct for them.

Resistance soldering is a type of reflow soldering. This method generates heat by the flow of an electric current through a series circuit. Either a pair of electrodes may be applied across the joint or one of the materials to be soldered can be connected as an electrode.

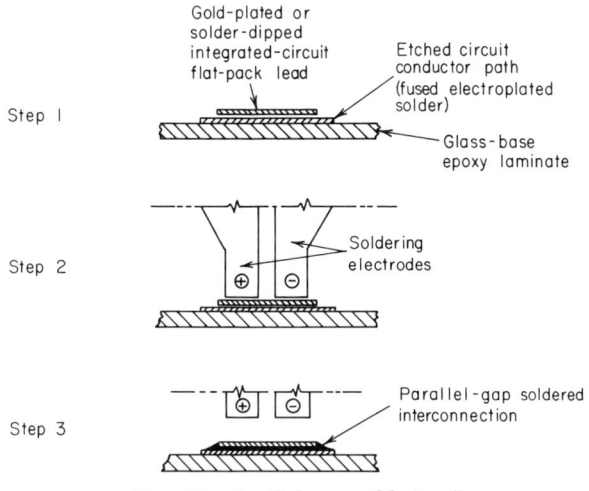

Fig. 54 Parallel-gap soldering.[47]

Parallel-gap welders are often utilized to effect resistance solder joints. The technique is referred to as *parallel-gap soldering*. It differs from other soldering methods because flux is used, and also additional solder, resistance welding equipment to accomplish the soldering process; the resultant solder joints are smaller than hand-soldered joints, can be made as fast as parallel-gap welded joints, are made without subjecting the integrated circuit to any appreciable temperature rise, and are repairable.

Figure 54 shows the basic parallel-gap soldering process. In step 1, the integrated-circuit flatpack lead is positioned over the etched circuit-conductor path, which has an outer layer of fused electroplated solder. In step 2, the electrodes are brought against the flatpack lead with a light pressure and a direct-voltage pulse of short duration (3 to 5 msec) causes current to flow through the joint, melting the existing solder on the joint surfaces. In step 3, the soldering electrodes have been raised, and the reflowed solder has been allowed to cool, forming the interconnection.

An electrode pressure of 1 to 2 lb and an electrode gap of 0.010 to 0.015 in. are recommended for use with gold-plated Kovar flatpack leads (0.004 in. thick by 0.015 to 0.020 in. wide). There are a number of different types of machines that may be used, such as constant-voltage dc parallel-gap welders; constant-voltage ac soldering machines; constant-voltage variable-pulse ac soldering machines; dual-head series welders. Some of these machines may be adapted so that multiple solder joints can be accomplished simultaneously.

The gold-plated Kovar leads are sometimes solder-coated. The leads are fluxed and dipped for a few seconds in $^{60}/_{40}$ solder at 450°F (solder pot temperature).

During parallel-gap soldering, it is possible to melt the printed-circuit board resin directly beneath the joint and under the copper foil. This resin flow could degrade the insulation resistance characteristics of the base laminate. Therefore, parallel-gap soldering schedules are adjusted so as not to cause an excessive amount of resin flow.

A flatpack having 14 leads can be gap-soldered in an average time of 0.53 min. Gap soldering to copper-clad circuit boards can be done equally as fast as gap-welding to nickel-clad circuit boards, with most of the time being spent in positioning the flatpack leads beneath the electrodes.

Fluxing of parallel-gap soldered joints is subject to some discussion. Keister[47] states that: "Fluxing of gap-soldered joints is unnecessary, leaves a residue which must later be removed, and produces joints of lower peel strength than those produced without flux. Fluxing is *not* recommended." However, Douglas[49] states that: "In every case, those (joints) made *with* flux were superior by a considerable margin. The only cases where fluxless joints were even marginally acceptable were those made with solder-plated circuitry and solder-tinned leads."

Parallel-gap soldering will produce acceptable joints from the standpoints of insulation resistance, electrical joint resistance, and mechanical strength based on the results of environmental tests, including (1) thermal shock, (2) humidity, (3) life at elevated temperature, (4) random and sinusoidal vibration, and (5) salt-water immersion.

Parallel-gap soldering joints using solder-dipped Kovar foil leads and gold-plated Kovar foil leads are *both* acceptable for forming integrated-circuit flatpack interconnections.

Gap-soldered joints had higher peel strengths than hand-soldered joints at all pull angles except 180°. Gap-soldered joints showed minimum peel strengths at pull angles from 45 to 90°. Peel-strength tests in shear (0° angle of pull) showed that lead failure would occur before joint failure. Figure 55 shows mechanical peel testing of parallel-gap solder joints at various angles of pull. Figure 56 is a plot of peel-strength values at various pull angles for parallel-gap-soldered gold-plated Kovar flatpack leads, and gold-plated solder-coated Kovar leads, and also manual soldering. Note that at 30 and 45° pull, gold-plated leads have higher strength, whereas, at 90 and 180° the solder-coated leads have higher pull strengths.[47] These data were taken from Keister.[47] However, others have indicated that gold-plated leads do not exhibit as high a pull strength as similar solder-coated leads (see Fig. 57).[49]

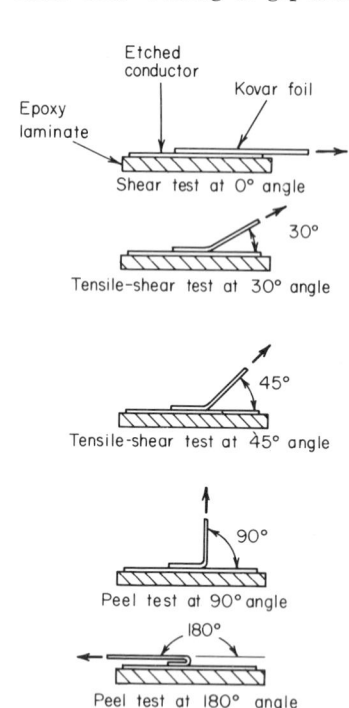

Fig. 55 Mechanical peel testing.[47]

At least two investigators have recently studied the effect of gold on solder. Bester[50] and Wild[51] detail this work. The summary of each of these reports is reprinted here:

The rapidity with which gold becomes a constituent in the ternary Sn-Pb-Au system when being soldered is well known. Large quiescent solder baths at 480 F take Au into solution at a rate of about 0.001 in. in 4 seconds. The acicular Au (Sn Pb)$_4$ phase identified by microprobe analysis is first apparent at 500× magnification with as little as 1 percent Au in Sn 60 and Sn 63 solders. The ultimate tensile strength of the ternary alloy remains close to that of Sn 63 and Sn 60 solders (8,000 psi) until the Au content exceeds 12 percent. At 7 percent Au, elongation and reduction in area drop below 10 percent. Bend ductility is considerably reduced at Au content above 5 percent, and a very significant transition in impact strength occurs when Au exceeds 4 percent. Failures occurring in

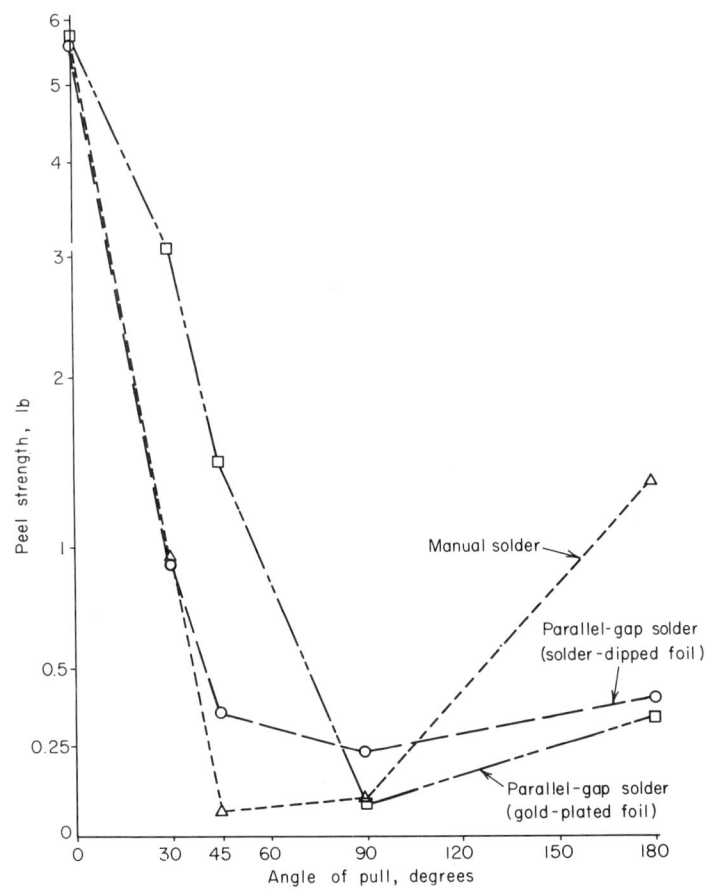

Fig. 56 Peel-strength values.[47]

solder-coated, Au-plated Kovar transistor leads and in lap-soldered joint were associated with massive and acicular Au (Sn Pb)$_4$ phases. The thickness of gold plating prior to soldering is, by itself, not an adequate criterion for predicting the quality of the solder joint. For severe vibration and shock environments, it is recommended that the Au concentration not be permitted to exceed 4 percent and that soldering parameters such as time, temperature, agitation, and relative mass of Au to solder be adjusted to assure that continuous intermetallic formations and concentration gradients that permit the Au content to exceed 4 percent be avoided.

The solderability of Au plating when unalloyed and not contaminated is good; however, contamination of the base metal by diffusion of oxidants into porous

13-84   Metals Joining of Electronic Circuitry

plating or oxidation of the base metal after it diffuses through the gold can reduce solderability considerably. Large amounts of acicular Au (Sn Pb)$_4$ crystals in solder, coupled with an increase in liquidus temperature of about 100°F when Au content reaches 15 percent can be expected to reduce fluidity and apparent wettability of Au plating. The addition of Au to solder baths to reduce the rate of solution of Au on parts being soldered is neither practical nor effective. Indium solders,

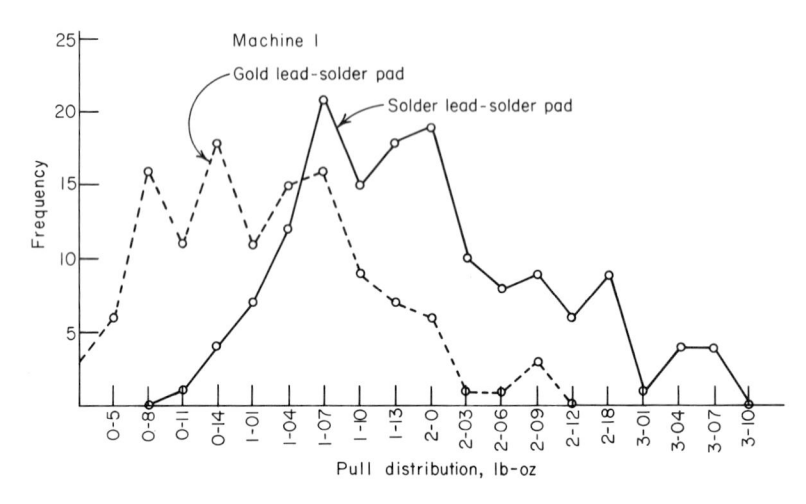

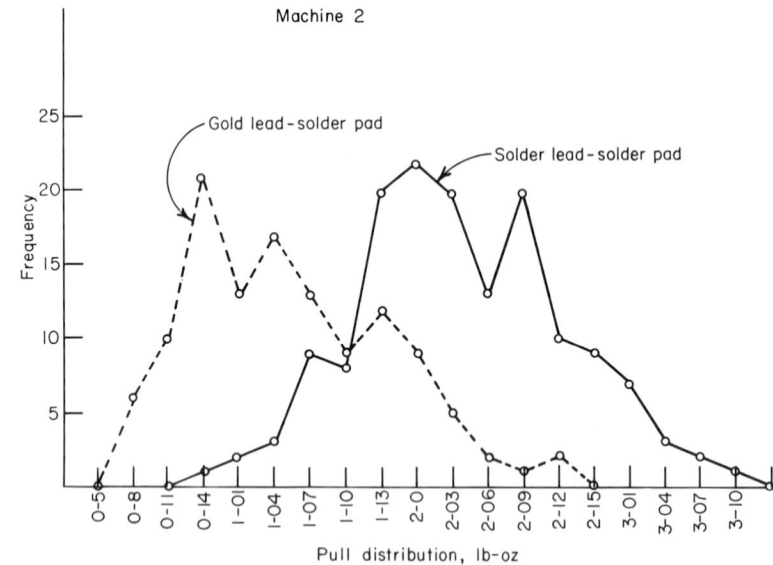

Fig. 57 Distribution of peel-strength values.[49]

particularly those containing small amounts of Zn, are of interest since their reaction rates with gold are reported to be considerably reduced. Electrical conductivity of Sn 63 was found to fall between 9.5 and 12.5 percent IACS as gold content increased to 10 percent.[50]

Various tests were made in this study to determine the effect of gold on the melting points, wettability, bulk properties, and solder joint properties of Sn 62 and Sn 63 solders. Metallographic changes in solder structure, with additions of gold are also discussed.

This study showed that it is feasible to allow the gold content in solder pots to accumulate ≈1.0 percent gold. The study also showed that there are no detrimental changes in the physical properties of solder joints with up to ≈1.0 percent gold. There is a gradual decrease in the overall solder joint strength with additions above 1.0 percent gold. Although the strength of the solder is at a maximum with ≈1 percent gold, the ductility and toughness of the solders were actually improved with additions of gold up to 2.5 percent. This paper also discusses precautionary measures which must be taken when increasing the allowable gold content in the solder pot or solder joints.

A significant cost savings can be realized by increasing the allowable gold content in high volume production type solder pots. This would be based primarily on: (1) using significantly less solder, (2) fewer interruptions of the production line, and (3) reduced cost of changing solder pots.[51]

Parallel-gap soldering will not produce any appreciable temperature rise within the flatpack case. Hand soldering of the individual flatpacks leads is suitable as long as the total volume of flatpacks being handled is low. It normally takes 3 to 6 min to hand-solder a single 14-lead flatpack to a printed circuit. The time can be reduced if the flatpack is prepositioned on the printed-circuit board. However, the prepositioning time must still be included in the overall analysis of the attachment rate.

As usage of flatpacks increases, hand soldering of one flatpack lead at a time will not meet production requirements. The equipment used in multilead reflow soldering consists of a suitable power supply, coupled with a reflow solder head. This combination must provide a precisely controlled pulse amplitude, pulse length, and tip pressure.

The design of the solder tip is critical in relation to the total reflow-solder process. The requirement for a very limited voltage across the flatpack leads is of prime importance.

Several reflow solder tips have been developed. The slotted tip was developed because of the control it appeared to provide in heat distribution across the tip. If any bar heats more than an adjacent bar, its resistance increases. The heating current is then forced to flow through the cooler bar. As a result, the heat of the individual bars is equalized across the tip. However, this held true only if every bar was on a lead. When flatpacks with leads omitted were soldered, the tip overheated at these points. The slotted reflow-solder tip presents some problems in alignment with the flatpack leads. The unslotted tip overcomes this problem.

An ac power supply provides an advantage in that long pulse lengths are easily obtained. As a result, a lower tip temperature can be used over a longer period of time. This results in longer tip life and provides an advantage in obtaining optimum solder joints. By using a longer heat pulse, the uneven solder heights that are sometimes experienced can be overcome. The longer heat pulse allows the highest-level solder to reflow, and enough pulse length still remains to reflow the balance of the leads. All the solder reflows equally, and consistent joints are obtained. The longer pulse length allows the use of an unslotted reflow solder tip. By combining two reflow solder tips in a special head, it is possible to reflow all 14 leads of a flatpack at one time. The flatpack is held in position by vacuum until the head is lowered. The downward stroke of the head is stopped just above the surface of the printed circuit board. The operator manually positions the board under the flatpack leads. The downward actuation of the head is continued, and upon reaching the preset pressure, the power supply is fired. From the time of initiation of the power-supply pulse, the control of the heat, pressure, and dwell time is automatic. As a result, more consistent joints are obtained.

A semiautomatic soldering machine has a number of flatpacks loaded in a nest that is mounted on a movable platform also containing the printed-circuit board. The operator moves a pantograph-type handle to a pickup position on a template. Depressing a switch on the handle causes the reflow-solder head to lower and pick up the proper flatpack from the nest. The operator then moves the handle to the desired reflow position on the template. Depressing the switch in this location lowers the reflow-solder head and the flatpack to the printed-circuit board and automatically initiates the complete solder cycle.

With this type of equipment, it is possible to position and reflow a flatpack in 8 sec. Typical user reports indicate that a rate of one flatpack each 12 sec can be achieved over an 8-hr period. This rate includes all material handling and operator relief time. The figures apply to a printed-circuit board with 14 flatpacks. A large number of flatpacks per board would decrease the average joining time per flatpack.

Several of the solder-reflow machines on the market are capable of being fully automated. It is possible to tape-control the positioning and all pickup and reflow cycles. A more sophisticated automatic control also would include removing the flatpacks from magazines and then positioning and reflow soldering.

Optimum solder joints are obtained if the flatpack leads are tinned. The plating on the printed circuit, when combined with that on the flatpack leads, produces optimum filleting around the flatpack leads.

With ultrasonic tinning, the immersion of the flatpack leads in the solder may be held to only 1 sec. The amount of solder build up on the flatpack lead depends on the total dwell time, the temperature, and the speed of withdrawal from the solder bath. The ultrasonic agitation removes all the gold from the flatpack lead. During the withdrawal of the lead from the bath, an even, consistent coating of solder is deposited on the Kovar lead material, in place of the gold plating. Ultrasonic tinning eliminates the requirement for flux in the tinning operation. Establishing a processing procedure when using the reflow-soldering technique is relatively simple. The reflow-solder tip pressure is adjusted so that sufficient pressure exists to assure intimate contact with all pads, but not enough pressure to damage the printed circuit. This pressure is normally 2 to 3 lb.

The pulse length on the power supply should be set to approximately 1 to 1½ sec. The pulse amplitude should then be adjusted so that solder reflow occurs. Then, it is only necessary to adjust the pulse time and energy levels for optimum reflow of the solder. On the semiautomatic reflow-soldering machines, the dwell time should be adjusted for complete resetting of the solder.

The multilead reflow-soldering technique for attaching flatpacks or similar packages to printed circuits has proved to be 18 times as rapid as hand soldering of individual flatpack leads. In large-scale production operations, this technique offers a suitable solution to maintaining an adequate production rate.

The reflow-soldering technique can be utilized with dual in-lines by forming the leads in the same manner as is done with flatpacks. It is possible to position and reflow-solder a dual in-line package every 8 sec. This time compares very favorably with the time required when using automatic insertion techniques and subsequent flow soldering. The printed-circuit board need be handled only one time during the attachment of the integrated circuit.[52]

Both parallel-gap soldering and parallel-gap welding can be done using identical equipment. The only equipment change necessary is the material from which the electrodes are machined. For parallel-gap welding, the electrodes are of RWMA-2 copper alloy, while parallel-gap soldering electrodes are of material not easily wet by solder (e.g., aluminum, tungsten, molybdenum, etc.).

Impulse soldering with a parallel-gap welder has been investigated. Although these bonds do not have strengths equivalent to those of welded connections, solder bonds are less sensitive to material and electrode pressure variations than welds, and components can be removed nondestructively. Bonding energies are about half those required for welding. Bonding energy tolerances are at least twice as great as those for parallel-gap welding.

A voltage-controlled welding power supply which was used can be set to deliver bonding pulses from 0 to 1.99 volts in amplitude (as measured at the electrodes) and from 0 to 9.9 sec in width. It cannot supply preheat or postheat pulses. The head was equipped with a compliant micrometer-type adjustable parallel-electrode holder. It has a 4½ in. throat. Its force can be varied from about 8 oz to 10 lb.

The electrodes must be of a material not wettable by solder. Both molybdenum and tungsten were used. Both made excellent bonds. The tungsten electrode, being harder, requires less dressing during use. The minimum electrode pressure is that which makes adequate contact for the welder to discharge, and the maximum is that pressure where deformation of electrode or bonding materials begins. No variation in the bonds was observed as a result of change of electrode pressure. For flatpack leads the range of force was about 1.5 to 6 lb using 0.015- by 0.025-in. molybdenum electrodes.

Soldering ranges were determined by bonding 0.016- by 0.004-in. solder-dipped Kovar flatpack leads to 0.020-in.-wide gold-plated 2-oz copper printed-circuit leads. The electrodes were 0.015- by 0.023-in. molybdenum. The results are shown in Fig. 58 for 0.035-in. electrode spacing and in Fig. 59 for 0.020-in. electrode spacing. Some representative test measurement points are indicated on the figures. The bonding regions shown at 40 msec remain constant as time is increased. The soldering range has as its lower boundary the lowest heat that will produce a satisfactory solder bond, and as its upper bond any welding no matter how slight. All bonds in the soldering range can be removed by the application of heat or by the

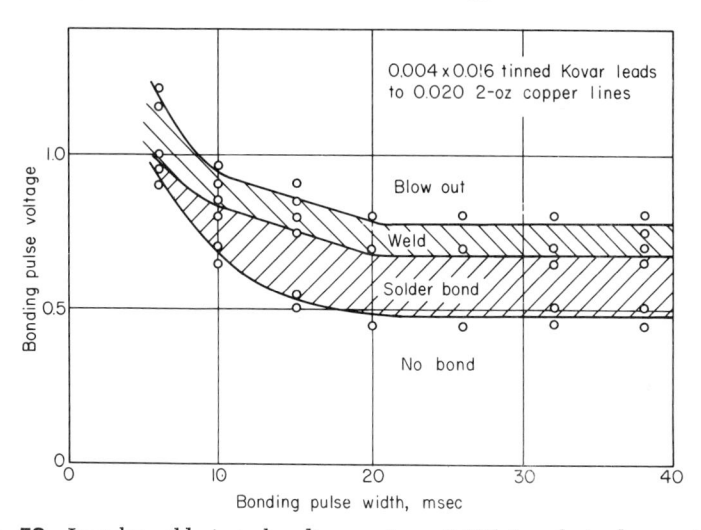

Fig. 58 Impulse soldering: bonding regions, 0.035 in. electrode spacing.[53]

use of a sharp knife. The welding range includes all welds short of blowout even though toward the lower boundary the welds would not be considered satisfactory as a bond. For comparison purposes, Fig. 60 shows the welding range for the same type of gold-plated leads but without solder welded to the same type of printed-circuit loads. The welding range indicated included all welded bonds without regard to quality.

Figures 58 and 59 show several of the important advantages of soldering over welding for exploratory development systems. The soldering range is at least twice as great as the welding range and is remote from the blowout region. Even if a bond is made at a heat somewhat above the soldering range, the only result would be a weld. Unless this bond later must be removed, no damage would be done.

Solder Preforms. Soldering parts in high-density electronic packages presents problems uncommon to other joining methods. The amount of solder and flux required for adequate joining is extremely small. Excessive solder and flux will result in a bridge that will cause short circuiting. The heat from the soldering operation and the residues from the soldering materials cannot be allowed to degrade

13-88 Metals Joining of Electronic Circuitry

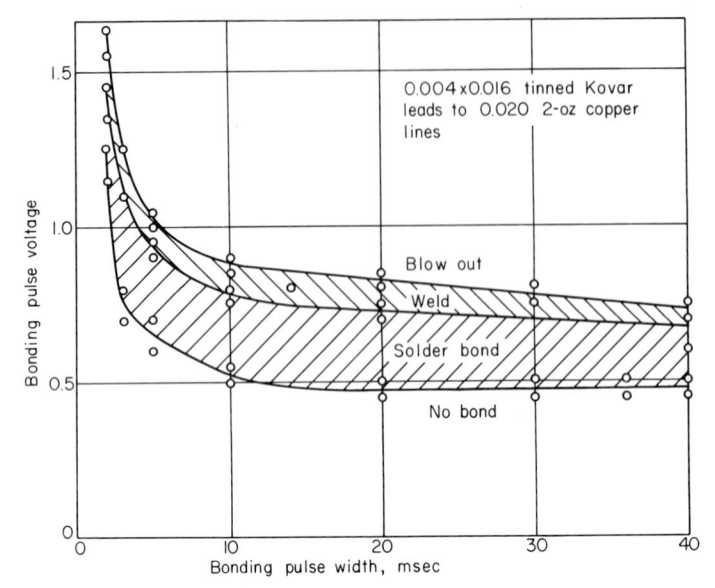

Fig. 59 Impulse soldering: bonding regions, 0.020 in. electrode spacing.[53]

the rest of the assembly. When soldering many joints on one assembly, remelt must not occur. Solder preforms are a solution to many of these problems.

One method of utilization of solder preforms is in hot-air soldering systems. The basic soldering conditions that the system must provide are shown in Table 38.

If these conditions are not provided, the solder joints are apt to be of poor quality. If the air velocity is too high, the solder is forced to one side of the pin, and if the air temperature is too high, the solder wicks up the pin excessively.

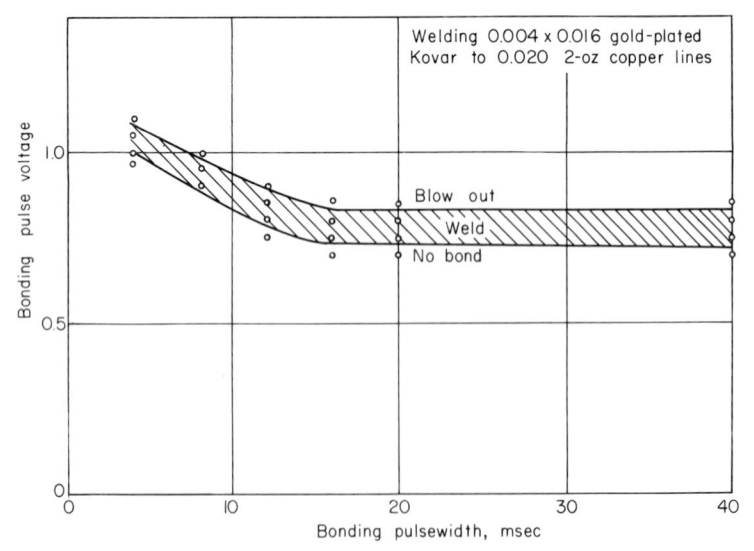

Fig. 60 Impulse soldering: bonding regions, 0.008 in. electrode spacing.[53]

Excessive wicking can also occur if the air delivery angle is nearly horizontal. However, since tests showed that good solder joints result from conditions listed in Table 38, the actual hot-air solder machine design is based on these conditions.

The solder machine layout, as shown in Fig. 61 was conceived to control the basic solder-joint variables and to control the hot-air leaks so that operator comfort is maintained without special enclosures. Air leaks are controlled by air curtains as shown and by careful design of the airflow path.

TABLE 38 Hot-air Soldering Parameters[54]

Variable	Desired range
Air temperature	300–325°C
Air velocity	1,300–1,600 ft/min
Air delivery angle	Approximately 25° with vertical

Three small standard blowers were used, rather than one large blower, to minimize machine size and to deliver a somewhat evenly distributed airstream into the mixing plenum. The air is heated by 16 Nichrome wire elements. Each element dissipates 1 kw.

Normal air-temperature spread across the machine at the air outlet to the soldering zone is 5°C or less for no-load conditions. Temperature spread across the outlet for a severely unbalanced load is 12°C. The average air outlet temperature to the soldering zone is 315°C without load and about 305°C with normal load. For a continuous load of boards 2 ft wide, the temperature at outlet to soldering

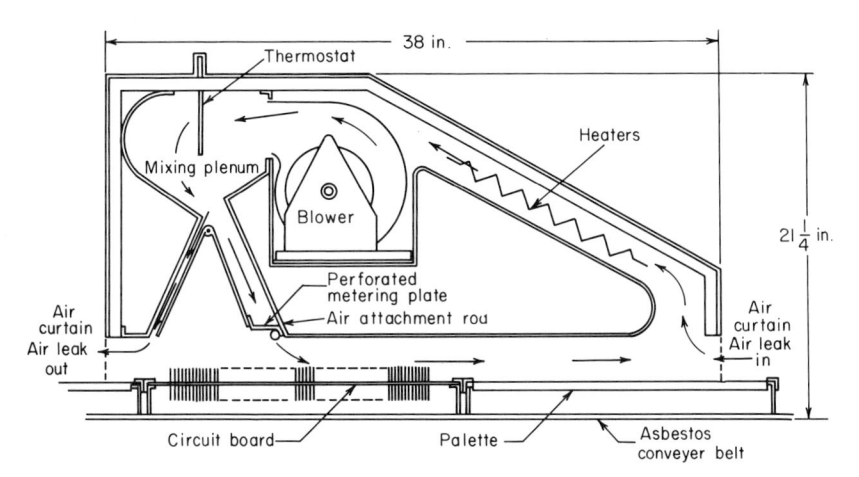

Fig. 61 Hot-air soldering machine.[54]

zone held above 300°C. Good uniform solder joints were obtained uniformly across the entire machine during this extreme load.

Solder is provided at the joint by a doughnut-shaped preform placed around each pin. Two preform sizes are used. The thicker preform is used for $\frac{1}{8}$-in.-thick boards having plated-through holes. The thinner preform is used on $\frac{1}{16}$-in.-thick boards having plated-through holes. It was found that the size and shape of preform was not as critical when proper air velocity and temperature were provided.

The hot-air soldering facility is capable of mass-soldering 7,500 pins/min on a continuous-flow basis. The pins pressed into boards and soldered are approximately 0.025 in. square and less than 2 in. long. The soldering variables are controlled so that joints are mass-soldered uniformly across the entire 24-in.-wide soldering zone. Solder wicking up the pins is controlled so that no appreciable wicking occurs above 0.055 in. from the board surface. Pins pressed into plated-through holes are soldered in one pass through the machine.

Infrared heating can also be used with solder preforms. Temperature-time curves comparing infrared heating and a typical soldering iron are shown in Fig. 62. The lower temperature with the infrared system was made possible partly because peaking is not a problem with infrared heating—nothing touches the work—and partly because of more precise control and more uniform heat transfer. With infrared heating, the workpiece temperature is the principal variable affecting the energy transfer rate. But, since the source-temperature is considerably greater than the work-piece temperature, the effect of workpiece temperature as a variable is negligible. Hence, the energy transfer rate is essentially constant. By comparison, heat transfer with a soldering iron depends on tip temperature and cross section of contact, both of which vary. The tip temperature varies with soldering speed and operator skill, while the cross section varies with the amount of solder used and the wet area on the soldering-iron tip.

One difficulty encountered with the infrared system was a tendency of the circuit board to burn if the center of heat concentration was not on the land area. Surface charring was noted when the deviation exceeded $\frac{1}{16}$ in. Also, the solder tended to run along the copper conductors if the heat was applied for too long a period. Both of these problems were eliminated as the operator gained proficiency.

An important consideration was the relative size of the infrared unit. Since the unit is much larger than a soldering iron, it was necessary to fix its position and to mount and manipulate the circuit board. This increased the time necessary for each operation and, the cost per operation.

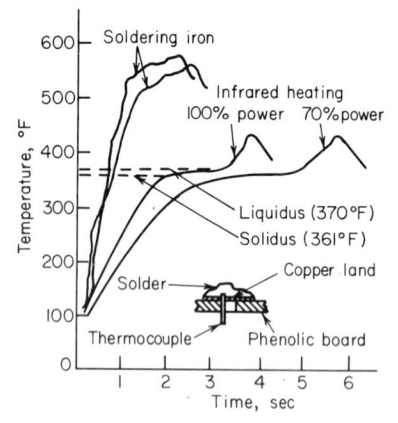

Fig. 62 Infrared soldering time-temperature curves.[55] The method of measurement is indicated by the sketch below the curves.

Another type of preform is prealloyed solder-flux creams. Solder creams are flux and solder combinations made from prealloyed solder powder and fully activated, mildly activated, or nonactivated flux. The selected composition of the prealloyed solder powder depends on the specific use to which it is to be put.

Unlike roll solders, exact amounts of solder pastes may be precisely located. A specific volume can be selectively screened or masked onto the assembly surface with an air syringe to provide close control over the volume of cream deposited.

Screening is usually appropriate for very thin deposits generally no greater than 0.001 in. thick. The screening process permits the deposition of solder creams in custom amounts and configurations to meet specific needs. Since heavier deposits may present registration problems, another application method should be employed when the required thickness exceeds 0.001 in.

Masking or extrusion is appropriate. Masking is used for heavier deposits of solder cream. The cream is applied in any configuration through a stainless steel mask or a preetched stainless steel template. The thickness of the stainless steel mask determines the thickness of the deposited layer of solder cream.

With solder cream, flatpacks can be uniformly soldered to printed-circuit boards and still be very easy to inspect. Two methods are available. By one, the solder cream is applied to the printed-circuit board pads immediately after board manufacture when the pads are still simple to solder. The flatpacks with solderable leads are simply placed where desired, and when heat is applied to the solder cream, a fillet is formed. The other approach involves placing the solder cream on the flatpack leads. The leads are then placed on the printed-circuit board pads. In this instance, the process is in effect reversed. Moreover, whereas in the first approach little solder is present on top of the lead, in the second the solder completely encompasses it. Thus, from an inspection standpoint, the first approach is more desirable.[56]

A recent development in solder preforms is the use of a plastic sheet with heat-shrinkable pockets formed on the same center distances as the terminations of a multilayer printed-circuit board. The solder balls in these pockets contain a precisely measured amount of solder. During heating, the heat-shrinkable pockets flatten and force the molten solder into the terminations. All surfaces have been raised to soldering temperature so that good wetting is obtained.

The most practical method of applying heat is in an oven. If low-temperature solder is used, such as 50 Sn 50 In (with a melting temperature of 243°F), oven temperatures as low as 300°F can be used for installation. For higher-temperature materials, Sn 63 solder can be used. In one application, soldering time was reduced from 20 to 1 hr on a previously hand-soldered circuit board, and the number of rejected boards was significantly reduced.

Ultrasonic Soldering. One of the main attributes of ultrasonic soldering is the elimination of precleaning, fluxing, flux residue removal, and postcleaning. The application of ultrasonics to the wave-solder process is being developed. The system employs ultrasonic energy applied below the crest of the flowing solder wave. Ultrasonic energy breaks down oxide films and thus eliminates the need for flux. Ultrasonic wave soldering has also been successful in the soldering of aluminum leads to printed-circuit-board copper conductors without the use of flux. The elimination of flux is important to wave-soldering applications. However, not all the effects of ultrasonics or the level of energy that can be tolerated by the circuitry are known.

Hand Soldering. In many cases individual joints have to be either soldered for the first time or resoldered in the printed-circuit board itself. This might be necessitated by repair work, touch-up, faulty connections, replacement of components, engineering changes, or the insertion of temperature-sensitive components. In some cases, this might even involve soldering the complete circuit board by hand because of low-volume production.

Soldering irons are normally classified by their wattage. In addition to the wattage, the efficiency of the iron, which is defined as "the quantity of the energy generated as heat which actually reaches the working tip and the solder area itself," should be given. The heat content of the iron and the recovery rate should be known. Because the heat source of the soldering iron is a resistance type of element, it is possible by changing the voltage of the input to change the caloric supply in the iron. Using this fact, a study was made of the input voltage. Figure 63 shows a typical TTV curve (tip temperature versus voltage).

Splashes[57] **and slivers**[58] Solder splashes are small globules, stringers, or spots of solder that splash or explode out from a terminal, pad, or conductor area and contact (or nearly contact) adjacent etched-circuit areas, resulting in electric short circuits or a reduction in the insulation spacing. Solder splashes can occur during initial soldering operations of modules, terminals, or components onto the circuit board or during repair or module replacement operations. They are also equally applicable to gold- and solder-plated through-hole circuit boards. Failures are usually not visually detectable because of the design of the circuit-board structure. X-ray examination of circuit boards, however, can reveal the presence of small solder splashes or splatters.

Solder splashes are caused by the expansion of trapped air pockets or flux gases which expand upon the application of heat and erupt in volcanic fashion, spraying minute particles of solder over unwanted areas of the circuit board.

Another cause is excess solder globules, which, upon becoming molten during the soldering operation, flow along paths that terminate in unwanted areas. If solder

13-92 Metals Joining of Electronic Circuitry

splashing does not occur during the initial soldering of a wire to a terminal or a component lead to the plated-through hole, it is unlikely that it will occur during any subsequent soldering or unsoldering operation.

Gold-plated circuit boards with hand-soldered terminals have more tendency toward solder splashing than solder-plated circuit boards. Solder-plated boards, having a minimum of fused electroplated solder on their surface, have less tendency to cause solder splashing than printed-circuit boards having thick solder-plated surfaces.

Solder-coated terminals have less tendency to cause solder splashing than either solder-plated or gold-plated terminals. Terminals that were fused in place (as opposed to hand soldering) showed considerably less solder splashing than terminals that were hand-soldered to the etched-circuit board. This is a result of using less solder in forming the fillet around the terminal by hot-oil fusing.

If the coating is adhesive-bonded tightly (no air bubbles) to the circuit board, solder splashing will be minimized. In addition, if repair or replacement operations are done quickly and with the coating removed from the area surrounding the

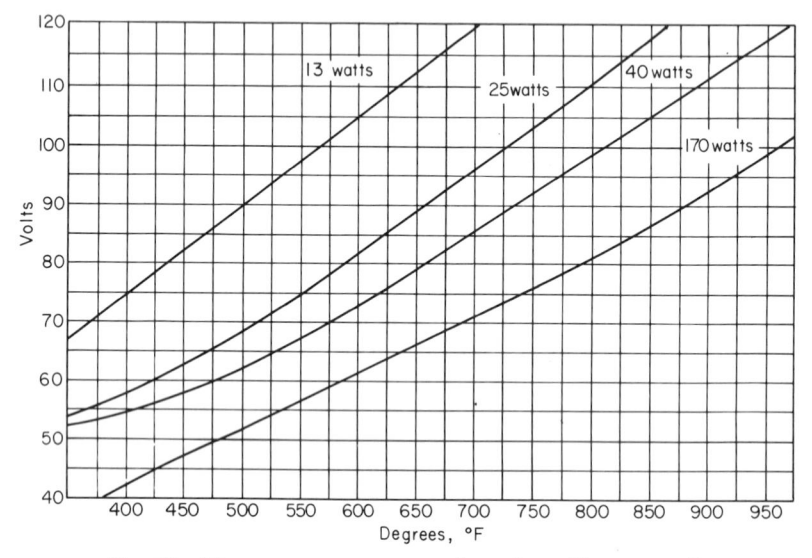

Fig. 63 Tip temperature versus voltage for soldering pen.[46]

solder joint, solder splashing will not occur. The reason is that the removal of organic material from around the solder joint allows an outlet for the expanding gases.

During the etching of solder-plated circuit boards (using the solder resist process) the copper is etched from beneath the solder. This undercutting action leaves an overhang of unetched solder plating. During fabrication of the circuit board, this is sometimes broken loose and removed by brushing; sometimes it is merely folded down over the edge of the conductor.

Subsequent handling of the circuit board during inspection, testing, assembly, etc., can break loose these overhang formations and create solder slivers. Extreme undercutting of solder-plated circuit boards has been reported by others, even to the extent of complete removal of the basic copper conductor lines. There are certain instances, however, where solder slivers do not exist. For example, if a circuit board has been flow-soldered, no slivers would exist on the bottom side of the circuit board, since they would all have been reflowed.

A way to avoid solder slivers is to do away with the solder overhang. The best technique found to date is a controlled hot-oil fusing (also known as flow melting or oil flowing). Controlled fusing is done in a bath with a time-temperature cycle selected to reflow the solder overhang without causing a heavy contoured surface on the conductors. The circuit board to be dipped is fluxed and clamped in a rigid fixture to prevent any warp or twist and dipped horizontally into the hot-oil bath. At the end of the proper immersion time, the board is removed from the hot-oil bath and allowed to cool. The oil and flux residues are then removed by immersion and brushing in solvent and detergent solutions. Controlled hot-oil fusing has been found to be a technique that will eliminate solder overhang and solder sliver problems on double-sided and multilayer plated-through-hole etched circuit boards, if done properly.

Resistance-welded Connections

In resistance welding the heat is produced within the material or leads being welded. This heat is produced by the resistance that the materials or leads themselves offer to the passage of an electric current. During the welding process, pressure or force is applied to the joint interface to actually produce the welded connection.

This combination of heat and pressure is used to produce either of two basic types of welds. Then if the temperature produced at the joint interface is high enough to actually cause the materials to melt, fusion welding takes place.

The second basic type of weld occurs when the temperature at the joint interface is high enough to bring the materials only to a plastic state without actually melting. When pressure is brought to bear on the materials while they are in this plastic condition, a forged weld results.

Fusion welds are the strongest joints. However, the electrical and thermal properties of the materials most often used in electronic joining are such that the most common joints are the forged-weld type.

Resistance welding may be classified into two basic types: opposed electrode and parallel electrode.

Pincer welding is a type of opposed-electrode welding, while gap, series, and series step welding are all types of parallel-electrode welding.

Opposed-electrode welding In this type of welding, the electrodes approach the parts to be welded from directly opposite sides. The electrodes have an included angle of 180°, and the tips or surfaces of the electrodes that come into contact with the work are directly opposed to each other, with the parts to be welded sandwiched in between.

A necessary requirement is access for the electrodes to approach the workpieces from opposite sides. The fact that the electrode tips bear directly against each other in opposite directions makes it possible to obtain relatively high forces. As a result, this process is capable of handling comparatively large wire sizes. The process also is one of the simplest to set up and perform. The weld heads and power supplies normally required are among the less costly types.

The disadvantage in opposed-electrode welding is the requirement for two-sided accessibility. The interconnections for some high-density modules cannot be made with this type of electrode setup.

Both ac and dc power supplies are used, although capacitance-discharge dc types are most common. Capacitance-discharge input is capable of providing high amperage and low-voltage welding energy to the joint interface. The disadvantage, however, with these power supplies is the time delay in recharging the capacitor bank and the factor of electrode polarity. Also, there tends to be some degradation in energy output from capacitor banks over extended periods of time.

Ac power supplies have the capability of medium to long pulse lengths and continuous output, and are not polarity-sensitive. In ac input the energy pulse can be "shaped" to have a waveform with an upslope, downslope, or both. This capability of ac power provides an advantage in welding materials that require preheat and postheat. The main disadvantage with ac supplies is the complexity of the controls needed.

13-94   Metals Joining of Electronic Circuitry

Figure 64 illustrates the setup for opposed-electrode welding. The approximate range of process parameters is shown in Table 39. This type of welding is used for two-sided work such as lap welding, crosswire welding, module welding, and seam welding by making a series of continuous overlapping welds. The advantages are that high weld forces can be obtained, equipment is relatively low in cost and is easy to set up and maintain, and only two welding variables are involved. The main disadvantage is that the interconnections must be accessible from opposite sides.

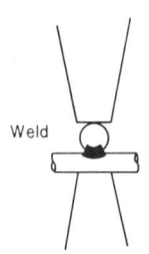

Fig. 64   Opposed-electrode welding.[59] (*Reprinted from Assembly Engineering, September, 1966. Copyright, 1966, by the Hitchcock Publishing Company, Wheaton, Illinois.*)

Selecting the electrode materials best suited for the materials being welded is one of the most important steps in obtaining reliable welds. These should be chosen on the basis of their suitability to the plating on the materials being welded, rather than the base metal itself. The electrode materials should not have the same thermal and electrical conductivity as the materials being welded.

Two problems in resistance welding are the deposition of minute particles of materials onto the electrodes and electrode pitting. Selecting the correct electrode materials is the best way to avoid these. However, the ideal electrode material

TABLE 39   Opposed-electrode Welding Process Parameters[59,*]

Power supplies:	
Ac............................	60 Hz; 0.5 watt to 1 kva
Dc............................	0.2–100 watts
Pulse length:	
Ac............................	Up to 0.1 sec
Dc............................	1.5–15 msec
Electrode force....................	3 oz to 20 lb
Electrode size.....................	0.005–0.1 in. diameter
Total thickness of welded materials....	0.060 in.
Maximum materials thickness ratio....	4:1

* Reprinted from *Assembly Engineering*, September, 1966. Copyright, 1966, by the Hitchcock Publishing Company, Wheaton, Illinois.

for each welding job cannot always be obtained. Therefore, electrode tips should be cleaned and refaced periodically.

Another problem is concentrating the welding heat at the joint interface. Heat balance must be controlled within acceptable limits to obtain reliable welds. If heat-balance problems are encountered, increase or decrease the size of the electrodes being used, change electrode materials, change the thickness ratio of the materials being welded, or change the size and shape of the electrodes for better thermal balance.

In the past, opposed-electrode welding was used, primarily for welding three-

Second-level Joining, Materials and Processes 13-95

dimensional modules such as wire matrixes and cordwood-type assemblies. However, recently the use of opposed-electrode welding in conjunction with flexible printed circuits has been proposed.[60]

Dc power sources provide a fast discharge (1 to 5 msec) at low voltage and high amperage. Dc welds are usually polarity-sensitive, requiring the negative electrode to be placed against the same material each time. Ac power sources

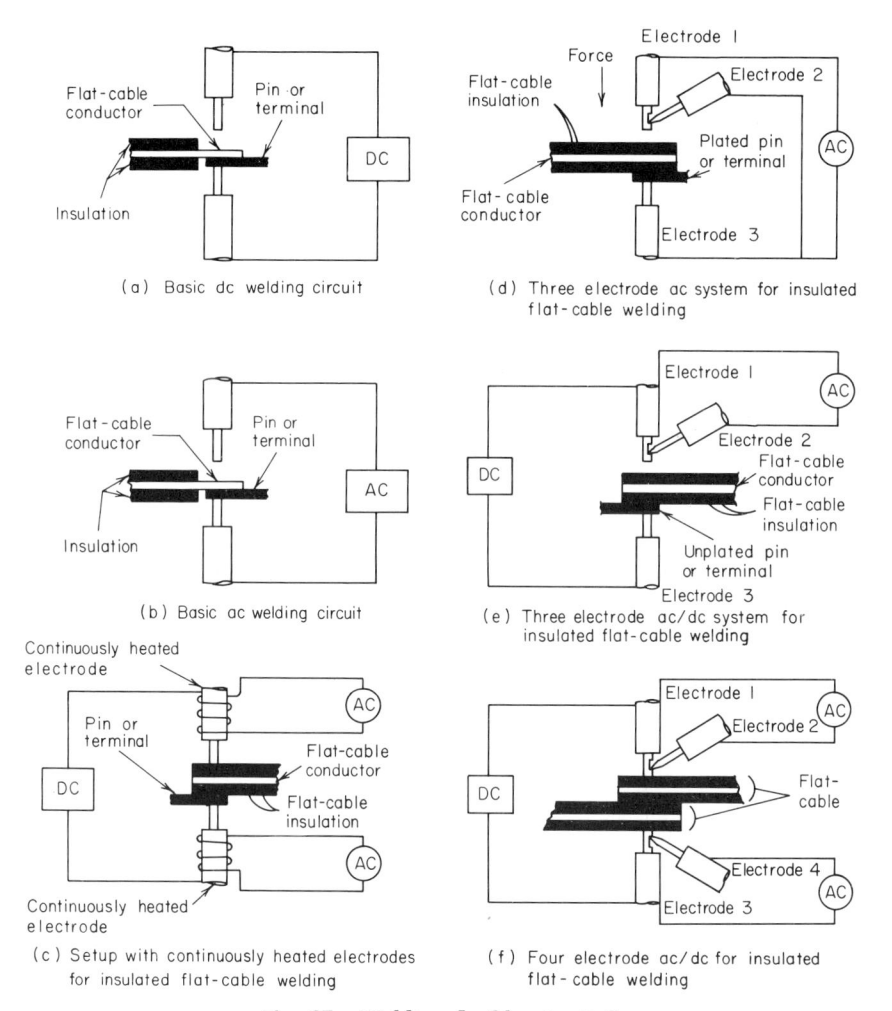

Fig. 65 Welding flexible circuits.[60]

tend to provide longer pulse lengths (8 msec to 3 sec) at a slightly higher voltage and lower amperage. Alternating current, of course, is not polarity-sensitive.

Copper-to-copper welds are best made using the dc approach shown in Fig. 65a because (1) the weld pulse is short enough to allow the weld to be made before the heat drains away from the weld area through the highly thermal conductive copper and (2) dc weld schedules are relatively easy to establish.

The ac approach shown in Fig. 65b might be utilized to weld the conductor

13-96 Metals Joining of Electronic Circuitry

to the terminal if one or the other were made of a material less conductive than copper—nickel, for example. Alternating current is also extremely useful for applying heat to the flat cable insulation during or before the actual welding process.

Figure 65c represents a continuously heated electrode setup for welding through flat cable insulation. The electrodes are kept in a heated condition at all times through heater elements powered by ac sources. The heated electrodes melt the insulation, exposing pure metal to the electrodes. A dc pulse is then applied to the electrodes to provide the actual weld. This setup can be used to weld flat cable to itself or to other pins or terminals, providing that the insulation itself has a relatively low melting point. The disadvantage of this type of setup is that when the electrodes are heated sufficiently to melt higher-temperature insulation, they will not easily pass direct current, and there is a sizable waste of energy.

Figure 65d shows the three-electrode, pure ac approach to insulated flat cable welding. Current initially flows through electrodes 1 and 2, building up intense heat at the tip of electrode 1. This heat melts the insulation on the flat cable conductor. As soon as the circuit "sees" a low-resistance path between electrodes 1 and 3, essentially all of the current flows through the parts to be joined, and a "weld" is produced. This is not a true weld, however, because the setup requires a "third" material in the form of a ductile plating on the terminal or pin.

Electrodes 1 and 3 are usually made of molybdenum while electrode 2 is made of tungsten. The entire welding process typically consumes less than 1 sec, with approximately 0.5 sec required for the vaporization of the insulation and another 0.1 or 0.2 sec for the joining operation.

The advantage of this setup is that the electrodes are heated only during the welding process. And, when they are heated, they reach a higher temperature than continuously heated tips and are therefore able to strip a higher-temperature insulation. The disadvantage is that the weld pulse does not always get its full portion of the total power-supply pulse owing to variations in material thicknesses.

Figure 65e shows an adaptation of the three-electrode process using both an ac and dc power supply. This setup, though more costly than a pure ac approach, overcomes the objections stated above and allows a true fusion weld to take place between the flat cable conductor and the pin or terminal. The necessity for the "third material" or plating on the pin is also eliminated.

The ac pulse requires approximately 0.4 sec and the dc pulse, triggered from 20 msec to 1 sec after the ac pulse, requires only 4 msec. The time delay allows the electrodes to cool down for easier passage of the direct current.

The most complete ac/dc setup—designed for welding flat cable to itself—consists of two shunt electrodes, each powered by an ac supply, and a dc supply to provide the actual weld current. This is called the "4-E process," and it is pictured in Fig. 65f.

When the two ac supplies are triggered (by the electrode force reaching a preset value), the shunt electrode pairs (electrodes 1 and 2, electrodes 3 and 4) build up heat, quickly vaporizing or melting the cable insulation adjacent to them. The alternating current runs for a preset time; then a time delay cuts in, which allows the electrodes to cool. This cooling period will be approximately 0.4 sec. As soon as the time delay ends, the dc power supply discharges, welding the conductors to each other. Multiple conductors, in a sandwich layer, can be welded at one time. Ground planes can be welded to conductors in the same manner.

Pincer welding Pincer resistance welding is similar to opposed electrode-welding except that the included angle between electrodes may vary from 20 to almost 180°. The included angle between electrodes, however, usually is 60°.

The reason for the pincer type of electrode setup is the accessibility advantages that it provides. The pincer setup is ideal for reaching into "tight" places such as those found in high-density modules. Pincer electrodes often have an insulated coating consisting of a dielectric varnish or similar material to prevent short-circuiting against parts of the assembly that they might come into contact with. This coating should be capable of withstanding a great deal of abrasive wear. Assembly operators will be able to maneuver the electrodes into very close quarters within an electronic assembly without short-circuiting or damaging any of the parts.

Because of the acute angle at which the electrodes sometimes are positioned, the squeezing action of the electrodes can impose high twisting forces on the holders and a spreading action at the electrode tips. For these reasons, "pin" the holders to the weld head, and limit the electrode force to 12 lb.

Figure 66 illustrates the technique, and Table 40 lists the approximate range of process parameters.

Pincer welding is used when access to interconnections is difficult. The process can be used for cross-wire welding, welding fine ribbon or wire, and making welded interconnections in high-density modules.

The advantages are that electrodes can be set up to operate either vertically or horizontally, and electrodes can reach into tight places.

One of the disadvantages is that as the included angle between electrodes is reduced, the weld forces must also be reduced to prevent spreading of the tips. Also, setup time usually is longer than that required in opposed-electrode welding, and refacing electrode tips is more difficult.

TABLE 40 Pincer Welding Process Parameters[59,*]

Power supplies:	
Ac	60 Hz; 0.5 watt to 1 kva
Dc	0.2–100 watts
Pulse length:	
Ac	Up to 0.1 sec
Dc	1.5–15 msec
Electrode force	3 oz to 12 lb
Electrode size	0.005–0.1 in. diameter
Total thickness of welded materials	0.045
Maximum materials thickness ratio	3:1

* Reprinted from *Assembly Engineering*, September, 1966. Copyright, 1966, by the Hitchcock Publishing Company, Wheaton, Illinois.

Parallel-gap welding Parallel-gap welding is a form of resistance welding in which both electrodes contact the workpieces on one side. The electrodes are positioned so close together that only one weldment is made between the two electrodes.

The electrodes are individually suspended in parallel-gap welding. The same force is applied to each electrode in practically all cases.

The parallel-gap process is commonly used for welding flatpack leads to printed-circuit boards, and component leads or interconnects to thin-film circuits. Welding ribbons as thin as 0.000125 in. and 0.005 in. wide has been successfully accomplished with the process. Normally, the upper limits for parallel-gap welding wires is about 0.012 in. in diameter, and for ribbon about 0.010 in. thick by 0.030 in. wide. The metallic-base materials being welded can range anywhere from 250 Å in thickness on up.

Split-tip welding is a variation of the parallel-gap process in which the two electrodes are mechanically locked together and separated by a thin dielectric spacer.

Figure 67 is an illustration of parallel-gap welding. Table 41 lists the approximate range of the various process parameters involved in parallel-gap welding.

Parallel-gap welding is used for welding applications such as in connecting flatpack leads to printed-circuit boards or in welding wire and ribbon leads to thin-film depositions. As discussed in the previous section, the process also can be used for soldering if component leads are solder-coated or if solder preforms are used.

The advantages are that welding can be accomplished from one side of the workpiece, visual inspection of the weld area is easy, electrode wear is minimal, and the process is not polarity-sensitive.

13-98   Metals Joining of Electronic Circuitry

The disadvantages are: the process cannot be used on material combinations that have high resistance, and weld parameters are more difficult to determine than for some of the other welding processes.

The gap between electrodes in parallel-gap welding is a critical factor in obtaining reliable welds. The electrode gap should be about two to three times the thickness of the wire or ribbon being welded. Also, the more conductive the wire or ribbon is, the narrower the electrode gap should be.

When welding leads to laminated printed-circuit boards, some "outgassing" of the boards can be expected. The amount of outgassing or size of the burned

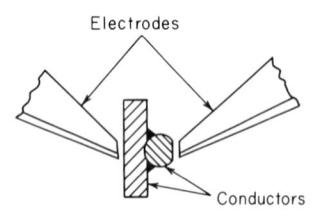

Fig. 66 Pincer welding.   **Fig. 67** Parallel-gap welding.

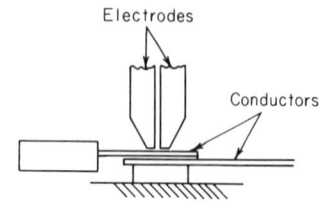

area under the weld is partly determined by the size of the electrode gap. Reducing electrode gap will reduce outgassing.

Both electrodes should be the same size, and the tips should be a few thousandths of an inch wider than the ribbon or wire being welded.

Regular dc capacitance-discharge power supplies can be used for parallel-gap welding; however, this type of supply usually requires a constant-voltage-control accessory for best results. If an ac power supply is used, it can be either 60 or 1,000 Hz, with a controlled voltage output. Such an ac power supply may

TABLE 41 Parallel-gap Welding Process Parameters[59,*]

Power supplies:	
Ac	1,000 Hz; 0.1–200 watts†
Dc	0.2–25 watts‡
Pulse length:	
Ac	Up to 1 sec
Dc	1.5–100 msec
Electrode force	2 oz to 8 lb.
Electrode size	0.005 in.² to 0.25− by 0.1 in.
Total thickness of welded materials	0.018 in.
Maximum materials thickness ratio	50:1

* Reprinted from *Assembly Engineering*, September, 1966. Copyright, 1966, by the Hitchcock Publishing Company, Wheaton, Illinois.
† May require upslope and downslope features and a constant-voltage source.
‡ Constant-voltage output may be desirable.

have a resistance feedback system so that the power supply will "lock out" if it senses a resistance that is too high.

Any of the commercially available flatpacks can be reliably welded to glass-epoxy copper printed-circuit boards if the conductors are properly designed. The guidelines for the design of the boards, whether multilayer, double, or single-sided, are as follows:

1. The width of the conductor pad must be as close as possible to the width of flatpack lead for optimum welding conditions. It should not be less than lead width by more than 0.002 in. or exceed it by more than 0.008 in.

2. The total copper thickness must be at least 0.0028 in., i.e., 2 oz in weight. For plated-through boards, 1 oz copper should be used with 1 oz of copper plating.

3. The finish of the boards must consist of 300 μin. nickel plating and 50 to 80 μin. of gold.

The metals involved are: leads: gold-plated Kovar; conductors: nickel- and gold-plated copper.

The melting points of nickel, Kovar, copper, and gold are 1455, 1450, 1083, and 1063°C, respectively. Most of the electric current passes through the Kovar lead since it is in direct contact with the welding electrodes. Hence, the Kovar lead reaches the highest temperature in the system, but because of its high melting point it does not melt. The Kovar is directly in contact with the gold which is plated to the lead and also the conductor pad. The gold then melts since it has the lowest melting point. If the weld pulse is of the proper amplitude, the gold is the only metal melting to form a braze bond between the Kovar of the lead and the nickel plating of the copper conductor. If the pulse is of higher amplitude than is necessary, then the copper also will melt. With even greater amplitude of pulse, the melted copper is expelled from the area beneath the electrodes so that voids result. The strength of a weld is a measure of its quality and reliability.

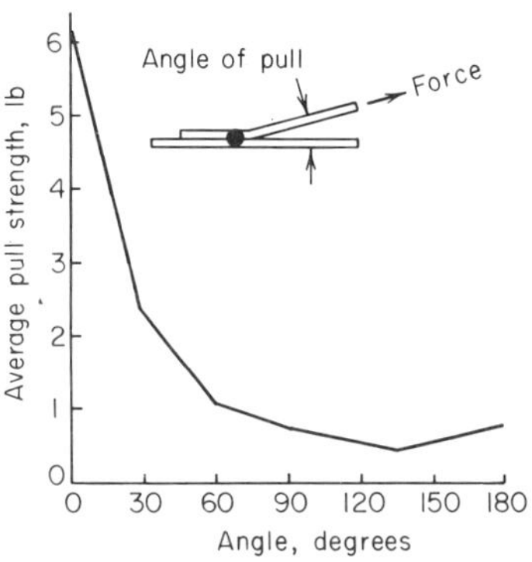

Figure 68 is a plot of the average pull strength versus angle of pull. From this figure it can be seen that the minimum pull strength of the joints occurs at 135° pull angle; in actual use, however, a lead that is part of a flatpack and is mounted on a board will never be subjected to any forces at an angle greater than 45°. The angle of pull in practice is determined by the height of the flatpack and the distance of the point of the weld from the root of the lead. For a typical flatpack lead welded ½ in. away from the glass and with the lead emerging 0.030 in. above the board, the angle of pull becomes $\tan^{-1} 30/125 = 13.5°$. At that angle the average pull strength is in excess of 4 lb. There are only a limited number of flatback header manufacturers, and all use Kovar leads to obtain a hermetic seal to the glass package. The standard finish is 75 to 150 μin. of gold.

Fig. 68 Variation in pull strength as a function of pull angle.[61]

The Joint Electron Device Engineering Council of the Electronic Industries Association (EIA) has acknowledged eight flatpack sizes as standard: they are registered at TO-84 through TO-91. The lead dimensions are all the same: 0.003 to 0.006 in. thick by 0.010 to 0.019 in. wide; i.e., the minimum lead size will be 0.003 by 0.010 in., and the maximum will be 0.006 by 0.019 in.

Table 42 illustrates specific resistance characteristics for four possible circuit-board

TABLE 42 Comparison of Resistances for Circuit-board Conductors
(Based on 0.003- by 0.010-in. Cross Section 1 in. in Length)[62]

Material	Resistivity, microhm-cm	Resistance, ohms/in.	Relative conductivity, %
Copper.	1.71	0.022	100
Nickel A	9.5	0.125	18
Kovar.	50.5	0.66	3.4
5% nickel A . . . . 90% iron . . . . 5% aluminum	5% Ni 6.84 ⎫ 90% Fe 9.71 ⎬ 9.21 5% Al 2.65 ⎭	5% Ni 1.8 ⎫ 90% Fe 0.143 ⎬ 0.12 5% Al 0.69 ⎭	18.6

materials using a representative circuit-run cross section of 0.003 by 0.010 in. The sandwich foil material shown in number 4 in the table consists of 5 percent nickel on the top surface metallurgically bonded with 90 percent iron and 5 percent aluminum, and represents a useful compromise of properties. The nickel provides a weldable surface with good storage life. The iron core tends to act as a thermal barrier preventing weld heat from degrading the bond to the glass-epoxy substrate, and the aluminum provides a good bond to the glass-epoxy.

Table 43 summarizes properties of resistivity, relative conductivity, and melting points for a group of metals and alloys that are representative of the most common

TABLE 43 Material Properties Comparison[62]

Material	Resistivity, microhm-cm	Relative conductivity, % LACS	Melting point, °C
Copper (ETP)	1.71	101	1083–1150
Nickel (Ni 99.95 + Co)	6.84	25.2	1453
Nickel A (Ni 99.4 + Co)	9.5	18	1435–1446
Kovar (Fe 54 Ni 28 Co 18)	50.5	3.7	1450
Nickel-iron (Ni 50 Fe 50)	48.0	3.5	
Phosphor bronze 1.25% (Cu 98.75 Sn 1.25)	3.6	48	930–1000
Phosphor bronze 5% (Cu 95 Sn 5)	11.0	15	950–1050
Phosphor bronze 8% (Cu 92 Sn 8)	13.0	13	880
Phosphor bronze 10% (Cu 90 Sn 10)	16.0	11	845–1000
Iron (99.99% pure)	9.71	17.75	1527–1537
Aluminum (99.99% pure)	2.65	64.94	660
Cupronickel 30% (Cu 70 Ni 30)	37.0	4.6	1170–1240
Cupronickel 10% (Cu 88.7 Ni 10 Fe 1.3)	14.7	9.1	1100–1150
Monel* (Ni 67 Cu 30)	48.2	3.58	1300–1350
Cobalt	6.24	27.6	1495
Commercial bronze (Cu 90 Zn 10)	3.9	44	1020–1045
Gilding 95% (Cu 95 Zn 5)	3.1	56	1275–1300
Beryllium copper (Cu 97.9 Be 1.9 Ni 0.2 or Co)	9.6–11.5	15–18, cold-worked 22–30, precipitation-hardened	870–980
Soft solder (Sn 63 Pb 37)	14.5	11.9	183
Constantan (Ni 45 Cu 55)	50.0	3.7	1210

* Trademark of Huntington Alloy Products Division, The International Nickel Co., Inc., Huntington, W. Va.

clad materials being offered by suppliers for weldable circuit applications at the present time. As a general statement, the higher-resistance materials offer improved weldability characteristics. A weldable multilayer circuit-board concept offers a unique design advantage. The use of multiple layers makes it possible to combine high conductivity etched copper-clad lower layers with a higher-resistance surface layer for weldability.

A second form of weldable multilayer circuit board which has been investigated does not use plated-through holes for interlayer conections. As shown in Fig. 69, this concept, known as step-weld interconnection, achieves interlayer connections by a welded ribbon in a "staircase" configuration formed by punching each layer with rectangular holes of progressively different lengths. This design concept required that a higher-resistance, weldable, clad material be used for all layers instead of for just the top layer, as with plated-through holes. The step-weld board manufacturing sequence includes etching individual circuit layers, punching the rectangular hole pattern in respective clad laminate and prepreg adhesive layers, registering circuit layers, welding interlayer step-weld connections, and laminating. Since lami-

Second-level Joining, Materials and Processes 13-101

nating is done after the interconnection welding is completed, the normal prepreg resin flow fills the holes and effectively encapsulates the interlayer welded connections. This design approach requires more board area to accomplish the interlayer connections and does not lend itself to the conventional copper-clad materials for any of the circuit layers.

Still other techniques that have been considered for achieving welded interlay

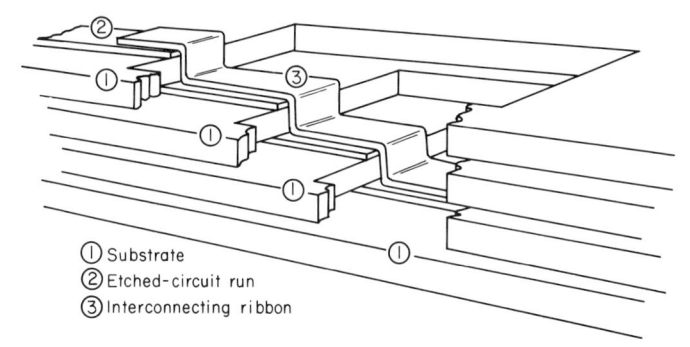

① Substrate
② Etched-circuit run
③ Interconnecting ribbon

Fig. 69 Step-weld interconnection technique.[62]

connections are summarized in Fig. 70. The tab-to-run interconnection illustrated is a special application of the step-weld for joining only two layers of circuitry.

The tab-to-tab interconnection uses opposed-electrode welding and requires clearance holes in both sides of the circuit board. For this reason it increases the board area for a given amount of circuitry. The approach also presents mechanical processing difficulties in properly aligning the tabs before welding.

The chimney or pushed-through interconnection illustrated in Fig. 70 is a variation of the tab-to-tab technique in which the weld joint is made at the top surface.

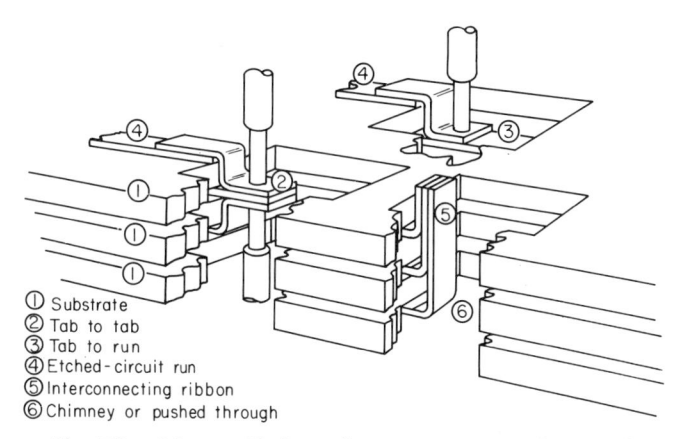

① Substrate
② Tab to tab
③ Tab to run
④ Etched-circuit run
⑤ Interconnecting ribbon
⑥ Chimney or pushed through

Fig. 70 Other welded interlayer-connection techniques.[62]

This design is adaptable to automated techniques. Although requiring higher resistances, this technique, when automated, offers a very fast turnaround cycle from circuit schematic diagram to finished board. A program sponsored by the Air Force Materials Laboratory has developed a dry process based on conventional G-10 epoxy-glass substrate, standard matrix conductors of preetched nickel, through connections of nickel ribbon, and welded joints. Starting with a schematic diagram and board mechanical specifications, interconnection signal routing is determined by computer; then the board is fabricated by a tape-controlled numerical machine.

Outputs from the computer design phase are automatically drawn assembly drawings and line drawings of the signal paths by layer, and fabrication information in the form of numerical-machine control tape. This concept offers a complete design and fabrication cycle of two weeks for high-density multilayer circuit boards.

Series welding Series welding is similar to parallel-gap welding in that both electrodes contact the work from one side. The principal difference between the two processes, however, is that series welding produces a separate weldment under each electrode. Also, in series welding the electrode gap, normally, is greater than 0.025 in.

The series welding process is used when it is necessary to make two welds simultaneously at an interconnection or a joint. Two welds may be needed for added reliability at the point of interconnection.

The electric current passes from one electrode through the base metal and parts being welded and back to the other electrode. The weld process is polarity-sensitive with dc power supplies, and one of the weldments usually is stronger than the other.

Wires up to 0.040 in. in diameter and ribbons up to 0.015 in. thick and 0.050 in. wide can be welded in the process. It is essential that the base metal have the capacity for carrying the electric current loads required for welding heavier wires and ribbons. Wide differences in material thickness cannot be tolerated.

Series welding is used to connect flatpack leads and other component leads to printed-circuit boards. The process is especially adapted for making two simultaneous welds at interconnections. However, there is a difference in the strength of the two weldments.

Normally the forces applied at the electrodes are the same for each electrode. However, if different-size weldments are desired, the forces at each electrode can be different.

As a general rule, the gap or spacing between electrodes must be greater than 0.025 in. Gap spacing beyond this dimension is not critical.

The most reliable welds are obtained when the base metal is more conductive than the wire or ribbon being welded.

A controlled-voltage output is not required in series welding. Regular dc capacitance discharge and ac power supplies can be used.

The process is polarity-sensitive. The approximate process parameters are contained in Table 44.

TABLE 44 Series Welding Process Parameters[59,*]

Power supplies:	
Ac	60 Hz; 0.5 watt to 0.5 kva
Dc	0.2–100 watts
Pulse length:	
Ac	Up to 0.1 sec
Dc	1.5–15 msec
Electrode force	3 oz to 10 lb
Electrode size	0.005–0.1 in. diameter
Total thickness of welded materials	0.020 in.
Minimum materials thickness ratio	2:1

* Reprinted from *Assembly Engineering*, September, 1966. Copyright, 1966, by the Hitchcock Publishing Company, Wheaton, Illinois.

Series step welding This is a type of resistance welding where one electrode is in contact with the base metal while the other electrode is in contact with the ribbon or wire being welded. It is used primarily on materials that cannot be welded by parallel-gap welding, and the process is especially well suited for welding high-resistance leads to base metals. A big advantage in this process is that high-resistance leads can be welded from one side of the workpiece. Wires

Second-level Joining, Materials and Processes   13-103

up to 0.030 in. in diameter can be welded to printed-circuit boards. However, base metals must be thick enough to carry the electric current needed for welding heavy leads. Welds on thin films are not very practical because of the low current-carrying capacity of the films.

Better heat-balance control is possible than can be achieved with parallel-gap welding. Establishing a good current path between the two electrodes is essential in obtaining reliable welds. This current flow can be improved in a number of different ways such as by using high electrode forces to make good electric contact or using different electrode materials or electrode sizes.

When welds are made to a base metal with limited current-carrying capacity, shape the electrode in contact with the base metal in the form of a crescent so that it surrounds the weld area. The approximate range of series step-welding process parameters are contained in Table 45.

TABLE 45 Series Step-welding Process Parameters[59,*]

Power supplies:	
Ac	60 Hz; 0.5 watt to 0.5 kva
Dc	0.2–100 watts
Pulse length:	
Ac	Up to 0.1 sec
Dc	1.5–15 msec
Electrode force	3 oz to 10 lb
Electrode size	0.005–0.1 in. diameter
Total thickness of welded materials	0.030 in.
Maximum materials thickness ratio	1:1

* Reprinted from *Assembly Engineering*, September, 1966. Copyright, 1966, by the Hitchcock Publishing Company, Wheaton, Illinois.

Other Welded Connections

Percussive arc welding[63] This type of welding is accomplished by bringing the two workpieces together during or immediately after the period of an electric arc between them. The arc melts the surfaces of both conductors; while the surfaces are molten, the pieces are jammed together and then allowed to cool.

Percussive arc welding can join a large variety of material combinations, from very fine wire to heavy power handling wires. Percussive arc welding is used to butt-weld leads and terminals.

There are four welding systems in use. One is a straight ac machine, usually 60 Hz. The basic components in this machine are a transformer, a mechanical or electronic switch, electric connections to the weldments, and usually a mechanically operated actuator. This type of machine is used a great deal in structural work for heavy rivets and studs.

It has been found limited to relatively few material combinations when applied to electronics. The advantage is its simplicity and capability of providing very high heating to the workpieces. The disadvantages to its use in electronics are overheating of materials and the fact that the weldments usually must be shaped on the end surfaces to provide material for melt and striking the arc.

The other systems are all of the dc capacitor-discharge type. The simplest type of machine has a dc power supply charging a capacitor for the weld energy, a switch to connect the capacitor bank to the weldments, and an electromechanical actuator to bring the two weldments together. The advantage of this is that it is very simple, rugged, and low priced; however, it is limited in the range of materials and sizes that it can weld, and the weldments must be shaped to strike the arc. Figure 71 is a schematic diagram of this type of machine.

Another system has a dc power supply charging a capacitor bank for energy storage with the addition of high-frequency rf pulse to initiate the arc. The rf pulse initiates the weld arc, and as the weld current flows through the electromechanical actuator, the two weldments are brought together to made the interconnection. The rf pulse initiating the arc eliminates the critical shaping of weldments and provides more uniform and reliable welding of a wide variety of materials.

In the fourth system, the electromechanical actuator is independently powered, and the rf pulse can be timed to occur at some definite point during the weldment travel. The weldments can attain velocities independent of weld power and provide optimum penetration of weld. By timing the initiation of the arc with the rf pulse, welding can be accomplished with less heat generated and less deformation of materials. With controls for the parameters of weld energy, actuator velocity, timing of rf pulse, and weldment gap, a wider variety of material combinations and sizes may be welded.

Percussive arc welding is capable of welding a variety of materials such as copper, steel, Alumel,* Chromel,* molybdenum, nickel, aluminum, and tantalum. These materials may also be welded to each other, such as copper to aluminum, copper to steel, nickel to tantalum, and aluminum to steel.

In the percussive arc-welding technique, the heat is concentrated at the weld interface only, and for an average of one millisecond only. The temperature rise beyond the intermediate weld has been found to be negligible for most applications.

The workpieces are used as an anode and a cathode. At the beginning of the weld cycle, the workpieces are separated by an air gap. When the power supply is triggered, a pulse of rf energy coupled to the output through a transformer ionizes the air gap. This creates an arc discharge of the direct current which heats the two surfaces to be joined to a welding temperature. Simultaneously, an electromagnetic actuator is energized and accelerates one workpiece toward the other. When the materials are in a proper welding state, the actuator percussively joins them together with sufficient force to fuse the molten surfaces, but does not damage materials such as small or stranded wire.

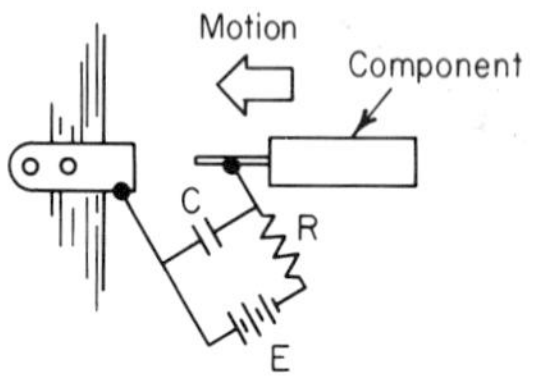

Fig. 71 Percussive arc welding.

The reliability of percussive arc-welded interconnections using radio frequency to initiate the welding arc is substantially greater than that with a machine using straight dc discharge. Critical wire shaping is eliminated, and the degree of melt on the weldments is more accurately controlled.

The tensile strength of weld joints is as high as the strength of the weakest of the parent material. The welded joint will stand up under vibration stresses as well as the wire. The electrical characteristics of the welded joint are not degraded under vibration stresses.

Ultrasonic welding[64] Ultrasonic welding is a process for joining similar and dissimilar metals by the introduction of high-frequency vibratory energy into the overlapping metals in the area to be joined. No fluxes or filler metals are used, no electric current passes through the weld metal, and usually no heat is applied. The workpieces are clamped together under moderately low static force, and ultrasonic energy is transmitted into them for a brief interval. A sound metallurgical bond is produced without arc or spark, without melting of the weld metal, without the cast structure resulting from melting.

The energy used in producing an ultrasonic weld passes into an electronic generator or rotating-machine device, which converts the 60-Hz line power to the desired frequency of operation, usually within the range of 1,000 to 100,000 Hz/sec. The high-frequency current is delivered to a transducer which converts the electric power into acoustic power at the same frequency. The acoustic power is then transmitted through a coupling member to the work-contacting tip and then into the material to be welded. Some of the energy traverses the weld zone and is dissipated elsewhere. Energy losses occur in the various links. The acoustic energy

---

* Trademarks of Hoskins Manufacturing Co., Detroit, Mich.

delivered to the weld is in the order of 15 to 30 percent of that originally drawn from the power line.

Welding occurs when the tip, clamped against the workpieces, oscillates in a plane parallel to the weld interface. The combined static and oscillating forces introduce dynamic stresses in the metal, producing elastoplastic deformations which effect a moderate temperature rise in the weld zone. The properties of the material in the weld zone are transiently altered.

There is an upper limit to the thickness of any given material that can be effectively welded, because of the power-handling capacity of the equipment. For a readily weldable material such as aluminum the maximum thickness in which reproducible welds can be produced is approximately 0.10 in.; for some of the harder metals the upper limit is in the range of 0.02 to 0.05 in. This limitation applies only to the thinnest member of the weldment; the other member may be of greater thickness. There appears to be no lower limit to weldable thickness. Fine wires of less than 0.0003 in. in diameter have been satisfactorily welded, and thin foils of 0.00017 in. thickness have been joined without rupture.

Among the most readily weldable materials are aluminum and its alloys. Copper and its alloys, and other soft metals, are also relatively easy to weld. High thermal conductivity of materials appear to be less of an adverse factor in ultrasonic weldability than in melting-type joining processes. Satisfactory bonds can be produced in iron and steel of various types. Nickel is readily weldable. Titanium and zirconium alloys can be welded most readily. Tungsten, tantalum, and molybdenum present certain problems, but thin gages of these materials have been satisfactorily joined. Beryllium has been welded experimentally.

No particular difficulties have been encountered with gold, silver, and platinum, as well as alloys of metals. These and other materials have been satisfactorily bonded to semiconductors such as germanium and silicon.

There is an optimum clamping force at which effective bonding of a given material combination is achieved with minimum vibratory energy.

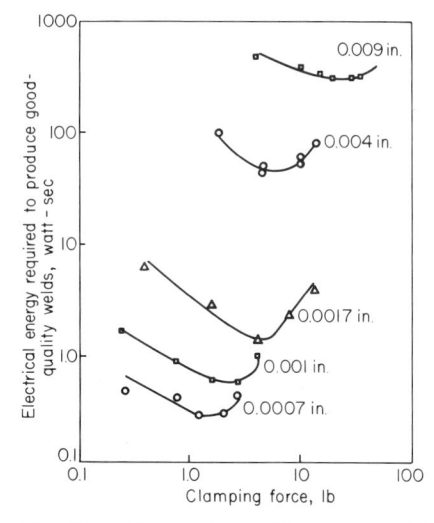

Fig. 72 Ultrasonic welding, threshold curves.[64]

The procedure used to establish this power–clamping-force relationship for spot-type welding is illustrated in Fig. 72. At a fixed weld time (in this case 1 sec) for the entire series of curves, each point was obtained by arbitrarily selecting a single clamping force and producing welds at decreasing values of power. Attempts to peel these specimens established a power value at which the bond ceased to exhibit weld button tear-out and could actually be peeled apart. The power setting corresponding to this transition is the threshold-power curve value at the selecting clamping force. The method is repeated for other arbitrarily selected clamping-force values, until sufficient data are obtained to form a curve of clamping force versus power which represents the threshold-power value for welding a given gage of material. Threshold curves for each of the gages noted in Fig. 72 were obtained in this manner.[64]

Ultrasonic welding can be used in the assembly of miniaturized components in the electronics industry, since the finest wires and the thinnest foils can be welded to pieces of almost any size or shape. The process is well suited to automated assembly procedures for such application because once the proper welding conditions are established, joints of high reproducibility can be obtained.

One application already discussed is the attachment of fine aluminum or gold lead wires to semiconductor materials. The leads can be precisely positioned and bonded in very small areas without contaminating the high-purity semiconductor material with arcing, sparking, or outgassing.

Copper, chromium, and silver leads can be attached to metallized glass surfaces as, for example, in the fabrication of high-temperature printed circuits. Delicate instrument components such as fine wire coils and aluminum pointers can be bonded by the technique. Ultrasonic welding is used for producing current-carrying junctions involving various combinations of aluminum, copper, silver, etc.

Electric wires of the same or dissimilar materials have been satisfactorily joined, even though certain insulation coatings and stranded aluminum wire have been joined to copper terminals.

Contact buttons of one material are readily welded to another material in sheet, plate, ribbon, or other form.

Air Force Contract AF33(600)-30839, Wafer Transformer Project, supported a study of ultrasonic-welded aluminum-copper junctions as a means of attaching leads to a wafer-type coil. A summary of the results obtained is shown in Table 46.

The tests were performed with all joints of equal cross section joined in series in ovens at 130 and 180°C under tungsten-lamp loads at various multiples of the basic transformer design standard of 1,000 amp/in.[2] of conductor cross section. All joints were tested under alternating current switched "on" 1½ hr and "off" ½ hr. Initially, tests were begun at 150 percent of the basic design current, but this was soon increased to 300 percent, then to 600 percent and finally 10 times in an attempt to force failure. Also initially only the current to the tests joints was switched, but this was soon changed to include temperature cycling of the ovens as well. However, that the ovens do not go below 30°C on the "off" cycle is a necessary precaution to exclude larger amounts of moisture, few of the test junctions being coated or protected.

Initial and final resistances were taken by both ac drop underload and by dc measurement on a Kelvin bridge, the condition of each joint being followed during the life testing by ac drop only.

Electron-beam welding One of the more recent welding techniques that is being used to join electronic devices is the electron-beam process. In this process, a high-power electron gun generates a narrow stream of high-velocity electrons which are beamed at the workpiece. As these high-velocity electrons strike the target, their kinetic energy is converted almost completely to thermal energy within the impact area by violently increasing the lattice vibration of the atoms. This energy absorption produces a corresponding increase in the temperature of the target material which is directly proportional to the power density of the incident electron beam. In the electron-beam welding process, sufficient beam power is applied to the parts to be joined to produce a narrow zone of molten metal which fuses the parts together upon solidification.

The guns used for this process are high-power versions of those employed in cathode-ray and x-ray tubes, and are composed of an electron-emitting cathode, a focusing electrode, and an accelerating anode. The gun systems employed for electron-beam welding differ in the emitter design, power output, and the focusing system for the output beam.

Since the emitting cathode must resist poisoning from metal vapors, released gas, and sputtering from the heated target, the emitter materials are pure refractory metals such as molybdenum, tantalum, and tungsten which necessitate a cathode operating temperature over 2000°C to produce sufficiently high electron emission. Tungsten is used for filament-type emitters, while tantalum is made into solid and stronger configurations and heated by electron bombardment from a filament placed behind the emitting surface. The constancy of the emitter's shape during operation and the excellent focusing characteristics produce a very high efficiency of emitted cathode current. Figure 73 illustrates an electron gun of this type in which 99 percent of the emitted cathode current can be focused through the anode aperture. The focusing electrode at the gun has only a small bias between it and the cathode for variable-focus control, and is eliminated in some electron-beam systems. The largest positive potential gradient that is maintained between the anode and the emitter accelerates the electrons toward the target. In some instances, the anode is merely the workpiece itself, which is at a positive potential difference from the cathode. This arrangement is called a *work-accelerated system*. In the self-accelerated gun, the accelerating anode is a separate disk structure with a hole in the center that allows the electron stream to pass through it. Since no potential gradient exists between the gun and the work in the self-accelerated gun, the electron beam is not disrupted by irregularities in the work surface or the workpiece geometry. Since the internal focusing system of the cathode converges the electron beam a very short distance away from the emitter, another auxiliary focusing system is required to refocus the beam a safe distance away from the cathode to prevent sputtering of metal vapor, etc., from deflecting the beam or damaging the emitter surface. Two methods, electrostatic and electromagnetic, are presently being used. In electrostatic focusing, the stream of electrons is focused into a narrow beam by the electric lines of force created around a centered hole in a plate. The electromagnetic focusing system utilizes a series of magnetic lense and deflection coils to align and focus the electron beam. The deflecting sys-

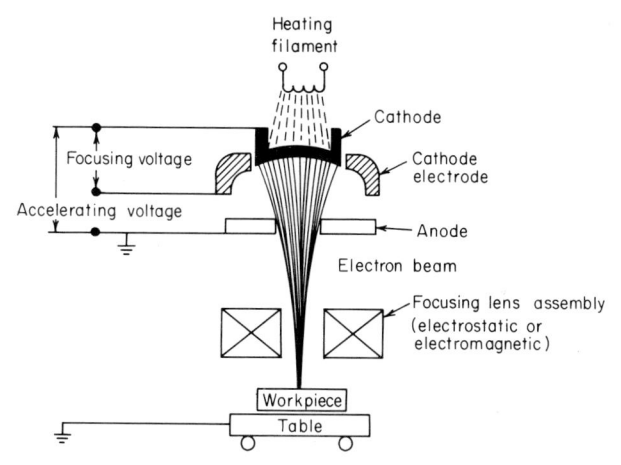

Fig. 73 Self-accelerating electron-beam welding system.[65]

tem enables the operator to move the beam a short distance over the target without moving the workpiece.

Since electrons are easily scattered by collisions with gas molecules, a good vacuum environment is required for precise control of the electron beam. Most systems operate at around 10^{-4} mm Hg with ultimate pump-down capabilities of about 10^{-6} mm Hg. The vacuum requirements for electron-beam processes present both advantages and disadvantages. The advantages gained are the ability to weld refractory metals such as tungsten, molybdenum, etc., without contamination from the air. On the other hand, the restriction placed on weld sizes and configurations by a leakproof vacuum chamber is a problem with larger workpieces.

The most important parameter of electron-beam welding is the accelerating voltage. Most electron-beam welders can be divided into low-voltage machines, which operate at 10 to 30 kv, and high-voltage welders which utilize 50 to 150 kv. Each type possesses unique advantages and applications. The high-voltage equipment has produced very high (25:1) depth-to-width ratios of welds with some metals. The minimum electron-beam spot size for these machines is about 0.001 in., as contrasted with about 0.010 in. for the low-voltage type. The high-voltage system is capable of achieving higher power densities on the target.

TABLE 46 Comparison of Aluminum-Copper Junctions[74]

Ultrasonic weld

Materials: gage and type	0.00025 Al, 0.0002 Cu (bare)	0.00025 Al, 0.002 Cu (bare)	0.0005 Al PYF,* 0.0028 Cu (bare)	0.0005 Al (bare), 0.002 at clad Cu, Ni plated	0.0005 Al (bare), 0.0028 Cu, Ni-Cd plated
Condition	....	Coated after	....	....	Chromated
Operating temperature, °C	180	180	180	180	180
Operating current,† asf	3×	3×	....	....	....
1st 1,000 hr	....	....	6×	6×	6×
Next 5,000 hr	....	....	10×	10×	10×
Number of samples	10	8	6	6	6
Type of change in resistance:‡					
1st 1,000 hr	S.D.	S.G.I. (1st 5,000 hr)	S.G.I.	S.G.I.	S.G.I.
Next 5,000 hr	N.F.C. (1st 5,000 hr)	N.S.C.	N.S.C.	See below.	S.G.D.
Special results or changes	Change in resistance confined largely to 3 samples (next 9,000 hr)			In next 5,000 hr some slightly up and some slightly down	

Capacitance discharge weld

Materials: gage and type	0.00025 Al, 0.002 Cu (bare)	0.0005 Al PYF,* 0.0025 Cu, Cu plated, 60:40 tin-lead	0.0005 Al PYF,* 0.0028 Cu, Cd plated	0.0005 Al PYF,* 0.0028 Cu, Cu plated, tinned	0.0005 Al PYF,* 0.0028 Cu (bare)	0.0005 Al (bare), 0.002 Cu, Al clad, Ni plated	0.0005 Al vinyl, 0.0028 Cu (bare)
Condition	Coated after						
Operating temperature, °C	180	180	180	180	180	180	180
Operating current,† asf	3×	....	....	....	....	....	....
1st 1,000 hr	....	6×	6×	6×	6×	6×	6×
Next 5,000 hr	....	10×	10×	10×	10×	10×	10×

Number of samples	7	6	6	6	6	6	6
Type of change in resistance:‡ 1st 1,000 hr	S.G.I. (1st 5,000 hr)	N.S.I.	S.G.I.	See below.	See below.	S.G.I.	See below.
Next 5,000 hr	N.S.C.	S.G.D.	S.G.D.	S.G.D.	See below.	S.G.D.	All show decrease.
Special results or changes			1 joint had original high dc resistance, and Cu was visibly corroded.		Low-resistance increase slight; high-resistance decrease slight 1st 1,000 hr	66% joints increased markedly in 1,000 hr. 100% increased in next 5,000 hr. Cu was visibly corroded.	All joints show radical increase in resistance; vinyl-coated Al used in error. All were badly corroded.

Soldered

Materials: gage and type	0.00025 Al, 0.002 Cu (bare)	0.00025 Al, 0.002 Cu (bare)
Condition	Not cleaned	Cleaned
Operating, temperature, °C	130	130
Operating current,† asf	3×	3×
Number of samples	7	3
Type of change in resistance:‡		
1st 5,000 hr	G.I.	S.G.I.
Next 5,000 hr	S.G.I.	S.G.I.

* PYF Polyvinyl formal resins.
† Design current = 1,000 asf (amperes per square inch).
‡ Total time includes 1½ hr on, ½ hr off.
S.D.   Slight decrease.
S.G.I. Slight general increase.
N.F.C. No further change.
N.S.C. No significant change.
S.G.D. Slight general decrease.
N.S.I. No significant increase.
G.I.   General increase.

13-110   Metals Joining of Electronic Circuitry

One of the advantages of the electron-beam process is that the incident electron beam penetrates deep into the target material with relatively little heat transfer by conduction to the surrounding metal area. This effect considerably reduces the hazard of thermal damage to adjacent components. This extremely localized heating produces a very limited fusion zone and an extremely narrow heat-affected zone. The heat-affected zone of many welds is the weakest area; the joint often exhibiting brittleness, lack of strength, and ductility. The thin but weak heat-affected zone of an electron-beam weld is mechanically supported by the surrounding stronger unaffected base metal. Although the electron-beam method has minimized the fundamental fusion-weld effects, it has not eliminated or changed the basic metallurgical principles that govern the final microstructure of any weld. The weld properties are dictated by the fundamental principles of thermodynamics, equilibria, etc., regardless of the welding process. Some problems are encountered with electron-beam welding.

The vacuum environment has apparently created a porosity problem in some electron-beam welds. Since many of the oxides and alloy constituents of the high-temperature alloys are volatile, the welding of these materials produces weld-metal porosity. Therefore, consideration must be given to the alloy composition and gas content when welding with electron-beam equipment. The extremely intense energy focused on a very concentrated area with electron beams has also created some problems. It may be necessary to stress-relieve most welds to remove any residual stresses present in the weld area.

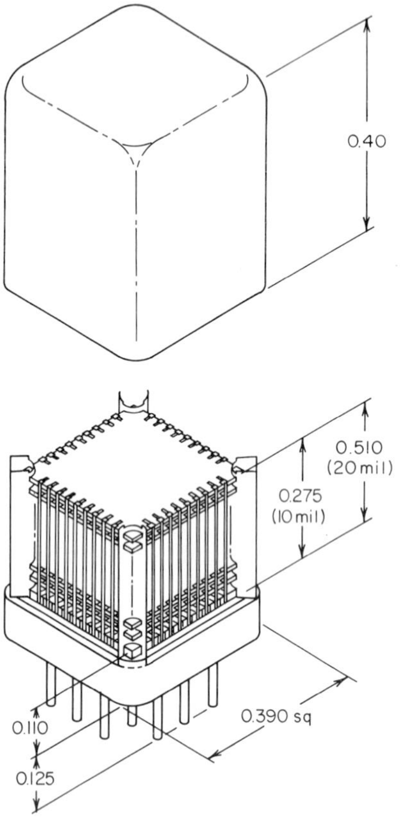

Hamilton Standard, Division of United Aircraft Corp., under sponsorship of the U.S. Army Electronic Command, has extensively investigated and developed electron-beam welding. The high-reliability enhanced micromodule is one example.[66] A conductor 0.002 by 0.010 in. OFHC copper ribbon is welded to the metallized edge of alumina ceramic wafers, as shown in Fig. 74. Welding is done in automated equipment. After each welding operation along the edge of any one wafer, the beam is deflected to the next welding station, is activated, and completes its programmed weld

Fig. 74 Enhanced micromodule.[66]

angle. After a row of microwelds is completed, the beam is deflected to the next microwafer, and the cycle continued.

Laser welding[67]   The laser (contraction of "light amplification by stimulated emission of radiation") is a device that emits collimated monochromatic light. It utilizes a ruby excited by a flash of light and consists of a rod of ruby with both ends silvered–one end partially transmitting. Surrounding it is a xenon flash lamp which excites or "pumps" the chromium ions in the ruby into higher-energy states. Normally, the ions will fall back to the ground state spontaneously and give up their energy in the form of fluorescent emission. When the proper conditions exist in the ruby, the excited ions are induced to give up energy of excitation simultaneously. This produces an avalanche of radiant energy rather than simply a fluorescent glow. This avalanche of radiant energy is characteristically monochromatic, coherent, collimated, and very intense.

A schematic diagram of a laser welder is shown in Fig. 75. The main components of a pulsed laser are a "pumping" source (usually a xenon flash lamp), the reflecting cavity to help direct radiation from the lamp onto the crystal (usually ruby), and the optical system for focusing the laser beam from the crystal onto the workpiece. Although there are many types of lasers, only the solid-state devices at present have sufficient power and intensity to perform welding tasks. Normally the output from such a laser has the wrong characteristic to be useful as a welding tool. The pulse duration is too short. This short pulse results in hole drilling instead of welding. To produce welds instead of holes it is necessary to design the unit to reproducibly obtain a long controlled pulse.

A variety of materials have been welded with the laser. These involve similar and dissimilar metal joints including copper, nickel, tantalum, stainless steel, Dumet,* Kovar, aluminum, tungsten, titanium, and columbium.

Four types of interconnection welds are shown in Fig. 76. With the butt configuration, two wires, preferably with squared-off ends, are placed end to end, and the laser beam is directed at the joint. The energy in the laser beam is

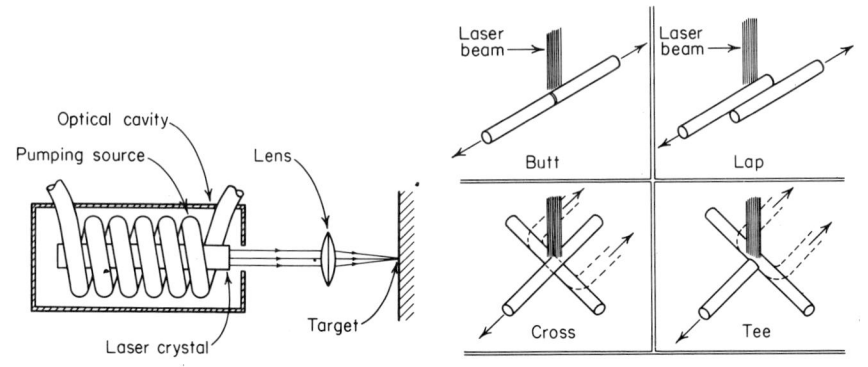

Fig. 75 Schematic diagram of laser welder.[67]

Fig. 76 Joint configuration for wire-to-wire welds.[67]

absorbed by the metal, creating a weld nugget. In the lap joint, the two pieces of wire are placed side by side, and the beam is directed at the area where they touch. The tee joint is a variation in which the one wire is at right angles to the other, whereas in the cross configuration one wire is on top of the other.

The preferred joint configuration for laser welding of wires is the lap joint, although sound welds can be produced in the other configurations. By comparison, the cross configuration is preferable for resistance welding so that pressure can be easily applied to the wires and the weld forms at the joint interface. With the lap joint, the laser energy is directed to the precise spot where the weld nugget is needed without stringent requirements for lead length, alignment, etc. Multiple spot welds can be made for added strength, as desired. Only a simple fixture or clamping arrangement that holds the wires together is necessary. When the cross configuration is used, direct the laser beam at an angle to the joint so that it strikes the interface between the wires. This eliminates the need to melt through the top wire in order to join it to the bottom one.

The optimum laser output energy for welding a particular wire depends on the physical properties of the metal (absorptivity, thermal conduction, density, heat capacity, and melting point) as well as on the laser pulse length. Materials with high surface reflectivity require greater laser energy output than those with low reflectivity or high absorptivity because a greater portion of the radiant energy is reflected. Metals such as copper and silver with high thermal conductivities

* Trademark of United Mineral and Chemical Corp., New York, N.Y.

require a greater output than low-conductivity materials like nickel or iron. An approximate correlation, which was derived empirically, is shown in Fig. 77.

The laser can produce fusion welds on printed-circuit boards without damage to the substrate. The board claddings tested include copper, nickel, and Kovar. For laser-welding to circuit boards, the usual procedure is to preform the leads so that they make contact with the pads on the board when the body of the device is placed next to the board, the device being held to the board by means of adhesive or double-sided tape. After alignment is checked, the leads are individually welded.

Figure 78 illustrates the end, center, and plug configurations investigated for welding integrated circuits to boards. With the end weld the lead is positioned next to the board and the laser directed so that the beam partially strikes the lead and partially strikes the board. In the center-weld configuration the beam wholly impinges on the lead. With plug-weld configuration the laser strikes a lead containing a drilled hole slightly smaller than the diameter of the focused laser spot. Thus,

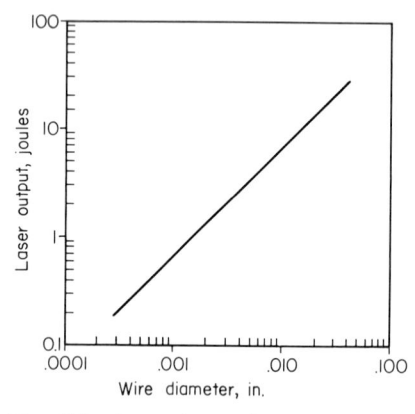

Fig. 77 Approximate laser output required.[67]

Fig. 78 Joint designs for integrated-circuit welding.[67]

the center portion of the laser beam strikes the board cladding, and the peripheral portion of the beam impinges on the lead.

The preferred board substrate for laser welding is epoxy–fiber glass although welds can be made on phenolic boards. Nickel allows the greatest latitude in process variables without producing unacceptable welds. However, consistently good welds can be obtained with copper boards. Kovar offers no welding advantages over nickel, and when conductor resistance and processing techniques are considered, becomes a less attractive choice.

The welding configurations in order of ease of production are plug, end, and center. With laser welding the energy for fusion comes from absorption of the laser beam by the material it strikes. With the center-weld configuration the pad on the board surface is shadowed by the lead being welded. The energy must be transferred through the lead and across the interface. This necessitates that there be intimate contact between the lead and the pad. The advantages of the plug weld are obvious. Although predrilled leads are required, the laser beam can strike simultaneously both pieces being joined and thus sidestep this heat-transfer problem. The end weld is another technique for achieving the same effect—putting energy into both halves of the weldment.

One important characteristic of laser-welded circuit-board joints is that they are true fusion welds. The laser microwelding method has some advantages over other methods of making small electrical joints. The technique does not require the high-vacuum environment necessary for the electron-beam technique. No foreign materials are introduced into the joint as in soldering. Laser welding does not require the general heating of the workpiece like that necessary to form thermal compression bonds and solder joints. In welding with a laser, nothing contacts the work. Parts are not physically displaced since pressure is not applied to the joint as in resistance welding. Having no electrodes, the laser is not subject to electrode erosion or oxidation which can cause deterioration of weld quality. The welding beam can reach into places that are inaccessible to the more conventional methods of joining.

Welding through transparent materials as well as in any desired atmosphere is possible. Precision welding can be done with a well-defined, focused spot. Spot sizes less than a thousandth of an inch in diameter can be achieved, along with accurate positioning.

Mechanical Connections

Pressure welding With this process a metallurgical bond is effected between two workpieces by exerting pressures 5 to 10 times the plastic deformation limit of the materials being welded. Copper to copper, aluminum to aluminum, and copper to aluminum have been welded in this fashion. The materials actually flow together at the high pressures involved (over 200,000 psi). The bonds are stronger than similar solder joints, but not as strong as similar arc-welded joints. The surface of the materials being pressure-bonded must be ultraclean. The method is used to hermetically seal certain types of packages.

Crimping and swaging are types of pressure welding. The deformed-tube configuration is used extensively for attaching wires to components. The advantages are that these methods are cheap and easy to install. However, such techniques usually introduce weak points, prone to fatigue failure.

Wire-wrap connections This type of connection consists of a post with a specific number of turns of solid wire tightly wrapped around it. The post wire, wire insulator, and wrapping tool must be matched. Contact force and friction are initially established by the controlled deformation of the corners of the post and the wire. After an extended period, diffusion of the post and wire material produces a cold weld. The technique is used for back-plane wiring of computers. Figure 79 is an illustration of a typical wire-wrapped joint. This type of connection has extremely high proved reliability. It is also a completely automated process. The greatest disadvantage is that it is bulky. Present systems require that post centers be spaced at $\frac{1}{10}$ in. apart, and that each post be $\frac{3}{10}$ in. long for each wire terminated to it.

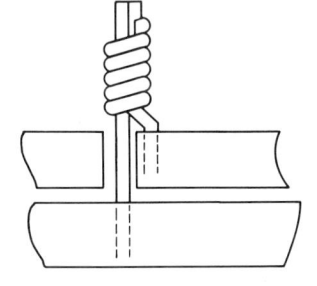

Fig. 79 Split-pin wire-wrap joint.

Conductive adhesives Epoxies heavily loaded with metal particles, generally copper, silver, gold, or aluminum, have been employed for some applications. Some semiconductor manufacturers make the header-to-die bond with this type of material. It is used occasionally in field-service tests. The main advantage is that less heat than soldering is required. The disadvantages are poor conduction and poor adhesion.

Comparison of Second-level Techniques[68]

In order to compare the types of connections, Bell Laboratories conducted tests of solderless wrap, soldered, percussive-welded, and resistance-welded connections. The tests measured the hours of survival under environment conditions of vibration, shock, thermal shock, corrosion, humidity, and bending. These tests based on fatigue life placed the percussive weld first, followed by soldered, wire wrap, and resistance weld as shown in Table 47.[68]

Information from high-temperature testing indicates that solder joints lose mechanical strength rapidly above 200°F. Crimp connections depend primarily on the strength characteristics of the wire itself, with the strength of the crimp approaching, but never quite attaining, the strength of the wire.

TABLE 47 Interconnection Ratings[68]

Environmental conditions	Number tested	Best	Second	Third	Poor
Vibration:					
90° bend	320	PW	S	WW	PW
Omitting 90° bend	160	WW	PW	S	RW
Laboratory shock	160	S	WW	PW	RW
Railroad shock	160	S	PW	WW	RW
Temperature	320	S	PW	WW	RW
Corrosion	160	PW	S	WW	RW
Humidity	160	PW	S	WW	RW
Bending:					
Lightly loaded	160	PW	WW	S	RW
Heavily loaded	160	WW	PW	S	RW

S Soldered. PW Percussive weld. WW Wire wrap. RW Resistance weld

Using time-to-failure curves with 5 years as a mean life goal, the following upper temperature limits are indicated:

Welded joints	540°F
Crimped connections	320°F
Wire wrap	320°F
Soldered contacts	140°F

Mechanical shock environment limits were also established. In order to set these limits, an average figure for shock rate was set at 1 impact 1 hr or a total of 40,000 shocks to failure. The following data were obtained:

Wire-wrap joints	8.4 g
Welded joints	7.6 g
Soldered contacts	2.7 g
Crimped connections	1.0 g

REFERENCES

1. Merrigan, M. A.: "Handbook of Design Criteria for Microelectronic System Packages," *RADC Tech. Rep.* 67-138, vol. 1 of AD 655 764, Hughes Aircraft Co., June, 1967.
2. Parks, M. S.: The Story of Microelectronics, *North American Rockwell Autonetics Div. Publ.* P3-75/317, 1966.
3. "Multilayer Printed Circuit Boards," a technical manual of the Institute of Printed Circuits, March, 1966.
4. MIL-STD-1313. Microelectronic Terms and Definitions, December, 1967.
5. Motorola Inc.: "Integrated Circuits: Design Principles and Fabrication," pp. 307–311, McGraw-Hill Book Company, New York, 1965.
6. Hansen, M.: "Constitution of Binary Alloys," 2d ed, McGraw-Hill Book Company, New York, 1958.
7. Farrell, J., and W. Gavin: In-house Deposition Techniques for Aluminum, *RADC Tech. Memo* EME-65-5, August, 1966.
8. Greenhouse, H. M., et al.: Thin Film Microcircuit Interconnections, Communications Division, *Bendix Corp. Tech. Rep.* ECOM-01482-F, January, 1967.
9. Calabrese, D.: Microphotographic and Photo Resist Techniques for Microelectronic Devices, *RADC Tech. Memo* EME-65-1, May, 1965.
10. Bogert, H., et al.: Large Scale Integrated Circuit Array, Philco Ford Corp., *USAF Contract* AF33(615)3620 *Quart. Rep.* 1, May, 1966.
11. Lathrop, J. W., et al.: Large Scale Integrated Circuit Array, Texas Instruments Corp., *USAF Contract* AF33(615)3546 *Quart. Rep.* 3, November, 1966.
12. Beadles, R. L.: "Interconnections and Encapsulation," AD 654-630, vol. 14

of "Integrated Silicon Device Technology," ASD-TDR-63-316, Research Triangle Institute, May, 1967.
13. Dicken, H. K., and D. B. Kret: Assembling Integrated Circuits, *Electron. Eng.*, October, 1966.
14. Ruggiero, E. M.: Aluminum Bonding Is Key to 40 Watt Microcircuits, *Electron.*, Aug. 23, 1965.
15. Larson, R. B.: Microjoining Processes for Electronic Packaging, *Assem. Eng.*, November, 1966. Copyright 1966. Reprinted by permission of Hitchcock Publishing Co., Wheaton, Ill.
16. Slemmons, J. W., and J. R. Howell: Microjoining Processes for Hybrid Thin Film Circuits, *Proc. WEPA/SAE Symp.*, October, 1964.
17. Arleth, J. M., and R. D. Demenus: New Test for Thermocompression Microbonds, *Electron. Prod.*, May, 1967.
18. Peterson, McKaig, and DePrisco: Ultrasonic Welding in Electronic Devices, part 6, *IRE Int. Conv. Rec.*, 1962.
19. O'Connell, E. P.: Development of a Microelectronics Capability and Facility at RADC, *RADC Tech. Rep.* 65-439, May, 1966.
20. Riben, A. R., and S. L. Sherman: Microbonds for Hybrid Microcircuits, *United Aircraft Hamilton Std. Div. Tech. Rep.* ECOM-03742-F, January, 1967.
21. Evaluation of Flip/Chip Integrated Circuit Interconnections, Honeywell, Inc., Computer Control Division, *Contract* NOBSR 95052, *Final Rep.*, January, 1967.
22. Moore, R. P.: Reliability Test Program of Ultrasonic Face Down Bonding Technique, Univac, *RADC Tech. Rep.* 67-138 (AD 655 781), June, 1967.
23. Scudder, N. F., et al.: Reliability of Ultrasonic Face Bonds, North American Rockwell Autonetics Division, *Contract* F30602-67-C-0083, *Interim Rep.* 2., January, 1968.
24. Wagner, S., et al.: Low Cost Integrated Circuit Techniques, *Philco Ford Corp. Tech. Rep.* ECOM-01424F (AD 652 699), May, 1967.
25. Mallery, P.: Bonding Beam Leads, *Workshop IV, Bonding Tech. for Hybrid Packag.*, 1967 Nat. Electron. Packag. and Prod. Conf., June, 1967.
26. Mallery, P., and M. Lepseller: Exploratory Study of Bonding Methods for Leads on 2.5 to 50 mil Centers, *Proc. Electron. Components Conf.*, 1966.
27. Uncased Semiconductors Task Group Report, Electronics Industry's Association Meeting, Nov. 13, 1967.
28. Knight, C. W. T. Economic Considerations in Design and Fabrication of Face Bonded Devices, *Workshop IV, Bonding Tech. for Hybrid Packag.*, 1967 Nat. Electron. Packag. and Prod. CONF., June, 1967.
29. MacAvoy, T. C., and S. A. Halaby: Substrates and Packages for Microelectronics," *Proc. 1964 Nat. Electron. Packag. and Prod. Conf.*, June, 1964.
30. Short, Q. A.: Conductor Composition for Fine-line Printing, *Electron. Packag. and Prod.*, February, 1968.
31. O'Connell, J. A.: Thick Film Technology, *Proc. 1966 Nat. Electron. Packag. and Prod. Conf.*, June. 1966.
32. Schwartz, B., and D. L. Wilcox: Laminated Ceramics, *Proc. 1967 Electron. Components Conf.*, May, 1967.
33. Abrams, H.: Development of Thin Film Circuits with Thick Film Conductor Networks and Crossovers, *Proc. 8th Int. Electron. Circuit Packag. Symp.*, August, 1967.
34. Black, J. R.: Mass Transport of Aluminum by Momentum Exchange with Conducting Electronics, *Proc. 6th Annu. Rel. Phys. Symp.*, November, 1967.
35. Bennett, R.: A Comparative Analysis of Ultrasonic Aluminum Wire Bonding versus Thermocompression Gold Wire Ball Bonding, Fairchild Semiconductor, Inc.
36. Rigling, W. S.: Designing and Making Multilayer Printed Circuits, *Electro-Technol.*, May, 1966.
37. Morrison, R.: Using the Clearance-hole Method for Multilayer Printed Wiring, *Electron. Packag. and Prod.*, August, 1965.
38. Design Specification for Rigid Multilayer Printed Wiring Boards, *Inst. Printed Circuits Spec.* IPC-ML-910.
39. Hayes, G. A.: The Packaging Revolution, part IV, Bigger and Better Multilayer Boards, *Electron.*, Nov. 29, 1965.
40. Specifications for Printed Wiring Boards Multilayer (Plated-through Hole), MIL-P-55640, February, 1969.
41. Burstein, E. B.: Materials and Processes for Multilayer Printed Wiring Boards, *Proc. 1966 Nat. Electron. Packag. and Prod. Conf.*, June 1966.
42. Broache, E., and J. Poch: Elimination of Fractures in Plated-through-hole Printed

Circuit Boards by the Use of Ductile Plating, *IEEE Trans. Pts., Mater, & Packag.*, vol. PMP-2, no. 4, December, 1966.
43. Take the Gamble Out of Plating of Printed Circuits, *Enthone, Inc.*, Form 48900-268-IOM.
44. Rhoades, W. T.: Microminiature Multiplane Interconnections, *Hughes Aircraft Co. Tech. Rep.* ECOM-01439-2, April, 1966.
45. Korb, R. W.: Multilayer Circuit Board Evaluation, TRW Systems, *NASA Contract* NAS9-4810 *Final Rep.*, December, 1965.
46. Manko, H. H.: Chaps. 11–14 in Clyde F. Coombs, Jr. (ed.), "Printed Circuits Handbook," McGraw-Hill Book Company, New York, 1967.
47. Keister, F. A.: Parallel-gap Soldering, an Advanced Technique for Interconnecting Integrated Circuits. *Proc. 1965 Nat. Electron. Packag. and Prod. Conf.*, June, 1965.
48. Cavasin, J.: Printed Wiring Protection, *Mach. Des.*, Dec. 21, 1967.
49. Douglas, R. R.: Manufacturing Process Development for Parallel-gap Soldering, *Proc. 1966 Nat. Electron. Packag. and Prod. Conf.*, June, 1966.
50. Bester, M. H.: Metallurgical Aspects of Soldering Gold and Gold Plating, *North American Rockwell Autonetics Div. Rep.*, T72368/301, October, 1967.
51. Wild, R. N.: Effects of Gold on the Properties of Solders, *Proc. 1968 Nat. Electron. Packag. and Prod. Conf.*, June, 1968.
52. Shultz, H. F.: Multiple-lead Reflow Soldering Techniques, *Proc. 1967 Nat. Electron. Packag. and Prod. Conf.*, June, 1967.
53. Mallery, P.: Impulse Microsoldering with a Parallel-gap Welder, *Microjoining Workshop, 1965 Nat. Electron. Packag. and Prod. Confer.*, June, 1965.
54. Weltha, M. D.: Design and Operation of a Hot Air Soldering Facility, *Proc. 1967 Nat. Electron. Packag. and Prod. Conf.*, June 1967.
55. Costello, B. J.: Soldering with Infrared Heating, *Electron. Packag. and Prod.*, March, 1965.
56. Manko, H. H.: Speed Microelectronic Soldering, *Electron. Des.*, no. 1, May 24, 1967.
57. Keister, F. A.: Packaging Engineers, Beware of Solder Splashes, *EDN*, Oct. 1966.
58. Keister, F. A.: Beware of Solder Slivers, *Electron. Packag. and Prod.*, November, 1966.
59. Larson, R. B.: Microjoining Processes for Electronic Packaging, part 1, *Assem. Eng.*, September, 1966.
60. Shultz, H. F.: Welded Interconnections for Flexible Circuits, *Proc. 1968 Nat. Electron. Packag. and Prod. Conf.*, June, 1968.
61. Koudounaris, A.: Parallel-gap Welding to Copper Printed Wiring Boards," *Douglas MSSD Rep.* SM49158, July, 1966.
62. Washer, R. B.: New Developments in Weldable Electronic Circuit Boards, *Proc. 1964 Nat. Electron. Packag. and Prod. Conf.*, June, 1964.
63. Erbe, A. R.: High Reliability Interconnection Systems with Percussive Arc Welding, *Proc. 1965 Nat. Electron. Packag. and Prod. Conf.*, June, 1965.
64. Ultrasonic Welding, sec. 3, chap. 49 in A. L. Phillips (ed.), "Welding Handbook," 5th ed., American Welding Society, New York, 1963.
65. Fraikor, F. J.: Developments in Electron-beam and Laser-beam Welding for Electron Tube Fabrication, *USAELRDL Tech. Rep.* 2341, January, 1963.
66. Riben, A. R., and R. E. Antolik: High Reliability Microcircuit Modules Interim Report, *United Aircraft Corp. Hamilton Std. Div. Tech. Rep.* ECOM-00131-1, August, 1966.
67. Jackson, J. E.: Packaging with Laser Welding, *Proc. 1965 Nat. Electron. Packag. and Prod. Conf.*, June, 1965.
68. Mills, G. W.: A Comparison of Permanent Electrical Connections, *BSTJ*, May, 1964.
69. Berkebile, M. J.: Investigation of Solder Cracking Problems on Printed Circuit Boards, *NASA Tech. Memo.* X53653, September, 1967.
70. Bryant, R. D., et al.: Reliability of Microelectronic Circuit Connections, *RADC Tech. Rep.* 67-221 (AD 820 993).
71. MIL-STD-883, Test Methods and Procedures for Microelectronics, May 1, 1968.
72. Herold, D. W.: Visual Inspection Guide for Welded Interconnections, *Electron. Packag. and Prod.*, November, 1967.
73. Deal, F. C.: Method For Nondestructive In-process Weld Evaluation, *Proc. 8th Int. Electron. Circuit Packag. Symp.*, August, 1967.
74. Ultrasonic Welded Aluminum-Copper Junctions as Electrical Connections, *USAF Contract* AF33(600)30839 *Tech. Rep.*, 1964.

Chapter **14**

Photofabrication

ROBERT J. RYAN
EDMUND B. DAVIDSON
HARVEY O. HOOK

RCA Laboratories, Princeton, New Jersey

Introduction	14–2
Glossary	14–3
Analysis of the Photofabrication Process and Applications	14–6
Process Description	14–6
Classification of Applications for Photofabrication	14–7
Comparison with Other Fabrication Methods	14–9
Photomask Fabrication	14–11
Master Artwork Design and Fabrication	14–11
Camera Work	14–19
Contact Printing	14–27
Macrocontact Printing	14–29
Microcontact Printing	14–29
Chemical Processing	14–32
Special Techniques	14–37
Registration for Exposure	14–40
Photosensitive Materials	14–41
Positive- and Negative-working Photoresists	14–42
Dichromated Resists	14–53
Three-dimensional Photosensitive Materials	14–55
Processes Used and Their Relation to Product Design	14–56
Pretreatment Processes	14–57

Etching	14–66
Plating	14–75
Vacuum Evaporation and Sputtering	14–79
Applications	14–81
Chemically Machined Parts	14–81
Printed Circuitry	14–90
Semiconductor Device Technology	14–100
Thin-film Applications	14–109
Screen Fabrication and Applications	14–114
Applications of Three-dimensional Photosensitive Materials	14–124
References	14–128

INTRODUCTION

Photofabrication is a processing method widely used in the manufacture of a variety of electronic components. In photofabrication the concepts of photography are employed along with photosensitive materials and chemical or physical processes, e.g., etching or plating to produce the required shape or pattern of the component. Its use in the electronics industry parallels the tremendous growth of electronics during the past 20 years. Continuing advances, especially in the area of solid-state devices, in color television, and in the packaging of devices, have been made possible through the use of improved photofabrication processes. The introduction of new products, such as large-scale integrated silicon circuits (LSI), will depend, to a high degree, on the refinement and implementation of this process by the design and development engineer.

Historically, photofabrication originated in the printing industry with the development of photoengraving. In its earliest form it involved the use of sensitized asphaltum and gelatin for the fabrication of etched printing plates and electrotypes. This was followed by the development of photoetched halftone screens that permit the reproduction of continuous-tone originals (photographs) as an array of "dots" of different size or spacing. This approach is still employed in both black and white and four-color printing. The use of photosensitive emulsions was also extended to the fabrication of silk screens for screen printing. In addition to the printing industry, photoengraving became an integral part of the decorative engraving and nameplate field.

The photofabrication process had its introduction to the electronics industry during the 1930s with the advent of screen printing for the processing of resistors and capacitors. During World War II, the printing of etch-resistant inks by the screening process was adapted to the fabrication of printed-circuit boards. Discovery of new photoresist materials with good acid-resistant properties during the 1950s led to the direct application of a photographically developed pattern to the processing of printed-circuit boards. Additional uses of photofabrication techniques evolved rapidly with the advent of television, solid-state devices, and computers. At present, photofabrication is considered a field in itself with application to all phases of the electronic industry.

Photofabrication encompasses a wide range of disciplines, materials, and processes. This chapter is intended to provide an understanding of the basic photofabrication method and its associated processes as they apply to the electronics industry. To accomplish this, the authors have organized the chapter into several main sections based on materials, processing, and applications. The basic terminology is defined and the process is described in the initial sections. These are followed by comprehensive sections on photomasks, photosensitive materials, and the processes employed in photofabrication. The final section presents the various areas of application along with typical examples of each.

GLOSSARY

Some terms commonly used in photofabrication are defined in Table 1.

TABLE 1 Terms and Definitions

Artwork. The master precision images on a film or glass support. The artwork is the starting point from which a mask or intermediate is made, e.g., by photoreduction or contact printing.

Artwork generator. A computer-controlled machine to produce artwork. An automatic plotter may be a large-scale artwork generator. See also *reticle generator*.

Bimetal mask. A mask formed by different metals combined by either electroforming or cladding. Apertures are selectively etched through one metal to form an image in a second metal.

Chemical blanking. Process of producing metallic and nonmetallic parts by chemical action.

Chemical machining. Fabrication of a material by chemical action as compared to mechanical methods.

Chemical milling. A chemical machining process involving large amounts of material removal such as in weight reduction.

Circle of confusion. Finite circle which is the image of a point. High-resolution images have small circles of confusion.

Conversion coating. A protective modification of a surface usually by chemical means such as the formation of a metal complex.

Coordinatograph. Machine for guiding a tool to cut on coordinate axis with high precision (used to generate artwork).

Cut and strip. Method of producing artwork by cutting the pattern and stripping away the unwanted areas of a two-layer system.

Densitometer. An instrument for measuring the optical density of a selected part of a partially transmitting medium such as a photographic film or plate.

Density. The logarithmic value of the ratio of flux transmitted I to that incident I_0 on a uniformly exposed and processed area:

$$D = \log_{10} \frac{I_0}{I}$$

Development. In photography: a chemical treatment used to render visible the latent image on an exposed film. In photoresist technology: formation of a photoresist pattern following exposure by chemical dissolution of the soluble portion.

Diffraction. The bending of light at the edge of an opaque object.

Diffusion. In photography: the spreading of light outside the intended area of exposure by multiple scattering. In semiconductor technology: the migration of dopant atoms into a semiconductor material to alter its resistivity.

Electroformed mask. Thin-metal patterns formed by plating in between a nonconducting image which has been placed on a conductive surface.

Electroforming. The fabrication of a part or a pattern through electrodeposition of a metal onto a permanent or temporary support.

Etch factor. The ratio of the depth of etch to the lateral etch or undercut distance.

Etch mask (stop). A protective pattern applied to material to prevent etching in unwanted areas.

Functional chemical machining. Fabrication of the shape of a part or a pattern by chemical etching through a protective mask image.

TABLE 1 Terms and Definitions (Continued)

Gamma. A term representing the contrast of a photographic process. The slope of the straight-line portion of the plot of density versus $\log_{10}$ of exposure. High gamma equals high contrast.

Halation. The spreading of light outside the intended area of exposure by reflection from the rear surface of the transparent base supporting the emulsion to be exposed. This is distinguished from diffusion, which takes place within the emulsion layer.

Image. (1) All geometric forms appearing in the functional pattern area. All images are identified by a reference or by actual dimension. (2) The replica of an object formed by an imaging process. An image is produced by a lens at its focus.

Keys. Patterns selectively placed within an array for testing and/or locating. On a single-pattern mask the key is outside the pattern areas.

Lateral reversal. Retains the negative or positive of the original, but has a mirror image of its geometric orientation. Contact printing and reversal processing produces a lateral reversal.

Mandrel. The surface that is used to replicate a part by electroforming and subsequently separated from it.

Mask. A coating or barrier used to prevent alteration of the underlying surface. Also called *resist*.

Metal etched mask. A mask formed by etching apertures through a metal protected by a photoresist.

Metal-on-glass mask. A mask that has thin-metal images on a glass base. Commonly *chrome mask*.

Mirror-image mask. Mask that rotates the geometric orientation 180°.

Overcutting. Scoring the peelable lacquer outside the desired area when cutting artwork.

Overhang. The inverted shell of plating formed when conductor material is selectively removed from under the plating.

Overlap. Area added to the die to ensure mask continuity. It is primarily used in step-and-repeat systems.

Pattern plating. Plating of a metal on a surface in between a plating mask pattern.

Peel strength. The force necessary to maintain steady separation of the metal cladding from the laminate to which it is bonded, when pulled at a prescribed rate at a right angle to the board. Given in pounds per inch of width of metal.

Peelable lacquer. A coating on a flexible support that will peel off after an area has been outlined with a cut.

Photochemical machining. Chemical machining using a photosensitive material to produce a mask pattern.

Photoengraving. Partial etching of a material using a photosensitive material to produce a mask pattern.

Photoetching. An etching process using a photosensitive material to produce a mask pattern.

Photoforming. An electroforming process using a photosensitive material to produce a mask pattern.

Photographic. Relating to, obtained by, or used in photography. A photographic coating or film is generally considered to be of the silver halide type.

Photomask. A photographically produced pattern containing the information to be transferred to a workpiece, usually by contact printing on photoresist.

TABLE 1 Terms and Definitions (Continued)

Photoresist. A film-forming material, sensitive to light or other radiation, such that later processing will result in changes in either the exposed or the unexposed areas exclusively.
  a. Negative working: The exposed areas are insolubilized on exposure. Development removes the unexposed areas, leaving a pattern corresponding to the clear or white areas of the original.
  b. Positive working: The exposed areas are solubilized on exposure, allowing to be removed in the development step. The remaining resist pattern corresponds to the opaque or block areas of the original.
Pinhole. A weakness in the photoresist layer, not generally visible, which allows etchants to penetrate the film locally, causing etching in areas which should be covered by the photoresist film.
Plated-metal etch resist (stop). An etch mask pattern formed by plating a dissimilar metal onto the metal to be processed and selectively etching the base metal (e.g., nickel on copper).
Plotter, automatic. A computer-controlled machine capable of producing precise artwork.
Process artwork. The artwork used to define the part to be processed. This could be a photomask, metal-on-glass mask, or metal etched mask.
Process materials. Materials used subsequently during a fabrication process as distinct from the material.
Registration. The accuracy of relative position of all patterns on any mask with the corresponding patterns of any other mask of a given device series when properly superimposed.
Resolution. The process or capability of making distinguishable the individual parts of closely adjacent images or sources of light.
Reticle. The master which is placed in the object plane of a final-size step-and-repeat camera.
Reticle generator. An artwork generator capable of sufficient precision to produce a step-and-repeat reticle as the original artwork.
Scattering. The diffraction of light by particles in a processed photographic emulsion. See also *Diffusion*.
Scribecoat. A material composed of a stable base, such as glass or film, with an opaque coating that can be scribed with a blade.
Scribing. The removing of the opaque film on scribe coat to generate artwork. This is usually done on a coordinatograph.
Shadow mask. A metal mask used to confine the projected electron beams of a color television kinescope to a particular phosphor dot pattern.
Single segment. A single functional pattern that has been reduced photographically. It is stepped and repeated to form an array.
Throwing power. The ability to deposit a plated metal in recesses and normally low-current-density areas.
Transmittance T. The ratio of flux transmitted to that incident on a uniformly dense area. For photographic emulsion the area should be large compared with the area of a grain. $T = I/I_0$ where I_0 is the incident intensity and I is the transmitted intensity.
Wafer. A semiconductor substrate onto which devices are fabricated, with the aid of photoresist mask patterns.

ANALYSIS OF THE PHOTOFABRICATION PROCESS AND APPLICATIONS

The chief aspect in photofabrication is the formation of a pattern or the shape of a part by photographic means. This is accomplished through the use of a photomask and a photosensitive material, along with subsequent chemical or physical processes. The adaptability of photofabrication to a particular application will be dependent upon the part design as well as the overall process characteristics and production requirements. All applications, however, regardless of complexity or number of parts to be processed, require the above essential steps. The ultimate choice should be determined by a thorough understanding of the factors involved in photofabrication. In this section, the process is broken down into its various steps and their purposes are defined. Types of applications are also summarized, along with a comparison to other fabrication methods. Details of these steps and application areas are presented in the later sections of this chapter.

Process Description

The essential steps in photofabrication are outlined in Fig. 1 and are summarized below.

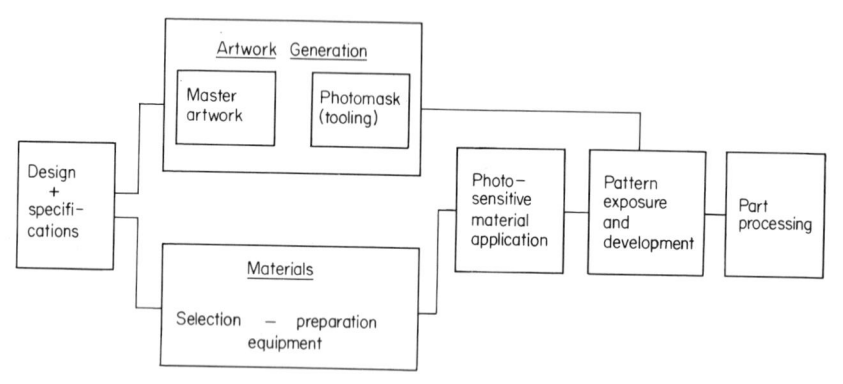

Fig. 1 Flow chart of photofabrication process.

Design and specification The design of a part is determined by its end use, i.e., by the interaction of the part with a larger system. The design data determine the configuration, specifications, and type of material required for the part. If it is to be prepared by photofabrication, these same design data dictate the type of artwork needed, the materials to be employed, and the processing involved in the subsequent fabrication sequence.

Artwork generation Under this wording is included the preparation of both a master pattern or "original" and the subsequent photomask. The methods and equipment to be used during photofabrication are determined by the design specification and overall precision required for the final product. Master artwork is generally prepared from photographic plate or scribing film (e.g., Rubilith*), whereas the process artwork can consist of a photomask such as a photographic transparency, or metal pattern on glass, or an etched metal mask made from the master. Good artwork is one of the most essential requirements in the photofabrication process because it is the "tool" used during processing and it sets an upper limit to the overall quality of the product.

Material selection and preparation A number of materials must be specified before a photofabrication process can be designed. These include: (1) the material of the part to be fabricated, (2) the photosensitive material, and (3) "process" materials. Process materials may include cleaners, solvents, developers, etching

* Trademark of Ulano Co., Brooklyn, N.Y.

Analysis of the Photofabrication Process and Applications

solutions, and plating solutions for various applications. Selection of part materials is determined by the design requirements and application. The part material guides the selection of the photosensitive and process materials, as each must be compatible with the others and engineered to produce a workable process. For most applications the photosensitive material is used as a thin protective film or coating and is applied from solution.

Material preparation steps involve pretreatment of the part, usually including a cleaning step, and monitoring and control of the photosensitive and processing chemicals.

The part design and materials employed also determine the equipment required. This can vary appreciably for different applications and determines to a large extent the economics of the overall process.

All these points will be discussed in detail below.

Photosensitive material application Coatings of photosensitive materials are applied by standard coating techniques such as spraying, dipping, flowing, whirling, and roller coating. The choice of a method will be determined in part by the particular materials used and the thickness and uniformity required. A detailed description of the various coating methods is presented later in this chapter.

Pattern exposure and development During exposure, light having a high intensity in the near-ultraviolet region (250 to 550 mμ), passed through a suitable photomask, is allowed to impinge on the photosensitive coating. This initiates a chemical reaction in the material, altering its chemical solubility properties. Photosensitive materials, e.g., photoresists, that are made less soluble under the influence of light are called *negative* photoresists, and those made more soluble are called *positive* photoresists.

The photosensitive film is then developed by dissolving the soluble portions in a suitable solvent. This may be followed by a postexposure or baking step to harden the remaining resist, make it insensitive to further illumination, and increase the chemical resistance.

Part processing In most applications, the pattern formed in the photosensitive material is used as a "resist" or protective coating during the actual part-fabrication step. Fabrication processes may include etching, plating, vacuum deposition, or other chemical treatment. After processing, the resist image is normally removed by chemical or physical methods to give the final part.

Classification of Applications for Photofabrication

Electronic applications can be classified into two general groups: those using a photosensitive material as a protective resist during processing and those using the photosensitive material as an integral part of the final product. The various applications are outlined in Table 2.

Most applications fall into the first category where the resist serves to delineate the shape of the part and to protect appropriate portions of the material during processing. These applications can be further divided according to the type of processing and parts obtained. Chemical machining involves the fabrication of individual parts by etching or plating, and open through-hole areas are usually produced during processing. In comparison, printed circuits are normally fabricated on a nonconducting substrate that is not modified during processing. For the processing of semiconductor devices, the artwork used differs from that for chemical machining and circuit-board fabrication mainly because of the resolution required; line widths and separations are one or two orders of magnitude smaller in semiconductor work. Processing steps for these devices nevertheless are very similar to those for printed circuitry in that patterns are formed on a supporting substrate. The semiconductor substrate, however, includes active electronic devices whose final properties are controlled to a large extent by the quality of the photofabrication processing. Similarly, thin-film applications require higher-resolution artwork than printed circuitry, and the materials processed have a thickness in the order of 5 μm and below.

The largest application in the second group involves patterns processed on woven screens. Here the photosensitive emulsion or film, when processed, forms the stencil for a subsequent screening operation. The formation of phosphor dot patterns for color kinescopes involves a completely different type of photofabrication. In this application the photosensitive material acts as a binder for the phosphor particles as well as a means for defining the required pattern in the dried coating. Although the binder is lost on subsequent baking of the color kinescopes, this application of photofabrication fits into the second of the two general groups. Other applications listed in Table 2 involve relatively thick layers of photosensitive materials that, after processing, function as a mechanical part. Their use is limited at present but represents a potential expansion for photofabrication processes in electronics.

TABLE 2 Applications of Photofabrication to Electronics

A. *Applications of photosensitive materials as masks*
   1. Chemically machined parts:
      a. Color kinescope shadow masks
      b. Lead frames for components
      c. Magnetic-recording head laminations
      d. Tube grids
      e. Switch contacts and assemblies
      f. Evaporation masks
      g. Masks for thick-film circuitry
      h. Perforated plastic insulator layers
   2. Printed circuitry:
      a. Etched circuits
      b. Plated circuits
   3. Semiconductor device processing
   4. Thin-film circuitry
B. *Applications of photosensitive materials as part of the product*
   1. Screens for printed circuits, thick-film circuits
   2. Color kinescope phosphor patterns
   3. Fluidic device prototypes
   4. Process jigs and fixtures
   5. Printing plates

Photofabrication processes can also be described as direct and indirect depending on their use in electronic processing. The direct process involves the direct application of a photosensitive material in the fabrication of the required part. In the indirect process, the photosensitive material is used to fabricate tools used in subsequent processing. Of the applications listed in Table 2, evaporation masks, stencil masks, screens, and photopolymer printing plates are examples of indirect processes. Screens, for example, are employed as a means of depositing materials in patterns useful in electronic applications, particularly in the preparation of thick-film circuitry and printed circuitry.

In applications requiring a chemical-resistant coating such as in printed circuitry, both direct or indirect methods can be used. The choice of method will depend chiefly on the pattern design and the tolerances required. Directly applied photoresists are capable of producing extremely fine patterns and are generally used for dimensions below 0.005 in. Better control of edge tolerances and definition is also attainable with photoresists. Table 3 presents a general comparison of the various image-transfer methods and shows the order of preference for their use.

Major electronic applications of photofabrication are discussed in detail later in the chapter.

Comparison with Other Fabrication Methods

Photofabrication offers many advantages over other fabrication methods for the applications listed in Table 2. Masks for transistors and integrated circuits could not be processed in a practical or economic way by any other method since the definition and tolerances required are beyond the capability of alternative patterning techniques. For other applications, the geometry of the part, its material composition, or economic considerations may be the determining factor in choosing photofabrication over alternative processes.

The photofabrication of thin-metal parts with perforations or cutout areas is generally classified as chemical machining or chemical blanking. Alternative fabrication methods include conventional mechanical blanking or punching and the more recent nontraditional[1] methods such as ultrasonic machining, abrasive jet machining, electrochemical machining, electric discharge machining, electron-beam machining, laser machining, and plasma-arc machining. A brief description of these processes is given in Table 4. In addition to photofabrication, only mechanical blanking has been used for the fabrication of large quantities of thin-metal parts. Some of the other methods may be preferred for a specific application, particularly where a small number of parts is involved. A detailed description and evaluation of these methods can be found in Ref. 1.

TABLE 3 Order of Preference for Various Resist Image-transfer Methods

Specification	Photo-resist	Screen printing	Stencil printing	Offset printing
Pattern definition	1	4	3	2
Control of tolerances	1	4	3	2
Applicability of large-area patterns	2	2	3	1
Applicability to double-sided printing	1	3	3	2
Thickness of coating attainable	4	2	1	3
Printing rate	3	1	1	2
Economics of process	4	1	2	3

Comparison of chemical and mechanical blanking In order to select either the chemical or mechanical blanking process for a particular application, an analysis must be made of the part design, material specification, and overall economics of the process. Any of these factors could determine the selection of one method over the other. A brief description of these factors is presented here to serve as a general guide to the use of the photofabrication process. Detailed information can be found under the section on Chemically Machined Parts (page 14-81).

Part Design. The design of the part and the tolerances required are the first consideration. Normally, part thickness is the main restriction of the chemical machining method because of the long etching times required and resultant undercutting of the material. A material thickness greater than 0.060 in. is rarely used, with most parts being in the range of 0.0005 to 0.020 in. Greater precision and maximum resolution are obtained by the chemical method when the thickness is minimum. In comparison, mechanical blanking becomes ineffective as material thickness is minimized, requiring greater precision and tighter tolerances for tool fabrication.

Minimum surface dimensions are also dependent on the thickness of the material, pattern resolution, and tolerances being inversely proportional to the thickness. This restriction holds for both the mechanical and chemical methods. The overall area of a particular sample has less of an effect on the process capabilities. It does, however, affect the tooling and equipment required and the economics of the process.

Pattern complexity of the part to be processed has little effect on the chemical method since two-dimensional artwork forms the tooling for the process. Tooling

14-10 Photofabrication

for mechanical blanking, however, such as required for a complex asymmetric pattern may be impossible to produce by presently known methods, or the cost may be prohibitive.

Material Specification. The mechanical properties of materials do not affect the chemical milling process. Parts can be produced from practically any metal or alloy with any degree of hardness with no change in the resultant material properties. Applications where these properties are critical favor the use of the chemical method, as the properties can be optimized before fabrication. This is particularly true for magnetic materials since the magnetic permeability is highly dependent upon mechanical stress in the material.

In comparison, the mechanical blanking method is dependent upon the mechanical properties of the part material, and control of these properties is difficult. Some materials, such as hardened metals and the superalloys, cannot be processed in

TABLE 4 Classification of Machining Processes

Mechanical blanking. Blanking of the workpiece by the shearing force of a shaped cutting tool using mechanical energy.

Ultrasonic machining. Removal of material from the workpiece with a shaped tool oscillating at approximately 20,000 cpm in an abrasive slurry.

Abrasive jet machining. Removal of material from the workpiece by the impingement of fine abrasive particles entrained in a high-velocity gas stream.

Electrochemical machining. Removal of material by anodic dissolution under the influence of an electric current between a shaped cathodic tool and an anodic workpiece.

Chemical machining. Removal of material by chemical dissolution of the workpiece, using a resistant masking material selectively applied to obtain the desired pattern.

Electric-discharge machining. Removal of metal through the vaporization of the workpiece by high-frequency electric sparks between the workpiece and a shaped electrode.

Electron-beam machining. Removal of material by thermal vaporization of the workpiece, using high-energy electron beam as an energy source.

Laser machining. Removal of material by thermal vaporization of the workpiece, using thermal energy generated from a focused monochromatic light beam with a high-power density.

Plasma-arc machining. Removal of metal by thermal evaporization of the workpiece, using a hot, partially ionized gas stream and an electric arc struck from a cathode within the plasma generator to the workpiece.

a practical manner by this method. In addition, many applications require controlled annealing steps after part fabrication to attain the required material properties.

Cost Factors. The factors that determine the overall cost of either process include the various costs for tooling, maintenance, overhead, production rate, and auxiliary processing required. In general, tooling costs for mechanical blanking are 20 to 60 times the cost of process artwork used in chemical milling. For this reason chemical milling is normally employed for small- to medium-size production runs and for applications where future design changes are required. The other costs, however, are in favor of mechanical blanking, particularly because of the larger production rates presently attainable. Costs in large-volume production amount to approximately one-half those of the chemical method.[2,3]

Miscellaneous Factors Affecting Selection of a Process. In addition to the above, time schedules required for tooling up for production may be an important consideration. A short time of one to three weeks is generally required to obtain process artwork for photofabrication. In comparison, dies used for mechanical blanking generally require much longer times.

PHOTOMASK FABRICATION

In all photofabrication processing some sort of mask is used to permit transfer of the desired pattern to the work, usually by contact printing. The nature of the product, the spectral response of the photosensitive material (e.g., photoresist), and the subsequent processing required are among the factors that determine the most useful form for this mask, and thereby the optimum procedure for its fabrication.

Accuracy and repeatability are important in photofabrication. In this section the accuracy will be stated in micrometers (μm), and the repeatability in parts per million (ppm). Some of the terms frequently used in photomask work are included in the glossary.

The flow chart of Fig. 2 outlines the steps common to the preparation of most masks for use in photofabrication.[4-8] The detailed contents of the blocks may vary widely. For example, pattern design may be as simple as a freehand soft-pencil sketch for a printed-circuit breadboard (which may be contact-printed to make the finished circuit board), or as complex as an elaborate computer-aided design for the several layers of a multilayer printed circuit or LSI circuit. The first example collapses the whole process into the first block, but a more typical mask will require at least a size change in the photographic manipulation stage. If multiple images are required on the mask, step-and-repeat or multi-image cameras can be used. In many processes, mask life is short. A number of identical masks may therefore be needed. Contact printing is usually the most economical way of producing these duplicates.

The examples of Table 5 are representative of mask-fabrication processes but by no means exhaustive. A full listing with recommended steps is beyond the scope of this chapter. Instead, each stage will be examined to establish principles on which reasonable choices can be made. Since the most demanding masks are for semiconductor production, these masks will be emphasized. For less demanding work, requirements can be relaxed.

Master Artwork Design and Fabrication

Methods and materials Master artwork originates in the mind of the designer. Mental designs may be set down as the starting point for a mask in several ways. Traditionally the designer makes a more or less careful drawing which becomes the master "pattern" from which the master pattern is "drawn" by a draftsman on a suitable medium for reproduction. This drawing may be done by (1) scribing through an opaque coat on a transparent film or on glass, (2) applying opaque black tape to transparent film or white paper (Fig. 3), or (3) cutting and peeling the ruby layer of a two-layer film (Fig. 4) as well as (4) the more traditional pen-and-ink means. For the last mentioned, inks compounded with a special plastic base are available and are usually applied to a stable polyester film drafting sheet. Cut and peel[8] is the favored means of manual artwork production when sharp, high-contrast lines are needed, and, when done on a coordinatograph, the precision may be as good as 30 ppm. From the data of Table 6 it is evident that the accuracy and precision of the initial pattern depend critically on the use of machine aids.

The choice of the medium is based on the stability required and the nature of the further processing.[9-17] Table 7 gives data on the stability of materials commonly used. The effects of temperature and humidity are shown in Figs. 5 and 6. The initial accuracy of hand-drawn artwork rarely justifies the stability of polyester film, let alone glass. Glass is the most stable base, with almost no humidity sensitivity and very low thermal expansion. Polyester is the most stable plastic, but

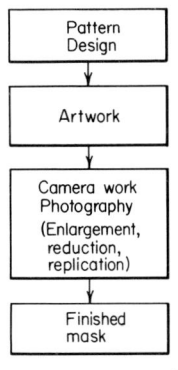

Fig. 2 Photomask fabrication flow chart.

TABLE 5   Representative Examples of Mask-fabrication Steps

Mask purpose	Single-layer printed-circuit etch, copper-clad board	Transistor fabrication — First method	Transistor fabrication — Second method	Etching Fotoform* glass
Pattern design	Sketch on gridded paper	Engineer's sketch fully dimensioned	Engineer's sketch with critical dimensions	Engineer's sketch fully dimensioned
Master pattern	Tape on polyester film	Cut 100× final size on manual coordinatograph on polyester-base ruby film	Program automatic drafting machine with computer aid. Draw 100×	Cut and peel on ruby 25×
First camera	4× reduction	10× reduction	10× reduction	25× reduction
Second camera	—	10× reduction in SAR (step and repeat) camera	1× step and repeat	—
Third camera	—	—	10× reduction	—
Other steps	Contact print or reversal process if desired to use positive photoresist	Contact print to make working masks. In some instances contact prints made on chrome	Contact print for production masks	Print on photoresist on Be-Cu foil. Plate up with Ni. Etch Be-Cu out of holes
Special requirement	±0.005-in. accuracy adequate	Registry of several layers to ±1 µm/in. over a 1- to 2-in. circle. Images 0.4 mm (0.015 in.) square. Smallest lines (0.0002 in.) typical	Registry of several layers to ±1 µm/in. over a 1- to 2-in. circle. Images 0.4 mm (0.015 in.) square. Smallest lines (0.0002 in.) typical	High contrast in the 3300 Å spectral range where glass is sensitive
Comments	—	Coverage of lens on step-and-repeat camera limits maximum device size	Large-scale step and repeat eases dust problems and allows retouching of master pattern. Final reduction lens is limiting factor in coverage and resolution	Most glass absorbs in this range. Dark areas of high-resolution photo-emulsions are partly transparent near 3000 Å

* Trademark of Corning Glass Works, Corning, N.Y.

Fig. 3 Hand taping of artwork.

temperature and humidity must be controlled in order to do the most precise work.

Acetate films are less stable than polyester but are less expensive. They are suitable for less precise work where the dimensional changes under the range of temperature and humidity encountered are acceptable.

Both glass and film bases have their advantages. Glass is more costly, heavier, harder to handle, and more fragile than film; but, even worse, when photoreduc-

TABLE 6 Typical Accuracy and Precision of Artwork

Drawing means	Accuracy μ	Precision, ppm	Finest line, μ	Basis
Manual..........................	±250	±250	250	Precision based on artwork 1 meter square (40 by 40 in.)
Machine-aided (manual coordinatograph).......................	±25	±25	500	
Machine-drawn (automatic plotter) 10 by 10 cm to 1 by 1 meter Scribe or cut photo	±25 ±12	±25 ±12	50	
Machine-drawn, (reticle generator), programmed step and flash with programmed rectangle 1 to 100 mm square	±1	±5	5	Precision based on 5 cm square (2 by 2 in.)
Machine-drawn (mask generator), drawn actual size for semiconductor masks 25 to 100 mm square	±1	±5	0.3	Projection based on state of the art. Not yet in existence. Precision based on 5 cm square (2 by 2 in.)

14-14 Photofabrication

tions are to be made, the lack of flatness of glass plates can result in serious magnification errors. For overall error tolerance of 1 part in 50,000, which is needed in some semiconductor masks, a bow of only 0.05 mm (0.002 in.) can introduce the whole of the allowable error.[18] The best large photographic plates commercially available are guaranteed to be flat to 6×10^{-5} cm/m[17], which for a 0.7-meter (25-in.) plate amounts to one-half or more of the allowable error, especially if (as is very likely) one or more of the plates in a set is bowed opposite to others. With careful humidity and temperature control in the rooms where

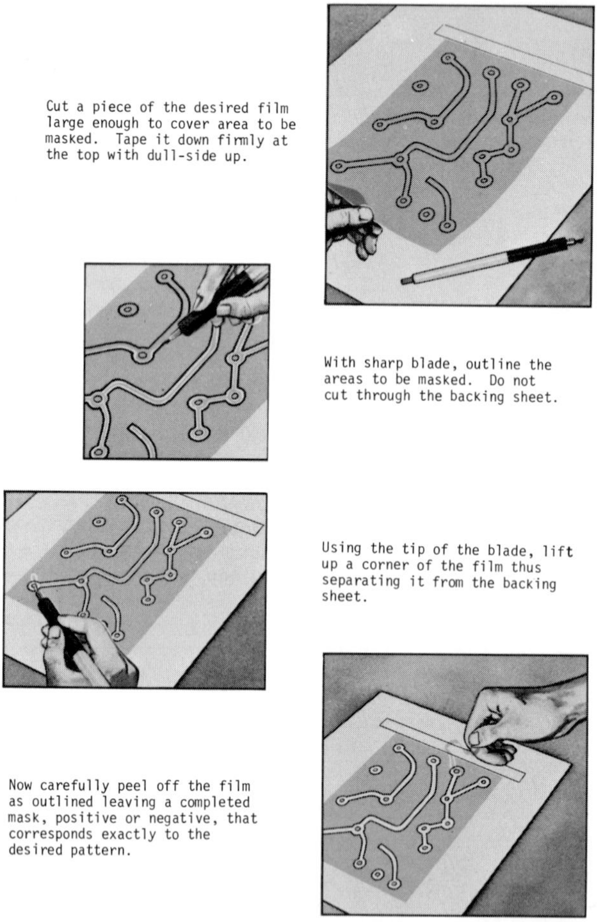

Cut a piece of the desired film large enough to cover area to be masked. Tape it down firmly at the top with dull-side up.

With sharp blade, outline the areas to be masked. Do not cut through the backing sheet.

Using the tip of the blade, lift up a corner of the film thus separating it from the backing sheet.

Now carefully peel off the film as outlined leaving a completed mask, positive or negative, that corresponds exactly to the desired pattern.

Fig. 4 Cut-and-peel artwork preparation. (*Ulano Co.*)

polyester film is used, the dimensional changes can match extremely closely for the several patterns of a set. Even if the copyboard of the camera is warped, the distortion is the same in all patterns. Copyboard glass can be selected from plate glass to achieve a flatness of 0.02 mm (0.001 in.) over a 1-meter (40-in.) square; so the distortion is lower than with glass plates.

Table 8 is an evaluation of the errors in fabricating masks to a demanding but not extreme specification in a fairly well-controlled laboratory. It is evident that no set of choices can assure consistent maintenance of tolerances. The humidity

Photomask Fabrication 14-15

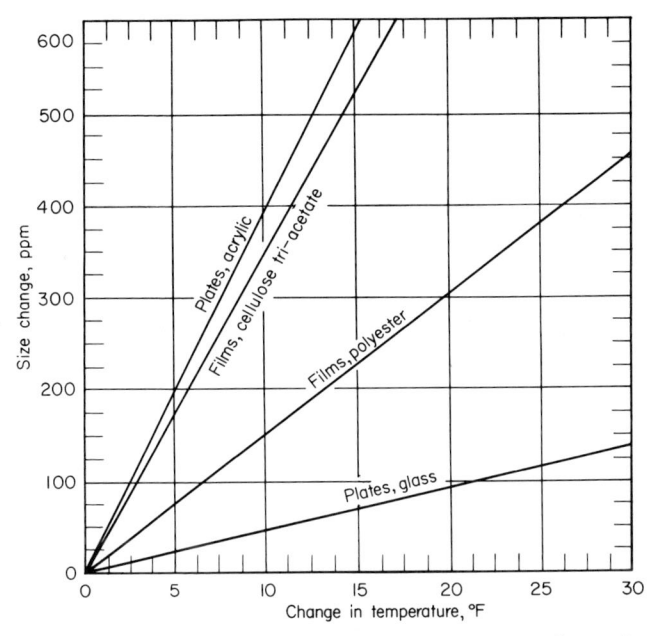

Fig. 5 Size changes of artwork due to temperature change.[16]

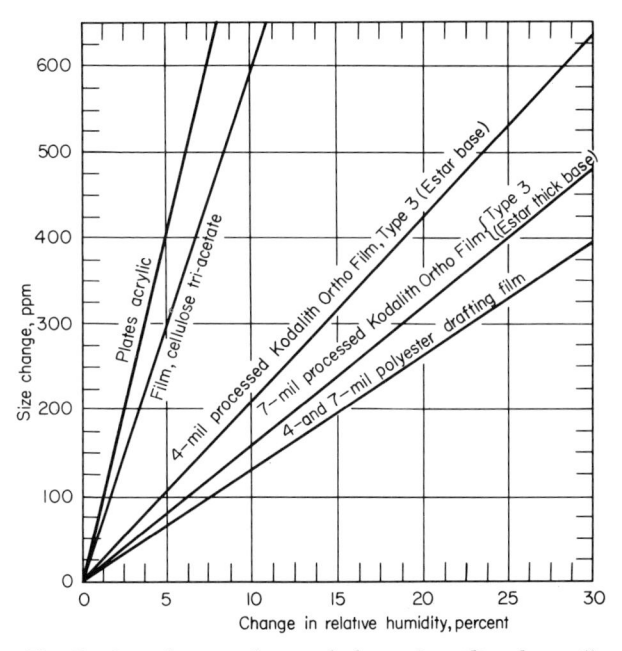

Fig. 6 Size changes of artwork due to humidity change.[16]

14-16 Photofabrication

TABLE 7 Stability of Materials Used as Bases for Artwork

Material	Temperature ppm/°C	Temperature ppm/°F	Relative humidity, ppm/%	Aging in 5 years 78°F, 60% RH	Aging in 5 years 90°F, 90% RH
Triacetate film	63	35	60	−300	−4700
Vinyl film	54	....	?		
Acrylic plate	70	40	80		
Estar* polyester film, 0.004 in	27	15	21	−250	
Estar polyester film, 0.007 in	27	15	16	−250	−100 +200
Polyester film, drafting	27	15	13		
Photographic glass	4.5	2.5	Nil		
Fused quartz	0.5	0.3	Nil		
Celanar† polyester film	20	11	11		

* Trademark of Eastman Kodak Co., Rochester, N.Y.
† Trademark of Celanese Corp., New York, N.Y.

range is too wide for polyester-based masters, and glass plates are not flat enough, although microflat plates come close. Four alternative means of getting required dimensional control are: (1) Decrease the range of humidity variation. (This is both difficult and costly.) (2) Secure a large supply of twin-ground or float-process plate glass, and select the 1 out of 4 to 10 sheets that are flat enough and free from defects. Coat them with peel coat or photographic emulsion, use them, strip them, and recoat for reuse. (3) Work harder. The humidity is likely to stay relatively constant for days or weeks at a time. If all the masters of a set are made while the humidity is constant and photographed under constant conditions, the results will match. But take the precaution of putting a large

TABLE 8 Error Analysis for Photomask Fabrication*

Specification	Errors, ppm Film, 0.007 in.	Errors, ppm Ultraflat glass	Errors, ppm Microflat glass
Artwork, 25X, (±0.0005 in. in 25 in.)	±20	±20	±20
Temperature error at 75°F ± 1°F	±15	±4	±4
Humidity error at 35–45% RH	±80†		
Processing size change	+50 to −80†		
Total artwork error	+165 to −215	±24	±24
Reduction camera, 75 in. to lens:			
Copyboard, ±0.001 in	13‡		
Large glass flatness	....	±160†	±20
2- by 2-in. glass flatness	±14	±34	±14
Total copying error	±14	±184	±34
Total overall error	+179 to −229	±208	±58

* Requirement: registry to ±1 μm in a 2.5-cm circle (±40 ppm total error).
† Trouble: Far out of tolerance.
‡ Repeating error which may be neglected if only requirement is registry within a set of masks.

scale mark on the masters so that the camera reduction can be adjusted if conditions do change. (Note that scale marks cannot help correct for warped glass artwork.) (4) Use a copyboard thick enough to vacuum-flatten glass plates without bowing the copyboard. The copyboard glass needs to be thicker than the glass plates in the ratio of the cube root of the ratio of the initial warp to final warp. A flat copyboard consisting of three sheets of glass cemented together will reduce the warp in a glass plate the thickness of one sheet by a factor of 27.

Computer-controlled, machine-generated artwork As masks become more complex, particularly for multilayer printed circuits and large-scale integrated circuits, the feasibility of using hand-cut or hand-drawn artwork vanishes. The probability of getting a defect-free set of masters is too low. The assistance of a computer in preparing the instructions to an artwork generation machine becomes essential. Figure 7 is a photograph of a Gerber plotter, a large-scale artwork generator. Various computer programs have been developed based on the needs and inclinations of the designers.[19-26] The basic varieties of programs can be characterized by how

Fig. 7 Large plotter type of artwork generator.

they perform in three parts of the operation. These parts are (1) data input, (2) data processing, (3) form of output to pattern generator.

Data Input. For multilayer printed circuits and discrete transistor designs a digitizer is a convenient form of input. A digitizer records in computer-readable form, for example on punched cards, the coordinates to which a cursor is moved on a master drawing. The digitizer can round the numerical data to the nearest grid point to assure registry of several layers. The data must be supplied to the digitizer operator in the form of an accurate, but not necessarily dimensioned, drawing. This is in contrast to the situation with a manual coordinatograph where the operator needs a carefully dimensioned but not necessarily accurate drawing. Since most designers are accustomed to thinking graphically, these forms of input seem natural. However, the artwork generator needs more information than just the locations of a sequence of points to produce a finished pattern. The digitizer operator must indicate whether a line is to connect two or more points, how wide the line should be, and whether a pattern from a "library" or "font" of stored patterns should be flashed with a given point as a center. Some form of digital keyboard is provided for these inputs. The cathode-ray-tube, light-pen combination used as a computer input may be considered a form of digitizer. The main alternative to digitizer methods is direct digital input to the computer. If the designer prepares this input, he may not need a drawing at all. At most, a rudimentary sketch with crucial coordinates designated should suffice. With sufficiently flexible and sophisticated computer programs, direct digital input is probably the most efficient way of supplying the data. With the possibility of seeing the resulting pattern almost immediately by using a time-shared computer with a plotting device or oscilloscope attached, it is almost certainly the best method of supplying the input data.

Data Processing by Computer. What the computer does with the input data depends partly on the nature of the data and partly on the end use of the artwork, but mostly on the nature of the artwork generator. The nature of the data affects only the input format which the computer must read and recognize. The examples below illustrate how the end use and artwork generator affect the computer's task. At present there is virtually no standardization in artwork generator input.

The large plotter type of artwork generator such as the one in the photograph of Fig. 7 may scribe to provide outlines only, may be equipped with blade to cut coated polyester film for hand stripping, or for greatest flexibility may be equipped with a photohead to expose photographic film or plates directly. These three modes of operation will serve to illustrate the differences in computer processing that may be required. Consider the pattern of Fig. 8 which is to be produced on different media. The designer has told the computer the location of the corners of the pattern and the desired result. The desired result of Fig. 8a is a line inscribed in the pattern so that its outer edge has the correct pattern dimensions. Such a pattern might be used for an etch master—the center portion drops out when the line etches through. Suppose an 0.13-mm (0.005-in.) line is required. The computer would compute a path 0.065 mm (0.0025 in.) inside the pattern and instruct the plotter to select the correct line width and draw along the computed path.

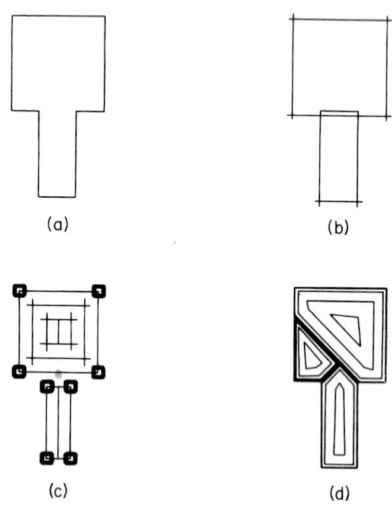

Fig. 8 Artwork generator motions to generate a simple pattern.

If the plotter is using a cut-and-strip film (see Fig. 8b), then the cuts must be made a little past the intersections, say 0.25 mm (0.010 in.), to assure clean peeling. Cuts through the area to be peeled are of no consequence, but cuts outside the peeled area are to be avoided. The computer now determines the boundary lines and starts the plotter 0.25 mm before each intersection and causes it to proceed 0.25 mm beyond the next intersection. For the patterns of Fig. 8c and d the computer is required to fill in the interior of the pattern solid. In Fig. 8c the plotter has available square apertures which may draw and flash in an overlapping fashion to fully expose the film. The necessary locations for flashing are computed, and the appropriate instructions prepared for the plotter. The pattern in Fig. 8c is drawn with special apertures to illustrate this. Reticle generators derived from step-and-repeat machines require this kind of instruction. Figure 9 is a photograph of one of these machines. The filling of the pattern of Fig. 8d is to be done with circular apertures by tracing around the inside of the pattern, repeatedly moving the path toward the center, and selecting larger line widths as permissible. So that the computer can keep track of the interior, all exterior angles smaller than 180° must be eliminated. In the particular routine used, all exterior angles smaller than 180° are bisected, each bisection producing an additional polygon. The selection of larger line widths was inhibited for this example so that the path taken by the plotter would be demonstrated. Of course, such a simple pattern would most likely be done by manual coordinatograph. The power of the automatic computer-controlled machines is most helpful when patterns with some form of repetition are to be drawn (e.g., a fan-out pattern and an encoder disk are repetitive patterns), or when part or all of the pattern is computer-generated, or stored in a library of patterns.

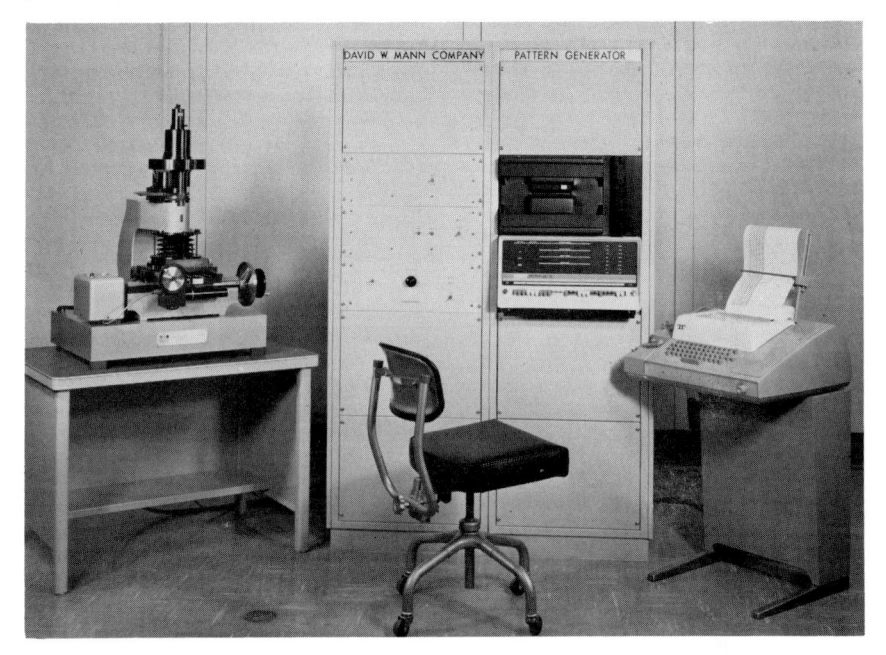

Fig. 9 Reticle generator. (*D. W. Mann Co., Div. GCA.*)

Camera Work

Camera work can be generally classified in three categories. These may be called *first reduction, step-and-repeat,* and *second reduction.* In micromask production the order of performance of step-and-repeat and second reduction may be interchanged or merged. In some work, for example, in printed-circuit work or sheet-metal layout, the master may be smaller than the final work.[24] For this work the first reduction is an enlargement. In general, each step of the mask-making process degrades the end result somewhat; therefore, the less camera work, the better. A one-step 25× reduction will almost always be better than two successive 5× reductions. Multiple reductions should be used only if the artwork generation precision is inadequate for single reduction.

Optical "facts of life" Figure 10 illustrates a lens system which identifies the symbols used in Table 9. The notation follows that of Morgan,[18] who gives some good examples of the use of the equations. Although the geometric-optics approach is adequate for determining magnification, it is wholly inadequate for describing the resolution to be expected from the highly corrected lenses required for photomask work. Many lenses are available which approach within a few percent the theoretical diffraction limit of resolution. In addition the medium, e.g., film or photoresist, which "senses" the pattern affects the useful resolution. The classical resolution criteria are generally satisfactory when the eye is the sensor. The Abbe criterion is based on the assumption that two points can barely be distinguished when the peaks of their respective diffraction patterns coincide with the first zero-energy ring of the adjacent diffraction pattern. This criterion results in Eq. (7) of Table 9. The Rayleigh criterion, which can also accommodate lens aberrations, requires that the wavefront be deformed from its theoretically perfect position by no more than one-quarter wave. In air Rayleigh's criterion is expressed by

$$\Delta = \frac{\lambda}{NA^2}$$

for diffraction-limited optics. Hopkins[27] has proposed a criterion based on the measured response to a sinusoidal grating,

$$\Delta = \frac{0.40}{R(NA)^2}$$

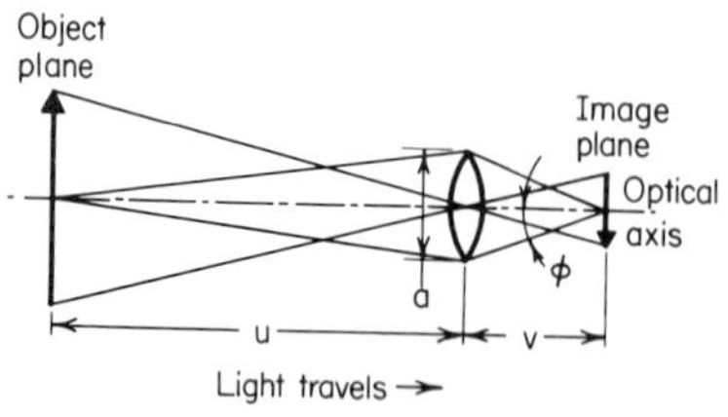

Fig. 10 Simple lens system.

where R is the spatial frequency to which the lens gives 80 percent response. A relation of Hardy and Perrin[28] seems to fit experimental results on high-resolution plates. Altman[29] reports the following results using 0.53-μm light (medium green):

Numerical aperture	Δ
0.1	35
0.16	14
0.32	4.5
0.65	0.75
0.90	0.4

TABLE 9 Formulas of Geometric Optics

$$1/f = 1/u + 1/v \qquad (1)$$
$$M = v/u \qquad (2)$$
$$v = f(1 + M) \qquad (3)$$
$$u = f(1 + 1/M) \qquad (4)$$
$$\Delta v = -M^2 \Delta u \qquad (5)$$
$$f/A = a \qquad (6)$$
$$d = 1.22 \lambda \sin \phi \qquad (7)$$
$$\Delta v = vd/a \qquad (8)$$
$$\Delta v = vdA/f \qquad (9)$$
$$\Delta M = -\Delta uv/u^2 \qquad (10)$$
$$\Delta M = \Delta v/u \qquad (11)$$
$$\phi = \arctan a/2v \qquad (12)$$

Symbols

- u Object-to-lens distance, mm
- v Image-to-lens distance, mm
- f Focal length of lens, mm
- M Magnification ratio
- d Diameter of circle of confusion, mm
- A Relative aperture of a lens, usually called f number or f stop
- a Lens-opening diameter, mm

One difficulty in applying any of these criteria is the inability to arrive at an adequate description of "satisfactory" resolution. For example, in MOS transistors uniform line spacing is more important than smooth edge contour; in some bipolar devices the length of the boundary is a crucial factor; for precision resistors in integrated circuits average line width must be accurate.

Spatially coherent illumination Another difficulty is illustrated in Fig. 11, which represents the spatial sine-wave response of a diffraction-limited lens under different conditions of illumination. The spatial frequency for zero response with spatially coherent illumination (point source focused in the lens) is just one-half that obtained with diffuse illumination; however, spatially coherent illumination maintains full response up to the cutoff, whereas with diffuse illumination the response decreases linearly with increasing spatial frequency. The transition between these conditions is continuous as coherence is varied. Thus a lens tested with different lighting conditions may give very different results. With spatially coherent illumination and high-contrast objects significant nonlinearities can occur in the *optical system,* in addition to the nonlinearities of the photosensitive material.

Diffuse illumination Harwick Johnson[31] has investigated the use of photographic plate nonlinearities with linear optical systems to improve edge sharpness. Edge sharpness can indeed be improved over the image but only by precise exposure and processing control. If small lines and small spaces are both to be improved, the exposures must be different for different areas of the pattern. Such spatial exposure modulation is used in graphic arts processes where it is called *masking.* Altman[29] treats masking in microphotography conceptually as well as indicating

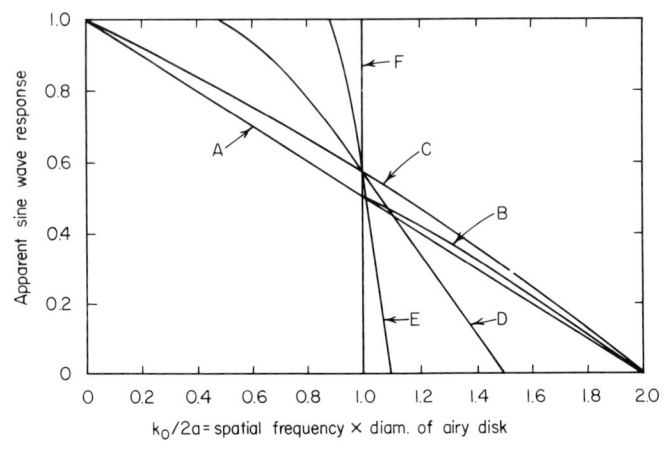

Fig. 11 Apparent sine-wave response versus spatial coherence of illumination.[30]

the theoretical masking required. However, the problem of applying masking for microphotography is difficult when many closely spaced areas must be treated differently. The curves of Fig. 12 show the relative image intensity for clear slits of different widths in the object.[16] It is from these curves that the criterion for minimum "useful" line width for a given lens has been derived. The narrowest lines or spaces of equal width that may be *arbitrarily* placed in a pattern and reproduced at their proper widths without masking, that is with equal exposures, will be called *minimum useful lines.** The correct exposure for minimum useful lines will be the correct exposure for all larger lines or spaces also, and turns out to be the exposure that centers the effective exposure range of the film about the 50 percent intensity level of the image. This criterion is easy to remember since it happens that for diffraction-limited optics the minimum useful line width in micrometers equals the f number of the lens (see Fig. 13).

Adjacent equal lines and spaces always approach a 50 percent response as they

* Strictly speaking this definition assumes an emulsion of infinite gamma, in other words one that changes from unexposed to fully exposed for an infinitesimal change of exposure. High-contrast emulsions used in photomask work have high enough gamma (about 15) to make this a useful definition.

14-22 Photofabrication

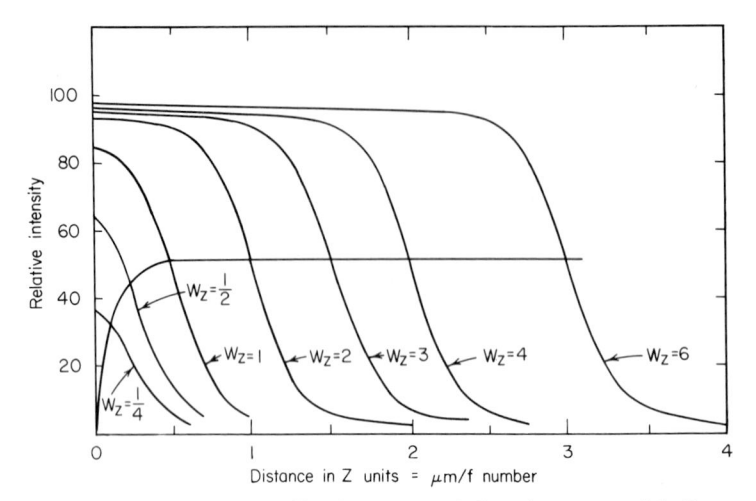

Fig. 12 Intensity profiles for images of slits of various widths.[29]

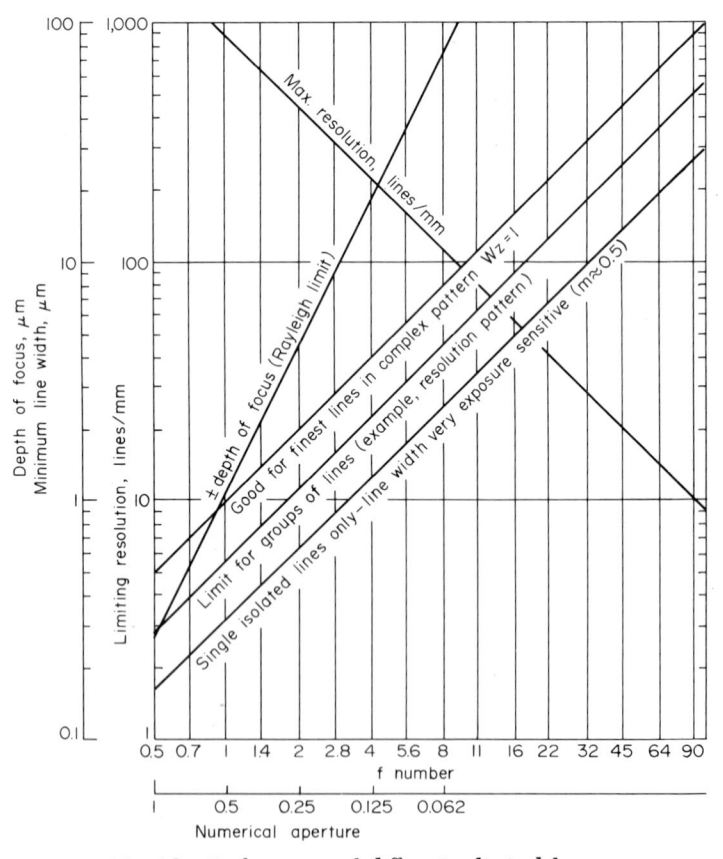

Fig. 13 Performance of diffraction-limited lenses.

get finer (as in a resolution chart), and so the same no-masking criterion allows one to produce an infinite grating with finer lines than indicated. The lines at the edge of a finite grating will be improperly reproduced, and exposure will be more critical if one attempts to take advantage of this effect. In general, finer lines than indicated by the "minimum-line-width" criterion will require the cut-and-try approach. If all parts of the mask-making process are under excellent control, the artwork can have the finest lines made slightly oversize, or the finer lines can be given more exposure to compensate for the reduced peak response. Altman[29] discusses masking as a means of compensation but does not estimate the limit that one can expect to achieve. Taking the case of a linearly falling spatial frequency response such as achieved with diffuse illumination, H. Johnson[31] estimates this limit for one processing condition for Kodak high-resolution plates. He suggests that, with careful control, lines one-fourth of the minimum useful line width can be achieved. Whether these lines would have adequate edge definition depends on the ultimate use of the masks.

When seeking the finest resolution possible, one can increase the apparent gamma of a photographic emulsion by exposing the whole plate uniformly to some low density. This expedient results in gray instead of clear areas which increases subsequent exposures. To understand how this works, consider the transfer curve of Fig. 14. If the plate is given a uniform exposure to ND 0.5, then the additional exposure to achieve ND 2 is about that required on nonpreexposed plate. The resultant edge sharpening may be worth the quadrupled exposure required to print the resulting mask.

First reduction For single-layer printed circuits, panel layouts, silk-screen masters, and similar work a graphic arts process camera can produce adequate results. However, multilayer printed circuits, semiconductor masks, and two-sided plated evaporation masks require a precision of registry difficult or impossible to achieve without a camera capable of adjustments of a few parts per million even for first reduction. For a 25× first reduction to make a step-and-repeat reticle with a 2.5-cm (1-in.) square pattern one might use a 75-mm focal-length lens. The lens-to-copy distance then would be $25(75 + 75/25)$ mm ≈ 1,950 mm. If a precision of 1 μm is required in the image, the precision of the lens-to-copy distance must be 40 ppm, or about 0.078 mm (0.003 in.). The lens-to-plate distance must be controlled to $1/25$ of this value, approximately 0.003 mm (~0.0001 in.)! The depth of focus is much greater than 0.003 mm.

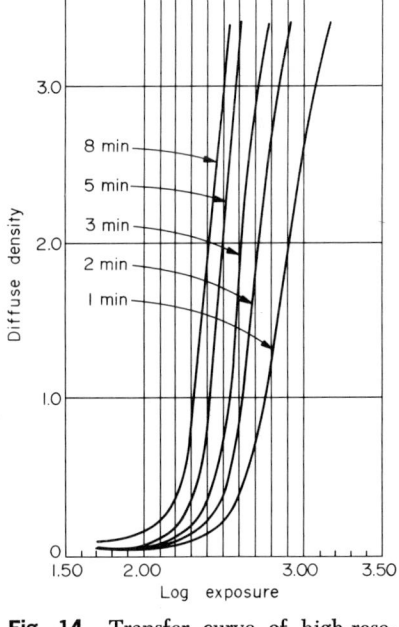

Fig. 14 Transfer curve of high-resolution plate.

Process Camera Work. Many types of photomasks such as single-layer printed circuits may be done with almost any graphic arts process lens, on a process camera. Although process lenses are not free of distortion, they may also be used for more critical work such as multilayer printed circuits if all the layers are made with the same lens in the same part of the field. If registry with other operations (such as automatic drilling) is required, low-distortion process lenses are available and should be used. Process lenses, especially the apochromats, have been carefully developed, and many manufacturers produce lenses of near theoretical resolution at $f16$ to $f22$.

Process cameras are traditionally used for making halftone and line engravings for printing. For black-and-white work modest accuracy is sufficient. However, color work requires superior registry, and accurate process cameras were developed for color work. These process cameras were the forerunners of what are now called *printed-board cameras* and *multilayer printed-board cameras*, respectively. Where large film and plate copies are to be made, a process camera is appropriate. Ingenious designs and massive structures are used to achieve impressive accuracy and repeatability. Figure 15a is a photograph of a large, precision process-type camera.

Precision Camera Work. Micromasks frequently require registry to 10 to 20 ppm over maximum dimensions generally less than 10 cm (4 in.). Cameras capable of such precision, and lenses to use with them, have been developed in response to the micromask needs of the semiconductor industry.[32] Since the required images are smaller than the usual capacity of process cameras, a smaller camera provides a more economical approach to the required precision. Figure 15b is a photograph of a precision micromask camera. The required precision is impressive. For a camera to be capable of a precision of ±10 ppm at 10 times reduction with a lens of 250-mm focal length, the lens-to-copy distance must be repeatable to ±25 μm (0.001 in.), and the lens-to-plate distance repeatable to ±2.7 μm (0.00011 in.). Furthermore, the parallelism of copyboard, lens board, and film plane must not deviate more than ±20 seconds of arc. The formulas of Table 9 can be used to compute the requirements for other conditions.

Lenses to achieve the low distortion and high resolution required over reasonably large fields are mostly products of modern computer designs using recently developed high-index-of-refraction glasses with low absorption in the portion of the spectrum for which they are corrected. In order to achieve the best resolution, lowest distortion, and largest field possible, these lenses are designed for use with a narrow-wavelength band of light and a single magnification. Performance deteriorates rapidly if other than design conditions apply. One lens designer estimates that no more than 5 percent deviation from design magnification can be tolerated without significant deterioration of resolution.*

Table 10 is a list of reduction-camera lenses with their expected performance for micromask work. The Bausch and Lomb Super Baltars are designed and corrected for 35-mm motion picture photography. These color-corrected lenses are more complex, and have smaller angular fields, poorer resolution, or higher distortion than the lenses designed for narrow-band light. In the table the minimum useful line width is theoretical on axis, assuming diffraction-limited response with diffuse illumination except where data are available to indicate that a lower resolution should be used. As noted above, spatially coherent illumination will give different results. The illumination in step-and-repeat cameras may be partially spatially coherent. The figures for data capacity are useful for comparing lenses within groups with similar properties. Large data capacity is easier to achieve with long-focal-length lenses with wide angular coverage but lower resolution.

Step-and-Repeat Cameras. The patterns on many micromasks are repetitive. A transistor mask may have 1,600 identical patterns in a square array; means of making multiple images are therefore essential. Many ingenious ways have been devised to get multiple images (see the following section), but the workhorses of multi-image repetitive patterns are step-and-repeat machines.[22,31,33-39] Two categories of machines might be called the *intermediate-size* and *final-size* step-and-repeat machines. The multiple-image pattern produced by an intermediate-size step-and-repeat machine is further reduced by a precision camera to make the final mask. Intermediate size step-and-repeat machines may produce the images by contact printing or projection printing repeatedly on a large film or plate which is translated (stepped) and stopped during exposure.

* B. E. Day of Wray Ltd., London, private communication.

Photomask Fabrication   14-25

(a)

(b)

Fig. 15 Cameras for photomask processing. (a) Precision process camera. (R. W. Borrowdale Co.) (b) Precision micromask camera.

Final-size step-and-repeat machines are all projection type and hence are often called step-and-repeat cameras. For the finest resolution and best precision, final-sized step-and-repeat is to be preferred. As a result, most of the masks for transistor production are made by final-size step-and-repeat machines. The precision of motion of modern step-and-repeat machines does not limit the precision of the

TABLE 10 Reduction Camera Lenses for Microcircuit Masks

Manufacturer	Focal length, mm	f number	Corrected magnification	Field diameter, mm	Corrected wavelength λ, μm	Minimum working line, μm	Data capacity, megabits	Approximate price
Nikon....	155	4.0	0.1	56	0.546	4	155	$5,300
	125	2.8	0.04	28	0.546	2.8	79	5,000
	105	2.8	0.033	24	0.546	2.8	58	4,800
Wray....	45	4.0	0.02	18	0.546	4	16	330
	75	4.0	0.04, 0.1, 0.2, 0.5	35	0.546	4	64	400
	127	4.0	0.025, 0.05, 0.1, 0.2	60	0.546	4	180	1,044
	254	4.0	0.1, 0.2	116	0.546	4	670	2,760
Kodak...	93	3.5	0.05	33	Visible	3.5	69	7,500
	50	2.8	0.1	20	Visible	2.8	40	4,000
Bausch & Lomb Super Baltars	25	2.0	....	14	Visible	6	4.8	403
	35	2.0	....	15	Visible	5	7	234
	50	2.0	....	23	Visible	5	17	220
	150	2.8	....	26	Visible	6	15	285
	225	4.0	....	26	Visible	6	15	400

Data capacity = $(\pi/4) \times$ (field diameter, mm)$^2 \times$ (minimum working line width, μm)$^{-2}$

final masks. The precision is limited by the lens distortions and magnification mismatch. If these were corrected, the limit would become the wavelength of light. Figure 16 is a photograph of a six-barrel step-and-repeat machine.

Lenses suitable for step and repeat at final size are listed in Table 11. Note that resolution is generally higher than that of the precision camera lenses but the fields are smaller. The Leitz microscope objectives are included for comparison. They are apochromatic; that is, they are color-corrected at three wavelengths in

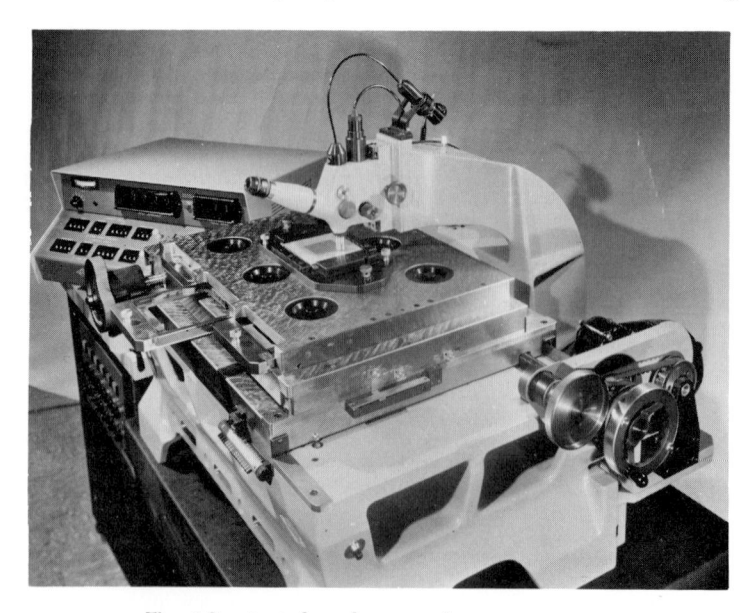

Fig. 16 A six-barrel step-and-repeat camera.

the red, green, and blue, and can be used with any visible light. All the other lenses are corrected for a narrow-band spectrum and, as a result, can have larger fields. Note that larger data capacity appears in lenses with larger fields and lower resolution. In integrated circuits the yield of good devices increases as the total device size decreases. The most desirable step-and-repeat lens therefore should have the necessary data capacity in the *smallest* field.

Step-and-repeat exposure of photoresist to make chrome master masks has become practical with the advent of lenses designed for the blue-violet portion of the spectrum.[40,41] The exposures required are 1 or 2×10^3 larger than for high-resolution silver halide emulsion; therefore the process is slower and more expensive, and its use should be reserved for those masks that really need resolution of lines less than 2.5 μm wide.

Other Methods of Forming Multiple Images. Many other methods of multiple-image production are possible. Some are straightforward; others are rather ingenious. A multipinhole camera has many attractive features, e.g., simplicity, freedom from geometric distortion, and wide acceptance angle. The optimum selection of pinhole size and magnification results in a minimum line width of about 0.2 mm, which is adequate for many purposes. If one needs 2-μm lines in the final work of a 50-mm (2-in.) square, at least 100× final reduction would be required from a master 5-meter (16½ ft) square, a very large copyboard.

Better resolution can be obtained by using simple lenses in an array. The job of arranging them to achieve uniform image size and spacing is formidable. If really accurate arrays are needed, it is doubtful whether this approach is less costly than step-and-repeat.

Carrying the lens-array concept to its logical conclusion has led to the "fly's eye" camera with, say, 200:1 reduction, which would make a final mask from a single-device master in one step. One such device uses two sets of cylindrical planoconvex lenses at right angles to each other with the convex faces nearly in contact. At best the resolution can be very good, but uniform resolution and magnification are very difficult to achieve, and the good portion of the field is only a small part of the total area.

Arrays of small mirrors or prisms may be arranged in an optical system to provide multiple images. The resolution is limited by the single lens used and can be excellent. Of all the multiple-imaging systems the multimirror type is probably capable of the best results and is surely the most expensive.

All these multiple-image systems have limitations that account for their lack of popularity in mask making. Probably the fatal limitation is inflexibility. Once the repeat distance for the array is established, it can be changed only with difficulty. For example, redesigning to reduce the linear size of a transistor by 40 percent should allow twice as many transistors to be made on a wafer since they can be formed closer together. The fixed array of a multi-image camera would not allow one to realize this gain since the center-to-center repeat distance is a constant. Another disadvantage is that where reductions are small, e.g. 10:1, the COS^4 falloff in light intensity toward the edge of the array becomes very serious, requiring some sort of exposure compensation. At present it seems unlikely that multi-image cameras will make any serious inroads into the step-and-repeat business, particularly for integrated-circuit mask fabrication.

Contact Printing

In the section on camera work the concept of data capacity was used. Contact printing allows an almost arbitrarily large amount of data to be replicated in a single step. If intimate contact is obtained, the resolution of the copy is limited only by the wavelength of light, the resolving power of the photosensitive material, and the pattern in the master. For electronic applications a vacuum printing frame is used to assure intimate contact. The usual pressure-type printing frames or contact printers do not provide adequate contact to prevent resolution loss or linewidth variations. Contact printing is generally used with either silver halide emulsion or photoresist as the sensitive medium for electronics work. Other sensitive

TABLE 11 Lenses for Step-and-Repeat Cameras

Manufacturer	Focal length, mm	Numerical aperture	f number	Corrected magnification	Field diameter, mm	Corrected wavelength λ, μm	Data capacity, megabits	Minimum line width, μm	Object-to-image distance, meters	Approximate price
Nikon	55	....	2.0	0.25	10	0.546	20	2.0	0.315	$2,000
	28	....	1.8	0.1	4	0.546	3.9	1.8	0.315	1,300
	29.5	....	1.2	0.04	2	0.546	2.1	1.2	0.810	5,000
	55	....	2.0	0.25	10	0.405	35	1.5	0.35	
	28	....	1.8	0.1	4	0.406	6.4	1.0	0.29	
Leitz, Canada	25	....	1.5	0.1	8	0.400–0.480	35	1.20	0.6	
		0.25	2.0	0.1	12–14	0.405–0.436	50	1.5	0.4	
Fairchild	50	....	1.26	0.1	7	0.416–0.466	39	1.0	0.63	5,000
Wray	75	....	2.8	0.1	11.2	0.436	20	2.25	0.83	1,300
Tropel	26.5	0.285	1.6	0.1	7	0.405–0.436	23	1.3	0.315	5,000
Leitz, Germany; metallurical microscope objectives		0.95	....	0.0125	0.4	0.4–0.7	0.5	0.5	0.2	411
		0.50	....	0.0312	1.1	0.4–0.7	0.72	1.0	0.2	369
		0.25	....	0.0625	2.2	0.4–0.7	0.72	2.0	0.2	240
		0.18	....	0.125	4.4	0.4–0.7	1.7	3.0	0.2	217
Cerco	30	0.452	1.0	0.1	4.7	0.436	27	0.8	0.4	5,500
	20	0.324	1.5	0.1	3.25	0.436	5.7	1.2	2.04	3,900

The lenses corrected for 0.546 μm are intended to be used with high-resolution silver halide emulsions. Those corrected for 0.405 to 0.436 μm are intended for exposing photoresist using the mercury g and h line radiation. At the time of writing, the Nikon, Wray, Tropel, and Cerco lenses appear to be commercially available. The Leitz microscope objectives are available as replacements for their large metallograph. The commercial production of the other lenses is in some doubt. They are in the preproduction or phototype stage. The data capacity is computed on the basis of the working minimum line width squared and divided into the area of the circular field.

media such as diazo and blueprint can also be used, but they are not common in mask making for electronics applications. Although there is much similarity between macro- and microcontact printing, they will be discussed separately to emphasize the techniques important to success.

Macrocontact Printing

For convenience, macrocontact printing will be defined as including any work for which lithographic film has adequate resolution. This choice limits the finest lines to about 25 µm (0.001 in.). Lithographic film is available in many sizes, base materials, and a variety of thicknesses.[43,44] The selection of base material was discussed previously. Naturally the vacuum contact-printing frame should be large enough to accommodate the largest work anticipated. Other than some awkwardness in operation, a large vacuum frame has no significant disadvantages for small work. The vacuum pump, lines, and valves should be large enough to permit pumping down to less than 5 torr absolute in a minute or so. Vacuum print frames have one glass side and one flexible side. For two-side exposure, as required for double-sided printed circuits, the flexible side is transparent, usually a vinyl film. Machines are available that use two light sources to expose both sides simultaneously, thus avoiding the necessity for turning the work in the frame, with possible loss of registry. For macrowork adequate registry can be maintained by providing registration holes in the film. Pins are put through the holes to register the masks mechanically during exposure. For some work taping the film masks in registry will suffice. A fuller discussion of macroregistry appears in Ref. 16.

Although a vacuum frame is necessary for good contact printing, it is not sufficient in itself. Strauss[32] has discussed the effect of the vacuum in warping glass plates and the support glass on copyboards. The same problems appear in contact printing with rigid materials thicker than a fraction of a millimeter. The curve

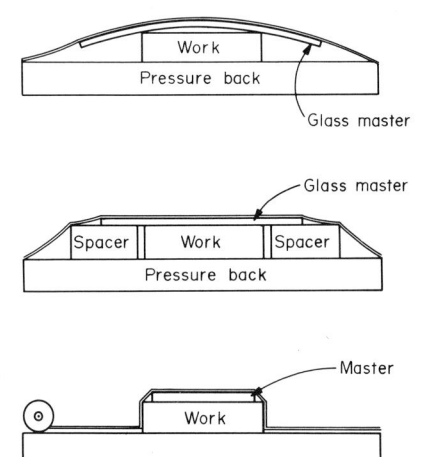

Fig. 17 Warping of copy in vacuum print frame or copyboard.[32]

involves using spacers to take the stresses which would otherwise bend the work out of contact. The causes of the problems and their cures are illustrated in Fig. 17.

Microcontact Printing

The sensitive materials used for microcontact printing are almost always high-resolution plates or photoresist films. Few other materials have the resolving power to reproduce the micrometer-sized lines required in photomasks for solid-state device and integrated-circuit fabrication. Only contact printing is capable of the nearly 2,000-megabit data transfer necessary to achieve micrometer-level resolution across a 50-mm (2-in.) diameter wafer. For slightly lower data-transfer capacity, projection printing may be used.[44] Table 12 lists some lenses for projection-wafer printing.

Such high-resolution contact printing requires the very best technique. Even minute spacings between the mask and the sensitive emulsion will spoil the result.[45] Stevens[46] illustrates the degradation in resolution due to the small spacing caused by coating pinhole areas in a negative with an opaque ink. He also shows the improvement obtained when a contact fluid is used. The photographs of Fig. 18 illustrate the effect. The thickness of the emulsion on high-resolution plates is about 6 µm before exposure. After exposure and processing, the thickness is

TABLE 12 Lenses for Projection Wafer Exposure

Manufacturer	Focal length, mm	Numerical aperture	f number	Corrected magnification	Field diameter, mm	Corrected wavelength λ, μm	Object-to-image distance, mm	Minimum working line width, μm	Data capacity, megabits	Approximate price
Leitz, Canada	....	0.20	2.5	0.5	40	0.405–0.436	0.5	2.0	314	
Fairchild	150	....	3.0	0.05	50	0.416–0.466	3.2	2.4	340	
Wray	250	....	2.8	0.1	35	0.436	2.75	2.2	200	$4,300
Zeiss	....	....	1.5	1.0	37	0.41–0.44	....	1.2	750	
Wray	75	....	2.8	1.0	35	0.436	0.3	8*	15	
Nikon	250	....	1.0	1.0	50	....	0.489	2.5*	314	$15,000

* From manufacturing data. The wide field results in a lower resolution than would be computed from the diffraction limit.

about 2 μm in the clear areas and 4 μm in the opaque areas. (Altman[29] has used these thickness differences to infer the density of fine lines, probably more accurately than they can be measured directly.) As a result of the emulsion thickness, the maximum distance between a part of the master mask and its image in the sensitive emulsion is 10 μm. For line widths approaching 1 μm that is very poor contact. Collimation of the light source would seem to be an obvious

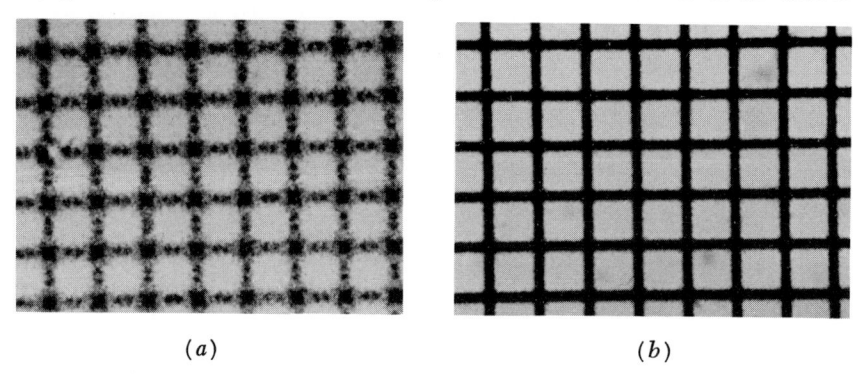

Fig. 18 Effect of lack of contact in contact printing; improvement due to contact fluid. (a) 1,800-line/in. grid at 1,000× without contact fluid. (b) 1,800-line/in. grid at 1,000× with contact fluid.

answer to lack-of-contact problems, but the improvement is illusory. For relatively coarse patterns and gross lack of contact, some improvement can be obtained, but when the resolution required approaches the wavelength of light, geometric optics no longer suffice, and Fresnel diffraction becomes significant. Lack of intimate contact is usually the problem when the pattern in the original can be seen in a microscope but the reproduction is poor. Another cause of poor resolution in contact printing is the lens effect at the boundary between clear and opaque areas of the negative. Figure 19 illustrates the lens effect for collimated light and geometric optics. This is illustrative only because, as consideration of the figure will reveal, the lens effect is opposite in sense to the Fresnel and Fraunhofer diffraction effects, both of which are present; and the distances are small enough so that interference patterns are significant even with fairly diffuse light of a fairly broad spectral range. Although the situation is quantitatively complex, minimizing the separation of the master mask

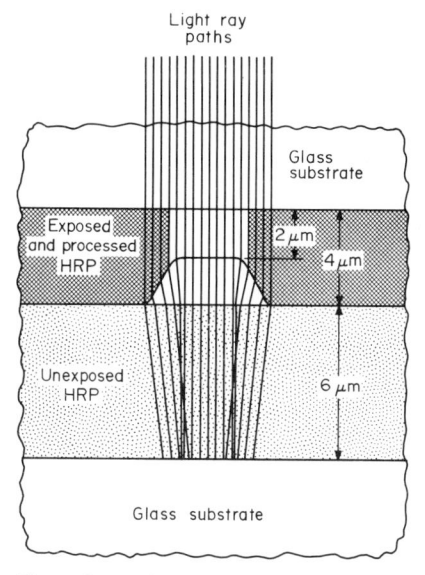

Fig. 19 Refraction at density transition in high-resolution emulsion.

from the sensitive medium always improves the reproduced image. The corollaries are: (1) the thinnest sensitive medium that can be used will yield the best result and (2) the thinner the opaque layer in the master and the nearer its surface the better.

Consideration of the mechanics of dispersion of microparticles in a fluid medium indicates that a substantial improvement in the density per unit thickness of silver

14-32 Photofabrication

halide high-resolution plates is very unlikely. Thinner coatings than the present 6 μm do not offer much improvement. A 2-μm coating would have a maximum optical density when developed of about 1, corresponding to a maximum contrast ratio of 10. Such a low-contrast ratio is not useful for most applications. Physical intensification may increase the contrast to some extent but probably at a cost in edge definition. The 6-μm coating is within the Rayleigh depth of field for a lens with 2.5-μm resolution; the 2-μm coating would allow a 1.4-μm resolution for the same criterion. Even thinner sensitive layers are required if submicron resolution is to be achieved.

At present the best results are accomplished with "chrome masks," i.e., glass plates coated with a thin layer of chromium, usually by vacuum evaporation. A photoresist layer on the chromium film provides the sensitive medium for exposure.[47] The developed photoresist pattern serves to define a pattern in the chromium layer by suitable etching. The chromium layer is about 0.1 μm or less thick, and the photoresist is usually about 0.5 μm thick. The sensitive material is thus only $\frac{1}{12}$ the thickness of the silver halide emulsion on a high-resolution plate. On a depth-of-focus basis, a resolution improvement of $\sqrt{12} = 3.5$ is expected. This improvement indicates a potential for producing 0.7-μm lines in a second photoresist layer by contact printing. Current experience substantiates the expected improvement. The improvement due to the "grainlessness" of photoresist is probably at least partly offset by a lower effective "contrast" compared to that of high-resolution plates. This last statement cannot be proved since the "contrast" of a photoresist is difficult to define. The edge transition is always abrupt in the etched chromium film. For convenience, the contrast might be expressed as the reciprocal of the minimum exposure increment within which it is uncertain whether or not the photoresist, after development, will prevent etching of the substrate. Even using such a definition, exposure data from different individuals and even different batches of photoresist exposed by the same individuals, give results that disagree by several factors of 10.

Registry in microcontact printing is a subject in itself and is treated in a section on "special techniques." Chemical processing of photographic emulsions is treated in the next section. Photoresist processing is described under Photosensitive Materials.

Chemical Processing

The development of photographic emulsions produces a silver deposit where the silver halide has been exposed to light; this process is called *direct development* and gives a *negative* image of the original artwork. *Reversal development* produces clear areas wherever the sensitive emulsion has been exposed and gives a *positive* image. Both types of development are useful for macro- and micromask work. For special purposes other processes may be used which provide dyed or stained images or images formed in a material other than reduced silver.

Direct processing Reduced to its simplest form, direct processing consists of chemical reduction of the exposed silver halide to metallic silver, and then dissolving away any unreduced silver halide. These steps are called *developing* and *fixing*, respectively. Usually several more steps are included to improve the image or decrease the time required for processing. Any textbook on photography includes a discussion of development.[48] Figure 20 is a flow chart for typical direct development as applied to the preparation of photomasks. Depending on the results desired, the development process can be modified in a variety of ways by including small quantities of special reagents. Most photomask work aims at maximum contrast with high-density black areas and clear areas of low density. These aims are best met using (1) long developing times just short of significantly increasing the background fog level, (2) very active, fast-working developers capable of producing high optical density in the exposed regions, and (3) minimum exposure to produce maximum density. In all development processing, constant temperature must be employed if reproducible results are to be achieved. Temperature changes increase the graininess and degrade edge definition. The choice of the developing

Photomask Fabrication 14-33

temperature is a compromise between higher temperatures for fast processing and lower temperatures for reduced swelling of the gelatin emulsion. The usual choice is between 18 and 20°C (65 and 70°F).

Chemically a developer reduces the silver halide to silver while in the process the developer itself is oxidized. The most active developers are easily oxidized on standing in air. Therefore, they have the shortest useful life after preparation. Frequently more consistent results are obtained by using a less active developer having a longer tank life, and accepting the reduction in contrast and maximum optical density. Nitrogen-burst agitation[49] and deep-tank processing prolong useful life by reducing air oxidation of the developer. For example, the useful life for one developer went from less than 4 hr for 1 gal of developer in a tray to over 2 weeks for 10 gal in a deep tank with nitrogen-burst agitation.

Table 13 is a compilation comparing several developers suitable for photomask work, the most active being listed first. The formulations are those recommended by Kodak,[50] but very similar formulas are recommended by most makers of photographic material. D-76, a general-purpose fine-grain developer for continuous-tone work, is included for comparison.

All developing baths contain basically four constituents: (1) the developer or reducing agent, (2) an accelerator or "promoter," (3) a preservative, and (4) a stabilizer. Other constituents sometimes used are (5) an antifog agent, and (6) silver halide solvents.

The activity of the reducing agent must be sufficient to reduce the exposed silver halide in a reasonable time without developing the unexposed portions. The most useful developing agents are organic compounds which ionize to a very limited extent in solution, and therefore act extremely slowly unless an ionization promoter is included. Usually an alkali, such as sodium hydroxide, or an alkaline salt, such as sodium carbonate or borax, is employed, although a weakly ionized acid such as boric acid is sometimes used as promoter (also called accelerator). Sodium sulfite is the most common preservative, although sodium bisulfite and potassium metabisulfite are sometimes used. The preservative reduces the rate of oxidation of the developing agent. A soluble halide, which is almost always potassium bromide, serves to stabilize the performance of the developing solution. Soluble halides are produced by the action of the developer on the silver bromide of the emulsion; this "by-product" retards the developing process. By starting with a relatively large amount of soluble halide, the change in its concentration during development is small, and the developer remains nearly constant with use. The soluble halide also inhibits development of unexposed areas, thus reducing the background fog level.

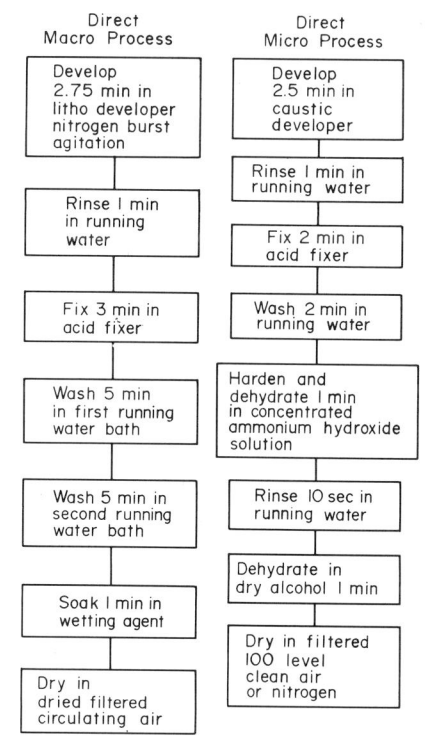

Direct Macro Process
- Develop 2.75 min in litho developer nitrogen burst agitation
- Rinse 1 min in running water
- Fix 3 min in acid fixer
- Wash 5 min in first running water bath
- Wash 5 min in second running water bath
- Soak 1 min in wetting agent
- Dry in dried filtered circulating air

Direct Micro Process
- Develop 2.5 min in caustic developer
- Rinse 1 min in running water
- Fix 2 min in acid fixer
- Wash 2 min in running water
- Harden and dehydrate 1 min in concentrated ammonium hydroxide solution
- Rinse 10 sec in running water
- Dehydrate in dry alcohol 1 min
- Dry in filtered 100 level clean air or nitrogen

Fig. 20 Flow chart for direct photographic processing.

Referring to Table 13, it can be seen why the developers behave as they do. D-8 uses a high concentration of hydroquinone, a high-contrast developing agent that tends to develop the more highly exposed areas faster. A large amount of very ionizable sodium hydroxide is included which allows the developer to work rapidly.

14-34 Photofabrication

This combination is so active that a large quantity of potassium bromide is needed to restrain the development of unexposed areas, and even the relatively high concentration of sodium sulfite preservative gives a life of only a few hours. D-11 and D-19 are similar, differing significantly only in the proportions of hydroquinone and Metol, a low-contrast developer, and in the amount of alkali. D-76 with no potassium bromide and much preservative depletes rapidly but keeps well if not used. Borax is a weakly ionized alkaline material which, together with the smaller amounts of developing agent, results in long developing times. The balance between low- and high-contrast developing agents results in long tonal range and high maximum density. Similar reasoning can be employed to predict the characteristics of other developer formulations.

D-85 is an example of an "infectious" developer which enhances edge contrast by drawing silver from adjacent unexposed areas to increase the edge density in exposed areas.

TABLE 13   Developers Suitable for Photomask Work*

Agent	D-8	D-11	D-85	D-19	D-76
High-contrast developer:					
Hydroquinone	45	9	22.5	8	5
Paraformaldehyde	....	...	7.5		
Low-contrast developer:					
Metol† (Elon‡)	....	1	....	2	2
Promoter accelerator:					
Sodium hydroxide	37.5				
Sodium carbonate	....	25	....	52.5	
Borax	....	...	....	....	2
Boric acid (crystalline)	....	...	7.5		
Preservative:					
Sodium sulfite	90	75	30	90	100
Sodium bisulfite	....	...	2.2		
Soluble halide:					
Potassium bromide	30	5	1.6	5	

* Quantities in grams to make 1 liter of working solution.
† Trademark of the Harshaw Chemical Co., Cleveland, Ohio.
‡ Trademark of Eastman Kodak Co., Rochester, N.Y.

The antifog agents act in various ways to suppress the conversion of unexposed silver halide to silver. The silver halide solvents can also act as antifog agents by dissolving unexposed silver bromide faster than it can develop. These silver halide solvents can also enhance edge resolution by dissolving slightly exposed grains before they develop. With proper adjustment of solvent activity one obtains the "monobath" process where development and fixing are completed simultaneously.

After a water rinse or an acid stop bath to terminate the development process, the film is fixed. Fixing removes the undeveloped silver halide from the emulsion. Thorough fixing and washing afterward are necessary if permanence is required. For a great deal of photomask work the master film or plate is discarded after photomasks are prepared, and permanence is not required. Therefore, the fixing times can be reduced to just the time necessary to clear the background, and small amounts of fixer may be present in the emulsion after washing. In quantity production the time saved may be worthwhile. The active ingredient in the fixer is usually sodium or ammonium thiosulfate, the latter being used for rapid fixing. Fixers are usually acid, to neutralize the alkaline component in the developer. They

Photomask Fabrication 14-35

also often include a hardener to "tan" the gelatin and improve the scratch resistance of the emulsion. Table 14 compares the formulations of several fixers applicable to photomask work.

Reversal processing Reversal processing[50] has long been used for home movies; and color slide film is reversal-processed. Reversal processing produces clear areas where the exposure was high, and dense areas where the initial exposure was low. In mask making, reversal processing has two main applications: (1) where a mask of the same sense as the original is wanted, one step can be omitted; and (2) where a "right-handed" mask is available and a "left-handed" mask is required, it can be formed by contact printing. The quality of the masks formed as in application 1 depends critically on the reversal process. Reversing the "handedness" of a mask by techniques other than reversal processing, application 2 requires exposure to the back of a mask or to the back of the artwork. Use of such "reversed" artwork results in little loss of definition if the camera is adjusted to allow for the thickness of the film. If the film is 0.18 mm (0.007 in.) thick,

TABLE 14 Fixers for Photomask Work*

Agent	Formulation			
	F-5	F-7	F-9	F-24
Silver halide solvent:				
Sodium thiosulfate..........	240	360	360	240
Ammonium chloride.........	....	50	....	....
Ammonium sulfate..........	....	....	60	....
Acid:				
Acetic acid, 28%............	48 ml	48	48	....
Boric acid.................	7.5	7.5	7.5	....
Preservative:				
Sodium bisulfite............	....	....	....	25
Sodium sulfite..............	15	15	15	10
Hardner:				
Potassium alum............	15	15	15	....

* Quantities in grams to make 1 liter.
F-5, general purpose; F-7, rapid; F-9, rapid, lower corrosion on stainless steel; F-24, nonhardening.

the whole camera must be moved toward the copyboard $0.007 \times n$, where n is the index of refraction of the film base. Usually $n = 1.5$ is an adequate approximation for copyboard allowance. Therefore, the lens and plate back must be moved about 0.27 mm (0.0105 in.) toward the copy to keep the magnification the same as if the artwork were facing normally. The second and much less desirable solution is to make a contact print of the artwork from the back. Even with a point light source, a loss in resolution will be observed owing to diffraction. With reversal processing the least costly and best results are achieved using a contact print made on high-resolution emulsion, or alternatively, direct-positive film can be used to make lateral reversal copies of large-scale artwork. Figure 21 is a flow chart for reversal processing of high-resolution plates. Table 15 lists formulations of the etch and bleach baths for this reversal process. Reversal-process kits intended for processing of home movies may be used with high-resolution emulsion and give very satisfactory results, especially if high-activity developers such as D-8 replace those in the kit. More exposure is usually recommended for reversal processing than for direct processing; however, exposure and development interact so strongly with high-contrast emulsions that no general rules can be applied.

14-36 Photofabrication

Other chemical processing The other principal chemical processes are reduction, intensification, and stencil processing. Usually corrective processes such as these are best avoided by redoing the work. However, an expensive mask can sometimes be rescued through a corrective chemical process. For example, overexposure of a master due to malfunction in a photoplotter might be corrected by judicious reduction. As another example, Kodak, Ltd., in England supplies maximum-resolution emulsion in a form that may be reconstituted and coated to give layers thinner than normal for high-resolution plates. The maximum density of very thin coatings may not be adequate without intensification. Since mask work almost always requires the highest possible contrast and the sharpest edges possible, reducers that preferentially reduce the less dense (often called shadow) areas are employed. These are called cutting reducers. Similarly the most useful intensifiers are those that add density most rapidly to the areas which are already most dense, without increasing the density of the clear areas. Table 16 gives the formulation for Farmer's reducer which is effective in clearing low-optical-density areas and a formula for a chromium intensifier which is relatively easy to use and keeps well. Also included is the prehardener bath required. The flow charts for using Farmer's reducer and the chromium intensifier are given in Fig. 22. Any photographic formulary will provide other processes which may suit particular needs. These two were selected for presentation here because they give good results and are relatively easy to use. Mercury intensifier is also easy to use but is toxic and not very permanent. Silver intensifier is tricky to use but matches the original color well. Intensifiers using thiocyanate are hazardous because thiocyanate is a powerful fogging agent which can spoil unexposed film or cause spots during development if the thiocyanate dust is not very carefully controlled. Other reducers such as the permanganate–sulfuric acid type are either more corrosive or toxic than Farmer's reducer, and many are more difficult to control.

Stencil processing, which completely removes the gelatin from clear areas of the emulsion, may be used for either macro- or micromasters. Film macromasters are less sensitive to humidity changes if the gelatin is removed from most of the surface. Micromasters are less subject to scratching if glass is exposed in the clear areas rather than gelatin. The line edges produced by stencil processing

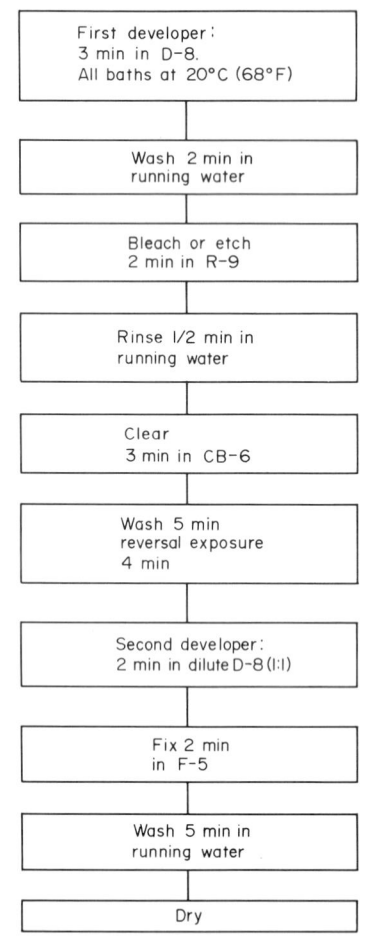

Fig. 21 Flow chart for reversal processing.

First developer:
3 min in D-8.
All baths at 20°C (68°F)
↓
Wash 2 min in running water
↓
Bleach or etch 2 min in R-9
↓
Rinse 1/2 min in running water
↓
Clear 3 min in CB-6
↓
Wash 5 min
reversal exposure 4 min
↓
Second developer: 2 min in dilute D-8 (1:1)
↓
Fix 2 min in F-5
↓
Wash 5 min in running water
↓
Dry

TABLE 15 Bleach and Clearing Baths for Reversal Processing

Bleach bath R9 (1 liter):
  Potassium dichromate .... 9.5 g
  Sulfuric acid .... 12 ml
Clearing bath CB-6 (1 liter):
  Calgon* .... 0.5 g
  Sodium bisulfite .... 15.0 g

* Trademark of Calgon Corporation, Pittsburgh, Pa.

TABLE 16   Reducer and Intensifier for Photomask Work

Farmer's reducer

Solution 1:
  Hypo (sodium thiosulfate) 240
  Water to make .... 1 liter
Solution 2:
  Potassium ferricyanide .... 10 g
  Water to make .... 250 ml
For use mix one part of solution 2 to 4 parts of solution 1 immediately before use.
Keeps only a short time after mixing.

Chromium intensifier (stock solution)

Potassium bichromate .... 90 g
Hydrochloric acid .... 74 ml
Water to make .... 1 liter

---

tend to appear ragged when examined microscopically but may still print well. Stencil processing is described by Neblett.[51]

Special Techniques

The following section on photoresists provides information on their choice and use. This section will not go into detail on the photoresist techniques used in making some of these special masks.

For many purposes photomasks using silver halide emulsions are less satisfactory than masks produced by other techniques. For example, chromium-on-glass masks are re-

Reducer

- Mix 1 part solution A and 4 parts solution B just before use
- Treat until sufficient reduction is obtained
- Wash 5 min and dry

Intensifier

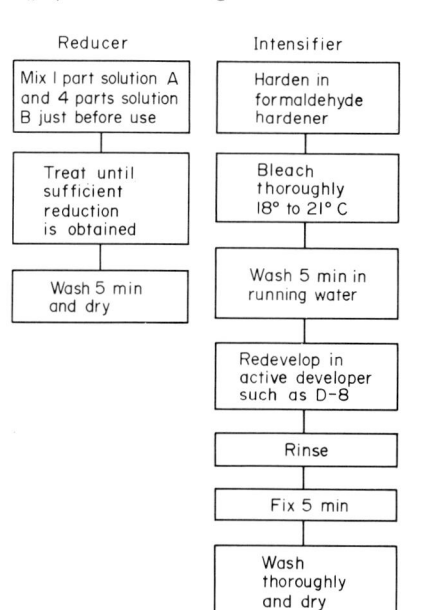

- Harden in formaldehyde hardener
- Bleach thoroughly 18° to 21° C
- Wash 5 min in running water
- Redevelop in active developer such as D-8
- Rinse
- Fix 5 min
- Wash thoroughly and dry

Fig. 22 Flow chart for reduction and intensification.

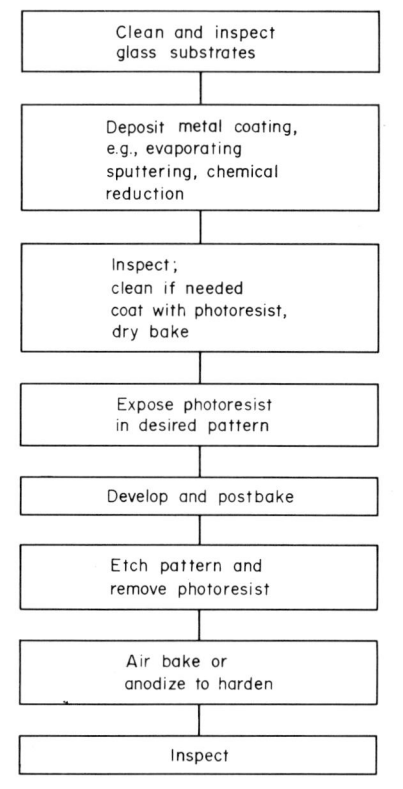

- Clean and inspect glass substrates
- Deposit metal coating, e.g., evaporating sputtering, chemical reduction
- Inspect; clean if needed coat with photoresist, dry bake
- Expose photoresist in desired pattern
- Develop and postbake
- Etch pattern and remove photoresist
- Air bake or anodize to harden
- Inspect

Fig. 23 Flow chart for metal-on-glass masks (chrome masks).

proted to give 10 to 100 times the lifetime of silver halide masks when used for contact printing in semiconductor processing. They also yield more repeatable results. In some applications, such as vacuum evaporation masking and screen printing, the clear areas of the mask must be physically open. The processes used to make these photomasks replace the cameras, developers, and processing tanks used with silver halide emulsions by such equipment as precision-registry vacuum-contact printing frames, organic solvents, vapor degreasers, spray etchers, electroplating equipment, spin coaters, ovens, and vacuum evaporators. The processing steps are usually more numerous and less well controlled than for silver halide masks; mask quality depends more on the skill and experience of the mask maker.

Metals on glass Figure 23 is a flow chart for the fabrication of metal-on-glass masks. Chromium is almost universally used as the metal. The metal should adhere firmly to the glass, be opaque in very thin layers over the entire actinic spectrum, have abrasion and scratch resistance comparable to that of glass, be etchable in etchants that do not destroy the photoresist and be easy to apply with uniform thickness, optical density, and etch properties. Table 17 lists comparative properties of several metals which might be used for metal-on-glass marks. This comparison shows why chromium is usually the preferred material. Other deposition methods such as sputtering can improve the adhesion of materials like copper, which might be preferred for ease of etching, but the abrasion resistance is still

TABLE 17 Properties of Metals for Metal-on-Glass Masks

Metal	Deposition method	Adhesion	Opacity	Abrasion resistance	Ease of etching	Uniformity
Aluminum	Vacuum evaporation	Good	Good	Fair	Good-poor	Good
Copper	Vacuum evaporation	Fair	Good	Poor	Good	Good
Chromium	Vacuum evaporation	Good	Good	Good	Good-fair	Good
Gold	Vacuum evaporation	Poor	Good	Poor	Poor	Good
Nichrome*	Vacuum evaporation	Good	Good	Good	Poor	Fair
Nickel	Vacuum evaporation	Fair	Good	Fair	Fair	Good
Inconel†	Vacuum evaporation	Good	Good	Good	Poor	Fair
Silver	Vacuum evaporation	Poor	Fair	Poor	Fair	Good
Silver	Chemical reduction	Fair	Good	Poor	Fair	Poor

* Trademark of Driver-Harris Co., Harrison, N.J.
† Trademark of Huntington Alloy Products Division, The International Nickel Co., Inc., Huntington, W. Va.

inferior. The nickel-chromium alloys are excellent for adhesion and wear resistance but are very difficult to etch. Chromium etches easily and uniformly if it is deposited under conditions that give a high-purity metallic deposit of uniform thickness. These coatings are softer than coatings that are oxidized slightly during deposition. The increased difficulty of etching the oxidized chromium and the nonuniformity sometimes introduced by the oxidizing conditions make it preferable to produce the highest-purity chromium deposit possible. After defining a pattern in the chromium film, it can be hardened by oxidizing the surface. Aluminum may also be hardened by oxidizing the surface, but, unlike chromium oxide, aluminum oxide is transparent to actinic light. Chromium should be deposited in a good vacuum, free from water vapor, oil, and oxygen contamination and from high-purity chromium source material. Chromium plated heater rods is a common source for vacuum evaporation. The as-plated purity depends on the plating process, but careful vacuum heat treatment is usually needed to attain adequate source purity. The heat treatment may be performed in the evaporation chamber, but since it is a lengthy process and leaves at least some of the impurities in the vacuum system, a separate vacuum furnace treatment is usually preferred.

Evaporation should be started while a shutter covers the substrates and brought to a constant rate before the shutter is opened. This initial evaporation gathers impurities from the residual gas in the evaporation chamber and covers possible gas-emitting surfaces with a coating of chromium. The optimum rate of evaporation is not critical. At very low rates there is time for impurities to land on the coating; at very high rates the atoms do not have time to adjust their positions in the film before more material arrives to lock them in place where they landed. The practical upper bound on rate is usually set by the stability of the source. Too high a rate results in explosive boiling and splattered material. Heating the substrates before and during deposition helps remove contaminants from the surface and anneals the metal film for better etchability. A program for vacuum evaporation of chromium is outlined in Fig. 24. A coating thickness that gives about 1 percent light transmission is usually preferred. Thicker coatings take longer to etch; thinner coatings have inadequate contrast. Complete freedom from pinholes is needed, a condition that places stringent requirements on cleanliness every step of the process.

The optimum photoresist coating is the thinnest one that has adequate etch resistance. The thinner the coating, the better the resolution, the shorter the required exposure, but the greater the likelihood of pinholes and inadequate etch resistance. For chrome mask work photoresist coatings of 0.2 to 0.5 µm are usually used. (Note that this thickness is $\frac{1}{30}$ to $\frac{1}{12}$ of the thickness of high-resolution silver halide emulsions.) Application of the photoresist is usually done on a high-speed whirler with rapid acceleration. The properly diluted and filtered photoresist is applied to the chrome-coated plate on the whirler while it is stationary. The whirler is then rapidly accelerated to throw off most of the resist, leaving a coating whose thickness is determined by the centrifugal force, viscosity, and surface tension of the resist. There are probably many combinations of whirler speed, acceleration, and resist dilution that will give adequate results.[52] Since other factors such as the time lapse between depositing the resist and spinning, the temperature, humidity, and airflow affect the results, each operator must determine by trial his best conditions. Resist exposure is similarly subject to a large number of variables and must be determined by experiment. The manufacturer's instructions are satisfactory as a starting point. Resist speed increases during drying, and so adequate drying and prebake are essential. Postexposure baking increases etch resistance and the difficulty of eventual resist removal. Up to the temperature that results in deterioration of the resist more postbake produces more etch resistance. For overall operating ease and speed most operators use the minimum postbake that gives adequate etch resistance. When akali-resistant photoresists are used, a sodium hydroxide–ferricyanide chromium etchant is fast and gives good results.

Positive-working resists, based on azide photochemistry are not alkali-resistant, and acid-type etchants are required. Generally these are proprietary formulations available from their vendors as liquids. The azide resists are easier to remove than negative-working resists; acetone is generally completely effective. Many pro-

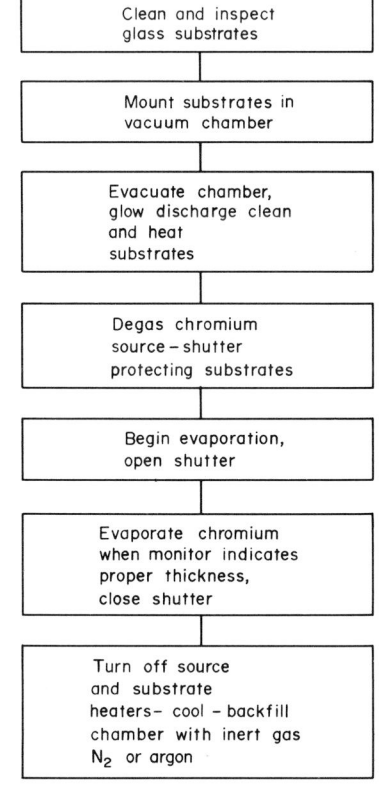

Fig. 24 Flow chart for vacuum evaporation of chromium on glass.

- Clean and inspect glass substrates
- Mount substrates in vacuum chamber
- Evacuate chamber, glow discharge clean and heat substrates
- Degas chromium source – shutter protecting substrates
- Begin evaporation, open shutter
- Evaporate chromium when monitor indicates proper thickness, close shutter
- Turn off source and substrate heaters – cool – backfill chamber with inert gas N_2 or argon

14-40   Photofabrication

prietary resist strippers are available, varying widely in their effectiveness, toxicity, flammability, stability, volatility, effect on substrate, and cost. They are usually mixtures of organic solvents, alcohols, organic acids, phenolic compounds, and detergents. With some substrates very powerful reagents such as chromic acid may be used. Glow discharge in oxygen is a very effective stripper but is usually a more costly process than solvent stripping.

Self-supporting metal   Masks consisting of metal foils containing apertures are employed sometimes in place of photomasks. Such masks are fabricated using a photoresist to define a pattern and etching the exposed metal. The processing of these masks is described on page 14-86 under Chemically Machined Parts.

Registration for Exposure

Where more than one photomask is used to produce a pattern on a substrate, registration of the several patterns is necessary. For double-sided printed circuits simple registry means may suffice, but in fabrication of large-scale integrated devices the best registration possible is barely good enough. Registration may be classified, as before, into macro- and microcategories. Macroregistration includes registry for which 25-μm (0.001 in.) error or greater is permissible. Microregistry includes the range from 25 μm downward to 0.2 μm (10^{-5} in.), which is about the limit for optional registry.

For all macroregistry, mechanical devices suffice. For some two-sided printed circuits the masks may be taped together in registry and the substrate slipped between for exposure in a vacuum frame which permits exposure from both sides. With care registry within 125 μm (0.005 in.) may be obtained with film masters and thin substrates. If the substrates are thick, more accurate registry is required; if registry must be obtained with previously determined features, more elaborate mechanical registry is needed. Registry holes punched in polyester-base film photomasks or drilled in glass photomasks may be aligned with holes in the substrate, using close-fitting pins in the holes. Registry errors can be kept as low as 25 μm (0.001 in.) or lower.

Mechanical means of microregistry can attain errors as low as 2.5 μm (0.0001 in.) but more accurate registry requires optical alignment of registration keys.

For accurate mask registration where mechanical holding fixtures are used, glass masks may be cemented in frames which are held against three pins by spring arrangements. If the pins are accurately set, perpendicular to the faceplate, and the springs are strong enough to overcome all frictional effects, registration errors can be as small as to 2 to 5 μm. The difficulty of aligning the two masks accurately on their frames and cementing them in place may be avoided by making only one of the masks, cementing it in place on one frame, and cementing an unexposed photographic plate to the mating frame. The plate is then exposed in contact with the first mask in the registry fixture and reversal-processed to produce the matching mask for the other side of the work. Of course, if the two masks are different, this means of production may not be applicable. Where it is applicable, it provides the best mechanical registry of photomasks. Figure 25 is a photograph of a registry fixture for 4- by 5-in. glass plate photomasks. If the patterns differ only in the dimensions of the pads or lines, sometimes over- or underexposure can be utilized to change some of the dimensions and contact printing can still be used to produce the mask.

For the optical registration of semiconductor patterns the first pattern put on the wafer may be located more or less arbitrarily. Before subsequent exposures registry keys or registry marks that were applied in the first pattern are used to align each mask before the photoresist is exposed. With sufficient magnification and near contact between the wafer and the mask which are being registered and where the geometry is favorable, registering as close as 0.1 μm (4 μin.) can be attained. Since the smallest line that can be produced optically is two or three times this dimension, this registry is adequate for any optical pattern. In alignment machines the mask and the wafer are both held firmly in a mechanism able to produce a very fine relative translation between the two. During this translation, the two are held out of contact to avoid damaging either the mask or the wafer. After at least approximate registry has been obtained, the mask and wafer are moved into contact for printing. For coarser geometries the printing may be done while they are slightly out of contact. These machines include the light sources and optics necessary to produce a light intensity that allows reasonably short exposure. In the semiconductor industry this type of equipment, although expensive, is economical to operate in production because accurate registering and contact printing can be completed rapidly and repeatably. The registration figures

Fig. 25 Mechanical mask registration fixture.

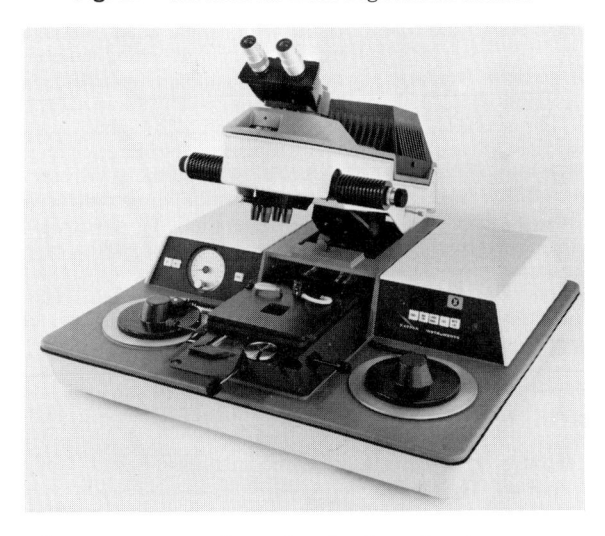

Fig. 26 Optical registration machine (mask aligner). (*Kasper Instruments Co.*)

quoted will not be achieved unless the optical viewing system of the machine has sufficient magnification and resolution, the motions are smooth and precise, and adequate contact is obtained during the printing. Figure 26 is a photograph of a mask-alignment machine for semiconductor processing.

PHOTOSENSITIVE MATERIALS

Photosensitive materials can be defined, within the context of this chapter, as having the ability to form a continuous film which is sensitive to light or other

Photofabrication

radiation so that the exposed (or unexposed) areas of the film can be processed without affecting the unexposed (or exposed) areas. The processing referred to above is generally treatment with a solvent to give selective dissolution. The net result, therefore, of applying a photoresist film, irradiating it through a suitable mask, and dissolving the unwanted areas is to form a pattern in the resist. The pattern can be described as two-dimensional since the thickness of the photoresist is generally small in comparison to the other dimensions of the pattern.

The essential feature of a photoresist is that the irradiation must produce a change in it that will enable later operations to discriminate between the exposed areas and the unexposed areas. For all common applications of photoresists, light-induced alteration in solubility is used.

Positive- and Negative-working Photoresists

The alteration in solubility might be one from a soluble to an insoluble species, or one from solubility in one class of solvents to solubility in a different class of solvents. The differentiation between the above alternatives may seem trivial at first glance; however, it provides the distinction between the two broad classes of photoresists, namely, negative and positive working. A negative-working photoresist is one that produces a resist pattern corresponding to the transparent areas of the mask through which it is exposed. In other words, the pattern is produced by removing the unexposed areas of photoresist, leaving material in areas corresponding to the negative of the mask pattern.

A positive-acting photoresist, on the other hand, gives a pattern that corresponds to the unexposed area of the photoresist. Since the exposed areas are removed during the development step, it is clear that the action of the light is to alter the solubility properties of the exposed areas rather than to insolubilize them. A positive-acting photoresist has the possibility of being made negative working by the proper choice of developing solvents. A negative-acting photoresist can never be anything but negative working, however.

The principles underlying the operation of a positive and negative photoresist are sufficiently different so that they will be described separately.

A negative-working photoresist is composed generally of a polymeric film-forming material, together with a light-sensitive agent in a suitable solvent. The polymer has the ability to react further with itself to form a three-dimensional crosslinked network incapable of being dissolved without degradation. The polymeric substance may or may not itself be light-sensitive. In any event, light-sensitive chemicals are added to increase the efficiency of the absorption of light energy and of the transfer of the energy to the reactive polymer, e.g., KPR.* The light-sensitive material in some cases decomposes to yield new reactive species which then interact with the polymer causing the crosslinking, e.g., in KTFR.* Another mode of operation of the sensitizing chemicals is the direct production of a crosslinking agent from the sensitizer itself, e.g., PVA-dichromate.

A positive-acting photoresist, on irradiation, suffers a change in chemical constitution, e.g., decomposition or isomerization, so that the final product is still soluble; however, its solubility characteristics are sufficiently different from those of the starting material that the proper choice of solvent will allow selective removal while leaving the unexposed areas untouched. Positive photoresists could be made negative by choosing for the development step a selective solvent which will dissolve the unexposed, rather than the exposed, portion.

Photoresists, as a prime requirement, must be light-sensitive. A good photoresist will be sufficiently sensitive so that it will be economically attractive to use. Another requirement of a good photoresist is that it produce a pinhole-free film at film thicknesses consistent with a short exposure time and the desired resolution. A rule of thumb suggested for negative photoresists is that the thickness of the photoresist layer should be no more than one-third the width of the finest geometry line to be resolved.[53] The limit of resolution routinely obtained in microelectronic

* Trademarks of Eastman Kodak Co., Rochester, N.Y.

applications is 0.1 mil (2.5 µm). To achieve this resolution the photoresist layer must theoretically be below approximately 8000 Å in thickness, and indeed microelectronic applications do use layers from 5000 to 10,000 Å thick. Resolution of 0.5 µm has been claimed for positive photoresists in resist coatings 19 µm thick.[54] The difference between positive and negative photoresists with regard to resolution versus photoresist thickness has been attributed to the absence in the former of swelling of the image during the developing step.

It is desirable from the standpoint of resolution to use photoresist films that

TABLE 18 Suppliers of Photoresist Materials

Supplier	Name	Remarks
Du Pont[56]	Riston	Supplied as a film; for plating and etching.
Dynachem[57]	DCR 3140	General-purpose resist.
	DCR 3154	Improved adhesion to aluminum and improved resistance to alkaline etches.
	DCR 3118, 3118H	Roller-coating formulation.
	DCR 3116	Provides heavy resist layer.
	DCR 3170	Microelectronic formulation.
G.A.F.[58]		Positive resist in field testing.
Philip A. Hunt[59]	Waycoat No. 10	General-purpose etching and plating resist.
	Waycoat No. 20	Especially useful as a plating resist.
Kodak[60]	KPR	Used on copper and copper-based alloys; used on clear and light-colored anodized aluminum.
	KPR2	Used on copper and copper-based alloys; electroplating resist.
	KPR3	Formulated for dip-coating systems.
	KPR4	Formulated for roller-coating systems; good for plated-through holes.
	KOR	Similar to KPR 2; possesses greater spectral sensitivity than other products.
	KMER	Used on all surfaces except copper and copper alloys and clear and light-colored anodized aluminum.
	KTFR	Microelectronic formulation.
	KPL	Used to increase viscosity of KPR; rarely used alone.
Norland[61]	Photoresist 30	One-part water-based resist for stainless regular steel, nickel, copper, brass.
	Photoresist 22	Two-part water-based system for Kovar-type metals.
Shipley[54] (positive photoresists)	AZ-111	General-purpose photoresist.
	AZ-119	As above but formulated for roller coating.
	AZ-340	Used for circuit boards and plated-through holes.
	AZ-345	Higher solids and viscosity than AZ-340 to provide lands around plated-through holes.
	AZ-1350	Used for microelectronic applications.
	AZ-1350H	As above, but formulated for roller coating.

are as thin as possible; however, as the films decrease in thickness, the number of voids or pinholes in the film increases.[55] After the substrate is processed, these pinholes will manifest themselves as small etch pits in areas that should have been completely protected by the resist. Each user of photoresists must choose a suitable thickness of photoresist that balances the needed resolution against the deleterious effect of pinholes.

Commercially available photoresists Listed in Table 18 are the suppliers of photosensitive materials applicable to the subject matter of this section. The photoresists are negative-working unless otherwise noted. Where a supplier has formula-

14-44 Photofabrication

tions for specific uses or methods of application, this has been noted. Materials that are being field-tested are so designated. It is not possible to list all the products that are on the market, and so the list is comprised of those materials that are most widely used, or those that are new but possess some interesting feature. It is apparent from the table that the major uses of photoresists are as etch and plating resists for the chemical milling of specialized metal parts, and as etch resists for microelectronic device and circuit fabrication.

Manipulation of photoresists The sequence of steps involved in the use of photoresists is presented in Fig. 27. Each step will be described separately. The preexposure bake and the postbake are described in conjunction with the application step and development step, respectively, because a particular photoresist may or may not require either or both of these baking operations. The need for the preexposure bake is dependent on the volatility of the solvent for the photoresist; the need for the postbake is dependent on the type of photoresist being used and also on the particular substrate processing involved.

Application In photoresist processing the application step is critical since it will influence the final performance of the resist. Cleanliness is important, both of the substrate surface and of the facility in which the application is performed. Clean room equipment is available from a number of suppliers.[62] Items such as laminar-flow hoods incorporating filters capable of removing airborne particles down to 0.3 μm and static-free gowns and caps are widely used in microelectronics processing.

Metallic substances should be scrub-cleaned to remove all traces of grease and contaminants, well rinsed, and thoroughly dried. Scrub cleansers as well as chemical cleaning solutions are available from some suppliers of photoresists.[54,61]

Semiconductor device processing requires exceptionally careful attention to the surface cleanliness, and one supplier of photoresist has recommended the procedure given in Table 19 for cleaning oxidized silicon wafers.[57] The cleaning procedure ensures the optimum contact and adhesion of the photoresist layer to the substrate. Even this, however, is insufficient in many microelectronic applications. For this reason, efforts have been directed at finding a surface treatment that will promote the adhesion of the photoresist to the oxide substrate. Reactive alkoxy silanes have been used with varying degrees of success.[63]

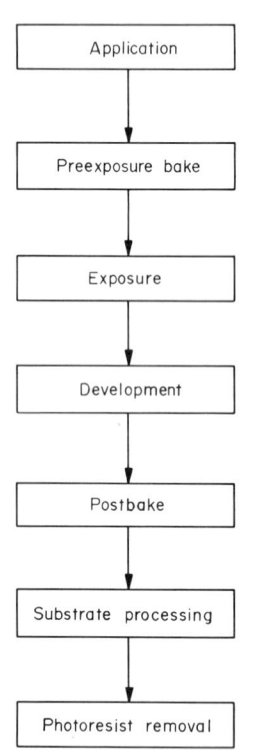

Fig. 27 Flow chart for photoresist processing.

Photoresists are generally sold prefiltered; however, a final filtration at the point of use is advisable. Many of the impurities present in photoresists are deformable. That is, pressure filtration can force particles having equilibrium sizes larger than those of the filter pores through the filter. For this reason, filtration should be conducted under the lowest practical pressure head. Equipment is available for the removal of submicrometer particles from solution on both a laboratory and a production scale.[64]

Photoresist can be applied by a number of methods, each of which has its own advantages. The method to be used will be determined by the thickness and the uniformity desired, by the geometry of the piece to be coated, and by the number of pieces to be coated simultaneously. Suppliers of photoresist application equipment can be found in the trade literature.[65]

Flow Coating. The easiest method of coating is flow coating. In this method the photoresist is poured onto the piece, and the piece is tilted and rotated so that the photoresist solution eventually covers the entire surface. The surface tension of the photoresist is generally high enough to prevent the photoresist from spilling over the edge of the piece. Alternatively, a glass rod can be placed parallel to the substrate surface and touching the photoresist surface. Movement of the rod over and parallel to the substrate will give rapid covering of the substrate. Once the substrate is covered, it can be air-dried on a flat surface, in which case the coating will be thickest at the center and will thin out toward the edge. Alternatively, the substrate can be tilted or suspended so that the excess runs off from a corner. Either of these methods gives a nonuniform coating thickness. With the latter method, repeating the process and allowing the drainage to occur at the opposite corner tends to even out the thickness across the surface. This method is useful for small parts in noncritical and short-run applications.

Roller Coating. An obvious extension of the above method is to doctor-blade a photoresist onto the substrate. This can be done for noncritical work on a short-run basis; however, for larger runs and better control, roller coating is advisable. A more uniform coating is achieved in roller coating over simple doctor blading owing to conformance of the transfer roller to variations of stock thickness.

TABLE 19 Procedure for Cleaning Oxidized Silicon Wafers Prior to Photoresist Application[57]

The following precoating cleaning procedure assumes that the oxidized wafer to be coated is visibly contaminated. Wafers that are not visibly contaminated do not require steps 1 through 4.
1. Using a cotton swab, scrub the wafer with a solution of Isonox or other comparable cleaning compound. (*Note:* Isonox is a surface cleaning compound manufactured by Dynachem Corporation.)
2. Rinse the wafer with D. I. (deionized) water, and place in a 50-ml beaker which has also been scrubbed with the cleaning compound and rinsed.
3. Flush the beaker with D. I. water.
4. After pouring off the last water, cover the wafer with methyl alcohol, and ultrasonically agitate for 30 sec.
5. Solvent-degrease by heating gently in trichlorethylene until boiling has been evident for 30 sec.
6. Wash with methyl alcohol.
7. Flush with D. I. water.
8. Boil wafer in c.p. grade nitric acid at 80°C for 20 min.
9. Flush with D. I. water.
10. Rinse wafer with methyl alcohol.
11. Store in trichlorethylene.

In roller coating, the photoresist is applied to the stock in a printing type of operation from a rubber transfer roller. A controlled quantity of photoresist is continually supplied to the transfer roller by the wringing action of a second doctor roller. Careful control of the roller settings, transfer pressure, conveyer speed, and the photoresist solutions must be maintained if uniformity and reproducibility are needed. The thickness of the applied coating is proportional to the photoresist viscosity, surface tension, and roll speed, and inversely proportional to the doctor-roll contact pressure. Excess roller pressure can result in skips or beads in the coating and a thin photoresist layer. Low pressures may produce too thick a coating and plugged holes in open-hole stock.

Photoresist viscosity is an important parameter in determining the transfer of photoresist to the substrate so that a uniform even coat is obtained and in preventing the filling of open holes, for example, in through-hole printed-circuit boards. If the viscosity is too low, inadequate chemical masking and filled holes will result. Too high a viscosity can result in excessive photoresist thickness and bubbles in the dried film. The use of double wet or dry coating produces greater uniformity and is preferred for critical applications.

In operation, the stock to be coated is passed through the coater at a conveyer speed of 4 to 10 fpm. For the double-coating technique, the stock is rotated

90° after the first pass and passed through a second time. The coated sheets are then air-dried 10 to 15 min and prebaked according to the recommended schedule for the particular photoresist. Drying in between the first and second coat will increase the thickness approximately 40 percent over a double wet pass.

The roller-coating method of application is suitable for large-scale production of parts of small or large areas and containing open holes. It can provide a uniform coating of ±5 percent thickness variation and very efficient utilization of photoresist as compared to other methods. Since a resilient rubber roller is employed under pressure to transfer the photoresist to the stock, compensation is made for thickness variations, and small voids in the surface tend to be force-filled. This technique is particularly advantageous in printed circuitry where thickness variations of ±0.005 in. are not uncommon for conventional board laminates. It also prevents photoresist from coating the inside hole walls in through-hole circuit boards. Typical roller-coating application data for various resists are presented in Table 20.

Dip Coating. Dip coating is an extremely simple concept that may be complicated in practice, depending on the restrictions that are imposed upon it. It is the easiest and quickest way of applying photoresist coatings to both sides of a board simultaneously and is suitable for many applications. In its simplest form, the stock is immersed in the photoresist solution, withdrawn, and held in a vertical position. Excess photoresist is allowed to drain, and the film is then dried. Thin,

TABLE 20 Typical Roller-coating Application Data for Various Resists*

Specification	DCR 3118	AZ-119	KPR4
Resist viscosity at 77°F...............	No. 4 Zahn cup, 25–35 sec	No. 2 Zahn cup, 55–75 sec	360–460 cp
Conveyor speed, fpm.................	4	4	7
Approximate film thickness (2 wet passes), in........................	0.00025	0.00015	0.00010
Resist temperature..................	75 ± 5°F		
Panel adjustment dial setting..........	0.020–0.030 in. less than stock thickness		
Roller dial..........................	8–12 psi		

* Data for roller coater of Gyrex types 630 and 730.

uniform films of photoresist are obtainable when withdrawal speed is regulated. This can be understood from the following considerations. The amount of photoresist remaining on the substrate as it is withdrawn from the reservoir is a function of the surface tension and viscosity of the photoresist. As the substrate is advanced out of the photoresist reservoir, photoresist adheres. As the level of the adhering liquid photoresist film rises above the level of the reservoir, gravity will tend to make it flow back into the reservoir. The viscosity of the photoresist impedes this motion. With time, the solvent evaporates, and eventually the solvent loss is such that the photoresist ceases to flow. A slow withdrawal speed will give a thinner coating than a fast withdrawal speed, because solvent evaporation takes place more slowly. More viscous photoresist solutions will give a thicker coat than less viscous solutions because flow is slower with the more viscous material. Temperature and rate of air movement affect the deposition of the photoresist layer. Temperature affects the viscosity and the rate of evaporation of the solvent from the photoresist film. A rise in the temperature would tend to reduce thickness by reducing the photoresist viscosity; however, it would have the opposite effect on film thickness from the standpoint of evaporation. Controlled withdrawal is capable of giving a uniform, thin coating provided attention is paid to the above variables.

The main disadvantage of this technique is that the coating will tend to be wedge-shaped or thicker at the bottom. Also, it is not applicable to perforated stock, as a nonuniform thickness is obtained owing to runout from the holes. The thickness of the photoresist layer obtained is typically in the order of 0.0001 to 0.00015 in. and may not be suitable for applications requiring deep etching or long electroplating cycles. The use of a double dip does not significantly increase the thickness owing to partial solution of the first coat and may lead to nonuniform film. Representative dip-coating data for various resists are included in Tables 21 and 22.

Spray Coating. Photoresists can be applied by spraying, and this process is applicable to general all-around work for either small- or large-scale production. The simplest example of a spray process is the application of photoresist from a commercially available aerosol dispenser.[66] This technique is solely manual in operation, and its success is dependent on the dexterity of the operator. It is convenient for coating of small pieces on a sample or short-run basis. However, the evenness of

TABLE 21 Typical Dip-coat Film Thickness at a Withdrawal Rate of 4 in./min

Photoresist material	Film thickness measured		
	Top*	Middle*	Bottom*
KPR	0.07	0.10	0.10
KPR2	0.12	0.13	0.15
DCR 3140	0.13	0.15	0.16

* Average of three readings, in mils.

TABLE 22 Effect of Withdrawal Rate on Dip-coated Photoresist Thickness

Withdraw rate, in./min	KPR2 film thickness measured		
	Top*	Middle*	Bottom*
2	0.12	0.13	0.13
4	0.12	0.13	0.15
10	0.12	0.15	0.17

* Average of three readings, in mils.

the coating and the reproducibility are generally poor. Photoresist can be applied also with compressed air as the propellant and conventional manual spray equipment. Greater control of the spray can be achieved with this type of equipment, but it still suffers from the problems noted above. Semiautomatic and automatic spray equipment is available which can provide maximum efficiency and uniformity. Photoresist is fed to the airstream by either a gravity feed or a pressure pot. Airless or vapor-spray equipment is also available.[67] This method makes use of an inert organic vapor as the propellant. The advantages claimed for such a system are cleanliness and a finer, more easily controlled spray at the low pressures provided by the higher-molecular-weight propellants. Lower-pressure operation reduces overspray and minimizes air entrapment.

Spray coating has the advantage over other methods in being able to deposit a wide range of film thicknesses by adjusting the thinner that is used and the number of spray passes. The thickness of the applied photoresist layer is dependent upon the mechanical setup of the equipment, namely: the size of the nozzle opening, the distance of the gun from the work, the speed with which the gun travels across the work, and the degree of overlap of the sprayed pattern. These factors are controllable with mechanized spray equipment. In such equipment, an automatic spray gun is held at a fixed distance from the workpiece and traversed back and forth by a reciprocating mechanism at a constant speed during the spray cycle. A work-holding fixture is also employed to advance the work the required distance. Equipment of this type is essential for production runs to obtain good reproducibility. The above variables must be held constant, and rigid cleaning procedures for the spray gun must be adhered to in order to prevent clogging of the spray nozzles. The operation must also be carried out in a spray booth with sufficient exhaust to carry away solvent vapors. Suitable air filtration and control of temperature and humidity are also required, particularly for work of high resolution. The method is useful for production runs of chemically machined parts and printed circuits with and without through holes. Representative data for various resists are shown in Table 23.

TABLE 23 Typical Mechanized Spray-coating Application Data for Various Photoresists

Photoresist material	Film thickness, mils			Spray data, in.		Thinner and reduction used
	One coat	Two coats	Three coats	Orifice	Overlap	
KPR2	0.13–0.18	....	....	0.052	4	Blend 33% KPR thinner and 6% KOR thinner, 2:1.
	0.05–0.07	0.10–0.15	....	0.031	4	
	0.10–0.12	0.19–0.22	0.27–0.32	0.031	2	KOR thinner, 5:1.
DCR 3140	0.10–0.15	0.20–0.28	....	0.031	4	None
	0.18–0.21	....	....	0.031	2	None
	0.08–0.10	0.18–0.21	....	0.031	4	Xylene, 5:1.
	0.07–0.10	0.15–0.18	....	0.031	4	Xylene, 4:1.
	0.08–0.10	0.16–0.21	....	0.031	4	Toluene, 4:1.
AZ-340	0.16–0.22	0.30–0.40	....	0.031	4	None

Whirl Coating. The application of photosensitive emulsions by slow-speed whirling in the range of 75 to 100 rpm is in common use in the graphic arts industry for the coating of plates. This method has only limited use today in the photofabrication of electronic parts. This principle, however, has been extended to high-speed whirling equipment and represents the most widely used method in the microelectronics industry for photoresist-coating the small-sized silicon wafers normally encountered. Larger pieces such as found in thin-film applications have also been successfully coated in this manner. The piece to be coated is mounted on a spinner shaft; normally smaller pieces are held by a vacuum chuck. The surface of the piece is flooded with photoresist, and the piece is then spun at a constant rpm, using a high initial acceleration rate. Centrifugal force causes the coating to move out to the edges and produces a thin coat on the surface with good uniformity. It is reproducible if careful control of resist viscosity, solids content, whirler acceleration, and whirler speed is exercised.[52] Because the linear velocity and acceleration will differ along a radius as one moves from the center to the periphery, the resist will normally be thicker at the center than near the edge. A thick rim of resist may be present owing to surface-tension effects at the periphery of the wafer.

The problems associated with this method are no different from those associated with other methods. Unless proper control of the above parameters is exercised wafer coatings will be nonuniform. Improper adjustment of the solid content and solvent composition will cause "stringers" at the edge of the wafer. Dust or surface imperfections will cause depressions in and rippling of the resist as it flows around the obstacle. Adequate exhaust design must be maintained to prevent spun-off resist particles from redepositing on the wafer as contamination. The thickness of the photoresist layer is inversely proportional to the whirler speed. As the spin speed and acceleration are decreased, flow defects will be minimized; however, the rimming effect will be more pronounced. Figure 28 shows the effect of whirler speed on the thickness obtained for a typical thin-film photoresist at various dilutions. In practice, a whirler speed in the range of 2,000 to 8,000 rpm and a total spin time of 20 to 30 sec are used for the photoresists commonly employed in semiconductor work. A typical procedure includes mounting the silicon wafer on the vacuum chuck, covering it with a layer of photoresist from a filtered syringe, and then initiating the cycle. The cycle may be repeated a second time to reduce possible defects in the coating. The wafer is then removed, inspected, and baked before further processing.

Special Methods. There are some photoresists that require special application methods. Riston,* a photopolymer resist film made by Du Pont, is supplied as a solid photoresist film sandwiched in between sheets of polyester film and polyethylene film. This photoresist film is applied to a substrate by a lamination technique on equipment available from the photoresist manufacturer. During application, the polyethylene film is automatically removed immediately before lamination of the photoresist film to the substrate by a heated roller (110 to 120°C). Since the photoresist is applied dry, baking prior to exposure is not required. A 30-min "holding delay" period is used before exposure, however, to allow stress relieving of the photoresist film on the substrate. The polyester cover sheet remains in place during pattern exposure and acts as a protective mechanical cover-up to the development step. This type of resist is well suited to large-area printed-circuit-board manufacturing since the large areas are difficult to cover uniformly by liquid application techniques. Control by the user of many of the variables affecting the uniformity and pinhole density characteristics of the photoresist layer are obviated since these properties are controlled by the photoresist manufacturer.

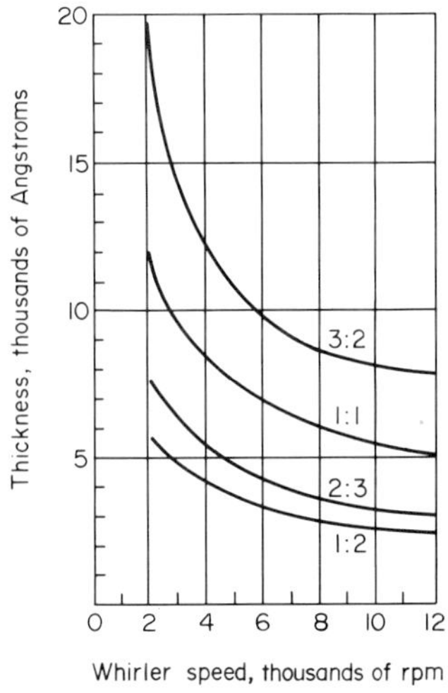

Fig. 28 Effect of whirler speed on photoresist thickness.[52]

There is only one material available that can be applied from the vapor state.[68] This has not found widespread use since other features of the material make it difficult to work with. The concept of a vapor-deposited resist film has merits from the standpoint of ease of mass-coating substrates with thin, uniform pinhole-free layers.

Exposure The exposure of a photoresist can be performed by the means which will supply the necessary energy in a reasonable period of time and which has the ability, inherent or otherwise, to deliver this energy to selected areas of the substrate. Systems that satisfactorily meet these requirements are ultraviolet and visible light sources used in conjunction with suitable transparency masks, and scanning electron beams and modulated or driven lasers. The last of these systems has received little attention in the published literature, although it is an area that will probably attract increasing attention in the future. The electron-beam exposure of photoresists is advanced to the stage where equipment for this purpose is commercially available, and end products are being made which use electron-beam expo-

---

* Trademark of E. I. du Pont de Nemours & Co., Wilmington, Del.

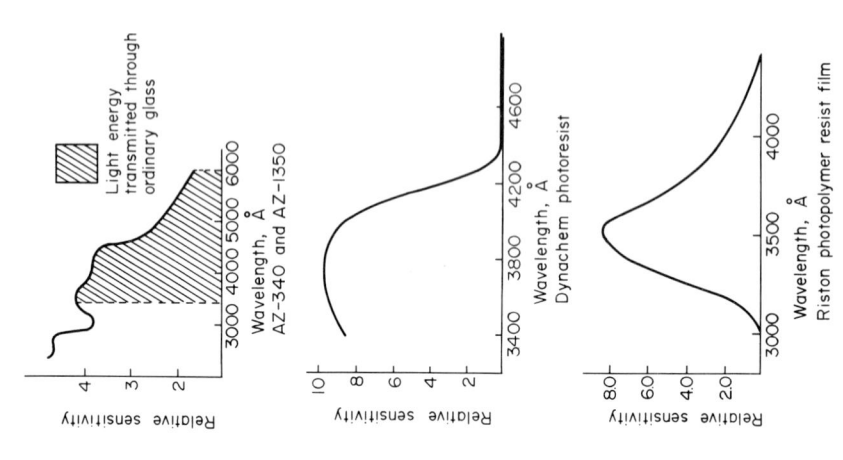

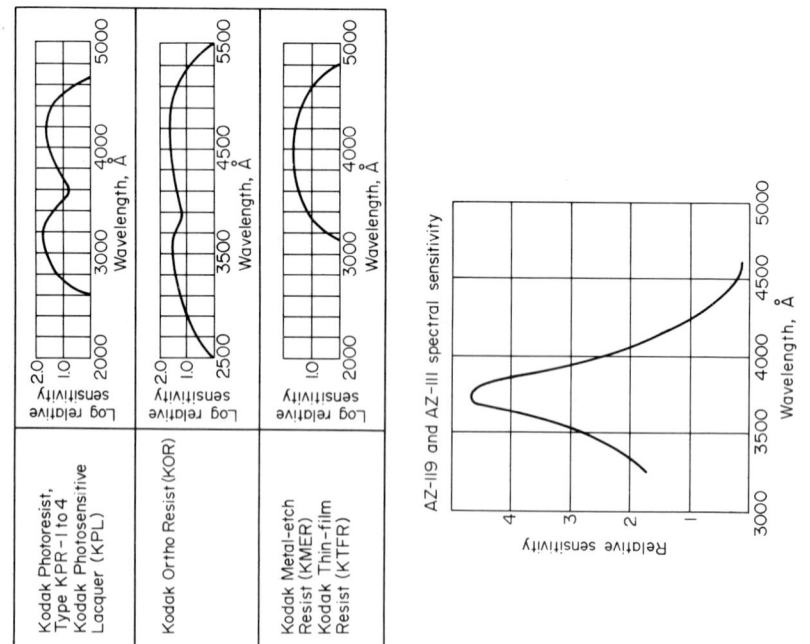

Fig. 29 Spectral response curves for various photoresists.[56-60]

sure for part of the processing.[69] The most important exposure means in terms of usage is optical exposure through masks. This method will be discussed in detail.

It is a fundamental law of photochemistry that only light energy which is absorbed by a system can be effective in causing a chemical reaction. Therefore, in evaluating a light source, the important feature to consider is the available energy in the spectral region absorbed by the photoresist, and not the integrated output of the source. For this reason a 200-watt mercury lamp is often a better exposure source than a 650-watt tungsten lamp. The spectral response curves for some commercial photoresists are given in Fig. 29. Inspection of these curves reveals that most photoresists are sensitive to the highest-energy radiation to which glass is transparent. The reasons for this are twofold. First, sensitivities in this region allow Pyrex* or soft glass flats to be used as mask substrates, avoiding the higher cost of quartz or silica substrates. Second, since the photochemistry of photoresists involves the breaking of chemical bonds, the highest-energy photons are most effective.

As discussed earlier, the photochemical reactions which the various photoresists undergo are not the same. The reaction of photoresists that crosslink by a free-radical process is inhibited by oxygen (air). This will result in a thinner film after exposure and development than was applied to the wafer. The optimum exposure time may also be increased in an oxygen environment because of the inhibition. Oxygen inhibition can be overcome by conducting the exposure in a vacuum chuck so that the concentration of oxygen at the surface of the substrate is low, or else by flooding the surface of the substrate with an inert gas, such as nitrogen or carbon dioxide, before the substrate and the mask are brought into contact.

Intimate contact between mask and substrate can be obtained by forcing them together in a vacuum frame, by forcing one against the other by gas pressure, or by the use of a compression spring. The size of the substrate will determine to a large extent the means required to obtain good contact. The necessity for good contact is apparent when one considers that any abrupt change in the optical density in the artwork, such as obtained by moving across any boundary, will act like a scattering center. The closer the mask is to the surface of the resist, the shorter the path length of the scattered light, and hence the less the degradation of the image in the reproduction. The scattering of light is much more noticeable when a collimated light source is used. The resolution capability of a photoresist is affected by the light-scattering process during exposure. This produces a partial polymerization at the bottom corners of a negative-working photoresist due to diffusion of light under the film image from side-angle light. With positive photoresists, the effect is noted by rounding of the top corners and sharper definition at the bottom. Reflections off the bottom of the substrate also contribute to the effect. Thin photoresist films minimize these effects. This phenomenon is generally referred to as *coving* and is illustrated in Fig. 30.

The correct exposure time for a particular photoresist or a given substrate is something that must be experimentally determined[70] and periodically reverified. Parameters to be considered are: (1) the distance of the exposure source from the substrate (more important for uncollimated sources than for collimated ones), (2) the intensity of the source (this will, in general, change slowly with time), (3) the spectral distribution of the source, (4) the thickness of the photoresist layer, (5) the pattern features of the mask (small features require longer exposures than larger ones), (6) the transmission of supporting members (glass or plastic films), and (7) the optical density of the opaque areas (to avoid "burning" through these areas).

Development After exposure, the photoresist layer is developed in a suitable solvent to remove the soluble portion. The procedure and the chemicals for developing the exposed photoresist image are generally available from the photoresist manufacturer. The manufacturer, however, cannot offer the ideal combination of

---

* Trademark of Corning Glass Works, Corning, N.Y.

14-52 Photofabrication

materials and processing for all possible substrates, photoresist thicknesses, pattern sizes, etc., that a given user may require. Therefore, if the recommended procedures do not give satisfactory results, the recommended development procedure must be altered somewhat until it suits the particular application. A number of pertinent remarks on methods of development and the importance of developing properly follows.

Negative-acting photoresists are crosslinked in those areas that are exposed to light. These photoresists undergo a change which is more physical than chemical. Chemically, the solubility for solvents that dissolve in the resist is unaltered in the unexposed areas; what keeps the exposed areas from dissolving is that the exposure step has caused individual polymer chains to be crosslinked or joined into a three-dimensional network somewhat analogous to a set of monkey bars. No solvent will dissolve this network since the various segments are constrained to each other. This is an example of physical rather than chemical insolubility. The analogy is not entirely a good one since the polymer chains are coiled and flexible, not

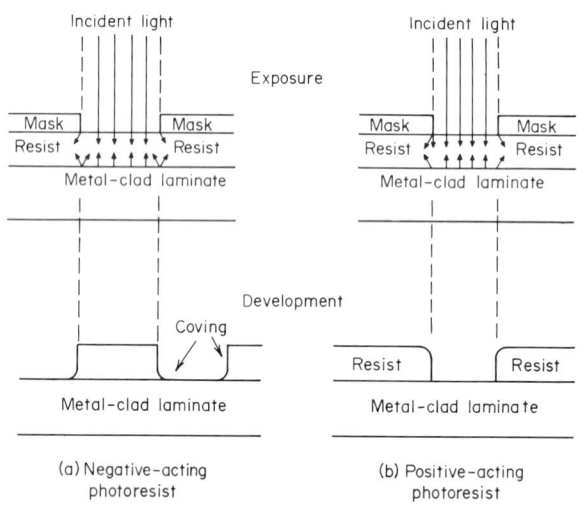

Fig. 30 Coving of negative- and positive-acting photoresists.[72]

rigid and fully extended as are the rungs on a monkey bar. As a result, when a crosslinked polymer is exposed to a good solvent, i.e., one that would dissolve it rapidly if it were not crosslinked, it will swell until the chains are extended as far as the crosslinks will allow. There is a considerable driving force or pressure associated with this process. This pressure will tend to break the bonds that hold the photoresist to the substrate surface since these bonds are much weaker than the chemical bonds holding the photoresist together. As a result of the swelling pressure, therefore, the photoresist-substrate interface is weakened, and the weakness may manifest itself during subsequent processing steps. The greater the number of photochemically induced crosslinks present at the developing step, the less the tendency for swelling. This is why the exposure step must be optimized to provide the largest number of crosslinkages in the exposed regions without making the unexposed regions insoluble. Several techniques can be used for the development step, namely: simple immersion in the solvent solution, immersion with mechanical or ultrasonic agitation, spraying, or a combination of these. Development time is decreased and resolution generally improved with increase in the degree of agitation. Spray systems have the advantage of producing mechanical force and solvent flow on the surface to be developed to dislodge photoresist particles. Systems comprising a combination of immersion and spray rinsing have proved satisfactory for general use. The optimum developing time is dependent upon the particular resist and thickness, pattern resolution, developing equipment, and environment, and must be determined experimentally.

After the use of the developing solution, a fixing solution consisting of a second solvent or solvent mixture is employed to rinse the coated surface and restore the photoresist pattern to the proper size. This solution must be a sufficiently good solvent so that it will not precipitate the dissolved polymer still on the surface of the substrate, thereby redepositing polymer in "opened" areas. On the other hand, it must not be good enough to continue the developing process to any considerable extent. Since this solution has less of an affinity for the photoresist than the developing solution, it will shrink the photoresist pattern and "squeeze out" the developing solution that is imbibed in it.

The final step in the process is to bake the photoresist pattern at an elevated temperature. This process removes the last traces of imbibed solvent, restores the photoresist pattern to its equilibrium size, further crosslinks and strengthens the photoresist, and reestablishes bonds at the photoresist-substrate interface. The photoresist pattern is now ready for processing.

Removal Once the pattern has been transferred to the underlying substrate, e.g., by etching or plating, the photoresist has served its purpose and is generally removed. There are three methods that can be used to effect removal: one is to use commercial "strippers" which swell the photoresist and render it easily removed by mechanical means such as swabbing or brushing; the second is to use liquid oxidizing reagents which decompose the photoresist; the third is to use reactive gaseous species to decompose and volatilize it.

The first of the above methods needs little clarification. Many strippers are available: suppliers can be found in trade journals.[65] They all work, some better than others; however, they do differ in their purity and the extent to which they affect the electrical properties of the devices.

The second method employs known oxidizing solutions such as sulfuric-dichromate, ammoniacal hydrogen peroxide, metachloroperbenzoic acid, etc. Such agents may be effective where the photoresists are resistant to or incompletely removed by the strippers mentioned above.

The third method utilizes oxygen which has been made reactive by passing it through an rf field. The reactive oxygen oxidizes the substrate, giving volatile oxides of carbon, hydrogen, and nitrogen. There is evidence that the process is detrimental to certain devices.[71]

Troubleshooting From the point of view of the user, the successful use of photoresists is one based on art rather than science. Kodak has published a brief compilation of commonly occurring problems along with their probable causes and their cures. This serves as an invaluable aid to the novice in the photoresist area; it is reproduced here in Table 24.

Dichromated Resists

Dichromated resists are negative-working, water-soluble systems which for many years were based on naturally occurring materials such as gelatin, fish glue, shellac, and starch. The polymeric base must have easily oxidized groups in it since the photoreaction of the dichromate yields a reduced form of the chromate ion while the polymeric component is simultaneously oxidized. The reduced form of the dichromate then binds the polymer, and crosslinking results. At present, there are now a number of synthetic water-soluble polymers capable of taking part in the photoreaction with dichromate, notably polyvinyl alcohol.

These systems often suffer from limited shelf life owing to the possible thermal reaction between the dichromate and the polymer. For this reason, the resists are generally sold as a two-part system. The speed of a dichromated resist depends on the polymer component, the particular dichromate salt being used, the pH of the dried film, and the moisture content of the film.[73]

These types of resists are not generally used in the microelectronics industry since the very fine detail required in microelectronic applications cannot be main-

14-54 Photofabrication

tained by the dichromated-type resists in the aqueous etchants employed. However, the resists have found wide application in the color television industry. In the production of shadow masks, a dichromated fish glue is employed as the metal etch resist. Dichromated polyvinyl alcohols are also used in the defining of color

TABLE 24   Kodak Photosensitive Resist Troubleshooting Chart

Problem	Cause	Correction
1. Insufficient resist coverage	Insufficient cleaning Resist diluted too much ⎫ (spraying) Resist applied too fast ⎭	Improve cleaning methods. Check for surface recontamination. Decrease resist dilution. Decrease rate of travel of spray gun. Adjust spray gun to obtain even coverage.
2. Coating has "orange peel."	Too-rapid drying Poor leveling of the coating Too much resist applied to surface (spraying)	Decrease drying temperatures (if used). Dilute resist with recommended thinner. Decrease amount of resist passing through spray nozzle.
3. Image wash-off during development.	Insufficient cleaning Insufficient exposure Resist too thick Resist not thoroughly dry before exposure	Improve cleaning methods. Increase exposure. (Make step-tablet exposure determination.) Decrease coating thickness. Increase prebaking times.
4. Developed image has resist in nonimage areas. (Resist will not develop even after prolonged soaking.)	Resist is fogged (heat). Resist is fogged (light). Transparency has low density.	Check preexposure drying temperatures. Do not exceed 250°F (120°C). Check safelights for UV leakage and level of illumination. Check density of black areas on transparency (should be at least 1.0).
5. Developed image is "streaky."	Runback of unexposed resist from edges of material Clamp used to hold work unsuitable	Extend development time. Clamps should be capable of rapid drain-off of developer. Clamps should be light enough so that very little heat transfer takes place (vapor degreasers only).
6. Developed image "puckers."	Resist coating too thick Development time too long Insufficient exposure Resist not thoroughly dry before exposure	Decrease resist coating thickness. Decrease development time (vapor degreaser only). Increase exposure. (Make step-tablet exposure determination.) Increase prebaking times.
7. Developed image has pinholes.	Dust and dirt coating Dust and dirt on transparency Dust and dirt on vacuum frame Poor housekeeping	Be sure surface is free from dust before coating with resist. Clean transparency before using. Clean glass and plastic (if any) on vacuum frame at least once a day. Coating and exposing areas must be clean. Etching and other areas should be clean.
8. Developed image has "halos" around holes (through-hole printed-circuit boards).	Improper thinner used with resist Wrong resist	Use only recommended thinners (see product data sheets). KPR2 or KPR3 (both diluted with KOR thinner) will minimize this effect.
9. Developed image has a scum of dye after water wash-off.	Insufficient rinsing Resist fogged Insufficient development	Aerated sprays of water should be used after dye step. (A 10-sec rinse with isopropyl alcohol can also be used.) See problem 4. Increase development time.
10. Image breaks down during etching.	Insufficient coating thickness Poor adhesion High concentration of mineral acids in etchant	Increase coating thickness. Improve surface cleaning before coating. Decrease amounts of acid in etchant.

TABLE 24 Kodak Photosensitive Resist Troubleshooting Chart (Continued)

Problem	Cause	Correction
11. Severe undercutting of image when etched	Lack of or improper conversion coating (if needed) Conversion coating too thick Etchant exhausted Wrong etchant for material in process No etchant agitation (noncopper metals)	Apply conversion coating as per recommendations. Decrease treatment time. Regenerate or replace etchant. Use recommended etchant for material in process. Use spray, splash, or bubble agitation.
12. Fine threads of metal in etched portion of image	Resist fogged Insufficient development	See problem 4.
13. Etched portions of metal are rough or severely pitted (rough etch or "cratering").	Copper ion concentration too high in inhibited ferric chloride (noncopper metals only)	See problems 4 and 9. In inhibited ferric chloride etch systems, noncopper metals should be etched first. Excessive copper ion concentrations will cause "cratering" of noncopper metals if this is not done. If etchant has this contamination, discard it.
14. Deep channels cut into non-etched portions of metal	Deep scratches on surface Surface finish too rough	Use different metal (if possible). Surface finish should be about 15 μin. Increase coating thickness (emergency measure only).
15. Heavy haze of nodules on plated work	Pits or peaks on metal surface	Dry buffing or abrasive slurry buffing will relieve this condition.
16. Plating confined by resist image is burned, cracked, or peeling.	Insufficient development Plating current density too high Plating bath contaminated	See problems 4 and 9. Calculate current density on the basis of exposed metal. Analyze plating bath.

SOURCE: Eastman Kodak Company.

phosphor dots.[74] Recently, Bell Labs has announced the use of dichromated gelatin as a hologram recording medium.[75]

Three-dimensional Photosensitive Materials

Photoresist materials and dichromated emulsions are generally used in thin coatings of 0.001 in. or less and can be considered as producing "two-dimensional images" when exposed and developed. Photosensitive materials are available in rigid layers, however, which can be utilized in the forming of three-dimensional structures by photofabrication. These materials include photosensitive glasses and plastics.

Photosensitive glass Photosensitive glass was developed by Corning Glass Works in 1947.[76] In 1953, Corning announced the commercial availability of this material for precision chemical machining and its processing details.[77] Photosensitive glasses are similar to conventional glasses in composition except for the addition of various photosensitive metals and sensitizers. For chemical machining, a lithium silicate glass modified by potassium oxide and aluminum oxide has been found suitable.[77] Traces of cerium and silver compounds are added as the photosensitive materials. The glass is exposed to ultraviolet light in a similar manner to other photosensitive materials and developed by heat treatment at 550 to 600°C. The essential reactions occurring within the exposed glass during exposure and development consist of the formation of silver crystals followed by nucleation and growth of lithium metasilicate crystals. Cerium acts as the photosensitizer during the exposure. The spectral sensitivity of a cerium-sensitizer glass is shown in Fig. 31. After heat treatment the crystalized portion exhibits a solubility in dilute hydrofluoric acid solutions of 15 times that of the unexposed glass.

The properties of Fotoform glass and Fotoform glass ceramics are included in Table 25. Processing and application data are presented later in the chapter.

14-56   **Photofabrication**

Photosensitive plastics Photosensitive polymer materials were developed for the fabrication of "relief" printing plates in the printing industry. These consist of Dycril* photopolymer printing plates[78] developed by Du Pont and a photosensitive polyamide plate developed by Time-Life, Inc.[79] Du Pont Dycril plates were first made available in 1960. Du Pont later marketed a series of photopolymer sheet materials called Templex† for the fabrication of various jigs and fixtures and other components.[80] The later material has been investigated for possible use in the electronics industry and will be considered further.

Templex photopolymer material can be "formed" by photofabrication into intricate and precise three-dimensional configurations. Exposure to light between 310 and 420 mμ results in a polymerization of the underlying photosensitive material. Maximum sensitivity is exhibited at a wavelength of 330 mμ, and polyester film photomasks can be used for exposure. The exposed pattern is developed to form the three-dimensional structure by washing away the unexposed portions in a dilute caustic solution.

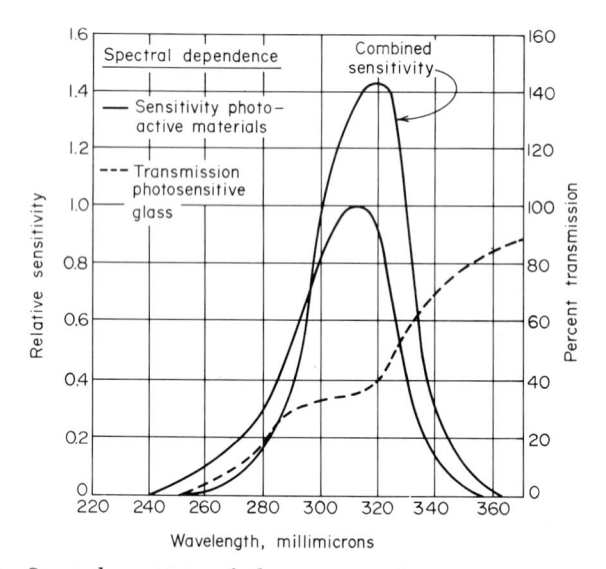

Fig. 31   Spectral sensitivity of photosensitive glass with cerium sensitizer.[77]

The physical and electrical properties of processed Templex photopolymer are shown in Table 26. Chemically, Templex is affected by lower-molecular-weight alcohols and ketones. Aromatic and saturated hydrocarbons and chlorinated solvents show little effect. The compatibility of specific chemicals must be determined under service conditions for a particular application. The processing characteristics are discussed under the application section later in the chapter, and typical examples are outlined.

PROCESSES USED AND THEIR RELATION TO PRODUCT DESIGN

Several processes must be specified for materials that are to undergo photofabrication. These can be divided into the pretreatment processes used before application

---

* Trademark for photopolymer printing plates, E. I. du Pont de Nemours & Co., Wilmington, Del.

† Trademark for photopolymer material, E. I. du Pont de Nemours & Co., Wilmington, Del.

of the photosensitive materials and the various fabrication processes such as etching or plating used on the material once a resist pattern has been formed. The type of processing utilized and the materials involved will determine the characteristics of the photosensitive resist and overall product conformance to the design specification. Also, dimensional changes may occur in parts during processing, depending on the process used. A knowledge of these factors is needed before part and process design in order that suitable compensation can be made. A general treatment of the various processes is presented in the following sections, along with the factors to be considered in product and process design.

TABLE 25 Properties of Fotoform Glass and Fotoform Glass Ceramics[77b]

Property	Unceramed Type B	Unceramed Type C	Ceramed
Mechanical:			
Specific gravity	2.36	2.37	2.46
Modulus of rupture, psi $\times 10^{-3}$ (abraded samples)	>8.7	>16.0	>20.0
Modulus of elasticity, psi $\times 10^{-6}$	11.0	12.0	13.5
Modulus of shear, psi $\times 10^{-6}$	4.6	5.1	5.7
Working tensile strength, psi $\times 10^{-3}$	1.0	....	3.0
Knoop hardness (100 g), kg/mm²	507	566	581
Softening temperature, °C	....	....	700
Specific heat, cal/(g)(°C):			
At 25°C	....	....	0.209
At 200°C	....	....	0.256
Thermal conductivity, cal/(sec-cm²)(°C/cm):			
At 25°C	....	....	0.0056
At 200°C	....	....	0.0050
Surface finish (as abraded), μin	....	....	80–120
Poisson's ratio	0.20	0.18	0.20
Electrical:			
Power factor at 1 MHz:			
20°C	0.005	0.003	0.006
200°C	0.130	0.021	0.014
Dielectric constant at 1 MHz:			
20°C	6.5	5.7	5.6
200°C	8.3	6.3	6.3
Dissipation factor at 1 MHz and 20°C	....	....	0.0062
Loss factor at 1 MHz and 20°C	0.033	0.017	0.034
Dielectric strength, volts/mil	....	....	>450
Surface resistivity, ohms/sq:			
Untreated surface	....	....	>10⁸
Silicone-treated surface	....	....	>4 × 10¹²

Pretreatment Processes

Surface preparation of the materials to be fabricated is an important step in determining the quality and yield of finished parts. All materials contain surface contaminants resulting from their manufacture, handling, and contact with their surroundings. These must be removed effectively before application of resist layers and part processing.

Two objectives are required in any pretreatment process, namely, (1) removal of all contamination that would interfere with part fabrication and (2) proper

Photofabrication

TABLE 26 Properties of Templex Photosensitive Plastic[a,80]

ASTM Method	Property	Value

D 638	Tensile strength, psi:
	  −40°F
	  73°F
	  173°F
	Elongation, %:
	  −40°F
	  73°F
	  173°F
	Yield stress at 73°F, psi
	Tensile modulus, psi:
	  −40°F
	  73°F
	  173°F
		
		14,200
		5,080
		3,920
		
		11
		19.7
		21.8
		4,980
		
		256,000
		201,000
		192,000

D 256	Impact strength (Izod):[b,c]
	  Notched bar, ft-lb/in. of notch
	  Unnotched bar, ft-lb/in.
		
		0.66
		6.6

D 1044 D 1706 D 785 D 789	Taber abrasion (CS-17 wheel/1,000-g load), g/kHz Durometer hardness, D scale[b,d] Rockwell hardness, R scale[b] Melting point (Fisher-Johns), °C	0.0373 83 D 101 R 300; darkens at 210

D 696	Coefficient of linear thermal expansion, in./(in.)(°C) $\times 10^5$:
	  0–23°C
	  23–66°C
	  0–66°C
		
		10.35
		7.58
		8.55

Cenco-Fitch Nonstandard D 149 D 495	Thermal conductivity at 50°C,[c] cal/(cm)(sec)(°C) $\times 10^4$ Specific heat at 50°C, cal/g[c] Dielectric strength at 68°F, volts/mil[f] Arc resistance, sec	4.95 0.89 435 69

D 257	Surface resistivity at 23°C, ohms/sq $\times 10^{-11}$
	Volume resistivity, ohm-cm:
	  Mercury cell:[e,g]
	    23°C
	    50°C
	    75°C
	  Balsbough cell:[e,h]
	    23°C
	    50°C
	    75°C
		1.6
		
		5.8×10^{10}
		7.1×10^{8}
		1.6×10^{8}
		
		6.3×10^{10}
		1.6×10^{9}
		2.1×10^{8}

D 150	Dielectric constant:[i]
	  At 30°C:
	    100 Hz
	    1 kHz
	    10 kHz
	    100 kHz
	    1 MHz
	  At 75°C:
	    100 Hz
	    1 kHz
	    10 kHz
	    100 kHz
	    1 MHz
		
		11.0
		8.6
		7.6
		6.9
		6.3
		
		85
		20
		10.5
		8.5
		7.4

Footnotes appear at end of table.

TABLE 26 Properties of Templex Photosensitive Plastic[c,80] **(Continued)**

ASTM Method	Property	Value

	Dissipation factor:[i]
	 At 30°C:
	   100 Hz
	   1 kHz
	   10 kHz
	   100 kHz
	   1 MHz
		
		0.31
		0.76
		0.30
		0.14
		0.09

D 792	Specific gravity	1.323
D 542	Index of refraction n_D at 25°C	1.495
D 568-61	Flammability, in./min[b]	13
D 635		8

[a] Data shown are average values obtained under standard ASTM conditions and should not be used as minima for material specifications. Except where otherwise noted, tests were carried out at 73°F (23°C) with unsupported 0.047-in.-thick specimens obtained by 90-sec exposure to 140-amp carbon-arc lamp at 30 in. distance from sample surface, followed by 11-min washout under standard conditions. Samples were conditioned for 30 days at 73°F and 50% RH to equilibrium moisture content before testing, unless other test conditions are specified.

[b] Test sample not moisture-conditioned, but had come to thermal equilibrium at 73°F (23°C) ambient atmosphere.

[c] 0.148-in.-thick specimen (15-min exposure to 140-amp carbon arc at 50-in. distance, 28-min washout under standard conditions).

[d] 0.012-in.-thick specimen (90-sec exposure to 140-amp carbon arc at 30-in. distance, 5-min washout under standard conditions).

[e] 0.019-in.-thick specimen (90-sec exposure to 140-amp carbon arc at 30-in. distance, 7-min washout under standard conditions).

[f] Samples tested in oil. Values for 0.012- and 0.019-in.-thick specimens are 925 and 750 volts/mil, respectively.

[g] Electrode diam. 0.90 in., voltage 500.

[h] Electrode diam. 1.5 in., voltage 500.

[i] Balsbough cell with 1-in. electrodes; samples oven-conditioned for 30 min before test.

preparation of the surface to provide maximum adhesion of the resist layer. Improper removal of contaminants can result in poor resist adhesion, and nonuniform etching or plating. In addition to surface cleaning, conversion coatings may be employed to promote better resist adhesion and reduce undercutting during etching.

Cleaning Factors to be considered in the selection and use of a cleaning process include the type of contamination or soil to be removed, characteristics and shape of the material to be cleaned, and the required degree of cleanliness.

Soils can be classified simply into organic and inorganic types. Organic soils usually consist of various oils and greases resulting from material fabrication and handling. Inorganic soils consist chiefly of surface oxides, cleaning abrasives, and other solid particles. Organic soils may contain polar or nonpolar groups that influence their adhesion to the surface. Polar groups, for example, will form strong bonds to metal surfaces, and in some cases form metal compounds which are extremely difficult to remove. The various types of soils common to metals and their removal have been reviewed extensively in the literature.[81,82] The data and cleaning methods discussed for metals can often be applied to other materials.

Generally, chemical and mechanical cleaning techniques are used alone or in combination, depending on the application, material, and size of the part to be processed. Mechanical methods involve the use of abrasives along with sanding, polishing, brushing, and blasting techniques. The hardness and size of any abrasives used must be selected so as to prevent excessive scratching of the surface, which may interfere with the adhesion of thin resist layers. Deep unidirectional scratches should be avoided with resists to avoid possible "wicking" of processing solutions under the resist layer.[83]

Mechanical Cleaning. Mechanical cleaning methods are used to remove heavy soils, scale, and surface burrs. The mechanical force required will vary with the type of abrasive employed and the method of applying the force. Brushing is used in most applications with various scrubbing cleansers, mild alkali cleaners, and fine abrasives. Heavy soils require coarse abrasives with brushing or the use of sanding methods. Sandblasting and vapor-blasting techniques employ abrasives

TABLE 27 Typical Applications of Various Vapor-degreasing Solvents[81]

Application	Solvent	Approximate vapor temperature, °F	Factors affecting selection
Removal of soils from parts	Trichloroethylene	188	Most commonly used degreasing solvent.
Removal of slightly soluble (high melting) soils	Perchloroethylene	250	Used where higher operating temperature is desirable.
Removal of water films from metals	Perchloroethylene	250	Rapid and complete drying in one operation.
Cleaning coils and components for electric motors	Methyl chloroform	165	Solvent must not damage wire coating or sealing agents. Requires special equipment design. Selection should be based on preliminary trials.
	Trichlorotrifluoroethane	118	
Cleaning temperature-sensitive materials	Methylene chloride	104	Used where parts must not be exposed to higher vapor temperatures during cleaning. Special corrosion-resistant equipment is required.
	Trichlorotrifluoroethane	118	
Cleaning components for rockets or missiles	Trichloroethylene	188	Cleaned parts must be free of soils or residues which might react with oxidizers.
Cleaning with ultrasonics	Trichloroethylene	188	For cleaning efficiency beyond that obtained from standard vapor degreasing. Solvent must be kept clean by continuous distillation and filtration during use. Selection should be based on preliminary trials.
	Perchloroethylene	250	
	Methylene chloride	104	
	Fluorinated hydrocarbon	118	

carried in an air or liquid stream and directed against the surface to be cleaned. These methods are effective in cleaning surfaces containing recesses that could not be cleaned by other methods. A wide variety of cleaning equipment and materials designed for specific photofabrication applications is available commercially. Detailed information on the use of mechanical cleaning methods can be found in the literature.[84]

In addition to surface cleaning, abrasive methods are used to deglaze highly polished surfaces to promote better bonding of resist layers and plated coatings. Most applications of plating on nonconductors require a certain degree of surface roughness in order to obtain good adhesion. As previously noted, care must be taken to avoid deep unidirectional scratches, especially when thin photoresist layers are employed.

Processes Used and Their Relation to Product Design   14-61

Chemical Cleaning. Chemical cleaning is effective in the surface preparation of most materials used in photofabrication and related processes. Chemical cleaning methods include (1) solvent cleaning, (2) emulsion cleaning, (3) alkaline cleaning, (4) electrolytic cleaning, and (5) acid cleaning. Adequate cleaning may be obtained using just one of these methods, or a combination may be required where heavy contamination is involved.

Solvent cleaning involves the removal of soluble oil or grease or other contaminants through contact with a solvent solution or vapor. This method is used extensively as an initial cleaning procedure prior to mechanical or aqueous chemical cleaning methods. Trichloroethylene, perchloroethylene, and other halogenated solvents are commonly employed, and a wide variety of equipment designed for their use in the vapor phase is available. Chlorinated solvents are also used for the development of many types of photoresists and are effective in removing residual resist films in the developed out areas prior to etching or plating. A typical vapor degreaser includes a heated sump solvent compartment and an upper compartment with water cooling in the walls to provide a continuous layer of fresh solvent vapors at the top of the compartment. In use, parts are immersed in the chamber above

TABLE 28 Typical Applications Using Solvent as the Final Cleaner[82]

Type of work	Type of soil removed	Solvent	Method of cleaning	Subsequent operations
Silicon and germanium pellets	Wax	Trichlorethylene	Spray or boiling solvent	Diffusion, etching, alloying
Electrical chassis	Finger marks at final touch-up	Naphtha	Hand wipe	Assemble into cabinets
Cabinets, 7 by 1½ by 2½ ft	Shop soil	Aromatic paint thinner	Spray and hand wipe	Paint priming
Printed-circuit boards	Soldering flux	1,1,1-trichloroethane or Freon* alcohol	Soak in trays	Assemble into units
Aluminum boxes, 4 by 4 by 6 in.	Shop soil	Trichlorethylene	Soak in bucket	Painting
Steel panels, 3½ by 2 ft	Machine oil	Kerosene containing safety solvent	Hand wipe	Phosphating

* Trademark of E. I. du Pont de Nemours & Co., Wilmington, Del.

the solvent where solvent condenses on the surface until an equilibrium temperature is reached. Some equipment includes a spray of solvent during this step. The part is then removed through the vapor layer at the top. The vapor method is more effective than simple immersion methods in that fresh solvent vapor comes in contact with the part continuously. Most contamination is carried with the condensate into the sump. Typical applications of various vapor-degreasing solvents are presented in Table 27. The choice of solvent is determined by its boiling range and its effect on the materials to be cleaned.

Solvent cleaning by simple immersion spraying or wiping is used with chlorinated and other organic solvents or their mixtures at room or slightly elevated temperatures. Cleaning action is not as effective as with the vapor-degreasing method because of the lower temperatures employed and the possibility that contaminants from the bath may be deposited on the surface as the solvent dries. Treatment should be followed by spray rinsing with fresh solvent to remove residual films. Solvent cleaning often forms the final cleaning procedure prior to further processing. Some typical applications are listed in Table 28. Agitation of immersion baths with ultrasonics produces a very effective cleaning. A mechanical scrubbing action is set up by cavitation of the solvent on the surface, aiding dissolution of contaminants and dislodgement of solid particles. This method is very effective for use with porous samples or samples containing holes or cutout areas.

14-62   Photofabrication

Emulsion cleaning is a variation of solvent cleaning where organic solvents are dispersed in an aqueous medium with the aid of an emulsifying agent. Various detergents and surface-active agents are added to the water phase. This method is used with soak or spray equipment to remove light soils.

Alkaline cleaning with soaking or spray methods is effective in removing oils, grease, fats, and solid soils from metals. Soil removal is accomplished by saponification or emulsification processes. Solutions are normally used with agitation and at a sufficiently high temperature to decrease cleaning time without adversely affecting the material. Typical alkaline cleaner formulations for various metals are shown in Table 29. The use of spray equipment or ultrasonic agitation is more effective than soaking because of the increased mechanical action at the surface. A wide variety of proprietary solutions are available with light, medium, and heavy cleaning action. In general, light and medium cleaners are sufficient to remove most light soils. Heavy-duty cleaners are required for heavy contamination. These, however, must not be used for aluminum or other materials adversely affected by strong alkaline solutions. Nonetch and neutral soaking solutions are available from many suppliers for specific applications.

Electrolytic cleaning involves passage of a direct current through an alkaline electrolyte, the metal to be cleaned forming the anode or cathode of the cell with an inert metal serving as the counter electrode. Cleaning is accomplished through chemical reaction and through the mechanical agitation set up by the generation of gas at the surface. Hydrogen is evolved at the cathode and oxygen at the anode in a 2:1 ratio for a given current density. Cathodic cleaning is more effective owing to the greater quantity of gas evolved. Also, negatively charged colloidal particles normally found in alkaline solutions are repelled from the cathode surface. Cathodic cleaning is subject, however, to deposition of positively charged metals (metal smuts) as contaminants. Also, since hydrogen is evolved at the surface, cathodic cleaning should not be used with metals that adsorb or react with this element. Periodic reverse-current methods provide better cleaning efficiency. Both cathodic and anodic cleaning are involved, and the beneficial qualities of each are obtained.

Acid cleaning with a variety of mineral and organic acids and solutions of acid salts is used to remove metal oxides and light soils and to neutralize surfaces after alkaline cleaning methods. A wide variety of proprietary compositions is available, ranging from weak to strong. With most metals, weak solutions of acid etchants, e.g., concentrations of 10 percent, can be used. In general, solutions should be of sufficient strength to remove the surface oxide but no so strong that the metal is appreciably attacked.

Generally, cleaning procedures will vary considerably with various materials and applications. An outline of metal cleaning processes selected for various purposes is presented in Table 30. All cleaning steps should be followed by rinsing in fresh solvent or water to avoid drag-out of contaminants from the cleaning solutions. A sufficient rinse time should be used to completely remove all dissolved and loose soils, residues, and cleaning-solution films. Water-rinse baths should be of the water overflow type to avoid buildup of contaminants. Suitable test procedures for evaluating cleanliness should be established for particular applications. These can consist of a simple water-break test in the final rinsing step or a standardized spray or atomizer test. They are adequate for most applications. Ultimately the effectiveness of any cleaning method must be judged by the subsequent processability of the cleaned part. Detailed cleaning tests are described in the literature references.

After the cleaning process, parts should proceed to the next processing step without delay, to minimize recontamination. Where resist application follows, parts are forced-air-dried and heated to remove surface moisture. Parts to be plated or etched can proceed immediately to the various processes after the final rinse. It should be emphasized, however, that in all processes involving surface coatings, etching, or plating, a uniform, clean surface will ensure reproducibility of the process and high quality. Detailed information on physical and chemical cleaning and procedures for specific applications can be found in the literature.[81-87]

TABLE 29 Typical Alkaline Cleaner Formulations for Various Metals[82]

Formulation and conditions	Aluminum Soak	Aluminum Spray	Copper Soak	Copper Spray	Copper Electrolytic	Copper plate electrolytic	Iron and Steel Soak	Iron and Steel Spray	Iron and Steel Electrolytic	Magnesium Soak	Magnesium Spray	Zinc Soak	Zinc Spray	Zinc Electrolytic
						Composition of cleaner, % by weight								
Builders:														
Sodium hydroxide, ground	...	...	20	15	15	55	20	20	55	20	20	...	15	15
Sodium carbonate, dense	...	...	18	...	...	8	18	29	8.5	18	29	...	...	...
Sodium bicarbonate	21	24	...	34	34	...	...	...	...	...	...	...	35	34
Sodium tripolyphosphate	30	30	...	10	10	10	...	...	...	...	...	90	10	10
Tetrasodium pyrophosphate	...	...	20	...	...	...	20	20	10	20	20	...	...	...
Sodium metasilicate, anhydrous	45	45	30	40	40	25	30	30	25	30	30	...	40	40
Surface-active (wetting) agents:														
Sodium resinate	...	...	5	...	...	...	5	...	...	5	...	5	...	...
Alkyd aryl sodium sulfonate	3	...	5	...	...	...	5	...	1	5	...	5	...	...
Alkyl aryl polyether alcohol	...	...	2	...	...	1	2	...	...	2	...	...	...	...
Nonionics high in ethylene oxide	1	1	...	1	1	1	...	1	0.5	...	1	...	...	1
						Other conditions								
Operating temperature of solution, °F	160	160	180	170	160	180	200	170	180	200	170	180	170	180
Concentration of cleaner, oz/gal H$_2$O	4	1	8	1	8	8	8	1	8	8	1	4	1	6
Relative cost of chemicals*	136	121	148	102	104	118	150	107	111	150	107	194	100	104

* The cost index is based on the lowest-cost formulation (spray zinc) as 100. The base value of 100 corresponds to a cost of $4 to $8 per 100 lb of the cleaner formulation, depending on quantity and local purchasing conditions.

14-63

TABLE 30 Metal Cleaning Processes Typically Selected for Various Purposes[82]

(Processes listed in order of decreasing preference)

Type of production	In-process cleaning	Preparation for painting	Preparation for phosphating	Preparation for plating
	Removal of pigmented drawing compounds[a]				
Occasional or intermittent	(1) Hot-emulsion hand slush, spray emulsion in single stage; or (2) vapor slush degrease[b]	(1) Boiling alkaline, blow off, hand wipe; or (2) vapor slush degrease, hand wipe; or (3) acid clean[c]	Hot-emulsion hand slush, spray emulsion in single stage, hot rinse, hand wipe	Hot alkaline soak, hot rinse (hand wipe if possible), electrolytic alkaline, cold-water rinse
Continuous high production	Conveyorized spray-emulsion washer	Alkaline soak, hot rinse, alkaline spray, hot rinse	Alkaline soak, hot rinse, alkaline spray, hot rinse	Hot emulsion or alkaline soak, hot rinse, electrolytic alkaline, hot rinse
	Removal of unpigmented oil and grease				
Occasional or intermittent	(1) Emulsion dip or spray; (2) vapor degrease; (3) cold solvent dip; or (4) alkaline dip, rinse, dry (or dip in rust preventive)	Vapor degrease or phosphoric acid clean	(1) Emulsion dip or spray, rinse; or (2) vapor degrease	Emulsion soak, barrel rinse, electrolytic alkaline, rinse, hydrochloric acid dip, rinse
Continuous high production	(1) Automatic vapor degrease; or (2) emulsion, tumble, spray, rinse, dry	Automatic vapor degrease	(1) Emulsion power-spray, rinse; (2) vapor degrease; or (3) acid clean[c]	Automatic vapor degrease, electrolytic alkaline, rinse, hydrochloric acid dip, rinse[d]

Removal of chips and cutting fluids

Occasional or intermittent	(1) Alkaline dip and emulsion surfactant; (2) Stoddard solvent or trichlorethylene; or (3) steam	(1) Alkaline dip and emulsion surfactant; or (2) solvent or vapor	Alkaline dip, rinse, electrolytic alkaline[f] rinse, acid dip, rinse[a]
Continuous high production	Alkaline (dip or spray) and emulsion surfactant	(1) Alkaline dip and emulsion surfactant[e]; or (2) solvent or vapor	Alkaline soak, rinse, electrolytic alkaline,[f] rinse, acid dip and rinse[a]
	Alkaline (dip or spray) and emulsion surfactant[c]	Alkaline (dip or spray) and emulsion surfactant	

Removal of polishing and buffing compounds

Occasional or intermittent	Seldom required	(1) Surfactant (agitated soak), rinse; or (2) emulsion soak, rinse	(1) Surfactant (agitated soak), rinse, electroclean[h]
Continuous high production	Seldom required	(1) Surfactant alkaline spray, spray rinse; or (2) emulsion spray, rinse	(1) Surfactant alkaline soak and spray, alkaline soak, spray rinse, electrolytic alkaline[h]
	(1) Surfactant alkaline spray, spray rinse; (2) agitated soak or spray, rinse[i]		

[a] For complete removal of pigment, parts should be cleaned immediately after the forming operation, and all rinses should be of spray type, where practical.
[b] Used only when pigment residue can be tolerated in subsequent operations.
[c] Phosphoric acid cleaner-coaters are often sprayed onto parts to clean the surface and leave a thin phosphate coating.
[d] Some plating processes may require additional cleaning dips.
[e] Neutral emulsion or solvent should be used before manganese phosphating.
[f] Reverse-current cleaning may be necessary to remove chips from parts having deep recesses.
[g] For cyanide plating, acid dip and water rinse are followed by alkaline and water rinses.
[h] Second preference: stable or diphase emulsion spray or soak, rinse, alkaline spray or soak, rinse, electroclean. Third preference: solvent presoak, alkaline soak or spray, electroclean.
[i] Third preference: emulsion spray, rinse.

14-65

Photofabrication

Conversion coatings A conversion coating is one formed chemically on the surface of a metal by the reaction of the surface with a chemical solution. The reaction can result in the formation of a chemical compound or a "complex" between the base metal or moisture absorbed on the surface and the reagent. The film formed on the surface changes its nature and produces a continuous protective layer on the base metal. Principal uses of these coatings in photofabrication include: promotion of adhesion of resists to the surface, reduction of undercutting during etching, and protection against corrosion. In addition to the treating of metals, nonmetallic surfaces containing reactive groups or hydrogen-bonded water molecules can be treated to produce complexes with similar improvements.

Acidic solutions containing chromates and phosphates are used to form conversion coatings with metals.[84,85] Solutions of hexavalent chromium with sulfates or chlorides react with aluminum, magnesium, zinc, cadmium, silver, copper, and copper alloys to produce chrome complex films. Initially, a gel-like film is formed which transforms into an amorphous hard surface on drying. Concentrated phosphoric acid solutions containing an accelerator such as a nitrate react with nickel, zinc, manganese, magnesium, iron, and various steels to produce uniform corrosion-resistant nonconductive metal phosphates. Conversion coatings are recommended for metals such as aluminum, magnesium, and nickel which oxidize readily after cleaning. Such coatings can produce better bonding of the resist and help prevent chemical attack of the part through thin resist areas. The degree of improvement realized will vary with the different photoresists, processing chemicals, and processing steps involved. Generally, such films are more resistant to the etchants used, resulting in a reduction of undercutting at the edge of the pattern (see Fig. 32).

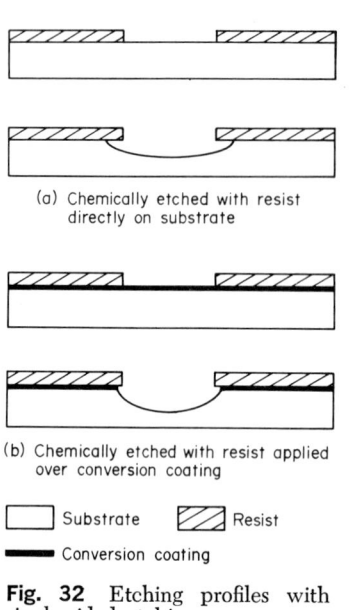

Compounds such as the titanium esters and siloxanes react with surface hydroxyl groups and adsorbed water to form surface complexes and are used as adhesion promoters on metals, glass, semiconductors, and plastics. Treatment consists of coating of the surface with a dilute solvent solution of the complexing agent, drying, and heat curing of the resultant film. The complex formed produces a hydrophobic surface and promotes better adhesion of organic resists and coatings.

(a) Chemically etched with resist directly on substrate

(b) Chemically etched with resist applied over conversion coating

Substrate   Resist   Conversion coating

Fig. 32 Etching profiles with single-sided etching.

A conversion coating consisting of a thin continuous layer is desirable. Heavy coatings tend to be powdery and noncontinuous and should be avoided. The use of a conversion coating must be evaluated in the overall fabrication process. If an improvement is realized, adequate control of the treating times or coating application must be carried out to ensure reproducibility.

Etching

Most photofabrication applications in the electronics industry involve chemical etching of the part after a resist pattern has been formed. Chemical etching is the removal of material by dissolution or chemical reaction in a suitable etchant. With metals, dissolution occurs as a result of the formation of a salt soluble in the etchant. Plastics are etched by either solvent action of the etchant or chemical breakdown of the polymer. Factors to be considered in the etching process include: (1) the material to be etched, (2) part design and specifications, (3) type of mask or resist used, (4) type of etchant, (5) effect of the physical process parameters, e.g., temperature, flow rate, etc., and (6) economics. The composition of the part to be processed, part design, and specifications will dictate the choice of etchants and the process to be used. A brief discussion of these factors is presented here as a general guide, along with information relating to the etching of various materials.

Role of materials The characteristics of the material to be etched will affect the quality of the etched part. Materials of uniform composition and free of defects will exhibit a regular etch rate and produce good pattern reproducibility. For example, metals having a homogeneous composition and uniform grain size and distribution can be expected to produce high yields of etched parts. Random etching in metals is attributed to nonuniform grain size, to segregation or inclusion of metal contaminants, and to contaminants from milling operations. Manufacturers of metals for electronic applications produce special grades of exceptionally homogeneous glass-sealing and magnetic alloys for chemical machining processes.

Single-crystal materials will exhibit different etch rates along specific crystal planes in certain etchants. This characteristic can be used effectively to produce high etch factors and parts having minimum spacing between etched-out areas. Silicon wafers, for example, can be etched preferentially through the (100) lattice plane in caustic alcohol solutions with a minimum of undercutting.

Desirable etchant properties include a fast uniform etch rate, high etch factor, simple control, good compatibility with mask materials and equipment, and low cost. To determine the suitability of an etchant, the chemical reactivity of the material to be etched will serve as a guide. For example, acids and bases that form salts soluble in aqueous solutions are generally used for metals. In addition to reactivity, the chemical and physiological properties associated with the handling and disposal of etchants must be considered, particularly in large-volume operations. As a consequence, an etchant having optimum etching properties for a given material may not be suitable in a production process.

Table 31 lists a variety of materials and suitable etchants for use with them. Additional data for the etching of thin-film materials are presented in a later section. Applications may exist where photoetching of a specific material not included in the tables is required. The other chapters of this handbook should be consulted for detailed information relating to the chemical properties of materials to aid in the selection of an etchant. Proprietary etchants for specific materials are available from many commercial sources. These usually consist of common etchant reagents modified by the incorporation of additives including catalysts, organic additives to improve etch factor, and surface-active agents. These formulations offer good control of the etching process.

Etch rate is the quantity of material dissolved by a particular etchant for a specific process in a given time (usually expressed in thickness dissolved per unit of time, e.g., mils per minute). Process parameters affecting etch rate include: concentration of etchant and reaction products, temperature, agitation, chemical additives and impurities, and the area of exposed material. In the etching of a particular part, the etch rate is calibrated as a function of etchant concentration and temperature, using controlled agitation and equipment. While the rate increases with temperature, the optimum temperature for a given process will be determined by the material characteristics and the equipment to be employed in the etching process. Etch rate at the desired operating temperature is then determined as a function of by-product concentration to establish operating limits for the process. Curves showing the etch characteristics of ferric chloride in the etching of copper are given in Figs. 33 and 34.[94] Similar curves can be generated for the etching of other metals with ferric chloride and for other systems.[87,91] Once the optimum etch rate is determined for a particular metal and etchant, methods must be worked out to control the process and the total amount of material removed.

Effect of etching on part design The term *etch factor* refers to the degree of lateral etching or undercutting obtained under the edge of the resist image and is defined as the ratio of etch depth to undercut distance. This is illustrated in Fig. 35. Etch factor is dependent upon the material being etched, the etch rate, and the type of equipment used, and is approximately constant for a fixed

14-68   Photofabrication

TABLE 31   Etchants Commonly Used in the Processing of Materials [1,86,89-93]

Material	Etchant	Concentration	Temperature, °F	Etch rate, in./min (fresh solution)	Typical etch factor
Alfenol................	FeCl₃	42°Bé	120		
	HNO₃:HCl:H₂O	1:1:2	100-120		
Aluminum and aluminum alloys	NaOH or KOH	10-20%	140-190		
	FeCl₃	12 to 18°Bé	120	0.001+	1.5:1-2.0:1
	HCl:H₂O	1:4			
	HCl:HNO₃:H₂O[a]	10:1:9	120	0.001-0.002	2:1
	HCl:FeCl₃(42°Bé)	1:10	110		
	H₃PO₄(85%):HNO₃(70%):H₂O	10:1:2½ by vol.	113-131		
Chromium..............	HNO₃:H₂O	3:1	175		
	FeCl₃(42°Bé):HCl	2:1	175		
	33% NaOH:25% K₃Fe(CN)₆ (Kodak EB-5[b] bath)	1:3	70-75		
Cold-rolled steel........	FeCl₃	42-49°Bé	120-130	0.001	2:1
	HNO₃	10-15%	120	0.001	1.5-2.0:1
Constantin............	FeCl₃	42°Bé			
Copper and copper alloys	FeCl₃	42°Bé	120	0.002	2.5-3.0:1
	CuCl₂ solutions	2M CuCl₂ in 6N HCl (typical solution)			
	CuCl₂	35°Bé	130	0.00055	2.5-3.0:1
	Chromic-sulfuric	20-30% H₂SO₄[7,26] 10-20% chromate	120	0.0015	2-3:1
	(NH₄)₂S₂O₈	20%	90-120	0.001	2-3:1
	NH₄Cl sat. with NaCl[c]				
Germanium[f]...........	HF, HNO₃ mixtures	1:1; 2:1			
Gold..................	HCl:HNO₃	3:1	90-100	0.001-0.002	
	NaCN solutions with H₂O₂				
	Alkaline cyanide solutions[c]				
	Iodine-iodide solutions 60g KI, 15g I₂ 70-75° + 1-3l H₂O				
Hardened tool steel.....	HNO₃	10-15%	100-120	0.0005-0.001	1-2:1
HyMu 80 and other magnetic alloys..........	FeCl₃	42-49°Bé	110-130		
	FeCl₃(42°Bé):HCl(20°Bé)	9:1	110-120		
Inconels...............	FeCl₃	42-49°Bé	110-130		
Kovar[d]................	Chromic-sulfuric	40°Bé	120	0.001	2-1
Lead..................	FeCl₃	42°Bé	130		
	FeCl₃(42°Bé):HCl(20°Bé)	9:1	110-120		
Moly permalloy........	FeCl₃	42-49°Bé	130		
	FeCl₃(42°Bé):HCl(20°Bé)	9:1	110-120		
Molybdenum...........	H₂SO₄:HNO₃:H₂O	1:1:1-5	130	0.001 at 130°	
	HNO₃:HCl:H₂O	1:1:1-2			
	K₃Fe(CN)₆ (200 g/l):NaOH (20-25 g/l):Na₂C₂O₄ (3-3.5 g/l)				
	NaOH(10-20%),[e] Na₂C₂O₄(5%)				
Nickel and nickel-iron, alloys................	FeCl₃	42-49°Bé	110-130	0.0005-0.001	1-3:1
Nickel-silver, alloys......	FeCl₃	42°Bé	130		
	Chromic-sulfuric		120		
	(NH₄)₂S₂O₈	20%	90-120		
Phosphor bronze........	FeCl₃	42°Bé	80	0.0005	2-1
	Chromic-sulfuric		80	0.0005	2-1
	(NH₄)₂S₂O₈	20%	80	0.0003	2-1
	FeCl₃(42°Bé):HCl(2°Bé)	9:1	110-120		

[a] HNO₃ used when titanium parts are involved.
[b] For vacuum-deposited coatings.
[c] For electrolytic etching.
[d] Trademark of Westinghouse Electric Corporation, Pittsburgh, Pa.
[e] For electrolytic etching at 6 volts, using stainless steel cathode.
[f] For detailed information, see P. J. Holmes (ed.), "The Electrochemistry of Semiconductors", Academic Press, New York, 1962.

TABLE 31 Etchants Commonly Used in the Processing of Materials[1,86,89−93] **(Continued)**

Material	Etchant	Concentration	Temperature, °F	Etch rate, in./min (fresh solution)	Typical etch factor
Silicon[f]	HF-HNO₃ mixtures				
	HF-HNO₃-CH₃COOH mixtures	1:2:1			
	FeCl₃(42°Bé):HNO₃:HF	4:4:1			
Silicon dioxide	HF; NH₄F-HF mixtures				
Silicon steel	FeCl₃	42°Bé	130	0.001	1.5–2:1
Silver	FeNO₃	55% wt./vol.	110–120	0.0008	
	HNO₃:H₂O	50–90%	100–120	0.0005–0.001	
Stainless steel[g]					
Low 300–400, series	FeCl₃(42°Bé):HCl	3% HCl by vol.	130		
High 300–400, series	FeCl₃	36–49°Bé	130	0.0008	1.5–2:1
	HNO₃:HCl:H₂O	1:1:1–3			
	HCl:H₂O[h]	1:3			
Tin	FeCl₃	42°Bé	90–130		
Titanium	HF	10–50% by vol.	90–120	0.0005 at 90° (10% soln.)	
	HF:HNO₃:H₂O	1:2:7	90	0.00075	
	NH₄HF₂:HCl:H₂O	Various	90–120		
Zinc	HNO₃	10–15% by vol.	100–120	0.001	

[g] The presence of copper ions produce non-uniform etching of steels and should be avoided.
[h] For electrolytic etching at 6 volts with stainless steel cathode.

process (material, mask, etchant, and equipment). For a particular material the etch factor is lower with etchants of highest reaction rate. The etch factor is generally improved by increasing the flow of etchant directed against the surface of the sample and increasing the etching rate. For example, etching rate is increased and the etch factor raised in going from immersion etching to splash etching to spray etching.

The etch factor must be considered in artwork preparation to obtain precise dimensional control of the part. This can be understood from the following example, assuming an etch factor of 2 as shown in Fig. 35. For each 0.001 in. of etch depth, a corresponding undercutting equal to 0.0005 in. occurs under the

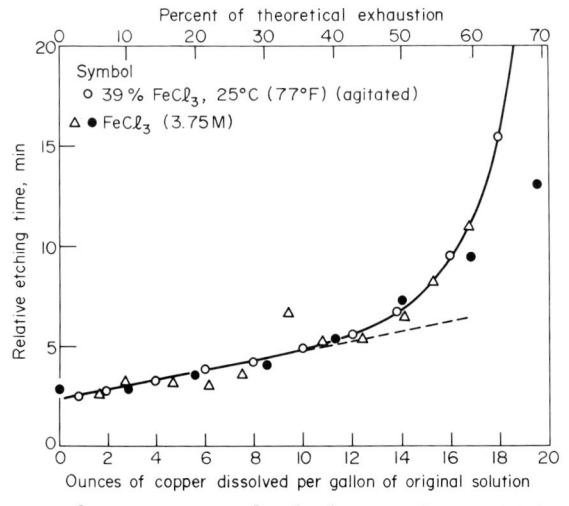

Fig. 33 Relative etching time versus dissolved copper for a 42° Bé FeCl₃ solution.[94]

14-70 Photofabrication

resist image. If a total etch depth of 0.010 in. is required, the surface dimensions of the part etched would be reduced by 0.005 in. at each edge. Thus, the pattern dimensions of the resist must be oversize by an equivalent amount, or 0.005 in. per edge, in order to obtain the designed part size. Artwork dimensions must be adjusted similarly. Where negative photoresists are employed, the negative pattern of the part on the photomask (clear image) must be increased by 0.005 in.

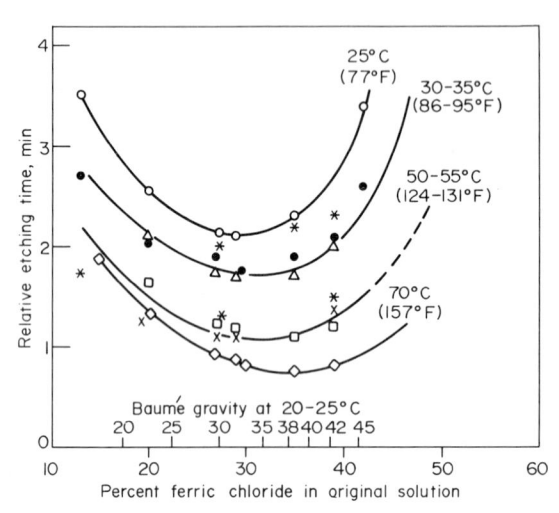

Fig. 34 Relative etching time versus $FeCl_3$ concentration for various temperatures.[94]

per edge; positive photoresists would require an increase in the positive pattern of the part on the photomask (opaque image). In two-sided etching only one-half the above compensation, or 0.0025 in. per edge, is needed since the part is etched to half the total thickness from each side.

In practice, etch factors must be determined experimentally for a particular material and application under controlled processing conditions. This value can then be used for artwork compensation and layout as described above. The etch factors listed in Table 31 are typical for those materials with spray etching and can be used as a general guide to the material processing characteristics.

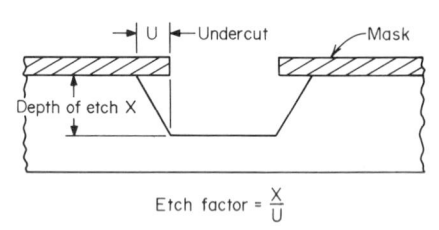

Fig. 35 Illustration of etch factor. Etch factor = $\frac{X}{U}$

The type of resist used and the method of applying it will be determined by the chemical resistance required, pattern resolution and tolerances, part design, and overall economics. Generally, selection of a resist material must be considered along with selection of materials, etchants, and other process parameters in order to optimize a particular process. Detailed properties of various photoresists were discussed previously. A general summary of their use in the etching process is presented here.

Resist characteristics The etch resist used must be capable of reproducing the required pattern and be sufficiently inert in the etchant to prevent breakdown or lifting during the etching process. Pattern definition, accuracy, and the thickness of the material to be etched will dictate the thickness of resist layer used. For example, in semiconductor applications a resist thickness in the order of 0.5 μm (0.00002 in.) is required in order to produce patterns of lines and spaces under

10 µm (0.0004 in.). For patterns that do not require highest resolution, thicker resist layers can be used.

Most photoresists and other organic-resist materials are reasonably inert chemically to the various acid etchants. Their ability to stand up to the etchant, however, is affected by the thickness of the resist layer, the temperature of the etching process, and mechanical agitation at the material-solution interface. Chemical resistance decreases as resist thickness is decreased. Also, the number of pinholes increases, and problems associated with dust and other contaminants become more acute. Higher etching temperatures lead to a swelling of the resist, a weakening of the mechanical properties, and greater chemical attack. Chemical attack of the resist-material interface also increases with increase in temperature, and lifting of the resist image may occur. Generally, the operating temperature of the etching process should be chosen to provide a good etch rate but should be low enough to minimize detrimental effects to the resist layer.

Mechanical agitation at the surface must also be considered. Increase in agitation can lead to greater chemical attack of the resist as well as fracturing of the resist image, particularly at the pattern edges.

Chemical attack of resists is generally greater with alkaline etchants, particularly with positive photoresists. The use of thicker resist layers, lower processing temperatures, and conversion coatings should be considered in applications involving alkaline solutions.

In addition to the chemical and physical properties required of a resist, it should be capable of being easily removed, preferably by simple chemical dissolution without affecting the processed part. Many resists (in particular, where high baking temperatures are used for maximum resistance to etching) require mechanical action to remove them completely. This may limit their use where thin parts are being processed.

Other considerations Economics must be considered in the selection and processing of a resist mask for a given application. Material, equipment, and processing costs vary widely among photoresists as well as among other types of resists. Photoresists generally cost more than other types of resists. Both photoresists and printed resists (screen printing, offset printing, etc.) require high equipment costs but afford high production rates. Selection should be determined according to the number and complexity of parts being processed.

The physical parameters of the etching process will have a decided effect on the process materials used and quality of the finished parts. As previously mentioned, fluid agitation at the etchant-part material interface varies with the type of equipment used. This will affect the etch rate, etch factor, and types of etchants and mask layers used in the process. Generally, increase in agitation at the solution-part interface will increase the etch rate by removing gas and other reaction products from the surface and will minimize local effects due to temperature. Increase in the mechanical force of the etchant on the surface, however, may affect the adherence of the mask layer and must be considered in process design. Where mask adhesion is marginal, only very mild agitation should be considered.

Process parameters such as heat generation, gas evolution, and foaming must be controlled to obtain good-quality etching. Most etching reactions are highly exothermic; for example, the reaction of copper with persulfate liberates 92.5 kcal/mole of copper etched. This necessitates the removal of heat by cooling coils or continuous circulation of the etchant through a heat exchanger for any large operation. Good control of temperature is essential to guarantee a uniform etch rate and to prevent attack on the resist image. Proper safeguards must also be included to prevent damage to plastic etching tanks due to an uncontrolled temperature rise.

Etching methods Three methods of chemical etching are used: immersion etching, splash etching, and spray etching. Simple immersion etching involves immersion of the part to be etched in a suitable etchant until etching is complete. Etching times are slow, and poor etch factors are obtained; hence, very little use is made of this method in production facilities. The addition of solution agitation in the form of stirring, air bubbling, or ultrasonics affords considerable improvement. Production equipment has been designed based on immersion etching with bubble agitation.

Splash etching involves splashing of the etchant against the surface to be etched, using rotating concave paddles. A greater mechanical motion of the fluid etchant against the surface and better removal of reaction products are obtained. This type of equipment is still in wide use in the photoengraving industry.

Spray etching is widely used at present for most photofabrication applications and offers the best control of the etching process. Equipment includes single- and double-sided designs with either vertical or horizontal positioning of spray nozzles. Fast etch rates and good etch factors are obtained through control of spray velocity and the spray pattern. Uniformity of the spray pattern is improved in many models by rotation of the workpiece and oscillation of the spray nozzles.

The materials used in all etching equipment must be compatible with the etchants employed. Most units use plastics such as polyvinyl chloride and polycarbonate. Metals include titanium and alloys of the Hastelloy* C type. Care must be taken with plastics to avoid noncompatible corrosive solvents and high temperatures. The type and construction of process equipment must be considered in product and process design. Normally, the physical parameters of the equipment and the etchant will determine the etch rate and etch factor as a function of pattern design. Such data must be determined for a particular application and correlated with the part design and processing of artwork.

The pattern on the part to be etched will also affect the etching parameters. The etch rate will generally decrease as the spacing between unetched portions is narrowed. This result is due to reduced mechanical agitation at the solution interface in these narrow regions which enables higher concentrations of reaction products to build up. In spray etching, this effect is noted with spaces below 0.005 in. for metals and spaces below 0.010 in. for plastics. This factor must be taken into account in designing artwork, particularly where patterns including both large and small spacing are required.

In addition to the process parameters required of an etchant and the equipment needed for a particular application, the toxicological and disposal problems must be considered. Most etching chemicals are highly corrosive acids which must be confined to the etchant system. Adequate safety precautions should be maintained to protect personnel from any direct contact. Similarly, proper methods of disposal must be used to minimize detrimental effects to sewer systems.

The economics of the etching process are an important consideration in production facilities. Factors determining process costs include capital equipment, etchant cost as a function of production rate (e.g., per weight of metal removed), reclamation, and disposal. There is a considerable difference in the costs of etchants and associated processes; costs will vary depending upon the type of facility and location. Where choice of etchants is possible, all aspects must be evaluated, including the type of mask required and the quality and yield of parts produced. Regenerative systems should be considered for large-production requirements in the chemical milling and printed-circuit fields, as these methods offer promise of substantially reduced etching costs.[95-97]

Operation and control The operation and control of etchants will vary according to the type of the process and facility. For small-scale laboratory processes using immersion etching with stirring or bubble agitation, fresh etchants can be utilized, visually monitored, and discarded as etching times become excessive. Similar techniques can be employed with laboratory and pilot-line etching equipment. The problem with this method is that continuous monitoring is required by process personnel, and the quality of the finished parts is largely determined by the skill of the operator. In production operations, a more positive process control method is required that gives good-quality product at high production rates and low cost. Such processes are usually based on replenishment or regeneration of etchants at specified times or after a predetermined quantity of product has been made. The

* Trademark of Union Carbide Corporation, New York, N.Y.

process may be controlled either by an operator or automatically. Etchants are replaced on schedule before the etching time becomes excessive, and production rates are lowered. A second more attractive method of control is through sensing and feedback control of process variables to replenish or regenerate the etchant so that a constant etch rate is maintained. Etching control methods and systems have been reviewed extensively by Benton.[97]

TABLE 32 Types and Characteristics of Etching Control Systems[97]

Type of system	Monitoring method	Control parameters	Effectiveness
Analog..........	Monitor physical and/or chemical properties related to etch rate	pH, redox potentials, specific gravity, chemical composition	Generally crude control—poor correlation of single parameter with overall process variables
Analog-direct....	Monitor etch rate under static condition and correlate to operating process	Etch times measured for material in a separate immersion or smaller spray system and correlated to production system	Fair control, provided operating parameters are not varied widely
Direct..........	Monitor rate under conditions identical to operating process	Etch times measured in special monitoring equipment using a stream of etchant from the main etching equipment under conditions identical with the production system	Good to excellent control, depending on how well parameters are matched

A comparison of various control systems is presented in Table 32. Direct systems offer the best control. However, excellent control can be obtained with other systems for applications where process parameters vary uniformly and contamination is minimum. For example, etching of steel in ferric chloride is effectively controlled by measuring specific gravity and ferrous iron content. The methods commonly used to adjust etch rates in conjunction with the above control systems include automatic chemical makeup feed, automatic time cycle controls for batch etching, and automatic conveyer speed control for conveyerized etching. Both so-called

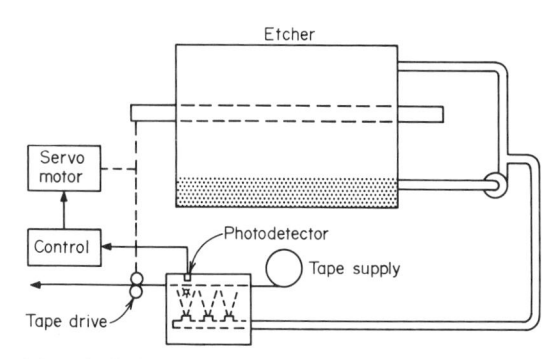

Fig. 36 Block diagram of direct-feedback etchant controller.[97]

analog-direct and direct control systems are used in conveyerized etchers for chemical milling and for printed-circuit fabrication to control the conveyer rate. The conveyer rate is adjusted to maintain the amount of material removed as constant, regardless of etchant strength.

Figure 36 shows a block diagram of a direct servo feedback controller used with spray etchers in printed-circuit manufacturing. In operation, a laminated copper-Mylar* test tape is fed into a spray cell that uses a portion of the etchant from the main etching equipment under conditions that duplicate the physical and chemical characteristics of the spray etcher. The etching of copper from the plastic tape is monitored by a photocell and compared with an etch time predicted for the process. Actual etching times faster or slower than the predetermined time indicate correspondingly too strong or too weak etchant. Tape movement is slowed or speeded up accordingly while the conveyer speed is adjusted simultaneously through a mechanical interlock mechanism. With this type of controller, tolerances on the order of 0.0003 in. in the removal of copper can be maintained over an etchant composition range of 0 to 14 oz of copper per gal of $FeCl_3$. The control automatically compensates for temperature variations of $\pm 20°F$, pressure differentials of ± 25 percent, and specific-gravity variations of $\pm 5°$ Baumé. Etchants are disposed of, once metal contamination reaches a specified level.

Both etchant replenishment and replenishment-regeneration systems have been devised that maintain a uniform etch rate and give extended etchant life. Continuous-replenishment systems are based on the addition of fresh etchant and overflow of spent etchant. This approach has proved practical for continuous-production applications using multiple-compartment etchers with progressive sections operating at higher and higher by-product metal ion concentration levels. The economics of this approach are satisfactory provided a high enough metal ion concentration level is attained before dumping. Replenishment-recovery systems and replenishment-regeneration systems have been recently developed. These systems are referred to as *closed-loop etching*. The spent etchant is sent to an adjacent facility for metal recovery and/or replenishment before returning to the etcher in a continuous cycle.

The Caper system (continuous ammonium persulfate etching and recovery) was developed by The Food Machinery Corporation[96,98] for the etching of copper. In the process, the ammonium persulfate with dissolved copper is pumped to a crystallizer where the double salt of copper sulfate–ammonium sulfate $CuSO_4 \cdot (NH_4)SO_4 \cdot 6H_2O$ is crystallized by cooling to $40°F$. The salt crystals are filtered out, and the remaining weak persulfate solution is pumped to a storage tank for replenishment with fresh ammonium persulfate prior to return to the etcher. This system eliminates disposal of spent etchant and increases the persulfate use efficiency from 50 to 95 percent. In addition, the crystallized double salt can be resold, lowering the overall process cost considerably.

Replenishment-regeneration systems are in use for the etching of iron and iron alloys with ferric chloride and copper with cupric chloride based on the regeneration of metal-loaded etchant with chlorine gas. The chemical reactions involved in the ferric chloride system are:

$$2FeCl_3 + Fe \rightarrow 3FeCl_2 \qquad \text{Etching}$$

$$2FeCl_2 + Cl_2 \xrightarrow{H^+} 2FeCl_3 \qquad \text{Regeneration}$$

The bath is controlled by measuring the ferrous iron content and regulating the chlorine gas feed accordingly to hold that concentration constant. The specific gravity is also measured and maintained within a preset range by addition of water. Etchant temperature, spray pressure, and work feed are maintained constant to aid in uniform removal of metal.

The cupric chloride system for copper[95,99] works similarly with the following reactions taking place:

$$CuCl_2 + Cu \rightarrow 2CuCl \qquad \text{Etching}$$

$$2CuCl + Cl_2 \xrightarrow{H^+} 2CuCl_2 \qquad \text{Regeneration}$$

Automatic control is accomplished by continually monitoring the ratio of cupric chloride to cuprous chloride, using a modified colorimeter. Chlorine gas is fed

* Trademark of E. I. du Pont de Nemours & Co., Wilmington, Del.

in as required to maintain a constant ratio. Water is added to maintain the specific gravity of the etchant between 34 and 36° Baumé. Excess etchant is accumulated and stored for disposal or recovery.

Systems of the above type offer many advantages for large-production facilities, namely: continuous process control, higher production rates, uniform quality, reduction in disposal problems, and lower costs. In the electronics industry automatic etching systems developed to date are widely applied to the etching of copper printed circuits and the etching of steel aperture masks for color kinescopes. Details of these systems can be obtained from the literature and respective system manufacturers. The characteristics of various etchants used in printed-circuit processing are discussed in a later section.

Plating

Plating processes are used extensively in photofabrication, directly in combination with photoresists (electroforming) as well as indirectly in the formation of base layers for subsequent fabrication. Some specific applications are listed in Table 33.

TABLE 33 Photofabrication Applications Utilizing Plating of Metals

A. *With Direct Use of Photofabricated Masks*
  1. Printed circuits
  2. Evaporation masks
  3. Fine-mesh screens
  4. Resistor patterns
  5. Plated etch resists
  6. Fine-dimension printing stencils
B. *Indirect Uses*
  1. Thin-film components
  2. Semiconductor metallization

Of the applications listed, only fine-mesh screens are fabricated by electroforming as separate discrete parts. The screens are electroplated on a temporary supporting conductive plate called a *mandrel* and subsequently stripped off. The various types of mandrels and their fabrication by photoresist processing have been described by Kodak.[100] The remaining applications involve plating of a metal on the part to be processed, and functions as a portion of the component.

The types of resist layers, their selection, and the preparation of base materials used with these processes are similar to those used in etching, as previously discussed. The choice of resist layer and method of application will depend upon the resolution and accuracy of the pattern required and the chemical and physical properties needed for the plating process. Precision parts such as fine-mesh screens and evaporation masks require the use of photoresists. Printed-circuit applications can often utilize resist patterns formed by printing methods rather than by exposure of photoresist.

Plating processes can be divided into three basic types: electroplating, electroless plating, and immersion or replacement plating.

Electroplating Electroplating is the deposition of a metal from solution on passage of an electric current. In the process an electrochemical reduction of the metal ion occurs on the cathodic surface. Deposition rate is a direct function of the applied current and is the fastest of the plating methods. Most electronic plating applications requiring thick metallic layers use this deposition method with a variety of standard metal-plating baths. Resist layers must be chosen based on the chemical corrosiveness and operating conditions of the bath. The resist must be sufficiently inert to prevent unwanted plating due to breakdown through to the base metal or lifting of the edge of the image.

The precautions required for resist-patterned electroplating are different from those needed with etching. In etching processes small traces of contaminants are normally removed from the surface, owing to the strong chemical and physical forces exerted. Consequently, these impurities are not important with respect to the ultimate use of the etched part. However, surfaces to be plated must be wetted by the plating solution in order for deposition to occur. Thus, cleaning is required after development of the resist pattern so that all resist residues, films, and surface oxides are removed. Generally, rinsing in fresh solvent after development and the use of a mild surface-etch solution immediately prior to plating are sufficient.

Secondly, to carry the current and to distribute it uniformly, good electric contact must be made to the part to be plated. Good contact is normally achieved by providing extra material outside the useful sample area, material that can later be discarded. This also provides a more uniform current density over the required pattern since higher current densities are normally obtained at the outer edges. Uniform distribution of current is dependent on the surface structure of the part to be plated, as well as voltage gradients in the plating bath, and is more difficult to provide. High current densities are obtained on projections and edges, resulting in thicker deposits in these areas.

Because of their corrosiveness, long plating times, and hydrogen evolution, many plating baths impose more stringent requirements on the resist layer than etching baths do. Alkaline plating solutions, particularly the cyanide type, may soften and lift the resist layer and should be avoided where possible. The majority of photoresists can be used with the various acid-plating baths and, with care, in mild alkaline baths. Thicker photoresist layers, however, are generally required. Plating voltages should be controlled to prevent hydrogen evolution and possible lifting of the resist film.

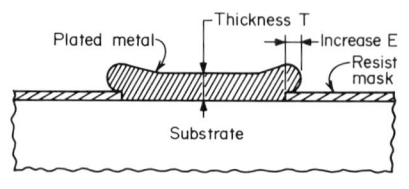

Fig. 37 Profile of electroplated pattern.

Effect of Plating on Part Design. Plating is an additive process and requires different considerations in part design and generation of photomasks from those required for etching. The thickness of metal deposited is directly proportional to current density on the surface plated. Plating of a pattern on a plane surface will result in an increase in the thickness of the pattern and a corresponding increase in the lateral dimensions of the image over the resist layer due to plating at the edges. This phenomenon is illustrated in Fig. 37. The amount of lateral growth of the plated pattern is approximately equal to the plated thickness. Thus, a line plated to a thickness of 0.001 in. will result in a corresponding increase in line width of approximately 0.002 in. (0.001 in. per side). This effect varies with the potential gradient in the plating bath at the surface of the part, as previously noted, and will be dependent upon the type of bath, the current density, solution agitation, the position and movement of the part, and the bath temperature. Higher potential gradients are set up on the edges of the plated image, causing a higher current density in these areas and a corresponding increase in thickness. Thus, a slight crowning occurs at line edges, as noted in Fig. 37. The ability of a bath to produce uniform deposits over irregular surfaces is usually termed *throwing power*. This characteristic is an important consideration in choosing baths for nonuniform surface plating such as in plated-through-hole printed circuits.

The increase in dimensions obtained with plating must be taken into account in part design and in the generation of artwork. As a general rule, dimensions between edges in the resist pattern should be smaller by a factor equal to twice the thickness to be plated. Compensation should also be made for the thickness of the resist used. These factors will limit the resolution capabilities and the thickness of plated metal permissible. For example, high-line-resolution patterns will require a thin photoresist layer; the plated layer must therefore be thin to avoid short-circuiting across the resist image.

A listing of commonly electroplated metals and their electrochemical equivalents appears in Table 34. Detailed data on plating processes, composition, and operational data for various metal baths can be obtained in the literature.[84,101-103]

TABLE 34 Electrochemical Equivalents for Various Metals[84]
(Calculated on basis of 100% cathode efficiency)

Metal	Symbol	Valence	Atomic weight	Specific gravity	Weight for 0.001 in., oz/ft²	Thickness of 1 oz/ft², in.	Grams deposited per amp-hr	Oz/amp-hr	Amp-hr/ft² to deposit 0.001 in.
Antimony	Sb	3	121.76	6.68	0.56	0.00180	1.514	0.053	10.4
Arsenic	As	3	74.91	5.73	0.47	0.00213	0.932	0.033	14.4
Cadmium	Cd	2	112.41	8.65	0.71	0.00139	2.097	0.074	9.73
Chromium	Cr	6	52.01	7.1	0.59	0.00169	0.323	0.011	51.8
Cobalt(ous)	Co	2	58.94	8.9	0.74	0.00135	1.100	0.039	19.0
Copper(ous)	Cu	1	63.54	8.93	0.74	0.00134	2.371	0.084	8.89
Copper(ic)	Cu	2	63.54	8.93	0.74	0.00134	1.186	0.042	17.8
Gold(ous)	Au	1	197.2	19.3	1.47*	0.00068*	7.356	0.236*	6.2
Gold(ic)	Au	3	197.2	19.3	1.47*	0.00068*	2.450	0.079*	18.6
Indium	In	3	114.76	7.31	0.56*	0.00182*	1.427	0.045*	12.0
Iron(ous)	Fe	2	55.85	7.87	0.65	0.00153	1.042	0.037	17.9
Lead	Pb	2	207.21	11.35	0.94	0.00106	3.865	0.136	6.9
Nickel	Ni	2	58.69	8.90	0.74	0.00135	1.095	0.039	19.0
Palladium	Pd	2	106.7	11.40	0.86*	0.00116*	1.990	0.064*	13.5
Platinum	Pt	4	195.23	21.45	1.60*	0.00062*	1.821	0.058*	27.8
Rhodium	Rh	3	102.91	12.5	0.95*	0.00106*	1.280	0.041*	22.9
Silver	Ag	1	107.88	10.5	0.79*	0.00126*	4.025	0.129*	6.2
Tin(ous)	Sn	2	118.70	7.33	0.61	0.00164	2.214	0.078	7.8
Tin(ic)	Sn	4	118.70	7.3	0.61	0.00164	1.107	0.039	15.6
Zinc	Zu	2	65.38	7.14	0.59	0.00168	1.219	0.043	14.3

* These figures are for 1 troy oz/ft².²

Electroless plating Electroless plating is the controlled autocatalytic chemical reduction of a metal ion by a suitable reducing agent. No external current is involved in the process, and metals can be deposited on properly activated nonconductors as well as on conductor surfaces. Electroless baths normally consist of a metal salt, metal complexing agent, pH buffer, and reducing agent. The reduction of the metal is initiated by a catalytic agent on the surface of the part to be plated. Compounds that have been used for this purpose include the salts of tin, palladium, platinum, silver, gold, and titanium. Once the reduction is initiated, the plating continues to build up by autocatalytic reaction on the surface.

In contrast to electroplating, only a limited number of metals can be plated electrolessly. These are listed in Table 35. Of the metals listed, copper and nickel

TABLE 35 Electroless Plated Metals

Metals	Alloys
Cu, Ag, Au, Pd	Ni-W
Fe, Co, Ni	Ni-P*
Cr, Sb, As	Ni-B*

* Phosphorus and boron are codeposited with nickel from the chemical reduction of the reducing agent.

have been used extensively in the electronics industry for the fabrication of printed circuits and contacts. Electroless plated alloys are used for magnetic resistive films. Nickel-phosphorus alloys have been used in the processing of resistor networks on copper printed-circuit boards, both the copper and resistor patterns being defined by photofabrication. The capability of plating on nonconductors by simple immer-

14-78   Photofabrication

sion has led to the exclusive use of electroless baths for plated-through-hole printed circuits.

In practice, the surface to be plated is first activated by treating in a catalytic solution, e.g., in solutions of tin chloride and palladium chloride. Then it is immersed in the electroless plating solution for the required time. Deposition rate is dependent upon the bath composition and the concentration of constituents, pH, and temperature. Rates for nickel baths are as high as 0.001 in./hr; for copper the rates are generally lower by an order of magnitude. In contrast to electrodeposition, metals deposited by electroless deposition are of uniform thickness on all sensitized surfaces in contact with the solution, regardless of the shape of the part.

The adhesion of electroless deposited metals to nonconductors is dependent upon physical and mechanical forces and varies with the nature of the surface. Highly polished, defect-free surfaces will exhibit poor adhesion and must be deglazed before plating. Suitable surface preparation usually entails mechanical abrasion or chemical etching. Primers and adhesive films may also be used. Chemical surface treatments have been developed for a variety of plastics that produce peel strengths for plated-metal layers as high as 20 lb/in.[104]

In most applications only a thin layer of metal—in the order of 100 μin.—is deposited. Thicker layers may be mechanically weak and may lift from the surface, particularly if adhesion is minimal. Applications requiring heavier deposits are generally built up by electroplating. Electroless copper baths have been developed which show good ductility for thick deposits. These have been used for printed-circuit fabrication.[105] In the selection of the type of resist to be used, similar criteria apply to electroless baths and to the electroplating baths. Artwork must be adjusted for increases in the dimensions of the plated image where thick deposits are required.

Data are available in the literature on the application of electroless baths, including bath composition, operation, and pretreatment processes.[106] Most laboratory and production facilities use one of the many proprietary solutions available on the market.

Immersion plating Immersion plating (displacement or contact plating) is the galvanic displacement of one metal by a more noble metal (higher electrode potential) in solution. This type of plating is exemplified by the well-known coating of iron with copper in a copper-salt solution (e_0 of iron = -0.44 volt; e_0 copper = $+0.34$ volt). Immersion plating is similar to electroless plating in that no external current is involved. However, the thickness of metal obtainable with this method is limited to less than 100 μin. since plating stops once all of the basis metal is coated and isolated from the solution.

The use of immersion-plating baths in electronic photofabrication applications is restricted almost entirely to immersion tin and gold plating in printed-circuit applications. It is also used as part of the electroless plating process in the activation of noncatalytic metals with palladium solutions. For this use, only a "seeding" of the surface is required to initiate the subsequent electroless plating.

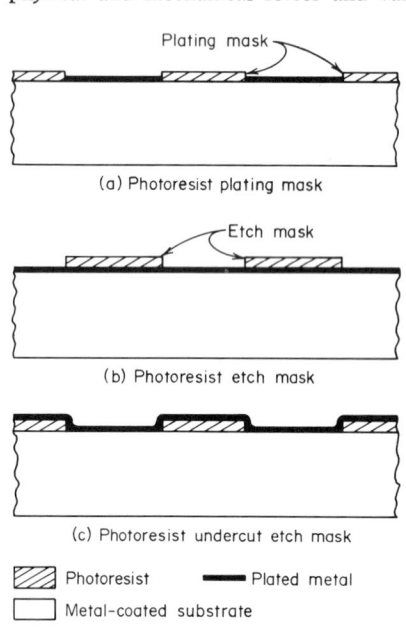

Fig. 38 Typical uses of photoresists with plating processes.

(a) Photoresist plating mask

(b) Photoresist etch mask

(c) Photoresist undercut etch mask

Application techniques In the electronics industry, three fabrication techniques are employed, comprising a combination of photoresist patterning and plating (see Fig. 38):
 1. Photoformation of a mask pattern over a metal or metal-coated substrate, followed by plating, e.g., bimetallic metal evaporation masks. See page 14-87.
 2. Photoformation of a positive mask pattern over a plated metal, followed by etching, e.g., conventional etched-down printed-circuit boards. See page 14-90.
 3. Photoformation of a negative mask pattern over an insulating substrate, followed by plating and dissolution of the negative mask, e.g., certain thin-film resistor patterns.

The considerations noted previously for the use of photoresists and plating processes apply also to these techniques. In the first method, a photoformed image is used to define the pattern plating of a metal for use as, for example, a mandrel in the electroforming of metal parts. Applications requiring electroplated layers above 0.0005 in. involve plating times of 20 min or more and good chemical resistance of the resist layer. The requirements are less stringent where only thin plated layers are deposited and short plating times are employed.

The second method utilizes the resist image as an etch mask in the etching of a plated metal. Applications involving the third method, plating of metals over a photoformed image, are limited to the deposition of thin-metal layers. Electroless plating is used to deposit a metal layer over the entire surface of the part, followed by dissolution of the resist layer and its plating by undercutting in a suitable resist solvent. Electroplating, if used over the electroless coating, is restricted to a thin layer to permit removal of the plated resist layer. Edge definition of the final image is not as good as that obtained with the other methods. This technique is discussed further in the processing of thin films using vacuum techniques.

Vacuum Evaporation and Sputtering

Vacuum evaporation and sputtering processes are considered in detail elsewhere in this book. Photochemical processes are often used directly and indirectly in these processes. Here, we shall consider only those evaporation and sputtering parameters that bear on photochemical processes. The main use of photochemical processes in evaporation and sputtering is for pattern generation in thin films.

Evaporation Evaporation is a film deposition process performed in a vacuum system at pressures less than 10^{-5} torr. The material to be deposited is heated until its vapor pressure reaches 10^{-2} to 10^{-1} torr. The vapor is then condensed onto the substrates, as well as all other surfaces in the vacuum system that can be reached by a direct path. The material arriving at the substrate is generally in a low energy state, typically 0.001 to 0.1 ev.

This is, of course, a very simple view of a rather complex process.[107-110] The essential features of the process as they bear on photochemical processes are:
 1. The energy of the depositing material as it arrives at the substrate is low.
 2. The background gas pressure is low so that evaporated material has a long mean free path.

Sputtering Sputtering is defined as the removal of material from a solid surface when the surface is bombarded by ions or atoms. The dominant mode by which material is removed from the surface is momentum transfer from high-energy ions or atoms to surface atoms. There are numerous second-order phenomena as well. In sputtering, the energy required to remove atoms from the source is not supplied thermally, but usually by energetic ion bombardment from a plasma. Several reviews of the basic physical phenomena and deposition techniques exist.[108,111-114]

The essential features of sputtering as they relate to photochemical processes are:
 1. The film material is in a very highly energetic state as it arrives at the substrate (5 to 50 ev).
 2. The background gas pressure is high, resulting in an extensive amount of scattering by collisions between sputtered material and background gas.

Delineation of patterns in evaporated and sputtered films There are, in essence, three ways used to form patterns in thin films (see Fig. 39):
1. Deposition through a mechanical mask.
2. Photoformation of a film (usually photoresist) on a substrate, followed by deposition of the film of interest. The photoformed film is then removed, along with the deposited film lying on it.
3. Deposition of the desired film, followed by deposition and photoformation of a suitable etch resist (usually photoresist). The desired film material is then etched through the resist pattern, and the resist is subsequently removed.

Each of these techniques has its uses and limitations, as described below.

Mechanical Masks. In general, mechanical masking offers the poorest definition, but has two advantages that weigh heavily in its favor: (1) No chemical reagents need be used on the film and (2) sequential masking operations can be accomplished in one pump-down if a suitable mask changer is used.

The ultimate cause of poor definition when mechanical masks are used is the fact that truly intimate contact between the mask and the substrate cannot be achieved in a vacuum system. Gray and Weimer[115] have described the effects introduced by this poor contact. Their investigations were performed using a wire-grill mask, but the same effects occur when photoengraved (chemically milled) masks are used. Figure 40 illustrates the effect of scattering by the mask itself. Material may be scattered by the mask into both the exposed regions of the substrate and the shadow of the mask. In the shadow, material is deposited where none is desired. In the exposed areas, the addition of scattered material produces films that differ physically from those formed by direct deposit.

The magnitude of the scattering depends on the material deposited, the type of mask, the deposition rate, the pressure, and the temperature of the mask and substrate. Scattering of most materials increases with:
1. Substrate temperature
2. Background gas pressure
3. Slower deposition rates

In addition to scattering, surface migration of metallic films, a well-known phenomenon,[116,117] affects the dimensions of conducting patterns. Surface migration is enhanced by the same parameters that cause increased scattering.[116-119] The lack

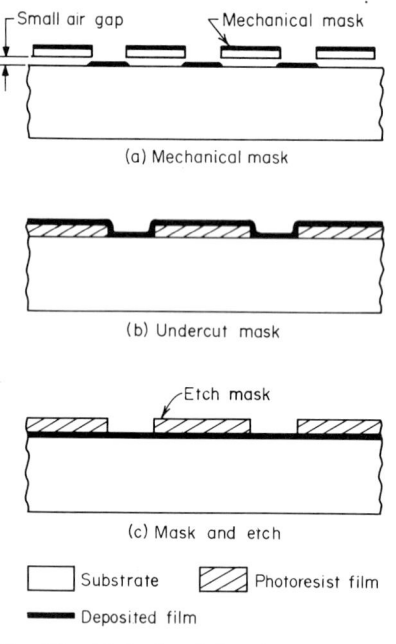

Fig. 39 Methods of forming patterns in thin films.

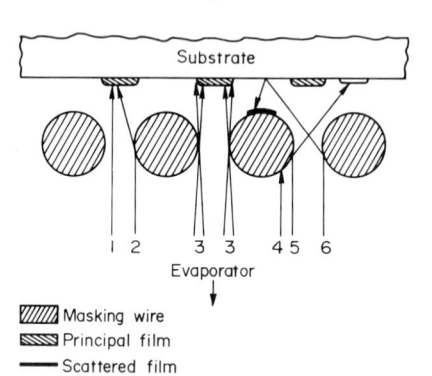

Fig. 40 Cross section of a target with a wire-masking grill during evaporation.[115]

of intimate contact between mask and substrate allows free migration in all directions.

Because of the higher pressures and higher-energy incoming atoms used in sputtering, as compared to evaporation, the definition of sputtered films deposited through mechanical masks is very poor.

Undercut Masking Techniques. The use of photoresists to undercut deposited films (Fig. 39b) eliminates the scattering and surface-migration problems because the masking film is intimately bonded to the substrate. When the photoresist is washed out, the edges of the film near the resist are literally torn, often leaving a fairly ragged edge with chips of film material sticking up from the substrate. Despite this, pattern definition is substantially better than with mechanical masking.

This technique is often used for film materials that are difficult to etch chemically without destroying the photoresist patterns. Undercut techniques using photoresists cannot be used with sputter deposition. The energy of arrival of sputtered material is high enough so that material can penetrate the photoresist and render it insoluble in the usual photoresist strippers.

Materials other than photoresists can be employed as the layer to be undercut.[120] In this case, a resist pattern is first defined by photoresist and etch techniques. Then the photoresist is removed, leaving the secondary resist. The desired film is deposited, and the pattern is then undercut. This technique is usually employed when the desired film is soluble in photoresist strippers. If the undercut layer is hard and dense (i.e., does not allow deep penetration), this technique can be used with sputter deposition.

Photomask and Etch Techniques. The most common technique for pattern delineation in thin films is chemical etching through a photoresist mask (Fig. 39c). The definition obtained depends on the quality of the resist, the adhesion of the resist to the film surface, the nature of the material to be etched, and the etchant used. With materials that are difficult to etch, strong chemicals are needed. This often results in a degradation of the resist adhesion, leading to a large amount of undercutting (Fig. 32a). This loss of adhesion can sometimes be reduced by surface treatment[121] or by applying a conversion coating[88] (Fig. 32b). There is always some undercutting since chemical etching proceeds in all directions.

Radio-frequency sputter etching,[122–126] which utilizes sputtering rather than chemical attack, as an etchant gives substantially improved pattern definition. In this process, the photoresist is removed at the same time as the material to be etched. In theory, any solid material can be sputter-etched with definition limited only by the quality of the photoresist pattern[123–126] and the thickness of the original photoresist layer. Figure 41 illustrates the quality of patterns produced by this technique.

APPLICATIONS

The processing sequence involved in any specific application of photofabrication is the same as that presented in Fig. 1. The individual processing steps will vary according to the part design, the material, and the subsequent processing required. A brief description of the types of applications found in the electronics industry was given on page 14-7. A number of applications are listed in Table 2.

A general discussion was presented on page 14-56 regarding the processes involved in part fabrication that can be used as a guide in the selection of a suitable process for a given application. In this section, applications of photofabrication are examined, and important parameters that must be specified by the engineer or fabricator for the design and processing of parts are defined. Emphasis will be placed on those aspects of photofabrication related to part design and to the quality of the finished product. Examples of typical processes for each type of application will be given.

Chemically Machined Parts

Chemical machining is used extensively for the production of flat metal parts for a variety of electronic applications. This method is also referred to as chemical blanking, functional chemical machining, and photochemical machining. Many ad-

14-82 Photofabrication

vantages are inherent in this method over conventional machining methods, resulting in both improved quality and lower cost of the parts. These advantages include: (1) the ability to fabricate practically any metal or alloy with no alteration of material properties, (2) an absence of burrs on the finished part, (3) freedom of part design, (4) good control of tolerance, and (5) minimum tooling costs and time.

Design considerations In designing parts to be processed by chemical machining, the tolerances associated with the etching process must be considered. These

(a)

(b)

Fig. 41 Typical sputter etched patterns. (a) Photomicrograph of tungsten sputter-etched from a silicon surface. (Dark lines are Si, 10 μm wide.) (b) 3° angle lapped section of a sputter-etched line. The interference fringes show that the walls are vertical. Angle lapping includes the effects of photoresist definition and etch profile.

will determine the corrections necessary in artwork preparation and the quality of the parts obtainable.

The etch factor or amount of undercutting will determine the artwork compensation required. The majority of production applications utilize double-sided etching to reduce the etching time and to obtain better control of part dimensions and shape. Typical processing steps are illustrated in Fig. 42. Etch factors are calculated for one-half the thickness for the material when two-sided etching is used. This factor is also dependent upon the type and thickness of material etched, the etchant, the photoresist, and the processing equipment. Since these will vary with each facility, the etch factor should be determined under controlled process conditions (etchant concentration and temperature and etching time), using test samples for a given application. Once determined, the factor will remain constant with good control of the etching process and can be used in part design. Typical etch factors obtained with various metals in spray etching are included in Table 31. The undercutting obtained will generally produce a slope or taper of the sidewalls of the etched pattern. For standard production facilities using spray etching, this is usually controlled to 25 to 50 percent of material thickness and is minimized by overetching the parts. This effect will vary with the pattern and with solution agitation. Increase in spray pressure will tend to minimize undercutting.

Variation in pattern design will also affect etch uniformity. Sharp inside corners cannot be etched out, owing to poor accessibility to the etchant. High-resolution

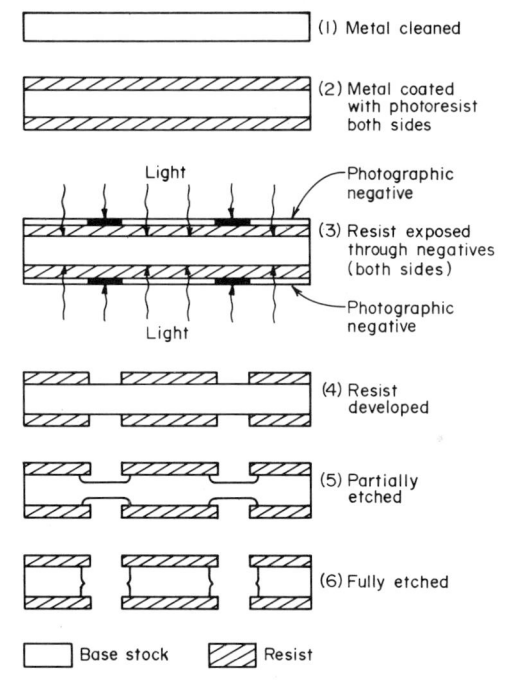

Fig. 42 Double-sided etching-processing sequence.

patterns will also exhibit a slower etch rate due to a similar effect and the resultant buildup of reaction-product concentration at the surface. Thus patterns of varying high and low resolution will be limited by the finer pattern. Etching times, undercutting, and artwork compensation must be determined accordingly. Sharp outside corners exhibit an opposite effect to inside corners as the etchant will have easy access to the edge surfaces. Generally, a radius of one-half to one times the thickness is obtained on inside corners, and one-fourth to one-third times the thickness on outside corners.

Because of undercutting during etching, there is a minimum dimension that can be obtained for holes and slots in the etched part while maintaining good sidewall slope. Benton[127] lists these limitations with spray etching as 0.7t for copper alloys, 1.0t for steel alloys and 1.4t for aluminum alloys and stainless steel, where t is the thickness of the metal stock. For good part design and reproducibility of a projecting piece, a minimum dimension equal to twice the metal thickness should

14-84   Photofabrication

be used when possible. Separation of holes and slots should be limited to one-half the metal thickness for metals up to 0.010 in. Thicker samples will require greater separations. The above figures should be used as a general guide, since the ability to etch fine patterns will vary considerably with the type of material and the etching facility. Metals showing high etch factors will enable finer patterns to be obtained and should be selected for applications requiring highest resolution. Thus, copper-base alloys would be preferred over nickel, and both preferred over aluminum for such parts.

In general, the surface tolerances attainable will be dependent upon the type and thickness of the material to be processed. Minimum tolerances are obtained with thin materials, owing to the short etching times and small amount of undercutting involved. Tolerances will increase as the thickness increases because of an increase in the amount of undercutting and a greater variation in etch uniformity down through the metal. As a general guide, a tolerance of ±20 percent of metal thickness can be expected for a production process for materials varying in thickness from 0.0005 to 0.030 in. Tighter tolerances can be obtained with greater process control, or in a limited area of the sample provided tolerances of other areas can be relaxed. Tolerances of metals of greater thickness will be proportionally greater. Typical tolerances for the commonly used metals are listed in Table 36.

TABLE 36   Representative Thickness-Tolerance Comparison of Etched Metals[128-130]

Metals	0.0005 in.	0.001 in.	0.002 in.	0.005 in.	0.010 in.	0.020 in.	0.030 in.
Aluminum alloys	±0.0005	±0.0005	±0.0008	±0.0015	±0.0025	±0.005	±0.008
Copper and copper alloys	±0.0002	±0.0005	±0.0005	±0.001	±0.0025	±0.004	±0.006
Invar	±0.0004	±0.0005	±0.0005	±0.001	±0.0025	±0.005	
Magnetic alloys (Ni-Fe)		±0.0005	±0.001	±0.002	±0.0025	±0.005	
Molybdenum, titanium		±0.0005	±0.001	±0.002			
Nickel		±0.001	±0.001	±0.0015	±0.0025	±0.005	±0.008
Steel alloys	±0.0004	±0.0005	±0.0005	±0.001	±0.002	±0.004	±0.006
Stainless steel	±0.0004	±0.0005	±0.0005	±0.0015	±0.0025	±0.005	±0.008

Processing.   The general guidelines presented in the above discussions are based upon the proper choice of artwork and materials and good process control. Engineering drawings should include registration holes or marks and tabs where required for processing or for the etched part. Compensation must be made for undercutting as established by the process guidelines. The artwork generation method is selected according to the tolerances required by the part. In most applications for electronic parts, artwork is initially prepared oversize to take advantage of the greater accuracies obtainable. Process artwork is then prepared by photoreduction, step and repeating, and contact printing as required.

The steps involved in chemical blanking employing a negative resist material and two-sided etching are shown in Fig. 42. Similar steps are involved with positive resists except that opposite polarity artwork is required. Process artwork consists of mirror images of the pattern to be etched with the emulsion side held against the surface of the stock. Registration is accomplished with indexing pins, taping, or various fixtures. Matching indexing holes in the artwork and the stock offer the best means of obtaining precise registration from front to back and are preferred for production runs. Pin bars and exposure jigs can also be employed to register artwork. These should be set up and the initial registration made with a spacer of the same thickness as the required stock in between to compensate for stock thickness during processing. Taping methods are suitable for sample fabrication and where registration tolerances are not critical. Artwork is aligned on a light table and taped along one edge to a spacer of equivalent stock thickness.

Negative and positive resists and various dichromated emulsions are presently used for chemical machining. Factors affecting choice of a resist were reviewed previously under etching processes. Resists are normally applied by spray or dipping methods with double-sided work. Films are then air-dried and baked for the required schedule. After resist application, the stock is positioned in between the registered artwork in a double-sided exposure frame, and the resist is exposed to a suitable light source. The exposed resist image is developed and the stock transferred to the etching station. A baking cycle is often required after development, particularly for the etching of thick stock, to impart greater chemical inertness. Spray etching is almost universally used for chemical blanking. The etch cycle in most production facilities consists of continuous travel through a spray chamber and directly into rinsing, resist removal, and drying stations. In some applications the resist is allowed to remain on the finished parts to act as a protective coating.

Part Handling. Three methods are used to handle parts during and after etching and are normally referred to as the drop-out, tabbing, and back-coating methods. The drop-out method involves etching out complete parts with the parts retained on a screen holder in the etching unit. Parts can be used directly after removal of resist. Handling becomes a problem with small parts after etching, and this method is seldom used.

The second method utilizes an interconnecting gridwork of tabs to hold the parts together in a strip or sheet. This provides easy handling of small parts during etching and can be designed to facilitate part assembly in the final application. For example, magnetic-recording head laminations are etched out into an array or fret, and the arrays are stacked with a suitable insulating resin and laminated to form the recording head. Lead frames for integrated-circuit packages are also designed in an array to provide simple handling during the package fabrication step. The major disadvantage with the tabbing method is that additional equipment and processing steps are required for cutting and finishing of individual parts.

Back coating involves partial etching of the parts, removing from the etcher, drying, and coating of one side with a continuous resin layer by spraying or by using a pressure-sensitive adhesive film. The coated sheets are then returned to the etcher, and the etch is continued from the other side until completed. Parts are allowed to remain on the backing until ready for use. The backing is readily dissolved away in organic solvents. This method is good for handling large quantities of small parts.

Each of the above methods has its advantages, and the choice among them will depend on the part size, the number of parts being etched, and the post-processing steps involved in the final application. Some typical examples of chemically machined parts are presented in the following section.

Processing of microcircuit lead frames Kovar is used for the fabrication of leads in many microcircuit packages because of its close thermal expansion match to glass and ceramics. Special etching grades having uniform grain size and structure are produced for chemical milling applications. These give parts having good edge definition and minimum undercutting. A typical processing schedule for the photofabrication of Kovar lead frames is outlined as follows:[93]

1. Vapor-degrease Kovar sheet in trichloroethylene; immerse in 50 percent sodium hydroxide at 180°F for 1 min followed by 5 percent nitric acid dip at room temperature. Force-air-dry with filtered air.

2. Apply photoresist by dip or roller coating to obtain a uniform thickness of approximately 200 μin.

3. Dry and bake-resist layer, e.g., 15 to 30 min air dry plus 10 to 30 min at 150 to 250°F (depending on type of resist).

4. Position coated stock in between photomasks in exposure frame, and apply vacuum.

5. Expose for required time. Exposure time is dependent upon type of resist and thickness and must be predetermined by test samples.

6. Develop resist pattern, and rinse.

7. Bake, e.g., 1 to 3 min to 300 to 350°F.

14-86 Photofabrication

8. Etch in ferric chloride or chromic acid solutions according to predetermined time schedule. A typical etch rate for 38° Baumé ferric chloride at 120°F is 0.0008 in./min (etch factor 2.0).

9. Strip resist, and clean parts to remove etch residues in commercial desmutting solution.

10. Inspect parts.

Figure 43 shows a chemical-etched lead-frame array for a dual in-line integrated-circuit package. The etching profile obtained in the two-sided etching process is shown in Fig. 43b.

Metal-masks—etched and electroformed When masks with physical openings are required, such as for masking in vacuum evaporation, etched (chemically ma-

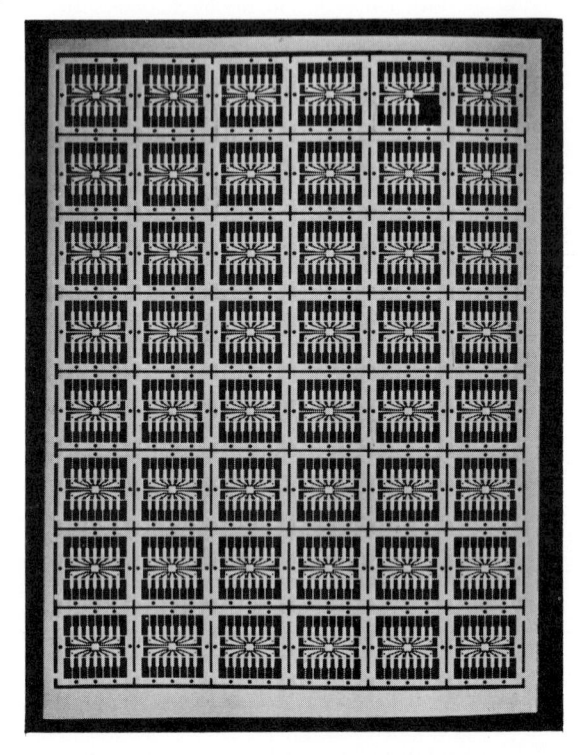

(a)

(b)

Fig. 43 Etched microcircuit-package lead-frame array and cross-sectional profile.

chined) or electroformed metal masks may be used. Electroformed masks can be made with more accurate openings than etched masks but require more steps in their fabrication. Fotoform glass can also be used to make masks of this type.

Depending on the resolution and physical properties required, many metals may be used for etched masks. The metals most commonly employed are beryllium-copper, copper, molybdenum, stainless steel and low-thermal-expansion nickel-iron alloys. Table 37 compares the salient properties of these metals with those of some other metals which are frequently chemically milled. Beryllium-copper is outstanding because it combines good strength with good etchability and its fine grain structure results in smoother edges on etched holes. Copper is economical and etches well but has modest strength. Molybdenum has high strength at high temperature and thus can be used where temperatures will rise considerably as, for example, in evaporation masks where the substrate is heated during evaporation. Of the metals listed, molybdenum is the most difficult to etch. The low-thermal-expansion nickel-iron alloys are available in specially purified form for good etchability. Although these sheet materials are intended for glass-to-metal seal applications, such as integrated-circuit lead frames, their properties commend them for making evaporation masks. The low thermal expansion helps avoid the problem that as evaporation proceeds, the mask expands more than its mounting frame and hence

TABLE 37 Properties of Metals Commonly Used for Etched Masks

Metals	Tensile strength, psi $\times 10^{-3}$	Yield strength at 0.2% offset, psi $\times 10^{-3}$	Thermal expansion at 25–500°C, in./(in.)(°C) $\times 10^{-6}$	Melting point, °C
Copper.	32	10	18.3	1083
Beryllium-copper.	60–120*	28–112*	17.8–300°C	
Nickel.	55	15	15.2	1440
Stainless steel (304).	105	45 at 0.5% offset	18.3	1475
Molybdenum.	70	56	5.7	2622
Tungsten.	260	. . . .	4.6	3410
Iron-cobalt-nickel alloys.	90	50	6.2	1450

* Properties dependent upon degree of temper of available grades.

sagging away from the substrate it is intended to mask. However, unless a low-expansion frame is used, the eventual heating of the frame may stretch the mask beyond its yield point, making it slack when it cools.

The technique for making etched masks is similar to the process outlined in Fig. 42 for two-sided etching. Figure 44 compares the cross-section contours of the holes produced by etching from one side, etching from both sides, and electroforming. When masks are etched from both sides, the photomasks for exposing the photoresist must be registered as previously discussed.

Electroformed masks are often called bimetal or trimetal masks. These masks have a core of one metal with a patterned electroformed surface layer, or layers, of a different metal. Since, as the flow chart of Fig. 45 shows, the core metal must be etched away without attacking the electroformed skin, the choice of materials is limited. The electroformed metal is almost always nickel-plated from a sulfamate bath. Few, if any, other metals or plating baths produce a deposit with low enough stress to avoid warping the mask. Copper and copper alloys such as beryllium-copper are almost always used as the substrate for the nickel plating. Complexing etchants rich in ammonia dissolve the copper alloys about 1,000 times faster than they do nickel. The trimetal mask with nickel on both sides of the center foil remains flat with temperature changes because the stresses due to differential thermal expansion are balanced. In addition, the premasking

14-88   Photofabrication

of the lower nickel seems to improve the edge resolution of the evaporated deposit by blocking material arriving at glancing angles to the surface of the substrate. Typical processing details for bimetallic masks processed with a positive-acting photoresist are presented in the following outline.

1. Prepare artwork at 10 to 20×, using a coordinatograph. Patterns must be made oversize to compensate for electroplated nickel thickness. (If negative resist is used, patterns are made smaller by a similar amount.) Photoreduce to required size on high-resolution photographic film or plate.

2. Clean scratch-free beryllium-copper sheet by vapor degreasing in trichloroethylene, followed by a 3- to 5-min soak in a neutral copper cleaner (e.g., Shipley Neutra-Clean 68). Rinse in cold water, and force-dry with filtered air. The metal sheet should be oversize by ½ to 1 in. on each side of the desired final mask size.

3. Coat with filtered Shipley photoresist in a laminar-flow hood by dipping or spraying to obtain uniform thin coating.

4. Air-dry in laminar-flow hood for 2 min.

5. Position photomask with emulsion side against photoresist coating in vacuum frame, apply vacuum, and expose for required time (determined by test samples).

6. Develop in Shipley developer; rinse in cold water.

7. Dry with filtered air, and inspect.

8. Coat backside with plating stop off.

9. Soak clean in Shipley Neutra-Clean 68 for 1 to 3 min; rinse in cold water followed by distilled water. Transfer directly to plating bath.

10. Immerse in nickel-plating tank with current in ON position. Electroplate at required current density to produce desired nickel thickness. (A nickel thickness of 0.5 to 1 mil is typical.) Rinse with water, and dry with filtered air.

11. Strip photoresist layer, and inspect.

12. Etch from the front side in a selective etchant for beryllium-copper.

13. Rinse in weak acetic acid followed by water.

14. Inspect for completion of etch.

15. Carefully remove plating stop off from back of mask.

16. Soak clean in Shipley Neutra-Clean 68 for 1 to 2 min, water-rinse, and dry carefully with filtered air.

17. Final-inspect and assemble in mask frame.

Etched masks for stencil printing applications are photofabricated with a fine grid structure on one side, prepared by electroforming or etching, and photofabricated with the desired printing pattern on the other side prepared by etching.[131] The fine grid structure extending partially through the thickness of the completed stencil adds strength to the stencil when large open areas are present on the pattern side. Masks of this type are capable of producing patterns of much finer definition and resolution than woven wire screens and offer better control of ink transfer.

Color television shadow masks   The shadow mask for color television cathode-ray tubes represents the largest-volume production application of a chemically machined part in the electronics industry today. It is an essential component of the tube and is utilized in both its processing and its operation. In principle, the mask functions to separate the three primary colors on the tube screen by permitting a particular color dot to be exposed only to the electron beam of its corresponding electron gun (the remaining two beams being shadowed from it). Details of the tube operation are reviewed elsewhere. A summary of the processing of the shadow mask is presented below.[132]

A typical shadow mask is constructed from 0.006-in.-thick cold-rolled steel sheet in which is etched from both sides approximately 400,000 holes varying in size from 0.010 to 0.012 in. and separated by about 0.015 in. In the manufacturing process steel coils are processed in a continuous manner. The sheet is fed into a machine which cleans the surface with caustic and acid solutions to remove all soils and is dried. Both sides are then coated with a thin layer of a dichromate-sensitized fish-glue resist; the resist is dried and the stock recoiled. The strips are next fed through a manually indexed exposure machine containing glass photomasks of the dot patterns held in vacuum frames, aligned, and exposed to high-intensity arc lamps. In the next machine, unexposed resist is removed by water spray, and the strip is continued through an oven to bake the remaining exposed resist image. The final process consists in etching the holes with ferric chloride spray, removing the hardened resist in caustic solution, rinsing, and drying.

The set of photomasks used in processing shadow masks is very critical. Hole diameters are different on each side in order to produce a tapered hole.[132] The hole diameter and the degree of tapering are also varied from the center of the

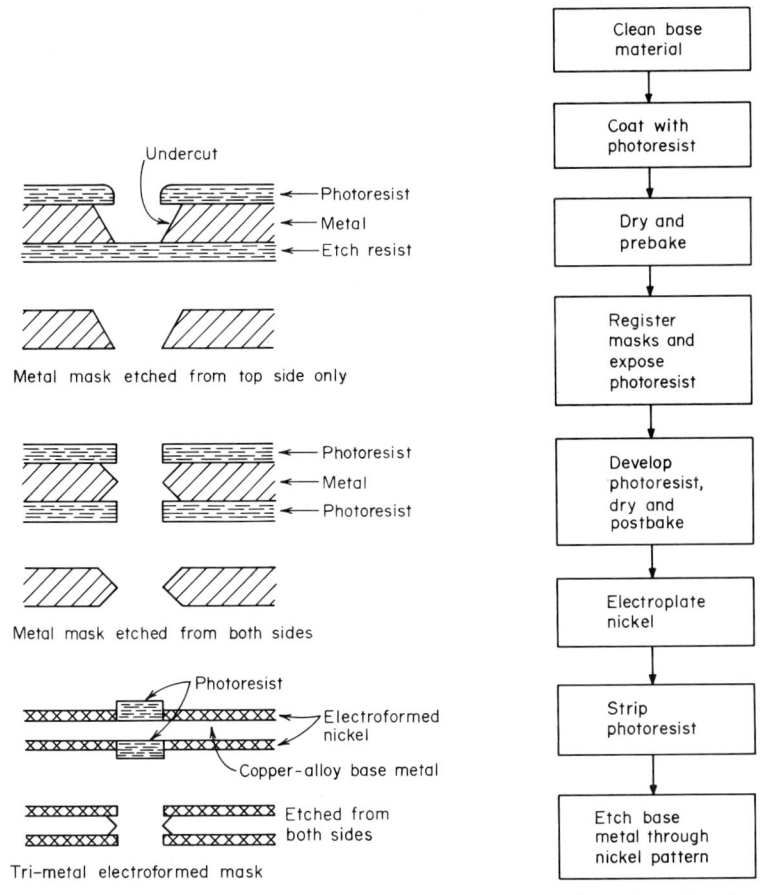

Fig. 44 Cross sections of masks produced by various methods.

Fig. 45 Flow chart for electroformed mask.

mask out to the periphery to produce a more uniform intensity of the transmitted electron beam. Improvements in mask quality in recent years have largely been due to improved photomask generation techniques.

After etching, the mask is formed into the required contour, blackened, welded to a rigid frame, and thoroughly inspected. Each shadow mask is then mated to a glass panel and used as the photomask for exposure of the three primary-color phosphor-dot patterns. The three dichromate-sensitized phosphor layers are separately applied and exposed in a sequential manner through the shadow mask, using a point light source positioned to expose only the one phosphor-dot array (cor-

Photofabrication

responding to the appropriate electron gun in the finished tube). The shadow mask thus serves an important function in the field of color television.

Printed Circuitry

The term printed circuitry generally refers to a pattern of a conducting material supported on a nonconducting substrate. Although this general term can be applied to a variety of metals and nonconductors, it is normally restricted to the use of copper conductors on plastic substrates. Other metals such as nickel and Kovar are used in applications requiring welded components, but they represent a very small fraction of the current market. The photofabrication of printed circuits utilizes the same general processes involved in the chemical machining and electroforming of parts, that is, etching and plating. Printed circuitry differs significantly, however, in that a relatively small number of metals are used, at thicknesses of less than 0.003 in. and always supported on a nonconducting substrate. A general summary of the various printed-circuit types is presented in the following sections, along with the photofabrication techniques used in their processing. The discussion will be restricted to printed circuits formed by etching or plating after application of a photoresist pattern. Other methods of generating resist patterns, such as screen printing and offset printing, will be discussed in a later section.

Types of printed circuits Printed circuits can be classified[133] into rigid and flexible types composed of single-, double-, or multiple-conductor layers, each conductor layer being separated from its neighbors by an insulating material. Rigid circuits are composed of relatively nonflexible insulating substrates, whereas flexible circuits utilize thin, free-flexing insulating layers. These can be subclassified into subtractive (or etched-foil circuits) and additive (or plated circuits). Rigid printed circuits using subtractive processes make up the bulk of circuits presently being manufactured. The processes used in circuit fabrication are the same for both rigid and flexible types; hence, a general discussion of processing will apply to each. Only base materials will differ, although these must still be considered in the selection of the overall process.

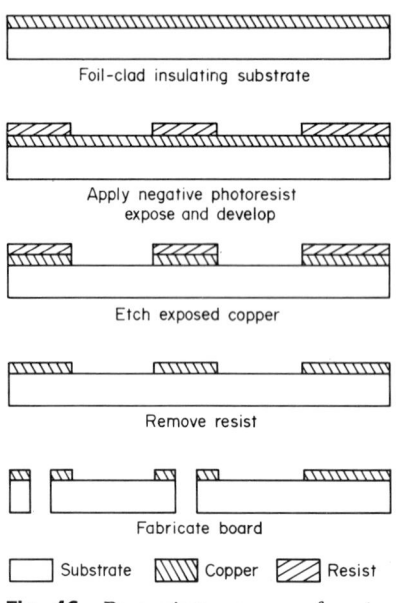

Fig. 46 Processing sequence for single-layer printed circuit, using a negative-photoresist etch mask.

Subtractive Circuits. Two methods are in common use for the fabrication of subtractive or etched-foil circuits: (1) etching of the metal foil, using a photoresist image as the etch mask, and (2) pattern plating of an etch-resistant metal, using a photoresist image as a plating resist followed by removal of the photoresist and etching of the exposed foil. Solder or gold are generally used as the plated etch resists. These methods are illustrated in Figs. 46 and 47. Method 1 is normally used for single-sided circuitry and method 2 for double and multilayer circuits using plated-through-hole techniques. The plated-through-hole process requires a plating step to build up sufficient metal for the through-hole interconnection prior to resist application.

Additive Circuits. The methods used for the fabrication of additive printed circuits using plating processes are illustrated in Figs. 48 and 49. These include: (1) pattern plating of the circuit directly onto an insulating substrate (or on an insulating substrate covered with a thin conducting metal film), using a photoresist image as a plating resist, and (2) pattern plating of the circuit on a temporary support layer, using a photoresist image as a plating resist followed by transfer of the circuit to a permanent insulating layer. The first method can be employed to process single- and double-layer circuits over any suitable insulating layer using electroless and electroplating techniques. Method 2 is generally used

(1) Drill and de-burr holes

(2) Apply electroless copper plate

(3) Apply electrodeposited copper plate

(4) Apply negative photoresist, expose and develop

(5) Plate with dissimilar metal (e.g., gold, lead-tin solder, etc.)

(6) Strip photoresist and etch copper

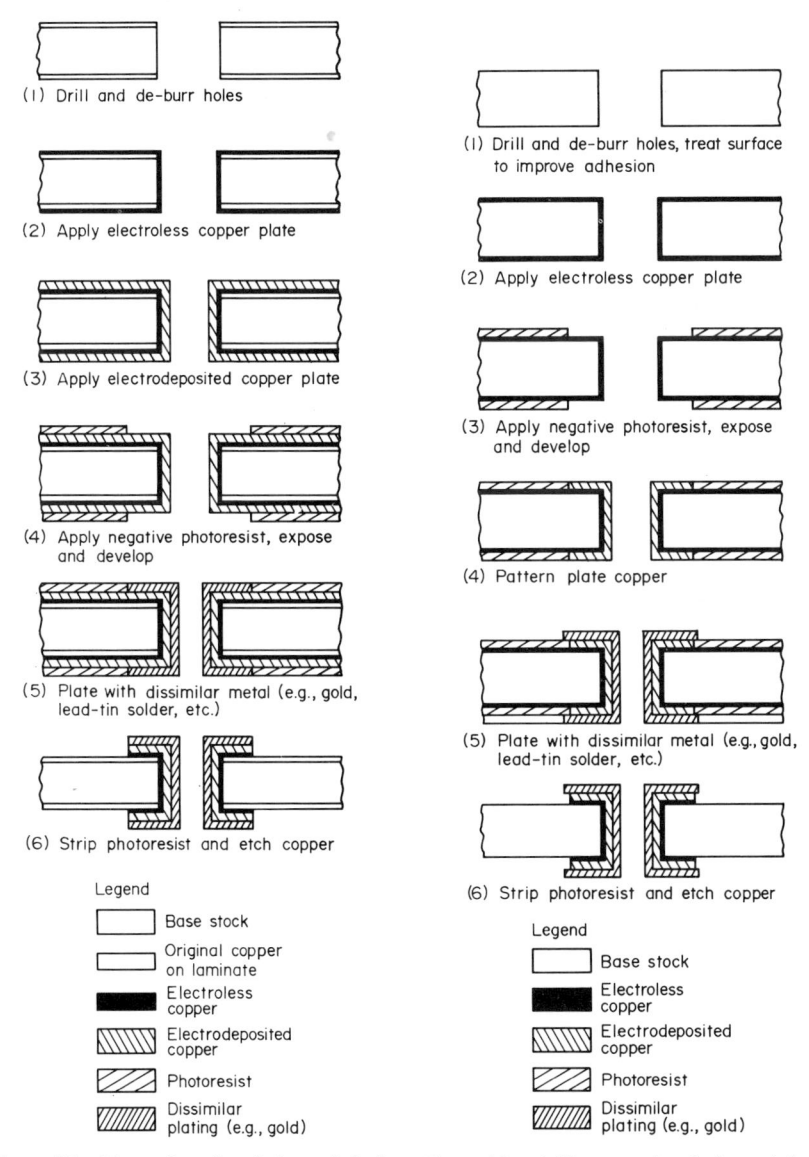

Legend
- Base stock
- Original copper on laminate
- Electroless copper
- Electrodeposited copper
- Photoresist
- Dissimilar plating (e.g., gold)

Fig. 47 Typical plated-through-hole processing sequence, using a negative photoresist and a plated-metal etch resist.

(1) Drill and de-burr holes, treat surface to improve adhesion

(2) Apply electroless copper plate

(3) Apply negative photoresist, expose and develop

(4) Pattern plate copper

(5) Plate with dissimilar metal (e.g., gold, lead-tin solder, etc.)

(6) Strip photoresist and etch copper

Legend
- Base stock
- Electroless copper
- Electrodeposited copper
- Photoresist
- Dissimilar plating (e.g., gold)

Fig. 48 Additive plated-through-hole processing sequence pattern plating, using a negative-photoresist plating mask.

14-92 Photofabrication

in applications requiring metal circuits flush with the surface such as with printed rotary switches.[91]

Design considerations for photofabrication of printed circuits The design of printed circuits is determined by the electrical requirements of the application, the packaging technique used, and reliability level needed. Printed circuits have been placed by the industry into four classes, depending on complexity and tolerances involved. These include:

    Class 1. Generally indicates maximum tolerances, low-cost tooling and/or materials.

    Class 2.
    Class 3. } Indicates progressive tightening of tolerances and upgrading of tooling and/or materials and increase in costs.
    Class 4.

Table 38 lists conductor-width and space-design dimensions and representative processing tolerances for the various classes. Information relating to electrical design and layouts of printed circuitry and the packaging concepts involved with their use can be found in the literature.[133,134] General design guidelines are summarized in Table 39.

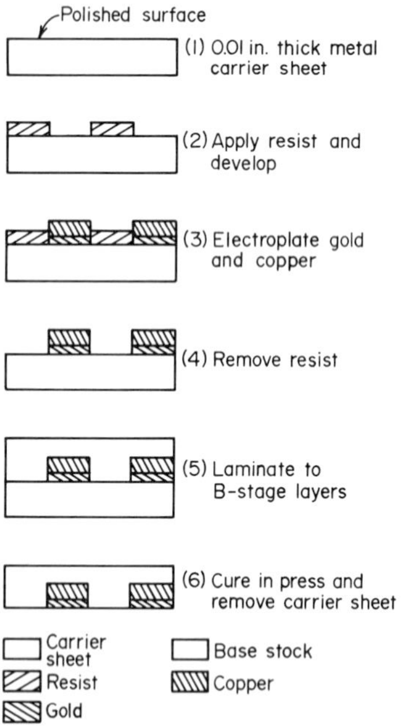

In the processing of printed circuits using photofabrication techniques, any deleterious effects of the processes on the conductor pattern and on the board material must be considered. With regard to the conductor pattern, changes occurring owing to undercutting or plating must be accounted for in layout of artwork, as noted previously. Typical etch factors obtained for copper in spray etching range from 2 to 3, as shown in Table 31. Undercutting is increased considerably with the use of plated-metal etch resists, however, because of the galvanic action set up by the copper-plated metal couple. The magnitude of this effect will increase with the increase in potential difference of the couple; thus a greater undercutting is obtained with the use of plated-gold etch resist as opposed to a solder etch resist. The artwork compensation required in the etching of copper foil of various thicknesses with different resists is presented in Table 40. This factor will have a decided effect on the design of fine-line circuitry and on the selection of processing materials. For example, circuits of the Class 4 type would be impossible to produce using standard etching techniques with a plated-gold resist.

Artwork compensation for additive circuit fabrication should be guided by the factors discussed previously under plating. Generally, the plated pattern will increase in dimension on each side by an amount equal to the plated-metal thickness. Thus, artwork must be prepared undersize or oversize by this amount, depending on the type of resist used (negative or positive).

Selection of artwork generation methods will be determined by the class of circuitry and the overall size. Printed circuits, for example, for home instrument applications generally fall in Classes 1 and 2, and artwork is prepared by drafting or taping methods. Artwork for a large computer multilayer "platter," however, must be prepared using an automatic coordinatograph.

Printed-circuit processing Selection of materials for printed-circuit applications has been narrowed down in the industry to a relatively small number of metals

Fig. 49 Additive printed-circuit processing sequence using the transfer method.

(1) 0.01 in. thick metal carrier sheet
(2) Apply resist and develop
(3) Electroplate gold and copper
(4) Remove resist
(5) Laminate to B-stage layers
(6) Cure in press and remove carrier sheet

Carrier sheet | Base stock
Resist | Copper
Gold

Applications   14-93

and plastics. These are summarized in Tables 41 and 42. Preference for a certain material is determined by one or more of their specific properties or cost. Generally, most of the board materials listed have been optimized for printed-circuit applications and can be used with the various photoresists and common etchants. The characteristics of these materials can be obtained elsewhere in this handbook and in the literature references.

The mechanical and chemical processes involved in the fabrication of printed circuits are outlined in Table 43. The general aspects of these processes in photofabrication were discussed previously. The important considerations for their use in printed circuitry will be reviewed in this section. Process details can be obtained in the literature references.

TABLE 38   Dimensional Data for Printed-circuit Conductors*,[134]

Conductor	Class 1	Class 2	Class 3	Class 4
Examples by classes using worst-case process reduction				

Conductor width:
  Design minimum
  Finished minimum†
	0.031
	
	0.021
		0.015
		
		0.010
			0.010
			
			0.006
				0.005
				
				0.003

Expected process tolerances‡				

Conductor width:
  No plating
 	
  With plating
 	
Conductor spacing:
  No plating
 	
  With plating
	+0.006
	−0.010
	+0.015
	−0.010
	
	+0.010
	−0.006
	+0.010
	−0.015
		+0.004
		−0.005
		+0.008
		−0.005
		
		+0.005
		−0.004
		+0.005
		−0.008
			+0.002
			−0.004
			+0.004
			−0.004
			
			+0.004
			−0.002
			+0.004
			−0.004
				+0.001
				−0.002
				+0.002
				−0.002
				
				+0.002
				−0.001
				+0.002
				−0.002

* Final product drawings and/or specifications should call out only minimums for conductors and spacings. Artwork should be done on a magnified scale suitable to produce required tolerances (normally 4:1). All dimensions in inches.

† These minimums do not make allowance for nicks, pinholes, and scratches. These imperfections, are normally acceptable providing the line is not reduced by more than 20 percent.

‡ Given as a guide only. Specific process tolerances should be ascertained by your supplier.

The above line and space tolerances can be applied to plated-through-hole boards where the laminate is not over 0.0625 thickness. In cases where thicker laminates are used, and hole-diameter-to-laminate thickness ratio is less than 2:3, plating may further expand conductor widths.

Stated tolerances are based on 1 oz copper. For each 0.001 in. additional copper thickness, an additional 0.001 in. reduction per conductor side can be expected.

Method of measurement:
  A   Pocket comparator—gap gage.
  B   40× microscope or optical projector comparator.
  CAUTION: Measuring plated conductor width does not give a true reading of the remaining copper width owing to undercutting in the process.

Of the mechanical processes listed, only hole fabrication has a direct effect on the use of photoresists. Plated-through-hole applications require holes that are clean and free of burrs to produce a continuous metallization of the hole wall. Cleanliness is critical in multilayer circuitry as all contamination and board resin smears must be removed from the exposed inner copper layers within the holes. For rigid reinforced laminates the best holes are produced by drilling, and this method is used for multilayer and most double-layer applications.

The choice of photoresist and method of application will be determined by the type of circuit, the plating process, and the etchant used. Most photoresists are compatible with the various plating baths and etchants employed in printed-circuit

14-94 Photofabrication

TABLE 39 Design Guidelines for Printed Circuits[134]

1. Conductor thickness and width are determined on the basis of the current-carrying capacity required and allowable temperature rise.
   Consider:
   (*a*) Safe operating temperature of the laminate.
   (*b*) Maximum ambient temperature in operating location.
   (*c*) Spacing of parallel conductors versus adjacent free panel area.
2. Conductors shall contain no exterior corners having less than a 90° included angle.
3. Conductor lengths shall be held to a minimum between various terminal areas.
4. A distance of not less than the board thickness is preferred between the edge of conductors and the edge of the board.
5. Use single-sided circuit boards where practical.
6. When conductors are required on both sides of the boards, parts are placed one one side only, avoiding moisture traps between components and bridged conductors. Electrical interference should be avoided.
7. Conductive patterns larger than 0.5 in. shall be relieved to prevent blistering or warpage in soldering.
8. A hole within a terminal area shall be provided for each component-part lead to be mounted on the printed wiring board.
9. A distance of not less than the board thickness is preferred between the edges of component or lead holes as well as between the edges of component or lead holes and the edge of the board.
10. Use heat sinks where necessary to avoid hot spots.
11. Check compatibility of electrical, mechanical, and chemical characteristics of materials in the specific application as well as United Laboratories and Military Standards where applicable.
12. Consider space required for support and mounting of the board as well as weight, size, and location of components to prevent fracture or loosening resulting from flexing, vibration, and shock.
13. The manufacturing process must be considered in the master drawing (artwork) in that most processes will affect line widths, spacing (Table 38), and terminal area.

processing. High alkaline baths, such as alkaline cyanide gold and alkaline etch, cannot be used with the positive and other alkaline-soluble resists. Generally, thicker photoresist layers and prebaking are required for plating applications. Table 44 illustrates the use of resists with various types of printed circuits. The majority of plated-through-hole circuits are processed with a negative photoresist and a plated, etch-resistant metal (Fig. 47). This is adequate for circuitry up to Class 3. Circuits of Class 4 type, however, require greater control of the etching process to avoid undercutting problems. Negative resists are seldom used for direct etching

TABLE 40 Required Artwork Dimensions for Etched Copper Foil

Resist system	½ oz (0.0007 in.)	1 oz (0.0014 in.)	2 oz (0.0028 in.)
Negative artwork:			
Photoresist:			
Lines and pads............	X in.*	$X + 0.0005$ in.	$X + 0.0015$ in.
Spaces...................	X in.	$X - 0.0005$ in.	$X - 0.0015$ in.
Positive artwork (*pattern plate*):			
Solder plate:			
Lines and pads............	$X + 0.00025$ in.	$X + 0.001$ in.	$X + 0.0025$ in.
Spaces...................	$X - 0.0025$ in.	$X - 0.001$ in.	$X - 0.0025$ in.
Gold plate:			
Lines and pads............	$X + 0.002$ in.	$X + 0.0035$ in.	$X + 0.0065$ in.
Spaces...................	$X - 0.0002$ in.	$X - 0.0035$ in.	$X - 0.0065$ in.

* X in. = desired etched dimension.

TABLE 41 Properties of Various Plastics Used in Printed Circuitry[87,135]

Base resin	Type of circuitry	Maximum* service temperatures, °C	Dielectric constant	Effect of processing chemicals under average conditions — Etchants	Effect of processing chemicals under average conditions — Resist, solvents and strippers	General chemical properties
Thermosetting:						
Allylic	Rigid	230	3.4–5	None	None	Attacked by oxidizing acids
Epoxy	Rigid and flexible	125–250	4.6–5.2	None	None	Dissolves in hot concentrated sulfuric
Fluorocarbon	Flexible	200–250	2.2–2.5	None	None	Decomposes in strong acids
Melamine	Rigid	100–200	7.5–7.8	None	None	Decomposed by strong oxidizing acids and strong alkalies
Phenolic	Rigid	120–260	4.6–5.5	None to 30 minutes	Attacked by long exposure to chlorinated solvents	
Polyester	Rigid and flexible	150	2–3.5 4.3 (reinforced)	None	None	Attacked by strong acids and strong alkalies
Polyimide	Flexible	400	3.5	None	None	Dissolves in hot caustic
Silicone	Rigid	250–315	4.2	Ferric chloride or ammonium persulfate recommended	May be attacked by organic strippers	
Thermoplastic:						
Polycarbonate	Flexible	120–150	3.0	Attacked by chromic acid above room temperature	None below 60°	Soluble in chlorinated and aromatic hydrocarbons—attacked by strong acids and alkalies
Polyethylene	Flexible	110–120	2.2	None	None below 60°	
Polypropylene	Flexible	125–160	2.0	None	None	
Polysulfone	Rigid and flexible	160	2.8	Attacked by chromic acid above room temperature	Attacked by chlorinated and aromatic hydrocarbons	Attacked by strong oxidizing acids

* Varies with type of resin and filler.

Photofabrication

TABLE 42 Properties of Metals Commonly Used in Printed Circuitry[133]

Metal	Symbol	Electromotive potential, volts	Resistivity, microhms/cm²	Relative conductivity (Cu = 100)	Oz/ft² for 0.001 in.	Coefficient of thermal expansion, in./(in.)(°F) × 10⁵	Modulus of elasticity in tension, psi × 10⁻⁶	Brinell hardness
Aluminum	Al	+1.67	2.665(20°C)	65	0.22	1.33	10	15
Nickel	Ni	+0.25	6.84 (20°C)	25	0.74	0.76	30	110
Tin	Sn	+0.14	11.50(20°C)	15	0.60	1.30	6	5.2
Lead	Pb	+0.13	20.65(20°C)	7.7	0.94	1.60	2.6	3.9
Copper	Cu	−0.34	1.726(23°C)	100	0.74	0.91	16	42
Silver	Ag	−0.80	1.59 (20°C)	104	0.79*	1.05	11	95
Rhodium	Rh	−0.82	9.83 (0°C)	33	0.95*	0.43	21	37
Gold	Au	−1.68	2.19 (0°C)	70	1.47*	0.80	12	28

* Troy ounces.

of through-hole boards since complete exposure and hardening of resist within the holes is difficult to achieve. This is not a problem with the positive-working resists and successful through-hole etching methods have been developed for their use.[136] The electroplating baths used in printed circuitry are listed in Table 45. Bath composition and operational details can be obtained in the literature.

The choice of etchant is guided by the resist, the processing equipment, the etch factor, operational characteristics, and cost. Table 46 lists the ethants used in the processing of copper printed circuits and their compatibility with various resists. Spray etching is preferred for all printed-circuit applications. Optimum etchant compositions and the effects of dissolved copper on the etching rate are shown in Fig. 50. Generally, the etch factors for spray etching with photoresist

TABLE 43 Printed-circuit Fabrication Processes

Process	Function	Application
Machining: Punching, Drilling	Hole fabrication	Circuit registration, component mounting, and interconnection
Cutting	Trimming and separation of circuits	Board finishing
Image transfer: Artwork preparation	Circuit pattern generation	Process tooling
Board cleaning	Contamination removal and surface preparation	Board processing
Photoresist imaging, Screen printing, Offset printing	Resist mask or etch mask formation	
Plating: Electroless	Plating of insulator surface	Plated-through-hole interconnection. Additive circuit processing
Electroplating	Buildup of conductor layer	Plated-through-hole interconnection. Additive circuit processing. Plated etch resists. Improving contacts and surface protection
Etching	Conductor circuit formation	Subtractive circuit application
Laminating	Lamination of individual circuit layers	Multilayer circuits. Insulating and protection of circuitry

Applications 14-97

TABLE 44 Application of Resists in Printed Circuitry

Type of circuitry	Hole fabrication	Type of resist	Application method
Classes 1 and 2 (½–2 oz copper):			
Single layer......	Punched after circuit etching	Screen resist	Screening roller, spray
		Negative and positive photoresists	
Double layer.....	Drilled or punched before plating	Screen resist	Screening
		Negative photoresist with plated etch resist	Roller coat, spray, dip
		Positive photoresist with plated etch resist	Roller coat
		Positive photoresist only	Dip coat
Class 3 (½–2 oz copper):			
Single layer......	Drilled or punched after circuit etching	Negative and positive photoresists	Roller, spray
		Screen resist	High-precision screening
Double layer.....	Drilled or punched before plating	Negative photoresist with plated etch resist	Roller coat, spray, dip
		Positive photoresist with plated etch resist	Roller coat
		Positive photoresist only	Dip coat
Multilayer.......	Drilled before plating	Same as double layer	
Class 4 (½–1 oz copper)..	Use same procedures as in Class 3 above except much greater control of the resist application, plating, and etching must be maintained. Plating resists should be avoided when possible with acid etchants (see Table 40).		

masks are highest for ferric chloride and cupric chloride and are slightly lower with ammonium persulfate and the other etchants. Highest etch factors for plated-metal resists are obtained with the alkaline etch owing to a lowering of the galvanic corrosion. Operational data for the various etchants is summarized in Table 47.

The processing of printed circuits by the subtractive and additive techniques is outlined in the section beginning on page 14-98.

TABLE 45 Electroplating Baths Used in Printed-circuit Applications[87]

Metal	Function	Main bath constituents	Operating pH and temperature
Copper..........	Base conductor metal	(A) Copper pyrophosphate	pH 8.1–8.8 with ammonia 122–140°F
		(B) Acid–copper sulfate	Strong acid
		Copper sulfate–sulfuric acid	68–122°F
		(C) Acid–copper fluoborate	pH 0.2–1.7
		Copper fluoborate–fluoboric acid–boric acid	70–120°F
Gold............	Plating resist electric contacts	(A) Acid gold (organic acid and salts)	pH 3.5–4.5 80–120°F
		(B) Alkaline bright cyanide gold	pH 12
		Gold–potassium cyanide–potassium carbonate	75–100°F
Solder (tin-lead)..	Plating resist	Stannous fluoborate–lead fluoborate	pH 0.5 or less
		Fluoboric acid–boric acid	60–100°F
Tin-nickel.......	Plating resist	(A) Stannous chloride–nickel chloride	pH 2–2.5
		Ammonium bifluoride	155 ± 5°F
Nickel..........	Undercoat for gold- and rhodium-plated contacts	Nickel sulfamate	pH 3.5–4.2
		Nickel chloride–boric acid	75–100°F
Rhodium........	Electric contacts	(A) Rhodium sulfate–sulfuric acid	40–45°C
		(B) Rhodium phosphate–phosphoric acid	
Tin.............	Solderability Electric contact	(A) Stannous fluoborate	pH 0.2 or less
		Fluoboric acid–boric acid	70–120°F
		(B) Stannous fluoborate–nickel fluoborate	pH 3.7–4.0 150°F
		Ammonium bifluoride	

TABLE 46 Etchants for Copper and Their Resist Compatibility[97]

Etchant	Screen resist	Photo-resist	Solder plate	Tin plate	Gold and nickel-gold nickel-rhodium
Ferric chloride............	X	X			X
Ammonium persulfate......	X	X	X	X	X
Chromic acid.............	X	X	X	X	X
Cupric chloride*...........	X	X		X	X
Alkaline etch.............	X†	X†	X	X	X

* Regenerated.
† May not be used with alkaline-soluble resists.

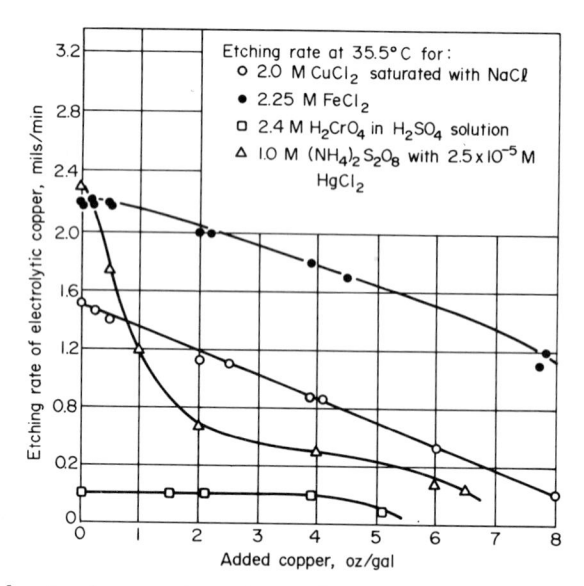

Fig. 50 Etch rate of copper in various etchants as a function of dissolved copper content.[91]

Etching rate at 35.5°C for:
○ 2.0 M $CuCl_2$ saturated with NaCl
● 2.25 M $FeCl_2$
□ 2.4 M H_2CrO_4 in H_2SO_4 solution
△ 1.0 M $(NH_4)_2S_2O_8$ with 2.5×10^{-5} M $HgCl_2$

Fabrication of Double-sided Circuits—Subtractive Method with Plated Etch Resist. The basic steps required for the fabrication of double-sided circuit boards with plated-through holes is illustrated in Fig. 47. A typical processing schedule is summarized in the following outline.

1. Drill double copper-clad laminate for through-hole interconnections, and deburr holes.
2. Clean by vapor degreasing in trichloroethylene, followed by scrubbing with fine pumice in alkali or copper cleaner. Spray water rinse.
3. Etch copper in dilute ammonium persulfate at room temperature (2 lb/gal water) for 1 to 2 min. Spray water rinse.
4. Immerse in sulfuric acid 25 percent by volume for 1 min at room temperature, followed by 15-sec deionized water rinse.
5. Electroless copper plate. A typical schedule involves:
    a. Dip in hydrochloric acid (25 percent by volume) for 1 to 2 min.
    b. Sensitize in stannous chloride solution 1 to 2 min—deionize water rinse.

c. Activate in palladium chloride solution 1 to 2 min—deionize water rinse.
  d. Deposit copper in electroless bath for 5 to 20 min (depending on bath characteristics)—deionize water rinse.
 6. Electroplate copper in pyrophosphate bath to deposit 0.001 to 0.002 in. of copper in holes, e.g., 50 to 60 min at a current density of 30 amp/ft^2. Spray rinse in deionized water, dry, and inspect.
 7. Coat both surfaces of board with negative photoresist by spray or roller coating to achieve dry-film thickness of 0.00015 to 0.00035 in. Air-dry and bake at recommended temperature.
 8. Register board to photomaster in exposure frame with registration pins (three minimum), apply vacuum, and expose for required time.
 9. Develop resist pattern, postbake, and inspect; touch up minor flaws.
 10. Electroplate metal etch resist (e.g., 0.0001-in. gold or 0.0005-in. tin-lead). Board surface is precleaned before plating (alkali dip, ammonium persulfate etch, and sulfuric acid dip with deionized water-rinse cycles in between).
 11. Strip photoresist, and etch exposed copper in ferric chloride for gold-plated circuits and chromic acid for tin-lead.
 12. Spray rinse and air dry.

TABLE 47 Operational Characteristics of Various Copper Etchants[97]

Etchant	Corrosiveness	Neutralization or disposal problems	Toxicity	Venting requirements	Postetch cleaning difficulty
Ferric chloride	H	M	L	L	M
Ammonium persulfate	L	L	L	L	L
Chromic acid	H	H	H	H	M
Cupric chloride	H	L	M	M	M
Alkaline etch	L	L	M	H	L

H High.
M Medium.
L Low.

The above procedure is also applicable to plated-through-hole multilayer structures.

Additive Processing of Double-sided Circuits—Pattern Plating Method. The processing sequence for the fabrication of additive double-sided circuit boards is illustrated in Fig. 48. In comparison to the subtractive method outlined above, the process differs only in the pretreatment and copper-plating steps. Substrates are initially treated to alter the surface to promote adhesion of the placed metal and then electrolessly plated over the entire surface and through all holes as in step 5 above. A photoresist pattern is applied, and the board is pattern plated with copper (circuit only) to the required thickness (0.0015 in. typical). This step is followed by plating of gold or solder over the pattern. The board is completed by removal of the resist pattern and flash etching of the thin electroless copper layer beneath it.

Good control of the line width is essential for fine-line circuitry or strip-line applications. In the pattern plating process, the width of a plated line increases by an amount equivalent to twice the plated-metal thickness, as noted previously. The photoresist image has little effect on the amount of spreading, since most resists are applied in a thickness of the order of 0.0002 in., and plating proceeds out over the resist image. Control of the spreading can be achieved through the use of a nonliquid photoresist film such as Riston. This resist can be applied in a controlled thickness from 0.001 to 0.003 in. and exposed with collimated light to achieve near-vertical sidewalls in the resist image after development. Subsequent plating of copper or other metals is then confined by the thick resist layer. As a result, finer line patterns with better definition can be processed. The thicker resist layer also affords greater protection during plating with a minimum of defects occurring because of pinholing.

Direct Etching of Printed Circuits with Photoresist Masks. The etching of printed-circuit patterns can be carried out by the direct application of a negative or positive photoresist mask applied to one or both sides of a copper-clad circuit laminate. Figure 46 illustrates the sequence for a single-sided board. The cleaning and photoresist processing steps are identical to those noted above for the subtractive process using a plated-metal etch resist. After image development, boards are etched in a spray etcher, the resist is removed, and the board is blown dry. Photoresists are normally used for all fine-line circuitry of the Class 3 and 4 types. Most other boards are processed by silk-screening an etch-resistant mask pattern.

Semiconductor Device Technology

Present-day semiconductor devices include a variety of rectifiers, transistors, and integrated circuits fabricated in thin sheets of silicon. Although the number of processing steps increases in going from simple diodes to a complex integrated circuit, the additional steps represent repeat operations rather than new processing. Processing steps common to all devices are outlined in Fig. 51.

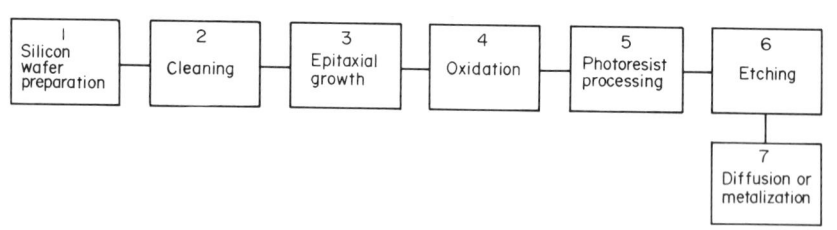

1 Silicon wafer preparation	2 Cleaning	3 Epitaxial growth	4 Oxidation	5 Photoresist processing	6 Etching

	7 Diffusion or metalization

Fig. 51 Flow chart for semiconductor processing.

The processing of simple diodes involves (steps 5 to 7, Fig. 51) a photoresist pattern formation, an oxide opening, and a diffusion step to form a junction. The oxide serves as a barrier to diffusion so that dopant atoms enter the silicon only where there are holes in the oxide. The photoresist pattern determines where these openings in the oxide will be after etching. A second photoresist and etching step serves to open up contact windows, and a third photoresist and etching step following metallization defines contact regions. For conventional integrated-circuit production six photoresist patterning steps are required for the various diffusions, contact openings, and intraconnections. Additional photoresist steps are required for circuits where dielectric isolation of circuit components or multilevel metallization is employed, and where thin-film resistors and capacitors are also formed on the chip.

For reasons of economy, reliability, and speed of operation, the silicon area needed for each device is made as small as possible. Although circuit design guidelines are usually dictated by the materials used, heat-transfer, and other factors, the lower limit in size is determined by the accuracy with which the physical dimensions of circuit elements can be defined. This depends on the control of the opening step, which depends in turn on the definition in the photoresist pattern, and ultimately on the original artwork used in the exposure of the photoresist. The factors that affect pattern definition and resolution in the silicon thus include mask quality, resist pattern quality, and oxide etching and diffusion process parameters. Since, at some stage, all patterns are defined optically in a photoresist layer, photofabrication represents a key process in the production of these devices.

In comparison to the majority of other photofabrication operations, semiconductor devices require much greater control of the photoresist processing. Pattern dimensions are more than an order of magnitude smaller, and registration of many masks is required. Circuits with patterns of 0.3- to 0.5-mil element size are typical of present-day devices. Figure 52 and Table 48 illustrate typical design rules for the layout of complementary MOS circuits.[137] Other devices in the development stage have element dimensions in the order of 1 to 3 μm.

Typical silicon wafers used in production have diameters ranging from ¾ to 2 in. The pattern complexity, element size, and the number of patterns in the array determine the registration accuracy that must be attained. More complex patterns with small elements require tighter tolerance during alignment. Individual bipolar devices and circuits may measure 30 to 200 mils across, with a total number of up to 3,000 devices or circuits per wafer. In the sequential processing of devices, each succeeding photoresist step positions one element of a device within another element of the device or makes a connection between elements. Thus, each photomask must give a photoresist pattern superimposing precisely on those prepared previously. Since six or more separate masks are used to prepare a device of an integrated circuit, each having a pattern such as in Fig. 52, the importance of mask quality and registration is easily understood. Production of devices of this complexity requires photomasks of high precision, dimensional control, and image quality, and a reproducible method of precise alignment. Photomask requirements were summarized in a recent review of integrated-circuit technology by Camenzind as follows:[138]

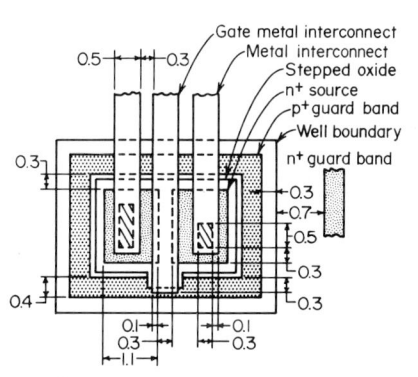

Fig. 52 Typical complementary MOS device layout.[137]

1. The master drawings of all masks must be made with great precision; this is why they are made several hundred times larger than the actual circuit.
2. The process by which a pattern is arrayed on the mask in rows and columns must be extremely precise. The major limitation here is the precision of the step-and-repeat mechanism which moves the projection of the pattern across the mask.
3. The mask material must be dimensionally stable; it must not contract or expand with variations in temperature and humidity.
4. The operator who positions the mask on top of the slice with the aid of a high-powered microscope must perform this task accurately.

TABLE 48 Typical MOS Device Dimensions[137]

Characteristic	Dimension
Channel length	0.3 mil
Channel width	As required (1.5 mil minimum)
Gate metallization	0.5 mil (0.1-mil overlap each side)
Contact opening	0.3 by 0.5 mils (The minimum dimension from the edge of diffusion is 0.3 mil.)
Contact metallization	0.5 mil (0.1-mil overlap each side)
Interconnect metallization	0.3 mil
Separation between metal conductors	0.3 mil
Guard bands (n+ or p+)	0.5-mil minimum (Expansion to fill nonactive area preferred.)
Bond pads	4 by 4 mils (The minimum distance from metal, diffused area, or dicing line is 2 mils. Center-to-center-bond-pad spacing is 8 mils.)
Dicing allowance	4 mils

The methods used for photomask fabrication were described in a previous section. This section will present the requirements of photoresists and processing of photoresist patterns for the fabrication of semiconductor devices. The alignment of photomasks and their application will also be included.

Photoresist requirements The photoresist must serve two main functions, namely, be capable of reproducing the photomask pattern with a high degree of accuracy and, secondly, provide satisfactory masking of the coated surface during subsequent etching or metallization. In order to achieve high resolution, a thin coating of uniform thickness is required. However, thin layers exhibit a poorer etch resistance and a larger pinhole density owing to dust and other airborne contamination.[55,139] A large number of pinholes can occur because of solid particles of dust or hardened resist suspended in the resist solution or embedded in the coating during application. These can be minimized with proper microfiltration of resist solutions prior to application, and work should be carried out in laminar-flow boxes.

Generally, the resist coating should be no greater than one-third the finest line dimensions to be reproduced in the oxide. For example, patterns of 0.3 to 0.5 mil as shown in Fig. 52 could be processed with a resist thickness in the order of 2 μm. A resist thickness of 0.5 to 1.5 μm is satisfactory for most integrated-circuit applications.

The techniques for processing photoresist patterns, namely, pretreatment of wafers and materials, methods of application, development and removal, were discussed in previous sections. Spinning techniques offer the best control for application of thin resist layers and are preferred over other methods in contemporary integrated-circuit processing. Even flow of resist and proper solvent evaporation during the coating step are mandatory. Rapid exposure and development are highly desirable. The developed pattern must have good adhesion to the wafer, show good stability during etching, and be easily removed in solvents that do not attack the substrate once the etching step has been completed. It should be emphasized that, because of the small dimensions involved, the degree of cleanliness and uniformity is of utmost importance throughout semiconductor processing in order to achieve good device yields. Wafers must be suitably protected to avoid contamination and handled only with tweezers.

Resist problems become more critical as pattern dimensions are decreased, and circuits more complex. Since many interdependent photofabrication steps are used, good yields can only be obtained by employing a high degree of process control during each individual step. Pinholes are much more critical in the processing of integrated circuits than in transistor processing because of the increased critical area in which a pinhole can lead to failure.

Photomask alignment After coating with photoresist, drying, and baking, the wafer must be aligned to a photomask and exposed by contact printing with a suitable highly collimated light source. The wafer is initially aligned centrally on a vacuum chuck, and the first pattern is exposed. Development and etching of the oxide completes the first photofabrication step and produces an array of openings over the wafer. The next photomask is then aligned to this initial pattern.

The registration of each subsequent photomask to the previous pattern is a critical operation in the photoengraving process for semiconductor devices. All previous patterns on the wafer must be positioned in an x, y, and θ (rotation) direction and aligned to the photomask pattern to within several millionths of an inch. Extreme care must be used to avoid mechanical damage to the resist coated surface and to the image surface of the photomask. Scratches and solid particles on either surface can lead to pinholes or, alternatively, to obstruction of some desired openings. Both lead to lower device yields. The aligned mask and wafer must then be clamped for contact printing without lateral displacement.

Mask alignment is generally achieved with an optical alignment fixture consisting of a special-purpose microscope and micropositioning table. A compound microscope of the metallurgical type is normally used with vertical illumination. Stereomicroscopes are seldom used because of losses arising from systematic misalignment.

Factors to be considered in the selection and use of a compound microscope include: the field of view decreases as the magnification increases; light is required with intensity in proportion to the square of the magnification; the lower limit of magnification is set by the resolution capability of the eye retina; the upper limit is determined by physical limitations of the various optics such as working distance and depth of field; as magnification is increased, a higher order of mechanical accuracy is required during scanning; the magnification should be sufficiently high to resolve the small dimensions of the devices being processed.

Typical data for a Bausch and Lomb microscope built for mask alignment work is presented in Table 49. Single-field microscopes require alignments of two or more widely spaced patterns on the wafer by scanning the different positions. Double-field or split-field systems are preferred as they permit simultaneous viewing of two positions on the wafer and provide rapid alignment.

TABLE 49 Data for Bausch and Lomb Mask Alignment Microscope[140]

Objective	N.A.*	Total magnification	Objective† resolution, μm	Field of view, mm	Working distance, mm	Depth of field, μm
2.6×	0.08	52	6 μ	3.2	43.5	150
4	0.1	80	3.5	2.2	38.0	100
8	0.2	160	1.7	1.5	14.2	25
10	0.25	200	1.3	0.9	10.6	16
20	0.4	400	0.8	0.6	2.8	6
25	0.5	500	0.65	0.45	1.8	4

* N.A. = numerical aperture = sine of half-angle of cone of light received by objective. For example, a 0.5 N.A. objective receives a 60° cone of light from a point on the wafer. The f number, as used in photography, would be $f/0.87$.

† Theoretical resolution of a microscope lens based on the Airy diffraction disk is

$$\frac{0.6\lambda}{N'x(\text{N.A.})}$$

where λ = wavelength of light
 N' = index of objective to object medium (1.0 in air)
 N.A. = numerical aperture

Attainable resolution varies with observer's eye and training, objective quality, lighting, color, and contrast of subject, etc.

Important considerations for microscope illumination are summarized in the following. The light system used should produce a light beam which fills the objective with light as if coming from a source at the eyepiece in order to obtain maximum resolution. Light intensity should provide normal opening of the pupil of the eye (diffraction can occur at high intensity and loss of resolution at low intensity). Incandescent illumination with a sharp yellow filter provides light equivalent to peak eye sensitivity and above the cutoff sensitivity for photoresists.

The alignment fixture must provide rigid holding of the wafer and mask in a flat parallel position, e.g., by a flat vacuum chuck. The mask holder should support the mask so that bowing is absent during contact with the wafer. The spacing and clamping system must bring the mask and wafer into intimate parallel contact, separate them at a minimum fixed parallel gap, and reclamp them for exposure with no lateral movement. The micropositioner must be capable of aligning the wafer under the mask in x, y, and θ directions within the required limits of resolution of the pattern and microscope.

Most alignment machines contain light sources for exposing the photoresist pattern. The exposure system must provide light of uniform intensity and well collimated over the entire surface of the wafer. A high-pressure mercury-vapor lamp is preferred as a source and provides minimum exposure times, as noted previously.

14-104 Photofabrication

Details of mask alignment and various alignment equipment have been presented by Schwartz.[140] Present-day machines are available that rapidly align wafers to an accuracy of 0.000010 in. in x-y and 0.000050 radian in angular alignment (see Fig. 26). Operators must be thoroughly familiar with the equipment, and control tests should be run to establish optimum procedures for a given application. Adequate control procedures must also be established to maintain proper microscope and illuminator performance and resist exposure.

Etching considerations The materials most commonly etched in the fabrication of semiconductor devices are silicon dioxide and aluminum. Discussion will be limited to the processing of these materials, as they are representative of the large

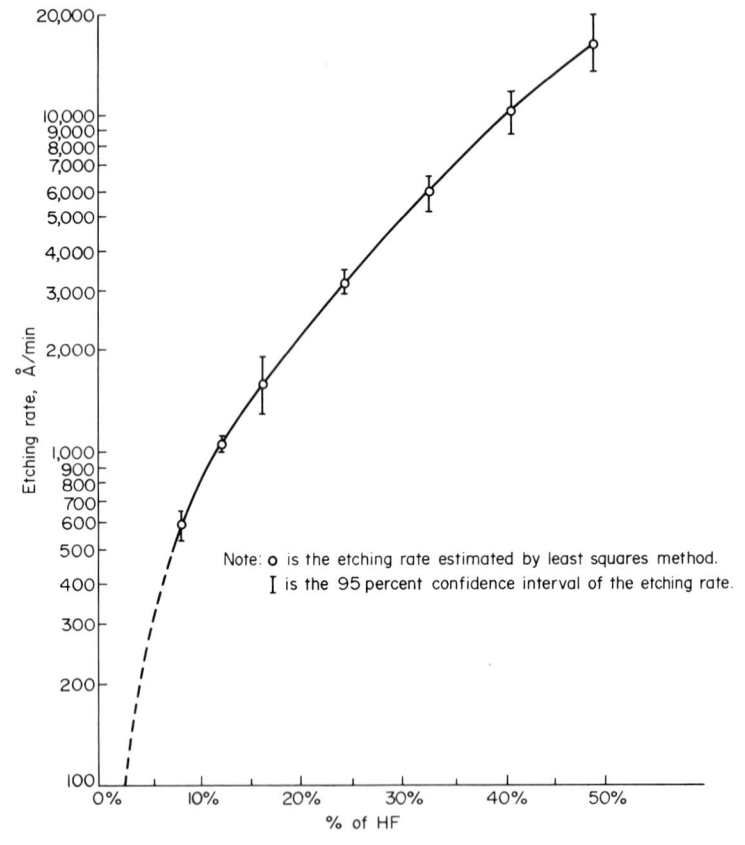

Fig. 53 Etching rate of silicon oxide film versus concentration of hydrofluoric acid at a temperature of 25°C and at a stirring rate of 100 rpm.[142]

majority of present-day devices. Similar process guidelines would be applicable to other semiconductor materials and metals using the appropriate etchants presented elsewhere in this chapter.

In the processing of semiconductor devices all acids, solvents, and water used should be of the highest chemical purity to avoid possible altering of the semiconductor properties.[141] Electronic-grade standards have been established and should be specified in the purchase of all chemicals. Deionized water with a low nonvolatile organic content should be used throughout processing. Impurities that can act as n or p dopants, such as arsenic, boron, copper, phosphorous, nickel, and the heavy metals, must specifically be excluded from the processing solutions.

Applications   14-105

In general, most commercially available photoresists can be used in the etching of the above materials. Special grades and/or dilutions which provide uniform thin coatings are available for semiconductor and thin-film etching. In comparison to other applications for photoresists, the thickness of the material to be etched is small except for deep etching of silicon (SiO_2 is typically 5,000 to 10,000 Å; aluminum is typically 1 μm). Etch factors must be considered, as previously noted. Undercutting may be a serious problem for small-geometry devices. Etch factors of 1 to 2 are considered acceptable for most applications.

Lifting of resist edges and surface etching between the resist and silicon dioxide interface (fringing) is generally more detrimental than undercutting. The majority of these problems are associated with moisture on the wafer surface and can be minimized through adequate control of baking cycles prior to the coating and etching steps. The factors influencing resist adherence were discussed previously.

Silicon Dioxide Etching. A number of etchants have been used successfully for the etching of silicon oxide films. Of these, solutions of hydrofluoric acid are the most common. The etch rate of silicon dioxide is a function of the hydrofluoric acid content and increases as this concentration is increased. Data for the etching of silicon dioxide (thermally grown in wet oxygen) are shown in Fig. 53. Optimum etching is generally achieved at etch rates of up to 2,000 Å/min. An increase in etch rate produces greater undercutting and problems associated with resist breakdown. Control of etch rate is necessary in order to obtain a high degree of dimensional control and good device yields. If all oxide is not removed, the pattern is not satisfactory for the next processing step. Overetching produces wider openings in the oxide and thins down fine oxide lines. The etch rate can be controlled by the addition of buffering compounds to hydrofluoric acid and by maintaining a relatively constant temperature during etching. Solutions buffered with ammonium fluoride are widely used in most applications. Figure 54 presents the etch rate of a typical buffered etchant in the etching of thermally grown oxide as a function of temperature.

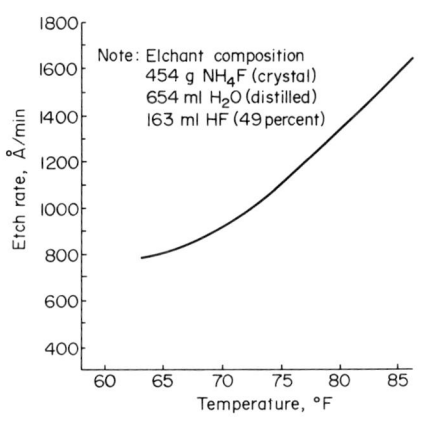

Fig. 54 Etching rate of silicon dioxide in buffered HF-NH$_4$F solution versus temperature.

Other factors affecting etch rate include the preparation method of impurities or dopants in the oxide and agitation of the etch.[143] Table 50 gives a comparison of SiO_2 films grown by various methods and with various heat treatments. The etch rate of vapor-grown SiO_2 (as deposited) in buffered etchant is much greater than that of thermally grown oxide. With high-temperature densification, the rate decreases to that for thermal oxide, 18 Å/sec for HF-NH$_4$F buffer solution at 25°C.

A typical procedure for etching patterns in silicon dioxide using Kodak KTFR photoresist is summarized as follows:

1. Clean wafer. If processing directly follows oxidation step, wafer can be coated directly. Wafers stored in a suitable clean container with no noticeable contamination can be cleaned with methylene chloride or trichloroethylene by applying fresh solvent (electronic grade) to the wafer surface and spinning dry (30 sec at 2,000 to 4,000 rpm). Bake wafer on a hot plate at 300 to 350°F for 1 min.
2. Place wafer on spinner chuck, and apply vacuum.
3. Apply photoresist from filter syringe to cover entire surface of wafer.
4. Start spin cycle, and spin 1 min at 4,000 to 10,000 rpm.
5. Visually inspect for flaws. The surface should appear uniform with a slight

rim around the edge. Strip resist on rejected wafers with methylene chloride, and repeat above process.

6. Dry in nitrogen box 15 to 30 min or in desiccator for 1 hr.
7. Bake on hot plate at 85°C for 10 to 20 min.
8. Place wafer on the vacuum chuck of the alignment fixture, and align to proper mask.
9. Expose 5 sec with 200-watt mercury-vapor source.
10. Develop by immersing in beaker with developing solution (90 percent Stoddard solvent, 10 percent xylene) for 1 min followed by spray of fresh solution. Immerse in fixing solution (90 percent Stoddard solvent, 10 percent denatured alcohol) for 15 sec, and spray with fresh solution.
11. Spin dry or blow dry with nitrogen.
12. Inspect for sharply defined pattern.
13. Bake on hot plate at 175 to 200°C for 10 to 15 min; allow to cool to room temperature.
14. Etch oxide regions in buffered hydrofluoric acid solution at controlled temperature in a polyethylene container. Etch time is determined by oxide thickness and etch rate.
15. Rinse in deionized water, and inspect. The complete removal of oxide is noted by a nonwetting surface in water and a white appearance. If oxide is still present, continue etching for a short interval, rinse, and reinspect.
16. Remove resist by immersing in commercial stripper solution for 10 min.
17. Rinse with deionized water, using final hot-water rinse.
18. Spin dry or blow dry with nitrogen and place in a suitable container, and transfer to next processing step.

NOTES: (1) Proper precautions for handling of hydrofluoric acid solutions should be enforced at all times.

(2) Techniques for determining the etch end point based on the wetting properties of the surface have been described by Kern.[145]

After openings (windows) are etched into the oxide, several processes may follow, namely, diffusion, oxide regrowth, or metallization. Diffusion processes form the various circuit elements in the wafer by selectively doping the silicon with a p

TABLE 50 Comparative Data for Deposited and Grown One-micrometer-thick SiO$_2$ Films on Silicon[144]

Method of oxide preparation	Thickness by Tolansky interferometry, Å	Refractive index	Dissolution rate* at 25°C, Å/sec	Dielectric constant at 1 kHz	Dielectric field maximum sustained, dc volts/cm × 10^{-6}
From SiH$_4$ at 325°C (760 Å/min):					
As deposited................	10,190	1.43	87.3	5.73	2.5
After 12 min at 770°C (argon)	9,560	1.46	30.3	4.75	3.7
After 5 min at 1250°C (argon)	9,670	1.46	21.3		
From SiH$_4$ at 475°C (2500 Å/min):					
As deposited................	10,340	1.46	81.1	4.30	3.4
After 12 min at 770°C (argon)	9,930	1.46	33.4	4.93	4.0
After 5 min at 1250°C (argon)	9,570	1.49	18.3		
Thermal oxidation at 1250°C (steam) 10,000 Å/hr:					
As grown....................	9,650	1.47	18.1	3.83	5.2

* Standard HF-NH$_4$F buffer solution: 454 g ammonium fluoride crystals; 654 cm^3 distilled water; 163 cm^3 hydrofluoric acid solution 49% (used after 24 hr of equilibration).

or an n type material. Typical junction depths are 7 to 10 μm for the collector-substrate, 0.15 to 0.25 for the base-collector, and 0.12 to 0.18 for the emitter-base. During diffusion the dopant diffuses laterally under the oxide so that the resultant edge of junction intersects the surface beneath the oxide film. Oxide regrowth may be carried out for several purposes, such as protective layers over previously formed regions, dielectric layers for capacitors or multilayer metallization, and channel regions in MOS devices. Metallization forms contacts to opened device regions in the oxide and intraconnections between various components.

In the fabrication of a semiconductor device a combination of the above processes is employed to form the various components and circuits. Figure 55 illustrates some of the various types of components formed for integrated circuits. A flow chart showing the steps in the processing of a typical integrated-circuit wafer is presented in Fig. 56.[146] In addition to forming a barrier layer during diffusion, silicon dioxide films are also used as etch masks for the etching of silicon nitride and silicon in the processing of beam-lead and mesa-type devices.

Etching of Patterns in Aluminum. After the last oxide etching step, a metallization step is needed to complete the device. After opening contact areas, a thin layer (e.g., aluminum) is deposited on the wafer by vacuum deposition. An aluminum thickness of 1 μm is satisfactory for most circuits; high-power circuits require up to an 8-μm-thick layer. The aluminum layer is then coated with photoresist, a metallization pattern is formed in the resist, and the aluminum is etched. A typical procedure for processing patterns in aluminum using Kodak KTFR photoresist is the following.

1. After coating with aluminum, the wafers should be resist-coated directly or stored in a dry nitrogen chamber. If they are stored for any length of time, heat on a hot plate at 150°C for 10 min before resist coating.
2. Place wafer on spinner chuck, and apply vacuum.
3. Apply photoresist from filter syringe to cover entire surface of wafer.
4. Start spin cycle, and spin 1 min at 4,000 to 10,000 rpm.
5. Repeat steps 3 and 4 for second coat.
6. Visually inspect for flaws; strip with methylene chloride, and repeat coating procedure if required.
7. Place wafer on hot plate at 85°C for 10 to 20 min.
8. Place wafer on the vacuum chuck of the alignment fixture, and align to proper metallization mask.

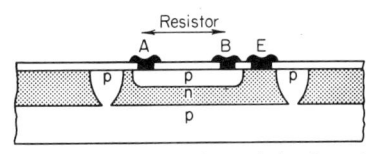

(a) Resistor: p-layer used as resistive element

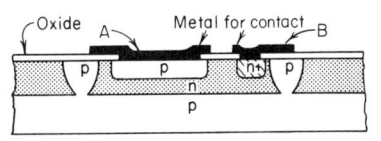

(b) Junction capacitor

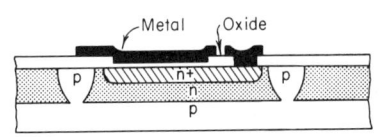

(c) Thin-film capacitor

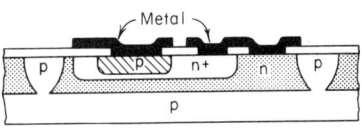

(d) High-speed diode (collector-base short)

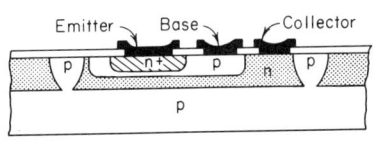

(e) Bipolar transistor

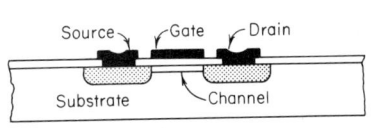

(f) MOSFET transistor

Fig. 55 Typical circuit elements formed during integrated-circuit processing.

14-108 Photofabrication

9. Expose 5 sec with 200-watt mercury-vapor source.

10. Develop by immersing in beaker with developing solution (90 percent Stoddard solvent, 10 percent xylene) for 5 min, followed by spray of fresh solvent. Immerse in fixing solution (90 percent Stoddard solvent, 10 percent denatured alcohol) for 15 sec, and spray with fresh solution.

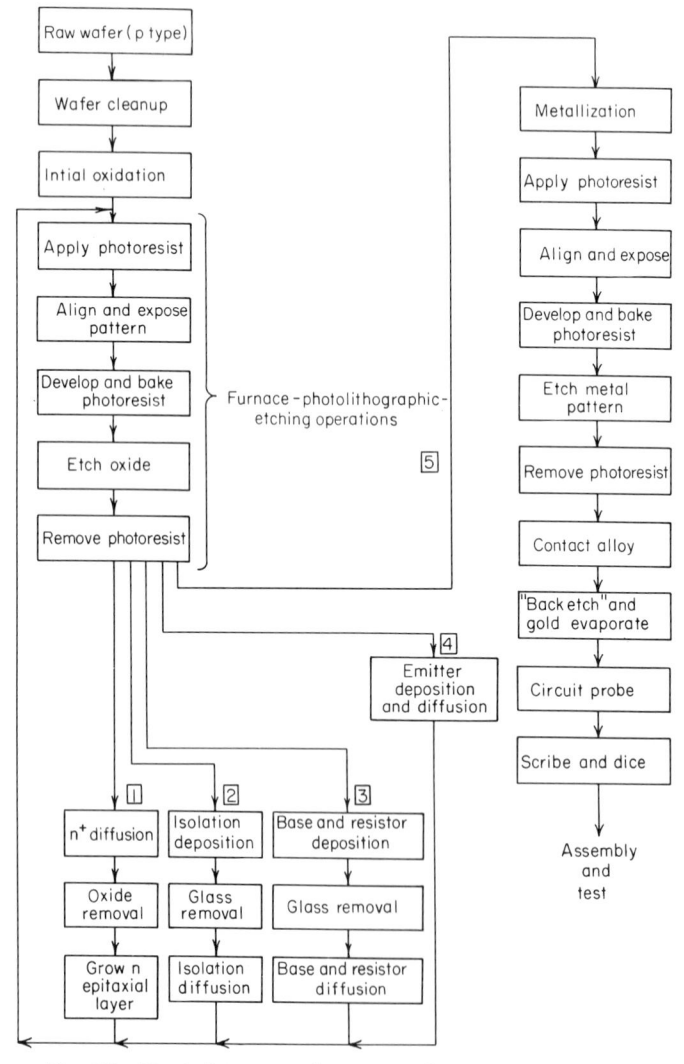

Fig. 56 Typical integrated-circuit wafer processing steps.[146]

11. Spin dry or blow dry with nitrogen.
12. Inspect for sharply defined pattern.
13. Bake on hot plate at 100 to 135°C for 30 min.
14. Etch exposed aluminum in phosphoric-nitric etchant (85 H_3PO_4 reagent grade, 70 HNO_3 reagent grade: deionized H_2O; 10:1:2½ by volume). Immerse wafer in etchant at 45 to 55°C until etching is complete, as determined by visual observation. Rinse in deionized water.

Applications   14-109

15. Spin dry or blow dry with nitrogen.
16. Examine under microscope for completeness of etching; re-etch if required.
17. Remove resist by immersing in hot commercial stripper solution in a hood (e.g., IRC Laboratory J-100 or B&A A-20 diluted, 1 part stripper to 3 parts methylene chloride; reflux boil). Rinse in deionized water; soak in hot methyl alcohol (electronic grade), and rinse in cold fresh methyl alcohol. Spin dry.
18. Inspect under microscope for complete removal of resist.
19. Place finished wafer in a chemically cleaned container.

After this processing, the aluminum will generally undergo an alloying step at a temperature of approximately 540°C to improve the aluminum-silicon contact to provide a good electric contact. Wafers are then tested for device electrical characteristics, scribed, and separated into individual circuits.

In addition to the above etchant, some use is made of electrolytic etching in KOH solutions. Caustic solutions should be avoided, however, where the alkali-metal ions could contaminate the device, for example, in MOS devices.

Thin-film Applications

The term thin films in electronics generally refers to electronic components and circuits applied by vacuum evaporation, sputtering, chemical vapor deposition, electroless plating, and electroplating techniques, and defined by photofabrication. Details of these processes and the factors related to photofabrication were presented on page 14-79. The major applications are passive microcircuits, ferromagnetic and cryoelectric memory arrays, and active devices. In addition to circuit applications, photoetched patterns of thin chromium or other metal layers on glass (see page 14-37) are used as photomasks in the processing of semiconductor and thin-film circuits.

A wide variety of materials have been used to form thin-film elements for electronic applications, as discussed in detail in Chap. 11 of this handbook. Of the applications listed above, passive circuits fabricated on various insulators and semiconductors are the most widespread. The processing of these circuits is representative of thin-film techniques and will be discussed in this section along with typical examples.

Passive thin-film microcircuits   Thin-film microcircuits are fabricated by a sequence of material deposition, photoresist pattern formation, and etching steps, in a manner similar to that employed in the processing of semiconductors. Anodization is also used, for example, in tantalum circuitry, to form oxide layers for capacitors and to adjust resistive components to tolerance. Alternatively, thin-film circuits can be fabricated from a series of photoetched-metal evaporation masks as noted previously. Patterns made from these masks have poorer definition and resolution than those obtained from direct photoresist masks and are more restricted in pattern layout. They do, however, offer lower-cost processing for high-volume production.

The design and layout of thin-film passive circuits are dependent upon the circuit specifications, the material parameters, and the process capabilities.[147-150] Resistor values, for example, are determined by the sheet resistivity (ohms per unit square*) of the deposited film and the geometry of the resistor pattern. Sheet resistivity is chosen so that the resistors will cover an area sufficient to provide the necessary heat dissipation and, in addition, will enable suitably high yields of devices to be obtained. Similarly, capacitance is directly proportional to the area and inversely proportional to the thickness of the dielectric film; capacitor patterns are designed accordingly. A survey of various thin-film applications showed minimum dimensions to be in the order of 0.5 mil for thin-film circuits applied to silicon devices, whereas the majority of circuits fabricated on ceramics or glass substrates have minimum dimensions of 5 mils or greater.

In the processing of thin-film circuits, several precision photomasks are required

* Sheet resistivity is a normalized measure of the resistance: $R(\text{ohms/sq}) = \rho(\text{ohm-cm})/t(\text{cm})$, where ρ = specific resistivity and t = film thickness.

to define the various conductor, resistor, and capacitor patterns. These are employed sequentially with the respective deposition and etching steps. To obtain maximum precision, the basic artwork is prepared oversize on a coordinatograph and photo-reduced onto high-resolution plates. Circuits are generally step and repeated to give an array. In this way optimum use is made of the available substrate area, and the cost per circuit due to the processing is thereby lowered.

Photoresist and etching Most photoresists will satisfy the requirements of thin-film processing, using techniques described previously (see Fig. 39). Generally, photoresists formulated for semiconductor processing are preferred, particularly where high-resolution patterns are involved. In applications where a continuous conductive film is deposited over a continuous resistive film, a positive photoresist can be used in such a manner that the conductor and resistor patterns are delineated and etched in a sequential manner with the one resist application and two photoresist exposure steps.

Substrates must be free of contamination and handled only with tweezers or at the edges before resist coating. Normally the photoresist should be applied to substrates immediately after removal from the deposition chamber. Samples stored for considerable periods or those that exhibit contamination of any kind should be cleaned in detergent solutions and degreaser solvents, and baked at a temperature of 150 to 200°C before coating with photoresist.

Spinning and spray methods of application are preferred. These give good uniformity and control of photoresist thickness. The optimum thickness for this film is dependent upon the material to be etched. Films of oxides and cermets that are difficult to etch require thicker photoresist layers than single-metal films and are often etched by the undercut masking technique. A photoresist thickness range of 0.5 to 1 µm is adequate for most applications.

A wide variety of aligning and exposure equipment is in use for thin-film applications ranging from standard graphic arts plate-making tables to custom-designed fixtures for large-volume production. In most applications, substrates are in the order of 2 by 2 in. or larger and are exposed in contact with a photomask in a vacuum frame or in a mechanical jig with a clamping mechanism. Alignment can be made visually with an optical microscope or in a jig, two substrate edges providing the registration. A registration of ±0.001 in. is adequate for most applications. For thin films on semiconductor wafers, alignment is carried out in a wafer alignment fixture, as discussed previously. Exposure is followed by development, and where additional etch resistance and adhesion are required, baking at temperatures of 150 to 200°C is employed.

The etching of thin films is performed by immersion etching with agitation or in spray-etch equipment. Spray etching produces a faster etch rate and better edge definition than other methods. Many materials and etchants have been applied to the fabrication of thin-film circuits. Table 51 lists the more common materials, along with typical etchants. Other etchants can be used successfully, such as those listed in Table 31. For specific applications the etchant must be optimized with suitable test samples to establish the process control required for good uniformity and yield. Where undercutting or lifting of photoresist occurs during etching, the cleaning and baking procedures should be examined. Materials that are difficult to etch may require the use of a conversion coating or a resist layer formed by printing or in some other manner. For example, silicon dioxide films are used as etch masks in the etching of silicon nitride for semiconductor-device passivation. Patterns of a metal such as copper are also used as secondary resists where high processing temperatures preclude the use of photoresists themselves. After etching, the photoresist or other mask layer is removed, and the circuits proceed to the next processing sequence.

The processing of thin-film aluminum conductors on silicon devices using a negative photoresist was outlined on page 14-107. Additional examples of thin-film applications will be presented here to illustrate this important area of photofabrication.

Nichrome-gold thin-film circuitry[152] Many pure metal systems are applicable to the formation of thin-film resistors. Of these, nickel-chromium alloys are the

TABLE 51 Etchants for Selected Deposited Films[91]

Deposited film	Etchant*	Comments
Aluminum	2.25–3.75 M ferric chloride (1) 2–3 N sodium hydroxide (2) 85% phosphoric acid, 70% nitric acid, water (10:1:2½ by vol.)	Tantalum, titanium, gold, and platinum unaffected by these etchants.
Chromium	6 N hydrochloric acid (3) 2 pts. etchant no. 1, 1 pt. concentrated hydrochloric acid (4)	6 N sulfuric acid also will work but is less desirable.
Chromium–silicon oxide cermet[151]	Hydrochloric acid–hydrofluoric acid mixtures at 33°C	HCl to HF ratio dependent upon Cr content. HCl increased as Cr increases.
Constantin	Etchant no. 1	
Copper	See Table 31	Also for high-copper alloys.
Gold	Aqua regia (5) 1 pt. concentrated hydrochloric acid, 1 pt. concentrated nitric acid, 2 pts. water (6) Potassium iodide–iodine[152]	Tantalum and titanium are unaffected by etchant no. 6. Best to bake the photosensitive resist.
Gold–palladium	2 pts. concentrated hydrofluoric acid, 1 pt. concentrated nitric acid, 3 pts. water (7) Potassium iodide–iodine[152] 30g KI + 7.5g I$_2$ + 1200 ml H$_2$O	For fired-on films. Will attack glass and organic resists in time.
Hafnium[153]	Hydrofluoric acid 1–2%	
Manganese	Etchant no. 3 or 4	
Molybdenum	1 pt. concentrated sulfuric acid, 1 pt. concentrated nitric acid, 3 pts. water	Can also be etched anodically in chromic acid (100 g/l).
Nichrome	4 pts. concentrated hydrochloric acid, 1 pt. water Stannous chloride–hydrochloric acid (70% by vol.)[152] Etchant no. 1	Suitable for nickel and nickel-base magnetic alloys. See also comments under aluminum. Etched at 205–220°F.
Palladium	Etchant no. 5 or 6	
Platinum	Etchant no. 5 or 6	Also for platinum alloys. Etchant no. 7 for fired films.
Silicon	4 pts. etchant no. 1, 4 pts. concentrated nitric acid, 1 pt. concentrated hydrochloric acid.	
Silicon dioxide	Saturated aqueous solution of ammonium fluoride Hydrofluoric–nitric acid mixtures	Removes about 1,000 Å of oxide per min.
Silver	2 M ferric nitrate	Dilute nitric acid may also be used but is less desirable.
Steel	Etchant no. 1	Also for stainless steel.
Tantalum	1 pt. concentrated hydrofluoric acid, 1 pt. nitric acid, 2 pts. water	Gold and Nichrome unaffected by this etchant, but on glass these may slough off if not protected.
Tellurium	6 N nitric acid 1 M ammonium persulfate	
Tin oxide	Powdered zinc in dilute hydrochloric acid with vigorous stirring. Then use etchant no. 1 or 2 M cupric chloride in 4 N hydrochloric acid	Baked resist needed. Nascent hydrogen reduces tin oxide to tin which is soluble in chloride etchants.
Titanium	1 pt. concentrated hydrofluoric acid, 20 pts. water	Gold, Nichrome, and tantalum unaffected by this etchant.

* The etchants listed are used at room temperature unless otherwise noted. The term pts. indicates parts by volume.

most common. A typical processing sequence for the fabrication of Nichrome-gold thin-film circuits on a ceramic substrate includes: (1) cleaning the ceramic substrate; (2) vacuum deposition of the resistor film (e.g., 75 Å Nichrome); (3) vacuum deposition of the conductor film (e.g., 2,000 Å gold); (4) application of a positive photoresist and development of a composite resistor-conductor photoresist pattern (the gold and Nichrome are etched sequentially in two steps); (5) the photoresist is exposed using the resistor pattern only, and the exposed gold areas are etched away to form the resistors; (6) active components are mounted and tested. Details of the photofabrication steps (4 and 5 above) are presented in Table 52. The

TABLE 52 Photoetch Process for Nichrome-Gold Thin-film Circuitry[152]

Step	Process	Solution or material	Temperature, °F	Time
1	Apply photoresist	Positive-acting resist. Syringe with filter attachment	Room	As required
2	Whirl 1,500–3,000 rpm	....	Room	15–30 sec
3	Dry photoresist	Fresh-air circulating oven	150 ±10	4–6 min
4	Expose photoresist	Phototransparency of composite circuit, ultraviolet light source	Room	2–4 min
5	Develop	Appropriate developer	Room	As required
6	Spray rinse	Deionized water	Room	15–30 sec
7	First gold etch	Potassium iodide–iodine etchant	Room	Minimum time required to remove gold film
8	Spray rinse	Deionized water	Room	15–30 sec
9	Nichrome etch	Hydrochloric acid–stannous chloride etchant	210 ±10	As required
10	Rinse, spray, and overflow immersion	Deionized water	Room	15–30 sec
11	Dry	Dry nitrogen	Room	As required
12	Reexpose photoresist	Phototransparency of resistor pattern, ultraviolet light source	Room	2–4 min
13	Develop and rinse	Repeat steps 5 and 6		
14	Second gold etch	Potassium iodide–iodine etchant	Room	See step 7
15	Spray rinse	Deionized water	Room	30–60 sec
16	Dry	Dry nitrogen	Room	As required
17	Remove resist	Acetone	Room	As required
18	Dry and store			

important features of this process are the application of selective etchants and the double-exposure capability of a positive resist. The majority of metal resistor films are deposited to give resistivities in the range of 25 to 100 ohms/sq. This leads to the formation of stable resistors that usually have a low thermal coefficient of resistance.

Chromium-silicon monoxide cermet resistors Chromium and silicon monoxide can be simultaneously vacuum-deposited to form stable, resistive films with sheet resistivities in the range 100 to 10,000 ohms/sq. This system is representative of cermet materials, i.e., a mixture of a metal and a ceramic dielectric. Cermet films are preferred over pure metal systems where high-resistance-value resistors are required. Etching is generally more difficult than with most metallic films because of the chemical inertness of the glass-metal composite, and undercut masking techniques or rf sputter-etching are often employed. Typical examples of the pro-

Applications 14-113

cessing of chromium–silicon monoxide resistor circuits are illustrated in Figs. 57 and 58. In these figures a simple photoresist mask and etch technique and an undercut technique with a metal contact mask are illustrated, respectively.

Tantalum thin-film circuitry The refractory metals, tantalum, tungsten, rhenium, and niobium, are attractive for the fabrication of thin-film circuit components for

Glazed alumina substrate, cleaned ultrasonically in detergent and degreaser solvents

Chromium-gold conductor pattern vacuum-deposited through a metal mask (e.g., 250Å Cr; 1500Å Au)

Chromium-silicon monoxide vacuum-deposited by flash evaporation with substrate at 350°C

Photoresist applied and resistor pattern, exposed, developed and baked

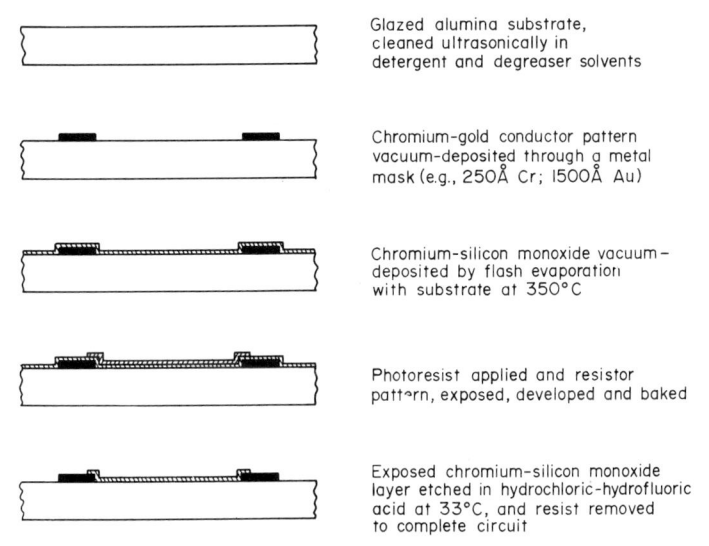

Exposed chromium-silicon monoxide layer etched in hydrochloric-hydrofluoric acid at 33°C, and resist removed to complete circuit

Fig. 57 Fabrication sequence for chromium–silicon monoxide/gold thin-film circuits.

Oxidized silicon device wafer

Aluminum vacuum-deposited over silicon oxide layer

Photoresist applied and negative resistor pattern exposed and developed; aluminum etched down to oxide layer

Photoresist removed and chromium-silicon monoxide film vacuum-deposited by flash evaporation

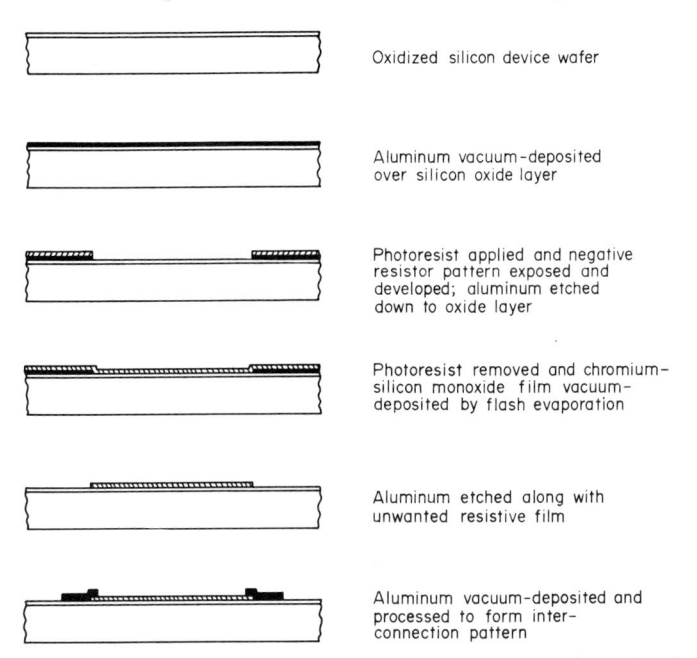

Aluminum etched along with unwanted resistive film

Aluminum vacuum-deposited and processed to form interconnection pattern

Fig. 58 Chromium–silicon monoxide resistors fabricated onto silicon by the undercut mask method.[154]

several reasons: thin films of these elements are highly stable, have high resistivities, and form stable anodic and thermal oxides that can be used for capacitor and insulating dielectrics. Thus, circuits of conductors, resistors, and capacitors can be fabricated, based on a single metal. Of these, the use of sputtered tantalum in photodefined patterns has been highly developed. This system is currently employed for many thin-film circuit applications.

In the sputtering process, material properties are controlled by variation in the deposition parameters, codeposition with other metals, and reactive sputtering with various gases to form the circuit elements.[155] Dielectric layers are formed, and resistors are adjusted by anodization. The processing of tantalum circuit elements is summarized below. The techniques outlined with various modifications are used for the fabrication of hybrid circuits on a mass-production basis.

Tantalum resistors are processed from three types of films: tantalum nitride, tantalum containing controlled quantities of interstitial oxygen, and low-density tantalum. These films can be fabricated into resistors by application of photoresist and chemical etching. The resistors can subsequently be trimmed to tolerance by selective anodization. Conductor films are then applied through a mechanical mask or by deposition and photoetching, as previously described. Similarly, resistor and conductor films can be deposited sequentially, followed by selective photoetching of the conductor and resistor patterns. The resistors can be trimmed by controlled anodization to precise values.

Capacitors are fabricated with the following sequence: (1) deposit a tantalum layer approximately 5,000 Å thick; (2) coat with photoresist, and define capacitor pattern; (3) anodize exposed tantalum to form tantalum pentoxide at a constant-forming voltage for the time necessary to achieve the required capacitance value; (4) strip photoresist; (5) recoat with resist and define conductor pattern, etch tantalum in hydrofluoric acid mixture, strip resist layer; (6) deposit aluminum or gold top electrode by deposition through a mechanical mask. (The initial tantalum layer acts as the bottom electrode.)

Other applications All thin-film applications use one or more of the processing steps outlined in the above examples. Chromium photomasks, for example, are prepared either by vacuum evaporation onto a flat glass plate, followed by photoresisting and etching, or by the undercut method. Similarly, cryogenic memory planes are fabricated by vacuum deposition and photoresist patterning of superconducting metals such as lead and tin, along with silicon monoxide or other insulating layers. Applications utilizing organic-base substrate materials are relatively few, owing to contamination problems that can occur in vacuum processing. One such application not involving vacuum deposition is the fabrication of thin-film nickel-phosphorus resistors on standard printed-circuit substrate materials by electroless plating, photoresist, and etching techniques.[156]

Screen Fabrication and Applications

Screen process printing is a printing method based on the use of a stencil pattern supported on a woven screen. With a squeegee, ink is forced through the open areas of the screen onto a suitable substrate. The process originated in the printing of decorative patterns in the fifth to seventh centuries A.D., using fine screens made from silk. The term *silk screening* is still in wide use today to describe screen process printing, regardless of the type of screen material used.

The process was adopted by the electronics industry in the 1940s for the printing of capacitors and carbon-film resistors. Its use was extended to the printing of conductor patterns; various modular forms of electronic circuits, comprising resistors, conductors, and capacitors; and protective resists for printed-circuit etching. With the development of high-alumina and beryllia ceramics, the process became important for fabricating packages for electronic devices. The process is currently used on a large scale for the printing of thick-film passive circuits on ceramic substrates (see Chap. 12) and in the preparation of printed-circuit etch masks.

The screen-supported stencil is the basic tool for the process and determines to a large extent the quality of the finished component. For electronic applications screens are prepared exclusively by photofabrication. The materials, design considerations, and processing of screens will be discussed in detail in this section along with a summary of typical applications. Details of the screen-printing process can be obtained in the literature references.

Materials and assembly A screen for screen process printing consists of a woven screen material mounted in a supporting frame with a photosensitive emulsion applied to the screen material. The frame serves to support the mesh with suitable tautness to provide flatness and registration of the pattern. It also provides a

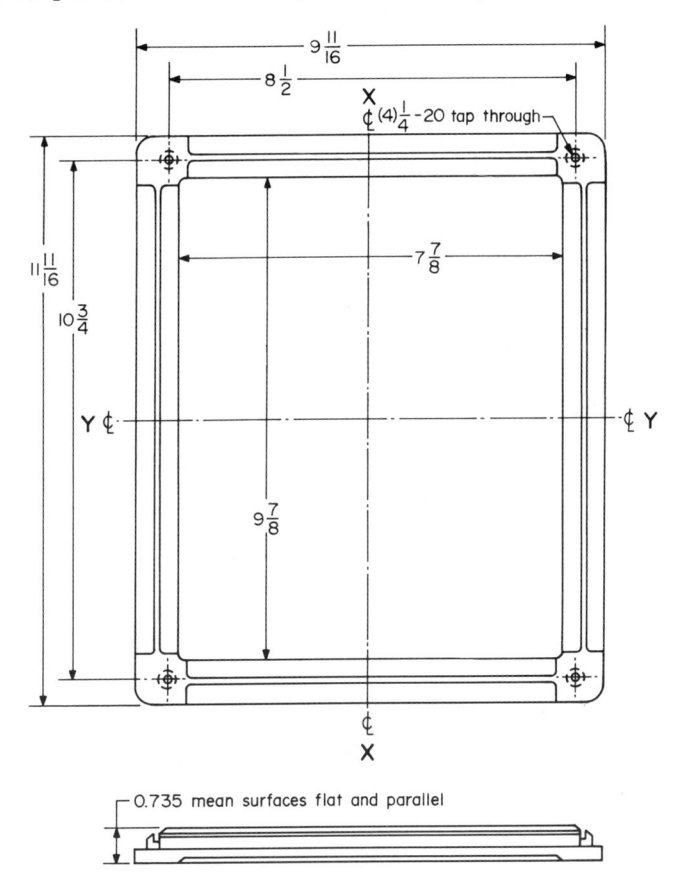

Fig. 59 Crimped attached screen frame. (*IRI, Nashua, N.H.*)

means for mounting the screen during the printing process. Frames must be sufficiently rigid and flat to maintain screen tension and proper alignment. Applications requiring high precision such as fine-line circuitry require frames machined to close tolerances. For electronic applications frames are made almost exclusively from cast aluminum and are available with mesh areas ranging from 2 by 3 in. to 36 in. square, in sizes to fit the various screen-printing equipment. The mesh is attached to the frame with an assembly fixture and is stretched by wedging plastic or metal tubing in slots on the frame. An improved version of this type of frame is illustrated in Fig. 59. The castings used are annealed and machined flat to ±0.002 in. Frames are also available with adjustable slotted aluminum bars. These permit positioning of untensioned screens on the frame and allow adjustment to proper tension. This type of frame produces equal tension on the

14-116   Photofabrication

screen via 100 percent gripping around the periphery and a pull parallel to the plane of the fabric. The functional parts of this frame are shown in Fig. 60.

Screen Mesh. Stainless steel and monofilament nylon mesh are used in making stencil screens for most electronic screen-processing applications. Stainless steel is two to three times stronger than nylon and has a high modulus of elasticity, permitting tighter stretching on the frames. Nylon, however, has a greater resiliency and shows less of an effect due to extreme squeegee pressures and accidental bumps with blunt objects. Nylon is essentially unaffected by the processing chemicals used in electronic applications, but does absorb water up to 6 percent and is highly sensitive to heat (235°C softening point). During exposure of the photosensitive stencil, light can travel along nylon fibers into the unexposed regions, causing unwanted exposure. Dyed nylon grades are available and should be used for fine detail work. Both stainless steel and nylon can produce satisfactory screens; stainless steel is usually preferred for long production runs.

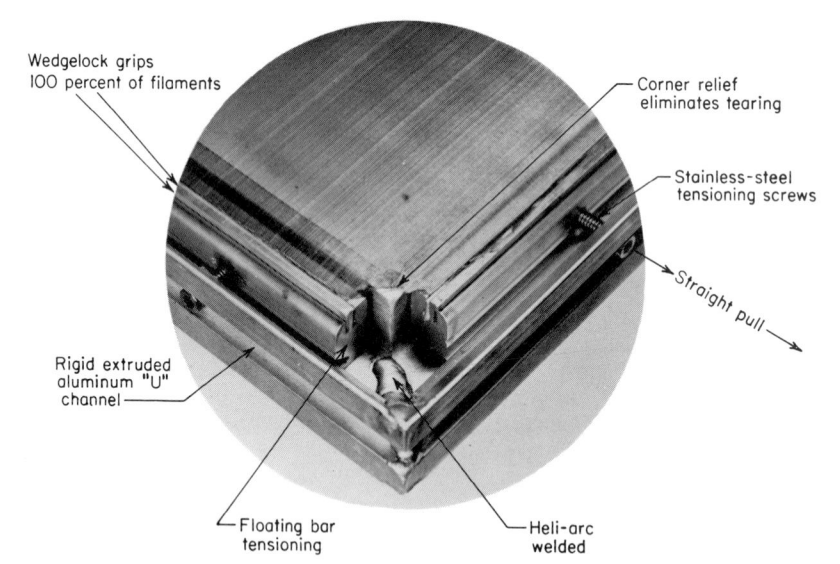

Fig. 60   Dia-print screen chase.   (*Kressilk Products Inc., Elmsford, N.Y.*)

Plain weave mesh is normally used and is available in a wide variety of sizes and wire diameters. Mesh specifications for typical mesh from 105 to 465 are listed in Table 53. Mesh count varies in manufacturing and is controlled to within ±6 percent. Wire diameter is altered in relation to mesh size so that the percent open area does not vary too widely from one size to another. It should be noted, however, that different manufacturers may use different wire diameters for a given mesh size, and the percent of open area can vary from that listed in Table 53.

Screens are normally aligned so that the mesh is square with the frame. The importance of good alignment of the mesh with the screen and the importance of a uniform screen tension have been reviewed by Coronis.[157] Square alignment permits easier pattern alignment and setup during printing. Screen tension should be uniform from screen to screen to permit good repeatability in printing. It was shown that the screen tension required is directly proportional to frame size and wire diameter. Less force is required to deflect large screens than smaller screens. Also, fine mesh with smaller wire diameters shows greater deflection for a constant applied force. Squeegee force must be controlled to provide good screen life. Normally, the applied force should be sufficient to bring the screen into contact with the substrate to be printed and to completely fill the screen cavities.

TABLE 53 Stainless Steel and Nylon Screen Mesh Specifications

Mesh	Stainless steel wire	Nitex	Wire diameter	Mesh opening	% open area
105	105	...	0.0030	0.0065	46.9
120	120	...	0.0026	0.0057	47.3
135	135	...	0.0023	0.0051	47.4
145	145	...	0.0022	0.0047	46.4
165	165	...	0.0019	0.0042	47.1
180	180	...	0.0018	0.0042	46.6
185	...	185	0.0017	0.0037	47.5
200	200	...	0.0016	0.0034	46.2
206	...	206	0.0015	0.0035	50.0
229	...	229	0.0014	0.0030	46.5
230	230	...	0.0014	0.0029	46.0
240	...	240	0.0014	0.0028	44.0
250	250	...	0.0018		
260	...	260	0.0014	0.0024	45.0
270	270	...	0.0016		
283	...	283	0.0012	0.0024	44.5
306	...	306	0.0012	0.0021	41.0
325	325	...	0.0011	0.0020	42.2
330	...	330	0.0012	0.0018	37.0
350	...	350	0.0012	0.0016	34.0
380	...	380	0.0012	0.0014	30.0
400	400	...	0.0010	0.0015	36.0
465	...	465	0.0012	0.00098	21.0

Recommended screen tensions for various stainless steel meshes, based on a constant applied force, are summarized in Table 54.

Screen Emulsions. Two types of photosensitive emulsions are used in screen fabrication: those directly applied and those indirectly applied or transferred to the screen. Both are of the negative-acting type. Figure 61 illustrates the cross sections of screens of each type.

Direct emulsions are water solutions of gelatin, polyvinyl alcohol, or polyvinyl acetate sensitized with potassium, sodium, or ammonium dichromate or diazo com-

TABLE 54 Recommended Tensions for Stainless Steel Screens on a 5- by 5-in. Frame[157]

Mesh size	Wire diameter	Mesh opening	Deflection*
105	0.0030	0.0065	0.035
150	0.0026	0.0041	0.035
165	0.0020	0.0042	0.050
200	0.0016	0.0034	0.055
200	0.0021	0.0029	0.045
230	0.0015	0.0029	0.055
250	0.0016	0.0024	0.050
325	0.0011	0.0020	0.065

* Deflection tolerance ± 0.012 in. Data measured on a Presco STG-3 tension test gage under applied force of 1 lb.

14-118 Photofabrication

pounds. Emulsions are usually supplied unsensitized and are mixed with sensitizer just before use. The dichromated sensitized solutions have a short shelf life because of oxidation of the emulsified resin and breakdown of the dichromate in acid solutions. Films made from aged solutions show a lowered light sensitivity and a decrease in acid resistance. Emulsions sensitized with diazo compounds exhibit a lower light sensitivity and improved shelf life. The chemistry and characterization of various dichromated colloid solutions have been reviewed extensively by Kosar.[73] Good screens can be processed in a reproducible manner by following the supplier's instructions for a particular emulsion. The reference literature should be consulted, however, to obtain a better understanding of the factors affecting emulsions prior to and after application to the screen and their use and storage properties.

In use, the emulsion solutions are applied directly to the screens from one or both sides, using a wide blade or squeegee, and allowed to partially dry. A second coat is preferred on the outside or printing surface to ensure complete filling of the screen. Double coating tends to build up the overall thickness of the screen. Increasing the thickness will result in a larger volume for each of the open areas of the washed-out pattern and, hence, thicker ink deposits. The application of a thick emulsion must be controlled to ensure that the thickness is uniform. This can be accomplished by forming a thin tape border on the screen outside the image area. This tape, which serves as a shim, determines the thickness of the emulsion when a straightedge blade is passed across the surface. During washout, thick emulsion coatings tend to be undercut. This leads to a printed ink pattern larger than the open areas of the artwork. Coated screens, when dried, contain residual moisture in equilibrium with the surrounding atmosphere. The presence of water has a marked effect on spectral sensitivity, which increases with an increase in moisture content, as shown in Fig. 62. As a result, processing of screens should be conducted under controlled temperature and humidity conditions so that the proper exposure time will be known.

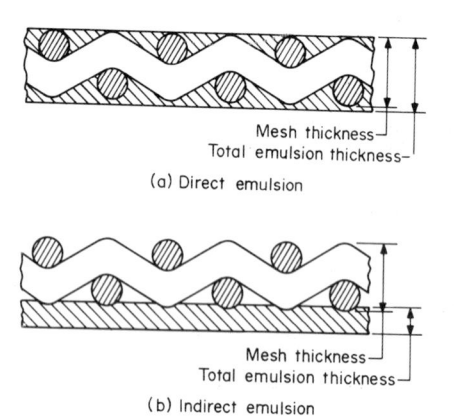

Fig. 61 Cross sections of direct- and indirect-emulsion screens.

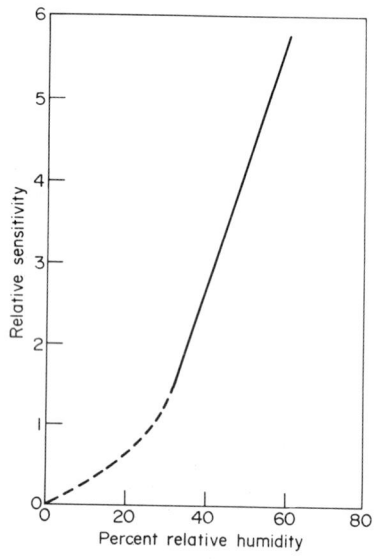

Fig. 62 Effect of relative humidity on dichromate-sensitized emulsion sensitivity.[73]

Indirect emulsions are supplied either as an unsupported film or as a supported film on a plastic backing sheet. Dry-film thickness for different types of film is usually in a range of 0.001 to 0.002 in. Both presensitized and unsensitized films are available commercially. Most indirect emulsions have a composition similar to that of the direct emulsions and are processed similarly. Recently, film compositions have become available that exhibit improved resistance to heat and better physical properties. Some of these types require a heat treatment or a chemical conversion step to provide optimum optical and physical properties before use.

The main difference between the indirect emulsion and the direct type is that the indirect is exposed and the pattern is developed before attachment to the screen. Improved pattern detail is obtained since the screen does not interfere with the development process. This pattern detail is maintained after transfer to the screen. After development, the soft film is pressed onto the screen, with a light pressure, so that it is firmly attached and flat. It is then allowed to dry. This produces a screen with the mesh slightly embedded in the emulsion, as shown in Fig. 61b.

In normal use, a direct-emulsion screen is capable of producing 100,000 good impressions. Indirect-emulsion screens are more subject to wear, and a maximum of about 25,000 impressions can be obtained with one of them. The actual life of a screen, however, is dependent upon the particular application. In thick-film printing, for example, a maximum of only 20,000 and 5,000 impressions, respectively, can be expected for the two screen types because of the abrasive nature of the inks used. Other applications requiring only a printed resist image can attain the normal maximum life if proper process controls are employed.

The following factors should be considered in choosing an emulsion for a given application:
1. Pattern resolution and definition required
2. Stability with respect to process reagents and ink solvents
3. Number of parts to be printed
4. Good flexibility to prevent flaking of the screen during printing
5. Minimum pinholing tendencies
6. Ease of preparation of the stencil pattern—light sensitivity, ease of washout of unexposed areas, ease of attaching emulsion to screen (indirect emulsion)
7. Reasonable shelf life

Design considerations The major factors to be considered in the design of screens are (1) the pattern and (2) the thickness of ink required by the application. Two general types of electronic applications can be defined: those in which the printed image will function as a protective mask and those in which the printed image forms an active part of the product. The first type includes the printing of etch resists and plating resists. The second includes the printing of various conductive, resistive, or dielectric patterns for passive components or circuitry. The ink thickness required is highly dependent upon the application. In the printing of protective resists, an ink layer of sufficient thickness to hold up during the subsequent processing of the parts is required. Thicknesses in the neighborhood of 0.0005 in. or more are usually satisfactory. For resistive or dielectric inks, however, the thickness must be carefully controlled since this parameter, along with pattern area, determines the characteristics of printed resistors and capacitors.

The pattern design, resolution, and tolerances obtained with a screen process are determined to a large extent by the mesh used in screen fabrication. An examination of mesh specifications will indicate design guidelines to be used in pattern layout. The important mesh parameters to be considered include the dimensions of open areas, wire diameter, and variations in mesh count. Specifications for a 200-mesh screen are illustrated in Fig. 63. If the edge of the pattern falls in between two wires, two effects can occur for direct-emulsion screens, namely, the unhardened portion of a partially exposed opening may remain in the mesh, or the hardened portion may be removed from the opening during development (or on subsequent printing). This creates a nonuniform edge in the screen pattern equivalent to plus or minus one-half the mesh opening and is commonly referred to as "stepping" (Fig. 63b). Most direct emulsions are subject to the above limitation; dichromate-sensitized emulsions show more of an effect than nondichromated types. Because of this factor, patterns of lines only one mesh opening wide or smaller should be avoided where continuous lines are to be printed. Also, the

14-120 Photofabrication

tolerance that can be held in the printed image is limited to approximately plus or minus one-half the mesh opening. Thus, in designing patterns to be printed with a 200-mesh screen, a minimum line width equivalent to two mesh openings or 0.0084 in. should be specified with an expected tolerance of ±0.001 in. With reasonable care, direct-emulsion screens can be used for patterns containing 0.005 in. lines on 0.010-in. centers.

Good design practice requires layout of patterns on an x-y grid, using the mesh wire spacing as a guide. The patterns are then aligned to the wires in the mesh during exposure. Registration of the total pattern is hard to attain, however, because of variations of mesh count, and some "stepping" may occur. It should also be noted that stepping is hard to avoid with nonlinear patterns.

(a) 200 mesh specifications

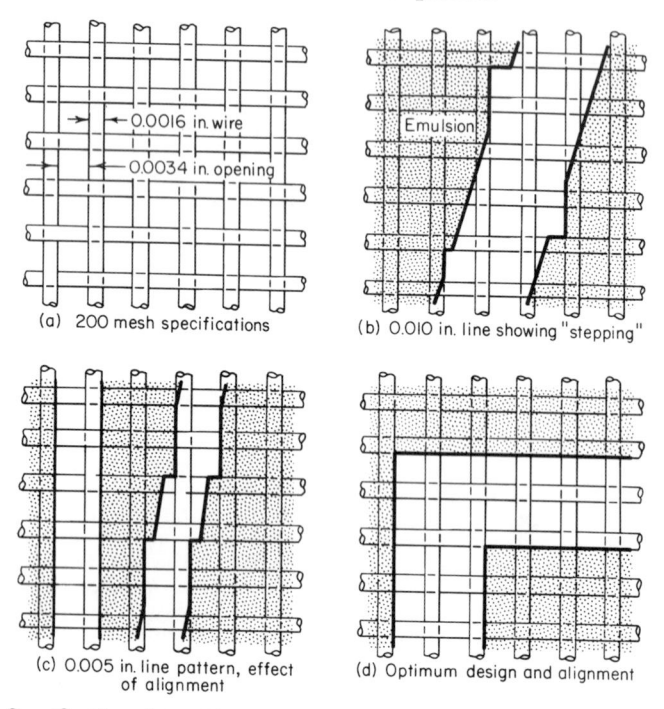

(b) 0.010 in. line showing "stepping"

(c) 0.005 in. line pattern, effect of alignment

(d) Optimum design and alignment

Fig. 63 Specifications for a 200-mesh stainless steel screen and the effect of pattern alignment.

Patterns with better edge definition and higher resolution can be obtained with indirect emulsions since development is not subject to screen interference, and therefore stepping is minimized. Line resolution is subject to the wire diameter in that a sufficient opening must be available for ink transfer. Patterns with 0.003-in. lines and 0.005-in. spaces are attainable with careful control of processing for indirect emulsions attached to screens of 325 mesh or higher.

Wire diameter in the mesh and overall screen-emulsion thickness have a direct effect on the thickness of the printed image. This factor is extremely important in the printing of electronic passive components, and most studies have centered in this area. The effect of mesh size on the printed-film thickness for several resistor inks using direct-emulsion screens is summarized in Tables 55 and 56. As the wire diameter is decreased with an increase in mesh size, the overall thickness of ink transferred is decreased proportionately.

The above data are based on typical thick-film resistor ink compositions having viscosities ranging from 150,000 to 300,000 cp (Brookfield, RTV, 10 rpm, no. 7

Applications 14-121

TABLE 55 Effect of Mesh Parameters on Film Thickness and Electrical Properties of Various Resistor Compositions[158]

E. I. du Pont composition	Mesh	Film thickness, mils Dry	Film thickness, mils Fired	Resistance, ohms/sq	Hot (125°C) TCR
7800	105	1.23	0.66	0.15	133
	165	1.00	0.64	0.16	421
	200	0.59	0.45	0.21	643
	325	0.32	0.29	0.29	461
7826	105	1.64	0.99	153	662
	165	0.88	0.66	313	646
	200	0.75	0.45	532	594
	325	0.39	0.36	1,034	547
7827	105	1.55	0.94	1,245	404
	165	1.13	0.77	1,480	405
	200	0.82	0.55	2,051	319
	325	0.56	0.28	3,953	229
7828	105	1.48	0.98	2,650	148
	165	1.09	0.73	3,520	124
	200	0.84	0.58	4,720	81
	325	0.57	0.31	7,860	2.8
7832	105	1.59	1.10	5,560	65
	165	1.29	0.75	6,430	58
	200	0.99	0.53	8,230	31
	325	0.73	0.35	13,180	19
7860	105	1.68	0.98	9,830	403
	165	1.27	0.76	14,380	365
	200	0.89	0.50	32,900	262
	325	0.62	0.28	75,500	205

TEST DATA: Four-square test patterns (0.25 by 0.75 in.) printed onto 1 by 1 by 0.025-in. AlSiMag 614 substrates with a Presco 100B printer. Terminations were Du Pont 7553 Pt-Au fired at 1000°C and a 10-min soak. Resistors fired at 760°C peak and 45-min cycle through a continuous-belt furnace. Dry films measured optically with a vernier optical gage. Fired films measured by a Starrett Model 652 gage.

TABLE 56 Effect of Mesh Characteristics (Analysis of Data from Table 55)

Stainless steel mesh	Wire diameter, in.	Mesh opening, in.	% open area	Average film thickness, in. Dry (T_D)	Average film thickness, in. Fired (T_F)	Ratio of thickness to wire diameter T_D/d	Ratio of thickness to wire diameter T_F/d
105	0.003	0.0065	46.9	0.0015	0.00094	0.51	0.31
165	0.0019	0.0042	47.1	0.0011	0.00072	0.58	0.38
200	0.0016	0.0034	46.2	0.00081	0.00051	0.51	0.32
325	0.0014	0.0017	30.5	0.00053	0.00031	0.38	0.22

14-122 Photofabrication

spindle, 25°C ± 1°C). Wet-film thickness for the above inks is approximately double the dry-film thickness, averaging 0.0015 in. for 200-mesh screens and 0.002 in. for 165-mesh. The inks used for masking applications have a still lower density and viscosity and transfer a thinner inked image. Data for the printing of a resist-type ink using various monofilament nylon screens are presented in Table 57. Little effect of mesh open area is noted on the thickness and the printed ink pattern because of the leveling and flow characteristics of this type of ink.

TABLE 57 Effect of Mesh Characteristics on Printed-ink Thickness[159]

Nitex mesh	Thread diameter, in.	Mesh opening, in.	% open area	Printed wet-paint thickness	Ratio of thickness to thread diameter
120-T	0.0028	0.0055	44.5	0.000925	0.33
185-T	0.0019	0.0035	43	0.00059	0.31
240-S	0.0014	0.0027	44	0.00053	0.38
240-T	0.0015	0.0026	39	0.00057	0.38
240-HD	0.0017	0.0024	35	0.00061	0.36
306-T	0.0014	0.0019	33.5	0.00049	0.35

TEST DATA: Ink: 37% binder, 44% pigment, 19% solvent, viscosity of 1,144 cp. 60–65% relative humidity at 68°F. Direct-emulsion screen with no overfill. Pattern printed on aluminum foil.

General guidelines for use in the layout and design of screen patterns for large-scale production are summarized as follows:

1. Pattern layout should be based on the mesh-wire spacing and openings in an x-y orientation.
2. Minimum line dimensions should be equal to two mesh openings.
3. A tolerance of plus or minus one mesh opening should be allowed.
4. Nonlinear designs should be avoided with fine-line patterns.
5. A minimum line width of 0.010 in. should be used for either direct-emulsion screens or indirect-emulsion screens.
6. Patterns should be specified on x and y axes with the y axis representing the direction of squeegee motion.
7. A minimum spacing of 1 in. outside the squeegee contact area on all sides should be provided to avoid undue strain at the edges of the screen.

The pattern design will vary considerably with different applications. Detailed guidelines for the design of thick-film patterns have been summarized by Keister and Auda.[160] Further details of the screen printing process for fabricating microcircuits can be found in the *Proceedings of the Electronic Components Conference*, the *Proceedings of the International Society of Hybrid Microelectronic Conference*, Ref. 158, and a recent book by Hughes.[161]

Requirements for the screening of mask patterns for printed-circuit plating and etching are less stringent since wide lines and bigger line separations are generally needed. Most applications are in home-instrument-type circuits of Classes 1 and 2 as listed in Table 38. The above general guidelines should be used in circuit design. Further details can be found in Ref. 87.

In addition to emulsion-screen stencils, two other types of stencils are used in the screen printing of fine-detailed patterns. These include photoetched-metal masks suspended on a metal screen and photoetched-metal masks directly attached to a standard screen frame. Both types are produced by photoetching, or photoetching in combination with photoforming, and are described in detail on page 14-86. Etched mask screens produce better definition than can be obtained with emulsion screens, and give better control of ink transfer. This type of mask should be considered for patterns having minimum line dimensions of 0.005 in. or less where maximum resolution is required.[162]

Processing of screen stencils Material and design guidelines for emulsion screen stencils were presented in the previous sections. Details for the fabrication of screen-stencil patterns are presented here, based on average processing schedules from various emulsion suppliers. Film or glass artwork can be used. Artwork is first generated with the accuracy necessary for the particular application. The procedures outlined below apply to stainless steel or nylon mesh screens assembled on cast aluminum frames.

All screens must be cleaned thoroughly to provide adequate wetting and adhesion of the emulsion. New stainless steel mesh can be adequately cleaned, using degreasing solvents such as detergent-trichloroethylene mixtures. Scrubbing with a stiff nylon brush will aid removal of other soils. When using reclaimed screens, two cycles of scrubbing with a fine nonchlorinated pumice cleanser or trisodiumphosphate powder is recommended. The cleaned screens are then rinsed in hot water and dried before emulsion coating.

Nylon screens can be cleaned by scrubbing as above; however, caution should be used to prevent mechanical injury to the threads. New nylon screens have a very smooth thread surface and must be pretreated to produce good adhesion of the emulsion. This can be accomplished by deglazing in a 20 percent caustic solution for 10 to 20 min, followed by neutralization in a weak acid solution. Proprietary treating solutions which promote good adhesion are also available from emulsion suppliers. After treatment, screens are rinsed in hot water and dried.

Coating and exposure procedures are summarized in the following. The coating process should be performed under low-voltage incandescent light or yellow lights.

Direct-emulsion Screens

1. Mix sensitizer and emulsion solutions thoroughly according to supplier's instructions, using plastic or wooden stirrer and plastic or glass container.

2. Pour a heavy strip of emulsion along one edge of the outside of the screen (printing contact side). Squeegee the emulsion with an aluminum, a plastic, or a stainless steel applicator blade across the screen with two or three even strokes. Repeat procedure to coat the other side.

3. Allow screen to dry. Drying is normally carried out in circulating airflow (120°F maximum) for a period of 20 to 60 min.

4. Repeat the coating procedure with a second thin coat on the outside of the screen to produce a smooth printing surface, and dry as above.

5. Then expose the screen to a photomask of the desired pattern, using a vacuum contact frame. The pattern is normally centered and aligned to the mesh wires in an x-y registration. Exposure times will vary according to the particular emulsion used and the thickness. With dichromate-sensitized emulsion, exposure times of 1 to 4 min are typical for carbon-arc light sources of 35 to 50 amp at a 3-ft distance from the lamp. Exposure times for diazo-sensitized emulsions require two to three times longer exposures for equivalent conditions.

6. Wash out the unexposed portions by soaking in warm water (100 to 120°F) for approximately 2 min, and then directing a forced spray against both sides of the screen to develop a sharp pattern. Ultrasonics have also been used to wash out the pattern in place of a spray.

7. Inspect for emulsion thickness, pattern definition and alignment, pinholes, and tension. Pinholes and open areas around the edge of the frame are filled with blockout lacquer.

Indirect-emulsion Screens

1. Presensitized films are cut to size and placed in a vacuum exposure frame with the artwork against the plastic support film. Exposure times for most films will vary from 1 to 4 min, for arc lamps of 35 to 50 amp. Optimum exposure times for a particular facility should be determined by running a series of test samples covering a range of exposures.

2. The exposed film is developed in the recommended developer solution for 1 to 6 min.

3. Wash out the film in warm water (110 to 120°F) until a sharp, clear image is obtained, and rinse in cold water.

4. Place the washed-out film with the emulsion side up on a flat support board. Position the screen over the film in correct alignment, and lower into contact with the emulsion. A light pressure should be used to give a uniform level contact across the screen.

5. Using a light pressure, blot excess water from screen with an absorbent paper such as newsprint, and allow film to dry slowly.

6. When the film is dry, peel off the plastic backing sheet, and wash surface with naphtha or toluene, if required, to remove any adhesive films present.

7. Inspect for pattern definition and alignment, pinholes, and tension.

The instructions presented above are typical for screen fabrication. Processing should be carried out under controlled temperature and humidity and clean conditions. Optimum exposure, development, and washout times should be determined for a particular emulsion, using test samples and recommended solutions. Subsequent processing should be controlled accordingly.

Applications of Three-dimensional Photosensitive Materials

Photosensitive resist materials, when used as protective masks, serve as essentially two-dimensional coatings to define an image in the part to be processed. The resist thickness is important only in relation to its protective value. Screening emulsions, on the other hand, utilize the emulsion thickness to control the quantity of ink transferred, and hence function in a three-dimensional sense. The capability of shaping a photosensitive material in three dimensions extends its usefulness to the direct photochemical production of parts previously requiring mechanical fabrication. Two such three-dimensional photosensitive systems have been applied to the electronics industry, namely, the photosensitive glasses developed by Corning, and the thick organic photopolymer materials developed by Du Pont. The properties of these materials were presented under Photosensitive Materials, page 14-56. The processing characteristics will be summarized here, and areas of application will be reviewed.

In the processing of a three-dimensional photosensitive material, exposure to light of suitable wavelength alters the chemical properties within the bulk of the material as well as at the surface. The shape of the exposed material is determined by the photomask pattern and the angle of incidence for the light hitting the surface. Thus, for a perfect collimated light source, a column of material is exposed down through the sample. Changing the angle of incidence creates a tapered exposure profile and a similar slope in the final etched part. The attainable exposure depth is limited only by the absorption of light in the material.

Photosensitive glass Photosensitive glass materials are produced by Corning in flat sheets in a thickness range of 0.052 to 0.235 in. Exposure of these materials to light, followed by suitable heat treatment, produces an increase in the solubility of the exposed glass regions. For hydrofluoric acid solutions this increase is a ratio of approximately 15:1 for the exposed over the unexposed portion. If collimated light is employed for exposure, a slight tapering in the sidewalls with depth of about 4° is obtained during subsequent etching because of the above etch factor. This creates a conical-shaped hole for single-sided etching and an hour-glass-shaped hole for double-sided etching. The limiting dimensions for the fabrication of holes and the tolerances expected are shown in Tables 58 and 59, for one- and two-sided etching. Closer hole spacing can be achieved where double-sided etching is used. It should be noted that control of etching with the photosensitive glasses is dependent upon the bulk properties of the materials and not just the surface image. Thus, much smaller hole diameters and closer spacing can be obtained for a given thickness than could be obtained with simple resist etching of a nonphotosensitive glass layer.

Processing. The following general processing conditions have been described by Stookey for chemical machining of photosensitive glass:[77] (1) A saturation exposure is used; (2) heat treatment should be restricted to the minimum time

TABLE 58 Allowable Dimensions for Etched Holes in Photosensitive Glass[77b]

One-sided etch

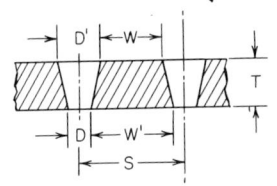

Min D = $T/20$ or 0.007 in., whichever is larger
Max T = $8D$ or 0.235 in., whichever is smaller
Min W = 0.004
D' = $D + T/20$
Min W' = $0.004 + T/10$
Min S = $D + 0.004 + T/20$ (for $D = T/8$)

Two-sided etch

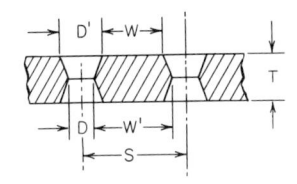

Min D = $T/20$ or 0.007 in., whichever is larger
Max T = $10D$ or 0.235 in., whichever is smaller
Min W = 0.004
D' = $D + T/40$
Min W' = $0.004 + T/20$
Min S = $D + 0.004 + T/40$ (for $D = T/10$)

TABLE 59 Tolerances for Photosensitive Glass Parts[77b]

Dimension (all ± tolerances), in.	Unceramed	Ceramed
Cut edge	1/16	1/16
Ground edge:		
Standard	0.005	0.005
Premium	0.002	0.002
Etched edge:		
0–1/2 in. long	0.001	0.0015
Over 1/2 in. long	0.002	0.0025
Hole ID, and slot length (or width):		
0–1 in	0.0015	0.002
1–3 in	0.003	0.004
Hole spacing (center to center):		
0–1/2 in	0.001	0.0015
1/2 in. and over	0.002	0.002
Out of flat	0.001	0.002
Pattern concentricity:		
Exposed separately	0.006	0.006
Exposed simultaneously	Same as hole-spacing tolerances	

required for complete development; (3) maximum etching efficiency is obtained with suitable agitation in a 5 percent aqueous hydrofluoric acid solution at 20°C.

During exposure, the number of crystal nuclei in the material defined by the mask reaches a maximum and saturates. Overexposure results in further crystallization outside the image area, owing to dispersion of the light in the glass, and an enlargement of the image results. Cerium-sensitized glass exhibits a maximum sensitivity at a wavelength of 3100 Å (see Fig. 31). Film or glass photomasks absorb strongly in this range and require long exposure times. Optimum results can be obtained by using an etched-metal mask to define the pattern or a quartz-base photomask.

Development time is proportional to the viscosity of the glass and decreases exponentially with increase in temperature. Complete development of the exposed material is obtained at 600°C in 1 hr. On continued heating, further crystal growth takes place, causing a decrease in the image accuracy.

Etching is normally performed in a stirred 2 to 10 percent aqueous hydrofluoric acid solution at room temperature. Etch rate of the exposed image in a 5 percent solution is approximately 40 mils/min. Two-sided etching is generally preferred for hole patterns.[77b] Smaller, closer-spaced holes can be obtained, and sharp-corner edges at the surface of the holes are avoided (see Table 58). This is particularly important in applications requiring plating of through holes such as is required for printed circuitry.

Electronic Applications. For photosensitive glasses these include various parts for vacuum tubes, digital converters, thermocouple holders and tubes, printed-circuit substrates, various jigs and fixtures, and evaporation masks. Corning has utilized the photosensitized glasses for the fabrication of fluid-circuit components. The required channel configurations are etched into a layer of glass and sealed with a cover layer of similar material. Complex multilayer fluid structures can also be produced. The details of these "fluidic" circuits have been reviewed elsewhere.[163]

Photopolymer systems Templex photosensitive plastic material[80] is available from Du Pont in supported and unsupported sheets up to 18 by 24 in., with thicknesses in the range of 0.008 to 0.047 in. A supported photopolymer material called Dycril is also available from Du Pont for use in the preparation of relief printing plates.[78] Both materials are applicable to the electronics industry.

Process Considerations. The processing of Templex involves exposure to light between 3200 and 4000 Å through a photographic film transparency, followed by washout of the unexposed portions in a dilute caustic solution. Film photomasks with a matte finish on the photographic emulsion side are preferred to avoid pockets of trapped air between the film and the photopolymer surface. Processing equipment includes a vacuum frame having a clear plastic cover, a high-intensity carbon-arc lamp, and a high-pressure spray washout unit.

Two factors affect the exposure of these photopolymers and the resultant shape of the processed material, namely, the presence of absorbed oxygen and the incident light angle. Oxygen present in the plastic from the atmosphere prevents polymerization in depth because of a preferential reaction of oxygen with the sensitizer materials used. Conditioning the photopolymer in carbon dioxide or nitrogen removes the oxygen and enables complete polymerization in depth of the exposed layer to be achieved during exposure.

The incident light affects the shape of the sidewalls in a manner similar to that noted for the photosensitive glasses. Near-vertical sidewalls can be obtained in two ways: by exposure of conditioned material to collimated light or by exposure of unconditioned material to polymerize the surface layer, followed by controlled washout to prevent excessive undercutting of unpolymerized material. The effects of variation of the incident light angle and of conditioning are illustrated in Fig. 64 for line patterns exposed in a 0.042-in.-thick photopolymer layer.

Pattern design must take both these factors into account as well as the washout characteristics of the material. A shrinkage of the polymer occurs during exposure as a result of the polymerization. This shrinkage amounts to approximately 0.001 in. for any edge when supported material is employed; corresponding compensation should be made during artwork preparation. Unsupported samples exhibit a greater amount of shrinkage and are more difficult to control dimensionally.

The washout or etching of the unexposed plastic is conducted in 0.036 N sodium hydroxide solution between 28.3 and 29.5°C with a spray pressure of 40 to 50 psi. Etch rate is 0.006 to 0.007 in./min and can be controlled to ±0.001 in. for patterns with openings above 0.010 in. The minimum pattern resolution that can be washed out is dependent upon the spray characteristics. Generally, patterns with open spaces below 0.010 in. are restricted in depth to the space dimension. For example, 0.005-in.-wide spaces can be washed out to a depth of 0.005 in. under the above spray conditions.

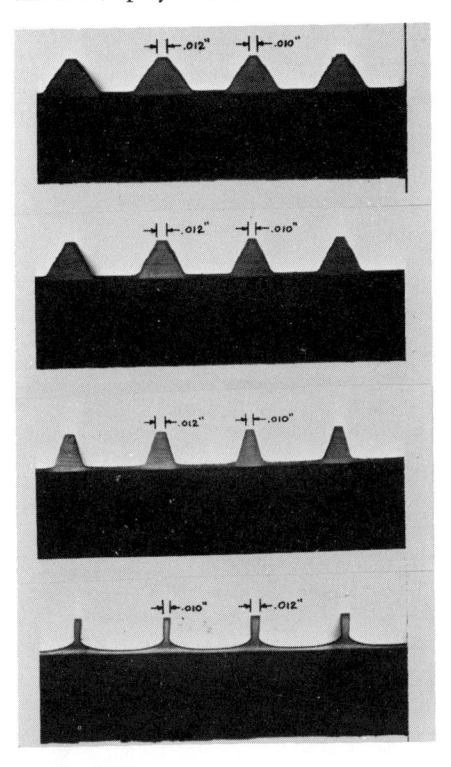

Fig 64 Type 42A101 Templex plates exposed at various lamp-to-sample distances. (a) 30-in. distance conditioned; (b) 40-in. distance conditioned; (c) 50-in. distance conditioned; (d) 50-in. distance unconditioned.

Fig. 65 Dycril dry offset printing plate for the printing of resist patterns in printed-circuit processing. Line dimensions are 0.004 in.

Applications. Du Pont photopolymer materials have been used to fabricate tooling jigs, alignment fixtures, pantograph masters, casting molds, and relief printing plates for electronic applications. Patterns for fluid-device components can also be readily formed in supported material for prototype and production work, using cemented plastic cover sheets. The processing of typical patterns is outlined below.

Processing of an Alignment Jig. An alignment jig with 0.040-in.-deep cavities and vertical sidewalls can be processed as follows:

1. Mount unconditioned Templex 42A101 plate in vacuum frame with film negative in place (e.g., Du Pont COD film)—emulsion side against photopolymer surface.

2. Expose: e.g., 2 to 2½ min with MacBeth 140-amp carbon arc at a distance of 50 in. or 3 to 5 min in a plate-making table with a 30-amp arc lamp.
3. Washout: 0.036 N NaOH at 29°C, 8 to 12 min in a Master Washout Unit (40 to 50 psi with nozzles located 5¾ in. from the plate). Alternate washout; 16 min in a Chemcut Model 610 etcher.
4. Rinse with water.
5. Air-dry.
6. Postexpose 2 to 4 min where added chemical resistance is required.

Processing of Printing Plates. Relief plates for the offset printing of plating or etching masks are processed according to the following schedule:
1. Condition Dycril plates in carbon dioxide 5 hr for 0.015-in. relief and 24 hr for 0.040-in. relief.
2. Mount in vacuum frame with film negative in place, emulsion side against photopolymer surface.
3. Expose 2 to 2½ min with MacBeth 140-amp carbon arc at a distance of 30 to 40 in. (this produces a tapered support to the printing relief pattern—see Fig. 64).
4. Wash out to produce minimum relief of 0.005 in.
5. Rinse with water.
6. Air-dry.
7. Postexpose 2 to 4 min where added chemical resistance is required.

Photopolymer printing plates can be used for the dry offset printing of plating or etching resist patterns in a manner similar to screen printing.[164] This type of printing is capable of defining patterns over large areas. Pattern dimensions down to 0.004 in. with a registration of 0.001 in. are easily formed by this technique. Figure 65 shows a photograph of a Dycril printing plate containing 0.004-in.-wide lines. This plate is suitable for the printing of a resist pattern and can be used in the fabrication of printed circuits for the interconnection of integrated-circuit chips.

At present the electronic applications of three-dimensional photopolymer systems are relatively limited compared with those for photoresist materials. Their usefulness and potential advantages have been established, however, and a large increase in the number of applications can be anticipated for the near future.

REFERENCES

1. R. K. Springborn (ed.): "Non-traditional Machining Processes," American Society of Tool and Manufacturing Engineers, 1967.
2. Woodring, G. Daniel: Production Chemical Machining: How It Operates and What It Costs, *Amer. Soc. Tool and Mfg. Eng. Tech. Paper* MR66-155, 1966.
3. Photofabrication, *Eastman Kodak Co. Pam.* 3-8.
4. Hook, H.: Automated Mask Production for Semiconductor Technology, *Solid State Technol.*, July, 1967.
5. Levine, J. E.: Process Analysis of Mask Making, *Solid State Technol.*, July, 1968.
6. Payne, P. D.: Photomask Technology in Integrated Circuits, *Solid State Technol.*, July, 1967.
7. Holthaus, D. J.: The Basic ABC's of Mask Making, *Proc. 2d Kodak Seminar on Microminiaturization*, Apr. 4–5, 1966.
8. Bell, E.: Recent Breakthroughs in Large Area, Extremely High-resolution Mask Making, *Proc. 2d Kodak Seminar on Microminiaturization*, April 4–5, 1966.
9. Kodak Plates and Films for Science and Industry, *Eastman Kodak Co. Period.* P1-66-Z, 1966.
10. Dimensional Stability of Kodak Acetate Films for the Graphic Arts, *Eastman Kodak Co. Publ.* Q-33.
11. Physical Properties of Kodak Ester Base Films for the Graphic Arts, *Eastman Kodak Co. Publ.* Q-34.
12. Physical Characteristics of Kodak Glass Base Plates, *Eastman Kodak Co. Publ.* Q-35.
13. Template Making with Cronaflex, *E. I. du Pont de Nemours & Co. Tech. Bull.* 1.

14. *Eastman Kodak Co. Pam.* P-34, 1968.
15. Celanar Polyester Film Properties, *Celanese Plastics Co. Bull.* DIA.
16. Artwork for Photofabrication, *Eastman Kodak Co. Period.* P-1-63-2, 1963.
17. Eastman Kodak Company: *Kodak Tech. Bits*, no. 3, 1968.
18. Morgan, R. A.: The Brightening Outlook for Optical Design, *Prod. Eng.*, Apr. 11, 1966.
19. Strickland and Crawford: Advances in Computer Generation of Master Artwork for Microminiature Circuits, *Solid State Technol.*, July, 1967.
20. Cook, P. W., W. E. Donath, G. A. Lemke, and A. E. Brennemann: An Automatic Integrated Circuit Mask Artwork Generating System, *IEEE J. Solid State Circuits*, vol. SC-2, no. 4, December, 1967.
21. Tong, J. B.: Mask Manufacture for Integrated Circuits, *Solid State Technol.*, July, 1968.
22. Beeh, R. C. M.: Automation and Motor Function Routines for Mask Making, *Solid State Technol.*, July, 1968.
23. Glendinning, W. B., and S. Marshall: Microcircuit Photomasks from Automatic Techniques, *IEEE Trans. Electron Devices*, vol. Ed-12, no. 12, December, 1965.
24. Fisk, C. J., D. L. Caskey, L. E. West: ACCEL: Automated Circuit Card Etching Layout, *Proc. IEEE*, vol. 55, no. 11, November, 1967.
25. Spitalny, A., and M. J. Goldberg: On-line Graphics Applied to Layout Design of Integrated Circuits, *Proc. IEEE*, vol. 55, no. 11, November, 1967.
26. Lathrop, J. W., R. S. Clark, J. E. Hull, and R. M. Jennings: A Discretionary Wiring System as the Interface between Design Automation and Semiconductor Array Manufacture, *Proc. IEEE*, vol. 55, no. 11, November, 1967.
27. Hopkins, H. H.: The Frequency Response of a Defocused Optical System, *Proc. Roy. Soc.*, 231A, p. 91, 1955.
28. Hardy, L. H., and F. H. Perrin: "The Principles of Optics," McGraw-Hill Book Company, New York, 1932.
29. Altman, J. H.: Photography of Fine Slits Near the Diffraction Limit, *Proc. 2d Kodak Seminar on Microminiaturization*, Apr. 4–5, 1966.
30. Swing, R. E., and J. R. Clay: Ambiguity of the Transfer Function with Partially Coherent Illumination, *J. Opt. Soc. Amer.*, October, 1967.
31. Johnson, H.: Restoration of High Frequencies in Photomask Line Patterns by Photographic Nonlinearity, *Solid State Technol.*, July, 1967.
32. Strauss, W. A., Jr.: Photofabrication Camera Performance: Random Image Distortion, *Reprographics*, October, 1966.
33. W. Watson & Sons, Ltd., Barnet Herts, U.K.: Provisional Information on the New Watson Step and Repeat Camera Mk III.
34. David W. Mann Company: Manufacturer's literature.
35. The Jade Corporation: Manufacturer's literature.
36. Opto Mechanisms, Inc.: Manufacturer's literature.
37. Beeh, R. C. M.: A High Accuracy Automated Microflash Camera, *Solid State Technol.*, July, 1967.
38. Hilton, E. A., and D. M. Cross: Laser Brightens the Picture for IC Mask-making Camera, *Electron.*, Aug. 7, 1967.
39. Staff Report: Boosting IC's, *Electron.*, Apr. 17, 1967.
40. Schuetze, H. J., and K. E. Hennings: Micron and Sub-micron Patterns for Semiconductor Devices and Integrated Circuits, Electron Devices Meeting, IEEE, Washington, D.C., Oct. 20–22, 1967.
41. Lovering, H. B.: Direct Exposure of Photoresist by Projection, *Solid State Technol.*, July, 1968.
42. Basic Photography for the Graphic Arts, *Eastman Kodak Co. Publ.* Q-1.
43. Photographic Materials for the Graphic Arts, *Eastman Kodak Co. Publ.* Q-2.
44. Buzwa, M. J., G. G. Milne, and A. M. Smith: Optical Systems for Direct Projection on Photoresist, *Proc. Seminar on Ultra-microminiaturization*, Society of Photographic Scientists and Engineers, Nov. 7–8, 1968.
45. Geihes, G. I., and B. D. Ables: Contact Printing Associated Problems, *Kodak Photoresist Seminar Proc. Publ.* P-192-B, 1968.
46. Stevens, G. W. W.: "Microphotography," 2d ed., John Wiley & Sons, Inc., New York, 1968.
47. *Eastman Kodak Co. Pam.* P174, 1968.
48. Mees, K. E., and T. H. James (eds.): "The Theory of the Photographic Process," Macmillan Co. The Macmillan Company, New York.
49. Gaseous Burst Agitation for the Graphic Arts, *Eastman Kodak Co. Publ.* Q-37.

50. Techniques of Microphotography, *Eastman Kodak Co. Publ.* P-52, 1967.
51. Neblett, C. B.: "Photography: Its Materials and Processes," 5th ed., D. Van Nostrand Company, Inc., Princeton, N.J., 1952.
52. Damon, G. F.: The Effect of Whirler Acceleration on the Properties of the Photoresist Film, *Proc. 2d Kodak Seminar on Microminiaturization,* April 4–5, 1966; *Photochem. Fabr.,* vol. 2, no. 1, February, 1969.
53. Martinson, L. E.: The Technology of Microimage Resists, *Proc. 2d Kodak Seminar on Microminiaturization,* Apr. 4–5, 1966.
54. Shipley Co., Inc.: Technical data sheets on Shipley photoresist.
55. Kelley, R. F.: Relationship between Resist Thickness and Pinholing, *Proc. Kodak Seminar on Microminiaturization,* June 3–4, 1965.
56. E. I. du Pont de Nemours & Co.: Photopolymer Resist System, technical bulletin.
57. Dynachem Corporation: Technical data sheets on Dynachem photoresists.
58. General Aniline and Film Corporation: technical data sheets.
59. Philip A. Hunt Chemical Corporation: technical data sheets on Hunt photoresists.
60. *Eastman Kodak Co. Pam.* P-81 to P-86 and P-137.
61. Norland Products Incorporated: Technical data sheets on Norland photosensitive materials.
62. See, for example, *Electron. Packag. and Prod.,* vol. 8, no. 7, 1968; *Circuits Mfg.,* vol. 8, no. 5, 1968.
63. *a.* Kodak Photoresist Seminar, May 20–21, 1968.
    b. Bortfeld, R.: Requirements of a Production Photoresist System, *Electrochem. Technol.,* vol. 1, p. 242, 1963.
64. *a.* Millipore Corporation: Technical data sheets.
    b. Bendix Filter Division: Technical data sheets.
65. See references cited in 62, also:
    a. Journal of Photochemical Etching.
    b. Photochemical Fabrication.
    c. Solid State Technology.
66. Miller-Stephenson Chemical Co., Inc.: Technical data sheets.
67. Zicon Corporation: Technical data sheets.
68. Union Carbide Corporation: Thin Films of Poly-para-xylylene Technology and Applications, technical bulletin.
69. *a.* JEOL Instruments: Technical data sheets.
    b. Perkins, K.: Electron Beam Exposure of Photoresists, Kodak Photoresist Seminar, May 20–21, 1968.
70. *a. Journal of Photochemical Etching,* p. 4, November, 1966.
    b. Htoo, M. S.: *Photogr. Sci. and Eng.,* vol. 12, p. 109, 1968.
71. *a.* Tracerlab: Technical data sheets.
    b. Irving, S.: A Dry Photoresist Removal Method, *Kodak Photoresist Seminar Proc. Publ.* P-192B, 1968.
72. Duffek, E. F., and E. Armstrong: in C. F. Coombs, Jr. (ed.), "Printed Circuits Handbook," McGraw-Hill Book Company, New York, 1967.
73. Kosar, J.: "Light Sensitive Systems," John Wiley & Sons, Inc., New York, 1965.
74. Holahan, J. F.: *Electron. World,* December, 1965.
75. *Electron. News,* p. 42, Aug. 5, 1968.
76. Stookey, S. D.: Photosensitive Glass, *Ind. Eng. Chem.,* April, 1949.
77. *a.* Stookey, S. D.: Chemical Machining of Photosensitive Glass, *Ind. Eng. Chem.,* January, 1953.
    b. Lazar, N.: Photosensitive Glass and Chemical Machining, *Prod. Eng.,* July 11, 1960.
78. E. I. du Pont de Nemours & Co.: *Du Pont Graphic Arts Regist.,* vol. 5, no. 3, September, 1964.
79. U.S. Patent 3,081,168, Polyamide Photographic Printing Plate, Mar. 12, 1963.
80. E. I. du Pont de Nemours & Co.: *Du Pont Templex Ind. Bull.* 1, April, 1965.
81. Spring, S.: "Metal Cleaning," Reinhold Publishing Corporation, New York, 1963.
82. Lisman, T. (ed.): "Metals Handbook: Heat Treating, Cleaning and Finishing," 8th ed., vol. 2, American Society for Metals, 1964.
83. An Introduction to Photofabrication Using Kodak Photosensitive Resists, *Eastman Kodak Co. Publ.* P-79, 1966.

84. Metals and Plastics Publications, Inc.: "Metal Finishing Guidebook," 1968.
85. Eastman Kodak Co. Period. P-200-67-2, 1967.
86. Eastman Kodak Co. Publ. P-91, 1967.
87. Coombs, C. F. Jr., (ed.): "Printed Circuits Handbook," McGraw-Hill Book Company, New York, 1967.
88. Eastman Kodak Co. Period. P-200-67-3, 1967.
89. Etching Considerations in Chemical Milling, *Chemcut Corp. Tech. Bull.* 167, 1967.
90. *Hunt Chemical Corp. Tech. Bulls.* 3, 11-13, 16, 18, and 19.
91. Schlabach, T. D., and D. K. Rider: "Printed and Integrated Circuitry," McGraw-Hill Book Company, New York, 1963.
92. Etching Metals with Ammonium Persulfate, *FMC Corp. Bull.* 111, 1967.
93. Westinghouse Electric Corporation; Chemical Etching of Kovar Lead Frames, technical bulletin.
94. Nehervis, W. F.: The Use of Ferric Chloride in the Etching of Copper, The Dow Chemical Company, Midland, Mich., 1962.
95. Cupric Chloride Regeneration System, *Chemlea Corp. Bull.* ERS-67.
96. FMC Corporation: Caper Process, technical bulletin.
97. Printed Circuit Etching Facilities and Processes, *Chemcut Corp. Tech. Bull.* 567.
98. Rudimer, K. J.: The Caper Process, Institute of Printed Circuits Seminar, Mar. 20, 1968.
99. Benton, Rufus R. C.: Cupric Chloride–Copper Regeneration, Institute of Printed Circuits Seminar, Mar. 20, 1968.
100. Chemical Milling with Kodak Photosensitive Resists, *Eastman Kodak Co. Publ.* P-131, 1968.
101. Lowenheim, F. A.: "Modern Electroplating," John Wiley & Sons, Inc., New York, 1963.
102. Graham, A. K.: "Electroplating Engineering Handbook," 2d ed., Reinhold Publishing Corporation, New York, 1962.
103. Brenner, A.: "Electrodeposition of Alloys," vols. 1 and 2, Academic Press, Inc., New York, 1963.
104. Staff Report: *Mod. Plast.*, November, 1967.
105. U.S. Patent 3,257,215, June 21, 1966, and U.S. Patent 3,361,580, Jan. 2, 1968.
106. Goldie, W.: "Metallic Coating of Plastics," vol. 1, Electrochemical Publications Limited, Middlesex, England, 1968.
107. Dushman, S.: "Scientific Foundations of Vacuum Technique," 2d ed., John Wiley & Sons, Inc., New York, 1962.
108. Holland, L.: "Vacuum Deposition of Thin Films," John Wiley & Sons, Inc., New York, 1958.
109. Hass, G. (ed.): "Physics of Thin Films," vol. 1, Academic Press, Inc., New York, 1963.
110. Hass, G., and R. E. Thun (eds.): "Physics of Thin Films," vols. 2 and 3, Academic Press, Inc., New York, 1964, 1966.
111. Wehner, G. K.: *Advan. Electron. and Electron Phys.*, vol. 7, p. 239, 1955.
112. Kay, E.: *Advan. Electron. and Electron Phys.*, vol. 17, p. 245, 1962.
113. Maissel, L. I.: The Deposition of Thin Films by Cathode Sputtering, in G. Hass and R. E. Thun (eds.), "Physics of Thin Films," vol. 3, Academic Press, Inc., New York, 1966.
114. Kaminsky, M.: "Atomic and Ionic Impact Phenomena on Metal Surfaces," Academic Press, Inc., New York, 1965.
115. Gray, S., and P. K. Weimer: *RCA Rev.*, vol. 20, p. 413, 1959.
116. Basset, Mentor, and Pashley: "Structure and Properties of Thin Films," John Wiley & Sons, Inc., New York, 1959.
117. Pashley, D. W.: *Advan. Phys.*, vol. 5, p. 173, 1956.
118. Schwarz, H.: *J. Appl. Phys.*, vol. 34, p. 2053, 1963.
119. Chopra, K. L.: *J. Appl. Phys.*, vol. 37, p. 3405, 1966.
120. Baker, A. G.: *Proc. 2d Symp. on Deposition of Thin Films by Sputtering*, University of Rochester, June 7, 1967.
121. Lussow, R. O.: *J. Electrochem. Soc.*, vol. 115, p. 660, 1968.
122. Anderson, G. S., W. N. Mayer, and G. K. Wehner: *J. Appl. Phys.*, vol. 33, p. 2991, 1962.
123. Davidse, P. D.: Extended Abstracts, *13th Nat. Vac. Symp.*, American Vacuum Society, San Francisco, Calif., 1966.

124. Tsui, R. T. C.: *Proc. 2d Symp. on Deposition of Thin Films by Sputtering,* pp. 48–52, University of Rochester, June 7, 1967; and *Semicond. Prod. and Solid State Technol.,* vol. 10, p. 33, 1967.
125. Vossen, J. L., and J. J. O'Neill, Jr.: *RCA Rev.,* vol. 29, p. 135, 1968.
126. Heil, R., S. Hurwitt, and W. Huss: Sputter Etching of Microcircuits and Components, *Solid State Technol.,* vol. 11, p. 12, 1968.
127. Benton, Rufus R. C.: Chemical Machining, *ASME Tech. Paper* 685, 1965.
128. Photofabrication, *Eastman Kodak Co. Pam.* H3-8, 1967.
129. Chemical Machining, *Chemcut Corp. Bull.* CM-65A.
130. Chemical Micro Milling Company: Tolerance Data Sheet for Chemical Machined Parts.
131. Dougherty, F.: The Production of Masks by Electroforming or Etching Methods, *Photochem. Fabr.,* October, 1968.
132. Holahan, J. F.: Manufacture of Color Picture Tubes, *Electron. World,* December, 1965.
133. Schnorr, D. P.: chap. 1, in C. A. Harper (ed.), "Handbook of Electronic Packaging," McGraw-Hill Book Company, New York, 1969.
134. The Institute of Printed Circuits: "Printed Circuits Technical Manual."
135. *Mod. Plast.,* encyclopedia issue, 1968.
136. Steinhoff, T. L.: Some Unique Applications of Positive-working Photo Resists in the Plating of Printed Wiring Boards, *Proc. Plating in Electron. Symp.,* American Electroplators Society, 1966.
137. Medwin, A. H.: Fabrication of Complementary MOS Circuits, *RCA Eng.,* vol. 13, no. 3, 1967.
138. Camerzind, H. R.: A Guide to Integrated Circuit Technology, *Electro-Technol.,* February, 1968.
139. Lawson, Jr., T. R.: A Prediction of the Photoresist Influence on Integrated Circuit Yield, *Solid State Technol.,* vol. 9, no. 7, July, 1966.
140. Schwartz, S.: "Integrated Circuit Technology," McGraw-Hill Book Company, New York, 1966.
141. Staff Report: Chemicals in Semiconductor Manufacturing, *Electron. Packag. and Prod.,* June, 1968.
142. Mai and Looney: Thermal Growth and Chemical Etching of Silicon Dioxide Films, *Solid State Technol.,* vol. 9, no. 1, January, 1966.
143. Dey, J., M. Lundgren, and S. A. Harrell: Parameters Affecting the Etching Edge of Silicon Dioxide, *Kodak Photoresist Seminar Proc. Publ.* P-192-B, 1968.
144. Kern, W., and N. Goldsmith: The Deposition of Vitreous Silicon Dioxide Films from Silane, *RCA Rev.,* vol. 28, no. 1, 1967.
145. Kern, W.: A Technique for Measuring Etch Rates of Dielectric Films, *RCA Rev.,* vol. 29, no. 4, 1968.
146. Troy, E. M.: Integrated Circuit Operations at RCA Somerville, *RCA Eng.,* vol. 13, no. 1, 1967.
147. Holland, L. (ed.): "Thin Film Microelectronics," John Wiley & Sons, Inc., New York, 1965.
148. Peek, J. R.: *Semicond. Prod. and Solid State Technol.,* vol. 10, no. 5, 1967.
149. Maissel, L. I.: *Semicond. Prod. and Solid State Technol.,* vol. 11, no. 5, 1968.
150. Gregor, L. V., and R. E. Jones: *Semicond. Prod. and Solid State Technol.,* vol. 11, no. 5, 1968.
151. Wagner, R., and R. R. Urlau: Miniature High Range Resistors, *Proc. Electron. Components Conf.,* 1967.
152. Guttenplan, J. D., D. Maslow, and W. S. DeForest: A Photo-etch Process for Fabrication of Hybrid Thin Film Circuitry, *Proc. 2d Symp. an Hybrid Microelectron.,* 1967.
153. Huber, F., W. Witt, and I. H. Pratt: Thin Film Hafnium Technology, *Proc. Electron. Components Conf.,* 1967.
154. Schwartz, N., and R. W. Berry: in G. Hass and R. E. Thun (eds.), "Physics of Thin Films," vol. 2, Academic Press, Inc., New York, 1964.
155. McLean, D. A., and W. H. Orr: Tantalum Integrated Circuits, *Bell Lab. Rec.,* October–November, 1966.
156. Foley, M. A.: Batch Fabrication of Thin Film Resistors for Hybrid Circuitry by Electroless Deposition of Nickel Phosphide, *Proc. Elec. Components Conf.,* 1967.
157. Coronis, H. L.: Screen Tension and Mesh Alignment for Control of the Off-contact Printing Process with Direct Emulsion Screens and Indirect Metal Mask,

2d Symp. on Hybrid Microelectron., International Society for Hybrid Microelectronics, October, 1967.
158. E. I. du Pont de Nemours and Co.: "Thick Film Handbook," sec. R-2, 1967.
159. Kressilk Products, Inc.: General Catalogue.
160. Keister, F. Z., and D. Auda: Design Guide for Thick Film Hybrid Microcircuits, *2d Symp. on Hybrid Microelectron.*, International Society for Hybrid Microelectronics, October, 1967.
161. Hughes, D. C., Jr.: "Screen Printing of Microcircuits," Dan Mar Publishing Co., Somerville, N.J., 1967.
162. Short, O. A.: Conductor Compositions for Fine Line Printing, *Electron. Packag. and Prod.*, February, 1968.
163. Shinners, S. M.: *Electro-Technol.*, March, 1968.
164. Ryan, R. J., T. E. McCurdy, and N. E. Wolff: Additive Processing Techniques for Printed Circuit Boards, *RCA Rev.*, vol. 29, no. 4, 1968.

Chapter 15

Materials for the Space Environment

JOHN B. RITTENHOUSE
Advanced Technology, Lockheed-California Company,
Burbank, California

JOHN B. SINGLETARY
Lockheed Missiles and Space Company,
Sunnyvale, California

CLAUS G. GOETZEL
Materials Sciences Laboratory, Lockheed Research Laboratory,
Palo Alto, California

The Space Environment	15–2
Ascent Environment	15–2
The Upper Atmosphere	15–6
Nonpenetrating Radiation	15–9
Penetrating Radiation	15–16
Meteoroids	15–28
Effect of Space Environment on Electronic Materials and Components	15–33
Principal Environmental Effects	15–33
Other Environmental Effects	15–49
Outlook	15–50
Acknowledgment	15–51
References	15–51

15-2   Materials for the Space Environment

THE SPACE ENVIRONMENT

Ascent Environment

The placement into earth or solar orbit of a spacecraft requires that a considerable amount of energy be expended in removing it from the gravitational field of the earth. During this removal process or ascent phase, much of the energy is unavoidably wasted, such as that used to heat the atmosphere, and some of the waste energy is transferred to the spacecraft or payload stage of the rocket vehicle. This may take the form either of thermal energy arising from aerodynamic interaction of the vehicle with the atmosphere or of mechanical energy sources such as aerodynamic pressure and shear forces, acceleration loading, and shock, and vibration. All these factors are here collectively termed the ascent environment of the spacecraft.

Since the ascent environment is not a natural one but a consequence of spacecraft launching, it will show considerable variation, depending on the booster rocket used, as well as the design of the spacecraft, particularly its coupling with the booster upper stage and its nose-fairing protective device. Because of this variability in the environment, it must be rather carefully defined in nature and effects for each combination of spacecraft and launch vehicle.

Aerodynamic heating   Aerodynamic heating refers to the thermal energy transfer between air and a body moving through it at high velocity. The sources of the thermal energy are the frictional forces generated by the relative movement of the air and the body.

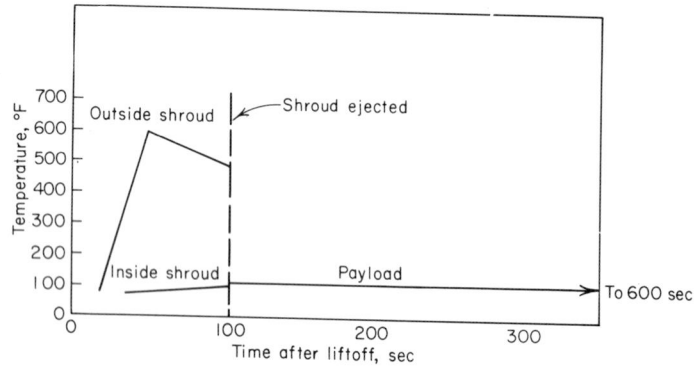

Fig. 1   Representative heating profile for Scout launch vehicle.

Fundamentally, air particles exchange energy with the skin element upon collision. The mechanisms giving rise to these collisions and influencing the attendant energy interchange may include molecular thermal conduction between the skin surface and the adjacent air layer as well as between layers of air, energy transport by diffusion of turbulent eddies of air from a hot portion to a cooler portion, and the formation of shock waves. Three types of molecular flow may be considered in discussing aerodynamic effects: continuum, transition, and free-molecule flow. Continuum flow is commonly encountered during ascent to altitudes of about 60 sm.[*] In this type of interaction, molecule-molecule and molecule-vehicle collisons are so numerous that the air may be treated as a continuous fluid. When a body moves through this fluid at a high velocity, kinetic energy of motion is converted to thermal energy both by compression, as through shock waves, and by friction between fluid particles.

As higher altitudes are reached, the average distance traveled by molecules between collisions (mean free path) becomes larger, as a consequence of the decreasing

[*] Statute miles.

air density. This causes deviations from continuum flow, at first in the immediate vicinity of the vehicle surface. With further decrease in air density, the mean free path of the gas molecules becomes greater by an order of magnitude greater than the dimensions of the space vehicle, resulting in free-molecule flow in which individual molecular collisions must be treated by kinetic theory.

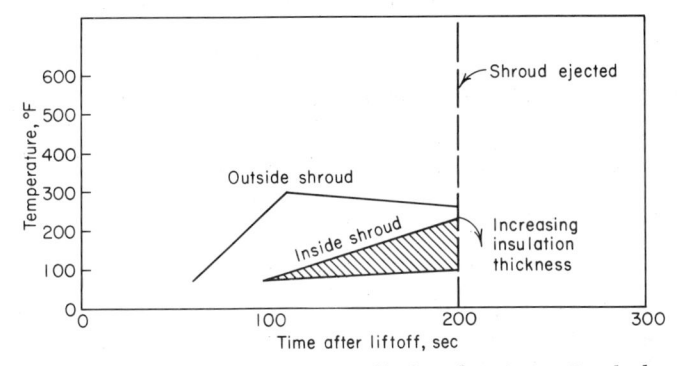

Fig. 2 Representative heating profile for Thor-Agena B vehicle.

The magnitude of aerodynamic heating on the exterior of the shroud may be quite intense during the launch phase, even though it is of short duration. Most of the heat absorbed by the shroud during launch is, however, carried away when the shroud is jettisoned. As representative illustrations of the sort of heating to

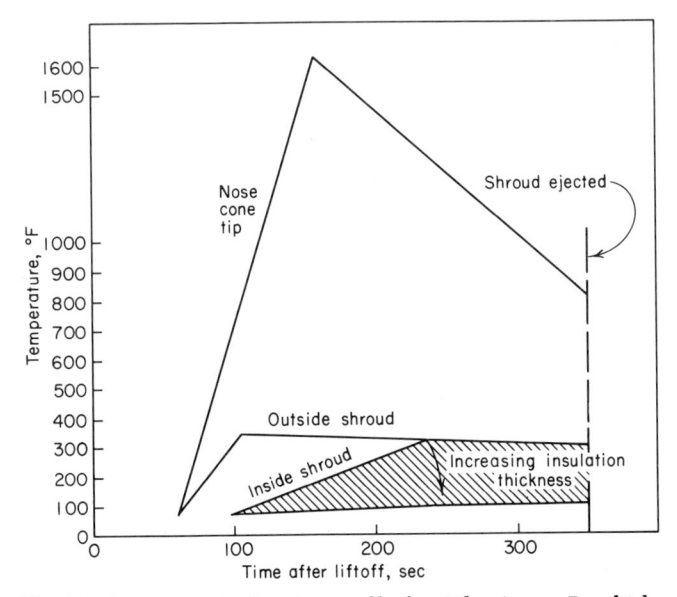

Fig. 3 Representative heating profile for Atlas-Agena B vehicle.

be expected, idealized launch heating profiles are shown in Figs. 1 to 3 for three typical rocket combinations.[1] Heating of these and other launch vehicles will generally be contained in the following profile: nose-cone stagnation temperature from ambient to 1600°F in 100 sec and shroud temperature near cylindrical section to 600°F in 100 sec.

15-4   Materials for the Space Environment

Peak internal temperatures of the shroud and of the spacecraft payload are much lower than peak external temperatures, as shown in the examples of Figs. 1 to 3. The maximum temperature of the inside of the forward cylindrical section of the shroud is no more than 300°F, reached in 200 sec.[1] After shroud ejection, the spacecraft is subjected to free molecular heating as well as direct solar radiation. For some trajectories, considerable heating may occur in free-molecule flow before the vehicle reaches altitudes at which aerodynamic heating is negligible. The magnitude of this effect must be examined for each spacecraft, trajectory, and launch-vehicle combination considered.

Vibration, shock, and random noise   The coupling of mechanical energy into a spacecraft payload from its booster rocket system may be accomplished in a variety of ways. It is termed vibration, shock, random noise, or sometimes aerodynamic or acoustic excitation, depending on the distribution of the excitation with time and with frequency. In a fundamental sense, all these terms refer to the same thing, insofar as they are all manifestations of mechanical energy; however, because of their variation in time and frequency distribution as well as their differing sources, it is convenient to give them separate treatment. Shock usually refers to a force or an acceleration represented by a discontinuous function of time, or in a practical sense, a force or acceleration that has a rise time very short compared with other characteristic times of the system under consideration. On the other hand, vibration refers to a force or acceleration which varies continuously with time above and below mean value. Usually, vibration is restricted to mean a periodic function of time, but it may also be used to include random noise or vibration that has no definite period.

The spacecraft vibration environment depends, as does the ascent heating environment, on the spacecraft size and construction as well as on the characteristics of the boost rockets and the efficiency of energy coupling between rocket and spacecraft. Vibration and shock inputs may arise from several sources, of which the major ones are probably engine ignition shocks, engine acoustic pressures, aerodynamic forces, and stage separation shocks. Such inputs may contribute the following effects to the spacecraft vibration environment:

   Fatigue of structural components

   Loss of temperature control surfaces through loosening of the surface from its substrate

   Degradation of shaped optical surfaces

At least three aspects of the vibration environmental problem immediately present themselves: prediction of loads, calculation of response, and design of simulation. Of these, the description of the environment or prediction of loads is probably the least advanced owing to both theoretical and experimental difficulties.[2] The various loads from different sources that are experienced by a spacecraft appear at different places during different periods of its trajectory, and their intensity and excitation characteristics depend on both trajectory and geometric parameters. An estimate of the environment for a particular space vehicle is often a composite of theoretical computation, laboratory studies on models, and flight measurements on vehicles.[3]

A typical pattern of vibration excitation of a spacecraft mounted on a boost vehicle may consist of several relatively widely separated but related parts. While the vehicle is still on the ground, the intense acoustic field created by the rocket-engine exhaust is reverberated from the ground to space-vehicle components. This source of excitation diminishes rapidly as the vehicle gains altitude. For a period, the excitation decreases to approximately the preignition hum level, indicating an essentially still condition. When the missile velocity approaches the speed of sound, the excitation increases sharply and attains a maximum level during the transonic period. This source of vibration decays at approximately the same rate at which it built up. The vibration level again diminishes to the preignition level and remains very low until separation, at which time shocks are sensed throughout the vehicle when the connecting bolts are exploded and the second-stage engine is ignited. These shocks constitute the last major excitation detected.

This typical pattern may be seen in Fig. 4, which traces the vibration history of a spacecraft lifted by a liquid-fueled booster.[4] The initial contribution of acoustic excitation is seen as the early peak to about 8g. The transonic vibration maximum is the large peak at about 50 sec after lift-off, while stage separation shock is seen as a narrow peak at about 165 sec. The actual time of the stage separation shock is much smaller than shown in Fig. 4, in which the peak is broadened by the characteristics of the measuring instrument. These values were taken from

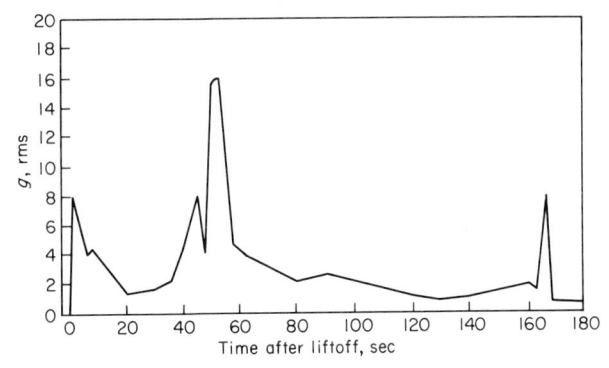

Fig. 4 Vibration history of spacecraft on liquid-fueled booster.

telemetered data on actual flights. The distribution with frequency of the acceleration energy density and the acoustic energy density for the same flight situation are shown in Figs. 5 and 6.

A considerable degree of protection from the hazards associated with the ascent environment may be afforded by a nose fairing which is jettisoned after ascent through the major part of the atmosphere. However, the process of jettisoning this nose fairing may in itself create new potential problem areas of the ascent

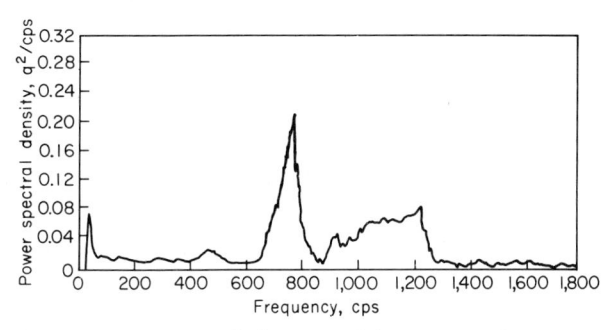

Fig. 5 Acceleration-power spectral density at lift-off versus frequency of spacecraft on liquid-fueled booster.

environment. For instance, in one study of shock loading due to a pyrotechnic nose-fairing separation device,[5] it was found that shock levels ranged from approximately 5,000 peak g adjacent to the separation charge to less than 100g at the lowest point observed on the spacecraft. In a similar study of shroud-separation shock spectra on the Ranger spacecraft, about 50g peak-to-peak amplitude was observed.[6] Pretesting of spacecraft components and subassemblies to the expected shock loading is thus indicated if pyrotechnic separation devices are used.

15-6   Materials for the Space Environment

The Upper Atmosphere

The main regions of the earth's atmosphere are listed in Table 1. Since the dominant process in the atmosphere is mixing, up to at least 56 sm (90 km), the composition of the air and the mean molecular weight remain constant from 0 to 56 sm, defining a region termed the *homosphere*. Above 56 sm, the molecular weight decreases, as the composition changes with altitude, because of molecular

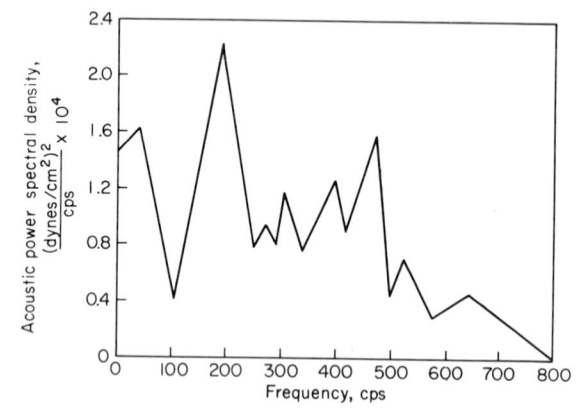

Fig. 6 Acoustic-power spectral density at lift-off versus frequency of spacecraft on liquid-fueled booster.

dissociation and diffusion. The "U.S. Standard Atmosphere"[7] is the best available reference for atmospheric properties in the homosphere. This atmosphere is a middle-latitude (approximately 45°) year-round mean over the range of solar activity between sunspot minima and maxima. Seasonal and latitudinal variations of properties in the homosphere can be taken into account, if desired, by using values from the supplemental atmospheres derived in Ref. 8.

The principal problem related to atmospheric structure is the calculation of accurate values for all primary properties in the heterosphere (above 56 sm) as a

TABLE 1   Main Regions of the Earth's Atmosphere

Atmospheric region	Region	Approximate altitude range, sm*	Characteristic features
Homosphere....	Troposphere	0–7	Mean molecular-weight constant; heat transfer by convection
	Stratosphere	7–30	Constant molecular weight; increasing temperatures; region strongly heated by both earth infrared and solar ultraviolet radiation
	Mesosphere	30–56	Constant molecular weight; decreasing temperature; mixing processes dominant throughout homosphere
Heterosphere....	Thermosphere	56–340	Frequent particle collisions; diffusion process dominant
	Exosphere	340–37,000	Collisions rare; temperature constant to about 5,300 sm; diffusion process dominant

* sm = statute miles.

function of time, location, and date by taking into account the relevant processes. The primary atmospheric properties are temperature, pressure, density, and mean molecular mass (or composition). To account for their variations, the following factors must be considered: (1) time (hour, day, sun-rotation period, season, year, sunspot cycle); (2) location (altitude, latitude, longitude); (3) solar activity (ultraviolet radiation, x-rays, solar plasma, and associated magnetic storms); and (4) processes (conduction, diffusion, mass transport, photoionization, dissociation, recombination, particle escape into space). The problem of describing upper-atmospheric behavior is difficult because many of these elements are interrelated.

Atmospheric variations The magnitude of the atmospheric properties in the upper atmosphere are derived from measurements made from satellites, rockets, meteor observations, sky emissions, and the propagation of sound and radio waves. The data are sparse and are very uncertain above 125 sm. The density (drag) data resulting from tracking satellites are the most precise and also the most numerous. Study of the orbital decay data has clearly established that two major systematic density variations occur: (1) a solar activity effect in which variations in atmospheric heating and density occur above 56 sm due to variations in solar ultraviolet radiation; and (2) a diurnal (time-of-day) effect, in which the solar heating results in the atmosphere bulging toward the sun, producing relatively large density increases at altitudes above 190 sm in the sunlit region of the earth. At 500 sm, owing to effect 1, the density can be 40 times greater during solar maximum conditions than during solar minimum; and owing to effect 2, 15 times greater during the day than during the night. The combination of effects 1 and 2 can result in densities 500 times greater at 500 sm during solar maximum (day) than during solar minimum (night). This extreme variability in density also applies to pressure.

Upper-atmosphere density variations with latitude and season are much smaller than the two preceding primary effects. The diurnal effect is one in which the density varies with local time, or the longitude difference between the given location and the subsolar point. From the strong diurnal effect, one would expect latitudinal and seasonal effects, and considerably lower density in the winter polar region. Until recently, it was not possible to prove definitely the existence of any significant latitudinal variation in density. Different investigators have reported slight latitudinal density variations, but the various findings were not in agreement. Now, the existence of an appreciable latitudinal density gradient has been demonstrated by May,[9] who analyzed data from certain recoverable satellites, the altitudes of which ranged between 118 and 162 sm and inclinations of about 80°. May concluded that the air density at a fixed height is a function of latitude and is about 30 percent smaller at the poles than at the equator. His conclusion is corroborated by an independent analysis by Anderson,[10] based on a different method of deriving the behavior of density with latitude. Theoretical support for the stated density variation is given by Lagos and Mahoney,[11] whose study of solar heating shows that seasonal variability is much more important than the diurnal variability at 45° latitude. Another conclusion from their study is that the meridional density gradients are significant at all times and are particularly large at solstice. It should be emphasized that circulation effects, neglected by Lagos and Mahoney, may introduce appreciable modifications in the latitudinal density variation. At higher altitudes—above 380 sm—this simple picture may not hold near the polar regions. Keating and Prior[12] conclude that there are two diurnal atmospheric density bulges, one due to atomic oxygen and the other due to helium. Thus, as the altitude increases to where helium replaces atomic oxygen as the principal constituent (380 to 440 sm), there is a shift from the atomic oxygen bulge on the sun side of the equator to a high-latitude "winter" helium bulge on the opposite side of the equator.

The concept that the atmosphere extending above 250 sm over a given location on the earth's surface is isothermal in the sense that the temperature does not vary with altitude is now well established. In the tenuous gas of the upper thermosphere, the thermal conductivity is independent of the pressure, whereas the heat capacity varies linearly with density. Consequently, the conductivity is very

15-8   Materials for the Space Environment

large compared with the heat capacity. Above 250 sm the absorption of energy is negligible, and the relatively high heat conductivity eliminates temperature differences; hence, the kinetic temperature is nearly constant with altitude for many thousands of miles. The kinetic temperature can be determined only for gas with a Maxwellian velocity distribution. As pointed out under the discussion of the accuracy of the hydrostatic assumption, the Maxwellian velocity distribution applies in the exosphere, provided that the escape of particles to space is negligible. For hydrogen, the escape of atoms is comparatively rapid so that the velocity distribution in the upper exosphere is not Maxwellian. Therefore, the hydrogen atoms in the upper exosphere have a non-Maxwellian distribution that becomes more pronounced with altitude.[10] Under these circumstances, the concept of kinetic temperature is not entirely applicable, although an effective temperature can be defined by considering the average energy of the hydrogen atoms. This effective temperature is not constant as a function of distance from the earth. The same altitude above which the use of the hydrostatic equation becomes questionable can also be taken to be where the deviation between the kinetic and effective temperature becomes significant. This altitude was at about 5,300 sm for average sunspot conditions.

Atmospheric composition   The neutral atmosphere above 60 sm consists almost completely of molecular nitrogen and oxygen and atomic argon, oxygen, helium, and hydrogen; the relative concentrations of these constituents depend strongly on altitude and temperature. The composition of the upper atmosphere can be explained, at least in a qualitative sense, by the types of photochemical reactions that can occur. In the following paragraphs, the reactions leading to the neutral constituents only will be discussed.

At about 60 sm, the absorption of solar radiation with wavelengths shorter than 1850 Å down to about 1300 Å leads to the dissociation of oxygen molecules into oxygen atoms. For wavelengths of less than 1026 Å, the oxygen molecule can be ionized; this ionization is normally followed by a dissociative recombination producing atomic oxygen. Although the oxygen atoms can recombine into molecules, photochemical equilibrium does not prevail because of the important role played by vertical transport processes in determining the atomic and molecular concentrations at various levels near 60 sm. More oxygen dissociates than recombines above 60 sm owing to the rapid decrease of the recombination processes with altitude. Below 60 sm, collisions occur frequently enough for recombination to prevail, and hence more oxygen recombines than dissociates. Consequently, there is a steady flux of molecular oxygen upward and atomic oxygen downward through the 60-sm level because of the effects of diffusion and mixing. Atomic oxygen is the most important constituent in the upper thermosphere.

The most active process leading to the dissociation of molecular nitrogen is ionization, followed by dissociative recombination, producing atomic nitrogen. Atomic nitrogen can react with molecular oxygen to form nitric oxide and atomic oxygen. The nitric oxide in turn reacts with atomic nitrogen to form molecular nitrogen and atomic oxygen. The effectiveness of these reactions, together with the slowness with which molecular nitrogen dissociates, causes atmospheric nitrogen to remain predominantly in molecular form.

Photodissociation of water vapor and methane near 50 sm constitutes the principal source of atomic hydrogen. Owing to the small mass of the hydrogen atom compared with other atmospheric constituents, the hydrogen concentration does not increase with altitude as rapidly as do the other atmospheric constituents in the altitude region where diffusion proceeds rapidly; hence, atomic hydrogen becomes an increasingly important atmospheric constituent with increasing altitude. However, atomic hydrogen is such a minor constituent in the thermosphere that it does not become the dominant constituent until an altitude of 800 to 1,900 sm is reached, remaining so until about 12,000 sm where the hydrogen ion becomes dominant. The source of atomic hydrogen near 50 sm can be expected to remain essentially constant through the sunspot cycle, but the rate of escape, depending on the temperature at the base of the exosphere, varies with the sunspot cycle. The escape will be relatively rapid when the temperature is high, and the concentration of hydrogen will be correspondingly low in the exosphere near sunspot maximum. The escape is relatively slow when the temperature is low, so that the concentration must be comparatively high near sunspot minimum.

Nicolet[13] showed that helium atoms are an important constituent in the lower exosphere. He explains the high densities derived from the rate of change of the period of the Echo satellite by the presence of helium. The slow density decrease between 470 and 930 sm cannot be attributed to atomic oxygen, nitrogen, or hydrogen. Although atomic oxygen is the most important constituent in the upper thermosphere, atomic helium dominates over atomic oxygen somewhere above 500 to 620 sm. Atomic hydrogen dominates over helium somewhere above 800 to 1,900 sm.

A model for atmospheric properties Upper-atmosphere measurements are not made on a regular enough basis, in kind, time, or space, to allow them to be used without the aid of a model to represent atmospheric conditions. The approach of most models used to derive atmospheric properties is to assume altitude profiles for some of the properties in order to calculate the remainder. Almost all the models deal with data referring only to density, pressure, or temperature. The altitude variation of the mean molecular mass is introduced somewhat arbitrarily, and therefore a physically consistent vertical distribution of the composition could not be obtained. To avoid some of the assumptions of the other methods and to attempt to take into account the factors mentioned above under atmospheric physics and composition, a new method was devised for computing atmospheric properties.[14] In this mode, no major assumptions are made regarding the property profiles. Instead, the primary properties are calculated by starting with an empirical density profile from a density model as the chief input, assuming diffusive equilibrium conditions above 68 sm and isothermal conditions with altitude above 250 sm. The density profile used as starting input for this new property model is represented in a previous empirical model as a function of local time and solar activity from 125 to 500 sm.[15] The results of the computations made from the model are given in Tables 2 and 3 and Fig. 7.

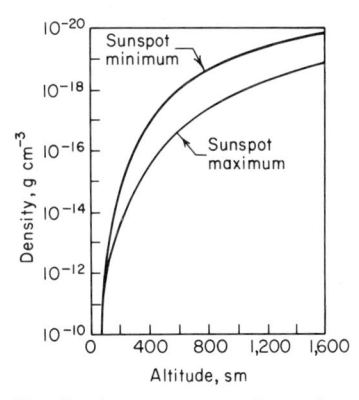

Fig. 7 Average atmospheric density at extremes of sunspot cycles.

The tables exhibit the neutral atmospheric properties and compositions (number densities) versus altitude for both sunspot maximum and sunspot minimum. The differential value of the extreme ultraviolet flux S' has been taken as 250×10^{-22} watt/(m^2)(Hz) for sunspot maximum and 50×10^{-22} watt/(m^2)(Hz) for sunspot minimum. Both tables are for $t = 2,100$ hr local time; the density for this time has been found to approximate closely the diurnally averaged density or the sum of the densities for every hour of the day divided by 24. Since secondary effects such as those due to latitude have been neglected, the model is most representative of conditions in the equatorial and midlatitude regions. The computations of the basic atmospheric properties—pressure, temperature, and mean molecular weight—are in good agreement with published values, the error being less than 5 percent throughout the region specified by the input data.[16] Also, the basic atmospheric concentrations of N_2, O_2, O, and He agree with published values within errors of 6, 20, 12, and 6 percent, respectively, at 120 km.

Nonpenetrating Radiation

Temperatures of artificial satellites or spacecraft orbiting about a planet (or a moon) in the solar system are determined mainly by (1) the direct solar radiation; (2) the reflected solar radiation, or albedo, of the planet (moon); (3) the emitted

TABLE 2 Neutral Properties of Upper Atmosphere at Sunspot Maximum for Various Altitudes

Altitude		Pressure*		Temperature, °K	Molecular weight	Scale height, km	Density, g/cc	Total, cm^{-3}	Concentration* Constituent, cm^{-3}					
km	sm	dynes/cm²	torr						$n(N_2)$	$n(O_2)$	$n(A)$	$n(O)$	$n(He)$	$n(H)$
100	62.1	3.3 (−1)	2.48 (−4)	223	27.76	7.0	5.00 (−10)	1.09 (13)	8.15 (12)	1.76 (12)	7.42 (10)	8.81 (11)	4.10 (8)	1.11 (6)
120	74.6	3.8 (−2)	2.85 (−5)	348	24.41	12.6	3.18 (−11)	7.83 (11)	4.54 (11)	6.89 (10)	1.92 (9)	2.59 (11)	1.18 (8)	3.53 (4)
140	87.0	1.1 (−2)	8.25 (−6)	569	22.44	22.5	5.42 (−12)	1.45 (11)	6.68 (10)	8.27 (9)	1.53 (8)	7.01 (10)	5.91 (7)	2.06 (4)
160	99.4	5.5 (−3)	4.13 (−6)	758	21.12	32.0	1.84 (−12)	5.24 (10)	1.95 (10)	2.11 (9)	2.99 (7)	3.07 (10)	3.88 (7)	1.49 (4)
180	111.9	3.1 (−3)	2.33 (−6)	826	20.09	36.9	9.00 (−13)	2.70 (10)	8.15 (9)	7.89 (8)	8.90 (6)	1.80 (10)	3.18 (7)	1.33 (4)
200	124.3	1.8 (−3)	1.35 (−6)	882	19.24	41.4	4.84 (−13)	1.51 (10)	3.68 (9)	3.21 (8)	2.94 (6)	1.11 (10)	2.68 (7)	1.22 (4)
240	149.1	7.7 (−4)	5.78 (−7)	984	18.01	49.9	1.69 (−13)	5.64 (9)	8.81 (8)	6.36 (7)	4.00 (5)	4.68 (9)	1.99 (7)	1.04 (4)
300	186.4	2.6 (−4)	1.95 (−7)	1,085	16.94	59.6	4.83 (−14)	1.72 (9)	1.38 (8)	7.77 (6)	2.96 (4)	1.56 (9)	1.41 (7)	8.85 (3)
340	211.3	1.3 (−4)	9.75 (−8)	1,108	16.50	63.2	2.41 (−14)	8.78 (8)	4.56 (7)	2.19 (6)	6.12 (3)	8.19 (8)	1.18 (7)	8.34 (3)
400	249	5.3 (−5)	3.98 (−8)	1,109	16.01	66.4	9.24 (−15)	3.48 (8)	9.24 (6)	3.54 (5)	6.26 (2)	3.29 (8)	9.38 (6)	7.87 (3)
500	311	1.2 (−5)	9.00 (−9)	1,109	15.15	72.2	2.06 (−15)	8.17 (7)	6.89 (5)	1.82 (4)	1.53 (1)	7.45 (7)	6.47 (6)	7.17 (3)
600	373	3.4 (−6)	2.55 (−9)	1,109	13.59	82.9	5.01 (−16)	2.22 (7)	5.53 (4)	1.02 (3)	...	1.76 (7)	4.51 (6)	6.51 (3)
700	435	1.2 (−6)	9.00 (−10)	1,109	10.93	106	1.37 (−16)	7.54 (6)	4.77 (3)	6.19 (1)	...	4.35 (6)	3.18 (6)	5.96 (3)
800	497	5.2 (−7)	3.90 (−10)	1,109	7.95	150	4.46 (−17)	3.38 (6)	4.40 (2)	4.06 (0)	...	1.11 (6)	2.26 (6)	5.48 (3)
900	559	2.9 (−7)	2.18 (−10)	1,109	5.84	210	1.87 (−17)	1.93 (6)	4.34 (1)	...	...	2.96 (5)	1.62 (6)	5.04 (3)
1,000	621	1.9 (−7)	1.42 (−10)	1,109	4.76	264	1.00 (−17)	1.26 (6)	4.55 (0)	...	...	8.17 (4)	1.18 (6)	4.64 (3)
1,200	746	9.9 (−8)	7.43 (−11)	1,109	4.11	323	4.40 (−18)	6.45 (5)	...	...	...	6.88 (3)	6.34 (5)	3.97 (3)
1,400	870	5.5 (−8)	4.13 (−11)	1,109	3.99	350	2.37 (−18)	3.57 (5)	...	...	...	6.58 (2)	3.53 (5)	3.43 (3)
1,600	994	3.1 (−8)	2.33 (−11)	1,109	3.96	372	1.35 (−18)	2.05 (5)	...	...	...	7.03 (1)	2.02 (5)	2.98 (3)
1,800	1,119	1.9 (−8)	1.42 (−11)	1,109	3.94	393	7.94 (−19)	1.22 (5)	...	...	...	8.50 (0)	1.19 (5)	2.60 (3)
2,000	1,243	1.1 (−8)	8.25 (−12)	1,109	3.91	416	4.81 (−19)	7.41 (4)	...	...	...	1.13 (0)	7.18 (4)	2.29 (3)
2,500	1,553	3.7 (−9)	2.78 (−12)	1,109	3.79	481	1.52 (−19)	2.42 (4)	...	...	...	...	2.24 (4)	1.70 (3)
3,000	1,864	1.4 (−9)	1.05 (−12)	1,109	3.58	569	5.49 (−20)	9.26 (3)	...	...	...	...	7.95 (3)	1.31 (3)
3,500	2,175	6.4 (−10)	4.80 (−13)	1,109	3.26	693	2.25 (−20)	4.15 (3)	...	...	...	...	3.12 (3)	1.03 (3)
4,000	2,486	3.3 (−10)	2.48 (−13)	1,109	2.86	872	1.03 (−20)	2.17 (3)	...	...	...	...	1.34 (3)	8.27 (2)
4,500	2,796	2.0 (−10)	1.50 (−13)	1,109	2.44	1,122	5.28 (−21)	1.30 (3)	...	...	...	...	6.25 (2)	6.78 (2)
5,000	3,107	1.3 (−10)	9.75 (−14)	1,109	2.07	1,450	3.00 (−21)	8.75 (2)	...	...	...	...	3.11 (2)	5.64 (2)
6,000	3,728	7.6 (−11)	5.70 (−14)	1,109	1.55	2,290	1.28 (−21)	4.99 (2)	...	...	...	...	9.11 (1)	4.08 (2)
7,000	4,350	5.2 (−11)	3.90 (−14)	1,109	1.28	3,226	7.23 (−22)	3.40 (2)	...	...	...	...	3.21 (1)	3.08 (2)
8,000	4,971	3.3 (−11)	2.93 (−14)	1,109	1.15	4,142	4.86 (−22)	2.53 (2)	...	...	...	...	1.30 (1)	2.40 (2)
9,000	5,592	3.0 (−11)	2.25 (−14)	1,109	1.09	5,019	3.60 (−22)	1.99 (2)	...	...	...	...	5.97 (0)	1.93 (2)
10,000	6,214	2.5 (−11)	1.88 (−14)	1,109	1.06	5,878	2.83 (−22)	1.62 (2)	...	...	...	...	3.00 (0)	1.59 (2)

* Numbers in parentheses denote powers of 10. Thus, 3.3 (−1) means 3.3×10^{-1}.

TABLE 3 Neutral Properties of Upper Atmosphere at Sunspot Minimum for Various Altitudes

Altitude		Pressure*		Temper- ature, °K	Molecu- lar weight	Scale height, km	Density,* g/cc	Total, cm⁻³	Concentration* Constituent, cm⁻³					
km	sm	dynes/cm²	torr						$n(N_2)$	$n(O_2)$	$n(A)$	$n(O)$	$n(He)$	$n(H)$
100	62.1	2.9 (−1)	2.18 (−4)	204	28.11	6.4	4.83 (−10)	1.04 (13)	8.14 (12)	1.74 (12)	6.65 (10)	4.80 (11)	1.44 (8)	1.83 (6)
120	74.6	2.1 (−2)	1.58 (−5)	287	26.77	9.4	2.40 (−11)	5.41 (11)	4.07 (11)	5.86 (10)	1.39 (9)	7.49 (10)	1.57 (7)	1.06 (5)
140	87.0	3.9 (−3)	2.92 (−6)	420	25.08	14.8	2.79 (−12)	6.71 (10)	4.43 (10)	4.90 (9)	6.86 (7)	1.79 (10)	8.28 (6)	6.77 (4)
160	99.4	1.3 (−3)	9.75 (−7)	588	23.53	22.3	6.23 (−13)	1.60 (10)	8.91 (9)	8.23 (8)	7.97 (6)	6.20 (9)	4.93 (6)	4.62 (4)
180	111.9	6.0 (−4)	4.50 (−7)	730	22.24	29.4	2.19 (−13)	5.93 (9)	2.79 (9)	2.25 (8)	1.63 (6)	2.91 (9)	3.47 (6)	3.60 (4)
200	124.3	3.2 (−4)	2.40 (−7)	824	21.15	35.1	9.91 (−14)	2.82 (9)	1.11 (9)	7.98 (7)	4.64 (5)	1.63 (9)	2.74 (6)	3.10 (4)
240	149.1	1.2 (−4)	9.00 (−8)	948	19.44	44.6	2.91 (−14)	9.01 (8)	2.41 (8)	1.43 (7)	6.06 (4)	6.43 (8)	1.95 (6)	2.56 (4)
300	186.4	3.4 (−5)	2.55 (−8)	996	17.75	52.2	7.39 (−15)	2.51 (8)	3.59 (7)	1.62 (6)	4.24 (3)	2.12 (8)	1.43 (6)	2.28 (4)
340	211.3	1.6 (−5)	1.20 (−8)	997	17.02	55.1	3.36 (−15)	1.19 (8)	1.08 (7)	4.12 (5)	7.74 (2)	1.07 (8)	1.20 (6)	2.18 (4)
400	249	5.7 (−6)	4.28 (−9)	997	16.27	58.7	3.12 (−15)	4.15 (7)	1.83 (6)	5.42 (4)	6.30 (1)	3.87 (7)	9.32 (5)	2.05 (4)
500	311	1.1 (−6)	8.25 (−10)	997	15.21	64.7	2.06 (−16)	8.16 (6)	1.02 (5)	1.99 (3)	1.06 (0)	7.42 (6)	6.17 (5)	1.85 (4)
600	373	2.6 (−7)	1.95 (−10)	997	13.34	75.9	4.27 (−17)	1.93 (6)	6.15 (3)	8.07 (1)		1.49 (6)	4.13 (5)	1.67 (4)
700	435	8.4 (−8)	6.30 (−11)	997	10.13	103	1.02 (−17)	6.10 (5)	4.02 (2)	3.58 (0)		3.14 (5)	2.80 (5)	1.51 (4)
800	497	3.8 (−8)	2.85 (−11)	997	6.87	156	3.13 (−18)	2.74 (5)	2.84 (1)			6.91 (4)	1.92 (5)	1.38 (4)
900	559	2.2 (−8)	1.65 (−11)	997	4.95	223	1.32 (−18)	1.61 (5)	2.16 (0)			1.58 (4)	1.33 (5)	1.25 (4)
1,000	621	1.5 (−8)	1.13 (−11)	997	4.10	276	7.35 (−19)	1.08 (5)				3.78 (3)	9.27 (4)	1.15 (4)
1,200	746	7.8 (−9)	5.86 (−12)	997	3.54	337	3.32 (−19)	5.65 (4)				2.41 (2)	4.66 (4)	9.64 (3)
1,400	870	4.5 (−9)	3.38 (−12)	997	3.25	387	1.75 (−19)	3.24 (4)				1.77 (1)	2.42 (4)	8.19 (3)
1,600	994	2.8 (−9)	2.10 (−12)	997	2.95	448	9.83 (−20)	2.00 (4)				1.48 (0)	1.30 (4)	7.00 (3)
1,800	1,119	1.8 (−9)	1.35 (−12)	997	2.64	528	5.80 (−20)	1.33 (4)					7.23 (3)	6.04 (3)
2,000	1,243	1.3 (−9)	9.75 (−13)	997	2.32	629	3.61 (−20)	9.36 (3)					4.12 (3)	5.24 (3)
2,500	1,553	6.8 (−10)	5.10 (−13)	997	1.69	969	1.38 (−20)	4.91 (3)					1.13 (3)	3.78 (3)
3,000	1,864	4.4 (−10)	3.30 (−13)	997	1.34	1,368	7.05 (−21)	3.18 (3)					3.56 (2)	2.82 (3)
3,500	2,175	3.2 (−10)	2.40 (−13)	997	1.17	1,741	4.43 (−21)	2.29 (3)					1.26 (2)	2.17 (3)
4,000	2,486	2.4 (−10)	1.80 (−13)	997	1.08	2,065	3.16 (−21)	1.75 (3)					4.94 (1)	1.70 (3)
4,500	2,796	1.9 (−10)	1.43 (−13)	997	1.05	2,354	2.41 (−21)	1.39 (3)					2.11 (1)	1.37 (3)
5,000	3,107	1.6 (−10)	1.20 (−13)	997	1.03	2,624	1.92 (−21)	1.13 (3)					9.68 (0)	1.12 (3)
6,000	3,728	1.1 (−10)	8.26 (−14)	997	1.01	3,156	1.32 (−21)	7.88 (2)					2.47 (0)	7.85 (2)
7,000	4,350	8.0 (−11)	6.00 (−14)	997	1.00	3,707	9.65 (−22)	5.78 (2)						5.78 (2)
8,000	4,971	6.1 (−11)	4.58 (−14)	997	1.00	4,290	7.36 (−22)	4.42 (2)						4.42 (2)
9,000	5,592	4.8 (−11)	3.60 (−14)	997	1.00	4,912	5.80 (−22)	3.49 (2)						3.49 (2)
10,000	6,214	3.9 (−11)	2.93 (−14)	997	1.00	5,574	4.70 (−22)	2.83 (2)						2.83 (2)

* Numbers in parentheses denote powers of 10. Thus, 2.9 (−1) means 2.9×10^{-1}.

radiation of the planet (moon); and (4) the surface characteristics of the artificial satellite or spacecraft, which depend on its composition, shape, and orientation. This radiation can be considered to be nonpenetrating radiation. With regard to the planet earth, its albedo radiation (reflected solar radiation) consists of ultraviolet, visible, and infrared emissions between 0.29 and 4.0 microns (Table 4), where 1 micron = 10^{-4} cm = 10^4 Å. The radiaton categories cited above will be discussed in turn in this chapter.

Solar radiation The total amount of electromagnetic radiation emitted from the sun is the same as that from a 5800°K blackbody. Solar radiation covers the spectrum from wavelengths shorter than 10^{-4} to longer than 10^8 microns (see Table 4).

TABLE 4 Classification of Electromagnetic Radiation by Wavelength

Wavelength, microns	Classification
<0.1	X-rays and gamma rays
0.01–0.2	Far ultraviolet
0.2 –0.32	Middle ultraviolet
0.32–0.38	Near ultraviolet
0.38–0.72	Visible
0.72–1.5	Near infrared
1.5 –5.6	Middle infrared
5.6 –1,000	Far infrared
>1,000	Microwaves and radio waves

The nonpenetrating portion of the solar spectrum lies between the wavelength limits 0.01 and 15 microns. About 99 percent of the energy of the solar spectrum lies between 0.3 and 4.0 microns. The solar constant is the total solar irradiation at the earth's mean distance from the sun; it has the value 0.140 watt/cm². Because the orbit of the earth around the sun is slightly elliptical, the distance of the earth from the sun—and therefore the value of the solar constant—changes throughout the year. Use of the solar constant at the mean distance cited above results in a maximum error of ±3.5 percent.

The sun is a very stable source of radiation in the visible and adjoining spectral regions. The continuum radiation probably does not change noticeably during the solar cycle, which averages 11 years, even in the ultraviolet. However, the line radiations exhibit variations that become more important at the shorter wavelengths. The largest solar cycle variation occurs in the far ultraviolet below 0.1 micron, at wavelengths sometimes designated as the extreme ultraviolet (Table 4). The 10.7-cm flux from the sun is an approximate index of this variation.[17]

The data and observations discussed above have been gathered into a single curve of the solar spectrum[18] in Fig. 8, which shows the spectral radiation versus wavelength. From 0.14 micron to the middle infrared, the spectrum is a continuum with superimposed absorption (Fraunhofer) lines. This portion of the curve has been smoothed to eliminate the fine detail. The solar spectrum below 0.14 micron consists entirely of sharp emission lines. These have been smoothed to intervals of 0.005 micron in order to make the continuous curve. They contribute only a small amount of energy compared with the continuum at all wavelengths longer than 0.085 micron, except in the region of the atomic-hydrogen Lyman-alpha line (0.1216 micron). At shorter wavelengths, it is also likely that the background continuum radiation contains more energy than all but the strongest lines. The intensities of the lines over the spectral region below 0.1 micron must still be regarded as tentative. Below about 0.01 micron, the coronal spectrum consists of x-rays.

Albedo The albedo of an object is the ratio of the emergent light flux scattered and reflected in a given spectral region to the incident light flux in the same spectral region. If the emergent spectral intensity is known, then the albedo a_1 for a source like the sun is

$$a_1(\lambda_1,\lambda_2) = \frac{\int_{\lambda_1}^{\lambda_2} \int_0^{2\pi} \int_0^{\pi/2} I_\lambda(\theta,\varphi) \cos\theta \sin\theta \, d\theta \, d\varphi \, d\lambda}{\int_{\lambda_1}^{\lambda_2} \pi F_\lambda \cos Z \, d\lambda} \qquad (1)$$

where I_λ = emergent spectral intensity
F_λ = incident spectral intensity
Z = solar zenith angle

The albedo a_1 is the albedo of a point. The albedo of an object would have to be obtained by averaging over the object.

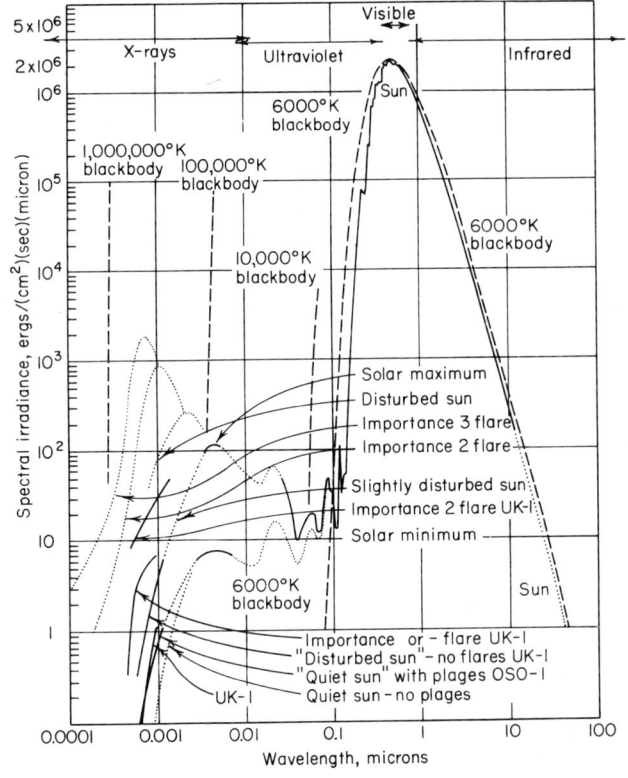

Fig. 8 The solar electromagnetic radiation spectrum. Solid lines represent measurements; dotted lines, estimates.

The albedo of a point on the earth's surface a_2, as measured from an orbiting satellite, is[19]

$$a_2 = \frac{F}{\cos Z \int S_\lambda \Phi(\lambda) \, d\lambda} \qquad (2)$$

where F = flux measured by the sensor
S_λ = incident spectral flux
$\Phi(\lambda)$ = effective spectral response of the sensor

In order to use Eq. (2) to determine the albedo of an area, an assumption must be made as to the scattering function of the area. One assumption is that the surface scatters isotropically. The albedo of an area measured in this way is dependent (especially for a portion of the earth where scattering from the atmosphere is the major contributor) on the exact shape of the spectral response of the sensor. Actually, the area neither reflects nor radiates according to Lambert's law (like a perfectly diffuse surface). Assumptions about the nonisotropic nature of the radiation can be made so that calculations of this radiation emerging from the atmosphere can be based on model atmospheres.[20]

For the earth, the visual albedo will not be observed directly until a lunar observatory is established on the surface of the moon or an orbital observatory is used at a distance comparable to that of the moon. Usually, the albedo of the earth is derived theoretically or inferred from a series of local observations, each of which refers to a limited area. Since the total of these observations covers only a small portion of the surface of the earth in a reasonable period of time, they leave considerable doubt as to the total albedo. The albedo of the earth has been computed from measurements made from artificial satellites. However, there is some uncertainty concerning the satellite data used because of instrumental degradation.[21]

Another, little used, method is the computation of the albedo of the entire earth from observations of earth light reflected from a new moon.[22] In this type of observation, a comparison is made between the brightness of the portion of the moon illuminated by the sun and the portion illuminated by the earth. A correction is applied for the geometry, and the brightness of the earth relative to that of the sun is derived. This then leads to the albedo of the earth. To calculate accurately the albedo of the earth by this method, it is necessary to know the phase curve of both the moon and the earth. Because of the long-term nature of these studies, it is important that observations be made continuously, and preferably they should be made from scattered localities over the earth to establish an albedo for the whole earth. Phase curves for the portions of the earth observable from each of these places are necessary. It will be some years before they are available for localities other than Europe.[23]

TABLE 5 Earth Albedo

Source	Albedo range, %	Ref.
Clouds	17–81	24
Terrain	7–28	25
Soils	5–43	26
Snow cover	29–86	26
Ice surface	12–36	26
Sea surface	2–70	26
Latitude and seasonal averages	28–50	26
Latitude variation under various sky conditions	12–92	27
Various surfaces for different wavelengths and cloudiness	1–37	26

The earth's albedo may be regarded as the sum of three components: (1) the albedo of the earth's surface, (2) the albedo associated with backscattering of the incident radiation by the atmosphere, and (3) the albedo of clouds. The mean albedo is about 50 percent in the ultraviolet range and 28 percent in the infrared range. In the visible region, it is 30 to 40 percent. For a cloud-covered earth, the albedo is 50 to 60 percent; for a cloudless earth, it is about 16 percent. Table 5 presents a list of the albedo ranges of the earth for various cases.

Most of the solar energy reflected to space is reflected from clouds. The albedo of clouds is so variable that it is impossible to specify an average value for the total reflection of clouds. Therefore, it is necessary to measure the albedo of the whole earth first and then subtract the contributions by the atmosphere and the ground. The remainder is the albedo of the clouds. One computation, based on moon measurements, indicates that the average albedo of clouds is about 50 to 55 percent of the energy incident on the clouds.[28]

Inasmuch as the data referenced in Table 5 represent experimental averages, individual measurements may depart considerably from these data. The values of the albedo are averages over various kinds of surface cover and also averages with respect to time. The term *soil* must be defined in terms of the amount of moisture, the grain size (roughness of surface), and the color of the surface. The albedo of all kinds of soil decreases with increasing moisture because the albedo of water is lower than that of soil. The albedo of snow depends on the amount of impurity it contains, on its surface roughness, and on the angle of incidence.

The values of the albedo for water surfaces are noticeably different for direct solar radiation and for scattered sky radiation. Robinson[26] presents computed and observed values of the albedo for direct radiation at various angles of incidence on a smooth water surface. The experimental values of the albedo are in good agreement with the theoretical calculations made with the Fresnel formula. Robinson also gives theoretical values of the albedo for smooth and rough water surfaces at various zenith distances from the sun. Experimental determinations of the albedo of water surfaces for scattered radiation have led to values between 5 and 8 percent. This albedo varies considerably with the degree of cloudiness of the sky. However, even during a clear sky, it varies markedly, owing to anisotropy of the sky's luminance. Since the albedo varies with the elevation angle of the sun, it exhibits a diurnal and an annual variation, increasing with decreasing elevation of the sun. Also, the albedo depends on the spectral composition of the incident radiation. The incident radiation contains a higher proportion of the short-wave radiation as the elevation of the sun increases. The effect of the wavelength on the albedo is toward an increase in the albedo with decreasing elevation of the sun, contributing toward a diurnal variation.

TABLE 6 Total Long-wave Radiation from the Earth and Atmosphere

Latitude, deg	Long-wave energy radiated, watts/cm^2				
	Jan.	Mar 21	July	Sept. 23	Annual
0–10	0.0203	0.0212	0.0209	0.0206	0.0225
10–20	0.0206	0.0210	0.0210	0.0211	0.0230
20–30	0.0203	0.0204	0.0213	0.0213	0.0228
30–40	0.0193	0.0194	0.0216	0.0213	0.0222
40–50	0.0175	0.0175	0.0202	0.0201	0.0210
50–60	0.0164	0.0164	0.0195	0.0185	0.0195
60–90	0.0156	0.0152	0.0189	0.0177	0.0183

Thermal radiation of the earth The thermal radiation of the earth and atmosphere is infrared radiation with a wavelength greater than 1.5 microns (Table 4). Table 6 gives values for the total outgoing radiation, as extracted by Johnson[29] from Baur and Phillips.[30]

The values have been averaged over a period of a day or longer. Houghton's annual-average values for the long-wave radiation[31] are shown in the Annual column of Table 6. His data are considered to be more accurate than Baur and Phillips' in terms of annual averages, but they do not show the seasonal trend indicated by Baur and Phillips. The quantities of interest for satellite thermal control are the long-term averages. Thermal time constants and the times for material degradation are great enough to mask the effects of short-term, localized variations in earth thermal emission. Only small errors ensue when the annual average emission is used for any season. Also, because of the effect of heat storage, the change in absorbed solar radiation causes only small seasonal variations in the average emission of the whole earth. Therefore, earth emission can be taken to be independent of season without introducing significant error. Houghton's average annual values listed in Table 6 are adequate for any time of the year.

Johnson[29] has presented curves for the earth and atmospheric emission spectra. These curves are shown in Fig. 9, where the solid curve is the approximate radiation from the earth and atmosphere; the 288°K blackbody curve approximates the radiation from the earth's surface; and the 128°K blackbody curve approximates the radiation from the atmosphere in spectral regions where the atmosphere is opaque. The temperature of a satellite depends on absorbed energy or on the spectral distribution of incident energy and the spectral radiation characteristics of the satellite surface. However, the total earth emissive power and total absorptance for long-wavelength energy can be used to calculate satellite heating. Absorptance for most materials shows only small variations with wavelength beyond 8 microns, and, as illustrated in Fig. 9, nearly all the energy emitted by the earth-atmosphere system is beyond 8 microns. Camack[27] presents a model that establishes extremes

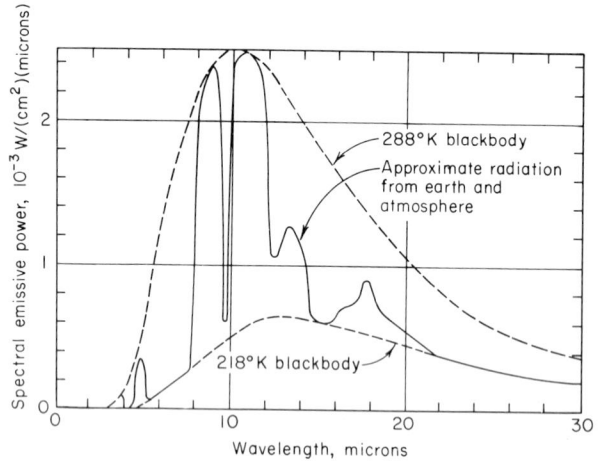

Fig. 9 Typical spectral emissive power curve for the thermal radiation leaving the earth.

in the variation of earth radiation and albedo and assumes a frequency distribution for occurrences between these extremes.

Penetrating Radiation

The components of the penetrating radiation environment for a given spacecraft which may arise from various sources are characterized by quite different spatial configurations and time variations. Thus, assessment of the accumulated fluxes of radiation encountered by a space vehicle over any length of time is, in general, a complex problem. These components, from natural and man-made sources, are discussed in the following sections, along with a description of the methods available for calculating the cumulative external radiation environment and the local environment within a spacecraft or other configuration.

Natural radiation environment The natural radiation environment, as differentiated from the radiations that are present only because of actions taken by man, consists of energetic particles both charged and neutral as well as high-energy electromagnetic radiations. The most important of these from the standpoint of damage to electronic materials in space are the charged particles, because of their relatively high intensities and damage coefficients.

Charged Particles. Although most of the known charged particles may be present in space, only electrons, protons, and helium nuclei exist in sufficient numbers to be of interest in their effects on electronic materials.

These particles display varying energy distributions and temporal fluctuations in different regions of space, so that it is convenient to assess the effects of a single particle type on a given spacecraft separately for the various spatial regions.

GEOMAGNETICALLY TRAPPED RADIATION: For orbits near the earth (up to approximately 20,000 nm* or 23,000 sm in altitude), geomagnetically trapped or Van Allen radiation is confined to the volume occupied by the earth's magnetic field which is bounded by the region of interaction between the magnetosphere and the solar plasma. The earth's magnetic field is very nearly a pure dipole field, so that the magnetic shell described by the set of field lines having the same integral invariant is an approximately toroidal surface. A charged particle trapped on a field line follows a helical path around the field line; this path is determined by the initial conditions at the time the particle is injected, such as the initial energy of the particle and its pitch angle with respect to the field direction. At some point on the field line, the particle "mirrors" or reverses its motion parallel to the field and bounces back and forth between mirror points of equal field strength, at the same time drifting in longitude on the shell. If the mirror points are in a low enough atmospheric density so that the probability of scattering is small, the particle will be stably trapped unless the field line on which it is trapped becomes distorted. If one or both of the mirror points are in a part of the atmosphere where the density is sufficiently great, the motion of the particle may be changed by scattering from atoms or electrons; and as a result of the energy loss or change in direction of the particle, it is eventually lost as a trapped radiation particle.

Since the spatial distribution of trapped particles is a function of the magnetic field, and since a dipole field possesses a high degree of symmetry, a convenient system for mapping the intensities of trapped particles is in terms of two coordinates that describe the magnetic shell upon which a particle remains trapped: the total field strength B, and the radial distance to the intersection of the field line with the magnetic equatorial plane L.[32] This system can be transformed for easier visualization to R, λ coordinates where R is the radial distance from the center of the earth, and λ is the geomagnetic latitude. In this system, R and L are approximately equal at the geomagnetic equator and are usually expressed in units of earth radii.

Trapped protons: The trapped proton belt is the region of the radiation environment in low to medium altitude of the high flux intensities, large damage coefficients, and deep penetration of materials.

A compilation of measurements of the trapped protons[33,34] made between 1961 and 1965 has resulted in a good model of this portion of the environment including spatial intensity distributions, energy spectra, and some assumptions as to the temporal variations of the first two parameters. The integral flux distributions above 0.4-, 4-, and 34-Mev energy thresholds are shown in Figs. 10 to 12 in the form of isoflux contours on R, λ maps. It is evident from the difference in spatial extent between the 0.4-Mev map and the two higher-energy maps that it is convenient to think of zones in the proton belt, one with virtually no protons with energies greater than 4 Mev. This may be called the *outer* radiation zone, which extends from L value of about 4 (in units of earth radii) to the outer boundary of particle trapping. This zone is characterized by time variations in flux intensities and corresponding changes in energy spectra with time.

Energy spectra at the magnetic equator for various L values in the inner and outer proton zones are presented in Figs. 13 and 14. Variations in the outer-zone spectra with time occur because of energy-selective changes in proton intensities. The spectra shown can generally be fit rather well by an exponential form: $\Phi(E) = \Phi_1 \exp[-(E - E_1)/E_0]$, where Φ_1 is the integral flux above the reference

* Nautical miles.

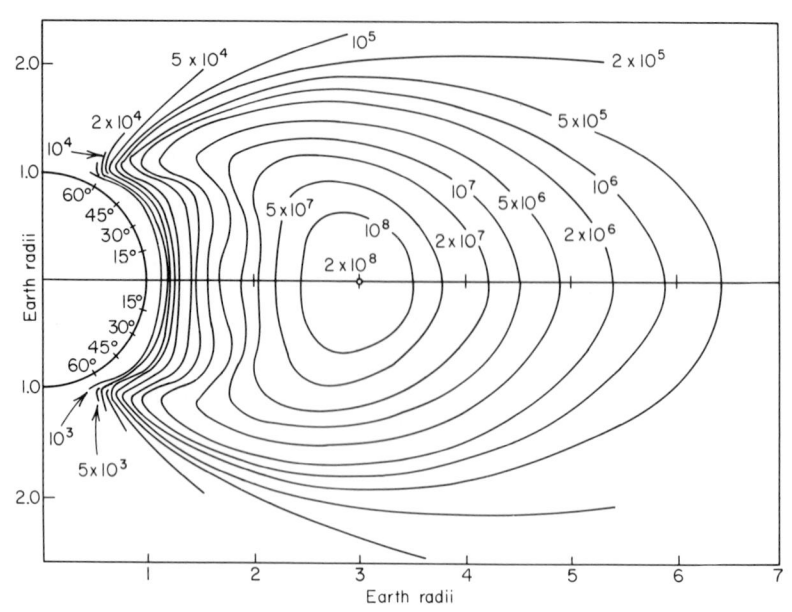

Fig. 10 Proton isoflux contours ($E > 0.4$ Mev). Contours are labeled in units of protons per cm^2-sec.

energy E_1 (0.4 Mev in this case), and E_0 is the spectral energy parameter. A power-law spectral fit can also represent the data over a limited energy range.

The intensities and energy spectra of protons for $L \lesssim 2$ in the inner zone are much more stable in time than those in the outer zone.[35,36] This is due to the

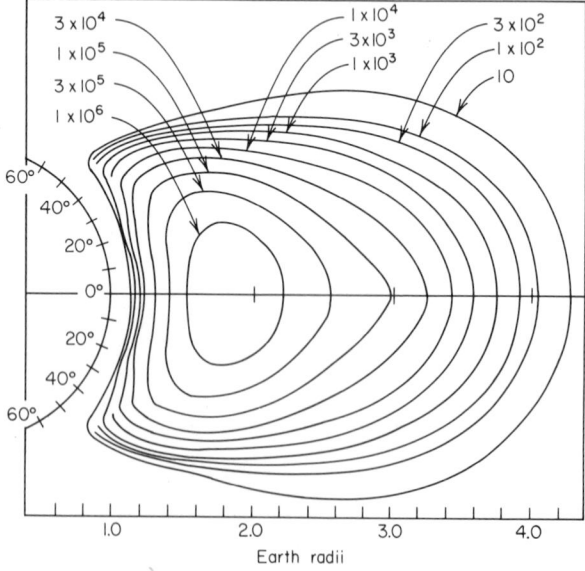

Fig. 11 Proton isoflux contours ($E > 4$ Mev). Contours are labeled in units of protons per cm^2-sec.

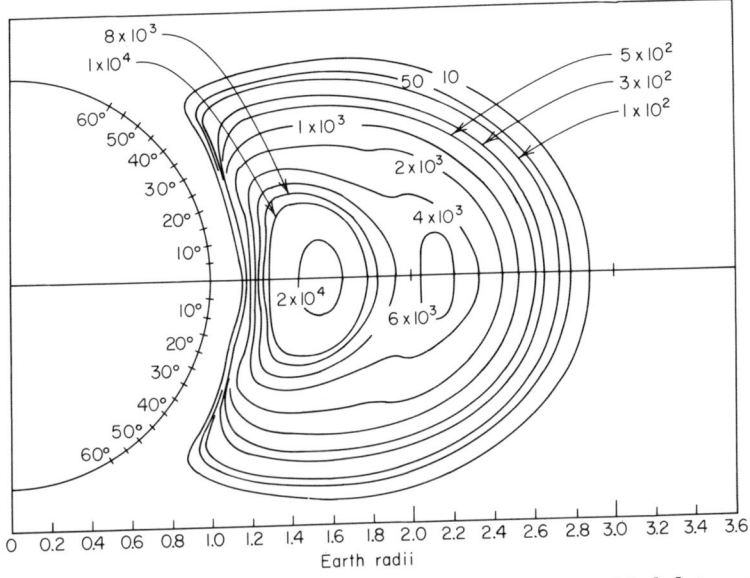

Fig. 12 Proton isoflux contours ($E > 34$ Mev). Contours are labeled in units of protons per cm²-sec.

fact that the higher rigidity of the magnetic field in this region makes it less susceptible to distortion from magnetic disturbances originating outside the magnetosphere. There is a variation in flux intensities in the lower edge of the inner zone over the 11-year period of the solar cycle.[37,38] This is due to an increase

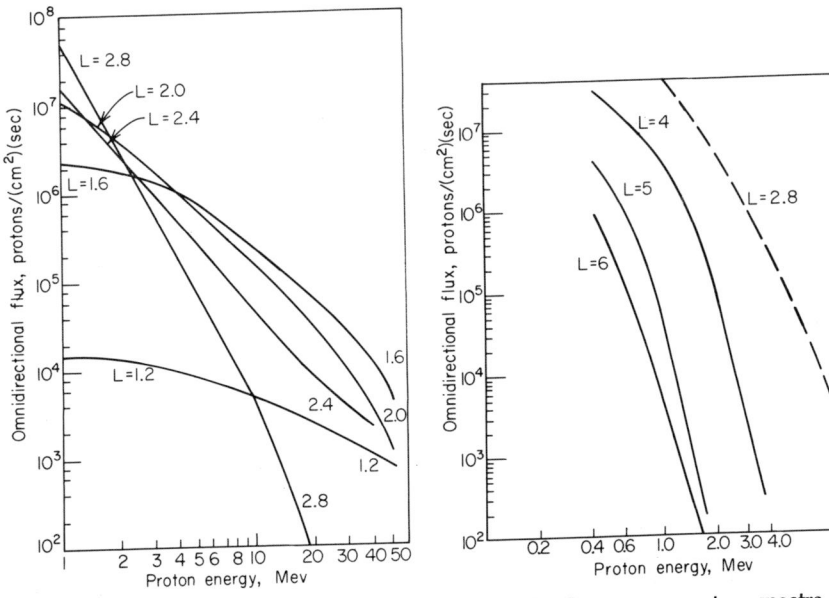

Fig. 13 Inner-zone proton spectra.

Fig. 14 Outer-zone proton spectra.

Materials for the Space Environment

in the effective density in the upper atmosphere during the period of maximum solar activity. The effect of this variation will be a decrease in flux intensities at altitudes up to a few hundred nautical miles during the solar maximum as compared with the solar minimum period. This reduction in fluxes is approximately a factor of 2 between 1966 and November, 1967, for altitudes up to about 300 nm.[38] The energy spectra of protons in this region would be expected to change with the variations in flux since particle loss from atmospheric scattering is an energy-dependent process. Few measurements of these spectral changes during solar maximum are available, but more should be forthcoming soon and through the current solar cycle maximum.

Trapped electrons: The trapped-electron belt coincides spatially with the proton belt but has different configurations in its intensity and energy spectrum distributions. A model of those distributions was derived from data[39] accumulated between late 1962 and 1965. All these measurements were made after the creation of

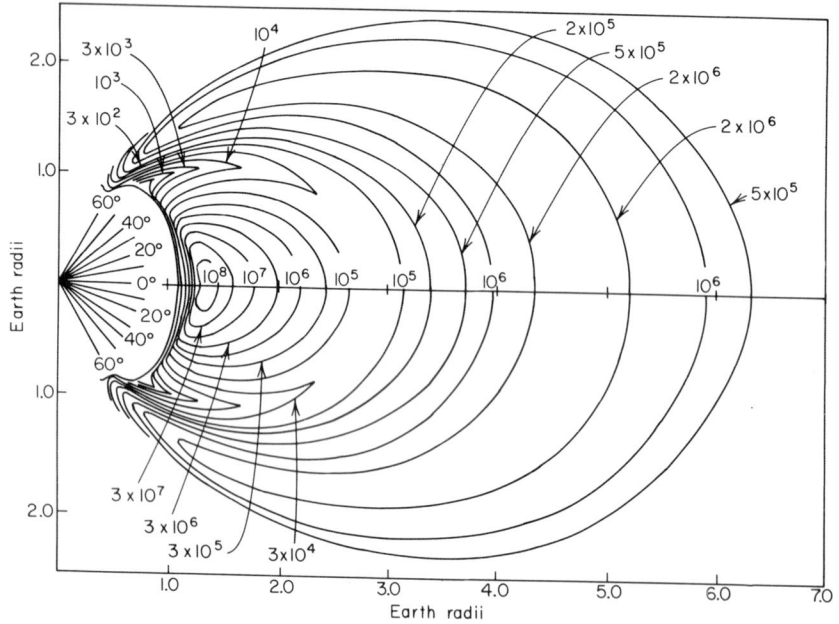

Fig. 15 Trapped-electron isoflux contours ($E > 0.5$ Mev) as of August, 1964. Contours are labeled in units of electrons per cm^2-sec.

the artificial electron belt by beta-decay electrons from the Starfish high-altitude nuclear explosion on July 9, 1962. A more detailed discussion of man-made portions of the radiation environment appears in a later section (page 15–25), and only the gross characteristics of the artificial electron belt as they pertain to the current environment will be covered here. Since trapped electrons of natural origin were not well measured before 1962, our present knowledge does not permit a clean separation in the inner radiation belt between naturally occurring electrons and those of artificial origin.

The integral flux distribution above 0.5 Mev electron energy as of August, 1964, is given in Fig. 15. As in the case of trapped protons, the electron belt is divided into an inner and an outer zone, with the zone boundary being taken at a minimum in the distribution of high-energy electrons at $L \sim 2.5$ to 3 earth radii. The inner zone in late 1964 was characterized by energy spectra generally similar to a fission beta spectrum (Fig. 16) and by monotonic losses in intensity, the loss rate being highest at very low L values and fairly uniform at about a factor of 3 decrease in intensities each year[40] for $L \lesssim 1.3$. Thus, for the main portion of the inner zone, the fluxes of artificially injected electrons will be about two orders of magnitude lower in late 1968 than those shown in Fig. 15.

The electron flux intensities in the outer zone ($L \gtrsim 2.5$) shown in Fig. 15 are approximate mean values from data taken from 1962 to 1964, near a period of minimum solar activity. Intensities throughout this zone show fluctuations of as much as two orders of magnitude over time periods of weeks or a few months. Since changes in spectral shape might be expected to accompany the intensity fluctuations, the spectra shown for $L = 3$, 4, and 5 in Fig. 16 are typical only. They may, however, be representative of longer time-averaged spectra, since selected data from which they were derived agreed relatively well over the period 1962 to 1964. For a period of maximum solar activity, it may not be unreasonable to assume that the average electron flux intensities in the outer zone may be closer to the peak fluxes measured in the quiet period 1962 to 1964, or nearly an order of magnitude higher than those shown in Fig. 15. This assumption, although it is not supported by experimental data, will result in more conservative radiation effects predictions for the solar maximum period 1968 to 1971.

Trapped alpha particles: Alpha particles trapped in the geomagnetic field have been observed[41] from the satellite Injun IV. The peak of the alpha flux intensity occurs at $L \sim 3.1$, and at this position the ratio of the flux of alpha particles with energies greater than 2.09 Mev to that of protons with energies greater than 0.52 Mev is 2.3×10^{-4}. The energy spectrum of the trapped alpha particles through the region of maximum intensity can be represented analytically by an exponential form with a spectral parameter energy $E_0 = 1.5$ Mev. This steep spectrum, with almost no particles above 10 Mev, coupled with the low integral intensities, allows one to ignore trapped alpha particles as compared with protons and electrons in radiation effects calculations.

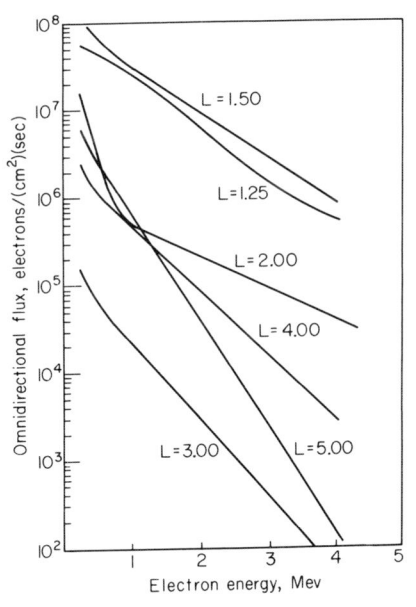

Fig. 16 Trapped-electron spectra.

Calculation of accumulated fluxes: It can be seen from the foregoing discussion of trapped radiation that the calculation of the particle fluxes accumulated by a particular spacecraft at a given time involves many variables and is not simple to perform. Computers are quite well suited to this type of calculation. Programs have been in use for some time which compute B and L coordinates from the spacecraft-position coordinates, interpolate the flux at that position from an input flux map, and sum the resultant fluxes over the spacecraft orbit for any required period of time. The program computes the instantaneous and summed integral fluxes of protons and electrons above five threshold energies for each type of particle. The electron map is adjusted by the program for losses from the inner zone and solar maximum increases in the outer zone for future proposed vehicles by putting in the elapsed time between the date of the original map and the orbiting date of the vehicle.

Since a large number of spacecraft in the past have been in low-altitude circular orbits and it is probable that this type of orbit will continue to be used in the future, calculations have been made for two orbital inclination angles, 30 and 90° at various altitudes. The electron flux per day of particles with energies greater

15-22 Materials for the Space Environment

than 0.5 Mev in December, 1968, versus orbital altitude is presented in Fig. 17. The fluxes shown in this figure can be taken as upper limits for later dates, barring any future artificial injections of electrons, since the inner-belt intensities are declining and a best estimate of the maximum outer-belt fluxes occurring at the solar maximum period has been used. Figure 18 shows the corresponding proton fluxes per day with energies greater than 4 Mev. The solar cycle changes in the lower edges of the belt have not been included in these calculations, since the magnitude of the effect has not been well established. For this reason, the figure is not dated for a specific time period, but the fluxes can still be used as conservative upper limits for the solar maximum period. Obviously, spacecraft with highly eccentric or unusual orbital configurations must be treated individually because variables such as the angle of the line of apsides will change the attitude of the orbit with respect to the trapped radiation belts.

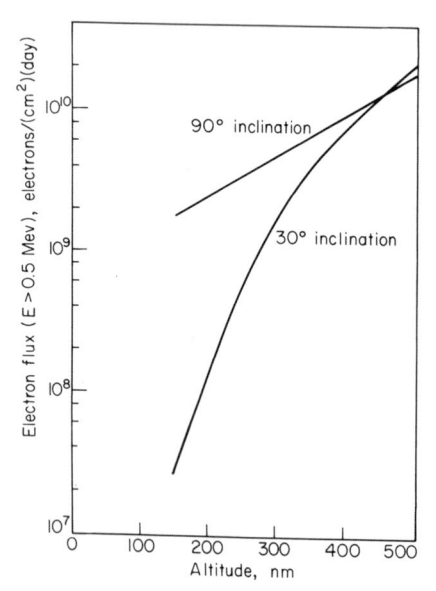

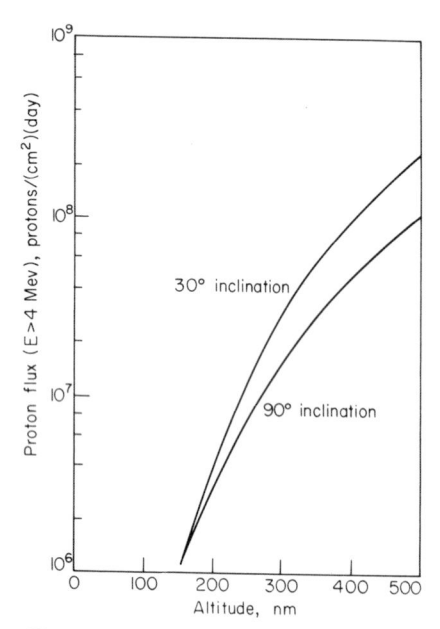

Fig. 17 Electron flux per day encountered in circular orbits for December, 1968.

Fig. 18 Proton flux per day encountered in circular orbits.

SOLAR PARTICLES: The geomagnetic field deflects charged particles incident on it from interplanetary space and thus provides very effective shielding to the region of space between about 60° north and south magnetic latitudes within the magnetosphere. Near the magnetic poles and in interplanetary space outside the boundary of the magnetosphere, the direct charged-particle radiation from the sun can be observed. This radiation consists of two components: high-energy particles that occur sporadically, usually in correlation with visible disturbances on the surface of the sun or solar flares, and low-energy protons and electrons, which are present more continuously.

Solar flare radiation: The charged particles in solar flare radiation events are electrons, protons, alpha particles or helium nuclei, and very small numbers of charged nuclei of mass greater than helium.

Electrons in the energy range 40 to 150 kev have been measured when accompanying a number of small solar flares during solar minimum.[42] One case of very high energy electrons with energies in the range of 100 to 1,000 Mev was observed after a succession of three fairly large solar flares[43] within four days during 1961. The fluxes of electrons observed in all cases were small from a damage standpoint.

Protons from solar flares present perhaps the most important source of damaging particles for many orbital configurations. Since solar proton events occur sporadically and vary widely in peak proton flux and duration, the total flux of protons expected within a particular time period is treated statistically. Information obtained on proton fluxes during the last solar cycle maximum has been incorporated into a statistical model which gives the probability of encountering a total flux of protons greater than Φ for various time periods.[44] Plots of this probability versus the total flux of protons with energies greater than 30 Mev for mission durations of two weeks and one year are shown in Fig. 19. The extrapolation of the 52-week curve follows from extending the measured distribution of sizes of individual flares. This allows for the possibility of the occurrence of proton events more intense than any observed during the last solar maximum period. The energy spectra of solar flare protons vary from event to event and also change with time during a single event.[45] A spectrum taken as most representative of the average of 29 solar flares[44] between 1956 and 1961 is shown in Fig. 20 as a differential energy spectrum normalized to an integral flux above 30 Mev or 1 proton/cm². The proton intensity at any energy is obtained by multiplying the value from Fig. 20 by the integral flux above 30 Mev for a selected probability from Fig. 19. The curve in Fig. 20 follows the same analytic form below 30 Mev proton energy as it does at higher energies, but is shown as a dashed line because the direct-particle measurements did not extend this low in energy and the uncertainties associated with other types of measurements in the low-energy region are much greater. The solar proton model described, although based on solar maximum conditions, can serve as a conservative upper limit for estimates during solar minimum.

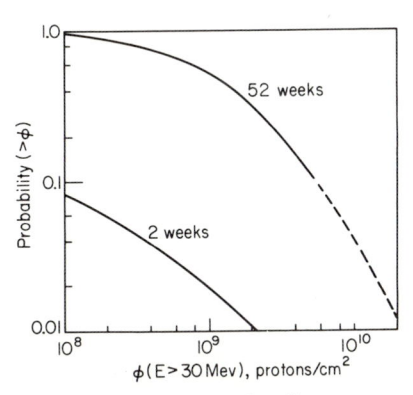

Fig. 19 Free-space solar flare proton environment for a 2-week and a 52-week mission.

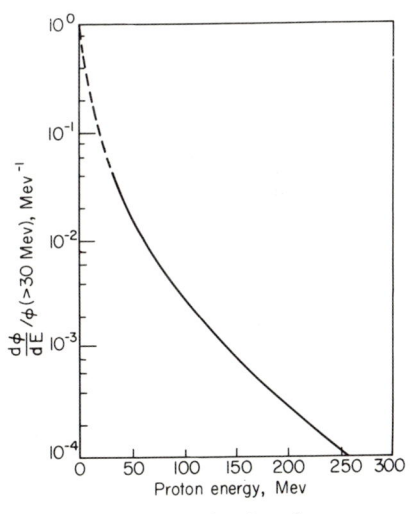

Fig. 20 Differential solar flare proton spectrum in free space.

Alpha particles and charged nuclei of higher atomic number accompany the fluxes of protons from solar flares. In several cases where both alphas and heavier nuclei have been observed, the ratio between their numbers has been constant at about 60. The ratio of protons to alphas within the same energy range appears to vary considerably, from about 10 to several hundred.[45,46]

Solar wind: The solar wind is a plasma consisting of protons, electrons, and alpha particles which continuously streams radially outward from the sun.[47] The particle velocity in the vicinity of the earth was found to vary with solar modulation between about 350 and 700 km/sec, which corresponds to energies of approximately 0.6 to 2.6 kev for protons. The particle flux intensity varied between about 3×10^7 and 1×10^9 cm^{-2}-sec^{-1}. These are large fluxes of particles, but since the energy per particle is small, the damage to materials from solar wind particles will be confined to surfaces.

Auroral region radiation: Intense fluxes of protons and electrons have been observed in auroral regions from about 60 to 70° geomagnetic latitude with somewhat lower fluxes at higher latitudes up to the magnetic poles.[48,49] The particle intensities fluctuate over several orders of magnitude but may always be present in these regions at altitudes up to at least 500 nm. The exact origin of these fluxes and the mechanisms of their trapping or storage and precipitation into the atmosphere are not well understood. They seem to be correlated with solar activity, however, and the most reasonable source with sufficient total energy to produce the observed fluxes is the solar wind.

The average energies of electrons observed in the auroral regions is of the order of a few kilovolts to tens of kilovolts.[50] Intensities for these electrons are difficult to estimate for assessing damage to materials because of their large variations, the relatively short time span covered by observations, and the fact that the fluxes may be highly anisotropic at times and nearly isotropic at other times. A rough estimate based on the highest activity data and assuming an average energy of 10 kev gives approximately 10^{12} electrons/(cm^2)(day) for a low-altitude polar orbiting satellite.

Observations of precipitating protons in the auroral regions in 1965 showed average particle energies of 10 to 20 kev and peak fluxes greater than 10^6 protons/(cm^2)(sec)(steradian) for energies greater than 20 kev.[51] A rough estimate similar to that provided above for the electrons gives approximately 10^{10} protons/(cm^2)(day), with an average energy of 15 kev.

GALACTIC COSMIC RAYS: Galactic cosmic rays have been studied for several decades, but more detailed data are still needed for a complete quantitative understanding of the subject. Within the solar system, cosmic rays consist mostly of protons (about 90 percent), 10 percent helium nuclei, and very small percentages of heavier nuclei having atomic numbers up to approximately 30.[52] There is also evidence of primary cosmic electrons and positrons,[53] constituting perhaps less than 1 percent of the proton flux at a given energy. The energies of cosmic protons may be as high as 10^{10} Gev, although most of the flux is in the energy range from approximately 1 to 10 Gev. The integrated flux of protons above 40 Mev is about 2 protons/(cm^2)(sec) during solar maximum and about 5 protons/(cm^2)(sec) during solar minimum. This solar modulation is energy-dependent, affecting the fluxes of protons with energies between a few tenths of a giga electron volt and a few giga electron volts strongly.

Neutral Particles and Electromagnetic Radiation. Since uncharged particles and electromagnetic radiation are unaffected by magnetic fields, they will be more uniformly distributed in space near the earth from primary sources than, for instance, the trapped charged-particle radiation. The spatial distribution of the radiation arising from the interaction of primary charged particles with matter such as the atmosphere of the earth and the spacecraft itself will be quite localized in nature.

PRIMARY SOURCES: *Solar origin:* An upper limit of 10^{-2} neutron/(cm^2)(sec) average flux of direct solar neutrons has been estimated as the solar minimum.[54] Similarly, only upper-limit estimates have been made for direct neutrons accompanying solar flares. In either case, the total flux is negligible compared with that from other sources.

The sun continuously emits fluxes of x-rays in addition to visible and ultraviolet light. The average intensity in the energy range 1.55 to 12.4 kev during a period of moderate solar activity in 1966 was 2×10^4 ergs/(cm^2)(sec).[55] The intensity of the x-ray flux in the energy range 1 to 6 kev from a major solar x-ray flare on July 7, 1966, was observed to rise from about 10^{-3} erg/(cm^2)(sec) before the flare to a peak of 3×10^{-2} erg/(cm^2)(sec) during the flare.[56] The time integral for this flare was 97 ergs/cm^2. The observations of solar flare x-rays are too limited to try to estimate the magnitude of their effect over a long time period. However, the energies of the x-rays are fairly low so that damage from them will generally be limited to surfaces.

An experiment to measure gamma rays with energies greater than 20 Mev did not show a significant increase in flux over background from the direction of the sun, but set an upper limit of 2.8×10^{-2} photon/(cm^2)(sec) for this source during a mildly active period of the sun in 1959.[57]

Galactic origin: If neutrons are produced in our galaxy or others, they would not be detectable as such except in areas relatively close to their source since free neutrons decay with a 12-min half-life into protons, which would then merely add to the cosmic-ray proton flux. Thermalized neutral atoms, mostly hydrogen, exist in interstellar space at an approximate average density of 1 atom/cm^2, although the regional density may vary by orders of magnitude.[58] To a space vehicle traveling through this medium, the particles would appear as a bombarding flux with a relative velocity equal to that of the vehicle. A vehicle with a velocity of 10^8 cm/sec would encounter an average flux of 10^8 atoms/(cm^2)(sec) with an apparent energy of about 5 kev.

Fluxes of x-rays from stellar sources have been observed to be of the order of 0.1 photon/(cm^2)(sec) in the energy range of about 20 to 40 kev.[59,60] Gamma rays with energies greater than 20 Mev from the directions of several stellar x-ray sources were not detectable above background level, and upper limits of 1×10^{-2} to 2×10^{-2} photon/(cm^2)(sec) were established for the period of observation.[57]

SECONDARY SOURCES: Numerous types of particles as well as electromagnetic radiation may arise from the interaction of primary particles from solar and galactic origin when incident on the atmosphere of the earth. The most significant of these secondary radiations are bremsstrahlung x-rays produced by trapped electrons stopping in vehicle materials, and to a lesser degree by solar flare electrons stopping in the vehicle and atmosphere and albedo neutrons produced by nuclear reactions between high-energy solar and galactic protons and the vehicle and atmosphere. The bremsstrahlung intensity and energy spectrum is a function of the electron flux intensity and energy spectrum and the material in which the electrons are stopped. In general, the effect of the bremsstrahlung will be much lower in magnitude than effects from primary radiation. The flux of albedo neutrons from atmospheric reactions is approximately 1 neutron/(cm^2)(sec) at altitudes up to a few thousand nautical miles.

Man-made radiation The most intense man-made radiations in space have originated from high-altitude nuclear device detonations. In this discussion these radiations will be divided into two categories: (1) charged and neutral particles and electromagnetic radiations arising from the detonation itself and from the fission debris, and (2) the charged particles injected by the explosion into trapped orbits in the earth's magnetosphere. Other possible sources of direct and trapped radiation such as leakage from satellite-borne reactors and radioactive isotope sources or proposed experimental accelerators could be significant, but to date have not presented appreciable radiation levels in space.

Direct Radiation from Nuclear Devices. A large proportion of the energy of a nuclear detonation is released in the form of thermal radiation.[61] In the exploding device, the temperature is several tens of million degrees Kelvin. Most of the (primary) thermal radiation is then in the wavelength range of about 0.1 to 100 Å, i.e., 120 to 0.12 kev energy, corresponding roughly to the soft x-ray region.[61] For altitudes above 70 sm, the air density is less than 10^{-7} of that at sea level. The mean free path of the dominant thermal x-rays produced in the nuclear detonation is then several hundred miles.[61]

It is convenient, for practical purposes, to consider the nuclear radiations from an explosion as being divided into two time categories, initial or prompt and residual. Most of the neutrons and some of the gamma rays are emitted in the fission and/or fusion processes simultaneously with the explosion. Residual radiations (gamma rays, electrons, positrons, neutrons, and alpha particles) are liberated over an extended period of time as the fission products undergo radioactive decay. At any point of interest in space the relative densities of both the prompt and residual radiations from a nuclear detonation depend on its altitude, the particular characteristics of the nuclear device, and the location of the point of interest relative to the position of the detonation. It is an extremely complex system to assess, even when all the parameters for a given case are specified, so that a quantitative treatment of the subject is beyond the scope of this handbook.

Particles Trapped in the Geomagnetic Field. The charged-particle radiations from nuclear devices which may become trapped in the geomagnetic field include electrons and perhaps positrons from beta decay of the fission fragments, protons from the decay of neutrons emitted by the device, and heavier-charged particles. In addition, significant perturbations in the trapped populations may also occur at the time of a nuclear detonation. Electrons have been observed to remain trapped in large numbers for various lengths of time after these tests. The trapping of the other kinds of particles in conjunction with nuclear detonations has not been well established.

Three detonations in the Argus series in 1958 were designed to investigate the injection of electrons into the geomagnetic field using nuclear devices as a source.[62] The devices had nominal yields of 1.4 kilotons of TNT equivalent and were detonated at ~200- to 400-km altitudes on L shells of 1.7 to 2.1 in the Southern Hemisphere. The observed lifetime of electron fluxes was approximately three to four weeks, with initial maximum fluxes of 10^5 to 10^6 electrons/(cm^2)(sec).[63]

Three detonations by the U.S.S.R. in late 1962 on L shells of 1.8, 1.9, and 2.0 resulted in initial maximum fluxes of about 10^7 electrons/(cm^2)(sec) with lifetimes of approximately one month for each injection.[63] The altitudes of the detonations and exact yields are not available. However, it was estimated that two of the series were submegaton yield and the third was of the order of a megaton yield.[63]

The Starfish nuclear burst occurred on July 9, 1962, at an altitude of about 400 km near Johnston Island which corresponds[63,64] to an L shell of 1.12. The yield was 1.4 megatons, which could produce approximately 5×10^{26} fission-product decay electrons. The initial electron fluxes were very high—of the order of 10^9 electrons/(cm^2)(sec) in the first day. Many of the particles were in very unstable trapping configurations, as evidenced by the rapid decrease in intensity in the first few weeks following the detonation. At later times, the decrease in intensity appeared to slow to rates that could be described by exponential functions in which the exponential parameter, the mean lifetime, is a function of L. Measurements of mean lifetime made in late 1962 and 1963 showed a peak value of about 1½ years for $L = 1.5$ earth radii. This corresponds to the slowest decay rate, so that the maximum in flux intensity shown in Fig. 15 will shift toward $L = 1.5$ as the Starfish electron belt decays.

From the foregoing examples, it is evident that the intensities of electron fluxes and the length of time they remain trapped after injection depend on the yield of the devices and the altitude and geomagnetic location of the detonation. In particular, as a result of a nuclear detonation, high fluxes of electrons can be injected into low-altitude regions of space where the fluxes of naturally trapped electrons and protons are rather low. Thus, the injection of electrons from nuclear detonations can present a serious hazard for years to many satellite programs designed to take advantage of the normally low radiation levels at altitudes of ~150 nm.

Internal environments Because of the presence of shielding provided by the spacecraft structure and to the possible existence of an on-board nuclear reactor or radioactive isotopes, the internal radiation environment may be quite different from the external environments discussed in the previous section.

Existing procedures for calculating modifications of the intensities and energy spectra by the shielding are rather complex. Accordingly, the subject will be discussed here only in rather general terms. In addition, radiation levels will be given for typical on-board reactors or radioactive sources.

Attenuation by Shielding. As they penetrate matter, radiations lose energy by a variety of interactions, some leading to secondary radiations. Accordingly, calculations of transmissions may involve many complex procedures. In certain special cases, various assumptions that lead to rather simple calculations can be made. For example, with thin, low-atomic-number shielding, the secondary radiations may not make a significant contribution. Thus, the environment at the center of a uniform thin spherical shell in an omnidirectional flux can be calculated rather well with simple techniques. For thicker, more complex shielding configuration, it may be necessary to employ more sophisticated time-consuming computational procedures. Computer programs have been developed which take into account complex geometries while using simplified interaction models.[65-67] For a more detailed and realistic treatment of the environment and its interaction with shielding materials, Monte Carlo programs are used. Various computer codes of this type, in which individual particles of photons are followed as they interact with the shielding material, have been developed for electron, nucleon, and electromagnetic radiations.[68-70]

Energy Deposit in Materials. For certain radiation damage assessments, direct correlation of damage coefficients with fluxes and energy spectra of the incident radiations are most satisfactory. This is true of cases such as semiconductor damage, which involves perturbation of atoms in a crystal lattice by particle radiations, particularly protons, neutrons, and heavier particles where the total damage may be more than that produced by ionization energy deposit. For other types of radiation damage, it is convenient to convert the radiation environment incident on material to the energy deposited per unit mass of dose in the material.[61,71,72] This applies particularly to materials in which the main damage is from breakage of chemical bonds (e.g., organic spacecraft materials and animal tissue). The currently used unit of dose is the rad, defined as 100 ergs of energy absorbed per gram of material on which the radiation is incident.

Particle radiations will produce damage in materials that is disproportionate to the linear energy transfer from the particles through ionization. The damage per unit energy transfer becomes greater as the particle slows down and its ionization rate becomes greater. This excess damage is specified for tissue dose as a relative biological effectiveness (RBE) factor which is a function of particle energy and is used as a multiplier to convert doses from lower-energy particles to what would be produced by high-energy particles only. For biological doses, the dose unit is the rem, which is equal to a rad multiplied by the RBE for the radiation.

The charged particle dose in biological tissue thick enough to stop all the incident radiation is given by

$$D = \int_0^R \frac{dE}{dx}(E) \cdot \text{RBE}(E)\, dx \qquad (3)$$

where RBE(E) = relative biological effectiveness
dE/dx = linear energy transfer of the particle or, conversely, the stopping power of the material
R = range of distance the particle travels before it is stopped

The integral is evaluated over the range of the particle, and the quantities dE/dx and RBE will be functions of the remaining particle energy. A simplification of this relationship holds when the thickness of the tissues is such that a small amount of the particle energy is lost in penetrating the material. Thus, surface dose or skin dose from particles of energy E can be expressed as

$$D_s = \frac{dE}{dx}(E) \cdot \text{RBE}(E)\, \Delta t \qquad (4)$$

where Δt is the material or skin thickness.

Environment from On-board Sources and Nuclear Reactors. The use of nuclear reactors for propulsion or for power generation and the use of radio isotope power sources on board space vehicles can pose damage problems from the leakage fluxes.

Isotope sources have a relatively low potential power output per unit mass. Their use would probably be constrained to modest power requirements from weight considerations such that the radiation escaping the power source and contributing to the vehicle internal environment may not be a serious problem compared with that caused by external radiation. The principal leakage radiation from such sources would be gamma rays and bremsstrahlung, the relative intensities depending upon the choice of isotope.

Nuclear reactors for either power or propulsion have the capability of very high power output, and for efficient utilization would probably range from hundreds of kilowatts to many megawatts. The primary leakage radiations from reactor cores are neutrons and gamma rays from the fission process. Since a significant fraction of the neutrons have energies in excess of 1 Mev and the fission gamma-ray spectrum peaks at around 1 Mev, these radiations are difficult to shield against. The unshielded radiation rates 100 ft from a 1,000-Mw reactor might be as high as 10^{11} neutrons/(cm^2)(sec) and 10^6 rad/hr from gamma rays. Thus, serious damage to many components would occur in a short time in this environment unless very effective shielding were provided.

Meteoroids

One of the environments in which the spacecraft must function successfully is that of the extraterrestial solid debris or meteoroids. Until the advent of the space age in 1957, our understanding of this environment has stemmed from terrestial observations of the interaction of the debris with the upper atmosphere, studies of the solar corona and zodiacal light, and collection of dust particles at high altitudes or recovered from deep sea or ocean sediments. Starting with Explorer 1 in 1958, several spacecraft have carried some form of meteoroid detector, culminating with the penetration devices carried on board Explorer 16 and 23 and Pegasus 1, 2, and 3. Each of the observational techniques—astronomical, photographic, radio, dust collection, and spacecraft impact detectors—has a sensitivity for a particular range of particles mass population.[73]

A meteor is an extraterrestial particle originating in the solar system, the light of which, generated by interaction with the atmosphere, is detectable visually or by radio. A meteoroid is a particle traveling in space. A micrometeoroid is a small meteoroid, usually of mass smaller than 10^{-6} g. A meteorite is a meteor that has survived the passage through the atmosphere and may be recovered on the earth's surface. A micrometeorite is a small meteorite, a small fragment of a larger meteorite, or actual microscopic particles that have survived the passage through the atmosphere. The term *dust* is used to describe any extraterrestial particles of microscopic size that originate in the solar system and may be applied to micrometeoroids or micrometeorites. The particles of debris are also termed *sporadic meteoroids*, which are particles of close proximity having a similar but independent solar orbit.[74]

The space debris is believed to originate primarily in the solar system, caused by the following: cometary disintegration, which contributes more than 90 percent of the total; fragmentation of asteroids (solid bodies concentrated in interplanetary space between the orbits of Mars and Jupiter), which contributes between 2 and 10 percent of the total; material ejected from the surface of the moon by meteoroid impact; interstellar capture, which contributes about 1 percent of the total; and condensation of interplanetary gas.[73,74]

Meteoroid density The density of the asteroids is about 9 g/cm³. They are composed of iron, iron-nickel alloys, or mixtures of these with metallic oxides (stones). Meteorite particles of this origin will be of relatively high density.

Meteorites of cometary origin are conglomerates of mineral particles and have low density, high porosity, and high frangibility. The conglomerate structure, with voids, will have an average density of about 0.4 g/cm³, which will increase to the density of a mixture of the heavier elements, iron and nickel and their oxides, which is about 3.5 g/cm³. A mass density of 0.5 g/cm³ for meteoroids has been tentatively agreed upon.

Meteoroid velocity The minimum theoretical velocity of a meteoroid near the earth is the escape velocity of 11 km/sec. The maximum velocity of a heliocentric particle at 1 astronomical unit (1 au = 93,000,000 sm), the distance of the earth from the sun, is about 42 km/sec. The earth has an orbital velocity of 30 km/sec, and so the maximum velocity of a meteoroid in parabolic retrograde orbit in the ecliptic plane at 1 au will be the summation of these velocities, or 72 km/sec. Very few meteoroids have hyperbolic trajectories, and so very few will have velocities greater than 72 km/sec.

The reflection of electromagnetic radiation causes a change in the momentum of a particle in solar orbit which imparts an outward force opposing the inward solar gravitational force. This effect of solar electromagnetic radiation pressure can cause a particle either to be forced out of the solar system or to spiral inward toward the sun.

The latter is the Poynting-Robertson effect. This effect tends to cause meteoroids to assume a circular orbit (gradually) rather than an eccentric orbit and thus influences the size and velocity in orbit. A particle of 1 micron diameter and having the density of aluminum (2.79 g/cm^3) in the vicinity of the earth is of the proper size to just balance the solar pressure and gravitational field. A particle 200 microns in diameter would spiral into the sun in 4,000 years from the earth's orbit under the influence of the Poynting-Robertson effect. The solar wind also influences the size and velocity of particles orbiting the sun. Thus, there can be a sorting out of particles according to their size, the smaller particles being faster than the larger.

Flux-mass relationship All these factors have an influence on the flux-mass and size-velocity relationships of meteors observed. The velocity variation with visual magnitude assumes a velocity of 28 km/sec for magnitude 0 to 7. The increasing influence of radiation pressure with decreasing size reduces the velocity by 1 km/sec for each visual magnitude to a constant velocity of 15 km/sec at visual magnitude 20. The velocity of 15 km/sec is assumed constant for all larger visual magnitudes. A normal impact velocity for meteoroids of 25 km/sec has been tentatively agreed upon.[74,75]

Increases in the average hourly rate of meteor influx are observed at regular intervals during the calendar year. These increases are caused by the passage of the earth in its orbit through the orbit of a stream of particles, probably of cometary origin, that are traveling in similar heliocentric orbits. The ratios of the meteoroidal stream flux to the sporadic background can increase by as much as a factor of 20 during a meteor shower. In some streams the peak flux will increase to 15 percent of the maximum in about 1¾ hr, then decrease in the next 1½ hr, and reduce to the sporadic background in another 1¼ hr for a total stream lifetime of 4½ hr. Some of these streams have a period of about 1.5 years, whereas others recur in about 100 years. The flux generally decreases with increasing mass of the meteoroid in the stream fluences. The sporadic fluence also decreases in flux with increasing mass.

The tentatively established flux-mass relationship[75] is shown in Table 7. These data are plotted as the log of the flux versus the log of the mass in Fig. 21. Photographic data, penetration data from Explorers 16 and 23 and Pegasus 1, 2, and 3, and Gemini window crater data were used to establish the curve.[75-77] Influx

TABLE 7 Tentative Annual Average Sporadic Meteoroid Environment Model
(at 1 astronomical unit)

Mass meteoroid, grams	Flux, particles/(m²)(sec)
1.0	3.89×10^{-15}
10^{-6}	8.17×10^{-8}
4.46×10^{-9}	2.86×10^{-6}
5.0×10^{-11}	1.52×10^{-5}

rates based on radio observation or spacecraft microphone sensors were considered for the derivation of this model but were not included.

Meteoroid impingement The impingement of high-velocity particles on thin-skinned pressurized structures can cause small punctures that could result in gradual loss of pressure of the space vehicle or actual rupture by explosive decompression. The kinetic energy of the impinging particle is absorbed by the particle and the skin to cause fragmentation and vaporization of both. Interaction of these fragments with the spacecraft atmosphere can result in oxidative explosions to produce high-temperature and high-pressure fluctuations. The penetration, impingement, or erosion of other spacecraft components could result in nonstandard operation, malfunction, or failure. The flux-mass relationship then should be converted to a flux-penetration relationship in some standard approved manner so that the meteoroid penetration hazard to spacecraft can be estimated.

Perforation of Materials. Terminal ballistics of the penetration of armor by projectiles has been studied for many years. The physical phenomenon of perforation, cratering, and spalling under hypervelocity impact simulated in the laboratory has yielded some empirical equations relating the velocity of the projectiles and their mechanical and physical properties to the mechanical and physical properties of the target plate.[73,78-82]

The major limitation of the ground-based simulation techniques of meteoroid impact has been inability to achieve the high velocity (25 km/sec) estimated for meteoroids in space. Also the projectiles used have not been able to simulate the estimated density of meteoroids (0.5 g/cm³).

By the use of simulation techniques and comparison of the empirical equations developed from them, a tentative relationship between thickness of material just perforated (sometimes called the ballistic limit), its properties, the properties of the meteoroid, and its velocity have been established as shown in the following equation:[75]

$$t = 0.65 \left(\frac{1}{\epsilon_t}\right)^{1/8} \left(\frac{\rho_m}{\rho_t}\right)^{1/2} (V_m)^{7/8} (d_m)^{19/18} \tag{5}$$

where t = thickness of sheet just perforated, cm
ϵ_t = percentage elongation of sheet material, %
ρ_t = mass density of sheet material, g/cm³
ρ_m = mass density of meteoroid, 0.5 g/cm³
V_m = normal impact velocity, 25 km/sec
d_m = meteoroid diameter, cm, assuming a spherical particle

Inserting the material constants for 2024-T3 aluminum alloy and the meteoroid constants into Eq. (5) results in

$$t = 3.30(d_m)^{19/18} \tag{6}$$

A plot of the logarithm of the thickness of 2024-T3 aluminum alloy calculated from Eq. (6) as a function of the logarithm of the flux of meteoroid penetration per square foot per day is shown in Fig. 22. For comparison, the data of the Gemini window crater, the Explorer 23, and the Pegasus spacecraft meteoroid penetration experiments as evaluated by Nauman[77] are also shown in Fig. 22.

Single-wall Perforation. The meteoroid environment in terms of penetration flux and the thickness of sheet material just perforated permits the selection of the single-wall thickness of a spacecraft based upon the desired reliability for no perforations in a given area in a given time. A Poisson distribution of the number of penetrations n permits the calculation of the probability $p(n)$ for penetration of area A in time τ by the meteoroid flux ϕ from the following equation:

$$p(n) = e^{-\phi A \tau} \sum_{n=0}^{n} \frac{(\phi A \tau)}{n^n} \tag{7}$$

The flux corresponding to a probabilty $p(0)$ of no penetrations is then

$$\phi = \frac{A\tau}{-\ln p(0)} \quad (8)$$

The probability of 0.99, 0.95, 0.90 for no perforations by meteoroids through a given single-sheet thickness of 2024-T3 aluminum alloy for a given area-mission duration product is shown in Fig. 23. These data were derived from the flux-penetration curve of Fig. 22 and Eqs. (7) and (8).

Multiple-wall Penetration. To design a single-wall spacecraft for a high probability of no perforations for a large area-duration product may impose an unacceptable weight penalty. To avoid this and achieve optimum weight, the spaced armor principle has been studied to afford protection of the main spacecraft wall by a thin sheet spaced away from it a given distance. The distance of the main wall sheet, the front wall sheet (sometimes called the bumper or shield), and the spacing of the front wall from the main wall depend upon the meteoroid velocity,

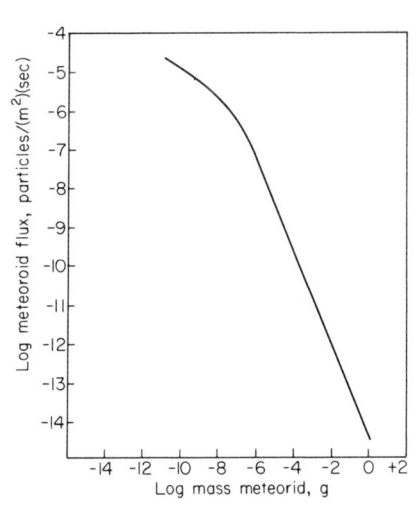

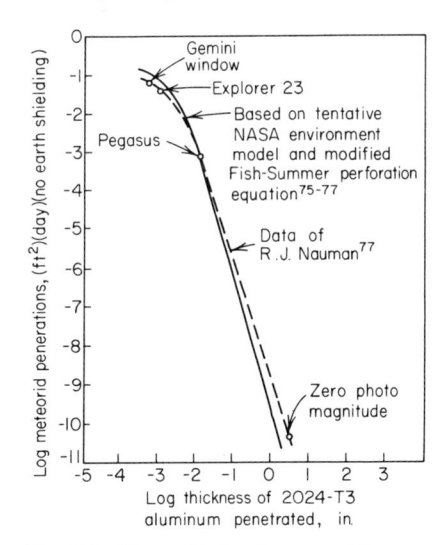

Fig. 21 Tentative NASA annual average sporadic meteoroid environment model for 1 au.[75,76]

Fig. 22 Upper limit of meteoroid penetrations through 2024-T3 aluminum alloy.

its mass, and the material of the spaced sheets. Structural and load considerations also enter into the configuration.

In simulation tests, it was found that more than two sheets did not provide as much protection as two properly designed spaced sheets.[83] It was also found that honeycomb or other filler material between the two walls was not as protective as two walls with proper spacing and thickness distribution between them without filler.

Three possibilities for the behavior of a particle upon impacting the front sheet are: it is stopped; it perforates without damage to the particle to travel on to impact the second sheet; or it can pass through the sheet and become fractured, molten, or vaporized. The meteoroid velocities are large enough to cause melting and vaporization.

If the meteoroid should penetrate the front sheet and impact the second sheet, a shock wave is generated within the second sheet that can cause formation of an internal fracture or spall. If the shock is intense enough, material may fracture from the inner surface of the main wall and be propelled about the inside of the spacecraft to damage components or occupants. The impact of the meteoroid debris on the second sheet can produce an impulsive load (for a few microseconds) that can cause deformation of the material and failure in tension or shear.

Theoretically, it is believed that at meteoroid velocities the material behaves hydrodynamically and the meteoroid becomes molten or vaporized as it passes through the shield. The hole in the sheet has the same diameter as the meteoroid; the shock is absorbed in the small volume about the hole; the shield debris, although traveling at slower velocities than the meteoroid fragments, can be ignored; and the meteoroid fragments are distributed in a gaussian manner.[82,84]

Investigators have shown experimentally that the spacing between sheets can be from 1 to 10 in. The outer and inner sheets can be configured to give the

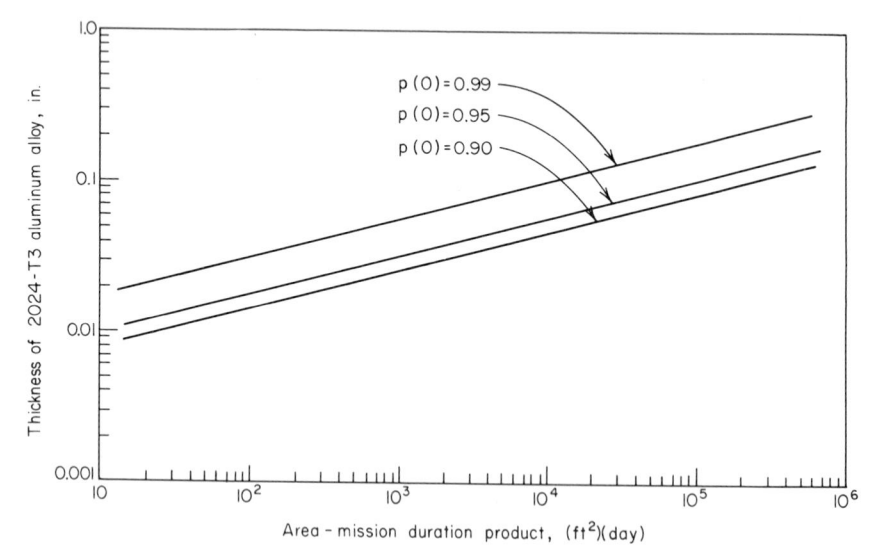

Fig. 23 Probability $p(0)$, of no meteoroid penetrations through a single sheet of 2024-T3 aluminum alloy.

best distribution for protection of the inner sheet. Two empirical equations have been proposed,[82,84] shown below as Eqs. (9) and (10).

$$V = \left[\frac{4\pi\sigma}{E} - \sqrt{\frac{1-\nu}{3(1+\nu)}} \, C \right] \left(\frac{\rho_0 d}{m_p} \right)^2 S^2 t_0 t_1 \qquad (9)$$

where V = velocity of meteoroid, km/sec
m = mass of meteoroid, g
d = diameter of meteoroid, cm
ν = Poisson's ratio of sheet material
ρ_0 = density of sheet material, g/cm^3
σ = critical stress in sheet, dynes/cm^2
C = velocity of sound in sheet, cm/sec
E = modulus of elasticity of sheet, dynes/cm^2
S = sheet spacing, cm
t_0 = thickness of shield, cm
t_1 = thickness of spacecraft sheet, cm

$$t_f = \frac{m_p}{0.045} \left(\frac{5.08^2}{S} \right) \left[\frac{0.0102}{(t_s/d)^2} + 0.079 \right] \qquad (10)$$

where t_f = second sheet fracture thickness, cm
 s = sheet spacing, cm
 t_s = shield thickness, cm
 m_p = meteoroid mass, g
 d = meteoroid diameter, cm

This treatment assumes a ratio of $t_s/d = 0.1$. Under these conditions, an efficiency factor of about 7 of the spaced double-wall over the single-wall configuration is achieved. This configuration is for the case where the wall spacing is 2 in., and the thickness of the inner wall is about 5 times the thickness of the shield.

For the Apollo program, the empirical equation being used to estimate the double-wall configuration makes the assumption that in actual meteoroid penetration through the sheet (the shield), the particle is not vaporized. It is assumed that since there is vacuum in the spacing between the shield and the spacecraft wall (the second sheet), some meteoroid particulate matter will strike the second sheet and some impulsive loading on it will occur.

To account for this behavior, the tentatively accepted Eq. (11) is used, where it is assumed that t_s/d is equal to 0.1 and that t_b remains constant when s/d is greater than 25.

$$t_b = \frac{K(\rho_m)^{0.15}(M_m)^{0.35}V_m}{\sqrt{S}} \qquad (11)$$

where t_b = thickness of second sheet, cm
 K = constant of sheet material (for aluminum alloy $K = 0.06$)
 ρ_m = density of meteoroid, g/cm³
 M_m = mass of meteoroid, g
 V_m = velocity of meteoroid, km/sec
 S = sheet spacing, cm

This equation is also a specialized one and has certain limitations. Thus, like the other equations for double-wall penetration, Eq. (11) does not completely cover all conditions. Equations (9) to (11) should be used with complete awareness of their limitations.

To correct the material parameters of Eq. (11) for other materials, Eq. (12) is tentatively proposed.

$$t_b = \frac{K(\rho_m)^{0.15}(M_m)^{0.35}V_m}{\sqrt{S}} \sqrt{\frac{70{,}000}{\sigma_y}} \qquad (12)$$

where σ_y is the 0.2 percent offset yield strength of the substituted material in pounds per square inch, and all other parameters are as in Eq. (11).[85]

EFFECT OF SPACE ENVIRONMENT ON ELECTRONIC MATERIALS AND COMPONENTS

Penetrating radiation constitutes the most important direct space environment parameter affecting the operation of electronic materials and components in space. The nature of applications of electronic components generally results in their being protected from other aspects of the space environment.

Principal Environmental Effects

The designer of electronic components and devices for space applications is interested in how the radiation environment will change the properties of these components and devices and at what level of radiation dose and under what environmental conditions the components and devices will cease to operate satisfactorily. Data needed to answer these questions fall into two general categories: those concerning the materials of which a component or device is built and those concerning the entire component or device. Data on radiation effects on components and devices are of greater usefulness to the aerospace designer, but unfortunately, these data are still incomplete, partially because the many types and varieties of electronic

15-34 Materials for the Space Environment

components and devices would require a much greater investment of time and equipment for testing than has heretofore been expended. Furthermore, advances in electronic technology are introducing novel and modified varieties of components at a fast rate, and test data for these new items are necessarily scarce at best.

Radiation response of electronic materials Since direct data on radiation effects on components and devices will frequently be unavailable to the designer of aerospace equipment, he must then resort to the other category of data mentioned—the individual materials of which the component is made. On the basis of radiation effects on materials, some extrapolation can be made as to the behavior of components and the limiting conditions that a component can withstand.

This section presents a discussion of the general effects of space radiation on materials used in the manufacture of electronic components.[80-95] Available information on the components and devices themselves will be presented in a subsequent section, and finally the effects of other environmental factors will be summarized.

Metals and Alloys. Of all the materials used in electronics, metals and their alloys are probably the most resistant to the effects of radiation. Changes in their mechanical properties are of little concern in electronics because they are comparatively small and a large percentage of these changes anneal at rather low temperature.

Effects of radiation on electrical resistivity have been studied extensively. Changes in this property are essentially due to lattice imperfections, which are more efficiently produced by heavy particles than by electrons or gamma radiation. Neutron radiation at temperatures between 30 and 300°C causes small permanent changes in most metals. At room temperature, increases in resistivity range from 0 to a few percent for an integrated slow neutron flux of 10^{19} neutrons/cm^2 for metals commonly used in electronics such as copper, silver, nickel, platinum, and their alloys. Substantial effects can be observed, however, in metals having a high melting point: about 10 percent for the same neutron dose in tungsten and molybdenum at room temperature. During exposure to an integrated fast neutron flux of 2×10^{18} neutrons/cm^2, copper-nickel and nickel-chrome, commonly used in wire-wound resistors, showed a maximum increase in resistance of 0.8 percent. However, no permanent effect was observed. The effects of radiation on the work function and emitting property of metals and alloys used for thermionic and photocathodes have not been studied.

Ceramic Materials. In general, ceramic materials used in capacitors, including those containing some organic binder, can withstand appreciable doses of radiation. Only negligible degradation of electrical and physical characteristics occurs up to 10^{15} to 10^{18} neutrons/cm^2.

In thermistors made of metal oxides, the most susceptible constituent is the organic binder. Gas evolution may cause some decay of mechanical characteristics. However, exposure to fast neutrons up to 5.5×10^{10} rads produces no change in the negative temperature coefficient. In one test with slow neutrons, a dose of 2×10^{15} neutrons/cm^2 did not produce any significant effect.

Silicon thermistors are preferred in some applications because of their positive temperature coefficient. Their resistance, however, is very sensitive to radiation. At a dose of 10^{17} neutrons/cm^2, the resistance increases by a factor of 10^6, of which only a negligible fraction can be attributed to irradiation temperature.

Prolonged irradiation in all oxides leads to permanent displacement effect and loss of oxygen, especially near the surface.

In glasses, radiation-induced changes of concern in electronics are electric conductivity, mechanical degradation, and absorption of visible light. No quantitative information is available on increased conductivity, although some test data seem to indicate that this effect is unimportant in lead glass irradiated with 1-Mev electrons. Coloration in most glasses saturates at 1×10^{10} rads or less. High-purity vitreous silica remains transparent at 1×10^8 rads. Addition of 1 or 2 percent CeO$_2$ increases resistance to coloration. The cerium ion is efficient in removing radiation-produced free electrons which would otherwise form color centers.

Electrons or gammas are more efficient in creating color centers than in displacing atoms. Consequently, degradation in mechanical properties occurs at higher doses than does coloration. Bombardment with neutrons and heavy particles leads to mechanical failure at rather low doses. The presence of boron in borosilicate glasses makes them particularly susceptible to neutrons because of the transmutation reaction in boron during which its nucleus absorbs a neutron, then breaks up into an alpha particle and a lithium atom. Cracking and chipping are observed in these glasses at doses ranging from 10^{15} to 10^{20} neutrons cm^2, depending on the boron content, the initial annealing, and the particular device.

Semiconductor Materials. Nuclear radiations produce permanent effects on semiconductors, mainly through displacement of individual atoms of the bombarded material. The minimum energy which a bombarding particle would require to displace one atom is usually estimated at 30 ev for germanium and 25 ev for silicon. Most displaced-atoms defects anneal out at room temperature or form secondary associations with each other and with impurities initially present in the lattice. Those that are thermally stable, however, can affect electronic properties of semiconductors by increasing the concentration of trapping, scattering, and recombination centers. Trapping centers remove carriers that would otherwise be available for conduction. Silicon of n- and p-type gradually changes to the intrinsic material; n-type germanium may convert to p-type. Creation of scattering centers shortens the mean free path of free carriers, whereas additional recombination centers reduce the minority-carrier lifetime.

Solar cells suffer two main types of damage: decrease in minority-carrier lifetime and interaction of vacancies with impurities such as oxygen and phosphorus. Altering the concentration and nature of impurities can reduce the extent of damage appreciably.

Silicon n-p cells are more radiation-resistant than silicon p-n cells. The reason is probably that the recombination length in the p region of the first category is larger than the recombination length in the n region of the second category after prolonged bombardment. For proton irradiation the threshold for 25 percent degradation is about 10^{11} protons/cm^2 at a primary energy of 240 Mev. Silicon lithium-drift p-i-n diodes suffer the usual degradation in minority-carrier lifetime. In addition, however, a redistribution of lithium ions occurs in the drifted region. The increased lithium gradient at the junction leads to a narrower depletion width, causing the lithium drift p-i-n diodes to be more vulnerable to radiation than the p-n type.

Dielectric Materials. At a total gamma dose of 1×10^8 rads, polytetrafluoroethylene, polytrifluorochloroethylene, polystyrene, and polyethylene decrease in volume resistivity by a factor of about 10, and polyethylene terephthalate (Mylar*) by a factor of about 6.

Changes in mechanical properties are generally more serious. At gamma doses as low as 8.7×10^3 rads the melt viscosity of Teflon* decreases by several orders of magnitude (due to predominant scission). Approximately 3.6 µg of corrosive fluorine gas per gram of Teflon is evolved at a dose of 4.35×10^6 rads.

Teflon capacitors exposed to 10^{14} neutrons/cm^2 increase in capacitance by about 10 percent, and the leakage resistance drops from 6×10^{11} to 6×10^8 ohms. In capacitors, the oriented film Mylar shows no objectionable degradation in electrical properties up to 10^8 rads and in mechanical properties up to 10^9 rads of gamma radiation. During irradiation, the dielectric constant undergoes some changes but recovers after irradiation.

Mica used in capacitors behaves satisfactorily up to a neutron dose of 5×10^{18} neutrons/cm^2. Epoxy binders used in carbon-composition resistors become increasingly conductive under irradiation, with no recovery. A fast neutron dose of 2.4×10^{17} neutrons/cm^2 decreases the resistance of a 100-ohm resistor by 4 percent.

Damage in metallized paper capacitors is primarily due to gas evolution from the impregnants (mineral oil, wax, castor oil, petroleum jelly, chlorinated compounds), causing pressure buildup, bursting, and short circuit. For fast neutrons, the threshold of damage is around 10^{13} neutrons/cm^2.

The main radiation effects on circuit boards are increased leakage current and

---

* Trademarks of E. I. du Pont de Nemours & Co., Wilmington, Del.

mechanical damage in the form of blistering and warping. The extent of damage depends critically on the manufacturing process, but test data seem to indicate that the glass-melamine type is more susceptible to blistering and warping than the glass-epoxy, paper-phenolic, steatite, and nylon-phenolic types.

Increase in leakage current is believed to be due to radiation-induced ionization in air, establishing conductive paths between exposed copper conductors. Coatings reduce this leakage, and the most efficient one seems to be silicone varnish. The leakage current in an uncoated board was less than 10^{-12} amp before exposure and 1.09×10^{-7} amp during exposure to gamma radiation at a rate of 70 rads/sec. Under the same irradiation condition, a board coated with silicone varnish showed a leakage current of less than 10^{-12} amp before exposure and 5.6×10^{-10} amp during exposure.

Insulating Materials. Organic dielectrics, encapsulants, and other types of insulating materials are relatively sensitive to radiation damage with respect to their physical strength and dimensional stability as well as their electrical properties. The most sensitive organic insulation possesses a radiation tolerance of up to 1×10^7 rads.[97] Radiation-resistant organic materials of certain types have been known to perform satisfactorily in an environment up to 10^{10} rads, whereas inorganic insulation and dielectrics may be usuable to radiation levels up to 5×10^{12} rads. Table 8 presents[97,98]

TABLE 8 Changes in Electrical Properties of Electrical Insulators Due to Radiation and Vacuum Exposure

Material	Radiation exposure, rads	Dissipation factor* Before	Dissipation factor* After	Dielectric constant* Before	Dielectric constant* After	Volume resistivity, ohm-cm $\times 10^{-17}$ Before	Volume resistivity, ohm-cm $\times 10^{-17}$ After
Polytetrafluoroethylene, TFE-6	4×10^7	<0.001	0.081	2.08	2.12	>10.0	0.31
Polytetrafluoroethylene, TFE-7	4×10^6	<0.001	0.019	2.09	2.13	>10.0	0.28
Fluorinated propylene, FEP-100	1.6×10^6	....	....	....	.....	>10.0	0.048
Monochlorotrifluoroethylene, K-4	2.5×10^6	0.008	0.005	....	.....	>10.0	0.38
Monochlorotrifluoroethylene, K-5	2.5×10^6	0.011	0.004	....	.....	>10.0	2.30
Kynar 400†	$6.6 > 10^7$	0.0042	0.0100	1.00	0.988	0.04	0.0005

* Measurements taken at 60 Hz.
† Trademark of Pennsalt Chemical Corporation, Philadelphia, Pa.

some of the changes in electrical properties of fluorocarbon materials on irradiation. Silicone dielectrics are generally resistant to particle radiation.[99] In one study, it was found that although crosslinking occurred at 10^6 rads in dimethyl-based silicones and at 10^8 rads in methyl phenyl-based silicones, with accompanying brittleness, no significant changes were observed in the electrical properties.

Wire insulations of alternate laps of polyimide, H-film,* and Teflon FEP of 3-, 4-, and 6-mil (0.006 in.) total thickness, and polyolefin with a polyvinylidene fluoride jacket of about 9-mil total thickness were irradiated for 10 hr at 6,000 rads/hr of x-rays in vacuum at 150°C (300°F). The voltage breakdown and insulation resistance of twisted pairs of the wires were measured before and after exposure. No significant changes in these properties occurred as a result of the x-ray radiation. The insulation resistance of some combinations increased about one order of magnitude after irradiation. For example, the resistance of the poly-

* Trademark of E. I. du Pont de Nemours & Co., Inc.

olefin–polyvinylidene fluoride combination increased from 6.3×10^{13} ohms before to 3.1×10^{14} ohms after irradiation.[100]

Several encapsulating compounds have been found serviceable for space applications. The cure is an important factor in space stability, a higher-temperature cure being preferred to room-temperature cure. Solvent systems are generally not satisfactory as they tend to dissolve the insulation of embedded wires. Some of the classes of encapsulating materials and the temperature and radiation environments in which they can be used[101] are shown in Table 9.

TABLE 9 Electrical Encapsulant Materials

Material	Continuous temperature range for successful use, °F	Radiation level at which electrical properties may begin to change, rads
Polyurethanes	−65 to 165	10^8–10^9
Silicone rubbers	−85 to 500	5×10^6
Polysulfide	−70 to 300	10^6
Epoxies	−65 to 200	10^7–5×10^8

Magnetic Materials. The two main classes of magnetic materials used in the construction of magnetic devices and circuits are metallic and nonmetallic. Metallic magnetic materials are specialized conductors which at higher operating frequencies show eddy-current effects. This then leads to the lamination of metallic sheets to form useful magnetic devices. To prevent magnetic short circuits, the thin metallic sheets are separated by an insulator such as paper, mica, lacquer, or metallic oxides.

Metallic conductors are little affected by radiation effects. The insulation may be affected if exposed to high fluences of energetic penetrating radiation. Above the Curie temperature, metallic magnetic materials lose their intrinsic magnetic capabilities, and the ordered domain structure of the materials is changed to unordered or random arrangement of the magnetic domains.

For many ferromagnetic materials, the Curie temperature is above 500°C (932°F). The specialized magnetic materials such as permanent magnets and magnetic materials used to store information operate only in the small temperature range −10 to +100°C and may lose their magnetic capability well below their intrinsic Curie temperature.

Local magnetic heating which develops hot spots can produce localized demagnetization. Such a change in the local order of the domains results in a pronounced change of the electromagnetic behavior of the magnetic material.

Nonmetallic magnetic materials are separated into two classes: powder cores and solid-state ferric oxide cores. Powdered cores overcome the frequency limitations of metallic magnetic materials. Fine magnetic metallic or magnetic ferric oxide powders are pressed with a binder into a suitable shape, and fired to form a hard compact. This powder core is mounted into a nonconducting container and wound with the required windings. Changes in the properties of these powdered cores are mainly due to changes in the binder.

Solid-state ferric oxides or ferrites are widely used for high-frequency inductors and with a specialized magnetic nonlinear characteristics as information storage elements. Ferrite materials are not affected by penetrating radiation. The square-loop ferrite cores used to store information are temperature-sensitive. This is a secondary temperature effect, which can be induced by the penetrating radiation.

The effects of reactor neutron and gamma radiation on some components containing magnetic materials[102] are shown in Table 10. The majority of the failures occurred at high neutron fluences and were attributed to short circuits caused by insulation embrittlement or loss in integrity of a powder metallurgically produced compact. Only transient changes in magnetic properties of the materials occurred.

TABLE 10 Effect of Reactor Radiation on Electronic Components Containing Magnetic Materials

Component	Neutrons/cm^2	Gamma radiation	Remarks
Plate transformer....	4×10^{17}	2×10^9	Windings failed at 4×10^{17} neutrons/cm^2. Irradiated at 932°F (500°C). Failure attributed to effect of elevated temperature.
Filament transformer	4×10^{17}	2×10^9 2×10^6/hr	Of two units irradiated, one did not fail, and one developed an open circuit in primary winding at 1×10^{17} neutrons/cm^2. Irradiated at 932°F (500°C).
Inductor (reactor)...	4×10^{17}	2×10^9	Of two units irradiated, one did not fail, and one developed an open circuit at about 5×10^{16} neutrons/cm^2. Irradiated at 932°F (500°C).
Magnetic coils.......	2.5×10^{18}	....	Polymeric insulation materials, Bakelite* varnish, polystyrene, rubber, and friction tape; embrittled.
Magnetic core materials 2V Permendur laminated	2.5×10^{18}	3×10^8	Little change in magnetic properties at direct current, 60 and 400 Hz a-c frequencies. Cobalt-containing alloy, Permendur, had induced radioactivity.
2–81 Mo Permalloy (powder metallurgy process); 50–50 Ni-ferrite ceramic, linear	2.5×10^{18}	3×10^8	Changes in magnetic properties of 2–81 Mo Permalloy because of radiation effects on binder increased air gaps with loss of magnetic flux density. No change in ceramic material.
Nickel ferrite, yttrium-iron garnet	1×10^{17}	5×10^9	No change in magnetic properties of square-hysteresis-loop materials.
Ferrite core.........	4×10^{17}	2×10^{19}	No change in magnetic properties.

* Trademark of Union Carbide Corporation, New York, N.Y.

Threshold damage to iron and 5-molybdenum permalloy magnetic components occurs at about 10^{16} protons/cm^2 and 1.5 Mev. At 10^{17} protons/cm^2, the maximum permeability was decreased 22 and 36 percent, respectively. At 10^{17} protons/cm^2 and 4 Mev, the maximum permeability of the 5 Mo permalloy was reduced 49 percent.

An optically polished yttrium-iron garnet single crystal for a ferromagnetic power limiter was irradiated with 10-Mev protons, 1-Mev electrons, and low-intensity cobalt-60 gamma radiation. The changes in electrical and magnetic properties detected were attributed to the electron- and proton-induced radiation damage of the Teflon holder which supported the magnetic device in the resonant transmission cavity.[97]

Radiation response of electronic devices *Semiconductor Devices.* Over the past decade, the amount of radiation effects data on semiconductor devices has increased by about three orders of magnitude, and the degree of sophistication in radiation effects information has also increased to a large extent. Yet the designer is still not able to achieve reliably hardened designs by selecting components and materials from various compendia of radiation effects information. The reason for this apparent paradox lies in the basic limitation in the component selection approach to hardening, an approach that is only one of the major factors in evaluation and circumvention of system and subsystem vulnerability. The complexities that broaden the scope of the problem of radiation vulnerability and circumvention result from (1) multiplicity in degradation modes and corresponding multiplicity in tolerance levels, (2) possible interaction between modes, (3) dependence on secondary environments, (4) damage to circuits or degradation of system performance by transient effects, (5) dependence on nonradiation ambients such as temperature, pressure, mechanical stress, and electrical stress, (6) wide variations in radiation tolerance of devices of similar electrical characteristics fabricated by different manufacturers or even by the same manufacturer, (7) very rapid developments and changes in electronic device technology and application, and (8) proximal or interface effects related to configuration rather than basic transducer properties.

To cope with this broad scope of problems, it is necessary to adopt a broad approach which seeks to utilize indirect as well as direct methods of hardening and circumvention. The designer must incorporate radiation vulnerability—its as-

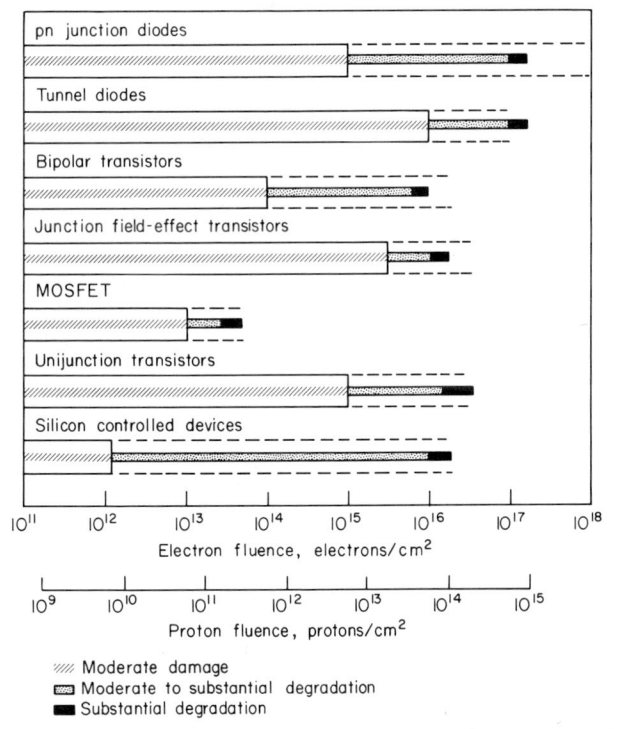

Fig. 24 Useful ranges for semiconductor devices exposed to space radiation (correlated with Table 11).

sessment and circumvention—into his planning and design at very early stages in order to maximize his flexibility in circumvention through circuit design, programming, device selection, packaging, and location. He must understand the manner in which radiation affects the functional parameters of interest; he must know how these changes will affect his subsystem and the overall system; he must think of alternatives, trade-offs, possible relocations, and, finally, possible device selection or hardening. In short, the designer must consider radiation as an added ambient environment along with and in context with temperature, pressure, etc. Hardening by comparing environmental specifications with previously reported data on specific components is very risky unless such information is taken as a guideline and in specific context with application and service conditions.

DIODES, TRANSISTORS, AND SOLAR CELLS: In context with the above discussion

15-40 Materials for the Space Environment

of the nature of radiation vulnerability and the approach to circumvention and hardening, the radiation tolerance information in this section is presented in terms of "useful range" in Figs. 24 and 25 and by specification of dominant modes of degradation in Table 11. The generic categories have been selected on the basis of conventional usage and transducer configuration. The dominant failure modes are based on radiation response and parameters of importance in common applications. It is intended that the designer use Table 11 as a guide to the useful range or the range of radiation in which a particular generic class of devices is expected to suffer significant changes in one or more of the commonly used characteristics but not to the extent of degradation beyond usefulness in conventional application. If it can be ascertained from Fig. 24 that a particular device under consideration belongs to a class well below the useful range (significant change, but still useful), perhaps the vulnerability problem can be regarded as minor or nil unless an unusually tight requirement is imposed. If the application is a critical one or if the radiation level is expected to fall into or close to the range of significant damage, Table 11 should be consulted to find the dominant mode of degradation and to help assess the systems implications of these anticipated effects.

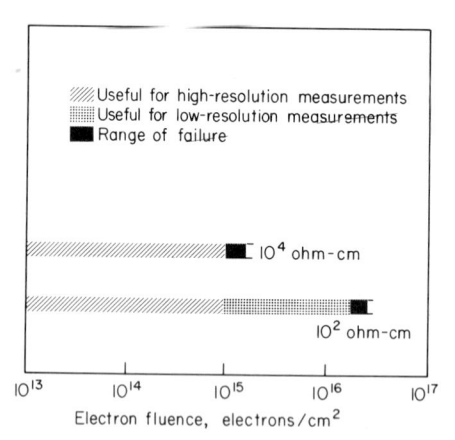

Fig. 25 Useful ranges for silicon charged-particle detectors exposed to space radiation (phosphorus-diffused devices).

Although the format chosen gives only approximate information and some devices in a given category may fall outside the average envelope indicated for the group, it is clear that the measured radiation tolerance of a specific device in a particular test is subject to wide variation in other applications and under different conditions of use. Furthermore, specifying the measured radiation tolerance or performance in a test to the observed degree of precision often tends to mislead because there is a natural tendency to infer the spread of the tolerance range from the precision of a set of measurements.

The specific intent of the tabular information is only to provide the designer with a warning signal of situations of potentially marginal performance and reliability and to give indications of the nature of the malfunction. When a potentially marginal situation arises, the designer may choose to circumvent by alternate design or functional requirement, to circumvent by temporary overrides or disconnects, to modify the design to allow substantial degradation in pertinent parameters without major system compromise, to relocate or repackage or provide additional shielding if reasonable, or to undertake a systematic test program to establish and proof-test a given level of system performance under appropriate test conditions.

The best conditions for such testing, would, of course, be those of actual usage—the space environment. One example of such tests carried out in space is the extensive study of radiation damage to solar cells and transistors performed by the Applied Physics Laboratory[103] beginning in 1961. In particular, the influence of shielding of various types of thicknesses was studied in detail. Shielding used on the solar cells studied ranged from none up to 0.125-in. quartz (0.700 g/cm^2) and up to 0.3-in. aluminum (2.057 g/cm^2) for the transistors. The results of these flight experiments indicate that optimum power-to-weight-ratio solar arrays will be obtained by use of n-on-p solar cells with 6-mil quartz covers. In one experiment, for example, n-on-p solar cells with 0.125-in. quartz covers decreased in current output by 10.5 percent in 500 days, whereas the same type of cells

TABLE 11 Trends of Radiation Damage in Semiconductor Devices

Device type	Operating parameter	Parameter change with increasing fluence	Comments
Diodes, p-n junction	Forward voltage V_F Reverse breakdown voltage V_R Reverse leakage current (I_R) Rise time T_R Storage time T_S	Increase Slight increase Increase Increase Decrease	Fast diodes with low breakdown voltages are less sensitive to permanent effects because of doping profiles.
Diodes, tunnel	Peak current I_P Valley current I_v	Slight decrease Increase	Because of high doping, this device is not affected until relatively high dose levels.
Transistors, bipolar	Current gain h_{FE} Reverse leakage current I_{co} Saturation voltage $V_{CE, \text{sat}}$	Decrease Increase Increase	High-frequency, low-breakdown voltages, characteristic of epitaxial devices, provide better radiation resistance. Some manufacturers advertise radiation-resistant devices incorporating these features.
Transistors, junction field effect	Transconductance, mhos Drain current with ($V_{gs} = 0$), I_{DSS} Pinchoff voltage V_p	Decrease Decrease Decrease	Gain parameter (transconductance) is not appreciably degraded until relatively high radiation dose levels, making the device less sensitive than comparable bipolar devices.
Transistors, MOS field effect*	Transconductance, grams Threshold voltage V_τ Channel resistance r_{DS}	Decrease Increase Increase	Very easily degraded in an ionizing radiation environment via charge storage in the oxide layer.
Transistors, unijunction	Interbase resistance R_{BB} Valley current I_v Valley voltage V_v Intrinsic standoff ratio	Increase Decrease Increase Remains same	Interbase resistance very sensitive to neutron-induced permanent radiation damage due to doping concentrations.
Silicon controlled devices	Gate current for triggering I_{GF} Gate voltage (required for I_{GF}), V_{GF}	Increase Increase	New device designs, employing selective current gains in the analog of the device, claim greatly increased radiation tolerance.
Solar cells	Short-circuit current I_{SC} Open-circuit voltage V_{oc} Maximum available power ratio $P_{\text{max}}/P_{\text{max},0}$	Decrease Decrease Decrease	n-on-p are approximately one order of magnitude more resistant than the p-on-n devices. Cadmium sulfide thin-film cells look promising, but extensive degradation data are not available at this time.

* MOS Metal oxide semiconductor.

Materials for the Space Environment

TABLE 12 Effect of Space Radiation in Orbit on Transistors

Transistor type	Shielding	\multicolumn{4}{c}{Percent of original value of transistor β retained after indicated days in orbit}
		10 days	50 days	100 days	200 days
2N2222*	None	93	85	80	74
	0.1 in Al (0.685 g/cm^2)	99	98	97.5	97
2N2586*	None	100	98	97.5	97
	0.1 in Al (0.685 g/cm^2)	98	96	94	93
2N2907-A*	0.1 in Al (0.685 g/cm^2)	100	100	99	98
2N1711† (unaltered)	Epoxy-glass honeycomb (0.28 g/cm^2)	92	72	68	66
2N1711† (punctured case)	Epoxy-glass honeycomb (0.28 g/cm^2)	98	91	87	83

* Exposed to environment of satellite 1964-83-C, apogee 1,070 km, perigee 1,027 km, inclination 89.99°, launch date Dec. 12, 1964.

† Exposed to environment of satellite 1963-38C, apogee 1,120 km, perigee 1,070 km, inclination 89.9°, launch date Sept. 28, 1963.

decreased by only 8 percent in the same time under 0.006-in. covers. This difference is attributed to discoloration of the thick cover slide.

The transistor reliability studies, performed on satellite 1964-83C, are summarized in Table 12.

The lower damage shown by the 2N1711 transistor with punctured case as compared with transistors with unpunctured cases is ascribed to ionization of gases with consequent surface damage in the unventilated transistors.

Solar cells have been considered separately in Fig. 26 and Table 11 because of their importance in space-vehicle applications in the foreseeable future, and because they are by function and by mechanism substantially different from the circuit component devices. Experimental data on actual exposure to the space environment have been obtained for solar cells on a number of spacecraft flights. The early data from ATS-1 given in Table 13 show response of a variety of cell types and shielding combinations.[104]

Some serious thought is now being given to preirradiation of solar cells before flight. Since most serious loss of solar cell output occurs in the first few days or weeks in orbit, preirradiation would result in cell outputs more constant with time (although lower). Since solar-cell arrays must now be designed to give sufficient power after radiation degradation, their excess power during the early times in orbit must be dissipated; preirradiation would reduce this requirement.

The term "threshold of failure" or "degradation" was avoided in the previous discussion and is avoided in the illustrations because these terms and their variations have been used in a variety of different ways in the radiation effects literature and because the concept of the word "threshold" implies a sharp boundary which is inappropriate in most applications of radiation damage data. However, the commonly used meanings of threshold are described and compared in the following paragraphs for purposes of clarification and to indicate the context in which tabular information is presented.

In the simplest case, failure behavior can be described in terms of a step function; i.e., zero failures occur up to a certain critical level of radiation, and 100 percent failure is encountered beyond this level. This behavior can thus be characterized by a discrete threshold. However, failure data conforming to such go or no-go response are rather rarely encountered in radiation-induced failure processes. More often, the failure behavior is characterized by a frequency distribution in the failure probability versus dose. Accordingly, the failure threshold can be defined in various arbitrary ways:

1. The maximum permissible radiation constraint which, within a set of nominally identical specimens, produces an average absolute percent change in an operating characteristic such that the latter is still within prescribed limits of tolerance. (This definition neglects consideration of the statistical spread in degradation behavior. As a consequence, some outlying members of the set may actually exceed the design specifications.)

2. A definition similar to the above but superior in its usefulness assumes the thresholds to be given by the maximum permissible radiation constraint which, among a set of nominally identical specimens, produces a given maximum but still acceptable change in an operating characteristic such that this maximum change is not exceeded by more than some specified small proportion of nominally identical

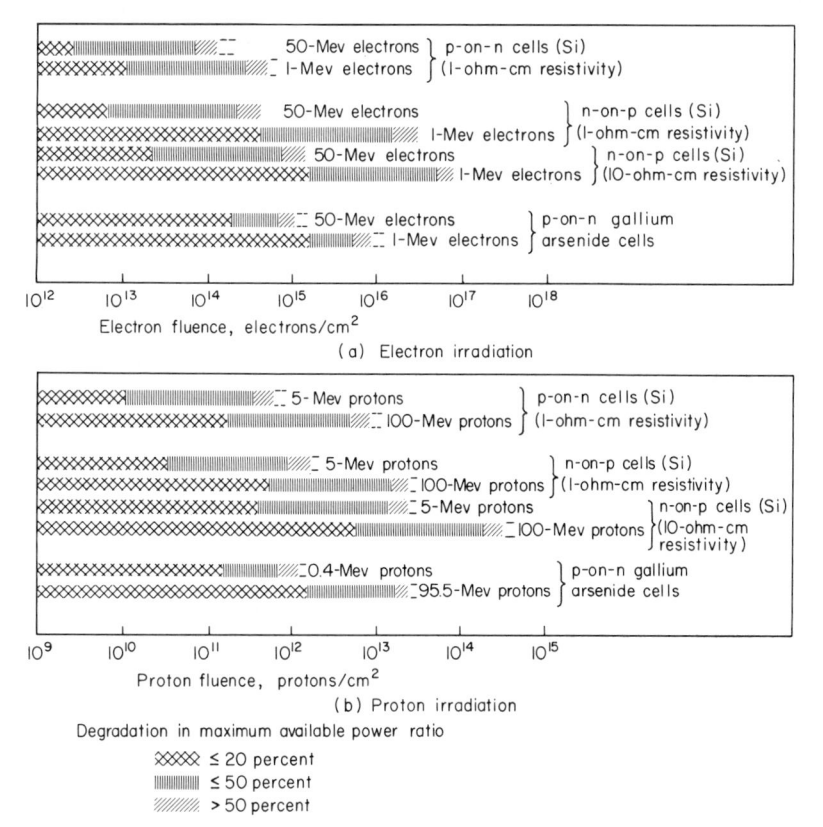

Fig. 26 Useful ranges for solar cells exposed to space radiation.

specimens. This is the concept used in establishing the approximate end points of the useful ranges in Fig. 24.

3. The maximum permissible radiation producing a still acceptable proportion of defective specimens in a given set of nominally identical specimens.

4. The radiation level at which a given radiation-induced effect becomes barely detectable by means of a specified technique of measurement.

INTEGRATED CIRCUITS: Owing to the wide variability in the structural and functional design of integrated circuits or microcircuits, their response to radiation is discussed here only in general terms. Moreover, there is to date a paucity of experimental data on specific integrated circuits to permit a comprehensive evaluation of the behavior of generically related systems. However, one study of commercially

15-44 Materials for the Space Environment

TABLE 13 Changes in Characteristics of Experimental Solar Cells Aboard ATS-1 Spacecraft

Solar cell type	Base resistivity, ohm-cm	Dopant	Shield material	Shield thickness, mils	Cell short-circuit current, ma At lift-off	After 3.3 days in orbit	% change	Cell voltage, mv At lift-off	After 3.3 days in orbit	% change
n on p	10	Al	Sapphire	30	46.4	46.8	+ 0.9	531	522	− 1.7
n on p	13	B	7940 Silica	6	63.4	63.8	+ 0.6	554	547	− 1.3
n on p	10	B	7940 Silica	6	67.5	67.0	− 0.7	567	565	− 0.4
n on p	7	B	7940 Silica	6	62.7	62.4	− 0.5	573	566	− 1.2
n on p	3	B	7940 Silica	6	62.4	62.4	0	586	580	− 1.0
n on p	Graded	B	7940 Silica	6	55.5	55.9	+ 0.7	600	597	− 0.5
p on n	1	P	None	0	58.7	21.9	− 62.7	563	435	− 22.7
n on p	1	B	None	0	53.7	40.0	− 25.5	569	474	− 16.7
n on p	10	B	7740 Glass	1	62.0	60.9	− 1.8	554	545	− 1.6
n on p	10	B	0211 Glass	6	66.4	67.2	+ 1.2	575	567	− 1.4
n on p	10	B	7940 Silica	60	68.0	68.5	+ 0.7	576	570	− 1.0
n on p	10	B	7940 Silica	30	68.4	68.9	+ 0.7	570	565	− 0.9
n on p	10	B	7940 Silica	15	67.0	67.6	+ 0.9	572	575	+ 0.5
n on p	10	B	None	0	69.5	60.2	− 13.4	559	431	− 22.9
n on p	10	Al	7940 Silica	30	66.9	67.5	+ 0.9	570	561	− 1.6
n on p	10	Al	7940 Silica	6	64.0	63.0	− 1.4	568	557	− 1.9

available circuits reveals that the better-constructed circuits are immune to transient radiation disturbances to over 10^9 rad and to permanent damage up to at least 10^{13} neutrons/cm^2.[105]

In a space radiation environment, the principal modes of degradation expected will arise from (1) cumulative displacement damage, (2) leakage currents, (3) surface and interface effects, and (4) effects of charge storage in the oxide layer of metal-oxide-silicon-type structures. These modes of damage are, to a greater or lesser degree, also in evidence upon irradiation of discrete components, but they are less accessible to study at the locus of their occurrence in microcircuits because of their small scale and peculiar geometries. The high surface-to-volume ratio of the constituent circuit elements generally contributes a predominance of the role of surface and interfacial phenomena. As a consequence, the radiation response of these structures is very sensitive to subtle configurational, topological, and compositional variations—factors currently not adequately understood or adequately controlled during the fabrication process. These difficulties are compounded by the fact that the performance of microcircuits and hence their response to radiations are of necessity a function of the circuit parameters.

One approach to predicting microcircuit performance upon irradiation has been to treat the constituent circuit elements as though they were discrete devices. This practice yields, at best, crude estimates of performance for preliminary design purposes and definitely does not eliminate the need for testing. Two general rules for maximizing radiation stability of microcircuits can be given:

- The circuit selected should be relatively insensitive to transistor gain degradation.

Effect of Space Environment on Electronic Materials and Components

- The transistor elements should have the highest circuit-compatible gain-bandwidth product.

However, the trend toward development of large-scale monolithic arrays will progressively preclude the application of this approach and will pose new problems in the formulation of effective and economical techniques for predicting and testing circuit response.

Electron Tubes. Until about a decade ago the electron tube was the "workhorse" in active electronic circuitry while the proportionate use of transistors was increasing but remained relatively low. In the past four years, the solid-state electronic device for general circuit application has virtually replaced electron tubes in ground-based equipment and in space-vehicle instrumentation except for special-purpose applications. This trend has had a strong effect on the manufacturing and development of electron tubes and on the radiation testing programs. The radiation damage data on electron tubes remain fragmentary, and in a sense have been abandoned as a result of the rapid increase in the use of semiconductors. Moreover, most of the tests on electron tubes were performed during a period when the use of nuclear reactor facilities was prevalent, and many experimenters reported results in total dose of neutrons integrated over time and large portions of the energy spectrum plus the associated reactor gamma environment. Because of their greater physical dimensions, very few data are available on the response of electron tubes in electron and proton accelerator beams. The combination of these factors makes it difficult to correlate precisely to obtain highly reliable information on the tolerance of electron tubes to natural space radiation.

As a rule of thumb, it is probably safe and conservative to assume that electron-tube devices as a class are at least two orders of magnitude higher in tolerance to permanent effects of penetrating radiation than the semiconductor component group. It is recognized that there is a wide variation in both categories, and this comment is made only to indicate a general trend. Reactor experiments indicate usable ranges of 10^{15} to 10^{17} neutrons/cm^2 (plus associated gamma doses of 10^5 to 10^7 rads) for general-purpose tubes of conventional construction. The General Electric TIMMS (thermionic integrated micromodules) units, developed for high-temperature application, were found to be about two orders of magnitude more resistant than those of conventional construction.

The permanent failure of hard vacuum tubes is generally due to cracking of the glass envelope or separation at the Kovar*-to-glass seals. The optical darkening of the glass in the ultraviolet and visible range after relatively low exposures is of little consequence except for tubes with optical applications. Cracking in the envelope and breaks in the seal occur in fairly wide statistical distributions and at levels well below the range of serious catastrophic damage to the individual materials, glass and Kovar. This leads to the conclusion that these failures are due in part to mechanical stresses and to interface stresses which are probably aggravated by variations in fabrication control. There are indications that tube life (active) is diminished after irradiations in the range of 10^{15} to 10^{17} neutrons/cm^2 plus associated gammas, but the statistics are poor. Transient effects on the operating characteristics are not serious at 10^6 rad/hr and 10^{11} neutrons/cm^2.

Gas tubes are similar to vacuum tubes in most respects and can be expected to suffer permanent damage due to glass envelope failure and seal failure in about the same range. The characteristic differences in gas tubes are the partial pressure of filling gas and the somewhat different electrode configurations. The presence of the filling gas and its function give rise to two additional failure modes. At high ionization dose rates, transient currents are produced, but no permanent damage is expected. Neutron-induced transmutations in the gas are possible and would tend to produce a change in gas pressure at a rate dependent on the half-life of the unstable isotope produced. However, this mode is relatively unimportant in space application unless reactor power units are used in the immediate vicinity.

Light-sensitive tubes are probably the most important class of vacuum tubes remaining in extensive use in space vehicles, although the solid-state electrooptical

---

* Trademark of Westinghouse Electric Corporation, Pittsburgh, Pa.

devices are also gaining rapidly in use and performance. The basic construction is closely related to that of the hard vacuum tube, and the mechanical failure problems are expected to be similar in kind and in exposure range. Glass darkening occurs in the range of 10^3 to 10^8 rads and tends to saturate at about 10^{10} to 10^{11} protons/cm². Quartz and fused silica are about two decades more resistant. Most of the coloration induced by radiation can be annealed by moderate-temperature immersions but might prove to be inconvenient in operational equipment. Another problem encountered in photosensitive tubes is the radiation-induced transient response because many applications involve low signal-to-noise ratios which are close to the capability of the device and its ancillary equipment. Among the suspect modes are luminescence of the glass envelope, secondary electrons produced in the photocathode region, recoil electrons produced in the dynode region, and leakage across insulators. A recent report[106] indicates that glass luminescence appears to dominate under exposure to 2.6-Mev electrons (or bremsstrahlung formed in aluminum by 2.6-Mev electrons). The dark current attributable to luminescence was about four times that due to photocathode recoil electrons; and the effects due to recoil electrons produced in the dynode section were another half-decade below the photocathode effects. No leakage across insulators was indicated, presumably because of the geometric configuration of the electron beam.

Except for the filamentless TIMMS construction, which is probably practical for applications on reactor-powered vehicles, the radiation response of electron tubes is characterized by their general construction, which involves a glass envelope and vacuum seals, insulation, filling gas in the case of gas tubes, photosensitive surfaces, and dynode structures in the case of light-sensitive tubes. This range of general characteristics covers most conventional and special-purpose tubes, including microwave power tubes which may have specialized applications in space. Therefore, the behavior of tube types not specifically discussed can probably be inferred from their general construction and functions.

In summation, it appears that electron tubes in space applications will be relegated to highly specialized functions only, and that solid-state devices will continue to dominate until important breakthroughs are achieved in other approaches as fluidics. The continued domination by solid-state systems is guaranteed by the low power requirements and the rapid trend toward LSI (large-scale integration) and other forms approaching the ultimate limits in semiconductor device compactness.

Resistive Components. Comparatively little work has been done recently concerning radiation effects on resistors and capacitors, probably because of the greater radiation resistance of these elements compared with that of the semiconductors with which they are used. The resistive components examined fall into three general categories: carbon composition, film, and wire-wound. The resistors in each category may be either fixed or variable.

Of the fixed resistors, the wire-wound resistors are the least radiation-sensitive, requiring fluences of 10^{15} to 10^{16} neutrons/cm² for significant changes. Oxide film resistors are the most sensitive, showing degradation at fluences of 10^{12} neutrons/cm². Resistances of carbon composition begin to show damage at fluences of about 10^{13} neutrons/cm². In general, the larger the value of resistance, the greater the percentage of change in resistance under radiation, with 1-megohm resistors changing by a factor of 3 to 4 over 100-ohm resistors.

The variable resistors are the potentiometer and trimmer resistance type and are either wire-wound or composition film. In general, these resistive elements show performance similar to that of their fixed resistor counterparts.

The effects of gamma radiation on resistive components is generally insignificant compared with the effects on semiconductors and capacitors. Gamma-rate effects are momentary, lasting only a few microseconds, and are usually due to ionization causing a shunt leakage path. Therefore, these effects will be more significant for high-value rather than low-value resistors. Although gamma-rate effects are small in resistors, these effects may become significant for rates as low as 10^7 R/sec.

A summary of radiation effects on resistive components[89] is shown in Fig. 27. These data are of a general nature, of course, so that if information is desired

Effect of Space Environment on Electronic Materials and Components 15-47

on a particular resistive component, it is advisable to run laboratory tests on that component.

Because of the radiation insensitivity of resistors, all wire-wound, carbon, and composition resistors and potentiometers tested are suitable for all one-year circular-orbit space missions.

The trends of radiation influences on resistors[107] are summarized in Table 14.

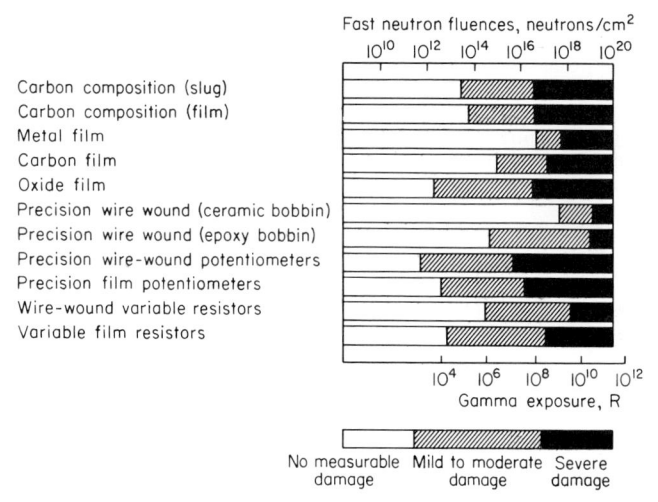

Fig. 27 Estimated effects of a radiation environment on commonly used resistance devices.

Capacitive Elements. Comparatively little recent work has been reported on capacitors, again probably because of their relative insensitivity compared with that of semiconductors. Capacitors are divided into six categories, according to the type of dielectric used: ceramic, glass-porcelain, mica, paper, plastic, and electrolytic. The parameters affected by radiation are capacitance, dissipation factor, and leakage resistance. Although changes in capacitance may be either positive or

TABLE 14 Trends of Radiation Damage in Resistive Elements

Resistor parameter	Dependence on radiation	Specific examples
Transient leakage resistance R_s	Inversely as gamma dose rate	2×10^3 kilohms at 10^{10} rads/sec for carbon composition
Compton replacement current I_R	Directly as gamma dose rate	5 ma at 10^{10} rads/sec for carbon composition
Permanent resistance changes ΔR	Directly as neutron fluence	2% reduction at 10^{14} neutrons/cm² (fast) for carbon composition

negative, the dissipation factor invariably increases and the leakage resistance invariably decreases. In addition, oil-impregnated paper capacitors are subject to failure by case rupture due to evolved gas from the oil during radiation.

Paper and paper-plastic capacitors appear to be the most sensitive to radiation of the types tested, being worse than the inorganic dielectric types by a factor of about 1,000. Of all capacitors, it appears that glass capacitors have the greatest

15-48   Materials for the Space Environment

radiation resistance, followed by mica and ceramic. Both tantalum and aluminum electrolytic capacitors have relatively low radiation thresholds; however, they appear capable of withstanding extended exposure to radiation before suffering severe damage.

Transient radiation is principally manifested as a transient drop in the dielectric resistance of the capacitors. The capacitors invariably recover to their initial state. The relative radiation sensitivity of the six basic categories of capacitors[108] is shown in Fig. 28. The data given are approximate only. If detailed information concerning radiation effects on a particular capacitor is desired, it is advisable to make radiation tests on that particular type.

Because of the relative insensitivity of capacitances, all capacitors of the types discussed are suitable for all one-year circular-orbit space missions.

Inductive Components and Electromechanical Devices. INDUCTORS: Very few data are available concerning radiation effects on inductors, and these data appear to be confined to high-frequency chokes, with an inductance range[86] of 0.1 to 10,000 μh. The chokes are of the molded type, having phenolic cores for low inductances, powdered iron cores for medium inductances, and ferrite cores for higher inductances.

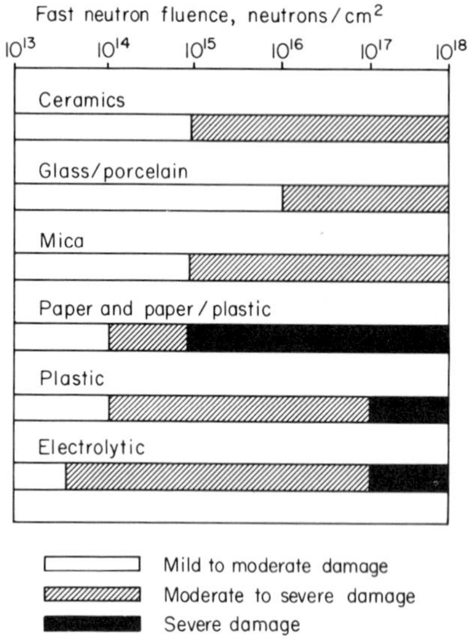

The extent of radiation damage to these components depends on the type of resin used for coil encapsulation. For an integrated fast neutron exposure of up to 4×10^{16} neutrons/cm^2, radio-frequency chokes encapsulated in either diallyl phalate or epoxy have inductance changes of about 5 percent and resistance changes of about 10 percent. For an integrated fast flux of 10^{14} neutrons/cm^2 and a gamma exposure of 5×10^6 rads, the changes are insignificant.

No data appear to be available for air coils or for low-frequency inductors. However, it is anticipated that, at least for air coils, very little sensitivity would be noticed except for perhaps inductance variations due to dimensional changes and for transient effects due to ionization of the air surrounding the coil. Low-frequency inductors in the range of 0.001 to 200 henrys are generally constructed similar to transformers, and it is expected that their radiation characteristics would be similar.

Fig. 28 Relative radiation sensitivity of capacitors.

TRANSFORMERS: The principal effect of radiation on transformers appears to be physical damage rather than a direct effect on their electrical operation. The usual physical damage is case rupture of hermetically sealed units caused by gas evolved from the potting compound. In some instances, the case rupture caused short circuiting of the terminals of the transformers. The experimental data, however, show that no significant degradation of the electrical characteristics of transformers occurs as a result of irradiation, at least up to 5×10^{17} neutrons/cm^2 and 2×10^9 rads.

RELAYS AND SWITCHES: The primary damage occurring in relays and switches due to nuclear radiation is damage to the insulating and construction materials used in these devices. Their radiation resistance is considerably enhanced by the use of such materials as Mycalex* and fiber glass. With proper materials, relays

---

* Trademark of Mycalex Corp. of America.

have operated satisfactorily at fast neutron fluxes of up to 6.5×10^{14} neutrons/cm^2. Some microswitches have shown case damage at gamma-ray exposures as low as 4 to 6×10^6 rads or neutron fluxes of 10^{15} neutrons/cm^2. However, in general, these switches can withstand gamma-radiation exposures of 10^7 rads with only minor deterioration, but major case embrittlement occurs at gamma-radiation exposures of 1.2×10^8 rads. Since a major factor in the radiation resistance of relays and switches is insulation and construction materials, additional information concerning the effects of radiation on these devices should be obtained.

Batteries. One aspect of electrical equipment that has been of considerable interest is the operating temperature of spacecraft batteries. Generally, these batteries are made up of several cells of 1.2 volts (20 to 30 amp) each. The active elements of the cells are enclosed in a case of acrylonitrile-butadiene-styrene for silver-zinc or nylon or nickel for nickel-cadmium. For primary cells, the active elements are silver, silver oxide, and zinc electrodes with cellophane spacers and insulators; the electrolyte is potassium hydroxide. For rechargeable cells, the active elements are nickel, nickel oxide, and cadmium; the other elements are similar to those in the primary silver-zinc cell. Cells generate gas during their operation (hydrogen when in use, and oxygen when completely discharged), and so they are vented to prevent excess internal pressure. The charge and discharge rates of the cells, and consequently of the battery, are sensitive to the operating temperature. To maintain proper operating temperatures in the range of 0 to 38°C, batteries require thermal control in the form of insulation or heaters to keep them warm, or radiating elements or surfaces to reduce the operating temperature. If the battery runs too cold (below $-7°$C), insufficient power will be developed. If the battery gets too hot (above 55°C), irregular charge-discharge characteristics are developed which, if not corrected, can force the battery into irregular cycles that can become cumulative in their effect and ultimately destroy the battery. One reason for this behavior is that at higher temperature the internal resistance decreases and the electric heating of the cell increases, which causes a runaway increase in the battery temperature.

Other Environmental Effects

Penetrating radiation constitutes the most direct space-environmental parameter of significance to the operation of electronic components and materials in space. Some other environmental factors of less direct consequence are enumerated and discussed briefly below.

Vacuum Exposed insulators, such as wire insulation, connector insulation, and encapsulating materials, may be subject to outgassing, sublimation, and perhaps vacuum decomposition in some cases. Very few cases would have serious consequences because low vapor pressure and vacuum-stable materials for electrical functions are readily available.

High-voltage breakdown problems were reported during the thermal vacuum testing of satellites Ariel 1 and 2, Explorer 17, Nimbus, OAO, and OGO. In the OSO 1 and 2, there was no evidence of this problem in thermal vacuum test, but it did occur in orbit. This behavior indicates inability to completely simulate the space environment. The major difference is the ionized elements, energetic particles, and ultraviolet and gamma radiation that are not normally present in thermal vacuum acceptance testing of spacecraft subsystems or systems. It is probable that adequate space simulation could be provided at the subsystem level to detect the potential problem area before flight.

Humidity and other contaminants In most cases, the operation of electric equipment is unimpaired or even improved by lowering the water content and other contaminants of the atmospheres. In exceptional cases, such as carbon-brush contacts, the component can be sealed with an optimized atmosphere, or another brush material can be substituted.

Zero gravity The number of electronic devices depending on a gravitational field is very small. Such devices are usually specialized equipment for experimental purposes and are not generally included among the electronic components of missiles.

Micrometeoroids Most electronic components are protected mechnically from physical impact with micrometeoroids, but exposed insulators and optical surfaces may sustain damage. Solar-cell cover plates may be damaged optically in time, but sandblasted plates do not seem to reduce the cell output seriously. Light shields on phototubes and other light-sensitive devices may be subject to puncture, but such applications are usually of a special nature, and the problem is treated as a prime one.

The effect of simulated exposure to micrometeoroids has been studied in connection with specific space programs. In one such study,[96] it was concluded that the damage to protected solar cells would be very low after 55 days in space, with the extent of damage strongly dependent upon the meteoroid model chosen.

Magnetic fields The magnitude of the magnetic intensity and the variations anticipated are not expected to have serious effects on electronic components. Although some electrical equipment does depend upon magnetic fields, the magnitude of these fields is very high compared with those in space. Possibly the magnetic field changes in the vicinity of a nuclear detonation would be sizable, but it appears that other catastrophic effects would be present and possibly overwhelming. Sensitive scientific instruments such as magnetometers are exceptions to this discussion, but these are used for special purposes and should not be classified with normal engineering components.

Visible light Except for special devices such as phototubes, fluorescent devices, and photovoltaic systems, light in the visible region is not of serious consequence. Light shields and baffles for these types are generally considered as primary design parameters.

Ultraviolet radiation Ultraviolet radiation can affect electrical components by inducing a spurious photoresponse or by degrading such surface materials as organic dielectrics and optical filters. Data on the combined effects of ultraviolet, vacuum, and penetrating radiation are still very scarce, but it should be emphasized that ultraviolet radiation must be considered carefully in the selection of surface insulators and dielectrics for long-term operation. The production of spurious responses by direct interaction or through intermediate mechanisms such as fluorescence is subject to considerations similar to those for visible light.

Temperature Not rightfully a primary parameter of space, temperature is a phenomenon in a spacecraft that results from a number of other factors. Electrical components are sensitive to temperature, and many components such as transistors and resistors have temperature-dependent parameters. These effects are well known and are generally compensated in circuit design and by means of spacecraft thermal control. That this has not always been completely achieved in the past, however, is indicated by the experience with transistors on the Relay 1 Satellite which were mounted so that they received no transferred heat from the rest of the power supply and thus were at about −20°C. Since this was a critical temperature for the material, a power drain resulted because of a change in transistor parameters. After several days in space, these transistors cooled further, to −30°C, and at this less critical temperature their characteristics became favorable and offset the power drain. At warmer temperatures, this power drain would not occur. The solution to the problem was to provide a heat-conduction path from the power supply to the transistor mounting to keep the transistors at a temperature above the critical temperature.

OUTLOOK

In summary it may safely be said that the most damaging environment in space to electronic components and materials is that of penetrating radiation. As described above, the other space environments cause minimal degradation due to factors of location and function of electronic components.

As missions calling for longer and longer lifetimes are planned, the long-term damage to electronic materials and components may become a limiting factor in such planning. Some communications satellites are even now being proposed for lifetimes of five or more years in space. At the opposing extreme in lifetime considerations are the short-time but high-intensity environments associated with nuclear explosions. These environments have not been considered in this chapter but may become important in future planning. As the requirement for long lifetimes exhausts the available radiation resistance of present electronic components, a deliberate program of component improvement with respect to radiation may be needed. Although such a program would be difficult owing to uncertainties in the efficiency of accelerated life testing, the needs of advanced space exploration will undoubtedly require this improvement in the long run.

ACKNOWLEDGEMENT

The authors wish to acknowledge that certain sections of this chapter were extracted and revised from chapters in the "Space Materials, Handbook," 3d edition. The chapters in the "Space Materials Handbook" were developed by A. E. Anderson, P. C. Castro, E. Gaines, R. A. Glass, W. Imhof, J. C. Lee, P. V. Phung, and F. F. Stucki. The "Space Materials Handbook," 3d edition, was produced by John B. Rittenhouse and John B. Singletary at the Materials Sciences Laboratory, Palo Alto Research Laboratory, Lockheed Missiles and Space Company, under a contract administered by Mr. E. L. Horne, Materials Information Branch, Wright-Patterson Air Force Base, Ohio, and Mr. George C. Deutsch, Assistant Director, Materials, Office of Advanced Research and Technology, National Aeronautics and Space Administration, Washington, D.C.

REFERENCES

1. Neff, W. J., and R. A. Montes de Oca: Launch Environment Profiles for Sounding Rockets and Spacecraft, *NASA Tech. Note* D-1916, January, 1964.
2. Lyon, R. H.: Random Noise and Vibration in Space Vehicles, *Shock and Vibration Information Center Rep.* SVM-1, U.S. Department of Defense, 1967.
3. Flight-loads Measurements during Launch and Exit, *NASA Rep.* SP-8002, National Aeronautics and Space Administration, Washington, D.C., December, 1964.
4. Osgood, C. C.: "Spacecraft Structures," Prentice-Hall, Inc., Englewood Cliffs, N.J., 1966.
5. Otera, J. M: Spacecraft Shock Environment Induced by Nose Fairing Separation Pyrotechnics, *Space Tech. Lab. Rep.* 2411-6001-RU-000, Redondo Beach, Calif., April, 1963.
6. Wiksten, D. B.: Dynamic Environment of the Ranger Spacecraft I through IX (final report), *Jet Propulsion Lab. Tech. Rep.* 32-909, Pasadena, Calif., May, 1966.
7. "U.S. Standard Atmosphere, 1962," Government Printing Office, Washington, D.C., 1962.
8. "U.S. Standard Atmosphere Supplements, 1966," Government Printing Office, Washington, D.C., 1966.
9. May, R. R.: Upper Air Density Derived from Orbits of [Recoverable] Satellites and Its Variations with Latitude," *Planet. Space Sci.,* vol. 12, pp. 1179–1185, 1964.
10. Anderson, A. D.: Existence of a Significant Latitudinal Variation in Density from 200 to 800 Kilometers, *Nature,* vol. 209, pp. 656–661, 1966.
11. Lagos, C. P., and J. R. Mahoney: Numerical Studies of Seasonal and Latitudinal Variability in a Model Thermosphere, *J. Atmos. Sci.* vol. 24, pp. 89–94, 1967.
12. Keating, G. M., and E. J. Prior: The Winter Helium Bulge, 1967 IQSY/COSPAR Assembly, London, England, July, 1967, pp. 17–29.
13. Nicolet, M.: Helium: an Important Constituent in the Lower Thermosphere, *J. Geophys. Res.* vol. 66, pp. 2263–2264, 1961.
14. Anderson, A. D., and W. E. Francis: The Variation of the Neutral Atmospheric Properties with Local Time and Solar Activity from 100 to 10,000 km," *J. Atmos. Sci.,* vol. 23, pp. 110–124, 1966.
15. Anderson, A. D.: On the Inexactness of the 10.7-cm Flux from the Sun as an Index of the Total Extreme Ultraviolet Radiation, *J. Atmos. Sci.,* vol. 21, pp. 1–14, 1964.

15-52 Materials for the Space Environment

16. Francis, W. E.: A Least Squares Method of Computing Density-based Model Atmospheres, *J. Atmos. Sci.*, vol. 23, pp. 431–442, 1965.
17. Anderson, A. D.: Long-term (Solar Cycle) Variation of the Extreme Ultraviolet Radiation and 10.7-cm Flux from the Sun, *J. Geophys. Res.*, vol. 70, pp. 3231–3234, 1965.
18. "Solar Electromagnetic Radiation," *NASA Rep.* SP-8005, National Aeronautics and Space Administration, Washington, D.C., June, 1965.
19. Whitehill, L. P.: "Survey on Earth Albedo," *Mass. Inst. Technol. Tech. Note* 1966-53, October, 1966.
20. Bandeen, W., et al.: Infrared and Reflected Radiation Measurements from the Tiros II Meteorological Satellite, *J. Geophys. Res.*, vol. 66, p. 3171, 1961.
21. Nordberg, W.: Research with Tiros Radiation Measurements, *Astronaut. and Aerosp. Eng.*, vol. 1, p. 82, April, 1963.
22. Danjon, A.: Albedo, Color, and Polarization of the Earth, in G. P. Kuiper (ed.), "The Earth as a Planet," pp. 726–738, University of Chicago Press, Chicago, 1954.
23. Murcray, W. B.: Study of the Albedo of the Earth, Geophysical Institute, University of Alaska, *Contract* Nonr 3010(03)(AD 642 356), *Final Rep.*, September, 1966.
24. Fritz, S.: Solar Radiant Energy and Its Modification by the Earth and Its Atmosphere, in T. F. Malone (ed.), "Compendium of Meteorology," pp. 13–33, American Meteorological Society, Boston, 1951.
25. List, R. J.: "Smithsonian Meteorological Tables," vol. 114, pp. 442–444, Smithsonian Institution, Washington, D. C., 1958.
26. Robinson, N.: "Solar Radiation," American Elsevier Publishing Company, Inc., New York, 1966.
27. Camack, W. G.: Albedo and Earth Radiation, chap. 6 in C. Goetzel, J. Rittenhouse, and J. Singletary (eds.), "Space Materials Handbook," Addison-Wesley Publishing Company, Inc., Reading, Mass., 1965.
28. Fritz, S.: The Albedo of the Planet Earth and of Clouds, *J. Meteorol.*, vol. 6, pp. 277–282, 1949.
29. Johnson, F. S.: Solar Radiation, chap. 4 in F. S. Johnson (ed.), "Satellite Environment Handbook," 2d ed., Stanford University Press, Stanford, Calif., 1965.
30. Baur, F., and H. Phillips: Der Wärmenhanshalt der Lufthülle der Nordhalbkugel in Januar und Juli and zur Zeit der Äquinoktien und Solstien, part 1, *Gerlands Beitr. Geophys.*, vol. 42, pp. 160–207, 1934.
31. Houghton, H. H.: On the Annual Heat Balance of the Northern Hemisphere, *J. Meteorol.*, vol. 11, pp. 1–9, 1954.
32. McIlwain, C. E.: Coordinates for Mapping the Distribution of Magnetically Trapped Particles, *J. Geophys. Res.*, vol. 66, p. 3681, 1961.
33. Vette, J. I.: "Models of the Trapped Radiation Environment," vol. 1, "Inner Zone Protons and Electrons, *NASA Rep.* SP-3024, 1966.
34. King, J. H.: "Models of the Trapped Radiation Environment," vol. 4, "Low Energy Protons," *NASA Rep.* SP-3024, 1967.
35. Pizzella, G., C. E. McIlwain, and J. A. Van Allen: Time Variations of Intensity in the Earth's Inner Radiation Zone, October 1959 through December 1960, *J. Geophys. Res.*, vol. 67, p. 1235, 1962.
36. Fillius, R. W.: Trapped Protons of the Inner Radiation Belt, *J. Geophys. Res.*, vol. 71, p. 97, 1966.
37. Blanchard, R. C., and W. N. Hess: Solar Cycle Changes in Inner Zone Protons, *J. Geophys. Res.*, vol. 69, p. 3927, 1964.
38. Nakano, G. H., and H. H. Heckman: Evidence for Solar Cycle Changes in the Inner Belt Protons (manuscript).
39. Vette, J. I., A. B. Lucero, and J. A. Wright: "Models of the Trapped Radiation Environment," vol. 2, "Inner and Outer Zone Electrons," *NASA Rep.* SP-3024, 1966.
40. Bostrom, C. O., and D. J. Williams: Time Decay of the Artificial Radiation Belt, *J. Geophys. Res.*, vol. 70, p. 240, 1965.
41. Krimigis, S. M., and J. A. Van Allen: Geomagnetically Trapped Alpha Particles, *J. Geophys. Res.*, vol. 72, p. 5779, 1967.
42. Schardt, A. W., and A. G. Opp: Particles and Fields: Significant Achievements, *Rev. Geophys.*, vol. 5, p. 411, 1967.
43. Meyer, P., and R. Vogt: High-energy Electrons of Solar Origin, *Phys. Rev. Lett.*, vol. 8, p. 387, 1962.

References

44. Modisette, J. L., T. M. Vinson, and A. C. Hardy: Model Solar Proton Environments for Manned Spacecraft Design, *NASA Tech. Note* D-2746, April, 1965.
45. Webber, W. R.: An Evaluation of the Radiation Hazard Due to Solar-particle Events, *Boeing Co. Rep.* D2-90469, Seattle, Wash., December, 1963.
46. Biswas, S., and C. E. Fichtel: Composition of Solar Cosmic Rays, *Space Sci. Rev.*, vol. 4, p. 709, 1965.
47. Snyder, C. W., and M. Neugebauer: Interplanetary Solar Wind Measurements by Mariner II, *Space Res.*, vol. 4, p. 89, 1964.
48. Evans, J. E., et al.: Recent Results from Satellite Measurements of Low Energy Particles Precipitated at High Latitudes, *Space Sci. Rev.* vol. 7, p. 263, 1967.
49. O'Brien, B. J.: Precipitation of Energetic Particles into the Atmosphere, chap. 5 in Martin Walt (ed.), "Auroral Phenomena" Stanford University Press, Stanford, Calif., 1965.
50. Sharp, R. D., and R. G. Johnson: Some Average Properties of Auroral Electron Precipitation as Determined from Satellite Observations, *J. Geophys. Res.*, vol. 73, 969, 1968.
51. Sharp, R. D., et al.: Satellite Measurements of Precipitating Protons in the Auroral Zone, *J. Geophys. Res.*, vol. 72, p. 227, 1967.
52. Vernov, S. N., and A. E. Chudahov: Terrestrial Corpuscular and Cosmic Rays, *Space Res.*, vol. 1, p. 751, 1960.
53. Meyer, P.: Primary Electrons and Positrons in the Cosmic Radiation, *Proc. Int. Conf. on Cosmic Rays*, London, 1965.
54. Bame, S. J., and J. R. Asbridge: A Search for Solar Neutrons near Solar Minimum, *Los Alamos Sci. Lab. Rep.* LA-DC-7973, 1965.
55. Gregory, B. N., and R. W. Kreplin: Observations of Solar X-ray Activity below 20 Angstroms, *J. Geophys. Res.*, vol. 72, p. 4815, 1967.
56. Van Allen, J. A.: Solar X-ray Flare of July 7, 1966, *J. Geophys. Res.*, vol. 72, p. 5903, 1967.
57. Frye, G. M., Jr., F. Reines, and A. H. Armstrong: Search for Solar and Cosmic Gamma Rays, *J. Geophys. Res.*, vol. 71, p. 3119, 1966.
58. Goldberg, L., and E. R. Dyer, Jr.: Galactic and Extragalactic Astronomy, chap. 18 in L. V. Berkner and H. Odishaw (eds.), "Science in Space," McGraw-Hill Book Company, New York, 1961.
59. Haymes, R. C., and W. L. Craddock, Jr.: High Energy X-rays from the Crab Nebula, *J. Geophys. Res.*, vol. 71, p. 3261, 1966.
60. Edwards, P. J., and K. G. McCracken: Upper Limits to the Hard X-ray Flux from the Quiet Sun and Jupiter, *J. Geophys. Res.*, vol. 72, p. 1809, 1967.
61. Glasstone, S.: The Effects of Nuclear Weapons, U.S. Atomic Energy Commission, 1962.
62. Christofilos, N. C.: Sources of Artificial Radiation Belts, pp. 565–574 in B. M. McCormac (ed.), "Radiation Trapped in the Earth's Magnetic Field," D. Reidel Publishing Co., Dordrecht, Holland, 1966.
63. Van Allen, J. A.: Spatial Distribution and Time Decay of the Intensities of Geomagnetically Trapped Electrons from the High Altitude Nuclear Burst of July 1962, in B. M. McCormac (ed.), "Radiation Trapped in the Earth's Magnetic Field," pp. 575–592, D. Reidel Publishing Co., Dordrecht, Holland, 1966.
64. Imhof, W. L., et al.: Analysis and Evaluation of Measurements of Geomagnetically Trapped Electrons from High-altitude Nuclear Explosions, *NASA Rep.* 1540, Lockheed Missiles & Space Co., Palo Alto, Calif., 1964.
65. Dye, D. L.: Space Proton Doses at Points within the Human Body, *Proc. Symp. on Prot. against Radiat. Hazards in Space*, NASA Tech. Note D-7652, November, 1962.
66. Mar, B. W.: Electron Shielding Codes for Evaluation of Space Radiation Hazards, *Boeing Co. Rep.* D2-90414, Seattle, Wash., June, 1963.
67. Simpson, K. M., C. W. Hill, and C. C. Douglass: A Space Radiation Shielding Code for Realistic Vehicle Geometries, *2d Symp. on Prot. against Radiat. in Space*, NASA Rep. SP-71, 1965.
68. Berger, M. J.: "Methods in Computational Physics," vol. 1, Academic Press, Inc., New York, 1963.
69. Computer Codes for Space Radiation Environment and Shielding, *Air Force Weapons Lab. Rep.* TDR-64-71, vol. 1, 1964.
70. Kinney, W. E.: The Nucleon Transport Code, NTC, *Oak Ridge Nat. Lab. Rep.* 3610, August, 1964.
71. Hine, G. J., and G. L. Brownell (eds.): "Radiation Dosimetry," Academic Press, Inc., New York, 1956.

72. Radiation Quantities and Units, *Int. Comm. on Radiol. Units and Meas. Rep.* 10a, *Nat. Bur. Stand. Handb.*, vol. 84, November, 1962.
73. Cosby, W. A., and R. G. Lyle: The Meteoroid Environment and its Effects on Materials and Equipment, *Nat. Res. Counc. Rep.* ASP-78, Prevention of Deterioration Center, National Academy of Sciences–National Research Council, 1965.
74. Burbank, P. B., B. G. Cour-Palais, and W. E. McAllum: A Meteoroid Environment for Near-earth, Cislunar and Near-lunar Operations, *NASA Tech. Note* D-2747, Manned Spacecraft Center, Houston, Tex., April, 1965.
75. Charak, Mason: Personal communication, NASA Design Criteria Branch, Office of Advanced Research Technology, Washington, D.C., Apr. 8, 1968.
76. Charak, M. T., and S. A. Mills: The NASA Vehicle Design Criteria Program: Engineering Models of the Environment, *3d Nat. Conf. on Aerosp. Meteorol.*, New Orleans, La., May 6–9, 1968.
77. Nauman, R. J.: The Near Earth Meteoroid Environment, *NASA Tech. Note* D-3717, George C. Marshal Space Flight Center, Huntsville, Ala., November, 1966.
78. Summers, J. L.: Investigation of High-speed Impact: Region of Impact and Impact at Oblique Angles, *NASA Tech. Note* D-94, Ames Research Center, Moffet Field, Calif., October, 1959.
79. Kruszewski, E. T.: Meteoroids: Need for Penetration Sealing Laws and the Potentials of Simulation Techniques, Conference on the Role of Simulation in Space Technology, Blacksburg, Va., August 17–21, 1964.
80. Thomson, R. G., and E. T. Kruszewski: Effect of Target Material Yield Strength on Hypervelocity Perforation and Ballistic Limit, Hypervelocity Impact Symposium, Tampa, Fla., Nov. 17–19, 1964.
81. Thomson, R. G.: Analysis of Hypervelocity Perforation of a Visco-plastic Solid including the Effects of Target-material Yield Strength, *NASA Tech. Rep.* R-221, Langley Research Center, Langley Station, Hampton, Va., April, 1965.
82. Kruszewski, E. T., and R. T. Hayduk: Implications of the Meteoroid Environment on the Design of Spacecraft, *8th AIAA-ASME Struct., Struct. Dyn. and Mater. Conf.*, Palm Springs, Calif., Mar. 29–31, 1967.
83. McMillan, A. R.: Experimental Investigations of Simulated Meteoroid Damage to Various Spacecraft Structures, *NASA Rep.* CR-915, General Motors Corp., Santa Barbara, Calif., January, 1968.
84. Madden, R.: Ballistic Limit of Double-walled Meteoroid Bumper Systems, *NASA Tech. Note.* D-3916, Langley Research Center, Langley Station, Hampton, Va., April, 1967.
85. Cour-Palais, B. G.: Personal communications, NASA Manned Spacecraft Center, Houston, Tex., May 22, 1968.
86. Hamman, D. J., et al.: "Space Environment Effects on Materials and Components" (AD 601 876), vol. 2, "Electronic and Mechanical Components," Redstone Arsenal, Ala., Apr. 1, 1964.
87. Hulten, W. C., et al.: Irradiation Effects of 22 and 240 MeV Protons on Several Transistors and Solar Cells, *NASA Tech Note* D-718, Langley Research Center, Langley Field, Hampton, Va., April, 1961.
88. Lamond, P., and P. Berman: High Efficiency Silicon Solar Cells, ARDA-80-59 *Rep.* 3, Transition Electronic Corp., Wakefield, Mass., July 1, and Dec. 31, 1960.
89. Spradin, B. C.: *Nuclear Radiation Effects on Resistive Elements*, Radiation Effects Information Center, Battelle Memorial Institute, Columbus, Ohio, July, 1966.
90. Kennedy, B. W.: Effects of Gamma on Selected Potting Compounds and Insulating Materials, NASA, Marshall Space Flight Center, Huntsville, Ala., November, 1963.
91. Gri, N. J.: Radiation Damage in Silicon p-i-n Lithium-drift Solar Cells, *IEEE Trans. on Nucl. Sci.*, vol. NS-12, pp. 464–471, February, 1965.
92. "TREE Handbook" (Transient-radiation Effects on Electronics), *DASA Rep.* 1420, Battelle Memorial Institute, Columbus, Ohio, August, 1967.
93. Kircher, J. F., and R. E. Bowman (eds.): "Effects of Radiation on Materials and Components," Reinhold Publishing Corporation, New York, 1964.
94. "Radiation Effects: Survey of Soviet-bloc Scientific and Technical Literature, 1960–1963, Aerospace Technological Division, Library of Congress, Washington, D.C., Mar. 24, 1965.
95. Hamman, D. J., et al.: Radiation Effects: State of the Art, 1961–1965, *REIC Rep.* 38, Battelle Memorial Institute, Columbus, Ohio, June 30, 1965.

96. Mirtich, M. J., and R. L. Bowman: Effect of Simulated Micrometeoroid Exposure on Performance of n/p Silicon Solar Cells, *AIAA J.*, vol. 5, pp. 1364–1466, July, 1967.
97. Drennan, J. E., and D. J. Hamman: Space Radiation Damage to Electronic Components and Materials, *REIC Rep.* 39, Battelle Memorial Institute, Columbus, Ohio, January, 1966.
98. Kerlin, F. E., and E. T. Smith: Measured Effects of the Various Combinations of Nuclear Radiation, Vacuum, and Cryotemperatures on Engineering Materials, *General Dynamics Corp. Rep.* FZK-290, Fort Worth, Tex., July, 1966.
99. Ringwood, A. F.: Behavior of Plastics in Space Environment, *Mod. Plast.*, vol. 41, p. 173, January, 1964.
100. Frisco, L. J., and K. N. Mathes: Evaluation of Thin-wall Spacecraft Electrical Wiring, General Electric Co., *Contract* NAS 9-4549, vol. 1, *Final Rep.* Schenectady, N.Y., September, 1965.
101. "Program 461 Spacecraft Materials Handbook," *Lockheed Missiles & Space Co. Rep.* A327227, Sunnyvale, Calif., May, 1964.
102. Goetzel, C. F., J. B. Rittenhouse, and J. B. Singletary (eds.): "Space Materials Handbook," 2d ed., *Lockheed Missiles & Space Co. Reps.* ML-TDR-64-40 (AD 460 399), X65-14878, Sunnyvale, Calif., January, 1965.
103. Fischell, R. E., et al.: Radiation Damage to Orbiting Solar Cells and Transistors, TG-886 (AD 657 155) (N67-38181), Applied Physics Laboratory, Johns Hopkins University, Silver Spring, Md., March, 1967.
104. Waddel, R. C.: Early Results from the Solar Cell Radiation Damage Experiment on ATS-1, *NASA Tech. Memo.* X-55772 (N67-26568), Goddard Space Flight Center, Greenbelt, Md., April, 1967.
105. Perkins, C. W., R. W. Marshall, and A. M. Liebschultz: Radiation Effects on (Monolithic) Microelectronic Circuits, *Hughes Aircraft Co. Rep.* 5 (AD 642 301)(N67-15500), Fullerton, Calif., November, 1966.
106. Favale, A. J., F. J. Kuehne, and M. D. D'Agostino: Electron Induced Noise in Star Tracker Photomultiplier Tubes, *IEEE Trans. on Nucl. Sci.*, vol. NS-14, p. 190, December, 1967.
107. Finnell, J. T., and F. W. Karpowich: Skipping the Hard Part of Radiation Hardening, *Electron.*, pp. 122–127, Mar. 4, 1968.
108. Hanks, C. L., and D. J. Hamman: The Effect of Nuclear Radiation on Capacitors, *REIC Rep.* 44, Battelle Memorial Institute, Columbus, Ohio, Dec. 30, 1966.

Index

Index

ABS (*see* Thermoplastics)
Adhesives:
   ASTM standards for, **2**-99
   epoxy, properties of, **2**-101, **2**-102
   joining of laminates, **2**-96, **2**-97
   MIL specifications for, **2**-98
   pullout strength, **2**-96
   shear strength, **2**-103
   temperature effects of, **2**-103, **2**-104
   types and properties of, **2**-101
Age hardening, **9**-5
Alkyds (*see* Coatings; Laminates;
   Thermosetting plastics)
Alloys (*see* specific metal)
Alumina (*see* Ceramics; Substrates)
Aluminum:
   alloy compositions, **9**-16
   alloy groups, **9**-13
   bend radii, **9**-32, **9**-33
   brazing, **9**-33 to **9**-35
   cast alloys, **9**-37, **9**-38
   clad products, **9**-18
   densities, **9**-30
   electrical properties, **9**-12, **9**-28, **9**-29
   fabrication characteristics, **9**-31 to **9**-34
   finishing (*see* Finishes)
   general properties, **9**-12, **9**-13
   heat treatment, **9**-36, **9**-37
   mechanical properties of, **9**-14, **9**-19, **9**-20
   melting points, **9**-30
   specifications for, **9**-21 to **9**-23
   temperature designations for, **9**-14, **9**-15

Aluminum (*Cont.*):
   temperature effects of, **9**-17
   tensile properties of, **9**-24 to **9**-27
   thermal properties of, **9**-28, **9**-29
   weldability of, **9**-34 to **9**-36
AMS specifications for coatings, **5**-71
Anodizing, **10**-48
Artwork:
   automatically generated, **12**-14
   coordinatographs, **12**-13
   cut-and-strip method, **12**-12
   tape method, **12**-12
   (*See also* Photofabrication)
ASTM specifications:
   for adhesives, **2**-99
   for elastomers, **3**-2, **3**-3, **3**-39
   for ferrous metals, **8**-24
   for laminates, **2**-3, **2**-92
   for micas, **6**-36
   for printed circuits, **2**-105
   for thermoplastics, **1**-79
   for tubing, **4**-80

Beryllia (*see* Ceramics; Substrates)
Beryllium and Lockalloy:
   availability of, **9**-39
   chemical compositions, **9**-40
   corrosion resistance, **9**-41
   cost, **9**-43
   fabrication, **9**-42
   finishing, **9**-41, **9**-42
   (*See also* Finishes)
   impact data, **9**-39

Beryllium and Lockalloy (*Cont.*):
  joining, **9**-43
  mechanical properties of, **9**-40, **9**-41
  physical properties of, **9**-41, **9**-42
  toxicity, **9**-42
Beryllium-copper:
  availability of, **9**-48
  compositions, **9**-48
  mechanical properties of, **9**-49
Bonds and bonding:
  adhesive (*see* Adhesives)
  ball (nailhead), **13**-20
  beam-lead pulse-soldered, **13**-51, **13**-52
  bird-beak, **13**-21, **13**-22
  bond-site damage, **13**-41
  ceramic- and glass-to-metal, **6**-49, **8**-25
  chip, **13**-35, **13**-43, **13**-49
    thermal, **13**-53
  combination, **13**-23
  electrical continuity, **13**-41
  epoxy-glass, **13**-34
  experimental cell, **13**-42
  face, **13**-33, **13**-37, **13**-38
  glass substrates, **13**-39
  green ceramic process, **13**-63
  heated-tip, **13**-23
  hole cleaning, **13**-72
  hole drilling, **13**-72
  KMER, **13**-51
  Kovar, **13**-50
  lead, **13**-17
  liquid fluxes, comparison of, **13**-77
  microbonder, **13**-30
  microbonds weld schedule, **13**-28, **13**-31
  50-micron-diameter, **13**-46
  microwelding, **13**-28 to **13**-30
  MLB fabrication, **13**-74
  molybdenum, **13**-50
  pad areas, **13**-46
  plated-up MLB, **13**-70
  registration mirror, **13**-48
  registration system, **13**-48
  screen and fire process, **13**-62
  shear strength, **13**-45, **13**-49
  silicon chips, **13**-39
  sintering, **13**-67
  solder-glass chip, **13**-17
  solder pads, **13**-47
  soldering (*see* Soldering)
  split-tip welder, **13**-32
  stitch (scissors), **13**-21, **13**-22
  strain-gage, **13**-40
  substrate internal connections, **13**-55
  substrate thermal resistance, **13**-54
  thermal resistance, **13**-54, **13**-55
  thermocompression, **13**-18
    process parameters, **13**-24

Bonds and bonding (*Cont.*):
  thermocompression, stitch-bond, **13**-24
  through-hole plating thickness, **13**-75
  tip, **13**-36
  tip designs, comparison, **13**-44
  ultrasonic, **13**-25, **13**-43
    electric conductors, **13**-27
    pull strength, **13**-29
  wedge, **13**-19
  welding (*see* Welding)
  (*See also* Soldering; Welding)
Brasses:
  compositions, **9**-46
  leaded, **9**-46
Brazing (*see* Aluminum, brazing)

Capacitors (*see* Ceramics; Glasses; Micas; etc.)
Casting (*see* Embedding)
Ceramics:
  capacitors, **6**-56
    aging, **6**-69, **6**-71
    barrier-layer, **6**-66, **6**-70
    color code, **6**-64
    construction, **6**-57
    degradation, **6**-71
    disc, **6**-63
    general purpose, **6**-63
    multilayer, **6**-63
    stability, **6**-68
    summary chart, **6**-58
    TC, characteristics of, **6**-62
    temperature-compensating, **6**-60
    value ranges, **6**-60
  dielectrics, **6**-10, **6**-56, **6**-61, **6**-65
    EIA specifications for, **6**-61
    representative formulations for, **6**-61
  differential expansion, **6**-50 to **6**-53
  fine-line printing, **13**-65
  glass substrate, **13**-64
  green, **13**-63 to **13**-66
  insulators, **6**-63
    alumina, **6**-12 to **6**-20
    beryllia, **6**-13 to **6**-20
    carbides, **6**-22
    cordierite, **6**-12
    electrical porcelain, **6**-3, **6**-7
    forsterite, **6**-12
    magnesia, **6**-21
    nitrides, **6**-24, **6**-25
    silicon nitride, **6**-26
    steatite, **6**-8
    zirconia, **6**-22
  magnetic, **6**-87
    characteristics of, **6**-91
    classification of, **6**-101

Ceramics (*Cont.*):
  magnetic, hysteresis, **6**-91 to **6**-93
  magnetism, **6**-87
  magnetostriction, **6**-92
  magnets, permanent, **6**-107
  melting temperatures, **6**-7
  piezoelectric, **6**-85
  Pyroceram, comparison properties of, **6**-46
  seals: bonding, **6**-49
    ceramic-to-metal, **6**-49
  semivitreous and refractory products, **6**-11
  sequentially fired, **13**-63
  sintering, **13**-67
  substrate materials, **13**-64
  substrates: ceramic and glass, **6**-38 to **6**-40
    dielectric constant, **6**-42
    electric conductivity, **6**-40
    surface smoothness, **6**-38
    thermal conductivity, **6**-41
    and glazes, **6**-42
  tests and measurements, **6**-2, **6**-4 to **6**-6
  thermal expansion, **6**-7
  thermal-stress resistance, **6**-9
  thick-film, **13**-63
    (*See also* Thick films)
Cermet materials:
  conductors: bulk, **12**-44
    criteria, **12**-43
    dissipation, **12**-45
    gold, **12**-41
    palladium-gold, **12**-42
    palladium-silver, **12**-42
    platinum-gold, **12**-42
    surface, **12**-44
  dry-transfer tapes, **12**-46
  insulators, **12**-40 to **12**-45
  thick-film, active, **12**-45
    (*See also* Thick films)
  thick-film conductors, **12**-37 to **12**-39
    load-life comparisons, **12**-39
    (*See also* Thick films)
  thick-film resistors, **12**-31 to **12**-36
    resistive pastes, variation of, **12**-40
    (*See also* Thick films)
Chemical machining:
  applications, **14**-81 to **14**-89
  color television shadow masks, **14**-88
  design considerations, **14**-82
  metal-masks: electroformed, **14**-86 to **14**-88
    etched, **14**-86 to **14**-88
  metal tolerances, **14**-84
  microcircuit lead frames, **14**-85
  part handling, **14**-85

Chip and wire assembly:
  aluminum wire, **12**-73
  ball and wedge, **12**-70
  eutectic die bonding, **12**-66
  flip-chip bonding, **12**-67
  infrared microscope, **12**-68
  paste bonding, **12**-67
  pure gold wire, **12**-69
  thermal compression, **12**-69
  ultrasonic, **12**-71, **12**-72
  (*See also* Bonds and bonding)
Circuit boards:
  clearance-hole, **13**-69
  conductors, **13**-99
  copper-plating, **13**-75
  electroless plating, **13**-75
  hole cleaning, **13**-72
  hole drilling, **13**-72
  interconnections, **13**-69
  IPC-ML-910, **13**-72
  MLB, **13**-70
  multilayer-board fabrication, **13**-73 to **13**-75
  plated-through hole (PTH), **13**-70
  plated-up, **13**-71
  soldering, **13**-92
  strain on copper, **13**-74
  (*See also* Printed circuits)
Cleaning:
  nature of particles, **5**-52
  potential contaminants, **5**-51
  pulsating-spray, **5**-55
  solvent purity, **5**-52
  solvents and solutions, **5**-52
  ultrasonic, **5**-54
  vapor degreasing, **5**-54
Coatings:
  aluminum, **10**-47
  aluminum protection for, **5**-44
  AMS specifications for, **5**-71
  anodized, **10**-45
  application methods for, **5**-55 to **5**-64
    brush, **5**-64
    dip coating, **5**-59
    electrostatic spray coating, **5**-58
    fluidized bed, **5**-60 to **5**-63
    rollercoating, **5**-63
    screening, **5**-63
    spray coating, **5**-57, **5**-58
    vacuum impregnation, **5**-63
  barrier and junction, **5**-24
  chromate, **10**-48
  cleaning methods: nature of particles, **5**-52
    potential contaminants, **5**-51
    pulsating-spray, **5**-55
    solvent purity, **5**-52
    solvents and solutions, **5**-52

6 Index

Coatings, cleaning methods (*Cont.*):
    ultrasonic, **5**-54
    vapor degreasing, **5**-54
  coil varnishes, **5**-16
  for conductors, **5**-28, **5**-29
    compositions analysis of, **5**-29
    metal-filled plastics, data for, **5**-29
  conversion, **10**-48
  corrosion protection for, **5**-40
    modes for, **5**-41
  dielectric properties of, **5**-29
    breakdown voltage, **5**-32
    of polymer coatings, **5**-30
    strength, **5**-32
  for electronic equipment: electrical protection, **5**-25
    insulation resistance, **5**-25
    properties of, **5**-25
  for electronic packaging, **5**-19 to **5**-21
  fluidized bed, **5**-60 to **5**-63
    plastics, summary of, **5**-62
    typical examples of, **5**-62
  hardness values of, Sward, **5**-50
  humidity testing, **5**-39
  magnesium-lithium alloy protection, **5**-42
  magnesium protection, **5**-42
    chemical and anodic treatments, **5**-43
  maskants, **5**-11 to **5**-13
  mechanical protection, **5**-48
    abrasion resistance, **5**-48
    coefficient of friction, **5**-49
    frictional resistance, **5**-48
  microorganism protection, **5**-44, **5**-45
  MIL specifications for, **5**-3, **5**-9, **5**-66 to **5**-71
  moisture protection for, **5**-36
  moisture-vapor transmission rates, **5**-40
  nature of, **5**-2
  NEMA specifications for, **5**-14, **5**-16, **5**-67
  nonmetallic, **10**-45
  phosphate, **10**-49
  polymer: arc resistance, **5**-34, **5**-37
    dielectric strength, **5**-35, **5**-36
    dissipation factors, **5**-33
    thermal conductivity data for, **5**-46, **5**-47
    volume resistivities, **5**-26
    water absorption, **5**-38
  polymer-type, **5**-4 to **5**-7
  for printed-circuit boards, **5**-3 to **5**-8
    coating thickness, **5**-8
    coating types, **5**-3
    humidity effect, **5**-11
    inspection, **5**-9

Coatings, for printed-circuit boards (*Cont.*):
    reliability, **5**-3
    reworkability of coated assemblies, **5**-10
    salt-spray effects, **5**-10
    spacings between conductors, **5**-12
  resins, **5**-4 to **5**-7
  resistivity: effects of variables, **5**-27
    surface, **5**-28
  for semiconductors: adhesion, **5**-23
    corrosivity, **5**-22
    MIL specifications for, **5**-23
    moisture resistance of, **5**-21
    purity of, **5**-21
    stresses, **5**-23
  solder resists, **5**-11
  specifications: for cleaning and surface treatments, **5**-71
    for coatings, paints, and primers, **5**-69, **5**-70
    for electrical requirements, **5**-66
    for electronic requirements, **5**-66
    for finishing systems, **5**-68
    for solvents, resins, and other coating ingredients, **5**-71
    for test methods, **5**-73
    for test procedures, **5**-72
  thermal protection, **5**-45
    heat resistance, **5**-45
    test methods, **5**-47
    thermal conductivity, **5**-45 to **5**-47
  thick-film circuits, **5**-13
    (*See also* Thick films)
  thin-film circuits, **5**-13
    deposition methods, **5**-64, **5**-65, **5**-67
    general properties, **5**-70
    (*See also* Thin films)
  tin-lead protection, **5**-44
  water absorption, **5**-38
  water-vapor permeability, **5**-39
  wire varnishes, **5**-15, **5**-16
  (*See also* Electrodeposition; Nonelectrolytic deposition)
Coaxial cables:
  attenuation ratings, **4**-52, **4**-53
  design with, **4**-46 to **4**-49
  impedance, **4**-48
  MIL specifications for, **4**-49
  power ratings of, **4**-50, **4**-51
  TFE, **4**-47
  x-ray irradiation, **4**-48
Conductivity, thermal (*see* Thermal conductivity)
Conductors:
  bare, **4**-7
  coated, **4**-7
  coatings for, **4**-9

Conductors (*Cont.*):
  construction of, **4**-12
  copper-clad steel wire, **4**-9
  materials, **4**-2, **4**-6, **4**-8, **4**-12
  solid, properties of, **4**-10
  stranded, details of, **4**-13
Connections:
  chimney, **13**-101
  circuit boards, **13**-68
  etching, **13**-15
  flash evaporation, **13**-12
  masking, **13**-13
  module, **13**-56
  multiple reduction, **13**-13
  photomask, **13**-13
  photoresist techniques, **13**-13
  resistance-welded, **13**-93
  solder, **13**-75
  sputtering, **13**-15
  substrate internal, **13**-55
  tab-to-tab, **13**-101
  thick-film, **13**-68
  thin-film, **13**-55, **13**-68
  welding, **13**-93, **13**-113
Copper:
  beryllium-copper, **9**-47 to **9**-49
  brasses (*see* Brasses)
  bronzes, **9**-46, **9**-47
  characteristics of, **9**-43
  electrical conductivity of, **9**-48, **9**-49
  high-conductivity alloys, **9**-43 to **9**-45
  silver bearing, **9**-45
Corrosion:
  designing for, resistant finish, **10**-55
  environments and constituents, **5**-41
  galvanic, **10**-50 to **10**-52
  metal, comparisons of, **5**-41
  MIL specifications for, resistance, **10**-50 to **10**-53
  organics, **10**-53
  protective coatings for, **5**-43

Definitions (*see* Terms and definitions)
Depositions:
  catalytic, **10**-37
  chemical-and-vacuum-deposited metals, **10**-40
  electroplated buildup, **10**-44
  metallic substrates, **10**-36
  nonmetallic substrates, **10**-39 to **10**-43
  platable plastics, **10**-43
  (*See also* Electrodeposition; Nonelectrolytic deposition)
Dielectric constant:
  elastomers, **3**-20, **3**-22, **3**-27
  glasses, **6**-32
  plastics (*see* specific plastic)

Dielectric constant (*Cont.*):
  (*See also* specific material)
Dielectric loss:
  glasses, **6**-34, **6**-35
  (*See also* specific material)
Dielectric strength:
  glasses, **6**-34, **6**-35
  (*See also* specific material)
Dielectrics (*see* specific material)
Differential expansion:
  ceramics, **6**-50 to **6**-53
  (*See also* Thermal expansion)

EIA specifications for ceramics, **6**-61
Elastomers:
  acrylic, **3**-9
  ASTM specifications for, **3**-2, **3**-3, **3**-39
  butadiene, **3**-10
  butyl, **3**-8, **3**-26, **3**-27
  carboxylic, **3**-10
  cellular sponge, **3**-36
  chemical properties of, **3**-15, **3**-19
    swell, **3**-16, **3**-17
  compounding, **3**-33, **3**-34
    curing systems effects, **3**-36
    silicone compounds, **3**-35
    wet electrical properties, **3**-34, **3**-36
  costs, **3**-11
  electrical properties of, **3**-18 to **3**-25
    dielectric strength, **3**-20
    electrical values, **3**-24
    insulating, **3**-24
    polymers, **3**-25
    power factor, **3**-23
    resistance, **3**-21
    rubber, **3**-23
    specific inductive capacity (SIC), **3**-22
    water absorbing, **3**-26
  EPR, **3**-26 to **3**-28
  EPT, **3**-8
  fluoroelastomers, **3**-10
  foam rubber, **3**-36 to **3**-39
  general properties, **3**-4, **3**-5
  Hydrin, **3**-10
  Hypalon, **3**-9
  isoprene rubber, **3**-7
  manufacturing methods, **3**-31, **3**-32
    millable gums, **3**-31
  natural rubber, **3**-6
  neoprene, **3**-8
  nitrile, **3**-8
  nomenclature, **3**-4, **3**-5
  oxidation rate, **3**-26
  physical properties, **3**-12 to **3**-15
    radiation stabilities, **3**-15
    thermal stabilities, **3**-15

8 Index

Elastomers (*Cont.*):
  Polyblend, **3**-11
  polyethylene, **3**-26 to **3**-28
  polysulfide, **3**-9
  rubber compounds, **3**-14
  Rubber Manufacturers Association, **3**-29
  silicones, **3**-9, **3**-16, **3**-35
  specification sources, **3**-30
  specifications, general, **3**-28
    for materials, **3**-29
    for products, **3**-29
  sponge rubber, **3**-36 to **3**-38
  styrene-butadiene copolymer, **3**-7
  test methods for, **3**-30
Electrical properties:
  descriptions of, **1**-8, **1**-9
  (*See also* specific material)
Electrodeposition:
  anodizing, **10**-48
  basic theory, **10**-5
    Faraday's laws, **10**-5
  bath constituents, **10**-11
    anions, **10**-11
  bath operation, **10**-12, **10**-25
    agitation, **10**-13
    anodes, **10**-13
    chromium, **10**-14
    current density, **10**-14
    metal concentration, **10**-13
    metal costs, **10**-15
    operating temperature, **10**-13
    solution control, **10**-12
  brush plating, **10**-30
  coating properties, **10**-27 to **10**-30
  current distribution, **10**-8 to **10**-10
  design, **10**-15, **10**-17 to **10**-19
    angles, **10**-15
    contacts, **10**-17
    corners, **10**-15
    edges, **10**-15
    gas traps, **10**-17
    holes, **10**-15
    joints, **10**-15
    special shapes, **10**-15
  electroforming, **10**-32 to **10**-35
    matrix materials, **10**-33
    parting, **10**-34
  electromachining, **10**-35
  electropolishing, **10**-23
  field theory, **10**-5
  mechanical properties of, range of, **10**-31
  metal distribution of, **10**-9, **10**-10
  physical properties of, **10**-31
  postplating, **10**-27
  rinsing, **10**-25
  strike plating, **10**-23

Electrodeposition (*Cont.*):
  substrate metals, **10**-19
    alloys, **10**-26, **10**-27
    aluminum, **10**-20, **10**-21
    magnesium, **10**-20, **10**-22, **10**-23
    other metals, **10**-26
  substrates, **10**-2
  testing of, **10**-58 to **10**-61
Electroless plating (*see* Nonelectrolytic deposition)
Embedding:
  epoxies, **1**-95
    curing agents, **1**-95
    flexibilized, **1**-96
    modified, **1**-96
    novolak, **1**-95
    polysulfide, **1**-97
  processes, **1**-87, **1**-88
    casting, **1**-87
    encapsulating, **1**-87
    impregnating, **1**-87
    mold selection, **1**-89
    potting, **1**-87
    transfer molding, **1**-87, **1**-89
  resins, **1**-91 to **1**-102
    epoxies, **1**-93 to **1**-97
    exothermic properties, **1**-91
    fillers, **1**-102 to **1**-104
    low-density foams, **1**-101
    polybutadienes, **1**-99, **1**-100
    polyesters, **1**-99
    polysulfides, **1**-101
    processing, **1**-91
    silicones, **1**-97
    stresses, **1**-104
    urethanes, **1**-99
    viscosity, **1**-91
Etching:
  material, characteristics of, **14**-67 to **14**-69
  methods, **14**-71
  operation and control, **14**-72 to **14**-79
    comparisons, **14**-73
    direct-feedback etchant controller, **14**-73
  part design, **14**-67
  resist characteristics, **14**-70
  (*See also* Chemical machining; Photo fabrication)

Ferrites:
  applications for, **6**-102
  ceramic magnets, **6**-107
  commercial materials, **6**-103
  core materials, **6**-104, **6**-105
  garnets, **6**-99, **6**-100
  hard, **6**-104

Index 9

Ferrites (*Cont.*):
  hexagonal, **6**-96
  memory core data, **6**-108
  microwave, **6**-109 to **6**-113
  planar, **6**-97, **6**-98
  polycrystalline, **6**-109
  soft, **6**-101
  spinel, **6**-93 to **6**-95
    cobalt, **6**-95
  square-loop, **6**-106
Ferrous metals:
  ASTM specifications for, **8**-24
  calculating values, **8**-31
  Ceramvar, **8**-30
  electrical-resistance alloys, **8**-32
  expansion alloys, **8**-30
  ferrous alloys, **8**-33
  glass-to-metal seal, **8**-25
  high-permeability alloys, **8**-2, **8**-18
  magnetic alloys: Armco ingot iron, **8**-17
    chemical compositions, **8**-19
    machining, **8**-18
    magnetic properties, **8**-19
    mechanical properties, **8**-13, **8**-16, **8**-17
    MMPA standards for, **8**-20
    permanent, **8**-17, **8**-20
    stainless steels, **8**-21, **8**-24
      austenitic, **8**-29
      ferritic, **8**-28
    temperature-compensator, **8**-22, **8**-23
    thermal expansion, **8**-18
  thermal-expansion properties of, **8**-23, **8**-24
Filament winding (*see* Laminates)
Films:
  cellulose, **2**-115
  definition of, **2**-115
  electrical, properties of, **2**-115 to **2**-123
  fluorocarbon, **2**-115
  Kapton, **2**-127, **2**-128
  Mylar, **2**-125, **2**-129
  plastic, **2**-127
  polyamide, **2**-115
  polyester, **2**-124
  polyimide, **2**-125
  polystyrene, **2**-124
  selector chart, **2**-124
  (*See also* Photofabrication; Thick films; Thin films)
Finishes:
  coatings (*see* Coatings; Electrodeposition; Nonelectrolytic deposition)
  MIL specifications for, **10**-61
  organic (*see* Coatings)
Flexible insulations:
  definition of, **2**-127

Flexible insulations (*Cont.*):
  inorganic paper, **2**-128 to **2**-130
  MIL specifications for, **2**-128
  varnished cloth, **2**-128
  (*See also* Films)
Flexible wiring (*see* Multiconductor cables; Printed circuits; Wiring and cabling)

Glasses:
  capacitors, **6**-75
    construction of, **6**-75
    operating characteristics of, **6**-76
  commercial, **6**-27
    chemical compositions, **6**-27
  dielectric constant, **6**-32
  dielectric loss, **6**-34, **6**-35
  dielectric strength, **6**-29 to **6**-32
  differential expansion, **6**-50 to **6**-53
  glass-ceramics, **6**-42
  Pyroceram, comparison properties of, **6**-46
  resistivity: electrical, **6**-28
    surface, **6**-29
    volume, **6**-28
  seals: bonding, **6**-44, **6**-49
    ceramic-to-metal, **6**-49
    glass-metal combinations, **6**-48
    glass-to-metal, **6**-42
    stresses, **6**-45
  substrates: ceramic and glass, **6**-38 to **6**-40
    dielectric constant, **6**-42
    differential thermal expansion, **6**-41
    electric conductivity, **6**-40
    properties of glass substrates, **6**-43
    surface smoothness, **6**-38
    thermal conductivity, **6**-41
    thermal shock resistance, **6**-41
  thick film: cermets, **6**-55
    dielectrics, **6**-56
    resistors, **6**-54
Glossary (*see* Terms and definitions)

Hardness:
  deposits, **10**-60
  elastomers, **3**-12
  laminates, **2**-7
Hookup wire, **4**-33
  abrasion resistance, **4**-42
  automated-termination data, **4**-45
  copper conductor data, **4**-39, **4**-40
  data, **4**-33, **4**-36 to **4**-40

Hookup wire (*Cont.*):
  Kapton film, **4**-43
  MIL-W-81044/4 wire, **4**-37
  outer-space, **4**-41 to **4**-44
  ultrahigh vacuum, **4**-46
  wire weight comparisons, **4**-41

Insulation:
  application of, **4**-21, **4**-22
  construction of, **4**-21, **4**-22
  materials, **4**-15 to **4**-21
    chemical properties of, **4**-18
    electrical properties of, **4**-17
    mechanical properties of, **4**-16
    physical properties of, **4**-16
  selections, **4**-32
  (*See also* specific material)
Insulation resistance:
  definition of, **1**-8
  (*See also* specific material)
Insulators (see specific material)
Interconnections:
  crossover, **13**-60, **13**-61
  definitions of, **4**-31
  hookup wire (*see* Hookup wire)
  interfacial, **13**-57
  MIL specifications for, **4**-36
  screen and fire, **13**-62
  tantalum-aluminum, **13**-48
  thin-film, **13**-56, **13**-59
  welding, **13**-114
  wire data, **4**-33 to **4**-36
    copper conductors, **4**-34
  (*See also* Intraconnections; specific type of interconnections)
Intraconnections:
  aluminum-silicon phase, **13**-10
  crossover defect, **13**-15
  flash evaporation, **13**-12
  mechanical assembly levels, **13**-8
  metallization, **13**-11
  multilevel, **13**-15
  photoresist techniques, **13**-13
  thin-film aluminum, **13**-12
  (*See also* Thin films)
  thin-film conductors, **13**-13
  (*See also* Interconnections; specific type of interconnections)
IPC classes of circuit boards, **13**-72

Joining:
  flash evaporation, **13**-12
  intraconnections, **13**-8
  (*See also* Adhesives; Bonds and bonding; Interconnections; specific type of bonding)

Kapton (*see* Films)

Laminates:
  ASTM specifications for, **2**-3, **2**-92
  cost, **2**-53
  design criteria for, **2**-12
  filament winding, **2**-83
    hoop strength data, **2**-84
    internal-pressure vessel, **2**-84
  glass fiber finishes, **2**-14, **2**-39
  industrial, **2**-62
  insulating, **2**-63
  joining, **2**-93 to **2**-96
    screws, holding forces of, **2**-93 to **2**-96
  material types, **2**-17 to **2**-48
    diallyl phthalate, **2**-32
    diphenyl oxide, **2**-47, **2**-50
    epoxy, **2**-17, **2**-30
    melamine, **2**-17, **2**-28
    phenolics, **2**-16
    phenylsilane, **2**-41
    polybenzimidazole, **2**-46, **2**-49
    polyester, **2**-29, **2**-34, **2**-36 to **2**-38
    polyimide, **2**-42, **2**-44, **2**-47
    polyimide-glass, **2**-43
    silicone, **2**-33, **2**-40
    Teflon, **2**-41
    thermoplastics, **2**-48
      glass-mat-reinforced, **2**-51
  NEMA classes, **2**-48, **2**-54 to **2**-58, **2**-62 to **2**-74
    of rods, **2**-61
    of thickness, **2**-76, **2**-78, **2**-79
    of tubing: molded, **2**-59
      rolled, **2**-60
  prepreg, **2**-66 to **2**-69
  property chart, **2**-18 to **2**-27
  prototypes, **2**-70, **2**-82
    autoclave molding, **2**-81
    hand lay-up, **2**-71
    molding methods, **2**-82
    pressure-bag, **2**-77
    spray-up, **2**-72
    vacuum bag, **2**-73, **2**-74
    vacuum injection, **2**-82
  reinforcing fibers, **2**-13
  selector chart, **2**-15
  space-age, **2**-84
    beryllium, **2**-87, **2**-89
    beryllium-wire, **2**-89
    boron, **2**-84
    boron polyimide, **2**-86
    carbon, **2**-86
    epoxy-boron, **2**-86
    graphite, **2**-87
    polyimide-graphite, **2**-88

Laminates, space-age (*Cont.*):
SiC, **2**-90
whiskers, **2**-85, **2**-88
test methods, **2**-3 to **2**-11
thermal aging, **2**-63
thermal rating, **2**-53
vulcanized fiber, **2**-63, **2**-76, **2**-80
(*See also* Printed circuits)
Leads:
alloys, **9**-68
lead frames (*see* Bonds and bonding; Chemical machining; Ferrous metals)
Lockalloy (*see* Beryllium and Lockalloy)

Magnesium:
alloy compositions, **9**-53
alloys, **9**-50 to **9**-59
availability, **9**-52
castings, **9**-53
designations, **9**-52
fabrication, **9**-58
finishing (*see* Finishes)
forms, **9**-51
joining, **9**-59
mechanical properties of, **9**-54 to **9**-58
specific gravities, **9**-51
specifications for, **9**-51
thermal strength properties of, **9**-52
Magnet wire:
conductors, **4**-68, **4**-70
high-temperature applications, **4**-70
design considerations, **4**-74, **4**-75
film-insulated wire characteristics, **4**-75
insulations, **4**-69, **4**-71 to **4**-75
MIL specifications for, **4**-70
NEMA standards for, **4**-70
types of, **4**-68, **4**-70
Magnetic ceramics (*see* Ceramics)
Magnetic metals (*see* Ferrous metals)
Materials (*see* specific material)
Metal bonding (*see* Aluminum, brazing; Bonds and bonding; Soldering)
Metals (*see* Electrodeposition; Ferrous metals; Nonferrous metals)
Micas:
ASTM specifications for, **6**-36
capacitors: characteristics of, **6**-78
color code for, **6**-80
encapsulation, **6**-78
mica-paper, **6**-78
operating characteristics of, **6**-79
silvered-mica, **6**-77
stacked mica-foil, **6**-77
glass-bonded, **6**-38, **6**-40
muscovite-ruby mica, **6**-35
phlogopite-amber, **6**-37

Micas (*Cont.*):
properties of, **6**-37
reconstituted, **6**-38
synthetic mica-fluorophlogopite, **6**-37
Microelectronics:
classifications of, **11**-4, **11**-5
(*See also* Bonds and bonding; Photofabrication: Thin films
MIL specifications:
for adhesives, **2**-98
for coatings, **5**-3, **5**-9, **5**-23, **5**-66 to **5**-71
for coaxial cables, **4**-49
for corrosion resistance, **10**-50 to **10**-53
for finishes, **10**-61
for flexible insulation, **2**-128
for hookup wire, **4**-37, **4**-38
for interconnections, **4**-36
for magnet wire, **4**-70
for multiconductor cables, **4**-58
for printed circuits, **2**-105
for thick film, **12**-75
for tubing, **4**-80, **4**-83, **4**-85
Modulus:
elastomers, **3**-12
(*See also* specific material)
Molded products:
economics of, **1**-83 to **1**-85
guidelines for, **1**-83 to **1**-85
Molding materials (*see* Thermoplastics; Thermosetting plastics)
Multiconductor cables:
airborne, **4**-49
building-wire, **4**-66, **4**-68
cabling-concept comparisons, **4**-56
comparisons, **4**-56
copper conductors, **4**-62
design considerations, **4**-59, **4**-66
flat-conductor, **4**-55 to **4**-61
flat-flexible-cable, **4**-65
flexible printed wiring, **4**-61
ground electronics, **4**-54
ground support, **4**-54
IPCEA standards for, **4**-66
MIL specifications for, **4**-58
NAS standards for, **4**-61
NEMA standards for, **4**-66
rubber-insulated, **4**-66
shipboard, **4**-66
types, **4**-49 to **4**-68
Mylar (*see* Films)

NAS standards for multiconductor cables, **4**-61

12 Index

NEMA standards:
  for coatings, **5**-14, **5**-67
  for laminates, **2**-48 to **2**-74
  for magnet wire, **4**-70
  for multiconductor cables, **4**-66
  for printed circuits, **2**-106, **2**-112
  for tubing, **4**-80
Nickel:
  alloy compositions, **9**-60
  fabrication, **9**-60 to **9**-62
  forms, **9**-59
  mechanical properties, **9**-62, **9**-64
  physical properties, **9**-62
Nonelectrolytic deposition:
  conversion coatings, **10**-48
  metallic substrates, **10**-36
  reasons for, **10**-2 to **10**-5
  testing of, **10**-58 to **10**-61
    adhesion, **10**-60
    mechanical property, **10**-60
    porosity, **10**-59
    solderability, **10**-60
    stress, **10**-60
    thickness, **10**-58
Nonferrous metals:
  alloy selection, **9**-3 to **9**-11
  aluminum (*see* Aluminum)
  brasses (*see* Brasses)
  comparative data, **9**-11
  copper (*see* Copper)
  heat absorption, **9**-11
  joining, **9**-8
  magnesium (*see* Magnesium)
  mechanical properties of, **9**-6
  nickel (*see* Nickel)
  physical properties of, **9**-7
  refractory metals, **9**-68 to **9**-70
  resistivity, **9**-11
  stiffness, **9**-11
  strength-density ratios, **9**-11
  strengthening mechanisms, **9**-5
  titanium (*see* Titanium)
  wrought, **9**-4
  (*See also* specific metal)

Photofabrication:
  applications, **14**-7
    chemical machining (*see* Chemical machining)
    classifications of, **14**-7
    to electronics, **14**-8
    printed circuits (*see* Printed circuits)
    screen fabrication (*see* Screen fabrication)
    semiconductors (*see* Semiconductors)

Photofabrication, applications (*Cont.*):
    thin films (*see* Thin films)
    three-dimensional photosensitive materials (*see* Photosensitive materials)
  artwork, **14**-6
  cleaning, **14**-59
    alkaline, **14**-62
    chemical, **14**-61 to **14**-64
    mechanical, **14**-60
    metal, **14**-64
  comparisons, **14**-9
  conversion coatings, **14**-66
  design, **14**-6
  material selection, **14**-6
  processes, **14**-6, **14**-56
    machining, **14**-10
    pretreatment, **14**-57
  resist image-transfer, **14**-9
  (*See also* Photomasks; Photosensitive materials; Printed circuits)
Photomasks:
  artwork, **14**-11 to **14**-19
    accuracy and precision, **14**-13
    computer controlled, **14**-17
    cut-and-peel, **14**-14
    materials, stability of, **14**-16
    size changes: due to humidity, **14**-15
      due to temperature, **14**-15
  camera work, **14**-19, to **14**-27
    diffuse illumination, **14**-21
    geometric optics, **14**-20
    lens system, **14**-19, **14**-23, **14**-24
    photomask processing, **14**-25
    precision, **14**-24
    precision process, **14**-25
    spatial sine-wave, **14**-21
    step-and-repeat, **14**-24, **14**-28
  chemical processing, **14**-32
    developers, **14**-34
    direct, **14**-32
    Farmer's reducer, **14**-37
    fixers, **14**-35
    other, **14**-36
    reversal processing, **14**-35
  contact printing, **14**-27, **14**-31
  fabrication, **14**-11
    error analysis, **14**-16
    mask-fabrication steps, **14**-12
  macrocontact printing, **14**-29
  microcontact printing, **14**-29
  projection-wafer printing, **14**-30
  registration for exposure, **14**-40
  special techniques, **14**-37 to **14**-39
    metals on glass, **14**-38

Index 13

Photoresists (*see* Photosensitive materials)
Photosensitive materials:
  alignment jig, processing of, **14**-127
  applications of, **14**-124 to **14**-128
  dichromated resists, **14**-53
  glasses, **14**-124
  negative resists, **14**-42 to **14**-53
  photopolymer systems, **14**-126
  photoresists: application of, **14**-44
    coving of, **14**-52
    development of, **14**-51
    dip coating, **14**-46
    exposure, **14**-49, **14**-51
    flow coating, **14**-44
    removal, **15**-52
    roller coating, **14**-45
    special methods, **14**-49
    special response, **14**-50
    spray coating, **14**-47
    suppliers, **14**-43
    whirl coating, **14**-48
  positive resists, **14**-42 to **14**-53
  Templex, **14**-126 to **14**-128
  three-dimensional materials, **14**-55, **14**-124 to **14**-128
    glass, **14**-55, **14**-57
    plastics, **14**-56, **14**-58, **14**-59
  troubleshooting chart, **14**-54, **14**-55
Piezoelectricity, **6**-79, **6**-81
  ceramics, **6**-85
    applications of, **6**-86, **6**-87
    commercial, properties of, **6**-88, **6**-89, **6**-90
  coupling factor, **6**-85
  elastopiezodielectric matrices of crystal glasses, **6**-82, **6**-83
  ferroelectricity, **6**-85
  properties of, **6**-84
  pyroelectricity, **6**-85
Plastics:
  adhesives (*see* Adhesives)
  coatings (*see* Coatings)
  embedding materials (*see* Embedding)
  laminates (*see* Laminates)
  parts design, **1**-86
  processing guidelines for, **1**-83 to **1**-85
  processing methods for, **1**-83 to **1**-85
  thermoplastics (*see* Thermoplastics)
  thermosetting plastics (*see* Thermosetting plastics)
Plating:
  applications of, **14**-75
    techniques, **14**-79
  electroless, **14**-77
  electroplating, **14**-75
    electrochemical equivalents, **14**-77

Plating (*Cont.*):
  immersion, **14**-78
  (*See also* Electrodeposition; Non-electrolytic deposition)
Polymers (*see* Plastics)
Precious metals:
  mechanical properties of, **9**-73
  physical properties of, **9**-73
Printed circuits:
  applications of, **14**-90 to **14**-99
  artwork, **14**-94
  ASTM specifications for, **2**-105
  coatings for, **5**-3 to **5**-12
  copper-clad, **2**-106 to **2**-111
    solder float test for, **2**-108
    testing of, **2**-107
  copper-foil, **2**-109
  design considerations of, **14**-92
    guidelines for, **14**-94
  dimensional data, **14**-93
  dimensions, **2**-111
  double-sided circuits, **14**-98
    additive processing of, **14**-99
    fabrication of, **14**-98
  electroplating baths, **14**-97
  fabrication of, **14**-96
  flexible, **2**-113
  metals, properties of, **14**-96
  microwave strip line, **2**-114
    electrical properties, **2**-114
  MIL specifications for, **2**-105
  multilayer, **2**-111 to **2**-113
  NEMA standards for, **2**-106, **2**-112
  plastics, properties of, **14**-95
  processing, **14**-92
  resists: application of, **14**-97
    copper, **14**-98
  selector chart, **2**-109
  thickness tolerances, **2**-110
  tolerances, **2**-111
  types of, **14**-90
    additive, **14**-90
    subtractive, **14**-90
Printed wiring (*see* Printed circuits)

Radiation (*see* Space effects)
Refractory metals, properties of, **9**-69, **9**-70
Resins (*see* Plastics)
Resistivity (*see* specific materials)
Resists (*see* Photosensitive materials)
Rubber (*see* Elastomers)

SAE specifications for tubing, **4**-83
Screen fabrication, applications of, **14**-114 to **14**-123

14   Index

Screens:
    design considerations, **14**-119, **14**-122
    Dia-print, **14**-116
    emulsions: direct, **14**-117
        indirect, **14**-118, **14**-123
    mesh, **14**-116, **14**-121
    nylon screen mesh, **14**-117
    stainless steel, **14**-117
        tensions for, **14**-117
    stencils, **14**-123
    (*See also* Thick films)
Seals:
    ceramic-to-metal, **6**-49
    glass, **6**-44 to **6**-49
    glass-to-metal, **8**-25
    (*See also* Adhesives; Ceramics; Ferrous metals; Glasses)
Semiconductors:
    applications of, **14**-100 to **14**-108
    band gaps, **7**-22
    Bausch and Lomb microscope, **14**-103
    bismuth telluride, **7**-17
    boron, **7**-17
    cadmium sulfide, **7**-18
    carrier concentration measurement, **7**-57
    conductivity type, **7**-33
    crystal growth: closed-tube sublimation, **7**-85
        closed-tube transport, **7**-85, **7**-86
        Czochralski, **7**-77
        doping, **7**-89
        dynamic vacuum evaporation, **7**-84
        environment, **7**-75 to **7**-77
        epitaxy, **7**-89
        float zoning, **7**-78
        flow systems, **7**-89
        flux growth, **7**-80
        melt growth, **7**-81
        open-tube transport, **7**-86 to **7**-88
        purification, **7**-82
        ribbon growth, **7**-79
        uniform distribution, **7**-83
        vapor-phase, **7**-87
        zone leveling, **7**-79
    crystal structure, **7**-4 to **7**-7
    crystal symmetry, **7**-11
        tensor properties, **7**-12
    crystallographic defects, **7**-11 to **7**-16
        device problems, **7**-13
        dislocations, **7**-12
        grain boundaries, **7**-15
        hexagonal lattice spacing, **7**-37
        inclusions, **7**-16
        lattice defects, **7**-13
        lattice spacing, **7**-35
        lineage, **7**-15
        placement errors, **7**-11

Semiconductors, crystallographic defects (*Cont.*):
        separate phases, **7**-16
        stacking faults, **7**-15
        twinning, **7**-14, **7**-16
    defect observation, **7**-28
    diamond, **7**-18
    diffusion, **7**-99 to **7**-108
        bilateral, **7**-102
        concentration step, **7**-104
        error-function algebra, **7**-105
        error-function values, **7**-100
        impurity sources for open-tube diffusion: of boron, **7**-107
            of phosphorus, **7**-108
        infinite source, **7**-102
        limited source, **7**-103
        three-dimensional solution, **7**-105
        time-varying temperature, **7**-104
        transfer of impurities, **7**-105
    diffusion coefficients, impurities in III-V compounds, **7**-30
    dislocation etches, **7**-52 to **7**-54
    effects of organic materials, **1**-27 to **1**-29, **5**-21 to **5**-24
    elastic constants, **7**-31
    etching: aluminum, **14**-107
        silicon dioxide, **14**-105
    finished devices measurements, **7**-67 to **7**-75
        delineating layers, **7**-70 to **7**-73
        interferometry, **7**-69
        layer thickness, **7**-68
        VAMFO, **7**-75
        visual determination, **7**-74
    gallium arsenide, **7**-18
    general properties of, **7**-3
    germanium, **7**-19, **7**-27
    gray tin, **7**-19
    impurity concentration, **7**-58
    impurity sources, **7**-107
    index of properties of, **7**-23
    indium antimonide, **7**-20
    Knoop hardness, **7**-32
    lead sulfide, **7**-20
    lifetime measurements, **7**-60 to **7**-67
        carrier measuring methods, **7**-63
        diffusion length, **7**-66
        drift method, **7**-65
        MOS capacitor, **7**-65
        photoconductive decay, **7**-62
        range and limitations, **7**-64
        surface photovoltage, **7**-67
    melting points, **7**-38
    mobility of carriers, **7**-39
    MOS, **14**-101
    orientation determination, **7**-24 to **7**-28

Semiconductors (*Cont.*):
  photofabrication operations, **14**-100 to **14**-109
  photomask equipment, **14**-102
  photoresist, **14**-102
  piezoresistance coefficients, **7**-40
  plane indices, **7**-8
    angles, **7**-10
    equations, **7**-9
  refractive indices, **7**-41
  resisitivity: capacitance-voltage, **7**-51
    direct method, **7**-36
    linear four-point probe, **7**-40
    measurement methods, **7**-55
    profiling, **7**-56 to **7**-59, **7**-62
    sheet resistance, **7**-44, **7**-48
    spreading resistance, **7**-51
    square four-point array, **7**-43
    three-point probe, **7**-56
    two-point probe, **7**-37
  segregation coefficients, **7**-43
  selenium, **7**-20
  shaping: abrasives, **7**-94
    chemical shaping and polishing, **7**-98
    cleaving, **7**-97
    diamond-saw data, **7**-92
    grinding, **7**-96
    grinding-wheel data, **7**-92
    mechanical damages, **7**-93
    polishing, **7**-93
    sandblasting, **7**-94
    sieve sizes, **7**-95
    slicing, **7**-90
    smoothing, **7**-91
    ultrasonic cutting, **7**-94
  silicon, **7**-21, **7**-27
  surface energies, **7**-45
  tellurium, **7**-22
  thermal conductivity, **7**-49
  (*See also* Bonds and bonding; Thick films; Thin films)
Shield jackets:
  abrasion resistance, **4**-25
  materials, **4**-24, **4**-28
  properties, **4**-26
Shielding:
  effectivity, **4**-23
  materials, **4**-22, **4**-24
  (*See also* specific type of wiring)
Sleeving (*see* Tubing)
Soldering:
  connections, **13**-75
  extension, **13**-90
  flatpacks, **13**-85
  fluxes, **13**-76
  hand, **13**-91
  hot-air, **13**-89

Soldering (*Cont.*):
  impulse, **13**-86 to **13**-88
  infrared, **13**-90
  masking, **13**-90
  parallel-gap, **13**-81
  peel, **13**-82 to **13**-84
  plating, **13**-80
  preforms, **13**-87
  processes, **13**-77
  reflow, **13**-80, **13**-85
  resistance, **13**-81
  slivers, **13**-91
  soft solder-alloy, **13**-78
  solder slinging, **13**-80
  splashes, **13**-91
  ultrasonic, **13**-91
  wave, **13**-78
Solders:
  compositions, **9**-70
  mechanical properties of, **9**-71
  melting points of, **9**-70
  shear strength, **9**-72
Space effects:
  albedo, **15**-12, **15**-15
    earth, **15**-14
  ascent, **15**-2
  atmosphere: earth's, **15**-6
    upper, **15**-6, **15**-10
  atmospheric composition, **15**-8
  atmospheric properties, **15**-9
  atmospheric variations, **15**-7
  batteries, **15**-49
  booster, liquid-fueled, **15**-5
  electron tubes, **15**-45
  electrons, trapped, **15**-20
  environments, **15**-33
    effects of, **15**-33
    internal, **15**-26 to **15**-28
    nuclear reactors, **15**-27
    on-board sources, **15**-27
    radiation, **15**-34 to **15**-44
  flux-mass, **15**-29
  fluxes, **15**-21
  galactic cosmic rays, **15**-24
  heating, aerodynamic, **15**-2
  humidity, **15**-49
  magnetic fields, **15**-50
  materials: energy deposits, **15**-27
    perforation, **15**-30
  meteoroids, **15**-28 to **15**-33
    density, **15**-28
    impingement, **15**-30
    penetrations, **15**-32
    sporadic, **15**-29, **15**-31
    velocity, **15**-28
  micrometeoroids, **15**-49
  outlook, **15**-50

16 Index

Space effects (*Cont.*):
  particles: changed, **15**-17
    geomagnetic, **15**-26
    neutral, **15**-24
    trapped alpha, **15**-21
  penetration, multiple-wall, **15**-31
  perforation, single-wall, **15**-30
  proton flux, **15**-22
  proton isoflux contours, **15**-20
  protons: inner-zone, **15**-19
    trapped, **15**-17
  radiation: auroral region, **15**-24
    effect on ceramics, **15**-34
    effect on dielectrics, **15**-35
    effect on electronic components, **15**-38
    effect on electronic devices, **15**-38
    effect on inductors, **15**-48
    effect on insulating materials, **15**-36
    effect on integrated circuits, **15**-43
    effect on magnetics, **15**-37
    effect on metals and alloys, **15**-34
    effect on relays, **15**-48
    effect on resistive devices, **15**-47
    effect on resistive elements, **15**-47
    effect on semiconductor devices, **15**-41
    effect on semiconductors, **15**-35, **15**-39
    effect on solar cells, **15**-43
    effect on switches, **15**-48
    effect on transistors, **15**-42
  electromagnetic, **15**-12, **15**-13, **15**-24
  geomagnetically trapped, **15**-17
  long-wave, **15**-15
  man-made, **15**-25
  natural, **15**-16
  nonpenetrating, **15**-9
  nuclear devices, **15**-25
  penetrating, **15**-16 to **15**-18
  response of electronic materials, **15**-34
  solar, **15**-12
  thermal, **15**-15
  ultraviolet, **15**-50
  random noise, **15**-4
  resistive components, **15**-46
  shielding, **15**-26
  shock, **15**-4
  silicon charged-particle detectors, **15**-40
  solar particles, **15**-22
    solar flare: free-space, **15**-23
      radiation, **15**-22
  temperature, **15**-50
  TIMMS, **15**-45
  vacuum, **15**-49
  vibration, **15**-4

Space effects (*Cont.*):
  visible light, **15**-50
  zero gravity, **15**-49
Sputtering, **14**-79 to **14**-82
  (*See also* Thin films)
Standards and specifications:
  for adhesives, **2**-98
  for aluminum, **9**-21 to **9**-23
  AMS, for coatings, **5**-71
  ASTM (*see* ASTM specifications)
  EIA, for ceramics, **6**-61
  for elastomers, **3**-28 to **3**-30, **3**-39
  for electroless finishes, **10**-61
  for electrolytic finishes, **10**-61
  IPC, for circuit boards, **13**-72
  for laminates, **2**-48, **2**-58
  for magnesium alloys, **9**-51
  MIL (*see* MIL specifications)
  NAS, for multiconductor cables, **4**-61
  NEMA (*see* NEMA standards)
  SAE, for tubing, **4**-83
Strain hardening, **9**-5
Substrates:
  ceramic and glass, **6**-38 to **6**-42
  dielectric constant, **6**-42
  electrical conductivity, **6**-40
  surface smoothness, **6**-38
  thermal conductivity, **6**-41
  (*See also* specific material)

Terminations, wire and cable:
  allowable wire gage, **4**-77
  hardware for, **4**-76, **4**-80
  identification for, **4**-79
  shielding, **4**-78
  terminals, **4**-77 to **4**-79
  voltage between terminals, volts, **4**-76
Terms and definitions:
  finishes, **10**-3 to **10**-5
  laminates, **2**-11
  metals joining, **13**-3 to **13**-6
  photofabrication, **14**-3 to **14**-5
  plastics for electronics, **1**-3 to **1**-7
  thick film, **12**-4
  wires and cables, **4**-3 to **4**-6
Thermal conductivity:
  ceramic substrates, **11**-19
  coatings, **5**-45 to **5**-47
  epoxies, **1**-103
  glass substrates, **11**-19
  metal comparisons, **9**-10
  nickel alloys, **9**-62, **9**-63
  (*See also* specific material)
Thermal expansion:
  aluminum alloys, **9**-30
  ceramic substrates, **11**-19

Index

Thermal expansion (*Cont.*):
  ceramics, **6**-7
  epoxies, **1**-102
  glass substrates, **11**-19
  metals, **8**-23, **8**-24
  nickel alloys, **9**-62
  titanium alloys, **9**-65
  (*See also* specific material)
Thermal properties:
  aluminum alloys, **9**-28 to **9**-30
  metals, **9**-12
  nickel alloys, **9**-62
  titanium alloys, **9**-65
Thermoplastics:
  ABS plastics, **1**-29, **1**-38
    electrical properties of, **1**-38
    mechanical properties of, **1**-29
    property comparisons of, **1**-39
  ABS-polycarbonate alloy, **1**-39
  acetals, **1**-40
    electrical properties of, **1**-41
    fluorocarbon fiber-filled, **1**-41
    mechanical properties of, **1**-40
  acrylics, **1**-41
    electrical properties of, **1**-42
    mechanical properties of, **1**-42
    property comparisons of, **1**-43
    TFE fiber-filled, **1**-43
  application information, **1**-30 to **1**-33
  ASTM specifications for, **1**-79
  cellulosics, **1**-43
  chlorinated polyether, **1**-45
    electrical properties of, **1**-45
    mechanical properties of, **1**-45
  electrical properties of, typical, **1**-37
  ethylene-vinyl acetates, **1**-45
  fluorocarbons, **1**-45
    creep modulus, **1**-47
    electrical properties of, **1**-49
    FEP, **1**-52
    mechanical properties of, **1**-46
    space environment, **1**-51
    tensile strength, **1**-50
    TFE, **1**-52
  glass-filled, **1**-79
  glass-reinforced, **1**-81
  ionomers, **1**-53
    electrical properties of, **1**-54
    mechanical properties of, **1**-54
  Kapton, **1**-62
  mechanical properties of, typical, **1**-34 to **1**-36
  Mylar, **1**-62
  nylons, **1**-54
    electrical properties of, **1**-57
    mechanical properties of, **1**-55
    Nomex, **1**-58
    temperature effects of, **1**-55, **1**-58

Thermoplastics (*Cont.*):
  parylenes, **1**-58
  phenoxies, **1**-59
  physical properties of, typical, **1**-34 to **1**-36
  plastic processing, **1**-82
    guidelines for, **1**-83 to **1**-85
    part design for, **1**-86
  polyallomers, **1**-61
  polyamide-imides, **1**-61
  polycarbonates, **1**-66 to **1**-69
    electrical properties of, **1**-69
    mechanical properties of, **1**-67
  polyesters, **1**-69, **1**-72
    Mylar, **1**-72
    types of, **1**-71
  polyethylenes, **1**-60
  polyolefins, crosslinked, **1**-61
  polyphenylene oxides, **1**-71
  polypropylenes, **1**-60
  polystyrenes, **1**-74
    expanded styrene, **1**-76
    radar band frequencies, **1**-74
  polysulfones, **1**-76
    glass contents effects, **1**-80
    raw, **1**-79
  vinyls, **1**-78
Thermosetting plastics:
  alkyds, **1**-12 to **1**-14
  aminos, **1**-15
    electrical properties of, **1**-17
    mechanical properties of, **1**-17
    physical properties of, **1**-16
  application of, **1**-10
  diallyl phthalates, **1**-18, **1**-20
  epoxies, **1**-20
  insulation, electrical, **1**-8
  phenolics, **1**-21
    properties of, **1**-24
  polyesters, **1**-22
  silicones, **1**-26
    properties of, **1**-27
    water-extract resistivity, **1**-27, **12**-9
Thick films:
  artwork, **12**-11 to **12**-15, **12**-17
  cermets (*see* Cermet materials)
  cleanliness, **12**-80 to **12**-82
    ultrasonic, **12**-81
      selection guide for, **12**-81
  comparisons, **12**-4
  crossovers, **12**-9
  drying process, infrared, **12**-53
  encapsulation, **12**-76
    hermetic sealing, **12**-78
    leak detection, **12**-80
  exposure, **12**-25
  firing, **12**-54 to **12**-56
    furnace, **12**-54

18   Index

Thick films, firing (*Cont.*):
   resistor pastes, **12**-56
   glasses, **6**-54 to **6**-56
   hybrid circuit, **12**-74, **12**-77
   hermetic package specification for, **12**-76
   layout criteria for, **12**-5 to **12**-8
   layout suggestions for, **12**-8
   MIL specifications for, **12**-75
   photoreduction, **12**-15
      artwork film flatness, **12**-17
      extensive distortion, **12**-17
      inspection systems, **12**-19
      lenses, **12**-18
      optical distortions, **12**-17
      processing problems, **12**-18
      reduction cameras, **12**-19
      trapezium distortion, **12**-17
      trapezoidal effect, **12**-16
   power dissipation, **12**-8
   printing processes, **12**-47 to **12**-52
      film thickness measurements, **12**-50
      hybrid printer, **12**-48
      squeegee parameters, **12**-47
      surfanalyzer, **12**-50 to **12**-52
   probing equipment, **12**-73
   resistor patterns, **12**-10
   resistor trimming, **12**-56 to **12**-61
      air abrasion, **12**-57
      laser, **12**-60, **12**-61
      pulse, **12**-58
      scribe, **12**-59
   resistor value calculations, **12**-6
   resistors, effects of firing temperatures on, **11**-8
   screens, **12**-21
      emulsion, **12**-23
      inspection, **12**-26
      mesh sizes, **12**-21, **12**-25
      metal mask, **12**-26
      tension, **12**-21
   solder assembly, **12**-61 to **12**-65
      fluxing, **12**-62
      reflow, **12**-63 to **12**-65
      tinning, **12**-62
   substrate materials, **12**-27 to **12**-31
      ceramics, **12**-29 to **12**-31
      machinability, **12**-28
      properties, **12**-28
Thin films:
   aluminum-tantalum, **11**-35
   application comparisons, **11**-3
   applications for, **14**-109 to **14**-113
   capacitors, **11**-9, **11**-22
      characteristics of, **11**-24
      silicon monoxide, **11**-20
   cermet, **11**-30
   chemical deposition, **11**-17

Thin films (*Cont.*):
   chromium-resistive, **11**-29
   chromium-silicon monoxide, **14**-112
   deposition guidelines for, **11**-10 to **11**-12
   deposition rates, **11**-14 to **11**-16
   dielectric film properties of, **11**-9, **11**-21, **11**-23
   diffused resistors, **11**-24 to **11**-26
   diffusion, **11**-17
   electrical properties of, **11**-31
   electrode configuration, **11**-16
   etching, **14**-110
   fabrication, **11**-8
   flip-flop devices, **11**-37
   getter sputtering, **11**-17
      metal film properties, **11**-18
   hot-stage metallograph, **11**-53
   instrument use, **11**-40 to **11**-51
   instrumentation for microelectronics, **11**-38 to **11**-59
   interconnections, **11**-31 to **11**-36
   metal film conductors, **11**-32
   metallization systems, **11**-32
      chromium-gold, **11**-33
      Kovar, **11**-33
      molybdenum-gold, **11**-34
   microcircuits, **14**-109
   microscopes: electron, **11**-56
      radiometric, **11**-57
      scanning electron, **11**-55
   modifiers, **11**-7
   nichrome-gold, **14**-110
      photoetch, **14**-112
   nickel chromium, **11**-28
   photoresists, **14**-110
   reliability, **11**-36
   resistive lengths, **11**-26
   resistors, **11**-23, **11**-27 to **11**-30
   RF sputtering, **11**-9, **11**-13
   screen deposition, **11**-3
   sensitivities of instruments, **11**-38
   sputtered oxide properties, **11**-20
   substrates, **11**-18
   tantalum oxide, **11**-21, **11**-28, **14**-113
   thick film versus, **11**-2
   titanium, **11**-29
   vacuum evaporation, **11**-9
   x-ray: diffraction, **11**-54
      electron-probe microanalyzer, **11**-58
      fluorescence, **11**-54
Titanium:
   alloys, **9**-63
   bend radii, **9**-67
   electrical properties of, **9**-65
   forms of, **9**-67
   gas contamination, **9**-66
   mechanical properties of, **9**-66

Titanium (*Cont.*):
  physical properties of, **9**-65
Tubing:
  ASTM specifications for, **4**-80
  braided sleeving, **4**-82
  extruded, **4**-81
  heat shrinkable properties of, **4**-84
  MIL specifications for, **4**-80, **4**-83, **4**-85
  NEMA standards for, **4**-80
  SAE specifications for, **4**-83
  shrinkable devices, **4**-83
  support spacing, **4**-85
  (*See also* Laminates; Nonferrous metals)

Vacuum:
  delineation of patterns, **14**-80
  evaporation, **14**-79
  (*See also* Bonds and bonding; Space effects; Thin films)

Welding:
  aluminum-copper, **13**-108
  conductive adhesives, **13**-113
  copper-to-copper, **13**-95
  electron-beam, **13**-106
  of flexible circuits, **13**-95
  interlayer-connections, **13**-101 to **13**-112
  lap joint, **13**-111
  laser, **13**-110
  material comparisons, **13**-100

Welding (*Cont.*):
  melting points, **13**-99
  micromodule, **13**-110
  opposed-electrode, **13**-93
  parallel-gap, **13**-97
  percussive arc, **13**-103
  pincer, **13**-96 to **13**-98
  power clamping force, **13**-105
  pressure, **13**-113
  pull-angle, **13**-99
  resistance, **13**-93
  second-level techniques, **13**-113
  self-accelerating electron-beams, **13**-107
  series, **13**-102
  series step, **13**-102
  step-weld, **13**-101
  ultrasonic, **13**-104
  wire-to-wire, **13**-111
  wire-wrap connections, **13**-113
  (*See also* Bonds and bonding)
Wiring and cabling:
  conductors (*see* Conductors)
  current capacity, **4**-28 to **4**-31
  design considerations for, **4**-28
    wire-gage selection of, **4**-28
  flexible, **4**-55 to **4**-66
  insulation (*see* Insulation)
  jackets, **4**-24 to **4**-28
  magnet wire (*see* Magnet wire)
  shielding (*see* Shielding)
  terminations (*see* Terminations)
  terms for, **4**-3 to **4**-6
  (*See also* Coaxial cables; Hookup wire; Magnet wire; Multiconductor cables; Shielding)